OBJECTIVE NCERT PUNCH 4.0

MCQs Extracted from NCERT Line by Line

AIIMS PYQs along with NEET to level up

PHYSICS

UPDATED WITH HIGH ORDER TIME INTENSIVE MCQs
AS PER 2025 NEET PAPER

Dr. Manish Raj (M.R. Sir)

EDITION: Fourth

Published By: Physicswallah Limited

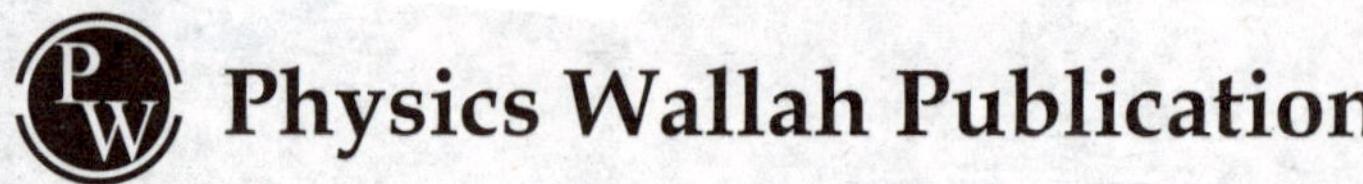

ISBN: 978-93-6897-472-7

MRP: 749/-

Edited by : Dr. Manish Raj (M.R. Sir)

Mobile App: Physics Wallah (Available on Play Store)

Website: www.pw.live

Youtube Channel: Physics Wallah - Alakh Pandey
Competition Wallah
NCERT Wallah

Email: publication@pw.live

SKU Code: 50a88cf3-6001-4c33-83d4-448f52df04bd

Printed at First Impression, Noida

Preface

A dedicated team at Physics Wallah (PW) works diligently to provide students with quality content for the NEET exam.

From the beginning, the whole content team comprising Associate faculties, DTP operators, Proofreaders and others are involved in shaping the material to their best knowledge and experience to produce powerful content for the students.

Faculties have adopted a new style of presenting the content in easy-to-understand language and have provided the team with their guidance and supervision throughout the creation of this Study Material.

Physics Wallah (PW) strongly believes in conceptual and fun-based learning. PW provides exam-focused content designed to offer clarity and value to students.

A plethora of NEET Study Material is available in the market but PW professionals are continuously Working to provide comprehensive study material for our NEET students.

This Study Material adopts a multi-faceted approach to master and understanding the concepts by having a rich diversity of questions asked in the examination and equip the students with the knowledge for the competitive exam.

The main objective of the study material is to provide a large number of quality problems with varying cognitive levels to facilitate the teaching-learning of concepts that are presented through the book.

It is widely used by aspirants because of its easy-to-understand language and clear illustrations with diagrams, mnemonics, flow charts and tables.

Students can benefit themselves by attempting the exercise given in this Study Material for the self-assessment and also mastering the basic techniques of problem-solving.

The questions are strictly designed in accordance with the exam relevant topics that helps to develop examination temperament within the students.

By mastering the Physics Wallah (PW) study material curated by the PW team, students can improve their chances of qualifying for the NEET exam and achieving a strong rank.

In each chapter, for better understanding, questions have been classified according to the latest syllabus of NEET

- The nature and diversity of the questions help students to ace the examination
- Quality questions to strengthen the concept of the topic at the zenith level

This Book Consists of

The following features are included in each chapter for systematic preparation:

- **NEET PYQs Analysis:** Year wise number of questions and topicwise number of questions.
- **NCERT-Picks:** Provides formulas, numerical profiles and statements asked in NEET exams.
- **NCERT Topic-wise MCQs:** Aligned with the latest NCERT and NEET syllabus.
- **High-Order Time Intensive MCQs:** Assists in practicing high order time consuming questions with detailed explanations.
- **Statement-based, Assertion & Reason, Match the Column MCQs:** Provides a variety of question types for practice.
- **NCERT Exemplar MCQs:** Includes questions from the NCERT Exemplar with explanations.
- **NEET & AIIMS Past Year MCQs and NEET 2025 Paper:** Provides Past Years Questions for Practice
- **Rationalised Content *[RC]*:** We have highlighted the content or chapters that have been rationalized by the NCERT.

About the Editor

Dr. Manish Raj, from **Bihar**, Popularly known as **M.R. sir** amongst his students is **renowned faculty of physics at PW** with **11 years** of teaching experience. He received his education from university of Delhi.

Till now he has taught **1.5 million** students and still counting.

He is recognized nationally for his teaching style and has great influence as a professor of physics. Mentor of AIR 24, 28, 39, 42, 59, 67 in NEET and many more.

INKZOID FOUNDATION honours him as **'The Star Teacher Of The Year 2022'** for teaching complete NEET Physics on alternate days for the **longest time duration** of 24 hours on YouTube.

Topper's Testimonials

Reading and understanding NCERT text books is a must do for NEET aspirants. Many students fail to realise the importance of NCERT. Once they realise, they find it challenging.To retain the information given in NCERT, one needs to practice a lot of questions. NCERT punch offers a variety of questions which helps in buliding grip on core concepts for NEET ,boards and JEE main

Samuel Harshith
NEET AIR-24

NEET PAST YEAR WEIGHTAGE ANALYSIS

Physics-XI

Chapters Name	Year Wise Number of Questions														
	2025	2024 Re	2024	2023	2023 Manipur	2022	2022 Re	2021	2020	2020 Covid	2019	2018	2017	2016 II	2016 I	2015
Units & Dimensions	2	1	3	2	1	3	1	3	3	2	1	1	1	1	0	1
Motion in a Straight Line	2	1	1	2	1	2	2	0	0	0	0	1	2	1	1	1
Motion in a Plane	0	1	0	2	1	1	1	2	1	1	3	0	0	3	0	2
Law of Motion	3	2	3	1	1	1	2	1	1	1	2	3	2	0	1	2
Work, Energy, and Power	2	3	2	2	3	2	2	1	0	1	3	1	2	3	4	3
Systems of Particles and Rotational Motion	2	2	3	2	3	3	1	1	2	2	2	4	3	3	2	3
Gravitation	2	2	2	2	3	2	2	3	1	1	2	2	2	2	2	1
Mechanical Properties of Solids	0	0	1	1	1	1	1	0	1	0	1	1	1	0	0	1
Mechanical Properties of Fluids	2	1	1	2	2	2	2	0	1	2	2	1	1	2	1	1
Thermal Properties of Matter	1	2	1	0	0	0	1	1	1	0	2	1	2	1	2	1
Thermodynamics	1	2	1	1	1	1	1	0	2	0	1	3	2	1	2	3
Kinetic Theory	2	1	1	1	1	1	1	1	3	2	1	1	2	2	0	2
Oscillations	2	1	2	1	1	1	2	2	1	1	3	1	1	0	0	2
Waves	1	1	0	1	1	1	1	0	1	1	0	2	2	3	3	1

Physics-XII

Chapters Name	Year Wise Number of Questions															
	2025	2024 Re	2024	2023	2023 Manipur	2022	2022 Re	2021	2020	2020 Covid	2019	2018	2017	2016 II	2016 I	2015
Electric Charges and Fields	1	1	0	2	3	1	2	1	1	1	3	0	1	0	1	1
Electrostatic Potential and Capacitance	3	3	5	2	1	2	2	4	3	2	0	1	2	1	1	1
Current Electricity	3	5	4	5	4	3	4	4	3	4	3	3	2	2	2	3
Moving Charges and Magnetism	3	0	2	2	2	3	4	3	1	0	2	2	1	3	2	2
Magnetism and Matter	0	4	2	1	1	1	0	1	1	1	1	2	2	0	1	0
Electromagnetic Induction	0	1	2	1	1	2	1	1	0	2	2	1	2	1	1	1
Alternating Current	2	3	1	3	4	2	3	4	2	2	0	1	0	3	2	1
Electromagnetic Waves	1	3	1	2	1	2	2	1	1	2	2	1	1	1	0	2
Ray Optics and Optical Instruments	2	1	2	3	5	3	2	5	3	5	3	2	2	3	3	2
Wave Optics	2	3	2	1	0	1	2	0	2	1	1	3	3	1	1	1
Dual Nature of Radiation and Matter	2	2	2	2	2	2	2	1	3	2	1	2	2	1	2	2
Atoms	2	2	2	2	3	1	1	0	1	1	1	1	1	2	1	1
Nuclei	0	2	1	1	0	2	2	3	2	2	1	1	1	1	0	1
Semiconductor Electronics	2	3	3	3	3	3	3	2	4	4	2	3	3	3	2	2

Contents

CLASS-XI

CLASS-XII

Class-XI

CHAPTER 0

Mathematical Tools

NCERT-PICKS

Quadratic Equation

An equation of second degree is called a quadratic equation. It is of the form

$ax^2 + bx + c = 0$

The roots of quadratic equation are

$$x = \frac{-b \pm (b^2 - 4ac)^{1/2}}{2a}$$

Binomial Theorem

If n is any integer, or a fraction and x is any real number, then

$$(1 + x)^n = 1 + nx + \frac{n(n-1)x^2}{2!} + \ldots$$

If $|x| << 1$, then $(1 + x)^n \simeq 1 + nx$.

Mensuration

- Area of circle $= \pi r^2 = \frac{\pi D^2}{4}$ $\left(\because r = \frac{D}{2}\right)$
- Surface area of a sphere $= 4\pi r^2 = \pi D^2$
- Volume of sphere $= \frac{4}{3}\pi r^3$
- Surface area of a cylinder $= 2\pi r(r + h)$
- Volume of a cylinder $= \pi r^2 h$
- Curved surface area of a cone $= \pi rl$
- Volume of cone $= \frac{1}{3}\pi r^2 h$
- Surface area of a cube $= 6(\text{side})^2$
- Volume of a cube $= (\text{side})^3$

Four Quadrants

- I^{st} quadrants $\Rightarrow$ All trigonometric ratios are positive
- IInd quadrants $\Rightarrow$ Only $\sin\theta$ and $\text{cosec}\,\theta$ are positive
- IIIrd quadrants $\Rightarrow$ Only $\tan\theta$ and $\cot\theta$ are positive
- IVth quadrants $\Rightarrow$ Only $\cos\theta$ and $\sec\theta$ are positive

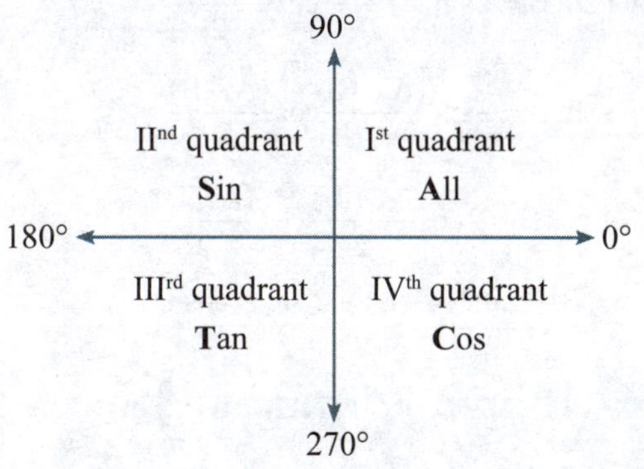

Position of a Point in xy Plane

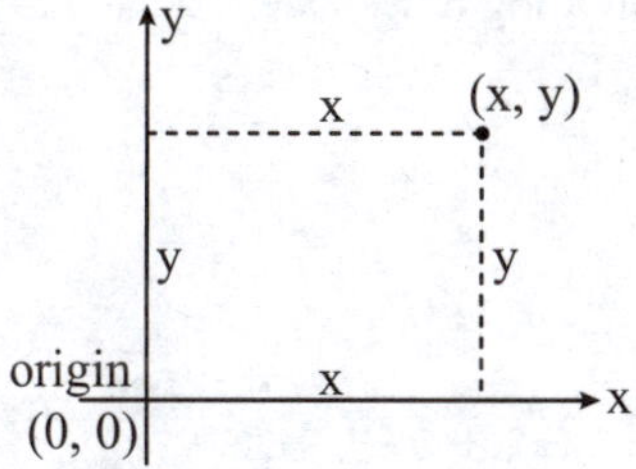

Distance Formula

- The distance (d) between two points P (x_1, y_1) and Q (x_2, y_2) is given by $d = \sqrt{(x_2 - x_1)^2 + (y_2 - y_1)^2}$

Fundamental Trigonometric Relations

- $\text{cosec}\,\theta = \frac{1}{\sin\theta}$
- $\sec\theta = \frac{1}{\cos\theta}$
- $\cot\theta = \frac{\cos\theta}{\sin\theta} = \frac{1}{\tan\theta}$
- $\tan\theta = \frac{\sin\theta}{\cos\theta}$

- $\sin^2\theta + \cos^2\theta = 1$
- $1 + \tan^2\theta = \sec^2\theta$
- $1 + \cot^2\theta = \text{cosec}^2\theta$
- $\sin(A + B) = \sin A \cos B + \cos A \sin B$
- $\cos(A + B) = \cos A \cos B - \sin A \sin B$
- $\sin(A - B) = \sin A \cos B - \cos A \sin B$
- $\cos(A - B) = \cos A \cos B + \sin A \sin B$
- $\tan(A+B) = \dfrac{\tan A + \tan B}{1 - \tan A \tan B}$
- $\sin 2A = 2 \sin A \cos A$
- $\cos 2A = 2\cos^2 A - 1 = 1 - 2\sin^2 A = \cos^2 A - \sin^2 A$
- $\sin A + \sin B = 2\sin\dfrac{(A+B)}{2}\cos\dfrac{(A-B)}{2}$
- $\sin A - \sin B = 2\sin\dfrac{(A-B)}{2}\cos\dfrac{(A+B)}{2}$
- $\cos A + \cos B = 2\cos\dfrac{(A+B)}{2}\cos\dfrac{(A-B)}{2}$
- $\sin(A + B) + \sin(A - B) = 2 \sin A \cos B$

Logarithmic Formulas

- $\log_a mn = \log_a m + \log_a n$
- $\log_a\left(\dfrac{m}{n}\right) = \log_a m - \log_a n$
- $\log_a m = \log_b m \times \log_a b$
- $\log_a b^c = c \log_a b$
- $\log_a 1 = 0$
- $\log_a a = 1$
- $x = e^{\log x}$

Approximate values

If angle (θ) is small then

- $\sin\theta \approx \theta$
- $\cos\theta \approx 1$
- $\tan\theta \approx \theta$
- $\sin\theta \approx \tan\theta$

Differential Formulas

- $\dfrac{dc}{dx} = 0,$ [c = constant]
- $\dfrac{d(cy)}{dx} = c\dfrac{dy}{dx}$
- $\dfrac{d(x^n)}{dx} = nx^{n-1}$
- $\dfrac{d[f(x) \pm g(x)]}{dx} = \dfrac{df(x)}{dx} \pm \dfrac{dg(x)}{dx}$
- $\dfrac{d\{f(x)g(x)\}}{dx} = f(x)\dfrac{dg(x)}{dx} + g(x)\dfrac{df(x)}{dx}$
- $\dfrac{d}{dx}\left\{\dfrac{f(x)}{g(x)}\right\} = \dfrac{g(x)\dfrac{df(x)}{dx} - f(x)\dfrac{dg(x)}{dx}}{\{g(x)\}^2}$
- $\dfrac{dy}{dx} = \dfrac{dy}{du}\dfrac{du}{dx}$
- $\dfrac{d}{dx}e^x = e^x$
- $\dfrac{d}{dx}u^n = nu^{n-1}\dfrac{du}{dx}$
- $\dfrac{d}{dx}\log x = \dfrac{1}{x}$
- $\dfrac{d}{dx}(\sin x) = \cos x$
- $\dfrac{d}{dx}(\cos x) = -\sin x$
- $\dfrac{d}{dx}(\tan x) = \sec^2 x$
- $\dfrac{d}{dx}(\cot x) = -\text{cosec}^2 x$
- $\dfrac{d}{dx}(\sec x) = \sec x \tan x$
- $\dfrac{d}{dx}(\text{cosec } x) = -\text{cosec } x \cot x$
- $\dfrac{d}{dx}\sin^{-1} x = \dfrac{1}{\sqrt{1-x^2}}$
- $\dfrac{d}{dx}\cos^{-1} x = \dfrac{-1}{\sqrt{1-x^2}}$

Slope of a Line

It is the tan of angle made by a line with the positive direction of *x*-axis, measured in anticlockwise direction.

- Slope = $\tan\theta$
- In figure (*a*)slope is positive
- In figure (*b*)slope is negative

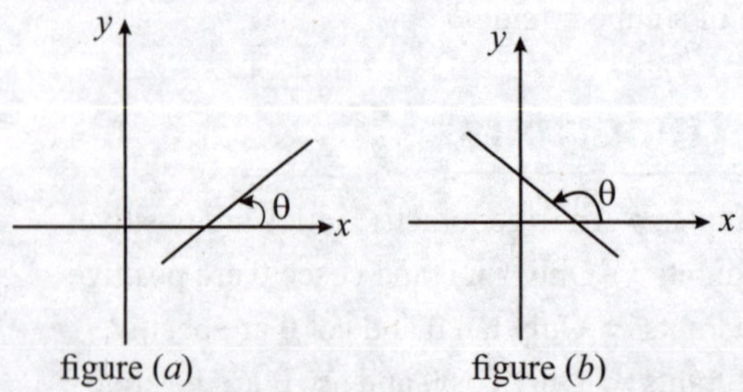

figure (*a*) figure (*b*)

Maxima and Minima

- Condition for Minima: $\frac{dy}{dx} = 0$ and $\frac{d^2y}{dx^2} > 0$
- Condition for Maxima: $\frac{dy}{dx} = 0$ and $\frac{d^2y}{dx^2} < 0$

Integral Formulas

- $\int dx = x + c$ where c = constant
- $\int x^n dx = \frac{x^{n+1}}{n+1} + c$
- $\int \left(\frac{1}{x}\right) dx = \log_e x + c$
- $\int \sin x \, dx = -\cos x + c$
- $\int \sin ax \, dx = -\frac{\cos ax}{a} + c$
- $\int \cos x dx = \sin x + c$
- $\int \sec^2 x dx = \tan x + c$
- $\int \operatorname{cosec}^2 x dx = -\cot x + c$
- $\int \tan x dx = \log |\sec x| + c$
- $\int e^x dx = e^x + c$

Area Under a Curve and Definite Integration

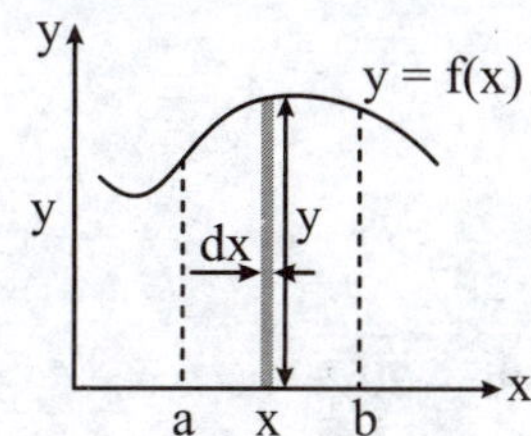

- $\int_a^b f(x)dx$ = Shaded area between curve and x-axis

[Where $\int_a^b f(x)dx$: Definite integral, a: Lower limit, b: Upper limit]

Some Standard Graphs and their Equations

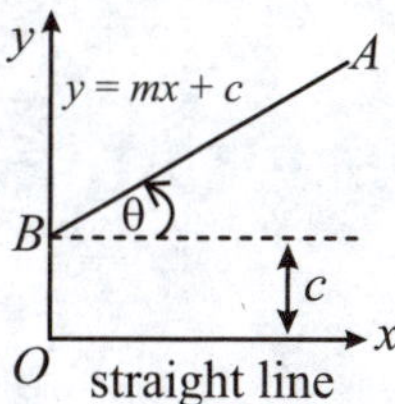

straight line

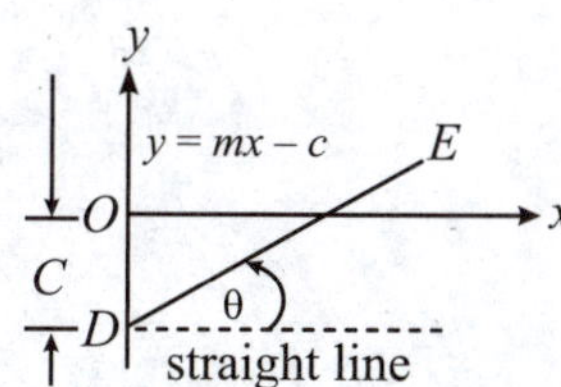

straight line

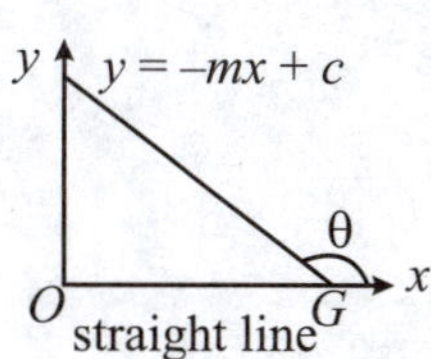

straight line

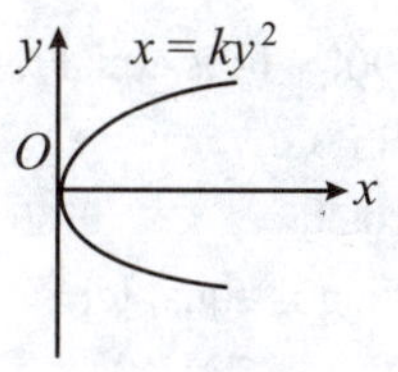

Parabola

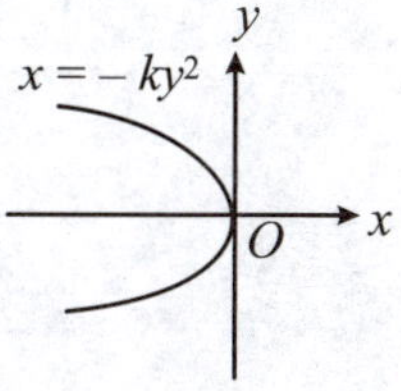

Parabola

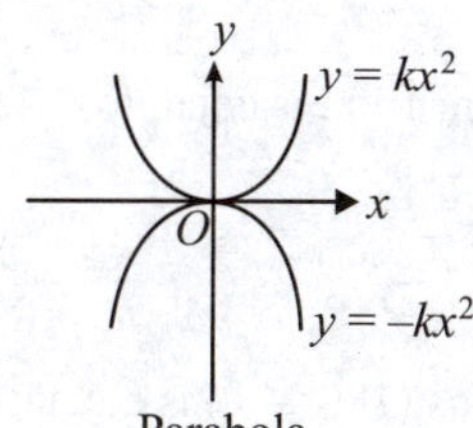

Parabola

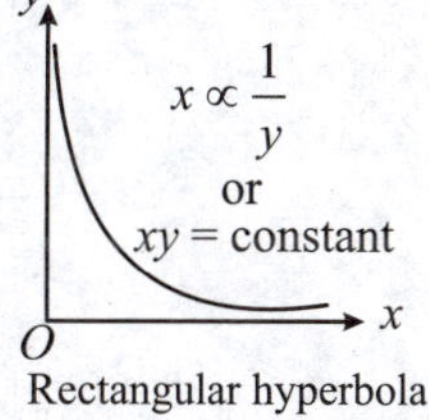

Rectangular hyperbola

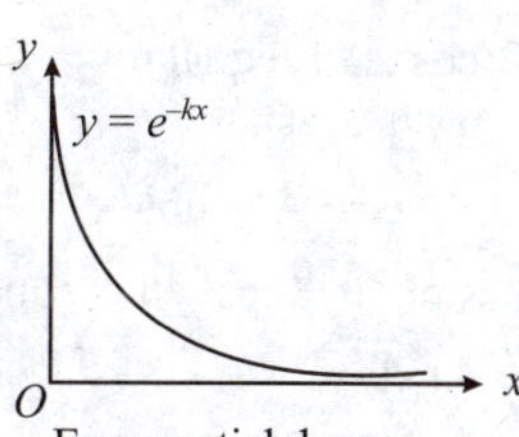

Exponential decay

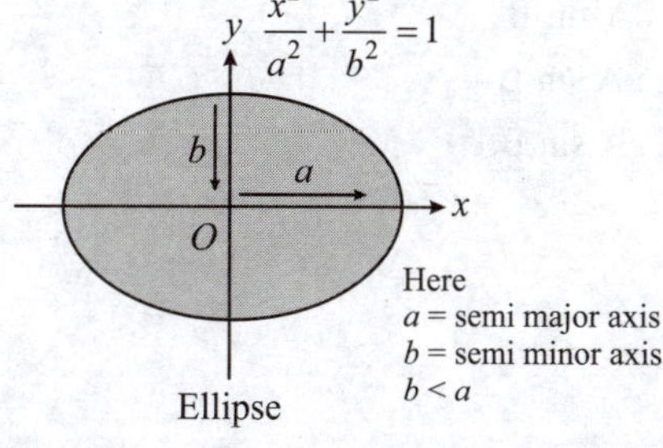

Ellipse

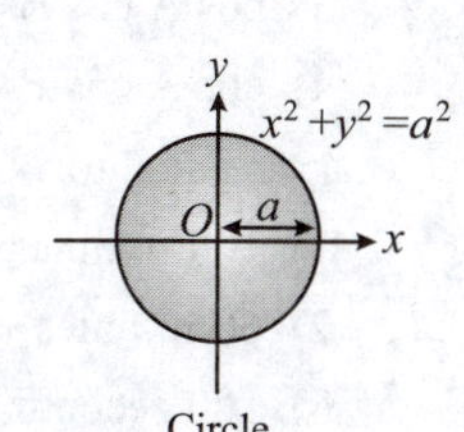

Circle

MCQs ?

NCERT Topic-wise MCQs

Trigonometry

1. $\sin(90° - \theta)$ is equal to:
   1) $\sin\theta$ 2) $-\sin\theta$
   3) $\cos\theta$ 4) $-\cos\theta$
2. 1 radian is equal to:
   1) $\frac{\pi}{180}$ degree 2) $\frac{180}{\pi}$ degree
   3) $\frac{90}{\pi}$ degree 4) $\frac{18}{\pi}$ degree
3. $\sin 100\pi$ is equal to:
   1) 1 2) 100
   3) Zero 4) $\frac{1}{2}$
4. $\cos(180° - \theta)$ is equal to:
   1) $-\cos\theta$ 2) $\cos\theta$
   3) $\sin\theta$ 4) $-\sin\theta$
5. cos 2A is equal to:
   1) $1 - 2\sin^2 A$ 2) $2\cos^2 A - 1$
   3) $\cos^2 A - \sin^2 A$ 4) All
6. $\sin^2 4\theta + \cos^2 4\theta$ is equal to:
   1) 4 2) 2
   3) -1 4) 1
7. $\cos(A + B)$ is equal to:
   1) cos A cos B + sin A cos B
   2) cos A sin B – sin A sin B
   3) cos A cos B + sin A sin B
   4) cos A cos B – sin A sin B
8. 1° is equal to:
   1) 100 minutes
   2) 60 minutes
   3) 100 seconds
   4) 60 seconds
9. $1 + \tan^2\theta$ is equal to:
   1) $\sec\theta$ 2) $\sec 2\theta$
   3) $\sec^2\theta$ 4) $2\sec\theta$
10. sin 300° is equal to:
    1) $-\cos 30°$ 2) $\cos 60°$
    3) $\sec 30°$ 4) $\sec 60°$
11. Evaluate:
    1) cos 15° 2) cos 53°
    3) tan 37° 4) sin53° – cos 37°
12. Evaluate:
    1) 2 sin 15° cos 15° 2) sin 22.5° cos 22.5°
    3) tan 75° 4) $\sin^2 22.5°$
13. Evaluate:
    1) $\cos\left(\frac{5\pi}{4}\right)$ 2) $\sin\left(\frac{2\pi}{3}\right)$
    3) $\sin\left(\frac{3\pi}{4}\right)$ 4) $\tan\left(\frac{7\pi}{6}\right)$
14. $\sqrt{1+\sin\theta}$ is equal to
    1) $(\sin\theta + \cos\theta)$ 2) $\sin\theta - \cos\theta$
    3) $\sin\frac{\theta}{2}+\cos\frac{\theta}{2}$ 4) $\sin\frac{\theta}{2}-\cos\frac{\theta}{2}$
15. If $\frac{\sin\theta+\cos\theta}{\sin\theta-\cos\theta}=\frac{7}{3}$, then $\tan\theta =$
    1) $\frac{3}{5}$ 2) $\frac{5}{2}$
    3) $\frac{5}{3}$ 4) $\frac{2}{5}$

Algebra

16. The roots of $100x^2 - 20x + 1 = 0$ is:
    (1) $\frac{1}{20}$ and $\frac{1}{20}$ (2) $\frac{1}{10}$ and $\frac{1}{20}$
    (3) $\frac{1}{10}$ and $\frac{1}{10}$ (4) None of these
17. If ½ is a root of the quadratic equation $x^2 - mx - 5/4 = 0$, then value of m is:
    1) 2 2) -2
    3) -3 4) 3
18. Find the sum of first 20 natural Number.
    1) 210 2) 200
    3) 220 4) 230
19. The sum of the first five multiples of 3 is:
    1) 45 2) 55
    3) 65 4) 75

20. Find sum of $1-\frac{1}{2}+\frac{1}{4}-\frac{1}{8}+\frac{1}{16}-\frac{1}{32}$ ...... ∞.

1) $\frac{3}{4}$ 2) $\frac{4}{3}$

3) $\frac{2}{3}$ 4) $\frac{3}{2}$

21. Find sum of $1+\frac{1}{3}+\frac{1}{9}+\frac{1}{27}+$ .... up to ∞.

1) $\frac{3}{2}$ 2) $\frac{2}{3}$

3) $\frac{4}{3}$ 4) $\frac{3}{4}$

22. The sum of 1 + 3 + 5 + 7 + ... upto n terms is:

1) $(n+1)^2$ 2) $(2n)^2$

3) n^2 4) $(n-1)^2$

23. Find approximate value of the $\sqrt{0.95}$:

1) 1 2) 0.595

3) 0.60 4) 0.975

24. Find approximate value of the $\sqrt{104}$:

1) 10.2 2) 12

3) 13.5 4) 15

25. Find approximate value of the $(4.04)^3$

1) 60.05 2) 75.63

3) 65.92 4) 55.72

26. Find approximate value of the $(9.6)^4$

1) 4200 2) 3600

3) 2100 4) 8400

27. log 25 + log 4 – log 5 is equal to

1) log 20 2) log 25

3) log 15 4) log 10

28. $\log_e 8$ is equal to

1) $\log_e 2$ 2) $2\log_e 2$

3) $3\log_e 2$ 4) $4\log_e 2$

29. $\log_e 15$ is equal to

1) $\log_e 3 + \log_e 5$ 2) $\log_e 5 - \log_e 3$

3) $\log_e 10 + \log_e 5$ 4) $\log_e 10 - \log_e 5$

30. $\log_5 x - \log_5(y) = 2$, find the value of $\frac{x}{y}$

1) 100 2) 25

3) 50 4) 75

Calculus

31. If $y = \sin^2 x - 2\tan^2 x$, then $\frac{dy}{dx}$ at $x = \frac{\pi}{4}$ is:

1) –11 2) –7

3) –13 4) –15

32. If $y = x^3 + 2x + 1$ then $\frac{dy}{dx}$ at $x = 1$ is:

1) 6 2) 7

3) 8 4) 5

33. $y = \frac{1+x}{e^x}$ then $\frac{dy}{dx}$ is equal to:

1) $\frac{x}{e^x}$ 2) $-\frac{x}{e^x}$

3) $\frac{(x+1)}{e^x}$ 4) None of these

34. $\frac{d}{dr}\left(\pi r^2\right)$ is equal to:

1) $2\pi r$ 2) πr

3) $2r$ 4) $\frac{\pi r}{2}$

35. If $y = x^2 + x - 1$ then $\frac{dy}{dx}$ at $x = 1$ is equal to:

1) 3 2) –3

3) 0 4) None

36. Find derivative of $y = x \sin x$ with respect to independent variable x.

$y = x \sin x$

1) $\sin x + x\cos x$ 2) $\sin x - x\cos x$

3) $\cos^2 x - x\sin^2 x$ 4) $\sin^2 x - x\cos^2 x$

37. Find derivative of $y = e^x \ell n x$ with respect to x.

1) $e^x \ell n x - \frac{e^x}{x}$ 2) $e^x \ell n x - \frac{e^x}{x^2}$

3) $e^x \ell n x + \frac{e^x}{x^2}$ 4) $e^x \ell n x + \frac{e^x}{x}$

38. Find derivative of $y = (x-1)(x^2+x+1)$ with respect to independent variable x.

1) $\frac{dy}{dx} = 3x$ 2) $\frac{dy}{dx} = 3x^2$

3) $\frac{dy}{dx} = 2x^2$ 4) $\frac{dy}{dx} = 2x$

39. If $y = a(1 - \cos\theta)$ and $x = a(\theta + \sin\theta)$, then find the slope of tangent of y verses x at $\theta = \frac{\pi}{2}$.

1) 1 2) – 1

3) 2 4) – 2

40. The slope of the normal to the curve $y = x^2 - \frac{1}{x^2}$ at (– 1, 0) is:

1) $\frac{1}{4}$ 2) $-\frac{1}{4}$

3) 4 4) – 4

41. $y = \sin u, u = \cos x$. Find $\frac{dy}{dx}$.

1) $-\cos u.\sin x$
2) $-\cos u.\sqrt{1-u^2}$
3) $-\sin x.\sqrt{1-y^2}$
4) All of above

42. $y = 3t^2 - 1, x = t^2$. Find $\frac{dy}{dx}$.

1) 3
2) 2
3) 1/3
4) 1/2

43. The value of $\int_0^{\pi/2} \sin 2\theta d\theta$ is:

1) 0
2) 1
3) –1
4) 2

44. Value of $\int_0^1 \frac{1}{(3-2x)^2} dx$ is:

1) $-\frac{1}{9}$
2) $\frac{1}{3}$
3) $-\frac{4}{9}$
4) None of these

45. $\int (4\cos t + t^2) dt$ is equal to:

1) $-4\sin t + \frac{t^3}{3} + C$
2) $-4 \sin t + t^2 + C$
3) $4\sin t + \frac{t^3}{3} + C$
4) $4 \sin t + 2t^3 + C$

46. $\int \frac{1}{x} dx$ is:

1) $\log_e x + c$
2) $e^x + c$
3) $x^{3/2} + c$
4) $5 + c$

47. $\int \frac{1}{\sqrt{x}} dx$ is

1) $\frac{\sqrt{x}}{2} + C$
2) $2\sqrt{x} + C$
3) $\frac{2}{\sqrt{x}} + C$
4) None of these

48. If $y = \sin(2x^2)$, then $\frac{dy}{dx}$ is:

1) $4x \cos(2x^2)$
2) $2 \cos (2x^2)$
3) $4 \cos (2x^2)$
4) $-4 \cos (2x^2)$

49. Maximum and minimum values of function $2x^3 - 15x^2 + 36x + 11$ is

1) 39, 38
2) 93, 83
3) 45, 42
4) 59, 58

50. $\int \left(\frac{3}{2\sqrt{x}}\right) dx$ will be

1) $2\sqrt{x^3} + C$
2) $3\sqrt{x} + C$
3) $\sqrt{x^3} + C$
4) $\sqrt{x^4} + C$

51. $\int \left(\frac{1}{3\sqrt[3]{x}}\right) dx$ will be?

1) $\frac{x^{\frac{3}{4}}}{2} + C$
2) $\frac{x^{\frac{2}{3}}}{3} + C$
3) $x^{\frac{2}{3}} + C$
4) $\frac{x^{\frac{2}{3}}}{2} + C$

52. $\int_{-4}^{-1} \frac{\pi}{2} d\theta$ will be?

1) $\frac{3\pi}{2}$
2) $\frac{2\pi}{3}$
3) 3π
4) 2π

53. Use a definite integral to find the area of the region between the given curve $y = 2x$ and the x-axis on the interval $[0, b]$

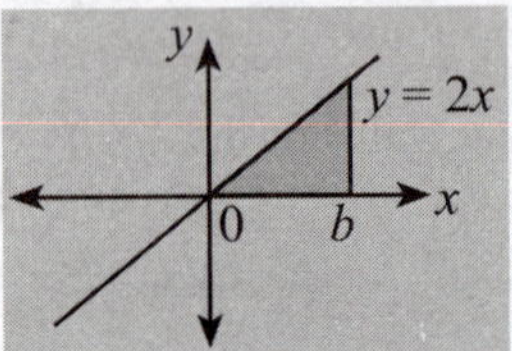

1) b^2
2) b^3
3) $2b^2$
4) $\frac{b^3}{3}$

54. For function $y = f(x) = x^2$, find average value of y from $x = 1$ to $x = 3$

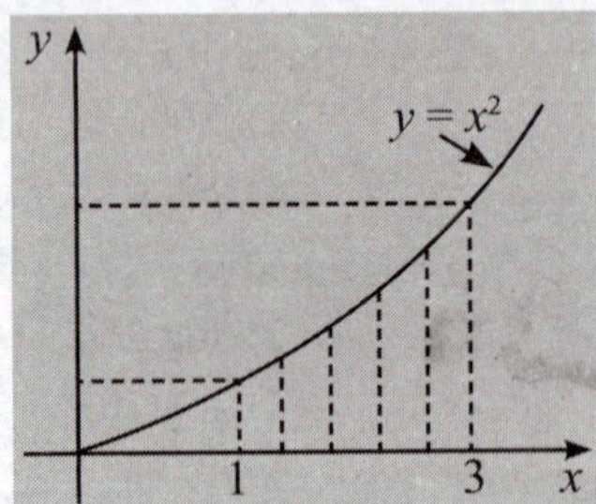

1) $\frac{13}{3}$
2) $\frac{2}{13}$
3) $\frac{13}{4}$
4) $\frac{4}{13}$

55. Find the area of the curve $y = \sin x$ between 0 and π.

1) 2 sq. unit
2) 5 sq. unit
3) 4 sq. unit
4) 10 sq. unit

56. For function $y = 2 \sin\theta$
Find average of y from $\theta = 0$ to $\theta = \pi$.

1) $\frac{4}{\pi}$
2) $\frac{2}{\pi}$
3) $\frac{\pi}{4}$
4) $\frac{\pi}{2}$

57. Particle's position as a function of time is given by $x = -t^2 + 4t + 4$, find the maximum value of position co-ordinate of particle.

1) 2
2) 4
3) –8
4) 8

58. Find minimum value of the function: $y = 25x^2 + 5 - 10x$

1) 4 2) 3

3) 2 4) 1

59. Determine the position where potential energy will be minimum if $U(x) = 100 - 50x + 1000x^2$.

1) 0.25×10^{-2} 2) 2.5×10^{-2}

3) 2.5×10^{-1} 4) 250×10^{-2}

60. Find out minimum/maximum value of $y = 4x^2 - 2x + 3$ also find out those points where value is minimum/maximum.

1) $\text{min} = \frac{11}{4}, x = \frac{1}{2}$ 2) $\text{max} = \frac{11}{4}, x = \frac{1}{4}$

3) $\text{min} = \frac{11}{4}, x = \frac{1}{4}$ 4) $\text{max} = \frac{11}{4}, x = \frac{1}{2}$

61. $x + y = 10$ then find the maximum value of $f = xy$?

1) 25 2) 30

3) 15 4) 35

62. If $\ell + r = 12$ here ℓ is length of cylinder and r is radius of cylinder then find maximum value of volume of cylinder

1) 156 π 2) 350 π

3) 256 π 4) 250 π

Coordinate Geometry & Graphs

63. Correct graph of $y = |3x + 4|$ is:

1) 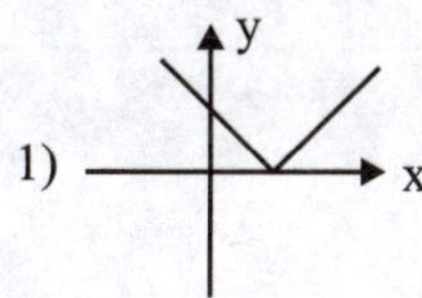
2) 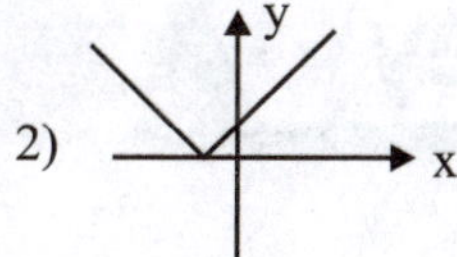

3) 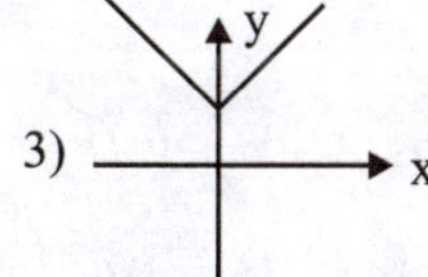
4) 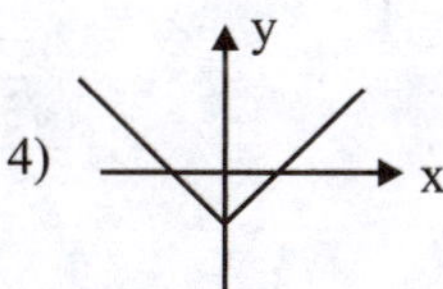

64. Correct graph of $y = -(x + 2)^2$ is:

1) 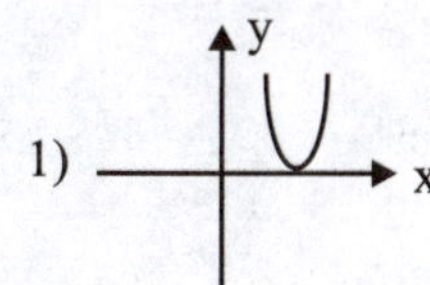
2) 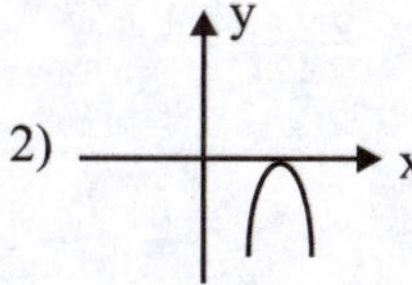

3) 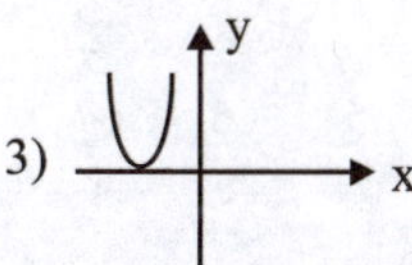
4) 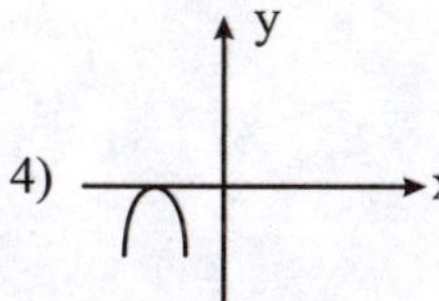

65. Graph of $y = 2(x + 1)^2 + 2$ is:

1) 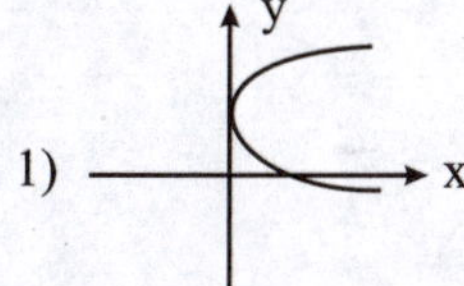
2) 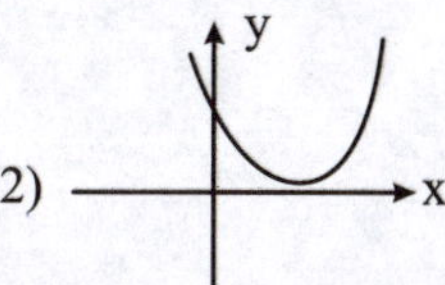

3) 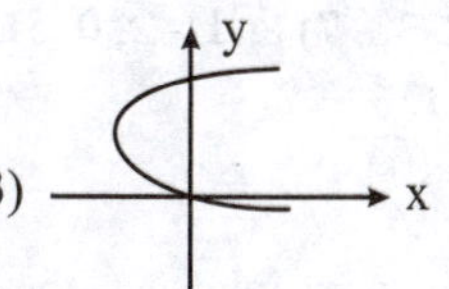
4) 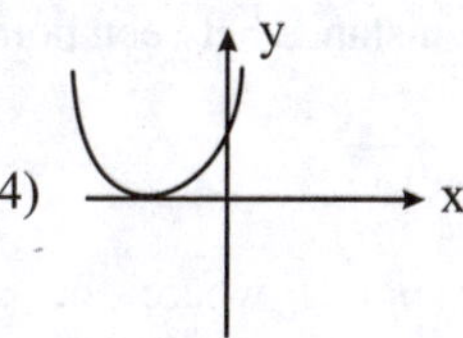

66. Graph of $y = 3x^2 - 4x + 1$ is:

1) 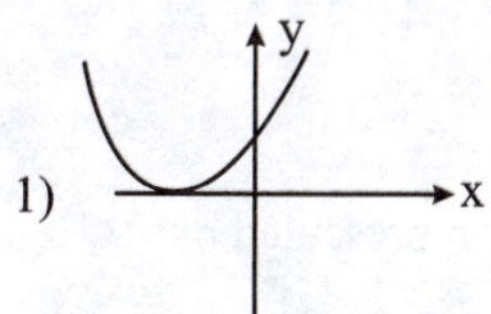
2) 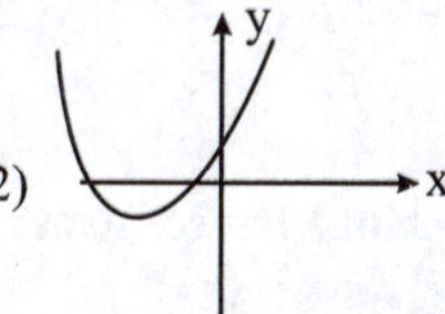

3) 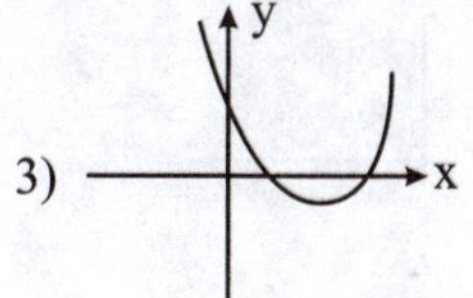
4) 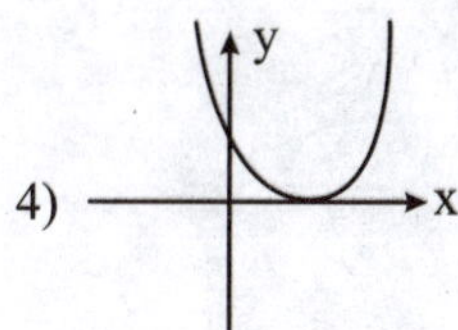

67. Graph of $x = 2(y - 1)^2$ is:

1) 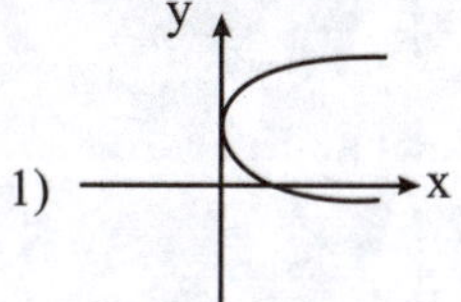
2) 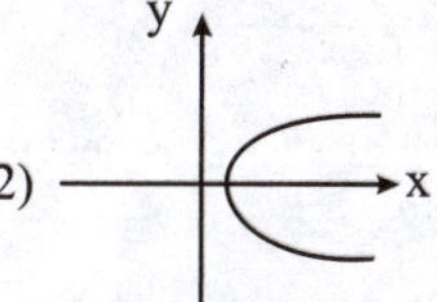

3) 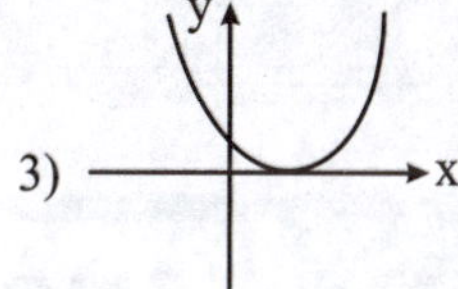
4) 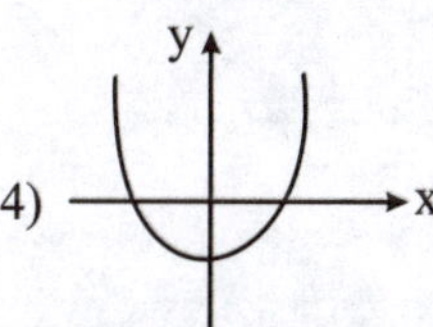

68. Find the slope of the line $-3x + 4y = 11$.

1) $\frac{-4}{3}$ 2) $\frac{-3}{4}$

3) $\frac{3}{4}$ 4) $\frac{11}{3}$

69. Graph of $y = 1 - e^{-x}$ is best represented by (for $x > 0$):

1) 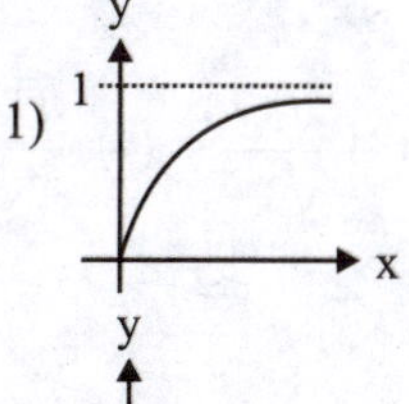

2) 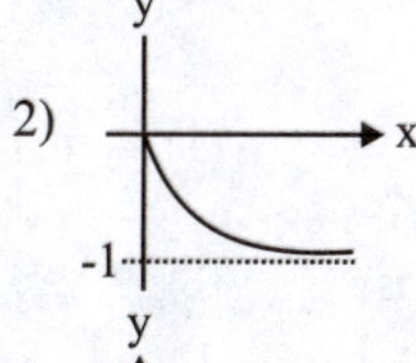

3) 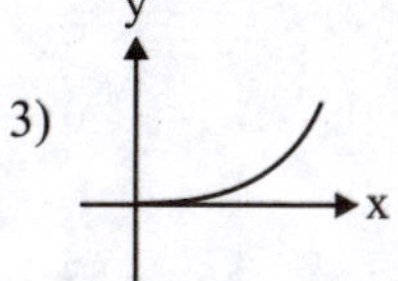
4) 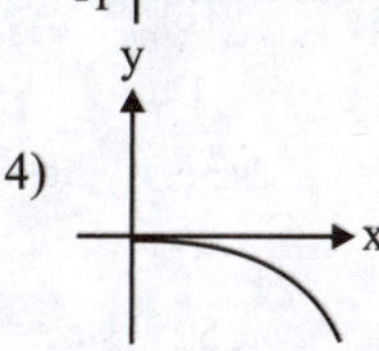

70. The equation of straight line having slope $\sqrt{3}$ and y intercept of –2 will be:

1) $y = \sqrt{3}x + 2$ 2) $y = \sqrt{3}x - 2$

3) $y = -\sqrt{3}x - 2$ 4) $y = -\sqrt{3}x + 2$

71. The equation of straight line is $2y = 3x - 6$ its x and y intercepts are

1) –3, 2 2) 3, –2

3) 2, –3 4) –2, 3

72. The distance between points (2, 3, –7) and (–2, 0, 5) is

1) 5
2) 13
3) $\sqrt{145}$
4) $\sqrt{119}$

73. The points where the line $y = x$ intersect the curve $x^2 + y^2 = 32$ is

1) $4\sqrt{2}, 4\sqrt{2}$
2) 4, 4
3) $-4\sqrt{2}, -4\sqrt{2}$
4) –2, –2

74. Graph of PV = constant is best represented by:

1) 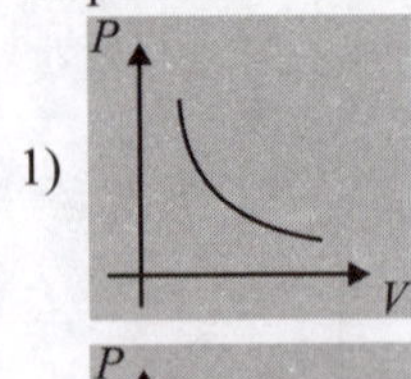

2) 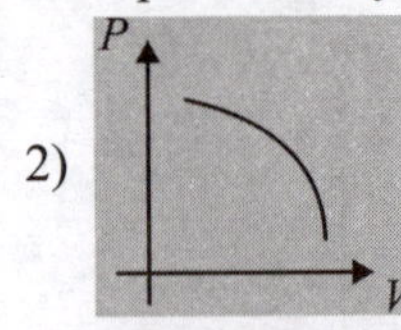

3) 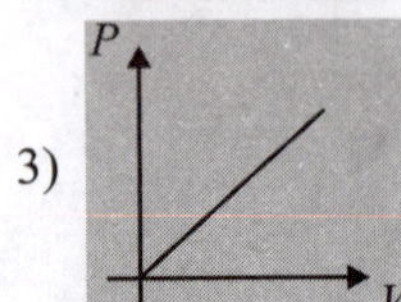

4) 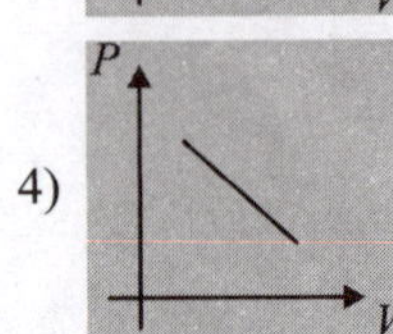

75. Graph of $x^2y = 2$ is best represented by:

1) 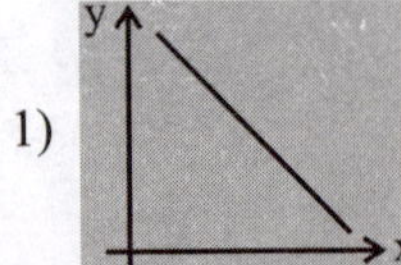

2) 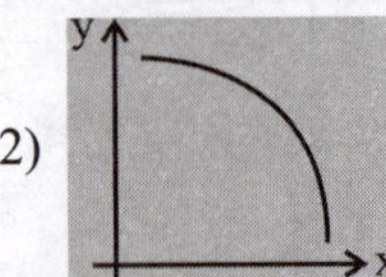

3) 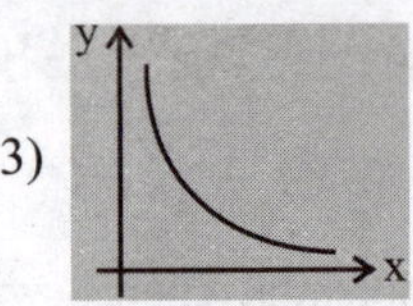

4) 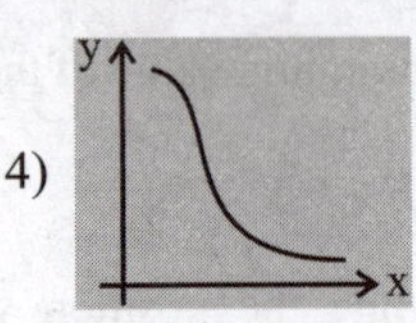

76. If $y = (x^2 + x - 2)$, $y - x$ graph is:

1) 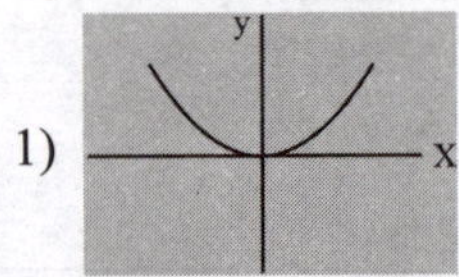

2) 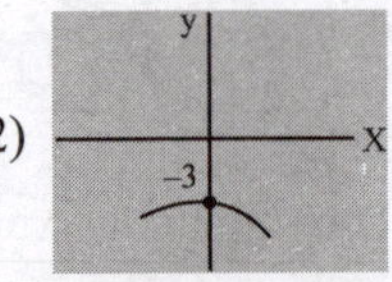

3) 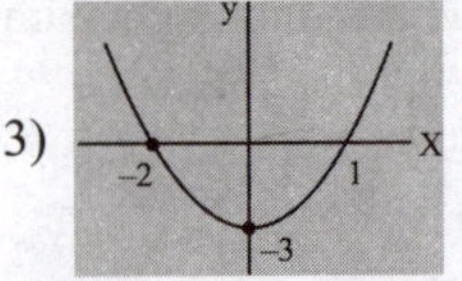

4) 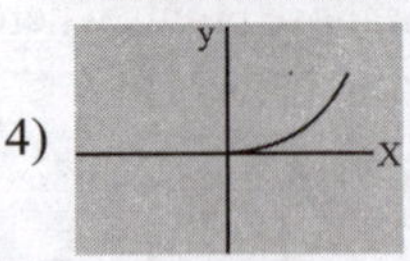

77. If $y = x^2 + 2x - 3$, y-x graph is:

1) 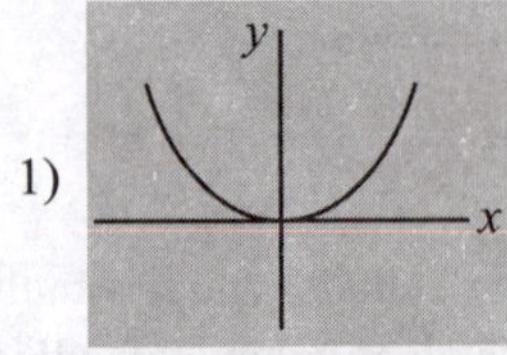

2) 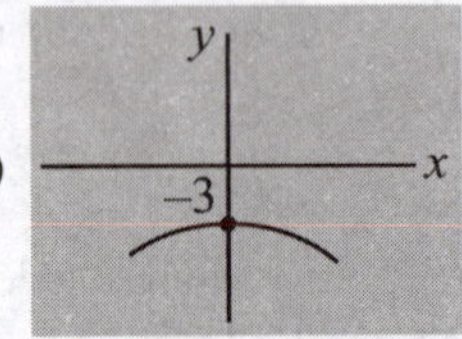

3) 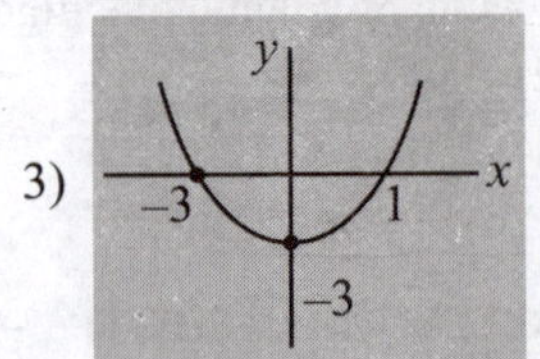

4) 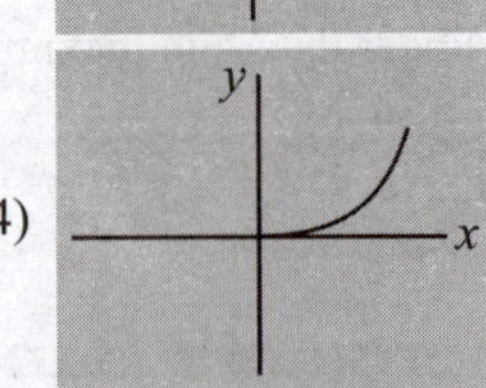

Answer Key

NCERT Topic-wise MCQs

1	2	3	4	5	6	7	8	9	10	11
3	2	3	1	4	4	4	2	3	1	(1) $\frac{\sqrt{3}+1}{2\sqrt{2}}$, (2) $\frac{3}{5}$, (3) $\frac{3}{4}$, (4) 0

12	13	14	15	16	17
(1) $\frac{1}{2}$, (2) $\frac{1}{2\sqrt{2}}$, (3) $\frac{\sqrt{3}+1}{\sqrt{3}-1}$, (4) $\frac{\sqrt{2}-1}{2\sqrt{2}}$	(1) $-\frac{1}{\sqrt{2}}$, (2) $\frac{\sqrt{3}}{2}$, (3) $\frac{1}{\sqrt{2}}$, (4) $\frac{1}{\sqrt{3}}$	3	2	3	2

18	19	20	21	22	23	24	25	26	27	28	29	30	31	32	33	34
1	1	3	1	3	4	1	3	4	1	3	1	2	2	4	2	1
35	**36**	**37**	**38**	**39**	**40**	**41**	**42**	**43**	**44**	**45**	**46**	**47**	**48**	**49**	**50**	**51**
1	1	4	2	1	1	4	1	2	2	3	1	2	1	1	2	4
52	**53**	**54**	**55**	**56**	**57**	**58**	**59**	**60**	**61**	**62**	**63**	**64**	**65**	**66**	**67**	**68**
1	1	1	1	1	4	1	2	3	1	3	2	4	2	3	1	3
69	**70**	**71**	**72**	**73**	**74**	**75**	**76**	**77**								
1	2	3	2	2	1	3	3	3								

CHAPTER

1 Units & Dimensions

Topicwise Number of Questions Analysis (2025-2016)

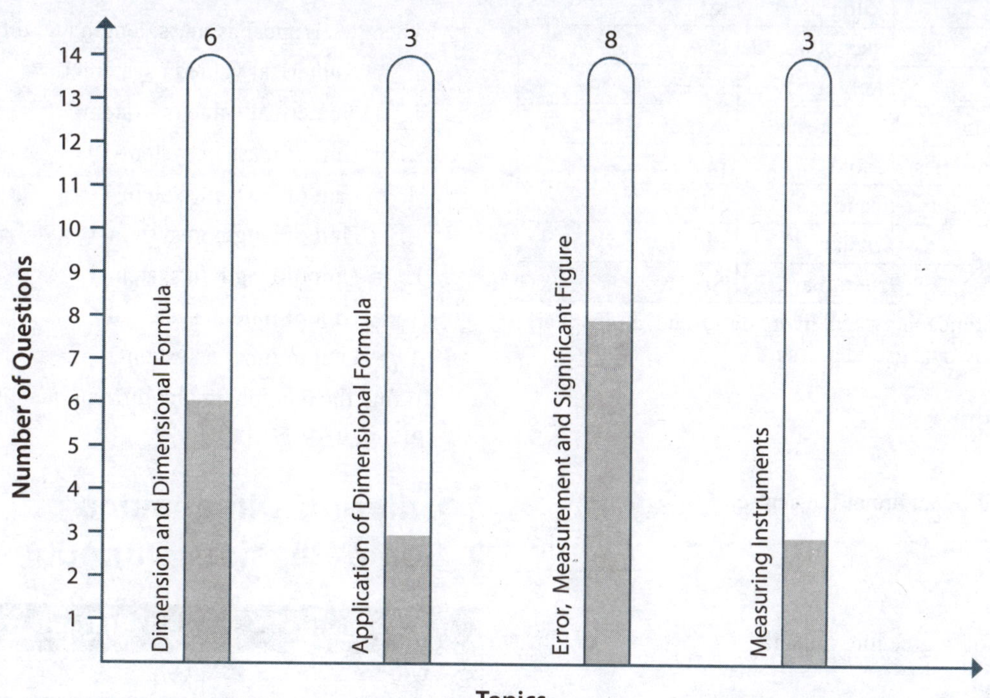

NCERT-Picks

Physical Constants

- Mass of electron (m_e) = 9.109×10^{-31} kg
- Mass of proton (m_p) = 1.6726×10^{-27} kg
- Mass of a neutron (m_n) = 1.6749×10^{-27} kg
- Charge of an electron (e) = -1.602×10^{-19}C
- Speed of light in vacuum (c) = 3×10^8 m/s
- Planck's constant (h) = 6.626×10^{-34} J -s
- Universal Gravitational constant (G) = 6.67×10^{-11} Nm^2/kg^2
- Avogadro's Number (N_A) = 6.022×10^{23} mol^{-1}
- Boltzmann constant (k_B) = 1.38×10^{-23} J/K
- Stefan's constant (σ) = 5.67×10^{-8} W m^{-2} K^{-4}
- Wien's constant (b) = 2.898×10^{-3} m K
- Solar constant (S) = 1.366×10^3 W m^{-2}
- Mass of the sun (M_S) = 1.989×10^{30} kg
- Mass of the earth (M_E) = 5.97×10^{24} kg
- Radius of the earth (R_e) = 6371 km $\approx 6.4 \times 10^6$ m
- Density of earth = 5.513×10^3 kg/m^3
- Average angular velocity of the earth = 7.29×10^{-5} rad/s
- Average distance between the Sun and the Earth (AU) = 1.496×10^{11} m
- Average distance between moon and the earth = 3.844×10^8m
- Magnetic Moment of the earth = 8.22×10^{22} A - m^2

Conversion Coefficients

- 1 Light year = 9.46×10^{15} m.
- 1 A.U. = 1.496×10^{11} m.
- 1 Å = 10^{-10} m.
- 1 Fermi = 10^{-15} m.
- 1 eV = 1.6×10^{-19} Joule.
- 1 Horse Power = 746 Watt.

System of Units

Formulas

- System of Units:
  - (a) C.G.S = Centimeter-Gram-Second
  - (b) F.P.S = Foot-Pound-Second
  - (c) M.K.S = Meter-Kilogram-Second
  - (d) M.K.S.A = Meter-Kilogram-Second-Ampere

Quantity	Unit	Symbol of Unit
Length	metre	m
Mass	kilogram	kg
Time	second	s
Temperature	kelvin	k
Electric current	ampere	A
Number of particles	mole	mol
Luminous intendity	candela	cd
Plane angle	radian	rad
Solid angle	steradian	Sr

Pressure difference between hydrostatic and atmospheric pressure is known as gauge pressure.

Numerical Profile

Plane Angle:

Angle substended by an arc at the centre of a circle

$$d\theta = \frac{ds}{r} = \frac{\text{arc}}{\text{radius}}$$

(i) Conversion of degree into radian:

$$1° = \frac{\pi}{180}\text{ rad}$$

(ii) Conversion of degree into minutes or seconds:

$$1° = \frac{\pi}{180}\text{ rad} = (60)' = (3600)''$$

where,' = minute, " = second

Solid Angle:

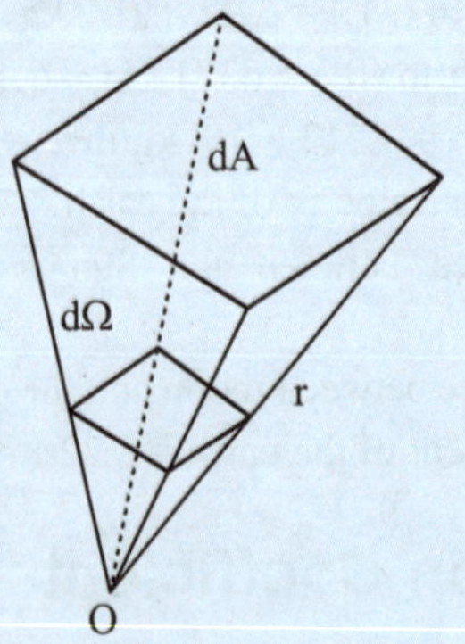

Solid angle, $d\Omega = \frac{dA}{r^2}$

Where, dA = intercepted area of spherical surface

r = radius of the sphere

dΩ = solid angle

Applications of Dimensions

Formulas

1. To check the correctness of an equation
   Dimensions on LHS = Dimensions on RHS
2. Conversion of one system of units into another system, we use

$$n_2 = n_1\left(\frac{M_1}{M_2}\right)^a\left(\frac{L_1}{L_2}\right)^b\left(\frac{T_1}{T_2}\right)^c$$

[n_2 and n_1 is the numerical values in the two systems]

a, b, c is constant

[where M, L and T is mass, length and time]

n_1 = Numerical value in system-1

n_2 = Numerical value in system-2

M_1 = Unit of mass in system-1

M_2 = Unit of mass in system-2

L_1 = Unit of length in system-1

L_2 = Unit of length in system-1

T_1 = Unit of time in system-1

T_2 = Unit of time in system-1

3. To find the relation among diffrent quantities,
   $[P] \propto [Q]^a [R]^b [S]^c$

Unit and Dimensions of Some Physical Quantities

Quantity	SI Unit	Dimension
Density	kg/m^3	M/L^3
Force	newton (N)	ML/T^2
Work	joule (J)(= N m)	ML^2/T^2
Energy	joule(J)	ML^2/T^2
Power	Watt (W) (= J/s)	ML^2/T^3
Momentum	kg m/s	ML/T
Gravitational constant	$N\ m^2/kg^2$	L^3/MT^2
Angular velocity	radian/s	T^{-1}
Angular acceleration	radian/s^2	T^{-2}
Angular momentum	kg m^2/s	ML^2/T
Moment of inertia	kg m^2	ML^2
Torque	N m	ML^2/T^2
Angular frequency	radian/s	T^{-1}
Frequency	hertz (Hz)	T^{-1}
Period	s	T
Surface Tension	N/m	M/T^2
Coefficient of viscosity	N s/m^2	M/LT

Quantity	SI Unit	Dimension
Wavelength	m	L
Intensity of wave	W/m^2	M/T^3
Specific heat capacity	J/kg K	L^2/T^2K
Stefan's constant	$W/m^2\ K^4$	M/T^3K^4
Heat	J	ML^2/T^2
Thermal conductivity	W/m-K	ML/T^3K
Current density	A/m^2	I/L^2
Electrical conductivity	$1/\Omega$ m(= mho/m)	I^2T^3/ML^3
Electric dipole moment	C m	LIT
Electric field	V/m (=N/C)	ML/IT^3
Potential (voltage)	Volt (V) (=J/C)	ML^2/IT^3
Electric flux	V m	ML^3/IT^3
Capacitance	Farad (F)	I^2T^4/ML^2
Electromotive force	Volt (V)	ML^2/IT^3
Resistance	ohm (Ω)	ML^2/I^2T^3
Permittivity of space	$C^2/N\ m^2$ (=F/m)	I^2T^4/ML^3
Permeability of space	N/A^2	ML/I^2T^2
Magnetic field	Tesla (T) (= Wb/m^2)	M/IT^2
Magnetic dipole moment	A m^2	IL^2
Inductance	Henry (H)	ML^2/I^2T^2

Errors & Measurement, Significant Figures

Formulas

Different types of Errors

1. **Absolute error**

   Absolute error = |True value - measured value|

   $\Delta a_i = |a_{mean} - a_i|$

   The absolute error is always taken as positive.

2. **Mean absolute error** - The arithmetic mean of all the absolute errors is known as the mean absolute error of the physical quantity.

$$\Delta a_{mean} = \frac{|\Delta a_1| + |\Delta a_2| + --- |\Delta a_n|}{n}$$

$$\therefore \Delta a_{mean} = \frac{1}{n}\sum_{i=1}^{\infty}|\Delta a_i|$$

   The mean absolute error is always positive.

3. **Relative error** $= \dfrac{\text{mean absolute error}}{\text{mean value}} = \dfrac{\Delta a_{mean}}{a_{mean}}$

4. **Percentage error** $= \dfrac{\Delta a_{mean}}{a_{mean}} \times 100\%$

Numerical Profile

Propagation of errors under different cases

(i) Addition of the quantities

Let, $Z = A + B$

$\Rightarrow Z \pm \Delta Z = (A \pm \Delta A) + (B \pm \Delta B)$

Max. Possible error in Z is given by

$\Delta Z = \Delta A + \Delta B$

Relative error in Z $= \dfrac{\Delta Z}{Z} = \dfrac{\Delta A + \Delta B}{A + B}$

Percentage error in Z $= \dfrac{\Delta A + \Delta B}{A + B} \times 100$

(ii) Subtraction of the quantities

Let $Z = A - B$

$Z \pm \Delta Z = (A \pm \Delta A) - (B \pm \Delta B)$

Maximum Possible error in Z is given by

$\Delta Z = \Delta A + \Delta B$

Relative error in Z $= \dfrac{\Delta A + \Delta B}{A - B}$

Percentage error in Z $= \dfrac{\Delta A + \Delta B}{A - B} \times 100$

(iii) Multiplication of the quantities

Let, $Z = AB$

Maximum possible relative error in Z is given by

$$\frac{\Delta Z}{Z} = \frac{\Delta A}{A} + \frac{\Delta B}{B}$$

Percentage error in Z is given by

$$\frac{\Delta Z}{Z} \times 100 = \left(\frac{\Delta A}{A} \times 100\right) + \left(\frac{\Delta B}{B} \times 100\right)$$

(iv) Division of the quantities:

Let, $Z = \dfrac{A}{B}$

Maximum possible relative error in Z is given by

$$\frac{\Delta Z}{Z} = \frac{\Delta A}{A} + \frac{\Delta B}{B}$$

Percentage error in division is

$$\frac{\Delta Z}{Z} \times 100 = \left(\frac{\Delta A}{A} \times 100\right) + \left(\frac{\Delta B}{B} \times 100\right)$$

(v) If $Z = \dfrac{A^p B^q}{C^r}$

then maximum fractional error in Z is

$$\frac{\Delta Z}{Z} = p\frac{\Delta A}{A} + q\frac{\Delta B}{B} + r\frac{\Delta C}{C}$$

Error in Z is given by

$$\frac{\Delta Z}{Z} \times 100 = p\frac{\Delta A}{A} \times 100 + q\frac{\Delta B}{B} \times 100 + r\frac{\Delta C}{C} \times 100$$

Significant Figures

(i) All the non-zero digits in a given number are significant irrespective of the location of the decimal point if any.
E.g.: 194.53, 1945.2 or 194.52 all have five of significant figures.

(ii) All zeros occurring between two non zero digits are significant. E.g.: 107008, 107.008 or 1.07008 has six significant figures.

(iii) Zeros to the left of a non zero digit and right of decimal are not significant. 0.00045 has two significant figures.

(iv) For a number ending in zeros which are not to the right of a decimal, the zeros are not significant e.g., 12000 has two significant figures.

Rule for Rounding off the Number

I Rule: If the digit which is to be rounded off is more than 5, then the preceding digit is increased by one.

e.g., $3.87 \approx 6.9$

II Rule: If the digit which is to be rounded off is less than 5, then the preceding digit is left unchanged.

e.g., $3.94 \approx 3.9$

Note: If the immediate insignificant digit to be dropped is 5 then there will be two different cases

(a) If the preceding digit is even, it is left unchanged and 5 is dropped.
Ex: 4.7252 is to be rounded off to two decimal places. The digit to be dropped here is 5 (along with 3) and the preceding digit 2 is even and hence to be retained as two only
$4.7252 = 4.72$

(b) If the preceding digit is odd, it is raised by 1
Ex: 4.7153 is to be rounded off to two decimal places. As the preceding digit 1 is odd, it is to be raised by 1 as 2.
$4.7153 = 4.72$

Measuring Instruments

Vernier Calliper

It is a device used to measure accurately upto 0.1 mm. There are two scales in the vernier calliper, vernier scale and main scale. The main scale is fixed whereas the vernier scale is movable along the main scale.

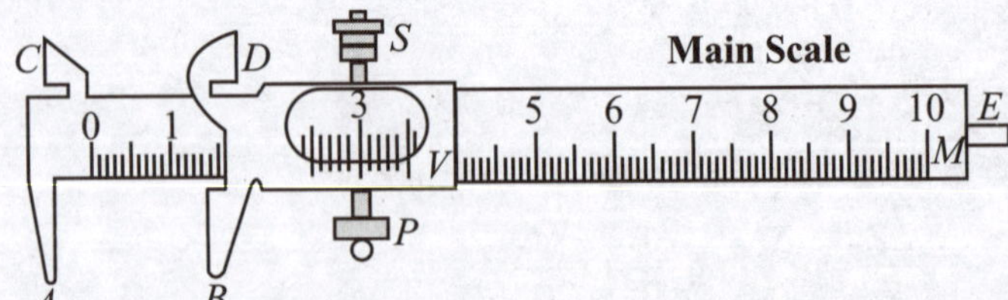

Screw Gauge

Screw gauge is a mechanical tool that allows precise measurement of the diameter, radius, or thickness of a thin wire or a thin metal sheet. It is also known as a micrometer screw gauge.

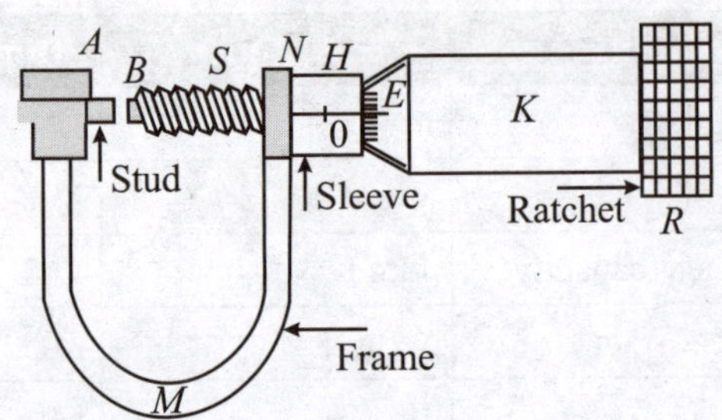

Least Count

Least count of Varnier Calipers:

L.C. = 1 MSD – 1 VSD

or

$$\text{L.C.} = \left(\frac{b-a}{b}\right)M$$

a = No. of division on Main scale.
b = No. of division on Vernier scale.
1 MSD = 1 Main Scale division.
1 VSD = 1 Vernier scale division

Least count of Screw Gauge:

$$\text{Least Count} = \frac{\text{pitch of the screw}}{\text{no. of rotations on the circular scale}}$$

$$\text{Pitch} = \frac{\text{Distance moved by the screw on the linear scale}}{\text{Number of full rotations given}}$$

Statements Asked in Exam

- Length, Mass and Time are chosen as base quantities.
- Mere dimensional correctness of an equation does not ensure its physical correctness. For eg, torque $= \frac{1}{2}mv^2$, is dimensionally correct, but not Physically correct.
- All constants are not dimensionless. For e.g., universal gravitational constant, Plank's constant, etc., do have dimensions.
- 1 parsec = 3.08×10^{16} m
(Parsec is the distance at which average radius of earth's orbit subtends of an angle of 1 arc second)
- Parallax method can be used for measuring distances of stars which are less than 100 light years away.
- The cylinder used in defining kg is made of platinum iridium alloy.
- Systematic error can be minimised.
- In random error, we take large number of observation so that random error reduces to $\frac{1}{n}$ times.
- The random errors are those errors, which occur irregularly and hence are random in to sign and size.
- Strain, angle, solid angle, trignometrical ratios, refractive index, relative density, relative permeability and relative permitivity have no dimensions.

MCQs

NCERT Topic-wise MCQs

Units, System of Units

1. Which of the following is not the unit of time?
   1) Solar day 2) Parallactic second
   3) Leap year 4) Lunar month
2. A unit less quantity
   1) Never has a non zero dimension
   2) Always has a non zero dimension
   3) May have a non zero dimension
   4) Does not exit
3. Which of the following is not the name of a physical quantity?
   1) Kilogram 2) Impulse
   3) Energy 4) Density
4. Parsec is a unit of
   1) Time 2) Angle
   3) Distance 4) Velocity
5. Which of the following system of units is not based on the unit of mass, length and time alone
   1) FPS 2) SI
   3) CGS 4) MKS

Dimensions & Dimensional Formulas

6. The dimensions of magnetic moment are:
   1) L^2A^{-1} 2) L^2A^1
   3) LA^2 4) L^2A^{-3}
7. The velocity V of a particle is given in terms of time t as $V = at + \frac{b}{t+C}$.
   The dimensions of a, b, c are:
   1) L^2; MLT^{-2} 2) LT^2; LT; L
   3) LT^{-2}; L; T 4) L; LT; T^2
8. In a particular system, the unit of length, mass and time are chosen to be 10 cm, 10 g and 0.1 s respectively. The unit of force in this system will be equivalent to:
   1) 0.1 N 2) 1 N
   3) 10 N 4) 100 N
9. The time dependence of a physical quantity P is given by $P = P_0 \exp(-\alpha t^2)$, where α is a constant and t is time. The constant α:
   1) Is dimensionless 2) Has dimensions T^{-2}
   3) Has dimensions of P 4) Has dimensions T^2
10. A unitless quantity:
    1) May have a dimensions
    2) Always has a dimensions
    3) Never has a dimensions
    4) Does not exist
11. Which of the following sets cannot enter into the list of fundamental quantities in any system of units?
    1) Length, time and velocity
    2) Length, mass and velocity
    3) Mass, time and velocity
    4) Length, time and mass
12. A dimensionless quantity:
    1) May have a unit 2) Never has a unit
    3) Always has a unit 4) Doesn't exist
13. The position x of a particle at time t is given by-
    $$x = \frac{v_0}{a}\left(1 - e^{-at}\right)$$
    where v_0 is a constant and a > 0.
    The dimensions of v_0 and a are:
    1) $M^0 L T^{-1}$ and T^{-1} 2) $M^0 L T^0$ and T^{-1}
    3) $M^0 L T^{-1}$ and LT^{-2} 4) $M^0 L T^{-1}$ and T
14. The dimensions of intensity are:
    1) $M^1 L^0 T^{-3}$ 2) $M^2 L^1 T^{-2}$
    3) $M^1 L^2 T^{-2}$ 4) $M^2 L^0 T^{-3}$
15. Which one of the following are the dimensions of pressure?
    1) $\frac{ML}{T^2}$ 2) $\frac{M}{L^2T^2}$
    3) $\frac{M}{LT^2}$ 4) $\frac{M}{LT}$
16. The dimensions of RC, where R is resistance and C is capacitance are same as that of:
    1) Inverse of time 2) Time
    3) Square of time 4) Square root of time

17. Dimensions of gravitational constant are:

1) $M^{-1}L^3T^{-2}$ 2) $M^{-2}L^3T^{-1}$
3) $M^3L^{-1}T^{-2}$ 4) $M^{-1}L^2T^{-3}$

18. The dimensional formula of magnetic flux is:

1) $[M^1L^2T^{-2}A^{-1}]$ 2) $[M^1L^0T^{-2}A^{-2}]$
3) $[M^0L^{-2}T^{-2}A^{-2}]$ 4) $[M^1L^2T^{-1}A^3]$

19. The dimensions of Planck's constant and angular momentum are, respectively:

1) $M^1L^2T^{-1}$ and M^1LT^{-1} 2) $M^1L^2T^{-1}$ and $M^1L^2T^{-2}$
3) M^1LT^{-1} and $M^1L^2T^{-1}$ 4) $M^1L^2T^{-1}$ and $M^1L^2T^{-1}$

20. Dimensions of resistance are same as that of (h is Planck's constant and e is electronic charge):

1) $\frac{h}{e}$ 2) $\frac{h}{e^2}$
3) $\frac{h^2}{e}$ 4) $\frac{h^2}{e^2}$

21. The pair having the same dimensions is:

1) Angular momentum, work
2) Work, torque
3) Potential energy, linear momentum
4) Kinetic energy, velocity

22. $ML^3T^{-1}A^{-2}$ are the dimensions of:

1) Resistivity 2) Conductivity
3) Resistance 4) None of these

23. $[ML^2T^{-3}]$ are the dimensions of:

1) Work 2) Power
3) Force 4) Momentum

24. If R and L represent respectively resistance and self inductance, which of the following combinations has the dimensions of frequency?

1) $\frac{R}{L}$ 2) $\frac{L}{R}$
3) $\sqrt{\frac{R}{L}}$ 4) $\sqrt{\frac{L}{R}}$

25. The dimensions of interatomic force constant are:

1) MT^{-2} 2) MLT^{-1}
3) MLT^{-2} 4) $ML^{-1}T^{-1}$

26. The dimensions of emf in MKS is:

1) $ML^{-1}T^{-2}A^{-2}$ 2) $ML^2T^{-2}A^{-2}$
3) $MLT^{-2}A^{-1}$ 4) $ML^2T^{-3}A^{-1}$

27. The physical quantities not having same dimensions are:

1) Speed and $(\mu_0\varepsilon_0)^{-1/2}$
2) Torque and work
3) Momentum and Planck's constant
4) Stress and Young's modulus

28. The dimensions of physical quantity X in the equation force $= \frac{X}{\sqrt{\text{Density}}}$, is given by:

1) $M^1L^4T^{-2}$ 2) $M^2L^{-2}T^{-1}$
3) $M^{3/2}L^{-1/2}T^{-2}$ 4) $M^1L^{-2}T^{-1}$

29. The dimensions of $\frac{1}{\sqrt{\varepsilon_0\mu_0}}$ is that of:

1) Velocity 2) Time
3) Capacitance 4) Distance

30. A quantity X is given as, $X = \frac{\varepsilon_0 LV}{t}$; ε_0 is the permittivity of free space, L is length, V is potential difference and t is time. The dimensions of X are same as that of:

1) Resistance 2) Charge
3) Voltage 4) Current

31. Given that V is speed, r is the radius and g is the acceleration due to gravity, which of the following is dimensionless?

1) V^2/rg 2) V^2r/g
3) V^2g/r 4) V^2rg

32. The dimensions of electric potential are the same as that of:

1) Work 2) Electric field per unit charge
3) Work per unit charge 4) Force per unit charge

33. The dimensional formula for latent heat is:

1) $[M^0L^2T^{-2}]$ 2) $[MLT^{-2}]$
3) $[ML^2T^{-2}]$ 4) $[ML^2T^{-1}]$

34. If $V = \sqrt{\frac{\gamma P}{\rho}}$, then dimensions of γ are (V = speed, P = pressure, ρ = density):

1) $[M^0L^0T^0]$ 2) $[M^0L^0T^{-1}]$
3) $[M^1L^0T^0]$ 4) $[M^0L^1T^0]$

35. In the relation:

$y = a \sin(\omega t - kx)$,

the dimensional formula for k is (symbols have usual meaning):

1) $[M^0LT]$ 2) $[M^0L^{-1}T^0]$
3) $[M^0LT^{-1}]$ 4) $[M^0L^{-1}T^{-1}]$

36. Displacement, y of a particle on a point of a rope placed along x-axis, at time t, is given as relation $y = a\cos(\omega t + kx)$. The dimensional formula for kx is same as that of:

1) a/ω 2) a/y
3) $\omega t/a$ 4) $ya/\omega t$

37. Suppose refractive index μ is given as:

$$\mu = A + \frac{\beta}{\lambda^2}$$

where A and B are constants and λ is the wavelength, then dimensions of B are same as that of:

1) Wavelength 2) Volume
3) Pressure 4) Area

38. The modulus of elasticity is dimensionally equivalent to:

1) Strain 2) Force
3) Stress 4) Coefficient of viscosity

39. In the relation: $\frac{dy}{dt} = 2\omega\sin(\omega t + \phi_0)$, the dimensional formula for $(\omega t + \phi_0)$ is:

1) [MLT] 2) $[MLT^0]$
3) $[ML^0T^0]$ 4) $[M^0L^0T^0]$

40. Dimensions of Stefan's constant are:

1) $[M L^2 T^{-2}]$ 2) $[ML^2 T^{-2} K^{-4}]$
3) $[M T^{-3} K^{-4}]$ 4) $[M L^0 T^{-2}]$

41. With usual notation, amongst the following, the one which does not represent the dimensions of time is:

1) $\left[\frac{L}{R}\right]$ 2) [RC]
3) $\left[\sqrt{LC}\right]$ 4) $\left[\frac{1}{\sqrt{LC}}\right]$

42. The dimensions of angular momentum are:

1) $[ML^2 T^{-2}]$ 2) $[ML^{-2} T^{-1}]$
3) $[MLT^{-1}]$ 4) $[ML^2 T^{-1}]$

43. The dimensions of K in equation $W = \frac{1}{2}Kx^2$ is (W = work done, x = extension in spring):

1) $[M^1 L^0 T^{-2}]$ 2) $[M^0 L^1 T^{-1}]$
3) $[M^1 L^1 T^{-2}]$ 4) $[M^1 L^0 T^{-1}]$

44. The velocity of a particle (V) at a instant (t) is given by $V = at + bt^2$. The dimensions of b are:

1) L 2) LT^{-1}
3) LT^{-2} 4) LT^{-3}

45. If h = Planck's constant, then $\frac{h}{2\pi}$ has the dimension of:

1) Velocity 2) Momentum
3) Energy 4) Angular momentum

46. Surface tension has the same dimensions as that of:

1) Coefficient of viscosity 2) Impulse
3) Momentum 4) Spring constant

47. The equation of a wave is given by:

$$y = A\sin\omega\left(\frac{x}{v} - k\right)$$

where ω is the angular velocity and v is the linear velocity. The dimensions of k are:

1) LT 2) T
3) T^{-1} 4) T^2

Applications of Dimensions

48. If units of length, mass and force are chosen as fundamental units, the dimensions of time would be:

1) $M^{1/2} L^{-1/2} F^{1/2}$ 2) $M^{1/2} L^{1/2} F^{1/2}$
3) $M^{1/2} L^{1/2} F^{-1/2}$ 4) $M^1 L^{-1/2} F^{-1/2}$

49. If speed of light (c), acceleration due to gravity (g) and pressure (P) are taken as fundamental quantities, the dimensions of gravitational constant (G) are:

1) $C^0 g P^{-3}$ 2) $C^2 g^3 P^{-2}$
3) $C^0 g^2 P^{-1}$ 4) $C^2 g^2 P^{-2}$

50. If energy (E), velocity (V) and force (F) be taken as fundamental quantities, then what are the dimensions of mass?

1) EV^2 2) EV^{-2}
3) FV^{-1} 4) FV^{-2}

51. If area (A), velocity (V) and density (ρ) are taken as fundamental quantities, what is the dimensional formula for force?

1) $AV^2\rho$ 2) $A^2V\rho$
3) $AV\rho^2$ 4) $AV\rho$

52. A student writes the escape velocity as:

$$V_e = \sqrt{\frac{GM}{R^2}}$$

The equation is:

1) Dimensionally incorrect 2) Dimensionally correct
3) Numerically correct 4) Both (1) and (3)

53. If force (F), length (l), current (I) and time (T) are taken as base quantities then the dimensions of ε_0 are:

1) $[FL^2 A^2 T^{-2}]$ 2) $[F^{-1} L^2 A^2 T^2]$
3) $[F^{-1}L^{-2}T^2A^2]$ 4) $[F^2L^2T^2A^2]$

54. E, M, J and G denote energy, mass, angular momentum and gravitational constant respectively. Then the dimensions of EJ^2/M^5G^2 are:

1) Angle 2) Length
3) Mass 4) Time

55. Planck's constant (h), speed of light in vacuum (c) and unitersal gravitational constant (G) are three fundamental constants. Which of the following combinations has the dimensions of length?

1) $\frac{\sqrt{hc}}{G}$ 2) $\frac{\sqrt{G}}{h^3}$
3) $\frac{\sqrt{hG}}{C^5}$ 4) $\frac{\sqrt{hG}}{C^5}$

56. A book with many printing errors contains four different formulas for the displacement y of a particle undergoing a certain periodic motion.

I. $y = a\sin\frac{2\pi t}{T}$

II. $y = a \sin vt$

III. $y = \left(\frac{a}{T}\right)\sin(t/a)$

IV. $y = \left(\frac{a}{\sqrt{2}}\right)\left(\sin\frac{2\pi t}{T} + \cos\frac{2\pi t}{T}\right)$

(where, a = maximum displacement of the particle, v = speed of the particle, T = time period of motion). Which is the wrong formula on dimensional grounds?

1) I, II and III 2) III and II
3) I and IV 4) II and I

57. The SI unit of energy is kg $m^2 s^{-2}$, that of speed v is ms^{-1} and of acceleration a is ms^{-2}. Which of the formula for kinetic energy (K) given below can you rule out on the basic of dimensional arguments (m stands for the mass of the body)?

I. $K = m^2v^2$ II. $K = (1/2)\ mv^2$
III. $K = ma$ IV. $K = (3/16)\ mv^2$
V. $K = \left(\frac{1}{2}\right)mv^2 + ma$

1) I and II 2) Only II
3) II and IV 4) I, III and V

Errors & Measurement, Significant Figures

58. The position of the particle moving along Y-axis is given as $y = At^2 - Bt^3$, where y is measured in metre and t in second. Then, the dimensions of B are:

1) $[LT^{-2}]$ 2) $[LT^{-1}]$
3) $[LT^{-3}]$ 4) $[MLT^{-2}]$

59. In a vernier calipers, one main scale division is x cm and n divisions of the vernier scale coincide with (n – 1) divisions of the main scale. The least count (in cm) of the calipers is:

1) $\left(\frac{n-1}{n}\right)x$ 2) $\left(\frac{nx}{n-1}\right)$
3) $\frac{x}{n}$ 4) $\left(\frac{x}{n-1}\right)$

60. If the length of rod A is 3.25 ± 0.01 cm and that of B is 4.19 ± 0.01 cm, then the rod B is longer than rod A by:

1) 0.94 ± 0.00 cm 2) 0.94 ± 0.01 cm
3) 0.94 ± 0.02 cm 4) 0.94 ± 0.005 cm

61. A physical quantity P is given by $P = \frac{A^3b^{1/2}}{C^{-4}D^{3/2}}$. The quantity which brings in the maximum percentage error in P is:

1) A 2) B
3) C 4) D

62. A physical quantity A is related to four observations a, b, c and d as $A = \frac{a^2b^3}{c\sqrt{d}}$. The percentage errors in measurements of a, b, c and d are 1%, 3%, 2% and 2% respectively. What is the percentage error in the quantity A?

1) 12% 2) 7%
3) 5% 4) 14%

63. Error in the measurement of radius of a sphere is 1%. The error in the calculated value of its volume is:

1) 1% 2) 3%
3) 5% 4) 7%

64. A wire has a mass 0.3 ± 0.003g , radius 0.5 ± 0.005 mm and length 6 ± 0.06 cm. The maximum percentage error in the measurement of density is:

1) 1 2) 2
3) 3 4) 4

65. A body travels uniformly a distance of (13.8 ± 0.2)m in a time (4.0 ± 0.3)sec. The velocity of the body within error limits is:

1) (3.45 ± 0.2)m/sec 2) (3.45 ± 0.3)m/sec
3) (3.45 ± 0.4)m/sec 4) (3.45 ± 0.5)m/sec

66. The values of two resistors are R_1 = (6 ± 0.3) KΩ and R_2 = (10 ± 0.2)kΩ. The percentage error in the equivalent resistance when they are connected in parallel is:

1) 5.125% 2) 2%
3) 3.125% 4) 10.125%

67. If $x = a^n$, then fractional error $\frac{\Delta x}{x}$ is equal to:

1) $\pm\left(\frac{\Delta a}{a}\right)^n$ 2) $\pm n\left(\frac{\Delta a}{a}\right)$
3) $\pm n \log_e \frac{\Delta a}{a}$ 4) $\pm n \log \frac{\Delta a}{a}$

Least Count

68. The main scale division of vernier calliper is equal to 1 mm. 19 division of main scale are equal in length to 20 division of vernier scale. In measuring diameter, the main scale read 35 division and 4th division of vernier scale coincide with main scale. find the least Count

1) 0.02 mm 2) 0.03 mm
3) 0.04 mm 4) 0.05 mm

69. 1 cm on the main scale of the vernier callipers is divided into ten equal parts. If 20 division of the vernier scale coicide with 18 small divisions of the main scale. determine the lease count of the callipers?

1) 0.01 cm 2) 0.02 cm
3) 0.03 cm 4) 0.04 cm

70. A spherometer has 100 equal divisions marked along the periphery of its disc, and one full rotation of the disc advances on the main scale by 0.01 cm. Fing the least count of the system

1) 10^{-2} cm 2) 10^{-3} cm
3) 10^{-4} cm 4) 10^{-5} cm

High Order Time Intensive MCQs

1. The relation $p=\frac{\alpha}{\beta}e^{\frac{-\alpha Z}{k\theta}}$ where p is pressure, Z is distance, k is Boltzmann constant and θ is temperature. The dimensional formula of β will be:
   1) $[M^0L^2T^0]$ 2) $[ML^2T]$
   3) $[ML^0T^{-1}]$ 4) $[M^0L^2T^{-1}]$

2. From the dimensional consideration, which of the following equations is correct?
   1) $T=2\pi\sqrt{\frac{R^3}{GM}}$ 2) $T=2\pi\sqrt{\frac{GM}{R^3}}$
   3) $T=2\pi\sqrt{\frac{GM}{R^2}}$ 4) $T=2\pi\sqrt{\frac{R^2}{GM}}$

3. A uniform wire of length L, diameter D and density ρ is stretched under a tension T. The correct relation between its fundamental frequency f, the length L and the diameter D is:
   1) $f\propto\frac{1}{LD}$ 2) $f\propto\frac{1}{L\sqrt{D}}$
   3) $f\propto\frac{1}{D^2}$ 4) $f\propto\frac{1}{LD^2}$

4. The speed of light c, gravitational constant G and Planck's constant h are taken as fundamental quantities in a system. The dimensions of time in this new system should be:
   1) $[G^{1/2}h^{1/2}c^{-5/2}]$ 2) $[G^{-1/2}h^{1/2}c^{1/2}]$
   3) $[G^{-1/2}h^{1/2}c^{-3/2}]$ 4) $[G^{-1/2}h^{1/2}c^{1/2}]$

5. If E = energy, G = gravitational constant, I = impulse and M = mass, then dimensions of $\frac{GIM^2}{E^2}$ are same as that of
   1) Time 2) Mass
   3) Length 4) Force

6. The dimensions of $\frac{e^2}{4\pi\varepsilon_0 hc}$ are (where e, ε_0, h and c are electric charge, electric permittivity, Planck's constant and velocity of light in vacuum respectively):
   1) $[M^0L^0T^0]$ 2) $[ML^0T^0]$
   3) $[M^0LT^0]$ 4) $[M^0L^0T]$

7. When a metal block is hanged from a spring balance, it reads 5.00 ± 0.05 N. While in water, it reads 4.00 ± 0.05 N. The relative density would be quoted as:
   1) 5.00 ± 0.05 2) 5.00 ± 11%
   3) 5.00 ± 0.10 4) 5.00 ± 6%

8. A gas bubble formed from an explosion under water oscillates with a period T proportional to $p^a d^b E^c$, where p is pressure, d is the density of water and E is the total energy of explosion. The values of a, b and c are
   1) a = 1, b = 1, c = 2 2) a = 1, b = 2, c = 1
   3) $a=\frac{5}{6}, b=\frac{1}{2}, c=\frac{1}{3}$ 4) $a=-\frac{5}{6}, b=\frac{1}{2}, c=\frac{1}{3}$

9. A physical quantity x is given by $x=\frac{2k^3l^2}{m\sqrt{n}}$. The percentage error in the measurements of k, l, m and n are 1%, 2%, 3% and 4% respectively. The value of x is uncertain by:
   1) 8% 2) 10%
   3) 12% 4) None of these

10. A normal human eye can see an object making an angle of 1.8° at the eye. What is the minimum height of object which can be seen by an eye placed at a distance of 1 m from the object?

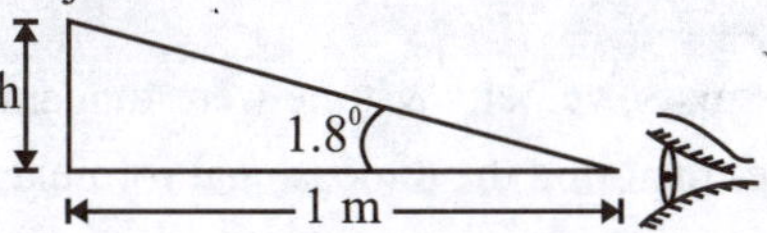

   1) π cm 2) 2π cm
   3) 4π cm 4) 3π cm

11. Consider the equation $\frac{d}{dt}\left(\int\vec{F}\cdot d\vec{S}\right)=A\left(\vec{F}\cdot\vec{p}\right)$ where $\vec{F}$ = force, $\vec{s}$ = displacement, t = time and $\vec{p}$ = momentum. The dimensional formula of A will be :
   1) $M^0L^0T^0$ 2) ML^0T^0
   3) $M^{-1}L^0T^0$ 4) $M^0L^0T^{-1}$

12. A uniform wire of length L and mass M is stretched between two fixed points, keeping a tension F. A sound of frequency μ is impressed on it. Then the maximum vibrational energy is existing in the wire when μ=
   1) $\frac{1}{2}\sqrt{\frac{ML}{F}}$ 2) $\sqrt{\frac{FL}{M}}$
   3) $2\sqrt{\frac{FM}{L}}$ 4) $\frac{1}{2}\sqrt{\frac{F}{ML}}$

13. If P, Q, R are physical quantities, having different dimensions, which of the following combinations can never be a meaningful quantity?
   1) $\frac{(P-Q)}{R}$ 2) $PQ-R$
   3) $\frac{PQ}{R}$ 4) $\frac{(PR-Q^2)}{R}$

14. In an experiment with Vernier callipers of least count 0.1 mm, when two jaws are joined together the zero of Vernier scale lies right to the zero of the main scale and 6th division of Vernier scale coincides with the main scale division. While measuring the diameter of a spherical bob, the zero of vernier scale lies in between 3.2 cm and 3.3 cm marks, and 4th division of vernier scale coincides with the main scale division. The diameter of bob is measured as:
   1) 3.18 cm 2) 3.25 cm
   3) 3.26 cm 4) 3.22 cm

15. The vernier scale used for measurement has a positive zero error of 0.2 mm. If while taking a measurement it was noted that zero on the vernier scale lies between 8.5 cm and 8.6 cm. and vernier coincidence is 6, then the correct value of measurement is ______ (least count = 0.01 cm)
1) 8.58 cm
2) 8.54 cm
3) 8.56 cm
4) 8.36 cm

16. If the units of length, velocity and force are halved, then the units of power will be:
1) Doubled
2) Halved
3) $\frac{1}{4}$th
4) Unaffected

17. Suppose mass, velocity & time were fundamental physical quantities then find the dimensional formula of pressure.
1) $[M^1V^{-1}T^{-2}]$
2) $[M^1V^{-1}T^{-3}]$
3) $[M^1V^{-2}T^{-3}]$
4) $[M^2V^{-2}T^{-3}]$

18. Force F and density d are related as $F = \frac{\alpha}{\beta + \sqrt{d}}$ then find the dimensions of α:
1) $[M^{1/2}L^{-1/2}T^{-2}]$
2) $[M^{3/2}L^{1/2}T^{2}]$
3) $[M^{3/2}L^{-1/2}T^{-2}]$
4) $[M^{2}L^{-1/2}T^{2}]$

19. Frequency is the function of density (ρ), length (a) and surface tension (T). Frequency is equal to:
1) $\frac{k\rho^{1/2}a^{3/2}}{\sqrt{T}}$
2) $\frac{k\rho^{3/2}a^{3/2}}{\sqrt{T}}$
3) $\frac{k\rho^{1/2}a^{3/2}}{T^{3/4}}$
4) None of these

20. According to Newton, the viscous force acting between liquid layers of area A and velocity gradient $\frac{\Delta V}{\Delta Z}$ is given by $F = -\eta A \frac{\Delta V}{\Delta Z}$, where η is a constant called as coefficient of viscosity. The dimensional formula of η is:
1) $[ML^{-2}T^{-2}]$
2) $[M^0 L^0 T^0]$
3) $[ML^2 T^2]$
4) $[ML^{-1} T^{-1}]$

21. A liquid of coefficient of viscosity η is flowing steadily through a capillary tube of radius r and length l. If V is volume of liquid flowing per second, the pressure difference P between the ends of tube is given by:
1) $P = \frac{8\pi l V}{\eta r^4}$
2) $P = \frac{8\eta r^4 l}{\pi V}$
3) $P = \frac{8\eta l V}{\pi r^4}$
4) $P = \frac{8\eta r^4 V}{\pi l}$

22. Given that $\int \frac{dx}{\sqrt{2ax - x^2}} = a^n \sin^{-1}\left(\frac{x-a}{a}\right)$
where a = constant. Using dimensional analysis, the value of n is-
1) 1
2) –1
3) 0
4) None of the above

23. There are 100 divisions on the circular scale of a screw gauge of pitch 1 mm. With no measuring quantity in between the jaws, the zero of the circular scale lies 5 divisions below the reference line. The diameter of a wire is then measured using this screw gauge. It is found the 4 linear scale divisions are clearly visible while 60 divisions on circular scale coincide with the reference line. The diameter of the wire is:
1) 4.65 mm
2) 4.55 mm
3) 4.60 mm
4) 3.35 mm

Statement Based MCQs

Directions: These questions consist of two statements each, printed as Statement-I and Statement-II. While answering these questions, you are required to choose any one of the following four responses.

1) Both Statement-I and Statement-II are correct.

2) Both Statement-I and Statement-II are incorrect.

3) Statement-I is correct & Statement-II is incorrect.

4) Statement-I is incorrect & Statement-II is correct.

1. **Statement-I:** The units of some physical quantities can be expressed as combination of the base units.
Statement-II: We need only a limited number of units for expressing the derived physical quantities.

2. **Statement-I:** The number 1.202 has four significant figures and the number 0.0024 has two significant figures.
Statement-II: All the non zero digits are significant.

3. **Statement-I:** All physical quantities can be added or subtracted.
Statement-II: The physical quantities should not have same dimensions for addition and subtraction.

4. **Statement-I:** Let us consider an equation $1/2\ mv^2 = mgh$ where, m is the mass of the body, v is velocity, g is the acceleration due to gravity and h is the height. This equation is dimensionally correct.
Statement-II: All dimensionally correct equations are correct.

5. **Statement-I:** The given equation $x = x_0 + u_0 t + \frac{1}{2}at^2$ is dimensionally correct, where x is the distance travelled by a particle in time t, x_0 is initial position, u_0 is initial velocity and a is uniform acceleration along the line of motion.
Statement-II: Dimensional analysis can be used for checking the dimensional consistency or homogeneity of the equation.

Assertion & Reason MCQs

Directions: These questions consist of two statements each, printed as Assertion and Reason. While answering these questions, you are required to choose any one of the following four responses.

1) **Both Assertion and Reason are True and the Reason is a correct explanation of the Assertion.**
2) **Both Assertion and Reason are True but Reason is not a correct explanation of the Assertion.**
3) **Assertion is True but the Reason is False.**
4) **Assertion is False but Reason is True.**

1. **Assertion:** Parallax method can be used for measuring distances of heavenly bodies more than 100 light years away.

   Reason: For a distance more than 100 light years, parallax angle is reduced to such an extent that it cannot be measured accurately.

2. **Assertion:** Light year is the distance that light travels with velocity of 3×10^8 m s^{-1} in one year.

   Reason: Light year is the unit for measuring time.

3. **Assertion:** Dimensions of Planck's constant is equal to the dimensions of angular momentum.

   Reason: Dimensions of work are ML^2T^{-2}.

4. **Assertion:** A dimensionally wrong or inconsistent equation must be wrong.

   Reason: A dimensionally consistent equation is a correct equation.

5. **Assertion:** When the units of measurement of a quantity are changed, its numerical value changes.

   Reason: Smaller the unit of measurement, smaller is its numerical value.

Match the Columns MCQs

1. If R is resistance, L is inductance, C is capacitance, H is latent heat and s is specific heat, then match the quantity given in column-I with the dimensions given in column-II

	Column-I		Column-II
(a)	LC	(i)	$[L^2T^{-2}]$
(b)	LR	(ii)	$[L^2T^{-2}K^{-1}]$
(c)	H	(iii)	$[T^{-2}]$
(d)	s	(iv)	$[M^2L^4T^{-5}A^{-4}]$

1) (a)-(iii) (b)-(i) (c)-(iv) (d)-(ii)
2) (a)-(iii) (b)-(iv) (c)-(i) (d)-(ii)
3) (a)-(i) (b)-(iii) (c)-(ii) (d)-(iv)
4) (a)-(iv) (b)-(ii) (c)-(iii) (d)-(i)

2. There are four vernier scales; whose specifications are given in column I and the least count is given in column II. Match the column-I and II with correct specification and corresponding least count (s = value of main scale division, n = number of marks on vernier)

	Column-I		Column-II
(a)	s = 1 mm, n = 10	(i)	0.05 mm
(b)	s = 0.5 mm, n = 10	(ii)	0.01 mm
(c)	s = 0.5 mm, n = 20	(iii)	0.1 mm
(d)	s = 1 mm, n = 100	(iv)	0.025 mm

1) (a)-(i) (b)-(iii) (c)-(ii) (d)-(iv)
2) (a)-(iv) (b)-(i) (c)-(iii) (d)-(ii)
3) (a)-(ii) (b)-(iv) (c)-(iii) (d)-(i)
4) (a)-(iii) (b)-(i) (c)-(iv) (d)-(ii)

NCERT Exemplar MCQs

1. The number of significant figures in 0.06900 is:

   1) 5 2) 4
   3) 2 4) 3

2. The sum of the numbers 436.32, 227.2 and 0.301 in appropriate significant figures is:

   1) 663.821
   2) 664
   3) 663.8
   4) 663.82

3. The mass and volume of a body are 4.237 g and 2.5 cm^3, respectively. The density of the material of the body in correct significant figures is:

   1) 1.6048 g cm^{-3} 2) 1.69 g cm^{-3}
   3) 1.7 g cm^{-3} 4) 1.695 g cm^{-3}

4. The numbers 2.745 and 2.735 on rounding off to 3 significant figures will give:

   1) 2.75 and 2.74 2) 2.74 and 2.73
   3) 2.75 and 2.73 4) 2.74 and 2.74

5. The length and breadth of a rectangular sheet are 16.2 cm and 10.1 cm, respectively. The area of the sheet in appropriate significant figures and error is:
   1) 164 ± 3 cm^2
   2) 163.62 ± 2.6 cm^2
   3) 163.6 ± 2.6 cm^2
   4) 163.62 ± 3 cm^2

6. Which of the following pairs of physical quantities does not have same dimensional formula?
   1) Work and torque
   2) Angular momentum and Planck's constant
   3) Tension and surface tension
   4) Impulse and linear momentum

7. You measure two quantities as A = 1.0 m ± 0.2 m, B = 2.0 m ± 0.2 m. We should report correct value for $\sqrt{AB}$ as:
   1) 1.4 m ± 0.4 m
   2) 1.41 m ± 0.15 m
   3) 1.4 m ± 0.3 m
   4) 1.4 m ± 0.2 m

8. Which of the following measurements is most precise?
   1) 5.00 mm 2) 5.00 cm
   3) 5.00 m 4) 5.00 km

9. The mean length of an object is 5 cm. Which of the following measurements is most accurate?
   1) 4.9 cm
   2) 4.805 cm
   3) 5.25 cm
   4) 5.4 cm

10. Young's modulus of steel is 1.9×10^{11} N m^2. When expressed in CGS units of dyne/cm^2, it will be equal to: (1 N = 10^5 dyne, 1 m^2 = 10^4 cm^2)
   1) 1.9×10^{10}
   2) 1.9×10^{11}
   3) 1.9×10^{12}
   4) 1.9×10^{13}

11. If momentum (p), area (A) and time (T) are taken to be fundamental quantities, then energy has the dimensional formula:
   1) $[pA^{-1}T^1]$
   2) $[p^2AT]$
   3) $[pA^{-1/2}T]$
   4) $[pA^{1/2}T^{-1}]$

12. Photon is quantum of radiation with energy E = hν, where ν is frequency and h is Planck's constant. The dimensions of h are the same as that of:
   1) Linear impulse
   2) Angular impulse
   3) Linear momentum
   4) Angular momentum

13. If Planck's constant (h) and speed of light in vacuum (c) are taken as two fundamental quantities, which one of the following can, in addition, be taken to express length, mass and time in terms of the three chosen fundamental quantities?
   1) Mass of electron (m_e)
   2) Universal gravitational constant (G)
   3) Charge of electron (e)
   4) Mass of proton (m_p)

14. Which of the following ratios express pressure?
   1) Force/Area
   2) Energy/Volume
   3) Energy/Area
   4) Force/Volume

15. Which of the following is not a unit of time?
   1) Second
   2) Parsec
   3) Year
   4) Light year

NEET & AIIMS Past Year MCQs

1. The quantities which have the same dimensions as those of solid angle are: **(2024)**
   1) strain and arc 2) angular speed and stress
   3) strain and angle 4) stress and angle

2. A force defined by $F = \alpha t^2 + \beta t$ acts on a particle at given time t. The factor which is dimensionless, if α and β are constants, is: **(2024)**
   1) $\alpha\beta t$ 2) $\alpha\beta/t$
   3) $\beta t/\alpha$ 4) $\alpha t/\beta$

3. The pitch of an error free screw gauge is 1 mm and there are 100 divisions on the circular scale. While measuring the diameter of a thick wire, the pitch scale reads 1 mm and 63rd division on the circular scale coincides with the reference line. The diameter of the wire is: **(2024 Re)**
   1) 1.63 cm
   2) 0.163 cm
   3) 0.163 m
   4) 1.63 m

4. In a vernier calipers, $(N + 1)$ divisions of vernier scale coincide with N divisions of main scale. If 1 MSD represents 0.1 mm, the vernier constant (in cm) is: **(2024)**
   1) 100 N 2) $10(N+1)$
   3) $\frac{1}{10N}$ 4) $\frac{1}{100(N+1)}$

5. The errors in the measurement which arise due to unpredictable fluctuations in temperature and voltage supply are: **(2023)**

1) Random errors
2) Instrumental errors
3) Personal errors
4) Least count errors

6. A metal wire has mass (0.4± 0.002) g, radius (0.3± 0.001) mm and length (5 ± 0.02) cm. The maximum possible percentage error in the measurement of density will nearly be: **(2023)**

1) 1.4%
2) 1.2%
3) 1.3%
4) 1.6%

7. Plane angle and solid angle have: **(2022)**

1) Both units and dimension
2) Units but no dimensions
3) Dimensions but no units
4) No units and no dimensions

8. The dimension $[MLT^{-2}A^{-2}]$ belong to the : **(2022)**

1) Electric permittivity
2) Magnetic flux
3) Self inductance
4) Magnetic permeability

9. The area of a rectangular field (in m^2) of length 55.3 m and breadth 25 m after rounding off the value for correct significant digits is : **(2022)**

1) 14×10^2
2) 138×10^1
3) 1382
4) 1382.5

10. If E and G respectively denote energy and gravitational constant, then E/G has the dimensions of: **(2021)**

1) $[M][L^{-1}][T^{-1}]$
2) $[M][L^0][T^0]$
3) $[M^2][L^{-2}][T^{-1}]$
4) $[M^2][L^{-1}][T^0]$

11. A screw gauge gives the following readings when used to measure the diameter of a wire **(2021)**

Main scale reading : 0 mm

Circular scale reading : 52 divisions

Given that 1 mm on main scale corresponds to 100 divisions on the circular scale. The diameter of the wire from the above data is:

1) 0.026 cm
2) 0.26 cm
3) 0.052 cm
4) 0.52 cm

12. If force [F], acceleration [A] and time [T] are chosen as the fundamental physical quantities. Find the dimensions of energy. **(2021)**

1) $[F] [A] [T^2]$
2) $[F] [A] [T^{-1}]$
3) $[F] [A^{-1}] [T]$
4) $[F] [A] [T]$

13. The density of a cube is measured by $T = 2\pi\sqrt{\ell/g}$ Given fractional errors in T and l are + x and + y respectively: **(AIIMS 2014)**

1) x + y
2) x − y
3) 2x + y
4) 2x − y

14. What is the dimension of magnetic field B in terms of C = coulomb, M, L, and T: **(AIIMS 2011)**

1) $M^1L^1T^{-2}C^1$
2) $M^1L^0T^{-1}C^1$
3) $M^1L^0T^{-2}C^1$
4) $M^1L^0T^{-1}C^{-1}$

15. If C and L denote the capacitance and inductance respectively, the units of LC are: **(AIIMS 2011)**

1) $[M^0L^0T^2]$
2) $[M^0L^2T^{-2}]$
3) $[MLT^{-2}]$
4) $[M^0L^0T^0]$

16. The dimensions of specific resistance are: **(AIIMS 2009)**

1) $[M L^2T^{-2}A^{-1}]$
2) $[M L^3T^{-3}A^{-2}]$
3) $[M L^3T^{-2}A^{-1}]$
4) $[M L^2T^{-2}A^{-2}]$

17. Which of the following physical quantities do not have same dimensions? **(AIIMS 2007)**

1) Pressure and stress
2) Tension and surface tension
3) Strain and angle
4) Energy and work.

18. Length cannot be measured by: **(AIIMS 2002)**

1) Fermi
2) Micron
3) Debye
4) Light year

Answer Key

NCERT Topic-wise MCQs

1	2	3	4	5	6	7	8	9	10	11	12	13	14	15	16	17
2	1	1	3	2	2	3	1	2	3	1	1	1	1	3	2	1

18	19	20	21	22	23	24	25	26	27	28	29	30	31	32	33	34
1	4	2	2	1	2	1	1	4	3	3	1	4	1	3	1	1

35	36	37	38	39	40	41	42	43	44	45	46	47	48	49	50	51
2	2	4	3	4	3	4	4	1	4	4	4	2	3	3	2	1

52	53	54	55	56	57	58	59	60	61	62	63	64	65	66	67	68
1	3	1	4	2	4	3	3	3	3	4	2	4	2	4	2	4

69	70
1	3

High Order Time Intensive MCQs

1	2	3	4	5	6	7	8	9	10	11	12	13	14	15	16	17
1	1	1	1	1	1	2	4	3	1	3	4	1	1	2	3	2

18	19	20	21	22	23
3	4	4	3	3	2

Statement Based MCQs

1	2	3	4	5
1	1	2	3	1

Assertion & Reason MCQs

1	2	3	4	5
4	3	2	3	3

Match the Column MCQs

1	2
2	4

NCERT Exemplar MCQs

1	2	3	4	5	6	7	8	9	10	11	12	13	14	15
2	3	3	4	1	3	4	1	1	3	4	2,4	1,2,4	1,2	2,4

NEET & AIIMS Past Year MCQs

1	2	3	4	5	6	7	8	9	10	11	12	13	14	15	16	17
3	4	2	4	1	4	2	4	1	4	3	1	3	4	1	2	2

18
3

CHAPTER

2 Motion in a Straight Line

Topicwise Number of Questions Analysis (2025-2016)

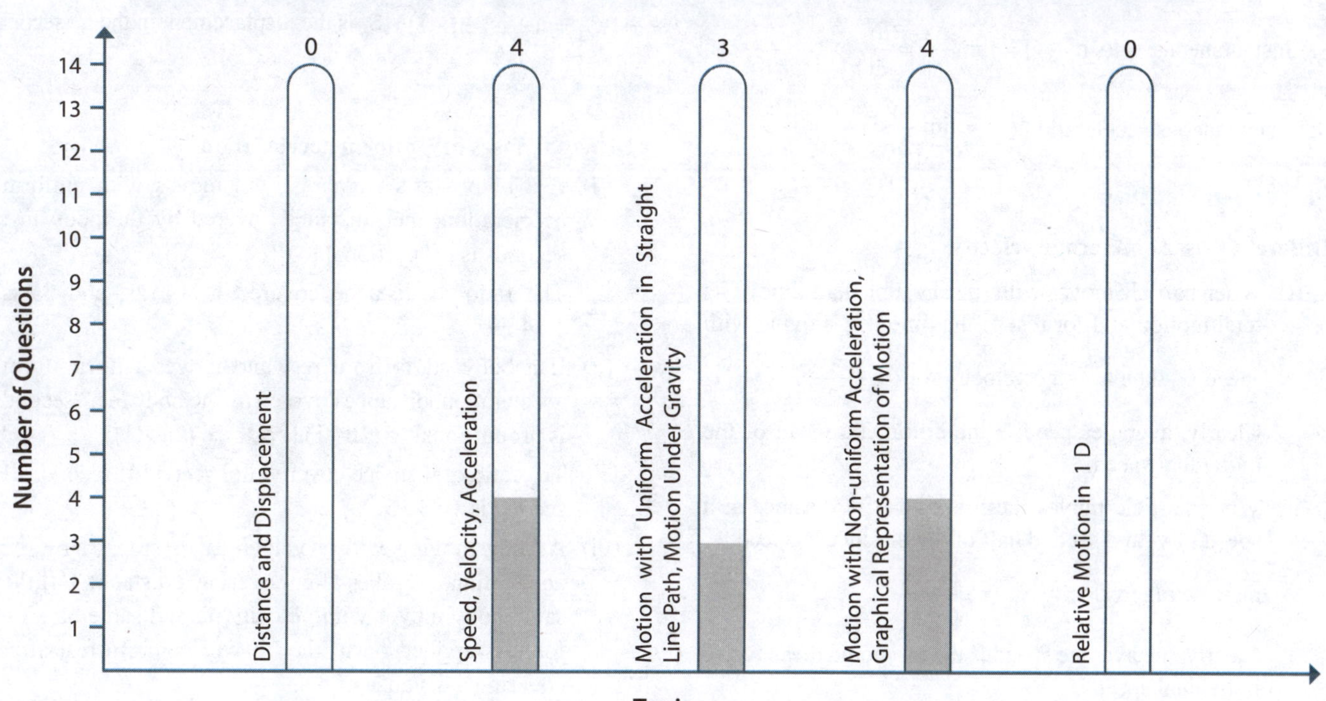

NCERT-Picks

Introduction

- When an object changes its location in relation to its surroundings throughout time, it is said to be in motion.
- Motion in straight line is also known as linear motion.

Position, Path Length and Displacement

Formulas

1. A position indicates the location of object with respect to arbitrary reference point.
2. Distance = Length of the path travelled by an object.
3. Displacemnet = $\vec{x}_f - \vec{x}_i$
   Displacement is the shortest distance between the initial point and final point.

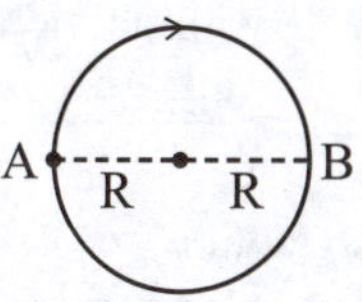

Let an object moves along the circular path from point A to point B,

Then, distance travelled = πR

and displacement = 2R

Speed, Velocity and Acceleration

Formulas

1. $\text{Speed} = \dfrac{\text{Distance}}{\text{Time}}$
2. $\text{Velocity} = \dfrac{\text{Displacement}}{\text{Time}}$

3. Acceleration $= \dfrac{\text{Change in Velocity}}{\text{Time taken}}$
4. Average speed $= \dfrac{\text{Total distance travelled}}{\text{Total time taken}}$
5. Average velocity $= \dfrac{\text{Total displacement}}{\text{Total time taken}}$
6. Average acceleration $= \dfrac{\text{Change in Velocity}}{\text{Change in time}}$
7. Instantaneous speed $= \lim_{\Delta t \to 0} \dfrac{\Delta x}{\Delta t} = \dfrac{dx}{dt}$
8. Instantaneous velocity (v) $= \lim_{\Delta t \to 0} \dfrac{\Delta x}{\Delta t} = \dfrac{dx}{dt}$
9. Instantaneous acceleration (a) $= \lim_{\Delta t \to 0} \dfrac{\Delta v}{\Delta t} = \dfrac{dv}{dt} = \dfrac{d^2x}{dt^2}$

Numerical Profile

Different Cases of Average velocity

(i) When particle moves with speed v_1 upto half time of its total motion and for rest of the time it is moving with speed v_2, then average velocity $v_{av} = \dfrac{v_1 + v_2}{2}$

Clearly, average speed is the arithmetic mean of the individual speed.

(ii) When particle moves the first half of a distance at a speed of v_1 and second half of the distance at speed v_2, then average velocity $v_{av} = \dfrac{2v_1v_2}{v_1 + v_2}$

Clearly, the average speed is the harmonic mean of the individual speeds.

(iii) When particle covers one third distance at speed v_1, next one third at speed v_2 and last one third at speed v_3, then average velocity $v_{av} = \dfrac{3v_1v_2v_3}{v_1v_2 + v_2v_3 + v_1v_3}$.

(iv) If a particle accelerated for a time t_1 with acceleration a_1 and for time t_2 with acceleration a_2 then average acceleration is $a_{av} = \dfrac{a_1t_1 + a_2t_2}{t_1 + t_2}$

(v) For non-uniform motion

(a) Instantaneous velocity v is given by

$$v = \frac{dx}{dt}$$

$$x_f - x_i = \int v\,dt$$

(x_f is the final position and x_i is the initial position)

(b) Instantaneous acceleration is given by

$$a = \frac{dv}{dt}$$

$$v_f - v_i = \int a\,dt$$

(v_f is the final velocity and v_i is the initial velocity)

Kinematic Equations for Uniformly Accelerated Motion

Formulas

Equations of motion

1. $v = u + at$ [where, u is initial velocity, v is final velocity and a is acceleration]
2. $s = ut + \dfrac{1}{2}at^2$ [s is the displacement of the object]
3. $v^2 = u^2 + 2as$
4. $S_n = u + \dfrac{a}{2}(2n - 1)$ [S_n is the displacement in the nth second]

Numerical Profile

Different cases of Uniform acceleration

(i) If a body starts from rest and moves with uniform acceleration, then distance covered by the body in t seconds is proportional to t^2(i.e. $s \propto t^2$).

The ratio of distance covered in 1s, 2s, and 3s is 1 : 4 : 9.

(ii) If a body starts from rest and moves with uniform acceleration, distance covered by the body in n^{th} second is proportional to (2n –1) [i.e., $s_n \propto (2n - 1)$].

The ratio of distance covered in I second II second, III second is 1 : 3 : 5.

(iii) A body moving with a velocity u is stopped by the application of brakes after covering a distance s. If the same body moves with velocity nu and same braking force is applied on it, then it will come to rest after covering a distance of n^2s.

As $v^2 = u^2 - 2as$

$\Rightarrow 0 = u^2 - 2as \Rightarrow s = u^2/2a \Rightarrow s \propto u^2$ [since a is constant] So we can say that if u becomes n times, then s becomes n^2 times that of previous value.

(iv) A particle moving with uniform acceleration from A to B along a straight line has velocities v_1 and v_2 at A & B respectively. If C is the mid-point between A & B then velocity of the particle at C is equal to-

$$v = \sqrt{\frac{v_1^2 + v_2^2}{2}}$$

Different cases of Motion under gravity

(i) A body is dropped from the top of a tower of height '*h*' and at the same time another body is projected vertically up from the foot of the tower with a velocity '*u*'.

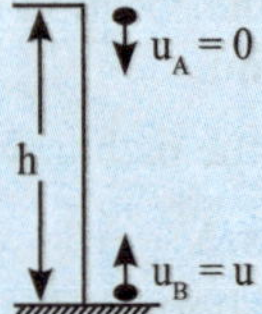

(a) The time after which they meet $t = \frac{h}{u}$

(b) The separation between the two bodies after 't' seconds is = (h – ut)

(c) The height at which they meet above the ground $= \left(h - \frac{gh^2}{2u^2}\right)$

(d) The time after which their velocities are equal in magnitudes is $t = \frac{u}{2g}$

(e) If they meet at mid point, velocity of the body thrown upwards is u = gt

(ii) A body projected vertically up crosses a point P at a height 'h' above the ground at time 't_1' seconds and thereafter at time 't_2' seconds the body returns to the same point while coming down.

(a) Then total time of flight $T = t_1 + t_2$

(b) Height of point P is $h = \frac{1}{2}gt_1t_2$

(c) Maximum height reached above the ground $H = \frac{1}{8}g(t_1 + t_2)^2$

(d) Speed of throw, $u = \frac{t_1 + t_2}{2}$

(iii) Let, a body is projected vertically upwards with a velocity 'u' from the top of a tower of height 'h'

(a) If time taken t to reach the ground is given, then height of the tower is

$$h = -ut + \frac{1}{2}gt^2$$

(b) If height h of the tower is given then time taken by the body to reach the ground

$$t = \frac{u + \sqrt{u^2 + 2gh}}{g}$$

(c) The velocity of the body at the foot of the tower

$$v = \sqrt{u^2 + 2gh}$$

(d) Velocity of the body after 't' sec. is $v = u - gt$

(iv) Let a body projected vertically down from the top of a tower with a velocity 'u' reaches the foot of the tower after a time 't_1' with a velocity 'v_1'. Another body projected vertically up from the top of tower with the same velocity reaches the foot of the tower after a time 't_2' with a velocity 'v_2'. A third body is dropped from the top of the tower reaches the foot of the tower after a time 't' with a velocity 'v', then

(a) $t = \sqrt{t_1 t_2}$

(b) $h = \frac{1}{2}gt_1t_2$ [h: Height of tower]

(c) $u = \frac{1}{2}g(t_1 - t_2)$

(d) $v_1 = v_2 = \sqrt{u^2 + 2gh}$

(v) An elevator is accelerating upwards with an acceleration a. If a person inside the elevator throws a particle vertically up with velocity u relative to the elevator, then the time of flight of the body is $t = \frac{2u}{g + a}$

In the above case if elevator accelerates down, time of flight is $t = \frac{2u}{g - a}$

Graphical Representation of Motion

Formulas

1. Velocity of a particle = Slope of displacement time graph

2. Let v_1 and v_2 be the velocities of two particles having displacement time graph with slopes θ_1 and θ_2 respectively then-

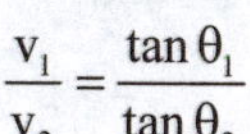

$$\frac{v_1}{v_2} = \frac{\tan\theta_1}{\tan\theta_2}$$

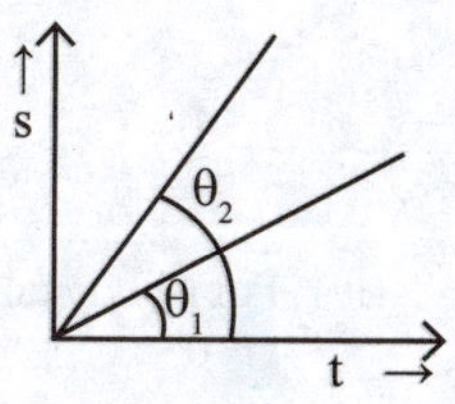

3. Area under velocity time graph = Displacement

4. Slope of velocity time graph = Acceleration

5. Speed of particle = Slope of distance time graph

6. Area under speed time graph = Distance

7. A straight line graph has a single slope. So if the displacement-time graph is a straight line, it represents a constant velocity. If the velocity-time graph is a straight line, it represents a constant acceleration.

8. Curved graph has multiple slopes. As the displacement-time graph bends upwards with the passage of time, the value of θ increases, slope (= tan θ) of the curve increases, consequently the velocity increases and hence the motion is accelerated.

Displacement →

Time →

9. As the displacement-time graph bends downwards with the passage of time, the value of θ decreases, the slope of the curve decreases, consequently the velocity decreases and hence the motion is decelerated.

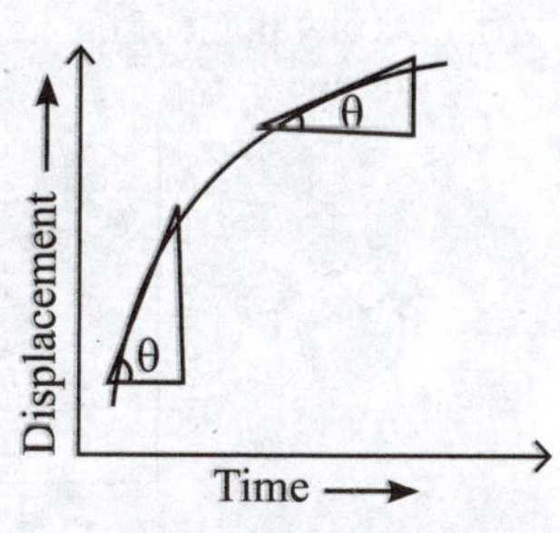

Graphs

Position vs time graph

(i) For a body at rest.

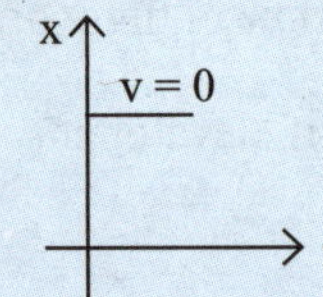

(ii) For a body in uniform motion.

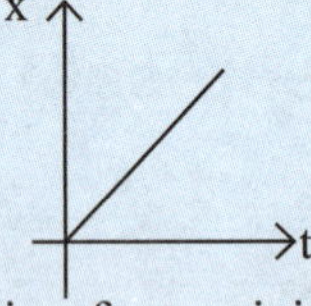

(iii) For a body starting from origin with zero speed and accelerating uniformly.

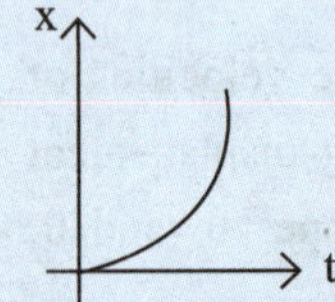

(iv) For a body uniformly accelerated with $u \neq 0$ and $x \neq 0$ at $t = 0$.

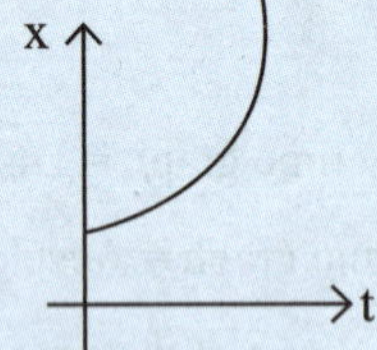

(v) For a uniformly retarded body.

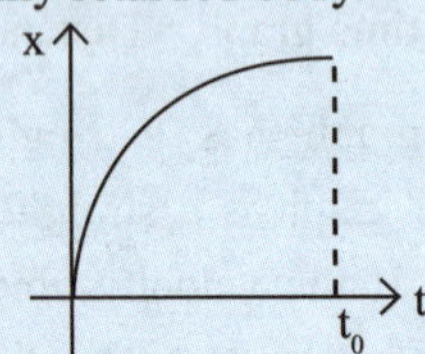

Distance vs time graph

(i) The distance covered by a body cannot decrease with the increase of time. So the distance-time graph shown below is not possible.

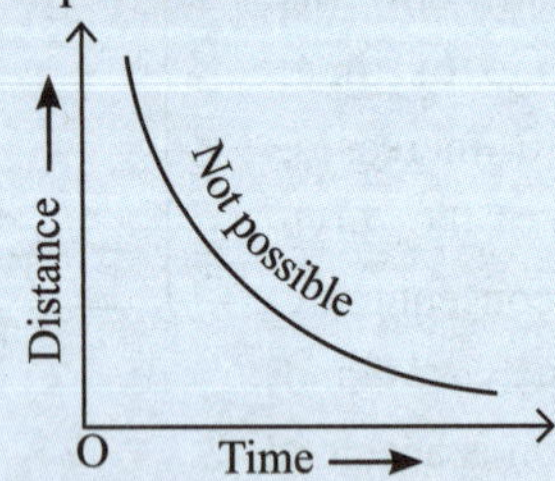

(ii) At any point of time a body cannot have two values of distance.

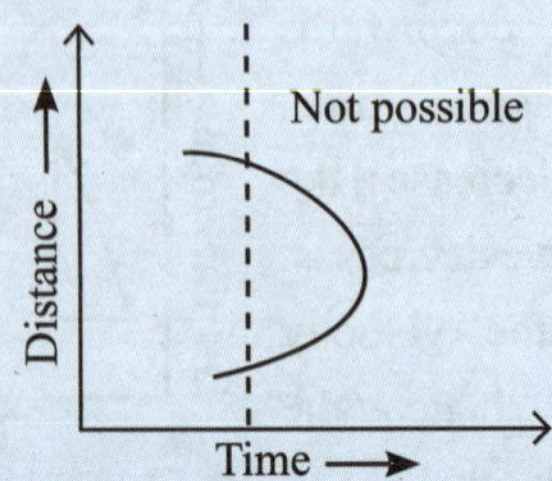

Velocity vs time graph

(i) For a body in uniform motion.

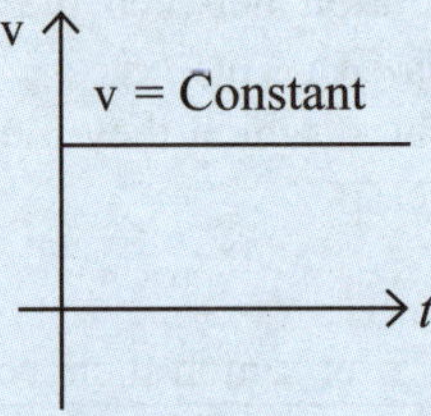

(ii) For a uniformly accelerated body with initial velocity $u = 0$.

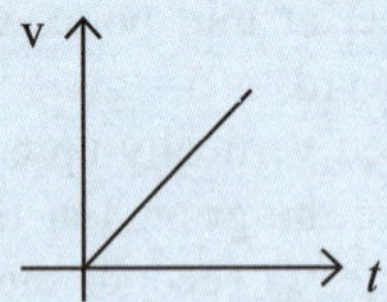

(iii) For a uniformly accelerated body with initial velocity $= u$.

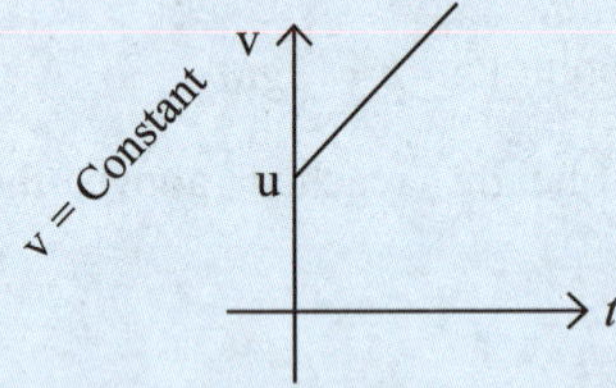

(iv) For a uniformly retarded body till its velocity becomes zero.

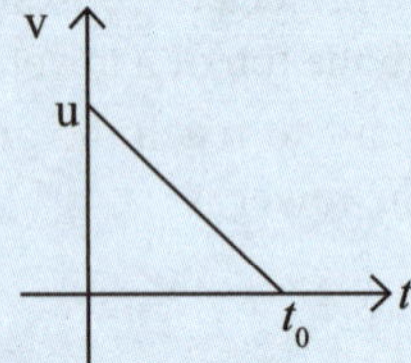

(v) For a body thrown upwards.

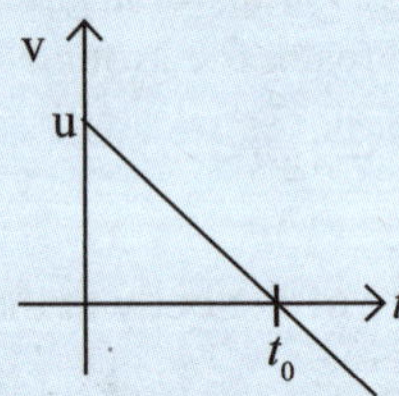

(vi) When the acceleration decreases and increases.

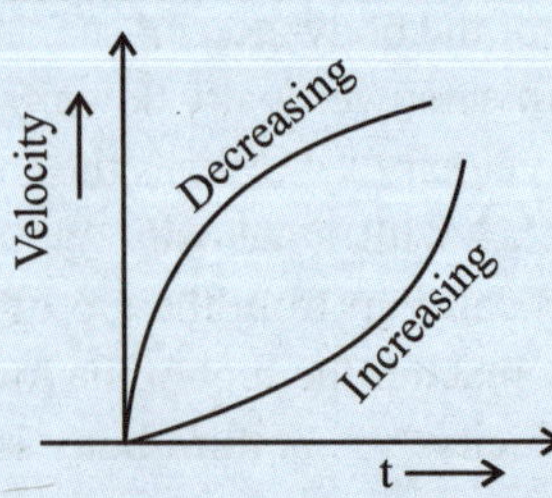

(vii) Area under v – t curve gives displacement.

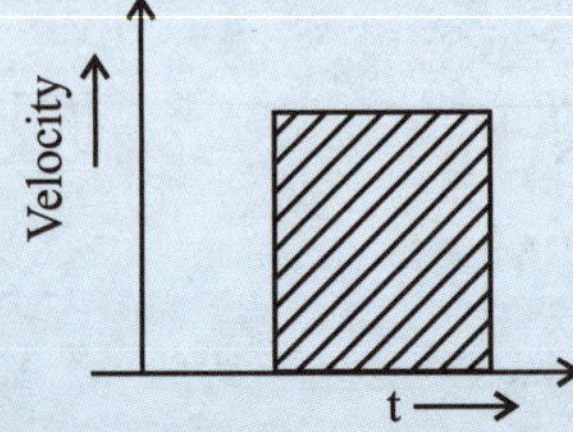

Acceleration vs time graph

(i) When a = 0

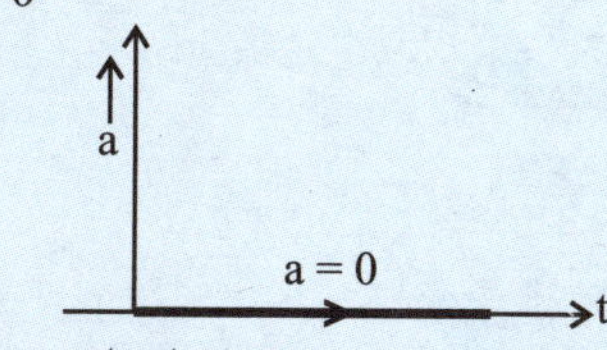

(ii) When a = constant

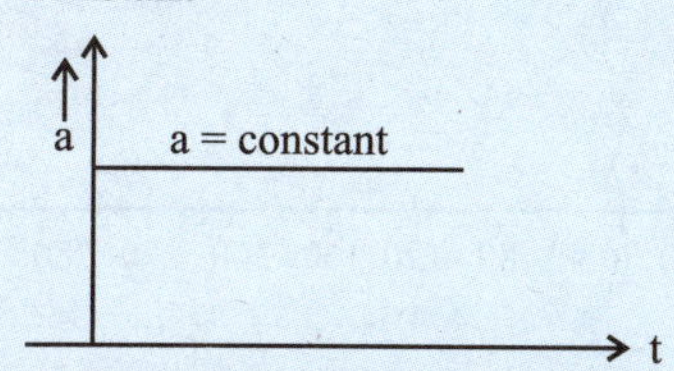

(iii) The change in velocity of the particle = area enclosed by the acceleration-time graph.

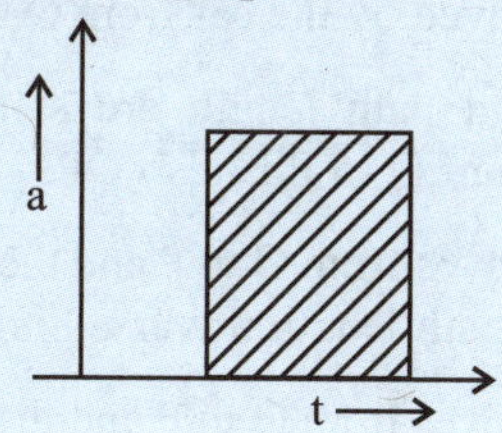

Relative Motion in 1-D

Formulas

1. 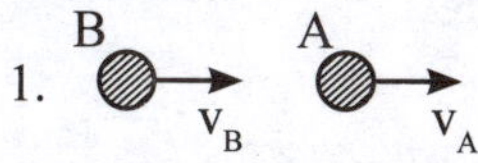

If two objects A & B move along parallel lines in the same direction, then the relative velocity of A w.r.t B is given by $v_{AB} = v_A - v_B$

and the relative velocity of B w.r.t A is given by $v_{BA} = v_B - v_A$.

2. If two objects A and B move along parallel lines in the opposite direction, then the relative velocity of A w.r.t B is given by $v_{AB} = v_A - (-v_B) = v_A + v_B$ and the relative velocity of B w.r.t A is given by $v_{BA} = -v_B - v_A$.

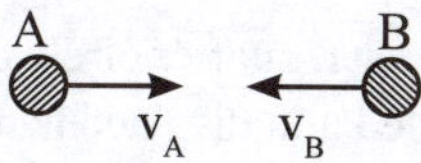

Numerical Profile

Position time graph of relative velocity under different cases

(i) Case-1: When the two objects move with same velocity in the same direction. That is $v_1 = v_2$ and relative velocity, $v_2 - v_1 = 0$

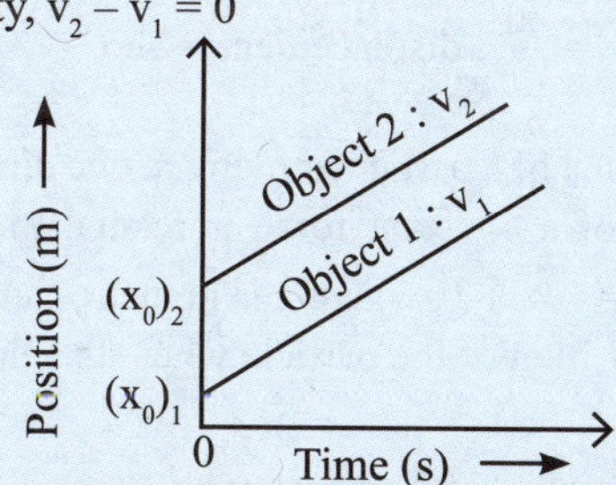

(ii) Case-2: When $v_2 > v_1$ or relative velocity $(v_2 - v_1)$ is positive.

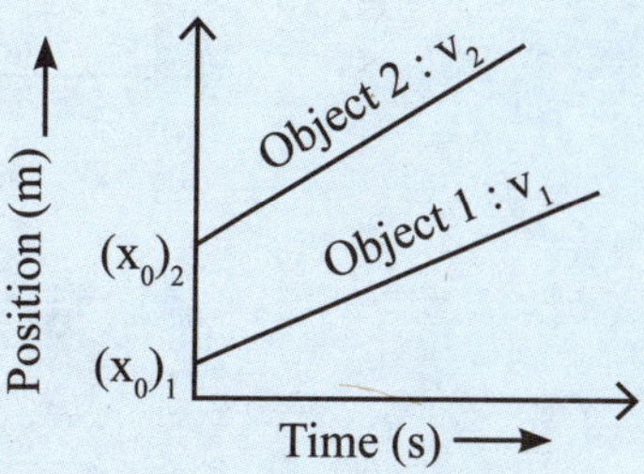

(iii) Case-3: When $v_2 < v_1$ or relative velocity $(v_2 - v_1)$ is negative.

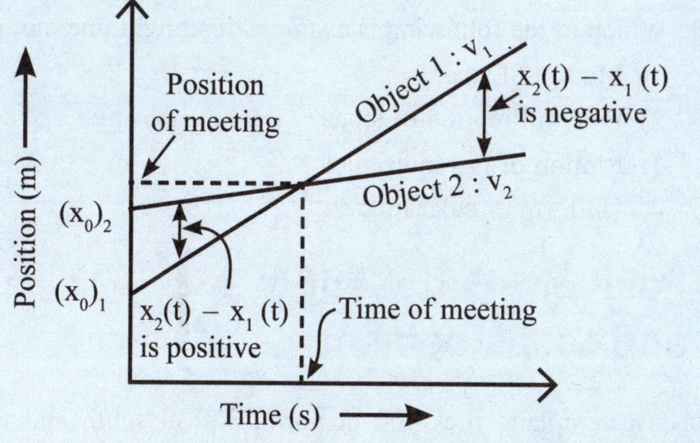

Statements Asked in Exam

- The magnitude of the displacement for a motion between two points may be zero but the corresponding path length is not zero.
- The magnitude of displacement may or may not be equal to the path length traversed by an object.
- When only magnitude of velocity changes then acceleration will be either parallel or anti-parallel to velocity, e.g., motion under gravity.
- The zero velocity of particle at any instant does not necessarily imply zero acceleration at the instant.
- The direction of velocity of a body can change even when acceleration is constant.
- The average speed of a body is equal to its instantaneous speed if the body moves with a constant speed.
- Average speed of an object is greater than or equal to magnitude of average velocity.
- An object with zero speed may have non zero acceleration at that instant.
- A body with positive value of acceleration may not be speeding up.
- Speedometer measures the instantaneous speed of a vehicle.
- The direction in which an object moves is given by the direction of velocity of the object and not by the direction of acceleration.
- The body thrown vertically upwards returns to its point of projection with the same magnitude of the velocity with which it was thrown, provided air resistance is negligible.
- The acceleration of the falling body does not depend on the mass of the body.

NCERT Topic-wise MCQs

Introduction

1. Which of the following is example of straight line motion?
   1) Motion of snake
   2) Free fall motion of stone
   3) Motion of gas molecules
   4) Uniform circular motion

Position, Path Length and Displacement

2. An aeroplane flies 400 m north, 300 m south and then 1200 m upwards. The net displacement is:
   1) 1204 m   2) 1304 m
   3) 1404 m   4) 1504 m

3. The numerical ratio of magnitude of distance to displacement is:
   1) Always equal to one
   2) Always less than one
   3) Always greater than one
   4) Equal to or more than one

4. An athlete completes one round of a circular track of radius R in 80 sec. What will be his displacement at the end of 5 minutes?
   1) $2\pi R$ 2) $3\pi R$
   3) $\sqrt{2}R$ 4) Zero

5. A wheel of radius 3 m rolls forward half a revolution on a horizontal ground. The magnitude of the displacement of the point of the wheel initially in contact with the ground is:
   1) 2π m
   2) $\sqrt{2}\pi$ m
   3) $\sqrt{\pi^2+4}$ m
   4) $3\sqrt{\pi^2+4}$m

   r = 3m
   P
   P

6. An aeroplane flies from (–4m, –5m, + 8m) to (7m, –2m, –3m) in the xyz coordinates. Displacement of the aeroplane in co-ordinate form is given by:
   1) (3m, –4m, 5m)   2) (4m, –5m, 11m)
   3) (11m, 3m, –11m)   4) (11m, –6m, 7m)

7. Let P, Q and R represent the positions of the car at different instants of time.

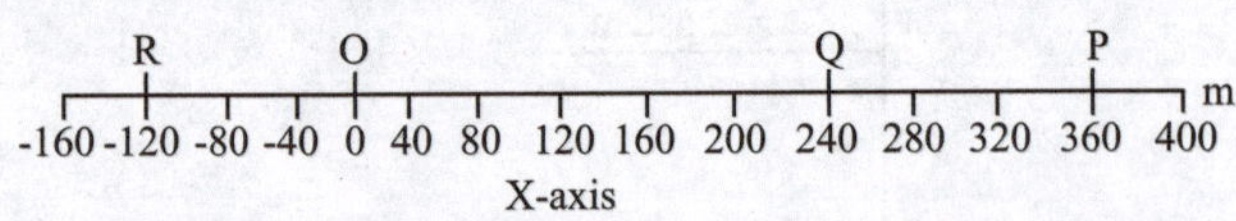

Consider the following statements with reference to the above situation:

I. Distance moved by the car from O to P is OP = 360 m.

II. Distance and path length are same, so path length traversed from O to P is also OP = 360 m.

III. If the car moves from O to P and then moves back from P to Q, the path length traversed is OP + PQ = 480 m.

IV. If the car moves from O to P and then moves back from P to Q, the path length traversed is OQ = 240 m.

Choose the correct option
   1) Statements I, II and III are correct
   2) Statements I, II and IV are correct
   3) Statements I and III are correct
   4) Statement IV is correct

8. Let x_1 and x_2 be the positions of an object at time t_1 and t_2. Then, its displacement, denoted by A, in time Δt = B, is given by the difference of position C from position D. Here A, B, C and D refer to
   1) $A \rightarrow \Delta x$, $B \rightarrow t_1 - t_2$, $C \rightarrow$ final, $D \rightarrow$ initial
   2) $A \rightarrow \Delta x$, $B \rightarrow t_2 - t_1$, $C \rightarrow$ final, $D \rightarrow$ initial
   3) $A \rightarrow x$, $B \rightarrow t_2 - t_1$, $C \rightarrow$ initial, $D \rightarrow$ final
   4) $A \rightarrow \Delta x$, $B \rightarrow t_2 - t_1$, $C \rightarrow$ initial, $D \rightarrow$ final

9. An athlete moves on a quarter circle of radius r. The values of distance travelled and displacement are
   1) distance = πr, displacement = r
   2) distance = $\pi r\sqrt{2}$, displacement = $\pi r\sqrt{2}$
   3) distance = $\frac{\pi r}{2}$, displacement = $r\sqrt{2}$
   4) distance = $\frac{\pi r}{2}$, displacement = zero

10. The position x of a particle moving in one dimension, under the action of a constant force is related to time t by the equation, $t = \sqrt{x} + 3$, where x is in metres and t in seconds. Find the position of the particle when its velocity is zero.
    1) 0   2) 3
    3) 2   4) 9

Speed, Velocity and Acceleration

11. In 1.0 s, a particle goes from point A to point B, moving in a semicircle (see figure). The magnitude of the average velocity is:

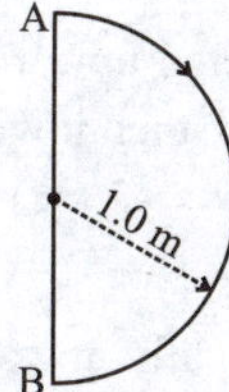

1) 3.14 m/s 2) 2.0 m/s
3) 1.0 m/s 3) zero

12. The displacement of a body along x - axis depends on time as $\sqrt{x} = 3t + 5$. Then the velocity of body:
1) Increases with time 2) is independent of time
3) Decreases with time 4) None of these

13. A truck travels a distance A to B at a speed of 40 km/h and returns to A at a speed of 50 km/h, then what is the average velocity of the whole journey?
1) 34.5 km/h 2) Zero
3) 35 km/hr 4) 40 km/hr

14. An athlete participates in a race. He is moving on a circular track of radius 80 m and completes half a revolution in 20s. His average velocity is:
1) 8 m/s 2) 16 m/s
3) 10 m/s 4) 12 m/s

15. The motion of a particle is described by the equation $x = a + bt^2$, where $a = 10$ cm, $b = 15$ cm/s². Its instantaneous velocity at t = 3 second will be?
1) 10 cm/s 2) 20 cm/s
3) 60 cm/s 4) 90 cm/s

16. A car moving on a straight road covers one third of distance with a speed of 20 m/s, other one third with speed of 40 m/s and next one third with speed 60 m/s, then the average speed of the car is:
1) 32.7 m/s 2) 40 m/s
3) 31 m/s 4) 33 m/s

17. A train moves with a speed of 30 km/h in the first 15 minutes, with the speed of 40 km/h in the next 15 minutes, and then with a speed of 60 km/h in the last 30 minutes. Calculate the average speed of the train for this journey.
1) 75 km/h 2) 47.5 km/h
3) 60 km/h 4) 40 km/h

18. A particle is in motion along a square path ABCD of side length 10 meters. If it starts at point A and reaches point C. What is the magnitude of the displacement of the particle?
1) 10 m 2) 20 m
3) $10\sqrt{2}$ m 4) $5\sqrt{2}$ m

19. A particle is moving with a constant speed v in a circle. What is the magnitude of average velocity after one-fourth rotation?
1) $\frac{\pi v}{\sqrt{2}}$ 2) $\frac{\sqrt{2}v}{\pi}$
3) $\frac{2\sqrt{2}v}{\pi}$ 4) $\frac{\pi V}{2\sqrt{2}}$

20. A body travels a distance s_1 with velocity v_1 and distance s_2 with velocity v_2 in the same direction. Calculate the average velocity of the body.
1) $\frac{(s_1+s_2)v_1v_2}{s_1v_2+s_2v_1}$ 2) $\frac{v_1v_2}{s_1v_2+s_2v_1}$
3) $\frac{(s_1-s_2)v_1v_2}{s_1v_2-s_2v_1}$ 4) $\frac{v_1+v_2}{2}$

21. The velocity of the particle at any time t is given by $v = 2t(3 - t)$ ms⁻¹. At what time is its velocity maximum?
1) 2 seconds 2) 1.5 seconds
3) 1 second 4) 5 seconds

22. The velocity of an object changes from 50 m/s to 60 m/s in 20 seconds. What is the average acceleration of the object over that interval?
1) 5 m/s² 2) 0.5 m/s²
3) 0.2 m/s² 4) 1.5 m/s²

23. A particle moves in a straight line so that its displacement x at any time t is given by $x^2 = 1 + t^2$. Its acceleration at any time t is x^{-n} where n = ______.
1) 1 2) 2
3) 3 4) 4

24. A particle is moving with speed $v = b\sqrt{x}$ along positive x-axis. Calculate the speed of the particle at time $t = \tau$ (assume that the particle is at origin at $t = 0$)
1) $\frac{b^2\tau}{4}$ 2) $\frac{b^2\tau}{2}$
3) $b^2\tau$ 4) $\frac{b^2\tau}{\sqrt{2}}$

25. The relation between time t and distance x for a moving body is given as $t = mx^2 + nx$, where m and n are constants. The retardation of the motion is: (Where v stands for velocity)
1) $2n^2v^2$ 2) $2mnv^3$
3) $2mv^3$ 4) $2nv^3$

Paragraph for Q. 26 to 28

Given questions are based on following passage. Choose the correct option from those given below.

The position of an object moving along X-axis is given by $x = a + bt^2$, where a = 8.5 m, b = 2.5 ms⁻² and t is measured in seconds.

26. The velocity at t = 0 s is
1) 0 m s⁻¹ 2) 10 m s⁻¹
3) 8.5 m s⁻¹ 4) 2.5 m s⁻¹

27. The velocity at t = 2.0 s is
1) 10 m s⁻¹ 2) 0 m s⁻¹
3) 5 m s⁻¹ 4) 2.5 m s⁻¹

28. The average velocity between t = 2.0 s and t = 4.0 s is:
1) 30 m s^{-1}
2) 15 m s^{-1}
3) 10 m s^{-1}
4) None of these

29. The position of an object moving along X-axis is given by $x - a - bt^2$, where a = 8.5 m, b = 2.5 ms^{-2} and t is measured in seconds.
For the above situation, match the situations in Column-I with the values of Column-II and choose the correct option from the codes given below:

	Column-I		Column-II
(a)	Velocity of object at t = 2.0 s	(i)	– 15 ms^{-1}
(b)	Velocity of object at t = 0s	(ii)	– 10 ms^{-1}
(c)	Instantaneous speed of object at t = 2.0 s	(iii)	0 ms^{-1}
(d)	Average velocity between t = 2.0 s and t = 4.0 s	(iv)	10 ms^{-1}

Codes:
1) (a)-(i) (b)-(ii) (c)-(iii) (d)-(iv)
2) (a)-(ii) (b)-(iii) (c)-(iv) (d)-(i)
3) (a)-(iv) (b)-(iii) (c)-(ii) (d)-(i)
4) (a)-(iii) (b)-(ii) (c)-(i) (d)-(iv)

30. The x-t graph for motion of a car is given below. With reference to the graph which of the given statement(s) is/are correct?

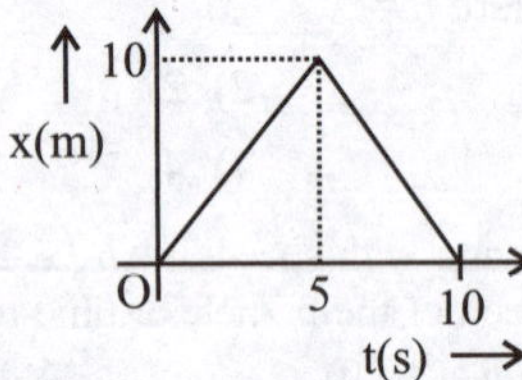

I. The instantaneous speed during the interval t = 5 s to t = 10 s is negative at all time instants during the interval.
II. The velocity and the average velocity for the interval t = 0 s to t = 5s are equal and positive.
III. The car changes its direction of motion at t = 5 s
IV. The instantaneous speed and the instantaneous velocity is positive at all time instants during the interval t = 0 s to t = 5 s

Choose the correct option:
1) I, II and III
2) II and III
3) II, III and IV
4) I, II, III and IV

Kinematic Equations for Uniformly Accelerated Motion

31. A bullet emerges from a barrel of length 1.2 m with a speed of 640 ms^{-1}. Assuming constant acceleration the approximate time that it spends in the barrel after firing the bullet is:
1) 75 ms
2) 3.75 ms
3) 4 second
4) 0.1875 second

32. A bullet fired into a fixed target loses half of its velocity after penetrating 4 cm. How much further it will penetrate before coming to rest assuming that it faces constant resistance to motion?
1) 0.2 cm
2) 5 cm
3) 2 cm
4) 1.33 cm

33. An object is projected upwards with a velocity of 100 m/sec. If g = 10m/sec^2, then it will strike the ground in,
1) 10 sec
2) 20 sec
3) 15 sec
4) 5 sec

34. A ball thrown vertically upwards with a speed of 19.6 m/s from the top of a tower returns to the earth in 6 seconds. What is the height of the tower?
1) 40 m
2) 58.8 m
3) 50 m
4) 70 m

35. A particle starts moving from rest with acceleration 2 m/s^2. Distance traveled by it in 5th half second is:
1) 1.25 m
2) 2.25 m
3) 6.25 m
4) 30.25 m

36. A juggler maintains four balls in motion making each of them rise to a height of 40m from his hand. What time interval should he maintain for the proper distance between them?
1) 1.71 seconds
2) 2.14 seconds
3) 1.41 seconds
4) 4 seconds

37. A body starting from rest moves with constant acceleration. The ratio of distance covered by the body during 8th second to that covered in 8 seconds is:
1) $\frac{15}{60}$
2) $\frac{15}{64}$
3) $\frac{12}{15}$
4) 1

38. A car moving with a velocity of 10 m/s can be stopped by the application of a constant force F in a distance of 20 m. If the velocity of the car is 30 m/s it can be stopped by this force in:
1) 100 m
2) 120 m
3) 180 m
4) 160 m

39. A particle moves for 50 seconds. It first accelerates from rest and then retards to rest. If the retardation be 5 times the acceleration then the time for retardation is:
1) 25/3 second
2) 50/3 second
3) 25 second
4) 100/3 second

40. A body is projected vertically upward from the surface of earth. If upward direction is taken as positive, then acceleration of body during its upward and downward journey are respectively:
1) Positive, negative
2) Negative, negative
3) Positive, positive
4) Negative, positive

41. A ball is thrown upward with a velocity of 100 m/s it will reach the ground after:

1) 10 s
2) 20 s
3) 5 s
4) 40 s

42. A stone falls from a cliff and travels 24.5 m in the last second before it reaches the ground at the foot of the cliff. Find the height of the cliff.

1) 48.2 m
2) 44.1 m
3) 40 m
4) 28 m

43. A ball is dropped from a height h above ground. Neglect the air resistance, its velocity (v) varies with its height y above the ground as:

1) $\sqrt{2g(h-y)}$
2) $\sqrt{2gy}$
3) $\sqrt{2g(h+y)}$
4) $\sqrt{2gh}$

44. A body thrown vertically upwards with a speed of 19.6 ms^{-1} from the top of a tower returns to the earth in 10 seconds. What is the height of the tower? ($g = 10ms^{-2}$)

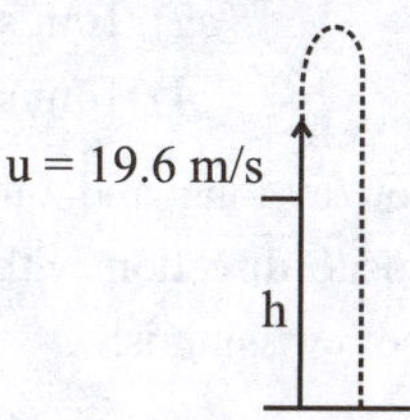

1) 304 m
2) 308 m
3) 310 m
4) 312 m

45. Water drops fall at regular intervals from a tap which is 8 m above the ground. The third drop is leaving the tap at the instant the first drop touches the ground. The second drop at that instant, is at height (from ground):

1) 2 m
2) 5 m
3) 4 m
4) 6 m

46. When a ball is thrown up vertically with velocity v_0, it reaches a maximum height of h. If one wishes to triple the maximum height then the ball should be thrown with velocity:

1) $\sqrt{3}v_0$
2) 3 v_0
3) 9 v_0
4) 3/2 v_0

47. A balloon is rising vertically up with a velocity of 30 m/s. A stone is dropped from it reaches the ground in 8 seconds. The height of the balloon when the stone was dropped from it is:

1) 80 m
2) 100 m
3) 85 m
4) 95 m

48. A man throws a ball vertically upward and it rises upto 40 m and returns to his hands. What was the initial velocity of the ball and for how much time (T) it remained in the air?

1) $T = 5$ sec, $u = 20\sqrt{3}$ m/s
2) $T = 5.64$ sec, $u = 10\sqrt{2}$ m/s
3) $T = 5.66$ sec, $u = 20\sqrt{2}$ m/s
4) None of these

49. A ball is projected vertically upwards. The times to cross a height *h* while ascending and while descending are t_1 and t_2 respectively. Then the velocity of projection is:

1) $\frac{g(t_1+t_2)}{2}$
2) $t_1 + t_2$
3) $2g(t_1 + t_2)$
4) $\frac{3}{2}g(t_1 + t_2)$

Graphical Representation of Motion

50. The slope of the tangent to the v – t curve gives the value of

1) Instantaneous acceleration
2) Instantaneous velocity
3) Average acceleration
4) None of the above

51. For a body moving with uniform acceleration along straight line, the variation of its velocity (v) with position (x) is best represented by

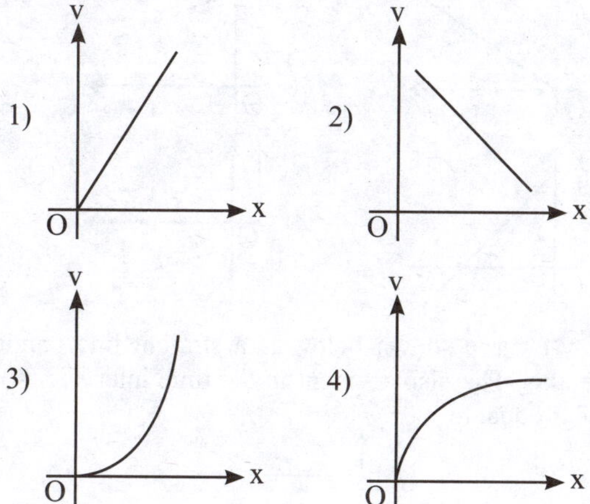

52. The position-time graph for a particle moving along a straight line is shown in figure. The total distance travelled by it in time t = 0 to t = 10 s is:

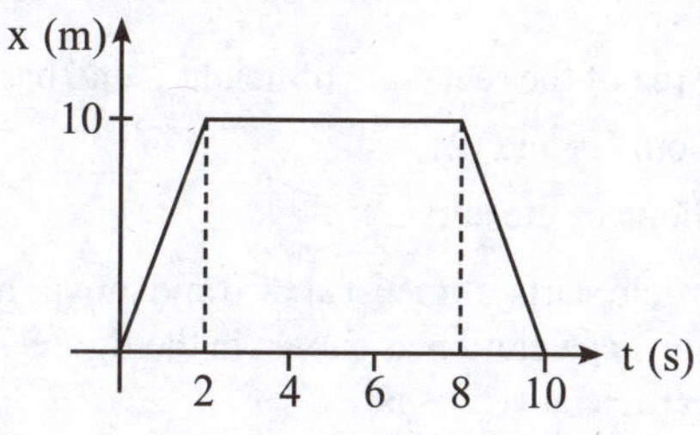

1) Zero
2) 10 m
3) 20 m
4) 80 m

53. The position-time graph for a body moving along a straight line between O and A is shown in figure. During its motion between O and A, how many times body comes to rest?

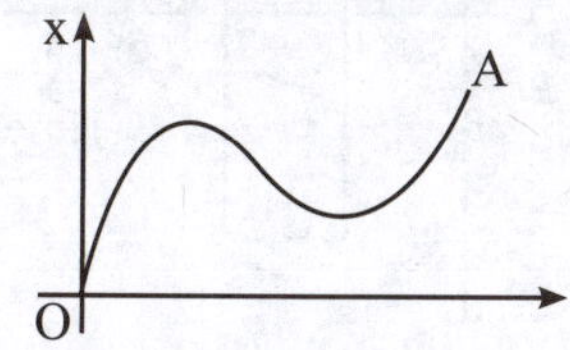

1) Zero
2) 1 time
3) 2 time
4) 3 time

54. What kind of motion is rectilinear motion?
1) One dimensional 2) Two dimensional
3) Three dimensional 4) Zero dimensional

55. An object is moving in positive direction with a positive acceleration. The velocity-time graph with constant acceleration which represents the above situation is:

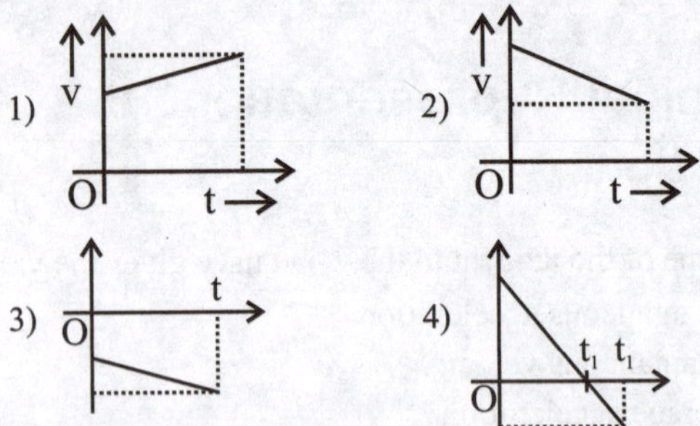

56. An object is moving in negative direction with a negative acceleration. The velocity-time graph with constant acceleration which represents the above situation is:

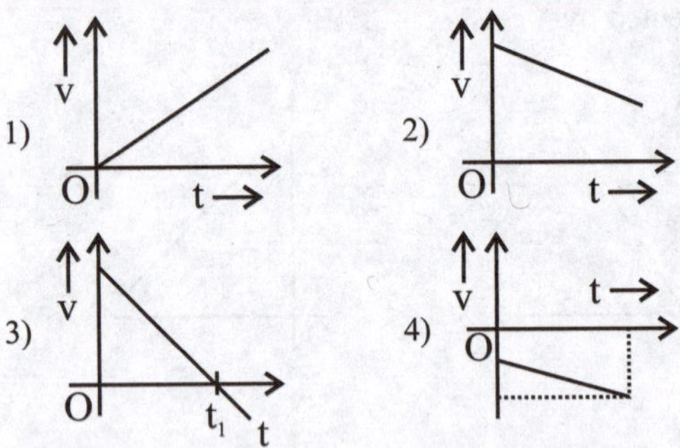

57. The v-t curve shown below is a straight line parallel to time-axis. The displacement in the time interval t = 0 and t = T is equal to

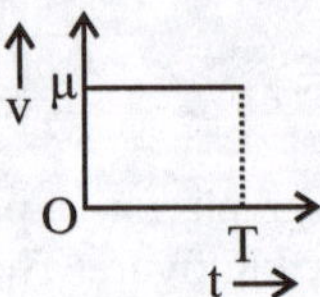

1) μT
2) Area of the rectangle of height μ and base T
3) Both (1) and (2)
4) Slope of the curve

58. A particle starts from rest at t = 0 and moves in a straight line with an acceleration as shown in the figure. The velocity of the particle at t = 3 s is:

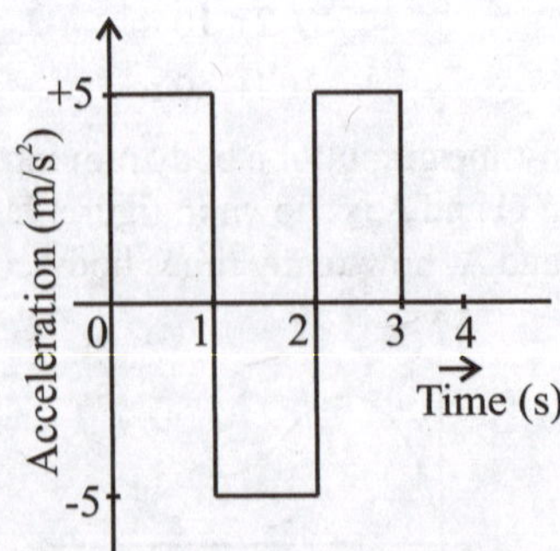

1) 5 m/s 2) 6 m/s
3) 10 m/s 4) 15 m/s

Relative Motion in 1D

59. Two cars are moving along a straight line in opposite direction with the same speed v. The relative velocity of two cars w.r.t. each other is
1) 2v 2) v
3) $\frac{v}{2}$ 4) Zero

60. Two objects A and B are moving with speeds v_A and v_B respectively in the same direction. The magnitude of relative velocity of A w.r.t. B is
1) $v_A - v_B$ 2) $v_A + v_B$
3) $v_B - v_A$ 4) v_A / v_B

61. A small block is placed over another block which is moving with a velocity of 5m/s. What is the absolute velocity of the small block?
1) 5m/s 2) 10m/s
3) 0m/s 4) 14m/s

62. Two trains one 60 m long and other is 80 m long are travelling in opposite direction with velocity 10 m/s and 25 m/s. The time of crossing is:
1) 4 seconds 2) 5 seconds
3) 6 seconds 4) 3 seconds

63. The distance between two trucks moving towards each other is decreasing at the rate of 10 m/s. If these trucks travel with same speeds in same direction the separation increases at the rate of 5 m/s. The velocity of the trucks are:
1) $v_1 = 8.5$ m/s , $v_2 = 1.5$ m/s
2) $v_1 = 7.5$ m/s , $v_2 = 2.5$ m/s
3) $v_1 = 5$ m/s , $v_2 = 5$ m/s
4) None of these

64. A bus is moving with a speed of 10 m/s on a straight road. A person on a scooter wishes to overtake the bus in 50 seconds. If the bus is at a distance of 1 km from the person, with what speed should the person chase the bus?
1) 50 m/s 2) 60 m/s
3) 80 m/s 4) 30 m/s

65. Two balls are dropped from same height at a gap of 1 second between them. The separation between the two balls after 4 seconds of the drop of the 1st ball is:
1) 30 m 2) 35 m
3) 40 m 4) 48 m

66. A rocket travelling at a speed of 200 m/s ejects its products of combustion at the speed of 1200 m/s relative to the rocket, then the speed of escaping vapours with respect to a person on the ground is:
1) 1000 ms^{-1}
2) 1200 m/s
3) 1400 m/s
4) 200 m/s

High Order Time Intensive MCQs

1. A car, starting from rest, accelerates at the rate f through a distance S, then continues at constant speed for time t and then decelerates as the rate $f/2$ to come to rest. If the total distance travelled is 15 S, then:
   1) $S = ft$
   2) $S = \frac{1}{6}ft^2$
   3) $S = \frac{1}{72}ft^2$
   4) $S = \frac{1}{4}ft^2$

2. A particle moving along X-axis has acceleration f, at time t, given by f = $f_0\left(1-\frac{t}{T}\right)$, where f_0 and t are constants between t = 0. The instant when f = 0, the particle's velocity (v_x) is:
   1) Zero
   2) f_0T
   3) $\frac{1}{4}f_0T$
   4) $\frac{1}{2}f_0T$

3. Velocity of a particle changes with position according to following curve. Acceleration of the particle at S = 1 m

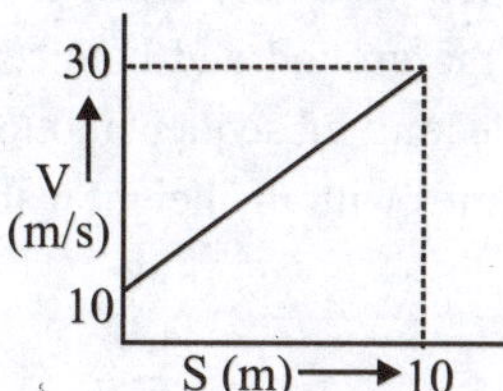

   1) 24 m/s²
   2) 2 m/s²
   3) 20 m/s²
   4) 3 m/s²

4. A stone starts to fall under gravity. The distances covered by it in first, second and third second are in ratio:
   1) 1 : 3 : 5
   2) 1 : 2 : 3
   3) 1 : 4 : 9
   4) 1 : 5 : 9

5. A bird is tossing (flying to and fro) between two cars moving towards each other on a straight road. One car has speed of 27 km h^{-1} while the other has the speed of 18 km h^{-1}. The bird starts moving from first car towards the other and is moving with the speed of 36 km h^{-1} when the two cars were separated by 36 km. The total distance covered by the bird is:
   1) 28.8 km
   2) 38.8 km
   3) 48.8 km
   4) 58.8 km

6. On a long horizontally moving belt, a child runs to and fro with a speed 9 km h^{-1} (with respect to the belt) between his father and mother located 50 m apart on the moving belt. The belt moves with a speed of 4 km h^{-1}. For an observer on a stationary platform, the speed of the child running in the direction of motion of the belt is:
   1) 4 km h^{-1}
   2) 5 km h^{-1}
   3) 9 km h^{-1}
   4) 13 km h^{-1}

7. A ball is dropped from the top of a building. The ball takes 0.5 s to fall past the 3 m length of window some distance below from the top of building. With what speed does the ball pass the top of window?
   1) 6 ms^{-1}
   2) 12 ms^{-1}
   3) 7 ms^{-1}
   4) 3.5 ms^{-1}

8. A particle is released from rest from a tower of height 3h. The ratio of the intervals of time to cover three equal heights h is:
   1) $t_1 : t_2 : t_3 = 3 : 2 : 1$
   2) $t_1 : t_2 : t_3 = 1 : (\sqrt{3}-2) : (2-\sqrt{5})$
   3) $t_1 : t_2 : t_3 = \sqrt{3} : \sqrt{2} : 1$
   4) $t_1 : t_2 : t_3 = 1 : (\sqrt{2}-1) : (\sqrt{3}-\sqrt{2})$

9. A ball is released from the top of a tower of height h. It takes T seconds to reach the ground. What is the position of the ball at T/3 seconds?
   1) h/9 from ground
   2) 7h/9 form ground
   3) 8h/9 from ground
   4) 17h/18 from ground

10. From a building two balls A and B are thrown such that A is thrown upwards and B downwards with same velocity. v_A & v_B are the velocities on reaching the ground then:
   1) $v_B > v_A$
   2) $v_A = v_B$
   3) $v_A > v_B$
   4) Velocity depends upon mass

11. Which one of the following represents the x-t graph of two objects A and B moving with zero relative speed?

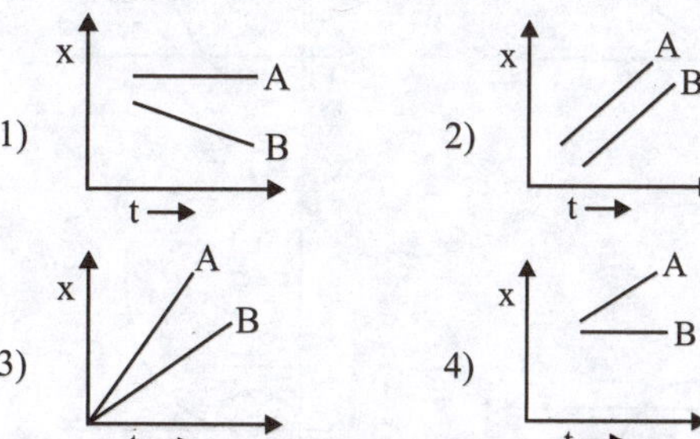

12. A car accelerates from rest at constant rate α for some time & decelerates to rest at rate β. Find the maximum velocity attained if total time of frovd is t.
   1) $v = t\frac{\alpha\beta}{\alpha-\beta}$
   2) $v = t\left(\frac{\beta^2}{\alpha-\beta}\right)$
   3) $v = t\frac{\alpha^2}{\alpha+\beta}$
   4) $v = t\frac{\alpha\beta}{\alpha+\beta}$

13. A particle travelling along a straight line traverses one third of the total distance with a velocity v_0. The remaining part of the distance was covered with a velocity v_1 for half the time and velocity v_2 for the other half of the time. Find the mean velocity of the particle averaged over the whole motion of time:

1) $\dfrac{3v_0(v_1+v_2)}{v_1+v_2+4v_0}$ 2) $\dfrac{v_0(v_1+v_2)}{v_1+v_2+12v_0}$

3) $\dfrac{v_0^2}{v_1+v_2+v_0}$ 4) $\dfrac{(v_1+v_2+v_0)^2}{v_0}$

14. An object falling through a fluid is observed to have acceleration given by $a = g - bv$ where g = gravitational acceleration and b is constant. After a long time of release, it is observed to fall with constant speed. The value of constant speed is

1) $\dfrac{g}{b}$ 2) $\dfrac{b}{g}$

3) bg 4) b

15. A train of 150 m length is going toward north direction at a speed of 10 m/s. A parrot flies at a speed of 5 ms^{-1} towards south direction parallel to the railway track. The time taken by the parrot to cross the train is equal to

1) 12 s 2) 15 s

3) 8 s 4) 10 s

16. A body is at rest at x = 0. At t = 0, it starts moving in the positive x-direction with a constant acceleration. At the same instant another body passes through x = 0 moving in the positive x-direction with a constant speed. The position of the first body is given by x_1 (t) after time t and that of the second body by x_2 (t) after the same time interval. Which of the following graphs correctly describes $(x_1 - x_2)$ as a function of time 't' ?

1) 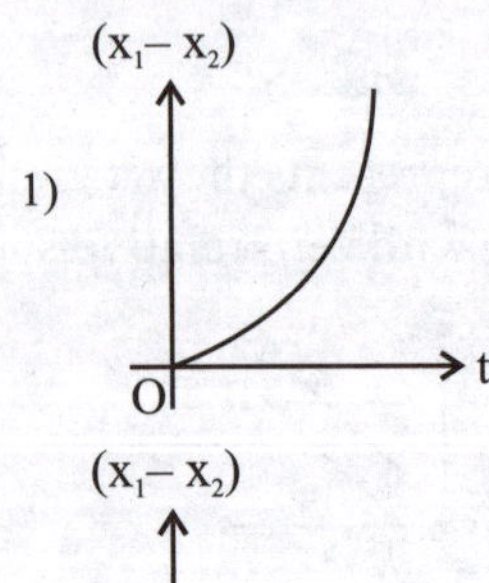

2) 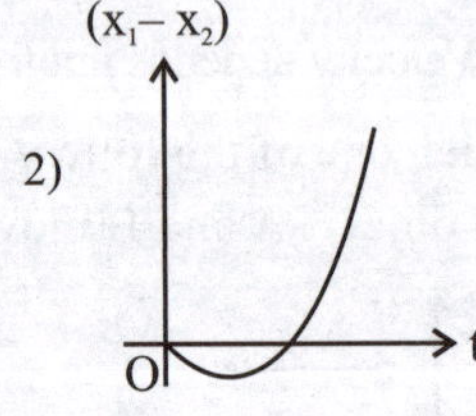

3) 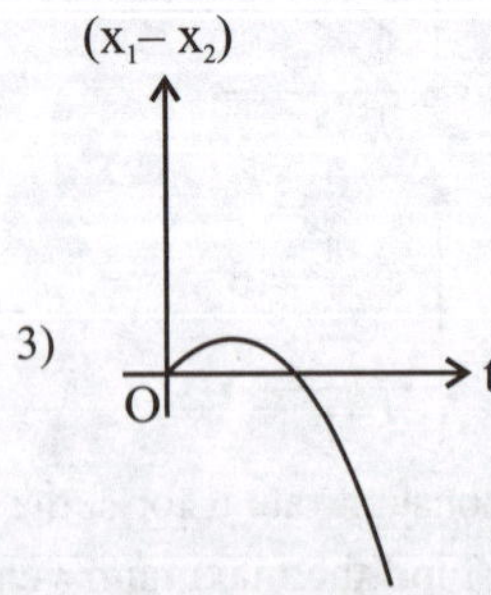

4) 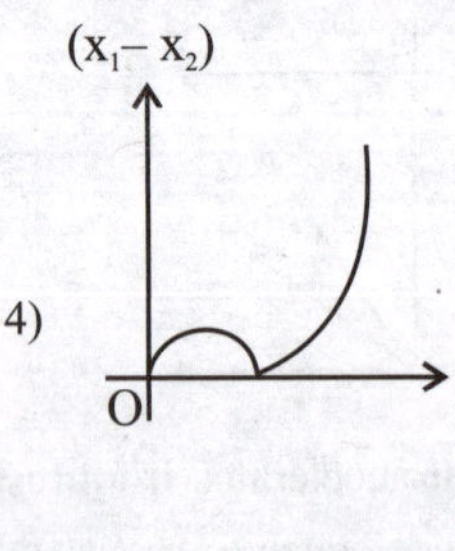

17. A person driving a car with a speed 72 km/h, suddenly sees a boy crossing the road. If the distance moved by a car, before the person applies brakes is 5 m, the reaction time of the person is:

1) 0.5 s 2) 0.10 s

3) 10 s 4) 0.25 s

18. The velocity v and displacement r of a body are related as $v^2 = kr$, where k is a constant. What will be the velocity after 1 second?

(Given that the displacement is zero at t = 0).

1) $\sqrt{kr}$ 2) $kr^{3/2}$

3) $\dfrac{k}{2}r^0$ 4) Data is not sufficient

19. A particle moving with uniform acceleration has average velocities v_1, v_2 and v_3 over the successive intervals of time t_1, t_2 and t_3 respectively. The value of $\dfrac{(v_2 - v_1)}{v_3 - v_2}$ will be:

1) $\dfrac{t_1 - t_2}{t_2 - t_3}$ 2) $\dfrac{t_1 - t_2}{t_2 + t_3}$

3) $\dfrac{t_1 + t_2}{t_2 - t_3}$ 4) $\dfrac{t_1 + t_2}{t_2 + t_3}$

20. The relation between time and displacement is $t = \alpha x^2 + \beta x$, where α, β are constants. If v is the velocity then the retardation is:

1) $2\alpha v^3$ 2) $2\beta v^3$

3) $2\alpha\beta v^3$ 4) $2\beta^2 v^3$

21. A ball is dropped vertically from a height d above the ground. It hits the ground and bounces up vertically to a height d/2. Neglecting subsequent motion and air resistance, its velocity v varies with the height h above the ground as

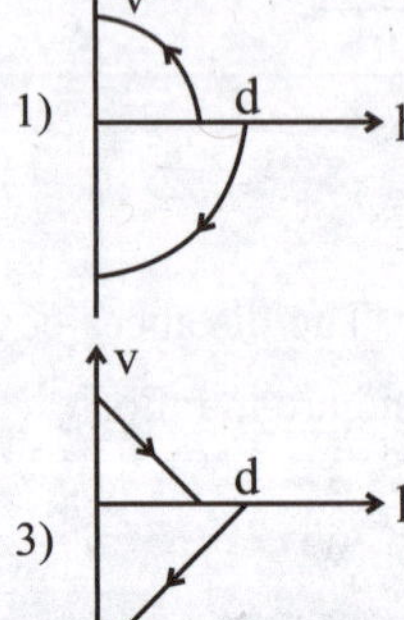

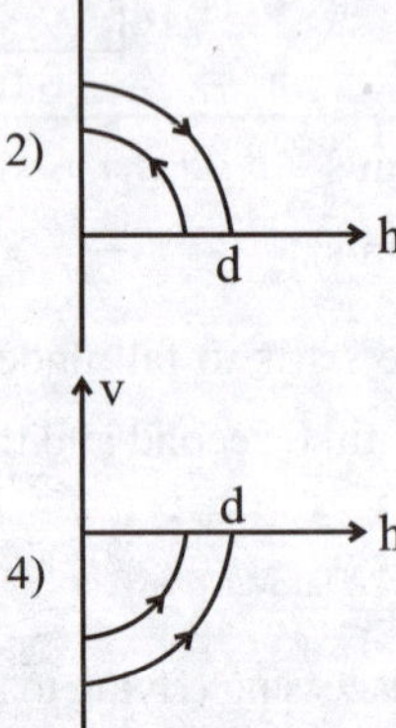

22. A stone is dropped from the top of a tower of height *h*. After 1 second another stone is dropped from the balcony 20 m below the top. Both reach the bottom simultaneously. What is the value of *h*?

1) 31.25 m 2) 100 m

3) 120 m 4) 130 m

23. A crazy ball is dropped on to the floor from the hand of the children from the height of 2 m. It rebounds to the height of 1.5 m. If the ball was in contact with the floor for 0.02 second, what was the average acceleration during contact?

1) 545 ms^{-2}

2) 600 ms^{-2}

3) 590 ms^{-2}

4) 400 ms^{-2}

Statement Based MCQs

Directions: Each question consists of two statements, Statement-I and Statement-II. Choose any one of the following four responses.

1) **Both Statement-I and Statement-II are correct.**
2) **Both Statement-I and Statement-II are incorrect.**
3) **Statement-I is correct & Statement-II is incorrect.**
4) **Statement-I is incorrect & Statement-II is correct.**

1. **Statement-I:** The speed of the particle remains constant when the angle between $\vec{a}$ and $\vec{v}$ is equal to 90°
   Statement-II: The distance covered by a particle never decrease with time, it always increases.
2. **Statement-I:** If a body moves in a straight line in one direction then the average speed of the body over a given interval of time is equal to the average velocity of the body in the same interval of time.
   Statement-II: When a body moves without change of direction the distance travelled by the body is equal to the displacement of the body.
3. **Statement-I:** All bodies fall freely with the same acceleration.
   Statement-II: The acceleration of the falling bodies depend on the mass of the body.
4. **Statement-I:** The area under a – t graph gives the change in velocity.
   Statement-II: For uniformaly accelerated motion ($a \neq 0$), v – t graph is a straight line whose slope gives the acceleration of the particle.
5. **Statement-I:** For a given time interval, average velocity is single valued while average speed may have many values.
   Statement-II: Velocity is a vector quantity and speed is a scalar quantity.

Assertion & Reason MCQs

Directions: Each question consists of two statements each, Assertion and Reason. Choose any one of the following four responses.

1) **Both Assertion and Reason are True and the Reason is the correct explanation of the Assertion.**
2) **Both Assertion and Reason are True but Reason is not the correct explanation of the Assertion.**
3) **Assertion is True but the Reason is False.**
4) **Assertion is False but Reason is True.**

1. **Assertion:** Positive acceleration of a body can be associated with 'slowing down' of the body.
   Reason: Acceleration is a vector quantity.
2. **Assertion:** A body can have acceleration even if its velocity is zero at a given instant of time.
   Reason: A body moving in a straight line is momentarily at rest when it reverses its direction of motion.
3. **Assertion:** 'Speeding up' of a body can be associated with negative acceleration of the body .
   Reason: Increase in speed of a moving body is independent of its direction of motion.
4. **Assertion:** Position-time graph of a stationary object is a straight line parallel to time axis.
   Reason: For a stationary object, position does not change with time.
5. **Assertion:** A body falling freely may do so with constant velocity.
   Reason: The body fall freely, when acceleration of a body is equal to acceleration due to gravity.

Match the Column MCQs

1. A particle moves along a straight line such that its displacements S varies with time t as $s = \alpha + \beta t + \gamma t^2$.

	Column-I		Column-II
(a)	Acceleration at t = 2s	(i)	$\beta + 5\gamma$
(b)	Average velocity during third second	(ii)	2γ
(c)	Velocity at t = 1s	(iii)	α
(d)	Initial displacement	(iv)	$\beta + 2\gamma$

1) (a)-(ii); (b)-(i); (c)-(iv); (d)-(iii)
2) (a)-(iii); (b)-(iv); (c)-(i); (d)-(ii)
3) (a)-(i); (b)-(ii); (c)-(iv); (d)-(iii)
4) (a)-(i); (b)-(ii); (c)-(iii); (d)-(iv)

2. For a body projected vertically up with a velocity $\vec{v}_0$ from the ground, match the following.

	Column-I		Column-II
(a)	$\vec{v}_{av}$ (Average velocity)	(i)	Zero for round trip
(b)	u_{av} (Average speed)	(ii)	$\frac{\vec{v}_1 + \vec{v}_2}{2}$ over any time interval
(c)	T_{ascent}	(iii)	$\frac{v_0}{2}$ over the total time of its flight
(d)	$T_{descent}$	(iv)	$\frac{v_0}{g}$

1) (a)-(i, iv); (b)-(iii); (c)-(iv); (d)-(i)
2) (a)-(i, ii); (b)-(iv); (c)-(iii); (d)-(i)
3) (a)-(i, ii); (b)-(iii); (c)-(iv); (d)-(iv)
4) (a)-(i, iv); (b)-(iv); (c)-(i); (d)-(iii)

3. A particle is moving along x-axis. Its x-coordinate varies with time as: $x = -20 + 5t^2$
For the given equation match the following two columns:

	Column-I		Column-II
(a)	Particle will cross the origin at	(i)	Zero second
(b)	At what time velocity and acceleration are equal	(ii)	1s
(c)	At what time particle changes its direction of motion	(iii)	2s
(d)	At what time velocity is zero	(iv)	None of these

1) (a)-(iv); (b)-(ii); (c)-(iv); (d)-(iii)
2) (a)-(iii); (b)-(ii); (c)-(iv); (d)-(i)
3) (a)-(ii); (b)-(i); (c)-(iii); (d)-(iv)
4) (a)-(iv); (b)-(ii); (c)-(iii); (d)-(i)

4. A particle moves along x-axis. Its position (x) is shown as function of time on following graph

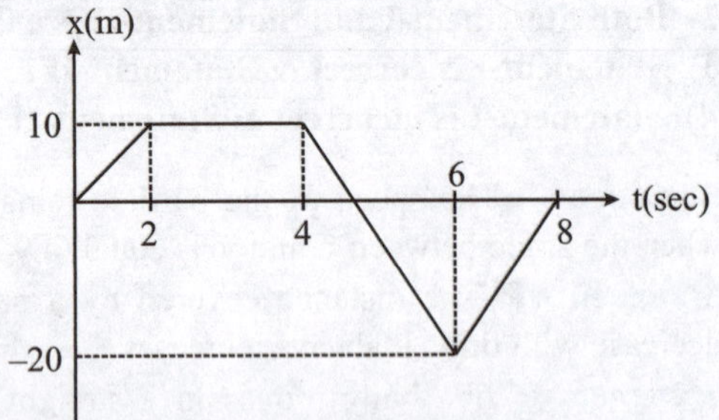

	Column-I		Column-II
(a)	Displacement of the particle at t = 1 sec.	(i)	60 m
(b)	Total distance covered by the particle.	(ii)	5m
(c)	Total displacement of the particle.	(iii)	4.67 or 14/3 sec
(d)	The instant i.e. time when particle crosses is initial position.	(iv)	zero

1) (a)-(i,ii); (b)-(ii,iv); (c)-(i,iii); (d)-(i,iii)
2) (a)-(i,iii); (b)-(ii,iii); (c)-(iii); (d)-(i)
3) (a)-(ii); (b)-(i); (c)-(iv); (d)-(iii)
4) (a)-(ii); (b)-(iii); (c)-(iv); (d)-(ii)

NCERT Exemplar MCQs

1. Among the four graphs shown in the figure there is only one graph for which average velocity over the time interval (0, T) can vanish for a suitably chosen T. Which one is it?

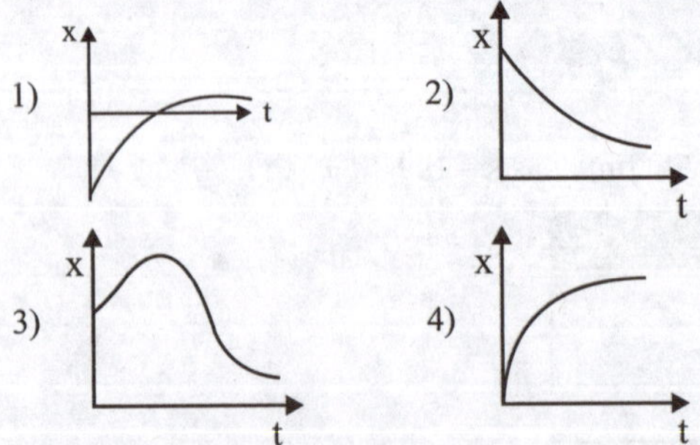

2. A lift is coming from 8th floor and is just about to reach 4th floor. Taking ground floor as origin and positive direction upwards for all quantities, which one of the following is correct?

1) $x < 0, v < 0, a > 0$
2) $x > 0, v < 0, a < 0$
3) $x > 0, v < 0, a > 0$
4) $x > 0, v > 0, a < 0$

3. In one dimensional motion, instantaneous speed v satisfies $0 \le v < v_0$.

1) The displacement in time T must always take non-negative values
2) The displacement x in time T satisfies $-v_0 T < x < v_0 T$
3) The acceler ation is always a non-negative number
4) The motion has no turning points

4. A vehicle travels half the distance L with speed v_1 and the other half with speed v_2, then its average speed is:

1) $\frac{v_1 + v_2}{2}$
2) $\frac{2v_1 + v_2}{v_1 + v_2}$
3) $\frac{2v_1 v_2}{v_1 + v_2}$
4) $L\frac{(v_1 + v_2)}{v_1 v_2}$

5. The displacement of a particle is given by $x = (t - 2)^2$ where x is in metre and t in second. The distance covered by the particle in first 4 seconds is:
1) 4 m
2) 8 m
3) 12 m
4) 16 m

6. At a metro station, a girl walks up a stationary escalator in time t_1. If she remains stationary on the escalator, then the escalator takes her up in time t_2. The time taken by her to walk up on the moving escalator will be:
1) $(t_1 + t_2)/2$
2) $t_1t_2/(t_2 - t_1)$
3) $t_1t_2/(t_2 + t_1)$
4) $t_1 - t_2$

7. The variation of quantity A with quantity B, plotted in figure, describes the motion of a particle in a straight line.

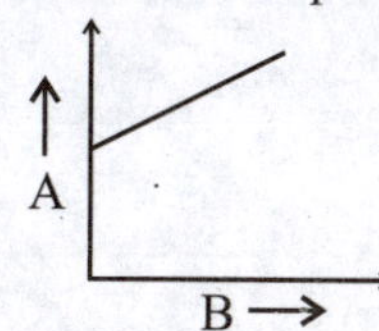

1) Quantity B may represent time
2) Quantity A is velocity if motion is uniform
3) Quantity A is displacement if motion is uniform
4) Quantity A is velocity if motion is uniformly accelerated

8. A graph of x versus t is shown in figure. Choose correct alternatives given below:

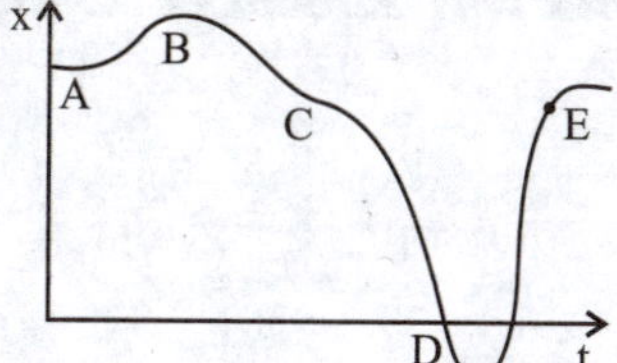

1) The particle was released from rest at t = 0
2) At B, the acceleration a > 0
3) Velocity vanishes at point C
4) The speed at D exceeds that at E

9. For the one-dimensional motion, described by $x = t - \sin(t)$:
1) $x(t) > 0$ for all $t > 0$
2) $v(t) > 0$ for all $t > 0$
3) $a(t) > 0$ for all $t > 0$
4) $v(t)$ lies between 0 and 2

10. A spring with one end attached to a mass and the other to a rigid support is stretched and released:
1) Magnitude of acceleration, when just released is maximum
2) Magnitude of acceleration, when at equilibrium position, is maximum
3) Speed is maximum when mass is at equilibrium position
4) Magnitude of displacement is always maximum whenever speed is minimum

11. A ball is bouncing elastically with a speed 1 m/s between walls of a railway compartment of size 10 m in a direction perpendicular to walls. The train is moving at a constant velocity of 10 m/s parallel to the direction of motion of the ball. As seen from the ground:
1) The direction of motion of the ball changes every 10 seconds
2) Speed of ball changes every 10 seconds
3) Average speed of ball over any 20 seconds interval is fixed
4) The acceleration of ball is the same as from the train

NEET & AIIMS Past Year MCQs

1. A particle is moving along x-axis with its position (x) varying with time (t) as $x = \alpha t^4 + \beta t^2 + \gamma t + \delta$. The ratio of its initial velocity to its initial acceleration, respectively, is: **(2024 Re)**
1) $2\alpha : \delta$
2) $\gamma : 2\delta$
3) $4\alpha : \beta$
4) $\gamma : 2\beta$

2. The velocity (v) – time (t) plot of the motion of a body is shown below: **(2024)**

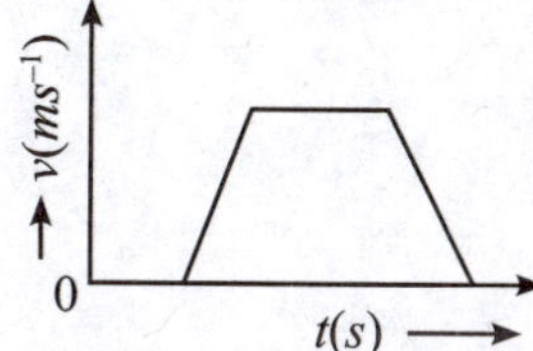

The acceleration (a) – time (t) graph that best suits this motion is:

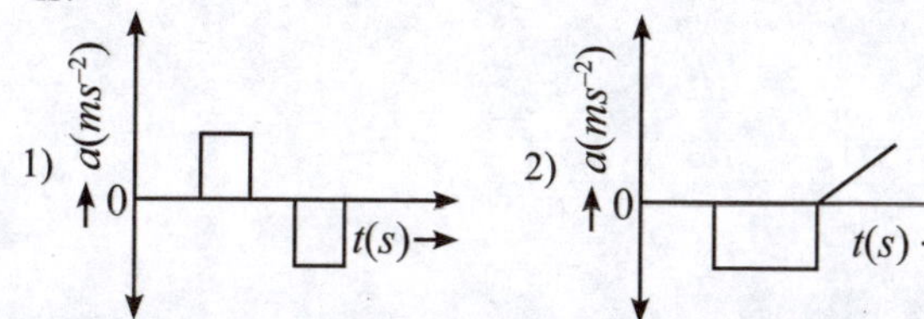

3)

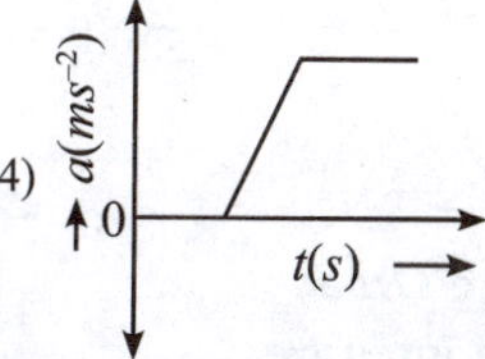

3. A vehicle travels half the distance with speed v and the remaining distance with speed 2v. Its average speed is: **(2023)**
1) $\frac{3v}{4}$
2) $\frac{v}{3}$
3) $\frac{2v}{3}$
4) $\frac{4v}{3}$

4. A horizontal bridge is built across a river. A student standing on the bridge throws a small ball vertically upwards with a velocity 4 ms^{-1}. The ball strikes the water surface after 4 s. The height of bridge above water surface is (Take g = 10 ms^{-2}): **(2023)**
1) 68 m
2) 56 m
3) 60 m
4) 64 m

5. The ratio of the distance traveled by a freely falling body in the 1st, 2nd, 3rd and 4th second : **(2022)**

1) 1 : 1 : 1 : 1 2) 1 : 2 : 3 : 4
3) 1 : 4 : 9 : 16 4) 1 : 3 : 5 : 7

6. The displacement time graphs of two moving particle make angles of 30° and 45° with the x-axis as shown in the figure. The ratio of their respective velocity is : **(2022)**

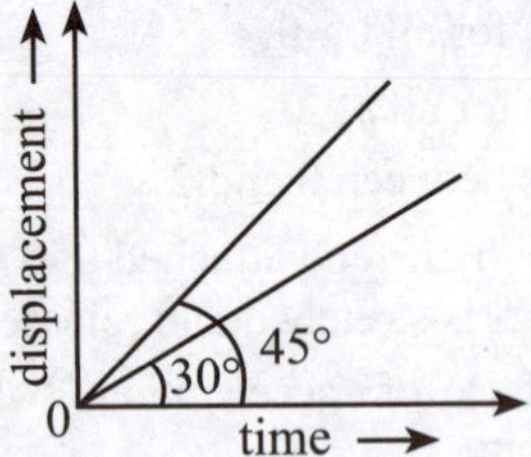

1) $1:\sqrt{3}$
2) $\sqrt{3}:1$
3) 1 : 1
4) 1 : 2

7. A man is at a distance of 6 m from a bus. The bus begins to move with a constant acceleration of 3 m s^{-2}. In order to catch the bus, the minimum speed with which the man should run towards the bus is: **(AIIMS 2016)**

1) 2 m s^{-1}
2) 4 m s^{-1}
3) 6 m s^{-1}
4) 8 m s^{-1}

Answer Key

NCERT Topic-wise MCQs

1	2	3	4	5	6	7	8	9	10	11	12	13	14	15	16	17
2	1	4	3	4	3	1	2	3	1	2	1	2	1	4	1	2
18	19	20	21	22	23	24	25	26	27	28	29	30	31	32	33	34
3	3	1	2	2	3	2	3	1	1	2	2	4	2	4	2	2
35	36	37	38	39	40	41	42	43	44	45	46	47	48	49	50	51
2	3	2	3	1	2	2	2	1	1	4	1	1	3	1	1	4
52	53	54	55	56	57	58	59	60	61	62	63	64	65	66		
3	3	1	1	4	3	1	1	1	1	1	2	4	2	1		

High Order Time Intensive MCQs

1	2	3	4	5	6	7	8	9	10	11	12	13	14	15	16	17
3	4	1	1	1	4	4	4	3	2	2	4	1	1	4	2	4
18	19	20	21	22	23											
3	4	1	1	1	3											

Statement Based MCQs

1	2	3	4	5
1	1	3	1	1

Assertion & Reason MCQs

1	2	3	4	5
2	2	2	1	4

Match the Column MCQs

1	2	3	4
1	3	2	3

NCERT Exemplar MCQs

1	2	3	4	5	6	7	8	9	10	11
3	3	2	3	2	3	1	1,3,4	1,4	1,3	2,3,4

NEET & AIIMS Past Year MCQs

1	2	3	4	5	6	7
4	1	4	4	4	1	3

CHAPTER 3

Motion in a Plane

Topicwise Number of Questions Analysis (2025-2016)

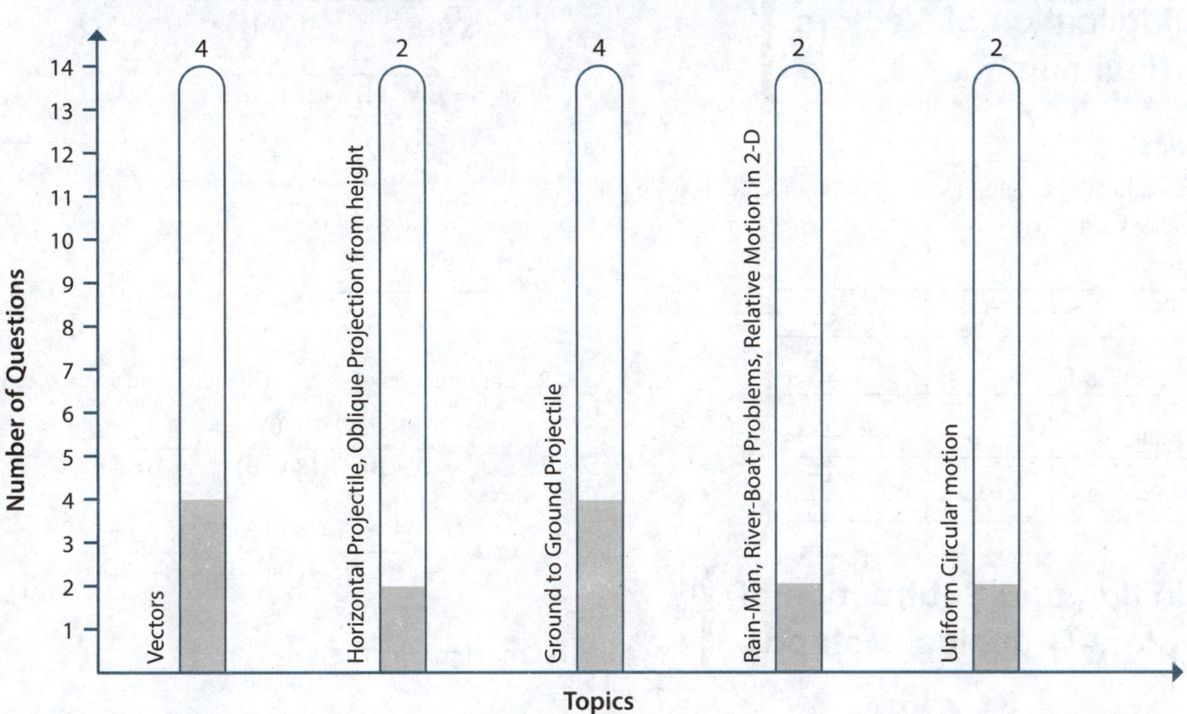

NCERT-Picks

Introduction

- Motion in a plane is also referred to as motion in two dimensions (2-D).
- In order to deal with motion of an object in more than one dimension, we need to introduce the concept of physical quantity, called vector.

Scalar and Vector

- Scalar has only magnitude and no direction.
- Vector has both magnitude and direction.

Formulas

1. Representation of vector

   $\vec{A} = \overrightarrow{PQ}$

   Tail ——→ Head
   P Q

2. Unit vector is the vector having unit magnitude and direction parallel to the given vector.

   $$\hat{A} = \frac{\vec{A}}{|\vec{A}|} = \frac{\text{vector}}{\text{modulus of vector}}$$

3. Magnitude of a vector $\vec{A} = A_x\hat{i} + A_y\hat{j} + A_z\hat{K}$ is given by

   $$|\vec{A}| = \sqrt{A_x^2 + A_y^2 + A_z^2}$$

Numerical Profile

(i) $\overrightarrow{OP}$ is the position vector of the object point P.

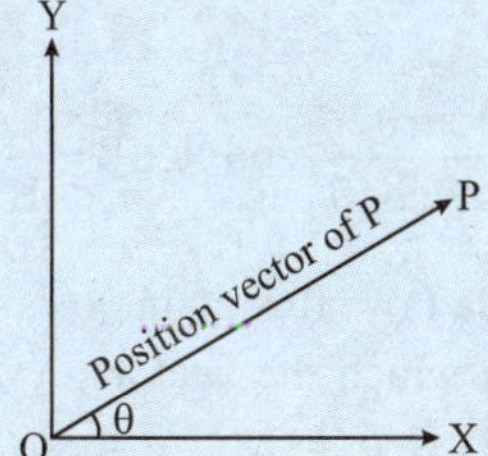

(ii) $\overline{PQ}$ is the displacement vector of the object in time t to t'.

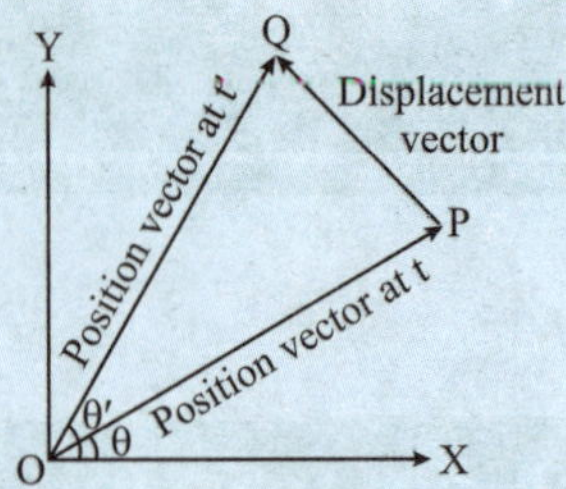

Multiplication of Vectors by Real number

Formulas

1. If a vector $\vec{A}$ is multiplied by a real number k, then we get another vector $\vec{B}$.

$\vec{B} = k\vec{A}$

k	$\vec{A}$	$\vec{B} = k\vec{A}$
Less than 1	$\vec{A}$ →	$(\vec{B} < \vec{A})$ →
Equal to 1	$\vec{A}$ →	$\vec{B} = \vec{A}$ →
More than 1	$\vec{A}$ →	$(\vec{B} > \vec{A})$ →

Addition and Subtraction of Vectors-Graphical Method

Formulas

Addition

1. **The parallelogram law:** Resultant $\vec{R}$ is the diagonal of the parallelogram of which $\vec{A}$ and $\vec{B}$ are the adjacent sides as shown in figure. Magnitude of $\vec{R}$ is given by

$$R = \sqrt{A^2 + B^2 + 2AB\cos\theta}$$

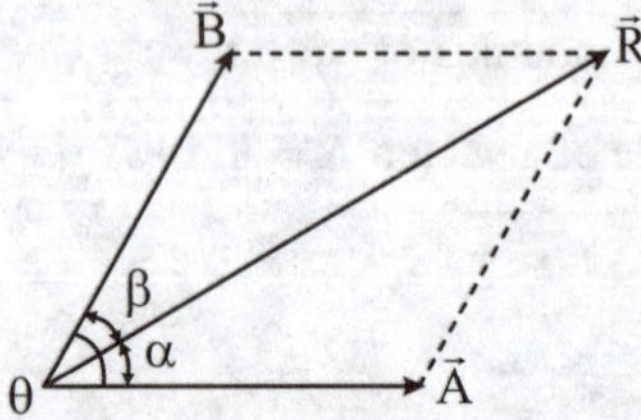

The direction of $\vec{R}$ can be found by angle α or β of $\vec{R}$ with $\vec{A}$ or $\vec{B}$.

$$\tan\alpha = \frac{B\sin\theta}{A + B\cos\theta} \text{ and } \tan\beta = \frac{A\sin\theta}{B + A\cos\theta}$$

2. **The triangle law:** Resultant $\vec{R}$ is the same irrespective of the order in which the vectors $\vec{A}$ and $\vec{B}$ are taken, Thus, $\vec{R} = \vec{A} + \vec{B} = \vec{B} + \vec{A}$

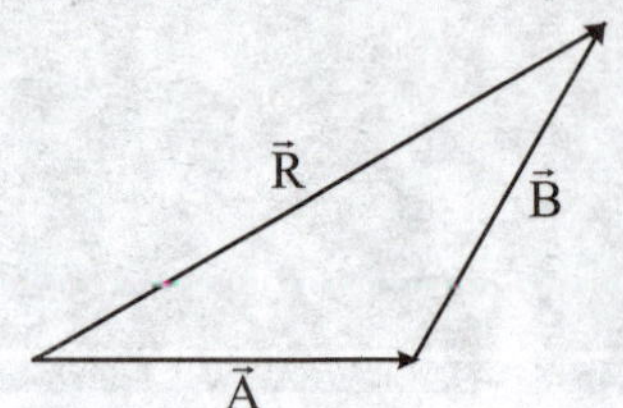

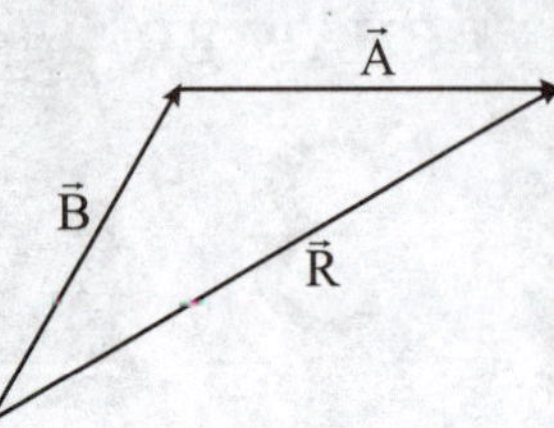

Subtraction

1. Magnitude of $\vec{S} = \vec{A} - \vec{B}$ will be thus given by

$$|\vec{S}| = |\vec{A} - \vec{B}| = |\vec{A} + (-\vec{B})|$$

$$S = \sqrt{A^2 + B^2 + 2AB\cos(180 - \theta)}$$

$$\text{or } S = \sqrt{A^2 + B^2 - 2AB\cos\theta}$$

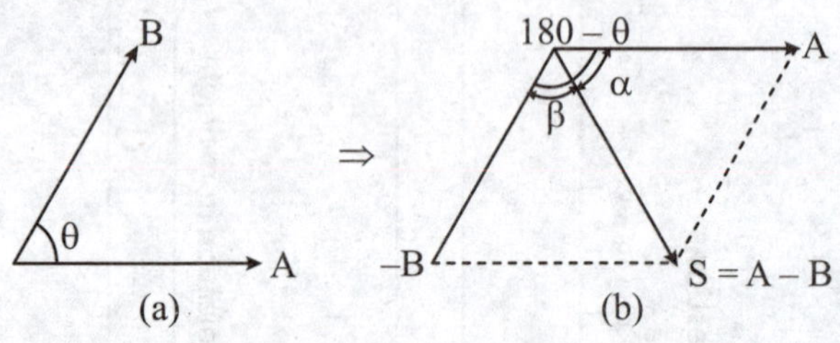

2. For direction of $\vec{S}$ we will either find angle α or β.

$$\tan\alpha = \frac{B\sin(180 - \theta)}{A + B\cos(180 - \theta)} = \frac{B\sin\theta}{A - B\cos\theta}$$

$$\text{or } \tan\beta = \frac{A\sin(180 - \theta)}{B + A\cos(180 - \theta)} = \frac{A\sin\theta}{B - A\cos\theta}$$

Numerical Profile

(i) Vector addition is commutative: $\vec{A} + \vec{B} = \vec{B} + \vec{A}$

(ii) Vector addition is distributive: $n(\vec{A} + \vec{B}) = n\vec{A} + n\vec{B}$

(iii) Vector addition is associative:

$\vec{A} + (\vec{B} + \vec{C}) = (\vec{A} + \vec{B}) + \vec{C}$

(iv) When $\vec{A}$ and $\vec{B}$ in same direction:

Resultant: $\vec{R} = \vec{A} + \vec{B}$ $(\because \theta = 0°)$

(v) Vector subtraction is not commutative: $\vec{A} - \vec{B} \neq \vec{B} - \vec{A}$

(vi) Vector subtraction is distributive: $n(\vec{A} - \vec{B}) = n\vec{A} - n\vec{B}$

(vii) Vector subtraction is not associative:

$\vec{A} - (\vec{B} - \vec{C}) \neq (\vec{A} - \vec{B}) - \vec{C}$

(viii) When $\vec{A}$ and $\vec{B}$ in opposite direction $(\vec{A} > \vec{B})$:

Resultant: $\vec{S} = \vec{A} - \vec{B}$ $(\because \theta = 180°)$

Resolution of Vector

- It is a process of splitting a vector into two or more vectors in such a way that their combined effect is same as that vector.

Formulas

1. Rectangular components of a vector in a plane

$\vec{A}_X = \vec{A}\cos\theta$

$\vec{A}_Y = \vec{A}\sin\theta$

$|\vec{A}| = \sqrt{A_x^2 + A_y^2}$

$\tan\theta = \frac{A_Y}{A_X}$

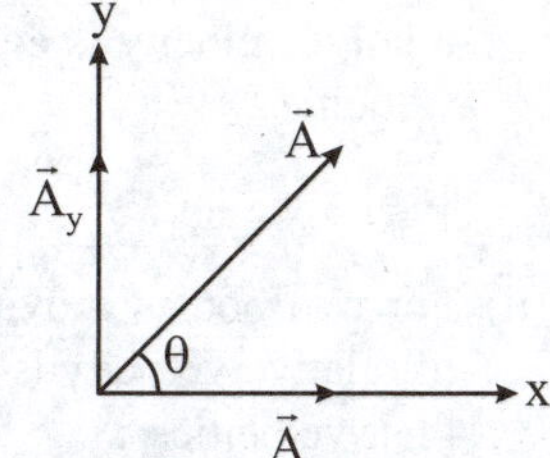

2. Rectangular compounds of vector in a space.

$\vec{A}_X = \vec{A}\cos\alpha$

$\vec{A}_Y = \vec{A}\cos\beta$

$\vec{A}_Z = \vec{A}\cos\gamma$

$\vec{A} = \sqrt{A_X^2 + A_Y^2 + A_Z^2}$

$\cos\alpha = \frac{A_X}{A}, \cos\beta = \frac{A_Y}{A}, \cos\gamma = \frac{A_Z}{A}$

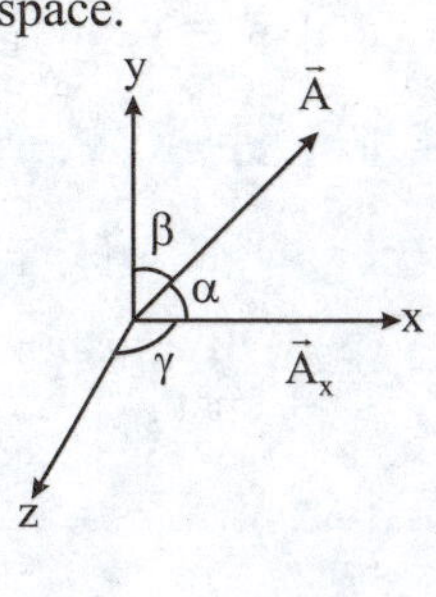

Numerical Profile

(i) Projection of a vector $\vec{A}$ in the direction of vector $\vec{B}$

$= \frac{\vec{A}.\vec{B}}{|\vec{B}|}$

[dot product of $\vec{A}$ and $\vec{B}$ divided by magnitude of $\vec{B}$]

(ii) Projection of a vector $\vec{B}$ in the direction of vector $\vec{A}$

$= \frac{\vec{A}.\vec{B}}{|\vec{A}|}$

[dot product of $\vec{A}$ and $\vec{B}$ divided by magnitude of $\vec{A}$]

Vector Addition – Analytical Method

Formulas

1. If we have two vectors $\vec{A}$ and $\vec{B}$.

$\vec{A} = A_X\hat{i} + A_Y\hat{j} + A_Z\hat{k}$

$\vec{B} = B_X\hat{i} + B_Y\hat{j} + B_Z\hat{k}$

Then resultant $\vec{R} = \vec{A} + \vec{B} = R_X\hat{i} + R_Y\hat{j} + R_Z\hat{k}$

$R_X = A_X + B_X$

$R_Y = A_Y + B_Y$

$R_Z = A_Z + B_Z$

2. Analytical method of addition (or subtraction) can be extended for any number of vectors.

Motion in a Plane

Formulas

1. **Expression for position vector:** Position vector $\overrightarrow{OP}$ of a particle located at P with respect to the origin O. If (x, y) are the coordinates of point P, then

$\overrightarrow{OP} = \overrightarrow{OA} + \overrightarrow{OB}$

or, $\vec{r} = x\hat{i} + y\hat{j}$

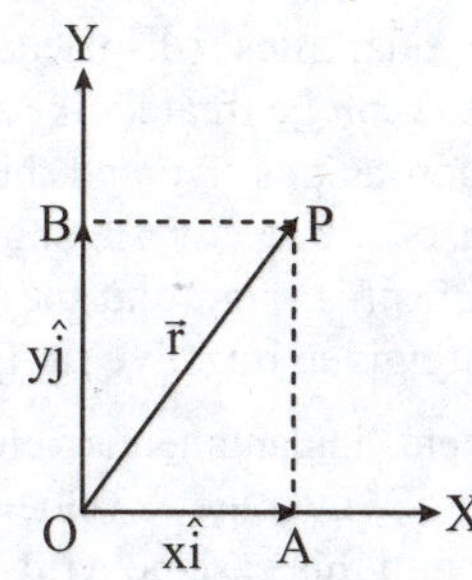

2. **Expression for displacement vector:** The particle is at point P(x, y) at time t and at P'(x', y') at time t'.

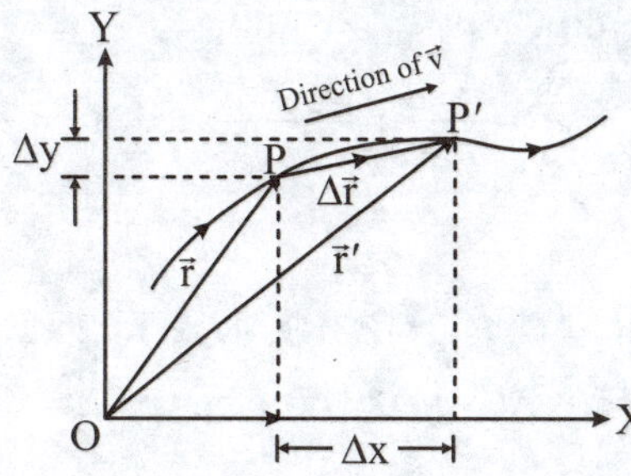

Using triangle law of vectors addition,

$\overrightarrow{OP'} = \overrightarrow{OP} + \overrightarrow{PP'}$

So the displacement vector is

$\overrightarrow{PP'} = \overrightarrow{OP'} - \overrightarrow{OP}$

or, $\Delta\vec{r} = \vec{r}' - \vec{r}$

3. **Average velocity:** The ratio of the displacement and the corresponding time interval.

$$\vec{v} = \frac{\Delta\vec{r}}{\Delta t} = \frac{\Delta x\hat{i} + \Delta y\hat{j}}{\Delta t} = \frac{\Delta x}{\Delta t}\hat{i} + \frac{\Delta y}{\Delta t}\hat{j}$$

[Here $\Delta t = t' - t$]

or, $\vec{v} = v_x\hat{i} + v_y\hat{j}$.

4. **Instantaneous velocity:** The limiting value of its average velocity when the time interval approaches zero.

$$\vec{v} = \lim_{\Delta t\to 0}\frac{\Delta\vec{r}}{\Delta t} = \frac{d\vec{r}}{dt}$$

5. **Average acceleration:** The ratio of the change in velocity and the corresponding time interval.

$$\vec{a} = \frac{\vec{v}' - \vec{v}}{\Delta t} = \frac{\Delta\vec{v}}{\Delta t} = \frac{\Delta(v_x\hat{i} + v_y\hat{j})}{\Delta t} = \frac{\Delta v_x}{\Delta t}\hat{i} + \frac{\Delta v_y}{\Delta t}\hat{j}$$

or $\vec{a} = a_x + a_y\hat{j}$

6. **Instantaneous acceleration:** The limiting value of its average acceleration when the time interval approaches zero.

$$\vec{a} = \lim_{\Delta t\to 0}\frac{\Delta\vec{v}}{\Delta t} = \frac{d\vec{v}}{dt}.$$

Numerical Profile

(i) If motion is one dimensional (let along x-axis) then all vector quantities (displacement, velocity and acceleration) can be treated like scalars by assuming one direction as positive and the other as negative. In this case, all vectors along positive direction are given positive sign and the vectors in negative direction are given negative sign.

(ii) Displacement, instantaneous velocity, instantaneous acceleration, average velocity and average acceleration in this case be written as

$s = \Delta r = r_f - r_i$ or $x_f - x_i$

$v = \frac{ds}{dt}$ or $\frac{dx}{dt}$ or $a = \frac{dv}{dt}$

$v_{av} = \frac{\Delta s}{\Delta t} = \frac{r_f - r_i}{\Delta t} = \frac{x_f - x_i}{\Delta t}$ and $a_{av} = \frac{\Delta v}{\Delta t} = \frac{v_f - v_i}{\Delta t}$

Motion in a Plane with Constant Acceleration

Formulas

1. Velocity of a particle moving with constant acceleration $\vec{a}$ at any time t.

   $\vec{v} = \vec{v}_0 + \vec{a}t$, where $\vec{v}_0$ is the velocity of particle at t = 0.

2. Average velocity over time interval t, $\vec{v}_{av} = \frac{\vec{v} + \vec{v}_0}{2}$

   In terms of components: $v_x = v_{0x} + a_x t$; $v_y = v_{0y} + a_y t$

3. Displacement of the particle,

   $$\vec{r} - \vec{r}_0 = \left(\frac{\vec{v} + \vec{v}_0}{2}\right)t = \left(\frac{(\vec{v}_0 + \vec{a}t) + \vec{v}_0}{2}\right)t$$

   $$\vec{r} = \vec{r}_0 + \vec{v}_0 t + \frac{1}{2}\vec{a}t$$

   In terms of components:

   $$x = x_0 + v_{0x}t + \frac{1}{2}a_x t^2 \text{ and } y = y_0 + v_{0y}t + \frac{1}{2}a_y t^2$$

Relative Velocity in Two Dimensions

Formulas

1. It v_A and v_B are the velocities of two bodies, then relative velocity of A with respect to B

   $\vec{v}_{AB} = \vec{v}_A - \vec{v}_B$

2. Let two objects A and B are moving with velocities v_A and v_B, the magnitude of the relative velocity v_{AB} is given by

   $v_{AB} = \sqrt{v_A^2 + v_B^2 - 2v_A v_B \cos\theta}$

   (θ is the angle between v_A and v_B)

3. Suppose the relative velocity v_{AB} makes an angle β with v_A. Then

   $\beta = \tan^{-1}\left(\frac{v_B \sin\theta}{v_A - v_B \cos\theta}\right)$

Note:

(i) For two bodies moving in the same direction, magnitude of relative velocity is equal to the difference of magnitude of velocities.

$|\vec{v}_R| = v_A - v_B$ [v_R is the relative velocity]

(ii) For two bodies moving in opposite directions, magnitude of relative velocity is equal to the sum of the magnitude of their velocities.

$|\vec{v}_R| = v_A + v_B$

Numerical Profile

River Boat Concept

(i) If a man wants to cross the river along the shortest path

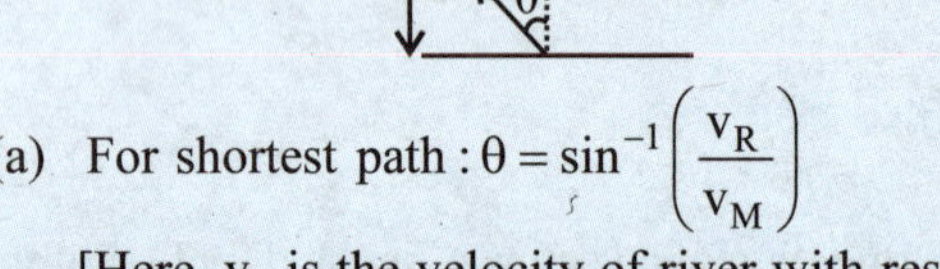

(a) For shortest path : $\theta = \sin^{-1}\left(\frac{v_R}{v_M}\right)$

[Here, v_R is the velocity of river with respect to ground]

(b) time of crossing $= \frac{d}{\sqrt{v_M^2 - v_R^2}}$

[v_M is the velocity of man with respect to river]

(ii) If the man wants to cross the river in the shortest time

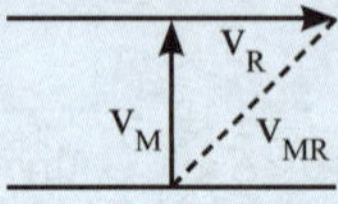

(a) For shortest time

$t_{cross} = \frac{d}{v_M}$

(b) Distance covered by the man along the river

$s = v_R \times \frac{d}{v_M}$

Rain Umbrella Concept

(i) If rain is falling vertically with a velocity $\vec{v}_R$ and an observer is moving horizontally with velocity $\vec{v}_M$, the velocity of rain relative to observer will be:

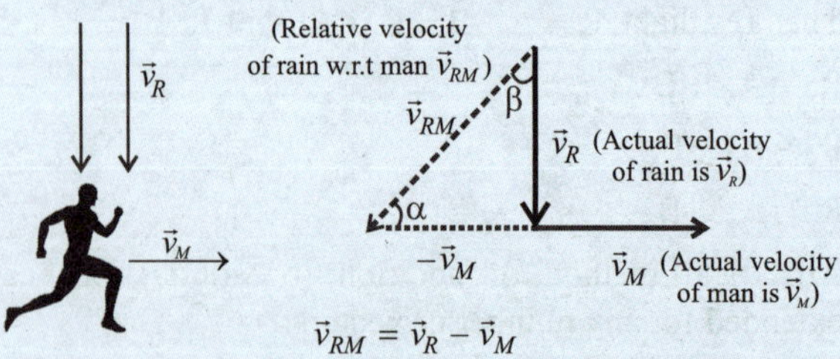

(a) The magnitude of velocity of rain relative to man is $v_{RM} = \sqrt{v_R^2 + v_M^2}$

(b) If β is the angle made by the umbrella with vertical, then, $\tan\beta = \frac{v_M}{v_R}$

(ii) If a man is moving with a velocity v_M relative to ground towards east(positive x-axis), and the rain is falling with a velocity $\vec{v}_R$ relative to ground by making an angle θ with vertical(negative z-axis). Then the velocity of rain relative to man $\vec{v}_{RM}$ is as shown in figure.

$\vec{v}_R = v_{R_x}\hat{i} - v_{R_y}\hat{k}$; $\vec{v}_M = v_M\hat{i}$

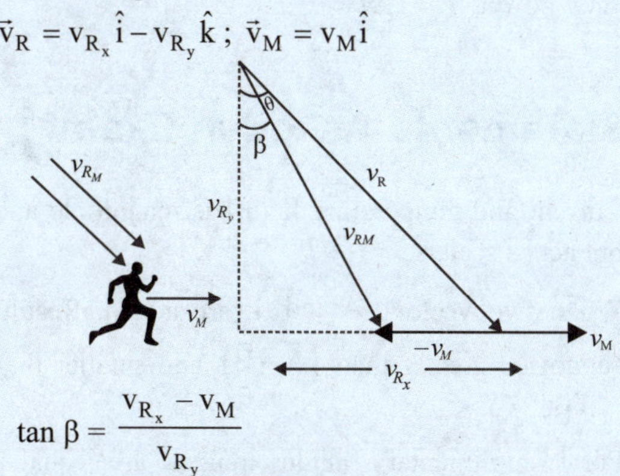

$$\tan\beta = \frac{v_{R_x} - v_M}{v_{R_y}}$$

Projectile Motion

Formulas

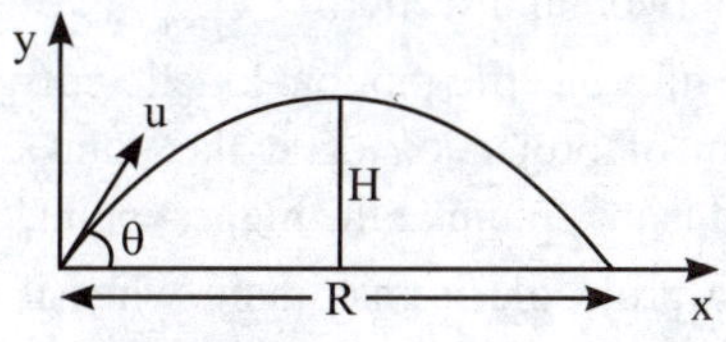

1. Time of flight, $t = \dfrac{2u\sin\theta}{g}$
2. Range of the projectile, $R = \dfrac{u^2\sin 2\theta}{g}$
3. Maximum height $= \dfrac{u^2\sin^2\theta}{2g}$
4. Equation of Trajectory, $y = x\tan\theta - \dfrac{1}{2u^2}\dfrac{gx^2}{\cos^2\theta}$
5. Let t_1 and t_2 are the time of flights for complementary angles then

$$\frac{t_1}{t_2} = \frac{2u\sin\theta}{g} \times \frac{g}{2u\sin(90-\theta)}$$

$$\frac{t_1}{t_2} = \tan\theta$$

6. Product of time of flights for complimentary angles:

$$t_1t_2 = \frac{2u\sin\theta 2u\cos\theta}{g^2} = \frac{2R}{g}$$

Numerical Profile

Horizontal Projectile

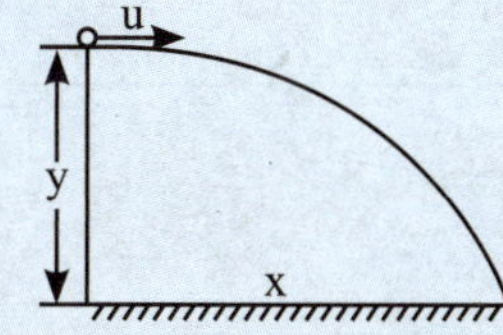

1. Distance covered along horizontal direction, $x = ut$
   [u is the speed in horizontal direction, t is the time taken to hit the ground]
2. Distance covered along vertical direction, $y = \dfrac{1}{2}\dfrac{gx^2}{u^2}$
   [g is the acceleration due to gravity]
3. Time taken to reach the ground, $t = \sqrt{\dfrac{2y}{g}}$
4. Distance covered along horizontal direction, $x = u\sqrt{\dfrac{2y}{g}}$
   [y is the height at which ball is projected]
5. Resultant velocity $(\vec{v})$
   $\vec{v} = \sqrt{v_x^2 + v_y^2}$
   [v_x and v_y is the velocity in x and y direction]
6. Angle made by the resultant velocity with the horizontal direction, $\tan\beta = \dfrac{v_y}{v_x} = \left(\dfrac{gt}{u}\right)$

Horizontal Projectile under Different Cases

(i) Projectile thrown up from the top of tower making an angle θ with the horizontal.

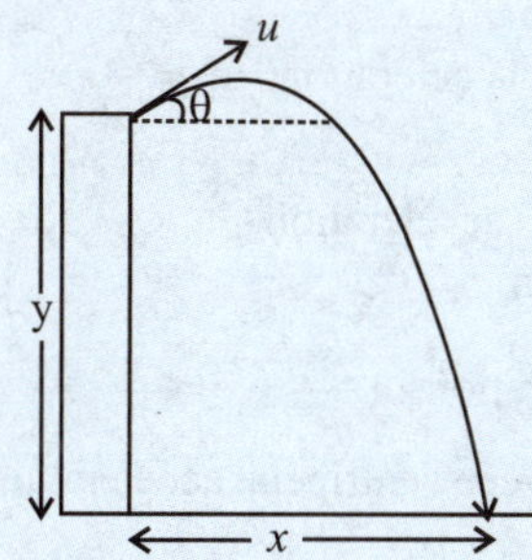

(a) $y = (-u\sin\theta)t + \frac{1}{2}gt^2$
[t is the time taken to reach the ground]

(b) $x = u\cos\theta \times t$

(c) Velocity with which it strikes the ground
$v = \sqrt{(u\sin\theta)^2 + 2gy}$
[v is the magnitude of final velocity]

(d) The angle α with the horizontal at which it strikes the ground

$$\alpha = \tan^{-1}\left[\frac{-u\sin\theta + gt}{u\cos\theta}\right] = \tan^{-1}\left[\frac{\sqrt{u^2\sin^2\theta + 2gy}}{u\cos\theta}\right]$$

(ii) Projectile thrown down from the top of a tower making an angle θ with the horizontal.

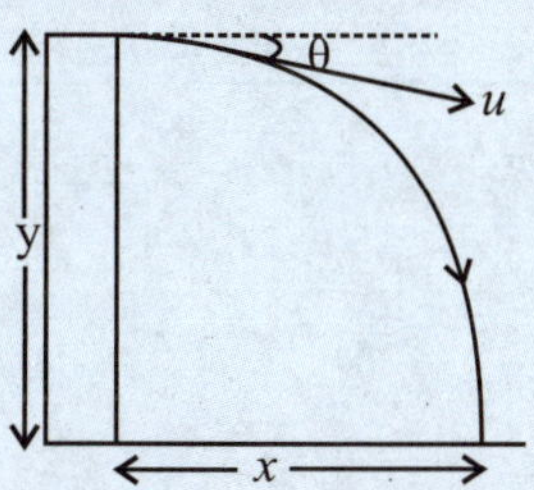

(a) $y=(u\sin\theta)t+\frac{1}{2}gt^2$

(b) $x = u\cos\theta.t$

(c) The velocity with which it strikes the ground $v=\sqrt{u^2+2gy}$

(d) The angle α with the horizontal with which it strikes the ground

$$\alpha=\tan^{-1}\left[\frac{\sqrt{u^2\sin^2\theta+2gy}}{u\cos\theta}\right]$$

Uniform Circular Motion

Formulas

1. Angular speed, $\omega=\frac{\Delta\theta}{\Delta t}$ [θ is the angular displacement]
2. Linear speed in circular motion, $v = \omega R$.
   [R is the radius of the circle, ω is the angular velocity]
3. Centripetal acceleration, $a_c=\frac{v^2}{R}$
4. Tangential acceleration, $a_t = R\alpha$ [α is the angular acceleration]
5. Net acceleration, $a=\sqrt{a_c^{\,2}+a_t^{\,2}}$
6. Angle between centripetal acceleration and net acceleration,
   $\tan\beta=\frac{a_t}{a_c}=\frac{R\alpha}{v^2/R}=\frac{R^2\alpha}{v^2}$

Numerical Profile

(i) Vectorially, linear velocity $\vec{v}=\vec{\omega}\times\vec{r}$

(ii) $F_t = 0; F_{net} = F_c=\frac{mv^2}{r}=m\omega^2 r$

(iii) $F_c=\frac{mv^2}{r}$, Since the speed v is constant, the magnitude of centripetal force is constant.

$\vec{F}_c$ is always perpendicular to velocity.

$\therefore$ Direction of $\vec{F}_c$ must change.

In other words $\vec{F}_c$ is a variable force.

(iv) Work done by $\vec{F}_c$ is always zero.

$\therefore$ Centripetal force cannot change the kinetic energy of a particle. It provides acceleration but does no work.

(v) Power, $P=\vec{F}_c.\vec{v}=0$

Statements Asked in Exam

- Current and temperature both has magnitude and direction but act as scalars.
- When two vectors $\vec{A}$ and $\vec{B}$ are added. Resultant vector cannot be greater than $(\vec{A}+\vec{B})$ and smaller than $(\vec{A}-\vec{B})$ or $(\vec{B}-\vec{A})$
- For complimentary angles ranges are equal, therefore, horizontal range of projectile is same when angle of projection is (i) θ and $90°-\theta$, (ii) $45°+\theta$ and $45°-\theta$.
- At the highest point the projectile possesses velocity only along the horizontal direction.
- In the case of projectile motion, kinetic energy is maximum at the point of projection or at the point of reaching the ground and is minimum at the highest point.
- Range of a projectile is maximum when it is projected at angle 45°.
- Time of flight, horizontal range, maximum height are independent of the mass of a projectile.
- The maximum height attained by a projectile is equal to one fourth of its maximum range.
- Without centripetal force, a body cannot move on a circular path.
- The acceleration vector of a particle in uniform circular motion averaged over one cycle is null.
- A circular motion is an accelerated motion.
- Minimum number of coplanar vectors of unequal magnitude which can give zero resultant is three.
- A vector subtraction does not obey commutative and associative laws.
- At the highest point of a projectile the velocity and acceleration are perpendicular to each other.
- In the case of oblique projection of a projectile momentum is never zero throughout the motion.
- In the case of non-uniform circular motion of an object, is the object possesses both centripetal and tangential acceleration.

MCQs ?

NCERT Topic-wise MCQs

Introduction of Vectors, Vectors Addition & Subtraction

1. Two equal forces have their resultant equal to either of the force, at what angle are they inclined?

1) 120° 2) 360°
3) 90° 4) 60°

2. A particle starting from the origin (0,0) moves in a straight line in the (x,y) plane, its coordinates at a later time are $(\sqrt{3},3)$. The path the particle makes with x-axis, is inclined at an angle of:

1) 45° 2) 60°
3) 30° 4) 90°

3. A unit vector parallel to the resultant of the vectors $\vec{A}=4\hat{i}+3\hat{j}+6\hat{k}$ and $\vec{B}=-\hat{i}+8\hat{j}-8\hat{k}$

1) $\frac{3\hat{i}+11\hat{j}-2\hat{k}}{2}$ 2) $\frac{\hat{i}+2\hat{j}-3\hat{k}}{\sqrt{166}}$
3) $\frac{3\hat{i}+11\hat{j}-2\hat{k}}{\sqrt{134}}$ 4) $\frac{4\hat{i}+6\hat{j}+8\hat{k}}{\sqrt{11}}$

4. Two force whose magnitudes are in the ratio of 3:5 given a resultant of 35 N. If the angle of inclination be 60°, calculate the magnitude of the largest force.

1) 30 N 2) 45 N
3) 25 N 4) 20 N

5. Two equal forces act at a point and are inclined to each other at an angle of 60°. The magnitude of their resultant is:

1) $\sqrt{3}$A 2) 2A
3) 3A 4) $\sqrt{2}$A

6. If vectors $\hat{i}-3\hat{j}+5\hat{k}$ and $\hat{i}-3\hat{j}-a\hat{k}$ are equal vectors then the value of a is:

1) 5 2) 4
3) –4 4) –5

7. If the magnitudes of vectors $\vec{A}$, $\vec{B}$ and $\vec{C}$ are 4, 3 and 5 units respectively and $\vec{A}+\vec{B}=\vec{C}$, the angle between vectors $\vec{A}$ and $\vec{B}$ is:

1) 90° 2) $\cos^{-1}\left(\frac{5}{16}\right)$
3) 30° 4) $\tan^{-1}\left(\frac{12}{5}\right)$

8. Two vectors $\vec{A}$ and $\vec{B}$ are such that $|\vec{A}+\vec{B}|=|\vec{A}-\vec{B}|$ then the angle between the two vectors $\vec{A}$ and $\vec{B}$ will be:

1) 0° 2) $\frac{\pi}{3}$
3) $\frac{\pi}{6}$ 4) $\frac{\pi}{2}$

9. The vector sum of two forces is perpendicular to their vector differences. In that case, the forces:

1) Cannot be predicted
2) Are perpendicular to each other
3) Are equal to each other in magnitude
4) Are not equal to each other in magnitude

10. Find the resultant of three vectors $\overrightarrow{OA}, \overrightarrow{OB}$ and $\overrightarrow{OC}$ shown in the following figure. Radius of the circle is *R*.

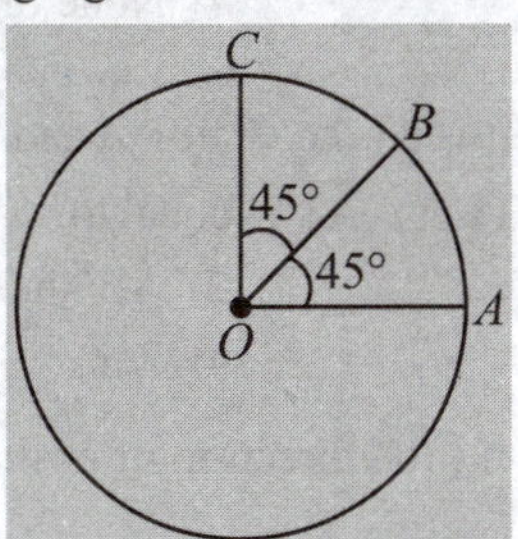

1) $2R$ 2) $R(1+\sqrt{2})$
3) $R\sqrt{2}$ 4) $R(\sqrt{2}-1)$

Vector Products & Resolution

11. The vector projection of a vector $3\hat{i}+5\hat{k}$ on y-axis is:

1) 5 2) 4
3) 3 4) Zero

12. The component of a vector is:

1) Always less than its magnitude
2) Always greater than its magnitude
3) Always equal to its magnitude
4) None of the above

13. If $|\vec{A}\times\vec{B}|=\frac{(\vec{A}.\vec{B})}{\sqrt{3}}$ then the value of angle between $\vec{A}$ and $\vec{B}$ is:

1) 0^0 2) π
3) $\frac{\pi}{2}$ 4) $\frac{\pi}{6}$

14. If a vector $2\hat{i}+4\hat{j}+6\hat{k}$ is perpendicular to the vector $5\hat{i}+6\hat{j}-a\hat{k}$. Then the value of a is:
1) 14/3
2) 17/9
3) 14/5
4) 17/3

15. Consider two vectors $\vec{F_1}=2\hat{i}+5\hat{k}$ and $\vec{F_2}=3\hat{j}+4\hat{k}$. The magnitude of the scalar product of these vectors is:
1) 26 2) 28
3) 30 4) 20

16. Two vectors $\vec{A}$ and $\vec{B}$ have equal magnitudes. If magnitudes of $\vec{A}+\vec{B}$ is equal to n times the magnitude of $\vec{A}-\vec{B}$, then the angle between $\vec{A}$ and $\vec{B}$ is:
1) $\cos^{-1}\left(\frac{n+n^2}{n-n^2}\right)$ 2) $\cos^{-1}\left(\frac{n^2+1}{n^2-1}\right)$
3) $\cos^{-1}\left(\frac{n^2-1}{n^2+1}\right)$ 4) $\cos^{-1}\left(\frac{2+n}{2-n}\right)$

17. What is the angle between $\hat{i}+\hat{j}+\hat{k}$ and $\hat{i}$?
1) $\frac{\pi}{6}$ 2) $\frac{\pi}{4}$
3) $\frac{\pi}{2}$ 4) None of these

18. If the angle between the vector $\vec{A}$ and $\vec{B}$ is 0, the value of the product $(\vec{B}\times\vec{A}).(\vec{A})$ is equal to:
1) $BA^2\cos\theta$ 2) $AB^2\sin\theta$
3) $A^2B\cos\theta$ 4) Zero

19. If $\vec{A}=3\hat{i}+5\hat{j}-7\hat{k}$, the direction of cosines of the vector $\vec{A}$ are:
1) $\frac{2}{\sqrt{83}},\frac{5}{\sqrt{83}},\frac{7}{\sqrt{83}}$ 2) $\frac{3}{\sqrt{83}},\frac{5}{\sqrt{83}},\frac{7}{\sqrt{83}}$
3) $\frac{1}{\sqrt{83}},\frac{2}{\sqrt{83}},\frac{5}{\sqrt{83}}$ 4) $\frac{3}{\sqrt{83}},\frac{5}{\sqrt{83}},\frac{-7}{\sqrt{83}}$

20. The angle between vectors $(\hat{i}+\hat{j})$ and $(\hat{j}+\hat{k})$ is;
1) 90° 2) 180°
3) 0° 4) 60°

21. Let $\vec{A}=A\cos\theta\,\hat{i}+A\sin\theta\,\hat{j}$, be any vector. Another vector $\vec{B}$ which is normal to $\vec{A}$ is:
1) $B\cos\theta\,\hat{i}+B\sin\theta\,\hat{j}$ 2) $B\sin\theta\,\hat{i}+B\cos\theta\,\hat{j}$
3) $B\sin\theta\,\hat{i}-B\cos\theta\,\hat{j}$ 4) $A\cos\theta\,\hat{i}-A\sin\theta\,\hat{j}$

22. If $|\vec{a}\cdot\vec{b}|=|\vec{a}\times\vec{b}|$, then angle between $\vec{a}$ and $\vec{b}$ will be;
1) 30° 2) 45°
3) 90° 4) 180°

23. The adjacent sides of a parallelogram are $\vec{A}=2\hat{i}-3\hat{j}+\hat{k}$ and $\vec{B}=-2\hat{i}+4\hat{j}-\hat{k}$. What is the area of parallelogram?
1) 4 units 2) 7 units
3) $\sqrt{5}$ units 4) $\sqrt{8}$ units

24. Find a unit vector perpendicular to both the vectors $(2\hat{i}+3\hat{j}+\hat{k})$ and $(\hat{i}-\hat{j}+2\hat{k})$.
1) $\frac{1}{\sqrt{8}}(4\hat{i}+3\hat{j}-5\hat{k})$ 2) $\frac{1}{\sqrt{83}}(7\hat{i}-3\hat{j}-5\hat{k})$
3) $\frac{1}{\sqrt{8}}(4\hat{i}+3\hat{j}+1\hat{k})$ 4) $\frac{1}{\sqrt{3}}(7\hat{i}-3\hat{j}-5\hat{k})$

25. A unit vector parallel to the resultant of the vectors $\vec{A}=4\hat{i}+3\hat{j}+6\hat{k}$ and $\vec{B}=-\hat{i}+8\hat{j}-8\hat{k}$
1) $\frac{3\hat{i}+11\hat{j}-2\hat{k}}{2}$ 2) $\frac{\hat{i}+2\hat{j}-3\hat{k}}{\sqrt{166}}$
3) $\frac{3\hat{i}+11\hat{j}-2\hat{k}}{\sqrt{134}}$ 4) $\frac{4\hat{i}+6\hat{j}+8\hat{k}}{\sqrt{11}}$

26. If $\vec{A}=3\hat{i}+4\hat{j}$ and $\vec{B}=6\hat{i}+8\hat{j}$ and A and B are the magnitudes of $\vec{A}$ and $\vec{B}$, then which of the following is not true?
1) $\vec{A}\times\vec{B}=\vec{0}$ 2) $\frac{A}{B}=\frac{1}{2}$
3) $\vec{A}.\vec{B}=48$ 4) $A=5$

27. Two vectors $\vec{P}$ and $\vec{Q}$ are inclined to each other at angle θ. Which of the following is the unit vector perpendicular to $\vec{P}$ and $\vec{Q}$?
1) $\frac{\vec{P}\times\vec{Q}}{P.Q}$ 2) $\frac{\hat{P}\times\hat{Q}}{\sin\theta}$
3) $\frac{\vec{P}\times\vec{Q}}{PQ\sin\theta}$ 4) $\frac{\hat{P}\times\hat{Q}}{PQ\sin\theta}$

28. The angle between vector $(\vec{A}\times\vec{B})$ and $(\vec{B}\times\vec{A})$ is:
1) π rad 2) $\frac{\pi}{2}$ rad
3) $\frac{\pi}{4}$ rad 4) zero

29. If three vector satisfy the relation $\vec{A}.\vec{B}=0$ and $\vec{A}.\vec{C}=0$ then $\vec{A}$ can be parallel to
1) $\vec{C}$ 2) $\vec{B}$
3) $\vec{B}\times\vec{C}$ 4) $\vec{B}.\vec{C}$

Horizontal Projectile, Oblique Projection From Height

30. Which of the following is true regarding projectile motion?
1) Horizontal velocity of projectile is constant
2) Vertical velocity of peojectile is constant
3) Acceleration is not constant
4) Momentum is constant

31. A stone is just released from the window of a train moving along a horizontal straight track. The stone will hit the ground following:
1) Strainght path
2) circulat path
3) Parabolic path
4) Hyperbolic path

32. The angle of projection at which the horizontal range and maximum height of projectile are equal is:
1) 45°   2) $\theta = \tan^{-1}(0.25)$
3) $\theta = \tan^{-1} 4$ 4) 60°

33. At the top of the trajectory of a projectile, the directions of its velocity and acceleration are
1) perpendicular to each other
2) Parallel to each other
3) Inclined to each other at an angle of 45°
4) Antiparallel to each other

34. An object is thrown along a direction inclined at an angle of 45° with the horizontal direction. The horizontal range of the particle is equal to
1) Vertical height
2) Twice the vertical height
3) Thrice the vertical height
4) Four times the vertical height

35. If R and H represent horizontal range and maximum height of the projectile, then the angle of projection with the horizontal is
1) $\tan^{-1}\left(\frac{H}{R}\right)$ 2) $\tan^{-1}\left(\frac{2H}{R}\right)$
3) $\tan^{-1}\left(\frac{4H}{R}\right)$ 4) $\tan^{-1}\left(\frac{4R}{H}\right)$

36. A ball is thrown horizontally from the top of a tower of height 8 m. It touches the ground at a distance of 90 m from the foot of the tower. The initial velocity of the body is:
1) 60 m/s   2) 75 m/s
3) 69.75 m/s   4) None

37. A stone is dropped from the window of a bus moving at 80 km/h. If the window is 150 cm high, find the distance along the track which the stone moves before striking the ground.
1) 15.26 m   2) 13 m
3) 14.16 m   4) 12 m

38. A particle is projected horizontally with a speed 20 m/s from the top of a tower. After what time will the velocity of particle be at 45° angle from the initial direction of projection ? [Let $g = 10\ ms^{-2}$]
1) 2s   2) 5s
3) 4s   4) 7s

39. A body is thrown horizontally with a velocity $\sqrt{2gh}$ from the top of a tower of height h. It strikes the level ground through the foot of the tower at a distance x from the tower. The value of x is-
1) h   2) $\frac{h}{2}$
3) 2h   4) $\frac{2h}{3}$

40. The relation between the time of flight of a projectile, T_f and time to reach the maximum height, t_m is
1) $T_f = 2t_m$ 2) $T_f = t_m$
3) $T_f = \frac{t_m}{2}$ 4) $T_f = \sqrt{2}(t_m)$

Ground to Ground Projectile

41. The maximum range of a gun on horizontal terrain is 15 km. If $g = 10\ m/s^2$. What must be the muzzle velocity of the shell?
1) 387.2 m/s   2) 390 m/s
3) 380 m/s   4) 350 m/s

42. A person can throw a stone to a maximum distance of 500 m. The greatest height to which he can throw the stone is:
1) 50 m   2) 250 m
3) 260 m   4) 1000 m

43. A cricketer can throw a ball to a maximum horizontal distance of 180 m. The speed with which he throws the ball is:
1) 35 m/s   2) 30 m/s
3) 40 m/s   4) 42 m/s

44. A stone is projected in air. Its time of flight is 3 seconds and range is 150 m. the horizontal component of velocity of projection of the stone is ($g = 10\ ms^{-2}$)
1) $22.5\ ms^{-1}$ 2) $15\ ms^{-1}$
3) $50\ ms^{-1}$ 4) $30\ ms^{-1}$

45. A bullet is fired with a velocity u making an angle of 30° with the horizontal plane. The horizontal component of the velocity of the bullet when it reaches the maximum height is:
1) $\frac{u}{\sqrt{2}}$ 2) u
3) $u\sqrt{3}$ 4) $\frac{\sqrt{3}}{2}u$

46. For an object thrown at 30° to horizontal, the maximum height (H) and horizontal range (R) are related as:
1) $R = 16H$ 2) $R = 4\sqrt{3}H$
3) $R = 2\sqrt{2}H$ 4) $R = 2H$

47. The trajectory of a projectile near the surface of the earth is given as $y = 2x - 9x^2$.

If it were launched at an angle θ_0 with speed v_0, then (take, $g = 10\ ms^{-2}$)

1) $\theta_0 = \sin^{-1}\left(\frac{1}{\sqrt{5}}\right)$ and $v_0 = \frac{5}{3} ms^{-1}$

2) $\theta_0 = \cos^{-1}\left(\frac{2}{\sqrt{5}}\right)$ and $v_0 = \frac{3}{5} ms^{-1}$

3) $\theta_0 = \cos^{-1}\left(\frac{1}{\sqrt{5}}\right)$ and $v_0 = \frac{5}{3} ms^{-1}$

4) $\theta_0 = \sin^{-1}\left(\frac{2}{\sqrt{5}}\right)$ and $v_0 = \frac{3}{5} ms^{-1}$

48. Two projectiles A and B thrown with speeds in the ratio $1:\sqrt{2}$ acquired the same heights. If A is thrown at an angle of 60° with the horizontal, the angle of projection of B will be:

1) $\sin^{-1}\left(\frac{\sqrt{3}}{2\sqrt{2}}\right)$ 2) 30°
3) 45° 4) 90°

49. A cricketer hits a ball and it possess kinetic energy E, ball making an angle 30° with horizontal, then its K.E at the highest point is:

1) $\frac{3E}{2}$ 2) E
3) $\frac{\sqrt{3}E}{4}$ 4) $\frac{3E}{4}$

50. A gun fires two bullets at 60° and 30° with the horizontal. The bullets strikes at some horizontal distance. The ratio of maximum heights for the two bullets is in the ratio:

1) 1/4 2) 3/4
3) 3 4) 2

51. A particle is projected from the ground at an angle such that it just clears the top of a pole after time t_1 in its path. It takes further time t_2 to reach the ground. What is the height of the pole ?

1) $\frac{t_1 - t_2}{2g}$ 2) $\frac{1}{2}gt_1t_2$
3) $\frac{g(t_1+t_2)}{2}$ 4) $\frac{g}{t_1t_2}$

Rain-Man, River-Boat Problems, Relative Motion in 2D

52. Suppose that two objects A and B are moving with velocities $\vec{v}_A$ and $\vec{v}_B$ (each with respect to some common frame of reference). Let $\vec{v}_{AB}$ represent the velocity of A with respect to B. Then

1) $\vec{v}_{AB} + \vec{v}_{BA} = 0$ 2) $\vec{v}_{AB} - \vec{v}_{BA} = 0$
3) $\vec{v}_{AB} = \vec{v}_A + \vec{v}_B$ 4) $|\vec{v}_{AB}| \neq |\vec{v}_{BA}|$

53. A car is going in south with a speed of 5 m/s. To a man sitting in car a bus appears to move towards west with a speed of $2\sqrt{6}$ m/s. What is the actual speed of the bus?

1) 4 ms^{-1} 2) 3 ms^{-1}
3) 7 ms^{-1} 4) none of these

54. A flag is mounted on a car moving due North with velocity of 20 km/hr. Strong winds are blowing due East with velocity of 20 km/hr. The flag will point in direction

1) East 2) North-East
3) South-East 4) South-West

55. A man wishes to cross a river in a boat. If he crosses the river in minimum time he takes 10 minutes with a drift of 120 m. If he crosses the river taking shortest route, he takes 12.5 minutes, find velocity of the boat with respect to water.

1) 20 m/min 2) 12 m/min
3) 10 m/min 4) 8 m/min

56. Suppose that two objects A and B are moving with velocities $\vec{v}_A$ and $\vec{v}_B$ (each with respect to some common frame of reference). Let $\vec{v}_{AB}$ represents the velocity of A with respect to B then:

1) $\vec{v}_{AB} + \vec{v}_{BA} = 0$ 2) $\vec{v}_{AB} - \vec{v}_{BA} = 0$
3) $\vec{v}_{AB} \neq \vec{v}_A + \vec{v}_B$ 4) $|\vec{v}_{AB}| \neq |\vec{v}_{BA}|$

57. Rain is falling vertically with speed of 35 m/s. A woman rides a bicycle with a speed of 15 m/s in East to West direction. What is the direction in which she should hold her umbrella?

1) $\tan^{-1}\left(\frac{3}{7}\right)$ with the vertical towards west
2) $\tan^{-1}\left(\frac{4}{7}\right)$ with the vertical towards east
3) $\tan^{-1}\left(\frac{5}{7}\right)$ with the vertical towards east
4) Towards North downward

58. The speed of boat is 18 km/h in still water. It crosses a river of width 2 km along the shortest path in 7 minutes. The velocity of the river is:

1) 5.5 km/h
2) 6 km/h
3) 7 km/h
4) 10 km/h

59. A man walks in rain with a velocity of 10 km/h. The rain drops strikes at him at an angle of 45° with the horizontal. What will be the downward velocity of the raindrops and relative velocity of rain w.r.t man?

1) $\vec{v}_r = 10$ km/h, $\vec{v}_{rm} = 10$ m/s
2) $\vec{v}_r = 20$ km/h, $\vec{v}_{rm} = 10\sqrt{2}$ km/h
3) $\vec{v}_r = 10$ km/h, $\vec{v}_{rm} = 10\sqrt{2}$ km/h
4) $\vec{v}_r = 1$ km/h, $\vec{v}_{rm} = 1$ km/h

60. A boat man could row his boat with a speed 10m/sec. He wants to take his boat from P to a point Q just opposite on the other bank of the river flowing at a speed 4m/sec. He should row his boat:
1) At right angle to the stream
2) At an angle of $\sin^{-1}$ (2/5) with PQ up the stream
3) At an angle $\sin^{-1}$ (2/5) with PQ down the stream
4) At an angle $\cos^{-1}$ (2/5) with PQ down the stream

61. A boat which has a speed of 5 km/hr in still water crosses a river of width 1 km along the shortest possible path in 15 minutes. The velocity of the river water in km/hr is:
1) 1 2) 3
3) 4 4) $\sqrt{41}$

62. A man standing on a road has to hold his umbrella at 30° with the vertical to keep the rain away. He thrown the umbrella and starts running at 10 km/h. He finds that rain drop are hitting his head vertically. Find the speed of rain w.r.t. road:
1) 10 km/s 2) 20 km/h
3) $10\sqrt{3}$ km / s 4) $20\sqrt{3}$ km / s

63. The position vector of particle changes with time according to the relation $\vec{r}(t)=15t^2\hat{i}+(4-20t^2)\hat{j}$. What is the magnitude of the acceleration (in ms^{-2}) at t = 1s?
1) 50 2) 100
3) 25 4) 40

64. A man moves on a cycle with a velocity of 4 km/hr. The rain appears to fall to him with a velocity of 3 km/hr vertically. The actual velocity of rain is
1) 7 km hr^{-1} 2) $\frac{4}{3}$ km hr^{-1}
3) $\frac{3}{4}$ km hr^{-1} 4) 5 km/hr

Circular Motion Kinematics

65. A particle is moving on a circular path with constant speed. Then its acceleration will be
1) Zero
2) External radial acceleration
3) Internal radial acceleration
4) Contant acceleration

66. Certain neutron stars are believed to be rotating at about 1 rev/sec. If such a star has a radius of 20 km, the acceleration of an object on the equator of the star will be
1) 20×10^8 m/sec^2 2) 8×10^5 m/sec^2
3) 120×10^5 m/sec^2 4) 4×10^8 m/sec^2

67. The length of second's hand in a watch is 1 cm. The change in velocity of its tip in 15 seconds is
1) Zero 2) $\frac{\pi}{30\sqrt{2}}$ cm/sec
3) $\frac{\pi}{30}$ cm/sec 4) $\frac{\pi\sqrt{2}}{30}$ cm/sec

68. A particle moves in a circle of radius 25 cm at two revolutions per second. The acceleration of the particle in m/s^2 is
1) π^2 2) $8\pi^2$
3) $4\pi^2$ 4) $2\pi^2$

69. A particle is performing circular motion of radius 1 m. Its speed is v = ($2t^2$) m/s. What will be the magnitude of its acceleration at t = 1s ?
1) $4\sqrt{2}$ 2) $2\sqrt{2}$
3) $5\sqrt{2}$ 4) $7\sqrt{2}$

70. In a ground, a cyclist follows a track that turn to his right by an angle of 60° after every 100 m. The displacement of the cyclist from the starting point to the fourth turn:
1) $200\sqrt{10}$ 2) $100\sqrt{2}$
3) $2\sqrt{2}$ 4) $4\sqrt{6}$

71. If a body is moving in a circle of radius r with a constant speed v, its angular velocity is:
1) $\frac{v^2}{r}$ 2) vr
3) $\frac{v}{r}$ 4) $\frac{r}{v}$

72. The angular speed of seconds needle in a mechanical watch is:
1) π rad/s 2) $\pi \times 30$ rad/s
3) $\frac{\pi}{2}$ rad / s 4) $\frac{\pi}{30}$ rad / s

73. If a_r and a_t represents radial and tangential accelerations, the motion of a particle will be uniformly circular if:
1) $a_r = 0$ and $a_t = 0$ 2) $a_r = 0$ and $a_t \neq 0$
3) $a_r \neq 0$ and $a_t = 0$ 3) $a_r \neq 0$ and $a_t \neq 0$

74. Which of the following statements is not correct in uniform circular motion?
1) The speed of the particle remains constant
2) The linear velocity remains constant
3) The acceleration always points towards the centre
4) The angular speed remains constant

75. The angle turned by a body undergoing circular motion depends on time as $\theta = \theta_0 + \theta_1 t + \theta_2 t^2 + \theta_3 t^3$. The angular acceleration will be:
1) θ_3 2) $2\theta_2 + 6\theta_3 t$
3) $6\theta_3 t$ 4) $\theta_2 + 6\theta_3$

76. A particle moving along the circular path with a speed v and its speed increases by g in one second. If the radius of the circular path be r, then the net acceleration of the particle is:

1) $\frac{v^2}{r}+g$
2) $\frac{v^2}{r^2}+g^2$
3) $\left(\frac{v^2}{r}+g\right)^{1/2}$
4) $\left(\frac{v^4}{r^2}+g^2\right)^{1/2}$

77. An object of mass 4 kg is being rotated with 120 rev. per minute in a horizontal circular path of radius 4m. Its kinetic energy is:

1) 5120 J
2) 4126 J
3) 6000 J
4) 346 J

78. A car is moving with speed 40 m/s on a circular path of radius 500 m. Its speed is increasing at the rate of 4 m/s^2. What is the acceleration of the car:

1) 6.12 m/s^2
2) 4×10^2 m/s^2
3) 5.122 m/s^2
4) 3.9 m/s^2

79. A body moving along a circular path of radius R with velocity v, has centripetal acceleration a. If its velocity is made equal to 2v. What will be the centripetal acceleration?

1) 4a
2) 2a
3) $\frac{a}{4}$
4) $\frac{a}{2}$

High Order Time Intensive MCQs

1. What is the magnitude of average velocity after half rotation, if a particle is moving with constant speed v in a circle?

1) 2v
2) $2\frac{v}{\pi}$
3) $\frac{v}{2}$
4) $\frac{v}{2\pi}$

2. The equation of a projectile is $y=\sqrt{3}x-\frac{gx^2}{2}$. the angle of projection is

1) $\tan^{-1}\left(\frac{1}{\sqrt{3}}\right)$
2) $\tan^{-1}\left(\sqrt{3}\right)$
3) $\frac{\pi}{2}$
4) Zero

3. A body is projected horizontally from the top of a tower with a velocity of 10 m/s. If it hits the ground at an angle of 45°, then the vertical component of velocity when it hits ground (in m/s) is

1) $10\sqrt{2}$
2) $5\sqrt{2}$
3) 5
4) 10

4. A body is projected with an angle θ. The maximum height reached is h. If the time of flight is 4s and g is 10 m/s^2, then value of h is

1) 40 m
2) 20 m
3) 5 m
4) 10 m

5. An aeroplane is flying horizontally with a velocity of 600 km/h and at a height of 1960 m. When it is vertically above a point A on the ground, a bomb is released from it. The bomb strikes the ground at point B. The distance AB is

1) 1200 m
2) 0.33 km
3) 33.3 km
4) 3.33 km

6. The equation of trajectory of a projectile is $y=10x-\left(\frac{5}{9}\right)x^2$. If we assume g = 10 ms^{-2}. What will be the range of projectile ?

1) 36 m
2) 24 m
3) 18 m
4) 9 m

7. A body is projected with a speed u m/s at an angle β with the horizontal. The kinetic energy at the highest point is $\frac{3}{4}$th of the initial kinetic energy. The value of β is

1) 30°
2) 45°
3) 60°
4) 120°

8. A particle is projected with a velocity v such that its range on the horizontal plane is twice the greatest height attained by it. What is the range of the projectile ?

1) $\frac{4v^2}{5g}$
2) $\frac{4g}{5v^2}$
3) $\frac{v^2}{g}$
4) $\frac{4v^2}{\sqrt{5}g}$

9. A projectile can have the same range R for two angles of projection. If t_1 and t_2 be the times of flights in the two cases, then the product of the two times of flights is proportional to

1) R^2
2) $\frac{1}{R^2}$
3) $\frac{1}{R}$
4) R

10. A ball is projected from the ground at angle θ with horizontal. After 1 s it is moving at angle 45° with the horizontal and after 2 s it is moving horizontally. What is the velocity of projection of the ball?

1) $10\sqrt{3}ms^{-1}$ 2) $20\sqrt{3}ms^{-1}$

3) $10\sqrt{5}ms^{-1}$ 4) $20\sqrt{2}ms^{-1}$

11. A projectile is projected at an angle 60° with horizontal with speed 10 ms^{-1}. The minimum radius of curvature of trajectory is:

1) 2.55 m 2) 2 m

3) 10 m 4) None of these

12. Two stones are projected with same velocity v at an angle θ and (90° – θ). If H and H_1 are the greatest heights in the two paths, then what is the relation between R, H and H_1?

1) $R = 4\sqrt{HH_1}$ 2) $R = \sqrt{HH_1}$

3) $R = HH_1$ 4) None of these

13. A stone thrown at an angle θ to the horizontal reaches maximum height H. Then, the time of flight of stone will be

1) $\sqrt{\frac{2H}{g}}$ 2) $2\sqrt{\frac{2H}{g}}$

3) $2\frac{\sqrt{2H\sin\theta}}{g}$ 4) $\frac{\sqrt{2H\sin\theta}}{g}$

14. Given, $\vec{A} = 2\hat{i} + 3\hat{j}$ and $\vec{B} = \hat{i} + \hat{j}$. The component of vector A along vector B is

1) $\frac{1}{\sqrt{2}}$ 2) $\frac{3}{\sqrt{2}}$

3) $\frac{5}{\sqrt{2}}$ 4) $\frac{7}{\sqrt{2}}$

15. A B C D E F is a regular hexagon with point O as centre. The value of $\overrightarrow{AB} + \overrightarrow{AC} + \overrightarrow{AD} + \overrightarrow{AE} + \overrightarrow{AF}$ is

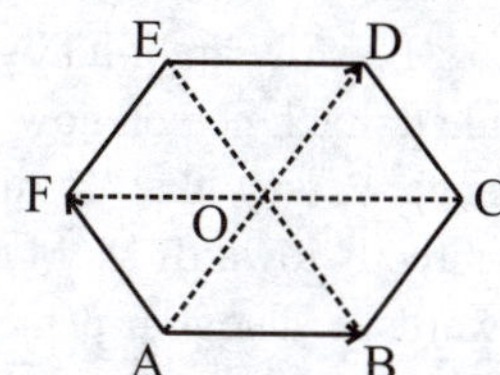

1) $2\overrightarrow{AO}$ 2) $4\overrightarrow{AO}$

3) $6\overrightarrow{AO}$ 4) Zero

16. An insect trapped in a circular groove of radius 12 cm moves along the groove steadily and completes 7 revolutions in 100 s. The linear speed of the insect is

1) 4.3 cm s^{-1} 2) 5.3 cm s^{-1}

3) 6.3 cm s^{-1} 4) 7.3 cm s^{-1}

17. A car travels 6 km towards the north at an angle 45° to the east and then travels distance of 4 km towards the north at an angle 135° to the east. How far is the point from the starting point? What angle does the straight line joining its initial and final positions makes with the east?

1) $\sqrt{50}$ km and $\tan^{-1}(5)$

2) 10 km and $\tan^{-1}(\sqrt{50})$

3) $\sqrt{52}$ km and $\tan^{-1}(5)$

4) $\sqrt{52}$ km and $\tan^{-1}(\sqrt{5})$

18. A proton in a cyclotron changes its velocity from 30 km/h north to 45 km/h east in 20 s. What is the magnitude of average acceleration during the same?

1) 2.5 km/s^2 2) 12.5 km/s^2

3) 22.5 km/s^2 4) 32.5 km/s^2

19. Given, two vectors $\vec{A} = -4\hat{i} + 4\hat{j} + 2\hat{k}$ and $\vec{B} = 2\hat{i} - \hat{j} - \hat{k}$. The angle made by $(\vec{A} + \vec{B})$ with $\hat{i} + 2\hat{j} - 4\hat{k}$ is

1) 30° 2) 45°

3) 60° 4) 90°

20. Vector which is perpendicular to $(a\cos\theta\hat{i} + b\sin\theta\hat{j})$ is

1) $b\sin\theta\hat{i} - a\cos\theta\hat{j}$ 2) $\frac{1}{a}\sin\theta\hat{i} - \frac{1}{b}\cos\theta\hat{j}$

3) $5\hat{k}$ 4) All of the above

21. If $\vec{A} = 2\hat{i} + 3\hat{j} + 6\hat{k}$ and $\vec{B} = -\hat{i} + 3\hat{j} + 4\hat{k}$, then projection of $\vec{A}$ on $\vec{B}$ will be

1) $\frac{31}{\sqrt{13}}$ 2) $\frac{31}{\sqrt{26}}$

3) $\sqrt{\frac{31}{26}}$ 4) $\sqrt{\frac{31}{13}}$

22. The position vector of a particle is r = (a cos ωt) $\hat{i}$ + (as in ωt) $\hat{j}$.

The velocity vector of the particle is

1) Parallel to position vector

2) Perpendicular to position vector

3) Directed towards the origin

4) Directed away from the origin

23. Angle (in rad) made by the vector $\sqrt{3}\hat{i} + \hat{j}$ with the X-axis is

1) $\frac{\pi}{6}$ 2) $\frac{\pi}{4}$

3) $\frac{\pi}{3}$ 4) $\frac{\pi}{2}$

24. Raindrops are falling vertically with a velocity of 10 m/s. To a cyclist moving on a straight road the raindrops appear to be coming with a velocity of 20 m/s. The velocity of cyclist is:

1) 10 m/s 2) $10\sqrt{3}$ m/s

3) 20 m/s 4) $20\sqrt{3}$ m/s

25. A man starts running along a straight road with uniform velocity $u\hat{i}$ observes that the rain is falling vertically downward. If he doubles his speed, he finds that the rain is coming at an angle θ to the vertical. The velocity of rain with respect to the ground is:

1) $u\hat{i} - u\tan\theta\hat{j}$ 2) $u\hat{i} - u\cot\theta\hat{j}$

3) $u\hat{i} + u\cot\theta\hat{j}$ 4) $\frac{u}{\tan\theta}\hat{i} - u\hat{j}$

26. A body of mass 3 kg is at rest. If a force of $\vec{F} = 6t^2\hat{i} + 3\hat{j}$ is applied on the body then the velocity of body at t = 3 sec is

1) $18\hat{i} + 3\hat{j}$ 2) $18\hat{i} + 6\hat{j}$

3) $3\hat{i} + 18\hat{j}$ 4) $18\hat{i} + 4\hat{j}$

27. A projectile is projected at an angle (α > 45°) with an initial velocity u. The time t, at which its magnitude of horizontal velocity will equal the magnitude of vertical velocity is:

1) $t = \frac{u}{g}(\cos\alpha - \sin\alpha)$ 2) $t = \frac{u}{g}(\cos\alpha + \sin\alpha)$

3) $t = \frac{u}{g}(\sin\alpha - \cos\alpha)$ 4) $t = \frac{u}{g}(\sin^2\alpha - \cos^2\alpha)$

28. A projectile is thrown from a point in a horizontal plane such that the horizontal and vertical velocities are 9.8 ms^{-1} and 19.6 ms^{-1}. It will strike the plane after covering distance of

1) 39.2 m 2) 19.6 m

3) 9.8 m 4) 4.9 m

29. Rain water is falling vertically downwards with velocity v, when the velocity of wind is zero, water is collected at a rate R. When the wind starts blowing horizontally at a speed u, the rate of collection of water in the same vessel is:

1) $\sqrt{u^2 + v^2}R$ 2) $\frac{v}{u}R$

3) $\frac{uR}{v}$ 4) R

30. The magnitude of the displacement of a particle moving in a circle of radius a with constant angular speed ω varies with time t as:

1) 2asinωt 2) $2a\sin\frac{\omega t}{2}$

3) 2acosωt 4) $2a\cos\frac{\omega t}{2}$

31. Two guns A and B can fire bullets at speeds 1 km/s and 2 km/s, respectively. From a point on a horizontal ground they are fired in all possible directions. The ratio of maximum areas covered by the bullets on the ground fired by the two guns is

1) 1 : 4 2) 1 : 16

3) 1 : 8 4) 1 : 2

32. A particle moves along the positive part of the curve $y = \frac{x^2}{2}$, where $x = \frac{t^2}{2}$. What will be velocity of particle at t = 2 sec.

1) $2\hat{i} - 4\hat{j}$ 2) $2\hat{i} + 4\hat{j}$

3) $4\hat{i} + 2\hat{j}$ 4) $4\hat{i} - 2\hat{j}$

33. A boat move along the flow of river between two fixed point A and B. It take t_1 time when going downstream and takes t_2 time when going upstream between these two points. What time it will take in still water to cover the distance equal to AB.

1) $\frac{2t_1t_2}{t_1 + t_2}$ 2) $t_1 + t_2$

3) t_1/t_2 4) $\frac{2t_1t_2}{t_1 - t_2}$

34. A boat can be rowed at 5 m/s in still water. It is used to cross a 200 m wide river from south bank to the north bank. The river current has uniform velocity of 3 m/s due east.

In which direction must it be steered to cross the river perpendicular to current ?

1) 37° 2) 30°

3) 60° 4) 90°

35. A particle moves in xy plane with a velocity given by $\vec{v} = (8t - 2)\hat{i} + 2\hat{j}$. If it passes through the point (0, 0) at t = 0 sec., then give equation of the path.

1) y(y – 1) = x 2) y(x – 1) = y

3) x(y – 1) = x 4) x(x – 1) = y

36. A ball is projected so as to pass a wall at a distance *a* from the point of projection at an angle of 45° and falls at a distance *b* on the other side of the wall. If *h* is height of wall then:

1) $h = a\sqrt{2}$ 2) $h = b\sqrt{2}$

3) $h = \frac{\sqrt{2}ab}{a + b}$ 4) $h = \frac{ab}{a + b}$

37. Four persons K, L, M, N are initially at the four corners of a square of side d. Each person now moves with a uniform speed v in such a way that K always moves directly towards L, L directly towards M, M directly towards N and N directly towards K. The four persons will meet at a time

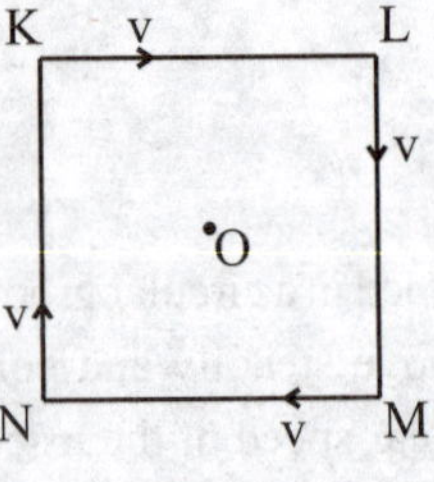

1) 2d/v 2) d/v

3) 3d/ v 4) d/3v

38. The x and y co-ordinates of a particle are

$x = A\sin\omega t$ and $y = 2A\sin\left(\omega t + \frac{\pi}{2}\right)$, then the motion of the particle is

1) Circular
2) Parabolic
3) Elliptical anti clockwise
4) Elliptical clockwise

39. A ball *B* is at 300 cm distance from origin on a line 37° above horizontal. Another ball *A* is projected directly aiming *B* with initial velocity 700 cm/s. At the same instant *B* is released from its position. How far will *B* have fallen when it is hit by *A*?

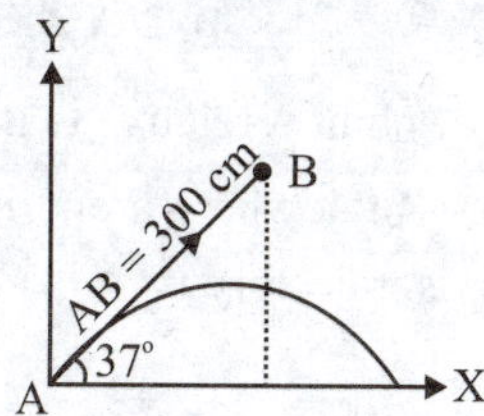

1) 90 cm
2) 80 cm
3) 70 cm
4) 60 cm

40. A particle starts with initial velocity of $(2\hat{i}+\hat{j})$ m/s. Uniform acceleration $(-\hat{i}+3\hat{j})$ m/s². What is *y* component of velocity at the instant when *x* component of velocity becomes zero?

1) 5 m/s
2) 7 m/s
3) 6 m/s
4) 10 m/s

41. A particle is projected in such a way that, taking point of projection as origin, $y = 8t - 5t^2$ and $x = 6t$. What is the range of projectile?

1) 48 m
2) 4.8 m
3) 9.6 m
4) 24 m

42. A body falling freely from a given height H hits an inclined plane in its path at a height 'h'. As a result of this impact the direction of the velocity of the body becomes horizontal. Find the total time the body will take to reach the ground.

1) $\sqrt{\frac{2}{g}}\left(\sqrt{h}+\sqrt{H-h}\right)$
2) $\sqrt{2gh}$
3) $\sqrt{\frac{2}{g}}\left(\sqrt{H-h}\right)$
4) None of the above

43. A particle is projected from the ground with an initial speed of v at an angle θ with horizontal. The average velocity of the particle between its point of projection and highest point of trajectory is

1) $\frac{v}{2}\sqrt{1+2\cos^2\theta}$
2) $\frac{v}{2}\sqrt{1+\cos^2\theta}$
3) $\frac{v}{2}\sqrt{1+3\cos^2\theta}$
4) $v\cos\theta$

44. The horizontal range and maximum height attained by a projectile are R and H respectively. If a constant horizontal acceleration a = g/4 is imparted to the projectile due to wind, then its horizontal range and maximum height will be

1) $(R+H), \frac{H}{2}$
2) $\left(R+\frac{H}{2}\right), 2H$
3) $(R + 2H), H$
4) $(R + H), H$

45. The velocity vectors of a point A varies with time as $\vec{v} = a\hat{i} + bt\hat{j}$, where a and b are positive constants. Find the time dependence of magnitude of velocity vector.

1) $a^2 + b^2t^2$
2) $\sqrt{a+b^2t^2}$
3) $\sqrt{a^2+b^2t^2}$
4) $\sqrt{a^2+bt}$

46. Two bodies fall freely from the same height, but the second body starts falling T seconds after the first. The time (after which the first body begins to fall) when the distance between the bodies equals L is:

1) $\frac{1}{\sqrt{gh}}$
2) $\frac{1}{\sqrt{2gh}}$
3) $\frac{1}{gh}$
4) None of these

47. A body is projected at t = 0 with a velocity 10 ms^{-1} at an angle of 60° with the horizontal. The radius of curvature of its trajectory at t = 1s is R. Neglecting air resistance and taking acceleration due to gravity g = 10 ms^{-2}, the value of R is

1) 10.3 m
2) 2.8 m
3) 5.1 m
4) 2.5 m

48. Two particles A and B are moving with equal speed v but in opposite direction such that distance between them remains constant which is D. After how much time A will reach the initial position?

1) $\frac{2D}{v}$
2) $\frac{\pi D}{v}$
3) $\frac{2\pi D}{v}$
4) $\frac{\pi D}{2v}$

49. A point moves along a circle with a velocity v = at, where a = 0.50 m/s². Find the total acceleration of the point at the moment when it has covered the nth (n = 0.10) fraction of the circle after beginning of the motion.

1) 0.8 m/s²
2) 0.6 m/s²
3) 0.7 m/s²
4) 0.9 m/s²

50. A particle rotates along a circle of radius $R = \sqrt{2}$m with an angular acceleration $\alpha = \frac{\pi}{4}\text{rad/s}^2$ starting from rest. Calculate the magnitude of average velocity of the particle over the time it rotates a quarter circle.

1) 0.5 m/s
2) 1 m/s
3) 2 m/s
4) 1.5 m/s

51. The velocity-time graph of a particle in one dimensional motion is shown in the figure. Which of the following formula is correct for describing the motion of the particle over the time interval t_1 to t_2?

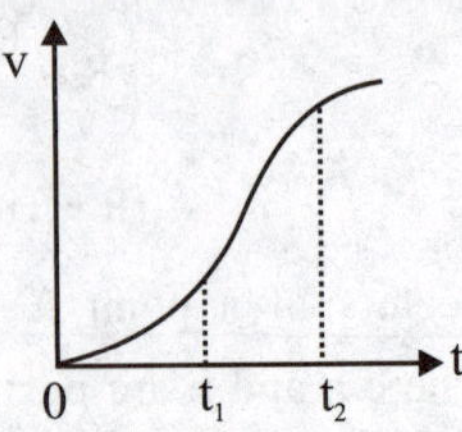

1) $x(t_2) = x(t_1) + v(t_1)(t_2 - t_1) + \left(\frac{1}{2}\right)a(t_2 - t_1)^2$

2) $v(t_2) = v(t_1) + a(t_2 - t_1)$

3) $v_{average} = \frac{[x(t_2) + x(t_1)]}{(t_2 - t_1)}$

4) $a_{average} = \frac{[v(t_2) - v(t_1)]}{(t_2 - t_1)}$

52. A particle moves from the point $(2.0\hat{i} + 4.0\hat{j})$m at t = 0 with an initial velocity $(5.0\hat{i} + 4.0\hat{j})\text{ms}^{-1}$. It is acted upon by a constant force which produces a constant acceleration $(4.0\hat{i} + 4.0\hat{j})\text{ms}^{-2}$. What is the distance of the particle from the origin at time 2 s?

1) 5m 2) $20\sqrt{2}$ m
3) $10\sqrt{2}$ m 4) 15 m

53. The vector $\hat{i} + x\hat{j} + 3\hat{k}$ is rotated through an angle θ and doubled in magnitude, then it becomes $4\hat{i} + (4x - 2)\hat{j} + 2\hat{k}$. The values of x is

1) $-\frac{2}{3}$ 2) $\frac{1}{3}$
3) $\frac{2}{3}$ 4) None

54. Three vectors as shown in the fig. have magnitudes $|\vec{a}| = 3, |\vec{b}| = 4$ and $|\vec{c}| = 10$. Find the numbers p and q such that $\vec{c} = p\vec{a} + q\vec{b}$.

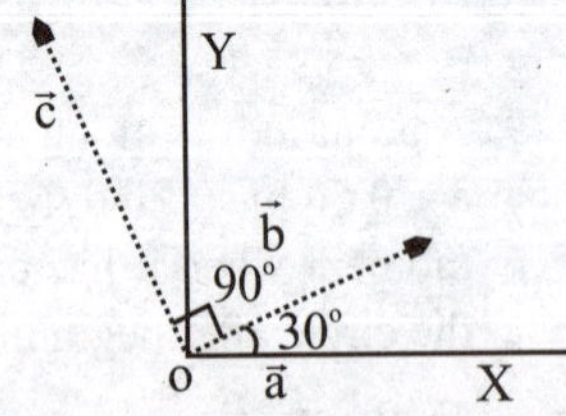

1) $-\frac{20}{3}, \frac{5\sqrt{3}}{2}$

2) $\frac{20}{3}, \frac{5\sqrt{3}}{2}$

3) $-\frac{10}{3}, \frac{\sqrt{3}}{2}$

4) None

55. Find a unit vector perpendicular to both the vectors $(2\hat{i} + 3\hat{j} + \hat{k})$ and $(\hat{i} - \hat{j} + 2\hat{k})$.

1) $\frac{1}{\sqrt{8}}(4\hat{i} + 3\hat{j} - 5\hat{k})$

2) $\frac{1}{\sqrt{83}}(7\hat{i} - 3\hat{j} - 5\hat{k})$

3) $\frac{1}{\sqrt{8}}(4\hat{i} + 3\hat{j} + 1\hat{k})$

4) None of these

56. The resultant of $\vec{A} + \vec{B}$ is $\vec{R}_1$. On reversing the vector $\vec{B}$, the resultant becomes $\vec{R}_2$. What is the value of $R_1^2 + R_2^2$?

1) $A^2 + B^2$ 2) $A^2 - B^2$
3) $2(A^2 + B^2)$ 4) $2(A^2 - B^2)$

57. A radius vector of point A relative to the origin varies with time t as $\vec{r} = at\hat{i} - bt^2\hat{j}$ where a and b are constants. The equation of point's trajectory is

1) $y = -\frac{b}{a^2}x^2$

2) $y = \frac{b}{a^2}x^2$

3) $y = -\frac{2b}{a^2}x^2$

4) $y = \frac{2b}{a^2}x^2$

58. A rigid body is spinning with an angular velocity of 4 rad/sec about an axis $(3\hat{j} + \hat{k})$ passing through the point A (1, 3, –1). Then the velocity of the particle at the point P (4, –2, 1) with respect to point A is:

1) $\frac{4}{\sqrt{10}}(\hat{i} - 3\hat{j} - 9\hat{k})$

2) $-\hat{i} + 3\hat{j} + 9\hat{k}$

3) $13(3\hat{i} - 5\hat{j} + 2\hat{k})$

4) None of above

59. Find the resultant of three vectors shown in figure. Radius of circle is R:

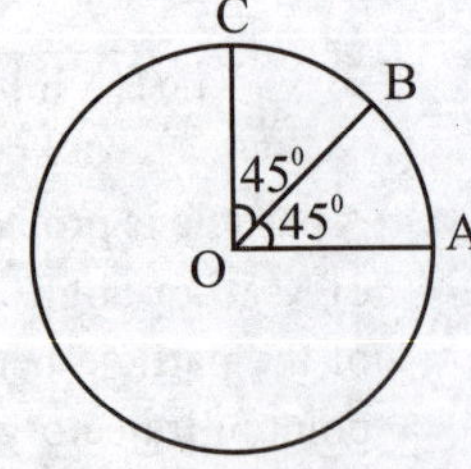

1) $R(1 + \sqrt{2})$

2) $R(\sqrt{2} - 1)$

3) $R^2(\sqrt{2} - 1)$

4) $R(\sqrt{3} - \sqrt{2})$

Statement Based MCQs

Directions: These questions consist of two statements each, printed as Statement-I and Statement-II. While answering these questions, you are required to choose any one of the following four responses.

1) **Both Statement-I and Statement-II are correct.**
2) **Both Statement-I and Statement-II are incorrect.**
3) **Statement-I is correct & Statement-II is incorrect.**
4) **Statement-I is incorrect & Statement-II is correct.**

1. **Statement-I:** Horizontal velocity of a projectile remains constant throughout the motion of the projectile.

   Statement-II: There is an acceleration acting on the projectile along the horizontal direction.

2. **Statement-I:** Equal vectors must have same magnitude as well as same direction.

   Statement-II: Negative vectors have equal magnitude and are in the same direction.

3. **Statement-I:** In the presence of air resistance, the range and maximum height attained by projectile are reduced but the time of flight and the angle with which the projectile strikes the ground are increased.

   Statement-II: Air resistance is like a drag force which always oppose the motion (relative motion).

4. **Statement-I:** A body thrown horizontally strikes the ground with the same velocity with which it was thrown.

   Statement-II: Vertical velocity of a projectile thrown horizontally remains same throughout the motion.

5. **Statement-I:** When a particle moves in a circular path with a uniform speed, its velocity and acceleration both remains same.

   Statement-II: The centripetal acceleration in circular motion is independent of the angular velocity of the body.

Assertion & Reason MCQs

Directions: These questions consist of two statements each, printed as Assertion and Reason. While answering these questions, you are required to choose any one of the following four responses.

1) **Both Assertion and Reason are True and the Reason is a correct explanation of the Assertion.**
2) **Both Assertion and Reason are True but Reason is not a correct explanation of the Assertion.**
3) **Assertion is True but the Reason is False.**
4) **Assertion is False but Reason is True.**

1. **Assertion:** For large angular displacements, average angular velocity is a scalar quantity.

   Reason: Large angular displacement is a scalar.

2. **Assertion:** A man can cross river of width d in time t. On increasing river velocity, time taken to cross the river by man will remain unchanged.

   Reason: Velocity of river is perpendicular to width of river. So time to cross the river is independent of velocity of river.

3. **Assertion:** When a body is dropped or thrown horizontally from the same height, it reaches the ground at the same time.

   Reason: They have same acceleration and same initial speed in vertical direction.

4. **Assertion:** In the presence of air friction, if the ball is thrown vertically upwards, the time of ascent is less than the time of descent.

   Reason: Force due to air friction always acts opposite to the motion of the body.

5. **Assertion:** Range of projectile is maximum at $\theta = 45°$.

   Reason: For two projectiles ranges are not equal at complimentary angles.

Match the Columns MCQs

1. A particle is projected from level ground. Assuming projection point as origin, x-axis along horizontal and y-axis along vertically upwards. If particle moves in x-y plane and its path is given by $y = ax - bx^2$ where a, b are positive constants. Then match the physical quantities given in column-I with the values given in column-II. (g in column-II is acceleration due to gravity.)

	Column-I		Column-II
(a)	Horizontal component of velocity	(i)	a/b
(b)	Time of flight	(ii)	$\frac{a^2}{4b}$
(c)	Maximum height	(iii)	$\sqrt{\frac{g}{2b}}$
(d)	Horizontal range	(iv)	$\sqrt{\frac{2a^2}{bg}}$

1) (a)-(iii) (b)-(iv) (c)-(ii) (d)-(i)
2) (a)-(i) (b)-(iv) (c)-(ii) (d)-(iii)
3) (a)-(ii) (b)-(iii) (c)-(i) (d)-(iv)
4) (a)-(iv) (b)-(iii) (c)-(ii) (d)-(i)

2. It $\vec{v}_{mw}$ is the velocity of a man relative to water, $\vec{v}_w$ is the velocity of water, and $\vec{v}_m$ is the velocity of man relative to ground, match the following:

	Column-I		Column-II
(a)	Minimum distance for $v_{mw} > v_w$	(i)	$\theta = \sin^{-1}\left(\frac{v_w}{v_m}\right)$
(b)	Minimum time for $v_{mw} \geq v_w$	(ii)	$\vec{v}_m \perp \vec{v}_w$
(c)	Minimum distance for $v_{mw} < v_w$	(iii)	$\vec{v}_{mw} \perp \vec{v}_w$
(d)	Minimum time for $v_{mw} < v_w$	(iv)	$\theta = \sin^{-1}\frac{v_w}{v_{mw}}$

where θ is the angle between $\vec{v}_{mw}$ and the width of the river.

1) (a)-(i) (b)-(ii) (c)-(iii) (d)-(iv)
2) (a)-(iv) (b)-(i) (c)-(iii) (d)-(ii)
3) (a)-(ii) (b)-(iii) (c)-(i) (d)-(iv)
4) (a)-(iii) (b)-(iv) (c)-(i) (d)-(ii)

3. A body is projected with a velocity of 60 ms^{-1} at 30° to horizontal.

	Column-I		Column-II
(a)	Initial velocity	(i)	$60\sqrt{3}\hat{i}+40\hat{j}$
(b)	Velocity after 3 s	(ii)	$30\sqrt{3}\hat{i}+10\hat{j}$
(c)	Displacement after 2 s	(iii)	$30\sqrt{3}\hat{i}+30\hat{j}$
(d)	Velocity after 2s	(iv)	$30\sqrt{3}\hat{i}$

1) (a)-(iii) (b)-(iv) (c)-(i) (d)-(ii)
2) (a)-(ii) (b)-(iii) (c)-(i) (d)-(iv)
3) (a)-(iv) (b)-(i) (c)-(ii) (d)-(iii)
4) (a)-(i) (b)-(ii) (c)-(iii) (d)-(iv)

4. In projectile motion, if vertical component of velocity is increased to two times, keeping horizontal component unchanged, then

	Column-I		Column-II
(a)	Time of flight	(i)	will remain same
(b)	Maximum height	(ii)	will become two times
(c)	Horizontal range	(iii)	will become four times
(d)	Angle of projection with horizontal	(iv)	None

1) (a)-(iii) (b)-(iv) (c)-(i) (d)-(ii)
2) (a)-(ii) (b)-(iii) (c)-(i) (d)-(iv)
3) (a)-(iv) (b)-(ii) (c)-(ii) (d)-(iii)
4) (a)-(ii) (b)-(iii) (c)-(ii) (d)-(iv)

NCERT Exemplar MCQs

1. The angle between $\vec{A} = \hat{i}+\hat{j}$ and $\vec{B} = \hat{i}-\hat{j}$ is:
1) 45° 2) 90°
3) –45° 4) 180°

2. Which one of the following statements is true?
1) A scalar quantity is the one that is conserved in a process
2) A scalar quantity is the one that can never take negative values
3) A scalar quantity is the one that does not vary from one point to another in space
4) A scalar quantity has the same value for observers with different orientation of the axes

3. Figure shows the orientation of two vectors u and v in the xy-plane:

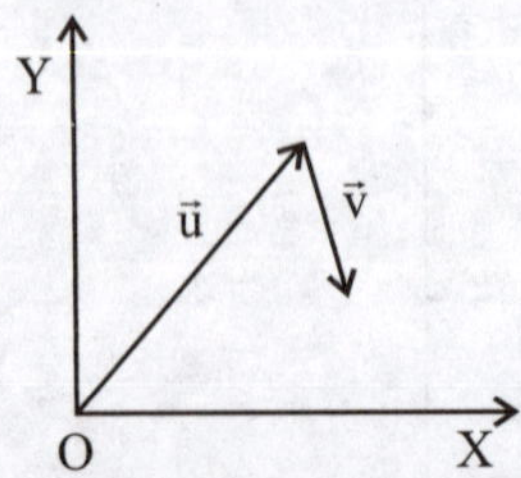

If $\vec{u} = a\hat{i}+b\hat{j}$ and $\vec{v} = p\hat{i}+q\hat{j}$
1) a and p are positive while b and q are negative
2) a, p and b are positive while q is negative
3) a, q and b are positive while p is negative
4) a, b, p and q are all positive

4. The component of a vector r along X-axis will have maximum value if:
1) r is along positive Y-axis
2) r is along positive X-axis
3) r makes an angle of 45° with the X-axis
4) r is along negative Y-axis

5. The horizontal range of a projectile fired at an angle of 15° is 50 m. If it is fired with the same speed at an angle of 45°, its range will be:
1) 60 m
2) 71 m
3) 100 m
4) 141 m

6. Consider the quantities pressure, power, energy, impulse, gravitational potential, electrical charge, temperature, area. Out of these, the only vector quantities are:
   1) Impulse, pressure and area
   2) Impulse and area
   3) Area and gravitational potential
   4) Impulse and pressure

7. In a two dimensional motion, instantaneous speed v_0 is a positive constant. Then, which of the following are necessarily true?
   1) The average velocity is not zero at any time
   2) Average acceleration must always vanish
   3) Displacements in equal time intervals are equal
   4) Equal path lengths are traversed in equal intervals

8. In a two dimensional motion, instantaneous speed v_0 is a positive constant. Then, which of the following are necessarily true?
   1) The acceleration of the particle is zero
   2) The acceleration of the particle is bounded
   3) The acceleration of the particle is necessarily in the plane of motion
   4) The particle must be undergoing a uniform circular motion

NEET & AIIMS Past Year MCQs

1. The two-dimensional motion of a particle, described by $\vec{r} = (\hat{i} + 2\hat{j})A\cos\omega t$ is a/an:
   A. parabolic path B. elliptical path
   C. periodic motion D. simple harmonic motion
   Choose the correct answer from the options given below: **(2024 Re)**
   1) B, C and D only 2) A, B and C only
   3) A, C and D only 4) C and D only

2. A football player is moving southward and suddenly turns eastward with the same speed to avoid an opponent. The force that acts on the player while turning is: **(2023)**
   1) along south-west
   2) along eastward
   3) along northward
   4) along north-east

3. A bullet is fired from a gun at the speed of 280 m/s in the direction 30° above the horizontal. The maximum height attained by the bullet is **(2023)**
   ($g = 9.8\ ms^{-2}$, $\sin 30° = 0.5$)
   1) 3000 m 2) 2800 m
   3) 2000 m 4) 1000 m

4. A ball is projected with a velocity, $10\ ms^{-1}$, at an angle of 60° with the vertical direction. Its speed at the highest point of its trajectory will be: **(2022)**
   1) $10\ ms^{-1}$ 2) Zero
   3) $5\sqrt{3}\ ms^{-1}$ 4) $5\ ms^{-1}$

5. A particle moving in a circle of radius R with a uniform speed takes a time T to complete one revolution.
   If this particle were projected with the same speed at an angle 'θ' to the horizontal, the maximum height attained by it equals 4R. The angle of projection, θ is then given by: **(2021)**
   1) $\theta = \cos^{-1}\left(\frac{\pi^2 R}{gT^2}\right)^{1/2}$
   2) $\theta = \sin^{-1}\left(\frac{\pi^2 R}{gT^2}\right)^{1/2}$
   3) $\theta = \sin^{-1}\left(\frac{2gT^2}{\pi^2 R}\right)^{1/2}$
   4) $\theta = \cos^{-1}\left(\frac{gT^2}{\pi^2 R}\right)^{1/2}$

6. A car starts from rest and accelerates at $5\ m/s^2$. At t = 4 s, a ball is dropped out of a window by a person sitting in the car. What is the velocity and acceleration of the ball at t = 6s? **(2021)**
   1) 20 m/s, 0 2) $20\sqrt{2}$ m/s, 0
   3) $20\sqrt{2}$ m/s, $10\ m/s^2$ 4) 20 m/s, $5\ m/s^2$

7. An arrow is projected into air. Its time of flight is 5s and range 200m. What is the maximum height reached by it? (Take $g = 10\ m\ s^{-2}$): **(AIIMS 2014)**
   1) 31.25 m 2) 24.5 m
   3) 18.25 m 4) 46.75 m

8. A particle is projected from the ground with an initial speed of 'v' at angle θ with horizontal. The average velocity of the particle between its point of projection and highest point of trajectory is: **(AIIMS 2013)**
   1) $\frac{v}{2}\sqrt{1+2\cos^2\theta}$ 2) $\frac{v}{2}\sqrt{1+\cos^2\theta}$
   3) $\frac{v}{2}\sqrt{1+3\cos^2\theta}$ 4) $v\cos\theta$

9. Ratio of initial kinetic energies of two projectiles of equal masses is 4 : 1 and their ratio of maximum heights is 4 : 1. The ratio of their horizontal ranges is: **(AIIMS 2010)**
   1) 4 : 1 2) 1 : 4
   3) 2 : 1 4) None

10. Which of the following velocity-time graphs shows a realistic situation for a body in motion? **(AIIMS 2007)**

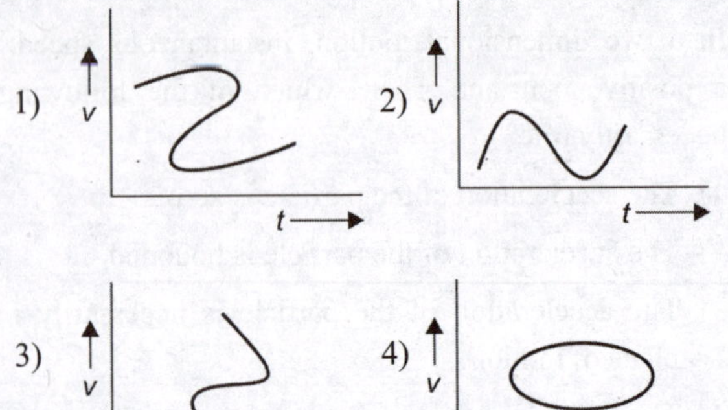

11. The direction of the angular velocity vector is along: **(AIIMS 2004)**

1) The tangent to the circular path
2) The inward radius
3) The outward radius
4) The axis of rotation

Answer Key

NCERT Topic-wise MCQs

1	2	3	4	5	6	7	8	9	10	11	12	13	14	15	16	17
1	2	3	3	1	4	1	4	3	2	4	4	4	4	4	3	4
18	**19**	**20**	**21**	**22**	**23**	**24**	**25**	**26**	**27**	**28**	**29**	**30**	**31**	**32**	**33**	**34**
4	4	4	3	2	3	2	3	3	2	1	3	1	3	3	1	4
35	**36**	**37**	**38**	**39**	**40**	**41**	**42**	**43**	**44**	**45**	**46**	**47**	**48**	**49**	**50**	**51**
3	3	4	1	3	1	1	2	4	3	4	2	3	1	4	3	2
52	**53**	**54**	**55**	**56**	**57**	**58**	**59**	**60**	**61**	**62**	**63**	**64**	**65**	**66**	**67**	**68**
1	3	3	1	1	1	1	3	2	2	2	1	4	3	2	4	3
69	**70**	**71**	**72**	**73**	**74**	**75**	**76**	**77**	**78**	**79**						
1	2	3	4	3	2	2	4	1	3	1						

High Order Time Intensive MCQs

1	2	3	4	5	6	7	8	9	10	11	12	13	14	15	16	17
2	2	4	2	4	3	1	1	4	3	1	1	2	3	3	2	3
18	**19**	**20**	**21**	**22**	**23**	**24**	**25**	**26**	**27**	**28**	**29**	**30**	**31**	**32**	**33**	**34**
1	4	4	2	2	1	2	3	2	3	1	4	2	2	2	1	1
35	**36**	**37**	**38**	**39**	**40**	**41**	**42**	**43**	**44**	**45**	**46**	**47**	**48**	**49**	**50**	**51**
1	4	2	4	1	2	3	1	3	4	3	4	2	2	1	2	4
52	**53**	**54**	**55**	**56**	**57**	**58**	**59**									
2	1	1	2	3	1	1	1									

Statement Based MCQs

1	2	3	4	5
3	3	1	2	2

Assertion & Reason MCQs

1	2	3	4	5
1	1	1	1	3

Match the Columns MCQs

1	2	3	4
1	3	1	4

NCERT Exemplar MCQs

1	2	3	4	5	6	7	8
2	4	2	2	3	2	4	3

NEET & AIIMS Past Year MCQs

1	2	3	4	5	6	7	8	9	10	11
4	4	4	3	3	2	1	3	1	2	4

CHAPTER

4 Laws of Motion

Topicwise Number of Questions Analysis (2025-2016)

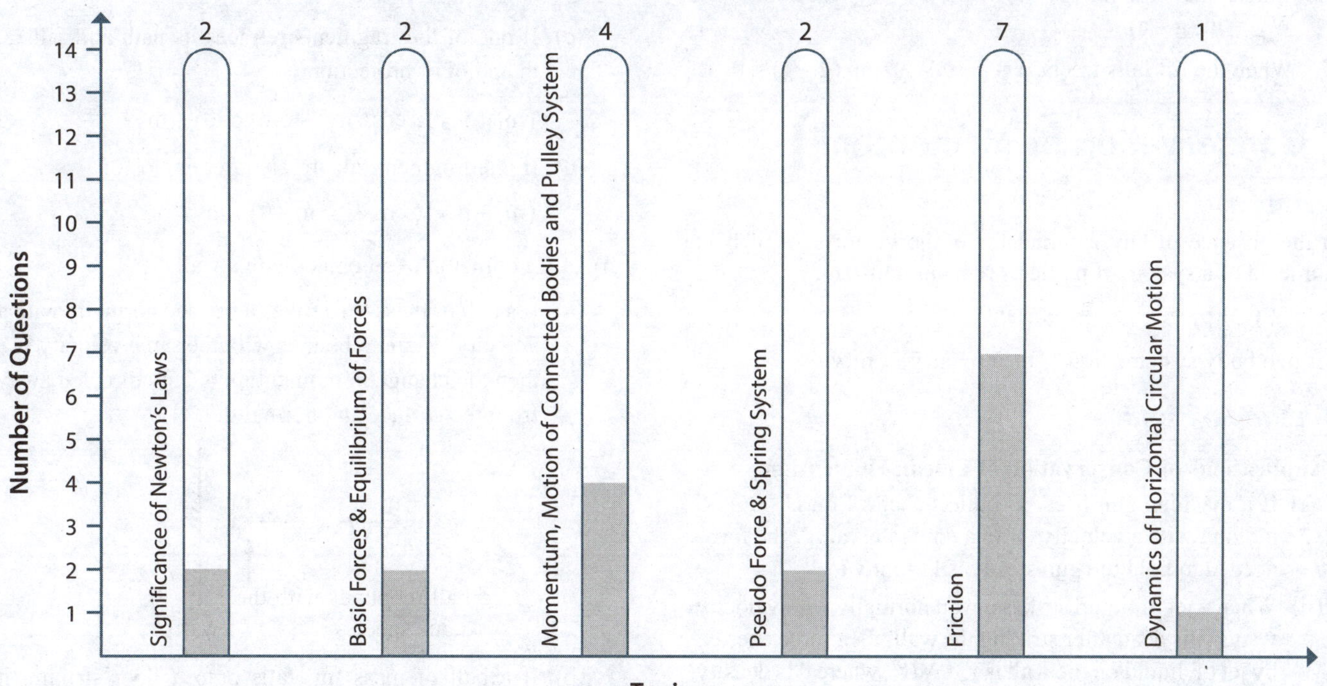

NCERT-PICKS

Introduction

Aristotle's Fallacy: An external force is required to keep a body in motion.

Inertia: A property of matter by which it continues in its existing state of rest or uniform motion in a straight line, unless that state is changed by an external force.

Newton's First Law of Motion: It states that a body remains in the state of rest or uniform motion in a straight line unless and until an external force acts on it.

Newton's Second Law of Motion

It states that the rate of change of linear momentum of a body is directly proportional to the applied force and the change in momentum takes place in the direction of the applied force.

Formulas

Linear momentum $\vec{p} = m \times \vec{v}$

Force $\vec{F} = \dfrac{d\vec{p}}{dt} = \dfrac{d(m\vec{v})}{dt} = m\vec{a}$

Impulse $\vec{I} = \vec{p}_f - \vec{p}_i = \Delta\vec{p} = \int_{t_1}^{t_2} \vec{F}\,dt$ = area under force – time graph

Numerical Profile

If all the bodies of a system have a common acceleration then that common acceleration can be given by

$$a = \frac{\text{Net pulling/pushing force}}{\text{Total mass}} = \frac{F_{net}}{m_{total}}$$

Newton's Third Law of Motion

It states that to every action there is an equal and opposite reaction.

Formulas

Reaction force = – Action force

Numerical Profile

The apparent weight of a man in a lift:

1. When the lift moves upwards with acceleration a, $W_{app} = m(g + a)$.
2. When the lift moves downward with acceleration a, $W_{app} = m(g - a)$.
3. When the lift falls freely, a = g, so $W_{app} = m(g - g) = 0$.

Conservation of Momentum

Formulas

In the absence of any external force, the vector sum of linear momenta of a system of particles remains constant.

$\Rightarrow \vec{p}_1 + \vec{p}_2 + \vec{p}_3 + ... + \vec{p}_n = \text{constant}$

For two body system, $m_1\vec{u}_1 + m_2\vec{u}_2 = m_2\vec{v}_2 + m_2\vec{v}_2$

Numerical Profile

Applications of Conservation of Linear Momentum

(i) If a machine gun fires 'n' bullets each of mass 'm' and moving with a velocity 'v' in a time interval 't', the force needed to hold the gun steady is F = mnv/t.

(ii) When a jet of liquid strikes a wall normally with velocity v and falls dead after striking the wall, then force exerted by jet of liquid on the wall is $F = Adv^2$, where d is density of the liquid.

However, if the liquid bounces back with the same velocity then force exerted by the liquid on the wall $F = 2Adv^2$

(iii) When a jet of liquid strikes a wall by making an angle 'θ' with the wall with velocity 'v' and rebounds with the same velocity then force exerted by water jet on the wall is $F = 2dAv^2\sin\theta$

(iv) A metallic plate of mass 'M' is held in mid air by firing 'n' bullets in 't' seconds each of mass 'm' with a velocity 'v' from below, then

(a) If the bullets fall dead after hitting the plate then

$$\frac{mnv}{t} = Mg$$

(b) If the bullets rebound after hitting the plate with same velocity then, $\frac{2mnv}{t} = Mg$

(v) Explosion of a shell

(a) Consider a shell of mass *M* as a projectile with velocity *u* and angle of projection θ. Suppose the shell breaks into two fragments of masses m_1 and m_2 at the maximum height

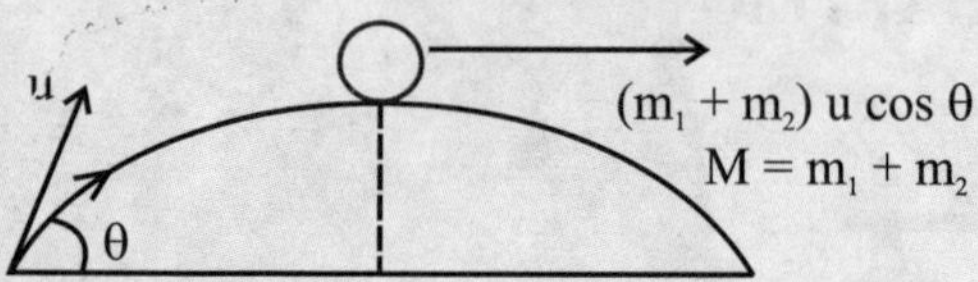

As the momentum is conserved just before and just after the explosion.

$$(m_1 + m_2)u\cos\theta\vec{i} = m_1\vec{v}_1 + m_2\vec{v}_2$$

(b) If the fragments travel in opposite directions after explosion then

$$(m_1 + m_2)u\cos\theta\vec{i} = m_1v_1\vec{i} - m_2v_2\vec{i}$$

(c) If one of the fragments retraces its path and falls at the point of projection

$$(m_1 + m_2)u\cos\theta\vec{i} = -m_1u\cos\theta\vec{i} + m_2\vec{v}$$

(d) If one fragment falls freely after explosion

$$(m_1 + m_2)u\cos\theta\vec{i} = m_1(0) + m_2\vec{v}_2$$

(vi) In the collision of an object with a wall

(a) If a ball of mass 'm' strikes a surface normally with a velocity 'v' and rebounds with the same velocity 'v', then the change in momentum is 2mv directed away from the surface and normal to it.

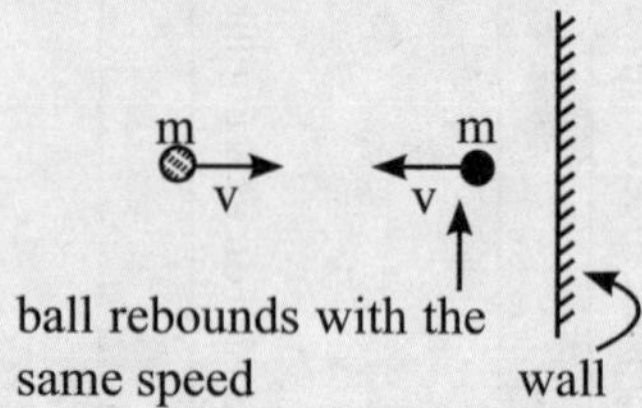

(b) If a ball of mass 'm' falls onto a floor striking it normally with a velocity 'u' and bounces again normal to the surface with a velocity 'v', then the magnitude of change in momentum is m (v + u) and is directed vertically upward.

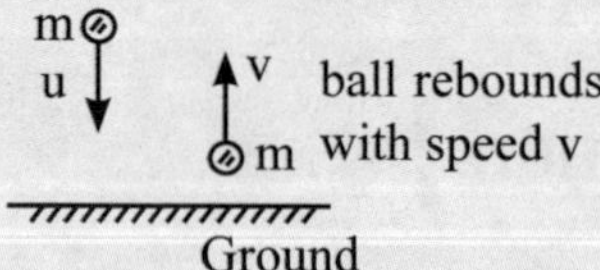

(c) If a ball of mass 'm' strikes a wall with a velocity 'v' making an angle 'θ' with the wall and bounces with the same speed 'v' at same angle with the wall, then the change in momentum of the ball is 2 mv sinθ and is directed away from the surface along the normal.

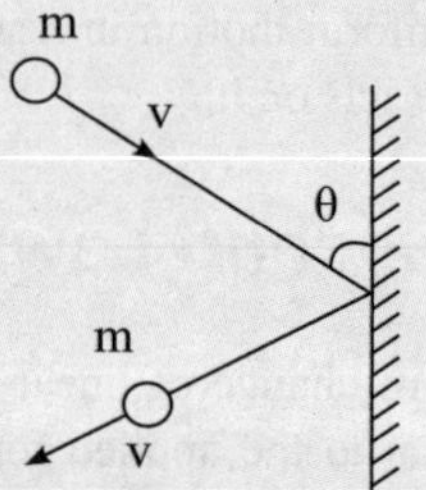

Equilibrium of a Particle

Equilibrium of a particle in mechanics refers to the situation when the net external force on the particle is zero. This means that the particle is either at rest or in uniform motion.

Formulas

A number of concurrent forces are said to be in equilibrium if their resultant is zero.

$\vec{F} = \vec{F_1} + \vec{F_2} + \vec{F_3} + \ldots + \vec{F_n} = \vec{0}$

Numerical Profile

If $\vec{F_1}, \vec{F_2}$ and $\vec{F_3}$ are three concurrent forces in equilibrium

1. $\vec{F_1} + \vec{F_2} + \vec{F_3} = \vec{0}$
2. $\frac{F_1}{\sin\alpha} = \frac{F_2}{\sin\beta} = \frac{F_3}{\sin\gamma}$ (Lami's Theorem)

Friction

Formulas

1. Force of limiting friction acting on a body $f_{lim} = \mu_s N$
2. Force of kinetic friction $f_k = \mu_k N$
3. Angle of Repose: It is the angle of the inclined plane at which a body placed on it just begins to slide

$\therefore \quad mg\sin\theta = f$

$mg\sin\theta = \mu mg\cos\theta$

$\therefore \quad \mu = \tan\theta$

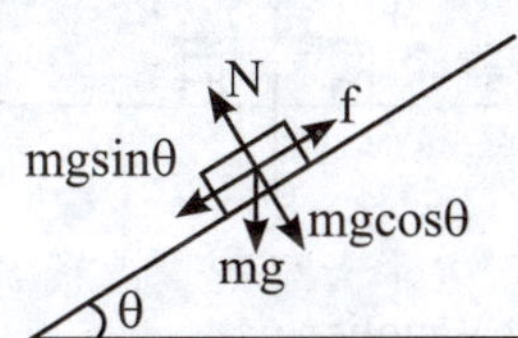

Numerical Profile

Motion of a body on an inclined plane

(i) If a body starts from rest from the top of an inclined plane of length l, time taken by the body to reach the bottom of the plane

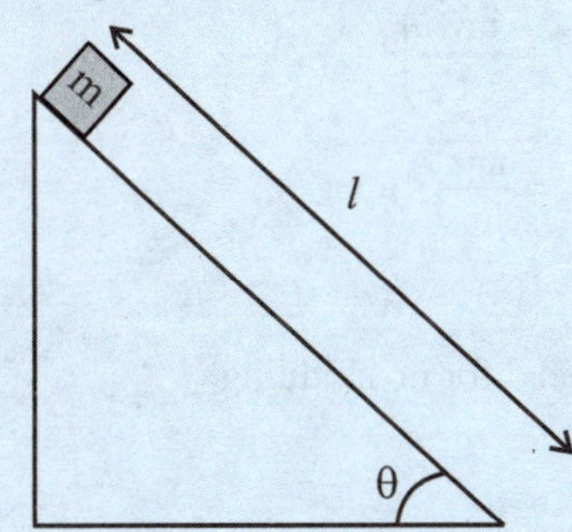

$s = ut + \frac{1}{2}at^2$

[where t is the time taken by the mass to reach the bottom and a is the acceleration of the mass]

$\ell = 0 + \frac{1}{2}at^2$ $\quad [\because s = \ell \text{ and } u = 0]$

$\Rightarrow \quad t = \sqrt{\frac{2\ell}{a}}$

$\therefore \quad t = \sqrt{\frac{2l}{g(\sin\theta - \mu_k\cos\theta)}}$

[where μ_k is the coefficient of kinetic friction]

(ii) Velocity of the body at the bottom of the plane.

$\because v^2 = u^2 + 2a\ell$

$v = \sqrt{0^2 + 2a\ell}$

$v = \sqrt{2a\ell}$

$v = \sqrt{2gl\,(\sin\theta - \mu_k\cos\theta)}$

(iii) When a body is pushed upwards along an inclined plane and it just reaches the top of the inclined plane of length l, then

Time of travel $t = \sqrt{\frac{2\ell}{g\,(\sin\theta + \mu_k\cos\theta)}}$

(iv) Force required to drag the block with an acceleration 'a' is

$F = (\mu_k\, mg\cos\theta + mg\sin\theta + ma)$

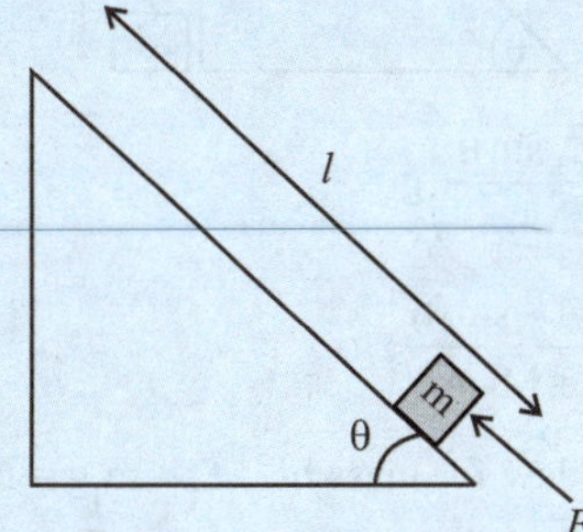

Motion of Connected Bodies

Formulas

1.

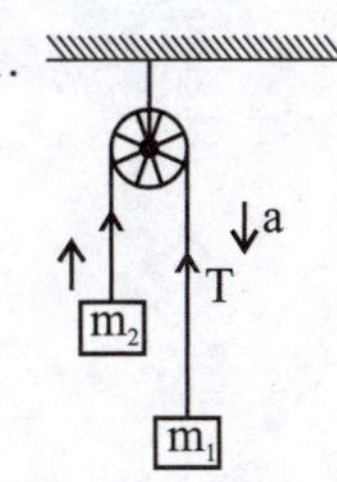

Acceleration of the blocks, $a = \frac{(m_1 - m_2)g}{(m_1 + m_2)}$

Tension in the string, $(T) = \frac{2m_1m_2g}{m_1 + m_2}$

2. $a = \left(\frac{m_2}{m_1 + m_2}\right)g$

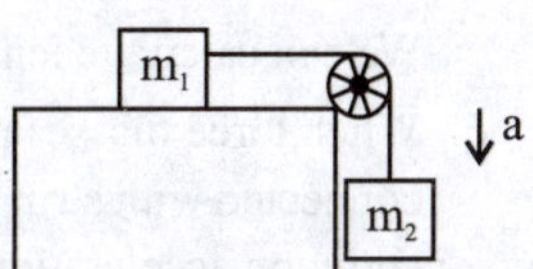

3. In the system shown below if $m_1 > m_2$, then acceleration of the system is given by

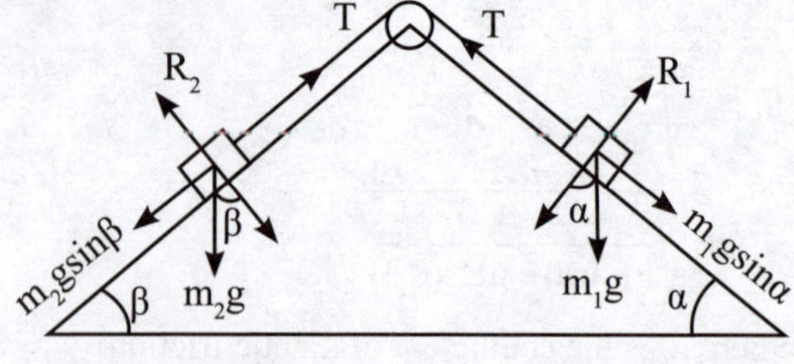

$$a = \frac{(m_1 \sin\alpha - m_2 \sin\beta)g}{m_1 + m_2}$$

[a is the acceleration of the system]

Tension in the string is given by

$$T = \frac{m_1 m_2 (\sin\alpha + \sin\beta)g}{(m_1 + m_2)}$$

4. Masses attached to a string passing through the pulley attached to the edge of an inclined plane

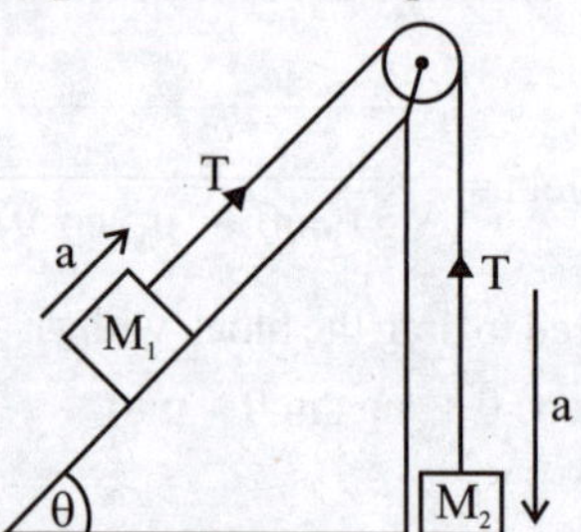

$$a = \left(\frac{M_2 - M_1 \sin\theta}{M_1 + M_2}\right) g$$

$$T = \left\{\frac{M_1 M_2 (1 + \sin\theta)}{(M_1 + M_2)}\right\} g$$

5. Newton's 2nd law for mass m_1 : $T_1 - m_1 g = m_1 a$ ...(i)

Newton's 2nd law for mass m_2 : $m_2 g + T_2 - T_1 = m_2 a$ ...(ii)

Newton's 2nd law for mass m_3 : $m_3 g - T_2 = m_3 a$ ...(iii)

Solving equ. (i), (ii) and (iii)

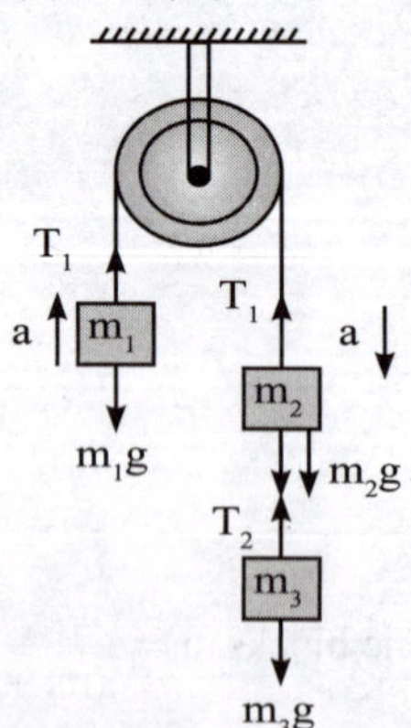

$$\Rightarrow a = \frac{(m_2 + m_3 - m_1)}{(m_1 + m_2 + m_3)} g$$

We can calculate tension T_1 and T_2 from above equations.

6. When three masses placed on a smooth horizontal table are connected with strings and mass m_1 is pulled with a force F_1, common acceleration produced is

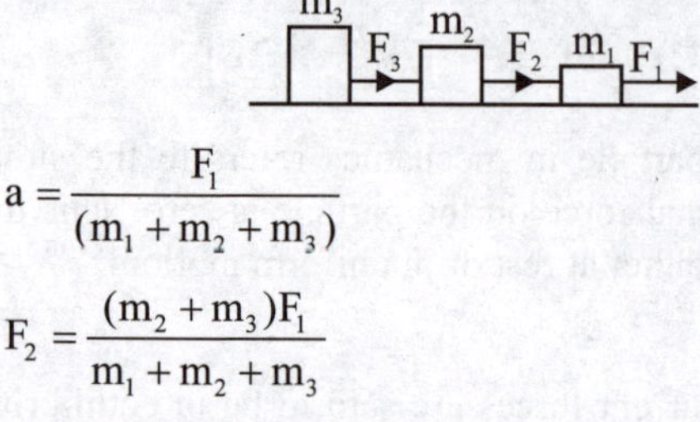

$$a = \frac{F_1}{(m_1 + m_2 + m_3)}$$

$$F_2 = \frac{(m_2 + m_3)F_1}{m_1 + m_2 + m_3}$$

$$F_3 = \frac{m_3 F_1}{m_1 + m_2 + m_3}$$

Circular Motion

Formulas

Centripetal force, $F_C = \frac{mv^2}{R} = mR\omega^2$

R = Radius of the circle.

Numerical Profile

Motion on a Levelled Curved Road

Suppose a car of mass *m* is taking a circular turn on a level road. It has a tendency of skidding sideways i.e., away from the centre of the circle. Due to this, the static friction '*f*' acts towards the centre and provides the necessary centripetal force for the motion along the curve.

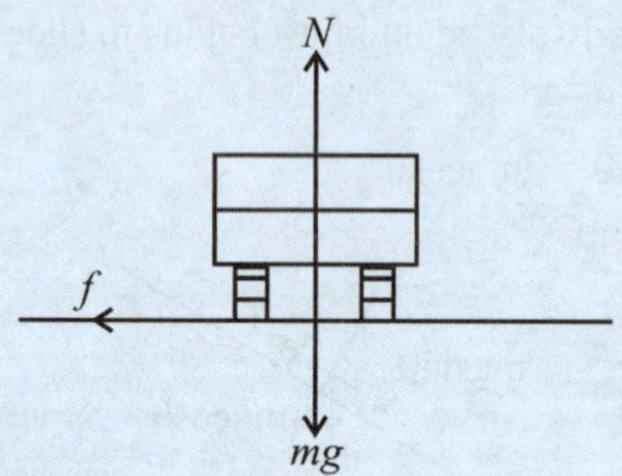

As there is no vertical motion

$\therefore N = mg$

where, N is normal reaction.

Centripetal force required for circular motion is provided by the frictional force.

For no slipping

$$f \le f_{lim} \quad \left(\text{Here, } f = \frac{mv^2}{r}\right)$$

$$\Rightarrow \frac{mv^2}{r} \le \mu_s N \Rightarrow \frac{mv^2}{r} \le \mu_s mg$$

$$v \le \sqrt{\mu_s rg}$$

$\therefore$ Maximum speed for no skidding

$$v_{max} = \sqrt{\mu_s rg}$$

Banking of Roads

(i) In case of frictionless road, the road is banked at the turn such that the outer part of the road is some what lifted compared to the inner part.

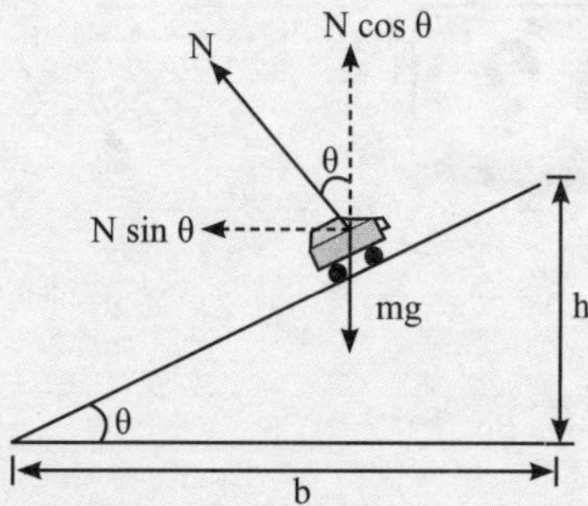

For equilibrium

$N \sin\theta = \frac{mv^2}{r}$ and $N \cos\theta = mg$

[where, v is the velocity of the car and r is the radius of the circular path]

$\Rightarrow \tan\theta = \frac{v^2}{rg} \Rightarrow v = \sqrt{rg \tan\theta}$

(ii) When friction is present, the direction of frictional force is determined by the velocity of the vehicle v for a given banking angle θ

(a) If $v = 0$, then frictional force is up the plane

(b) If $v > \sqrt{gr \tan\theta}$, then frictional force is down the plane.

(c) If $v = \sqrt{gr \tan\theta}$, then frictional force $f = 0$

(d) If $v < \sqrt{gr \tan\theta}$, then frictional force f is up the plane.

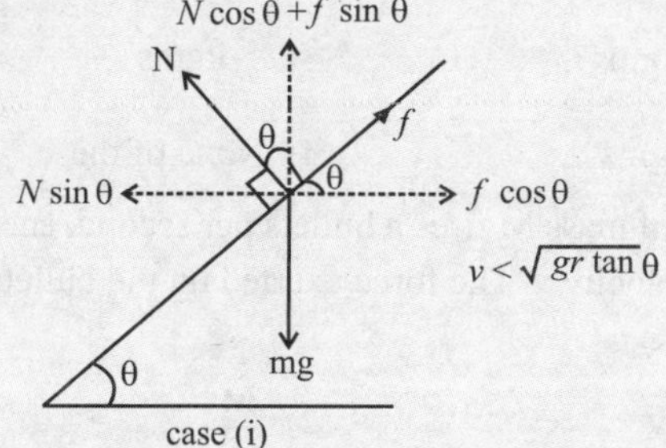

case (i)

To find minimum safe speed we use fig (i), and

$$N \cos\theta + f \sin\theta = mg \quad \text{...(i)}$$

$$N \sin\theta - f \cos\theta = \frac{mv^2_{min}}{r} \quad \text{...(ii)}$$

Also, $f = \mu N$ (for $v = min$) ...(iii)

Solving (i), (ii) and (iii)

$$v_{min} = \sqrt{\frac{rg(\sin\theta - \mu\cos\theta)}{(\cos\theta + \mu\sin\theta)}}$$

To find maximum safe speed, we have to consider figure (ii).

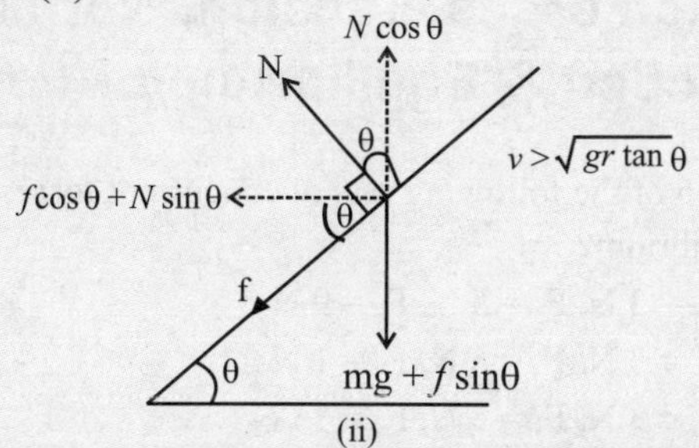

(ii)

$$N \sin\theta + f \cos\theta = \frac{mv^2_{max}}{r} \quad \text{...(i)}$$

$$N \cos\theta = f \sin\theta + mg \quad \text{...(ii)}$$

$$f = \mu N \text{ (for } v = max) \quad \text{...(iii)}$$

From (i), (ii) and (iii) we get

$$v_{max} = \sqrt{\frac{rg(\sin\theta + \mu\cos\theta)}{(\cos\theta - \mu\sin\theta)}}$$

Bending of cyclist

A cyclist is taking a turn in a circle of radius r with speed v, To avoid slipping he should bend through an angle θ with the vertical such that $\tan\theta = \frac{v^2}{rg}$

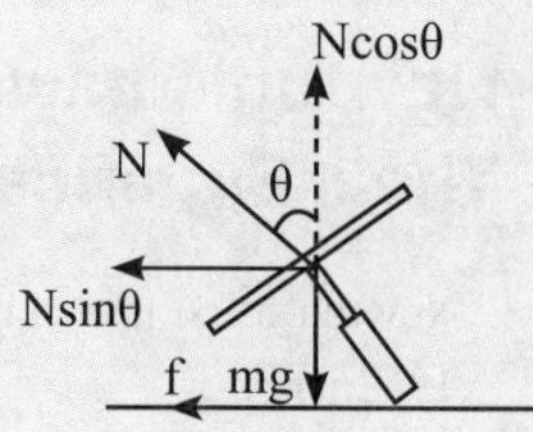

Statements Asked In Exam

- Newton's first and third law can be derived from second law. Hence, Newton's second law is the most fundamental law out of three.
- All central forces are conservative forces.
  i. Gravitational force ii. Electric force iii. Elastic force
- Common forces in mechanics are non-conservative forces.
  i. Normal ii. Tension iii. Friction.
- Forces in nature occur in pairs. A single isolated force is not possible.
- A frame of reference which is at rest or which is moving with a uniform velocity along a straight line is called an inertial frame of reference. In inertial frame of reference Newton's laws of motion hold good.
- Inertial frame of reference are also called unaccelerated frame of reference or Newtonian or Galilean frame of reference.
- Accelerated frame of reference are called non-inertial frame of reference. Newton's laws of motion are not applicable in non-inertial frame of reference.
- Law of conservation of linear momentum is independent of frame of reference, though linear momentum depends on frame of reference.
- Linear momentum of a body can change only in the direction of applied force.
- Kinetic friction does not depend upon the velocity of the body.
- It is easier to maintain the motion of an object than to start it, because force of limiting friction is more than the kinetic friction.
- Frictional force gets increased when two surfaces in contact are polished beyond a certain limit.
- Sand is thrown on the tracks covered with snow, so as to increase friction.
- It is easier to pull a lawn roller than push it because effective weight of the lawn roller while pulling is less than the effective weight while pushing it.
- A shell is fired from a canon, it explodes in mid air, its total K.E increases.

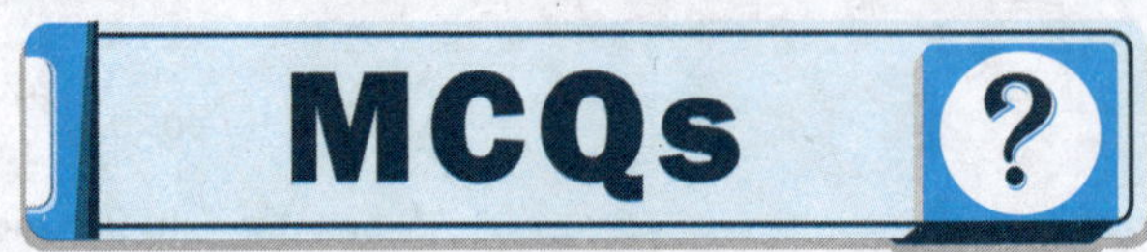

NCERT Topic-wise MCQs

Newton's Laws & Their Significance

1. Newton's first law of motion describes the following
   1) Energy 2) Work
   3) Inertia 4) Moment of Inertia

2. Which is the type of inertia?
   1) Inertia of rest 2) Inertia of direction
   3) Inertia of motion 4) All the above options

3. A person sitting in an open car moving at constant velocity throws a ball vertically up into air. The ball falls
   1) Outside the car
   2) In the car ahead of the person
   3) In the car to the side of the person
   4) Exactly in the hand which threw it up

4. A man is standing in a bus. He suddenly falls ahead on a woman when bus suddenly applies brakes. Woman filed a complaint on him. What will be the correct reason man should explain?
   1) Man is taken back
   2) Man is sleeping
   3) Inertia of rest keeps the lower part of the body at rest whereas the upper part of body remains in motion due to inertia of motion
   4) None of the above

5. A particle is moving with a constant speed along a straight line path. A force is not required to
   1) Increases its speed
   2) Decrease the momentum
   3) Change the direction
   4) Keep it moving with uniform velocity.

6. A mass (M) of 1kg is suspended by a string A. Another string C is connected to its lower end. If a sudden jerk is given to C then

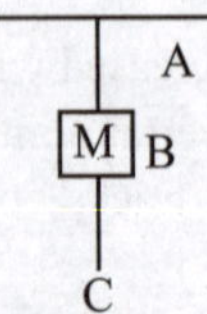

   1) The string AB will break
   2) The string BC will break
   3) None of the strings will break
   4) The mass will start rotating

7. Same force acts on two bodies of different masses 8 kg and 10 kg initially at rest. The ratio of times required to acquire same final velocity is
   1) 1 : 1 2) 8 : 11
   3) 4 : 5 4) 2 : 3

8. A car is moving with a velocity 8m/sec whose mass is 150 kg. To stop it in $\frac{1}{5}$sec, the required force in opposite direction is
   1) 300 N
   2) 100 N
   3) 200 N
   4) 6000 N

9. A body of mass 2 kg moves with an acceleration 3 ms^{-2}. The change in momentum in one second is
   1) $\frac{2}{3}kgms^{-1}$ 2) $\frac{3}{2}kgms^{-1}$
   3) 6 $kgms^{-1}$ 4) None of these

10. A gun of mass M fires n bullets per second, each of mass m with velocity v. The force exerted by the bullets on the gun is
    1) nmv 2) $\frac{Mv}{n}$
    3) n^2mv 4) $\frac{Mv^2}{n}$

11. A wagon of mass 2000 kg is separated from a train travelling at 9 km/h and comes to rest in 2 minutes. What is the resistance between the wheels of the wagon and the track?
    1) 42 N 2) 41.7 N
    3) 45 N 4) 47 N

Basic Forces, Tension, Normal Reaction & Equilibrium of Forces

12. Which of the following sets of concurrent forces, may be in equilibrium?
    1) $F_1 = 3$ N, $F_2 = 5$ N, $F_3 = 9$ N
    2) $F_1 = 3$ N, $F_2 = 5$ N, $F_3 = 1$ N
    3) $F_1 = 3$ N, $F_2 = 5$ N, $F_3 = 15$ N
    4) $F_1 = 3$ N, $F_2 = 5$ N, $F_3 = 6$ N

13. There are four forces acting at a point P produced by strings as shown in figure, which is at rest. The forces F_1 and F_2 are

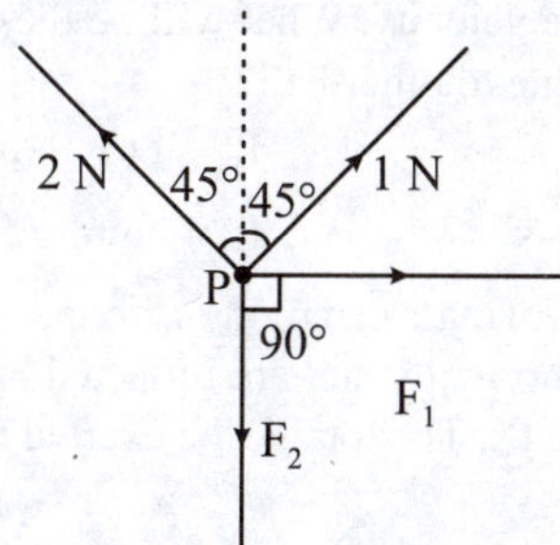

1) $\frac{1}{\sqrt{2}}N, \frac{3}{\sqrt{2}}N$ 2) $\frac{3}{\sqrt{2}}N, \frac{1}{\sqrt{2}}N$

3) $\frac{1}{\sqrt{2}}N, \frac{1}{\sqrt{2}}N$ 4) $\frac{3}{\sqrt{2}}N, \frac{3}{\sqrt{2}}N$

14. A rope of length 20 m is kept on frictionless surface and a force of 5 N is applied to one of its end. Find the tension in the rope at 5m from this end.

1) 5 N   2) 2 N
3) 3.75 N   4) 4 N

15. If rope of lift breaks suddenly the normal force exerted by the surface of lift will be: (a = acceleration of lift)

1) m (g + a)   2) mg
3) m (g – a)   4) Zero

16. With what minimum acceleration can an athlete climb a rope whose breaking strength is $\frac{1}{3}$rd of his weight?

1) 6.67 m/s²   2) 7 m/s²
3) 8 m/s²   4) 7.23 m/s²

17. A car of mass 1000 kg is moving with a velocity of 15 m/s and is acted upon by a forward force of 500 N due to engine and retarding force of 120 N due to friction. What will be its velocity after 10 seconds.?

1) 15 m/s   2) 16 m/s
3) 18.8 m/s   4) 20 m/s

18. A monkey of mass 40 kg climbs on a massless rope of breaking strength 600N. The rope will break if the monkey... (Take g = 10 m/s²)

1) Climbs up with uniform velocity of 6m/s
2) Climbs up with an acceleration of 6 m/s²
3) Climbs down with an acceleration of 4 m/s²
4) Climbs down with an acceleration 5 m/s²

19. A body of weight 2kg is suspended as shown in figure. The tension T_1 in the horizontal string (in kg wt) is

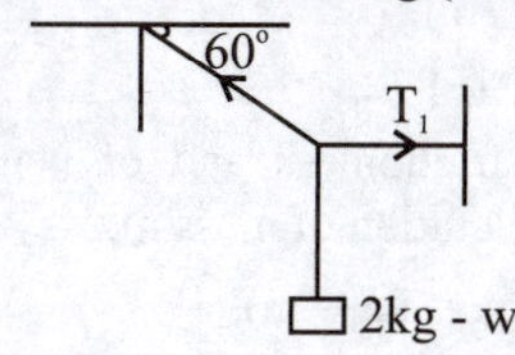

1) $\frac{2}{\sqrt{3}}$ 2) $\frac{\sqrt{3}}{2}$

3) $\frac{2}{3}$ 4) $\frac{3}{2}$

Impulse, Momentum & Pulley Systems

20. Figures I, II, III and IV depict variation of force with time

1) 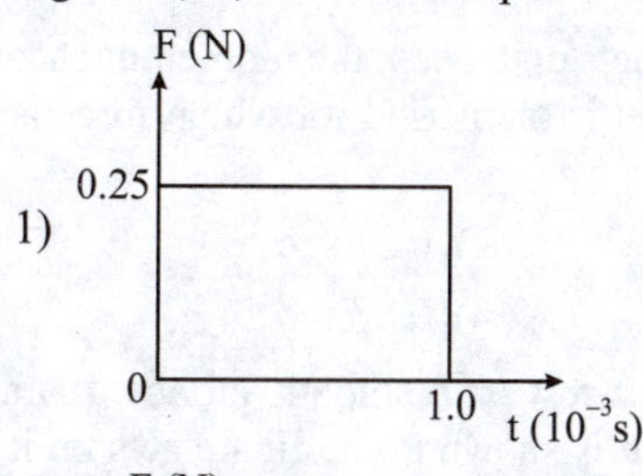

2) 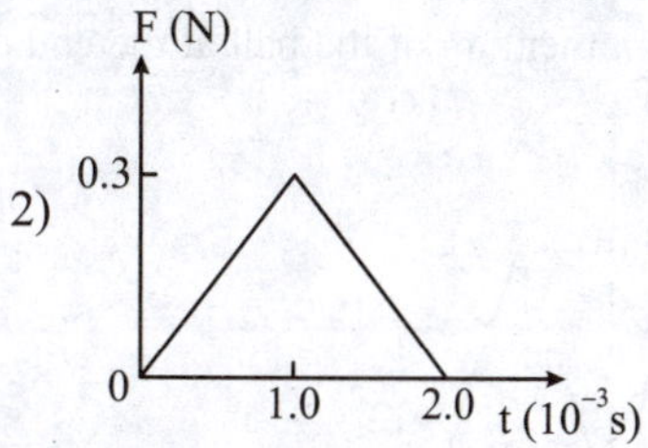

3) 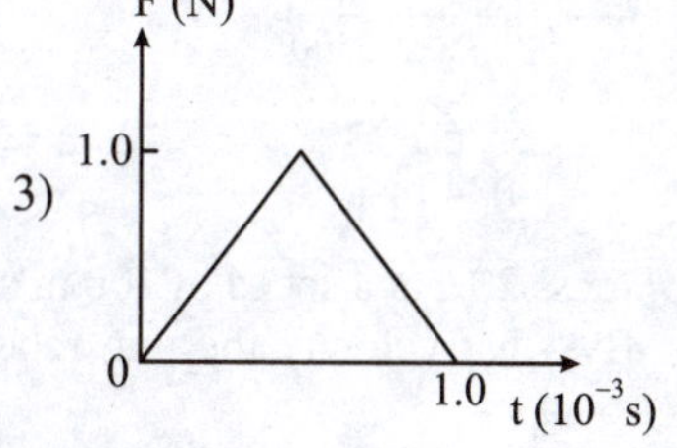

4) 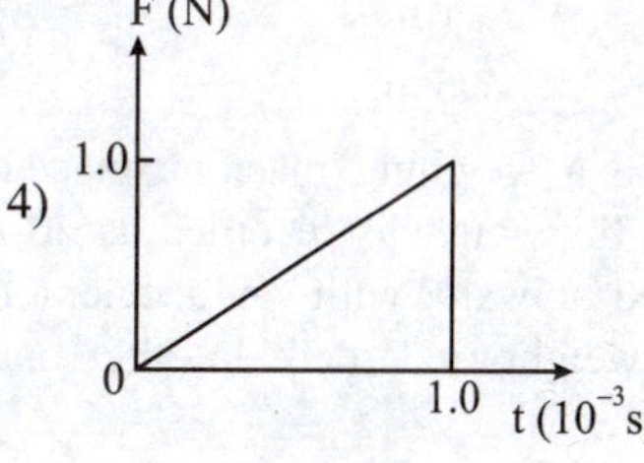

21. A force of 50 dyne is acted on a body of mass 5 g which is at rest for an interval of 3 s. Then the impulse is

1) 0.15×10^{-3} Ns   2) 0.98×10^{-3} Ns
3) 1.5×10^{-3} Ns   4) 2.5×10^{-3} Ns

22. If a bullet of mass 10 gm moving with velocity 150 m/sec penetrates the wooden block upto 6 cm. Then the average force imposed by the bullet on the block is

1) 1875 N   2) 1900 N
3) 2000 N   4) 1500 N

23. A force time graph for the motion of a body is given in figure. Change in linear momentum between 0 and 6s is

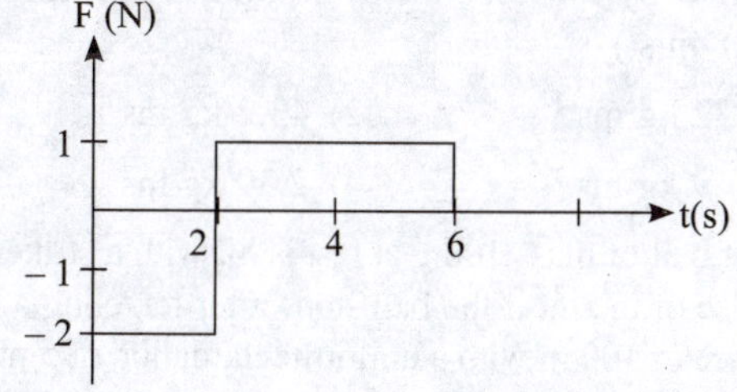

1) Zero   2) 8 Ns
3) 4 Ns   4) 2 Ns

24. A cricket ball of mass 200 g collides with a bat with velocity 30 m/s and returns with the same velocity within 0.01 sec. The force acted on the bat is:

1) 1200 N 2) 1400 N
3) 1600 N 4) None of these

25. If a force of 500 N acts on a body, the momentum acquired is 135 kg -m/s. What is the period for which force acts on the body?

1) 0.22 sec 2) 0.21 sec
3) 0.25 sec 4) 0.27 sec

26. A body is initially at rest on a smooth surface. A force F, whose time variation is shown in the figure acts on it for a duration of 4 s. The momentum of the ball at the end of the 4 s (in Ns) is

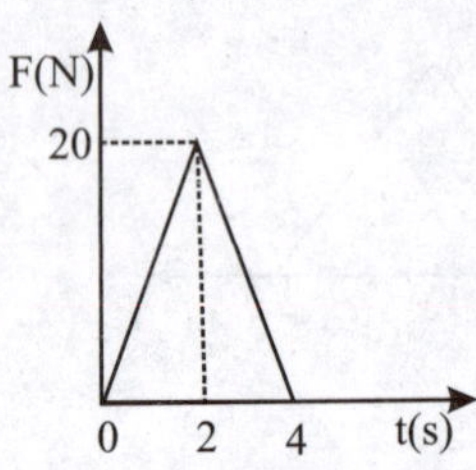

1) 10 Ns 2) 20 Ns
3) 30 Ns 4) 40 Ns

27. A man fires a bullet of mass 20 g at a speed of 100 m/s. The gun is of 2 kg mass. By what velocity the gun rebounds backward?

1) 2 m/s 2) 1 m/s
3) 5 m/s 4) 6 m/s

28. A man is standing on a weighing machine placed in a lift. When stationary, his weight is recorded as 40 kg. If the lift is accelerated upward with an acceleration of 2 m/s^2, then the weight recorded in the machine ($g = 10\ m/s^2$) will be

1) 48 kg wt 2) 32 kg wt
3) 42 kg wt 4) 40 kg wt

29. A machine gun is mounted on a 2800 kg vehicle on a horizontal smooth rod (negligible friction). The gun fires 10 bullets per sec with a velocity of 200 m/s. If the mass of each bullet is 50 g. What is the acceleration produced in the vehicle?

1) 3.6 cm/s^2 2) 4 cm/s^2
3) 5.2 cm/s^2 4) 4.9 cm/s^2

30. A spherical body of mass 100 g is dropped from a height of 10 m from the ground. After hitting the ground, the body rebounds to a height of 5m. The impulse of force imparted by the ground to the body is given by: (given $g = 9.8\ m/s^2$)

1) 4.32 kg ms^{-1} 2) 43.2 kg ms^{-1}
3) 23.9 kg ms^{-1} 4) 2.39 kg ms^{-1}

31. A golf ball of mass 100 g at rest is hit with a striker. Find the impulse of the hit if the ball stops after traveling a horizontal distance of 100 m with a uniform retardation of 5 m/s^2?

1) 0.3 Ns 2) 0.1 Ns
3) 3.16 Ns 4) 0.9 Ns

32. Two billiard balls each of mass 50 g moving in opposite directions with a speed of 40 km/h collide and rebound with the same velocity. What will be the impulse imparted to each ball due to other ball?

1) 2.22 kg m/s 2) 1.11 kg m/s
3) 3.33 kg m/s 4) 4 kg m/s

33. 100 balls each of mass m moving with speed v simultaneously strike a wall normally and are reflected back with the same speed, in time t s. The total force exerted by the balls on the wall is

1) $\frac{100mv}{t}$ 2) $\frac{200mv}{t}$
3) 200 mvt 4) $\frac{mv}{100t}$

34. Two masses of 4 kg and 11kg are connected by a string passing through a frictionless pulley and are kept on a frictionless table. The acceleration of 11 kg mass is

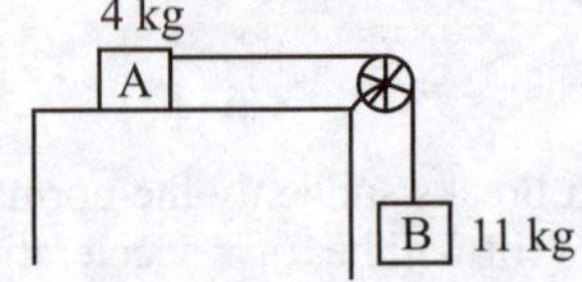

1) 7.33 m/s^2 2) 49 m/s^2
3) 5.44 m/s^2 4) 19.5 m/s^2

35. In the arrangement shown in the figure, the pulley has a mass 3m. Neglecting friction on the contact surface, the force exerted by the supporting rope AB on the ceiling is

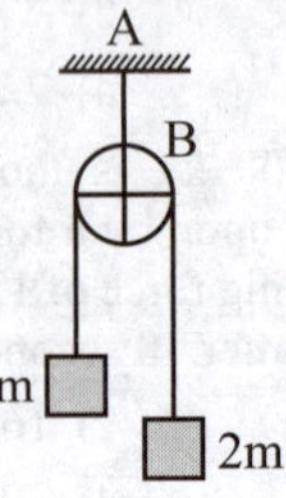

1) 6 mg 2) 3 mg
3) 4 mg 4) None of these

36. A system consists of three masses m_1, m_2 and m_3 connected by a string passing over a pulley P. The mass m_1 hangs freely and m_2 and m_3 are on a rough horizontal table (the coefficient of friction = μ).

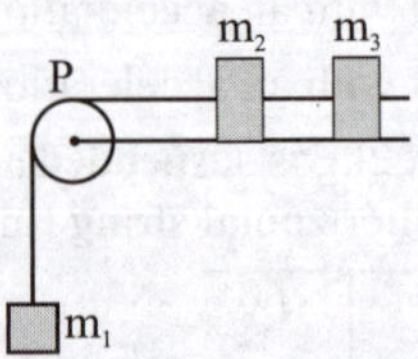

The pulley is frictionless and of negligible mass. The downward acceleration of mass m_1 is

(Assume $m_1 = m_2 = m_3 = m$)

1) $\frac{g(1-2\mu)}{9}$ 2) $\frac{2g\mu}{3}$
3) $\frac{g(1-2\mu)}{3}$ 4) $\frac{g(1-2\mu)}{2}$

37. Three blocks of masses 2 kg, 3 kg and 5 kg are connected to each other with light string and are then placed on a smooth frictionless surface. The system is pulled with force F from side of the lighter mass so that it moves with an acceleration of 1 ms^{-2}, T_1 and T_2 denote the tension in the strings. The value of F is

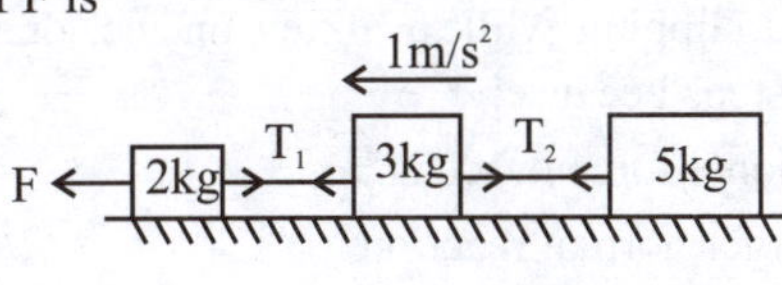

1) 30 N
2) 20 N
3) 10 N
4) 40 N

38. A light string passes over a frictionless pulley. To one of its ends a mass of 10 kg is attached to its other end a mass of 15 kg is attached. The tension in the thread will be

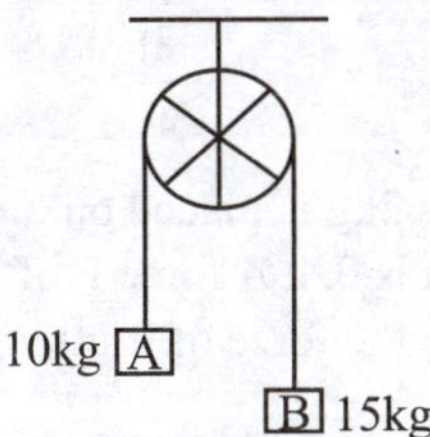

1) 120 N
2) 118 N
3) 110 N
4) None

39. Three blocks with masses m, 2 m, 3 m are connected by strings as shown in the figure. After an upward force F is applied on block m, the masses move upward at constant speed v. What is the net force on the block of mass 2m? (g is the acceleration due to gravity)

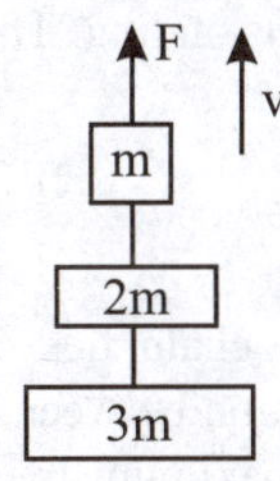

1) Zero
2) 2mg
3) 3mg
4) 6mg

40. The pulleys and strings shown in the figure are smooth and of negligible mass. For the system to remain in equilibrium, the angle θ should be

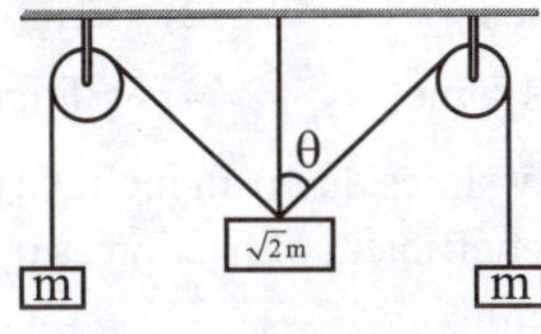

1) 0°
2) 30°
3) 45°
4) 60°

41. A block of mass 'm' is connected to another block of mass 'M' by a string (massless). The blocks are kept on a smooth horizontal plane. Initially the blocks are at rest and the string is unstretched. Then a constant force 'F' starts acting on the block of mass 'm' to pull it. Find the force on the block of mass 'M'.

1) $\frac{Mm}{M+m}$
2) $\frac{F}{M+m}$
3) $\frac{mF}{M+m}$
4) $\frac{MF}{M+m}$

42. Two masses of 8 kg and 4 kg are connected by a string as shown in figure over a frictionless pulley. The acceleration of the system is

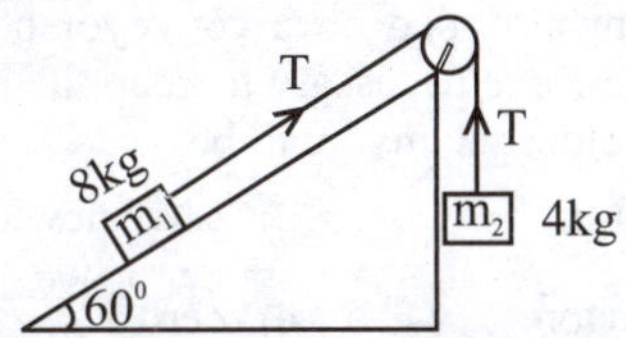

1) –0.1 m/s^2
2) 4 m/s^2
3) 3 m/s^2
4) –2.44 m/s^2

43. Two masses M_1 and M_2 are accelerated uniformly on a frictionless surface. The ratio of the tensions T_1/T_2 is

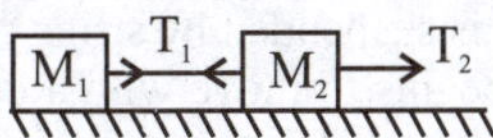

1) $\frac{M_1}{M_1+M_2}$
2) $\frac{M_2}{M_1+M_2}$
3) $\frac{M_2}{M_1}$
4) $\frac{M_1M_2}{M_1+M_2}$

44. If the resultant of all the external forces acting on a system of particles is zero then from an inertial frame, one can surely say that

1) Linear momentum of the system does not change with time
2) Kinetic energy of the system does not change with time
3) Angular momentum of the system does not change with time
4) Potential energy of the system does not change in time

45. A string of negligible mass going over a clamped pulley of mass m supports a block of mass M as shown in the figure. The force on the pulley by the clamp is given by

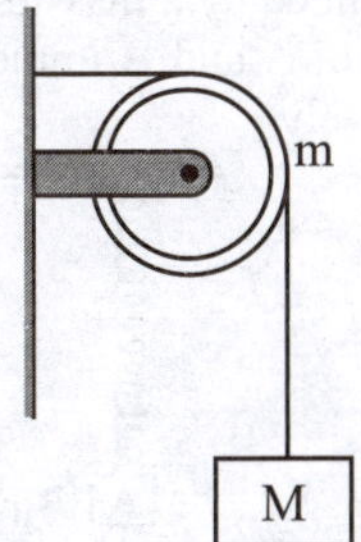

1) $\sqrt{2}$ Mg
2) $\sqrt{2}$ mg
3) $g\sqrt{(M+m)^2+m^2}$
4) $g\sqrt{(M+m)^2+M^2}$

46. A ball of mass m moves with speed v and it strikes normally with a wall and is reflected back normally. If its time of contact with wall is t, then find force exerted by ball on the wall.

1) $\frac{2mv}{t}$ 2) $\frac{mv}{t}$
3) mvt 4) $\frac{mv}{2t}$

47. Sand is being dropped on a conveyor belt at the rate of M kgs^{-1}. The force necessary to keep the belt moving with a constant velocity v ms^{-1} will be

1) Mv newton 2) 2 Mv newton
3) $\frac{Mv}{2}$ newton 4) Zero

Inertial, Non-Inertial Frames, Pseudo Force & Spring Systems

48. A small sphere is suspended by string from the ceiling of a car. If the car begins to move with a constant acceleration a, the tension generated in the string is

1) $m\sqrt{g^2+a^2}$ 2) $m\sqrt{g^2-a^2}$
3) $m\sqrt{g^2+2a^2}$ 4) Mg

49. The tension in the spring is

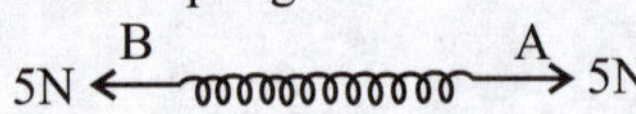

1) Zero 2) 3 N
3) 5 N 4) 1 N

50. A body of mass 8 kg is suspended through two light springs X and Y connected in series as shown in figure. The readings in X and Y respectively are

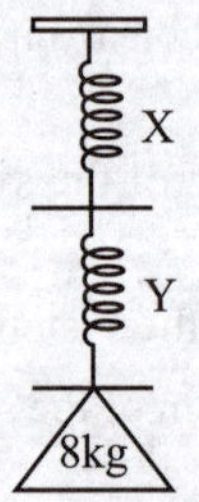

1) 8 kg, zero 2) Zero, 8 kg
3) 6 kg, 2 kg 4) 8 kg, 8 kg

51. Two blocks A and B of masses 3m and m respectively are connected by a massless and inextensible string. The whole system is suspended by a massless spring. The magnitudes of acceleration of A and B immediately after the string is cut are respectively

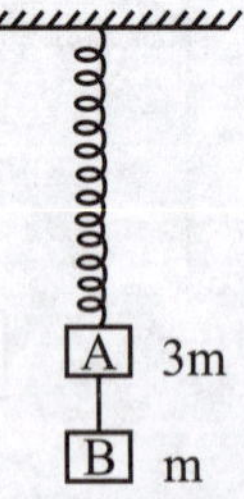

1) g, g 2) g/3, g/3
3) g/2, g/2 4) g/3, g

Friction: Static, Limiting of Kinetic, Angle of Friction & Angle of Repose

52. To avoid slipping while walking on ice, one should take smaller steps because:

1) frictional force of ice is large
2) of larger normal reaction
3) frictional force of ice is small
4) of smaller normal reaction

53. A block is thrown on ice with a velocity of 6 m/s and it comes to rest after traveling 10m. Find the coefficient of friction.

1) 1 2) 0.18
3) 0.17 4) 2

54. A block of mass 2kg is placed on the floor. The coefficient of static friction is 0.4. A force F of 2.5 N is applied on the block. Calculate the force of friction between the block and the floor.

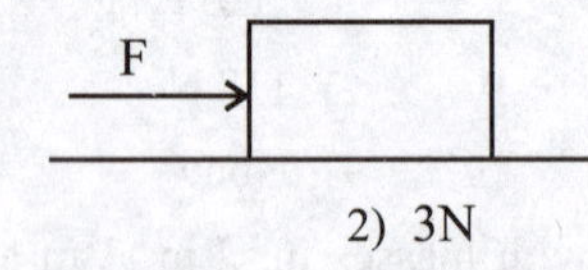

1) 1.5 N 2) 3N
3) 4N 4) 2.5 N

55. A uniform metal chain is placed on a rough table such that one end of chain hangs down over the edge of the table. When two- third of its chain length hangs over the edge the chain just starts sliding. Then the coefficient of static friction is

1) 1 2) 2
3) 3 4) 4

56. A heavy uniform chain lies on a horizontal top. If the coefficient of friction between the chain and table surface is 0.50 then the maximum fraction of length of the chain that can hang over one edge of the table is

1) 33.3% 2) 40%
3) 50% 4) 60%

57. If the normal reaction is doubled, the coefficient of friction

1) is Doubled 2) is Halved
3) Remains same 4) is Tripled

58. When a body slides down an inclined plane of inclination θ and having coefficient of friction as μ_K, then its acceleration is given by.

1) $g(\mu_K \sin\theta + \cos\theta)$
2) $g(\mu_K \sin\theta - \cos\theta)$
3) $g(\sin\theta + \mu_K \cos\theta)$
4) $g(\sin\theta - \mu_K \cos\theta)$

59. In the figure shown, the frictional coefficient between table and block is 0.2. Find the ratio of tensions in the right and left strings.

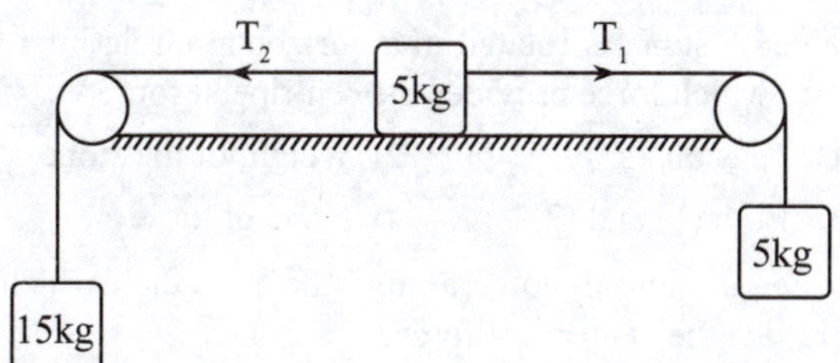

1) 17 : 24
2) 34 : 12
3) 2 : 3
4) 3 : 2

60. A block rests on a rough inclined plane making an angle of 30° with horizontal. The coefficient of static friction between the block and inclined plane is 0.8. If the frictional force on the block is 10 N, the mass of the block (in kg) is (Take, $g = 10\ m/s^2$)

1) 2.0
2) 4.0
3) 1.6
4) 2.5

61. A block of mass 20 kg is sliding on a surface is inclined at an angle of 45° with the horizontal. What will be the acceleration of the block if the coefficient of kinetic friction between the block and the surface is 0.7?

1) $2.01\ m/s^2$
2) $2.1\ m/s^2$
3) $3.08\ m/s^2$
4) None of these

62. What is the maximum value of the force F such that the block shown in the arrangement, does not move ?

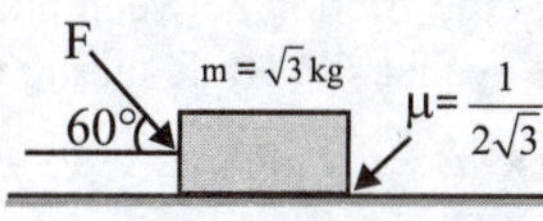

1) 20 N
2) 10 N
3) 12 N
4) 15 N

63. A block slides down an inclined plane of 45° with the horizontal. Starting from rest it covers 10 m in the first 3 seconds. Find the coefficient of kinetic friction between the block and plane.

1) 0.58
2) 0.74
3) 0.69
4) 0.61

64. A block of mass 4 kg is placed on a rought horizontal plane. A time dependent horizontal force F = kt acts on the block. Here $k = 2\ Ns^{-1}$. The frictional force between the block and plane at time t = 2s is ($\mu = 0.2$)

1) 4 N
2) 8 N
3) 12 N
4) 10 N

65. A block of mass 0.1 kg is held against a wall applying a horizontal force of 5 N on the block. If the co-efficient of friction between the block and the wall is 0.5, the magnitude of the friction force acting on the block is

1) 2.5 N
2) 0.98 N
3) 4.9 N
4) 0.49 N

66. An insect crawls up a hemispherical surface very slowly as shown in the figure. The coefficient of friction between the surface and the insect is 1/3. If the line joining the centre of the hemispherical surface to the insect makes an angle α with the vertical, the maximum possible value of α is given by

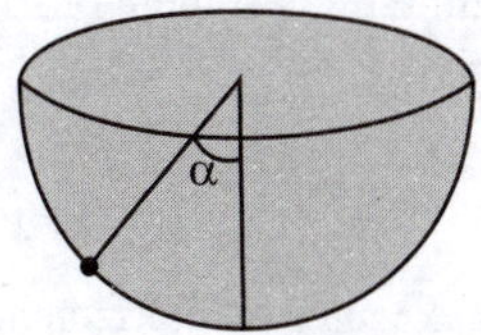

1) $\cot\alpha = 3$
2) $\tan\alpha = 3$
3) $\sec\alpha = 3$
4) $\operatorname{cosec}\alpha = 3$

67. If coefficient of friction is 1.73 then what will be the minimum angle of inclined plane to just start motion of objects?

1) 60°
2) 30°
3) 45°
4) 90°

68. A block of mass m is in contact with the cart C as shown in the figure.

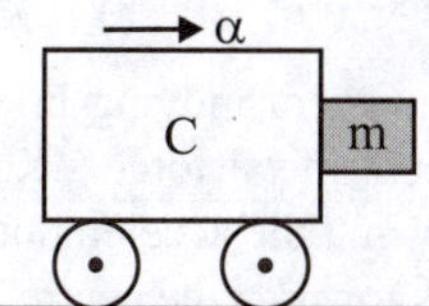

The coefficient of static friction between the block and the cart is μ. The acceleration α of the cart that will prevent the block from falling satisfies

1) $\alpha > \frac{mg}{\mu}$
2) $\alpha > \frac{g}{\mu m}$
3) $\alpha \geq \frac{g}{\mu}$
4) $\alpha < \frac{g}{\mu}$

69. A body of mass 2 kg just begins to slide down on inclined plane at an angle of 45° with the horizontal. The friction force will be

1) 19.6 tan 45°
2) 9.8 sin 45°
3) 9.8 cos 45°
4) 19.6 cos 45°

70. A body of mass M starts sliding down on the inclined plane, where the critical angle is $\angle ACB = 30°$ as shown in figure. The coefficient of friction will be

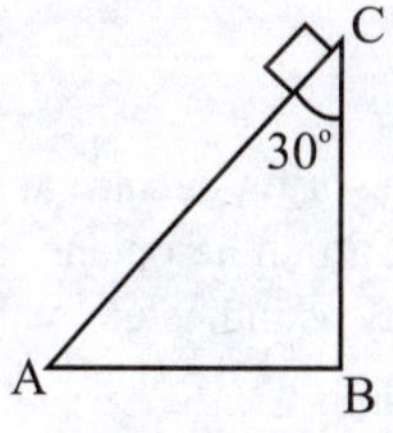

1) $Mg/\sqrt{3}$
2) $\sqrt{3}Mg$
3) $\sqrt{3}$
4) None of these

Dynamics of Horizontal Circular Motion / Uniforms Circular Motion

71. A bend on a level road has a radius of 180 m. Find the maximum speed which a car turning this bend may have without skidding if coefficient of friction between the tyres and the road is 1.5

1) 52 m/s
2) 62.3 m/s
3) 48 m/s
4) 49.2 m/s

72. An aircraft executes a horizontal loop at a speed of 800 km/h with the wings banked at an angle of 30° with the horizontal. What is the radius of the loop?

1) 7.2×10^2 m
2) 9.615×10^3 m
3) 8.543×10^3 m
4) None

73. A car of mass 2000 kg is moving with a speed of 15 m/s on a circular path of radius 20m on a level road. What should be the frictional force between the car and the road so that the car does not slip and what should be the value of coefficient of friction to attain this force?

1) 1.11×10^4 N, 0.86
2) 3.25×10^4 N, 1.41
3) 1.16×10^4 N, 0.67
4) 2.25×10^4N, 1.125

74. If a body of mass m moving in circle of radius r, then a force called centripetal force ...X... act towards the centre with a speed v and the acceleration ...Y... called centripetal acceleration is towards the centre. Here, X and Y are equal to

1) $X = \frac{mv^2}{r}, Y = \frac{v^2}{r}$
2) $X = \frac{2mv^2}{r}, Y = \frac{v^2}{r}$
3) $X = \frac{mv^2}{2r}, Y = \frac{2v^2}{r}$
4) None of the above

75. When a stone is rotated in a horizontal circle by a string, then which force provides the centripetal force?

1) Tension
2) Weight of the stone
3) Both (1) and (2)
4) None of these

76. When a planet revolve around the Sun, then which force provide the centripetal force?

1) Gravitational force on the planet due to Sun
2) Gravitational force on the planet due to Moon
3) Gravitational force on the planet due to other planet
4) None of the above

77. When a car is taking a circular turn on a horizontal road the centripetal force is the force of ...X.... Here, X refer to

1) Friction
2) Weight of the car
3) Weight of the tyres
4) None of these

78. As car is moving along a circle on a level road, then which force balance the weight of the car?

1) Friction force
2) Weight of the tyres
3) Normal reaction (N)
4) None of these

79. When a car is moving along a circle on a level road the centripetal force is provided by f, where f is given as:

1) $f < \mu_s N = \frac{mv^2}{r}$
2) $f \le \mu_s N = \frac{mv^2}{r}$
3) $f = \mu_s N = \frac{mv^2}{r}$
4) $f = \mu_k N = \frac{mv^2}{r}$

80. A cyclist speeding at 18 km/h on a level road takes a sharp circular turn of radius 3 m without reducing the speed. The coefficient of static friction between the tyres and the road is 0.1. Will the cyclist slip while taking the turn?

1) Yes, because $v > \sqrt{\mu_s gR}$
2) Yes, because $v < \sqrt{\mu_s gR}$
3) No, because $v \le \sqrt{\mu_s gR}$
4) None of these

High Order Time Intensive MCQs

1. If force on a rocket having exhaust velocity of 300 m/sec is 210 N, then rate of combusion of the fuel is

1) 0.7 kg/s
2) 1.4 kg/s
3) 0.07 kg/s
4) 10.7 kg/s

2. If a person with a spring balance and a body hanging from it goes up and up in an aeroplane, then the reading of the weight of the body as indicated by the spring balance will

1) Go on increasing
2) Go on decreasing
3) First increase and then decrease
4) Remain the same

3. A force vector applied on a mass is represented as $\vec{F} = 6\hat{i} - 8\hat{j} + 10\hat{k}$ and accelerates with 1 m/s². What will be the mass of the body?

1) $10\sqrt{2}$ kg
2) $2\sqrt{10}$ kg
3) 10 kg
4) 20 kg

4. The average force necessary to stop a bullet of mass 20 g moving with a speed of 250 m/s, as it penetrates into the wood for a distance of 12 cm is

1) 2.2×10^3 N
2) 3.2×10^3 N
3) 4.2×10^3 N
4) 5.2×10^3 N

5. Two masses of 4 kg and 5 kg are connected by a string passing through a frictionless pulley and are kept on a frictionless table as shown in the figure. The acceleration of 5 kg mass is:

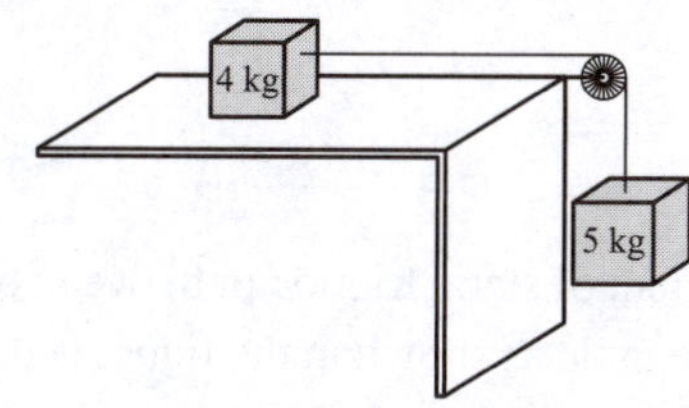

1) 49 m/s² 2) 5.44 m/s²
3) 19.5 m/s² 4) 2.72 m/s²

6. Three blocks of masses 2 kg, 3 kg and 5 kg are connected to each other with light string and are then placed on a frictionless surface as shown in the figure. The system is pulled by a Force F = 10 N, then tension T_1 =

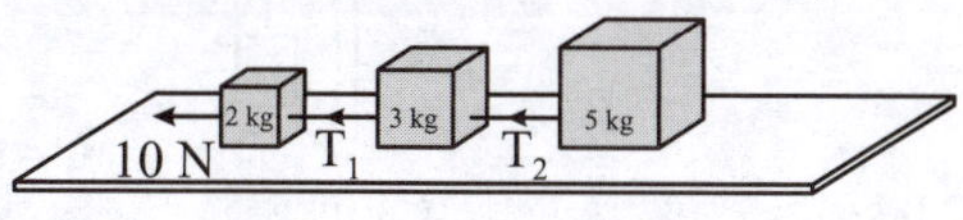

1) 1 N 2) 5 N
3) 8 N 4) 10 N

7. A body of 2 kg has an initial speed 5 ms⁻¹. A force acts on it for some time in the direction of motion. The force-time graph is shown in figure. The final speed of the body.

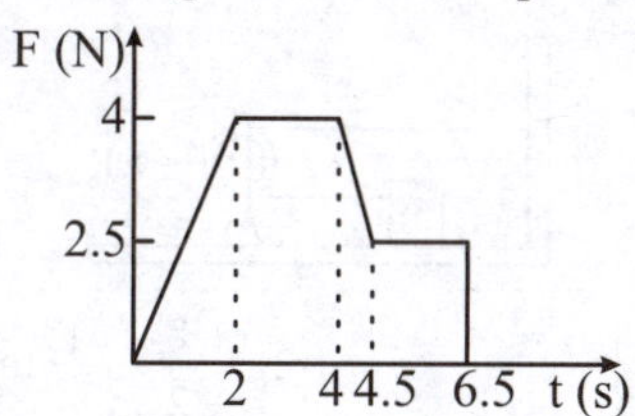

1) 9.25 ms⁻¹ 2) 5 ms⁻¹
3) 14.25 ms⁻¹ 4) 4.25 ms⁻¹

8. Match Column-I with Column-II.

Column-I		Column-II	
(A)	Definition of force	(p)	Newton's third law
(B)	Measure of force	(q)	Impulse
(C)	Effect of force	(r)	Newton's second law
(D)	Recoiling of gun	(s)	Newton's first law

1) A-q, B-p, C-r, D-s 2) A-p, B-q, C-r, D-s
3) A-s, B-r, C-q, D-p 4) A-s, B-q, C-r, D-p

9. A bird is sitting on stretched telephone wired. If its weight is W, then the additional tension produced by it in the wires will be

1) T = W 2) T > W
3) T < W 4) T = 0

10. When a body is acted by a constant force, then which of the following quantities remains constant?

1) Velocity 2) Acceleration
3) Momentum 4) None of these

11. A body of mass 200 g is accelerated upon by a constant force, travels a distance of 10 m in 1st second and 30 m in 2nd second. The foce acting on the body is

1) 1 N 2) 2 N
3) 3 N 4) 4 N

12. A massless string is passing over a frictionless pulley and is carrying two masses m_1 & m_2 ($m_1 > m_2$) at its ends. If g is the acceleration due to gravity then the thrust on the pulley by the string is

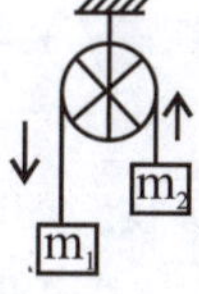

1) $\left(\frac{4m_1m_2}{m_1+m_2}\right)g$ 2) Zero
3) $\left(\frac{2m_1m_2}{m_1+m_2}\right)g$ 4) $\left(\frac{2m_1m_2}{m_1m_2}\right)g$

13. A balloon with mass m is descending down with an acceleration a (where a < g). How much mass should be removed from it so that it starts moving up with an acceleration a ?

1) $\frac{2ma}{g+a}$ 2) $\frac{ma}{g+a}$
3) $\frac{2ma}{g-a}$ 4) $\frac{ma}{g-a}$

14. A machine gun is mounted on a 200 kg vehicle on a horizontal smooth road (friction negligible). The gun fires 10 bullets/s with a velocity of 500 ms⁻¹. If the mass of each bullet is 10 g, what is the acceleration produced in the vehicle?

1) 25 cm/s²
2) 30 cm/s²
3) 50 cm/s²
4) 50 cm/s²

15. The acceleration of system of two bodies over the wedge as shown in figure is (m_1 = 3 kg) (m_2 = 5 kg)

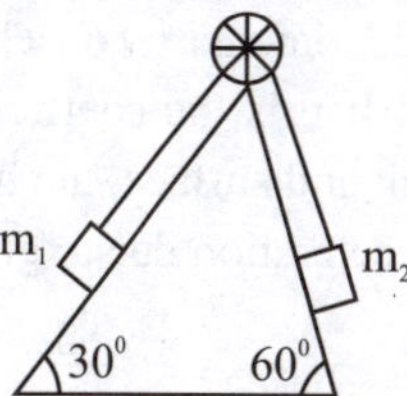

1) 3.5 m/s² 2) 4 m/s²
3) 1 m/s² 4) 2.5 m/s²

16. If the coefficient of friction between an insect and bowl is μ and the radius of the bowl is r, the maximum height to which the insect can crawl in the bowl is :

1) $\frac{r}{\sqrt{1+\mu^2}}$ 2) $r\left[1-\frac{1}{\sqrt{1+\mu^2}}\right]$
3) $r\sqrt{1+\mu^2}$ 4) $r\left[\sqrt{1+\mu^2}-1\right]$

17. A 0.5 kg ball moving with a speed of 12 ms^{-1} strikes a hard wall at an angle of 30° with the wall. It is reflected with the same speed and at the same angle. If the ball is in contact with the wall for 0.25 s, the average force acting on the wall is

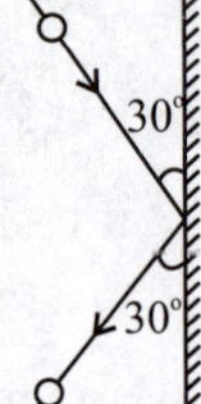

1) 48 N
2) 24 N
3) 12 N
4) 96 N

18. A disc of mass 100 g is kept floating horizontally in air by firing bullets, each of mass 5 g with the same velocity at the same rate of 10 bullets per second. The bullets rebound with the same speed in opposite direction. The velocity of each bullet at the time of impact is

1) 196 cm/s
2) 9.8 cm/s
3) 98 cm/s
4) 980 cm/s

19. In the figure given the position- time graph of a particle of mass 0.1 kg is shown. The impulse at t = 2s is

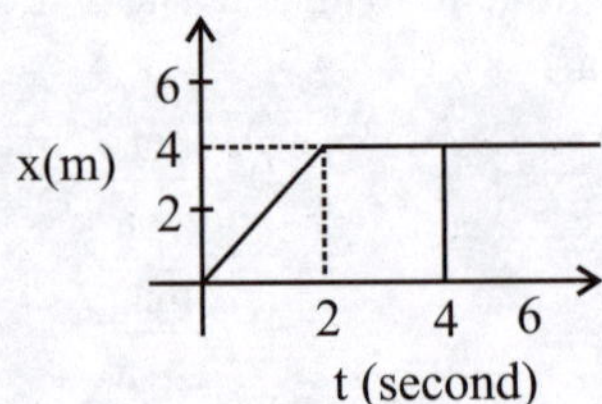

1) 0.2 kg ms^{-1}
2) – 0.2 kg ms^{-1}
3) 0.1 kg ms^{-1}
4) – 0.4 kg ms^{-1}

20. A 5000 kg rocket is set for vertical firing. The exhaust speed is 800 ms^{-1}. To given an initial upward acceleration of 20 ms^{-2}, the amount of gas ejected per second to supply the needed thrust will be

1) 137.5 kg/s
2) 185.5 kg/s
3) 127.5 kg/s
4) 187.5 kg/s

21. A block is pushed momentarily on a horizontal surface with initial velocity v. If μ is the coefficient of sliding friction between the block and surface, the block will come to rest after time (g = acceleration due to gravity)

1) $\frac{v}{\mu g}$
2) $\frac{vg}{\mu}$
3) $\frac{v\mu}{g}$
4) $\frac{\mu g}{v}$

22. A smooth block is released at rest on a 45° incline and then slides a distance d. The time taken to slide is n times as much to slide on rough incline, than on a smooth incline. The coefficient of friction is

1) $\mu_k = 1 - \frac{1}{n^2}$
2) $\mu_k = \sqrt{1 - \frac{1}{n^2}}$
3) $\mu_s = 1 - \frac{1}{n^2}$
4) $\mu_s = \sqrt{1 - \frac{1}{n^2}}$

23. Mass m and M on pulley move 0.6 m in 4 s. What is the ratio of $\frac{m}{M}$?

1) $\frac{55}{11}$
2) $\frac{113}{117}$
3) $\frac{57}{55}$
4) $\frac{397}{403}$

24. The coefficient of static friction μ, between block A of mass 2 kg and the table as shown in the figure is 0.2. What would be the maximum mass of block B so that the two blocks do not move? The string and the pulley are assumed to be smooth and massless (take, g = 10 ms^{-2}).

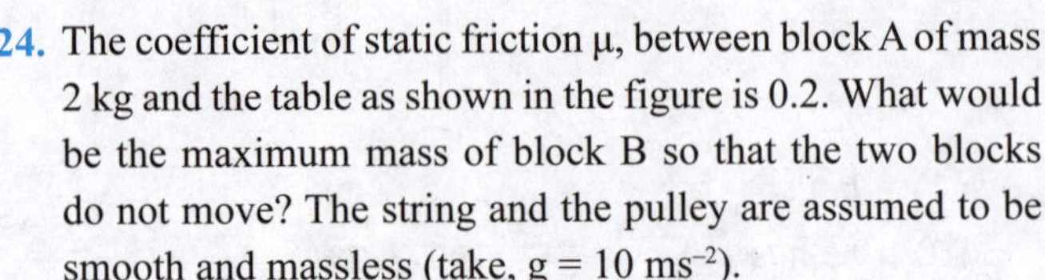

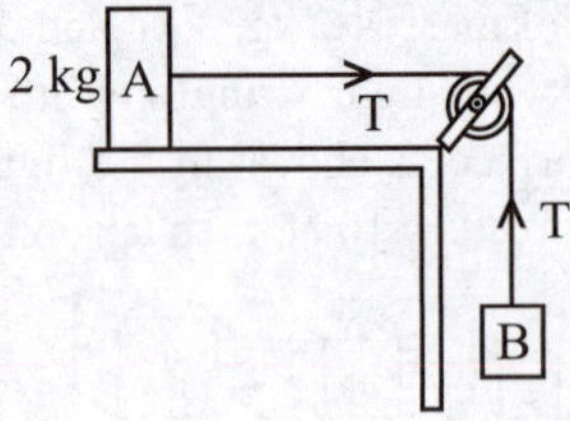

1) 2.0 kg
2) 4.0 kg
3) 0.2 kg
4) 0.4 kg

25. In the given figure the pulley is assumed massless and frictionless. If the friction force on the object of mass m is f, then its acceleration in terms of the force F will be equal to

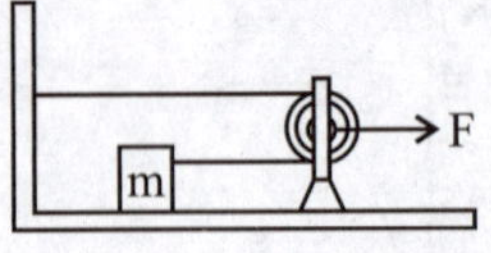

1) $\left(F - \frac{f}{m}\right)$
2) $\left(\frac{F}{2} - f\right)/m$
3) $\frac{F}{2m}$
4) None of these

26. To determine the coefficient of friction between a rough surface and a block, the surface is kept inclined at 45° and the block is released from rest. The block takes a time t in moving a distance d. The rough surface is then replaced by a smooth surface and the same experiment is repeated. Block now takes a time t/2 in moving down the same distance d. The coefficient of friction is

1) 3/4
2) 5/4
3) 1/2
4) $1/\sqrt{2}$

27. A block of mass m is placed on a surface with a vertical cross-section given by $y = \frac{x^3}{6}$. If the coefficient of friction is 0.5. The maximum height above the ground at which the block can be placed without slipping is

1) $\frac{1}{6}$m
2) $\frac{2}{3}$m
3) $\frac{1}{3}$m
4) $\frac{1}{2}$m

28. A smooth inclined plane is inclined at an angle θ with the horizontal. A body starts from rest and slides down the inclined surface, then the time taken by the body to reach the bottom is

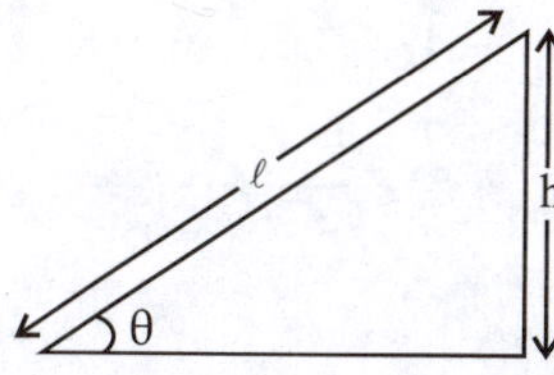

1) $\sqrt{\frac{2h}{g}}$
2) $\sqrt{\frac{2\ell}{g}}$
3) $\frac{1}{\sin\theta}\sqrt{\frac{2h}{g}}$
4) $\sin\theta\sqrt{\frac{2h}{g}}$

29. An object is kept on a smooth inclined plane of length ℓ and of height 1 m. The horizontal acceleration to be imparted to the inclined plane so that the object is stationary relative to incline is

1) $g\sqrt{\ell^2-1}$
2) $g(\ell^2-1)$
3) $\frac{g}{\sqrt{\ell^2-1}}$
4) $\frac{g}{\ell^2-1}$

30. A body takes time t to reach the bottom of an inclined plane of angle θ with the horizontal. If the plane is made rough, time taken now is 2t. The coefficient of friction of the rough surface is

1) $\frac{3}{4}\tan\theta$
2) $\frac{2}{3}\tan\theta$
3) $\frac{1}{4}\tan\theta$
4) $\frac{1}{2}\tan\theta$

31. For the system shown in the figure, the pulleys are light and frictionless. The tension in the string will be:

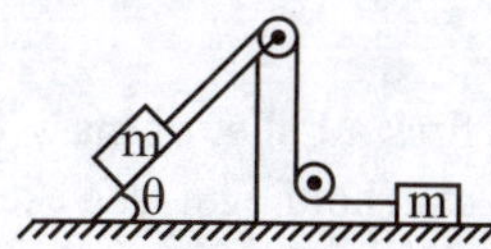

1) $\frac{2}{3}mg\sin\theta$
2) $\frac{3}{2}mg\sin\theta$
3) $\frac{1}{2}mg\sin\theta$
4) 2 mg sin θ

32. Two identical particles A and B, each of mass m, are interconnected by a spring of stiffness k. If the particle B experiences a force F & the elongation of the spring is x, the relative acceleration between the particles is equal to:

1) F/2m
2) $\frac{F-kx}{m}$
3) $\frac{F-2kx}{m}$
4) $\frac{kx}{m}$

33. Two masses M and m are connected by the arrangement shown in figure. What is the downward acceleration of mass M?

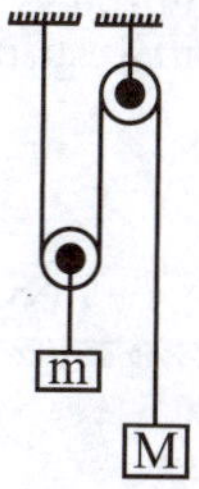

1) $\left(\frac{2M-m}{3M+m}\right)g$
2) $\left(\frac{2M-m}{4M+m}\right)2g$
3) $\left(\frac{M-m}{M+m}\right)g$
4) $\left(\frac{M}{2M+m}\right)g$

34. For the arrangement shown in the figure the extension in the spring, for which the block remains at rest is ($g = 10$ m/s²)

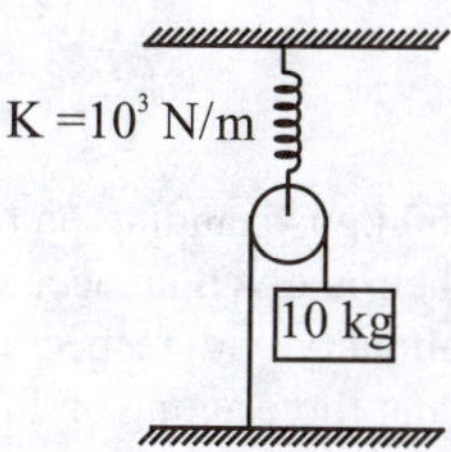

1) 20 cm
2) 25 cm
3) 15 cm
4) 30 cm

35. A pendulum with a bob of mass 2 kg is hanging from the ceiling of a trolley as shown in the figure. The trolley is uniformly accelerating with an acceleration a, then ($g = 10$ m/s²)

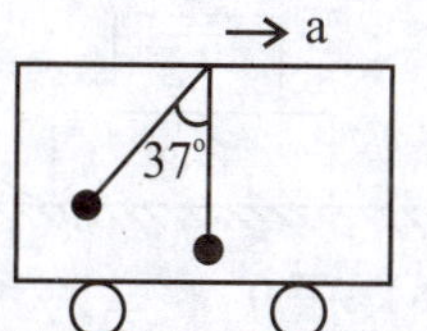

1) $a = 6$ m/s²
2) $a = 7.5$ m/s²
3) $a = 4$ m/s²
4) $a = 2.5$ m/s²

36. The acceleration of blocks of mass 2 kg and 4 kg are respectively (Pulleys and threads are massless) ($g = 10$ m/s²)

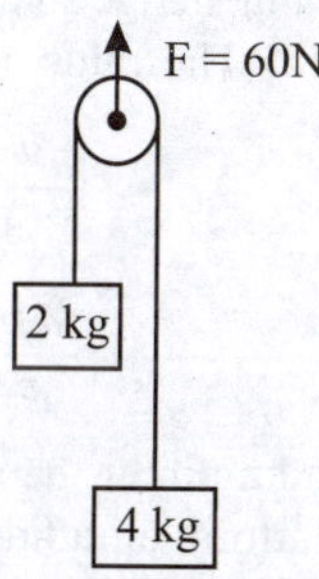

1) $a_1 = 5$ m/s²; $a_2 = 2.5$ m/s²
2) $a_1 = a_2 = 0$
3) $a_1 = a_2 = \frac{20}{6}$ m/s²
4) $a_1 = 20$ m/s²; $a_2 = 5$ m/s²

37. A block of mass m is placed on smooth triangular block as shown in the figure. The triangular block is moving horizontally with uniform speed $2\sqrt{3}$ m/s. The acceleration of block of mass m with respect to the triangular block is:

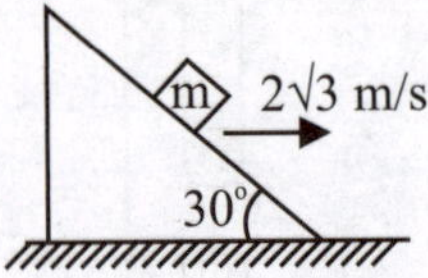

1) $5\ m/s^2$ 2) $8\ m/s^2$
3) $4\ m/s^2$ 4) $6\ m/s^2$

38. A block slides down an inclined plane of slope θ with constant velocity. It is then projected up the plane with an initial speed v_0. How far up the incline will it move before coming to rest?

1) $\frac{v_0^2}{4g\sin\theta}$ 2) $\frac{v_0^2}{g\sin\theta}$
3) $\frac{v_0^2}{2g\sin\theta}$ 4) $\frac{v_0^2}{2g}$

39. A man of mass 60 kg is standing on a horizontal conveyor belt. When the belt is given an acceleration of $1\ ms^{-2}$, the man remains stationary with respect of the moving belt. If $g = 10\ ms^{-2}$, the net force acting on the man is

1) 0.6 N 2) 6 N
3) 60 N 4) 600 N

40. Two blocks of mass 4 kg and 6 kg are placed one over the other. If a force of 10 N is applied on the block of mass 4 kg, then the accelerations of the two blocks at the instant shown is ($g = 10\ m/s^2$)

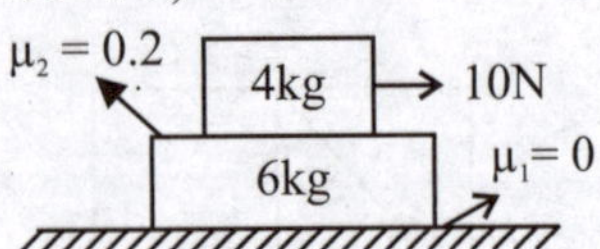

1) $a(4\text{ kg}) = a(6\text{ kg}) = 0$
2) $a(4\text{ kg}) = 0.5\ m/s^2$, $a(6\text{ kg}) = 0$
3) $a(4\text{ kg}) = 2\ m/s^2$, $a(6\text{ kg}) = 0.5\ m/s^2$
4) $a(4\text{ kg}) = 1\ m/s^2$, $a(6\text{ kg}) = 1\ m/s^2$

41. Radius of the curved road on a national highway is R. Width of the road is b. The outer edge of the road is raised by h with respect to inner edge so that a car with velocity v can pass safely over it. The value of h is :

1) $\frac{v^2b}{Rg}$ 2) $\frac{v}{Rgb}$
3) $\frac{v^2R}{bg}$ 4) $\frac{v^2b}{R}$

42. A boy holds a pendulum in his hand while standing at the edge of a circular platform of radius r rotating at an angular speed ω. The pendulum will hang at an angle θ with the vertical so that:

1) $\tan\theta = 0$ 2) $\tan\theta = \frac{\omega^2 r^2}{g}$
3) $\tan\theta = \frac{r\omega^2}{g}$ 4) $\tan\theta = \frac{g}{\omega^2 r}$

43. Two forces are simultaneously applied on an object. What third force would make the net force on the particle zero?

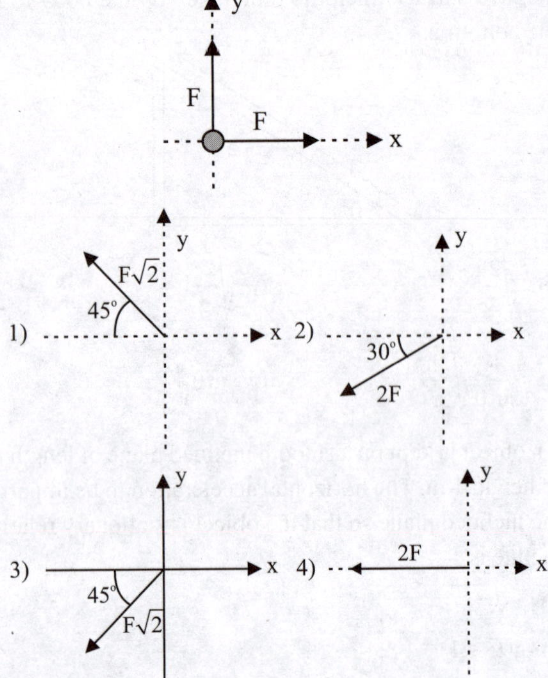

44. If masses are released from the position shown in figure then time elapsed before mass m_1 collides with the floor will be:

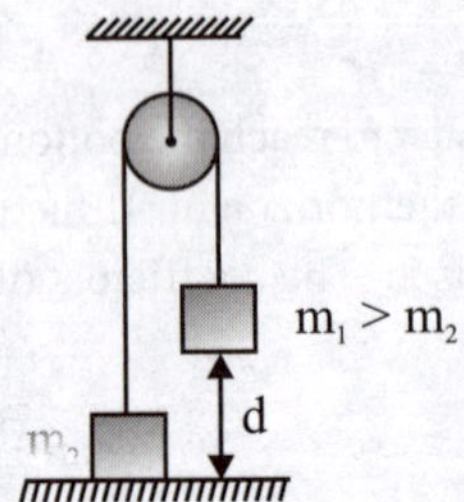

1) $\sqrt{\frac{2m_1gd}{m_1+m_2}}$ 2) $\sqrt{\frac{2(m_1+m_2)d}{(m_1-m_2)g}}$
3) $\sqrt{\frac{2(m_1-m_2)d}{(m_1+m_2)g}}$ 4) None of these

45. A machine gun fires a bullet of mass 40 g with a velocity $1{,}200\ ms^{-1}$. The man holding it, can exert a maximum force of 144 N on the gun. How many bullets can he fire per second at the most?

1) One 2) Four
3) Two 4) Three

46. A player caught a cricket ball of mass 150 g moving at a rate of 20 m/s. If the ball catching process is completed in 0.1 s, the force of blow exerted by the ball on the hand of the player is equal to

1) 30 N
2) 300 N
3) 150 N
4) 3 N

47. An elevator accelerates upwards at a constant rate. A uniform string of length L and mass m supports a small block of mass M that hangs from the ceiling of the elevator. The tension at distance l from the ceiling is T. The acceleration of the elevator is:

1) $\dfrac{T}{M+m-\dfrac{ml}{L}}-g$

2) $\dfrac{T}{2M+m-\dfrac{ml}{L}}+g$

3) $\dfrac{T}{M+\dfrac{ml}{L}}-g$

4) $\dfrac{T}{2M-m+\dfrac{ml}{L}}-g$

48. If the coefficient of friction of a plane inclined at 45° is 0.5, then acceleration of a body sliding freely on it is

1) $4.9\ m/s^2$ 2) $\dfrac{9.8}{\sqrt{2}}\ m/s^2$

3) $9.8\ m/s^2$ 4) $\dfrac{9.8}{2\sqrt{2}}\ m/s^2$

49. A block of mass m lying on a rough horizontal plane is acted upon by a horizontal force P and another force Q inclined at an angle θ to the vertical. The block will remain in equilibrium, if the coefficient of friction between it and the surface is

1) $(P + Q\sin\theta)/(mg + Q\cos\theta)$

2) $(P\cos\theta + Q)/(mg - Q\sin\theta)$

3) $(P + Q\cos\theta)/(mg + Q\sin\theta)$

4) $(P \sin\theta - Q)/(mg - Q\cos\theta)$

Statement Based MCQs

Directions: Each question consist of two statements, Statement-I and Statement-II. Choose any one of the following four responses.

1) **Both Statement-I and Statement-II are correct.**
2) **Both Statement-I and Statement-II are incorrect.**
3) **Statement-I is correct & Statement-II is incorrect.**
4) **Statement-I is incorrect & Statement-II is correct.**

1. **Statement-I:** A bird alights on a stretched wire depressing it slightly. The increase in tension of the wire is more than the weight of the bird.

   Statement-II: The tension must always be more than the weight as it needs to balance weight.

2. **Statement-I:** Motion of a rocket is based on conservation on linear momentum.

   Statement-II: Frictional forces are conservative forces.

3. **Statement-I:** Gravitational forces are conservative forces.

   Statement-II: Work done in a round trip in a gravitational field is never zero.

4. **Statement-I:** Conservation of momentum is valid when the external force on the system is zero.

   Statement-II: Impulse is product of force and displacement.

5. **Statement-I:** Newton's first law is the real law of motion.

   Statement-II: Newton's second law is the real law of motion.

Assertion & Reason MCQs

Directions: Each question consists of two statements Assertion and Reason. Choose any one of the following four responses.

1) **Both Assertion and Reason are True and the Reason is a correct explanation of the Assertion.**
2) **Both Assertion and Reason are True but Reason is not a correct explanation of the Assertion.**
3) **Assertion is True but the Reason is False.**
4) **Assertion is False but Reason is True.**

1. **Assertion:** When a sphere rolls on a horizontal table it slows down and eventually stops.

   Reason: When the sphere rolls on the table, both the sphere and the surface are deformed near the contact. As a result, the sphere slows down.

2. **Assertion:** It is difficult to move a cycle along a road with its brakes on.

   Reason: Sliding friction coefficient is greater than the rolling friction coefficient.

3. **Assertion:** Two teams having a tug of war always pull equally hard on one another.

   Reason: The team that pushes harder against the ground, in a tug of war, wins.

4. **Assertion:** The shafts of motors are provided with ball bearings to reduce friction.

   Reason: The rolling friction is less than sliding friction.

5. **Assertion:** In Karate, a brick can be broken with a bare hand by applying the force swiftly.

   Reason: In this process the impulse is sharp.

Match the Columns MCQs

1. Match the following with correct response.

Column-I		Column-II	
(a)	Inertia	(i)	Product of mass and velocity
(b)	Friction	(ii)	Mass of the object
(c)	Momentum	(iii)	Rate of change of momentum
(d)	Force	(iv)	Necessary evil

1) (a)-(iv) (b)-(i) (c)-(iii) (d)-(ii)
2) (a)-(i) (b)-(iii) (c)-(ii) (d)-(iv)
3) (a)-(iii) (b)-(ii) (c)-(iv) (d)-(i)
4) (a)-(ii) (b)-(iv) (c)-(i) (d)-(iii)

2. Match the following with correct response.

Column-I		Column-II	
(a)	Newton's first law of motion	(i)	The acceleration produced is directly proportional to product of mass & force applied
(b)	Newton's second law of motion	(ii)	When no force is exerted on an object it stays at rest or it moves in a straight line with constant speed
(c)	Newton's third law of motion	(iii)	To every action there is an equal & opposite reaction
(d)	Galileo's law of inertia	(iv)	An object at rest or in uniform motion, will remain at rest or uniform motion unless an unbalanced force is applied on it

1) (a)-(ii) (b)-(iv) (c)-(i) (d)-(iii)
2) (a)-(i) (b)-(iii) (c)-(ii) (d)-(iv)
3) (a)-(iv) (b)-(i) (c)-(iii) (d)-(ii)
4) (a)-(iii) (b)-(ii) (c)-(iv) (d)-(i)

3. Match the following with correct response.

Column-I		Column-II	
(a)	Friciton	(i)	When one object rolls over another
(b)	Limiting friction	(ii)	Force just sufficient to move the object
(c)	Sliding friction	(iii)	Force which opposes motion
(d)	Rolling friction	(iv)	Force sufficient to slide one object over another

1) (a)-(iii) (b)-(ii) (c)-(iv) (d)-(i)
2) (a)-(iv) (b)-(i) (c)-(iii) (d)-(ii)
3) (a)-(ii) (b)-(iv) (c)-(i) (d)-(iii)
4) (a)-(i) (b)-(iii) (c)-(ii) (d)-(iv)

4. Match the following with correct response.

Column-I		Column-II	
(a)	F	(i)	$m \times v$
(b)	P	(ii)	Momentum after collision are equal
(c)	ΔP	(iii)	$\frac{P_2 - P_1}{t}$
(d)	Momentum before collision	(iv)	$m \times a$

1) (a)-(iv) (b)-(i) (c)-(iii) (d)-(ii)
2) (a)-(ii) (b)-(iv) (c)-(i) (d)-(iii)
3) (a)-(iii) (b)-(ii) (c)-(iv) (d)-(i)
4) (a)-(i) (b)-(iii) (c)-(ii) (d)-(iv)

NCERT Exemplar MCQs

1. A ball is traveling with uniform translatory motion. This means that:
   1) It is at rest
   2) The path can be a straight line or circular and the ball travels with uniform speed
   3) All parts of the ball have the same velocity (magnitude and direction) and the velocity is constant
   4) The center of the ball moves with constant velocity and the ball spins about its center uniformly

2. A meter scale is moving with uniform velocity. This implies:
   1) The force acting on the scale is zero, but a torque about the center of mass can act on the scale
   2) The force acting on the scale is zero and the torque acting about center of mass of the scale is also zero
   3) The total force acting on it need not be zero but the torque on it is zero
   4) Neither the force nor the torque need to be zero

3. A cricket ball of mass 150 g has an initial velocity $\vec{u} = (3\hat{i} + 4\hat{j})ms^{-1}$ and a final velocity $\vec{v} = -(3\hat{i} + 4\hat{j})ms^{-1}$, after being hit. The change in momentum (final momentum - initial momentum) is (in kg ms^{-1}):
   1) Zero
   2) $-(0.45\hat{i} + 0.6\hat{j})$
   3) $-(0.9\hat{i} + 1.2\hat{j})$
   4) $-5(\hat{i} + \hat{j})$

4. In the previous problem (Q.3), the magnitude of the momentum transferred during the hit is:
   1) Zero
   2) 0.75 kg-m s^{-1}
   3) 1.5 kg-m s^{-1}
   4) 14 kg-m s^{-1}

5. Conservation of momentum in a collision between particles can be understood from:
   1) Conservation of energy
   2) Newton's first law only
   3) Newton's second law only
   4) Both Newton's second and third law

6. A hockey player is moving northward and suddenly turns westward with the same speed to avoid an opponent. The force that acts on the player is:
   1) Frictional force along westward
   2) Muscle force along southward
   3) Frictional force along south-West
   4) Muscle force along south-West

7. A body of mass 2 kg travels according to the law $x(t) = pt + qt^2 + rt^3$ where, $q = 4\ ms^{-2}$, $p = 3\ ms^{-1}$ and $r = 5\ ms^{-3}$. The force acting on the body at t = 2s is:
   1) 136 N 2) 134 N
   3) 158 N 4) 68 N

8. A body with mass 5 kg is acted upon by a force $\vec{F} = (-3\hat{i} + 4\hat{j})N$. If its initial velocity at t = 0 is $\vec{u} = (6\hat{i} - 12\hat{j})ms^{-1}$, the time at which it will just have a velocity along the Y-axis is:
   1) Never 2) 10 s
   3) 2 s 4) 15 s

9. A car of mass m starts from rest and acquires a velocity along east, $\vec{v} = v\hat{i}\ (v > 0)$ in two seconds. Assuming the car moves with uniform acceleration, the force exerted on the car is:
   1) $\frac{mv}{2}$ eastward and is exerted by the car engine
   2) $\frac{mv}{2}$ eastward and is due to the friction on the tyres exerted by the road
   3) More than $\frac{mv}{2}$ eastward exerted due to the engine and overcomes the friction of the road
   4) $\frac{mv}{2}$ exerted by the engine

10. The motion of a particle of mass m is given by x = 0 for t < 0s x(t) = A sin 4 πt for 0 < t < (1/4)s (A > 0), and x = 0 for t > (1/4) s. Which of the following statements are true?
   1) The force at t = (1/8)s on the particle is $-16\pi^2$A-m
   2) The particle is acted upon by an impulse of magnitude 4 π^2A-m at t = 0 and t = (1/ 4) s
   3) The particle is not acted upon by any force
   4) The particle is not acted upon by a constant force

11. In figure the coefficient of friction between the floor and the body B is 0.1. The coefficient of friction between the bodies B and A is 0.2. A force F is applied as shown on B. The mass of A is m/2 and of B is m. Which of the following statements are true?

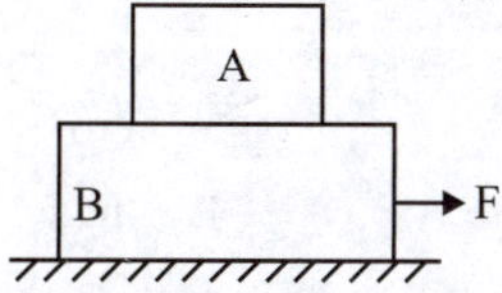

   1) The bodies will move together if F = 0.25 mg
   2) The body A will slip with respect to B if F = 0.5 mg
   3) The bodies will move together if F = 0.5 mg
   4) The bodies will be at rest if F = 0.1 mg

12. Mass m_1 moves on a slope making an angle θ with the horizontal and is attached to mass m_2 by a string passing over a frictionless pulley as shown in figure. The coefficient of friction between m_1 and the sloping surface is μ. Which of the following statements are true?

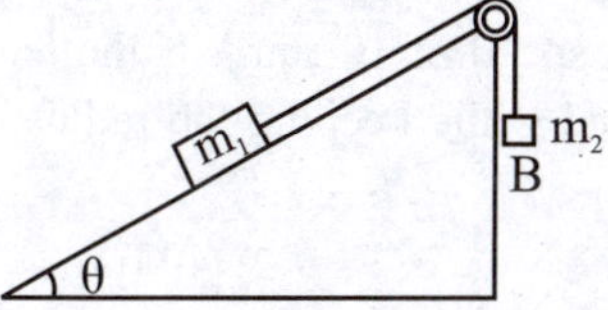

   1) If $m_2 > m_1 \sin\theta$, the body will move up the plane
   2) If $m_2 > m_1 (\sin\theta + \mu\cos\theta)$, the body will move up the plane
   3) If $m_2 < m_1 (\sin\theta + \mu\cos\theta)$, the body will move up the plane
   4) If $m_2 < m_1 (\sin\theta - \mu\cos\theta)$, the body will move down the plane

13. In figure, a body A of mass m slides on plane inclined at angle θ_1 to the horizontal and μ is the coefficient of friction between A and the plane. A is connected by a light string passing over a frictionless pulley to another body B, also of mass m, sliding on a frictionless plane inclined at an angle θ_2 to the horizontal. Which of the following statements are true?

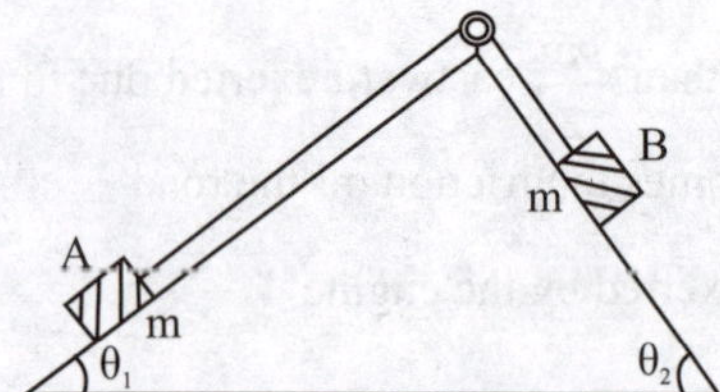

1) A will never move up the plane
2) A will just start moving up the plane when $\mu = \frac{\sin\theta_2 - \sin\theta_1}{\cos\theta_1}$
3) For A to move up the plane, θ_2 must always be greater than θ_1
4) B will always slide down with constant speed

14. Two billiard balls A and B, each of mass 50 g and moving in opposite directions with speed of 5 ms^{-1} each, collide and rebound with the same speed. If the collision lasts for 10^{-3} s, which of the following statements are true?

1) The impulse imparted to each ball is 0.25 kg-ms^{-1} and the force on each ball is 250 N
2) The impulse imparted to each ball is 0.25 kg-ms^{-1} and the force exerted on each ball is 25×10^{-5} N-s
3) The impulse imparted to each ball is 0.5 N-s
4) The impulse and the force on each ball are equal in magnitude and opposite in directions

NEET & AIIMS Past Year MCQs

1. A box of mass 5 kg is pulled by a cord, up along a frictionless plane inclined at 30° with the horizontal. The tension in the cord is 30 N. The acceleration of the box is (Take $g = 10$ ms^{-2}) **(2024 Re)**

1) 2 m s^{-2} 2) Zero
3) 0.1 m s^{-2} 4) 1 m s^{-2}

2. A horizontal force 10 N is applied to a block *A* as shown in figure. The mass of blocks *A* and *B* are 2 kg and 3 kg, respectively. The blocks slide over a frictionless surface. The force exerted by block *A* on block *B* is: **(2024)**

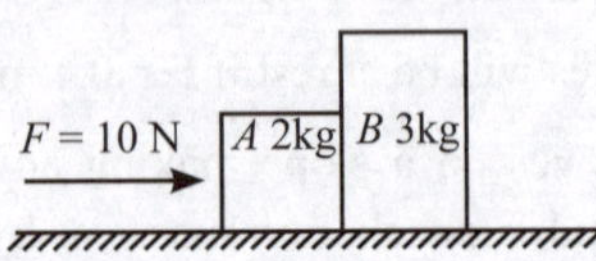

1) 6 N 2) 10 N
3) Zero 4) 4 N

3. A bob is whirled in a horizontal circle by means of a string at an initial speed of 10 rpm . If the tension in the string is quadrupled while keeping the radius constant, the new speed is: **(2024 Re)**

1) 20 rpm 2) 40 rpm
3) 5 rpm 4) 10 rpm

4. A particle moving with uniform speed in a circular path maintains **(2024)**

1) constant velocity by varying acceleration.
2) varying velocity and varying acceleration.
3) constant velocity.
4) constant acceleration.

5. A bob is whirled in a horizontal plane by means of a string with an initial speed of ω rpm. The tension in the string is *T*. If speed becomes 20 ω while keeping the same radius, the tension in the string becomes: **(2024)**

1) $\frac{T}{4}$ 2) $\sqrt{2}T$
3) T 4) $4T$

6. Calculate the maximum acceleration of a moving car so that a body lying on the floor of the car remains stationary. The coefficient of static friction between the body and the floor is 0.15 (g = 10 ms^{-2}). **(2023)**

1) 50 ms^{-2} 2) 1.2 ms^{-2}
3) 150 ms^{-2} 4) 1.5 ms^{-2}

7. A shell of mass m is at rest initially. It explodes into three fragments having mass in the ratio 2 : 2 : 1. If the fragments having equal mass fly off along mutually perpendicular directions with speed v, the speed of the third (lighter) fragment is: **(2022)**

1) $3\sqrt{2}$ v 2) v
3) $\sqrt{2}$ v 4) $2\sqrt{2}$ v

8. A small block slides down on a smooth inclined plane, starting from rest at time t = 0. Let S_n be the distance travelled by the block in the interval t = n – 1 to t = n. Then, the ratio $\frac{S_n}{S_{n+1}}$ is: **(2021)**

1) $\frac{2n-1}{2n+1}$ 2) $\frac{2n+1}{2n-1}$ 3) $\frac{2n}{2n-1}$ 4) $\frac{2n+1}{2n}$

9. A man (mass = 50 kg) is in an elevator that is moving with an acceleration 0.49 m/s^2 upwards. Find normal reaction exerted by man on floor of the elevator. **(AIIMS 2018)**

1) 214.5 N 2) 314.5 N
3) 414.5 N 4) 514.5 N

10. The figure shows a horizontal force $\vec{F}$ acting on a block of mass m on an inclined plane (angle θ). What is the normal reaction *N* on the block? **(AIIMS 2017)**

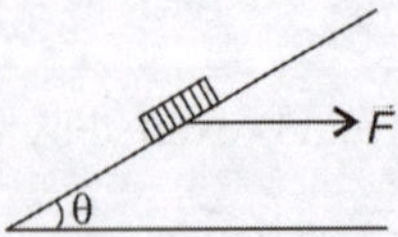

1) mgsinθ + Fcosθ 2) mgsinθ – Fcosθ
3) mgcosθ – Fsinθ 4) mgcosθ + Fsinθ

11. A ball is thrown vertically upwards. Which of the following plots represents the speed-time graph of the ball during its flight if the air resistance is not ignored: **(AIIMS 2003)**

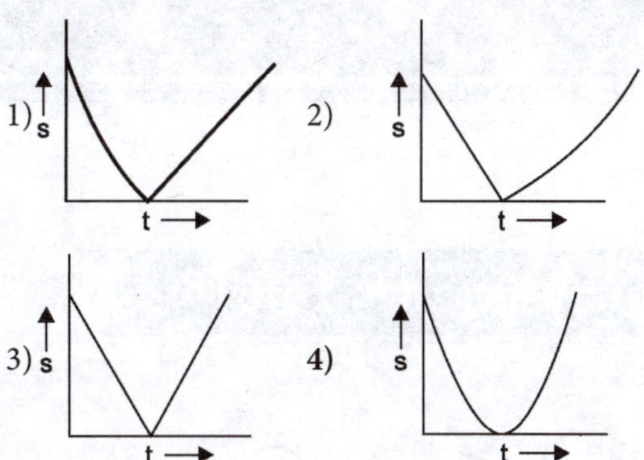

12. The forces, which meet at one point but their lines of action do not lie in one plane, are called: **(AIIMS 1996)**

1) Non-coplanar concurrent forces

2) Coplanar concurrent forces

3) Non-coplanar non-concurrent forces

4) Coplanar non-concurrent forces.

Answer Key

NCERT Topic-wise MCQs

1	2	3	4	5	6	7	8	9	10	11	12	13	14	15	16	17
3	4	4	3	4	2	3	4	3	1	2	4	1	3	4	1	3
18	19	20	21	22	23	24	25	26	27	28	29	30	31	32	33	34
2	1	3	3	1	1	1	4	4	2	1	1	4	3	2	2	1
35	36	37	38	39	40	41	42	43	44	45	46	47	48	49	50	51
4	3	3	1	1	3	4	4	1	1	4	1	1	1	3	4	4
52	53	54	55	56	57	58	59	60	61	62	63	64	65	66	67	68
2	2	4	2	1	3	4	1	1	2	1	3	1	2	1	1	3
69	70	71	72	73	74	75	76	77	78	79	80					
4	3	1	3	4	1	1	1	1	3	2	1					

High Order Time Intensive MCQs

1	2	3	4	5	6	7	8	9	10	11	12	13	14	15	16	17
1	3	1	4	2	3	3	3	2	2	4	1	1	1	1	2	2
18	19	20	21	22	23	24	25	26	27	28	29	30	31	32	33	34
4	2	4	1	1	4	4	2	1	1	3	3	1	3	3	2	1
35	36	37	38	39	40	41	42	43	44	45	46	47	48	49		
2	1	1	1	3	4	1	3	3	2	4	1	1	4	1		

Statement Based MCQs

1	2	3	4	5
3	3	3	3	4

Assertion & Reason MCQs

1	2	3	4	5
1	1	2	1	1

Match the Columns MCQs

1	2	3	4
4	3	1	1

NCERT Exemplar MCQs

1	2	3	4	5	6	7	8	9	10	11	12	13	14
3	2	3	3	4	3	1	2	2	1,2,4	1,2,4	1,4	2,3	3,4

NEET & AIIMS Past Year MCQs

1	2	3	4	5	6	7	8	9	10	11	12
4	1	1	2	4	4	4	1	4	4	4	1

CHAPTER

5 Work, Energy and Power

Topicwise Number of Questions Analysis (2025-2016)

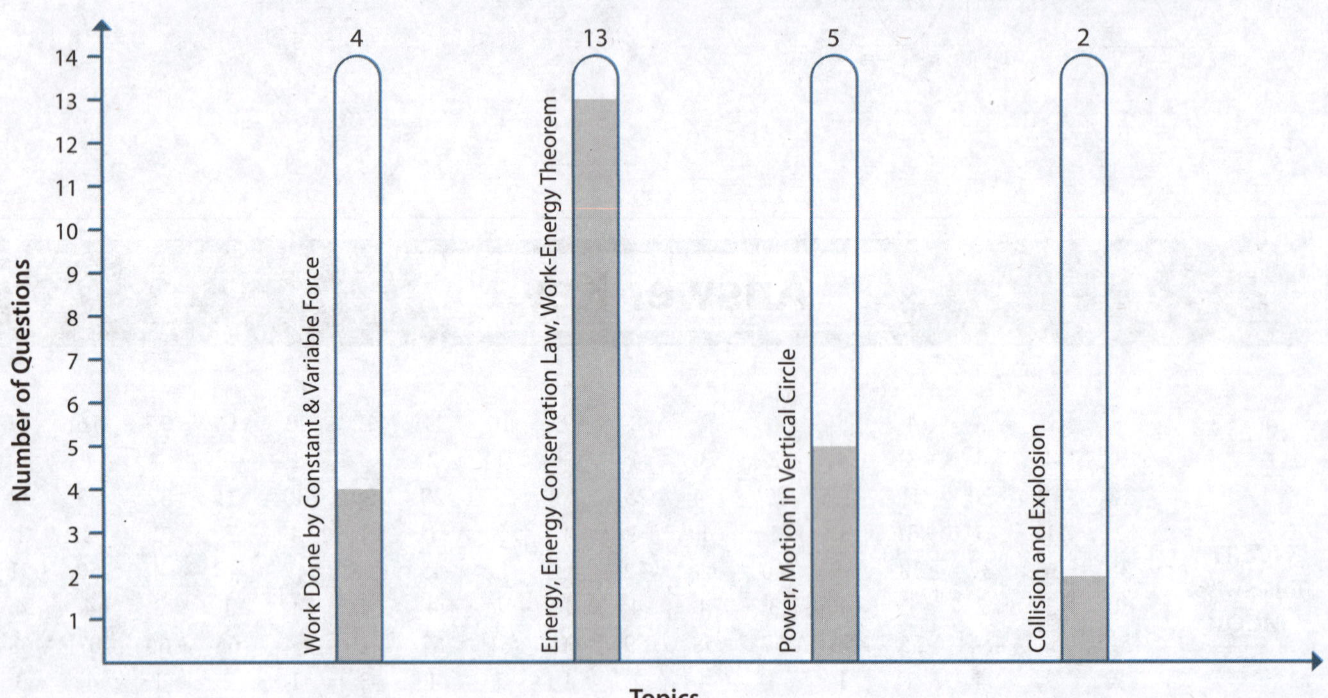

NCERT-PICKS

Work Done

Formulas

1. Work done, $W = \vec{F}.\vec{s} = Fs\cos\theta$

   Its S.I. unit is Joule (J).

2. In the case of a variable force

   $W = \int_{s_1}^{s_2} \vec{F}.d\vec{s}$ [s_1 and s_2 are the displacement]

   Work is a scalar quantity.

Numerical Profile

Work done under different cases

(i) If a force changes linearly from F_1 to F_2 over a displacement *S*, then

$$W = \left(\frac{F_1 + F_2}{2}\right)S$$

(ii) An object of mass m is lifted to a height h above the ground.

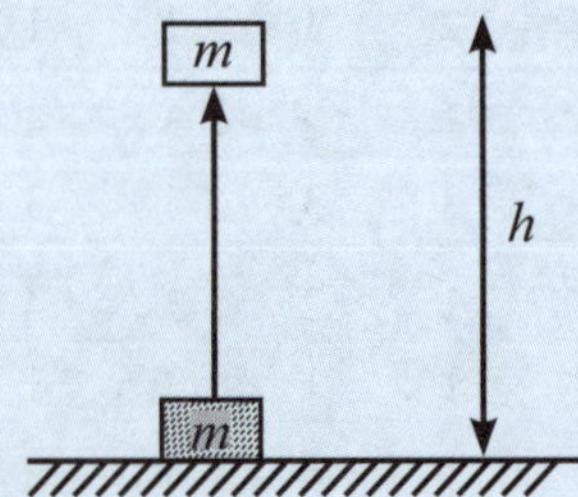

Work done by the gravitational force is given by

$W_g = -mgh$

(iii) Work done in pulling the bob of a simple pendulum of length L through an angle θ with the vertical by means of a horizontal force F.

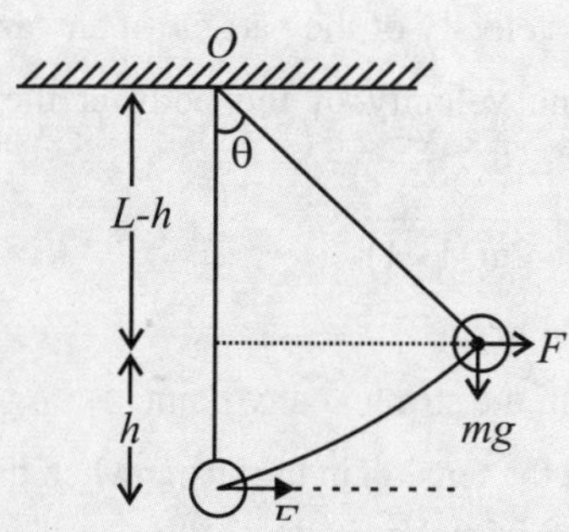

$$\cos\theta = \frac{L-h}{L} = 1 - \frac{h}{L}$$

$$\frac{h}{L} = 1 - \cos\theta$$

$h = L(1 - \cos\theta)$

Work done by gravitational force

$W_g = -mgh = -mgL(1 - \cos\theta)$

Work done by horizontal force F is $W_F = FL \sin\theta$

Work done in accelerated frame

(i) A block of mass m is suspended vertically using a rope of negligible mass. If the rope is used to lift the block vertically up with uniform acceleration '*a*', work done by tension in the rope in lifting the block through a height h.

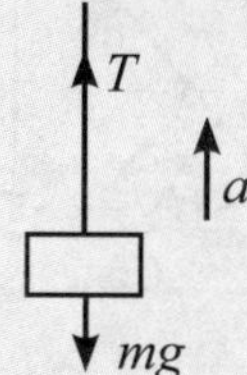

$W = T \times h$

$W = m(g + a)h$ (h = height)

(ii) If block is going down with acceleration '*a*'

$W = m(g - a)h$

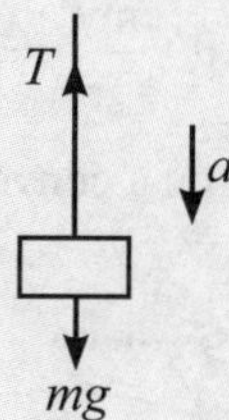

Case Based on Chain

(i) A uniform chain of mass m and length L is kept on smooth horizontal table such that $\frac{1}{n^{th}}$ of its length is hanging over the edge of the table. The work done by the pulling force to bring the hanging part on to the table is

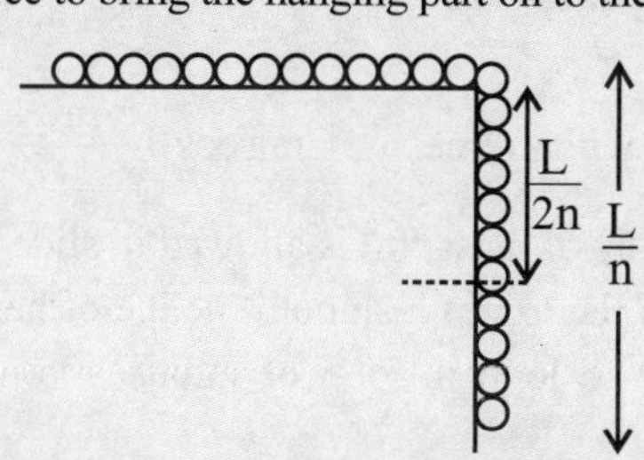

Weight of hanging part, $\frac{m}{n}g$

$$\text{Work} = \frac{m}{n}gh = \frac{mg}{n}\frac{L}{2n} = \frac{mgL}{2n^2}$$

(ii) A uniform chain of mass m and length L rests on a smooth horizontal table with $\frac{1}{n_1}$ part of its length is hanging from the edge of the table. Work done in pulling the chain partially such that $\frac{1}{n_2}$ part is hanging from the edge of the table is given by

$$W = \frac{mgL}{2}\left[\frac{1}{n_1^2} - \frac{1}{n_2^2}\right]$$

Energy

Formulas

1. Kinetic Energy(KE) = $\frac{1}{2}mv^2$.
2. Gravitational Potential Energy(PE) = mgh.
3. Potential Energy in a spring = $\frac{1}{2}kx^2$.
4. K.E = $\frac{P^2}{2m}$

Work-Energy Theorem

Formulas

Work energy theorem: Change in the kinetic energy of a body is equal to the amount of work done on it by all the forces.

$$W_{all} = \frac{1}{2}mv^2 - \frac{1}{2}mu^2 \Rightarrow W_{all} = \Delta K$$

Numerical Profile

Applications of work-energy theorem

(i) A particle of mass 'm' is thrown vertically up with a speed 'u'. Neglecting air friction, work done by the gravitational force, as the particle reaches maximum height is

$W_g = \Delta K = K_f - K_i$

$$W_g = -\frac{1}{2}mu^2 \Rightarrow -mgh = -\frac{1}{2}mu^2 \Rightarrow u = \sqrt{2gh}$$

(ii) A block of mass 'm' slides down a frictionless smooth incline of inclination 'θ' to the horizontal. If h is the height of the incline, the velocity with which the body reaches the bottom of the incline is

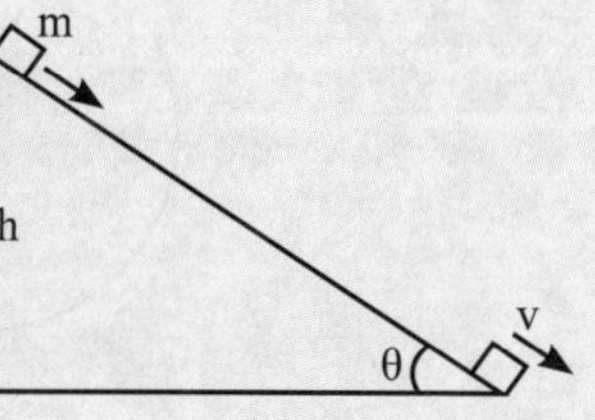

$W_g = \Delta K$

$$mgh = \frac{1}{2}mv^2 - 0$$

$$mgh = \frac{1}{2}mv^2$$

$$v = \sqrt{2gh}$$

(iii) A body of mass 'm' starts sliding from the top of a rough inclined plane of inclination 'θ' and length 'L'. The velocity 'v' with which it reaches the bottom of the incline is

$$W_g + W_f = \Delta K$$

$$(mg\sin\theta)L + (-\mu_k mg\cos\theta)L = \frac{1}{2}mv^2 - 0$$

where μ_k is the coefficient of friction

$$\Rightarrow v = \sqrt{2gL(\sin\theta - \mu_k\cos\theta)}$$

(iv) Consider a simple pendulum of length L. If it is given a horizontal speed u at its lowest position then the speed of the bob when it makes an angle θ with the vertical is v, then

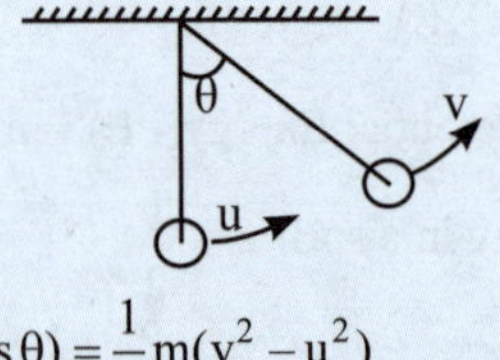

$$-mgL(1-\cos\theta) = \frac{1}{2}m(v^2 - u^2)$$

$$\Rightarrow v = \sqrt{u^2 - 2gL(1-\cos\theta)}$$

Motion in a Vertical Circle

Formulas

The minimum velocity(v) of a body in verticle plane at any point to complete full circle is $\sqrt{gR(3+2\cos\theta)}$

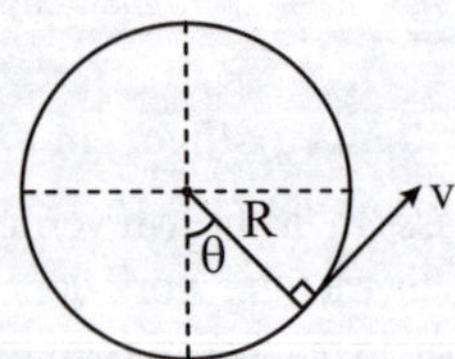

Numerical Profile

Velocity and Tension in Vertical Circular Motion

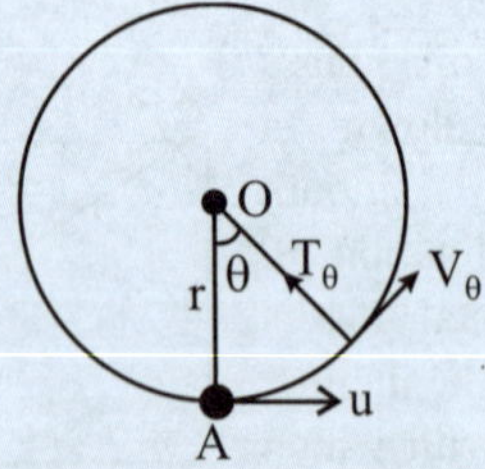

(i) Let u be the velocity of the body at the lowest point. Velocity of the body at any point on the vertical circle

$$\Rightarrow v_\theta = \sqrt{u^2 - 2gr(1-\cos\theta)}$$

[v_θ is the velocity of the particle at any angle θ]

If v is the velocity of the body at the highest point (θ = 180°)

$$v = \sqrt{u^2 - 2gr(1+1)}$$

$$v^2 = u^2 - 4\,gr$$

(ii) Tension in the string at any point:

Let T_θ be the tension in the string when the string makes an angle θ with the vertical.

$$T_\theta = \frac{mv_\theta^{\,2}}{r} + Mg\cos\theta$$

(iii) The minimum speed of the particle at the highest point to just complete the vertical circle $v = \sqrt{gr}$

(iv) If $u \ge \sqrt{5gR}$, particle completes the circle.

(v) If $\sqrt{2gr} < u < \sqrt{5gr}$ the particle is not able to complete the vertical circle, it goes to a certain height and leaves the circular path.

Motion of an object on a curved surface

(i) When a car moves on a concave bridge of radius r

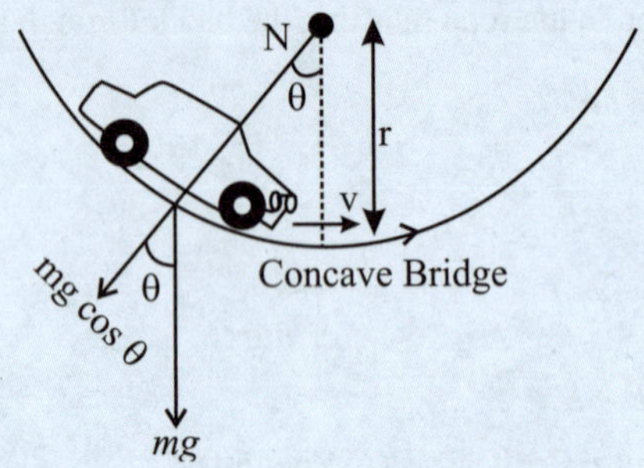

Centripetal force $= N - mg\cos\theta = \frac{mv^2}{r}$ and reaction force, $N = mg\cos\theta + \frac{mv^2}{r}$

(ii) When car moves on a convex bridge of radius r

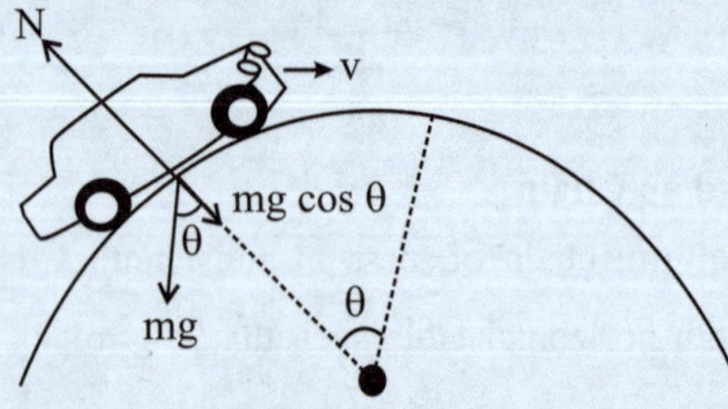

Centripetal force $= mg\cos\theta - N = \frac{mv^2}{r}$

and reaction force, $N = mg\cos\theta - \frac{mv^2}{r}$

(iii) A body of mass 'm' is allowed to slide down from rest, from the top of a smooth incline of height 'h'. For the body to loop in loop of radius 'r' on arriving at the bottom.

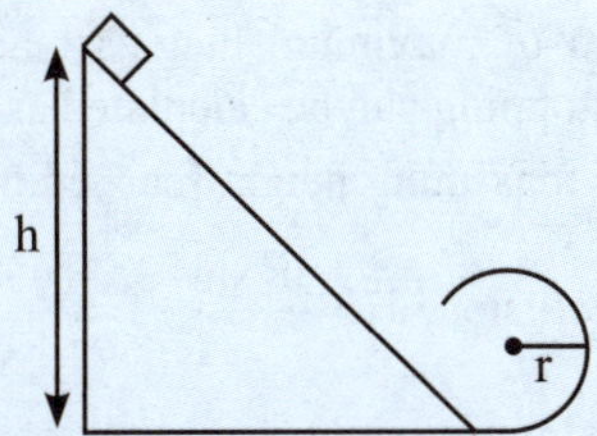

Minimum height of smooth incline so that the particle completes the vertical circle, $h = \left(\frac{5r}{2}\right)$

(iv) A small block is sliding down from the top of a smooth hemisphere of radius r, placed on a table with its flat face on the table then

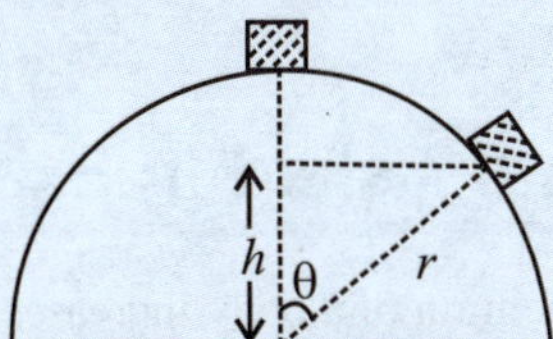

(a) the vertical height h from the table at which the block leaves the hemisphere is h = 2r/3

(b) If the position vector of the block with respect to the centre of curvature makes an angle q with vertical when the block leaves the hemisphere, in that case $\cos\theta = 2/3$

(c) Velocity of the block at the instant it leaves the hemisphere, $v = \sqrt{\frac{2gr}{3}}$.

Power

Formulas

1. $P_{av} = \frac{\text{Total work done}}{\text{time}} = \frac{W}{t}$

S.I unit of power is watt [P_{av} is the average power delivered]

2. $P = \frac{dW}{dt} = \frac{d}{dt}\vec{F}.d\vec{r}$ [P is the instantaneous power]

3. $P = \vec{F}\cdot\vec{v}$ [F is the force and v is the velocity]

Numerical Profile

Calculation of Power under different Cases

(i) The power of a machine gun firing 'n' bullets each of mass 'm' with a velocity 'v' in a time interval 't' is given by $P = \frac{nmv^2}{2t}$

(ii) A crane lifts a body of mass m with a constant velocity v from the ground, its power is $P = Fv = mgv$

(iii) A conveyor belt is moving with a constant speed 'v' horizontally and gravel is fed on it at a rate of $\frac{dm}{dt}$ then $F = v\frac{dm}{dt}$. Then additional power required to drive the belt is, $P = Fv = v^2\frac{dm}{dt}$

(iv) When a liquid of density 'ρ' coming out of a hose pipe of area of cross section 'A' with a velocity 'v' and strikes the wall normally. Then power exerted by the liquid is $P = \frac{1}{2}\frac{mv^2}{t} = \frac{1}{2}\rho A v^3$

(v) A vehicle of mass 'm' is driven with constant acceleration along a straight level road against a constant external resistance 'R' when the velocity is 'v' power of engine is

$P = Fv$

$P = (R + ma)v$

(vi) If a motor lifts water from a well of depth 'h' and with a velocity 'v' then power of the pump

$$P = \frac{mgh + \frac{1}{2}mv^2}{t}$$

Collision

Formulas

1. Elastic collision in one dimension

Before collision After collision

(a) $v_1 = \left(\frac{m_1 - m_2}{m_1 + m_2}\right)u_1 + \left(\frac{2m_2}{m_1 + m_2}\right)u_2$

(b) $v_2 = \left(\frac{2m_1u_1}{m_1 + m_2}\right) - \left(\frac{m_1 - m_2}{m_1 + m_2}\right)u_2$

Special Cases

(i) If colliding particles are of equal masses i.e., $m_1 = m_2 = m$

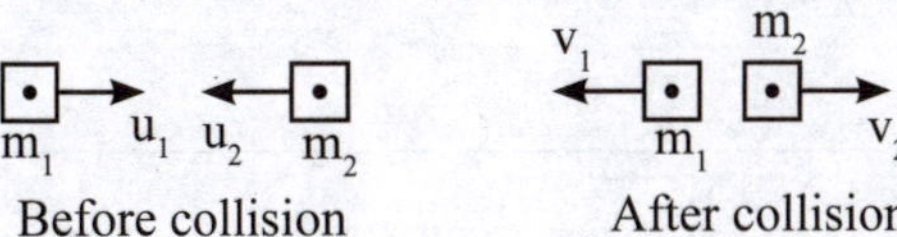

Before collision After collision

Here speed will be interchanged, $v_1 = u_2$ and $v_2 = u_1$

(ii) A lighter particle collides with heavier particle which is at rest i.e., $m_1 \lll m_2$ and $u_2 = 0$

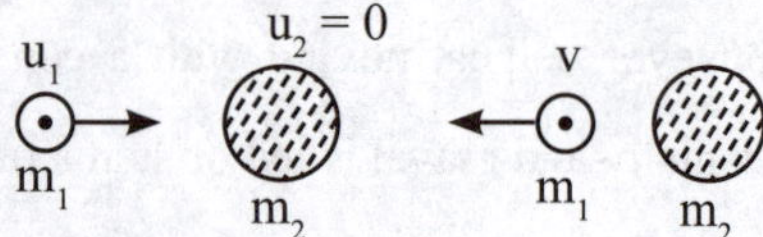

here, $v = -u_1$

2. Coefficient of Restitution (e)

$$= \frac{\text{Relative velocity of separation}}{\text{Relative velocity of approach}} = \frac{(v_2 - v_1)}{(u_1 - u_2)}$$

(a) For a perfectly elastic collision e = 1

(b) For an inelastic collision $0 < e < 1$

(c) For completely inelastic collision e = 0

Numerical Profile

Collision under different cases

(i) A ball is projected with an initial velocity u and at an angle θ with the horizontal surface. If 'e' is the coefficient of restitution between the ball and the surface then

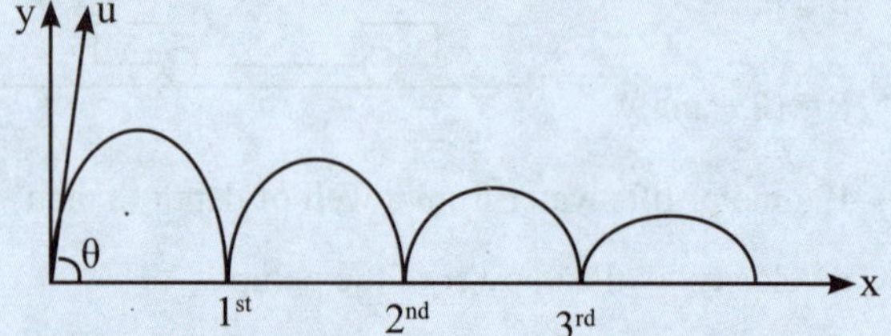

The total time taken by ball to come to rest can be calculated as

(a) Time it takes for 1st collision, $T = \frac{2u\sin\theta}{g}$

(b) Time interval between 1st and 2nd collision

$$T_1 = \frac{2v_1\sin\theta}{g} \qquad (\because v_1 = eu)$$

$$T_1 = \frac{2(eu)\sin\theta}{g} = eT$$

(c) The total time of flight, $T_{total} = \frac{T}{1-e}$

(ii) The total horizontal range covered before the ball comes to rest can be calculated as

(a) The horizontal distance covered by the ball in 1st collision

$$R = \frac{u^2 \sin 2\theta}{g}$$

(b) The horizontal distance covered by it between 1st and 2nd collisions $R_1 = u\cos\theta \times eT = eR$

(c) Total horizontal distance or range by the ball

$$R_{total} = \frac{R}{1-e}$$

(iii) The sum of maximum heights reached by the ball before stopping can be calculated as follows

(a) The maximum height reached by the ball before 1st collision, $H = \frac{u^2\sin^2\theta}{2g} = \frac{(u\sin\theta)^2}{2g}$

(b) Maximum height it reaches between 1st and 2nd collision, $H_1 = \frac{(eu\sin\theta)^2}{2g} = e^2H$

(c) The sum of maximum heights reached by the ball,

$$H_{total} = \frac{H}{1-e^2}$$

Statements Asked In Exam

- When the momentum of a body increases by a factor n, then its kinetic energy increases by factor n^2.
- A body can have momentum even when its energy is negative.
- When a light body and heavy body are moving with same kinetic energy and same retarding force is applied on each, both the bodies will stop after travelling the same distance.
- Kinetic energy of a system can be increased or decreased without applying any external force on the system. As in the case of explosion, the kinetic energy increases even without the application of external forces.
- Remember that work done in moving a body over a smooth inclined plane does not depend upon slope of inclined plane W = mgh, and it depends only on height h of the inclined plane.
- Potential energy of a system increases if conservative force does work on it.
- When a spring is compressed its potential energy increase.
- An air bubble rises in water, its potential energy decreases.
- In elastic collision, forces involved are conservative.
- In case of collisions as the impulsive force acting during collision is internal, the total momentum of the system always remains conserved.
- In an oblique elastic collision, if $m_1 = m_2$ and m_2 is initially at rest, then after the collision the two masses will move in directions inclined at 90° to each other.

MCQs

NCERT Topic-wise MCQs

Work Done by Constant & Variable Force

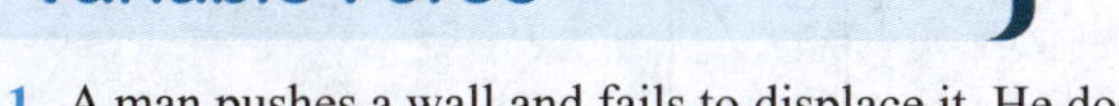

1. A man pushes a wall and fails to displace it. He does
1) Negative work
2) Positive but not maximum work
3) No work at all
4) Maximum work.

2. A force of 10 N acts on a 10 kg body initially at rest. The work done by the force during the first 5 seconds and 5th second of motion of body is. (in joule)
1) 3, 1.5 2) 9, 4
3) 11.5, 5 4) 125, 45

3. A cord is used to lower vertically a block of mass m by a distance d with constant downward acceleration $\frac{g}{4}$. Work done by the cord on the block is
1) $\frac{3}{4}mgd$ 2) $-\frac{3}{4}mgd$
3) mgd 4) $\frac{mgd}{4}$

4. The work done in pulling up a block of wood weighing 200N for length of 10 m on a smooth plane inclined at an angle of 30° with the horizontal is
1) 18 kJ 2) 20 kJ
3) 1 kJ 4) None

5. If $\vec{F} = 2\hat{i} + 3\hat{j} + 4\hat{k}$ acts on a body and displaces it by $\vec{S} = 3\hat{i} + 2\hat{j} + 5\hat{k}$ then the work done by the force is
1) 12 J 2) 20
3) 32 J 4) 64 J

6. When a force vector $\vec{F} = (\hat{i} + 2\hat{j} + \hat{k})N$ acts on a body and produces a displacement of $\vec{S} = (4\hat{i} + \hat{j} + 7\hat{k})m$, then the work done is
1) 9J 2) 13J
3) 5J 4) 1J

7. A force of 5 N acts on a 15 kg body initially at rest. The work done by the force during the first second of motion of the body is
1) 5 J 2) $\frac{5}{6}$ J
3) 6 J 4) 75 J

8. Figure shos the F-x graph, where F is the force applied and x is the distance covered by the body along a straight line path.

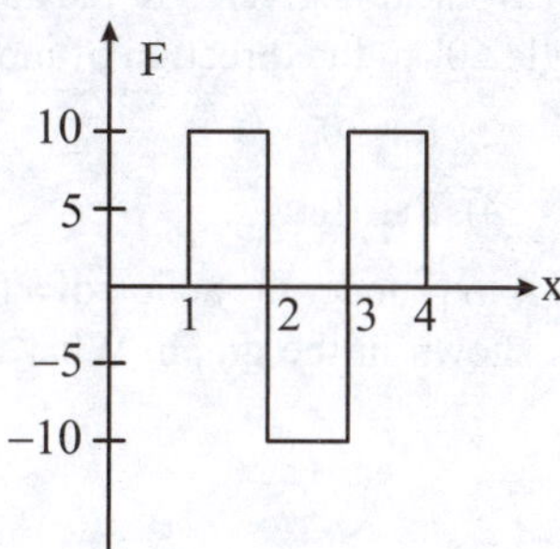

Given that F is in newton and x in metre, what is the work done.
1) 10 J 2) 20 J
3) 30 J 4) 40 J

9. If W_1, W_2, and W_3 represent the work done in moving a particle from A to B along three 1, 2, and 3 respectively (as shown) in the gravitational field of a point mass m, find the correct relation between W_1, W_2 and W_3.

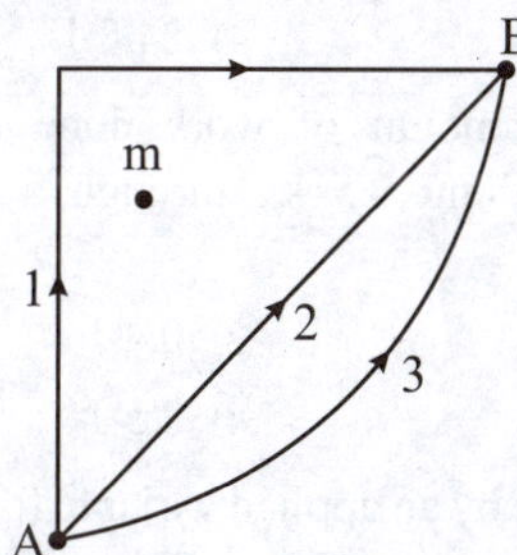

1) $W_1 > W_2 > W_3$ 3) $W_1 = W_2 = W_3$
3) $W_1 < W_2 < W_3$ 4) $W_2 > W_1 > W_3$

10. A body of mass 5 kg is moved up over 10 m along the line of greatest slope of a smooth inclined plane of inclination of 30° with the horizontal. If g = 10 m/s^2, the work done will be
1) 500 J 2) 2500 J
3) 250 J 4) 25 J

11. A bicycle chain of length 1.6 m and of mass 1 kg is lying on a horizontal floor. If g = 10 ms^{-2}, the work done in lifting it with one end touching the floor and the other end 1.6 m above the floor is
1) 10 J 2) 3.2 J
3) 8 J 4) 16 J

12. A particle is moving under the influence of a force given by F = Kx
where K is a constant and x is the distance (in m) moved. The energy (in kJ) gained by the particle in moving from x = 0 to x = 5 is
1) 0.125K kJ
2) 0.185K kJ
3) 0.625K kJ
4) 0.0125K kJ

13. A gardner moves a lawn roller through a distance of 100 m with a force of 60 N. Calculate his wages, if he is to be paid 10 paise for doing 15 J of work. It is given that the applied force is inclined at an angle 60° to the direction of motion.
1) Rs. 40 2) Rs. 30
3) Rs. 20 4) Rs. 10

14. A body moves from point A to B under the action of a force, varying in magnitude, as shown in the graph. What is the work done?

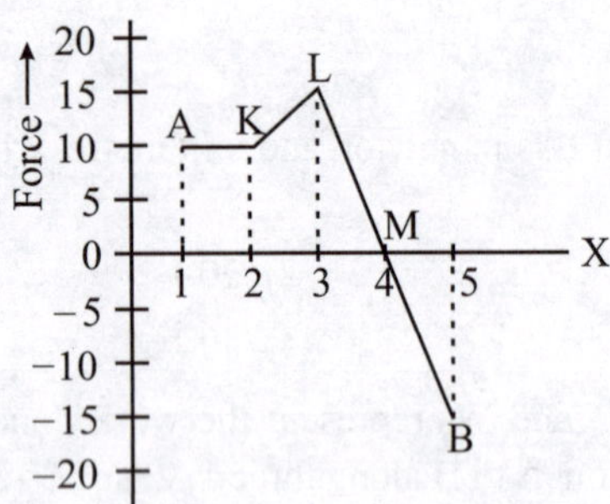

1) 22.5 J 2) 24 J
3) 20 J 4) 24.5 J

15. What is the amount of work done in raising a glass of water weighing 0.5 kg through a height of 50 cm? ($g = 10\ m/s^2$)
1) 1 J 2) 0.4 J
3) 0.20 J 4) 2.5 J

16. The work done by an applied variable force, $F = x + x^3$ from x = 0 m to x = 2m, where x is displacement, is
1) 6 J 2) 8 J
3) 10 J 4) 12 J

17. A rain drop of mass (1/10) gram falls vertically at constant speed under the influence of the forces of gravity and viscous drag. In falling through 100 m, the work done by gravity is
1) 0.98 J 2) 0.098 J
3) 9.8 J 4) 98 J

18. A force $\vec{F} = (6\hat{i} - 8\hat{j})N$, acts on a particle and displaces it over 4 m along the X-axis and then displaces it over 6 m along the Y-axis. The total work done during the two displacements is
1) 72 J 2) 24 J
3) – 24 J 4) Zero

19. A lawn roller is pulled along a horizontal surface through a distance of 20 m by a rope with a force of 200 *N*. If the rope makes an angle of 60° with the vertical while pulling, the amount of work done by the pulling force is
1) 4000 J 2) 1000 J
3) $2000\sqrt{3}$ J 4) 2000 J

20. A force acts on a body and displaces it in its direction. The graph shows the relation between the force and displacement, the work done by the force is

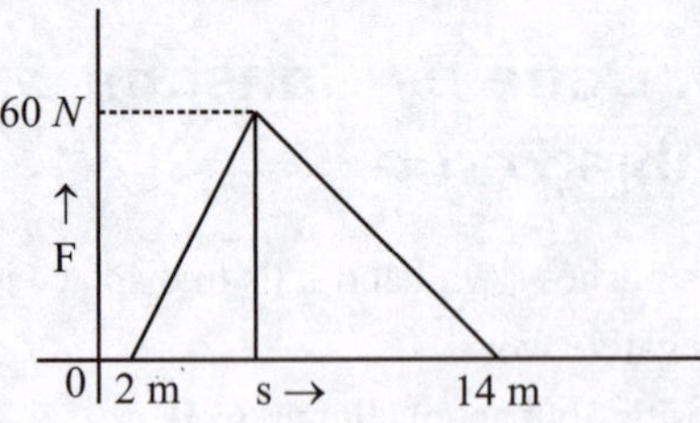

1) 420 J 2) 360 J
3) 840 J 4) 720 J

21. A string of length L and force constant K is stretched to obtain extension ℓ. It is further stretched to obtain extension ℓ_1. The work done in second stretching is
1) $\frac{1}{2}K\ell_1(2\ell + \ell_1)$ 2) $\frac{1}{2}K\ell_1^2$
3) $\frac{1}{2}K(\ell^2 + \ell_1^2)$ 4) $\frac{1}{2}K(\ell_1^2 - \ell^2)$

Energy, Work-Energy Theorem

22. The work done by a body against friction always results in
1) loss of kinetic enrgy
2) loss of potential energy
3) gain of kinetic energy
4) gain of potential energy

23. A body is falling freely under the action of gravity alone in vacuum. Which of the following quantities remain constant during the fall?
1) Kinetic energy
2) Potential energy
3) Total mechanical energy
4) Total linear momentum

24. A ball of mass 2 kg and another of mass 4 kg are dropped together from a 60 feet tall building. After a fall of 30 feet each towards earth, their respective kinetic energies will be in the ratio of
1) $\sqrt{2}:1$ 2) 1 : 4
3) 1 : 2 4) $1:\sqrt{2}$

25. A 50 g bullet moving with velocity 10 m/s strikes a block of mass 950 g at rest and gets embedded in it. The loss in kinetic energy will be
1) 100 % 2) 95 %
3) 5 % 4) 50 %

26. If the unit of force and length each be increased by four times, then the unit of energy is increased by

1) 16 times 2) 8 times
3) 2 times 4) 4 times

27. A body of mass 4 kg is thrown upward with a velocity of 40m/s it comes to rest after attaining maximum height of 30 m. How much energy is lost due to air friction?

1) 2000 J 2) 118 J
3) 1800 J 4) 1810 J

28. A shot is fired at 30° with the vertical from a point on the ground with kinetic energy *K*. If air resistance is ignored, the kinetic energy at the top of the trajectory is

1) $3K/4$ 2) $K/2$
3) K 4) $K/4$

29. A body starts from rest and is acted on by a constant force. The ratio of kinetic energy gained by it in the first five seconds to that gained in the next five second is

1) 2 : 1 2) 1 : 1
3) 3 : 1 4) 1 : 3

30. The mass of a simple pendulum bob is 100 gm. The length of the pendulum is 1 m. The bob is drawn aside from the equilibrium position so that the string makes an angle of 60° with the vertical and let go. The kinetic energy of the bob while crossing its equilibrium position will be

1) 0.49 J 2) 0.94 J
3) 1 J 4) 1.2 J

31. A body starts from rest and moves with uniform acceleration. What is the ratio of kinetic energies at the end of 1st, 2nd and 3rd seconds of its journey?

1) 1 : 8 : 27 2) 1 : 2 : 3
3) 1 : 4 : 9 4) 3 : 2 : 1

32. A spring when stretched by 2 mm its potential energy becomes 4J. If it is stretched by 6 mm its potential energy becomes.

1) 32 J 2) 36 J
3) 30 J 4) 40 J

33. A body of mass 2 kg moves with velocity of 18 m/s hits a spring fixed at other end of force constant 1000 N/m and comes to rest after compressing the spring. The compression of the springs is

1) 0.80 m 2) 0.60 m
3) 0.79 m 4) 0.100 m

34. A uniform chain of length 3m is kept on a table such that a length of 70 cm hangs freely from the edge of the table. The total mass of the chain is 4 kg. What is the work done in pulling the entire chain on the table?

1) 1.713 J
2) 3.174 J
3) 2.750 J
4) 2.500 J

35. Two bodies A and B have masses 20 kg and 5 kg respectively. Each one is acted upon by a force of 8 kg wt. If they acquire the same kinetic energy in time t_A and t_B then the ratio $\frac{t_A}{t_B}$ is

1) 1 2) 2
3) 3 4) 4

36. Two stones of masses m and 2 m are projected vertically upwards so as to reach the same height. The ratio of the kinetic energies of their projection is

1) 2 : 1 2) 1 : 2
3) 4 : 1 4) 1 : 4

37. A 150 gm mass has a velocity $(2i + 6j)$ m/s at a certain instant. Its kinetic energy is

1) 6 J 2) 2 J
3) 3 J 4) 8 J

38. A body of mass 2 kg is projected with an initial velocity of 5 ms^{-1} along a rough horizontal table. The work done on the body by the frictional forces before it is brought to rest is

1) 250 J 2) 2.5 J
3) 10 J 4) 25 J

39. An object is acted on by a retarding force of 10 N and at a particular instant its kinetic energy is 6 J. The object will come to rest after it has travelled a distance of

1) 3/5 m 2) 5/3 m
3) 4 m 4) 16 m

40. A bullet fired into a trunk of a tree loses 1/4 of its kinetic energy in traveling a distance of 5 cm. Before stopping it travels a further distance of

1) 150 cm 2) 1.5 cm
3) 1.25 cm 4) 15 cm

41. Two bodies of masses 3m and 6m have same momentum. Their respective kinetic energies E_1 and E_2 are in the ratio

1) 1 : 2 2) 2 : 1
3) $\sqrt{2}:1$ 4) 1 : 1

42. Two bodies of masses 3 m and m have their K.E in the ratio 27 : 1, then the ratio of their momenta is

1) 9 :1 2) 1 : 9
3) $1:3\sqrt{3}$ 4) 1 : 1

43. A 2 kg ball is thrown up with an initial speed 20 m/s and reaches a maximum height of 10 m. How much energy is dissipated by air drag acting on the ball during the ascent?

1) 180 J 2) 160 J
3) 150 J 4) 200 J

44. A tank of size 10 m × 10 m × 10 m is full of water and built on the ground. If g = 10 ms^{-2}, the potential energy of the water in the tank is

1) 5×10^7 J 2) 1×10^8 J
3) 5×10^4 J 4) 5×10^5 J

45. AB is a frictionless inclined surface making an angle of 30° with the horizontal. A is 6.3 m above the ground while B is 3.8 m above the ground. A block slides down from A, initially starting from rest. Its velocity of block on reaching B is

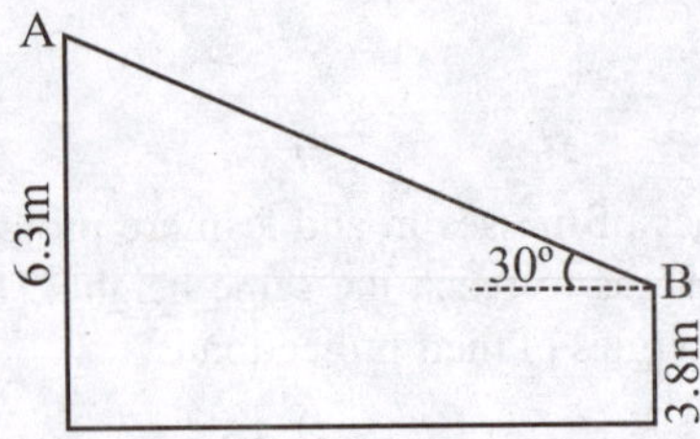

1) 7 m s^{-1}
2) 14 m s^{-1}
3) 7.4 m s^{-1}
4) 4.9 m s^{-1}

46. A spring when compressed by 4 cm has 2 J energy stored in it. The force required to extend it by 8 cm will be

1) 20 N
2) 2 N
3) 200 N
4) 2000 N

47. The elastic potential energy of a stretched spring is given by $E = 50x^2$ where x is the displacement in meter and E is in joule, then the force constant of the spring is

1) 50 Nm
2) 100 Nm^{-1}
3) 100 N/m^2
4) 100 Nm

48. A particle of mass m at rest is acted upon by a force F for a time t. Its kinetic energy after the interval t is

1) $2\,Ftm$
2) $\frac{2Ft}{m}$
3) $\frac{Ft}{m}$
4) $\frac{F^2t^2}{2m}$

49. If a bullet losses one-third of its velocity on penetrating 4 cm in a wooden block, then how much will it penetrate more before coming to rest?

1) 1 cm
2) 3.2 cm
3) 3 cm
4) 2.7 cm

50. A shooter fires a bullet with a speed of 200 m/s can just penetrate 3 planks of equal thickness x. Then the number of such planks penetrated by this bullet if its velocity becomes 3 times initial. (consider retardation is constant)

1) 11
2) 15
3) 27
4) 30

51. A block of mass 1 kg moving with a speed of 4 ms^{-1}, collides with another block of mass 2 kg which is at rest. The lighter block comes to rest after collision. The loss in the K.E of the system.

1) 8 J
2) 4×10^{-7} J
3) 4 J
4) 0 J

52. A ball of mass 2 kg and another of mass 6 kg are dropped together from a 80 feet tall building. After a fall of 50 feet each towards the earth, their respective kinetic energies will be in the ratio of

1) 1 : 3
2) $1:\sqrt{2}$
3) 2 : 1
4) $\sqrt{3}:1$

53. Work-energy theorem is valid in the presence of

1) External forces only
2) Internal forces only
3) Conservative forces only
4) All types of forces

54. In a ballistic demonstration a police officer fires a bullet of mass 100 g with a speed of 250 m/s on wooden block of thickness 4 cm. The bullet emerges with only 20% of its initial kinetic energy. What will be the emergent speed of the bullet?

1) 121 m/s
2) 141 m/s
3) 112 m/s
4) 131 m/s

55. A body is dropped from a height h. If it acquires a momentum p, then the mass of the body is

1) $\frac{P}{\sqrt{2gh}}$
2) $\frac{P^2}{2gh}$
3) $\frac{2gh}{P}$
4) $\sqrt{2gh/P}$

56. Which of the following graph is correct between kinetic energy (E) and potential energy (U) with height (h) from the ground of the particle?

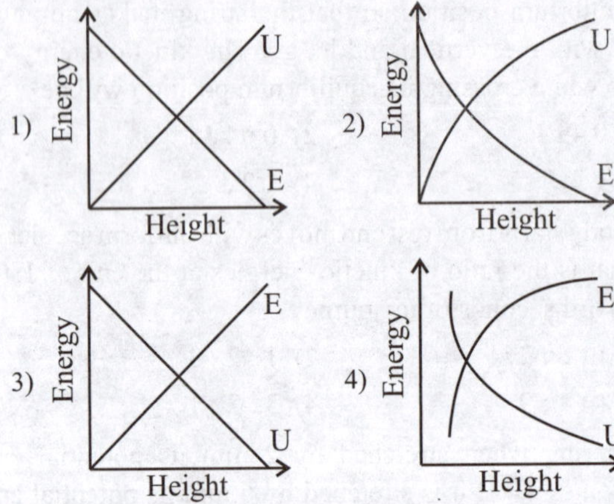

57. A particle of mass 100 g is thrown vertically upwards with a speed of 5 ms^{-1}. The work done by the force of gravity during the time, the particle goes up is

1) – 0.5 J
2) – 1.25 J
3) 1.25 J
4) 0.5 J

58. A spring of spring constant 5×10^3 Nm^{-1} is stretched initially by 5 cm from the unstretched position. Then, the work required to stretch it further by another 5 cm is

1) 12.50 Nm
2) 18.75 Nm
3) 25.00 Nm
4) 6.25 Nm

59. A simple pendulum is oscillating with an angular amplitude 60°. If mass of bob is 50 gram, the tension in the string at mean position is (g = 10 ms^{-2})

1) 0.5 N
2) 1 N
3) 1.5 N
4) 2N

60. A body of mass 2 kg is thrown up vertically with kinetic energy of 490 J. If g = 9.8 ms^{-2}, the height at which the kinetic energy of the body becomes half of the original value, is

1) 50 m 2) 25 m
3) 12.5 m 4) 19.6 m

Power, Motion in Vertical Circle

61. An elevator can carry a maximum load of 1800 kg (elevator + passengers) is moving up with a constant speed of 2 ms^{-1}. The frictional force opposing the motion in 4000 N. What is minimum powe delivered by the motor to the elevator?

1) 22 kW 2) 44 kW
3) 66 kW 4) 88 kW

62. An electric pump lifts 1000 litre of water to a height of 10 m in 980 s. The power of the pump in kilowatt is (g = 9.8 ms^{-2}, density of water = 1 kg $litre^{-1}$)

1) 100 2) 125
3) 0.1 4) 0.01

63. A force $(4\hat{i}+\hat{j}-2\hat{k})N$ acting on a body maintains its velocity at $(2\hat{i}+2\hat{j}+3\hat{k})ms^{-1}$. The power exerted is

1) 4 W 2) 5 W
3) 2 W 4) 8 W

64. The human heart discharges 80 ml of blood at each beat against pressure of 0.1 mm of Hg. What is the power of the heart assuming that the pulse frequency is 75 beats per minute? (Density of mercury = 13.6×10^3 kg/m^3)

1) 1.33 watt 2) 121 watt
3) 1330 k watt 4) None

65. A body of mass m accelerates uniformly from rest to v_1 in time t_1. As a function of time t, the instantaneous power delivered to the body is

1) $\frac{mv_1 t}{t_1}$ 2) $\frac{mv_1^2 t}{t_1}$
3) $\frac{mv_1 t_1^2}{t}$ 4) $\frac{mv_1^2 t}{t_1^2}$

66. An electric motor is used to fill an overhead tank of capacity $10m^3$ kept at a height of 20 m above the ground. If the pump takes 6 minutes to fill the tank by consuming 20 kW power the efficiency of the pump should be (g = 10 m/s^2)

1) 27.5% 2) 30%
3) 32% 4) 40%

67. An engine develops 20 kW power. How much time will it take to lift a mass of 200 kg to a height of 50 m? (g = 10 m/s^2)

1) 1 sec 2) 2 sec
3) 5 sec 4) 7 sec

68. A particle moves with a velocity $\left(5\hat{i}-3\hat{j}+6\hat{k}\right)m/s$ under the influence of a constant force $\left(10\hat{i}+10\hat{j}+20\hat{k}\right)$ N. The instantaneous power applied to the particle is

1) 200 W 2) 40 W
3) 140 W 4) 170 W

69. A motor boat is going in a river with a velocity $\vec{v}=\left(4\hat{i}-2\hat{j}+\hat{k}\right)ms^{-1}$. If the resisting force due to stream is $\vec{F}=\left(5\hat{i}-10\hat{j}+6\hat{k}\right)N$, then the power of the motor boat is

1) 100 *W* 2) 50 *W*
3) 46 *W* 4) 23 *W*

70. It is said the power of a human are typically 10W for a 70 kg adult. In one year time the total energy required in kWh will be

1) 87.6 *kWh*
2) 100 *kWh*
3) 1000 *kWh*
4) 1 *kWh*

71. A car weighing 1000 kg is going up an incline with a slope of 2 in 25 at a steady speed of 18 km/h. If g = 10 ms^{-2}, the power of its engine is

1) 4 *kW* 2) 50 *kW*
3) 625 *kW* 4) 25 *kW*

72. A man carries a load of 50 kg through a height of 40 m in 25 seconds. If the power of the man is 1568 W, his mass is

1) 5 kg 2) 1000 kg
3) 200 kg 4) 50 kg

73. A force applied by an engine of a train of mass 2.05×10^6 kg changes its velocity from 5 m/s to 35 m/s in 15 minutes. The power of the engine is

1) 1.36 hp 2) 1.36 MW
3) 1.36 W 4) None

74. A force of $(3\hat{i}+4\hat{j}+5\hat{k})$ N acts on a body for 5 seconds, produces a displacement of $(2\hat{i}+3\hat{j}+4\hat{k})$m. The power used is:

1) 5 watt 2) 7.6 watt
3) 8 watt 4) 8.4 watt

75. An engine pump up 200 kg of water through a height of 20 m in 10 second. If the efficiency of the engine is 80%. Calculate the power of the engine.

1) 3800 W 2) 5000 W
3) 4900 W 4) 4800 W

76. Water falling from a 60 m high is to be used for generating electric power. If 9×10^5 kg of water falls per hour and half the gravitation potential energy can be converted into electric energy. How many 100 watt bulbs can be lit ? (g = 10 m/s^2)

1) 750 2) 770
3) 550 4) 900

77. The power of a water pump is 5 kW, the amount of water it can raise in 5 minutes to a height of 50 m is ($g = 10\ m/s^{-2}$)

1) 5×10^4 L
2) 10^6 L
3) 3×10^5 L
4) 3×10^3 L

78. A body slides from an initial height of 10 m, what should be the radius of the circular loop so that the body just completes the loop.

1) 2 m 2) 3 m
3) 4 m 4) 5 m

79. A body crosses the top most point of a vertical circle with critical speed. Its centripetal acceleration, when the string at the lowest point of circle/loop:

1) 3 g 2) g
3) 5 g 4) 6 g

80. One end of the string of length 1.0 m is tied to a body of mass 0.5 kg. It is whirled in a vertical circle with angular frequency 4 rad/s. The tension in the string when the body is at the lower most point of its motion will be equal to (take, $g = 10\ m/s^2$)

1) 3 N 2) 5 N
3) 8 N 4) 13 N

81. An electric motor creates a tension of 4500 newton in a hoisting cable and reels it in at the rate of 2m/s. What is the power of the motor?

1) 15 kW 2) 9 kW
3) 225 W 4) 9000 kW

82. A water pump driven by petrol raises water at a rate of 0.5 m^3/min from a depth of 30 m. If the pump is 70% efficient, what power is developed by the engine?

1) 1750 W 2) 2450 W
3) 3500 W 4) 7000 W

83. A body of mass m is rotated at uniform speed along vertical circle with help of light string. If T_1, T_2 are tensions in the string when the body is crossing highest and lowest point of vertical circle respectively then which of following expression is correct.

1) $T_2 - T_1 = 6\ mg$
2) $T_2 - T_1 = 4mg$
3) $T_2 - T_1 = 2\ mg$
4) $T_2 - T_1 = mg$

84. A stone of mass m is tied to a string and is moved in a vertical circle of radius r making n rev/min. The total tension in the string when the stone is at the lowest point is

1) mg 2) $m(g + \pi n r^2)$
3) $m(g + nr)$ 4) $m\left(g + \frac{\pi^2 n^2 r}{900}\right)$

85. A stone is attached to one end of a string and rotated in a vertical circle. If string breaks at the position of maximum tension, it will break at

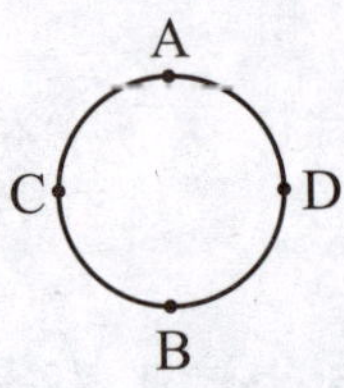

1) A 2) B
3) C 4) D

86. A ball of mass 0.6 kg attached to a light inextensible string rotates in a vertical circle of radius 0.75 m such that it has speed of 5 ms^{-1} when the string is horizontal. Tension in the string when it is horizontal on the other side is (take, g 10 ms^{-2})

1) 30 N 2) 26 N
3) 20 N 4) 6 N

87. A particle is moving in a vertical circle. The tensions in the string when passing through two positions at angle 30° and 60° from vertical (lowest position) are T_1 and T_2 respectively, then

1) $T_1 = T_2$
2) $T_1 > T_2$
3) $T_1 < T_2$
4) Tension in the string always remains the same

Collision

88. What is the coefficient of restitution (e)?

1) Ratio of final to initial relative velocities
2) Ratio of initial to final relative velocities
3) Ratio of final to initial velocity of a body
4) Product of initial to final relative velocities

89. When two same masses travelling in opposite directions with different velocities collide perfectly elastically, then their velocities ________________.

1) Exchange 2) Remain unchanged
3) Increase 4) Decrease

90. Two base balls A and B having masses 200 g and 150 g respectively are moving in opposite direction. Velocity of A is 3 m/s. After the collision both the balls come to rest then speed of other ball was.

1) 3 m/s 2) –2 m/s
3) –4 m/s 4) 1 m/s

91. A ball of mass 10 kg is moving with velocity of 12 m/s. It strikes another ball of mass 6 kg which is moving in the same direction with a velocity of 8 m/s. If the collision is elastic, their respective velocities (in m/s) after the collision will be:

1) 10.5, 10.5 2) 4, 5
3) 10.9, 11 4) 9, 13

92. A body falls on a surface of coefficient of restitution 0.6 from a height of 3 m. The body rebounds to a height of

1) 1 m 2) 1.08 m
3) 2 m 4) 2.5 m

93. A particle falls from a height h on a fixed horizontal plane and rebounds. If e is the coefficient of restitution, the total distance travelled by the particle before the rebounding stops is.

1) $2h\left[\frac{1+e^2}{1+e}\right]$ 2) $h\left[\frac{1+e^2}{1-e^2}\right]$

3) $2h\left[\frac{1-e}{1+e}\right]$ 4) $h\left[\frac{1-e^2}{1-2e^2}\right]$

94. A neutron makes a head on elastic collision with a stationary deuteron. The fractional energy loss of the neutron in the collision is

1) 8/9 2) 9/8
3) 3/4 4) 5/2

95. Two masses of 2 kg and 5 kg each move towards each other with speed 3 m/s and 4 m/s collide and stick together. What is the final velocity?

1) 5.2 m/s 2) 4 m/s
3) 2 m/s 4) 6 m/s

96. Two identical bodies of mass w moving with velocities u_1 and u_2 collide perfectly inelastically. The loss in energy is

1) $w(u_1 + u_2)$ 2) $\frac{w}{4}(u_1 - u_2)^2$

3) $2w(u_1 - u_2)^2$ 4) None

97. A body of mass 8 kg makes an elastic collision with another body at rest and continues to move in the original direction with a speed equal to $\frac{1}{4}$ of its original speed. What is the mass of the second body?

1) 5.4 kg 2) 4.8 kg
3) 3 kg 4) 2.4 kg

98. A sphere of mass m moving with velocity u hits another stationary sphere of same mass. If e is coefficient of restitution. What is the ratio of velocities of two spheres after collision?

1) $\frac{1-e}{e}$ 2) $\frac{1+e}{e}$

3) $\frac{1+e}{1-e}$ 4) $\frac{1-e}{1+e}$

99. During inelastic collision of two particles:

1) $(K.E)_{final} = (K.E)_{initial}$
2) $(K.E)_{final}$ must be greater than $(K.E)_{initial}$
3) $(K.E)_{final}$ must be less than $(K.E)_{initial}$
4) $(K.E)_{final}$ may be greater or less than $(K.E)_{initial}$

100. A 100 kg mass is travelling at a speed of 5 m/s. Another 150 kg mass is travelling at a speed of 10 m/s in the same direction, strikes the first mass. After the collision the 100 kg mass is travelling with a speed of 6 m/s. The coefficient of restitution of the collision is

1) 0.66 2) 0.77
3) 0.78 4) 0.68

101. A body of mass m is moving towards east and another body of equal mass is moving towards the north. If, after collision both stick together, their speed after collision would be

1) v 2) v/2
3) $\sqrt{2}$ 4) $v/\sqrt{2}$

102. A ball falling freely from a height of 4.9 m hits a horizontal surface. If $e = \frac{3}{4}$, then the ball will hit the surface second time after

1) 0.5 s 2) 1.5 s
3) 3.5 s 4) 3.4 s

High Order Time Intensive MCQs

1. A cord is used to lower vertically a block of mass M by a distance d with constant downward acceleration $\frac{g}{4}$. Work done by the cord on the block is

1) $Mg\frac{d}{4}$ 2) $3Mg\frac{d}{4}$

3) $-3Mg\frac{d}{4}$ 4) Mgd

2. A body of mass 4 kg is moving with momentum of 8 kg ms^{-1}. a force of 0.2 N acts on it in the direction of motion of the body for 10 s. The increase in kinetic energy is

1) 10 J 2) 8.5 J
3) 4.5 J 4) 4 J

3. If the force acting on a body is inversely proportional to its speed, then its kinetic energy is

1) linearly related to time
2) inversely proportional to time
3) inversely proportional to the square of time
4) a constant

4. The potential energy of a spring when stretched through a distance x is 10 J. What is the amount of work done on the same spring to stretch it through an additional distance x?

1) 10 J
2) 20 J
3) 30 J
4) 40 J

5. A vertical spring with force constant k is fixed on a table. A ball of mass m at a height h above the free upper end of the spring falls vertically on the spring, so that the spring is compressed by a distance d. The net work done in the process is

1) $mg(h+d)+\frac{1}{2}kd^2$ 2) $mg(h+d)-\frac{1}{2}kd^2$

3) $mg(h-d)-\frac{1}{2}kd^2$ 4) $mg(h-d)+\frac{1}{2}kd^2$

6. A stone of mass *m*, tied to a string is being whirled in a vertical circle with a uniform speed. The tension in the string is

1) The same throughout the motion
2) Minimum at the highest position of the circle path
3) Minimum at the lowest position of the circular path
4) Minimum when the rope is the horizontal position

7. A block of mass 2 kg is free to move along the x-axis. It is at rest and from t = 0 onwards it is subjected to a time-dependent force F(t) in the x-direction. The force F(t) varies with t as shown in the figure. The kinetic energy of the block after 4.5 s is

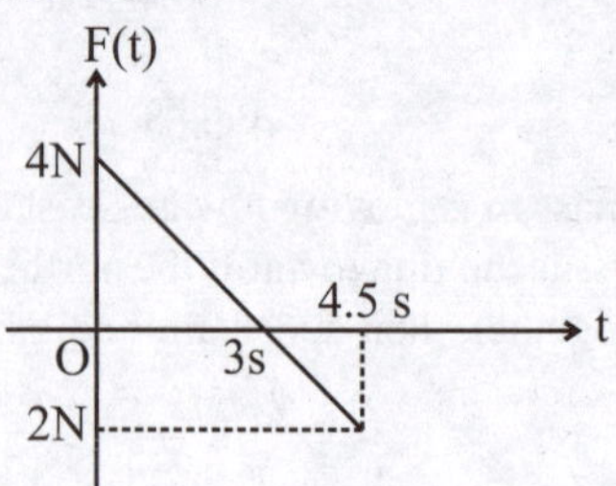

1) 4.50 J 2) 7.50 J
3) 5.06 J 4) 14.06 J

8. A bullet of mass m moving with a velocity v strikes a suspended wooden block of mass M remains inside the block. If the block rises to a height h. The initial velocity of the block just after the bullet enters it, will be

1) $\frac{M+m}{m}\sqrt{2gh}$ 2) $\frac{M+m}{m}\sqrt{gh}$

3) $\frac{m}{M+m}2gh$ 4) $\frac{M+m}{M}\sqrt{2gh}$

9. A boy of mass 40 kg gets an energy of 21 kJ by eating a bread. If the efficiency is 28%, then the height that can be climbed by him using this energy, is

1) 22.5 m 2) 15 m
3) 10 m 4) 5 m

10. A stone is tied to a string of length l is whirled in a vertical circle with the other end of the string at the centre. At a certain instant of time, the stone is at its lowest position and has a speed u. The magnitude of the change in velocity of string when it is at horizontal position (g being acceleration due to gravity) is:

1) $\sqrt{2u^2-2gl}$ 2) $\sqrt{u^2-gl}$

3) $u-\sqrt{u^2-2gl}$ 4) $\sqrt{2gl}$

11. A block of mass 10 kg is moving in x-direction with a constant speed of 10 m/s. It is subjected to a retarding force F = – 0.1x J/m during its travel from x = 20 m to x = 30 m. Its final kinetic energy will be

1) 475 J 2) 450 J
3) 275 J 4) 250 J

12. A body is initially at rest. It undergoes one dimensional motion with constant acceleration. The power delivered to it at time t is proportional to

1) $t^{1/2}$ 2) t
3) $t^{3/2}$ 4) t^2

13. If a machine gun fires n bullets per second each with kinetic energy, K, then the power delivered to the bullets is

1) nK^2 2) $\frac{K}{n}$
3) n^2K 4) nK

14. A car of mass m starts from rest and accelerates so that, the instantaneous power delivered to the car has a constant magnitude P_0. The instantaneous velocity of this car is proportional to

1) t^2P_0 2) $t^{1/2}$
3) $t^{-1/2}$ 4) $t/\sqrt{m}$

15. A box is moved along a straight line by a machine delivering constant power. The distance moved by the body in time t is proportional to

1) $t^{1/2}$ 2) $t^{3/4}$
3) $t^{3/2}$ 4) t^2

16. A motor is used to deliver water at a certain rate through a given horizontal pipe. To deliver n-time the water through the same pipe in the same time the power of the motor must be increased as follows

1) n times 2) n^2 times
3) n^3 times 4) n^4 times

17. A particle of mass m is moving in a circular path of constant radius r such that centripetal acceleration, a_c varying with time is $a_c = k^2rt^2$, where k is a constant. What is the power delivered to the particle by the force acting on it?

1) $2mkr^2t$ 2) mkr^2t^2
3) mk^2r^2t 4) mk^2rt

18. A spherical ball A of mass 4 kg, moving in a straight line strikes another spherical ball B of mass 1 kg at rest. After the collision, A and B move with velocities v_1 ms^{-1} and v_2 ms^{-1}, respectively making angles of 30° and 60° with respect to the original direction of motion of A. The ratio v_1/v_2 will be

1) $\frac{\sqrt{3}}{4}$ 2) $\frac{4}{\sqrt{3}}$

3) $\frac{1}{\sqrt{3}}$ 4) $\sqrt{3}$

19. A body of mass 4 kg moving with velocity 12 ms^{-1} collides with another body of mass 6 kg at rest. If two bodies stick together after collision, then the loss of kinetic energy of system is

1) Zero 2) 288 J
3) 172.8 J 4) 144 J

20. A force, $\vec{F} = 2x\vec{i} + 2\vec{j}$, acts on a particle placed at origin. What is the work done by this force in moving a particle from origin to P(1, 1)?

1) 3 J 2) 4 J
3) 5 J 4) 6 J

21. A body is acted upon by a force which is inversely proportional to square root of the distance covered. The work done will be proportional to:

1) s 2) s^2
3) $\sqrt{s}$ 4) log s

22. The kinetic energy acquired by a mass m in travelling a certain distance d starting from rest under the action of a constant force is directly proportional to

1) $\sqrt{m}$ 2) Independent of m
3) $1/\sqrt{m}$ 4) m

23. During inelastic collision between two bodies, which of the following quantities always remain conserved?

1) Total kinetic energy
2) Total mechanical energy
3) Total linear momentum
4) Speed of each body

24. The potential energy of a body is given by, $U = A - Bx^2$ (where x is the displacement). The magnitude of force acting on the particle is

1) Constant 2) Proportional to x
3) Proportional to x^2 4) Inversely proportional to x

25. A body is moved along a straight line by a machine delivering constant power. The distance moved by the body in time t is proportional to

1) $t^{1/2}$ 2) $t^{3/4}$
3) $t^{3/2}$ 4) t^2

26. The power of a water pump is 2kW. If g = 10 ms^{-2}, the amount of water it can raise on one minute to a height of 10 m is

1) 2000 litre 2) 1000 litre
3) 100 litre 4) 1200 litre

27. An engine is working at a constant power draws a load of mass m against a resistance r. Find maximum speed of load.

1) $\frac{P}{r}$ 2) $\frac{P}{2r}$
3) $\frac{r}{P}$ 4) $\frac{2r}{P}$

28. A particle of mass m is moving in a circular path of constant radius r such that tangential acceleration a_t is varying with time t as $a_t = k^2 r t^2$ where k is constant. What is the power delivered to the particle by the forces acting on it?

1) $\frac{mk^4r^2t^2}{4}$ 2) $\frac{mk^4r^3t^4}{5}$
3) $\frac{mk^4r^2t^5}{3}$ 4) $\frac{mk^4r^2t^5}{9}$

29. Two equal billiard balls are in contact on the table. A third equal ball strikes them symmetrically and remains at rest after impact. The coefficient of restitution is

1) $\frac{1}{3}$ 2) $\frac{2}{3}$
3) $\frac{1}{6}$ 4) $\frac{\sqrt{3}}{2}$

30. The potential energy of a 1 kg particle free to move along the x-axis is given by $U(x) = \left(\frac{x^4}{4} - \frac{x^2}{2}\right)$J. The total mechanical energy of the particle is 2 J. Then, the maximum speed (in m/s) is-

1) $3/\sqrt{2}$ 2) $\sqrt{2}$
3) $1/\sqrt{2}$ 4) 2

31. With what minimum velocity v_0 should block be projected from left end A towards end B such that it reaches the other end B of conveyor belt moving with constant velocity v? Friction coefficient between block and belt is μ.

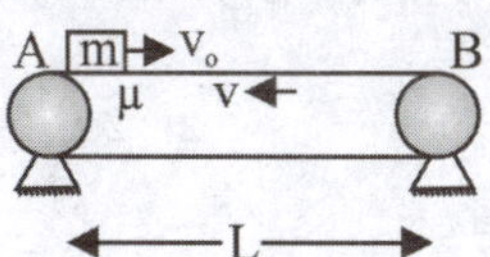

1) $\sqrt{\mu gL}$ 2) $\sqrt{2\mu gL}$
3) $\sqrt{3\mu gL}$ 4) $2\sqrt{\mu gL}$

32. A body of mass 5 kg strikes a body of mass 2.5 kg initially at rest, which after the collision begins to move with a kinetic energy of 5J. Assuming the collision to be central and elastic, the K.E. of the first body before impact is

1) 0.62 J 2) 5.62 J
3) 7.28 J 4) 4.20 J

33. The power of a heart which pumps 5×10^3 cc of blood per minute at a pressure of 120 mm of mercury (g = 10 ms^{-2} and density of Hg = 13.6×10^3 kg/m^3) is :

1) 1.36 W 2) 13.6 W
3) 0.136 W 4) 136 W

34. Work done in time t on a body of mass m which is accelerated from rest to a speed v in time t_1, as a function of time t is given by:

1) $\frac{m}{2}\left[\frac{v}{t_1}\right]t^2$ 2) $m\left[\frac{v}{t_1}\right]t^2$
3) $\frac{1}{2}m\left[\frac{vt}{t_1}\right]^2$ 4) $\frac{1}{2}\left[\frac{v^2}{t_1^2}\right]t^2$

35. Force acting on a particle moving in a straight line varies with the velocity of the particle as $F = Kv^{-2}$, where K is constant. The work done by this force in time t is:

1) Kvt
2) $K^2v^2t^2$
3) K^2vt
4) $\frac{3Kt}{2v}$

36. A block of mass m moving with speed v compresses a spring through distance x before its speed is halved. What is the value of spring constant?

1) $\frac{3mv^2}{4x^2}$
2) $\frac{mv^2}{4x^2}$
3) $\frac{mv^2}{2x^2}$
4) $\frac{2mv^2}{x^2}$

37. System shown in the figure is released from rest when spring is unstretched. Pulley and spring are massless and friction is absent everywhere. The speed of 5 kg block when 2 kg block leaves the contact with ground is (Take force constant of spring k = 40 N/m and g = 10m/s²)

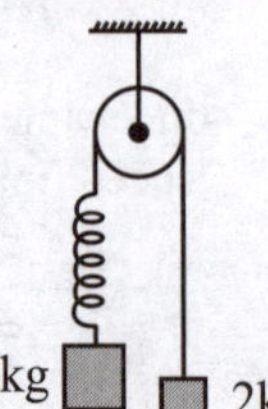

1) $\sqrt{2}$m/s
2) $2\sqrt{2}$m/s
3) 2 m/s
4) $4\sqrt{2}$m/s

38. The kinetic energy of a particle moving along a circle of radius 'R' depends on the distance covered 'x' as $K = \alpha x^2$, where α is a constant, find the net force acting on the particle as a function of 'x' is

1) $2\alpha x\sqrt{1+\frac{x^2}{R^2}}$
2) $2\alpha\sqrt{1-\frac{x^2}{R^2}}$
3) $2x\sqrt{\alpha+\frac{x^2}{R^2}}$
4) None

39. An object of mass m is tied to a string of length l and a variable horizontal force is applied on it which is initially zero and gradually increases until the string makes an angle θ with the vertical. Work done by the force F is:

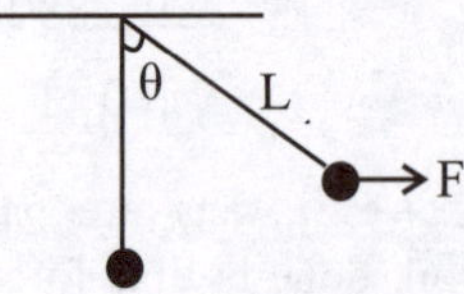

1) $Fl(1 - \sin\theta)$
2) mgl
3) $Fl \sin\theta$
4) $mgl(1 + \cos\theta)$

40. A particle initially at rest on a frictionless horizontal surface is acted upon by a horizontal force which is constant in size and direction. A graph is plotted between the work done (W) on the particle, against the speed of the particle. (v). If there are no other horizontal forces acting on the particle the graph would look like

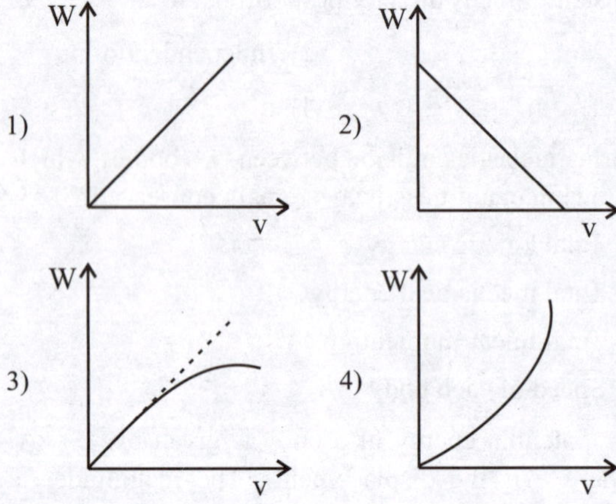

Statement Based MCQs

Directions: These questions consist of two statements each, printed as Statement-I and Statement-II. While answering these questions, you are required to choose any one of the following four responses.

1) **Both Statement-I and Statement-II are correct.**
2) **Both Statement-I and Statement-II are incorrect.**
3) **Statement-I is correct & Statement-II is incorrect.**
4) **Statement-I is incorrect & Statement-II is correct.**

1. **Statement-I:** If the internal forces are conservative, the work done by the external force is equal to the change in mechanical energy.

   Statement-II: Work done on a system by all the (external and internal) force is equal to the change in its kinetic energy and the change in the potential energy of a system corresponding to conservative internal forces is equal to negative of the work done by these forces.

2. **Statement-I:** An air bubble rises in water, its potential energy decreases.

   Statement-II: When a spring is compressed its potential energy decreases.

3. **Statement-I:** eV and joule are the SI units of energy used in modern physics and mechanics respectively.

   Statement-II: Different types of energies have different SI units.

4. **Statement-I:** A heavy weight is suspended from a spring. A person raises the weight slowly till the spring become slack. The work done by person is W. The energy stored on the stretched spring was E. The gain in gravitational potential energy is (W + E).

   Statement-II: Work done by the spring force is always zero.

5. **Statement-I:** Mass and energy are not conserved separately, but are conserved as a single entity called mass-energy.

   Statement-II: Mass and energy are inter-convertible in accordance with Einstein's relation, $E = mc^2$.

Assertion & Reason MCQs

Directions: These questions consist of two statements each, printed as Assertion and Reason. While answering these questions, you are required to choose any one of the following four responses.

1) **Both Assertion and Reason are True and the Reason is a correct explanation of the Assertion.**
2) **Both Assertion and Reason are True but Reason is not a correct explanation of the Assertion.**
3) **Assertion is True but the Reason is False.**
4) **Assertion is False but Reason is True.**

1. **Assertion:** Work done can be positive, negative and zero.

   Reason: Work done is a scalar quantity.

2. **Assertion:** In uniform circular motion of a particle, sum of power delivered to it by centripetal force acting on the particle is zero.

   Reason: Dot product of two perpendicular vectors is always zero.

3. **Assertion:** A body cannot have energy without having momentum but it can have momentum without having energy.

   Reason: Momentum and energy have different dimensions.

4. **Assertion:** Mechanical energy of a particle is conserved if it is acted upon by a conservative force only.

   Reason: Work done by a conservative force in a closed path is zero.

5. **Assertion:** Kinetic energy of a system can be increased without applying any external force on the system.

   Reason: Single external force acting on a particle necessarily changes its kinetic energy.

Match the Columns MCQs

1. Match the column-I with column-II.

	Column-I		Column-II
(a)	A stone at a height	(i)	Power
(b)	A moving ball	(ii)	joule
(c)	Energy	(iii)	work done in 1 s
(d)	Power	(iv)	potential energy
(e)	watt	(v)	kinetic energy

1) (a)-(iv); (b)-(v); (c)-(ii); (d)-(iii); (e)-(i)
2) (a)-(iii); (b)-(iv); (c)-(ii); (d)-(i); (e)-(v)
3) (a)-(ii); (b)-(iv); (c)-(iii); (d)-(i); (e)-(v)
4) (a)-(iv); (b)-(iii); (c)-(ii); (d)-(i); (e)-(v)

2. Match the column-I with column-II.

	Column-I		Column-II
(a)	Kinetic energy	(i)	mgh
(b)	Power	(ii)	$1/2\ mv^2$
(c)	Potential energy	(iii)	mv
(d)	Momentum	(iv)	W/t

1) (a)-(ii); (b)-(iv); (c)-(i); (d)-(iii)
2) (a)-(iii); (b)-(ii); (c)-(iv); (d)-(i)
3) (a)-(iii); (b)-(iv); (c)-(ii); (d)-(i)
4) (a)-(i); (b)-(iv); (c)-(iii); (d)-(ii)

3. Match the column-I with column-II.

	Column-I		Column-II
(a)	Joule	(i)	Power
(b)	Watt	(ii)	1hp
(c)	1 kWh	(iii)	Energy
(d)	746 W	(iv)	3.6×10^6 J

1) (a)-(i); (b)-(ii); (c)-(iii); (d)-(iv)
2) (a)-(iii); (b)-(ii); (c)-(iv); (d)-(iv)
3) (a)-(iii); (b)-(i); (c)-(iv); (d)-(ii)
4) (a)-(i); (b)-(iv); (c)-(iii); (d)-(ii)

4. Match the column-I with column-II.

	Column-I		Column-II
(a)	Mass of 1 kg, pulled through 1 m horizontally, force applied 8N	(i)	Work done = 0
(b)	Suitcase on head, person moves upstairs	(ii)	Work done = mgh J
(c)	No component of force in the direction of motion	(iii)	fs cos θ
(d)	Work done	(iv)	Work done = 8 J

1) (a)-(i); (b)-(ii); (c)-(iii); (d)-(iv)
2) (a)-(iv); (b)-(ii); (c)-(i); (d)-(iii)
3) (a)-(iii); (b)-(iv); (c)-(ii); (d)-(i)
4) (a)-(i); (b)-(iv); (c)-(iii); (d)-(ii)

5. Match the column-I with column-II.

	Column-I		Column-II
(a)	Force	(i)	$kg\ m^2s^{-2}$
(b)	Work	(ii)	$kg\ m\ s^{-1}$
(c)	Momentum	(iii)	$kg\ m^2s^{-3}$
(d)	Power	(iv)	$kg\ m\ s^{-2}$

1) (a)-(iv); (b)-(i); (c)-(ii); (d)-(iii)
2) (a)-(iii); (b)-(ii); (c)-(iv); (d)-(i)
3) (a)-(iii); (b)-(iv); (c)-(ii); (d)-(i)
4) (a)-(i); (b)-(iv); (c)-(iii); (d)-(ii)

NCERT Exemplar MCQs

1. In a shot put event, an athlete throws the shot put of mass 10 kg with an initial speed of 1 ms^{-1} at 45° from a height 1.5 m above ground. Assuming air resistance to be negligible and acceleration due to gravity to be 10 ms^{-2}, the kinetic energy of the shot put when it just reaches the ground will be:
   1) 2.5 J
   2) 52.5 J
   3) 5.0 J
   4) 155.0 J

2. Which of the diagrams in figure correctly shows the change in kinetic energy of an iron sphere falling freely in a lake having sufficient depth to impart it a terminal velocity?

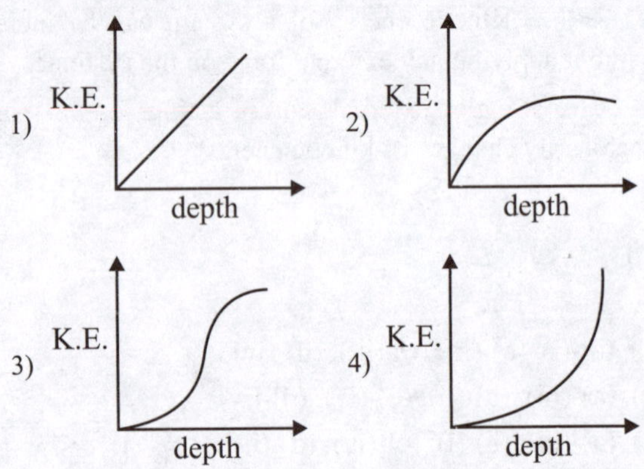

3. A man squatting on the ground gets straight up and stand. The force of reaction of ground on the man during the process is:
   1) Constant and equal to mg in magnitude
   2) Constant and greater than mg in magnitude
   3) Variable but always greater than mg
   4) At first greater than mg and later becomes equal to mg

4. A bicyclist comes to a skidding stop in 10 m. During this process, the force on the bicycle due to the road is 200 N and is directly opposed to the motion. The work done by the cycle on the road is:
   1) + 2000 J
   2) – 200 J
   3) Zero
   4) – 20,000 J

5. A body is falling freely under the action of gravity alone in vacuum. Which of the following quantities remain constant during the fall?
   1) Kinetic energy
   2) Potential energy
   3) Total mechanical energy
   4) Total linear momentum

6. Two inclined frictionless tracks, one gradual and the other steep meet at A from where two stones are allowed to slide down from rest, one on each track as shown in figure. Which of the following statement is correct?

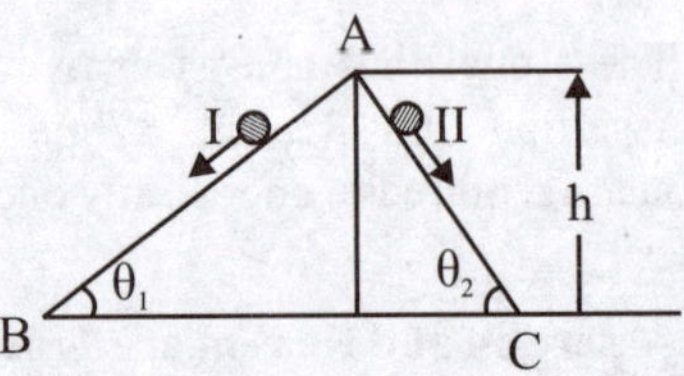

   1) Both the stones reach the bottom at the same time but not with the same speed
   2) Both the stones reach the bottom with the same speed and stone-I reaches the bottom earlier than stone-II
   3) Both the stones reach the bottom with the same speed and stone-II reaches the bottom earlier than stone-I
   4) Both the stones reach the bottom at different times and with different speeds

7. A body of mass 0.5 kg travels in a straight line with velocity $v = a\,x^{3/2}$ where $a = 5\ m^{-1/2}s^{-1}$. The work done by the net force during its displacement from x = 0 to x = 2 m is:
   1) 1.5 J
   2) 50 J
   3) 10 J
   4) 100 J

8. A body is moving unidirectionally under the influence of a source of constant power supplying energy. Which of the diagrams shown in figure correctly shows the displacement-time curve for its motion?

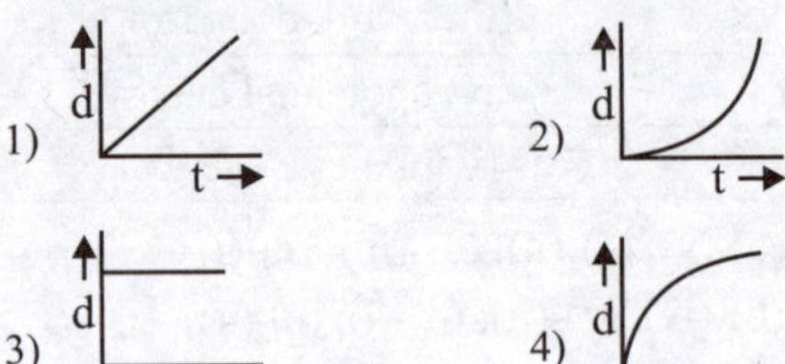

9. Which of the diagrams shown in figure represents variation of total mechanical energy of a pendulum oscillating in air as function of time?

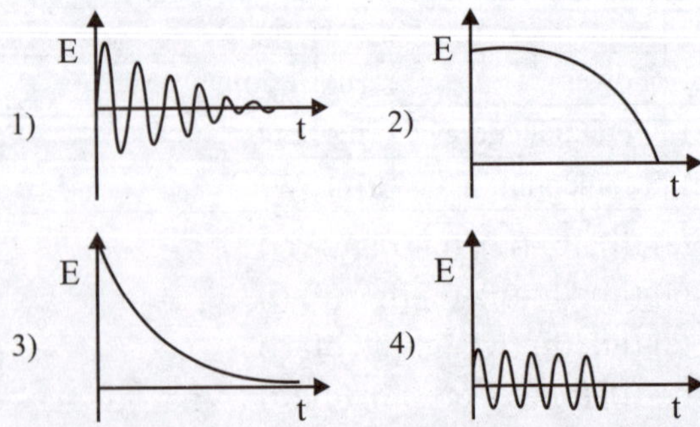

10. A mass of 5 kg is moving along a circular path of radius 1 m. If the mass moves with 300 rev/min, its kinetic energy would be:
    1) $250\pi^2$
    2) $100\pi^2$
    3) $5\pi^2$
    4) 0

NEET & AIIMS Past Year MCQs

1. The potential energy of a particle moving along x-direction varies as $V = \frac{Ax^2}{\sqrt{x}+B}$. The dimensions of $\frac{A^2}{B}$ are: **(2024 Re)**
   1) $[M^{3/2} L^{1/2} T^{-3}]$ 2) $[M^{1/2} L T^{-3}]$
   3) $[M^2 L^{1/2} T^{-4}]$ 4) $[ML^2 T^{-4}]$

2. An object falls from a height of 10 m above the ground. After striking the ground it loses 50% of its kinetic energy. The height upto which the object can rebounce from the ground is: **(2024 Re)**
   1) 7.5 m 2) 10 m 3) 2.5 m 4) 5 m

3. An object moving along horizontal x-direction with kinetic energy 10 J is displaced through $x = (3\hat{i})m$ by the force $\vec{F} = (-2\hat{i} + 3\hat{j})$N. The kinetic energy of the object at the end of the displacement x is **(2024 Re)**
   1) 10 J 2) 16 J 3) 4 J 4) 6 J

4. At any instant of time t, the displacement of any particle is given by $2t - 1$ (SI unit) under the influence of force of 5 N. The value of instantaneous power is (in SI unit): **(2024)**
   1) 7 2) 6 3) 10 4) 5

5. Two bodies A and B of same mass undergo completely inelastic one dimensional collision. The body A moves with velocity v_1 while body B is at rest before collision. The velocity of the system after collision is v_2. The ratio $v_1 : v_2$ is: **(2024)**
   1) 4 : 1 2) 1 : 4 3) 1 : 2 4) 2 : 1

6. The potential energy of a long spring when stretched by 2 cm is U. If the spring is stretched by 8 cm, potential energy stored in it will be: **(2023)**
   1) 16U 2) 2U 3) 4U 4) 8U

7. A bullet from a gun is fired on a rectangular wooden block with velocity u. When bullet travels 24 cm through the block along its length horizontally, velocity of bullet become $\frac{u}{3}$. Then it further penetrates into the block in the same direction before coming to rest exactly at the other end of the block. The total length of the block is: **(2023)**
   1) 30 cm 2) 27 cm 3) 24 cm 4) 28 cm

8. An electric lift with a maximum load of 2000 kg (lift + passengers) is moving up with a constant speed of 1.5ms^{-1}. The frictional force opposing the motion is 3000 N. The minimum power delivered by the motor to the lift in watts is : (g = 10 ms^{-2}) **(2022)**
   1) 23500 2) 23000
   3) 20000 4) 34500

9. The energy that will be ideally radiated by a 100 kW transmitter in 1 hour is: **(2022)**
   1) 1×10^5 J 2) 36×10^7 J
   3) 36×10^4 J 4) 36×10^5 J

10. Water falls from a height of 60 m at the rate of 15 kg/s to operate a turbine. The losses due to frictional force are 10% of the input energy. How much power is generated by the turbine? (g = 10 m/s^2) **(2021)**
    1) 8.1 kW 2) 12.3 kW
    3) 7.0 kW 4) 10.2 kW

11. A body moves from rest with a constant acceleration. Which one of the following graphs represents the variation of its kinetic energy K with the distance travelled (x)? **(AIIMS 2016)**

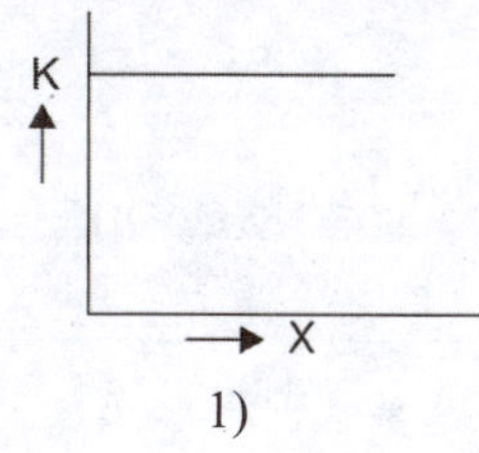

1)

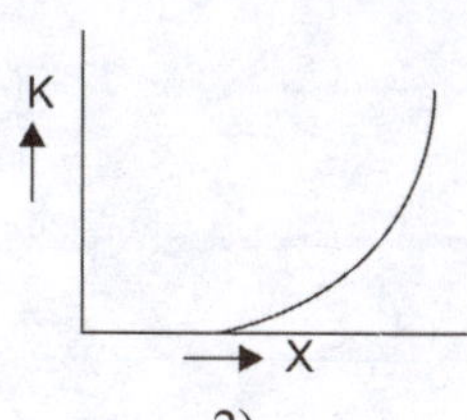

2)

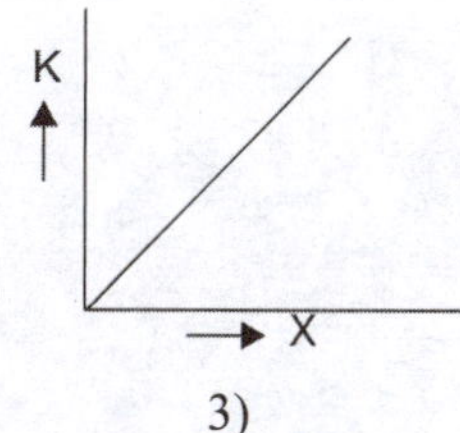

3)

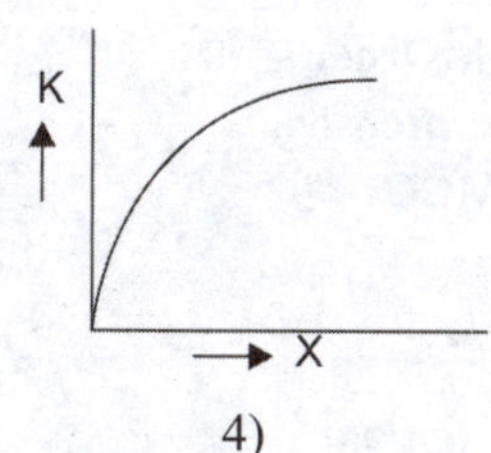

4)

12. Which of the following graph depicts spring constant k versus length l of the spring correctly? **(AIIMS 2009)**

1) 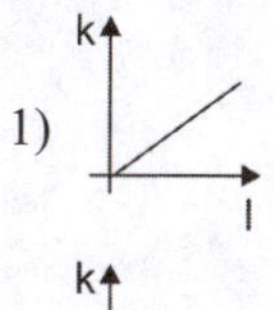

2) 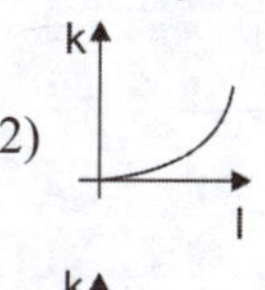

3) 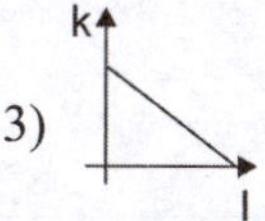

4) 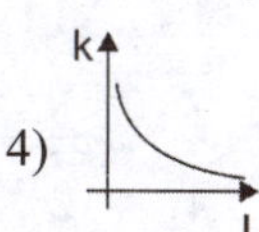

13. A ball is bouncing down a flight of stairs. The coefficient of restitution is e. The height of each step is d and the ball descends one step each bounce. After each bounce it rebounds to a height 1 above the next lower step. The height is large enough compared with the width of step so that the impacts are effectively head-on. Find the relationship between h and d: **(AIIMS 2009)**
    1) $h = d/1 - e^2$ 2) $h = d/1 + e^2$
    3) $h = d/1 + e$ 4) $h = d/1 - e^2$

14. A body of mass 5 kg moving with a speed of 1.5 m/s on a horizontal smooth surface collides with a nearly weightless spring of force constant k = 5 N/m. The maximum compression of the spring would be: **(AIIMS 2009)**
    1) 0.5 m 2) 0.15 m
    3) 1.5 m 4) 0.12 m

Answer Key

NCERT Topic-wise MCQs

1	2	3	4	5	6	7	8	9	10	11	12	13	14	15	16	17
3	4	2	3	3	2	2	1	2	3	3	4	3	1	4	1	2
18	19	20	21	22	23	24	25	26	27	28	29	30	31	32	33	34
3	3	2	4	1	3	3	2	1	1	4	4	1	3	2	1	2
35	36	37	38	39	40	41	42	43	44	45	46	47	48	49	50	51
2	2	3	4	1	4	2	1	4	1	1	3	2	4	2	3	3
52	53	54	55	56	57	58	59	60	61	62	63	64	65	66	67	68
1	4	3	1	1	2	2	2	3	2	3	1	1	4	1	3	3
69	70	71	72	73	74	75	76	77	78	79	80	81	82	83	84	85
3	1	1	4	2	2	3	1	4	3	3	4	2	3	3	4	2
86	87	88	89	90	91	92	93	94	95	96	97	98	99	100	101	102
3	2	1	1	3	4	2	2	1	3	2	2	4	4	1	4	2

High Order Time Intensive MCQs

1	2	3	4	5	6	7	8	9	10	11	12	13	14	15	16	17
3	3	1	3	2	2	3	1	2	1	1	2	4	2	3	3	3
18	19	20	21	22	23	24	25	26	27	28	29	30	31	32	33	34
1	3	1	4	2	3	2	3	4	1	3	2	1	2	2	1	3
35	36	37	38	39	40											
4	1	2	1	3	4											

Statement Based MCQs

1	2	3	4	5
1	3	3	3	1

Assertion & Reason MCQs

1	2	3	4	5
2	1	4	2	3

Match the Column MCQs

1	2	3	4	5
1	1	3	2	1

NCERT Exemplar MCQs

1	2	3	4	5	6	7	8	9	10
4	2	4	3	3	3	2	2	3	1

NEET & AIIMS Past Year MCQs

1	2	3	4	5	6	7	8	9	10	11	12	13	14
3	4	3	3	4	1	2	4	2	1	3	4	1	3

CHAPTER 6

System of Particles and Rotational Motion

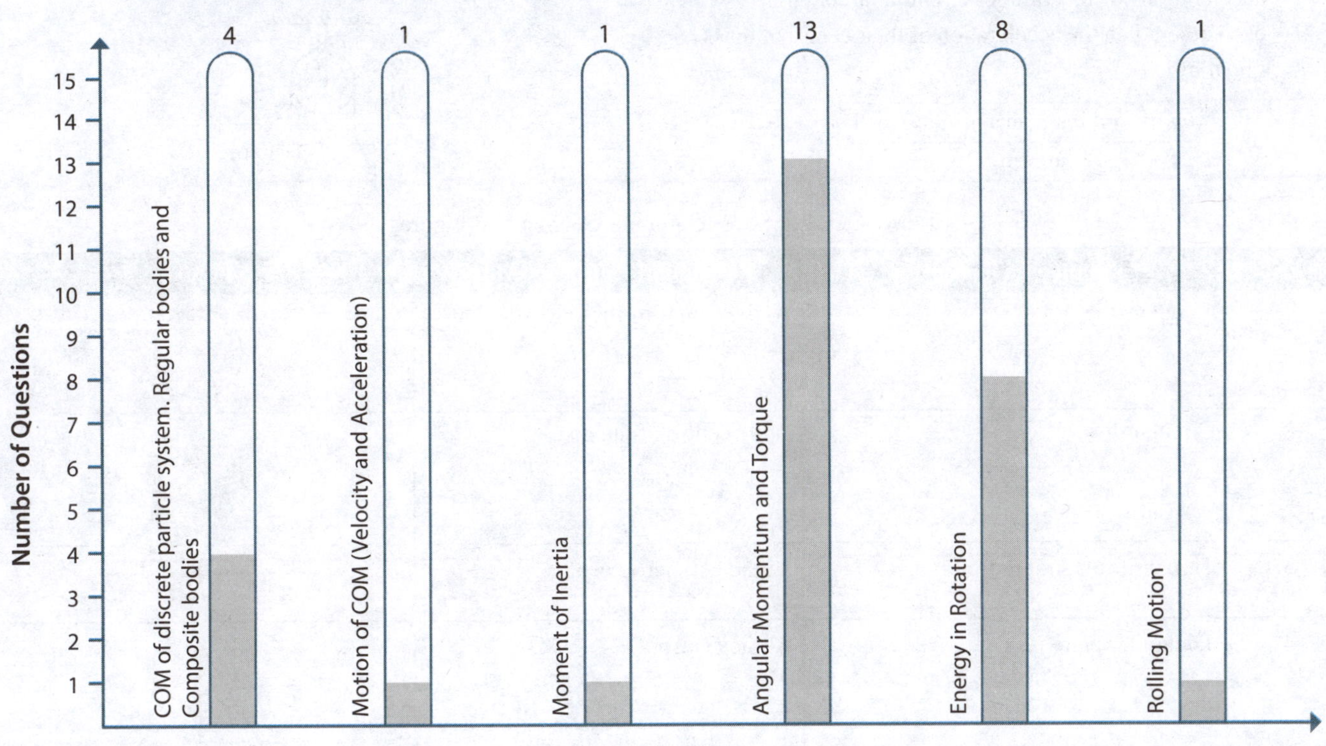

NCERT-Picks

Centre of Mass

Formulas

1. Coordinates of the centre of mass of a system of particles is given by

$$x_{cm} = \frac{m_1x_1 + m_2x_2 + ... + m_nx_n}{m_1 + m_2 + ... + m_n} = \frac{\sum m_ix_i}{\sum m_i}$$

$$y_{cm} = \frac{m_1y_1 + m_2y_2 + ... + m_ny_n}{m_1 + m_2 + ... + m_n} = \frac{\sum m_iy_i}{\sum m_i}$$

$$z_{cm} = \frac{m_1z_1 + m_2z_2 + ... + m_nz_n}{m_1 + m_2 + ... + m_n} = \frac{\sum m_iz_i}{\sum m_i}$$

2. Velocity of centre of mass of a system of particles

$$v_{cm} = \frac{m_1\vec{v}_1 + m_2\vec{v}_2 + ... + m_n\vec{v}_n}{m_1 + m_2 + ... + m_n}$$

3. Acceleration of centre of mass,

$$a_{cm} = \frac{m_1\vec{a}_1 + m_2\vec{a}_2 + ... + m_n\vec{a}_n}{m_1 + m_2 + ... + m_n}$$

Numerical Profile

Velocity and Acceleration of Centre of Mass

(i) If two particles of masses m_1 and m_2 are moving with velocities of magnitudes v_1 and v_2 making an angle 'θ' with each other, then the velocity of their centre of mass, $v_{cm} = \frac{\sqrt{m_1^2v_1^2 + m_2^2v_2^2 + 2(m_1v_1)(m_2v_2)\cos\theta}}{(m_1 + m_2)}$

(a) If the particles move at right angles to each other, then $\theta = 90^\circ$ and $v_{cm} = \dfrac{\sqrt{m_1^2 v_1^2 + m_2^2 v_2^2}}{m_1 + m_2}$

(b) If the particles move in opposite directions, then $\theta = 180^\circ$ and $v_{cm} = \dfrac{m_1 v_1 - m_2 v_2}{m_1 + m_2}$

(c) If the particles move in the same direction, then $\theta = 0^\circ$ and $v_{cm} = \dfrac{m_1 v_1 + m_2 v_2}{m_1 + m_2}$

(ii) If two particles of masses m_1 and m_2 are moving with accelerations of magnitudes a_1 and a_2 at an angle θ with each other, then the acceleration of their centre of mass is given by

$$a_{cm} = \frac{\sqrt{m_1^2 a_1^2 + m_2^2 a_2^2 + 2(m_1 a_1)(m_2 a_2)\cos\theta}}{m_1 + m_2}$$

(a) If the particles move in the same direction, then $\theta = 0^\circ$ and $a_{cm} = \dfrac{m_1 a_1 + m_2 a_2}{m_1 + m_2}$

(b) If the particles move in opposite directions, then $\theta = 180^\circ$ and $a_{cm} = \dfrac{m_1 a_1 - m_2 a_2}{m_1 + m_2}$

(iii) The acceleration of C.M. of a system of two blocks of mass m_1, m_2 ($m_1 > m_2$) joined by a massless string and passing over a pulley of negligible mass, is given by $\left(\dfrac{m_1 - m_2}{m_1 + m_2}\right)^2 g$.

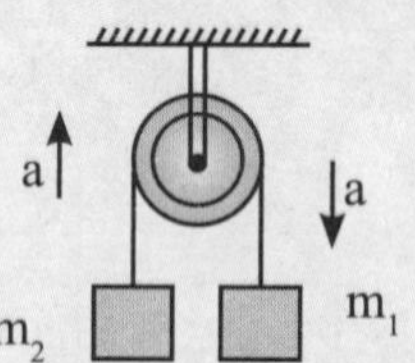

Table: COM of Some Common Geometrical Figures

S. No.	Shape of The Body	Position of Centre of Mass	Figure
1.	Circular ring	At the centre of the ring	
2.	Circular disc	A the centre of the disc	
3.	Thin uniform straight rod	At the geometric centre	
4.	Triangular plate	At the centroid	
5.	Square plate	At the point of intersection of the diagonals	
6.	Rectangular plate	At the point of intersection of the diagonals	
7.	Hollow or solid sphere	At the centre of the sphere	
8.	Hollow cone	At a height of $\dfrac{h}{3}$, from the base	

9.	Solid cone or Pyramid	At a height of $\frac{h}{4}$ from the base	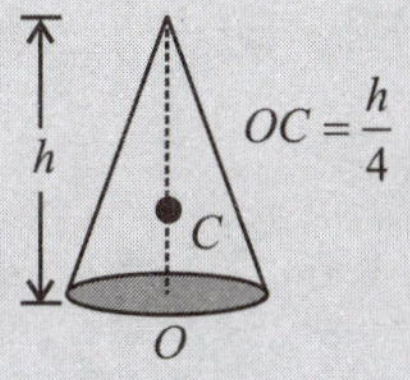 $OC = \frac{h}{4}$
10.	Solid (or) hollow cylinder	At the mid-point of its own axis	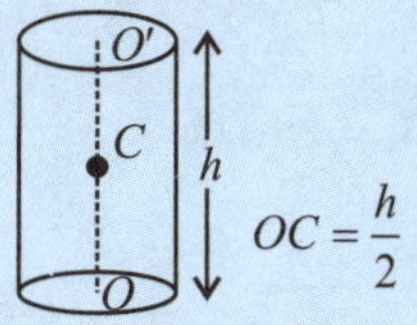 $OC = \frac{h}{2}$
11.	An arc of radius R subtending an angle α at its centre of curvature	At a distance of $\frac{2R}{\alpha}\sin\left(\frac{\alpha}{2}\right)$ from its centre of curvature on the axis of symmetry	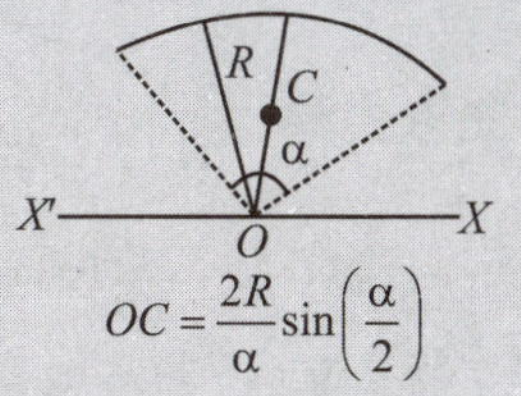 $OC = \frac{2R}{\alpha}\sin\left(\frac{\alpha}{2}\right)$
	(i) A semi-circle of radius 'R'	At a distance of $\frac{2R}{\pi}$ from its centre on the axis of symmetry	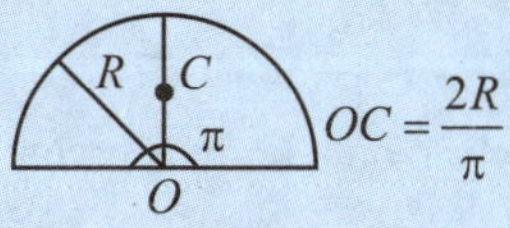 $OC = \frac{2R}{\pi}$
	(ii) A quadrant of a circle of radius 'R'	At a distance of $\frac{4R}{\pi\sqrt{2}}$ from its centre 'O' on the axis of symmetry	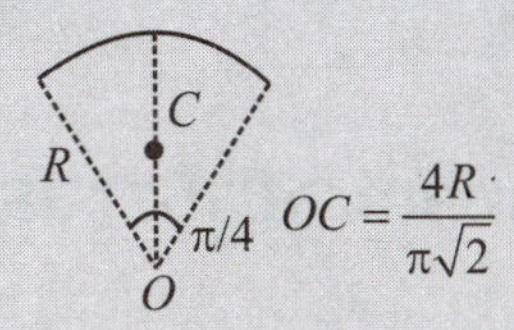 $OC = \frac{4R}{\pi\sqrt{2}}$
12.	Semi-circular disc	At a distance of $\frac{4R}{3\pi}$ from its centre 'O' on the axis of symmetry	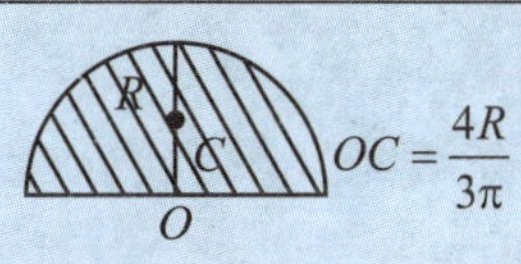 $OC = \frac{4R}{3\pi}$
13.	Solid hemi-sphere	At a distance of $\frac{3R}{8}$ from its centre 'O' on the axis of symmetry	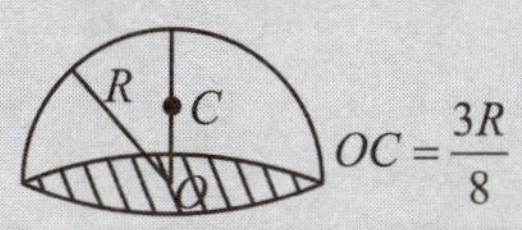 $OC = \frac{3R}{8}$
14.	Hollow hemi-sphere (or) Hemi-spherical shell	At a distance of $\frac{R}{2}$ from its centre 'O' on the axis of symmetry	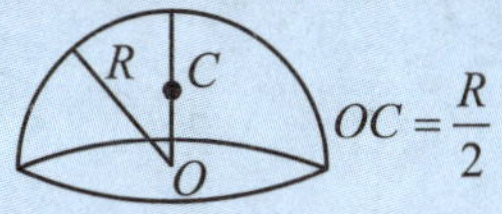 $OC = \frac{R}{2}$
15.	Horse-shoe magnet	At its centre within the boundary limits	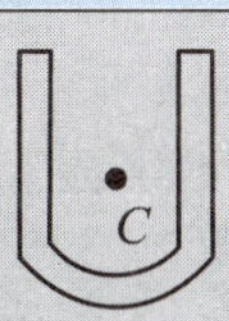 C
16.	Semi-Circular annular plate	At a distance of $OC = \frac{4\left(R_1^2 + R_1R_2 + R_2^2\right)}{3\pi\left(R_1 + R_2\right)}$ from its centre of symmetry	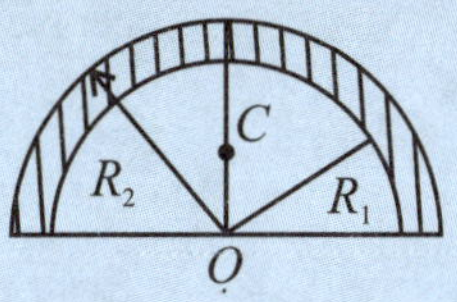 R_2, R_1, C, O

Rotational Parameters

Formulas

1. $\omega = \omega_0 + \alpha t$ [where, ω_0 is initial angular velocity, α is angular acceleration and t is time]
2. $\theta = \omega_0 t + \frac{1}{2}\alpha t^2$ [θ is angular displacement]
3. $\omega^2 - \omega_0^2 = 2\alpha\theta$
4. $\omega = \frac{v}{r}$
5. $\omega = \frac{d\theta}{dt}$
6. $\alpha = \frac{d^2\theta}{dt^2}$
7. $a_t = r \times \alpha$ [where, a_t is tangential acceleration]

Torque

Formulas

Torque acting on a body is given by

$\vec{\tau} = \vec{r} \times \vec{F}$ [$\vec{\tau}$ is torque, $\vec{r}$ is the position vector and $\vec{F}$ is force]

$\Rightarrow \tau = rF\sin\theta$ (θ is the angle between $\vec{r}$ and $\vec{F}$)

Moment of Inertia

Formulas

S.No.	Body	Axis of rotation	Moment of inertia (I)
1.	Ring	about its own axis	MR^2
2.	Disc	about its own axis	$\frac{MR^2}{2}$
3.	Solid cylinder of length L, radius R and mass M	(i) about its own axis	$\frac{1}{2}MR^2$
		(ii) about an axis passing through centre of cylinder and ⊥ to its length	$M\left[\frac{L^2}{12}+\frac{R^2}{4}\right]$
4.	Thin rod of length L	(i) about an axis through centre of rod and ⊥ to rod.	$\frac{ML^2}{12}$
		(ii) about an axis through one end and ⊥ to rod	$\frac{ML^2}{3}$
5.	Hollow sphere	About its own axis	$\frac{2}{3}MR^2$
6.	Solid sphere	About its own axis	$\frac{2}{5}MR^2$

Parallel Axis Theorem

$I = I_C + Mh^2$

Where,

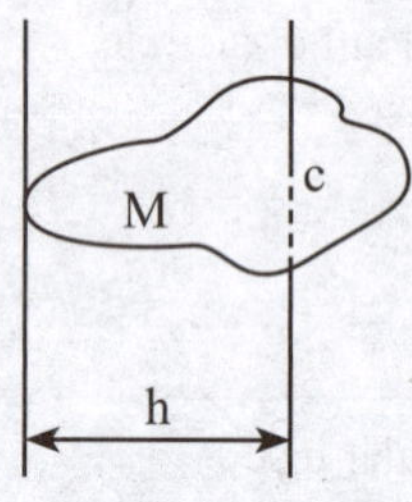

- I is the moment of inertia of the body about the axis parallel to an axis passing through centre of mass.
- I_C is the moment of inertia about the center of mass.
- M is the mass of the body
- h is the distance between the two axes

Perpendicular Axis Theorem

$I_a = I_b + I_c$

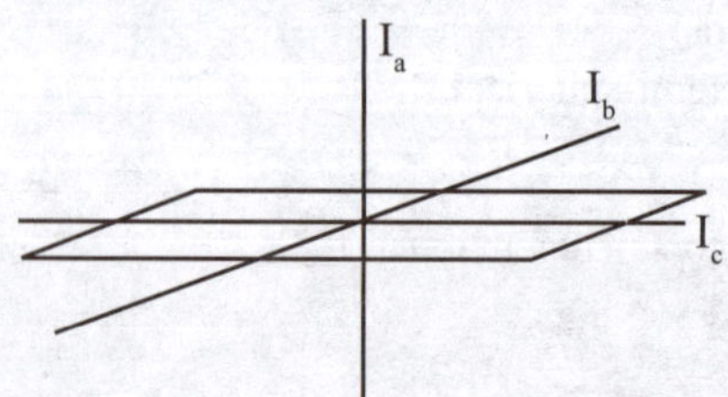

Numerical Profile

Work done in Rotational Motion

(i) Work done by external torque on rotating body is $W = \int \tau d\theta$

If τ is constant $W = \tau\theta$

(ii) Work done by external torque on rotating rigid body is equal to change in rotational kinetic energy.

$$\int \tau d\theta = \frac{1}{2}I\omega^2 - \frac{1}{2}I\omega_0^2$$

Angular Momentum

Formulas

- $\vec{L} = \vec{r} \times \vec{p}$
- $\vec{L} = L = r\,p \sin\theta$

Conservation of Angular Momentum

Formulas

$$\vec{\tau}_{ext} = \frac{d\vec{L}}{dt}$$

If $\vec{\tau}_{ext} = 0$

$\Rightarrow \vec{L} =$ constant

Rolling Motion

Formulas

1. For pure rolling on ground

   $v_{cm} = \omega r$
2. For rolling with skidding

   $v_{cm} > \omega r$
3. For rolling with slipping

   $v_{cm} < \omega r$
4. Total kinetic energy of a rolling body

   $$K = \frac{1}{2}mv^2 + \frac{1}{2}I\omega^2$$

Numerical Profile

Velocity at a point on the rim of a rolling wheel

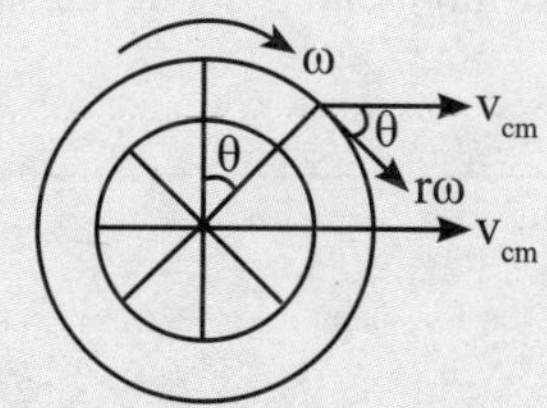

For pure rolling $v_{cm} = r\omega$

As v_{net} is the vector sum of v_{cm} and ωr

$$\Rightarrow v_{net} = 2v_{cm}\cos\frac{\theta}{2}$$

Total K.E of a rolling body

Total kinetic energy is given by

$$K.E_{total} = K.E_{translational} + K.E_{rotational}$$

$$K.E_{total} = \frac{1}{2}mv_{cm}^2\left(1+\frac{k^2}{r^2}\right)$$ [v_{cm} is the velocity of C.M and k is radius of gyration]

(i) KE of a rolling solid cylinder, $K = \frac{3}{4}mv_{cm}^2$

(ii) KE of a rolling solid sphere, $K = \frac{7}{10}mv_{cm}^2$

(iii) KE of a rolling hollow sphere, $K = \frac{5}{6}mv_{cm}^2$

Pure Rolling of a body on an inclined plane (without slipping)

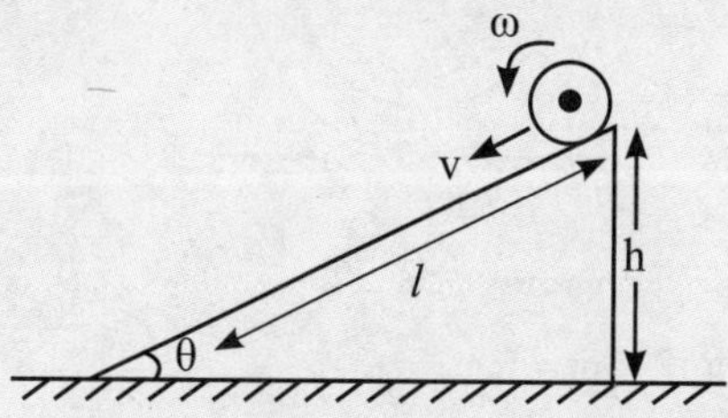

(i) Let a body is rolling down on an inclined plane, velocity of the body at the bottom of the inclined plane is given by

$$v = \sqrt{\frac{2gh}{1+\frac{k^2}{r^2}}} = \sqrt{\frac{2gl\sin\theta}{1+\frac{k^2}{r^2}}} \text{(since } h = l\sin\theta)$$

(ii) Acceleration of the body is given by $a = \dfrac{g\sin\theta}{1+\frac{k^2}{r^2}}$

(iii) Time taken by the body to reach the bottom is given by

$$t = \sqrt{\frac{2l(1+k^2/r^2)}{g\sin\theta}} = \frac{1}{\sin\theta}\sqrt{\frac{2h}{g}\left(1+\frac{k^2}{r^2}\right)}$$

A body rolling up on an inclined plane

Using conservation of energy $\frac{1}{2}mv^2 + \frac{1}{2}I\omega^2 = mgh$

Height attained by the body, $h = \dfrac{v^2}{2g}\left[1+\dfrac{k^2}{r^2}\right]$

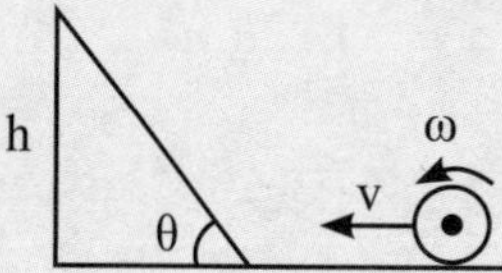

(i) For a solid sphere, $h = \dfrac{7v^2}{10g}$

(ii) For a hollow sphere $h=\frac{5v^2}{6g}$

(iii) For a disc (or) solid cylinder $h=\frac{3v^2}{4g}$

(iv) For a ring (or) hollow cylinder, $h=\frac{v^2}{g}$

Angular Momentum of Rolling Wheel

Angular momentum of a rolling wheel about an axis passing through the point of contant P and perpendicular to the plane of wheel:

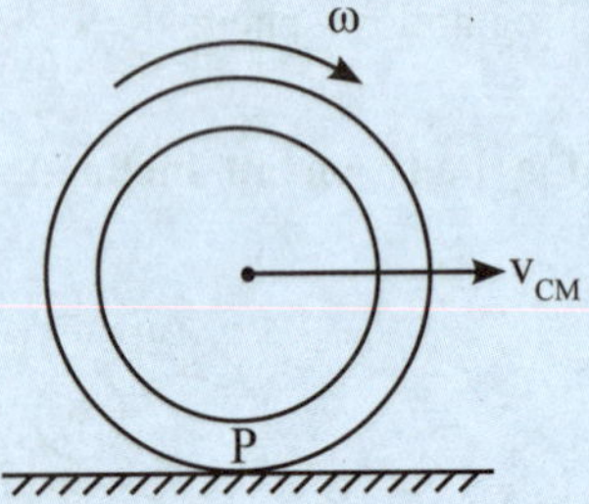

$\vec{L}=\vec{L}_{translation}+\vec{L}_{rotation}$

$\vec{L}=\left(I_{CM}+mR^2\right)\omega=I_P\omega$

Statements Asked in Exam

- Motion of a body is said to be translational, when at any instant of time all particles of the body have the same velocity.
- The forces leading to explosion are internal forces. They contribute nothing to the motion of centre of mass.
- If one of the particles is heavier than the other, the centre of mass of two particle system will shift towards the heavier particle.
- The centre of mass of a body may lie within or outside the material of the body.
- A person is sitting in the compartment of a train moving with uniform velocity on a smooth track. The velocity of centre of mass of the system (train + man) does not change if the person begins to run in the compartment.
- An isolated system is that on which no external forces are acting.
- Rotational analogue of mass is moment of inertia.
- A wheel moving down a perfectly frictionless inclined plane will undergo slipping not rolling motion.
- Rolling motion is a combination of rotatory motion as well as translatory motion parallel to the plane on which the body rolls.
- The instantaneous speed of the point of contact during rolling on ground is zero.
- For pure rolling motion, work done against friction is zero.
- For a body to roll down an inclined plane the force of friction is necessary, otherwise it will slide.
- There are two spheres of same mass and same radius, one is solid and other is hollow. Hollow has a larger moment of inertia about its diameter.
- Angular momentum of a particle about a given axis is twice the product of mass of the particle and its areal velocity.

  $\left|\vec{L}\right|=2m\left(\frac{dA}{dt}\right)$
- Sometimes, a rigid body may be in partial equilibrium i.e., it may be in translational equilibrium and not in rotational equilibrium or vice versa.
- The centre of gravity (CG) of a body is a point where the weight of the body acts and torque due to gravity about this point is zero on the body is zero.
- If the body is so extended that $\vec{g}$ varies from part to part of the body, then the centre of gravity shall not coincide with the centre mass of the body.
- Coefficient of friction for rolling of solid cylinder without slipping down the rough inclined plane is

  $\mu=\frac{1}{3}\tan\theta$.
- In a rigid body, the distance between any pair of particles does not change.
- Raw egg will spin with smaller angular acceleration than a hard boiled egg for same applied torque.
- A couple produces rotation without translation.
- We prefer to use a wrench with long arm because it produces large torque even with a small force.
- When the total external force on the system is zero, the velocity of centre of mass remains constant.
- A planet revolves around a massive star in a highly elliptical orbit. Its angular momentum about the star is constant over the entire orbit.

MCQs ?

NCERT Topic-wise MCQs

Center of Mass of Discrete Particle System, Regular Bodies & Composite Bodies

1. The centre of mass of a body:
   1) Depends on the choice of co-ordinate system
   2) Is independent of the choice of co-ordinate system
   3) May or may not depend on the choice of co-ordinate system
   4) None of the above

2. A system consists of two masses 9 kg & 11 kg which are connected by a massless rod that lies along the z-axis. The 9 kg mass is at a distance z = 5 m while the 11 kg mass is at a distance z = 9 m. The z-coordinate and x-coordinate of system are (in m)
   1) 5, 4 2) 7.2, 7.2
   3) 7.2, 0 4) 2, 3

3. Mass is non-uniformly distributed on the circumference of a ring of radius r, centred at origin. Let a be the distance of the centre of mass of the ring from origin. Then
   1) $r = a$ 2) $0 < a \le r$
   3) $a < r$ 4) $a > r$

4. Centre of mass of the given system of particles will be at

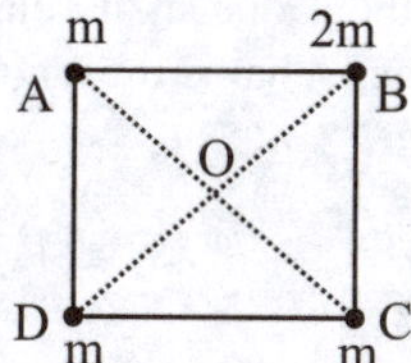

   1) OD 2) OC
   3) OB 4) AO

5. A body A of mass M while falling vertically downward under gravity breaks into two parts, a body B of mass $\frac{M}{2}$ and a body C of mass $\frac{M}{2}$. The centre of mass B and C taken together shifts, compared to that of body A, towards
   1) Body C
   2) Body B
   3) Depends on height of breaking
   4) Does not shift

6. A small disc of radius 2 cm is cut from a disc of radius 6 cm. If the distance between their centers is 3.2 cm, what is the shift in the centre of mass of the disc?
   1) 0.4 m 2) 2.4 cm
   3) 1.8 cm 4) 1.2 cm

7. Find the position of the centre of mass of the T-shaped plate from O

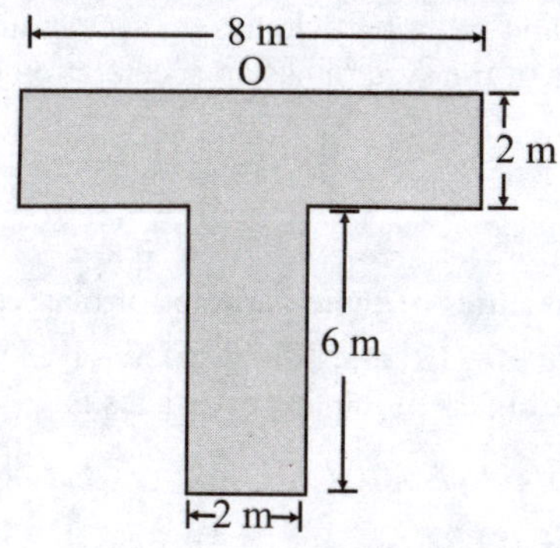

   1) 2.71 m 2) 3.5 m
   3) 2 m 4) 2.50 m

8. The distance between the centres of carbon and oxygen atoms in the carbon monoxide gas molecule is 1.13 Å. Locate the centre of mass of the gas molecule relative to the carbon atom.
   1) 0.646 Å 2) 1.534 Å
   3) 0.857 Å 4) 2.785 Å

9. The centre of mass of three particles of mass 1 kg, 2 kg and 3 kg is at (3, 3, 3) with reference to a fixed coordinate system. Where should a fourth particle of mass 4 kg be placed, so that the centre of mass of the system of all particles shifts to a point (1, 1, 1)?
   1) (–1, –1, –1) 2) (–2, –2, –2)
   3) (2, 2, 2) 4) (1, 1, 1)

Motion of Center of Mass (Velocity & Acceleration)

10. Two objects of masses 300 g and 700 g possess velocities $15\hat{i}$ m/s and $4\hat{i}+6\hat{j}$ m/s respectively. The velocity of their centre of mass in m/s is
    1) $7\hat{i}+8\hat{j}$ 2) $7.3\hat{i}+4.2\hat{j}$
    3) $5\hat{i}+4\hat{j}$ 4) $2\hat{i}+6\hat{j}$

11. Two masses m_1 and m_2 ($m_1 > m_2$) are connected by a massless and inextensible string passing over a massless and frictionless pulley. The acceleration of the centre of mass of the system is:

1) $\left(\frac{m_1 - m_2}{m_1 + m_2}\right)^2 g$ 2) $\left(\frac{m_1 + m_2}{m_1 - m_2}\right)^2 g$

3) $\left(\frac{m_1 - m_2}{m_1 + m_2}\right) g$ 4) g

12. A system of two blocks A and B of masses 20 kg and 10 kg are connected by a string which hangs over a frictionless massless pulley. What will be the acceleration of the centre of mass?

1) 1.09 m/s^2 2) 2 m/s^2

3) 0.3 m/s^2 4) 0.9 m/s^2

13. A system consists of two identical particles. One particle is at rest and other particle has an acceleration a. The centre of mass of the system has an acceleration of

1) 2a 2) a/2

3) a 4) $\frac{2a}{3}$

14. The velocities of three particles of masses 20 g, 30 g, and 50 g are $10\hat{i}$, $10\hat{j}$ and $10\hat{k}$ respectively. The velocity of the centre of mass of the three particles is

1) $2\hat{i}+3\hat{j}+5\hat{k}$ 2) $-\hat{i}-3\hat{j}-2\hat{k}$

3) $-\hat{i}+3\hat{j}+2\hat{k}$ 4) $-\hat{i}+3\hat{j}-2\hat{k}$

15. Two identical particles move towards each other with velocities 2v and v respectively. What is the velocity of centre of mass ?

1) 2v 2) v

3) $\frac{3v}{2}$ 4) $\frac{v}{2}$

Rotational Parameters, Moment of Inertia, Torque, Angular Momentum

16. The angular speed of a motor wheel is increased from 1200 rpm to 3120 rpm in 16 seconds. The angular acceleration of the motor wheel 1s

1) 2π rad s^{-2} 2) 4π rad s^{-2}

3) 6π rad s^{-2} 4) 8π rad s^{-2}

17. A grindstone has a moment of inertia of 6 kgm^2. A constant torque is applied and the grindstone is found to have a speed of 150 rpm, 10 seconds after starting from rest. The torque is

1) 3π Nm 2) 3 Nm

3) $\frac{\pi}{3}$Nm 4) 4π Nm

18. Angular momentum of the particle rotating with a central force is constant due to

1) constant torque

2) constant force

3) constant linear momentum

4) zero torque

19. The position of a particle is give by $\vec{r} = \hat{i} + 2\hat{j} - \hat{k}$ and its linear momentum is given by $\vec{p} = 3\hat{i} + 4\hat{j} - 2\hat{k}$. Then its angular momentum about the origin is perpendicular

1) x-axis 2) y-axis

3) z-axis 4) yz-plane

20. Match Column-I with Column-II.

Column-I		Column-II	
(A)	For translational equilibrium	(p)	Mk^2
(B)	For rotational equilibrium	(q)	Angular acceleration
(C)	Moment of inertia of a body	(r)	$\sum \vec{F} = 0$
(D)	Torque is required to produce	(s)	$\sum \vec{\tau} = 0$

1) A-p, B-q, C-r, D-s 2) A-q, B-r, C-s, D-p

3) A-r, B-q, C-p, D-s 4) A-r, B-s, C-p, D-q

21. The angular velocity of a wheel increases from 150 rps to 200 rps in 18 sec. The number of revolutions made during that time is

1) 2000 2) 3150

3) 3000 4) 1570

22. A fly wheel gains a speed of 680 rpm in 9 sec. Its angular acceleration will be

1) 4π rad/s^2 2) 2.5π rad/s^2

3) 2π rad/s^2 4) π rad/s^2

23. The direction of angular velocity vector is along

1) The tangent to the circular path

2) The inward radius

3) The outward radius

4) The axis of rotation

24. A uniform circular disc of mass 200 g and radius 5 cm is rotated about one of its diameters at an angular speed of 15 rad/sec. What is the angular momentum about the axis of rotation?

1) 1.6×10^{-3} Js

2) 18.75×10^{-4} Js

3) 19.1×10^{-1} Js

4) 22 Js

25. Calculate the angular momentum of a particle of mass 5 mg revolving about a circle of radius 60 cm with a speed of 2 km/s.

1) $1\times10^{-3}\,kgm^2s^{-1}$ 2) $4\times10^{-3}\,kgm^2s^{-1}$

3) $6\times10^{-3}\,kgm^2s^{-1}$ 4) $6\times10^{-4}\,kgm^2s^{-1}$

26. A rope is wound round a hollow cylinder of mass 5 kg and radius 60 cm. If the rope is pulled with a force of 80 N. What is the angular acceleration of the cylinder?

1) 25 rad/s^2 2) 26.7 rad/s^2

3) 30 rad/s^2 4) 38 rad/s^2

27. A grindstone has a moment of inertia of 10 kgm^2. A constant torque is applied and the grindstone is found to have a speed of 200 rpm, 20 sec after starting from rest. What is the value of the torque?

1) 2.14 Ns
2) 6 Ns
3) 4 Ns
4) 10.47π Ns

28. A constant torque of 2000 N-m, turns a wheel of moment of inertia 300 kg-m^2 about an axis through the centre. Angular velocity of the wheel after 5 s will be

1) 33.3 rad/s^2
2) 4.44 rad/s^2
3) 50 rad/s^2
4) None

29. The angular momentum of a wheel changes from 3 L to 7 L in 4 sec. What is the magnitude of torque acting on it?

1) L
2) 2 L
3) 3 L
4) 4 L

30. A flywheel of moment of inertia 9×10^5 kg m^2 is rotating with uniform angular speed of 5 rad s^{-1} If a torque of 9×10^4 Nm retards the wheel, then time in which the wheel comes to rest is

1) 25 s
2) 30 s
3) 50 s
4) 60 s

31. A 5 kg mass is rotating on a circular path of radius 1 m with angular velocity of 50 rad/sec. If radius of path becomes 5 m. The value of angular velocity will be

1) 7 rad/s
2) 3 rad/s
3) 4 rad/s
4) 2 rad/s

32. A particle of mass m is rotating in a circular path of radius R. Its angular momentum is L. The centripetal force acting on the particle is

1) $\frac{m^2L^2}{r}$
2) $\frac{L^2}{r^3m}$
3) $\frac{mL}{r}$
4) None

33. Two discs have radii 0.1 m and 0.4 m. Their moment of inertia about their axes will be in the ratio of (assume same thickness and material)

1) 1:144
2) 1:225
3) 1:256
4) 1:169

34. A wheel of mass 15 kg has a moment of inertia of 200 kg-m^2 about its own axis. The radius of gyration will be:

1) 3.65 m
2) 4.25 m
3) 4.67 m
4) 1.21 m

35. The moment of inertia of a uniform ring of mass M and radius r about a tangent lying in its own plane is

1) 2 Mr^2
2) $\frac{3}{2}Mr^2$
3) Mr^2
4) $\frac{Mr^2}{2}$

36. Radius of gyration of a body depends upon

1) Mass and size of body
2) Shape of body
3) Radius or Area of body
4) Mass distribution and axis of rotation

37. Three rings each of mass M and radius R are arranged as shown in the figure. The moment of inertia of the system about yy′ will be:

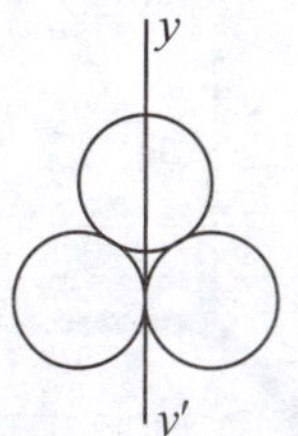

1) $\frac{3}{2}MR^2$
2) 3 MR^2
3) 5 MR^2
4) $\frac{7}{2}MR^2$

38. Two discs have their moments of inertia about their own axes in the ratio 1:2 and their diameters are in ratio 2:1. The ratio of their masses will be

1) 1 : 4
2) 1 : 8
3) 3 : 4
4) 5 : 4

39. From a uniform circular disc of radius R and mass 9 M, a small disc of radius $\frac{R}{3}$ is removed as shown in the figure. The moment of inertia of the remaining disc about an axis perpendicular to the plane of the disc and passing through centre of disc is:

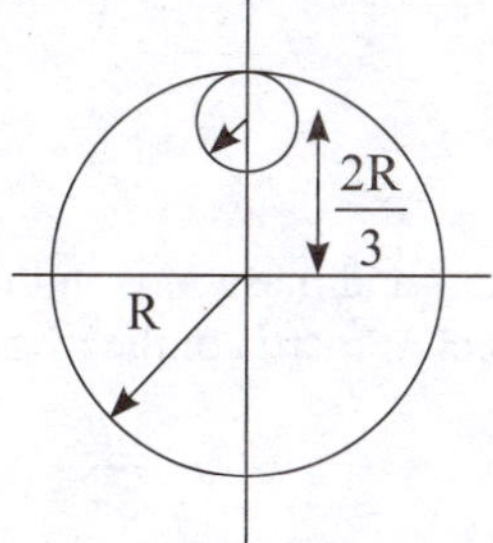

1) $\frac{37}{9}MR^2$
2) 4 MR^2
3) $\frac{40}{9}MR^2$
4) 10 MR^2

40. A heavy iron bar of weight 12 kg is having its one end on the ground and the other on the shoulder of a man. The rod makes an angle 60° with the horizontal, the weight experienced by the man is:

1) 6 kg
2) 12 kg
3) 3 kg
4) $6\sqrt{3}$ kg

41. A heavy iron bar, of weight W is having its one end on the ground and the other on the shoulder of a person. The bar makes an angle θ with the horizontal. The weight experienced by the person is:

1) $W/2$
2) W
3) $W \cos\theta$
4) $W \sin\theta$

42. The diameter of wheel increases by 1%. What will be percentage increases in moment of inertia about axis of symmetry?

1) 1% 2) 4%
3) 3% 4) 2%

43. Three thin rods each of length L and mass M are placed along x, y, z-axes in such a way that one end of each of the rods is at the origin. The moment of inertia of this system about z-axis is:

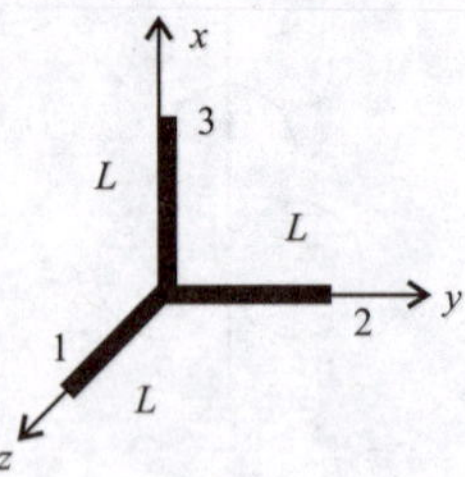

1) $\frac{2ML^2}{3}$ 2) $\frac{4ML^2}{3}$
3) $\frac{5ML^2}{3}$ 4) $\frac{ML^2}{3}$

44. Three identical thin rods each of length l and mass M are joined together to form a letter H. The moment of inertia of the system about one side of H is:

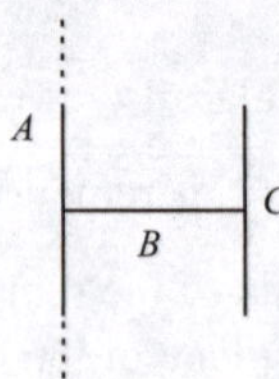

1) ML^2 2) $\frac{4}{3}ML^2$
3) $\frac{ML^2}{2}$ 4) $\frac{3}{4}ML^2$

45. A wire of length l and mass M is in the form of a circular ring. The moment of inertia of the ring about its axis is

1) $\frac{Ml^2}{2\pi^2}$ 2) $\frac{Ml^2}{2\pi}$
3) $\frac{Ml^2}{4\pi}$ 4) $\frac{Ml^2}{4\pi^2}$

46. The moment of inertia of a solid sphere about an axis passing through centre of gravity is $\frac{2}{5}MR^2$, then its radius of gyration about a parallel axis at a distance 4R from first axis is

1) $R\sqrt{\frac{82}{5}}$ 2) $R\sqrt{\frac{22}{5}}$
3) $R\sqrt{\frac{10}{3}}$ 4) $R^2\sqrt{10}$

47. For the same total mass which of the following will have the largest moment of inertia about an axis passing through the centre of gravity and perpendicular to the plane of the body?

1) A disc of radius a
2) A ring of radius a
3) a square lamina of side 2a
4) Four rods forming square of side 2a

48. The angular momentum of a rotating body changes from A_0 to 4 A_0 in 4 sec. The torque acting on the body is

1) $\frac{3}{4}A_0$ 2) 4 A_0
3) 3 A_0 4) $\frac{3}{2}A_0$

Conservation of Angular Momentum, Motion of Connected Bodies, Energy in Rotation

49. A body in rotational motion possesses rotational kinetic energy given by ________.

1) $KE = \frac{1}{2}I^2\omega$ 2) $KE = \frac{1}{2}I\omega^2$
3) $KE = 2I^2\omega$ 4) $KE = I\omega$

50. Two rotating bodies have same angular momentum but their moments of inertia are I_1 and I_2 respectively ($I_1 > I_2$). Which body will have higher kinetic energy of rotation?

1) First
2) Second
3) Both will have same rotational kinetic energy
4) Not possible to predict

51. A round disc having moment of inertia I_2 about an axis perpendicular to its plane and passing through its centre is placed over another disc having moment of inertia I_1 rotating with an angular velocity ω about the same axis. The final angular velocity of the combination of discs is:-

1) ω 2) $\frac{I_1\omega}{I_1+I_2}$
3) $\frac{(I_1+I_2)\omega}{I_1}$ 4) $\frac{I_2\omega}{I_1+I_2}$

52. If the earth loses its atmosphere suddenly, then the duration of the day will:-

1) Increase
2) Decrease
3) Remain unchanged
4) Nothing can be definitely said

53. Two rigid bodies A and B rotate with rotational kinetic energies E_A and E_B respectively. The moments of inertia of A and B about the axis of rotation are I_A and I_B respectively. If $I_A = I_B/4$ and $E_A = 100\ E_B$ the ratio of angular momentum (L_A) of A to the angular momentum (L_B) of B is

1) 25 2) 5/4
3) 5 4) 1/4

54. A swimmer while jumping into water from a height begins to rotate fastly in the air if
   1) He pulls his arms and legs in
   2) He spread his arms and legs
   3) He keeps himself straight
   4) None of the above

55. A uniform metre stick of mass M is hinged at one end and held in horizontal direction by a string attached to the other end. What should be the initial acceleration (in rad/s^2) of the stick if the string is cut?
   1) $\frac{3g}{2L}$ 2) g/L
   3) 3 g/L 4) 4 g/L

56. For a system of particles,
   1) $\frac{d\vec{P}}{dt} = \vec{F}_{ext}$ 2) $\frac{d\vec{L}}{dt} = \vec{\tau}_{ext}$
   3) $\frac{d\vec{L}}{dt} = \vec{\tau}_{ext}$ 4) Both (1) and (2)

57. For a system of particles if $\tau_{ext} = 0$, then choose the correct option:
   1) $\frac{d\vec{L}}{dt} \neq 0$ 2) $\vec{L}$ is a variable
   3) $\vec{L}$ = constant 4) none of these

58. The torque applied to a ring revolving about its own axis so as to change its angular momentum by 2 J-s in 5 s, is
   1) 10 N-m 2) 2.5 N-m
   3) 0.1 N-m 4) 0.4 N-m

59. When a mass is moving in a circle about a fixed point P, its angular momentum about P is directed along
   1) Radius
   2) The tangent to the orbit
   3) A line perpendicular to the plane of rotation
   4) None of the above

60. If a gymnast, sitting on a rotating stool with his arms outstretched, suddenly pulls his hands inwards
   1) The angular velocity decreases
   2) His moment of inertia decreases
   3) The angular velocity stays constant
   4) The angular momentum increases

61. A uniform heavy disc is rotating at constant angular velocity ω about a vertical axis through its centre and perpendicular to the plane of the disc. Let L be its angular momentum. A lump of plasticine is dropped vertically on the disc and sticks to it. Which of the following will be constant
   1) ω 2) ω and L both
   3) L only 4) Neither ω nor L

62. If the radius of the earth suddenly contracts to half of its present value, then the duration of the day will be of
   1) 6 hours 2) 12 hours
   3) 18 hours 4) 24 hours

Paragraph (Q. No. 63-66)

Answer these questions on the basis of given figure and information.

A cord of negligible mass is wound round the rim of a fly wheel (disc) of mass 20 kg and radius 20 cm. A steady pull of 25 N is applied on the cord as shown in figure. The flywheel is mounted on a horizontal axle with frictionless bearings.

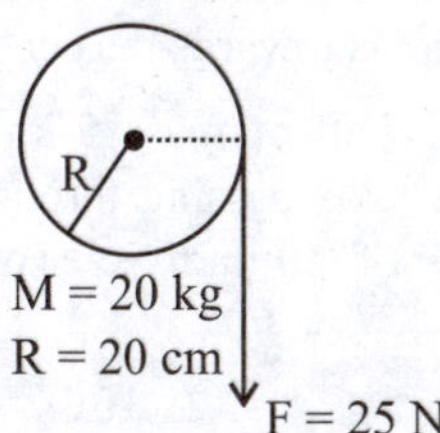

63. Compute the angular acceleration of the wheel.
   1) 12.50 rad/s^2 2) 6 rad/s^2
   3) 10 rad/s^2 4) 8 rad/s^2

64. Find the work done by the pull, when 2 m of the cord is unwound
   1) 40 J 2) 50 J
   3) 30 J 4) 20 J

65. Find the kinetic energy of the wheel when 2 m of chord is unwound (Assume that the wheel starts from rest)
   1) 10 J 2) 30 J
   3) 50 J 4) 20 J

66. A rope of negligible mass is wound round a hollow cylinder of mass 3 kg and radius 40 cm. What is the angular acceleration of the cylinder, if the rope is pulled with a force of 30 N? What is the linear acceleration of rope? (Assume that there is no slipping)
   1) 25 rad/s^2, 10 m/s^2
   2) 25 rad/s^2, 5 m/s^2
   3) 5 rad/s^2, 20 m/s^2
   4) 5 rad/s^2, 15 m/s^2

67. A flywheel rotating about a fixed axis has a kinetic energy of 360 joules when its angular speed is 30 rad/ sec. The moment of inertia of the wheel about the axis of rotation is
   1) 0.6 kgm^2 2) 0.15 kgm^2
   3) 0.8 kgm^2 4) 0.75 kgm^2

68. The position of a particle is given by $\vec{r} = (\hat{i} + 2\hat{j} - \hat{k})$ and momentum $\vec{P} = (3\hat{i} + 4\hat{j} - 2\hat{k})$. The angular momentum about the origin is
   1) Perpendicular to X-axis
   2) Perpendicular to Y-axis
   3) Perpendicular to Z-axis
   4) A line at equal angles to all the three axes

69. Which of the following (if mass and radius are assumed to be same) have maximum percentage of total K.E. in rotational form while pure rolling?
   1) Disc 2) Sphere
   3) Ring 4) Hollow sphere

70. A particle is confined to rotate in a circular path with decreasing linear speed. Then which of the following is correct ?

1) $\vec{L}$ (angular momentum) is conserved about the centre
2) only direction of angular momentum $\vec{L}$ is conserved
3) it spirals towards the centre
4) its acceleration is towards the centre.

71. A spherical solid ball of mass 5 kg and radius 5 cm is rotating about an axis passing through its centre with an angular velocity of 50 rad/sec. The kinetic energy of rotation is

1) 12 J 2) 10 J
3) 7 J 4) 6.25 J

72. A uniform rod AB of length l and mass m is free to rotate about A. The rod is released from rest in the horizontal position. The initial angular acceleration of the rod will be:

1) $\frac{3g}{2\ell}$ 2) $\frac{2g}{3\ell}$
3) $mg\frac{\ell}{2}$ 4) $\frac{3}{2}g\ell$

73. One end of a uniform rod of mass m and length l is clamped. The rod lies on a smooth horizontal surface and rotates on it about the clamped end at a uniform angular velocity ω. The force exerted by the clamp on the rod has a horizontal component

1) $m\omega^2 l$ 2) Zero
3) mg 4) $\frac{1}{2}m\omega^2 l$

74. The M.I. of a body about a given axis is 1.2 kg m^2. Initially the body is at rest. In order to produce a rotational kinetic energy of 1500 joules an angular acceleration of 25 rad/sec^2 must be applied about the axis for a duration of

1) 4 sec 2) 2 sec
3) 8 sec 4) 10 sec

75. A wheel having moment of inertia 2 kg-m^2 about its vertical axis, rotates at the rate of 60 rpm about this axis. The torque which can stop the wheel's rotation in one minute would be:

1) $\frac{2\pi}{15}$ *N-m* 2) $\frac{\pi}{12}$ *N-m*
3) $\frac{\pi}{15}$ *N-m* 4) $\frac{\pi}{18}$ *N-m*

76. A person is standing on the edge of a circular platform, which is moving with constant angular speed about an axis passing through its centre and perpendicular to the plane of platform. If person is moving along radius towards axis of rotation then the angular velocity will :

1) Decrease 2) Remain unchanged
3) Increase 4) Data is insufficient

77. A thin circular ring of mass M and radius 'r' is rotating about its axis with a constant angular velocity ω. Four objects each of mass m, are kept gently at the opposite ends of two perpendicular diameters of the ring. The angular velocity of the ring will become:

1) $\frac{M\omega}{4m}$ 2) $\frac{M\omega}{M+4m}$
3) $\frac{(M+4m)\omega}{M}$ 4) $\frac{(M+8m)\omega}{M+4m}$

Rolling Motion

78. Choose the correct statement.

1) A sphere can do pure rolling on smooth horizontal surface
2) A sphere can't do pure rolling on a fixed smooth wedge
3) Friction can act parallel or antiparallel to the direction of motion
4) All of these

79. Two solid spheres of different mass, radii and density roll down a rough inclined plane under identical situation .their time to come down is independent of their

1) mass 2) radius
3) density 4) all of these

80. The speed of a uniform solid cylinder after rolling down an inclined plane of vertical height H, from rest without sliding is

1) $\frac{\sqrt{gh}}{3}$ 2) $\sqrt{\frac{2gh}{3}}$
3) $\sqrt{gh}$ 4) $\sqrt{\frac{4gh}{3}}$

81. A thin uniform circular ring is rolling down an inclined plane of inclination 30° without slipping. Its linear acceleration along the inclined plane will be

1) $g/4$ 2) $g/2$
3) $2g/3$ 4) $g/3$

82. A solid cylinder and a solid sphere of same mass and radius roll without slipping on a rough inclined plane.
The force of friction is

1) Same for both the bodies
2) More for sphere
3) More for cylinder
4) Double for sphere than for cylinder

83. A solid sphere is rolling without slipping on a horizontal surface. The ratio of its rotational kinetic energy to its translational kinetic energy is:

1) 2/9 2) 2/7
3) 2/5 4) 7/2

84. A disc rolls on an inclined plane without slipping. Ratio of rotational and total kinetic energy is

1) 3 : 1 2) 1 : 3
3) $\sqrt{3}:1$ 4) $1:\sqrt{3}$

85. An object is rolling without slipping on a horizontal surface and its rotational kinetic energy is two-third of translational kinetic energy. The body may be.
1) Disc
2) Solid sphere
3) Hollow cylinder
4) Spherical shell

86. An inclined plane is making an angle of 60° with the vertical. A ring rolling down this inclined plane from rest without slipping has an acceleration equal to
1) g/4 2) g/2
3) $\frac{2}{5}g$ 4) $\frac{3}{2}g$

87. A ring of mass 1 kg and diameter 1 m is rolling on a plane road with a speed 2 m/s. Its kinetic energy would be
1) 1 joule 2) 4 joules
3) 2 joules 4) 0.5 joule

88. A disc of mass m and radius r rolls on a horizontal plane without slipping with the speed u. If it starts climbing up an inclined plane, the maximum height it would attain will be
1) $\frac{3u^2}{4g}$ 2) $\frac{4u^2}{3g}$
3) $\frac{2u^2}{3g}$ 4) $\frac{u^2}{3g}$

89. The centre of a wheel rolling on a plane surface moves with a speed v_0. A particle on the rim of the wheel at the same level as the centre will be moving at speed
1) Zero 2) v_0
3) $\sqrt{2}v_0$ 4) $2v_0$

90. For pure rolling, the relation between v_P and v_Q is:

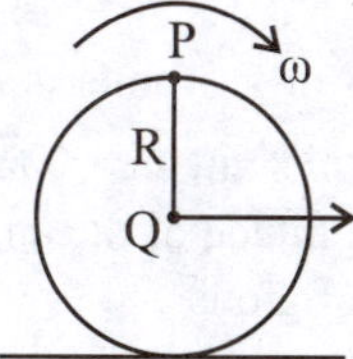

1) $v_P = v_Q$ 2) $v_P = 2v_Q$
3) $v_P = \frac{v_Q}{2}$ 4) None of these

91. If rotational kinetic energy is 50% of total kinetic energy then the body will be:
1) Ring
2) Cylinder
3) Hollow sphere
4) Solid sphere

92. The velocity of centre of mass of a disc rolling on an inclined plane changes from V to 2V. If M is the mass of disc then increase in its kinetic energy will be
1) (9/4) MV^2 2) $MV^2/2$
3) MV^2 4) $3MV^2$

93. A solid sphere and a solid cylinder having the same mass and radius, roll down the same incline. The ratio of their accelerations will be
1) 15 : 14 2) 21 : 10
3) 5 : 3 4) 3 : 5

94. A uniform thin ring of mass 0.4 kg rolls without slipping on a horizontal surface with a linear velocity of 10 cm/s. The kinetic energy of the ring is
1) 4×10^{-3} joules
2) 4×10^{-2} joules
3) 2×10^{-3} joules
4) 2×10^{-2} joules

95. A uniform sphere is released on a smooth inclined plane from the top. When it moves down, its angular momentum is:
1) Conserved about every point
2) Conserved about the point of contact only
3) Conserved about the centre of the sphere only
4) Conserved about any point on a fixed line parallel to the inclined plane and passing through the centre of the ball.

96. A force F is applied on a sphere as shown in the figure. For pure accelerated rolling, the force of friction will be towards

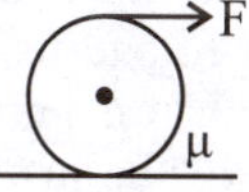

1) Forward direction
2) Backward direction
3) Normal to the surface
4) Zero

97. If KE of rolling motion of a body is K then its rotational KE about its CM of a body is
1) $K + Mv_{CM}^2/2$ 2) $K - Mv_{CM}^2/2$
3) $K + Mv_{CM}^2$ 4) $K - Mv_{CM}^2$

98. When a sphere having moment of inertia I about its centre of gravity and mass 'm' rolls from rest down an inclined plane without slipping. its kinetic energy is (symbols have usual meaning)
1) $\frac{1}{2}I\omega^2$ 2) $\frac{1}{2}mv^2$
3) $I\omega + mv$ 4) $\frac{1}{2}I\omega^2 + \frac{1}{2}mv^2$

High Order Time Intensive MCQs

1. The moment of inertia of two equal masses each of mass m at separation L connected by a rod of mass M, about an axis passing through centre and perpendicular to length of rod is

1) $\frac{(M+3m)L^2}{12}$ 2) $\frac{(M+6m)L^2}{12}$

3) $\frac{ML^2}{4}$ 4) $\frac{ML^2}{12}$

2. The moment of inertia of a circular disc of radius 2m and mass 1 kg about an axis passing through the centre of mass but perpendicular to the plane of the disc is 2 kg-m². Its moment of inertia about an axis parallel to this axis but passing through the edge of the disc is (see the given figure).

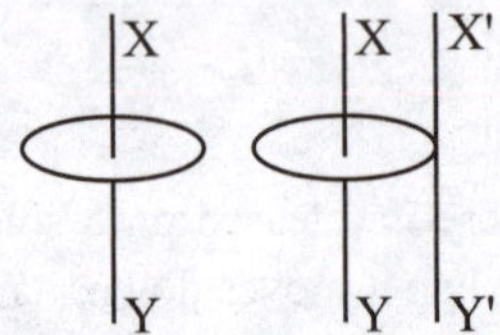

1) 8 kg-m² 2) 4 kg-m²

3) 10 kg-m² 4) 6 kg-m²

3. Two identical concentric rings each of mass m and radius R are placed so that they have their planes perpendicular to each other. What is the moment of inertia about the axis of one of the rings?

1) $\frac{3}{2}MR^2$ 2) $2MR^2$

3) $3\ MR^2$ 4) $\frac{1}{4}MR^2$

4. A circular platform is mounted on a frictionless vertical axle. Its radius is 2 m and its moment of inertia about the axle is 200 kg-m². It is initially at rest. A 50 kg man stands on the edge of the platform and begins to walk along the edge at the speed of 1 ms⁻¹ relative to the ground. Time taken (in sec) by the man to complete one revolution is

1) π 2) $\frac{3\pi}{2}$

3) 2π 4) $\frac{\pi}{2}$

5. A particle of mass m is projected with a velocity v making an angle of 45° with the horizontal. The magnitude of angular momentum of the projectile about the point of projection when the particle is at maximum height h is

1) Zero 2) $\frac{mv^3}{4\sqrt{2}g}$

3) $mv(2gh^3)$ 4) $m\ (2gh^3)$

6. Two persons of masses 55 kg and 65 kg respectively, are at the opposite ends of a boat. The length of the boat is 3.0 m and weight is 100 kg. The 55 kg man walks up to the 65 kg man and sits with him. If the boat is in still water, the centre of mass of the system shifts by

1) 3.0 m 2) 2.3 m

3) Zero 4) 0.75 m

7. A particle of mass m moves in the XY-plane with a velocity v along the straight line AB. If the angular momentum of the particle with respect to origin O is L_A when it is at A and L_B when it is at B, then

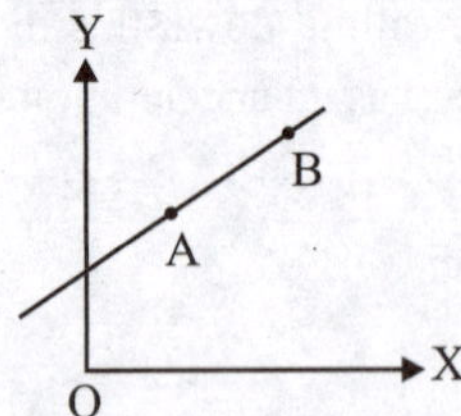

1) $L_A > L_B$

2) $L_A = L_B$

3) The relationship between L_A and L_B depends upon the slope of the line AB

4) $L_A < L_B$

8. A solid sphere is rotating about a diameter at an angular velocity ω. If it cools so that its radius reduces to 1/n of its original value, its angular velocity becomes

1) $\frac{\omega}{n}$ 2) $\frac{\omega}{n^2}$

3) $n^2\omega$ 4) $n\omega$

9. A thin rod of mass m and length 2ℓ is made to rotate about an axis passing through its centre and perpendicular to it. If its angular velocity changes from 0 to ω in time t, the torque acting on it is

1) $\frac{m\ell^2\omega}{12t}$ 2) $\frac{m\ell^2\omega}{3t}$

3) $\frac{m\ell^2\omega}{t}$ 4) $\frac{4m\ell^2\omega}{3t}$

10. A particle of mass m = 5 units is moving with a uniform speed, $v = 3\sqrt{2}$ units in the XOY plane along the line y = x + 4. The magnitude of the angular momentum of the particle about the origin is

1) 60 units 2) $40\sqrt{2}$units

3) Zero 4) 30 Units

11. A uniform disc of mass M and radius R is mounted on an axle supported on frictionless bearings. A light cord is wrapped around the rim of the disc and a steady downward pull, T is exerted on the cord. The angular acceleration of the disc is

1) $\frac{MR}{2T}$ 2) $\frac{2T}{MR}$

3) $\frac{T}{MR}$ 4) $\frac{MR}{T}$

12. A cubical block of side 30 cm is moving with velocity 2 ms^{-1} on a smooth horizontal surface. The surface has a bump at a point O as shown in the figure.

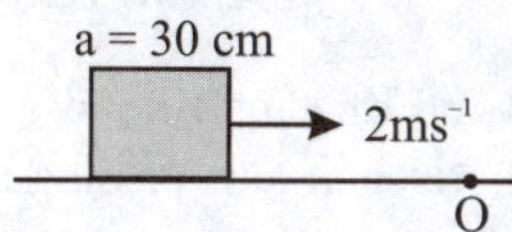

The angular velocity (in rad/s) of the block immediately after it hits the bump is (assume the collision to be perfectly inelastic)

1) 9.4
2) 6.7
3) 5.0
4) 13.3

13. Three bodies, a ring, a solid cylinder and a solid sphere roll down the same inclined plane without slipping. They start from rest. The radii of the bodies are identical. Which of the bodies reach the ground with maximum velocity?

1) Solid sphere
2) Ring
3) Solid cylinder
4) All will have same velocity

14. An object of mass m is attached to light string which passes through a hollow tube. The object is set into rotation in a horizontal circle of radius r_1. If the string is pulled, shortening the radius to r_2, the ratio of new kinetic energy to the original kinetic energy is

1) $\left(\frac{r_2}{r_1}\right)^2$
2) $\left(\frac{r_1}{r_2}\right)^2$
3) $\frac{r_1}{r_2}$
4) $\frac{r_2}{r_1}$

15. A solid cylinder is rolling down on an inclined plane of angle θ. The coefficient of static friction between the plane and cylinder is μ_s, then condition for the cylinder not to slip is

1) $\tan\theta \geq 3\mu_s$
2) $\tan\theta > 4\mu_s$
3) $\tan\theta \leq 3\mu_s$
4) $\tan\theta < \mu_s$

16. A cylinder is rolling down on an inclined plane of inclination 60°. What is its acceleration?

1) g/3
2) $g/\sqrt{3}$
3) $\sqrt{\frac{2g}{3}}$
4) None of the above

17. Two spheres of unequal mass but same radius are released on inclined plane. They roll down with slipping. Which one will reach the ground first?

1) Lighter sphere
2) Heavier sphere
3) Both will reach at the same time
4) None of the above

18. A solid cylinder rolls down an inclined plane of height 3 m and reaches the bottom of plane with angular velocity of $2\sqrt{2}rad^{-1}$. The radius of cylinder must be (take g = 10 ms^{-2})

1) 5 m
2) 0.5 m
3) $\sqrt{10}$ m
4) $\sqrt{7.5}$ m

19. For the given uniform square lamina ABCD, whose centre is O,

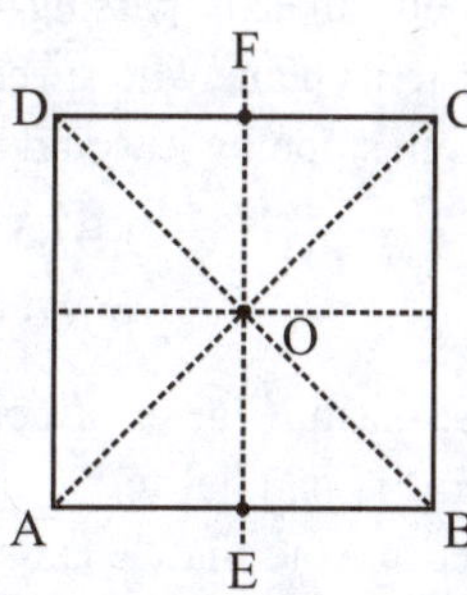

1) $I_{AC} = \sqrt{2}I_{EF}$
2) $\sqrt{2}I_{AC} = I_{EF}$
3) $I_{AC} = 3I_{EF}$
4) $I_{AC} = I_{EF}$

20. A solid cylinder of mass M and radius R rolls without slipping down an inclined plane making an angle θ with the horizontal. Then its acceleration is:

1) $\frac{1}{3}g\sin\theta$
2) $\frac{2}{3}g\sin\theta$
3) $\frac{2}{5}g\sin\theta$
4) $\frac{2}{7}g\sin\theta$

21. When a solid sphere rolls without slipping down an inclined plane making an angle θ with the horizontal, the acceleration of its centre of mass is a. If the same sphere slides without friction, its acceleration a′ will be:

1) $\frac{7}{2}a$
2) $\frac{5}{7}a$
3) $\frac{7}{5}a$
4) $\frac{5}{2}a$

22. A ring of radius R is rotating with an angular speed ω_0 about a horizontal axis. It is placed on a rough horizontal table. The coefficient of kinetic friction is μ_k. The time after which it starts rolling is:

1) $\frac{\omega_0\mu_k R}{2g}$
2) $\frac{\omega_0 R}{2\mu_k R}$
3) $\frac{2\omega_0 R}{\mu_k g}$
4) $\frac{\omega_0 R}{2\mu_k g}$

23. A uniform rod of mass M and length l is pivoted at one end so that it can rotate in a vertical plane (see figure). There is negligible friction at the pivot. The free end is held vertically above the pivot and then released. The angular acceleration of the rod when it makes an angle θ with the vertical is

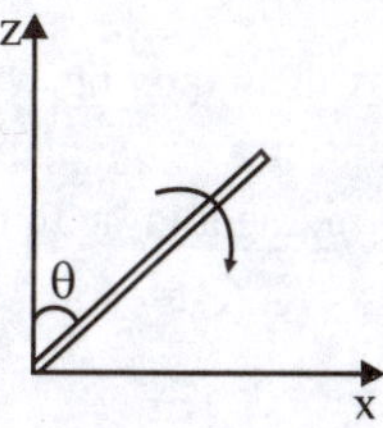

1) $\frac{3g}{2l}\sin\theta$
2) $\frac{2g}{3l}\sin\theta$
3) $\frac{3g}{2l}\cos\theta$
4) $\frac{2g}{3l}\cos\theta$

24. A circular disc has a mass of 1 kg and radius 40 cm. It is rotating about an axis passing through its centre and perpendicular to its plane with speed of 10 rev/s. The work done in joules in stopping it would be ($\pi^2 \approx 10$):

1) 4 2) 47.5
3) 79 4) 160

25. A 100 cm rod is moving on a horizontal surface. At an instant, when it is parallel to the x-axis its ends A and B have velocities 30 cm/s and 20 cm/s as shown in the figure. Find velocity of its center.

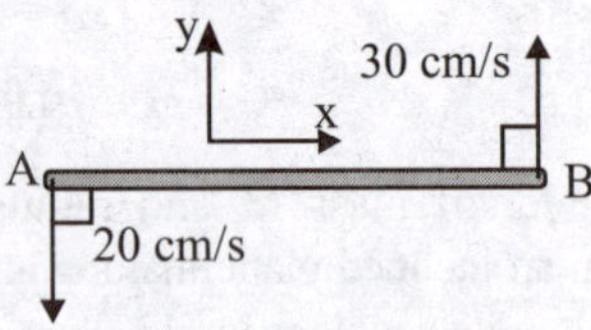

1) $0.5\hat{j}$ cm/s 2) $50\hat{j}$ cm/s
3) $0.05\hat{j}$ cm/s 4) $5.0\hat{j}$ cm/s

26. A hoop of radius r and mass m rotating with an angular velocity ω_0 is placed on a rough horizontal surface. The initial velocity of the centre of the hoop is zero. What will be the velocity of the centre of the hoop when it ceases to slip?

1) $\frac{r\omega_0}{4}$ 2) $\frac{r\omega_0}{3}$
3) $\frac{r\omega_0}{2}$ 4) $r\omega_0$

27. A mass m is supported by a massless string wound around a uniform hollow cylinder of mass m and radius R. If the string does not slip on the cylinder, with what acceleration will the mass fall on release ?

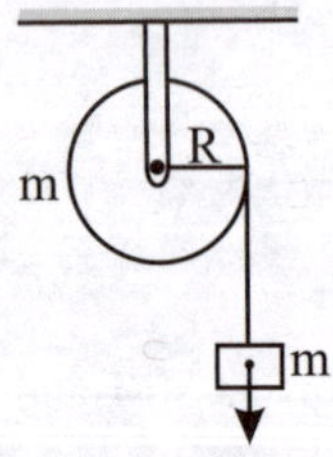

1) $\frac{2g}{3}$ 2) $\frac{g}{2}$
3) $\frac{5g}{6}$ 4) g

28. A particle of mass m is moving with a constant velocity v parallel to x-axis in an x–y plane as shown in figure. Calculate angular momentum with respect to origin at any instant.

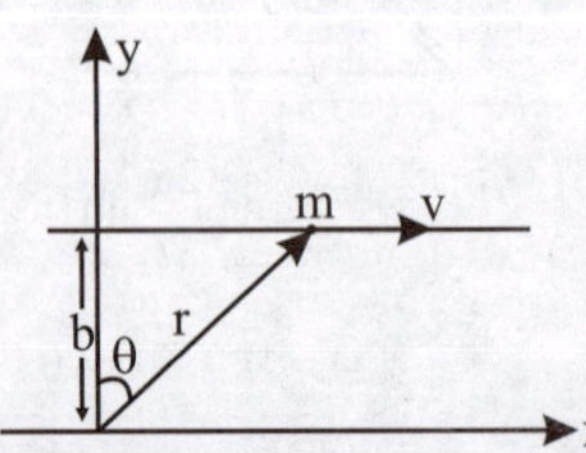

1) $-mvb\hat{k}$ 2) Zero
3) $\frac{mvb}{2}\hat{k}$ 4) $mvb\cos\theta\hat{k}$

29. A thin uniform rod of length l and mass m is swinging freely about a horizontal axis passing through its end. Its maximum angular speed is ω. It centre of mass rises to a maximum height of .

1) $\frac{l^2\omega^2}{6g}$ 2) $\frac{l^2\omega^2}{2g}$
3) $\frac{l^2\omega^2}{3g}$ 4) $\frac{l\omega}{g}$

30. A circular disc of mass M and radius R is rotating with an angular velocity ω about an axis passing through its centre and perpendicular to the plane of the disc. A small point like object of mass m detaches from the rim of the disc and continues to move with same angular speed. The angular velocity of remaining disc now will become

1) $\left(\frac{M-2m}{M+m}\right)\omega$ 2) $\left(\frac{M+2m}{M+m}\right)\omega$
3) ω 4) $\left(\frac{M+2m}{M-m}\right)\omega$

31. Two blocks of masses 10 kg and 4 kg are connected by a spring of negligible mass and placed on a frictionless horizontal surface. An impulse gives a velocity of 14 m/s to the heavier block in the direction of the lighter block. The velocity of the centre of mass is

1) 30 m/s 2) 20 m/s
3) 10 m/s 4) 5 m/s

32. A uniform rod hinged at its one end (O) is allowed to rotate in vertical plane. Rod is given an angular velocity ω in its vertical position as shown in figure. The value of ω for which the force exerted by the hinge on rod is zero in this position is:

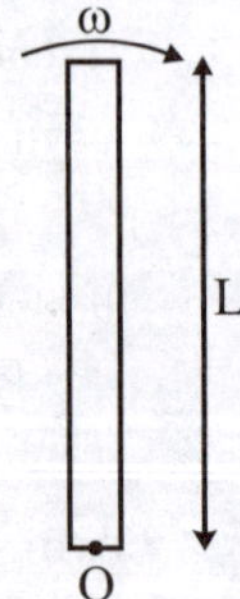

1) $\sqrt{\frac{g}{L}}$ 2) $\sqrt{\frac{2g}{L}}$
3) $\sqrt{\frac{g}{2L}}$ 4) $\sqrt{\frac{3g}{L}}$

33. A body is rolling without slipping on a moving plank as shown in figure. Relation of v and v_1 is .

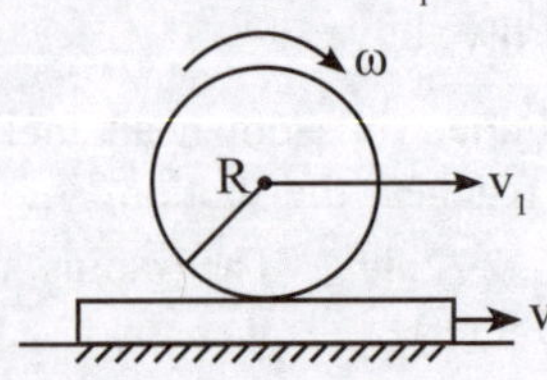

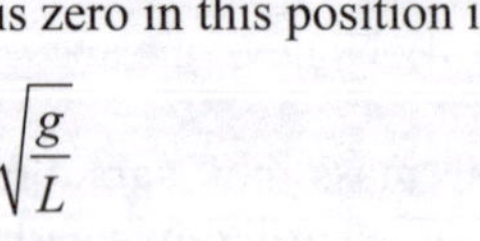

1) $v_1 - v = R\omega$ 2) $v_1 + v = R\omega$
3) $v_1 + v = -R\omega$ 4) None of these

Statement Based MCQs

Directions: Each question consist of two statements, Statement-I and Statement-II. Choose any one of the following four responses.

1) **Both Statement-I and Statement-II are correct.**
2) **Both Statement-I and Statement-II are incorrect.**
3) **Statement-I is correct & Statement-II is incorrect.**
4) **Statement-I is incorrect & Statement-II is correct.**

1. **Statement-I:** Torque acting on a body is the product of force on the body and the perpendicular distance of the force from the axis of rotation.

   Statement-II: In the absence of any external torque, the total angular momentum of the system remains conserved.

2. **Statement-I:** A ballet dancer folds her arms and legs to decrease her moment of inertia.

   Statement-II: Moment of inertia of a uniform solid sphere is $\frac{2}{5}MR^2$.

3. **Statement-I:** To unscrew a rusted nut, we need a wrench with longer arm.

   Statement-II: Wrench with longer arm increases the force needed to rotate on the arm.

4. **Statement-I:** Moment of inertia of a rod about an axis passing through its centre is $ml^2/3$, where m is mass of the rod and l is length of the rod.

   Statement-II: Torque on rod about centre of mass due to gravitational force is zero.

5. **Statement-I:** A helicopter must necessarily have two propellers.

   Statement-II: Two propellers are provided in helicopter in order to conserve linear momentum.

Assertion & Reason MCQs

Directions: Each question consist of two statements, Assertion and Reason. Choose any one of the following four responses.

1) **Both Assertion and Reason are True and the Reason is correct explanation of the Assertion.**
2) **Both Assertion and Reason are True but Reason is not correct explanation of the Assertion.**
3) **Assertion is True but the Reason is False.**
4) **Assertion is False but Reason is True.**

1. **Assertion:** The spokes near the top of a rolling bicycle wheel look more blurred than those near the bottom of the wheel.

   Reason: The spokes near the top of wheel move faster than those near the bottom of the wheel.

2. **Assertion:** A ladder is more likely to slip when a person is near the top than when he is near the bottom.

   Reason: The friction between the ladder and the floor decreases as he climbs up.

3. **Assertion:** The hard boiled egg and raw egg can be distinguished by spinning them.

   Reason: While spinning the moment of inertia of hard boiled egg is more as compared to raw egg.

4. **Assertion:** Many great rivers flow towards the equator. The sediments that they carry, decreases the angular speed of the the earth about its own axis.

   Reason: The angular momentum of the earth about its rotation axis is conserved, therefore due to deposition of sediments moment of inertia of the earth increases and angular speed decreases.

5. **Assertion:** In the case of pure rolling, force of friction acting on the body must be zero.

   Reason: In case of sliding motion on a rough surface, force of friction acts on the body.

Match the Columns MCQs

1. A uniform sphere of mass M and radius R is placed on a rough horizontal surface (Fig.). The sphere is struck horizontally at a height h from the floor. Match the Column-I with Column-II.

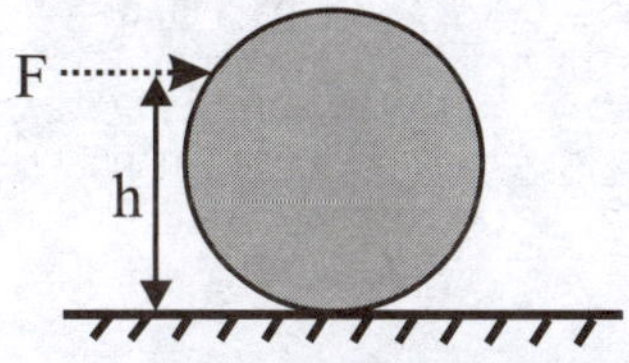

	Column-I		Column-II
(a)	$h = \frac{R}{2}$	(i)	Sphere rolls without slipping with a constant velocity and no loss of energy
(b)	$h = R$	(ii)	Sphere spins clockwise, loses energy by friction.
(c)	$h = \frac{3}{2}R$	(iii)	Sphere spins anti-clockwise, loses energy by friction.
(d)	$h = \frac{7}{5}R$	(iv)	Sphere has only translational motion, loses energy by friction.

1) (a)-(iii) (b)-(iv) (c)-(ii) (d)-(i)
2) (a)-(iv) (b)-(i) (c)-(iii) (d)-(ii)
3) (a)-(ii) (b)-(iii) (c)-(i) (d)-(iv)
4) (a)-(i) (b)-(ii) (c)-(iv) (d)-(iii)

2. Match the rotational motion quantities to their translation motion counterparts:

	Column-I		Column-II
(a)	Torque	(i)	Mass
(b)	Moment of Inertia	(ii)	Linear Momentum
(c)	Angular momentum	(iii)	Velocity
(d)	ω	(iv)	Force

1) (a)-(i) (b)-(iv) (c)-(iii) (d)-(ii)
2) (a)-(iv) (b)-(i) (c)-(ii) (d)-(iii)
3) (a)-(i) (b)-(ii) (c)-(iii) (d)-(iv)
4) (a)-(iii) (b)-(iv) (c)-(ii) (d)-(i)

3. Match the entries of Column-I with their analogues in column-II:

	Column-I		Column-II
(a)	$\frac{1}{2} mv^2$	(i)	$\frac{1}{2} I\omega^2$
(b)	$a = dv/dt$	(ii)	$\tau = (dL/dt)$
(c)	$F = dp/dt$	(iii)	$\alpha = (d\omega/dt)$

1) (a)-(i) (b)-(iii) (c)-(ii)
2) (a)-(iii) (b)-(i) (c)-(ii)
3) (a)-(i) (b)-(ii) (c)-(iii)
4) (a)-(ii) (b)-(iii) (c)-(i)

4. Match the entries of Column-I with Column-II.

	Activity		Type of Motion
(a)	Movement of the tip of the minute hand of a clock	(i)	Rotational motion
(b)	A stone falling from a height in a straight line path	(ii)	Circular motion
(c)	Moving wheel of a sewing machine	(iii)	Linear motion

1) (a)-(i) (b)-(ii) (c)-(iii)
2) (a)-(ii) (b)-(iii) (c)-(i)
3) (a)-(i) (b)-(ii) (c)-(iii)
4) (a)-(iii) (b)-(i) (c)-(ii)

NCERT Exemplar MCQs

1. When a disc rotates with uniform angular velocity, which of the following is not true?
1) The sense of rotation remains same
2) The orientation of the axis of rotation remains same
3) The speed of rotation is non-zero and remains same
4) The angular acceleration is non-zero and remains same

2. A uniform square plate has a small piece Q of an irregular shape removed and glued to the centre of the plate leaving a hole behind in figure. The moment of inertia about the z-axis is then:

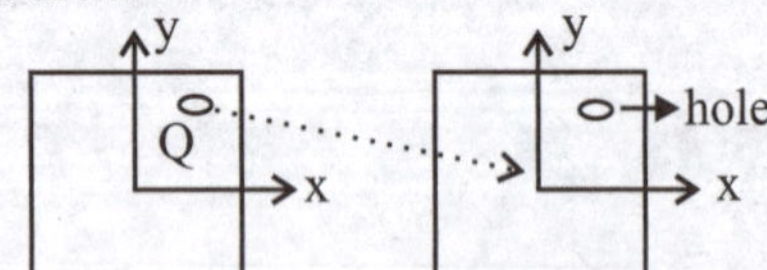

1) Increased
2) Decreased
3) The same
4) Changed in unpredicted manner

3. In problem 2, the C.M of the plate is now in the following quadrant of x – y plane:
1) I 2) II 3) III 4) IV

4. A merry-go-round, made of a ring-Like platform of radius R and mass M, is revolving with angular speed ω. A person of mass M is standing on it. At one instant, the person jumps off the round, radially away from the centre of the round (as seen from the round). The speed of the round afterwards is:
1) 2 ω 2) ω
3) $\frac{\omega}{2}$ 4) 0

NEET & AIIMS Past Year MCQs

1. Let ω_1, ω_2 and ω_3 be the angular speed of the second hand, minute hand and hour hand of a smoothly running analog clock, respectively. If x_1, x_2 and x_3 are their respective angular distances in 1 minute then the factor which remains constant (k) is **(2024 Re)**

1) $\frac{\omega_1}{x_1} = \frac{\omega_2}{x_2} = \frac{\omega_3}{x_3} = k$ 2) $\omega_1 x_1 = \omega_2 x_2 = \omega_3 x_3 = k$
3) $\omega_1 x_1^2 = \omega_2 x_2^2 = \omega_3 x_3^2 = k$ 4) $\omega_1^2 x_1 = \omega_2^2 x_2 = \omega_3^2 x_3 = k$

2. The radius of gyration of a solid sphere of mass 5 kg about XY is 5 m as shown in figure. The radius of the sphere is $\frac{5x}{\sqrt{7}}$ m, then the value of x is: **(2024 Re)**

X

Y

1) 5
2) $\sqrt{2}$
3) $\sqrt{3}$
4) $\sqrt{5}$

3. The moment of inertia of a thin rod about an axis passing through its mid point and perpendicular to the rod is 2400 g cm^2. The length of the 400 g rod is nearly: **(2024)**

1) 20.7 cm 2) 72.0 cm
3) 8.5 cm 4) 17.5 cm

4. A wheel of a bullock cart is rolling on a level road as shown in the figure below. If its linear speed is v in the direction shown, which one of the following options is correct (P and Q are any highest and lowest points on the wheel, respectively)? **(2024)**

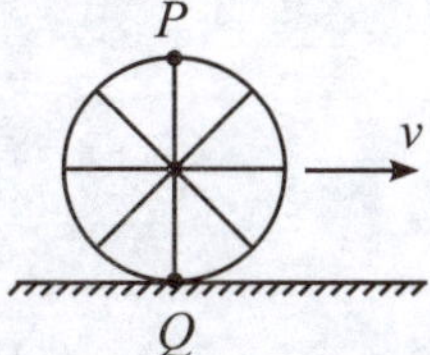

1) Both the points P and Q move with equal speed.
2) Point P has zero speed.
3) Point P moves slower than point Q.
4) Point P moves faster than point Q.

5. The angular acceleration of a body, moving along the circumference of a circle, is: **(2023)**

1) along the axis of rotation
2) along the radius, away from centre
3) along the radius towards the centre
4) along the tangent to its position

6. The ratio of radius of gyration of a solid sphere of mass M and radius R about its own axis to the radius of gyration of the thin hollow sphere of same mass and radius about its axis is: **(2023)**

1) 5 : 2 2) 3 : 5 3) 5 : 3 4) 2 : 5

7. The angular speed of a flywheel moving with uniform angular acceleration changes from 1200 rpm to 3120 rpm in 16 seconds. The angular acceleration in rad/s^2 is: **(2022)**

1) 104π 2) 2π 3) 4π 4) 12π

8. The ratio of the radius of gyration of a thin uniform disc about an axis passing through its centre and normal to its plane to the radius of gyration of the disc about its diameter is: **(2022)**

1) $1:\sqrt{2}$ 2) 2 : 1
3) $\sqrt{2}:1$ 4) 4 : 1

9. Two objects of mass 10 kg and 20 kg respectively are connected to the two ends of a rigid rod of length 10 m with negligible mass. The distance of the centre of mass of the system from the 10 kg mass is: **(2022)**

1) 5m 2) $\frac{10}{3}$m
3) $\frac{20}{3}$m 4) 10 m

10. A uniform rod of length 200 cm and mass 500 g is balanced on a wedge placed at 40 cm mark. A mass of 2 kg is suspended from the rod at 20 cm and another unknown mass 'm' is suspended from the rod at 160 cm mark as shown in the figure. Find the value of 'm' such that the rod is in equilibrium. (g = 10 m/s^2) **(2021)**

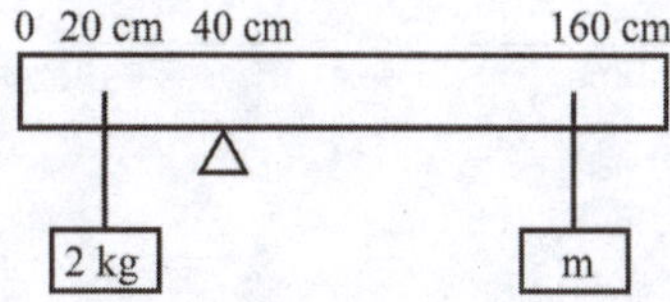

1) $\frac{1}{3}$kg 2) $\frac{1}{6}$kg 3) $\frac{1}{12}$kg 4) $\frac{1}{2}$kg

11. What will be the position of centre of mass of a half disc as shown : **(AIIMS 2018)**

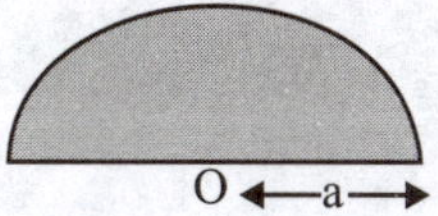

O ←a→

1) $\frac{2a}{\pi}$ 2) $\frac{4a}{3\pi}$ 3) $\frac{a}{\pi}$ 4) $\frac{2a}{3\pi}$

12. In the diagram shown below all three rods are of equal length L and equal mass M. The system is rotated such that rod B is the axis. What is the moment of inertia of the system? **(AIIMS 2007)**

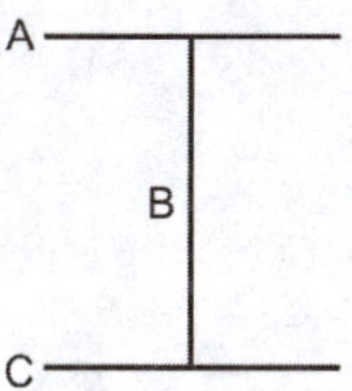

1) ML^2/6 2) 4/3 ML^2
3) ML^2/3 4) 2/3 ML^2

13. If a street light of mass M is suspended from the end of a uniform rod of lenght L in different possible patterns as shown in figure, then: **(AIIMS 2006)**

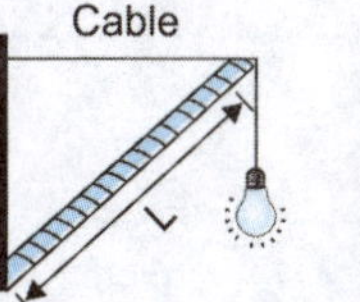

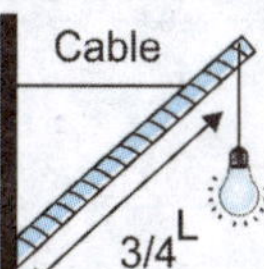

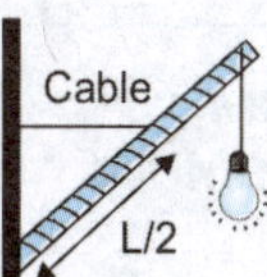

1) Pattern A is more sturdy
2) Pattern B is more sturdy
3) Pattern C is more sturdy
4) All will have same sturdiness

14. A solid sphere is rolling on a frictionless surface, shown in figure with a translational velocity v m/s. If it is to climb the inclined surface then v should be: **(AIIMS 2005)**

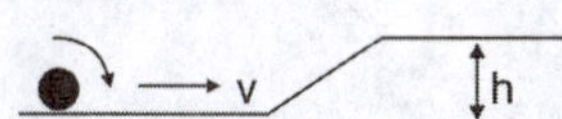

1) $\geq \sqrt{10/7gh}$ 2) $\geq \sqrt{2gh}$

3) 2 gh 4) 10/7 gh

15. A ladder is leaned against a smooth wall and it is allowed to slip on a frictionless floor. Which figure represents trace of its centre of mass: **(AIIMS 2005)**

1) 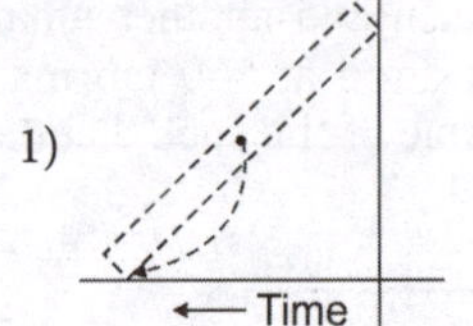

2) 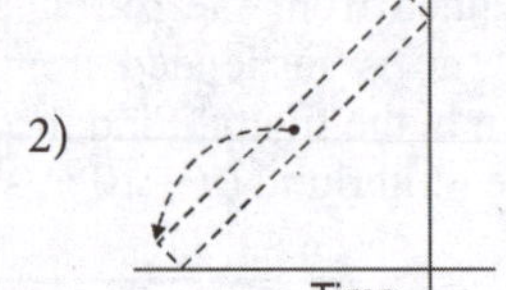

3) 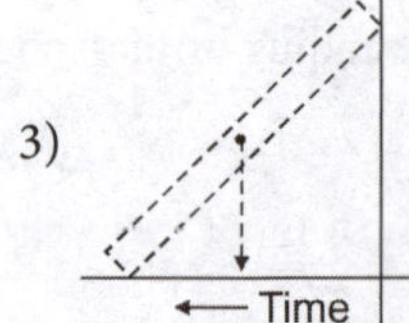

4) 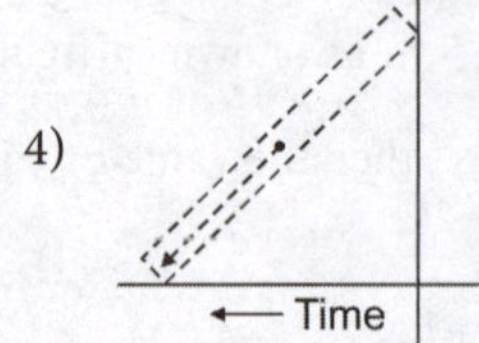

16. Two masses are attached to a rod end to end. If torque is applied, they rotate with angular acceleration α. If their distances are doubled and same torque is applied, then they move with angular acceleration: **(AIIMS 2001)**

1) 4α 2) α

3) 3α 4) $\alpha/4$

Answer Key

NCERT Topic-wise MCQs

Q	1	2	3	4	5	6	7	8	9	10	11	12	13	14	15	16	17
Ans	2	3	2	3	4	1	1	1	2	2	1	1	2	1	4	2	1
Q	18	19	20	21	22	23	24	25	26	27	28	29	30	31	32	33	34
Ans	4	1	4	2	2	4	2	3	2	4	1	1	3	4	2	3	1
Q	35	36	37	38	39	40	41	42	43	44	45	46	47	48	49	50	51
Ans	2	4	4	2	2	3	1	4	1	2	4	1	4	1	2	2	2
Q	52	53	54	55	56	57	58	59	60	61	62	63	64	65	66	67	68
Ans	2	3	1	1	4	3	4	3	2	3	1	1	2	3	1	3	1
Q	69	70	71	72	73	74	75	76	77	78	79	80	81	82	83	84	85
Ans	3	2	4	1	4	2	3	3	2	4	4	4	1	3	3	2	4
Q	86	87	88	89	90	91	92	93	94	95	96	97	98				
Ans	1	2	1	3	2	1	1	1	1	4	1	2	4				

High Order Time Intensive MCQs

Q	1	2	3	4	5	6	7	8	9	10	11	12	13	14	15	16	17
Ans	2	4	1	3	2	3	2	3	2	1	2	3	1	2	3	2	3
Q	18	19	20	21	22	23	24	25	26	27	28	29	30	31	32	33	
Ans	4	4	2	3	4	1	4	4	3	2	1	1	3	3	2	1	

Statement Based MCQs

Q	1	2	3	4	5
Ans	1	1	3	4	3

Assertion & Reason MCQs

Q	1	2	3	4	5
Ans	1	3	3	1	4

Match the Column MCQs

Q	1	2	3	4
Ans	1	2	1	2

NCERT Exemplar MCQs

Q	1	2	3	4
Ans	4	2	3	1

NEET & AIIMS Past Year MCQs

Q	1	2	3	4	5	6	7	8	9	10	11	12	13	14	15	16
Ans	1	4	3	4	1	None	3	3	3	3	2	1	1	1	1	4

CHAPTER 7

Gravitation

Topicwise Number of Questions Analysis (2025-2016)

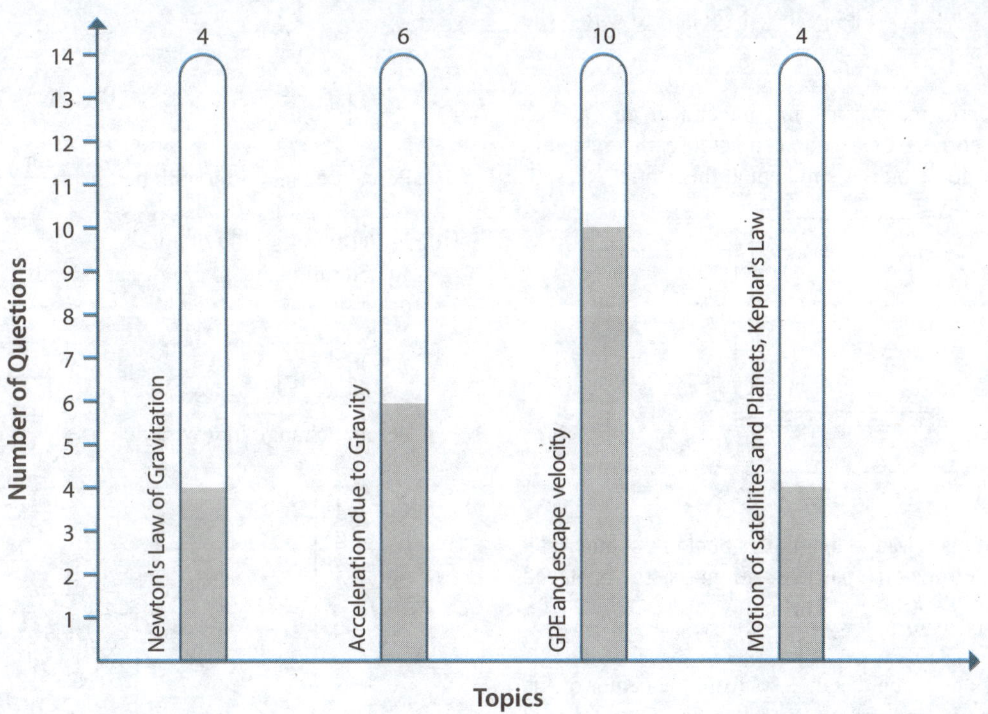

NCERT-PICKS

Newton's Law of Gravitation

Formulas

1. According to Newton's law of gravitation,

$$|\vec{F}| = \frac{GM_1M_2}{R^2}$$

M_1 ——— M_2, separated by R

Note: Gravitational force F acting between two bodies is always attractive. It is independent of the medium between the bodies.

2. The net gravitational force on any particle is the vector sum of individual gravitational forces on it by all other particles in the systems.

$$\vec{F}_{net} = \vec{F}_{12} + \vec{F}_{13} + \vec{F}_{14} + \ldots\ldots \vec{F}_{1n}$$

Numerical Profile

Gravitational force due to the masses

(i) If two particles of equal masses 'm' are placed at the two vertices of an equilateral triangle of side 'b', then the resultant gravitational force on mass 'm' placed at the third vertex is given by

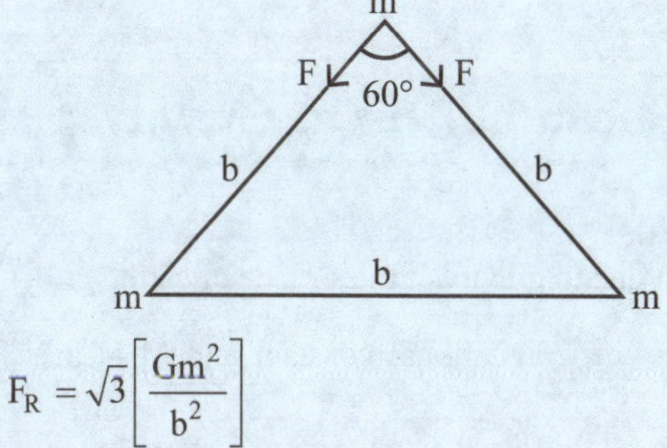

$$F_R = \sqrt{3}\left[\frac{Gm^2}{b^2}\right]$$

(ii) If equal masses each of mass m, are kept at the vertices of a square of length b, the gravitational force of attraction on any one of the particles is

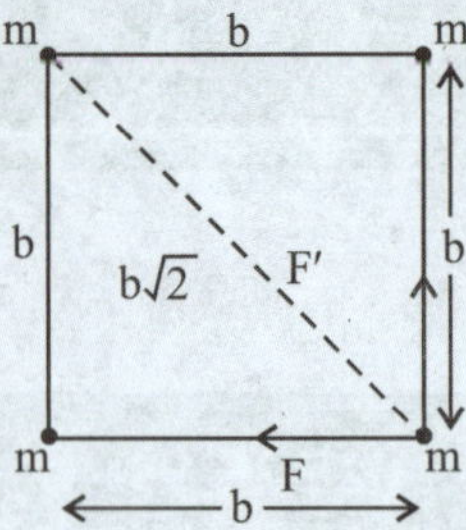

$$F_R = \frac{Gm^2}{b^2}\left(\sqrt{2}+\frac{1}{2}\right)$$ along the diagonal towards the opposite corner.

(iii) If four different masses m_1, m_2, m_3 and m_4 are placed at the four corners of a square of side b the resultant gravitational force on mass m kept at the centre.

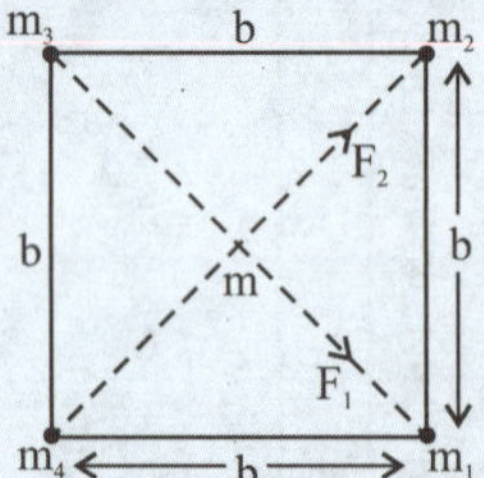

$$F = \frac{2Gm}{b^2}\sqrt{(m_1-m_3)^2+(m_2-m_4)^2}$$

(iv) A thin rod of mass M and length L is bent in a semi circle. Gravitational force on a particle with mass "m" is placed at center of curvature, $F = \frac{GMm}{L^2}2\pi$

(v) A point mass m is at a distance 'y' from the center of the ring of mass M and radius R on its axis. Gravitational force between two is

$$F = \frac{GMmy}{(R^2+y^2)^{\frac{3}{2}}}$$

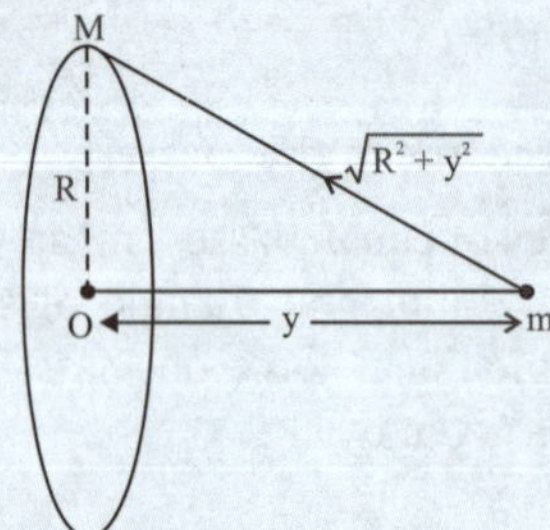

Acceleration Due to Gravity

Formulas

Acceleration due to gravity at the surface of earth, $g = \frac{GM}{R^2}$

[where, G is the gravitational constant and M is mass of the Earth]

The value of g on the earth is taken to be 9.8 ms^{-2}

Numerical Profile

(i) Relation between g and G

$$g = \frac{GM_e}{R_e^2} = \frac{G\frac{4}{3}\pi R_e^3\rho}{R_e^2} = \frac{4}{3}\pi G R_e \rho$$

(ii) Variation of g with altitude h

Acceleration due to gravity at height h from the surface of earth

$$g' = \frac{g}{\left(1+\frac{h}{R}\right)^2}$$

If h << R

$$g' = g\left[1-\frac{2h}{R}\right]$$ [R is the radius of the planet]

Percent change in g with height $= \frac{2h}{g} \times 100\%$

(iii) Variation of g with depth d

Acceleration due to gravity at a depth d below the surface of earth

$$g' = g\left[1-\frac{d}{R}\right]$$

Percent change in g with depth $= \left(\frac{d}{R}\times 100\right)\%$

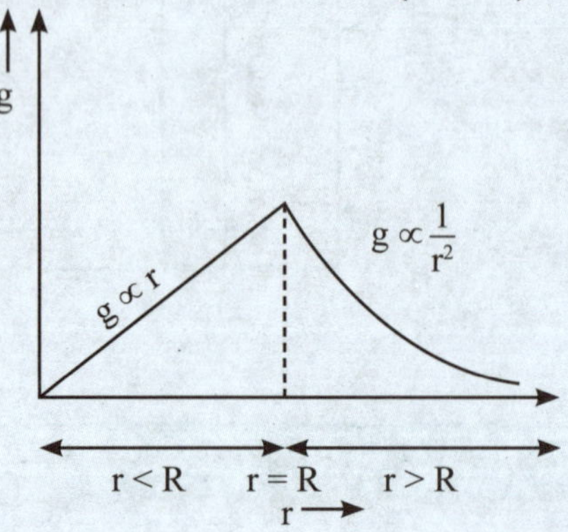

Gravitational Filed Strength (or) Intensity of Gravitational Field

Formulas

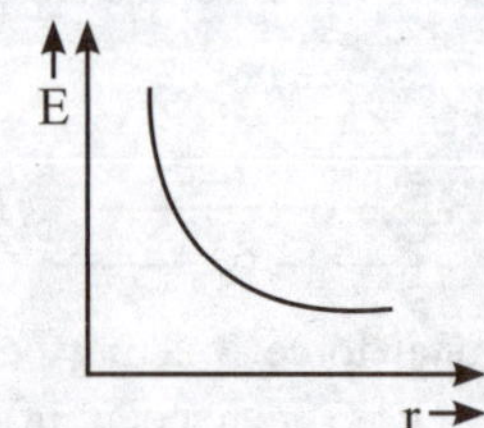

1. Gravitation field strength $\vec{E}_g = \frac{\vec{F}}{m_0}$
2. The intensity of gravitational field at a distance r form a point mass 'M' is given by

$$E_G = \frac{GM}{r^2}$$

Numerical Profile

(i) Gravitational field intensity due to a circular ring at any point on its axis is given by

$$E_g = \frac{GMx}{(x^2+R^2)^{3/2}} \text{ along } \overrightarrow{PO}$$

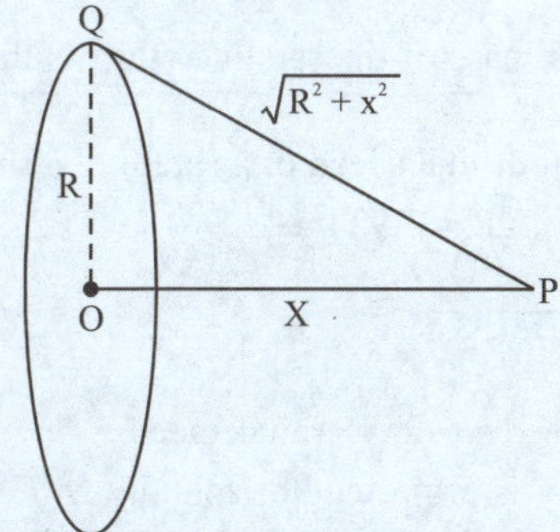

- Gravitational field intensity is directed towards the centre of the circular ring.
- At the centre of the circular ring, $E_g = 0$
- E_g is maximum, at $x = \frac{R}{\sqrt{2}}$ and

$$E_{max} = \frac{2GM}{3\sqrt{3}R^2}$$

(ii) Gravitational field intensity due to a circular disc at any point on the axial line

$$E = \frac{2GM}{R^2}\left(1 - \frac{x}{\sqrt{x^2+R^2}}\right) = \frac{2GM}{R^2}(1-\cos\alpha)$$

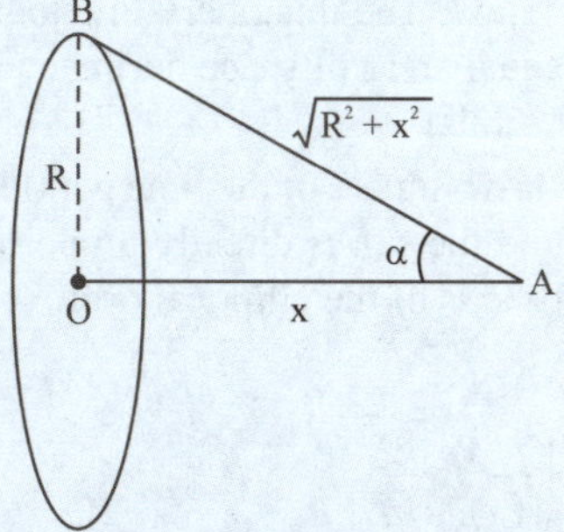

(iii) Gravitational Field Due to Straight Rod

Intensity due to rod of length $2l$, density ρ, placed along x-axis, such that mid point of rod is coincides with origin. The gravitational field intensity at a point $P(a, o)$ is

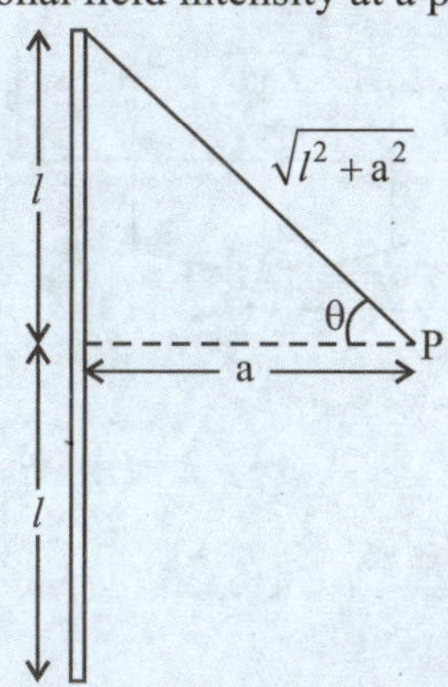

$$E_g = \frac{2G\rho}{a}\frac{l}{\sqrt{l^2+a^2}} \Rightarrow E_g = \frac{2G\rho}{a}\sin\theta$$

Above expression can be written as

$$E_g = \frac{2G\rho}{a}\frac{1}{\sqrt{1+\frac{a^2}{l^2}}}$$

$$E_g = \frac{2G\rho}{a}\left(1 - \frac{a^2}{2l^2} + high\ powers\ of\ \frac{a^2}{l^2} + - - - - -\right)$$

If l is ∞ then $E_g = \frac{2G\rho}{a}$

(iv) Gravitational field due to a thin uniform spherical shell

(a) At a point inside the spherical shell, E = 0

(b) At a point on the surface of spherical shell $E = \frac{GM}{R^2}$

(c) At a point outside the spherical shell, $E = \frac{GM}{r^2}$

(v) Field due to Solid Sphere(uniform mass density):

(a) At a point inside surface of solid sphere

$$E = \frac{GMx}{R^3} \text{ (for } x < R)$$

(b) At a point on the surface of the solid sphere,

$$E = \frac{GM}{R^2} \text{ (for } x = R)$$

(c) At a point outside the solid sphere,

$$E = \frac{GM}{x^2} \text{ (for } x > R)$$

Gravitational Potential and Energy

Formulas

1. Potential due to a point mass at a distance r, $V = \frac{-GM}{r}$
2. Potential due to uniform ring on its axis r, $V = \frac{-GM}{\sqrt{(a^2+r^2)}}$

[r is the distance of the point from the centre of the ring, V is the potential due to the ring at a point on its axis and a is the radius of the ring].

Numerical Profile

(i) Potential due to uniform thin spherical shell of mass M and radius R

(a) At a point outside the shell, $V = -\frac{GM}{r}$ [if r > R]

(b) At a point inside the shell, $V = -\frac{GM}{R}$ [if r < R]

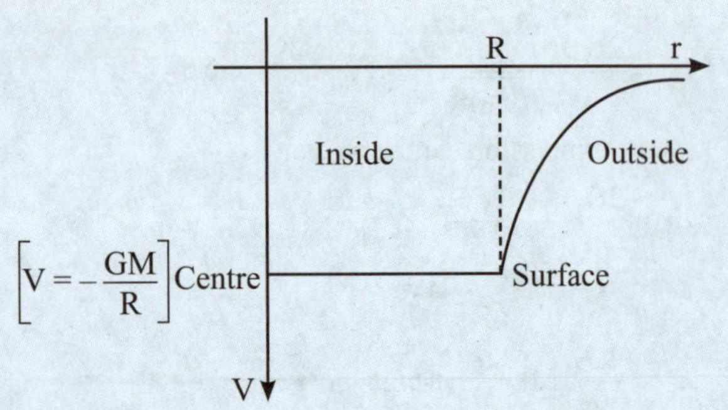

(ii) Potential due to uniform solid sphere of mass M and radius R

(a) At a point outside the sphere, $V = -\frac{GM}{r}$ [if $r > R$]

(b) At a point inside the sphere, $V = -\frac{GM}{2R^3}[3R^2 - r^2]$ [if $r < R$]

[r is the distance of the point from the centre]

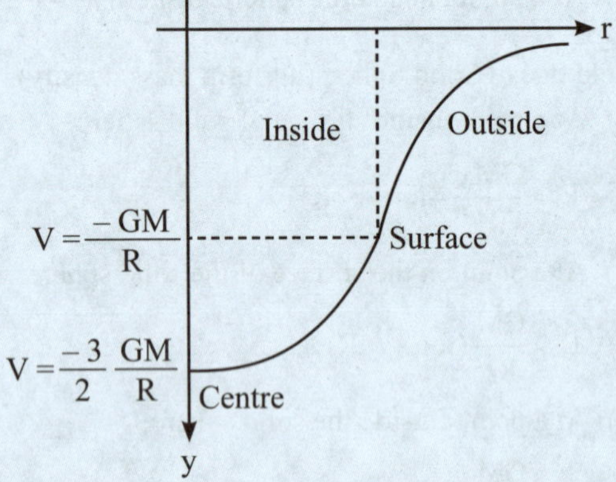

(iii) Relation between Gravitational field and potential

$\vec{E} = -\text{grad } V$

(iv) Gravitational potential energy $U = -\frac{Gm_1m_2}{r}$

[m_1 and m_2 are the masses of the object]

(v) Gravitational potential energy at a height 'h' above the surface of earth

$U_h = \frac{-GMm}{R+h}$

Satellite, Escape velocity

Formulas

1. Escape velocity, $v_e = \sqrt{\frac{2GM}{R}} = \sqrt{2gR}$

[where M is mass of the plannet]

2. Orbital speed of satellite, $v = R\sqrt{\frac{g}{R+h}} = \sqrt{\frac{GM}{R+h}}$

[h is height of satellite from the surface of earth]

3. Time period of revolution of satellite, $T = \frac{2\pi}{R}\sqrt{\frac{(R+h)^3}{g}}$

4. Energies of a satellite

Kinetic energy $(E_K) = \frac{GMm}{2r}$

Potential energy $(E_P) = \frac{-GMm}{r}$

Total energy $(E) = \frac{-GMm}{2r}$

[where m is mass of the satellite and r is the radius of the orbit]

5. The value of orbital speed of a satellite, near the surface of earth is $v = \sqrt{gR} = 7.92 \text{ kms}^{-1}$

Numerical Profile

(i) Escape speed = $\sqrt{2}$ × orbital speed.

(ii) Let, a body is projected horizontally with speed v, from any height from the surface of earth.

If $v > v_e$, the body will escape, following a hyperbolic, path. The speed of body v′ moving in interstellar space is

$v' = \sqrt{v^2 - v_e^2}$

Kepler's Laws of Planetary Motion

Formulas

1. **Kepler's 1st Law:** Every planet revolves around the sun in an elliptical orbit and the sun is situated at one focus of the ellipse.
2. **Kepler's 2nd Law:** The imaginary line joining a planet and sun sweeps equal areas of space during equal time intervals as the planet orbits.
3. **Kepler's 3rd law:** Square of the time period of revolution of a planet around the sun is directly proportional to the cube of semi-major axis of the elliptical orbit.

$T^2 \propto r^3$

$\Rightarrow \left(\frac{T_1}{T_2}\right)^2 = \left(\frac{r_1}{r_2}\right)^3$

Numerical Profile

Analysis of Satellite's Motion in Orbit:

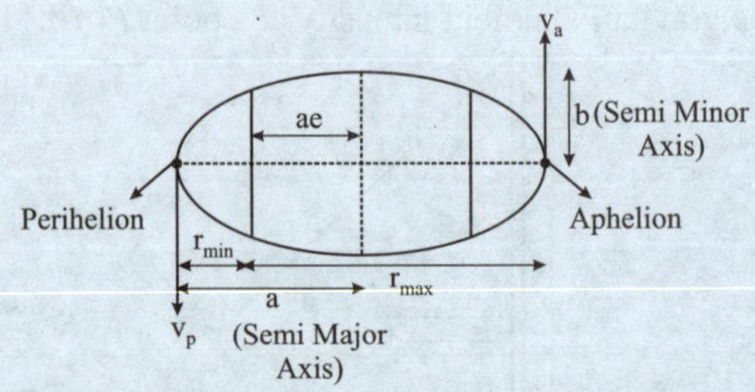

(i) $r_{max} = a(1 + e)$ (ii) $r_{min} = a(1 - e)$

(iii) $\frac{v_p}{v_a} = \frac{v_{max}}{v_{min}} = \frac{1+e}{1-e}$ (where, e is eccentricity.)

Statements Asked in Exam

- Universal Gravitational constant G is a scalar quantity. It is independent of nature and size of the bodies and has same value throughout the universe. Value G in S.I system $= 6.67 \times 10^{-11}\ Nm^2kg^{-2}$
- Escape velocity of a body is independent of mass of the body and the angle of projection.
- Escape velocity is equal for an ant as well as for an elephant.
- If by some miracle G becomes one tenth, the gravitational force will become one tenth and then we will be able to jump ten times higher than the present value.
- Gravitational intensity at a point inside a hollow spherical shell is zero.
- In case of a solid sphere ,the gravitational intensity at an internal point (r < R) varies linearly with r, according to relation

$$E = \frac{GM'}{r^2} = \frac{G}{r^2} \times \frac{4}{3}\pi r^3 p = \frac{G}{R^2} \times \frac{4\pi}{3} r^3 \times \frac{M}{\frac{4}{3}\pi R^3}$$

$$= \frac{GM}{R^3} r; \text{ i.e., } E \propto r$$

At the centre of earth r = 0; therefore. $E_{centre} = 0$

- Work done on a satellite at any instant is zero as the gravitational force is always perpendicular to the displacement.
- When the total energy of a satellite is zero or positive it will escape away from its orbit and its path becomes parabolic.
- The direction of orbital velocity of a satellite at an instant is along the tangent to the orbital path at that instant.
- The orbital speed of a satellite decreases with an increase in the radius of orbit or increase in height from the surface of earth/planet.
- If the speed of a satellite orbiting the earth is increased by 41.4% or its kinetic energy is doubled, it will escape away from the gravitational field of earth.
- For h << R the rate of decrease of the acceleration due to gravity with height is twice as compared to that with depth.
- The acceleration due to gravity increases from equator to pole due to rotation of earth.
- The value of acceleration due to gravity $[g = Gm/(R^2)]$ is minimum at mercury and maximum at Jupiter.
- The value of g at a location gives the value of intensity of gravitational field at that point
- For a satellite, orbiting close to the surface of earth (h << R), the time period of revolution is,

$$T = 2\pi\sqrt{R/g} = 84.6 \text{ minutes.}$$

- Total energy of satellite = E

$$E = -E_K = \frac{1}{2}E_P$$

[Where E_K: kinetic energy of satellite and E_P: Potential energy of satellite]

- The energy required to remove the statellite from its orbit around the earth to infinity is called Binding Energy.

$$\text{Binding Energy} = \frac{GMm}{2r}$$

- When the total energy of satellite is negative. It will be moving in either a circular or an elliptical orbit.
- When the total energy of satellite is positive, it will escape away from the orbit following a hyperbolic path.
- When the speed of satellite is increased, its total energy will increase and it will start orbiting in a circular path of larger radius.

NCERT Topic-wise MCQs

Newton's Law of Gravitation

1. If there was a reduction in gravitational effect, then which of the following do you think would change in some respect?
   1) Magnetic fore 2) Electrostatic force
   3) Viscous force 4) Archimedes uplift

2. If the distance between sun and earth is increased by 3 times, then attraction between the two will be
   1) Remains constant 2) Decrease by 63%
   3) Increase by 63% 4) Decreases by 89%

3. One can easily weight the earth by calculating the mass of earth by using formula
   1) $\frac{GR_E^2}{g}$ 2) $\frac{gR_E^2}{G}$
   3) $\frac{gR_E}{G}$ 4) $\frac{GR_E^3}{g}$

4. Two spherical bodies of mass M and 5M and radii R and 2R respectively are released in free space with initial separation between their centres equal to 12R. If they attract each other due to gravitational force only, then the distance covered by the smaller body just before collision is
   1) 2.5R 2) 4.5 R
   3) 7.5R 4) 1.5 R

5. Two spheres of mass m and M are situated in air and the gravitational force between them is F. The space around the masses is now filled with a liquid of specific gravity 3. The gravitational force will now be
   1) $\frac{F}{3}$ 2) $\frac{F}{9}$
   3) 3F 4) F

Acceleration Due to Gravity; Variation of g With Height &Depth

6. Determine the speed with which the earth would have to rotate on its axis so that a person on the equator would weigh 3/5th as much as at present. Take the equatorial radius as 6400 km.
   1) 6.7×10^{-4} rad s^{-1} 2) 7.8×10^{-4} rad s^{-1}
   3) 8.7×10^{-4} rad s^{-1} 4) 4.5×10^{-4} rad s^{-1}

7. A planet has same density and same acceleration due to gravity as of earth and universal gravitational constant G is twice of earth. The ratio of their radius
   1) 1 : 4
   2) 1 : 5
   3) 1 : 2
   4) 3 : 2

8. Find the percentage decrease in the weight of a body when taken to a height of 32 km above the surface of the earth. Radius of the earth is 6400 km
   1) 1% 2) 5%
   3) 2.5% 4) 8%

9. What will happen to the weight of the body at the south pole if the earth stops rotating about its polar axis?
   1) No change 2) Increases
   3) Decreases but no zero 4) Reduces to zero

10. The height at which weight of the body become $\frac{1}{16}$ its weight on the surface of earth, is
    1) 5R
    2) 15 R
    3) 3R
    4) 4R

11. What will be the acceleration due to gravity at a depth d, where g is acceleration due to gravity on the surface of earth?
    1) $\frac{g}{\left[1+\frac{d}{R}\right]^2}$
    2) $g\left[1-\frac{2h}{R}\right]$
    3) $\frac{g}{\left[1-\frac{d}{R}\right]^2}$
    4) $g\left[1-\frac{d}{R}\right]$

12. The height vertically above earth's surface at which the acceleration due to gravity becomes 1% of its value at the surface is
    1) 8R 2) 9R
    3) 10R 4) 20R

13. Which of the following graphs shows the variation of acceleration due to gravity g with depth h from the surface of the earth?

1) 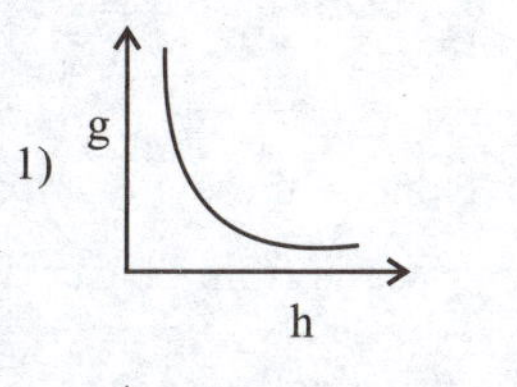

2) 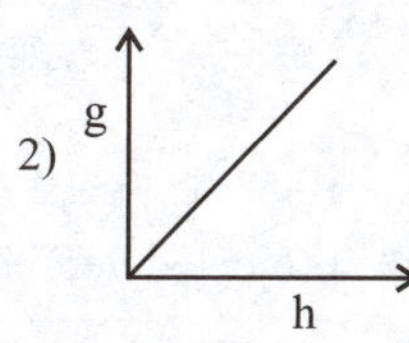

3) 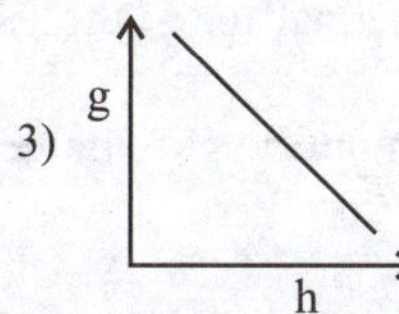

4) 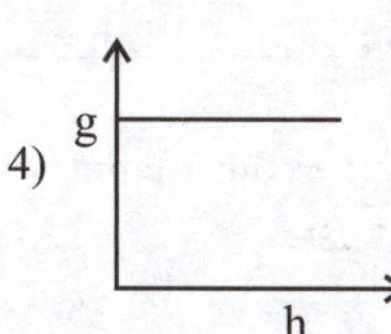

14. At what altitude (h) above the earth's surface would the acceleration due to gravity be 1/4 of its value at the earth's surface?

1) h = R 2) h = 4R
3) h = 2R 4) 16R

15. At what height above the surface of the earth the value of acceleration due to gravity would be half of its value on the surface of earth? [radii of earth = 6400 km]

1) 2561 km 2) 2650 km
3) 3200 km 4) 9800 km

16. Two planets have radii r_1 and r_2 and densities d_1 and d_2 respectively. Then the ratio of acceleration due to gravity on them is

1) $r_1d_1 : r_2d_2$ 2) $r_1d_2 : r_2d_1$
3) $r_1^2d_1 : r_2^2d_2$ 4) $r_2^2d_1 : r_1^2d_2$

17. Find the ratio of acceleration due to gravity g at depth d and height h, where d = 2h.

1) 1 : 1 2) 1 : 2
3) 2 : 1 4) 1 : 4

18. How far away from the surface of earth does the acceleration due to gravity become 4% of its value on the surface of earth? Radius of earth = 6400 km.

1) 30,500 km 2) 28,600 km
3) 25,600 km 4) 30,400 km

19. Density of newly discovered planet is twice that of earth. The acceleration due to gravity at the surface of the planets is equal to that at the surface of the earth. If the radius of earth is R, then the radius of the planet would be

1) 2R 2) 4R
3) $\frac{R}{4}$ 4) $\frac{R}{2}$

20. The acceleration due to gravity on the planet A is 9 times the acceleration due to gravity on planet B. A man jumps to a height of 2 m on the surface of the planet A, then the maximum height he can attain on planet B would be:

1) 18 m 2) 16 m
3) 20 m 4) 30 m

21. Find the percentage decrease in weight of a body, when taken 16 km below the surface of the earth. Take radius of the earth as 6400 km.

1) 1% 2) 0.25%
3) 0.5% 4) None of these

22. At what height above the earth's surface, the value of g is same as in a mine 80 km deep?

1) 40 km 2) 30 km
3) 20 km 4) 10 km

23. If the rotational speed of earth increases, then weight of the body at the equator

1) Increases 2) Decreases
3) No effect 4) Becomes double

24. The acceleration due to gravity near the surface of a planet of radius R and density d is proportional to

1) $\frac{d}{R^2}$
2) dR^2
3) dR
4) $\frac{d}{R}$

25. Calculate that imaginary angular velocity of the earth (in rad s^{-1}) for which effective acceleration due to gravity at the equator becomes zero? Given radius of the earth = 6400 km and g = 10 ms^{-2}.

1) 2×10^{-3} rad/s 2) 5.5×10^{-3} rad/s
3) 1.25×10^{-3} rad/s 4) 1.5×10^{-3} rad/s

GPE and Escape Velocity; Gravitational Intensity and Gravitational Potential & Escape Velocity

26. A body of mass m is raised to a height 10R from the surface of the earth, the increase in potential energy is

1) $\frac{GMm}{11R}$ 2) $\frac{GMm}{10R}$
3) $\frac{mgR}{11G}$ 4) $\frac{10\,GMm}{11R}$

27. A body of mass m taken from the earth's surface to the height equal to the twice the earth radius R of the earth. The change in potential energy of the body will be

1) 2mgR 2) $\frac{2}{3}$ mgR
3) 3 mg R 4) $\frac{mgR}{3}$

28. A particle is fired with a speed of 20 km/s. The speed with which it will move in interstellar space is

1) 16.5 km/s 2) 11.2 km/s
3) 10 km/s 4) 8.8 km/s

29. Which one of the following plots represents the variation of gravitational field of a particle with distance r due to thin spherical shell of radius R?

1) 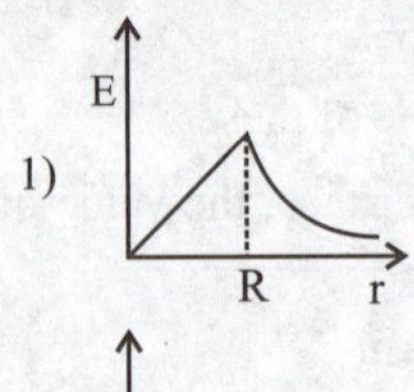
2) 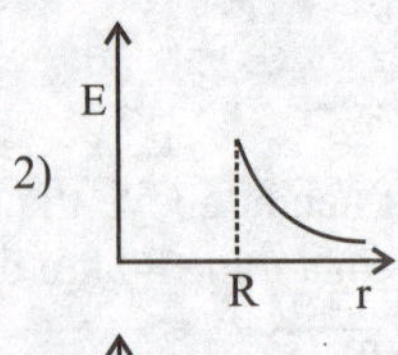
3) 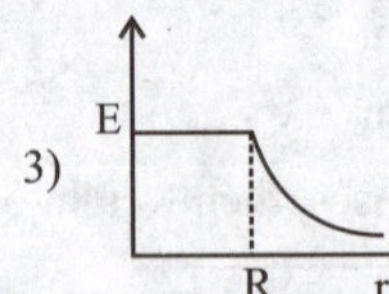
4) 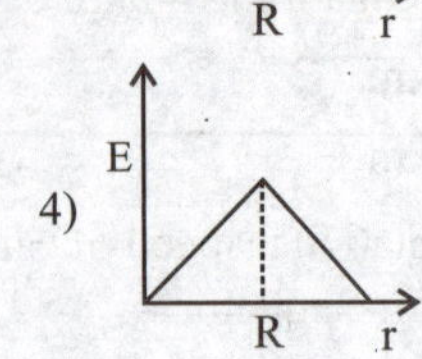

30. Two planets A and B have the same material density. If the radius of A is twice that of B, then the ratio of escape velocity $\frac{v_A}{v_B}$ is

1) 2
2) $\sqrt{2}$
3) $\frac{1}{\sqrt{2}}$
4) $\frac{1}{2}$

31. The escape velocity for the earth is v_e. The escape velocity for a planet whose radius $\frac{1}{4}$th radius of earth and mass half that of earth is

1) $\frac{v_e}{\sqrt{2}}$
2) $\sqrt{2}\,v_e$
3) $2v_e$
4) $\frac{v_e}{2}$

32. Change in the gravitational potential energy when a body of mass m is raised to a height nR above the surface of the earth is

1) $\left[\frac{n}{n+1}\right]$ mgR
2) $\left[\frac{n}{n-1}\right]$ mgR
3) nmgR
4) $\frac{GMm}{R(n+1)}$

33. Energy required to move a satellite of mass m from an orbit of radii 2R to 3R around earth of mass M

1) $\frac{GMm}{3R}$
2) $\frac{GMm}{6R}$
3) $\frac{GMm}{R}$
4) $\frac{GMm}{2R}$

34. The binding energy of a satellite of mass m in an orbit of radius r is

1) $\frac{mgR^2}{r}$
2) $\frac{mgR^2}{2r}$
3) $\frac{-mgR^2}{r}$
4) $\frac{-mgR^2}{2r}$

35. Escape velocity from a planet is v_e. If its mass is increased to 8 times and its radius is increased to 2 times, then new escape velocity would be

1) v_e
2) $\sqrt{2}$
3) $2v_e$
4) $2\sqrt{2}\,v_e$

36. The radius of a planet is double that of the earth but their average densities are the same. If the escape velocities at the planet and at the earth are v_p and v_e respectively, then v_p is equal to.

1) v_e
2) $2v_e$
3) $3v_e$
4) None of these

37. A body is projected vertically upwards from the surface of planet of radius r with a velocity equal to $\frac{1}{3}$ the escape velocity for the planet. The maximum height attained by the body is

1) $\frac{R}{2}$
2) $\frac{R}{3}$
3) $\frac{R}{5}$
4) $\frac{R}{8}$

38. An artificial satellite moving in a circular orbit around the earth has a total (kinetic + potential) energy E_0. Its potential energy is:

1) $-E_0$
2) $1.5E_0$
3) $2E_0$
4) E_0

Motion of Satellites, Planets and Kepler's Law

39. Calculate the value of orbital velocity for an artificial satellite of earth orbiting at height equal to the radius of the earth (6400 km).

1) 5.6 km/s
2) 4.3 km/s
3) 2.8 km/s
4) None of these

40. At a height H from the surface of the earth, the total energy of a satellite is equal to the potential energy of a body of equal mass at a height 3R from the surface of the earth. The value of H is

1) R
2) $\frac{4R}{3}$
3) 3R
4) $\frac{R}{3}$

41. Two satellites A and B go around a planet P in circular orbit having radius 4R and R, respectively. If the speed satellite A is 3v, then speed of satellite B will be

1) 6v
2) 9v
3) 3v
4) None

42. A spherical planet has a mass M_p and diameter D_p. A particle of mass m falling freely near the planet surface will experience an acceleration due to gravity, equals to

1) $\frac{4\,GM_p}{D_p^2}$
2) $\frac{4\,GM_p m}{D_p^2}$
3) $\frac{GM_p m}{D_p^2}$
4) $\frac{4\,GM_p m}{D_p^2}$

43. Average distance of the earth from the sun is L_1. If one year of earth = D days, one year of another planet whose average distance from the sun is L_2 will be

1) $D\left[\frac{L_2}{L_1}\right]^{1/2}$

2) $D\left[\frac{L_2}{L_1}\right]^{3/2}$

3) $\frac{GM_p m}{L_2^2}.D$

4) $\frac{GM_p m}{L_2^2}\frac{D}{L_1}$

44. The Earth moves around the Sun in an elliptical orbit. The ratio $\frac{OA}{OB} = x$. The ratio of speeds at B and A of Earth is:

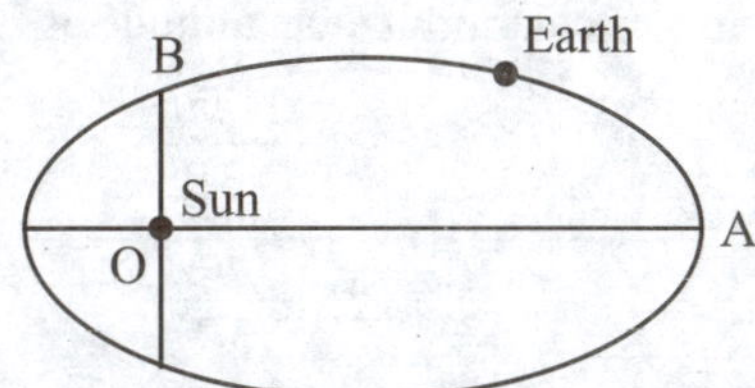

1) $\sqrt{x}$ 2) x

3) x^2 4) $x\sqrt{x}$

45. A space-ship is launched into a circular orbit close to the earth's surface. What additional velocity has now to be imparted to the space-ship in the orbit to overcome the gravitational pull? (Radius of the earth = 6400 km, g = 9.8 ms^{-2}).

1) 3.278 km/s 2) 4.5 km/s

3) 6.5 km/s 4) None of these

46. One planet is orbiting around the sun, At a point P it is at minimum distance d_1 from the sun and at that time speed is v_1. If at an another point Q it is at maximum distance d_2 from the sun then at this point the speed of planet will be:

1) $\frac{d_1^2 v_1}{d_2^2}$ 2) $\frac{d_2 v_1}{d_1}$

3) $\frac{v_1 d_1}{d_2}$ 4) $\frac{d_2^2 v_1}{d_1^2}$

47. The ratio of KE of a planet at the points 1 and 2 is

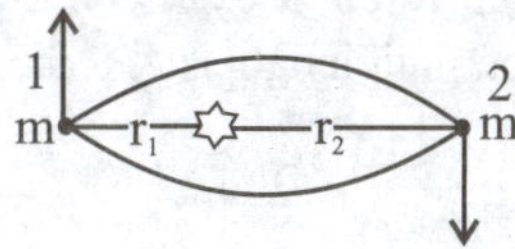

1) $\left(\frac{r_1}{r_2}\right)^2$ 2) $\left(\frac{r_2}{r_1}\right)^2$

3) $\frac{r_1}{r_2}$ 4) $\frac{r_2}{r_1}$

High Order Time Intensive MCQs

1. A satellite is rotating around a planet in the orbit of radius *r* with time period *T*. If gravitational force changes according to $r^{-5/2}$ the T^2 will be:

1) r^3 2) $r^{7/2}$

3) $r^{9/2}$ 4) $r^{3/2}$

2. Two particles of equal mass m go around a circle of radius *R* under the action of their mutual gravitational attraction. The speed of each particle is:

1) $\frac{1}{2R}\sqrt{\frac{1}{Gm}}$ 2) $\sqrt{\frac{Gm}{2R}}$

3) $\frac{1}{2}\sqrt{\frac{Gm}{R}}$ 4) $\sqrt{\frac{4Gm}{R}}$

3. If the gravitational force had varied as $r^{-5/2}$ instead of r^{-2}, the potential energy of a particle at a distance *r* from the centre of the Earth would be proportional to:

1) $r^{3/2}$ 2) $r^{-3/2}$

3) r^{-1} 4) r^{-2}

4. The angular velocity of rotation of a star (of mass *M* and radius *R*), at which the matter starts to escape from its equator, is:

1) $\sqrt{2GM^2/R}$ 2) $\sqrt{2GM/R^3}$

3) $\sqrt{2GM/R}$ 4) $\sqrt{2GR/M}$

5. A mass m is at a distance x from one end of a uniform rod of length L and mass M. The gravitational force on the mass due to the rod is:

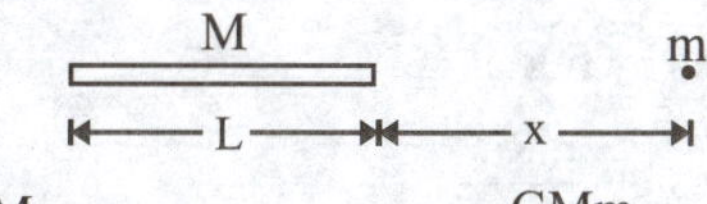

1) $\frac{GMm}{(x+L)^2}$ 2) $\frac{GMm}{x(L+x)}$

3) $\frac{GMm}{x^2}$ 4) $\frac{GmM}{\left(\frac{L}{2}+x\right)^2}$

6. An asteroid of mass m is approaching earth, initially at a distance of 10 R_e with speed v_i. It hits the earth with a speed v_f (R_e and M_e are radius and mass of earth), then:

1) $v_f^2 = v_i^2 + \frac{2Gm}{M_e R}\left(1-\frac{1}{10}\right)$

2) $v_f^2 = v_i^2 + \frac{2GM_e}{R_e}\left(1+\frac{1}{10}\right)$

3) $v_f^2 = v_i^2 + \frac{GM_e}{R_e}\left(\frac{1}{10}-1\right)$

4) $v_f^2 = v_i^2 + \frac{2GM_e}{R_e}\left(1-\frac{1}{10}\right)$

7. The escape velocity of a body is v_e. A body is projected with velocity $2v_e$. With what constant velocity will it move in the inter planetary space?

1) v_e
2) $\sqrt{2}v_e$
3) $\sqrt{3}v_e$
4) $\sqrt{5}v_e$

8. The condition for a uniform spherical mass m of radius r to be a black hole is:
(G = gravitational constant and g = acceleration due to gravity):

1) $\left(\frac{2Gm}{r}\right)^{1/2} \le c$
2) $\left(\frac{2gm}{r}\right)^{1/2} = c$
3) $\left(\frac{2Gm}{r}\right)^{1/2} \ge c$
4) $\left(\frac{gm}{r}\right)^{1/2} \ge c$

9. A straight rod of length L extends from x = a to x = L + a. The gravitational force, it exerts on a point of mass m at x = o, if mass per unit length is $A + Bx^2$:

1) $Gm\left[A\left(\frac{1}{a+L}-\frac{1}{a}\right)+BL\right]$
2) $Gm\left[A\left(\frac{1}{a}-\frac{1}{a+L}\right)+BL\right]$
3) $Gm\left[A\left(\frac{1}{a+L}-\frac{1}{a}\right)-BL\right]$
4) $Gm\left[A\left(\frac{1}{a}-\frac{1}{a+L}\right)-BL\right]$

10. If earth, while rotating with angular velocity ω becomes stand still, what will be the effect on the weight of a body of mass m at a latitude of 45°:

1) Remain Unchanged
2) Decreases by $mR\omega^2$
3) Increases by $mR\omega^2$
4) Increases by $\frac{mR\omega^2}{2}$

11. The distance of Venus from the sun is 0.72 AU. Find the orbital period of Venus.

1) 240 days
2) 223 days
3) 213 days
4) 230 days

12. Suppose that gravitational force varies inversely as the nth power of distance. Then the time period of a planet in circular orbit of Radius R around the Sun will be proportional to:

1) $R^{\frac{(n+1)}{2}}$
2) $R^{\frac{(n-1)}{2}}$
3) R^n
4) $R^{\frac{(n-2)}{2}}$

13. A satellite goes along an elliptical path around Earth. The rate of change of area swept by the line joining Earth and satellite is proportional to:

1) $r^{1/2}$
2) r
3) $r^{3/2}$
4) r^2

14. A body is released from a point distance r from the centre of Earth. If R is the radius of the earth and r > R, then velocity of the body at the time of striking the earth will be:

1) $\sqrt{gR}$
2) $\sqrt{2gR}$
3) $\sqrt{\frac{2gR_r}{r-R}}$
4) $\sqrt{\frac{2gR(r-R)}{r}}$

15. A solid sphere is of density J and radius R. The gravitational field at a distance R from the centre of the sphere, when r < R is:

1) $\frac{J\pi GR^3}{r}$
2) $\frac{4\pi GJr^2}{3}$
3) $\frac{4\pi GJR^2}{3r^2}$
4) $\frac{4\pi GJr}{3}$

16. What is the minimum energy required to launch a satellite of mass m from the surface of a planet of mass M and Radius R in a circular orbit at an altitude of 2R?

1) $\frac{5GmM}{6R}$
2) $\frac{2GmM}{3R}$
3) $\frac{GmM}{2R}$
4) $\frac{GmM}{3R}$

17. The time period of a satellite of the Earth is 5h. If the separation between the Earth & satellite is increased to 4 times the previous value, the new time period will become:

1) 10h
2) 80h
3) 40h
4) 20h

18. The curve for potential energy (E_p) and kinetic energy (E_k) of two particle system are shown in the figure. At what points the system will be bound energy?

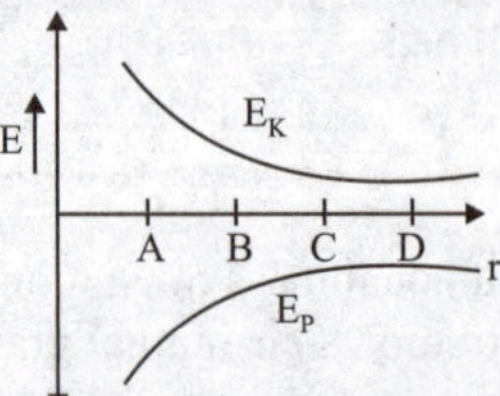

1) Only at A
2) Only At D
3) At A & D
4) At A, B, C and D

19. The mass of a spaceship is 1000 kg. It is to be launched from Earth into space. The required energy will be:

1) 6.4×10^{11}J
2) 6.4×10^{8}J
3) 6.4×10^{9}J
4) 6.4×10^{10}J

20. A research satellite of mass 200 kg circle the Earth in an orbit of average radius $\frac{3R}{2}$. Assuming the gravitational pull of mass of 1 kg on earth's surface to be 10 N, the pull on the satellite will be:

1) 880 N
2) 889 N
3) 885 N
4) 892 N

21. Imagine the acceleration due to gravity on earth is 10 m/s^2 and on mars is 4 m/s^2. A traveller of mass 60 kg goes from earth to mars by a rocket moving with constant velocity. If effect of other planets is assumed to be negligible, which one of the following graphs shows the variation of weight of traveller with time.

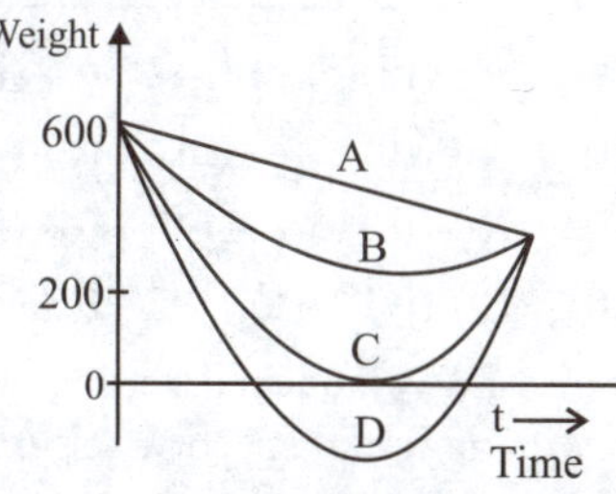

1) A 2) B
3) C 4) D

22. The value of g at any point is 9.8 m/s^2. If earth is reduced to half of its present size by shrinking without loss of mass but the point remains at the same position, then the value of g at that point will be :
1) 4.9 m/s^2 2) 3.1 m/s^2
3) 9.8 m/s^2 4) 19.6 m/s^2

23. The loss in weight of a body taken from earth's surface to a height h is 1%. The percentage change in weight taken into a mine of depth h will be :
1) 1% loss 2) 1% gain
3) 0.5 gain 4) 0.5% loss

24. The gravitational field in a region is given by $(2\hat{i}+2\hat{j})$ N/kg. What is the work done by an external agent to slowly shift a particle of mass 10 kg from the point (0, 0) to a point (5m, 4m)?
1) 180 J 2) –180 J
3) 90 J 4) –90 J

25. Two bodies of masses m_1 and m_2 are initially at rest at infinite distance apart. They are then allowed to move towards each other under mutual gravitational attraction. Their relative velocity of approach at a separation distance r between them is
1) $\left[2G\frac{(m_1-m_2)}{r}\right]^{1/2}$ 2) $\left[\frac{2G}{r}(m_1+m_2)\right]^{1/2}$
3) $\left[\frac{r}{2G(m_1m_2)}\right]^{1/2}$ 4) $\left[\frac{2G}{r}m_1m_2\right]^{1/2}$

26. The escape velocity from the earth is about 11 km/s. The escape velocity from a planet having twice the radius and the same mean density as the earth is:
1) 22 km/s 2) 11 km/s
3) 5.5 km/s 4) 15.5 km

27. The density of core of a planet is ρ_1 and that of outer shell is ρ_2. The radius of core is R and that of planet is 2R. Gravitational field at outer surface of planet is same as at the surface of core. What is the ratio ρ_1/ρ_2.

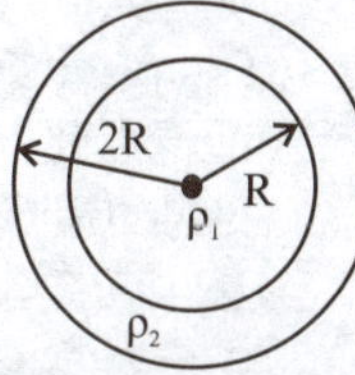

1) $\frac{3}{4}$ 2) $\frac{5}{3}$
3) $\frac{7}{3}$ 4) $\frac{3}{5}$

28. Work done by an external agent in bringing three particles of mass m at the vertices's of an equilateral triangle of length ℓ is
1) $\frac{3Gm^2}{\ell}$ 2) $-\frac{3Gm^2}{\ell}$
3) $\frac{3Gm^2}{2\ell}$ 4) None of these

29. The orbital velocity of a satellite at point B with radius r_B is v. The radius of point A is r_A. If the orbit is increased in radial distance so that r_A becomes $1.2r_A$ find the orbital velocity at $(1.2r_A)$

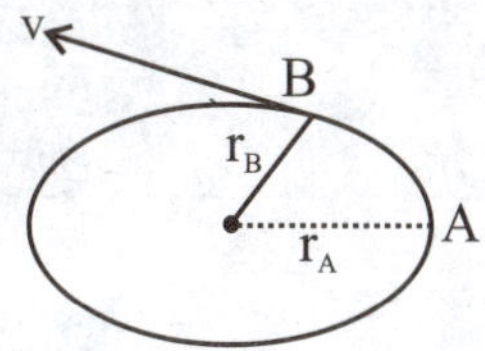

1) $\frac{vr_B}{r_A\sqrt{1.2}}$ 2) $\frac{vr_A}{1.2r_B}$
3) $\frac{vr_B}{1.2r_A}$ 4) $\frac{vr_A}{r_B\sqrt{1.2}}$

30. A satellite revolves in elliptical orbit around a planet of mass M. Its time period is T and M is at the centre of the path. The length of the major axis of the path is : (neglect the gravitational effect of other objects in space)
1) $2\left[\frac{GMT^2}{4\pi^2}\right]^{1/3}$ 2) $\left[\frac{GMT^2}{4\pi^2}\right]^{1/3}$
3) $\frac{1}{2}\left[\frac{GMT^2}{4\pi^2}\right]^{1/3}$ 4) None of these

31. A satellite is orbiting with areal velocity v_a. At what height from the surface of the earth, it is rotating, if the radius of earth is R?
1) $\frac{4v_a^2}{gR^2}-R$ 2) $\frac{2v_a^2}{gR^2}-R$
3) $\frac{v_a^2}{gR^2}-R$ 4) $\frac{v_a^2}{2gR^2}-R$

32. A spherical planet has uniform density $\frac{\pi}{2}\times10^4$ kg/m^3. Find out the minimum period for a satellite in a circular orbit around it in seconds.
$\left(\text{Use } G=\frac{20}{3}\times10^{-11}\frac{N-m^2}{kg^2}\right)$
1) 7500 2) 3000
3) 4500 4) 6000

33. P(r) is magnitude of physical quantity as a function of r (distance from centre of spherical distribution of radius R). Match the following columns and select the correct option from the codes given below:

	Column-I		Column-II
(a)	P (r) o r r = R	(i)	P(r) is variation of gravitational potential for a uniform volumetric spherical distribution of mass
(b)	P (r) o r r = R	(ii)	P(r) is variation of gravitational potential for a uniform spherical shell.
(c)	P (r) o r r = R	(iii)	P(r) is variation or gravitational field intensity for a uniform volumetric spherical distribution of mass
(d)	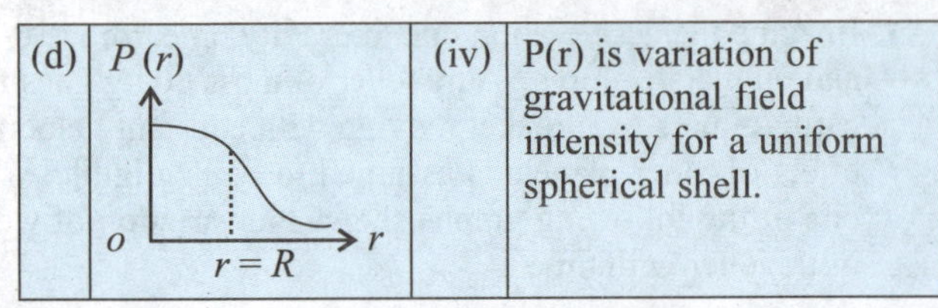	(iv)	P(r) is variation of gravitational field intensity for a uniform spherical shell.

1) (a)-(iv) (b)-(iii) (c)-(ii) (d)-(i)
2) (a)-(ii) (b)-(iv) (c)-(iii) (d)-(i)
3) (a)-(ii) (b)-(iii) (c)-(iv) (d)-(i)
4) (a)-(iii) (b)-(iv) (c)-(i) (d)-(ii)

34. A rocket of mass M is launched vertically from the surface of the earth with an initial speed v. Assuming the radius of the earth to be R and negligible air resistance, the maximum height attained by the rocket about the surface of the earth is

1) $R/\left(\frac{gR}{2v^{-2}}-1\right)$
2) $R\left(\frac{gR}{2v^{-2}}-1\right)$
3) $R/\left(\frac{2gR}{v^{2}}-1\right)$
4) $R\left(\frac{2gR}{2v^{-2}}-1\right)$

Statement Based MCQs

Directions: These questions consist of two statements each, printed as Statement-I and Statement-II. While answering these questions, you are required to choose any one of the following four responses.

1) **Both Statement-I and Statement-II are correct.**
2) **Both Statement-I and Statement-II are incorrect.**
3) **Statement-I is correct & Statement-II is incorrect.**
4) **Statement-I is incorrect & Statement-II is correct.**

1. **Statement I:** The square of the period of revolution of a planet is proportional to the cube of its distance from the sun.
   Statement II: Revolution of moon around the earth is due to gravitational force.
2. **Statement I:** Acceleration due to gravity increases as we go towards the centre of the earth.
   Statement II: Acceleration due to gravity is maximum near the earth's equator.
3. **Statement I:** Escape velocity is different for two bodies of different masses.
   Statement II: Escape velocity is maximum when angle of projection is 90°.
4. **Statement I:** We cannot move even a finger, without disturbing all the stars.
   Statement II: Everybody in this universe attracts every other body with a force which is inversely proportional to the square of distance between them.
5. **Statement I:** Temperature of mercury is more than that of Earth.
   Statement II: Escape velocity of body on the surface of earth is less than that on the moon.

Assertion & Reason MCQs

Directions: These questions consist of two statements each, printed as Assertion and Reason. While answering these questions, you are required to choose any one of the following four responses.

1) **Both Assertion and Reason are True and the Reason is a correct explanation of the Assertion.**
2) **Both Assertion and Reason are True but Reason is not a correct explanation of the Assertion.**
3) **Assertion is True but the Reason is False.**
4) **Assertion is False but Reason is True.**

1. **Assertion:** The stars twinkle, while the planets do not.
   Reason: The stars are much bigger in size than the planets.
2. **Assertion:** Escape velocity of an elephant from the surface of the earth is equal to the escape velocity of an ant.
   Reason: Escape velocity of a body is independent of the mass of the body.
3. **Assertion:** Although gravitational pull of sun on earth is more than that of moon, yet moon's contribution towards tide formation on earth is greater than that due to sun.
   Reason: Because the tidal effect is inversely proportional to the cube of the distance.

4. Assertion: A planet is a heavenly body revolving round the sun.

Reason: Star is a self-luminous body made of gaseous material.

5. Assertion: When planet moves in elliptical orbit around sun. Its angular momentum about sun remains conserved.

Reason: Total mechanical energy of planet- sun system remains conserved.

Match the Columns MCQs

1. | Potential | on the surface of a solid sphere is x and radius is y. Match the following two columns.

	Column-I		Column-II
(a)	Field strength at distance 2y from centre	(i)	$\frac{x}{2y}$
(b)	| Potential | at distance y/2 from centre	(ii)	$\frac{x}{2}$
(c)	Field strength at distance y/2 from centre	(iii)	$\frac{x}{4y}$
(d)	| Potential | at distance 2y from centre	(iv)	None

1) (a)-(iii) (b)-(iv) (c)-(i) (d)-(ii)
2) (a)-(iv) (b)-(iii) (c)-(ii) (d)-(i)
3) (a)-(i) (b)-(iii) (c)-(iv) (d)-(ii)
4) (a)-(iii) (b)-(ii) (c)-(iv) (d)-(i)

2. Match the following two columns.

	Column-I		Column-II
(a)	Work done in raising a mass m to a height = R	(i)	$\frac{1}{4}$mgR
(b)	Kinetic energy of a satellite of mass m at height	(ii)	mgR
(c)	Difference in energies of two satellites each of mass m but one at height h_1 = R and another of height h_2 = 2R	(iii)	$\frac{1}{2}$mgR
(d)	Kinetic energy required to raise a particle of mass m to a height h = R. if projected vertically from surface of earth.	(iv)	None

1) (a)-(ii) (b)-(iii) (c)-(iv) (d)-(i)
2) (a)-(iii) (b)-(i) (c)-(iv) (d)-(iii)
3) (a)-(ii) (b)-(iv) (c)-(iii) (d)-(i)
4) (a)-(iv) (b)-(ii) (c)-(i) (d)-(iii)

3. Match the following two columns.

	Column-I		Column-II
(a)	Gravitational field strength is maximum at	(i)	r = 0
(b)	Gravitational field strength is zero at	(ii)	r = R
(c)	Gravitational potential is minimum at	(iii)	$r = \frac{R}{\sqrt{2}}$
(d)	Gravitational potential is zero at	(iv)	None of these

Here, r is the distance form centre of a solid sphere or distance from centre of a ring along its axis.

1) (a)-(ii, iii) (b)-(iii) (c)-(ii) (d)-(i)
2) (a)-(iv) (b)-(ii, iii) (c)-(i) (d)-(ii)
3) (a)-(iv) (b)-(ii) (c)-(iii) (d)-(i, iv)
4) (a)-(ii) (b)-(iii, i) (c)-(iv) (d)-(ii)

4. Let V and E denote the gravitational potential and gravitational field respectively at a point due to certain uniform mass distribution described in four different situations of column-I. Assume the gravitational potential at infinity to be zero. The value of E and V are given in column-II. Match the statement in column-I with results in column-II.

	Column-I		Column-II
(a)	At centre of thin spherical shell	(i)	E = 0
(b)	At centre of solid sphere	(ii)	E ≠ 0
(c)	A solid sphere has a non-concentric spherical cavity. At the centre of the spherical cavity	(iii)	V ≠ 0
(d)	At centre of line joining two point masses of equal magnitude	(iv)	V = 0

1) (a)-(ii, iii) (b)-(iv, iii) (c)-(i, iii) (d)-(i, iii)
2) (a)-(iv, iii) (b)-(i, iii) (c)-(ii, iii) (d)-(ii, iii)
3) (a)-(i, iii) (b)-(i, iii) (c)-(ii, iii) (d)-(i, iii)
4) (a)-(iii, ii) (b)-(ii, iv) (c)-(iv, ii) (d)-(iv, iii)

5. A particle is taken to a distance r (> R) from centre of the earth. R is radius of the earth. It is given velocity V which is perpendicular to radius. With the given values of V in column-I you have to match the resultant path of particle in column-II. Here 'G' is the universal gravitational constant and 'M' is the mass of the earth.

	Column-I (Velocity)		Column-II (Path)
(a)	$v = \sqrt{GM/r}$	(i)	Elliptical
(b)	$v = \sqrt{2GM/r}$	(ii)	Parabolic
(c)	$v > \sqrt{2GM/r}$	(iii)	Hyperbolic
(d)	$\sqrt{GM/r} < v < \sqrt{2GM/r}$	(iv)	Circular

1) (a)-(iv) (b)-(ii) (c)-(iii) (d)-(i)
2) (a)-(i) (b)-(iv) (c)-(ii) (d)-(iii)
3) (a)-(ii) (b)-(iii) (c)-(i) (d)-(iv)
4) (a)-(iv) (b)-(ii) (c)-(ii) (d)-(i)

NCERT Exemplar MCQs

1. The Earth is an approximate sphere. If the interior contained matter which is not of the same density everywhere, then on the surface of the Earth, the acceleration due to gravity:
   1) Will be directed towards the centre but not the same everywhere
   2) Will have the same value everywhere but not directed towards the centre
   3) Will be same everywhere in magnitude directed towards the centre
   4) Cannot be zero at any point
2. As observed from the Earth, the Sun appears to move in an approximate circular orbit. For the motion of another planet like Mercury as observed from the Earth, this would:
   1) Be similarly true
   2) Not be true because the force between the Earth and mercury is not inverse square law
   3) Not be true because the major gravitational force on mercury is due to the Sun
   4) Not be true because mercury is influenced by forces other than gravitational forces
3. Different points in the Earth are at slightly different distances from the Sun and hence experience different forces due to gravitation. For a rigid body, we know that if various forces act at various points in it, the resultant motion is as if a net force acts on the CM (centre of mass) causing translation and a net torque at the CM causing rotation around an axis through the C. M. for the Earth-Sun system (approximating the Earth as a uniform density sphere):
   1) The torque is zero
   2) The torque causes the earth to spin
   3) The rigid body result is not applicable since the earth is not even approximately a rigid body
   4) The torque causes the earth to move around the sun
4. Satellites orbiting the Earth have finite life and sometimes debris of satellites fall to the Earth. This is because:
   1) The solar cells and batteries in satellites run out
   2) The laws of gravitation predict a trajectory spiraling inwards
   3) Of viscous forces causing the speed of satellite and hence height to gradually decrease
   4) Of collisions with other satellites
5. Both the Earth and the Moon are subject to the gravitational force of the Sun. As observed from the Sun, the orbit of the Moon:
   1) Will be elliptical
   2) Will not be strictly elliptical because the total gravitational force on it is not central
   3) Is not elliptical but will necessarily be a closed curve
   4) Deviates considerably from being elliptical due to influence of planets other than the Earth
6. In our solar system, the inter-planetary region has chunks of matter (much smaller in size compared to planets) called asteroids. They:
   1) Will not move around the Sun, since they have very small masses compared to the Sun
   2) Will move in an irregular way because of their small masses and will drift away into outer space
   3) Will move around the Sun in closed orbits but not obey Kepler's laws
   4) Will move in orbits like planets and obey Kepler's laws
7. Choose the wrong option:
   1) Inertial mass is a measure of difficulty of accelerating a body by an external force whereas the gravitational mass is relevant in determining the gravitational force on it by an external mass
   2) The gravitational mass and inertial mass are equal is an experimental result
   3) The acceleration due to gravity on the Earth is the same for all bodies is due to the equality of gravitational mass and inertial mass
   4) Gravitational mass of a particle like proton can depend on the presence of neighbouring heavy objects but the inertial mass cannot
8. Particles of masses 2M, m and M are respectively at points A, B and C with $AB = \frac{1}{2}(BC)$. m is much-smaller than M and at time t = 0, they are all at rest as given in figure: At subsequent times before any collision takes place.

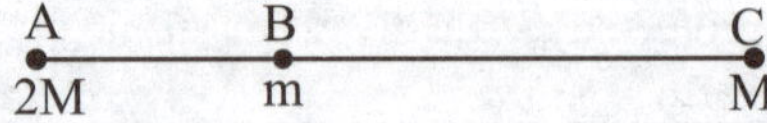

   1) m will remain at rest
   2) m will move towards M
   3) m will move towards 2M
   4) m will have oscillatory motion
9. Which of the diagrams shown in figure most closely shows the variation in kinetic energy of the earth as it moves once around the sun in its elliptical orbit?

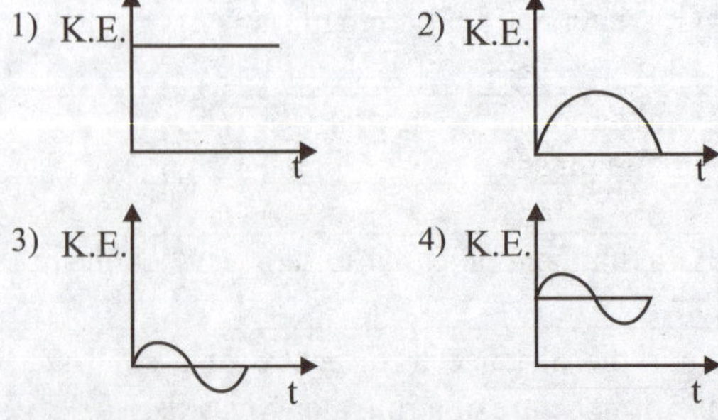

NEET & AIIMS Past Year MCQs

1. The mass of a planet is $\frac{1}{10}$th that of the earth and its diameter is half that of the earth. The acceleration due to gravity on that planet is: **(2024)**
   1) 4.9 m s^{-2} 2) 3.92 m s^{-2}
   3) 19.6 m s^{-2} 4) 9.8 m s^{-2}

2. An object of mass 100 kg falls from point A to B as shown in figure. The change in its weight, corrected to the nearest integer is (R_E is the radius of the earth) **(2024 Re)**

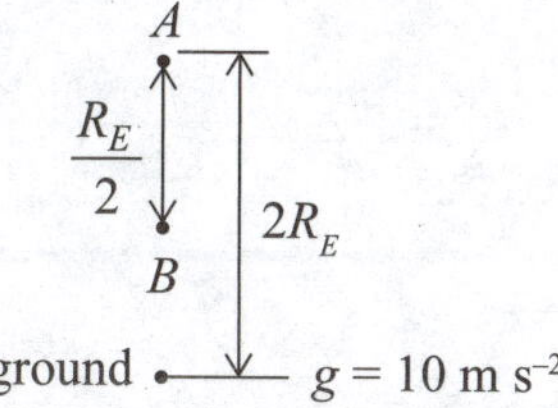

   1) 49 N 2) 89 N
   3) 5 N 4) 10 N

3. The escape velocity for earth is v. A planet having 9 times mass that of earth and radius, 16 times that of earth, has the escape velocity of: **(2024 Re)**
   1) $\frac{V}{3}$ 2) $\frac{2v}{3}$ 3) $\frac{3v}{4}$ 4) $\frac{9v}{4}$

4. The minimum energy required to launch a satellite of mass m from the surface of earth of mass M and radius R in a circular orbit at an altitude of 2R from the surface of the earth is: **(2024)**
   1) $\frac{GmM}{2R}$ 2) $\frac{GmM}{3R}$ 3) $\frac{5GmM}{6R}$ 4) $\frac{2GmM}{3R}$

5. Two bodies of mass m and 9m are placed at a distance R. The gravitational potential on the line joining the bodies where the gravitational field equals zero, will be (G = gravitational constant): **(2023)**
   1) $-\frac{20Gm}{R}$ 2) $-\frac{8Gm}{R}$
   3) $-\frac{12Gm}{R}$ 4) $-\frac{16Gm}{R}$

6. A satellite is orbiting just above the surface of the earth with period T. If d is the density of the earth and G is the universal constant of gravitation, the quantity $\frac{3\pi}{Gd}$ represents: **(2023)**
   1) $\sqrt{T}$ 2) T
   3) T^2 4) T^3

7. A body of mass 60 g experiences a gravitational force of 3.0 N, when placed at a particular point. The magnitude of the gravitational field intensity at that point is: **(2022)**
   1) 180 N/kg 2) 0.05 N/kg
   3) 50 N/kg 4) 20 N/kg

8. Match Column-I and Column-II **(2022)**

	Column-I		Column-II
(a)	Gravitational constant (G)	(i)	$[L^2T^{-2}]$
(b)	Gravitational potential Energy	(ii)	$[M^{-1}L^3T^{-2}]$
(c)	Gravitational potential	(iii)	$[LT^{-2}]$
(d)	Gravitational Intensity	(iv)	$[ML^2T^{-2}]$

   Choose the correct answer from the options given below:
   1) (a)-(iv) (b)-(ii) (c)-(i) (d)-(iii)
   2) (a)-(ii) (b)-(i) (c)-(iv) (d)-(iii)
   3) (a)-(ii) (b)-(iv) (c)-(i) (d)-(iii)
   4) (a)-(ii) (b)-(iv) (c)-(iii) (d)-(i)

9. A particle is released from height S from the surface of the Earth. At a certain height its kinetic energy is three times its potential energy. The height from the surface of earth and the speed of the particle at that instant are respectively: **(2021)**
   1) $\frac{S}{4}, \sqrt{3gS}$ 2) $\frac{S}{2}, \frac{\sqrt{3gS}}{2}$
   3) $\frac{S}{4}, \sqrt{\frac{3gS}{2}}$ 4) $\frac{S}{4}, \frac{3gS}{2}$

10. The escape velocity from the Earth's surface is v. The escape velocity from the surface of another planet having a radius, four times that of Earth and same mass density is: **(2021)**
   1) 2v 2) 3v
   3) 4v 4) v

11. A particle of mass 'm' is projected with a velocity $v = KV_e (k < 1)$ from the surface of the earth.
   The maximum height above the surface reached by the particle is: **(2021)**
   1) $R\left(\frac{K}{1+K}\right)^2$ 2) $\frac{R^2K}{1+K}$
   3) $\frac{RK^2}{1-K^2}$ 4) $R\left(\frac{K}{1-K}\right)^2$

12. A satellite which is revolving around earth has minimum distance from earth equal to r_1 and maximum distance equal to r_2 then time period of the satellite will be **(AIIMS 2018)**
   1) $2\pi\sqrt{\frac{r_1r_2}{Gm}}$ 2) $\frac{2\pi\left(\frac{r_1+r_2}{2}\right)^{3/2}}{\sqrt{Gm}}$
   3) $\frac{2\pi}{\sqrt{Gmr_1r_2}}$ 4) $\frac{\sqrt{Gmr_1r_2}}{2\pi}$

13. The change in the gravitational potential energy when a body of mass m is raised to a height nR above the surface of the earth is (here R is the radius of the earth): **(AIIMS 2017)**

1) $\left(\frac{n}{n+1}\right)mgR$ 2) $\left(\frac{n}{n-1}\right)mgR$

3) nmgR 4) $\frac{mgR}{n}$

14. A wire of length l and mass m is bent in the form of a semicircle. The gravitational field intensity at the centre of semicircle is: **(AIIMS 2009)**

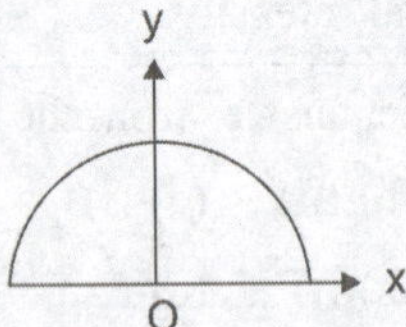

1) $\frac{Gm}{\pi l}$ along x-axis

2) $\frac{Gm}{\pi l}$ along y-axis

3) $\frac{2\pi Gm}{l^2}$ along y-axis

4) $\frac{2\pi Gm}{l^2}$ along x-axis

15. Suppose the gravitational force varies inversely as the nth power of distance. Then the time period of a planet in circular orbit of radius R around the sun will be proportional to: **(AIIMS 2008)**

1) $R^{(n+1)/2}$ 2) $R^{(n-1)/2}$

3) R^n 4) $R^{(n-2)/2}$

Answer Key

NCERT Topic-wise MCQs	1	2	3	4	5	6	7	8	9	10	11	12	13	14	15	16	17
	4	4	2	3	4	2	3	1	1	3	4	2	3	1	2	1	1
	18	19	20	21	22	23	24	25	26	27	28	29	30	31	32	33	34
	3	4	1	2	1	2	3	3	4	2	1	2	1	2	1	2	2
	35	36	37	38	39	40	41	42	43	44	45	46	47				
	3	2	4	3	1	1	1	1	2	2	1	3	2				

High Order Time Intensive MCQs	1	2	3	4	5	6	7	8	9	10	11	12	13	14	15	16	17
	2	3	2	2	2	4	3	3	2	4	2	1	1	4	4	1	3
	18	19	20	21	22	23	24	25	26	27	28	29	30	31	32	33	34
	4	4	2	3	3	4	2	2	1	3	2	1	1	1	2	3	3

Statement Based MCQs	1	2	3	4	5
	1	2	2	1	3

Assertion & Reason MCQs	1	2	3	4	5
	2	1	1	2	2

Match the Columns MCQs	1	2	3	4	5
	1	2	1	3	1

NCERT Exemplar MCQs	1	2	3	4	5	6	7	8	9
	4	3	1	3	2	4	4	3	4

NEET & AIIMS Past Year MCQs	1	2	3	4	5	6	7	8	9	10	11	12	13	14	15
	2	1	3	3	4	3	3	3	3	3	3	2	1	3	1

CHAPTER

8 Mechanical Properties of Solids

Topicwise Number of Questions Analysis (2025-2016)

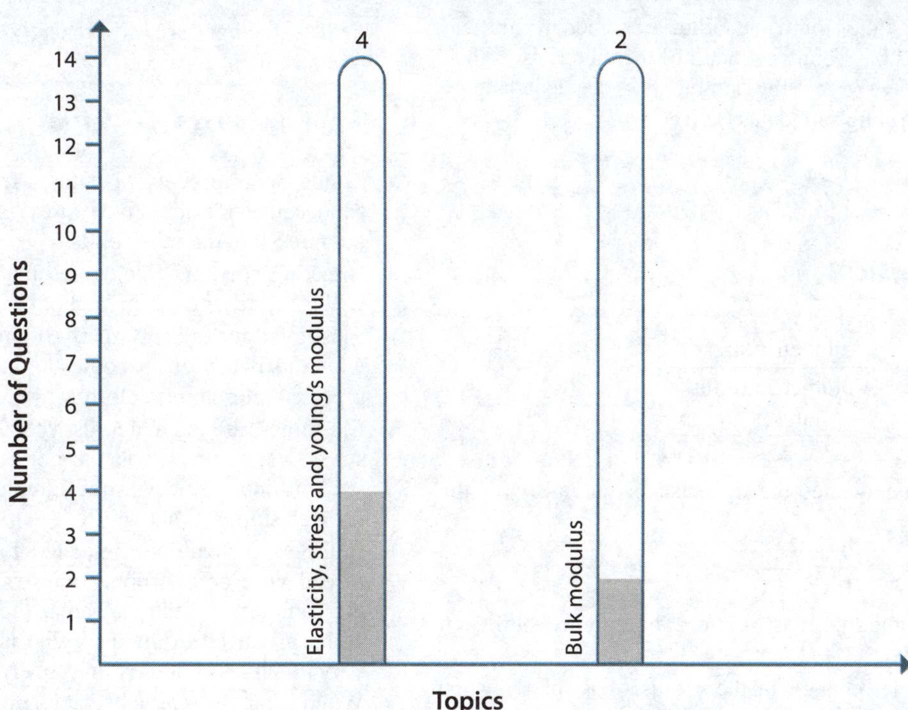

NCERT-Picks

Stress, Strain

Formulas

1. Stress $= \dfrac{\text{Restoring force}}{\text{Area}}$
2. Longitudinal strain $= \dfrac{\text{change in length}}{\text{original length}}$
3. Volumetric strain $= \dfrac{\text{change in volume}}{\text{original volume}}$

Young's Modulus

Formulas

1. Young's modulus, $Y = \dfrac{\text{Normal Stress}}{\text{Longitudinal Strain}}$

$$Y = \frac{F/A}{\Delta l/l} = \frac{Fl}{A\Delta l}$$

Note: Steel is more elastic than rubber

$\because Y_{steel} > Y_{rubber}$

2. Percentage increase in length, $\dfrac{\Delta l}{l} \times 100 = \dfrac{F}{AY} \times 100$

Where, Force F is in Newton, area A is in m^2, stress in Nm^{-2}, Young's modulus Y in Nm^{-2} or Pa.

Numerical Profile

(i) Elongation of a wire under its own weight

$$e = \frac{\ell^2 dg}{2Y}$$

(where l is length of the wire d is density and Y is the young's modulus of the wire).

(ii) If l_1, l_2 are the lengths of a wire under tensions T_1 & T_2, then the length of the unstretched wire.

$$\ell = \frac{\ell_1 T_2 - \ell_2 T_1}{T_2 - T_1}$$

(iii) The elongation produced in a wire when a body of mass M hung at the free end of the wire is immersed in a liquid $= \frac{(Mg - V\rho g)L}{AY}$

[where, L is Length of the wire, V is volume of the body immersed in the liquid, A is area of cross section of the wire, ρ-density of the liquid]

(iv) Two wires of same lengths and radii are joined end to end and loaded. If Young's moduli of the materials are Y_1 & Y_2, the Young's modulus of the combination is given as

$$\frac{2}{Y} = \frac{1}{Y_1} + \frac{1}{Y_2} \Rightarrow Y = \frac{2Y_1Y_2}{Y_1 + Y_2}$$

(v) Two wires of same length and radius are joined in parallel and loaded. If the Young's moduli of the materials of the wires are Y_1 & Y_2 and the combination acts as a single wire then its Young's modulus is given as

$$Y = \frac{Y_1 + Y_2}{2}$$

Bulk-modulus

Formulas

1. Bulk modulus $= \frac{\text{Volumetric stress}}{\text{Volumetric strain}}$

or, $B = -\frac{F/A}{\Delta V/V} = -\frac{p}{\Delta V/V} = -V\frac{p}{\Delta V}$

Negative sign indicates the decrease in volume with the increase in stress

2. Compressibility $= \frac{1}{B} = -\frac{\Delta V}{pV}$

where, Bulk modulus B is in Nm^{-2} and compressibility in $N^{-1}\ m^2$ or Pa^{-1}.

Note: Fractional change in the radius of a sphere when subjected to volumetric stress is $\frac{\Delta V}{V} = \frac{3\Delta R}{R}$

[Where, R is radius of the sphere]

Modulus of Rigidity

Formulas

1. Shearing stress $= \frac{\text{Tangential Force}}{\text{Area}} = \frac{F}{A}$

2. Shearing strain $= \theta = \frac{\Delta l}{l}$

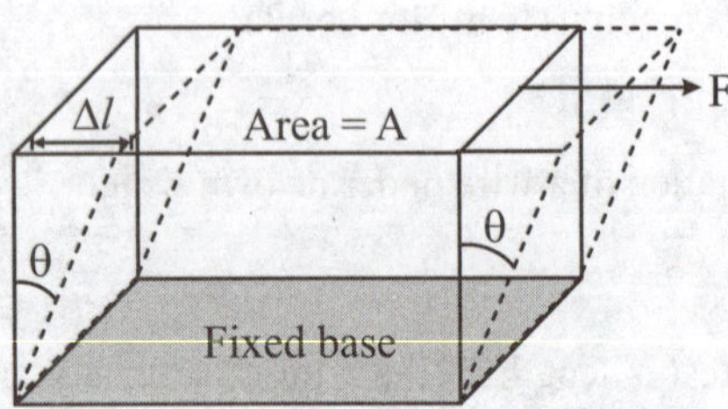

3. Modulus of rigidity $= \frac{\text{Shearing stress}}{\text{Shearing strain}}$

$$\Rightarrow G = \frac{F/A}{\theta} = \frac{F/A}{\Delta l/l} = \frac{Fl}{A\Delta l}$$

Where, Modulus of rigidity G is in Nm^{-2}.

Thermal Stress

Formulas

1. Potential energy of a stretched wire

$U = \frac{1}{2} \times$ stretching Force $\times$ extension $= \frac{1}{2}F\Delta l$

$U = \frac{1}{2} \times$ stress $\times$ strain $\times$ volume of the wire.

$U = \frac{1}{2} \times Y \times (\text{Strain})^2 \times V$ [Where, V is the volume]

2. Thermal Stress $= Y\,\alpha\,\Delta\theta$

($\Delta\theta$ = temperature difference, α is coefficient of linear expansion)

Statements Asked in Exam

- Solids are more elastic and gases are least elastic. It is so because for the given stress applied gases are more compressible than the solids.
- Breaking stress is fixed for a material, but breaking force will vary, depending on area of cross section of the wire.
- Elastic fatigue is the property of an elastic body by virtue of which its behaviour becomes less elastic under the action of repeated alternating deforming forces.
- Elastomers are the materials whose stress vs strain graph is not a straight line. Stress vs strain graph for a elastomer is shown fig,

- If the same loads are attached to a lead wire and a rubber wire of the same cross-sectional area, the strain produced in the lead wire is very much less than that in rubber wire.
- When a load W is attached at its middle point, the depression δ produced at middle point in the bar is

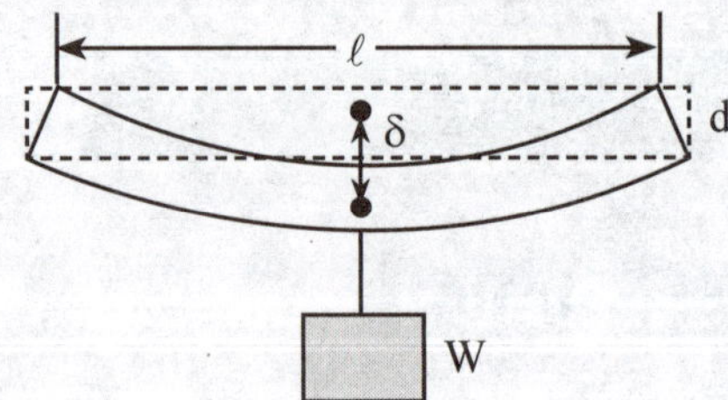

$$\delta = \frac{Wl^3}{4bd^3Y}$$

l = length, b = breadth, d = thickness

- Theoretical value of poisson's ratio range from –1 to 0.5 but practical value ranges from 0 to 0.5.
- Hooke's law is valid only in the linear part of stress - strain curve.
- In daily life, we feel that a material which stretches more is more elastic, but it is a misnomer. In fact material which is less stretched for a given load is considered to be more elastic.
- Maximum height of a mountain on the surface of earth $\approx 10^4$ m.
- For the same size and same material hollow shafts are stronger than solid shafts.

NCERT Topic-wise MCQs

Elasticity, Stress and Young's Modulus

1. A rod has a radius of 100 mm and a length of 10 cm. A 100 N force compress along its length. Calculate the longitudinal stress developed in the rod.

1) 2000 N/m^2 2) 2422 N/m^2
3) 3182 N/m^2 4) 5500 N/m^2

2. The Young's modulus of wire of length L and radius r is Y. If the length and radius are reduced to $\frac{L}{3}$ and $\frac{r}{2}$, then its Young's modulus will be

1) Y 2) $\frac{3}{4}Y$
3) 3Y 4) 4Y

3. On increasing the length by 0.5 mm in a steel wire of length 4m and area of cross section 5 mm^2, the force required is ($Y = 11.90 \times 10^{12}$ N/m^2)

1) 4×10^3 N 2) 2×10^3 N
3) 7.4×10^3 N 4) 10^3 N

4. The Young's modulus of brass and steel are respectively 1.0×10^{11} Nm^{-2} and 2.0×10^{11} Nm^{-2}. A brass wire and a steel wire of the same length are extended by 1 mm each under the same force. If radii of brass and steel wires are R_B and R_S respectively, then

1) $R_S = \sqrt{2}R_B$ 2) $R_S = \frac{R_B}{\sqrt{2}}$
3) $R_S = 4R_B$ 4) $R_S = \frac{R_B}{2}$

5. A and B are the two wires, the radius of A is thrice that of B. They are stretched by the same load. Then the stress on B is

1) Remains same 2) 2 times of A
3) 3 times of A 4) 9 times of A

6. A block of mass 500 kg is suspended by wire of length 70 cm. The area of cross-section of wire is 10 mm^2. When the load is removed, the wire contracts by 0.5 cm. The Young's modulus of the material of wire will be ($g = 10$ m/s^2)

1) 10×10^{14} N/m^2 2) 4×10^{14} N/m^2
3) 8×10^{11} N/m^2 4) 7×10^{10} N/m^2

7. Two wires A and B are of same materials. Their length are in the ratio 3 : 4 and diameters are in the ratio 5 : 1 when stretched by force F_A and F_B respectively they get equal increase in their lengths. The ratio of F_B/F_A is.

1) 0.01 2) 0.03
3) 0.04 4) 0.05

8. In plotting stress versus strain curves for two materials P and Q, a student by mistake puts strain on the Y-axis and stress on the X-axis as shown in the figure. The the correct statements is/are

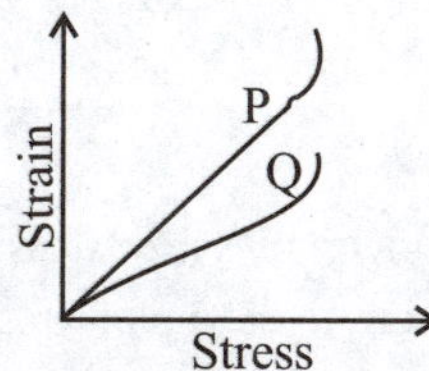

1) P has less tensile strength then Q
2) P is more ductile than Q
3) P is more brittle than Q
4) The Young's modulus of P is more than that of Q

9. A wire stretches by l under a load. If the load and radius both are increased to four times, the stretch in the wire will be

1) $\frac{l}{2}$ 2) $\frac{l}{3}$
3) $\frac{l}{4}$ 4) $\frac{l}{6}$

10. A wire can be broken by applying a load of 30 kg-wt. The force required to break the wire of twice the diameter is

1) 240 kg-wt 2) 100 kg-wt
3) 120 kg-wt 4) 150 kg-wt

11. The following four wires are made of the same material. Which of these will have the largest extension when the same tension is applied ?

1) Length = 50 cm, diameter = 0.5 mm
2) Length = 100 cm, diameter 1 mm
3) Length = 200 cm, diameter = 2 mm
4) Length = 300 cm, diameter = 3 mm

12. Two wires P and Q are of same material. Their length are in the ratio 4 : 3 and the diameters are in the ratio 2 : 3. If they are pulled by the same force their increase in length will be in the ratio :

1) 3/1 2) 2/3
3) 5/4 4) 3/5

13. A rubber cord of length L is suspended vertically. Density of rubber is D and Young's modulus is Y. If the cord extends by a length l under its own weight then

1) $l = \frac{L^2 Dg}{Y}$ 2) $l = \frac{L^2 Dg}{2Y}$

3) $l = \frac{L^2 Dg}{4Y}$ 4) $l = \frac{L^2 Dg}{8Y}$

14. The extension produced in a wire by the application of load is 5 mm. The extension in a wire of the same material and length but half the radius, by the same load is:

1) 20 mm 2) 30 mm

3) 40 mm 4) 50 mm

15. The metal ring of initial radius r and cross-sectional area A is fitted onto a wooden disc of radius R > r. If Young's modulus of the metal is Y, then the tension in the ring is

1) $\frac{AYR}{r}$ 2) $\frac{AY(R-r)}{r}$

3) $\frac{Y}{A}\left(\frac{R-r}{r}\right)$ 4) $\frac{Yr}{AR}$

16. The graph below shows the extension ($\Delta\ell$) of a wire of length 1 m suspended from the top of a roof at one end with a load W connected to the other end. If the cross-sectional area of the wire is 10^{-6} m^2, what is Young's modulus of the wire?

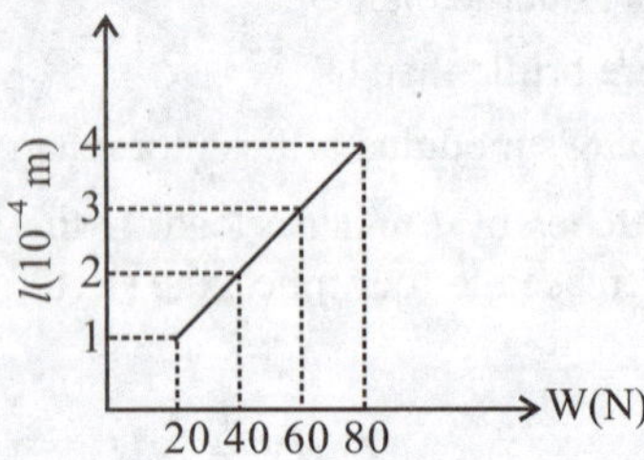

1) 2×10^{11} Nm^{-2} 2) 2×10^{-11} Nm^{-2}

3) 2×10^{-12} Nm^{-2} 4) 2×10^{-14} Nm^{-2}

Bulk-modulus

17. Isothermal elasticity of a gas is equal to

1) Density 2) Pressure

3) Volume 4) Specific heat

18. To what depth below the surface of sea should a rubber ball be taken as to decrease its volume by 0.4% [density of sea water = 1000 kg m^{-3}, Bulk modulus of rubber = 9×10^8 Nm^{-2}, g = 10 m/s^2]

1) 5×10^2 m 2) 30×10^6 m

3) 40×10^6 m 4) 36×10^1 m

19. A uniform cube is subjected to volume compression. If each side is decreased by 3%, the bulk strain is

1) 27% 2) 8%

3) 9% 4) 1%

20. The fractional change in volume per unit increase in pressure is called

1) Pressure coefficient 2) Volume coefficient

3) Bulk modulus 4) Compressibility

21. Bulk modulus of a perfectly rigid body is

1) Zero 2) Unity

3) Infinity 4) less than 1

22. If the bulk modulus of lead is 8.0×10^9 Nm^{-2} and the initial density of lead is 11.4 g/c.c., then under the pressure of 2.0×10^8 N/m^2, the density of the lead is

1) 11.3 g/cm^3 2) 11.5 g/cm^3

3) 11.6 g/cm^3 4) 11.7 g/cm^3

23. The pressure on an object of bulk modulus B undergoing hydraulic compression due to a stress exerted by surrounding fluid having volume strain $\left(\frac{\Delta V}{V}\right)$ is

1) $B^2\left(\frac{\Delta V}{V}\right)$ 2) $B\left(\frac{\Delta V}{V}\right)^2$

3) $\frac{1}{B}\left(\frac{\Delta V}{V}\right)$ 4) $B\left(\frac{\Delta V}{V}\right)$

24. A ball falling in a lake of depth 300 m shows decrease of 0.3% in its volume at the bottom. What is the bulk modulus of material of the ball ? (g = 10 m/s^2)

1) 10^6 N/m^2 2) 10^9 N/m^2

3) 3×10^2 N/m^2 4) None

25. If the pressure of the gas is increased from 1.01×10^5 pa to 2×10^5 Pa and volume is decreased by 25% at constant temperature, then the bulk modulus of the gas is

1) 3.96×10^5 N/m^2 2) 4.8×10^5 N/m^2

3) 10^{-5} N/m^2 4) 10^5 N/m^2

26. A spherical ball contracts in volume by 0.08% when subjected to a normal uniform pressure of 500 atm. What is the bulk modulus of ball.

1) 4.7×10^5 N/m^2 2) 5×10^7 N/m^2

3) 2×10^7 N/m^2 4) 6.3×10^{10} N/m^2

27. The average depth of Indian ocean is about 5000 m. What will be the percentage fractional compression of water at the bottom of the ocean, given that the bulk modulus of water is 3×10^9 Pa and g = 9.8 m/s^2 (1Pa = 1 N/m^2)

1) 17% 2) 1.6%

3) 19% 4) 18%

Modulus of Rigidity

28. A 8 cm cube has its upper face displaced by 0.5 mm by a tangential force of 10 kN. What is the shear modulus of the cube?

1) 2×10^{10} N/m^2 2) 2.5×10^8 N/m^2

3) 1×10^9 N/m^2 4) 1 N/m^2

29. A square lead slab of side 1 metre and thickness 1 metre is subjected to a shearing force on its narrow edge of 15×10^4 N. The lower edge is riveted to the floor. How much will the upper edge be displaced?
($G = 5.6 \times 10^9$ N/m^2)
1) 27 μm 2) 30 μm
3) 21 μm 4) None

30. The upper end of a wire of radius 4 mm and length 100 cm is clamped and its other end is twisted through an angle of 30°. The angle of shear is
1) 12° 2) 1.2°
3) 0.12° 4) 0.012°

Poisson Ratio and Relation Between Moduli of Elasticity

31. Which of the following is right for a metal?
1) $Y < G$ 2) $Y = G$
3) $Y > G$ 4) $Y < \frac{1}{G}$

32. If the Young's modulus of the material is 6 times its modulus of rigidity, then its volume elasticity will be
1) $\frac{2Y}{3}$ 2) ∞
3) $\frac{Y}{9}$ 4) $\frac{-Y}{9}$

33. Minimum and maximum value of poisson ratio for a metal lies between
1) 0 to 1 2) 0 to 0.5
3) $-\infty$ to 1 4) $-\infty$ to ∞

34. Calculate the poisson ratio for silver, given its Young's modulus is 7.25×10^{10} N/m^2 and bulk modulus(B) is 11×10^{10} N/m^2
1) 0.39 2) 0.29
3) 0.49 4) 0.59

35. For a given material the Young's modulus is 3 times the Bulk modulus of material then its modulus of rigidity is
1) 2 B 2) 1.5 B
3) B 4) B/2

36. A material has Poisson ratio 0.30. If a uniform rod suffers a longitudinal strain of 3×10^{-4}, then the percentage change in volume.
1) 0.05% 2) 0.02%
3) 0.01% 4) 0.08%

High Order Time Intensive MCQs

1. When metal wire is elongated by hanging load on it, then fractional change in volume $\frac{\Delta V}{V}$ is proportional to
1) $l\Delta l$ 2) $\frac{\Delta l}{l}$
3) $\frac{2\Delta l}{l}$ 4) $\sqrt{\frac{\Delta l}{l}}$

2. A stress of 6×10^6 Nm^{-2} is required for breaking a material. The density ρ of the material is 3×10^3 kg m^{-3}. If the wire is to break under its own weight, the length of the wire made of that material should be (take, $g = 10$ ms^{-2})
1) 20 m 2) 200 m
3) 100 m 4) 2000 m

3. If the angle of shear is 30° for a cubical body and the change in length is 250 cm, then what must be the volume of this cubical body.
1) 100.3 m^3 2) 81.2 m^3
3) 50.6 m^3 4) 20.4 m^3

4. The length of a thin metallic wire is l_1 when the tension in it is T_1 and l_2 when the tension in it is T_2. What is the original length of the metallic wire?
1) $\frac{T_2 l_1 - T_1 l_2}{T_2 - T_1}$ 2) $\frac{T_2 l_2 - T_1 l_1}{T_1 - T_2}$
3) $\frac{T_2 l_1 - T_1 l_2}{l_1 - l_2}$ 4) $(T_2 l_1 - T_1 l_2)(l_1 - l_2)$

5. A material has Poisson's ratio $\frac{1}{2}$. If a uniform rod of it suffers a longitudinal strain of 3×10^{-2}, then the percentage change in volume is
1) 1% 2) 0%
3) 5% 4) 7%

6. A uniform rod of length L and density ρ is being pulled along a smooth floor with a horizontal acceleration α (see figure). The magnitude of the stress at the transverse cross-section through the mid-point of the rod is .

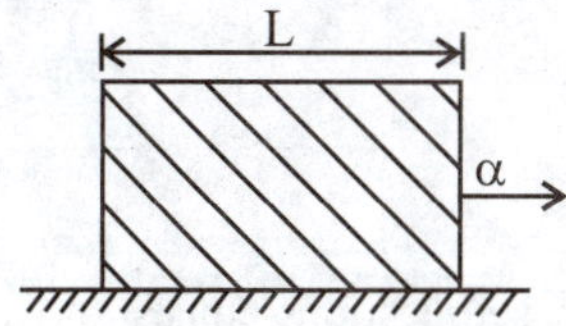

1) $\frac{1}{2}\rho\alpha L$ 2) $\frac{3}{2}\rho\alpha L$
3) $\frac{5}{2}\rho\alpha L$ 4) $2\rho\alpha L$

7. One end of a horizontal thick wire of length 2L and radius 2R is welded to another wire of same material of length L and radius R. When the arrangement is stretched by applying forces at two ends, find the ratio of the elongation in the thin wire to that in the thick wire
1) 0.25 2) 0.50
3) 2.00 4) 4.00

8. A bar of cross section A is subjected to equal and opposite tensile forces at the ends as shown below:

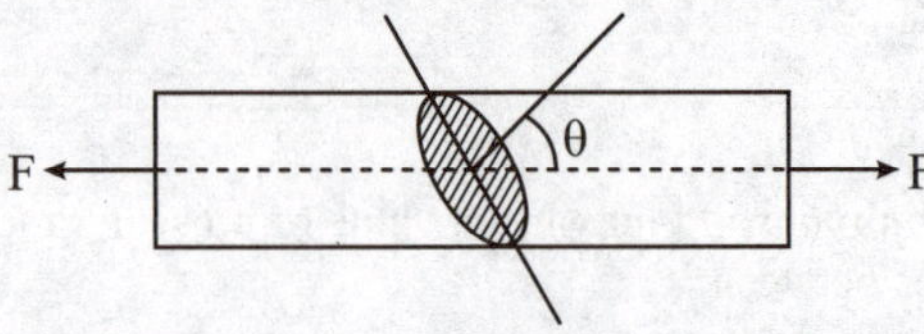

Let the normal to the cross section of the bar makes an angle θ with the axis of the bar. The tensile stress on this plane is

1) $\frac{F\cos^2\theta}{A}$ 2) $\frac{F}{A\cos^2\theta}$

3) $\frac{F\cos\theta}{A}$ 4) $\frac{F}{A\cos\theta}$

9. A sphere of radius 3 cm is subjected to a pressure of 100 atm. Its volume decreases by 0.3 cm³. Bulk modulus of the sphere is

1) $4\pi \times 10^7$ atm 2) $4\pi \times 3\times 10^3$ atm

3) $4\pi \times 10^5$ atm 4) $4\pi \times 5\ 10^4$ atm

10. Two wires A and B are of same material. Their lengths are in the ratio 1 : 2 and diameters are in the ratio 2 : 1. When stretched by forces F_A and F_B respectively, they get equal increase in their lengths. Then, the ratio $F_A : F_B$ should be

1) 1 : 2 2) 1 : 1

3) 2 : 1 4) 8 : 1

11. A man grows into a giant such that his linear dimensions increase by a factor of 9. Assuming that his density remains same, the stress in the leg will change by a factor of

1) $\frac{1}{9}$ 2) 81

3) $\frac{1}{81}$ 4) 9

12. A cable that can support a load of 800 N is cut into two equal parts. The maximum load that can be supported by either part is:

1) 100 N 2) 400 N

3) 800 N 4) 1600 N

13. The relation between Young's modulus (Y), bulk modulus (K) and modulus of rigidity (η) is

1) $\frac{9}{Y}=\frac{1}{K}+\frac{3}{\eta}$ 2) $\frac{1}{Y}=\frac{3}{\eta}+\frac{1}{3K}$

3) $\frac{1}{Y}=\frac{1}{K}+\frac{3}{\eta}$ 4) $\frac{1}{\eta}=\frac{1}{3K}+\frac{3}{Y}$

14. If longitudinal strain for a wire is 0.03 and its poisson ratio is 0.6, then its lateral strain is equal to

1) 0.018 2) 0.015

3) 0.005 4) 0.02

15. A solid sphere of radius R made of material of bulk modulus B is surrounded by a liquid in a cylindrical container. A massless piston of area A floats on the surface of a liquid when a mass m is placed on the piston to compress the liquid, the fractional change in the radius of the sphere $\frac{dR}{R}$ is

1) $\frac{mg}{4AB}$ 2) $\frac{mg}{3AB}$

3) $\frac{mg}{2AB}$ 4) $\frac{mg}{AB}$

16. Two wires of same length and radius are joined in parallel and loaded. The Young's modulii of the material of the wires are Y_1 & Y_2. If the combination is taken as a single wire then its Young's modulus is

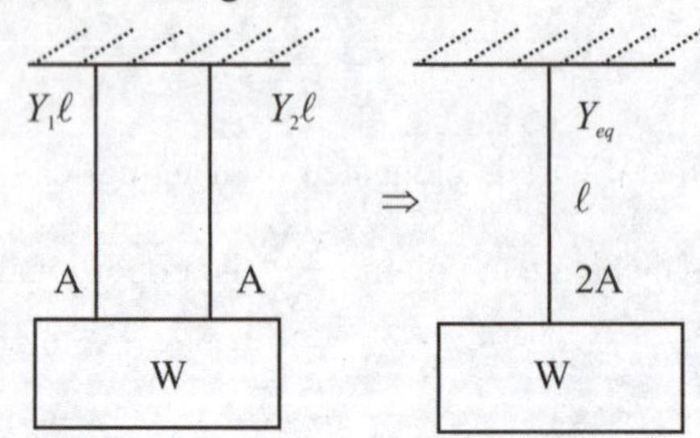

1) $\frac{Y_1Y_2}{Y_1+Y_2}$ (2) $\frac{Y_1+Y_2}{2}$

3) $\frac{2Y_1Y_2}{Y_1+Y_2}$ (4) $\frac{Y_1+Y_2}{Y_1Y_2}$

Statement Based MCQs

Directions: These questions consist of two statements each, printed as Statement-I and Statement-II. While answering these questions, you are required to choose any one of the following four responses.

1) **Both Statement-I and Statement-II are correct.**
2) **Both Statement-I and Statement-II are incorrect.**
3) **Statement-I is correct & Statement-II is incorrect.**
4) **Statement-I is incorrect & Statement-II is correct.**

1. **Statement-I:** Identical springs of steel and copper are equally stretched. More work will be done on the steel spring.

Statement-II: Steel is more elastic than copper.

2. **Statement-I:** Strain is a unit less quantity.

Statement-II: Strain is equivalent to force.

3. **Statement-I:** For an elastomer, stress vs strain graph is a straight line.

Statement-II: Steel is an example of elastomer.

4. **Statement-I:** Within the elastic limit, stress vs strain graph is a straight line.

Statement-II: Compressibility is the reciprocal of bulk modulus.

5. **Statement-I:** Bulk modulus is the ratio of normal stress to the volumetric strain.

Statement-II: Shear modulus is the ratio of tangential stress to the shearing strain.

Assertion & Reason MCQs

Directions: These questions consist of two statements each, printed as Assertion and Reason. While answering these questions, you are required to choose any one of the following four responses.

1) **Both Assertion and Reason are True and the Reason is a correct explanation of the Assertion.**
2) **Both Assertion and Reason are True but Reason is not a correct explanation of the Assertion.**
3) **Assertion is True but the Reason is False.**
4) **Assertion is False but Reason is True.**

1. **Assertion:** (i) Bridges are declared unsafe after a long time of their use.

   Reason: This is due to the loss of elastic properties because of the action of repeated alternating deforming forces.

2. **Assertion:** If a wire can bear maximum force F, then wire of same material but double thickness can bear maximum force 4F.

   Reason: Breaking force $\propto \frac{1}{\pi r^2}$.

3. **Assertion:** When a wire is stretched by a force along its length then its length increases and the radius decreases

   Reason: Poison's ratio = lateral strain/longitudinal strain

4. **Assertion:** A hollow shaft is found to be stronger than a solid shaft.

   Reason: The torque required to produce a given twist in hollow cylinder is greater than the torque required to produce same twist in a solid cylinder of same size and same material.

5. **Assertion:** Steel is more elastic than rubber.

   Reason: Young's modulus of steel is more than the Young's modulus of rubber.

Match the Columns MCQs

1. Match the following type questions.

	Column-I		Column-II
(a)	Longitudinal stress	(i)	Volume changes
(b)	Shear stress	(ii)	Shape changes
(c)	Volumetric stress	(iii)	Volume does not change
(d)	Tensile stress	(iv)	Shape does not change only size changes

1) (a)-(i,iv) (b)-(ii,iii) (c)-(i,iv) (d)-(i,iv)
2) (a)-(ii,iii) (b)-(i,iv) (c)-(i,iv) (d)-(i,iv)
3) (a)-(i,iv) (b)-(i,iv) (c)-(ii,iii) (d)-(i,iv)
4) (a)-(i,iv) (b)-(i,iv) (c)-(i,iv) (d)-(ii,iii)

2. With regard to dependence of quantities given in column-I and column-II, match the following:

	Column-I		Column-II
(a)	Young's modulus of a substance	(i)	Depends on temperature
(b)	Bulk modulus of a substance	(ii)	Depends on length
(c)	Modulus of rigidity of substance	(iii)	Depends on area of cross section
(d)	Volume of a substance for a given mass	(iv)	Depends on the nature of material

1) (a)-(i,iv) (b)-(i,iv) (c)-(i,iv) (d)-(i,ii,iii)
2) (a)-(i,iv) (b)-(i,iv) (c)-(i,iv) (d)-(i,iv)
3) (a)-(i,ii,iii) (b)-(i,iv) (c)-(i,iv) (d)-(i,ii,iii)
4) (a)-(i,ii,iii) (b)-(i,iv) (c)-(i,iv) (d)-(i,iv)

3. Match the column-I with column-II

	Column-I		Column-II
(a)	A body which regains its original shape after the removal of external force.	(i)	Elasticity
(b)	A body which does not regain its original shape after the removal of external force.	(ii)	Elastic body
(c)	A body which does not show any deformation on applying external force.	(iii)	Plastic body
(d)	The property of the body to regain its original configuration when the deforming forces are removed.	(iv)	Rigid body

1) (a)-(i) (b)-(iii) (c)-(iv) (d)-(ii)
2) (a)-(ii) (b)-(iv) (c)-(iii) (d)-(i)
3) (a)-(ii) (b)-(iii) (c)-(iv) (d)-(i)
4) (a)-(i) (b)-(iv) (c)-(ii) (d)-(iii)

NCERT Exemplar MCQs

1. Modulus of rigidity of ideal liquids is:
   1) Infinity
   2) Zero
   3) Unity
   4) Some finite small non-zero constant value
2. The maximum load a wire can withstand without breaking, when its length is reduced to half of its original length, will:
   1) Be half 2) Be double
   3) Be four times 4) Remain same
3. The temperature of a wire is doubled. The Young's modulus of elasticity:
   1) Will also double 2) Will become four times
   3) Will remain same 4) Will decrease
4. A spring is stretched by applying a load to its free end. The strain produced in the spring is:
   1) Volumetric
   2) Shear
   3) Longitudinal and shear
   4) Longitudinal
5. A rigid bar of mass M is supported symmetrically by three wires each of length *l*. Those at each end are of copper and the middle one is of iron. The ratio of their diameters, if each is to have the same tension, is equal to:
   1) $\frac{Y_{copper}}{Y_{iron}}$ 2) $\sqrt{\frac{Y_{iron}}{Y_{copper}}}$
   3) $\frac{Y^2_{iron}}{Y^2_{copper}}$ 4) $\frac{Y_{iron}}{Y_{copper}}$
6. A rectangular frame is to be suspended symmetrically by two strings of equal length on two supports (figure). It can be done in one of the following three ways:

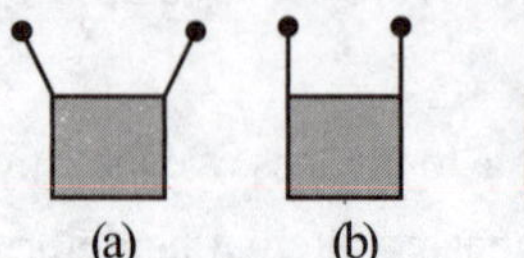

The tension in the string will be:
   1) The same in all cases
   2) Least in a
   3) Least in b
   4) Least in c

NEET & AIIMS Past Year MCQs

1. The maximum elongation of a steel wire of 1 m length if the elastic limit of steel and its Young's modulus, respectively, are 8×10^8 N m^{-2} and 2×10^{11} N m^{-2}, is: **(2024)**
   1) 40 mm 2) 8 mm
   3) 4 mm 4) 0.4 mm
2. Let a wire be suspended from the ceiling (rigid support) and stretched by a weight W attached at its free end. The longitudinal stress at any point of cross-sectional area A of the wire is: **(2023)**
   1) Zero 2) $\frac{2W}{A}$
   3) $\frac{W}{A}$ 4) $\frac{W}{2A}$
3. Given below are two statements: One is labelled as Assertion (A) and the other is labelled as Reason (R)

   Assertion (A): The stretching of a spring is determined by the shear modulus of the material of the spring

   Reason (R): A coil spring of copper has more tensile strength than a steel spring of same dimensions.

   In the light of the above statements, choose the most appropriate answer from the options given below: **(2022)**
   1) (A) is false but (R) is true
   2) Both (A) and (R) are true and (R) is the correct explanation of (A)
   3) Both (A) and (R) are true and (R) is not the correct explanation of (A)
   4) (A) is true but (R) is false
4. For a constant hydraulic stress on an object, the fractional change in the object's volume $\left(\frac{\Delta V}{V}\right)$ its bulk modulus (B) are related as **(AIIMS 2005)**
   1) $\frac{\Delta V}{V} \propto B$ 2) $\frac{\Delta V}{V} \propto \frac{1}{B}$
   3) $\frac{\Delta V}{V} \propto B^2$ 4) $\frac{\Delta V}{V} \propto B^{-2}$

Answer Key

NCERT Topic-wise MCQs

1	2	3	4	5	6	7	8	9	10	11	12	13	14	15	16	17
3	1	3	2	4	4	2	2	3	3	1	1	2	1	2	1	2
18	19	20	21	22	23	24	25	26	27	28	29	30	31	32	33	34
4	3	4	3	4	4	2	1	4	2	2	1	3	3	4	2	1
35	36															
2	3															

High Order Time Intensive MCQs

1	2	3	4	5	6	7	8	9	10	11	12	13	14	15	16
2	2	2	1	2	1	3	1	2	4	4	3	1	1	2	2

Statement Based MCQs

1	2	3	4	5
1	3	2	4	1

Assertion & Reason MCQs

1	2	3	4	5
1	3	2	1	1

Match the Column MCQs

1	2	3
1	2	3

NCERT Exemplar MCQs

1	2	3	4	5	6
2	4	4	3	2	3

NEET & AIIMS Past Year MCQs

1	2	3	4
3	3	4	2

CHAPTER 9

Mechanical Properties of Fluids

Topicwise Number of Questions Analysis (2025-2016)

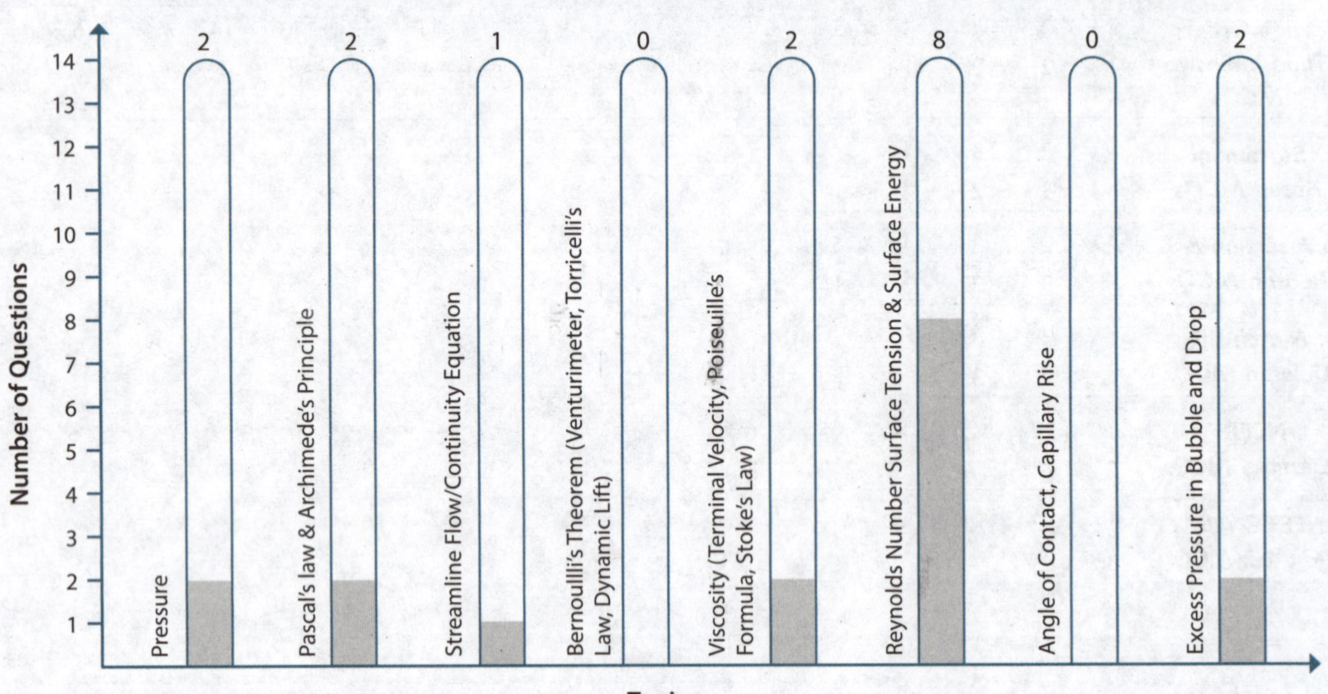

NCERT-Picks

Pressure

Formulas

1. $P = \frac{F}{A}$ where F is force and P is pressure exerted by the liquid and A is area.
2. Variation of pressure with depth

   Pressure difference between two positions inside a fluid that are separated at a depth of h are related by the expression:

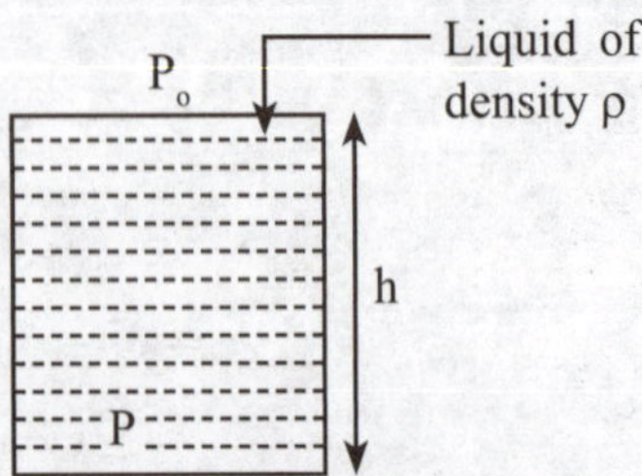

$P - P_o = h\rho g$

Pressure difference between hydrostatic and atmospheric pressure is known as gauge pressure.

Numerical Profile

Variation of Pressure with acceleration:

(i) When a vessel containing liquid moves in the upward direction with uniform acceleration 'a', $P = hd(g + a)$ where g is acceleration due to gravity.

(ii) When a vessel containing liquid is moving in the downward direction with uniform acceleration 'a', $P = hd(g - a)$ where, g is acceleration due to gravity

(iii) If a U shape tube is moving horizontally with an acceleration 'a' as figure shows. then $\tan\theta = \frac{h}{l} = \frac{a}{g}$

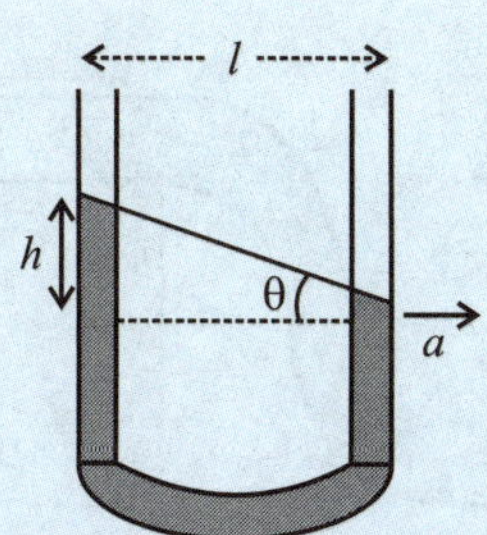

(iv) If container is accelerated 'a' at some angle with the horizontal,

$$\tan\alpha = \frac{a_x}{a_y + g}$$

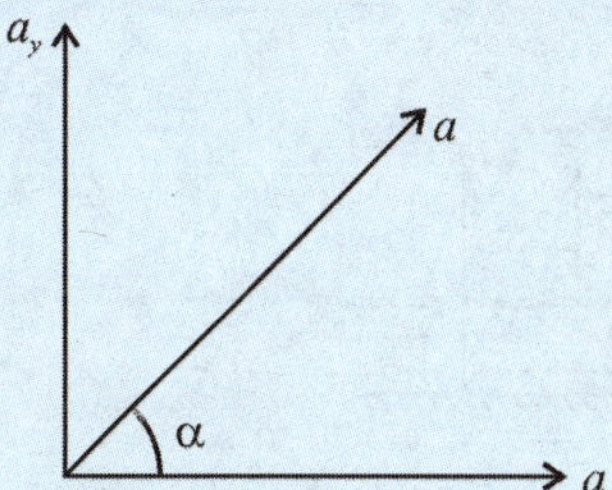

a_y = vertical component of acceleration of container $= a\sin\alpha$

a_x = horizontal component of acceleration of container $= a\cos\alpha$

α = angle of inclination of free surface of the liquid with horizontal.

Pascal's Law and Archimede's Principle

Formulas

1. Hydraulic Lift:

If d_1 and d_2 are the distances covered by the smaller and larger piston.

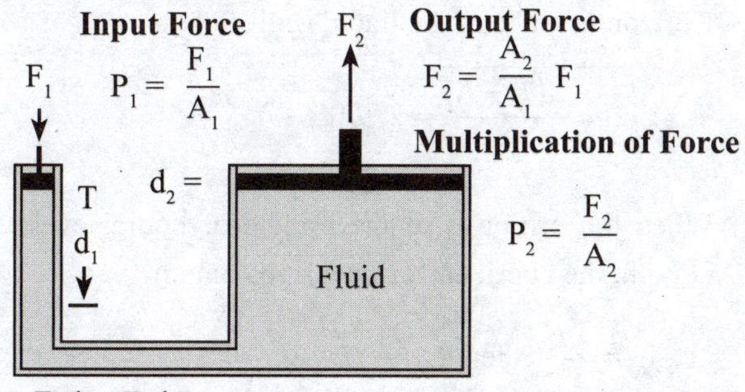

$$F_1 d_1 = F_2 d_2$$

$$d_1 = \frac{F_2}{F_1} d_2 = \frac{A_2}{A_1} d_2$$

2. Archimedes Principles

(a) Apparent weight = Actual weight – upthrust

$$W_{app} = W\left(1 - \frac{\sigma}{\rho}\right)$$

ρ = body density

σ = liquid density

(b) If a body of volume V is immersed in liquid of density σ

W_1 = Weight of the body in air

W_2 = Weight of the body in water

Apparent weight loss $= W_1 - W_2 = V\sigma g$

$$V = \frac{W_1 - W_2}{\sigma g}$$

(c) When a body floats then the weight of body = upthrust

$$V\rho g = V_{in}\sigma g \Rightarrow V_{in} = \left(\frac{\rho}{\sigma}\right)V$$

$$V_{out} = V - V_{in} = \left(1 - \frac{\rho}{\sigma}\right)V$$

$$\frac{V_{out}}{V} = \left(1 - \frac{\rho}{\sigma}\right)$$

Numerical Profile

Density and Relative density

$$\text{Density} = \frac{\text{Mass of body}}{\text{Volume of body}} = \frac{M}{V}$$

$$\text{Relative density} = \frac{\text{Density of body}}{\text{Density of water}}$$

$$\text{R.D} = \frac{\text{Weight of body}}{\text{loss of weight in water}} = \frac{w_1}{w_1 - w_2}$$

Density of a mixture:

(i) When two liquids of masses m_1, m_2 and densities ρ_1, ρ_2 are mixed then the effective density of the mixture is

$$\rho = \frac{M}{V} = \frac{m_1 + m_2}{V_1 + V_2} = \frac{m_1 + m_2}{\left(\frac{m_1}{\rho_1} + \frac{m_2}{\rho_2}\right)} = \frac{(m_1 + m_2)\rho_1\rho_2}{m_1\rho_2 + m_2\rho_1}$$

(ii) If masses of the two liquids are equal i.e., $m_1 = m_2$ then density of the mixtures $\rho = \frac{2\rho_1\rho_2}{\rho_1 + \rho_2}$

(iii) If two liquids having volumes V_1, V_2 and densities ρ_1, ρ_2 respectively are mixed together, then the resultant density of the mixture is

$$\rho = \frac{m}{V} = \frac{m_1 + m_2}{V_1 + V_2} = \frac{V_1\rho_1 + V_2\rho_2}{V_1 + V_2}$$

Note: If $V_1 = V_2$ then, $\rho = \frac{\rho_1 + \rho_2}{2}$

Variation of Density with Pressure

With rise in pressure, volume reduces, density increases:

$$\rho_0 = \rho\left[1 - \frac{\Delta p}{K}\right]$$

$$\Rightarrow \rho = \rho_0\left[1 - \frac{\Delta p}{K}\right]^{-1} = \rho_0\left(1 + \frac{\Delta p}{K}\right)$$

where ρ_0 = density at NTP

ρ = Density at desired pressure

Δp = Increase in pressure

K = Bulk modulus

Streamline Flow

Formulas

1. **Sreamline flow**

   At any point of the streamline flow every particle has the same velocity in magnitude and direction as that of its proceeding particle.

   ∴ At any point of streamline, velocity = const

   Note: Two streamlines never intersect each other.

2. **Equation of Continuity**

   (a) For the streamline flow of an ideal fluid the rate of flow of fluid remains constant.

   $\Rightarrow a_1v_1 = a_2v_2$

   Where a_1, a_2 are the area of cross sections and v_1 and v_2 are the velocities of flow.

   (b) Rate of flow of a liquid through a tube $Q = \frac{Volume}{time}$

   S.I unit of Q is m^3s^{-1}

   Equation of continuity represents the conservation of mass in case of moving fluids.

Bernoulli's Theorem, Torricelli's Law

Formulas

1. **Bernoulli theorem:**

   $P + \rho gh + \frac{1}{2}\rho v^2 = \text{constant}$

   where P is pressure, ρ is density of the liquid, h is height of the tube and v is speed of the liquid through the tube.

2. **Toricelli theorem:**

   Speed of efflux, $v = \sqrt{2gh}$

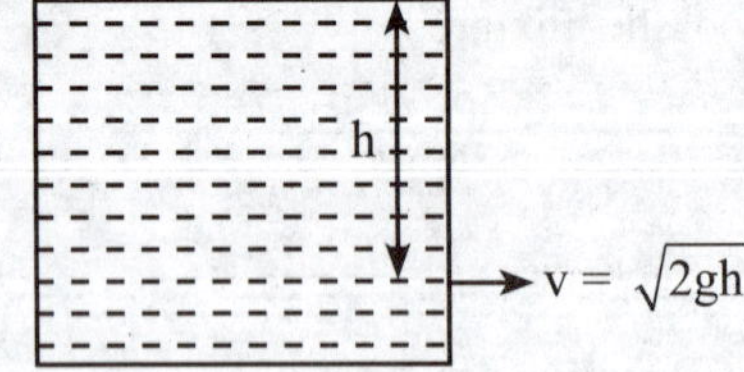

Numerical Profile

According to Bernoulli

$$p_1 + \frac{1}{2}\rho v_1^2 + \rho g y_1 = p_2 + \frac{1}{2}\rho v_2^2 + \rho g y_2$$

If the two ends of a tube are at the same height then $y_1 = y_2$, then

$$p_1 + \frac{1}{2}\rho v_1^2 = p_2 + \frac{1}{2}\rho v_2^2$$

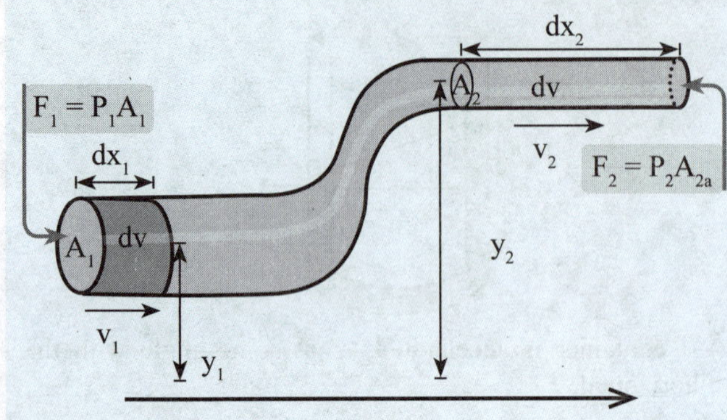

The speed of efflux:

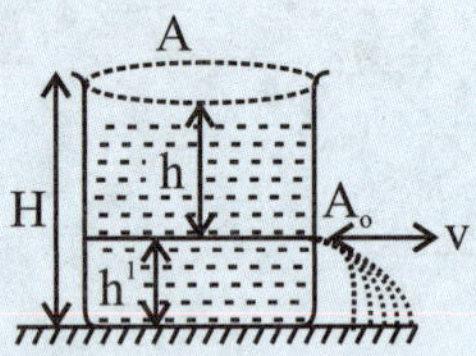

(a) The speed of efflux of a fluid through a hole at the bottom of a tank filled to a depth 'h' is given by: $\boxed{v = \sqrt{2gh}}$ where the letter g is the acceleration.

(b) Time taken by the efflux liquid to reach the ground is given as $t = \sqrt{\frac{2(H-h)}{g}}$

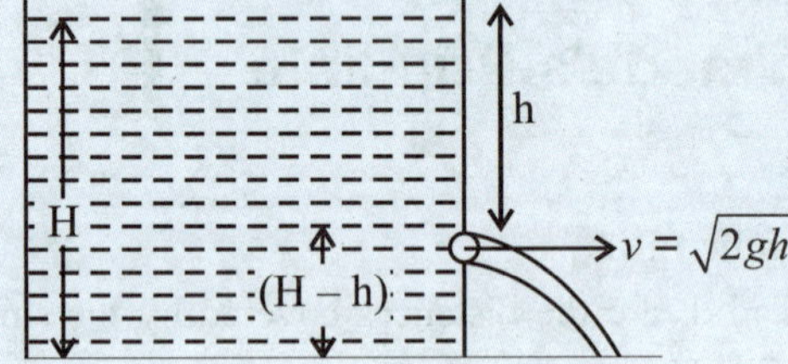

(c) Horizontal range of the liquid is given by $R = V \times t$

$$R = \sqrt{2gh.\frac{2(H-h)}{g}} = 2\sqrt{h(H-h)}$$

(d) When the orifice is at the middle of liquid level and bottom, then horizontal range is maximum

i.e., if $h = \frac{H}{2}$ then $R_{Max} = H = 2h$

(e) If the hole is at the bottom of the tank & A and A_o are the area of cross section of tank and orifice. Then time taken to empty the tank $t = \frac{A}{A_o}\sqrt{\frac{2H}{g}}$

Time taken by the water to fall from level h to (H – h)

$$t = \frac{A}{A_o}\sqrt{\frac{2h}{g}} - \frac{A}{A_o}\sqrt{\frac{2(H-h)}{g}}$$

Viscosity, (Terminal Velocity, Poiseuille's Formula, Stoke's Law)

Formulas

1. Viscous Drag

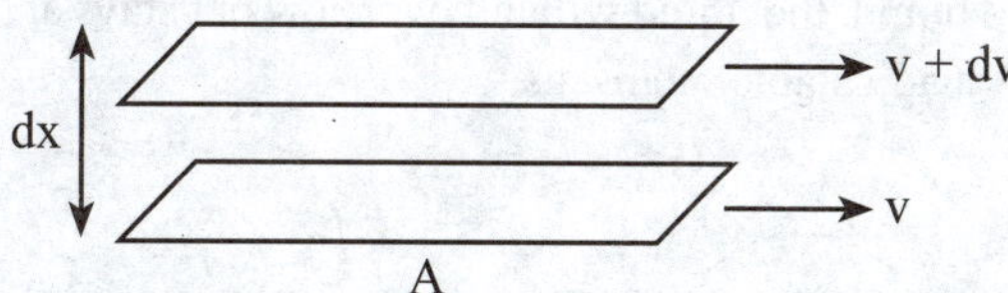

Viscous drag between two layers of liquid is given by

$$F = -\eta \frac{Adv}{dx}$$

2. **Stoke's Law**

The force of viscosity on a small sphere moving through a viscous fluid is given by:

$F = 6\pi\eta rv$

Where, F is the viscous force

η is the coefficient of viscosity

r is the radius of the spherical object

v is the flow velocity relative to the object

3. **Terminal Velocity**

It is the maximum velocity acquired by a body while falling through a viscous liquid.

Terminal velocity, $v = \frac{2}{9}\frac{(\rho - \sigma)r^2 g}{\eta}$

4. **Poiseuille's Formula**

Rate of flow of a liquid through a tube of radius r

$$V = \frac{\pi P r^4}{8\eta l}$$

Where, η is the coefficient of viscosity of the liquid

l is the length of the tube

P is the difference is pressure

Critical Velocity & Reynolds Number

Formulas

1. **Critical Velocity**

Critical velocity of liquid $v_c = \frac{R_e \eta}{\rho D}$

R_e = Reynold number

D = Diameter of tube

2. **Reynold's Number**

$$R_e = \frac{\rho v d}{\eta}$$

Where, ρ = density of the fluid

v = speed of the fluid

d = diameter of pipe

η = coefficient of viscosity of the fluid

For $R_e < 1000$ flow remains streamline or laminar flow

For $R_e > 2000$ for turbulent flow

For $1000 < R_e < 2000$ flow is unsteady

Surface Tension

Formulas

1. Surface Tension (S) is given by $S = \frac{F}{\ell}$

Force due to surface tension:

(a) On a needle

$F = 2lS$

(b) On a square frame

$F = 2 \times 4l \times S$

(c) On a square plate

$F = 4lS$

(d) On a solid disc

$F = 2\pi r \times S$

(e) On a thin ring

$F = 2\pi(r + r)S$

2. Work done or energy stored due to surface tension

$W = S \times \Delta A$

Where ΔA is increase in area

Numerical Profile

Work done under different cases

(i) Work done in forming a liquid drop is, W = Change in surface Area × Surface tension = $\Delta A \times S = 4\pi r^2 S$.

(ii) Work done in increasing the size of liquid drop from radius r_1 to r_2 is $W = 4\pi S(r_2^2 - r_1^2)$

(iii) Work done in blowing a soap bubble is $w = 8\pi r^2 S$.

(iv) Work done in increasing the size of a soap bubble from radius r_1 to r_2 is, $W = 8\pi S(r_2^2 - r_1^2)$

(v) Work done to split a liquid drop of radius 'R' into 'n' no. of identical droplets is

$W = 4\pi R^2 S(n^{1/3} - 1)$.

(vi) Work done to form a big drop from 'n' identical droplets each of radius 'r' is $W = 4\pi r^2 S(n - n^{2/3})$.

(vii) If a number of identical smaller drops each of radius 'r' merge to form a single big drop of radius R, the rise in temperature of the drop is $\frac{3S}{dJS}\left[\frac{1}{r} - \frac{1}{R}\right]$

(viii) A large number of identical smaller drops each of radius r merge to form a single drop of radius R. The energy released in the process is converted into KE of bigger drop formed. Then speed of the bigger drop formed is $\sqrt{\frac{6S}{d}\left[\frac{1}{r} - \frac{1}{R}\right]}$.

Angle of Contact

It is the angle which the tangent to the liquid surface makes with the solid surface which is inside the liquid.

θ is the angle of contact

Formulas

1. If angle of contact $\theta < 90°$ = Liquid wets the walls of the container, meniscus concave upward

   $\theta > 90°$ = Liquid does not wet the walls of the container, meniscus convex upward.
2. The value of angle of contact for pure water and glass (without grease) is zero.
3. Angle of contact decreases on adding soluble impurity to a liquid.
4. Angle of contact increases with increase in temperature of liquid.

Capillarity

Formulas

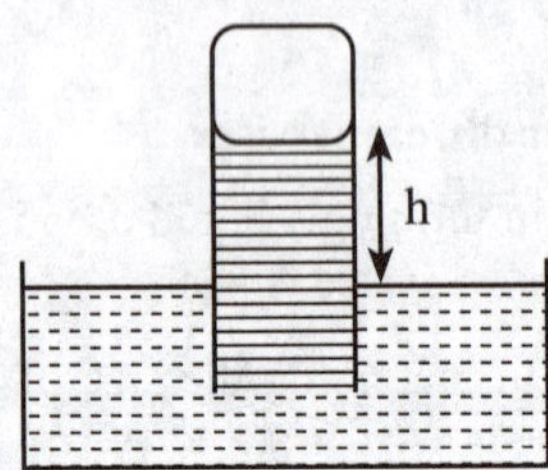

$$h = \frac{2S\cos\theta}{\rho r g} \text{ (Ascent formula)}$$

Where, h is the rise of liquid in the tube

ρ is the density of the liquid

S is the surface tension of the liquid

θ is the angle of contact

Numerical Profile

Rise of liquid in a capillary tube

(i) If the radii of the two limbs of a U-tube are r_1 and r_2, the difference between the levels of a liquid poured

$$h = \frac{2S}{dg}\left[\frac{1}{r_1} - \frac{1}{r_2}\right] \text{ here } r_1 < r_2$$

(ii) When the lift is moving down with an acceleration 'a' then the height of the liquid in the tube is given by

$$h' = h\left[\frac{g}{g-a}\right]$$

(iii) When the lift is moving up with an acceleration 'a' then the height of the liquid in the tube is given by

$$h' = h\left[\frac{g}{g+a}\right]$$

(iv) When a capillary tube of insufficient length is dipped in the liquid, the liquid will not overflow but stays at the top with adjustable meniscus.

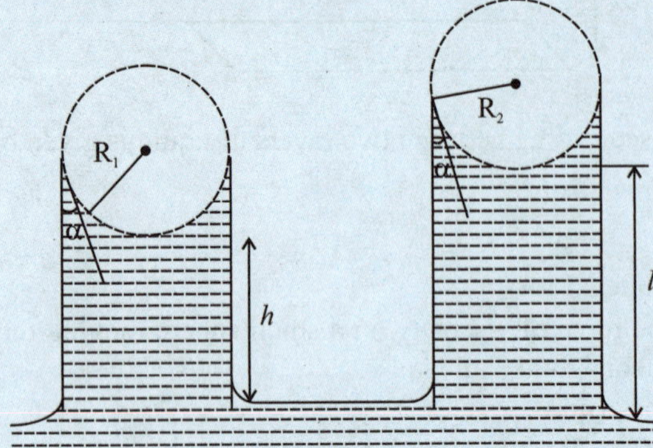

$\because$ hR = constant

For a tube of insufficient length h is less therefore R increases.

Excess Pressure in Soap Bubble and Liquid Drop

Formulas

1. Excess pressure inside a liquid drop

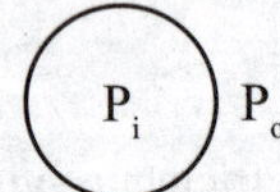

$$P_i - P_o = \frac{2S}{R}$$

$$\Delta P = \frac{2S}{R}$$

2. Excess pressure inside a soap bubble

$$P_i - P_o = \frac{4S}{R}$$

$$\Delta P = \frac{4S}{R}$$

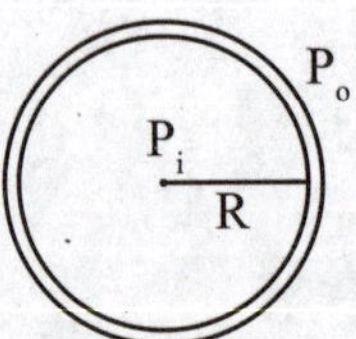

3. Total pressure inside the air bubble at a depth h below the surface of a liquid of density ρ is

$$P = P_o + h\rho g + \frac{2S}{R}$$

(where P_0 = atmospheric pressure)

Numerical Profile

Combination of drops and Excess pressure

(i) Two spherical drops are formed in vacuum. If they coalesce to form a single spherical bubble at constant temperature then the radius of the bubble so formed is $r = \sqrt{r_1^2 + r_2^2}$

(ii) As the bubbles are in vacuum. the pressure of air inside them are $p_1 = \frac{4S}{r_1}$ and $P_2 = \frac{4S}{r_2}$.

(iii) If the bubbles come together, the difference of the excess pressure on either side of the common surface must be same $\left(\frac{4S}{r_1} - \frac{4S}{r_2} = \frac{4S}{r}\right)$

(iv) When a soap bubble of radius r_1 and another of radius r_2 are brought together, then the radius of curvature 'R' on the common interface is

$$R = \frac{r_1 r_2}{r_2 - r_1} (\text{If } r_2 > r_1)$$

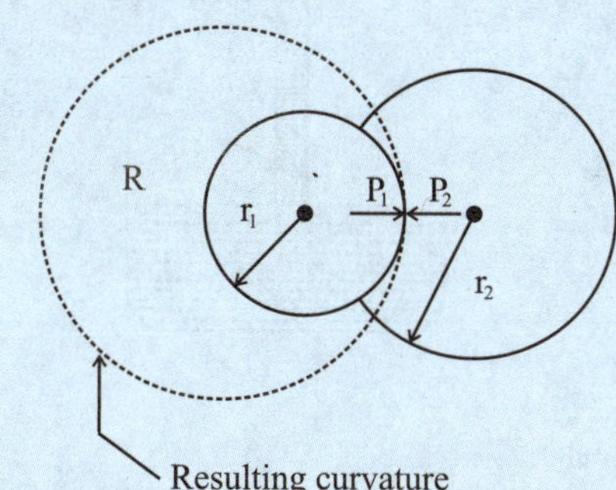

(v) If p_1 and p_2 be the excess pressure in soap bubbles of radii r_1 and r_2 respectively, then the resultant excess pressure is given as $p = p_1 - p_2$

If r be the radius of the interface, then

$$\frac{4S}{r} = \frac{4S}{r_1} - \frac{4S}{r_2}$$

or $\frac{1}{r} = \frac{1}{r_1} - \frac{1}{r_2}$ or $\frac{1}{r} = \frac{r_2 - r_1}{r_1 r_2}$

or $r = \frac{r_1 r_2}{r_2 - r_1}$

Statements Asked in Exam

- When a liquid is in equilibrium the force acting on its surface is perpendicular every where.
- Hydrometer device for measuring some characteristics of a liquid, such as its density (weight per unit volume) or specific gravity (weight per unit volume compared with water)
- Hydrostatic pressure is a scalar quantity even though pressure is force divided by area and force is a vector.
- If a body just float in a liquid (density of the body is equal to the density of liquid) then the body sinks if it is pushed downwards.
- The force required by a man to move his limbs immersed in water is smaller than the force for the same movement in air. This is due to the upthrust by water.
- Water flows faster than honey due to less viscosity
- The velocity of fall of a man jumping with a parachute first increases and then becomes constant.
- We cannot sip a drink with a straw on the moon because there is no atmosphere on the moon.
- Surface tension does not depend on the area of the surface.
- When the liquid drops merge with each other to form a larger drop, energy is released.
- Liquid meniscus is concave in the tube containing water.
- Height of the liquid Column-In a capillary tube on the moon is six times than that of earth because gravitational force on moon is less than that of earth's gravitational field.
- Angle of contact is independent of the angle of inclination of the walls.
- A stream line is a map of fluid flow. In a steady flow two streamlines do not intersect as it means that the fluid particle will have two possible velocities at the point of intersection.
- The blood pressure in human is greater at the feet than at the brain.
- Bernoulli's principle does not hold in presence of viscous drag on the fluid. The work done by this dissipative force must be taken into account in this case.
- Water with detergent dissolved in it has small angles of contact (θ) this is because for a small θ, there is a fast capillary rise of the detergent in the cloth. The capillary rise of a liquid is directly proportional to the cosine of the angle of contact (θ).
- Surface tension of a liquid is zero at its boiling point and it vanishes at the critical temperature.
- Surface tension of a liquid is independent of the area of liquid surface.
- Small drops of mercury are spherical and large drops become flat.
- It is better to wash the clothes in hot soap solution.
- Surface tension concept is only holds for liquids but not for gases which are also fluids.
- The tip of the nib of a pen is split in order to provide a capillary which helps the ink to rise to the end of the nib and enables it to write continuously.
- A fluid flowing out of a small hole in a vessel, results in a backward thrust on the vessel.
- When a shaving brush is taken out of water its hair cling together.
- Deep water runs slow.
- At narrower portions where the streamlines are closely spaced, velocity increases and its vice versa.

NCERT Topic-wise MCQs

Pressure & Density

1. If pressure at half of depth of a lake is equal to the $\frac{1}{4}$ of the pressure at the bottom of the lake then what is the depth of the lake?

1) 30.6 m 2) 41.6 m
3) 53.4 m 4) 56 m

2. A uniformly tapering vessel is filled with a liquid of density of 1000 kg/m³. What is the force acting on the base of the vessel due to the liquid?

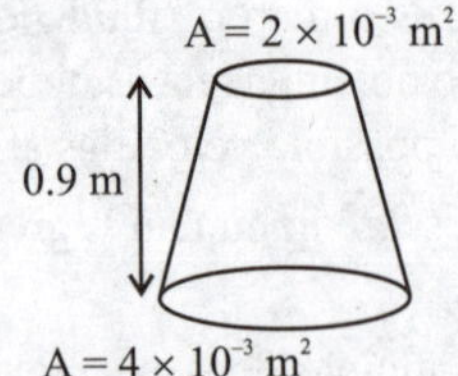

1) 15 N 2) 30 N
3) 35 N 4) 40 N

3. A beaker is filled with two non-mixing liquids 1 and 2 having densities ρ_1 and ρ_2 respectively. A solid ball made of a material of density ρ_3 is dropped in the beaker. It comes in equilibrium position after few seconds. Which of the following relation is true?

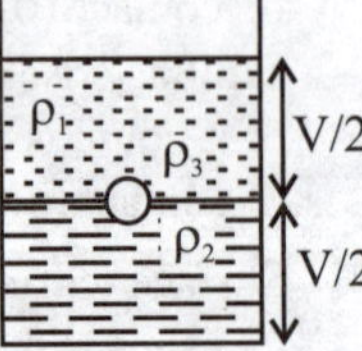

1) $\rho_2 > \rho_3 > \rho_1$ 2) $\rho_3 > \rho_2 > \rho_1$
3) $\rho_1 > \rho_2 > \rho_3$ 4) None

4. The volume of an air bubble becomes two times as it rises from the bottom of a lake to its surface. Assuming atmospheric pressure 75 cm of Hg and the density of water is $\frac{1}{11}$ times the density of mercury. The depth of the lake is:

1) 11 m 2) 5.6 m
3) 7 m 4) 6.8 m

5. A barometer tube reads 76 cm of mercury. If the tube is gradually inclined at an angle of 30° with the vertical, keeping the open end immersed in the mercury reservoir, the length of the mercury column will be

1) 70 cm 2) 89 cm
3) 88 cm 4) 100 cm

6. The pressure on a swimmer 40 m below the surface of water at sea level is.

1) 4.08×10^5 N/m² 2) 5.01×10^5N/m²
3) 0.4×10^5 N/m² 4) None

7. A barometer made of a very narrow tube (see figure) is placed at normal temperature and pressure. The coefficient of volume expansion of mercury is 0.00018/°C and that of the tube is negligible. The temperature of mercury in the barometer is now raised by 1°C but the temperature of the atmosphere does not change. Then, the mercury height in the tube remains unchanged.

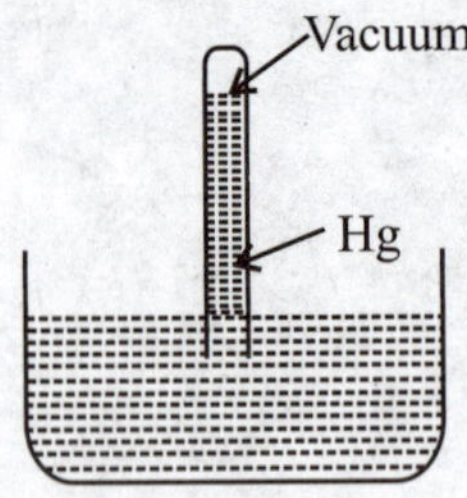

1) Increase
2) Decrease
3) No change
4) First decrease then increase

8. In a wind tunnel experiment, the pressure on the upper and lower surface of the wings are 5×10^5 Pa and 10×10^5 Pa respectively. If the area of the wing is 80 m², the net lifting force on the wing is.

1) 4×10^7N 2) 5×10^7 N
3) 3×10^7 N 4) 2×10^7N

9. The height of a mercury barometer is 90 cm at sea level and 60cm at the top of a hill. The ratio of density of mercury to that of air is 10^4. The height of the hill is

1) 2 km 2) 3 km
3) 4 km 4) 5 km

Pascal's Law & Archimedes Principle

10. An iceberg of density 700 kg/m³ is floating in water of density 1000 kg/m³. The percentage of volume of ice-cube outside the water is.

1) 30% 2) 40%
3) 50% 4) 60%

11. A cylinder is floating in a beaker which contains the liquids. The height h_1 and h_2 of the cylinder is submerged in the liquids of densities ρ_1 and ρ_2 respectively as shown in the figure. What will be change in submerged length of cylinder in fluids-2, if we increase the height of fluid-1 above fluid-2?

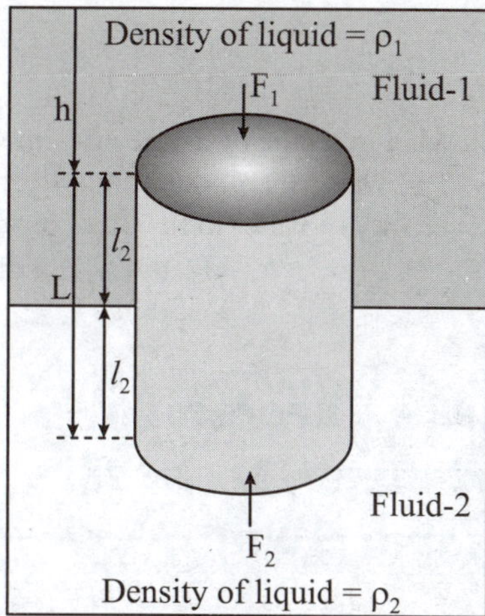

1) $l_1 + l_2$
2) l_1
3) Zero
4) l_2

12. A block floats in water with two-third of its volume above the surface of water. If it is placed in a liquid, it floats with half of its volume above the surface of the liquid. The specific gravity of the liquid is

1) 4 : 1
2) 3 : 2
3) 1 : 4
4) 2 : 3

13. A solid uniform ball of volume V floats on the interface of two immiscible liquids (Fig.) The specific gravity of the upper liquid is ρ_1 and that of lower one is ρ_2 whilst the specific gravity of ball is $\rho(\rho_1 < \rho < \rho_2)$. The fraction of the volume of the ball in the upper liquid is

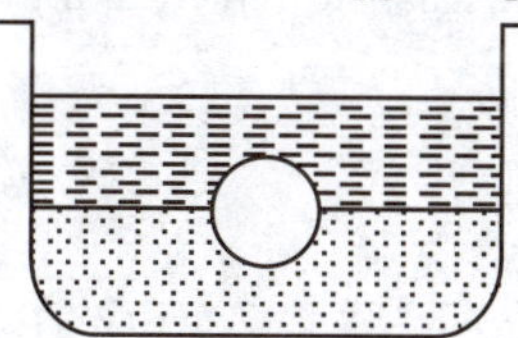

1) $\frac{\rho_2}{\rho_1}$
2) $\frac{\rho_2 - \rho}{\rho_2 - \rho_1}$
3) $\frac{\rho - \rho_1}{\rho_2 - \rho_1}$
4) $\frac{\rho_1}{\rho_2}$

14. A motor boat having a length of 10m and breadth 5m is floating on a lake. The boat sinks by 5cm when load gets on it. What is the mass of load?

1) 3000 g
2) 2500 Kg
3) 4010 Kg
4) None

15. In the hydraulic car lift as shown in the figure, the input piston on the left has a radius of r_1 = 0.010 m and a negligible weight. The output plunger on the right has a radius of r_2 = 0.20 m. The combined weight of the car and the plunger is 20,000 N. Suppose that the bottom surfaces of the piston and plunger are at the same level, so that h = 0 m in figure. What is the magnitude F_1 of the input force needed?

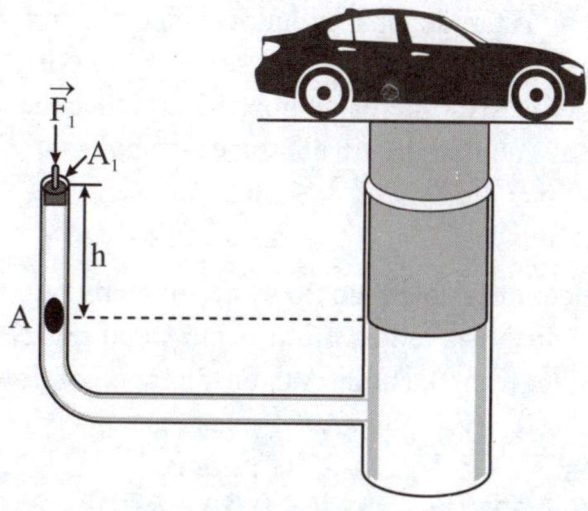

1) 10 N
2) 500 N
3) 50 N
4) 20 N

16. A log of wood floats in water with $\frac{1}{5}$ of its volume above the surface. What is the density of wood?

1) 600 kg/m³
2) 800 kg/m³
3) 1200 kg/m³
4) 500 kg/m³

17. A U-tube of uniform cross section as shown in figure is partially filled with a liquid I. Another liquid II which does not mix with liquid I is poured into one side. It is found that the liquid levels of the two sides of the tube are the same, while the level of liquid I has risen by 2 cm. If the specific gravity of liquid I is 1.1 the specific gravity of liquid II must be

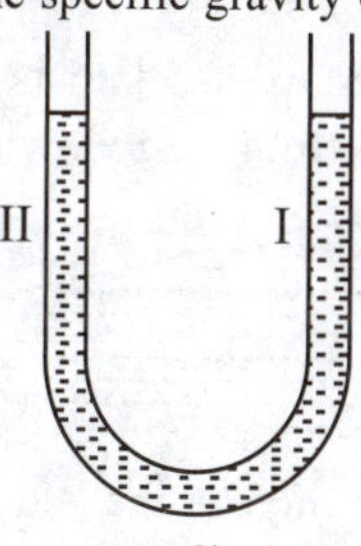

1) 1.12
2) 1.1
3) 1.05
4) 1.0

18. A body floats in a liquid contained in a beaker. The whole system is falling under gravity. The upthrust on the body due to liquid is

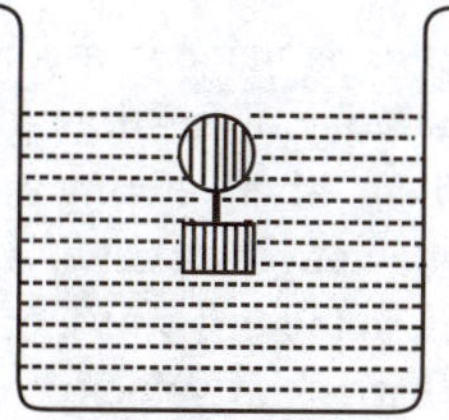

1) Zero
2) Equal to weight of liquid displaced
3) Equal to weight of the body in air
4) Equal to weight of the immersed body

Streamline Flow

19. A fluid is flowing through a tube of length L. The radius of the tube is r and the velocity of the fluid is v. If the radius of the tube is increased to 2r then what will be the new velocity?

1) 4 v
2) v/4
3) v/2
4) 2v

20. A liquid is flowing in a cylindrical pipe of internal diameter 4 cm with a velocity of 5 m/s. If this tube is joined with another tube of internal diameter 2 cm, then the velocity of flow of liquid in the smaller tube will be

1) 10 m/s 2) 20 m/s
3) 40 m/s 4) 5 m/s

21. An incompressible fluid flows steadily through a cylindrical pipe which has radius 4R at point A and radius R at point B farther from the point A in the direction of flow of water. If the velocity at point A is 5 m/s then velocity at B is

1) 5m/s 2) 20m/s
3) 40 m/s 4) 80 m/s

22. Water is flowing through a tube of non uniform cross-section. If the radius of the tube at the entrance and exit is 3 : 2. Then the ratio of velocity of liquid entering and leaving the tube is

1) 8 : 27 2) 4 : 9
3) 1 : 1 4) 9 : 4

23. Water is flowing through a tube of non-uniform cross-section. If the radii of the tube at the entrance and the exit points are in the ratio of 5 : 4, then the ratio of the velocities of flow of water at the entrance and exit is.

1) 25 : 16 2) 16 : 25
3) 12 : 15 4) 10 : 13

24. A liquid is allowed to flow into a cone shape pipe. Choose the correct option from the following.

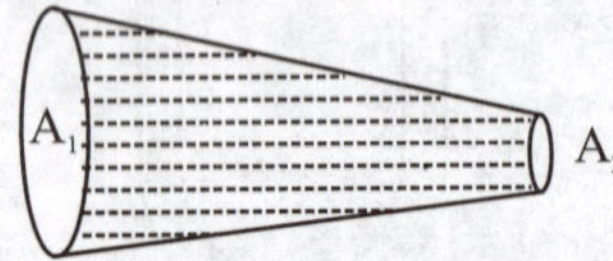

1) The velocity is high at the wider end and low at narrow end
2) The velocity low at the wider end and high at the narrow end
3) The velocity is same at both the ends
4) The liquid/fluid flows with uniform velocity in pipe

25. A water connection is made in a colony from main water supply plant. Water is flowing through a pipe line of area of cross-section 70 cm² after entering into colony it is divided into five pipe line network of area A_1, A_2, A_3, A_4 in which water flows with velocities 5 m/s, 5 m/s, 2 m/s and 1 m/s respectively. What is the velocity of water in main pipe line? (in m/s)

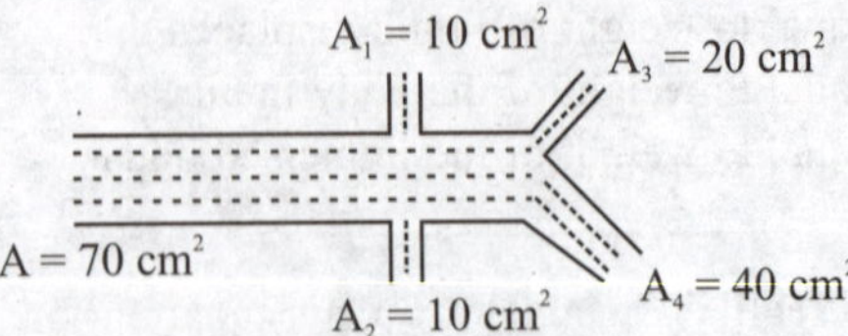

1) 2.57 2) 3.4
3) 5.1 4) 6.2

26. An ideal liquid flows through the pipe, which is of uniform cross - section area. The speed V_A and V_B, and pressure P_A and P_B at points A and B respectively are

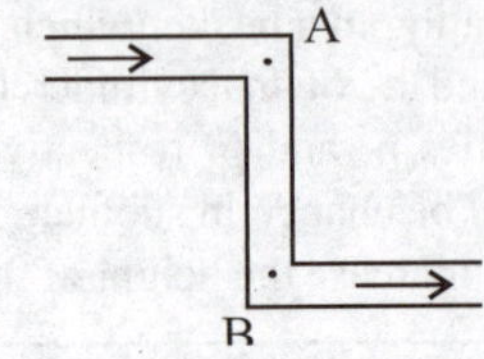

1) $V_A = V_B$ 2) $V_B > V_A$
3) $P_A = P_B$ 4) $P_B > P_A$

27. Water is flowing in a pipe of diameter 4 cm with a velocity of 20 m/s. The water then enters into a tube of diameter 2 cm. The velocity of water in the other pipe is:

1) 40 m/s 2) 10 m/s
3) 80 m/s 4) 50 m/s

Bernoulli's Theorem (Venturi-meter, Torricelli's Law Dynamic Lift)

28. A diagram shows a venturi-meter, through which water is flowing. The speed of water at X is 2 cm/s. The speed of water at Y ($g = 1000\ cm/s^2$) is

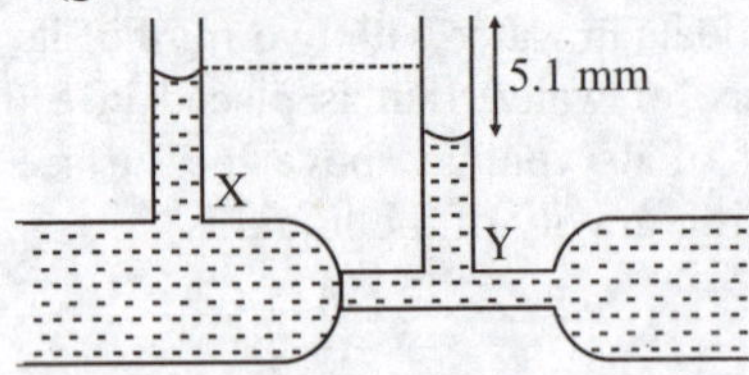

1) 23 cm/s 2) 32 cm/s
3) 101 cm/s 4) 1024 cm/s

29. A large tank filled with water to a height 'h' is to be emptied through a small hole at the bottom. The ratio of time taken for the level of water to fall from h to $\frac{h}{2}$ and from $\frac{h}{2}$ to bottom is

1) $\sqrt{2}$ 2) $\frac{1}{2}$
3) $\sqrt{2}-1$ 4) $\frac{1}{\sqrt{2}-1}$

30. Three tubes A, B and C are connected to a horizontal pipe in which liquid is flowing. The radii of pipe at the joints of A, B and C are 2 cm, 1 cm and 2 cm respectively. The height of liquid

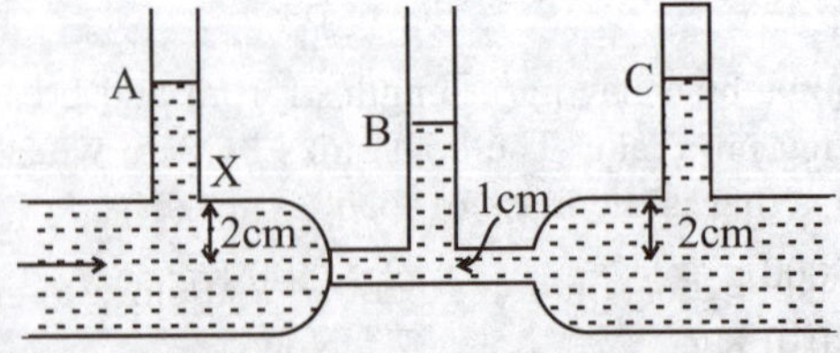

1) In A is maximum 2) In A and B is equal
3) Is same in all the three 4) In A and C is same

31. In a test experiment on a model aeroplane in a wind tunnel, the flow speeds on the upper and lower surfaces of the wing are 70 m/s and 63 m/s respectively. What is the lift on the wing if its area is 2.5 m²? The density of air is 1.3 kg/m³.

1) 1000 N 2) 2500 N
3) 1500 N 4) 2000 N

32. The weight of an aeroplane flying in the air is balanced by
    1) Vertical component of thrust created by air currents striking the lower surface of the wing.
    2) Force due to the pressure difference between the upper and lower surface of the wings created by different air speeds on the surfaces.
    3) Force due to reaction of gases ejected by the revolving propeller.
    4) Upthrust of the air which will be equal to the weight of the air having the same volume as the plane.

33. Bernoulli's equation is applicable in the case of:
    1) Streamlined flow of compressible fluids
    2) Streamlined flow of incompressible fluids
    3) Turbulent flow of compressible fluids
    4) Turbulent flow of incompressible fluids

34. At what speed, the velocity head of the water is twice than that of pressure head of 75 cm of Hg?
    1) 10 m/s 2) 20 m/s
    3) 30 m/s 4) 40 m/s

35. Water is flowing with a velocity of 5 m/s in a horizontal pipe with cross-sectional area decreasing from 5×10^{-2} m^2 to 2×10^{-2} m^2 at pressure 8×10^4 pascal. What will be the pressure at smaller cross-section?
    1) 130 kPa 2) 400 kPa
    3) 146 kPa 4) 200 kPa

36. A small hole is made at a height of $\frac{1}{\sqrt{3}}$m from the bottom of the cylindrical water tank and at a depth of h = 1.71 m from the upper level of water in the tank. What is the distance, where the water emerging from the hole strike the ground?
    1) 6 m 2) 4 m
    3) 3 m 4) 2 m

37. A tank containing liquid has an orifice in one vertical side. If the center of orifice is 19.6 m below the surface level in the tank. Find the velocity of discharge assuming there is no wastage of energy.
    1) 18.4 m/s 2) 19.6 m/s
    3) 15 m/s 4) 9.8 m/s

38. Dynamic lift caused by spinning is called
    1) Torricelli's law 2) Archimedes Principle
    3) Pascal's law 4) Magnus effect

39. Two motor boats moving parallel and close to each other they will be:
    1) Pulled towards each other
    2) Repel from each other
    3) Keep moving uniformly
    4) Both (1) & (2) depends upon velocity

40. Which of the following is conserved in Bernoulli's theorem?
    1) Mass 2) Momentum
    3) Energy 4) Density

41. The velocity of outflow of liquid through an orifice does not depends upon.
    1) Radius of orifice
    2) Height of the liquid column
    3) Acceleration due to gravity
    4) Density of the liquid

42. A drum of radius R full of liquid of density d is rotated at ω rad/s. The increase in pressure at the outer edge of the drum will be
    1) $\frac{\omega^2 R^2 d}{2}$ 2) $\frac{\omega^2 R d}{2}$
    3) $\frac{\omega R d^2}{2}$ 4) $\frac{\omega^2 R^2 d^2}{2}$

43. A vessel containing water has area of cross-section 16 cm^2. A small square hole of side 1 cm is at the bottom. What will be the time taken by the water to empty the vessel if vessel is filled up to the height of 32 cm? ($g = 9.8$ m/s^{-2})
    1) 3 sec 2) 5 sec
    3) 4 sec 4) 8 sec

Viscosity (Terminal Velocity, Poiseuille's Formula Stoke's Law)

44. Ten drops of a liquid of density ρ and each of radius r are falling through air with a constant velocity 12 m/s. When the 8 drops out of 10 drops coalesce to from a single drop, the terminal velocity of the new drop will be
    1) 48 cm s^{-1} 2) 52 cm s^{-1}
    3) 68 cm s^{-1} 4) 70cm s^{-1}

45. Water flows in a streamline manner through a capillary tube of radius a. The pressure difference being P and the rate of flow is Q. If the radius is reduced to a/2 and the pressure is increased to 2P then the rate of flow becomes
    1) 4Q 2) Q/2
    3) Q 4) Q/8

46. For any layer of liquid, its upper layer pulls it forward while lower layer pulls it backward, which results in force between the layer. What is the types of flow?
    1) Streamline 2) Turbulent
    3) Laminar 4) None

47. Which of the following statement is correct?
    1) The viscosity of liquid increases with temperature while it decreases in the case of gases
    2) The viscosity of liquid decreases with temperature while it is increases in the case of gases.
    3) Viscosity of both decreases
    4) Viscosity of both increase

48. In Poiseuille's method of determination of coefficient of viscosity the physical quantity that requires greater accuracy in measurement is
    1) Pressure difference
    2) Volume of the liquid collected
    3) Length of capillary tube
    4) Inner radius of the capillary tube

49. The velocity of the surface layer of water in a river of depth 20 m is 5 m/s. The shearing stress between the surface layer and the bottom layer is ($\eta = 10^{-3}$ poise)
1) 2.5×10^{-4} N/m^2 2) 3×10^{-4} N/m^2
3) 3.5×10^{-5} N/m^2 4) 4×10^{-3} N/m^2

50. A ball of mass m and radius r is released in viscous liquid. The value of its terminal velocity is proportional to
1) $\frac{1}{r}$ only 2) m/r
3) $(m/r)^{1/2}$ 4) m only

51. More viscous oil is used in summer than in winter in motors due to:
1) Rise in temperature in summer the viscosity of oil increases
2) Rise in temperature in summer the viscosity of oil decreases
3) Surface tension of oil decreases
4) Surface tension of oil increases

52. The value of coefficient of viscosity, in comparison to coefficient of friction is:
1) Very large 2) Very small
3) Nearly same 4) 8-10 times more

53. A spherical body falling through a viscous liquid of infinite extent ultimately attains a constant value. when
1) Upthrust + weight = viscous drag
2) Weight + viscous drag = upthrust
3) Viscous drag + upthrust = weight
4) Viscous drag + upthrust > weight

54. If the terminal speed of a sphere of gold (density = 19.5×10^3 kg/m^3) is 0.5 m/s in a viscous liquid (density = 3 kg/m^3). Find the terminal velocity of a sphere of silver (density = 10.5×10^3 kg/m^3) of the same size in the same liquid.
1) 0.30 m/s 2) 0.20 m/s
3) 0.51 m/s 4) 0.22 m/s

55. A layer of glycerin of thickness 3 mm is present between a large surface area and a surface area of 0.2 m^2, with what force the small surface is to be pulled, so that it can move with a velocity of 5 m/s? ($\eta = 0.09$ kgm^{-1}s^{-1}).
1) 10 N 2) 30N
3) 40 N 4) 50N

56. A small solid ball is released in a viscous liquid. The velocity of ball is given by

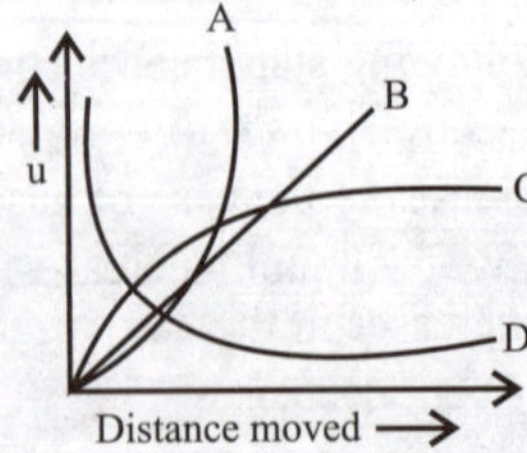

1) Curve D 2) Curve C
3) Curve B 4) Curve A

57. Two small spherical drops having radii in the ratio 4 : 3 fall from a height through the atmosphere. Their momenta on reaching the earth are in the ratio
1) 4 : 4 2) 1 : 2
3) 4 : 1 4) 5 : 1

58. There is a 1 mm thick layer of glycerine between a plate of area 100 cm^2 and a large plate. If the coefficient of viscosity of glycerine is 1.0 kg/ms, then what force is required to move the smaller plate with a velocity of 7 cm/s.
1) 7 N 2) 77 N
3) 70 N 4) 0.7 N

Reynolds Number Surface Tension & Surface Energy

59. What is the force required to take away a flat plate of radius 10 cm from the surface of water, if surface tension of water is 72×10^{-3} Nm^{-1}?
1) 5×10^{-2}N 2) 4.5×10^{-2}N
3) 3×10^{-2} N 4) 3×10^{3}N

60. The surface tension of soap solution is 30 dyne/cm. The radius of a spherical soap bubble is increased from 1 cm to 5 cm. The work done will be
1) 2880 π erg 2) 5760 π erg
3) 1920 π erg 4) 3840 π erg

61. What is the work done in breaking a drop of water of 4 mm diameter into 1000 drops of same size, surface tension of water is 72×10^{-3} Nm^{-1}?
1) 1.125×10^{-4}J 2) 1.125×10^{-5} J
3) 3.25×10^{-5}J 4) 3.25×10^{-6}J

62. Antiseptics (dettol. savlon, etc.) spread over the surface faster due to
1) Less density 2) Low specific heat
3) Low viscosity 4) Low surface tension

63. Writing on a black board with a piece of chalk is possible by the property of
1) Adhesive force 2) Cohesive forced
3) Surface tension 4) Viscosity

64. Soap helps in cleansing clothes because
1) Chemicals of soap changes
2) It increase the surface tension of the solution
3) It absorbs the dirt
4) It lowers the surface tension of the solution

65. A drop of oil is placed on the surface of water. Which of the following statement is correct?
1) It will remain on it as a sphere
2) It will spread as a thin layer
3) It will be partly as spherical droplets and party as thin film
4) It will float as a distorted drop on the water surface

66. Due to which property of water, tiny particles of camphor dance on the surface of water?
1) Viscosity 2) Surface tension
3) Weight 4) Floating force

67. The surface tension of a liquid is 10 N/m. If a thin film of the area 0.05 m^2 is formed on a loop. Then the surface energy will be
1) 5 J 2) 3 J
3) 2 J 4) 1 J

68. If two soap bubbles of equal radii r coalesce then the radius of curvature of interface between two bubbles will be

1) 2 2) 1

3) 0 4) ∞

69. The work done in increasing the size of a soap film from 11 cm × 7 cm to 20 cm × 8 cm is 8×10^{-2}J. The surface tension of the film is

1) 4.8 N/m 2) 5 N/m

3) 6.2 N/m 4) None

70. 27 mercury drops coalesce to form one mercury drop, the energy changes by a factor of

1) 6 times 2) 10 times

3) 9 times 4) 2 times

71. The work done in blowing a bubble of radius R is W, then the work done in making a bubble of radius 2R from that solution is

1) W/2 2) 2W

3) 4W 4) $2^{1/3}$ W

72. The surface tension does not depend upon:

1) Presence of impurities 2) Temperature

3) Atmospheric pressure 4) Nature of the liquid

73. A long cylindrical glass vessel has a small hole of radius r at its bottom. the depth to which the vessel can be lowered vertically in a deep water bath without any water entering inside is (surface tension = T and d = density of water)

1) $\frac{4T}{rdg}$ 2) $\frac{5T}{rdg}$

3) $\frac{3T}{rdg}$ 4) $\frac{2T}{rdg}$

74. At critical temperature the surface tension of a liquid becomes:

1) Zero

2) Infinity

3) Same as that at any other temperature

4) Cannot be determined

Angle of Contant, Capillary Rise, Excess Pressure in Bubble and Drop

75. A capillary tube of 2mm diameter. What is the height to which water rises, if surface tension of water is 7.4×10^{-3} Nm^{-1} and angle of contact is 0°?

1) 0.25 mm 2) 1.51 mm

3) 0.4 mm 4) 0.5 mm

76. What is the diameter of a capillary tube in which mercury is depressed by 1.21 cm and surface tension for mercury is 540×10^{-3} Nm^{-1} the angle of contact is 120° and density of mercury is 13.6×10^{2} kg m^{-3}?

1) 0.06 mm 2) 10 mm

3) 2 mm 4) 6 mm

77. Angle of contact of liquid with a solid depends on

1) Solid only

2) Liquid only

3) Both solid and liquid only

4) Orientation of the solid surface in liquid

78. Due to capillary action, a liquid will rise in a tube, if the angle of contact is

1) Acute 2) Obtuse

3) 90° 4) Zero

79. Two capillaries made of same material but of different radii are dipped in a liquid. The rise of liquid in one capillary is 4 cm and that in second tube is 5cm . The ratio of their radii is

1) 1/2 2) 2/3

3) 1/4 4) 5/4

80. Two capillary tubes of same diameter are put vertically one each in two liquids whose relative densities are 0.8 and 0.6 and surface tensions are 70 and 40 dyne/ cm respectively. The ratio of heights of liquids in the tubes $\frac{h_1}{h_2}$ is

1) 5/4 2) 2/3

3) 21/16 4) 16/21

81. When a capillary tube is dipped in a liquid, the capillary rise is h_1, when the inner surface is coated with wax, the capillary rise is h_2 then

1) $h_1 = h_2$ 2) $h_1 < h_2$

3) $h_1 > h_2$ 4) None

82. If a capillary tube is dipped into liquid and the levels of the liquid inside and outside are same, then the angle of contact is

1) 0° 2) 90°

3) 45° 4) 30°

83. If h is the height of capillary rise and r is the radius of the capillary tube, then which of the following relation will be correct?

1) hr = constant 2) h/r² = constant

3) hr² = constant 4) h/r = constant

84. Figure shows a capillary tube C dipped in a liquid that wets it. The liquid rises to a point A. If we blow air through the horizontal tube H, what will be happen to the liquid Column-In the capillary tube?

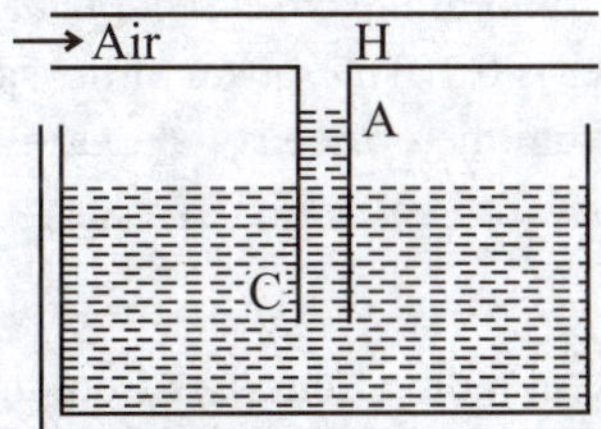

1) Level will rise above A

2) Level will fall below A

3) Level will remain at A

4) It is difficult to predict

85. The soap bubble formed at the end of a tube is blown very slowly with time. Which graph is best suited for the excess pressure inside the bubble with time.

1) 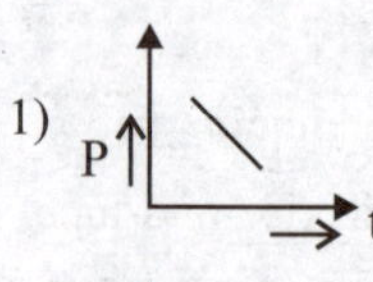

2) 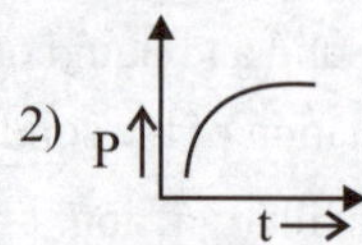

3) 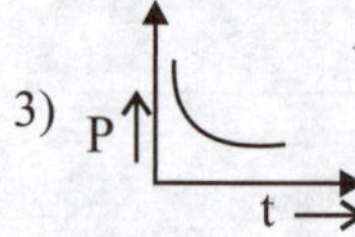

4) 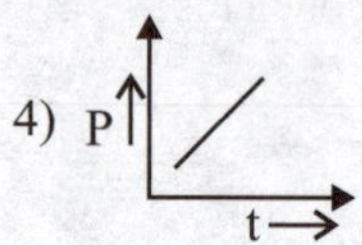

86. The excess pressure inside a spherical drop of water is four times that of another drop. Then their respective mass ratio is

1) 1 : 16 2) 8 : 1
3) 1 : 4 4) 1 : 64

87. Pressure inside two soap bubbles are 1.06 and 1.08 atm. Ratio between their volumes is

1) 16 : 25 2) 64 : 27
3) 16 : 9 4) 3 : 10

88. When a large bubble rises from the bottom of a lake to the surface, its radius triples. If atmospheric pressure is equal to that of column of water height H, then the depth of the lake (h) is

1) 28 H 2) 25 H
3) 27 H 4) 26 H

89. If excess of pressure inside a soap bubble of radius 2 cm is balanced by that due to column of oil of (specific gravity = 0.8) 2mm high. What is the surface tension of soap bubble?

1) 12×10^2 N/m 2) 6×10^{-2} N/m
3) 8.12×10^{-2} N/m 4) 7.84×10^{-2} N/m

High Order Time Intensive MCQs

1. A vertical tank with depth H is full with water. A hole is made on one side of the walls at a depth h below the water surface. At what distance from the foot of the wall does the emerging stream of water strike the foot?

1) $\sqrt{h(H-h)}$ 2) $2\sqrt{h(H-h)}$
3) $2(H-h)\sqrt{h/(H-h)}$ 4) $\sqrt{2h/(H-h)}$

2. If a capillary tube of radius R is immersed in water; the mass of water risen in capillary is M. If the radius of capillary be doubled, the mass of water risen in the capillary will be:

1) $\frac{M}{2}$ 2) M
3) 2M 4) 4M

3. Water flows along a horizontal pipe whose cross-section is not constant. The pressure is 1 cm of Hg where the velocity is 35 cms^{-1}. At a point where the velocity is 65 cms^{-1}, the pressure will be

1) 0.89 cm of Hg 2) 8.9 cm of Hg
3) 0.5 cm of Hg 4) 1 cm of Hg

4. Water is filled in a cylindrical container to a height of 3 m. An orifice is made at a height 52.5 cm from bottom of cylinder. The ratio of the cross-sectional area of the orifice and the beaker is 0.1. The square of the speed of the liquid coming out from the orifice is (take, g = 10 ms^{-2})

1) 50 m^2 s^{-2} 2) 50.5 m^2 s^{-2}
3) 51 m^2 s^{-2} 4) 52 m^2 s^{-2}

5. An air bubble of radius 1 cm rises from the bottom portion through a liquid of density 1.5 gcc^{-1} at a constant speed of 0.25 cms^{-1}. If the density of air is neglected, the coefficient of viscosity of the liquid is approximately (in Pas).

1) 1300 2) 1300
3) 130 4) 13

6. The cylindrical tube of spray pump has a cross-section of 8 cm^2, one end of which has 40 fine holes each of area 10^{-8} m^2. If the liquid flows inside the tube with a speed of 0.15 m min^{-1}, the speed with which the liquid is ejected through the holes is

1) 50 ms^{-1} 2) 5 ms^{-1}
3) 0.05 ms^{-1} 4) 0.5 ms^{-1}

7. A hole is in the bottom of the tank having water. If total pressure at the bottom is 3 atm (1 atm = 10^5 Nm^{-2}), then velocity of water flowing from hole is

1) $\sqrt{400}$ ms^{-1} 2) $\sqrt{600}$ ms^{-1}
3) $\sqrt{60}$ ms^{-1} 4) None of the above

8. The speeds of air-flow on the upper and lower surfaces of a wing of an aeroplane are v_1 and v_2, respectively. If A is the cross-sectional area of the wing and ρ is the density of air, then the upward lift is

1) $\frac{1}{2}\rho A(v_1 - v_2)$ 2) $\frac{1}{2}\rho A(v_1 + v_2)$
3) $\frac{1}{2}\rho A(v_1^2 - v_2^2)$ 4) $\frac{1}{2}\rho A(v_1^2 + v_2^2)$

9. Water flows steadily through a horizontal pipe of variable cross-section. If the pressure of water is p at a point where flow speed is v, the pressure at another point where the flow of speed is 2v, is (take density of water as ρ)

1) $p-\frac{3\rho v^2}{2}$ 2) $p-\frac{\rho v^2}{2}$
3) $p-\frac{3\rho v^2}{4}$ 4) $p - \rho v^2$

10. A given shaped glass tube having uniform cross-section is filled with water and is mounted on a rotatable shaft as shown in the figure. If the tube is rotated with a constant angular velocity ω, then

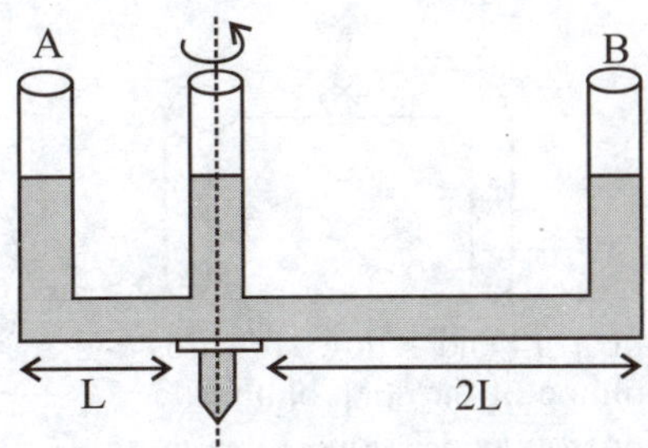

1) Water levels in both sections A and B go up
2) Water level in section A goes up and that in B comes down
3) Water level in section A comes down and that in B it goes up
4) Water levels remain same in both sections

11. A drop of some liquid of volume 0.04 cm^3 is placed on the surface of a glass slide. Then another glass slide is placed on it in such a way that the liquid forms a thin layer of area 20 cm^2 between the surfaces of the two slides. To separate the slides, a force of 16×10^5 dyne has to be applied normal to the surfaces. The surface tension of the liquid is (in dyne-cm^{-1})
1) 60 2) 70
3) 80 4) 90

12. The rate of flow of water in a capillary tube of length ℓ and radius r is V. The rate of flow in another capillary tube of length 4ℓ and radius 2r for same pressure difference would be
1) V 2) 4V
3) 2V 4) 8V

13. The radius of spherical drop of water is 1 mm. If surface tension of water be 70×10^{-3} Nm^{-1}, the pressure difference inside and outside the drop will be
1) 70 Nm^{-2} 2) 140 Nm^{-2}
3) 280 Nm^{-2} 4) Zero

14. What should be the pressure inside a small air bubble of 0.1 mm radius situated just below the water surface? (surface tension of water = 7.2×10^{-2} Nm^{-1} and atmospheric pressure = 1.013×10^5 Nm^{-2})
1) 2.012×10^5 Nm^{-2} 2) $2.0.12 \times 10^4$ Nm^{-2}
3) 1.027×10^5 Nm^{-2} 4) 1.027×10^4 Nm^{-2}

15. If the length of tube is less and cannot accommodate the maximum rise of liquid, then
1) Liquid will form fountain
2) Liquid will not rise
3) The meniscus will adjust itself so that the water does not spill
4) None of the above

16. Water rises in a capillary tube to a certain height such that the upward force due to surface tension is balanced by 75×10^{-4} N force due to the weight of the liquid. If the surface tension of the water is 6×10^{-2} Nm^{-1}, the inner circumference of the capillary must be
1) 1.25×10^{-2} m 2) 0.50×10^{-2} m
3) 6.5×10^{-2} m 4) 12.5×10^{-2} m

17. Water rises upto height h in a capillary tube of certain diameter. This capillary tube is replaced by similar tube of half the diameter. Now, the water will rise to the height of
1) 4h 2) 3h
3) 2h 4) h

18. A bubble is at the bottom of the lake of depth h. As the bubble comes to sea level, its radius increases three times. If atmospheric pressure is equal to l metres of water column, then h is equal to:
1) 26 l 2) l
3) 25 l 4) 30 l

19. Energy needed in breaking a drop of radius R into n drops of radii r is given by
1) $4\pi T(nr^2 - R^2)$ 2) $\frac{4}{3}\pi\left(r^3 n - R^2\right)$
3) $4\pi T(R^2 - nr^2)$ 4) $4\pi T(nr^2 + R^2)$

20. A space 2.5 cm wide between two large plane surfaces is filled with oil. Force required to drag a very thin plate of area 0.5 m^2 just midway the surfaces at a speed of 0.5 m/sec is 1N. The coefficient of viscosity in kg-s/m^2 is:

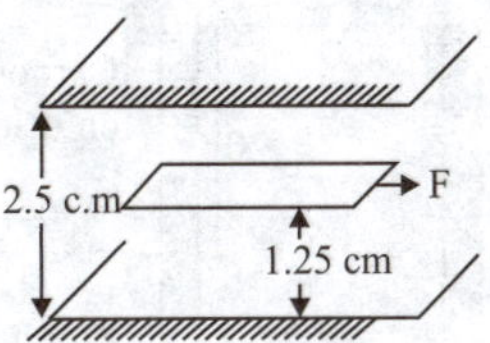

1) 5×10^{-2} 2) 2.5×10^{-2}
3) 1×10^{-2} 4) 7.5×10^{-2}

21. A homogeneous solid cylinder of length L(L <H/2) and cross sectional area A/5 is immersed such that it floats with its axis vertical at the liquid-liquid interface with length L/4 in the denser liquid as shown in the figure. The lower density liquid is open to atmosphere having pressure P_0. Then density D of solid is given by:

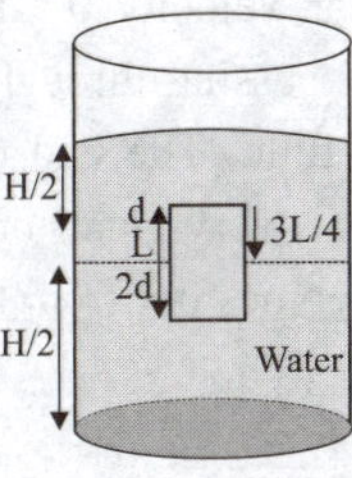

1) $\frac{5}{4}d$ 2) $\frac{4}{5}d$
3) $\frac{d}{5}$ 4) None of these

22. Find out $\sin\theta$ in the diagram given below:

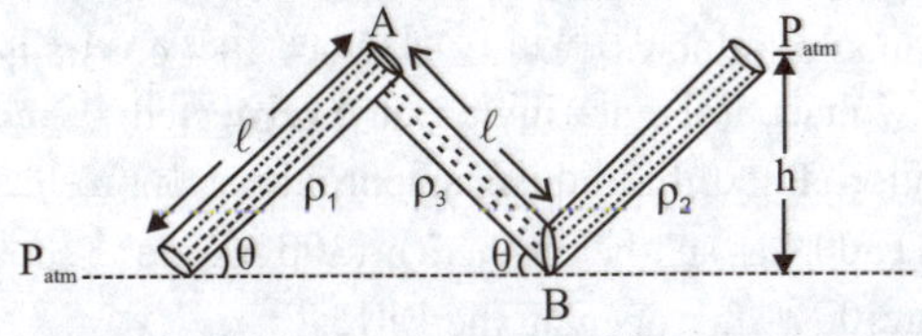

1) $\frac{\rho_2 h}{(\rho_3 - \rho_1)\ell}$ 2) $\frac{\rho_1 h}{(\rho_3 - \rho_2)\ell}$

3) $\frac{\rho_3 h}{(\rho_2 - \rho_1)\ell}$ 4) None of these

23. A submarine experiences a pressure of 5.05×10^6 Pa at a depth of d_1 in a sea. When it goes further to a depth of d_2, it experiences a pressure of 8.08×10^6 Pa, then $d_2 - d_1$ is approximately (density of water = 10^3 kg/m³ and g = 10 m/s²)

1) 500 m 2) 400 m
3) 600 m 4) 300 m

24. The rate of steady volume flow of water through capillary tube of length l and radius r under a pressure difference of P is V. The tube is connected with another tube of the double of length and half the radius in series. Then the rate of steady volume flow through them is (pressure difference across the combination is p)

1) V/18 2) V/17
3) V/2 4) V/33

25. Calculate the value of h in U-tube shown in the following figure.

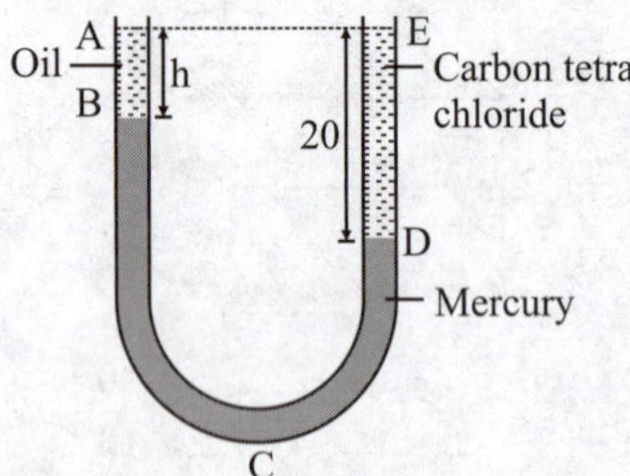

Density of oil = 0.9 g/cm³
Density of carbon tetra chloride = 1.6 g/cm³
Density of mercury = 13.6 g/cm³

1) 18.9 cm 2) 20.9 cm
3) 16.9 4) 14.9

26. A horizontal oriented tube AB of length 5 m rotates with a constant angular velocity 0.5 rad/s about a stationary vertical axis OO′ passing through the end A. The tube is filled with ideal fluid. The end of the tube is open, the closed end B has a very small orifice. The velocity with which the liquid comes out from the hole (in m/s) is:

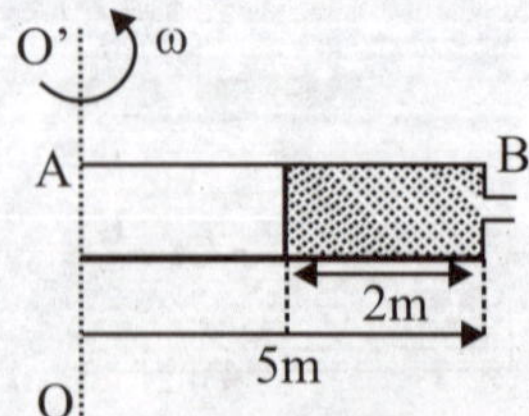

1) 1 m/s 2) 2 m/s
3) 4 m/s 4) 3 m/s

27. A glass tube of uniform internal radius r has a valve separating the two identical ends. Initially, the valve is in a tightly closed position. End 1 has a hemispherical soap bubble of radius r. The end 2 has sub-hemispherical soap bubble as shown in the figure. Just after opening the valve:

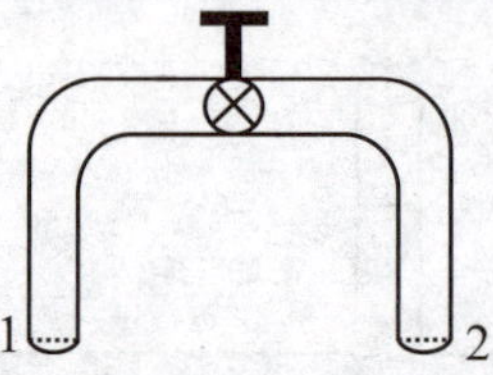

1) The air from end 1 flows towards end 2. No change in the volume of the soap bubbles.
2) The air from end 1 flows towards end 2. Volume of the soap bubble at end 1 decreases.
3) The no change occurs.
4) The air from end 2 flows towards end 1. Volume of the soap bubble at end 1 increases.

28. A wooden ball of density D is immersed in water of density d to a depth h below the surface of water. Up to which height will the ball jump out of water, when it is released?

1) $\frac{d}{D}h$ 2) $\left(\frac{d}{D} - 1\right)h$
3) h 4) Zero

29. A soap bubble having radius $\frac{1}{\sqrt{\pi}}$ cm is expanded to a bubble of radius $\frac{2}{\sqrt{\pi}}$ cm. If the surface tension of soap solution is 30 dyne/cm, the work done is

1) 180 erg 2) 360 erg
3) 720 erg 4) 960 erg

30. A wooden block, with a coin placed on its top, floats in water as shown in the figure. The distance l and h are shown there. After some time the coin falls into the water. Then:

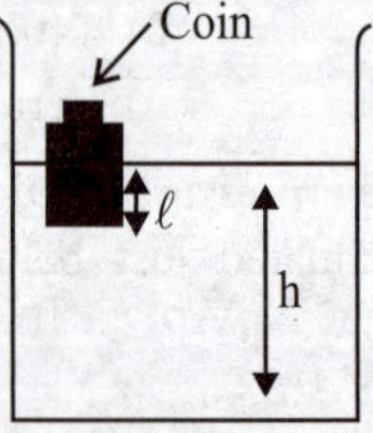

1) l decreases and h increases
2) l increases and h decreases
3) Both l and h increases
4) Both l and h decreases

31. A large open tank has two holes in the wall. One is a square hole and another is circular. Square hole has a side L at a depth h from the top and circular hole has radius r at a depth 6h from the top. When the tank is completely filled with water the quantity of water flowing out per second from both the holes are the same. What is the value of r?

1) $L^2/6$ 2) $\left(\frac{L}{6\pi}\right)^{1/2}$

3) $\left(\frac{L^2}{\pi\sqrt{6}}\right)^{1/2}$ 4) None

32. An inverted bell lying at the bottom of a lake 50 m deep has 100m^2 of air trapped in it. The bell is brought to the surface of lake. The volume of the trapped air will be (atmospheric pressure = 70 cm of Hg and density of Hg = 13.6 g/cc)
1) 515 cm^3
2) 600cm^3
3) 625.2 cm^3
4) 416.8 cm^3

33. A ball whose density is 5×10^3 kg/m^3 falls into the water from a height of 5m. To what depth the ball will go?
1) 7.8 m
2) 5 m
3) 6.25 m
4) 9 m

34. Two soap bubbles of radii r_1 and r_2, $r_1 > r_2$ get attached to each other to have a common interface. The radius of this interface is
1) $r_1 - r_2$
2) $r_1 + r_2$
3) $\frac{r_1 r_2}{r_1 + r_2}$
4) $\frac{r_1 r_2}{r_1 - r_2}$

35. A vessel containing water is given a constant acceleration a towards the right along a straight horizontal path. Which of the following diagrams represents the surface of the liquid?

1) 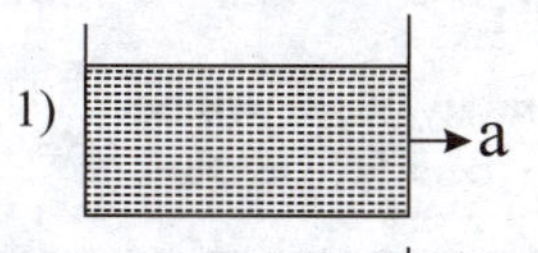

2) 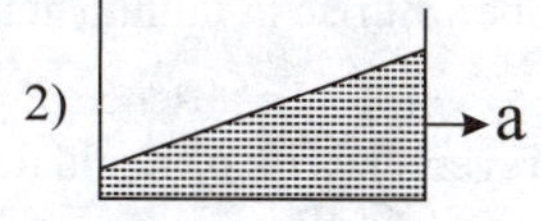

3) 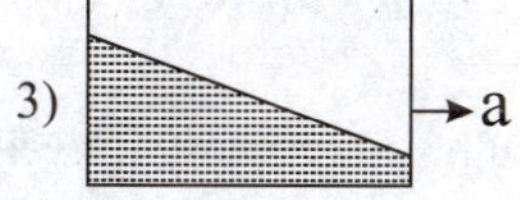

4) None of these

36. Water flow through a horizontal pipe of varying cross section at the rate of 50 litres per minute then what will be the velocity of water at a point where diameter is 10cm?
1) 0.301 ms^{-1}
2) 0.629 ms^{-1}
3) 0.0424 ms^{-1}
4) 0.112 ms^{-1}

37. A vessel containing water stand on a table of height H. A small hole is punched in the side of vessel at its base. The steam of water strikes the ground at a horizontal distance R from the table. Then the depth of water in the cylinder is
1) H
2) R
3) $\sqrt{RH}$
4) $\frac{R^2}{4H}$

38. A pipe is running full of water. At a certain point A it tapers from 80 cm to 10 cm diameter at B. The pressure difference at A and B is 120 cm of water column. What will be the rate of flow of water through the pipe? (in m^3 /sec).
1) 3×10^{-2}
2) 4×10^{-2}
3) 6×10^{-2}
4) 7×10^{-2}

39. A container with square base of side a is filled up to a height H with a liquid. A hole is made at a depth h from the free surface of water. With what acceleration the container must be accelerated, so that the water does not come out?
1) g
2) $\frac{g}{2}$
3) $\frac{2gH}{2}$
4) 2gH

40. A vertical U-shaped tube having a liquid rotates about the vertical axis. The value of ω is:

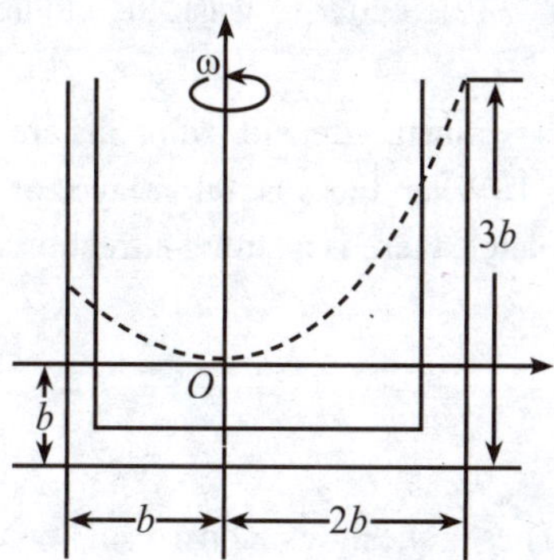

1) $\sqrt{\frac{g}{b}}$
2) $2\sqrt{\frac{g}{3b}}$
3) $\sqrt{\frac{g}{2b}}$
4) None of these

41. A glass capillary tube is of the shape of truncated cone with an apex angle α so that its two ends have cross sections of different radii. when dipped in water vertically, water rises in it to a height h, where the radius of its cross section is b. If the surface tension of water is S, its density is ρ, and its contact angle with glass is θ, the value of h will be (g is the acceleration due to gravity)

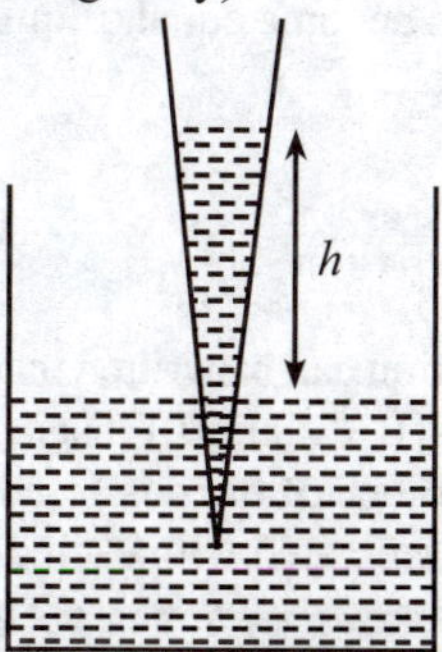

1) $\frac{2S}{b\rho g}\cos(\theta - \alpha)$
2) $\frac{2S}{b\rho g}\cos(\theta + \alpha)$
3) $\frac{2S}{b\rho g}\cos(\theta - \alpha/2)$
4) $\frac{2S}{b\rho g}\cos(\theta + \alpha/2)$

Statement Based MCQs

Directions: These questions consist of two statements each, printed as Statement-I and Statement-II. While answering these questions, you are required to choose any one of the following four responses.

1) **Both Statement-I and Statement-II are correct.**
2) **Both Statement-I and Statement-II are incorrect.**
3) **Statement-I is correct & Statement-II is incorrect.**
4) **Statement-I is incorrect & Statement-II is correct.**

1. **Statement-I:** It is better to wash the clothes in cold soap solution.

   Statement-II: Surface tension of paints are kept low.

2. **Statement-I:** When there is a thin layer of water between two glass plates, there is a strong attraction between them.

   Statement-II: The pressure between the plates becomes less than atmospheric pressure due to presence of the liquid.

3. **Statement-I:** When a large soap bubble and a small soap bubble are connected by a capillary tube, the large bubble expands and while the small bubble shrinks.

   Statement-II: The excess pressure inside a bubble is inversely proportional to radius.

4. **Statement-I:** Body of fish is streamlined.

   Statement-II: The streamlined body helps the fish to cut its way through the water.

5. **Statement-I:** Density of ice is less than the density of sea water.

   Statement-II: Upthrust always acts in the downward direction.

Assertion & Reason MCQs

Directions: These questions consist of two statements each, printed as Assertion and Reason. While answering these questions, you are required to choose any one of the following four responses.

1) **Both Assertion and Reason are True and the Reason is a correct explanation of the Assertion.**
2) **Both Assertion and Reason are True but Reason is not a correct explanation of the Assertion.**
3) **Assertion is True but the Reason is False.**
4) **Assertion is False but Reason is True.**

1. **Assertion:** At critical temperature, surface tension of a liquid becomes zero.

   Reason: At this temperature, intermolecular forces for liquids and gases become equal. Liquid can expand without any restriction.

2. **Assertion:** In gravity free space, the liquid in a capillary tube will rise to infinite height.

   Reason: In the absence of gravity, there will be no force to prevent the rise of liquid due to surface tension.

3. **Assertion:** Pressure is low in a region where velocity of air is high.

   Reason: According Bernoulli's Theorem if velocity increases, pressure decreases.

4. **Assertion:** Hot soup tastes better than the cold soup.

   Reason: Hot soup spread properly on our tongue due to lower surface tension.

5. **Assertion:** Roof of buildings are blown off during a strong storm.

   Reason: Roof of buildings becomes lighter during storm.

Match the Columns MCQs

1. Bernoulli's equation can be written in the following different forms (Column-I). Column-II lists certain units each of which pertains to one of the possible forms of the equation. Match the unit associated with each of the equations.

	Column-I		Column-II
(a)	$\frac{v^2}{2g}+\frac{p}{\rho g}+z=\text{constant}$	(i)	Total energy per unit mass
(b)	$\frac{\rho v^2}{2}+p+\rho gz=\text{constant}$	(ii)	Total energy per unit weight
(c)	$\frac{v^2}{2}+\frac{p}{\rho}+gz=\text{constant}$	(iii)	Total energy per unit volume

1) (a)-(ii) (b)-(i) (c)-(iii)
2) (a)-(ii) (b)-(iii) (c)-(i)
3) (a)-(iii) (b)-(i) (c)-(ii)
4) (a)-(i) (b)-(iii) (c)-(ii)

2. Match the Column-I with Column-II.

	Column-I		Column-II
(a)	Bernoulli's theorem	(i)	Narrower pipes have less pressure
(b)	Ball moving with spin	(ii)	Paint gun
(c)	Artificial high pressure	(iii)	Non-viscous fluids
(d)	Streamlined flow	(iv)	Conservation of energy
		(v)	Uplift due to pressure difference

1) (a)-(iii, iv) (b)-(v) (c)-(ii) (d)-(i, iii)
2) (a)-(iii) (b)-(i) (c)-(v) (d)-(ii)
3) (a)-(iv) (b)-(ii) (c)-(iii) (d)-(v)
4) (a)-(ii, iv) (b)-(v) (c)-(iii) (d)-(i)

3. Match the Column-I with Column-II.

	Column-I		Column-II
(a)	Magnus energy	(i)	Pascal's Law
(b)	Loss of energy	(ii)	Archimede's principle
(c)	Pressure is same at the same level in a liquid	(iii)	Viscous force
(d)	Hydraulic machines	(iv)	Lifting of as bestos roofs

1) (a)-(iv) (b)-(i) (c)-(iii) (d)-(ii)
2) (a)-(ii) (b)-(iv) (c)-(i) (d)-(iii)
3) (a)-(iv) (b)-(iii) (c)-(i) (d)-(i)
4) (a)-(i) (b)-(iv) (c)-(iii) (d)-(ii)

4. A cylinder of weight W is floating in two liquids as shown in figure. Net force on cylinder from top is F_1 and force on cylinder from the bottom is F_2. Match the following two columns.

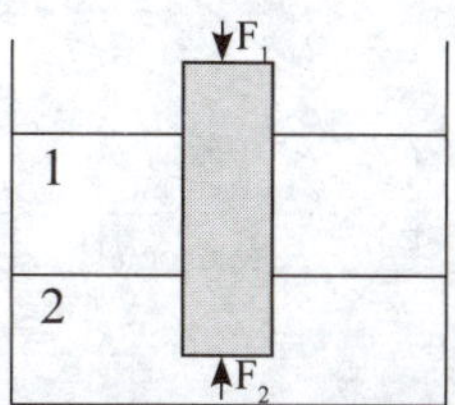

	Column-I		Column-II
(a)	Net force on cylinder from liquid-1	(i)	Zero
(b)	$F_2 - F_1$	(ii)	W
(c)	Net force on cylinder from liquid-2	(iii)	Net upthrust
(d)	Net force on cylinder from atmosphere	(iv)	none of these

1) (a)-(iv) (b)-(iii) (c)-(i) (d)-(ii)
2) (a)-(i) (b)-(ii, iii) (c)-(iv) (c)-(iv)
3) (a)-(ii) (b)-(iii) (c)-(i) (c)-(iv)
4) (a)-(iii) (b)-(ii, iii) (c)-(i) (c)-(iv)

NCERT Exemplar MCQs

1. A tall cylinder is filled with viscous oil. A round pebble is dropped from the top with zero initial velocity. From the plot shown in figure, indicate the one that represents the velocity (v) of the pebble as a function of time (t):

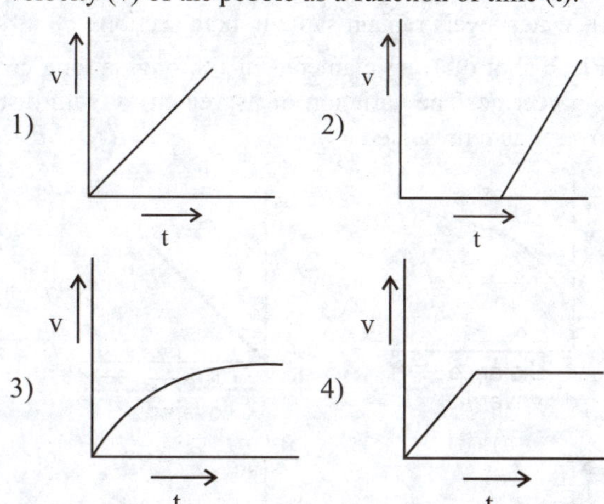

2. Which of the following diagrams does not represent a streamline flow?

1) 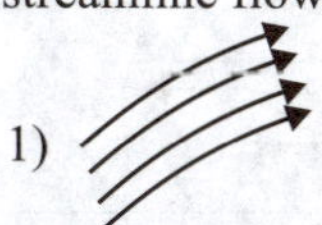

2) 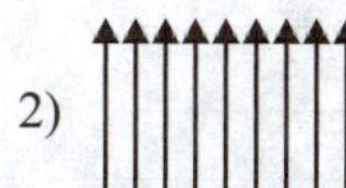

3)

4) 

3. Along a streamline:
1) The velocity of a fluid particle remains constant
2) The velocity of all fluid particles crossing a given position is constant
3) The velocity of all fluid particles at a given instant is constant
4) The speed of a fluid particle remains constant

4. An ideal fluid flows through a pipe of circular cross-section made of two sections with diameters 2.5 cm and 3.75 cm. The ratio of the velocities in the two pipes is:
1) 9 : 4
2) 3 : 2
3) $\sqrt{3}:\sqrt{2}$
4) $\sqrt{2}:\sqrt{3}$

5. The angle of contact at the interface of water-glass is 0°, ethyl alcohol-glass is 0°, mercury-glass is 140° and methyl-iodide-glass is 30°. A glass capillary is put in a trough containing one of these four liquids. It is observed that the meniscus is convex. The liquid in the trough is:
1) Water
2) Ethyl alcohol
3) Mercury
4) Methyl iodide

NEET & AIIMS Past Year MCQs

1. An ideal fluid is flowing in a non-uniform cross-sectional tube *XY* (as shown in the figure) from end *X* to end *Y*. If K_1 and K_2 are the kinetic energy per unit volume of the fluid at *X* and *Y* respectively, then the correct option is: **(2024 Re)**

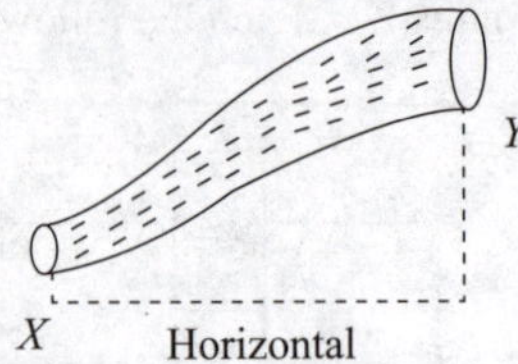

1) $K_1 = K_2$ 2) $2K_1 = K_2$
3) $K_1 > K_2$ 4) $K_1 < K_2$

2. A thin flat circular disc of radius 4.5 cm is placed gently over the surface of water. If surface tension of water is 0.07 Nm^{-1}, then the excess force required to take it away from the surface is: **(2024)**
1) 1.98 mN 2) 99 N
3) 19.8 mN 4) 198 N

3. The amount of energy required to form a bubble of radius 2 cm from a soap solution is nearly: (surface tension of soap solution = 0.03 N m^{-1}) **(2023)**
1) 50.1×10^{-4} J 2) 30.1×10^{-4} J
3) 5.06×10^{-4} J 4) 3.01×10^{-4} J

4. The venturi-meter works on: **(2023)**
1) The principle of perpendicular axes
2) Huygen's principle
3) Bernoulli's principle
4) The principle of parallel axes

5. A spherical ball is dropped in a long column of a highly viscous liquid. The curve in the graph shown which represents the speed of the ball (v) as a function of time (t) is: **(2022)**

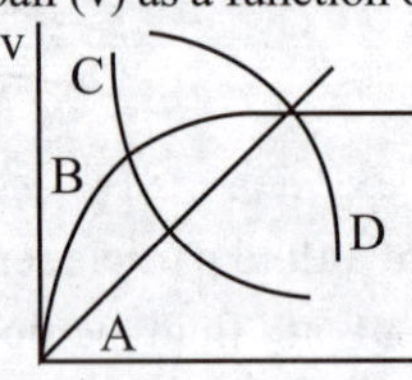

1) D 2) A 3) B 4) C

6. If a soap bubble expands, the pressure inside the bubble: **(2022)**
1) Is equal to the atmospheric pressure
2) Decreases
3) Increases
4) Remains the same

7. If there were no gravity, which of the following will not be there for a fluid? **(AIIMS 2011)**
1) Viscosity 2) Surface tension
3) Pressure 4) Archimedes' upward thrust

8. A thread is tied slightly loose to a wire frame as in figure and the frame is dropped into a soap solution and taken out. The frame is completely covered with the film. When the portion *A* is punctured with a pin, the thread: **(AIIMS 2008)**

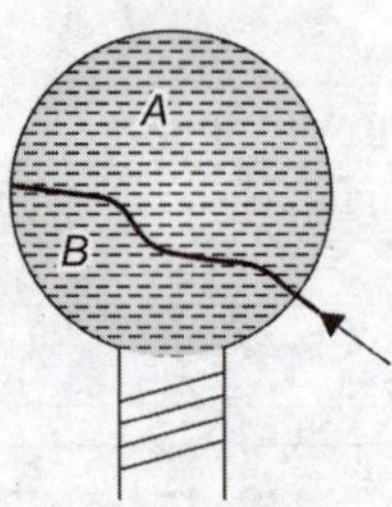

1) Becomes concave towards *A*
2) Becomes convex towards *A*
3) Either a. or b. depending on the size of *A* with respect to *B*
4) Remains in the initial position.

9. In old age arteries carrying body become narrow resulting in an increase in the blood pressure. This follows from: **(AIIMS 2007)**
1) Pascal's law
2) Stoke's law
3) Bernoulli's principle
4) Archimedes principle

10. A given shaped glass tube having uniform cross section is filled with water and is mounted on rotatable shaft as shown in figure. If the tube is rotated with a constant angular velocity then: **(AIIMS 2005)**

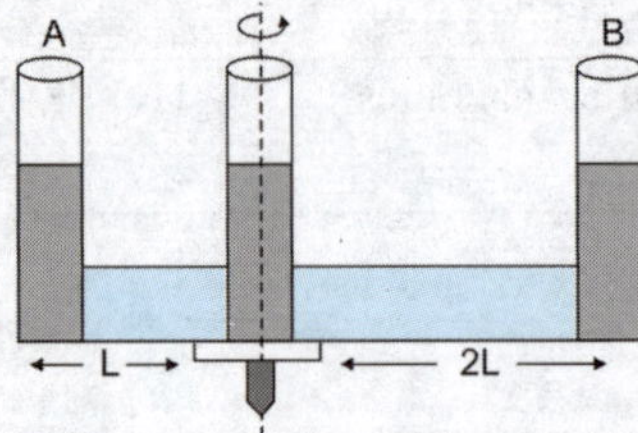

1) Water levels in both sections A and B go up
2) Water level in section A goes up and that in B comes down
3) Water level in section A comes down and that in B it goes up
4) Water levels remain same in both sections

11. A lead shot of 1mm diameter falls through a long column of glycerine. The variation of its velocity v with distance covered is represented by: **(AIIMS 2003)**

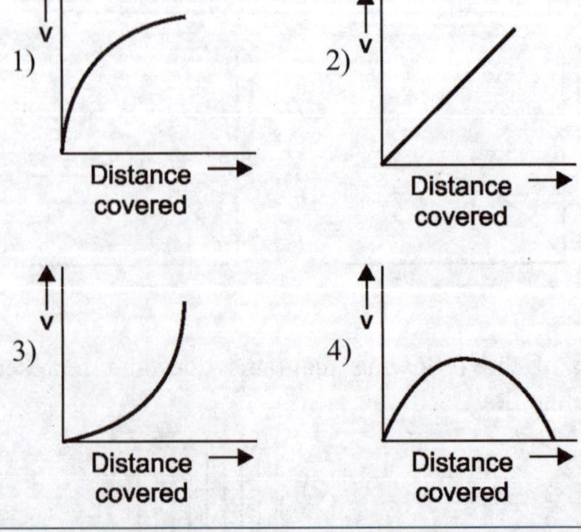

Answer Key

NCERT Topic-wise MCQs

1	2	3	4	5	6	7	8	9	10	11	12	13	14	15	16	17
1	3	1	4	3	2	1	1	2	1	3	4	2	2	3	2	2
18	19	20	21	22	23	24	25	26	27	28	29	30	31	32	33	34
1	2	2	4	2	2	2	1	4	3	2	3	4	3	2	2	2
35	36	37	38	39	40	41	42	43	44	45	46	47	48	49	50	51
3	4	2	4	1	3	1	1	3	1	4	3	2	4	1	2	2
52	53	54	55	56	57	58	59	60	61	62	63	64	65	66	67	68
2	3	4	2	2	3	4	2	2	3	4	1	4	2	2	4	4
69	70	71	72	73	74	75	76	77	78	79	80	81	82	83	84	85
1	3	3	3	4	1	2	1	3	1	4	3	3	2	1	1	3
86	87	88	89													
4	2	4	4													

High Order Time Intensive MCQs

1	2	3	4	5	6	7	8	9	10	11	12	13	14	15	16	17
2	3	1	1	3	2	1	3	1	4	3	2	2	3	3	4	3
18	19	20	21	22	23	24	25	26	27	28	29	30	31	32	33	34
1	1	2	1	1	4	4	1	2	2	2	3	4	3	3	3	4
35	36	37	38	39	40	41										
3	3	4	1	1	2	4										

Statement Based MCQs

1	2	3	4	5
4	1	1	1	3

Assertion & Reason MCQs

1	2	3	4	5
1	1	1	1	3

Match the Column MCQs

1	2	3	4
2	1	3	2

NCERT Exemplar MCQs

1	2	3	4	5
3	4	2	1	3

NEET & AIIMS Past Year MCQs

1	2	3	4	5	6	7	8	9	10	11
3	3	4	3	3	2	4	2	3	1	1

CHAPTER

10 Thermal Properties of Matter

Topicwise Number of Questions Analysis (2025-2016)

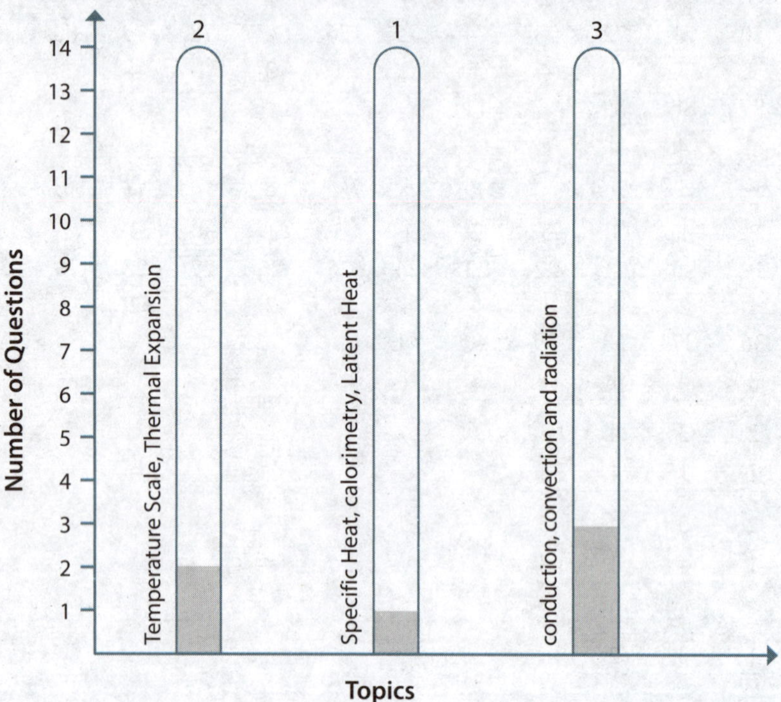

NCERT-Picks

Temperature Scale

Formulas

1. $\frac{T_C - 0}{100} = \frac{T_F - 32}{180} = \frac{T_K - 273.15}{100}$

   $\Rightarrow \frac{T_C}{5} = \frac{T_F - 32}{9} = \frac{T_K - 273.15}{5}$

   Where, T_C is temperature in °C

   T_F is temperature in °F

   T_K is temperature is K

2. To measure temperature t of a mercury column in a thermometer.

   Temperature of the mercury column,

   $$t = \left(\frac{l_t - l_0}{l_{100} - l_0}\right) \times 100°C$$

   [l_0 is length at 0°C, l_{100} is length at 100°C, l_t is length at t °C]

3. Temperature of the gas, $t = \frac{P_t - P_0}{P_{100} - P_0} \times 100°C$

   [P_t is pressure at t°C, P_0 is pressure at 0°C and P_{100} is pressure at 100°C]

Thermal Expansion

Formulas

1. Length at temperature t °C,

   $L_t = L_0 (1 + \alpha\Delta T)$

   [Where, L_0 is initial length, α is coefficient of linear expansion, and ΔT is change in temperature]

2. Area at temperature t°C,

   $A_t = A_0(1 + \beta\Delta T)$

   [A_0 is initial area and β is coefficient of areal expansion]

3. Volume at t° C,

   $V_t = V_0(1 + \gamma\Delta T)$

   [V_0 is initial volume and γ is coefficient of volume expansion]

4. Relation between α, β and γ

   $\alpha : \beta : \gamma = 1 : 2 : 3$

Numerical Profile

Coefficient of linear expansion

(i) If numerical value of coefficient of linear expansion of a solid is α_c when the temperature is measured in celsius scale and its value is α_F if the temperature is measured in Fahrenheit scale then

(a) $\alpha_F < \alpha_C$.

(b) $\alpha_F = \left(\frac{5}{9}\right)\alpha_C$

(ii) A composite rod is made by joining two rods of different materials and of same cross section. If l_1, l_2 are their initial lengths at t_1 °C, then

(a) The effective coefficient of linear expansion of the composite rod, $\alpha = \frac{l_1\alpha_1 + l_2\alpha_2}{l_1 + l_2}$.

(b) The increase in length of composite rod at t_2 °C,

$\Delta l = (l_1\alpha_1 + l_2\alpha_2)(t_2 - t_1)$

(iii) If two metal rods of coefficients of linear expansions α_1 and α_2 have same length at t_1°C and t_2°C respectively and the common temperature at which they have again the equal length is t, then

$$t = \frac{\alpha_1 t_1 - \alpha_2 t_2}{\alpha_1 - \alpha_2}.$$

Calorimetry

Formulas

1. According to the principle of calorimetry,

   Heat gain = Heat loss

   $m_1C_1(T_1 - T_{mix}) = m_2C_2(T_{mix} - T_2)$

   $$T_{mix} = \frac{m_1C_1T_1 + m_2C_2T_2}{m_1C_1 + m_2C_2}$$

   [m_1, m_2 = masses of the substances, C_1, C_2 = specific heat capacities, T_1, T_2 = Temperature, T_{mix} = Temperature of mixture at equilibrium]

2. To change the state of a substance

   Amount of heat, $Q = mL$ [where L is the latent heat]

Conduction

Formulas

1. Rate of heat flow through a conductor is given by

   $$\frac{\Delta Q}{\Delta t} = \frac{KA(T_1 - T_2)}{\ell}$$

   [K is thermal conductivity, ℓ is the length and T_1 and T_2 are the temperature of both ends.

2. Thermal resistance,

   $$\because H = \frac{\Delta Q}{\Delta t} = \frac{KA}{\ell}\Delta T = \frac{\Delta T}{\ell / KA} = \frac{\Delta T}{R_H}$$

   where,

   $R_H = \frac{\ell}{KA}$ = Thermal resistance

   (a) For series connection thermal resistance, $R_{eq} = R_1 + R_2$

   [R_1 and R_2 is the thermal resistance of individual rod]

   (b) For parallel connection thermal resistance,

   $$\frac{1}{R_{eq}} = \frac{1}{R_1} + \frac{1}{R_2}$$

3. Equivalent thermal conductivity of two metal rods of same cross-section area combined in series-

K_1	K_2
←ℓ_1→	←ℓ_2→

   $$K_s = \frac{\ell_1 + \ell_2}{\frac{\ell_1}{K_1} + \frac{\ell_2}{K_2}}$$

4. Equivalent thermal conductivity of two metal rods of same length combined in parallel

   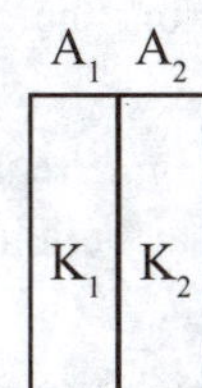

   $$K_p = \frac{K_1A_1 + K_2A_2}{A_1 + A_2}$$

Numerical Profile

If two rods of same length having thermal conductivities K_1 and K_2 are connected in series, then their equivalent thermal conductivity is $K_{eq} = \frac{2K_1K_2}{K_1 + K_2}$

and in parallel combination $K_{eq} = \frac{K_1 + K_2}{2}$

Radiation

Formulas

1. Kirchhoff's law of Radiation

$$E = \frac{e}{a} = \frac{\text{Emissive power}}{\text{Absorptive power}}$$

Where, $e = \frac{1}{A}\frac{\Delta Q}{\Delta t}$

$$a = \frac{\text{energy absorbed}}{\text{energy incident}}$$

Statements Asked In Exam

- Temperature is a relative measure, it is an indication of hotness or coldness of a body.
- Heat is the form of energy which is transferred between two (or more) systems or a system and its surroundings by virtue of temperature difference.
- Mercury and alcohol are the liquids used in most liquid-in-glass thermometers.
- The ice and steam point have values 32°F and 212°F respectively on the fahrenheit scale and 0°C and 100°C on the celcius scale.
- The triple point of water has a unique value of 273.16 K. At particular values of volume and pressure, the triple point of water is always 273.16 K.
- On the fahrenheit scale, there are 180 equal intervals between two reference points, and on the Celsius scale there are 100.
- The change of state from solid to liquid is called melting, from liquid to gas is called vaporisation and from liquid to solid is called freezing.
- Both the solid and liquid states of a substance co-exist in thermal equilibrium during the change of state from solid to liquid.
- Heating system based on circulation of steam are more efficient in warming a building than those based on circulation of hot water.
- The temperature at which the solid and liquid states of substance are thermal equilibrium with each other is known as the melting point of the substance. It also depends on pressure.
- Melting point of a substance changes due to increase in pressure. This phenomenon is known as regelation.
- A piece of red glass when heated and taken out. It glows with green light.
- The temperature at which the liquid and the vapour states of a substance coexist is called its boiling point.
- The temperature of a substance remains constant during its change of state.
- The triple point of water occurs at temperature 273.16 K and pressure 6.11×10^{-3} Pa.
- The amount of heat per unit mass transfered during change of state of the substance is called latent heat of the substance for the process.
- Conduction is the mechanism of transfer of heat between two adjacent parts of a body because of their temperature differences.
- A perfect black body emits radiation of all possible wavelength.
- A body with large reflectivity is a poor emitter.
- The vaporisation line is called steam line in case of water and sublimation line is called the frost line in case of water.
- Trade wind is the steady surface wind on the earth blowing in from north-east towards the equator.
- Animals curl into a ball, when they feel very cold.
- Mercury, though a liquid is heated by conduction and not by convection.
- Even though air is a bad conductor of heat, we don't feel warm without clothes.
- Conduction and convection are processes where heat is transferred by the collision of molecules. Hence, for these processes to take place we need a material medium.
- Convection and conduction are slow phenomenon but radiation takes place at the speed of light.
- Path of heat transfer for conduction and convection are irregular but for radiation it is a straight line.
- Stainless steel cooking pans are preferred with extra copper bottoms because of high thermal conductivity of copper.
- An ink dot on a cup of porcelain appears dark. When the same cup is heated to a high temperature, the dot becomes brighter than the rest of the cup.

NCERT Topic-wise MCQs

Temperature Scale, Thermal Expansion

1. Oxygen boils at –183°C. This temperature is approximately.
   1) 215°F
   2) –297°F
   3) 329°F
   4) 361°F

2. The absolute zero is the temperature at which
   1) Water freezes
   2) All substance exist in solid state
   3) Molecular motion ceases
   4) None of the above

3. The scale on a steel meter stick is calibrated at 15°C. What is the error in the reading of 60 cm at 27°C? $\alpha_{steel} = 1.2 \times 10^{-5}\ (°C)^{-1}$.
   1) 0.001 cm
   2) 0.00864 cm
   3) 0.013 cm
   4) 0.1 cm

4. How much should the temperature of a brass rod be increased so as to increase its length 3% (α for brass is $2 \times 10^{-5}/°C^{-1}$)?
   1) 1500°C
   2) 150°C
   3) 2000°C
   4) 2500°C

5. The volume of a metal block is 4000 cm³ at 273K. What would be its volume at 373 K? (give α of the metal is $18 \times 10^{-6} K^{-1}$)
   1) 5.001 m³
   2) 4×10^{-3} m³
   3) 4.002×10^{-3} m³
   4) 1.002×10^{-2} m³

6. A sheet of brass is 100 cm long and 40 cm broad at 273 K. The area of the surface increase by 2 cm² at 100°C. What is the coefficient of linear expansion of brass?
   1) $5 \times 10^{-6}\ °C^{-1}$
   2) $2.5\ 10^{-6}\ °C^{-1}$
   3) $3 \times 10^{-6}\ °C^{-1}$
   4) $2 \times 10^{-5}\ °C^{-1}$

7. A metal rod is used as a bar pendulum. If the room temperature rise by 10°C and coefficient of linear expansion of the metal of the rod is 3×10^{-5} per °C, the period of pendulum will have percentage increases of
   1) 0.816%
   2) 0.15%
   3) 0.01%
   4) 0.015%

8. The graph AB shown in the figure is a plot of the temperature of a body in degree Celsius and degree Fahrenheit. Then

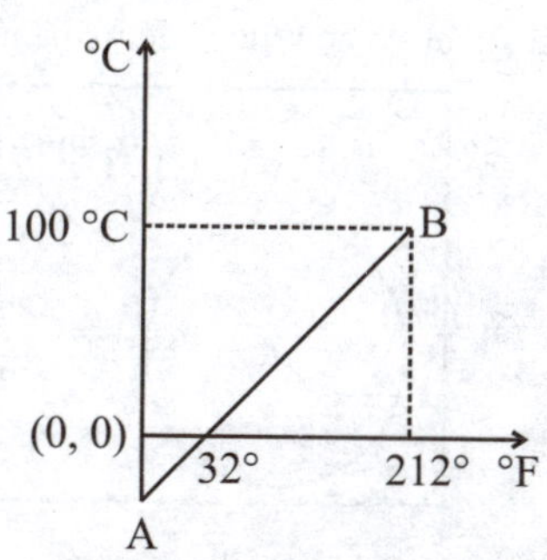

   1) Slope of the line AB is 9/5
   2) Slope of the line AB is 5/9
   3) Slope of the line AB is 1/9
   4) Slope of the line is 3/9

9. A faulty thermometer has its fixed points marked as 5° and 95°. This thermometer reads the temperature of a body as 59°. Then the correct temperature of the body on the Celsius scale is
   1) 59°
   2) 60°
   3) 48.6°
   4) 58°

10. The coefficient of volumetric expansion of lamina is $5 \times 10^{-5}/°C$. Its coefficient of linear expansion will be:
    1) $1.6 \times 10^{-5}/°C$
    2) $1.1 \times 10^{5}/°C$
    3) $2 \times 10^{-6}/°C$
    4) $3 \times 10/°C$

11. A lamina copper is 100 cm long and 15 cm broad at 273 K. The area of cross-section increases by 2 cm² at 100°C. What will be the coefficient of linear expansion of sheet material?
    1) $10^{-6}/°C$
    2) $2 \times 10^{-6}/°C$
    3) $6.6 \times 10^{-6}/°C$
    4) $5 \times 10^{-6}/°C$

12. If on heating liquid from 273K to 353K, the mass expelled is $\left(\frac{1}{50}\right)^{th}$ of mass still remaining, the coefficient of apparent expansion of liquid is
    1) $3.2 \times 10^{-4}/°C$
    2) $4.4 \times 10^{-2}/°C$
    3) $3 \times 10^{-3}/°C$
    4) $2.5 \times 10^{-4}/°C$

13. A faulty thermometer has its lower fixed point marked as –10° and upper fixed point market as 110°. If the temperature of the body shown in this scale is 62°, the temperature shown on the Celsius scale is
    1) 72°C
    2) 82°C
    3) 60°C
    4) 42°C

14. When temperature of a rod increases from t to t + Δt the moment of inertia of the rod about an axis increases from I to I + ΔI. If α be the coefficient of linear expansion of the rod, then the value of $\frac{\Delta I}{I}$ is

1) $2\alpha\Delta t$ 2) $\alpha\Delta t$

3) $\frac{\alpha\Delta t}{2}$ 4) $\frac{\Delta t}{\alpha}$

15. Two holes of unequal diameters d_1 and d_2, ($d_1 > d_2$) are cut in a metal sheet. When the sheet is heated,

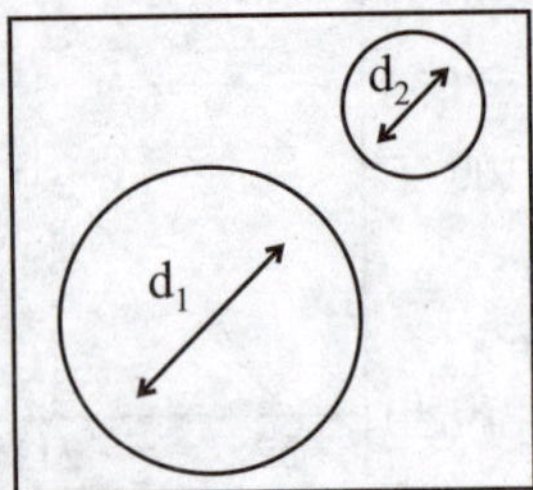

1) d_1 will increase, d_2 will decrease
2) d_1 will decrease, d_2 will increase
3) Both d_1 and d_2 will decrease
4) Both d_1 and d_2 will increase

Specific Heat, Calorimetry

16. When 0.20 kg of ice at 0°C mixed with 0.30 kg of water at 70°C in a container, the resulting temperature is 10°C. What is the latent heat of fusion of ice? (S_{water} = 4186 J kg^{-1}°C^{-1})

1) 4×10^5 J kg^{-1} 2) 3.34×10^5 J kg^{-1}

3) 6×10^5 J kg^{-1} 4) 7×10^5 J kg^{-1}

17. 80 gm of water at 30°C are poured on a large block of ice at 0°C. The mass of ice that melts is

1) 30 gm 2) 80 gm

3) 1600 gm 4) 150 gm

18. If specific heat of substance is infinite it means

1) Heat is emitted
2) Heat is absorbed
3) No change in temperature takes place whether heat is taken in or given out
4) All of the above

19. Two spheres made of same substance have diameters in the ratio 1 : 4. Their thermal capacities are in the ratio of

1) 16 : 1 2) 32 : 1

3) 64 : 1 4) 1 : 64

20. One kilogram of steam at 100°C can melt how much ice at 0°C?

1) 8.0 kg 2) $\frac{8}{54}$kg

3) $\frac{54}{8}$kg 4) 540 kg

21. Amount of heat required to raise the temperature of a body through 1° unit is called its

1) Water equivalent 2) Thermal capacity

3) Entropy 4) Specific heat

22. A bubble of 8 mole of helium is submerged at a certain depth in water. The temperature of water increase by 60°C. How much heat is added approximately to helium during expansion ?

1) 1000 cal 2) 7000 cal

3) 2400 cal 4) 1500 cal

23. Water is boiled to convert into stream. At this stage the specific heat of water is

1) < 1 2) ∞

3) 1 4) 0

24. The sprinkling of water reduces the temperature of a closed room because

1) The temperature of water is less than that of the room
2) Water is the bad conductor of heat
3) Water drops absorbs heat of room
4) Water drop looses some heat in the room

Conduction & Convection

25. The coefficient of thermal conductivity of a material depends upon

1) Temperature gradient between two surfaces
2) Area of the material
3) Thickness of the material
4) Nature of the material

26. Two identical plates of different metals are joined to form a single plate whose thickness is 2 times the thickness of each plate. If the coefficients of conductivity of each plate are 4 and 6 respectively, then the conductivity of composite plate will be

1) 5 2) 4.8

3) 4.9 4) 4

27. The freezer in a refrigerator is located at the top section so that

1) Conduction take place
2) The heat gained from the environment is low
3) The motor is not heated
4) Convection can take place so that entire refrigerator is cooled quickly

28. In which of the following convection will not take place primarily?

1) Sea and land breeze
2) Trade wind
3) Boiling of water
4) Warming of glass of bulb due to filament.

29. Two identical square rods of metal are welded end-to-end as shown in Fig. (1). 20 calories of heat will flow through it in 4 minutes. If the rods are welded as shown in Fig. (2), the same amount of heat flows through the rods in

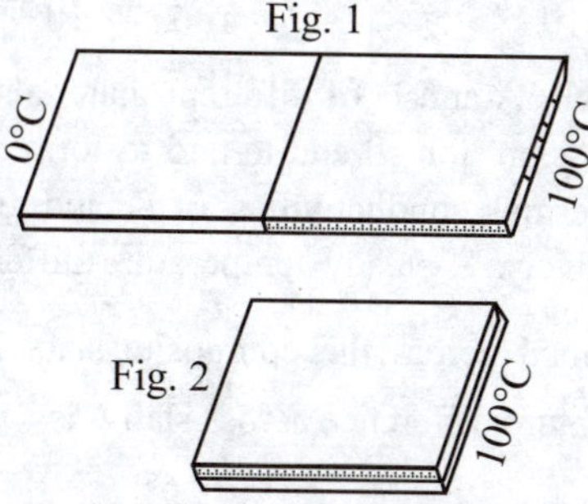

1) 1 minute
2) 2 minutes
3) 4 minutes
4) 16 minutes

30. Two identical rods with different thermal conductivities k_1 & k_2 and different temperatures are first placed along length and then along area, then the ratio of rates of heat flow in both cases is

1) $\frac{k_1k_2}{k_1+k_2}$
2) $\frac{4k_1^2k_2}{(k_1+k_2)^2}$
3) $\frac{(k_1+k_2)^2}{4k_1}$
4) $\frac{4k_1k_2}{(k_1+k_2)^2}$

31. Three identical rods A, B and C are placed end to end. A temperature difference is maintained between the free ends of A and C. The thermal conductivity of B is thrice of that of C and half of that of A. The effective thermal conductivity of the system will be (K_A is the thermal conductivity of rod A).

1) $\frac{1}{3}K_A$
2) $3K_A$
3) $2K_A$
4) $\frac{2}{3}K_A$

32. Three rods of the same dimension have thermal conductivities 3K, 2K and 4K. They are arranged as shown in diagram. There ends kept at 150°C, 20°C, 80°C. The temperature of the junction is

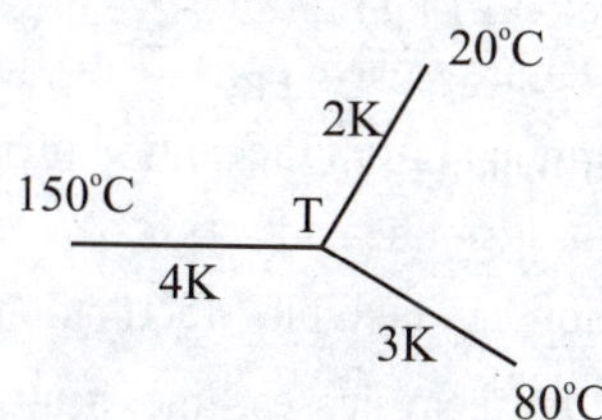

1) 90°C
2) 97.7°C
3) 112°C
4) 50°C

33. Can we boil egg inside the earth satellite by convection?

1) Yes
2) No
3) Nothing can be said
4) Data insufficient

34. Two identical conducting rods are first connected independently to two vessels, one containing water at 100°C and the other containing ice at 0°C. In the second case, the rods are joined end to end and connected to the same vessels. Let q_1 and q_2 be the rate of melting of ice in two respective case. The ratio $\frac{q_1}{q_2}$ is

1) $\frac{1}{2}$
2) $\frac{2}{1}$
3) $\frac{4}{1}$
4) $\frac{1}{4}$

35. Two walls of thickness d_1 and d_2 and thermal conductivities K_1 and K_2 are in contact. In the steady state, if the temperatures at the outer surface are T_1 and T_2, the temperature at the common wall is

1) $\frac{(K_1T_1d_2+K_2T_2d_1)}{(K_1d_2+K_2d_1)}$
2) $\frac{(K_1T_1+K_2T_2)}{(K_1d_2+K_2d_1)}$
3) $\left(\frac{K_1d_1+K_2d_2}{T_1+T_2}\right)T_1T_2$
4) $K_1A_2=K_2A_1$

36. A cylinder of radius R made of material of thermal conductivity K_1 is surrounded by a cylindrical shell of inner radius R and outer radius 3 R made of a material of thermal conductivity K_2. The two ends of the combined system are maintained at two different temperatures. There is no loss of heat across the cylindrical surface and the system is in steady state. The effective thermal conductivity of the system is

1) K_1+K_2
2) $\frac{(K_1+8K_2)}{9}$
3) $\frac{K_1K_2}{(K_1+K_2)}$
4) $K_1A_1=K_2A_2$

37. On heating one end of a rod, the temperature of whole rod will be uniform when

1) $K=1$
2) $K=0$
3) 0.74 K/W
4) $K=\infty$

38. The coefficient of thermal conductivity of copper is 9 times that of steel. In the composite cylindrical bar shown in the figure, what will be the temperature at the junction of copper and steel?

100°C	0°C
Copper	Steel
←18 cm→	←6 cm→

1) 75°C
2) 67°C
3) 25°C
4) 33°C

39. Two rods of same length and area of cross-section A_1 and A_2 have their ends at the same temperature. If K_1 and K_2 are their thermal conductivities, then the rate of flow of heat is the same in both the rods if

1) $K_1A_2=K_2A_1$
2) $\frac{(K_1+8K_2)}{9}$
3) $\frac{K_1K_2}{(K_1+K_2)}$
4) $K_1A_1=K_2A_2$

High Order Time Intensive MCQs

1. The ratio of densities of two bodies is 5 : 6 and specific heat in the ratio 3 : 5. The ratio of their thermal capacities for unit volume is

1) 25 : 18 2) 1 : 2
3) 2 : 1 4) 1 : 1

2. Two uniform metal rods of lengths ℓ_1 and ℓ_2 and coefficients of linear expansion α_1 and α_2, respectively are connected to form a single rod of length $(\ell_1 + \ell_2)$. When the temperature of the combined rod is raised by t°C, the length of each rod increases by the same amount. Then $\left(\frac{\alpha_2}{\alpha_1+\alpha_2}\right)$ is

1) $\frac{\ell_1}{(\ell_1+\ell_2)}$ 2) $\frac{(\ell_1+\ell_2)}{\ell_1}$
3) $\frac{\ell_2}{(\ell_1+\ell_2)}$ 4) $\frac{(\ell_1+\ell_2)}{\ell_2}$

3. A block of ice of mass 50 kg is sliding on a horizontal plane. It starts with speed 5 m/s and stops after moving through some distance. The mass of ice that has melted due to friction between the block and the surface is (assuming that no energy is lost and latent heat of fusion of ice is 80 cal/g, 1 cal = 4.2 J)

1) 2.86 g 2) 3.86 g
3) 0.86 g 4) 1.86 g

4. One end of a copper rod of length 2m and area of cross section 10^{-3}m² is immersed in boiling water and the other in ice. If the coefficient of thermal conductivity of copper is 92 cal/m-s-°C and the latent heat of ice is 9×10^4 cal/kg (assumed) then the amount of ice which will melt in three minute is

1) 15×10^{-1} kg 2) 40×10^{-2} kg
3) 92×10^{-4} kg 4) 25×10^{-2} kg

5. 19 g of water at 30°C and 5 g of ice at – 20°C are mixed together in a calorimeter. What is the final temperature of the mixture?(Given, specific heat of ice = 0.5 cal g^{-1} $(°C)^{-1}$ and latent heat of fusion of ice = 80 cal g^{-1})

1) 0°C 2) –5°C
3) 5°C 4) 10°C

6. A lead bullet, of initial temperature 27°C and speed v km/h penetrates into a solid object and melts. If 50% of the kinetic energy is used to heat it, the value of v in km h^{-1} is (for lead melting point = 600 K, latent heat of fusion = 2.5×10^4 J/kg, specific heat = 125 $Jkg^{-1} K^{-1}$)

1) 3600 2) 1800
3) 1200 4) 1000

7. A 42 kg block of ice moving on rough horizontal surface stops due to friction, after sometime. If the initial velocity of the decelerating block is 4 ms^{-1}, the mass of ice (in kg) that has melted due to the heat generated by the friction is (Latent heat of ice is 3.36×10^5 J kg^{-1})

1) 1×10^{-3} 2) 1.5×10^{-3}
3) 2×10^{-3} 4) 2.5×10^{-3}

8. Two slabs A and B of different materials but of the same thickness are joined end to end to form a composite slab. The thermal conductivities of A and B are K_1 and K_2 respectively. A steady temperature difference of 12°C is maintained across the composite slab. If $K_1 = \frac{K_2}{2}$, the temperature difference across slab A is

1) 4°C 2) 6°C
3) 8°C 4) 10°C

9. Three rods made of same material and having the same cross-section have been joined. Each rod is of the same length. The left and right ends are kept at 0°C and 100°C respectively. The temperature at the junction of the three rods will be.

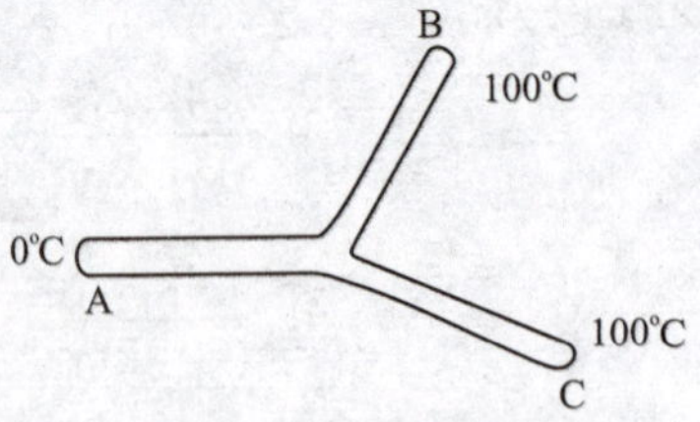

1) 50°C 2) 60°C
3) 66.6°C 4) 40°C

10. A beaker contains 200 g of water. The heat capacity of the beaker is equal to that of 20 g of water. The initial temperature of water in the beaker is 20°C. If 440g of hot water at 92°C is poured in it the final temperature (neglecting radiation loss), will be nearest to

1) 58°C 2) 68°C
3) 73°C 4) 78°C

11. A spherical iron ball is placed on a large block of dry ice at 0°C. The ball sinks into the ice until it is half submerged. What was the temperature of iron ball?

Density of iron = 7675 kg m^{-3}

Density of ice = 921 kg m^{-3}

Specific heat of iron = 504 J kg^{-1} K^{-1}

Latent heat of fusion of ice = 336×10^3 J kg^{-1}

1) 30°C 2) 35°C
3) 40°C 4) 45°C

12. If there are no heat losses, the heat released by the condensation of x gm of steam at 100°C into water at 100°C can be used to convert y gm of ice at 0°C can be used to convert y gm of ice at 0°C into water at 100°C. Then the ratio y : x is nearly

1) 5 : 1 2) 2 : 1
3) 2.5 : 1 4) 3 : 1

13. A lead bullet hits a steel plate with a velocity 400 m/s. If the impact is perfectly inelastic and heat produced is equally shared between the bullet and the target then the rise in temperature of bullet is (specific heat capacity of lead = 125J $kg^{-1} K^{-1}$)

1) 320°C
2) 321°C
3) 221°C
4) 210°C

14. A space object has the shape of a sphere of radius R. Heat sources ensure that the heat evolution at a constant rate is distributed uniformly over its volume. The amount of heat liberated by an unit surface area is proportional to the fourth power of thermodynamic temperature. In what proportion would the temperature of the object change if its radius is decreased to half?

1) 2.19
2) 5.19
3) 1.19
4) 4.19

15. The room temperature is +20°C when outside temperature is –20°C and room temperature is + 10°C when outside temperature is –40°C. Find the temperature of the radiator heating the room:

1) 30°C
2) 60°C
3) 90°C
4) 45°C

16. Some water at 0°C is placed in a large insulated enclosure (vessel). The water vapour formed is pumped out continuously. What fraction of the water will ultimately freeze, if the latent heat of vaporization is seven times the latent heat of fusion?

1) 7/8
2) 8/7
3) 3/8
4) 5/8

17. A substance is in the solid form at 0°C. The amount of heat added to this substance and its temperature are plotted in the following graph. If the relative specific heat capacity of the solid substance is 0.5, find from the graph, the mass of the substance:

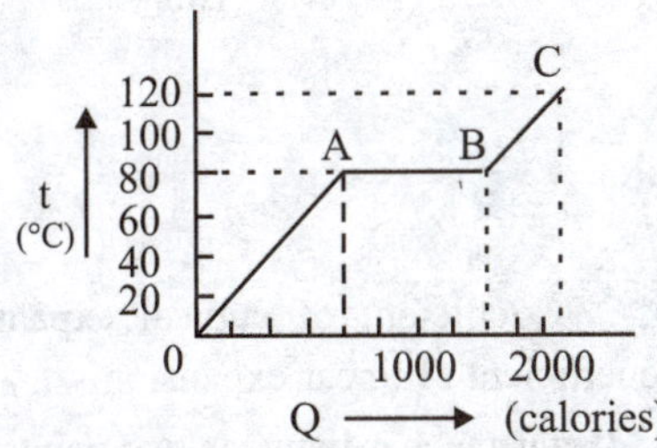

(Specific heat capacity of water = 1000 cal $kg^{-1} K^{-1}$)

1) 0.02 kg
2) 2 kg
3) 0.04 kg
4) 0.05 kg

18. An iron ball has a diameter of 6 cm and is 0.010 mm too large to pass through a hole in a brass plate when the ball and plate are at a temperature of 30°C. At what temperature, the same for ball and plate, will the ball just pass through the hole? (α_{brass} = 19 × 10^{-6} °C^{-1} and α_{iron} = 12 × 10^{-6} °C^{-1})

1) 23.8°C
2) 13.8°C
3) 53.8°C
4) 83.8°C

19. What should be the lengths of steel and copper rod so that the length of steel rod is 5 cm longer then the copper rod at all the temperatures? Coefficients of linear expansion for copper and steel are 1.7 × 10^{-5} °C^{-1} and 1.1 × 10^{-5} °C^{-1}:

1) 2.17, 14.17 cm
2) 9.17, 14.17 cm
3) 9.17, 18.17 cm
4) 3.17, 5.17 cm

20. A copper rod 2 m long has a circular cross section of radius 1 cm. One end is kept at 100°C and the other at 0°C, and the surface is insulated so that negligible heat is lost through the surface. Thermal conductivity of copper is 401 W/m–K. Find the thermal resistance of the bar.

1) 15.9 K/W
2) 20.9 K/W
3) 40.9 K/W
4) 50.9 K/W

21. A copper rod 2 m long has a circular cross section of radius 1 cm. One end is kept at 100°C and the other at 0°C, and the surface is insulated so that negligible heat is lost through the surface. Thermal conductivity of copper is 401 W/m–K. Find the temperature gradient $\frac{dT}{dx}$.

1) –10°C/m
2) –40°C/m
3) –50°C/m
4) –90°C/m

22. Three rods of same dimensions have thermal conductivities 3K, 2K and K. They are arranged as shown below

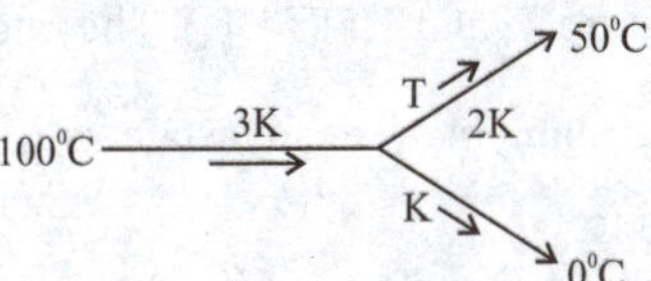

What will be the temperature T of the junction?

1) $\frac{200}{3}$°C
2) $\frac{100}{3}$°C
3) 75°C
4) $\frac{50}{3}$°C

23. Two spheres P and Q of same colour having radii 8 cm and 2 cm are maintained at temperature 127°C and 527°C respectively. The ratio of energy radiated by P and Q is

1) 0.054
2) 0.0034
3) 1
4) 2

24. The following figure shows tow air-filled bulbs connected by a U-tube partly filled with alcohol. What happens to the levels of alcohol in the limbs X and Y when an electric bulb placed midway between the bulbs is lighted?

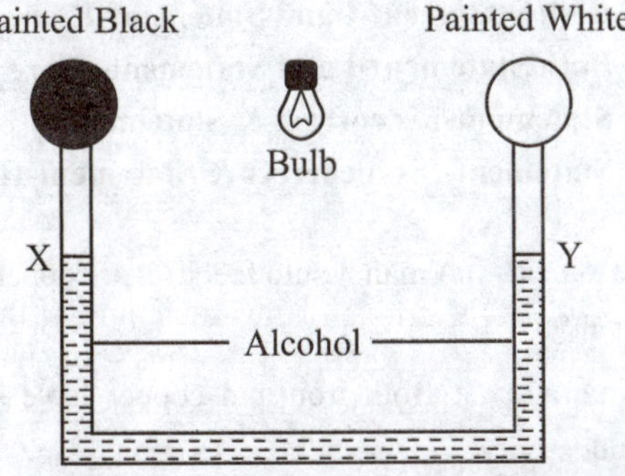

1) The level of alcohol in limb X falls while that in limb Y rises
2) The level of alcohol in limb X rises while that limb Y falls
3) The level of alcohol falls in both limbs
4) There is no change in the levels of alcohol in the two limbs

25. The graphs gives the variation of temperature of two bodies having the same surface area with time where A_x and A_y represent absorptivity and ε_x and ε_y represent emissivity then:

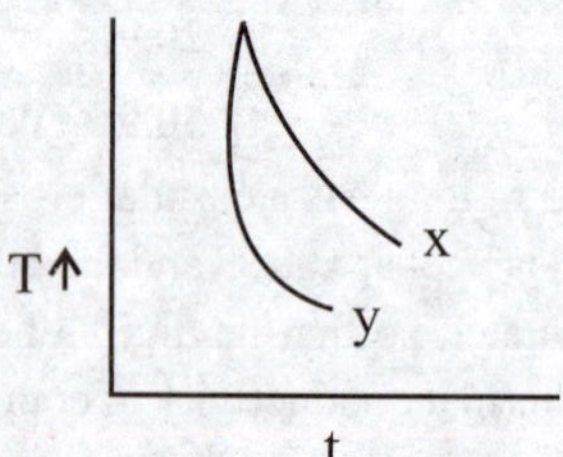

1) $\varepsilon_x > \varepsilon_y$ and $A_y > A_x$
2) $\varepsilon_x < \varepsilon_y$ and $A_x > A_y$
3) $\varepsilon_x > \varepsilon_y$ and $A_x > A_y$
4) $\varepsilon_y > \varepsilon_x$ and $A_y > A_x$

26. The temperature of the two outer surfaces of a composite slab, consisting of two materials having coefficients of thermal conductivity K and 2 K and thickness x and 4x respectively are T_2 and $T_1 (T_2 > T_1)$. The rate of heat transfer through the slab, in a steady state is $\left(\frac{A(T_2 - T_1)K}{x}\right)f$, with f equal to:

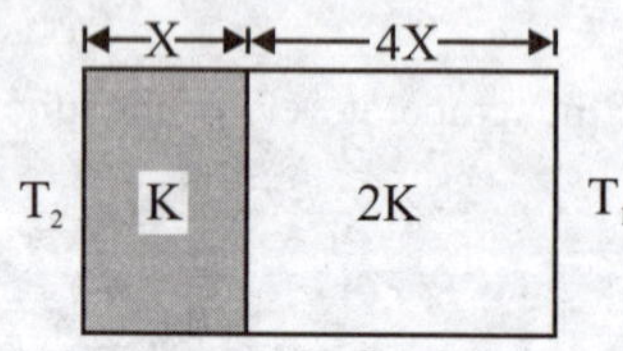

1) 1
2) 1/2
3) 2/3
4) 1/3

27. Two holes of unequal diameters d_1 and d_2 $(d_1 > d_2)$ are cut in a metal sheet. If the sheet is heated :
1) Both d_1 and d_2 will decrease
2) Both d_1 and d_2 will increase
3) d_1 will increase, d_2 will decrease
4) d_1 will decrease, d_2 will increase

28. The temperature of a body is increased from 27°C to 127°C. The radiation emitted by it increases by a factor of
1) $\frac{256}{81}$
2) $\frac{15}{9}$
3) $\frac{4}{5}$
4) $\frac{12}{27}$

29. A metal sphere of radius r and specific heat S is rotated about an axis passing through its centre at a speed of f rotations per second. It is suddenly stopped and 50% of its energy is used in increasing its temperature. Then the rise in temperature of the sphere is :
1) $\frac{2\pi^2 f^2 r^2}{5S}$
2) $\frac{\pi^2 f^2}{10 r^2 S}$
3) $\frac{7}{8}\pi r^2 f^2 S$
4) $\frac{5(\pi r f)^2}{14S}$

30. The density of a substance at 0°C is 10 g/cc and 100°C, its density is 9.7 g/cc. The coefficient of linear expansion of the substance is
1) 10^{-4}
2) 10^{-3}
3) 10^{-2}
4) 10^{2}

31. A copper ball of mass 0.2 kg falls freely from height of 15m and bounces to a height of 5 m from the ground. If the half of the dissipated energy in this process is absorbed by the ball. The rise in temperature is (specific-heat of copper = 386.04 $Jkg^{-1} K^{-1}$)
1) 0.129°C
2) 0.130°C
3) 0.113°C
4) 0.100°C

Statement Based MCQs

Directions: These questions consist of two statements each, printed as Statement-I and Statement-II. While answering these questions, you are required to choose any one of the following four responses.
1) **Both Statement-I and Statement-II are correct.**
2) **Both Statement-I and Statement-II are incorrect.**
3) **Statement-I is correct & Statement-II is incorrect.**
4) **Statement-I is incorrect & Statement-II is correct.**

1. **Statement-I:** A man would feel iron or copper balls equally hot at 98.4°F

Statement-II: Both iron and copper have same thermal conductivity.

2. **Statement-I:** Coefficient of cubical expansion is three times the coefficient of linear expansion.

Statement-II: Distance is between two points on a metallic sheet increases with rise in temperature.

3. **Statement-I:** Blue star is at lower temperature than red star.

Statement-II: Wein's displacement law states that $T \propto (1/\lambda_m)$.

4. **Statement-I:** Rate of flow of heat in a metallic rod is directly proportional to the difference in temperature between the two ends of the rod.

Statement-II: At –40° both Celsius and Fahrenheit thermometer give the same reading.

Assertion & Reason MCQs

Directions: These questions consist of two statements each, printed as Assertion and Reason. While answering these questions, you are required to choose any one of the following four responses.

1) **Both Assertion and Reason are True and the Reason is a correct explanation of the Assertion.**
2) **Both Assertion and Reason are True but Reason is not a correct explanation of the Assertion.**
3) **Assertion is True but the Reason is False.**
4) **Assertion is False but Reason is True.**

1. **Assertion:** In aircraft manufacture, rivets and other fasteners are often cooled in dry ice before insertion and then allowed to expand to a tight fit.

   Reason: In a solid, heat flows due to convection.

2. **Assertion:** Woolen clothes keep the body warm in winter.

   Reason: Woolen fibers enclose a large amount of air in them. Both air and wool are the bad conductors of heat.

3. **Assertion:** An ice skater can slide over ice smoothly if the skate blades are sharp.

   Reason: Melting point of ice decreases with increase in pressure.

4. **Assertion:** On the sea shore cool breeze flows in the evening.

   Reason: Convection currents are set up from sea to the land since land cools slower than water.

5. **Assertion:** Dental materials used for fillings must be matched in their thermal expansion properties to tooth enamels.

   Reason: If the material used for tooth enamel filling has thermal expansion different from that of tooth then it shall cause pain.

Match the Columns MCQs

1. A metal piece of density ρ_1 floats on liquid of density ρ_2. The coefficients of volume expansion of the metal and liquid are γ_1 and γ_2, respectively. The temperatures of both liquid and metal are increased by ΔT. Then match the following:

	Column-I		Column-II
(a)	If $\gamma_2 > \gamma_1$	(i)	the solid lifts up
(b)	$\gamma_2 = \gamma_1$	(ii)	fraction of the volume of metal submerged in liquid increases
(c)	If $\gamma_2 < \gamma_1$	(iii)	the solid sinks
		(iv)	no effect on fraction of solid submerged in liquid

1) (a)-(i) (b)-(iv) (c)-(iii)
2) (a)-(ii) (b)-(iv) (c)-(i)
3) (a)-(i) (b)-(ii) (c)-(iii)
4) (a)-(ii) (b)-(iii) (c)-(i)

2. Three liquids P, Q and R having same specific heat and mass m, 2m and 3m have temperature 20°C, 40°C and 60°C respectively. Temperature of the mixture when:

	Column-I		Column-II
(a)	P and Q are mixed	(i)	35°C
(b)	P and R are mixed	(ii)	52°C
(c)	Q and R are mixed	(iii)	50°C
(d)	P, Q and R all three are mixed	(iv)	45°C
		(v)	None

1) (a)-(i) (b)-(iv) (c)-(iii) (d)-(ii)
2) (a)-(iii) (b)-(i) (c)-(iv) (d)-(ii)
3) (a)-(i) (b)-(ii) (c)-(iii) (d)-(iv)
4) (a)-(v) (b)-(iii) (c)-(ii) (d)-(v)

3. The surface of a household radiator has an emissivity of 0.55 and an area of 1.5 m^2. Its equilibrium temperature is 50°C and the surroundings are at 22°C ($\sigma = 5.67 \times 10^{-8}$ W/m^{2}K^4).

	Column-I		Column-II
(a)	Rate at which radiation is emitted by the radiator is	(i)	155
(b)	Rate at which radiation is absorbed by the radiator is	(ii)	509
(c)	Net value of radiation from the radiator is	(iii)	354

1) (a)-(ii) (b)-(iii) (c)-(i)
2) (a)-(i) (b)-(iii) (c)-(ii)
3) (a)-(ii) (b)-(i) (c)-(iii)
4) (a)-(i) (b)-(iii) (c)-(ii)

NCERT Exemplar MCQs

1. A bimetallic strip is made of aluminum and steel ($\alpha_{Al} > \alpha_{steel}$). On heating, the strip will:
   1) Remain straight
   2) Get twisted
   3) Will bend with aluminum on concave side
   4) Will bend with steel on concave side
2. A uniform metallic rod rotates about its perpendicular bisector with constant angular speed. If it is heated uniformly to raise its temperature slightly:
   1) Its speed of rotation increases
   2) Its speed of rotation decreases
   3) Its speed of rotation remains same
   4) Its speed increases because its moment of inertia increases
3. An aluminum sphere is dipped into water. Which of the following is true?
   1) Buoyancy will be less in water at 0°C than that in water at 4°C
   2) Buoyancy will be more in water at 0°C than that in water at 4°C
   3) Buoyancy in water at 0°C will be same as that in water at 4°C
   4) Buoyancy may be more or less in water at 4°C depending on the radius of the sphere
4. As the temperature is increased, the period of a pendulum:
   1) Increases as its effective length increases even though its centre of mass still remains at the centre of the bob
   2) Decreases as its effective length increases even though its centre of mass still remains at the centre of the bob
   3) Increases as its effective length increases due to shifting to centre of mass below the centre of the bob
   4) Decreases as its effective length remains same but the centre of mass shifts above the centre of the bob
5. Heat is associated with:
   1) Kinetic energy of random motion of molecules
   2) Kinetic energy of orderly motion of molecules
   3) Total kinetic energy of random and orderly motion of molecules
   4) Kinetic energy of random motion in some cases and kinetic energy of orderly motion in other
6. The radius of a metal sphere at room temperature T is R and the coefficient of linear expansion of the metal is α. The sphere heated a little by a temperature ΔT so that its new temperature is T + ΔT. The increase in the volume of the sphere is approximately:
   1) $2\pi R\alpha\Delta T$
   2) $\pi R^2\alpha\Delta T$
   3) $4\pi R^3\alpha\Delta T/3$
   4) $4\pi R^3\alpha\Delta T$

NEET & AIIMS Past Year MCQs

1. A metallic bar of Young's modulus, 0.5×10^{11} Nm^{-2} and coefficient of linear thermal 10^{-5}°C^{-1} length 1m and area of cross-section 10^{-3} m^2 is heated from 0°C to 100°C without expansion or bending. The compressive force developed in it is: **(2024)**
   1) 100×10^3N
   2) 2×10^3N
   3) 5×10^3N
   4) 50×10^3N
2. Water is used as a coolant in a nuclear reactor because of its **(2024 Re)**
   1) high thermal expansion coefficient
   2) high specific heat capacity
   3) low density
   4) low boiling point
3. Given below are two statements: One is labelled as Assertion A and the other is labelled as Reason R.

   Assertion A: Houses made of concrete roofs overlaid with foam keep the room hotter during summer.

   Reason R: The layer of foam insulation prohibits heat transfer, as it contains air pockets.

   In the light of the above statements, choose the correct answer from the options given below. **(2024 Re)**
   1) A is true but R is false.
   2) A is false but R is true.
   3) Both A and R are true and R is the correct explanation of A.
   4) Both A and R true but R is NOT the correct explanation of A.
4. A bimetallic strip consists of metals X and Y. It is mounted rigidly at the base as shown. The metal X has a higher coefficient of expansion compound to that for metal Y. When bimetallic strip is placed in a cold bath. **(AIIMS 2006)**

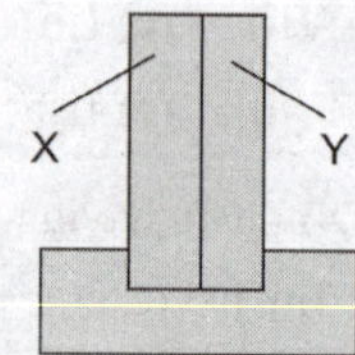

   1) It will bend towards the right
   2) It will bend towards the left
   3) It will not bend but shrink
   4) It will neighter bend nor shrink

5. Black body radiation curves shown below are at temperatures T_1 and T_2 ($T_2 > T1$). Which of the following plots is correct:- (AIIMS 2003)

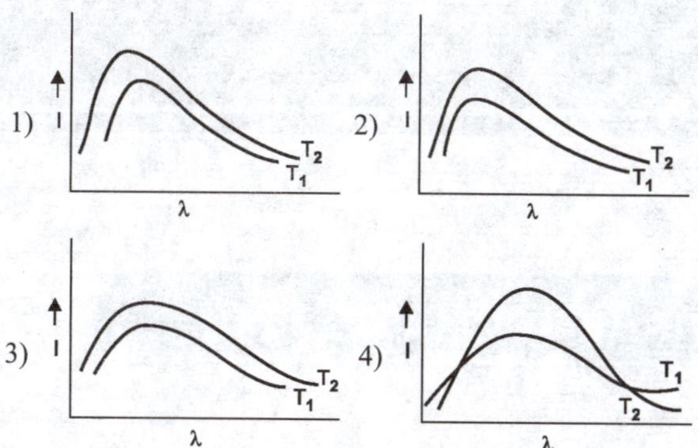

6. For an enclosure maintained at 1000 K, the maximum radiation occurs at wavelength λ_m. If the temperature is raised to 2000 K, the peak will shift to (AIIMS 2001)

1) $\frac{5}{2}\lambda_m$

2) $\frac{1}{2}\lambda_m$

3) $\frac{7}{2}\lambda_m$

4) $\frac{3}{2}\lambda_m$

Answer Key

NCERT Topic-wise MCQs

1	2	3	4	5	6	7	8	9	10	11	12	13	14	15	16	17
2	3	2	1	3	2	4	2	2	1	3	4	3	1	4	2	1
18	**19**	**20**	**21**	**22**	**23**	**24**	**25**	**26**	**27**	**28**	**29**	**30**	**31**	**32**	**33**	**34**
3	3	1	2	3	2	3	4	2	4	4	1	4	1	2	2	3
35	**36**	**37**	**38**	**39**												
1	2	4	1	4												

High Order Time Intensive MCQs

1	2	3	4	5	6	7	8	9	10	11	12	13	14	15	16	17
2	3	4	3	3	2	1	3	3	2	3	4	1	3	2	1	1
18	**19**	**20**	**21**	**22**	**23**	**24**	**25**	**26**	**27**	**28**	**29**	**30**	**31**			
3	2	1	3	1	3	1	4	4	2	1	1	1	1			

Statement Based MCQs

1	2	3	4
3	1	4	1

Assertion & Reason MCQs

1	2	3	4	5
3	1	1	3	1

Match the Column MCQs

1	2	3
2	4	1

NCERT Exemplar MCQs

1	2	3	4	5	6
4	2	1	1	1	4

NEET & AIIMS Past Year MCQs

1	2	3	4	5	6
4	2	2	2	1	2

CHAPTER

11 Thermodynamics

Topicwise Number of Questions Analysis (2025-2016)

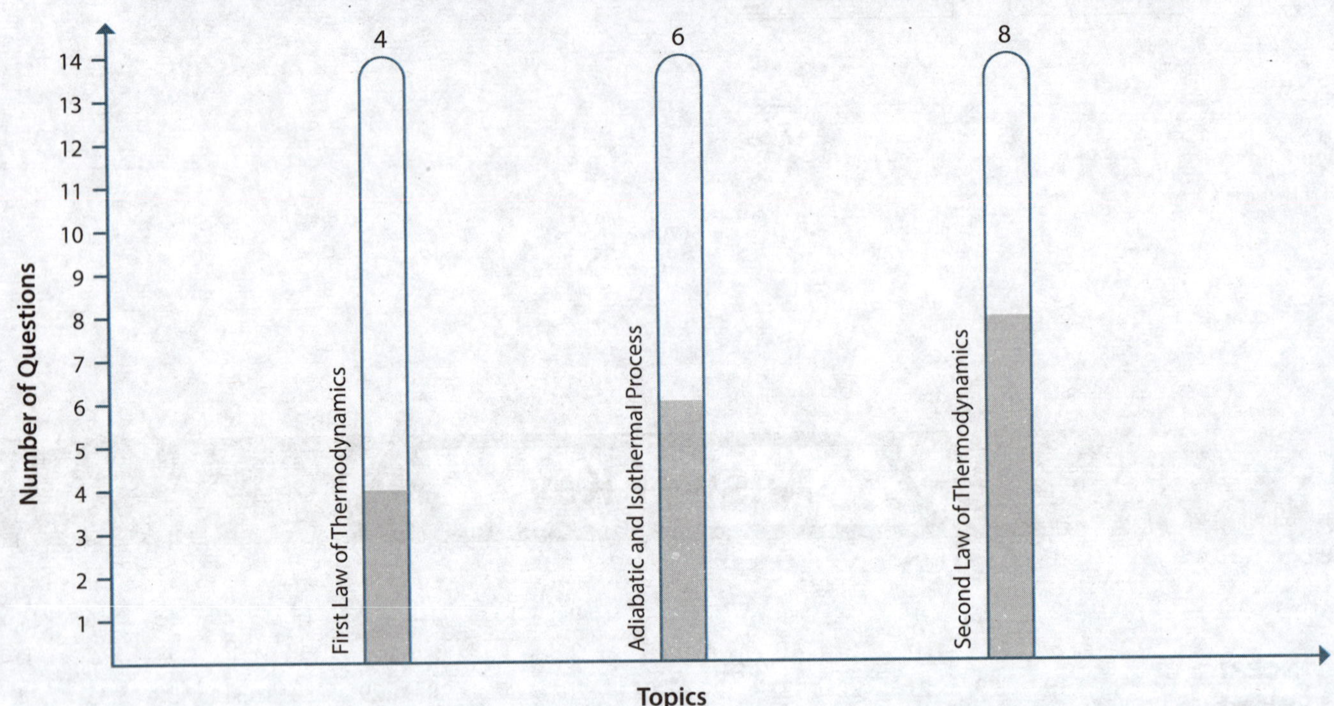

NCERT-Picks

Heat and Law of Thermodynamics

Formulas

1. **Zeroth law of thermodynamics**

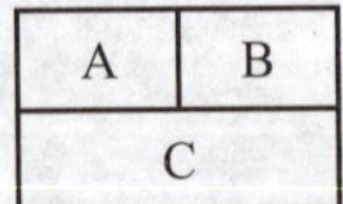

If A and C are in thermal equilibrium, B and C are in thermal equilibrium, then A and B must be in thermal equilibrium

2. **First law of thermodynamics:**

$dQ = dU + dW$

Where dU is the change in internal energy, dQ is the heat given to the system and dW is the work done by gas.

$dQ = +Ve$, when heat is supplied

$= -Ve$, when heat is rejected

$dU = +Ve$, when temperature increases

$= -Ve$, when temperature decreases

$dW = +Ve$, when work is done by the system (expansion).

$= -Ve$, when work is done on the system (compression).

3. If W is the work done, P is the pressure and dV is the change in volume, then $W = \int PdV$

4. Change in internal energy of a system is independent of the path on P – V diagram.

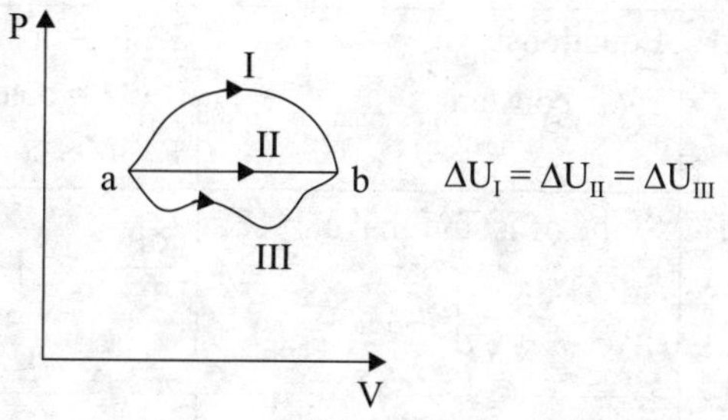

$\Delta U_I = \Delta U_{II} = \Delta U_{III}$

Numerical Profile

Work Done by a Gas due to Change in Volume

(i) If the volume changes from V_1 to V_2, the total external work done is

$$W = \int_{V_1}^{V_2} P\,dV = \text{Area under P-V curve.}$$

Sign Convention

(i) If the work is done by the system, then work done is (+) ve.

$$(\because dV > 0 \Rightarrow W > 0)$$

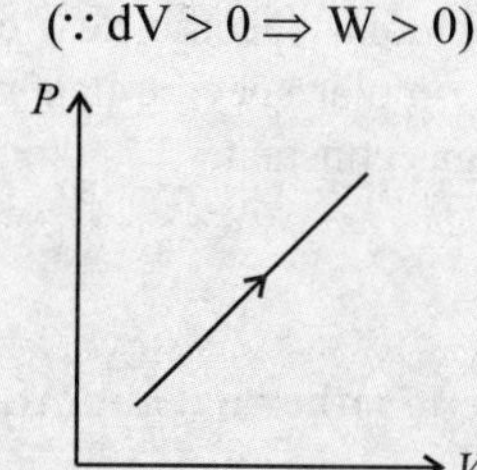

(ii) When work is done on the system, then it is negative.

$$(\because dV < 0 \Rightarrow W < 0)$$

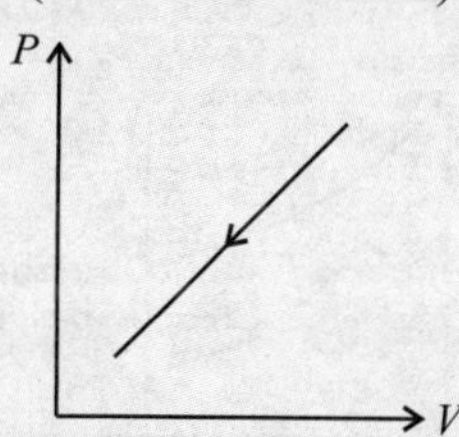

(iii) In a cyclic process (the system finally reaches the same initial state), value of work done is equal to the area enclosed by the cycle.

It is positive if the cycle is clockwise.

It is negative if the cycle is anticlockwise.

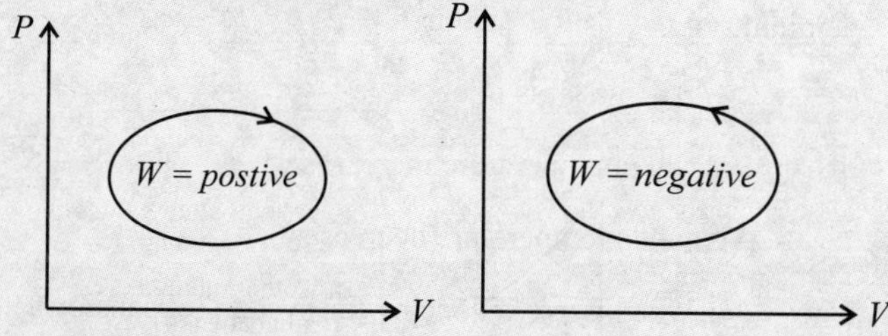

Application of Joule's Constant

(i) The height at which ice is released to melt completely is

$$h = \frac{JL}{g} \qquad (\because mgh = JmL)$$

[J is joule's constant, g is acceleration due to gravity and L is Latent heat of ice.]

Application of Joule's Constant

(i) The height at which ice is released to melt completely is

$$h = \frac{JL}{g} \qquad (\because mgh = JmL)$$

[J is joule's constant, g is acceleration due to gravity and L is Latent heat of ice.]

(ii) The increase in temperature of water when it falls from a height h to the ground is,

$$\Delta\theta = \frac{gh}{Js} \qquad (\because Jms\,\Delta\theta = mgh)$$

where 's' is specific heat of water

(iii) If a bullet of mass m moving with a velocity v is stopped abruptly by a target and all of its heats energy liberated is retained by bullet, then the rise in temperature is.

$$\Delta\theta = \frac{v^2}{2\,Js} \qquad (\because Jms\,\Delta\theta = \frac{1}{2}mv^2)$$

(iv) If a block of ice of mass M is dragged with constant velocity on a rough horizontal surface of coefficient of friction μ, through a distance *d*, then the mass of ice that melts is,

$$m = \frac{\mu Mgd}{JL}$$

(v) If a bullet at a temperature lesser than its melting point just melts when abruptly stopped by an obstacle and if all the heat produced is absorbed by the bullet then, we have

$$J(ms\Delta\theta + mL) = \frac{1}{2}mv^2$$

Where L is called Latent heat of the material of the bullet, m and v is the mass and velocity of the bullet,

Thermodynamics Processes

Formulas

1. Adiabatic process

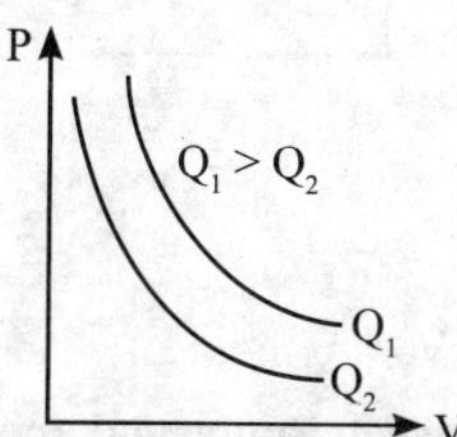

(a) $P_1V_1^{\gamma} = P_2V_2^{\gamma}$

(b) $P_2^{1-\gamma}T_2^{\gamma} = P_1^{1-\gamma}T_1^{\gamma}$

(c) $T_2V_2^{\gamma-1} = T_1V_1^{\gamma-1}$ Where $\gamma = \frac{C_p}{C_v}$

(d) Work done in adiabatic expansion $W = \frac{nR}{1-\gamma}(T_2 - T_1)$

(e) Slope of an a adiabatic curve at a point $= \frac{dP}{dV} = -\gamma\frac{P}{V}$

2. Isothermal process

(a) PV = constant

$P_1V_1 = P_2V_2$

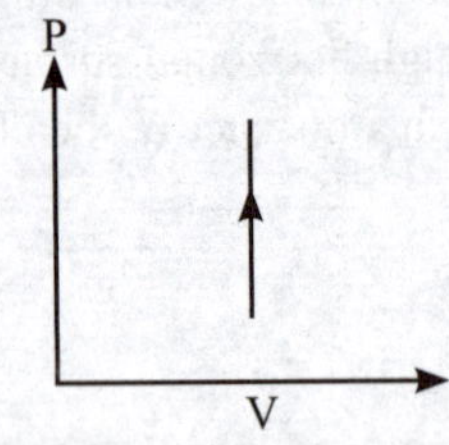

(b) Work done by the gas in isothermal expansion

$$W = 2.303\,nRT\log\frac{V_2}{V_1}$$

$$W = 2.303\,nRT\log\frac{P_1}{P_2}$$

(c) Slope of isothermal curve at any point, $\frac{dP}{dV} = -\frac{P}{V}$

3. Isochoric Process

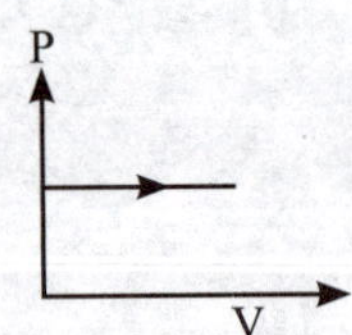

(a) $\frac{P}{T} = \text{constant} \Rightarrow \frac{P_1}{T_1} = \frac{P_2}{T_2}$

(b) $W = \int pdV = 0$

(c) Change in internal energy

$$\Delta U = \mu C_V \Delta T = \frac{\mu R}{\gamma - 1}\Delta T$$

4. Isobaric Process

(a) $\frac{V_1}{T_1} = \frac{V_2}{T_2}$

(b) $W = P(V_2 - V_1)$

5. Comparison between isothermal and adiabatic changes

	Isothermal	Adiabatic
(i)	$\Delta T = 0$	$\Delta Q = 0; \Delta S = 0$
(ii)	Changes are slow	Sudden changes
(iii)	System is thermally conducting	Thermally insulated
(iv)	$\Delta U = 0$	$\Delta U \neq 0$
(v)	Specific heat, $C = \infty$	$C = 0$
(vi)	Equations PV = constant	PV^{γ} = constant, $TV^{\gamma-1}$ = constant $P^{1-\gamma}T^{\gamma}$ = constant
(vii)	Slope of isothermal curve $\frac{dP}{dV} = -\left(\frac{P}{V}\right)$	$\frac{dP}{dV} = -\gamma\left(\frac{P}{V}\right)$
(vii)	Work done in expansion $W = 2.303\ nRT\log\frac{V_2}{V_1}$ $W = 2.303\ P_1V_1\log\frac{V_2}{V_1}$ $W = 2.303\ nRT\log\frac{P_1}{P_2}$	$W = nR\frac{(T_2 - T_1)}{1-\gamma}$ $W = \frac{P_2V_2 - P_1V_1}{1-\gamma}$ $W = C_V(T_1 - T_2)$

6. Mayer's formula

$C_P - C_V = R$

Where, C_P is the molar specific heat capacity at constant pressure C_V is the molar specific heat capacity at constant volume and R is gas constant.

Numerical Profile

Comparison between Isothermal and Adiabatic Curves

(i) When expanded to the same volume from the same initial state.

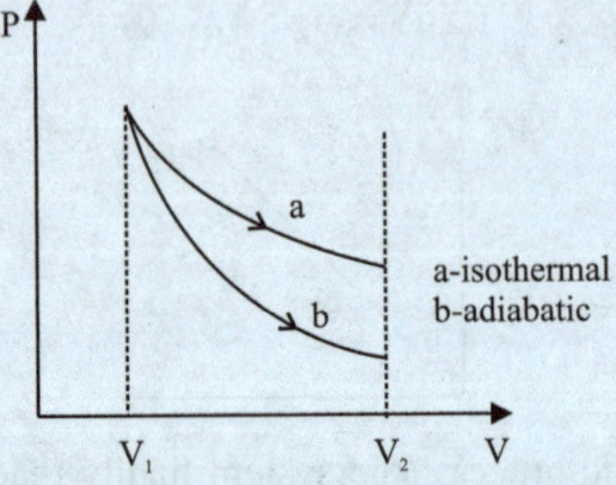

Work done: $W_a > W_b$

Final pressure: $P_a > P_b$

Final temperature: $T_a > T_b$

(ii) When compressed to the same volume from the same initial state.

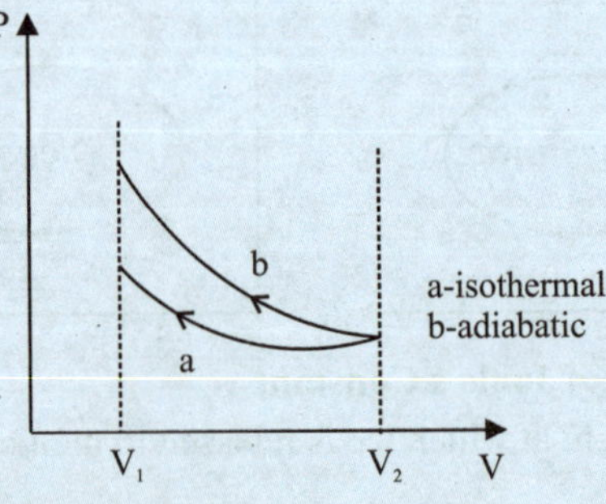

Work done: $W_a < W_b$

Final pressure: $P_a < P_b$

Final temperature: $T_a < T_b$

Statements Asked in Exam

- Celsius scale can never have the same reading as the kelvin scale, at any temperature.
- The direction of flow of heat is determined by the temperature alone and not by the total heat energy.
- Internal energy of a gas consists of energy of translation rotation of the molecules of gases.
- A gas does not possess a unique or single specific heat.
- In case of an ideal gas, internal energy depends only on the temperature; in case of a real gas it depends on volume and temperature.
- The temperature beyond which a gas cannot be liquified regardless of the amount of pressure applied to it is critical temperature.
- In an isothermal process, T = constant
  $\therefore$ dT = 0 and dU = 0
  Therefore, dQ = dW = P(dV)
- In an isochoric process, also called constant volume process
  V = constant $\therefore$ dV = 0, dW = PdV = 0
- In an isothermal quasi-static process the gas has the same temperature throughout the change of volume.
- Heat gained or lost and work done by or on a gas are path dependent.
- First law of thermodynamics tells about the conversion of mechanical energy into heat energy.
- Temperature of a body can be raised without heating it. For example during adiabatic compression temperature increases.
- Second law of thermodynamics is Kelvin Planck statement. It is impossible to construct a heat engine that takes heat from a hot body and converts all of the energy into useful work without losing heat to the cold reservoir.
- When a cyclic process is represented clockwise, work done in the process is positive and vice-versa.
- It is not possible for a system, unaided by external agency to transfer heat from a body at low temperature to another at higher temperature.
- An ideal gas is compressed at a constant temperature, its internal energy will not increase or decrease.
- In an isothermal process and in a cyclic process, there is no change in internal energy.
- Graphs for expansion of gas

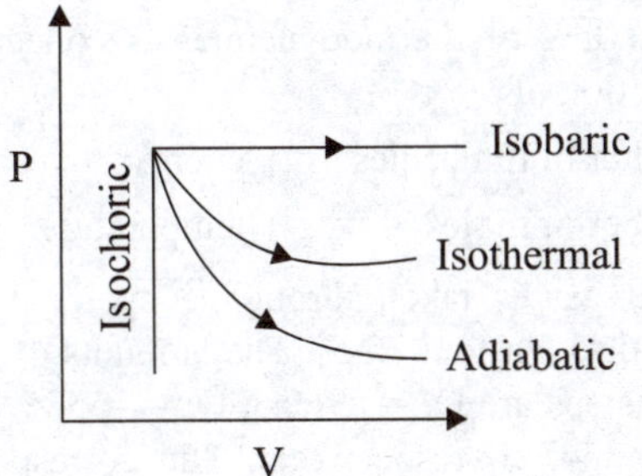

- Graph for compression of gas

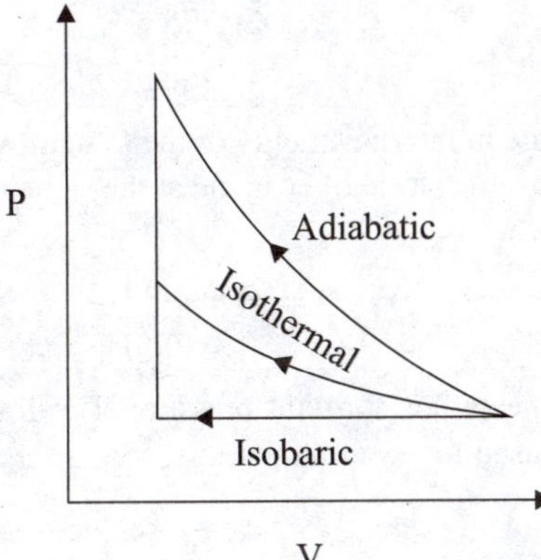

MCQs

NCERT Topic-wise MCQs

First Law of Thermodynamics

1. The first law of thermodynamics is concerned with the conservation of
   1) number of molecules 2) energy
   3) number of moles 4) temperature

2. An ideal gas is taken through a cyclic thermodynamic process through four steps. The amounts of heat involved in these steps are $Q_1 = 5960$ J, $Q_2 = -5585$ J, $Q_3 = -2980$ J, $Q_4 = 3645$ J, respectively. The corresponding works involved are $W_1 = 2200$ J, $W_2 = -825$ J, $W_3 = -1100$ J, and W_4 respectively, the value of W_4 is:
   1) 780 2) 775
   3) 765 4) 760

3. The change in internal energy of the system when a system absorbs 2 kilo calorie of heat and at the same time does 500 joule of work is
   1) 7900 J 2) 8200 J
   3) 5600 J 4) 6400 J

4. A gas is heated at constant pressure. The fraction of heat supplied used for external work is:
   1) $1-\frac{2}{\gamma}$ 2) $1-\frac{3}{\gamma}$
   3) $1-\frac{2}{3\gamma}$ 4) $1-\frac{1}{\gamma}$

5. 1 cm³ of water at its boiling point absorbs 540 cal of heat to become steam with a volume of 1671 cm³ at the atmospheric pressure = 1.013×10^5 Nm^{-2} and the mechanical equivalent of heat = 4.19 J/cal, the energy spent in this process in overcoming intermolecular forces is:
   1) 500 cal 2) 499 cal
   3) 502 cal 4) 490 cal

6. A gas is compressed at a constant pressure of 50 N/m² from a volume of 10 m³ to volume of 4 m³. Energy of 100 joule is thus added to gas by heating. Its internal energy is:
   1) 450 J 2) 400 J
   3) 350 J 4) 410 J

7. A cylinder of fixed capacity 44.8L contains 2 moles of He gas at STP. What is the amount of heat needed to raise the temperature of the gas in the cylinder by 20°C?
   1) 496 J 2) 495 J
   3) 502 J 4) 499 J

8. If temperature is increased by 1 K at constant volume then work done on the gas is:
   1) Positive 2) Negative
   3) Zero 4) Data Inadequate

9. The P – V diagram of a system undergoing thermodynamic transformation is shown in figure. The work done by the system in going from A → B → C is 30 joule and 40 joule heat is given to the system. The change in Internal energy between A and C is:

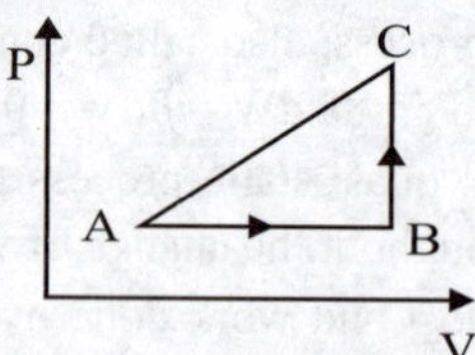

   1) 8 2) 9
   3) 11 4) 10

10. Which of the following is not a path function?
    1) ΔW 2) ΔQ + ΔW
    3) ΔQ – ΔW 4) ΔQ

11. If 150 joule of heat is added to a system and the work done by the system is 110 Joule, then change in internal energy
    1) 35 2) 38
    3) 42 4) 40

12. An ideal gas is taken through a cycle ABCA as shown in the P – V diagram The work done during the cycle is:

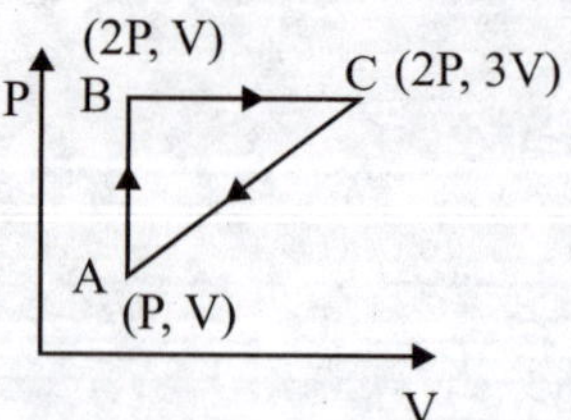

    1) 2 PV 2) 3 PV
    3) PV 4) PV/3

13. The state of thermodynamical system is represented by:
    1) P and V only 2) T and V only
    3) P, V, T 4) None

14. If Q, E and W denote respectively the heat added, change in internal energy and the work done in a closed cyclic process then which of these will follow:
    1) ΔW = 0 2) ΔQ = 0
    3) ΔE = 0 4) None of these

15. A cyclic process is shown in the figure. Work done during the cyclic process ABCDA

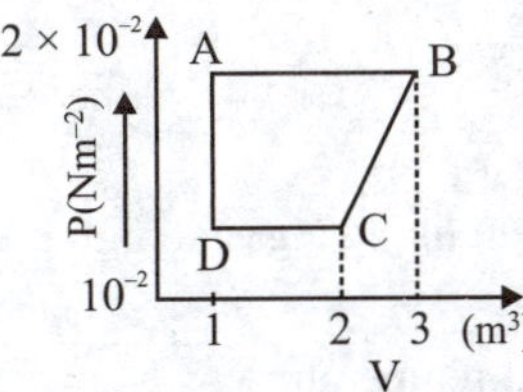

1) 150 J
2) 163 J
3) 145 J
4) 166 J

16. A system is taken from state A to state B along two different paths 1 and 2. The heat absorbed and work done by the system along these two paths are Q_1, Q_2 and W_1, W_2 respectively then

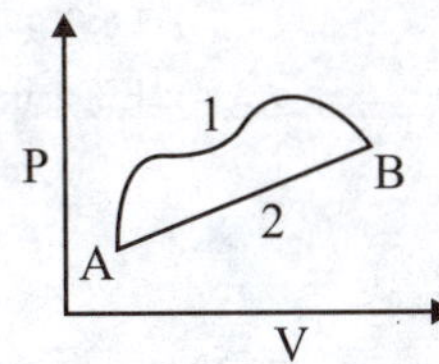

1) $Q_1 - W_1 = Q_2 - W_2$
2) $Q_1 = Q_2$
3) $W_1 = W_2$
4) None

17. An ideal gas is taken around the cycle ABCA as shown in the figure then work done by the gas during the cycle is

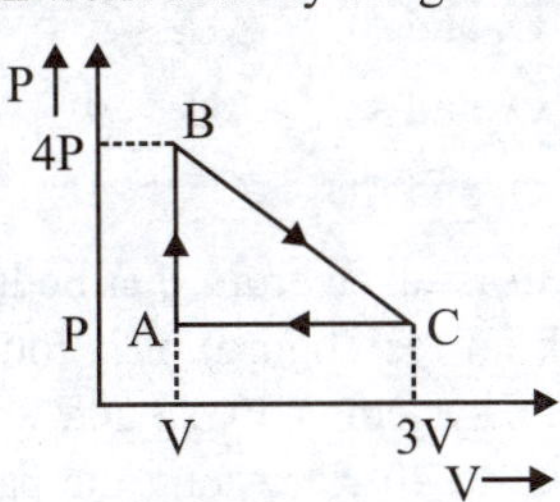

1) 3PV
2) 2 PV
3) PV
4) $\frac{2}{3}$PV

18. In a cyclic process, work done by the system is
1) $\Delta Q = \Delta U - \Delta W$
2) $\Delta Q = \Delta W$
3) $\Delta Q \neq \Delta W$
4) $\Delta Q = \Delta U$

Adiabatic and Isothermal Processes

19. An ideal mono atomic gas at 27°C is compressed adiabatically to $\frac{8}{27}$ times to its present volume. The increase in temperature of the gas is
1) 402°C
2) 375°C
3) 475°C
4) 175°C

20. No heat flows between systems and surrounding, then process is
1) Adiabatic
2) Isothermal
3) Isobaric
4) Isochoric

21. Work done per mol in an isothermal change is
1) $RT \log_e \frac{V_1}{V_2}$
2) $RT \log_e \frac{P_2}{P_1}$
3) $RT \log_e \frac{V_2}{V_1}$
4) $P\Delta V$

22. 10 moles of an ideal gas at constant temperature 600 K is compressed from 100L to 10L. The work done in the process is
1) 12×10^4 J
2) -12×10^4 J
3) 11.4×10^4 J
4) -11.4×10^4 J

23. For an adiabatic process, the relation between V and T is given by
1) $TV^{\gamma-1} = K$
2) $PV^3 = K$
3) $T^1V^3 = K$
4) $T^{\gamma}V^{\gamma} = K$

24. The temperature of the system decreases in the
1) Adiabtic
2) Isobaric
3) Isothermal
4) Isochoric expansion

25. In which one of the following, does the internal energy of the system remain constant?
1) Isothermal
2) Isochoric
3) Isobaric
4) Adiabatic

26. Isobaric modulus of elasticity is equal to zero when the process is:
1) Isochoric
2) Isothermal
3) Adiabatic
4) Isobaric

27. A sample of gas expands from volume V_1 to V_2. The amount of work done by the gas is greatest when the expansion is:

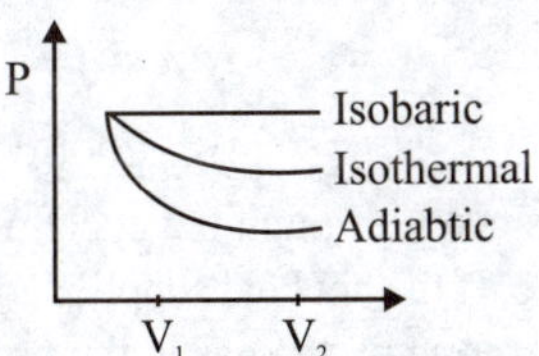

1) Isochoric
2) Isobaric
3) Isothermal
4) None

28. In an adiabatic changes the pressure and temperature of a mono atomic gas are related with relation $P \propto T^C$, where C is equal to
1) $\frac{3}{2}$
2) $\frac{5}{2}$
3) $\frac{7}{9}$
4) $\frac{9}{7}$

29. At 27°C a gas is suddenly compressed such that its pressure becomes $\frac{1}{8}$ of its original pressure. The temperature of the gas will be, if $\gamma = \frac{5}{3}$
1) −172°C
2) −168°C
3) 4K
4) −142°C

30. If there is a straight line parallel to volume axis in a P – V diagram, then it is a graph of:

1) Isothermal 2) Isobaric
3) Isochoric 4) Adiabatic

31. A gas at NTP is suddenly compressed to one – fourth of its original volume.
If $\frac{C_P}{C_V}$ is supposed to be $\frac{3}{2}$, then the final pressure is

1) 4 atmosphere 2) 3/2 atmosphere
3) 8 atmosphere 4) 1/4 atmosphere

32. The molar specific heats of an ideal gas at constant pressure and volume are denoted by C_p and C_V respectively. If $\gamma = \frac{C_P}{C_V}$ and R is the universal gas constant, then C_V is equal to

1) $\frac{\gamma R}{\gamma - 1}$ 2) $\frac{R}{\gamma - 1}$
3) $\frac{(\gamma - 1)R}{\gamma}$ 4) None

33. If a cylinder containing a gas at high pressure explodes, the gas undergoes.

1) Decrease in temperature
2) Increase in temperature
3) Temperature does not alter
4) None

34. A thermo-flask made of stainless steel contains several tiny lead shots. If flask is quickly shaken up and down several times, the temperature of lead shots.

1) Increase 2) Decrease
3) Remains same 4) First increase then decrease

High Order Time Intensive MCQs

1. A gas is taken through the cycle A → B → C → A, as shown. What is the net work done by the gas?

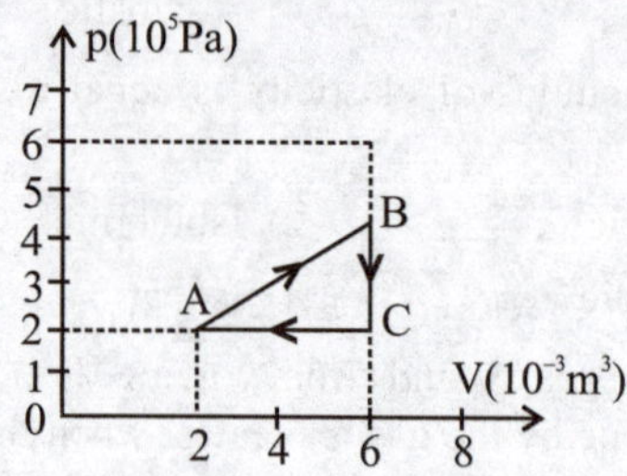

1) 2000 J 2) 400 J
3) Zero 4) – 2000 J

2. An ideal gas is taken through the cycle A → B → C → A, as shown in figure. If the net heat supplied to the gas in the cycle is 5 J, the work done by the gas in the process C → A is

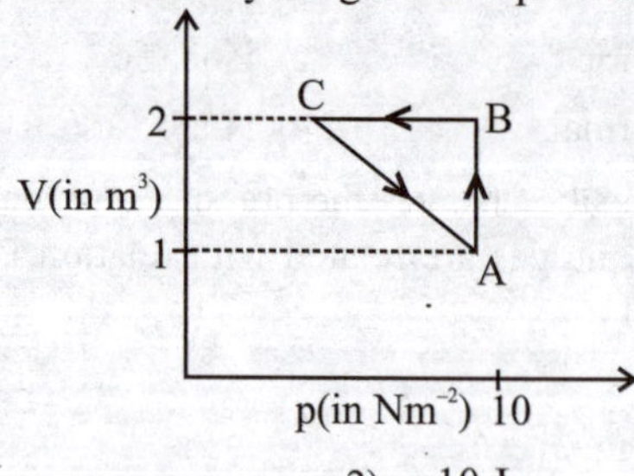

1) – 5 J 2) – 10 J
3) – 15 J 4) – 20 J

3. An ideal gas goes from state A to state B via three different processes as indicated in the p-V diagram.

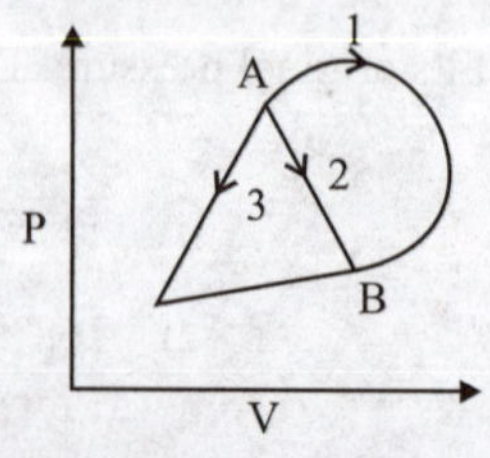

If Q_1, Q_2 and Q_3 indicate the heat absorbed by the three processes and ΔU_1, ΔU_2, and ΔU_3 indicate the change in internal energy along the three processes respectively, then

1) $Q_1 > Q_2 > Q_3$ and $\Delta U_1 = \Delta U_2 = \Delta U_3$
2) $Q_1 > Q_2 > Q_3$ and $\Delta U_1 < \Delta U_2 < \Delta U_3$
3) $Q_1 = Q_2 = Q_3$ and $\Delta U_1 > \Delta U_2 > \Delta U_3$
4) $Q_1 > Q_2 > Q_3$ and $\Delta U_1 > \Delta U_2 > \Delta U_3$

4. When a system is taken from thermodynamic state i to f along the path iaf (see figure), it is found that the heat Q absorbed by the system is 50 cal and work W done by the system is equal to 20 cal. Along the path ibf Q = 36 cal. What is W along the path ibf?

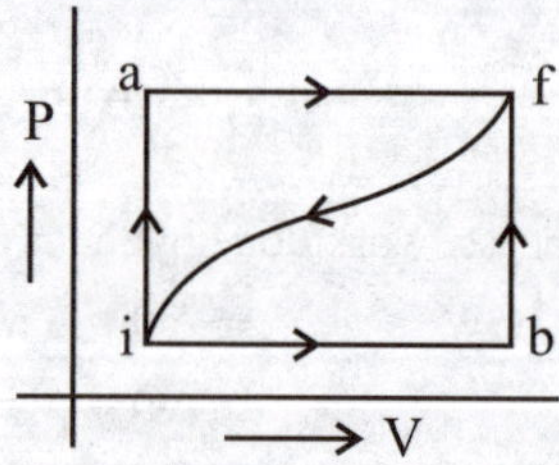

1) 6 cal 2) 18 cal
3) 40 cal 4) – 43 cal

5. p-V plots for two gases during adiabatic process are shown in the figure. Plots 1 and 2 should correspond respectively to

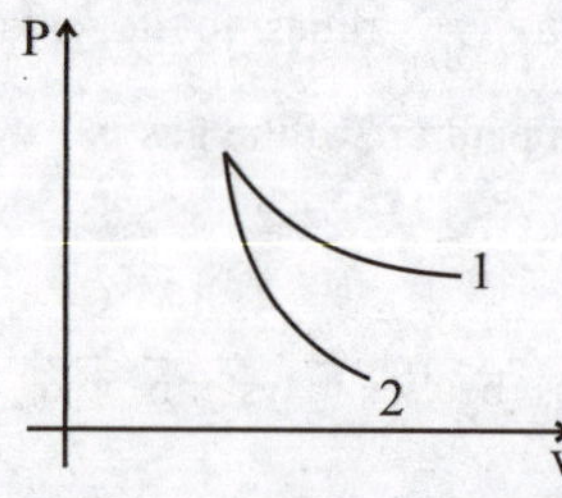

1) He and O_2 2) O_2 and He
3) He and Ar 4) O_2 and N_2

6. An ideal gas after going through a series of four thermodynamic states in order, reaches the initial state against (cyclic process). The amounts of heat and work involved in these states are

$Q_1 = 6000$ J, $Q_2 = -5500$ J

$Q_3 = -3000$ J, $Q_4 = 3500$ J

$W_1 = 2500$ J, $W_2 = -1000$ J

$W_3 = -1200$ J, $W_4 = x$J

The ratio of net work done by the gas to the total heat absorbed by the gas is η. The value of x and η are nearly

1) 500, 7.5%
2) 700, 10.5%
3) 1000, 21%
4) 1500, 15%

7. P-V diagram of an ideal gas is as shown in figure. Work done by the gas in the process ABCD is:

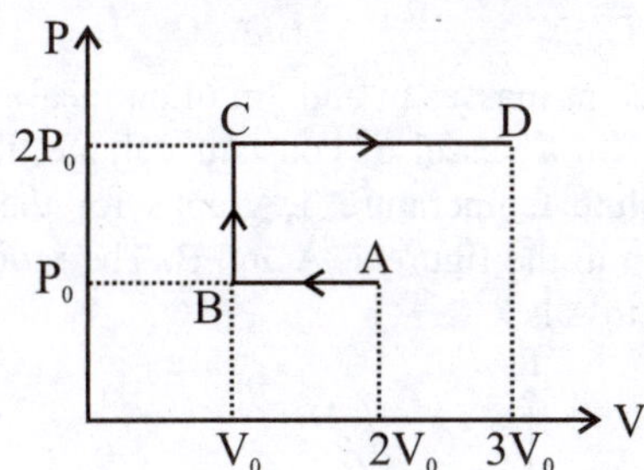

1) $4P_0V_0$ 2) $2P_0V_0$
3) $3P_0V_0$ 4) P_0V_0

8. A mono atomic gas at a pressure P, having a volume V expands isothermally to a volume 2V and then adiabatically to a volume 16V. What is the final pressure of the gas? (take, $\gamma = \frac{5}{3}$)

1) 64 P 2) 32 P
3) $\frac{P}{64}$ 4) 16 P

9. Heat energy absorbed by a system in going through a cyclic process shown in figure is

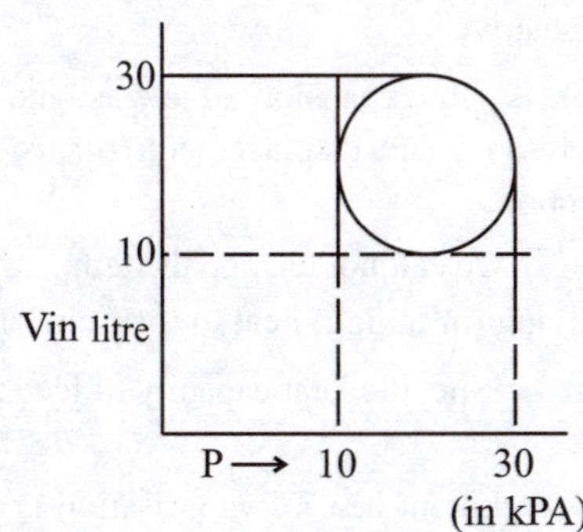

1) 10^7 πJ 2) 10^4 πJ
3) 10^2 πJ 4) 10^{-3} πJ

10. During an adiabatic process, the pressure of a gas is found to be proportional to the cube of its temperature. The ratio of γ for the gas is

1) $\frac{43}{3}$ 2) 2
3) $\frac{5}{3}$ 4) $\frac{3}{2}$

11. One mole of an ideal gas goes from an initial state A to final state B via two processes. It first undergoes isothermal expansion from volume V to 3V and their volume is reduced from 3V to V at constant pressure. The correct P-V diagram representing the two processes is

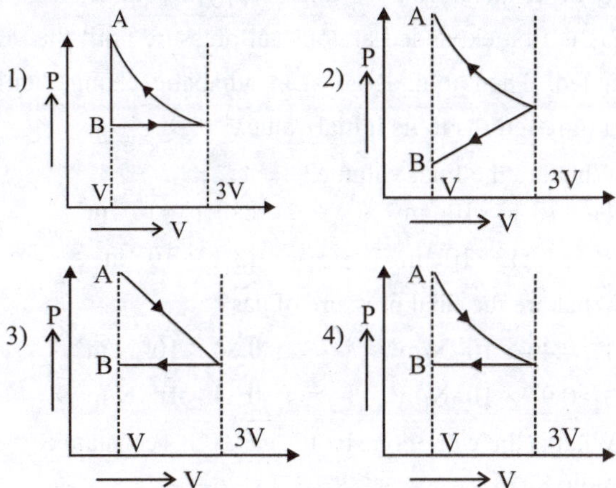

12. If $Q_{iaf} = 80$ cal, $W_{iaf} = 60$ cal and $W = -30$ cal for the curved path f → i, then value of Q for path f → i, will be

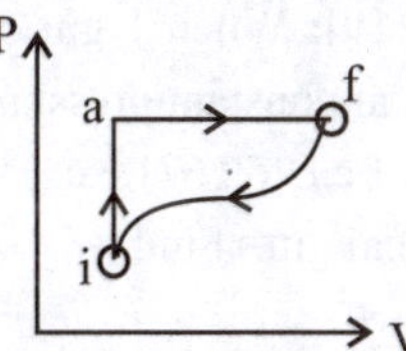

1) 50 cal 2) 30 cal
3) –30 cal 4) –50 cal

13. The P – V diagram of 2 g of Helium gas for a certain process A → B is shown in figure. Heat given to the body during process A → B is:

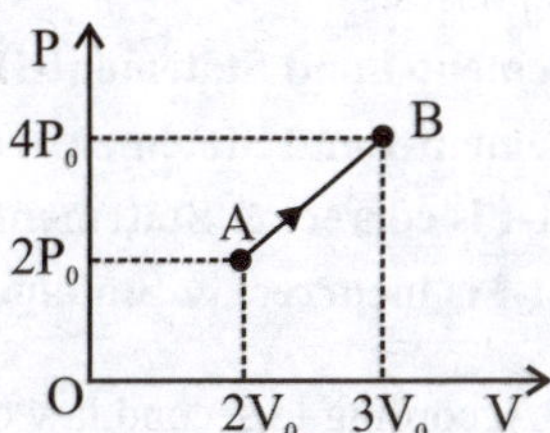

1) $4P_0V_0$
2) $12P_0V_0$
3) $6P_0V_0$
4) $15P_0V_0$

14. For the process as shown below the heat supplied to the system is 30 kJ per cycle. The efficiency of the cycle is:

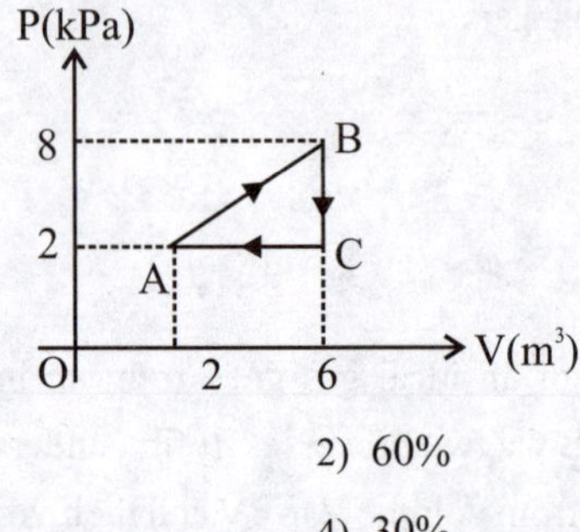

1) 40%
2) 60%
3) 50%
4) 30%

Paragraph (Q. 15 to 17): Two moles of Helium gas ($\gamma = 5/3$) are initially at temperature 27°C and occupy a volume of 20 litres. The gas is first expanded at constant pressure until the volume is doubled. Then it undergoes an adiabatic change until the temperature returns to its initial value.

15. What are the final volume?
1) 113.13×10^{-3} m^3
2) 213.13×10^{-3} m^3
3) 313.13×10^{-3} m^3
4) 13.13×10^{-3} m^3

16. What are the final pressure of gas?
1) 0.44×10^5 N/m^2
2) 0.84×10^5 N/m^2
3) 0.94×10^5 N/m^2
4) 0.34×10^5 N/m^2 9

17. What is the work done by the gas? (Gas constant R = 8.3 T/ mole K)
1) 13450 J
2) 14450 J
3) 16450 J
4) 12450 J

Paragraph (Q. 18 to 19): When 1 gm of water changes from liquid to vapour phase at constant pressure of 1 atmosphere, the volume increases from 1 c.c to 1671 c.c. The heat of vaporization at this pressure is 540 cal/gm. Find

18. The work done (in J) in change of phase
1) 170.78 Joule
2) 200.67 Joule
3) 190.78 Joule
4) 168.67 Joule

19. Increase in internal energy of water.
1) 2099.33 J
2) 3099.33 J
3) 4099.33 J
4) 5099.33 J

20. For an ideal gas:
A. The change in internal energy in a constant pressure process from temperature T_1 to T_2 is equal to $nC_v(T_2 - T_1)$, where C_v is the molar specific heat at constant volume and n the number of moles of the gas.
B. The change in internal energy of the gas and the work done by the gas are equal in magnitude in an adiabatic process.
C. The internal energy does not change in an isothermal process.
D. No heat is added or removed in an adiabatic process.
Which of the combination are correct?
1) A, B
2) A, B, C
3) A, B, C, D
4) A, C

21. Two different masses m and 3m of an ideal gas are heated separately in a vessel of constant volume, the pressure p and absolute temperature T, graphs for these two cases are shown in the figure as A and B. The ratio of slopes of curves B to A is

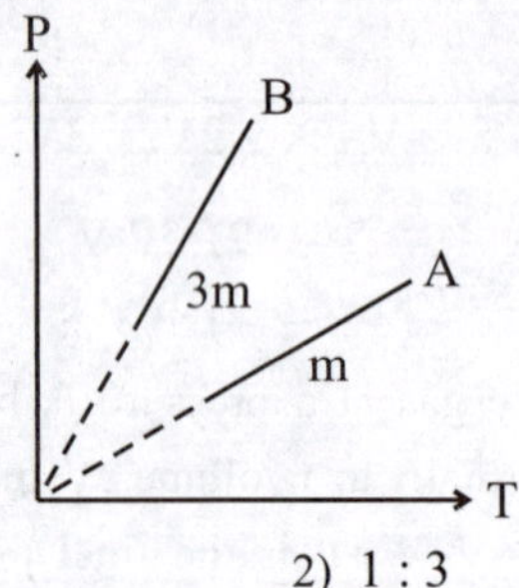

1) 3 : 1
2) 1 : 3
3) 9 : 1
4) 1 : 9

Statement Based MCQs

Directions: These questions consist of two statements each, printed as Statement-I and Statement-II. While answering these questions, you are required to choose any one of the following four responses.

1) **Both Statement-I and Statement-II are correct.**
2) **Both Statement-I and Statement-II are incorrect.**
3) **Statement-I is correct & Statement-II is incorrect.**
4) **Statement-I is incorrect & Statement-II is correct.**

1. Statement I: According to second law of thermodynamics heat can not flow from a body at low temperature to a body at higher temperature on its own.

Statement II: Efficiency of a real engine can not be 100%.

2. Statement I: In a cyclic process change in internal energy is never zero.

Statement II: Pressure reamains constant in isochoric process.

3. Statement I: When a gas is allowed to expand, work done by gas is positive.

Statement II: In expansion of a gas, the force due to gaseous pressure and displacement (of piston) are in the same direction.

4. Statement I: We can not change the temperature of a body without giving (or taking) heat to (or from) it.

Statement II: Specific heat capacity of ice is less than that of water.

5. Statement I: Latent heat of vaporisation is more than the latent heat of fusion.

Statement II: During isobaric process temperature of the system remains constant.

Assertion & Reason MCQs

Directions: These questions consist of two statements each, printed as Assertion and Reason. While answering these questions, you are required to choose any one of the following four responses.

1) **Both Assertion and Reason are True and the Reason is a correct explanation of the Assertion.**
2) **Both Assertion and Reason are True but Reason is not a correct explanation of the Assertion.**
3) **Assertion is True but the Reason is False.**
4) **Assertion is False but Reason is True.**

1. **Assertion:** If temperature remains constant throughout, the process is said to be isothermal process.

   Reason: Thermodynamic system are always perfectly conducting to the surrounding .

2. **Assertion:** $Q = \Delta W$ during an isothermal process.

   Reason: For isothermal process, temperature T = constant; $\Delta U = 0$; where, U is the internal energy of the system.

3. **Assertion:** The instant expansion of the enclosed air into the atmosphere when a motor tyre bursts, is approximately adiabatic and thus the tyre is cooled.

   Reason: The actual 'pop' is so quick that there is really no time for any significant heat to be gained or lost by the rubber.

4. **Assertion:** If the working substance is taken in a expanding chamber in which the pressure is kept constant, the process is called isobaric process

   Reason: In these processes the gas either expands or contract to maintain a constant pressure.

5. **Assertion:** At constant temperature, pressure of a gas is inversely proportional to the volume.

   Reason: At constant pressure, volume of a gas is directly proportional to its temperature..

Match the Columns MCQs

1. When a sample of a gas is taken from state i to state f along path iaf, heat supplied to the gas is 60 cal and work done by the gas is 25 cal. If it is taken by path ibf, then heat supplied is 42 cal.

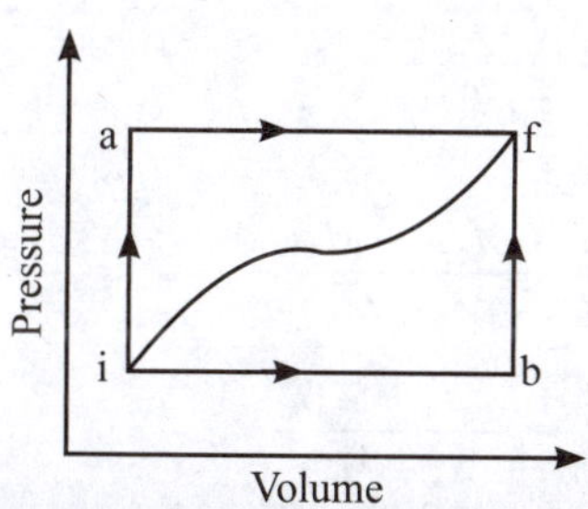

	Column-I		Column-II
(a)	Work done by the gas along path ibf is	(i)	7 cal
(b)	If work done upon the gas is 13 cal for the return path fi, then heat rejected by the gas along path if is	(ii)	19 cal
(c)	If internal energy of the gas at state i is 10 cal, then internal energy at state f is	(iii)	45 cal
(d)	If internal energy at state b is 22 cal and at i is 10 cal then heat supplied to the gas along path is	(iv)	48 cal

1) a-(i); b-(iv); c-(iii); d-(ii)
2) a-(iii); b-(ii); c-(iv); d-(i)
3) a-(iv); b-(i); c-(ii); d-(iii)
4) a-(ii); b-(iii); c-(i); d-(iv)

2. Match the change in the internal energy of the system given in Column-II for each situation in given Column-I.

	Column-I		Column-II
(a)	A thermodynamic system absorbs 400 cal of heat and at the same time does 300 J of work.	(i)	–3347 J
(b)	A thermodynamic system absorbs 300 cal and at the same time 320 J of work is done on it	(ii)	1374 J
(c)	Eight eight hundred calories is removed from a gas held at constant volume.	(iii)	1157 J

1) a-(ii); b-(iii); c-(i)
2) a-(ii); b-(i); c-(iii)
3) a-(iii); b-(ii); c-(i)
4) a-(ii); b-(iii); c-(i)

3. Math the column-I and column-II and select the correct choice out of the four options.

	Column-I		Column-II
(a)	Zeroth law of thermodynamics	(i)	Give us the concept of temperature
(b)	First law of thermodynamics	(ii)	Defines internal energy
(c)	Second law of thermodynamics	(iii)	Change in internal energy is zero
(d)	In a complete cycle of changes	(iv)	The efficiency of on engine is always less than unity

1) a-(iii); b-(ii); c-(i, iv); d-(iv)
2) a-(i); b-(iii); c-(iv); d-(ii)
3) a-(ii); b-(i); c-(iii); d-(iv)
4) a-(i); b-(ii); c-(iv); d-(iii)

NCERT Exemplar MCQs

1. If an average person jogs, he produces 14.5×10^3 cal / min. This is removed by the evaporation of sweat. The amount of sweat evaporated per minute (assuming 1 kg requires 580×10^3 cal for evaporation) is:

1) 0.25 kg
2) 2.25 kg
3) 0.05 kg
4) 0.20 kg

2. Consider P-V diagram for an ideal gas shown in figure:

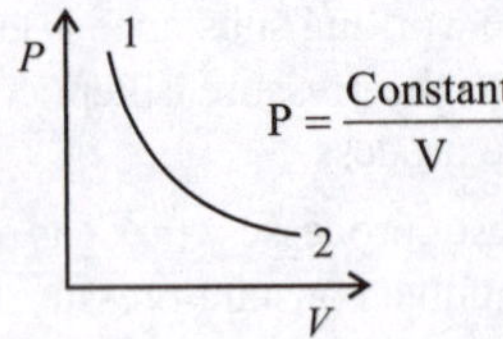

Out of the following diagrams, which figure represents the T-P diagram?

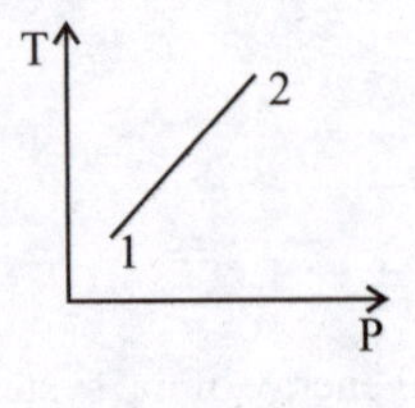

(i)

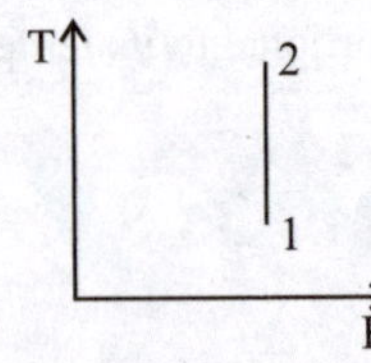

(ii)

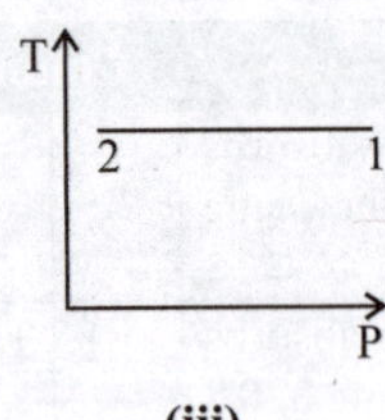

(iii)

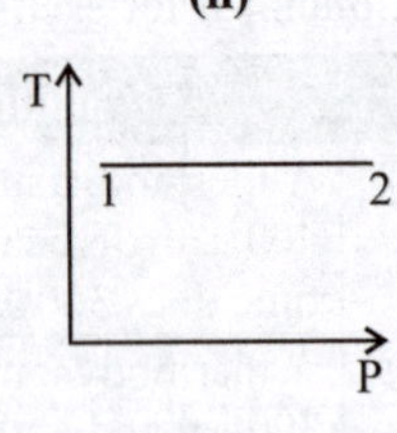

(iv)

1) (iv)
2) (ii)
3) (iii)
4) (i)

3. An ideal gas undergoes cyclic process ABCDA as shown in given P-V diagram. The amount of work done by the gas is:

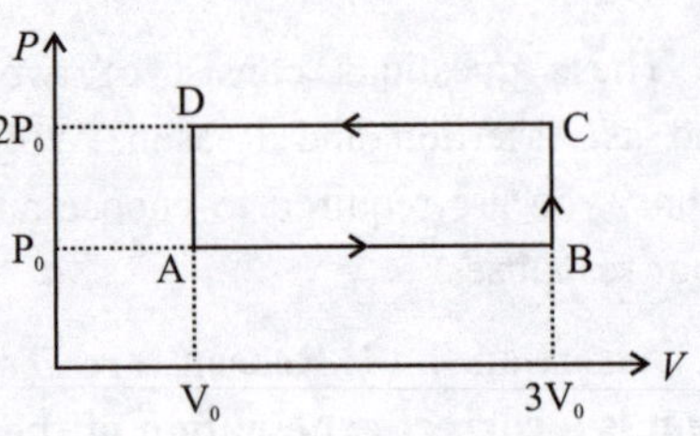

1) $6 P_0V_0$
2) $-2 P_0V_0$
3) $+2 P_0V_0$
4) $+4 P_0V_0$

4. Consider two containers A and B containing pressure, volume and temperature. The gas in container A is compressed to half of its original volume isothermally while the gas in container B is compressed to half of its original value adiabatically. The ratio of final pressure of gas in B to that of gas in A is:

1) $2^{\gamma-1}$
2) $\left(\frac{1}{2}\right)^{\gamma-1}$
3) $\left(\frac{1}{1-\gamma}\right)^2$
4) $\left(\frac{1}{\gamma-1}\right)^2$

5. Three copper blocks of masses M_1, M_2 and M_3 kg respectively are brought into thermal contact till they reach equilibrium. Before contact, they were at T_1, T_2, T_3 ($T_1 > T_2 > T_3$). Assuming there is no heat loss to the surroundings, the equilibrium temperature T is (s is specific heat of copper):

1) $T = \frac{T_1 + T_2 + T_3}{3}$
2) $T = \frac{M_1T_1 + M_2T_2 + M_3T_3}{M_1 + M_2 + M_3}$
3) $T = \frac{M_1T_1 + M_2T_2 + M_3T_3}{3(M_1 + M_2 + M_3)}$
4) $T = \frac{M_1T_1s + M_2T_2s + M_3T_3s}{M_1 + M_2 + M_3}$

NEET & AIIMS Past Year MCQs

1. The equilibrium state of a thermodynamic system is described by

A. Pressure
B. Total heat
C. Temperature
D. Volume
E. Work done

Choose the most appropriate answer from the options given below. **(2024 Re)**

1) A, B and E only
2) B, C and D only
3) A, B and C only
4) A, C and D only

2. A thermodynamic system is taken through the cycle *abcda*. The work done by the gas along the path *bc* is: **(2024)**

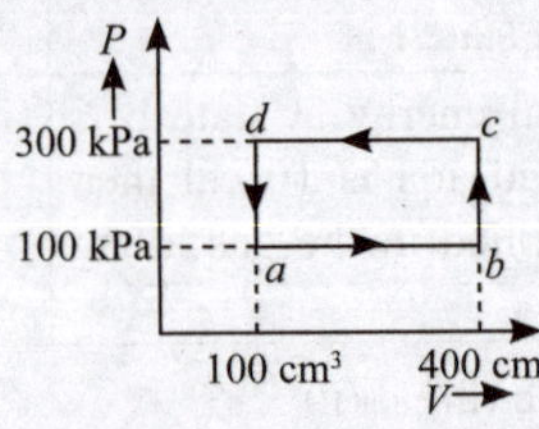

1) – 90J
2) – 60J
3) Zero
4) 30J

3. A Carnot engine has an efficiency of 50% when its source is at a temperature 327°C. The temperature of the sink is: **(2023)**

1) 200° C
2) 27° C
3) 15° C
4) 100° C

4. An ideal gas undergoes four different processes from the same initial state as shown in the figure below. Those processes are adiabatic, isothermal isobaric and isochoric. The curve which represents the adiabatic process among 1, 2, 3 and 4 is : **(2022)**

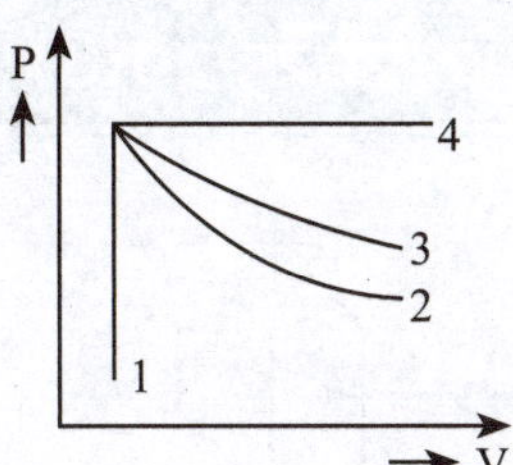

1) 4
2) 1
3) 2
4) 3

5. If 1 cm^3 of water is vaporized (latent heat of vaporization = 540 cal/g°C) at P = 1 atm. If the volume of steam formed is 1671 cm^3, calculate increase in internal energy. **(AIIMS 2018)**

1) 540 cal
2) 167 cal
3) 373 cal
4) 707 cal

6. An engine has an efficiency of 1/6. When the temperature of sink is reduced by 62°C, its efficiency is doubled. The temperature of source will be: **(AIIMS 2013)**

1) 37°C
2) 62°C
3) 99°C
4) 124°C

Answer Key

NCERT Topic-wise MCQs

1	2	3	4	5	6	7	8	9	10	11	12	13	14	15	16	17
2	3	1	4	1	2	4	3	4	3	4	3	3	3	1	1	1
18	19	20	21	22	23	24	25	26	27	28	29	30	31	32	33	34
2	2	1	3	4	1	1	1	4	2	2	4	2	3	2	1	1

High Order Time Intensive MCQs

1	2	3	4	5	6	7	8	9	10	11	12	13	14	15	16	17
2	1	1	1	2	2	3	3	3	4	4	4	4	1	1	1	4
18	19	20	21													
4	1	3	1													

Statement Based MCQs

1	2	3	4	5
1	2	1	4	3

Assertion & Reason MCQs

1	2	3	4	5
3	1	1	1	2

Match the Columns MCQs

1	2	3
1	1	4

NCERT Exemplar MCQs

1	2	3	4	5
1	3	2	1	2

NEET & AIIMS Past Year MCQs

1	2	3	4	5	6
4	3	2	3	3	3

CHAPTER

12 Kinetic Theory

Topicwise Number of Questions Analysis (2025-2016)

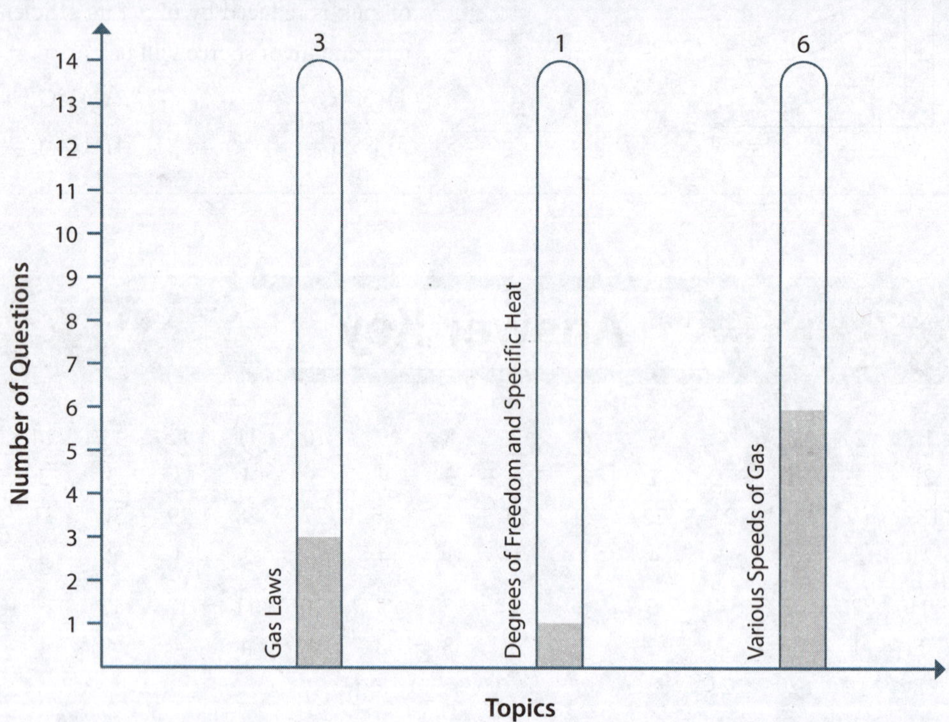

NCERT-Picks

Gas Laws and Kinetic Theory

Formulas

1. Ideal gas equation PV = nRT

   [P is the pressure, V is the volume and T is the temperature. n is the number of moles and R is gas constant, R is equal to 8.314 JK^{-1} mol^{-1}]

2. N is the total number of molecules and m is mass of each molecule, then pressure exerted by each molecule.

   $$P = \frac{1}{3}\frac{mN}{V}v_{rms}^2 = \frac{1}{3}\rho v_{rms}^2$$

   Where, v_{rms} = root mean square speed of the each molecule

3. Average kinetic energy per molecule = $\frac{3}{2}KT$

   [where K is Boltzmann constant = 1.38×10^{-23} Jk^{-1} and T is the temperature]

4. Total Kinetic energy of all the molecules = $\frac{3}{2}nRT$

5. $v_{rms} = \sqrt{\frac{3KT}{m}}$ [m is the mass per molecule]

6. $v_{rms} = \sqrt{\frac{3RT}{M_0}}$ [$M_0 = mN_A$ where N_A is Avogadro Number]

Numerical Profile

Deduction from kinetic theory

(i) Boyle's Law: At constant temperature

$\Rightarrow P_1V_1 = P_2V_2$

(ii) Charles's Law: At constant pressure

$V \propto T$

$\therefore \frac{V_1}{T_1} = \frac{V_2}{T_2}$

(iii) Gay Lussac's law: Volume is kept constant

$$\frac{P}{T} = \text{constant} \Rightarrow \frac{P_1}{T_1} = \frac{P_2}{T_2}$$

(iv) Graham's law of Diffusion:

Rate of diffusion of the gas $\propto \frac{1}{\sqrt{M}}$

Where, M is the molecular mass of the gas.

$$\Rightarrow \frac{R_1}{R_2} = \frac{\sqrt{M_2}}{\sqrt{M_1}}$$

R_1 is the rate of diffusion of the first gas.

R_2 is the rate of diffusion of the second gas.

M_1 is the molecular mass of the first gas.

M_2 is the molecular mass of the second gas

(v) Dalton's Law of partial pressure: Total pressure exerted by a mixture of non reacting gases occupying a given volume is

$P = P_1 + P_2 + P_3 + ...$

where P_1, P_2, P_3 ... are the partial pressure of the gases.

Types of Velocities of gas molecules

(i) Average velocity: $\bar{v} = \frac{v_1 + v_2 + v_3 + ... + v_n}{n}$

$$\bar{v} = \sqrt{\frac{8P}{\pi\rho}} = \sqrt{\frac{8RT}{\pi M}} = \sqrt{\frac{8kT}{\pi m}}$$

(ii) Root Mean Square Velocity:

$$v_{rms} = \sqrt{\frac{v_1^2 + v_2^2 + v_3^2 + ... + v_n^2}{n}} = \sqrt{\overline{v^2}}$$

$$\because P = \frac{1}{3}\rho v_{r.m.s}^2$$

$$v_{rms} = \sqrt{\frac{3P}{\rho}}$$

(iii) Most Probable Velocity: $v_{mp} = \sqrt{\frac{2P}{\rho}}$

(iv) $v_{rms} : \bar{v} : v_{mp} = 1.73 : 1.60 : 1.41$

Clearly, $v_{rms} > \bar{v} > v_{mp}$.

Degree of Freedom

Formulas

1.

Atomicity	no. of degree of freedom	C_P	C_V	$\gamma = \frac{C_P}{C_V}$
Monoatomic	3	$\frac{5R}{2}$	$\frac{3R}{2}$	$\frac{5}{3}$
Diatomic	5	$\frac{7R}{2}$	$\frac{5R}{2}$	$\frac{7}{5}$
Triatomic:				
Linear	7	$\frac{9R}{2}$	$\frac{7R}{2}$	$\frac{9}{7}$
Non-Linear	6	4R	3R	$\frac{4}{3}$

2. Heat supplied at constant volume $Q = nC_V\Delta T$

Heat supplied at constant pressure $Q = nC_P\Delta T$

[C_p is specific heat at constant pressure, C_V is the specific heat at constant volume]

Numerical Profile

Specific heat of mixture of gases

(i) If n_1 moles of C_{v_1} is mixed with the n_2 moles of C_{v_2} then by conservation of energy $U_1 + U_2 + ... = U$

$$n_1C_{V_1}\Delta T + n_2C_{V_2}\Delta T + ... = (n_1 + n_2 + ...)C_{V_{mix}}\Delta T$$

$$\therefore C_{V_{mix}} = \frac{n_1C_{V_1} + n_2C_{V_2} + ...}{n_1 + n_2 + ...} \text{ and } C_{P_{mix}} = C_{V_{mix}} + R$$

(ii) If n_1 moles of γ_1 and n_2 moles of γ_2.... are mixed then by conservation of energy we have $U_1 + U_2 + ... = U$

$$n_1C_{V_1}\Delta T + n_2C_{V_2}\Delta T + ... = (n_1 + n_2 + ...)C_{V_{mix}}\Delta T$$

$$n_1\left(\frac{R}{\gamma_1 - 1}\right) + n_2\left(\frac{R}{\gamma_2 - 1}\right) + ... = (n_1 + n_2 + ...)\frac{R}{\gamma_{mix} - 1}$$

$$\frac{n_1}{\gamma_1 - 1} + \frac{n_2}{\gamma_2 - 1} + ... = \frac{(n_1 + n_2 + ...)}{\gamma_{mix} - 1}$$

For two gases we have $\frac{n_1}{\gamma_1 - 1} + \frac{n_2}{\gamma_2 - 1} = \frac{n_1 + n_2}{\gamma_{mix} - 1}$

Statements Asked in Exam

- One mole of any substance at any pressure temperature or volume always contains 6.022×10^{23} particles.
- Real gases behave like perfect gases at low pressure and high temperature also at higher temperature and low density.
- At high temperature and low pressure real gases behave as an ideal gas as potential energy due to intermolecular forces becomes less significant compared with particle kinetic energy.
- Gases such as helium; hydrogen and oxygen etc. which cannot be liquified easily are called permanent gases.
- If C_1, C_2 are the rms velocities of the gases of molecular weights M_1 and M_2 then at the given temperature.

$$\frac{C_1}{C_2} = \sqrt{\frac{M_2}{M_1}}$$

- In an ideal gas and real gas both internal energy depends on different states function.

- Average speed of the gas molecule of mass m is given by

$$C_{av} = \sqrt{\frac{8KT}{m}}$$

- Real gases do not obey Boyle's law at high pressure and low temperature.
- Air pressure in a car tyre increases during driving in summer.
- On driving a vehicle for a long time, air pressure inside the tyres slightly increases.
- Root mean square velocity of a gas molecule is given by

$$C_{rms} = \sqrt{\frac{3\,RT}{m}} = \sqrt{\frac{3\,KT}{m}}$$

- Root mean square velocity and average velocity of gas molecules, both increase with rise in temperature
- Although the R.M.S speed of gas molecules is of the order of the speed of sound in that gas, yet on opening bottle of perfume in one corner of a room, takes time in reaching the other corner.
- If μ_1 mole of a non-reactive gas in thermodynamical state (P_1, V_1, T_1) are mixed with μ_2 moles of another non-reactive gas at (P_2, V_2, T_2) and the resultant mixture is at (P, V, T), then by conservation of mass

$$\mu_1 \frac{P_1V_1}{T_1} + \mu_2 \frac{P_2V_2}{T_2} = \frac{PV}{T}\mu$$

- If hydrogen and oxygen gases are at the same temperature T, then the ratio of kinetic energies of Oxygen molecule and hydrogen molecule is one.
- Kinetic interpretation of pressure

$$P = \frac{1}{3}\frac{MC^2}{V}, \text{ with } C^2 \propto T$$

(i) If volume V and temperature T are constant, $P \propto M$ i.e., if mass of a gas is increased, number of molecules increases, number of collisions/sec increases, therefore the pressure of the gas increases

(ii) If mass and temperature are constant, $P \propto \frac{1}{V}$

(iii) If mass and volume are constant $P \propto C^2 \propto T$

NCERT Topic-wise MCQs

Gas Laws

1. Two vessels A and B having equal volume contain equal masses of hydrogen in A and helium in B at 300 K. Then mark the correct statement.
   1) The pressure exerted by hydrogen is half that exerted by helium.
   2) The pressure exerted by hydrogen is equal to that exerted by helium.
   3) The pressure exerted by hydrogen is twice that exerted by helium.
   4) Average K.E. of the molecules of hydrogen is half the average K.E. of the molecules of helium.

2. In the given V – T diagram, what is the relation between pressure P_1 and P_2?

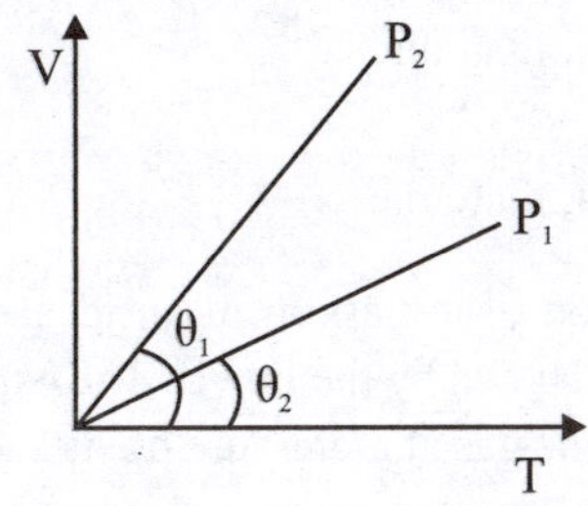

   1) $P_1 = P_2$ 2) $P_2 > P_1$
   3) $P_1 > P_2$ 4) Data insufficient

3. Critical temperature of CO_2 is 31.2°C. In summer, the room temperature is 40°C, then
   1) CO_2 cannot be liquefied
   2) CO_2 can be liquefied with increase of pressure
   3) CO_2 can be liquefied with decrease of pressure
   4) CO_2 can be liquefied if temperature of CO_2 is decreased below 31.2°C

4. A balloon is filled at 27°C and 1 atm pressure by 500 m^3 He. At –3°C and 0.5 atm pressure, the volume of He gas contained in balloon will be
   1) 500 m^3 2) 850 m^3
   3) 900 m^3 4) 895 m^3

5. Air inside a closed container is saturated with vapour. The air pressure is ρ and the saturated vapour pressure of water is $\bar{P}$. If the mixture is compressed to one-half of its volume by maintaining temperature constant, the pressure becomes
   1) $2\rho + \bar{P}$ 2) $2(\rho + \bar{P})$
   3) No effect   4) Quadrupled

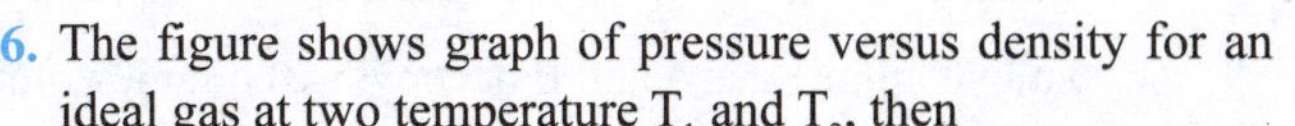

6. The figure shows graph of pressure versus density for an ideal gas at two temperature T_1 and T_2, then

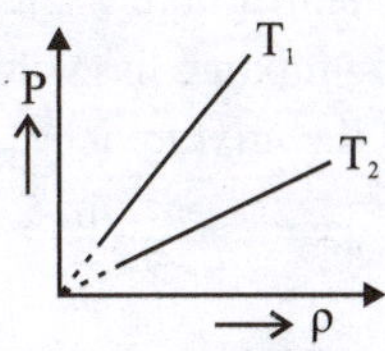

   1) $T_1 > T_2$ 2) $T_1 = T_2$
   3) $T_1 < T_2$ 4) None

7. A closed vessel contains 8 g of oxygen and 7g of N_2 Total pressure is 10 atm at a given temperature. If now oxygen is absorbed by introducing a suitable absorbent, the pressure of the remaining gas will be
   1) 2 atm   2) 10atm
   3) 4 atm   4) 5 atm

8. Air is filled at 60°C in a vessel of open mouth. The vessel is heated to a temperature T so that 1/4th of air escapes. Assuming the volume of the vessel remaining constant the value of T is
   1) 80°C   2) 444°C
   3) 333°C   4) 171°C

9. By what percentage should the pressure of a given mass of a gas be increased, so as to decreases its volume by 10% at a constant temperature?
   1) 5%   2) 7.2%
   3) 14.8%   4) 11.1%

10. If pressure of a gas contained in a closed vessel is increased by 0.4% when heated by 1°C, its initial temperature must be
    1) 250 K   2) 250°C
    3) 2500 K   4) 25°C

11. The figure shows the pressure p versus volume V graphs for a certain mass of a gas at two constant temperatures T_1 and T_2. Which of the following inferences is correct?

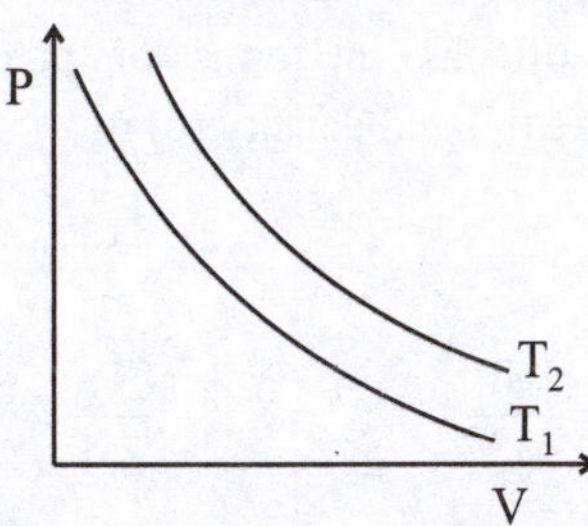

1) $T_1 = T_2$
2) $T_1 > T_2$
3) $T_1 < T_2$
4) no inference can be drawn due to insufficient information.

12. Average momentum of an ideal gas depends upon

1) Temperature 2) Mass
3) Volume 4) It is Zero

13. Two mono atomic gas A and B occupying the same volume V are at the same temperature T and pressure P. If they are mixed, the resulting mixture has volume V and temperature T. The pressure of the mixture is

1) P 2) $\frac{P}{2}$
3) 4P 4) 2P

14. How much heat energy in joules must be supplied to 14g of N_2 at room temperature to raise its temperature by 40°C at constant pressure?

1) 50R 2) 60R
3) 70R 4) 80R

15. The phenomenon of Brownian movement may be taken as evidence of

1) Kinetic theory of matter
2) EMT of radiation
3) Corpuscular theory of light
4) Photoelectric phenomenon

16. A vessel has 6g of oxygen at pressure P and temperature 400 K. A small hole oxygen leaks out if the final pressure is P/2 and temperature is 300 K?

1) 3g 2) 2g 3) 4g 4) 5g

17. If P is the pressure, V the volume, R the gas constant, K the Boltzmann constant and T the absolute temperature then the number of molecules in the given mass of the gas is given

1) $\frac{PV}{RT}$ 2) $\frac{PV}{KT}$
3) $\frac{PR}{T}$ 4) pV

18. What is the mass of 2ℓ of N_2 at 22.4 atm pressure and 273 K?

1) 28 g 2) 14×22.4 g
3) 56 g 4) 15 g

19. Equation of gas in terms of pressure (P), absolute temperature (T) and density (d) is

1) $\frac{P_1}{T_1 d_1}$ 2) $\frac{P_1 T_1}{d_1} = \frac{P_2 T_2}{d_2}$
3) $\frac{P_1 d_2}{T_2} = \frac{P_2 d_1}{T_1}$ 4) $\frac{P_1}{T_1 d_1} = \frac{P_2}{T_2 d_2}$

Degrees of Freedom and Specific Heat

20. Mean free path of molecules of a gas is inversely proportional to

1) r 2) r^3
3) r^2 4) $r^{1/2}$

21. The value of C_V for O_2 is $\frac{5R}{2}$ with increase in temperature. It becomes $\frac{7R}{2}$ due to

1) Translational motion
2) Rotational motion
3) Vibrational motion
4) None

22. At ordinary temperature, the molecules of an ideal gas have only translational and rotational kinetic energies. At high temperature they may also have vibrational energy as a result of this at high temperature

1) $C_V < \frac{3R}{2}$, monoatomic
2) $C_V = \frac{3R}{2}$, monoatomic
3) $C_V < \frac{5R}{2}$, diatomic
4) $C_V = \frac{5R}{2}$, diatomic

23. The expansion of a mass m of an ideal gas at a constant pressure p is shown by the line H. The expansion of a mass 2m of the same gas at a pressure p/2 is shown by

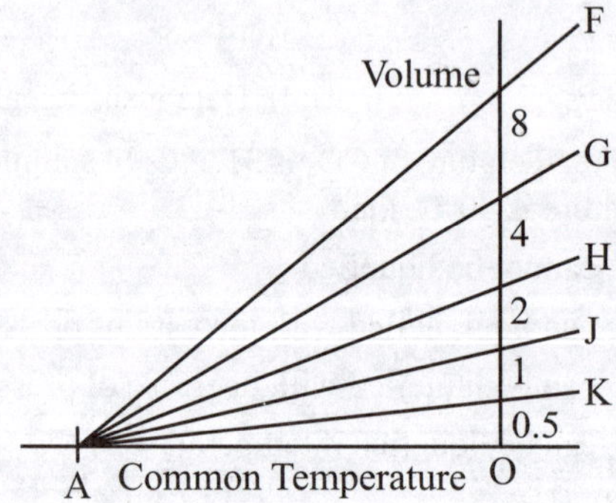

1) line F 2) line G
3) line H 4) line J

24. Which of the following statement given below are correct for mono atomic gas?

I. $C_V = \frac{3}{2}R$ II. $C_P = \frac{5}{2}R$
III. $C_P - C_V = 2R$ IV. $\frac{C_P}{C_V} = \frac{5}{3}$

1) I only 2) I, II and IV
3) III only 4) None

25. If the pressure in a closed vessel is reduced by drawing out some of the gas, the mean free path of two molecules
1) is increased
2) is decreased
3) remains unchanged
4) increases or decreases according to the nature

26. If quantity of heat 1163.4J supplied to 1 mole of N_2 gas, at room temperature at constant pressure, then rise in temperatures
1) 20K 2) 30K
3) 40K 4) 58K

27. The molar specific heat at constant pressure of an ideal gas is $\frac{7}{2}R$. The ratio of specific heat at constant pressure to that at constant volume is
1) $\frac{7}{5}$ 2) $\frac{8}{7}$
3) $\frac{5}{7}$ 4) $\frac{9}{7}$

28. For a gas molecule with 6 degrees of freedom, the law of equipartition of energy gives the following relation between the molecular specific heat (C_V) and gas constant (R)
1) $C_V = \frac{R}{2}$ 2) $C_V = R$
3) $C_V = 2R$ 4) $C_V = 3R$

29. A polyatomic gas with a n degrees of freedom has a mean energy per molecule given by
1) $\frac{nkT}{N}$ 2) $\frac{nkT}{2N}$
3) $\frac{nkT}{2}$ 4) $\frac{3kT}{N}$

30. If γ is the ratio of specific heats and R is the universal gas constant, then the molar specific heat at constant pressure C_p is given by
1) γR 2) $\frac{(\gamma-1)R}{\gamma}$
3) $\frac{R}{\gamma-1}$ 4) $\frac{\gamma R}{\gamma-1}$

Various Speeds of Gas

31. A gas is compressed isothermally. The rms velocity of its molecules
1) Increases 2) Decreases
3) Remains same 4) First increase then decrease

32. The velocity of 4 gas molecules are 1 km/s, 3 km/s, 5 km/s and 7 km/s. Calculate the difference between average and rms velocity (in km/s):
1) 0.338 2) 0.438
3) 0.583 4) 0.683

33. At what temperature will the oxygen molecules have the same rms as H_2 at 300K?
1) 4800 K 2) 3600 K
3) 5000 K 4) 8000 K

34. Average velocity of an ideal gas molecule is
1) Proportional to $\sqrt{T}$ 2) Proportional to T_2
3) Proportional to T_3 4) Zero

35. The temperature of an ideal gas is increased from 120 K to 480 K. If at 120 K, the root mean square speed of gas molecules is V, then at 480, it will be
1) 4V 2) 2V
3) $\frac{V}{2}$ 4) $\frac{V}{4}$

36. 10 mole of an ideal mono atomic gas at 10°C is mixed with 20 mole of another mono atomic gas at 20°C then the temperature of the mixture is
1) 15.5°C 2) 15°C
3) 16°C 4) 16.6°C

37. At what temperature is the rms velocity of a hydrogen molecule equal to that of an O_2 at 47°C?
1) 80K 2) –73K
3) 3K 4) 20 K

38. Root mean square velocity of a particle is V at pressure P. If pressure is increased two times, then the rms velocity becomes
1) 2V 2) 3V
3) 0.5V 4) V

39. The ratio of root mean square velocities of O_3 and O_2 is
1) 1 : 1 2) 2 : 3
3) 3 : 2 4) $\sqrt{2} : \sqrt{3}$

40. The amount of heat energy required to raise the temperature of 1 gm of He at NTP, from T_1 K to T_2 K is
1) $3/4\ K_b(T_1 - T_2)$ 2) $3/5\ K_b(T_1 - T_2)$
3) $3/8\ K_bN_a(T_1 - T_2)$ 4) $3/8\ K_bN_a(T_2 - T_1)$

41. Heat is supplied to diatomic gas at constant pressure the ratio between heat energy supplied and work done is
1) $\frac{5}{2}$ 2) $\frac{9}{7}$
3) $\frac{7}{2}$ 4) $\frac{8}{9}$

42. The kinetic energy of 1 g molecules of a gas at normal temperature and pressure is
1) 1.3×10^2J 2) 2.7×10^2 J
3) 0.56×10^4 J 4) 3.4×10^3 J

43. When the container stopped suddenly, ordered kinetic energy gets converted into disordered kinetic energy which increases the temperature of gas. KE per unit volume is E. The pressure exerted by the gas is given by

1) $\frac{E}{3}$
2) $\frac{2E}{3}$
3) $\frac{3E}{2}$
4) $\frac{E}{2}$

44. A closed compartment containing gas is moving with some acceleration in horizontal direction. Then the pressure in the compartment is

1) Higher in front side
2) Lower in front side
3) Lower in rear side
4) Lower in upper side

45. A sealed container with negligible coefficient of volumetric expansion, contains He gas. When it is heated from 300 K to 600 K, the average KE of He atom will be:

1) Halved
2) Unchanged
3) Doubled
4) Increased by factor $\sqrt{2}$

46. If at the same temperature and pressure, the densities of two diatomic gases are d_1 and d_2 respectively. The ratio of mean kinetic energy per molecules of gases will be

1) 1 : 1
2) $d_1 : d_2$
3) $\sqrt{d_1} : \sqrt{d_2}$
4) $\sqrt{d_2} : \sqrt{d_1}$

High Order Time Intensive MCQs

1. The temperature at which the velocity of oxygen will be half that of hydrogen at NTP is:

1) 400°C
2) 421°C
3) 500°C
4) 819°C

2. When temperature of an ideal gas is increased from 27°C to 227°C, its rms speed is changed from 400 ms^{-1} to v_s, Then v_s is:

1) 516 ms^{-1}
2) 4150 ms^{-1}
3) 310 ms^{-1}
4) 746 ms^{-1}

3. How many degrees of freedom the gas molecules have if under STP the gas density $\rho = 1.3$ kg/m^3 and the velocity of sound propagation in it is 330 ms^{-1}?

1) 3
2) 5
3) 7
4) 8

4. Two vessels A and B having equal volume contain equal masses of hydrogen in A and helium in B at 300 K. Then, mark the correct statement?

1) The pressure exerted by hydrogen is half that exerted by helium
2) The pressure exerted by hydrogen is equal to that exerted by helium
3) Average KE of the molecule of hydrogen is half the average KE of the molecules of helium
4) The pressure exerted by hydrogen is twice that exerted by helium

5. The average kinetic energy of a gas molecule is

1) Proportional to pressure of gas
2) Inversely proportional to volume of gas
3) Inversely proportional to absolute temperature of gas
4) Directly proportional to absolute temperature of gas

6. Pressure of an ideal gas is increased by keeping temperature constant. What is the effect on kinetic energy of molecules?

1) Increases
2) Decreases
3) No change
4) Cannot be determined

7. Two perfect gases at absolute temperatures T_1 and T_2 are mixed. There is no loss of energy. The temperature of mixture of masses of molecules are m_1 and m_2 and the number of molecules in the gases are n_1 and n_2 respectively, is

1) $\frac{T_1 + T_2}{2}$
2) $\frac{n_1T_1 + n_2T_2}{n_1 + n_2}$
3) $\frac{n_1T_2 + n_2T_1}{n_1 + n_2}$
4) $\sqrt{T_1T_2 / n_1n_2}$

8. For a gas at 400 K temperature energy is 300 J and speed is 400 m/sec. If temperature is made 3 times of initial then the final energy and speed are

1) 600 J, 300 $\sqrt{3}$ m/s
2) 900 J, 400 $\sqrt{3}$ m/s
3) 1200 J, 500 $\sqrt{3}$ m/s
4) 1500 J, 600 $\sqrt{3}$ m/s

9. A vessel contains 1 mole of O_2 (M = 32) at temperature T. The pressure of the gas is P. An identical vessel containing 1 mole of He gas (m = 4) at temperature 2T has a pressure of:

1) $\frac{P}{8}$
2) P
3) 2P
4) 8P

10. A certain amount of an ideal mono atomic gas needs 200 J of heat energy to raise its temperature by 10°C at constant pressure. The heat needed for the same temperature rise of the same gas but at constant volume will be:

1) 300 J
2) 120 J
3) 2000 J
4) 400 J

11. A gas mixture contains 8 moles of oxygen and 2 moles of argon at room temperature T. The total internal energy of the mixture is:

1) 11 RT
2) 23 RT
3) 22 RT
4) 26 RT

12. Two cylinders contain ideal diatomic gas. As shown in figure same amount of heat is given to the two cylinders. If temperature rise in cylinder B is T_0 then temperature rise in cylinder A will be

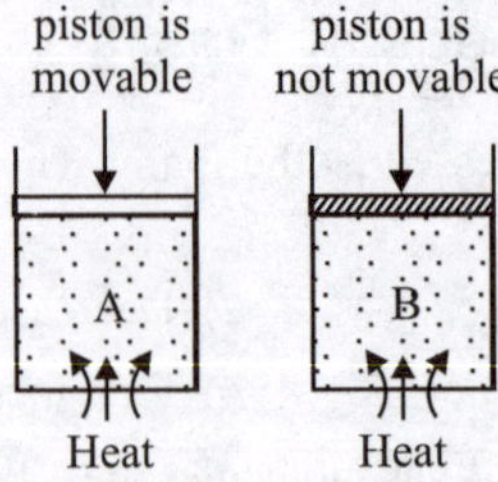

1) $\frac{7}{5}T_0$
2) $2T_0$
3) $\frac{3}{5}T_0$
4) $\frac{5}{7}T_0$

13. The relation between r.m.s. velocity, v_{rms} and the most probable velocity, v_{mp} of a gas is

1) $v_{rms} = v_{mp}$
2) $v_{rms} = \sqrt{\frac{3}{2}} v_{mp}$
3) $v_{rms} = \sqrt{\frac{2}{3}} v_{mp}$
4) $v_{rms} = \frac{2}{3} v_{mp}$

14. Two identical vessels A & B contain equal amount of ideal monoatomic gas. The piston of A is fixed but that of B is free. Same amount of heat is absorbed by A & B. If B's internal energy increases by 100 J the change in internal energy of A is

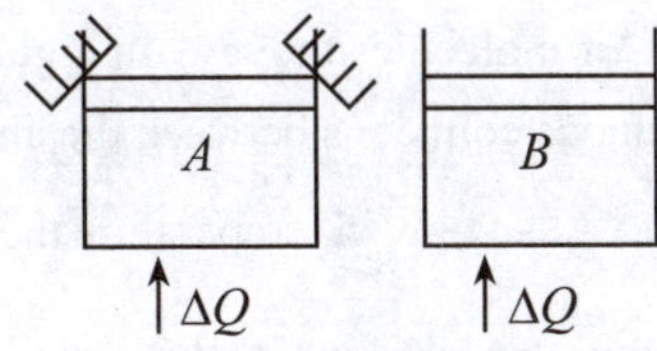

1) 100J
2) –90°C
3) 250 J
4) None of these

15. The P-V diagrams of two different masses m_1 and m_2 for an ideal gas at constant temperature T is given in figure. Then:

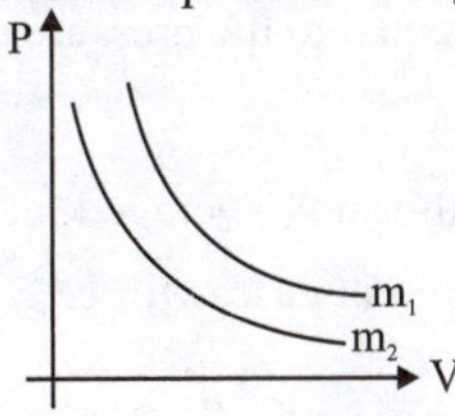

1) $m_1 = m_2$
2) $m_1 > m_2$
3) $m_1 < m_2$
4) Data is insufficient

16. The root mean square speed of hydrogen molecule is v_1 at temperature T K. The most probable speed of Oxygen molecule is v_2 at same temperature. Then $\frac{V_1}{V_2}$ is equal to:

1) $2\sqrt{6}$
2) 4
3) $3\sqrt{6}$
4) 2

17. During an experiment an ideal gas is found to obey an additional law VP^2 = constant. The gas is initially at temp T and volume V. What will be the temperature of the gas when it expands to a volume 2V?

1) $T' = \sqrt{4}T$
2) $T' = \sqrt{2}T$
3) $T' = \sqrt{5}T$
4) $T' = \sqrt{6}T$

18. The gas in a vessel is subjected to a pressure of 20 atmosphere at a temperature 27°C. The pressure of the gas in a vessel after one half of the gas is released from the vessel and the temperature of the remainder is raised by 50°C is

1) 8.5 atm
2) 10.8 atm
3) 11.7 atm
4) 17 atm

19. The fastest moving gas molecules are those of

1) Nitrogen
2) Oxygen
3) Chlorine
4) Hydrogen

20. An oxygen cylinder of volume 30 litres has an initial gauge pressure of 15 atm and a temperature of 27°C. After some oxygen is withdrawn from the cylinder, the gauge pressure drops to 11 atm and its temperature drops to 17°C. Estimate the mass of oxygen taken out of the cylinder ($R = 8.31 J$ $mol^{-1} K^{-1}$, molecular mass of $O_2 = 32$, u)

1) 0.121 Kg
2) 0.140 Kg
3) 0.131 Kg
4) Data insufficient

21. A container contain 0.1 mole of H_2 and 0.1 mole of O_2. If the gases are in thermal equilibrium then

1) Only the average kinetic energy of the molecule of H_2 and O_2 is same
2) Average speed of the molecule of H_2 and O_2 is same
3) Only the specific heat at constant pressure of two gases is same
4) The specific heat at constant pressure and the kinetic energy are same for both the gases

22. Two gases A and B having the same temperature T, same pressure p and same volume V are mixed. If the mixture is at the same temperature T and occupies a volume V, the pressure of the mixture is

1) 2p
2) p
3) p/2
4) 4p

Statement Based MCQs

Directions: These questions consist of two statements each, printed as Statement-I and Statement-II. While answering these questions, you are required to choose any one of the following four responses.

1) **Both Statement-I and Statement-II are correct.**
2) **Both Statement-I and Statement-II are incorrect.**
3) **Statement-I is correct & Statement-II is incorrect.**
4) **Statement-I is incorrect & Statement-II is correct.**

1. **Statement-I:** RMS velocity of a gas molecule is directly proportional to the square root of temperature.

   Statement-II: RMS velocity is more than the average velocity.

2. **Statement-I:** For a mono-atomic gas, degree of freedom is 5.

   Statement-II: Energy associated with each molecule per degree of freedom is $\frac{1}{2}k_B T$.

3. **Statement-I:** According to the assumptions of kinetic theory of gases, collisions between the gas molecules are perfectly elastic.

   Statement-II: Graham's law of diffusion states that the rate of diffusion is inversly proportional to the square root of the molecular mass of the gas.

4. **Statement-I:** The root mean square velocity of molecules of a gas is higher than their most probable velocity, at any temperature.

   Statement-II: Most probable velocity is possessed by least number of the gas molecules.

5. **Statement-I:** Real gases show ideal behaviour at low pressure.

   Statement-II: Real gases show ideal gas behaviour at high temperature.

Assertion & Reason MCQs

Directions: These questions consist of two statements each, printed as Assertion and Reason. While answering these questions, you are required to choose any one of the following four responses.

1) **Both Assertion and Reason are True and the Reason is a correct explanation of the Assertion.**
2) **Both Assertion and Reason are True but Reason is not a correct explanation of the Assertion.**
3) **Assertion is True but the Reason is False.**
4) **Assertion is False but Reason is True.**

1. **Assertion:** For mono-atomic gas molecules the value of $\frac{C_P}{C_V} = 1.67$

   Reason: For diatomic gas molecules the value of $\frac{C_P}{C_V} = 1.2$

2. **Assertion:** Molecules of air in a container do not all fall and settle on the ground due gravity.

   Reason: Air molecules move with high velocity and there are continuous collisions between the air molecules.

3. **Assertion:** At any given temperature the ratio of rms speed and average speed of a gas molecules is $\sqrt{3} : \sqrt{\frac{8}{\pi}}$

   Reason: As v_{rms} is more than v_{av}.

4. **Assertion:** For a mixture of non reactive ideal gases, the total pressure is equal to the pressure due to each gas in the mixture.

   Reason: In equilibrium, average kinetic energy of the gas molecules of different gases will be equal.

5. **Assertion:** The ratio of $\frac{C_P}{C_V}$ for a diatomic gas is more than that for a mono-atomic gas.

   Reason: The molecules of a mono-atomic gas have more degrees of freedom than the molecules of a diatomic gas.

Match the Columns MCQs

1. Match the columns and select the correct option:

	Column-I		Column-II
(a)	Avg. speed	(i)	$\sqrt{\frac{3RT}{M}}$
(b)	R.M.S speed	(ii)	$\sqrt{\frac{8RT}{\pi M}}$
(c)	Most probable speed	(iii)	$\frac{3RT}{M}$
(d)	Mean square speed	(iv)	$\sqrt{\frac{2RT}{M}}$

1) (a)-(i) (b)-(iv) (c)-(iii) (d)-(ii)
2) (a)-(ii) (b)-(iii) (c)-(i) (d)-(iv)
3) (a)-(ii) (b)-(i) (c)-(iv) (d)-(iii)
4) (a)-(ii) (b)-(iv) (c)-(i) (d)-(iv)

2. Match the columns and select the correct option:

	Column-I		Column-II
(a)	$\frac{C_P}{C_v}$	(i)	$\frac{nRT}{V}$
(b)	Energy associated with each degrees of freedom	(ii)	$1+\frac{2}{n}$ for n degree of freedom
(c)	Pressure	(iii)	$P \propto T$
(d)	Gay Lussac's Law (at constant volume)	(iv)	$\frac{kT}{2}$

1) (a)-(i) (b)-(iv) (c)-(iii) (d)-(ii)
2) (a)-(ii) (b)-(iv) (c)-(i) (d)-(iii)
3) (a)-(i) (b)-(ii) (c)-(iii) (d)-(iv)
4) (a)-(iii) (b)-(iv) (c)-(ii) (d)-(i)

NCERT Exemplar MCQs

1. A cubic vessel (with face horizontal + vertical) contains an ideal gas at NTP. The vessel is being carried by a rocket which is moving at a speed of 500 m s^{-1} in vertical direction. The pressure of the gas inside the vessel as observed by us on the ground:
   1) Remains the same because 500 ms^{-1} is very much smaller than v_{rms} of the gas
   2) Remains the same because motion of the vessel as a whole does not affect the relative motion of the gas molecules and the walls
   3) Will increase by a factor equal to $(v_{rms}^2 + (500^2)) / v_{rms}^2$ where v_{rms} was the original mean square velocity of the gas
   4) Will be different on the top wall and bottom wall of the vessel

2. 1 mole of an ideal gas is contained in a cubical volume V, ABCDEFGH at 300K (Figure). One face of the cube (EFGH) is made up of a material which totally absorbs any gas molecule incident on it. At any given time:

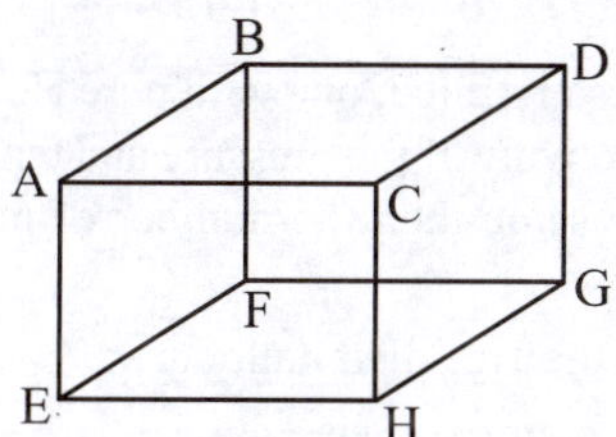

   1) The pressure on EFGH would be zero
   2) The pressure on all the faces v will the equal
   3) The pressure of EFGH would be double the pressure on ABCD
   4) The pressure on EFGH would be half that on ABCD

3. Boyle's law is applicable for an:
   1) Adiabatic process
   2) Isothermal process
   3) Isobaric process
   4) Isochoric process

4. A cylinder containing an ideal gas is in vertical position and has a piston of mass M that is able to move up or down without friction (figure). If the temperature is increased:

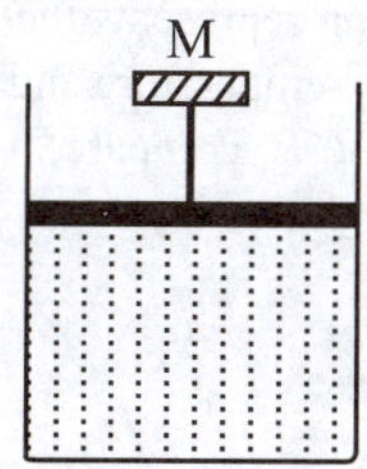

   1) Both p and V of the gas will change
   2) Only p will increase according to Charles' law
   3) V will change but not p
   4) p will change but not V

5. 1 mole of H_2 gas is contained in a box of volume V = 1.00 m^3 at T = 300 K. The gas is heated to a temperature of T = 3000 K and the gas gets converted to a gas of hydrogen atoms. The final pressure would be: (considering all gases to be ideal)
   1) Same as the pressure initially
   2) 2 times the pressure initially
   3) 10 times the pressure initially
   4) 20 times the pressure initially

6. A vessel of volume V contains a mixture of 1 mole of hydrogen and 1 mole oxygen (both considered as ideal). Let $f_1(v)dv$, denote the fraction of molecules with speed between v and (v + dv) with $f_2(v)dv$, similarly for oxygen. Then:
   1) $f_1(v) + f_2(v) = f(v)$ obeys the Maxwell's distribution law
   2) $f_2(v)$ will obey the Maxwell's distribution law separately
   3) Neither f_1 (v), nor $f_2(v)$ will obey the Maxwell's distribution law
   4) $f_2(v)$ and $f_1(v)$ will be the same

7. An inflated rubber balloon contains one mole of an ideal gas, has a pressure p, volume V and temperature T. If the temperature rises to 1.1 T, and the volume is increased to 1.05 V, the final pressure will be:
   1) 1.1 p
   2) p
   3) Less than p
   4) Between p and 1.1p

NEET & AIIMS Past Year MCQs

1. The following graph represents the *T-V* curves of an ideal gas (where *T* is the temperature and *F* the volume) at three pressures P_1, P_2 and P_3 compared with those of Charles s law represented as dotted lines. **(2024)**

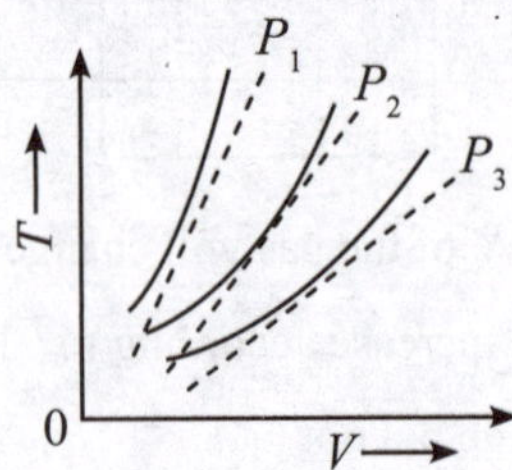

1) $P_2 > P_1 > P_3$
2) $P_1 > P_2 > P_3$
3) $P_3 > P_2 > P_1$
4) $P_1 > P_3 > P_2$

2. According to the law of equipartition of energy, the number of vibrational modes of a polyatomic gas of constant $\gamma = \frac{C_p}{C_v}$ is (C_P where C_V are the specific heat capacities of the gas at constant pressure and constant volume, respectively): **(2024 Re)**

1) $\frac{4+3\gamma}{\gamma-1}$
2) $\frac{3+4\gamma}{\gamma-1}$
3) $\frac{4-3\gamma}{\gamma-1}$
4) $\frac{3-4\gamma}{\gamma-1}$

3. The temperature of a gas is –50 °C. To what temperature the gas should be heated so that the rms speed is increased by 3 times? **(2023)**

1) 223 K
2) 669 °C
3) 3295 °C
4) 3097 K

4. The volume occupied by the molecules contained in 4.5 kg water at STP, if the intermolecular forces vanish away is: **(2022)**

1) 5.6 m^3
2) 5.6×10^6 m^3
3) 5.6×10^3 m^3
4) 5.6×10^{-3} m^3

5. Match Column-I and Column-II and choose the correct match from the given choices. **(2021)**

	Column-I		Column-II
A.	Root mean square speed of gas molecules	p.	$\frac{1}{3}nmv^2$
B.	Pressure exerted by ideal gas	q.	$\sqrt{\frac{3RT}{M}}$
C.	Average kinetic energy of a molecule	r.	$\frac{5}{2}RT$
D.	Total internal energy of 1 mole of a diatomic gas	s.	$\frac{3}{2}k_BT$

1) A-(q); B-(r); C-(s); D-(p)
2) A-(q); B-(p); C-(s); D-(r)
3) A-(r); B-(q); C-(p); D-(s)
4) A-(r); B-(p); C-(s); D-(q)

6. The root mean square velocity of hydrogen molecule at 27°C is V_H and that of oxygen at 402°C is V_o then: **(AIIMS 2017)**

1) $V_o > V_H$
2) $4\ V_o = 9\ V_H$
3) $2\ V_o = 3\ V_H$
4) $9\ V_o = 134\ V_H$

7. When an ideal mono atomic gas is heated at constant pressure, fraction of heat energy supplied which increases the internal energy of gas, is: **(AIIMS 2017)**

1) 2/5
2) 3/5
3) 3/7
4) 3/4

8. In a cylinder there are 60 g Ne and 64 g O_2. If pressure of mixture of gases in cylinder is 30 bar then in this cylinder partial pressure of O_2 is (in bar): **(AIIMS 2013)**

1) 30
2) 20
3) 15
4) 12

9. Two balloons are filled, one with pure He gas and the other by air, respectively. If the pressure and temperature of these balloons are same then the number of molecules per unit volume is: **(AIIMS 2006)**

1) More in the He filled balloon
2) Same in both balloons
3) More in air filled balloon
4) In the ratio of 1 : 4

10. The temperature of a gas is held constant, while its volume is decreased. The pressure exerted by the gas on the walls of the container increases, because its molecules: **(AIIMS 1996)**

1) Strike the walls more frequently
2) Strike the walls with higher velocities
3) Are in contact with the walls for a shorter time
4) Strike the walls with larger force.

11. If C_p and C_v are the specific heats for a gas at constant pressure and at constant volume respectively, then the relation $C_p - C_v = R$ is exact for: **(AIIMS 1994)**

1) Ideal gas and nearly true for real gases at high pressure
2) Ideal and real gases at all pressures
3) Ideal gas and nearly true for real gases at moderate pressure
4) Ideal gas at all pressure and real gas at moderate pressure.

Answer Key

NCERT Topic-wise MCQs

1	2	3	4	5	6	7	8	9	10	11	12	13	14	15	16	17
3	3	1	3	1	1	4	4	4	1	3	4	4	3	1	2	1
18	19	20	21	22	23	24	25	26	27	28	29	30	31	32	33	34
3	4	3	3	2	1	2	1	3	1	4	3	4	3	3	1	4
35	36	37	38	39	40	41	42	43	44	45	46					
2	4	4	4	4	4	3	4	2	2	3	1					

High Order Time Intensive MCQs

1	2	3	4	5	6	7	8	9	10	11	12	13	14	15	16	17
4	1	2	4	4	3	2	2	3	2	2	4	2	2	2	2	2
18	19	20	21	22												
3	4	2	4	1												

Statement Based MCQs

1	2	3	4	5
1	4	1	3	1

Assertion & Reason MCQs

1	2	3	4	5
3	1	2	2	4

Match the Column MCQs

1	2
3	2

NCERT Exemplar MCQs

1	2	3	4	5	6	7
2	4	2	3	4	2	4

NEET & AIIMS Past Year MCQs

1	2	3	4	5	6	7	8	9	10	11
2	3	3	1	2	3	2	4	2	1	3

CHAPTER 13

Oscillations

Topicwise Number of Questions Analysis (2025-2016)

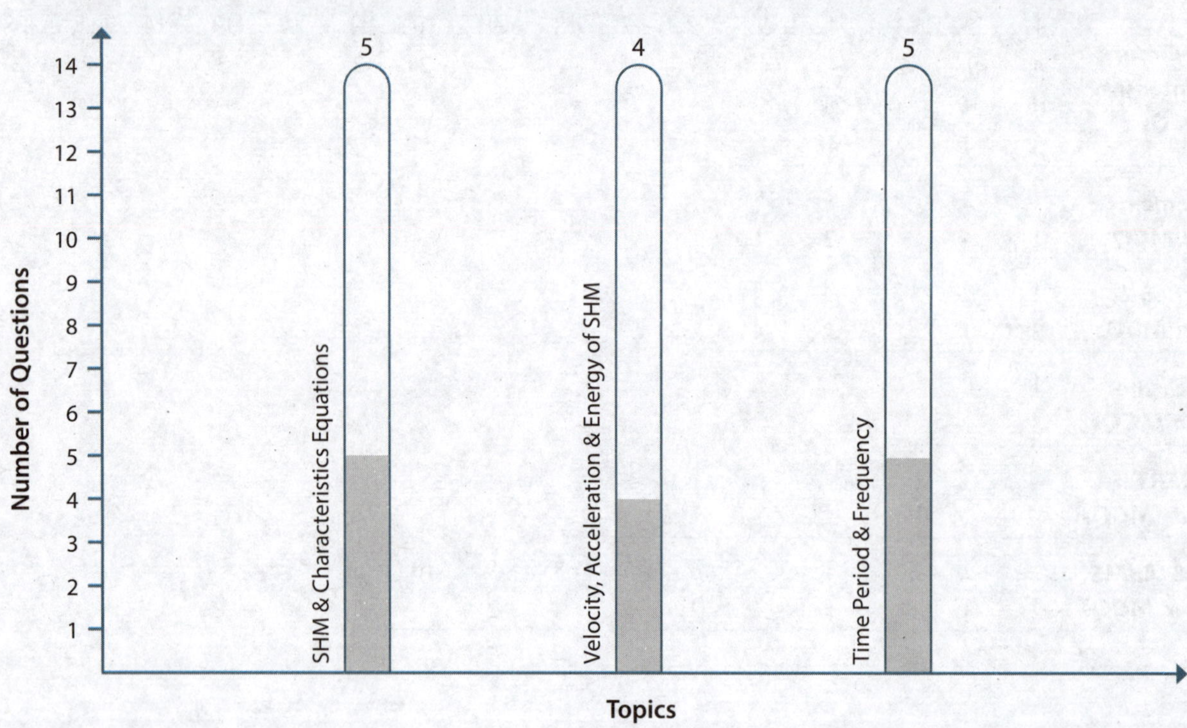

NCERT-Picks

SHM: Characteristic Equation

Formulas

1. Displacement in SHM at any time t is

   $y = A\sin(\omega t \pm \phi)$

2. **Amplitude:** It is the maximum value of displacement of the particle from its equilibrium position

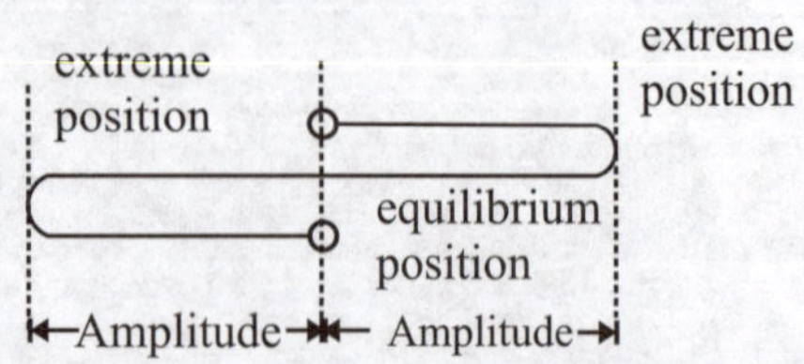

Amplitude = 1/2 (distance between extreme points/position)

3. **Angular Frequency (ω):** $\omega = \frac{2\pi}{T} = 2\pi f$ and its unit is rad/sec.

Numerical Profile

Necessary Condition for executing S.H.M.

(i) In linear S.H.M.:

$\therefore F \propto -x \Rightarrow F = -kx$, [k is restoring force constant]

(–) ve sign shows that direction of force is towards mean position and x is displacement of particle from the equilibrium position.

(ii) In angular S.H.M.:

$\therefore \tau \propto -\theta \Rightarrow \tau = -C\theta$ [θ is the angular displacement and C is restoring torque constant]

Velocity, Acceleration & Energy of SHM

Formulas

1. Velocity in SHM at any time t is given by

$$v = A\omega\cos(\omega t \pm \phi) \Rightarrow v = \omega\sqrt{A^2 - y^2}$$

(i) Velocity is maximum at the mean position i.e., at the mean position $v_{max} = \omega A$

(ii) Velocity is minimum when $x = \pm A$ i.e., at the position of maximum displacement velocity is zero $v_{min} = 0$.

2. Acceleration in SHM, $a = -\omega^2 y$

(i) In the case of SHM, acceleration is maximum when $x = \pm A$ i.e., $a_{max} = \pm \omega^2 A$ (towards mean position)

(ii) In the case of SHM acceleration is minimum when $x = 0$ i.e., $a_{min} = 0$

3. Potential Energy, $PE = \frac{1}{2}m\omega^2 y^2 = \frac{1}{2}ky^2$

[where m is the mass of the particle]

4. Kinetic Energy, $KE = \frac{1}{2}m\omega^2(A^2 - y^2) = \frac{1}{2}k(A^2 - y^2)$

5. Total Energy, $E = \frac{1}{2}m\omega^2 A^2 = \frac{1}{2}kA^2$

Numerical Profile

Energy conservation in S.H.M.

(i) Kinetic Energy (KE): The kinetic energy

$$= \frac{1}{2}m\omega^2 A^2\cos^2(\omega t + \phi)$$

(ii) Potential Energy(PE) $= \frac{m\omega^2 y^2}{2} = \frac{m\omega^2}{2}A^2\sin^2(\omega t + \phi)$

(iii) Total energy = Kinetic energy + Potential energy

$$= \frac{1}{2}m\omega^2 A^2$$

(iv) If initial potential energy is considered to be equal to zero i.e. $U_0 = 0$

then $KE_{max} = PE_{max} = TE = \frac{1}{2}kA^2$

(v) If $U_0 \neq 0$ then, $PE_{max} = TE = \frac{1}{2}kA^2 + U_0$

& $KE_{max} = \frac{1}{2}kA^2$

(vi) If frequency of S.H.M. is f then frequency of oscillation of KE = Frequency of oscillation of PE = 2f

Spring Mass System

Formulas

1. System, given in figure, is in equilibrium condition. Mass m is displaced by x distance in + x direction.

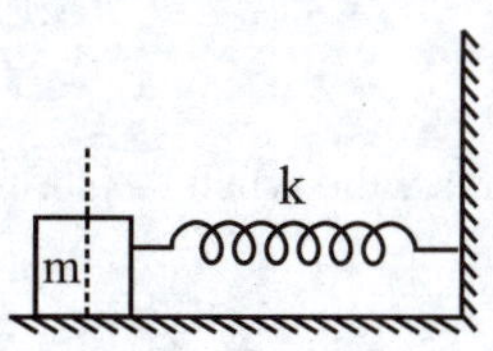

$$\therefore \nu = \frac{1}{2\pi}\sqrt{\frac{k}{m}} \text{ or } T = 2\pi\sqrt{\frac{m}{k}}$$

2. In figure (a) only spring is hanging. Here one mass is hanged as shown in figure. It displaces spring to x_0 distance. At condition of balance or equilibrium

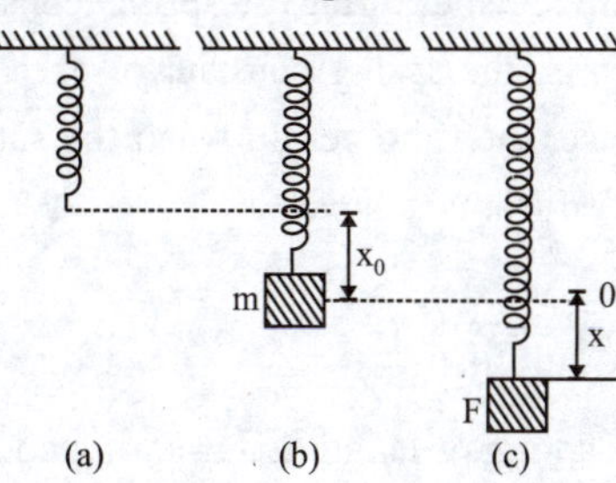

$$a = -\frac{k}{m}x$$

$$a = -\omega^2 x$$

$$\omega = \sqrt{\frac{k}{m}} = \frac{2\pi}{T} = 2\pi\nu$$

3. When springs are connected in series $\frac{1}{k} = \frac{1}{k_1} + \frac{1}{k_2}$

$$\therefore \quad T = 2\pi\sqrt{\frac{m}{k}} \Rightarrow T = 2\pi\sqrt{\frac{m(k_1 + k_2)}{k_1 k_2}}$$

$$\nu = \frac{1}{2\pi}\sqrt{\frac{k}{m}}, \omega = \sqrt{\frac{k}{m}}$$

4. When springs are connected in parallel extension is same for both spring but force acting will be different.

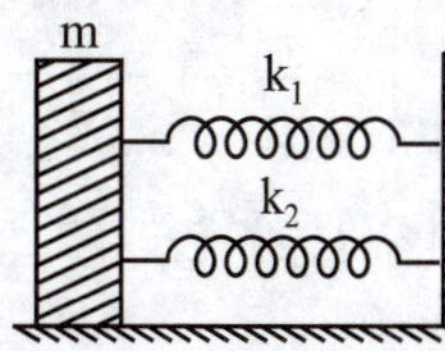

$$\Rightarrow F = -kx$$

$$\therefore k = k_1 + k_2$$

$$\therefore \omega = \sqrt{\frac{k_1 + k_2}{m}}, \nu = \frac{1}{2\pi}\sqrt{\frac{k_1 + k_2}{m}}$$

$$T = 2\pi\sqrt{\frac{m}{k_1 + k_2}}$$

Numerical Profile

Cases of equivalent spring constant

(i) If a mass M is suspended from a spring of force constant K. Then the time period of oscillation will be

$$T = 2\pi\sqrt{\frac{M}{K}}$$

This relation is valid when the spring is assumed to be massless

(ii) If the length of spring of spring constant K is made n times then its effective spring constant = K/n and time period when the same mass M is suspended from it becomes $T' = 2\pi\sqrt{\frac{M}{K/n}} = \sqrt{n}T$

(iii) If we divide the spring of spring constant K into n equal parts, the spring constant of each part becomes nk. Hence, the time period when the same mass M is suspended from each part is

$$T' = 2\pi\sqrt{\frac{M}{nK}} = \frac{T}{\sqrt{n}}.$$

(iv) When a spring of length L and spring constant K, is cut into two pieces of length L_1 and L_2 of spring constant K_1 and K_2 such that $L_1 = nL_2$ (where n is an integer), then

$$K_1 = \frac{K(n+1)}{n} \text{ and } K_2 = (n+1)K.$$

(v) If two blocks of masses m_1 and m_2 are connected at the two ends of a spring of spring constant K, then time period of their oscillation is given by $T = 2\pi\sqrt{\frac{\mu}{K}}$

$\mu = \frac{m_1 m_2}{m_1 + m_2}$, is called reduced mass.

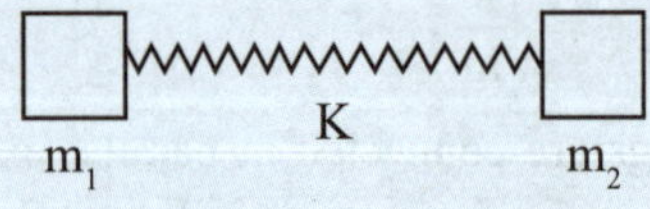

Simple Pendulum and Superposition of SHM

Formulas

1. Time period of simple pendulum, $T = 2\pi\sqrt{\frac{\ell}{g}}$

   [g is the acceleration due to gravity, *l* is the length of simple pendulum]

2. Angular frequency, $\omega = \sqrt{\frac{g}{\ell}}$

Numerical Profile

Time period of simple pendulum in different cases

(i) Time period of simple pendulum of infinite length is 84.6 min which is maximum and is equal to

$$T = 2\pi\sqrt{\frac{R}{g}} = 84.6 \text{ min.}$$

(ii) If g remains constant and $\Delta\ell$ is change in length, then percentage change in time period is

$$\frac{\Delta T}{T}\times 100 = \frac{1}{2}\frac{\Delta\ell}{\ell}\times 100 = 50\alpha\Delta\theta$$

[where α is coefficient of linear expansion of wire and $\Delta\theta$ is change in temperature]

(iii) If ℓ remains constant and Δg is change in acceleration then, $\frac{\Delta T}{T}\times 100 = -\frac{1}{2}\frac{\Delta g}{g}\times 100$

negative sign indicates time period decreases as g increases.

(iv) If $\Delta\ell$ is change in length and Δg is change in acceleration due to gravity, then

$$\frac{\Delta T}{T}\times 100 = \left[\frac{1}{2}\frac{\Delta\ell}{\ell} + \frac{1}{2}\frac{\Delta g}{g}\right]\times 100$$

Simple pendulum in Accelerating Frame

Condition for applying this formula: $|\vec{g} - \vec{a}|$ = constant

(i) If reference system is lift,

(a) If velocity of lift is constant,

v = constant

$\Rightarrow a = 0 \Rightarrow g_{eff} = g$ [a is the acceleration of the lift]

$\therefore T = 2\pi\sqrt{\frac{\ell}{g}}$

(b) If lift is moving downwards with acceleration a,

$g_{eff} = g - a$

$\therefore T = 2\pi\sqrt{\frac{\ell}{g-a}}$

(c) If lift is moving upwards with acceleration a,

$g_{eff} = g + a$

$\therefore T = 2\pi\sqrt{\frac{\ell}{g+a}}$

(d) If lift falls downwards freely,

$g_{eff} = 0$

$T = \infty$

(ii) If reference system is a vehicle moving with an acceleration "a",

$g_{eff} = \sqrt{g^2 + a^2}$

$T = 2\pi\sqrt{\frac{\ell}{\sqrt{g^2 + a^2}}}$

(iii) If electric field is horizontal; $g_{eff} = \frac{\sqrt{(mg)^2 + (qE)^2}}{m}$

$$T = 2\pi\sqrt{\frac{\ell}{(g)^2 + \left(\frac{qE}{m}\right)^2}} \quad \text{[E is the electric field]}$$

(iv) Let, a charged sphere is oscillating in a vertical electric field E. For it,

$$g_{eff} = \frac{mg \pm qE}{m}$$

$$T = 2\pi\sqrt{\frac{\ell}{g \pm \frac{qE}{m}}}$$

Superposition of two SHM's

(i) In same direction and of same frequency

$x = x_1 + x_2$

$x = A_1 \sin \omega t + A_2 \sin (\omega t + \theta)$

[A_1 and A_2 are the amplitudes of SHM]

$x = A \sin (\omega t + \phi)$

where $A = \sqrt{A_1^2 + A_2^2 + 2A_1A_2 \cos\theta}$

& $\phi = \tan^{-1}\left[\frac{A_2 \sin\theta}{A_1 + A_2 \cos\theta}\right]$

(ii) In same direction but are of different frequencies.

$x_1 = A_1 \sin \omega_1 t$ and $x_2 = A_2 \sin \omega_2 t$

then resultant displacement $x = x_1 + x_2$

$= A_1 \sin \omega_1 t + A_2 \sin \omega_2 t$

(iii) In two different directions

Let $x = A_1 \sin \omega t$ and $y = A_2 \sin(\omega t + \phi)$

(a) If $\phi = 0$ or π, then $y = \pm\left(\frac{A_2}{A_1}\right)x$

In this case path will be a straight line.

(b) If $\phi = \frac{\pi}{2}$ then

$x = A_1 \sin \omega t$

$y = A_2 \sin (\omega t + \pi/2)$

so, resultant will be $\frac{x^2}{A_1^2} + \frac{y^2}{A_2^2} = 1$. i.e., equation of an ellipse and if $A_1 = A_2$, then superposition will be an equation of circle.

Damped Harmonic Oscillations

[RC]

Formulas

1. Damping force $\vec{F}_d = -b\vec{v}$, where $\vec{v}$ is the velocity of oscillator and b is a damping constant, then displacement of oscillator is given by

$$x(t) = A_0 e^{-bt/2m} \cos(\omega' t + \phi)$$

where ω' is angular frequency of the damped oscillator

$$\omega' = \sqrt{\frac{k}{m} - \frac{b^2}{4m^2}}$$

2. The velocity amplitude of the system is greatest when $\omega_d = \omega$

This condition is called resonance.

Statements Asked in Exam

- A particle will be executing S.H.M. when the force acting on the particle at an instant is directly proportional to the displacement and is directed towards the mean position.
- In S.H.M. the bob of simple pendulum moves faster initially after crossing the mean position but the becomes slower later on. Therefore, the time taken by bob of simple pendulum for the displacement $y = 0$ to $y = \frac{r}{2}$ is $\frac{T}{12}$ and time taken to move from $y = \frac{r}{2}$ to $y = r$ is $\frac{T}{6}$, where T is the time period of oscillation of the pendulum.
- Restoring force on the particle in S.H.M. at an instant is $F = -M\omega^2 y$. It is maximum when the particle is at the extreme position and is minimum when the particle is at the mean position.
- Kinetic energy of a particle at an instant in S.H.M. is

$$E_K = \frac{1}{2} m\omega^2(A^2 - y^2) = \frac{1}{2} m\omega^2 A^2 \cos^2 \omega t = \frac{kA^2}{2} \cos^2 \omega t$$

which is maximum at the mean position and is minimum at the extreme position.
- Potential energy of a particle at an instant in S.H.M. is

$$E_p = \frac{1}{2} m\omega^2 y^2 = \frac{1}{2} m\, \omega^2 A^2 \sin^2 \omega t = \frac{1}{2} kA^2 \sin^2 \omega t$$

which is maximum at the extreme position and is minimum at the mean position.
- Thus maximum value of time period of vibration of simple pendulum is 84.6. minutes.
- When the bob of a simple pendulum describes a horizontal circle and length of simple pendulum makes an angle, θ with the vertical, then time period of rotatory motion is

$$T = 2\pi\sqrt{\frac{L \cos\theta}{g}}$$

- If a simple pendulum of length L is suspended from the roof of compartment of a train moving down an inclined plane of inclination θ, then time period of oscillation is

$$T = 2\pi\sqrt{\frac{L}{(g \cos\theta)}}$$

- In case of a simple pendulum of length L and time period T,
  - (i) The graph between L and T is a parabola
  - (ii) The graph between L and T^2 is a straight line.
- A pendulum clock runs slow due to increase in its time period and becomes fast due to decrease in its time period.
- If the acceleration due to gravity decreases by x%, then time period of simple pendulum increase by $\frac{x}{2}\%$.
- If the length of simple pendulum increases by y% then its time period increases by $\frac{y}{2}\%$.
- If a body of mass m is lying on a plate which is oscillating vertically with an amplitude a, then the body will not leave the contact with the platform, if time period of oscillation of platform is $T = 2\pi\sqrt{\frac{a}{g}}$.
- If a ball of radius r oscillates in a bowl of radius R (where R > r) then the time period of oscillation of the ball is

$$T = 2\pi\sqrt{\frac{R-r}{g'}}$$

- If a body oscillates with time periods T_1 and T_2 under the influence of forces F_1 and F_2 respectively, then the time period of oscillation of the bob under the influence of force $\vec{F} = \vec{F_1} + \vec{F_2}$ is given by

$$T = \frac{T_1 T_2}{(T_1^2 + T_2^2)^{1/2}}$$

- When a liquid in a U tube is executing S.H.M., then time period of oscillation.

$$T = 2\pi\sqrt{\frac{L}{2g}} = 2\pi\sqrt{\frac{2h}{2g}} = 2\pi\sqrt{\frac{h}{g}} \quad [\because L = 2h]$$

where L = total length of liquid column in a U-tube, h = height of liquid column in one limb of U-tube.

MCQs

NCERT Topic-wise MCQs

SHM: Characteristic Equation

1. A body oscillates with SHM according to the equation $x = 5\cos(2\pi t + \frac{\pi}{4})$. Its instantaneous displacement at t = 1 sec is

1) $\frac{\sqrt{2}}{5}$ 2) $\frac{1}{\sqrt{3}}$

3) $\frac{1}{\sqrt{2}}$ 4) $\frac{5}{\sqrt{2}}$

2. Which of the following expressions does not represent SHM?

1) $A\cos\omega t$ 2) $A\sin 2\omega t$

3) $A\sin\omega t + B\cos\omega t$ 4) $A\sin^2\omega t$

3. A particle executing a simple harmonic motion has a period of 6 sec. The time taken by the particle to move from the mean position to half the amplitude, starting from the mean position is

1) $\frac{3}{2}$s 2) $\frac{1}{2}$s

3) $\frac{3}{4}$s 4) $\frac{1}{4}$s

4. Two simple harmonic motions A and B are given respectively by the following equations.

$$y_1 = a\sin\left[\omega t + \frac{\pi}{6}\right]$$

$$y_2 = a\sin\left[\omega t + \frac{3\pi}{6}\right]$$

Phase difference between the waves is

1) $\frac{\pi}{2}$ 2) $\frac{\pi}{6}$

3) $\frac{\pi}{3}$ 4) Zero

5. A block is resting on a piston which is moving vertically with SHM of period 1 sec. At what amplitude of motion, will the block and piston separate?

1) 0.2 m 2) 0.25 m

3) 0.3 m 4) 0.35 m

6. A particle is executing SHM. Then, the graph of velocity as a function of displacement is

1) Straight line 2) Circle

3) Ellipse 4) Hyperbola

7. The displacement equation of a simple harmonic oscillator is given by

$y = A\sin\omega t - B\cos\omega t$. The amplitude of oscillator will be:

1) $A - B$ 2) $A + B$

3) $\sqrt{A^2 + B^2}$ 4) $A^2 + B^2$

8. A particle executes SHM, its time period is 16 sec. If it passes through the centre of oscillation, then its velocity is 2 m/s at time 2 sec. The amplitude will be:

1) 7.2 m 2) 4 cm

3) 6 cm 4) 0.72 cm

9. If a particle executes SHM with angular velocity 3.5 rad/s and maximum acceleration 7.5 m/s^2, then the amplitude of oscillation will be

1) 0.8 m 2) 0.69 m

3) 0.41 m 4) 0.61 m

10. A body executing SHM has velocity 10 cm/s and 7 cm/s, when its displacements from the mean position are 3 cm and 4 cm respectively. Length of path

1) 10 cm 2) 4.8 cm

3) 4 cm 4) 11.36 cm

11. The displacement of a particle varies according to the relation:

$x = 4(\cos\pi t + \sin\pi t)$

The amplitude of the particle is:

1) 8 2) –4

3) 4 4) $4\sqrt{2}$

12. How long after the beginning of motion is the displacement of a harmonically oscillating point equal to one half of its amplitude, if the period is 24 sec and initial phase is zero?

1)12 sec 2) 2 sec

3) 4 sec 4) 6 sec

13. A particle executes simple harmonic motion between $x = -A$ and $x = +A$. The time taken for it to go from 0 to A/2 is T_1 and to go from A/2 to A is T_2. Then:

1) $T_1 < T_2$ 2) $T_1 > T_2$

3) $T_1 = T_2$ 4) $T_1 = 2T_2$

14. A simple harmonic motion having an amplitude A and time period T is represented by the equation :

$y = 5\sin\pi(t + 4\pi)$ m

Then, the values of A (in m) and T (in sec) are :

1) A = 2; T = 2 2) A = 10; T = 1

3) A = 5 ; T = 2 4) A =10; T = 2

15. Which one of the following equations of motion represents simple harmonic motion?

1) Acceleration $= -k_0x + k_1x^2$
2) Acceleration $= -k(x + a)$
3) Acceleration $= k(x + a)$
4) Acceleration $= kx$

Velocity, Acceleration & Energy of SHM; PE, KE and TE in SHM & Spring

16. When the displacement of a particle executing simple harmonic motion is half its amplitude, the ratio of its kinetic energy to potential energy is

1) 1 : 3 2) 2 : 1
3) 3 : 1 4) 1: 2

17. A mass M, attached to a horizontal spring, executes SHM with a amplitude A_1. When the mass M passes through its mean position then a smaller mass m is placed over it and both of them move together with amplitude A_2. The ratio of $\left(\frac{A_1}{A_2}\right)$ is:

1) $\frac{M}{M+m}$ 2) $\frac{M+m}{M}$
3) $\left(\frac{M}{M+m}\right)^{1/2}$ 4) $\left(\frac{M+m}{M}\right)^{1/2}$

18. The amplitude of a particle executing SHM is 4 cm. At the mean position, the speed of the particle is 16 cm/s. The distance of the particle from the mean position at which the speed of the particle becomes $8\sqrt{3}$ cm/sec will be

1) $2\sqrt{3}$ 2) $\sqrt{3}$
3) 1 4) 2

19. If a body is executing simple harmonic motion and its current displacements is $\frac{\sqrt{3}}{2}$ times the amplitude from its mean position, then the ratio between potential energy and kinetic energy is

1) 3 : 2 2) 2 : 3
3) $\sqrt{3}:1$ 4) 3 : 1

20. A 10 kg metal block is attached to a spring of spring of spring constant 1000 Nm^{-1}. A block is displaced from equilibrium by 10 cm and released, the maximum acceleration of block is

1) 10 m/s^2 2) 100 m/s^2
3) 200 m/s^2 4) 0.1 m/s^2

21. The KE and PE of a particle executing SHM of amplitude a will be equal when displacement is

1) $\frac{A}{2}$ 2) $A\sqrt{2}$
3) 2A 4) $\frac{A}{\sqrt{2}}$

22. A particle is executing SHM with amplitude A and has maximum velocity v_0. Its speed at displacement A/2 will be:

1) $(\sqrt{3})v_0/2$ 2) $v_0/\sqrt{2}$
3) v_0 4) $v_0/4$

23. A long spring, when stretched by a distance x, has potential energy U. On increasing the stretching to nx, the potential energy of the spring will be

1) $\frac{U}{n}$ 2) nU
3) n^2U 4) $\frac{U}{n^2}$

24. When a spring is stretched by 10 cm, the potential energy stored is E. When the spring is stretched by 10 cm more, the potential energy stored in the spring becomes

1) 2E 2) 4E
3) 6E 4) 10E

25. U is the PE of an oscillating particle and F is the force acting on it at a given instant. Which of the following is true?

1) $\frac{U}{F}+x=0$ 2) $\frac{2U}{F}+x=0$
3) $\frac{F}{U}+x=0$ 4) $\frac{F}{2U}+x=0$

26. The displacement of a body executing SHM is given by $x = A \sin(2\pi t + \pi/3)$. The first time from t = 0 when the velocity is maximum is :

1) 0.33 sec 2) 0.16 sec
3) 0.25 sec 4) 0.5 sec

27. The displacement of a particle executing SHM is given by $y=10\sin\left(6t+\frac{\pi}{3}\right)$, where y is in meter at time t in seconds. The initial displacement and velocity of the particle are respectively:

1) $5\sqrt{3}$ m and 30m / sec 2) 15 m and $5\sqrt{3}$m / sec
3) $15\sqrt{3}$ m and 30m / sec 4) $\frac{10\sqrt{3}}{2}$ m and 30m / sec

28. A body of mass 1 kg is executing simple harmonic motion. Its displacement y(cm) at t seconds is given by $y=6\sin\left(100t+\frac{\pi}{4}\right)$. Its maximum kinetic energy is:

1) 6 J 2) 18 J
3) 24 J 4) 36 J

29. The average acceleration of a particle performing SHM over one complete oscillation is

1) $\frac{\omega^2 A}{2}$ 2) $\frac{\omega^2 A}{\sqrt{2}}$
3) Zero 4) $A\omega^2$

Time Period and Frequency of Various Oscillating System

30. What will be time period of the displaced body of mass m?

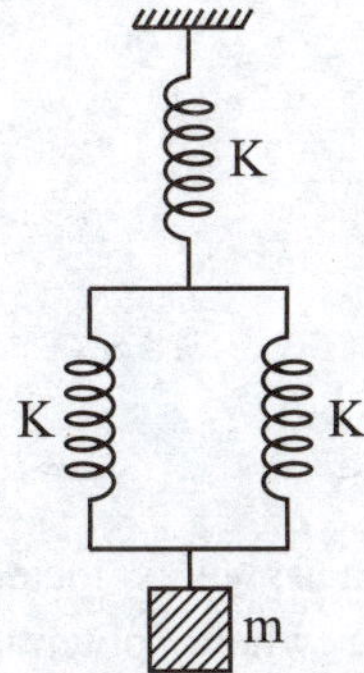

1) $2\pi\sqrt{\frac{3k}{m}}$ 2) $2\pi\sqrt{\frac{m}{3k}}$ 3) $2\pi\sqrt{\frac{3m}{2k}}$ 4) $2\pi\sqrt{\frac{2m}{3k}}$

31. If the maximum velocity and acceleration of a particle executing SHM are equal in magnitude, the time period will be

1) 1.57 2) 3.17 3) 6.28 4) 12.56

32. A simple pendulum of frequency n falls freely under gravity from certain height from the ground level. Its frequency of oscillation.

1) Remain unchanged 2) > n
3) Becomes zero 4) Becomes infinite

33. Equation of a particle executing SHM is $\frac{2d^2x}{dt^2} + 32x = 0$, then the time period of the body is

1) Zero 2) π
3) 2π 4) $\frac{\pi}{2}$

34. The mass and diameter of a planet are twice that of the earth. The period of oscillation of pendulum on this planet will be (if it is a second's pendulum on earth)

1) $1/\sqrt{2}$ 2) $2\sqrt{2}$
3) 2 4) 1/2

35. A spring-mass system consists of a platform attached to a spring, and a block of ice is placed on the platform. The system starts oscillating vertically. As time progresses, the ice melts and drips away without affecting the vertical motion directly. Assuming no external damping forces, how do the frequency $\omega(t)$ and amplitude $A(t)$ of oscillation change over time?

1) $\omega(t)$ increases, $A(t)$ decreases
2) $\omega(t)$ increases, $A(t)$ remains constant
3) $\omega(t)$ decreases, $A(t)$ increases
4) $\omega(t)$ increases, $A(t)$ decreases

36. A pendulum has time period T in air when it is made to oscillate in water, it acquired a time period $T = \sqrt{2}T$. The density of the pendulum bob is equal to

1) $\sqrt{2}$ 2) 2
3) $2\sqrt{2}$ 4) None

37. Two pendulums have time period T and 5T/4. They starts SHM at the same time from the mean position. What will be the phase difference between them after the bigger pendulum completed one oscillation?

1) 45° 2) 90°
3) 60° 4) 30°

38. A simple pendulum has a time period T_1 on the surface of earth of radius R. When taken to a height of R above the earth's surface, its time period is T_2, ratio $\frac{T_2}{T_1}$ is

1) $\frac{1}{\sqrt{2}}$ 2) $\sqrt{2}$
3) 2 4) 4

39. A simple pendulum has a time period T in vacuum. Its time period when it is completely immersed in a liquid of density 1/8 of density of material of the bob is

1) $\sqrt{\frac{8}{7}}T$ 2) $\sqrt{\frac{5}{8}}T$
3) $\sqrt{\frac{3}{2}}T$ 4) $\sqrt{\frac{3}{8}}T$

40. The potential energy of a particle of mass 1 kg in motion along the x-axis is given by: $U = 4(1 - \cos 2x)$J, where x is in metres. The period of small oscillations (in sec) is:

1) 2π 2) π
3) $\frac{\pi}{2}$ 4) $\sqrt{2}\pi$

41. A mass-spring system oscillates vertically with an initial mass m_0. A small hole causes it to lose mass uniformly at a slow rate. Which of the following statements correctly describes the system's natural frequency and energy over time?

1) Natural frequency increases; total mechanical energy remains constant
2) Natural frequency decreases; total mechanical energy increases
3) Natural frequency increases; total mechanical energy decreases
4) Both natural frequency and total mechanical energy increase

42. A particle of mass 1 kg is undergoing S.H.M., for which graph between force and displacement (from mean position) is shown. Its time period, in seconds, is:

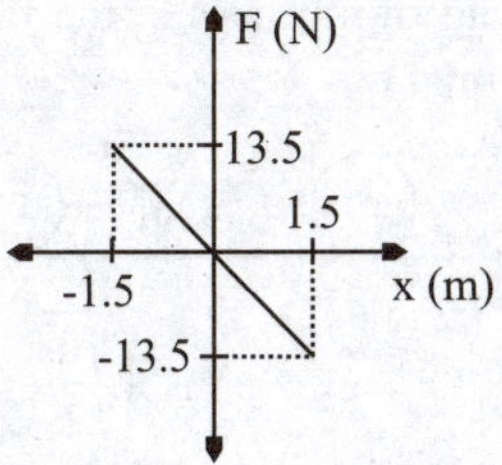

1) $\pi/3$ 2) $2\pi/3$
3) $\pi/6$ 4) $3/\pi$

43. The period of oscillation of simple pendulum of length L suspended from the roof of a vehicle which moves without friction down an inclined plane of inclination α is given by:

1) $2\pi\sqrt{\frac{L}{g\cos\alpha}}$ 2) $2\pi\sqrt{\frac{L}{g\sin\alpha}}$

3) $2\pi\sqrt{\frac{L}{g}}$ 4) $2\pi\sqrt{\frac{L}{g\tan\alpha}}$

44. The particle is executing S.H.M. on a line 4cm long. If its velocity at mean position is 12 cm/sec, its frequency in Hertz will be:

1) $2\pi/3$
2) $3/2\pi$
3) $\pi/3$
4) $3/\pi$

High Order Time Intensive MCQs

1. A heavy brass sphere is hung from a weightless inelastic spring and as a simple pendulum its time period of oscillation is *T*. When the sphere is immersed in a non-viscous liquid of density 1/10 that of brass, it will act as a simple pendulum of period

1) T 2) $\frac{10}{9}T$

3) $\sqrt{\left(\frac{9}{10}\right)}T$ 4) $\sqrt{\left(\frac{10}{9}\right)}T$

2. A block of mass 100 g attached to a spring of spring constant 100 N/m is lying on a frictionless floor as shown. The block is moved to compress the spring by 10 cm and then released. If the collisions with the wall in front are elastic, then the time period of the motion is:

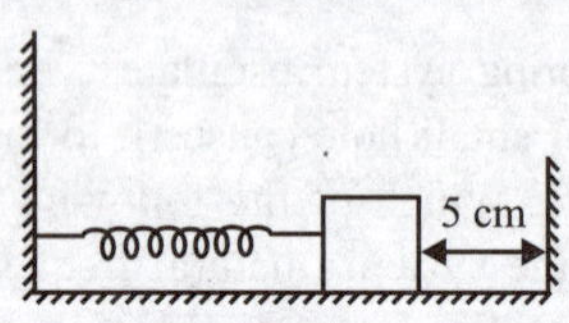

1) 0.2 s 2) 0.1 s
3) 0.15 s 4) 0.132 s

3. A block is resting on a rough horizontal surface, which is executing SHM of amplitude A. The frequency of oscillation of horizontal surface for which the block just starts to slip is

1) Zero 2) ∞

3) $\frac{1}{2\pi}\sqrt{\frac{Kx^2}{m}}$ 4) $\frac{1}{2\pi}\sqrt{\frac{\mu g}{A}}$

4. Two identical particles each of mass m are interconnected by a light spring of stiffness k, the time period for small oscillation is equal to:

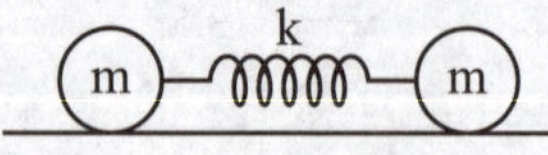

1) $2\pi\sqrt{\frac{m}{k}}$ 2) $\pi\sqrt{\frac{m}{k}}$

3) $2\pi\sqrt{\frac{m}{2k}}$ 4) $\pi\sqrt{\frac{2m}{k}}$

5. A particle of mass *m* is located in a one-dimensional potential field, where potential energy is given by $V(x) = A(1 - \cos px)$, where *A* and *p* are constants. The period of small oscillations of the particle is

1) $2\pi\sqrt{\frac{m}{Ap}}$ 2) $2\pi\sqrt{\frac{m}{Ap^2}}$

3) $2\pi\sqrt{\frac{m}{A}}$ 4) $\frac{1}{2\pi}\sqrt{\frac{Ap}{m}}$

6. A point mass *m* is suspended at the end of a massless wire of length *L* and cross-section area *A*. If *Y* is the Young's modulus for the wire, then the frequency of oscillations for the SHM along the vertical line is

1) $\frac{1}{2\pi}\sqrt{\frac{YA}{mL}}$ 2) $2\pi\sqrt{\frac{mL}{YA}}$

3) $\frac{1}{\pi}\sqrt{\frac{YA}{mL}}$ 4) $\pi\sqrt{\frac{mL}{YA}}$

7. The equation of a damped simple harmonic motion is $m\frac{d^2x}{dt^2}+b\frac{dx}{dt}+kx=0$. Then, the angular frequency (ω) of oscillation is **[RC]**

1) $\omega=\left(\frac{k}{m}-\frac{b^2}{4m^2}\right)^{1/2}$ 2) $\omega=\left(\frac{k}{m}-\frac{b}{4m}\right)^{1/2}$

3) $\omega=\left(\frac{k}{m}-\frac{b^2}{4m}\right)^{1/2}$ 4) $\omega=\left(\frac{k}{m}-\frac{b^2}{4m^2}\right)$

8. A simple pendulum performs simple harmonic motion about $x = 0$ with an amplitude *a* and time period *T*. The speed of the pendulum at $x=\frac{a}{2}$ will be

1) $\frac{\pi a\sqrt{3}}{2T}$ 2) $\frac{\pi a}{T}$

3) $\frac{3\pi^2 a}{T}$ 4) $\frac{\pi a\sqrt{3}}{T}$

9. Two masses m_1 and m_2 are suspended together by a massless spring of force constant *k*, as shown in figure. When the masses are in equilibrium, mass, m_1 is removed without disturbing the system. The angular frequency of oscillation of mass m_2 is

1) $\sqrt{\frac{k}{m_2}}$ 2) $\sqrt{\frac{k}{m_1}}$

3) $\sqrt{\frac{k\,m_1}{m_2^2}}$ 4) $\sqrt{\frac{k\,m_2}{m_1^2}}$

10. A rectangular block of mass m and area of cross-section A floats in a liquid of density ρ. If it is given a small vertical displacement from equilibrium it undergoes oscillation with a time period T. Then,

1) $T \propto \sqrt{\rho}$ 2) $T \propto \frac{1}{\sqrt{A}}$

3) $T \propto \frac{1}{\rho}$ 4) $T \propto \frac{1}{\sqrt{m}}$

11. A body performs S.H.M. The K.E, (K) varies with T as indicated in the graph:

1) 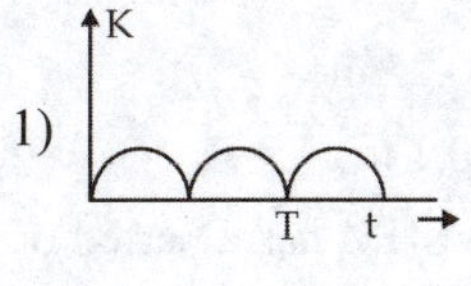

2) 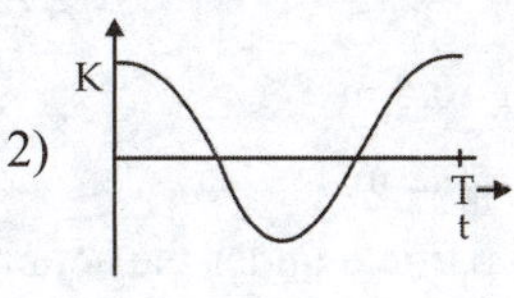

3) 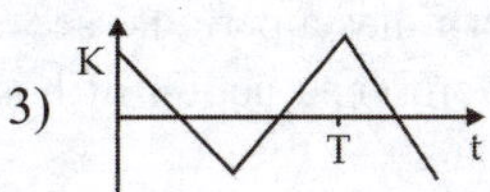

4) 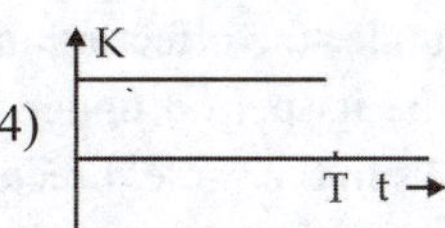

12. The system shown in the figure can move on a smooth surface. The spring is initially compressed by 6 cm and then released. The amplitude of 3 kg block will be

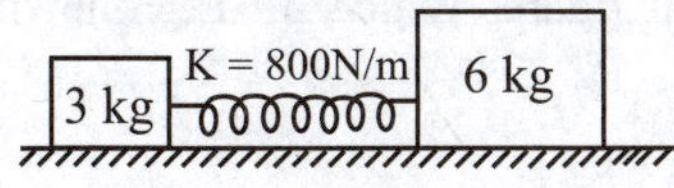

1) 2 cm 2) 4 cm

3) 6 cm 4) None of these

13. A heavy small-sized sphere is suspended by a string of length l. The sphere rotates uniformly in a horizontal circle with the string making an angle θ with the vertical. Then, the time period of this conical pendulum is

1) $t = 2\pi\sqrt{\frac{l}{g}}$ 2) $t = 2\pi\sqrt{\frac{l\sin\theta}{g}}$

3) $t = 2\pi\sqrt{\frac{l\cos\theta}{g}}$ 4) $t = 2\pi\sqrt{\frac{l}{g\cos\theta}}$

14. A mass M is suspended from a spring of negligible mass. The spring is pulled a little and then released so that the mass executes SHM of time period T. If the mass is increased by m, the time period becomes $5T/3$, then the ratio of $\frac{m}{M}$ is

1) $\frac{3}{5}$ 2) $\frac{24}{9}$

3) $\frac{16}{9}$ 4) $\frac{5}{3}$

15. The angular amplitude of a simple pendulum is θ_0. The maximum tension in its string will be

1) $mg(1-\theta_0)$ 2) $mg(1+\theta_0)$

3) $mg(2-3\cos\theta_0)$ 4) $mg(3-2\cos\theta_0)$

16. In the given diagram S_1 and S_2 are identical springs. The frequency of oscillation of the mass m is f. If one of the springs is removed, then frequency will be

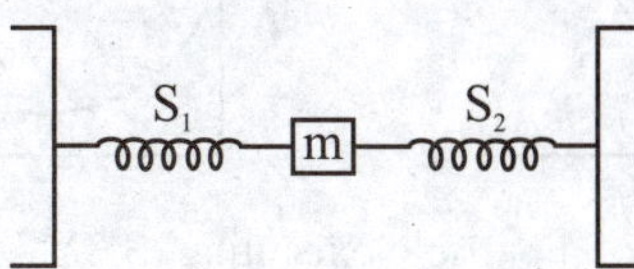

1) f 2) $2f$

3) $\sqrt{2}f$ 4) $\frac{f}{\sqrt{2}}$

17. Time period of a spring mass system is T. If this spring is cut into two parts whose lengths are in the ratio 1 : 3 and same mass is attached to the longer part, the new period will be

1) $\sqrt{\frac{3}{2}}T$ 2) $\frac{T}{\sqrt{3}}$

3) $\frac{\sqrt{3}T}{2}$ 4) $\sqrt{3}T$

18. The displacement of a particle executing SHM is given by:

$$y = 5\sin\left(4t + \frac{\pi}{3}\right)$$

If T is the time period and the mass of the particle is 2 g, the kinetic energy of the particle when $t = \frac{T}{4}$ is given by:

1) 0.4 J 2) 0.5 J

3) 3 J 4) 0.3 J

19. A simple harmonic motion has an amplitude A and time period T. The time required by it to travel from x = A to x = A/2 is:

1) T/6 2) T/4

3) T/3 4) T/2

20. A uniform spring whose unstressed length is l, has a force constant K. The spring is cut into two pieces of unstressed lengths l_1 and l_2, where $l_2 = nl_1$, n being an integer. Now a mass m is made to oscillate with first spring. The time period of its oscillation would be:

1) $T = 2\pi\sqrt{\frac{mn}{K(n+1)}}$ 2) $T = 2\pi\sqrt{\frac{m}{K(n+1)}}$

3) $T = 2\pi\sqrt{\frac{m}{nK}}$ 4) $T = 2\pi\sqrt{\frac{m(n+1)}{Kn}}$

21. A box filled with grains is oscillating vertically on a spring. The grains leak out gradually during the motion. Which of the following graphs best represents the variation of angular frequency $\omega(t)$ with time?

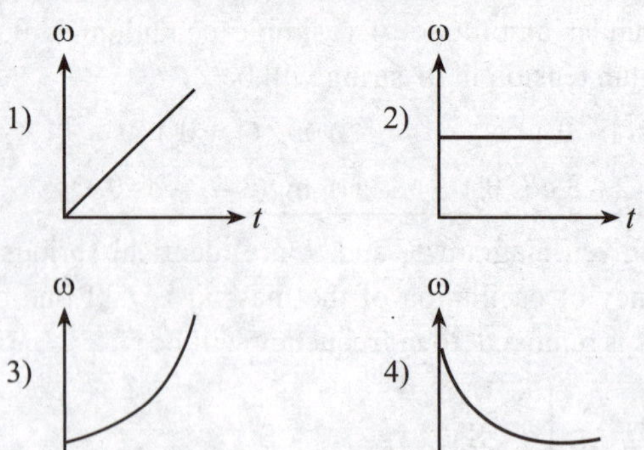

22. The period of a particle executing SHM is 8 s. At t = 0, it is at the mean position. The ratio of the distance covered by the particle in the 1st second to the 2nd second is :

1) $\frac{1}{\sqrt{2}+1}$ 2) $\sqrt{2}$

3) $\frac{1}{\sqrt{2}}$ 4) $\sqrt{2}+1$

23. The time period of a simple pendulum inside a stationary lift is $\sqrt{5}$. What will be the time period when the lift moves upwards with an acceleration $\frac{g}{4}$?

1) $\sqrt{5}$ s 2) $2\sqrt{5}$ s

3) $(2+\sqrt{5}\,s)$ 4) 2 s

24. Five identical springs are used in the three configurations as shown in figure. The time periods of vertical oscillations in configurations (a), (b) and (c) are in the ratio :

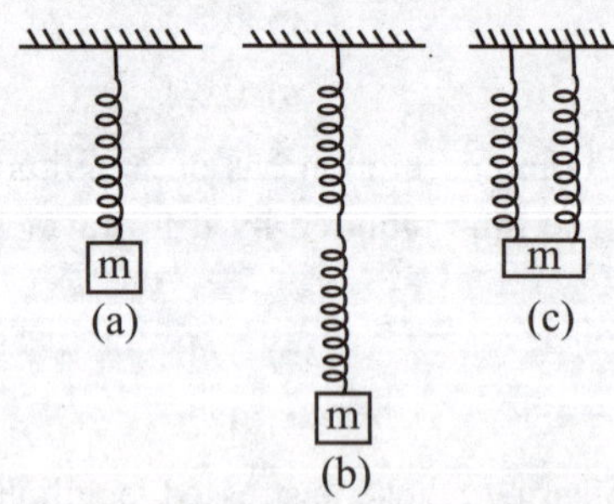

1) $1:\sqrt{2}:\frac{1}{\sqrt{2}}$ 2) $2:\sqrt{2}:\frac{1}{\sqrt{2}}$

3) $\frac{1}{\sqrt{2}}:2:1$ 4) $2:\frac{1}{\sqrt{2}}:1$

25. When the displacement of a particle executing SHM is one-fourth of its amplitude, what fraction of the total energy is the kinetic energy?

1) $\frac{16}{15}$ 2) $\frac{15}{16}$

3) $\frac{3}{4}$ 4) $\frac{4}{3}$

26. A mass M = 5 kg is attached to a spring as shown in the figure and held in position so that the spring remains unstretched. The spring constant is 200N/m. The mass M is then released and begins to undergo small oscillations. The amplitude of oscillation is:

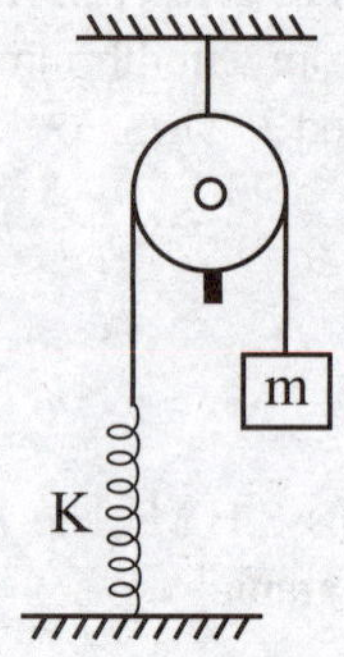

1) .0.5 m 2) 0.25 m

3) 0.2 m 4) 0.1 m

27. A particle under the action of a SHM has a period of 3 sec and under the effect of another it has a period 4 sec. What will be its period under the combined action of both the SHM's in the same direction?

1) 7 sec 2) 5 sec

3) 2.4 sec 4) 0.4 sec

28. The displacement-time graph of a particle executing SHM is shown in figure. Which of the following statement is true?

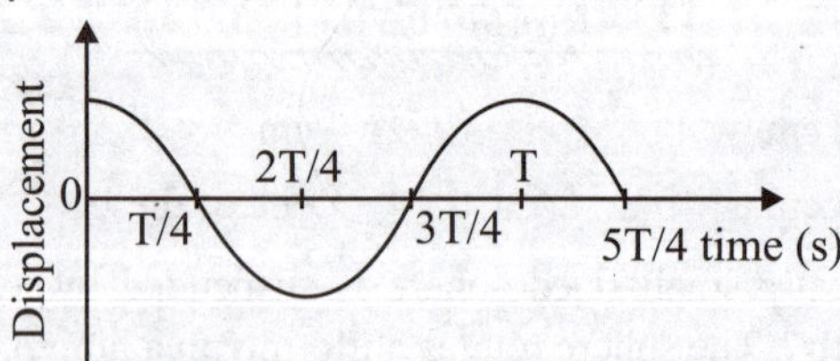

1) The force is zero at $t=\frac{3T}{4}$

2) The acceleration is maximum at $t=\frac{4T}{4}$

3) The velocity is maximum at $t=\frac{T}{4}$

4) All of these

Statement Based MCQs

Directions: These questions consist of two statements each, printed as Statement-I and Statement-II. While answering these questions, you are required to choose any one of the following four responses.

1) **Both Statement-I and Statement-II are correct.**
2) **Both Statement-I and Statement-II are incorrect.**
3) **Statement-I is correct & Statement-II is incorrect.**
4) **Statement-I is incorrect & Statement-II is correct.**

1. **Statement-I:** For an oscillating simple pendulum, the tension in the string is maximum at the mean position and minimum at the extreme position.

   Statement-II: The velocity of oscillating bob in simple harmonic motion is maximum at the mean position.

2. **Statement-I:** The graph of total energy of a particle in S.H.M is a straight line.

   Statement-II: Total energy of a particle in S.H.M. remains constant.

3. **Statement-I:** Acceleration of a particle executing S.H.M. at mean position is zero.

   Statement-II: The frequency of a second pendulum does not depend upon acceleration due to gravity.

4. **Statement-I:** The restoring force acting on the pendulum is towards the extreme position.

   Statement-II: The total energy is directly proportional to the amplitude of vibration of the harmonic oscillator.

Assertion & Reason MCQs

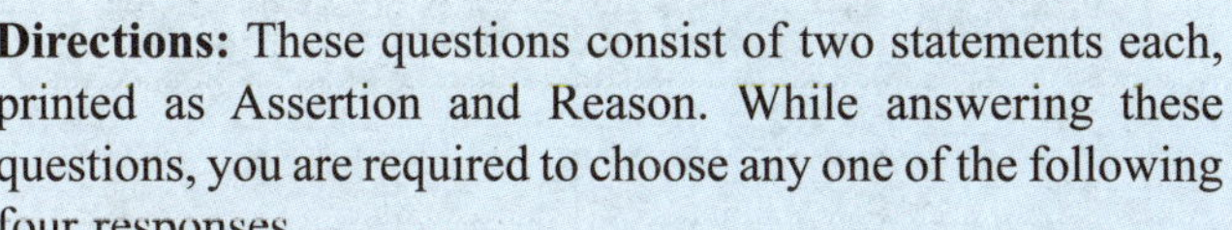

Directions: These questions consist of two statements each, printed as Assertion and Reason. While answering these questions, you are required to choose any one of the following four responses.

1) **Both Assertion and Reason are True and the Reason is a correct explanation of the Assertion.**
2) **Both Assertion and Reason are True but Reason is not a correct explanation of the Assertion.**
3) **Assertion is True but the Reason is False.**
4) **Assertion is False but Reason is True.**

1. **Assertion:** Damped oscillation indicates loss of energy.

   Reason: The energy loss in damped oscillation may be due to friction, air resistance etc. *[RC]*

2. **Assertion:** In Simple harmonic motion, acceleration is always directed towards the mean position.

   Reason: In Simple harmonic motion, the body stops momentarily at the extreme position and moves back to mean position.

3. **Assertion:** In a S.H.M, kinetic and potential energies become equal when the displacement is $1/\sqrt{2}$ times the amplitude.

   Reason: In S.H.M, kinetic energy is zero when potential energy is maximum.

4. **Assertion:** Amplitude of the particle is minimum at resonance.

   Reason: If a hole were drilled through the centre of earth and a ball dropped into the hole at one end, it will execute S.H.M. *[RC]*

5. **Assertion:** The motion of the earth around the sun is periodic but not oscillatory.

   Reason: Oscillatory motion is necessarily periodic but periodic motion is not necessarily oscillatory.

Match the Columns MCQs

1. Match the following:

	Column-I		Column-II
(a)	A constant force acting along the line of SHM affects	(i)	the time period
(b)	A constant torque acting along the arc of angular SHM affects	(ii)	the frequency
(c)	A particle falling on the block executing SHM when the later crosses the mean possition affects	(iii)	the mean position
(d)	A particle executing SHM when kept on a uniformly accelerated car affects	(iv)	the amplitude

1) (a)-(i,ii,iv) (b)-(iv,i) (c)-(iii,ii) (d)-(i,iv)
2) (a)-(iii,iv) (b)-(iii,iv) (c)-(i,ii,iv) (d)-(iii,iv)
3) (a)-(ii,i) (b)-(iii,i) (c)-(iv,ii) (d)-(iii,i,iv)
4) (a)-(iii,i) (b)-(ii,iii) (c)-(iv,ii) (d)-(ii,i)

2. In simple harmonic motion, match the following graphs:

	Column-I		Column-II
(a)	Position (y) vs. time (x)	(i)	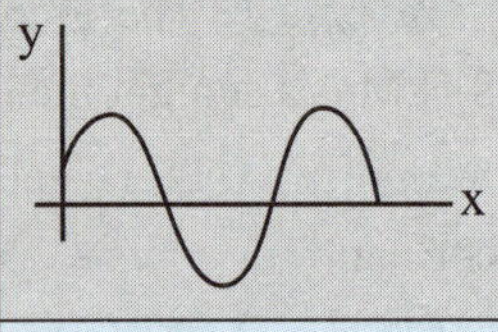
(b)	Velocity (y) vs. time (x)	(ii)	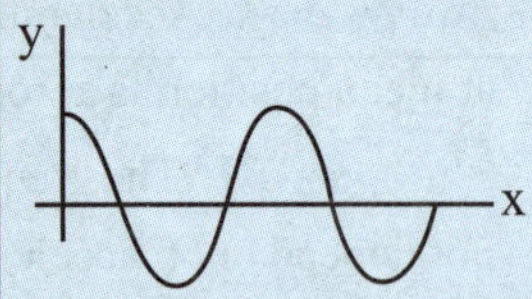
(c)	Potential energy (y) vs. time (x)	(iii)	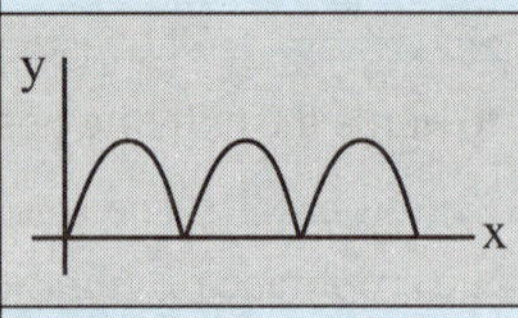
(d)	Total energy (y) vs. time (x)	(iv)	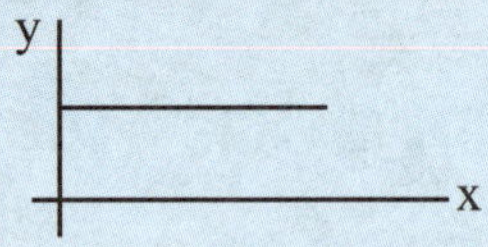

1) (a)-(i,ii) (b)-(i,ii) (c)-(iii) (d)-(iv)
2) (a)-(iv) (b)-(i,ii) (c)-(iii) (d)-(i,ii)
3) (a)-(iv,i) (b)-(iii,ii) (c)-(i) (d)-(ii)
4) (a)-(iii) (b)-(i,iv) (c)-(ii) (d)-(iv)

3. Match the following:

	Column-I		Column-II
(a)	Linear combination of two SHMs	(i)	$T = 2\pi\sqrt{\frac{R}{g}}$ (R is radius of the earth)
(b)	$y = A \sin\omega_1 t + A \sin(\omega_2 t + \phi)$	(ii)	SHM for equal frequencies and amplitude
(c)	Time period of a pendulum of infinite length	(iii)	Superposition may not always be an SHM
(d)	Maximum value of time period of an oscillating pendulum	(iv)	Amplitude will be $\sqrt{2}A$ for $\omega_1 = \omega_2$ and phase difference of $\pi/2$

1) (a)-(i,ii) (b)-(i,ii) (c)-(iii) (d)-(iv)
2) (a)-(iv) (b)-(i,ii) (c)-(iii) (d)-(i,ii)
3) (a)-(iv,i) (b)-(iii,ii) (c)-(i) (d)-(ii)
4) (a)-(ii,iii) (b)-(ii,iii,iv) (c)-(i) (d)-(i)

4. Match the terms in column-I with the most suitable options in column-II.

	Column-I		Column-II
(a)	$\sin\omega t - \cos\omega t$	(i)	SHM with time period $\frac{\pi}{\omega}$
(b)	$\cos\left(\frac{\pi}{4} - 2\omega t\right)$	(ii)	SHM with time period $\frac{2\pi}{\omega}$
(c)	$1 + \omega t + \omega^3 t^3$	(iii)	Periodic but not SHM
(d)	$\sin^3 \omega t$	(iv)	Non periodic

1) (a)-(i) (b)-(iii) (c)-(iv) (d)-(ii)
2) (a)-(iii) (b)-(iv) (c)-(ii) (d)-(i)
3) (a)-(ii) (b)-(i) (c)-(iv) (d)-(iii)
4) (a)-(i) (b)-(ii) (c)-(iv) (d)-(iii)

5. Column-I shows the different physical situations and lists column-II lists their time periods. Match the situations in column-I to their time periods in column-II.

	Column-I		Column-II
(a)	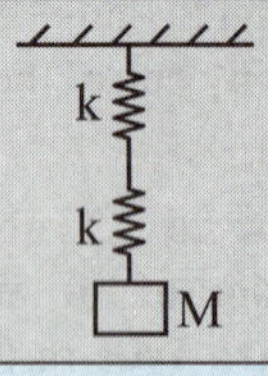 k, k, M	(i)	$T = 2\pi\sqrt{\frac{M}{2k}}$
(b)	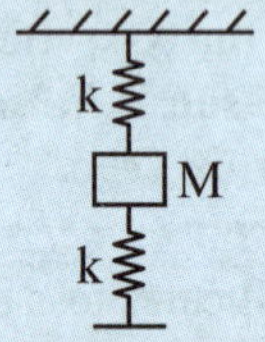 k, M, k	(ii)	$T = 2\pi\sqrt{\frac{M}{k}}$
(c)	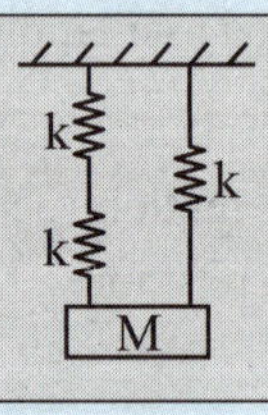 k, k, k, M	(iii)	$T = 2\pi\sqrt{\frac{2M}{3k}}$
(d)	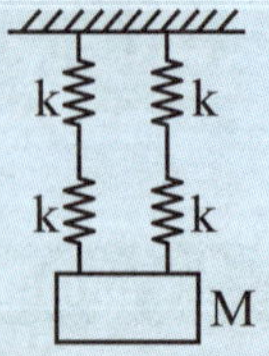 k, k, k, k, M	(iv)	$T = 2\pi\sqrt{\frac{2M}{k}}$

1) (a)-(iv) (b)-(i) (c)-(ii) (d)-(iii)
2) (a)-(iv) (b)-(i) (c)-(iii) (d)-(ii)
3) (a)-(i) (b)-(ii) (c)-(iii) (d)-(iv)
4) (a)-(iii) (b)-(ii) (c)-(i) (d)-(iv)

NCERT Exemplar MCQs

1. The displacement of a particle is represented by the equation $y = 3\cos\left(\frac{\pi}{4} - 2\omega t\right)$. The motion of the particle is:
   1) Simple harmonic with period $2\pi/\omega$
   2) Simple harmonic with period π/ω
   3) Periodic but not simple harmonic
   4) Non-periodic
2. The displacement of a particle is represented by the equation $y = \sin^3\omega t$. The motion is:
   1) Non-periodic
   2) Periodic but not simple harmonic
   3) Simple harmonic with period $2\pi/\omega$
   4) Simple harmonic with period π/ω
3. The relation between acceleration and displacement of four particles are given below:
   1) $a_x = +2x$
   2) $a_x = +2x^2$
   3) $a_x = -2x^2$
   4) $a_x = -2x$

   Which, one of the particle is exempting simple harmonic motion?
4. Motion of an oscillating liquid column in a U-tube is:
   1) Periodic but not simple harmonic
   2) Non-periodic
   3) Simple harmonic and time period is independent of the density of the liquid
   4) Simple harmonic and time period is directly proportional to the density of the liquid
5. A particle is acted simultaneously by mutually perpendicular simple harmonic motion $x = a\cos\omega t$ and $y = a\sin\omega t$. The trajectory of motion of the particle will be:
   1) An ellipse
   2) A parabola
   3) A circle
   4) A straight line
6. The displacement of a particle varies with time according to the relation $y = a\sin\omega t + b\cos\omega t$:
   1) The motion is oscillatory but not SHM
   2) The motion is SHM with amplitude $a + b$
   3) The motion is SHM with amplitude $a^2 + b^2$
   4) The motion is SHM with amplitude $\sqrt{a^2 + b^2}$
7. Four pendulums A, B, C and D are suspended from the same elastic support as shown in figure. A and C are of the same length, while B is smaller than A and D is larger than A. If A is given a transverse displacement:

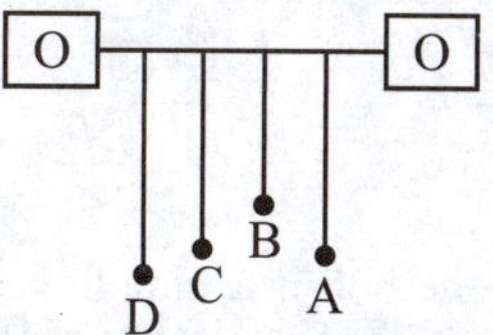

   1) D will vibrate with maximum amplitude
   2) C will vibrate with maximum amplitude
   3) B will vibrate with maximum amplitude
   4) All the four will oscillate with equal amplitude
8. Figure shows the circular motion of a particle. The radius of the circle, the period, sense of revolution and the initial position are indicated on the figure. The simple harmonic motion of the x-projection of the radius vector of the rotating particle P is:

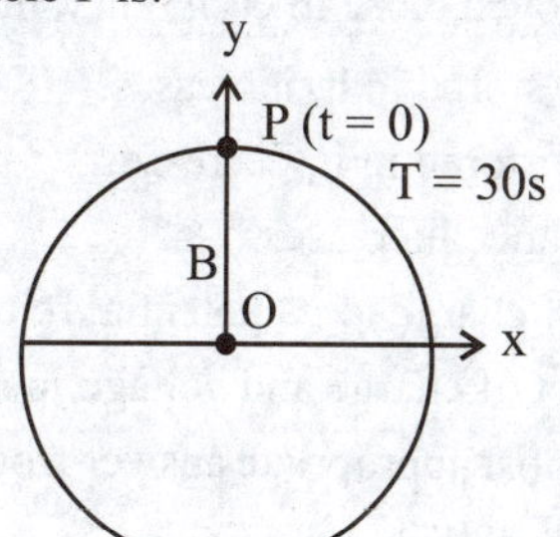

   1) $x(t) = B\sin\left(\frac{2\pi t}{30}\right)$
   2) $x(t) = B\cos\left(\frac{\pi t}{15}\right)$
   3) $x(t) = B\sin\left(\frac{\pi t}{15} + \frac{\pi}{2}\right)$
   4) $x(t) = B\cos\left(\frac{\pi t}{15} + \frac{\pi}{2}\right)$
9. The equation of motion of a particle is $x = a\cos(\alpha t)^2$. The motion is:
   1) Periodic but not oscillatory
   2) Periodic and oscillatory
   3) Oscillatory but not periodic
   4) Neither periodic nor oscillatory
10. A particle executing SHM has a maximum speed of 30 cm/s and a maximum acceleration of 60 cm/s². The period of oscillation is:
    1) π sec
    2) $\frac{\pi}{2}$ sec
    3) 2π sec
    4) $\frac{\pi}{t}$ sec

NEET & AIIMS Past Year MCQs

1. A particle executing simple harmonic motion with amplitude A has the same potential and kinetic energies at the displacement **(2024 Re)**

1) $2\sqrt{A}$ 2) $\frac{A}{2}$

3) $\frac{A}{\sqrt{2}}$ 4) $A\sqrt{2}$

2. If the mass of the bob in a simple pendulum is increased to thrice its original mass and its length is made half its original length, then the new time period of oscillation is $\frac{x}{2}$ time its original time period. Then the value of x is: **2024)**

1) $2\sqrt{3}$ 2) 4

3) $\sqrt{3}$ 4) $\sqrt{2}$

3. If the plate of a parallel plate capacitor connected to a battery are moved close to each other, then **(2024)**

A. the charge stored in it. increases.
B. the energy stored in it. decreases.
C. its capacitance increases.
D. the ratio of charge to its potential remains the same.
E. the product of charge and voltage increases.

Choose the most appropriate answer from the given below:

1) B. D and E only
2) A, B and C only
3) A, B and E only
4) A, C and E only

4. The x–t graph of a particle performing simple harmonic motion is shown in the figure. The acceleration of the particle at t = 2 s is: **(2023)**

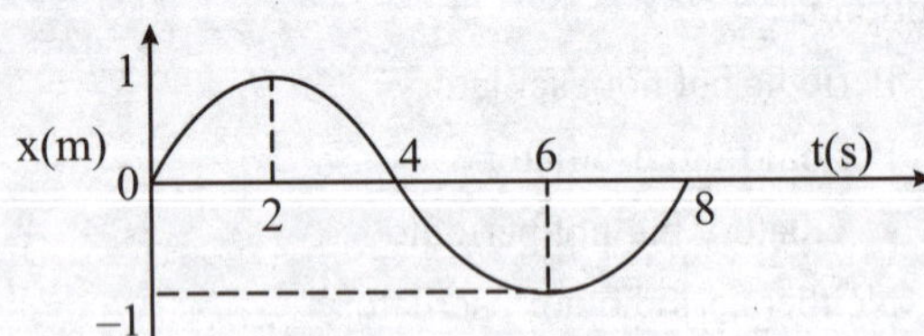

1) $-\frac{\pi^2}{16}ms^{-2}$ 2) $\frac{\pi^2}{8}ms^{-2}$

3) $-\frac{\pi^2}{8}ms^{-2}$ 4) $\frac{\pi^2}{16}ms^{-2}$

5. Two pendulums of length 121 cm and 100 cm start vibrating in phase. At some instant, the two are at their means position in the same phase. The minimum number of virbrations of the shorter pendulum after which the two are again in phase at the means position is: **(2022)**

1) 8 2) 11 3) 9 4) 10

6. A body is executing simple harmonic motion with frequency 'n', the frequency of its potential energy is: **(2021)**

1) 2n 2) 3n 3) 4n 4) n

7. A spring is stretched by 5 cm by a force 10 N. The time period of the oscillations when a mass of 2 kg is suspended by it is: **(2021)**

1) 6.28 s 2) 3.14 s

3) 0.628 s 4) 0.0628 s

8. A particle performing SHM with angular frequency ω = 5000 radian/second and amplitude A = 2 cm and mass of 1 kg. Find the total energy of oscillation. **(AIIMS 2018)**

1) 2 kJ 2) 5 kJ

3) 7 kJ 4) 15 kJ

9. The angular amplitude of a simple pendulum is θ_0. The maximum tension in its string will be: **(AIIMS 2016)**

1) $mg(1-\theta_0)$ 2) $mg(1+\theta_0)$

3) $mg(1-\theta_0^2)$ 4) $mg(1+\theta_0^2)$

10. A mass M is suspended from a spring of negligible mass. The spring is pulled a little and then released so that the mass executes simple harmonic oscillations with a time period T. If the mass is increased by m, then the time period becomes $\frac{5}{4}T$. The ratio of $\frac{m}{M}$ is: **(AIIMS 2016)**

1) 9/16 2) 5/4

3) 25/16 4) 4/5

11. In simple harmonic motion, loss of K.E. is proportional to: **(AIIMS 2011)**

1) e^x 2) x^3

3) $\log x$ 4) x^2

12. Which of the following functions represents a simple harmonic oscillation: **(AIIMS 2005)**

1) $\sin \omega t - \cos \omega t$ 2) $\sin^2 \omega t$

3) $\sin \omega t + \cos 2\omega t$ 4) $\sin \omega t - \cos 2\omega t$

Answer Key

NCERT Topic-wise MCQs

1	2	3	4	5	6	7	8	9	10	11	12	13	14	15	16	17
4	4	2	3	2	3	3	1	4	1	4	2	1	3	2	3	4
18	**19**	**20**	**21**	**22**	**23**	**24**	**25**	**26**	**27**	**28**	**29**	**30**	**31**	**32**	**33**	**34**
4	4	1	4	1	3	2	2	1	1	2	3	3	3	3	4	2
35	**36**	**37**	**38**	**39**	**40**	**41**	**42**	**43**	**44**							
4	2	2	3	1	3	3	2	1	4							

High Order Time Intensive MCQs

1	2	3	4	5	6	7	8	9	10	11	12	13	14	15	16	17
4	4	4	4	2	1	1	4	1	2	1	2	3	3	4	4	3
18	**19**	**20**	**21**	**22**	**23**	**24**	**25**	**26**	**27**	**28**						
4	1	2	3	4	4	1	2	2	3	4						

Statement Based MCQs

1	2	3	4
1	1	3	2

Assertion & Reason MCQs

1	2	3	4	5
1	2	2	4	2

Match the Column MCQs

1	2	3	4	5
2	1	4	3	2

NCERT Exemplar MCQs

1	2	3	4	5	6	7	8	9	10
2	2	4	3	3	4	2	1	3	1

NEET & AIIMS Past Year MCQs

1	2	3	4	5	6	7	8	9	10	11	12
3	4	4	1	2	1	3	2	4	1	4	1

CHAPTER

14 Waves

Topicwise Number of Questions Analysis (2025-2016)

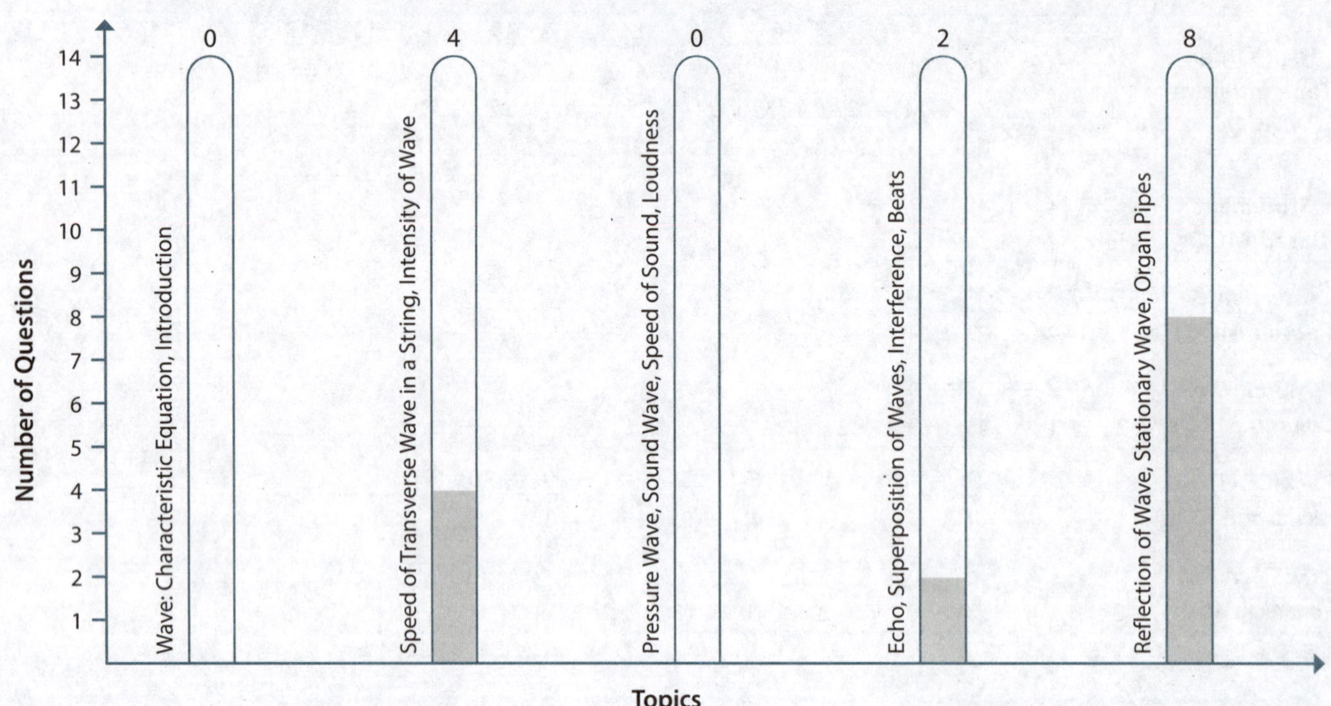

NCERT-PICKS

Equation of Waves

Formulas

1. If A is amplitude, k is wave number and t is the time, then Equation of a plane progressive wave is
$y = A \sin(kx - \omega t)$
2. If λ is the wavelength, then $\lambda = \frac{2\pi}{k}$
3. If ν is the frequency, then $\nu = \frac{\omega}{2\pi}$
4. Phase difference between two points on a wave $\Delta\phi = \frac{2\pi}{\lambda}$ (Path difference).
5. (a) For a wave travelling along positive X-axis,
$v_{particle} = -v_{Wave}\left(\frac{dy}{dx}\right)$
(b) For a wave travelling along negative X-axis,
$v_{particle} = v_{wave}\left(\frac{dy}{dx}\right)$
6. Wave velocity $v = \nu\lambda$
7. Particle velocity $v = \omega\sqrt{A^2 - x^2}$, x is displacement of the wave.

Numerical Profile

(i) **Superposition of waves:** If displacement of two waves reaching at a point
$x_1 = a_1 \sin \omega t$
$x_2 = a_2 \sin(\omega t + \phi)$
Then Resultant amplitude $A = \sqrt{a_1^2 + a_2^2 + 2a_1a_2\cos\theta}$
and Resultant intensity $I = I_1 + I_2 + 2\sqrt{I_1I_2}\cos\theta$.

(ii) When a wave is reflected back from a free end and there is no phase change.

y_1 = Asin(ωt – kx) is the incident wave propagating along + x-direction and y_2 = Asin(ωt + kx) is the reflected wave along –x direction.

(iii) When a wave is reflected from a rigid surface, a phase difference of π is introduced in the reflected wave. If y_1 = Asin(ωt – kx) is the wave propagating along + ve 'x' direction

The reflected wave is $y_2 = A\sin[(\omega t + kx) + \pi]$

Equation of Stationary Wave

(i) If the wave is reflected at free boundary

y = 2A cos kx sin ωt (where 2 A cos kx is known as the amplitude of the stationary wave)

(ii) If the wave is reflected at fixed boundary

y = 2A sin kx cos ωt (where 2 A sin kx is the amplitude)

Speed of Wave in a String and Sound Waves

Formulas

1. Velocity of wave propagating through a string, $v = \sqrt{\frac{T}{\mu}}$

[μ is mass per unit length of the string, T is the tension in the string]

Note: T remains unchanged as the part of the string vibrates up and down this is valid only for small amplitude.

2. Velocity of sound through a medium $v = \sqrt{\frac{E}{\rho}}$,

E = modulus of elasticity and ρ = density.

In solids, e is repleced by Y = young's modulus.

As modulus of elasticity is more for solids,

So $v_{solids} > v_{liquids} > v_{gases}$.

3. Speed of sound in gases, $v = \sqrt{\frac{\gamma P}{\rho}}$

$\gamma = \frac{C_p}{C_v}$ (ratio of two specific heats)

[Where ρ is density of the gas and P is the pressure]

4. Factors affecting velocity of sound v

a. Density,

$v \propto \frac{1}{\sqrt{\rho}}$ [Here, ρ shown density]

b. Pressure,

$v \propto P^0$ i.e., velocity of sound is indepenedent of the change in pressure provided temperature is constant.

c. Temperature,

$v \propto \sqrt{T}$ [Here, T is the temperature]

d. Humidity,

Velocity of sound increase with increase in humidity.

$\rho_m < \rho_d \Rightarrow v_m > v_d$ { m = moist air, d = dry air}

Vibration of Stretched Strings

When sonometer wire vibrates in 1 loop, then its frequency of vibration is called I[st] harmonic (n_1) or fundamental note $\lambda_1 = 2\ell$;

$$n_1 = \frac{1}{2\ell}\sqrt{\frac{T}{\mu}} \qquad \left[\because \mu = \frac{m}{\ell}\right]$$

Formulas

1. When the sonometer wire vibrates in 2 loops then its frequency of vibration is called 2nd harmonic (n_2) or 1st overtone

$$\lambda_2 = \frac{2\ell}{2};\ n_2 = 2\left(\frac{1}{2\ell}\right)\sqrt{\frac{T}{\mu}} = 2n_1$$

2. When the sonometer wire vibrates in 3 loops, then its frequency of vibration is called 3rd harmonic (n_3) or 2nd overtone.

$$\lambda_3 = \frac{2\ell}{3};\ n_3 = 3\left(\frac{1}{2\ell}\right)\sqrt{\frac{T}{\mu}} = 3n_1$$

[λ is the wavelength, ℓ is the length of the wire and T is the tension]

3. When sonometer wire is vibrating in 'x' loops then its frequency of vibration is called x[th] harmonic or (x – 1)[th] overtone.

$$\lambda_x = \frac{2\ell}{x};\ n_x = x\left(\frac{1}{2\ell}\right)\sqrt{\frac{T}{\mu}} = xn_1$$

Numerical Profile

Fundamental frequency

(i) $n_1 = \frac{1}{2\ell}\sqrt{\frac{T}{\mu}} = \frac{1}{2\ell}\sqrt{\frac{T}{A\rho}} = \frac{1}{2\ell}\sqrt{\frac{T}{\pi r^2\rho}} = \frac{1}{\ell d}\sqrt{\frac{T}{\pi\rho}}$, d is the diameter of wire

Laws of transverse vibrations of strings.

(i) Law of length: $n \propto \frac{1}{l}$

$$\frac{n_1}{n_2} = \frac{l_2}{l_1}$$

(ii) Law of tension: $n \propto \sqrt{T}$

$$\frac{n_1}{n_2} = \sqrt{\frac{T_1}{T_2}}$$

(iii) Law of linear density: $n \propto \frac{1}{\sqrt{\mu}}$

$$\frac{n_1}{n_2} = \sqrt{\frac{\mu_2}{\mu_1}}$$

Organ Pipes

Formulas

1. Closed organ pipe

a. Fundamental note $n_1 = \frac{v}{4L}$

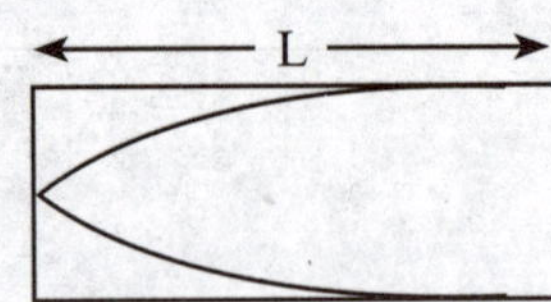

b. First overtone or 3rd harmonic, $n_2 = 3n_1 = \frac{3v}{4L}$

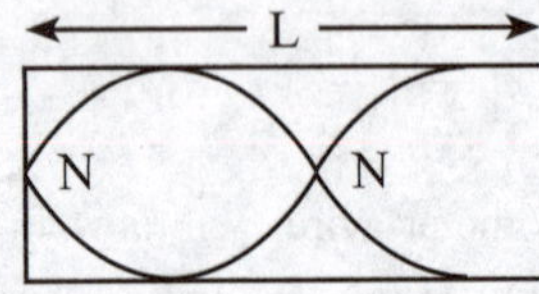

c. Second overtone or 5Th harmonic, $n_3 = 5n_1 = \frac{5v}{4L}$

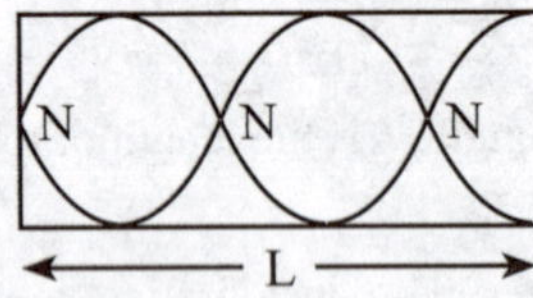

2. Open organ pipe

a. Fundamental note $n_1 = \frac{v}{2L}$

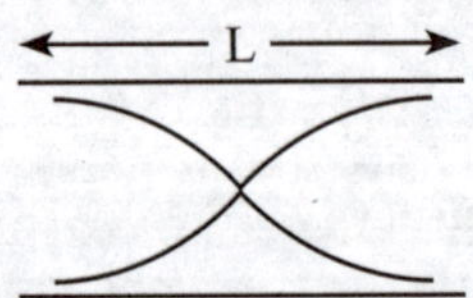

b. First overtone or 2nd harmonic, $n_2 = 2n_1 = \frac{v}{L}$

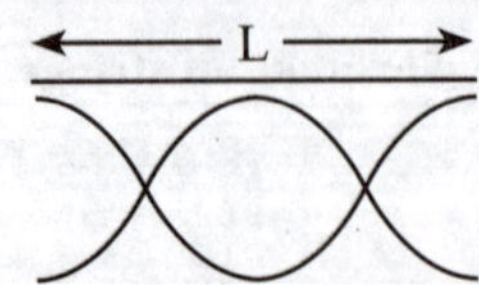

c. Second overtone or 3rd harmonic, $n_3 = 3n_1 = \frac{3v}{2L}$

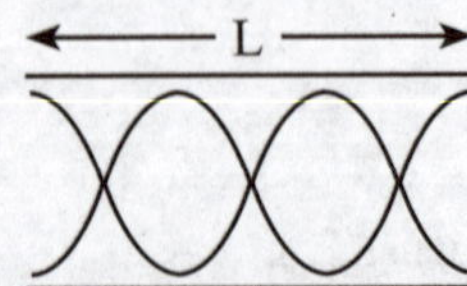

[n_1 is fundamental frequency and v is speed of the wave]

Numerical Profile

Organ Pipes

(i) For open organ pipe, the harmonics excitable in the tube are all integral multiples of its fundamental frequency.

$\therefore \lambda = \frac{2L}{n}$, where n = 1, 2, ...

Frequency $= \frac{v}{\lambda} = \frac{nv}{2L}$, where n = 1, 2, ...

(ii) For closed organ pipe

$\lambda = \frac{4L}{(2n+1)}$, where n = 0, 1, 2, ...

Frequency $= \frac{(2n+1)v}{4L}$, where n = 0, 1, 2, ...

(iii) Beat frequency $= n_1 - n_2$, where n_1 and n_2 are the frequencies of two sound waves.

(iv) The time interval between a minimum and immediate maximum is $\frac{1}{2(n_1 - n_2)}$.

Statements Asked in Exam

- A material medium is a must for propagation of waves. It should possess the properties of inertia and elasticity.
- Transverse wave motion that travels in the form crests and troughs
- Longitudinal wave motion that travels in the form of compressions and rarefactions.
- Power per unit area at a distance r from the source is $\frac{p}{4\pi r^2}$. Thus $I \propto \frac{1}{r^2}$.
- Speed of longitudinal waves in a long solid rod is

$$v = \sqrt{\frac{Y}{\rho}}$$

Where, Y is Young's modulus of the material of solid rod and ρ is density of the material. The speed of longitudinal waves in a liquid is given by $v = \sqrt{\frac{K}{\rho}}$, where K is bulk modulus of elasticity of the liquid.

- The expression for speed of longitudinal waves in a gas, as suggested by Newton and modified later by Laplace is

$$v = \sqrt{\frac{B_a}{\rho}} = \sqrt{\frac{\gamma P}{\rho}}$$

Where B_a is coefficient of adiabatic elasticity of the gas $B_a = \gamma P$, where P is the pressure exerted by the gas and $\gamma = \frac{C_P}{C_V}$

- The speed of transverse waves over a string is given by $v = \sqrt{\frac{T}{m}}$ where T is tension in the string and m is mass of unit length of the string.
- Equation of plane progressive waves travelling with a velocity v along positive directions of X-axis is $y = a \sin \frac{2\pi}{\lambda}(vt - x)$
here λ is wave length of the wave, a is amplitude of particle, and x is the distance from the origin.
- On a string transverse stationary waves are formed due to superposition of direct and the reflected transverse waves.
In the 1st normal mode of vibration, $\lambda_1 = 2L$
$$\nu_1 = \frac{v}{\lambda_1} = \frac{v}{2L}$$
In the 2nd mode of vibration, $\lambda_2 = L$
$$\therefore \nu_2 = \frac{v}{\lambda_2} = \frac{v}{L} = 2\nu_1$$
[This note is called 2nd harmonic.]
In the 3rd normal mode of vibration $\lambda_3 = \frac{2L}{3}$
$$\nu_3 = \frac{v}{\lambda_3} = \frac{v}{2L/3} = \frac{3v}{2L} = 3\nu_1$$
[This note is called 3rd harmonic.]
- The wavelength of nth mode of vibration of a stretched string is
$\lambda_n = \frac{2L}{n}$ and its frequency $\nu_n = n\nu_1$
[This note is called nth harmonic.]
- Node are the points, where amplitude of vibration is zero. In the nth mode of vibration, there are (n + 1) nodes located at distances (from one end)
$$x = 0, \frac{L}{n}, \frac{2L}{n} \ldots L$$
- Antinodes are the point, where amplitude of vibration is maximum. In the nth mode of vibration, there are n antinodes, located at distances (from one end)
$$x = \frac{L}{2n}, \frac{3L}{2n}, \frac{5L}{2n}, \ldots \frac{(2n-1)}{2n}L.$$
- Distance between successive nodes or antinodes = $\frac{\lambda}{2}$
- Distance between adjacent nodes or antinodes = $\frac{\lambda}{4}$
- Beats: Beats is the phenomenon of regular variation in the intensity of sound with time when two sources of nearly equal frequencies are sounded together. One minimum of sound followed by one maximum constitute one beat.
- If ν_1 and ν_2 are the frequencies of two sources producing beats, then time interval between two-successive maxima $= \frac{1}{\nu_1 - \nu_2}$
- Number of beats/sec = $(\nu_1 - \nu_2)$ = difference in frequencies of two sources.
- For hearing distinct beats, difference in frequencies of two sources producing beats should be equal to or less than 10.
- If m is number of beats/sec, ν_1 is known frequency of one source, then the unknown frequency ν_2 of other sources will be $\nu_2 = \nu_1 \pm m$.
The plus/minus sign of m is decided either by loading or by filing one of the sources.
Note: Loading increasing the mass of the source and filing decreases the mass of the source.

NCERT Topic-wise MCQs

Wave: Introduction Characteristic Equation

1. When a wave undergoes refraction:
   1) Frequency changes
   2) Amplitude changes
   3) Velocity unchanged
   4) Both amplitude and frequency remains unchanged

2. If the equation of a progressive wave is given by:
   $$y = 4\sin\pi\left[\frac{t}{5} - \frac{x}{9} + \frac{\pi}{6}\right]$$
   where y is in cm
   then, which of the following is correct?
   1) v = 7 cm/sec 2) $\lambda = 9$ m
   3) A = 4 cm 4) f = 20 Hz

3. A boat at anchor is rocked by waves of velocity 25 m/s having crests 100 m apart. They reach the boat once every
   1) 4 sec 2) 8 sec
   3) 2 sec 4) 0.25 sec

4. The displacement y of a particle in a medium can be expressed as $y = 10^{-6}\sin\left[100t + 20x + \frac{\pi}{4}\right]$. The speed of wave is
   1) 2000m/s 2) 5m/s
   3) 20 m/s 4) 5πm/s

5. Which of the following equations represents a wave travelling along y-axis?
   1) $y = A\sin(kx + \omega t)$
   2) $y = A\sin ky\cos\pi t$
   3) $x = A\sin(ky - \omega t)$
   4) $y = A\cos kx\sin\omega t$

6. A progressive wave moving along x axis is represented by $y = A\sin\left[\frac{2\pi}{\lambda}(vt - x)\right]$. The wavelength at which the maximum particle velocity is 3 times the wave velocity is
   1) $\frac{A}{3}$ 2) $\frac{2A}{3\pi}$
   3) $\frac{3}{4}\pi A$ 4) $\frac{2}{3}\pi A$

7. The equation of a simple harmonic progressive wave is given by $y = A\sin(100\pi t - 3x)$. Find the distance between 2 particles having a phase difference of $\frac{\pi}{3}$.
   1) $\frac{\pi}{9}$ 2) $\frac{\pi}{18}$
   3) $\frac{\pi}{6}$ 4) $\frac{\pi}{3}$

8. In sine wave, minimum distance between 2 particles which always have same speed is
   1) $\frac{\lambda}{2}$ 2) $\frac{\lambda}{4}$
   3) $\frac{\lambda}{3}$ 4) λ

9. A wave motion is described by $y(x, t) = A\sin(Kx - \omega t)$ then, the ratio of the maximum particle velocity to the wave velocity is
   1) KA 2) $\frac{\omega}{K}$
   3) $\frac{d\omega}{K}$ 4) $\frac{x}{1}$

10. The equation of a wave travelling on a string $y = 4\sin\left[\frac{\pi}{2}\left(8t - \frac{x}{8}\right)\right]$ the velocity of the wave is
    1) 64 cm/s in – x direction
    2) 32 cm/s in –x direction
    3) 32 cm/s in + x direction
    4) 64 cm/s in + x direction

11. If the phase difference between two sound waves of wavelength λ is 60°, the corresponding path difference is
    1) $\frac{\lambda}{6}$ 2) $\frac{\lambda}{2}$
    3) 2λ 4) $\frac{\lambda}{4}$

12. The phase difference between two corks A and B on the water wave is:

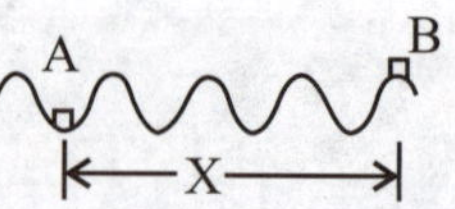

    1) Zero 2) 3π
    3) 6π 4) 7π

13. An observer standing at the seacoast observes 54 waves reaching the coast per minute. If the wavelength of the waves is 10 m, its velocity is -

1) 90 m/s 2) 90 cm/s
3) 9 m/s 4) 900 m/s

14. Two corks are 10 m apart in a lake. Each goes up and down with period 0.5s. It is observed that when one is at its highest point, other one is at lowest point. The possible speed of wave is :

1) 2.5 m/s 2) 5 m/s
3) 40 m/s 4) 4 m/s

Speed of Transverse Waves in a String, Intensity of Wave

15. A metallic wire of 1 m length has a mass of 10×10^{-3} kg If a tension of 100 N is applied to a wire, what is the speed of transverse wave?

1) 100 m/s 2) 10 m/s
3) 200 m/s 4) 0.1 m/s

16. A string has tension T for tripling the frequency the tension in the string will be:

1) 3T 2) $\frac{T}{9}$
3) $\frac{T}{\sqrt{3}}$ 4) 9T

17. A wire is 4 m long and has a mass 0.2 kg. The wire is kept horizontally. A transverse pulse is generated by plucking one end of the taut (tight) wire. The pulse makes four trips back and forth along the cord in 0.8 sec. The tension is the cord will be-

1) 80 N 2) 160 N
3) 240 N 4) 320 N

18. Two strings of copper are stretched to the same tension. If their cross-section area are in the ratio 1 : 4, then the respective wave velocities will be-

1) 4 : 1 2) 2 : 1
3) 1 : 2 4) 1 : 4

19. A transverse sinusoidal wave moves along a string in the positive X-direction at a speed of 10 cm/s. The wavelength of the wave is 0.5 m and its amplitude is 10 cm. At a particular time t, the snap-shot of the wave is shown in the figure. The velocity of point P when its displacement is 5 cm, is:

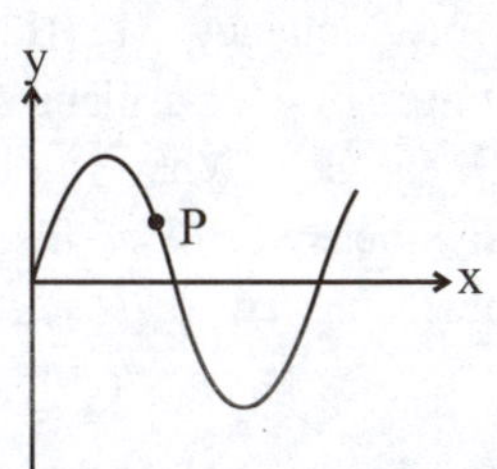

1) $\frac{\sqrt{3}\pi}{50}\hat{j}$ m/s 2) $-\frac{\sqrt{3}\pi}{50}\hat{j}$ m/s
3) $\frac{\sqrt{3}\pi}{50}\hat{i}$ m/s 4) $-\frac{\sqrt{3}\pi}{50}\hat{i}$ m/s

20. A uniform rope of length 10 m and mass 15 kg hangs vertically from a rigid support. A block of mass 5 kg is attached to the free end of the rope. A transverse pulse of wavelength 0.08 m is produced at the lower end of the rope. The wavelength of the pulse when it reaches the top of the rope will be-

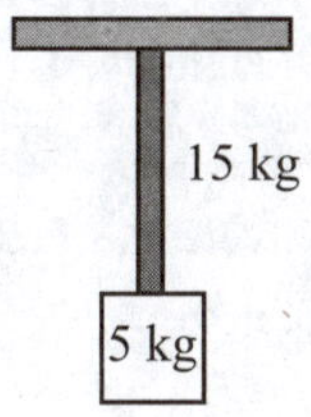

1) 0.08 m 2) 0.04 m
3) 0.16 m 4) 0 m

21. Two stretched strings have lengths l and $2l$ while tensions are T and 4T respectively. If they are made of same material the ratio of their frequency is:

1) 2 : 1 2) 1 : 2
3) 1 : 1 4) 1 : 4

22. A string fixed at both ends oscillates in 5 segments, length 10 m and velocity of wave is 20 m/s. What is the frequency?

1) 5 Hz 2) 15 Hz
3) 10 Hz 4) 2 Hz

23. Two uniform strings A and B made of steel are made to vibrate under the same tension. If the first overtone of A is equal to the second overtone of B and if the radius of A is twice that of B, the ratio of the lengths of the strings is:

1) 2 : 1 2) 3 : 2
3) 3 : 4 4) 1 : 3

24. A heavy rope is suspended from a rigid support. A wave pulse is set up at the lower end, then-

1) The pulse will travel with uniform speed
2) The pulse will travel with increasing speed
3) The pulse will travel with decreasing speed
4) The pulse cannot travel through the rope

Sound Wave, Speed of Sound, Loudness

25. Velocity of sound in a gaseous medium is 330 ms^{-1}. If the pressure is increased by 4 times without change in volume, the velocity of sound in the gas is -

1) 330 ms^{-1} 2) 660 ms^{-1}
3) 156 ms^{-1} 4) 990 ms^{-1}

26. Speed of sound wave in a gas is v_1 and rms speed of molecules of the gas at the same temperature is v_2. Then

1) $v_1 = v_2$ 2) $v_1 < v_2$
3) $v_1 > v_2$ 4) $v_1 \le v_2$

27. The ratio of intensities between two coherent sound source is 4 : 1. The difference of loudness in decibels (dB) between maximum and minimum intensities, when they interfere in space is -

1) 10 log 2 2) 20 log 3
3) 10 log 3 4) 20 log 2

28. The nature of sound waves in gases is-

1) Transverse
2) Longitudinal
3) Transverse and longitudinal
4) None of these

29. Sound waves of wavelength greater than that of audible sound are called-

1) Seismic waves 2) Sonic waves
3) Ultrasonic waves 4) Infrasonic waves

30. The wavelength of ultrasonic waves in air is of the order of-

1) 5×10^{-5} cm 2) 5×10^{3} cm
3) 5×10^{5} cm 4) 5×10^{8} cm

31. Sound waves have the following frequencies that are audible to human beings-

1) 5 c/s 2) 27000 c/s
3) 5000 c/s 4) 50,000 c/s

32. A big explosion on the moon cannot be heard on the earth because-

1) The explosion produces high frequency sound waves which are inaudible
2) Sound waves require a material medium for propagation
3) Sound waves are absorbed in the atmosphere of moon
4) Sound waves are absorbed in the earth's atmosphere

33. The intensity of a sound wave gets reduced by 20% on passing through a slab. The reduction in intensity on passage through two such consecutive slabs is

1) 40% 2) 36%
3) 30% 4) 50%

34. When a certain volume of water is subjected to increase of 100kPa pressure, the volume of water decrease by 5×10^{-3} %. The speed of sound in water must be

1) 140m/s 2) 300m/s
3) 1400m/s 4) 5000 m/s

35. Oxygen is 16 times heavier than hydrogen. Equal volume of hydrogen and oxygen are mixed. The ratio of speed of sound in the mixture to that in hydrogen is

1) $\sqrt{8}$ 2) $\sqrt{\frac{2}{17}}$
3) $\sqrt{\frac{1}{8}}$ 4) $\sqrt{\frac{32}{17}}$

36. Pressure variation in the propagation of sound waves in gaseous medium are:

1) Adiabatic 2) Isothermal
3) Isobaric 4) Isochoric

37. Temperature at which the speed of sound in air becomes doubled its value at 0°C is:

1) 1092°C 2) 819°C
3) 819K 4) 546°C

38. Stationary point source emits sound uniformly in all directions in non-absorbing medium two points P and Q are at a distance of 4 m and 9 m respectively from the source. The ratio of amplitudes of the waves at P and Q is

1) $\frac{3}{2}$ 2) $\frac{4}{9}$
3) $\frac{2}{3}$ 4) $\frac{9}{4}$

39. The intensity of sound increases at night due to

1) Increase in air density 2) Decrease in air density
3) Low temperature 4) None

Echo, Superposition of Waves, Interference, Beats

40. A source of unknown frequency gives 4 beat/s when sounded with a source of unknown frequency 250 Hz. The second harmonic of the source of unknown frequency gives 5 beats per sec when sounded with a source of frequency 513 Hz. The unknown frequency is (in Hz).

1) 254 2) 246
3) 240 4) 260

41. If v_1 and v_2 are the frequencies of two turning forks, then beat frequency is

1) $\frac{v_1}{v_2}$ 2) $v_1 + v_2$
3) $\frac{v_2}{v_1}$ 4) $v_1 - v_2$

42. Two tuning forks A and B, produce notes of frequencies 258 Hz and 262 Hz. An unknown note sounded with A produces certain beats. When the same note is sounded with B, the beat frequency gets doubled. The unknown frequency is

1) 250 Hz 2) 252 Hz
3) 254Hz 4) 245 Hz

43. A tuning fork of frequency 512 Hz makes 4 beats/sec with the vibrating string of a piano. The beat frequency decreases to 2 beats/sec. When the tension in the piano string is slightly increased. The frequency of the piano string before increasing tension is.

1) 510 2) 514
3) 516 4) 508

44. A disc of a siren containing 60 holes rotates at a constant speed of 360 rpm. The emitted sound is in unison with a turning fork of frequency is

1) 10 Hz 2) 360 Hz
3) 216 Hz 4) 60 Hz

45. Beats are produced by two waves given by y_1 = a sin 2000 πt and y_2 = a sin 2008 πt. The number of beats heard per second

1) Zero 2) One
3) Four 4) Eight

46. Three sound waves of equal amplitudes have frequencies ν – 1, ν, ν + 1, They superpose to produce beats. The number of beats produced per sec will be

1) ν 2) $\frac{\nu}{2}$
3) 2 4) 1

47. Beats are produced by frequencies f_1 and f_2 ($f_1 > f_2$). The duration of time between two successive maxima or minima is equal to

1) $\frac{1}{f_1 + f_2}$ 2) $\frac{2}{f_1 - f_2}$
3) $\frac{2}{f_1 + f_2}$ 4) $\frac{1}{f_1 - f_2}$

48. Number of beats between A and B are 5 and between B and C are 3. Beat frequency between A and C may be

1) 1 2) 2
3) 8 4) None

49. Three waves of equal frequency having amplitudes 10mm, 4mm and 7mm arrive at a given point with successive phase difference π/2. The amplitude of the resulting wave in mm is given by

1) 7 2) 6
3) 5 4) 4

50. A fork *A* has frequency 2% more than the standard fork and *B* has a frequency 3% less than the frequency of same standard fork. The forks *A* and *B* when sounded together produced 6 beats/s. The frequency of fork *A* is

1) 116.4 Hz 2) 120 Hz
3) 122.4 Hz 4) 238.8 Hz

51. The frequencies of two tuning forks *A* and *B* are respectively 1.5% more and 2.5% less than that of the tuning fork *C*. When *A* and *B* are sounded together, 12 beats are produced in 1 s. The frequency of the tuning fork *C* is

1) 200 Hz 2) 240 Hz
3) 360 Hz 4) 300 Hz

52. A man fired a bullet in front of a mountain and he heard its echo after 2 seconds. After traveling a distance of 85 m towards the mountain, he fired another bullet and heard its echo after 1.5 seconds. The velocity of sound and distance between the mountain and the man when the first bullet was fired are

1) 340 m/s, 340m 2) 340 m/s, 140 m
3) 140 m/s, 340 m 4) 140 m/s, 140 m

Reflection of Wave, Stationary Wave, Organ Pipes; Normal Modes of Vibration of Strings & Organ Pipe

53. The number of possible natural oscillations of air column in a pipe closed at one end of length 85 cm whose frequencies lie below 1250 Hz (v = 340 m/s)

1) 3 2) 4
3) 5 4) 6

54. A whistle whose air column is open at both ends has a fundamental frequency of 5100 Hz. If the speed of sound in air 340 m/s, the length of the whistle is

1) $\frac{5}{3}$ cm 2) $\frac{10}{3}$ cm
3) 5 cm 4) $\frac{20}{3}$ cm

55. What is the fractional change in tension necessary in a sonometer of fixed length to produce a note one octave lower than before?

1) $\frac{1}{4}$ 2) $\frac{1}{2}$
3) $\frac{2}{3}$ 4) $\frac{3}{4}$

56. In the case of a travelling wave, the reflection at a right boundary will take place with a phase change of

1) $\frac{\pi}{2}$ 2) $\frac{\pi}{4}$
3) π 4) $\frac{\pi}{6}$

57. In the fundamental mode time taken by the wave to reach the closed end of the air filled pipe is 10^{-2} sec. The fundamental frequency

1) 25 2) 12.5
3) 20 4) 15

58. If the length of an open organ pipe is 33.3 cm, then the frequency of fifth overtone (v_s = 333m/s)

1) 3000 Hz 2) 1500 Hz
3) 2500 Hz 4) 1250 Hz

59. In 1m long open pipe, what is the harmonic of resonance obtained with a tuning fork of frequency 480 Hz

1) First 2) Second
3) Third 4) Fourth

60. A standing wave having 3 nodes and 2 antinodes is formed between two atoms having a distance 1.21 Å between them. The wavelength of the standing wave is

1) 1.21Å
2) 1.42Å
3) 6.05Å
4) 3.63Å

61. A segment of wire vibrates with fundamental frequency of 450 Hz under a tension of 9 Kg wt then, tension at which the fundamental frequency of the same wire becomes 900 Hz is

1) 36 kg wt
2) 27 kg wt
3) 18 kg wt
4) 72 kg wt

62. If λ_1, λ_2 and λ_3 are the wavelengths of the waves giving resonance with the fundamental, first and second overtones respectively of a closed organ pipe. Then, the ratio of wavelengths $\lambda_1 : \lambda_2 : \lambda_3$ is:

1) 1 : 3 : 5
2) 1 : 2 : 3
3) 5 : 3 : 1
4) $1 : \frac{1}{3} : \frac{1}{5}$

High Order Time Intensive MCQs

1. Sound waves are passing through two routes-one in straight path and the other along a semicircular path of radius *r* and are again combined into one pipe and superposed as shown in the figure. If the velocity of sound waves in the pipe is *v*, then frequencies of resultant waves of maximum amplitude will be multiples integral of

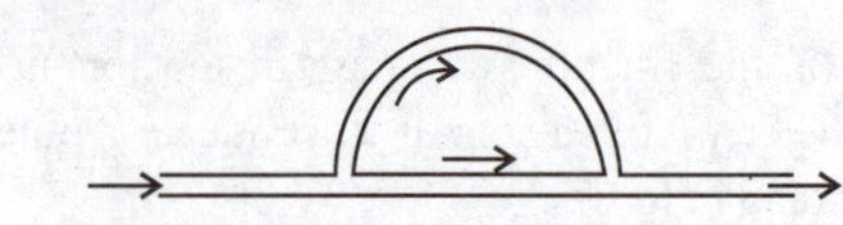

1) $\frac{v}{r(\pi-2)}$
2) $\frac{v}{r(v-1)}$
3) $\frac{2v}{r(\pi-1)}$
4) $\frac{v}{r(\pi+1)}$

2. Two sound waves, each of amplitude *A* and frequency ω superpose at a point with phase difference of π/2. The amplitude and frequency of the resultant wave are respectively

1) $\frac{A}{\sqrt{2}}, \frac{\omega}{2}$
2) $\frac{A}{\sqrt{2}}, \omega$
3) $\sqrt{2}A, \frac{\omega}{2}$
4) $\sqrt{2}A, \omega$

3. 5 beats/s are produced on blowing together two closed organ pipes of the same diameter but of different lengths. If shorter pipe is of 10 cm length and speed of sound in air is 300 m/s, length of other pipe is

1) 10.06 cm
2) 11.22 cm
3) 16 cm
4) 14 cm

4. Each of the two strings of length 51.6 cm and 49.1 cm are tensioned separately by 20 N force. Mass per unit length of both the strings is same and equal to 1 gm^{-1}. When both the strings vibrate simultaneously, the number of beats is

1) 4
2) 7
3) 8
4) 3

5. A string of density 7.5 g cm^{-3} and area of cross-section 0.2 mm^2 is stretched under a tension of 20 N. when it is plucked at the mid-point, the speed of the transverse waves on the wire is:

1) 116 ms^{-1}
2) 40 ms^{-1}
3) 200 ms^{-1}
4) 80 ms^{-1}

6. A transverse wave is described by the equation $y = y_0 \sin 2\pi\left(ft - \frac{x}{\lambda}\right)$. The maximum particle velocity is equal to four times the wave velocity if:

1) $\lambda = \frac{\pi y_0}{4}$
2) $\lambda = \frac{\pi y_0}{2}$
3) $\lambda = \pi y_0$
4) $\lambda = 2\pi y_0$

7. A sinusoidal wave moving along a string is shown twice in the figure as crest A travels in the positive direction of an x-axis by distance d = 6.0 cm in 4.0 ms. The tick marks along the axis are separated by 10 cms. height H = 6.00 mm. The wave equation is:

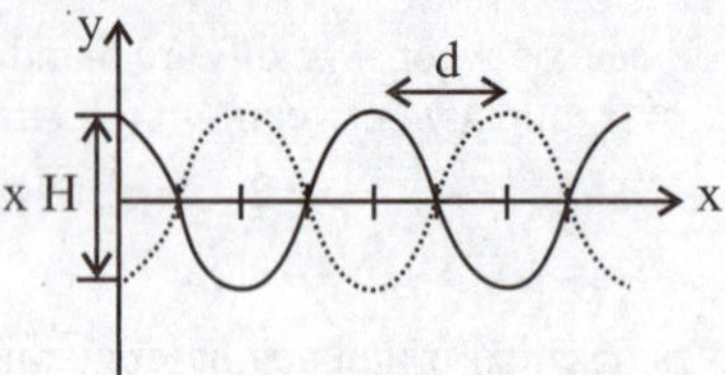

1) y = (3 mm) sin (16x – 2.4 × 10^2t)
2) y = (3 mm) sin (16x + 2.4 × 10^2t)
3) y = (3 mm) sin (8x + 2.4 × 10^2t)
4) y = (3 mm) sin (8x – 2.4 × 10^2t)

8. $y(x,t) = \frac{1.0}{[(ax+bt)^2 + (cx+dt)^2 + 5]}$ represents a moving pulse where x and y are in metres and t in seconds. It has speed of pulse b/a. Find the correct option.

1) bc = ad
2) bd = ac
3) ab = cd
4) None of these

9. A tuning fork of frequency n is held near the open end of a tube which is closed at the other end and the length of the tube is adjusted until resonance occurs. If the two shortest lengths that produce resonance are L_1 and L_2, the speed of the sound is:

1) $n(L_2 - L_1)$
2) $n(L_2 - L_1)/2$
3) $2n(L_2 + L_1)$
4) $2n(L_2 - L_1)$

10. A transverse wave propagating on a stretched string of linear density 3 × 10^{-3} kg/m is represented by the equation y = 0.2 sin (1.5x + 60t). The tension in the string

1) 0.24
2) 0.187 × 10^{-5} N
3) 1.20
4) 1.80

11. A transverse progressive wave is established on a 100 m long rope of mass 4 kg under tension of 100 N. If 25 wave pulses can be accommodated on the string and maximum particle velocity of the string is 5 cm/s and at t = 0, particle at x = 0 is at its mean position and going down, find out the transverse wave equation. Assume sinusoidal wave going towards right.

1) $y=\left(\frac{\pi}{100}\right)\sin\left(12.5\pi t-\frac{\pi x}{2}+\pi\right)$

2) $y=\left(\frac{1}{500\pi}\right)\sin\left(25\pi t-\frac{\pi x}{2}+\pi\right)$

3) $y=\left(\frac{1}{50\pi}\right)\sin\left(25\pi t-\frac{\pi x}{2}\right)$

4) $y=\left(\frac{\pi}{10}\right)\sin\left(25\pi t-\frac{\pi x}{2}+\pi\right)$

12. The displacement-time graph for two sound waves A and B are shown in the figure, then the ratio of their intensities I_A / I_B is equal to

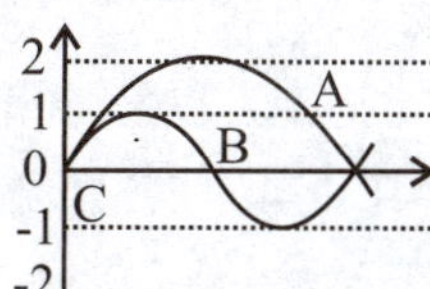

1) 1 : 4 2) 1 : 16
3) 1 : 2 4) 1 : 1

13. Two mechanical waves $y_1 = 2 \sin 2\pi (50t - 2x)$ & $y_2 = 4 \sin 2\pi (ax + 100t)$ propagate in a medium with same speed. Choose the correct statement. ***[RC]***

1) The ratio of their intensities is 1 : 16
2) The ratio of their intensities is 1 : 4
3) The value of 'a' is 3 units
4) The value of 'a' is 2 units

14. Two sources vibrates with different amplitudes and frequencies. If two identical strings having equal tension are attached to the source individually, they send harmonic transverse waves. Find the incorrect option.

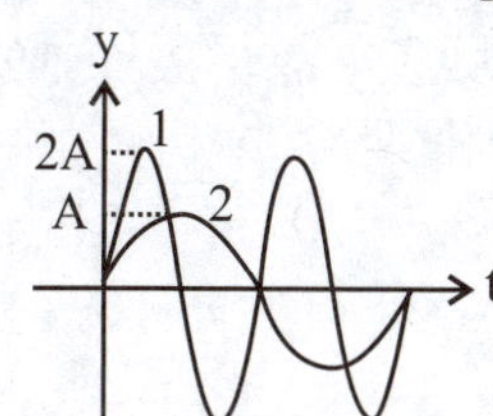

1) The ratio of power delivered in the strings is 1 : 1
2) The ratio of power delivered by the sources is 16 : 1
3) The ratio of speeds of the waves is 1 : 1
4) The ratio of wavelengths of the waves is 1 : 2

15. A person is talking in a small room and the sound intensity level is 60 dB everywhere within the room. If there are eight people talking simultaneously in the room, what is the sound intensity level ?

1) 60 dB 2) 69 dB
3) 74 dB 4) 81 dB

16. The equation of stationary wave along a stretched string is given by:

$$y = 5\sin\frac{\pi x}{3}\cos 40\pi t.$$

Where x and y are in cm and t in second. The separation between two adjacent nodes is:

1) 1.5cm 2) 3 cm
3) 6 cm 4) 4 cm

17. A sonometer wire resonates with a given tuning fork forming standing waves with five antinodes between the two bridges when a mass of 9 kg is suspended from the wire. When this mass is replaced by a mass M, the wire resonates with the same turning fork forming three antinodes for the same position of the bridges. The value of M is:

1) 25 kg
2) 5 kg
3) 12.5 kg
4) (1/25) kg

18. Two sources of sound A and B produces the wave of 350 Hz, they vibrate in the same phase. The particle P is vibrating under the influence of these two waves, if the amplitudes at the point P produced by the two waves is 0.3 mm and 0.4 mm, then the resultant amplitude of the point P will be, (when AP – BP = 25 cm and the velocity of sound is 350 m/sec)

1) 0.7 mm 2) 0.1 mm
3) 0.2 mm 4) 0.5 mm

19. Two vibrating strings of same length, same cross section area and stretched to same tension are made of materials with densities ρ & 2ρ. Each string is fixed at both ends. If v_1 represents frequency of the fundamental mode of vibration of the one made with density ρ and v_2 for another, then v_1/v_2 is

1) 1/2 2) 2
3) $\sqrt{2}$ 4) $\frac{1}{\sqrt{2}}$

20. A chord attached about an end to a vibrating fork divides it into 6 loops, when its tension is 36 N. The tension at which it will vibrate in 4 loops is.

1) 24 N
2) 36 N
3) 64 N
4) 81 N

21. A closed pipe and an open pipe of same length produce 2 beats, when they are set into vibration simultaneously in their fundamental mode. If the length of the open pipe is halved and that of closed pipe is doubled, and if they are vibrating in the fundamental mode, then the number of beats produced is

1) 4 2) 7
3) 2 4) 8

22. The vibrations of four air columns are represented in the figure. The ratio of frequencies $n_p : n_q : n_r : n_s$ is:

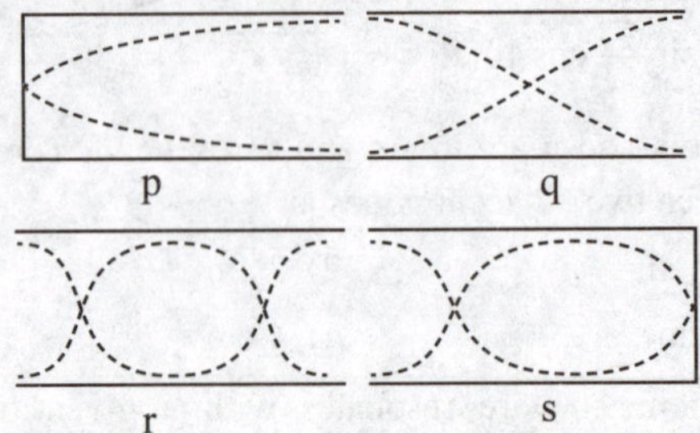

1) 12 : 6 : 3 : 4
2) 1 : 2 : 4 : 3
3) 4 : 2 : 3 : 1
4) 6 : 2 : 3 : 4

23. A two-fold increase in intensity of a wave implies an increase of (Given : $\log_{10} 2 = 0.3010$.)
1) 2 dB
2) 10 dB
3) 3.01 dB
4) 0.5 dB

24. A person is standing at a distance D from an isotropic point source of sound. He walks 50.0 m towards the source and observes that the intensity of the sound has doubled, hits initial distance D from the source is:
1) $50\sqrt{2}$ m
2) $\frac{50\sqrt{2}}{\sqrt{2}-1}$ m
3) $\frac{50}{\sqrt{2}-1}$ m
4) $100\sqrt{2}$ m

25. A certain transverse sinusoidal wave of wavelength 20 cm is moving in the positive x direction. The transverse velocity of the particle at x = 0 as a function of time is shown the amplitude of the motion is:

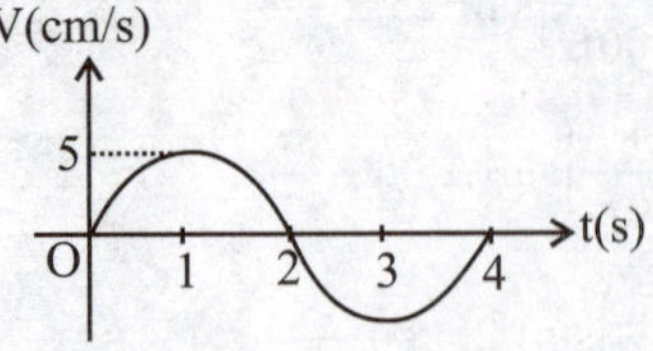

1) $\frac{5}{\pi}$ cm
2) $\frac{\pi}{2}$ cm
3) $\frac{10}{\pi}$
4) 2π cm

Statement Based MCQs

Directions: These questions consist of two statements each, printed as Statement-I and Statement-II. While answering these questions, you are required to choose any one of the following four responses.

1) **Both Statement-I and Statement-II are correct.**
2) **Both Statement-I and Statement-II are incorrect.**
3) **Statement-I is correct & Statement-II is incorrect.**
4) **Statement-I is incorrect & Statement-II is correct.**

1. **Statement-I:** In longitudinal wave propagation, the distance between two consecutive compression is equal to wavelength of wave.

Statement-II: Standing wave is not a wave as it does not transport energy.

2. **Statement-I:** A wind is different from the sound wave in air.

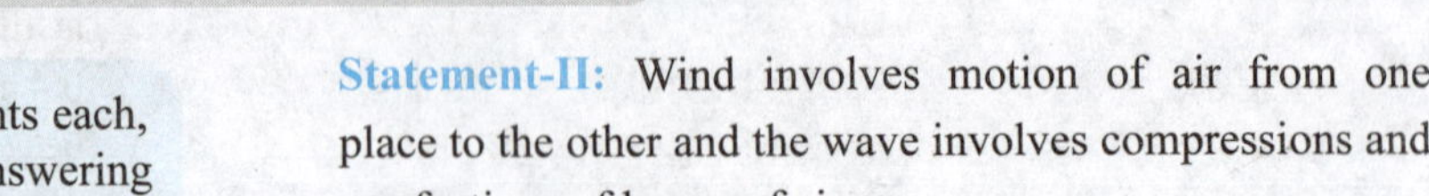

Statement-II: Wind involves motion of air from one place to the other and the wave involves compressions and rarefactions of layers of air.

3. **Statement-I:** SONAR is employed in detection and location of objects under the hills.

Statement-II: SONAR makes use of reflection of ultrasonic waves.

4. **Statement-I:** Sound travels faster in air than in water.

Statement-II: Air is a rarer medium as compared to water medium.

5. **Statement-I:** Fundamental frequency of open organ pipe is twice the fundamental frequency of closed organ pipe.

Statement-II: First overtone of open organ pipe is twice its fundamental frequency.

Assertion & Reason MCQs

Directions: These questions consist of two statements each, printed as Assertion and Reason. While answering these questions, you are required to choose any one of the following four responses.

1) **Both Assertion and Reason are True and the Reason is a correct explanation of the Assertion.**
2) **Both Assertion and Reason are True but Reason is not a correct explanation of the Assertion.**
3) **Assertion is True but the Reason is False.**
4) **Assertion is False but Reason is True.**

1. **Assertion:** Sound waves cannot travel through vacuum but light waves can.

Reason: Sound waves cannot be polarised but light waves can be.

2. **Assertion:** The presence of moisture increases the velocity of sound in air.

Reason: Density of moist air is more than the density of dry air.

3. **Assertion:** Explosions on the other planets are not heard on the earth.

Reason: To hear different beats, difference in frequencies of two sources should be less than 10.

4. **Assertion:** Ultrasonics is the acoustic analogue of ultraviolet radiation.

Reason: Ultraviolet rays do not produce visual sensations similarly ultrasonic waves are not heard by the human ear.

5. **Assertion:** Infrasonic waves have frequency range that lies below 20 Hz.

Reason: Pitch of female voice is more than the pitch of male voice.

Match the Columns MCQs

1. A closed organ pipe of length L vibrating in second overtone, then match the following:

Column-I		Column-II	
(a)	Displacement node	(i)	Closed end
(b)	Displacement antinode	(ii)	open end
(c)	Pressure node	(iii)	4L/5 from closed end
(d)	Pressure antinode	(iv)	L/5 from closed end

1) (a)-(i,iii) (b)-(ii,iv) (c)-(ii,iv) (d)-(i,iii)

2) (a)-(ii,iv) (b)-(i,iv) (c)-(ii,iii) (d)-(i,iii)

3) (a)-(iii,iv) (b)-(ii,iii) (c)-(i,iv) (d)-(i,ii)

4) (a)-(i,iii) (b)-(iii,iv) (c)-(i,iv) (d)-(ii,iii)

2. Three successive resonance frequencies in a an organ pipe are 1310, 1834 and 2358 Hz. Velocity of sound in air is 340 m/s, then match the items given in column I with that in column II:

Column-I		Column-II	
(a)	Length of the pipe in cm	(i)	262
(b)	Fundamental frequency (Hz)	(ii)	786
(c)	Frequency of fifth harmonic (Hz)	(iii)	32.4
(d)	Frequency of 1 overtone (Hz)	(iv)	1310

1) (a)-(iii) (b)-(i) (c)-(iv) (d)-(ii)

2) (a)-(ii) (b)-(iii) (c)-(i) (d)-(iv)

3) (a)-(i) (b)-(ii) (c)-(iii) (d)-(iii)

4) (a)-(i) (b)-(ii) (c)-(iv) (d)-(iii)

3. In each of the four situations of column-I a stretched string or an organ pipe is given aloong with the required data. In case of strings the tension in string is T = 102.4 N and the mass per unit length of string is 1 g/m. Speed of sound in air is 320 m/s. Neglect end corrections. The frequencies of resonance are given in Column-II. Match each situation in column-I with the possible resonance frequencies given in column-II.

Column-I		Column-II	
(a)	Strin fixed at both ends 0.5 m Fixed Fixed	(i)	320 Hz
(b)	String fixed at one end and free at other end 0.5 m Fixed end Free end	(ii)	480 Hz
(c)	Open organ pipe 0.5m	(iii)	640 Hz
(d)	Closed organ pipe 0.5m	(iv)	800 Hz

1) (a)-(i,ii) (b)-(i,ii) (c)-(iii) (d)-(iv)

2) (a)-(iv) (b)-(i,ii) (c)-(iii) (d)-(i,ii)

3) (a)-(i,iii) (b)-(ii,iv) (c)-(i,iii) (d)-(ii,iv)

4) (a)-(ii,iii) (b)-(ii,iii,iv) (c)-(i) (d)-(i)

4. Match the following.

Column-I		Column-II	
(a)	Pitch	(i)	Waveform
(b)	Quality	(ii)	Frequency
(c)	Loudness	(iii)	Intensity
(d)	Nodes	(iv)	Position of zero amplitude

1) (a)-(i) (b)-(iii) (c)-(iv) (d)-(ii)

2) (a)-(ii) (b)-(i) (c)-(iii) (d)-(iv)

3) (a)-(ii) (b)-(i) (c)-(iv) (d)-(iii)

4) (a)-(i) (b)-(ii) (c)-(iv) (d)-(iii)

5. Match the following.

	Column-I		Column-II
(a)	$y = \sin(5x - 4t) + 3\cos(4t - 5x + \pi/6)$	(i)	Particles at every position are performing SHM
(b)	$y = 10\cos\left(t - \frac{x}{330}\right)\sin(100)\left(t - \frac{x}{330}\right)$	(ii)	Equation of traveling wave
(c)	$y = 10\sin(2\pi x - 120t) + 10\cos(120t + 2\pi x)$	(iii)	Equation of standing wave
(d)	$y = 10\sin(2\pi x - 120t) + 8\cos(118t - 59/30\pi x)$	(iv)	Equation of Beats

1) (a)-(iv) (b)-(i) (c)-(ii) (d)-(iii)
2) (a)-(iv) (b)-(ii) (c)-(iii) (d)-(ii)
3) (a)-(i) (b)-(ii) (c)-(iii) (d)-(iv)
4) (a)-(i,ii) (b)-(iv) (c)-(i,iii) (d)-(iv)

NCERT Exemplar MCQs

1. Water waves produced by a motorboat sailing in water are:
1) Neither longitudinal nor transverse
2) Both longitudinal and transverse
3) Only longitudinal
4) Only transverse

2. Sound waves of wavelength λ travelling in a medium with a speed of v m/s enter into another medium where its speed in 2v m/s. Wavelength of sound waves in the second medium is:
1) λ
2) $\frac{\lambda}{2}$
3) 2λ
4) 4λ

3. Speed of sound wave in air:
1) Is independent of temperature
2) Increases with pressure
3) Increases with increase in humidity
4) Decreases with increase in humidity

4. Change in temperature of the medium changes:
1) Frequency of sound waves
2) Amplitude of sound waves
3) Wavelength of sound waves
4) Loudness of sound waves

5. With propagation of longitudinal waves through a medium, the quantity transmitted is:
1) Matter
2) Energy
3) Energy and matter
4) Energy, matter and momentum

6. Which of the following statements are true for wave motion?
1) Mechanical transverse waves can propagate through all mediums
2) Longitudinal waves can propagate through solids only
3) Mechanical transverse waves can propagate through solids only
4) Longitudinal waves can propagate through vacuum

7. A sound wave is passing through air column in the form of compression and rarefaction. In consecutive compressions and rarefactions:
1) Density remains constant
2) Boyle's law is obeyed
3) Bulk modulus of air oscillates
4) There is no transfer of heat

8. Equation of a plane progressive wave is given by $y = 0.6\sin 2\pi\left(t - \frac{x}{2}\right)$. On reflection from a denser medium its amplitude becomes $\frac{2}{3}$ of the amplitude of the incident wave. The equation of the reflected wave is:
1) $y = 0.6\sin 2\pi\left(t + \frac{x}{2}\right)$
2) $y = -0.4\sin 2\pi\left(t + \frac{x}{2}\right)$
3) $y = 0.4\sin 2\pi\left(t + \frac{x}{2}\right)$
4) $y = -0.4\sin 2\pi\left(t - \frac{x}{2}\right)$

9. A string of mass 2.5 kg is under tension of 200 N. The length of the stretched string is 20.0 m. If the transverse jerk is struck at one end of the string, the disturbance will reach the other end in:
1) 1s
2) 0.5 s
3) 2s
4) Data given is insufficient

NEET & AIIMS Past Year MCQs

1. The displacement of a travelling wave $y = C\sin\frac{2\pi}{\lambda}(at - x)$ where t is time, x is distance and λ is the wavelength, all in S.I. units. Then the frequency of the wave is **(2024 Re)**

1) $\frac{2\pi\lambda}{a}$
2) $\frac{2\pi a}{\lambda}$
3) $\frac{\lambda}{a}$
4) $\frac{a}{\lambda}$

2. The ratio of frequencies of fundamental harmonic produced by an open pipe to that of a closed pipe having the same length is **(2023)**

1) 3 : 1
2) 1 : 2
3) 2 : 1
4) 1 : 3

3. If the initial tension on a stretched string is doubled, then the ratio of the initial and final speed of a transverse wave along the string is: **(2022)**

1) 1 : 2
2) 1 : 1
3) $\sqrt{2}:1$
4) $1:\sqrt{2}$

4. A string is clamped at its midpoint and its fundamental frequency is given. Calculate velocity of wave: **(AIIMS 2015)**

1) νL
2) ν/L
3) νL^2
4) $2\nu L$

5. If two waves of wavelengths 50 cm and 51 cm produced 12 beats/s, the velocity of sound is: **(AIIMS 2015)**

1) 360 ms^{-1}
2) 306 ms^{-1}
3) 331 ms^{-1}
4) 340 ms^{-1}

Answer Key

NCERT Topic-wise MCQs

1	2	3	4	5	6	7	8	9	10	11	12	13	14	15	16	17
4	3	1	2	3	4	1	1	1	4	1	4	3	3	1	4	1
18	19	20	21	22	23	24	25	26	27	28	29	30	31	32	33	34
2	1	3	3	1	4	2	1	2	2	2	4	1	3	2	2	3
35	36	37	38	39	40	41	42	43	44	45	46	47	48	49	50	51
2	1	2	4	1	1	4	3	4	2	3	3	4	2	3	3	4
52	53	54	55	56	57	58	59	60	61	62						
1	4	2	4	3	1	1	3	1	1	4						

High Order Time Intensive MCQs

1	2	3	4	5	6	7	8	9	10	11	12	13	14	15	16	17
1	4	1	2	1	2	1	1	4	2	2	4	1	1	2	2	1
18	19	20	21	22	23	24	25									
4	3	4	2	2	3	2	3									

Statement Based MCQs

1	2	3	4	5
3	1	4	4	1

Assertion & Reason MCQs

1	2	3	4	5
2	3	2	2	2

Match the Column MCQs

1	2	3	4	5
1	1	3	2	4

NCERT Exemplar MCQs

1	2	3	4	5	6	7	8	9
2	3	3	3	2	3	4	2	2

NEET & AIIMS Past Year MCQs

1	2	3	4	5
4	3	4	1	2

Class-XII

CHAPTER 1

Electric Charges and Fields

Topicwise Number of Questions Analysis (2025-2016)

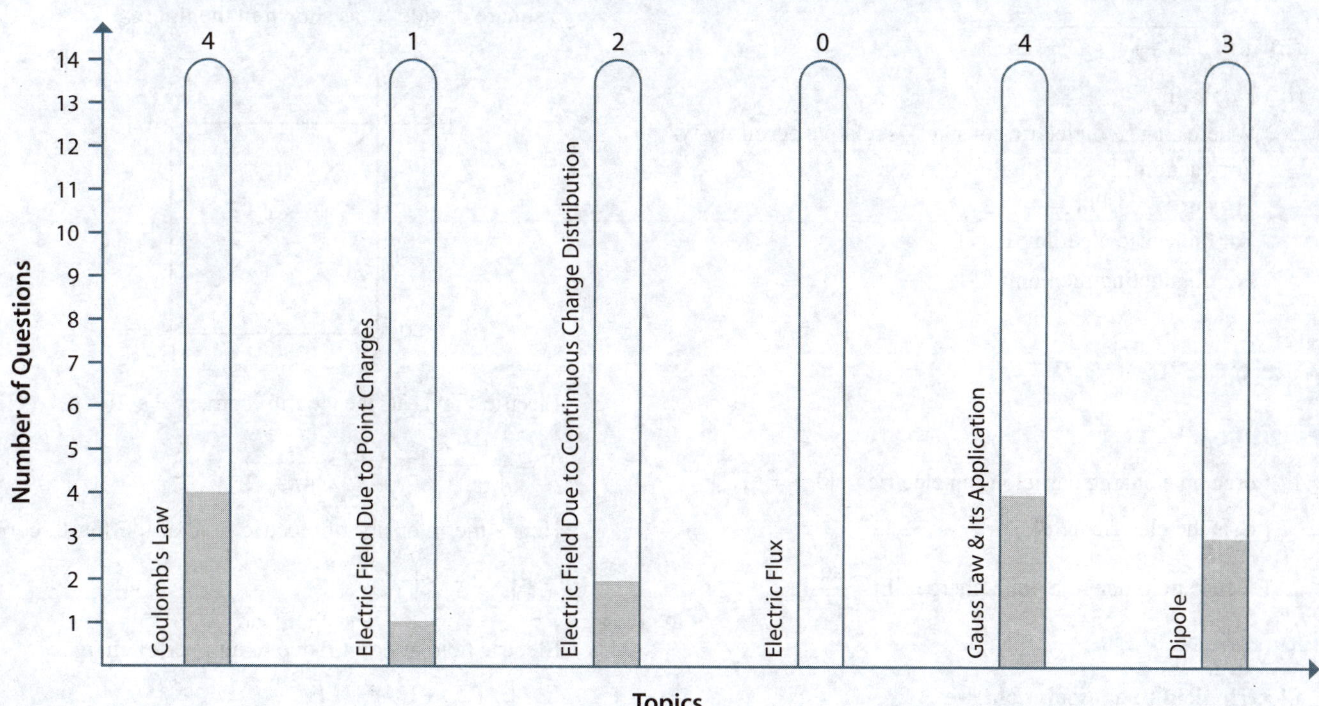

NCERT-PICKS

Introduction

Electrostatics is the branch of physics that deals with the study of forces, fields and potentials arising from static charges.

Coulomb's Law

Formulas

1. Coulomb's law, $|\vec{F}| = \frac{kq_1q_2}{r^2}$

   where, $k = \frac{1}{4\pi\varepsilon_o} = 9\times10^9\ Nm^2C^{-2}$

   (q_1, q_2 are the charges on the particles, r is the separation between them and F is the force between the charges)

2. Coulomb's law in vector form:

   Electric force on charge q_1 due to q_2 is: $\vec{F}_{12} = \frac{kq_1q_2}{|\vec{r_1}-\vec{r_2}|^3}(\vec{r_1}-\vec{r_2})$

   Electric force on charge q_2 due to q_1 is: $\vec{F}_{21} = \frac{kq_1q_2}{|\vec{r_2}-\vec{r_1}|^3}(\vec{r_2}-\vec{r_1})$

3. $\vec{F}_{12} = -\vec{F}_{21}$ and $|\vec{F}_{12}| = |\vec{F}_{21}|$

4. Quantisation of charge, q = ne

   [n is the number of charges and e is the charge on one electron]

Numerical Profile

(i) **Principle of super position:** The total force $\vec{F_1}$ on the charge q_1, due to all other charges is vector sum of all the forces.

$\vec{F_1} = \vec{F_{12}} + \vec{F_{13}} + ... + \vec{F_{1n}}$

$\vec{F_1} = k\left(\frac{q_1q_2}{r_{12}^2}\hat{r}_{12} + \frac{q_1q_3}{r_{13}^2}\hat{r}_{13} + ... + \frac{q_1q_n}{r_{1n}^2}\hat{r}_{1n}\right)$

$\vec{F_1} = kq_1\sum_{i=2}^{n}\frac{q_i}{r_{1i}^2}\hat{r}_{1i}$

(ii) $F = \sqrt{F_1^2 + F_2^2 + 2F_1F_2\cos\theta}$

(iii) $F_{air} = \varepsilon_r F_{medium}$

Where ε_r (= K, dielectric constant) is relative permittivity.

For vacuum: K = 1

For air: K = 1.00054

For dielectric medium: K > 1

For conducting medium: K = ∞

Electric Field

Formulas

1. Force on a charge particle is an electric field $\vec{F} = q\vec{E}$

   [$\vec{E}$ is the electric field]

2. Electric field due to a point charge, $|\vec{E}| = \frac{kq}{r^2}$

Numerical Profile

Electric field due to point charges

(i) Two charges +q and –q are separated by a distance 'r'. Electric field at the mid-point of the line joining the charges, $E = \frac{1}{4\pi\varepsilon_0}\frac{8q}{r^2}$

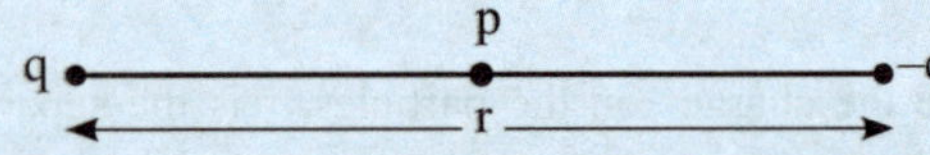

(ii) Two charges +q, –q are kept at the two vertices of an equilateral triangle of side 'r', then electric field at the third vertex is

$E' = 2E\cos\frac{\theta}{2} = \sqrt{3}E$ [θ = 60°]

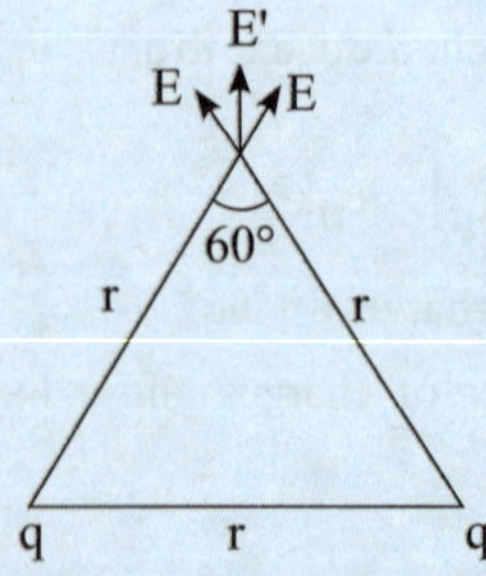

$E' = \sqrt{3}\frac{1}{4\pi\varepsilon_0}\frac{q}{r^2}$

(iii) If three charges +q each are kept at the three vertices of an equilateral triangle of side 'r' then the electric field at the centroid is zero.

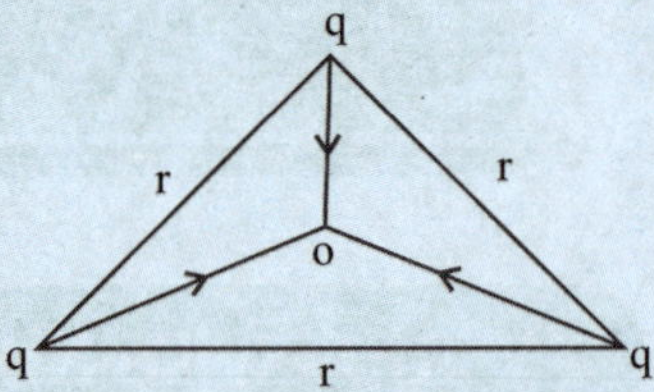

(iv) Let, three charges +q each are kept at the three corners of a square of side 'r' as shown in the figure.

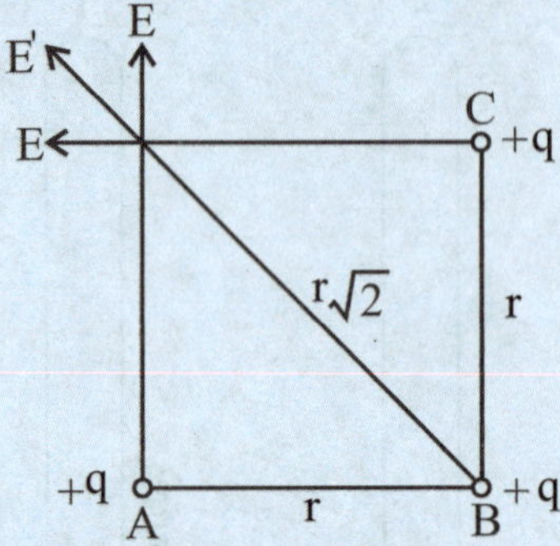

Electric field at the fourth corner $= \sqrt{2}E + E'$. Here $E = \frac{1}{4\pi\varepsilon_0}\frac{q}{r^2}$ and $E' = \frac{1}{4\pi\varepsilon_0}\frac{q}{2r^2} = \frac{E}{2}$.

Hence the intensity of electric field at the fourth corner $= E\left(\sqrt{2} + \frac{1}{2}\right)$

(v) Electric field at point P due to number of charges.

$\vec{E} = \vec{E}_1 + \vec{E}_2 + \vec{E}_3 + ... + \vec{E}_n$

$\vec{E} = k\sum_{i=1}^{n}\frac{q_i}{r_{ip}^2}\hat{r}_{ip}$

Gauss Law and Electric field due to Uniform Bodies

Formulas

1. Electric flux, $\phi = \vec{E}.\vec{A}$ [A is the area]

2. Gauss law, $\int\vec{E}.d\vec{s} = \frac{q_{in}}{\varepsilon_0}$

   [q_{in} is the charge enclosed by the gausian surface]

3. Electric field due to uniformly charged sphere at a point outside it, $|\vec{E}| = \frac{q}{4\pi\varepsilon_0 r^2}$

   [where r is the distance of the point from the center of the sphere]

4. Electric field due to uniformly charged sphere at a point inside the sphere, $|\vec{E}| = \frac{qr}{4\pi\varepsilon_0 R}$

   [R is radius of the sphere]

Numerical Profile

Electric Field under different conditions

(i) Electric field due to an infinitely long charged wire.

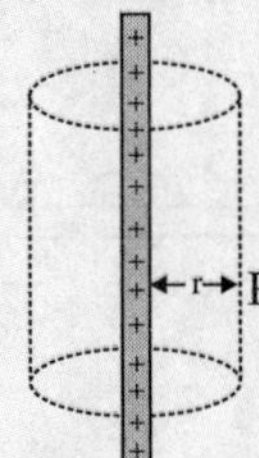

Electric field at point P, $E = \dfrac{\lambda}{2\pi\varepsilon_0 r}$

[λ is the charge per unit length]

Note: Direction of the field is away from the wire if charge is positive and towards the wire if the charge is negative.

(ii) Electric field due to a uniformly charged infinite plane sheet.

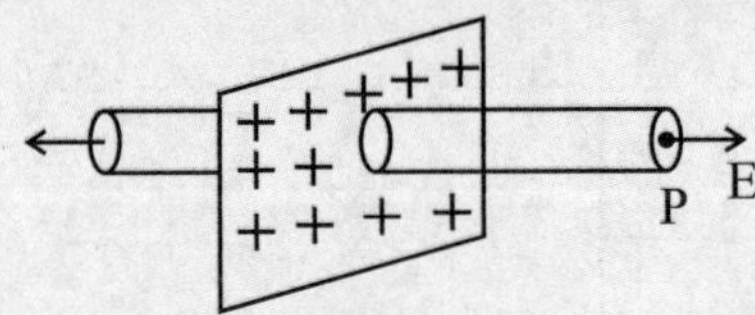

Electric field at point P, $E = \dfrac{\sigma}{2\varepsilon_0}$

[σ is charge per unit area]

(iii) Electric field intensity due to two equally and oppositely charged parallel plane sheets of charge.

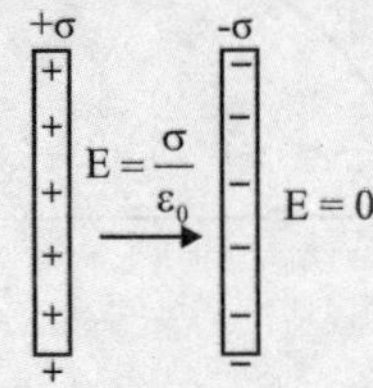

$E = \dfrac{\sigma}{\varepsilon_0}$ (between two plates)

$E = 0$ (outside the plates)

(iv) Electric field intensity due to two positively charged parallel plane sheets of charge

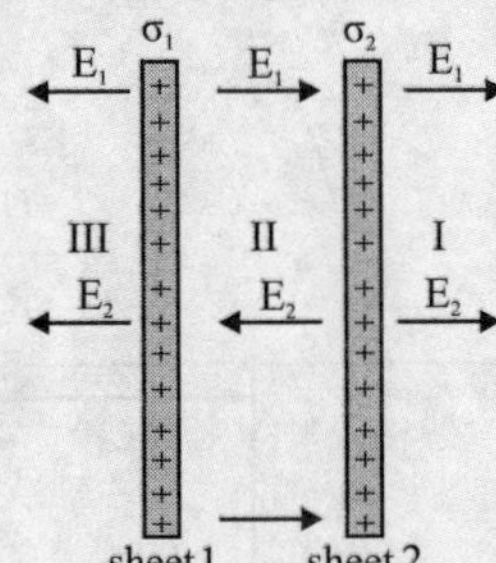

Let $\sigma_1 > \sigma_2$

[E_1 and E_2 is the electric field due to sheet 1 and sheet 2]

In region I: $E_{net} = E_1 + E_2 = \dfrac{(\sigma_1 + \sigma_2)}{2\varepsilon_0}$ (towards right)

In region II: $E_{net} = E_1 - E_2 = \dfrac{(\sigma_1 - \sigma_2)}{2\varepsilon_0}$ (towards right)

In region III: $E_{net} = E_1 + E_2 = \dfrac{(\sigma_1 + \sigma_2)}{2\varepsilon_0}$ (towards left)

(v) Electric field due to a thin spherical shell of surface charge density σ and radius R.

$E = \dfrac{1}{4\pi\varepsilon_0} \cdot \dfrac{q}{r^2}$ {For r > R i.e., Outside}

$E = 0$ {For r < R i.e., Inside}

$E = \dfrac{1}{4\pi\varepsilon_0} \cdot \dfrac{q}{R^2}$ {For r = R i.e., At surface}

Surface charge density: $\sigma = \dfrac{q}{4\pi A^2}$

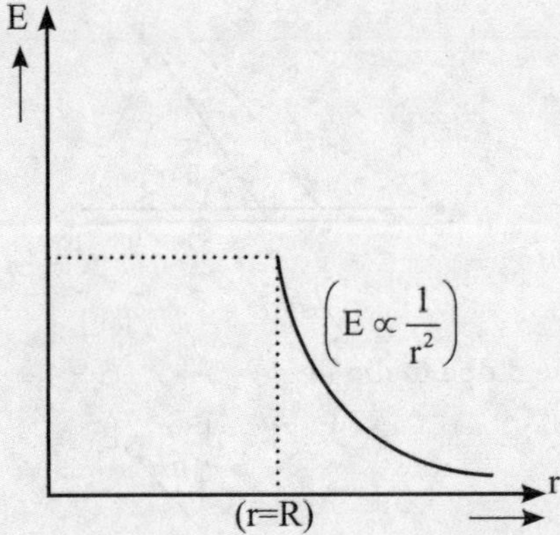

(vi) Electric field due to a solid sphere shell of uniform charge density ρ and radius R.

$E = \dfrac{1}{4\pi\varepsilon_0} \cdot \dfrac{q}{r^2}$ {For r > R i.e., Outside}

$E = \dfrac{1}{4\pi\varepsilon_0} \cdot \dfrac{qr}{R^3}$ {For r < R i.e., Inside}

$E = \dfrac{1}{4\pi\varepsilon_0} \cdot \dfrac{q}{R^2}$ {For r = R i.e., At surface}

Surface charge density: $\rho = \dfrac{q}{\frac{4}{3}\pi R^3}$

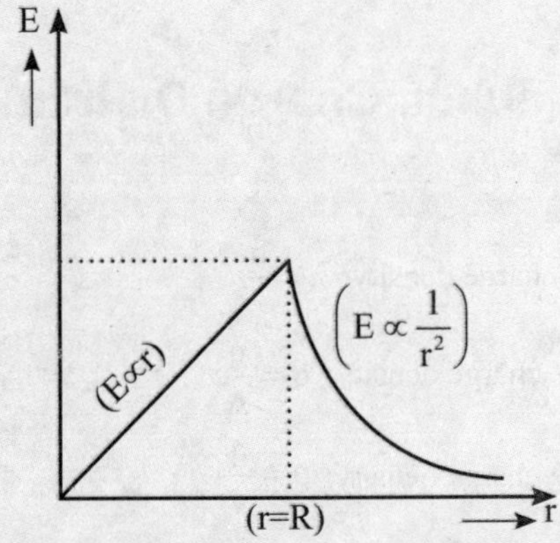

Dipole

Formulas

1. Electric dipole moment, $\vec{p} = q \times \vec{d}$

   [q is the charge on the particle and d is the distance between the two charged particles]
2. Torque acting on an electric dipole placed in a uniform electric field, $\vec{\tau} = \vec{p} \times \vec{E}$
3. Potential energy of a dipole, $U = \vec{p} \cdot \vec{E}$

Numerical Profile

Electric Potential due to dipole,

Potential at point P, $V = \frac{1}{4\pi\varepsilon_0} \frac{p\cos\theta}{r^2}$

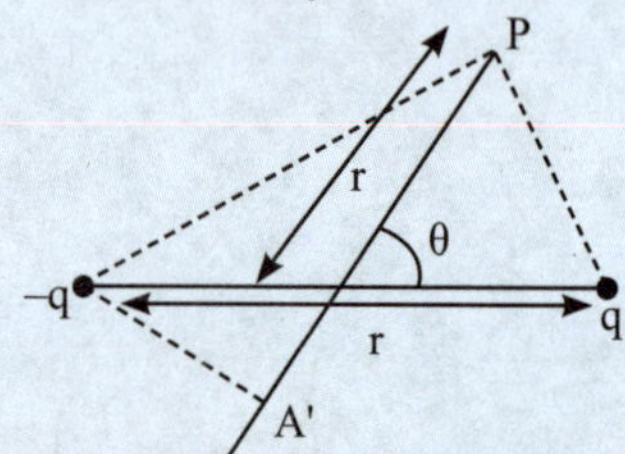

Electric field due to dipole

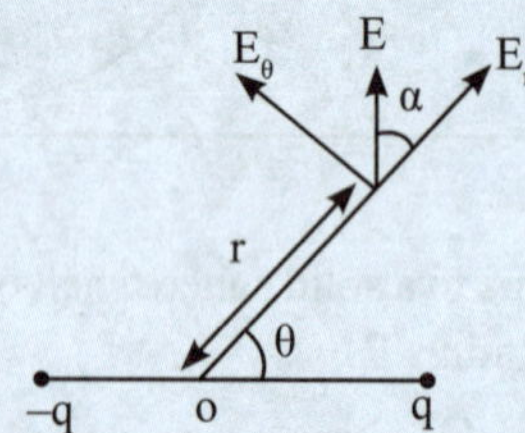

(i) Electric field in radial direction, $E_r = \frac{1}{4\pi\varepsilon_0} \frac{2p\cos\theta}{r^3}$

(ii) Electric field in transverse direction, $E_\theta = \frac{1}{4\pi\varepsilon_0} \frac{p\sin\theta}{r^3}$

(iii) $E_{net} = \sqrt{E_r^2 + E_\theta^2} = \frac{1}{4\pi \epsilon_0} \frac{p}{r^3}\left(\sqrt{3\cos^2\theta + 1}\right)$

If E_{net} makes angle α with OP, then

$\alpha = \tan^{-1}\left(\frac{1}{2}\tan\theta\right)$

Continuous Charge Distribution

Formulas

1. Linear charge density: $\lambda = \frac{q}{\ell}$
2. Surface charge density: $\sigma = \frac{q}{A}$
3. Volume charge density: $\rho = \frac{q}{V}$

[Where:- q: Charge, ℓ: length, A: Area, V: Volume]

Numerical Profile

Electric field due to finite wire

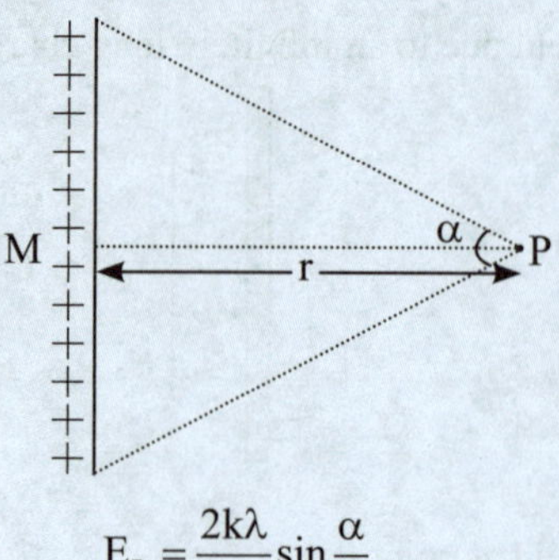

$$E_P = \frac{2k\lambda}{r}\sin\frac{\alpha}{2}$$

Electric field due to infinite wire

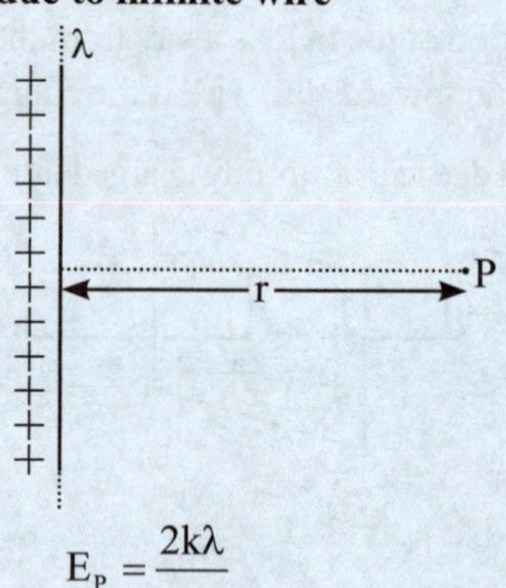

$$E_P = \frac{2k\lambda}{r}$$

Electric field due to Charged arc

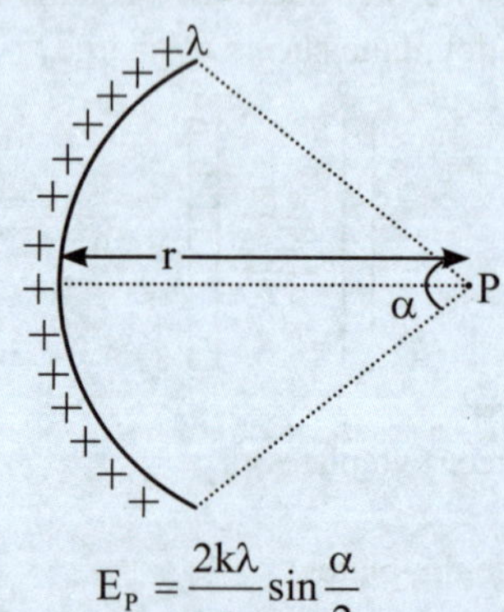

$$E_P = \frac{2k\lambda}{r}\sin\frac{\alpha}{2}$$

Electric field due to charged ring

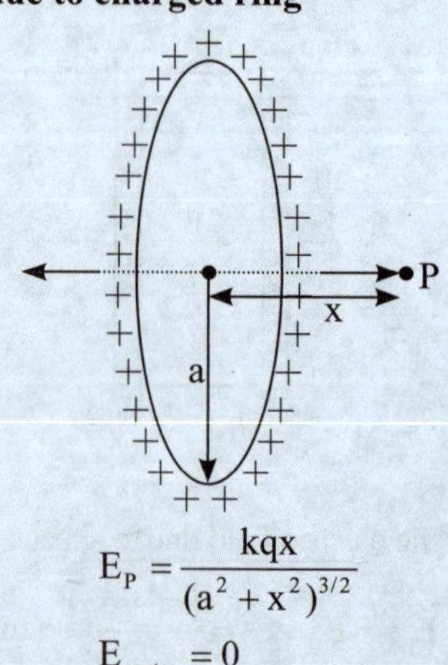

$$E_P = \frac{kqx}{(a^2 + x^2)^{3/2}}$$

$$E_{center} = 0$$

Statements Asked in Exam

- It is observed that if two glass rods rubbed with wool or silk clothes are brought close to each other, they repel each other.
- On rubbing two bodies, the mass of one body slightly gets lowered.
- While calculating the force between the two charges using Coulomb's law, never take into account, the sign of two charges. It just indicates nature of the force.
- The test charge used to measure electric field at a point is taken as unit in magnitude or very small.
- Field lines are always directed from higher potential to lower potential.
- Neutral point is a point where resultant electric field is zero.
- When a moving charged particle enters a uniform electric field at right angle to the field the trajectory is parabola.
- A surface having same electric potential at every point is an equipotential surface.
- Work done in moving a charge along an equipotential surface is always zero.
- Relative permitivity of the material is also known as specific inductive capacitance.
- In air if intensity of electric field exceeds the value 3×10^6 N/C, air ionizes.
- Force characteristic of field is intensity and work characteristic of field is potential.
- 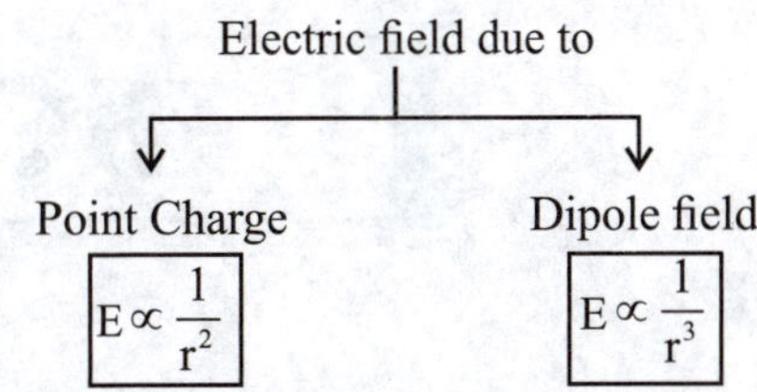

- Inducting body neither gain nor loses charge.
- Quantisation of charge means that charge can exist only as an integral multiple of charge on one electron.
- At the macroscopic level, the quantisation of charge has no practical consequences and can be ignored
- Conservation of charge means that in an isolated system total electric charge always remains constant.
- For metals, dielectric constant K = ∞ and induced charge is opposite to the inducing charge.
- Coulomb's law is valid at a distance greater than 10^{-15} m (1 fermi).
- Decreasing order of fundamental forces is
- $F_{Nuclear} > F_{Electromagnetic} > F_{Weak\ nuclear} > F_{Gravitational}$
- At the center of the line joining two equal & similar charges $V \neq 0, E = 0$
- Charges can be created only in equal and unlike pairs.
- Coulomb's law of electric force obey Newton's third law of motion.
- A positively charged particle is free to move in an electric field along the line of force.

MCQs ?

NCERT Topic-wise MCQs

Coulomb's Law, Electric Field Due to Point Charges

1. Four charged particles are placed at the vertices of square as shown in the figure. An electron that is free to move is placed at the exact centre of the square. In which direction will the electron move?

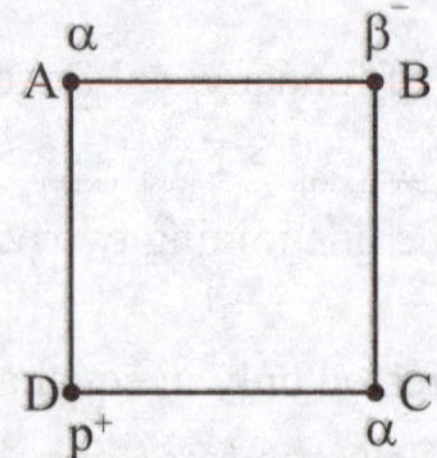

1) It will move toward A 2) It will move toward B
3) It will move toward C 4) It will move toward D

2. Three charges 4q, Q and q are in a straight line in the position of 0, $\ell/2$ and ℓ respectively. The resultant force on q will be zero, if Q =

1) –q 2) –2q
3) $-\frac{q}{2}$ 4) 4q

3. Two charged particles having charge 2×10^{-8} C each are joined by an insulating string of length 1m and the system is kept on a smooth horizontal table, what is the tension in the string?

1) 3.6×10^{-6}N 2) 3.4×10^{-6}N
3) 4×10^{-7}N 4) 3×10^{-4}N

4. A charge q is placed at the center of the line joining two equal charges of magnitude Q. All the three charges are in equilibrium then what will be the value of 'q'?

1) $\frac{Q}{4}$ 2) $\frac{-Q}{4}$
3) Q 4) $\frac{-Q}{8}$

5. How many electrons must be removed from a piece of metal to give it a positive charge of 1×10^{-7}C ?

1) 6.25×10^{-11} 2) 6.25×10^{-12}
3) 6.25×10^{11} 4) 6.25×10^{13}

6. Two charged spheres separated at a distance R exert a force F on each other. If they are immersed in a liquid of dielectric constant 5 then what is the new force between them

1) $\frac{F}{5}$ 2) F
3) 5F 4) $\frac{F}{2}$

7. Charges on two spheres are +10μC and –5μC respectively. They experience a force F. If each of them is given an additional charge + 2μC then new force between them keeping the same distance is

1) 18F 2) $\frac{F}{25}$
3) $\frac{18F}{25}$ 4) $\frac{25}{18}$F

8. Three charges q_1, q_2, q_3 each equal to q are placed at the vertices of an equilateral triangle of side ℓ. What will be the force on a charge Q placed at the centroid of triangle?

1) 4N 2) 5N
3) 10N 4) Zero

9. Total charge Q is broken in two parts Q_1 & Q_2 and they are placed at a distance R from each other. The maximum force of repulsion between them will occur when

1) $Q_2 = \frac{Q}{R}, Q_1 = Q - \frac{Q}{R}$
2) $Q_2 = \frac{Q}{4}, Q_1 = Q - \frac{2Q}{R}$
3) $Q_2 = \frac{Q}{4}, Q_1 = Q - \frac{3Q}{4}$
4) $Q_2 = \frac{Q}{2}, Q_1 = Q - \frac{Q}{2}$

10. Figure shows the electric lines of force emerging from a charged body. If the electric field at A and B are E_A and E_B respectively and if the displacement between A and B is r then

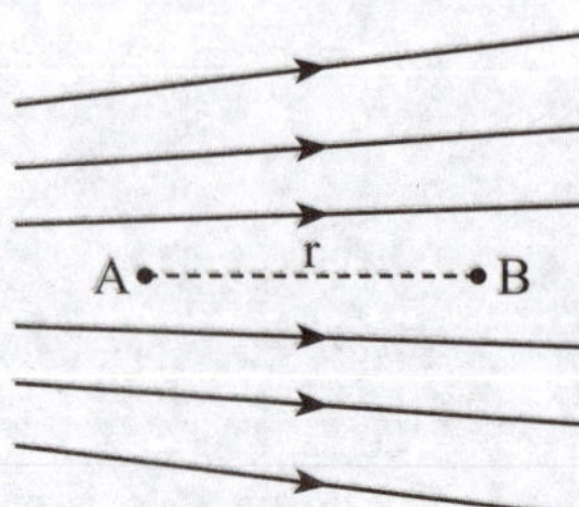

1) $E_A > E_B$ 2) $E_A < E_B$
3) $E_A = \frac{E_B}{r}$ 4) $E_A = \frac{E_B}{r^2}$

11. If there are 100 free electrons excess on a body, then what will be net charge on it?

1) 1.6×10^{-17} C
2) 1.6×10^{-21} C
3) -1.6×10^{-17} C
4) -1

12. The magnitude of force, exerted by a uniform electric field on an electron having mass m_e and a proton of mass m_p are represented as F_e and F_p respectively, are related as

1) $F_p = F_e$
2) $\frac{F_e}{F_p} = \frac{m_e}{m_p}$
3) $\frac{F_e}{F_p} = \frac{m_p}{m_e}$
4) $\frac{F_e}{F_p} = \frac{m_e^2}{m_p^2}$

13. The minimum electrostatic force between two charged particles placed at a distance of 1 m is,

1) 2.3×10^{-28} N
2) 6.2×10^{-34} N
3) 1.02×10^{-26} N
4) 4.2×10^{-27} N

14. In which of the following cases the electric field at the centre is not zero?

1) 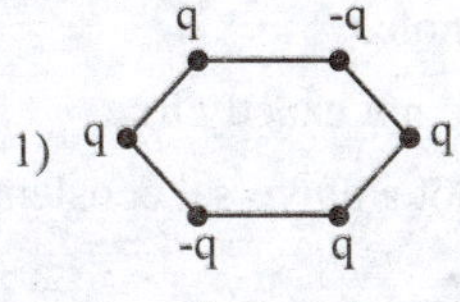

2) 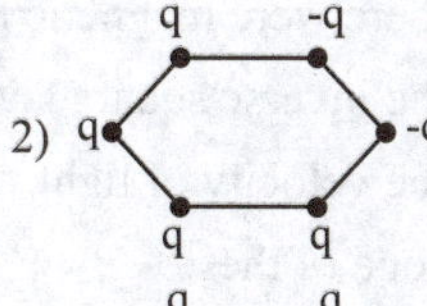

3) 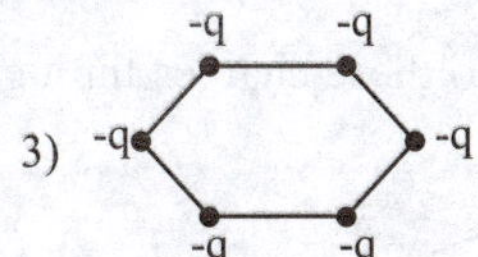

4) q q q q q q

15. Two charge q and $-3q$ are placed fixed on x-axis separated by a distance 'd'. Where should a third charge 2q be placed such that it will not experience any force?

1) $\frac{d}{2}(1+\sqrt{3})$ to the left of q
2) $\frac{d}{2}(1+\sqrt{3})$ to the right of q
3) $\frac{d}{2}(1+\sqrt{5})$ to the left of q
4) $\frac{d}{2}(1+\sqrt{5})$ to the right of q

Electric Field Due to Uniform Bodies, Field, Motion of Charge

16. Which one of the following is not a property of electrostatic field lines?

1) Two lines cannot cross each other
2) Field line start from negative and ends at positive charge
3) They cannot form closed loop
4) None of the above

17. A particle of mass m and charge q is thrown at a speed u against a uniform electric field E. How much distance will it travel before coming to rest?

1) $mu^2/2qE$
2) $2mu^2/qE$
3) $mqE/2u^2$
4) mu/qE

18. A uniform electric field of 10 N/C is created between two parallel charged plates (as shown in figure). An electron enters the field symmetrically between the plates with a kinetic energy 0.5eV. The length of each plate is 10 cm. The angle (θ) of deviation of the path of electron as it comes out of the field is (in degree).

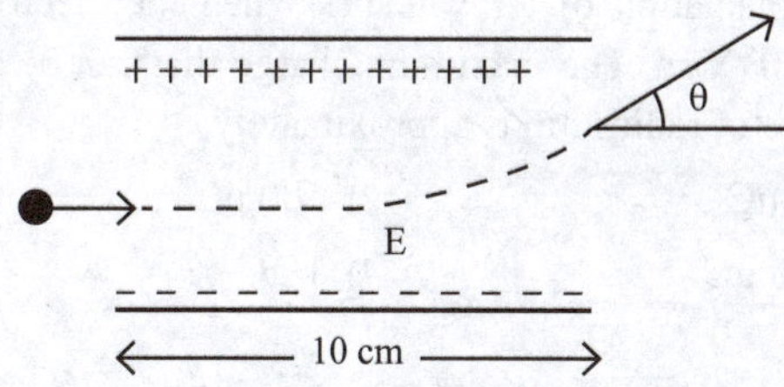

1) 30
2) 45
3) 60
4) 90

19. Two equal and opposite charges of masses m_1 & m_2 are accelerated in a uniform electric field through the same distance. What is the ratio of magnitudes of their accelerations if their ratio of masses is $\frac{m_1}{m_2} = 0.5$.

1) $\frac{a_1}{a_2} = 0.5$
2) $\frac{a_1}{a_2} = 1$
3) $\frac{a_1}{a_2} = 2$
4) $\frac{a_1}{a_2} = 3$

20. The electric field intensity at a point in vacuum is equal to

1) Zero
2) Force a proton would experience there
3) Force an electron would experience there
4) Force a unit positive charge would experience there

21. In a uniform electric field if a charge is fired in a direction different from the line of electric field then the trajectory of the charge will be a

1) Straight line
2) Circle
3) Parabola
4) Ellipse

22. Two spheres of radii R_1 & R_2 respectively are charged and joined by a wire. The ratio of electric field of spheres at any point equal distant from both the spheres is:

1) $\left(\frac{R_2}{R_1}\right)^2$
2) $\left(\frac{R_1}{R_2}\right)^2$
3) $\frac{R_2}{R_1}$
4) $\frac{R_1}{R_2}$

23. A uniform electric field $E = (8m/e)$ V/m is created between two parallel plates of length 1 m as shown in figure, (where m = mass of electron and e = charge of electron). An

electron enters the field symmetrically between the plates with a speed of 2 m/s. The angle of the deviation (θ) of the path of the electron as it comes out of the field will be.

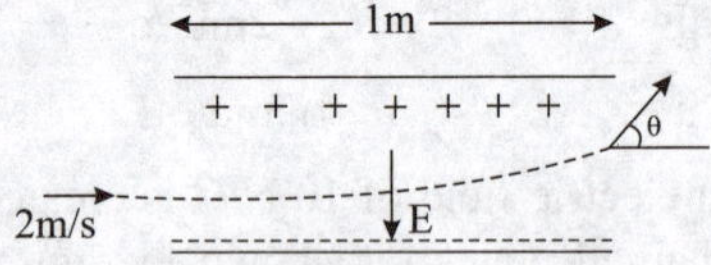

1) $\tan^{-1}(4)$ 2) $\tan^{-1}(2)$
3) $\tan^{-1}\left(\frac{1}{3}\right)$ 4) $\tan^{-1}(3)$

24. The insulation of air vanishes when the electric field is 5×10^5 V/m. The maximum charge that can be given to a sphere of radius 3m is approximately

1) 5 μC 2) 500 μC
3) 10 μC 4) 1 μC

Gauss Law & Its Application, Conductors

25. In a uniform electric field E if we consider an imaginary cubical closed surface of side a, then find the net flux through the cube?

1) $\sqrt{2}Ea^2$ 2) 0
3) $2Ea^2$ 4) Ea^2

26. A charge q is situated at the centre of a cube. Electric flux through one of the faces of the cube is

1) $\frac{q}{\varepsilon_0}$ 2) $\frac{q}{3\varepsilon_0}$
3) $\frac{q}{6\varepsilon_0}$ 4) Zero

27. The value of electric flux leaving out from a unit positive charge in vacuum is:

1) $4\pi\varepsilon_0$ 2) ε_0
3) ε_0^{-1} 4) $(4\pi\varepsilon_0)^{-1}$

28. A ball having a charge –100e is placed at the centre of a hollow spherical shell which has a net charge of –20e. What is the charge on the shell's outer surface?

1) + 80e 2) –20e
3) –100e 4) –120e

29. The electric field inside a spherical shell of uniform surface charge density is

1) Zero
2) Constant less than zero
3) Directly proportional to the distance from the centre
4) None of the above

30. Electric field intensity at a point in between two parallel sheets with like charges of same surface charge densities (σ) is

1) $\frac{\sigma}{2\varepsilon_o}$ 2) $\frac{3\sigma}{\varepsilon_o}$
3) $\frac{2\sigma}{\varepsilon_o}$ 4) Zero

31. For a given surface the Gauss's law is stated as $\int \vec{E}.d\vec{S} = 0$. From this we can conclude that

1) E is necessarily zero on the surface
2) E is perpendicular to the surface at every point
3) The total flux through the surface is zero
4) The flux is only going out of the surface

32. An electric charge q is placed at the center of a cube of side ℓ. The electric flux on one of its faces will be

1) $\frac{6q}{\varepsilon_o}$ 2) $\frac{q}{\varepsilon_o}$
3) $\frac{q}{6\varepsilon_o}$ 4) $\frac{2}{3}\frac{q}{\varepsilon_o}$

33. Gauss's law should be invalid if

1) There were magnetic monopoles
2) The inverse square law were not exactly true
3) The velocity of light were not a universal constant
4) None of these

34. The flux of electric field due to these charges through the surface S is

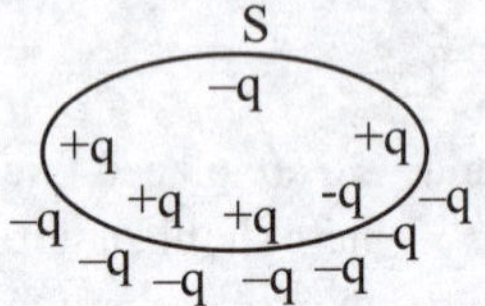

1) $\frac{2q}{\varepsilon_o}$ 2) $\frac{3q}{\varepsilon_o}$
3) $\frac{-5q}{\varepsilon_o}$ 4) $q\varepsilon_o$

35. Gauss law is valid for

1) Closed loops 2) Open loops
3) Both (1) and (2) 4) none of the above

36. A circular disc of radius 'r' is placed along the plane of paper. A uniform electric field $\vec{E}$ is also present and is parallel to the plane of the paper. What amount of electric flux is associated with it?

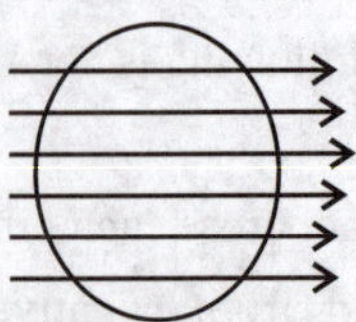

1) $E\pi r^2$ 2) Zero
3) $2E\pi r$ 4) $\frac{\pi r^2 q}{E}$

37. Five electric dipoles are placed inside a shell. What will be the amount of electric flux associated with the shell?

1) $5q\varepsilon_o$
2) $\frac{10q}{\varepsilon_o}$
3) Zero
4) $\frac{5q}{\varepsilon_o}$

38. What is the flux linked through a cube of side 'ℓ' if a charge 2q is placed at any one vertex of the cube?

1) $\frac{q}{\varepsilon_o}$
2) $\frac{q}{8\varepsilon_o}$
3) $\frac{2q}{\varepsilon_o}$
4) $\frac{q}{4\varepsilon_o}$

39. In a region, the intensity of an electric field is given by $\vec{E} = 8\hat{i} + 8\hat{j} + \hat{k}$ in NC^{-1}. The electric flux through a surface $\vec{A} = 10\hat{i}\,m^2$ in the region is

1) 10 Nm^2C^{-1}
2) 80 Nm^2C^{-1}
3) 8 Nm^2C^{-1}
4) None of these

40. A sphere S_1 of radius R, encloses a total charge q. If there is another concentric sphere S_2 of radius 2R and there is no additional charges between S_1 and S_2. What is the ratio of electric flux through S_1 and S_2?

1) 1.5
2) 2
3) 1
4) 0.5

41. Two parallel line charges $+\lambda$ and $-\lambda$ are placed with a separation distance R in free space. The net electric field exactly mid way between the two line charge is

1) Zero
2) $\frac{2\lambda}{\pi\varepsilon_o R}$
3) $\frac{\lambda}{\pi\varepsilon_o R}$
4) $\frac{1}{2\pi\varepsilon_o R}$

42. A charge q is located at the center of a cube. The electric flux through any two consecutive faces will be

1) $\frac{\pi q}{6(4\pi\varepsilon_o)}$
2) $\frac{q}{6(4\pi\varepsilon_o)}$
3) $\frac{2\pi q}{6(4\pi\varepsilon_o)}$
4) $\frac{4\pi q}{3(4\pi\varepsilon_o)}$

43. If the electric flux entering and leaving an enclosed surface respectively are ϕ_1 & ϕ_2 the electric charge inside the surface will be

1) $\frac{\phi_2 - \phi_1}{\varepsilon_o}$
2) $\frac{\phi_2 + \phi_1}{\varepsilon_o}$
3) $(\phi_1 - \phi_2)\,\varepsilon_o$
4) $\varepsilon_o(\phi_2 - \phi_1)$

44. Electric flux emanating through a surface element $d\vec{s} = 5\hat{i}$ placed in an electric field $E = 4\hat{i} + 4\hat{j} + 4\hat{k}$ is

1) 10 units
2) 20 units
3) 4 units
4) 16 units

Dipole

45. Force of interaction between two co-axial short electric dipoles whose centres distance apart varies as

1) $\frac{1}{R}$
2) $\frac{1}{R^2}$
3) $\frac{1}{R^3}$
4) $\frac{1}{R^4}$

46. The angle between the dipole moment and electric field at any point on the equatorial plane is

1) 180°
2) 0°
3) 45°
4) 90°

47. The torque acting on a dipole of moment $\vec{P}$ in an electric field $\vec{E}$ is

1) $\vec{P} \cdot \vec{E}$
2) $\vec{P} \times \vec{E}$
3) Zero
4) $\vec{E} \times \vec{P}$

48. An electric dipole consists of a positive charge and a negative charge of magnitude 4μC separated by a distance of 5mm. What is the dipole moment?

1) 2×10^{-6} C-m
2) 2×10^{-8} C-m
3) 4×10^{-5} C-m
4) 5×10^{-8} C-m

49. An electric dipole consists of two charges of magnitude 0.1 μC separated by a distance of 2 cm. The dipole is in an external field of 10^5 N/C. What maximum torque does the field exert on the dipole?

1) 10^{-4}N–m
2) 2×10^{-8}N–m
3) 2×10^{-4}N–m
4) 3×10^{-4}N–m

50. Electric charges q, q, and –2q are placed at the corners of an equilateral triangle ABC of side ℓ. The magnitude of electric dipole moment of the system is

1) $\sqrt{3}q\ell$
2) $\sqrt{2}q\ell$
3) $q\ell$
4) $2q\ell$

51. An electric dipole consist of a positive and negative charge of 8μC each placed at a distance of 5mm. What is the magnitude of dipole moment?

1) 10^{-8} C-m
2) 2×10^{-8} C-m
3) 4×10^{-8} C-m
4) None of these

52. What is the magnitude of electric field intensity due to a dipole of moment 4×10^{-8} C-m at a point distant 0.5m from the center of dipole when line joining the point to the center of dipole makes an angle of 60° with dipole axis?

1) 4×10^{-5} N/C
2) 5×10^{-11} N/C
3) 2×10^{7} N/C
4) 3.8×10^{3} N/C

53. A water molecule has an electric dipole moment 7×10^{-30} C-m when it is in vapour state. The distance between the center of positive and negative charge of the molecule is

1) 5×10^{-11} m
2) 2×10^{-11} m
3) 4.1×10^{-1} m
4) 4.375×10^{-11} m

54. An electric dipole, when held at 60° with respect to a uniform electric field of 10^7 NC^{-1} experience a torque of 12×10^{-20} N–m. What is the value of dipole moment?

1) 1.4×10^{-26} C–m
2) 8×10^{-27} C–m
3) 3×10^{-26} C–m
4) 1.4×10^{-27} C–m

55. An electric dipole is placed at an angle of 45° with an electric field of intensity 10^7 NC^{-1}. It experiences a torque equal to 15×10^2 N – m. What will be magnitude of charge on dipole if it is 5 cm long?

1) 1 mC
2) 2 mC
3) 5 mC
4) 4 mC

56. An electric dipole of moment p is lying along a uniform electric field E. The work done in rotating the dipole by 180° is

1) $\sqrt{2}pE$
2) pE
3) 2 pE
4) $\frac{pE}{\sqrt{2}}$

High Order Time Intensive MCQs

1. Two pith balls carrying equal charges are suspended from a common point by strings of equal length, the equilibrium separation between them is r. Now, the strings are rigidly clamped at half the height. The equilibrium separation between the balls now become

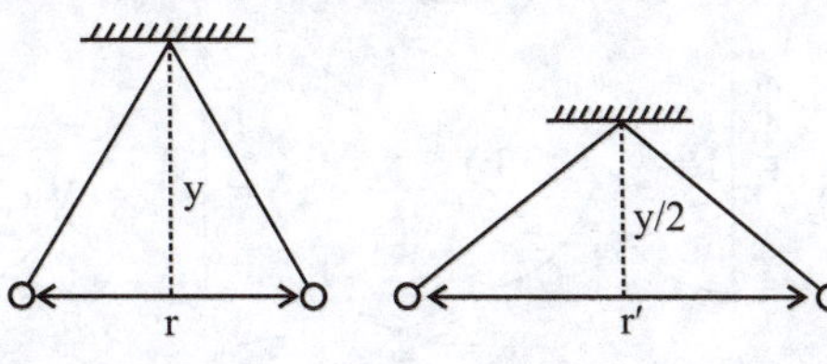

1) $\left(\frac{1}{\sqrt{2}}\right)^2$ 2) $\left(\frac{r}{\sqrt[3]{2}}\right)$
3) $\left(\frac{2r}{\sqrt{3}}\right)$ 4) $\left(\frac{2r}{3}\right)$

2. Consider a non-spherical conductor shown in the figure which is given a certain amount of positive charge. The charge distributes itself on the surface such that the charge densities are σ_1, σ_2 and σ_3 at the region 1, 2 and 3 respectively. Then

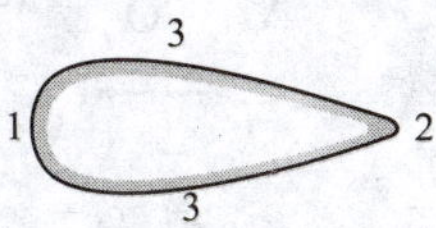

1) $\sigma_1 > \sigma_2 > \sigma_3$ 2) $\sigma_2 > \sigma_3 > \sigma_1$
3) $\sigma_3 > \sigma_1 > \sigma_2$ 4) $\sigma_2 > \sigma_1 > \sigma_3$

3. There are two charge +1μc and +2μc kept at certain separartion, the ratio of electrostatic forces acting on them will be in the ratio
1) 1 : 2 2) 2 : 1
3) 1 : 1 4) 1 : 4

4. A conductor has 14.4×10^{-19} C positive charge. The conductor has (charge on electron = 1.6×10^{-19} C)
1) 9 electrons in excess 2) 27 electrons in short
3) 27 electrons in excess 4) 9 electrons in short

5. A deuteron and an α-particle are placed in an electric field.The forces acting on them are F_1 and F_2 and their accelerations are a_1 and a_2 respectively.
(*i*) $F_1 = F_2$
(*ii*) $F_1 \neq F_2$
(*iii*) $a_1 = a_2$
(*iv*) $a_1 \neq a_2$
1) (*i*), (*ii*) 2) (*ii*), (*iii*)
3) (*ii*), (*iv*) 4) (*ii*), (*iv*)

6. Three concentric metallic spherical shells of radii R, 2 R, 3 R are given charges Q_1, Q_2, Q_3, respectively. It is found that the surface charge densities on the outer surfaces of the shells are equal. Then, the ratio of the charges given to the shells, $Q_1 : Q_2 : Q_3$, is
1) 1 : 2 : 3 2) 1 : 3 : 5
3) 1 : 4 : 9 4) 1 : 8 : 18

7. Two identical charged spheres of material density ρ, suspended from the same point by inextensible strings of equal length make an angle θ between the strings. When suspended in a liquid of density σ the angle θ remains the same. The dielectric constant *k* of the liquid is
1) $\frac{\rho}{\rho-\sigma}$ 2) $\frac{\rho-\sigma}{\rho}$
3) $\frac{\rho}{\rho+\sigma}$ 4) $\frac{\rho+\sigma}{\rho}$

8. An electron moving with the speed 5×10^6 m/s is shooted parallel to the electric field of intensity 1×10^3 N/C. Field is responsible for the retardation of motion of electron. Now evaluate the distance travelled by the electron before coming to rest for an instant (mass of $e = 9 \times 10^{-31}$ kg, charge = 1.6×10^{-19} C)
1) 7 m 2) 0.7 m
3) 7 cm 4) 0.7 cm

9. The electric charges are distributed in a small volume. The flux of the electric field through a spherical surface of radius 10 cm surrounding the total charge is 20 Vm. The flux over a concentric sphere of radius 20 cm will be
1) 20 Vm 2) 25 Vm
3) 40 Vm 4) 200 Vm

10. The electric intensity outside a charged sphere of radius *R* at a distance $r (r > R)$ is
1) $\frac{\sigma R^2}{\varepsilon_0 r^2}$ 2) $\frac{\sigma r^2}{\varepsilon_0 R^2}$
3) $\frac{\sigma r}{\varepsilon_0 R}$ 4) $\frac{\sigma R}{\varepsilon_0 r}$

11. If a positively charged pendulum is oscillating in a uniform electric field as shown in diagram. Its frequency compared to that when it was uncharged-

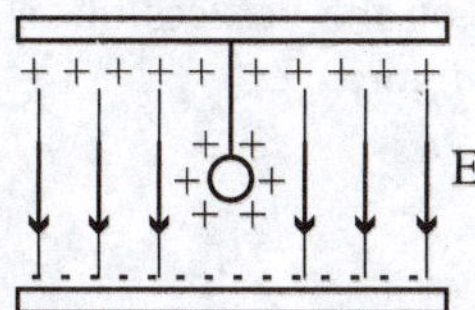

1) Will increase
2) Will decrease
3) Will not change
4) Will first increase and then decrease

12. The figure shows some of the electric field lines corresponding to an electric field. The figure suggests

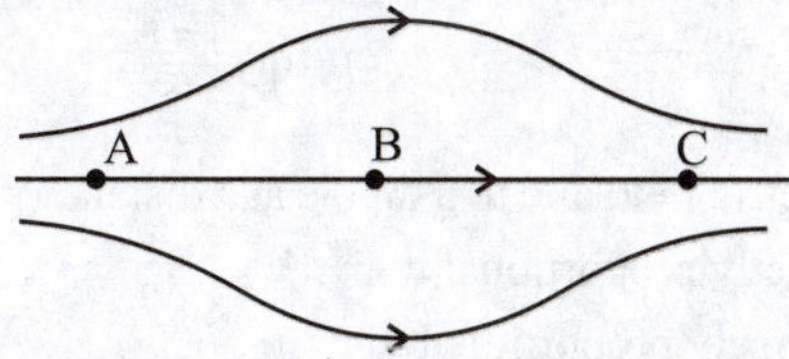

1) $E_A > E_B > E_C$ 2) $E_A = E_B = E_C$
3) $E_A = E_C > E_B$ 4) $E_A > E_C > E_B$

13. Two spheres A and B are charged with the charges of +10 and +20 coulomb respectively and separated by a distance of 80 cm. The electric field at point on the line joining the centres of the two spheres will be zero at a distance from the sphere A:
1) 20 cm 2) 33 cm
3) 45 cm 4) 60 cm

14. An infinite line charge produces a field of 18×10^4 N/C at a distance of 2 metre from it. The linear charge density is?
1) 10^{-5} 2) 2×10^{-5}
3) Zero 4) 4×10^{-5}

15. Five point charges (+q each) are placed at the five vertices of a regular hexagon of side 2 a. What is the magnitude of the net electric field at the centre of the hexagon?
1) $\frac{1}{4\pi\varepsilon_0}\frac{q}{a^2}$ 2) $\frac{q}{16\pi\varepsilon_0 a^2}$
3) $\frac{\sqrt{2}}{4\pi\varepsilon_0}\frac{q}{a^2}$ 4) $\frac{5q}{16\pi\varepsilon_0 a^2}$

16. A hemispherical surface of radius R is kept in a uniform electric field E as shown in figure. The flux through the curved surface is

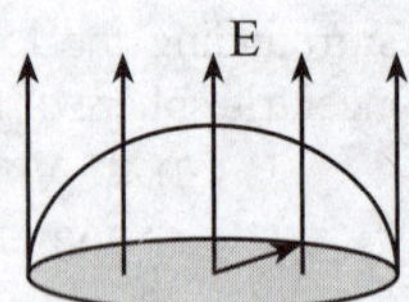

1) $E2\pi R^2$ 2) $E\pi R^2$
3) $E4\pi R^2$ 4) Zero

17. If an electric field is given by $10\hat{i}+3\hat{j}+4\hat{k}$. Calculate the electric flux through a surface of area 10 units lying in yz plane.
1) 100 units 2) 10 units
3) 30 units 4) 40 units

18. Determine the electric dipole moment of the system of three charges, placed on the vertices of an equilateral triangle, as shown in the figure:

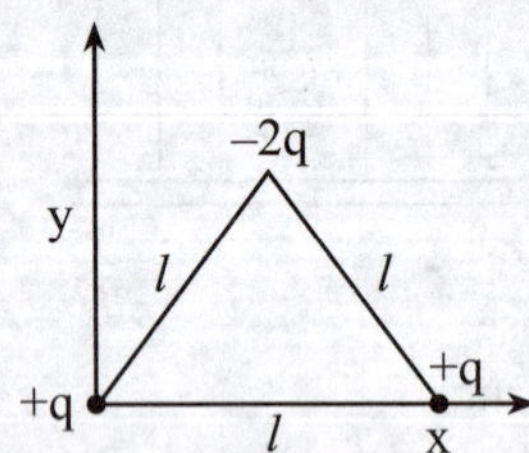

1) $\sqrt{3}ql\frac{\hat{j}-\hat{i}}{\sqrt{2}}$ 2) $2ql\hat{j}$
3) $-\sqrt{3}ql\hat{j}$ 4) $(ql)\frac{\hat{j}-\hat{i}}{\sqrt{2}}$

19. The electric field due to a dipole at a distance r on its axis is
1) Directly proportional to r^3
2) Inversely proportional to r^3
3) Directly proportional to r^2
4) Inversely proportional to r^2

20. Which of the following represents the variation of the electric field with distance from the centre of a uniformly charged non-conduction sphere of radius R?

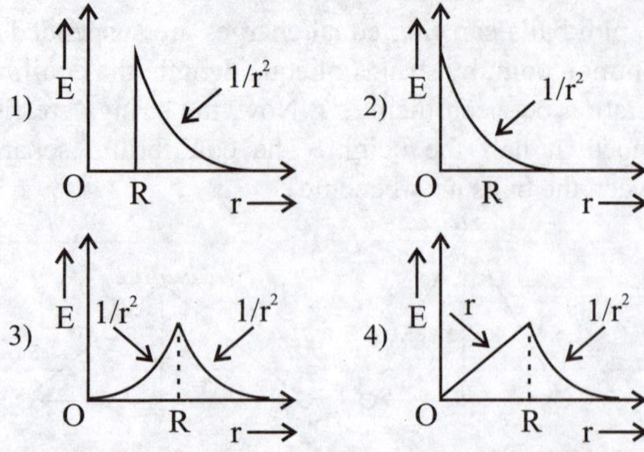

21. Six charges, three positive and three negative of equal magnitude are to be placed at the vertices of a regular hexagon, such that the electric field, at the point O is double the electric field, when only one positive charge of same magnitude is placed at R. Which of the following arrangements of charges is possible for P, Q, R, S, T and U respectively?

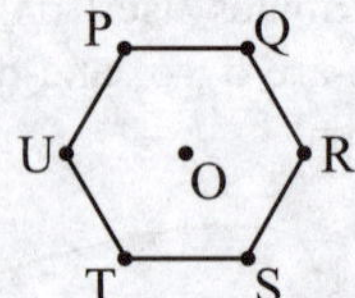

1) +, –, +, –, –, + 2) +, –, +, –, +, –
3) +, +, –, +, –, – 4) –, +, +, –, +, –

22. A hollow cylinder has a charge q C within it. If ϕ is the electric flux in unit of volt meter associated with the curved surface B, the flux linked with the plane surface A in unit of volt meter will be

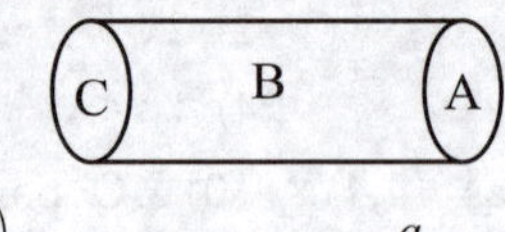

1) $\frac{1}{2}\left(\frac{q}{\varepsilon_0}-\phi\right)$ 2) $\frac{q}{2\varepsilon_0}$
3) $\frac{\phi}{3}$ 4) $\frac{q}{\varepsilon_0}-\phi$

23. Three infinitely long charge sheets are placed as shown in the figure.

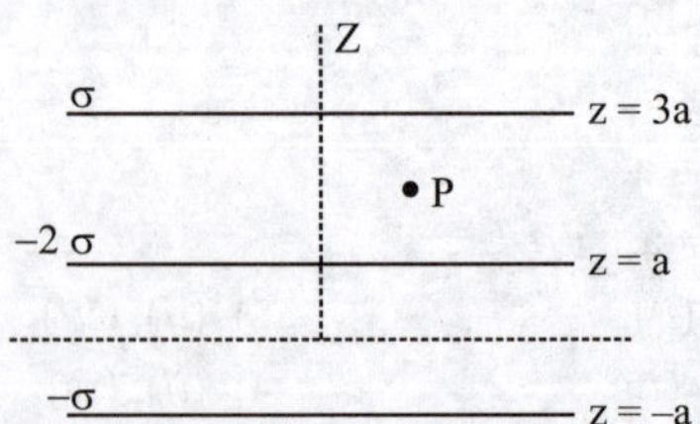

The electric field at point P is
1) $\frac{2\sigma}{\varepsilon_0}\hat{k}$ 2) $-\frac{2\sigma}{\varepsilon_0}\hat{k}$
3) $\frac{4\sigma}{\varepsilon_0}\hat{k}$ 4) $-\frac{4\sigma}{\varepsilon_0}\hat{k}$

Paragraph for Q. 24 to 25

A particle of mass m carrying charge q_1 is revolving around a fixed charge $-q_2$ in a circular path of radius r. Calculate the period of revolution and its speed also.

24. Speed of particle

1) $v = \sqrt{\frac{q_1 q_2}{4\pi\varepsilon_0 mr}}$ 2) $v = \sqrt{\frac{4\pi\varepsilon_0 mr}{q_1 q_2}}$

3) $v = \sqrt{\frac{q_1 q_2}{2\pi\varepsilon_0 mr}}$ 4) $v = \sqrt{\frac{2\pi\varepsilon_0 mr}{q_1 q_2}}$

25. Calculate the period of revolution

1) $2\sqrt{\frac{\pi^3\varepsilon_0 r^3 m}{q_1 q_2}}$ 2) $\sqrt{\frac{\pi^3\varepsilon_0 r^3 m}{q_1 q_2}}$

3) $4\sqrt{\frac{\pi^3\varepsilon_0 r^3 m}{q_1 q_2}}$ 4) None of the above

26. A particle of mass m and charge q is located midway between two fixed charged particles each having a charge q and a distance 2ℓ apart. If it is displaced along the line connecting them and released. Also find its time period for SHM.

1) $4\pi\sqrt{\frac{m\pi\varepsilon_0\ell^3}{q^2}}$ 2) $\pi\sqrt{\frac{m\pi\varepsilon_0\ell^3}{2q^2}}$

3) $2\pi\sqrt{\frac{m\pi\varepsilon_0\ell^3}{q^2}}$ 4) $\pi\sqrt{\frac{2m\pi\varepsilon_0\ell^3}{q^2}}$

27. A conductor *PQ* as shown in figure is charged. σ_A, σ_B, σ_C and σ_D are surface charge densities of points *A*, *B*, *C* and *D* respectively

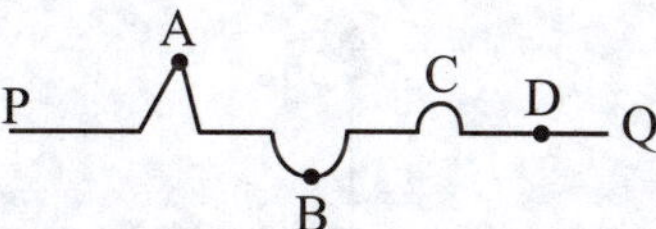

1) $\sigma_A < \sigma_B < \sigma_C < \sigma_D$ 2) $\sigma_A = \sigma_B = \sigma_C > \sigma_D$

3) $\sigma_A > \sigma_C > \sigma_B > \sigma_D$ 4) $\sigma_A = \sigma_B = \sigma_C = \sigma_D$

28. Two spherical conductors B and C having equal radii and carrying equal charges in them repel each other with a force F when kept apart at some distance. A third spherical conductor having same radius as that of B but uncharged, is brought in contact with B, then brought in contact with C and finally removed away from both. The new force of repulsion between B & C is

1) $\frac{F}{4}$ 2) $\frac{3F}{4}$

3) $\frac{F}{8}$ 4) $\frac{3F}{8}$

29. Two identical charged shperes suspended from a common point by two massless strings of lengths ℓ, are initially at a distance $d(d << \ell)$ apart because of their mutual repulsion. The charges begin to leak from both The spheres at a constant rate. As a result, the spheres approach each other with a velocity v. Then v varies as a function of the distance x between the spheres, as

1) $v \propto x^{\frac{1}{2}}$ 2) $v \propto x$

3) $v \propto x^{-\frac{1}{2}}$ 4) $v \propto x^{-1}$

30. Two small conducting spheres of equal radius have charges +10 μC and –20 μC respectively and placed at a distance R from each other. They experience force F_1. If they are brought in contact and separated to the same distance, they experience force F_2. The ratio of F_1 to F_2 is

1) 1 : 8 2) –8 : 1

3) 1 : 2 4) –2 : 1

31. Eight charges, 1μC, –7μC, –4μC, 10μC, 2μC, –5μC, –3μC and 6μC are situated at the eight corners of a cube of side 20 cm. A spherical surface of radius 80 cm encloses this cube. The centre of the sphere coincides with the centre of the cube. Then the total outgoing flux from the spherical surface (in unit of volt meter) is:

1) $36\pi \times 10^3$ 2) $684\pi \times 10^3$

3) Zero 4) None of the above

32. A positive charge Q is uniformly distributed along a circular ring of radius R. A small test charge q is placed at the centre of the ring as shown in figure. Then

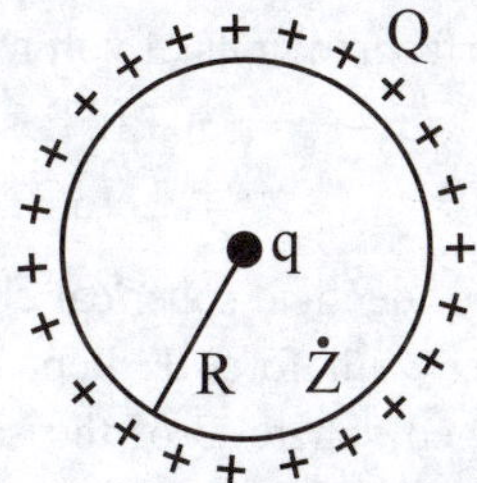

1) If q > 0, and is displaced away from the centre in the plane of the ring, it will be pushed back towards the centre.

2) If q < 0 and is displaced away from the centre in the plane of the ring, it will never return to the centre and will continue moving till it hits the ring.

3) If q < 0 it will perform SHM for small displacement along the axis.

4) All of the above

33. A metallic spherical shell has an inner radius R_1 and outer radius R_2. A charge is placed at the centre of the spherical cavity. The surface charge density on the inner surface is

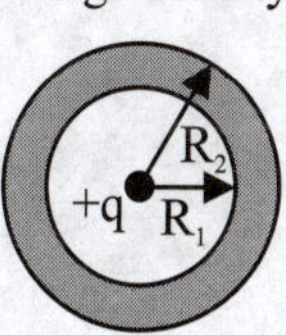

1) $\frac{q}{4\pi R_1^2}$ 2) $\frac{-q}{4\pi R_1^2}$

3) $\frac{q^2}{4\pi R_2^2}$ 4) $\frac{q}{4\pi R_2^2}$

34. The electric field components in the given figures are $E_x = \alpha x^{1/2}$, $E_y = E_z = 0$ in which $\alpha = 800 N C^{-1} m^{-1/2}$. The charge within the cube is if net flux through the cube is 1.05 N m²C⁻¹ (assume a = 0.1 m)

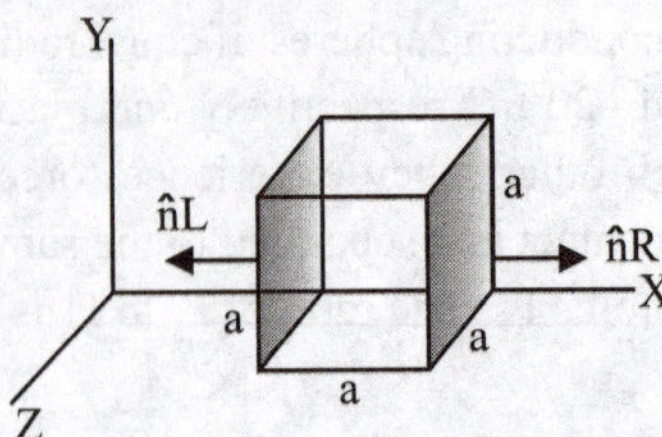

1) 9.27×10^{-12} C
2) 9.27×10^{12} C
3) 6.97×10^{-12} C
4) 6.97×10^{12} C

35. A square surface of side L metre in the plane of the paper is placed in a uniform electric field E (volt/m)

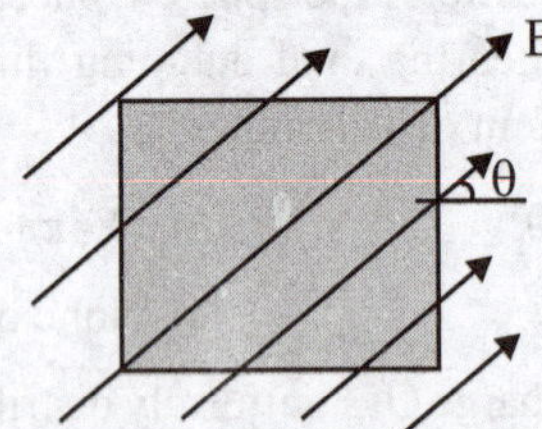

acting along the same plane at an angle θ with the horizontal side of the square as shown in the figure. The electric flux linked to the surface, in units of volt m, is

1) EL^2
2) $EL^2 \cos\theta$
3) $EL^2 \sin\theta$
4) zero

36. A and B are two identical spherical charged bodies which repel each other with force F, kept at a finite distance. A third uncharged sphere C of the same size is brought in contact with sphere B and removed. It is then kept at midpoint of A and B. Find the magnitude of force on C.

1) F/2
2) F/8
3) F
4) zero

37. A body has charge of $-2\mu C$. If it has 2.5×10^{13} protons, then how many electrons the body has ?

1) 1.25×10^{13}
2) 2.5×10^{13}
3) 3.75×10^{13}
4) None of these

38. A uniformly charged then spherical shell of radius R carries uniform surface charge density of σ per unit area. It is made of two hemispherical shells, held together by pressing them with force F (see figure).

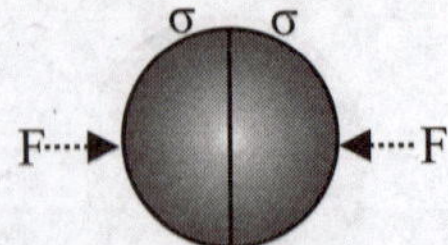

F is proportional to:

1) $\frac{1}{\varepsilon_o}\sigma^2 R^2$
2) $\frac{1}{\varepsilon_o}\sigma^2 R$
3) $\frac{1}{\varepsilon_o}\frac{\sigma^2}{R}$
4) $\frac{1}{\varepsilon_o}\frac{\sigma^2}{R^2}$

39. q, 2q, 3q and 4q charges are placed at the four corners A, B, C and D of a square. The field at the centre P of the square has the direction along

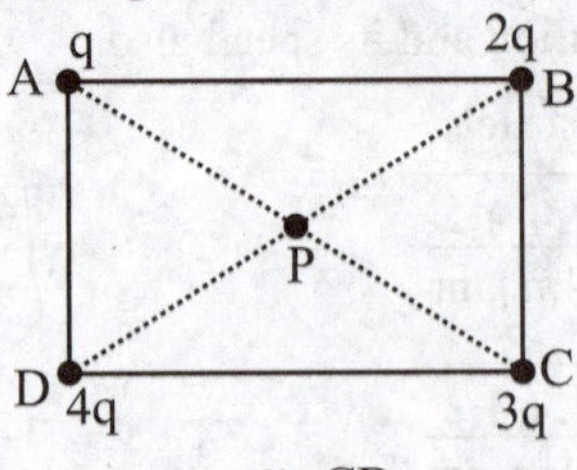

1) AB
2) CB
3) AC
4) BD

40. A solid conducting sphere of radius a has a net positive charge 2Q. A conducting spherical shell of inner radius b and outer radius c is concentric with the solid sphere and has a net charge –Q. The surface charge density on the inner and out surface of th spherical shell will be

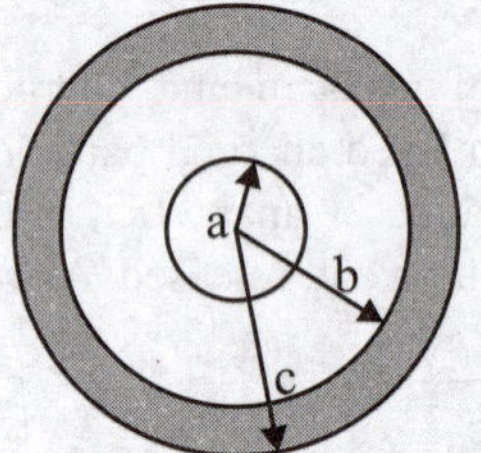

1) $-\frac{2Q}{4\pi b^2}, \frac{Q}{4\pi c^2}$
2) $-\frac{Q}{4\pi b^2}, \frac{Q}{4\pi c^2}$
3) $0, \frac{Q}{4\pi c^2}$
4) None of these

41. 4 charges are placed each at a distance 'a' from origin. The dipole moment of configuration is:

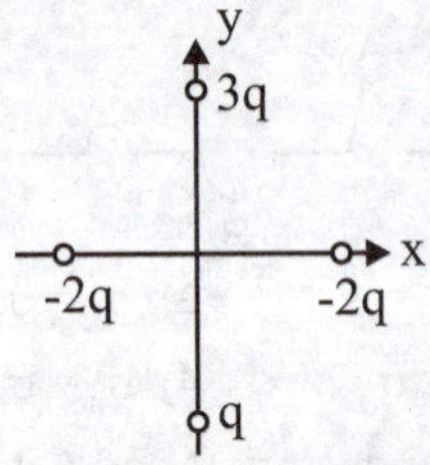

1) $2qa\hat{j}$
2) $3qa\hat{j}$
3) $2aq[\hat{i}+\hat{j}]$
4) None of these

42. Two metal spheres of same mass are suspended from a common point by a light insulating string. The length of each string is same. The sphere are given electric charges +q on one end and +4q on the other. Which of the following diagrams best shows the resulting positions of sphere?

1) 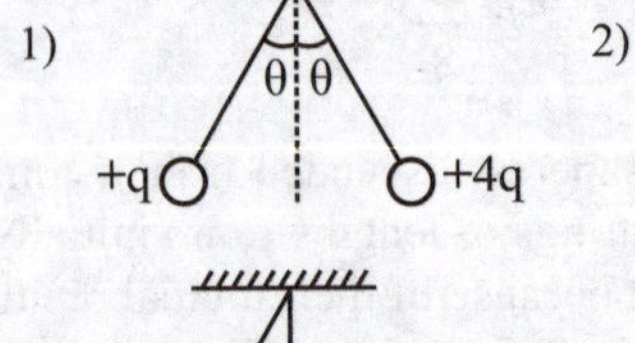

2) 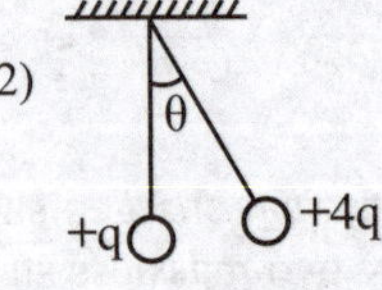

3) 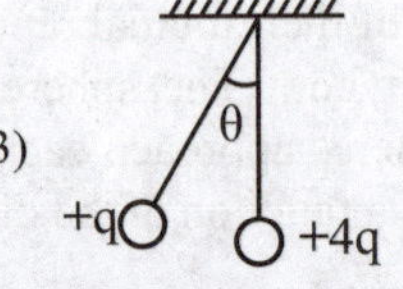

4) 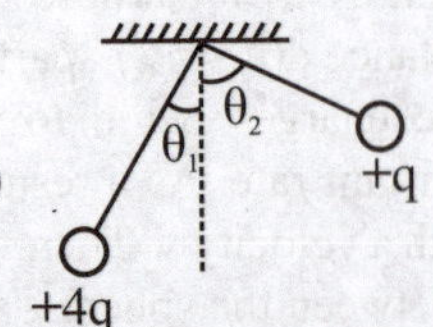

Statement Based MCQs

Directions: These questions consist of two statements each, printed as Statement-I and Statement-II. While answering these questions, you are required to choose any one of the following four responses.

1) **Both Statement-I and Statement-II are correct.**
2) **Both Statement-I and Statement-II are incorrect.**
3) **Statement-I is correct & Statement-II is incorrect.**
4) **Statement-I is incorrect & Statement-II is correct.**

1. **Statement-I:** If a point charge q is placed in front of an infinite grounded conducting plane surface, the point charge will experience a force.

   Statement-II: The force on the point charge in front of the conducting plane surface is due to the induced charge on the conducting surface which is at zero potential.

2. **Statement-I:** When charges are shared between two bodies, there occurs no loss of charge, but there does occur a transformation of energy.

   Statement-II: In case of sharing of charges conservation of energy fails.

3. **Statement-I:** On going away from a point charge or a small electric dipole, electric field decreases at the same rate in both the cases.

   Statement-II: Tangent at any point of electric line of force gives the direction of electric field

4. **Statement-I:** The surface charge densities of two spherical conductors of different radii are equal. Then the electric field intensities near their surface are also equal.

   Statement-II: Surface charge density is equal to charge per unit area.

5. **Statement-I:** If a conducting medium is placed between two charges, then electric force between them becomes zero.

   Statement-II: When a dielectric medium is placed between two charges, then force between the two charges is reduced.

Assertion & Reason MCQs

Directions: These questions consist of two statements each, printed as Assertion and Reason. While answering these questions, you are required to choose any one of the following four responses.

1) **Both Assertion and Reason are True and the Reason is a correct explanation of the Assertion.**
2) **Both Assertion and Reason are True but Reason is not a correct explanation of the Assertion.**
3) **Assertion is True but the Reason is False.**
4) **Assertion is False but Reason is True.**

1. **Assertion:** Charge is invariant.

   Reason: Charge does not depend on speed or frame of reference.

2. **Assertion:** The lightning conductor at the top of a building has sharp pointed ends.

   Reason: The surface charge density at sharp points is very high resulting in of electric wind.

3. **Assertion:** A metallic shield in form of a hollow shell may be built to block an electric field.

   Reason: In a hollow spherical shield, the electric field inside it is zero at every point.

4. **Assertion:** Two protons repel each other.

   Reason: Like charges repel each other while unlike charges attract each other.

5. **Assertion:** In a cavity within a conductor, the electric field is non-zero.

   Reason: Charges in a conductor reside only its surface.

Match the Columns MCQs

1. The Column-I gives the two point charge system separated by 2a and the column-II gives the variation of magnitude of electric field intensity along x-axis. Match the situation in Column-I with the results in Column-II.

	Column-I		Column-II
(a)	q at (–a, 0), q at (a, 0); x′ ← → x; (0, 0); a	(i)	Increases as x increases in the interval $0 \le x < a$
(b)	q at (–a, 0), –q at (a, 0); x′ ← → x; (0, 0); a	(ii)	Decreases as x increases in the interval $0 \le x < a$
(c)	y; q at (0, + a); x; (0, 0); q at (0, – a)	(iii)	Zero at x = 0
(d)	y; –q at (0, + a); x; (0, 0); q at (0, – a)	(iv)	Decreases as x increases in the interval $a < x < \infty$

1) (a)-(i, ii) (b)-(ii) (c)-(i, ii) (d)-(iii)
2) (a)-(i, iii, iv) (b)-(i, iv) (c)-(iii, iv) (d)-(ii, iv)
3) (a)-(i, ii) (b)-(iii) (c)-(ii) (d)-(iv)
4) (a)-(iii, iv) (b)-(iv) (c)-(iii, i) (d)-(ii)

2. Column-I gives certain situations involving two thin conducting shells connected by a conducting wire via a key K. In all situations one sphere has net charge +q and other sphere has no net charge. After the key K is pressed, Column-II gives some resulting effect. Match the figures in Column-I with the statements in Column-II.

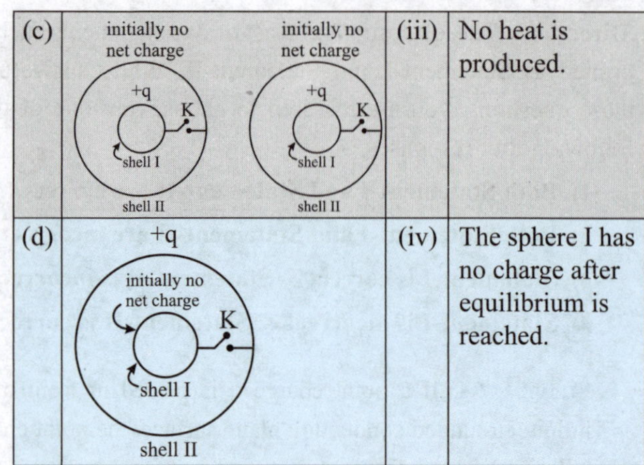

	Column-I		Column-II
(a)	+q; K; initially no net charge; shell I; shell II	(i)	Charge flows through connecting wire
(b)	+q; K; initially no net charge; shell I; shell II	(ii)	Potential energy of system of spheres decreases.
(c)	initially no net charge; +q; K; shell I; shell II — initially no net charge; +q; K; shell I; shell II	(iii)	No heat is produced.
(d)	+q; initially no net charge; K; shell I; shell II	(iv)	The sphere I has no charge after equilibrium is reached.

1) (a)-(i, ii) (b)-(ii, iv) (c)-(i, iii) (d)-(i, iii)
2) (a)-(i, ii) (b)-(i, ii) (c)-(i, ii, iv) (d)-(iii, iv)
3) (a)-(i, ii) (b)-(iv) (c)-(iv, ii) (d)-(iv)
4) (a)-(iii, iv) (b)-(iv) (c)-(iii, i) (d)-(ii)

3. Five identical charges are kept at five vertices of a regular hexagon. If in the given situation electric field at centre is E. Then match the column-I and column-II.

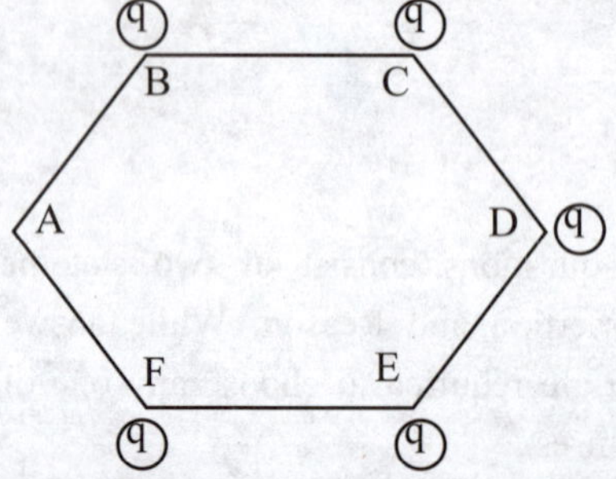

	Column-I		Column-II
(a)	If charge at B is removed, then electric field will become	(i)	2E
(b)	If charge at C is removed, then electric field will become	(ii)	E
(c)	If charge at D is removed, then electric field will become	(iii)	Zero
(d)	If charge at B and C both are removed, then electric field will become	(iv)	$\sqrt{3}E$

1) (a)-(iv) (b)-(ii) (c)-(iii) (d)-(i)
2) (a)-(ii) (b)-(iv) (c)-(i) (d)-(iii)
3) (a)-(iii) (b)-(iv) (c)-(i) (d)-(ii)
4) (a)-(iv) (b)-(iii) (c)-(ii) (d)-(i)

4. The electric field E is measured at a point P (0, 0, d) generated due to various charge distributions and the dependence of E on d is found to be different for different charge distributions. Column-I contains different relations between E and d. Column-II describes different electric charge distributions, along with their locations. Match the functions in Column-I with the related charge distributions in Column-II.

Column-I		Column-II	
(a)	E is independent of d	(i)	A point charge Q at the origin
(b)	$E \propto \frac{1}{d}$	(ii)	A small dipole with point charges Q at (0, 0, ℓ) and – Q at (0, 0, – ℓ). Take $2\ell << d$
(c)	$E \propto \frac{1}{d^2}$	(iii)	An infinite line charge coincident with the x-axis, with uniform linear charge density λ
(d)	$E \propto \frac{1}{d^3}$	(iv)	Ininite plane charge coincident with the xy-plane with uniform surface charge density.

1) (a)-(iii) (b)-(iv) (c)-(ii) (d)-(i)
2) (a)-(iv) (b)-(iii) (c)-(i) (d)-(ii)
3) (a)-(i) (b)-(ii) (c)-(iii) (d)-(iv)
4) (a)-(ii) (b)-(i) (c)-(iv) (d)-(iii)

NCERT Exemplar MCQs

1. In figure two positive charges q_2 and q_3 fixed along the y-axis, exert a net electric force in the + x-direction on a charge q_1 fixed along the x-axis. If a positive charge Q is added at (x, 0), the force on q_1:

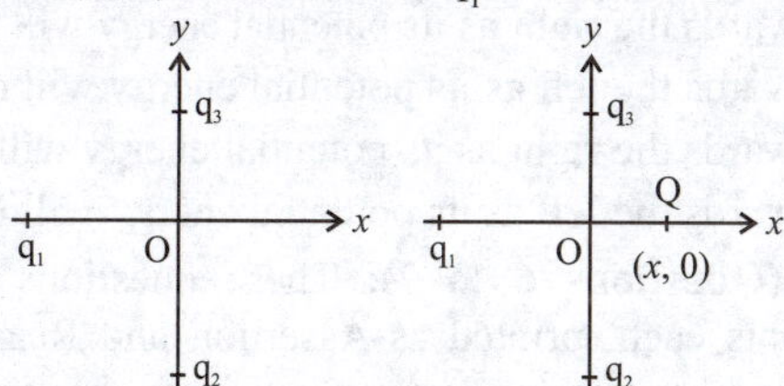

1) Shall increase along the positive x-axis
2) Shall decrease along the positive x-axis
3) Shall point along the negative x-axis
4) Shall increase but the direction changes because of intersection of Q with q_2 and q_3

2. A point positive charge is brought near an isolated conducting sphere (figure). The electric field is best given by:

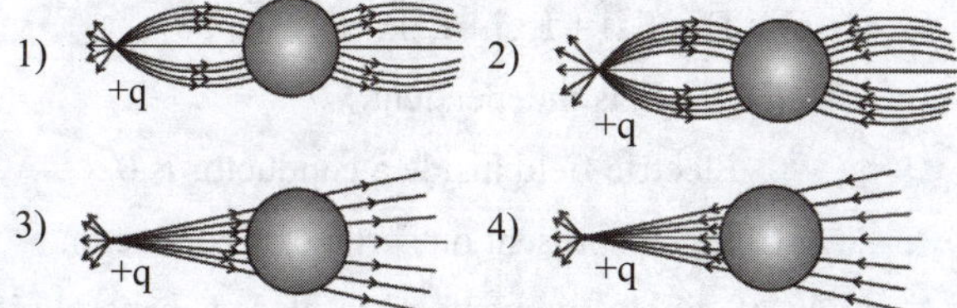

3. The electric flux through the surface:

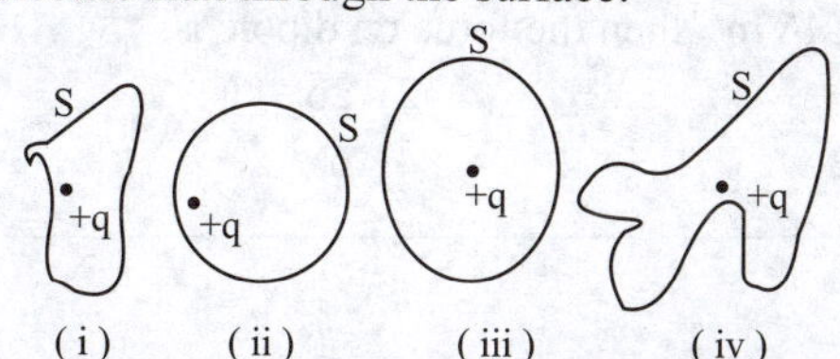

1) In Fig. (iv) is the largest
2) In Fig. (iii) is the least
3) In Fig. (ii) is same as Fig. (iii) but is smaller than Fig. (iv)
4) Is the same for all the figures

4. Five charges q_1, q_2, q_3, q_4, and q_5 are fixed at their positions as shown in Figure, S is a Gaussian surface. The Gauss's law is given by $\int_S E.dS = \frac{q}{\varepsilon_0}$. Which of the following statements is correct?

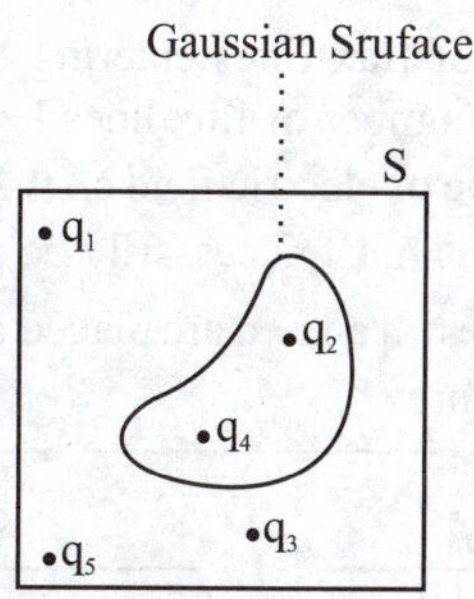

1) E on the LHS of the above equation will have a contribution from q_1, q_5 and q_1, q_5 and q_3 while q on the RHS will have a contribution from q_2 and q_4 only
2) E on the LHS of the above equation will have a contribution from all charges while q on the RHS will have a contribution from q_2 and q_4 only
3) E on the LHS of the above equation will have a contribution from all charges while q on the RHS will have a contribution from q_1,q_3 and q_5 only
4) Both E on the LHS and q on the RHS will have contributions from q_2 and q_4 only

5. Figure shows electric field lines in which an electric dipole P is placed as shown. Which of the following statements is correct?

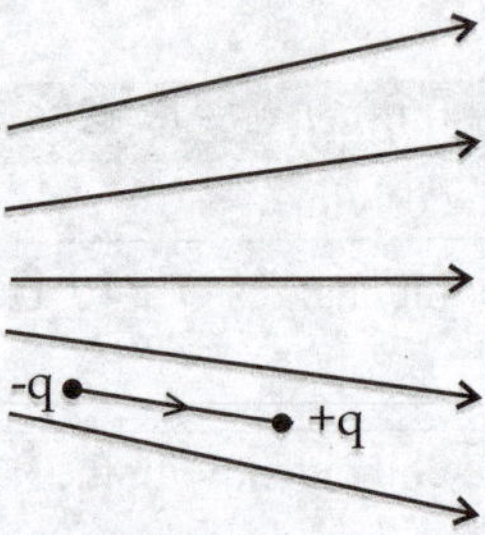

1) The dipole will not experience any force
2) The dipole will experience a force towards right
3) The dipole will experience a force towards left
4) The dipole will experience a force upwards

6. A point charge + q is placed at a distance d from an isolated conducting plane. The field at a point P on the other side of the plane is:
1) Directed perpendicular to the plane and away from the plane
2) Directed perpendicular to the plane but towards the plane
3) Directed radially away from the point charge
4) Directed radially towards the point charge

7. A hemisphere is uniformly charged positively. The electric field at a point on a diameter away from the centre is directed:
1) Perpendicular to the diameter
2) Parallel to the diameter
3) At an angle tilted towards the diameter
4) At an angle tilted away from the diameter

NEET & AIIMS Past Year MCQs

1. A metal cube of side 5 cm is charged with 6 μC. The surface charge density on the cube is **(2024 Re)**
1) 0.125×10^{-3} C m^{-2} 2) 0.25×10^{-3} C m^{-2}
3) 4×10^{-3} C m^{-2} 4) 0.4×10^{-3} C m^{-2}

2. An electric dipole is placed at an angle of 30° with an electric field of intensity 2×10^5 NC^{-1}. It experiences a torque equal to 4 Nm. Calculate the magnitude of charge on the dipole, if the dipole length is 2 cm. **(2023)**
1) 2 mC 2) 8 mC
3) 6 mC 4) 4 mC

3. If $\oint_s \vec{E} \cdot \vec{dS} = 0$ over a surface, then: **(2023)**
1) the electric field inside the surface is necessarily uniform.
2) the number of flux lines entering the surface must be equal to the number of flux lines leaving it.
3) the magnitude of electric field on the surface is constant.
4) all the charges must necessarily be inside the surface.

4. Two point charges –q and +q are placed at a distance of L, as shown in the figure

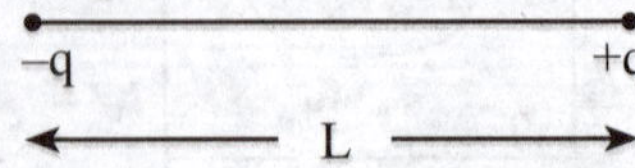

The magnitude of electric field intensity at a distance R (R >> L) varies as **(2022)**
1) $\frac{1}{R^6}$ 2) $\frac{1}{R^2}$
3) $\frac{1}{R^3}$ 4) $\frac{1}{R^4}$

5. A dipole is placed in an electric field as shown. In which direction will it move? **(2021)**

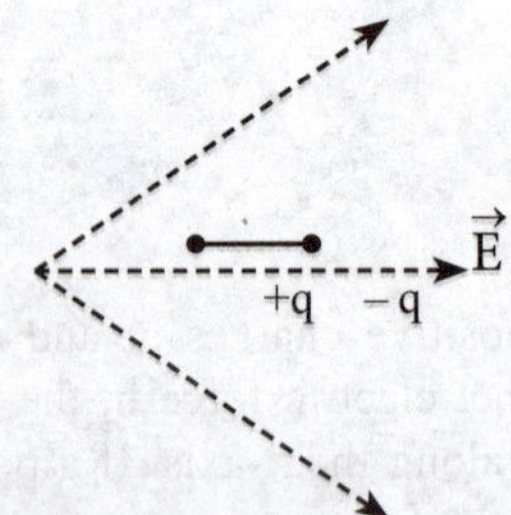

1) Towards the right as its potential energy will decrease.
2) Towards the left as its potential energy will decrease.
3) Towards the right as its potential energy will increase.
4) Towards the left as its potential energy will increase.

Directions (Questions 6 to 7): These questions consist of two statements each, printed as Assertion and Reason. While answering these questions, you are required to choose any one of the following four responses. **(AIIMS 2018)**
1) If both Assertion & Reason are True & the Reason is a correct explanation of the Assertion.
2) If both Assertion & Reason are True but Reason is not a correct explanation of the Assertion.
3) If Assertion is True but the Reason is False.
4) If **both Assertion & Reason are False.**

6. **Assertion:** $\vec{E} = E_x\hat{i} + E_y\hat{j} + E_z\hat{k}, \quad \vec{\nabla} \times \vec{E} = 0$
Reason: E_x, E_y, E_z is independent.

7. **Assertion:** Electric field inside a conductor is 0.
Reason: Charge is present on surface of conductor.

8. A dipole of dipole moment 'p' is placed in non-uniform electric field along x-axis. Electric field is increasing at the rate of 1Vm^{-1} then the force on dipole is: **(AIIMS 2013)**
1) 0 2) 2p
3) p/2 4) p

Answer Key

NCERT Topic-wise MCQs

1	2	3	4	5	6	7	8	9	10	11	12	13	14	15	16	17
4	1	1	2	3	1	3	4	4	1	1	1	1	2	1	2	1
18	19	20	21	22	23	24	25	26	27	28	29	30	31	32	33	34
2	3	4	3	3	2	2	2	3	3	4	1	4	3	3	2	1
35	36	37	38	39	40	41	42	43	44	45	46	47	48	49	50	51
1	2	3	4	2	3	2	4	4	2	4	1	2	2	3	1	3
52	53	54	55	56												
4	4	1	2	3												

High Order Time Intensive MCQs

1	2	3	4	5	6	7	8	9	10	11	12	13	14	15	16	17
2	4	3	4	2	2	1	3	1	1	1	3	2	2	2	2	1
18	19	20	21	22	23	24	25	26	27	28	29	30	31	32	33	34
3	2	4	4	1	2	1	3	3	3	4	3	2	3	4	2	1
35	36	37	38	39	40	41	42									
4	3	3	1	2	1	1	1									

Statement Based MCQs

1	2	3	4	5
1	3	4	1	1

Assertion & Reason MCQs

1	2	3	4	5
1	1	1	1	4

Match the Column MCQs

1	2	3	4
2	2	1	2

NCERT Exemplar MCQs

1	2	3	4	5	6	7
1	1	4	2	3	1	1

NEET & AIIMS Past Year MCQs

1	2	3	4	5	6	7	8
4	1	2	3	1	1	1	2

CHAPTER 2

Electrostatic Potential and Capacitance

Topicwise Number of Questions Analysis (2025-2016)

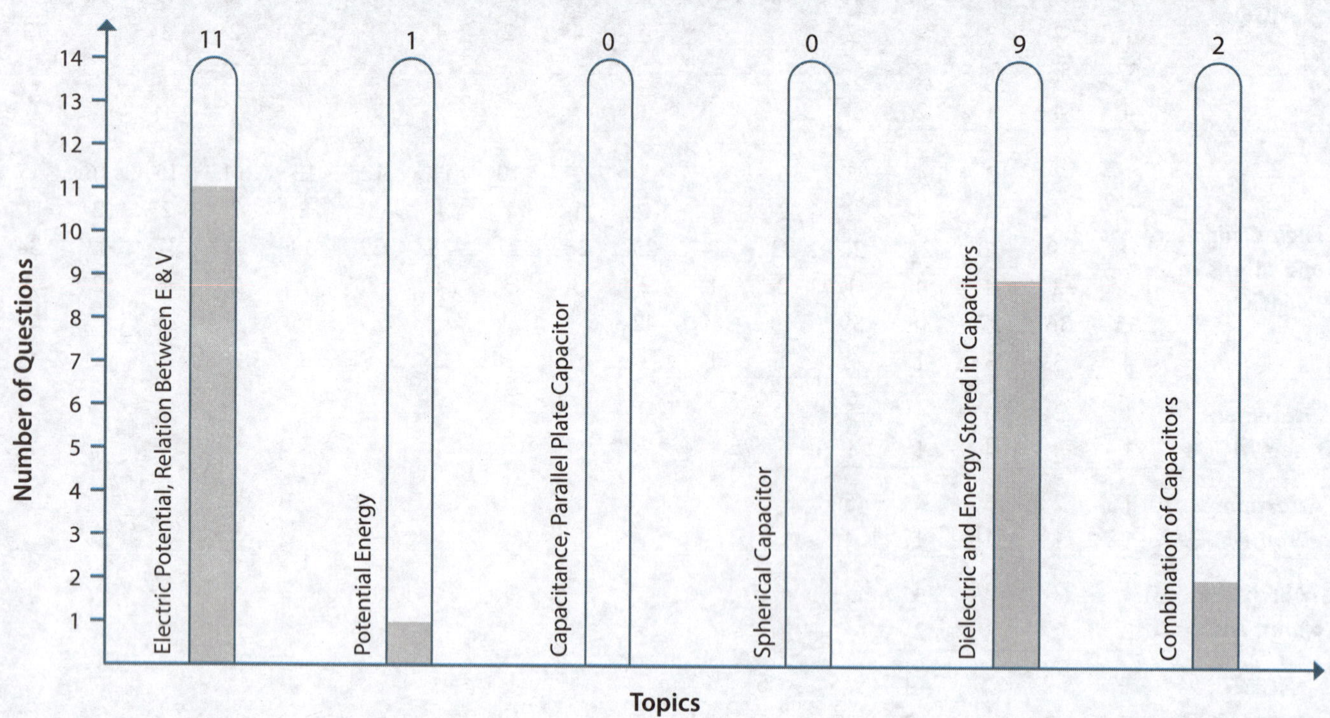

NCERT-PICKS

Introduction

Electrostatic potential at a point is defined as the work done by an external agent in bringing a unit positive charge from infinite to that point.

Any arrangement of two isolated conductors carrying equal and opposite charge can be called as capacitor.

Electric Potential and Potential Difference

Formulas

1. Potential difference between two points A and B

   $V_B - V_A = \frac{W_{AB}}{q}$ [V_B is potential at B and V_A is potential at A, W_{AB} is the work done to take the charge q from point A to B]

2. Potential at point B, if point A lies at infinity

   $V = V_B - V_\infty = V_B = \frac{W_B}{q}$ [W_B is the work done to take the charge q from ∞ to point B]

3. Electric Potential $= \frac{\text{Work done}}{\text{Charge}}$

Numerical Profile

Potential due to a point charge q at distance r from it

$$V = \frac{1}{4\pi\varepsilon_0} \cdot \frac{q}{r}$$

Potential at a point due to N point charges

$$V = \frac{1}{4\pi\varepsilon_0} \sum_{i=1}^{N} \frac{q_i}{r_i}$$

Potential at a point due to a dipole

$$V = \frac{1}{4\pi\varepsilon_0} \frac{p\cos\theta}{r^2} = \frac{1}{4\pi\varepsilon_0} \cdot \frac{\vec{p}.\vec{r}}{r^3}$$

(i) When the point lies on the axial line of the dipole, = 0° or 180°

$$V = \pm \frac{1}{4\pi\varepsilon_0} \cdot \frac{p}{r^2}$$

i.e., the potential has greatest positive or the greatest negative value.

(ii) When the point lies on the equatorial line of the dipole, = 90°, and V = 0, i.e., the potential at any point on the equatorial line of the dipole is zero. However, the electric field at such points is non-zero.

Potential due to spherical shell

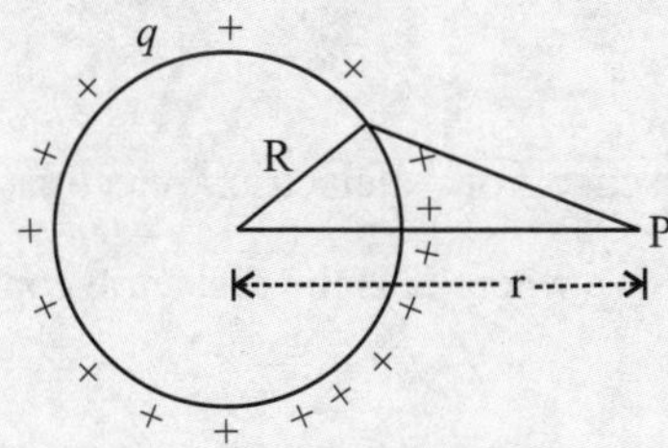

(i) When the point P lies outside the shell: $V = \frac{1}{4\pi\varepsilon_0} \cdot \frac{q}{r}$

(ii) When P lies on the surface of the spherical shell:

In this case r = R

Potential at the surface = $\frac{1}{4\pi\varepsilon_0} \cdot \frac{q}{R}$

(iii) When point P lies inside the shell, $V = \frac{1}{4\pi\varepsilon_0} \cdot \frac{q}{R}$

Potential Due to a Charged Ring

A charge q is distributed over the circumference of ring (radius = R)

(i) Electric potential at the centre of the ring, $V = \frac{1}{4\pi\varepsilon_o} \frac{q}{R}$.

(ii) At a distance 'r' from the centre of the ring on its axis would be $V = \frac{1}{4\pi\varepsilon_o} \cdot \frac{q}{\sqrt{R^2 + r^2}}$

Equipotential Surface

Formulas

1. No work is done in moving a test over an equipotential surface.

   W_{AB} = Charge × Potential Difference

   $= q(V_B - V_A)$

   As the surface is equipotential, So $V_B - V_A = 0$. Hence $W_{AB} = 0$.

2. Equipotential surfaces are closer together in the regions of strong field and farther apart in the regions of weak field.

   $dr \propto \frac{1}{E}$

Numerical Profile

Equipotential surfaces of a positive point charge

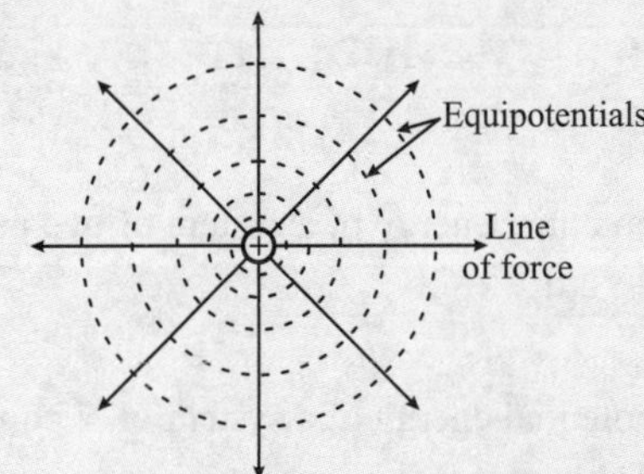

Equipotential surfaces of two equal and opposite point charges

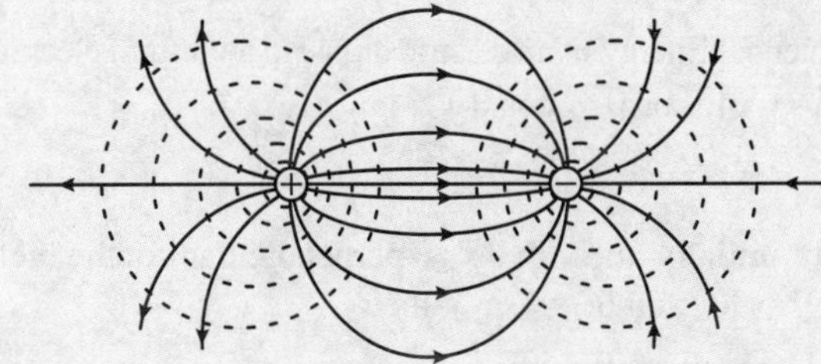

Equipotential surfaces of two equal positive charges

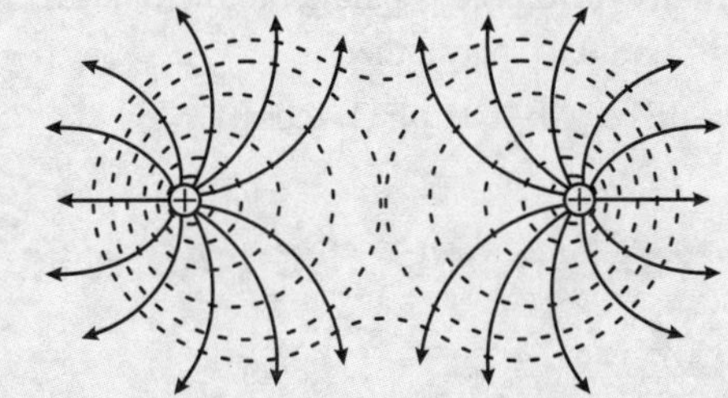

Equipotential Surfaces for a uniform electric field

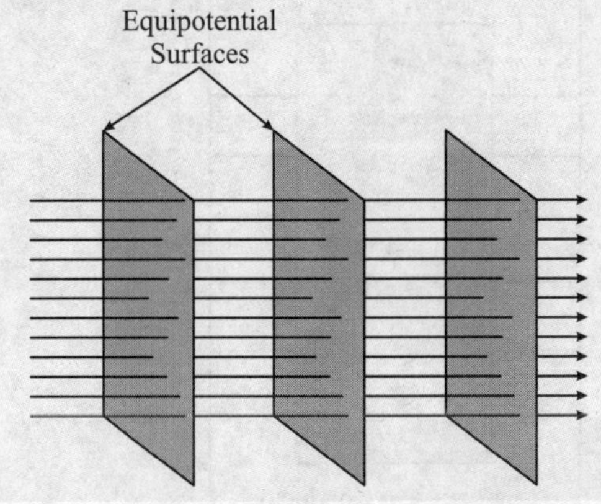

Relation between field and potential

(i) Electric field in a region can be determined from the electric potential by using relation,

$$\vec{E} = -\frac{dV}{\vec{dr}}$$

or $E_x = -\frac{\partial V}{\partial x}$, $E_y = -\frac{\partial V}{\partial y}$, $E_z = -\frac{\partial V}{\partial z}$

(ii) Electric field between two parallel conductors,

$$E = \frac{V}{d}$$

(iii) Electric potential in a region can be determined from the electric field by using the relation,

$$V = -\int_{\infty}^{r} \vec{E}.\vec{dr}$$

Electric Potential Energy

Formulas

1. Electric potential energy of a system of two point charges,

$$U = \frac{1}{4\pi\varepsilon_0} \cdot \frac{q_1 q_2}{r_{12}}$$

2. Electric potential energy of a system of N point charges,

$$U = \frac{1}{4\pi\varepsilon_0} \sum_{\text{all pairs}} \frac{q_i q_j}{r_{ij}}$$

3. Potential Energy of an electric dipole in a uniform electric field, $U = -pE(\cos\theta_2 - \cos\theta_1)$

Numerical Profile

(i) If initially the dipole is perpendicular to the field E, $\theta_1 = 90°$ and $\theta_2 =$ (say), then

$$U = -pE\cos\theta = -\vec{p}\cdot\vec{E}$$

(ii) If initially the dipole is parallel to the field E, $\theta_1 = 0°$ and $\theta_2 =$ (say), then

$$U = -pE(\cos\theta - 1) = pE(1 - \cos\theta)$$

Electrostatics of Conductors

Formulas

1. Net electostatic field is zero in the interior of a conductor.

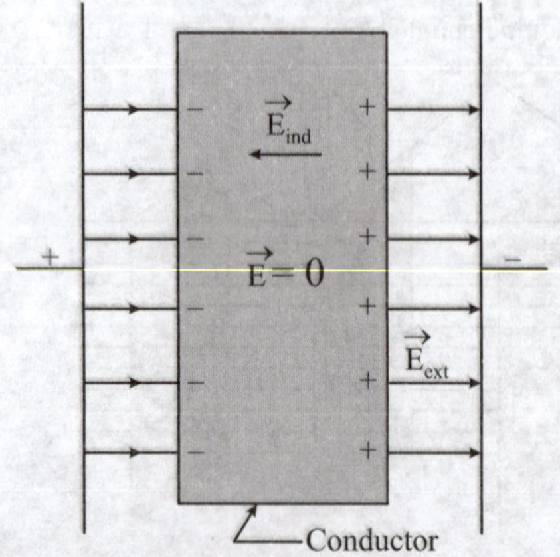

2. The net charge in the interior of a conductor is zero and any excess charge resides at its surface.

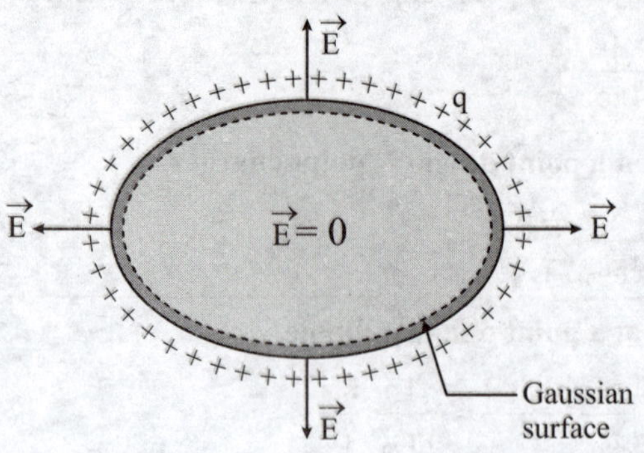

Numerical Profile

Electric field at the surface of a charged conductor

$$E = \frac{\sigma}{\varepsilon_0}$$

Potential is constant within and on the surface of a conductor

$$\frac{dV}{dr} = 0 \text{ or } V = \text{constant}$$

Electrostatic pressure on the surface of a charged conductor

$$P = \frac{\sigma^2}{2\varepsilon_0} = \frac{1}{2}\varepsilon_0 E^2$$

Note: In presence of dielectric, electrostatic pressure $= \frac{1}{2}\varepsilon_0 KE^2$ where K is the dielectric constant.

Capacitor and Capacitance

Formulas

1. Charge on the capacitor, $Q = CV$

   [V is voltage and C is capacitance]

2. Capacitance $= \frac{\varepsilon_0 A}{d}$

   [A is the area and d is the distance between plates]

3. Capacitance of the spherical capacitor, $C = \frac{4\pi\varepsilon_0 R_1 R_2}{R_2 - R_1}$

   [R_2 and R_1 is the radius of outer and inner sphere]

4. Capacitance of the Isolated sphere, $C = 4\pi\varepsilon_0 R$

5. Voltage on the Isolated sphere, $V = \frac{Q}{4\pi\varepsilon_0 R}$

6. Energy stored in a capacitor, $U = \frac{Q^2}{2C} = \frac{CV^2}{2} = \frac{QV}{2}$

7. Energy density $= \frac{1}{2}\varepsilon_0 E^2$

Numerical Profile

Cylindrical capacitor

(i) Potential difference between the cylinders,

$$V = \frac{Q}{2\pi\varepsilon_0 \ell}\ln\frac{R_2}{R_1}$$

[R_2 & R_1 are the radius and l is the length of the cylinder]

(ii) Capacitance of the cylindrical capacitor,

$$C = \frac{2\pi\varepsilon_0 \ell}{\ln\frac{R_2}{R_1}}$$

Parallel plate capacitor

(i) Electric field due to positive plate of capacitor, $E = \frac{Q}{2A\varepsilon_0}$

(ii) Force of attraction on the negative plate by the positive plate,

$$F = \frac{Q^2}{2A\varepsilon_0}$$

Dielectric

Formulas

1. Electric field after insertion of dielectric,

$$E = \frac{Q}{\varepsilon_0 AK}$$ [K is dielectric constant]

2. Potential difference between the plates,

$$V = \frac{Qd}{\varepsilon_0 AK}$$ [d is the distance between the plate]

3. When a dielectric slab is inserted capacitance of the parallel plate capacitor becomes, $C = \frac{K\varepsilon_0 A}{d}$

Numerical Profile

If a large number of dielectric slabs inserted between the plate as shown, capacitance

$$C' = \frac{\varepsilon_0 A}{d-(t_1+t_2+t_3+......)+\left(\frac{t_1}{K_1}+\frac{t_2}{K_2}+\frac{t_3}{K_3}+........\right)}$$

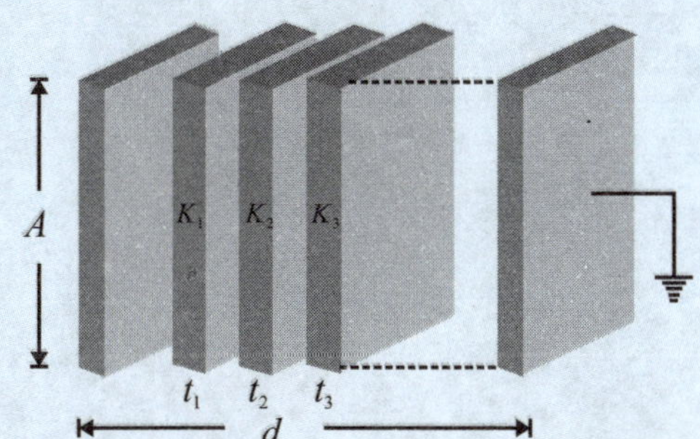

When a metallic slab is inserted between the plates

$$C' = \frac{\varepsilon_0 A}{(d-t)}$$

Grouping of Capacitors

Formulas

1. Series Grouping

(i) $C_{eq} = \frac{C_1 C_2}{C_1 + C_2}$

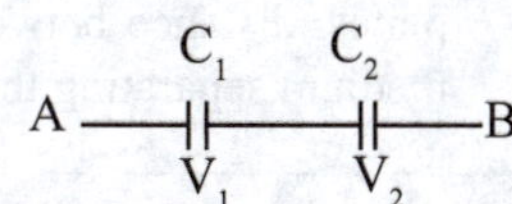

[C_{eq} is the equivalent capacitance]

(ii) $V = V_1 + V_2$

(iii) $V_1 = \left(\frac{C_2}{C_1 + C_2}\right)V$

(iv) $V_2 = \left(\frac{C_1}{C_1 + C_2}\right)V$

2. Parallel Grouping

(i) $C_{eq} = C_1 + C_2$

(ii) $Q = Q_1 + Q_2$

(iii) $Q_1 = \left(\frac{C_1}{C_1 + C_2}\right)Q$

(iv) $Q_2 = \left(\frac{C_2}{C_1 + C_2}\right)Q$

Numerical Profile

Connection of Capacitors

(i) If n identical capacitors each having capacitances C are connected in series with supply voltage V then Equivalent capacitance $C_{eq} = \frac{C}{n}$ and potential difference across each capacitor $V' = \frac{V}{n}$

(ii) If n identical plates are arranged as shown below, they constitute (n – 1) capacitors in series. If each capacitor has capacitance $\frac{\varepsilon_0 A}{d}$ then $C_{eq} = \frac{\varepsilon_0 A}{(n-1)d}$

(iii) If n identical capacitors are connected in parallel, then Equivalent capacitance C_{eq} = nC and Charge on each capacitor $Q' = \frac{Q}{n}$

Statements Asked in Exam

- With rise in temperature dielectric constant of liquid decreases.
- Dielectric constant (K) of an insulator can't be ∞.
- Two plates of unequal area can also form a capacitor, effective overlapping area is considered.
- If n identical capacitors are connected in parallel which are charged to a potential V. If these are connected in series then potential difference of combination will be nv.
- Capacity of earth is 711 μF. ($C = 4\pi\varepsilon_0 R$)
- Total energy stored in any set up of condensers is sum of the energies stored in individual condensers whatever be the mode of combination.
- Capacitance of a capacitor depends on geometry of the plates, distance between the plates and nature of dielectric medium separating the plates.
- To store charge and energy is the basic use of capacitor.
- An equipotential surface is that surface, at every point of which electric potential is the same. Equipotential surface are always perpendicular to the electric field lines.
- Introduction of dielectric (K) slab in between the plates of charged capacitor.

	With battery	Without battery
Potential (V)	V	$\frac{V}{K}$
Capacity (C)	KC	KC
Charge (Q)	KQ	Q
Energy (U)	KU	$\frac{U}{K}$
Electric field (E)	E	$\frac{E}{K}$

MCQs ?

NCERT Topic-wise MCQs

Electric Potential, Potential Energy and Relation Between E & V

1. A small conducting sphere of radius r is lying concentrically inside a bigger hollow conducting sphere of radius R. The bigger and smaller spheres are charged with Q and q (Q > q) and are insulated from each other. The potential difference between the spheres will be

1) $\frac{1}{4\pi\varepsilon_0}\left(\frac{q}{r}-\frac{q}{R}\right)$ 2) $\frac{1}{4\pi\varepsilon_0}\left(\frac{q}{R}-\frac{Q}{r}\right)$

3) $\frac{1}{4\pi\varepsilon_0}\left(\frac{q}{r}-\frac{Q}{R}\right)$ 4) $\frac{1}{4\pi\varepsilon_0}\left(\frac{Q}{R}+\frac{q}{r}\right)$

2. The potential of the electric field produced by point charge at any point (x, y, z) is given by $V = 3x^2 + 5$ where x, y are in meter and V is in volt. The intensity of electric field at (–2, 1, 0) is

1) +17 Vm^{-1} 2) –17 Vm^{-1}

3) +12 Vm^{-1} 4) –12 Vm^{-1}

3. An electric dipole of dipole moment 2×10^{-12} C–m is oriented at an angle of 60° with electric field of intensity 3NC^{-1}. What work need to be done to turn the dipole so that it becomes anti-parallel to the field?

1) 12×10^{-12} J 2) -9×10^{-12}J

3) 9×10^{-12}J 4) -12×10^{-12}J

4. Work done in carrying an electric charge Q_1 once round a circle of radius R with a charge Q_2 at the center of the circle is

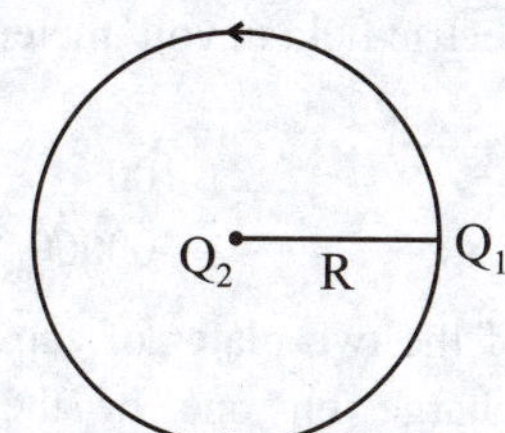

1) $\frac{Q_1Q_2}{4\pi\varepsilon_o R}$ 2) ∞

3) $\frac{Q_1Q_2}{4\pi\varepsilon_o R}$ 4) Zero

5. The kinetic energy of an electron which is accelerated in the potential difference of 100V is

1) 1.6×10^{-17} J 2) 1.6×10^{-14} J

3) 1.6×10^{-10} J 4) 1.6×10^{-8} J

6. Three charges are placed at the vertex of an equilateral triangle as shown in figure. For what value of Q the electrostatic potential energy of the system is zero?

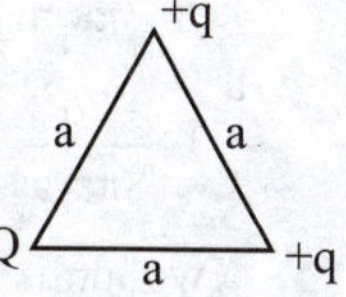

1) –q 2) q/2

3) –2q 4) $\frac{-q}{2}$

7. Two charged spheres of radii R_1 & R_2 have equal surface charge density. The ratio of their potential will be:

1) R_1/R_2 2) R_2/R_1

3) $(R_1/R_2)^2$ 4) $(R_2/R_1)^2$

8. The electric potential at any point x, y, z in metres is given by $V = 8x^3$ the electric field at a point (0, –3, 1) is

1) 12 Vm^{-1} 2) –6 Vm^{-1}

3) 6 Vm^{-1} 4) 0

9. If 20 J of work has to be done to move an electric charge of 4 C from a point where potential is 10 V to another point where potential is V volt find the value of V:

1) 2 V 2) 70 V

3) 5 V 4) 15 V

10. The potential of large liquid drop when eight liquid drops are combined is 20V. The potential of each drop was:

1) 10 V 2) 7.5 V

3) 5 V 4) 2.5 V

11. Electric potential at the center of a charged hollow metal sphere is

1) Zero

2) Twice as that on surface

3) Half of that on the surface

4) Same as that on the surface

12. The electric potential at the surface of an atomic nucleus (Z = 50) of radius 9×10^{-15} m is

1) 80 V 2) 8×10^6 V

3) 9 V 4) 9×10^5 V

13. If 40 J of work is to be done in moving a charge of 500C from A to B. What is the magnitude of potential difference?

1) 0.08 V 2) 0.10 V

3) 0.05 V 4) 0.02 V

14. Two positive point charges of 18μC and 15 μC are 1m apart. What is the work done in bringing them 0.4m apart?

1) 15×10^{-10} J 2) 3.6×10^{-6} J

3) 4×10^{-2} J 4) 2×10^{-14} J

15. Two points A and B are maintained at the potentials of 18 V and –5V respectively. What will be the work done in moving 150 α-particles from A to B?

1) 4×10^{-12} J
2) 8×10^{-8} J
3) 2×10^{-15} J
4) 1.1×10^{-15} J

16. Two equal charges q are placed at a distance of 2a and a third charge –2q is placed at the midpoint. The potential energy of the system is

1) $\frac{8q^2}{7\pi\varepsilon_o a}$
2) $\frac{7q^2}{8\pi\varepsilon_o a}$
3) $\frac{-7q^2}{8\pi\varepsilon_o a}$
4) None of these

17. Two unlike charge of magnitude q are separated by a distance 5d. The potential at a point midway between them

1) Zero
2) $\frac{1}{4\pi\varepsilon_o}\frac{q}{d}$
3) $\frac{3}{2\pi\varepsilon_o}\frac{q}{d}$
4) $\frac{5}{4\pi\varepsilon_o}\frac{q}{d}$

18. An electric dipole of magnitude 17×10^{-8}C–m is ligned in a uniform electric field of 2×10^5 N/C. What is value of potential energy of the dipole to hold the dipole at 30° with the direction of electric field?

1) -1.7×10^{-3} J
2) -1.7×10^{-2} J
3) -2.9×10^{-2} J
4) 5.1×10^{-2} J

19. Two charges of 10μC are placed at the corners A and B of an equilateral triangle of side length 0.5 m in air. The electric potential at C is

1) 36×10^4 V
2) 5×10^6 V
3) 7×10^9 V
4) 10^7 V

20. In the rectangle, the two corners have charges q_1 = –8 μC and q_2 = +4 μC. The work done in moving a charge +3 μC from B to A

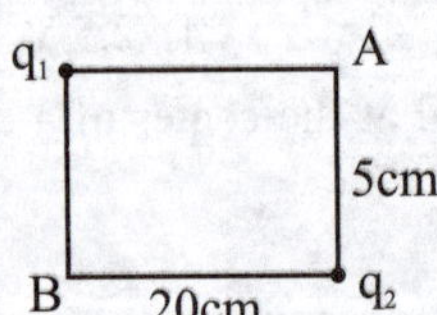

1) 3.2 J
2) 4.9 J
3) 5.6 J
4) 4.2 J

21. A thin spherical conducting shell of radius R has a charge q. Another charge Q is placed at the center of the shell. The electric potential at a point P at distance $\frac{R}{2}$ from the center of the shell is (Q = 10 μC, q = –2 μC)

1) $\frac{1.6\times10^5}{R}$ volt
2) 1.6×10^5 R volt
3) $\frac{1.4\times10^5}{R}$ volt
4) $\frac{4\times10^{-6}}{R}$ volt

22. The electrostatic potential energy between proton and α-particle separated by a distance 2Å is:

1) 20×10^{-18} J
2) 2.2×10^{-18} J
3) 34×10^{-19} J
4) None of these

Parallel Plate Capacitor and Capacitance

23. What is the value of capacitance, if the thin metallic plate is introduced between two parallel plates of area A and separated at distance d?

1) $\frac{\varepsilon_o A}{d-t}$
2) $\frac{\varepsilon_o A}{d}$
3) $\frac{2\varepsilon_o A}{d}$
4) $\frac{3\varepsilon_o A}{2d}$

24. The capacity of parallel plate capacitor depends upon

1) Separation between the plates
2) Nature of element / metal
3) Potential difference across plates
4) All of these

25. The capacitance of a parallel plate capacitor is 60 μF. If the distance between the plates is tripled and area doubled then new capacitance will be

1) 10 μF
2) 40 μF
3) 80 μF
4) 100 μF

26. A 500 μF capacitor is charged at a rate of 125 μC/sec. The potential difference across the capacitor will be 10 V after an interval of

1) 80 sec
2) 50 sec
3) 20 sec
4) 40 sec

27. Two capacitors of capacitances C_1 and C_2 are connected in parallel. If a charge q is given to the assembly, the charge gets shared. The ratio of the charge on the capacitor C_1 to the charge that on C_2 is

1) $\frac{C_1}{C_2}$
2) $\frac{C_2}{C_1}$
3) C_1C_2
4) $\frac{1}{C_1C_2}$

28. Between the plates of parallel plate condenser there is 1mm thick medium shoot of dielectric constant 4. It is charged at 100 volt. The electric field in volt/ meter between the plates of capacitor is:

1) 100000
2) 100
3) 25000
4) 400000

29. The potential of the two plates of capacitor are +10 V & –10 V. The charge on one of the plates is 40 C. The capacitance of the capacitor is

1) 2 F
2) 4 F
3) 0.5 F
4) 0.25 F

30. If the distance between parallel plates of a capacitor is halved and dielectric constant is doubled then the capacitance-

1) Decreases two times
2) Increases two times
3) Increases four times
4) Remains the same

31. If plates of charged parallel plate capacitor are pulled apart:
1) The capacitance increases
2) Charge increases
3) Potential difference remains same
4) Potential difference increases

32. The capacity of an isolated conducting sphere of radius R is proportional to
1) R^2
2) R
3) $R^{1/2}$
4) $R^{1/3}$

33. A spherical conductor with radius of 2m. What will be the value of capacitance?
1) 10^{-2} F
2) 4×10^{-2} F
3) 2.2×10^{-10} F
4) 5×10^{-5} F

34. In the arrangement $C_1 = 10\ \mu F$, $C_2 = 10\ \mu F$ and $C_3 = 20\ \mu F$. What is the equivalent capacitance (in μF) between points A and B?

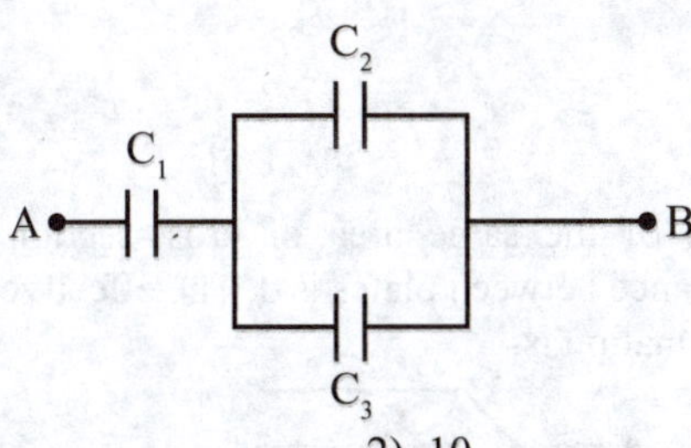

1) 7.5
2) 10
3) 12
4) 15

Dielectric Slab Inside the Capacitor and Energy Stored in Capacitor

35. A parallel plate air capacitance C. Now half of the space is filled with a dielectric material with dielectric constant K as shown in figure. The new capacitance is C'. Then

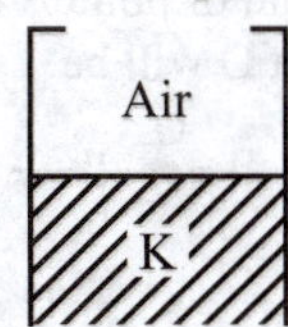

1) $C' = \frac{C}{2}[1+K]$
2) $C' = C[1+K]$
3) $C' = C\left[1+\frac{K}{2}\right]$
4) $C' = C\left[\frac{1}{2}+K\right]$

36. If the charge of 10 μC and –2 μC are given to two plates of a capacitor which are connected across a battery of 12V, find the capacitance of the capacitor.
1) 0.33 μF
2) 0.5 μF
3) 0.41 μF
4) 0.66 μF

37. A parallel plate capacitor is charged. If the plates are pulled apart,
1) The capacitance increases
2) The potential difference increases
3) The total charge increases
4) The charge and potential difference remain the same

38. In a capacitor of capacitance 20 μF the distance between the plates is 5 mm. If a dielectric slab of width 2 mm and dielectric constant 3 is inserted between the plates, then the new capacitor will be
1) 22.3μF
2) 25 μF
3) 19 μF
4) 27.2 μF

39. A 80 μF capacitor in a defibrillator is charged to 1000V. The energy stored in the capacitor is sent through the patient during a pulse of duration 3 ms. The power delivered to the patient is
1) 13.3 kW
2) 14.5 kW
3) 15 kW
4) 9 kW

40. A condenser of capacitance 12μF was initially charged to 20V. Now potential difference is made 40V. The increase in potential energy is
1) 7.2×10^{-3} J
2) 4×10^{-3} J
3) 3×10^{-4} J
4) 5×10^{-6} J

41. Two parallel plates of area A are separated by two different dielectric. The net capacitance is

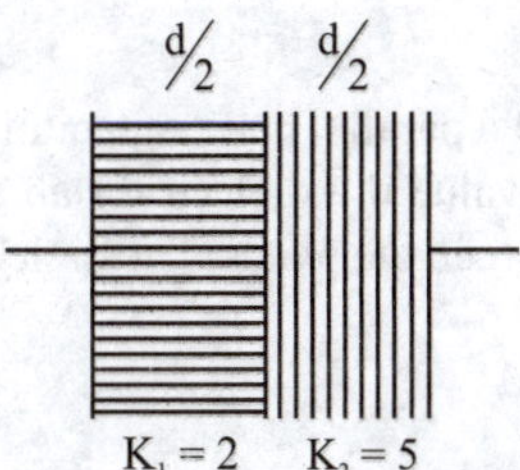

1) $\frac{\varepsilon_0 A}{d}$
2) $\frac{7}{20}\frac{\varepsilon_0 A}{d}$
3) $\frac{20}{7}\frac{\varepsilon_0 A}{d}$
4) $\frac{2\varepsilon_0 A}{d}$

42. A capacitance of an air capacitor is 25 μF the separation between the parallel plates is 8 mm. An aluminum plate of thickness 2mm thickness is introduced symmetrically between the plates. The new capacitance will be.
1) 10 μF
2) 33.3 μF
3) 44 μF
4) 50 μF

43. A parallel plate capacitor with air as the dielectric has capacitance C. A slab of dielectric constant K and having the same thickness as the separation between the plates is introduced. So as to fill $\frac{1}{3}$ of the capacitor. The new capacitance will be

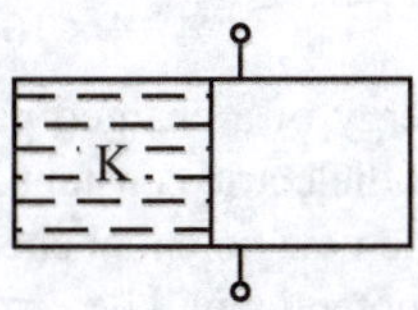

1) $\frac{(K+1)}{C}$
2) $\frac{C}{3}(K+2)$
3) 2C
4) $\frac{C}{3}(2K+3)$

44. A parallel plate capacitor is filled with dielectrics. What is its capacitance?

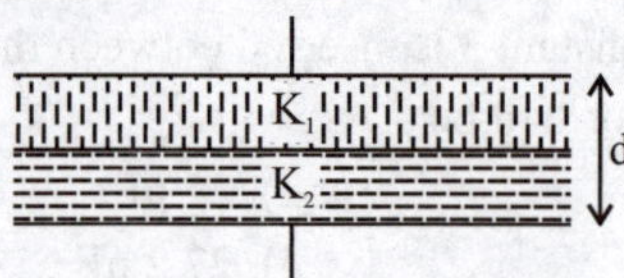

1) $\frac{2K_1K_2}{K_1+K_2}$ 2) $\frac{K_1K_2}{K_1+K_2}$

3) $\frac{2\varepsilon_0 A}{d}\left(\frac{K_1K_2}{K_1+K_2}\right)$ 4) $\frac{2\varepsilon_0 A}{d}\left(\frac{2K_1+K_2}{2K_1K_2}\right)$

45. A 2 μF capacitor is charged to 100 volt and then it plates are connected by a conducting wire. The heat produced is

1) 1 J 2) 0.1 J
3) 0.01 J 4) 0.001 J

46. An air filled parallel plate condenser has a capacity of 2 pF. The separation of the plates is doubled and the interspace between the plates is filled with wax. If the capacity is increased to 6pF, the dielectric constant of wax is

1) 2 2) 3
3) 4 4) 6

47. Capacitance of a parallel plate capacitor becomes 4/3 times of its original value if a dielectric slab of thickness t = d/2 is inserted between the plates. The dielectric constant of the slab is

1) 8 2) 4
3) 6 4) 2

48. A parallel plate capacitor with air as the dielectric has capacitance C. A slab of dielectric constant K and having the same thickness as the separation between the plates, is introduced so as to fill one-fourth of the capacitor as shown in the figure. The new capacitance will be:

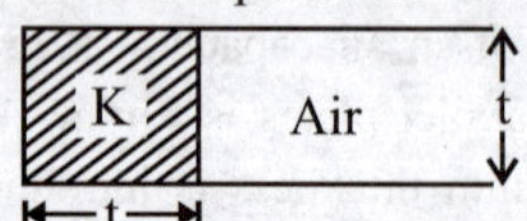

1) $(K+3)\frac{C}{4}$ 2) $(K+2)\frac{C}{4}$

3) $(K+1)\frac{C}{4}$ 4) $\frac{KC}{4}$

49. A capacitor having capacity of 2μF is charged to 200 V and then the plates of the capacitor are connected to a resistance wire. The heat produced in joule will be:

1) 2×10^{-2} 2) 4×10^{-2}
3) 4×10^{4} 4) 4×10^{10}

50. The potential energy of a charged parallel plate capacitor is U0. If a slab of dielectric constant K is inserted between the plates, then the new potential energy when capacitor is charged to same potential will be:

1) $\frac{U_0}{K}$ 2) U_0K

3) $\frac{U_0}{K^2}$ 4) U_0^2

Combination of Capacitors

51. Four capacitors 4 μF, 6 μF, 5 μF, 3 μF are connected in series to a source of 150V. The potential difference in volt, across the 5 μF capacitor will be

1) 30 volt 2) 40 volt
3) 50 volt 4) 60 volt

52. Five capacitor each of capacitance value C are connected as shown in figure. The ratio of capacitance between P to R, and the capacitance between P and Q is

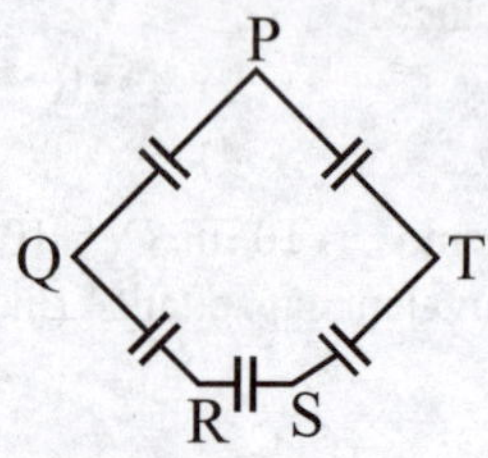

1) 1 : 4 2) 2 : 3
3) 2 : 5 4) 1 : 1

53. 4 plates of the same area of cross-section are joined. The distance between plates is d. The effective capacitance of combination is:

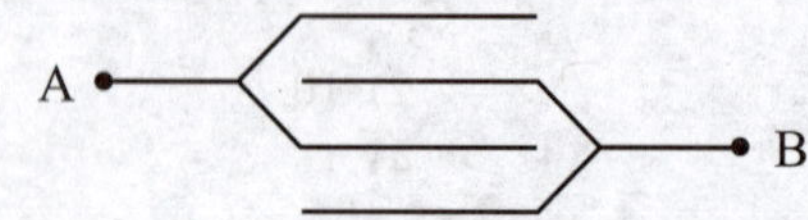

1) $\frac{3\varepsilon_0 A}{d}$ 2) $\frac{\varepsilon_0 A}{d}$

3) $\frac{4\varepsilon_0 A}{d}$ 4) None of these

54. Two condensers C_1 and C_2 in a circuit are joined as shown in figure. The potential of point A is V_1 and that of B is V_2. The potential of point O will be

A O B
V_1 C_1 C_2 V_2

1) $C_1V_1 + C_2V_2$ 2) $\frac{C_2V_1 + C_1V_1}{C_1 + 2C_2}$

3) $\frac{C_1V_1 + C_2V_2}{C_1 + C_2}$ 4) $\frac{V_1 + V_2}{2}$

55. Find the capacitance of the infinite ladder between point A and B

5μF 5μF 5μF 5μF
A
2μF 2μF 2μF 2μF
B

1) 4.3 μF 2) 10 μF
3) 8.6 μF 4) 3 μF

56. Three capacitors of capacitances 2μF, 5μF, 8μF are connected first in parallel combination and then in series combination, The ratio of their equivalent capacitances will be

1) 15 : 1 2) 20 : 1
3) 13 : 1 4) 12 : 1

57. For circuit the equivalent capacitance between A and B is (each capacitor is of 3μF)

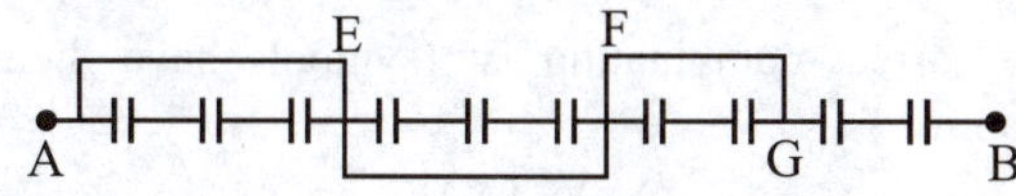

1) 1 μF
2) 5 μF
3) 6 μF
4) 8 μF

58. If each capacitor of 18μF arranged as shown in figure below, then the equivalent capacitance between the points P and Q

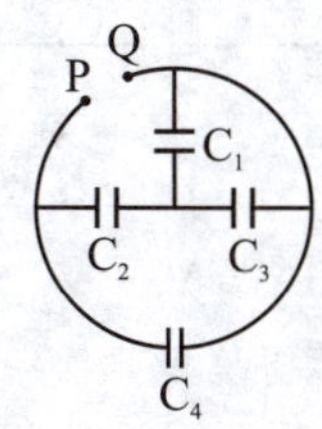

1) 30 μF
2) 90 μF
3) 20 μF
4) 18 μF

59. Three equal capacitors each with capacitance C are connected as shown in figure then the equivalent capacitance is

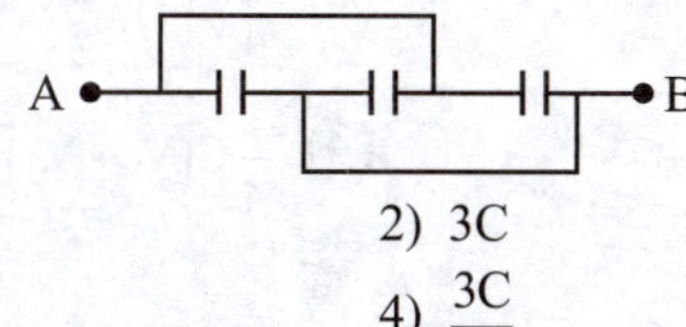

1) C
2) 3C
3) C/3
4) $\frac{3C}{2}$

60. The following arrangements consists of five identical metal plates parallel to each other. Area of each plate is A and separation between the successive plates is d. The equivalent capacitance between P and Q is -

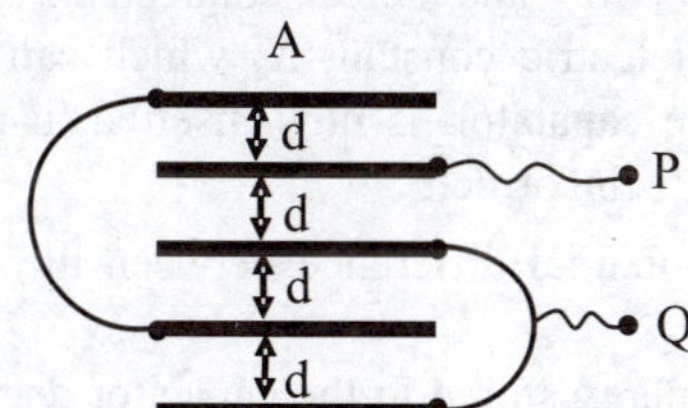

1) $\frac{5\varepsilon_0 A}{d}$
2) $\frac{7}{3}\frac{\varepsilon_0 A}{d}$
3) $\frac{4}{3}\frac{\varepsilon_0 A}{d}$
4) $\frac{5}{3}\frac{\varepsilon_0 A}{d}$

61. The ratio of equivalent capacitances between points P and Q in the following cases is –

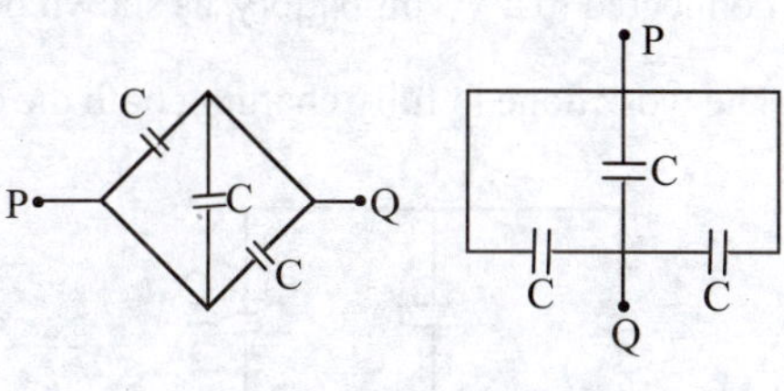

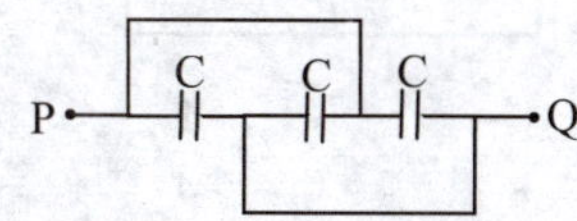

1) 1 : 2 : 3
2) 3 : 2 : 1
3) 1 : 1 : 1
4) 1 : 1 : 2

Sharing of Charge and Capacitor Circuits

62. The capacitors of capacitance 4 μF and 6 μF are connected in series. A potential difference of 500 V is applied to the outer plates of two capacitor system. Then the change on each plate of each capacitor is numerically.

1) 6000 μC
2) 1200 μC
3) 1200 C
4) 6000 C

63. A capacitor of 20 μF and charged to 500 V is connected in parallel with another capacitor of 10 μF charged to 200 V. The common potential will be

1) 400 V
2) 300 V
3) 200 V
4) 100 V

64. A 3 μF capacitor is charged to a potential of 300 V and 2 μF capacitor is charged to 200 V. The capacitors are then connected in parallel with plates of opposite polarity joined together. What amount of charge will flow, when the plates are so connected?

1) 600 μC
2) 800 μC
3) 700 μC
4) 300 μC

65. In the given circuit, a charge of +80 μC is given to the upper plate of the 4 μF capacitor. Then in the steady state, the charge on the upper plate of the 3 μF capacitor is

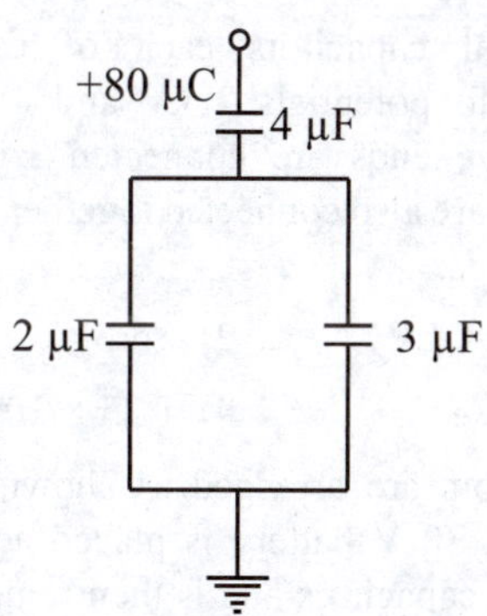

1) +32 μC
2) +40 μc
3) +48 μC
4) +80 μC

66. Two spherical conductors each of capacity C are charged to potential 25 V and 10 V. These are connected by a wire then the loss energy will be (C = 5 F)

1) 125 J
2) 425 J
3) 700 J
4) 281.25 J

67. A capacitor of capacity C_1 is charged upto V volt and then connected to an uncharged capacitor of capacity C_2. Then final potential difference across each will be

1) $\frac{C_1V_1 + C_2V_2}{C_1 + C_2}$
2) $\left(\frac{C_1}{C_1 + C_2}\right)V$
3) $C_1 + 2C_2$
4) $\left(\frac{C_1 + C_2}{C_1C_2}\right)V$

68. In the circuit the potential difference across the 4.5μF capacitor is

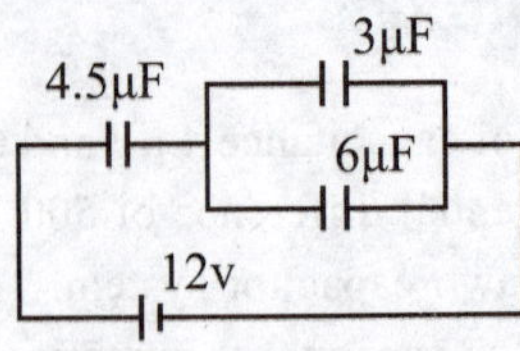

1) 8/3 volt
2) 4 volt
3) 6 volt
4) 8 volt

69. Three capacitors 3 μF, 6 μF and 6 μF are connected in series to a source of 120 V. The potential difference, in volt, across the 3 μF capacitor will be:

1) 24 V
2) 30 V
3) 40 V
4) 60 V

70. Two capacitors of 10 pF and 20 pF are connected to 200V and 100 V sources, respectively. If they are connected by the wire, then what is the common potential of the capacitors?

1) 133.3 V
2) 150 V
3) 300 V
4) 400 V

71. 64 small drops of water having same charge and same radius are combined to form one big drop. The ratio of capacitance of big drop to small drop is:

1) 2 : 1
2) 1 : 2
3) 4 : 1
4) 1 : 4

72. The correct combination is: (symbols have their usual meanings)

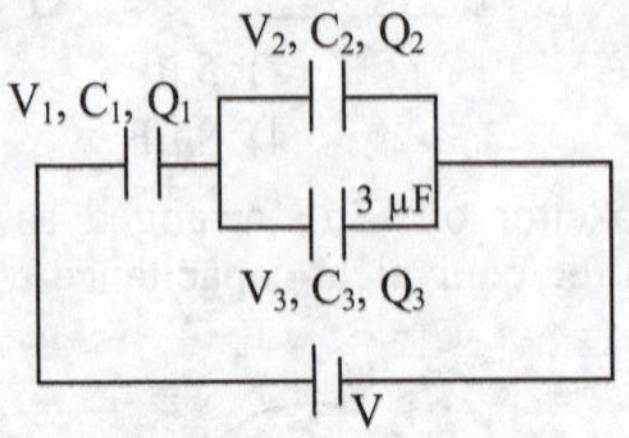

1) $Q_1 = Q_2 = Q_3$ and $V_1 = V_2 = V_3 = V$
2) $Q_1 = Q_2 + Q_3$ and $V = V_1 + V_2 + V_3$
3) $Q_1 = Q_2 + Q_3$ and $V = V_1 + V_2$
4) $Q_2 = Q_3$ and $V_2 = V_3$

73. Two identical capacitors have the same capacitance C. One of them is charged to potential V_1 and the other to V_2. The negative ends of the capacitors are connected, the decrease in energy of the combined system is

1) $\frac{1}{4}C(V_1^2 - V_2^2)$
2) $\frac{1}{4}C(V_1^2 + V_2^2)$
3) $\frac{1}{4}C(V_1 - V_2)^2$
4) $\frac{1}{4}C(V_1 + V_2)^2$

High Order Time Intensive MCQs

1. Two identical capacitors each of capacitance 5 μF are charged to potentials 2 kV and 1 kV, respectively. Their negative ends are connected together. When the positive ends are also connected together, the loss of energy of the system is:

1) 160 J
2) Zero
3) 5 J
4) 1.25×10^6 J

2. Four capacitors are arranged as shown. All are initially uncharged. A 30 V battery is placed across terminal PQ to charge the capacitors and is then removed. The voltage across the terminals RS is then (in volt)

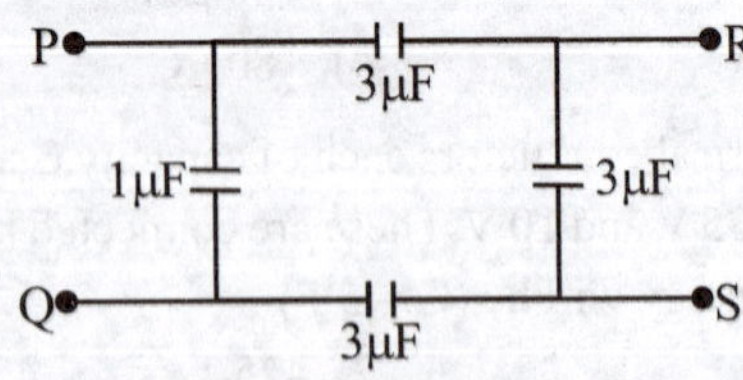

1) 10
2) 20
3) 30
4) 40

3. A charged oil drop of mass 2.5×10^{-7} kg is in space between the two plates, each of area 2×10^{-2} m² of a parallel plate capacitor. When the upper plate has a charge of 5×10^{-7} C and the lower plate has an equal negative charge, the oil remains stationary. The charge of the oil drop is (take, g = 10 m/s²):

1) 9×10^{-1} C
2) 9×10^{-6} C
3) 8.85×10^{-13} C
4) 1.8×10^{-14}

4. A parallel plate air capacitor of capacitance C_0 is connected to a cell of emf *V* and then disconnected from it. A dielectric slab of dielectric constant *K*, which can just fill the air gap of the capacitor, is now inserted in it. Which of the following is incorrect?

1) The potential difference between the plates decreases K times
2) The energy stored in the capacitor decreases *K* times
3) The change in energy $\frac{1}{2}C_0V^2(K-1)$
4) The change in energy $\frac{1}{2}C_0V^2(\frac{1}{K}-1)$

5. Two condensers, one of capacity C and the other of capacity $\frac{C}{2}$ are connected to a V volt battery, as shown below in the figure. The work done in fully charging both the condensers is:

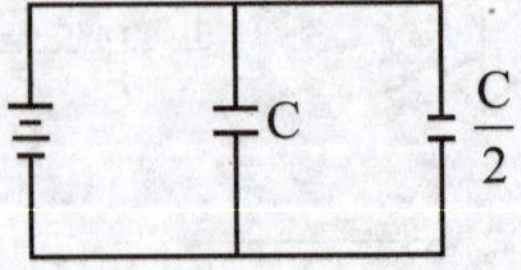

1) $2CV^2$
2) $\frac{1}{4}CV^2$
3) $\frac{3}{4}CV^2$
4) $\frac{1}{2}CV^2$

6. An insulator plate is passed between the plates of a capacitor. Then, current:

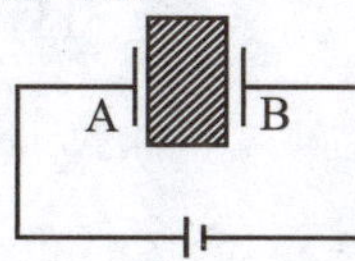

1) First flown from A to B and then from B to A
2) First flown from B to A and then from A to B
3) Always flows form B to A
4) Always flows from A to B

7. The capacitance of a parallel plate capacitor when a very thin metallic sheet is placed in the space between the plates, parallel to the plates, will

1) Increase
2) Decrease
3) Remain same
4) Nothing can be said

8. A capacitor having capacitance 1 μF with air, is filled with two dielectrics as shown below. How many times capacitance will increase?

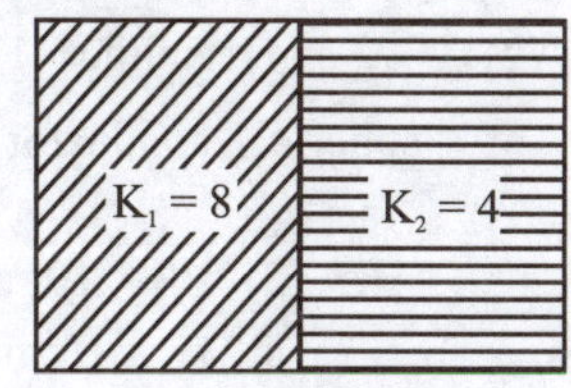

1) 4C
2) 6C
3) 5C
4) 3C

9. For the given arrangement of capacitors C_1 and C_2, which of the following statement is true?

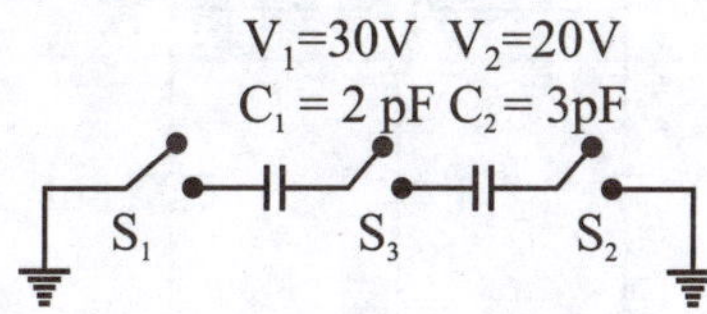

1) When S_1 and S_3 closed, $V_1 = 30$ V, $V_2 = 20$ V
2) When S_1 and S_3 closed, $V_1 = V_2 = 0$
3) When S_1 is closed, $V_1 = 15$ V, $V_2 = 20$ V
4) When S_3 is closed, $V_1 = V_2 = 20$ V

10. Two plates (area = S) charged to + q_1 and + q_2 ($q_2 < q_1$) are brought closer to form a capacitor of capacitance C. The potential difference across the plates is:

1) $\frac{q_1 - q_2}{2C}$
2) $\frac{q_1 - q_2}{C}$
3) $\frac{q_1 - q_2}{4C}$
4) $\frac{2(q_1 - q_2)}{C}$

11. A series combination of n_1 capacitors, each of value C_1 is charged by a source of potential difference 4 V. When another parallel combination of n_2 capacitors, each of value C_2 is charged by a source of potential difference V, it has the same (total) energy stored in it, as the first combination has. The value of C_2, in terms of C_1 is, then:

1) $\frac{2C_1}{n_1 n_2}$
2) $16\frac{n_1}{n_2}C_1$
3) $2\frac{n_2}{n_1}C_1$
4) $\frac{16C_1}{n_1 n_2}$

12. A slab of material of dielectric constant K has the same area as the plates of a parallel plate capacitor but has a thickness (3/4)d, where d is the separation of the plates. The ratio of the capacitance C (In the presence of the dielectric) to the capacitance C_0 (in the absence of the dielectric) is:

1) $\frac{3K}{K+4}$
2) $\frac{3K}{4}$
3) $\frac{4K}{K+3}$
4) $\frac{4}{3}K$

13. An air filled parallel plate condenser has a capacity of 2 pF. The separation of the plates is doubled and the interspace between the plates is filled with wax. If the capacity is increased to 6pF, the dielectric constant of wax is

1) 2
2) 3
3) 4
4) 6

14. The plate separation in a parallel plate condenser is d and plate area is A. If it is charged to V volt and battery is disconnected then the work done in increasing the plate separation to 2d will be–

1) $\frac{3}{2}\frac{\varepsilon_0 AV^2}{d}$
2) $\frac{\varepsilon_0 AV^2}{d}$
3) $\frac{2\varepsilon_0 AV^2}{d}$
4) $\frac{\varepsilon_0 AV^2}{2d}$

15. Initially K_1 is closed, now if K_2 is also closed, find the heat dissipated in the resistances of connecting wires.

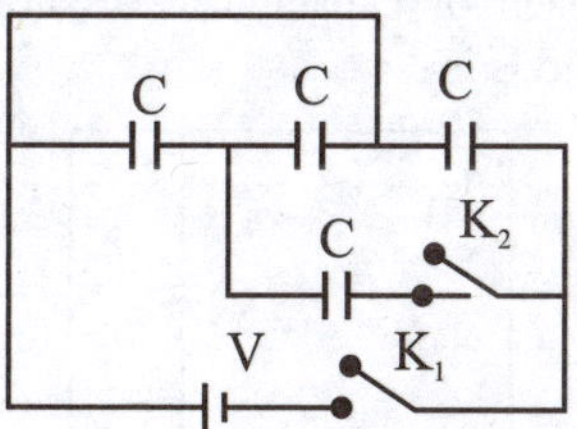

1) $\frac{1}{2}CV^2$
2) $\frac{2}{3}CV^2$
3) $\frac{1}{3}CV^2$
4) $\frac{1}{4}CV^2$

16. What is the dielectric constant of a metal?

1) –1
2) 0
3) 1
4) Infinite

17. Two spheres of capacitances 3μF and 5μF are charged to 300 V and 500 V respectively and are connected together. The common potential in steady state will be:

1) 400 V
2) 425 V
3) 350 V
4) 375 V

18. A parallel plate condenser with a dielectric constant K between the plates has a capacity C and is charged to a potential V volt. The dielectric slab between the plates is slowly removed and then reinserted. The net work done by the system in this process is

1) Zero
2) $\frac{1}{2}(K-1)CV^2$
3) $\frac{CV^2(K-1)}{K}$
4) $(K-1)\,CV^2$

19. The ratio of potential difference between 1 μF and 5 μF capacitors –

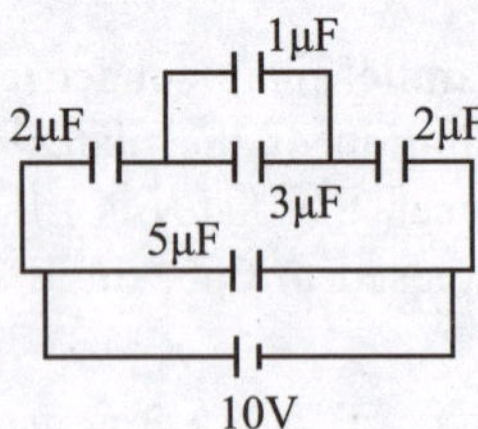

1) 1 : 2
2) 3 : 1
3) 1 : 5
4) 10 : 1

20. Two parallel plate capacitors whose capacities are C and 2 C respectively, are joined in parallel. These are charged by V potential difference. If the battery is now removed and a dielectric of dielectric constant K is filled in between the plates of the capacitor C, then what will be the potential difference across each capacitor:

1) $\frac{V}{K+2}$
2) $\frac{2V}{K+2}$
3) $\frac{3V}{K+2}$
4) $\frac{2+K}{3V}$

21. The equivalent capacitance of the combination of three capacitors, each of capacitance C shown in the figure, between A and B is

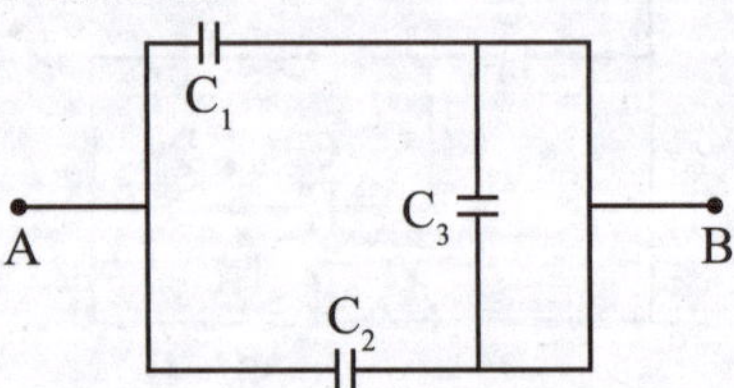

1) $\frac{C}{2}$
2) $\frac{3C}{2}$
3) $\frac{1}{3C}$
4) 2 C

22. The potential to which a conductor is raised, depends on

1) The amount of charge
2) Geometry of the conductor
3) Size of the conductor
4) All of these

23. In the given circuit, find potential at point 'O'.

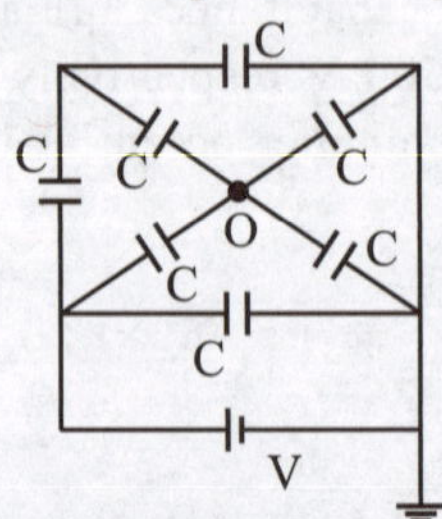

1) $\frac{4V}{11}$
2) $\frac{3V}{11}$
3) $\frac{9V}{11}$
4) None of these

24. For a spherical capacitor shown, at which points the electric field will be zero? (r is the distance of point from centre O)

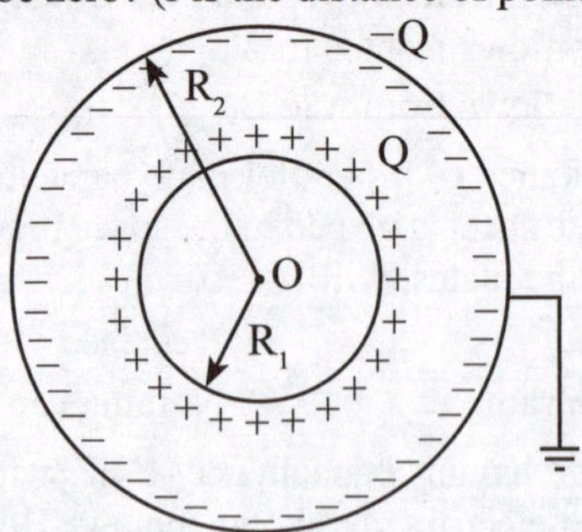

1) At point $r > R_2$
2) At point $r < R_1$
3) Both (1) & (2)
4) Neither (1) or (2)

25. If the capacity of a spherical conductor is 1 μF, then its diameter is

1) 18×10^3 m
2) 1.8×10^7 m
3) 1.8×10^3 m
4) 18×10^4 m

26. A 2μF capacitor is charged as shown in the figure. The percentage of its stored energy dissipated after the switch S is turned to position 2 is

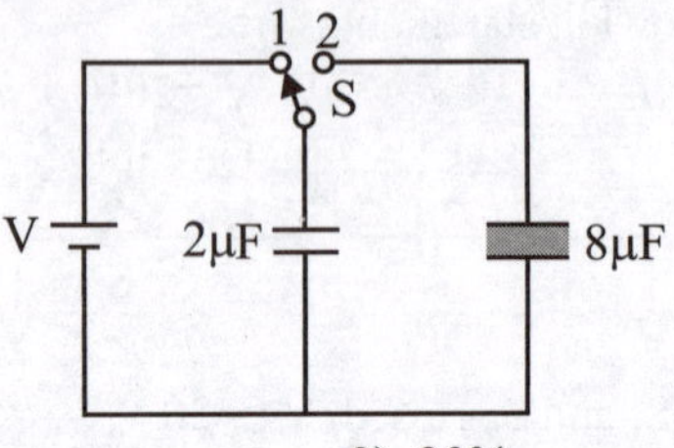

1) 0%
2) 20%
3) 75%
4) 80%

27. A charged ball B hangs from a silk thread S, which makes an angle with a large charged conducting sheet P, as shown in Figure. The surface charge density of the sheet is proportional to

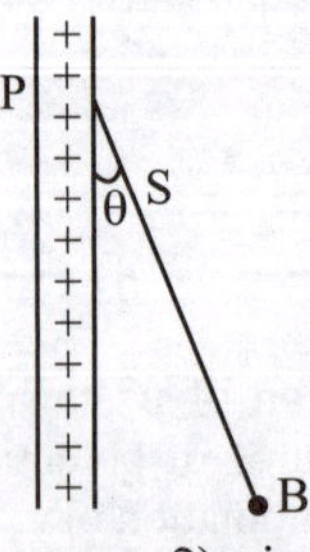

1) tan
2) sin
3) cot
4) cos

28. A ball of mass 1 g and charge 10^{-8} C moves from a point A, where potential is 600 V to the point B where potential is zero. Velocity of the ball at the point B is 20 cms^{-1}. The velocity of the ball at the point A will be

1) 22.8 cms^{-1}
2) 228 cms^{-1}
3) 16.8 ms^{-1}
4) 168 ms^{-1}

29. There is a uniform electric field of intensity E which is as shown. How many labeled points have the same electric potential as the fully shaded point?

E

1) 2 2) 3
3) 8 4) 11

30. If a charged spherical conductor of radius 10 cm has potential V at a point distant 5 cm from its centre, then the potential at a point distant 15 cm from the centre will be

1) $\frac{1}{3}V$ 2) $\frac{2}{3}V$
3) $\frac{3}{2}V$ 4) 3V

31. Three charges –q, + Q and + q are placed in a straight line as shown. If the total potential energy of the system is zero, then the ratio $\frac{q}{Q}$ is

−q +Q −q

x x

1) 2 2) 5.5
3) 4 4) 1.5

32. Two charges q_1 and q_2 are placed 30 cm apart, as shown in the figure. A third charge q_3 is moved along the arc of a circle of radius 40 cm from C to D. The change in the potential energy of the system is $\frac{q_3}{4\pi\varepsilon_0}$ k, where k is

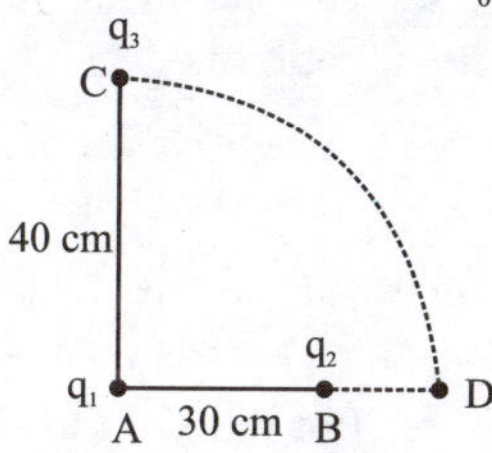

1) $8q_2$ 2) $8q_1$
3) $6q_2$ 4) $6q_1$

33. The radius of the circular plates of a parallel plate capacitor is R. Air is dielectric medium between the plates. If the capacitance of the capacitor is equal to the capacitance of a sphere of radius R, then the distance between the plates is

1) R/4 2) R/2
3) R 4) 2R'

34. The resultant capacity between the points P and Q of the given figure is

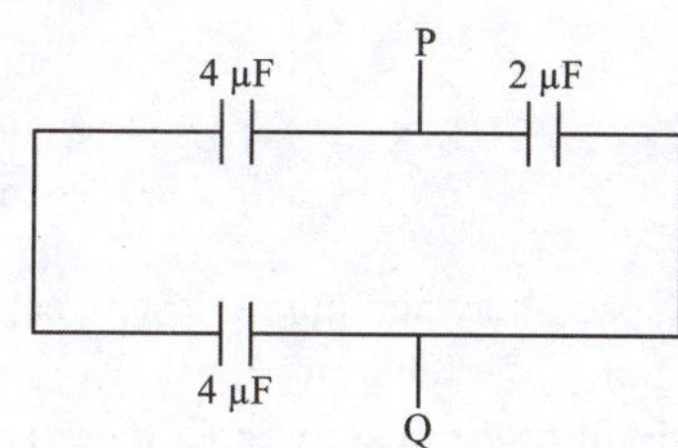

1) 4 µF 2) $\frac{16}{3}$ µF
3) 1.6 µF 4) 1 µF

35. If 3 capacitors of values 1, 2 and 3 µF are available. The maximum and minimum values of capacitances one can obtain by different combinations of the three capacitors together are respectively

1) 6µF, $\frac{6}{11}$µF
2) 6µF, $\frac{11}{6}$µF
3) 3µF, 1µF
4) 4µF, 2µF

36. The plates of a parallel plate capacitor have an area of 90 cm^2 each and are separated by 2 mm. The capacitor is charged by connecting it to a 400 V supply. Then the density of the energy stored in the capacitor is ($\varepsilon_0 = 8.8 \times 10^{-12}$ F/m)

1) 0.113 Jm^{-3} 2) 0.176 Jm^{-3}
3) 0.152 Jm^{-3} 4) 0.226 Jm^{-3}

37. Two spheres of radii 1 cm and 2 cm have been charged with 1.5×10^{-8} and 0.3×10^{-7} coulomb of positive charge. When they are connected with a wire, charge:

1) Will flow from the first to the second
2) Will flow from the second to the first
3) Will not flow at all
4) May flow either from first to second, or from the second to first, depending upon the length of the connecting wire

38. The capacitance of a capacitor is

1) Directly proportional to the dielectric constant of the medium between the plates
2) Inversely proportional to the dielectric constant of the medium between the plates
3) Proportional to the square of the dielectric constant of the medium between the plates
4) Independent of the dielectric constant of the medium between the plates

39. A parallel plate capacitor is charged and then isolated. On increasing the plate separation–

	Charge	Potential	Capacitance
1)	remains constant	remains constant	decreases
2)	remains constant	increases	decreases
3)	remains constant	decreases	increases
4)	increases	increases	decreases

40. The equivalent capacitance between A and B of the combination, shown in the figure, will be

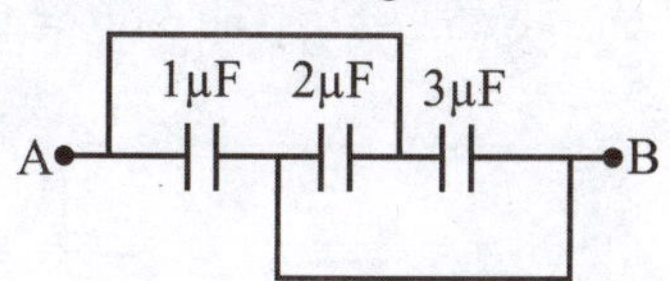

1) 1.5 µF 2) 3.0 µF
3) $\frac{6}{11}$ µF 4) 6 µF

41. The effective capacity in the following figure between the points P and Q will be

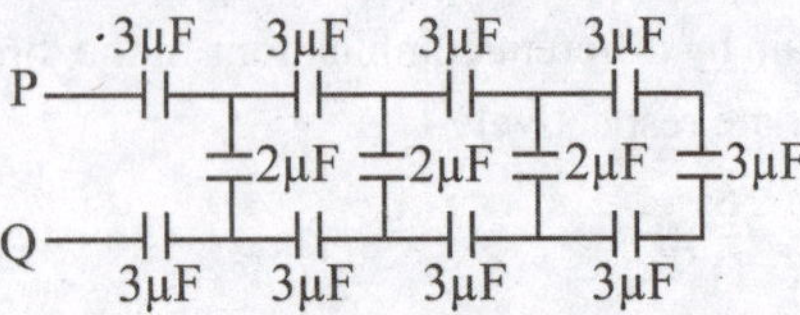

1) 3μF
2) 5μF
3) 2μF
4) 1μF

42. The energy and capacity of a charged parallel plate capacitor are U and C respectively. Now a dielectric slab of $_r$ = 6 is inserted in it then energy and capacity becomes (Assuming charge on plates remains constant)

1) 6U, 6C
2) U, C
3) $\frac{U}{6}$, 6C
4) U, 6C

43. Two capacitors C_1 = 2 μF and C_2 = 6 μF in series, are connected in parallel to a third capacitor C_3 = 4 μF. This arrangement is then connected to a battery of e.m.f. = 1V, as shown in the figure. How much energy is given by the battery in charging the capacitors:

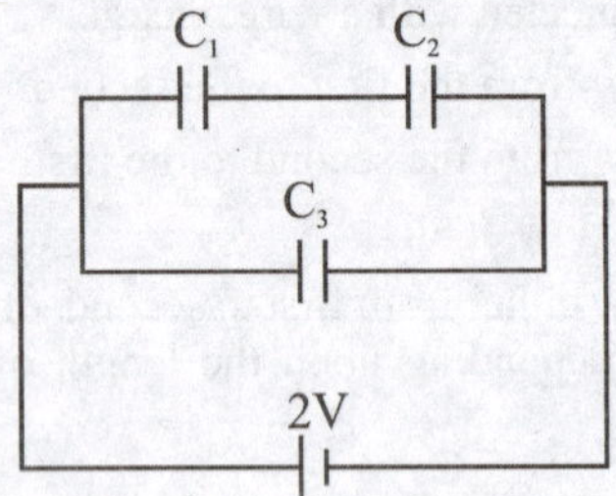

1) 22×10^{-6} J
2) 11×10^{-6} J
3) $\frac{32}{3} \times 10^{-6}$ J
4) $\frac{16}{3} \times 10^{-6}$ J

44. Two identical parallel plate capacitors are connected in series and then joined with a battery of 100V. A sheet of dielectric constant 4.0 is inserted between the plates of second capacitor. The potential difference across the capacitors will be respectively

1) 50 V, 50V
2) 80 V, 20 V
3) 20V, 80V
4) 75 V, 25 V

45. A capacitor is connected to 10 V battery. The charge on plates is 40μC. If medium between plates is air. The charge on plates will become 100C if the space between the plates is filled with oil. The dielectric constant of oil is:

1) 2.5
2) 4
3) 6.25
4) 0

46. In the given circuit switch K is open. The charge on the capacitor is 'C' in steady state is q_1. Now key is closed and steady state charge on C is q_2. The ratio of charges q_1/q_2 is -

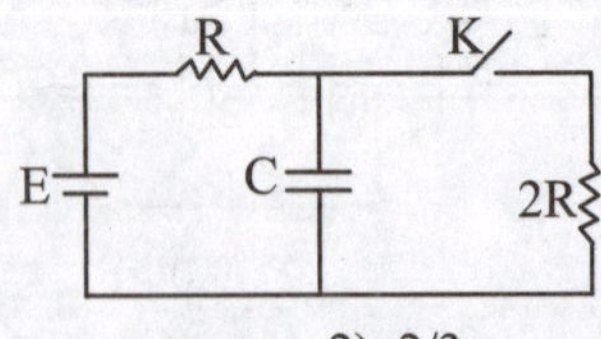

1) 3/2
2) 2/3
3) 1
4) 1/2

47. To reduce the capacitance of a parallel plate capacitor, the space between the plates is

1) Filled with dielectric material
2) Reduced and area of the plates is increased
3) Increased and area of the plates is decreased
4) Increased and area is increased relatively

48. A parallel plate condenser with plate separation d is charged with the help of a battery so that U_0 energy is stored in the system. A plate of dielectric constant K^2 and thickness d is placed between the plates of condenser while battery remains connected. The new energy of the system will be -

1) KU_0
2) K^2U_0
3) $\frac{U_0}{K}$
4) $\frac{U_0}{K^2}$

49. A slab of copper of thickness d/2 is introduced between the plates of a parallel plate capacitor where d is the separation between its two plates. If the capacitance of the capacitor without copper slab is C and with copper slab is C then $\frac{C'}{C}$ is:

1) $\sqrt{2}$
2) 2
3) 1
4) $\frac{1}{\sqrt{2}}$

50. A conducting sphere of radius 10 cm is charged with 10 mC. Another uncharged sphere of radius 20 cm is allowed to touch it for some time. After that if the sphere are separated, then surface density of charged on the spheres will be in the ratio of:

1) 1 : 4
2) 1 : 3
3) 2 : 1
4) 1 : 1

51. A network of four capacitors of capacity equal to C_1 = C, C_2 = 2C, C_3 = 3C and C_4 = 4C are connected to a battery as shown in the figure. The ratio of the charges on C_2 and C_4 is:

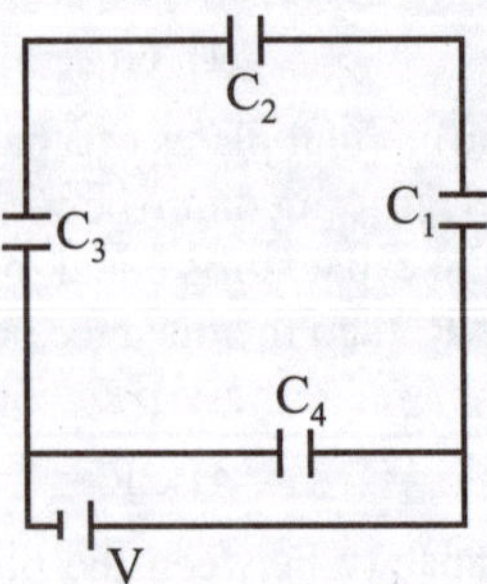

1) $\frac{22}{3}$
2) $\frac{3}{22}$
3) $\frac{7}{4}$
4) $\frac{4}{7}$

52. Half of the space between parallel plate capacitor is filled with a medium of dielectric constant K parallel to the plates. If initially the capacity is C, then the new capacity will be:

1) 2KC/(1+K)
2) C (K+1)/2
3) CK/(1+K)
4) KC

53. Consider the diagram,

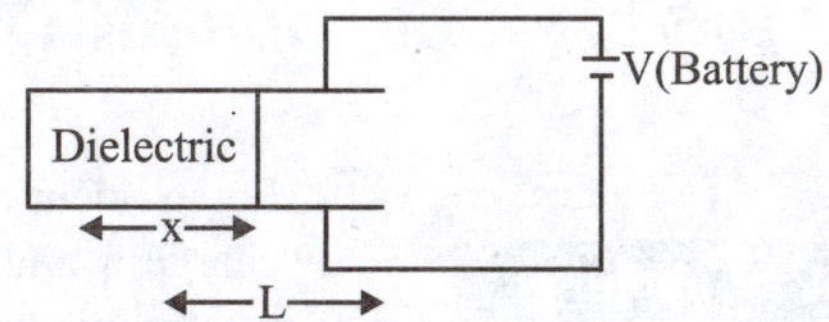

A parallel plate capacitor has the plate width t and length L while the separation between the plates is d. The capacitor is connected to a battery of voltage rating V. A dielectric which carefully occupy, the space between the plates of the capacitor is slowly inserted between the plates. When length x of the dielectric slab is introduced into the capacitor, then energy stored in the system is

1) $\dfrac{\varepsilon_0 tV^2}{2d}L$ 2) $\dfrac{\varepsilon_0 tV^2}{2d}L[x+1]$ 3) $\dfrac{\varepsilon_0 tV^2}{2d}[L+x(K-1)]$ 4) $\dfrac{\varepsilon_0^2 t^2 V^2}{2d^2}[L^2+x+1]$

Statement Based MCQs

Directions: These questions consist of two statements each, printed as Statement-I and Statement-II. While answering these questions, you are required to choose any one of the following four responses.

1) **Both Statement-I and Statement-II are correct.**
2) **Both Statement-I and Statement-II are incorrect.**
3) **Statement-I is correct & Statement-II is incorrect.**
4) **Statement-I is incorrect & Statement-II is correct.**

1. **Statement-I:** When some charge is given to an irregular shaped conductor, it distributes itself so that charge density is same everywhere.

   Statement-II: A conductor has to be equipotential under static condition.

2. **Statement-I:** If two protons are brought near one another, the potential energy of the system will increase.

   Statement-II: When a capacitor is charged by a battery, both the plates receive charge equal in magnitude, no matter sizes of plates are identical or not.

3. **Statement-I:** Lightning conductor protects a building by neutralising or conducting the charge of cloud to the ground.

   Statement-II: When a soap bubble is charged, it contracts.

4. **Statement-I:** Potential difference between two concentric charged metal spherical shells is directly proportional to the charge of inner sphere.

   Statement-II: The field inside the space between the two spheres described above will be only due to charge of inner sphere.

5. **Statement-I:** Capacity of a parallel plate capacitor increases when distance between the plates is decreased.

   Statement-II: This follows from $C=\dfrac{\varepsilon_0 A}{d}$.

Assertion & Reason MCQs

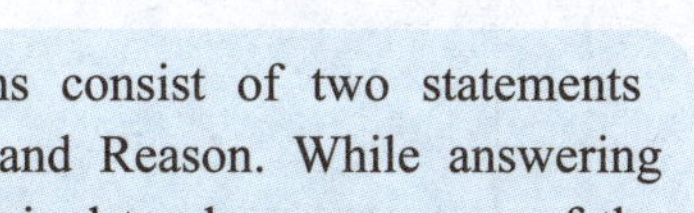

Directions: These questions consist of two statements each, printed as Assertion and Reason. While answering these questions, you are required to choose any one of the following four responses.

1) **Both Assertion and Reason are True and the Reason is a correct explanation of the Assertion.**
2) **Both Assertion and Reason are True but Reason is not a correct explanation of the Assertion.**
3) **If Assertion is True but the Reason is False.**
4) **Assertion is False but Reason is True.**

1. **Assertion:** If two concentric conducting spheres are connected by a conducting wire, no charge can exist on the inner sphere.

   Reason: When charge on outer sphere then potential of inner shell and outer shell will be same.

2. **Assertion:** Two concentric spherical shell of different radii are at potentials V_A and V_B. If outer shell is earthed then potential difference will not be changed.

   Reason: Potential difference between the surfaces of two concentric spherical shells does not depends on the charge on the outer shell.

3. **Assertion:** When a capacitor is charged by a battery, both the plates receive charges which are equal in magnitude, no matter whether the sizes of plates are identical or not.

   Reason: The charge distribution on the plates of capacitor is in accordance with charge conservation principle.

4. **Assertion:** A charged metallic shell may have charged particle inside it.

   Reason: There can't exist electric field lines inside the conductor.

5. **Assertion:** The tyres of aircrafts are made slightly conducting.

   Reason: If a conductor is connected to ground by a wire, the extra charge induced on conductor will flow to ground.

Match the Columns MCQs

1. Match the column-I with column-II.

	Column-I		Column-II
(a)	Potential due to a point charge varies as	(i)	$1/r^2$
(b)	Potential due to an electric dipole varies as	(ii)	Constant
(c)	Potential difference between any two points on an equipotential surface is	(iii)	$1/r$
(d)	Potential inside a charged conductor is	(iv)	Zero

1) (a)-(ii) (b)-(i) (c)-(iv) (d)-(iii)
2) (a)-(iii) (b)-(iv) (c)-(i) (d)-(ii)
3) (a)-(iii) (b)-(ii) (c)-(iv) (d)-(i)
4) (a)-(iii) (b)-(i) (c)-(iv) (d)-(ii)

2. In the figure shown, C_1 = 4μF (without dielectric) and C_2 = 4μF (with a dielectric slab of dielectric constant K = 2). Now, the same slab after removing from C_2 is filled in C_1. Then, match the following column-I and column-II.

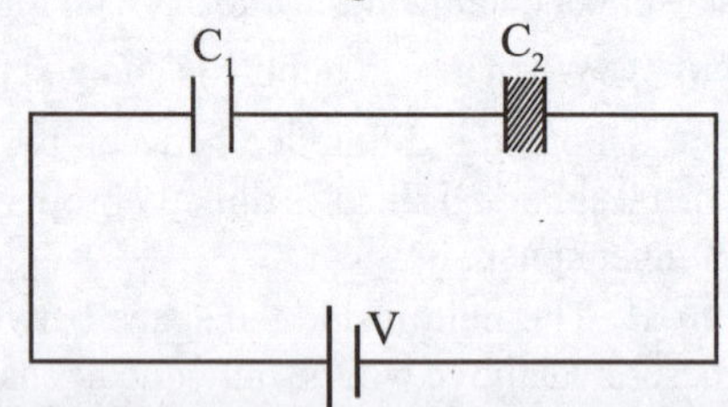

	Column-I		Column-II
(a)	Charge on C_2	(i)	will increase
(b)	Energy stored in C_2	(ii)	will decrease
(c)	Potential difference across C_2	(iii)	will remain same
(d)	Electric field between the plates of C_2	(iv)	data indufficient

1) (a)-(ii) (b)-(i) (c)-(i) (d)-(i)
2) (a)-(ii) (b)-(iii) (c)-(i) (d)-(i)
3) (a)-(i) (b)-(i) (c)-(ii) (d)-(iii)
4) (a)-(iii) (b)-(i) (c)-(ii) (d)-(iv)

3. In an electric field $\vec{E} = (2\hat{i} + 4\hat{j})\,N/C$, electric potential at origin is 0V. Match the following column-I and column-II.

	Column-I		Column-II
(a)	Potential at (4m, 0)	(i)	8 V
(b)	Potential at (–4m, 0)	(ii)	–8 V
(c)	Potential at (0, 4m)	(iii)	16 V
(d)	Potential at (0, –4m)	(iv)	–16 V

1) (a)-(i) (b)-(ii) (c)-(iv) (d)-(iii)
2) (a)-(ii) (b)-(i) (c)-(iv) (d)-(iii)
3) (a)-(i) (b)-(ii) (c)-(iii) (d)-(iv)
4) (a)-(iv) (b)-(ii) (c)-(i) (d)-(iii)

4. Let n identical drops each of radius r, capacitance C', charge q and potential V' are combined to form a larger drop. Radius of big drop is R, capacitance C, charge Q and potential V. Match the following column-I and column-II.

	Column-I		Column-II
(a)	Ratio $\frac{R}{r}$	(i)	$n^{1/3}$
(b)	Ratio $\frac{C}{C'}$	(ii)	$n^{2/3}$
(c)	Ratio $\frac{V}{V'}$	(iii)	$n^{5/3}$

1) (a)-(i) (b)-(ii) (c)-(ii)
2) (a)-(i) (b)-(iii) (c)-(i)
3) (a)-(i) (b)-(i) (c)-(ii)
4) (a)-(ii) (b)-(i) (c)-(iii)

5. Match the column-I with column-II.

	Column-I		Column-II
(a)	$V = \frac{C_1V_1 + C_2V_2}{C_1 + C_2}$	(i)	V_1 C_1 V C_2 V_2
(b)	$\Delta U = \frac{1}{2}\frac{C_1C_2}{C_1 + C_2}(V_1 - V_2)^2$	(ii)	C_1, V_1 K C_2, V_2 → C_1, V C_2, V
(c)	$\frac{(V - V_1)}{C_1} = \frac{V_2 - V}{C_2}$	(iii)	V C_1 V_1 C_2 V_2
(d)	Total energy $= \frac{1}{2}C_1(V - V_1)^2 + \frac{1}{2}C_2(V_1 - V)^2$	(iv)	V_1 C_1 C_2 V_2 V
		(v)	V_1 C_2 V C_1 V_2

1) (a)-(i, iv) (b)-(v) (c)-(i, ii) (d)-(iii)
2) (a)-(iv, v) (b)-(ii) (c)-(iv, ii) (d)-(ii)
3) (a)-(i, ii) (b)-(ii) (c)-(i, ii) (d)-(ii)
4) (a)-(i, v) (b)-(iii) (c)-(i, ii) (d)-(iv)

NCERT Exemplar MCQs

1. A capacitor of 4 μF is connected as shown in the circuit. The internal resistance of the battery is 0.5 Ω. The amount of charge on the capacitor plates will be:

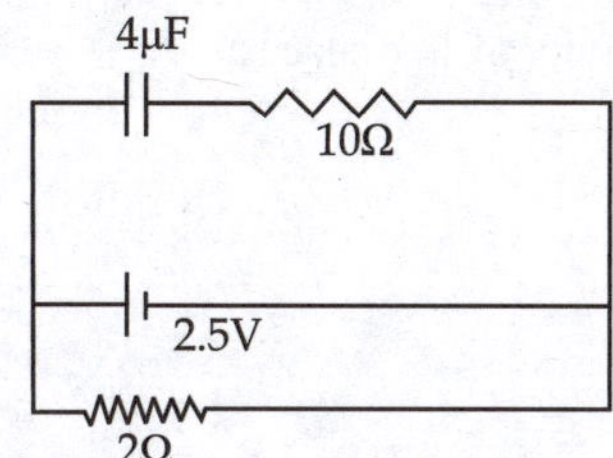

1) 0
2) 4 μC
3) 16 μC
4) 8 μC

2. A positively charged particle is released from rest in a uniform electric field. The electric potential energy of the charge:
1) Remains constant because the electric field is uniform
2) Increases because the charge moves along the electric field
3) Decreases because the charge moves along the electric field
4) Decreases because the charge moves opposite to the electric field

3. Figure shows some equipotential lines distributed in space. A charged object is moved from point A to point B.

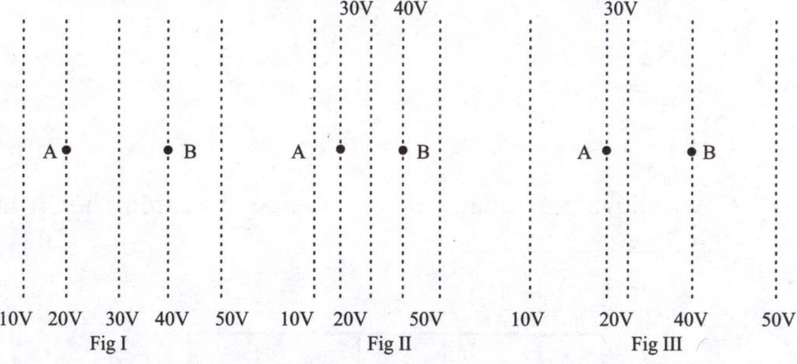

1) The work done in Fig. (I) is the greatest
2) The work done in Fig. (II) is least
3) The work done is the same in Fig. (I), Fig. (II) and Fig. (III)
4) The work done in Fig. (III) is greater than Fig. (II) but equal to that work done in fig (I)

4. The electrostatic potential on the surface of a charged conducting sphere is 100V. Two statements are made in this regard:

S_1 : At any point inside the sphere, electric intensity is zero.

S_2 : At any point inside the sphere, the electrostatic potential is 100V.

Which of the following is a correct statement?
1) S_1 is true but S_2 is false
2) Both S_1 and S_2 are false
3) S_1 is true, S_2 is also true and S_1 is the cause of S_2
4) S_1 is true, S_2 is also true but the statements are independent

5. Equipotentials at a great distance from a collection of charges whose total sum is not zero are approximately:
1) Spheres
2) Planes
3) Paraboloids
4) Ellipsoids

6. A parallel plate capacitor is made of two dielectric blocks in series. One of the blocks has thickness d_1 and dielectric constant K_1 and the other has thickness d_2 and dielectric constant K_2 as shown in figure. This arrangement can be thought as a dielectric slab of thickness d (= d_1 + d_2) and effective dielectric constant K. The K is:

1) $\frac{K_1d_1+K_2d_2}{d_1+d_2}$
2) $\frac{K_1d_1+K_2d_2}{K_1+K_2}$
3) $\frac{K_1K_2(d_1+d_2)}{K_1d_2+k_2d_1}$
4) $\frac{2K_1K_2}{K_1+K_2}$

NEET & AIIMS Past Year MCQs

1. The value of electric potential at a distance of 9 cm from the point charge 4×10^{-7} C is

$\left[\text{Given } \frac{1}{4\pi\varepsilon_0} = 9\times10^9 \text{ Nm}^2\text{C}^{-2}\right]$: **(2024 Re)**

1) 4×10^2 V
2) 44.4 V
3) 4.4×10^5 V
4) 4×10^4 V

2. Given below are two statements: one is labelled as **Assertion A** and the other is labelled as **Reason R.** **(2024)**

Assertion A: The potential (V) at any axial point, at 2 m distance (r) from the centre of the dipole of dipole moment vector $\vec{P}$ of magnitude, 4×10^{-6}C m, is $\pm 9 \times 10^3$ V.

$\left(\text{Take } \frac{1}{4\pi \epsilon_0} = 9\times10^9 \text{ SI Units}\right)$

Reason R: $V = \pm\frac{2P}{4\pi \epsilon_0 r^2}$, where r is the distance of any axial point, situated at 2 m from the centre of the dipole.

In the light of the above statements, choose the *correct* answer from the options given below:
1) A is true but R is false.
2) A is false but R is true.
3) Both A and R are true and R is the correct explanation of A.
4) Both A and R are true and R is NOT the correct explanation of A.

3. A thin spherical shell is charged by some source. The potential difference between the two points *C* and *P* (in *V*) shown in the given is: $\left(\text{Take } \frac{1}{4\pi \epsilon_0} = 9\times10^9 \text{ SI units}\right)$ **(2024)**

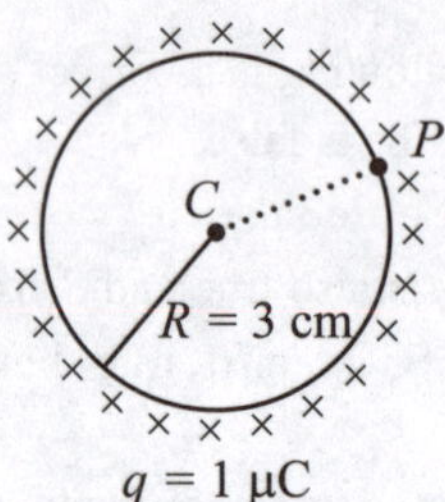

1) 0.5×10^5 2) zero
3) 3×10^5 4) 4×10^5

4. The capacitance of a capacitor with charge *q* and a potential difference *V* depends on **(2024 Re)**
1) both *q* and *V*
2) the geometry of the capacitor
3) q only
4) *V* only

5. In the following circuit, the equivalent capacitance between terminal *A* and terminal *B* is: **(2024)**

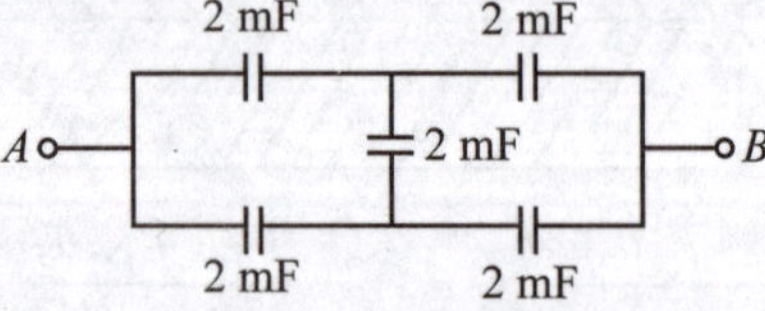

1) 0.5 mF 2) 4 mF
3) 2 mF 4) 1 mF

6. A 12 pF capacitor is connected to a 50 V battery, the electrostatic energy stored in the capacitor in nJ is **(2024 Re)**
1) 15 2) 7.5
3) 0.3 4) 150

7. The equivalent capacitance of the system shown in the following circuit is: **(2023)**

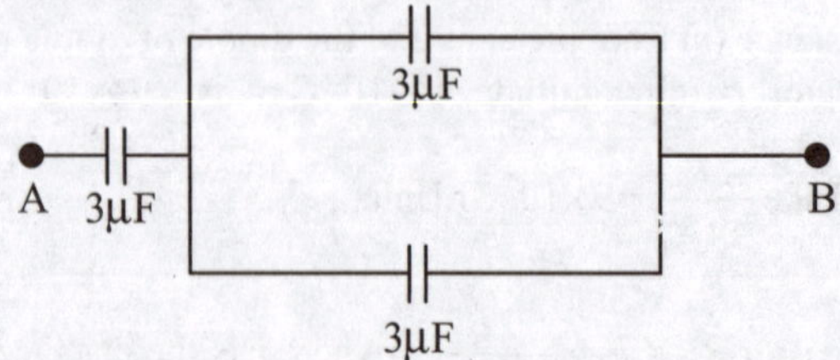

1) 9 μF 2) 2 μF
3) 3 μF 4) 6 μF

8. An electric dipole is placed as shown in the figure.

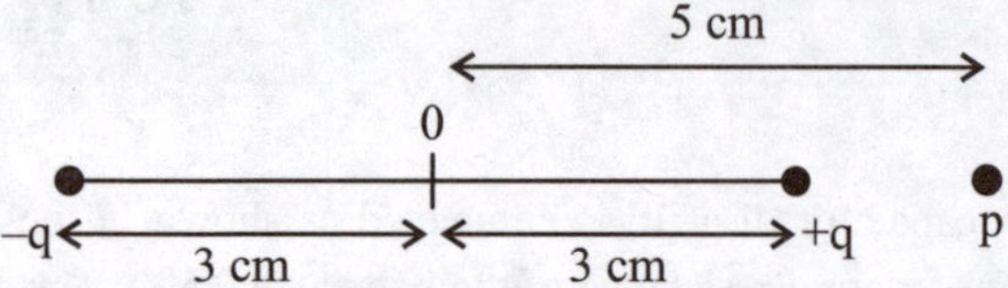

The electric potential (in 10^2 V) at point P due to the dipole is (ϵ_0 = permittivity of free space and $\frac{1}{4\pi\epsilon_0} = K$): **(2023)**
1) $\left(\frac{8}{3}\right)$qK 2) $\left(\frac{3}{8}\right)$qK
3) $\left(\frac{5}{8}\right)$qK 4) $\left(\frac{8}{5}\right)$qK

9. Two hollow conducting sphere of radii R_1 and R_2 ($R_1 >> R_2$) have equal charges. The potential would be **(2022)**
1) Dependent on the material property of the sphere
2) More on bigger sphere
3) More on smaller sphere
4) Equal on both the sphere

10. The angle between the electric lines of force and the equipotential surface is **(2022)**
1) 180° 2) 0°
3) 45° 4) 90°

11. Two charged spherical conductors of radius R_1 and R_2 are connected by a wire. Then the ratio of surface charge densities of the spheres (σ_1/σ_2) is **(2021)**
1) $\frac{R_2}{R_1}$ 2) $\sqrt{\left(\frac{R_1}{R_2}\right)}$
3) $\frac{R_1^2}{R_2^2}$ 4) $\frac{R_1}{R_2}$

12. The equivalent capacitance of the combination shown in the figure is **(2021)**

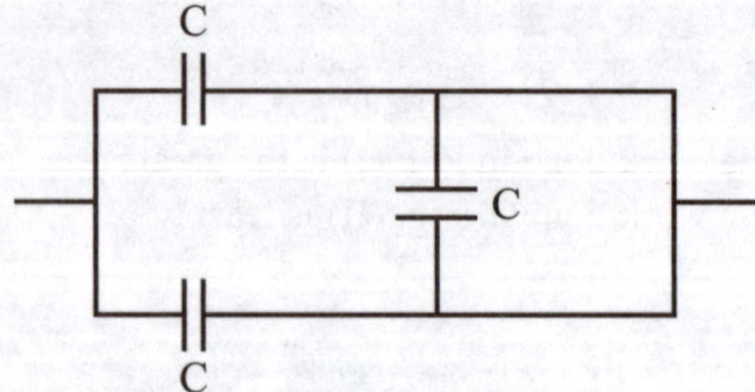

1) 2C 2) C/2
3) 3C/2 4) 3C

13. A parallel plate capacitor has a uniform electric field '$\vec{E}$' in the space between the plates. If the distance between the plates is 'd' and the area of each plate is 'A', the energy stored in the capacitor is **(2021)**
(ε_0 = permittivity of free space)
1) ε_0EAd 2) $\frac{1}{2}\varepsilon_0 E^2 Ad$
3) $\frac{E^2 Ad}{\varepsilon_0}$ 4) $\frac{1}{2}\varepsilon_0 E^2$

14. Twenty seven drops of same size are charged at 220 V each. They combine to form a bigger drop. Calculate the potential of the bigger drop. **(2021)**

1) 1320 V
2) 1520 V
3) 1980 V
4) 660 V

15. **Assertion:** All the charge in a conductor gets distributed on whole of its outer surface.

Reason: In a dynamic system, charges try to keep their potential energy minimum **(AIIMS 2018)**

1) If both Assertion & Reason are True & the Reason is a correct explanation of the Assertion.
2) If both Assertion & Reason are True but Reason is not a correct explanation of the Assertion.
3) If Assertion is True but the Reason is False.
4) If both Assertion & Reason are False.

16. Equipotential surfaces associated with an electric field which is increasing in magnitude along the x-direction are: **(AIIMS 2017)**

1) Planes parallel to yz-plane
2) Planes parallel to xy-plane
3) Planes parallel of xz-plane
4) Co-axial cylinders of increasing radii around the x-axis

17. A conducting sphere of radius 10 cm is charged with 10 μC. Another uncharged sphere of radius 20 cm is allowed to touch it for some time. After that if the spheres are separated, then surface density of charges on the spheres will be in the ratio of: **(AIIMS 2002)**

1) 1 : 4
2) 2 : 1
3) 1 : 3
4) 1 : 1

18. Electric potential V at any point x, y, z in space is given by $V = 6z^2$. The value of the electric field at the point (2, –1, 3) is: **(AIIMS 2000)**

1) 24
2) –12
3) –36
4) 12

19. If an electron is brought toward another electron, the electrostatic potential energy of the system: **(AIIMS 1995)**

1) Becomes zero
2) Increases
3) Remains the same
4) Decreases

Answer Key

NCERT Topic-wise MCQs

1	2	3	4	5	6	7	8	9	10	11	12	13	14	15	16	17
1	3	3	4	1	4	1	4	4	3	4	2	1	2	4	3	1
18	19	20	21	22	23	24	25	26	27	28	29	30	31	32	33	34
3	1	2	3	2	2	1	2	4	1	1	1	3	4	2	3	1
35	36	37	38	39	40	41	42	43	44	45	46	47	48	49	50	51
1	2	2	4	1	1	3	2	2	3	3	4	4	1	2	2	1
52	53	54	55	56	57	58	59	60	61	62	63	64	65	66	67	68
2	1	3	1	4	2	1	3	4	3	2	1	1	3	4	2	4
69	70	71	72	73												
4	1	3	3	3												

High Order Time Intensive MCQs

1	2	3	4	5	6	7	8	9	10	11	12	13	14	15	16	17
4	1	3	4	3	2	3	2	1	1	4	3	4	4	3	4	2
18	19	20	21	22	23	24	25	26	27	28	29	30	31	32	33	34
1	3	3	4	4	1	3	1	4	1	1	1	2	3	1	1	1
35	36	37	38	39	40	41	42	43	44	45	46	47	48	49	50	51
1	2	3	1	2	4	4	3	2	2	1	1	3	2	2	3	2
52	53															
1	3															

Statement Based MCQs

1	2	3	4	5
4	1	3	1	1

Assertion & Reason MCQs

1	2	3	4	5
1	1	1	2	2

Match the Column MCQs

1	2	3	4	5
4	1	2	3	3

NCERT Exemplar MCQs

1	2	3	4	5	6
4	3	3	3	1	3

NEET & AIIMS Past Year MCQs

1	2	3	4	5	6	7	8	9	10	11	12	13	14	15	16	17
4	1	2	2	3	1	2	2	3	4	1	1	2	3	1	1	2
18	19															
3	2															

CHAPTER

3 Current Electricity

Topicwise Number of Questions Analysis (2025-2016)

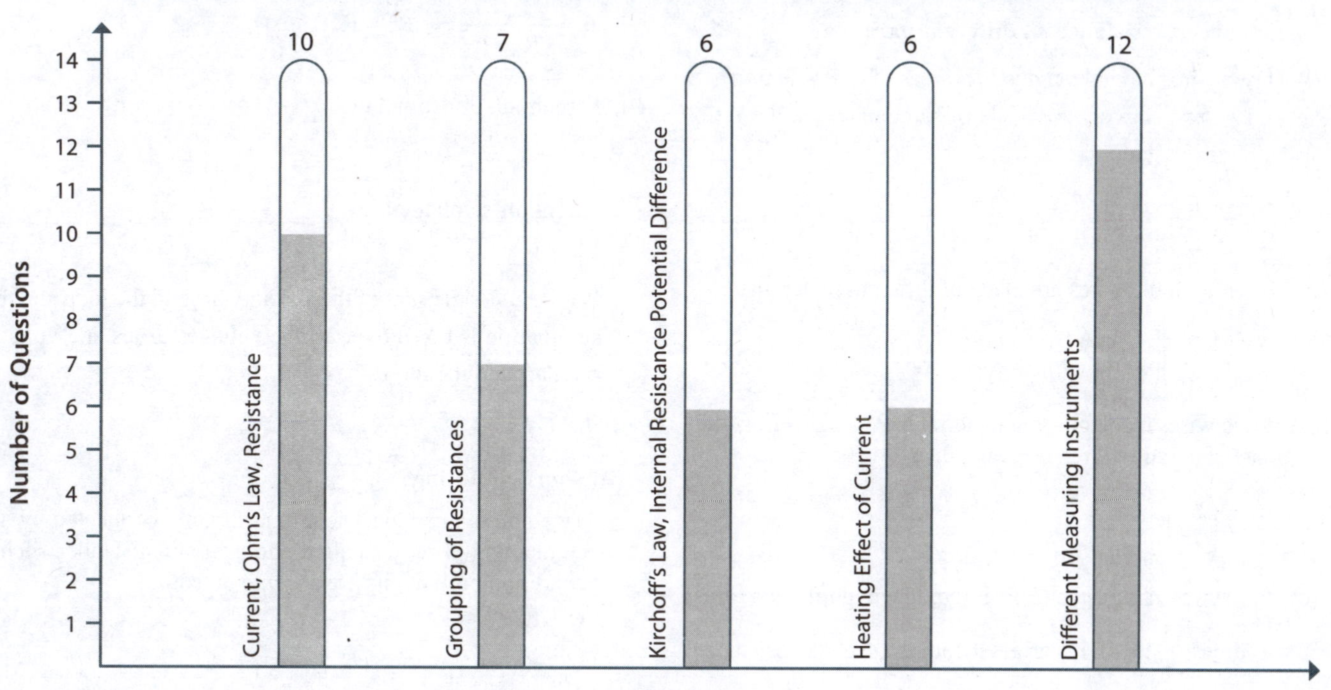

NCERT-PICKS

Current and Ohm's law

Formulas

1. Electric current $= \frac{\text{Charge}}{\text{Time}} \Rightarrow I = \frac{q}{t} = \frac{ne}{t}$, as q = ne
2. In case of an electron revolving in a circle of radius r with speed v, period of revolution of the electron is

   $T = \frac{2\pi r}{v}$

   Frequency of revolution, $\nu = \frac{1}{T} = \frac{v}{2\pi r}$

   Current at any point of the orbit,

   I = Charge flowing in 1 revolution × No. of revolutions per second

   $\Rightarrow I = e\nu = \frac{ev}{2\pi r}$.
3. Ohm's law, $R = \frac{V}{I}$ or $V = IR$
4. Resistance of a uniform conductor, $R = \rho\frac{l}{A}$
5. Resistivity or specific resistance, $\rho = \frac{RA}{l}$
6. Conductance $= \frac{1}{\text{Resistance}} = \frac{1}{R}$
7. Conductivity $= \frac{1}{\text{Resistivity}} \Rightarrow \sigma = \frac{1}{\rho} = \frac{l}{RA}$
8. Current density $= \frac{\text{Current}}{\text{Area}} \Rightarrow J = \frac{I}{A}$

Formulas

Current produced under different cases

(i) If there are n particles per unit volume each having a charge q and moving with velocity v, the current through cross section A, $I = nqvA$

(ii) If the point charge 'q' is moving in a circle of radius 'r' with speed 'v' with angular speed ω, then current will be,

$$I = \frac{q}{T} = \frac{qv}{2\pi r} = \frac{q\omega}{2\pi}$$

Dependence of resistance on different parameters

(i) If two wires having lengths l_1, l_2 areas of cross-section A_1 and A_2 respectively are made of same material, the ratio of their resistances.

$$\frac{R_1}{R_2} = \left(\frac{l_1}{l_2}\right)\left(\frac{A_2}{A_1}\right)$$

(ii) When both the wires are made of different materials.

$$\frac{R_1}{R_2} = \left(\frac{\rho_1}{\rho_2}\right)\left(\frac{l_1}{l_2}\right)\left(\frac{A_2}{A_1}\right)$$

(iii) If two wires made of same material have lengths l_1, l_2 and masses m_1 and m_2 respectively, then

$$\frac{R_1}{R_2} = \left(\frac{l_1^2}{l_2^2}\right)^2\left(\frac{m_2}{m_1}\right)$$

(iv) If a wire of resistance R is stretched until its radius becomes $\frac{1}{n}$ th of its original radius then its resistance becomes n^4R.

(v) If two wires of resistivities ρ_1 and ρ_2, lengths l_1 and l_2 are connected in series, the equivalent resistivity

$$\rho = \frac{\rho_1 l_1 + \rho_2 l_2}{l_1 + l_2}$$

(vi) If two wires of resistivities ρ_1 and ρ_2, Areas of cross section A_1 and A_2 are connected in parallel, the equivalent resistivity, $\rho = \frac{\rho_1\rho_2(A_1 + A_2)}{\rho_1 A_2 + \rho_2 A_1}$

Drift of Electrons, Resistivity and Kirchhoff's Law

Formulas

1. Current in terms of drift velocity (v_d) is $I = neAv_d$
2. Current density, $J = env_d$
3. In terms of relaxation time τ,

$$R = \frac{ml}{ne^2\tau A} \text{ and } \rho = \frac{m}{ne^2\tau}$$

4. Relation between current density and electric field,

$$J = \sigma E \quad \Rightarrow \quad E = \rho J$$

5. Mobility, $\mu = \frac{v_d}{E} = \frac{q\tau}{m}$
6. Electric current, $I = enA\mu E$
7. Conductivity of metallic conductor, $\sigma = ne\mu_e$
8. Conductivity of a semiconductor, $\sigma = ne\mu_e + pe\mu_h$
9. Temperature coefficient of resistance, $\alpha = \frac{R_2 - R_1}{R_1(t_2 - t_1)}$

If $t_1 = 0$ °C and $t_2 = t$ °C, then

$$\alpha = \frac{R_t - R_0}{R_0 \times t} \Rightarrow R_t = R_0(1 + \alpha t)$$

10. Kirchhoff's Junction law: $\sum_{i=1}^{n} i_i = 0$

$$\Rightarrow i_1 + i_2 + \ldots i_n = 0$$

11. Kirchhoff's voltage Law:

$$\Sigma V = 0$$

Where ΣV is the sum of the voltages around the loop and the summation is taken over all the voltages drops and voltage rises around the loop.

Numerical Profile

(i) Short circuiting

Two points in an electric circuit directly connected by a conducting wire are called short circuited. Under such condition both points are at same potential.

$$V_A = V_B$$

$$V_{R_1} = V_A - V_B = 0 \quad [\because I_{R_1} = 0]$$

$$I_{R_2} = \frac{E}{R_2}$$

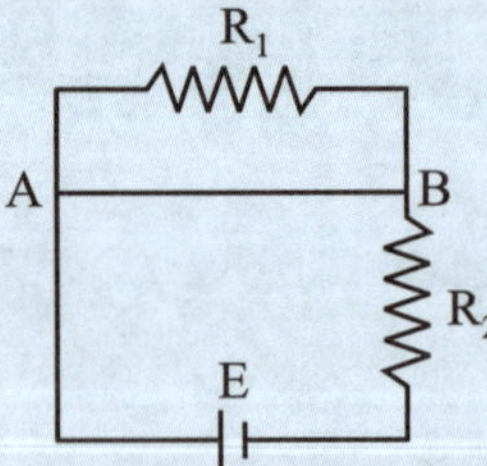

(ii) Distribution of current in parallel connections

When more than one resistances are connected in parallel, the potential difference across them is equal and the current is distributed among them in inverse ratio of their resistance as

$$i = \frac{V}{R} \Rightarrow i \propto \frac{1}{R} \text{ for same value of V}$$

For example in the figure,

$$i_1 : i_2 : i_3 = \frac{1}{R} : \frac{1}{2R} : \frac{1}{3R} = 6:3:2$$

$$\therefore i_1 = \left(\frac{6}{6+3+2}\right)i = \frac{6}{11}i$$

$$i_2 = \left(\frac{3}{6+3+2}\right)i = \frac{3}{11}i$$

$$i_3 = \left(\frac{2}{6+3+2}\right)i = \frac{2}{11}i$$

Note: In case of only two resistances, $\frac{i_1}{i_2} = \frac{R_2}{R_1}$

(iii) Distribution of potential in series connections

When more than one resistances are connected in series, the current through them is same and the potential is distributed in the direct ratio of their resistance as

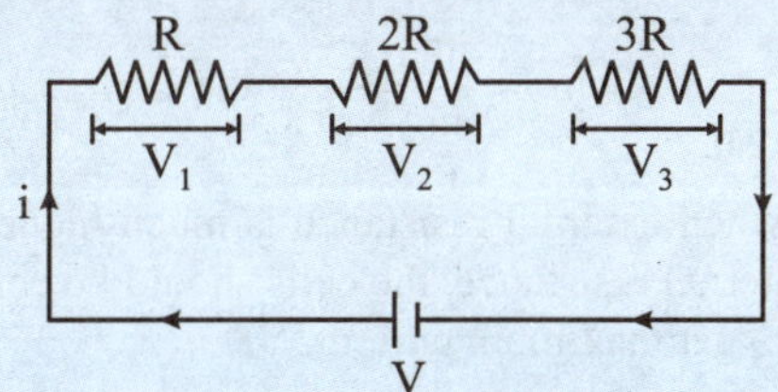

$V = iR$ or $V \propto R$ for same value of i.

For example in the figure,

$$V_1 : V_2 : V_3 = R : 2R : 3R = 1:2:3$$

$$\therefore V_1 = \left(\frac{1}{1+2+3}\right)V = \frac{V}{6}$$

$$V_2 = \left(\frac{2}{1+2+3}\right)V = \frac{V}{3}$$

$$V_3 = \left(\frac{3}{1+2+3}\right)V = \frac{V}{2}$$

Combination of Resistors

Formulas

1. The equivalent resistance R_s of a number of resistances connected in series is given by

   $$R_s = R_1 + R_2 + R_3 + ...$$

2. The equivalent resistance R_p of a number of resistances connected in parallel is given by

   $$\frac{1}{R_p} = \frac{1}{R_1} + \frac{1}{R_2} + \frac{1}{R_3} + ...$$

3. For two resistances in parallel,

   Currents through the two resistors will be

   $$I_1 = \frac{R_2 I}{R_1 + R_2} \text{ and } I_2 = \frac{R_1 I}{R_1 + R_2}$$

Numerical Profile

Resistance under different cases

(i) A wire of resistance 'R' is cut into 'n' equal parts and all of them are connected in parallel, equivalent resistance becomes $\frac{R}{n^2}$.

(ii) In 'n' wires of equal resistances are given, the number of combinations that can be made to give different resistances is 2^{n-1}.

(iii) A wire of resistance 'R' is bent in the form of circle, effective resistance across diameter is R/4

(iv) A wire of resistance 'R' is bent in the form of circle, effective resistance between two points A and B which makes an angle 'α' at the centre is

$$R_{AB} = \frac{R\alpha(2\pi - \alpha)}{4\pi^2}$$

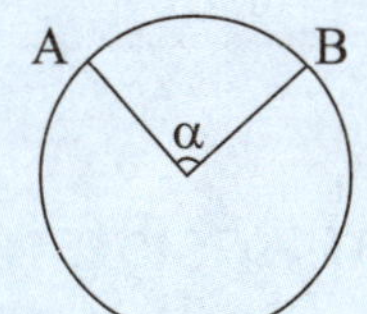

Electrical Energy, Power

Formulas

1. Work done, $W = I^2Rt$ (joule) $= \frac{I^2Rt}{4.18}$ cal [t is the time]
2. Power, $\frac{W}{t} = VI = I^2R = \frac{V^2}{R}$
3. Electrical energy, $W = Pt = VIt = I^2Rt$
4. Efficiency of cell, $\eta = \left(\frac{R}{R+r}\right)$
5. Maximum power is dissipated in circuit when R = r, then power dissipation is maximum about 50%.
6. Power output of cell, $P = \left(\frac{E}{R+r}\right)^2 \times R$
7. Charging current $= \frac{\text{emf of charger} - \text{emf of cell}}{\text{total resistace of circuit}}$
8. Power in series, $\frac{1}{P_{eq}} = \frac{1}{P_1} + \frac{1}{P_2} + \frac{1}{P_3} ...... \frac{1}{P_n}$
9. Power in parallel, $P_{eq} = P_1 + P_2 + P_3 .... P_n$

Numerical Profile

Calculation of power under different cases

(i) If 'n' appliances, each of resistance 'R' are connected in parallel with a voltage source 'V', the power dissipated 'P_P' will be

$$P_P = \frac{V^2}{(R/n)} = \frac{nV^2}{R}$$

For series combination, $P_S = \frac{V^2}{nR}$

$\Rightarrow \frac{P_P}{P_S} = n^2 \Rightarrow P_P = n^2 P_S$

This shows that power consumed by 'n' equal resistances in parallel is n^2 times that of power consumed in series if the voltage remains same.

(ii) For a given voltage V, if resistance is changed from 'R' to $\left(\frac{R}{n}\right)$, power consumed changes from 'P' to 'nP' where

$P' = \frac{V^2}{R'}$, then

$P' = \frac{V^2}{(R/n)} = \frac{nV^2}{R} = nP$

EMF, Internal Resistance, Terminal Potential Difference and Grouping of Cells

Formulas

1. EMF of a cell, $E = \frac{W}{q}$
2. For a cell of internal resistance r, the emf is $E = V + Ir = I(R + r)$
3. Terminal potential difference of a cell, $V = IR = \frac{ER}{R + r}$
4. Terminal potential difference when a current is being drawn from the cell, $V = E - Ir$
5. Terminal potential difference when the cell is being charged, $V = E + Ir$
6. Internal resistance of a cell, $r = R\left[\frac{E - V}{V}\right]$

Numerical Profile

Condition for maximum current from a series combination of cells

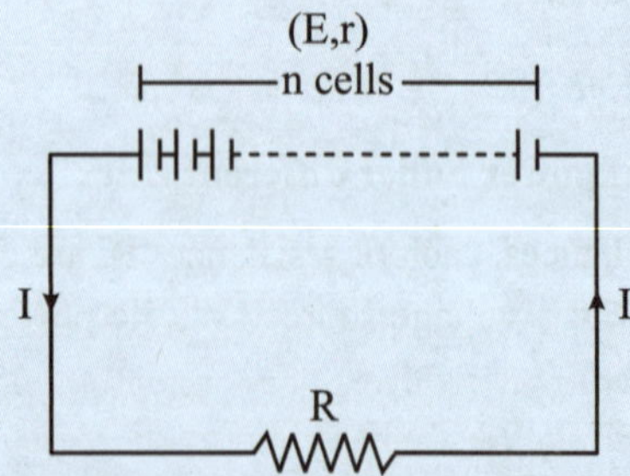

Fig.: A series combination of n cells.

(i) Total emf of n cells in series = Sum of emfs of all cells = nE

(ii) Total internal resistance of n cells in series = $r + r + r + \ldots$ n terms = nr

(iii) Total resistance in the circuit = R + nr

(iv) The current in the circuit is, $I = \frac{\text{Total emf}}{\text{Total resistance}} = \frac{nE}{R + nr}$

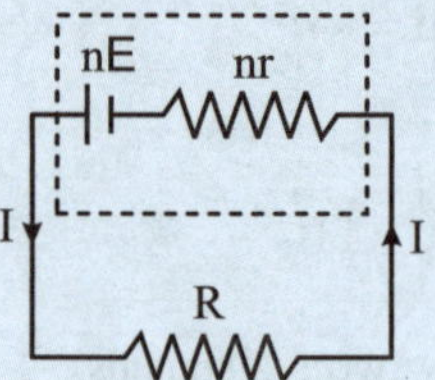

Fig.: Equivalent circuit.

Special Case

(i) If $R >> nr$, then $I = \frac{nE}{R}$ = n times the current (E/R) that can be drawn from one cell.

(ii) If $R << nr$, then $I = \frac{nE}{nr} = \frac{E}{r}$ = the current given by a single cell

Thus, when external resistance is much higher than the total internal resistance, the cells should be connected in series to get maximum current.

Condition for maximum current from a parallel combination of cells

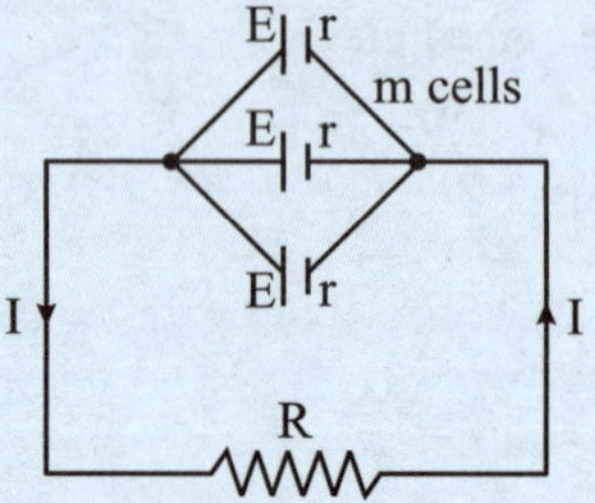

Fig.: A parallel combination of m cells.

(i) Equivalent resistance R′ is given by

$\frac{1}{R'} = \frac{1}{r} + \frac{1}{r} + \frac{1}{r} + \ldots$ m terms $= \frac{m}{r}$

or $R' = \frac{r}{m}$

(ii) Total resistance in the circuit $= R + R' = R + \frac{r}{m}$

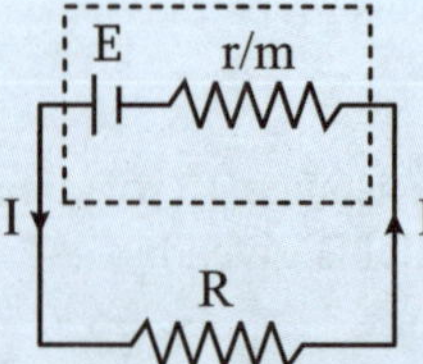

Fig.: Equivalent circuit.

(iii) Total emf of parallel combination = emf due to single cell = E

(iv) ∴ The current in the circuit is

$I = \frac{E}{R + r/m} = \frac{mE}{mR + r}$

Special Case

(i) If $R << \frac{r}{m}$, then

$$I = \frac{mE}{r} = m \text{ times the current due to a single cell.}$$

(ii) If $R >> \frac{r}{m}$, then

$$I = \frac{E}{R} = \text{ the current given by a single cell.}$$

Thus, when external resistance is much smaller than the net internal resistance, the cells should be connected in parallel to get maximum current.

We have n cells each of emf E. Of these, polarity of m cells (where n > 2m) is reversed. Then, net emf in the circuit is (n – 2m)E and resistance of the circuit is R. therefore,

$$i = \frac{(n-2m)E}{R}$$

Mixed Grouping

The situation is shown in figure.

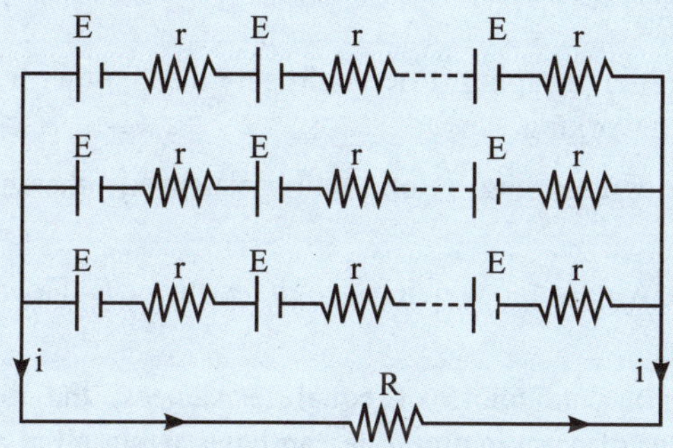

Fig.: Mixed Grouping of Cells

There are n identical cells in a row and number of rows are m. Emf of each cell is E and internal resistance is r. Treating each row as a single cell of emf nE and internal resistance nr, we have

(i) Net emf = nE

(ii) Total internal resistance = $\frac{nr}{m}$

(iii) Total external resistance = R

(iv) Current through the external resistance R is

$$i = \frac{nE}{R + \frac{nr}{m}}$$

This expression after some rearrangements can also be written as

$$i = \frac{mnE}{(\sqrt{mR} - \sqrt{nr})^2 + 2\sqrt{mnrR}}$$

(v) i is maximum when, $\sqrt{mR} = \sqrt{nr}$ or $R = \frac{nr}{m}$

or total external resistance = total internal resistance

Thus, we can say that the current and hence power transferred to the load is maximum when load resistance is equal to internal resistance. This is known as **maximum power transfer theorem**.

Different Measuring Instruments

[RC]

Formulas

1. For measuring internal resistance of a cell,

$$r = \frac{l_1 - l_2}{l_2} \times R$$

2. Potential gradient of the potentiometer wire,

$$k = \frac{V}{l}$$

3. Unknown emf balanced against length l, $E = kl$
4. For a balanced Wheatstone bridge,

$$\frac{P}{Q} = \frac{R}{S}$$

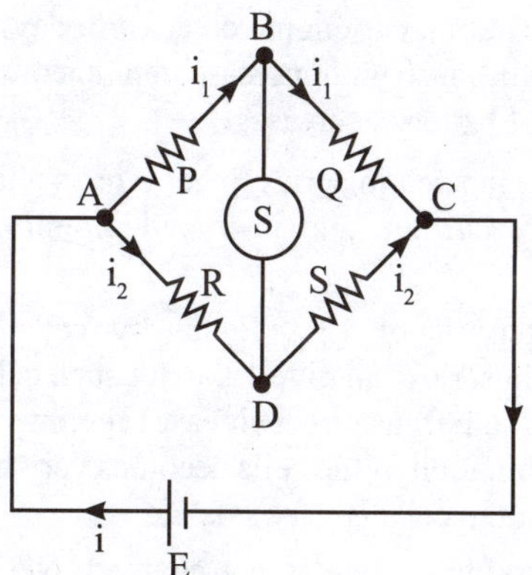

If X is the unknown resistance

$$\frac{P}{Q} = \frac{R}{X} \text{ or } X = \frac{RQ}{P}$$

5. In slide wire bridge, if balance point is obtained at l cm from the zero end, then

$$\frac{P}{Q} = \frac{R}{X} = \frac{l}{100-l} \text{ or } X = \left(\frac{100-l}{l}\right)R$$

Statements Asked in Exam

- The direction of electronic current is opposite to that of conventional current. Conventional current is due to flow of positive charge and electronic current is due to negative charge.
- Although a direction is associated with electric current yet it is a scalar quantity. It doesn't obey vector law of addition.

- Free electrons in conductors are always in continuous random motion and arises due to thermal velocities of the electrons which is usually very large.
- Thermal energy increases the velocity of electron in continuous motion.
- If a wire is stretched to double its original length without loss of mass, there is no effect on the resistivity of the wire. Resistivity of a wire depends on the nature of its material.
- Ohm-metre is the unit of resistivity whereas ohm meter is an instrument to measure the resistance.
- Ohm's law fails when V/I is not constant
- If the potential difference across a given copper wire is increased, drift velocity will also get increased. Drift velocity is directly proportional to the potential difference applied between the ends of the wire.
- Resistance of a conductor does not depend upon the potential difference applied across it. Resistance depends upon nature of material of the conductor and dimensions of the conductor.
- Connecting wires are made up of copper. Copper has low resistivity, high conductivity, and are diamagnetic material.
- If a steady current flows in a conductor, electric field exist within the conductor. It is the electric field which imparts acceleration to electrons for the flow of current.
- When n cells each of internal resistance r are connected in series the net internal resistant = nr. i.e there is no effect on the total internal resistance of the cells.
- It is easy to start a car engine on a warm day than on a chill day. This is because with increase in temperature the internal resistance of battery decreases.
- Cells are connected in series to get more voltage, in parallel to get more current, and in mixed grouping to get more power.
- A car battery is of 12V. Eight dry cells each of 1.5V connected in series can give 12V. But such cells are not used in car. This is because in such cases the internal resistance of the combination of the cells becomes very high, so current drawn from the cells is very small.
- A wire carrying current is not charged. Net charge on the wire is zero.
- Relaxation time decreases with rise in temperature.
- A meterbridge cannot be used to measure very high or very low resistances.
- Balanced position of meterbridge is not affected on interchanging the positions of the battery and the galvanometer.
- Potentiometer is based on null deflection method.
- Electric current should not be passed through a potentiometer wire for long time continuously as this will increase the resistance of the wire due to heating effect of current. So, fall of potential drop will not remain constant throughout the length of the potentiometer wire.
- A Potentiometer is preferred over a voltmeter for determining the emf of a cell. This is because a potentiometer doesn't draw any current from the cell whereas a voltmeter draws small current from the cell.
- Kirchoff's law are equally applicable to a.c as well as d.c circuits.
- Bending or reorienting the wire does not change the validity of Kirchhoff's junction rule. Kirchhoff's junction rule is based on conservation of charge and the outgoing currents add up and are equal to incoming current at a junction.
- Joule's heating effect of current is common to both a.c & d.c. That is why the instruments or electrical appliances such as heater, iron, geyser, toaster etc. work on both a.c and d.c
- A domestic electrical appliance , working on a three pin, will continue working even if the top pin is removed. The top pin is used only as a safety device.
- Heater wires carrying current becomes very hot while the wires carrying current remains cold. This is because the element of heater is made of nichrome which has high resistance, whereas the current carrying wires are made of copper (or aluminium) which has low resistance.
- In series combination of resistors, the potential difference and power consumption will be more on the resistors having high resistance as $P = I^2R$, $V = IR$
- In parallel combination, the current & power consumed will be more in the resistors having low resistances, as $P = \frac{V^2}{R}$. $I = \frac{V}{R}$
- In parallel grouping if one bulb gets fused, other bulbs will remain working.
- In series grouping, if one bulb gets fused other bulb will work.
- A fuse wire is one which has high resistance and low melting point.
- Using n conductors of equal resistances, the maximum number of combination one can have, using all at a time is 2^{n-1}
- Using n conductors of different resistances, the number of possible combinations are 2^n
- If the radius of the metallic wire becomes n times, the resistance becomes $\frac{1}{n^4}$ times.
- In parallel combination of electric bulb, the resistance of 100W bulb is less than that of 50W bulb.
- In series combination of electric bulb, the brightness of 50W bulb is more than that of 100W bulb.
- When two bulls of different wattages but of same rated voltages are connected in series and if the voltage supplied is twice the rated voltage then the bulb with the least wattage will fuse off.
- Current is a scalar quantity.
- $J = \sigma E$ is the microscopic form of ohm's law
- The electrical conductance of electrolytes is less than that of metals. Reason is that metals have a larger number of free charges carriers (electrons) as compared to the positive and negative ions in the electrolytes.

MCQs ?

NCERT Topic-wise MCQs

Current, Electric Conduction Ohm's law, Resistance

1. Electric field (E) and current density (J) have relation

1) $E \propto J^{-1}$ 2) $E \propto J$

3) $E \propto \frac{1}{J^2}$ 4) $E^2 \propto \frac{1}{J}$

2. The current flowing through a wire depends on time as $I = 3t^2 + 2t + 5$. The charge flowing through a cross section of the wire from time t = 0 to t = 2 sec is

1) 22 C 2) 20 C

3) 18 C 4) 5 C

3. Ohm's law is true

1) for metallic conductors at constant temperature.

2) for metallic conductors at variable temperature.

3) for electrolytes when current passes through them.

4) for diode when current flows.

4. Match Column-I with Column-II:

	Column-I		Column-II
(a)	$\frac{\rho \ell}{A}$	(i)	Ohm's Law
(b)	$\frac{RA}{\ell}$	(ii)	Conductivity
(c)	$\frac{1}{\rho}$	(iii)	Resistance
(d)	V = IR	(iv)	Resistivity

1) (a)-(iii) (b)-(iv) (c)-(ii) (d)-(i)

2) (a)-(ii) (b)-(i) (c)-(iv) (d)-(iii)

3) (a)-(iv) (b)-(iv) (c)-(i) (d)-(ii)

4) (a)-(ii) (b)-(i) (c)-(iii) (d)-(iv)

5. Two conductors of the same material of the same cross-sectional area of length 10 m and 20 m respectively. If the resistance of the first conductor is 100 ohm then the resistance of the second conductor is?

1) 50 ohm 2) 100 ohm

3) 150 ohm 4) 200 ohm

6. The specific resistance of wire is ρ. Its volume is $3m^3$ and its resistance is 3Ω, then its length will be

1) $\frac{1}{\sqrt{\rho}}$ 2) $\frac{3}{\sqrt{\rho}}$

3) $\frac{\sqrt{3}}{\rho}$ 4) $\frac{\rho}{\sqrt{3}}$

7. The drift velocity of electron in a copper wire of length 2m under the application of a potential difference of 200V is $0.5\ ms^{-1}$. Their mobility is (in $m^2\ V^{-1}\ s^{-1}$)

1) 2.5×10^{-3} 2) 2.5×10^{-2}

3) 5×10^{-2} 4) 5×10^{-3}

8. 5A of current is passed through a metallic wire/conductor. The charge flowing in one minute in coulomb will be

1) 5 2) 12

3) 1/12 4) 300

9. The resistance of a wire of uniform diameter d and length L is R. The resistance of another wire of the same material but diameter 2d and length 4L will be

1) 2R 2) R

3) R/2 4) R/4

10. The resistance of a wire is 10 Ω. Its length is increased by 10% by stretching. The new resistance will now be

1) 12 ohm 2) 1.2 ohm

3) 13 ohm 4) 11 ohm

11. In a neon discharge tube, 2.9×10^{18} Ne^+ ions move to the right in each second, while 1.2×10^{18} electrons move to the left per second. Charge on each electron is 1.6×10^{-19}C. The current in the tube is

1) 1A towards right 2) 0.66A towards right

3) 0.66 A towards left 4) Zero

12. A uniform wire of resistance R is uniformly compressed along its length until its radius becomes n times the original radius. Now resistance of the wire becomes

1) $\frac{R}{n^4}$ 2) $\frac{R}{n^2}$

3) $\frac{R}{n}$ 4) nR

13. An electron (charge = 1.6×10^{-19} coulomb) is moving in a circle of radius 5.1×10^{-11} m at a frequency of 6.8×10^{16} revolution/sec. The equivalent current is approximately

1) 5.1×10^{-3}A 2) 6.8×10^{-3} A

3) 1.1×10^{-2}A 4) 2.2×10^{-2}A

14. The resistance of a wire is 5Ω at 50°C and 6Ω at 100°C. The resistance of the wire at 0°C will be

1) 3Ω 2) 2Ω
3) 1Ω 4) 4Ω

15. A wire of diameter 0.02m contains 10^{28} free electrons per cubic-metre. An electrical current of 100 A flows through the wire. The drift velocity of the free electrons in the wires is nearly

1) 10^{-19} m/s 2) 5×10^{-10} m/s
3) 2×10^{-4} m/s 4) 8×10^{3} m/s

16. The identical conductors maintained at same temperature are given potential difference in the ratio 1 : 2. Then the ratio of their drift velocities is

1) 1 : 2 2) 3 : 2
3) 1 : 1 4) 1 : 4

17. The length of the wire is doubled without changing its other dimensions. Then its conductance will be

1) Remain unchanged 2) Halved
3) Quadrupled 4) 1/4 of original

18. The graph between resistivity and temperature, for a limited range of temperature is a straight line for a material like

1) Copper 2) Nichrome
3) Silicon 4) Mercury

19. If a wire is stretched to make it 0.1% longer then its resistance will

1) Increase by 0.05% 2) Increase by 0.2%
3) Decrease by 0.2 % 4) Decrease by 0.05%

Grouping of Resistances, Kirchoff's Law, Internal Resistance, Potential Difference

20. If each resistor in a parallel circuit has a resistance of R, what is the total resistance of N identical resistors connected in parallel?

1) $\frac{R}{N}$ 2) R
3) NR 4) N^2R

21. n number of resistors, each of resistance R, are connected in series. What is the total resistance?

1) $\frac{R}{N}$ 2) R
3) nR 4) n^2R

22. The lowest resistance which can be obtained by connecting 10 resistors each of $\frac{1}{10}$ ohm is [RC]

1) $\frac{1}{250}$ ohm 2) $\frac{1}{200}$ ohm
3) $\frac{1}{100}$ ohm 4) $\frac{1}{10}$ ohm

23. A metallic wire of resistance 12 Ω is bent to form a square. The resistance between two diagonal points would be [RC]

1) 12 Ω 2) 24 Ω
3) 6 Ω 4) 3Ω

24. Resistance of 6 ohm each are connected in the manner shown in adjoining figure with the current 0.5 ampere as shown in figure. The potential difference $V_P - V_Q$ is

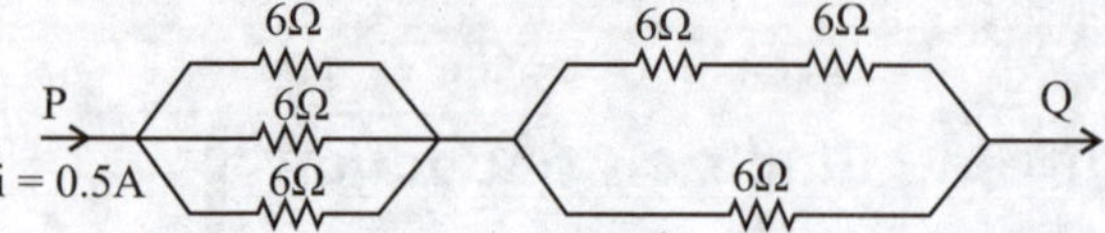

1) 3.6 V 2) 6 V
3) 3 V 4) 7.2 V

25. The current passing through the ideal ammeter in the circuit given below is

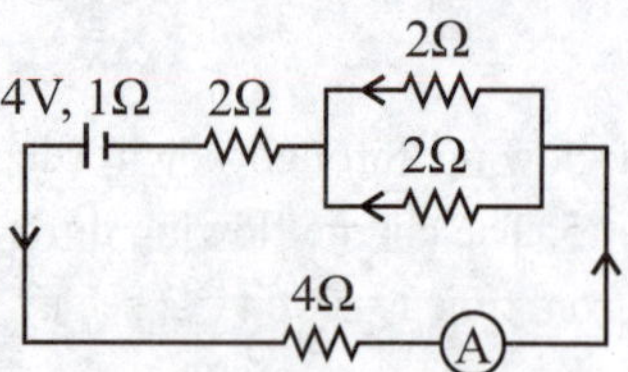

1) 1.25 A 2) 1 A
3) 0.75 A 4) 0.5 A

26. The equivalent resistance between points a and b of a network shown in figure is [RC]

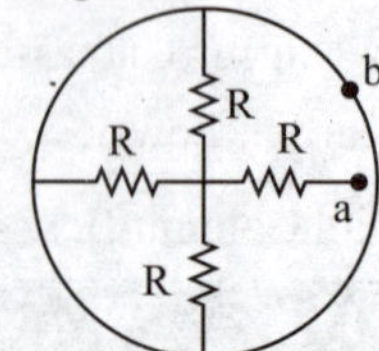

1) $\frac{3}{4}R$ 2) $\frac{4}{3}R$
3) $\frac{5}{4}R$ 4) $\frac{4}{5}R$

27. A wire of resistance R is cut into 'n' equal parts. These parts are then connected in parallel. The equivalent resistance of the combination will be [RC]

1) nR 2) $\frac{R}{n}$
3) $\frac{n}{R}$ 4) $\frac{R}{n^2}$

28. The current (I) in the given circuit is

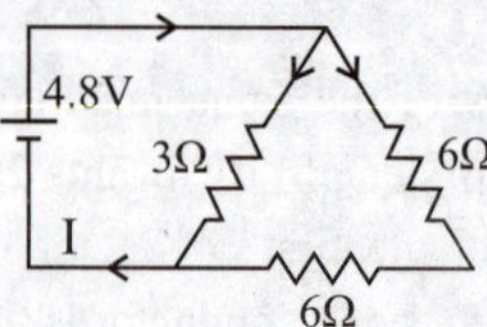

1) 8.31 A 2) 6.82 A
3) 4.92 A 4) 2 A

29. The figure below shows current in a part of electric circuit. The current I is

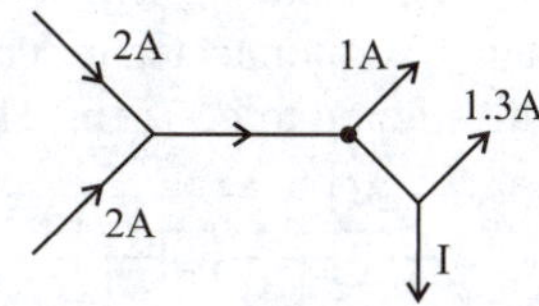

1) 1.7 A
2) 3.7 A
3) 1.3 A
4) 1 A

30. Two identical cells send the same current in 2 Ω resistance whether connected in series or in parallel. The internal resistance of the cell should be
1) 1 Ω
2) 2 Ω
3) 0.5 Ω
4) 2.5 Ω

31. 100 cells each of emf 5 volt and internal resistance 1ohm are to be arranged so as to produce maximum current in a 25 ohm resistance. Each row is to contain equal numbers of cells. The number of rows should be
1) 2
2) 4
3) 5
4) 10

32. When a resistor of 11 ohm is connected in series with an electric cell, the current flowing in it is 0.5 A. Instead, when 5Ω resistor is connected to the same electric cell in series the current increases by 0.4 A. The internal resistance of the cell is
1) 1.5 ohm
2) 2 ohm
3) 2.5 ohm
4) 3.5 ohm

33. The equivalent resistance between points A and B of an infinite network of resistances, each of 1Ω, connected as shown, is **[RC]**

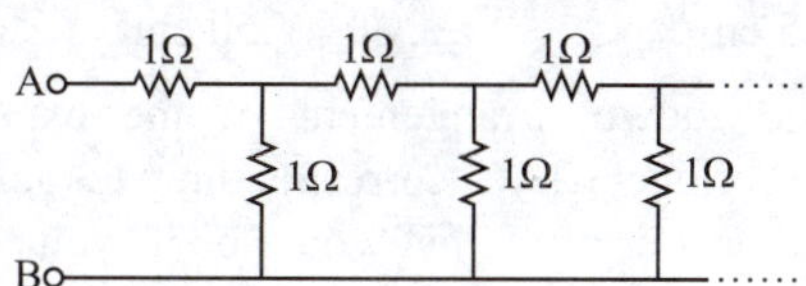

1) infinite
2) 2Ω
3) $\frac{1+\sqrt{5}}{2}\Omega$
4) Zero

34. n rows each containing m cells in series are joined in parallel. Maximum current is taken from this combination across and external resistance of 3ohm resistance. If the total number of cells used are 24 and internal resistance of each cells is 0.5ohm then
1) m = 8, n = 3
2) m = 6, n = 4
3) m = 12, n = 2
4) m = 2, n = 12

35. Two batteries, one of emf 18volt and internal resistance 2 ohm, the other of emf 12 volt and internal resistance 1 ohm are connected as shown, The voltmeter V will record a reading of

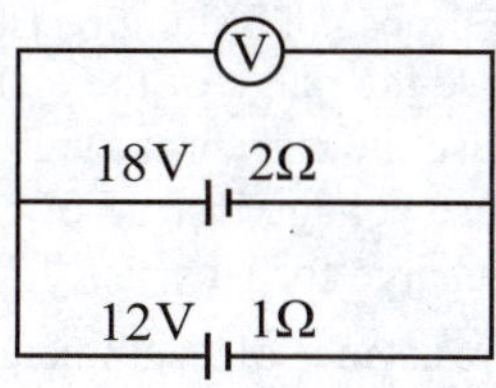

1) 15 V
2) 30 V
3) 45 V
4) 14 V

36. Two sources of equal e.m.f are connected to an external resistance R in series. The internal resistance of the two sources are R_1 and R_2 ($R_2 > R_1$). If the potential difference across the source having internal resistance R_2 is zero, then
1) $R = R_1R_2/(R_1 + R_2)$
2) $R = R_1R_2/(R_2 - R_1)$
3) $R = R_2(R_1 + R_2)/(R_2 - R_1)$
4) $R = R_2 - R_1$

37. When the resistance of 9 ohm is connected at the ends of a battery its terminal voltage decreases from 40 volt to 30 volt. The internal resistance of the battery is
1) 6 ohm
2) 3 ohm
3) 9 ohm
4) 15 ohm

38. A cell of emf E is connected with an external resistance R, then potential difference across cell is V. The internal resistance of a cell will be
1) $\frac{(E-V)R}{E}$
2) $\frac{(E-V)R}{V}$
3) $\frac{(V-E)R}{V}$
4) $\frac{(V-E)R}{E}$

Heating Effect of Current & Chemical Effect of Current

39. If voltage across a bulb rated 200 V, 100 W drops by 2.5% of its rated value, the percentage of the rated value, by which the power would decrease is
1) 20%
2) 2.5%
3) 5%
4) 10%

40. The power dissipated in the circuit shown in the figure is 30 W, the value of R is

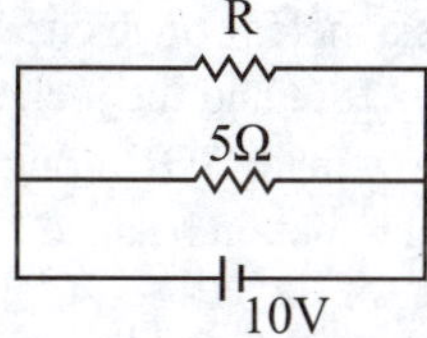

1) 20 Ω
2) 15 Ω
3) 10 Ω
4) 30 Ω

41. If two bulbs of wattages 25 W and 100 W respectively each rated by 220 V are connected in series with the supply of 440 V then which bulb will fuse?
1) 100 W bulb
2) 25 W bulb
3) None of these
4) Both of these

42. There are the two electric bulbs of 40 W and 100 W. Which one will glow brighter when first connected in series and then in parallel?
1) 40W in series and 100W in parallel
2) 100W in series and 40W in series
3) 40W both in series and parallel will be uniform
4) 100W both in series and parallel will be uniform

43. Two bulb consume same power when operated at 200 V and 300V respectively. When these bulbs are connected in series across a DC source of 500 V then
1) Ratio of potential differences across them is 3/2
2) Ratio of potential difference across them is 9/4
3) Ratio of power consumed across them is 4/9
4) Ratio of power consumed across them is 2/3

44. Two similar heater coils separately takes 10 minutes to boil a certain amount of water. If both coils are connected in series, time taken to boil the same amount of water will be
1) 15 min
2) 20 min
3) 7.5 min
4) 25 min

45. One filament takes 10 minute to heat a kettle and another takes 15min. If connected in parallel after how much time (in minutes) they combinedly take to heat the same kettle?
1) 6
2) 12.5
3) 25
4) 7.5

46. Two wires have resistances R and 2R, when both are joined in series and in parallel, then ratio of heats generated in these situations on applying the same voltage is
1) 2 : 1
2) 1 : 2
3) 2 : 9
4) 9 : 2

47. A resistor R has power dissipation P with cell voltage E. The resistor is cut in n equal parts and all parts are connected in parallel with same cell. The new power dissipation is
1) nP
2) nP^2
3) n^2P
4) n/P

Different Measuring Instruments

48. The figure shows a meter bridge circuit with AB = 100 cm, X = 12Ω and R = 18Ω and the jockey J in the position of balance. If R is now made 8Ω, through what distance will J have to be moved to obtain balance?

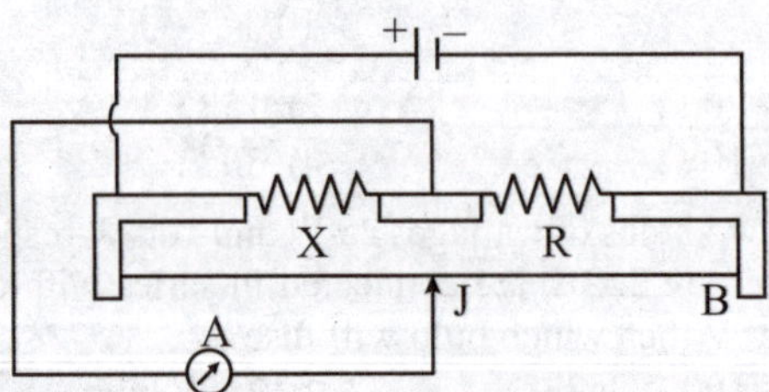

1) 10 cm
2) 20 cm
3) 30 cm
4) 40 cm

49. In a meter bridge experiment, resistance are connected as shown in figure, The balancing length l, is 55cm. Now an unknown resistance x is connected in series with P and new balancing length is found to be 75cm. The value of x is

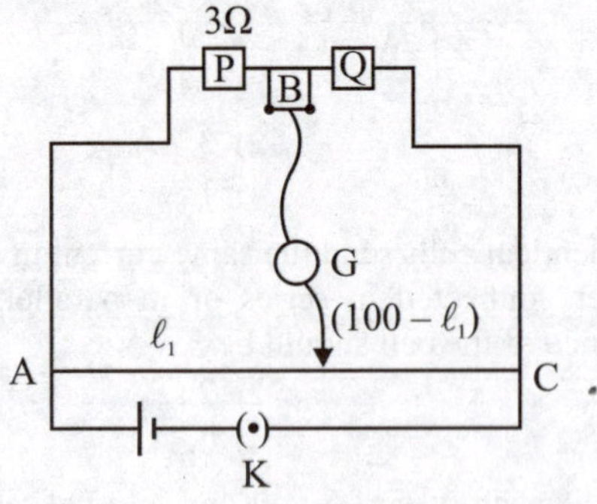

1) $\frac{54}{13}\Omega$
2) $\frac{20}{11}\Omega$
3) $\frac{48}{11}\Omega$
4) $\frac{11}{48}\Omega$

50. A cell of internal resistance 3 Ω and emf 10 volt is connected to a uniform wire of length 500 cm and resistance 3 ohm. The potential gradient in the wire is
1) 30 mV/cm
2) 2 mV/cm
3) 10 mV/cm
4) 5 V/m

51. Resistance in the two gaps of a meter bridge are 10 ohm and 30 ohm respectively. If the resistances are interchanged the balance point shifts by
1) 33.3 cm
2) 66.67 cm
3) 25 cm
4) 50 cm

52. In the shown arrangement of the experiment of the meter bridge if AC corresponding to null deflection of galvanometer is x, what would be its value if the radius of the wire AB is doubled?

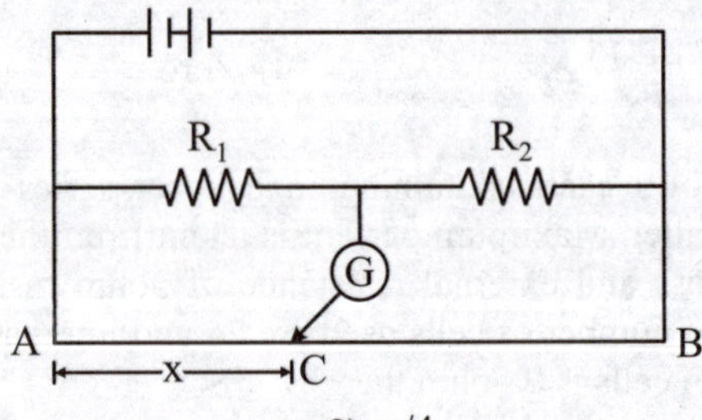

1) x
2) x/4
3) 4x
4) 2x

High Order Time Intensive MCQs

1. Two conducting wire of identical dimensions are connected in parallel. If 10 $Ohm^{-1}m^{-1}$ and 15 $Ohm^{-1}m^{-1}$ are conductivities of the wires then effective conductivity of the combination is:
1) 25 Ω^{-1} m^{-1}
2) 12 Ω^{-1} m^{-1}
3) 12.5 Ω^{-1} m^{-1}
4) 150 Ω^{-1} m^{-1}

2. What is the purpose of a shunt resistor in an ammeter?
1) To increase the range of the ammeter.
2) To decrease the current through the ammeter.
3) To measure voltage instead of current.
4) To store charge for later use.

3. Find the net resistance of an ammeter having coil resistance 100 Ohm and shunt resistance 10 Ohm.
1) 10 ohm
2) 9.1 Ohm
3) 99 Ohm
4) 110 Ohm

4. If a battery with an electromotive force (emf) 'E' and internal resistance 'Ri' is connected to a resistor of resistance 'R', what is the current 'I' through the circuit according to Ohm's Law?

1) $\frac{E}{R}$ 2) $\frac{E}{R_i + R}$

3) $\frac{E}{R_i - R}$ 4) $\frac{E}{2R}$

5. Three identical bulbs are connected in series and these together dissipates a power P. If now the bulbs are connected in parallel then the power dissipated will be:

1) $\frac{P}{3}$ 2) 3P 3) 9P 4) $\frac{P}{9}$

6. A wire of resistance R is elongated n-fold to make a new uniform wire. The resistance of new wire is:

1) nR 2) n^2R

3) $2nR$ 4) $2n^2R$

7. 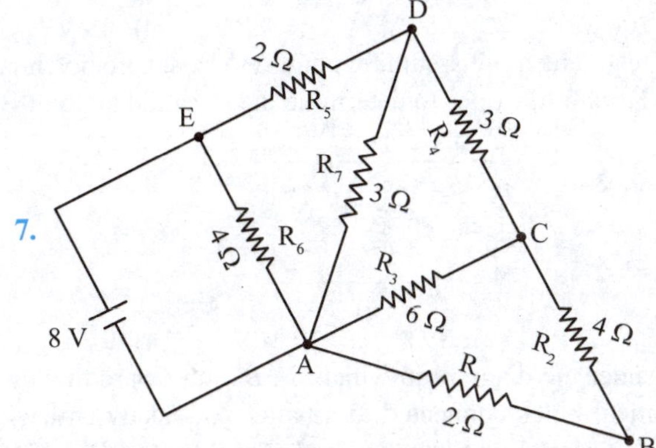

The current flowing through R_2 is:

1) $\frac{2}{3}A$ 2) $\frac{1}{4}A$ 3) $\frac{1}{2}A$ 4) $\frac{1}{3}A$

8. In the given circuit diagram, a wire is joining points B and D. The current in this wire is

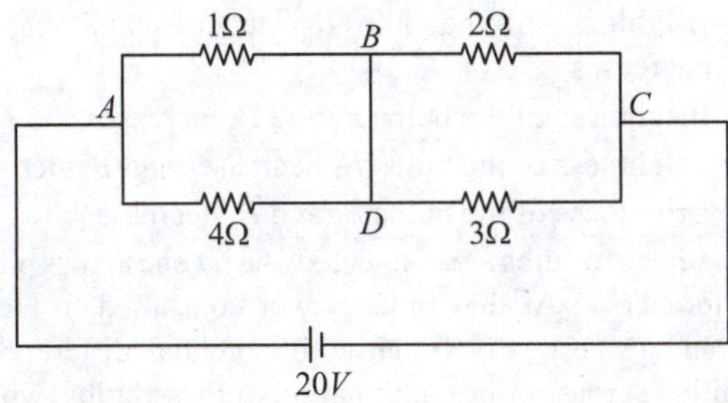

1) Zero 2) 2 A

3) 0.4 A 4) 4 A

9. Which part represents the negative dynamic resistance?

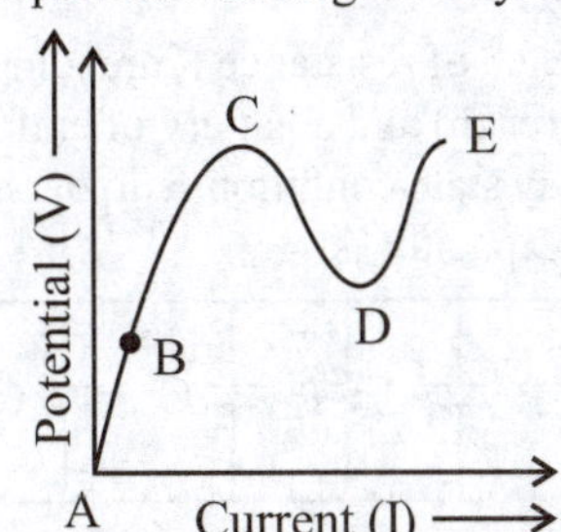

1) AB 2) BC 3) CD 4) DE

10. A ring is made of a wire having aresistance $R_0 = 12\ \Omega$. Find the points A and B as shown in the figure, at which a current carrying conductor should be connected so that the resistance R of the sub-circuit between these points is equal to $8/3\ \Omega$.

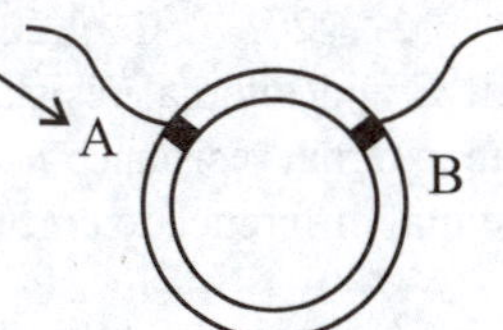

1) $\frac{l_1}{l_2} = \frac{5}{8}$ 2) $\frac{l_1}{l_2} = \frac{1}{3}$ 3) $\frac{l_1}{l_2} = \frac{3}{8}$ 4) $\frac{l_1}{l_2} = \frac{1}{2}$

11. A wire of resistance per unit length 12 Ω-m^{-1} is bents to form a complete circle of radius 10 cm. The resistance between its two diametrically opposite points A and B as shown in the figure, is: *[RC]*

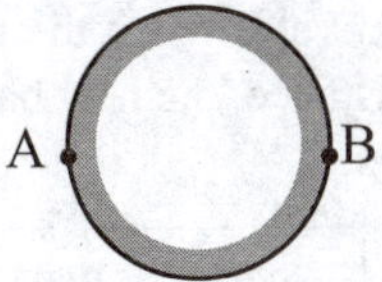

1) 0.6 πΩ 2) 3 Ω 3) 6 πΩ 4) 6 Ω

12. In the circuit shown below, the cells A and B have negligible resistance. For $V_A = 12$ V, $R_1 = 500\ \Omega$ and $R = 100\ \Omega$, the galvanometer (G) shows no deflection. The value of V_B is:

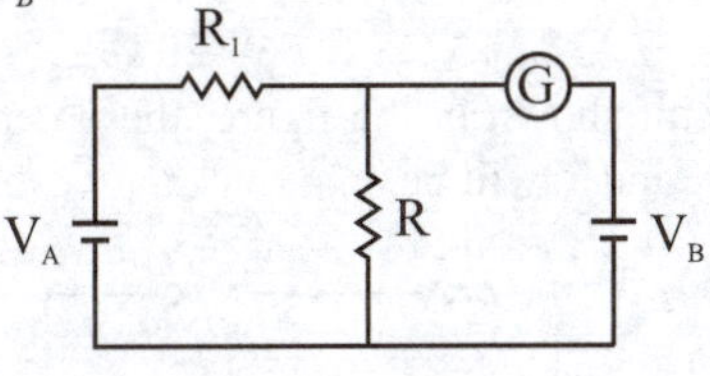

1) 4 V 2) 2 V 3) 12 V 4) 6 V

13. When a resistance R_1 is connected across the terminals of a cell, the current is I_1. When the resistance is changed to R_2, the current is I_2. The internal resistance of the cell is:

1) $\frac{I_1R_1 + I_2R_2}{I_1 + I_2}$ 2) $\frac{I_1R_2 + I_2R_1}{I_1 + I_2}$

3) $\frac{I_1R_2 - I_2R_1}{I_1 - I_2}$ 4) $\frac{I_2R_2 - I_1R_1}{I_1 - I_2}$

14. The current-voltage graphs for a given metallic wire at two different temperatures T_1 and T_2 are shown in the figure. The temperatures T_1 and T_2 are related as:

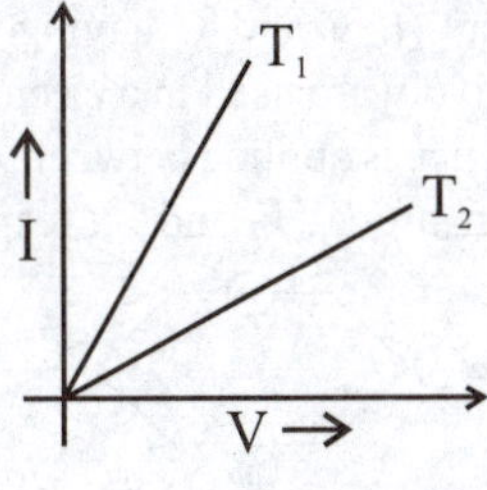

1) $T_1 > T_2$ 2) $T_1 < T_2$

3) $T_1 = T_2$ 4) $T_1 = 2T_2$

15. If 3 A of current is flowing between points P and Q in the circuit, then the potential difference between P and Q is

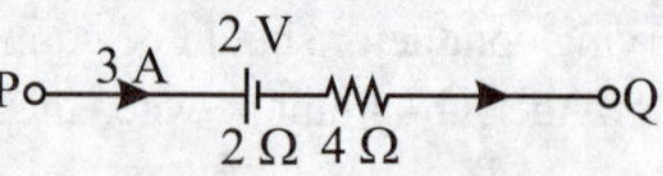

1) 30 V 2) 22 V 3) 20 V 4) 15 V

16. A cell of emf E and internal resistance r is connected in series with an external resistance nr. Then, the ratio of the terminal potential difference to emf is:

1) $\left(\frac{1}{n}\right)$ 2) $\frac{1}{(n+1)}$ 3) $\frac{n}{(n+1)}$ 4) $\frac{(n+1)}{n}$

17. The potential difference between A and B in the following figure is:

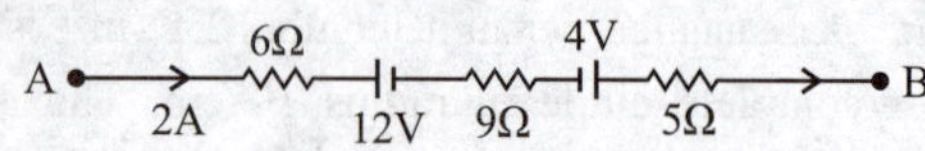

1) 32 V 2) 48 V 3) 24 V 4) 14 V

18. The resistances in the two arms of the meter bridge are 5 Ω and RΩ, respectively. When the resistance R is shunted with an equal resistance, the new balance point is at 1.6 l_1. The resistance R is: **[RC]**

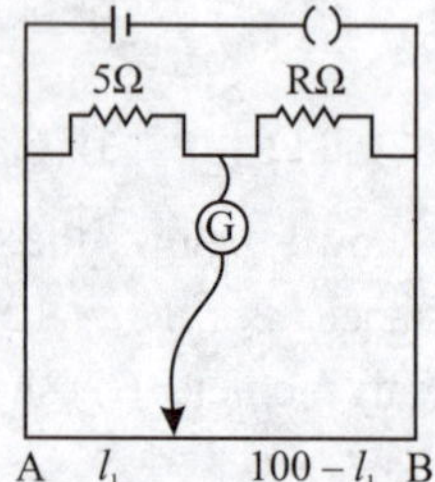

1) 10 Ω 2) 15 Ω 3) 20 Ω 4) 25 Ω

19. In the circuit shown in the figure, the potential difference between X and Y will be

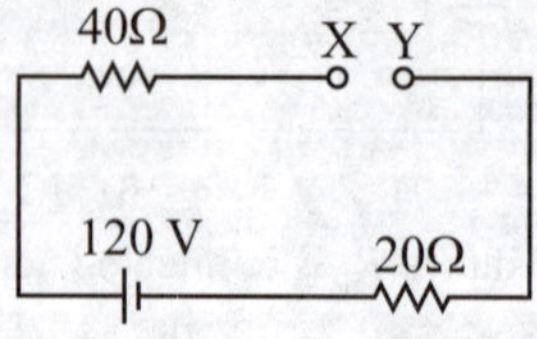

1) zero 2) 20 V 3) 60 V 4) 120 V

20. If the ammeter in the given circuit reads 2 A, the resistance R is

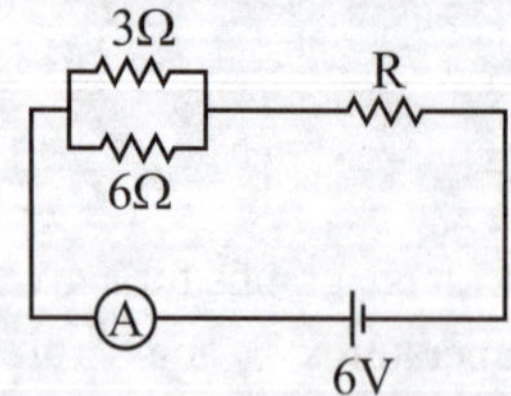

1) 1 Ω 2) 2 Ω 3) 3 Ω 4) 4 Ω

21. Three voltmeters A, B and C having resistances R, $1.5R$ and $3R$, respectively are used in a circuit as shown. When a potential difference is applied between X and Y, the readings of the voltmeters are V_1, V_2 and V_3, respectively then.

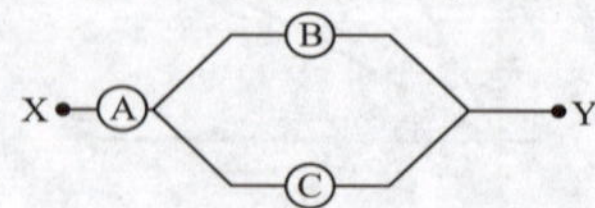

1) $V_1 = V_2 = V_3$ 2) $V_1 < V_2 = V_3$
3) $V_1 > V_2 > V_3$ 4) $V_1 > V_2 = V_3$

22. A 6 V battery is connected to the terminals of a 3 m long wire of uniform thickness and resistance of 100 Ω. The difference of potential between two points on the wire separated by a distance of 50 cm will be:

1) 2 V 2) 3 V 3) 1 V 4) 1.5 V

23. In the series combination of n cells each having emf ε and internal resistance r. If three cells are wrongly connected, then total emf and internal resistance of this combination will be:

1) nε, (nr – 3r) 2) (nε – 2ε), nr
3) (nε – 4ε), nr 4) (nε – 6ε), nr

24. Bulb B_1(100 W – 250 V) and bulb B_2(100 W – 200 V) are connected across 250 V. What is potential drop across B_2?

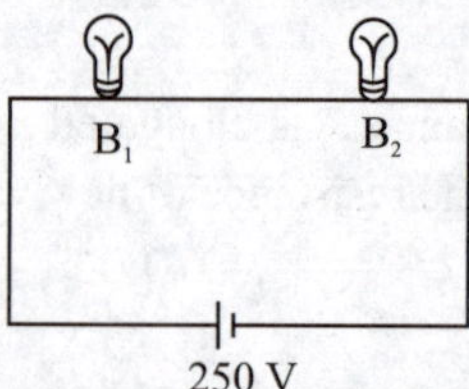

1) 200 V 2) 250 V 3) 98 V 4) 48 V

25. In the given circuit assuming point A to be at zero potential, use Kirchhoff's rules to determine the potential at point B.

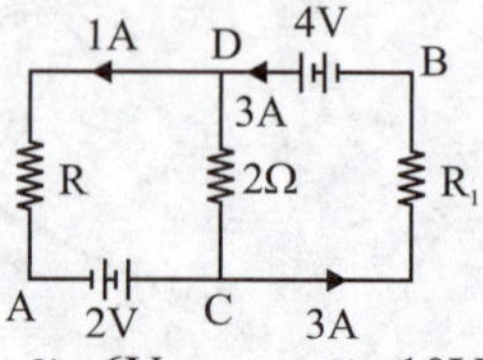

1) 2V 2) 6V 3) 10V 4) 0V

26. Consider the diagram in which B_1, B_2 and B_3 are the three identical bulbs connected to a battery of steady emf with key k closed. What happens to the brightness of the bulbs B_1 and B_2 when the key is opened?

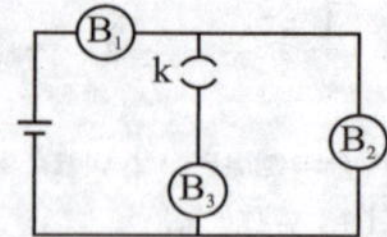

1) Brightness of the bulb B_1 increases and that of B_2 decreases
2) Brightness of the bulbs B_1 and B_2 increases
3) Brightness of the bulbs B_1 decreases and B_2 increases
4) Brightness of the bulbs B_1 and B_2 decreases

27. Two electric bulbs are connected one by one across potential difference V. At that time, power consumed in them are P_1 and P_2 respectively. Now, if potential difference V is applied across series combination of these bulbs, what will be total power consumed?

1) $\frac{P_1P_2}{P_1+P_2}$ 2) P_1P_2 3) $P_1 + P_2$ 4) $\sqrt{P_1P_2}$

28. n resistances each of resistance R are joined with capacitors of capacity C (each) and a battery of emf E as shown in the figure. In steady state condition ratio of charge stored in the first and last capacitor is:

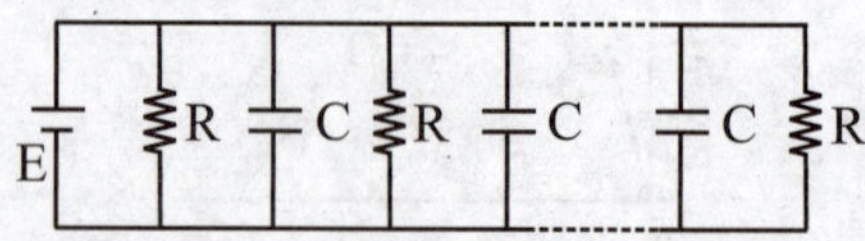

1) n : 1 2) (n – 1) : (n + 1)
3) $(n^2 + 1) : (n^2 - 1)$ 4) 1 : 1

29. In the circuit shown in the figure, the charge on the upper plate A is

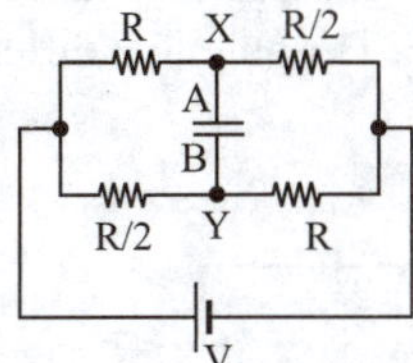

1) Positive
2) Negative
3) Zero
4) Oscillating

30. In the circuit shown, the ratio $\frac{I_1}{I_2}$ is equal to

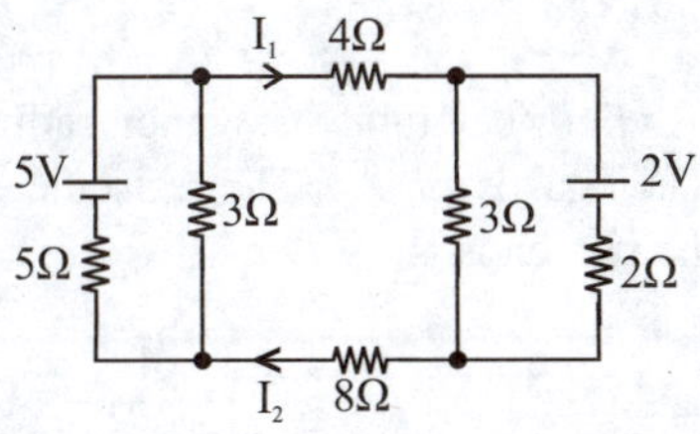

1) 4.0
2) 3.0
3) 2.5
4) 1.0

31. A resistance wire has a resistance R. Half of this wire is stretched to double its length and half is twisted to double its thickness, then its new resistance becomes

1) $\frac{17R}{8}$
2) $\frac{17R}{16}$
3) $\frac{65R}{32}$
4) $\frac{65R}{16}$

32. Two batteries of different e.m.f.'s and internal resistances are connected in series with each other and with an external load resistor. The current is 2.0 amp. When the polarity of one battery is reversed, the current becomes 1.0 amp. The ratio of the e.m.f.'s of the two batteries is

1) 2.5
2) 2.0
3) 3.0
4) 1.0

33. Consider the figure, cross-sectional area of conductor at A and B is $2a$ and a respectively. Current density at A and B are given by J_A and J_B then J_A/J_B is **[RC]**

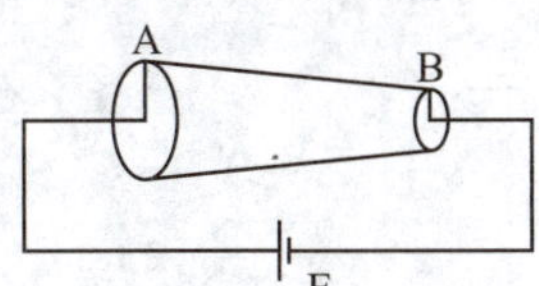

1) $\frac{1}{2}$
2) 2
3) 1
4) 4

34. A part of a circuit is shown in figure. $V_B - V_C$ is equal to 12 V. I_1 is equal to

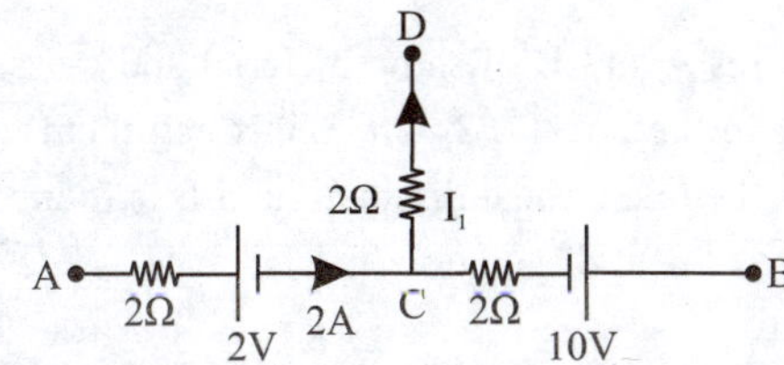

1) 1 A
2) 2 A
3) 3 A
4) 4 A

35. In the circuit shown, the total power output is P when S is closed. As the switch S is opened, the power output becomes (all bulbs are identical)

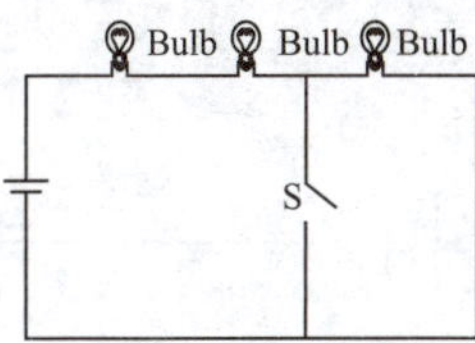

1) $\frac{2}{3}P$
2) $\frac{3}{2}P$
3) $\frac{4}{9}P$
4) $\frac{9}{4}P$

36. Null point with 1 V cell comes out to be 55 cm and with R = 10 Ω it is 50 cm. What is the internal resistance of the cell? **[RC]**

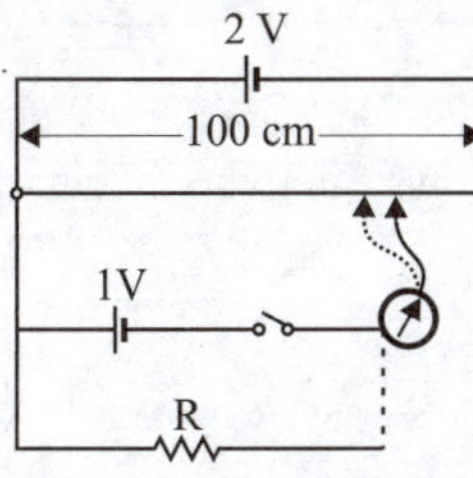

1) 0·5 Ω
2) 0·4 Ω
3) 1Ω
4) 0·2 Ω

37. A battery of emf 1.4 V and internal resistance 2Ω is connected to a 100 Ω resistor through an ammeter. The resistance of the ammeter is 4/3 Ω. A voltmeter is also connected to find the potential difference across the resistor. The ammeter reads 0.02A. What is the resistance of the voltmeter?

1) 200 Ω
2) 300 Ω
3) 400 Ω
4) 150 Ω

38. A battery of emf 1.4 V and internal resistance 2Ω is connected to a 100 Ω resistor through an ammeter. The resistance of the ammeter is 4/3 Ω. A voltmeter is also connected to find the potential difference across the resistor. The voltmeter reads 1.10 V. What is the error in reading?

1) 0. 3 V
2) 0.43 V
3) 0.53 V
4) 0.23 V

39. When two resistances X and Y are put in the left hand and right hand gaps in a wheatstone meter bridge, the null point is at 60cm. If X is shunted by a resistance equal to half of itself then find the shift in the null point. **[RC]**

1) 26.7 cm
2) 36.7 cm
3) 46.7 cm
4) 96.7 cm

40. In a gas discharge tube if 3×10^{18} electrons are flowing per sec from left to right and 2×10^{18} protons are flowing per second from right to left through a given cross section. The magnitude and direction of current through the cross section

1) 0.48, left to right
2) 0.48 A, right to left
3) 0.80A, left to right
4) 0.80 A, right to left

41. In the given circuit, $R_1 = 10Ω$, $R_2 = 6Ω$ and E = 10 V. Then reading of A_2 is:

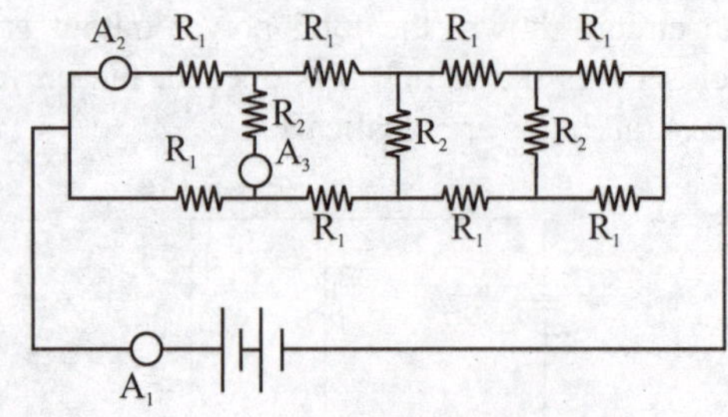

1) 1 amp
2) 2 amp
3) 1/4 amp
4) 3 amp

42. A current of 2 A flows in resistances as shown. The potential difference $V_A - V_B$ will be

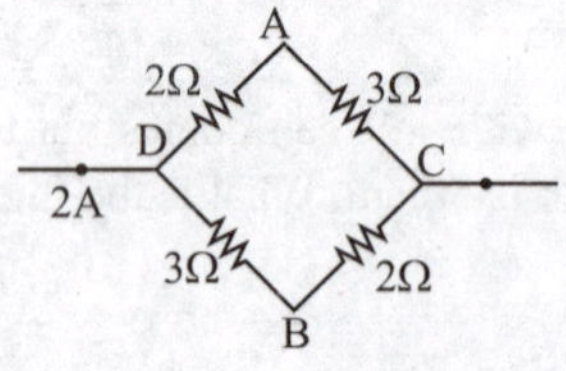

1) +4 V
2) +2 V
3) +1 V
4) −1 V

43. The figure shows a network of current. The current I will be

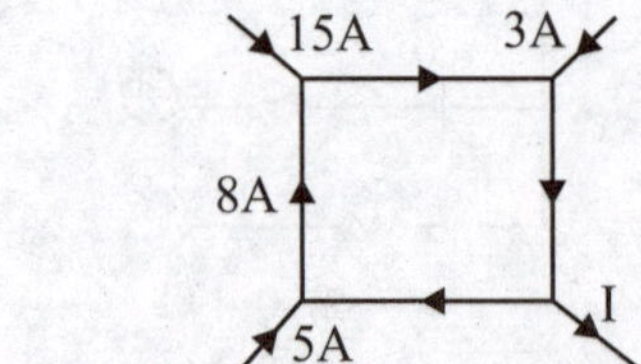

1) 3 A
2) 13 A
3) 23 A
4) −3 A

44. Two resistors R and 2R are connected in series in an electric circuit. The thermal energy developed in R and 2R are in the ratio

1) 1 : 2
2) 2 : 1
3) 1 : 4
4) 4 : 1

45. The steady state current in a 2Ω resistor shown in fig will be – (The internal resistance of the battery is negligible and the capacitance of the condenser C is 0.2μF)

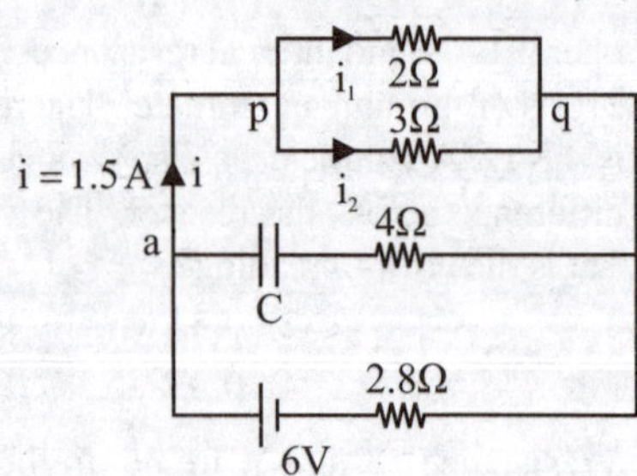

1) 1.5 A
2) 0.9 A
3) 1.2 A
4) 1.3 A

46. A and B are two points on a uniform ring of resistance 15 Ω. The ∠AOB = 45°. The equivalent resistance between A and B is:

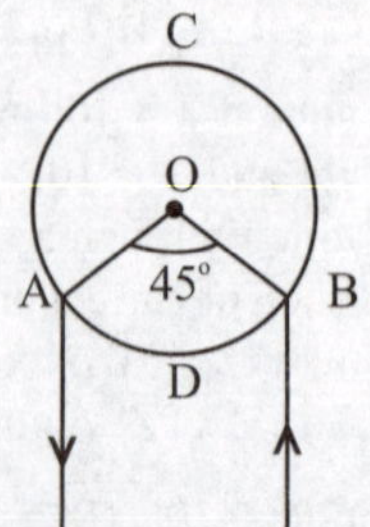

1) 1.64 Ω
2) 2.84 Ω
3) 4.57 Ω
4) 2.64 Ω

47. Four resistors are connected as shown in the figure. A 6V battery of negligible resistance is connected across terminals A and C. The potential difference across terminals B and D will be:

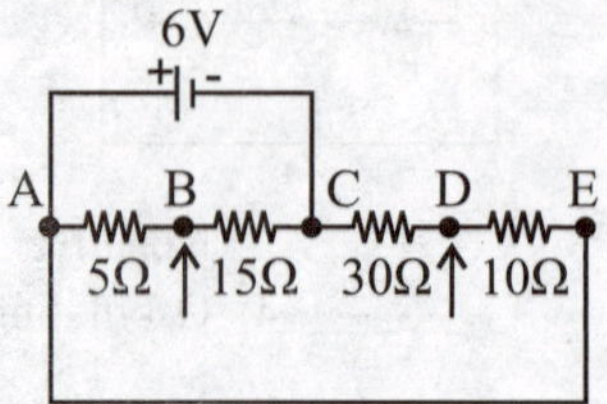

1) Zero
2) 1.5 V
3) 2 V
4) 3 V

48. A copper cylindrical tube has inner radius a and outer radius b. The resistivity is ρ. The resistance of the cylinder between the two ends is:

1) $\frac{\rho l}{b^2 - a^2}$
2) $\frac{\rho l}{2\pi(b-a)}$
3) $\frac{\rho l}{\pi(b^2 - a^2)}$
4) $\frac{\pi(b^2 - a^2)}{\rho l}$

49. The charge flowing through a resistance R varies with time according to $q = \alpha t - \beta t^2$. Calculate the total heat produced in R.

1) $\frac{\alpha^2 R}{2\beta}$
2) $\frac{\alpha^3 R}{6\beta}$
3) $\frac{\alpha^3 R}{\beta}$
4) None of these

50. In the circuit shown in figure, the value of resistance x, when the potential difference between the points B and D is zero, will be **[RC]**

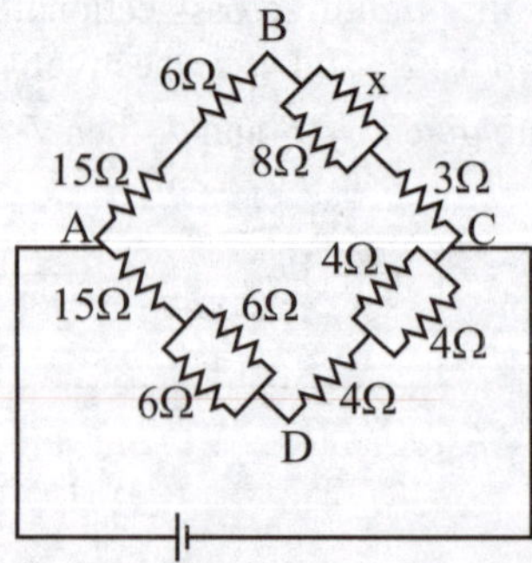

1) 9 Ω
2) 8 Ω
3) 6 Ω
4) 4 Ω

51. Two wires A and B of same material and mass have their lengths in the rate 1 : 2. On connecting them to the same source, the rate of heat dissipation in B is found to be 5 W. The rate of heat dissipation in A is

1) 10 W
2) 5 W
3) 20 W
4) None of these

Statement Based MCQs

Directions: These questions consist of two statements each, printed as Statement-I and Statement-II. While answering these questions, you are required to choose any one of the following four responses.

1) **Both Statement-I and Statement-II are correct.**
2) **Both Statement-I and Statement-II are incorrect.**
3) **Statement-I is correct & Statement-II is incorrect.**
4) **Statement-I is incorrect & Statement-II is correct.**

1. **Statement-I:** The standard resistance coils are made of manganin.

   Statement-II: For manganin, the temperature coefficient of resistance is almost zero and its resistivity is high.

2. **Statement-I:** An electric bulb glows immediately when it is switched on.

   Statement-II: The drift velocity of electrons in a metallic wire is very high (near speed of light).

3. **Statement-I:** In a conductor, not carrying any current free electrons keep on moving but no magnetic force acts on that conductor in a magnetic field.

   Statement-II: In a conductor, average thermal velocity of electrons is zero. Hence no current flows through the conductor.

4. **Statement-I:** Bulbs always fuse when they are switched off.

   Statement-II: When a bulb is switched off, at that instant its resistance increases.

5. **Statement-I:** Electrolytes are more conducting than metals.

   Statement-II: Electrons moves faster than Ions.

Assertion & Reason MCQs

Directions: These questions consist of two statements each, printed as Assertion and Reason. While answering these questions, you are required to choose any one of the following four responses.

1) **Both Assertion and Reason are True and the Reason is a correct explanation of the Assertion.**
2) **Both Assertion and Reason are True but Reason is not a correct explanation of the Assertion.**
3) **Assertion is True but the Reason is False.**
4) **Assertion is False but Reason is True.**

1. **Assertion:** A metal has resistance and it gets heated due to flow of current.

   Reason: When free electrons drift through a metal they make collisions with the lattice and there is transfer of energy to the lattice.

2. **Assertion:** Nichrome is used for making the heating wire of electrical appliances like electric irons, room heaters, water heaters and toasters etc.

   Reason: Nichrome has very high resistivity and it does not burn ever when red hot.

3. **Assertion:** If lightning strikes a wet tree then most of the current passes through the water contained by the tree.

   Reason: Water has more electrical conductivity than wood.

4. **Assertion:** Consider the two situations shown in the figure. Potential difference between points A and B, in case I is more as compared to case II.

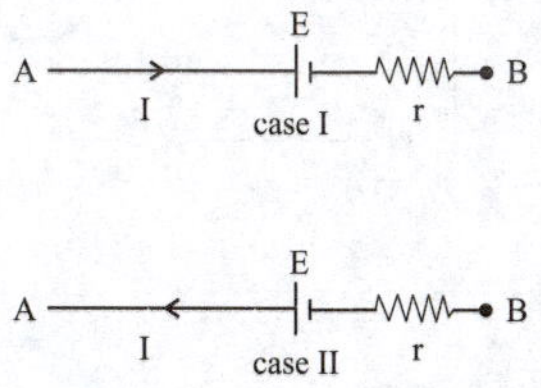

   A —→— I —|⊢— E —WWW— r —• B
   case I

   A —←— I —|⊢— E —WWW— r —• B
   case II

   Reason: In case I: $V_A - V_B = E + Ir$

   In case II: $V_A - V_B = E - Ir$

5. **Assertion:** Though the same current flows through the line wires and the filament of the bulb but heat produced in the filament is much higher than that in line wires.

   Reason: The filament of bulbs is made of a material of high resistance and low melting point.

Match the Columns MCQs

1. Match Column-I with Column-II:

Column-I		Column-II	
(a)	Electrical conductivity of conductor depends on	(i)	dimensions (length, area of cross section etc.)
(b)	Conductance of a conductor depends on	(ii)	temperature
(c)	For a conductor of given dimensions and at a given temperature, current density depends on	(iii)	nature of conductor
(d)	For a given potential difference applied across a conductor of given length, current in it will depend on	(iv)	electric field strength

1) (a)-(i) (b)-(ii, iii) (c)-(iv) (d)-(ii)
2) (a)-(ii, iii) (b)-(i, ii, iii) (c)-(iii, iv) (d)-(i, ii, iii)
3) (a)-(v) (b)-(ii) (c)-(iii) (d)-(iv)
4) (a)-(iii) (b)-(v) (c)-(ii) (d)-(i)

2. For the circuit shown in figure, all the three batteries are ideal. Match the column.

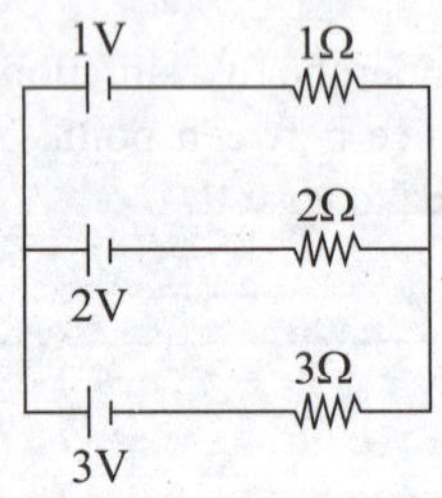

Column-I		Column-II	
(a)	The value of R such that current through 3 V battery is zero.	(i)	1Ω
(b)	The value of R such that current through 1 V battery is zero.	(ii)	2Ω
(c)	The value of R such that current through 2 V battery is zero.	(iii)	3Ω
(d)	The value of R for which current through 3 V battery is 5/8 A.	(iv)	not possible for any value of R

1) (a)-(i) (b)-(ii) (c)-(iii) (d)-(iv)
2) (a)-(ii) (b)-(i) (c)-(iv) (d)-(iii)
3) (a)-(iv) (b)-(iv) (c)-(i) (d)-(ii)
4) (a)-(ii) (b)-(i) (c)-(iii) (d)-(iv)

3. Match the Column-I with Column-II.

Column-I		Column-II	
(a)	Average velocity of free electrons in metals when $E = 0$	(i)	10^5 ms^{-1}
(b)	Random velocity of free electrons in metals when $E = 0$	(ii)	10^{-5} ms^{-1}
(c)	Drift velocity of free electrons in metals when $E \neq 0$	(iii)	0 ms^{-1}

1) (a)-(iii) (b)-(i) (c)-(ii)
2) (a)-(i) (b)-(ii) (c)-(iii)
3) (a)-(ii) (b)-(i) (c)-(iii)
4) (a)-(iii) (b)-(i) (c)-(i)

NCERT Exemplar MCQs

1. Consider a current carrying wire (current I) in the shape of a circle. Note that as the current progresses along the wire, the direction of J (current density) changes in an exact manner, while the current I remain unaffected. The agent that is essentially responsible for is:
   1) Source of emf
   2) Electric field produced by charges accumulated on the surface of wire
   3) The charges just behind a given segment of wire which push them just the right way by repulsion
   4) The charges ahead

2. Two batteries of emf ε_1 and ε_2 ($\varepsilon_2 > \varepsilon_1$) and internal resistances r_1 and r_2 respectively are connected in parallel:
   1) The equivalent emf ε_{eq} of the two cells is between ε_1 and ε_2, *i.e.*, $\varepsilon_1 < \varepsilon_{eq} < \varepsilon_2$
   2) The equivalent emf ε_{eq} is smaller than ε_1
   3) The ε_{eq} is given by $\varepsilon_{eq} = \varepsilon_1 + \varepsilon_2$ always
   4) ε_{eq} is independent of internal resistances r_1 and r_2

3. A resistance R is to be measured using a meter bridge, student chooses the standard resistance S to be 100Ω. He finds the null point at $l_1 = 2.9$ cm. He is told to attempt to improve the accuracy: Which of the following is a useful way? **[RC]**
   1) He should measure l_1 more accurately
   2) He should change S to 1000 Ω and repeat the experiment
   3) He should change S to 3 Ω and repeat the experiment
   4) He should give up hope of a more accurate measurement with a meter bridge

4. A metal rod of length 10 cm and a rectangular cross-section of 1 cm × $\frac{1}{2}$ cm is connected to a battery across opposite faces. The resistance will be:
   1) Maximum when the battery is connected across 1 cm × $\frac{1}{2}$ cm faces
   2) Maximum when the battery is connected across 10 cm × 1 cm faces
   3) Maximum when the battery is connected across 10 cm × $\frac{1}{2}$ cm faces
   4) Same irrespective of the three faces

5. Which of the following characteristics of electrons determines the current in a conductor?
   1) Drift velocity alone
   2) Thermal velocity alone
   3) Both drift velocity and thermal velocity
   4) Neither drift nor thermal velocity

NEET & AIIMS Past Year MCQs

1. A uniform wire of diameter d carries a current of 100 mA when the mean drift velocity of electrons in the wire is v. For a wire of diameter $\frac{d}{2}$ of the same material to carry a current of 200 mA, the mean drift velocity of electrons in the wire is **(2024 Re)**
   1) $4v$ 2) $8v$ 3) v 4) $2v$

2. In an electrical circuit, the voltage is measured as $V = (200 \pm 4)$ volt and the current is measured as $I = (20 \pm 0.2)A$. The value of the resistance is: **(2024 Re)**
   1) $(10 \pm 4.2)\ \Omega$ 2) $(10 \pm 0.3)\ \Omega$
   3) $(10 \pm 0.1)\ \Omega$ 4) $(10 \pm 0.8)\ \Omega$

3. The steady state current in the circuit shown below is: **(2024 Re)**

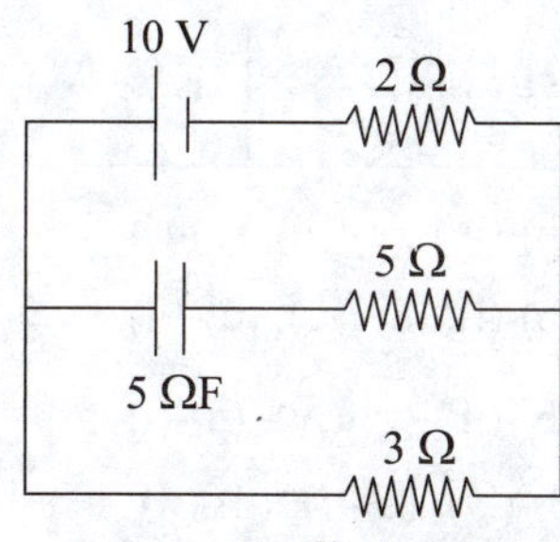

   1) 0.67 A 2) 1.5 A
   3) 2 A 4) 1 A

4. A uniform metal wire of length l has 10 Ω resistance. Now this wire is stretched to a length $2l$ and then bent to form a perfect circle. The equivalent resistance across any arbitrary diameter of that circle is **(2024 Re)**
   1) 10 Ω 2) 5 Ω
   3) 40 Ω 4) 20 Ω

5. A wire of length 'l' and resistance 100Ω is divided into 10 equal parts. The first 5 parts are connected in series while the next 5 parts are connected in parallel. The two combinations are again connected in series. The resistance of this final combination is: **(2024)**
   1) 55 Ω 2) 60 Ω
   3) 26 Ω 4) 52 Ω

6. The terminal voltage of the battery, whose emf is $10V$ and internal resistance 1Ω, when connected through an external resistance of 4Ω as shown in the figure is : **(2024)**

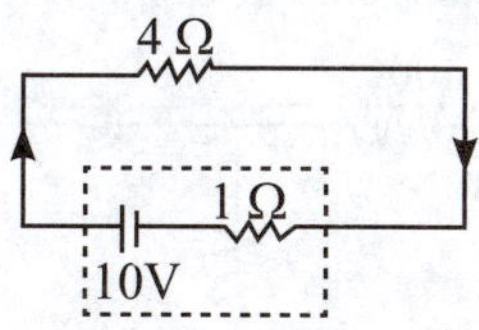

   1) 8 V 2) 10 V 3) 4 V 4) 6 V

7. The given circuit shows a uniform straight wire AB of 40 cm length fixed at both ends. In order to get zero reading in the galvanometer G, the free end of J is to be placed from B at: **(2024 Re)**

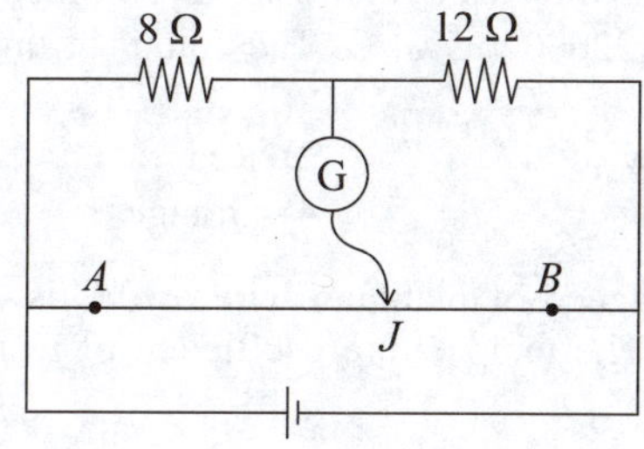

   1) 32 cm 2) 8 cm
   3) 16 cm 4) 24 cm

8. Choose the correct circuit which can achieve the bridge balance. **(2024)**

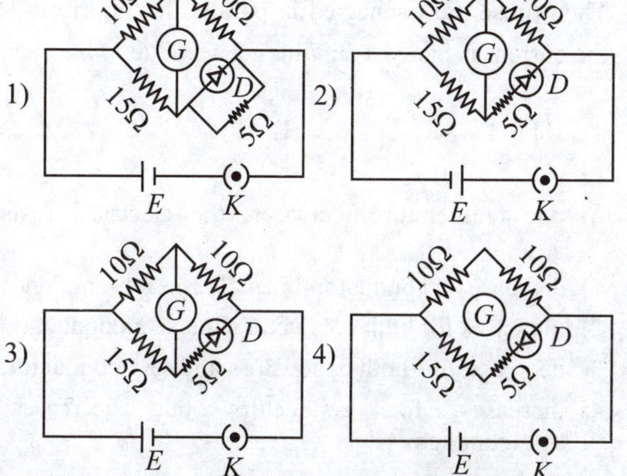

9. Two heaters A and B have power rating of 1 kW and 2 kW, respectively. Those two are first connected in series and then in parallel to a fixed power source. The ratio of power outputs for these two cases is: **(2024)**
   1) 1 : 2 2) 2 : 3 3) 1 : 1 4) 2 : 9

10. The magnitude and direction of the current in the following circuit is (2023)

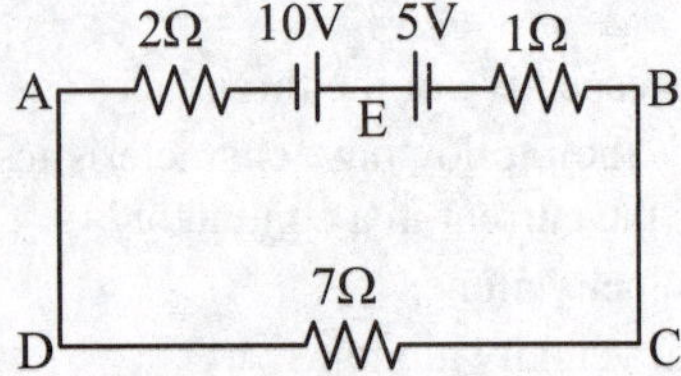

1) 1.5 A from B A through E
2) 0.2 A from B to A through E
3) 0.5 A from A to B through E
4) $\frac{5}{9}$ A from A to B through E

11. If the galvanometer G does not show any deflection in the circuit shown, the value of R is given by: (2023)

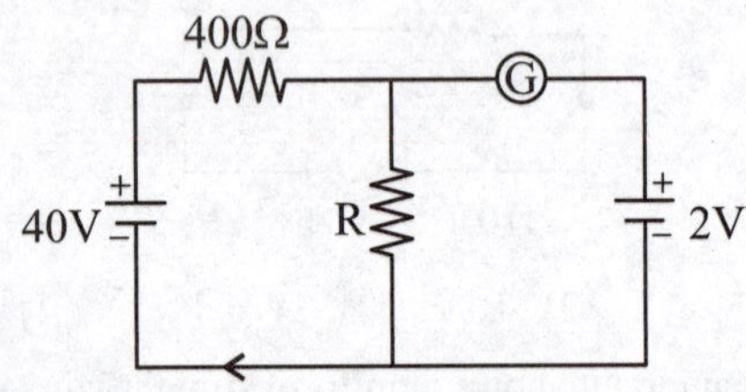

1) 400 Ω
2) 200 Ω
3) 50 Ω
4) 100 Ω

12. Resistance of a carbon resistor determined from colour codes is (22000 ± 5%) Ω. The colour of third band must be: (2023)

1) Yellow
2) Red
3) Green
4) Orange

13. The resistance of platinum wire at 0°C is 2Ω and 6.8Ω at 80°C. The temperature coefficient of resistance of the wire is: (2023)

1) $3 \times 10^{-1}\,°C^{-1}$
2) $3 \times 10^{-4}\,°C^{-1}$
3) $3 \times 10^{-3}\,°C^{-1}$
4) $3 \times 10^{-2}\,°C^{-1}$

14. 10 resistors, each of resistance R are connected in series to a battery of emf E and negligible internal resistance. Then those are connected in parallel to the same battery, the current is increased n times. The value of *n* is: (2023)

1) 1000
2) 10
3) 100
4) 1

15. As the temperature increases, the electrical resistance (2022)

1) decrease for conductors but increases for semiconductors
2) increases for both conductors and semiconductors
3) decreases for both conductors and semiconductors
4) increases for conductors but decreases for semiconductors

16. A copper wire of length 10 m and radius $(10^{-2}/\sqrt{\pi})$ m has electrical resistance of 10 Ω. The current density in the wire for an electric field strength of 10 V/m is (2022)

1) 10^5 A/m²
2) 10^4 A/m²
3) 10^6 A/m²
4) 10^{-5} A/m²

17. Two resistors of resistance, 100 Ω and 200 Ω are connected in parallel in an electrical circuit. The ratio of the thermal energy developed in 100 Ω to that in 200 Ω in a given time is (2022)

1) 4 : 1
2) 1 : 2
3) 2 : 1
4) 1 : 4

18. The effective resistance of a parallel connection that consists of four wires of equal length, equal area of cross-section and same material is 0.25 Ω. What will be the effective resistance if they are connected in series? (2021)

1) 0.5 Ω
2) 1 Ω
3) 4 Ω
4) 0.25 Ω

19. Column-I gives certain physical terms associated with flow of current through a metallic conductor. Column-II gives some mathematical relations involving electrical quantities. Match Column-I and Column-II with appropriate relations. (2021)

Column-I		Column-II	
(a)	Drift Velocity	(i)	$\frac{m}{ne^2\rho}$
(b)	Electrical Resistivity	(ii)	nev_d
(c)	Relaxation time	(iii)	$\frac{eE}{m}\tau$
(d)	Current Density	(iv)	$\frac{E}{J}$

1) (a)-(iii), (b)-(iv), (c)-(ii), (d)-(i)
2) (a)-(iii), (b)-(i), (c)-(iv), (d)-(ii)
3) (a)-(iii), (b)-(ii), (c)-(iv), (d)-(i)
4) (a)-(ii), (b)-(iv), (c)-(i), (d)-(ii)

20. Three resistors having resistances r_1, r_2 and r_3 are connected as shown in the given circuit. The ratio $\frac{i_3}{i_1}$ of currents in terms of resistances used in the circuit is (2021)

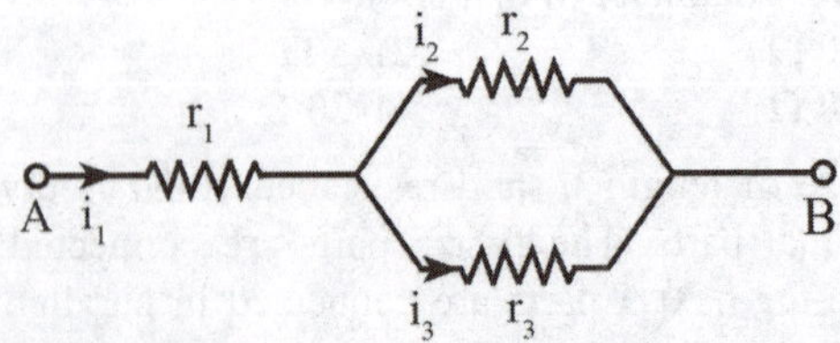

1) $\frac{r_2}{r_2 + r_3}$
(2) $\frac{r_1}{r_1 + r_2}$
3) $\frac{r_2}{r_1 + r_3}$
4) $\frac{r_1}{r_2 + r_3}$

21. Range of the ammeter is 5 ampere and full scale deflection current is 0.5 μA. If resistance of galvanometer is 50 Ω then find shunt resistance. (AIIMS 2018)

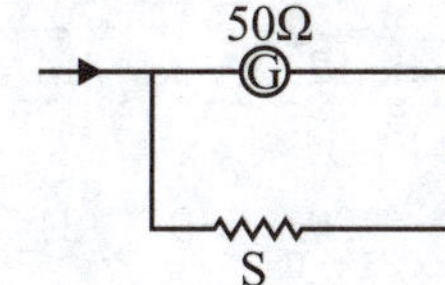

1) $6 \times 10^{-6}\Omega$
2) $5 \times 10^{-4}\Omega$
3) $5 \times 10^{-6}\Omega$
4) $6 \times 10^{-4}\Omega$

22. In the diagram find out the current through 2Ω: (AIIMS 2018)

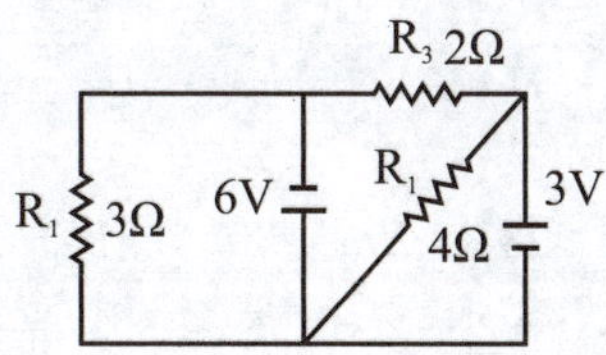

1) $\frac{5}{2}$A
2) 9A
3) 6A
4) $\frac{9}{2}$A

23. The resistance of an ammeter is 13 Ω and its scale is graduated for a current up to 100 A. After an additional shunt has been connected to this ammeter it becomes possible to measure currents up to 750 A by this meter. The valve of shunt resistance is: (AIIMS 2017)

1) 20 Ω
2) 2 Ω
3) 0.2 Ω
4) 2 kΩ

24. The temperature (T) dependence of resistivity (ρ) of a semiconductor is represented by: (AIIMS 2004)

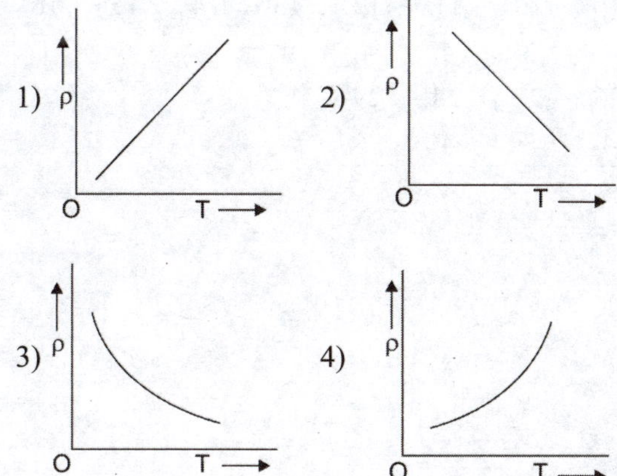

25. A metal rod consumes power P on passing current. If it is cut into two half and joined in parallel, it will consume power: (AIIMS 2001)

1) P
2) $2P$
3) 4P
4) $P/4$.

26. Given a current carrying wire of non uniform cross section. Which of the following is constant throughout the length of the wire? (AIIMS 2000)

1) Current, electrtic field and drift speed
2) Drift speed only
3) Current and drift speed
4) Current only

27. The resistance between the terminal points A and B of the given infinitely long circuit will be: (AIIMS 2000)

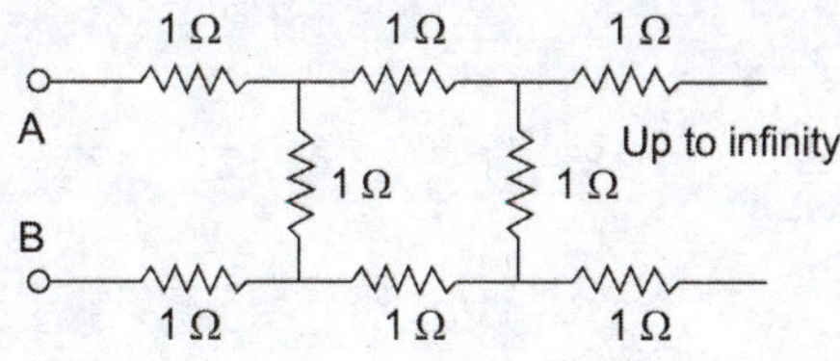

1) $(\sqrt{3}-1)\Omega$
2) $(1-\sqrt{3})\Omega$
3) $(1+\sqrt{3})\Omega$
4) $(2+\sqrt{3})\Omega$

28. What is the equivalent resistance between A and B in the given figure: (AIIMS 1995)

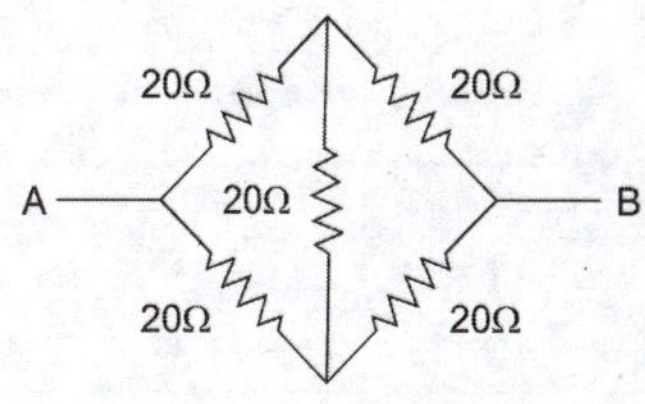

1) 40 Ω
2) 10 Ω
3) 50 Ω
4) 20 Ω

29. The heat produced by a 100 W heater in 2 min is equal to: (AIIMS 1995)

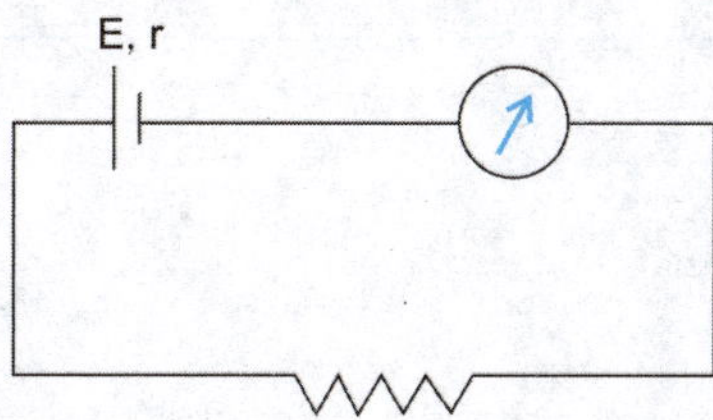

1) 10.5 kcal
2) 16.3 kcal
3) 2.8 kcal
4) 14.2 kcal

Answer Key

NCERT Topic-wise MCQs

1	2	3	4	5	6	7	8	9	10	11	12	13	14	15	16	17
2	1	1	1	4	2	4	4	2	1	2	1	3	4	3	1	2
18	**19**	**20**	**21**	**22**	**23**	**24**	**25**	**26**	**27**	**28**	**29**	**30**	**31**	**32**	**33**	**34**
2	2	1	3	3	4	3	4	2	4	4	1	2	1	3	3	3
35	**36**	**37**	**38**	**39**	**40**	**41**	**42**	**43**	**44**	**45**	**46**	**47**	**48**	**49**	**50**	**51**
4	4	2	2	3	3	2	1	3	2	1	3	3	2	3	3	4
52																
1																

High Order Time Intensive MCQs

1	2	3	4	5	6	7	8	9	10	11	12	13	14	15	16	17
3	1	2	2	3	2	4	2	3	4	1	2	4	2	3	3	2
18	**19**	**20**	**21**	**22**	**23**	**24**	**25**	**26**	**27**	**28**	**29**	**30**	**31**	**32**	**33**	**34**
2	4	1	1	3	4	3	1	3	1	4	2	4	1	3	1	3
35	**36**	**37**	**38**	**39**	**40**	**41**	**42**	**43**	**44**	**45**	**46**	**47**	**48**	**49**	**50**	**51**
1	3	1	4	1	4	3	3	3	1	2	1	1	3	2	2	3

Statement Based MCQs

1	2	3	4	5
1	3	1	2	4

Assertion & Reason MCQs

1	2	3	4	5
1	1	1	1	3

Match the Column MCQs

1	2	3
2	3	1

NCERT Exemplar MCQs

1	2	3	4	5
2	1	3	1	1

NEET & AIIMS Past Year MCQs

1	2	3	4	5	6	7	8	9	10	11	12	13	14	15	16	17
2	2	3	1	4	1	4	3	4	3	4	4	4	3	4	1	3
18	**19**	**20**	**21**	**22**	**23**	**24**	**25**	**26**	**27**	**28**	**29**					
3	4	1	3	4	2	3	3	4	3	4	3					

CHAPTER

4

Moving Charges and Magnetism

Topicwise Number of Questions Analysis (2025-2016)

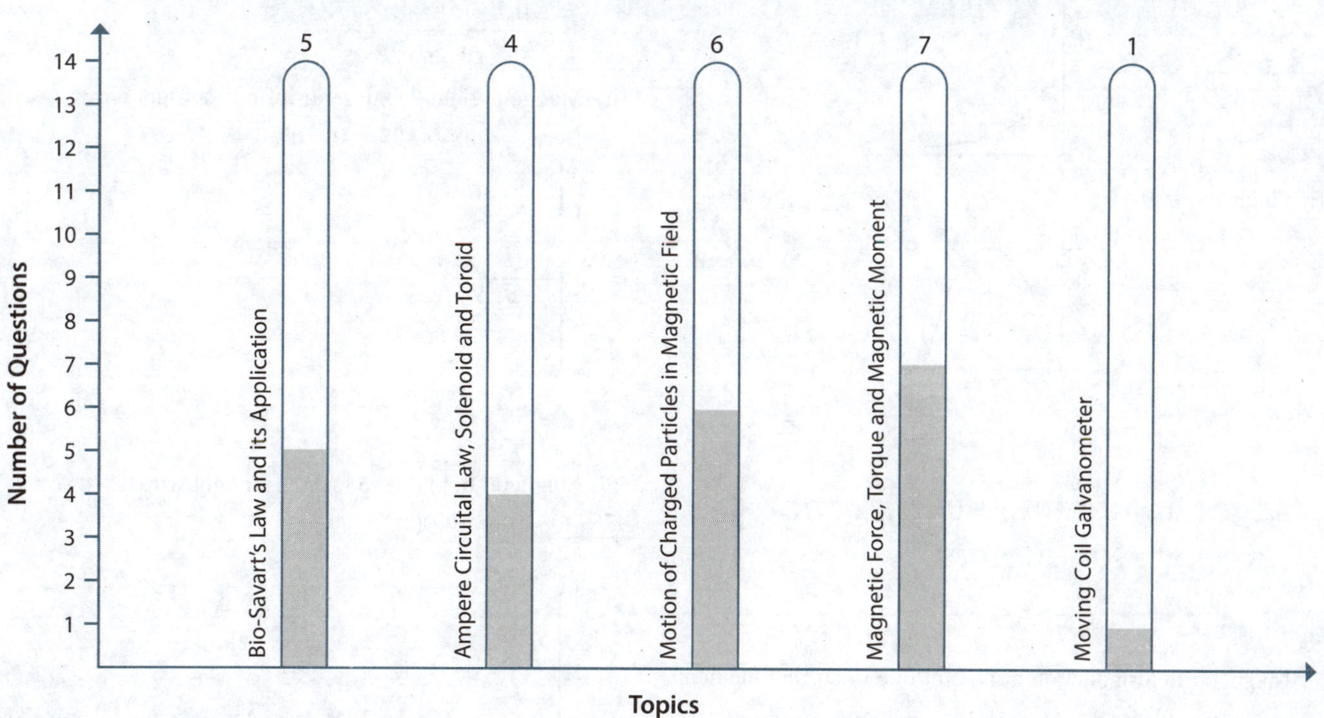

NCERT-Picks

Introduction

The branch of physics which deals with magnetism due to electric current or moving charge (i.e. electric current is equivalent to the charges or electrons in motion) is called electromagnetism.

Biot-Savart Law

Formulas

1. Magnetic field induction $d\vec{B}$ produced by an element $d\vec{\ell}$ carrying a current I at a distance r

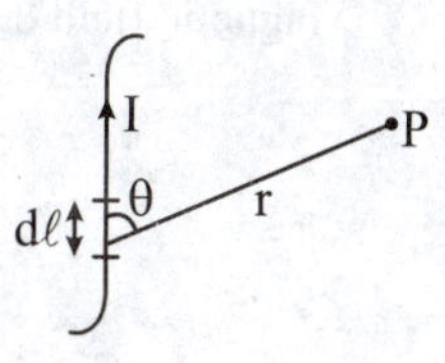

$$d\vec{B} = \frac{\mu_0 I}{4\pi} \times \frac{d\vec{\ell} \times \vec{r}}{r^3} = \frac{\mu_0 I}{4\pi} \times \frac{Id\ell \sin\theta}{r^2}\hat{n}$$

[Where μ_0 is permeability of free space ($4\pi \times 10^{-7}$ T mA^{-1})]

2. Intensity of magnetic field due to entire length of the conductor.

$$\vec{B} = \int d\vec{B}$$

Numerical Profile

Variation of magnetic field induction

(i) with current (I)

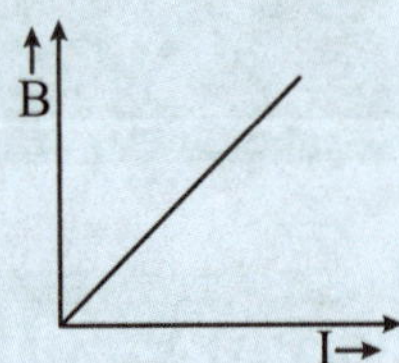

(ii) with distance (r)

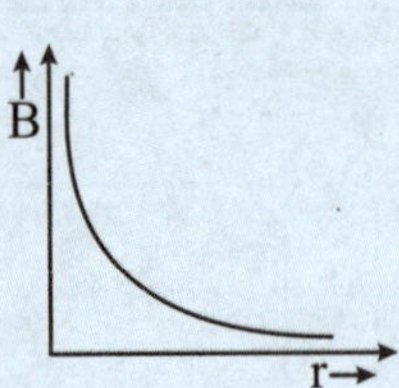

(iii) with angle (θ)

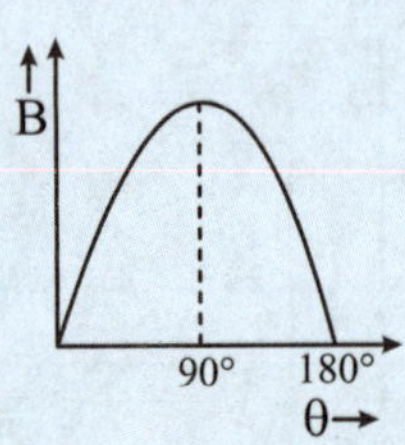

Magnetic field due to current Carrying Conductors

Formulas

1. Magnetic field induction at a point due to current element

$$dB = \frac{\mu_o}{4\pi} \times \frac{Id\ell \sin\theta}{r^2}$$

2. Magnetic field induction at a point due to linear conductor carrying current is

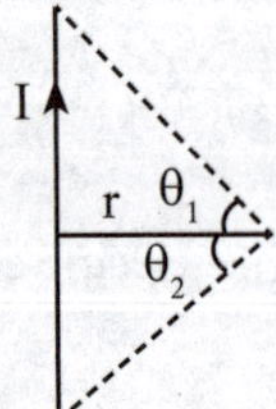

$$B = \frac{\mu_o I}{4\pi r}[\sin\theta_1 + \sin\theta_2]$$

3. Magnetic field due to a current carrying arc at its centre

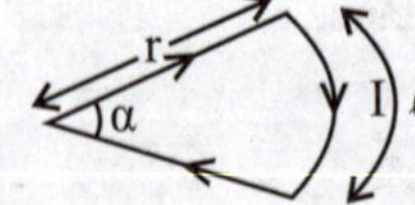

$$B = \frac{\mu_o I}{4\pi r}(\alpha)$$

4. Magnetic field induction at a point on the axis of the circular coil radius (r) carrying current I is

$$B = \frac{\mu_o}{4\pi} \times \frac{2\pi N I r^2}{(r^2 + x^2)^{3/2}} = \frac{\mu_o N I r^2}{2(r^2 + x^2)^{3/2}}$$

[Where x is distance of the point on the axis of circular coil from its center and N is number of turns.]

Numerical Profile

(i) Magnetic field on the perpendicular bisector of a finite straight wire of length (2L).

$$B = \frac{\mu_0 I}{2\pi a} \frac{L}{\sqrt{L^2 + a^2}}$$

(ii) Magnetic field due to semi-infinite straight wire:

Here θ_1=90°, θ_2=0° or θ_1= 0°, θ_2=90°

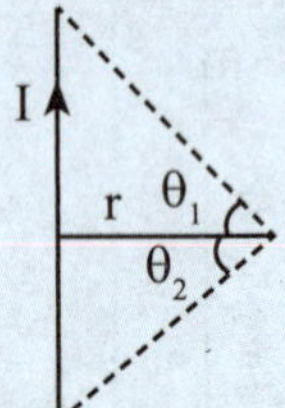

$$B = \frac{\mu_0 I}{4\pi a}$$

(iii) Magnetic field due to infinite straight wire

Here $\theta_1 = \theta_2 = 90°$

$$B = \frac{\mu_0 I}{2\pi a}$$

[where a is the distance of the point from the wire.]

(iv) For a current carrying circular coil with N no. of turns

$$B_{centre} = N\frac{\mu_0 I}{2r}$$

(v) For current carrying semicircular coil with N no. to turns

$$B_{centre} = N\frac{\mu_0 I}{4r}$$

(vi) Resultant magnetic field due to two current carrying circular coils.

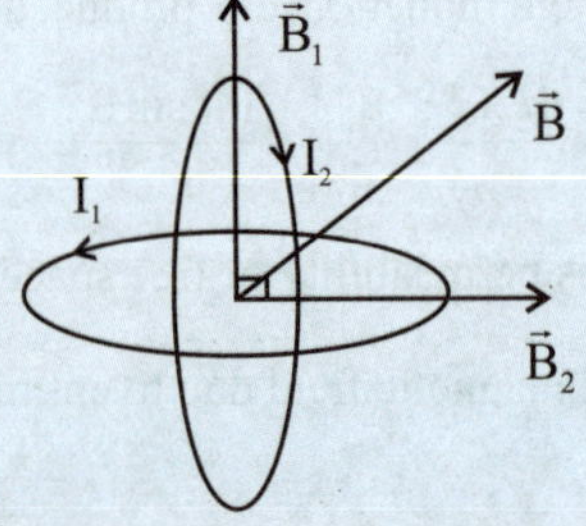

$$\vec{B} = \sqrt{B_1^2 + B_2^2 + 2B_1B_2\cos\theta}$$

When $B_1 = B_2,\ I_1 = I_2,\ r_1 = r_2$

$$\vec{B} = \sqrt{2}\,B_1 \text{ or } \sqrt{2}\,B_2$$

(vii) Magnetic field at the center of two concentric circular coils:

(a) Carrying current in same direction

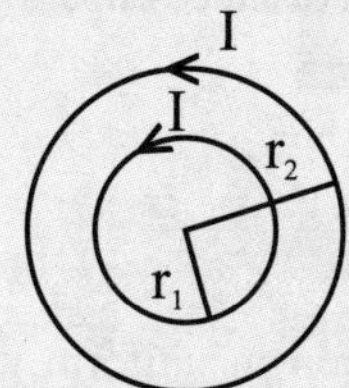

$$B = \frac{\mu_o n_1 I}{2r_1} + \frac{\mu_o n_2 I}{2r_2}$$

$$B = \frac{\mu_o I}{2}\left[\frac{n_1}{r_1} + \frac{n_2}{r_2}\right]$$

(b) Carrying current in opposite directions.

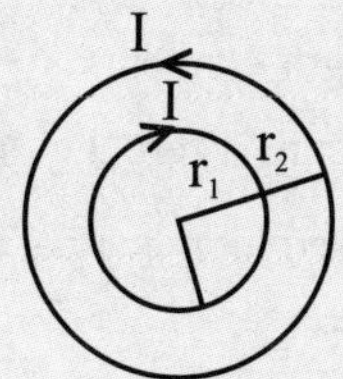

$$B = \frac{\mu_o n_1 I}{2r_1} + \frac{\mu_o n_2 I}{2r_2}$$

$$B = \frac{\mu_o I}{2}\left[\frac{n_1}{r_1} + \frac{n_2}{r_2}\right]$$

(viii) Magnetic field at a point from current carrying conductors carrying current in same direction.

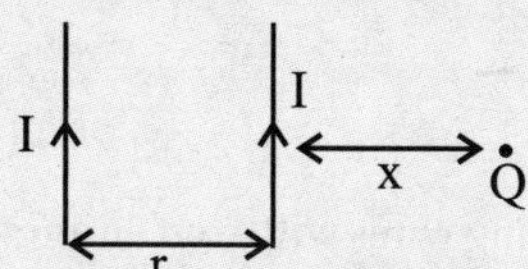

(a) $B = \frac{\mu_o}{4\pi} \times 2I\left[\frac{1}{x} + \frac{1}{r+x}\right]$

(b) Current carrying in opposite direction

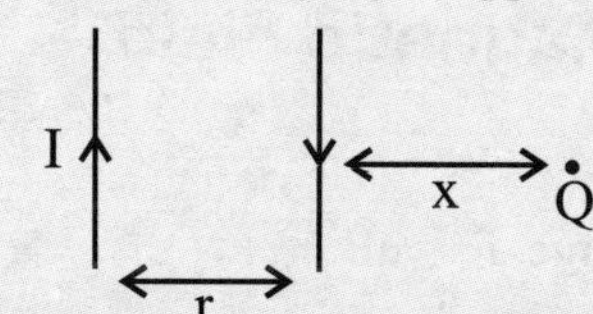

$$B = \frac{\mu_o}{4\pi} \times 2I\left[\frac{1}{x} - \frac{1}{r+x}\right]$$

(ix)

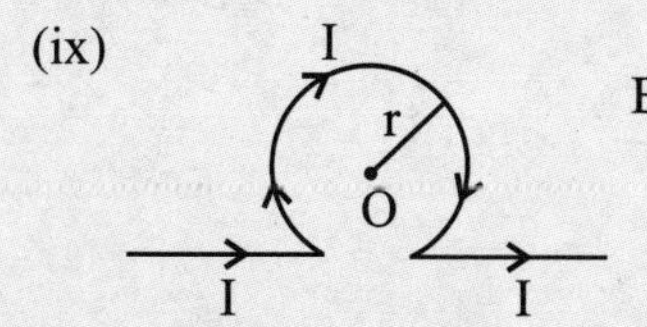

$$B = \frac{\mu_o}{4\pi}\left[\frac{2\pi I}{r} - \frac{2I}{r}\right]$$

(x)

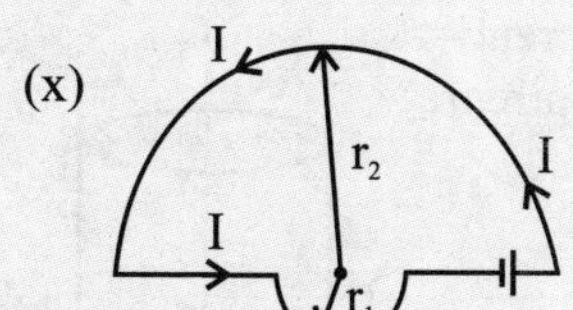

$$B = \frac{\mu_o}{4\pi}\left[\frac{\pi I}{r_1} + \frac{\pi I}{r_2}\right]$$

Ampere's Circuital Law

Formulas

1. The line integral of the magnetic field along any closed path is equal to μ_0 times of the net current bounded by the close path.

$$\oint \vec{B} \cdot d\vec{\ell} = \mu_0 I_{en}$$

2. It $I_1, I_2, I_3 \ldots I_n$ are currents in closed loop then

$$\oint \vec{B} \cdot d\vec{\ell} = \mu_0 \sum_{i=1}^{n} I_i$$

Numerical Profile

(i) Magnetic field due to a current carrying thin long hollow pipe

(a) $r > R$

$$\oint \vec{B} \cdot d\vec{\ell} = \oint B \cdot d\ell = B\oint d\ell = B(2\pi r)$$

$$\sum I_{en} = +I \Rightarrow B(2\pi r) = \mu_0 I \Rightarrow B = \frac{\mu_0 I}{2\pi r}$$

(b) $r < R$

$$\oint \vec{B} \cdot d\vec{\ell} = \oint B \cdot d\ell = B\oint d\ell = B(2\pi r)$$

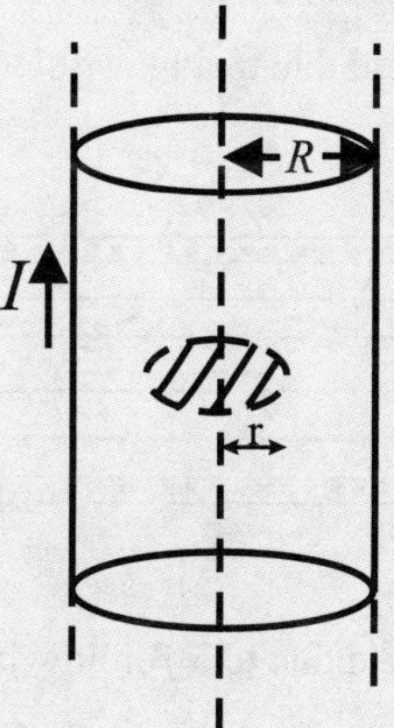

As pipe is hollow,

$$\sum I_{en} = 0$$

$$\Rightarrow B(2\pi r) = \mu_0(0) \Rightarrow B = 0$$

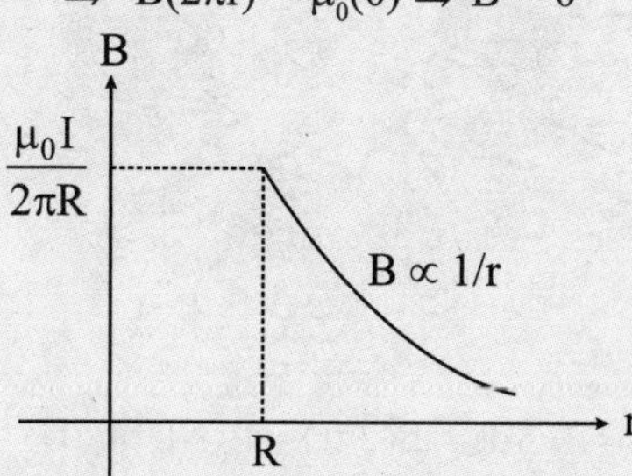

(ii) Magnetic field due to a current carrying rod having uniform current density

(a) $r > R$

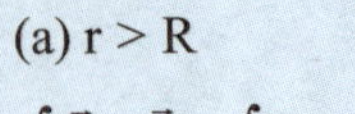

$$\oint \vec{B}\cdot d\vec{\ell} = \oint B\cdot d\ell = B\oint d\ell = B(2\pi r)$$

$$\sum I_{en} = +I \Rightarrow B(2\pi r) = \mu_0 I$$

$$\Rightarrow B = \frac{\mu_0 I}{2\pi r}$$

(b) $r < R$

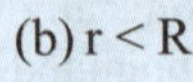

$$\oint \vec{B}\cdot d\vec{\ell} = \oint B\cdot d\ell = B\oint d\ell = B(2\pi r)$$

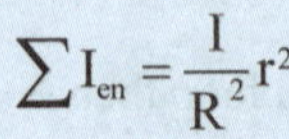

$$\sum I_{en} = \frac{I}{R^2}r^2$$

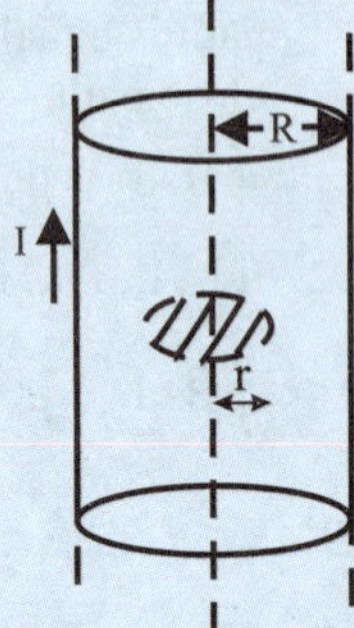

$$\Rightarrow B(2\pi r) = \mu_0\frac{I}{R^2}r^2$$

$$\Rightarrow B = \frac{\mu_0 I}{2\pi R^2}r$$

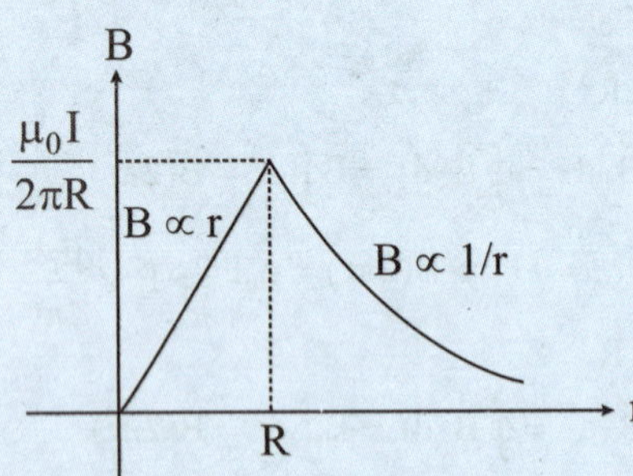

(iii) Magnetic field due to a long Solenoid

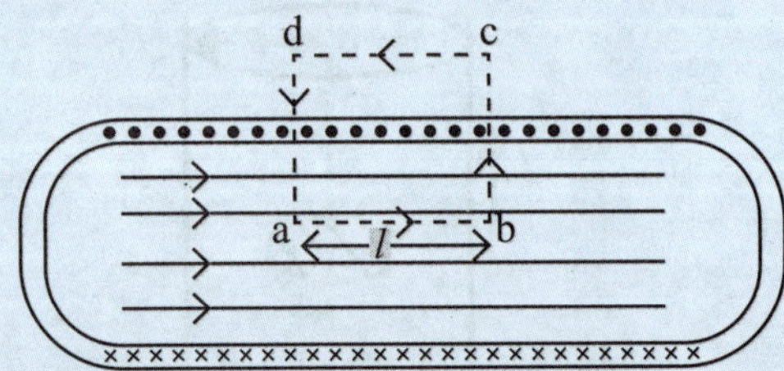

$B = \mu_o nI$

(iv) Magnetic field due to a Toroid with N turns *[RC]*

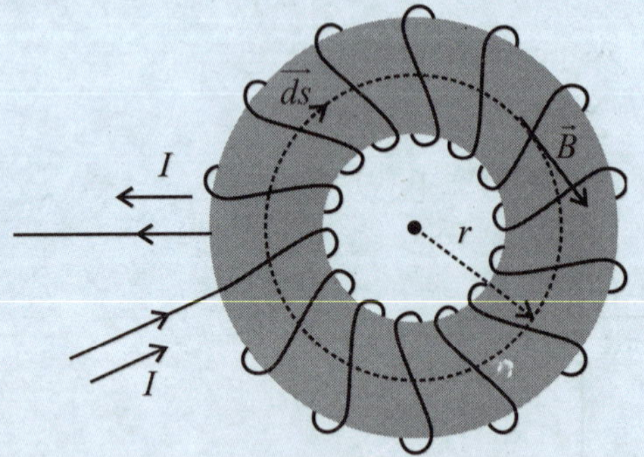

$$\oint \vec{B}.d\vec{s} = \oint B ds = B\oint ds = B(2\pi r) = \mu_0 NI \text{ or } B = \frac{\mu_0 NI}{2\pi r}$$

Magnetic Force

Formulas

1. Magnetic Force on a Current-Carrying Wire

$$\vec{F} = I(\vec{\ell}\times\vec{B})$$

$$F = I\ell B \sin\theta$$

2. Force per unit length on each of the two long parallel conductors carrying currents I_1 & I_2 placed at a distance r apart.

$$F = \frac{\mu_o}{4\pi}\times\frac{2I_1I_2}{r}$$

Numerical Profile

(i) Magnetic Force on a closed loop in uniform magnetic field

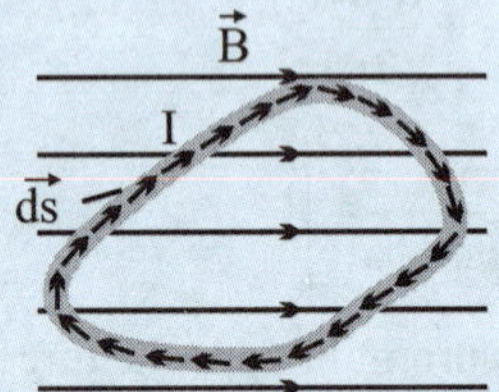

Since the set of differential length elements $d\vec{s}$ form a closed polygon, and their vector sum is zero, i.e. $\oint d\vec{s} = 0$.

The net magnetic force on a closed loop is $\vec{F}_B = 0$

(ii) Force Between Two Parallel Wires

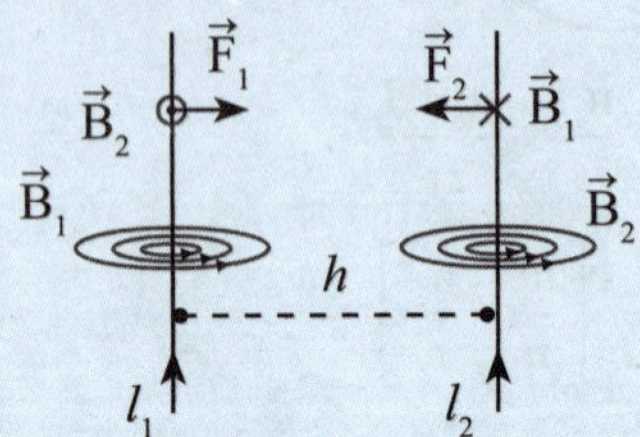

$$F_1 = I_1\left|\vec{\ell}\times\vec{B}_2\right| = \frac{\mu_0 I_1 I_2 \ell}{2\pi h}$$

$$F_2 = I_2\left|\vec{\ell}\times\vec{B}_1\right| = \frac{\mu_0 I_2 I_1 \ell}{2\pi h}$$

Note: Two parallel current-carrying wires will exert an attractive force on each other, if their currents are in the same direction.

Motion of charged Particle in Electric and Magnetic Field

Formulas

1. Electric force on a charge, $F_e = qE$
2. Magnetic force on a chage, $F_m = qvB\sin\theta$
3. In a perpendicular magnetic field, the charge follows a circular path.

$$qvB = \frac{mv^2}{r} \text{ or } r = \frac{mv}{qB}$$

$$T = \frac{2\pi m}{qB} \text{ and } f = \frac{qB}{2\pi m}$$

4. When $\vec{v}$ makes angle θ with $\vec{B}$, the charge follows helical path.

$$r = \frac{mv_{\perp}}{qB} = \frac{mv\sin\theta}{qB}$$

$$T = \frac{2\pi r}{v_{\perp}} = \frac{2\pi m}{qB}$$

5. Pitch of helix, $h = v_{\parallel}T = v\cos\theta \cdot T = \frac{2\pi mv\cos\theta}{qB}$

6. K.E. gained by an electron when accelerated through a potential difference V,

$$\frac{1}{2}mv^2 = eV \Rightarrow v = \sqrt{\frac{2eV}{m}}$$

Numerical Profile

1. Force on a moving charged particle in a region where both electric and magnetic fields are present

$$\vec{F} = q\vec{E} + q\left(\vec{v}\times\vec{B}\right) \quad \text{\{Lorentz Force\}}$$

2. Maximum K.E of accelerated charged particle is

$$E_{max} = \frac{B^2q^2r^2}{2m}$$

Torque on Current Loop and Magnetic Dipole

Formulas

1. Torque on current loop, $\vec{\tau} = \vec{\mu}\times\vec{B}$

[Where $|\vec{\mu}| = IA$, magnetic moment]

2. Magnitude of torque for a loop consisting N turns $\tau = NIAB\sin\theta$

Numerical Profile

(i) When $\theta = 0°$, $\tau = 0$, i.e., the torque is minimum when the plane of the loop is perperndicular to the magnetic field.

$$\tau_{min} = 0$$

(ii) When $\theta = 90°$, $\tau = NIBA$, i.e., the torque is maximum when the plane of the loop is parallel to the magnetic field. Thus

$$\tau_{max} = NIBA$$

The Moving Coil Galvanometer

Formulas

1. Current sensitivity $I_s = \frac{\theta}{I} = \frac{nBAI}{IK} = \frac{nBA}{k}$

(Unit = $radA^{-1}$ or div A^{-1})

2. Voltage sensitivity $V_s = \frac{\theta}{V} = \frac{nBAI}{IRK} = \frac{nBA}{RK}$

(Unit = rad V^{-1} or div V^{-1})

Numerical Profile

(i) **Galvanometer to Ammeter:** A small resistance called shunt S is connected in parallel with the galvanometer.

i_g = Full scale deflection current

$$i_g \times G = (i - i_g)S$$

Ammeter

$$i_g = \left(\frac{S}{S+G}\right)i$$

$$S = \frac{G}{\left(\frac{i}{i_g} - 1\right)}$$

(ii) **Galvanometer to Voltmeter:** A high resistance R is connected in series with the galvanometer.

$$V = i_g(G + R)$$

$$R = \frac{V}{i_g} - G$$

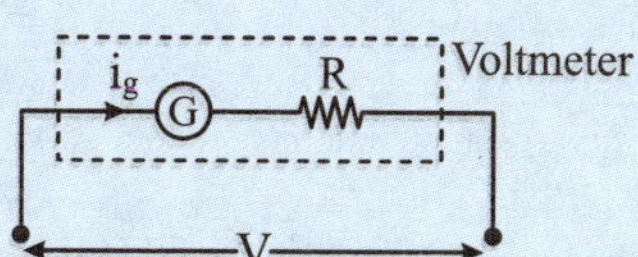

(iii) To increase the range of voltmeter n times the resistance to be connected in series $R = (n-1)G$

(iv) To increase the range of ammeter n times the value of shunt to be connected in parallel $S = \frac{G}{(n-1)}$

Statements Asked in Exam

- If the length of a solenoid is more than five or six times of its diameter then the solenoid is considered as infinitely long.
- A linear solenoid carrying current is equivalent to a bar magnet.
- The magnetic field induction at a point just outside the curved face of current carrying solenoid is zero.
- A solenoid tends to contract when current flows through it.
- If a current carrying circular loop (number of turns n = 1) is turned into a coil having n identical turns, then magnetic field at the center of the coil becomes n^2 times the previous value of the magnetic field i.e. $B' = n^2B_0$.
- Energy of a charged particle moving in a uniform magnetic field does not change because it experiences a force in a direction perpendicular to the direction of motion of the charged particle. If the magnetic field is not uniform it will change the energy of the charged particle.
- Main function of the soft iron core used in a galvanometer is to make the magnetic field lines radial.

- Range of ammeter can be increased by using a suitable value of shunt in parallel with the ammeter.
- Range of voltmeter can be decreased or increased by using suitable resistance in series or parallel with the galvanometer.
- Resistance of ideal voltameter and ideal ammeter is ∞ & zero respectively.
- An ammeter is connected in series, whereas a voltmeter is connected in parallel.
- Graduation on scale in ac ammeter and voltmeter are not equally distant.
- A dead-beat galvanometer is one in which the coil comes to rest at once when the passage of current through it is stopped.
- Nature of magnetic field is radial in a galvanometer.
- The amount of current which produces one scale deflection in the galvanometer is called "figure of merit of galvanometer.
- A moving coil galvanometer can't be used to measure the alternating current in a circuit since it measures average value of current and average value of A.C. is zero over a complete cycle.
- When the plane of the coil becomes perpendicular to the direction of the magnetic field. It will be in stable equilibrium.
- Magnetic field intensity at the centre of circular coil carrying current will be four times of its initial value, if current through the circular coil is doubled and the radius is halved.

MCQs

NCERT Topic-wise MCQs

Biot - Savart Law and Magnetic Field Due to Current Carrying Straight Conductors, Circular Coil

1. Magnetic field at a distance d from an infinite current carrying wire is
   1) $\frac{\mu_0 i}{2d}$ 2) $\frac{\mu_0 i}{2\pi d}$
   3) $\frac{\mu_0 i}{4d}$ 4) $\frac{\mu_0 i}{4\pi d}$

2. Magnetic field due to current element can be found by
   1) Ampere Circuital Law
   2) Biot- Savart's Law
   3) Gauss Law
   4) Coulomb's Law

3. The magnetic field outside a long straight current-carrying conductor varies with distance r as:
   1) $B \propto r$ 2) $B \propto \frac{1}{r}$
   3) $B \propto r^2$ 4) $B \propto \frac{1}{r^2}$

4. A wire of length L metre carrying a current I ampere is bent in the form of a circle. The magnitude of the magnetic moment is
   1) $\frac{L^2I^2}{4\pi}$ 2) $\frac{LI}{4\pi}$
   3) $\frac{L^2I}{4\pi}$ 4) $\frac{LI^2}{4\pi}$

5. Work done on an electron moving in a solenoid along its axis is equal to

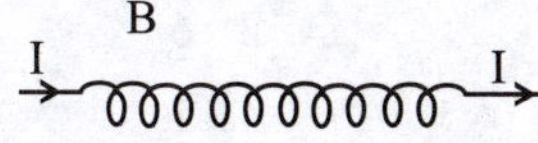

   1) Zero 2) evB
   3) $i\ell B$ 4) None of these

6. A circular coil carrying a certain current produces a magnetic field B_0 at its center. The coil is now rewound so as to have 3 turns and the same current is passed through it. The new magnetic field at the centre is
   1) $B_0/2$ 2) $9B_0$
   3) $B_0/3$ 4) $3B_0$

7. Two similar coils of radius R and number of turns N are lying concentrically with their planes at right angles to each other. The currents flowing in them are I & $I\sqrt{3}$ respectively. Find the magnetic field induction at the center of the coils.
   1) $\frac{2\mu_o NI}{R}$ 2) $\frac{3\mu_o NIR}{2}$
   3) $\frac{\mu_o NIR}{2}$ 4) $\frac{\mu_o NI}{R}$

8. An electric current passes through a long straight copper wire. At a distance 5cm from the straight wire the magnetic field is B. The magnetic field at 20cm from the straight wire would be:
   1) B/6 2) B/4
   3) B/3 4) 4B

9. A wire of mass 200g and length 1.5m carries a current of 2A. It is suspended in mid air by a uniform horizontal magnetic field B. The magnitude of B (in tesla) is ($g = 10\ ms^{-2}$)
   1) 2 2) 1.5
   3) 0.55 4) 0.66

10. A and B are two concentric circular coils having centre at O and carrying currents I_1 and I_2 as shown in the adjacent figure. If ratio of their radii is 1:2 and ratio of the magnetic flux densities at O due to A and B is 1 : 3 then the value of I_1/I_2 is

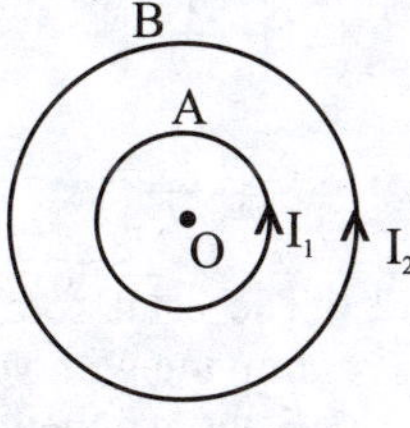

   1) 1/6 2) 1/4
   3) 1/3 4) 1/2

11. The dimension of the magnetic field intensity B is
   1) $[MLT^{-2}A^{-1}]$ 2) $[MT^{-2}A^{-1}]$
   3) $[ML^2TA^{-2}]$ 4) $[M^2LTA^{-1}]$

12. Two wires of same length are shaped into a square and a circle. If they carry same current. The ratio of the magnetic moment is
   1) $2 : \pi$ 2) $\pi : 2$
   3) $\pi : 4$ 4) $4 : \pi$

13. Magnetic field due to current carrying circular loop of radius 3cm at a point on the axis at a distance of 4cm from the center is 54 μT. What will be its value at the center of the loop?
   1) 250 μT 2) 150 μT
   3) 125 μT 4) 75 μT

14. The magnetic field due to a long straight wire carrying a current I is proportional to:

1) I 2) I^3
3) $I^{1/2}$ 4) 1/I

15. A current flows in a conductor from east to west. The direction of the magnetic field at a point above the conductor is

1) Towards north 2) Towards south
3) Towards east 4) Towards west

16. A long solenoid has n turns per meter and current 1 A is flowing through it. The magnetic field at the ends of the solenoid is

1) $\frac{\mu_o nI}{2}$ 2) $\mu_o nI$
3) Zero 4) $2\mu_o nI$

Motion of Charged Particle in Electric, Magnetic Field and Magnetic Force

17. The radius of the path of a charged particle in a magnetic field is directly proportional to:

1) The particle's charge.
2) The particle's mass.
3) The magnetic field strength.
4) The particle's mass and its velocity.

18. Two particles X and Y having equal charges are being accelerated through the same potential difference. Thereafter they enter normally in a region of uniform magnetic field and describes circular paths of radii R_1 and R_2 respectively. The mass ratio of X and Y is:

1) $\left(\frac{R_2}{R_1}\right)^2$ 2) $\left(\frac{R_1}{R_2}\right)^2$
3) $\left(\frac{R_1}{R_2}\right)$ 4) $\left(\frac{R_2}{R_1}\right)$

19. The total force F on a charged particle with charge q, moving with velocity v in the presence of both electric field E and magnetic field B is known as:

1) Gravitational force. 2) Centripetal force.
3) Lorentz force. 4) Frictional force.

20. A proton ($m_p = 1.67 \times 10^{-27}$ kg and $q_p = 1.6 \times 10^{-19}$C) enters perpendicular to the magnetic field of intensity $2Wb/m^2$ with a velocity 3.4×10^7m/sec. The acceleration of the proton should be

1) 6.5×10^{15} m/sec^2 2) 6.5×10^{13} m/sec^2
3) 6.5×10^{11} m/sec^2 4) 6.5×10^{9} m/ sec^2

21. A moving charge will gain kinetic energy due to the application of

1) Electric field 2) Magnetic field
3) Both of these 4) None of these

22. Two straight parallel wires both carrying 10A in the same direction attract each other with a force of 1×10^{-3} N. If both currents are doubled, the force of attraction will be

1) 10^{-3}N 2) 2×10^{-3}N
3) 4×10^{-3}N 4) 0.25×10^{-3}N

23. The radius of a circular loop is r and a current I is flowing in it. The equivalent magnetic moment will be

1) Ir 2) $2\pi Ir$
3) $I\pi r^2$ 4) $\frac{1}{r^2}$

24. A current of 5 A is flowing in a wire of length 1.5 m. A force of 7.5 N acts on it when it is placed in a uniform magnetic field of 2 tesla . The angle between the magnetic field and the direction of the current is

1) 30° 2) 45° 3) 60° 4) 90°

25. Two parallel beams of protons moving in the same direction produce a mutual force: (magnetic force)

1) Force of attraction in the plane of paper
2) Force of repulsion in plane of paper
3) Upward perpendicular to the plane of paper
4) Downward perpendicular to the plane of paper

26. Current I is carried in a wire of length L. If the wire is turned into a circular coil, the maximum magnitude of torque in a given magnetic field B will be

1) $\frac{LIB^2}{2}$ 2) $\frac{LI^2B}{2}$
3) $\frac{L^2IB}{2}$ 4) $\frac{L^2IB}{4\pi}$

27. A straight conductor carries a current of 5A. An electron travelling with a speed of 5×10^6 ms^{-1} parallel to the wire at a distance of 0.1 m from the conductor, experiences a force of

1) 8×10^{-20} N 2) 8×10^{-18} N
3) 8×10^{-10} N 4) 1.6×10^{-19} N

28. If a proton, deuteron and α - particle on being accelerated by the same potential difference enters perpendicular to the magnetic field then the ratio of their kinetic energies is

1) 1 : 2 : 2 2) 2 : 2 : 1
3) 1 : 2 : 1 4) 1 : 1 : 2

29. A charge particle is moving in a uniform magnetic field $(2\hat{i}+3\hat{j})$T. If it has an acceleration of $(\alpha\hat{i}-4\hat{j})$m/s^2, then the value of α is.

1) 3 2) 6
3) 12 4) 2

30. A particle with charge q is moving along a circle of radius R with uniform speed v. The associated magnetic moment μ is given by: ***[RC]***

1) $\frac{1}{2}v^2R$ 2) $\frac{1}{4}qvR$
3) $\frac{1}{2}qvR$ 4) $\frac{1}{2}q^2vR$

31. A charged particle enters a magnetic field B with its initial velocity making angle of 45° with B. The path of the particle will be

1) straight line 2) a circle
3) an ellipse 4) a helix

32. A proton enters a magnetic field of flux density 1.5 Wbm^{-2} with a speed of 2×10^7m/s at angle of 30° with the field. The force on a proton will be

1) 0.24×10^{-12}N
2) 2.4×10^{-12}N
3) 24×10^{-12}N
4) 0.024×10^{-12}N

33. A proton, a deuteron and α - particle with the same K.E enter a region of uniform magnetic field moving at right angle to B. What is the ratio of the radii of their circular path?

1) $1:\sqrt{2}:1$ 2) $1:\sqrt{2}:\sqrt{2}$
3) $\sqrt{2}:1:1$ 4) $\sqrt{2}:\sqrt{2}:1$

34. An electron and a proton enters in a magnetic field perpendicularly. Both have same kinetic energy. Which of the following is correct?

1) Trajectory of electron is less curved
2) Trajectory of proton is less curved
3) Both trajectories are equally curved
4) Both move on straight line path

35. A particle of charge q and mass m is moving with a velocity $-v\hat{i}\,(v \neq 0)$ towards a large screen placed in the $Y-Z$ plane at a distance d. If there is a magnetic field $\vec{B} = B_0\hat{k}$, the minimum value of v for which the particle will not hit the screen is:

1) $\frac{qdB_0}{3m}$ 2) $\frac{2qdB_0}{m}$
3) $\frac{qdB_0}{m}$ 4) $\frac{qdB_0}{2m}$

The Moving Coil Galvanometer

36. If a hypothetical galvanometer has a current sensitivity of 100 div/A and a voltage sensitivity of 25 div/V, what is the resistance of the galvanometer?

1) 4 Ω 2) 2 Ω
3) 4.5 Ω 4) 0.25 Ω

37. To use the galvanometer as an ammeter, what additional component should be connected, and how should it be connected?

1) A resistor in series to limit the current.
2) A high-resistance in parallel to measure high currents.
3) A low-resistance (shunt) in parallel to allow most of the current to bypass the galvanometer.
4) An inductor in series to increase current sensitivity.

38. An ammeter design requires that only 0.1% of the main current pass through the galvanometer. How much current passes through the shunt resistance?

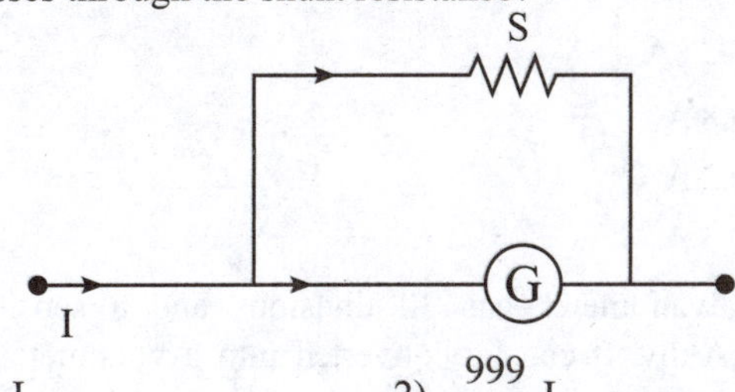

1) I 2) $\frac{999}{1000}$I
3) $\frac{1}{1000}$I 4) $\frac{9}{1000}$I

39. A Galvanometer coil is replaced by another coil of diameter 1/4 of its original diameter and the total number of turns is ten times the original number. What will be the new deflection if the same current is passed through it?

1) $\frac{5}{8}$ times the original deflection
2) $\frac{4}{5}$ times the original deflection
3) $\frac{4}{3}$ times original deflection
4) $\frac{5}{3}$ times original deflection

40. A resistance of 900 Ω is connected in series with a galvanometer of resistance 100 Ω. A potential difference of 1 volt produces 100 division deflection in the galvanometer. Find the figure of merit of galvanometer?

1) 10^{-6} A/div 2) 10^{-4}A/div
3) 10^{-7}A/div 4) 10^{-5}A/div

41. The deflection in moving coil galvanometer is

1) Directly proportional to the torsional constant.
2) Directly proportional to the number of torsional coil
3) Inversely proportional to the area of coil
4) Inversely proportional to the current flowing

42. A moving coil galvanometer has N number of turns in a coil of effective area A it carries a current I. The magnetic field B is radial. The torque acting on the coil is

1) NA^2BI 2) $NABI^2$
3) N^2ABI 4) NABI

43. The coil of galvanometer consists of 100 turns and effective area of 1 square cm. The restoring couple is 10^{-8} N-m/rad. The magnetic field between the pole pieces is 5T. The current sensitivity of this galvanometer will be

1) 5×10^4 rad/μA 2) 5×10^{-6} rad/A
3) 2×10^{-7} rad/A 4) 5 rad/μA

44. The relation between voltage sensitivity (σ_v) and current sensitivity (σ_i) of moving coil galvanometer is (Resistance of galvanometer = G)

1) $\frac{\sigma_i}{G} = \sigma_v$ 2) $\frac{\sigma_v}{G} = \sigma_i$
3) $\frac{G}{\sigma_v} = \sigma_i$ 4) $\frac{G^2}{\sigma_i} = \sigma_v$

45. A galvanometer having a resistance of 8 Ω is shunted by a wire of resistance 2 Ω. If the total current is 1 A the fraction of it passing through the shunt will be

1) 0.25 A

2) 0.8 A

3) 0.2 A

4) 0.5 A

46. A galvanometer has 30 divisions and a sensitivity of 16 μA/div. It can be converted into a voltameter to read 3V by connecting

1) Resistance of nearly 6 kΩ in series

2) 6kΩ in parallel

3) 500Ω in series

4) It cannot be converted

47. If resistance of galvanometer is 2Ω then find R when voltmeter reads 12 V and galvanometer reads 0.1 A?

1) 118 Ω 2) 120 Ω

3) 124 Ω 4) 114 Ω

48. A galvanometer has a resistance of 400 Ω and deflects full scale for current of 0.2 mA through it. The shunt resistance required to convert it into 3A ammeter is

1) 0.027 Ω 2) 0.0135 Ω

3) 0.054 Ω 4) None of these

49. The deflection in a moving coil galvanometer falls from 50 divisions to 10 divisions when a shunt of 12 Ω is connected with it. The resistance of galvanometer coil is

1) 20 Ω 2) 40 Ω

3) 50 Ω 4) 48 Ω

High Order Time Intensive MCQs

1. A wire carrying current i and other carrying 2i in the same direction produce a magnetic field B at the mid point. What will be the field when 2i current is switched off?

1) B/2 2) 2B

3) B 4) 4B

2. A wire loop PQRSP is constructed by joining two semi-circular coils of radii r_1 and r_2, respectively as shown in the figure. If the current flowing in the loop is i, then the magnetic induction at the point O is:

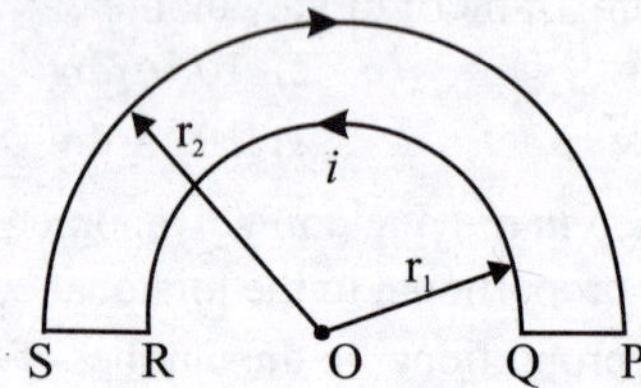

1) $\frac{\mu_0 i}{4}\left[\frac{1}{r_1}-\frac{1}{r_2}\right]$ 2) $\frac{\mu_0 i}{4}\left[\frac{1}{r_1}+\frac{1}{r_2}\right]$

3) $\frac{\mu_0 i}{2}\left[\frac{1}{r_1}-\frac{1}{r_2}\right]$ 4) $\frac{\mu_0 i}{2}\left[\frac{1}{r_1}+\frac{1}{r_2}\right]$

3. An electric current passes through a long straight copper wire. At a distance 5 cm from the straight wire, the magnetic field is B. The magnetic field at 25 cm from the straight wire would be:

1) $\frac{B}{6}$ 2) $\frac{B}{5}$

3) $\frac{B}{3}$ 4) $\frac{B}{2}$

4. A long straight wire of a circular cross-section (radius a) carries a steady current I and the current I is uniformly distributed across this cross-section. Which one of the following plots represents the variation of magnitude of magnetic field B with distance r from the centre of the wire.

1) 2)

3) 4)

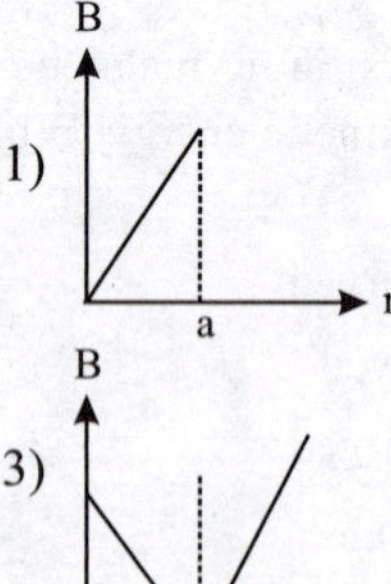

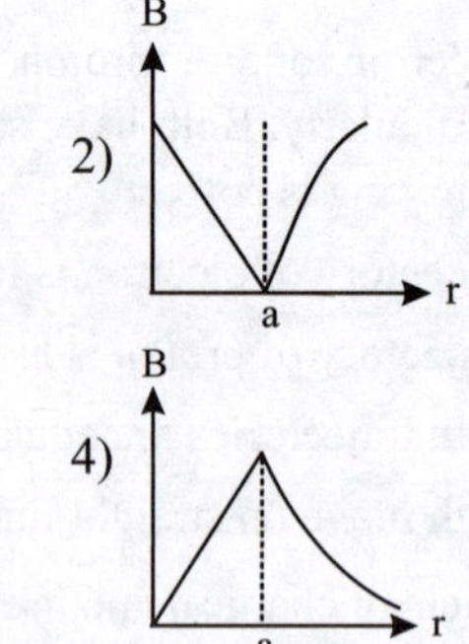

5. A square wire of each side ℓ carries a current I. What is the magnetic field at the midpoint of the square?

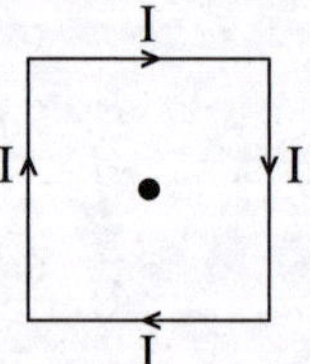

1) $4\sqrt{2}\frac{\mu_0}{4\pi}\frac{I}{\ell}$ 2) $8\sqrt{2}\frac{\mu_0}{4\pi}\frac{I}{\ell}$

3) $16\sqrt{2}\frac{\mu_0}{4\pi}\frac{I}{\ell}$ 4) $32\sqrt{2}\frac{\mu_0}{4\pi}\frac{I}{\ell}$

6. A proton of mass m and charge q is moving in a plane with kinetic energy E. If there exists a uniform magnetic field B, perpendicular to the plane of the motion, the proton will move in a circular path of radius:

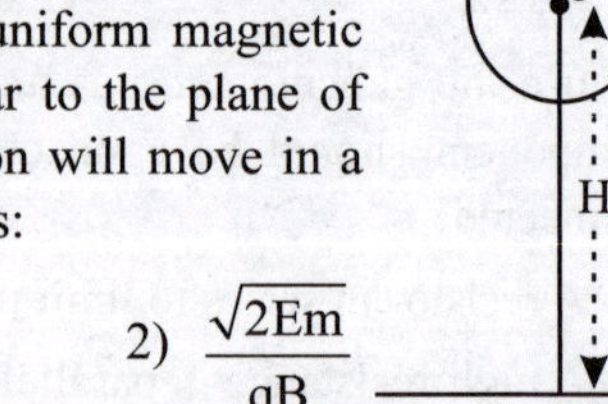

1) $\frac{2Em}{qB}$ 2) $\frac{\sqrt{2Em}}{qB}$

3) $\frac{\sqrt{Em}}{2qB}$ 4) $\frac{\sqrt{2Eq}}{qB}$

7. A proton carrying 1 MeV kinetic energy is moving in a circular path of radius R in uniform magnetic field. What should be the energy of an α-particle to describe a circle of same radius in the same field?

1) 2 MeV
2) 1 MeV
3) 0.5 MeV
4) 4 MeV

8. A conductor carrying current I is in the shape as shown in figure. Find the magnetic field induction at the common center O of all the three area.

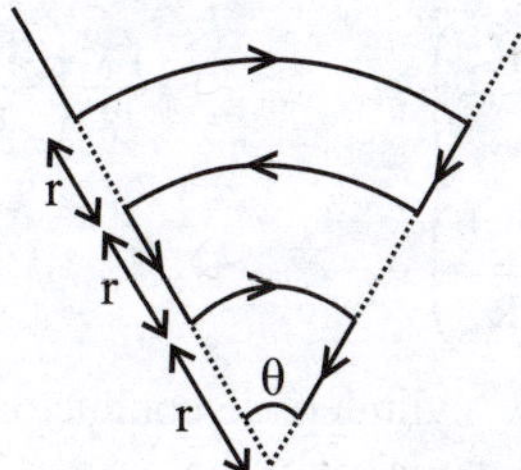

1) $\frac{3\mu_o I\theta}{4\pi r}$
2) $\frac{5\mu_o I\theta}{24\pi r}$
3) $\frac{\mu_o I\theta}{8\pi r}$
4) $\frac{\mu_o I\theta}{\pi r}$

9. An electric current of I ampere is flowing in a long conductor CG. Find the magnitude and direction of magnetic induction at the center O of circular part

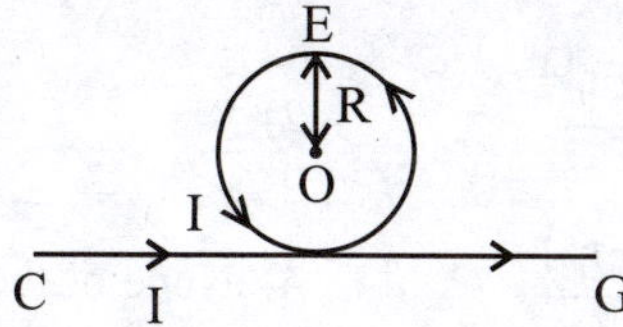

1) $\frac{\mu_o I}{2R}\left(\frac{1}{\pi}+1\right)$ acting ⊥ to the loop upward
2) $\frac{\mu_o I}{4R}\left(\frac{1}{\pi}+\pi\right)$ acting ⊥ to the loop upward
3) $\frac{3\mu_o I}{R}\left(\pi+\frac{1}{\pi}\right)$ acting ⊥ to the loop upward
4) $\frac{3\mu_o I}{R}\left(\frac{1}{\pi}+\pi\right)$ acting ⊥ to the loop upward

10. A proton moving with a constant velocity passes through a region of space without any change in its velocity. If $\vec{E}$ and $\vec{B}$ represent the electric and magnetic fields respectively, then the region of space may have:

(A) $E = 0, B = 0$
(B) $E = 0, B \neq 0$
(C) $E \neq 0, B = 0$
(D) $E \neq 0, B \neq 0$

Choose the most appropriate answer from the options given below:

1) (A), (B) and (C) only
2) (A), (C) and (D) only
3) (A), (B) and (D) only
4) (B), (C) and (D) only

11. A rectangular loop carrying a current I is situated near a long straight wire such that the wire is parallel to one of the sides of the loop and is in the plane of the loop. If a steady current I is established in the wire as shown in the figure, the loop will:

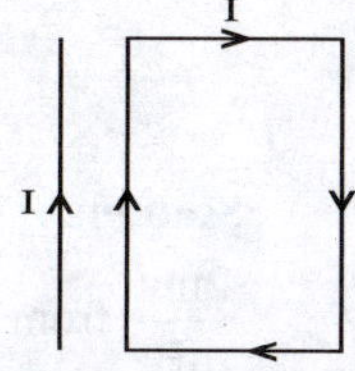

1) Rotate about an axis parallel to the wire
2) Move away form the wire
3) Move towards the wire
4) Remain stationary

12. A square current carrying loop is suspended in a uniform magnetic field acting in the plane of the loop. If the force on one arm of the loop is F, the net force on the remaining three arms of the loop is:

1) 3F
2) –F
3) –3F
4) F

13. A and B are two infinitely long straight parallel conductors. C is another straight conductor of length 1 m kept parallel to A and B as shown in the figure. Then, the force experienced by C is:

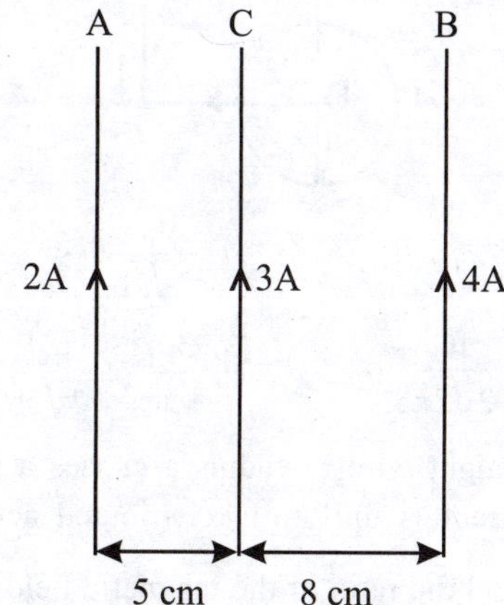

1) Towards A equal to 0.6×10^{-5} N
2) Towards B equal to 5.4×10^{-5} N
3) Towards A equal to 5.4×10^{-5} N
4) Towards B equal to 0.6×10^{-5} N

14. The magnetic force per unit length on a wire carrying a current of 10 A and making an angle of 45° with the direction of a uniform magnetic field of 0.20 T is

1) $2\sqrt{2}\text{Nm}^{-1}$
2) $\frac{2}{\sqrt{2}}\text{Nm}^{-1}$
3) $\frac{\sqrt{2}}{2}\text{Nm}^{-1}$
4) $4\sqrt{2}\text{Nm}^{-1}$

15. An equilateral triangle of side length ℓ is formed from a piece of wire of uniform resistance. The current I is fed as shown in the figure. Then the magnitude of the magnetic field at its centre O is:

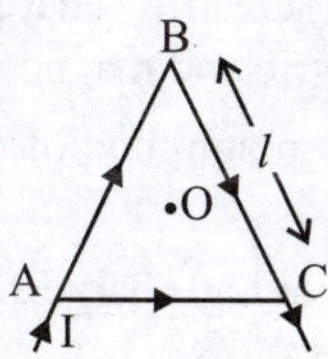

1) $\frac{\sqrt{3}\mu_0 I}{2\pi l}$
2) $\frac{3\sqrt{3}\mu_0 I}{2\pi l}$
3) $\frac{\mu_0 I}{2\pi l}$
4) Zero

16. Find the magnitude and direction of magnetic field at point O.

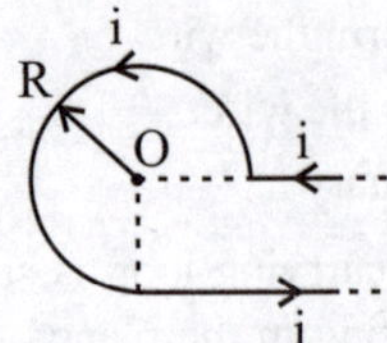

1) $\frac{\mu_0 i}{4R}\left(\frac{1}{\pi}+\frac{3}{2}\right)$, $\odot$
2) $\frac{\mu_0 i}{2R}\left(\frac{1}{\pi}+\frac{3}{2}\right)$, $\odot$
3) $\frac{\mu_0 i}{R}\left(\frac{1}{\pi}+\frac{3}{2}\right)$, $\odot$
4) $\frac{\mu_0 i}{2R}\left(\frac{1}{\pi}+\frac{3}{2}\right)$, $\otimes$

17. Evaluate the magnitude and direction of magnetic field at point O.

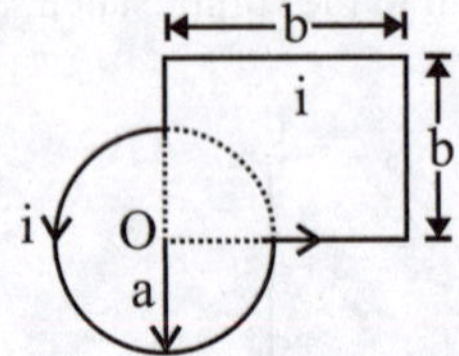

1) $\frac{3\mu_0 i}{8a}$, $\odot$
2) $\frac{\mu_0 i}{2\sqrt{2}b}$, $\odot$
3) $\frac{3\mu_0 i}{8a}+\frac{\mu_0 i}{2\sqrt{2}\pi b}$ $\odot$
4) $\frac{3\mu_0 i}{a}+\frac{\mu_0 i}{2\sqrt{2}b}$, $\otimes$

18. A long straight wire of radius a carries a steady current I. The current is uniformly distributed across its cross-section. Find the ratio of the magnetic field at $\frac{a}{2}$ and 2a.

1) 1
2) 2
3) 3
4) 4

19. In the loop shown, the magnetic induction at the point 'O' is

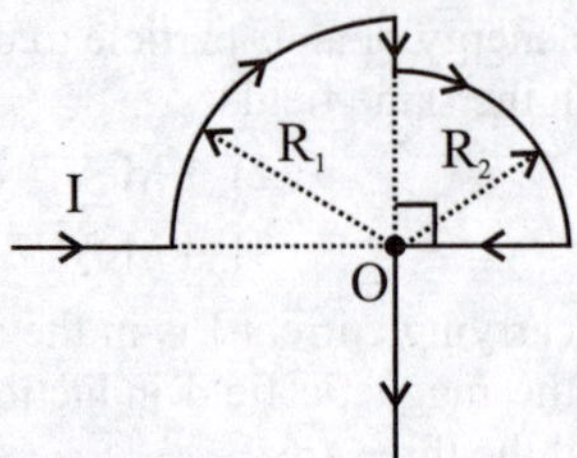

1) $\frac{\mu_0 I}{8}\left(\frac{R_1-R_2}{R_1R_2}\right)$
2) $\frac{\mu_0 I}{8}\left(\frac{R_1+R_2}{R_1R_2}\right)$
3) $\frac{\mu_0 I}{8}\left(\frac{R_1R_2}{R_1+R_2}\right)$
4) 0

20. A long hollow cylinderical conductor of radii r_1 and r_2 carries a current such that the current density is given by $J = ar^2$ over its cross-section. The magnetic field in the region $0 \le r \le r_1$.

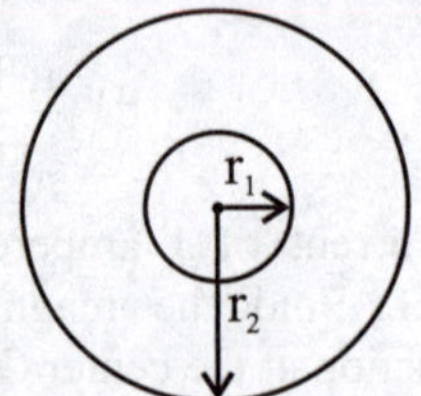

1) $\frac{\mu_0 a\left(r_2^4-r_1^4\right)}{4r}$
2) 0
3) $\frac{\mu_0 a\left(r_2^3-r_1^3\right)}{3r}$
4) None of these

Statement Based MCQs

Directions: These questions consist of two statements each, printed as Statement-I and Statement-II. While answering these questions, you are required to choose any one of the following four responses.

1) **Both Statement-I and Statement-II are correct.**
2) **Both Statement-I and Statement-II are incorrect.**
3) **Statement-I is correct & Statement-II is incorrect.**
4) **Statement-I is incorrect & Statement-II is correct.**

1. Statement-I: An electric charge in uniform motion produces both electric and magnetic field.

Statement-II: The magnitude of charge increase with the velocity of charge.

2. Statement-I: A moving charge produces magnetic field, but no electric field.

Statement-II: For Ampere's law $\oint \vec{B}.d\vec{\ell} = \mu_0 I$.

3. Statement-I: A solenoid tends to expand when a current passes through it.

Statement-II: Two straight parallel metallic wires carrying current in same direction repel each other.

4. Statement-I: When a charged particle moves in a magnetic field, the kinetic energy of charged particle remains constant.

Statement-II: Work done by magnetic force on moving charged particle is zero.

5. Statement-I: A positive charge q of mass, moving with a speed v, enters a region with uniform magnetic flux density B in a direction perpendicular to the field. It subsequently moves in a path of radius $\frac{mv}{qB}$.

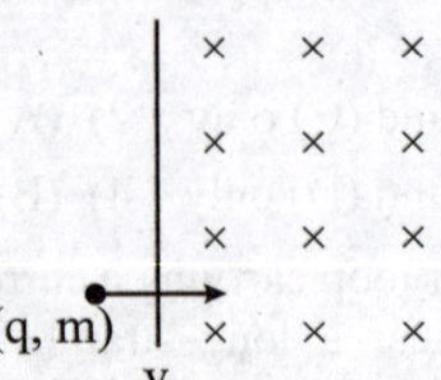

Statement-II: It comes out of the region at a distance of $\frac{2mv}{qB}$ from its entry point.

Assertion & Reason MCQs

Directions: These questions consist of two statements each, printed as Assertion and Reason. While answering these questions, you are required to choose any one of the following four responses.

1) **Both Assertion and Reason are True and the Reason is a correct explanation of the Assertion.**
2) **Both Assertion and Reason are True but Reason is not a correct explanation of the Assertion.**
3) **Assertion is True but the Reason is False.**
4) **Assertion is False but Reason is True.**

1. **Assertion:** An ammeter is connected in series in an electrical circuit and it should have a low resistance.

   Reason: The introduction of the ammeter must not affect the main current.

2. **Assertion:** A circular loop carrying current lies in XY plane with its center at origin having a magnetic flux in negative z-axis.

   Reason: Magnetic flux direction is independent of the direction of current in the conductor.

3. **Assertion:** If a proton and α-particle enter a uniform magnetic field perpendicularly, with the same speed, then the time period of revolution of the α-particle is double than that of proton.

   Reason: In a magnetic field, the time period of revolution of a charged particle is directly proportional to mass.

4. **Assertion:** In Ampere's circuital law the current outside the amperian loop is not included in the equation.

   Reason: Magnetic field calculated using Ampere's law is due to the current flowing inside the closed loop.

Match the Columns MCQs

1. An electron is moving towards positive x-direction. Match the following two columns for deflection of electron just after the fields are switched on. (E_0 & B_0 are positive constants)

	Column-I		Column-II
(a)	If magnetic field $B = B_0\hat{j}$ is switched on	(i)	Negative y-axis
(b)	If magnetic field $B = B_0\hat{k}$ is switched on	(ii)	Positive y-axis
(c)	If magnetic field $B = B_0\hat{i}$ and electric field $E = E_0\hat{j}$ is switched on	(iii)	Negative z-axis
(d)	If electric field $E = E_0\hat{k}$ is switched on	(iv)	Positive z-axis

1) (a)-(iii) (b)-(ii) (c)-(iv) (d)-(iii)
2) (a)-(iii) (b)-(ii) (c)-(i) (d)-(iii)
3) (a)-(i) (b)-(ii) (c)-(i) (d)-(iii)
4) (a)-(iv) (b)-(ii) (c)-(i) (d)-(iii)

2. In column-I, a current carrying loop and a uniform magnetic field are shown. Match this with column-II.

	Column-I		Column-II
(a)	B, I	(i)	Force = 0
(b)	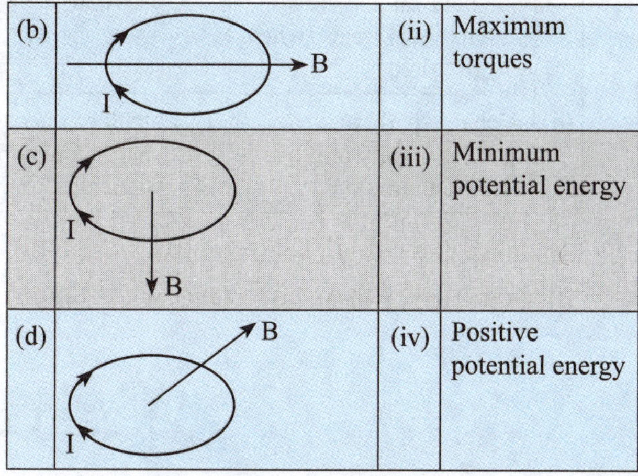 B, I	(ii)	Maximum torques
(c)	I, B	(iii)	Minimum potential energy
(d)	B, I	(iv)	Positive potential energy

1) (a)-(i, iii) (b)-(i, iii) (c)-(i, iii) (d)-(i, ii)
2) (a)-(ii, iii) (b)-(i, iii) (c)-(i, ii) (d)-(i, iv)
3) (a)-(ii, iii) (b)-(i, ii) (c)-(ii, iii) (d)-(i, ii)
4) (a)-(i, iv) (b)-(i, ii) (c)-(i, iii) (d)-(i, iv)

3. Column-I shows the state of motion of a charge particle . Column-II shows the possible combination of electric field and magnetic filed under which the path in Column-I is possible. Match Appropriately

	Column-I		Column-II
(a)	Charge at rest experiences a force.	(i)	$E = 0, B = 0$
(b)	A charge in motion goes undeviated with same velocity.	(ii)	$E \neq 0, B \neq 0$

(c)	A charge in motion goes undeviated with varying speed.	(iii)	$E = 0, B \neq 0$
(d)	A charged particle undergoes helical motion	(iv)	$E \neq 0, B = 0$

1) (a)-(i, ii) (b)-(ii) (c)-(iv) (d)-(iii, iv)
2) (a)-(ii, iv) (b)-(i, ii, iii) (c)-(ii, iv) (d)-(ii, iii)
3) (a)-(i, iv) (b)-(ii, iii) (c)-(iii, iv) (d)-(iv)
4) (a)-(iv) (b)-(ii, iii) (c)-(i, ii) (d)-(ii, iii)

4. Match the Column-I with Column-II

	Column-I		Column-II
(a)	A charge particle is moving in uniform electric and magnetic field in gravity free space.	(i)	Velocity of the particle may be constant.
(b)	A charge particle is moving in uniform electric, magnetic and gravitational fields.	(ii)	Path of the particle may be straight line.
(c)	A charge particle is moving in uniform magnetic and gravitational fields (where electric field is zero)	(iii)	path of the particle may be circular.
(d)	A charge particle is moving in only uniform electric field	(iv)	Path of the particle may be helical.

1) (a)-(i, ii, iv) (b)-(i, ii, iii) (c)-(i, ii, iv) (d)-(ii)
2) (a)-(i, ii, iv) (b)-(ii, iii) (c)-(i, iv) (c)-(iii)
3) (a)-(i, ii) (b)-(i, iv) (c)-(i, ii, iii) (c)-(i, ii, iv)
4) (a)-(i, iv) (b)-(i, ii) (c)-(iv) (c)-(i, iv)

5. Consider a closed loop in the form of a trapezium carrying current I. Match the following regarding the magnitude of magnetic field at point P.

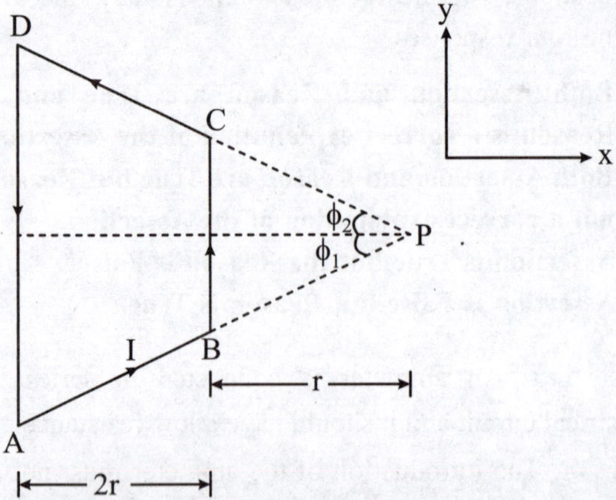

	Column-I		Column-II
(a)	Magnetic field due to AB	(i)	is greater than that due to DA
(b)	Magnetic field due to BC	(ii)	is greater than that due to CD
(c)	Magnetic field due to DA	(iii)	is not equal to zero
(d)	Magnetic field due to complete figure	(iv)	is zero

1) (a)-(i, iv) (b)-(ii) (c)-(i, ii) (d)-(i, ii, iii)
2) (a)-(iv) (b)-(i, ii, iii) (c)-(iv, ii) (d)-(i, ii)
3) (a)-(i, ii) (b)-(ii) (c)-(i, ii) (d)-(ii)
4) (a)-(iv) (b)-(i, ii, iii) (c)-(ii, iii) (d)-(i, ii, iii)

NCERT Exemplar MCQs

1. Two charged particles traverse identical helical paths in a completely opposite sense in a uniform magnetic field $\vec{B} = B_0\hat{k}$:
   1) They have equal z-component of momenta
   2) They must have equal charges
   3) They necessarily represent a particle and anti-particle pair
   4) The charge to mass ratio satisfy $\left(\frac{e}{m}\right)_1 + \left(\frac{e}{m}\right)_2 = 0$

2. Biot-Savart's law indicates that the moving electrons (velocity v) produce a magnetic field $\vec{B}$ such that:
   1) $\vec{B}$ is perpendicular of velocity
   2) $\vec{B}$ is parallel to velocity
   3) It obeys inverse cube law
   4) It is along the line joining the electron and point of observation

3. A current carrying circular loop of radius R is placed in the x-y plane with centre at the origin. Half of the loop with $x > 0$ is now bent so that it now lies in the y-z plane.
   1) The magnitude of magnetic moment now diminishes
   2) The magnetic moment does not change
   3) The magnitude of $\vec{B}$ at (0, 0, z), $z > R$ increases
   4) The magnitude of $\vec{B}$ at (0, 0, z), $z >> R$ is unchanged

4. An electron is projected with uniform velocity along the axis of a current carrying long solenoid. Which of the following is true?
   1) The electron will be accelerated along the axis
   2) The electron path will be circular about the axis
   3) The electron will experience a force at 45° to the axis and hence execute-a helical path
   4) The electron will continue to move with uniform velocity along the axis of the solenoid

5. A circular current loop of magnetic moment M is in an arbitrary orientation in an external magnetic field B. The work done to rotate the loop by 30^0 about an axis perpendicular to its plane is:

1) MB
2) $\sqrt{3}\frac{MB}{2}$
3) $\frac{MB}{2}$
4) Zero

NEET & AIIMS Past Year MCQs

1. A tightly wound 100 turns coil of radius 10 cm carries a current of 7 A. The magnitude of the magnetic field at the centre of the coil is (Take permeability of free space as $4\pi \times 10^{-7}$ SI units): **(2024)**

1) 4.4 mT
2) 44 T
3) 44 mT
4) 4.4 T

2. An iron bar of length L has magnetic moment M. It is bent at the middle of its length such that the two arms make an angle 60° with each other. The magnetic moment of this new magnet is: **(2024)**

1) 2 M
2) $\frac{M}{\sqrt{3}}$
3) M
4) $\frac{M}{2}$

3. A very long conducting wire is bent in a semi-circular shape from A to B as shown in figure. The magnetic field at point P for steady current configuration is given by: **(2023)**

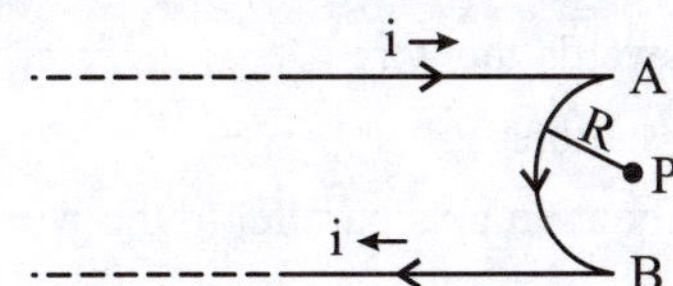

1) $\frac{\mu_0 i}{4R}\left[1-\frac{2}{\pi}\right]$ pointed into the page
2) $\frac{\mu_0 i}{4R}$ pointed into the page
3) $\frac{\mu_0 i}{4R}$ pointed away from the page
4) $\frac{\mu_0 i}{4R}\left[1-\frac{2}{\pi}\right]$ pointed away from page

4. A wire carrying a current I along the positive x-axis has length L. It is kept in a magnetic field $\vec{B} = (2\hat{i} + 3\hat{j} - 4\hat{k})\,T$. The magnitude of the magnetic force acting on the wire is: **(2023)**

1) $\sqrt{3}$ IL
2) 3 IL
3) $\sqrt{5}$ IL
4) 5 IL

5. Given below are two statements: **(2022)**

Statement I: Biot-Savart's law gives us the expression for the magnetic field strength of an infinitesimal current element ($Id\vec{\ell}$) of the current carrying conductor only.

Statement II: Biot-Savart's law is analogous to Coulomb's inverse square law of charge q, with the former being related to the field produced by a scalar source, $Id\vec{\ell}$ while the latter being produced by a vector source, q.

In light of above statements choose the most appropriate answer from the options given below

1) Statement I is incorrect and statement II is correct
2) Both Statement I and Statement II are correct
3) Both statement I and Statement II are incorrect
4) Statement I is correct and Statement II is incorrect.

6. A long solenoid of radius 1 mm has 100 turns per mm. If 1A current flows in the solenoid, the magnetic field strength at the centre of the solenoid is **(2022)**

1) 6.28×10^{-4} T
2) 6.28×10^{-2} T
3) 12.56×10^{-2} T
4) 12.56×10^{-4} T

7. From Ampere's circuital law for a long straight wire of circular cross section carrying a steady current, the variation of magnetic field in the inside and outside region of the wire is **(2022)**

1) A linearly decreasing function of distance upto the boundary of the wire and then a linearly increasing one for the outside region.
2) Uniform and remains constant for both the regions.
3) A linearly increasing function of distance upto the boundary of the wire and then decreasing outside region.
4) A linearly increasing function of distance r upto the boundary of the wire 1/r dependence for the outside region.

8. An infinitely long straight conductor carries a current of 5 A as shown. An electron is moving with a speed of 10^5 m/s parallel to the conductor. The perpendicular distance between the electron and the conductor is 20 cm at an instant. Calculate the magnitude of the force experienced by the electron at that instant. **(2021)**

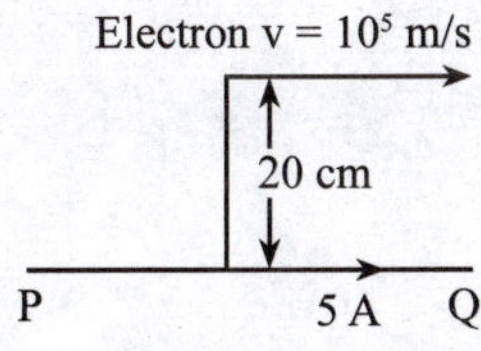

1) $8\pi \times 10^{-20}$ N
2) $4\pi \times 10^{-20}$ N
3) 8×10^{-20} N
4) 4×10^{-20} N

9. A thick current carrying cable of radius 'R' carries current 'I' uniformly distributed across its cross-section. The variation of magnetic field B(r) due to the cable with the distance 'r' from the axis of the cable is represented by: **(2021)**

1) 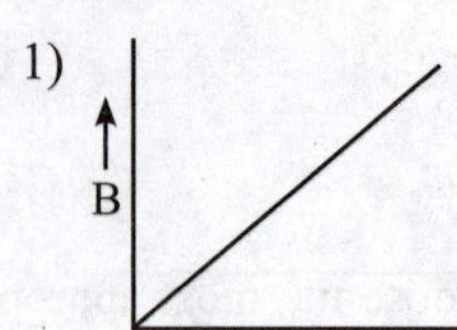

2) 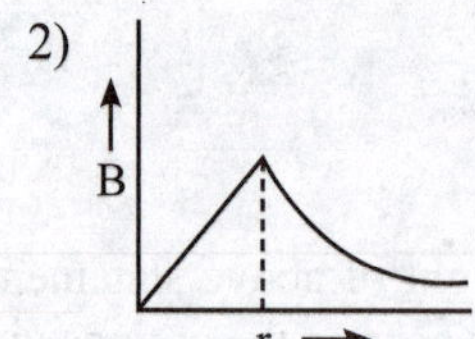

3) 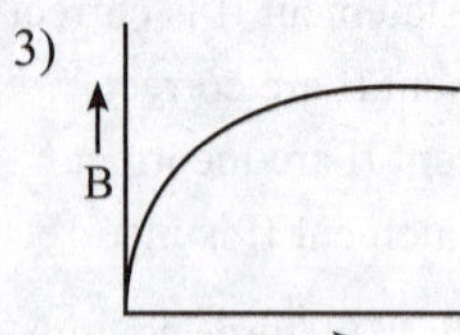

4) 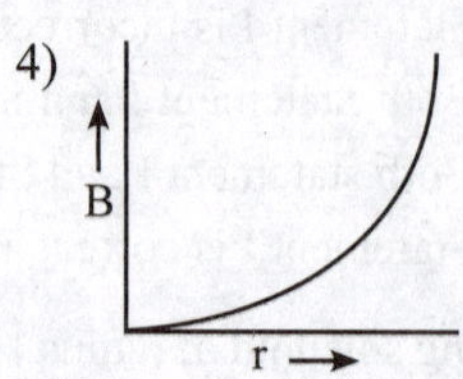

10. In the product **(2021)**

$\vec{F} = q(\vec{v} \times \vec{B})$

$= q\vec{v} \times (B\vec{i} + B\vec{j} + B_0\vec{k})$

For q = 1 and $\vec{v} = 2\hat{i} + 4\hat{j} + 6\hat{k}$ and

$\vec{F} = 4\vec{i} - 20\vec{j} + 12\vec{k}$

What will be the complete expression for $\vec{B}$?

1) $-6\hat{i} - 6\hat{j} - 8\hat{k}$
2) $8\hat{i} + 8\hat{j} - 6\hat{k}$
3) $6\hat{i} + 6\hat{j} - 8\hat{k}$
4) $-8\hat{i} - 8\hat{j} - 6\hat{k}$

11. Two changes $+q$ and $-q$ are attached to the two ends of a light rod of length L, as shown in figure. The system is given a velocity v perpendicular to magnetic field B. The magnetic force on the system of charges and magnitude of force on one charge by the rod, are respectively: **(AIIMS 2008)**

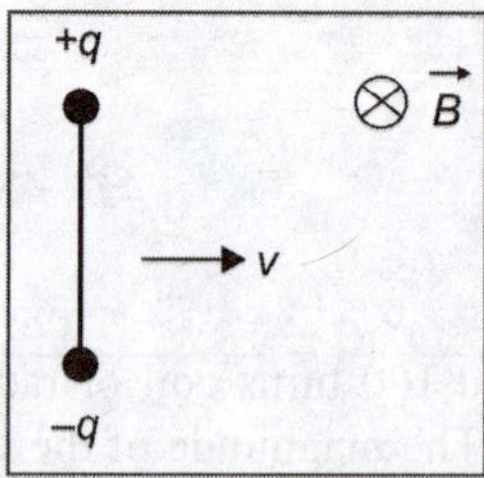

1) Zero, zero
2) Zero, qvB
3) $2qvB$, 0
4) $2qvB$, qvB

12. The ratio of magnetic fields on the axis of a circular current carrying coil of radius a to the magnetic field at its centre will be: **(AIIMS 2008)**

1) $1/(1 + x^2/a^2)^{3/2}$
2) $1/(1 + a^2/x^2)^{1/2}$
3) $1/(1 + a^2/x^2)^2$
4) $1/(1 + a^2/x^2)^3$

13. A rectangular loop carrying a current i_1, is situated near a long straight wire carrying a steady current i_2. The wire is parallel to one of the sides of the loop and is in the plane of the loop as shown in the figure. Then the current loop will:- **(AIIMS 2003)**

1) Move away from the wire
2) Move towards the wire
3) Remain stationary
4) Rotate about an axis parallel to the wire

Answer Key

NCERT Topic-wise MCQs

1	2	3	4	5	6	7	8	9	10	11	12	13	14	15	16	17
2	2	2	3	1	2	4	2	4	1	2	3	1	1	1	1	4
18	**19**	**20**	**21**	**22**	**23**	**24**	**25**	**26**	**27**	**28**	**29**	**30**	**31**	**32**	**33**	**34**
2	3	1	1	3	3	1	1	4	2	4	2	3	4	2	1	2
35	**36**	**37**	**38**	**39**	**40**	**41**	**42**	**43**	**44**	**45**	**46**	**47**	**48**	**49**		
3	1	3	2	1	4	2	4	4	1	2	1	1	1	4		

High Order Time Intensive MCQs

1	2	3	4	5	6	7	8	9	10	11	12	13	14	15	16	17
3	1	2	4	2	2	2	2	1	3	3	2	4	2	4	1	3
18	**19**	**20**														
1	2	2														

Statement Based MCQs

1	2	3	4	5
3	4	2	1	1

Assertion & Reason MCQs

1	2	3	4
1	3	2	2

Match the Column MCQs

1	2	3	4	5
2	4	2	1	4

NCERT Exemplar MCQs

1	2	3	4	5
4	1	1	4	4

NEET & AIIMS Past Year MCQs

1	2	3	4	5	6	7	8	9	10	11	12	13
1	4	4	4	4	3	4	3	2	1	2	1	2

CHAPTER 5

Magnetism and Matter

Topicwise Number of Questions Analysis (2025-2016)

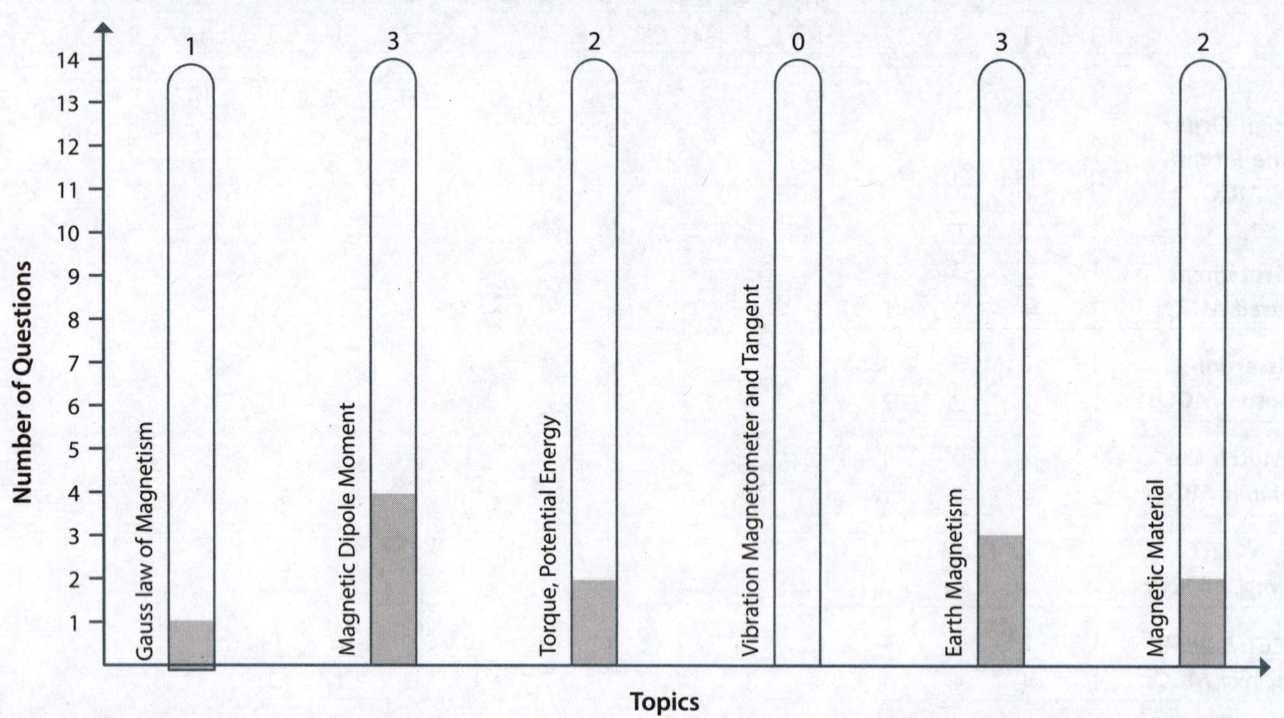

NCERT-PICKS

Introduction

A magnet is a material that has both attractive and directive properties.

Magnet attracts small pieces of iron, nickel, cobalt etc. This property of attraction is called magnetism.

The Bar Magnet

Formulas

1. Magnetic field at any point on the axis of a bar magnet $\vec{B} = \frac{\mu_0}{4\pi}\frac{2\vec{M}}{r^3}$
2. Magnetic field on equatorial point of the bar magnet, $\vec{B} = \frac{\mu_0}{4\pi}\frac{\vec{M}}{r^3}$ [$\vec{M}$ is magnetic dipole moment]
3. Torque $\vec{\tau} = \vec{M} \times \vec{B}$ [$\vec{M}$ is magnetic moment and $\vec{B}$ is uniform magnetic field]
4. Potential Energy = $-\vec{M} \cdot \vec{B} = -MB\cos\theta$
5. Magnetic moment of the bar magnet, $M = ml$ [m is the pole strength of the bar magnet and l is the length of the bar magnet]
6. $\frac{\text{Magnetic Length}}{\text{Geometrial Length}} = 0.84$

Numerical Profile

Magnetic field due to bar magnet

(i) Magnetic field on the axial line

$$\vec{B}_S = \frac{\mu_0}{4\pi}.\frac{m}{(r+\ell)^2} \text{ along PS} \qquad \text{[m is the pole strength]}$$

$$\vec{B}_N = \frac{\mu_0}{4\pi}\frac{m}{(r-\ell)^2} \text{ along NP}$$

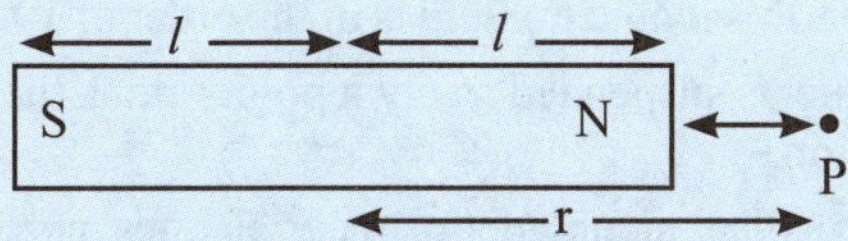

$$\vec{B}_{ax} = \vec{B}_N - \vec{B}_S = \frac{\mu_0}{4\pi}\frac{2Mr}{\left(r^2-\ell^2\right)^2}$$

(ii) Magnetic field on the equatorial line

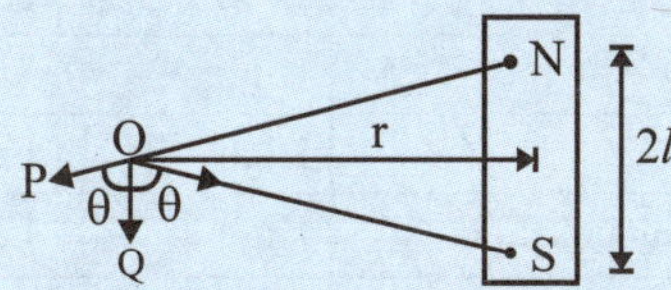

Net magnetic field at point O,

$$\vec{B} = \frac{\mu_0}{4\pi}\frac{\vec{M}}{(r^2+\ell^2)^{3/2}}$$

(iii) Magnetic field at a general position

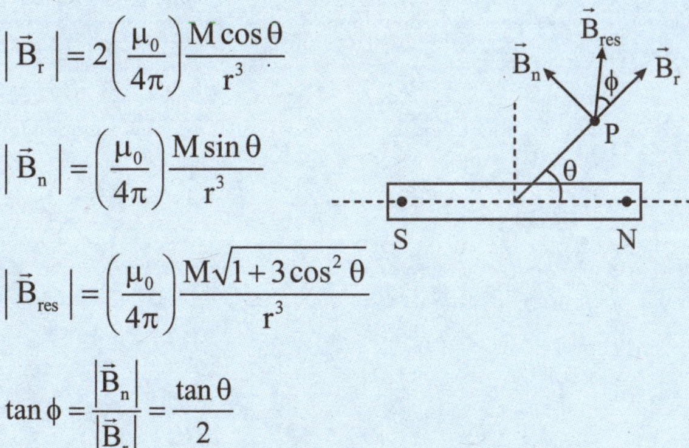

$$\left|\vec{B}_r\right| = 2\left(\frac{\mu_0}{4\pi}\right)\frac{M\cos\theta}{r^3}$$

$$\left|\vec{B}_n\right| = \left(\frac{\mu_0}{4\pi}\right)\frac{M\sin\theta}{r^3}$$

$$\left|\vec{B}_{res}\right| = \left(\frac{\mu_0}{4\pi}\right)\frac{M\sqrt{1+3\cos^2\theta}}{r^3}$$

$$\tan\phi = \frac{\left|\vec{B}_n\right|}{\left|\vec{B}_r\right|} = \frac{\tan\theta}{2}$$

Magnetism and Gauss's Law

Formulas

1. Horizontal component, $B_H = B\cos\delta$

   [where, δ is the angle of dip and B is earth's magnetic field]

2. Vertical component, $B_V = B\sin\delta$

3. Resultant magnetic field, $B = \sqrt{B_H^2 + B_V^2}$

4. Tangent of angle of dip, $\tan\delta = \dfrac{B_V}{B_H}$

5. The net magnetic flux through any closed surface

   $$\phi_B = \oint \vec{B}\cdot d\vec{S} = 0$$

Numerical Profile

Incoming flux = Outgoing flux.

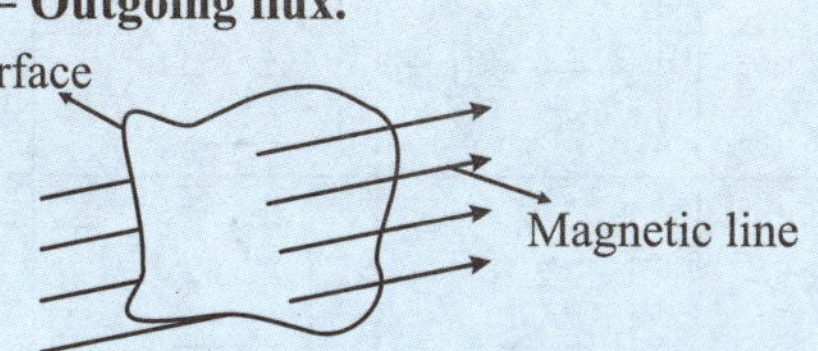

Magnetisation and Magnetic Intensity

Formulas

1. Magnetising field, $\vec{B} = \mu\vec{H}$

2. Intensity of magnetisation $\vec{I} = \dfrac{\text{Magnetic moment}}{\text{Volume}} = \dfrac{\vec{M}}{V}$

3. Magnetic permeability, $\mu_0 = \dfrac{\vec{B}}{\vec{H}}$

   [$\vec{B}$ is magnetic intensity and $\vec{H}$ is magnetising field]

4. Susceptibility, $\chi = \dfrac{\vec{I}}{\vec{H}}$ ($\vec{I}$ = intensity of Magnetization)

5. Relative permeability, $\mu_r = \dfrac{\mu}{\mu_0}$

   [μ_0 is the permeability of the vaccum, μ is the permeability of the material]

6. Permeability of the material, $\mu = \mu_0(1+\chi)$

7. Magnetic susceptibility, $\chi_m = \dfrac{C}{T}$

   [C is Curie constant and T is temperature]

8. $\vec{B} = \mu_0\left(\vec{H} + \vec{M}\right)$

Magnetic Properties of Materials

Formulas

1. Susceptibilities value,
   (i) Diamagnetic, $\chi < 1$
   (ii) Paramagnetic, $\chi > 1$
   (iii) Ferromagnetic, $\chi >>> 1$
2. Relative permeability value,
   (i) Diamagnetic, $0 \le \mu_r < 1$
   (ii) Paramagnetic, $1 < \mu_r < 1 + \varepsilon$
   (iii) Ferromagnetic, $\mu_r > 1000$

Numerical Profile

Property	Diamagnetic	Paramagnetic	Ferromagnetic
M – H curve	M, H	M, H	M, H

Property	Diamagnetic	Paramagnetic	Ferromagnetic
I – H curve	I, H	I, H	I, H
χ_m – T curve	χ_m, T	χ_m, T	χ_m, T_C, T

Statements Asked In Exam

- Electric lines of forces are discontinuous whereas magnetic lines of forces are closed continuous curves.
- Knowledge of declination at a place help in navigation.
- Earth's magnetic field extends upto a distance of 32000 km, which is about five times the radius of earth.
- Diamagnetic substances can be compared to non-polar dielectrics.
- Diamagnetism is a universal property of all substances.
- The susceptibility of a diamagnetic substance is independent of temperature
- For making electromagnet, coercivity should be low.
- The core of electromagnets are made up of ferromagnetic materials, which have high permeability and low retentivity. Soft iron is suitable for this purpose.
- Soft iron is used to make an electromagnet.
- Ferromagnetic materials have been divided into two types:
  - (i) Soft magnetic material:
    - (a) low retentivity, low coercivity
  - (ii) Hard magnetic material:
    - (a) High retentivity, High coercivity
- Strength of earth's magnetic field is of the order of 10^{-5} Tesla.
- When a magnet is suspended freely it points along the North-South direction.
- Susceptibility of a diamagnetic material does not depend upon temperature.

	Parameter	Soft Iron	Steel
1.	Retentivity	High	Low
2.	Coercivity	Low	High
3.	Permeability	High	Low
4.	Susceptibility	High	Low
5.	Intensity of magnetisation	High	Low
6.	Hystersis loss	Low	High

NCERT Topic-wise MCQs

Bar Magnet, Magnetic Dipole Moment, Torque and Potential Energy

1. A magnetised steel wire 31.4 cm long has pole strength of 0.2 Am. It is then bent in the form of a semicircle. What is the magnetic moment of the semicircle?

   1) 0.04 Am^2 2) 0.05 Am^2
   3) 0.06 Am^2 4) 0.07 Am^2

2. A short bar magnet has magnetic moment of 50 Am^2. Calculate the magnetic field intensity at a distance of 0.2m from its centre on axial line.

   1) 1.5×10^{-4} T 2) 1.25×10^{-4} T
   3) 1.5×10^{-3} T 4) 1.25×10^{-3} T

3. A magnetised needle of magnetic moment 4.8×10^{-2} Am^2 is placed at 30^0 with the direction of uniform magnetic field of magnitude 3×10^{-2} T. What is the torque acting on the needle?

   1) 7.2×10^{-1} N–m 2) 7.2×10^{-2} N–m
   3) 7.2×10^{-4} N–m 4) 7.2×10^{-3} N–m

4. A circular coil of 50 turns and radius 20 cm carries a current of 12A. Find magnetic moment associated with it:

   1) 72.5 Am^2 2) 75.4 Am^2
   3) 76.6 Am^2 4) 70 Am^2

5. A bar magnet is placed in the position of stable equilibrium in a uniform magnetic field of induction B. If it is rotated through an angle 180°, then the work done is:

   1) MB 2) 2 MB
   3) MB/2 4) Zero

6. A bar magnet has a coercivity 4 × 103 Am–1. It is desired to demagnetize it by inserting it inside a solenoid 12 cm long and having 60 turns. The current carried by the solenoid should be

   1) 8 A 2) 6 A
   3) 4.5 A 4) 2 A

7. A thin bar magnet of length 2L is bent at the mid point so that the angle between them is 60^0. The new effective length of the magnet is:

   1) $\sqrt{2}L$ 2) $\sqrt{3}L$
   3) 2L 4) L

8. Torque acting on a magnet of length 10cm and pole strength 15 A–m kept in a field of $B = 2 \times 10^{-5}$T at an angle of 30^0 is:

   1) 1.5×10^{-5} Nm 2) 1.5×10^{-3} Nm
   3) 1.5×10^{-2} Nm 4) 1.5×10^{-6} Nm

9. The magnetic field at a distance x on the axis of a small bar magnet is equal to the field at a distance y on the equator of the same magnet. The ratio of the distances x and y from the center of the magnet is:

   1) 2^{-3} 2) $2^{-1/3}$
   3) 2^3 4) $2^{1/3}$

10. A short bar magnet placed with its axis at 30° with a uniform external magnetic field of 0.35 T experiences a torque of magnitude equal to 4.5×10^{-2} N m. the magnitude of magnetic moment of the given magnet is

    1) 26 $J T^{-1}$ 2) 2.6 $J T^{-1}$
    3) 0.26 $J T^{-1}$ 4) 0.026 $J T^{-1}$

11. A magnetic needle lying parallel to a magnetic field requires W units of work to turn it through 60^0. The torque required to maintain the needle in this position will be:

    1) $\sqrt{3}W$ 2) W
    3) $\frac{\sqrt{3}W}{2}$ 4) 3W

12. Two magnets of equal magnetic moments m each are placed as shown in the figure. The resultant magnetic moment is

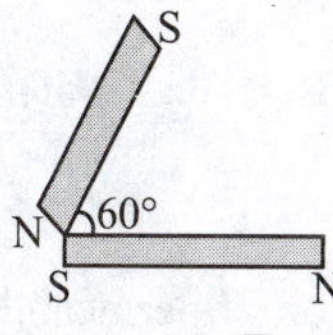

    1) M 2) $\sqrt{3}$ M
    3) $\sqrt{2}$ M 4) M/2

13. A straight wire carrying current I is turned into a circular loop. If the magnitude of magnetic moment associated with it in M.K.S unit is M, then length of the wire will be:

    1) $4\pi IM$ 2) $\sqrt{4\pi IM}$
    3) $\sqrt{\frac{4\pi M}{I}}$ 4) $\frac{M\pi}{4I}$

14. Two magnets each of magnetic moment 'M' are placed so as to form a cross at right angles to each other. The magnetic moment of the system will be:

    1) 2M 2) $\sqrt{2}M$
    3) $\frac{M}{2}$ 4) M

15. Two like magnetic poles of strength 10Am and 40Am are separated by a distance 30cm. The intensity of magnetic field is zero on the line joining them:
1) At a point 10 cm from the stronger pole
2) At a point 20 cm from the stronger pole
3) At the mid-point
4) At infinity

16. A bar magnet of magnetic moment 3.0 Am^2 is placed in a uniform magnetic field of 2×10^{-5}T. If each pole of the magnet experience a force of 6×10^{-4} N, the length of the magnet is:
1) 0.5 m 2) 0.3 m
3) 0.2 m 4) 0.1 m

17. A bar magnet of length 3 cm has a point A and B along its axis at distance of 24 cm and 48 cm from the mid-point of bar magnet. Ratio of magnetic fields at these points will be:

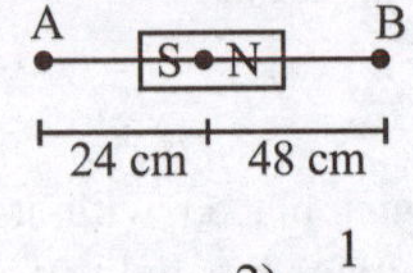

1) 8 2) $\frac{1}{\sqrt{2}}$
3) 3 4) 4

18. A current carrying loop is placed in a uniform magnetic field in four different orientations as shown in figure. Arrange them in the decreasing order of potential energy.

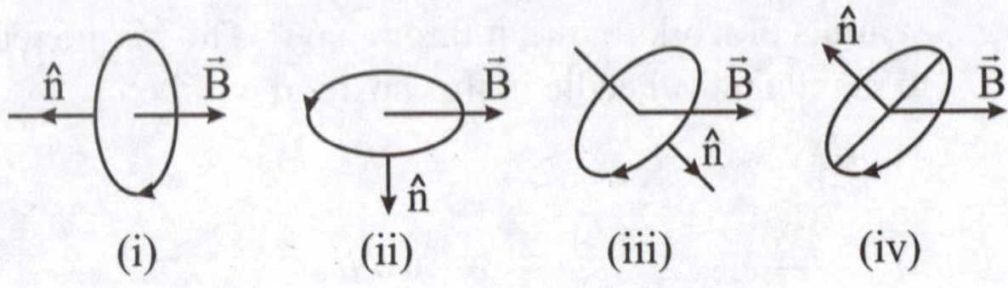

1) (iv), (ii), (iii), (i) 2) (i), (iv), (ii), (iii)
3) (iv), (iii), (ii), (i) 4) (i), (ii), (iii), (iv)

Magnetisation and Magnetic Intensity

19. A specimen of iron of permeability 8×10^{-3} weber/A m is placed in a magnetic field of strength 160 A/m. Magnetic induction in this iron is
1) 0.8 Wb/m^2 2) 5×10^{-5} Wb/m^2
3) 1.28 Wb/m^2 4) 20×10^3 Wb/m^2

20. A magnetising field of 1600 Am^{-1} produces a magnetic flux 2.4×10^{-5} weber in an iron bar of cross-sectional area 0.2 cm^2. The susceptibility of the iron bar will be
1) 1788 2) 1192
3) 596 4) 298

21. The magnetic moment produced in a substance of 1 g is 6×10^{-7} Am^2. If its density is 5g/cm^3, then the intensity of magnetisation (in A/m) will be:
1) 8.3×10^6 2) 3
3) 1.2×10^{-7} 4) 3×10^{-6}

22. The paramagnetic substance of susceptibility 3×10^{-4} is placed in a magnetic field of 4×10^{-4} Am^{-1}. Then the intensity of magnetisation in the unit of Am^{-1} is:
1) 1.33×10^8 2) 0.75×10^{-8}
3) 12×10^{-8} 4) 14×10^{-8}

23. A magnetising field of 2×10^3 amp/m produces a magnetic flux density of 8π testla in an iron rod. The relative permeability of the rod will be
1) 10^4 2) 10^3
3) 10^0 4) 10^2

24. The relation connecting magnetic susceptibility χ_m and relative permeability μ_r is
1) $\chi_m = \mu_r + 1$ 2) $\chi_m = \mu_r - 1$
3) $\chi_m = \frac{1}{\mu_r}$ 4) $\chi_m = 3(1+\mu_r)$

25. Point out the best representation of relation between magnetic susceptibility (χ) and temperature (T) for a paramagnetic material.

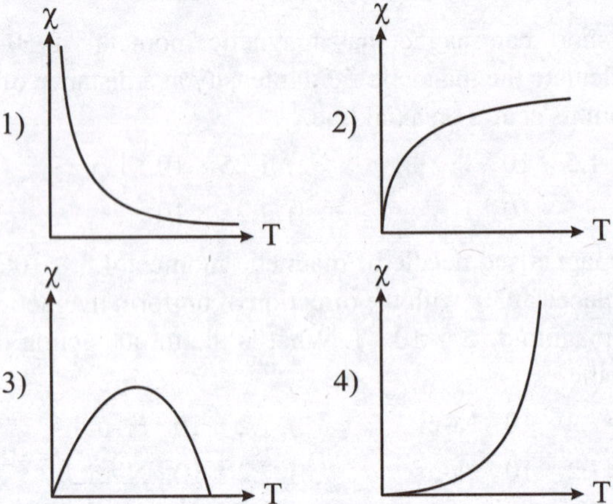

Magnetic Materials

26. The susceptibility of magnesium at 300K is 1.2×10^{-5}. At what temperature will the susceptibility be equal to 1.44×10^{-5}?
1) 260K 2) 255K
3) 250K 4) 400K

27. What is the dimensions of susceptibility?
1) (MLT^{-2}A^{-2}) 2) (ML^{0}T^{-3}A^{-1})
3) No dimension 4) (T^{-3}A^2)

28. The work done in rotating a bar magnet of magnetic moment M from its unstable equilibrium position to its stable equilibrium position in a uniform magnetic field B is:
1) MB 2) 2 MB
3) – MB 4) – 2 MB

29. The susceptibility of annealed iron at saturation is 5500. Find the permeability of annealed iron at saturation.
1) 7×10^{-3} TmA^{-1} 2) 6.9×10^{-3} TmA^{-1}
3) 3.2×10^{-3} TmA^{-1} 4) 2×10^{-4} TmA^{-1}

30. The magnetic field B and the magnetic intensity H in a material are found to be 1.6T and 1000 A/m respectively. Calculate the relative permeability μ_r and the susceptibility χ of material.

1) 1.3×10^{-3} each
2) 1.3×10^{3} each
3) 3.1×10^{-3} each
4) 3.1×10^{3} each

31. Which of the following is true?

1) Diamagnetism is temperature dependent
2) Paramagnetism is temperature dependent
3) Paramagnetism is temperature independent
4) None of these

32. For an isotropic medium B, μ, H and M are related as:

1) $(B-M) = \mu_0 H$
2) $M = \mu_0 (H+M)$
3) $H = \mu_0 (H+M)$
4) $B = \mu_0 (H+M)$

33. The universal property of all substances is:

1) Diamagnetism　2) Paramagnetism
3) None　4) Ferromagnetism

34. A curve between magnetic moment and temperature of magnet is:

1) 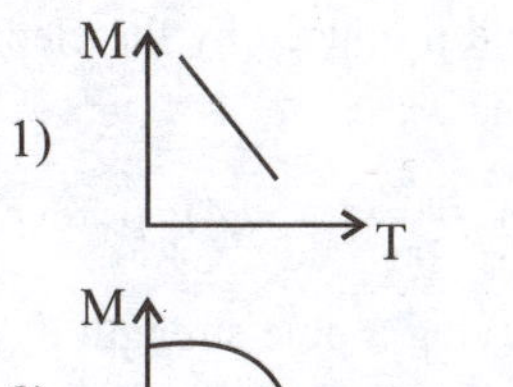

2) 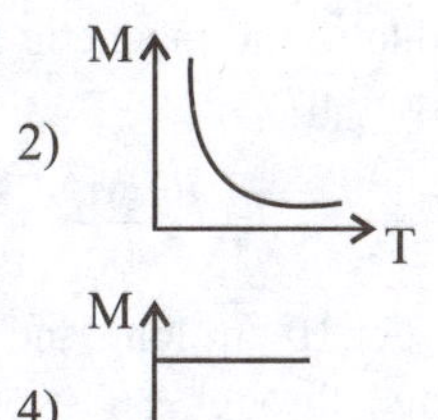

3) M, T

4) M, T

35. The magnetic susceptibility of a paramagnetic substance at -73^0C is 0.0060, then its value at -173^0C will be:

1) 0.0030　2) 0.0012
3) 0.0120　4) 0.041

36. The relative permeability of iron is 6000. Its magnetic susceptibility is:

1) 5999　2) 6001
3) 6000×10^{-7}　4) 6000×10^{7}

37. The magnetic susceptibility of paramagnetic material is:

1) Positive but very high　2) Negative but small
3) Negative but very high　4) Positive but small

38. Which one is not diamagnetic substance?

1) Cu　2) Sb
3) Bi　4) Cr

High Order Time Intensive MCQs

1. The bar magnet of magnetic moment 1.5 J/T is aligned with the direction of a uniform magnetic field of 0.22 T. The work done in turning the magnet so as to align its magnetic moment opposite to the field and the torque acting on it in this position are respectively.

1) 0.66 J, 0　2) 0.33 J, 0
3) 0.66 J, 6.66 N-m　4) 0.33 J, 6.66 N-m

2. The work done in turning a magnet of magnetic moment M by an angle of 90° from the meridian is n times the corresponding work done to turn it through an angle of 60°:

1) $n = 1/2$　2) $n = 2$
3) $n = 1/4$　4) $n = 1$

3. Work done in rotating a bar magnet from 0 to angle θ is:

1) $MH(1 - \cos\theta)$　2) $\frac{M}{H}(1-\cos\theta)$
3) $\frac{M}{H}(\cos\theta-1)$　4) $MH(\cos\theta - 1)$

4. A magnetic needle suspended parallel to a magnetic field requires $\sqrt{3}$ J of work to turn it through 60°. The torque needed to maintain the needle in this position will be:

1) $2\sqrt{3}$ J　2) 3 J
3) $\sqrt{3}$ J　4) $\frac{3}{2}$ J

5. A magnetic needle is kept in a non-uniform magnetic field. It experiences:

1) A force only but not a torque
2) A force and torque both
3) A torque only but not a force
4) Neither a torque nor a force

6. A bar magnet of magnetic moment M_1 is axially cut into two equal parts. If these two pieces are arranged perpendicular to each other, the resultant magnetic moment is M_2. Then, the value of M_1/M_2 is.

1) $\frac{1}{2\sqrt{2}}$　2) 1
3) $\frac{1}{\sqrt{2}}$　4) $\sqrt{2}$

7. If the magnetic susceptibility of a material is large and positive. The material is:

1) Diamagnetic　2) Ferromagnetic
3) Paramagnetic　4) Perfect diamagnetic

8. A paramagnetic substance is placed in a weak magnetic field and its absolute temperature T is increased. As a result, its magnetization:

1) Increases in proportion to T
2) Decreases in proportion to T
3) Increases in proportion to T^2
4) Decreases in proportion to $1/T^2$

9. An electron revolves in a circular orbit of radius r with constant speed v. What is the magnetic flux through the orbit due to the magnetic field produced by the electron's own motion?

1) $\frac{\mu_0 evr^2}{2\pi}$ 2) $\frac{\mu_0 evr^2}{4\pi}$ 3) $\frac{\mu_0 ev}{4\pi r}$ 4) $\mu_0 \pi evr$

10. A magnet 10 cm long and having a pole strength 2 Am is deflected through 30° from the magnetic meridian. The horizontal component of earth's induction is 0.32×10^{-4} T, then the value of deflecting couple is:

1) 16×10^{-7} Nm 2) 64×10^{-7} Nm
3) 48×10^{-7} Nm 4) 32×10^{-7} Nm

11. A magnet oscillating in a horizontal plane has a time period of 2s at a place where the angle of dip is 30° and 3s at another place where the angle of dip is 60°. The ratio of resultant magnetic fields at the two places is:

1) $\frac{4\sqrt{3}}{7}$ 2) $\frac{4}{9\sqrt{3}}$ 3) $\frac{9}{4\sqrt{3}}$ 4) $\frac{9}{\sqrt{3}}$

12. The magnetic moment of a bar magnet is 0.256 amp.m². Its pole strength is 400 milli amp.m. It is cut into two equal pieces and these two pieces are arranged at right angles to each other with their unlike poles in contact (or like poles in contact). The resultant magnetic moment of the system is

1) $\sqrt{2} \times 256 \times 10^{-3}$ Am² 2) 250×10^{-3}Am²
3) $\frac{256}{\sqrt{2}} \times 10^{-3}$ Am² 4) $\frac{128}{\sqrt{2}} \times 10^{-3}$ Am²

13. A magnet of length 2L and moment 'M' is axially cut into two equal halves 'P' and 'Q'. The piece 'P' is bent in the form of semi circle and 'Q' is attached to it as shown. Its moment is

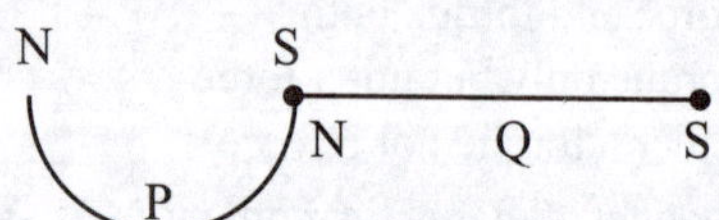

1) $\frac{M}{\pi}$ 2) $\frac{M}{2\pi}$
3) $\frac{M(2+\pi)}{2\pi}$ 4) $\frac{M\pi}{(2+\pi)}$

14. A bar magnet of magnetic moment 'M' is bent in the form of an arc which makes angle 60°. The percentage change in the magnetic moment is

1) 9% Increase 2) 9% Decrease
3) 4.5% Decrease 4) 4.5% Increase

15. A bar magnet of magnetic moment 3.0 amp.m² is placed in a uniform Magnetic induction field 2×10^{-5} T. If each pole of the magnet experience a force of 6×10^{-4} N, the length of the magnet is

1) 0.5 m 2) 0.3 m
3) 0.2 m 4) 0.1 m

16. The magnetic induction at a distance 'd' from the magnetic pole of unknown strength 'm' is B. If an identical pole is now placed at a distance of 2d from the first pole, the force between the two poles is

1) mB 2) $\frac{mB}{2}$
3) $\frac{mB}{4}$ 4) 2mB

17. Two identical north poles each of strength m are kept at vertices A and B of an equilateral triangle ABC of side a. The mutual force of repulsion between them has a magnitude of F. The magnitude of magnetic induction at C is

1) F/m 2) $F/\sqrt{3}m$
3) F/3m 4) $\sqrt{3}F/m$

18. A bar magnet of magnetic moment M_1 is suspended by a wire in a magnetic field. The top of the wire is twisted through 180°, then the magnet is rotated through 45°. Under similar conditions another magnet of magnetic moment M_2 is rotated through 30°, the ratio M_1:M_2 is

1) $9:10\sqrt{2}$ 2) $1:\sqrt{2}$
3) 1 : 1 4) 1 : 3

19. Two wires of same length are shaped into square and a circle. If they carry same current, ratio of the magnetic moments is:

1) 2 : π 2) π : 2
3) π : 4 4) 4 : π

20. Two identical bar Magnets each having magnetic moment of 'M' are kept at a distance of 2d with their axes perpendicular to each other in a horizontal plane. The Magnetic induction at midway between them is

1) $\frac{\mu_o}{4\pi}.(\sqrt{2})\frac{M}{d^3}$ 2) $\frac{\mu_o}{4\pi}.(\sqrt{3})\frac{M}{d^3}$
3) $\frac{\mu_o}{4\pi}.\frac{M}{d^3}$ 4) $\frac{\mu_o}{4\pi}.(\sqrt{5})\frac{M}{d^3}$

21. A dipole of magnetic moment $\vec{M} = 30\hat{j}$ A m² is placed along the y-axis in a uniform magnetic field $\vec{B} = (2\hat{i} + 5\hat{j})$ T. The torque acting on it is

1) $-40\hat{k}$ N m 2) $-50\hat{k}$ N m
3) $-60\hat{k}$ N m 4) $-70\hat{k}$ N m

22. Two short bar magnets of equal dipole moments 'M' each are fastened perpendicular at their centers as shown in figure. The magnitude of the magnetic field at 'P' at a distance d from their common center as shown in figure is

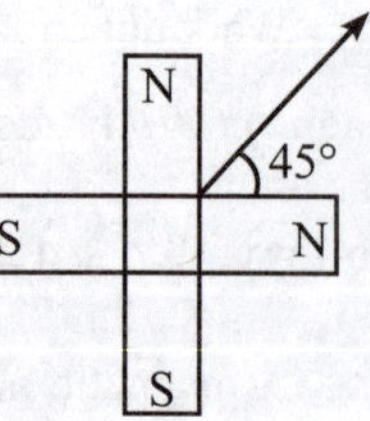

1) $\frac{\mu_0}{4\pi}\frac{M}{d^3}$ 2) $\frac{\mu_0}{4\pi}\frac{2\sqrt{2}M}{d^3}$
3) $\frac{\mu_0}{4\pi}\frac{2M}{d^3}$ 4) $\frac{\mu_0}{2\pi}\frac{M}{d^3}$

23. A short bar magnet is placed with its south pole facing geographic south and the distance between the null points is found to be 16 cm. When the magnet is turned pole to pole at the same place then the distance between the null points will be

1) Will be same along the axial line
2) Will be same along the equatorial line
3) Will be $16 \times 2^{1/3}$ times on the axial line
4) Will be $16 \times 2^{1/3}$ times on the equatorial line

24. A magnetic dipole is under the influence of two magnetic fields. The angle between the field directions is 60° and one of the fields has a magnitude of 1.2×10^{-2} T. If the dipole comes to stable equilibrium at an angle of 30° with this field, then the magnitude of the fields is

1) 1.2×10^{-4} T
2) 2.4×10^{-4} T
3) 1.2×10^{-2} T
4) 2.4×10^{-2} T

25. A bar magnet is placed with its north pole pointing north. Neutral point is at 12 cm. Another magnet is now placed on the first magnet, then the neutral point is found to be at 8 cm. The ratio of their magnetic moments is

1) 3:2
2) 27:19
3) 9:4
4) 9:5

26. The dipole moment of each molecule of paramagnetic gas is 1.5×10^{-23}Am^2 .The temperature of gas is 27°C and the number of molecules per unit volume in it is 2×10^{26}m^{-3}. The maximum possible intensity of magnetisation in the gas will be (in A/m) is

1) 3×10^3
2) 4×10^{-3}
3) 5×10^5
4) 6×10^{-4}

27. The correct value of dip angle at a place is 45°. If the dip circle is rotated by 45° out of the meridian, then the tangent of the angle of apparent dip at the place is

1) 1
2) 1/2
3) $1/\sqrt{2}$
4) $\sqrt{2}$

Statement Based MCQs

Directions: These questions consist of two statements each, printed as Statement-I and Statement-II. While answering these questions, you are required to choose any one of the following four responses.

1) Statement-I and Statement-II both are correct
2) Statement-I and Statement-II both are incorrect
3) Statement-I is correct and Statement-II is incorrect
4) Statement-I is incorrect but Statement-II is correct

1. **Statement-I:** A compass needle which is allowed to move in a horizontal plane is taken to a geomagnetic pole. It will stay in any position.

Statement-II: At geomagnetic pole $B_H = 0$.

2. **Statement-I:** The true magnetic north direction is found by using a compass needle.

Statement-II: The magnetic meridian of the earth is along the axis of rotation of the earth.

3. **Statement-I:** Earth's magnetic field inside a closed iron box is less as compared to the outside.

Statement-II: The magnetic permeability of iron is low.

4. **Statement-I:** If a compass needle be kept at magnetic north pole of the earth, the compass needle may stay in any direction.

Statement-II: Dip needle will stay vertical at the north pole of earth.

5. **Statement-I:** For situation shown a small body A, can be levitated in mid-air, if it is made of paramagnetic material.

Statement-II: In non-uniform external magnetic field, paramagnetic material is attracted to regions of greater magnetic field.

Assertion & Reason MCQs

Directions: These questions consist of two statements each, printed as Assertion and Reason. While answering these questions, you are required to choose any one of the following four responses.

1) Both Assertion and Reason are True and the Reason is the correct explanation of the Assertion.
2) Both Assertion and Reason are True but Reason is not the correct explanation of the Assertion.
3) If Assertion is True but the Reason is False.
4) If Assertion is False but the Reason is True.

1. **Assertion:** A dip circle is taken to geomagnetic equator. The needle is allowed to move in a vertical plane perpendicular to the magnetic meridian. The needle will stay in any direction if it is released.

Reason: At geomagnetic equator $B_V = 0$.

2. **Assertion:** A solenoid carrying current is equivalent to a bar magnet.

Reason: The magnetic field lines due to a current carrying solenoid resemble exactly field lines of a bar magnet.

3. **Assertion:** A charged particle is projected parallel to the earth surface toward north at magnetic equator then no magnetic force acts on it.

Reason: Magnetic field of the earth at the magnetic equator is horizontal and in the north direction.

4. **Assertion:** When a magnet is brought near iron nails, only translatory force act on it.

Reason: The field due to a magnet is generally uniform.

5. **Assertion:** For strong enough magnetic fields or low enough temperatures Curie's law fails.

Reason: At very low temperature atomic dipoles in paramagnetic material approach complete alignment and produce a maximum magnetization.

Match the Columns MCQs

1. Match the column-I with column-II.

Column-I		Column-II	
(a)	Magnets	(i)	Occurs when a magnet is heated.
(b)	Horse shoe magnet	(ii)	Should be kept in pairs with their opposite poles on the same side.
(c)	Bar magnet	(iii)	It is U-shaped.
(d)	Demagne-tisation	(iv)	Must always be kept away from mobiles, T.V., Music system, computers, etc.

1) (a)-(iii); (b)-(i); (c)-(ii); (d)-(iv)
2) (a)-(iii); (b)-(iv); (c)-(ii); (d)-(i)
3) (a)-(ii); (b)-(iv); (c)-(iii); (d)-(i)
4) (a)-(iv); (b)-(iii); (c)-(ii); (d)-(i)

2. Match the column-I with column-II.

Column-I (Quantity)		Column-II (Dimensions)	
(a)	Magnetic moment	(i)	A-m
(b)	Pole strength	(ii)	Am^{-1}
(c)	Intensity of magnetisation	(iii)	Weber
(d)	Magnetic flux	(iv)	Am^2

1) (a)-(iii); (b)-(i); (c)-(ii); (d)-(iv)
2) (a)-(iv); (b)-(i); (c)-(ii); (d)-(iii)
3) (a)-(ii); (b)-(iv); (c)-(iii); (d)-(i)
4) (a)-(iv); (b)-(iii); (c)-(ii); (d)-(i)

3. Match the column-I (material) with column-II (Magnetic susceptibility)

Column-I		Column-II	
(a)	Diamagnetic material	(i)	Small and positive
(b)	Paramagnetic material	(ii)	High and positive
(c)	Ferromagnetic material	(iii)	Small and negative

1) (a)-(i); (b)-(iii); (c)-(ii)
2) (a)-(iii); (b)-(i); (c)-(ii)
3) (a)-(iii); (b)-(ii); (c)-(i)
4) (a)-(i); (b)-(ii); (c)-(iii)

4. Match the column-I (Quantity) and column-II (Dimension)

Column-I (Quantity)		Column-II (Dimensions)	
(a)	Pole strength	(i)	$[L^{-1}A]$
(b)	Intensity of magnetisation	(ii)	$[L^0A^0]$
(c)	Magnetic moment	(iii)	[LA]
(d)	Relative permeability	(iv)	$[L^2A]$

1) (a)-(iii); (b)-(iv); (c)-(i); (d)-(ii)
2) (a)-(i); (b)-(ii); (c)-(iii); (d)-(iv)
3) (a)-(iii); (b)-(i); (c)-(iv); (d)-(ii)
4) (a)-(iv); (b)-(iii); (c)-(ii); (d)-(i)

NCERT Exemplar MCQs

1. A paramagnetic sample shows a net magnetisation of 8 A m^{-1} when placed in an external magnetic field of 0.6 t at a temperature of 4 K. When the same sample is placed in an external magnetic field of 0.2 T at a temperature of 16 K, the magnetisation will be

1) $\frac{32}{3}$ A m^{-1}
2) $\frac{2}{3}$ A m^{-1}
3) 6 A m^{-1}
4) 2.4 A m^{-1}

2. The magnetic field of the earth can be modelled by that of a point dipole placed at the centre of the earth. The dipole axis makes an angle of 11.3° with the axis of the earth. At Mumbai, declination is nearly zero. Then: *[RC]*

1) The declination varies between 11.3° W to 11.3° E
2) The least declination is 0°
3) The plane defined by dipole axis and the earth axis passes through Greenwich
4) Declination averaged over the earth must be always negative

3. Consider the two idealised systems (i) a parallel plate capacitor with large plates and small separation and (ii) a long solenoid of length L >> R, radius of cross-section. In (i) E is ideally treated as a constant between plates and zero outside. In (ii) magnetic field is constant inside the solenoid and zero outside. These idealised assumptions, however, contradict fundamental laws as below:

1) Case (i) contradicts Gauss' law for electrostatic fields
2) Case (ii) contradicts Gauss' law for magnetic fields
3) Case (i) agrees with $\oint E.dl = 0$.
4) Case (ii) contradicts $\oint H.dl = I_{en}$.

NEET & AIIMS Past Year MCQs

1. A sheet is place on a horizontal surface in front of a strong magnetic pole. A force is needed to: **(2024)**

A. hold the sheet there if it is magnetic.
B. hold the sheet there if it is non-magnetic.
C. move the sheet away from the pole with uniform velocity if it is conducting.
D. move the sheet away from the pole with uniform velocity if it is both, non-conducting and non-polar.

Choose the correct statement(s) from the options given below:

1) A, C and D only
2) C only
3) B and D only
4) A and C only

2. Match List-I with List-II. **(2024)**

List-I (Material)		List-II (Susceptibility (χ))	
(a)	Diamagnetic	(i)	$\chi = 0$
(b)	Ferromagnetic	(ii)	$0 > \chi \geq -1$
(c)	Paramagnetic	(iii)	$\chi >> 1$
(d)	Non-magnetic	(iv)	$0 < \chi < \varepsilon$ (a small positive number)

Choose the correct answer from the options given below:

1) (a)-(iii); (b)-(ii); (c)-(i); (d)-(iv)
2) (a)-(iv); (b)-(iii); (c)-(ii); (d)-(i)
3) (a)-(ii); (b)-(iii); (c)-(iv); (d)-(i)
4) (a)-(ii); (b)-(i); (c)-(iii); (d)-(iv)

3. The net magnetic flux through any closed surface is: **(2023)**

1) Negative
2) Zero
3) Positive
4) Infinity

4. A uniform conducting wire of length 12 a and resistance 'R' is wound up as a current carrying coil in the shape of,

(*i*) an equilateral triangle of side 'a'.
(*ii*) a square of side 'a'. **(2021)**

The magnetic dipole moments of the coil in each case respectively are:

1) 3 Ia^2 and Ia^2
2) 3 Ia^2 and 4 Ia^2
3) 4 Ia^2 and 3 Ia^2
4) $\sqrt{3}Ia^2$ and $3Ia^2$

Directions (Q. 5 to 7): In each of the following questions, a statement of assertion (A) is given followed by a corresponding statement of reason (R). Mark the correct answer.

1) If both Assertion & Reason are True & the Reason is a correct explanation of the Assertion.
2) If both Assertion & Reason are True but Reason is not a correct explanation of the Assertion.
3) If Assertion is True but the Reason is False.
4) If both Assertion & Reason are False.

5. **Assertion:** In water, value of magnetic field decreases.
Reason: Water is a diamagnetic substance. **(AIIMS 2016)**

6. **Assertion:** Magnetic field lines are continuous and closed.
Reason: Magnetic monopole does not exist. **(AIIMS 2016)**

7. **Assertion:** Magnetic susceptibility is a pure number.
Reason: The value of magnetic susceptibility for vacuum is one. **(AIIMS 2009)**

8. When a magnetic substance is heated, then it: **(AIIMS 1999)**

1) Remains the same
2) Loses its magnetism
3) Becomes a strong magnet
4) Either (1) or (3)

9. Domain formation is the necessary feature of: **(AIIMS 1998)**

1) Ferromagnetism
2) Paramagnetism
3) Diamagnetism
4) All of these

10. A permanent magnet attracts: **(AIIMS 1996)**

1) Some substances and repels others
2) All substances
3) Ferromagnetic substances and repels all other
4) Only ferromagnetic substances

Answer Key

NCERT Topic-wise MCQs

1	2	3	4	5	6	7	8	9	10	11	12	13	14	15	16	17
1	4	3	2	2	1	4	1	4	3	1	1	3	2	2	4	1
18	19	20	21	22	23	24	25	26	27	28	29	30	31	32	33	34
2	3	3	2	3	1	2	1	3	3	4	2	2	2	4	1	3
35	36	37	38													
3	1	4	4													

High Order Time Intensive MCQs

1	2	3	4	5	6	7	8	9	10	11	12	13	14	15	16	17
1	2	1	2	2	4	2	2	2	4	3	3	3	3	4	3	4
18	19	20	21	22	23	24	25	26	27							
1	3	4	3	2	3	3	2	1	4							

Statement Based MCQs

1	2	3	4	5
1	1	3	1	1

Assertion & Reason MCQs

1	2	3	4	5
1	1	1	4	1

Match the Columns MCQs

1	2	3	4
4	2	2	3

NCERT Exemplar MCQs

1	2	3
2	1	2

NEET & AIIMS Past Year MCQs

1	2	3	4	5	6	7	8	9	10
4	3	2	4	1	1	3	2	1	4

CHAPTER

6 Electromagnetic Induction

Topicwise Number of Questions Analysis (2025-2016)

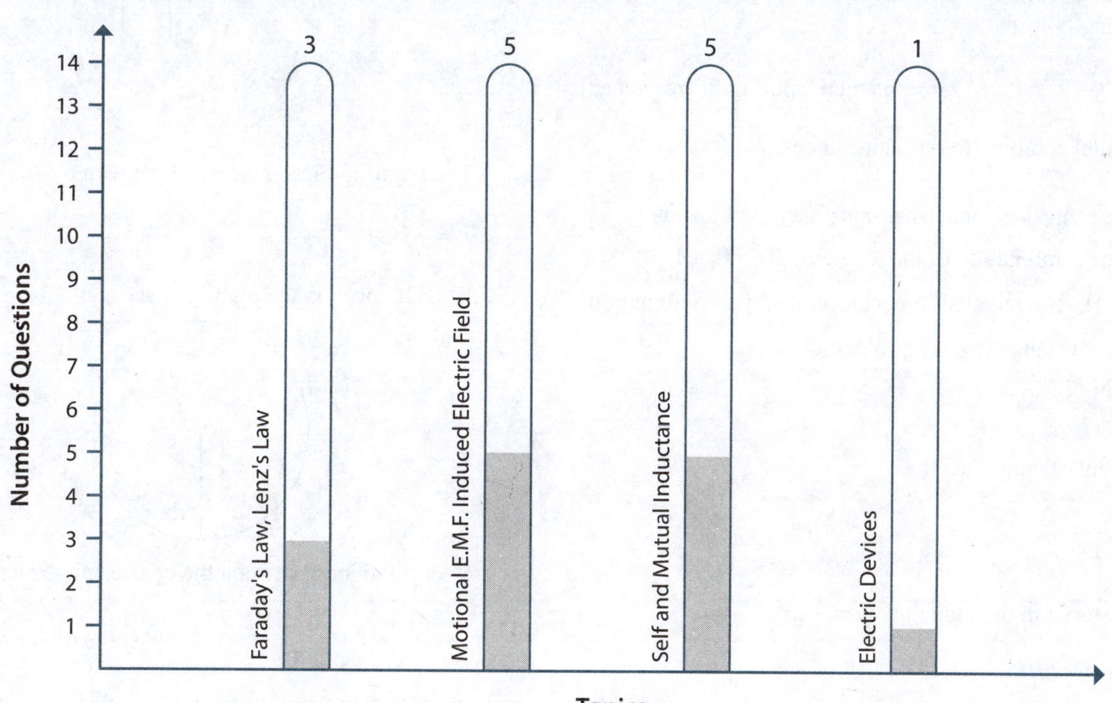

NCERT-PICKS

Magnetic Flux, Faraday's law

Formulae

1. Magnetic flux, $\phi = (\vec{B}.\vec{A}) = BA\cos\theta$

   [$\vec{B}$ is the magnetic field, $\vec{A}$ is the area vector and θ is the angle between magnetic field and area vector.]

2. Induced emf, $e = -N\frac{d\phi}{dt}$

   Where N is the number of turns in coil.

3. Magnitude of average induced emf, $e = -N\frac{\Delta\phi}{\Delta t} = N\frac{\phi_f - \phi_i}{\Delta t}$

   [ϕ_f and ϕ_i are the final and initial magnetic flux, Δt is change in time]

Motional emf, Self and Mutual Induction

Formulae

1. Induced emf of a rod, $e = Blv$

   [Here, v is the velocity and l is the length of the rod]

2. The charge induced in the circuit in time Δt,

   $$\Delta q = \frac{\Delta\phi}{R} = \frac{\text{Net change in magnetic flux}}{\text{Resistance}}$$

3. Force on the conductor, $F = BIl = B \times l \times \frac{Blv}{R} = \frac{B^2l^2v}{R}$

   [Here, R is the resistance and l is the length of the conductor]

4. Power, $P = F \times v = \frac{B^2l^2v^2}{R}$

   [P is the power and R is the resistance]

5. Magnetic flux, $\phi = LI$
[I is the magnitude of current and L is the coefficient of self induction]

6. Emf induced, $e = -L\dfrac{dI}{dt}$

7. Inductance of long solenoid, $L = \dfrac{\mu_0 N^2 A}{l}$

[N is the number of turns, A is the area of cross-section and l is the length of the solenoid]

8. Self inductance of a coil: Consider a circular coil of radius r and having N-turns.

$L = \dfrac{\mu_0 \pi N^2 r}{2}$ [μ_0 absolute permeability of free space]

9. For parallel combination of inductances, $\dfrac{1}{L_P} = \dfrac{1}{L_1} + \dfrac{1}{L_2}$

[L_p is the equivalent inductance in parallel combination.]

10. For series combination of inductances, $L_s = L_1 + L_2$

[L_s is equivalent inductance in series combination]

11. Mutual inductance two long solenoids,

$m = \dfrac{\mu_0 N_1 N_2 A}{l}$

Coefficient of coupling

$k = \dfrac{m}{\sqrt{L_1 L_2}}$

12. Energy stored in the inductor, $U = \dfrac{1}{2}LI^2$

Numerical Profile

Motional emf in different cases

(i) Motional EMF in Translatory Motion

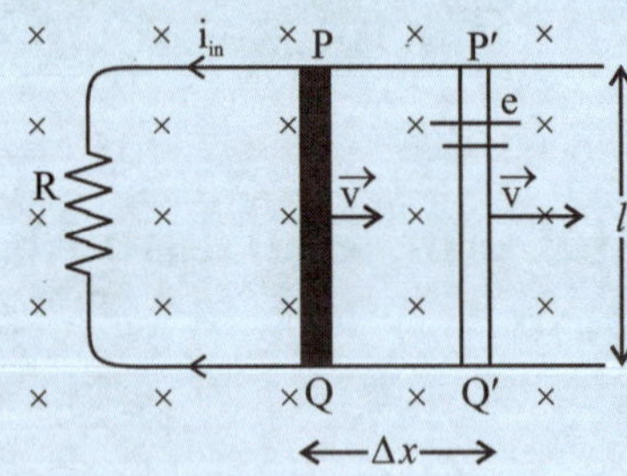

$$|e| = \frac{\Delta\phi}{\Delta t} = Bl\left(\frac{\Delta x}{\Delta t}\right) = Blv$$

(a) Induced current, $i_{in} = \dfrac{e}{R} = \dfrac{Bvl}{R}$.

(b) Magnetic force on the conductor

$$F_m = Bi_{in}l = \frac{B^2vl^2}{R}$$

(c) Power required to move conductor

$$P_{agent} = \vec{F}_{agent}\vec{v} = \frac{B^2v^2l^2}{R}$$

(d) Power dissipated through resistor

$$P_{Thermal} = (i_{in})^2 R = \frac{B^2v^2l^2}{R}$$

(ii) Motional EMF in Rotatory Motion: A conducting bar of length l rotates with a constant angular speed 'ω' about a pivot at one end. A uniform magnetic field B is at + 90° to the plane of rotation as shown in the figure.

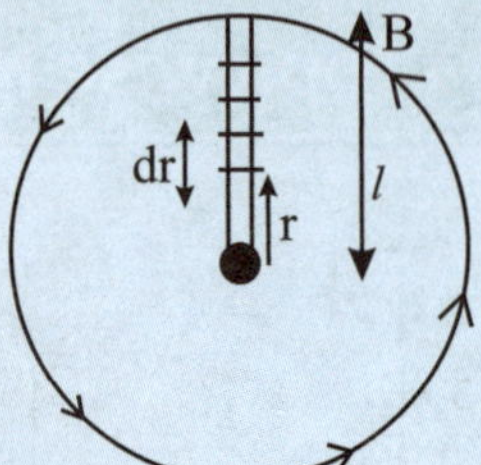

Then, emf induced in the rod is given by

$$e = \frac{1}{2}B\omega l^2$$

(iii) EMF Induced in a Rotating Disc:

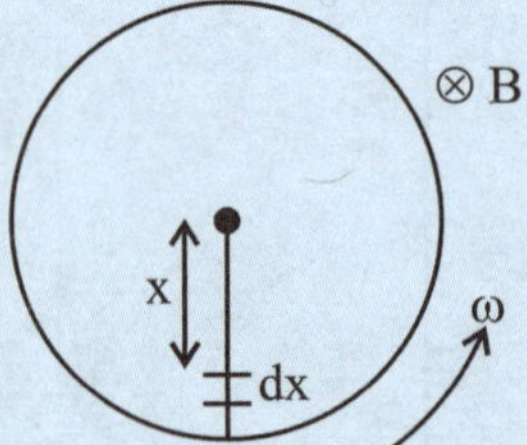

∴ emf induced between the centre and the edge of disc.

$e = \dfrac{B\omega r^2}{2}$ [r is the radius of the disc]

(iv) Induced e.m.f due to change in relative orientation of coil

$$e = \frac{-d\phi}{dt} = NBA\,\omega \sin \omega t$$

$$i = \frac{e}{R} = \frac{NBA\omega \sin \omega t}{R}$$

RL DC Circuit

Case-I: Growth of current

At any instant, Kirchoff's voltage law for the loop gives

$$E - L\frac{dI}{dt} = RI$$

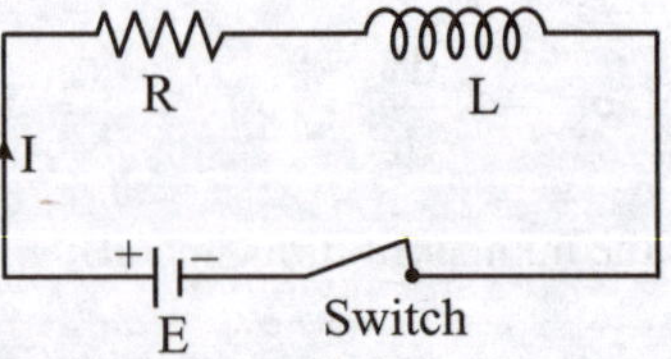

On rearranging the equation, we get

$$\frac{dI}{\frac{E}{R} - I} = \frac{R}{L}dt$$

On integrating both the sides we get

$$I = I_0\left(1 - e^{-t/\lambda}\right), \text{ where } I_0 = \frac{E}{R} \text{ and } \lambda = \frac{L}{R}$$

1. Just after the closing of the key inductance behaves like open circuit and current in circuit is zero.

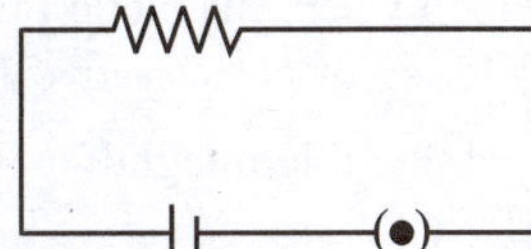

(Open circuit, $t = \infty, I = 0$)
(Inductor provide infinite resistance)

2. Some time after closing of the key inductance behaves like simple connecting wire (short circuit) and current in circuit is constant.

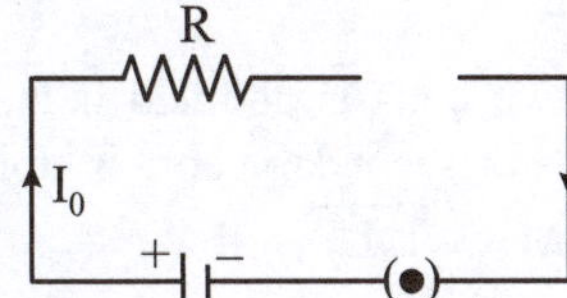

(Short circuit, $t \to \infty, I \to I_0$)
(Inductor provide zero resistance)

$I_0 = \frac{E}{R}$ (Final, steady, maximum or peak value of current)

Note: Peak value of current in circuit does not depends on self inductance of coil.

3. **Time constant of circuit (λ):** $\lambda = \frac{L}{R}$. Its SI unit is second(s)

It is a time in which current increases up to 63% or 0.63 times of peak current value.

4. **Half life (T):** It is a time in which current increases upto 50% or 0.50 times of peak current value.

$I = I_0(1 - e^{-t/\lambda})$

$$t = T, I = \frac{I_0}{2}, \frac{I_0}{2} = I_0\left(1 - e^{-T/\lambda}\right) \Rightarrow e^{-T/\lambda} = \frac{1}{2} \Rightarrow e^{T/\lambda} = 2$$

$$\frac{T}{\lambda}\log_e e = \log_e 2$$

$$T = 0.693\lambda$$

$$T = 0.693\frac{L}{R}\text{sec}$$

5. **Rate of growth of current at any instant:**

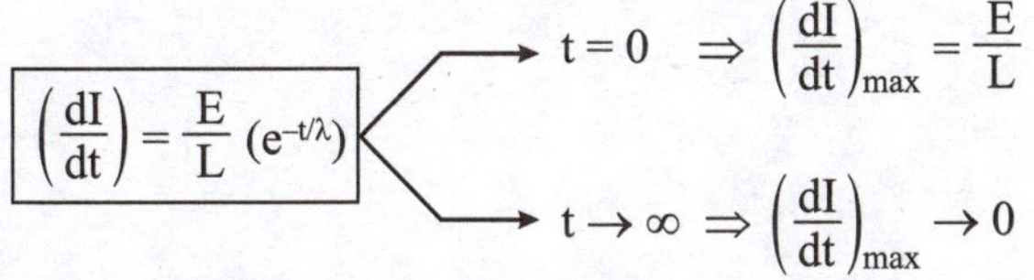

Note: Maximum or initial value of rate of growth of current does not depends upon resistance of coil.

Case-II: Decay of current:

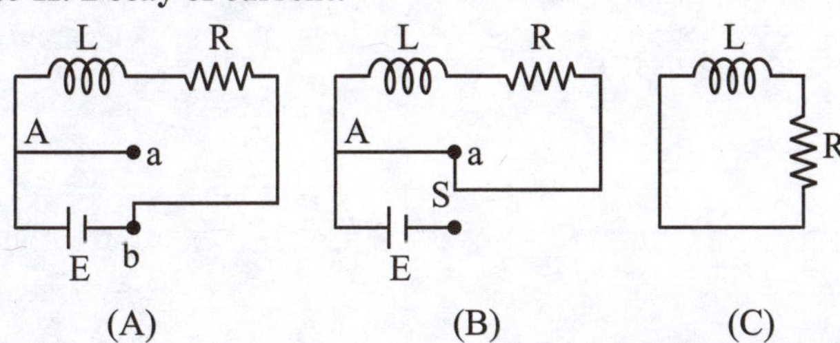

(A) (B) (C)

As the battery is disconnected, the current decreased in the circuit. This induced as emf is the inductor. As this is only emf in the circuit, we have

$$-L\frac{dI}{dt} = RI \quad \text{or} \quad \frac{dI}{I} = -\frac{R}{L}dt$$

on integrating both the sides, we get

$$\boxed{I = I_0 e^{\frac{-Rt}{L}} = I_0 e^{-t/\lambda}}$$

where $\lambda = L/R$ is the time constant of the circuit.

(Just after opening of key) $t = 0 \Rightarrow I = I_0 = \frac{E}{R}$

(Some time after opening of key) $t \to \infty \Rightarrow I \to 0$

1. **Time constant (λ):** It is a time in which current decreases up to 37% or 0.37 times of peak current value. $\boxed{\lambda = \frac{L}{R}\text{sec}}$

2. **Half life (T):** It is a time in which current decreases upto 50% or 0.50 times of peak current value. $\boxed{T = (0.693)\lambda \text{ sec}}$

3. **Rate of decay of current at any instant:**

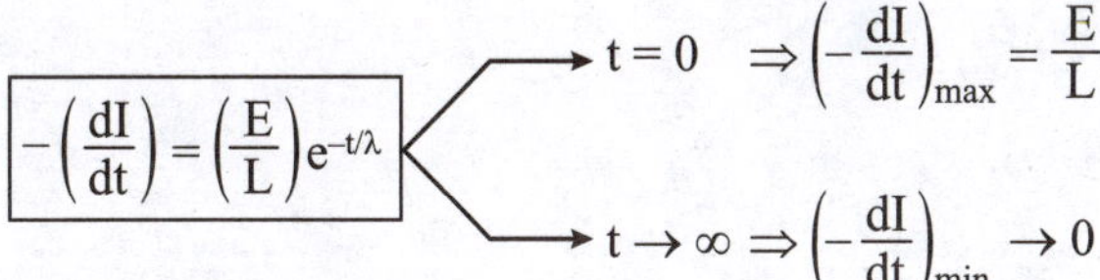

Special graph for R-L circuit:

Current Growth:

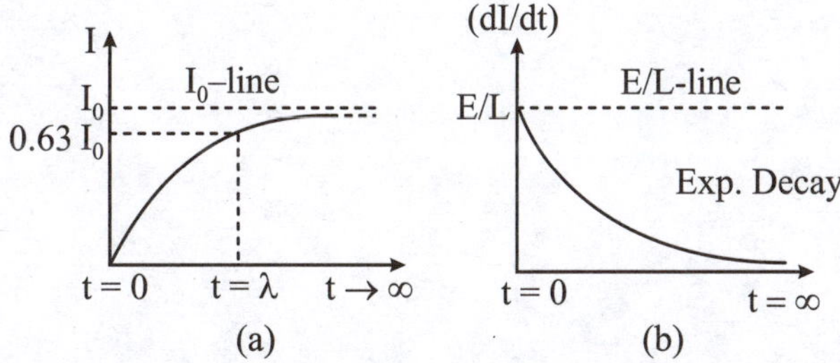

(a) (b)

Statements Asked in Exam

- Mechanical work done in moving the magnet w.r.t the coil changes into electrical energy producing induced current.
- Eddy current is produced in the conductor due to change in magnetic flux linked with it.
- We can reduce eddy current by cutting slots in the metal plate but we cannot eliminate them.
- Dimensional formula for the mutual inductance (M) is same as that of self-inductance.
- Mutual inductance of two coil depends upon,
  (i) Geometry of two coils, i.e., size, number of turns, nature of material
  (ii) Distance between two coils
  (iii) Relative placement of two coils (orientation)

- Coefficient of coupling (K) of two coils is a measure of the coupling between the two coils and it is given by

$$K = \frac{M}{\sqrt{L_1 L_2}}$$

L_1 and L_2 are the self inductance of two coils

M = Mutual inductance of two coil

- To calculate mutual inductance of two co-axial solenoid, area & length are considered of smaller and larger solenoid respectively.
- E.m.f is always induced whenever there is a change in magnetic flux, but current will be induced only when the circuit is complete,
- Induced e.m.f is sometimes named as back e.m.f.
- It is not possible to have mutual inductance without self inductance but it may be possible to have self inductance without mutual inductance
- Transformer has the highest possible efficiency out of all the electrical machines.
- Transformers are used only with an A.C source and not with the D.C source.
- A transformer can not be used to step up or step down dc voltage
- Core of a transformer is laminated to reduce eddy currents.
- An emf is induced in a closed loop where magnetic flux is varied. The induced electric field is not conservative field because for induced electric field, the line integral $\oint \vec{E}.d\vec{\ell}$ around a closed path is non-zero.
- Acceleration of a magnet falling through a long solenoid decreases because the current induced in the circuit always flows in such a direction that it opposes the change or the cause that produces it.
- The mutual inductance of two coils is doubled if the self inductance of the primary and secondary coil is doubled because mutual inductance $M \propto \sqrt{L_1 L_2}$
- An induced current has no direction of its own.

NCERT Topic-wise MCQs

Faraday's and Lenz's Law

1. A coil having an area A_0 is placed in a magnetic field which changes from B_0 to $4B_0$ in a time interval t. The e.m.f induced in coil will be:

1) $\frac{3A_0B_0}{t}$ 2) $\frac{4A_0B_0}{t}$
3) $\frac{3B_0}{A_0t}$ 4) $\frac{B_0A_0}{t}$

2. A coil having 500 square loops each of side 10 cm is placed normal to a magnetic flux which is increasing at the rate of 1 T/s. The magnitude of induced emf (in volts) is:

1) 0.1 2) 0.5
3) 1 4) 5

3. A coil of area 100cm^2 has 500 turns. Magnetic field of 0.1 Wb/m^2 is perpendicular to the coil. The field is reduced to zero in 0.1 second. The induced emf in the coil is

1) 1 volt 2) 5 volt
3) 50 volt 4) Zero

4. A 50 turns circular coil has a radius of 3cm, it is kept in a magnetic field acting normal to the area of the coil. The magnetic field B increased from 0.10 to 0.35 tesla in 2ms. The average induced emf in the coil is:

1) 1.77 volts 2) 17.7 volts
3) 177 volts 4) 0.177 volts

5. In the circuit with a coil of resistance 2Ω, the magnetic flux changes from 2 Wb to 10 Wb in 0.2 second. The charge that flows in the coil during this time is:

1) 5 coulomb 2) 4 coulomb
3) 1 coulomb 4) 0.25 coulomb

6. A magnetic field of 2×10^{-2} T acts at right angles to a coil of area 100 cm^2 with 50 turns. The average emf induced in the coil is 0.1 volt, when it is removed from the field in time t. The value of t is

1) 0.1 s 2) 0.01 s
3) 1 s 4) 20 s

7. The magnetic flux linked with a coil in Wb is given by the equation $\phi = 3t^2 + 4t + 9$. Then the magnitude of induced emf at t = 2 sec will be:

1) 2 V 2) 4 V
3) 8 V 4) 16 V

8. Choose the incorrect statement regarding Lenz law.

1) It gives the direction of induced emf.
2) It is based on the law of conservation of energy.
3) It is based on the law of conservation of momentum.
4) None of these

9. Magnetic flux through a cuircuit of resistance 20 ohm changes from 40 wb to 20wb in 5 ms/ charge passing through the circuit during this time is?

1) 1 C 2) zero
3) 2 C 4) 0.5 C

10. In a magnetic field of 0.05 T, area of a coil changes from 101cm^2 to 100cm^2 without changing the resistance which is 2Ω. The amount of charge that flow during this period is:

1) 2.5×10^{-6} C 2) 2 μC
3) 10^{-6} C 4) 8 μC

11. When a bar magnet falls through a long hollow metal cylinder fixed with its axis vertical, the final acceleration of the magnet is:

1) Zero 2) <g
3) >g 4) g

12. The dimensional formula of magnetic flux is:

1) $[ML^2T^{-2}A^{-2}]$ 2) $[MLT^{-1}A^{-1}]$
3) $[ML^2T^{-2}A^{-1}]$ 4) $[MLTA^{-3}]$

13. A metal disc of radius 200cm is rotated at a constant angular speed of 60 rad s^{-1} in a plane at right angles to an external field of magnetic induction of 0.05 Wbm^{-2}. Find the emf induced between the center and a point on the rim.

1) 12 V 2) 8 V
3) 2 V 4) 6 V

14. The magnetic flux associated with a closed loop is $\phi = 6t^2 + 7t + 1$, where ϕ is in milliweber and t in seconds. What will be the value of induced emf after 2 second ?

1) 29 mV 2) 60 mV
3) 31 mV 4) 22 mV

15. A coil of mean area 500cm^2 and having 1000 turns is held perpendicular to uniform field of 0.4 G. The coil is turned through 180^0 in 1/10 second. Calculate the average induced emf.

1) 0.02 V 2) 0.05 V
3) 0.06 V 4) 0.07 V

16. The flux associated with coil changes from 1.35 Wb to 0.79 Wb with in 1/10 s. Then the charge produced by the coil, (if resistance of coil is 7Ω) is:
1) 0.08 C 2) 0.8 C
3) 0.008 C 4) 8 C

17. A conducting circular loop is placed in a uniform magnetic field of induction B tesla with its plane normal to the field. Now, the radius of the loop starts shrinking at the rate $\left(\frac{dr}{dt}\right)$. Then the induced emf at the instant when the radius is r, is
1) $\pi rB\left(\frac{dr}{dt}\right)$ 2) $2\pi rB\left(\frac{dr}{dt}\right)$
3) $\pi r^2\left(\frac{dB}{dt}\right)$ 4) $\left(\frac{\pi r^2}{2}\right)^2 B\left(\frac{dr}{dt}\right)$

Electro Magnetic Induction, Motional emf, Self Induction, Mutual Inductance

18. If a 10m long metallic bar moves in a direction at right angles to the magnetic field with a speed of $5ms^{-1}$, 25V emf is induced in it. Find the value of magnetic field intensity.
1) 0.6 T 2) 0.4 T
3) 0.5 T 4) 0.8 T

19. The current passing through an inductor coil of 5 H is decreasing at the rate of 2 A/s, the emf developed across the coil is:
1) –10.0 V 2) + 2.5 V
3) – 2.5 V 4) + 10.0 V

20. An average induced emf of 0.20V appears in a coil when the current in it is changed from 5A in one direction to 5A in the opposite direction in 0.20 sec. Find the self induction of the coil.
1) 4×10^{-3} H 2) 2×10^{-3} H
3) $2\sqrt{2}\times10^{-3} H$ 4) $3\sqrt{3}\times10^{-1} H$

21. A coil of wire of certain radius has 600 turns and a self inductance of 108 MH. What will be the self inductance of a similar coil which has 500 turns?
1) 600 MH 2) 75 MH
3) 70 MH 4) None of these

22. When a wire loop is rotated in a magnetic field such that it continuously cuts the magnetic field lines, then the direction of induced emf changes once in every:
1) 1/4 revolution 2) 1/2 revolution
3) 1 revolution 4) 2 revolution

23. A coil of area 80 sq.cm and 50 turns is rotating with 2000 revolutions per minute about an axis perpendicular to a magnetic field of 0.05 tesla. The maximum value of the emf developed in it is:
1) 200 π volt 2) $\frac{10\pi}{3}$ volt
3) $\frac{4\pi}{3}$ volt 4) $\frac{2\pi}{3}$ volt

24. A long horizontal metallic rod with length along the east west direction is falling under gravity. The potential difference between its two ends will:
1) Be zero 2) Be constant
3) Increase with time 4) Decrease with time

25. A wheel with ten metallic spokes each 0.50 m long is rotated with a speed of 120 rev/min in a plane normal to the earth's magnetic field at the place. If the magnitude of the field is 0.4 gauss, the induced emf between the axle and the rim of the wheel is equal to:
1) 1.256×10^{-3} V 2) 6.28×10^{-4} V
3) 1.256×10^{-4} V 4) 6.28×10^{-5} V

26. A 50 mH coil carries a current of 2A. The energy stored in joules is:
1) 1 2) 0.1
3) 0.05 4) 0.5

27. Two coils of self inductance L_1 and L_2 are placed closer to each other so that total flux in one coil is completely linked with other. If M is mutual inductance between them then:
1) $M = L_1 L_2$ 2) $M_2 = L_1/L_2$
3) $M = \sqrt{L_1L_2}$ 4) $M = (L_1 L_2)^2$

28. An emf of 15 volts is applied in a circuit coil containing 5H inductance and 10Ω resistance. The ratio of the currents at time t = ∞ and t = 1 s is
1) $\frac{e^{1/2}}{e^{1/2}-1}$ 2) $\frac{e^2}{e^2-1}$
3) $1 - e^{-1}$ 4) e^{-1}

29. Self - inductance of a coil is 50mH. A current of 1A passing through the coil reduces to zero at steady rate in 0.1 sec, the self induced emf is:
1) 5 volt 2) 0.05 volt
3) 50 volt 4) 0.5 volt

30. 5cm long solenoid having 10 ohm resistance and 5mH inductance is joined to a 10 volt battery. At steady state the current through the solenoid in ampere will be:
1) 5 A 2) 1 A
3) 2 A 4) Zero

31. A metallic square loop ABCD is moving in its own plane with velocity *v* in a uniform magnetic field perpendicular to its plane as shown in Fig. An electric field is induced.

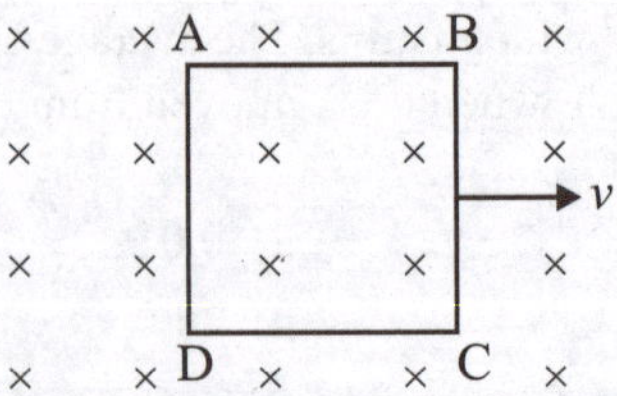

1) In AD, but not in BC
2) In BC, but not in AD
3) Neither in AD nor in BC
4) In both AD and BC

32. The self inductance of a coil is L, keeping the length and area same, the number of turns in the coil is increased to four times. The self inductance of coil will:
1) 1/4L 2) L
3) 4L 4) 16L

33. If the current is halved in a coil, then the energy stored is how much times the previous value:
1) 1/2 2) 1/4
3) 2 4) 4

34. Two circuits have mutual inductance of 0.09H. Average e.m.f. induced in the secondary by a change of current from 0 to 20 A in 0.006 s in primary will be
1) 120 v 2) 200 v
3) 180 v 4) 300 v

35. Two coils of self inductances 2mH and 8mH are placed so close together that the effective flux in one coil is completely linked with the other. The mutual inductance between these coils are:
1) 4 mH 2) 16 mH
3) 10 mH 4) 6 mH

36. An air core solenoid has 1000 turns and is one meter long. Its cross - sectional area is 10 cm^2. Its self inductance is:
1) 0.1256 mH 2) 12.56 mH
3) 1.256 mH 4) 125.6 mH

37. A coil of N = 100 turns carries a current I = 5A and creates a magnetic flux $\phi = 10^{-5}$ Tm^2 per turn. The value of its inductance will be:
1) 0.05 mH 2) 0.10 mH
3) 0.15 mH 4) 0.20 mH

38. A copper rod of length 'l' m is rotated about the end perpendicular to a uniform magnetic field of flux density 'B' with constant angular velocity 'ω'. The induced e.m.f. between the two ends is:
1) $2Bl\omega^2$ 2) $\frac{1}{4}(Bl^2\omega)$
3) $\frac{1}{8}(Bl\omega^2)$ 4) $\frac{1}{2}(Bl^2\omega)$

39. In figure magnetic energy stored in the coil is:

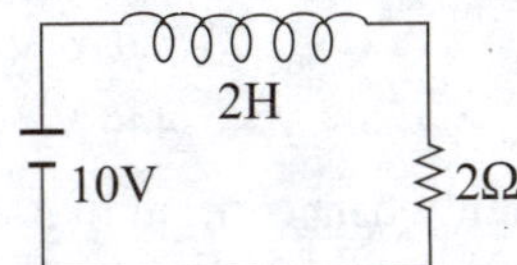

1) Zero 2) ∞
3) 25 J 4) None of these

40. Two solenoids of equal number of turns have their lengths and the radii in the same ratio 1 : 2. The ratio of their self inductances $\left(\frac{L_2}{L_1}\right)$ will be:
1) 1 : 2 2) 2 : 1
3) 1 : 1 4) 1 : 4

41. If self inductance of air core solenoid increases from 0.04 mH to 16 mH on introducing an iron core into it, then what is relative magnetic permeability of the core?
1) 400 2) 500
3) 600 4) 700

42. A jet plane is travelling west at the speed of 1800 km/h. What is the voltage difference developed between the ends of the wings 25m long, if the earth's magnetic field at the location has a magnitude of 5.0×10^{-4} T and the dip angle is 30°?
1) 2.2 volt 2) 4.2 volt
3) 3.1 volt 4) 1.1 volt

43. A 1m long conducting rod rotates with an angular frequency of 400 rad s^{-1} about an axis normal to the rod passing through its one end. The other end of the rod is in contact with a circular metallic ring. A constant magnetic field of 0.5T parallel to the axis exists everywhere. Calculate the e.m.f developed between the center and the ring?
1) 100 V 2) 101 V
3) 102 V 4) 103 V

Electric Devices (Application of EMI)

44. If rotational velocity of a dynamo armature is doubled, then induced e.m.f will become
1) Half 2) Two times
3) Three times 4) None of these

45. A flat square coil of 10 turns has sides of length 12cm. The coil rotates in a magnetic field whose flux density is 0.025 Wb/m^2. What is the angular velocity of the coil, if max. e.m.f. produced is 20mV?
1) 2.51 rad/s 2) 4.52 rad/s
3) 3.54 rad/s 4) 5.55 rad/s

46. An A.C generator of 220V having internal resistance r = 10 Ω and external resistance R = 100 Ω. What is the power developed in the external resistor?
1) 484 W 2) 400 W
3) 441 W 4) 369 W

47. Calculate the maximum emf induced when a square frame is making 240 r.p.m each side of square being 20cm and magnetic field intensity is 500 gauss?
1) 5.01×10^{-2} V 2) 5.02×10^{-2} V
3) 5.03×10^{-2} V 4) 8×10^{-3} V

48. A six pole generator with fix field excitation developed an emf of 100V, when operating at 1500 rpm. At what speed must it rotate to develop 120V?
1) 1200 rpm 2) 1800 rpm
3) 1500 rpm 4) 400 rpm

49. Which one is not an application of eddy current?
1) Induction furnace 2) Electric power meter
3) Induction meter 4) None of these

50. Which of the following is constructed on the principle of electromagnetic induction?
1) Galvanometer 2) Electric motor
3) Electric Generator 4) Voltmeter

High Order Time Intensive MCQs

1. A coil of resistance 400Ω is placed in a magnetic field. If the magnetic flux ϕ (Wb) linked with the coil varies with time t (s) as $\phi = 50t^2 + 4$. Current at 2 s is:
   1) 0.5 A 2) 0.1 A
   3) 2 A 4) 1 A

2. In a coil of resistance 10 Ω, the induced current developed by changing magnetic flux through it, is shown in figure as a function of time. The magnitude of change in flux through the coil (in weber) is:

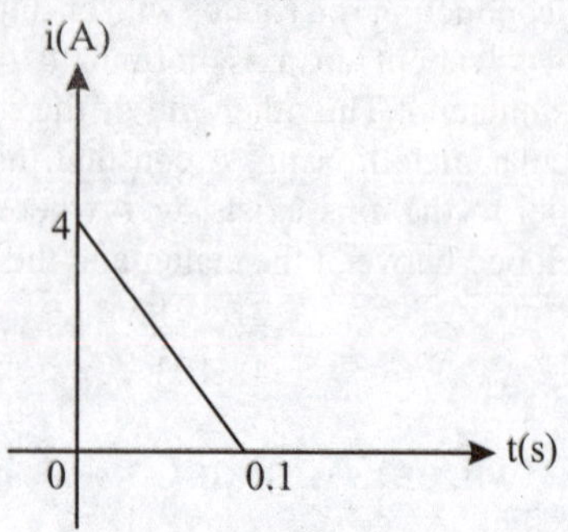

   1) 8 2) 2
   3) 6 4) 4

3. In figure, final value of current in 10 Ω resistor, when plug of key K is inserted is

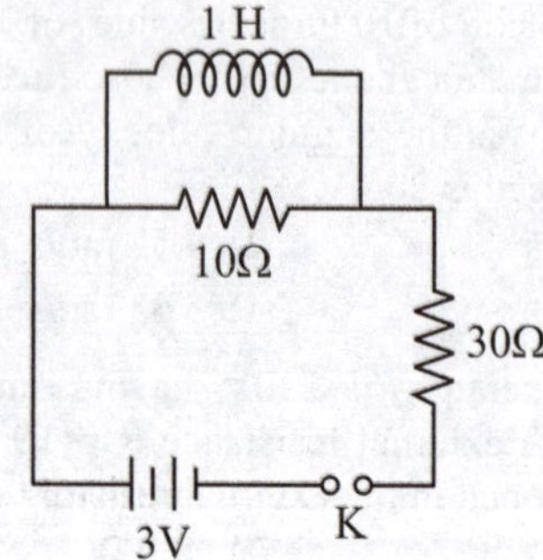

   1) $\frac{3}{10}$A 2) $\frac{3}{20}$A
   3) $\frac{3}{11}$A 4) zero

4. A conducting ring of radius r is placed perpendicularly inside a time varying magnetic field given by $B = B_0 + \alpha t$. B_0 and α are positive constants. Find the emf produced in the ring.
   1) $-\pi \alpha r^2$ 2) $-\pi \alpha r$
   3) $-\pi \alpha^2 r^2$ 4) $-\pi \alpha^2 r$

5. In the circuit shown in figure, L = 10 H, R = 5Ω, E = 15 V. the switch S is closed at t = 0. At t = 2s, the current in the circuit is

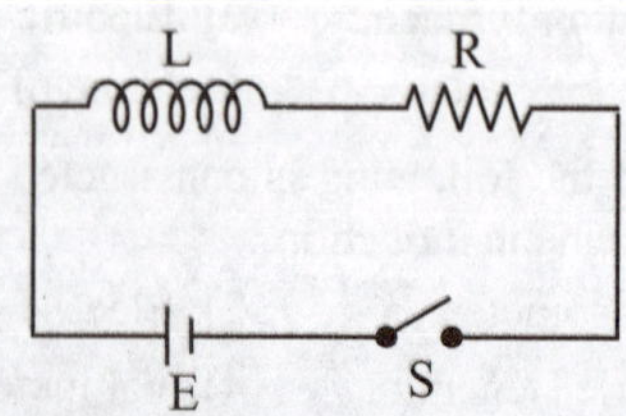

   1) $3\left(1-\frac{1}{e}\right)$A 2) $3\left(1-\frac{1}{e^2}\right)$A
   3) $3\left(\frac{1}{e}\right)$A 4) $3\left(\frac{1}{e^2}\right)$A

6. The north-pole of a long horizontal bar magnet is being brought closer to a vertical conducting plane along the perpendicular direction. The direction of the induced current in the conduction plane will be:
   1) Horizontal 2) Vertical
   3) Clockwise 4) Anti-clockwise

7. A coil of 1200 turns and mean area of 500 cm² is held perpendicular to a uniform magnetic field of induction 4×10^{-4}T. The resistance of the coil is 20 Ω. When the coil is rotated through 180° in the magnetic field in 0.1 s, the average electric current (in mA) induced is:
   1) 12 2) 24
   3) 36 4) 48

8. Two resistance of 10 Ω and 20 Ω and an inductor of inductance 5 H are connected to a battery of 2 V through a key K as shown in the figure. At time t = 0 when the key K is closed, the initial current through the battery is:

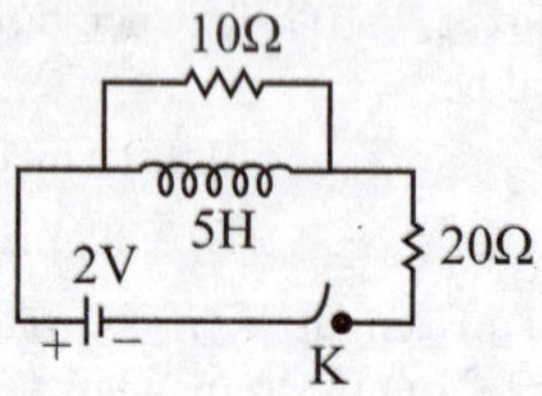

   1) 0.2 A 2) $\frac{2}{15}$A
   3) $\frac{1}{15}$A 4) 0A

9. The current in a self-inductance $L = 40$ mH is to be increased uniformly from 1 A to 11 A in 4 ms. The emf induced in the inductor during the process is:
   1) 100 V 2) 0.4 V
   3) 40 V 4) 440 V

10. When the current changes from + 2A to –2A in 0.05 s, an emf of 8V is induced in a coil. The coefficient of self-induction of the coil is:
   1) 0.2 H 2) 0.4 H
   3) 0.8 H 4) 0.1 H

11. The inductance of a coil is L = 10 H and resistance $R = 5\Omega$. If applied voltage of battery is 10 V and it switches off in 1 ms, then find induced emf of inductor.
   1) 2×10^4 V 2) 1.2×10^4 V
   3) 2×10^{-4} V 4) None of these

12. In a solenoid, if number of turns is doubled, then self-inductance will become:
1) Half 2) Double
3) $\frac{1}{4}$ times 4) Quadruple

13. A copper rod of length l is rotated about one end, perpendicular to the uniform magnetic field B with constant angular velocity ω. The induced emf between two ends of the rod is:
1) $\frac{1}{2}B\omega l^2$ 2) $B\omega l^2$
3) $\frac{3}{2}B\omega l^2$ 4) $2B\omega l^2$

14. An aeroplane having a wing space of 35 m flies due North with the speed of 90 ms^{-1}, given vertical component of earth's magnetic field is $B = 4 \times 10^{-5}$ T. The potential difference between the tips of the wings will be:
1) 0.013 V 2) 1.26 V
3) 12.6 V 4) 0.126 V

15. A straight conductor of length 4 m moves at a speed of 10 ms^{-1}. When the conductor makes an angle of 30° with the direction of magnetic field of induction of 0.1 Wb/m², then induced emf is:
1) 8 V 2) 4 V
3) 1 V 4) 2 V

16. For a solenoid having a primary coil of N_1 turns and a secondary coil of N_2 turns, the coefficient of mutual inductance is:
1) $\mu_0\mu_r\frac{N_1N_2}{l}$ 2) $\frac{\mu_0\mu_r N_1N_2}{Al}$
3) $\mu_0\mu_r N_1N_2 Al$ 4) $\frac{\mu_0\mu_r N_1N_2A}{l}$

17. In the circuit shown in the figure, what is the value of I_1 just after pressing the key K?

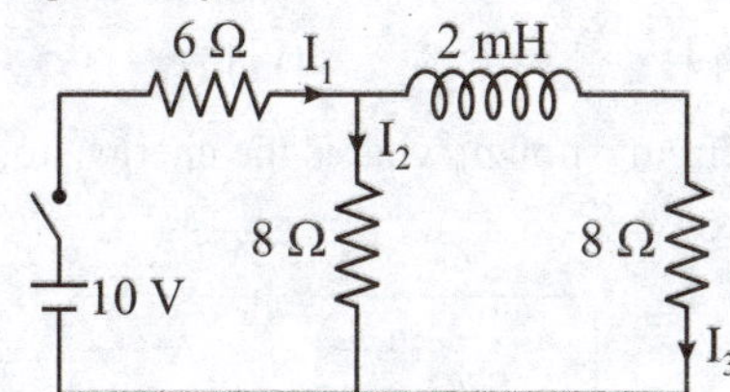

1) $\frac{5}{7}$A 2) $\frac{5}{11}$A
3) 1 A 4) None of these

18. Figure shows an L-R circuit. When the switch S is closed, the currents through resistors R_1, R_2 and R_3 are I_1, I_2 and I_3 respectively. the values of I_1, I_2 and I_3 at t = 0 are

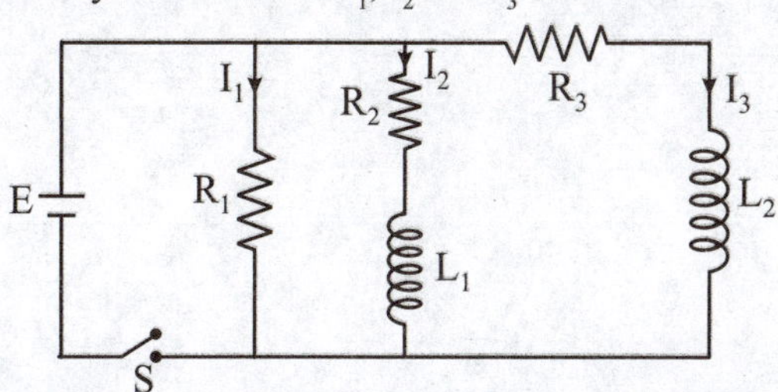

1) $I_1 = I_2 = I_3 = 0$
2) $I_1 = \frac{E}{R_1}, I_2 = I_3 = 0$
3) $I_1 = 0, I_2 = \frac{E}{R_2}, I_3 = \frac{E}{R_3}$
4) $I_1 = \frac{E}{R_1}, I_2 = \frac{E}{R_2 + L_1}, I_3 = \frac{E}{R_3 + L_3}$

19. A frame CDEF is placed in a region where a magnetic field B is present. A rod of length one metre moves with constant velocity 20 m/s and strength of magnetic field is one tesla. The power spent in the process is (Take R = 0.2 Ω and all other wires and rod have zero resistance)

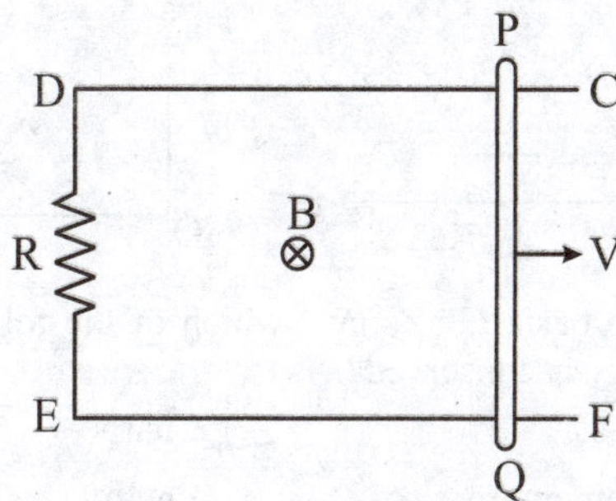

1) 1 kW 2) 2 kW
3) 3 kW 4) 4 kW

20. If 10 A current is flowing through a solenoid of inductance 5H, find the magnetic energy stored in the solenoid.
1) 50 J 2) 250 J
3) 500 J 4) 125 J

21. Two coils are wound on the same iron so that the flux generated by one passes through the other. The primary coil has N_p turns in it and when a current 2 A flows through it, the flux in it is 2.5×10^{-4} Wb. If the secondary coil has 12 turns, the mutual inductance of the coils is (assume the secondary coil is in open circuit)
1) 10×10^{-4} H 2) 15×10^{-4} H
3) 20×10^{-4} H 4) 25×10^{-4} H

22. The figure shows an L-R circuit, the time constant for the circuit.

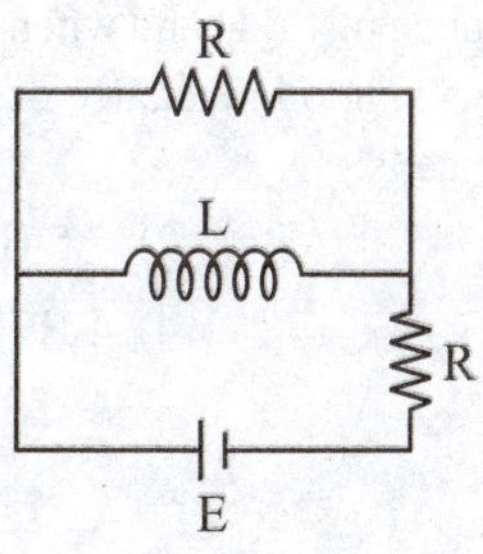

1) $\frac{L}{2R}$ 2) $\frac{2L}{R}$
3) $\frac{2R}{L}$ 4) $\frac{R}{2L}$

23. The current i in an induction coil varies with time t according to the graph shown in the figure. Which of the following graphs shows the induced emf(ε) in the coil with time?

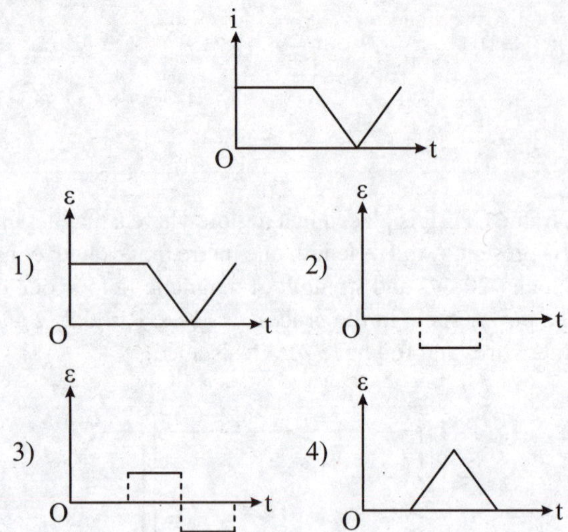

24. On the basis Lenz's law, which of the following physical quantity is conserved?

1) Mass
2) Charge
3) Momentum
4) Energy

25. A square loop of side b is rotated in a constant magnetic field B at angular frequency ω as shown in the figure. What is the emf induced in it?

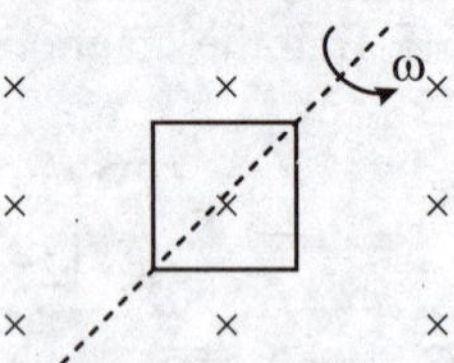

1) $b^2B\omega \sin \omega t$
2) $bB\omega \sin^2 \omega t$
3) $bB^2\omega \cos \omega t$
4) $b^2B\omega$

26. A coil having 500 sq loops of side 10 cm is placed normal to magnetic flux which increases at a rate of 1 T/s. The induced emf is

1) 0.1 V
2) 0.5 V
3) 1 V
4) 5 V

27. Two inductances connected in parallel are equivalent to a single inductance of 1.5 H and when connected in series are equivalent to a single inductance of 8 H. The difference in their inductance is:

1) 3H
2) 7.5 H
3) 3 H
4) 4 H

28. A conducting square loop of side l and resistance R is moving with a uniform velocity v perpendicular to one of its sides. A uniform and constant magnetic field B exists along the perpendicular to the plane of the loop as shown in figure:

The current induced in the loop is

1) $\frac{Blv}{R}$ clockwise
2) $\frac{Blv}{R}$ anti - clockwise
3) $\frac{2Blv}{R}$ anti - clockwise
4) Zero

29. A square loop of wire with side length 10 cm is placed to angle of 45° with a magnetic field that changes uniformly from 0.1 T to zero in 0.7s. The induced current in the loop (its resistance is 1Ω) is :

1) 1.0 mA
2) 2.5 mA
3) 3.5 mA
4) 4.0 mA

30. The current flowing in a coil of self inductance 0.4 mH is increased by 250 mA in 0.1 sec. The e.m.f. induced will be

1) +1 V
2) – 1 mV
3) +1 mV
4) – 1 V

31. A wire in the form of a circular loop of radius 10 cm lies in a plane normal to a magnetic field of 100 T. If this wire is pulled to take a square shape in the same plane in 0.1 s, average induced emf in the loop is:

1) 4.75 volt
2) 6 volt
3) 6.75 volt
4) 5.75 volt

32. In a coil when current changes from 10 A to 2 A in time 0.1 s, induced emf is 3.28 V. What is self-inductance of coil?

1) 4 H
2) 0.4 H
3) 0.04 H
4) 5 H

33. In the circuit shown, what is the energy stored in the coil at steady state ?

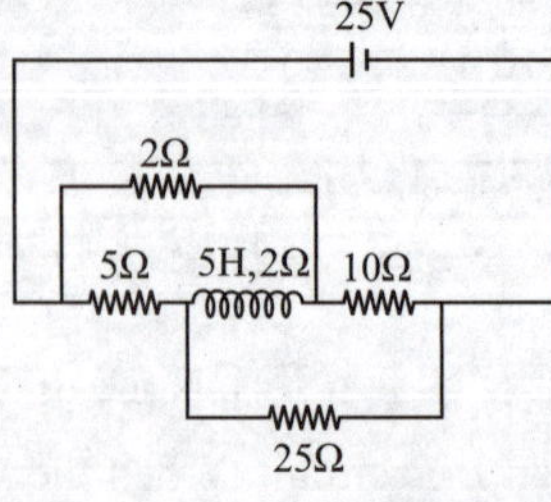

1) 21.3 J
2) 42.6 J
3) Zero
4) 213 J

Statement Based MCQs

Directions: These questions consist of two statements each, printed as Statement-I and Statement-II. While answering these questions, you are required to choose any one of the following four responses.

1) **Both Statement-I and Statement-II are correct.**
2) **Both Statement-I and Statement-II are incorrect.**
3) **Statement-I is correct & Statement-II is incorrect.**
4) **Statement-I is incorrect & Statement-II is correct.**

1. **Statement-I:** An electrical motor converts mechanical energy into electrical energy.

   Statement-II: An electrical motor works on the principle of electromagnetic induction

2. **Statement-I:** The mutual inductance of two coils is doubled, if the self-inductance of the primary and secondary coil is doubled. Given $M = k\sqrt{L_1L_2}$

   Statement-II: Mutual inductance is inversely proportional to the self-inductance of primary and secondary coils.

3. **Statement-I:** When two coils are wound on each other, the mutual induction between the coils is maximum.

   Statement-II: Mutual induction does not depend on the orientation of the coils.

4. **Statement-I:** An emf can be induced by moving a conductor in a magnetic field.

   Statement-II: An emf can be induced by changing the magnetic field.

Assertion & Reason MCQs

Directions: These questions consist of two statements each, printed as Assertion and Reason. While answering these questions, you are required to choose any one of the following four responses.

1) **Both Assertion and Reason are True and the Reason is a correct explanation of the Assertion.**
2) **Both Assertion and Reason are True but Reason is not a correct explanation of the Assertion.**
3) **Assertion is True but the Reason is False.**
4) **Assertion is False but Reason is True.**

1. **Assertion:** The direction of the induced electric field is always perpendicular to the direction of the (changing) magnetic field.

   Reason: The induced electric field is a non-conservative field.

2. **Assertion:** If electric current flowing through a circuit changes in magnitude, then eddy currents are induced in a nearby iron piece.

   Reason: Due to change of electric current, the magnetic flux linked with the iron piece changes, so eddy currents are induced in the iron piece.

3. **Assertion:** Induced e.m.f. is proportional to rate of change of magnetic field while induced current depends on resistance of the wire.

   Reason: The induced e.m.f. and the current will be same in two identical loops one of copper and other of aluminium, when rotated with same speed in the same magnetic field.

4. **Assertion:** Acceleration of a magnet falling through a long solenoid decreases.

   Reason: The induced current in a circuit always flow in such direction that it opposes the change or the cause that produced it.

5. **Assertion:** A spark occurs between the poles of a switch when the switch is opened.

   Reason: Current flowing in a conducting wire produces a magnetic field.

Match the Columns MCQs

1. In the circuit shown in figure, switch remains closed for long time. It is opened at time t = 0. Match the following two columns at t = (ln 2) second.

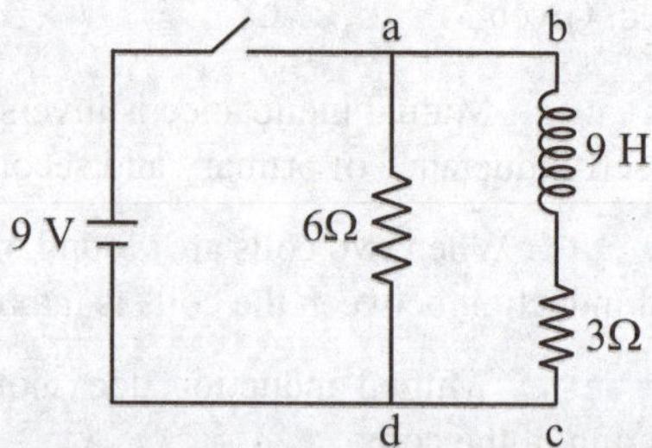

Column I		Column II	
(a)	Potential differences across inductor	(i)	9 V
(b)	Potential difference across 3 Ω resistance	(ii)	4.5 V
(c)	Potential difference across 6 Ω resistance	(iii)	6 V
(d)	Potential difference between points b and c	(iv)	None of these

1) (a)-(iv) (b)-(ii) (c)-(i) (d)-(i)
2) (a)-(i) (b)-(ii) (c)-(iii) (d)-(iv)
3) (a)-(i) (b)-(iii) (c)-(iv) (d)-(ii)
4) (a)-(ii) (b)-(iv) (c)-(i) (d)-(iii)

2. Magnetic flux passing through a coil of resistance 2 Ω is as shown in figure. Match the following two columns. In column II all physical quantities are in SI units.

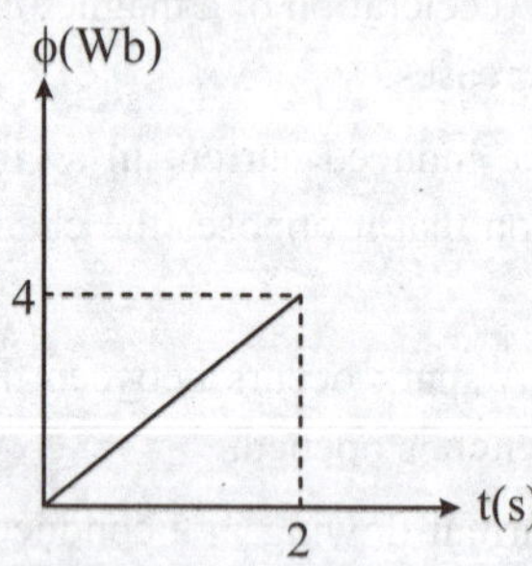

Column I		Column II	
(a)	Induced emf produced	(i)	4
(b)	Induced current	(ii)	1
(c)	Charge flow in 2 s	(iii)	8
(d)	Heat generation in 2 s	(iv)	2

1) (a)-(i) (b)-(iii) (c)-(iv) (d)-(iv)
2) (a)-(iv) (b)-(ii) (c)-(iv) (d)-(i)
3) (a)-(i) (b)-(iii) (c)-(iv) (d)-(ii)
4) (a)-(ii) (b)-(iv) (c)-(i) (d)-(iii)

3. A square loop is placed near a long straight current carrying wire as shown. Match the following two columns.

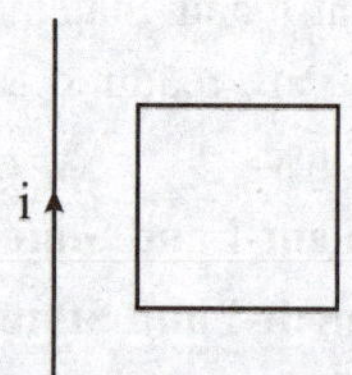

Column I		Column II	
(a)	If current is increased	(i)	Induced current in loop is clockwise
(b)	If current is decreased	(ii)	induced current in loop is anti-clockwise
(c)	If loop is moved away from the wire	(iii)	wire will attract the loop
(d)	If loop is moved towards the wire	(iv)	wire will repel the loop

1) (a)-(i) (b)-(iii) (c)-(iv) (d)-(iv)
2) (a)-(i) (b)-(ii) (c)-(iii) (d)-(iv)
3) (a)-(ii, iv) (b)-(i, iii) (c)-(i, iii) (d)-(ii, iv)
4) (a)-(ii) (b)-(iv) (c)-(i) (d)-(iii)

4. Magnetic flux in a circuit in a circular coil of resistance 10Ω changes with time as shown in the figure.

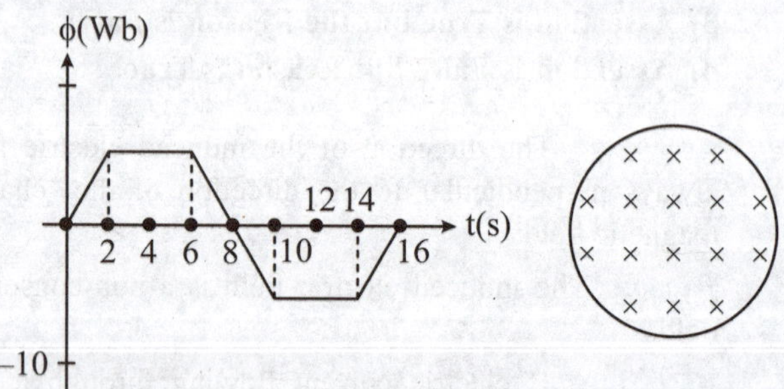

Column I		Column II	
(a)	At 1 s induced current is	(i)	Clockwise
(b)	At 5 s induced current is	(ii)	Anti-clockwise
(c)	At 9 s induced current is	(iii)	Zero
(d)	At 15 s induced current is	(iv)	2A

1) (a)-(i) (b)-(iii) (c)-(iv) (d)-(iv)
2) (a)-(i) (b)-(ii) (c)-(iii) (d)-(iv)
3) (a)-(i) (b)-(iii) (c)-(iv) (d)-(ii)
4) (a)-(ii) (b)-(iii) (c)-(i) (d)-(ii)

NCERT Exemplar MCQs

1. A square of side L metres lies in the xy-plane in a region, where the magnetic field is given by $B = B_0(2\hat{i}+3\hat{j}+4\hat{k})T$, where B_0 is constant. The magnitude of flux passing through the square is:
   1) $2B_0L_2$ Wb
   2) $3B_0L^2$ Wb
   3) $4B_0L^2$ Wb
   4) $\sqrt{29}B_0L^2Wb$

2. A loop, made of straight edges has six corners at A (0, 0, 0), B (L, 0, 0), C (L, L, 0), D (0, L, 0), E (0, L, L) and F (0, 0, L). A magnetic field $B = B_0(\hat{i}+\hat{k})T$ is present in the region. The flux passing through the loop ABCDEFA (in that order) is:
   1) B_0L^2 Wb
   2) $2B_0L^2$ Wb
   3) $\sqrt{2}B_0L^2Wb$
   4) $4B_0L^2$Wb

3. A cylindrical bar magnet is rotated about its axis. A wire is connected from the axis and is made to touch the cylindrical surface through a contact. Then:
   1) A direct current flows in the ammeter A
   2) No current flows through the ammeter A
   3) An alternating sinusoidal current flows through the ammeter A with a time period $T = \frac{2\pi}{\omega}$
   4) A time varying non-sinusoidal current flows through the ammeter A

4. There are two coils A and B as shown in the figure. A current starts flowing in B as shown, when A is moved towards B and stops when A stops moving. The current in A is counterclockwise. B is kept stationary when A moves. We can infer that

   A $\vec{v}$ B

   1) There is a constant current in the clockwise direction of A.
   2) There is a varying current in A.
   3) There is no current in A.
   4) There is a constant current in the counterclockwise direction in A.

5. The coil A is made to rotate about a vertical axis (figure). No current flows in B if A is at rest. The current in coil A, when the current in B (at t = 0) is counter-clockwise and the coil A is as shown at this instant, t = 0, is:

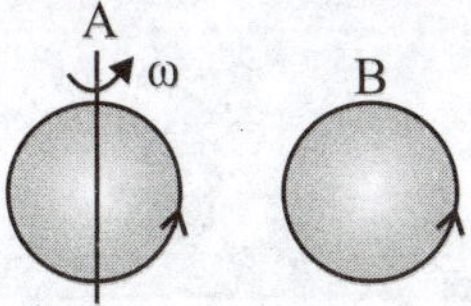

   1) Constant current clockwise
   2) Varying current clockwise
   3) Varying current counter clockwise
   4) Constant current counter clockwise

6. The self-inductance L of a solenoid of length l and area of cross-section A, with a fixed number of turns N increases as:
   1) l and A increase
   2) l decreases and A increases
   3) l increases and A decreases
   4) Both l and A decrease

NEET & AIIMS Past Year MCQs

1. **(2024)**

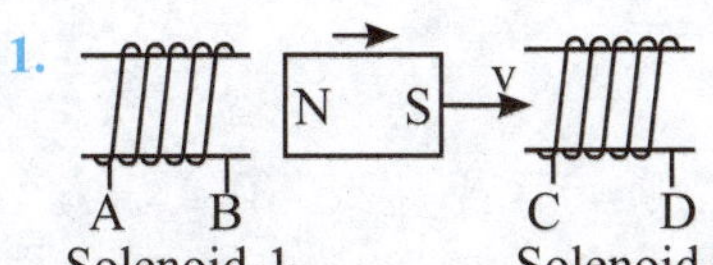

   In the above diagram, a strong bar magnet is moving towards solenoid-2 from solenoid-1. The direction of induced current in solenoid-1 and that in solenoid-2, respectively, are through the directions:
   1) *AB* and *CD*
   2) *BA* and *DC*
   3) *AB* and *DC*
   4) *BA* and *CD*

2. The magnetic energy stored in an inductor of inductance 4 μH carrying a current of 2 A is: **(2023)**
   1) 8 μJ
   2) 4 μJ
   3) 4 mJ
   4) 8 mJ

3. A square loop of side 1 m and resistance 1 Ω is placed in a magnetic field of 0.5T. If the plane of loop of perpendicular to the direction of a magnetic field, the magnetic flux through the loop is: **(2022)**
   1) zero weber
   2) 2 weber
   3) 0.5 weber
   4) 1 weber

4. A big circular coil of 100 turns and average radius 10 m is rotating about its horizontal diameter at 2 rad s^{-1}. If the vertical component of earth's magnetic field at that place is 2×10^{-5} T and electrical resistance of the coil is 12.56 Ω, then the maximum induced current in the coil will be: **(2022)**
   1) 2A
   2) 0.25 A
   3) 1.5 A
   4) 1 A

5. Two conducting circular loops of radii R_1 and R_2 are placed in the same plane with their centres coinciding. If $R_1 >> R_2$, the mutual inductance M between them will be directly proportional to: **(2021)**

1) $\frac{R_2}{R_1}$ 2) $\frac{R_1^2}{R_1}$

3) $\frac{R_2^2}{R_1}$ 4) $\frac{R_1}{R_2}$

6. If in an inductor of inductance L, radius r, current changes from I_1 to I_2. Find work done. (AIIMS 2018)

1) $\frac{1}{2}L\left(I_2^2 - I_1^2\right)$ 2) $\frac{1}{2}L\left(I_1^2 - I_2^2\right)$

3) $L\left(I_1^2 + I_2^2\right)$ 4) $L\left(I_2^2 - I_1^2\right)$

7. A metallic ring is dropped down, keeping its plane perpendicular to a constant and horizontal magnetic field. The ring enters the region of magnetic field at t = 0 and completely emerges out at t = T sec. The current in the ring varies as: (AIIMS 2006)

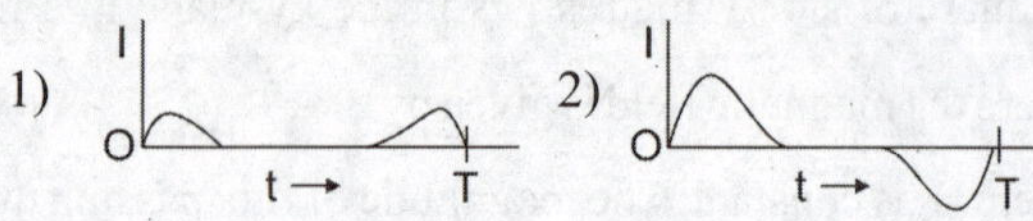

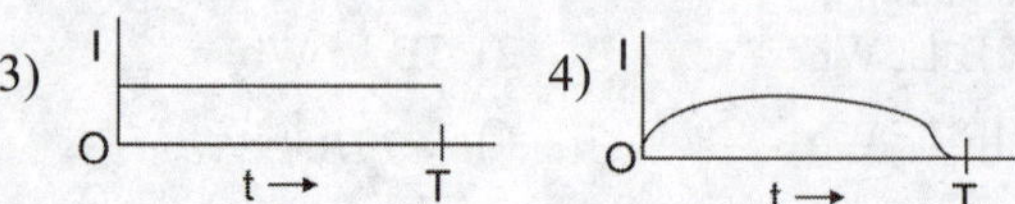

8. The coefficient of mutual inductance, when magnetic flux changes by 2×10^{-2} Wb and current changes by 0.01 A is: (AIIMS 2002)

1) 2 H 2) 4 H
3) 3 H 4) 8 H

Answer Key

NCERT Topic-wise MCQs

1	2	3	4	5	6	7	8	9	10	11	12	13	14	15	16	17
1	4	2	2	2	1	4	3	1	1	1	3	4	3	1	1	2
18	**19**	**20**	**21**	**22**	**23**	**24**	**25**	**26**	**27**	**28**	**29**	**30**	**31**	**32**	**33**	**34**
3	4	1	2	2	3	3	4	2	3	2	4	2	4	4	2	4
35	**36**	**37**	**38**	**39**	**40**	**41**	**42**	**43**	**44**	**45**	**46**	**47**	**48**	**49**	**50**	
1	3	4	4	3	2	1	3	1	2	4	2	2	2	4	3	

High Order Time Intensive MCQs

1	2	3	4	5	6	7	8	9	10	11	12	13	14	15	16	17
1	2	4	1	1	4	2	3	1	4	1	4	1	4	4	4	1
18	**19**	**20**	**21**	**22**	**23**	**24**	**25**	**26**	**27**	**28**	**29**	**30**	**31**	**32**	**33**	
2	2	2	2	2	3	4	1	4	4	4	1	2	3	3	3	

Statement Based MCQs

1	2	3	4
2	3	3	1

Assertion & Reason MCQs

1	2	3	4	5
2	1	3	1	2

Match the Columns MCQs

1	2	3	4
1	2	3	4

NCERT Exemplar MCQs

1	2	3	4	5	6
3	2	2	4	1	2

NEET & AIIMS Past Year MCQs

1	2	3	4	5	6	7	8
3	1	3	4	3	1	2	1

CHAPTER 7

Alternating Current

Topicwise Number of Questions Analysis (2025-2016)

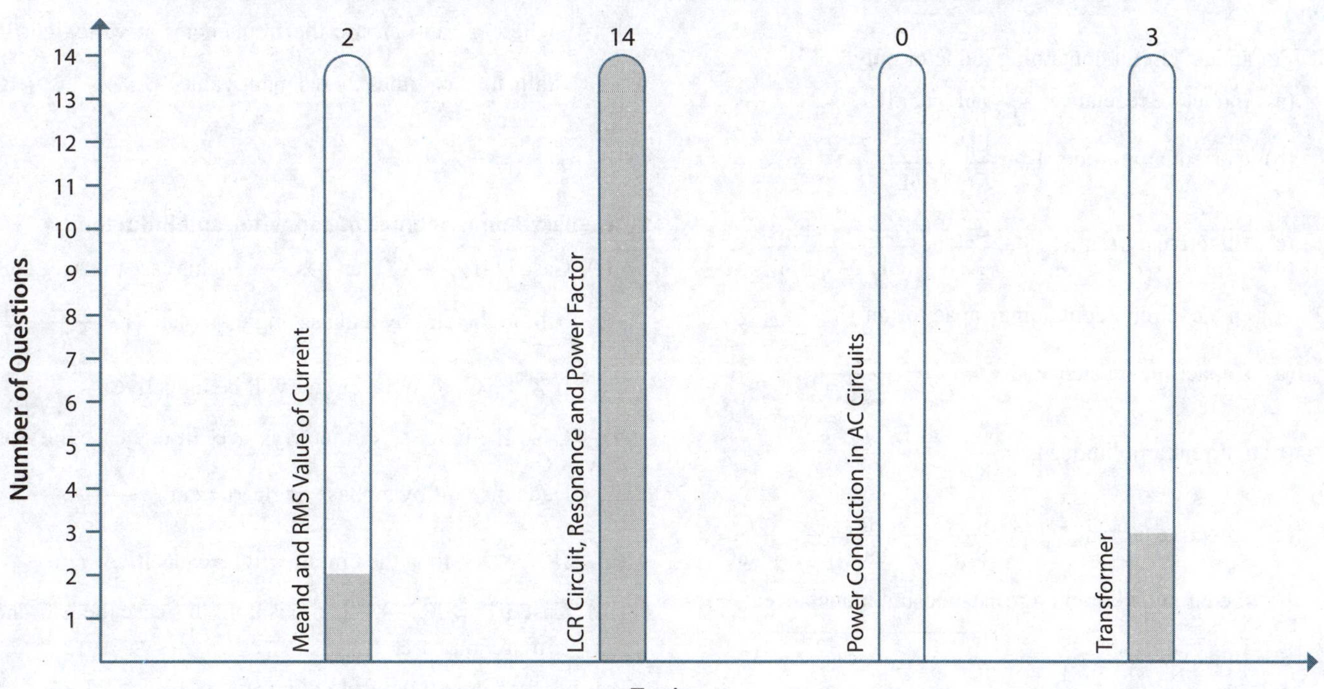

NCERT-PICKS

AC Circuits

Formulas

1. Instantaneous value of a.c., $I = I_0 \sin \omega t$, where I_0 is the peak or maximum value of a.c.

2. Average or mean value of a.c. over half cycle,

$$I_{av} = \frac{2}{\pi} I_0 = 0.637\ I_0$$

3. Effective or rms or virtual value of a.c.,

$$I_{eff} \text{ or } I_{rms} \text{ or } I_v = \frac{1}{\sqrt{2}} I_0 = 0.707\ I_0$$

4. For alternating voltages, we have

$$E = E_0 \sin \omega t, E_{av} = 0.637 E_0, E_{rms} = \frac{1}{\sqrt{2}} E_0$$

Numerical Profile

AC Throught A resistor

(i) Average value of emf can be found by the formula:

$$E_{av} = \frac{\int_{t_1}^{t_2} E(t)dt}{t_2 - t_1}$$

(ii) Average value of alternating current or emf over a long period of time will be zero, because for this time interval $t_2 - t_1$ will be large whereas quantity in the numerator will be finite.

(iii) Ordinary dc ammeters and voltmeters cannot measure alternating current or alternating emf. They will show zero reading. For this purpose, we use ac ammeters and voltmeters. They are known as hot wire instruments. They measure virtual values of current or voltage. For example, if we measure $I = 10 \sin \omega t$ A from ac ammeter, it will show $I_v = \frac{10}{\sqrt{2}} A$.

(iv) The symbol for an ac source is shown below.

AC Circuit Containing Inductor, Capacitor, Resistance. Resonance and Power Factor

Formulas

1. For an a.c. circuit containing inductor only,
   (a) Inductive reactance, $X_L = \omega L = 2\pi fL$
   (b) Current amplitude, $I_0 = \frac{E_0}{X_L} = \frac{E_0}{\omega L}$
   (c) Effective current, $I_{rms} = \frac{E_{rms}}{X_L} = \frac{E_{rms}}{\omega L} = \frac{E_0}{\sqrt{2}.\omega L}$
2. For an a.c. circuit containing capacitor only,
   (a) Capacitive reactance, $X_C = \frac{1}{\omega C} = \frac{1}{2\pi f C}$
   (b) Current amplitude, $I_0 = \frac{E_0}{X_C} = \frac{E_0}{1/\omega C}$
   (c) Effective current, $I_{rms} = \frac{E_{rms}}{X_C} = \frac{E_{rms}}{1/\omega C} = \frac{E_0}{\sqrt{2}.1/\omega C}$
3. For the circuit containing resistance and inductance:
   (a) Impedance, $Z = \frac{E_{rms}}{I_{rms}} = \sqrt{R^2 + X_L^2} = \sqrt{R^2 + \omega^2 L^2}$
   (b) Current, $I_{rms} = \frac{E_{rms}}{Z}$
   (c) Phase angle ϕ is given by
   $$\tan\phi = \frac{X_L}{R} = \frac{\omega L}{R} \text{ or } \cos\phi = \frac{R}{Z}.$$
   (d) Instantaneous current, $I = I_0 \sin(\omega t - \phi)$
4. For the circuit containing resistance and capacitance.
   (a) Impedance, $Z = \frac{E_{rms}}{I_{rms}} = \sqrt{R^2 + X_C^2} = \sqrt{R^2 + \frac{1}{\omega^2 C^2}}$
   (b) Current, $I_{rms} = \frac{E_{rms}}{Z}$
   (c) Phase angle ϕ is given by
   $$\tan\phi = \frac{X_C}{R} = \frac{1/\omega C}{R} \text{ or } \cos\phi = \frac{R}{Z}$$
   (d) Instantaneous current, $I = I_0 \sin(\omega t + \phi)$
5. For the circuit contacting resistance, inductance and capacitance.
   (a) Impedance of a series LCR-circuit,
   $$Z = \frac{E_{rms}}{I_{rms}} = \sqrt{R^2 + (X_L - X_C)^2} = \sqrt{R^2 + \left(\omega L - \frac{1}{\omega C}\right)^2}$$
   (b) Phase angle between current and voltage is given by
   $$\tan\phi = \frac{X_L - X_C}{R} \text{ or } \cos\phi = \frac{R}{Z}$$
   (c) Resonant frequency of LCR-series circuit (when $X_L = X_C$),
   $$f_r = \frac{1}{2\pi\sqrt{LC}}$$
   (d) Q - Factor $= \frac{\omega_r}{\omega_2 - \omega_1} = \frac{\omega_r L}{R} = \frac{1}{R}\sqrt{\frac{L}{C}}$
   where ω_1 and ω_2 are the frequencies at which current falls to $\frac{1}{\sqrt{2}}$ times its resonant value. *[RC]*

Numerical Profile

Case based on reactance of capacitor and inductor

(i) Case I : If $X_L > X_C$ then ϕ is +ve. In this case the A.C. lags behind the emf by a phase angle $\phi = \tan^{-1}\left(\frac{X_L - X_C}{R}\right)$.
If $X_L > X_C$, then the circuit will be inductive

(ii) Case II : If $X_L < X_C$ then ϕ is –ve. In this case the A.C leads the emf by a phase angle $\phi = \tan^{-1}\left(\frac{X_L - X_C}{R}\right)$.
If $X_L < X_C$, then the circuit will be capacitive

(iii) Case III : If $X_L = X_C$ then ϕ is 0. In this case the A.C and emf are in phase.

Graph based on Alternating Current

(i) Graph of X_L (in Ω) vs f (in Hz):

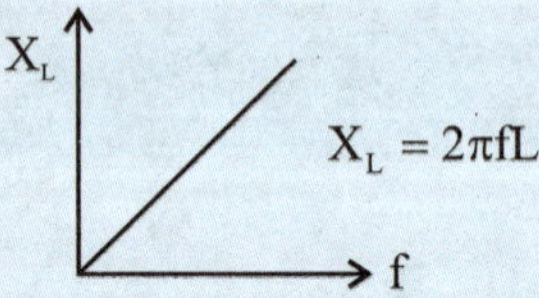

(ii) Graph of X_C (in Ω) vs f (in Hz):

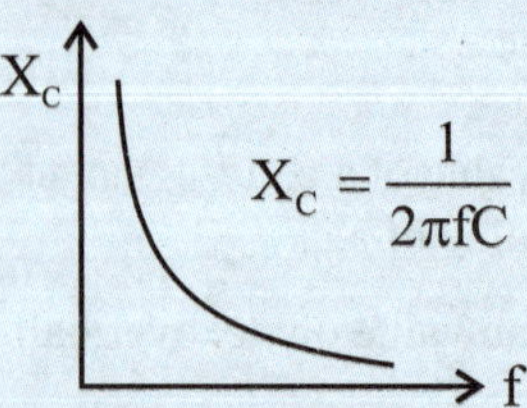

(iii) Graph showing growth of current in L-R circuit

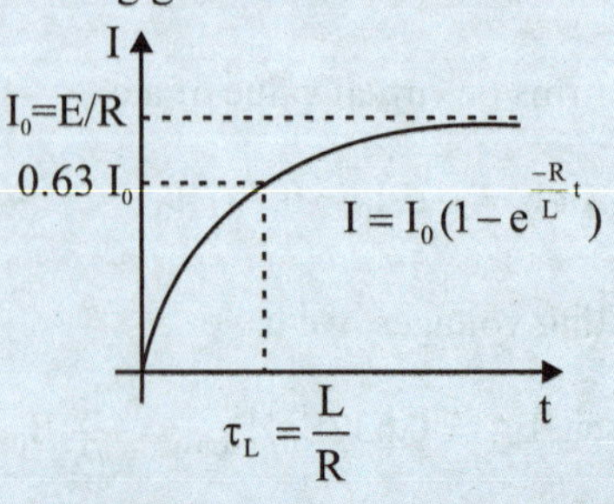

(iv) Decay of Current

The equation gives the value of current at any time t during decay of current in LR-circuit.

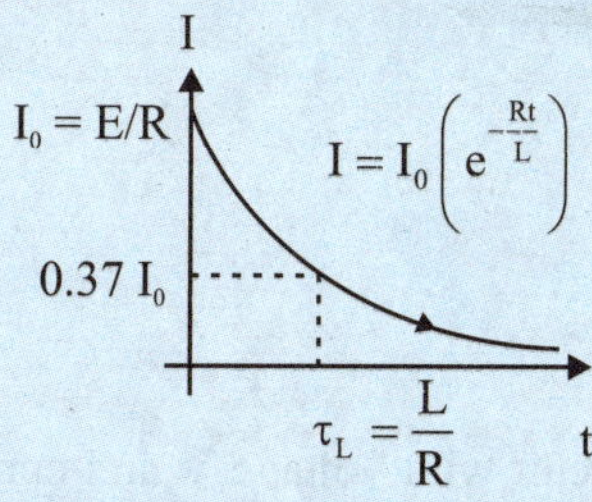

(v) Curves represents the variation of impedance (Z) with frequency f in series LCR circuit

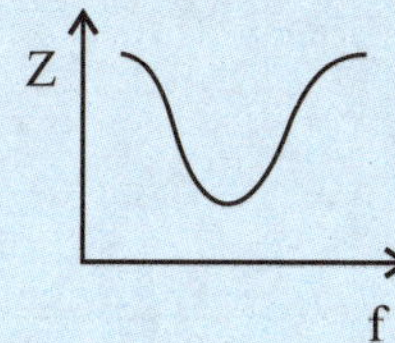

Power Consumption in AC Circuits

Formulas

1. Average Power = $I_{rms}\ E_{rms}\ \cos\phi$
2. Power factor = $\cos\phi = R/Z$
3. For RL circuit, $\cos\phi = \frac{R}{\sqrt{R^2 + X_L^2}}$
4. For RC circuit, $\cos\phi = \frac{R}{\sqrt{R^2 + X_C^2}}$
5. For LCR circuit, $\cos\phi = \frac{R}{\sqrt{R^2 + (X_L - X_C)^2}}$
6. For pure inductive & capacitive circuit, $\cos\phi = 0$
   $\therefore \phi$ between the I & V is $\pi/2$
7. For pure resistive circuit, $\cos\phi = 1 = \frac{R}{R}$

Transformer from EMI

Formulas

1. Emf induced, $e = e_0 \sin \omega t$
   [$\omega = 2\pi/T$ and T is the time period, e_0 is the maximum emf induced]
2. Efficiency, $\eta = \frac{\text{Output power}}{\text{Input power}} = \frac{E_s I_s}{E_P I_P}$
   [E_s = secondary voltage, E_p = primary voltage, I_s = current in secondary coil and I_p = current in primary coil]
3. When efficiency of transformer is 100%
   $\frac{E_s}{E_p} = \frac{N_s}{N_P} = \frac{I_P}{I_s} = K$ (Transformer ratio)
   [N_s = Number of secondary turns and N_p = number of primary turns]

Statements Asked in Exam

- Shock of A.C is attractive, whereas that of D.C is repulsive.
- For electroplating, electrorefining, electrotyping etc., A.C cannot be used, in such cases D.C is used.
- If a lamp and a capacitor are connected in series with an A.C source the lamp will shine, however, if the combination is connected with a D.C source the lamp will not shine.
- Capacitor gives infinite resistance to D.C ($\because X_c = 1/\omega C$ as, $\omega = 0 \Rightarrow X_c = \infty$)
- R.M.S value of A.C is that value of steady current, which would generate the same amount of heat in a given resistance in a given time as is done by the A.C when passed through the same resistance for the same time.
- $\therefore I_{rms} = \frac{I_0}{\sqrt{2}} = 0.707\ I_0$ = 70% of the peak value of A.C
- Inductor gives smooth pass to D.C as $X_L = \omega L$. For dc $\omega = 0 \Rightarrow X_L = 0$
- An inductor acts as a conductor for D.C supply.
- Inductor offers an easy path to D.C and resistive path to A.C.
- Self inductance plays a role of inertia (opposes both decay & growth of current)
- Work needs to be done against the back e.m.f E in establishing the current. This work done is stored as magnetic potential energy.
- Series resonance circuit is also known as acceptor circuit because it admits maximum current.
- Average power/cycle associated each with an inductor and a capacitor is zero. But average power per cycle associated with a resistor is not zero.
- $P = I_{rms}\ E_{rms}$
- Average power associated with pure inductor and pure capacitor is zero.
- A.C cannot produce chemical effects of currents e.g., electrolysis, electroplating etc. due to the large inertia of heavy ions which cannot follow the frequency of A.C.
- The divisions marked on a.c ammeter scale are not equally spaced because an a.c ammeter is based on heating effect of current.
- We can measure D.C by an D.C ammeter but not A.C because full cycle of A.C is zero
- The dimensional formula of $\sqrt{LC}$ and RC is equal to ($M^0L^0T^1$).
- A series L.C.R circuit have minimum impedance at resonance and hence admits maximum current.
- Whatever be the circuit, power dissipation always occurs in the resistor.
- Choke coil (inductor) is preferred in controlling A.C than the rheostat because no power or energy is dissipated in the case of choke coil. In rheostat some energy is dissipiated in the form of heat.
- There is no device that can control D.C. without any loss of power, even a choke coil cannot do so.
- Wattless current is one for which average power as well as power factor is zero.
- A hot wire ammeter can measure both alternating current and direct current.

NCERT Topic-wise MCQs

Power Factor, Power Consumption in AC circuits

1. A sinusoidal A.C current flows through a resistor of resistance R. If the peak current is I_p then the power dissipated is:
   1) $I_p^2 R\cos\theta$ 2) $\frac{1}{2}I_p^2 R$
   3) $\frac{4}{\pi}I_p^2 R$ 4) $\frac{I_p^2 R}{\pi}$

2. A 100Ω electric iron is connected to 200V, 50Hz ac source. Calculate average power delivered to iron, peak power and energy spent in one minute?
   1) 400W, 800W, 12×10^5 J
   2) 400W, 900W, 1.2×10^5 J
   3) 500W, 800W, 6×10^5 J
   4) 400W, 900W, 60×10^5 J

3. If a current I given by $I_O \sin(\omega t - \pi/2)$ flows in an a.c circuit across which an ac potential of $E = E_O \sin \omega t$ has been applied then the power consumption P in the circuit will be:
   1) $P = \frac{E_O I_O}{\sqrt{2}}$ 2) $P = \sqrt{2}E_O I_O$
   3) $P = \frac{E_O I_O}{2}$ 4) $P = 0$

4. An A.C voltage E = 200 sin 300t is applied across a series combination of R = 10Ω and L = 800 mH. Calculate the power factor of the circuit.
   1) 0.0315 2) 0.416
   3) 0.0416 4) 4.17

5. An alternating voltage e = 200 sin 100t is applied to a series combination of R = 30Ω and an inductor of 400 mH. The power factor of the circuit is:
   1) 0.01 2) 0.2
   3) 0.05 4) 0.6

6. Find the capacity of the capacitor which when put in series with a 10Ω resistor makes the power factor equal to 0.5. Assume an 80V-100Hz ac supply:
   1) 9.0×10^{-5} F 2) 9.1×10^{-5} F
   3) 9.2×10^{-5} F 4) 9.3×10^{-5} F

7. In an a.c circuit with voltage V and current I. The power dissipated is:
   1) VI
   2) $\frac{VI}{2}$
   3) $\frac{VI}{\sqrt{2}}$
   4) Can't find with this in formation

8. Calculate the value of inductance which should be connected in series with a capacitance of 5μF and a resistance of 10Ω and an a.c source of 50Hz, so that the power factor of the circuit is unity.
   1) 3.028 H 2) 4.029 H
   3) 2.026 H 4) 1.01 H

9. An ac supply gives $(30V)_{r.m.s}$ which passes through a 100Ω resistance. The power dissipated in it is:
   1) 30W 2) 90W
   3) 40W 4) 9W

10. Power dissipated in an LCR series circuit connected to an A.C source of emf ε is
    1) $\frac{\varepsilon^2\sqrt{R^2+\left(\omega L-\frac{1}{\omega C}\right)^2}}{R}$ 2) $\frac{\varepsilon^2\left[R^2+\left(\omega L-\frac{1}{\omega C}\right)^2\right]}{R}$
    3) $\frac{\varepsilon^2 R}{\sqrt{R^2+\left(\omega L-\frac{1}{\omega C}\right)^2}}$ 4) $\frac{\varepsilon^2 R}{R^2+\left[\left(\omega L-\frac{1}{\omega C}\right)^2\right]}$

11. Power delivered by the source of the circuit becomes maximum when:
    1) $\omega L = \omega C$ 2) $\omega L = \frac{1}{\omega C}$
    3) $\omega L = \left(\frac{1}{\omega C}\right)^2$ 4) $\omega L = \sqrt{\omega C}$

12. The power factor of LCR circuit at resonance is:
    1) 0.707 2) 1
    3) Zero 4) 0.5

13. The coefficient of induction of a choke coil is 0.1H and resistance is 12Ω. If it is connected to an a.c source of frequency 60Hz. Then the power factor will be:
    1) 0.35 2) 0.30
    3) 0.28 4) 0.24

14. There is a 5Ω resistance in an a.c circuit. Inductance of 0.1H is connected with it in series. If equation of A.C e.m.f is 5 sin 50t then the phase difference between current and e.m.f is:
1) $\frac{\pi}{2}$ 2) $\frac{\pi}{4}$
3) $\frac{\pi}{6}$ 4) Zero

15. The average power dissipation in a pure capacitance A.C circuit is:
1) $\frac{CV^2}{2}$ 2) CV^2
3) $\frac{CV^2}{4}$ 4) Zero

AC circuits

16. In an A.C circuit, peak value of voltage is 423 volts, its effective voltage is:
1) 400V 2) 323V
3) 299V 4) 340V

17. The peak value of 220 volts of A.C mains is:
1) 155.6 2) 220V
3) 311V 4) 440V

18. The r.m.s value of an ac of 50Hz is 10A. The time taken by the A.C in reaching from zero to maximum value and peak value of current will be:
1) 2×10^{-2} sec and 14.14A
2) 10^{-2} sec and 7.07A
3) 5×10^{-3} sec and 7.07A
4) 5×10^{-3} sec and 14.14A

19. An A.C is given by the equation $I = i_1\cos\omega t + i_2\sin\omega t$. The r.m.s current is given by:
1) $\frac{i_1 + i_2}{\sqrt{2}}$ 2) $\frac{(i_1 + i_2)^2}{\sqrt{2}}$
3) $\frac{(i_1^2 + i_2^2)^{1/2}}{\sqrt{2}}$ 4) $\frac{(i_1 + i_2)^{1/2}}{2}$

20. The frequency of an alternating voltage is 50 cycles/sec and its amplitude is 120V. Then the r.m.s value of voltage is:
1) 101.37V 2) 84.8V
3) 70.7V 4) 56.5V

21. Voltage and current in an A.C circuit are given by $V = 5\sin(100\pi t - \pi/6)$ and $I = 4\sin(100\pi t + \pi/6)$.
1) Voltage leads the current by 30°
2) Current leads the voltage by 30°
3) Current leads the voltage by 60°
4) Voltage leads the current by 60°

22. The voltage supplied to a circuit is given by $V = V_0 t^{3/2}$, where t is time in second. Find the rms value of voltage for the period, t = 0 to t = 1s.
1) $V_0/2$ 2) $3V_0/2$
3) $V_0/3$ 4) $V_0/5$

23. Calculat rms value of current for the following cases $i = 4 + 3\sin\omega t$
1) 4.52 units 2) 7.47 units
3) 6.49 units 4) 5.98 units

24. An AC voltage is given by $E = E_0 \sin\frac{2\pi t}{T}$
Then, the mean value of voltage calculated over any time interval of T/2 second.
1) is always zero 2) is never zero
3) is always $(2E_0/\pi)$ 4) may be zero

25. If the effective current in a 50 cycle per second a.c circuit is 5A, what is the peak value of current? What is the current 1/600 second after it was zero?
1) 6.07A, 3.535A 2) 7.07A, 3.645A
3) 5.05A, 5.05A 4) 7.07A, 3.535A

AC Circuit Containing Inductor, Capacitor, Resistance, Resonance

26. A 10 μF capacitor is charged to a 25V potential. The battery is then disconnected and pure 100 mH coil is connected across the capacitor so that LC oscillations are set up. Calculate maximum current in the coil:
1) 0.3A 2) 0.25A
3) 0.65A 4) 0.4A

27. Calculate the value of capacity in picofarad, which will make 101.4 microhenry inductance to oscillate with frequency of one megahertz.
1) 250 pF 2) 249.6 pF
3) 240 pF 4) 242 pF

28. A 100mH inductor, a 25 μF capacitor and 15Ω resistor are connected in series to a 120V, 50Hz a.c source. Calculate the resonant frequency:
1) 100.7 Hz 2) 100 Hz
3) 100.2 Hz 4) 1000 Hz

29. An alternating current of 1.5mA and angular frequency 300 rad/sec flows through a 10kΩ resistor and a 0.50μF capacitor in series. Find the rms voltage across the capacitor and also calculate the impedance of the circuit.
1) 10V; 1.2×10^4Ω 2) 10^4V; 10^4Ω
3) 50V; 1.0×10^4Ω 4) 10V; 10Ω

30. A bulb of resistance 10Ω, connected to an inductor of inductance L is in series with an a.c source marked 100V, 50Hz. If the phase angle between voltage and current is π/4 radian. What is the value of L?
1) 0.032 H 2) 0.032 μH
3) 0.032 mH 4) 0.032 MH

31. A coil of resistance 300Ω and inductance 1H is connected across an alternating voltage of frequency 300/2π Hz. Calculate the phase difference between voltage and current in the circuit.
1) 60° 2) 30°
3) 45° 4) 90°

32. L, C, R represents physical quantities inductance capacitance and resistance respectively. The combination representing dimensions of frequency is:

1) LC 2) $(LC)^{-1/2}$

3) $\left(\frac{L}{C}\right)^{-1/2}$ 4) $\left(\frac{L}{C}\right)^{1/2}$

33. In a L–R circuit, the value of L is $\left(\frac{0.4}{\pi}\right)$ henry and the value of R is 30Ω. If in the circuit, an alternating e.m.f of 200 volt at 50 cycle per sec is connected, the impedance of the circuit and current will be:

1) 11.4Ω; 17.5A 2) 30.7 Ω; 6.5A

3) 40.4Ω; 5A 4) 50Ω; 4A

34. A 3μF capacitor is connected to a 220V, 50Hz a.c source. Calculate the rms value of current through the circuit. Also, find the peak value of voltage across the capacitor.

1) 0.207A, 311.1V

2) 0.307A, 311.1V

3) 0.208A, 301V

4) 0.206A, 208V

35. What should be the frequency of alternating 200V supply so as to pass a maximum current of 0.9A through an inductance of 1 henry?

1) 40 Hz 2) 50 Hz

3) 60 Hz 4) 70 Hz

36. An alternating current of amplitude 0.25A and frequency 60Hz is fed to a coil of inductance 2H. Calculate the r.m.s potential difference across the inductor.

1) 130V 2) 134.4V

3) 133.4V 4) 132.4V

37. Light bulb is rated at 100W for a 220V supply. What is the rms current through the bulb?

1) 0.32A 2) 0.38A

3) 0.25A 4) 0.45A

38. A 100Ω iron is connected to a 110V–60Hz wall plug. What is the average potential difference over a half cycle?

1) 90V 2) 94.4V

3) 99.1V 4) 98.8V

39. A coil has an inductance of 5H and resistance of 20Ω. Peak emf of 100V is applied to it. What is the energy stored in magnetic form, when current has reached its final value?

1) 62.5J 2) 14.8 cal

3) Both (1) and (2) 4) None of these

40. A coil of inductance 0.4 mH is connected to a capacitor of capacitance 400 pF. To what wavelength is the circuit tuned?

1) 750m 2) 751m

3) 752m 4) 753m

41. A capacitor of 100μF and a coil of resistance 50Ω and inductance 0.5H are connected in series with a 110V–50Hz source. Calculate the r.m.s value of current in the circuit.

1) 0.916A 2) 0.816A

3) 0.10A 4) 0.707A

42. How much inductance should be connected to 200V 50Hz A.C supply so that a maximum current of 0.9A flows through it?

1) 1H 2) 2H

3) 3H 4) 4H

43. An electric lamp which runs at 80 volt D.C and consumes 10A is connected to 100 volt, 50Hz A.C mains. Calculate the inductance of the choke required.

1) 4×10^{-2} H 2) 5×10^{-2} H

3) 1.9×10^{-2} H 4) 3×10^{-2} H

44. At what frequency will a 0.5H inductor have a reactance of 1000Ω?

1) 320Hz 2) 318.5Hz

3) 319Hz 4) 319.5Hz

45. If $i_1 = 3 \sin \omega t$ and $i_2 = 4 \cos \omega t$, then i_3 is

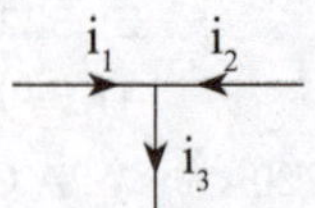

1) $5 \sin(\omega t + 53°)$ 2) $5 \sin(\omega t + 37°)$

3) $5 \sin(\omega t + 45°)$ 4) $5 \sin(\omega t + 35°)$

46. In an A.C circuit, the rms voltage is $100\sqrt{2}$ V. Find the peak value of voltage and its mean value during a positive half cycle.

1) 200V, 127.4V 2) 100V, 127.4V

3) 120V, 150V 4) 200V, 150V

47. An LC circuit contains a 20mH inductor and a 50μF capacitor with an initial charge of 10mC. The resistance of the circuit is negligible. At the instant the circuit is closed be t = 0. At what time, the energy stored is completely magnetic? **[RC]**

1) t = 1.6ms 2) t = 5ms

3) t = 7ms 4) t = 6ms

48. The capacity of a pure capacitor is 1 farad. In dc circuit its effective resistance will be:

1) Zero 2) Infinite

3) 1Ω 4) 1/2Ω

49. A resistance of 40Ω and an inductance of 95.5mH are connected in series in a 50 cycle/second ac circuit. The impedance of this combination is very nearly:

1) 30Ω 2) 40Ω

3) 50Ω 4) 60Ω

50. An inductive reactance of an inductor of $\frac{1}{\pi}$H at 50Hz frequency is:

1) $\frac{50}{\pi}\Omega$ 2) $\frac{\pi}{50}\Omega$

3) 100Ω 4) 50Ω

51. A coil of 200Ω resistance and 1H inductance is connected to an a.c source of frequency $\frac{100}{\pi}$Hz, phase angle between potential and current will be:

1) 30° 2) 45°
3) 60° 4) 90°

52. In LCR circuit the potential difference between the terminals of the inductance is 60V, between the terminals of the capacitor is 30V and that between the terminals of resistance is 40V. The supply voltage will be equal to:

1) 50V 2) 70V
3) 130V 4) 10V

53. A transistor - oscillator using a resonant circuit with an inductor L (of negligible resistance) and a capacitor C in series produce oscillation of frequency f. If L is doubled and C is changed to 4C the frequency will be: ***[RC]***

1) $\frac{f}{2\sqrt{2}}$ 2) $\frac{f}{2}$
3) $\frac{f}{4}$ 4) 8f

54. In a series resonant LCR circuit, the voltage across R is 100V and R = 1kΩ with C = 2μF. The resonant frequency ω is 200 rad/s. At resonance, the voltage across L is:

1) 40V 2) 250V
3) 4×10^{-3}V 4) 25mV

55. An ac source is connected in given series LCR circuit. The rms potential difference across the capacitor of 20 μF is .... V.

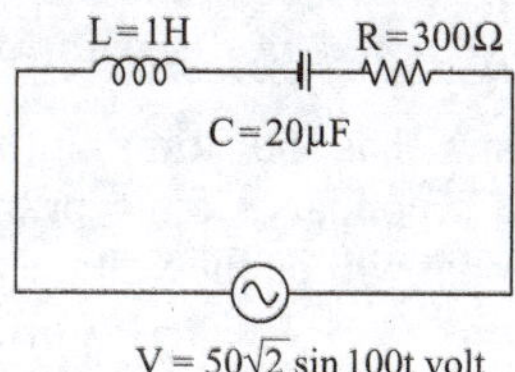

1) 20 V 2) 45 V
3) 50 V 4) 60 V

56. The self inductance of the motor of an electric fan is 10H. In order to impart maximum power at 50Hz, it should be connected to a capacitance of:

1) 1μF 2) 2mF
3) 4mF 4) 8mF

57. In the figure shown, three AC voltmeters are connected at resonance.

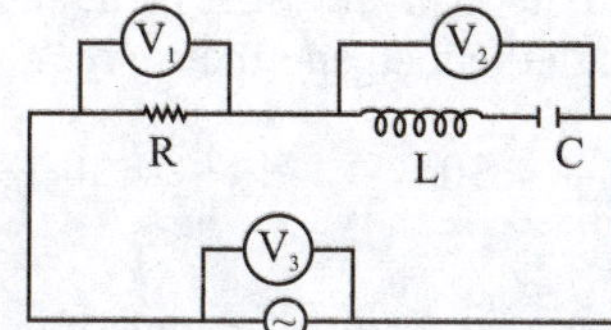

1) $V_2 = 0$ 2) $V_1 = 0$
3) $V_3 = 0$ 4) $V_1 = V_2 \neq 0$

58. A pure inductive coil of 30mH is connected to a source of 220V, 50Hz. The rms current in the coil is:

1) 50.35A 2) 23.4A
3) 30.4A 4) 12.3A

59. A transmitter transmit at a wavelength of 300m. A condenser of capacitance 2.4μF is being used. The value of the inductance for the resonant circuit is approximately:

1) 10^{-4}H 2) 10^{-6}H
3) 10^{-8}H 4) 10^{-10}H

60. A capacitor of capacitance 1μF is charged to a potential of 1V. It is connected in parallel to an inductor of inductance 10^{-3}H. The maximum current that will flow in the circuit has the value:

1) $\sqrt{1000}$mA 2) 1A
3) 1mA 4) 1000mA

61. In AC series circuit, the resistance, inductive reactance and capacitive reactance are 3Ω, 10Ω, 14Ω respectively. The impedence of the circuit is:

1) 5Ω 2) 4Ω
3) 7Ω 4) 10Ω

62. One 10V, 60W bulb is to be connected to 100V line. The required induction coil has self inductance of value (f = 50Hz)

1) 0.052H 2) 2.42H
3) 16.2mH 4) 1.62mH

63. An ac voltage source of variable angular frequency and fixed amplitude V is connected in series with a capacitance C and an electric bulb of resistance R (inductance zero), when ω is increased:

1) The bulb glows dimmer
2) The bulb glows brighter
3) Total impedance of the circuit is unchanged
4) Total impedance of the circuit increases

64. In the adjoining figure, the impedance of the circuit is:

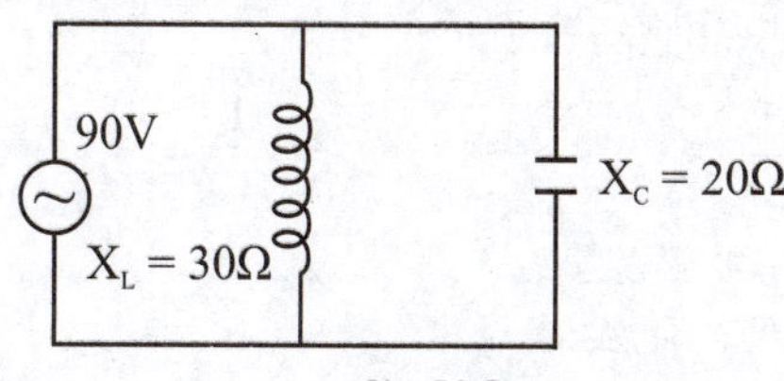

1) 120Ω 2) 50Ω
3) 60Ω 4) 90Ω

65. The variation of the instantaneous current (I) and the instantaneous emf (E) in a circuit is as shown in figure. Which of the following statement is correct.

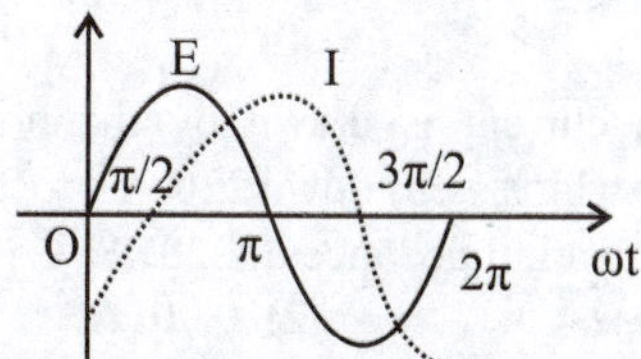

1) The voltage lag behind the current by π/2
2) The voltage leads the current by π/2
3) The voltage and the current are in phase
4) The voltage leads the current by π

66. An a.c source of variable frequency f is connected to an LCR series circuit. Which of the graphs in figure represents the variation of current I in the circuit with frequency f?

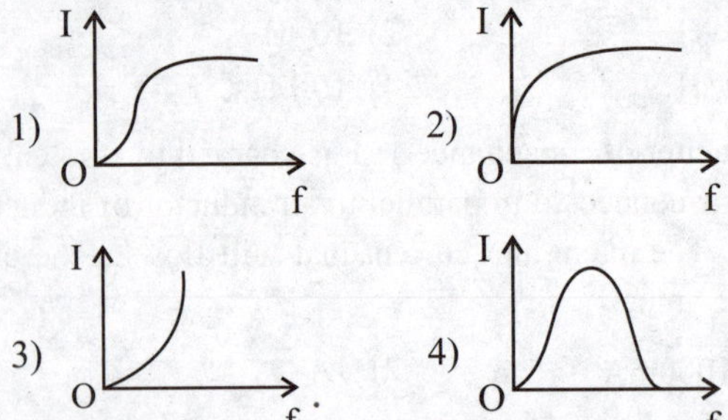

67. The power factor of the circuit as shown in figure is

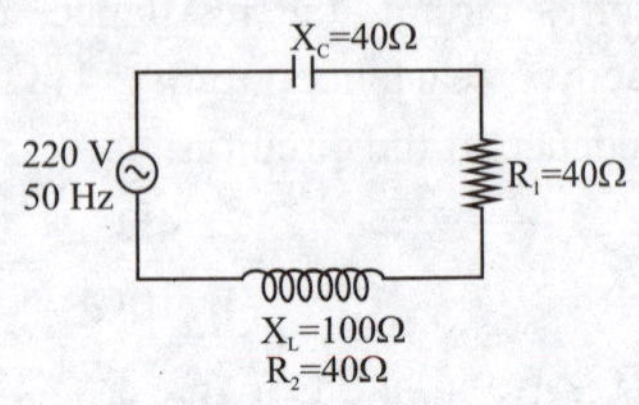

1) 0.2
2) 0.4
3) 0.8
4) 0.6

68. In an a.c circuit the potential difference across an inductance and resistance joined in series are respectively 16V and 20V. The total potential difference of the source is:
1) 20V
2) 25.6V
3) 31.9V
4) 53.5V

69. In series LCR circuit, the phase difference between applied voltage and current is:
1) Positive when $X_L > X_C$
2) Positive when $X_C > X_L$
3) 90^0
4) 0^0

70. What is the mechanical equivalent of spring constant K in LC oscillating circuit?
1) $\frac{1}{L}$
2) $\frac{1}{C}$
3) $\frac{L}{C}$
4) $\frac{1}{LC}$

71. In an oscillating system, a restoring force is a must. In an L-C circuit restoring force is provided by:
1) Capacitor
2) Inductor
3) Resistance
4) Both (1) and (2)

Transformer

72. How much current is drawn by the primary coil of a transformer which steps down 220 volt to 22 volt to operate a device with an impedance of 220Ω?
1) 0.01A
2) 0.001A
3) 0.0001A
4) 1A

73. A flat square coil of 10 turns has sides of length 12cm. The coil rotates in a magnetic field whose flux density is 0.025 Wb/m^2. What is the angular velocity of the coil, if max. e.m.f. produced is 20mV?
1) 2.51 rad/s
2) 3.52 rad/s
3) 4.54 rad/s
4) 5.55 rad/s

74. An A.C generator of 220V having internal resistance r = 10 Ω and external resistance R = 100 Ω. What is the power developed in the external resistor?
1) 484 W
2) 400 W
3) 441 W
4) 369 W

75. A transformer rated at 10 kW is used to connect a 5 kV transmission line to 240V circuit. The ratio of turns in the windings of a transformer?
1) 5
2) 20.8
3) 104
4) 40

76. Calculate the maximum emf induced when a square frame is making 240 r.p.m each side of square being 20cm and magnetic field intensity is 500 gauss?
1) 5.01×10^{-2} V
2) 5.02×10^{-2} V
3) 5.03×10^{-2} V
4) 8×10^{-3} V

77. The turn ratio of a transformer is given as 2 : 3. If the current through the primary coil is 3A, thus calculate the current through load resistance?
1) 1A
2) 4.5A
3) 2A
4) 1.5A

78. The primary winding of a transformer has 100 turns and its secondary winding has 200 turns. The primary is connected to an a.c supply of 120V and the current flowing in its is 10A. The voltage and the current in the secondary are:
1) 240V, 5A
2) 240V, 10A
3) 60V, 20A
4) 120V, 20A

79. A transformer has 50 turns in the primary and 100 turns in the secondary, if the primary is connected to a 220V d.c. What will be the voltage across the secondary?
1) 220V
2) 440V
3) 110V
4) Zero

80. A power transmission line feeds input power at 2300V to a step down transformer having 4000 turns in its primary. What should be the number of turns in secondary to get output power at 230V?
1) 200
2) 300
3) 400
4) 500

81. An a.c source of internal resistance 9000Ω is supply current to a load resistor of 10Ω. Here should the source be matched to the load and what is the ratio of the currents passing through the load and the source?
1) $\frac{n_s}{n_p} = \frac{1}{30}, \frac{I_s}{I_p} = 30$
2) $\frac{n_s}{n_p} = 30, \frac{I_s}{I_P} = \frac{1}{30}$
3) $\frac{n_s}{n_p} = 30, \frac{I_P}{I_S} = \frac{1}{30}$
4) None of these

82. A transformer with efficiency 80% works at 4kW and 100V. If the secondary voltage is 200V. Then the primary and secondary currents are respectively: (in ampere)
1) 40, 16
2) 16, 140
3) 16, 40
4) 20, 40

83. A step down transformer reduces the voltage of a transmission line from 2200V to 220V. The power delivered by it is 880W and its efficiency is 88%. The input current is:
1) 4.65 mA 2) 0.465 A
3) 0.4545 A 4) 4.65 A

84. What is increased in step down transformer?
1) Voltage 2) Current
3) Power 4) Current density

85. The ratio of secondary to the primary turns in a transformer is 3 : 2. If the power output be P, then the input power neglecting all losses must be equal to:
1) 5P 2) 1.5P
3) P 4) 2/5P

86. A transformer of efficiency 90% draws an input power of 4 kW. An electrical appliances connected across the secondary draws a current of 6A. The impedance of the device is:
1) 60Ω 2) 50Ω
3) 80Ω 4) 100Ω

87. An ideal transformer has 500 and 5000 turn in primary and secondary windings respectively. If the primary voltage is connected to a 6V battery then the secondary voltage is:
1) 0V 2) 60V
3) 0.6V 4) 6V

High Order Time Intensive MCQs

1. In an AC circuit, V and I are given by V = 150 sin (150 t) volt and I = 150 sin $\left(150t + \frac{\pi}{3}\right)$ amp.
The power dissipated in the circuit is
1) 106 W 2) 150 W
3) 5625 W 4) Zero

2. If net reactance in a circuit is $\sqrt{3}$ times of resistance, then find phase difference.
1) Zero 2) 30°
3) 60° 4) Data is incomplete

3. In an electrical circuit R, L, C and AC voltage source are all connected in series. When L is removed from the circuit, the phase difference between the voltage and the current in the circuit is π/3. If instead C is removed from the circuit, the phase difference is again π/3. The power factor of the circuit is:
1) 1/2 2) $1\sqrt{2}$
3) 1 4) $\sqrt{3}/2$

4. A coil of inductive reactance 31 Ω has a resistance of 8 Ω. It is placed in series with a condenser of capacitive reactance 25 Ω. The combination is connected to an AC source of 110 V. The power factor of the circuit is:
1) 0.56 2) 0.64
3) 0.80 4) 0.33

5. An AC voltage is applied to a resistance R and an inductor L in series. If R and the inductive reactance are both equal to 3 Ω, the phase difference (in rad) between the applied voltage and the current in the circuit is:
1) π/4 2) π/2
3) Zero 4) π/6

6. A resistor and a capacitor are connected in series with an AC source. If the potential drop across the capacitor is 5 V and that across resistor is 12 V, the applied voltage is:
1) 13 V 2) 17 V
3) 5 V 4) 12 V

7. The reactance of capacitor with capacitance C is X. If both frequency and capacitance become double,then new reactance will be
1) X 2) 2X
3) 4X 4) X/4

8. An inductor of inductance L and resistor of resistance R are joined in series and connected by a source of voltage V and frequency ω. Power dissipated in the circuit is:
1) $\left(\frac{R^2+\omega^2L^2}{V}\right)$ 2) $\frac{V^2R}{(R^2+\omega^2L^2)}$
3) $\left(\frac{V}{R^2+\omega^2L^2}\right)$ 4) $\frac{\sqrt{R^2+\omega^2L^2}}{V^2}$

9. A resistor and a capacitor are connected in series with an AC source. If the potential drop across the capacitor is 5 V and that across resistor is 12 V, the applied voltage is:
1) 13 V 2) 17 V
3) 5 V 4) 12 V

10. In a circuit L, C and R are connected in series with an alternating voltage source of frequency f. The current leads the voltage by 45°. The value of C is
1) $\frac{1}{2\pi f(2\pi fL+R)}$ 2) $\frac{1}{\pi f(2\pi fL+R)}$
3) $\frac{1}{2\pi f(2\pi fL-R)}$ 4) $\frac{1}{\pi f(2\pi fL-R)}$

11. The average current of a sinusoidally varying alternating current of peak value 5A with initial phase zero, between the instants t = T/8 to t = T/4 is (Where 'T' is time period)
1) $\frac{10}{\pi}\sqrt{2}A$ 2) $\frac{5}{\pi}\sqrt{2}A$
3) $\frac{20\sqrt{2}}{\pi}A$ 4) $\frac{10}{\pi}A$

12. An alternating current 'i' is given by $i = i_0 \sin 2\pi(t/T + 1/4)$. Then the avg current in the first one quarter time period is

1) $\frac{2i_0}{\pi}$ 2) $\frac{I_0}{\pi}$

3) $\frac{I_0}{2\pi}$ 4) $\frac{3I_0}{\pi}$

13. A 100 Ω resistance is connected in series with a 4H inductor. The voltage across the resistor is $V_R = 2\sin(1000t)V$. The voltage across the inductor is

1) $80\sin\left(1000t + \frac{\pi}{2}\right)$ 2) $40\sin\left(1000t + \frac{\pi}{2}\right)$

3) $80\sin\left(1000t - \frac{\pi}{2}\right)$ 4) $40\sin\left(1000t - \frac{\pi}{2}\right)$

14. The reading of voltmeter and ammeter in the following figure will respectively be

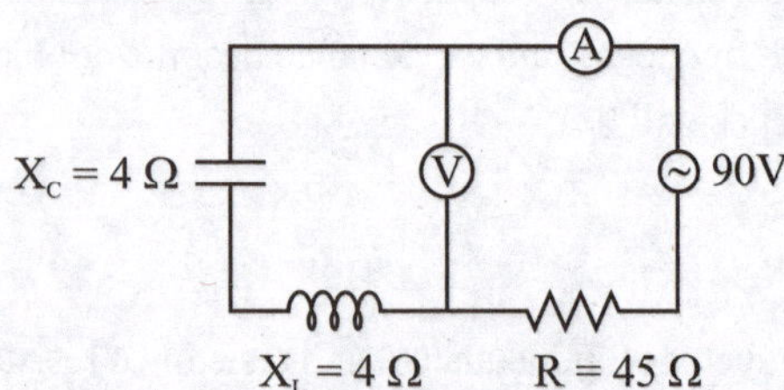

1) 0 and 2A 2) 2A and 0V

3) 2V and 2A 4) 0V and 0A

15. In the following circuit, the values of current flowing in the circuit at f = 0 and f = ∞ will respectively be

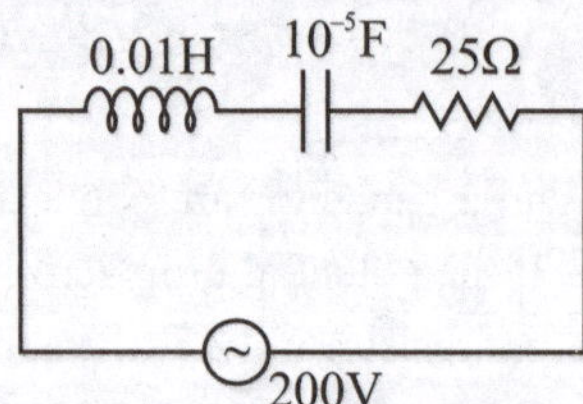

1) 8A and 0A 2) 0A and 0A

3) 8A and 8A 4) 0A and 8A

16. In the series LCR circuit figure the voltmeter and ammeter readings are

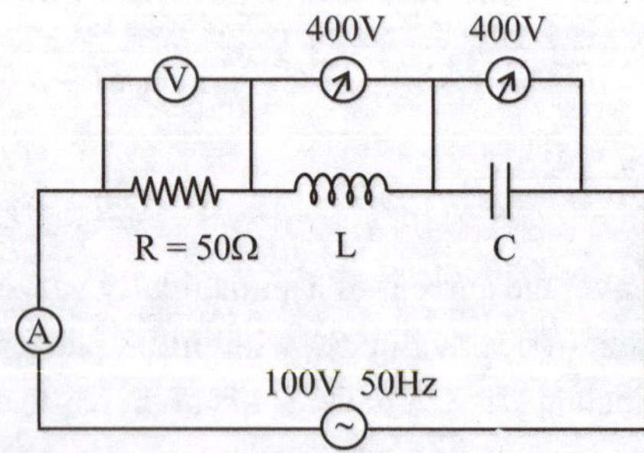

1) V = 100 volt, I = 2 amp

2) V = 100 volt, I = 5 amp

3) V = 1000 volt, I = 2 amp

4) V = 300 volt, I = 1 amp

17. The natural frequency of an LC - circuit is 1,25,000 cycles per second. Then the capacitor C is replaced by another capacitor with a dielectric medium of dielectric constant k. In this case, the frequency decreases by 25 kHz. The value of k is

1) 3.0 2) 2.1

3) 1.56 4) 1.7

18. In the given figure, the instantaneous value of alternating e.m.f. is e = 14.14 sin ωt. The reading of voltmeter in volt will be

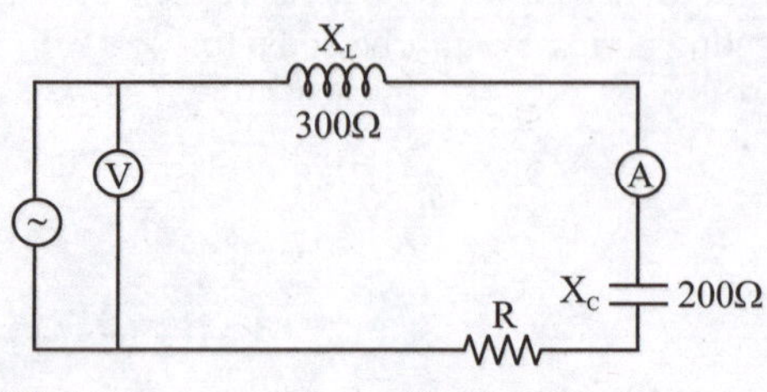

1) 141.4 2) 10

3) 200.0 4) 70.7

19. A 220V, 50Hz A.C generator is connected to an inductor and a 50Ω resistance in series. The current in the circuit is 1.0A. The potential difference across inductor is

1) 102.2V 2) 186.4V

3) 214.2V 4) 302V

20. The voltage of A.C. source varies with time according equation. V = 120 sin100πt cos100πt. Then frequency of source is

1) 50Hz 2) 100Hz

3) 150Hz 4) 200Hz

21. The current in a coil of self inductance 5H is increasing according to i = $2\sin^2 t$. The amount of energy spent during the period when current changes from 0 to 2 ampere is

1) 10J 2) 5J

3) 100J 4) 2J

22. The rms value of potential difference V shown in the figure is

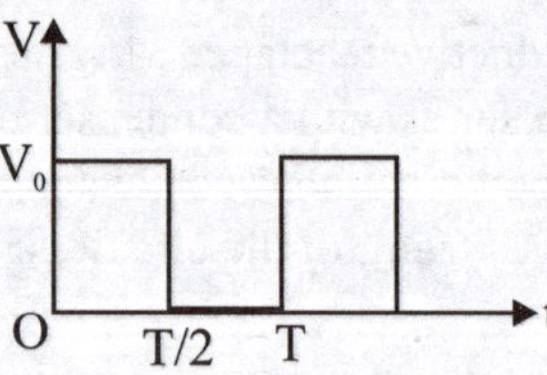

1) $\frac{V_0}{\sqrt{3}}$ 2) V_0

3) $\frac{V_0}{\sqrt{2}}$ 4) $\frac{V_0}{2}$

23. A transformer with efficiency 80% works at 2 kW and 100 V. If the secondary voltage is 200 V. Then, the primary and secondary currents are respectively:

1) 20 A, 8 A

2) 16 A, 40 A

3) 20 A, 40 A

4) 40 A. 20 A

24. The turns ratio of a transformer is given as 2 : 3. If the current through the primary coil is 6 A, thus calculate the current through load resistance.
1) 1 A
2) 4.5 A
3) 4 A
4) 1.5 A

25. For the LCR circuit, shown here, the current is observed to lead the applied voltage. An additional capacitor C', when joined with the capacitor C present in the circuit, makes the power factor of the circuit unity. The capacitor c', must have been connected in

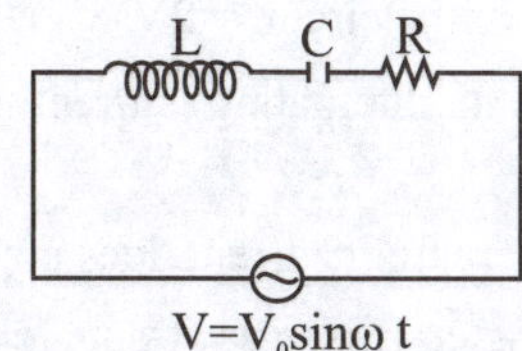

1) series with C and has a magnitude $\frac{1-\omega^2LC}{\omega^2L}$
2) series with C and has a magnitude $\frac{C}{(\omega^2LC-1)}$
3) parallel with C and has a magnitude $\frac{C}{(\omega^2LC-1)}$
4) parallel with C and has a magnitude $\frac{1-\omega^2LC}{\omega^2L}$

26. A current of 5 A is flowing at 220 V in the primary coil of a transformer. If the voltage produced in the secondary coil is 2200 V and 50% of power is lost, then the current in secondary will be:
1) 2.5 A
2) 5 A
3) 0.25 A
4) 0.5 A

27. The primary and secondary coils of a transformer have 50 and 1500 turns, respectively. If the magnetic flux ϕ linked with the primary coil is given by $\phi = \phi_0 + 4t$, where ϕ is in weber, t is time in second and ϕ_0 is a constant, then the output voltage across the secondary coil is:
1) 90 V
2) 120 V
3) 220 V
4) 30 V

28. An ac source is rated at 220V, 50Hz. The time taken for voltage to change from its peak value to zero is:
1) 50 sec
2) 0.02 sec
3) 5 sec
4) 5×10^{-3} sec

29. If an alternating voltage is represented as $E = 141\sin(628t)$, then the r.m.s value of the voltage and the frequency are respectively
1) 141V, 628 Hz
2) 100V, 50 Hz
3) 100V, 100Hz
4) 141V, 100Hz

30. A transformer has 500 primary turns and 10 secondary turns. If the secondary has a resistive load of 15Ω, the currents in the primary and secondary coils respectively, are:
1) 0.16A, 3.2×10^{-3} A
2) 3.2×10^{-3} A, 0.16 A
3) 0.16A, 0.16 A
4) 3.2×10^{-3} A, 3.2×10^{-3} A

31. A transformer of 100% efficiency has 200 turns in the primary and 40000 turns in secondary. It is connected to a 220 V main supply and secondary feeds to a 100 kΩ resistance. The potential difference per turn is:
1) 1.1 V
2) 25 V
3) 0.11 V
4) 11 V

32. The r.m.s voltage of the wave form shown is:

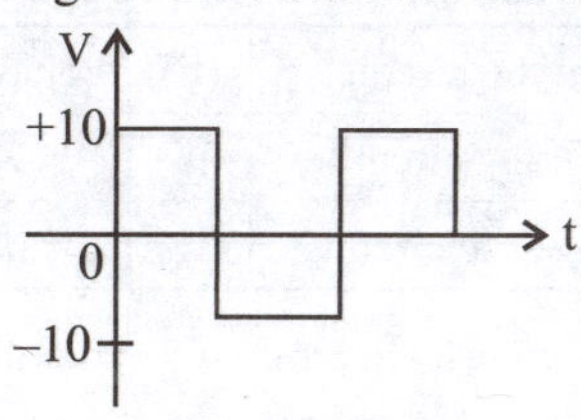

1) 10V
2) 7V
3) 6.37V
4) 11V

Statement Based MCQs

Directions: These questions consist of two statements each, printed as Statement-I and Statement-II. While answering these questions, you are required to choose any one of the following four responses.

1) **Statement-I and Statement-II both are correct**
2) **Statement-I and Statement-II both are incorrect**
3) **Statement-I is correct and Statement-II is incorrect**
4) **Statement-I is incorrect but Statement-II is correct**

1. **Statement-I:** An inductor acts as perfect conductor for D.C.

Statement-II: D.C remains constant in magnitude and direction.

2. **Statement-I:** The capacitive reactance limits the amplitude of the current in a purely capacitive circuit.

Statement-II: Capacitive reactance is proportional to the frequency and the capacitance.

3. **Statement-I:** A.C. is more dangerous than D.C.

Statement-II: For 120 V rms voltage, the peak voltage is 170V.

4. **Statement-I:** The inductive reactance limits amplitude of the current in a purely inductive circuit.

Statement-II: The inductive reactance is independent of the frequency of the current.

5. **Statement-I:** Average value of ac is always defined over half cycle.

Statement-II: Average value of A.C over a complete cycle is always zero.

Assertion & Reason MCQs

Directions: These questions consist of two statements each, printed as Assertion and Reason. While answering these questions, you are required to choose any one of the following four responses.

1) **Both Assertion and Reason are True and the Reason is the correct explanation of the Assertion.**
2) **Both Assertion and Reason are True but Reason is not the correct explanation of the Assertion.**
3) **Assertion is True but the Reason is False.**
4) **Assertion is False but the Reason is True**

1. **Assertion:** A transformer cannot work on DC supply.

   Reason: DC changes neither in magnitude nor in direction.

2. **Assertion:** Kirchhoff's second law can be applied in AC circuit shown below.

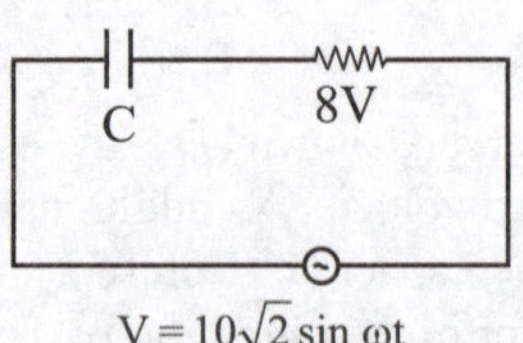

   Reason: V_C in the circuit = 2V.

3. **Assertion:** An alternating current does show magnetic effect.

   Reason: Alternating current does not vary with time.

4. **Assertion:** In a series LCR circuit at resonance, the voltage across the capacitor or inductor may be more than the applied voltage.

   Reason: At resonance in a series LCR circuit, the voltages across inductor and capacitor are out of phase.

Match the Column MCQs

1. Match the following two columns for a series AC circuit.

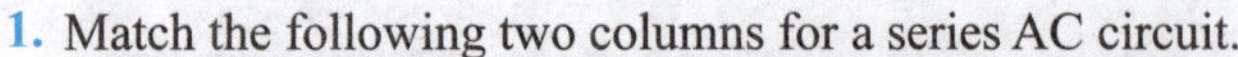

	Column-I		Column-II
(a)	Only C in the circuit	(i)	current will lead
(b)	Only L in the circuit	(ii)	voltage will lead
(c)	Only R in the circuit	(iii)	$\phi = 90°$
(d)	R and C in the circuit	(iv)	$\phi = 0°$

1) (a)-(i, iii); (b)-(ii, iii); (c)-(iv); (d)-(i)
2) (a)-(ii, i); (b)-(i,iv); (c)-(iii); (d)-(ii)
3) (a)-(i, iii); (b)-(ii,iii); (c)-(iv); (d)-(i)
4) (a)-(i, iii); (b)-(ii,iii); (c)-(iv); (d)-(i)

2. Applied AC voltage is given as, $V = V_0 \sin\omega t$.

   Corresponding to this voltage, match the following two columns.

	Column-I		Column-II
(a)	$I = I_0 \sin\omega t$	(i)	only R circuit
(b)	$I = -I_0 \cos\omega t$	(ii)	only L circuit
(c)	$I = I_0 \sin(\omega t + \pi/6)$	(iii)	may be C-R circuit
(d)	$I = I_0 \sin(\omega t - \pi/6)$	(iv)	may be L-C-R circuit

1) (a)-(i, iv); (b)-(ii); (c)-(iii, iv); (d)-(iv)
2) (a)-(iii, i); (b)-(iv); (c)-(i, iii); (d)-(ii)
3) (a)-(ii, iii); (b)-(ii); (c)-(iv, i); (d)-(iii)
4) (a)-(iv, i); (b)-(ii); (c)-(ii, ii); (d)-(i)

3. In series *R-L-C* circuit, $R = 100\ \Omega$, $C = \frac{100}{\pi}\mu F$. and $L = \frac{100}{\pi} mH$ are connected to an ac source as shown in figure. The rms value of ac voltage is 220 V and its frequency is 50 Hz. In column I, some physical quantities are mentioned, while in column II, information about quantities are provided. Match the entries of column I with the entries of column II.

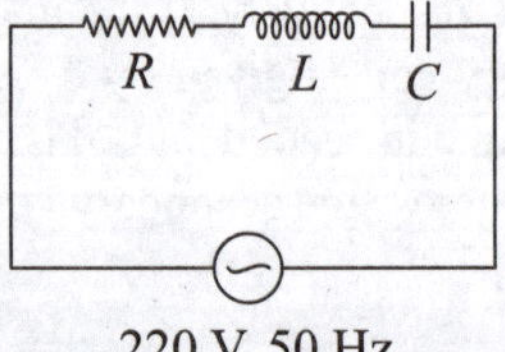

220 V, 50 Hz

	Column-I		Column-II
(a)	Average power dissipated in the resistor is	(i)	Zero
(b)	Average power dissipated in the inductor is	(ii)	Non-zero
(c)	Average power dissipated in the capacitor is	(iii)	163 SI units
(d)	The rms voltage across the capacitor is	(iv)	265.69 SI units

1) (a)-(iv,ii); (b)-(iii, i); (c)-(iv); (d)-(i)
2) (a)-(iii,iv); (b)-(ii,i); (c)-(ii); (d)-(iii)
3) (a)-(ii,iv); (b)-(i); (c)-(i); (d)-(ii,iii)
4) (a)-(i,iii); (b)-(ii,iii); (c)-(i); (d)-(iv)

NCERT Exemplar MCQs

1. If the rms current in a 50 Hz AC circuit is 5 A, the value of the current 1/300 s after its value becomes zero is:
   1) $5\sqrt{2}A$ 2) $5\sqrt{3/2}A$
   3) 5/6 A 4) $5/\sqrt{2}A$
2. An alternating current generator has an internal resistance R_g and an internal reactance X_g. It is used to supply power to a passive load consisting of a resistance R_g and a reactance X_L. For maximum power to be delivered from the generator to the load, the value of X_L is equal to:
   1) Zero 2) X_g
   3) $-X_g$ 4) R_g
3. When a voltage measuring device is connected to AC mains, the meter shows the steady input voltage of 220 V. This means:
   1) Input voltage cannot be AC voltage, but a DC voltage
   2) Maximum input voltage is 220 V
   3) The meter reads not v but $<v^2>$ and is calibrated to read $\sqrt{<v^2>}$
   4) The pointer of the meter is stuck by some mechanical defect
4. To reduce the resonant frequency in an LCR series circuit with a generator:
   1) The generator frequency should be reduced
   2) Another capacitor should be added in parallel to the first
   3) The iron core of the inductor should be removed
   4) Dielectric in the capacitor should be removed
5. Which of the following combinations should be selected for better tuning of an LCR circuit used for communication?
   1) R = 20 Ω, L = 1.5 H, C = 35 μF
   2) R = 25 Ω, L= 2.5 H, C = 45 μF
   3) R = 15 Ω, L = 3.5 H, C = 30 μF
   4) R= 25 Ω, L =1.5 H, C = 45 μF
6. An inductor of reactance 1Ω and a resistor of 2 Ω are connected in series to the terminals of a 6V (rms) AC source. The power dissipated in the circuit is:
   1) 8 W 2) 12 W
   3) 14.4 W 4) 18 W
7. The output of a step-down transformer is measured to be 24 V when connected to a 12 W light bulb. The value of the peak current is:
   1) $1/\sqrt{2}A$ 2) $\sqrt{2}A$
   3) 2 A 4) $2\sqrt{2}A$

NEET & AIIMS Past Year MCQs

1. In the circuit shown below, the inductance L is connected to an ac source. The current flowing in the circuit is $I = I_0 \sin\omega t$. The voltage drop (V_L) across L is **(2024 Re)**

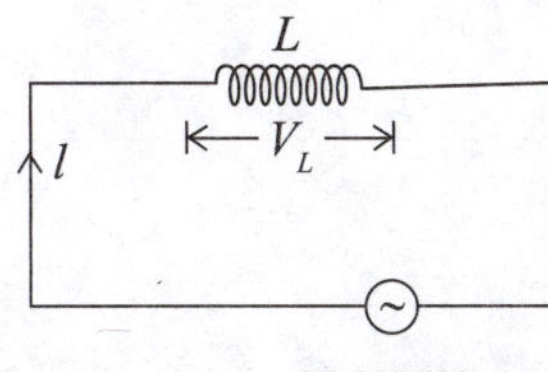

   1) $\omega L I_0 \sin\omega t$ 2) $\frac{I_0}{\omega L}\sin\omega t$
   3) $\frac{I_0}{\omega L}\cos\omega t$ 4) $\omega L I_0 \cos\omega t$
2. The amplitude of the charge oscillating in a circuit decreases exponentially as $Q = Q_0 e^{-Rt/2L}$, where Q_0 is the charge at t = 0 s. The time at which charge amplitude decreases to 0.50 Q_0 is nearly:
   [Given that $R = 1.5\ \Omega$, $L = 12$ mH, $ln(2) = 0.693$]**(2024 Re)**
   1) 19.01 ms 2) 11.09 ms
   3) 19.01 s 4) 11.09 s
3. A step up transformer is connected to an ac mains supply of 220 V to operate at 11000 V, 88 watt. The current in the secondary circuit, ignoring the power loss in the transformer, is **(2024 Re)**
   1) 8 mA 2) 4 mA
   3) 0.4 A 4) 4 A
4. A10mF capacitor is connected to a 210 V, 50 Hz source as shown in figure. The peak current in the circuit is nearly ($\pi = 3.14$): **(2024)**

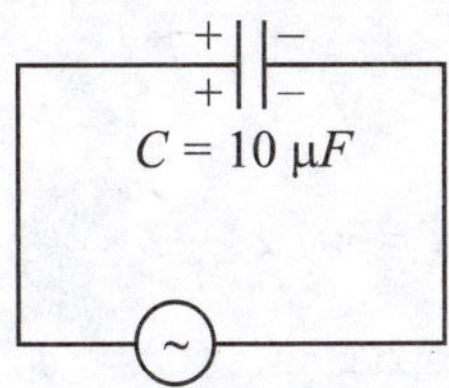

   1) 1.20 A 2) 0.35 A
   3) 0.58 A 4) 0.93 A
5. In an ideal transformer, the turns ratio is $\frac{N_P}{N_s} = \frac{1}{2}$ The ratio $V_s : V_p$ is equal to (the symbols carry their usual meaning): **(2024)**
   1) 1:1 2) 1:4
   3) 1:2 4) 2:1

6. In a series LCR circuit, the inductance L is 10 mH, capacitance C is 1μF and resistance R is 100Ω. The frequency at which resonance (2023)

1) 1.59 kHz
2) 15.9 rad/s
3) 15.9 kHz
4) 1.59 rad/s

7. A 12 V, 60 W lamp is connected to the secondary of a step down transformer, whose primary is connected to ac mains of 220 V. Assuming the transformer to be ideal, what is the current in the primary winding? (2023)

1) 0.37 A
2) 0.27 A
3) 2.7 A
4) 3.7 A

8. The net impedance of circuit (as shown in figure) will be: (2023)

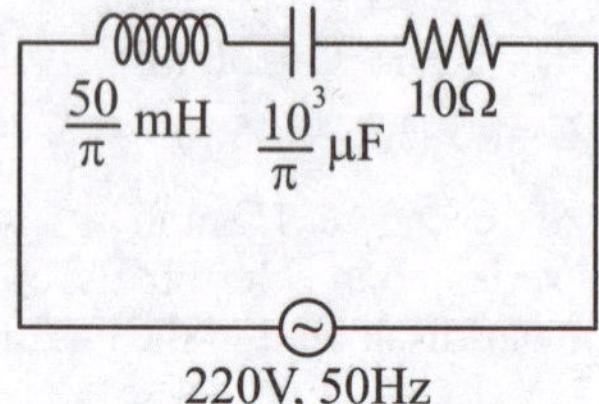

1) 25 W
2) $10\sqrt{2}\ \Omega$
3) 15 W
4) $5\sqrt{5}\ \Omega$

9. The peak voltage of the ac source is equal to (2022)

1) $\frac{1}{\sqrt{2}}$ times the rms value of the source
2) The value of voltage supplied to the circuit
3) The rms value of the ac source
4) $\sqrt{2}$ times the rms value of the ac source

10. A series LCR circuit with inductance 10 H, capacitance 10 μF, resistance 50Ω is corrected to an ac source of voltage, V = 200 sin(100t) volt. If resonant frequency of the LCR circuit is ν_0 and the frequency of the ac source is ν, then: (2022)

1) $\nu = 100\text{Hz}; \nu_0 = \frac{100}{\pi}\text{Hz}$
2) $\nu_0 = \nu = 50$ Hz
3) $\nu_0 = \nu = \frac{50}{\pi}\text{Hz}$
4) $\nu_0 = \frac{50}{\pi}\text{Hz}, \nu = 50\,\text{Hz}$

11. A capacitor of capacitance 'C', is connected across an ac source of voltage V, given by $V = V_0 \sin \omega t$

The displacement current between the plates of the capacitor, would then be given by: (2021)

1) $I_d = \frac{V_0}{\omega C}\cos \omega t$
2) $I_d = \frac{V_0}{\omega C}\sin \omega t$
3) $I_d = V_0 \omega C \sin \omega t$
4) $I_d = V_0 \omega C \cos \omega t$

12. An inductor of inductance L, a capacitor of capacitance C and a resistor of resistance 'R' are connected in series to an ac source of potential difference 'V' volts as shown in figure. Potential difference across L, C and R is 40V, 10V and 40V, respectively. The amplitude of current flowing through LCR series circuit is $10\sqrt{2}$A. The impedance of the circuit is: (2021)

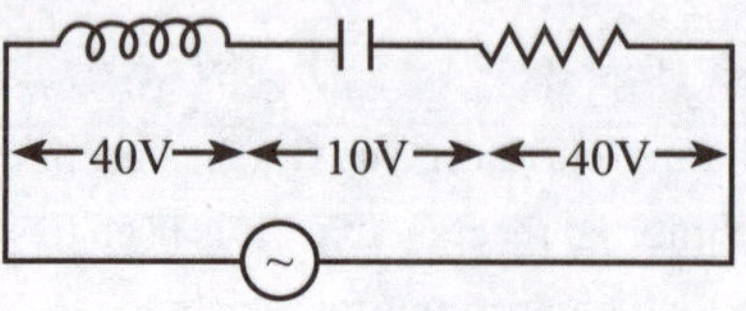

1) $5\sqrt{2}\Omega$
2) 4 Ω
3) 5 Ω
4) $4\sqrt{2}\Omega$

13. A step down transformer connected to an ac mains supply of 220 V is made to operate at 11 V, 44 W lamp. Ignoring power losses in the transformer, what is the current in the primary circuit? (2021)

1) 0.4 A
2) 2 A
3) 4 A
4) 0.2 A

14. A series LCR circuit containing 5.0 H inductor, 80 μF capacitor and 40 Ω resistor is connected to 230 V variable frequency ac source. The angular frequencies of the source at which power transferred to the circuit is half the power at the resonant angular frequency are likely to be: (2021) *[RC]*

1) 50 rad/s and 25 rad/s
2) 46 rad/s and 54 rad/s
3) 42 rad/s and 58 rad/s
4) 25 rad/s and 75 rad/s

15. A 50 Hz a.c. source of 20 volts is connected across R and C as shown in figure below. The voltage across R is 12 volts. The voltage across C is: (AIIMS 2005)

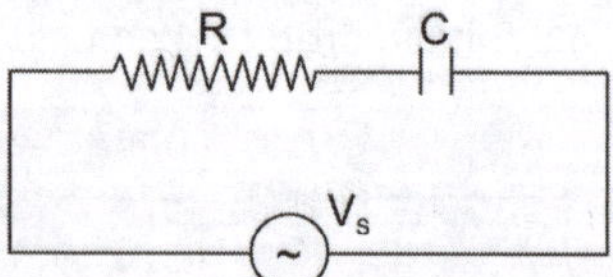

1) 8V
2) 16V
3) 10 V
4) Not possible to determine unless values of R and C are given

16. The circuit shown below acts as: (AIIMS 2005)

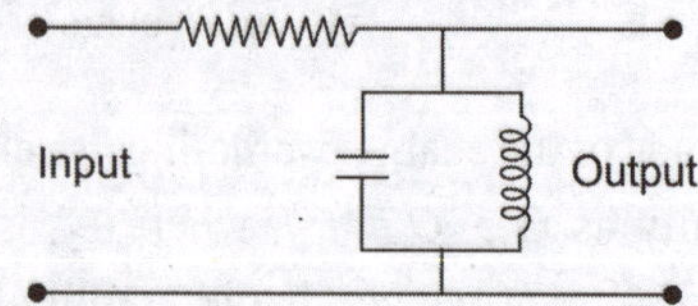

1) Tuned filter
2) Low pass filter
3) High pass filter
4) Rectifier

Answer Key

Topic-wise MCQs

1	2	3	4	5	6	7	8	9	10	11	12	13	14	15	16	17
2	1	4	3	4	3	4	3	4	4	2	2	2	2	4	3	3
18	**19**	**20**	**21**	**22**	**23**	**24**	**25**	**26**	**27**	**28**	**29**	**30**	**31**	**32**	**33**	**34**
4	3	2	3	1	1	3	4	2	2	1	1	1	3	2	4	1
35	**36**	**37**	**38**	**39**	**40**	**41**	**42**	**43**	**44**	**45**	**46**	**47**	**48**	**49**	**50**	**51**
2	3	4	3	3	4	2	1	3	2	1	1	1	2	3	3	2
52	**53**	**54**	**55**	**56**	**57**	**58**	**59**	**60**	**61**	**62**	**63**	**64**	**65**	**66**	**67**	**68**
1	1	2	3	1	1	2	3	1	1	1	2	3	2	4	3	2
69	**70**	**71**	**72**	**73**	**74**	**75**	**76**	**77**	**78**	**79**	**80**	**81**	**82**	**83**	**84**	**85**
1	2	4	1	4	2	2	2	3	1	4	3	1	1	3	2	3
86	**87**															
4	1															

High Order Time Intensive MCQs

1	2	3	4	5	6	7	8	9	10	11	12	13	14	15	16	17
3	3	3	3	1	1	4	2	1	1	1	1	1	1	2	1	3
18	**19**	**20**	**21**	**22**	**23**	**24**	**25**	**26**	**27**	**28**	**29**	**30**	**31**	**32**		
2	3	2	1	3	1	3	4	3	2	4	3	2	1	1		

Statement Based MCQs

1	2	3	4	5
1	3	1	3	1

Assertion & Reason MCQs

1	2	3	4
1	3	3	2

Match the Columns MCQs

1	2	3
1	1	3

NCERT Exemplar MCQs

1	2	3	4	5	6	7
2	3	3	2	3	3	1

NEET & AIIMS Past Year MCQs

1	2	3	4	5	6	7	8	9	10	11	12	13	14	15	16
4	2	1	4	4	1	2	4	4	3	4	3	4	2	2	1

CHAPTER

8 Electromagnetic Waves

Topicwise Number of Questions Analysis (2025-2016)

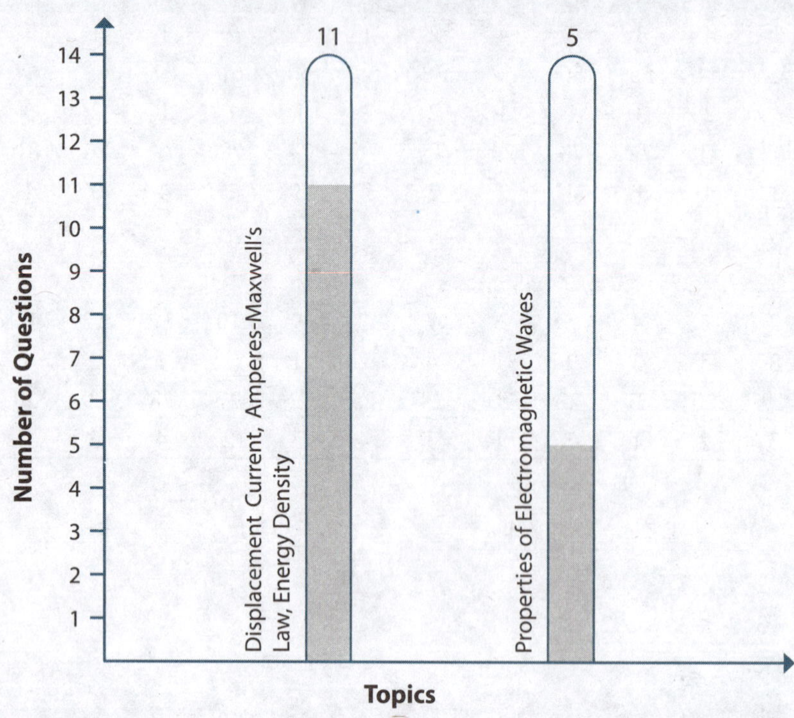

NCERT-PICKS

Displacement Current Ampere-Maxwell Law, Velocity, Amplitude Energy Density

Formulas

1. $I_D = \varepsilon_0 \frac{d\phi_E}{dt} = \varepsilon_0 A \frac{dE}{dt} = \frac{\varepsilon_0 A}{d}\frac{dV}{dt}$

   [I_D is displacement current, $_0$ is permittivity of free space, A is area of the plates of the capacitor, E is electric field between the plates of the capacitor and V is the potential difference between the plates of the capacitor]

2. $\oint \vec{B}.\vec{dl} = \mu_0(I+I_D)$

   [B is the magnetic field and I is the conduction current]

3. Maxwell's Equation:

   (a) Gauss law of electrostatics:

   $$\oint \vec{E}\cdot\vec{ds} = \frac{q}{\epsilon_0}$$

   (b) Gaus law of magnetism:

   $$\oint \vec{E}\cdot\vec{ds} = 0$$

   (c) Faraday's law of electromagnetic induction:

   $$\oint \vec{E}\cdot\vec{dl} = \frac{-d\phi_B}{dt}$$

   (d) Modified Ampere's law:

   $$\oint \vec{E}\cdot\vec{dl} = \mu_0[I_c + I_a] = \mu_0\left[I_c + \varepsilon_0\frac{d\phi_E}{dF}\right]$$

4. $u_E = \frac{1}{2}\varepsilon_0 E^2 = \frac{1}{2}\varepsilon_0\left(\frac{E_0}{\sqrt{2}}\right)^2 = \frac{1}{4}\varepsilon_0 E_0^2$

   [u_E is the electrical energy density and E_0 is peak value of the electric field]

5. $u_B = \frac{B^2}{2\mu_0} = \frac{(B_0/\sqrt{2})^2}{2\mu_0} = \frac{B_0^2}{4\mu_0}$

[u_B = magnetic energy density and B_0 is the peak value of magnetic field]

Intensity of electromagnetic wave,

$$I = \frac{1}{2}\varepsilon_0 E_0^2 c = \varepsilon_0 E_{rms}^2 c$$

$$I = \frac{1}{2\mu_0} B_0^2 c = \frac{1}{\mu_0} B_{rms}^2 c$$

6. Total average energy density $= u_E + u_B = 2\mu_E = 2\mu_B$

$$= \frac{1}{2}\varepsilon_0 E^2 + \frac{B^2}{2\mu_0} = \frac{1}{4}\varepsilon_0 E_0^2 + \frac{B_0^2}{4\mu_0}$$

Numerical Profile

Important Points

(i) The pressure exerted on the surface is force per unit area $\frac{F}{A}$. From Newton's second law, force is rate of change of momentum.

$$p = \frac{F}{A} = \frac{1}{A}\frac{d}{dt}\left(\frac{U}{c}\right) = \frac{1}{c}\frac{\left(\frac{dU}{dt}\right)}{A}$$

$\frac{(dU/dt)}{A}$ is the rate at which energy is coming at the surface per unit area.

(ii) If the surface is a perfect reflector and incidence is normal then the momentum carried to the surface is twice that given by Eq.(1). Therefore,

$P = \frac{2U}{c}$ (complete reflection)

(iii) The electric and magnetic fields of a sinusoidal plane electromagnetic wave propagating in the positive x-direction can also be written as

$E = E_m \sin(kx - t)$

$B = B_m \sin(kx - t)$

E_m, B_m are maximum values of electric and magnetic fields.

where is the angular frequency of the wave and k is wave number which is given by

$\omega = 2\pi f$ and $k = \frac{2\pi}{\lambda}$

$c = \frac{\omega}{k}$ and $c = \frac{E_m}{B_m}$

Statements Asked in Exam

- The electric and magnetic field vectors $\vec{E}$ and $\vec{B}$ are always to the direction of propagation of electromagnetic waves. Hence the electromagnetic waves is a transverse wave.
- The cross product $\vec{E}\times\vec{B}$ always gives the direction of propagation of electromagnetic wave.
- In an electromagnetic wave, the two fields (i.e., electric and magnetic) must vary sinusoidally, with the same frequency and in phase with each other.
- The ratio ω/K for a travelling wave is equal to its speed.
- Electromagnetic wave has linear momentum as well as energy. That is why electromagnetic waves exerts radiation pressure on the surface of object on which it falls.
- All the electromagnetic waves of different radiations move with the same speed in vacuum but they move with different speeds in a medium.
- Electromagnetic waves also transfer momentum to your hand but the amount of momentum transferred is extremely small and you do not feel the pressure.
- Infrared waves sometimes are called as Heat waves. It is also produced by hot bodies and molecules. Human eye can not detect them but snakes can detect them.
- Practically most of the radiations of wavelength less than 400 Å are absorbed by ozone layer and upper part of atmosphere and ionosphere.
- Electrical conductivity of earth's atmosphere increases with altitude.
- From the ionosphere layer of atmosphere, radio & microwaves are reflected back.
- Microwaves are the relevant waves in tele-communication.
- Microwaves are used in operating a RADAR.
- Accelerated charged particle produces electromagnetic waves.
- A radar using microwave can help in detecting the speed of tennis ball, cricket ball, automobile while in motion.
- Microwaves are used for observing the movements of trains, rails, while sitting in a microwave operated control rooms.
- Infrared waves are often known as heat waves.
- Satellites are used for long distance T.V transmission.

Type	Wavelength Range (λ)	Frequency Range (Hz)
1. Radiowaves	0.1 to 6×10^2 m	10^9– 5×10^5 Hz
2. Microwaves	10^{-3} to 0.1 m	3×10^{11}– 1×10^9 Hz
3. Infrared	7×10^{-7} to 10^{-3} m	4×10^{14}–3×10^{11} Hz
4. Visible light	4×10^{-7} to 7×10^{-7} m	8×10^{14}– 4×10^{14} Hz
5. Ultra violet	10^{-9} to 4×10^{-7}m	5×10^{17}–8×10^{14} Hz
6. X-rays	10^{-12} to 10^{-9} m	3×10^{21}–1×10^{16} Hz
7. γ - rays	< 10^{-12}	5×10^{22}– 3×10^{18} Hz

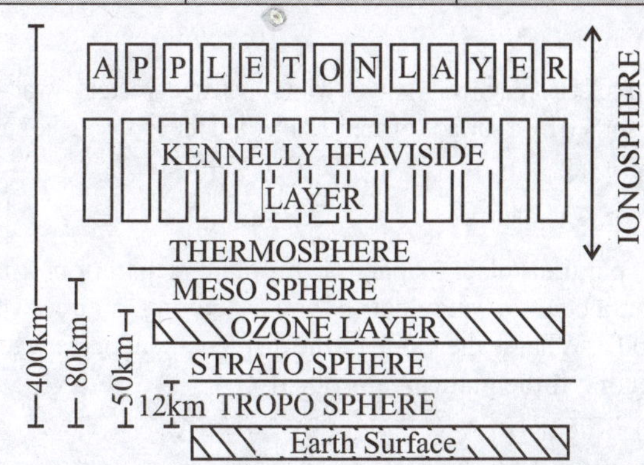

MCQs ?

NCERT Topic-wise MCQs

Displacement Current, Ampere-Maxwell Law, Velocity, Amplitude, Energy Density

1. How would you establish an instantaneous displacement current of 2A in the space between the two parallel plates of $1\mu F$ capacitor?
   1) $2\times10^{6}\,Vs^{-1}$ 2) $3\times10^{6}\,Vs^{-1}$
   3) $4\times10^{6}\,Vs^{-1}$ 4) $5\times10^{-6}\,Vs^{-1}$

2. Find the photon energy of photon in electron volts of electromagnetic waves of wavelength 40m?
   1) $3\times10^{-8}\,eV$ 2) $3.1\times10^{-8}\,eV$
   3) $3.2\times10^{-8}\,eV$ 4) $3.3\times10^{-8}\,eV$

3. Find the energy of photon in watt-hour for electromagnetic waves of wavelengths 3000Å.
   Given ($h = 6.6\times10^{-34}$ Js)
   1) 2×10^{-22} Wh 2) 2×10^{22} Wh
   3) 1.83×10^{-22} Wh 4) 1.83×10^{22} Wh

4. A beam of light traveling along *x*-axis is described by the magnetic field
   $B_X = 5\times10^{-9}\,T \sin\omega(t - y/c)$
   What is the maximum electric and magnetic forces on a charge i.e., α-particle moving along y-axis with a speed of 3×10^{8} m/s. (charge on electron $= 1.6\times10^{-19}$C).
   1) 4.80×10^{-17} N , 4.80×10^{-17} N
   2) 4.85×10^{-16} N , 5.6×10^{-17} N
   3) 9.2×10^{6} N , 9.2×10^{3} N
   4) 4.8×10^{-19} N , 4.8×10^{-19} N

5. If the intensity of the incident radiowave of 1watt/m² is reflected by the surface. Find the pressure exerted on the surface?
   1) $6.67\times10^{-9}N/m^2$ 2) $8\times10^{-9}N/m^2$
   3) $6\times10^{-10}N/m^2$ 4) $10^{-10}N/m^2$

6. An electromagnetic wave with intensity vector 6watt/m² is absorbed by a surface of area 12m². Find the force on the surface?
   1) 24N 2) 24×10^{10}N
   3) 24×10^{-10}N 4) 24×10^{-8}N

7. A parallel plate capacitor of plate separation 2mm is connected in an electric circuit having source voltage 400V. What is the value of the displacement current for 10^{-6} second if the plate area is 60cm²?
   1) 1.062×10^{-2}A 2) 2×10^{-2}A
   3) 2.1×10^{-4}A 4) None

8. A plane electromagnetic wave of frequency 25 MHz travel in free space along the *x*-direction. At a particular point in space and time, the electric vector is $\vec{E} = 6.3\,\hat{j}\ V/m$. What is the $\vec{B}$ at this point?
   1) $25\times10^{-8}\hat{k}\,T$ 2) $2.1\times10^{-8}\hat{k}\,T$
   3) $2.1\times10^{-8}\hat{j}T$ 4) $2\times10^{-8}\hat{j}T$

9. If $\vec{E}$ is an electric field and $\vec{B}$ is the magnetic induction, then the energy flow per unit area per unit time in an electromagnetic field is given by
   1) $\vec{E}\times\vec{B}$ 2) $\vec{E}/\vec{B}$
   3) $\vec{E}.\vec{B}$ 4) None of the above

10. An electromagnetic radiation has an energy 11 keV. To which region of electromagnetic spectrum does it belong? use : ($h = 6.6\times10^{-34}$Js , $e = 1.6\times10^{-19}$C).
    1) Infrared 2) Microwave
    3) X-ray 4) Y-rays

11. Which of the following equations is modified by Maxwell to include displacement current?
    1) Gauss's law for electricity
    2) Gauss's law for magnetism
    3) Ampere's circuital law
    4) Faraday's law of induction

12. Displacement current is produced when:
    1) Electrons move through vacuum
    2) Electric field between plates is constant
    3) Electric field changes with time
    4) Magnetic field is uniform

13. Total average energy density of an electromagnetic wave in free space is given by
    1) $\frac{1}{2}\varepsilon_0 E_{rms}^2 + \frac{B_{rms}^2}{2\mu_0}$ 2) $\frac{1}{2}\varepsilon_0 E_0^2 + \frac{1}{2}\varepsilon_0 B_0^2$
    3) $\frac{1}{2}\varepsilon_0 E_0^2$ 4) $\frac{E_0^2 + E_0^2}{2c}$

14. An electric field $\vec{E}$ and magnetic field $\vec{B}$ exist in a region. The field are not perpendicular to each other:
    1) This is not possible
    2) No electromagnetic wave is certainly passing through the region
    3) An electromagnetic wave may be passing through the region
    4) None of the above

15. The conduction current is same as displacement current when source is

1) a.c. only 2) d.c.only

3) Both a.c. and d.3) 4) Neither for d.c. nor for a.c.

16. A plane electromagnetic wave is incident on a material surface. The wave delievers momentum 'P' and energy 'E', then:

1) $P \neq 0 ; E \neq 0$ 2) $P = 0 ; E = 0$

3) $P = 0 ; E \neq 0$ 4) $P \neq 0 ; E = 0$

17. If the magnetic monopoles existed then which of the following Maxwell's equations would be modified ?

1) $\oint \vec{E}.\vec{ds} = \frac{q}{\epsilon_0}$

2) $\oint \vec{B}.\vec{ds} = 0$

3) $\oint \vec{B}.\vec{dl} = -\frac{d}{dt}\int \vec{B}.\vec{ds}$

4) $\oint \vec{B}.\vec{dl} = \mu_0 \epsilon_0 \frac{d}{dt}\int \vec{E}.\vec{ds} + \mu_0 I$

18. For television broadcasting, the frequency employed is normally:

1) 30-300 MHz 2) 30-300 GHz

3) 30-300 KHz 4) 30-300 Hz

19. A parallel plate capacitor is charged by a current of 2×10^{-7} A displaced between the plates of capacitor. When discharge of the capacitor takes place through a resistance, the rate of change of electric flux (in Wb/s) will be

1) 2.26×10^4 2) 4.26×10^8

3) 3.26×10^6 4) 6.26×10^9

20. In a plane e.m. wave the electric field oscillates sinusoidally at a frequency of 2.5×10^{10} Hz and amplitude 480 V/m. The amplitude of oscillating magnetic field will be:

1) 1.52×10^{-8} Wb/m^2 2) 1.52×10^{-7} Wb/m^2

3) 1.6×10^{-6} Wb/m^2 4) 1.6×10^{-7} Wb/m^2

21. If μ_0 be the permeability and K_0 the dielectric constant of a medium then refractive index given by:

1) $\frac{1}{\sqrt{\mu_0 K_0}}$ 2) $\frac{1}{\mu_0 K_0}$

3) $\sqrt{\mu_0 K_0}$ 4) $\mu_0 K_0$

22. If ε_0 and μ_0 represent the permitivity and permeability of vacuum and and μ represents the permitivity and permeability of medium then refractive index of the medium is given:

1) $\sqrt{\frac{\varepsilon_0 \mu_0}{\varepsilon \mu}}$ 2) $\sqrt{\frac{\varepsilon \mu}{\varepsilon_0 \mu_0}}$

3) $\sqrt{\frac{\varepsilon}{\mu_0 \varepsilon_0}}$ 4) $\sqrt{\frac{\mu_0 \varepsilon_0}{\varepsilon}}$

23. According to Maxwell's hypothesis, a changing electric field gives rise to:

1) An emf 2) Electric current

3) Magnetic field 4) Pressure radiant

24. The mean electric energy density between plates of a charged capacitor:

1) $\frac{q^2}{2\varepsilon_0 A^2}$ 2) $\frac{q}{2\varepsilon_0 A^2}$

3) $\frac{q^2}{2\varepsilon_0 A}$ 4) None of the above

25. If the wavelength of light is 4000Å, the number of waves in 1mm length will be:

1) 25 2) 0.25

3) 0.25×10^4 4) 2.5×10^4

26. The sun radiates electromagnetic energy at the rate of 3.9×10^{26} W. Its radius is 6.96×10^8 m. The intensity of sun light (in W/m^2) at the solar surface will be:

1) 5.6×10^6 2) 6.4×10^7

3) 4.2×10^6 4) 4.2×10^7

27. The speed of electromagnetic wave in a medium of dielectric constant 2.25 and relative permeability 4 is:

1) 10^8 m/s 2) 2.5×10^8 m/s

3) 2×10^8 m/s 4) 3×10^8 m/s

28. The amplitude of the sinusoidally oscillating electric field of plane wave is 60 V/m. Then the amplitude of the magnetic field is:

1) 2×10^7 T 2) 6×10^7 T

3) 6×10^{-7} T 4) 2×10^{-7} T

29. Which of the following are not electromagnetic waves?

1) Cosmic rays 2) γ-rays

3) Infrared 4) X-rays

30. Energy is not carried by which of the following wave?

1) Electromagnetic 2) Transverse

3) Stationary 4) Progressive

Uses of Electromagnetic Waves and Properties of Electromagnetic Waves

31. SONAR emits which of the following waves?

1) Ultrasonic 2) Radio

3) Light 4) None of these

32. In general the wavelength of microwaves is:

1) More than that of infrared waves

2) More than that of radio waves

3) Less than that of infrared waves

4) Less than that of ultraviolet waves

33. Which of the following wavelength of sun is used finally as electric energy?

1) Microwave 2) Infrared

3) UV 4) X-rays

34. If v_r, v_x, v_m are the speeds of γ-rays, X-rays and microwaves respectively in vacuum then:

1) $v > v_x > v_m$ 2) $v > v_x < v_m$
3) $v = v_x = v_m$ 4) $v \neq v_x < v_m$

35. The frequency order of γ-rays (A), X-rays (B), UV rays (C)?

1) A > B > C 2) B > A > C
3) C > A > B 4) B > C > A

36. A plane electromagnetic wave travelling along the x-direction has a wavelength of 3 mm. The variation in the electric field occurs in the y-direction with an amplitude 66 V m^{-1}. The equation for the electric and magnetic field as a function of x and t are respectively

1) $E_y = 33 \cos \times 10^{11}\left(t - \frac{x}{c}\right)$
$B_z = 1.1 \times 10^{-7} \cos \times 10^{11} (t - x/c)$
2) $E_y = 11 \cos 2 \times 10^{11} (t - x/c$
$B_z = 11 \times 10^{-7} \cos 2 \times 10^{11} (t - x/c)$
3) $E_x = 33 \cos \times 10^{11} (t - x/c)$
$B_y = 11 \times 10^{-7} \cos \times 10^{11} (t - x/c)$
4) $E_y = 66 \cos 2 \times 10^{11} (t - x/c)$
$B_z = 2.2 \times 10^{-7} \cos 2 \times 10^{11} (t - x/c)$

37. If $\vec{E}$ and $\vec{K}$ represent electric field and propagation vectors of the EM waves in vacuum, then magnetic field vector is given by: (ω – angular frequency):

1) $\frac{1}{\omega}(\vec{K} \times \vec{E})$ 2) $\omega(\vec{E} \times \vec{K})$
3) $\omega(\vec{K} \times \vec{E})$ 4) $\vec{K} \times \vec{E}$

38. Electric field in a plane electromagnetic wave is given by $E = 50 \sin (500x - 10 \times 10^{10} t)$V/m
The velocity of electromagnetic wave in this medium is: (Given c = speed of light in vacuum)

1) c 2) $\frac{c}{2}$
3) $\frac{2}{3}c$ 4) $\frac{3}{2}c$

39. A plane electromagnetic wave is propagating along the direction $\frac{\hat{i} + \hat{j}}{\sqrt{2}}$, with its polarization along the direction $\hat{k}$. The correct form of the magnetic field of wave would be (here B_0 is an appropriate constant)

1) $B_0 \frac{\hat{i} + \hat{j}}{\sqrt{2}} \cos\left(\omega t - k\frac{\hat{i} + \hat{j}}{\sqrt{2}}\right)$
2) $B_0 \frac{\hat{i} - \hat{j}}{\sqrt{2}} \cos\left(\omega t + k\frac{\hat{i} + \hat{j}}{\sqrt{2}}\right)$
3) $B_0 \frac{\hat{i} - \hat{j}}{\sqrt{2}} \cos\left(\omega t - k\frac{\hat{i} + \hat{j}}{\sqrt{2}}\right)$
4) $B_0 \hat{k} \cos\left(\omega t - k\frac{\hat{i} + \hat{j}}{\sqrt{2}}\right)$

40. Which of the following radiation in sunlight causes heating effect?

1) UV 2) IR
3) Visible 4) All

41. The wavelength of light visible to eye is of the order of:

1) 10^{-2}m 2) 10^{-10}m
3) 1m 4) 6×10^{-7}m

42. Which rays are not the part of em waves?

1) X-rays 2) Microwaves
3) α-rays 4) Radio-waves

43. Heat radiation propagate with the speed of:

1) α-rays 2) β-rays
3) Light waves 4) Sound waves

44. The electric and the magnetic field, associated with an e.m. wave, propagating along the +z-axis, can be represented by

1) $\left[\vec{E} = E_0 \hat{i}, \vec{B} = B_0 \hat{j}\right]$ 2) $\left[\vec{E} = E_0 \hat{k}, \vec{B} = B_0 \hat{i}\right]$
3) $\left[\vec{E} = E_0 \hat{j}, \vec{B} = B_0 \hat{i}\right]$ 4) $\left[\vec{E} = E_0 \hat{j}, \vec{B} = B_0 \hat{k}\right]$

45. The electric field in an electromagnetic wave is given by $\vec{E} = \hat{i}\, 40 \cos\omega\left(t - \frac{z}{c}\right) NC^{-1}$. The magnetic field induction of this wave is (in SI unit):

1) $\vec{B} = \hat{i}\frac{40}{c} \cos\omega\left(t - \frac{z}{c}\right)$ 2) $\vec{B} = \hat{j}\, 40 \cos\omega\left(t - \frac{z}{c}\right)$
3) $\vec{B} = \hat{k}\frac{40}{c} \cos\omega\left(t - \frac{z}{c}\right)$ 4) $\vec{B} = \hat{j}\frac{40}{c} \cos\omega\left(t - \frac{z}{c}\right)$

46. If λ_v, λ_x and λ_m represents the wavelength of visible light, x-ray and microwaves respectively, then:

1) $\lambda_m > \lambda_x > \lambda_v$ 2) $\lambda_v > \lambda_m \lambda_y$
3) $\lambda_m > \lambda_v > \lambda_x$ 4) $\lambda_v > \lambda_x > \lambda_m$

47. The refractive index and the permeability of a medium are respectively 1.5 and 5×10^{-7} Hm^{-1}. The relative permitivity of the medium is nearly

1) 25 2) 15
3) 81 4) 6

48. Which of the following rays are produced during nuclear reaction?

1) Microwave 2) UV
3) Infrared 4) γ-rays

49. Which of the following rays help us to detect the speed of tennis ball or cricket ball while in motion?

1) Radio wave
2) Infra red
3) Microwave
4) None of these

High Order Time Intensive MCQs

1. A plane electromagnetic wave propagating in the x-direction has wavelength of 6.0 mm. The electric field is in the y-direction and its maximum magnitude is 33 Vm^{-1}. The equation for the electric field as function of x and t is:
   1) $11 \sin \pi (t - x/c)$ 2) $33 \sin \pi \times 10^{11} (t - x/c)$
   3) $33 \sin \pi (t - x/c)$ 4) $11 \sin \pi \times 10^{11} (t - x/c)$

2. The sun delivers 10^4 W/m^2 of electromagnetic flux to the earth's surface. The total power that is incident on a roof of dimensions $(10 \times 10)m^2$ will be:
   1) 10^4 W 2) 10^5 W
   3) 10^6 W 4) 10^7 W

3. A perfectly reflecting mirror has an area of 1 cm^2. Light energy is allowed to fall on it for 1 hour at the rate of 10 W cm^{-2}. The force that acts on the mirror is:
   1) 3.35×10^{-8} N 2) 6.7×10^{-8} N
   3) 1.34×10^{-7} N 4) 2.4×10^{-4} N

4. The pressure exerted by an electromagnetic wave of intensity I (Wm^{-2}) on a non-reflecting surface is: (c is the velocity of light)
   1) Ic 2) Ic^2
   3) I/c 4) I/c^2

5. The condition under which a microwave oven heats up a food item containing water molecules most efficiently is:
   1) The frequency of the microwave must match the resonant frequency of the water molecules
   2) The frequency of the microwave has no relation with natural frequency of water molecules
   3) Microwaves are heat waves, so always produce heating
   4) Infrared waves produced heating in a microwave oven

6. Which of the following undergoes largest diffraction?
   1) Infrared light 2) Radio waves
   3) γ-rays 4) Ultraviolet light

7. A parallel plate capacitor is being charged so that the surface charge density increases uniformly with time. What is the nature of magnetic field between the plates due to displacement current?
   1) Zero everywhere
   2) Constant throughout the region
   3) Increases linearly with distance from center
   4) Maximum at center and zero at edges

8. We find that the temperature of air decreases as one goes up from the earth's surface, because:
   1) The atmospheric pressure drops with height
   2) The earth which radiates in the infrared region is the main heat source and temperature drops as we go away form it
   3) The density of air drops with height and the air therefore cannot hold stronger as we go up
   4) Winds are stronger as we go up

9. A plane electromagnetic wave travels in free space along X-direction. If the value of $\vec{B}$ (in tesla) at a particular point in space and time is $1.2 \times 10^{-8} \hat{k}$, the value of $\vec{E}$ (in V m^{-1}) at that point is:
   1) $1.2\hat{j}$ 2) $3.6\hat{k}$
   3) $1.2\hat{k}$ 4) $3.6\hat{j}$

10. The frequencies of X-rays, -rays and ultra-voilet rays are respectively a, b and c. Then
    1) a < b, b > c 2) a > b, b > c
    3) a > b, b < c 4) a < b, b < c

11. About 6% of the power of a 100 W bulb is converted to visible radiation. The average intensity of visible radiation at a distance of 8 m is (Assume that the radiation is emitted isotropically and neglect reflection):
    1) 3.5×10^{-3} W m^{-2} 2) 5.1×10^{-3} W m^{-2}
    3) 7.46×10^{-3} W m^{-2} 4) 2.3×10^{-3} W m^{-2}

12. The magnetic field of a beam emerging from a flood light is given by $B = 12 \times 10^{-8} \sin(1.20 \times 10^7 z - 3.60 \times 10^{15} t)$ T. The average intensity of the beam is:
    1) 1.71 Wm^{-2} 2) 2.1 W m^{-2}
    3) 3.2 W m^{-2} 4) 2.9 W m^{-2}

13. The electric field part of a electromagnetic wave in a medium is represented by $E_x = 0$.
    $$E_y = 2.5 \frac{N}{C} \cos\left[\left(2\pi \times 10^6 \frac{rad}{m}\right)t - \left(\pi \times 10^{-2} \frac{rad}{s}\right)x\right],$$
    $E_z = 0$. The wave is
    1) Moving along x direction with frequency 10^6 Hz and wavelength 100 m
    2) Moving along x direction with frequency 10^6 Hz and wavelength 200 m
    3) Moving along –x direction with frequency 10^6 Hz and wavelength 200 m
    4) Moving along y direction with frequency $2\pi \times 10^6$ Hz and wavelength 200 m

14. The magnetic field in a plane electromagnetic wave is given by
    $B_y = 2 \times 10^{-7} \sin (0.5 \times 10^3 x + 1.5 \times 10^{11} t)$.
    This electromagnetic wave is
    1) a visible light 2) a infrared wave
    3) a microwave 4) a radiowave

15. If a 100 watt bulb acts as a point source and has 2.5% efficiency, the intensity of this bulb at 3 m is
    1) 0.022 Wm^{-2} 2) 0.22 Wm^{-2}
    3) 4.4 Wm^{-2} 4) 0.033 Wm^{-2}

16. Consider the following types of electromagnetic radiation : radiowaves, infra-red, visible light. Which of the following statements are correct ?
    (i) Only radiowaves can be used to transmit audio information.
    (ii) Only infrared radiation is emitted by very hot object.
    (iii) Only visible light can be detected by human.
    1) Only (i) is correct
    2) Only (ii) is correct
    3) Only (iii) is correct
    4) None of the above is correct

Statement Based MCQs

Directions: These questions consist of two statements each, printed as Statement-I and Statement-II. While answering these questions, you are required to choose any one of the following four responses.

1) **Both Statement-I and Statement-II are correct.**
2) **Both Statement-I and Statement-II are incorrect.**
3) **Statement-I is correct & Statement-II is incorrect.**
4) **Statement-I is incorrect & Statement-II is correct.**

1. **Statement-I:** Induced electric field is not conservative (i.e. it forms closed loop).
   Statement-II: $\oint \vec{E}_{ind.} d\vec{\ell} \neq 0$
2. **Statement-I:** If earth did not have atmosphere, its average surface temperature would be lower than what it is now.
   Statement-II: Greenhouse effect of the atmosphere would be absent, if earth did not have atmosphere.
3. **Statement-I:** Induced electric field due to changing magnetic flux are more readily observed than induced magnetic field due to changing electric field.
   Statement-II: The induced field depends on change in flux.
4. **Statement-I:** Radio waves are better carriers of signals than microwaves.
   Statement-II: Electromagnetic waves require material medium for propagation.
5. **Statement-I:** Human eye is most sensitive for the light of frequency 5.36×10^{14} Hz.
   Statement-II: Frequency of light is directly proportional to wavelength of light.

Assertion & Reason MCQs

Directions: These questions consist of two statements each, printed as Assertion and Reason. While answering these questions, you are required to choose any one of the following four responses.

1) **Both Assertion and Reason are True and the Reason is a correct explanation of the Assertion.**
2) **Both Assertion and Reason are True but Reason is not a correct explanation of the Assertion.**
3) **Assertion is True but the Reason is False.**
4) **Assertion is False but Reason is True.**

1. **Assertion:** Accelerated charge radiate electromagnetic waves.
   Reason: As the EM wave propagate through the space, the oscillating electric and magnetic field regenerate each other.
2. **Assertion:** When variable frequency Alternating Current source is connected to a capacitor, displacement current increases with increase in frequency.
   Reason: If frequency increases conduction current will also increase.
3. **Assertion:** The differential equations of an electromagnetic wave can be given as
   $$\left(\frac{\partial^2 E}{\partial x^2}\right) = \mu_0\varepsilon_0\left(\frac{\partial^2 E}{\partial t^2}\right); \left(\frac{\partial^2 B}{\partial x^2}\right) = \mu_0\varepsilon_0\left(\frac{\partial^2 B}{dt^2}\right)$$
   Reason: The wave equation of en waves are similar to standing waves.
4. **Assertion:** For cooking in a microwave oven, food is always kept in metal containers.
   Reason: The energy of microwave is easily transferred to the food in metal container.
5. **Assertion:** Electromagnetic waves are transverse in nature.
   Reason: The magnetic and electric fields of an electromagnetic wave are perpendicular to each other and also perpendicular to the direction of propagation of waves

Match the Columns MCQs

1. Match the Column-I with Column-II.

	Column-I		Column-II
(a)	Hindrance	(i)	1/c
(b)	Poynting vector	(ii)	$\frac{1}{2}\varepsilon_0 E^2$
(c)	Energy density	(iii)	$\vec{E}\times\vec{H}$
(d)	$\frac{\lvert\vec{B}\rvert}{\lvert\vec{E}\rvert}$	(iv)	$\frac{E}{H}$

1) (a)-(i); (b)-(ii); (c)-(iv); (d)-(iii)
2) (a)-(iii); (b)-(iv); (c)-(ii); (d)-(i)
3) (a)-(iv); (b)-(iii); (c)-(ii); (d)-(i)
4) (a)-(i); (b)-(ii); (c)-(iii); (d)-(iv)

2. Match the Column-I with Column-II.

	Column-I		Column-II
(a)	Troposphere	(i)	Upto 12 km
(b)	Mesosphere	(ii)	10 km to 50 km
(c)	Ionosphere	(iii)	50 km to 80 km
(d)	Stratosphere	(iv)	Above 80 km

1) (a)-(i); (b)-(ii); (c)-(iii); (d)-(iv)
2) (a)-(i); (b)-(iii); (c)-(iv); (d)-(ii)
3) (a)-(iii); (b)-(ii); (c)-(i); (d)-(iv)
4) (a)-(i); (b)-(iii); (c)-(ii); (d)-(iv)

3. Match the Column-I with Column-II.

	Column-I Relative permeabilities & permitivities of the medium		Column-II Refractive index of the medium
(a)	$\varepsilon_r = 1, \mu_r = 2$	(i)	$n = 2$
(b)	$\varepsilon_r = 2, \mu_r = 1$	(ii)	$n = \sqrt{2}$
(c)	$\varepsilon_r = 2, \mu_r = 2$	(iii)	$n = \sqrt{3}$
(d)	$\varepsilon_r = 1, \mu_r = 3$	(iv)	$n = 4$

1) (a)-(*i*); (b)-(*ii*); (c)-(*iv*); (d)-(*iii*)
2) (a)-(*i*); (b)-(*ii*); (c)-(*i*); (d)-(*iii*)
3) (a)-(*ii*); (b)-(*ii*); (c)-(*i*); (d)-(*iii*)
4) (a)-(*ii*); (b)-(*ii*); (c)-(*iv*); (d)-(*iii*)

4. Match the Column-I with Column-II.

	Column-I EM waves		Column-II Association/Application
(a)	Infrared waves	(i)	To treat muscular & strain
(b)	Radio waves	(ii)	For broadcasting
(c)	X-rays	(iii)	To detect fracture of bones
(d)	Ultraviolet	(iv)	Absorbed by the ozone layer of the atmosphere

1) (a)-(iii); (b)-(ii); (c)-(i); (d)-(iv)
2) (a)-(i); (b)-(ii); (c)-(iii); (d)-(iv)
3) (a)-(iv); (b)-(iii); (c)-(ii); (d)-(i)
4) (a)-(i); (b)-(ii); (c)-(iv); (d)-(iii)

NCERT Exemplar MCQs

1. One requires 11 eV of energy to dissociate a carbon monoxide molecule into carbon and oxygen atoms. The minimum frequency of the appropriate electromagnetic radiation to achieve the dissociation lies in:

1) Visible region
2) Infrared region
3) Ultraviolet region
4) Microwave region

2. A linearly polarized electromagnetic wave given as $E = E_0\hat{i}\cos(kz-\omega t)$ is incident normally on a perfectly reflecting infinite wall at z = a. Assuming that the material of the wall is optically inactive, the reflected wave will be given as:

1) $E_r = E_0\hat{i}\cos(kz-\omega t)$
2) $E_r = E_0\hat{i}\cos(kz+\omega t)$
3) $E_r = E_0\hat{i}\cos(kz-\omega t)$
4) $E_r = E_0\hat{i}\sin(kz-\omega t)$

3. Light with an energy flux of 20 W/cm^2 falls on a non-reflecting surface at normal incidence. If the surface has an area of 30 cm^2, the total momentum delivered (for complete absorption) during 30 min is:

1) 36×10^{-5} kg-m/s
2) 36×10^{-4} kg-m/s
3) 108×10^{4} kg-m/s
4) 1.08×10^{7} kg-m/s

4. The electric field intensity produced by the radiations coming from 100 W bulb at a 3 m distance is E. The electric field intensity produced by the radiations coming from 50 W bulb at the same distance is:

1) $\frac{E}{2}$
2) 2E
3) $\frac{E}{\sqrt{2}}$
4) $\sqrt{2}E$

5. If $\vec{E}$ and $\vec{B}$ represent electric and magnetic field vectors of the electromagnetic wave, the direction of propagation of electromagnetic wave is along:
   1) $\vec{E}$ 2) $\vec{B}$
   3) $\vec{B} \times \vec{E}$ 4) $\vec{E} \times \vec{B}$
6. An electromagnetic wave travels in vacuum along z-direction
   $$\vec{E} = (E_1\hat{i} + E_2\hat{j})\cos(kz - \omega t)$$
   Choose the correct options from the following:
   1) The associated magnetic field is given as
   $$\vec{B} = \frac{1}{c}(E_1\hat{i} + E_2\hat{j})\cos(kz - \omega t)$$
   2) The associated magnetic field is given as
   $$\vec{B} = \frac{1}{c}(E_1\hat{i} - E_2\hat{j})\cos(kz - \omega t)$$
   3) The given electromagnetic field is circularly polarised
   4) The given electromagnetic wave is plane polarised

NEET & AIIMS Past Year MCQs

1. If the ratio of relative permeability and relative permittivity of a uniform medium is 1 : 4. The ratio of the magnitudes of electric field intensity (E) to the magnetic field intensity (H) of an EM wave propagating in that medium is
   $\left(\text{Given that } \sqrt{\frac{\mu_0}{\varepsilon_0}} = 120\pi\right)$ **(2024 Re)**
   1) $30\pi : 1$ 2) $1 : 120\,\pi$
   3) $60\pi : 1$ 4) $120\,\pi : 1$
2. The electromagnetic radiation which has the smallest wavelength are **(2024 Re)**
   1) X-rays 2) Gamma rays
   3) Ultraviolet rays 4) Microwaves
3. A parallel plate capacitor is charged by connecting it to a batter/through a resistor. If I is the current in the circuit, then in the gap between the plates: **(2024)**
   1) displacement current of magnitude equal to I flows in a direction opposite to that of I.
   2) displacement current of magnitude greater than I flows but can be in any direction.
   3) there is no current.
   4) displacement current of magnitude equal to I flows in the same direction as I.
4. The property which is not of an electromagnetic wave travelling in free space is that: **(2024)**
   1) they travel with speed equal to $\frac{1}{\sqrt{\mu_0 \epsilon_0}}$.
   2) they originate from charges moving with uniform speed.
   3) they are transverse in nature.
   4) the energy density in electric field is equal to energy density in magnetic field.
5. An ac source is connected to a capacitor C. Due to decrease in its operating frequency: **(2023)**
   1) capacitive reactance remains constant
   2) capacitive reactance decreases.
   3) displacement current increases.
   4) displacement current decreases.
6. In a plane electromagnetic wave travelling in free space, the electric field component oscillates sinusoidally at a frequency of 2.0×10^{10} Hz and amplitude 48 Vm^{-1}. Then the amplitude of oscillating magnetic field is:
   (Speed of light in free space = $3 \times 108\ ms^{-1}$) **(2023)**
   1) 1.6×10^{-6} T 2) 1.6×10^{-9} T
   3) 1.6×10^{-8} T 4) 1.6×10^{-7} T
7. When light propagates through a material medium of relative permitivity $_r$ and relative permeability μ_r, the velocity of light, v is given by : (c - velocity of light in vacuum) **(2022)**
   1) $v = \frac{c}{\sqrt{\epsilon_r \mu_r}}$ 2) $v = c$
   3) $v = \sqrt{\frac{\mu_r}{\epsilon_r}}$ 4) $v = \sqrt{\frac{\epsilon_r}{\mu_r}}$
8. Match List-I with List-II **(2022)**

List-I (Electromagnetic waves)		List-II (Wavelength)	
A.	AM radio waves	p.	10^{-10} m
B.	Microwaves	q.	10^{2} m
C.	Infraraed radiations	r.	10^{-2} m
D.	X-rays	s.	10^{-4} m

   Choose the correct answer from the options given below:
   1) A-(q); B-(r); C-(s); D-(p)
   2) A-(s); B-(r); C-(q); D-(p)
   3) A-(r); B-(q); C-(p); D-(s)
   4) A-(r); B-(s); C-(q); D-(p)
9. For a plane electromagnetic wave propagating in x-direction, which one of the following combination gives the correct possible directions for electric field (E) and magnetic field (B) respectively? **(2021)**
   1) $-\hat{j}+\hat{k}, -\hat{j}-\hat{k}$ 2) $\hat{j}+\hat{k}, -\hat{j}-\hat{k}$
   3) $-\hat{j}+\hat{k}, -\hat{j}+\hat{k}$ 4) $\hat{j}+\hat{k}, \hat{j}+\hat{k}$

10. Which wavelength of sun is used finally as electric energy? (AIIMS 2007)
   1) Radio waves
   2) Infra red waves
   3) Visible light
   4) Micro waves

11. The pressure exerted by an electromagnetic wave of intensity I (watt/m^2) on a nonreflecting surface is [c is the velocity of light]: (AIIMS 2005)
   1) Ic
   2) Ic^2
   3) I/c
   4) I/c^2

12. Characteristic X-rays are produced due to: (AIIMS 2003)
   1) Transfer of momentum in collision of electrons with target atoms
   2) Transition of electrons from higher to lower electronic orbits in an atom
   3) Heating of the target
   4) Transfer of energy in collision of electrons with atoms in the target

Directions: In each of the following questions, a statement of assertion (A) is given followed by a corresponding statement of reason (R). Mark the correct answer.

1) If both Assertion & Reason are True & the Reason is a correct explanation of the Assertion.
2) If both Assertion & Reason are True but Reason is not a correct explanation of the Assertion.
3) If Assertion is True but the Reason is False.
4) If both Assertion & Reason are False.

13. **Assertion:** Electromagnetic radiations exerts pressure.
   Reason: Electromagnetic-waves carry both momentum and energy. (AIIMS 2013)

14. **Assertion:** In electromagnetic waves electric field and magnetic field lines are perpendicular to each other.
   Reason: E and B are self sustaining. (AIIMS 2012)

Answer Key

NCERT Topic-wise MCQs

1	2	3	4	5	6	7	8	9	10	11	12	13	14	15	16	17
1	2	3	4	1	4	1	2	4	3	3	3	1	3	1	1	2
18	**19**	**20**	**21**	**22**	**23**	**24**	**25**	**26**	**27**	**28**	**29**	**30**	**31**	**32**	**33**	**34**
1	1	3	3	2	3	1	3	2	1	4	1	3	1	1	2	3
35	**36**	**37**	**38**	**39**	**40**	**41**	**42**	**43**	**44**	**45**	**46**	**47**	**48**	**49**		
1	4	1	3	3	2	4	3	3	1	4	3	4	4	3		

High Order Time Intensive MCQs

1	2	3	4	5	6	7	8	9	10	11	12	13	14	15	16
2	3	2	3	1	2	3	2	4	1	3	1	2	3	1	4

Statement Based MCQs

1	2	3	4	5
1	1	1	2	3

Assertion & Reason MCQs

1	2	3	4	5
2	1	3	4	1

Match the Columns MCQs

1	2	3	4
3	2	3	2

NCERT Exemplar MCQs

1	2	3	4	5	6
3	2	2	1	4	1, 4

NEET & AIIMS Past Year MCQs

1	2	3	4	5	6	7	8	9	10	11	12	13	14
3	2	4	2	4	4	1	1	1	2	3	2	1	2

CHAPTER 9

Ray Optics and Optical Instruments

Topicwise Number of Questions Analysis (2025-2016)

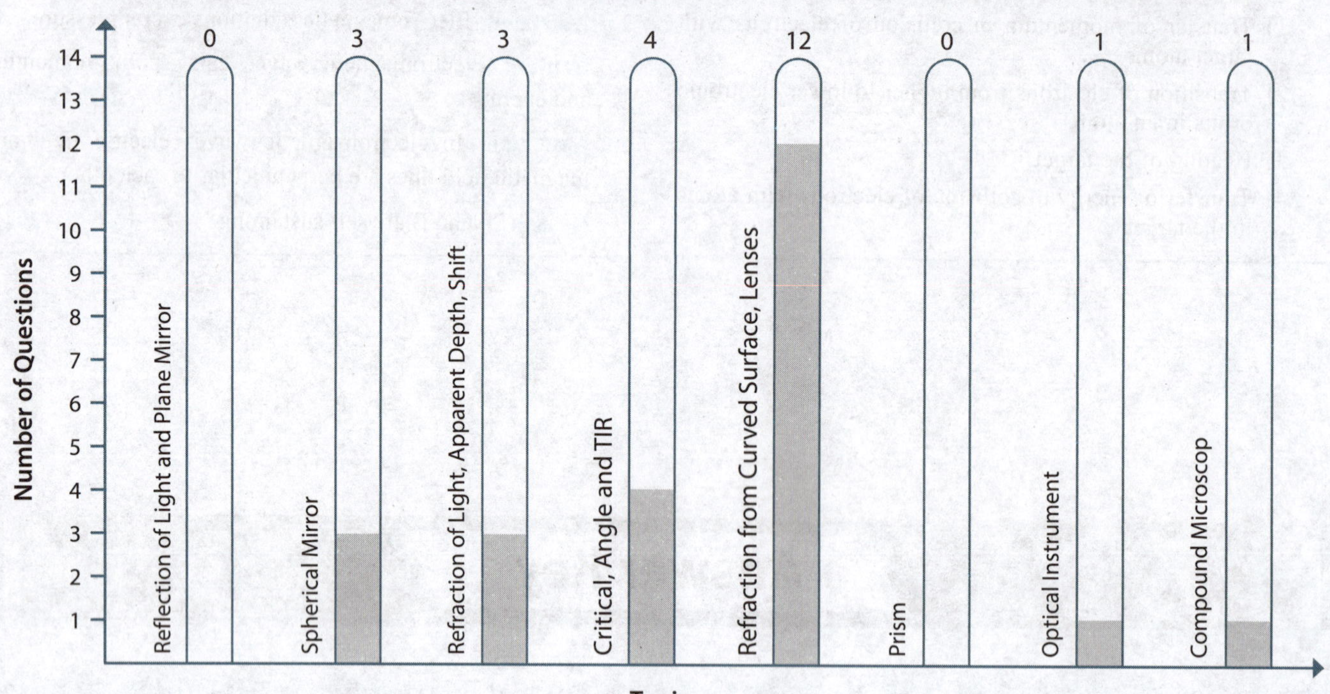

NCERT-Picks

Reflection of Light by a Plane Mirror

Formulas

1. Angle of incidence is equal to the angle of reflection i.e $\angle i = \angle r$

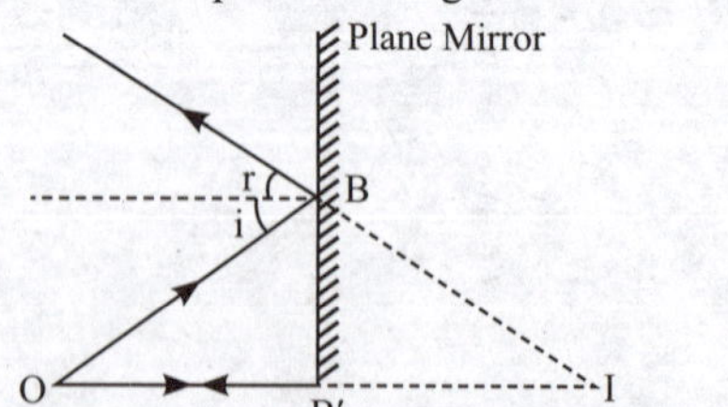

2. Number of images formed by two plane mirrors

(a) $n = \frac{360°}{\theta} - 1$, when $\frac{360°}{\theta}$ is even

(b) $n = \frac{360°}{\theta} - 1$, when $\frac{360°}{\theta}$ is odd and the object lies symmetrically

(c) $n = \frac{360°}{\theta}$, when $\frac{360°}{\theta}$ is odd and object lies unsymmetrically.

Numerical Profile

Reflection for plane mirror

(i) If two plane mirrors are inclined at 90°, the emergent ray is always anti parallel to incident ray if it suffers one reflection from each mirror whatever be the angle of incidence. The same is found to hold good for 3 plane mirrors forming the corner of a cube if the incident light suffers one reflection from each of them,

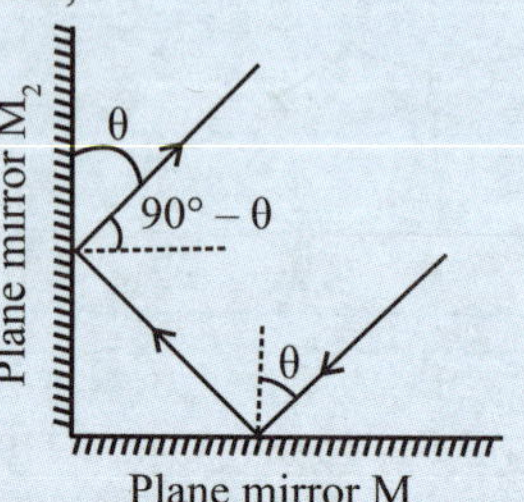

(ii) A person needs a plane mirror of minimum half of his height to see his full image.
(iii) A person standing in the middle of a room can see complete wall behind him if the mirror in front of him is ⅓rd of height of wall.
(iv) If an object move towards or away from a plane mirror at speed v, the image will also approach or recede at same speed v, i.e., the speed of image relative to object will be v- (-v) = 2v. Similarly if the mirror is moved towards or away from the object with speed v the image will move towards or away from the object with a speed 2v.
(v) When light gets reflected in the same medium, it suffers no change in frequency, speed and wavelength.

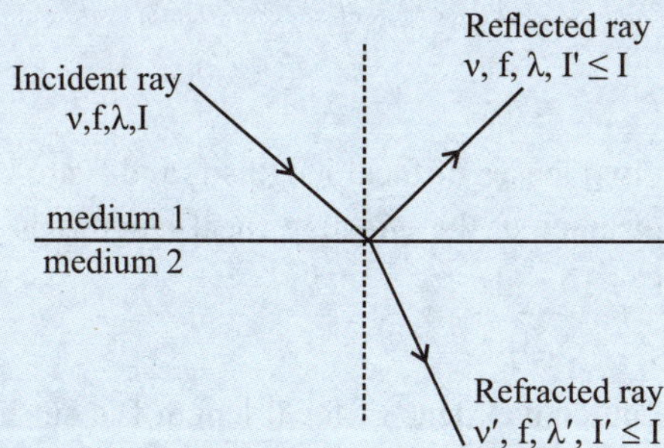

(vi) Frequency of light remains unchanged when it gets reflected or refracted.
(vii) Velocity, wavelength and intensity changes when it gets refracted.

Reflection of Light by a Spherical Mirror

Formulas

1. For paraxial rays
$f = R/2$
where, f is the focal length and R is the radius of curvature of the mirror
2. Mirror formula: $\frac{1}{v}+\frac{1}{u}=\frac{1}{f}$
where, u is the object distance, v is the image distance and f is the focal length of the mirror
Linear magnification $m=\frac{h_i}{h_o}=\frac{-v}{u}$
3. $u=\left(\frac{m-1}{m}\right)f$
4. $v = -(m-1)f$
5. Focal length: $f=\frac{uv}{v+v}$
6.

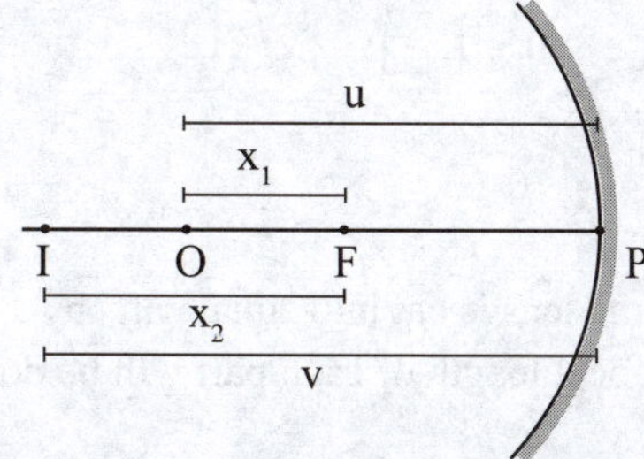

In case of spherical mirrors if object distance (x_1) and image distance (x_2) are measured from focus instead of pole $u = (f + x_1)$ and $v = (f + x_2)$, the mirror formula $\frac{1}{v}+\frac{1}{u}=\frac{1}{f}$ reduces to $\frac{1}{(f+x_2)}+\frac{1}{(f+x_1)}=\frac{1}{f}$

Which on simplification gives $x_1x_2 = f^2$

7. Axial or Longitudinal Magnification
$$m_{ax}=\frac{dv}{du}=\frac{x_2}{x_1}=\frac{v^2}{u^2}=m^2$$
where x_2 = size of image along principal axis.
x_1 = size of the object along principal axis.
8. Areal Magnification
$$m_{ar}=\frac{A_1}{A_0}=\frac{v^2}{u^2}=m^2$$
Where A_1 = Area of Image
A_0 = Area of Object

Refraction of light and TIR

Formulas

1. Snell's law: $\mu_a \sin\theta_a = \mu_b \sin\theta_b$
where μ_a and μ_b are the refractive indices of the two media, θ_a and θ_b are the angle of incidence and angle of refraction that the ray makes with the normal
2. Lateral shift: $d=\frac{t\sin(i-r)}{\cos r}$

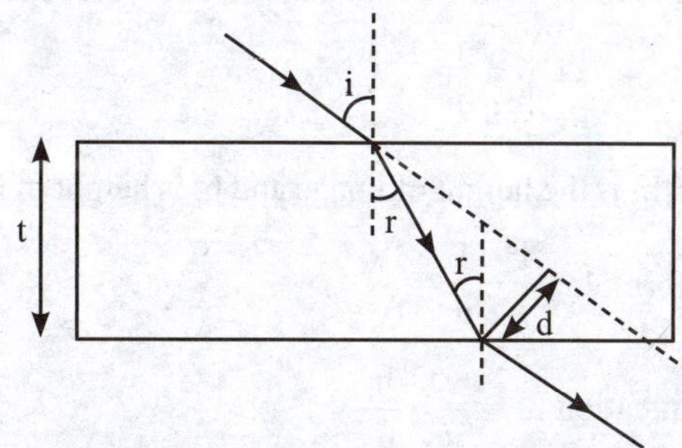

3. If x is the apparent shift in the position towards the observer and d is thickness of the medium over the object, then
$$x=\left(1-\frac{1}{\mu}\right)d$$
4. If θ_C is critical angle and μ is the refractive index of the denser medium with respect to the rarer, then $\theta_C=\sin^{-1}\left(\frac{1}{\mu}\right)$

Numerical Profile

Relation between object and image velocity

(i) When object is moving along principal axis
$$\frac{dv}{dt}=-\left(\frac{v^2}{u^2}\right)\frac{du}{dt}$$

(ii) If "h" is the depth of a fish from the surface of the water of refractive index μ, the radius of the circle on the surface of the water through which it can see the outer world is

$$R = h \tan C = \frac{h}{\sqrt{\mu^2 - 1}}$$

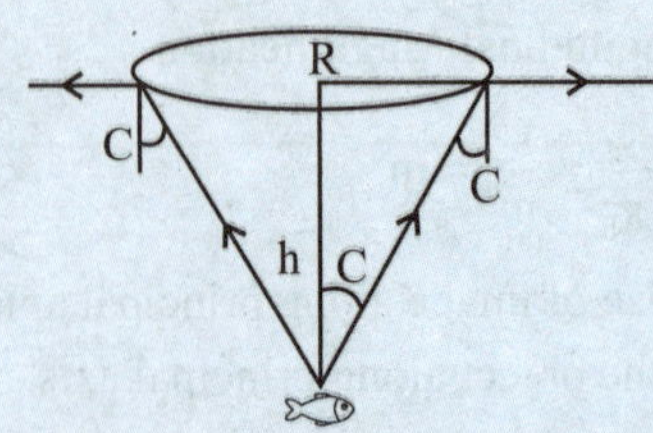

(iii) If there are r slabs of thickness $t_1, t_2, t_3, \ldots$ and refractive indices $\mu_1, \mu_2 \ldots$ then total apparent shift (t) is given by

$$t = t_1\left(1 - \frac{1}{\mu_1}\right) + t_2\left(1 - \frac{1}{\mu_2}\right) + \ldots$$

Refraction From Curved Surfaces and Lenses

Formulas

1. $\frac{\mu_2}{v} - \frac{\mu_1}{u} = \frac{(\mu_2 - \mu_1)}{R}$

   where, μ_1 and μ_2 are the refractive indices of the rarer and denser medium
2. Magnification for refraction from curved surface, $m = \frac{h_2}{h_1} = \left(\frac{\mu_1}{\mu_2}\right)\left(\frac{v}{u}\right)$

   where, h_2 is the height of image and h_1 is height of the object
3. Lens formula: $\frac{1}{v} - \frac{1}{u} = \frac{1}{f}$
4. Magnification $m = \frac{v}{u} = \frac{h_i}{h_o}$
5. Power of lens, $P = \frac{1}{f}$
6. Lens maker's formula

   $$\frac{1}{f} = (\mu - 1)\left(\frac{1}{R_1} - \frac{1}{R_2}\right)$$
7. Equivalent focal length

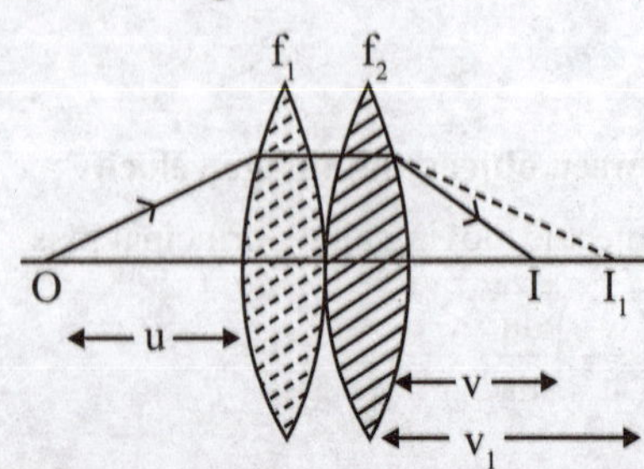

$$\frac{1}{f} = \frac{1}{f_1} + \frac{1}{f_2}$$

$$\Rightarrow P = P_1 + P_2$$

Numerical Profile

Lenses Separated by a Distance

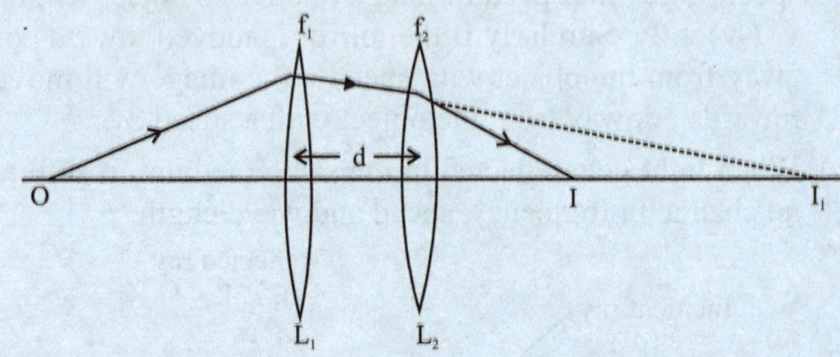

(i) When two lenses of focal lengths f_1 and f_2 are kept apart by a distance d, the effective focal length f is given by

$$\frac{1}{f} = \frac{1}{f_1} + \frac{1}{f_2} - \frac{d}{f_1 f_2}$$

(ii) If an equiconvex lens of focal length f is cut into equal parts by a horizontal plane AB then the focal length of each part will be equal to that of initial lens. Because μ, R_1 and R_2 will remain unchanged.

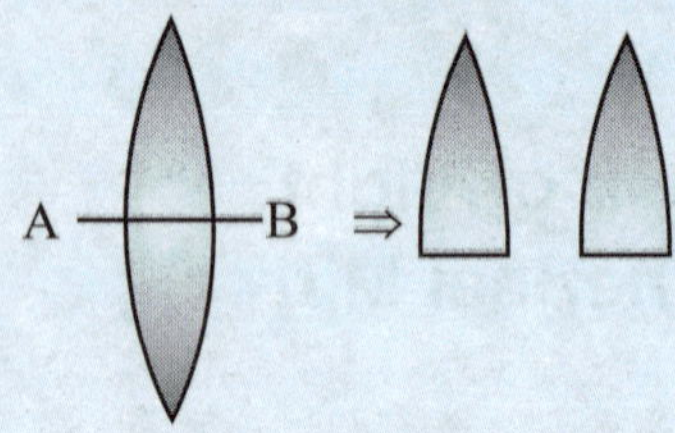

∵ Intensity, I ∝ $(\text{aperture})^2$

∴ Intensity through a single part will be reduced

(iii) Both lens act as a convex lens

f (positive) f (positive)

P′ = 2P

L_1 L_2

$$\frac{1}{F} = \frac{1}{f} + \frac{1}{f} = \frac{2}{f} \quad \text{so} \quad P' = 2P$$

(iv) One lens acts convex while another will act as concave

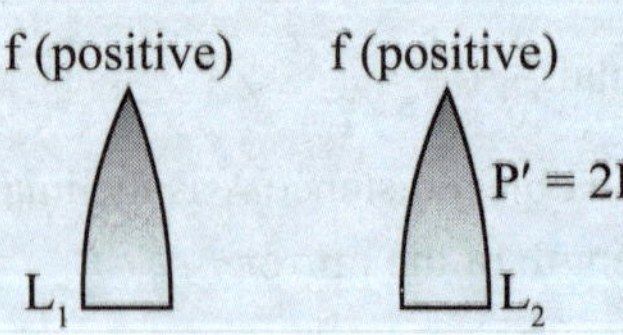

$$\frac{1}{F} = \frac{1}{f} - \frac{1}{f} = 0$$

(v) If the same lens is cut into equal parts by a vertical plane CD the focal length of each part will be double of initial value.

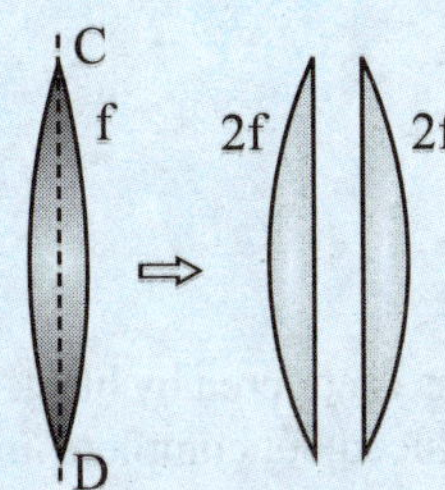

Focal length of each part become 2f

For equiconvex lens, $\frac{1}{f} = \frac{(\mu-1)\times 2}{R}$

For plane convex lens, $\frac{1}{f_1} = \frac{\mu-1}{R}$

so, $\frac{1}{f} = \frac{2}{f_1} \Rightarrow f_1 = 2f$

Prism

1. Angle of deviation (δ)

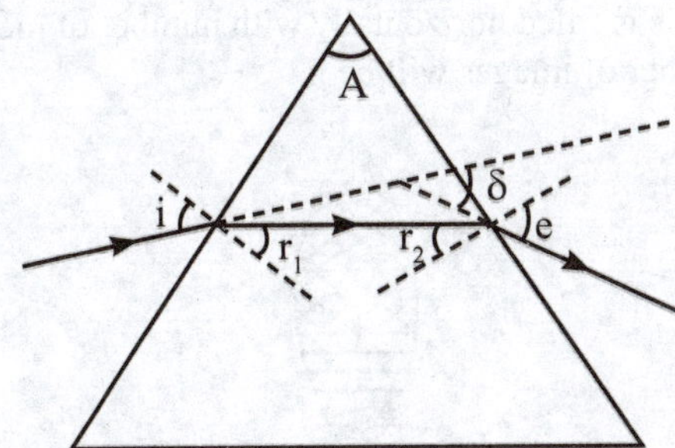

$\delta = i + e - A$

2. Condition of Grazing Emergence

$i = \sin^{-1}[\sin A\sqrt{\mu^2 - 1} - \cos A]$

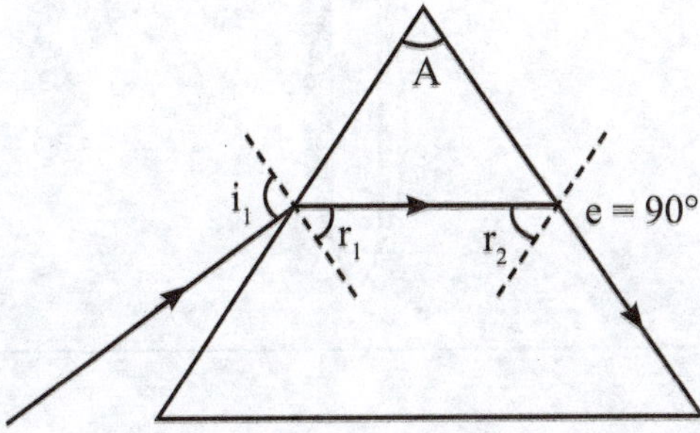

3. Condition of minimum deviation

For minimum deviation, i = e

$\Rightarrow \delta_{min} = 2i - A$

and $\mu = \frac{\sin\left(\frac{\delta_{min} + A}{2}\right)}{\sin\left(\frac{A}{2}\right)}$

4. Deviation produced by the prism

$\delta = (\mu - 1)A$

5. Dispersive Power

$\omega = \frac{\mu_v - \mu_r}{\mu_y - 1}$

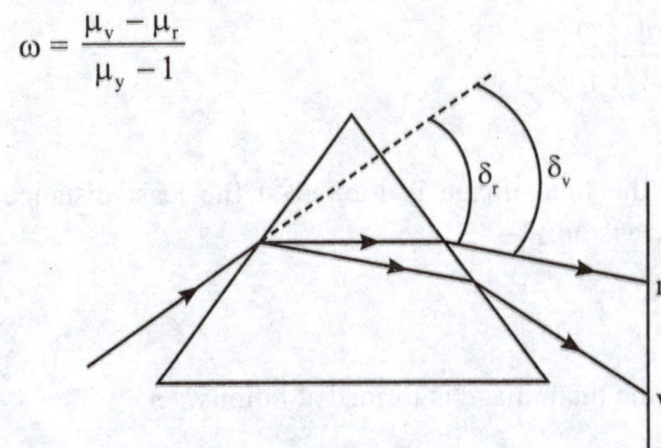

6. Deviation without Dispersion

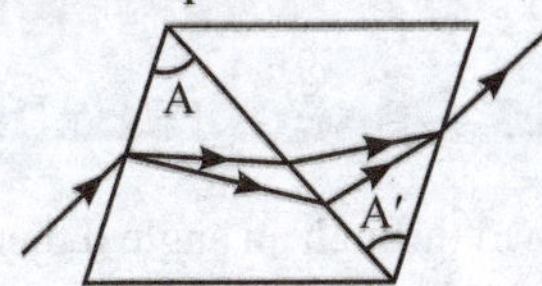

$(\mu_v - \mu_r)A + (\mu_v' - \mu_r')A' = 0 \Rightarrow A' = \frac{(\mu_v - \mu_r)A}{\mu_v' - \mu_r'}$

7. Dispersion without deviation

$(\mu - 1)A + (\mu' - 1)A' = 0 \Rightarrow A' = \frac{(\mu - 1)A}{(\mu' - 1)}$

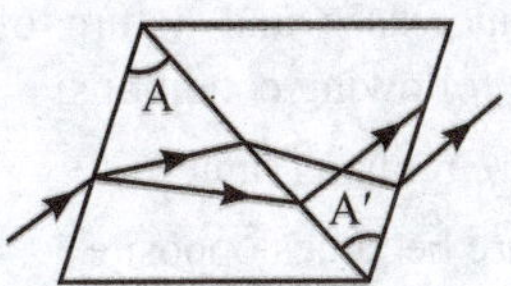

Numerical Profile

(i) For a light ray not to emerge from a prism

$\mu > \frac{1}{\sin\left(\frac{A}{2}\right)}$

(ii) For maximum deviation i = 90°

$\therefore \delta_{max} = 90° + e - A$

where, $e = \sin^{-1}[\mu \sin(A - \theta_C)]$

Simple Microscope

Formulas

1. Magnifying power of a simple microscope, $m = \left(1 + \frac{D}{f}\right)$
where, D is the least distance for clear vision

Compound Microscope and Telescope

Formulas

Compound Microscope:

(i) When the image is formed at least distance of distinct vision.

$m = \frac{v_o}{u_o}\left(1 + \frac{D}{f_e}\right)$

(ii) When the final image is formed at infinity.

$$m = -\frac{L}{f_o}\left(\frac{D}{f_e}\right)$$

Telescope:

(i) When the final image is formed at the least distance of distinct vision:

$$m = \frac{-f_o}{f_e}\left(1+\frac{f_e}{D}\right)$$

(ii) When the final image is formed at infinity.

$$m = -\left(\frac{f_o}{f_e}\right)$$ and distance between the lenses; $L = f_0 + f_e$

Statements Asked in Exam

- If a mirror is turned through an angle θ then the reflected ray turns through an angle 2θ.
- In total internal reflection, there is no loss of intensity because 100% of incident light is reflected back into the same medium, however there is always some loss of intensity of light in the case of reflection from a mirror.
- When a lens is immersed in a liquid or a medium its focal length always increases and nature of the lens changes according to the following conditions:

  If $n_o > n_m$ = nature remains same

  If $n_o < n_m$ = nature becomes opposite

  If $n_o = n_m$ lens disappears
- If a convex lens is dipped in a liquid whose refractive index is equal to the refractive index of lens. Then focal length will become infinite.
- Focal length of lens depends on wavelength

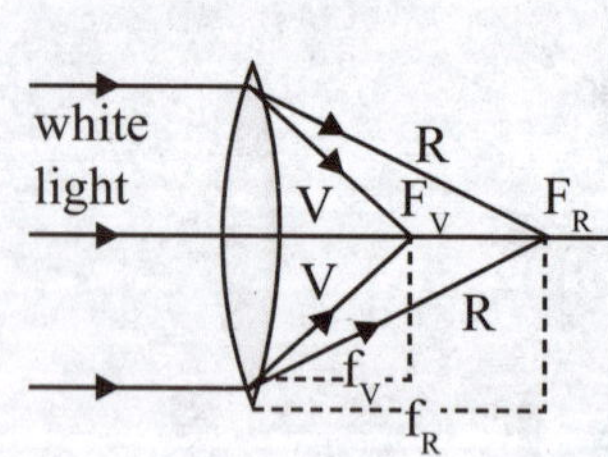

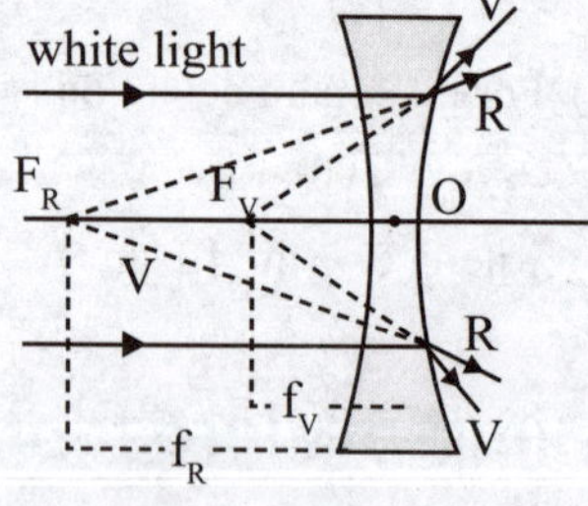

$$\because \frac{1}{f} \propto (\mu - 1) \propto \frac{1}{\lambda}$$

$$\therefore f \propto \lambda$$

$$\therefore f_R > f_V$$

- If half portion of lens is covered by black paper then intensity of image will be reduced but complete image will be formed.

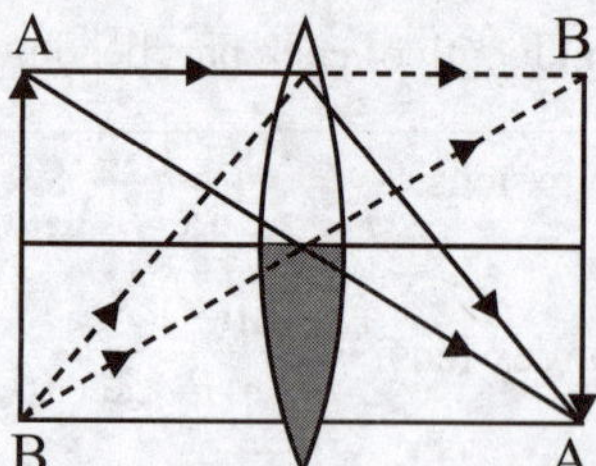

- A glass slab is placed over a page in which letters are printed in different colours, images of letters with violet colours are raised maximum.
- In a telescope, aperture of objective lens is made large to increase magnifying power and resolving power of the telescope.
- If the lens is filled horizontally with number of media n then the number of images will be n.

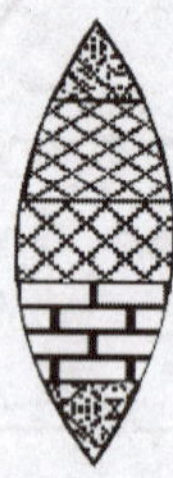

- If the lens is filled vertically with number of media n then the number of images will be 1.

NCERT Topic-wise MCQs

Reflection of Light Plane and Spherical Mirror

1. If an object is kept at a distance of 30 cm from a concave mirror of focal length 20 cm. Find the magnification of the image.

   1) – 1.5 2) – 2
   3) 2.5 4) – 3

2. Two plane mirrors are inclined at 70°. A ray incident on one mirror at angle θ after reflection falls on the second mirror and is reflected from there parallel to the first mirror, θ is:

   1) 50° 2) 45°
   3) 30° 4) 55°

3. Two plane mirrors are inclined to each other at an angle 72°, what is the number of images formed?

   1) 3 2) 5
   3) 9 4) 7

4. The image of a real object formed by a plane mirror is-

   1) Erect, real and of equal size
   2) Erect, virtual and of equal size
   3) Inverted, real and of equal size
   4) Inverted, virtual and of equal size

5. A light beam converges towards a convex mirror of focal length 20 cm as shown in the figure below, the position of the image is

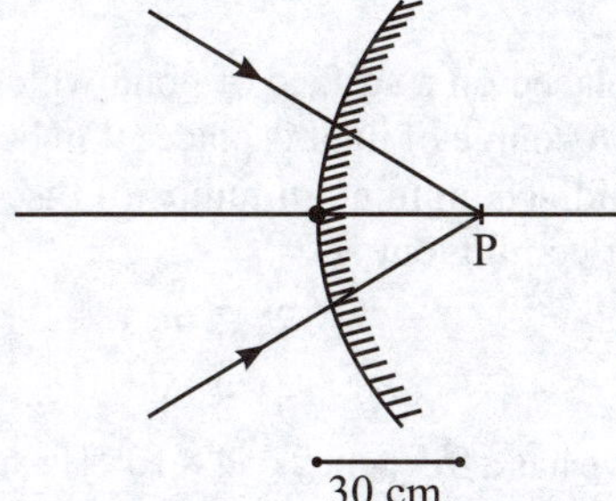

   1) 60 cm 2) 90 cm
   3) 75 cm 4) 45 cm

6. A point object at 15 cm from a concave mirror of radius of curvature 20 cm is made to oscillate along the principal axis with amplitude 2 mm. The amplitude of its image will be

   1) 2 mm 2) 4 mm
   3) 8 mm 4) 16 mm

7. A convex mirror of focal length f forms an image which is 1/n times the object. The distance of the object from the mirror is

   1) (n – 1) f
   2) $\sqrt{2}f$
   3) $\left(\frac{n+1}{n}\right)f$
   4) (n + 1) f

8. Two plane mirrors M_1 and M_2 are inclined at angle θ as shown. A ray of light 1, which is parallel to M_1 strikes M_2 and after two reflections, the ray 2 becomes parallel to M_2. Find the angle θ.

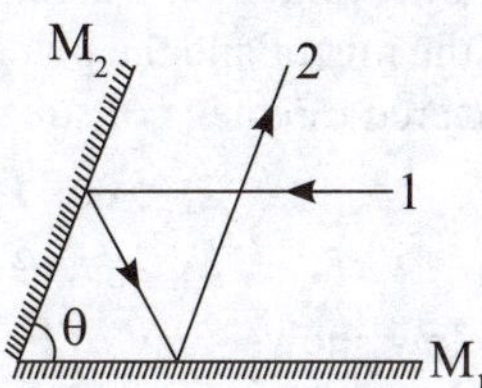

   1) 60° 2) 30°
   3) 40° 4) 50°

9. Two plane mirrors are lying perpendicular to each other, there is lamp in between the mirrors. Then number of images of lamp is

   1) 3 2) 7
   3) 9 4) 8

10. An object is at a distance of 0.5 m in front of a plane mirror. Distance between the object and image is

    1) 0.5 m 2) 1 m
    3) 0.25 m 4) 1.5 m

11. A virtual image three times the size of the object is obtained with a concave mirror of radius of curvature 36 cm. The distance of the object from the mirror is-

    1) 5 cm 2) 12 cm
    3) 10 cm 4) 20 cm

12. An object is placed 1m in front of a plane mirror. An observer stands 3m behind the object. For what distance must an observer focus his eyes in order to see the image of the object?

    1) 3m 2) 4m
    3) 5m 4) 8m

13. A concave mirror is placed on a horizontal table with its axis directed vertically upward. Let O be the pole of the mirror and C is centre of curvature. A point object is placed at C. It has a real image, also located at C. If the mirror is now filled with water, the image will be:

1) Real, and will remain at C
2) Real, and located at a point between C and ∞
3) Virtual and located at a point between C and O
4) Real, and located at a point between C and O

14. In an experiment to find focal length of a concave mirror, a graph is drawn between the magnitude of u and v. The graph looks like:

1) 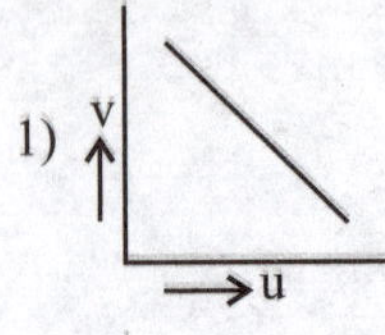

2) 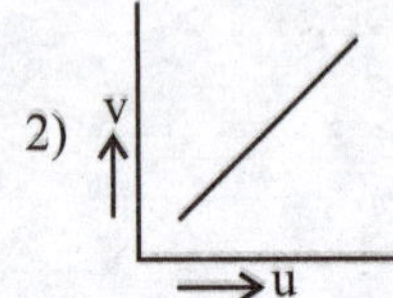

3) 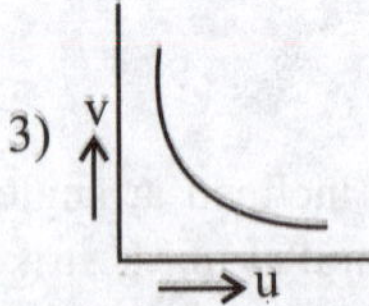

4) 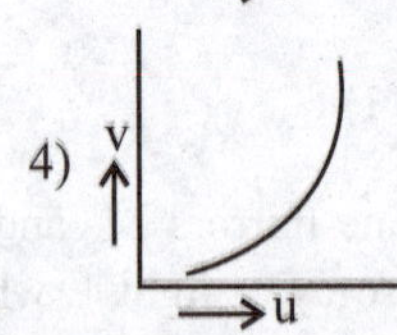

15. A square of side 3 cm is located at a distance 25 cm from a concave mirror of focal length 10 cm. The center of square is at the axis of the mirror and the plane is normal to axis of mirror. Area covered by image of square

1) 4 cm^2 2) 6 cm^2
3) 16 cm^2 4) 36 cm^2

16. The diagram below shows a ray of light travelling parallel to the principal axis of a concave spherical mirror. Point F is the principal focus and point C is the centre of curvature.

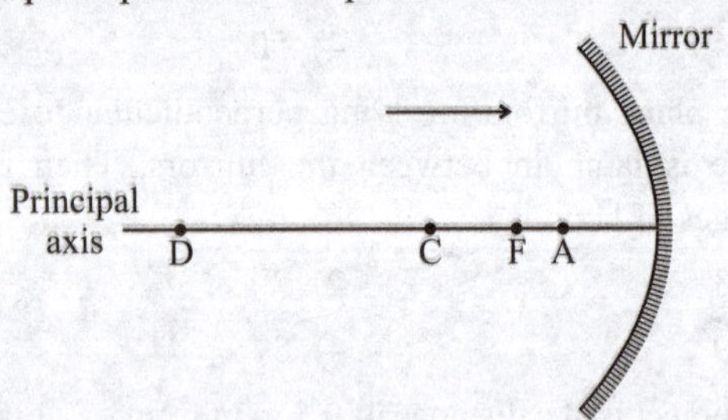

After striking the mirror, the ray of the light will be reflected through point

1) A 2) D
3) C 4) F

17. The graph in the below figure shows a plot of variation of v with change in u for a concave mirror. Points plotted above the point P on the curve are for values of v:

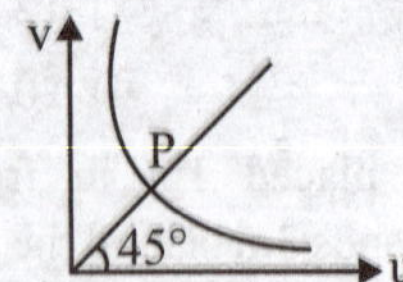

1) smaller than f
2) smaller than $2f$
3) larger than $2f$
4) larger than f but less than $2f$

18. If in the following figure, height of object O = +2.5cm, find the height of image I formed is:

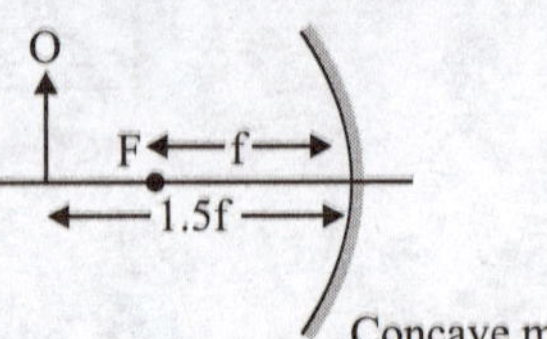

1) – 5cm 2) + 5cm
3) + 7.5cm 4) – 7.5cm

19. An infinitely long rod lies along the axis of a concave mirror of focal length f. The near end of the rod is at a distance u > f from the mirror. Its image will have a length

1) $\frac{f^2}{u+f}$ 2) $\frac{uf}{\mu-f}$
3) $\frac{f^2}{u-f}$ 4) uf/(u + f)

20. A luminous point object is moving along the principal axis of a concave mirror of focal length 12 cm towards it. When its distance from mirror is 20 cm its velocity is 4 cm/s. The velocity of the image in cm/s at that instant is:

1) 6 cm/s towards the mirror
2) 6 cm/s away from the mirror
3) 9 cm/s away from the mirror
4) 9 cm/s towards the mirror

21. A thin rod of length $f/3$ lies along the axis of a concave mirror of focal length f. One end of its magnified image touches an end of the rod. The length of the image is:

1) f 2) $\frac{1}{2}f$
3) $2f$ 4) $\frac{1}{4}f$

Refraction of Light, Apparent Depth, Shift, TIR

22. A disc is placed on a surface of pond which has refractive index 5/3. A source of light is placed 4 m below the surface of the liquid. The minimum radius of disc needed so that light is not coming out is

1) ∞ 2) 3 m
3) 6 m 4) 4 m

23. A light wave has a frequency of 4×10^{14} Hz and a wavelength of 5×10^{-7} m in a medium. The refractive index of the medium is:

1) 1.5 2) 1.33
3) 1.0 4) 0.66

24. How much water should be filled in container 21 cm in height, so that it appears half filled when viewed from the top of the container. (Given that ${}^a\mu_w = 4/3$)

1) 8 cm 2) 10.5 cm
3) 12 cm 4) None of the above

25. Monochromatic light is refracted from air into glass of refractive index μ. The ratio of the wavelength of incident and refracted waves is:

1) 1 : μ 2) 1 : μ^2

3) μ : 1 4) 1 : 1

26. For a colour of light the wavelength for air is 6000 Å and in water the wavelength is 4500 Å. Then the speed of light in water will be:

1) 5×10^{14} m/s 2) 2.25×10^8 m/s

3) 4×10^8 m/s 4) Zero

27. On heating a liquid the refractive index generally:

1) Decreases

2) Increase or decrease depending on the rate of heating

3) Do not change

4) Increase

28. A fish at a depth of 12 cm in water is viewed by an observer on the bank of a lake. To what height the image of the fish is raised? (${}_a\mu_w = 1.33$)

1) 9 cm 2) 12 cm

3) 3.8 cm 4) 3 cm

29. Each quarter of a vessel of depth H is filled of the refractive index n_1, n_2, n_3, n_4 from the bottom respectively. The apparent depth of the vessel when looked normally is:

1) H/4 $(n_1 + n_2 + n_3 + n_4)$ 2) H/4 $(n_1 + n_2 + n_3 + n_4)$

3) $(n_1 + n_2 + n_3 + n_4)/4H$ 4) $\frac{H}{4}\left(\frac{1}{n_1}+\frac{1}{n_2}+\frac{1}{n_3}+\frac{1}{n_4}\right)$

30. A ray of light strikes a transparent rectangular slab of refractive index $\sqrt{2}$ at an angle of incidence of 45°. The angle between the reflected and refracted ray is:

1) 75° 2) 90°

3) 105° 4) 120°

31. A transparent cube of 15 cm edge contains a small air bubble. Its apparent depth when viewed through one face is 6cm and when viewed through the opposite face is 4cm. Then the refractive index of the material of the cube is:

1) 2 2) 2.5

3) 1.6 4) 1.5

32. If ${}^i\mu_j$ represent refraction index when a light ray goes from medium $\hat{i}$ to $\hat{j}$ then the product ${}^2\mu_1 \times {}^3\mu_2 \times {}^4\mu_3$ will be:

1) ${}^3\mu_1$ 2) ${}^3\mu_2$

3) $\frac{1}{{}^1\mu_4}$ 4) ${}^3\mu_1$

33. The refractive index and the permeability of a medium are respectively 1.5 and 5×10^{-7} Hm^{-1}. The relative permittivity of the medium is nearly:

1) 25 2) 15

3) 81 4) 6

34. A large glass slab (μ = 5/3) of thickness 8 cm is placed over a point source of light on a plane surface. It is seen that light emerges out of the top surface of the slab from a circular area of radius R cm. What is the value of R?

1) 2 cm 2) 3 cm

3) 6 cm 4) 12 cm

35. Refractive index of water and glass are 4/3 & 3/2 respectively. A ray of light travelling in water is incident on the water glass interface at 30°. Calculate the angle of refraction.

1) 30° 2) 25°

3) 26.38° 4) 27°

36. A fish is a little away below the surface of a lake. If the critical angle is 49°, then the fish could see things above the water surface within an angular range of θ° where:

1) θ = 49° 2) θ = 90°

3) θ = 98° 4) $\theta = 24\frac{1}{2}^\circ$

37. A ray of light is incident at an angle *i* from denser to rarer medium. The reflected and the refracted rays are mutually perpendicular. The angle of reflection and the angle of refraction are respectively r & r' then the critical angle will be:

1) $\sin^{-1}(\sin r)$ 2) $\sin^{-1}(\tan r')$

3) $\sin^{-1}(\tan i)$ 4) $\sin^{-1}(\sin i)$

38. A ray of light passes through two slabs of same thickness. In the first slab n_1 waves are formed and in the second slab n_2. Find refractive index of second medium with respect to first.

1) $\frac{n_1}{n_2}$ 2) $\frac{n_1 + n_2}{n_2}$

3) $\frac{n_2}{n_1}$ 4) $\frac{n_1 + n_2}{n_1}$

39. A ray of light propagates from glass of $\mu_g = \frac{3}{2}$ to water of $\mu_w = \frac{4}{3}$. The value of the critical angle is:

1) $\sin^{-1}\left(\frac{1}{2}\right)$

2) $\sin^{-1}\left(\sqrt{\frac{8}{9}}\right)$

3) $\sin^{-1}\left(\frac{8}{9}\right)$

4) $\sin^{-1}\left(\frac{5}{7}\right)$

40. A ray of light enters from a rarer to denser medium. The angle of incidence is i. Then the reflected and refracted rays are mutually perpendicular to each other. The critical angle for a pair of media is:

1) $\sin^{-1}(\cot i)$ 2) $\cos^{-1}(\tan i)$

3) $\sin^{-1}(\tan i)$ 4) $\tan^{-1}(\sin i)$

41. Glass has refractive index μ with respect to air and the critical angle for a ray of light going from glass to air is θ. If a ray of light is incident from air on the glass with the angle of incidence θ, the corresponding angle of refraction is:

1) $\sin^{-1}\left(\frac{1}{\sqrt{\mu}}\right)$ 2) 90°

3) $\sin^{-1}\left(\frac{1}{\mu^2}\right)$ 4) $\sin^{-1}\left(\frac{1}{\mu}\right)$

42. White light is incident on the interface of glass and air as shown in figure. If green light is just totally internally reflected then the emerging ray in air contains:

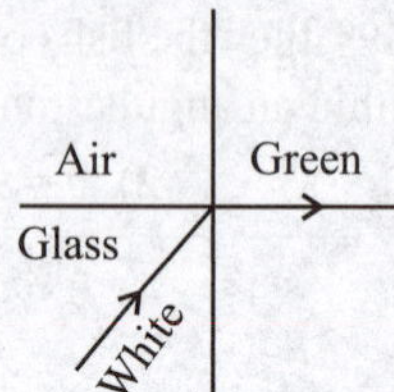

1) Yellow, Orange, Red 2) Indigo, Blue, Violet

3) All colours 4) All colours except green

43. In the figure for an angle of incidence 45°, at the top surface, what is the minimum refractive index needed for total internal reflection at vertical face:

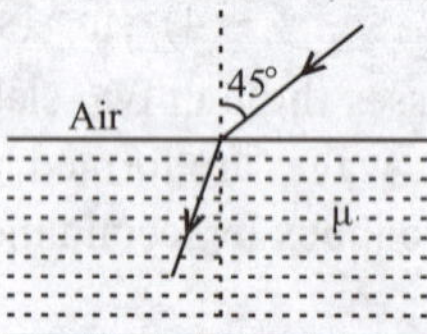

1) $\frac{\sqrt{2}+1}{2}$ 2) $\sqrt{3/2}$

3) $\sqrt{1/2}$ 4) $\sqrt{2}+1$

44. A normally incident ray reflected at an angle of 90°. The value of critical angle is:

1) 45° 2) 90°

3) 65° 4) 43.2°

45. A beaker contains water up to a height h_1 and kerosene of height h_2 above water so that the total height of (water + kerosene) is $(h_1 + h_2)$. Refractive index of water is μ_1 and that of kerosene is μ_2. The apparent shift in the position of the bottom of the beaker when viewed from above is

1) $\left(1-\frac{1}{\mu_1}\right)h_2+\left(1-\frac{1}{\mu_2}\right)h_1$

2) $\left(1+\frac{1}{\mu_1}\right)h_1-\left(1+\frac{1}{\mu_2}\right)h_2$

3) $\left(1-\frac{1}{\mu_1}\right)h_1+\left(1-\frac{1}{\mu_2}\right)h_2$

4) $\left(1+\frac{1}{\mu_1}\right)h_2-\left(1+\frac{1}{\mu_2}\right)h_1$

46. In vacuum, to travel distance d, light takes time t and in medium to travel 5d, it takes time T. The critical angle of the medium is:

1) $\sin^{-1}\left(\frac{5T}{t}\right)$ 2) $\sin^{-1}\left(\frac{5t}{T}\right)$

3) $\sin^{-1}\left(\frac{t}{5T}\right)$ 4) $\sin^{-1}\left(\frac{5t}{3T}\right)$

Prism

47. Deviation by a prism is lowest for:

1) Violet ray 2) Green ray

3) Red ray 4) Yellow ray

48. Which ray diagrams given below shows that a right angled isosceles prism can produce a deviation of 90° If the refractive index is more than $\sqrt{2}$.

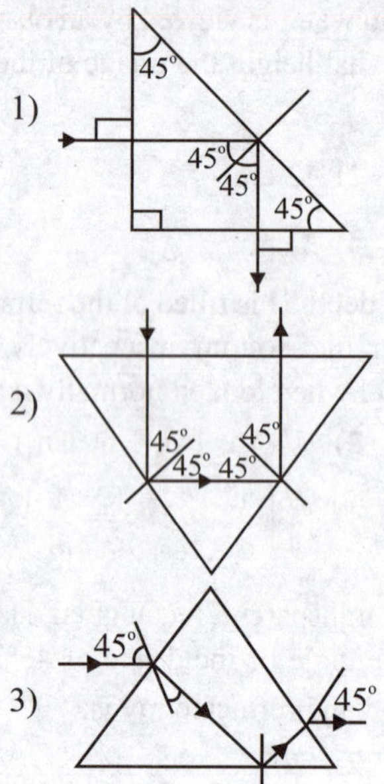

4) None of these

49. Dispersive power depends upon:

1) The shape of prism 2) Material of prism

3) Angle of prism 4) Height of the prism

50. The angle of minimum deviation for prism of angle π/3 is π/6. Calculate the velocity in the material of the prism if the velocity of light in vacuum is 3×10^8 m/s?

1) 2.12×10^8 m/s 2) 4×10^8 m/s

3) 5×10^8 m/s 4) 3.12×10^8 m/s

51. The refractive index of flint glass for red and blue colours are 1.664 & 1.644. What will be the dispersive power?

1) 0.305 2) 0.0305

3) 0.635 4) 0.0635

52. Deviation of 5° is observed from a prism whose angle is small and whose refractive index is 1.5. The angle of prism is

1) 7.5° 2) 10°

3) 5° 4) 3.3°

53. A ray incident at 15° on one refracting surface of a prism of angle 60°, suffers a deviation of 55°. What is the angle of emergence?

1) 95° 2) 45°
3) 30° 4) 100°

54. When light of wavelength λ is incident on an equilateral prism kept in its minimum deviation position, it is found that the angle of deviation equals the angle of the prism itself. The refractive index of the material of the prism for the wavelength λ is then:

1) $\sqrt{3}$
2) $\frac{\sqrt{3}}{2}$
3) 2
4) $\sqrt{2}$

55. A ray of light is incident at 60° on the one face of a prism of angle 30° and the emergent ray makes 30° with the incident ray. The refractive index of the prism is:

1) 1.732 2) 1.414
3) 1.5 4) 1.33

Refraction From Curved Surface, Lenses

56. The radius of curvature for a convex lens is 40 cm for each surface. Its refractive index is 1.5. The focal length will be:

1) 40 cm 2) 20 cm
3) 80 cm 4) 30 cm

57. Two lenses are placed in contact with each other and the focal length of combination is 80 cm. If the focal length of one is 20 cm, then the power of the other will be:

1) 1.66 D 2) 4.00 D
3) –1.00 D 4) –3.75 D

58. A thin convex lens of focal length 10 cm is placed in contact with the concave lens of same material and of same focal length. The focal length of combination will be:

1) Zero 2) Infinity
3) 10 cm 4) 20 cm

59. A needle placed 45 cm from a lens forms an image on the screen placed 90 cm on the other side of the lens. What is the size of image if the size of needle is 5 cm? And what is the focal length of lens?

1) 10 cm, 30cm 2) 20 cm, 30cm
3) 100cm, 10cm 4) 1 m, 100 cm

60. The distance between an object and a divergent lens is m times the focal length of the lens. The linear magnification produced by the lens is:

1) m 2) 1/m
3) m+1 4) $\frac{1}{m+1}$

61. A convex lens makes a real image 4 cm long on a screen when the lens is shifted to a new position without disturbing the object, we again get a real image on screen which is 16 cm tall. The length of the object must be:

1) 1/4 cm
2) 8 cm
3) 12 cm
4) 20 cm

62. An equiconvex lens is cut into two halves along (i) XOX' and (ii) YOY' as shown in the given figure. Let f, f' and f'' be the focal lengths of complete lens, of each half in case (i) and of each half in case (ii) respectively. Choose the correct statement from the following:

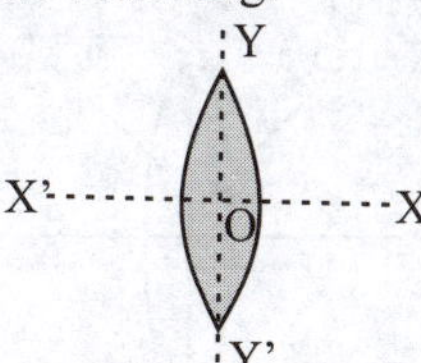

1) $f' = f, f'' = 2f$
2) $f' = 2f, f'' = f$
3) $f' = f, f'' = f$
4) $f' = 2f, f'' = 2f$

63. The focal length of a thin biconvex lens is 20 cm. When an object is moved from a distance of 25 cm in front of it to 50 cm, the magnification of its image changes from m_{25} to m_{50}. The ratio $\frac{m_{25}}{m_{50}}$ is

1) 4 2) 6
3) 1 4) 3

64. If the central portion of a convex lens is wrapped in black paper as shown in figure:

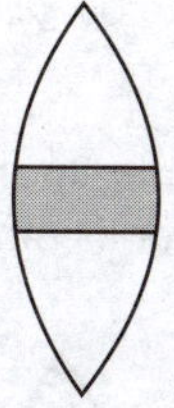

1) No image will be formed by the remaining portion of the lens
2) The full image will be formed but it will be less bright
3) The central portion of the image will be missing
4) There will be two images each produced by one of the exposed portion of the lens

65. A biconvex lens has a radius of curvature of magnitude 20 cm. Which one of the following options describe best the image formed of an object of height 2 cm placed 30 cm from the lens?

1) Virtual, upright, height = 1 cm
2) Virtual, upright height = 0.5 cm
3) Real, inverted, height = 4 cm
4) Real, inverted, height = 1 cm

66. A lens of refractive index n is put in a liquid of refractive index (n'), if focal length of lens in air is 'f', its focal length in liquid will be:

1) $\frac{-fn'(n-1)}{n'-n}$ 2) $\frac{-f(n'-n)}{n'(n-1)}$

3) $\frac{-n'(n-1)}{f(n'-n)}$ 4) $\frac{fn'n}{n-n'}$

67. A bi-convex lens is formed with two thin planoconvex lenses as shown in the figure. Refractive index n of the first lens is 1.5 and that of the second lens is 1.2. Both the curved surfaces are of the same radius of curvature R = 14 cm. For this bi-convex lens, for an object distance of 40 cm, the image distance will be

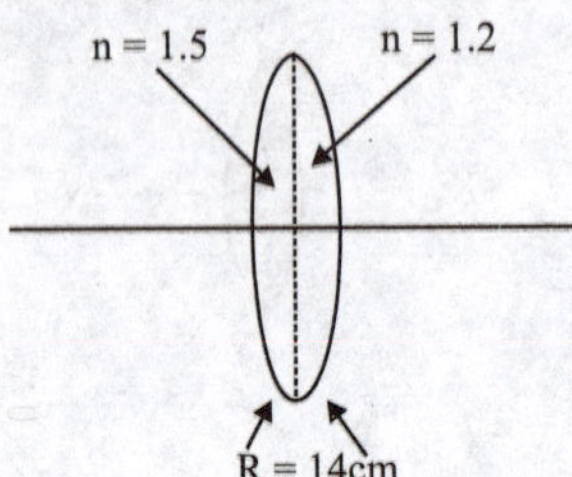

1) – 280.0 cm 2) 40.0 cm

3) 21.5 cm 4) 13.3 cm

68. A convex lens has a focal length f. It is cut into two parts along the dotted line as shown in figure. The focal length of each part will be:

1) $\frac{f}{2}$ 2) f

3) $\frac{3}{2}f$ 4) 2f

69. A convex lens is made up of three different material as shown in the figure, for a point object placed on its axis the number of image are:

1) 1 2) 5

3) 4 4) 3

70. A point object O is placed in front of a glass rod having spherical end of curvature is 30 cm. The image would be formed at

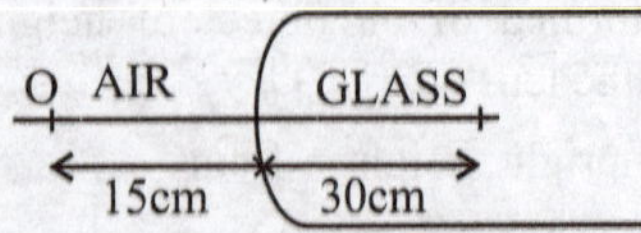

1) 30 cm to the left 2) Infinity

3) 1 cm to the right 4) 18 cm to the left

71. A mark placed on the surface of a sphere is viewed through glass from a position directly opposite. If the diameter of the sphere is 10 cm and refractive index of glass is 1.5, find the position of the image.

1) – 20 cm 2) 10 cm

3) 40 cm 4) –30 cm

72. In order to obtain a real image of magnification 2 using a converging lens of focal length 20 cm, where should an object be placed?

1) 50 cm 2) 30 cm

3) –50 cm 4) –30 cm

73. An effective power of a combination of 5 identical convex lenses which are kept in contact along the principal axis is 25 D. Focal length of each of the convex lens is:

1) 20 cm 2) 50 cm

3) 500 cm 4) 25 cm

74. There are two convex lenses L_1 and L_2 having focal lengths F_1 and F_2 respectively. The distance between L_1 and L_2 will be:

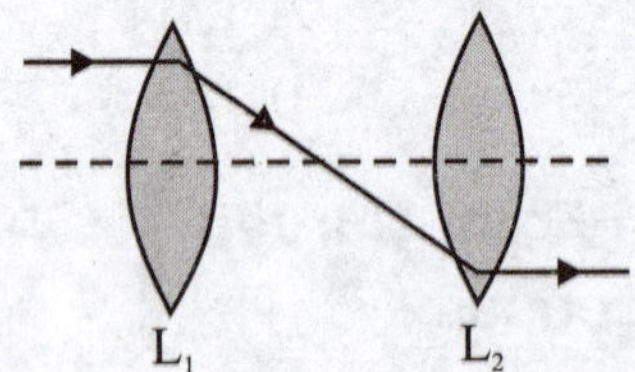

1) F_1 2) F_2

3) $F_1 + F_2$ 4) $F_1 - F_2$

75. A lens having focal length f and aperture of diameter d forms an image of intensity I. Aperture of diameter d/2 in central region of lens is covered by a black paper. Focal length of lens and intensity of image now will be respectively

1) f and $\frac{I}{4}$

2) $\frac{3f}{4}$ and $\frac{I}{2}$

3) f and $\frac{3I}{4}$

4) $\frac{f}{2}$ and $\frac{I}{2}$

76. The figure shows a thin lens with points C_1 and C_2 as the centres of curvature.

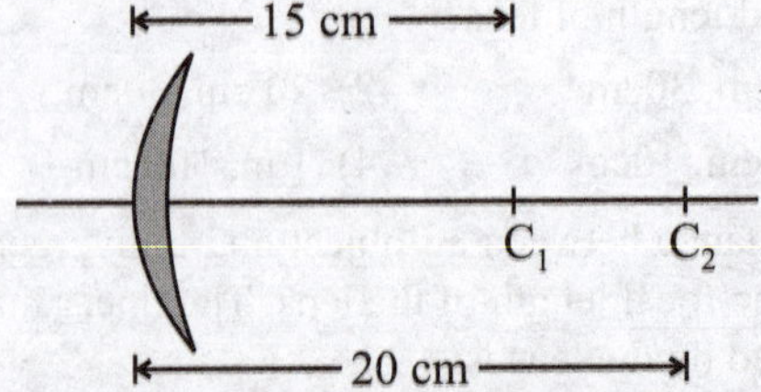

The focal length of lens will be (refractive index = 1.5)

1) 60 cm 2) 45 cm

3) 120 cm 4) 75 cm

Optical Instrument: Compound Microscope, Telescope

77. The magnifying power of a telescope is 9. When it is adjusted for parallel rays the distance between the objective and eyepiece is 20 cm. The focal length of lenses are:
1) 10 cm, 10 cm
2) 15 cm, 5 cm
3) 18 cm, 2 cm
4) 11 cm, 9 cm

78. Magnification produced by astronomical telescope for normal adjustment is 10 and length of telescope is 1.1 m. The magnification when the image is formed at least distance of distinct vision (D = 25 cm) is:
1) 14 2) 6
3) 16 4) 18

79. A telescope has focal length of objective and eye-piece as 200 cm and 5 cm. What is the magnification of telescope?
1) 40 2) 80
3) 50 4) 101

80. The sum of focal lengths of the two lenses of a refracting telescope is 105 cm. The focal length of one lens is 20 times that of the other. Determine the total magnification of the telescope when the final image is formed at infinity.
1) 20 2) 15
3) 30 4) 40

81. In a compound microscope, if the objective produces an image I_o and the eye piece produces an image I_e, then
1) I_o is virtual but I_e is real 2) I_o is real but I_e is virtual
3) I_o and I_e are both real 4) I_o and I_e are both virtual

82. In a compound microscope, the focal lengths of two lenses are 1.5 cm and 6.25 cm an object is placed at 2 cm from objective and the final image is formed at 25 cm from eye lens. The distance between the two lenses is:
1) 6.00 cm 2) 7.75 cm
3) 9.25 cm 4) 11.00 cm

83. An astronomical telescope has an angular magnification of magnitude 5 for distant objects. The separation between the objective and the eyepiece is 36 cm. The final image is formed at infinity. The focal lengths f_0 of the objective and f_e of the eyepiece are-
1) 45 cm and –9 cm respectively
2) 50 cm and 10 cm respectively
3) 7.2 cm and 5 cm respectively
4) 30 cm and 6 cm respectively

84. A compound microscope has magnifying power as 32 and magnifying power of eye-piece is 4, then the magnifying power of objective is-
1) 8 2) 10
3) 6 4) 12

85. A telescope has an objective lens of diameter 2m uses light of wavelength 5000Å for viewing stars. The minimum angular separation between two stars whose image is just resolved by the telescope is-
1) 4×10^{-4} rad 2) 0.25×10^{-6} rad
3) 0.31×10^{-6} rad 4) 5×10^{-3} rad

86. The focal lengths of the objective and eye piece of a compound microscope are 1 cm and 5 cm respectively. An object is placed at a distance of 1.1 cm from objective. If final image is formed at least distance of distinct vision, the magnifying power is-
1) 60 2) 50
3) 40 4) None of these

87. How should people wearing spectacles deals with a microscope?
1) They should keep on wearing their spectacles
2) They should take off their spectacles
3) They cannot use a microscope at all
4) They may keep on wearing or take off their spectacles. It makes no difference

88. If a compound microscope m_1 & m_2 be the linear magnification of the objective lens and eye lens respectively then magnifying power of the compound microscope will be:
1) $m_1 - m_2$ 2) $\sqrt{m_1 + m_2}$
3) $\frac{m_1 m_2}{2}$ 4) $m_1 m_2$

89. The maximum magnification that can be obtained with a convex lens of focal length 2.5 cm is (the least distance of distinct vision is 25 cm).
1) 10 2) 0.1
3) 62.5 4) 11

90. A compound microscope has an eye piece of focal length 10 cm and an objective of focal length 4 cm. Calculate the magnification. If an object kept at a distance of 5 cm from the objective so that image is formed at the least distance vision (20 cm).
1) 12 2) 11
3) 10 4) 13

91. A simple telescope consisting of an objective of focal length 60 cm and a single eye lens of focal length 5 cm is focused on a distant object in such a way that parallel rays come out from the eye lens. If the object subtends an angle 3° on objective. The angular width of the image will be
1) 40° 2) 36°
3) 80° 4) 50°

92. A compound microscope has an objective of focal length 1 cm and an eye piece of focal length 2.5 cm. An object has to be placed at a distance of 1.2 cm away from objective for normal adjustment. What will be the length of microscope tube?
1) 8.5 cm 2) 9.5 cm
3) 10.5 cm 4) 7.5 cm

93. The focal length of the objective and eye piece of an astronomical telescope are 25 cm & 2.5 cm respectively. The telescope is focused on an object 1.5 m from objective, the final image being formed 25 cm from eye of the observer. Calculate the length of telescope.

1) 30 cm 2) 31 cm
3) 32 cm 4) 33 cm

94. In compound microscope, the focal length of two lenses are 1.5 cm and 6.25 cm. An object is placed at 2 cm from objective and the final image is formed at 25 cm from eye lens. The distance between the two lenses is:

1) 6 cm 2) 7.75 cm
3) 9.25 cm 4) 11 cm

95. The length of a microscope tube is 20 cm. The focal length of the objective and eye lens are 0.5 cm and 1 cm. The magnifying power of the microscope is about:

1) 500 2) 10000
3) 1000 4) 1500

96. A telescope consists of two lenses of focal lengths 20 cm and 5 cm. Obtain its magnifying power when final image is at 25 cm from the eye.

1) 4.8 cm 2) – 4.8 cm
3) 2.4 cm 4) – 2.4 cm

97. In telescope power of an objective and eye lens are respectively + 1.25 D and – 20 D, then for relaxed vision the length and magnification will be:

1) 21.25 cm & 16 2) 75 cm & 20
3) 75 cm & 16 4) 8.5 cm & 21.25

98. A giant refracting telescope at an observatory has an objective lens of focal length 15 m. If an eye piece lens of focal length 1 cm is used. If this telescope is used to view the moon, what is the diameter of image of moon formed by objective lens ? The diameter of the moon is 3.42×10^6 m and radius of lunar orbit is 3.8×10^8 m

1) 13.5 cm 2) 3.5 cm
3) 11.5 cm 4) 15 cm

High Order Time Intensive MCQs

1. A microscope is focussed on a mark on a paper and then a slab of glass of thickness 3 cm and refractive index 1.5 is placed over the mark. How should the microscope be moved to get the mark in focus again?

1) 2 cm upward
2) 1 cm upward
3) 4.5 cm upward
4) 1 cm downward

2. A luminous object is placed at a distance of 30 cm from the convex lens of focal length 20 cm. On the other side of the lens, at what distance from the lens a convex mirror of radius of curvature 10 cm be placed in order to have an upright image of the object coincident with it?

1) 12 cm 2) 30 cm
3) 50 cm 4) 60 cm

3. A short linear object of length b lies along the axis of a concave mirror of focal length f at a distance u from the pole of the mirror, what is the size of image?

1) $\left(\frac{f}{u-f}\right)b$ 2) $\left(\frac{f}{u-f}\right)^2 b$
3) $\left(\frac{f}{u-f}\right)b^2$ 4) $\left(\frac{f}{u-f}\right)$

4. If angle of incidence is twice the angle of refraction in a medium of refractive index μ, then angle of incidence:

1) $2\cos^{-1}\frac{\mu}{2}$
2) $2\sin^{-1}\frac{\mu}{2}$
3) $2\cos^{-1}\mu$
4) $2\sin^{-1}\mu$

5. Light travels in two media A and B with speeds 1.8×10^8 ms^{-1} and 2.4×10^8 ms^{-1} respectively. Then, the critical angle between A and B:

1) $\sin^{-1}\left(\frac{2}{3}\right)$ 2) $\tan^{-1}\left(\frac{3}{4}\right)$
3) $\tan^{-1}\left(\frac{2}{3}\right)$ 4) $\sin^{-1}\left(\frac{3}{4}\right)$

6. A ray of light travelling in transparent medium of refractive index μ falls on a surface separating the medium from air at an angle of incidence of 45°. For which of the following value of μ the ray can undergo total internal reflection?

1) $\mu > 1.33$
2) $\mu > 1.4$
3) $\mu > 1.20$
4) $\mu > 1.25$

7. A combination of concave and convex lens in contact is placed over a plane mirror. When a pin P is placed as shown in the figure, the image is formed at the position of the pin itself. The focal length of the convex lens is 10 cm. Then the focal length of the concave lens is:

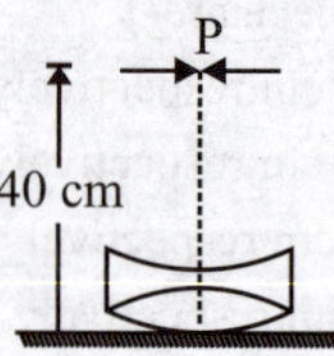

1) 40 cm 2) $\frac{40}{3}$ cm
3) 20 cm 4) $\frac{20}{3}$ cm

8. A small coin is resting on the bottom of a beaker filled with a liquid. A ray of light from the coin travels upto the surface of the liquid and moves along its surface.

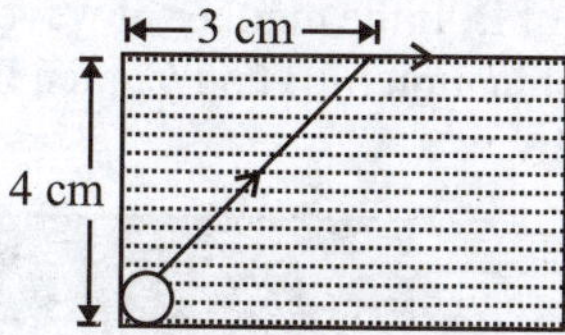

How fast is the light travelling in the liquid?

1) 1.8×10^8 ms^{-1}
2) 2.4×10^8 ms^{-1}
3) 3.0×10^8 ms^{-1}
4) 1.2×10^8 ms^{-1}

9. The angle of a prism is A. One of its refracting surfaces is silvered. Light rays falling at an angle of incidence $2A$ on the first surface returns back through the same path after suffering reflection at the silvered surface. The refractive index μ, of the prism is:

1) $2 \sin A$
2) $2 \cos A$
3) $\frac{1}{2}\cos A$
4) $\tan A$

10. A plano-convex lens fits exactly into a plano-concave lens. Their plane surfaces are parallel to each other. If lenses are made of different materials of refractive indices μ_1 and μ_2 and R is the radius of curvature of the curved surface of the lenses, What is the focal length of combination?

1) $\frac{R}{2(\mu_1+\mu_2)}$
2) $\frac{R}{2(\mu_1-\mu_2)}$
3) $\frac{R}{(\mu_1-\mu_2)}$
4) $\frac{2R}{(\mu_1-\mu_2)}$

11. Two plane mirrors A and B are parallel to each other and spaced 20 cm apart. An object is placed in between them at 15 cm from A. Out of the following at which point an image is not formed in mirror A:
(Distance measured from mirror A)

1) 15 cm
2) 25 cm
3) 45 cm
4) 55 cm

12. A thin plano-convex lens acts like a concave mirror of focal length 0.2 m, when silvered from its plane surface. The refractive index of the material of the lens is 1.5. The radius of curvature of the convex surface of the lens will be:

1) 0.1 m
2) 0.75 m
3) 0.4 m
4) 0.2 m

13. A driver at a depth 12 m inside water ($\mu = 4/3$) sees the sky in a cone of semi-vertical angle

1) $\sin^{-1}\left(\frac{4}{3}\right)$
2) $\tan^{-1}\left(\frac{4}{3}\right)$
3) $\sin^{-1}\left(\frac{3}{4}\right)$
4) 90°

14. A plano-convex lens has a maximum thickness of 6 cm. When placed on a horizontal table with the curved surface in contact with the table surface, the apparent depth of the bottommost point of the lens is found to be 4 cm. If the lens is inverted such that the plane face of the lens is in contact with the surface of the table, the apparent depth of the centre of the plane face is found to be $\frac{17}{4}$ cm. The radius of curvature of the lens is:

1) 68 cm
2) 75 cm
3) 128 cm
4) 34 cm

15. A boy is trying to start a fire by focussing sunlight on a piece of paper using an equi-convex lens of focal length 10 cm. The diameter of the sun is 1.39×10^9 m and its mean distance from the earth is 1.5×10^{11} m. What is the diameter of the sun's image on the paper ?

1) 9.2×10^{-4} m
2) 6.5×10^{-4} m
3) 6.5×10^{-5} m
4) 12.4×10^{-4} m

16. A double convex lens made of glass (refractive index $n = 1.5$) has both radii of curvature of magnitude 20 cm incident light rays parallel to the axis of the lens will converge at a distance L such that:

1) $L = 20$ cm
2) $L = 10$ cm
3) $L = 40$ cm
4) $L = \frac{20}{3}$ cm

17. The radii of curvature of the two surfaces of a lens are 20 cm and 30 cm and the refractive index of the material of the lens is 1.5. If the lens is concavo-convex, then the focal length of the lens is:

1) 24 cm
2) 10 cm
3) 15 cm
4) –120 cm

18. A thin prism P_1 with angle 4° and made from glass of refractive index 1.54 is combined with another prism P_2 made of glass of refractive index 1.72 to produce dispersion without deviation. The angle of prism P_2 is:

1) 5.33°
2) 4°
3) 2.6°
4) 3°

19. If one face of prism is silvered having prism angle 30° and $\mu = \sqrt{2}$. What will be the angle of incidence, so that the incident ray retraces its path?

1) 30°
2) 60°
3) 90°
4) 45°

20. For an angle of incidence θ on an equilateral prism of refractive index $\sqrt{3}$, the ray refracted is parallel to the base inside the prism. Then θ will be:

1) 30°
2) 45°
3) 60°
4) 75°

21. A plano-convex lens ($f = 20$ cm) is silvered at plane surface. Now, focal length of a plano-convex lens will be:

1) 20 cm
2) 40 cm
3) 30 cm
4) 10 cm

22. For the ray diagram shown in the adjoining figure, the false statement is:

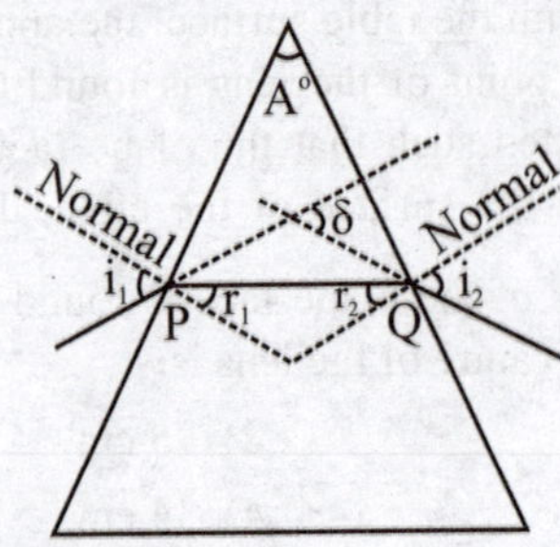

1) $\delta = (i_1+i_2) - (r_1 + r_2)$
2) $A = r_1 + r_2$
3) $(\sin i_1)(\sin r_2) = (\sin i_2)(\sin r_1)$
4) $(\sin i_1)(\sin r_1) = (\sin i_2)(\sin r_2)$

23. A prism of a certain angle deviates the red and blue rays by 8° and 12°, respectively. Another prism of the same angle deviates the red and blue rays by 10° and 14°, respectively. The prisms are small angled and made of different materials. The dispersive powers of the materials of the prisms are in the ratio:

1) 5 : 6 2) 9 : 11
3) 6 : 5 4) 11 : 9

24. A plane mirror is kept along positive x-axis facing along positive y-axis. The equation of a linear object is x = y. The equation of its image will be

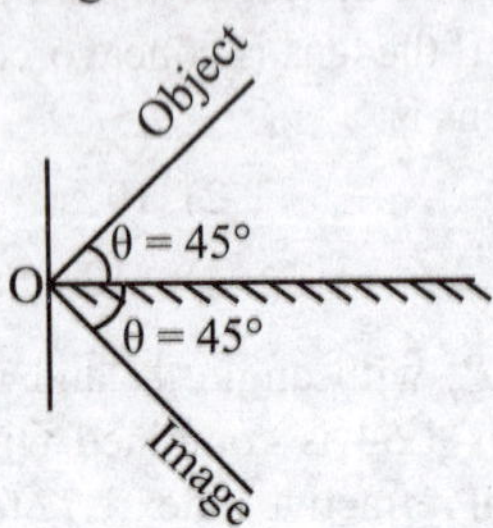

1) x = 2y
2) x + y = 0
3) 2x + 5y = 0
4) None of these

25. The effective focal length of the lens combination shown in the figure is – 60 cm. The radii of curvature of the curved surfaces of the plano-convex lenses are 12 cm each and refractive index of the material of the lens is 1.5. What would be the refractive index of the liquid?

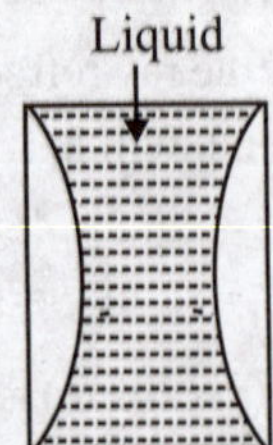

1) 1.33 2) 1.42
3) 1.53 4) 1.6

26. A slab of transparent material is made as shown in the given figure. Monochromatic parallel beams of light are normally incident on the slab. The thickness of C is twice the thickness of B. If the number of waves in A = number of waves in combination of B and C, then the refractive index of B is:

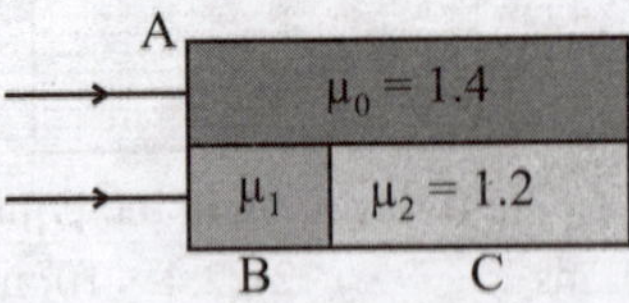

1) 1.33 2) 1.8
3) 1.6 4) 1.4

27. In a vessel two insoluble liquids are filled as shown in the figure. The effective refractive index of the combination for normal view is

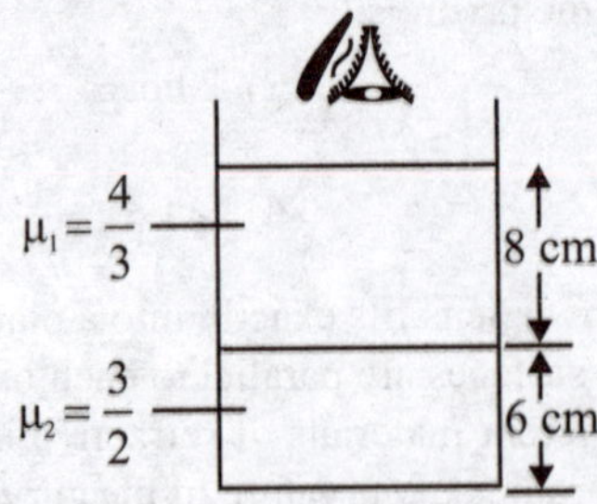

1) 1.33 2) 1.5
3) 1.67 4) 1.4

28. A beam is incident on a tube containing two convex lenses of focal length 15 cm and 20 cm as shown. If the diameter of incident beam is 10 cm, find the diameter of final beam.

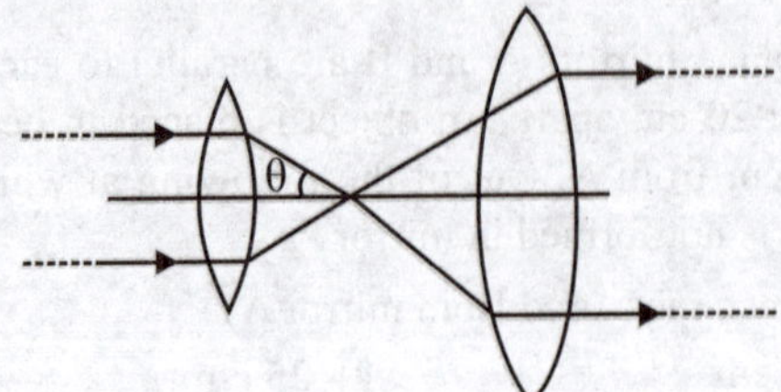

1) $\frac{40}{3}$ cm
2) $\frac{20}{3}$ cm
3) $\frac{50}{3}$ cm
4) $\frac{25}{3}$ cm

29. A lens is cut from the optical centre along the principal axis and the two parts of the lens are placed sideways as shown in figure. Then the new focal length of the combination:

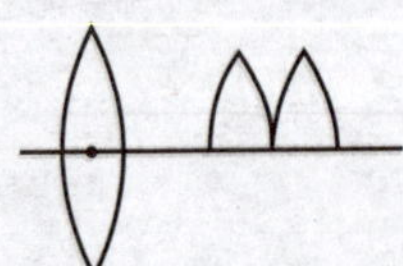

1) Remains the same 2) Becomes double
3) Becomes half 4) Becomes triple

30. Two plane mirrors A and B are aligned parallel to each other, as shown in the figure. A light ray is incident at an angle of 30° at a point just inside one end of A. The plane of incidence coincides with the plane of the figure. The maximum number of time the ray undergoes reflections (including the first one) before it emerges out is:

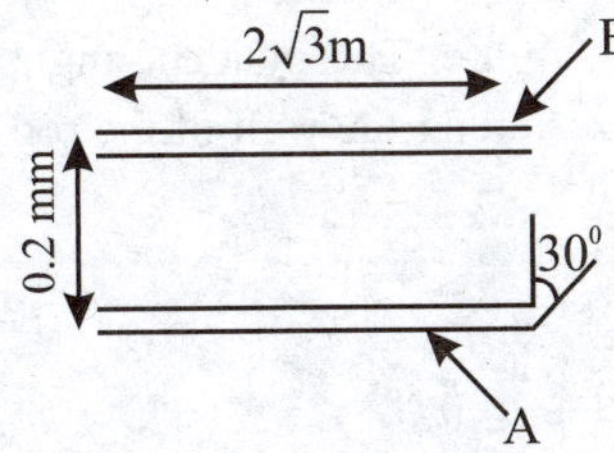

1) 28
2) 30
3) 32
4) 34

31. In Ramsden's eye-piece, the two plano-convex lenses each of focal length f, are separeted by a distance 12 cm. The equivalent focal length (in cm) of the eye lens :

1) 10.5
2) 12
3) 13.5
4) 15.5

32. The time taken by the light to cross a glass of thickness 4 mm and refractive index (μ = 3), will be

1) 4×10^{-11} sec
2) 16×10^{-11} sec
3) 8×10^{-11} sec
4) 24×10^{-10} sec

33. A hollow double concave lens is made of very thin transparent material. It can be filled with air or either of two liquids L_1 and L_2 having refractive indices n_1 and n_2 respectively ($n_2 > n_1 > 1$). The lens will diverge a parallel beam of light, if it is filled with :

1) air and placed in air
2) air and immersed in L_2
3) L_1 and immersed in L_2
4) L_2 and immersed in L_1

34. The graph in the below figure shows how the inverse of magnification 1/m produced by a convex thin lens varies with object distance u. What was the focal length of the lens used?

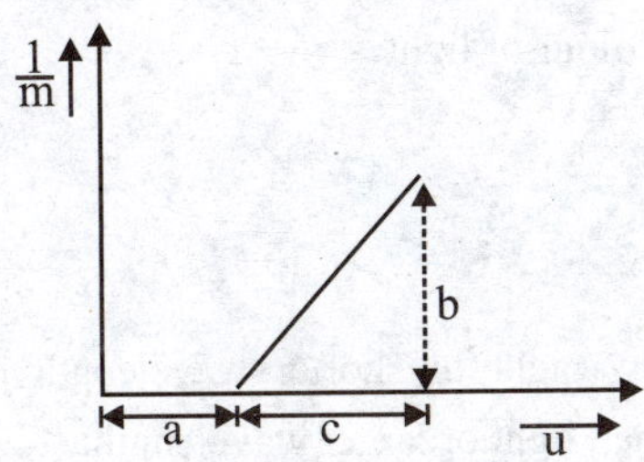

1) $\frac{b}{c}$
2) $\frac{b}{ca}$
3) $\frac{bc}{a}$
4) $\frac{c}{b}$

35. A spherical surface of curvature R separates air (μ = 1)from glass (μ = 1.5). The centre of curvature is in the glass. A point object P placed in air is found to have a real image Q in the glass. The line PQ cuts the surface at a point O and PO = OQ. The distance PO is equal to

1) 2R
2) 5R
3) 3R
4) 1.5R

36. A ray of light is incident on the left vertical face of a glass cube of refractive index μ_2, as shown in figure. The plane of incident is the plane of the page, and the cube is surrounded by liquid (μ_1). What is the largest angle of incidence θ_1 for which total internal reflection occurs at the top surface?

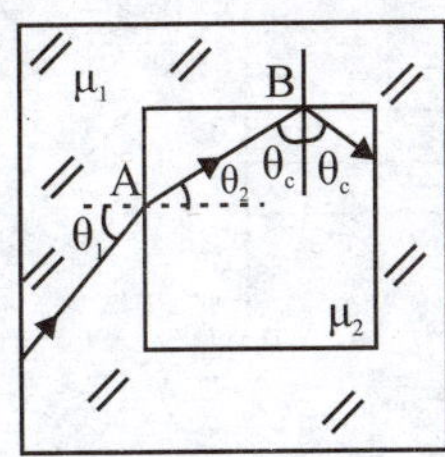

1) $\sin\theta_1 = \sqrt{\left(\frac{\mu_2}{\mu_1}\right)^2 - 1}$
2) $\sin\theta_1 = \sqrt{\left(\frac{\mu_2}{\mu_1}\right)^2 + 1}$
3) $\sin\theta_1 = \sqrt{\left(\frac{\mu_1}{\mu_2}\right)^2 + 1}$
4) $\sin\theta_1 = \sqrt{\left(\frac{\mu_1}{\mu_2}\right)^2 - 1}$

37. A concave lens of glass, refractive index 1.5 has both surfaces of radius of curvature R. On immersion in a medium of refractive index 1.75, it will behave as a :

1) convergent lens of focal length 3.5 R
2) convergent lens of focal length 3.0 R
3) divergent lens of focal length 3.5 R
4) divergent lens of focal length 3.0 R

38. The speed of light in media M_1 and M_2 are 1.5×10^8 m/s and 2.0×10^8 m/s respectively. A ray of light enters from medium M_1 to M_2 at an incidence angle i. If the ray suffers total internal reflection, the value of i is

1) Equal to $\sin^{-1}\left(\frac{2}{3}\right)$
2) Equal to or less than $\sin^{-1}\left(\frac{3}{5}\right)$
3) Equal to or greater than $\sin^{-1}\left(\frac{3}{4}\right)$
4) Less than $\sin^{-1}\left(\frac{2}{3}\right)$

39. Considering normal incidence of ray, the equivalent refractive index of combination of two slabs shown in the figure is

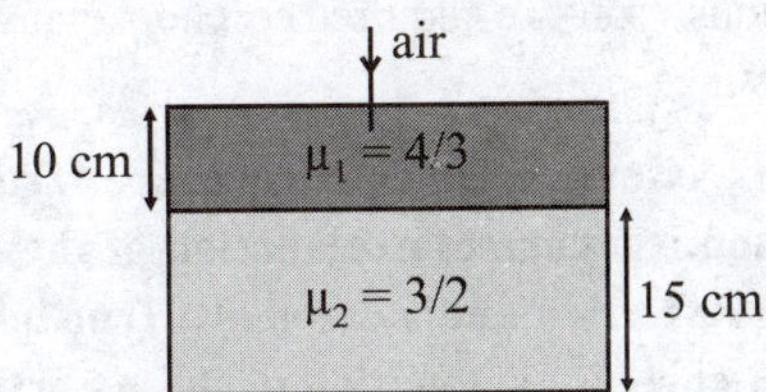

1) 1.8
2) 1.43
3) 2
4) None of these

40. A beam of light consisting of red, green and blue colours is incident on a right angled prism. The refractive index of the material of the prism for the above red, green and blue wavelengths are 1.39, 1.44 and 1.47, respectively.

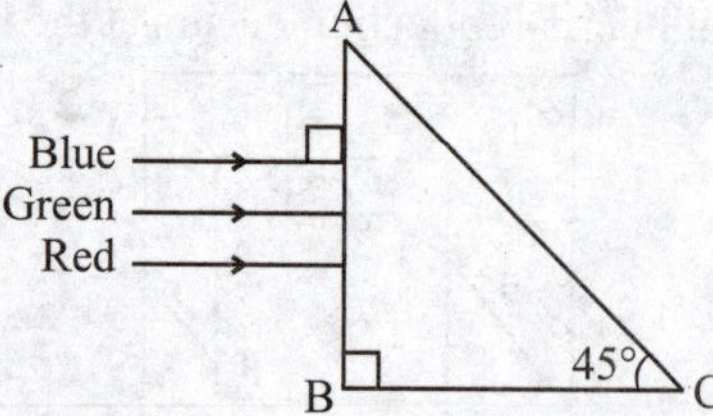

The prism will

1) separate the red colour part from the green and blue colours.
2) separate the blue colour part from the red and green colours.
3) separate all the three colours from one another.
4) not separate the three colours at all.

41. A transparent solid cylindrical rod has a refractive index of $2/\sqrt{3}$. If is surrounded by air. A light ray is incident at the mid point of one end of the rod as

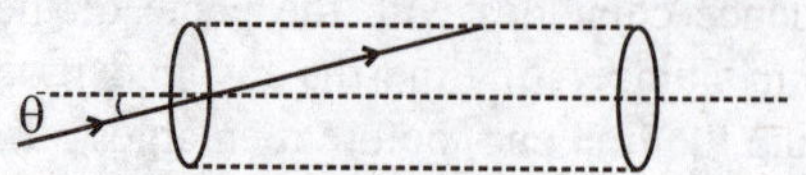

Shown in the figure. The incident angle θ for which the light ray grazes along the wall of the rod is

1) $\sin^{-1}\left(\frac{1}{2}\right)$
2) $\sin^{-1}\left(\frac{\sqrt{3}}{2}\right)$
3) $\sin^{-1}\left(\frac{2}{\sqrt{3}}\right)$
4) $\sin^{-1}\left(\frac{1}{\sqrt{3}}\right)$

Statement Based MCQs

Directions: These questions consist of two statements each, printed as Statement-I and Statement-II. While answering these questions, you are required to choose any one of the following four responses.

1) **Both Statement-I and Statement-II are correct.**
2) **Both Statement-I and Statement-II are incorrect.**
3) **Statement-I is correct & Statement-II is incorrect.**
4) **Statement-I is incorrect & Statement-II is correct.**

1. **Statement-I:** If objective and eye-piece of a microscope are interchanged, then it can work as a telescope.

   Statement-II: The objective lens of telescope has small focal length.

2. **Statement-I:** The velocity of light inside the glass slab is same for different colours.

   Statement-II: A ray of white light shows no dispersion on emerging from a rectangular glass slab.

3. **Statement-I:** The twinkling of stars is due to atmospheric reflection.

   Statement-II: In cold countries, the phenomenon of looming (i.e., ship appears inverted) takes place, due to total internal reflection.

4. **Statement-I:** Colour blind persons do not possess some cone shaped cells of retina.

   Statement-II: The cone shaped cells of our retina respond to the colour of light

Assertion & Reason MCQs

Directions: These questions consist of two statements each, printed as Assertion and Reason. While answering these questions, you are required to choose any one of the following four responses.

1) **Both Assertion and Reason are True and the Reason is a correct explanation of the Assertion.**
2) **Both Assertion and Reason are True but Reason is not a correct explanation of the Assertion.**
3) **Assertion is True but the Reason is False.**
4) **Assertion is False but Reason is True.**

1 **Assertion:** Rainbow is an example of the dispersion of sunlight by the water droplets.

   Reason: Light of shorter wavelength is scattered much more than light of larger wavelength.

2. **Assertion:** Well-cut diamond shines brilliantly.

   Reason: It happens due to total internal reflection.

3. **Assertion:** The focal length of a spherical mirror does not depend on the wavelength of light.

   Reason: In case of reflection of light from a denser medium the phase changes by π.

4. **Assertion:** A convex lens may be diverging.

   Reason: The nature of a lens depends upon the refractive indices of the material of lens and surrounding medium.

Match the Columns MCQs

1. For real objects, match the following tow Columns corresponding to linear magnification m given in Column-I.

	Column-I		Column-II
(a)	$m = -2$	(i)	Convex mirror
(b)	$m = -\frac{1}{2}$	(ii)	Concave mirror
(c)	$m = +2$	(iii)	Real image
(d)	$m = +\frac{1}{2}$	(iv)	Virtual image

1) (a)-(ii, iii) (b)-(ii, iii) (c)-(ii, iv) (d)-(i, iv)
2) (a)-(ii, iii) (b)-(i, iii) (c)-(i, iv) (d)-(ii, iv)
3) (a)-(i, iii) (b)-(ii, iii) (c)-(ii, iii) (d)-(i, iv)
4) (a)-(i, iii) (b)-(i, iii) (c)-(ii, iii) (d)-(i, iv)

2. Principal axis of a miror (AB), a point object O and its image I are shown in Column-I, match it with Column-II.

	Column-I		Column-II
(a)	O• A———B •I	(i)	Plane mirror
(b)	O• A———B •I	(ii)	Convex mirror
(c)	O• •I A———B	(iii)	Concave mirror
(d)	•I O•———	(iv)	Not possible

1) (a)-(iii) (b)-(iii) (c)-(i) (d)-(iii)
2) (a)-(ii) (b)-(iii) (c)-(iii) (d)-(iv)
3) (a)-(i) (b)-(ii) (c)-(iii) (d)-(iv)
4) (a)-(iii) (b)-(iii) (c)-(ii) (d)-(iv)

3. Match the following two columns corresponding to single refraction from plane surface. In an cases shown in Column I, $\mu_1 > \mu_2$.

	Column-I		Column-II
(a)	O, x, 1, 2	(i)	Image distance greater than x from plane surface
(b)	1, 2, x, O	(ii)	Image distance less than a from plane surface
(c)	1, 2, x, O	(iii)	Real image
(d)	O, x, 1, 2	(iv)	Virtual image

1) (a)-(ii, iv) (b)-(ii, iii) (c)-(i, iv) (d)-(i, iii)
2) (a)-(i, iii) (b)-(iii, ii) (c)-(i, iv) (d)-(iv)
3) (a)-(i, iii) (b)-(ii, iv) (c)-(iii) (d)-(iv)
4) (a)-(ii, iv) (b)-(ii) (c)-(iii) (d)-(ii, iii)

4. Consider the following diagram representing magnitude of magnification against image distance for a convex lens:

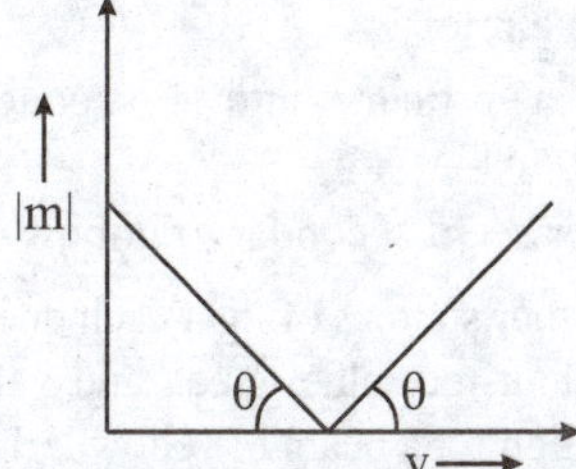

	Column-I		Column-II
(a)	Focal length of the lens	(i)	Inverse of slope of line
(b)	Intercept on x-axis	(ii)	Unity
(c)	Intercept on y-axis	(iii)	Focal length
(d)	Magnitude of slope of the line(s)	(iv)	Inverse of the focal length

1) (a)-(i) (b)-(ii) (c)-(iii) (d)-(iv)
2) (a)-(i) (b)-(iii) (c)-(ii) (d)-(iv)
3) (a)-(iii) (b)-(ii) (c)-(iii) (d)-(iv)
4) (a)-(iv) (b)-(iii) (c)-(ii) (d)-(iv)

NCERT Exemplar MCQs

1. A ray of light incident at an angle θ on a refracting face of a prism emerges from the other face normally. If the angle of the prism is 5° and the prism is made of a material of refractive index 1.5, the angle of incidence is:
   1) 7.5° 2) 5°
   3) 15° 4) 2.5°
2. A short pulse of white light is incident from air to a glass slab at normal incidence. After travelling through the slab, the first colour to emerge is:
   1) Blue 2) Green
   3) Violet 4) Red
3. An object approaches a convergent lens from the left of the lens with a uniform speed 5 m/s and stops at the focus. The image
   1) Moves away from the lens with an uniform speed 5 m/s
   2) Moves away from the lens with an uniform acceleration
   3) Moves away from the lens with a non-uniform acceleration
   4) Moves towards the lens with a non-uniform acceleration
4. A passenger in an aeroplane shall:
   1) Never see a rainbow
   2) May see a primary and a secondary rainbow as concentric circles
   3) May see a primary and a secondary rainbow as concentric arcs
   4) Shall never see a secondary rainbow
5. You are given four sources of light each one providing a light, of a single colour-red, blue, green and yellow. Suppose the angle of refraction for a beam of yellow light corresponding to a particular angle of incidence at the interface of two media is 90°. Which of the following statements is correct if the source of yellow light is replaced with that of other lights without changing the angle of incidence?
   1) The beam of red light would undergo total internal reflection
   2) The beam of red light would bend towards normal while it gets refracted through the second medium
   3) The beam of blue light would undergo total internal reflection
   4) The beam of green light would bend away from the normal as it gets refracted through the second medium
6. The radius of curvature of the curved surface of a plano-convex lens is 20 cm. If the refractive index of the material of the lens be 1.5, it will:
   1) Act as a convex lens only for the objects that lie on its curved side
   2) Act as a concave lens for the objects that lie on its curved side
   3) Act as a convex lens irrespective of the side on which the object lies
   4) Act as a concave lens irrespective of side on which the object lies
7. The phenomena involved in reflection of radio waves by ionosphere is similar to:
   1) Reflection of light by a plane mirror
   2) Total internal reflection of light in air during a mirage
   3) Dispersion of light by water molecules during the formation of a rainbow
   4) Scattering of light by the particles of air
8. The direction of ray of light incident on a concave mirror is shown by PQ while directions in which the ray would travel after reflection is shown by four rays marked 1, 2, 3 and 4 (figure). Which of the four rays correctly shows the direction of reflected ray?

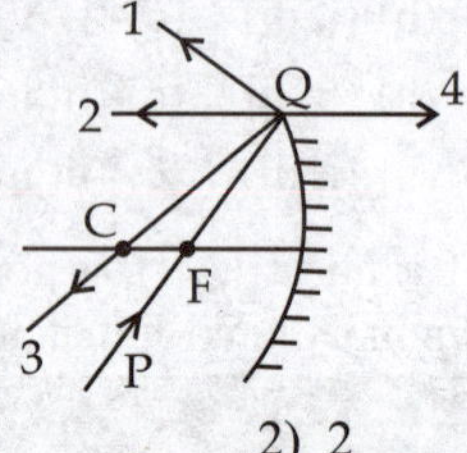

   1) 1 2) 2
   3) 3 4) 4
9. The optical density of turpentine is higher than that of water while its mass density is lower. Figure shows a layer of turpentine floating over water in a container. For which one of the four ray's incident on turpentine in figure, the path shown is correct?

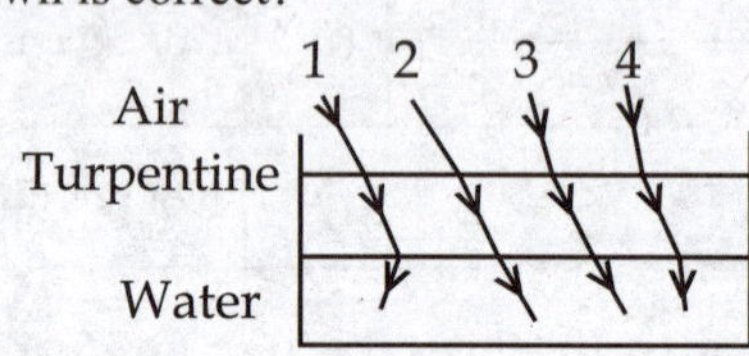

   1) 1 2) 2
   3) 3 4) 4
10. A car is moving with at a constant speed of 60 km h^{-1} on a straight road. Looking at the rear view mirror, the driver finds that the car following him is at a distance of 100 m and is approaching with a speed of 5 km h^{-1}.

    In order to keep track of the car in the rear, the driver begins to glance alternatively at the rear and side mirror of his car after every 2 s till the other car overtakes. If the two cars were maintaining their speeds, which of the following statement (s) is/are correct?
    1) The speed of the car in the rear is 65 km h^{-1}
    2) In the side mirror, the car in the rear would appear to approach with a speed of 5 km h^{-1} to the driver of the leading car
    3) In the rear view mirror, the speed of the approaching car would appear to decrease as the distance between the cars decreases
    4) In the side mirror, the speed of the approaching car would appear to increase as the distance between the cars decreases

11. There are certain material developed in laboratories which have a negative refractive index figure. A ray incident from air (Medium 1) into such a medium (Medium 2) shall follow a path given by:

1) 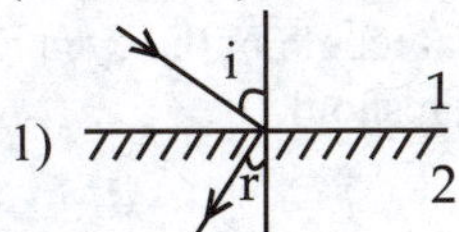

2) i, 1, r, 2

3) i, r, 1, 2

4) 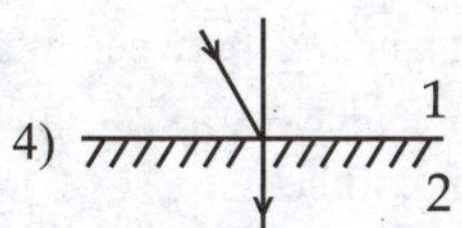

NEET & AIIMS Past Year MCQs

1. Given below are two statements:

Statement I: Image formation needs regular reflection and/or refraction.

Statement II: The variety in colour of objects we see around us is due to the constituent colours of the light incident on them.

In the light of the above statements, choose the most appropriate answer from the options given below: **(2024 Re)**

1) Statement I is correct but statement II is incorrect
2) Statement I is incorrect but Statement II is correct
3) Both Statement I and Statement II are correct
4) Both Statement I and Statement II are incorrect

2. A light ray enters through a right angled prism at point P with the angle of incidence 30° as shown in figure. It travels through the prism parallel to its base BC and emerges along the face AC. The refractive index of the prism is: **(2024)**

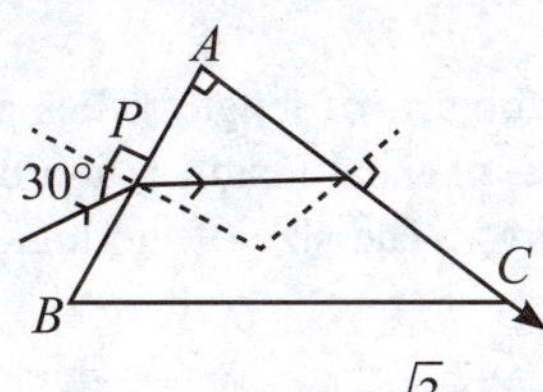

1) $\frac{\sqrt{3}}{4}$ 2) $\frac{\sqrt{3}}{2}$

3) $\frac{\sqrt{5}}{4}$ 4) $\frac{\sqrt{5}}{2}$

3. A small telescope has an objective of focal length 140 cm and an eye piece of focal length 5.0 cm. The magnifying power of telescope for viewing a distant object is: **(2024)**

1) 17 2) 32

3) 34 4) 28

4. Light travels a distance x in time t_1 in air and 10x in time t_2 in another denser medium. What is the critical angle for this **(2023)**

1) 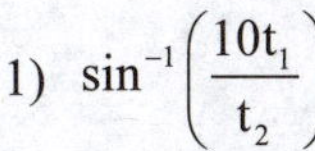 $\sin^{-1}\left(\frac{10t_1}{t_2}\right)$ 2) $\sin^{-1}\left(\frac{t_2}{t_1}\right)$

3) $\sin^{-1}\left(\frac{10t_2}{t_1}\right)$ 4) $\sin^{-1}\left(\frac{t_1}{10t_2}\right)$

5. In the figure shown here, what is the equivalent focal length of the combination of lenses (Assume that all layers are thin)? **(2023)**

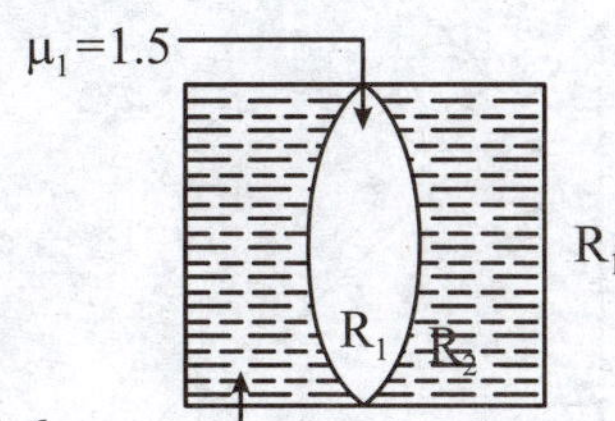

1) –50 cm 2) 40 cm

3) –40 cm 4) –100 cm

6. Two thin lenses are of same focal lengths (f), but one is convex and the other one is concave. When they are placed in contact with each other, the equivalent focal length of the combination will be: **(2023)**

1) Infinite 2) Zero

3) $\frac{f}{4}$ 4) $\frac{f}{2}$

7. A biconvex lens has radii of curvature, 20 cm each. If the refractive index of the material of the lens is 1.5, the power of the lens is: **(2022)**

1) Infinity 2) + 2D

3) + 20D 4) + 5D

8. A light ray falls on a glass surface of refractive index $\sqrt{3}$, at an angle 60°. The angle between the refracted and reflected rays would be: **(2022)**

1) 120° 2) 30°

3) 60° 4) 90°

9. Two transparent media A and B are separated by a plane boundary. The speed of light in those media are 1.5×10^8 m/s and 2.0×10^8 m/s, respectively. The critical angle for a ray of light for these two media is: **(2022)**

1) $\tan^{-1}(0.750)$ 2) $\sin^{-1}(0.500)$

3) $\sin^{-1}(0.750)$ 4) $\tan^{-1}(0.500)$

10. A lens of large focal length and large aperture is best suited as an objective of an astronomical telescope since: **(2021)**

1) A large aperture contributes to the quality and visibility of the images.
2) A large area of the objective ensures better light gathering power.
3) A large aperture provides a better resolution.
4) All of the above.

11. A convex lens 'A' of focal length 20 cm and a concave lens 'B' of focal length 5 cm are kept along the same axis with a distance 'd' between them. If a parallel beam of light falling on 'A' leaves 'B' as a parallel beam, then the distance 'd' in cm will be: **(2021)**

1) 15 2) 50
3) 30 4) 25

12. Find the value of the angle of emergence from the prism. Refractive index of the glass is $\sqrt{3}$. **(2021)**

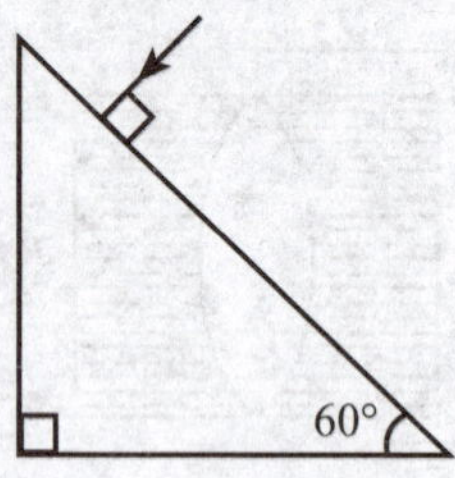

1) 30° 2) 45°
3) 90° 4) 60°

13. A point object is placed at a distance of 60 cm from a convex lens of focal length 30 cm. If a plane mirror were put perpendicular to the principal axis of the lens and at a distance of 40 cm from it, the final image would be formed at a distance of: **(2021)**

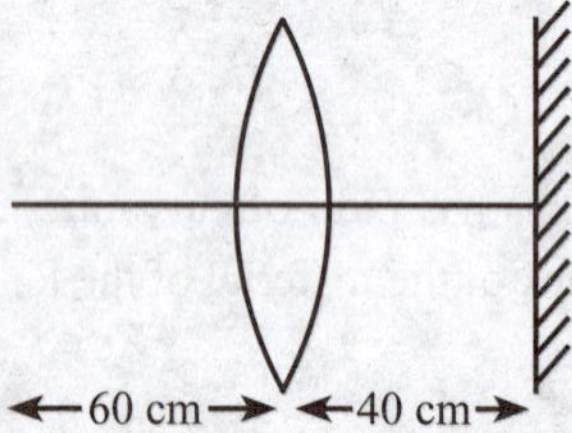

1) 30 cm from the lens, it would be a real image.
2) 30 cm from the plane mirror, it would be a virtual image.
3) 20 cm from the plane mirror, it would be a virtual image
4) 20 cm from the lens, it would be a real image.

14. Four lens having the focal length of +15 cm, 20 cm, +150 cm, and +250 cm respectively are provided to make an astronomical telescope. The focal length of the eye piece to produce the largest magnification should be: **(AIIMS 2017)**

1) +250 cm 2) +155 cm
3) 25 cm 4) +15 cm

15. Light wave enters from medium 1 to medium 2. Its velocity in 2nd medium is double from 1st. For total internal reflection the angle of incidence must be greater than: **(AIIMS 2016)**

1) 30° 2) 60°
3) 45° 4) 90°

16. Focal length of objective and eye piece of telescope are 200 cm and 4 cm respectively. What is the length of telescope for normal adjustment? **(AIIMS 2016)**

1) 196 cm 2) 204 cm
3) 250 cm 4) 225 cm

17. A thin prism P_1 of angle 4° and refractive index 1.54 is combined with another thin prism P_2 of refractive index 1.72 to produce dispersion without deviation. The angle of P_2 is: **(AIIMS 2011)**

1) 4° 2) 5.33°
3) 2.6° 4) 3°

18. Which of the following produces virtual image:

1) Simple microscope 2) Ordinary camera
3) Projector 4) Cinemascope

19. A short linear object, of length l, lies along the axis of a concave mirror, of focal length f, at a distance d from the pole of the mirror. The size of the image is then (nearly): **(AIIMS 2009)**

1) $\frac{If}{d-f}$ 2) $\frac{d-f}{if}$
3) $I\frac{f^2}{(d-f)^2}$ 4) $\frac{(d-f)^2}{f^2}I$

Answer Key

Topic-wise MCQs																	
	1	2	3	4	5	6	7	8	9	10	11	12	13	14	15	16	17
	2	1	2	2	1	3	1	1	1	2	2	3	4	3	1	4	3
	18	19	20	21	22	23	24	25	26	27	28	29	30	31	32	33	34
	1	3	3	2	2	1	3	3	2	1	4	4	3	4	3	4	3
	35	36	37	38	39	40	41	42	43	44	45	46	47	48	49	50	51
	3	3	3	3	3	1	3	1	2	2	3	2	3	1	2	1	2
	52	53	54	55	56	57	58	59	60	61	62	63	64	65	66	67	68
	2	4	1	1	1	4	2	1	4	2	1	2	2	3	1	2	4
	69	70	71	72	73	74	75	76	77	78	79	80	81	82	83	84	85
	4	1	1	4	1	3	3	3	3	1	1	1	2	4	4	1	3
	86	87	88	89	90	91	92	93	94	95	96	97	98				
	1	2	4	4	1	2	1	3	4	3	2	3	1				

High Order Time Intensive MCQs																	
	1	2	3	4	5	6	7	8	9	10	11	12	13	14	15	16	17
	2	3	2	1	4	2	2	1	2	3	3	4	3	4	1	1	4
	18	19	20	21	22	23	24	25	26	27	28	29	30	31	32	33	34
	4	4	3	4	4	3	2	4	2	4	1	3	2	3	1	4	4
	35	36	37	38	39	40	41										
	2	1	1	3	2	1	4										

Statement Based MCQs	1	2	3	4
	2	4	4	1

Assertion & Reason MCQs	1	2	3	4
	2	1	2	1

Match the Columns MCQs	1	2	3	4
	1	1	1	2

NCERT Exemplar MCQs	1	2	3	4	5	6	7	8	9	10	11
	1	4	3	2	3	3	2	2	2	4	1

NEET & AIIMS Past Year MCQs																	
	1	2	3	4	5	6	7	8	9	10	11	12	13	14	15	16	17
	3	4	4	1	4	1	4	3	3	4	1	4	3	4	1	2	4
	18	19															
	1	3															

CHAPTER

10 Wave Optics

Topicwise Number of Questions Analysis (2025-2016)

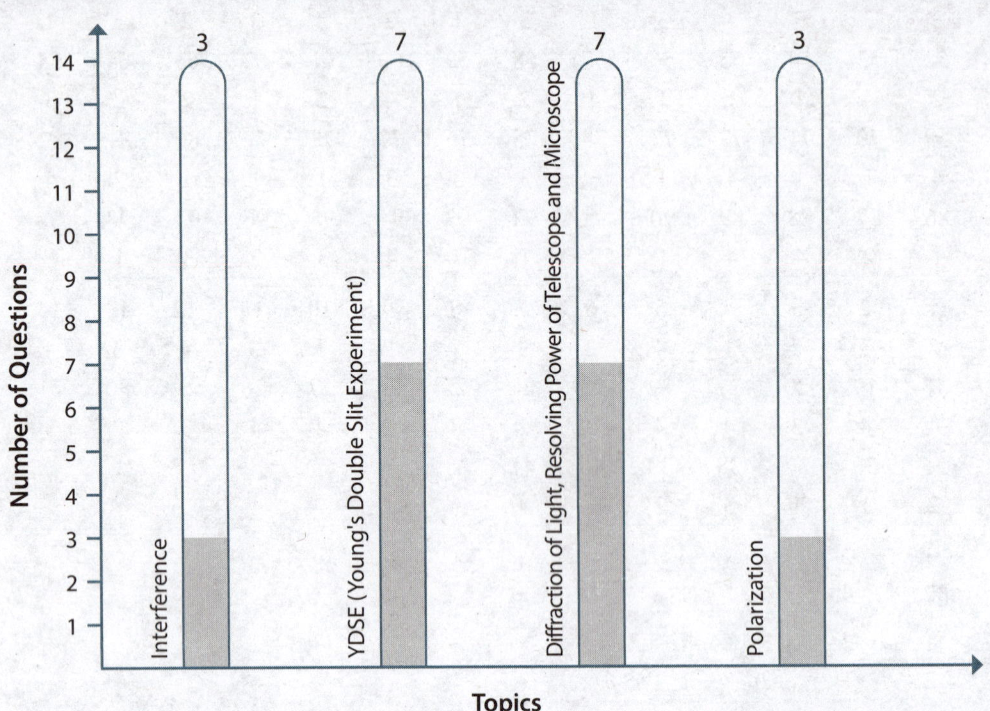

NCERT-Picks

Introduction

It is the branch of optics that studies interference, diffraction, polarization and other phenomenon for which the ray approximation of geometric optics is not valid.

It is also known as physical optics.

Huygens Principle

Formulas

1. Homogeneous medium, the wave front is always perpendicular to the direction of wave propagation.

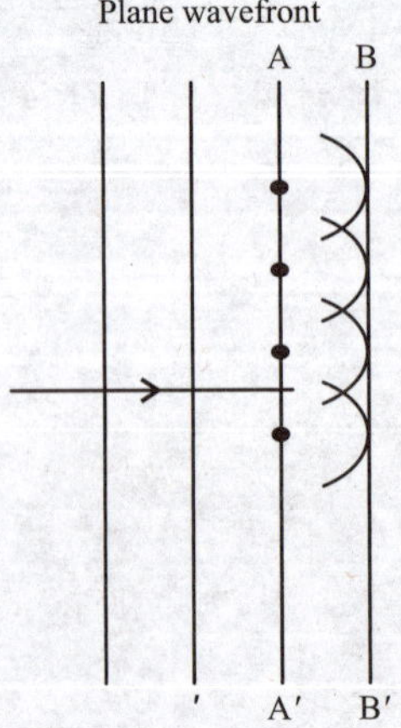

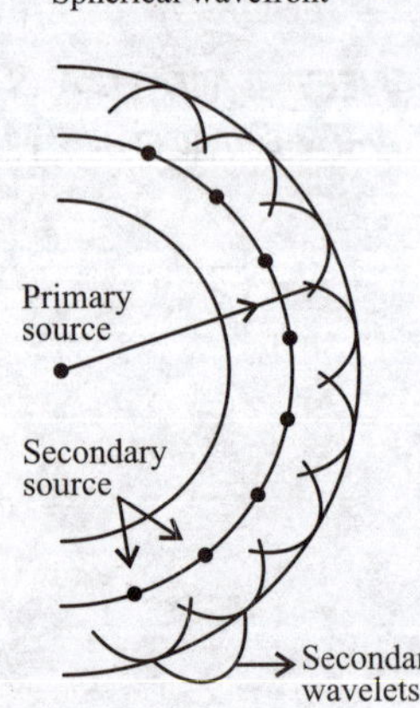

2. Time (t) remains same from one wavefront to another along the way.

Numerical Profile

(i) A point source of light produces a spherical wave front

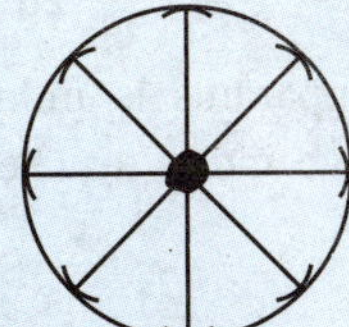
Point source

(ii) A linear source of light produces a cylindrical wave front.

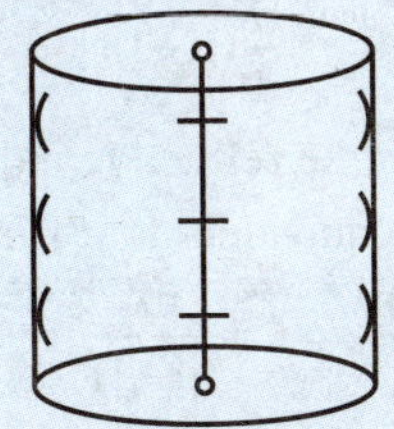
Line source

(iii) Any wave front from a distant source is a plane wave front.

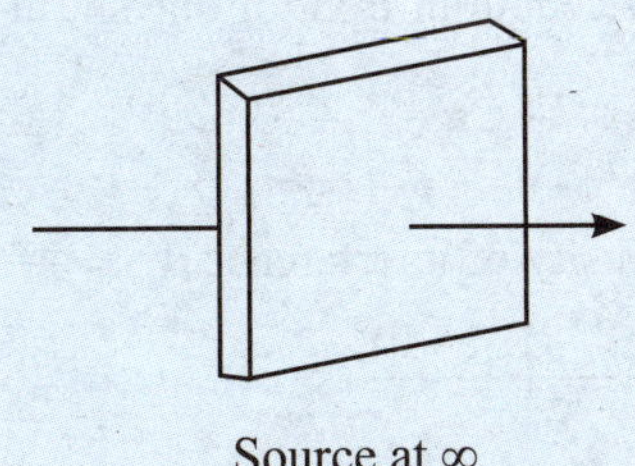
Source at ∞

Numerical Profile

Wavefront of Different Optical System

(i) Refraction of a plane wavefront by a thin prism

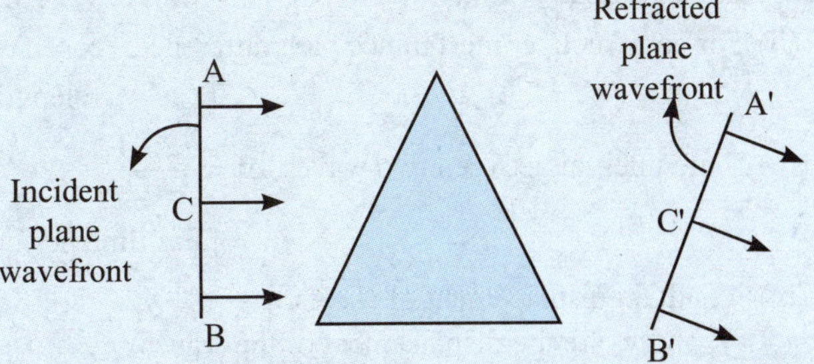

(ii) Refraction of a plane wavefront by a lens:

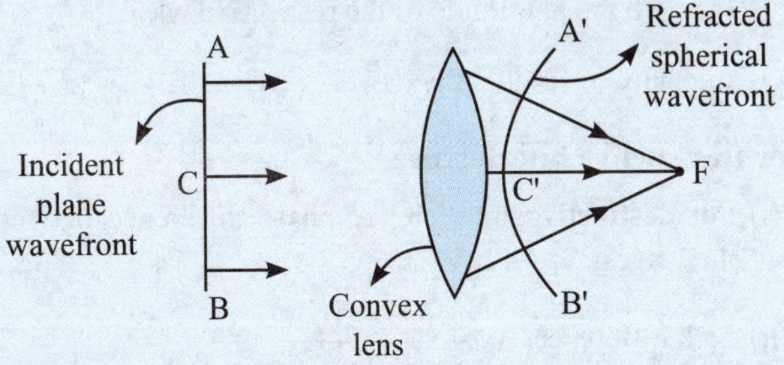

(iii) Reflection of a plane wavefront by a concave mirror

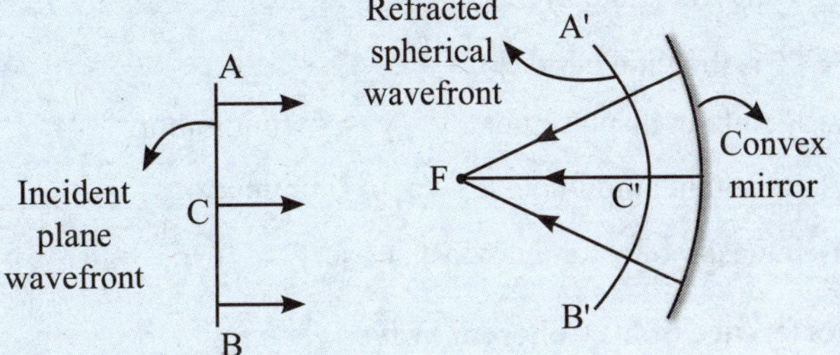

Refraction and Reflection of Plane Wave

Formulas

1. The angle of incidence is equal to angle of refelction.

$\angle i = \angle r$

2. Snell's law of refraction

$\frac{\sin i}{\sin r} = \mu$ (Refractive Index)

3. Relationship among refractive index, velocity and wave length.

$v \propto \lambda \propto \frac{1}{\mu}$

$\frac{v_1}{v_2} = \frac{\lambda_1}{\lambda_2} = \frac{\mu_2}{\mu_1}$

4. $\mu = \frac{\text{Speed of light in vacuum}}{\text{Speed of light in medium}} = \frac{c}{v}$

Coherent and Incoherent Addition of Waves

Formulas

1. The waves emitted by source 1 and source 2 start out in phase and arrive point at P in phase, leading to constructive interference at that point.

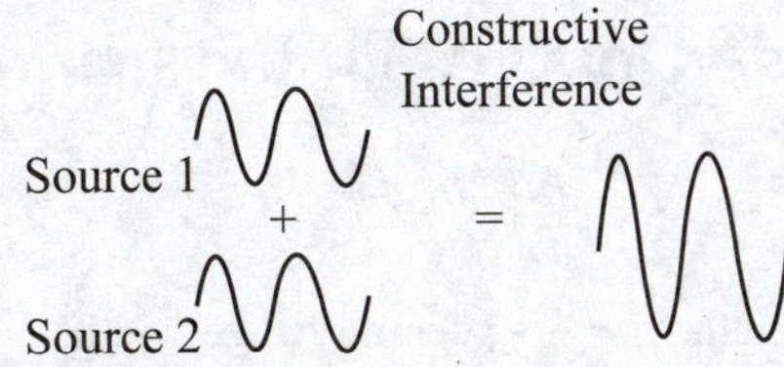

2. The waves emitted by two sources have the same phase to begin with, but they arrive at point P, out of phase. As a result, destructive interference occurs at P,

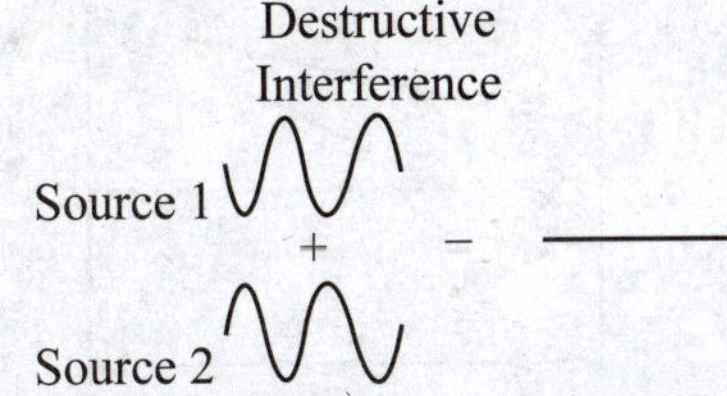

Numerical Profile

For Constructive interference

(i) For constructive interference phase differences between the waves, $\Delta\phi = 2n\pi$, [n = 0, 1, 2, 3...]

(ii) For constructive interference path difference, $\Delta x = n\lambda$ [λ is wavelength]

(iii) Time interval between two waves, $\Delta t = \frac{(2n)T}{2}$ [T is time period]

(iv) Resultant displacement of waves, $y = y_1 + y_2$ [y_1 and y_2 are the displacement of the waves)

(v) Resultant amplitude, R = (a + b) (maximum) [a and b are amplitude of the respective waves]

(vi) Intensity of resultant wave, $I = \left(\sqrt{I_1}+\sqrt{I_2}\right)^2 = (a+b)^2$

For Destructive Interference

(i) For destructive interference phase difference between the waves, $\Delta\phi = (2n-1)\pi$ [n = 1, 2, 3.....]

(ii) Path difference, $\Delta x = \frac{(2n-1)\lambda}{2}$

(iii) Time interval between two waves, $\Delta t = \frac{(2n-1)T}{2}$ [T is the time period]

(iv) Resultant displacement, $y = y_1 - y_2$ (minimum)

(v) Resultant amplitude, R = (a–b) (minimum)

(vi) Intensity of resultant wave, $I = (\sqrt{I_1}-\sqrt{I_2})^2 = (a-b)^2$

Interference of 'n' Coherent Waves

(i) If there are 'n' coherent waves each of intensity I_0 then the resultant maximum intensity is

$I = n^2 I_0$

(ii) If there are 'n' incoherent waves each of intensity I_0 then the resultant intensity is

$I = I_1 + I_2 + I_3 + ... + I_n$

$I = I_0 + I_0 + I_0 + ... + I_0 = nI_0$

Young's Double Slit Experiment

Formulas

1. Fringe width, $\beta = \frac{\lambda D}{d}$, central fringe width.

[D is separation between the slits and the screen, d is the separation between the slits]

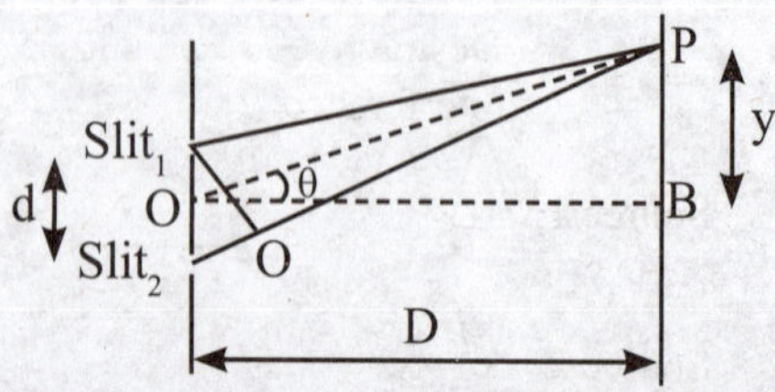

2. Position of bright fringe, $y = \frac{n\lambda D}{d}$ [n = 0, ±1, ±2, ±3,...]

3. Position of dark fringe, $y = \frac{(2n-1)\lambda D}{2d}$ [n = ±1, ±2, ±3,..]

4. When the entire apparatus is immersed in a transparent medium of refractive index μ. Then fringe width will be $\beta' = \frac{\lambda D}{\mu d} = \frac{\beta}{\mu}$

5. Angular width of interference fringes $= \frac{\beta}{D} = \frac{\lambda}{d}$

6. Shifting of fringe pattern

(a) Fringe shift = $\frac{D}{d}(\mu-1)t$

[Here, μ is the refractive index of the medium]

(b) If shift is equivalent to n fringes, then

$n = \frac{(\mu-1)t}{\lambda} \Rightarrow t = \frac{n\lambda}{(\mu-1)}$

7. Fringe visibility, $V = \frac{I_{max}-I_{min}}{I_{max}+I_{min}} = \frac{2\sqrt{I_1 I_2}}{(I_1+I_2)}$

8. Missing wavelength in front of one slit in YDSE

$\lambda = \frac{d^2}{(2n-1)D}$ n = 1, 2, 3....

9. Average intensity of interference $I_{av} = \frac{I_{max}+I_{min}}{2}$

$I_{av} = \frac{K(a+b)^2 + K(a-b)^2}{2}$

$I_{av} = K(a^2+b^2)$ [K is constant]

Numerical Profile

Analysis of Interference Pattern

(i) For maxima (constructive interference),

$\Delta x = \frac{d.y}{D} = n\lambda$

$\Rightarrow y = \frac{n\lambda D}{d}, n = 0, \pm 1, \pm 2, \pm 3$

Here $n = 0$ be similar to the central maxima

$n = \pm 1$ correspond to the 1st maxima

$n = \pm 2$ correspond to the 2nd maxima and so on.

(ii) For minima (destructive interference)

$\Delta x = \pm\frac{\lambda}{2}, \pm\frac{3\lambda}{2} \pm \frac{5\lambda}{2}$

$$\Rightarrow \Delta x = \begin{cases} (2n-1)\frac{\lambda}{2} & n = 1, 2, 3............ \\ (2n+1)\frac{\lambda}{2} & n = -1, -2, -3........ \end{cases}$$

$$\text{Therefore, } y = \begin{cases} (2n-1)\frac{\lambda D}{2d} & n = 1, 2, 3............ \\ (2n+1)\frac{\lambda D}{2d} & n = -1, -2, -3....... \end{cases}$$

Diffraction

Formulas

1. For diffraction at a single slit of width a,
   (a) Condition for n^{th} minimum is
   $a \sin \theta = n\lambda$, where n = 1, 2, 3,...
   (b) Condition of nth secondary maximum is
   $a \sin \theta = (2n+1)\frac{\lambda}{2}$, where n = 1, 2, 3,...
   (c) Angular position or direction of nth minimum,
   $\theta_n = \frac{n\lambda}{a}$
   (d) Distance of nth minimum from the centre of the screen,
   $x_n = \frac{nD\lambda}{a}$
   (e) Angular position of nth secondary maximum,
   $\theta_n' = (2n+1)\frac{\lambda}{2a}$
   (f) Distance of nth secondary maximum from the centre of the screen,
   $x_n' = (2n+1)\frac{D\lambda}{2a}$
   (g) Width of central maximum, $\beta_0 = 2\beta = \frac{2D\lambda}{a}$
   (h) Angular spread of central maximum on either side,
   $\theta = \pm\frac{\lambda}{a}$
   (i) Total angular spread of central maximum, $2\theta = \frac{2\lambda}{a}$
2. For diffraction at a circular aperture of diameter d,
   (a) Angular spread of central maximum, $\theta = \frac{1.22\lambda}{a}$
   (b) Linear spread, $x = D\theta$
   (c) Areal spread, $x^2 = (D\theta)^2$
   [where D is the distance at which the effect is considered]
3. Fresnel distance, $D_F = \frac{a^2}{\lambda}$
4. Size of Fresnel zone, $a_F = \sqrt{\lambda D}$.

Polarization

Formulas

1. Refractive index, $\mu = \tan i$ [Brewster's Law]
   [i is the angle of incidence] [RC]
2. Intensity of the transmitted light, $I = I_0 \cos^2\theta$

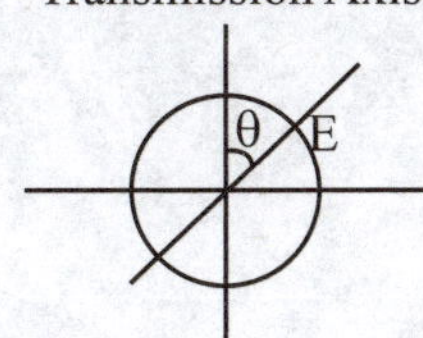

[I_0 is the intensity when E vector parallel to transmission axis, θ is angle between E vector and transmission axis]

Numerical Profile

Law of Malus

(i) If light passes through two successive polarisers, the emerging light has intensity $I = I_1 \cos^2\theta$.
[Where I_1 is the intensity that falls on the 2nd polariser and θ is the angle between the polarizing directions of two polarisers.]

(ii) If 'N' polarisers are arranged so that the first and last one are crossed. Let 'θ' is the angle between any two successive polarisers, 'I' is the intensity of incident light then the intensity of emerging light is
$I_1 = (I/2)\cos^n\theta$ where $n = 2(N-1)$

(iii) If P_1, P_2 and P_3 are three polarisers and angle between P_1 and P_2 is θ_1, P_2 and P_3 is θ_2 and I is the intensity of incident light, then the intensity of emerging light is $I_1 = \frac{I}{2}\cos^2\theta_1\cos^2\theta_2$

(iv) If $i = \theta_p$, the reflected light is completely polarised and the refracted light is partially polarised.
If $i < \theta_p$ or $i > \theta_p$, both reflected and refracted rays get partially polarised.
For glass $\theta_p = \tan^{-1}(1.5) \approx 57°$
For water $\theta_p = \tan^{-1}(1.33) \approx 53°$

ordinary light
Plane polarized light
i
air
glass
Partially polarized light

Statements Asked in Exam

- Corpuscular theory fails to explain the velocity of light in air and water.
- Polarization is used to check symmetry in crystals and molecules using the phenomenon of optical activity.
- Central maxima means the maxima formed with zero optical path difference. It may be formed anywhere on the screen.
- Fringes with red light are thicker than those for blue light.
- Ray optics is a limiting case of wave optics.
- Polaroids increase colour contrast in old paintings.
- Polaroids are used in three dimensional motion pictures and in optical stress analyses.
- Diffraction is common in light but not common in sound waves.
- The phenomena of interference and diffraction does not tell about transverse or longitudinal nature of light waves.
- To observe diffraction of light the size of obstacle / aperture should be of the order of 10^{-7} m.

Shape of light source	Variation of amplitude of distance	Variation of intensity with distance
Point source	$A \propto \frac{1}{r}$	$I \propto \frac{1}{r^2}$
linear/slit	$A \propto \frac{1}{\sqrt{r}}$	$I \propto \frac{1}{r}$
Extended light source and large distance	A = Constant	I = Constant

	Monochromatic light fringe		White light fringe
1.	Central fringe is of the same colour as that of the source.	1.	Central fringe is white.
2.	Dark and bright fringes are placed alternately about the central fringe.	2.	For small path difference, some coloured fringes are seen, outer edge is red and inner edge is violet.
3.	Width of each fringe is same.	3.	Fringe width is different $\beta_{red} > \beta_{violet}$
4.	Fringes of higher order also seen.	4.	Fringes of higher order not seen.

- The law of conservation of energy holds good in the case of interference of light.
- Fringes are neither image nor shadow of slit but locus of a point which moves in such a way that its path difference from the two sources remains constant.
- Young's double slit experiment is performed using a source of white light. The interference fringes will be coloured and the central maximum will be white, fringe next to central fringe will be violet and there will be no completely dark fringe.
- If the interference experiment is repeated with bichromatic light, the fringes of two wavelengths will be coincident for the first time when

  $$n(\beta)_{longer} = (n + 1)(\beta)_{shorter}$$

- No interference pattern is detected when two coherent sources are infinitely close to each other, because

  $$\beta \propto \frac{1}{d}$$

- Bubbles of colourless soap solution appear colourful in sun light.

NCERT Topic-wise MCQs

Huygens Principle, Interference and Young's Double Slit Experiment

1. If the two slit in young's experiment have width ratio 1 : 4. Deduce the ratio of intensity at maxima and minima in the interference pattern.

1) 9 : 1
2) 3 : 1
3) 9 : 2
4) 10 : 3

2. Two straight narrow parallel slits 3 mm apart are illuminated with monochromatic light of wavelength 5900×10^{-8} cm. Fringes are observed at a distance of 3 cm from the slits. Find the width of fringes.

1) 0.51 mm
2) 0.52 mm
3) 5.9 μm
4) 0.59 m

3. In young's double slit experiment, the width of the fringes obtained with light of wavelength 6000 Å is 2mm. What will be the fringe width, if the entire apparatus is immersed in a liquid of refractive index 1.33?

1) 5 mm
2) 3 mm
3) 1.5 mm
4) 1 mm

4. Two waves of amplitude 3 mm and 2 mm reach a point in the same phase. What is the resultant amplitude?

1) 2 mm
2) 5 mm
3) 7 mm
4) 8 mm

5. In Young's double slit experiment, the slits are 0.2 mm apart and the screen is 1.5 m away. It is observed that the distance between the central bright fringe and fourth dark fringe is 1.8 cm. Calculate wavelength of light used.

1) 6.86×10^{-7} m
2) 1×10^{-9} m
3) 7.34×10^{-7} m
4) 3.43×10^{-7} m

6. In Young's experiment, two coherent sources are 1.5 mm apart and the fringes are obtained at a distance of 2.5 m from them. If the sources produce light of wavelength 589.3 nm, find the number of fringes in the interference pattern, which is 4.9×10^{-3} m long.

1) 3
2) 5
3) 4
4) 7

7. In Young's experiment, what will be the phase difference and the path difference between the light wave reaching third bright fringe where λ = 5000 Å.

1) 3π, 15000Å
2) 2π, 7000 Å
3) 3π, 9000 Å
4) 6π, 15000 Å

8. Huygen's conception of secondary waves:

1) Allow us to find the focal length of a thick lens
2) Is a geometrical method to find a wave front
3) Is used to determine the velocity of light
4) Is used to explain polarisation

9. Find the maximum intensity in case of interference of n identical waves each of intensity I_0 if the interference is coherent.

1) nI_0
2) I_0
3) n^2I_0
4) n^nI_0

10. Two coherent monochromatic light beams of intensities I and 4I are superimposed. The maximum and minimum possible intensities in the resulting beam are:

1) 5I and I
2) 5I and 3I
3) 9I and I
4) 9I and 3I

11. Two identical light waves, propagating in the same direction have a phase difference δ. After they superpose the intensity of the resulting wave will be proportional to:

1) $\cos \delta$
2) $\cos \delta/2$
3) $\cos^2 \delta/2$
4) $\cos^2 \delta$

12. Find the ratio of the intensity at the centre of a bright fringe to the intensity at the point one quarter of the distance between two fringes from the centre.

1) 1 : 2
2) 2 : 1
3) 2 : 3
4) 4 : 3

13. The 6563 Å H_α line emitted by hydrogen in a star is found to be red-shifted by 15 Å. The speed with which the star is receding from the earth is

1) 3.21×10^5 ms^{-1}
2) 6.86×10^5 ms^{-1}
3) 2.12×10^5 ms^{-1}
4) 12.74×10^5 ms^{-1}

14. Two coherent sources of intensity ratio β interfere. Then the value of $(I_{max} - I_{min})/(I_{max} + I_{min})$ is :

1) $\frac{1+\beta}{\sqrt{\beta}}$
2) $\sqrt{\left(\frac{1+\beta}{\beta}\right)}$
3) $\frac{1+\beta}{2\sqrt{\beta}}$
4) $\frac{2\sqrt{\beta}}{1+\beta}$

15. Two waves of intensity ratio 1 : 9 cross each other at a point. Calculate the ratio of resultant intensities at a point, when (a) waves are incoherent (b) waves are coherent and differ in phase by 60°

1) 5 : 6
2) 13 : 10
3) 10 : 13
4) 6 : 5

16. Monochromatic green light of wavelength 5×10^{-7} m illuminates a pair of slits 1 mm apart. The separation of bright lines on the interference pattern formed on a screen 2 m away is:

1) 0.25 mm
2) 0.1 mm
3) 1 mm
4) 0.01 mm

17. In young's double slit experiment, the phase difference between the light waves reaching third bright fringe from the central fringe will be ($\lambda = 6000Å$)

1) Zero
2) 2π
3) 4π
4) 6π

18. In two separate set-ups of the young's double slit experiment fringes of equal width are observed when light of wavelengths in the ratio 1 : 2 are used. If the ratio of the slit separation in the two cases is 2 : 1 the ratio of the distances between the plane of the slits and the screen in the two set-ups is:

1) 4 : 1
2) 1 : 1
3) 1 : 4
4) 2 : 1

19. In a Young's double slit experiment, $\lambda = 500$ nm, d = 1.0 mm and D = 1.0 m. Find the minimum distance from the central maximum for which the intensity is half of the maximum intensity.

1) 1.25×10^{-4} m
2) 2×10^{-5} m
3) 12.5×10^{-4} m
4) 5.25×10^{-3} m

20. In young's double slit experiment angular width of fringes is 0.20 for sodium light of wavelength 5890 Å. If complete system is dipped in water then angular width of fringes becomes:

1) 0.11°
2) 0.15°
3) 0.22°
4) 0.30°

21. In a young's experiment, two coherent sources are placed 0.9 mm apart and the fringes are observed one metre away. If it produces the second dark fringe at a distance of 1 mm from the central fringe the wavelength of monochromatic light used would be:

1) 60×10^{-4} cm
2) 10×10^{-4} cm
3) 10×10^{-5} cm
4) 6×10^{-5} cm

22. In young's double slit experiment, the intensity of light coming from the first slit is double the intensity from the second slit. The ratio of the maximum intensity to the minimum intensity on the interference fringe pattern observed is:

1) 34
2) 40
3) 25
4) 38

23. Find the ratio of intensities at the two points X and Y on a screen in Young's double slit experiment, where waves from the two sources S_1 and S_2 have path differences of zero and $\lambda/4$ respectively.

1) $\frac{a+b}{a^2+b^2}$
2) $\frac{2(a+b)^2}{a^2+b^2}$
3) $\frac{(a+b)^2}{a^2+b^2}$
4) None of these

24. The light of wavelength 5890 Å falls normally on a thin air film. The minimum thickness of the film such that the film appears dark in reflected light is:

1) 2.945×10^{-7} m
2) 3.945×10^{-7} m
3) 4.95×10^{-7} m
4) 1.945×10^{-7} m

25. Young's double slit experiment is carried out by using green, red and blue light, one colour at a time. The fringe width recorded are β_G, β_R, β_B respectively. Then:

1) $\beta_G > \beta_B > \beta_R$
2) $\beta_B > \beta_G > \beta_R$
3) $\beta_R > \beta_B > \beta_G$
4) $\beta_R > \beta_G > \beta_B$

26. Two slits are separated by a distance of 0.5mm and illuminated with light of $\lambda = 6000$ Å. If the screen is placed 2.5m from the slits. The distance of the third bright fringe from the centre will be:

1) 1.5 mm
2) 3 mm
3) 6 mm
4) 9 mm

27. In young's double slit experiment the slits are 0.2 mm apart and the screen is 1.5 m away. It is observed that the distance between the central bright fringe and fourth dark fringe is 1.8 cm. Calculate wavelength of light used:

1) 6.86×10^{-7} m
2) 6.86×10^{-8} m
3) 6.86×10^{-18} m
4) 6.86×10^{-17} m

28. In a double slit experiment light has a frequency of 6×10^{14} s^{-1}. The distance between the centres of adjacent bright fringes is 0.75mm. What is the distance between the slits if the screen is 1.5m away?

1) 10^{-1} m
2) 10^{-3} m
3) 10^{-5} m
4) 1 m

29. In a young's double slit experiment $\lambda = 500$nm, d = 1.0mm and D = 1m. Find the minimum distance from the central maximum for which the intensity is half of the maximum intensity:

1) 1.25×10^{-3} m
2) 1.25×10^{-4} m
3) 2.25×10^{-1} m
4) 125 m

30. In Young's double slit experiment, the slits are horizontal. The intensity at a point P as shown in figure is $\frac{3}{4}I_0$, where I_0 is the maximum intensity.

Then the value of θ is

(Given the distance between the two slits S_1 and S_2 is 2λ.)

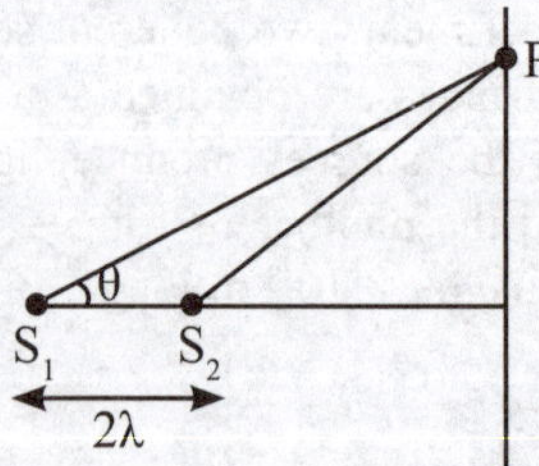

1) $\cos^{-1}\left(\frac{1}{12}\right)$
2) $\sin^{-1}\left(\frac{1}{12}\right)$
3) $\tan^{-1}\left(\frac{1}{12}\right)$
4) $\sin^{-1}\left(\frac{3}{5}\right)$

31. The wavelength of light used in young's double slit experiment is λ. The intensity at a point on the screen is I where the path difference is $\lambda/6$. If I_0 denotes the maximum intensity, then the ratio of I and I_0 is:
1) 0.866 2) 0.5
3) 0.707 4) 0.75

32. In young's double slit experiment, the distance between two slits is 0.6 mm and these are illuminated with a light of wavelength 4800Å. The angular width of first dark fringe on the screen distant 120 cm from slits will be:
1) 8×10^{-4} rad 2) 6×10^{-4} rad
3) 4×10^{-4} rad 4) 16×10^{-4} rad

33. In the phenomenon of interference, energy is:
1) Destroyed at bright fringes
2) Created at dark fringes
3) Conserved but it is redistributed
4) Same at all points

34. The young's double slit experiment is performed with light of wavelength 6000 Å where in 16 fringes occupy a certain region on the screen. If 24 fringes occupy the same region with another light of wavelength λ, then λ is:
1) 6000 Å 2) 4500 Å
3) 5000 Å 4) 4000 Å

35. Among two interfering sources, let A be ahead in phase by 54° relative to B. If the observations be taken from point P, such that PB – PA = 1.5 λ, deduce the phase difference between the waves from A and B reaching P.
1) 33π radian 2) 2.3π radian
3) 11π radian 4) 3.3π radian

Diffraction of Light, Resolving Power of Optical Instrument [RC]

36. Light of wavelength 600 nm is incident on an aperture of size 2mm. What will be the distance upto which the ray of light can travel, such that its spread is less than the size of the aperture?
1) 6.67 m 2) 7 m
3) 8 m 4) 6.2 m

37. A slit 4cm wide is irradiated with microwaves of wavelength 2cm. What will be angular spread of central maximum, assuming incidence normal to the plane of the slit?
1) 30° 2) 40°
3) 45° 4) 60°

38. What will be the resolving power of a microscope with cone angle of light falling on the objective equal to 30°? Take, λ = 600nm and μ for air = 1
1) 1.8×10^{6} 2) 1.67×10^{-6}
3) 1.6×10^{6} 4) 10^{6}

39. For what distance ray optics is a good approximation when the aperture is 3 mm wide and the wavelength is 500 nm?
1) 18 m 2) 20 m
3) 21 m 4) 25 m

40. Calculate the limit of resolution of microscope with cone angle of light falling on the objective equal to 60°. (Given λ = 5000Å, μ for air = 1)
1) 5×10^{-7} m 2) 0.5 μm
3) Both of these 4) None of these

41. A parallel beam of monochromatic light of wavelength 5000Å is incident normally on a single narrow slit of width 0.001mm. The light is focused by a convex lens on a screen placed on the focal plane. The first minima will be formed for the angle of diffraction equal to:
1) 0° 2) 15°
3) 30° 4) 60°

42. Angular width of central maxima in the Fraunhofer's diffraction pattern is measured, slit is illuminated by the light of wavelength 6000Å. If the slit is illuminated by light of another wavelength angular width decreases by 30%, wavelength of light used:
1) 3500 Å 2) 4200 Å
3) 4700 Å 4) 6000 Å

43. Two beams of light will not give rise to interference pattern if:
1) They are coherent
2) They have the same wavelength
3) They are lineraly polarised perpendicular to each other
4) They are not monochromatic

44. A single slit of width 0.20 mm is illuminated with light of wavelength 500nm. The observing screen is placed 80cm from the slit. The width of the central bright fringe will be:
1) 1 mm 2) 5 mm
3) 10 mm 4) 4 mm

45. A parallel beam of monochromatic light of wavelength 450 nm passes through a long slit of width 0.2 mm. Find the angular divergence in which most of the light is diffracted.
1) 1.5×10^{-3} rad 2) 4.5×10^{-3} rad
3) 3×10^{-3} rad 4) 6×10^{-3} rad

46. Direction of the first secondary maximum in the Fraunhoffer diffraction pattern at a single slit is given by (a is the width of the slit):
1) $a\sin\theta = \frac{\lambda}{2}$ 2) $a\cos\theta = \frac{3\lambda}{2}$
3) $a\sin\theta = \lambda$ 4) $a\sin\theta = \frac{3\lambda}{2}$

47. A beam of light of wavelength 600nm from a distant source falls on a single slit 1mm wide and the resulting diffraction pattern is observed on a screen 2 m away. What is the distance between the first dark fringe on either side of the central bright fringe?
1) 2 mm 2) 2.5 mm
3) 2.4 mm 4) 2.3 mm

48. Red light of wavelength 6500Å from a distant source falls on a slit 0.5 mm wide. What will be the distance between the two dark bands on each side of central bright band of the diffraction pattern observed on a screen placed 1.8 from the slit?

1) 4 mm 2) 4.68 mm
3) 5.06 mm 4) 6 mm

49. A parallel beam of monochromatic light of wavelength 450nm passes through a long slit of width 0.2mm. Find the angular divergence in which most of the light is diffracted:

1) 10^4 rad 2) 5.1×10^3 rad
3) 2.25×10^{-3} rad 4) 4.5×10^{-3} rad

50. When a compact disc is illuminated by a source of white light, coloured lines are observed. This is due to:

1) Dispersion 2) Diffraction
3) Interference 4) Refraction

51. A parallel beam of fast moving electrons is incident normally on a narrow slit. A screen is placed at a large distance from the slit. If the speed of the electrons is increased, which of the following statement is correct:

1) Diffraction pattern is not observed on the screen in the case of electrons
2) The angular width of the central maxima of the diffraction pattern will increase
3) The angular width of the central maxima will decrease
4) The angular width of the central maxima will remain the same

52. If the angular width of central maxima is $n \times 10^{-3}$ rad in Fraunhoffer diffraction when light of wavelength 12000nm falls on slit of width 5mm. Find the value of n.

1) 5 2) 4.8
3) 4.9 4) 2

53. Calculate wavelength of light used when limit of resolution of the microscope is 5×10^{-7} m and cone angle of light falling on the objective is equal to 90°. Given for air, $\mu = 1$. **[RC]**

1) 0.707×10^{-6} m 2) 7.07×10^{-6} m
3) 0.707×10^{-5} m 4) 6.06×10^{-4} m

Polarization

54. At what angle of incidence will the light reflected from liquid ($\mu = 1.732$) be completely polarised. Does this angle depend on the wavelength of light? **[RC]**

1) 53° 2) 60°
3) 30° 4) 42°

55. The refractive index of a medium is $\sqrt{3}$. What is the angle of refraction if the unpolarised light is incident on it at the polarising angle of the medium? **[RC]**

1) 30° 2) 45°
3) 6° 4) 90°

56. Polarisation of light is the only phenomenon that establishes:

1) Longitudinal nature of light
2) Transverse wave of light
3) Both of these
4) None of these

57. At what angle should the axes of two polaroids be placed so as to reduces the intensity of incident unpolarized light to 1/2?

1) 30° 2) 45°
3) 60° 4) 90°

58. The angle of polarisation for any medium is 60°. What will be critical angle for this? **[RC]**

1) $sin^{-1}\sqrt{3}$ 2) $\tan^{-1}\sqrt{3}$
3) $\cos^{-1}\sqrt{3}$ 4) $\sin^{-1}\frac{1}{\sqrt{3}}$

59. When the angle of incidence on a material is 60°, the reflected light is completely polarised. The velocity of the refracted ray inside the material is: **[RC]**

1) 3×10^8 2) $\frac{3}{\sqrt{2}} \times 10^8$
3) $\sqrt{3} \times 10^8$ 4) 0.5×10^8

60. When an unpolarized light of intensity, I_0 is incident on a polarizing sheet, the intensity of the light which does not get transmitted is:

1) Zero 2) I_0
3) $\frac{I_0}{2}$ 4) $\frac{I_0}{4}$

61. The Brewster angle for the glass-air interface is 53.74°. If ray of light going from air to glass strikes at an angle of incidence 45°, then the angle of refraction is ($\tan 53.74° \approx \sqrt{2}$) **[RC]**

1) 60° 2) 30°
3) 25° 4) 54.74°

62. 'n' polarizing sheets are arranged such that each makes an angle 45° with the preceeding sheet. An unpolarized light of intensity I is incident into this arrangement. The output intensity is found to be $\frac{I}{64}$. The value of n will be:

1) 3 2) 6
3) 5 4) 4

63. The graph showing the dependence of intensity of transmitted light on the angle between polariser and analyser is:

1) 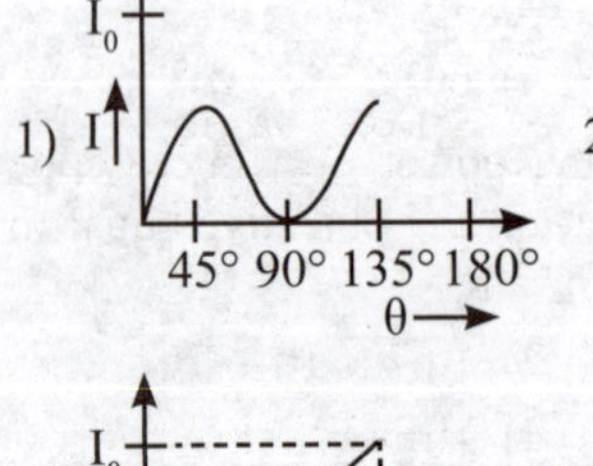

2) 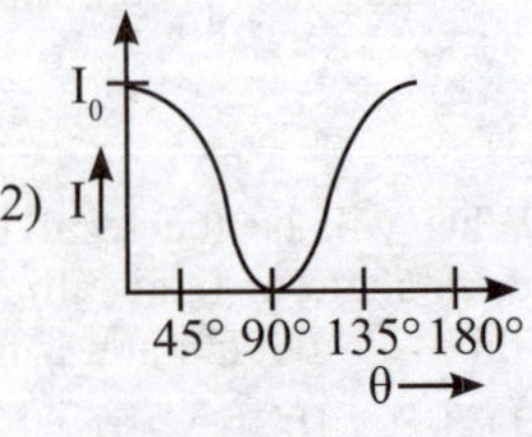

3) 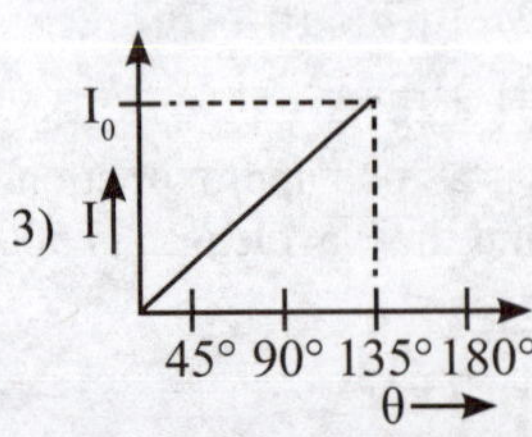

4) 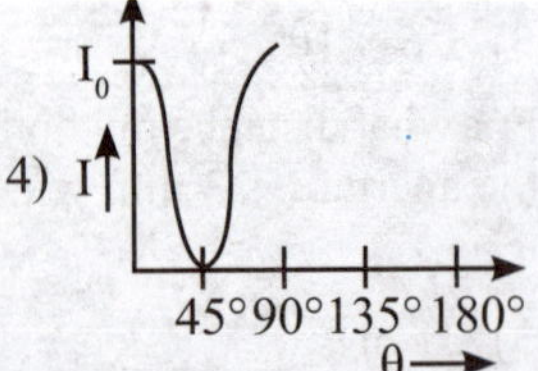

64. Which of the following phenomena is not common to sound and light waves?

1) Diffraction
2) Polarisation
3) Reflection
4) Interference

65. When unpolarised light beam is incident from air onto glass (n = 1.5) at the polarising angle: **[RC]**

1) Reflected beam is polarised 100%
2) Reflected and refracted beam are partially polarised
3) The reason for option (1) is that almost all the light is reflected
4) All of the above

66. A beam of light travelling in water falls on a glass plate immersed in water when the incidence angle is 51°, then the refracted beam of light is found polarised. Calculate μ of glass. Given μ of water is 4/3. ($\tan^{-1} 51° = 1.2349$) **[RC]**

1) 1.762
2) 1.647
3) 1.561
4) None of these

67. A light has amplitude A and angle between analyser and polariser is 60^0. Light is reflected by analyse has amplitude:

1) $A\sqrt{2}$
2) $\frac{A}{\sqrt{2}}$
3) $\frac{\sqrt{3}A}{2}$
4) $\frac{A}{2}$

68. As shown in the figures, three identical polaroids P_1, P_2 and P_3 are placed one after another. The pass axis of P_2 and P_3 are inclined at angle of 60° and 90° with respect to the axis of P_1. The source S has an intensity of $256\frac{W}{m^2}$. The intensity of light at point O is ________ $\frac{W}{m^2}$.

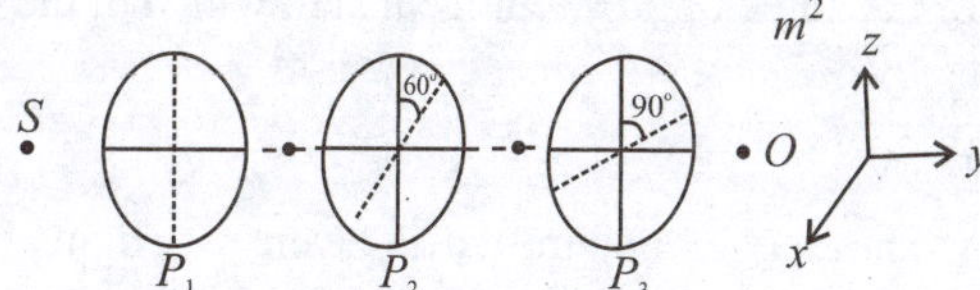

1) 24
2) 28
3) 32
4) 64

High Order Time Intensive MCQs

1. In the Young's double slit experiment, the intensity of light at a point on the screen (where the path difference is λ) is K, (λ being the wavelength of light used). The intensity at a point where the path difference is λ/4, will be:

1) K
2) $K/4$
3) $K/2$
4) Zero

2. In Young's double slit interference experiment, using two coherent waves of different amplitudes, the intensities ratio between bright and dark fringes is 3. Then, the value of the ratio of the amplitudes of the wave that arrive there is:

1) $\left(\frac{\sqrt{3}+1}{\sqrt{3}-1}\right)$
2) $\left(\frac{\sqrt{3}-1}{\sqrt{3}+1}\right)$
3) $\sqrt{3}:1$
4) $1:\sqrt{3}$

3. In a single slit diffraction experiment, first minimum for red light (660 nm) coincides with first maximum of some other wavelength λ'. Calculate λ'.

1) 440 nm
2) 210 nm
3) 340 nm
4) 220 nm

4. A Young's double slit set up for interference is shifted from air to water, then the fringe width:

1) Becomes infinite
2) Decreases
3) Increases
4) Remain unchanged

5. The wavelength of the light used in Young's double slit experiment is λ. The intensity at a point on the screen is I, where the path difference is $\frac{\lambda}{6}$. If I_0 denotes the maximum intensity, then the ratio of I and I_0 is:

1) 0.866
2) 0.5
3) 0.707
4) 0.75

6. A beam of light of wavelength 600 nm from a distant source falls on a single slit 1 mm wide and the resulting diffraction pattern is observed on a screen 2 m away. The distance between the first dark fringes on either side of the central bright fringe is:

1) 1.2 cm
2) 1.2 mm
3) 2.4 cm
4) 2.4 mm

7. The necessary condition for an interference by two sources of light is that:

1) Two light sources must have the same wavelength
2) Two light sources must have the different wavelength
3) Two sources should have the same wavelength, nearly the same amplitude and have a constant phase difference
4) Two point sources should have a randomly varying phase difference

8. Light from two coherent sources of the same amplitude A and wavelength λ illuminates the screen. The intensity, of the central maxima is I_0. If the sources were incoherent, the intensity at the same point will be:

1) $4I_0$
2) $2I_0$
3) I_0
4) $\frac{I_0}{2}$

9. Two light rays having the same wavelength λ in vacuum are in phase initially. Then , the first ray travels a path L_1 through a medium of refractive index n_1 while the second ray travels a path of length L_2 through a medium of refractive index n_2. The two waves are then combined to produce interference. Find the phase difference between two waves:

1) $\frac{2\pi}{\lambda}(L_2 - L_1)$
2) $\frac{2\pi}{\lambda}(n_1L_1 - n_2L_2)$
3) $\frac{2\pi}{\lambda}(n_2L_1 - n_1L_2)$
4) $\frac{2\pi}{\lambda}\left(\frac{L_1}{n_1} - \frac{L_2}{n_2}\right)$

10. Two waves originating from sources S_1 and S_2 having zero phase difference and common wavelength λ will show completely destructive interference at a point P if $(S_2P - S_1P)$ is:

1) 5λ 2) 2λ
3) $\frac{3\lambda}{4}$ 4) $11\frac{\lambda}{2}$

11. A parallel beam of monochromatic light of wavelength 5000 Å is incident normally on a single narrow slit of width 0.001 mm. The light is focused by a convex lens on a screen placed on the focal plane. The first minima will be formed for the angle of diffraction equal to which of the following?

1) 0° 2) 15°
3) 30° 4) 60°

12. A parallel monochromatic beam of light is incident normally on a narrow slit. A diffraction pattern is formed on a screen placed perpendicular to the direction of incident beam. At the first maxima of the diffraction pattern, the phase difference between the rays coming from the edges of the slit is:

1) 0 2) $\frac{\pi}{2}$
3) π 4) 2π

13. In a Fraunhofer diffraction at a single slit of width *d* with incident light of wavelength 5500 Å, the first minima is observed, at an angle 30°. The first secondary maxima is observed at an angle equals:

1) $\sin^{-1}\left(\frac{1}{\sqrt{2}}\right)$ 2) $\sin^{-1}\left(\frac{1}{4}\right)$
3) $\sin^{-1}\left(\frac{3}{4}\right)$ 4) $\sin^{-1}\frac{\sqrt{3}}{2}$

14. Single slit of width *d* is illuminated by violet light of wavelength 400 nm and the width of the diffraction pattern is measured as *y*. When half of the slit width is covered and illuminated by yellow light of wavelength 600 nm, Find the width of diffraction pattern:

1) The pattern vanishes and the width is zero
2) $\frac{y}{3}$
3) $3y$
4) None of the above

15. A ray of light of intensity I is incident on a parallel glass-slab at a point A as shown in below figure. It undergoes partial reflection and refraction. At each reflection 25% of incident energy is reflected. The rays AB and A'B' undergo interference. The ratio I_{max}/I_{min} is :

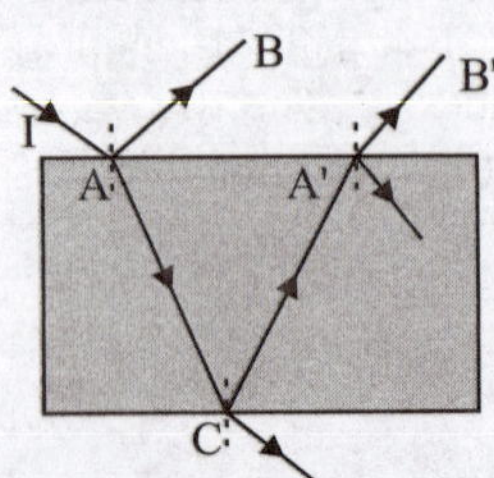

1) 4 : 1 2) 8 : 1
3) 7 : 1 4) 49 : 1

16. Two waves $y_1 = A_1 \sin(\omega t - \beta_1)$ and $y_2 = A_2 \sin(\omega t - \beta_2)$ superimpose to form a resultant wave whose amplitude is

1) $\sqrt{A_1^2 + A_2^2 + 2A_1A_2\cos(\beta_1 - \beta_2)}$
2) $\sqrt{A_1^2 + A_2^2 + 2A_1A_2\sin(\beta_1 - \beta_2)}$
3) $A_1 + A_2$
4) $|A_1 + A_2|$

17. Mixture of light consisting of wavelength 590 nm and an unknown wavelength illuminates young's double slit and gives rise to two overlapping interference patterns on the screen. The central maximum of both lights coincide. Further, it is observed that the third bright fringe of known light coincides with the 4th bright fringe of unknown light. From this data, the wave length of unknown light is

1) 885.0 nm 2) 442.5 nm
3) 776.8 nm 4) 393.4 nm

18. In a Young's double slit experiment, the two slits act as coherent sources of waves of equal amplitude A and wavelength λ. In another experiment with the same arrangement the two slits are made to act as incoherent sources of waves of same amplitude and wavelength. If the intensity at the middle point of the screen in the first case is I_1 and in the second case I_2, then the ratio I_1/I_2 is

1) 4 2) 2
3) 1 4) 0.5

19. White light is used to illuminate the two slits in Young's double slit experiment, separation between the slits is b and the screen is at a distance d(>>b) from the slits. At a point on the screen, directly in front of the slits, certain wavelengths are missing. Some of these missing wavelengths are

1) $\frac{b^2}{d}, \frac{b^2}{3d}$ 2) $\frac{2b^2}{d}, \frac{3b^2}{d}$
3) $\frac{2b^2}{3d}, \frac{4b^2}{5d}$ 4) $\frac{b^2}{5d}, \frac{7b^2}{5d}$

20. A screen is placed 50 cm from a single slit which is illuminate with 6000 Å light. If the distance between the first and third minima in the differaction pattern is 3.00 mm, what is the width of the slit ?

1) 0.3 mm 2) 0.5 mm
3) 0.2 mm 4) None of these

21. A plane wavefront ($\lambda = 6 \times 10^{-7}$ m) falls on a slit 0.4 mm wide. A convex lens of focal length 0.8 m placed behind the slit focusses the light on a screen. What is the linear diameter of second maximum ?

1) 9 mm 2) 15 mm
3) 57 mm 4) 6 mm

22. Determine the angular separation between the central maximum and first order maximum of the diffraction pattern due to a single slit of width 0.25 mm, when light of wavelength 5890 Å is incident on it normally.
    1) 35.3×10^{-3} rad
    2) 3.53×10^{-3} rad
    3) 0.353×10^{-3} rad
    4) None of these

23. In a Young's double slit experiment, two slits are illuminated by a mixture of two wavelengths 12000Å and 10000. At 6.0 mm from the common central bright fringe on a screen 2 m away from the slits, a bright fringe of one interference pattern coincides with the bright fringe of other. Distance between the slits should be:
    1) 1.0 mm 2) 1.5 mm
    3) 2.0 mm 4) 2.5 mm

24. Interference fringes from sodium light (λ = 5890Å) in a double slit experiment have an angular width 0.20°, To increase the fringe width by 10%, new wavelength should be
    1) 5896Å 2) 7321Å
    3) 6300Å 4) 6479Å

25. In Young's double slit experiment, the y-coordinates of central maxima and 10th maxima are 2 cm and 5 cm respectively. When the YDSE apparatus is immersed in a liquid of refractive index 1.5 the corresponding y-coordinates will be :
    1) 2 cm, 7.5 cm 2) 3 cm, 6 cm
    3) 2 cm, 4 cm 4) 4/3 cm, 10/3 cm

26. Two coherent monochromatic light sources are located at two vertices of an equilateral triangle. If the intensity due to each of the sources independently is 1 Wm^{-2} at the third vertex, the resultant intensity due to both the sources at that point (i.e., at the third vertex) is :
    1) Zero 2) $\sqrt{2}$
    3) 2 4) 4

Statement Based MCQs

Directions: These questions consist of two statements each, printed as Statement-I and Statement-II. While answering these questions, you are required to choose any one of the following four responses.

1) **Both Statement-I and Statement-II are correct.**
2) **Both Statement-I and Statement-II are incorrect.**
3) **Statement-I is correct & Statement-II is incorrect.**
4) **Statement-I is incorrect & Statement-II is correct.**

1. **Statement-I:** No interference pattern is detected when two coherent sources are infinitely close to each other.

   Statement-II: The fringe width is inversely proportional to the distance between the two slits.

2. **Statement-I:** The unpolarised light and polarised light can be distinguished from each other by using polaroid.

   Statement-II: A polaroid is capable of producing plane polarised beam of light.

3. **Statement-I:** In Young's double slit experiment energy conservation is not followed

   Statement-II: In Young's double slit experiment, if intensity of each source is I_0 then minimum and maximum intensity is zero and $4I_0$ respectively.

4. **Statement-I:** A point source of light gives rise to spherical wavefronts.

   Statement-II: A wave front always travels in forward direction in a medium.

5. **Statement-I:** Coloured spectrum is seen when we look through a muslin cloth.

   Statement-II: The coloured spectrum is due to diffraction of white light on passing through fines slits made by fine threads in the muslin cloth.

Assertion & Reason MCQs

Directions: These questions consist of two statements each, printed as Assertion and Reason. While answering these questions, you are required to choose any one of the following four responses.

1) **Both Assertion and Reason are True and the Reason is a correct explanation of the Assertion.**
2) **Both Assertion and Reason are True but Reason is not a correct explanation of the Assertion.**
3) **Assertion is True but the Reason is False.**
4) **Assertion is False but Reason is True.**

1. **Assertion:** Diffraction takes place for all types of waves mechanical or non-mechanical, transverse or longitudinal.
   Reason: Diffraction effects are perceptible only if wavelength of wave is comparable to dimensions of diffracting device.

2. **Assertion:** Light added to light can produce darkness.
   Reason: The destructive interference of two coherent light sources produce bright fringe.

3. **Assertion:** In YDSE, if a thin film is introduced in front of the upper slit, then the fringe pattern shifts in the downward direction.
   Reason: In YDSE independent of slit width, the minima will be completely dark.

4. **Assertion:** In Young's double slit experiment, interference pattern disappears when one of the slits is closed.

Reason: Interference occurs due to superimposition of light wave from two coherent sources.

5. **Assertion:** The phase difference between any two points on the same wave front is equal to 0.

Reason: Light from the source reaches every point of the wave front at the same time.

Match the Columns MCQs

1. In Young's double slit experiment, match the following two columns.

	Column-I		Column-II
(a)	When width of one slit is slightly increased	(i)	Maximum intensity will increase
(b)	When one slit is closed	(ii)	Maximum intensity will decrease
(c)	When both the sources are made incoherent	(iii)	Maximum intensity will remain same
(d)	When a glass slab is inserted in front of one of the slits	(iv)	Fringe pattern will disappear

1) (a)-(iii,ii) (b)-(iv) (c)-(i) (d)-(ii,iv)
2) (a)-(i) (b)-(ii,iv) (c)-(ii,iv) (d)-(iii)
3) (a)-(i,ii) (b)-(iii) (c)-(i) (d)-(iv,iii)
4) (a)-(iv,i) (b)-(iii,ii) (c)-(ii) (d)-(i)

2. Two waves from coherent sources meet at a point in a phase difference of ϕ. Both the waves have same intensities. Match the following two columns.

	Column-I		Column-II
(a)	If $\phi = 60°$	(i)	Resultant intensity will become four times
(b)	If $\phi = 90°$	(ii)	Resultant intensity will become two times
(c)	If $\phi = 0°$	(iii)	Resultant intensity will remain unchanged
(d)	If $\phi = 120°$	(iv)	Resultant intensity will become three times

1) (a)-(iv) (b)-(ii) (c)-(i) (d)-(iii)
2) (a)-(i) (b)-(ii) (c)-(iv) (d)-(iii)
3) (a)-(ii) (b)-(iii) (c)-(i) (d)-(iv)
4) (a)-(i) (b)-(iii) (c)-(ii) (d)-(iv)

3. In terms of fringe width ω, match the following two columns.

	Column-I		Column-II
(a)	Distance between central maxima and third order maxima	(i)	$2.5\ \omega$
(b)	Distance between central maxima and third order minima	(ii)	$3.0\ \omega$
(c)	Distance between first minima and fourth order maxima	(iii)	$3.5\ \omega$
(d)	Distance between second order maxima and fifth order minima	(iv)	None of these

1) (a)-(iii) (b)-(iv) (c)-(i) (d)-(ii)
2) (a)-(i) (b)-(iv) (c)-(ii) (d)-(iii)
3) (a)-(ii) (b)-(i) (c)-(iii) (d)-(i)
4) (a)-(iv) (b)-(iii) (c)-(ii) (d)-(i)

4. Match the following two columns.

	Column-I		Column-II
(a)	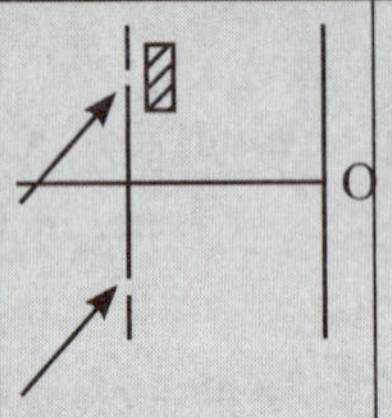 O	(i)	The zero order maxima will lie above point Q
(b)	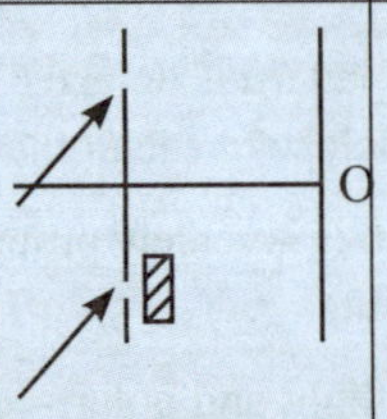 O	(ii)	The zero order maxima will lie below point Q
(c)	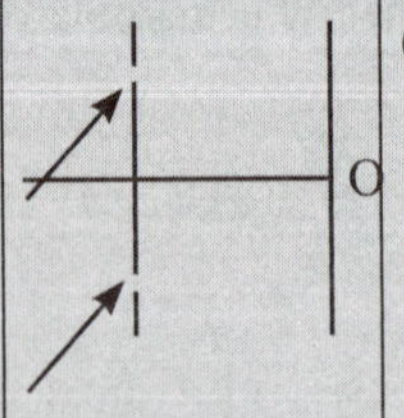 O	(iii)	The zero order maxima may lie above or below point O
(d)	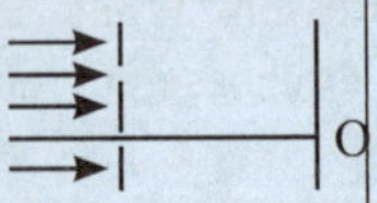 O	(iv)	The zero order maxima may lie at point O

1) (a)-(i) (b)-(iii,iv) (c)-(i) (d)-(i)
2) (a)-(i) (b)-(iii,iv) (c)-(ii) (d)-(iii)
3) (a)-(ii) (b)-(i) (c)-(iii) (d)-(i,iv)
4) (a)-(i,iv) (b)-(iii) (c)-(ii) (d)-(i)

NCERT Exemplar MCQs

1. Consider a light beam incident from air to a glass slab at Brewster's angle as shown in figure.

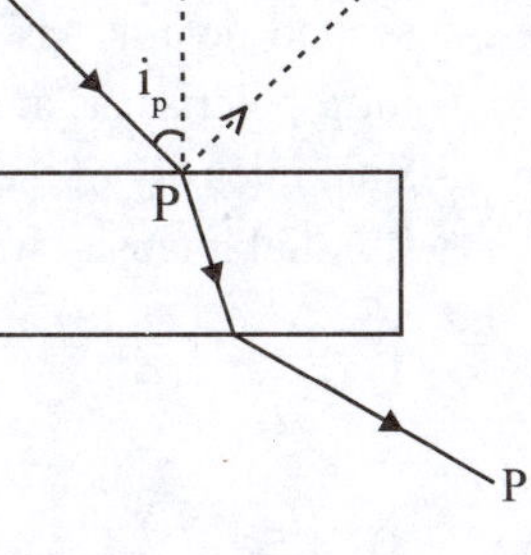

A polaroid is placed in the path of the emergent ray at point P and rotated about an axis passing through the centre and perpendicular to the plane of the polaroid:
   1) For a particular orientation, there shall be darkness as observed through the polaroid
   2) The intensity of light as seen through the polaroid shall be independent of the rotation
   3) The intensity of light as seen through the polaroid shall go through a minimum but not zero for two orientations of the polaroid
   4) The intensity of light as seen through the polaroid shall go through a minimum for four orientations of the polaroid

2. Consider sunlight incident on a slit of width 10^4Å. The image seen through the slit shall:
   1) Be a fine sharp slit white in colour at the centre
   2) A bright slit white at the centre diffusing to zero intensities at the edges
   3) A bright slit white at the centre diffusing to regions of different colours
   4) Only be a diffused slit white in colour

3. In a Young's double-slit experiment, the source is white light. One of the holes is covered by a red filter and another by a blue filter. In this case:
   1) There shall be alternate interference patterns of red and blue
   2) There shall be an interference pattern for red distinct from that for blue
   3) There shall be no interference fringes
   4) There shall be an interference pattern for red mixing with one for blue

4. Figure shows a standard two slit arrangement with slits S_1, S_2. P_1 and P_2 are the two minima points on either side of P (figure):

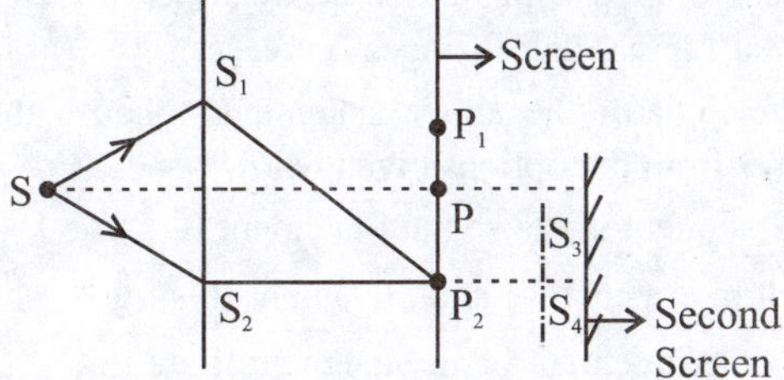

At P_2 on the screen, there is a hole and behind P_2 is a second 2-slit arrangement with slits S_3, S_4 and a second screen behind them:
   1) There would be no interference pattern on the second screen but it would be lighted
   2) The second screen would be totally dark
   3) There would be a single bright point on the second screen
   4) There would be a regular two slit pattern on the second screen

NEET & AIIMS Past Year MCQs

1. Two slits in Young's double slit experiment are 1.5 mm apart and the screen is placed at a distance of 1 m from the slits. If the wavelength of light used is 600×10^{-9} m then the fringe separation is **(2024 Re)**
   1) 4×10^{-5} m
   2) 9×10^{-8} m
   3) 4×10^{-7} m
   4) 4×10^{-4} m

2. Interference pattern can be observed due to superposition of the following waves:

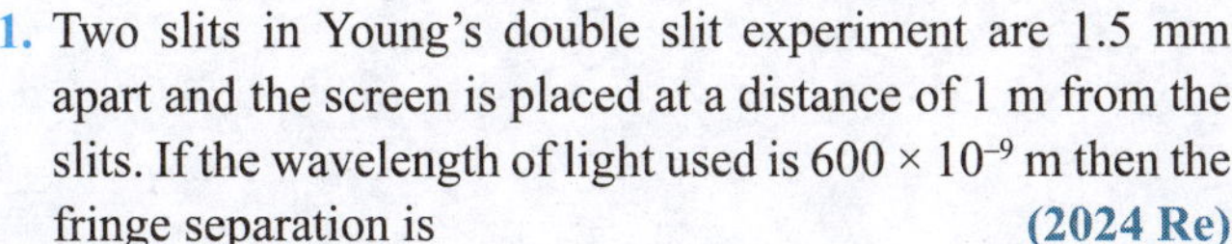
   1) $y = a \sin\omega t$
   2) $y = a \sin 2\omega t$
   3) $y = a \sin(\omega t - \phi)$
   4) $y = a \sin 3\omega t$

   Choose the correct answer from the options given below. **(2024 Re)**
   1) B and C
   2) B and D
   3) A and C
   4) A and B

3. A beam of unpolarized light of intensity I_0 is passed through a polaroid A, then through another polaroid B, oriented at 60° and finally through another polaroid C, oriented at 45° relative to B as shown. The intensity of emergent light is: **(2024 Re)**

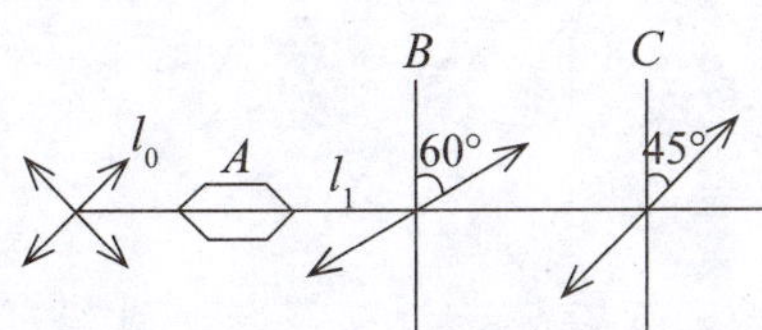

   1) $\frac{I_0}{16}$
   2) $\frac{I_0}{4}$
   3) $\frac{I_0}{2}$
   4) $\frac{I_0}{32}$

4. If the monochromatic source in Young's double slit experiment is replaced by white light, then **(2024)**
   1) there will be a central bright white fringe surrounded by a few coloured fringes.
   2) all bright fringes will be of equal width.
   3) interference pattern will disappear.
   4) there will be a central dark fringe surrounded by a few coloured fringes.

5. An unpolarised light beam strikes a glass surface at Brewster's angle. Then **(2024)**
   1) both the reflected and refracted light will be completely polarised.
   2) the reflected light will be completely polarised but the refracted light will be partially polarised.
   3) the reflected light will be partially polarised.
   4) the refracted light will be completely polarised.

6. For Young's double slit experiment, two statements are given below:

   Statement I: If screen is moved away from the plane of slits, angular separation of the fringes remains constant.

   Statement II: If the monochromatic source is replaced by another monochromatic source of larger wavelength, the angular separation of fringes decreases.

   In the light of the above statements, choose the correct answer from the options given below: **(2023)**
   1) Statement I is false but Statement II is true
   2) Both Statement I and Statement II is true
   3) Both Statement I and Statement II are true
   4) Statement I is true but Statement II are false

7. In a Young's double slit experiment, a student observes 8 fringes in a certain segment of screen when a monochromatic light of 600 nn wavelength is used. If the wavelength of light is changed to 400 nm, then the number of fringes he would observe in the same region of the screen is **(2022)**
   1) 12
   2) 6
   3) 8
   4) 9

8. An unpolarised light is passed through 3 polarisers. If the second polariser is at an angle 30° with the first and the third polariser is at an angle 60° with the second. Find the final intensity of the light passed through this combination if initial intensity was I_o. **(AIIMS 2018)**
   1) $\frac{3I_0}{32}$
   2) $\frac{3I_0}{8}$
   3) $\frac{3I_0}{16}$
   4) $\frac{5I_0}{16}$

9. If the intensity of light in YDSE is 50% of its maximum at a point. Calculate the path difference. **(AIIMS 2018)**
   1) $\frac{\lambda}{2}$
   2) $\frac{\lambda}{8}$
   3) $\frac{2\lambda}{3}$
   4) $\frac{\lambda}{4}$

10. In a single slit diffraction the distance between slit & screen is 1 m. The size of the slit is 0.7 mm & second maximum is formed at the distance of 2 mm from the centre of the screen, then find out the wavelength of light. **(AIIMS 2018)**
    1) 6500 Å
    2) 5600 Å
    3) 5000 Å
    4) 6000 Å

Answer Key

NCERT Topic-wise MCQs

1	2	3	4	5	6	7	8	9	10	11	12	13	14	15	16	17
1	3	3	2	1	2	4	2	3	3	3	2	2	4	3	3	4
18	**19**	**20**	**21**	**22**	**23**	**24**	**25**	**26**	**27**	**28**	**29**	**30**	**31**	**32**	**33**	**34**
1	1	2	4	1	3	1	4	4	1	2	2	1	4	1	3	4
35	**36**	**37**	**38**	**39**	**40**	**41**	**42**	**43**	**44**	**45**	**46**	**47**	**48**	**49**	**50**	**51**
4	1	4	3	1	3	3	2	3	4	2	4	3	2	3	2	3
52	**53**	**54**	**55**	**56**	**57**	**58**	**59**	**60**	**61**	**62**	**63**	**64**	**65**	**66**	**67**	**68**
2	1	2	1	2	2	4	3	3	2	2	2	2	1	2	4	1

High Order Time Intensive MCQs

1	2	3	4	5	6	7	8	9	10	11	12	13	14	15	16	17
3	1	1	2	4	4	3	4	2	4	3	4	3	3	4	1	2
18	**19**	**20**	**21**	**22**	**23**	**24**	**25**	**26**								
2	1	3	4	2	3	4	3	4								

Statement Based MCQs

1	2	3	4	5
1	1	4	1	1

Assertion & Reason MCQs

1	2	3	4	5
2	3	4	1	1

Match the Column MCQs

1	2	3	4
2	1	3	1

NCERT Exemplar MCQs

1	2	3	4
3	1	3	4

NEET & AIIMS Past Year MCQs

1	2	3	4	5	6	7	8	9	10
4	3	1	1	2	4	1	1	4	2

CHAPTER

11 Dual Nature of Radiation and Matter

Topicwise Number of Questions Analysis (2025-2016)

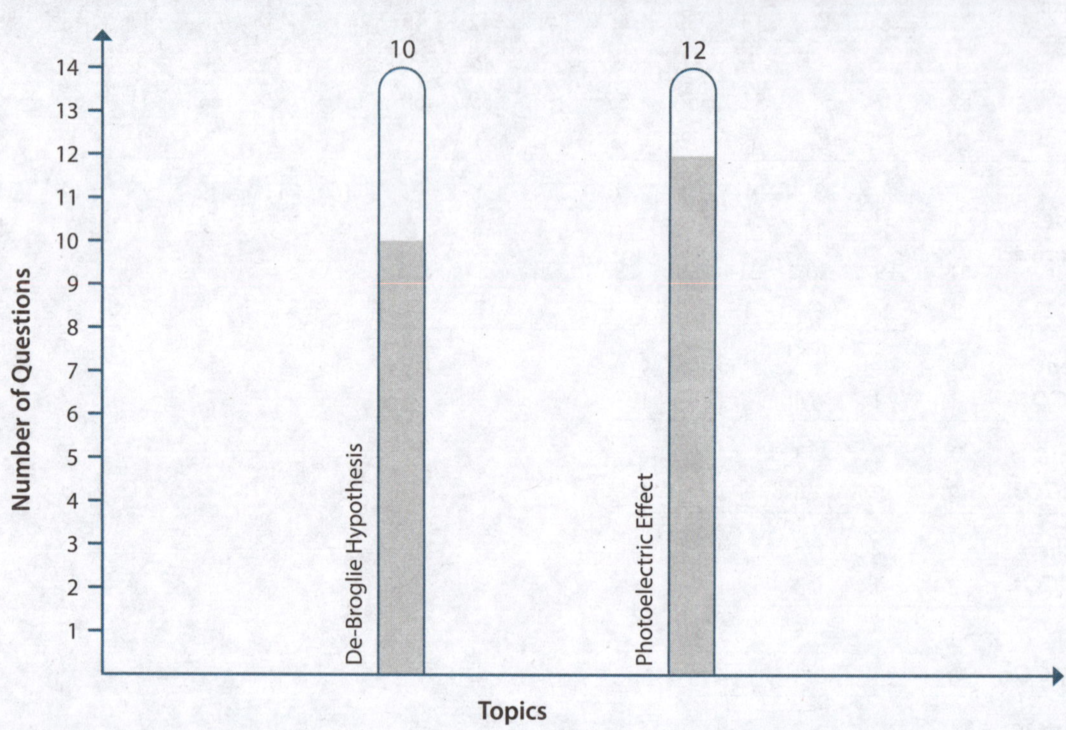

NCERT-Picks

Photoelectric Effect

Formulas

1. Maximum kinetic energy, $K_{max} = eV_0 = \frac{1}{2}mv_{max}^2$

[V_0 is the stopping potential]

2. $K_{max} = h\nu - W_0 = h\nu - h\nu_0$ [W_0 is the work function of metal, ν_0 is threshold frequency]

3. Kinetic energy, $eV_0 = hc\left(\frac{1}{\lambda} - \frac{1}{\lambda_0}\right)$ [λ_0 is threshold wavelength]

4. Ratio of the kinetic energy, $\frac{KE_2}{KE_1} = \frac{h\nu_2 - W_0}{h\nu_1 - W_0}$

[W_0 is the work function of the metal]

5. Energy of the particle, $E = \frac{p^2}{2m}$

[p is the momentum of the particle]

6. Momentum of the particle, $p = \sqrt{2mE}$

Numerical Profile

Graph

Stopping potential vs frequency:

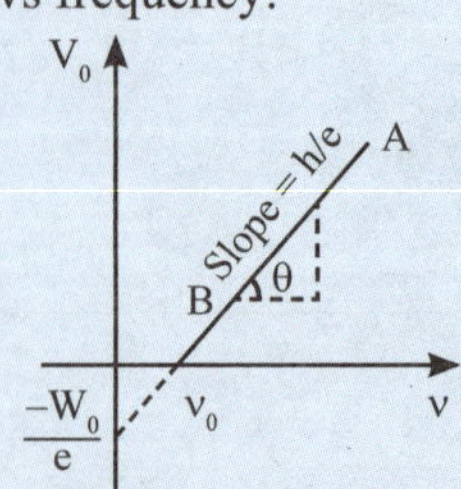

De-Broglie Hypothesis

Formulas

1. Wavelength of photon, $\lambda = \frac{h}{mv}$

   [h is planck's constant, m is the mass and v is the velocity of the particle]

2. Energy of photon, $E = h\nu = \frac{hc}{\lambda}$ [c is the speed of light]

3. Momentum of Photon, $p = \frac{E}{c} = \frac{h\nu}{c}$

   [ν is the frequency of photon]

4. Intensity of Photon,

   $I = \frac{E}{A \times t} = \frac{P}{t}$ [E = energy, P = Power]

5. Number of Photons emitted per second form a source of power P

   $n = \frac{P}{E} = \frac{P}{h\nu}$

6. Wavelength associated with electron $= \frac{h}{\sqrt{2mE}} = \frac{h}{\sqrt{2meV}}$

   [e is the charge of one electron]

Numerical Profile

De-broglie wave or Matter waves

When the material particles like neutrons are in thermal equilibrium at absolute temperature T, then their average kinetic energy is given by, $E = \frac{3}{2}kT$

Where k is Boltzmann's constant whose value is 1.38×10^{-23} joule/K

so that $\lambda = \frac{h}{\sqrt{2mE}} = \frac{h}{\sqrt{3mkT}}$ [T is temperature]

De-broglie Wavelength associated with uncharged particle

(i) For neutrons ($m_n = 1.67 \times 10^{-27}$ kg),

$$\lambda = \frac{h}{\sqrt{2mE}} = \frac{0.286}{\sqrt{E(eV)}} \text{Å} = \frac{1.23}{\sqrt{V}} \text{nm}$$

(ii) For gas molecules wavelength, $\lambda = \frac{h}{m \times v_{rms}}$

Statements Asked in Exam

- Millikan's experiment established that electric charge is quantised and he found the charge on an oil drop.
- The photon-particle collision is elastic collision because the momentum and total energy is conserved.
- If the frequency of incident radiation is equal or greater than that of threshold frequency photoelectric emission starts instantaneously in a time of the order of 10^{-9}s or less.
- Davisson and Germer performed an experiment to study the wave nature of electrons as suggested by the de-Broglie.
- An electron ,an alpha particle and a proton have the same kinetic energy, electron has largest de-Broglie wavelength.
- In an electron microscope a beam of electron is used in place of ordinary light and focussing is done by electrostatic
- Cathode ray was discovered by sir William Crooke. This is the stream of fast moving electrons.
- Matter wave represents the probability of finding a particle in space.
- Matter waves are not electromagnetic waves.
- de-Broglie or matter wave is independent of the charge on the material particle. It means matter wave or de-Broglie wave is associated with every moving particle whether charged or uncharged.
- De-Broglie waves with a moving football not are visible.
- Practical observation of matter waves is possible only when the de-Broglie wavelength is of the order of the size of the particles.
- An electron, an alpha-particle, and a proton have the same kinetic energy, alpha– particle has the shortest de Broglie wavelength.
- Electron microscope works on the phenomena of de-Broglie waves.
- Rest mass of photon is zero
- Work function of metal generally measured in eV.
- Photoelectric current is proportional to the number of photoelectrons emitted per second, so the number of photoelectrons emitted per second is directly proportional to the intensity of incident radiation.

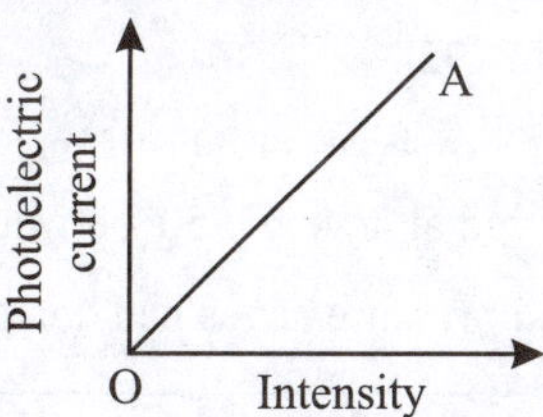

- The saturation current in photoelectric effect experiment with a light of frequency and intensity is independent of the anode potential.
- The minimum negative potential V_0 given to the anode plate E in figure for which the photoelectric current stops or become zero is called cut off or stopping potential.

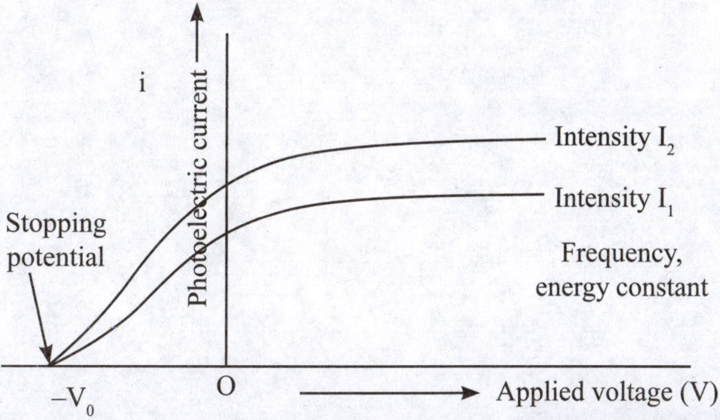

- Stopping potential is independent of the frequency of incident radiation.

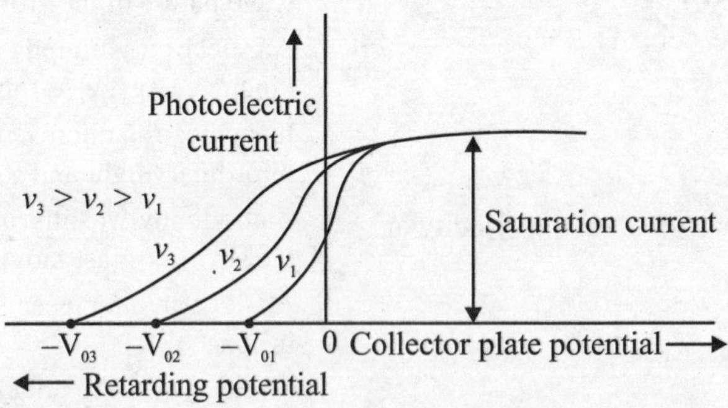

- For a given photosensitive material, there is a certain minimum cut off frequency ν_0 (called threshold frequency) for which the stopping potential is zero.

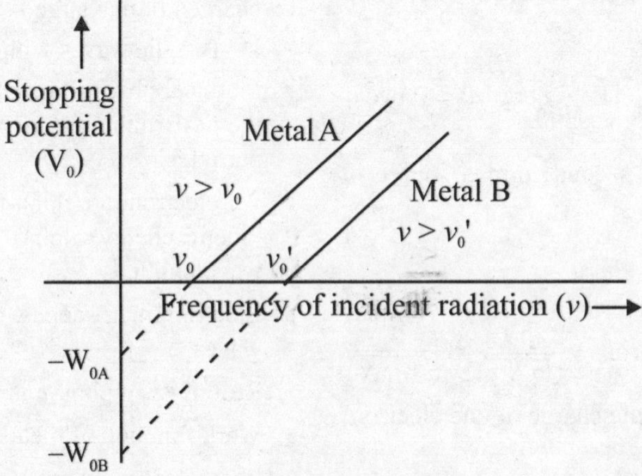

- In 1924, the French physicist Louis Victor de-Broglie (pronounced as de-Broglie) put forward the bold hypothesis of that moving particles of matter should display wave like properties under suitable condition.
- The wave nature of electrons was first experimentally verified by C.J Davisson and L.H Germer in 1927 and independently by G.P Thomson in 1928.
- Davisson and Germer observed diffraction effects with beams of electrons scattered by crystals and this peak diffraction of electrons observed at an angle 50^0 between the electron beam and diffracted electron beam and at 54V.
- Gases at normal pressure act as bad conductors and when subjected to low pressure and high voltage they act as a conductors.

MCQs ?

NCERT Topic-wise MCQs

De-Broglie Hypothesis

1. If the kinetic energy of an electron gets tripled then the de Broglie wavelength associated with it changes by a factor:
   1) 1/3
   2) $\sqrt{3}$
   3) $\frac{1}{\sqrt{3}}$
   4) 3

2. An electron, a proton and a uranium nucleus all have the same wavelength. The one with the most energy:
   1) Is the electron
   2) Is the proton
   3) Is the uranium nucleus
   4) Depends upon the wavelength and the properties of the particle

3. λ_e, λ_p and λ_α are the de-Broglie wavelengths of electron, proton and α particle. If all are accelerated by same potential, then
   1) $\lambda_e < \lambda_p < \lambda_\alpha$
   2) $\lambda_e < \lambda_p > \lambda_\alpha$
   3) $\lambda_e > \lambda_p < \lambda_\alpha$
   4) $\lambda_e > \lambda_p > \lambda_\alpha$

4. What will be the frequency associated with photon of energy 3.3×10^{-20}J?
   1) 6×10^{13} Hz
   2) 5×10^{11} Hz
   3) 5×10^{13} Hz
   4) None of these

5. What will be the numbers of photons emitted per second by a 25 watt source of monochromatic light of wavelength 6000 Å?
   1) 7.86×10^{-19}
   2) 7.55×10^{-16}
   3) 7.00×10^{2}
   4) 7.55×10^{19}

6. The de-Broglie wavelength associated with a material particle when it is accelerated through a potential difference of 150 volt is 1Å. What will be the de-Broglie wavelength associated with the same particle when it is accelerated through a potential difference of 4500V?
   1) $\frac{1}{\sqrt{3}}$Å
   2) $\frac{1}{\sqrt{30}}$Å
   3) $\frac{1}{\sqrt{300}}$Å
   4) $\frac{1}{3}$Å

7. Electrons are accelerated through a potential difference V and protons are accelerated through a potential difference 4V. The de-Broglie wavelength are λ_e and λ_p for electrons and protons respectively.

   The ratio of $\frac{\lambda_e}{\lambda_p}$ is given by (Given, m_e is mass of electrons and m_p is mass of proton).
   1) $\frac{\lambda_e}{\lambda_p} = \sqrt{\frac{m_p}{m_e}}$
   2) $\frac{\lambda_e}{\lambda_p} = \sqrt{\frac{m_e}{m_p}}$
   3) $\frac{\lambda_e}{\lambda_p} = \frac{1}{2}\sqrt{\frac{m_e}{m_p}}$
   4) $\frac{\lambda_e}{\lambda_p} = 2\sqrt{\frac{m_p}{m_e}}$

8. What is the de-Broglie wavelength of an atom of mass m at absolute temperature in T K?
   1) $\frac{h}{\sqrt{3mk_bT}}$
   2) $\frac{h}{\sqrt{3mk_b}}T$
   3) $\frac{hT}{\sqrt[3]{3k_b}}$
   4) $\frac{3K_bT}{\sqrt{h}}$

9. A radio transmitter operates at a frequency of 880 kHz and a power of 10 kW. Find the number of photons emitted per second.
   1) 2.50×10^{31}
   2) 1.72×10^{31}
   3) 2.90×10^{16}
   4) 2.71×10^{5}

10. Aparallelbeamofmonochromaticlightofwavelength660nm is incident on a totally reflecting plane mirror. The angle of incidence is 60^0 and the number of photons striking the mirror per second is 2×10^{19}. What will be the force exerted by the light beam on the mirror?
    1) 3×10^{8} N
    2) 3×10^{-8} N
    3) 9×10^{3} N
    4) 2×10^{-8} N

11. A proton is accelerated through one volt the increase in its kinetic energy is approximately:
    1) $\frac{1}{1870}$eV
    2) 1870 eV
    3) 1 eV
    4) 300 eV

12. An electron of mass m when accelerated through a potential difference V, has de-Broglie wavelength λ. The de-Broglie wavelength associated with a proton of mass M accelerated through the same potential difference will be:
    1) $\frac{\lambda m}{M}$
    2) $\lambda\sqrt{\frac{m}{M}}$
    3) $\frac{\lambda M}{m}$
    4) $\lambda\sqrt{\frac{M}{m}}$

13. Which of the following statement about photon is incorrect?
 1) Photons exerts no pressure
 2) Momentum of photon is $\frac{h\nu}{c}$
 3) Rest mass of photon is zero
 4) Energy of photon is hν

14. If m is the mass of an electron and c is the speed of light the ratio of the wavelengths of photon of energy E to that of the electron of the same energy:
 1) $c\sqrt{\frac{2m}{E}}$ 2) $\sqrt{\frac{2m}{E}}$
 3) $\sqrt{\frac{2m}{cE}}$ 4) $\sqrt{\frac{m}{E}}$

15. A proton and α-particle are accelerated through the same potential difference. The ratio of the de-Broglie wavelength λ_p to that λ_α is:
 1) $\sqrt{2}:1$ 2) $2:1$
 3) $\sqrt{6}:1$ 4) $\sqrt{8}:1$

16. The de-Broglie wavelength of a particle of kinetic energy K is λ. What will be the wavelength of the particle if its kinetic energy is K/4?
 1) λ 2) 2 λ
 3) $\frac{3\lambda}{4}$ 4) $\frac{5\lambda}{2}$

17. There are two sources of light each emitting a power of 100 W, one emits X-ray of wavelength 1 nm and the other visible light of 500 nm. The ratio of numbers of photons of X-rays to the photons of visible light of the given wavelength is:
 1) 1 : 500 2) 1 : 200
 3) 1 : 1000 4) 1 : 10

18. If the kinetic energy of the particle is increased by 16 times, the percentage change in the de-Broglie wavelength of the particle is:
 1) 25% 2) 75%
 3) 60% 4) 50%

19. Given below is the graph between frequency (ν) of the incident light and maximum kinetic energy (E_k) of the emitted photoelectrons. Find the value of work function from the graph.

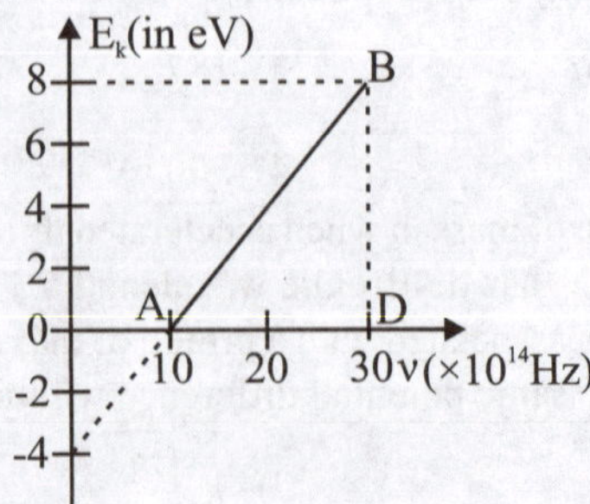

 1) 6 eV 2) 4 eV
 3) 3 eV 4) 2 eV

20. Photoelectric effect supports quantum nature of light because
 1) The energy of released electron is discrete
 2) The maximum kinetic energy of photoelectrons depends only on the frequency of light and not on its intensity
 3) Even when the metal surface is faintly illuminated, the photoelectrons leave the surface immediately
 4) Electric charge of the photoelectrons is quantized

21. The de-Broglie wavelength of a particle moving with a velocity 2.25×10^8 m/s is equal to the wavelength of photon. The ratio of kinetic energy of the particle to the energy of the photon is:
 1) 3/8 2) 5/8
 3) 7/8 4) 5/9

22. The kinetic energy of an electron with de-Broglie wavelength of 0.3 nanometer is:
 1) 0.168 eV 2) 16.8 eV
 3) 1.68 eV 4) 2.5 eV

23. Maximum kinetic energy of a photoelectron is E when the wavelength of incident light is. If energy becomes four times when wavelength is reduced to one-third, then work function of the metal is
 1) $\frac{3hc}{\lambda}$ 2) $\frac{hc}{3\lambda}$
 3) $\frac{hc}{\lambda}$ 4) $\frac{hc}{2\lambda}$

24. The de-Broglie wavelength of a neutron at 27°C is λ, what will be the wavelength at 927°C?
 1) λ/2 2) λ/8
 3) λ/10 4) λ/5

25. If a proton and electron have the same de-Broglie wavelength then:
 1) Kinetic energy of electron < Kinetic energy of proton
 2) Kinetic energy of electron = Kinetic energy of proton
 3) Momentum of electron > Momentum of proton
 4) Momentum of electron = Momentum of proton

26. A particle A of mass m and initial velocity v collides with a particle B of mass $\frac{m}{2}$ which is at rest. The collision is head on, and elastic. The ratio of the de-Broglie wavelength λ_A to λ_B after the collision is
 1) $\frac{\lambda_A}{\lambda_B} = 2$ 2) $\frac{\lambda_A}{\lambda_B} = \frac{2}{3}$
 3) $\frac{\lambda_A}{\lambda_B} = \frac{1}{2}$ 4) $\frac{\lambda_A}{\lambda_B} = \frac{1}{3}$

27. An electron beam is accelerated by a potential difference V to hit a metallic target to produce X-rays. It produces continuous as well as characteristic X-rays. If λ_{min} is the smallest possible wavelength of X-rays in the spectrum, the variation of log λ_{min} with log V is correctly represented in

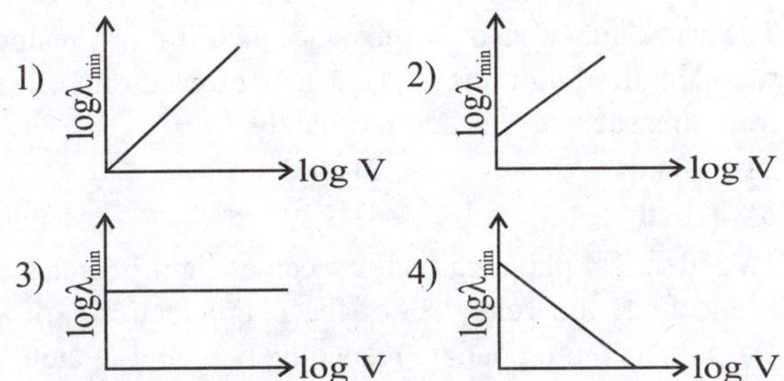

28. A beam of white light is incident normally on a plane surface absorbing 70% of the light and reflecting the rest. If the incident beam of light is of power P, what will be the force exerted by it on the surface?

1) $\frac{13P}{100c}$ 2) $\frac{130P}{10c}$
3) $\frac{13P}{10c}$ 4) $\frac{13c}{10P}$

29. The ratio of de-Broglie wavelength of molecules of hydrogen and helium in two gas jars kept separately at temperatures of 27°C and 127°C respectively is

1) $2/\sqrt{3}$ 2) 2/3
3) $\sqrt{3}/8$ 4) $\sqrt{8/3}$

Photoelectric Effect

30. Radiation of wavelength λ, is incident on a photocell. The fastest emitted electron has speed v. If the wavelength is changed to $\frac{3\lambda}{4}$, the speed of the fastest emitted electron will be

1) $> v\left(\frac{4}{3}\right)^{1/2}$ 2) $< v\left(\frac{4}{3}\right)^{1/2}$
3) $= v\left(\frac{4}{3}\right)^{1/2}$ 4) $= v\left(\frac{3}{4}\right)^{1/2}$

31. The threshold wavelength for photoelectric effect of a metal is 6500Å. The work function of the metal is approximately:

1) 2 eV 2) 1 eV
3) 0.1 eV 4) 3 eV

32. The photoelectric cut off voltage in a certain experiment is 1.5V. The maximum kinetic energy of photoelectrons emitted is:

1) 2.4 eV 2) 1.5 eV
3) 3.1 eV 4) 4.5 eV

33. In photoelectric effect the photo current:

1) Depends both on intensity and frequency of the incident light
2) Does not depend on the frequency of incident light
3) Decrease with increase in frequency of incident light
4) Increase with increase in frequency of incident light

34. When a surface 1 cm thick is illuminated with light of wavelength λ, the stopping potential is V_0. But when the same surface is illuminated by light of wavelength 3λ, the stopping potential is $V_0/6$, the threshold wavelength for metallic surface is

1) 4λ 2) 3λ
3) 5λ 4) 2λ

35. The work function for Al, K and Pt is 4.28eV, 2.30eV, 5.65eV respectively. Their respective threshold frequencies would be:

1) Pt > Al > K 2) Al > Pt > K
3) K > Al > Pt 4) Al > K > Pt

36. Which of the following statement is correct regarding the photoelectric experiment?

1) The photocurrent increases with intensity of light.
2) Stopping potential increases with increases in intensity of incident light.
3) The photo current increases with increases in frequency.
4) All of these

37. The photoelectric threshold frequency of a metal is ν. when light of frequency 4ν is incident on the metal, the maximum kinetic energy of the emitted photoelectron is:

1) 4 hν 2) 3 hν
3) 5 hν 4) $\frac{5h\nu}{2}$

38. Light of wavelength 0.6 μm from a sodium lamp falls on a photo cell and causes the emission of photo electrons for which the stopping potential is 0.5V. With light of wavelength 0.4μm from a sodium lamp, the stopping potential is 1.5V. With this data the value of h/e is:

1) 4×10^{-59} Vs 2) 0.25×10^{15} Vs
3) 4×10^{-15} Vs 4) 4×10^{-8} Vs

39. If the energy of a photon is 25 eV and work function of the material is 7 eV, then find the value of stopping potential:

1) 18V 2) 20V
3) 22V 4) 24V

40. Photons of energy 1eV and 3eV respectively are made incident on a metallic plate of work function 0.5eV one after the other. Find the ratio of maximum kinetic energy of photoelectrons emitted by them:

1) 1 : 10 2) 2 : 15
3) 1 : 5 4) 4 : 5

41. Calculate maximum kinetic energy of electrons emitted from a photo sensitive surface of work function 3.2eV, for the incident radiation of wavelength 300nm:

1) 0.1 eV 2) 0.2 eV
3) 0.5 eV 4) 0.925 eV

42. Find the difference of kinetic energy of photoelectrons emitted from a surface by light of wavelengths 2500Å and 5000Å:
1) 3.96×10^{-19} J
2) 4×10^{-9} J
3) 3×10^{9} J
4) 3 KJ

43. The threshold frequency of a certain metal is 3.3×10^{14} Hz. If light of frequency 8.2×10^{24} Hz is incident on the metal, then the cut off voltage for photoelectric emission is:
1) 2V
2) 4V
3) 6V
4) 8V

44. For a certain metal incident frequency v is five times of threshold frequency v_O and the maximum velocity of coming out photoelectrons is 8×10^6 ms^{-1}. If $v = 2v_o$,the maximum velocity of photoelectrons will be:
1) 4×10^6 m/s
2) 8×10^6 m/s
3) 4×10^{-6} m/s
4) 8 m/s

45. A photon of energy E ejects a photoelectrons from a metal surface whose work function is ϕ_o. If this electron enters into a uniform magnetic field B in a direction perpendicular to the field and describes a circular path of radius r then the radius r is:
1) $\frac{\sqrt{2m(E-\phi_o)}}{eB}$
2) $\sqrt{\frac{2m(E-\phi_o)}{eB}}$
3) $\sqrt{2m(E-\phi_o)eB}$
4) $\sqrt{\frac{2e(E-\phi_o)}{mB}}$

46. The photoelectric threshold wavelength for silver is λ_O, The energy of the electron ejected from the surface of silver by an incident wavelength λ ($\lambda < \lambda_O$) will be:
1) hc $(\lambda - \lambda_O)$
2) hc $(\lambda_O - \lambda)$
3) $\frac{hc}{\lambda_O - \lambda}$
4) $hc\left[\frac{\lambda_O - \lambda}{\lambda\lambda_O}\right]$

47. In Davisson and Germer experiment, the tungsten filament is coated with:
1) Aluminum oxide
2) Barium chloride
3) Titanium oxide
4) Barium oxide

48. The curve drawn between velocity and frequency of a photon in vacuum will be:
1) Straight line parallel to frequency axis
2) Straight line parallel to velocity axis
3) Straight line passing through origin and making an angle of 45^0 with frequency axis
4) Hyperbola

49. The work function of a metal is 4.2 eV, its threshold wavelength will be:
1) 4000 Å
2) 3500 Å
3) 2946 Å
4) 2500 Å

50. The work function for tungsten and sodium are 4.5 eV and 2.3 eV respectively. If the threshold wavelength λ for sodium is 5460Å, the value of λ for tungsten is:
1) 5893 Å
2) 10683 Å
3) 2791 Å
4) 528 Å

51. The work function of a metallic surface is 5.01 eV. The photo electrons are emitted when light of wavelength 2000Å falls on it. The potential difference applied to stop the fastest photo-electrons is:
1) 1.2V
2) 2.24V
3) 3.6V
4) 4.8V

52. The work function of a substance is 4.0 eV. The longest wavelength of light that can cause photoelectrons emission from this substance is approximately.
1) 540 nm
2) 400 nm
3) 310 nm
4) 220 nm

53. Two identical photo cathodes receives light of frequencies f_1 and f_2. If the velocities of the photoelectrons (of same mass 'm') coming out are respectively v_1 and v_2 then:
1) $v_1^2 - v_2^2 = \frac{2h}{m}(f_1 - f_2)$
2) $v_1 + v_2 = \left(\frac{2h}{m}(f_1 + f_2)\right)^{1/2}$
3) $v_1^2 + v_2^2 = \frac{2h}{m}(f_1 + f_2)$
4) $v_1 - v_2 = \left(\frac{2h}{m}(f_1 + f_2)\right)^{1/2}$

54. Maximum velocity of the photoelectrons emitted by a metal surface is 1.2×10^6 m/s. Assuming the specific charge of the electrons to be 1.8×10^{11} C Kg^{-1}, the value of the stopping potential in volts will be:
1) 2
2) 3
3) 4
4) 6

55. Threshold wavelength for photoelectric effect on aluminum is 5000Å. Its work function is:
1) 15J
2) 16×10^{-14}J
3) 4×10^{-19}J
4) 16×10^{-81}J

56. The photoelectric threshold wavelength of a certain metal is 3000Å. If the radiation of 2000Å is incident on the metal.
1) Electrons will be emitted
2) Positron will be emitted
3) Protons will be emitted
4) Electrons will not be emitted

57. A photocell stops emission if it is maintained at 2V negative potential. The energy of most energetic photoelectrons is:
1) 2 eV
2) 2J
3) 2 KJ
4) 2 KeV

58. When the energy of the incident radiation is increased by 20% the kinetic energy of the photoelectrons emitted from a metal surface increased from a 0.5eV to 0.8eV. The work function of the metal is:
1) 0.65 eV
2) 1 eV
3) 1.3 eV
4) 1.5 eV

59. Photoelectric emission is observed from a metallic surface for frequencies v_1 and v_2 of the incident light ($v_1 > v_2$). If the maximum value of kinetic energy of the photoelectrons emitted in the two cases are in the ratio 1 : n then the threshold frequency of the metallic surface is:
1) $\frac{v_1 - v_2}{n-1}$
2) $\frac{n v_1 - v_2}{n-1}$
3) $\frac{n v_2 - v_1}{n-1}$
4) $\frac{v_1 - v_2}{n}$

60. The threshold frequency for a certain photosensitive metal is v_0. When it is illuminated by light of frequency $v = 2v_0$, the maximum velocity of photoelectrons is V_0. What will be the maximum velocity of the photoelectrons when the same metal is illuminated by light of frequency $v = 5v_0$?

1) $\sqrt{2}V_o$ 2) $2V_o$
3) $2\sqrt{2}V_o$ 4) $4V_o$

61. Work function of nickel is 5.01 eV. When ultraviolet radiation of wavelength 2000Å is incident on it electrons are emitted. What will be the maximum velocity of emitted electrons?

1) 3×10^8 m/s 2) 6.46×10^5 m/s
3) 10.36×10^5 m/s 4) 8.54×10^6 m/s

62. Radiation of two photons energies twice and five times the work function of metal are incident successively on the metal surface. The ratio of the maximum velocity of the photoelectrons emitted in the two cases will be:

1) 1 : 1 2) 1 : 2
3) 1 : 3 4) 1 : 4

63. When a piece of metal is illuminated by monochromatic light of wavelength λ then the stopping potential for photoelectric current is 2.5 V_0. When the same surface is illuminated by light of wavelength 1.5 λ, then the stopping potential becomes V_0. Find the value of threshold wavelength for photoelectric emission.

1) 2.5 λ 2) 2.25 λ
3) 3.25 λ 4) 3.5 λ

64. Which type of collision is photon-electron collision in photoelectric effect?

1) Inelastic 2) Perfectly inelastic
3) Oblique 4) Elastic

65. The velocity of the most energetic electrons emitted from a metallic surface is doubled when the frequency **v** of the incident radiation is doubled. The work function of this metal is:

1) $\frac{hv}{4}$ 2) $\frac{hv}{3}$
3) $\frac{hv}{2}$ 4) $\frac{2hv}{3}$

66. A light of wavelength 5000Å falls on a sensitive plate with photoelectric work function 1.90eV, kinetic energy of the emitted electrons will be:

1) 0.1 eV 2) 2 eV
3) 0.58 eV 4) 1.581 eV

67. A photo sensitive surface is receiving light of wavelength 5000Å at the rate of 10^{-7} Js^{-1}. The number of photons ejected per second is:

1) 2.5×10^{12} 2) 2.5×10^{11}
3) 2.5×10^{10} 4) 2.5×10^9

68. Maximum velocity of the photo electrons emitted by a metal surface is 1.2×10^6 ms^{-1}. Assuming the specific charge of the electron to be 1.8×10^{11} C Kg^{-1}, the value of the stopping potential in volts will be:

1) 2 2) 3
3) 4 4) 6

High Order Time Intensive MCQs

1. The work function of a surface of a photo-sensitive material is 6.2 eV. The wavelength of the incident radiation for which the stopping potential is 5 V lies in the

1) infrared region 2) X-ray region
3) ultraviolet region 4) visible region

2. The moving proton and α particle are subjected to the same magnetic field so that the radii of their paths are equal to each other. Assuming the field induction B is perpendicular to the velocity vector of the α-particle and proton, the ratio of de-Broglie wavelength of α-particle to that of proton is

1) 1/4 2) 1/2
3) 1 4) 2

3. In an experiment on photoelectric effect, a student plots stopping potential V_0 against frequency v of the incident light for two different metals A and B. These are as shown in the figure.

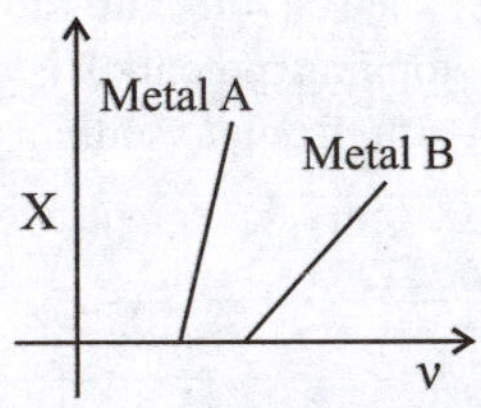

Looking at the graphs, you can most appropriately say that

1) Work function of metal B is greater than that of metal A
2) Work function of metal A is greater than that of metal B
3) Students data in not correct
4) None of the above

4. Photon and electron are given same energy (10^{-20} J) wavelength associated with photon and electron are λ_p and λ_e, the correct statement will be:

1) $\lambda_p > \lambda_e$ 2) $\lambda_p < \lambda_e$
3) $\lambda_p = \lambda_e$ 4) $\frac{\lambda_e}{\lambda_p} = c$

5. Light of two different frequencies whose photons have energies 1 eV and 2.5 eV respectively, illuminate a metallic surface whose work function is 0.5 eV successively. Ratio of maximum speeds of emitted electrons will be:

1) 1 : 2 2) 1 : 1
3) 1 : 5 4) 1 : 4

6. The de-Broglie wavelength λ of an electron accelerated through a potential V (in volts) is:

1) $\frac{1.227}{\sqrt{V}}$ nm 2) $\frac{0.1227}{\sqrt{V}}$ m

3) $\frac{0.01227}{\sqrt{V}}$ nm 4) $\frac{\sqrt{V}}{0.1227}$ nm

7. The de-Broglie wavelength of an electron and the wavelength of a photon are the same. The ratio between the energy of the photon and the momentum of the electron is (c = velocity of light, h = Planck's constant):

1) h 2) c

3) $\frac{1}{h}$ 4) $\frac{1}{c}$

8. Light of wavelength λ_A and λ_B falls on two identical metal plates A and B respectively. The maximum kinetic energy of photoelectrons is K_A and K_B respectively, then which one of the following relations is true? ($\lambda_A = 2\lambda_B$)

1) $K_A < \frac{K_B}{2}$ 2) $2K_A = K_B$

3) $K_A = 2K_B$ 4) $K_A > 2K_B$

9. For photoelectric emission for certain metal, the cut-off frequency is v. If radiation of frequency $2v$ impinges on the metal plate, the maximum possible velocity of the emitted electron will be (m is the electron mass):

1) $\sqrt{\frac{hv}{(2m)}}$ 2) $\sqrt{\frac{hv}{m}}$

3) $\sqrt{\frac{2hv}{m}}$ 4) $2\sqrt{\frac{hv}{m}}$

10. A 200 W sodium street lamp emits yellow light of wavelength 0.6 μm. Assuming it to be 25% efficient in converting electrical energy to light, the number of photon of yellow light it emits per second is:

1) 1.5×10^{20} 2) 6×10^{18}

3) 6.2×20^{20} 4) 3×10^{19}

11. In a photoelectric effect measurement, the stopping potential for a given metal is found to be V_0 volt when radiation of wavelength λ_0 is used. If radiation of wavelength $2\lambda_0$ is used with the same metal, then the stopping potential (in volt) will be

1) $\frac{V_0}{2}$ 2) $2\,V_0$

3) $V_0 + \frac{hc}{2e\lambda_0}$ 4) $V_0 - \frac{hc}{2e\lambda_0}$

12. If 5% of the energy supplied to a bulb is irradiated as visible light, how many quanta are emitted per second by a 100 W lamp? (Assume, wavelength of visible light as 5.6×10^{-5} cm):

1) 1.4×10^{19} 2) 3×10^{3}

3) 1.4×10^{-19} 4) 3×10^{4}

13. When 1 cm thick surface is illuminated with light of wavelength λ, the stopping potential is V. When the same surface is illuminated by light of wavelength 2λ, the stopping potential is $V/3$. Threshold wavelength for metallic surface is:

1) $4\lambda/3$ 2) 4λ

3) 6λ 4) $8\lambda/3$

14. Light of wavelength λ, strikes a photoelectric surface and electrons are ejected with an energy E. If E is to be increased to exactly twice its original value, the wavelength changes to λ', where:

1) λ' is less than $\frac{\lambda}{2}$

2) λ' is greater than $\frac{\lambda}{2}$

3) λ' is greater than $\frac{\lambda}{2}$ but less than λ

4) λ' is exactly equal to $\frac{\lambda}{2}$

15. Monochromatic light of frequency f is incident on emitter having threshold frequency f_0. The kinetic energy of ejected electron will be:

1) hf 2) $h(f - f_0)$

3) hf_0 4) $h(f + f_0)$

16. The energy that should be added to an electron, to reduce its de-Broglie wavelength from 2×10^{-9} m to 0.5×10^{-9} m will be

1) 1.1 MeV 2) 0.56 MeV

3) 0.56 keV 4) 5.6 eV

17. The threshold frequency for a certain metal is v_0. When a certain radiation of frequency $2v_0$ is incident on this metal surface the maximum velocity of the photoelectrons emitted is 2×10^6 ms^{-1}. If a radiation of frequency $3v_0$ is incident on the same metal surface, the maximum velocity of the photoelectrons emitted (in ms^{-1}) is:

1) 2×10^6 2) $2\sqrt{2} \times 10^6$

3) $4\sqrt{2} \times 10^6$ 4) $4\sqrt{3} \times 10^6$

18. A 5 W source emits monochromatic light of wavelength 5000 Å. When placed 0.5 m away, it liberates photoelectrons from a photosensitive metallic surface. When the source is moved to a distance of 1 m, the number of photoelectrons liberated will be reduced by a factor of:

1) 4 2) 8

3) 16 4) 2

19. Light of wavelength 0.6 μm from a sodium lamp falls on a photocell and caused the emission of photoelectrons for which the stopping potential is 0.5 V with light of wavelength 0.4 μm from a sodium lamp, the stopping potential is 1.5 V, with this data the value of $\frac{h}{e}$ is:

1) 4×10^{-19} V-s 2) 0.25×10^{-15} V-s

3) 4×10^{-15} V-s 4) 4×10^{-8} V-s

20. A free particle with initial kinetic energy E, de-Broglie wavelength λ, enters a region where in it has a potential energy V, what is the new de-Broglie wavelength?

1) $\lambda(1 - V/E)$ 2) $\lambda(1 + E/V)$

3) $\lambda/(1 - V/E)^{1/2}$ 4) $\lambda/(1 + V/E)^2$

21. The threshold wavelength for a metal having work function W_0 is λ_0. What is the threshold wavelength for a metal whose function is $\frac{W_0}{2}$?

1) $4\lambda_0$ 2) $2\lambda_0$

3) $\frac{\lambda_0}{2}$ 4) $\frac{\lambda_0}{4}$

22. If K_1 and K_2 are maximum kinetic energies of photoelectrons emitted when lights of wavelength λ_1 and λ_2 respectively incident on a metallic surface. If $\lambda_1 = 3\lambda_2$, then:

1) $K_1 > (K_2 / 3)$ 2) $K_1 < (K_2 / 3)$

3) $K_1 = 3K_2$ 4) $K_2 = 3K_1$

23. Monochromatic light frequency f_1 incident on a photocell and the stopping potential is found to be V_1. What is the new stopping potential of the cell, if the is radiated by monochromatic light of frequency f_2?

1) $V_1 - \frac{h}{e}(f_2 - f_1)$ 2) $V_1 + \frac{h}{e}(f_1 + f_2)$

3) $V_1 - \frac{h}{e}(f_1 + f_2)$ 4) $V_1 + \frac{h}{e}(f_2 - f_1)$

24. A photosensitive metallic surface has work function $h\nu_0$. If photons of energy $2h\nu_0$ fall on this surface, the electrons come out with a maximum velocity of 4×10^6 ms^{-1}. When the photon energy is increased to $5h\nu_0$, then maximum velocity of photoelectrons will be:

1) 2×10^6 ms^{-1} 2) 2×10^7 ms^{-1}

3) 8×10^5 ms^{-1} 4) 8×10^6 ms^{-1}

25. The graph between $KE\left(\frac{1}{2}mv^2_{max}\right)$ of ejected photoelectrons and the frequency ν of incident radiation for a material exhibiting photoelectric effect is shown in the figure. The work function of the material is:

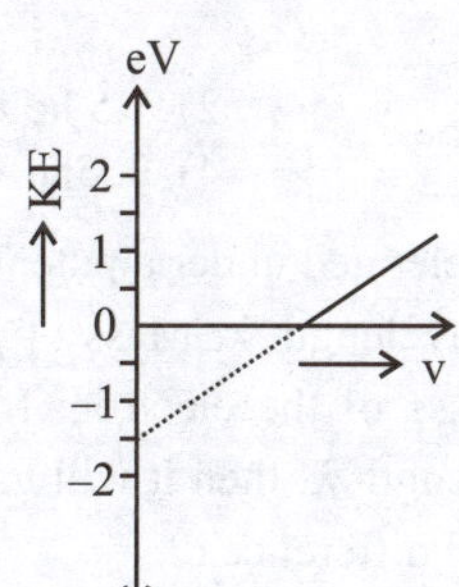

1) 1.5 eV 2) 1 eV

3) 3 eV 4) 2 eV

26. Light of wavelength 4000 Å is incident on a sodium surface for which the threshold wavelength of photoelectrons is 5420 Å. The work function of sodium is:

1) 4.58 eV 2) 2.29 eV

3) 1.17 eV 4) 0.57 eV

27. The PE of a particle of mass m is given by

$$U(x) = \begin{cases} E_0 & 0 \le x \le 1 \\ 0 & x = 1 \end{cases}$$

λ_1 and λ_2 an the de-Broglie wavelengths of the particle, when $0 \le x \le 1$ and $x > 1$ respectively. If the total energy of particle is $2E_0$, the ratio $\frac{\lambda_1}{\lambda_2}$ will be:

1) 2 2) 1

3) $\sqrt{2}$ 4) $\frac{1}{\sqrt{2}}$

28. The electric field of light wave is given as $E = 10^{-3}\cos\left(\frac{2\pi x}{5\times10^{-7}} - 2\pi\times6\times10^{14}t\right)\hat{x}NC^{-1}$. This light falls on a metal plate of work function 2 eV. The stopping potential of the photoelectrons is

Given, $E(\text{in eV}) = \frac{12375}{\lambda(\text{in Å})}$

1) 0.48 V 2) 0.72 V

3) 2.0 V 4) 2.48 V

29. Photoelectric effect experiment are per- formed using three different metal plates p, q and r having work function, $\phi_p = 2.0$ eV, $\phi_q = 2.5$ eV and $\phi_r = 3.0$ eV respectively. A light beam containing wavelength of 550 nm, 450 nm and 350 nm with equal intensities illuminates each of the plates. The correct I-V graph for the experiment is

(Take hc = 1240 eV nm)

1)

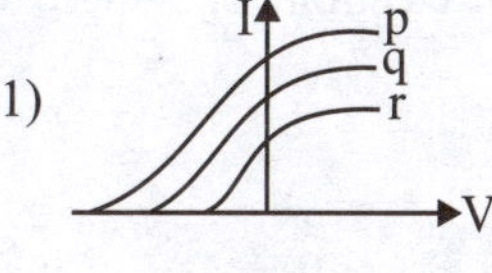

2)

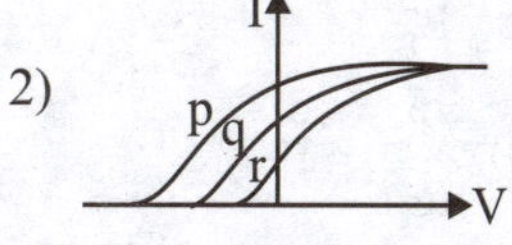

3)

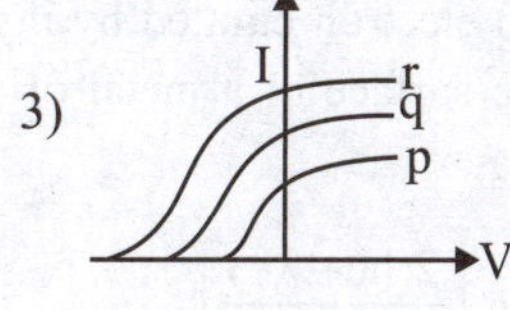

4)

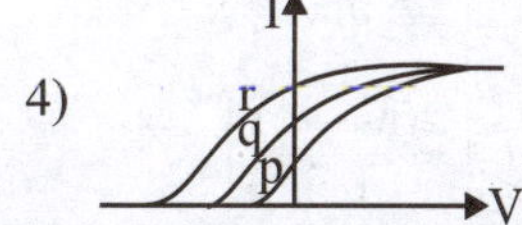

30. In a photoelectric effect experiment, the threshold wavelength of light is 380 nm. If the wavelength of incident light is 260 nm, the maximum kinetic energy of emitted electrons will be

Given, $E(\text{in eV}) = \frac{1237}{\lambda(\text{in nm})}$

1) 15.1 eV 2) 3.0 eV
3) 1.5 eV 4) 4.5 eV

31. The threshold frequency for a metallic surface corresponds to an energy of 6.2 eV and the stopping potential for a radiation incident on this surface is 5V. The incident radiation lies in

1) ultraviolet region 2) infrared region
3) visible region 4) X-ray region

32. The adjoining figure shows the variation of photo-current with anode potential for a photosensitive surface for three different radiations. Let I_a, I_b and I_c be the intensities and f_a, f_b and f_c be the frequencies for the curves *a, b* and *c* respectively. Then

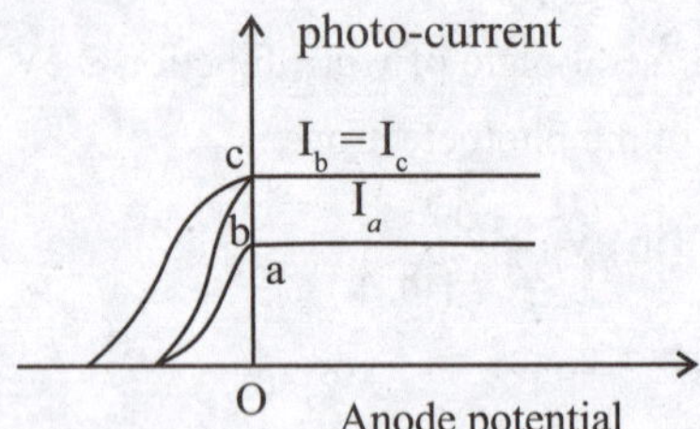

1) $f_a = f_b$ and $I_a = I_b$ 2) $f_a = f_b$ and $I_a \neq I_b$
3) $f_a = f_c$ and $I_a = I_c$ 4) $f_b = f_c$ and $I_b = I_c$

33. An α-particle moves in a circular path of radius 0.83 cm in the presence of a magnetic field of 0.25 Wb/m². The de-Broglie wavelength associated with the particle will be :

1) 1 Å 2) 0.1Å
3) 10 Å 4) 0.01 Å

34. The de-Broglie wavelength of a photon is twice the de-Broglie wavelength of an electron. The speed of the electron is $v_e = \frac{c}{100}$. Then:

1) $\frac{E_e}{E_p} = 10^{-4}$ 2) $\frac{E_e}{E_p} = 10^{-2}$

3) $\frac{P_e}{m_e c} = 10^{-1}$ 4) $\frac{P_e}{m_e c} = 10^{-4}$

35. The maximum velocity of an electron emitted by light of wavelength λ incident on the surface of a metal of work function ϕ is:

1) $\left[\frac{2(hc+\lambda\phi)}{m\lambda}\right]^{1/2}$ 2) $\left[\frac{2(hc-\lambda\phi)}{m\lambda}\right]^{1/2}$

3) $\frac{2(hc-\lambda\phi)}{m}$ 4) $\left[\frac{2(h\lambda-\phi)}{m}\right]^{1/2}$

36. When a centimeter thick surface is illuminated with light of wavelength λ, stopping potential is V. When the same surface is illuminated by light of wavelength 2λ, stopping potential is $\frac{V}{3}$. Threshold wavelength for metallic surface is:

1) $\frac{4\lambda}{3}$ 2) 4λ
3) 6λ 4) $\frac{8\lambda}{3}$

37. In a photoelectric experiment, the wavelength of the light incident on a metal is changed from 300 nm to 400 nm. The decrease in the stopping potential is close to $\left(\frac{hc}{e} = 1240 \text{ nmV}\right)$

1) 0.5 V 2) 2.0 V
3) 1.5 V 4) 1.0 V

38. The energy of photon is equal to the kinetic energy of a proton. The energy of photon is E. Let λ_1 be the de-Broglie wavelength of the proton and λ_2 be the wavelength of photon. The ratio λ_1/λ_2 is proportional to

1) E^0 2) $E^{1/2}$
3) E^{-1} 4) E^{-2}

39. A particle of mass M at rest decays into two particles of masses m_1 and m_2, having non-zero velocities. The ratio of the de-Broglie wavelengths of the particles, λ_1/λ_2, is

1) m_1/m_2 2) m_2/m_1
3) 1.0 4) $\sqrt{m_2}/\sqrt{m_1}$

40. The work functions of metals A and B are in the ratio 1 : 2. If light of frequencies f and 2f are incident on the surfaces of A and B respectively, the ratio of the maximum kinetic energies of photoelectrons emitted is (f is greater than threshold frequency of A, 2f is greater than threshold frequency of B)

1) 1 : 1 2) 1 : 2
3) 1 : 3 4) 1 : 4

41. When a metal surface is illuminated by light of wavelengths 400 nm and 250 nm, the maximum velocities of the photoelectrons ejected are v and 2v respectively. The work function of the metal is (h = Planck's constant, c = velocity of light in air)

1) $2hc \times 10^6$ J 2) $1.5\ hc \times 10^6$ J
3) $hc \times 10^6$ J 4) $0.5hc \times 10^6$ J

42. An electron accelerated under a potential difference V volt has a certain wavelength λ. Mass of proton is some 1840 times of the mass of the electron. If proton has to have the same wavelength λ, then it will have to be accelerated under a potential difference of

1) V volt 2) 1840 V volt
3) $\frac{V}{1840}$ volt 4) $\sqrt{1840}$ volt

43. A beam of electrons is used in a Young double slit experiment. The slit width is d. When the velocity of electron is increased, then
    1) No interference is observed
    2) Fringe width increases
    3) Fringe width decreases
    4) Fringe width remains same.

44. The de-Broglie wavelength associated with proton changes by 0.25% if its momentum is changed by p_0. The initial momentum was
    1) $100\ p_0$ 2) $p_0/400$
    3) $401\ p_0$ 4) $p_0/100$

45. If E_1 and E_2 are maximum kinetic energies of photoelectrons emitted when light of wavelength 2λ and λ respectively incident on a metallic surface. Then
    1) $E_1 > (E_2/2)$ 2) $E_1 < (E_2/2)$
    3) $E_1 = 2E_2$ 4) $E_2 = 2E_1$

46. e/m of electrons is 1.76×10^{11} C/kg and the stopping potential is 0.71 V, then the maximum velocity of photoelectrons is
    1) 150 km s^{-1} 2) 200 km s^{-1}
    3) 500 km s^{-1} 4) 250 km s^{-1}

Statement Based MCQs

Directions: These questions consist of two statements each, printed as Statement-I and Statement-II. While answering these questions, you are required to choose any one of the following four responses.

1) **Both Statement-I and Statement-II are correct.**
2) **Both Statement-I and Statement-II are incorrect.**
3) **Statement-I is correct & Statement-II is incorrect.**
4) **Statement-I is incorrect & Statement-II is correct.**

1. **Statement-I:** Though light of a single frequency (monochromatic) is incident on a metal, the energies of emitted photoelectrons are different.

   Statement-II: The energy of electrons emitted from inside the metal surface, is lost in collision with the other atoms in the metal.

2. **Statement-I:** A photon has no rest mass, yet it carries definite momentum.

   Statement-II: Momentum of photon is due to its energy and hence its equivalent mass.

3. **Statement-I:** In photoelectric emission, the velocity of electron ejected from near the surface is larger than that coming from interior of metal.

   Statement-II: The velocity of ejected electron will be zero.

4. **Statement-I:** A photocell is called an electric eye.

   Statement-II: When light is incident on some semiconductor, its electrical resistance is reduced.

5. **Statement-I:** The de Broglie equation has significance for any microscopic or sub-microscopic particle.

   Statement-II: The de Broglie wavelength is inversely proportional to the mass of the object if velocity is constant.

Assertion & Reason MCQs

Directions: These questions consist of two statements each, printed as Assertion and Reason. While answering these questions, you are required to choose any one of the following four responses.

1) **Both Assertion and Reason are True and the Reason is a correct explanation of the Assertion.**
2) **Both Assertion and Reason are True but Reason is not a correct explanation of the Assertion.**
3) **Assertion is True but the Reason is False.**
4) **Assertion is False but Reason is True.**

1. **Assertion:** The strength of photoelectric current depends upon the intensity of incident radiation.

   Reason: A photon of energy E possesses a mass equal to E/c^2 and momentum equal to E/c.

2. **Assertion:** As light travels from one medium to another, the frequency of light does not change.

   Reason: Frequency is the characteristic of the source.

3. **Assertion:** Visible light cannot eject photoelectrons from copper surface, whose work function is 4.4 eV.

   Reason: The threshold wavelength for copper surface is given by $\lambda_0 = \frac{hc}{\phi_0} = \frac{12400}{4.4} = 2818\,\mathring{A}$. This is less than the wavelength of visible light (3800 Å to 7800 Å).

4. **Assertion:** The ionising power of β-particle is less as compared to α-particle but its penetrating power is more.

   Reason: The mass of β-particle is less than the mass of α-particle.

5. **Assertion:** The relative velocity of two photons travelling in opposite directions is equal to c.

   Reason: The rest mass of the photon is always zero.

Match the Columns MCQs

1. Stopping potential versus frequency of incident light and maximum kinetic energy versus frequency of incident light graphs are shown in figure. Match the following two columns.

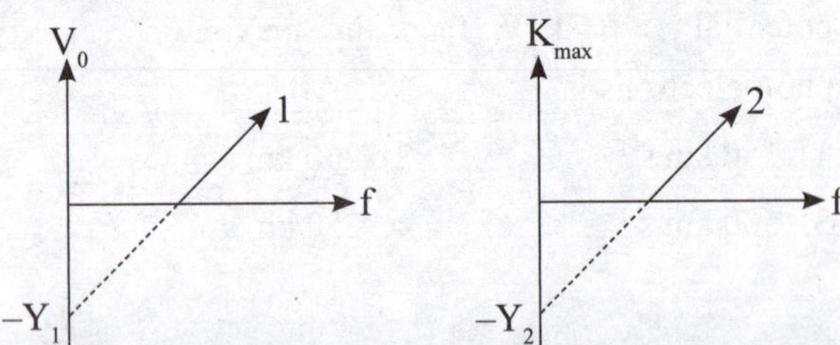

	Column-I		Column-II
(a)	Slope of line-1	(i)	h
(b)	Slope of line-2	(ii)	h/e
(c)	Y_2	(iii)	W
(d)	Y_1	(iv)	W/e

1) (a)-(i); (b)-(ii); (c)-(iii); (d)-(iv)
2) (a)-(iv); (b)-(iii); (c)-(ii); (d)-(i)
3) (a)-(ii); (b)-(i); (c)-(iii); (d)-(iv)
4) (a)-(i); (b)-(ii); (c)-(iv); (d)-(iii)

2. According to photoelectric effect, in Column-I, some physical quantities change while in Column-II effects of these changes are given. Match the entries of Column-I with the entries of Column-II.

	Column-I		Column-II
(a)	Intensity of incident light changes	(i)	Stopping potential changes
(b)	Frequency of incident light changes	(ii)	K_{max} of emitted photo-electrons changes
(c)	Target material changes	(iii)	Saturation current changes
(d)	Potential difference between the emitter and collector changes	(iv)	Time delay in emission of photoelectrons changes

1) (a)-(ii); (b)-(i); (c)-(iii); (d)-(iv)
2) (a)-(iii); (b)-(ii); (c)-(ii); (d)-(iv)
3) (a)-(iii); (b)-(i, ii); (c)-(i, ii); (d)-(iv)
4) (a)-(iv); (b)-(iii); (c)-(iv); (d)-(ii)

3. Match the column-I (Fundamental Experiment) with Column-II (its conclusion) and select the correct option from the choices given below the list:

	Column-I		Column-II
(a)	Franck-Hertz	(i)	Particle nature of light
(b)	Photo electric experiment	(ii)	Discrete energy levels of atom
(c)	Davision-Germer experiment	(iii)	Wave nature of electron

1) (a)-(i); (b)-(iv); (c)-(iii)
2) (a)-(ii); (b)-(iv); (c)-(iii)
3) (a)-(ii); (b)-(i); (c)-(iii)
4) (a)-(iv); (b)-(iii); (c)-(ii)

NCERT Exemplar MCQs

1. A particle is dropped from a height H. The de-Broglie wavelength of the particle as a function of height is proportional to:

1) H
2) $H^{1/2}$
3) H^0
4) $H^{-1/2}$

2. The wavelength of a photon needed to remove a proton from a nucleus which is bound to the nucleus with 1 MeV energy is nearly:

1) 1.2 nm
2) 1.2×10^{-3} nm
3) 1.2×10^{-6} nm
4) 1.2×10^{1} nm

3. Consider a beam of electrons (each electron with energy E_0) incident on a metal surface kept in an evacuated chamber. Then:

1) No electrons will be emitted as only photons can emit electrons
2) Electrons can be emitted but all with an energy, E_0
3) Electrons can be emitted with any energy, with a maximum of $E_0 - \phi$ (ϕ is the work function)
4) Electrons can be emitted with any energy, with a maximum of E_0

4. A proton, a neutron, an electron and an α-particle have same energy. Then, their de-Broglie wavelengths compare as:

1) $\lambda_p = \lambda_n > \lambda_e > \lambda_\alpha$
2) $\lambda_\alpha < \lambda_p = \lambda_n < \lambda_e$
3) $\lambda_e < \lambda_p = \lambda_n > \lambda_\alpha$
4) $\lambda_e = \lambda_p = \lambda_n = \lambda_\alpha$

5. An electron is moving with an initial velocity $v = v_0 \hat{i}$ and is in a magnetic field $B = B_0 \hat{j}$. Then, it's de-Broglie wavelength:

1) Remains constant
2) Increases with time
3) Decreases with time
4) Increases and decreases periodically

6. An electron (mass m) with an initial velocity $v = v_0\hat{i}$ ($v_0 > 0$) is in an electric field $E = -E_0\hat{i}$ (E_0 = constant > 0). It's de-Broglie wavelength at time t is given by:

1) $\dfrac{\lambda_0}{\left(1+\dfrac{eE_0}{m}\dfrac{t}{v_0}\right)}$ 2) $\lambda_0\left(1+\dfrac{eE_0 t}{mv_0}\right)$

3) λ_0 4) $\lambda_0 t$

7. An electron (mass m) with an initial velocity $v = v_0\hat{i}$ is in an electric field $E = E_0\hat{j}$. If $\lambda_0 = h/mv_0$, it's de-Broglie wavelength at time t is given by:

1) λ_0 2) $\lambda_0\sqrt{1+\dfrac{e^2E_0^2t^2}{m^2v_0^2}}$

3) $\dfrac{\lambda_0}{\sqrt{1+\dfrac{e^2E_0^2t^2}{m^2v_0^2}}}$ 4) $\dfrac{\lambda_0}{\left(1+\dfrac{e^2E_0^2t^2}{m^2v_0^2}\right)}$

NEET & AIIMS Past Year MCQs

1. If ϕ is the work function of photosensitive material in eV and light of wavelength of numerical value $\lambda = \dfrac{hc}{e}$ metre, is incident on it with energy above its threshold value at an instant then the maximum kinetic energy of the photo-electron ejected by it at that instant (Take h-Plank's constant, c-velocity of light in free space) is (in SI units): **(2023 Re)**

1) $e + 2\phi$ 2) $2e + \phi$

3) $e - \phi$ 4) $e + \phi$

2. An electron and an alpha particle are accelerated by the same potential difference. Let λ_e and λ_α denote the de-Broglie wavelengths of the electron and the alpha particle, respectively, then: **(2023 Re)**

1) $\lambda_e > \lambda_\alpha$ 2) $\lambda_e = 4\lambda_\alpha$

3) $\lambda_e = \lambda_\alpha$ 4) $\lambda_e < \lambda_\alpha$

3. The graph which shows the variation of $\left(\dfrac{1}{\lambda^2}\right)$ and its kinetic energy, E is (where λ is de Broglie wavelength of a free particle): **(2024)**

1) 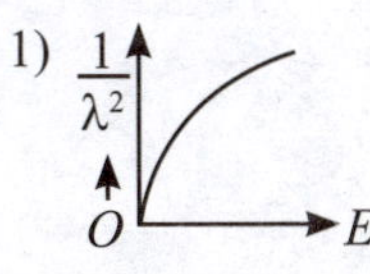

2) 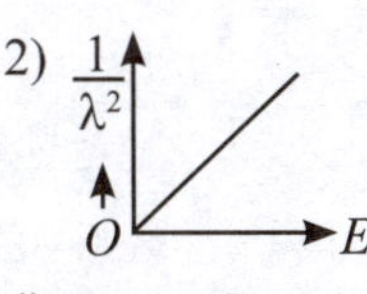

3) 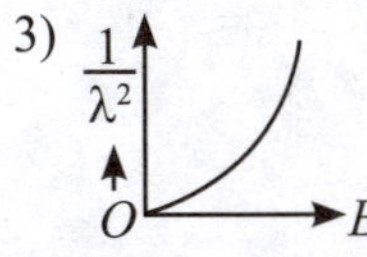

4) 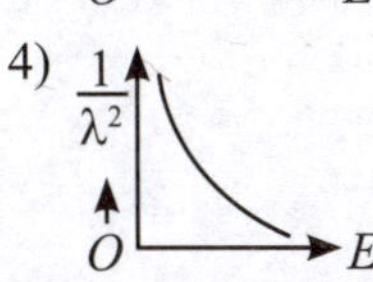

4. If c is the velocity of light in free space, the correct statements about photon among the following are: **(2024)**

A. The energy of a photon is $E = hv$

B. The velocity of a photon is c.

C. The momentum of a photon, $p = \dfrac{hv}{c}$

D. In a photon-electron collision, both total energy and total momentum are conserved.

E. Photon possesses positive charge.

Choose the correct answer from the options given below:

1) A, C and D only 2) A, B, D and E only

3) A and B only 4) A, B, C and D only

5. The minimum wavelength of X-rays produced by an electron accelerated through a potential difference of V volts is proportional to: **(2023)**

1) V^2 2) $\sqrt{V}$

3) $\dfrac{1}{V}$ 4) $\dfrac{1}{\sqrt{V}}$

6. The work functions of Caesium (Cs), Potassium (K) and Sodium (Na) are 2.14 eV, 2.30 eV and 2.75 eV respectively. If incident electromagnetic radiation has an incident energy of 2.20 eV, which of these photosensitive surfaces may emit photoelectrons? **(2023)**

1) Na only 2) Cs only

3) Both Na and K 4) K only

7. The graph which shows the variation of the de Broglie wavelength (l) of a particle and its associated momentum (p) is: **(2022)**

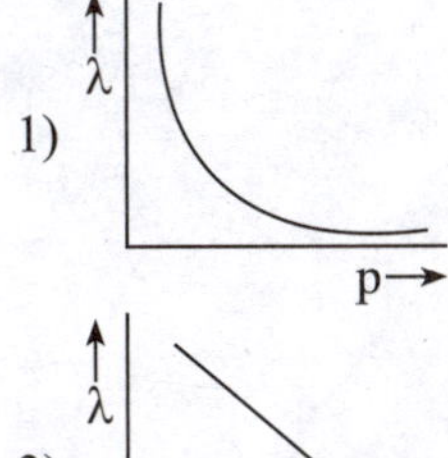

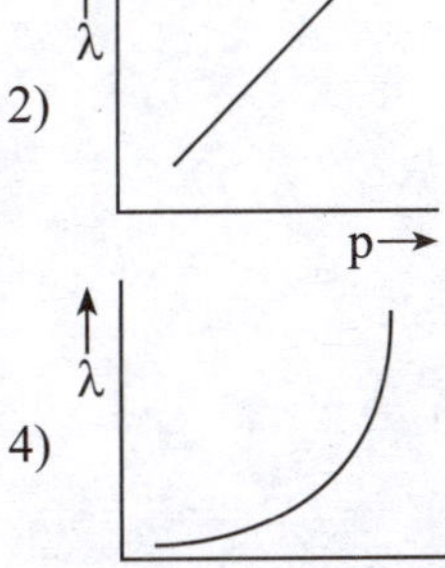

8. When two monochromatic light of frequency, ν and ν/2 are incident on a photoelectric metal, their stopping potential becomes $V_s/2$ and V_s respectively. The threshold frequency for this metal is: **(2022)**

1) $\dfrac{3}{2}\nu$ 2) 2ν

3) 3ν 4) $\dfrac{2}{3}\nu$

9. An electromagnetic wave of wavelength 'λ' is incident on a photosensitive surface of negligible work function. If 'm' mass is of photoelectron emitted from the surface has de-Broglie wavelength λ_d, then: (2021)

1) $\lambda_d = \left(\frac{2mc}{h}\right)\lambda^2$ 2) $\lambda = \left(\frac{2mc}{h}\right)\lambda_d^2$

3) $\lambda = \left(\frac{2h}{mc}\right)\lambda_d^2$ 4) $\lambda = \left(\frac{2m}{hc}\right)\lambda_d^2$

10. In a solar cell current is generated due to bond breakage in which region. (AIIMS 2018)

1) Depletion region 2) n-region

3) p-region 4) None of these

11. Which of the following has the longest de Broglie wavelength if they are moving with the same velocity? (AIIMS 2016)

1) Neutron 2) Proton

3) α particle 4) β-particle

12. If velocity of a particle is three times of that of electron and ratio of de Broglie wavelength of particle to that of electron is 1.814×10^{-4}. The particle will be: (AIIMS 2013)

1) Neutron 2) Deutron

3) Alpha 4) Tritium

13. Work function of nickel is 5.01 eV. When ultraviolet radiation of wavelength 2000 Å is incident on, electrons are emitted. What will be the maximum velocity of emitted electrons? (AIIMS 2010)

1) 3×10^8 m/s 2) 6.46×10^5 m/s

3) 10.36×10^5 m/s 4) 8.54×10^6 m/s

14. The de Broglie wavelength λ associated with neutrons is related with absolute temperature T as: (AIIMS 2009)

1) $\lambda \propto T$ 2) $\lambda \propto 1/T$

3) $\lambda \propto 1/T$ 4) $\lambda \propto T^2$

Answer Key

NCERT Topic-wise MCQs

1	2	3	4	5	6	7	8	9	10	11	12	13	14	15	16	17
3	1	4	3	4	2	4	1	2	4	3	2	1	1	4	2	1
18	**19**	**20**	**21**	**22**	**23**	**24**	**25**	**26**	**27**	**28**	**29**	**30**	**31**	**32**	**33**	**34**
2	2	2	1	2	2	1	4	1	4	3	4	1	1	2	2	3
35	**36**	**37**	**38**	**39**	**40**	**41**	**42**	**43**	**44**	**45**	**46**	**47**	**48**	**49**	**50**	**51**
1	1	2	3	1	3	4	1	1	1	1	4	4	1	3	3	1
52	**53**	**54**	**55**	**56**	**57**	**58**	**59**	**60**	**61**	**62**	**63**	**64**	**65**	**66**	**67**	**68**
3	1	3	3	1	1	2	2	2	2	2	2	4	4	3	2	3

High Order Time Intensive MCQs

1	2	3	4	5	6	7	8	9	10	11	12	13	14	15	16	17
3	2	3	1	1	1	2	1	3	1	4	1	2	3	2	4	2
18	**19**	**20**	**21**	**22**	**23**	**24**	**25**	**26**	**27**	**28**	**29**	**30**	**31**	**32**	**33**	**34**
1	3	3	2	2	4	4	1	2	3	1	1	3	1	2	4	2
35	**36**	**37**	**38**	**39**	**40**	**41**	**42**	**43**	**44**	**45**	**46**					
2	2	4	2	3	2	1	3	3	3	2	3					

Statement Based MCQs

1	2	3	4	5
1	1	3	1	1

Assertion & Reason MCQs

1	2	3	4	5
2	1	1	2	2

Match the Columns MCQs

1	2	3
3	3	3

NCERT Exemplar MCQs

1	2	3	4	5	6	7
4	2	4	2	1	1	3

NEET & AIIMS Past Year MCQs

1	2	3	4	5	6	7	8	9	10	11	12	13	14
3	1	2	4	3	2	1	None	2	1	4	1	2	3

CHAPTER

12 Atoms

Topicwise Number of Questions Analysis (2025-2016)

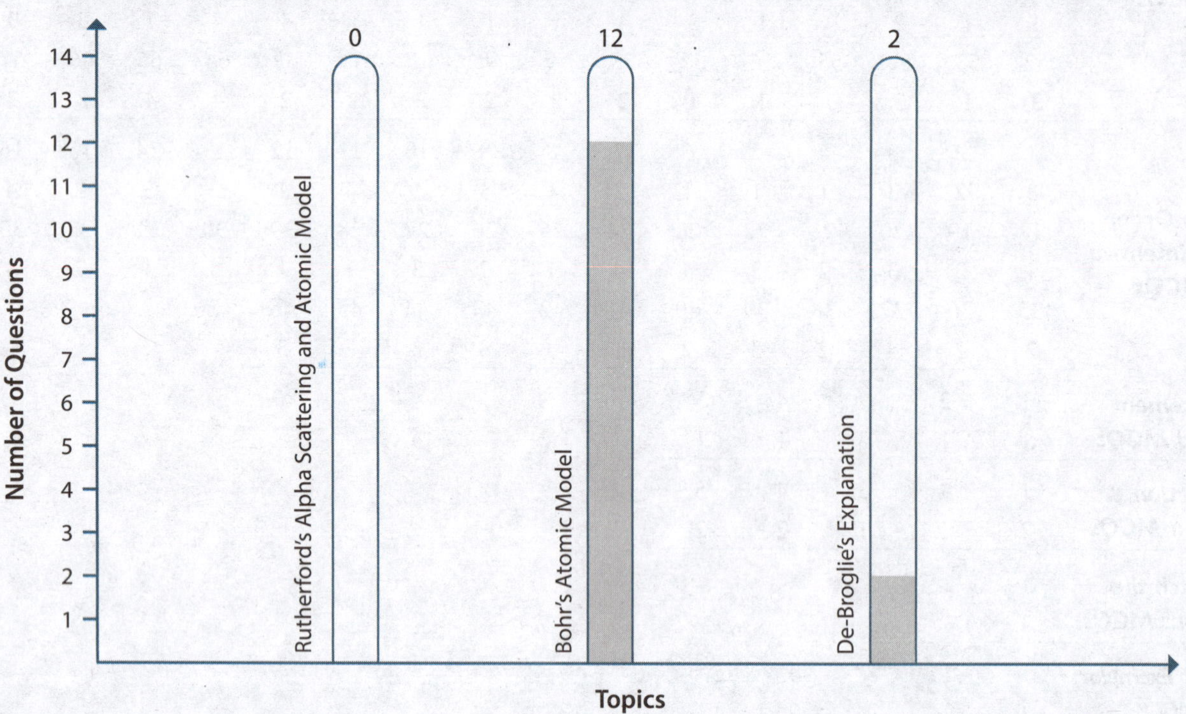

NCERT-PICKS

Rutherford's Alpha Scattering Experiment and Atomic Model

Formulas

1. Distance of closest approach, $r_0 = \dfrac{(Ze)(2e)}{4\pi\varepsilon_0 \times \left(\dfrac{1}{2}mv^2\right)}$

2. Impact parameter, $b = \dfrac{Ze^2 \cot\theta/2}{4\pi\varepsilon_o\left(\dfrac{1}{2}mv^2\right)}$

3. Nucleus size is of the order of 1 fermi = 10^{-15} m.

Bohr's Model

Formulas

1. Angular momentum of electron in stationary orbit,

   $L = mvr = \dfrac{nh}{2\pi}$

   m = mass of e^-

   v = velocity of e^-

   r = radius of orbit

   n = principal quantum number

2. Total path length of a stationary orbit = integral multiple of de-Broglie wavelength of the electron.

$2\pi r = nh$

where n = 1, 2, 3.... and $\lambda = \frac{h}{mv}$

$\Rightarrow 2\pi r = \frac{nh}{mv}$

3. The frequency of emitted radiation $\nu = \frac{E_2 - E_1}{h}$

   where, E_2 = energy of e^- in higher orbit

   E_1 = energy of e^- in lower orbit

4. Centripetal force for circular orbit is equal to the coulomb's force, i.e., $\frac{1}{4\pi\varepsilon_0}\frac{Ze^2}{r^2} = \frac{mv^2}{r}$

5. Kinetic energy of an electron in an atom having atomic number Z.

   $K.E = \frac{KZe^2}{2r}$

6. $P.E = \frac{-KZe^2}{r}$

7. $T.E = \frac{-KZe^2}{2r}$

8. Kinetic energy of electron in hydrogen atom $K.E = \frac{e^2}{8\pi\varepsilon_0 r}$

9. Potential energy of electron in hydrogen atom $P.E = \frac{-e^2}{4\pi\varepsilon_0 r}$

10. Total energy of electron in the case of hydrogen atom $T.E = \frac{-e^2}{8\pi\varepsilon_0 r}$

Numerical Profile

(i) Radius of orbit of electron is given by

$$r_n = \frac{\epsilon_0 n^2 h^2}{\pi m Z e^2} \Rightarrow r_n \propto \frac{n^2}{Z}$$

where, n = principal quantum number, h = Planck's constant, m = mass of an electron, Z = atomic number.

(ii) Velocity of electron in an orbit is given by

$$v = \frac{Ze^2}{2\epsilon_o hn} \Rightarrow v \propto \frac{Z}{n}$$

(iii) Kinetic energy of electron in any orbit is given by

$$E_k = \frac{13.6Z^2}{n^2}eV$$

(iv) Potential energy of electron in any orbit is given by

$$E_p = -27.2\frac{Z^2}{n^2}eV$$

$$\Rightarrow E_p \propto \frac{Z^2}{n^2}$$

(v) Total Energy of electron in any orbit is given by

$$E = \frac{-me^2Z^2}{8\epsilon_o^2 n^2h^2} = -13.6\frac{Z^2}{n^2}eV$$

(vi) Wavelength of radiation emitted when an electron jumps from orbit n_2 to n_1 is given by

$$\frac{1}{\lambda} = \frac{me^4}{8\epsilon_o^2 ch^3}\left[\frac{1}{n_1^2} - \frac{1}{n_2^2}\right]$$

$$\frac{1}{\lambda} = R\left[\frac{1}{n_1^2} - \frac{1}{n_2^2}\right]$$

$$R = \frac{me^4}{8\epsilon_o^2 ch^3} = 1.097 \times 10^7 m^{-1}$$

In quantum mechanics, the energies of a system are discrete or quantized. The energy of a particle of mass m is confined to a box of length L can have discrete values of energy given by the relation

$E_n = n^2 h^2/8mL^2; n < 1, 2, 3,...$

Spectral Line

Formulas

1. $\frac{1}{\lambda} = R\left(\frac{1}{n_1^2} - \frac{1}{n_2^2}\right)$, $R = 1.097 \times 10^7 m^{-1}$

Name	Spectral formula	No. of last orbit	Region
1. Lyman series	$\frac{1}{\lambda} = R\left(\frac{1}{(1)^2} - \frac{1}{(n)^2}\right)$	n = 2, 3, 4...	Ultraviolet
2. Balmer series	$\frac{1}{\lambda} = R\left(\frac{1}{(2)^2} - \frac{1}{(n)^2}\right)$	n = 3, 4, 5...	Visible
3. Paschen series	$\frac{1}{\lambda} = R\left(\frac{1}{(3)^2} - \frac{1}{(n)^2}\right)$	n = 4, 5, 6...	Infrared
4. Brackett series	$\frac{1}{\lambda} = R\left(\frac{1}{(4)^2} - \frac{1}{(n)^2}\right)$	n = 5, 6, 7...	Infrared
5. Pfund series	$\frac{1}{\lambda} = R\left(\frac{1}{(5)^2} - \frac{1}{(n)^2}\right)$	n = 6, 7, 8...	Infrared

2. Number of lines emitted when an electron comes back to ground state from the excited state = $\frac{n(n-1)}{2}$

Statements Asked in Exam

- Thomson's model of atom could not explain the origin of spectral series of hydrogen and other atoms, observed experimentally.
- Atoms of each element are stable and emit characteristic spectrum. The spectrum consists of a set of isolated parallel lines termed as line spectrum. It provides useful information about the atomic structure.
- While revolving in the permissible orbits, an electron does not radiate energy. These non-radiating orbits are called stationary orbits.
- The orbital picture in Bohr's model of the hydrogen atom was inconsistent with the uncertainty principle. It was replaced by modern quantum mechanics in which Bohr's orbits are regions where the electron may be found with high probability.

- Bohr's model is applicable only to hydrogenic atoms (single electron).
- An alpha particle is a helium nucleus containing 2 protons and 2 neutrons. Therefore, an alpha particle has 4 units of mass and 2 units of positive charge.
- In Rutherford's experiment, only about 0.14% of incident α-particles scatter by more than 1^0.
- About one α-particle in every 8000 α-particles deviate by more than 90^0.
- Distance of closest approach is that distance from the nucleus at which K.E of α-particle reduces to zero.
- Rutherford scattering experiment is a powerful way to determine the upper limit to the size of atomic nucleus.
- The centripetal force is provided by the electrostatic force of attraction between the electrons and the nucleus.
- Line spectra serve as fingerprints for identification of the gas.
- Every atom consists of a central core called nucleus, in which entire positive charge and almost entire mass of the atom is concentrated.
- Bohr model of hydrogen atom stresses that an electron in an atom can revolve only in certain stable orbits without the emission of energy.
- In the first orbit of hydrogen atom, orbital velocity of electron is 2.2×10^6 m/s which is roughly 1/137 of the velocity of light in vacuum.
- In the first orbit of hydrogen atom, the frequency of revolution of electron is 6.57×10^{15} rps.
- Total energy of an electron in a stationary orbit is negative which means the electron is bound to the nucleus and it is not free to leave.
- According to de Broglie, a stationary orbit is that which contains an integral number of de Broglie waves associated with the revolving electron. ($2\pi r = n\lambda$)
- If the electron in an atom were stationary they would fall into the nucleus on account of electrostatic force of attraction between electrons and the nucleus.

NCERT Topic-wise MCQs

Rutherford's Alpha Scattering and Atomic Model

1. In the Geiger - Marsden scattering experiment the number of scattered particles detected are maximum and minimum at the scattering angles respectively at:
   1) 0^0 and 180^0 2) 180^0 and 0^0
   3) 90^0 and 180^0 4) 45^0 and 90^0

2. In Geiger - Marsden scattering experiment, the trajectory traced by an α-particle depends on:
   1) Number of collision
   2) Number of scattered α-particles
   3) Impact parameter
   4) None of these

3. In a head on collision between an α-particle and gold nucleus, the closest distance of approach is 41.3 fm. Calculate the energy of α-particle (Z of gold = 79)
   1) 5.51 eV 2) 5.51 MeV
   3) 5.51 J 4) 5.51 KJ

4. What is the distance of closest approach when a 5 MeV proton approaches a gold nucleus (Z = 79)?
   1) 4×10^{-14} m 2) 2.3×10^{-14} m
   3) 3×10^{-14} m 4) None of these

5. If the electron in the hydrogen atom jumps from third orbit to second orbit, the wavelength of the emitted radiation in terms of Rydberg constant is
   1) $\frac{6}{5R}$ 2) $\frac{36}{5R}$
   3) $\frac{64}{7R}$ 4) None of these

6. The relation between the orbit radius and the electron velocity for a dynamically stable orbits in a hydrogen atom is (where all notations have their usual meanings):
   1) $v=\sqrt{\frac{4\pi\varepsilon_0}{me^2r}}$ 2) $r=\sqrt{\frac{e^2}{4\pi\varepsilon_0 v}}$
   3) $v=\sqrt{\frac{e^2}{4\pi\varepsilon_0 mr}}$ 4) $r=\sqrt{\frac{ve^2}{4\pi\varepsilon_0 m}}$

7. An α-particle of energy 5 MeV is scattered through 180º by a fixed uranium nucleus. The Closest distance is of the order of
   1) 1 Å 2) 10^{-10} cm
   3) 10^{-12} cm 4) 10^{-16} cm

8. In the Gieger - Marsden scattering experiment, in case of head - on collision the impact parameter should be:
   1) Maximum 2) Zero
   3) Infinite 4) Very large

9. An alpha nucleus of energy $\frac{1}{2}mv^2$ bombards a heavy target of charge Ze. Then, the distance of closest apporach for the alpha nucleus will be proportional to
   1) v^2 2) $1/m$
   3) $1/v^4$ 4) $1/Ze$

10. The volume occupied by an atom is greater than the volume of the nucleus by a factor of about:
    1) 10^1 2) 10^{-15}
    3) 10^{10} 4) 10^{15}

11. The Rutherford α-particle experiment shows that most of the α-particles pass through almost unscattered. While some are scattered through large angles. What information does it give about the structure of the atom?
    1) Atom is hollow
    2) The whole mass of the atom is concentrated in a small centre called nucleus
    3) Nucleus is positively charged
    4) All the above

12. Which radioactive source is used to produce α-particle beam?
    1) $^{214}_{83}Bi$ 2) $^{214}_{82}Pb$
    3) $^{14}_{6}C$ 4) $^{2}_{1}D$

13. An elementary particle of mass m and charge +e is projected with velocity v at a much more massive particle of charge Ze, where Z > 0. What is the closest possible approach of the incident particle ?
    1) $Ze^2/2\pi\varepsilon_0 mv^2$ 2) $Ze^2/4\pi\varepsilon_0 r_n$
    3) $Ze^2/8\pi\varepsilon_0 r_n$ 4) $-Ze^2/8\pi\varepsilon_0 r_n$

14. When an α-particle happens to come near a nucleus the intense electric field of these scatters it through a large angle then what is the effect of atomic electrons on α-particle?
    1) Repel α-particle lightly
    2) Deviate α-particle from their path
    3) Due to the electron cloud it blocks the path of α-particle
    4) Do not appreciably effect

15. Rutherford's atomic model could account for:
1) Stability of atom
2) Origin of spectra
3) The positive charged central core of an atom
4) Concept of stationary orbits

Spectral Lines

16. When an atomic gas or vapour is excited at low pressure by passing an electric current through it then:
1) Emission spectrum is observed
2) Absorption spectrum is observed
3) Band spectrum is observed
4) Both (2) and (3)

17. The first spectral series was discovered by:
1) Balmer 2) Lyman
3) Paschen 4) Pfund

18. In Balmer series of emission spectrum of hydrogen first four lines with different wavelength are H_α, H_β, H_γ, H_δ. Which among these has highest frequency?
1) H_α 2) H_β
3) H_γ 4) H_δ

19. Which of the following spectral series falls within the visible range of electromagnetic radiation?
1) Lyman series 2) Balmer series
3) Paschen series 4) Pfund series

20. If v_1 is the frequency of the series limit of Lyman series, v_2 is the frequency of the first line of Lyman series and v_3 is the frequency of the series limit of the Balmer series, then:
1) $v_1 - v_2 = v_3$ 2) $v_1 = v_2 - v_3$
3) $\frac{1}{v_2} = \frac{1}{v_1} + \frac{1}{v_3}$ 4) $\frac{1}{v_1} = \frac{1}{v_2} + \frac{1}{v_3}$

21. If the wavelength of the first line of the Balmer series of hydrogen is 6561Å, then the wavelength of the second line of the series should be:
1) 13122 Å 2) 3280 Å
3) 4860 Å 4) 2187 Å

22. If an electron in hydrogen atom jumps from an orbit of level n = 3 to an orbit of level n = 2 then emitted radiation has a frequency:
1) $\frac{Rc}{9}$ 2) $\frac{Rc}{25}$
3) $\frac{8Rc}{9}$ 4) $\frac{5Rc}{36}$

23. The ionization energy of the electron in the hydrogen atom in its ground state is 13.6 eV. The atoms are excited to higher energy level to emit radiations of 6 wavelengths. Maximum wavelength of emitted radiation corresponds to the transition between:
1) n = 3 and n = 2 states 2) n = 3 and n = 1 states
3) n = 2 and n = 2 states 4) n = 4 and n = 3 states

24. An electron is moving in an orbit of a hydrogen atom from which there can be a maximum of six transition. An electron is moving in an orbit of another hydrogen atom for which there can be a maximum of three transition. The ratio of the velocities of the electron in these two orbits is:
1) $\frac{1}{2}$ 2) $\frac{2}{1}$
3) $\frac{5}{4}$ 4) $\frac{3}{4}$

25. If λ_1 and λ_2 are the wavelengths of the first members of the Lyman and Paschen series respectively, then $\lambda_1 : \lambda_2$ is:
1) 1 : 3 2) 1 : 30
3) 7 : 50 4) 7 : 108

Bohr's Atomic Model and de-Broglie Explanation

26. Which one of the relations is correct between time period and number of orbits while an electron is revolving in an orbit?
1) $T \propto \frac{1}{n^2}$ 2) $T \propto n^2$
3) $T \propto n^3$ 4) $T \propto \frac{1}{n}$

27. The energy of an electron in an excited hydrogen atom is –3.4 eV. Calculate the angular momentum of the electron according to Bohr's theory. ($h = 6.6 \times 10^{-34}$ J-s):
1) 3×10^{-32} J-s 2) 2.1×10^{-34} Js
3) 4×10^{-34} Js 4) None of these

28. Bohr's model of atoms
1) Assumes that the angular momentum of electrons is quantised
2) Uses Einstein's photoelectric equation
3) Predicts continuous spectrum
4) Predicts the same emission spectra for all types of atoms

29. An electron in a hydrogen atom makes a transition from $n = n_1$ to $n = n_2$. The time period of the electron in the initial state is eight times that in the final state. The possible values of n_1 and n_2 are:
1) $n_1 = 4$, $n_2 = 2$ 2) $n_1 = 8$, $n_2 = 2$
3) $n_1 = 8$, $n_2 = 1$ 4) $n_1 = 6$, $n_2 = 2$

30. A hydrogen atom in its ground state absorbs 10.2 eV of energy. The angular momentum is increased by
1) 4.22×10^{-3} Js 2) 2.11×10^{-34} Js
3) 3.16×10^{-34} Js 4) 1.05×10^{-34} Js

31. In the Bohr model of the hydrogen atom, the lowest orbit corresponds to:
1) Infinite energy 2) Maximum energy
3) Minimum energy 4) Zero energy

32. A triply ionized beryllium (Be^{3+}) has the same orbital radius as the ground state of hydrogen. Then the quantum state n of Be^{3+} is:
1) n = 1 2) n = 2
3) n = 3 4) n = 4

33. Energy of an electron in the second orbit of hydrogen atom is E and the energy of electron in 3rd orbit of He will be
1) $E_3 = \frac{16E}{3}$ 2) $E_3 = \frac{16E}{9}$
3) $E_3 = \frac{4E}{9}$ 4) $E_3 = \frac{4E}{3}$

34. In accordance with the Bohr's model, the quantum number that characterises the earth's revolution around the sun in an orbit of radius 1.5×10^{11} m with orbital speed 3×10^4 ms^{-1} is: (Take mass of the earth = 6×10^{24} kg)
1) 5.98×10^{86} 2) 2.57×10^{38}
3) 8.57×10^{64} 4) 2.57×10^{74}

35. Consider an electron in the nth orbit of a hydrogen atom in the Bohr model. The circumference of the orbit can be expressed in terms of de Broglie wavelength λ of that electron as
1) $(0.529)n\lambda$ 2) $\sqrt{n}\,\lambda$
3) $(13.6)\,\lambda$ 4) $n\,\lambda$

36. The electric current I created by the electron in the ground state of H atom using Bohr model in terms of Bohr radius (a_o) and velocity (v_o) of electron in first orbit is:
1) $\frac{ev_o}{2\pi a_o}$ 2) $\frac{2\pi a_o}{ev_o}$
3) $\frac{2\pi a_o}{v_o}$ 4) $\frac{v_o}{2\pi a_o}$

37. Energy E of a hydrogen atom with principle quantum number n is given by $E = \frac{-13.6}{n^2}$eV. The energy of a photon ejected when the electron jumps from n = 3 state to n = 2 state of hydrogen is approximately
1) 1.5 eV 2) 0.85 eV
3) 3.4 eV 4) 1.9 eV

38. An electron is in an excited state in a hydrogen like atom. It has a total energy of –3.4 eV. The kinetic energy of the electron is E and its de-Bronglie wavelength is λ. Then: (Take $m_e = 9.1 \times 10^{-31}$ kg)
1) $E = 6.8$ eV, $\lambda = 6.6 \times 10^{-10}$ m
2) $E = 3.4$ eV, $\lambda = 6.6 \times 10^{-10}$ m
3) $E = 3.4$ eV, $\lambda = 6.6 \times 10^{-11}$ m
4) $E = 6.8$ eV, $\lambda = 6.6 \times 10^{-11}$ m

39. As the electron in Bohr's orbit of hydrogen atom passes from state n = 2 to n = 1, the KE (K) and the potential energy (U) change as:
1) K four fold, U also four fold
2) K two fold, U also two fold
3) K four fold, U two fold
4) K two fold, U four fold

40. The radius of the smallest electron orbit in H atom like ion is $\left(\frac{0.53}{4}\times 10^{-10}\right)$m, then it is:
1) Hydrogen atom 2) He^+
3) Li^{2+} 4) Be^{3+}

41. The electron in hydrogen atom revolves around the proton in a circular nth orbit of radius:
$$r_n = \frac{\varepsilon_0 n^2 h^2}{\pi m e^2} \text{ with speed, } v_n = \frac{e^2}{2\varepsilon_0 nh}$$
The current due to the circulating charge is proportional to:
1) e^2 2) e^3
3) e^5 4) e^6

42. In the Bohr model of the hydrogen atom, an electron moves in the n-th circular orbit of radius r with speed v, producing a magnetic field B at the center. The angular momentum is quantized as $mvr = \frac{nh}{2\pi}$. What is the magnetic flux linked with the orbit? (Given: $r = \frac{n^2h^2\in_0}{\pi me^2}$, where $\in_0$ is the permittivity of free space, e is the electron's charge, and other symbols have their usual meanings.)
1) $\frac{\mu_0 e^3}{8nh\in_0}$ 2) $\frac{\mu_0 e^3}{4nh\in_0}$
3) $\frac{\mu_0 e^3}{16nh\in_0}$ 4) $\frac{\mu_0 e^3}{4nh\in_0}$

43. In the Bohr model of a hydrogen atom, the centripetal force is furnished by the coulomb attraction between the proton and the electron. If a_0 is the radius of the ground state orbit, m is the mass and e is the charge on the electron and ε_0 is the vacuum permittivity, the speed of the electron is
1) $\frac{e}{\sqrt{4\pi\varepsilon_0 a_0 m}}$ 2) $\frac{e}{\sqrt{\varepsilon_0 a_0 m}}$
3) 0 4) $\frac{\sqrt{4\pi\varepsilon_0 a_0 m}}{e}$

44. The ratio of the speed of the electrons in the ground state of hydrogen to the speed of light in vacuum is
1) $\frac{1}{2}$ 2) $\frac{2}{137}$
3) $\frac{1}{137}$ 4) $\frac{1}{237}$

45. The ground state energy of hydrogen atom is –13.6 eV. The kinetic energy of the electron in this state is:
1) 27.2 eV 2) 13.6 eV
3) 6.8 eV 4) 3.4 eV

46. The de-Broglie wavelength of an electron in the first Bohr orbit is:
1) Equal to one fourth the circumference of the first orbit
2) Equal to half the circumference of the first orbit
3) Equal to twice the circumference of the first orbit
4) Equal to the circumference of the first orbit

47. The velocity of an electron in the second orbit of sodium atom is v. The velocity of an electron in its fifth orbit will be:
1) v 2) $\frac{22}{5}$v
3) $\frac{5}{2}$v 4) $\frac{2}{5}$v

48. Energy levels A, B, C of a certain atom corresponding to increasing value of energy, is, $E_A < E_B < E_C$. If λ_1, λ_2 and λ_3 are the wavelength of radiations corresponding to the transitions C to B, B to A and C to A respectively, which of the following statements is correct?

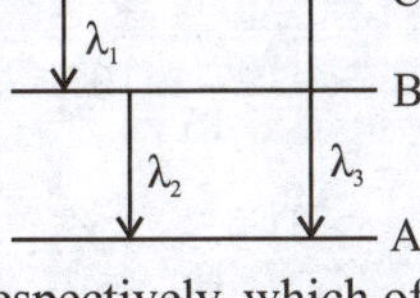

1) $\lambda_3 = \lambda_1 + \lambda_2$ 2) $\lambda_3 = \frac{\lambda_1\lambda_2}{\lambda_1 + \lambda_2}$
3) $\lambda_1 + \lambda_2 + \lambda_3 = 0$ 4) $\lambda_3^2 = \lambda_1^2 + \lambda_2^2$

49. The ionization energy of Li^{2+} is equal to:

1) 9 hcR 2) 6 hcR
3) 2 hcR 4) hcR

50. In hydrogen atom the electron is moving round the nucleus with velocity 2.18×10^6 m/s in an orbit of radius 0.528Å. The centripetal acceleration of the electron is:

1) 9×10^{18} m/s^2 2) 9×10^{22} m/s^2
3) 9×10^{-22} m/s^2 4) 9×10^{12} m/s^2

51. The ground state energy of hydrogen is –13.6 eV. What is the potential energy of the electron in this state?

1) 0 eV 2) –27.2 eV
3) 1 eV 4) 2 eV

52. The energy of electron in the nth orbit of H atom is expressed as $E_n = \frac{-13.6}{n^2}$ eV. The shortest and longest wavelength of Lyman series will be:

1) 910Å, 1213Å 2) 5463Å, 7858 Å
3) 1315Å, 1530Å 4) None of these

53. The ratio of the area of the orbits for the first excited state to the ground state for hydrogen atom is:

1) 16 : 1 2) 18 : 1
3) 4 : 1 4) 2 : 1

54. If the binding energy of the electron in a H atom is 13.6 eV, the energy required to remove the electron from the first excited state of Li^{2+} is:

1) 122.4 eV 2) 30.6 eV
3) 13.6 eV 4) 3.4 eV

55. If 13.6 eV energy is required to ionise the hydrogen atom, then energy required to remove an electron from n = 2 is

1) 10.2 eV 2) 0 eV
3) 3.4 eV 4) 6.8 eV

56. Minimum energy required to take out the electron from ground state of He^+ is:

1) 13.6 eV 2) 54.4 eV
3) 27.2 eV 4) 6.8 eV

57. The ratio of the wavelengths for 2 → 1 transition in Li^{++}, He^+ and H is:

1) 1 : 2 : 3 2) 1 : 4 : 9
3) 4 : 9 : 36 4) 3 : 2 : 1

58. The angular momentum (L) of an electron moving in a stable orbit around nucleus is:

1) Half integral multiple of $\frac{nh}{2\pi}$
2) Integral multiple of $\frac{n^2h}{2\pi}$
3) Integral multiple of $\frac{h}{2\pi}$
4) Half integral multiple of h

59. The orbit frequency of an electron in the hydrogen atom is proportional to:

1) n^3 2) n^{-3}
3) n 4) n^0

High Order Time Intensive MCQs

1. A stationary hydrogen atom emits a photon corresponding to first line of the Lyman series. The velocity acquired by the atom will be

1) 1.35 m/s 2) 3.25 m/s
3) 4.90 m/s 4) 7.31 m/s

2. An electron of a stationary hydrogen atom jumps from the 5th energy level to the ground level. The velocity that the atom acquired as a result of photon emission will be

1) $\frac{24hR}{25m}$ 2) $\frac{25hR}{24m}$
3) $\frac{25m}{24hR}$ 4) $\frac{24m}{25hR}$

3. An electron jumps from the 4th orbit to 2nd orbit of the hydrogen atom. Given, Rydberg's constant R = 10^5 cm^{-1}, the frequency in hertz of emitted radiation will be

1) $\frac{3}{16}\times10^5$ 2) $\frac{3}{16}\times10^{15}$
3) $\frac{9}{16}\times10^{15}$ 4) $\frac{3}{4}\times10^{15}$

4. The following diagram indicates the energy levels of a certain atom when the system moves from 4E level to E. A photon of wavelength λ_1 is emitted. The wavelength of photon produced during its transition from 7/3 E level to E is λ_2. The ration λ_1/λ_2 will be

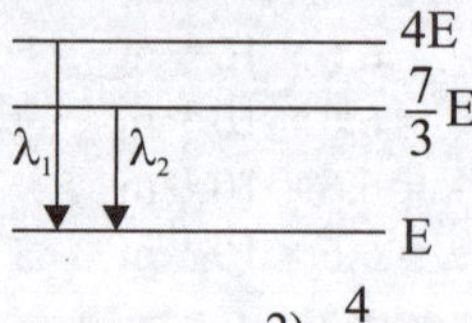

1) $\frac{9}{4}$ 2) $\frac{4}{9}$
3) $\frac{3}{2}$ 4) $\frac{7}{3}$

5. When in hydrogen like ion electron jumps from n = 3 to n = 1, the emitted photon has frequency 2.7×10^{15} Hz. When electron jumps from n = 4 to n = 1, the frequency is

1) 1.6×10^{15} 2) 2.8×10^{15}
3) 6.4×10^{15} 4) 4.8×10^{15}

6. The figure indicates the energy level diagram of an atom and the origin of six spectral lines in emission (e.g., line no. 5 arises from the transition from level B to A).

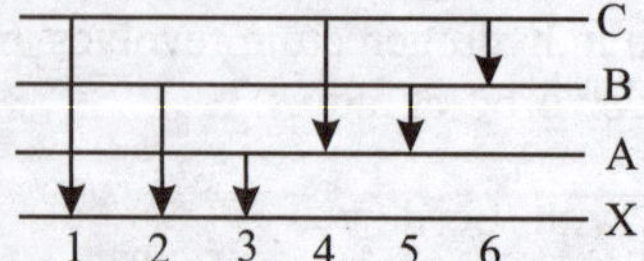

Which of the following spectral lines will occur in the absorption spectrum ?

1) 4, 5, 6 2) 1, 2, 3, 4, 5, 6
3) 1, 2, 3 4) 1, 4, 6

7. The shortest wavelength which can be obtained in hydrogen spectrum is ($R = 10^7\ m^{-1}$)
   1) 1000Å 2) 800Å
   3) 1300Å 4) 2100Å

8. Maximum energy is evolved during which of the following transition?
   1) n = 1 to n = 2 2) n = 2 to n = 6
   3) n = 2 to n = 1 4) n = 6 to n = 2

9. The diagram shows the energy levels for an electron in a certain atom. Which transition shown represents the emission of a photon with the most energy?

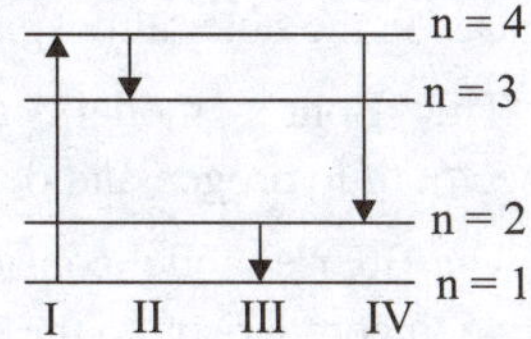

   1) III 2) IV
   3) I 4) II

10. The first member of the Balmer's series of the hydrogen has a wavelength λ, the wavelength of the second member of its series is
   1) $\frac{27}{20}\lambda$ 2) $\frac{20}{27}\lambda$
   3) $\frac{27}{10}\lambda$ 4) None of these

11. A H-atom in a state of binding energy 0.85 eV makes a transition to a state of excitation energy of 10.2 eV the initial state and final state of H-atoms are
   1) 4, 2 2) 2, 4
   3) 3, 2 4) 2, 3

12. The ratio of ionization energy of Bohr's hydrogen atom and Bohr's hydrogen like lithium atom is
   1) 1 : 1 2) 1 : 3
   3) 1 : 9 4) None of these

13. What is the angular momentum of an electron in Bohr's hydrogen atom whose energy is –0.544 eV ?
   1) $\frac{h}{\pi}$ 2) $\frac{2h}{\pi}$
   3) $\frac{5h}{2\pi}$ 4) $\frac{7h}{2\pi}$

14. Which of the following statements are true regarding Bohr's model of hydrogen atom ?
   A. Orbiting speed of electron decreases as it shifts to discrete orbits away from the nucleus.
   B. Radii of allowed orbits of electron are proportional to the square of principal quantum number
   C. Frequency with which electrons orbits around the nucleus in discrete orbits is inversely proportional to the cube of principal quantum number
   D. Binding force with which the electron is bound to the nucleus increases as it shifts to outer orbits
   1) A and C 2) B and D
   3) A, B and C 4) B, C and D

15. The following diagram indicates the energy levels of a certain atom when the system moves from 2E level to E, a photon of wavelength λ is emitted. The wavelength of photon produced during its transition from $\frac{4E}{3}$ level to E is:

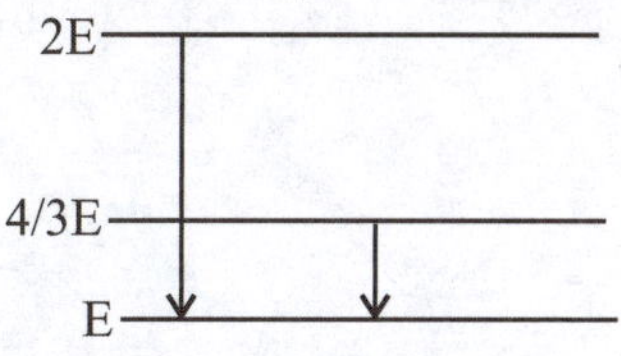

   1) $\frac{\lambda}{3}$ 2) $\frac{3\lambda}{4}$
   3) $\frac{4\lambda}{3}$ 4) 3λ

16. A double charged lithium atom is equivalent to hydrogen whose atomic number is 3. The wavelength of required radiation for emitting electron from first to third Bohr orbit in Li^{++} will be (Ionisation energy of hydrogen atom is 13.6 eV)
   1) 182.51 Å 2) 177.17 Å
   3) 142.25 Å 4) 113.74 Å

17. If the series limit of Lyman series for hydrogen atom is equal to the series limit of Balmer series for a hydrogen like atom, then atomic number of this hydrogen like atom will be
   1) 1 2) 2
   3) 3 4) 4

18. In terms of Rydberg constant R, the wave number of the first Balmer line is
   1) R 2) 3 R
   3) $\frac{5R}{36}$ 4) $\frac{8R}{9}$

19. A hydrogen atom emits a photon corresponding to an electron transition from n = 5 to n = 1. The recoil speed of hydrogen atom is almost (mass of proton = 1.6×10^{-27} kg).
   1) 10 ms^{-1} 2) $2 \times 10^{-2}\ ms^{-1}$
   3) 4 ms^{-1} 4) $8 \times 10^{2}\ ms^{-1}$

20. Some energy levels of a molecule are shown in the figure. The ratio of the wavelengths $r = \lambda_1/\lambda_2$ is given by

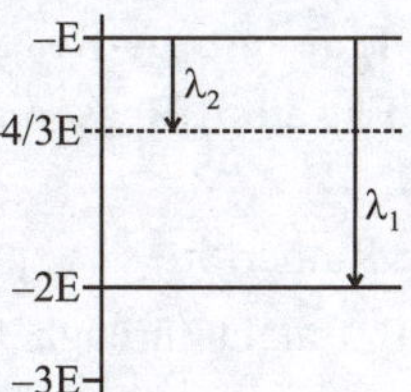

   1) $r = \frac{2}{3}$ 2) $r = \frac{3}{4}$
   3) $r = \frac{1}{3}$ 4) $r = \frac{4}{3}$

21. The ratio of the kinetic energy of the n = 2 electron for the H atom to that of He^+ ion is
   1) 1/4 2) 1/2
   3) 1 4) 2

22. A particle of mass m moves in a circular orbit in a central potential field $U(r)=\frac{1}{2}kr^2$. If Bohr's quantization conditions are applied, radii of possible orbitals and energy levels vary with quantum number n as

1) $r_n \propto n, E_n \propto n$ 2) $r_n \propto n^2, E_n \propto \frac{1}{n^2}$ 3) $r_n \propto \sqrt{n}, E_n \propto n$ 4) $r_n \propto \sqrt{n}, E_n \propto \frac{1}{n}$

Statement Based MCQs

Directions: These questions consist of two statements each, printed as Statement-I and Statement-II. While answering these questions, you are required to choose any one of the following four responses.

1) **Both Statement-I and Statement-II are correct.**
2) **Both Statement-I and Statement-II are incorrect.**
3) **Statement-I is correct & Statement-II is incorrect.**
4) **Statement-I is incorrect & Statement-II is correct.**

1. **Statement-I:** Electric conduction take place in gases at normal pressure.

Statement-II: Emission spectrum of hydrogen atom consists of only one line.

2. **Statement-I:** Dark lines are observed in the spectrum of light produced from the sun rays.

Statement-II: Dark lines are due to absorption of certain radiations by gases in the outer atmosphere of the sun.

3. **Statement-I:** The Bohr's model cannot differentiate between the spectra of hydrogen and deuterium.

Statement-II: The Bohr's model considers the nucleus as infinitely massive in comparison to the orbiting electrons.

4. **Statement-I:** Radius of the first orbit of hydrogen atom is 0.53Å.

Statement-II: Velocity of electron in the first orbit of hydrogen atom is 2.16×10^6 m/s.

5. **Statement-I:** The energy of an electron is largely determined by its principal quantum number.

Statement-II: The principal quantum number (n) is a measure of the most probable distance of finding the electron around the nucleus.

Assertion & Reason MCQs

Directions: These questions consist of two statements each, printed as Assertion and Reason. While answering these questions, you are required to choose any one of the following four responses.

1) **Both Assertion and Reason are True and the Reason is a correct explanation of the Assertion.**
2) **Both Assertion and Reason are True but Reason is not a correct explanation of the Assertion.**
3) **Assertion is True but the Reason is False.**
4) **Assertion is False but Reason is True.**

1. **Assertion:** In Rutherford's experiment, majority of α-particles went straight through the metal foil without being stopped or even appreciably deflected.

Reason: Almost all the space occupied in an atom is empty.

2. **Assertion:** Balmer series lies in the visible region of electromagnetic spectrum.

Reason: $\frac{1}{\lambda}=R\left(\frac{1}{2^2}-\frac{1}{n^2}\right)$, where n = 3, 4, 5.

3. **Assertion:** In the case of Lyman series electrons move from n > 1 level to n = 1 level.

Reason: In the case of Paschen series electrons move from n > 3 level to n = 3 level.

4. **Assertion:** For a hydrogen atom n = 1 level has energy –13.6 eV.

Reason: For hydrogen atom n = 2 level has energy – 3.4 eV.

5. **Assertion:** Bohr had to postulate that the electrons in stationary orbits around the nucleus do not radiate energy.

Reason: According to classical physics all accelerating electrons radiate energy.

Match the Column MCQs

1. For Bohr's atomic model, total energy (E), speed (v) of electron and radius of the orbit (r) are related with atomic number (z) and state of the electron (n), then match the column.

	Column-I		Column-II
(a)	v	(i)	$\frac{n^2}{z}$
(b)	E	(ii)	$\frac{z}{n}$
(c)	r	(iii)	$\frac{z^2}{n^2}$

1) (a)-(ii) (b)-(i) (c)-(iii)
2) (a)-(ii) (b)-(iii) (c)-(i)
3) (a)-(iii) (b)-(i) (c)-(ii)
4) (a)-(i) (b)-(ii) (c)-(iii)

2. For Bohr's atomic model $\frac{1}{\lambda} = Rz^2\left[\frac{1}{n_1^2} - \frac{1}{n_2^2}\right]$

	Column-I		Column-II
(a)	Lyman series	(i)	$n_1 = 3$
(b)	Balmer series	(ii)	$n_1 = 2$
(c)	Paschen series	(iii)	$n_1 = 4$
(d)	P fund series	(iv)	$n_1 = 1$
		(v)	$n_1 = 5$

1) (a)-(iv) (b)-(ii) (c)-(i) (d)-(v)
2) (a)-(i) (b)-(iii) (c)-(ii) (d)-(iv)
3) (a)-(ii) (b)-(iv) (c)-(i) (d)-(v)
4) (a)-(iv) (b)-(i) (c)-(ii) (d)-(iii)

3. If total energy is (1) in magnitude, potential energy is (2) in magnitude and kinetic energy is (3) then match the column for the Bohr's atomic model.

	Column-I		Column-II
A.	Total Energy	(i)	a
B.	PE	(ii)	–2a
C.	KE	(iii)	–a

1) (a)-(iii) (b)-(i) (c)-(ii)
2) (a)-(ii) (b)-(iii) (c)-(i)
3) (a)-(i) (b)-(iii) (c)-(ii)
4) (a)-(iii) (b)-(ii) (c)-(i)

NCERT Exemplar MCQs

1. Taking the Bohr radius as $a_0 = 53$ pm, the radius of Li^{++} ion in its ground state, on the basis of Bohr's model, will be about:
1) 53 pm
2) 27 pm
3) 18 pm
4) 13 pm

2. The binding energy of a H-atom, considering an electron moving around a fixed nuclei (proton), is

$$B = -\frac{me^4}{8n^2\varepsilon_0^2h^2} \text{ (M = electron mass)}$$

If one decides to work in a frame of reference where the electron is at rest, the proton would be moving around it. By similar arguments, the binding energy would be

$$B = -\frac{me^4}{8n^2\varepsilon_0^2h^2} \text{ (M = proton mass)}$$

This last expression is not correct, because:
1) n would not be integral
2) Bohr-quantisation applies only two electron
3) The frame in which the electron is at rest is not inertial
4) The motion of the proton would not be in circular orbits, even approximately.

3. The simple Bohr model cannot be directly applied to calculate the energy levels of an atom with many electrons. This is because:
1) Of the electrons not being subject to a central force
2) Of the electrons colliding with each other
3) Of screening effects
4) The force between the nucleus and an electron will no longer be given by Coulomb's law

4. For the ground state, the electron in the H-atom has an angular momentum = h, according to the simple Bohr model. Angular momentum is a vector and hence there will be infinitely many orbits with the vector pointing in all possible directions. In reality, this is not true:
1) Because Bohr model gives incorrect values of angular momentum
2) Because only one of these would have a minimum energy
3) Angular momentum must be in the direction of spin of electron
4) Because electrons go around only in horizontal orbits

5. O_2 molecule consists of two oxygen atoms. In the molecule, nuclear force between the nuclei of the two atoms:
   1) Is not important because nuclear forces are short-ranged
   2) Is as important as electrostatic force for binding the two atoms
   3) Cancels the repulsive electrostatic force between the nuclei
   4) Is not important because oxygen nucleus have equal number of neutrons and protons

6. Two H atoms in the ground state collide inelastically. The maximum amount by which their combined kinetic energy is reduced is:
   1) 10.20 eV   2) 20.40 eV
   3) 13.6 eV   4) 27.2 eV

7. A set of atoms in an excited state decays:
   1) In general to any of the states with lower energy
   2) Into a lower state only when excited by an external) electric field
   3) All together simultaneously into a lower state
   4) To emit photons only when they collide

NEET & AIIMS Past Year MCQs

1. Some energy levels of a molecule are shown in the figure with their wavelengths of transitions. Then: **(2024 Re)**

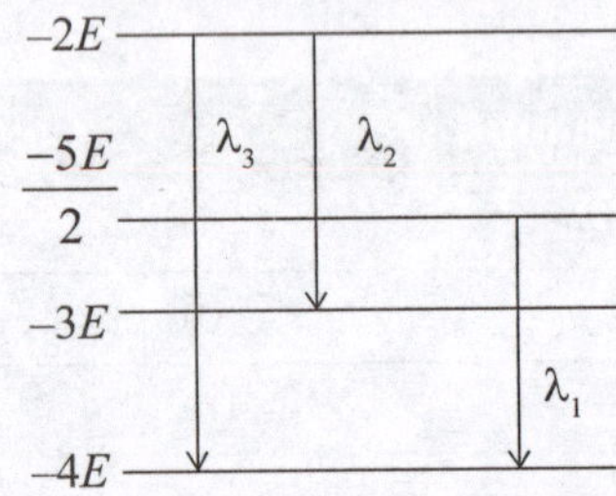

   1) $\lambda_3 > \lambda_2, \lambda_1 = 2\lambda_2$ 2) $\lambda_3 > \lambda_2, \lambda_1 = 4\lambda_2$
   3) $\lambda_1 > \lambda_2, \lambda_2 = 2\lambda_3$ 4) $\lambda_2 > \lambda_1, \lambda_2 = 2\lambda_3$

2. The spectral series which corresponds to the electronic transition from the levels $n_2 = 5, 6, \ldots$ to the level $n_1 = 4$ is **(2024 Re)**
   1) Pfund series   2) Brackett series
   3) Lyman series   4) Balmer series

3. Given below are two statements: **(2024)**

   Statement I: Atoms are electrically neutral as they contain equal number of positive and negative charges.

   Statement II: Atoms of each element are stable and emit their characteristic spectrum.

   In the light of the above statements, choose the *most appropriate* answer from the options given below:
   1) Statement I is correct but Statement II is incorrect.
   2) Statement I is incorrect but Statement II is correct.
   3) Both Statement I and Statement II are correct.
   4) Both Statement I and Statement II are incorrect.

4. Match List I with List II. **(2024)**

List I (Spectral Lines of Hydrogen for transitions from)		List II (Wavelengths (nm))	
A.	$n_2 = 3$ to $n_1 = 2$	I.	410.2
B.	$n_2 = 4$ to $n_1 = 2$	II.	434.1
C.	$n_2 = 5$ to $n_1 = 2$	III.	656.3
D.	$n_2 = 6$ to $n_1 = 2$	IV.	486.1

   Choose the correct answer from the options given below:
   1) A-IV, B-III, C-I, D-II   2) A-I, B-II, C-III, D-IV
   3) A-II, B-I, C-IV, D-III   4) A-III, B-IV, C-II, D-I

5. In hydrogen spectrum, the shortest wavelength in the Balmer series is λ. The shortest wavelength in the Bracket series is: **(2023)**
   1) 16 λ   2) 2 λ
   3) 4 λ   4) 9 λ

6. The radius of inner most orbit of hydrogen atom is 5.3×10^{-11} m. What is the radius of third allowed orbit of hydrogen atom? **(2023)**
   1) 4.77Å   2) 0.53Å
   3) 1.06Å   4) 1.59Å

7. Let T_1 and T_2 be the energy of an electron in the first and second excited states of hydrogen atom, respectively. According to the Bohr's model of an atom, the ratio $T_1 : T_2$ is: **(2022)**
   1) 9 : 4   2) 1 : 4
   3) 4 : 1   4) 4 : 9

8. What is the maximum wavelength for Balmer series in H atom. **(AIIMS 2018)**
   1) 6566 Å   2) 5665 Å
   3) 3574 Å   4) 6656 Å

9. **Assertion:** Balmer series lines in the visible region of electromagnetic spectrum. **(AIIMS 2017)**

   Reason: $\frac{1}{\lambda} = R\left(\frac{1}{(2)^2} - \frac{1}{(n)^2}\right)$, Where n = 3, 4, 5--------

   1. If both Assertion & Reason are True & the Reason is a correct explanation of the Assertion.
   2. If both Assertion & Reason are True but Reason is not a correct explanation of the Assertion.
   3. If Assertion is True but the Reason is False.
   4. If both Assertion & Reason are False.

Answer Key

NCERT Topic-wise MCQs

1	2	3	4	5	6	7	8	9	10	11	12	13	14	15	16	17
1	3	2	2	2	3	3	2	2	4	4	1	1	4	3	1	1
18	**19**	**20**	**21**	**22**	**23**	**24**	**25**	**26**	**27**	**28**	**29**	**30**	**31**	**32**	**33**	**34**
4	2	1	3	4	4	4	4	3	2	1	1	4	3	2	2	4
35	**36**	**37**	**38**	**39**	**40**	**41**	**42**	**43**	**44**	**45**	**46**	**47**	**48**	**49**	**50**	**51**
4	1	4	2	1	4	3	1	1	3	2	4	4	2	1	2	2
52	**53**	**54**	**55**	**56**	**57**	**58**	**59**									
1	1	2	3	2	3	3	2									

High Order Time Intensive MCQs

1	2	3	4	5	6	7	8	9	10	11	12	13	14	15	16	17
2	1	3	2	2	3	1	3	1	2	1	3	3	3	4	4	2
18	**19**	**20**	**21**	**22**												
3	3	3	1	3												

Statement Based MCQs

1	2	3	4	5
2	1	1	1	1

Assertion & Reason MCQs

1	2	3	4	5
1	1	2	2	2

Match the Columns MCQs

1	2	3
2	1	4

NCERT Exemplar MCQs

1	2	3	4	5	6	7
3	3	1	1	1	1	1

NEET & AIIMS Past Year MCQs

1	2	3	4	5	6	7	8	9
4	2	1	4	3	1	1	1	2

CHAPTER 13

Nuclei

Topicwise Number of Questions Analysis (2025-2016)

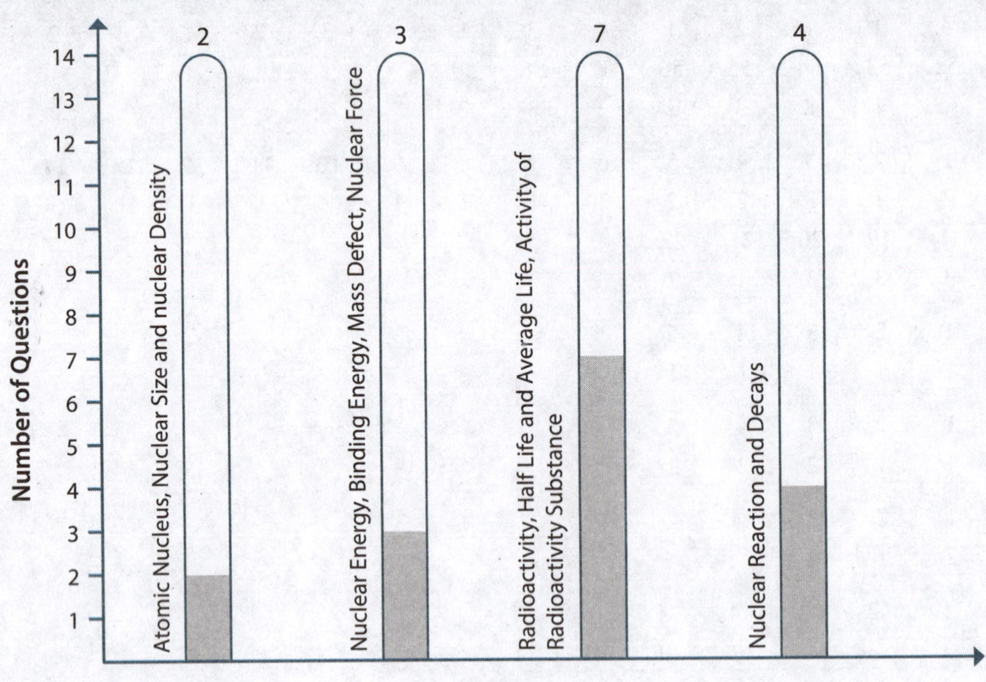

NCERT-Picks

Atomic Mass, Nucleus and Nuclear Density

Formulas

1. Radius of the nucleus, $R = R_0 A^{1/3}$

   [Where $R_0 = 1.1 \times 10^{-15}$ m and A is the mass number]

2. Volume of the nucleus, $V = \frac{4\pi R_0^3 A}{3}$

3. Density of the nucleus, $\rho = \frac{M}{V} = \frac{3M}{4\pi R_0^3 A}$

Binding Energy and Mass Defect

Formulas

1. Mass defect, $\Delta m = [Zm_p + (A - Z)m_n - M]$

   [Z is the atomic number, A is mass number, m_p and m_n is the mass of proton and neutron, M is the mass of nucleus]

2. Total Binding energy = $(\Delta m)\, c^2$ [c is the speed of light]

3. Average B.E / nucleon = $\frac{(\Delta m)c^2}{A}$

4. Binding energy = $\Delta m \times 931.5$ MeV

5. Packing fraction = $\frac{\text{Mass excess}}{\text{Mass number}} = \frac{\Delta m}{A}$

6. Einstein's mass-energy, $E = mc^2$

7. 1 a.m.u = $\frac{1}{12}$ Mass of C – 12 atom.

Radioactivity and Half Life

Formulas

1. Number of decay per unit time, $\frac{dN}{dt} = -\lambda N$

   [N is a number of active nuclei and λ is decay constant]

2. Number of active nuclei, $N = N_0\, e^{-\lambda t}$

   [N_0 is number of active nuclei at t = 0]

3. Time elapsed before the half active nuclei decay,

$$t_{1/2} = \frac{0.693}{\lambda}$$

4. Number of active nuclei at any time t, $N = \frac{N_0}{2^{t/t_{1/2}}}$
5. Number of decays per unit time,

$$A = \frac{A_0}{2^{t/t_{\frac{1}{2}}}}$$

6. Activity, $A = A_0 e^{-\lambda t}$
7. Average life of nucleus, $t_{av} = \frac{t_{1/2}}{0.693} = \frac{1}{\lambda}$

Numerical Profile

Law of radioactive decay

(i) For any radioactive substance, radioactive disintegration equation is

(a) $\lambda = \frac{2.303}{t} \log \frac{N_0}{N}$

(b) $\log \frac{N_0}{N} = 0.4342\lambda t$

N_0 = Initial number of radioactive atoms

t = time of decay

N = No. of radioactive atoms left after time '*t*'.

(ii) The number of half lives in the decay time '*t*' is given by

$$n = \frac{t}{t_{1/2}}$$

(iii) The fraction of the radioactive substance that remains undecayed after decay is equal to $\frac{1}{2^n}$

$$\frac{N}{N_0} = \frac{W}{W_0} = \frac{A}{A_0} = \frac{1}{2^n},\ n = \frac{t}{t_{1/2}}$$

[W_0 = initial weight of the radio active element,
W = Weight of the radio active element after the decay time 't']

(iv) For any radio active substance

(a) Activity (A) = λN

(b) Activity (A) $= \frac{0.693 \times N}{t_{\frac{1}{2}}}$

Nuclear Reaction and Decays

Formulas

1. Alpha decay, $_ZX^A \rightarrow {}_{Z-2}Y^{A-4} + {}_2He^4 + Q$
2. Beta$^-$ decay, $_ZX^A \rightarrow {}_{Z+1}Y^A + {}_{-1}e^0 + Q$
3. Beta$^+$ decay, $_ZX^A \rightarrow {}_{Z-1}Y^A + {}_1e^0 + Q$
4. Gamma decay, $_ZX^A \rightarrow {}_ZX^A + \gamma$
5. Alpha absorption, $_ZX^A + {}_2He^4 \rightarrow {}_{Z+2}C^{A+4} \rightarrow {}_{Z+1}Y^{A+3} + {}_1^1H + Q$

Numerical Profile

Q-value of nuclear reaction

Let m_x, m_y, m_{He} be the masses of parent nucleus X, daughter nucleus Y and alpha particle respectively then the value of Q is given as

$$Q = \left(m_x - m_y - m_{He}\right)c^2$$

This energy appears as the kinetic energy of the daughter nucleus Y and the kinetic energy of the alpha particle.

Nuclear Reactor

(i) The output power of a nuclear reactor is given by

P = nE

[n is No. of fissions taking place in 1 sec, E is Energy released per fission]

(ii) If X grams of a nuclear fuel of mass number 'A' undergo fission in a time of 't' sec. 'E' is energy released per fission, then the power output of the nuclear reactor is given by

$$P = \frac{NEX}{At}$$ [N is Avagadro number]

Statements Asked in Exam

- Neutrons were discovered by Chadwick, which is present in nucleus of the atom.
- Nucleon refers to a proton or a neutron.
- A free neutron, unlike a free proton, is unstable. It decays into a proton, an electron and an antineutrino (another elementary particle), and has a mean life of about 1000s.
- The density of nuclear matter is approximately 2.3×10^{17} Kgm^{-3}.
- The difference in mass of a nucleus and its constituents is called the mass defect.
- For nuclear fusion the light nuclei must have sufficient initial energy to overcome the coulomb potential barrier. That is why fusion requires a very high temperature.
- Nuclear forces are stronger than the electrostatic force.
- Nuclear forces act between a pair of neutrons, a pair of protons and also between a neutron proton pair, with the same strength. So, nuclear forces are independent of charge.
- Nuclear forces are negligible, when distance between nucleons is more than 10 fermi.
- Nuclear forces are non-central forces. This shows that the distribution of nucleons in a nucleus is not spherically symmetric.
- A japanese physicist Yukawa postulated in the year 1936 that nuclear forces between nucleons arise on account of continuous exchange of particles called mesons between the nucleons.

- S.I unit of radioacitvity is Becquerel (Bq)

  1Bq = 1 decay / sec

  1 millicurie = 3.7×10^7 decay / sec

  1 microcurie = 3.7×10^4 decay / sec

  1 rutherford = 10^6 decay / sec
- In nuclear reactions the following conservation laws are obeyed.
  - (i) Conservation of linear momentum
  - (ii) Conservation of total energy
  - (iii) Conservation of charge
  - (iv) Conservation of number of nucleons
- In α-decay, an α-particle is emitted from the parent nucleus and mass number of the atom decreases by 4 and atomic number decreases by 2.
- In β minus decay, there is a transformation of neutron into a proton, due to this mass number decreases by one, atomic number increases by one but atomic mass does not change
- In β^+ decay (β plus), orbital electron can combine with a proton in the nucleus to form a neutron and a neutrino, atomic number (z) decreases by one but atomic mass remains same.
- When a nucleus is in the excited state, either due to bombardment with high energy particles or by radioactive transformation it can decay to the ground state by emission of photons, neither mass number, atomic number and atomic mass change.
- Comparison of penetrating powers of α, β and γ

  $\gamma > \beta > \alpha$ (γ = 10000 times of α , β = 100 times of α)

NCERT Topic-wise MCQs

Atomic Nucleus, Nuclear Size and Nuclear Density

1. Two spherical nuclei have mass numbers 216 and 64 with their radii R_1 and R_2 respectively. The ratio, $\frac{R_1}{R_2}$ is equal to
   1) 3 : 2
   2) 1 : 3
   3) 1 : 2
   4) 2 : 3

2. How many neutrons are there in 14 g of $_6C^{14}$? (Avogadro's number = 6×10^{23})
   1) 48×10^{23}
   2) 64×10^{23}
   3) 56×10^{23}
   4) 80×10^{23}

3. If m, m_n and m_p are the masses of $_ZX^A$ nucleus, neutron and proton respectively, then:
   1) $m < (A - Z)m_n + Zm_p$
   2) $m = (A - Z)m_n + Zm_p$
   3) $m = (A - Z)m_p + Zm_p$
   4) $m > (A - Z)m_n + Zm_p$

4. The order of magnitude of density of uranium nucleus ($m_p = 1.67 \times 10^{-27}$ kg) is,
   1) 10^{20} kg m^{-3}
   2) 10^{17} kg m^{-3}
   3) 10^{14} kg m^{-3}
   4) 10^{11} kg m^{-3}

5. If the mass number of an atom is A = 40 and its electron configuration is $1s^2$, $2s^2$, $2p^6$, $3s^2$, $3p^6$, the number of neutrons and protons in its nucleus will be:
   1) 22, 18
   2) 18, 22
   3) 20, 20
   4) 18, 18

6. If the nucleus $^{125}_{32}Al$ has a nucleus radius of about 3.6 fm, then $^{125}_{32}Te$ would have its radius approximately as
   1) 9.6 fm
   2) 12.0 fm
   3) 4.8 fm
   4) 6.0 fm

7. The nucleus which has radius one third of the radius of Cs^{189} is:
   1) Be^a
   2) Li^7
   3) F^{19}
   4) $C1^2$

Nuclear Energy, Binding Energy, Mass Defect, Nuclear Force

8. Calculate binding energy of an alpha particle from the following data:
   Mass of free proton = 1.007825 a.m.u
   Mass of free neutron = 1.008665 a.m.u
   Mass of helium nucleus = 4.002800 a.m.u
   1 a.m.u = 931.5 Mev
   1) 28.5 Mev
   2) 28.3 Mev
   3) 28.1 Mev
   4) 28.7 Mev

9. The binding energy per nucleon of deuterium and helium nuclei are 1.1 MeV and 7 MeV respectively. When two deuterium nuclei fuse to form a helium nucleus the energy released in the fusion is:
   1) 24.6 MeV
   2) 28.5 MeV
   3) 22.3 MeV
   4) 23.6 Mev

10. Fission of nuclei is possible because the binding energy per nucleon in them:
    1) Increase with mass number at low mass number
    2) Decrease with mass number at low mass numbers
    3) Increase with mass number at high mass numbers
    4) Decrease with mass number at high mass numbers

11. If the binding energy per uncleon in $_3Li^7$ and $_2He^4$ nuclei are 5.60 MeV and 7.06 MeV respectovely, then in reaction
    $_1H^1 + {_3Li^7} \rightarrow 2\,{_2He^4}$,
    energy of proton must be
    1) 28.24 MeV
    2) 17.28 MeV
    3) 1.46 MeV
    4) 39.2 MeV

12. Calculate the binding energy per nucleon of $_{20}Ca^{40}$ nucleus $M(_{20}Ca^{40})$ = 39.962589 u, m_n = 1.008665 u, m_p = 1.007825 u:
    1) 8.34 MeV/N
    2) 9.54 MeV/N
    3) 8.46 MeV/N
    4) 10.56 MeV/N

13. Energy of 21g uranium is equal to:
    1) 9.0×10^{13} J
    2) 21×10^{-9} J
    3) 1.89×10^{15} J
    4) 1.9×10^{12} J

14. In a fission process, nucleus A divides into two nuclei B and C, their binding energies being E_a, E_b and E_c respectively. Then:
1) $E_b + E_c = E_a$ 2) $E_b + E_c > E_a$
3) $E_b + E_c < E_a$ 4) $E_b . E_c = E_a$

15. If 200 MeV energy is released in the fission of a single U^{235} nucleus, the number of fissions required per second to produce 1 kilowatt power shall be:
1) 3.125×10^{14} 2) 3.125×10^{13}
3) 3.125×10^{11} 4) 3.125×10^{16}

16. Energy released in the fission of a single $_{92}U^{235}$ nucleus is 200 MeV. The fission rate of a $_{92}U^{235}$ fueled reactor operating at a power level of 5W is:
1) $1.56 \times 10^{10}\ s^{-1}$ 2) $1.56 \times 10^{11}\ s^{-1}$
3) $1.56 \times 10^{16}\ s^{-1}$ 4) $1.56 \times 10^{17}\ s^{-1}$

17. If the energy released in the fission of one nucleus is 200 MeV, then the number of nuclei required per second in a power plant of 16 kW will be:
1) 0.5×10^{14} 2) 0.5×10^{12}
3) 5×10^{12} 4) 5×10^{14}

18. The energy in MeV released due to transformation of 1kg mass completely into energy:
1) 7.625×10 MeV
2) 10.5×10^{29} MeV
3) 2.8×10^{-28} MeV
4) 5.625×10^{29} MeV

19. Assume that a neutron breaks into a proton and an electron. The energy released during this process is:
1) 0.69 MeV 2) 0.73 MeV
3) 0.63 MeV 4) 0.81 MeV

20. If the speed of light were 2/3 of its present value, the energy released in a given atomic explosion will be decreased by a fraction:
1) 2/3 2) 4/9
3) 3/4 4) 5/9

21. For effective nuclear forces, the distance should be:
1) 1Å 2) 10 pm
3) 0.1 pm 4) 1 fm

22. Outside a nucleus:
1) Neutron is stable
2) Proton and neutron both are stable
3) Neutron is unstable
4) Neither neutron nor proton is stable

23. Nuclear forces are:
1) Short ranged attractive and charge independent
2) Short ranged attractive and charge dependent
3) Long ranged repulsive and charge independent
4) Long ranged repulsive and charge dependent

24. If m_0 is the mass of isotope $^{16}_{8}O$, m_p and m_n are the masses of a proton and neutron respectively, the nuclear binding energy of the isotope is
1) $(m_0 - 17m_n)c^2$ 2) $(m_0 - 8m_p)c^2$
3) $(m_0 - 8m_p - 8m_n)c^2$ 4) m_0c^2

25. Two nucleons are at a separation of 1×10^{-15} m. The nuclear force between them is F_1, if both are neutrons, F_2 if both are protons and F_3 if one is a proton and other is a neutron. In such a case:
1) $F_2 > F_1 > F_3$ 2) $F_1 = F_2 = F_3$
3) $F_1 = F_2 > F_3$ 4) $F_1 = F_3 > F_2$

Radioactivity, Half Life and Average Life, Activity of Radioactive Substance

26. In a radioactive substance at t = 0, the number of atoms is 8×10^4. Its half life period is 3 years. The number of atoms 1×10^4 will remain after interval:
1) 10 year 2) 9 year
3) 20 year 4) 3 year

27. The half life period of radium is 300 years. The fraction of sample of radium that would remain after 2700 year is:
1) 1/512 2) 1/128
3) 1/176 4) 1/200

28. The half life period of radium is 2000 years. Its average life time will be:
1) 1550 year 2) 3000 year
3) 2998 year 4) 2886 year

29. The half life of Bi^{210} is 5 days. What time is taken by $(7/8)^{th}$ part of the sample to decay?
1) 3.4 days 2) 10 days
3) 15 days 4) 20 days

30. Half life period of radioactive element depends upon:
1) Amount of element present
2) Temperature
3) Pressure
4) Nature of element

31. In which radioactive disintegration neutron dissociates into proton and electron?
1) He^{+1} emission 2) β-emisdion
3) γ-emission 4) Positron emission

32. The half life period of a radioactive substance is 5 min. The amount of substance decayed in 20 min will be:
1) 93.75% 2) 90%
3) 25% 4) 62.5%

33. If the radioactive decay constant of radium is 1.07×10^{-4} per year, then its half life period is approximately equal to:
1) 8,900 years 2) 7,000 years
3) 6,476 years 4) 2,520 years

34. ^{22}Ne nucleus, after absorbing energy, decays into two α-particles and an unknown nucleus. The unknown nucleus is

1) nitrogen 2) carbon
3) boron 4) oxygen

35. The decay constant of a radioactive sample is λ. The half-life and mean-life of the sample are respectively given by

1) 1/λ and (ln2)/λ 2) (ln 2) /λ and 1/λ
3) λ(ln2) and 1/λ 4) λ/(ln2) and /1λ

36. If half life of a radioactive element is 3 hours after 27 hours it activity becomes:

1) 1/32 2) 1/512
3) 1/8 4) 1/264

37. The half-life period of a radioactive element X is same as the mean-life time of another radioactive element Y. Initially both of them have the same number of atoms. Then

1) X and Y have the same decay rate initially
2) X and Y decay at the same rate always
3) Y will decay at a faster rate than X
4) X will decay at a faster rate than Y.

38. The activity of sample is 64×10^{-5} Ci. Its half-life is 3 days. The activity will become 5×10^{-6} Ci after:

1) 12 days 2) 7 days
3) 18 days 4) 21 days

39. 3.8 days is the half-life period of a sample. After how many days, the sample will become 1/8th of the original substance?

1) 11.4 2) 12.4
3) 16.4 4) 20.4

40. The half-life of ^{131}I is 8 day. Given a sample of ^{131}I at time t = 0, we can assert that

1) no nucleus will decay before t = 4 day
2) no nucleus will decay before t = 8 day
3) all nuclei will decay before t = 16 day
4) a given nucleus may decay at any time after t = 0

41. Three fourth of the active decays in a radioactive sample in 3/4 s. The half life of the sample:

1) 1/2 sec 2) 1 sec
3) 3/8 sec 4) 3/4 sec

42. During mean life of a radioactive element, the fraction that disintegrates is:

1) e 2) $\frac{1}{e}$
3) $\frac{e-1}{e}$ 4) $\frac{e}{e-1}$

43. 1 curie is equal to:

1) 3×10^{10} disintegration/sec
2) 3.7×10^{7} disintegration/sec
3) 5×10^{7} disintegration/sec
4) 3.7×10^{10} disintegration/sec

44. A 280 day old radioactive substance shows an activity of 6000 dps, 140 day later its activity becomes 3000 dps. What was its initial activity ?

1) 20000 dps 2) 24000 dps
3) 12000 dps 4) 6000 dps

45. The decay constant of a radioactive element is 1.5×10^{-9} per second. Its mean life in seconds will be:

1) 1.5×10^{9} 2) 4.62×10^{8}
3) 6.67×10^{8} 4) 10.35×10^{8}

46. 10g of radioactive material of half-life 15 year is kept in store for 20 years. The disintegrated material is given $(16)^{1/3} = 2.52$

1) 12.5 g 2) 10.5 g
3) 6.04 g 4) 4.03 g

47. A freshly prepared radioactive source of half life 2 hr emits radiation of intensity which is 64 times the permissible safe level. The minimum time after which it would be possible to work safely with this source is

1) 6 hr 2) 12 hr
3) 24 hr 4) 28 hr

48. A radioactive substance has an average life of 5 hr. In a time of 5 hours:

1) Half of the active nuclei decay
2) Less than half of the active nuclei decay
3) More than half of the active nuclei decay
4) All active nuclei decay

49. The half life of a sample of a radioactive substance is 1 hour. If 8×10^{10} atoms are present at t = 0, then the number of atoms decayed in the duration t = 2 hour to t = 4 hour will be:

1) 1.7×10^{7} 2) 1.5×10^{10}
3) 3×10^{7} 4) 2.8×10^{10}

50. After a certain lapse of time, the fraction of radioactive polonium undecayed is found to be 12.5% of the initial quantity. What is the duration of this time lapse if half life of polonium is 139 days?

1) 417 days 2) 500 days
3) 420 days 4) 512 days

51. Two nuclei P and Q have equal numbers of atoms at t = 0. Their half lives are 3 hours and 9 hours respectively. Compare their rates of disintegration after 18 hours from the start:

1) 3 : 1 2) 3 : 5
3) 3 : 10 4) 3 : 16

52. Find the number of atoms of $^{60}_{27}$Co necessary to provide a radioactive source of 8.0 mCi strength. The half-life of $^{60}_{27}$Co is 5.3 years.

1) 7.15×10^{16} atoms 2) 9.13×10^{14} atoms
3) 11.41×10^{16} atoms 4) 13.12×10^{14} atoms

53. The half life of $_{92}U^{238}$ against α-decay is 4.5 × 10^9 years. How many disintegrations occur per second in one gram of $_{92}U^{238}$?

1) 2.34 × 10^4 s^{-1}
2) 2.34 × 10^{-4} s^{-1}
3) 3.44 × 10^4 s^{-1}
4) 3.44 × 10^{-4} s^{-1}

54. The half life of radium Ra^{226} is 1620 years. How many atoms decay in 1 second in a 1 gram sample of radium?

1) 4 × 10^{10} s^{-1}
2) 3 × 10^{10} s^{-1}
3) 3 × 10^{-10} s^{-1}
4) 3 × 10^{10} s

55. The curve between the activity A of a radioactive sample and the number of active atoms N is:

1) 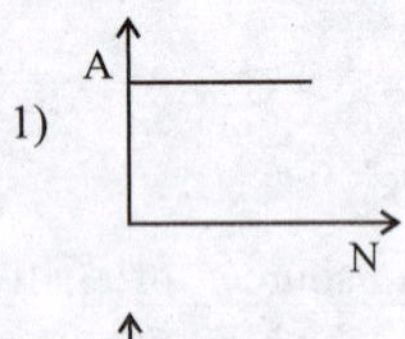

2) 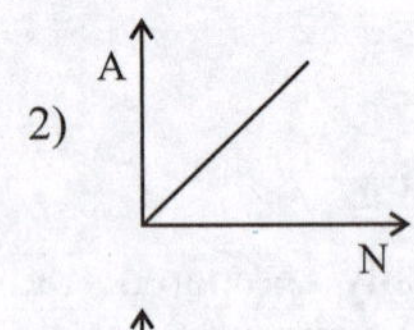

3) 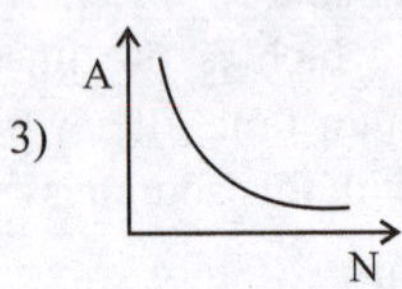

4) 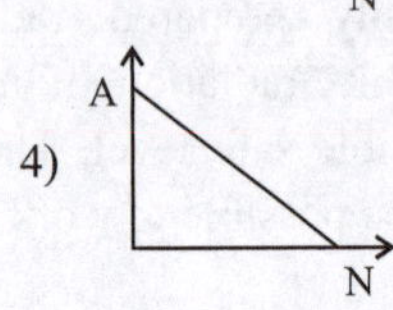

Nuclear Reaction and Decays

56. A deutron strikes $_8O^{16}$ nucleus with the subsequent emission of an alpha particle. What will be the atomic number, mass number and chemical name of the element so produced?

1) $_7N^{14}$
2) $_7N^{15}$
3) $_{13}Al^{27}$
4) None of these

57. A radioactive nucleus can decay by two different processes. The half-life for the first process is t_1 and that for the second process is t_2 then what will be the effective half-life t of the nucleus is

1) $t = t_1 + t_2$
2) $\frac{1}{t} = t_1 + t_2$
3) $t = \frac{t_1 t_2}{t_1 + t_2}$
4) $t = t_1 - t_2$

58. Consider α and β particles and γ - rays each having an energy of 0.5 MeV. In the increasing order of penetrating power, the radiation are respectively:

1) α, β, γ
2) α, γ, β
3) β, γ, α
4) γ, β, α

59. Pick out the incorrect statement from the following:

1) β-emission from the nucleus is always accompanied with a neutrino
2) The energy of the α-particle emitted from a given nucleus is always constant
3) γ-ray emission makes the nucleus more stable
4) Nuclear force is charge - independent

60. In the disintegration series:

$$^{238}_{92}U \xrightarrow{\alpha} X \xrightarrow{\beta^-} {}^{A}_{Z}Y$$

the value of Z and A respectively will be:

1) 92, 236
2) 88, 230
3) 90, 234
4) 91, 234

61. In a radioactive reaction $_{92}X^{232} \longrightarrow {}_{82}Y^{212}$, the number of α-particles emitted is:

1) 7
2) 6
3) 5
4) 4

62. In the given nuclear reaction A,B,C,D,E represent:

$$_{92}U^{238} \xrightarrow{\alpha} {}_BZ^A \xrightarrow{\beta} {}_DY^C \xrightarrow{e} {}_{92}X^{234}$$

1) A = 234 , B = 90 , C = 234 , D = 91 , e = β
2) A = 234 , B = 90 , C = 238 , D = 94 , e = α
3) A = 238 , B = 93 , C = 234 , D = 91 , e = β
4) A = 234 , B = 90 , C = 234 , D = 93 , e = α

63. A radioactive nucleus undergoes a series of decay according to the scheme:

$$A \xrightarrow{\alpha} A_1 \xrightarrow{\beta} A_2 \xrightarrow{\alpha} A_3 \xrightarrow{\gamma} A_4$$

If the mass number and atomic number of A are 180 and 72 respectively, then what are these number for A_4

1) 172 & 69
2) 174 & 70
3) 176 & 69
4) 176 & 70

64. The equation $4\,{}_1H^1 \rightarrow {}_2He^{4++} + Ze^+ + 26$ MeV represents:

1) γ - decay
2) β - decay
3) Fission
4) Fusion

65. In the following reaction the value of 'X' is:

$$_7N^{14} + {}_2He^4 \rightarrow X + {}_1H^1$$

1) $_8N^{17}$
2) $_8O^{17}$
3) $_7O^{16}$
4) $_7N^{16}$

66. In an atomic bomb, the energy is released due to:

1) Chain reaction of neutrons and $_{92}U^{235}$
2) Chain reaction of neutrons and $_{92}U^{238}$
3) Chain reaction of neutrons and $_{92}P^{240}$
4) Chain reaction of neutrons and $_{92}U^{236}$

67. Boron rods in nuclear reactor are used as a

1) Moderator
2) Control rods
3) Coolants
4) Protective shield

High Order Time Intensive MCQs

1. The mass of a ^7_3Li nucleus is 0.042u less than the sum of the masses of all its nucleons. The binding energy per nucleon of ^7_3Li nucleus is nearly? [use, 14 = 931 MeV]
   1) 46 MeV 2) 5.6 MeV
   3) 3.9 MeV 4) 23 MeV

2. An α-particle of mass m suffers one dimensional elastic collision with a nucleus of unknown mass. After the collision, the α-particle is scattered directly backwards losing 75% of its kinetic energy. Then, the mass of the nucleus is
   1) m 2) $\frac{2}{3}m$
   3) 3 m 4) $\frac{3}{2}m$

3. A radioactive nucleus (initial mass number A and atomic number Z) emits 3 α-particles and 2 positrons. The ratio of number of neutrons to that of protons in the final nucleus will be
   1) $\frac{A-Z-4}{Z-2}$ 2) $\frac{A-Z-8}{Z-4}$
   3) $\frac{A-Z-4}{Z-8}$ 4) $\frac{A-Z-12}{Z-4}$

4. Atomic mass of $^{13}_6C$ is 13.00335 amu and its mass number is 13. If, 1 amu = 931 MeV, binding energy of the neutrons presents in the nucleus is
   1) 0.24 MeV 2) 1.44 MeV
   3) 1.68 MeV 4) 3.12 MeV

5. The binding energies of the atoms of elements A and B are E_a and E_b respectively. Three atoms of the element B fuse to give one atom of element A. This fusion process is accompanied by release of energy e. Then, E_a , E_b and e are related to each other as.
   1) $E_a + e = 3E_b$
   2) $E_a = 3E_b$
   3) $E_a - e = 3E_b$
   4) $E_a + 3E_b + e = 0$

6. What is the Q - value of the reaction
   $p + {}^7Li \rightarrow {}^4He + {}^4He$
   The atomic masses of ^1H, ^4He and ^7Li are 1.007825 u, 4.002603 u and 7.016004u respectively
   1) 17.35 MeV 2) 18.06 MeV
   3) 177.35 MeV 4) 170.35 MeV

7. A nucleus disintegrates into two nuclear parts, which have their velocities in the ratio 2 : 1. The ratio of their nuclear sizes will be
   1) $2^{1/3} : 1$ 2) $1 : 3^{1/2}$
   3) $3^{1/2} : 1$ 4) $1 : 2^{1/3}$

8. In the reacion $^2_1H + {}^3_1H \rightarrow {}^4_2He + {}^1_0n$, if the binding energies of ^2_1H , ^3_1H and ^4_2He are respectively a,b and c (in MeV), then the energy (in MeV) released in this reaction is
   1) c + a – b 2) c – a – b
   3) a + b + c 4) a + b – c

9. Radio isotope X with a half - life 1.43×10^9 yr decays to Y which is stable. A sample of the rock from a cave was found to contain X and Y in the ratio 1 : 7. The age of the rock is
   1) 1.96×10^9 yr
   2) 3.96×10^9 yr
   3) 4.29×10^9 yr
   4) 8.40×10^9 yr

10. The half-life of a radioactive isotope X is 20 yr. It decays to another element Y which is stable. The two elements X and Y were found to be in the ration 1 : 7 in a sample of a given rock. The age of the rock is estimated to be
    1) 40 yr 2) 60 yr
    3) 80 yr 4) 100 yr

11. A mixture consists of two radioactive materials A_1 and A_2 with half lives of 20 s and 10 s respectively. Initially, the mixture has 40 g of A_1 and 160 g of A_2. The amount of the two in the mixture will become equal after
    1) 60 s 2) 80 s
    3) 20 s 4) 40 s

12. Two radioactive substances A and B have decay constants 5λ and λ respectively. At t = 0, they have the same number of nuclei. The ratio of number of nuclei of A to that of B will be $(1/e)^2$ after a time interval of
    1) $\frac{1}{\lambda}$ 2) $\frac{1}{2\lambda}$
    3) $\frac{1}{3\lambda}$ 4) $\frac{1}{4\lambda}$

13. A radioactive nucleus of mass M emits a photon of frequency ν and the nucleus recoils. The recoil energy will be
    1) $h^2 \nu^2 / 2Mc^2$ 2) Zero
    3) hν 4) $Mc^2 / h\nu$

14. The radioactivity of a sample is I_1 at a time t_1 and I_2 at a time t_2. If the half-life of the sample is $\tau_{1/2}$ then the number of nuclei that have disintegrated in the time $t_2 - t_1$ is proportional to
    1) $I_1 t_2 - I_2 t_1$ 2) $I_1 - I_2$
    3) $\frac{I_1 - I_2}{\tau_{1/2}}$ 4) $(I_1 - I_2) \tau_{1/2}$

15. The activity of a radioactive sample is measured as N_0 counts per minute at t = 0 and $\frac{N}{e}$ count per minute at t = 5 min. The time (in minute) at which the activity reduces to half its value is

1) $\log_e \frac{2}{5}$
2) $\frac{5}{\log_e 2}$
3) $5\log_e \frac{2}{5}$
4) $5 \log_e 2$

16. The graph that correctly represents the relation of frequency ν of a particular characteristic X-ray with the atomic number Z of the material is

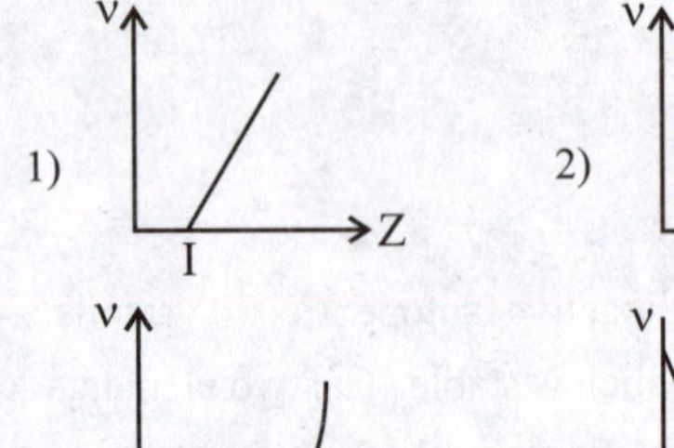

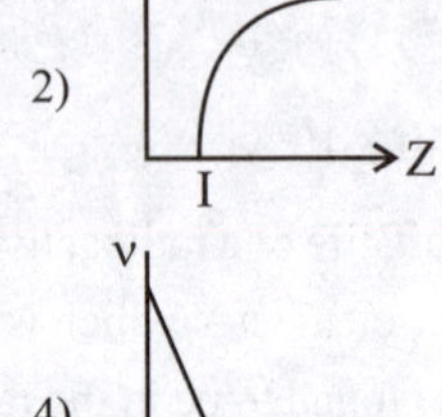

17. Two samples X and Y contain equal amount of radioactive substances. If $\frac{1}{16}$th of the sample X and $\frac{1}{256}$th of the sample Y, remain after 8 hours, then the ratio of half life periods of X and Y is:

1) 2 : 1
2) 1 : 2
3) 1 : 4
4) 4 : 1

18. If 200 MeV energy is released in the fission of a single nucleus of ${}^{235}_{92}U$, the fissions which are required to produce a power of 1 kW is :

1) 3.125×10^{13}
2) 1.52×10^{6}
3) 3.125×10^{12}
4) 3.125×10^{14}

19. The half life of a radioactive substance is 20 minutes. The approximate time interval $(t_2 - t_1)$ between the time t_2 when $\frac{2}{3}$ of it has decayed and time t_1 when $\frac{1}{3}$ of it had decayed is

1) 7 min
2) 14 min
3) 20 min
4) 28 min

20. The decay constant, for a given radioactive sample, is 0.3465 day^{-1}. What percentage of this sample will get decayed in a period of 4 days ?

1) 100%
2) 50%
3) 75%
4) 10%

21. The activity of a sample of radioactive material is R_1 at time t_1 and R_2 at time t_2 $(t_2 > t_1)$. If mean life of the radioactive sample is T, then:

1) $R_1 t_1 = R_2 t_2$
2) $\frac{R_1 - R_2}{t_2 - t_1} = \text{constant}$
3) $R_2 = R_1 \exp\left(\frac{t_1 - t_2}{T}\right)$
4) $R_2 = R_1 \exp\left(\frac{t_1}{T t_2}\right)$

Statement Based MCQs

Directions: These questions consist of two statements each, printed as Statement-I and Statement-II. While answering these questions, you are required to choose any one of the following four responses.

1) **Both Statement-I and Statement-II are correct.**
2) **Both Statement-I and Statement-II are incorrect.**
3) **Statement-I is correct & Statement-II is incorrect.**
4) **Statement-I is incorrect & Statement-II is correct.**

1. **Statement-I:** On α decay daughter nucleus shifts two place to the left from the parent nucleus.
Statement-II: An α particle have four units of mass.

2. **Statement-I:** Electrons of nuclear origin are called β-particles.
Statement-II: α-particles have a higher q/m (charge to mass ratio) than β -particles.

3. **Statement-I:** Fission reaction is used to initiate fusion process.
Statement-II: Fission produced large amount of heat inside the reactor and the temperature can be raised very high.

4. **Statement-I:** For nuclear fusion reactions very high temperature is required.
Statement-II: Only at high temperature the fusing nuclei get enough energy to overcome electrostatic repulsion between them.

5. **Statement-I:** The binding energy per nucleon decreases for a nucleus of large mass number.
Statement-II: The coulomb repulsion between the protons makes the nucleus increasingly less stable.

Assertion & Reason MCQs

Directions: These questions consist of two statements each, printed as Assertion and Reason. While answering these questions, you are required to choose any one of the following four responses.

1) **Both Assertion and Reason are True and the Reason is a correct explanation of the Assertion.**
2) **Both Assertion and Reason are True but Reason is not a correct explanation of the Assertion.**
3) **If Assertion is True but the Reason is False.**
4) **Assertion is False but Reason is True.**

1. **Assertion:** In β-decay, all the emitted electron do not have the equal amount energy.
   Reason: β-decay is not a two body decay process.

2. **Assertion:** Fe^{56} nucleus is more stable than U^{235} nucleus.
   Reason: Binding energy of Fe^{56} nucleus is more than binding energy of U^{235}

3. **Assertion:** Nuclear binding energy (or mass defect) of hydrogen nucleus is zero.
   Reason: Hydrogen nucleus contain only one nucleon.

4. **Assertion:** For the scattering of α-particles at a large angle only the nucleus of the atom is responsible.
   Reason: Nucleus is very heavy as compared to electrons.

5. **Assertion:** In a radioactive disintegration, an electron is emitted by the nucleus.
   Reason: Electrons are present inside the nucleus.

Match the Columns MCQs

1. In the following chain,

$$P \to Q \to R$$

P and Q are radioactive, while R is stable. Initially, we have only P and Q nuclei. There is no nucleus of R. As the time passes, match the two columns.

	Column-I		Column-II
(a)	Nuclei of (P + Q)	(i)	will increase continuously
(b)	Nuclei of Q	(ii)	will decrease continuously
(c)	Nuclei of (R + Q)	(iii)	will first increase then decrease
(d)	Nuclei of (P + R)	(iv)	data insufficient

1) (a)-(i) (b)-(iv) (c)-(iii) (d)-(ii)
2) (a)-(ii) (b)-(iii) (c)-(i) (d)-(iv)
3) (a)-(i) (b)-(ii) (c)-(iii) (d)-(iv)
4) (a)-(ii) (b)-(iv) (c)-(i) (d)-(iv)

2. There are some of the nuclear reactions given in column-I. Match this with the energy involved in these reactions in column-II.

	Column-I		Column-II
(a)	${}_1^2H + {}_1^2H \to {}_1^3H + {}_1^1H + E_1$	(i)	17.6 MeV
(b)	${}_1^3H + {}_1^2H \to {}_1^4He + {}_0^1n + E_2$	(ii)	18.3 MeV
(c)	${}_1^2H + {}_1^2H \to {}_2^3He + {}_0^1n + E_3$	(iii)	4 MeV
(d)	${}_2^3H + {}_1^2H \to {}_2^4He + {}_1^1H + E_4$	(iv)	13.3 MeV
		(v)	200 MeV

1) (a)-(i) (b)-(iv) (c)-(iii) (d)-(ii)
2) (a)-(iii) (b)-(i) (c)-(iv) (d)-(ii)
3) (a)-(i) (b)-(ii) (c)-(iii) (d)-(iv)
4) (a)-(iii) (b)-(iv) (c)-(ii) (d)-(i)

3. Match Column-I of the nuclear processes with Column-II containing parent nucleus and one of the end products of each process.

	Column-I		Column-II
(a)	Alpha decay	(i)	${}_8^{15}O \to {}_7^{15}N + ...$
(b)	β^+ decay	(ii)	$U^{238} \to U^{234} + ...$
(c)	Fission	(iii)	${}_{83}^{185}Bi \to {}_{82}^{184}Pb + ...$
(d)	Proton emission	(iv)	${}_{94}^{239}Pb \to {}_{57}^{140}La + ...$

1) (a)-(iv) (b)-(iii) (c)-(i) (d)-(ii)
2) (a)-(i) (b)-(iv) (c)-(ii) (d)-(iii)
3) (a)-(ii) (b)-(i) (c)-(iv) (d)-(iii)
4) (a)-(iv) (b)-(iii) (c)-(ii) (d)-(i)

4. Match the Column-I of properties with Column-II of reactions.

	Column-I		Column-II
(a)	Mass of products formed is less than the original mass of the system in	(i)	α-decay
(b)	Binding energy per nucleon increase in	(ii)	β-decay
(c)	Mass number is conserved in	(iii)	Nuclear fission
(d)	Charge number is conserved in	(iv)	Nuclear fusion

1) (a)-(i,ii,iii,iv) (b)-(i,ii,iii,iv) (c)-(i,ii,iii,iv) (d)-(i,ii,iii,iv)
2) (a)-(ii,i,iii,iv) (b)-(ii,i,iv) (c)-(i,ii,iv,iii) (d)-(i,ii,iii,iv)
3) (a)-(iv,ii,iii,i) (b)-(iii,ii,i,iv) (c)-(i,iii,ii) (d)-(i,ii,iii,iv)
4) (a)-(iii,ii,i) (b)-(iv,iii,i,iv) (c)-(iii,ii,iv) (d)-(i,ii,iii,iv)

NCERT Exemplar MCQs

1. Suppose we consider a large number of containers each containing initially 10000 atoms of a radioactive material with a half life of 1 yr. After 1 yr:

1) All the containers will have 5000 atoms of the material
2) All the containers will contain the same number of atoms of the material but that number will only be approximately 5000
3) The containers will in general have different numbers of the atoms of the material but their average will be close to 5000
4) None of the containers can have more than 5000 atoms

2. The gravitational force between a H-atom and another particle of mass m will be given by Newton's law:

$F = G\frac{M.m}{r^2}$, where r is in km and

1) $M = m_{proton} + m_{electron}$
2) $M = m_{proton} + m_{electron} - \frac{B}{C^2}(B = 13.6 \text{ eV})$
3) M is not relate to the mass of the hydrogen atom.
4) $M = m_{proton} + m_{electron} - \frac{|V|}{c^2}|V|$ = magnitude of the potential energy of electron in the H-atom

3. M_x and M_y denote the atomic masses of the parent and the daughter nuclei respectively in radioactive decay. The Q-value for a β^-decay is Q_1 and that for a β^+ decay is Q_2. If m_e denotes the mass of an electron, then which of the following statements is correct?

1) $Q_1 = (M_x - M_y)c^2$ and $Q_2 = [M_x - M_y - 2m_e]c^2$
2) $Q_1 = (M_x - M_y)c^2$ and $Q_2 = (M_x - M_y)c^2$
3) $Q_1 = (M_x - M_y - 2m_e)c^2$ and $Q_2 = (M_x - M_y + 2c_e)c^2$
4) $Q_1 = (M_x - M_y + 2m_e)c^2$ and $Q_2 = (M_x - M_y + 2m_e)c^2$

4. Tritium is an isotope of hydrogen whose nucleus triton contains 2 neutrons and 1 proton. Free neutrons decay into $p + \overline{e} + \overline{n}$. If one of the neutrons in Tritium decays, it would transform into He^3 nucleus. This does not happen. This is because:

1) Tritium energy is less than that of a He^3 nucleus
2) The electron created in the beta decay process cannot remain in the nucleus
3) Both the neutrons in triton have to decay simultaneously resulting in a nucleus with 3 protons, which is not a He_3 nucleus.
4) Free neutron decay due to external perturbations which is absent in triton nucleus

5. Heavy stable nuclei have more neutrons than protons. This is because of the fact that:

1) Neutrons are heavier than protons
2) Electrostatic force between protons are repulsive
3) Neutrons decay into protons through beta decay
4) Nuclear forces between neutrons are weaker than that between protons

6. In a nuclear reactor, moderators slow down the neutrons which come out in a fission process. The moderator used have light nuclei. Heavy nuclei will not serve the purpose, because:

1) They will break up
2) Elastic collision of neutrons with heavy nuclei will not slow them down
3) The net weight of the reactor would be unbearably high
4) Substances with heavy nuclei do not occur in liquid or gaseous state at room temperature

NEET & AIIMS Past Year MCQs

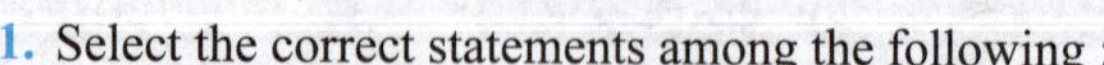

1. Select the correct statements among the following :

A. Slow neutrons can cause fission in $^{235}_{92}U$ than fast neutrons.

B. α-rays are Helium nuclei.

C. β-rays are fast moving electrons or positrons.

D. γ-rays are electromagnetic radiations of wavelengths larger than X-rays.

Choose the most appropriate answer from the options given below: **(2024 Re)**

1) A, B and C only
2) A, B and D only
3) A and B only
4) C and D only

2. The half life of a radioactive substance is 20 minutes. In how much time, the activity of substance drops to $(1/16)^{th}$ of its initial value? **(2023)**

1) 80 minutes
2) 20 minutes
3) 40 minutes
4) 60 minutes

3. In the given nuclear reaction, the element X is:

$^{22}_{11}Na \rightarrow X + e^+ + \nu$ **(2022)**

1) $^{22}_{12}Mg$
2) $^{23}_{11}Na$
3) $^{23}_{10}Ne$
4) $^{22}_{10}Ne$

4. A nucleus of mass number 189 splits into two nuclei having mass number 125 and 64. The ratio of radius of two daughter nuclei respectively is: **(2022)**

1) 25 : 16
2) 1 : 1
3) 4 : 5
4) 5 : 4

5. A nucleus with mass number 240 breaks into two fragments each of mass number 120, the binding energy per nucleon of unfragmented nuclei is 7.6 MeV while that of fragments is 8.5 MeV. The total gain in the Binding Energy in the process is: **(2021)**
   1) 9.4 MeV
   2) 804 MeV
   3) 216 MeV
   4) 0.9 MeV

6. The half-life of a radioactive nuclide is 100 hours. The fraction of original activity that will remain after 150 hours would be: **(2021)**
   1) $\frac{1}{2\sqrt{2}}$
   2) $\frac{2}{3}$
   3) $\frac{2}{3\sqrt{2}}$
   4) 1/2

7. A radioactive nucleus ${}_{Z}^{A}X$ undergoes spontaneous decay in the sequence ${}_{Z}^{A}X \rightarrow {}_{Z-1}B \rightarrow {}_{Z-3}C \rightarrow {}_{Z-2}D$, where Z is the atomic number of element X. The possible decay particles in the sequence are: **(2021)**
   1) $\alpha, \beta^{+}, \beta^{-}$
   2) $\beta^{+}, \alpha, \beta^{-}$
   3) $\beta^{-}, \alpha, \beta^{+}$
   4) $\alpha, \beta^{-}, \beta^{+}$

8. Find the minimum wavelength of X-rays tube emitted by X-ray tube, which is operating at 15 kv. Accelerating voltage will be? **(AIIMS 2018)**
   1) 1.2 Å
   2) 0.82 Å
   3) 1.88 Å
   4) 1.4 Å

9. An atom of mass number 15 and atomic number 7 captures an $\alpha-$particle and then emits a proton. The mass number and atomic number of the resulting atom will be respectively: **(AIIMS 2016)**
   1) 14 and 2
   2) 15 and 3
   3) 16 and 4
   4) 18 and 8

10. Total energy of the electron in hydrogen atom above 0 eV leads to: **(AIIMS 2012)**
    1) Continuous energy states
    2) Large number of discrete ionized states
    3) Balmer series
    4) Paschen series

11. The series corresponding to minimum wavelength transition in H-atom: **(AIIMS 2011)**
    1) Balmer
    2) Lyman
    3) Paschen
    4) Bracket

Answer Key

NCERT Topic-wise MCQs																	
	1	2	3	4	5	6	7	8	9	10	11	12	13	14	15	16	17
	1	1	1	2	1	4	2	3	4	4	2	1	3	2	2	2	4
	18	19	20	21	22	23	24	25	26	27	28	29	30	31	32	33	34
	4	2	2	4	3	1	3	2	2	1	4	3	4	2	1	3	2
	35	36	37	38	39	40	41	42	43	44	45	46	47	48	49	50	51
	2	2	3	4	1	4	3	3	4	2	3	3	2	3	2	1	4
	52	53	54	55	56	57	58	59	60	61	62	63	64	65	66	67	
	1	1	2	2	1	3	1	1	4	3	1	1	4	2	1	2	

High Order Time Intensive MCQs																	
	1	2	3	4	5	6	7	8	9	10	11	12	13	14	15	16	17
	2	3	3	4	1	1	4	2	3	2	4	2	1	4	4	3	1
	18	19	20	21													
	1	3	3	3													

Statement Based MCQs	1	2	3	4	5
	1	3	1	1	1

Assertion & Reason MCQs	1	2	3	4	5
	1	3	1	2	3

Match the Column MCQs	1	2	3	4
	4	2	3	1

NCERT Exemplar MCQs	1	2	3	4	5	6
	3	2	1	1	2	2

NEET & AIIMS Past Year MCQs	1	2	3	4	5	6	7	8	9	10	11
	1	1	4	4	3	1	2	2	4	1	2

CHAPTER

14 Semiconductor Electronics

Topicwise Number of Questions Analysis (2025-2016)

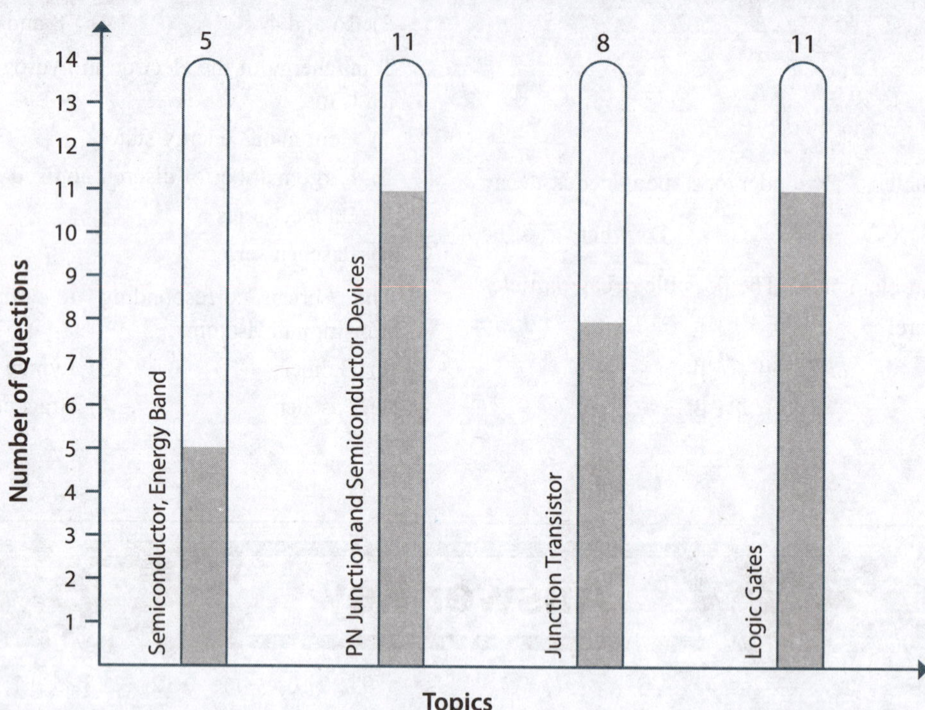

NCERT-PICKS

Intrinsic & Extrinsic Semiconductor

Formulas

1. Conductivity of a metal, $\sigma = \frac{nev}{E}$

   [n is the number density of the electrons, v is the drift speed, E is the electric field and e is the charge of one electron]

2. Current flowing in a semiconductor due to electrons and holes, $I = eA(n_e v_e + n_h v_h)$.

   [A is the area, μ_e and μ_h are the mobility of the electron and the holes]

3. The conductivity of n-type semiconductor, $\sigma_n = en_e\mu_e$

4. Conductivity of p-type semiconductor, $\sigma_p = en_h\mu_h$

5. Electrical conductivity, $\sigma = \frac{1}{\rho} = e(n_e \mu_e + n_h \mu_h)$

   [ρ is the resistivity of the material]

6. Conductivity of intrinsic semiconductor, $\sigma_i = n_i e(\mu_e + \mu_h)$

Numerical Profile

Density of electrons and holes

(i) The number of electrons and holes generated due to temperature are equal in the case of an intrinsic semiconductor. If n_e and n_h represents the number of electron and hole concentrations respectively then

$n_i = n_e = n_h$

For an extrinsic semiconductor,

$n_e n_h = n_i^2$

Where, n_i is number of intrinsic charge carriers.

(ii) In an n-type semiconductor,

$N_D \simeq n_e >> n_h$

(iii) In a p-type semiconductor,

$N_A \simeq n_h >> n_e$.

(iv) Minimum energy require to create a hole-electron pair,

$$E_g = h\nu_{min} = \frac{hc}{\lambda_{max}}$$

(v) Variation of conductivity with temperature,

$$\sigma = \sigma_o \exp\left(\frac{-Eg}{2k_B T}\right)$$

p–n Junction & Semiconductor Diode

Formulas

1. Dynamic resistance, $R_d = \frac{\Delta V}{\Delta I}$

   [ΔV is the change in voltage across the diode and ΔI is the change in current through the diode]

2. **Voltage across the divide circuit:**

   Applied voltage, $V = V_d + IR$

3. For a zener diode, $V_z = V_i - RI$

   Series resistor, $R = \frac{V_i - V_z}{R}$

4. Wattage of diode = Voltage drop × Current

5. The relation for the current I through the junction diode is

   $I = I_0 (e^{eV/kT} - 1)$

   [I_0 is called reverse saturation current, V is voltage across the diode, k is Boltzmann constant and T is absolute temperature]

Numerical Profile

p-n Junction Diode as Half·wave Rectifier

(i) Average output in one cycle

$$V_{dc} = \frac{V_0}{\pi}; I_0 = \frac{I_0}{\pi}$$

(ii) r.m.s. output: $I_{rms} = \frac{I_0}{2}, V_{rms} = \frac{V_0}{2}$

[I_0 is peak current]

(iii) The ratio of the effective alternating component of the output voltage current to the dc component is called as ripple factor.

$$r = \frac{I_{ac}}{I_{dc}} = \left[\left(\frac{I_{rms}}{I_{dc}}\right)^2 - 1\right]^{1/2} = 1.21$$

(iv) Efficiency: It is given by $\% \eta = \frac{P_{out}}{P_{in}} \times 100 = \frac{40.6}{1 + \frac{r_f}{R_L}}$

(a) If $R_L >> r_f$, then $\eta = 40.6\ \%$

(b) If $R_L = r_f$, then $\eta = 20.3\ \%$

(v) Form factor $= \frac{I_{rms}}{I_{dc}} = \frac{\pi}{2} = 1.57$

p-n Junction Diode as Full-Wave Rectifier

(i) Average output : $V_{av} = \frac{2V_0}{\pi}, i_{av} = \frac{2I_0}{\pi}$

(ii) r.m.s. output = $V_{rms} = \frac{V_0}{\sqrt{2}}, I_{rms} = \frac{I_0}{\sqrt{2}}$

(iii) Ripple frequency : The ripple frequency of full wave rectifier = 2 × (Frequency of input ac)

(iv) Efficiency: $\eta_{\%} = \frac{81.2}{1 + \frac{r_f}{R_L}}$ for $r_f << R_L$, $\eta = 81.2\%$

Junction Transistor (N – P – N, P – N – P Transistors)

Formulas

1. Emitter current, $I_E = I_B + I_C$

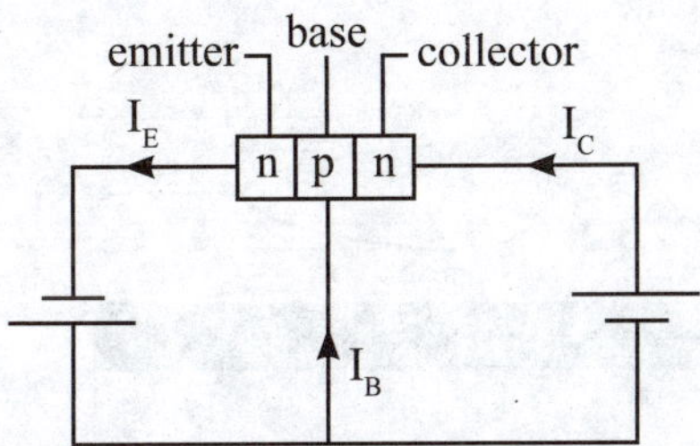

[I_B is base current, I_C is collector current]

2. D.C current gain for common base configuration, $\alpha = I_C / I_E$

3. A.C current gain, $\alpha = \frac{\Delta I_C}{\Delta I_E}$

4. A.C. power gain $= \frac{\text{Change in output power}}{\text{Change in input power}}$

   $$= \left(\frac{\Delta I_C}{\Delta I_E}\right)^2 \times \frac{R_L}{R_{BE}} = \alpha_{ac}^2 \times \text{resistance gain}$$

   (R_{BE} is the resistance offered by base emitter junction)

5. D.C current gain for common emitter configuration,

   $$\beta = \frac{I_C}{I_B} = \frac{I_C}{I_E - I_C} = \frac{\alpha}{1-\alpha} \qquad \alpha = \frac{\beta}{1+\beta}$$

6. A.C current gain for common emitter configuration,

   $$\beta_{a.c} = \frac{\Delta I_C}{\Delta I_B}$$

7. A.C voltage gain $(A_v) = \beta_{a.c} \times \frac{R_L}{R_{BE}}$

8. A.C Power gain $= \frac{\text{Change in output power}(P_o)}{\text{Change in input power}(P_i)} = \frac{\beta_{a.c}^2 \times R_L}{R_{BE}}$

Numerical Profile

(i) For common emitter configuration

(a) Dynamic input resistance $R_i = \left(\frac{\Delta V_{BE}}{\Delta I_B}\right)_{V_{CE} = \text{Constant}}$

(b) Dynamic output resistance $R_o = \left(\frac{\Delta V_{CE}}{\Delta I_C}\right)_{I_B = \text{constant}}$

(ii) For common base configuration

(a) Dynamic input resistance

$$R_i = \left(\frac{\Delta V_{EB}}{\Delta I_E}\right)_{V_{CB} = \text{constant}}$$

(b) Dynamic output resistance

$$R_o = \left(\frac{\Delta V_{CE}}{\Delta I_C}\right)_{I_E = \text{constant}}$$

Logic Gates

Formulas

1. For OR Gate

$X = A + B$

A	B	X
0	0	0
0	1	1
1	0	1
1	1	1

2. For AND Gate

$X = A.B$

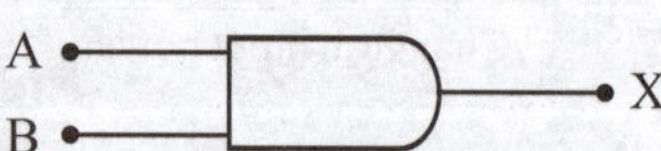

A	B	X
0	0	0
0	1	0
1	0	0
1	1	1

3. For NOT Gate

$X = \bar{A}$ Where X = output

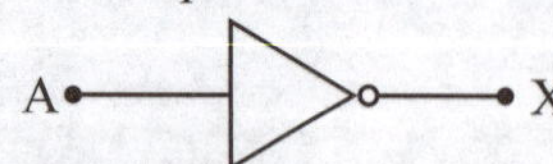

A	X
0	1
1	0

4. For NOR Gate

$X = \overline{A + B}$

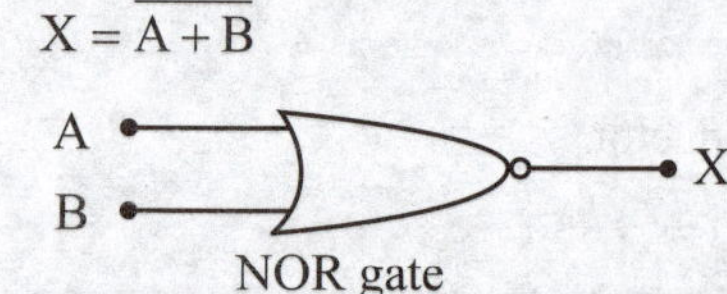

NOR gate

A	B	X
0	0	1
0	1	0
1	0	0
1	1	0

5. For NAND Gate

$X = \overline{A.B}$

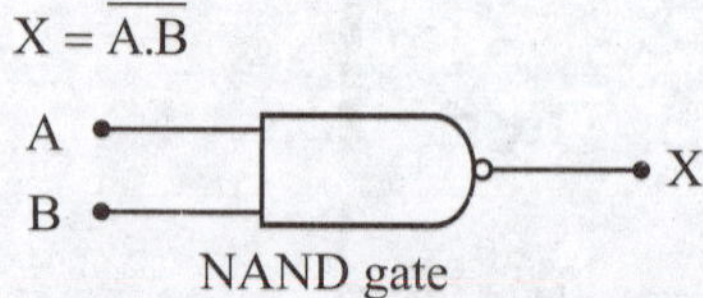

NAND gate

A	B	X
0	0	1
0	1	1
1	0	1
1	1	0

Statements Asked in Exam

- Fermi energy is the maximum possible energy possessed by free electrons of a metal at absolute zero temperature and its value is different for different elements.
- n - type semiconductor is better than p-type semiconductor as electrons have more mobility than holes.
- For a given applied voltage conduction current in n-type semiconductor is more than that of p-type semiconductor.
- For a silicon p-n junction the barrier potential is about 0.7V, whereas for a germanium p-n junction it is approximately 0.3V.
- An ideal diode permits only unidirectional conduction.
- Rectifier is a device which converts ac voltage to dc voltage. It is based on the fact that, a forward biased p-n junction diode conducts and a reverse bias p-n junction does not conduct.
- Solar cells are also used in artificial satellites to operate the various electrical instruments kept inside the satellite.
- A p-n junction with reverse bias can be used as a photodiode to measure intensity of light.
- We can't take one slab of p-type semiconductor and physically join it to another n-type semiconductor to get p-n junction.
- Zener diode is a special purpose diode in which the current suddenly increases in reverse bias condition after a certain voltage.

- Two p-n junction diodes placed back to back, cannot work as a pnp transistor.
- Breakdown voltage is that voltage at which breakdown of the junction of the diode occurs.
- Light emitting diodes give out light radiations when forward biased.
- Solar cells converts solar energy into electrical energy.
- In a transistor circuit, the reverse bias is high as compared to the forward bias. So that it may exert a large attractive force on the charge carriers to enter the collector region.
- Transistors are the current control devices.
- The semiconductors devices are current sensitive devices.
- Current due to minority carriers in the junction diode is independent of the applied voltage. It only depends upon the temperature of the diode.
- Transistor provides good power amplification when they are used in CE configuration.
- By increasing the load resistance we cannot increase the gain of transistor indefinitely.
- In a transistor base is made thin.
- NOR and NAND gates are universal gates because they can be used to perform the basic logic function of AND, OR and NOT gates.
- In modern day circuits, many logical gates or circuits are integrated in one single chip. These are known as integrated circuits (IC).

MCQs ?

NCERT Topic-wise MCQs

Intrinsic & Extrinsic Semiconductor

1. The mean free path of conduction electrons in copper is about 4×10^{-8} m. Find the electric field which can give, on an average, 2eV energy to a conduction electron in a block of copper.

1) 5×10^7 V/m
2) 5×10^6 V/m
3) 5×10^{-6} V/m
4) 5×10^{-7} V/m

2. In an intrinsic semiconductor, the number of conduction electron is 6×10^{19} per cubic meter. How many holes are there in a sample of size 1cm × 1cm × 2mm?

1) 5×10^2
2) 12×10^{-12}
3) 12×10^{12}
4) 4×10^8

3. An intrinsic semiconductor has a resistivity of 0.50 Ωm at room temperature. Find the intrinsic carrier concentration if the mobilities of electrons and holes are 0.39 $m^2V^{-1}s^{-1}$ and 0.11 $m^2V^{-1}s^{-1}$ respectively:

1) 1.2×10^{18} m^{-3}
2) 2.5×10^{19} m^{-3}
3) 1.9×10^{20} m^{-3}
4) 30×10^{-19} m^{-3}

4. The probability of electrons to be found in the conduction band of an intrinsic semiconductor of finite temperature:

1) Increases with increasing band gap
2) Decreases with increasing band gap
3) Decreases with increasing temperature
4) Is independent of the temperature and band gap

5. Pure Si at 300K has equal electron (n_e) and hole (n_h) concentration of 1.5×10^{16} m^{-3}. Doping by indium increases n_h to 4.5×10^{22} m^{-3}. Calculate n_e in the doped silicon:

1) 2×10^{10} m^{-3}
2) 3×10^{15} m^{-3}
3) 7×10^{15} m^{-3}
4) 5×10^9 m^{-3}

6. What are the safe limits of temperature for germanium and silicon?

1) 80°C to 200°C
2) 200°C to 300°C
3) 80°F to 20°F
4) 120°F to 390°F

7. A pure Si crystal has 5×10^{28} atoms m^{-3}. It is doped by 1 ppm concentration of pentavalent atoms. The number of holes is: ($n_i = 1.5 \times 10^{16}$ m^{-3})

1) 2.5×10^6 m^{-3}
2) 4.5×10^9 m^{-3}
3) 6×10^9 m^{-3}
4) 4×10^6 m^{-3}

8. In an n-type semiconductor when all donor states are filled, then the net charge density in the donor states becomes:

1) 1
2) >1
3) <1, but not zero
4) Zero

9. The forbidden energy gap is Ge is 0.72 eV. Given, hc = 12400 eV-Å. The maximum wavelength of radiation that will generate electron hole-pair is:

1) 172220 Å
2) 172.2 Å
3) 17222 Å
4) 1722 Å

10. In a semiconductor:

1) There are no free electrons at any temperature
2) The number of free electrons is more than that in a conductor
3) There are no free electrons at 0K
4) None of these

11. In the ratio of the concentration of electrons to that of holes in a semiconductor is 7/5 and the ratio of currents is 7/4 then what is the ratio of their drift velocities?

1) 4/7
2) 5/8
3) 4/5
4) 5/4

12. A piece of semiconductor is connected in series in an electric circuit. On increasing the temperature, the current in the circuit will be:

1) Decreased
2) Unchanged
3) Increased
4) Stop flowing

13. A pure semiconductor behaves slightly as a conductor at:

1) Room temperature
2) Low temperature
3) High temperature
4) Both (2) and (3)

14. If n_e and n_h are the number of electrons and holes in a semiconductor heavily doped with phosphorus, then:

1) $n_e >> n_h$
2) $n_e << n_h$
3) $n_e \leq n_h$
4) $n_e = n_h$

15. Semiconductor is damaged by the strong current due to:

1) Lack of free electron
2) Excess of electrons
3) Excess of protons
4) None of these

16. Energy bands in solids are a consequence of:

1) Ohm's law
2) Pauli's exclusion principle
3) Bohr's theory
4) Heisenberg's uncertainty principle

17. A semiconductor is cooled from T_1K to T_2K. Its resistance:
1) Will decrease
2) Will increase
3) Will first decrease and then increase
4) Will not change

18. The forbidden energy band gap in semiconductors, conductors and insulators are EG_1, EG_2 and EG_3 respectively. The relation among them is:
1) $EG_1 = EG_2 = EG_3$ 2) $EG_2 < EG_1 < EG_3$
3) $EG_1 > EG_2 > EG_3$ 4) $EG_1 < EG_2 > EG_3$

19. A piece of copper and the other of germanium are cooled from the room temperature to 100K, then which of the following would be a correct statement:
1) Resistance of each increases
2) Resistance of each decreases
3) Resistance of copper increases while that of germanium decreases
4) Resistance of copper decreases while that of germanium increases

20. Which of the following energy band diagram shows the n-type semiconductor:

1) 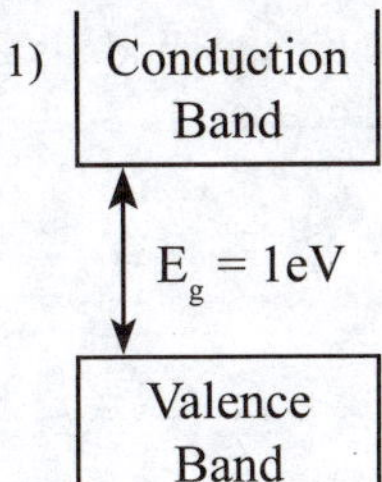

2) 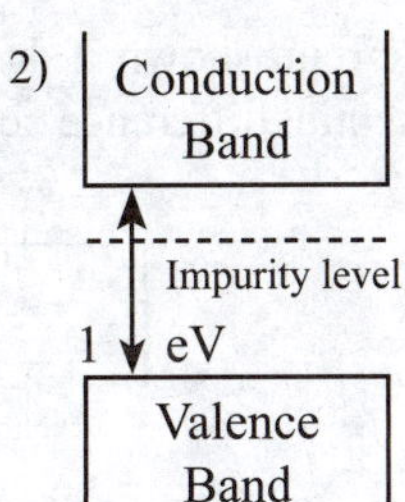

3) 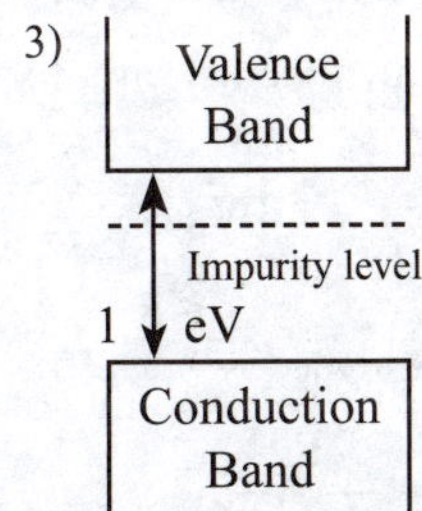

4) 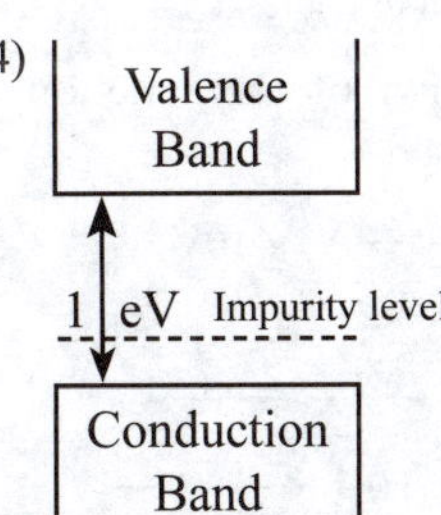

21. Intrinsic semiconductor is electrically neutral. Extrinsic semiconductor having large number of current carriers would be:
1) Positively charged
2) Negatively charged
3) Positively charged or negatively charged depending upon the type of impurity that has been added
4) Electrically neutral

22. In semiconducting materials, the mobilities of electrons and holes are μ_e and μ_h respectively. Which of the following is true ?
1) $\mu_e > \mu_h$ 2) $\mu_e < \mu_h$
3) $\mu_e = \mu_h$ 4) $\mu_e < 0$; $\mu_h > 0$

23. In sample of pure silicon 10^{13} atom/cm^3 is mixed of phosphorus. If all donar atoms are active then what will be resistivity at 20°C if mobility of electron is 1200 cm^2/volt-s
1) 0.5209 ohm cm 2) 5.209 ohm cm
3) 52.09 ohm cm 4) 520.9 ohm cm

24. In germanium crystal, the forbidden energy gap in joule is
1) 1.6×10^{-19} 2) zero
3) 1.12×10^{-19} 4) 1.76×10^{-19}

25. A N-type semiconductor is
1) Negatively charged 2) Positively charged
3) Neutral 4) None of these

26. If μ_e and μ_h are electron and hole mobility, E be the applied electric field, the current density J for intrinsic semiconductor is equal to
1) $n_i e(\mu_e + \mu_h)E$ 2) $n_i e(\mu_e - \mu_h)E$
3) $\dfrac{n_i e(\mu_e + \mu_h)}{E}$ 4) $\dfrac{E}{n_i e(\mu_e + \mu_h)}$

27. An electric field is applied to a semiconductor. Let the number of charge carries be n and the average drift speed be v. If the temperature is increased,
1) Both n and v will increase
2) n will increase but v will decrease
3) v will increases but n will decrease
4) Both n and v will decrease

28. When an impurity is doped into an intrinsic semiconductor, the conductivity of the semiconductor
1) Increases 2) Decreases
3) Remains the same 4) Become zero

29. In a P-type semiconductor, the acceptor level is 57 meV, above the valence band. The maximum wave length of light required to produce a hole will be-
1) 57 Å 2) 57×10^{-3} Å
3) 217100 Å 4) 11.61×10 Å

30. Which of the following when added as an impurity into silicon produces n-type semiconductor?
1) P 2) Al
3) B 4) Mg

31. The probability of electrons to be found in the conduction band of an intrinsic semiconductor at a finite temperature:
1) decreases exponentially with increasing band gap.
2) increases exponentially with increasing band gap.
3) decrease with increasing temperature.
4) is independent of the temperature and the band gap.

32. It is given $n_e = 7 \times 10^{11}$ per metre^3 and $n_h = 5 \times 10^{12}$ per metre^3. The semiconductor is
1) p-type 2) intrinsic
3) n-type 4) insulator

33. Regarding a semi-conductor which one of the following is wrong?

1) There are no free electrons at 0 K
2) There are no free electrons at room temperature
3) The number of free electrons increases with rise of temperature
4) The charge carriers are electrons and holes.

34. At absolute zero, Si acts as:

1) Non-metal 2) Metal
3) Insulator 4) None of these

35. By increasing the temperature, the specific resistance of a conductor and semiconductor:

1) Increases for both
2) Decreases for both
3) Increases, decreases respectively
4) Decreases, increases respectively

36. What is the energy gap in Si semiconductor?

1) 4.4 eV 2) 0.3 eV
3) 0.7 eV 4) 1.1 eV

37. In the energy band diagram of a material shown below, the open circles and filled circles denote holes and electrons respectively. The material is:

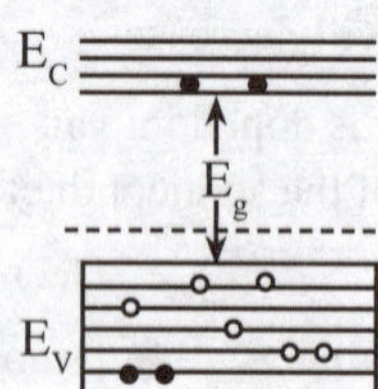

1) An n-type semiconductor
2) A p-type semiconductor
3) An insulator
4) A metal

38. Which one of the following statement is false ?

1) Pure Si doped with trivalent impurities gives a p-type semiconductor
2) Majority carriers in an n-type semiconductor are holes
3) Minority carriers in a p-type simiconductor are electrons
4) The resistance of intrinsic semiconductor decreases with increases of tempertaure

39. Carbon, Silicon and Germanium atoms have four valence electrons each. Their valence and conduction bands are separated by energy band gaps represented by $(E_g)_C$, $(E_g)_{si}$ and $(E_g)_{Ge}$ respectively. Which one of the following relationships is true on their case

1) $(E_g)_C < (E_g)_{Ge}$ 2) $(E_g)_C > (E_g)_{si}$
3) $(E_g)_C = (E_g)_{si}$ 4) $(E_g)_C < (E_g)_{si}$

p–n Junction and Semiconductor Diode / Semiconductor Devices

40. A diode connected to an external resistance and an e.m.f. Assuming that the barrier potential developed in diode is 0.5V. What will be the value of current in the circuit in milliampere?

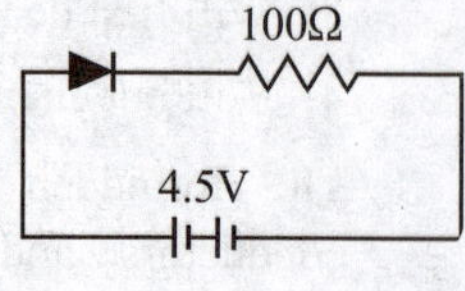

1) 40 mA 2) 4 mA
3) 0.4 mA 4) 400 mA

41. In an unbiased p–n junction, holes diffuse from the p-region to n-region because:

1) Free electrons in the n-region attract them
2) They move across the junction by the potential difference
3) Hole concentration in p-region is more as compared to n-region
4) All of these

42. Region without free electrons and holes in a p-n junction is:

1) n-region 2) p-region
3) Depletion region 4) None of these

43. In the circuit shown if drift current for the diode is 20μA, the potential difference across the diode is:

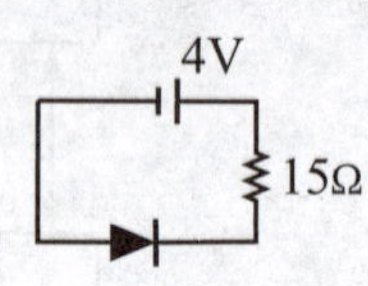

1) 2V 2) 4.5V
3) 4V 4) 2.5V

44. Which of the junction diode shown below are forward biased?

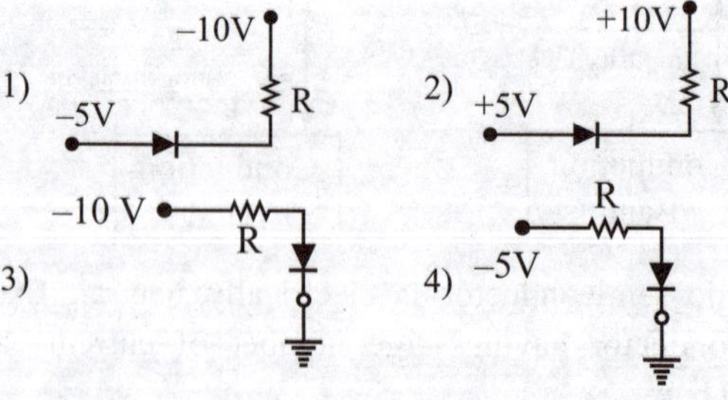

45. In a full wave junction diode rectifier the input ac has rms value of 20V. The transformer used is a step up transformer having primary and secondary turn ratio 1 : 2. The dc voltage in the rectified output is:

1) 12V 2) 24V
3) 36V 4) 42V

46. A Zener diode is specified as having a breakdown voltage of 9.1V, with a maximum power dissipation of 364 mW. What is the maximum current the diode can handle?

1) 40 mA 2) 50 mA
3) 60 mA 4) 70 mA

47. An ideal P–N junction diode shown in figure below. Find the current across the diode.

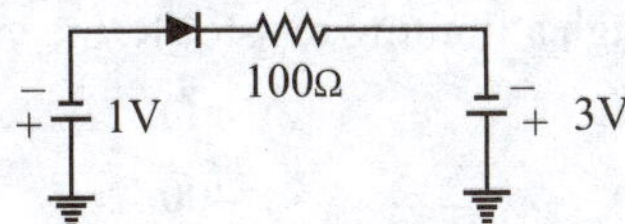

1) Zero
2) 10 mA
3) 20 mA
4) 50 mA

48. The electrical resistance of depletion region / layer is large because:

1) It has no charge carriers
2) It has a large number of charge carriers
3) It contains electrons as charge carriers
4) It has holes as charge carriers

49. The unidirectional flow of current through p-n junction makes it ideal to be used as

1) oscillator
2) rectifier
3) modulator
4) photoelectricity

50. On increasing the reverse bias to a large value in a p-n junction diode, current:

1) Increases slowly
2) Remains fixed
3) Suddenly increases
4) Decreased slowly

51. If a p–n junction diode is not connected to any circuit:

1) The potential is the same everywhere
2) The p-type is a higher potential than the n-type side
3) There is an electric field at the junction directed from the n-type side to the p-type side
4) None of the above

52. Which circuit will not show current in ammeter?

1) 2) 3) 4)

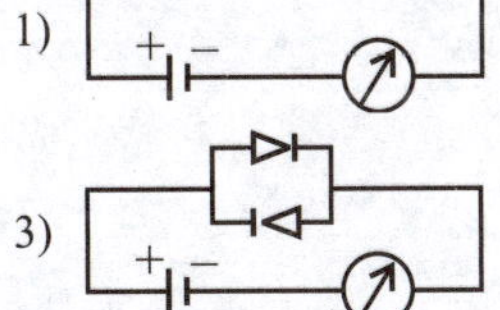

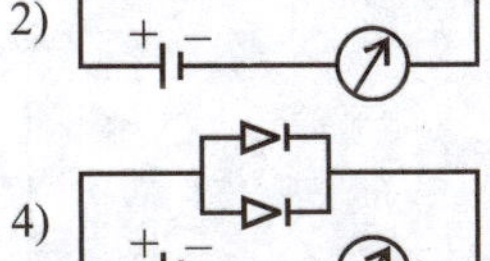

53. When a junction diode is reverse biased, the flow of current across the junction is mainly due to

1) Diffusion of charges
2) Depends upon the nature of material
3) Drift of charges
4) Both drift and diffusion of charges

54. Which is incorrect statement regarding reverse saturation current in p-n junction diode ?

1) Current is due to minority carriers
2) Current doubles for every 100 °C rise in temperature
3) Current carriers are produced by thermal agitation
4) Reverse saturation current is also known as leakage current

55. A potential barrier of 0.50V exist across a p–n junction. If the depletion region is 5×10^{-7} m wide, the intensity of the electric field in this region is:

1) 10^6 V/m
2) 10^5 V/m
3) 2×10^5 V/m
4) 2×10^6 V/m

56. Zener breakdown takes place if:

1) Doped impurity is low
2) Doped impurity is high
3) Less impurity in N-part
4) Less impurity in P-type

57. In a reverse bias diode when the applied voltage changes by 1V, the current is found to change by 0.5 μA. The reverse bias resistance of the diode is

1) 2×10^5 Ω
2) 2×10^6 Ω
3) 200 Ω
4) 2 Ω

58. In a half wave rectifier circuit operating from 50 Hz mains frequency, the fundamental frequency in the ripple would be:

1) 25 Hz
2) 50 Hz
3) 70.7 Hz
4) 100 Hz

59. A p-n photodiode is fabricated from a semiconductor with a band gap of 2.5 eV. The single wavelength is:

1) 5000Å
2) 5100Å
3) 5200Å
4) 500Å

60. Diffusion current in a p-n junction is greater than the drift current in magnitude:

1) if the junction is forward-biased
2) if the junction is reverse-biased
3) if the junction is unbiased
4) in no case

61. For the given circuit shown in fig, to act as full wave rectifier: –a.c. input should be connected across ________ and ________ the d.c. output would appear across ________ and ________.

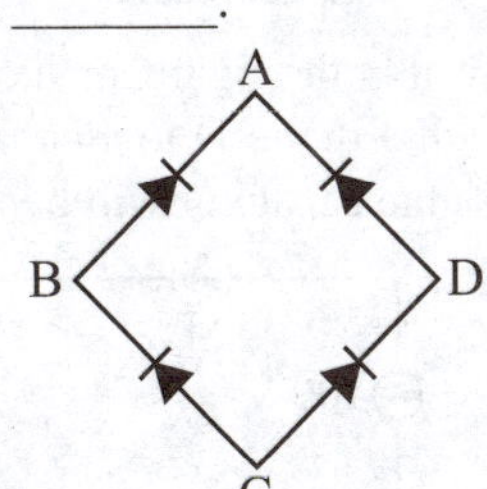

1) A, C, B, D
2) B, D, A, C
3) A, B, C, D
4) C, A, D, B

62. For the given circuit of P-N junction diode, which is correct?

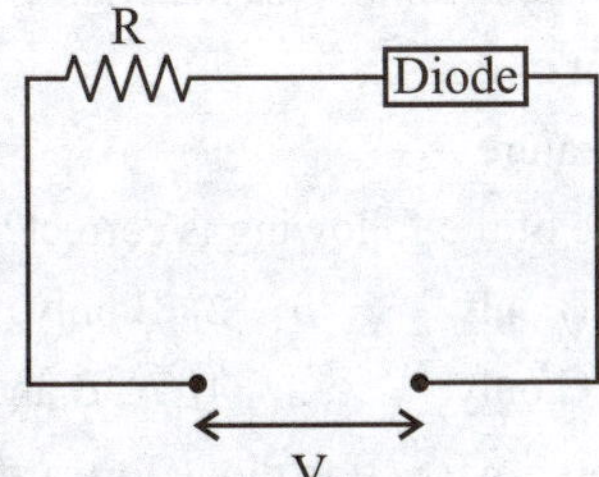

1) In F.B. the voltage across R is V
2) In R.B. the voltage across R is V
3) In F.B. the voltage across R is 2 V
4) In R.B. the voltage across R is 2 V

63. Reverse bias applied to a junction diode
1) Lowers the potential barrier
2) Raises the potential barrier
3) Increases the majority carrier current
4) Decreases the minority carrier current

64. In the middle of the depletion layer of a reverse biased p-n junction, the-
1) Electric field is zero
2) Potential is maximum
3) Electric field is maximum
4) Potential is zero

65. Reason for potential barrier in p-n junction is :
1) Excess of positive charge at junction
2) Deficiency of positive charge at junction
3) Deficiency of negative charge at junction
4) Excess of positive and negative charge at junction

66. For detecting intensity of light we use
1) photodiode in forward bias
2) photodiode in reverse bias
3) LED in forward bias
4) LED in reverse bias

67. In a reverse-biased Zener diode, when the applied bias voltage is equal to the breakdown voltage, then
1) Current remains constant while voltage increases sharply
2) Voltage remains constant while current increases sharply
3) Current and voltage increase
4) Current and voltage decrease

68. In the circuit given in the figure, an a.c. source of 100 V is connected through a diode D to a capacitor. The potential difference across the capacitor will be

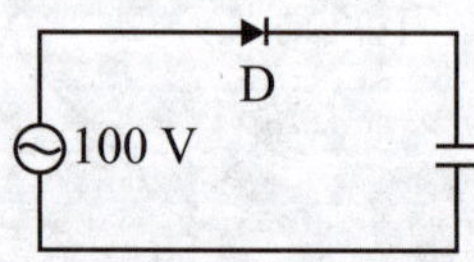

1) 100 V
2) 141 V
3) 200 V
4) 310 V

69. The barrier potential of a p-n junction depends on
A. Type of semi conductor material
B. Amount of doping
C. Temperature
Which one of the following is correct?
1) A and B only
2) B only
3) B and C only
4) A, B and C

70. For making p-n junction diode forward biased :
1) Same potential is applied
2) Greater potential is given to n compared to p
3) Greater potential is given to p compared to n
4) Unbalanced concentration

71. A light emitting diode has a voltage drop of 2 V across it and passes a current of 10 mA. When it operates with a 6 V battery through a limiting resistor R, the value of R is
1) 40 Ω
2) 4 Ω
3) 200 Ω
4) 400 Ω

72. Rectification is the process of conversion of
1) a.c. into d.3)
2) low a.c. into high a.c.
3) d.c. into a.3)
4) low d.c. into high d.c.

73. The current in the circuit shown in the figure considering ideal diode is

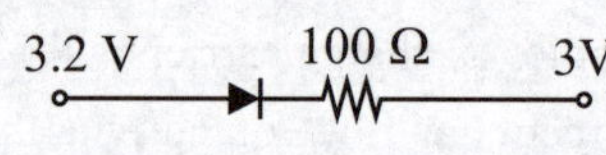

1) 20 A
2) 2×10^{-3} A
3) 200 A
4) 2×10^{-4} A

74. In the following, which one of the diodes is reverse biased?

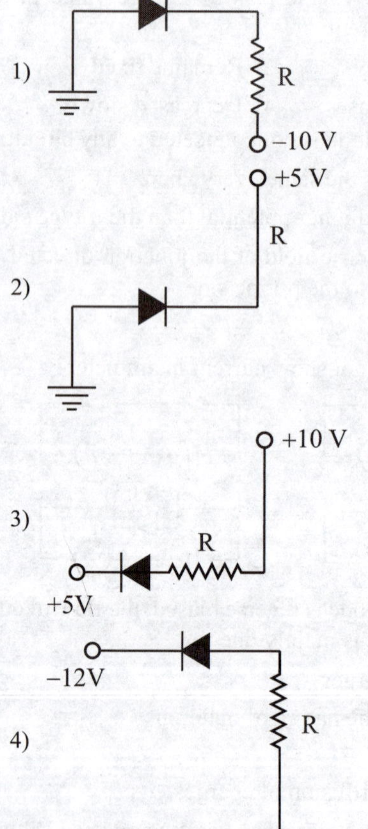

Logic Gates

75. Gates are shown in figure. The equivalent circuit is:

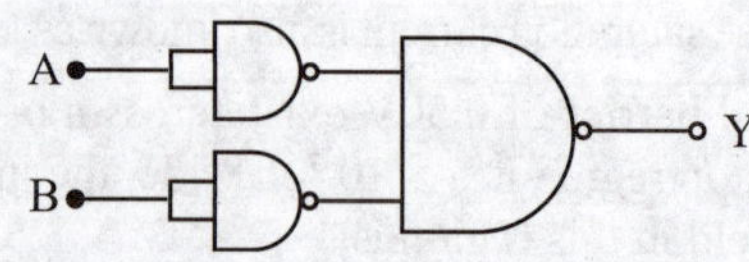

1) AND gate
2) NOR gate
3) OR gate
4) NOT gate

76. The truth table for the following logic circuit is:

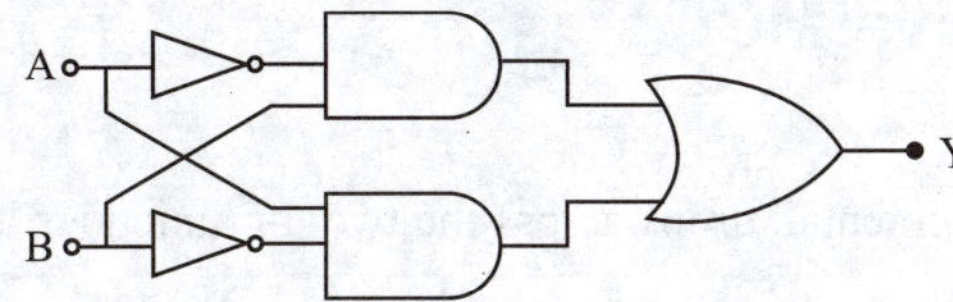

1)

A	B	Y
0	0	0
0	1	1
1	0	1
1	1	0

2)

A	B	Y
0	0	0
0	1	1
1	0	1
1	1	1

3)

A	B	Y
0	0	1
0	1	0
1	0	1
1	1	0

4)

A	B	Y
0	0	1
0	1	1
1	0	0
1	1	1

77. Truth table for the given circuit in figure is:

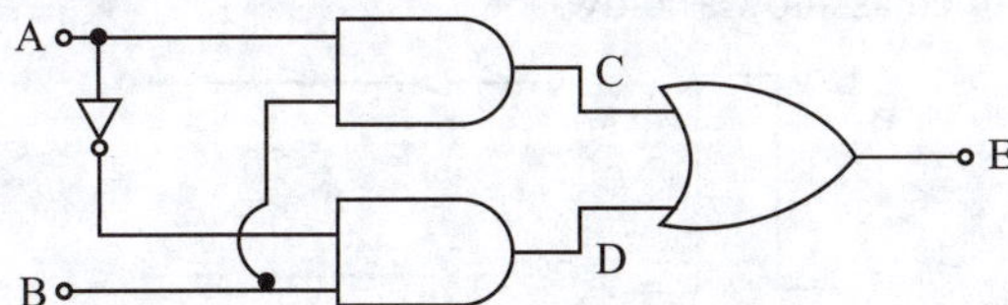

1)

A	B	E
0	0	1
0	1	0
1	0	1
1	1	0

2)

A	B	E
0	0	1
0	1	0
1	0	0
1	1	1

3)

A	B	E
0	0	0
0	1	1
1	0	0
1	1	1

4)

A	B	E
0	0	0
0	1	1
1	0	1
1	1	0

78. Minimum number of NAND gates used to form an AND gate:

1) 1
2) 2
3) 3
4) 4

79. What will be the input of A and B for the Boolean expression $\overline{(A+B)}.\overline{(A \cdot B)} = 1$

1) 0, 0
2) 0, 1
3) 1, 0
4) 1, 1

80. Which of the following logic gate is an universal gate?

1) OR
2) NOT
3) AND
4) NOR

81. Identify the operation performed by the circuit given below:

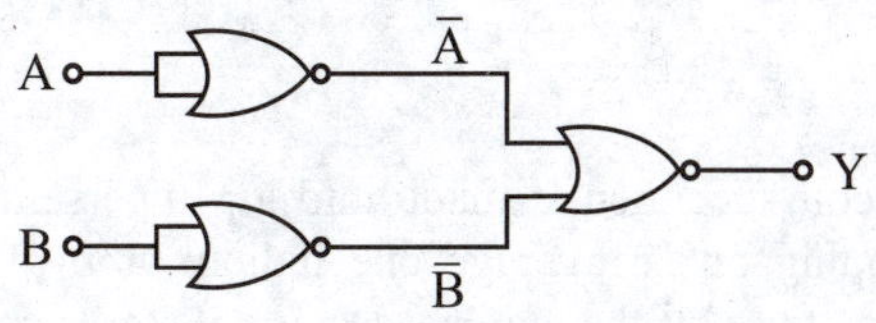

1) NOT
2) AND
3) OR
4) NAND

82. Identify the true statement of OR gate:

1) Output Y will be 1 when the input A or B or both are 1
2) Output Y will be 0 when either of the inputs A and B is 1
3) Output Y will be 1 only when both are inputs A and B are 1
4) Output Y will be 1 only when either of the inputs A and B is 1

83. How many gates are required to design; $y = A + \bar{A}.B$? Also name the gates.

1) 1, OR gate
2) 2, AND and OR gate
3) 3, AND, NOT and OR gate
4) 4, two NOT gates and two AND gates

84. Any digital circuit can be realised by repetitive use of only:

1) NOT gates
2) OR gates
3) AND gates
4) NOR gates

85. The following figure shows a logic gate circuit with two inputs A and B and the output Y. The voltage waveforms of A , B and Y are as given:

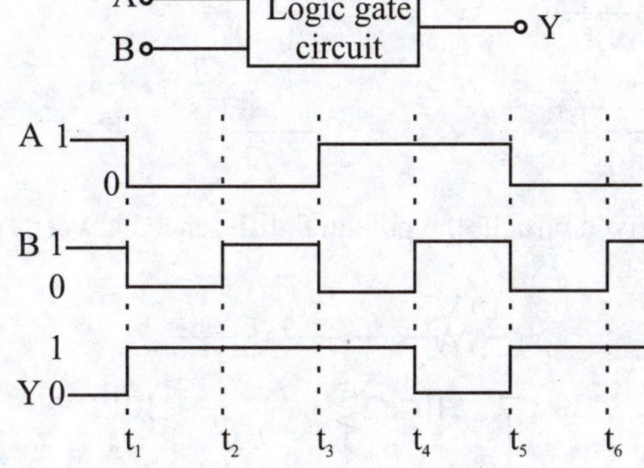

The logic gate is:

1) NOR gate
2) OR gate
3) AND gate
4) NAND gate

High Order Time Intensive MCQs

1. A silicon specimen is made into a *p*-type semiconductor by doping on an average one indium atom per 5×10^7 silicon atoms. If the number density of atoms in the silicon specimen is 5×10^{28} atom/m³, then the number of acceptor atoms in silicon per cubic centimeter will be:
   1) 2.5×10^{30} atom/cm³
   2) 2.5×10^{35} atom/cm³
   3) 1.0×10^{13} atom/cm³
   4) 1.0×10^{15} atom/cm³

2. A full wave rectifier circuit alongwith the output is shown in figure. The contribution(s) from the diode 1 is (are)

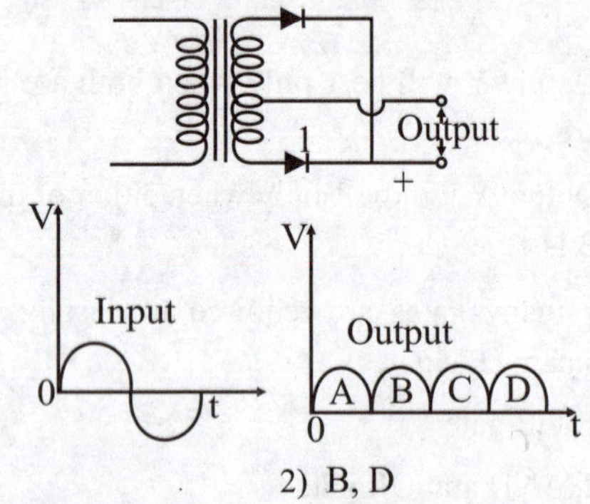

   1) C
   2) B, D
   3) A, C
   4) A, B, C, D

3. The value of the resistor R_S, needed in the de voltage regulator circuit shown here, equals

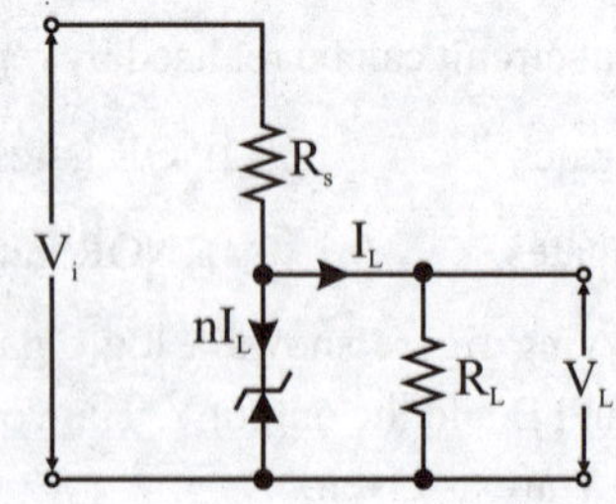

   1) $\frac{(V_i - V_L)}{(n+1)I_L}$
   2) $\frac{(V_i + V_L)}{(n+1)I_L}$
   3) $\frac{(V_i - V_L)}{nI_L}$
   4) $\frac{(V_i + V_L)}{nI_L}$

4. In the given circuit, the potential difference between A and B is

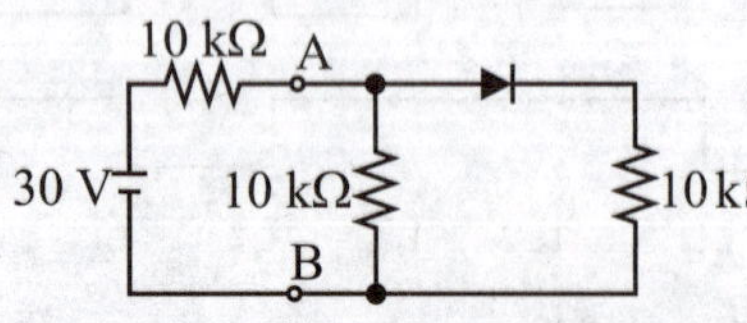

   1) 0
   2) 5 volt
   3) 10 volt
   4) 15 volt

5. Two identical p-n junctions may be connected in series with a battery in three ways, as shown in figure.

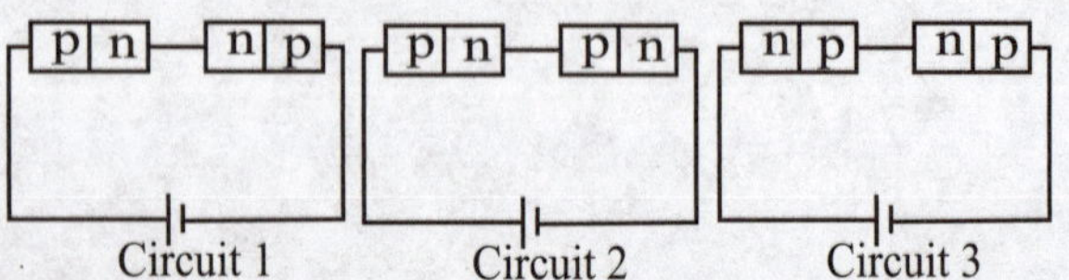

The potential drops across the two p-n junctions are equal in
   1) Circuit 1 and circuit 2
   2) Circuit 2 and circuit 3
   3) Circuti 3 and circuit 1
   4) Circuit 1 only

6. In the circuit shown below, assume the diode to be ideal. When V_i increases from 2 V to 6 V, the change in the current is (in mA)

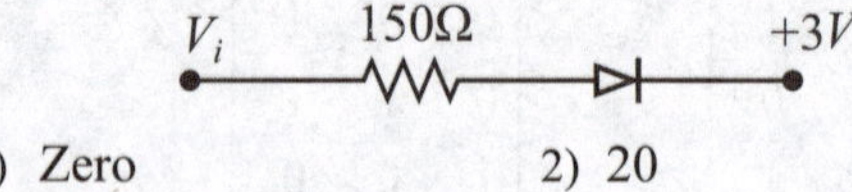

   1) Zero
   2) 20
   3) 80/3
   4) 40

7. If in a *p-n* junction diode, a square input signal of 10 V is applied as shown below.

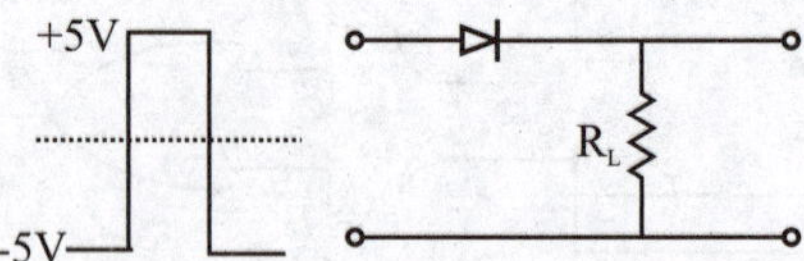

Then the output single across R_L will be

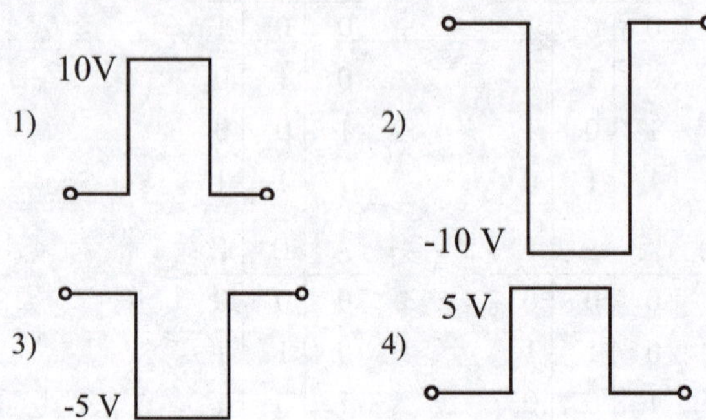

8. In the given circuit, the current through the resistor of 2 kΩ is:

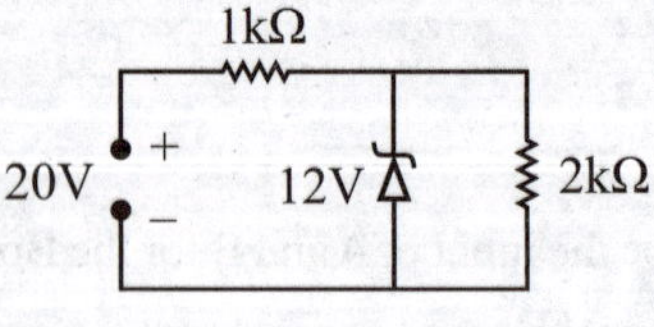

   1) 2 mA
   2) 4 mA
   3) 6 mA
   4) 1 mA

9. The logic gate having an output of 1 is

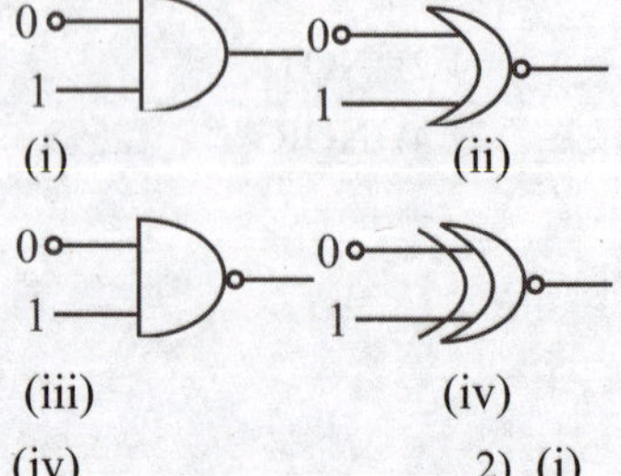

   1) (iv)
   2) (i)
   3) (ii)
   4) (iii)

10. In a common emitter configuration of a transistor, the voltage drop across a 500 Ω resistor in the collector circuit is 0.5 V when the collector supply voltage is 5 V. If the current gain in the common base mode is 0.96, the base current is:

1) $\frac{1}{20}\mu A$ 2) $\frac{1}{5}\mu A$
3) $\frac{1}{20}mA$ 4) $\frac{1}{24}mA$

11. In the following common emitter circuit, if $\beta = 100$, $V_{CE} = 7$ V, V_{BE} = negligible, $R_C = 2$ kΩ, then I_B = ?

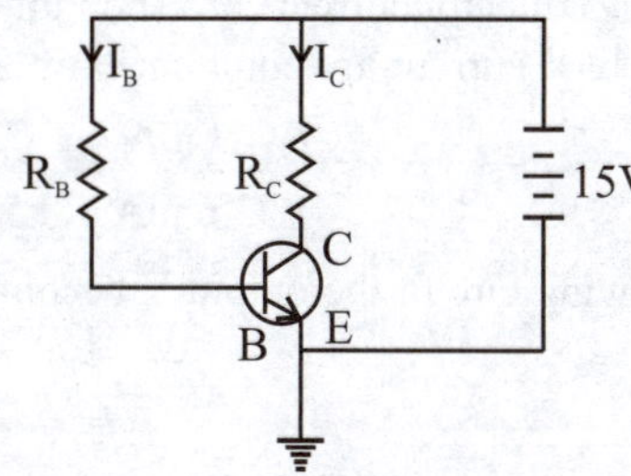

1) 0.01 mA 2) 0.04 mA
3) 0.02 mA 4) 0.03 mA

12. The input signal given to a CE amplifier having a voltage gain of 150 is $V_1 = 2\cos\left(15t+\frac{\pi}{3}\right)$. The corresponding output signal will be

1) $300\cos\left(15t+\frac{4\pi}{3}\right)$ 2) $300\cos\left(15t+\frac{\pi}{3}\right)$
3) $75\cos\left(15t+\frac{2\pi}{3}\right)$ 4) $2\cos\left(15t+\frac{5\pi}{6}\right)$

13. Identify the gate represented by the block diagram as shown in fig.

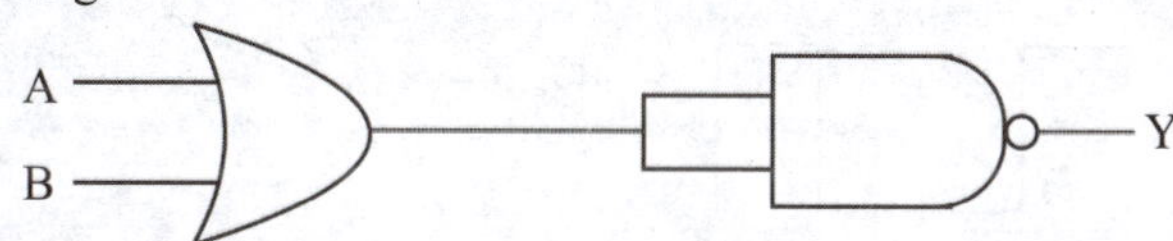

1) AND gate 2) NOT gate
3) NAND gate 4) NOR gate

14. To get an output, $Y = 1$ in given circuit, which of the following input will be correct?

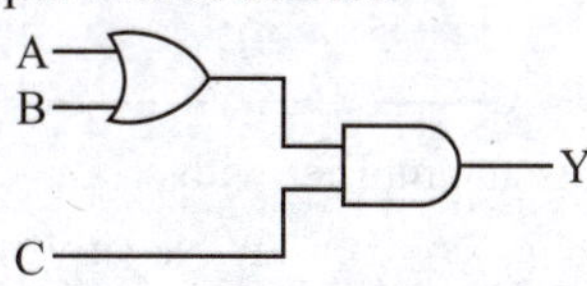

	A B C		*A B C*
1)	1 0 0	2)	1 0 1
3)	1 1 0	4)	0 1 0

15. In the circuit given *A*, *B* and *C* are inputs and *Y* is the output. The output of *Y* is:

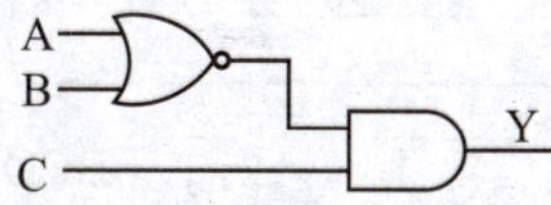

1) High for all high inputs
2) High for all the low inputs
3) High when $A = 0$, $B = 0$, $C = 1$
4) Low when $A = 0$, $B = 0$, $C = 1$

16. In the circuit below, A and B represent two inputs and C represents the output, the circuit represents

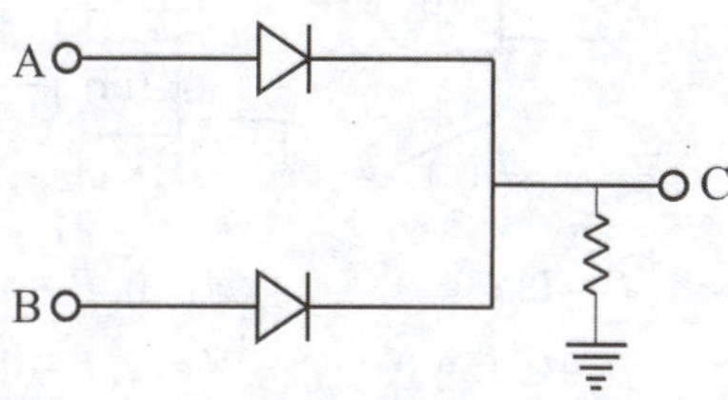

1) NOR gate 2) AND gate
3) NAND gate 4) OR gate

17. The given graph represents V-I characteristic for a semiconductor device.

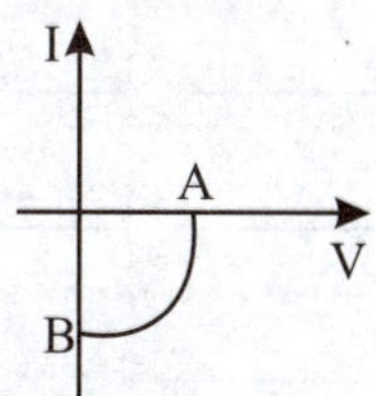

Which of the following statement is correct ?

1) It is V-I characteristic for solar cell where point A represents open circuit voltage and point B short circuit current.
2) It is for a solar cell and points A and B represent open circuit voltage and current, respectively.
3) It is for a photodiode and points A and B represent open circuit voltage and current, respectively.
4) It is for a LED and points A and B represents open circuit voltage and short circuit current respectively.

18. Which one of the entries given in the truth table is true for the following logic circuit?

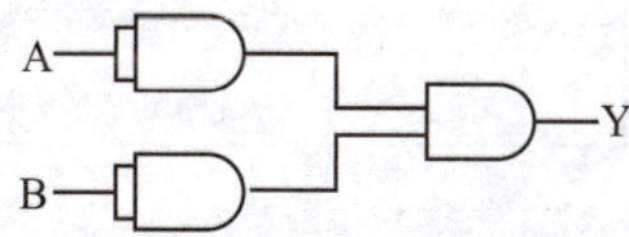

Entry No.	Input A	Input B	Output Y
1.	0	0	1
2.	0	1	0
3.	1	0	1
4.	1	1	0

1) 3 2) 4
3) 1 4) 2

19. The following configuration of gate is equivalent to

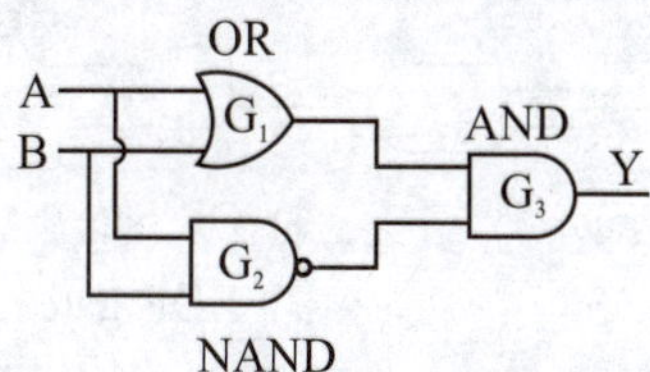

1) NAND 2) XOR
3) OR 4) None of the above

20. In the following circuit the output Y becomes 1 for the inputs:

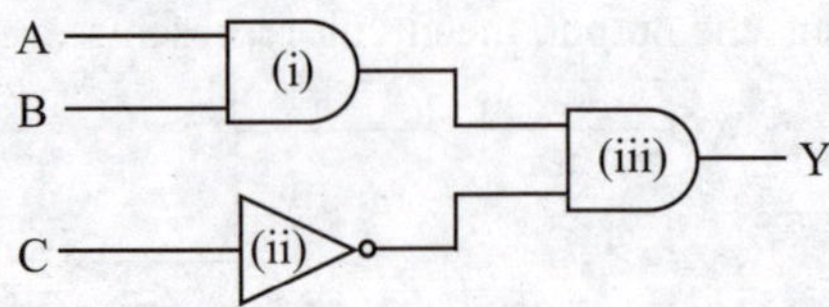

1) $A = 1, B = 0, C = 0$ 2) $A = 0, B = 1, C = 1$
3) $A = 0, B = 0, C = 0$ 4) $A = 1, B = 1, C = 0$

21. Two ideal junction diodes D_1, D_2 are connected as shown in the figure. A 3V battery is connected between A and B. The current supplied by the battery if its positive terminal is connected to A is

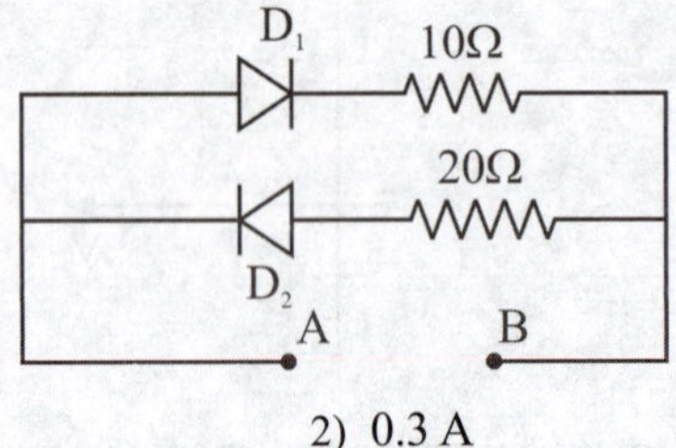

1) 0.1A 2) 0.3 A
3) 0.9 A 4) 90 A

22. A PN junction diode when forward biased has a drop of 0.5V which is assumed to be independent of current. The current in excess of 10mA through the diode produces large joule heating which damages the diode. If we want to use a 1.5V battery to forward bias the diode, the resistor used in series with the diode so that the maximum current does not exceed 5mA is

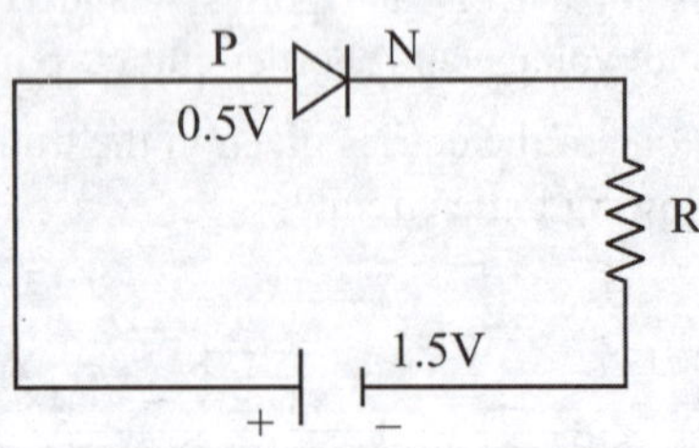

1) $2 \times 10^2 \Omega$ 2) $2 \times 10^5 \Omega$
3) $2 \times 10^3 \Omega$ 4) $2 \times 10^4 \Omega$

23. A cell of emf 4.5V is connected to a junction diode whose barrier potential is 0.7V. If the external resistance in the circuit is 190 Ω, the current in the circuit is

1) 20 mA 2) 2m A
3) 23mA 4) 200mA

24. The combination of gates shown below yields

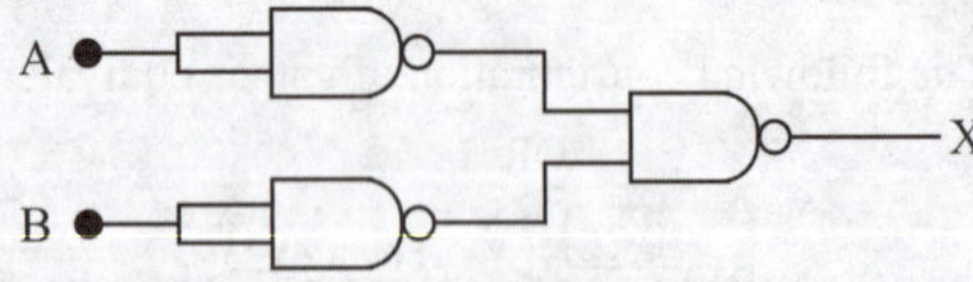

1) NAND gate 2) OR gate
3) NOT gate 4) XOR gate

25. Current amplification factor of a common base configuration is 0.88. Find the value of base current when the emitter current is 1 mA.

1) 0.12 mA. 2) 0.1 mA
2) 0.5 mA 4) 0.2 mA

26. For a transistor $\beta = 40$ and $I_B = 25\mu A$. Find the value of I_E.

1) 1mA 2) 1.025mA
3) 2mA 4) 1.2 mA

27. The constant α of a transistor is 0.9. What would be the change in the collector current corresponding to a change of 4 mA in the base current in a common emitter arrangement?

1) 30 mA 2) 63 mA
3) 36 mA 4) 3.6 mA

28. In the following circuit the output Y becomes zero for the inputs

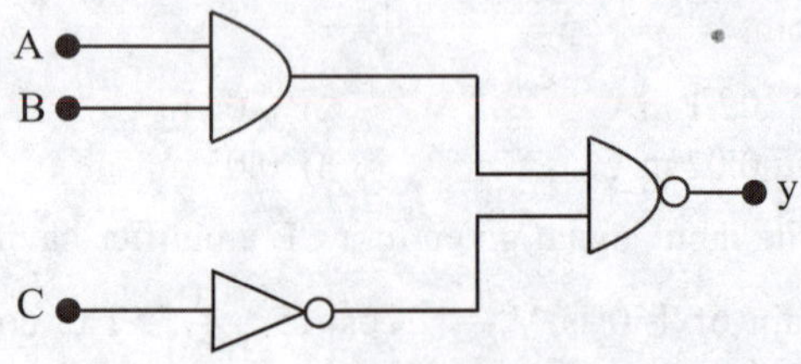

1) A = 1, B = 0, C = 0 2) A = 0, B = 1, C = 1
3) A = 0, B = 0, C = 0 4) A = 1, B = 1, C = 0

29. In a given circuit as shown the two input wave forms A and B are applied simultaneously

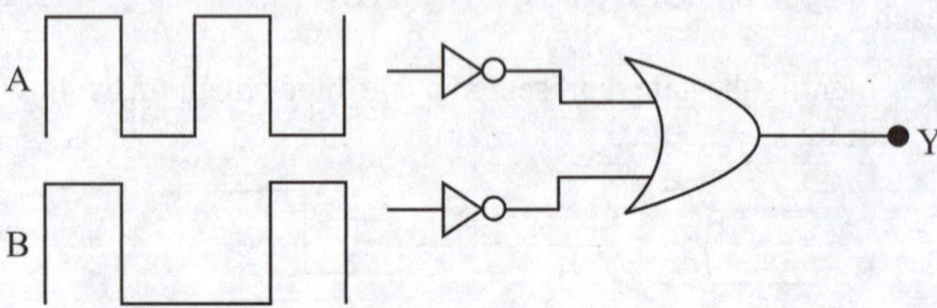

The resultant wave form at Y is

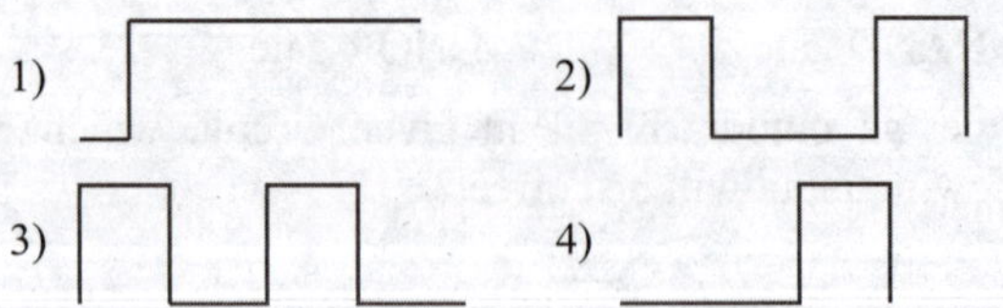

30. An oscillator is an amplifer with

1) A large gain 2) Negative feedback
3) Positive feedback 4) No feedback

31. The expression of Y in following circuit is

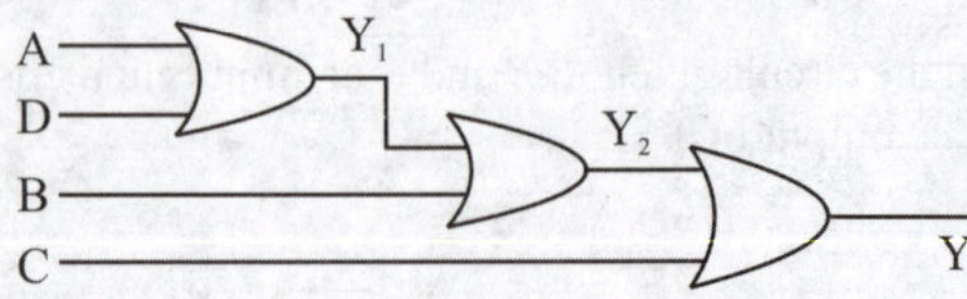

1) ABCD 2) A + BCD
3) A + B + C + D 4) AB + CD

Statement Based MCQs

Directions: These questions consist of two statements each, printed as Statement-I and Statement-II. While answering these questions, you are required to choose any one of the following four responses.

1) **Both Statement-I and Statement-II are correct.**
2) **Both Statement-I and Statement-II are incorrect.**
3) **Statement-I is correct & Statement-II is incorrect.**
4) **Statement-I is incorrect & Statement-II is correct.**

1. **Statement-I:** In an unbiased p-n junction, holes diffuse from the p-region to n-region.

   Statement-II: Hole concentration in p-region is more as compared to n-region.

2. **Statement-I:** At a fixed temperature, silicon will have a minimum conductivity when it has a smaller acceptor doping.

   Statement-II: The conductivity of an intrinsic semiconductor is slightly higher than that of a lightly doped p-type.

3. **Statement-I:** A p-type semiconductor when joined with n-type semiconductor does not form a p-n junction diode.

   Statement-II: n-type semiconductors are not electrically neutral as it has excess of electrons.

4. **Statement-I:** Two p-n junction diodes placed back to back, will work as a n-p-n transistor.

   Statement-II: The p-region of two p-n junction diodes placed back to back will form the base of n-p-n transistor.

5. **Statement-I:** In a common-emitter amplifier, the load resistance of the output circuit is 1000 times the load resistance of the input circuit. It $\alpha = 0.98$, then voltage gain is 49×10^3.

   Statement-II: $\alpha = \frac{\beta}{1-\beta}$ (symbols have their usual meaning.)

Assertion & Reason MCQs

Directions: These questions consist of two statements each, printed as Assertion and Reason. While answering these questions, you are required to choose any one of the following four responses.

1) **Both Assertion and Reason are True and the Reason is a correct explanation of the Assertion.**
2) **Both Assertion and Reason are True but Reason is not a correct explanation of the Assertion.**
3) **If Assertion is True but the Reason is False.**
4) **Assertion is False but Reason is True.**

1. **Assertion:** The depletion region of a junction diode is formed during the manufacturing process of junction.

   Reason: Zener breakdown is initiated through a direct rupture of covalent bonds due the existence of strong electric field.

2. **Assertion:** NOR and NAND gates are considered as universal gates.

   Reason: We can obtain all the gates by using NOR or NAND gates.

3. **Assertion:** p-n junction diode can be used as a rectifier. (AC $\rightarrow$ DC)

   Reason: p-n junction diode is one way device which offers a low resistance when forward biased and behaves like an insulator when reverse biased.

Match the Columns MCQs

1. Match the column-I with column-II.

	Column-I		Column-II
(a)	P-Type semiconductor	(i)	Mobility × Electric field.
(b)	N-Type semiconductor	(ii)	Impurity of IIIrd group.
(c)	Intrinsic semiconductor	(iii)	Impurity of Vth group
(d)	Drift velocity	(iv)	No-impurity

1) (a)-(ii) (b)-(iii) (c)-(i) (d)-(iv)
2) (a)-(i) (b)-(ii) (c)-(iv) (d)-(iii)
3) (a)-(iii) (b)-(iv) (c)-(i) (d)-(ii)
4) (a)-(ii) (b)-(iii) (c)-(iv) (d)-(i)

2. Match the column-I with column-II.

	Column-I		Column-II
(a)	Diode in forward-bias	(i)	↑↑↑
(b)	Diode in reverse-bias	(ii)	+ −
(c)	Light emitting diode	(iii)	
(d)	Zener diode	(iv)	− +

1) (a)-(ii) (b)-(iii) (c)-(i) (d)-(iv)
2) (a)-(i) (b)-(ii) (c)-(iv) (d)-(iii)
3) (a)-(ii) (b)-(iv) (c)-(i) (d)-(iii)
4) (a)-(ii) (b)-(iii) (c)-(iv) (d)-(i)

3. Match the following column-I with column-II of H.W.R. (Half wave rectifier) and F.W.R. (Full wave rectifier).

	Column-I		Column-II
(a)	Ripple factor of H.W.R.	(i)	1.57
(b)	Ripple factor of F.W.R.	(ii)	1.21
(c)	Form factor of H.W.R.	(iii)	1.11
(d)	Form factor of F.W.R.	(iv)	0.48

1) (a)-(ii) (b)-(iii) (c)-(i) (d)-(iv)
2) (a)-(ii) (b)-(iv) (c)-(i) (d)-(iii)
3) (a)-(iii) (b)-(iv) (c)-(iii) (d)-(i)
4) (a)-(i) (b)-(iii) (c)-(iv) (d)-(ii)

NCERT Exemplar MCQs

1. The conductivity of a semiconductor increases with increase in temperature, because:
1) Number density of free current carriers increases
2) Relaxation time increases
3) Both number density of carriers and relaxation time increase
4) Number density of carriers increases, relaxation time decreases but effect of decrease in relaxation time is much less than increase in number density

2. In figure given below V_0 is the potential barrier across a p-n junction, when no battery is connected across the junction:

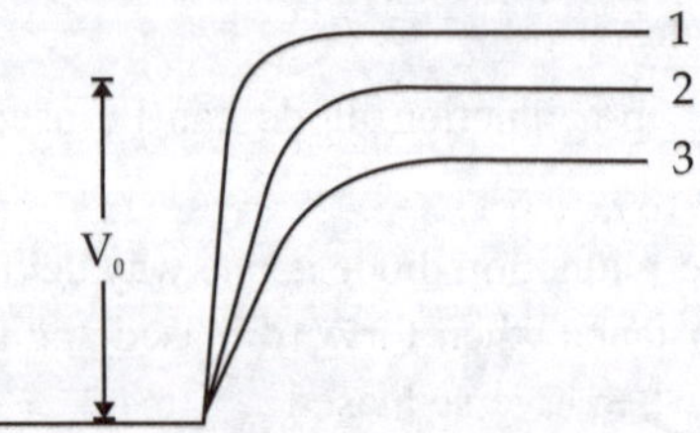

1) 1 and 3 both correspond to forward bias of junction
2) 3 corresponds to forward bias of junction and 1 corresponds to reverse bias of junction
3) 1 corresponds to forward bias and 3 corresponds to reverse bias of junction
4) 3 and 1 both correspond to reverse bias of junction

3. In figure given below, assuming the diodes to be ideal:

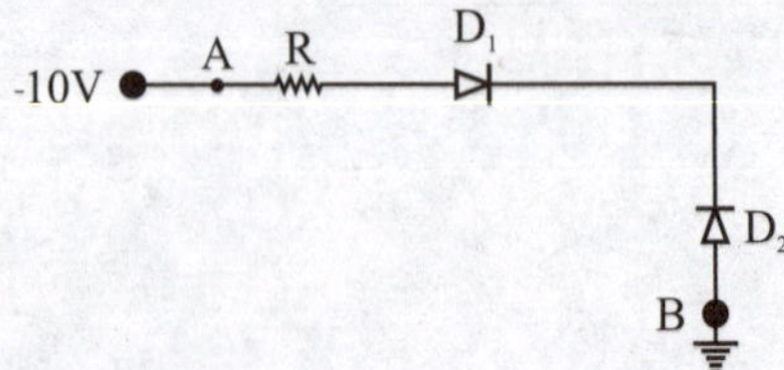

1) D_1 is forward biased and D_2 is reverse biased and hence current flows from A to B
2) D_2 is forward biased and D_1 is reverse biased and hence no current flows from B to A and vice-versa
3) D_1 and D_2 are both forward biased and hence current flows from A to B
4) D_1 and D_2 are both reverse biased and hence no current flows from A to B and vice-versa

4. A 220 V AC supply is connected between points A and B (figure). What will be the potential difference V across the capacitor?

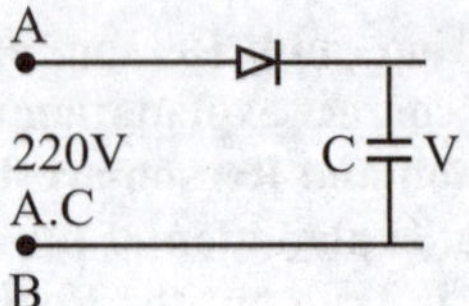

1) $200V$
2) $110V$
3) $0V$
4) $220\sqrt{2}V$

5. Hole is:
1) An anti-particle of electron
2) A vacancy created when an electron leaves a covalent bond
3) Absence of free electrons
4) An artificially created particle

6. In the circuit shown in figure given below, if the diode forward voltage drop is 0.3 V, the voltage difference between A and B is:

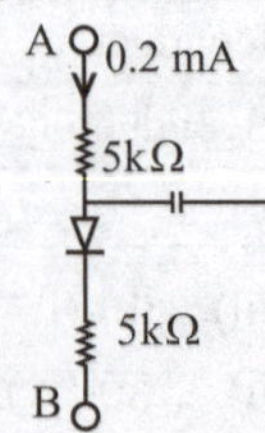

1) 1.3 V
2) 2.3 V
3) 0
4) 0.5 V

7. Truth table for the given circuit is:

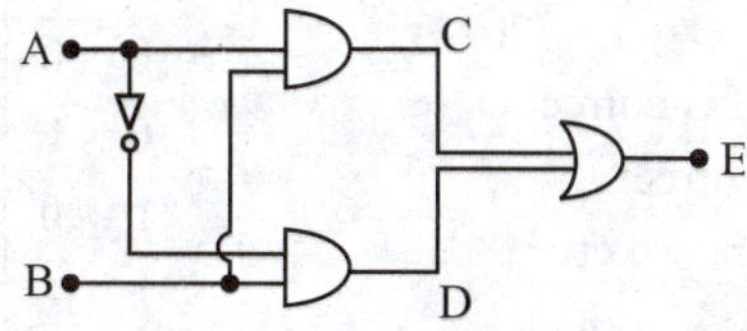

1)

A	B	E
0	0	1
0	1	0
1	0	1
1	1	0

2)

A	B	E
0	0	1
0	1	0
1	0	0
1	1	0

3)

A	B	E
0	0	0
0	1	1
1	0	0
1	1	1

4)

A	B	E
0	0	0
0	1	1
1	0	1
1	1	0

NEET & AIIMS Past Year MCQs

1.

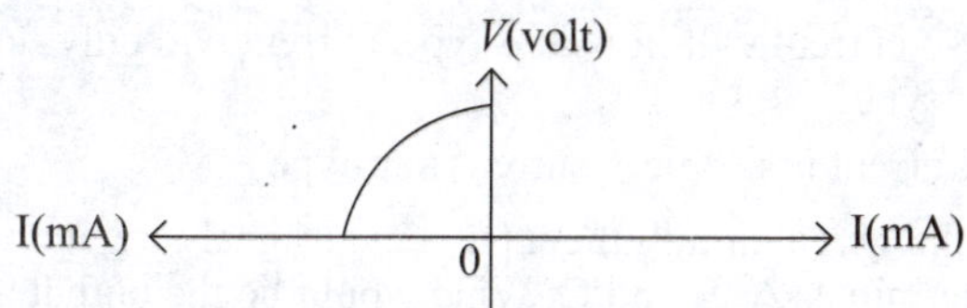

The I-V characteristics shown above are exhibited by a **(2024 Re)**

1) Light emitting diode 2) Zener diode
3) Photodiode 4) Solar cell

2. When the output of an OR gate is applied as input to a NOT gate, then the combination acts as a **(2024 Re)**

1) NAND gate 2) NOR gate
3) AND gate 4) OR gate

3. The output Y for the inputs A and B of the given logic circuit is: **(2024 Re)**

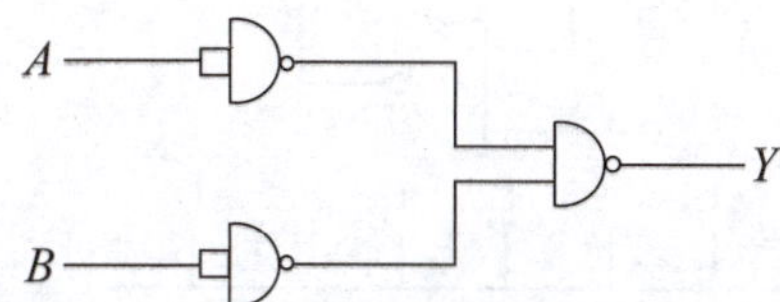

1) $A \cdot B$ 2) $\overline{A} \cdot \overline{B}$
3) $A + B$ 4) $\overline{A} + \overline{B}$

4. Consider the following statement A and B and identify the correct answer: **(2024)**

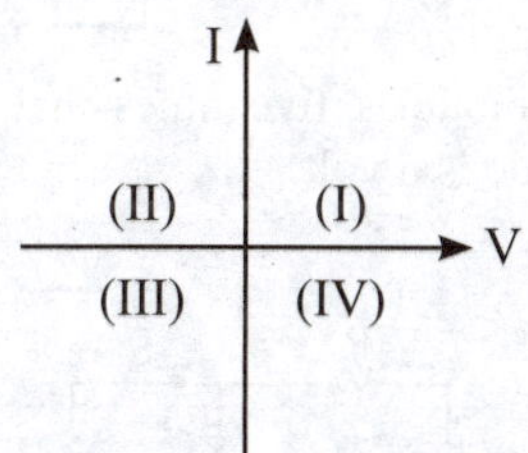

A. For a solar-cell, the I-V characteristics lies in the IV quadrant of the given graph.

B. In a reverse biased *pn* junction diode, the. current measured in (A), is due to majority charge carriers.

1) Both A and B are correct.
2) Both A and B are incorrect.
3) A is correct but B is incorrect.
4) A is incorrect but B is correct.

5. A logic circuit provides the output Y as per the following truth table: **(2024)**

A	B	Y
0	0	1
0	1	0
1	0	1
1	1	0

The expression for the output Y is:

1) $\overline{B}$ 2) B
3) $A.B+\overline{A}$ 4) $A.\overline{B}+\overline{A}$

6. The output (Y) of the given logic gate is similar to the output of an/a: **(2024)**

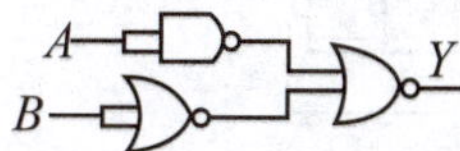

1) OR gate 2) AND gate
3) NAND gate 4) NOR gate

7. A full wave rectifier circuit consists of two p-n junction diodes, a centre-tapped transformer, capacitor and a load resistance. Which of these components remove the ac ripple from the rectified output **(2023)**

1) Load resistance
2) A centre-tapped transformer
3) p-n junction diodes
4) Capacitor

8. Given below are two statements:

Statement I: Photovoltaic devices can convert optical radiation into electricity.

Statement II: Zener diode is designed to operate under reverse bias in breakdown region.

In the light of the above statements, choose the most appropriate answer from the options given below: **(2023)**

1) Statement I is incorrect but Statement II is correct.
2) Both Statement I and Statement II are correct.
3) Both Statement I and Statement II are incorrect.
4) Statement I is correct but Statement II is incorrect.

9. For the following logic circuit, the truth table is: **(2023)**

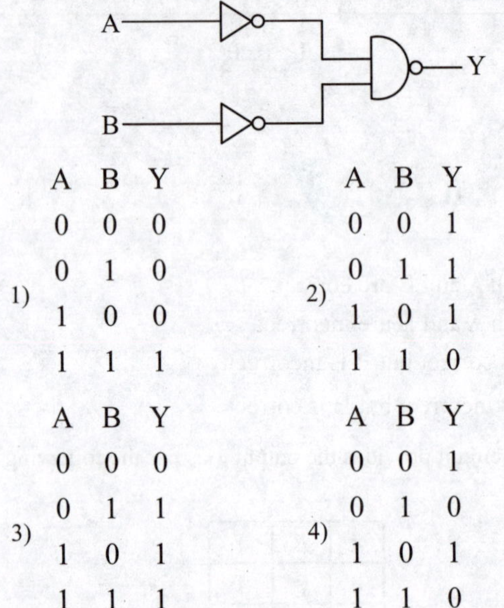

1)

A	B	Y
0	0	0
0	1	0
1	0	0
1	1	1

2)

A	B	Y
0	0	1
0	1	1
1	0	1
1	1	0

3)

A	B	Y
0	0	0
0	1	1
1	0	1
1	1	1

4)

A	B	Y
0	0	1
0	1	0
1	0	1
1	1	0

10. In half wave rectification, if the input frequency is 60 Hz, then the output frequency would be: **(2022)**

1) 120 Hz 2) 0
3) 30 Hz 4) 60 Hz

11.

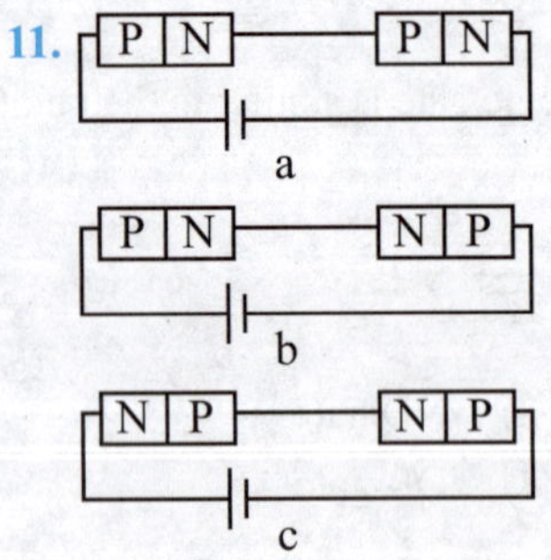

In the given circuits (a), (b) and (c), the potential drop across the two p-n junctions are equal in: **(2022)**

1) Both circuits (a) and (c) 2) Circuit (a) only
3) Circuit (b) only 4) Circuit (c) only

12. **(2022)**

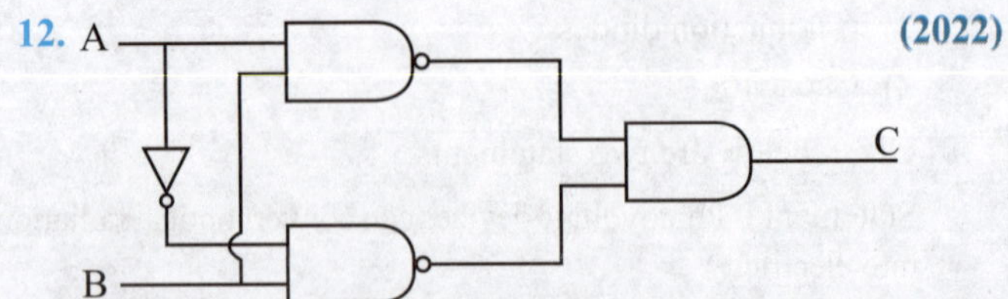

The truth table for the given logic circuit is:

1)

A	B	C
0	0	0
0	1	1
1	0	0
1	1	1

2)

A	B	C
0	0	0
0	1	1
1	0	1
1	1	0

3)

A	B	C
0	0	1
0	1	0
1	0	0
1	1	1

4)

A	B	C
0	0	1
0	1	0
1	0	1
1	1	0

13. The electron concentration in an n-type semiconductor is the same as hole concentration in a p-type semiconductor. An external field (electric) is applied across each of them. Compare the currents in them. **(2021)**

1) Current in p-type > current in n-type.
2) Current in n-type > current in p-type.
3) No current will flow in p-type, current will only flow in n-type.
4) Current in n-type = current in p-type

14. For the given circuit, the input digital signals are applied at the terminals A, B and C. What would be the output at the terminal y? **(2021)**

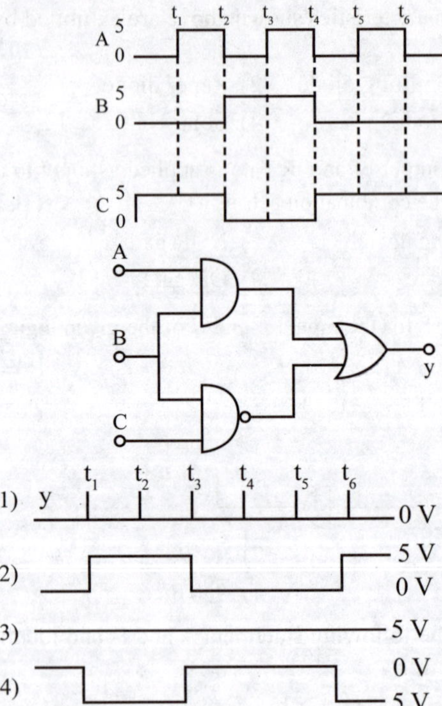

15. If current in diode is five times that in R_1. Breakdown voltage of diode is 6 volt. **(AIIMS 2018)**

Find R = ?

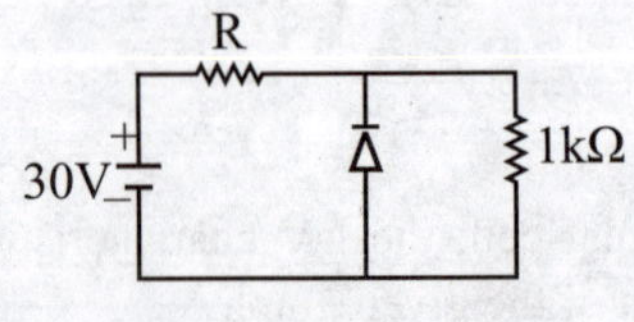

1) 2000/3 2) 1000/3
3) 200/3 4) 100/3

16. What is the out put of the given logic gate (AIIMS 2018)

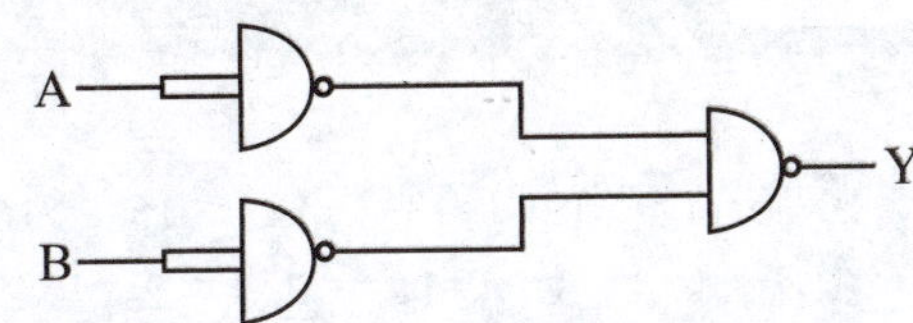

1) A.B
2) $\overline{A}.\overline{B}$
3) $\overline{A}+\overline{B}$
4) A + B

17. **Assertion:** The valve of current through p-n junction in the given figure will be 10 mA. (AIIMS 2017)

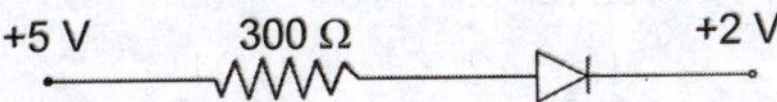

Reason: In the above figure, p-side is at higher potential than n-side.

1) If both Assertion & Reason are True & the Reason is a correct explanation of the Assertion.
2) If both Assertion & Reason are True but Reason is not a correct explanation of the Assertion.
3) If Assertion is True but the Reason is False.
4) If both Assertion & Reason are False.

18. **Assertion:** An *n*-type semiconductor has a large number of electrons but still it is electrically neutral. (AIIMS 2017)

Reason: An *n*-type semiconductor is obtained by doping an intrinsic semiconductor with a pentavalent impurity.

1) If both Assertion & Reason are True & the Reason is a correct explanation of the Assertion.
2) If both Assertion & Reason are True but Reason is not a correct explanation of the Assertion.
3) If Assertion is True but the Reason is False.
4) If both Assertion & Reason are False.

19. The output for the given, circuit is: (AIIMS 2016)

1) $(A+B).\overline{B}$
2) $(A\,.\,B).\overline{B}$
3) $(A+B).B$
4) $(A\,.\,B)+B$

20. A zener diode is specified as having a breakdown voltage of 9.1 V, with a maximum power dissipation of 364 mW. What is the maximum current the diode can handle? (AIIMS 2016)

1) 40 mA
2) 60 mA
3) 50 mA
4) 45 mA

21. The circuit as shown in figure: (AIIMS 2013)

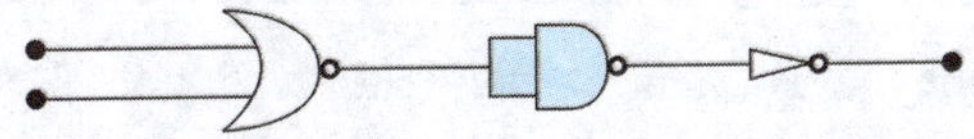

the equivalent gate is

1) NOR gate
2) OR gate
3) AND gate
4) NAND gate

Answer Key

NCERT Topic-wise MCQs

1	2	3	4	5	6	7	8	9	10	11	12	13	14	15	16	17
1	3	2	2	4	1	2	2	3	3	4	3	1	1	2	2	2
18	**19**	**20**	**21**	**22**	**23**	**24**	**25**	**26**	**27**	**28**	**29**	**30**	**31**	**32**	**33**	**34**
2	4	2	4	1	4	3	3	1	2	1	3	1	1	1	2	3
35	**36**	**37**	**38**	**39**	**40**	**41**	**42**	**43**	**44**	**45**	**46**	**47**	**48**	**49**	**50**	**51**
3	4	2	2	2	1	3	3	3	1	3	1	3	1	2	3	3
52	**53**	**54**	**55**	**56**	**57**	**58**	**59**	**60**	**61**	**62**	**63**	**64**	**65**	**66**	**67**	**68**
1	3	2	1	2	2	2	1	1	2	1	2	1	4	2	2	2
69	**70**	**71**	**72**	**73**	**74**	**75**	**76**	**77**	**78**	**79**	**80**	**81**	**82**	**83**	**84**	**85**
4	3	4	1	2	2	3	1	3	2	1	4	2	1	1	4	4

High Order Time Intensive MCQs

1	2	3	4	5	6	7	8	9	10	11	12	13	14	15	16	17
4	2	1	3	2	2	4	3	4	4	2	1	4	2	3	4	1
18	**19**	**20**	**21**	**22**	**23**	**24**	**25**	**26**	**27**	**28**	**29**	**30**	**31**			
4	2	4	2	1	1	2	1	2	3	4	1	3	3			

Statement Based MCQs

1	2	3	4	5
1	3	3	2	3

Assertion & Reason MCQs

1	2	3
2	1	1

Match the Column MCQs

1	2	3
4	3	2

NCERT Exemplar MCQs

1	2	3	4	5	6	7
4	2	2	4	2	2	3

NEET & AIIMS Past Year MCQs

1	2	3	4	5	6	7	8	9	10	11	12	13	14	15	16	17
4	2	3	3	1	2	4	2	3	4	1	4	2	3	1	4	2
18	**19**	**20**	**21**													
1	1	1	1													

CHAPTER

15 Experiment Skills in Physics

NCERT-Picks

Experiment 1: Vernier Calipers

Purpose: Measure internal and external diameters and depths of objects.

Setup: Main scale (M) and Vernier scale (V) that slides over the main scale.

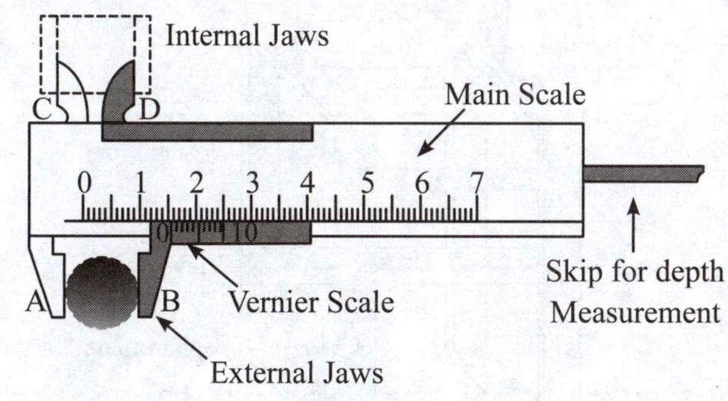

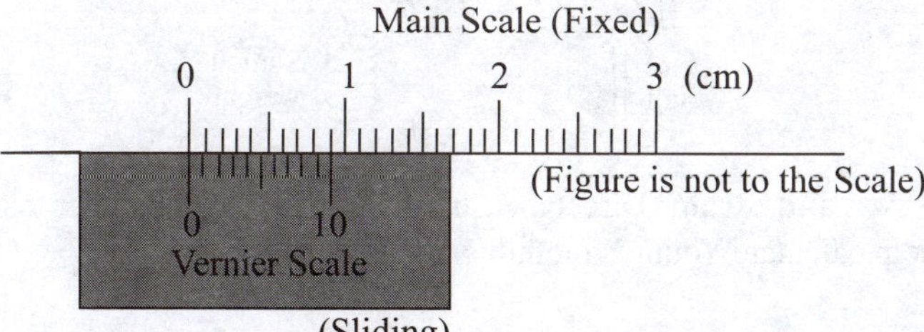

Formulas

Let N vernier scale divisions (V.S.D.) Coincide with (N – 1) main scale divisions (M.S.D.)

$\therefore$ N V.S.D. = (N – 1) M.S.D.

Least constant L.C = 1 M.S.D. – 1 V.S.D

$$\text{Least constant L.C.} = 1\text{ M.S.D.} - \left(\frac{N-1}{N}\right)\text{M.S.D} = \frac{1}{N}\text{ M.S.D}$$

$$\text{L.C} = \frac{\text{Value of one main scale division}}{\text{Total number of division on vernier scale}}$$

The vernier constant (V.C) or least count (L.C.) is the minimum distance that the vernier can reliably measure.

1. **Components:** Main scale (M) and Vernier scale (V) that slides over the main scale.
2. **Jaws:** Fixed jaws (A and C) for zero error checking, moving jaws (B and D) for measurement.
3. **Depth Measurement:** A thin metallic strip (E) attached to the back of the main scale for depth measurement.
4. **Zero Error and Correction:** Correct measurements require accounting for any misalignment between zero on the main scale and vernier scale when jaws A and B are closed.

Experiment 2: Screw Gauge

Purpose: Measure thickness or diameter of thin sheets or wires.

Setup: Consists of a screw that advances or retracts when rotated, providing precise measurements.

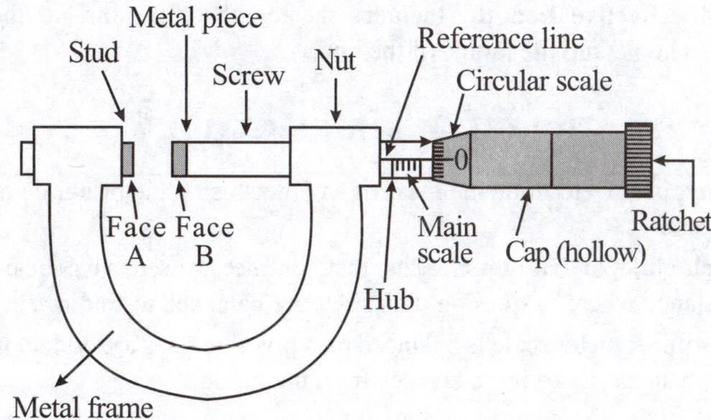

Pitch of a screw:

$$\text{Pitch} = \frac{\text{Distance moved by screw on the main scale}}{\text{Number of rotaion of circular scale}}$$

Least count of the screw gauge:

$$\text{Least count} = \frac{\text{Pitch}}{\text{Total number of division on the circular scale}}$$

The screw gauge provides higher precision compared to other measuring tools like a ruler due to its finer least count.

Measurement Calculation:

Total Reading = Main Scale Reading + (Circular Scale Reading × L.C)

Experiment 3: Simple Pendulum

Purpose: Investigate energy dissipation by plotting the square of amplitude versus time.

Setup: Consists of a pendulum bob suspended by a thread, attached to a clamp and stand.

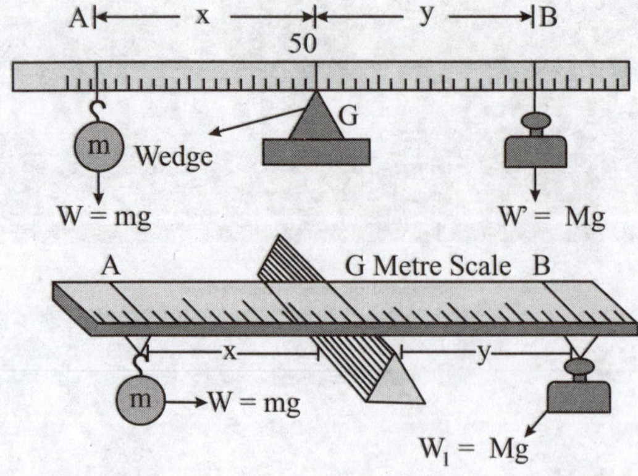

Formulas

1. **Force on Pendulum:**

$$F = -\frac{mg}{L}x$$

where m is mass of bob, L is effective length, and x is displacement.

2. **Energy Dissipation:**

$E \alpha A^2$

where A is amplitude.

3. **Energy Measurement:** Energy is proportional to the square of the amplitude of oscillation.
4. **Effective Length:** Includes the length of the thread, the hook, and the radius of the bob.

Experiment 4: Meter Scale

Purpose: Determine the mass of an object using the principle of moments.

Principle of Moments: The meter scale is used as a beam balance where torques on both sides are balanced to find mass.

Setup: A meter scale is balanced on a pivot; masses are added on both sides at varying distances from the pivot.

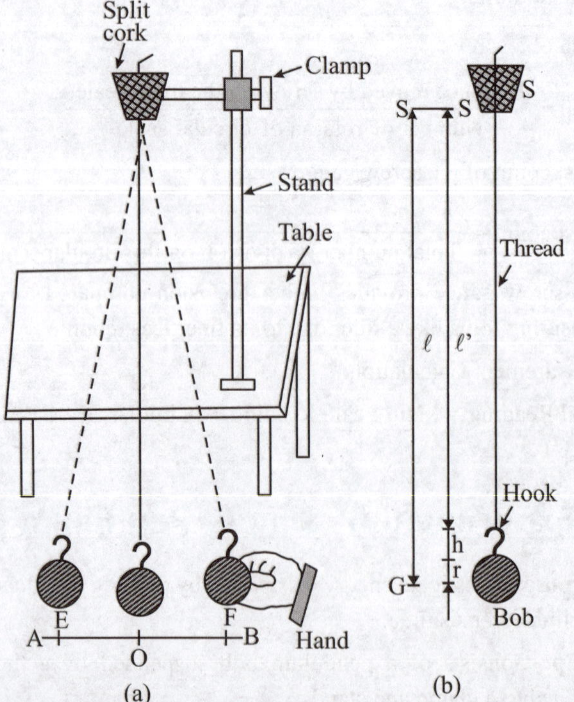

Formulas

$$mg \times x = Mg \times y \text{ or } m = M\left(\frac{y}{x}\right)$$

Where m is the unknown mass, M is the known mass, x and y are distances from the pivot.

Experiment 5: Young's Modulus of Elasticity

Purpose: Determine Young's modulus of the material of a metallic wire.

Setup: Searle's apparatus with a reference wire and a test wire where load is applied.

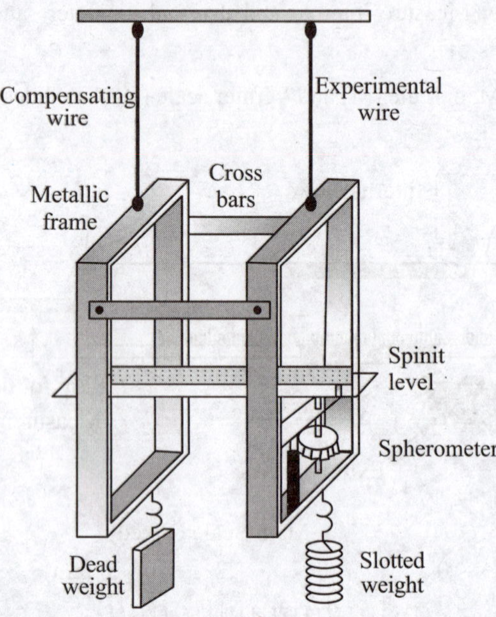

Stress and Strain: Direct measurements of extension under load help calculate Young's modulus.

Formulas

1. The material's Young's modulus for the given wire is given by

$$Y = \frac{MgL}{\pi r^2 \ell} \text{ N/m}^2$$

where

M = load applied at the end of the given wire

L = original length of the wire

r = radius of the wire

ℓ = extension produced in the wire

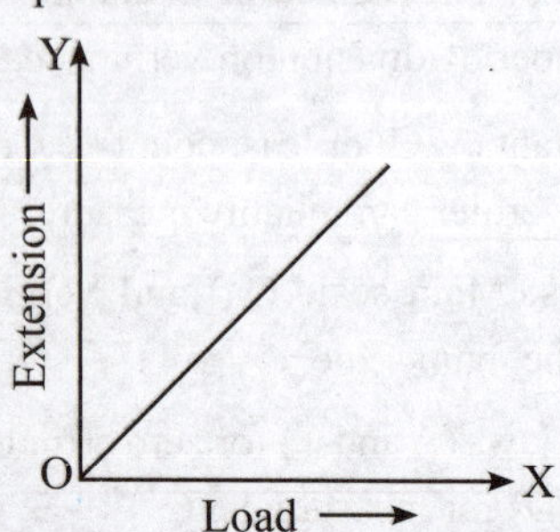

Experiment 6: Surface Tension

Purpose: Study capillary action and the effect of detergents on surface tension.

Setup:

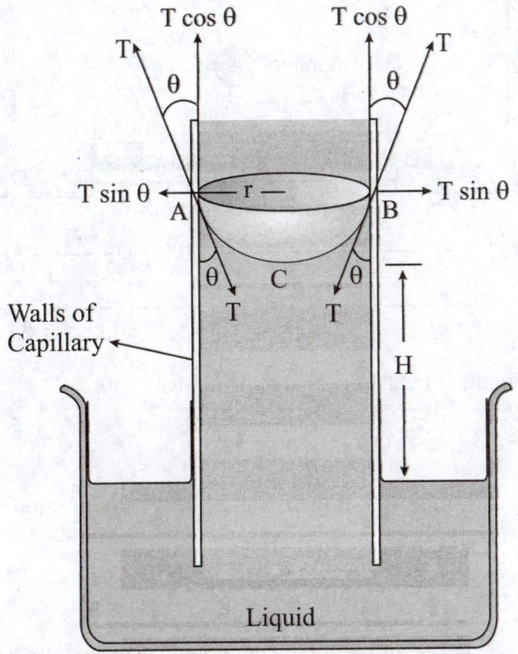

Capillary Rise: A tube is dipped in liquid to measure how high the liquid rises due to surface tension.

Effect of Detergents: Detergents decrease surface tension, affecting the height of capillary rise.

Formulas

Relation between capillary rise(h) and surface tension(T):

$$T = \frac{\left[H + \frac{r}{3}\right]\rho g}{2\cos\theta}$$

Where r is inner radius of capillary tube.

Experiment 7: Coefficient of Viscosity

Purpose: Determine the viscosity of a liquid by measuring the terminal velocity of a spherical body falling through the liquid.

Setup: A sphere is dropped in a viscous liquid contained within a graduated cylinder.

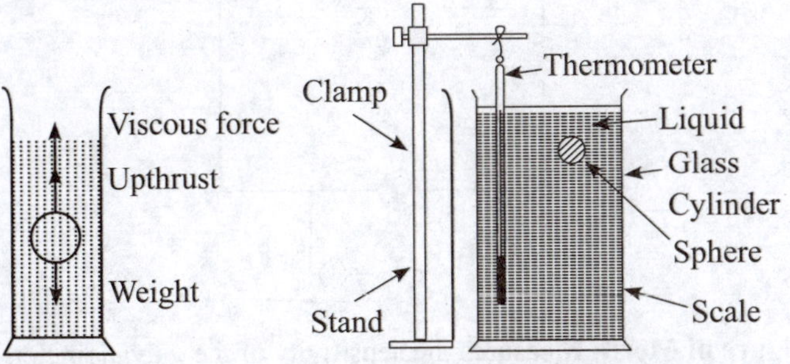

Formulas

1. **Viscous Force Calculation:**

   $F = 6\pi\eta r v$

   Where, η is the viscosity of the liquid, r is the radius of the sphere, and v is the terminal velocity.

   Viscosity (η): $\eta = \frac{2}{9} g(\sigma - \rho) . \frac{r^2}{v} = \frac{2}{9} \frac{(\sigma - \rho)g}{(\text{slope of line})}$

   Where, r radius, σ density of sphere, ρ density of liquid

2. **Terminal Velocity:** Reached when the gravitational force is balanced by the viscous drag and buoyant force.

Experiment 8: Speed of Sound in Air

Purpose: Measure the speed of sound in air using a resonance tube setup.

Setup: Consists of a tube partially filled with water, with a tuning fork vibrated above the opening.

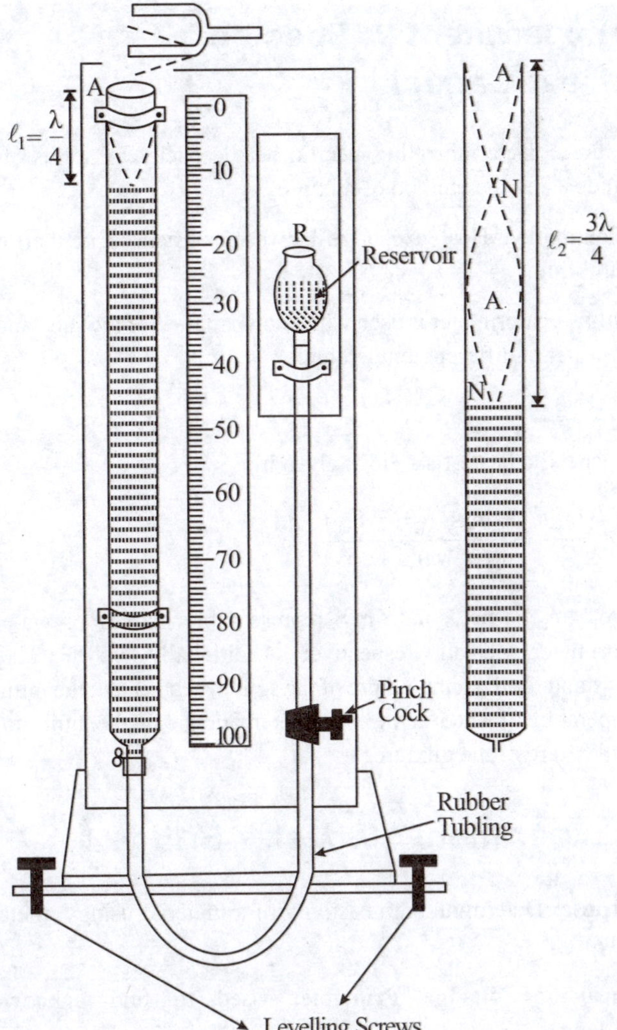

Resonance: Occurs when the frequency of sound waves matches the natural frequency of the air column.

Formulas

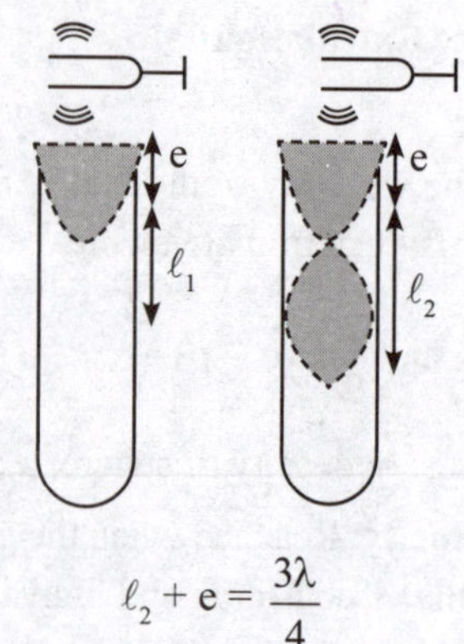

$\ell_1 + e = \frac{\lambda}{4}$, $\ell_2 + e = \frac{3\lambda}{4}$

After solving above equation:

Wavelength:

$\therefore \lambda = 2(l_2 - l_1)$

By applying the equation v = f λ, the velocity of sound is determined as

$v = 2f(l_2 - l_1)$ where, f is the frequency of the tuning fork.

Experiment 9: Specific Heat Capacity

Purpose: Determine the specific heat capacity of solids and liquids using the method of mixtures.

Calorimetry: Heat exchange between substances in thermal equilibrium.

Setup: A calorimeter is used with known masses of the substance and water at different temperatures.

Formulas

The specific heat of a solid is given by

$$s_s = \frac{(m_w s_w + m_c s_c)(\theta_e - \theta_w)}{m_s(\theta_s - \theta_e)}$$

Here, "m_w," "m_c," and "m_s" represent the masses of water, calorimeter, and solid, respectively. Additionally, "θ_s," "θ_w," and "θ_e" stand for the temperature of the solid when heated, the initial temperature of water in the calorimeter, and the final equilibrium temperature of the mixture.

Experiment 10: Metre Bridge

Purpose: Determine the resistivity of a material using a meter bridge.

Wheatstone Bridge Principle: Used to find unknown resistances.

Setup: Consists of a wire stretched along a ruler with known and unknown resistances placed in series.

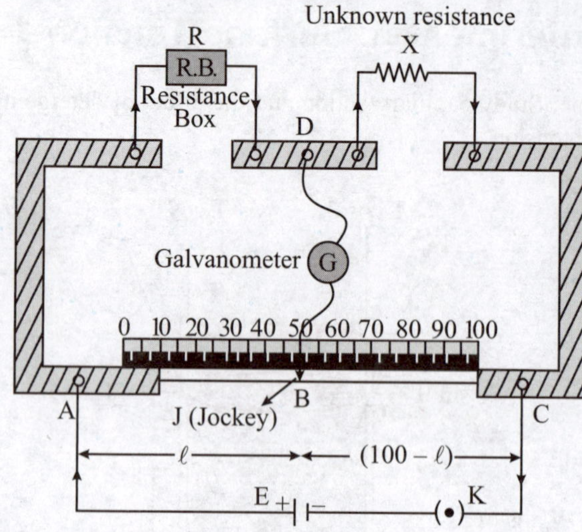

Formulas

According to Wheatstone bridge principal

$\frac{P}{Q} = \frac{R}{X}$ or $X = \frac{Q}{p}R$ or $X = \frac{(100-\ell)R}{\ell}$

Specific resistance (ρ) of the material of the given wire is given by

$$\rho = \frac{X\pi D^2}{4L}$$

Where, L is the length and D is the diameter of the given wire.

Experiment 11: Ohm's Law

Purpose: Determine the resistance of a wire using Ohm's law.

Setup: Consists of a voltage source connected to a resistor with an ammeter to measure current and a voltmeter to measure voltage across the resistor.

Principle: Current through a conductor between two points is directly proportional to the voltage across the two points.

Formulas

V = IR, where R is a constant called resistance.

Experiment 12: Figure of Merit of a Galvanometer

Purpose: Determine the resistance and figure of merit of a galvanometer using the half-deflection method.

Setup: A galvanometer is connected in series with a high resistance to limit current. The resistance is adjusted to reduce the deflection by half.

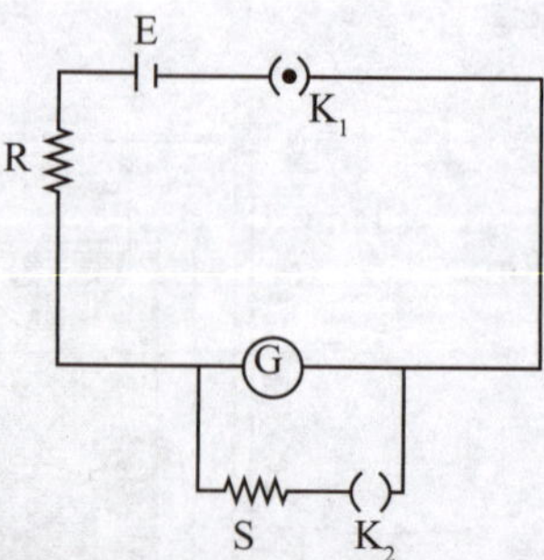

Figure of Merit: Measures the sensitivity of the galvanometer.

Formulas

1. Resistance of galvanometer: $G = \frac{RS}{R - S}$

2. Figure of merit $K = \frac{I_g}{\theta} = \frac{E}{(R+G)\theta}$

Experiment 13: Focal Length of Mirrors and Lenses

Purpose: Determine the focal lengths of mirrors and lenses using the parallax method.

Setup: Includes an optical bench with a source (object needle), a mirror or lens, and an image needle.

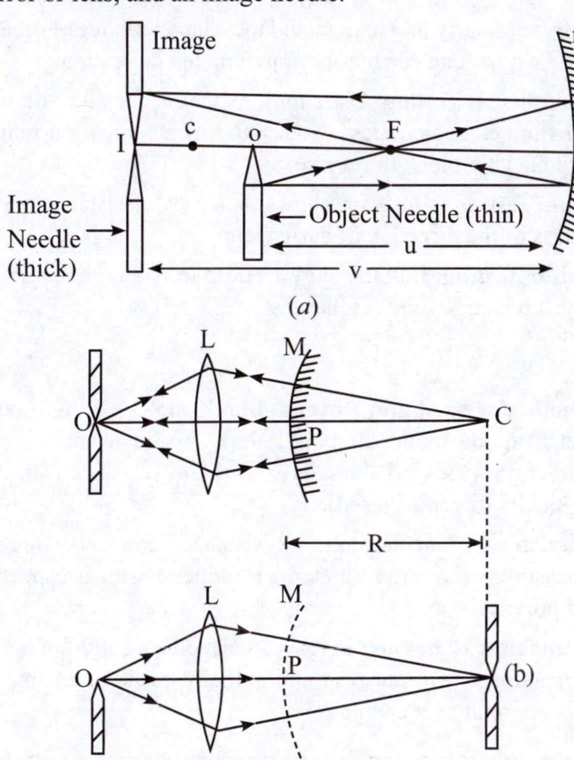

Parallax Method: Adjusts the position until no parallax (apparent shift) is observed between object and image.

Formulas

1. From mirror formula, $\frac{1}{f} = \frac{1}{v} + \frac{1}{u}$

   or $f = \frac{uv}{u+v}$

2. From lens formula $\frac{1}{f} = \frac{1}{v} - \frac{1}{u}$

Experiment 14: Angle of Deviation using a Prism

Objective: Plotting angle of deviation vs. Angle of incidence using parallax method).

Setup: A light source, a triangular prism, and a screen or detector to measure the angle of deviation.

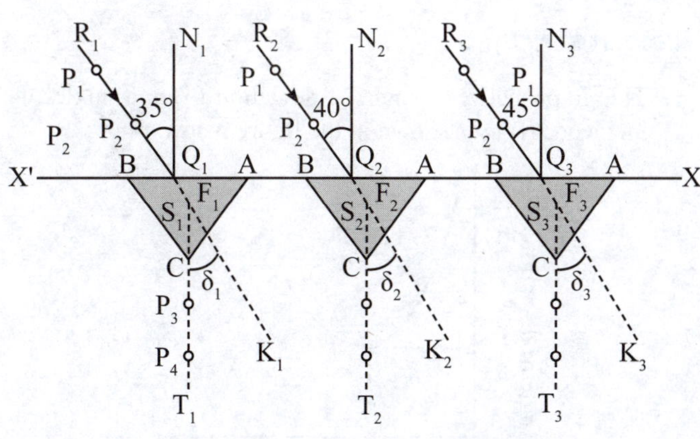

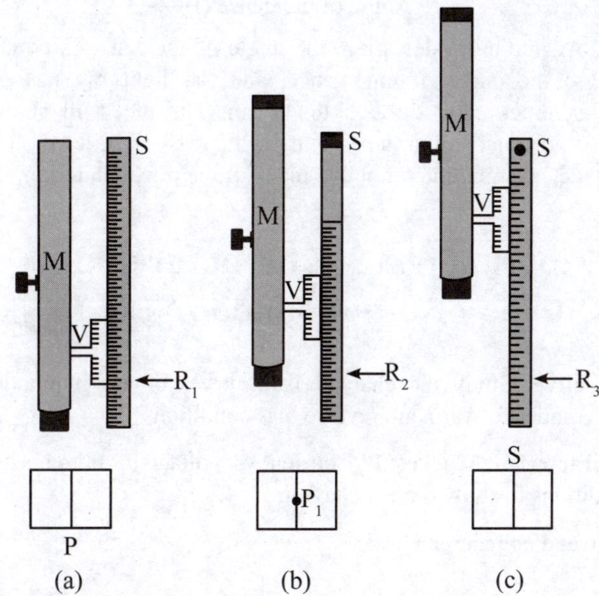

Refractive Index: Determines how much light is bent, or refracted, when it enters the prism.

Formulas

1. Refractive index of the material used in the prism:

$$n = \frac{\sin\left(\frac{A+\delta_m}{2}\right)}{\sin\left(\frac{A}{2}\right)}$$

   Where δ_m is the angle of minimum deviation and A is the prism angle.

2. The vertical scale reading when the microscope is focused on powder sprinkled on top of the slab is denoted as R_3. Real thickness = $(R_3 - R_1)$

   Apparent thickness = $(R_3 - R_2)$

   Refractive index a glass slab,

$$\mu = \frac{\text{Real thickness of slab}}{\text{Apparents thickness of slap}}$$

   Refractive index, $\mu = \frac{(R_3 - R_1)}{(R_3 - R_2)}$

Numerical Profile

- The graph between angle of deviation (δ) and angle of incidence (i) is as shown in the figure below.

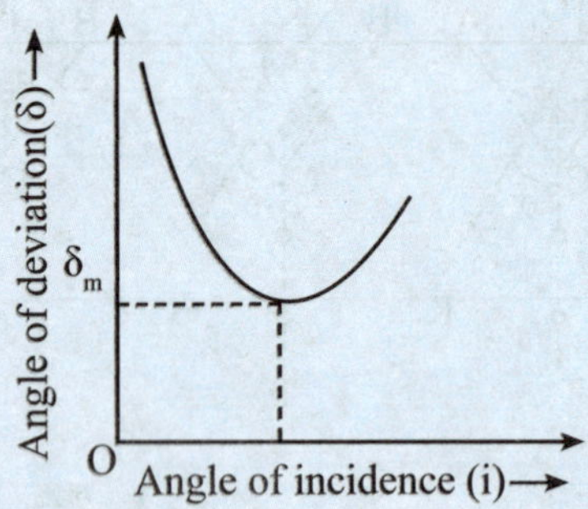

- At minimum deviation, the angle of incidence is equal to the angle of emergence, and the light ray passes symmetrically through the prism. The plot will show a distinct minimum point, which is significant for determining the refractive index of the prism material.

Experiment 15: Characteristic Curves of a p-n Junction Diode

Objective: Study the characteristic curves of a p-n junction diode under forward and reverse bias conditions.

Characteristic Curves: Plot current vs. voltage for both biasing conditions to study diode behavior.

Forward characteristics:

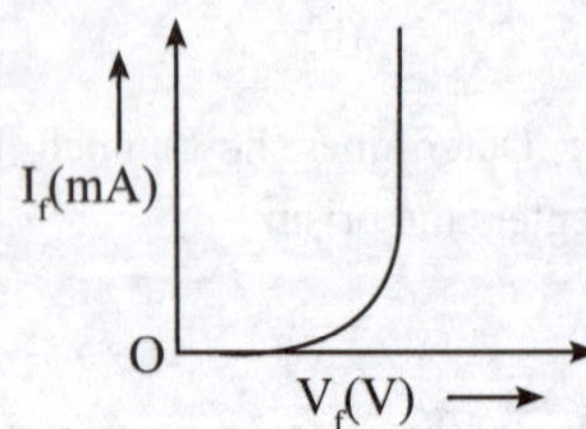

Reverse Characteristics:

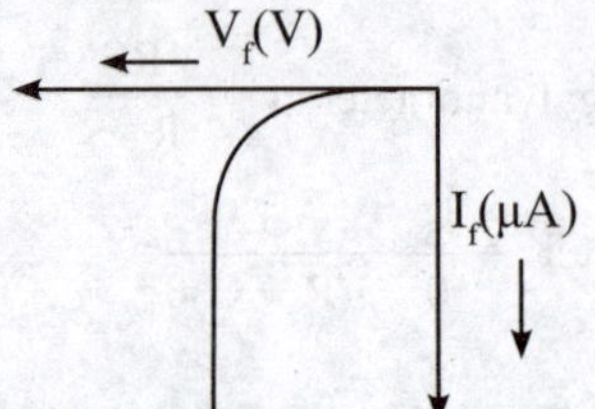

Zener Diode Breakdown: Specific focus on reverse bias breakdown properties.

Experiment 16: Diode, LED, Resistor, and Capacitor

Objective: Identify and understand the characteristics of diodes, LEDs, resistors, and capacitors from a mixed collection.

Diode and LED Testing: Use a multimeter to check for one-way current flow characteristics. Diodes and LEDs allow current in forward bias but block in reverse.

Resistor Testing: Resistor shows a steady resistance value regardless of the direction of current flow.

Capacitor Testing: Initially shows a drop from a high resistance to zero as it charges and discharges.

Formulas

1. **Diode Forward and Reverse Bias Test:** Use a multimeter set to diode mode; it should show low resistance in one direction (forward biased) and high resistance in the opposite (reverse biased).
2. **Resistor Identification:** Resistance can be directly measured with a multimeter across the resistor irrespective of polarity.
3. **Capacitor Behavior:** In resistance mode, a fully functional capacitor's resistance should initially peak then drop to open circuit as it charges.

MCQs ?

NCERT Topic-wise MCQs

Vernier Callipers

1. A vernier calliper has 20 divisions on the vernier scale which coincide with 19 divisions of the main scale the least count of the instrument is 0.1mm. The main scale are of
   1) 1 mm 2) 0.1 mm
   3) 2 mm 4) 0.1 mm
2. What is zero error in the vernier callipers?
   1) If the zeros of vernier scale and main scale do not coincide when the two jaws are in close contact without any gap
   2) When the Zero of the vernier scale is not marked correctly
   3) When the Zero of main scale is not marked correctly
   4) None of these
3. The depth of the vessel is measured by vernier callipers by the help of
   1) The thin rod projecting at the back
   2) The lower pair of jaws
   3) The upper pair of jaws
   4) None of these
4. A vernier calliper has 20 division on the vernier scale which coincide with 19 division of the main scale. what is the vernier constant of the instrument if the main scale division is of 2mm?
   1) 1mm 2) 0.1mm
   3) 0.2mm 4) 0.01mm
5. If a vernier calliper has least count 0.05mm, the accuracy of its measurement is:
   1) 0.05 mm 2) Less than 0.05mm
   3) Greater than 0.05 mm 4) None of these
6. What is the vernier – constant of the vernier calliper?
   1) The difference between one main scale division and one vernier scale division
   2) The value of one scale division
   3) The value of one vernier division
   4) $\dfrac{\text{One vernier division}}{\text{One scale division}}$

Screw Gauge

7. A student measured the diameter of a small steel ball using a screw gauge of least count 0.001 cm. The main scale reading is 5 mm and zero of circular scale division coincides with 25 divisions above the reference level. If screw gauge has a zero error of 0.004 cm, the correct diameter of ball is
   1) 0.521 cm 2) 0.525 cm
   3) 0.529 cm 4) 0.053 cm
8. If the pitch of the screw gauge is halved without changing the number of divisions on the circular scale, how does the accuracy of the instrument change?
   1) Accuracy increase
   2) Accuracy decrease
   3) Accuracy is not affected
   4) The data is insufficient to find the effect on accuracy

Simple Pendulum

9. The work done by a string of simple pendulum during one complete oscillation is equal to
   1) Total energy of pendulum
   2) Kinetic energy of pendulum
   3) Potential energy of pendulum
   4) Zero
10. The time period of oscillation of simple pendulum is given by $T = 2\sqrt{\dfrac{1}{g}}$, Where length 1 is about 100 cm and is known to have 1mm accuracy. The time period is 2 seconds. The time of 100 oscillations is measured by a stopwatch of least count 0.1 s, The percentage error in measurement of g is:
   1) 0.1% 2) 0.2%
   3) 0.8% 4) 1%

Young's Modulus of Elasticity of the Material of a Metallic Wire

11. The graph shows the change Δ1 in the length of a thin uniform wire used by the application of force F at two different temperature T_1 and T_2 the variation suggest that

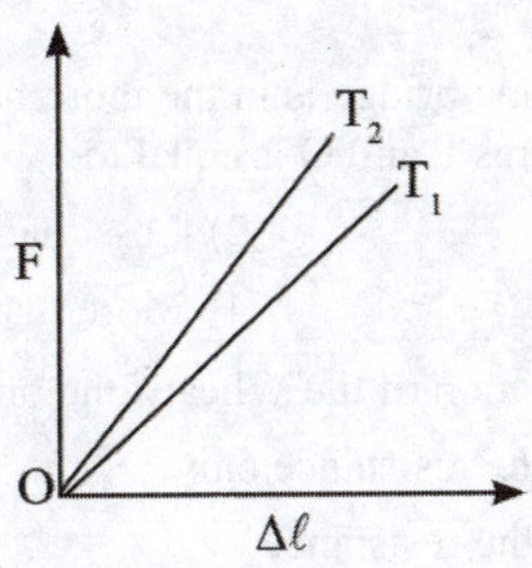

   1) $T_1 = T_2$ 2) $T_1 > T_2$
   3) $T_1 < T_2$ 4) None of these

12. When the stress is increased beyond the elastic limit the length of the wire starts increasing without increasing the force. the point is called
   1) Triple point
   2) Yield point
   3) Inverse point
   4) Breaking point

Surface Tension of Water by Capillary Rise and Effect of Detergents

13. In liquid capillaries of different diameters are dipped, then the graph between liquid rise (h) and diameter (d).

1) 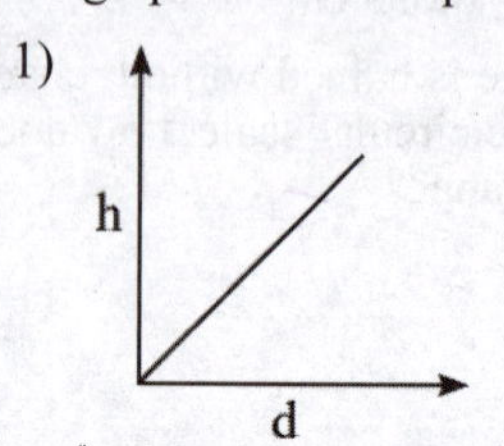

2) 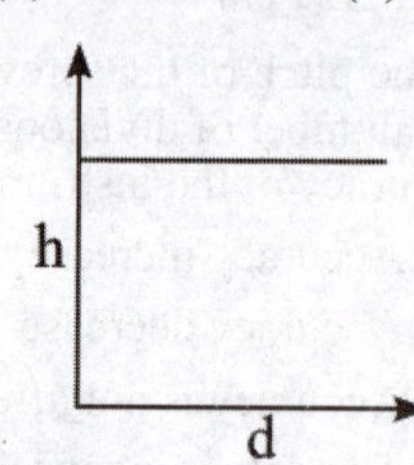

3) 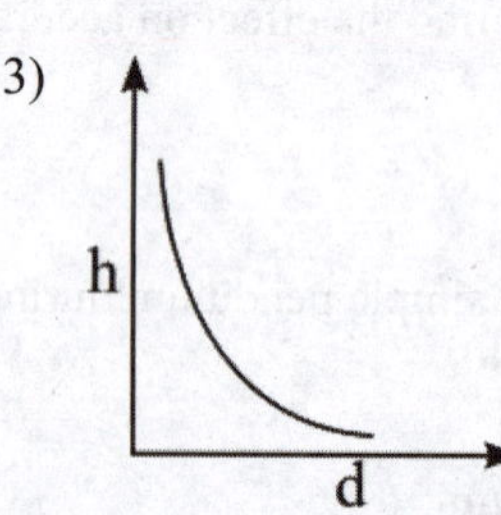

4) 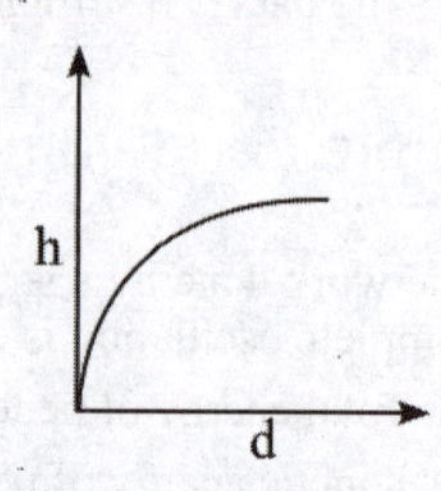

Resistivity of the Material of a Given Wire Using Metre Bridge

14. In a meter bridge, null point is at l = 33.7 cm, when the resistance S is shunted by 12 Ω resistance the null point is found to be shifted by a distance of 18.2 cm. The value of unknown resistance R should be
   1) 13.5 Ω
   2) 6.85 Ω
   3) 3.42 Ω
   4) None of these

15. The wire of meter-bridge needs to have
   1) High specific heat and low temperature coefficient of resistance
   2) Low specific heat and low temperature coefficient of resistance
   3) Low specific heat and high temperature coefficient of resistance
   4) High specific heat and high temperature coefficient of resistance

16. The Wheatstone bridge is in the more balanced state when the ratio of arms P and Q is equal to
   1) Zero
   2) Less than one
   3) One
   4) More than one

17. The main function of the Wheatstone bridge is to
   1) Measure the resistance only
   2) Compare the resistance
   3) Measure current and the resistance
   4) Measure the current only

18. The stripes fitted on a meter bridge are
   1) Thin copper strips
   2) Thick copper strips
   3) Thin constantan strips
   4) Thick constantan strips

The Resistance of a Given Wire Using Ohm's Law

19. If the position of ammeter and voltmeter are interchanged in ohm's law circuit then:
   1) Both instruments will be damaged for flow of large current
   2) No instrument will be harmed for flow of very small current
   3) No effect on reading of both instrument
   4) Both shows reading out of scale

20. A rheostat is used in an electrical circuit
   1) To change the resistance of the circuit
   2) To change the potential difference
   3) To change the emf
   4) To change the current through a particular instrument

21. Resistance of a conductor depends upon
   1) Applied potential difference across conductor and current passing through conductor
   2) Length and area of cross section of conductor
   3) Temperature of conductor
   4) Both (2) and (3)

22. The ohmic resistance can be measured by
   1) Ohm's law apparatus
   2) Wheatstone bridge
   3) AVO meter
   4) All of the above

Resistance and Figure of Merit of a Galvanometer by half Deflection Method

23. The sensitivity of a galvanometer depends on
   1) The magnetic field of cylindrical magnetic field used in the galvanometer
   2) Area of the coil
   3) Torsional constant of the spring
   4) All of these

24. The deflection (θ) produced in the coil of galvanometer is depends on
   1) Area of the coil
   2) Number of turns of coil
   3) Current passing through coil
   4) All of the above

The Focal Length of; (i) Convex Mirror (ii) Concave mirror, and (iii) Convex Lens, Using the Parallax Method

25. A convex lens of focal length F produce a real image n times than of size of object. The distance of object from lens is

1) $(n-1)F$ 2) $(n+1)F$

3) $\left(\frac{n+1}{n}\right)F$ 4) $\left(\frac{n-1}{n}\right)F$

Refractive Index of a Glass Slab Using a Traveling Microscope

26. In traveling microscope, focal length of objective is :
    1) Greater than focal length of eye piece
    2) Smaller than focal length of eye piece
    3) Equal to that of eye piece
    4) None of these
27. In a travelling microscope, the final image is formed at:
    1) Infinity
    2) At focus of eye piece
    3) At force of objective
    4) At least distance of distinct of vision from eye piece
28. In a traveling microscope, focal length of objective is:
    1) Greater than the focal length of eye piece
    2) Smaller than focal length of eye piece
    3) Equal to that of eye piece
    4) None of above

Characteristic Curves of a p-n Junction Diode in Forward and Reverse Bias

29. Which of the following statement is wrong regarding a p – n junction diode?
    1) When the p-type section is connected to the positive terminal and the n - type section to the negative terminal of the battery, the diode is called forward biased.
    2) When the p type section is connected to the positive terminal and the n type section to the negative terminal of the battery the diode is said to be reverse biased.
    3) The p-n junction conducts in forward biased mode
    4) The p-n junction has high resistance in reverse biased mode
30. What will be the resistance measured by an ohmmeter, if a p-n diode is reverse biased?
    1) Zero
    2) Low
    3) High
    4) Infinite
31. What will happen when a p-n diode is reversed-biased?
    1) No current flows
    2) The depletion region is increased
    3) The depletion region is reduced
    4) The height of the potential barrier is reduced

High Order Time Intensive MCQs

1. Read the screw gauge main scale has $\frac{1}{2}$ mm Marks. Circular scale has 50 division In complete rotation the screw advance by $\frac{1}{2}$ mm

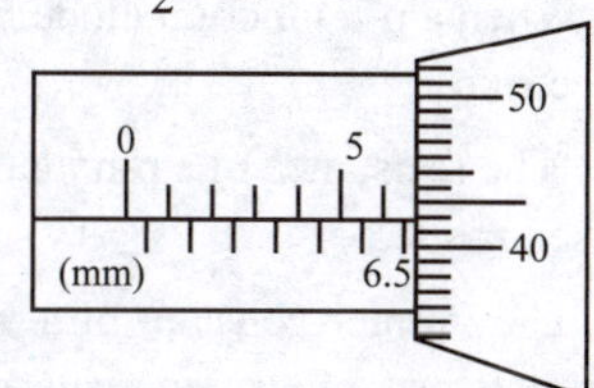

    1) 5.62 mm
    2) 6.19 mm
    3) 6.93 mm
    4) 6.83 mm
2. The vernier constant of two vernier calliper P and Q are 0.01cm and 0.01mm respectively. Which one can measure the length of an object more accurately?
    1) Both P and Q are equally accurate
    2) Vernier P
    3) Vernier Q
    4) Accuracy in measurement does not depend on vernier constant
3. The least count of the screw gauge is $\frac{1}{100}$ mm and the pitch of the screw is 1 mm. what is the maximum percentage error is of the instrument
    1) 5% 2) 2%
    3) 1% 4) 10%
4. Which of the following is not the cause of damping of oscillation of simple pendulum
    1) Friction due to two halves of split cork
    2) Air current due to fan
    3) Opening of doors and window of the room
    4) Closing of doors and window of the room

5. Total energy of particle executing S.H.M is proportional to
   1) Frequency of oscillation
   2) Displacement from mean position
   3) Square of amplitude
   4) Velocity in equilibrium

6. The meter bridge is most sensitive when
   1) Balance point in near 100 cm mark
   2) Balance point is near 50cm mark
   3) Balance point is near 150cm mark
   4) Balance point is near 75cm mark

7. In making an Ohm's law circuit, which of the following connection is correct?
   1) Voltmeter in series and ammeter in parallel
   2) Voltmeter in parallel and ammeter in series
   3) Voltmeter and ammeter both are in parallel
   4) Voltmeter and ammeter both are in series

8. If the position of an ideal voltmeter and the ammeter are interchanged in D.C circuit then:
   1) The voltmeter reads the emf of the source and the ammeter reading is zero
   2) The voltmeter reads the emf of the source and the ammeter shows maximum current.
   3) The voltmeter reading is zero and the ammeter reading is also zero
   4) Voltmeter reading to produce a deflection of one division

9. The figure of merit of a galvanometer is:
   1) The voltage required to produce a deflection of one division
   2) The current required to produce a deflection of 10 division
   3) The voltage required to produce a deflation of 10 division
   4) The current required to produce a deflection of one division

10. The traveling microscope is used for
    1) Determination of focal length of convex mirror
    2) Determination of focal length of convex lens
    3) Determination of refractive index of prism
    4) Determination of refractive index of glass slab

11. When the metre bridge is most sensitive
    1) $P = Q, R = X$
    2) $P = Q = R = X$
    3) $P = R\ Q = X$
    4) $P \neq Q \neq R \neq X$

12. The V-I graph for ohmic resistance is
    1) A straight line
    2) A rectangular hyperbola
    3) A parabola
    4) None of the above

13. Which of the following shows V-I straight line curve?
    1) NaCl solution 2) Junction diode
    3) Thyristor 4) Gold

14. The potential gradient of a potentiometer can be increased by which of the following operation?
    1) By decreasing the current through it
    2) By using a wire of material of low specific resistance
    3) By decreasing the area of cross section of the potentiometer wire
    4) By increasing the area of cross section of the potentiometer wire.

15. The current passing through galvanometer is 50 A. The pointer of galvanometer shows reading at tenth division. The figure of merit is
    1) 6A div^{-1} 2) 5A div^{-1}
    3) 15A div^{-1} 4) 500A div^{-1}

Statement Based MCQs

Directions: These questions consist of two statements each, printed as Statement-I and Statement-II. While answering these questions, you are required to choose any one of the following four responses.

1) Both Statement-I and Statement-II are correct.
2) Both Statement-I and Statement-II are incorrect.
3) Statement-I is correct & Statement-II is incorrect.
4) Statement-I is incorrect & Statement-II is correct.

1. **Statement-I:** The focal length of a convex lens can be determined using a traveling microscope.

   Statement-II: The focal length of a convex mirror cannot be accurately measured using a traveling microscope.

2. **Statement-I:** When a p-n junction diode is forward biased, it conducts electricity.

   Statement-II: The resistance of a p-n junction diode is low in reverse biased mode.

3. **Statement-I:** The vernier constant of a vernier calliper is the difference between one main scale division and one vernier scale division.

   Statement-II: The least count of a screw gauge is determined by the pitch of the screw.

4. **Statement-I:** The Wheatstone bridge is most balanced when the ratio of arms P and Q is equal to one.

   Statement-II: The Wheatstone bridge is used to measure resistance accurately.

Assertion & Reason MCQs

Directions: These questions consist of two statements each, printed as Assertion and Reason. While answering these questions, you are required to choose any one of the following four responses.

1) **Both Assertion and Reason are True and the Reason is a correct explanation of the Assertion.**
2) **Both Assertion and Reason are True but Reason is not a correct explanation of the Assertion.**
3) **Assertion is True but the Reason is False.**
4) **Assertion is False but Reason is True.**

1. **Assertion:** When a vernier calliper displays a positive error of 0.5 mm, the necessary adjustment is –0.5 mm to correct the measurement.

   Reason: A positive zero error situation leads to the measured value being higher than the true dimension of the object.

2. **Assertion:** A regular screw gauge is more accurate than a regular vernier calliper.

   Reason: The smallest measurement that a standard screw gauge can make is larger than the smallest measurement that a standard vernier calliper can make.

3. **Assertion:** The V-I characteristic of a P-N junction diode is identical to that of any other conductor.

   Reason: At room temperature, a P-N junction diode acts as a semi-conductor.

Match the Columns MCQs

1. Match the columns.

	Column-I		Column-II
(a)	Focal length of convex lens	(i)	Determined using a traveling microscope
(b)	Resistance of a p-n junction diode	(ii)	Most balanced when the ratio of arms P and Q is equal to one
(c)	Vernier constant of a vernier calliper	(iii)	Low in reverse biased mode
(d)	Wheatstone bridge	(iv)	Difference between one main scale division and one vernier scale division

1) (a)-(i) (b)-(iii) (c)-(iv) (d)-(ii)
2) (a)-(iii) (b)-(i) (c)-(iv) (d)-(ii)
3) (a)-(i) (b)-(ii) (c)-(iv) (d)-(iii)
4) None of these

2. Match the columns.

	Column-I		Column-II
(a)	Screw gauge	(i)	Used to measure the end correction
(b)	Resonance tube apparatus	(ii)	Determines the thickness/ diameter of a thin sheet/ wire
(c)	Meter bridge	(iii)	Measures the resistance of a wire
(d)	Potentiometer	(iv)	Determines the speed of sound in air

1) (a)-(ii) (b)-(iv) (c)-(i) (d)-(iii)
2) (a)-(ii) (b)-(i) (c)-(iii) (d)-(iv)
3) (a)-(ii) (b)-(iv) (c)-(iii) (d)-(i)
4) None of these

3. Match the columns.

	Column-I		Column-II
(a)	Least count of screw gauge	(i)	The difference between one main scale division and one vernier scale division
(b)	Zero error in vernier callipers	(ii)	Needs subtraction from the reading
(c)	Pitch of a screw	(iii)	Ratio of pitch to total number of divisions on the circular scale
(d)	Vernier constant of the vernier calliper	(iv)	Distance moved by screw per rotation

1) (a)-(ii) (b)-(i) (c)-(iv) (d)-(iii)
2) (a)-(iii) (b)-(ii) (c)-(i) (d)-(iv)
3) (a)-(ii) (b)-(iii) (c)-(iv) (d)-(i)
4) (a)-(iii) (b)-(ii) (c)-(iv) (d)-(i)

4. Match the columns.

	Column-I		Column-II
(a)	Resistor in a DC circuit	(i)	One-way current flow
(b)	Forward biased p-n junction	(ii)	Emits light when forward biased
(c)	Diode	(iii)	Conducts current
(d)	LED	(iv)	Exhibits a constant current

1) (a)-(ii) (b)-(i) (c)-(iv) (d)-(iii)
2) (a)-(iii) (b)-(ii) (c)-(i) (d)-(iv)
3) (a)-(ii) (b)-(iii) (c)-(iv) (d)-(i)
4) (a)-(iv) (b)-(iii) (c)-(ii) (d)-(i)

Answer Key

NCERT Topic-wise MCQs

1	2	3	4	5	6	7	8	9	10	11	12	13	14	15	16	17
3	1	1	2	1	1	3	1	4	2	2	2	3	2	1	3	2

18	19	20	21	22	23	24	25	26	27	28	29	30	31
2	1	1	4	4	4	4	3	2	4	4	2	3	2

High Order Time Intensive MCQs

1	2	3	4	5	6	7	8	9	10	11	12	13	14	15
3	3	3	4	3	2	2	1	4	4	2	1	4	3	2

Statement Based MCQs

1	2	3	4
3	1	4	3

Assertion & Reason MCQs

1	2	3
1	3	4

Match the Column MCQs

1	2	3	4
1	4	4	2

Explanations

CLASS-XI-Explanations

Mathematical Tools

NCERT Topic-wise MCQs

1. (3) $\sin(90° - \theta) = \cos\theta$

2. (2) $180° = \pi$ radian

3. (3) $\sin n\pi = 0, n = 0, 1, 2....$

4. (1) $\cos(180 - \theta) = -\cos\theta$

5. (4) **6. (4)** **7. (4)** **8. (2)** **9. (3)** **10. (1)**

11. (1) $\cos 15° = \cos(45° - 30°)$

$= \cos 45°\cos 30° + \sin 45°\sin 30°$

$= \frac{1}{\sqrt{2}} \times \frac{\sqrt{3}}{2} + \frac{1}{\sqrt{2}} \times \frac{1}{2} = \frac{\sqrt{3}+1}{2\sqrt{2}}$

(2) $\cos 53° = \frac{3}{5}$

(3) $\tan 37° = \frac{\sin 37°}{\cos 37°}$

$= \frac{3/5}{4/5} = \frac{3}{4}$

(4) $\sin 53° - \cos 37° = \frac{4}{5} - \frac{4}{5} = 0$

12. (1) $2\sin 15° \cos 15° = \sin 30° = \frac{1}{2}$

(2) $\sin 22.5° \cos 22.5° = \frac{1}{2}\sin 45° = \frac{1}{2\sqrt{2}}$

(3) $\tan 75° = \tan(45° + 30°)$

$= \frac{\tan 45° + \tan 30°}{1 - \tan 45° \tan 30°} = \frac{1+\frac{1}{\sqrt{3}}}{1-\frac{1}{\sqrt{3}}} = \frac{\sqrt{3}+1}{\sqrt{3}-1}$

(4) $\sin^2 22.5° = \frac{1-\cos 45°}{2} \quad \frac{1-1/\sqrt{2}}{2} = \frac{\sqrt{2}-1}{2\sqrt{2}}$

13. (1) $\cos\left(\frac{5\pi}{4}\right) = \cos\left(\pi + \frac{\pi}{4}\right) = -\cos\left(\frac{\pi}{4}\right) = -\frac{1}{\sqrt{2}}$

(2) $\sin\left(\frac{2\pi}{3}\right) = \sin\left(\pi - \frac{\pi}{3}\right) = \sin\left(\frac{\pi}{3}\right) = \frac{\sqrt{3}}{2}$

(3) $\sin\left(\frac{3\pi}{4}\right) = \sin\left(\pi - \frac{\pi}{4}\right) = \sin\left(\frac{\pi}{4}\right) = \frac{1}{\sqrt{2}}$

(4) $\tan\left(\frac{7\pi}{6}\right) = \tan\left(\pi + \frac{\pi}{6}\right) = \tan\frac{\pi}{6} = \frac{1}{\sqrt{3}}$

14. (3) $\sqrt{1 + 2\sin\frac{\theta}{2}\cos\frac{\theta}{2}}$

($\because \sin 2\theta = 2\sin\theta\cos\theta$)

$\sqrt{\sin^2\frac{\theta}{2} + \cos^2\frac{\theta}{2} + 2\sin\frac{\theta}{2}\cos\frac{\theta}{2}}$

$\sqrt{\left(\sin\frac{\theta}{2} + \cos\frac{\theta}{2}\right)^2} = \sin\frac{\theta}{2} + \cos\frac{\theta}{2}$

15. (2) $3\sin\theta + 3\cos\theta = 7\sin\theta - 7\cos\theta$

$\Rightarrow -4\sin\theta = -10\cos\theta \Rightarrow \tan\theta = \frac{5}{2}$

16. (3) $100x^2 - 20x + 1 = 0$

$\Rightarrow (10x - 1)^2 = 0$

$\therefore$ two roots are

$\frac{1}{10}, \frac{1}{10}$

17. (2) $\left(\frac{1}{2}\right)^2 - m\left(\frac{1}{2}\right) - \frac{5}{4} = 0 \Rightarrow m = -2$

18. (1) Sum of n natural numbers $= \frac{n(n+1)}{2}$

$= \frac{20}{2}(20+1) = 10 \times 21 = 210$

19. (1) The first five multiples of 3 is 3, 6, 9, 12 and 15

$a = 3$ and $d = 3$

$n = 5$

Sum, $S_n - \frac{n}{2}[2a + (n-1)d]$

$S_5 = \frac{5}{2}[2(3) + (5-1)3] = \frac{5}{2}[6+12] = \frac{5}{2}[18] = 5 \times 9 = 45$

20. (3) This series is G.P. of common ratio $r = -\frac{1}{2}$

$S_\infty = \frac{a}{1-r} = \frac{1}{1-\left(-\frac{1}{2}\right)} = \frac{1}{1+\frac{1}{2}} = \frac{2}{3}$

21. (1) This series is G.P. of common ratio $= \frac{1}{3}$

$S_\infty = \frac{a}{1-r} = \frac{1}{1-\frac{1}{3}} = \frac{1}{\frac{2}{3}} = \frac{3}{2}$

22. (3) $a = 1, d = 2$, number of terms $= n$

$\therefore S = \frac{n}{2}[2(1) + (n-1)2] = n^2$

23. (4) $(0.95)^{1/2}$

$(1 - 0.05)^{1/2}$

Binomial expression, if $x << 1$, then $(1 + x)^n = (1 + nx)$

$= \left(1 - \frac{0.05}{2}\right) = \left(1 - \frac{5}{200}\right) = \left(1 - \frac{1}{40}\right) = (1 - 0.025) = 0.975$

24. (1) $(100 + 4)^{1/2}$

$= \left[100\left(1+\frac{4}{100}\right)\right]^{1/2} = 10\left(1+\frac{2}{100}\right) = 10(1+0.02) = 10(1.02) = 10.2$

25. (3) $(4 + 0.04)^3$

$\left(4 + \frac{4}{100}\right)^3 = \left[4\left(1+\frac{1}{100}\right)\right]^3$

$= 4^3 \times \left(1 + \frac{3}{100}\right) = 64 \times \frac{103}{100} = 65.92$

26. (4) $(10 - 0.4)^4$

$$\left(10 - \frac{4}{10}\right)^4 = \left[10\left(1 - \frac{4}{100}\right)\right]^4 = 10^4\left(1 - \frac{16}{100}\right)$$

$= 10^4 (1 - 0.16) = 10^4 \times 0.84 = 8400$

27. (1) $\log 5^2 + \log 2^2 - \log 5$

$2 \log 5 + \log 2^2 - \log 5 \quad (\because \log m^n = n \log m)$

$\log (5 \times 4) = \log 20$

$(\because \log (m \times n) = \log m + \log n)$

28. (3) $\log 2^3 = 3 \log 2$

29. (1) $\log (3 \times 5) = \log 3 + \log 5$

30. (2) $\log_5\left(\frac{x}{y}\right) = \frac{2}{1} \Rightarrow \frac{x}{y} = 5^2 = 25$

31. (2) $\frac{dy}{dx} = 2 \sin x \cos x - 4 \tan x . \sec^2 x$

$= 2 \times \frac{1}{\sqrt{2}} \times \frac{1}{\sqrt{2}} - 4 \times 1 \times \sqrt{2} \times \sqrt{2} = 1 - 8 = -7$

32. (4) $\frac{dy}{dx} = 3x^2 + 2$

$f(x = 1) = 3 \times 1^2 + 2 = 5$

33. (2) $\frac{dy}{dx} = \frac{e^x - (1+x)e^x}{e^x . e^x} = \frac{-x}{e^x}$

34. (1) **35. (1)**

36. (1) $y = x \sin x$

$dy/dx = x \cos x + \sin x$

37. (4) $e^x \ell n x + \frac{e^x}{x}$

38. (2) $\frac{dy}{dx} = (x-1)(2x+1) + (x^2 + x + 1).1 = 3x^2$

39. (1) $\frac{dy}{d\theta} = a \sin \theta, \frac{dx}{d\theta} = a(1 + \cos \theta)$

Slope, $\frac{dy}{dx} = \frac{a \sin \theta}{a(1 + \cos \theta)} = \frac{\sin \frac{\pi}{2}}{1 + \cos \frac{\pi}{2}} = 1$

40. (1) Slope, of Tangent, $\frac{dy}{dx} = 2x - \frac{(-2)}{x^3} = 2x + \frac{2}{x^3}$

at point (–1, 0)

$\frac{dy}{dx} = 2 \times (-1) + \frac{2}{(-1)^3} = -2 - \frac{3}{1} = -4$

Slope of normal $= \frac{-1}{\text{Slope of Tangent}} = \frac{-1}{-4} = \frac{1}{4}$

41. (4) $\frac{dy}{dx} = \cos u \cdot \frac{du}{dx} = -\cos u \cdot \sin x$

42. (1) $\frac{dy}{dx} = \frac{\frac{dy}{dt}}{\frac{dx}{dt}} = \frac{6t}{2t} = 3$

43. (2) $\int_0^{\pi/2} \sin 2\theta \, d\theta = \left[\frac{-\cos 2\theta}{2}\right]_0^{\pi/2} = -\left[\frac{\cos \pi - \cos 0}{2}\right]$

$= -\left[\frac{-1-1}{2}\right] = 1$

44. (2) $\int_0^1 (3-2x)^{-2} dx = \left[\frac{(3-2x)^{-2+1}}{(-2+1) \times -2}\right]_0^1$

$= \frac{1}{2}\left(1 - \frac{1}{3}\right) = \frac{1}{3}$

45. (3) $\int (4 \cos t + t^2) dt = \int 4 \cos t \, dt + \int t^2 \, dt = 4 \sin t + \frac{t^3}{3} + C$

46. (1)

47. (2) $\int x^{\left(-\frac{1}{2}\right)} dx = \frac{x^{\left(-\frac{1}{2}+1\right)}}{1/2}$

$= 2\sqrt{x} + C$

48. (1) $\frac{dy}{dx} = \frac{d \sin(2x^2)}{d(2x^2)} \times \frac{d(2x^2)}{dx} = \cos (2x^2) \times 4x$

49. (1) For maxima and minima, $dy/dx = 0$

$6x^2 - 30x + 36 = 0$

$x^2 - 5x + 6 = 0 \Rightarrow x = 2, 3$

At $x = 2$, $y = 39$ (max.)

At $x = 3$, $y = 38$ (min.)

50. (2) $3\sqrt{x} + C$

51. (4) $\frac{x^{\frac{2}{3}}}{2} + C$

52. (1) $\frac{\pi}{2}[\theta]_{-4}^{-1} = \frac{\pi}{2}(-1 + 4) = \frac{3\pi}{2}$

53. (1) Area $= \int_0^b 2x \, dx = b^2$

54. (1) $y_{avg} = \frac{\int_1^3 x^2 dx}{3-1} = \frac{1}{2}\left[\frac{x^3}{3}\right]_1^3 = \frac{13}{3}$

55. (1) Given $y = \sin x$

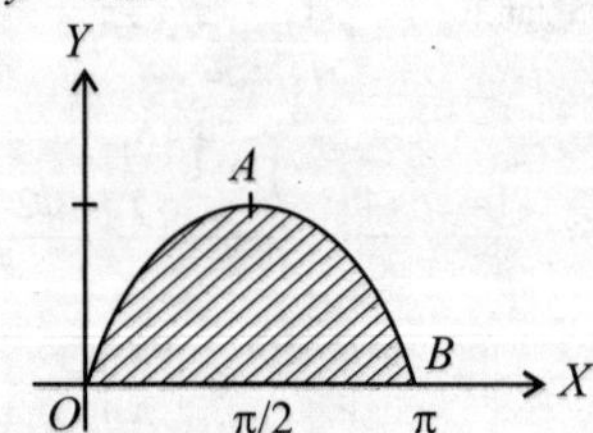

Area of OAB

$= \int_0^\pi y \, dx = \int_0^\pi \sin x \, dx = [-\cos x]_0^\pi$

$= -[\cos \pi - \cos 0] = -(-1 - 1) = 2$ sq. units

56. (1) $y_{avg} = \frac{2\int_0^\pi \sin \theta d\theta}{\pi - 0} = \frac{2}{\pi}[-\cos \theta]_0^\pi = \frac{4}{\pi}$

57. (4) $x = -t^2 + 4t + 4$

$\frac{dx}{dt} = -2t + 4 = 0$

$t = 2$

$x = -(2)^2 + 4(2) + 4$

$x = 8$

58. (1) $\frac{dy}{dx} = 50x - 10 \Rightarrow \frac{dy}{dx} = 0 \Rightarrow x = \frac{1}{5}$

$\frac{d^2y}{dx^2} > 0$

$\therefore\ y_{min} = 4$

59. (2) $U(x) = 100 - 50x + 1000x^2$

$\frac{dU}{dx} = -50 + 2000x = 0 \Rightarrow x = \frac{1}{40}$

$x = 2.5 \times 10^{-2}$

60. (3) $y = 4x^2 - 2x + 3$

$\frac{dy}{dx} = 8x - 2 = 0$

$x = \frac{1}{4}\quad y = 4\left(\frac{1}{4}\right)^2 - 2\left(\frac{1}{4}\right) + 3 = \frac{1}{4} - \frac{2}{4} + 3$

$y = \frac{11}{4} \Rightarrow \frac{d^2y}{dx^2} = 8 > 0 \rightarrow$ minimum

61. (1) $f = x\,(10 - x) = 10\,x - x^2$

$\frac{df}{dx} = -2x + 10 = 0$

$x = 5$

$\frac{d^2f}{dx^2} = -2$ max value at $x = 5, y = 5$

$f = xy = 25$

62. (3) Volume of cylinder = $\pi r^2 \ell$

$V = \pi r^2(12 - r)$

$V = \pi\,(12r^2 - r^3)$

$\frac{dV}{dr} = \pi\left[12(2r) - 3r^2\right] = 0$

$\frac{dV}{dr} = \pi\left[24r - 3r^2\right] = 0$

$24r = 3r^2$

$r = 8$

$\frac{d^2V}{dr^2} = \pi(24 - 6r)$

$= \pi\,(24 - 48) = -24\,\pi$ max.

$\therefore$ max value of $V = \pi \times 64 \times 4 = 256\,\pi$

63. (2) **64. (4)**

65. (2) Displaced point = (–1, 2), a > 0

66. (3) $h = -\frac{b}{2a} = \frac{4}{2\times 3} = \frac{2}{3}$

$k = -\left(\frac{b^2 - 4ac}{4a}\right) = -\left(\frac{16 - 4\times 1\times 3}{4\times 3}\right) = -\frac{1}{3}$

67. (1) Displace the graph of $x = 2y^2$ along positive y-axis by 1 unit.

68. (3) $-3x + 4y = 11$

$4y = 11 + 3x$

$y = \frac{11}{4} + \frac{3x}{4} \Rightarrow c = \frac{11}{4},\ m = \frac{3}{4}$

69. (1)

70. (2) $y = mx + c$

$y = \sqrt{3}x - 2$

71. (3) $y = \frac{3}{2}x - 3$

To get y – intercept, x term = 0

To get x – intercept, y term = 0

$x = 2$

$y = -3$

72. (2) Distance b/w two points (x_1, y_1, z_1) & (x_2, y_2, z_2) is

$= \sqrt{(x_2 - x_1)^2 + (y_2 - y_1)^2 + (z_2 - z_1)^2}$

$= \sqrt{(-2-2)^2 + (0-3)^2 + (5+7)^2}$

$= \sqrt{16 + 9 + 144} = \sqrt{169} = 13$

73. (2)

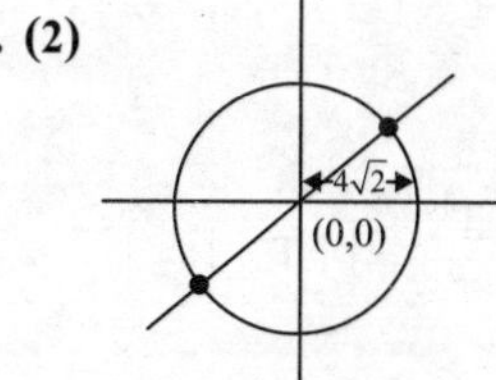

$x = y$ eq^n of straight line

$x^2 + y^2 = \left(4\sqrt{2}\right)^2$ eq^n of circle of centre (0, 0), radius $4\sqrt{2}$

St. line intersect circle at two points$(+a, +a)$ and $(-a, -a)$

$x^2 + x^2 = 32$

$2x^2 = 32$

$x = \pm 4$

$\therefore$ st line intersect circle at point (+4, +4) & (–4, –4)

74. (1) $PV = nRT$ at constant temp

$PV =$ const.

$P \propto \frac{1}{V}$

75. (3) $x^2y = 2$ graph is;

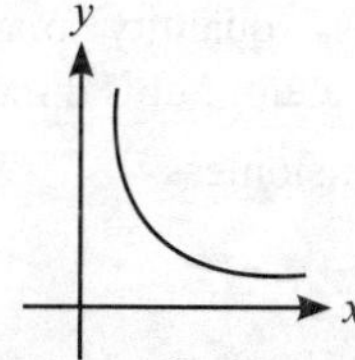

76. (3) D ≥ 0, real roots of quadratic equation (parabola)

77. (3)

D ≥ 0, real roots of quadratic equation (parabola)

Units & Dimensions

NCERT Topic-wise MCQs

1. (2) Solar day → Time for earth to take a complete rotation on its axis

Parallactic second [1 Parsec] → It is a distance corresponding to a parallax of one second of arc.

Leap year → A leap year is year (time) containing one extra day.

Lunar Month → A lunar month is the time between two identical view moons of full moons.

1 Lunar month = 29.53059 days.

2. (1) Fact based

3. **(1)** Kilogram is not a physical quantity, its a unit.

4. **(3)** PARSEC is a unit of distance.

It is used in astronomical science.

5. **(2)** SI system is NOT based on unit of mass, length and time alone.

This system is based on all 7 Fundamental physical quantities and 2 supplementary physical quantities.

6. **(2)** Magnetic moment = Current × Area

$\therefore [M] = [I]\,[A] = AL^2$

7. **(3)** As C is added to t, therefore, C has the dimensions of T.

As $\left[\frac{b}{t}\right] = V$,

$[b] = [V \times t] = LT^{-1} \times T = (L)$

From $[V] = [at]$, $[a] = \left[\frac{v}{t}\right] = \frac{LT^{-1}}{T} = \left[LT^{-2}\right]$

8. **(1)** $F = (M^1L^1T^{-2})$

$= [(10g)^1\,(10cm)^1\,(0.1s)^{-2}]$

$= [(10^{-2}kg)^1\,(10^{-1}m)^1\,(10^2s^{-2})]$

$F = 10^{-1}\,N = 0.1\,N$

9. **(2)** $P = P_0 \exp(-\alpha t^2)$

Since αt^2 is an exponent therefore it must be dimensionless for which $[\alpha] = \frac{1}{[T^2]} = T^{-2}$

10. **(3)** A unit-less quantity never has a non-zero dimension, i.e., it is dimensionless.

11. **(1)** Length, time and velocity can be deduced from one another. Therefore, they can not enter into the list of fundamental quantities in any system.

12. **(1)** A dimensionless quantity may have a unit. For example: angle has a unit but is dimensionless.

13. **(1)** As a × t is dimensionless

$[a] = \left[\frac{1}{t}\right] = \frac{1}{T} = \left[T^{-1}\right]$

Also, $[x] = \left[\frac{v_0}{a}\right]$

$[v_0] = [xa] = \left[LT^{-1}\right]$

14. **(1)** Intensity = Energy per second per unit area

$= \frac{ML^2T^{-2}}{TL^2} = \left[M^1L^0T^{-3}\right]$

15. **(3)** $P = \frac{F}{A} = \frac{MLT^{-2}}{L^2} = \left[ML^{-1}T^{-2}\right] = \frac{M}{LT^2}$

16. **(2)** $RC = \left(\frac{V}{I}\right)\left(\frac{q}{V}\right) = \frac{q}{I} = \frac{I \times t}{I} = t$

17. **(1)** From; $F = \frac{Gm_1m_2}{r^2}$

$G = \frac{Fr^2}{m_1m_2} = \frac{\left(MLT^{-2}\right)L^2}{M^2} = \left[M^{-1}L^3T^{-2}\right]$

18. **(1)** Magnetic flux $(\phi) = BA = \left(\frac{F}{I\ell}\right)A$

$= \frac{\left(MLT^{-2}\left(L^2\right)\right)}{AL} = \left[M^1L^2T^{-2}A^{-1}\right]$

19. **(4)** $h = \frac{E}{\nu} = \frac{ML^2T^{-2}}{T^{-1}} = \left[ML^2T^{-1}\right]$

$L = I\omega = \left(ML^2\right)\left(T^{-1}\right) = \left[M^1L^2T^{-1}\right]$

20. **(2)** $\frac{h}{e^2} = \frac{ML^2T^{-1}}{(AT)^2} = ML^2T^{-3}A^{-2}$ = Resistance (ohm)

21. **(2)** Dimensions of work and torque = $[ML^2T^{-2}]$

22. **(1)** $\text{Resistivity}(\rho) = \frac{[R][A]}{[L]}$ where; $R = [ML^2T^{-3}A^{-2}]$

23. **(2)** $\text{Power} = \frac{\text{Work}}{\text{Time}}$

$\therefore [\text{Power}] = \frac{[\text{Work}]}{[\text{Time}]} = \frac{\left[ML^2T^{-2}\right]}{[T]} = \left[ML^2T^{-3}\right]$

24. **(1)** $\frac{R}{L} = \frac{V/I}{V \times T/I} = \frac{1}{T}$ = Frequency

25. **(1)** $K = \frac{F}{x}$

$\Rightarrow [K] = \frac{[F]}{[x]} = \frac{[MLT^{-2}]}{[L]} = [MT^{-2}]$

26. **(4)** $e = L\frac{di}{dt}$

$[e] = \left[ML^2T^{-2}A^{-2}\right]\left[\frac{A}{T}\right] = \left[\frac{ML^2T^{-2}}{AT}\right] = \left[ML^2T^{-3}A^{-1}\right]$

27. **(3)** Momentum $[MLT^{-1}]$, Planck's constant $[ML^2T^{-1}]$

28. **(3)** $[X] = [F] \times [\rho]^{1/2}$

$= [MLT^{-2}] \times \left[\frac{M}{L^3}\right]^{1/2} = [M^{3/2}L^{-1/2}T^{-2}]$

29. **(1)** $\frac{1}{\sqrt{\mu_0\varepsilon_0}} = c$ = Velocity of light

30. **(4)** $[\varepsilon_0 \times L] = [C]$

$\therefore [X] = \frac{[\varepsilon_0 LV]}{[t]} = \frac{[CV]}{[t]} = \left[\frac{Q}{t}\right]$ = current

31. **(1)** Angle of banking ;

$\tan\theta = \frac{V^2}{rg}$ i.e., $\frac{V^2}{rg}$ is dimensionless

32. **(3)** An electric potential is the amount of work needed to move a unit of charge from a reference point to a specific point inside the field.

So $v = \frac{w}{q}$

33. (1) As $Q = mL$

$$[L] = \frac{[Q]}{[m]} = \frac{[ML^2T^{-2}]}{[M]} = [M^0L^2T^{-2}]$$

34. (1) $V = \sqrt{\frac{\gamma P}{\rho}}$ or $\gamma = \frac{V^2\rho}{P}$

$$[\gamma] = \frac{[LT^{-1}]^2[ML^{-3}]}{[ML^{-1}T^{-2}]} = [M^0L^0T^0]$$

35. (2) Here; kx is dimensionless. Hence,

$$[k] = \left[\frac{2\pi}{\lambda}\right] = [M^0L^{-1}T^0]$$

36. (2) Here, $[a] = [y]$.

So, $\frac{a}{y}$ is dimensionless. Same is the case with kx.

37. (4) As, $\mu = \frac{\text{Velocity of light in vaccum}}{\text{Velocity of light in medium}}$, hence μ is dimensionless.

Thus, each term on the RHS of given equation should be dimensionless.

$\therefore \frac{\beta}{\lambda^2}$ is dimensionless, i.e., β should have dimensions of λ^2, i.e., area.

38. (3) $E = \frac{\text{stress}}{\text{strain}}$

since, strain has no dimensions, therefore, dimension of E = dimension of stress

39. (4) Here, $(\omega t + \phi_0)$ is dimensionless because it is an argument of a trigonometric function.

40. (3) According to Stefan's law, energy emitted per second per unit area is:

$E = e\,\sigma\,T^4$

$$[\sigma] = \frac{[ML^2T^{-2}/L^2T]}{[K^4]} = [MT^{-3}K^{-4}]$$

41. (4) $e = L\left[\frac{dI}{dt}\right]$

$$\therefore [L] = \frac{[e][dt]}{[dI]} = \frac{[w/q][dt]}{[dI]} = \frac{[w][dt]}{[q][dI]} = \frac{[w]}{[dI]^2}$$

$$= \frac{[ML^2T^{-2}]}{[A]^2} = [ML^2T^{-2}A^{-2}]$$

$$[C] = \frac{[q]}{[v]}$$

$$= \frac{[q^2]}{[w]}$$

$$= \frac{[A^2T^2]}{[ML^2T^{-2}]}$$

$$= [M^{-1}L^{-2}T^4A^2]$$

$$[\sqrt{LC}] = \left([ML^2T^{-2}A^{-2}]\times[M^{-1}L^{-2}T^4A^2]\right)^{1/2}$$

$$= [T^2]^{1/2} = [T]$$

$$\therefore \left[\frac{1}{\sqrt{LC}}\right] = [T]^{-1}$$

42. (4) Angular momentum (L) = Moment of inertia (I) × Angular velocity (ω)

$\therefore [L] = [ML^2][T^{-1}] = ML^2T^{-1}$

43. (1) $W = \frac{1}{2}Kx^2 \Rightarrow [K] = \frac{[w]}{[x^2]} = \frac{[ML^2T^{-2}]}{[L^2]} = [MT^{-2}]$

44. (4) $V = at + bt^2$

$[V] = [bt^2]$

$[LT^{-1}] = [bT^2] \Rightarrow [b] = [LT^{-3}]$

45. (4) According to the relation

$L = \frac{nh}{2\pi}$ = Angular momentum

46. (4) Surface tension $= \frac{\text{Force}}{\text{Length}} = \frac{[MLT^{-2}]}{[L]} = [ML^0T^{-2}]$

Spring constant $= \frac{\text{Force}}{\text{Length}} = \frac{[MLT^{-2}]}{[L]} = [ML^0T^{-2}]$

47. (2) From the principle of homogeneity $\left(\frac{x}{v}\right)$ and K has dimensions of T.

48. (3) $F = M^1L^1T^{-2}$

$$\therefore T^2 = \frac{M^1L^1}{F}$$

$T = M^{1/2}L^{1/2}F^{-1/2}$

49. (3) Let $G = C^x g^y P^z$

$[M^{-1}L^3T^{-2}] = [LT^{-1}]^x[LT^{-2}]^y[ML^{-1}T^{-2}]^z$

$= M^z L^{x+y-z} T^{-x-2y-2z}$

Applying principle of homogeneity of dimensions, we get

$z = -1$, $x + y - z = 3$

$-x - 2y - 2z = -2$

On solving, we get,

$y = 2$, $x = 0$

$\therefore G = C^0 g^2 P^{-1}$

50. (2) Let $m \propto E^x V^y F^z$

By substituting the following dimensions:

$[E] = [ML^2T^{-2}]$, $[V] = [LT^{-1}]$, $[F] = [MLT^{-2}]$

and by equating the powers on both sides

$x = 1$, $y = -2$, $z = 0$, So $[m] = [EV^{-2}]$

51. (1) $F = A^\alpha V^\beta \rho^\gamma$

or $[MLT^{-2}] = [L^2]^\alpha [LT^{-1}]^\beta [ML^{-3}]^\gamma$

$= [M^\gamma L^{2\alpha+\beta-3\gamma} T^{-\beta}]$

This gives : $\gamma = 1$, $\beta = 2$, $2\alpha + \beta - 3\gamma = 1$

or $\alpha = 1$; $F = [AV^2\rho]$

52. (1) Dimensions of RHS $= \left[\dfrac{M^{-1}L^3T^{-2} \times M}{L^2}\right]^{1/2}$

$= [M^0LT^{-2}]^{1/2} = \dfrac{[L]^{1/2}}{T} \neq$ Dimensions of LHS

Hence, equation is dimensionally incorrect

Also; $V_e = \sqrt{\dfrac{2GM}{R}}$

Hence, given equation is dimensionally incorrect.

53. (3) $F = \dfrac{1}{4\pi\varepsilon_0}\dfrac{q_1q_2}{d^2}$

$\therefore [\varepsilon_0] = \dfrac{[q_1q_2]}{[Fd^2]} = \dfrac{[AT \times AT]}{[FL^2]} = [F^{-1}L^{-2}T^2A^2]$

54. (1) $[E] = [ML^2T^{-2}]$, $[J] = [ML^2T^{-1}]$, $[G] = [M^{-1}L^3T^{-2}]$

$$\left[\frac{EJ^2}{M^5G^2}\right] = \frac{[ML^2T^{-2}][ML^2T^{-1}]^2}{[M^5][M^{-1}L^3T^{-2}]^2} = [M^0L^0T^0] = [\text{Angle}]$$

55. (4) $\ell \propto h^x G^y c^z$

$[M^0L^1T^0] = [ML^2T^{-1}]^x [M^{-1}L^3T^{-2}]^y [LT^{-1}]^z$

$= [M]^{x-y}[L]^{2x+3y+z}[T]^{-x-2y-z}$

Equating on both sides: $x - y = 0$

$2x + 3y + z = 1$

$-x - 2y - z = 0$

Solving the above equations, $x = y = \dfrac{1}{2}, C = \dfrac{3}{2}$

$\ell \propto \dfrac{\sqrt{hG}}{C^{3/2}}$

56. (2) The dimension of LHS of each relation is [L].

I. As trigonometric term is dimensionless, therefore, dimensions of RHS = [L]. This formula is correct.

II. Dimension of RHS = $[L]\sin[LT^{-1}][T] = [L]\sin[L]$. As angle is not dimension less here. Therefore, this formula is wrong.

III. Dimension of RHS $= \dfrac{[L]}{[T]}\sin\dfrac{[T]}{[L]} = [LT^{-1}]\sin[TL^{-1}]$.

As angle is not dimensionless here, therefore this formula is wrong.

IV. Dimension of RHS $= [L]\left[\sin\dfrac{[T]}{[T]} + \cos\dfrac{[T]}{[T]}\right] = [L]$

As angle is dimensionless and dimension of RHS is equal to the dimension of LHS, therefore this formula is correct.

57. (4) The dimensions of the quantity on the right side are $[M^2L^2T^{-2}]$ for (I) : $[ML^2T^{-2}]$ for (II) and (IV) : $[MLT^{-2}]$ for (III). The quantity on the right side of (V) has no proper dimensions since, two quantities of different dimensions have been added. Since, the kinetic energy K has the dimensions of $[ML^2T^{-2}]$, formulas (I), (III) and (V) and ruled out. The correct dimensions for kinetic energy is given by (II) and (IV).

58. (3) As, $Y = At^2 - Bt^3$

$[Y] = [Bt^3]$

$[L] = [B][T^3]$

$\therefore [B] = [LT^{-3}]$

59. (3) One main scale division, 1 M.S.D = x cm

One vernier scale division, 1 V.S.D $= \dfrac{(n-1)x}{n}$

Least count = 1 M.S.D – 1 V.S.D

$= \dfrac{nx - nx + x}{n} = \dfrac{x}{n}$ cm

60. (3) $\Delta x = a - b \pm (\Delta a + \Delta b)$

$= 4.19 - 3.25 \pm (0.01 + 0.01) = 0.94 \pm 0.02$ cm

61. (3) Quantity C has maximum power, irrespective of sign. So; it brings maximum error in P.

62. (4) Percentage error in A =

$\left(2\times1 + 3\times3 + 1\times2 + \dfrac{1}{2}\times2\right)\% = 14\%$

63. (2) $\dfrac{\Delta V}{V} = 3\times\dfrac{\Delta r}{r} = 3\times\dfrac{1}{100} = \dfrac{3}{100} = 3\%$

64. (4) Since; $\rho = \dfrac{m}{\pi r^2 L}$

$\therefore \left(\dfrac{\Delta\rho}{\rho}\right)\times100 = \left(\dfrac{\Delta m}{m} + \dfrac{2\Delta r}{r} + \dfrac{\Delta L}{L}\right)\times100$

$= \left(\dfrac{0.003}{0.3} + 2\times\dfrac{0.005}{0.5} + \dfrac{0.06}{6}\right)\times100$

$= (0.01 + 0.02 + 0.01)\times100 = 4$

65. (2) Here ; $S = (13.8 \pm 0.2)$cm ; $t = (4.0 \pm 0.3)$sec

$\therefore V = \dfrac{13.8}{4} = 3.45\,\text{m/sec}$

Also; $\dfrac{\Delta V}{V} = \pm\left(\dfrac{\Delta S}{S} + \dfrac{\Delta t}{t}\right)$

$= \pm\left(\dfrac{0.2}{13.8} + \dfrac{0.3}{4.0}\right) = \pm0.0895$

$\Delta V = \pm0.0895 \times 3.45 = \pm0.308775 \approx 0.3$

$V = (3.45 \pm 0.3)$ m/sec

66. (4) $R_1 = (6 \pm 0.3)k\Omega$, $R_2 = (10 \pm 0.2)k\Omega$, $R_p = ?$

$R_p = \dfrac{R_1R_2}{R_1 + R_2}$

$\ln R_p = \ln R_1 + \ln R_2 - \ln$

Differentiating, $\dfrac{\Delta R_p}{R_p} = \dfrac{\Delta R_1}{R_1} + \dfrac{\Delta R_2}{R_2} + \left(-\dfrac{\Delta(R_1 + R_2)}{(R_1 + R_2)}\right)$

$\Rightarrow \dfrac{\Delta R_P}{R_P} = \dfrac{0.3}{6} + \dfrac{0.2}{10} + \dfrac{0.3}{16} + \dfrac{0.2}{16}$

$\therefore$ Total errors $= \dfrac{0.3}{6} + \dfrac{0.2}{10} + \dfrac{0.25}{8}$

$= 0.05 + 0.02 + 0.03125 = 0.10125$

67. (2) If $x = a^n$

then; $\frac{\Delta x}{x} = \pm n\left(\frac{\Delta a}{a}\right)$

68. (4) M = 1 mm

a = 19

b = 20

MSR = 35 division = 35 M = 35 mm

VSR = 4

$$LC = \frac{b-a}{b} \times M = \frac{20-19}{20} \times 35\text{mm} = 0.05\text{mm}$$

69. (1) 20 division of vernier scale = 18 division of main scale

$$\Rightarrow 1\text{VSD} = \left(\frac{18}{20}\right)\text{MSD} = 0.9\text{MSD}$$

Least count = 1 MSD – 1 VSD = 1 MSD – 0.9 MSD

= 0.1 MSD

$$= 0.1 \times 0.1 \text{ cm} = 0.01 \text{ cm} \qquad \left(\because 1\text{MSD} = \frac{1}{10}\text{cm} = 0.1\text{cm}\right)$$

70. (3) $\text{L.C.} = \frac{\text{Pitch}}{\text{Total no. of division on the circular scale}}$

$$= \frac{0.01}{100}\text{cm} = 10^{-4}\text{cm}$$

High Order Time Intensive MCQs

1. (1) Power of anything is dimensionless

$$\therefore \left[\frac{\alpha Z}{k\theta}\right] = [M^0L^0T^0]$$

$$\therefore [\alpha] = \left[\frac{k\theta}{Z}\right]$$

$$\Rightarrow [\alpha] = \frac{[ML^2T^{-2}K^{-1}][K]}{[L]} = [MLT^{-2}]$$

And $[p] = \left[\frac{\alpha}{\beta}\right]$

$$\Rightarrow [\beta] = \left[\frac{\alpha}{p}\right] = \frac{[MLT^{-2}]}{[ML^{-1}T^{-2}]} = [M^0L^2T^0]$$

2. (1) Taking, $T = 2\pi\sqrt{\frac{R^3}{GM}}$

Dimensions of LHS, = [T]

Dimensions of RHS,

$$= \left[\sqrt{\frac{R^3}{GM}}\right] = \sqrt{\frac{[L]^3}{[M^{-1}L^3T^{-2}][M]}} = \sqrt{[T]^2} = [T]$$

Thus, LHS = RHS for $T = 2\pi\sqrt{\frac{R^3}{GM}}$

3. (1) The fundamental frequency is given by $f = \frac{1}{2L}\sqrt{\frac{T}{\mu}}$,

μ = mass per unit length

$$\mu = \frac{M}{l} = \frac{PV}{l} = \rho\pi\frac{D^2}{4}$$

$$\Rightarrow f = \frac{1}{2L}\sqrt{\frac{T}{\rho\pi\frac{D^2}{4}}} = \frac{1}{LD}\sqrt{\frac{T}{\pi\rho}}$$

$\therefore f \propto \frac{1}{LD}$ (as T, π and ρ are constants)

4. (1) $T \propto c^xG^yh^z$

$$\Rightarrow \quad [M^0L^0T] = [LT^{-1}]^x[M^{-1}L^3T^{-2}]^y[ML^2T^{-1}]^z$$

$$\Rightarrow \quad [M^0L^0T] = [M^{-y+z}L^{x+3y+2z}T^{-x-2y-z}]$$

Comparing the powers of M, L and T, we get

$-y + z = 0$ ...(*i*)

$x + 3y + 2z = 0$ ...(*ii*)

$-x - 2y - z = 1$ ...(*iii*)

On solving Eqs. (*i*), (*ii*) and (*iii*), we get

$$x = -\frac{5}{2}, y = z = \frac{1}{2}$$

Hence, dimensions of time are $[G^{1/2}h^{1/2}c^{-5/2}]$.

5. (1) Substituting the dimensions of G,I, M and E, we get,

$$\left[\frac{GIM^2}{E^2}\right] = \frac{[M^{-1}L^3T^{-2}][MLT^{-1}][M^2]}{[ML^2T^{-2}]^2}$$

= [T] = dimensions of time

6. (1) We know that, $[e] = [AT]$, $[\varepsilon_0] = [M^{-1}L^{-3}T^4A^2]$, $[h] = [ML^2T^{-1}]$ and $[c] = [LT^{-1}]$

$$\therefore \left[\frac{e^2}{4\pi\varepsilon_0 hc}\right] = \left[\frac{A^2T^2}{M^{-1}L^{-3}T^4A^2 \times ML^2T^{-1} \times LT^{-1}}\right]$$

$$= [M^0L^0T^0]$$

7. (2) $\text{Relative density} = \frac{\text{Weight in air}}{\text{Loss of weight in water}}$

$$R = \frac{5.00}{1.00} = 5.00$$

$$\frac{dR}{R} = \frac{0.05}{5.00} + \frac{0.1}{1.00} = 0.11 = 11\%$$

R = 5.00 ± 11%

8. (4) According to question, $T \propto p^ad^bE^c$

$[M^0L^0T] = k\,[ML^{-1}T^{-2}]^a\,[ML^{-3}]^b\,[ML^2T^{-2}]^c$

Where, k is constant.

On comparing dimensions of similar terms, we have

$[M^0L^0T] = k[M^{a+b+c}\,L^{-a-3b+2c}\,T^{-2a-2c}]$

On comparing powers of M, L and T, we have

$0 = a + b + c$ ...(*i*)

$0 = -a - 3b + 2c$ ...(*ii*)

$1 = -2a - 2c$ ...(*iii*)

On solving Eqs. (*i*), (*ii*) and (*iii*), get

$$a = -\frac{5}{6}, b = \frac{1}{2}, c = \frac{1}{3}$$

9. (3) $x = \frac{2k^3 l^2}{m\sqrt{n}}$

Percentage error in x,

$$\frac{\Delta x}{x}\times 100 = \left(\frac{3\Delta k}{k}+\frac{2\Delta \ell}{\ell}+\frac{1}{2}\frac{\Delta n}{n}+\frac{\Delta m}{m}\right)\times 100$$

$$\frac{\Delta x}{x}\times 100 = \left[3\times 1 + 2\times 2 + \frac{1}{2}\times 4 + 3\right] = 12\%$$

10. (1) $\frac{h}{\ell} = \theta$

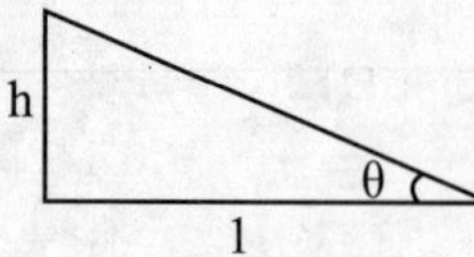

$$\frac{h}{\ell} = 1.8\times\frac{\pi}{180}m$$

$$h = \frac{1.8\times\pi\times 100}{180}cm$$

$h = \pi$ cm

11. (3) $\left[\frac{d}{dt}\left(\int \vec{F}\cdot d\vec{S}\right)\right] = \left[A\left(\vec{F}\cdot\vec{p}\right)\right]$

$$\Rightarrow \left[\frac{Fs}{t}\right] = [AFp]$$

$$\Rightarrow [A] = \left[\frac{s}{pt}\right] = \frac{L}{MLT^{-1}\times T} = M^{-1}$$

12. (4) $\mu \propto (F)^a (M)^b (L)^c$

$T^{-1} = (MLT^{-2})^a (M)^b (L)^c$

$T^{-1} = M^{a+b} L^{a+c} T^{-2a}$

$a + b = 0$ $\quad \boxed{b = \frac{-1}{2}}$

$a + c = 0$ $\quad \boxed{c = \frac{-1}{2}}$

$-2a = -1$ $\quad \boxed{a = \frac{1}{2}}$

$\mu = k F^{1/2} M^{-1/2} L^{-1/2}$

$$\mu = k\sqrt{\frac{F}{ML}}$$

13. (1) Physical quantities having different dimensions can not be added or subtracted.

As P, Q and R are physical quantities having different dimensions, therefore they can neither be added nor subtracted.

14. (1) ℓ Zero Error = 6 × Least Count

= 0.6 mm

∴ Reading = MSR + VSR × LC - Zero Error

= [32 *mm* + (0.1)4*mm*] – 0.6 *mm*

= 31.8 *mm* = 31.8×10^{-1} cm = 3.18 cm

= 3.18 *cm*

15. (2) Reading = *MSR* + *VSD* × *LC* – zero error

$$\text{Reading} = 8.5 + \frac{(0.1)\times 6}{10} - \frac{0.2}{10} = 8.54\text{cm}$$

16. (3) $P = L^a V^b F^c$

$[ML^2T^{-3}] = [L]^a [LT^{-1}]^b [MLT^{-2}]^c$

$c = 1$ $\quad 2 = a + b + c$ $\quad a + b = 1$

$-3 = -b - 2c$ $\quad a = 0$

$b = 1$

$P = V^1F^1$

⇒ P when V & F are halved

P becomes $\frac{1}{4}$

17. (2) $[P] = M^xV^yTz$

$$ML^{-1}T^{-2} = M^x\left(\frac{L}{T}\right)^y T^z$$

$ML^{-1}T^{-2} = M^xL^yT^{z-y}$

$\Rightarrow x = 1, \quad y = -1$

$z - y = -2$

$\Rightarrow z = -2 + y = -3$

18. (3) $[\alpha] = [F]\left[\sqrt{d}\right]$

$= [MLT^{-2}][ML^{-3}]^{1/2}$

$= M^{3/2}L^{-1/2}T^{-2}$

19. (4) Let $f = k\rho^a a^b T^c$ where $[f] = [T^{-1}][\rho] = [ML^{-3}]$, $[a] = [L]$ and $[T] = [MT^{-2}]$

Comparing dimensions on both sides, we get-

$a = -1/2$, $b = -3/2$ and $c = 1/2$

$\therefore \eta = k\rho^{-1/2} a^{-3/2} T^{-1/2}$

$$= \frac{k\sqrt{T}}{\rho^{1/2}a^{3/2}}$$

20. (4) Dimensions of force = $[MLT^{-2}]$

Dimensions of velocity gradient, $\frac{\Delta V}{\Delta Z} = \frac{[LT^{-1}]}{[L]} = [T^{-1}]$

Dimensions of Area, $A = [L^2]$

Given $F = -\eta A\frac{\Delta V}{\Delta Z}$

Dimensional formula for coefficient of viscosity

$$[\eta] = \frac{[F]}{[A]\left[\frac{\Delta V}{\Delta Z}\right]} = \frac{[MLT^{-2}]}{[L^2][T^{-1}]} = [ML^{-1}T^{-1}]$$

21. (3) L.H.S = P = $(M^1L^{-1}T^{-2})$

$$\text{R.H.S} = \frac{8\eta l v}{\pi r^4} = \frac{ML^{-1}T^{-1}(L)(L^3T^{-1})}{L^4} = (ML^{-1}T^{-2})$$

22. (3) The integral on LHS is in the form of log x, which is a number. Hence; a^n must be a number. So, n = 0.

23. (2) Least count of the screw gauge

$$= \frac{1}{100}mm = 0.01mm$$

zero error = +0.05 mm

Reading = 4 × 1 mm + 60 × 0.01 mm – 0.05 = 4.55 mm

Statement Based MCQs

1. (1) Both the statements are true
2. (1) Both the statements are true
3. (2) The recognition of concepts of dimensions, which guide the description of physical behavior is of basic importance as only those physical quantities can be added or subtracted which have the same dimensions.
4. (3) Statement I is true but statement II is false.
5. (1) Both the statements are true.

Assertion & Reason MCQs

1. (4) Due to parallax error the maximum distance of a star we can measure is limited to 100 light years. Parallax error increases due to increase in distance of star.
2. (3) Assertion is true but reason is false.
3. (2) Both assertion & reason are correct and reason is the not the correction explanation of the assertion.
4. (3) Assertion is true but reason is false.
5. (3) On changing the units of measurement of a physical quantity, the numerical value either decreases or increases.
If the system of measurement is large, then the numerical value will be smaller if the system of measurement is smaller, then numerical value will be higher.

Match the Column MCQs

1. (2) (a) - (iii), (b) - (iv), (c) - (i), (d) - (ii)

(i) $[LC] = \left[\frac{L}{R} RC\right] = T^2$

(ii) $[LC] = \left[\frac{L}{R} R^2\right] = T[R^2] = T\left[\frac{V}{I}\right]^2 = T\left[\frac{W}{QI}\right]$

$$= T\left(\frac{W}{I^2 t}\right)^2 = T\left(\frac{ML^2T^{-2}}{A^2T}\right) = M^2L^4T^{-5}A^{-4}$$

(iii) $[H] = \left[\frac{\text{Heat}}{\text{Mass}}\right] = \frac{ML^2T^{-2}}{M} = L^2T^{-2}$

(iv) $[s] = \left[\frac{\text{Heat}}{\text{Mass} \times \text{temperature}}\right] = \frac{ML^2T^{-2}}{MK} = L^2T^{-2}K^{-1}$

2. (4) (a) - (iii), (b) - (i), (c) - (iv), (d) - (ii)
We know that least count is given by s/n, so
(i) leat count = s/n = 1/10 = 0.1 mm
(ii) leat count = s/n = 0.5/10 = 0.05 mm
(iii) leat count = s/n = 0.5/20 = 0.025 mm
(iv) leat count = s/n = 1/100 = 0.01 mm

NCERT Exemplar MCQs

1. (2) In 0.06900, the underlined zeros are not significant. Hence, number of significant figures is four.
2. (3) The sum of the numbers can be calculated as 663.821 arithmetically. The number with least decimal places is 227.2 is correct to only one decimal place.
The final result should, therefore, be rounded off to one decimal place i.e., 663.8.
3. (3) In this question, density should be reported to two significant figures.

$$\text{Density} = \frac{4.237\text{ g}}{2.5\text{ cm}^3} = 1.6948\text{ g/cm}^3$$

After rounding off to two significatnt digits, we get density = 1.7g/cm^3

4. (4) Rounding off 2.745 to 3 significant figures it would be 2.74. Rounding off 2.735 to 3 significant figures it would be 2.74.
5. (1) Given, length $l = (16.2 \pm 0.1)$ cm
Breadth $b = (10.1 \pm 0.1)$ cm
Area $A = l \times b$
$= (16.2\text{ cm}) \times (10.1\text{ cm}) = 163.62\text{ cm}^2$
Rounding off to three significant digits, area A = 164 cm^2

$$\frac{\Delta A}{A} = \frac{\Delta l}{l} + \frac{\Delta b}{b} = \frac{0.1}{16.2} + \frac{0.1}{10.1}$$

$$= \frac{1.01 + 1.62}{16.2 \times 10.1} = \frac{2.63}{163.62}$$

$$\Delta A = A \times \frac{2.63}{163.62} = 163.62 \times \frac{2.63}{163.62} = 2.63\text{ cm}^2$$

$\Delta A = 3\text{ cm}^2$ (By rounding off to one significant figure)

Area, $A = A \pm \Delta A = (164 \pm 3)\text{ cm}^2$.

6. (3) (1) Work = force x distance= $[MLT^{-2}][L] = [ML^2T^{-2}]$
Torque = force × distance = $[ML^2T^{-2}]$
(2) Angular momentum = mvr = $[M][LT^{-1}][L] = [ML^2T^{-1}]$

$$\text{Planck's constant} = \frac{E}{\nu} = \frac{[ML^2T^{-2}]}{[T^{-1}]} = [ML^2T^{-1}]$$

(3) Tension = force = $[MLT^{-2}]$

$$\text{Surface tension} = \frac{\text{force}}{\text{length}} = \frac{[MLT^{-2}]}{[L]} = [ML^0T^{-2}]$$

(4) Impulse = force × time = $[MLT^{-2}][T] = [MLT^{-1}]$
Momentum = mass × velocity = $[M][LT^{-1}] = [MLT^{-1}]$

7. (4) Given, A = 1.0 m ± 0.2 m, B = 2.0 m ± 0.2 m
Let, $Y = \sqrt{AB} = \sqrt{(1.0)(2.0)} = 1.414\text{ m}$
Rounding off to two significant digits Y = 1.4

$$\frac{\Delta Y}{Y} = \frac{1}{2}\left[\frac{\Delta A}{A} + \frac{\Delta B}{B}\right] = \frac{1}{2}\left[\frac{0.2}{1.0} + \frac{0.2}{2.0}\right] = \frac{0.6}{2 \times 2.0}$$

$$\Rightarrow \quad \Delta Y = \frac{0.6Y}{2 \times 2.0} = \frac{0.6 \times 1.4}{2 \times 2.0} = 0.212$$

Rounding off to one significant digit ΔY = 0.2 m.
Thus,correct value for $\sqrt{AB} = Y + \Delta Y = 1.4 \pm 0.2\text{m}$

8. (1) All given measurements are correct upto two decimal places. As here 5.00 mm has the smallest unit and the error in 5.00 mm is least (commonly taken as 0.01 mm if not specified). Hence, 5.00 mm is most precise.

9. **(1)** Given length $l = 5$ cm

Now, checking the errors with each option one by one, we get

$\Delta l_1 = 5 - 4.9 = 0.1$ cm

$\Delta l_2 = 5 - 4.805 = 0.195$ cm

$\Delta l_3 = 5.25 - 5 = 0.25$ cm

$\Delta l_4 = 5.4 - 5 = 0.4$ cm

Error Δl_1 is least.

Hence, 4.9 cm is most precise.

10. **(3)** Given, Young's modulus $Y = 1.9 \times 10^{11}$ N/m²

1N = 10^5 dyne

Hence, $Y = 1.9 \times 10^{11} \times 10^5$ dyne/$(100)^2$ cm²

$\therefore Y = 1.9 \times 10^{11} \times 10^5$ dyne/$(100)^2$ cm²

$= 1.9 \times 10^{16-4}$ dyne/cm²

$Y = 1.9 \times 10^{12}$ dyne/cm²

11. **(4)** Given, fundamental quantities are momentum (p), area(A) and Time (T). We can write energy, $E \propto p^a A^b T^c$

$[E] = [ML^2T^{-2}]$

$[p] = [MLT^{-1}]$

$[A] = [L^2]$

$[T] = [T]$

$[E] = [k]\,[p]^a\,[A]^b\,[T]^c$

Putting all the dimensions, we get

$ML^2T^{-2} = [MLT^{-1}]^a\,[L^2]^b\,[T]^c$

$= M^aL^{2b+a}T^{-a+c}$

By principle of homogeneity of dimensions,

$a = 1,\ 2b + a = 2,\ -a + c = -2$

$\Rightarrow 2b + 1 = 2$

$\Rightarrow b = 1/2 - a + c = -2$

Now, $-a + c = -2$

$\Rightarrow \quad 2b+1=2$

$\Rightarrow \quad b=1/2-a+c=-2$

$\Rightarrow \quad c=-2+a=-2+1=-1$

Hence, $E = pA^{1/2}T^{-1}$

12. **(2, 4)** We know that energy of radiation, $E = h\nu$,

$$[h] = \frac{[E]}{[\nu]} = \frac{[ML^2T^{-2}]}{[T^{-1}]} = [ML^2T^{-1}]$$

As we know that linear impulse, $J = \Delta P$

Dimension of linear impulse = Dimensions of momentum

$= [MLT^{-1}]$

Angular impulse = τdt = ΔL = Change in angular momentum

Hence, dimension of angular impulse

= Dimension of angular momentum

$= [ML^2T^{-1}]$.

This is similar to the dimensions of Planck's constant h.

13. **(1, 2, 4)** We know the following dimensions

$[h] = [ML^2T^{-1}]$, $[c] = [LT^{-1}]$, $[m_e] = M$

$[G] = [M^{-1}L^3T^{-2}]$

$[e] = [AT]$, $[m_p] = [M]$

$$\left[\frac{hc}{G}\right] = \frac{[ML^2T^{-1}][LT^{-1}]}{[M^{-1}L^3T^{-2}]} = [M^2]$$

$$M = \sqrt{\frac{hc}{G}}$$

$$\frac{h}{c} = \frac{[ML^2T^{-1}]}{[LT^{-1}]} = [ML]$$

$$\Rightarrow L = \frac{h}{cM} = \frac{h}{c}\sqrt{\frac{G}{hc}} = \frac{\sqrt{Gh}}{c^{3/2}}$$

As, $c = LT^{-1}$

$$\Rightarrow [T] = \frac{[L]}{[c]} = \frac{\sqrt{Gh}}{c^{3/2}.c} = \frac{\sqrt{Gh}}{c^{5/2}}$$

Hence, a, b or d, any of these can be used to express L, M and T in terms of three chosen fundamental quantities.

14. **(1, 2)** We know that pressure $= \dfrac{\text{Force}}{\text{Area}}$

$$\text{Pressure} = \frac{\text{Force} \times \text{Distance}}{\text{Area} \times \text{Distance}} = \frac{\text{Work}}{\text{Volume}} = \frac{\text{Energy}}{\text{Volume}}$$

15. **(2, 4)** We know that 1 light year = 9.46×10^{11} m = distance that light travels in 1 year with speed 3×10^8 m/s.

1 parsec = 3.08×10^{16} m

= Distance at which average radius of earth's orbit subtends an angle of 1".

NEET & AIIMS Past Year MCQs

1. **(3)** Strain and angle are also dimensionless

Solid angle is dimensionless.

2. **(4)** $F = \alpha t^2 + \beta t$

$[MLT^{-2}] = \alpha\,[T^2]$

$\alpha = [MLT^{-4}]$

$[MLT^{-2}] = \beta\,[T]$

$\beta = [MLT^{-3}]$

$$\frac{\alpha t}{\beta} = \frac{[MLT^{-4}][T]}{[MLT^{-3}]} = [M^0L^0T^0]$$

3. **(2)** Least count $= \dfrac{\text{Pitch}}{\text{Divisions}} = \dfrac{1 \text{ mm}}{100} = 0.01$ mm.

Reading:

MSR = 1 mm

CSR = $63 \times 0.01 = 0.63$

Total reading = 1.63 mm = 0.163 cm.

4. **(4)** $(N + 1)$ vernier scale division (VSD) = (N) main scale division

$$1\,\text{VSD} = \left(\frac{N}{N+1}\right)\text{MSD}$$

L.C = Least count = 1 MSD – 1 VSD

$= \left(1 - \frac{N}{N+1}\right)$ MSD

$= \left(\frac{1}{N+1}\right)$ MSD

$= \left(\frac{1}{N+1}\right) \times \frac{0.1}{10}$ cm

$= \frac{1}{100(N+1)}$ cm

5. (1) As the factors controlling temperature and voltage supply are beyond prediction and control so the error occurred due to unpredictable fluctuations of temperature and voltage would be random errors.

6. (4) $[m = \rho\pi r^2 l]$

$\rho = \frac{m}{\pi r^2 l}$

$\frac{\Delta\rho}{\rho} = \frac{\Delta m}{m} + \frac{2\Delta r}{r} + \frac{\Delta l}{l}$

$\frac{\Delta\rho}{\rho} \times 100 = \frac{0.002}{0.4} \times 100 + \frac{2 \times 0.001}{0.3} \times 100 + \frac{0.02}{5} \times 100$

$= \frac{0.2}{0.4} + \frac{0.2}{0.3} + \frac{2}{5} = 0.5 + 0.67 + 0.4 = 1.57 = 1.6\%$

7. (2) Plane angle and solid angle are dimensionless physical quantities but have units.

8. (4) $[MLT^{-2}A^{-2}]$ is the dimensional formula for Magnetic permeability

9. (1) Area = L × B = 55.3 × 25 = 1382.5 = 14×10^2

Area should have 2 significant figures, as the minimum significant figure in the data given is two.

10. (4) E = energy = $[ML^2T^{-2}]$

G = Gravitational constant = $[M^{-1}L^3T^{-2}]$

So, $\frac{E}{G} = \frac{[E]}{[G]} = \frac{ML^2T^{-2}}{M^{-1}L^3T^{-2}} = \left[M^2L^{-1}T^0\right]$

11. (3) L.C. = $\frac{\text{Pitch}}{\text{CSD}} = \frac{1\text{mm}}{100} = 0.01\text{mm} = 0.001\text{cm}$

Radius = M.S. + n(L–1)

= 0 + 52 (0.001) = 0.052 cm

12. (1) $E \propto F^a A^b T^c$

$[M^1L^2T^{-2}] \propto [MLT^{-2}]^a [LT^{-2}]^b [T]^c$

a + b = 2; –2a – 2b + c = –2 and c = 2

a = 1 b = 1 c = 2

$E \propto [F]\,[A]\,[T^{+2}]$

13. (3) $T = 2\pi\sqrt{\frac{l}{g}} \Rightarrow g = 4\pi^2 \frac{l}{T^2}$

∵ Fractional error, $\frac{\Delta g}{g} = \frac{\Delta l}{l} + \frac{2\Delta T}{T}$

= y + 2x

14. (4) ∵ $F = q\,v\,B \Rightarrow [B] = \left[\frac{F}{qv}\right] = \frac{MLT^{-2}}{CLT^{-1}} = M^1L^0T^{-1}C^{-1}$

15. (1) Resonating frequency

$f = \frac{1}{2\pi\sqrt{LC}}$

$LC = \frac{1}{4\pi^2 f^2}$

Dimension of LC = Dimension of $\frac{1}{4\pi^2 f^2}$

= $[M^0\,L^0\,T^2]$

16. (2) **17. (2)** **18. (3)**

Motion in a Straight Line

NCERT Topic-wise MCQs

1. (2) Free fall motion of stone due to gravity is an example of straight line motion.

2. (1) An aeroplane flies 400 m north and 300 m south. Let the aeroplane starts its journey from point O.

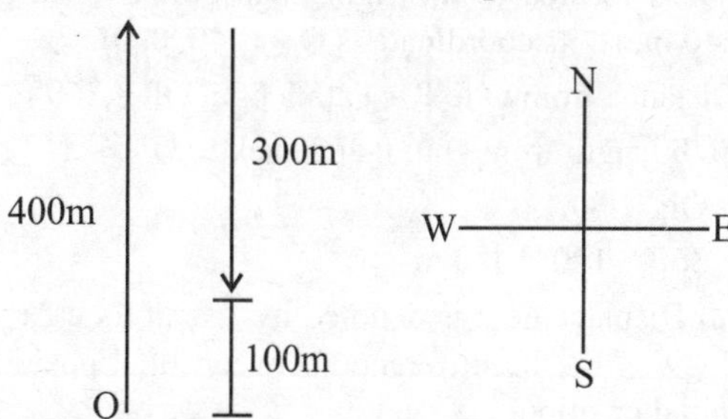

Displacement of the plane is 100 m. After it flies 1200 m upward, the net displacement is-

$= \sqrt{(100)^2 + (1200)^2} = \sqrt{10000 + 1440000}$

= 1204 m ≈ 1200 m

3. (4) Magnitude of displacement is the straight line distance between the initial and final position of the body which may be equal or less than the total distance covered by the body. So the ratio of distance to displacement may be equal or more than one.

$\frac{\text{Distance}}{\text{Displacement}} \geq 1$

4. (3) Initial position ← A

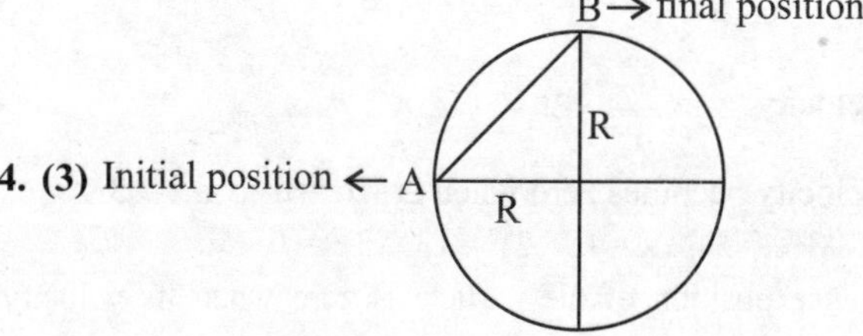

5 minutes = 300s

In 80s, he completes $\frac{300}{80} = 3\,\text{rounds} + \frac{3}{4}\,\text{round}$

∴ Displacement is line segment AB

$= \sqrt{R^2 + R^2} = \sqrt{2}\,R$

5. (4) Horizontal distance covered by the wheel in half revolution = $\pi R = 3\pi$

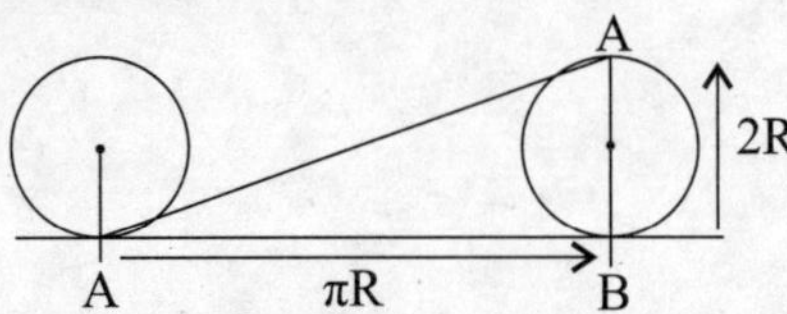

Net displacement of the point which was initially in contact with ground.

$= AA' = \sqrt{(\pi R)^2 + (2R)^2}$

$= \sqrt{(3\pi)^2 + (2\times 3)^2} = \sqrt{9\pi^2 + 36}$

$= 3\sqrt{\pi^2 + 4}$

6. (3) Displacement = (7m, –2m, –3m) – (–4m, –5m, +8m) = (11m, 3m, –11m)

7. (1) Point P is located on X-axis and to the right of origin 'O' at a distance of 360 m. Its position coordinate is P ≡ (+ 360, 0, 0).

'Q' is located to the right of origin 'O' at a distance of 240 m, so its coordinate is Q ≡ (240, 0, 0).

Distance from O to P = Path length OP = 360 m

Path length from O to P and back to Q

= OP + PQ

= 360 + 120 = 480 m

8. (2) Displacement is denoted by Δx, in time $\Delta t = t_2 - t_1$. It is given by the difference between final position (C) and initial position.

9. (3) Distance $= \dfrac{2\pi r}{4} = \dfrac{\pi r}{2}$

Displacement = $AB = \sqrt{2}\,r$

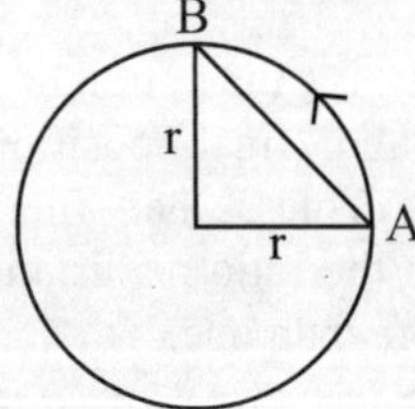

10. (1) Given : $t = \sqrt{x} + 3$ or $\sqrt{x} = t - 3$

or $x = (t-3)^2 = t^2 - 6t + 9$

Velocity, $v = \dfrac{dx}{dt} = 2t - 6$

Velocity becomes zero when $2t - 6 = 0$ or $t = 3$ s

∴ At t = 3s, $x = (t-3)^2 = (3-3)^2 = 0$

Hence position of the particle is zero when its velocity is zero.

11. (2) Average velocity $= \dfrac{\text{Total Displacement}}{\text{Total Time}}$

$= \dfrac{2\,m}{1\,s} = 2\,m/s$

12. (1) $\sqrt{x} = 3t + 5$

Squaring both side

$(\sqrt{x})^2 = (3t+5)^2$

$x = 9t^2 + 25 + 30t$

Differentiate both sides

$\dfrac{dx}{dt} = \dfrac{d(9t^2 + 30t + 25)}{dt}$

$v = 18t + 30$

Velocity will increase with time.

13. (2) ∵ Displacement is zero, hence average velocity will zero.

14. (1) Displacement = 2R

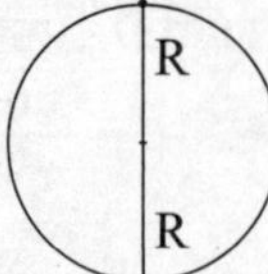

Time = 20 sec.

Average velocity $= \dfrac{2R}{20} = \dfrac{2\times 80}{20} = 8\,m/s$

15. (4) $x = a + bt^2$

Differentiating both sides

$\dfrac{dx}{dt} = \dfrac{d(a + bt^2)}{dt} \Rightarrow v = 0 + 2bt = 2bt$

After t = 3 sec,

$v = 0 + 2 \times 15 \times 3$

v = 90 cm/s

16. (1) Average speed $= \dfrac{\text{Total distance}}{\text{Total time}}$

Total time is $t_1 + t_2 + t_3$

$t_1 = \dfrac{x}{3\times 20}$, $t_2 = \dfrac{x}{3\times 40}$, $t_3 = \dfrac{x}{3\times 60}$

∴ Average speed $= \dfrac{x}{\dfrac{x}{3\times 20} + \dfrac{x}{3\times 40} + \dfrac{x}{3\times 60}}$

$= \dfrac{3}{\dfrac{9 + 4.5 + 3}{180}} = \dfrac{3\times 180}{16.5}$

Average speed = 32.7 m/s

Short trick:

Average speed $= \dfrac{3v_1v_2v_3}{v_1v_2 + v_2v_3 + v_3v_1}$

$= \dfrac{3\times 20\times 40\times 60}{20\times 40 + 40\times 60 + 60\times 20}$

$= \dfrac{144000}{800 + 2400 + 1200} = 32.7$ m/s

17. (2) Total distance covered

$= 30\times\dfrac{15}{60} + 40\times\dfrac{15}{60} + 60\times\dfrac{30}{60} = 47.5\,km$

Total time taken = 15 min + 15 min + 30 min = 60 min = 1 h

Average speed $= \dfrac{47.5}{1} = 47.5\,km\,h^{-1}$

18. (3) 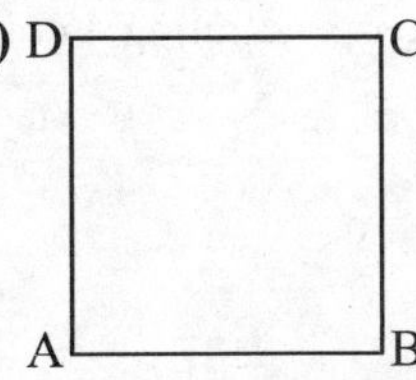

The points A and C are diagonally opposite each other in the square. Therefore, the displacement is the length of the diagonal of the square.

Using the Pythagorean theorem,

$$d = \sqrt{10^2 + 10^2} = \sqrt{100 + 100} = \sqrt{200} = 10\sqrt{2}$$

19. (3) Let R be the radius of the circle

Displacement of particle = r

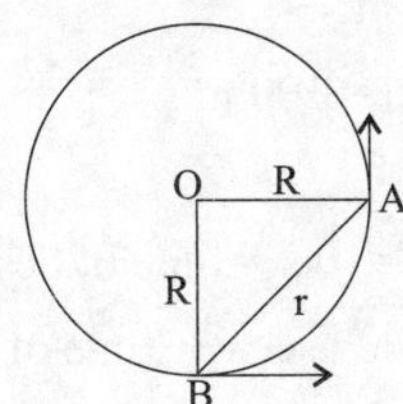

$$\text{Average velocity} = \frac{\text{displacement}}{\text{time}}$$

$$\text{Time} = \frac{\pi R}{2v} \quad (v = \text{velocity})$$

$$r = \sqrt{R^2 + R^2} = \sqrt{2}R$$

$$\text{Average velocity} = \frac{\sqrt{2}R}{\frac{\pi R}{2v}} = \frac{2\sqrt{2}v}{\pi}$$

20. (1) Total distance covered = $s_1 + s_2$

$$\text{Total time taken} = t_1 + t_2 = \frac{s_1}{v_1} + \frac{s_2}{v_2} = \frac{s_1 v_2 + s_2 v_1}{v_1 v_2}$$

$$\text{Average velocity} = \frac{s_1 + s_2}{\frac{s_1 v_2 + s_2 v_1}{v_1 v_2}} = \frac{(s_1 + s_2) v_1 v_2}{s_1 v_2 + s_2 v_1}$$

21. (2) At the time of maximum velocity, acceleration is zero.

$v = 2t(3 - t)$

$v = 6t - 2t^2$

Differentiating both sides

$$\frac{dv}{dt} = \frac{d(6t - 2t^2)}{dt}$$

$0 = 6 - 4t$

$t = 1.5$

22. (2) Average acceleration = $\dfrac{\text{Change in velocity}}{\text{Time in which the change occurred}}$

$$= \frac{60 - 50}{20} = \frac{10}{20} = 0.5\,\text{m/s}^2$$

23. (3) $x^2 = 1 + t^2$

Differenciating

$$2x\frac{dx}{dt} = 2t$$

$xv = t$

Again, differenciating

$$x\frac{dv}{dt} + v\frac{dx}{dt} = 1$$

$x.a + v^2 = 1$

$$a = \frac{1 - v^2}{x} = \frac{1 - t^2/x^2}{x}$$

$$a = \frac{1}{x^3} = x^{-3}$$

24. (2) $v = b\sqrt{x}$

$$\frac{dv}{dt} = \frac{b}{2\sqrt{x}}\frac{dx}{dt} \qquad a = \frac{bv}{2\sqrt{x}}$$

$$a = \frac{b(b\sqrt{x})}{2\sqrt{x}} \Rightarrow \frac{dv}{dt} = a = \frac{b^2}{2}$$

$$v = \frac{b^2}{2}\tau \qquad [\text{At } t = \tau]$$

25. (3) Given, $t = mx^2 + nx$

$$\frac{dt}{dx} = 2mx + n \Rightarrow \frac{dx}{dt} = \frac{1}{2mx + n}$$

$$v = \frac{1}{2mx + n}$$

Retardation of the motion.

$$a = v\frac{dv}{dx} = v\frac{-2m}{(2mx + n)^2} = -2mv^3$$

26. (1) Given, $x(t) = a + bt^2$; $(a = 8.5\text{ m and } b = 2.5\text{ ms}^{-2})$

$$v = \frac{dx}{dt} = \frac{d}{dt}(a + bt^2) = 2bt$$

At $t = 0$, $v = 2b \times 0 = 0\text{ ms}^{-1}$

27. (1) Since, $v = 2bt$

$\Rightarrow$ At t = 2s, velocity = $2 \times 2.5 \times 2$

$= 10\text{ ms}^{-1}$

28. (2) $x = a + bt^2$

$$\text{Average velocity} = \frac{x(t_2) - x(t_1)}{t_2 - t_1}$$

$$= \frac{(a + 16b) - (a + 4b)}{2.0} = 6.0\,b$$

$= 6.0 \times 2.5 = 15\text{ ms}^{-1}$

29. (2) Given, $x = a - bt^2$, a = 8.5m and b = 2.5 m/s²

$= 8.5 - 2.5t^2$

$$\text{Velocity of object} = \frac{dx}{dt} = 2bt$$

(a) velocity at $t = 2.0\,s = \left.\dfrac{dx}{dt}\right|_{t=2} = -4b$

$= -4 \times 2.5 = -10\text{ ms}^{-1}$

(b) velocity at $t = 0 = \left.\frac{dx}{dt}\right|_{t=0} = 0\,ms^{-1}$

(c) Instantaneous speed = Magnitude of instantaneous velocity

$= |-10\ ms^{-1}| = 10\ ms^{-1}$

(d) Average velocity $= \frac{x(t_2)-x(t_1)}{t_2-t_1}$

$= \frac{\left[a-b(4)^2\right]-\left[a-b(2)^2\right]}{2}$

$= \frac{4b-16b}{2} = -\frac{12b}{2} = -6b$

$= -6 \times 2.5\ ms^{-1} = -15\ ms^{-1}$

30. **(4)** I. The instantaneous speed is always positive as it is the magnitude of the velocity at an instant.

II. For t = 0s to t = 5s, the motion is uniform. So, the average velocity and the instantaneous velocity are equal.

III. During t = 0s to t = 5s. The slope of the graph is positive, hence the average velocity and the velocity both are positive. During t = 5s to t = 10s. The slope of the graph is negative, hence the velocity is negative.

Since, there is change in sign of velocity at t = 5s the car changes its direction at that instant.

IV. In figure, instantaneous velocity during t = 0s to t = 5s is positive Also, the speed is always a positive value.

31. **(2)** Length of the barrel = 1.2 m

Speed of the bullet = 640 ms^{-1}

According to the third equation of motion

$v^2 = u^2 + 2as$

$v^2 = 0 + 2as$

$640 \times 640 = 2 \times a \times 1.2$

$\therefore a = \frac{320\times640}{1.2}$

$640 = 0 + \frac{640\times320}{1.2}\times t$

$t = \frac{1.2\times640}{640\times320}$

t = 0.00375 s = 3.75 ms

32. **(4)** Let the initial velocity of the bullet = u

After penetrating its final velocity $= \frac{u}{2}$

From $v^2 - u^2 = 2as$

$\left(\frac{u}{2}\right)^2 - u^2 = 2\times a\times4$

$\frac{u^2}{4} - u^2 = 2\times a\times4$

$\frac{-3u^2}{4} = 8\times a \Rightarrow a = \frac{-3u^2}{32}$

Bullet will further penetrate after penetrating 4 cm.

Initial velocity $= \frac{u}{2}$

Final velocity = 0

From $v^2 - u^2 = 2as$

$0^2 - \left(\frac{u}{2}\right)^2 = 2\times\frac{-3u^2}{32}\times s$

$\frac{-u^2}{4} = \frac{-3u^2}{16}\times s$

$\Rightarrow s = \frac{4}{3} = 1.33\ cm$

33. **(2)** Apply v = u + at

v = u – gt, v = 0, u = 100 m/s

g = 10 m/s^2

Time of ascent, t = 10 s = time of descent

Total time = time of ascent + time of descent = 20s

34. **(2)** $u = -19.6\ ms^{-1}$, $a = 9.8\ ms^{-2}$, t = 6s

$s = ut + \frac{1}{2}at^2$

$s = -19.6\times6 + \frac{1}{2}\times9.8\times6^2$

$s = -19.6\times6 + 4.9\times36$

s = 58.8 m

35. **(2)** Distance travelled in 5th half second

= Distance thavelled in 2.5 – Distance travelled in 2

$= 0.5a\left(t_1^{\,2} - t_2^{\,2}\right)$

$= 0.5\times2\times\left(2.5^2 - 2^2\right)$

= 2.25 m

36. **(3)** If t is the total time of flight of ball in going up and coming back, then total displacement in time t is zero because ball comes back in hand of juggler.

When ball is at the highest point then v = 0

$v^2 - u^2 = 2as$

$0 - u^2 = 2 \times -g \times 40$

$-u^2 = -80 \times 10$

$u = 20\sqrt{2}$

Displacement of the ball is zero.

So, $s = ut + \frac{1}{2}(-g)t^2$

$0 = 20\sqrt{2}t - \frac{1}{2}\times10t^2$

$4\sqrt{2} = t$

$\therefore$ Time interval of each ball

$\frac{4\sqrt{2}}{4} = \sqrt{2} = 1.414\,sec.$

37. (2) Distance covered in 8th second

$$s_{8^{th}} = \frac{a}{2}(15)$$

Distance covered in 8 seconds-

$$s_8 = ut + \frac{1}{2}at^2 \Rightarrow 0 + \frac{1}{2}a \times 64 = 64\left(\frac{a}{2}\right)$$

$$\text{Ratio} = \frac{s_{8^{th}}}{s_8} = \frac{\frac{a}{2}(15)}{\frac{a}{2}(64)} = \frac{15}{64}$$

38. (3) According to the 3rd equation of motion

$v^2 - u^2 = 2as$

After applying the brakes car will come to rest. $v = 0$

$u^2 \propto s$

$$\therefore \frac{s_1}{s_2} = \frac{u_1^2}{u_2^2} \Rightarrow s_2 = \left(\frac{u_2}{u_1}\right)^2 \times s_1$$

$$s_2 = \left(\frac{30}{10}\right)^2 \times 20 = 9 \times 20 = 180\,\text{m}$$

39. (1) Let the particle accelerates with acceleration α for time t_1, and retard with retardation β for t_2.

$t_1 + t_2 = 50$

CASE -1	**CASE -2**
$v = u + \alpha t_1$	$v = u - \beta t_2$
$v = 0 + \alpha t_1$	$0 = u - \beta t_2$
$v = \alpha t_1$	$u = \beta t_2$

Initial velocity of case 2 is equal to the final velocity of Case 1.

$\alpha t_1 = \beta t_2$

$\alpha t_1 = 5\alpha t_2$

$\alpha t_1 = 5\alpha t_2 \Rightarrow 5t_2 + t_2 = 50$

$6t_2 = 50 \Rightarrow t_2 = 25/3$ sec.

40. (2) When body is moving up and down the gravitational force or g is acting on the body in only one direction i.e. downward so the acceleration is – ve. we are considering the upward direction as + ve, then the opposite direction, that is,downward direction is taken as – ve.

41. (2) Ball thrown vertically will suffer retardation of 'g'. Time taken by gravity to reduce balls speed to zero.

$0 = 100 + (-10)t$

$t = 10$ sec.

Ball will be at its peak at t = 10 sec. It will take same amount of time to comeback to ground total time T = 20 sec.

42. (2) Here $u = 0,\ g = 9.8\ \text{ms}^{-2}$

Let n^{th} second be the last second. Then

$$s_{n^{th}} = u + \frac{g}{2}(2n-1)$$

$$\text{or} \quad 24.5 = 0 + \frac{9.8}{2}(2n-1)$$

$$\text{or} \quad 2n - 1 = \frac{24.5}{4.9} = 5$$

or $n = 3$ or $t = 3s$

$$\text{Distance, } s = ut + \frac{1}{2}gt^2 = 0 + \frac{1}{2} \times 9.8 \times 3^2 = 44.1\,\text{m}$$

Hence the height of cliff is 44.1 m.

43. (1)

Using,

$v^2 - u^2 = 2as$

$v^2 = 2g(h - y)$

$v = \sqrt{2g(h-y)}$

44. (1) $h = ut + \frac{1}{2}gt^2$

$$h = -19.6 \times 10 + \frac{1}{2} \times 10 \times 100$$

$h = -196 + 5 \times 100$

$h = 500 - 196$

$h = 304$ m

45. (4) Time taken by first drop to reach the ground $t = \sqrt{\frac{2h}{g}}$

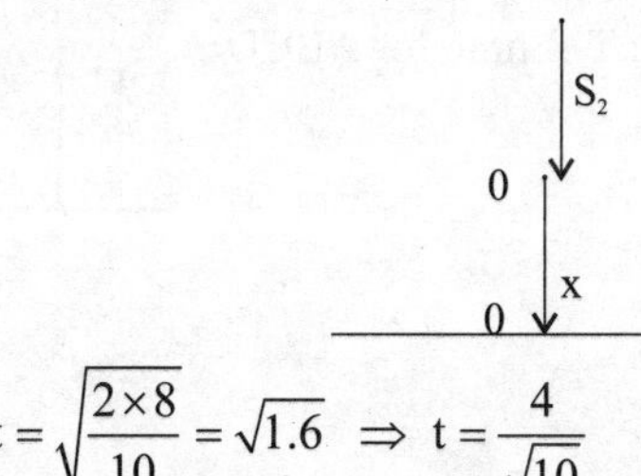

$$t = \sqrt{\frac{2 \times 8}{10}} = \sqrt{1.6} \Rightarrow t = \frac{4}{\sqrt{10}}$$

As the water drops falls at regular intervals of time. Let the time gap between successive drops be T.

$$\Rightarrow 2T = \frac{4}{\sqrt{10}} \Rightarrow T = \frac{2}{\sqrt{10}}$$

Distance of 2nd drop from the tap (u = 0)

$$s_2 = \frac{1}{2} \times 10 \times \left(\frac{2}{\sqrt{10}}\right)^2$$

= 2 m

Distance of drop from ground (x) = 8 – 2 = 6 m

46. (1) As we know that $v^2 = u^2 - 2gs$

v = 0 at maximum height

$-u^2 = -2gs$

$u^2 \propto s$

$$\frac{v_0^2}{v^2} = \frac{h}{3h} \Rightarrow v = \sqrt{3}v_0$$

47. **(1)** $u = 30$ m/s , $t = 8$ seconds, $a = -g$

$$s = ut + \frac{1}{2}at^2$$

$$-h = 30 \times 8 + \frac{1}{2} \times (-10) \times 64$$

$$= 240 - 64 \times 5$$

$$h = 80 \text{ m}$$

48. **(3)** $v = \sqrt{2 \times g \times h} = \sqrt{2 \times 10 \times 40} = \sqrt{800} = 20\sqrt{2}$ m/s

Time of ascent and descent are equal.

Let t_1 is the time of ascent and t_2 be that of descent

T = time of ascent + time of descent

$T = t_1 + t_2$

$\because v = u - gt$

For ascent

$0 = u - gt_1$

$$t_1 = \frac{u}{g}$$

$T = t_1 + t_2$

$$T = \frac{u}{g} + \frac{u}{g} = \frac{2u}{g} = \frac{2 \times \sqrt{800}}{10}$$

$T = 4 \times \sqrt{2}$ sec

$T = 5.66$ seconds

49. **(1)** Time for AB = t_1

Time for ABC = t_2 = Time for DCB

Total time of flight, T = time for ABCD

$= t_1 + t_2$

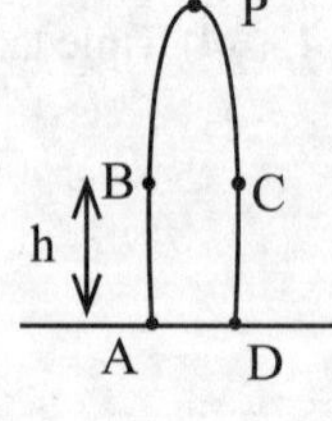

Now, between A&P,

$u = u$, $a = -g$, $v = 0$

$v = u - gt$

$$\Rightarrow u = g\left(\frac{t_1 + t_2}{2}\right)$$

50. **(1)** The acceleration at an instant is the slope of the tangent to the v – t curve at that instant.

51. **(4)** $a = v\dfrac{dv}{dx}$

$$\int a\,dx = \int v\,dv$$

$$ax = \frac{v^2}{2}$$

$$x = \left(\frac{1}{2a}\right)v^2$$

52. **(3)** From time t = 0 s to t = 2 s, distance travelled = 10 m

From time t = 2 s to t = 8 s, distance travelled = 0 m

From time t = 8 s to t = 10 s, distance travelled = 10 m

So total distance travelled = 10 + 10 = 20 m.

53. **(3)**

54. **(1)** Rectilinear motion happens along a straight line. A straight line is one dimensional. Hence, rectilinear motion is one dimensional.

55. **(1)** The velocity-time graph of a body having uniform acceleration (constant acceleration) is a straight line inclined to time axis. The above graph is for motion in positive direction as velocity is positive throughout the time interval and is also increasing, so the acceleration is positive. For positive acceleration, the slope of the graph must be positive.

56. **(4)** For negative direction, the velocity must be negative throughout the journey.

For negative acceleration, the slope of the v – t graph must be negative.

57. **(3)** For the graph shown, area under the v-t curve represents area of the rectangle of height μ and base T.

$\because$ Area under the v-t curve

= Displacement during t = 0 and t = T

$= \mu \times T = \mu T$

58. **(1)** Δv = Area under acceleration time graph

$v - o = (5 \times 1) - (5 \times 1) + (5 \times 1)$

$v = 5 - 5 + 5 = 5$ m/s

59. **(1)** When two bodies are moving towards each other, then their relative speed gets added.

Therefore, the magnitude of the relative speed of both the cars w.r.t each other is v + v = 2v.

60. **(1)**

61. **(1)** The absolute velocity of the small block is the same as that of the block moving with 5m/s. This is because absolute velocity is studied with respect to the ground frame. The relative velocity of the small block with respect to the other block is zero.

62. **(1)** Total length that has to be crossed = 60 + 80 = 140 m

Relative speed = 10 + 25 = 35 m/s

$$\text{Time} = \frac{\text{Displacement}}{\text{Velocity}}$$

$$\text{Time} = \frac{140}{35} = \frac{20}{5} = 4 \text{ seconds}$$

63. **(2)** According to the given equation

First case: Separation between the trucks decreases at the rate of 10 m/s due to the opposite motion of trucks towards each other-

$v_1 + v_2 = 10$ m/s ...(*i*)

Second case: Separation between the trucks increases due to the opposite relative motion of trucks in the same direction-

$v_1 - v_2 = 5$ m/s ...(*ii*)

From equation (*i*) and (*ii*), we get

$v_1 + v_2 = 10$

$v_1 - v_2 = 5$

$2v_1 = 15$

$v_1 = 7.5$ m/s

By putting the value of v_1 in equation (1)

$7.5 + v_2 = 10$

$v_2 = 2.5$ m/s

64. (4) Let the velocity of the scooter be v_S

Distance between the bus and scooter be 1 km = 1000 m

Velocity of bus = 10 m/s

Relative velocity of scooter w.r.t the bus = $v_S - 10$

Time taken to overtake, $t = \frac{1000}{v_S - 10}$

$50 = \frac{1000}{v_S - 10}$

$50\,v_S - 500 = 1000$

$50v_S = 1500$

$\Rightarrow v_s = 30$ m/s

65. (2) Motion of first ball u = 0 , a = –g

Let s_1 be the distance covered by the first ball in 4 seconds.

$-s_1 = ut + \frac{1}{2}gt^2$

$\Rightarrow s_1 = 0 + \frac{1}{2} \times 10 \times (4)^2 = 80\,\text{m}$

Let s_2 be the distance covered by the second ball in 3 seconds.

$-s_2 = 0 + \frac{1}{2} \times (-10) \times (3)^2 = 45\,\text{m}$

$\Rightarrow s_2 = 45\text{m}$

Separation between the two balls.

$s_1 - s_2 = 80$ m – 45 m = 35 m

66. (1) Relative velocity of combustion products of rocket w.r.t the motion of rocket

$\vec{v}_{cr} = +1200\,\text{m/s} \quad \vec{v}_r = -200\,\text{m/s}$

$\vec{v}_{cr} = \vec{v}_c - (\vec{v}_r)$

$1200 = \vec{v}_c - (-200)$

$\vec{v}_v = 1000\,\text{m/s}$

High Order Time Intensive MCQs

1. (3)

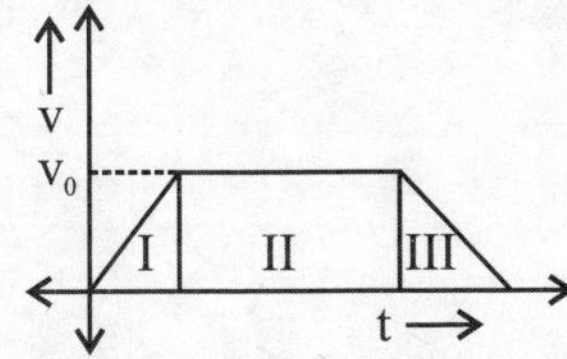

For I $\rightarrow$ $S = ut + \frac{1}{2}at_1^2$

u = O

$\therefore S = \frac{1}{2}at_1^2$

Now, $v_2 = u^2 + 2as$

$v_0^2 = 2fs \Rightarrow v_0 = \sqrt{2Sf}$

For III $0 = v_0^2 - 2\,(f/2)S_3 \Rightarrow S_3 = 2S$

For II,

$S_2 = 15\,S - (S + 2S) \Rightarrow 12\,S = v_0\,t$

$\Rightarrow S = \frac{v_0 t}{12}$

$\Rightarrow S = \frac{\sqrt{2Sf}\,t}{12}$

$\Rightarrow S = \frac{f\,t^2}{72}$

2. (4) $f = f_0 - \frac{f_0 t}{T} \Rightarrow f = 0$

t = T

Now

$\Rightarrow v = f_0 t - \frac{f_0 t^2}{2T}$

When, $f = 0 \Rightarrow t = T$

$v = f_0 T - \frac{f_0 T^2}{2T}$

$\Rightarrow v = \frac{f_0 T}{2}$

3. (1) From the graph

$v = \frac{20}{10}s + 10 \Rightarrow v = 2s + 10$

$a = v.\frac{dv}{ds} = 4s + 20$

At, s = 1 m

$\Rightarrow a = 24$ m/s²

4. (1) $S_{nth} = u + \frac{g}{2}(2n - 1)$

$S_1 = 0 + \frac{g}{2}(2 - 1) = \frac{g}{2}$

$S_2 = 0 + \frac{g}{2}(2 \times 2 - 1)$

$S_2 = \frac{g}{2}(3) = \frac{3}{2}g$

$S_3 = 0 + \frac{g}{2}(2 \times 3 - 1) = \frac{g}{2}(5) = \frac{5g}{2}$

$S_1 : S_2 : S_3 = \frac{g}{2} : \frac{3g}{2} : \frac{5g}{2} = 1:3:5$

5. (1) Total time taken by bird $\Rightarrow t = \frac{d}{v} = \frac{36}{(27+18)} = \frac{36}{45}hr$

Distance travelled by bird

s = v t

$s = 36 \times \frac{36}{45} = \frac{1296}{45} = 28.8$ km

6. **(4)** Speed of child w.r.t. belt = 9 km/hr

Speed of belt = 4 km/hr

From ground frame

Speed = (9 + 4) = 13 km/hr

7. **(4)** From $s = ut + \frac{1}{2}at^2$,

$$x = 0 + \frac{1}{2} \times 10t^2 = 5t^2 \quad \text{... }(i)$$

Also, $x + 3 = 0 + \frac{1}{2} \times 10(t + 0.5)^2$

$$\Rightarrow x + 3 = 5\left(t^2 + \frac{1}{4} + t\right) \quad \text{... }(ii)$$

Subtract Eq. (*ii*) from Eq. (*i*), we get

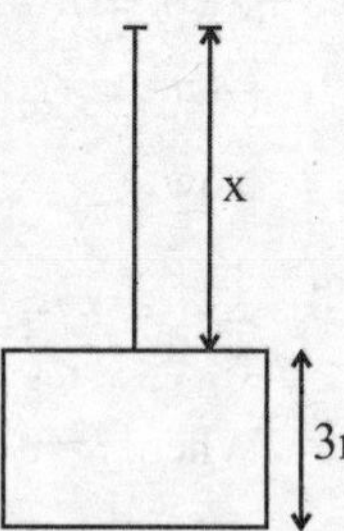

$$3 = 5\left(\frac{1}{4} + t\right) = \frac{5}{4} + 5t$$

$$3 - \frac{5}{4} = 5t$$

$$\frac{7}{4} = 5t \quad \text{or} \quad t = \frac{7}{20}\text{s}$$

From v = u + at,

$$v = 0 + 10 \times \frac{7}{20} = 3.5 \text{ ms}^{-1}$$

8. **(4)** t_1 – Cover h_1 distance

t_2 – Cover h_2 distance

t_3 – Cover h_3 distance

Now S = h, a = g, u = 0 ...(*i*)

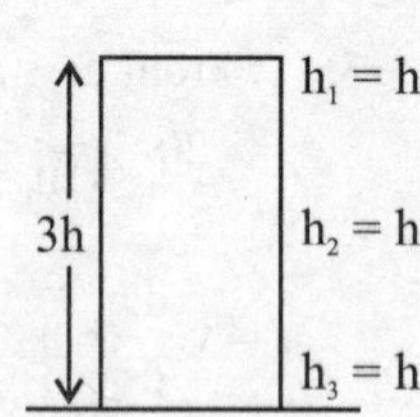

$$t_1 = \sqrt{\frac{2h}{g}}$$

Similarly, to cover S = 2h

$$\therefore 2h = \frac{1}{2}g(t'_2)^2$$

$$t'_2 = 2\sqrt{\frac{h}{g}}$$

$$\therefore t_2 = t'_2 - t_1$$

$$= 2\sqrt{\frac{h}{g}} - \sqrt{\frac{2h}{g}} = \sqrt{\frac{2h}{g}}\left(\sqrt{2} - 1\right)$$

$$t_2 = \sqrt{\frac{2h}{g}}\left(\sqrt{2} - 1\right) \quad \text{...}(ii)$$

For $S = 3h$, $u = 0$, $a = g$

$$t_3' = \sqrt{\frac{6h}{g}},$$

$$\Rightarrow t_3 = t_3' - t_2 - t_1$$

$$= \sqrt{\frac{2h}{g}}\left(\sqrt{3} - \sqrt{2} + 1 - 1\right)$$

$$t_3 = \sqrt{\frac{2h}{g}}\left(\sqrt{3} - \sqrt{2}\right)$$

$$t_1 : t_2 : t_3 = 1 : \left(\sqrt{2} - 1\right) : \left(\sqrt{3} - \sqrt{2}\right)$$

9. **(3)** Distance traveled in T/3 seconds

$$s = \frac{1}{2} \times g \times \left(\frac{T}{3}\right)^2$$

$$s = \frac{T^2}{9 \times 2}g$$

Now $T = \sqrt{\frac{2h}{g}}$

$$s = \frac{1}{18} \times \frac{2h}{g} \times g = \frac{h}{9}$$

Distance from ground $= h - \frac{h}{9} = \frac{8h}{9}$

10. **(2)** For, ball A

$S = -h$, $u = u$, $a = -g$

$v_A^2 = u^2 + 2as$

$\Rightarrow v_A^2 = u^2 + 2(-g)(-h)$

$\Rightarrow v_A = -\sqrt{2gh}$

For, ball B

$S = -h$, $u = -u$, $a = -g$

$v_B^2 = u^2 + 2as$

$\Rightarrow v_B^2 = (-u)^2 + 2(-g)(-h)$

$\Rightarrow v_B = -\sqrt{2gh}$

$v_A = v_B$

11. **(2)** For parallel lines, $v_1 = v_2$

$v_{rel} = v_1 - v_2 = 0$

12. **(4)** For acceleration $v_{max} = \alpha\, t_1$

$$\Rightarrow t_1 = \frac{v_{max}}{\beta}$$

For deceleration $0 = v_{max} - \beta\, t_2$

$$\Rightarrow t_2 = \frac{v_{max}}{\beta}$$

Now,

$t = t_1 + t_2$

$$\Rightarrow t = \frac{v_{max}}{\alpha} + \frac{v_{max}}{\beta}$$

$$\Rightarrow v_{max} = \frac{\alpha\beta t}{\alpha + \beta}$$

13. **(1)** Let the time taken for one third distance be t_1, then $t_1 = \frac{d}{3v_0}$ where d is the total length of the journey. Let the time taken for next 2d/3 distance be t_2,

$$\frac{2d}{3} = \frac{v_1t_2}{2} + \frac{v_2t_2}{2} = \frac{(v_1+v_2)t_2}{2}$$

$t_2 = \frac{4d}{3(v_1+v_2)}$. Thus, the total time taken for the journey is $\frac{d}{3v_0} + \frac{4d}{3(v_1+v_2)} = \frac{d(v_1+v_2+4v_0)}{3v_0(v_1+v_2)}$

Thus the average velocity $= \frac{\text{Total distance}}{\text{Total time}}$

$$= \frac{3v_0(v_1+v_2)}{v_1+v_2+4v_0}$$

14. (1) Here, $a = g - bv$

When an object falls with constant speed v_c, its acceleration becomes zero.

$\therefore\ g - bv_c = 0$ or $v_c = \frac{g}{b}$

15. (4) Relative velocity of the parrot w.r.t. the train

$[10 - (-5)] = 15$ m/s

Time taken by the parrot to cross the train

$= \frac{150}{15} = 10$ s.

16. (2) At t = 0 the first body starts moving with constant acceleration while the second body is already moving with certain constant speed. So the distance covered by the first body x_1 is smaller that covered by the second body x_2 i.e., $x_1 < x_2$ or $x_1 - x_2$ = negative till the first body attains the speed equal to that of second body. At that instant $x_1 = x_2$ or $x_1 - x_2 = 0$ and after that $x_1 > x_2$ i.e., Thereafter, the separation $x_1 - x_2$ increases till it $x_1 - x_2$ = positive and goes on increasing with increasing t.

17. (4) $v = 72\ \text{km/hr} = \frac{72\times5}{18} = 20$ m/s

Time taken to cover the distance of 5 m is:

$t = \frac{5}{20} = 0.25$sec

This time is called Reaction time

18. (3) $v^2 = kr$ or $v = \sqrt{kr}$

$$\frac{dv}{dt} = \sqrt{k}\frac{1}{2}r^{-1/2}\frac{dr}{dt} = \sqrt{k}\frac{1}{2}r^{-1/2}.v$$

$$= \sqrt{k}\frac{1}{2}r^{-1/2}.\sqrt{k}r^{1/2} = \frac{k}{2}r^0$$

Velocity after 1 sec $= 0 + \frac{k}{2}r^0 \times 1 = \frac{k}{2}r^0$

19. (4) Let u be initial velocity and a be the uniform acceleration

Average velocities in the intervals from-

0 to t_1, t_1 to t_2, t_2 to t_3 are

For constant acceleration

$$V_{avg} = \frac{u+v}{2}$$

$$v_1 = \frac{(u)+(u+at_1)}{2} = u + \frac{at_1}{2} \quad \text{... (1)}$$

$$v_2 = \frac{(u+at_1)+[u+a(t_1+t_2)]}{2} = u + at_1 + \frac{a}{2}t_2 \quad \text{... (2)}$$

$$v_3 = \frac{[u+a(t_1+t_2)]+u+a(t_1+t_2+t_3)}{2}$$

$$v_3 = u + at_1 + at_2 + \frac{a}{2}t_3 \quad \text{... (3)}$$

Subtracting (2) from (1), we get

$$v_2 - v_1 = \left(u + at_1 + \frac{a}{2}t_2\right) - \left(u + \frac{a}{2}t_1\right)$$

$$= u + at_1 + \frac{a}{2}t_2 - u - \frac{a}{2}t_1$$

$$v_2 - v_1 = \frac{a}{2}t_1 + \frac{a}{2}t_2 = \frac{a}{2}(t_1+t_2) \quad \text{... (4)}$$

Similarly subtracting (3) from (2), we get

$$(v_3 - v_2) = \frac{a}{2}t_2 + \frac{a}{2}t_3 = \frac{a}{2}(t_2+t_3) \quad \text{... (5)}$$

Divide (4) by (5) we get

$$\frac{v_2 - v_1}{v_3 - v_2} = \frac{\frac{a}{2}(t_1+t_2)}{\frac{a}{2}(t_2+t_3)} = \frac{(t_1+t_2)}{(t_2+t_3)}$$

20. (1) Given relation is $t = \alpha x^2 + \beta x$

Differentiating w.r.t x

$$\frac{dt}{dx} = \frac{d(\alpha x^2 + \beta x)}{dx}$$

$$\frac{1}{v} = 2\alpha x + \beta$$

$$\Rightarrow v = \frac{1}{2\alpha x + \beta}$$

Acceleration is $a = \frac{vdv}{dx}$

$$\Rightarrow a = \frac{-v.2\alpha}{(2\alpha x + \beta)^2} = -2\alpha.v.v^2$$

$$a = -2\alpha v^3$$

21. (1) When ball is dropped, initial velocity is zero, its velocity increases in its downward journey and is given by $v^2 = 2gy$ at any distance y from the point of release. When it hits the ground its velocity is $v = \sqrt{2gd}$. After hitting the ground its direction of motion reverses, it is again given by $0^2 - v^2 = -2gy$. The ball now rises to d/2, thus $v^2 = \frac{2gd}{2}$, $v = \sqrt{gd}$. Graph between v and d in the upward as well as in downward direction is parabola, the direction of velocity being reversed after bouncing graph (1) is correct choice.

22. (1) For 1st ball,

$$h = ut + \frac{1}{2}gt^2$$

$$\Rightarrow h = \frac{1}{2}gt^2 \quad ...(i)$$

For 2nd ball,

$$(h-20) = 0 + \frac{1}{2}g(t-1)^2$$

$$(h-20) = \frac{1}{2}g(t-1)^2 \quad ...(ii)$$

From the equations (*i*) and (*ii*) we get

$$h-(h-20) = \frac{1}{2}gt^2 - \frac{1}{2}g(t-1)^2$$

$$h-h+20 = \frac{1}{2}gt^2 - \frac{1}{2}gt^2 - \frac{1}{2}g + gt$$

$$20 = gt - \frac{g}{2}$$

$$\Rightarrow gt = 25 \quad (\because g = 10\text{ m/s}^2)$$

$$\Rightarrow t = 2.5\text{ second}$$

$$h = \frac{1}{2}\times 10\times(2.5)^2$$

$$= 31.25\text{ m}$$

23. (3) Average Acceleration $= \frac{\vec{v}_2 - \vec{v}_1}{\Delta t} = \frac{\text{Change in velocity}}{\text{Change in time}}$

When ball falls from height h_1 we get,

$$v_1^2 = u_1^2 + 2(-g)(-h_1)$$

$$v = -\sqrt{2gh_1} \quad ...(i)$$

The ball rises to height h_2 after striking the floor

$$\Rightarrow v_2^2 = u_2^2 + 2(-g)h_2$$

(After contact $v_2 = 0$ at rest highest point)

$$0 = (v')^2 - 2gh_2$$

$$(v')^2 = 2gh \Rightarrow v' = -\sqrt{2gh_2} \quad ...(ii)$$

$$\text{Average acceleration} = \frac{v_2 - v_1}{\Delta t}$$

$$\text{Average acceleration} = \frac{-\sqrt{2gh_1} - \sqrt{2gh_2}}{\Delta t}$$

$$= \frac{-\sqrt{2\times 10\times 2} - \sqrt{2\times 10\times 1.5}}{0.020}$$

$$\text{Average acceleration} = \frac{-\sqrt{40} - \sqrt{30}}{0.020} = \frac{-6.32 - 5.47}{0.020}$$

$$\text{Average acceleration} = -\frac{11.79\times 100}{2} = -\frac{1179}{2}$$

$$a_{av} = 589.5\text{ m/s}^2 \approx 590\text{ ms}^{-2}$$

Statement Based MCQs

1. (1) If angle between $\vec{v}$ and $\vec{a}$ is θ

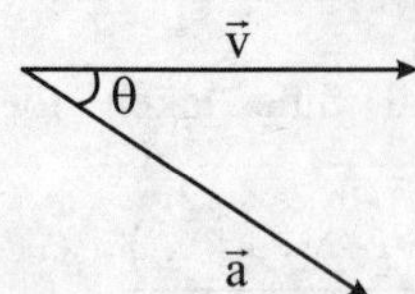

$a\sin\theta$ = Responsible for changing direction of velocity.

$a\cos\theta$ = Responsible for changing magnitude of velocity.

Here $\theta = 90°$

$|\vec{v}| \rightarrow$ Remain constant

2. (1) When a body moves on a straight path in one direction value of distance & displacement remains same.

3. (3) Any freely falling body hence acceleration = g

4. (1) Area → Integration

$$\frac{dv}{dt} = a$$

$$dv = adt$$

Change in velocity

$\vec{a}$ = constant (uniform)

$m = \tan\theta$ = constant

Curve will be straight line

5. (1) For a given interval of time, the body in motion may go from one position to another through different paths, where displacement in a given time has one value but distance travelled can have many values.

Assertion & Reason MCQs

1. (2) A body having positive acceleration can have decreasing speed if both are in opposite directions.

2. (2) When a body is moving vertically upwards then at the highest point of motion, its velocity is zero but its acceleration is equal to acceleration due to gravity.

3. (2) A body having negative acceleration can be associated with a speeding up. If object moves along negative X-direction with increasing speed then acceleration is also along –ve x.

4. (1) Position time graph for a stationary object is a straight line parallel to time axis showing that there is no change in position with time.

5. (4) Assertion is true but reason is false. A body moving along –ve x direction then slope of x – t graph will be –ve.

Match the Column MCQs

1. (1) Given, displacement, $s = \alpha + \beta t + \gamma t^2$

$$\therefore \text{Velocity } (v) = \frac{ds}{dt} = \beta + 2\gamma t$$

$$\therefore \text{Acceleration } (a) = \frac{dv}{dt} = 2\gamma$$

So, Acceleration at 2 second = 2γ (a → ii)

Average velocity during 3rd second,

$$v_{3^{rd}} = \frac{\int_2^3 v\,dt}{\int_2^3 dt} = \frac{\beta(3-2)+\gamma(9-4)}{1}$$

$v_{3^{rd}} = \beta + 5\gamma$ (b → i)

Velocity at 1 second = $\beta + 2\gamma(1)$
$= \beta + 2\gamma$ (c → iv)

Initial displacement (t = 0) = $\alpha + \beta(0) + \gamma(0)^2$
$= \alpha$ (d → iii)

2. (3) For a round trip, displacement is zero; hence $\vec{v}_{av} = 0$

$\vec{v}_{av} = \frac{\vec{v}_1 + \vec{v}_2}{2}$, when $\vec{v}_1$ is initial, $\vec{v}_2$ is final.

Hence a → i, ii

Average speed $(v_{av}) = \frac{\text{Total distance}}{\text{Time of flight}} = \frac{2(v_0^2/2g)}{2v_0/g} = \frac{v_0}{2}$

Hence b → iii

$T_{ascent} = T_{descent} = \frac{v_0}{g}$

Hence, b → iv, d → iv,

3. (2)

a. $x = 0$ at $t = 2s$

b. $v = \frac{dx}{dt} = 10t$

and $a = \frac{dv}{dt} = 10$

$v = a$ at 1s

c. $v = 10t$

Velocity is positive all the time.

d. $v = 0$ at $t = 0s$

4. (3)

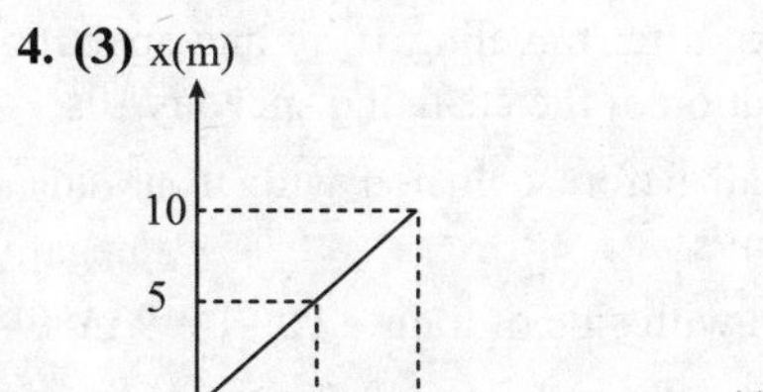

a. 5m

b. From t = 0s to t = 2s

$v = \frac{10}{2} = 5$ m/s (positive)

From t = 2s to t = 4s

$v = 0$

From t = 4 to t = 6s

$v = \frac{-20-10}{6-4} = \frac{-30}{2} = -15$ m/s

From t = 6s to t = 8s

$v = \frac{0+20}{8-6} = \frac{+20}{2} = +10$ m/s

Total Distance = (5 × 2) + (15 × 2) + (10 × 2)
= 10 + 30 + 20 = 60 m

c. at t = 8s x = 0 s = 0

d. x = 0 at $t = 0, \frac{14}{3}s, 8s$

$\therefore t = \frac{14}{3}s$

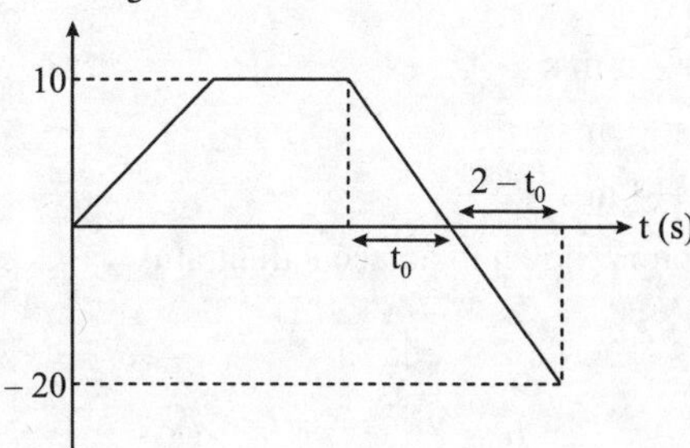

$$\frac{10}{t_0} = \frac{20}{2-t_0}$$

$$t_0 = \frac{2}{3}s$$

NCERT Exemplar MCQs

1. (3) In graph (c) for one value of position there are two different points of time. Hence, net displacement will be zero. So, $v_{avg} = 0$

2. (3) When the lift reaches 4th floor is about to stop hence, motion is retarding in nature. Hence, a > 0.

As displacement is in negative direction, velocity will also be negative i.e., v < 0.

Ground floor is origin and lift is in up words direction so its position is +v. x > 0

3. (2) For maximum and minimum displacement we have to keep the velocity of body maximum.

As maximum velocity in positive direction is v_0.

Maximum displacement in +ve direction = v_0T

Maximum velocity in opposite direction is also v_0.

Displacement can be only between $-v_0T$ and v_0T

Hence, $-v_0T < x < v_0T$.

4. (3) Time taken to travel first half distance $t_1 = \frac{l/2}{v_1} = \frac{l}{2v_1}$

Time taken to travel second half distance $t_2 = \frac{l}{2v_2}$

Total time = $t_1 + t_2$

$$= \frac{l}{2v_1} + \frac{1}{2v_2} = \frac{l}{2}\left[\frac{1}{v_1} + \frac{1}{v_2}\right]$$

We know that,

$$v_{av} = \text{Average speed} = \frac{\text{Total distance}}{\text{Total time}}$$

$$= \frac{l}{\frac{l}{2}\left[\frac{1}{v_1} + \frac{1}{v_2}\right]} = \frac{2v_1v_2}{v_1 + v_2}$$

5. (2) Given, $x = (t-2)^2$

Velocity, $v = \frac{dx}{dt} = \frac{d}{dt}(t-2)^2 = 2(t-2)\,\text{m/s}$

Acceleration, $a = \frac{dv}{dt} = \frac{d}{dt}\left[2(t-2)\right]$

$= 2[1 - 0] = 2\ \text{m/s}^2$

When,

t = 0, v = – 4 m/s

t = 2s; v = 0 m/s

t = 4s; v = 4 m/s

v – t graph is shown in adjacent diagram.

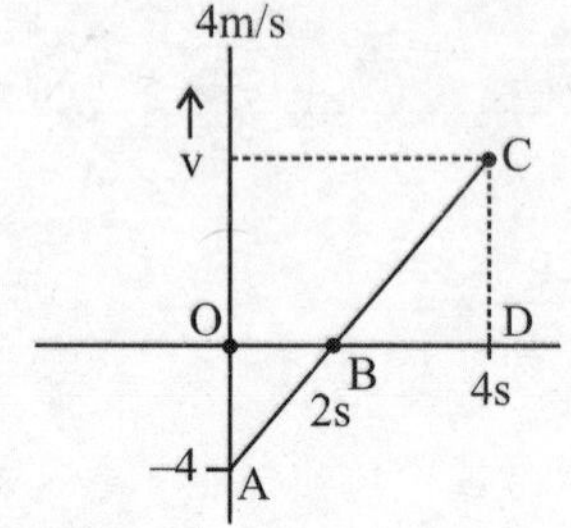

Distance travelled = area of the graph = area OAB + area CBD

$= \frac{4 \times 2}{2} + \frac{1}{2} \times 2 \times 4 = 8\ \text{m}$

6. (3) Let displacement is L, then

Velocity of girl, $v_g = \frac{L}{t_1}$

Velocity of escalator, $v_e = \frac{L}{t_2}$

Net velocity of the girl $= v_g + v_e = \frac{L}{t_1} + \frac{L}{t_2}$

If t is total time taken in covering distance L, then

$\frac{L}{t} = \frac{L}{t_1} + \frac{L}{t_2} \quad \Rightarrow \quad t = \frac{t_1 t_2}{t_1 + t_2}$

7. (1) When we are representing motion by a graph it may be displacement-time, velocity-time or acceleration-time. Hence, B may represent time. In the absence of any information of B, nothing can be said about A.

8. (1, 3, 4) As per the diagram, at point A the graph is parallel to time axis hence, $v = \frac{dx}{dt} = 0$. As the starting point is A hence, we can say that the particle is starting from rest.

At C, the graph changes slope, hence, velocity also changes. As graph at C is almost parallel to time axis hence, we can say that velocity vanishes.

At B the direction of velocity changes from +ve to –ve, therefore acceleration is negative.

From the graph it is clear that

|slope at D| > |slope at E|

Hence, speed at D will be more than at E.

9. (1,4) Given, $x = t - \sin t$

For t = 0, sin t = 0

For any t > 0, sin t < t

Hence, x(t) > 0 for all t > 0

Velocity $v = \frac{dx}{dt} = \frac{d}{dt}[t - \sin t] = 1 - \cos t.$

When, cos t = 1, velocity v = 0

$v_{max} = 1 - (\cos t)_{min} = 1 - (-1) = 2$

$v_{min} = 1 - (\cos t)_{max} = 1 - 1 = 0$

Hence, v lies between 0 and 2.

Acceleration $a = \frac{dx}{dt} = -\sin t$

When t = 0; x = 0, x = + 1, a = 0

When $t = \frac{\pi}{2}$; x = 1, v = 0, a = –1

When t = π; x = 0, x = – 1, a = 1

When t = 2 π; x = 0: x = 0, a = 0

So, a < 0 for some t > 0

10. (1, 3) When spring is stretched by x, restoring force will be F = – kx

Potential energy of the stretched spring $= PE = \frac{1}{2}kx^2$

When particle is released, it will execute S.H.M. about equilibrium position.

Acceleration will be $a = \frac{F}{m} = \frac{-kx}{m}$

At equilibrium position,

$x = 0 \quad \Rightarrow \quad a = 0$

Hence, when just released $x = x_{max}$

Hence, acceleration is maximum. Thus, option (1) is correct.

At equilibrium, whole PE will be converted to K.E. Hence, K.E will be maximum and hence, speed will be maximum.

11. (2, 3, 4) In this problem, we have to keep in mind the frame of the observer. Here we must be clear that we are considering the motion from the ground. Compared to velocity of train (10 m/s) speed of ball is less (1 m/s).

As speed is changing after traveling 10 m and speed is 1 m/s hence, time duration of the changing speed is 10s.

The speed of the ball before collision with front side of train is 10 + 1 = 11 m/s.

Speed after collision with side of train = 10 –1 = 9 m/s.

Speed of ball changes from 11 m/s in forward direction after every 10s and again back to 11 m/s in forward direction after another 10s.

In 20s, the ball repeats its behaviour again, so the average speed will be same in every 20s.

NEET & AIIMS Past Year MCQs

1. (4) Given position:

$x = \alpha t^4 + \beta t^2 + \gamma t + \delta$

Velocity:

$v = \frac{dx}{dt} = 4\alpha t^3 + 2\beta t + \gamma$

At $t = 0$: $v = \gamma$

Acceleration:

$$a=\frac{dv}{dt}=12\alpha t^2+2\beta$$

At $t=0$: $a=2\beta$

Ratio: $\frac{v}{a}=\frac{\gamma}{2\beta}$

2. (1)

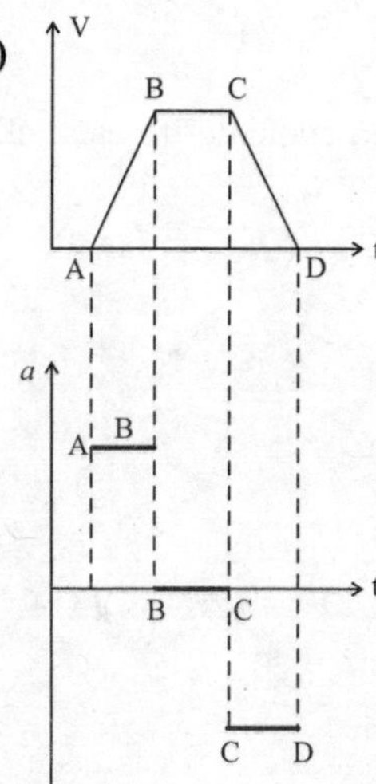

a = slope of $v-t$ graph

from A to B → slope is positive and constant → so acceleration is positive & constant

B to C →slope is zero ⇒ acceleration is zero.

C to D → slope is negative and constant → so acceleration is negative & constant

3. (4) Average speed $=\left(\frac{4v^2}{3v}\right)=\frac{4v}{3}$

4. (4) Let height of bridge = h

Displacement of ball, S = –h

$S=ut+\frac{1}{2}at^2 \quad -h=4\times4+\frac{1}{2}(-10)(4)^2 \Rightarrow h=64$ m

5. (4) Distance travelled during nth second is given by

$$S_{n^{th}}=u+\frac{a}{2}(2n-1)$$

$$=0+\frac{a}{2}(2n-1)$$

[As the body is starting from rest, ∴ u = 0]

$S_{n^{th}}=(2n-1)\Rightarrow S_{1st}:S_{2nd}:S_{3rd}:S_{4th}$

$=[2(1)-1]:[2(2)-1]:[2(3)-1]:[2(4)-1]$

$=1:3:5:7$

6. (1) ∵ Velocity is equal to slope of x-t graph

$\Rightarrow v\propto\tan\theta$

$$\frac{v_1}{v_2}=\frac{\tan30^\circ}{\tan45^\circ}=\frac{1}{\sqrt3}$$

7. (3) Man —— Bus

$a=3m/s^2$

Assuming man to be at rest 4 = 0

$v^2-u^2=2as$

$v^2=2\times3\times6$

$v^2=36$

$v=6m/s$

Motion in a Plane

NCERT Topic-wise MCQs

1. (1) As we know that-

$$R=\sqrt{A^2+B^2+2AB\cos\theta}$$

A = R = B = F (according to question)

$$F=\sqrt{F^2+F^2+2F^2\cos\theta}$$

$$F^2=2F^2(1+\cos\theta)$$

$$\frac{1}{2}-1=\cos\theta$$

$$\theta=120^\circ$$

2. (2) Position vector of a particle $\vec{r}=\sqrt3\hat{i}+3\hat{j}$

Angle with x-axis =

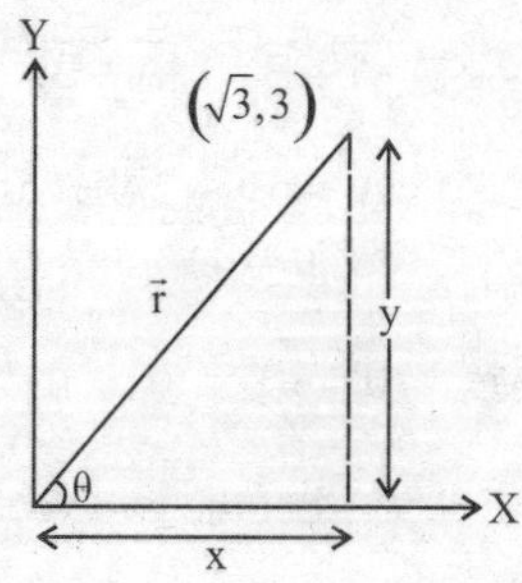

$$\theta=\tan^{-1}\left(\frac{y}{x}\right)=\tan^{-1}\left(\frac{3}{\sqrt3}\right)$$

$$\theta=\tan^{-1}\left(\sqrt3\right)$$

$$\theta=60^\circ$$

3. (3) Resultant of vectors $\vec{A}$ and $\vec{B}$

So, $\vec{R}=\vec{A}+\vec{B}$

$$\vec{R}=4\hat{i}+3\hat{j}+6\hat{k}-\hat{i}+8\hat{j}-8\hat{k}$$

$$\vec{R}=3\hat{i}+11\hat{j}-2\hat{k}$$

$$\vec{R}=\frac{3\hat{i}+11\hat{j}-2\hat{k}}{\sqrt{(3)^2+(11)^2+(-2)^2}}$$

$$\vec{R}=\frac{3\hat{i}+11\hat{j}-2\hat{k}}{\sqrt{9+121+4}}=\frac{3\hat{i}+11\hat{j}-2\hat{k}}{\sqrt{134}}$$

4. (3) Let P = 3x newton, Q = 5x newton, be the two forces

R = 35 N, $\quad\theta=60^\circ$

$$R=\sqrt{P^2+Q^2+2PQ\cos\theta}$$

or $35=\sqrt{(3x)^2+(5x)^2+2\times3x\times5x\cos60^\circ}$

or $35=7x$ or $x=\frac{35}{7}=5$

∴ $P=3\times5=15$ N and $Q=5\times5=25$ N.

5. (1) Let the two forces each of magnitude A then the resultant is

$$R=\sqrt{A^2+A^2+2A\times A\times\frac{1}{2}}=\sqrt{2A^2+A^2}=\sqrt{3A^2}$$

$$R=\sqrt3A$$

6. (4) Vectors are equal

$\hat{i}-3\hat{j}+5\hat{k}=\hat{i}-3\hat{j}-a\hat{k}$

so,

$5\hat{k}=-a\hat{k}$

$a=-5$

7. (1) Given condition is that

$\vec{C}=\vec{A}+\vec{B}$

5 is the resultant or modulus of a vector $|\vec{A}+\vec{B}|$

$|\vec{C}|^2=|\vec{A}+\vec{B}|^2$

$25 = 16 + 9 + 2 \times 4 \times 3 \cos\theta$

$\Rightarrow \cos\theta = 0$

$\Rightarrow \theta = 90°$

8. (4) $|\vec{A}+\vec{B}|=|\vec{A}-\vec{B}|$

Squaring both side

$|\vec{A}+\vec{B}|^2=|\vec{A}-\vec{B}|^2$

$\vec{A}^2+\vec{B}^2+2\vec{A}\vec{B}\cos\theta=\vec{A}^2+\vec{B}^2-2\vec{A}\vec{B}\cos\theta$

$\cos\theta = 0$

$\theta=\frac{\pi}{2}$

9. (3) Let the two vectors $\vec{A}=\vec{F_1}+\vec{F_2}$ and $\vec{B}=\vec{F_1}-\vec{F_2}$

As two vectors are perpendicular to each other so-

$\vec{A}.\vec{B}=0$

$(\vec{F_1}+\vec{F_2}).(\vec{F_1}-\vec{F_2})=0$

$(\vec{F_1})^2-(\vec{F_2})^2=0 \Rightarrow |\vec{F_1}|^2-|\vec{F_2}|^2=0$

$|\vec{F_1}|^2=|\vec{F_2}|^2$

$|\vec{F_1}|=|\vec{F_2}|$

10. (2)

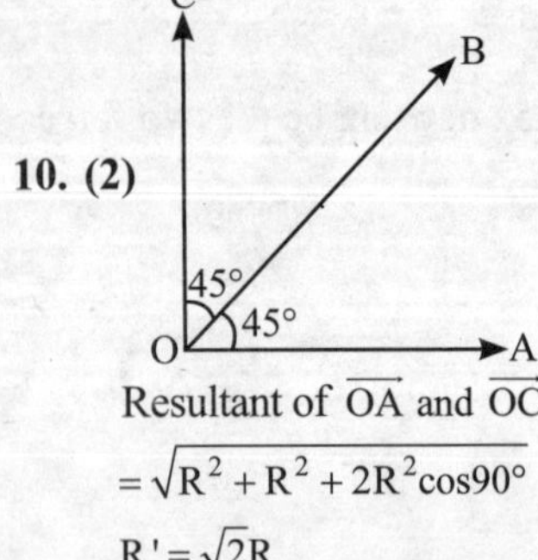

Resultant of $\overline{OA}$ and $\overline{OC}$

$=\sqrt{R^2+R^2+2R^2\cos 90°}$

$R'=\sqrt{2}R$

$R_{Net}=R'+R=R(1+\sqrt{2})$

11. (4) There is no y-component of the vector. So vector projection on y-axis is zero.

12. (4) Magnitude of vector depends upon the magnitudes of component vectors and their orientations with respect to each other, so that's why we can say the components of vectors may be equal to or less than or greater than the magnitude of vector.

13. (4) $|\vec{A}\times\vec{B}|=\frac{(\vec{A}.\vec{B})}{\sqrt{3}}$

$AB\sin\theta=\frac{AB\cos\theta}{\sqrt{3}}$

$\frac{\sin\theta}{\cos\theta}=\frac{1}{\sqrt{3}} \Rightarrow \theta=\tan^{-1}\left(\frac{1}{\sqrt{3}}\right) \quad \theta=\frac{\pi}{6}$

14. (4) When the vectors are perpendicular to each other then their dot product is equal to zero

$(2\hat{i}+4\hat{j}+6\hat{k}).(5\hat{i}+6\hat{j}-a\hat{k})=0$

$10 + 24 - 6a = 0$

$34 = 6a$

$a=\frac{17}{3}$

15. (4) For the scalar product, $F_1 . F_2 = (2\hat{i}+5\hat{k})(3\hat{j}+4\hat{k})$

$F_1 . F_2 = 2 \times 0 + 0 \times 3 + 5 \times 4 = 20$

16. (3) According to the question

$|\vec{A}+\vec{B}|=n|\vec{A}-\vec{B}|$

Squaring both side

$|\vec{A}+\vec{B}|^2=|\vec{A}-\vec{B}|^2$

$A^2 + B^2 + 2AB\cos\theta = n^2 (A^2 + B^2 - 2AB\cos\theta)$

$A^2 + B^2 + 2AB\cos\theta = n^2A^2 + n^2B^2 - n^2 \times 2AB\cos\theta$

$\because |\vec{A}|=|\vec{B}|$

$2A^2 + 2A^2\cos\theta = 2n^2A^2 - 2n^2A^2\cos\theta$

$2A^2(1 + \cos\theta) = 2n^2A^2(1 - \cos\theta)$

$1 + \cos\theta = n^2 - n^2\cos\theta$

$1 + \cos\theta + n^2\cos\theta = n^2$

$\cos\theta=\frac{n^2-1}{n^2+1}$

$\therefore \theta=\cos^{-1}\left(\frac{n^2-1}{n^2+1}\right)$

17. (4) Let $\vec{A}$ be $\hat{i}+\hat{j}+\hat{k}$ and $\vec{B}$ be $\hat{i}$

According to the relation $\frac{\vec{A}.\vec{B}}{|\vec{A}||\vec{B}|}=\cos\theta$

$\therefore \cos\theta=\frac{(\hat{i}+\hat{j}+\hat{k}).(\hat{i})}{\sqrt{1^2+1^2+1^2}\times\sqrt{1^2}}=\frac{1}{\sqrt{3}}$

$\cos\theta=\frac{1}{\sqrt{3}}$

18. (4) $\cos\theta=\frac{(\vec{B}\times\vec{A}).(\vec{A})}{|\vec{B}\times\vec{A}||\vec{A}|}$

$(\vec{B}\times\vec{A}).(\vec{A})=|\vec{B}\times\vec{A}||\vec{A}|\cos 90°$

$(\vec{B}\times\vec{A}).(\vec{A})=0$

19. (4) $\vec{A} = 3\hat{i} + 5\hat{j} - 7\hat{k}$

The direction cosines are

$$\cos\alpha = \frac{A_x}{|A|}, \quad \cos\beta = \frac{A_y}{|A|}$$

$$\cos\gamma = \frac{A_z}{|A|}$$

$$|\vec{A}| = \sqrt{(A_x)^2 + (A_y)^2 + (A_z)^2}$$

$$= \sqrt{9+25+49} = \sqrt{83}$$

$$\cos\alpha = \frac{3}{\sqrt{83}}, \cos\beta = \frac{5}{\sqrt{83}}, \cos\gamma = \frac{-7}{\sqrt{83}}$$

20. (4) $\vec{A} = \hat{i} + \hat{j}$

$\vec{B} = \hat{j} + \hat{k}$

$\vec{A}.\vec{B} = AB\cos\theta$

$\cos\theta = \frac{1}{2}$

$\theta = 60°$

21. (3) For perpendicular condition, $\vec{A}.\vec{B} = 0$

$\vec{A}.\vec{B} = (\hat{i}A\cos\theta + \hat{j}A\sin\theta)$

$(\hat{i}B\sin\theta - \hat{j}B\cos\theta)$

$= A\cos\theta\, B\sin\theta - A\sin\theta\, B\cos\theta$

$= AB\sin\theta\cos\theta - AB\sin\theta\cos\theta = 0$

$\vec{A}.\vec{B} = 0$ (condition satisfied)

22. (2) $|\vec{a}.\vec{b}| = |\vec{a} \times \vec{b}|$

$ab\cos\theta = ab\sin\theta$

$\tan\theta = 1$

$\theta = 45°$

23. (3) Area of parallelogram $= |\vec{A} \times \vec{B}|$

$\vec{A} = 2\hat{i} - 3\hat{j} + \hat{k}$

$\vec{B} = -2\hat{i} + 4\hat{j} - \hat{k}$

$|\vec{A} \times \vec{B}| = \sqrt{5}$ units

24. (2) Let $\vec{A} = 2\hat{i} + 3\hat{j} + \hat{k}$ and $\vec{B} = \hat{i} - \hat{j} + 2\hat{k}$ unit vector perpendicular to both $\vec{A}$ and $\vec{B}$ is

$$\hat{n} = \frac{\vec{A} \times \vec{B}}{|\vec{A} \times \vec{B}|}$$

$$\vec{A} \times \vec{B} = \begin{vmatrix} \hat{i} & \hat{j} & \hat{k} \\ 2 & 3 & 1 \\ 1 & -1 & 2 \end{vmatrix}$$

$$= \hat{i}(6+1) - \hat{j}(4-1) + \hat{k}(-2-3) = 7\hat{i} - 3\hat{j} - 5\hat{k}$$

$$\therefore |\vec{A} \times \vec{B}| = \sqrt{7^2 + (-3)^2 + (-5)^2}$$

$$= \sqrt{83} \text{ unit}$$

$$\therefore \hat{n} = \frac{1}{\sqrt{83}}(7\hat{i} - 3\hat{j} - 5\hat{k})$$

25. (3) $\vec{A} = 4\hat{i} + 3\hat{j} + 6\hat{k}, \vec{B} = -\hat{i} + 8\hat{j} - 8\hat{k}$

So, $\vec{R} = \vec{A} + \vec{B}$

$\vec{R} = 3\hat{i} + 11\hat{j} - 2\hat{k}$

$$\vec{R} = \frac{\vec{R}}{|\vec{R}|} = \frac{3\hat{i} + 11\hat{j} - 2\hat{k}}{\sqrt{134}}$$

26. (3) $\vec{A} = 3\hat{i} + 4\hat{j}, \vec{B} = 6\hat{i} + 8\hat{j}$

$$\vec{A} \times \vec{B} \begin{vmatrix} \hat{i} & \hat{j} & \hat{k} \\ 3 & 4 & 0 \\ 6 & 8 & 0 \end{vmatrix} = 0\hat{i} + 0\hat{j} + 0\hat{k} = \vec{0}$$

$$A = \sqrt{3^2 + 4^2} = 5 \;\&\; B = \sqrt{6^2 + 8^2} = 10$$

$$\vec{A}.\vec{B} = (3\hat{i} + 4\hat{j}).(6\hat{i} + 8\hat{j}) = 18 + 32 = 50$$

27. (2) $\vec{P} \times \vec{Q} = (PQ\sin\theta)\hat{n}$

$\hat{n}$ is a unit vector $\perp$ to plane of $\vec{P}$ and $\vec{Q}$

$$\hat{n} = \frac{\vec{P} \times \vec{Q}}{PQ\sin\theta} = \left(\frac{\vec{P}}{P}\right) \times \left(\frac{\vec{Q}}{Q}\right)\left(\frac{1}{\sin\theta}\right) = \frac{\hat{P} \times \hat{Q}}{\sin\theta}$$

28. (1) $(\vec{A} \times \vec{B})$ and $(\vec{B} \times \vec{A})$ are antiparallel vectors so angle between them 180° or π rad.

29. (3) $\vec{A} \cdot \vec{B} = 0 \Rightarrow \vec{A} \perp \vec{B}$

$\Rightarrow \vec{A} \cdot \vec{C} = 0 \Rightarrow \vec{A} \perp \vec{C}$

$\Rightarrow \vec{A} \parallel \vec{B} \times \vec{C}$

30. (1) Horizontal velocity of projectile is constant.

31. (3) Due to constant velocity along horizontal and vertical downward force of gravity stone will hit the ground following parabolic path.

32. (3) $R = 4H\cot\theta$

Where $R = H$ then $\cot\theta = 1/4 \Rightarrow \theta = \tan^{-1}(4)$

33. (1) Direction of velocity is always tangent to the path so at the top of trajectory it is in horizontal direction and acceleration due to gravity is always in vertically downward direction. It means angle between $\vec{v}$ and $\vec{g}$ are perpendicular to each other.

34. (4) $R = 4H\cot\theta$ if $\theta = 45°$ then $R = 4H\cot(45°) = 4H$

35. (3)

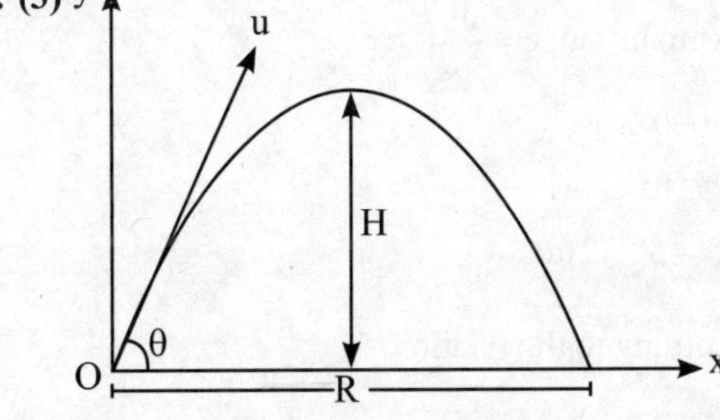

$$\text{Maximum height, } H = \frac{u^2\sin^2\theta}{2g} \quad ...(i)$$

$$\text{Horizontal range, } R = \frac{u^2\sin 2\theta}{g} = \frac{2u^2\sin\theta\cos\theta}{g} \quad ...(ii)$$

Divide (*i*) by (*ii*), we get

$$\frac{H}{R}=\frac{\tan\theta}{4} \text{ or } \tan\theta=\frac{4H}{R} \text{ or } \theta=\tan^{-1}\left(\frac{4H}{R}\right)$$

36. (3) According to the relation

$$R=u\sqrt{\frac{2h}{g}} \Rightarrow u=R\sqrt{\frac{g}{2h}}=90\sqrt{\frac{10}{2\times8}}$$

$$u=90\sqrt{\frac{10}{16}} \Rightarrow u=\frac{90}{4}\sqrt{10} \Rightarrow u=\frac{90}{4}\times3.1$$

$u = 69.75$ m/s

37. (4) As we know the relation,

$$u=80\times\frac{5}{18}=22.22\,\text{m/s}$$

$$R=u\sqrt{\frac{2h}{g}}$$

$h = 150 \times 10^{-2} = 1.5$ m

$$R=22.22\times\sqrt{\frac{2\times1.5}{10}}=22.22\sqrt{\frac{3}{10}}=22.22\times0.54$$

$= 11.99 \approx 12$ m

38. (1) Let x and y axis be adopted along horizontal and vertically downward direction respectively.

After time t, $v_x = u_x = 20\ \text{ms}^{-1}$, velocity in y-direction $v_y = u_y + a_y t = 0 + gt = gt$

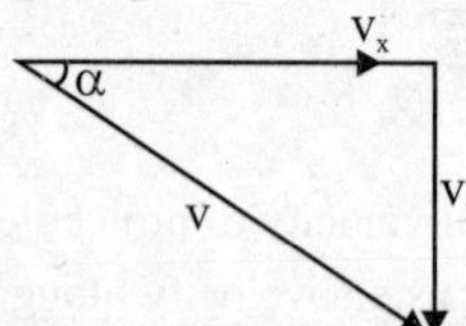

$$\tan\alpha=\frac{v_y}{v_x}=\frac{gt}{20} \quad \text{if } \alpha=45° \text{ than } 1=\frac{10\times t}{20} \Rightarrow t=2\text{s}$$

39. (3) $x=v\sqrt{\frac{2h}{g}}$

$$x=\sqrt{2gh}\times\sqrt{\frac{2h}{g}}=2h$$

40. (1) Time of flight is 2 times that of time taken to reach the maximum height, i.e., $T_f = 2t_m$

41. (1) Maximum range $=\frac{u^2}{g}$

$10 \times 15000 = u^2$

$150000 = u^2$

$100\sqrt{15}=387.2\,\text{m/s}$

42. (2) According to the relation, $H=\frac{u^2}{2g}$

$$R=\frac{u^2}{g}=500 \Rightarrow u^2=500\text{ g}$$

$$H=\frac{500g}{2g}=250\text{ m}$$

43. (4) Maximum horizontal distance (Range) $=\frac{u^2}{g}$

$180 \times 9.8 = u^2$

$\sqrt{1764}=u$

$u = 42$ m/s

44. (3) Time of flight (T) = 3 sec

$$\therefore \frac{2u\sin\theta}{g}=3$$

$$\text{Range}(R)=\frac{u^2\sin2\theta}{g}$$

$$\frac{u^2\sin2\theta}{g}=150=\frac{2(u\sin\theta)(u\cos\theta)}{g}$$

$150 = 3u\cos\theta$

Horizontal component of velocity, $u\cos\theta = 50\ \text{ms}^{-1}$

45. (4) Horizontal component of the velocity does not change in over all motion of projectile.

So, at the highest point it horizontal component of velocity is

$u\cos\theta = u\cos30°$

$$=\frac{u\sqrt{3}}{2}$$

46. (2) According to the relations

$$R=\frac{u^2\sin2\theta}{g},\ H=\frac{u^2\sin^2\theta}{2g}$$

$$H_{max}=\frac{u^2\sin^2 30^0}{2g}=\frac{u^2}{8g}$$

$$R=\frac{u^2\sin60°}{g}=\frac{u^2\sqrt{3}}{2g}$$

$$\therefore \frac{H}{R}=\frac{u^2}{8g}\times\frac{2g}{u^2\sqrt{3}} \Rightarrow R=4\sqrt{3}H$$

47. (3) Given, $g = 10\ \text{m/s}^2$

Equation of trajectory of the projectile,

$y = 2x - 9x^2$

In projectile motion, equation of trajectory is given by

$$y=x\tan\theta_0-\frac{gx^2}{2v_0^2\cos^2\theta_0}$$

Comparing these equations, we get

$\tan\theta_0 = 2$

$$\text{and} \quad \frac{g}{2v_0^2\cos^2\theta_0}=9$$

$$\text{or} \quad v_0^2=\frac{g}{9\times2\cos^2\theta_0}$$

$$\therefore v_0^2=\frac{10\times(\sqrt{5})^2}{2\times(1)^2\times9}=\frac{10\times5}{2\times9}$$

$\Rightarrow \quad v_0^2 = \frac{25}{9} \quad \text{or} \quad v_0 = \frac{5}{3}\ \text{m/s}$

and $\theta_0 = \cos^{-1}\left(\frac{1}{\sqrt{5}}\right)$

48. (1) For projectile A,

Max. Height $H_A = \frac{u_A^2 \sin^2 60°}{2g}$

For projectile B

$H_B = \frac{u_B^2 \sin^2 \theta}{2g}$

According to question

$\frac{u_A^2 \sin^2 60°}{2g} = \frac{u_B^2 \sin^2 \theta}{2g}$

$\frac{u_A^2}{u_B^2} = \frac{\sin^2 \theta}{\sin^2 60°} \Rightarrow \left(\frac{1}{\sqrt{2}}\right)^2 = \frac{\sin^2 \theta}{\left(\frac{\sqrt{3}}{2}\right)^2}$

$\Rightarrow \frac{3}{4} \times \frac{1}{2} = \sin^2 \theta$

$\theta = \sin^{-1}\left(\frac{\sqrt{3}}{2\sqrt{2}}\right)$

49. (4) Kinetic energy at the highest point is $K = K_0 \cos^2 \theta$

$K = E \cos^2 30° = \frac{3E}{4}$

50. (3) $H = \frac{u^2 \sin^2 \theta}{2g} \Rightarrow H \propto \sin^2 \theta$

$\frac{H_1}{H_2} = \frac{\sin^2 60°}{\sin^2 30°} = \frac{3}{4} \times \frac{4}{1} = 3$

51. (2) Height of the pole is equal to the vertical displacement of the particle at time t_1

Vertical displacement

$y = u_y t_1 + \frac{1}{2} a_y t_1^2 = u_y t_1 - \frac{1}{2} g t_1^2 \quad ...(i)$

and total flight time $t_1 + t_2 = \frac{2u_y}{g}$

$\Rightarrow u_y = \frac{g}{2}(t_1 + t_2)$

put value u_y in equation (*i*)

$y = \frac{g}{2}(t_1 + t_2)t_1 - \frac{1}{2} g (t_1)^2 = \frac{1}{2} g t_1 t_2$

so height of the pole $= \frac{1}{2} g t_1 t_2$

52. (1) Velocity of object A relative to that of B is

$\vec{v}_{AB} = \vec{v}_A - \vec{v}_B$

$\therefore \ \vec{v}_{AB} = -\vec{v}_{BA} \text{ and } |\vec{v}_{AB}| = |\vec{v}_{BA}|$

53. (3) $\vec{v}_c = -5\hat{j}$

$v_{bc} = -2\sqrt{6}\hat{i}$

$\vec{v}_b = \vec{v}_{bc} + \vec{v}_c = -2\sqrt{6}\hat{i} - 5\hat{j}$

$|\vec{v}_b| = \sqrt{4 \times 6 + 25} = 7\ \text{m/s}$

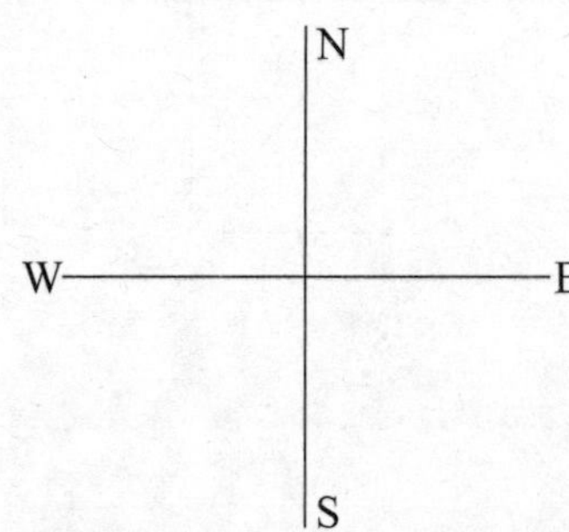

54. (3) $\vec{v}_w = 20\hat{i}, \vec{v}_c = 20\hat{j}$

Here we have to look for velocity of wind w.r.t. car.

So $\vec{v}_{w/c} = \vec{v}_w - \vec{v}_c = 20\hat{i} - 20\hat{j}$

This is in south-east direction.

55. (1)

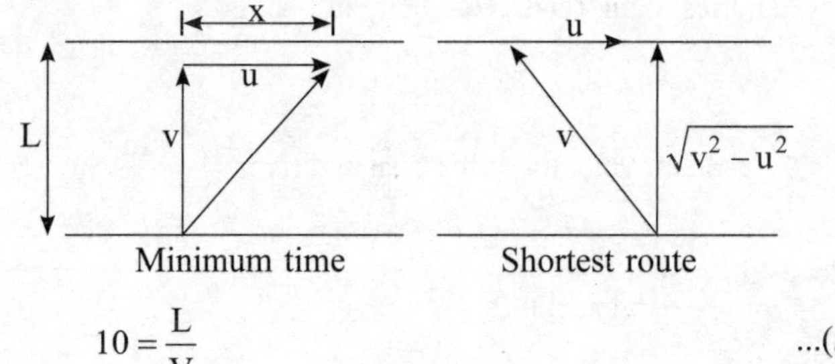

$10 = \frac{L}{V} \quad ...(i)$

$12.5 = \frac{L}{\sqrt{v^2 - u^2}} = \frac{L}{v\sqrt{1 - u^2/v^2}} \quad ...(ii)$

From (*i*) and (*ii*), $\frac{10}{12.5} = \frac{L}{v} \times \frac{v\sqrt{1 - u^2/v^2}}{L}$

$\frac{4}{5} = \sqrt{1 - \frac{12^2}{v^2}} \left[u = \frac{120}{10} = 12\right]$

$\Rightarrow \frac{12^2}{v^2} = 1 - \frac{16}{25} = \frac{9}{25}$

$\Rightarrow \frac{12}{v} = \frac{3}{5} \Rightarrow v = \frac{12 \times 5}{3} = 20\ \text{m/min}$

56. (1) Velocity of object A relative to that of object B is

$\vec{v}_{AB} = \vec{v}_A - \vec{v}_B$

Velocity of object B relative to that of A is

$\therefore \vec{v}_{BA} = \vec{v}_B - \vec{v}_A$

$\therefore \vec{v}_{BA} = -\vec{v}_{AB} \ \text{ and } \ |\vec{v}_{AB}| = |\vec{v}_{BA}|$

$\therefore \vec{v}_{AB} + \vec{v}_{BA} = 0$

57. (1)

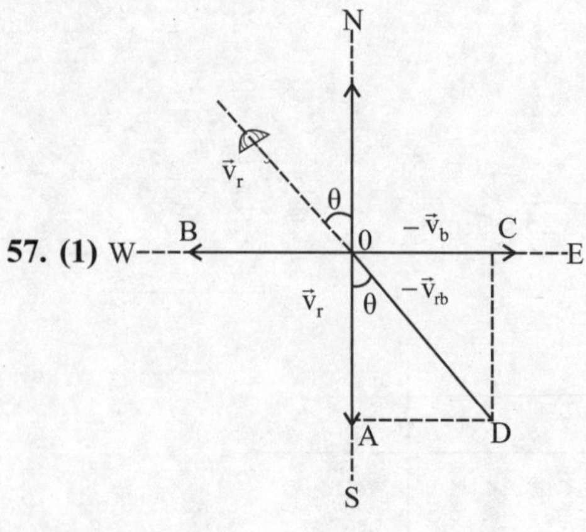

$\vec{v}_r = 35\,m/s$

$\vec{v}_b = 15\,m/s$

$\tan\theta = \frac{AD}{OA} \Rightarrow \tan\theta = \frac{\vec{v}_b}{\vec{v}_r} = \frac{15}{35} = \frac{3}{7}$

$\theta = \tan^{-1}\left(\frac{3}{7}\right)$ (Towards East)

58. (1) When a boat tends to cross a river of width along a shortest path, relative velocity of boat is-

$v_R = \sqrt{v_B^2 - v_r^2}$

Resultant velocity of the boat and river $= \dfrac{2\,km}{\frac{7}{60}\,hr}$

$= \frac{2\times 60}{7} = 17.14\ m/s$

$v_R^{\ 2} = v_B^{\ 2} - v_r^{\ 2} \Rightarrow v_r^{\ 2} = v_B^{\ 2} - v_R^{\ 2}$

$v_r^{\ 2} = \sqrt{v_B^{\ 2} - v_R^{\ 2}} = \sqrt{(18)^2 - (17.14)^2}$

$= \sqrt{30.3} = 5.5\,km/h$

59. (3) Velocity of man in rain = 10 km/h

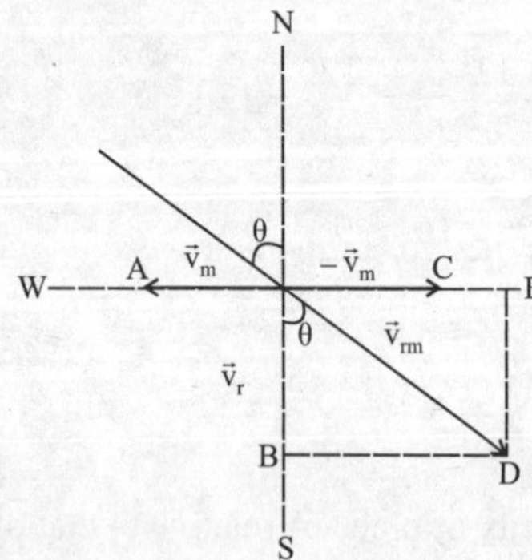

$\because\ \vec{v}_{rm} = \sqrt{v_r^{\ 2} + v_m^{\ 2} + 2v_r v_m \cos 90°}$

$\vec{v}_{rm}$ = relative velocity of rain w.r.t the man

where, $\vec{v}_m$ = velocity of man

$\tan\theta = \frac{BD}{OB} = \frac{v_m}{v_r}$

θ is the angle which v_{rm} makes with the vertical direction

$\tan 45° = \frac{10}{v_r} \Rightarrow v_r = 10\,km/h$

relative velocity of rain w.r.t the man

$\vec{v}_{rm} = \sqrt{(10)^2 + (10)^2} = \sqrt{100+100} = 10\sqrt{2}\,km/h$

60. (2) He should row his boat at an angle θ such that

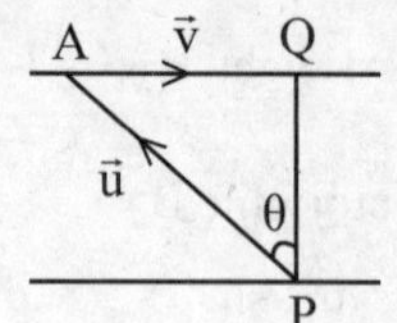

$\sin\theta = \frac{v}{u} = \frac{4}{10} = \frac{2}{5}$

$\therefore\ \theta = \sin^{-1}\left(\frac{2}{5}\right)$

61. (2)

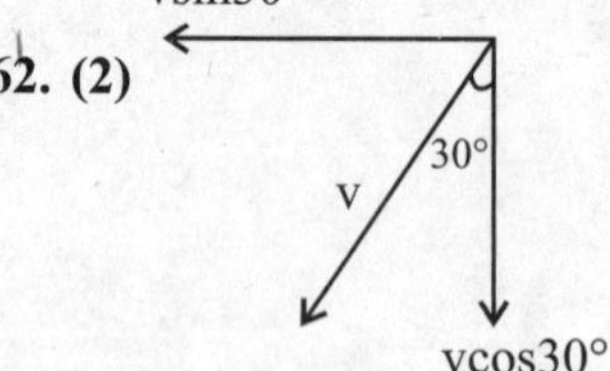

Velocity of the river is along AO

By applying triangle law of vector addition, value of AB comes to be 3 km/hr.

62. (2)

vsin30°
30°
v
vcos30°

Rain drop will hit vertical when

v sin 30° = 10 km/h

$\Rightarrow$ v = 20 km/h

63. (1) The position vector of particle is

$\vec{r}(t) = 15t^2\hat{i} + (4 - 20t^2)\hat{j}$

$\therefore$ Velocity of particle is

$\vec{v} = \frac{d\vec{r}}{dt} = 30t\,\hat{i} - 40t\,\hat{j}$

$\therefore$ Acceleration of particle is

$\vec{a} = \frac{d\vec{v}}{dt} = 30\hat{i} - 40\hat{j}$

So, At t = 1 sec.

$a_{(t=1)} = \sqrt{(30)^2 + (-40)^2} = 50\ ms^{-2}$

64. (4) Man velocity $(\vec{v}_m) = 4\hat{i}$

Rain velocity (apparent) $(\vec{v}_r) = -3\hat{j}$

$\therefore$ Velocity of rain : $|\vec{v}| = \sqrt{(\vec{v}_r)^2 + (\vec{v}_m)^2}$

Velocity of rain $= \sqrt{3^2 + 4^2} = 5\ km\ hr^{-1}$

65. (3) In uniform circular motion, acceleration causes due to change in direction and is directed radially towards centre.

66. (2) $a = \omega^2 r = 4\pi^2 n^2 r = 4\pi^2 \times 1^2 \times 20 \times 10^3$

$\therefore a \approx 8\times 10^5\ m/sec^2$

67. (4) In 15 seconds, the second's hand rotate through 90° angle.

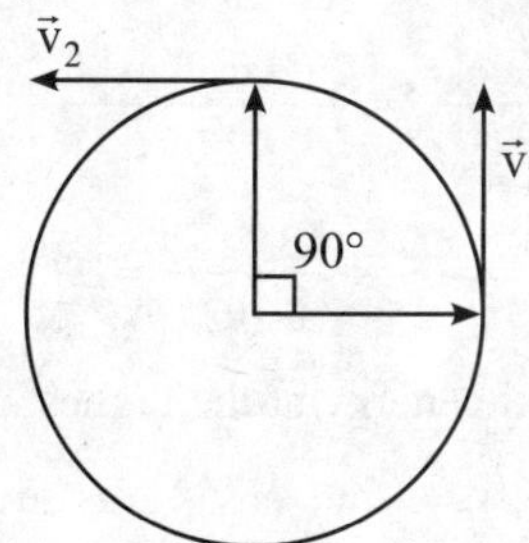

Change in velocity $|\vec{\Delta v}| = 2v\sin(\theta/2)$

$= 2(r\omega)\sin(90°/2) = 2\times1\times\frac{2}{T}\times\frac{1}{\sqrt{2}}$

$= \frac{4\pi}{60\sqrt{2}} = \frac{\pi\sqrt{2}}{30}\frac{cm}{sec}$ As [T = 60 sec]

68. (3) Since $n = 2, \omega = 2\pi\times2 = 4\pi\, rad/s^2$

So acceleration $= \omega^2 r = (4\pi)^2\times\frac{25}{100}\ m/s^2 = 4\pi^2$

69. (1) Tangential acceleration = $a_T = \frac{dv}{dt} = 4t$

At t = 1s $\Rightarrow a_T = 4\ m/s^2$

Centripetal acceleration $a_c = \frac{v^2}{r} = \frac{4t^4}{1} = 4t^4$

at t = 1 s $\Rightarrow a_c = 4\ m/s^2$

Net acceleration (a) = $\sqrt{a_T^2 + a_c^2} = \sqrt{4^2+4^2} = 4\sqrt{2}\ m/s^2$

70. (2) Cyclist starts from A and reaches a point E

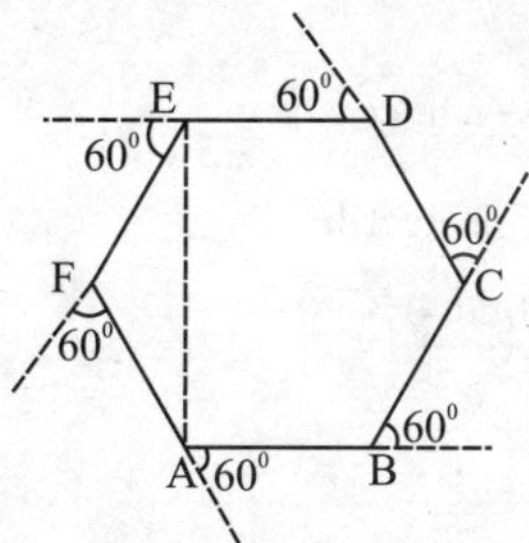

So their displacement is AE

$AE = \sqrt{AF^2 + FE^2}$ (According to Pythagoras theorem)

$AE = \sqrt{(100)^2 + (100)^2}$

$AE = 100\sqrt{2}$

71. (3) As we know that, $v = r\omega$

$\therefore \omega = \frac{v}{r}$

72. (4) $\omega = \frac{2\pi}{T} = \frac{2\times\pi}{60} = \frac{\pi}{30}$ rad/s

73. (3) In a uniform circular motion radial acceleration remains constant and tangential acceleration remains zero.

74. (2) In uniform circular motion speed and angular speed remains constant. Centripetal acceleration acts towards the center (centre seeking) because centripetal force acts towards the centre but velocity does not remains constant because direction is continuously changing.

75. (2) Angular acceleration $\alpha = \frac{d^2\theta}{dt^2}$

$\alpha = 2\theta_2 + 6\,\theta_3 t$

76. (4) Net acceleration = $\sqrt{a_c^2 + a_t^2}$

In circular motion magnitude of velocity change due to tangential acceleration.

Tangential acceleration = g

$\therefore a = \sqrt{a_c^2 + a_t^2}$

$a = \sqrt{\frac{v^4}{r^2} + g^2}$

77. (1) $K.E. = \frac{1}{2}mv^2 = \frac{1}{2}m\times(2\pi\nu R)^2$

$= \frac{1}{2}m\times4\times\pi^2\times\nu^2\times R^2$

$K.E = m\times2\times\pi^2\times\nu^2\times R^2$ $(3.14)^2 = 9.85 \approx 10$

$= 4\times2\times10\times\frac{120}{60}\times\frac{120}{60}\times4\times4$

$K.E = 4\times2\times10\times2\times2\times16 = 5120$ J

78. (3) Net acceleration of the car $a = \sqrt{(a_t)^2 + (a_c)^2}$

$\therefore a = \sqrt{(4)^2 + \frac{v^4}{r^2}} \Rightarrow a = \sqrt{16 + \frac{(40)^4}{(500)^2}}$

$\therefore a = \sqrt{16 + \frac{256\times10^4}{25\times10^4}} \Rightarrow a = \sqrt{16+10.24} = 5.122\ m/s^2$

79. (1) We know that centripetal acceleration is given by

$a = \frac{v^2}{R}$

For constant R .

$a \propto v^2$ or $\frac{a_1}{a_2} = \frac{v_1^2}{v_2^2}$

$\therefore \frac{a_1}{a_2} = \frac{v^2}{(2v)^2} = \frac{1}{4}$

$a_2 = 4\,a_1$ or $a_2 = 4\,a$

High Order Time Intensive MCQs

1. (2) The time taken to cover the complete circle would be equal to $T = \frac{2\pi r}{v}$,

Hence the time taken to cover half the rotation would be equal to $T_{1/2} = \frac{2\pi r}{2v} = \frac{\pi r}{v}$.

Hence the magnitude of average velocity after half rotation would be equal to $v_{av} = \frac{2rv}{\pi r} = \frac{2v}{\pi}$

2. (2) $x = u\cos\theta\, t$ ... *(i)*

$y = u\sin\theta t - \frac{1}{2}gt^2$ ... *(ii)*

Putting value of t from (*i*) and (*ii*)

$$y = u\sin\theta \times \frac{x}{u\cos\theta} - \frac{1}{2}g\frac{x^2}{u^2\cos^2\theta}$$

$$y = x\tan\theta - \frac{1}{2}\frac{gx^2}{u^2\cos^2\theta}$$

comparing it with the given relation

$$y = \sqrt{3}\times - g\frac{x^2}{2},$$

$\tan\theta = \sqrt{3}$

$\therefore \theta = \tan^{-1}\left(\sqrt{3}\right)$

3. (4) $\tan 45^\circ = \frac{v_y}{v_x}$

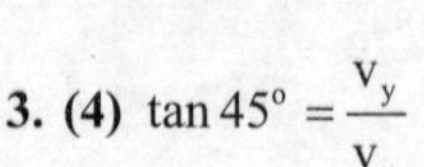

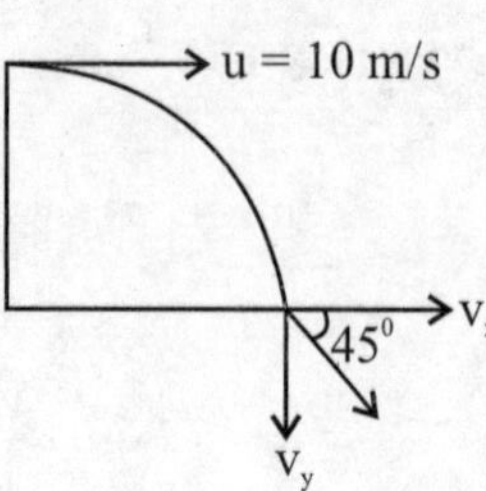

$1 = \frac{v_y}{v_x}$

$v_y = v_x = 10$ m/s

4. (2) The projectile is shown below

$$t = \frac{2u\sin\theta}{g} \Rightarrow 4 = \frac{2u\sin\theta}{g} \quad ...(i)$$

Also, $v^2 = u^2 + 2as \Rightarrow 0^2 = u^2\sin^2\theta - 2gh$

$$\frac{u^2\sin^2\theta}{2g} = h \quad ...(ii)$$

From equation (*ii*)

$$\frac{u^2\sin\theta}{g} = 2$$

$$\Rightarrow \frac{4g^2}{2g} = h$$

$\Rightarrow h = 20$ m

5. (4) From equation of motion where we have

$$h = \frac{1}{2}gt^2 \Rightarrow \quad t = \sqrt{\frac{2\times1960}{9.8}} = 20s$$

$$\therefore \quad s = AB = ut = 600\times\frac{20}{60\times60} = 3.33\text{km}$$

6. (3) Equation of projectile,

$$y = 10x - \left(\frac{5}{9}\right)x^2$$

Equation of trajectory is give by

$$y = x\tan\theta - \frac{g}{2u^2\cos^2\theta}.x^2$$

On comparing,

$$\tan\theta = 10 \text{ and } \frac{g}{2u^2\cos^2\theta} = \frac{5}{9}$$

or $10u^2\cos^2\theta = 9g$

$$\therefore \quad u^2\cos^2\theta = 9$$

Range of projectile,

$$R = \frac{2u^2\sin\theta\cos\theta}{g} = \frac{2u^2\tan\theta\cos^2\theta}{g} \quad (\because \sin\theta = \tan\theta\cos\theta)$$

$$= \frac{2(u^2\cos^2\theta).\tan\theta}{g} = \frac{2\times9\times10}{10} = 18\text{m}$$

7. (1) The kinetic energy at the highest point would be equal to $\frac{1}{2}m(u\cos\beta)^2 = \frac{1}{2}mu^2\cos^2\beta$ (as the vertical component of the velocity is zero).

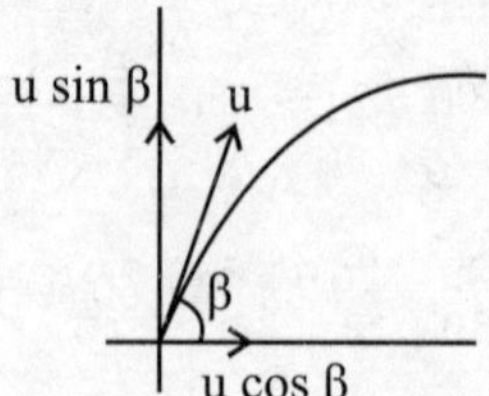

The initial kinetic energy is the maximum kinetic energy.

So, $(KE)_H = (KE)_i \cos^2\beta$ where, $(KE)_i = \frac{1}{2}mv^2$

Thus, $KE_i\cos^2\beta = \frac{3}{4}(KE)_i$

$$\Rightarrow \cos\beta = \frac{\sqrt{3}}{2} \Rightarrow \beta = 30^\circ$$

8. (1) Given, R = 2H

We know that, R = 4H cot θ

2H = 4H cot θ

$$\cot\theta = \frac{1}{2}$$

$$\Rightarrow \sin\theta = \frac{2}{\sqrt{5}} \text{ and } \cos\theta = \frac{1}{\sqrt{5}}$$

∴ Range of projectile,

$$R = \frac{2v^2\sin\theta\cos\theta}{g}$$

$$= \frac{2v^2}{g}\times\frac{2}{\sqrt{5}}\times\frac{1}{\sqrt{5}} = \frac{4v^2}{5g}$$

9. (4) $R = \frac{2u^2\sin\theta\cos\theta}{g}$

$$t_1 = \frac{2u\sin\theta}{g} \quad ...(i)$$

$$\text{And } t_2 = \frac{2u\sin(90^\circ-\theta)}{g} = \frac{2u\cos\theta}{g} \quad ...(ii)$$

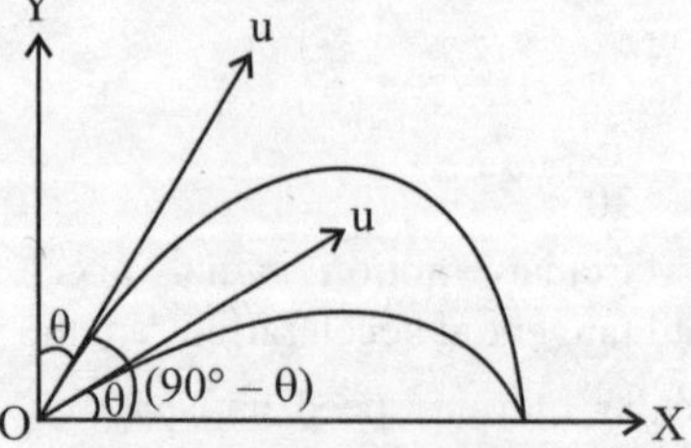

From Eqs. (*i*) and (*ii*), we get

$$t_1t_2 = \frac{4u^2\sin\theta\cos\theta}{g^2} = \frac{2R}{g} \Rightarrow t_1t_2 \propto R$$

10. (3) Let the angle made by the instantaneous velocity with the horizontal be α. Then,

$$\tan\alpha = \frac{v_y}{v_x} = \frac{u\sin\theta - gt}{u\cos\theta}$$

Given, $\alpha = 45°$, when t = 1s

$$\tan 45° = \frac{u\sin\theta - gt}{u\cos\theta}$$

$\Rightarrow u\cos\theta = u\sin\theta - g$ ...(*i*)

$\alpha = 0°$, when t = 2s

$\Rightarrow u\sin\theta - 2g = 0$ ... (*ii*)

$$\tan 0° = \frac{u\sin\theta - gt}{u\cos\theta}$$

On solving Eqs. (*i*) and (*ii*), we have

$u\sin\theta = 2g \Rightarrow u\cos\theta = g$

squaring and adding

$\Rightarrow \quad u = \sqrt{5}g = 10\sqrt{5}\,ms^{-1}$

11. (1) The radius of curvature is given by

$$r = \frac{mv^3}{|F\times v|} = \frac{mv^3}{Fv\sin\theta} = \frac{mv^3}{mgv\sin\theta}$$

$(\because F = mg)$

$$\Rightarrow \quad r = \frac{v^2}{g\sin\theta}$$

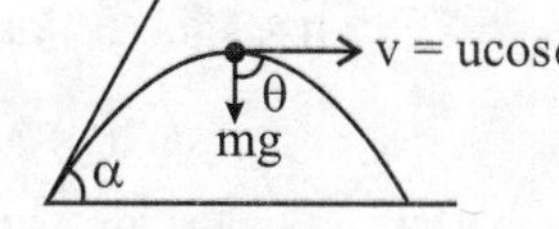

$$\therefore \quad r_{min} = \frac{v^2}{g(\sin\theta)_{max}}$$ [For P_{min}, θ should be maximum]

$\therefore \quad r_{min} = v^2/g \quad [\because (\sin\theta)_{max} = 1]$

i.e., θ = 90° is possible only when the projectile is at the maximum height.

At the maximum height the velocity,

$v = u\cos\alpha = 10\cos 60° = 5\ ms^{-1}$

$$\therefore \ r_{min} = \frac{5\times5}{9.8} = 2.55m$$

12. (1) Range of projectile is given by,

$$R = \frac{2u^2\sin\theta\cos\theta}{g} \quad ...(i)$$

$$\text{Height } H = \frac{u^2\sin^2\theta}{2g} \quad ...(ii)$$

$$\text{And, } H_1 = \frac{u^2\sin^2(90° - \theta)}{2g} = \frac{u^2\cos^2\theta}{2g} \quad ...(iii)$$

$$\text{Then, } HH_1 = \frac{u^2\sin^2\theta u^2\cos^2\theta}{2g\times2g} \quad ...(iv)$$

From Eq. (*i*), we get

$$R^2 = \frac{4u^2\sin^2\theta u^2\cos^2\theta\times4}{2g\times2g}$$

$R = \sqrt{16HH_1}$ [from Eq. (*iv*)]

$R = 4\sqrt{HH_1}$

13. (2) We know that maximum height is given by,

$$H = \frac{u^2\sin^2\theta}{2g} \quad ...(i)$$

and time of flight, $T = \frac{2u\sin\theta}{g}$

$$\Rightarrow u\sin\theta = \frac{Tg}{2} \quad ...(ii)$$

From Eqs. (*i*) and (*ii*), we get

$$H = \frac{1}{2g}\left(\frac{Tg}{2}\right)^2 \Rightarrow H = \frac{T^2g^2}{8g}$$

$$\Rightarrow \quad T^2 = \frac{8H}{g} \Rightarrow T = 2\sqrt{\left(\frac{2H}{g}\right)}$$

14. (3) The component of vector $\vec{A}$ along vector $\vec{B}$ would be equal to

$$\vec{A}.\frac{\vec{B}}{|\vec{B}|} = (2\hat{i} + 3\hat{j}).\frac{(\hat{i}+\hat{j})}{|(\hat{i}+\hat{j})|}$$

$$= (2\hat{i} + 3\hat{j}).\frac{(\hat{i}+\hat{j})}{\sqrt{2}} = \frac{(2+3)}{\sqrt{2}} = \frac{5}{\sqrt{2}}$$

15. (3)

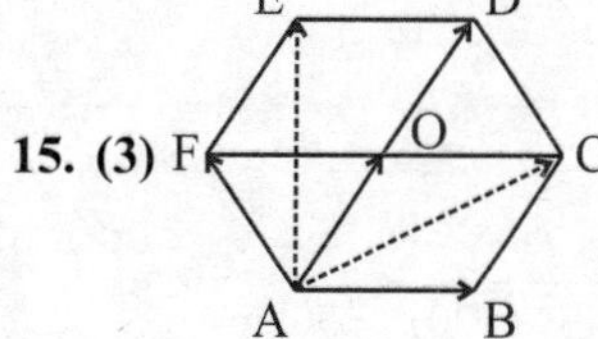

$\overrightarrow{AB} + \overrightarrow{AC} + \overrightarrow{AD} + \overrightarrow{AE} + \overrightarrow{AF}$

here $\overrightarrow{AD} = 2\overrightarrow{AO}$ ... (*i*)

$\overrightarrow{AB} + \overrightarrow{AF} = \overrightarrow{AO}$ ... (*ii*)

$\overrightarrow{AC} = \overrightarrow{AB} + \overrightarrow{AO}$

$\overrightarrow{AE} = \overrightarrow{AO} + \overrightarrow{AF}$

$\therefore \ \overrightarrow{AC} + \overrightarrow{AE} = \overrightarrow{AB} + \overrightarrow{AO} + \overrightarrow{AO} + \overrightarrow{AF} + 3\overrightarrow{AO}$ ... (*iii*)

Adding (*i*), (*ii*) and (*iii*) gives (as $\overrightarrow{AB} + \overrightarrow{AF} = \overrightarrow{AO}$)

$\overrightarrow{AB} + \overrightarrow{AC} + \overrightarrow{AD} + \overrightarrow{AE} + \overrightarrow{AF} = 6\overrightarrow{AO}$

16. (2) Here, r = 12 cm

$$\text{Frequency, } \nu = \frac{7}{100}\text{Hz}$$

The angular speed of the insect is

$$\omega = 2\pi\nu = 2\pi\times\frac{7}{100} = 4.44\ \text{rad s}^{-1}$$

The linear speed of the insect is

$v = \omega r = 0.44 \times 12 = 5.3\ cm\ s^{-1}$

17. (3) Net movement along x-direction

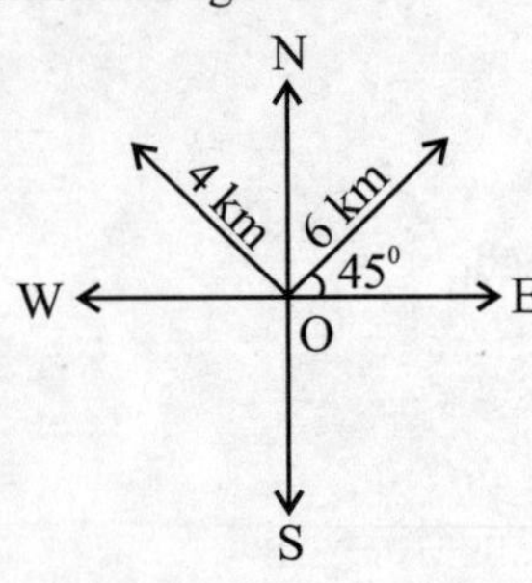

$s_x = 6\cos 45^\circ\hat{i} + 4\cos 45^\circ\left(-\hat{i}\right)$

$s_x = (6-4)\cos 45^\circ\hat{i}$

$= 2\times\frac{1}{\sqrt{2}} = \sqrt{2}\text{km}$

Net movement along y-direction

$s_y = 6\sin 45^\circ\hat{j} + 4\sin 45^\circ\hat{j}$

$s_y = (6+4)\sin 45^\circ\ \hat{j} = 10\times\frac{1}{\sqrt{2}} = 5\sqrt{2}\text{km}$

Net movement from starting point (O in the diagram)

$|s| = \sqrt{s_x^2 + s_y^2} = \sqrt{(\sqrt{2})^2 + (5\sqrt{2})^2} = \sqrt{52}\text{km}$

The tan of angle which resultant makes with the east direction,

$\tan\theta = \frac{\text{y-component}}{\text{x-component}} = \frac{5\sqrt{2}}{\sqrt{2}}$

$\Rightarrow \qquad \theta = \tan^{-1}(5)$

18. (1) Change in velocity $= \sqrt{(40)^2 + (30)^2} = 50\text{kms}^{-1}$

Average acceleration $= \frac{50}{20} = 2.5\text{kms}^{-2}$

19. (4) Given, $\vec{A} = -4\hat{i} + 4\hat{j} + 2\hat{k}$

And $\vec{B} = 2\hat{i} - \hat{j} - \hat{k}$

Let angle made by $(\vec{A} + \vec{B})$ with $\vec{C} = \hat{i} + 2\hat{j} - 4\hat{k}$ is θ, then

$\cos\theta = \frac{(\vec{A}+\vec{B}).\vec{C}}{|\vec{A}+\vec{B}|.|\vec{C}|}$

$= \frac{(-2\hat{i}+3\hat{j}+\hat{k}).(\hat{i}+2\hat{j}-4\hat{k})}{\sqrt{14}\times\sqrt{21}} = 0$

$\Rightarrow \qquad \theta = 90^\circ$

20. (4) For two vectors to be perpendicular, their dot product must be zero.

Given, $\vec{x} = a\cos\theta\hat{i} + b\sin\theta\hat{j}$

Let consider each option one by one.

$\vec{y} = b\sin\theta\hat{i} - a\cos\theta\hat{j}$

$\therefore\ \vec{x}.\vec{y} = (a\cos\theta\hat{i} + b\sin\theta\hat{j}) \Rightarrow (b\sin\theta\hat{i} - a\cos\theta\hat{j})$

$\Rightarrow \vec{x}.\vec{y} =$ ab sin θ cos θ – ab sin θ cos θ = 0

Also, $[a\cos\theta\hat{i} + b\sin\theta\hat{j}].[\frac{1}{a}\sin\theta\hat{i} - \frac{1}{b}\cos\theta\hat{i}] = 0$

And $[a\cos\theta\hat{i} + b\sin\theta\hat{j}].5\hat{k} = 0$

Thus, we see that all the vector given in the options are perpendicular to the vector $(a\cos\theta\hat{i} + b\sin\theta\hat{j})$.

21. (2) As, $|A| = \sqrt{2^2 + 3^2 + 6^2}$

$|A| = \sqrt{49} = 7$

And $|B| = \sqrt{(-1)^2 + 3^2 + 4^2} = \sqrt{26}$

$\therefore\ \vec{A}.\vec{B} = -2 + 9 + 24 = 31$

The projection of A on B (i.e., A cos θ)

$= \frac{31}{\sqrt{26}}$

22. (2) Position vector

$\hat{r} = (a\cos\omega t)\hat{i} + (a\sin\omega t)\hat{j}$

$\therefore$ Velocity, $\vec{v} = \frac{d\vec{r}}{dt} \Rightarrow \frac{d}{dt}[(a\cos\omega t)\hat{i} + (a\sin\omega t)\hat{j}]$

$= [(-a\sin\omega t)\hat{i} + (a\cos\omega t)\hat{j}]\omega$

$\therefore \vec{v}\vec{r} = [(-a\sin\omega t)\hat{i} + (a\cos\omega t)\hat{j}]\omega.[(a\cos\omega t)\hat{i} + (a\sin\omega t)\hat{j}]$

$= -a^2\cos\omega t\sin\omega t + a^2\omega\sin\omega t\cos\omega t = 0$

Here, $\vec{v}\perp\vec{r}$, i.e., velocity vector is perpendicular to position vector.

23. (1) Let $\vec{P} = \sqrt{3}\hat{i} + \hat{i}$

And $\vec{Q} = \hat{i}$ (unit vector along X-axis)

Now, $\vec{P}.\vec{Q} = (\sqrt{3}\hat{i} + \hat{j}).\hat{i}$

$= \sqrt{3} + 0 = \sqrt{3}\ (\because \hat{i}.\hat{i} = \hat{i}\ \hat{j}.\hat{i} = 0)$

Also, $|\vec{P}| = \sqrt{(\sqrt{3})^2 + (1)^2} = \sqrt{3+1} = 2$

$|\vec{Q}| = \sqrt{(1)^2} = 1$

$\because$ PQ cos θ = $\vec{P}.\vec{Q}$ where θ is the angle between vector $\vec{P}$ and X-axis.

$\cos\theta = \frac{\sqrt{3}}{2} \qquad \Rightarrow \qquad \theta = \frac{\pi}{6}$ rad

24. (2) By using addition of vector in vector form

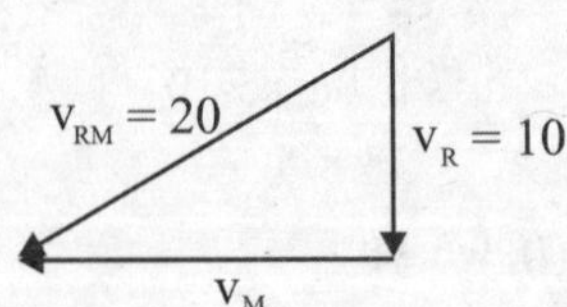

$v_m = \sqrt{(20)^2 - (10)^2}$

$= \sqrt{400 - 100}$

$= \sqrt{300} = 10\sqrt{3}$ m/s

25. (3) Velocity of rain along x-axis is v_0

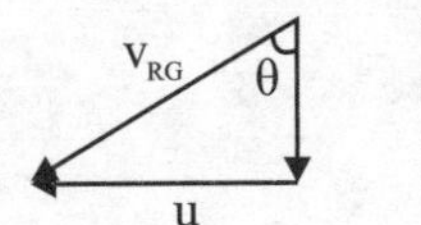

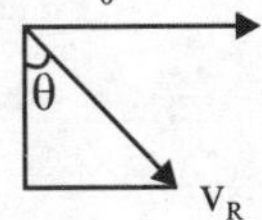

$$v_R = u\hat{i} + u\cot\theta\hat{j}$$

26. (2) $\vec{F} = m\vec{a} = \dfrac{md\vec{v}}{dt}$

$$\therefore\ 6t^2\hat{i} + 4t\hat{j} = \frac{md\vec{v}}{dt}$$

$$\frac{d\vec{v}}{dt} = \frac{6t}{3}2\hat{i} + \frac{4t}{3}\hat{j}$$

$$\text{or } d\vec{v} = \frac{\hat{i}6t^2dt}{3} + \hat{j}\frac{4t\,dt}{3}$$

Integrating on both sides:

$$\int\overrightarrow{dv} = \int(2t^2dt)\hat{i} + \left(\int\frac{4t}{3}dt\right)\hat{j}$$

$$\vec{v} = \frac{2t^3}{3}\hat{i} + \frac{2t^2}{3}\hat{j}$$

at t = 3 sec

$$\vec{v} = \frac{2}{3}(3)^3\hat{i} + \frac{2}{3}(3)^2\hat{j} = 18\hat{i} + 6\hat{j}$$

27. (3)

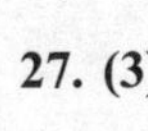

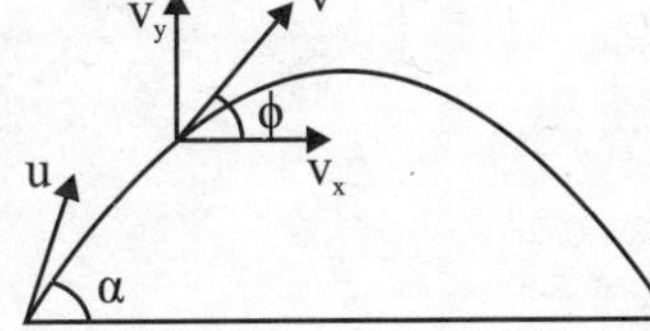

$$v_x = v_y \quad \Rightarrow \quad \phi = 45°$$

$$\tan\phi = \frac{v_y}{v_x} = \frac{u_y + a_y t}{v_x} = \frac{u\sin\theta - gt}{u\cos\theta}$$

$$\tan\phi = \tan\theta - \frac{gt}{u\cos\theta}$$

$$\tan 45^0 = \tan\alpha - \frac{gt}{u\cos\alpha} \qquad \text{[Given } \phi = 45°\text{]}$$

$$1 = \frac{\sin\alpha - \left(\dfrac{gt}{u}\right)}{\cos\alpha}$$

$$\cos\alpha = \sin\alpha - \frac{gt}{u}$$

$$\frac{gt}{u} = (\sin\alpha - \cos\alpha)$$

$$t = \frac{u}{g}(\sin\alpha - \cos\alpha)$$

28. (1) $R = \dfrac{2\times 9.8\times 19.6}{g(=9.8)} = 39.2$ m

29. (4) Rate of collection (R) = vA cos θ (θ is the angle between the velocity of the rain and area of vessel)

R = vA cos 0° = vA

When the wind blows R = v′A cos θ, (v is the new velocity of rain)

$$v' = \sqrt{v^2 + u^2}$$

$$\cos\theta = \frac{v}{\sqrt{v^2 + u^2}}$$

∴ R = v′A cos θ = vA = R

30. (2) In time t, particle has rotated an angle,

θ = ωt

$$s = PQ = \sqrt{QR^2 + PR^2}$$

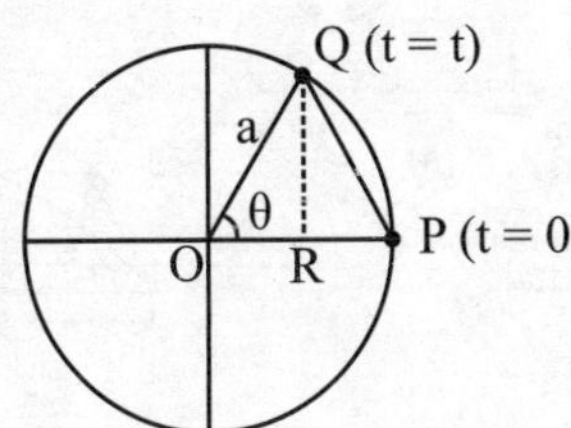

$$\text{or} \quad s = \sqrt{(a\sin\omega t)^2 + (a - a\cos\omega t)^2}$$

$$\text{or} \quad s = 2a\sin\frac{\omega t}{2}$$

31. (2) Bullet from guns can reach upto a distance of maximum range which occurs when projection is made at angle of 45°.

$$R_1 = \frac{u_1^2}{g}$$

$$R_2 = \frac{u_2^2}{g}$$

So, ratio of covered areas

$$\frac{A_1}{A_2} = \frac{\pi(R_1)^2}{\pi(R_2)^2} = \frac{u_1^4}{u_2^4}$$

Here, u_1 = 1 km/s and u_2 = 2 km/s

$$\text{So, ratio of areas} = \frac{1^4}{2^4} = \frac{1}{16} = 1:16$$

32. (2) $v = \dfrac{dx}{dt}\hat{i} + \dfrac{dy}{dt}\hat{j}$

$$\frac{dx}{dt} = t,\ \frac{dy}{dt} = x.\frac{dx}{dt} = \frac{t^3}{2}$$

$$\text{so, } v = 2\hat{i} + 4\hat{j}$$

33. (1) $t_1 = \dfrac{AB}{v_b + v_R}, t_2 = \dfrac{AB}{v_b - v_R}$ or $v_b + v_R = \dfrac{AB}{t_1}$

and $\quad v_b - v_R = \dfrac{AB}{t_2}$

$$\Rightarrow 2v_b = \frac{AB}{t_1} + \frac{AB}{t_2} = AB\left(\frac{t_1 + t_2}{t_1 t_2}\right)$$

or $\left(\dfrac{2t_1t_2}{t_1 + t_2}\right) = \dfrac{AB}{v_b}$ = time taken by the boat to cover AB

34. (1) To cross the river perpendicular to current i.e. along shortest path $\sin\theta = \frac{u}{v} = \frac{3}{5} \Rightarrow \theta = 37^\circ$

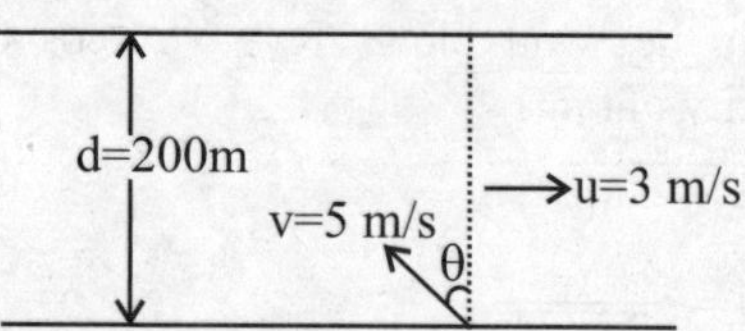

35. (1) $v_x = 8t - 2 \Rightarrow x = 4t^2 - 2t$

$v_y = 2 \Rightarrow y = 2t$

so, $x = y^2 - y = y(y-1)$

36. (4) For a projectile

$\tan\alpha + \tan\beta = \tan\theta$

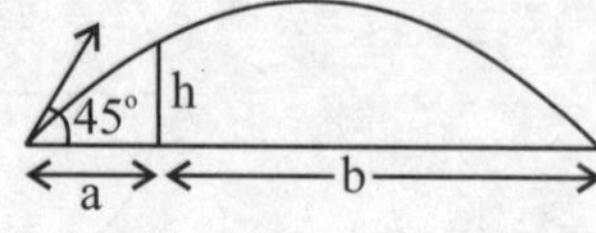

$\frac{h}{a} + \frac{h}{b} = \tan 45^\circ$

$h = \frac{ab}{a+b}$

37. (2) It is obvious from considerations of symmetry that at any moment of time all of the persons will be at the corners of square whose side gradually decreases and so they will finally meet at the centre of the square O.

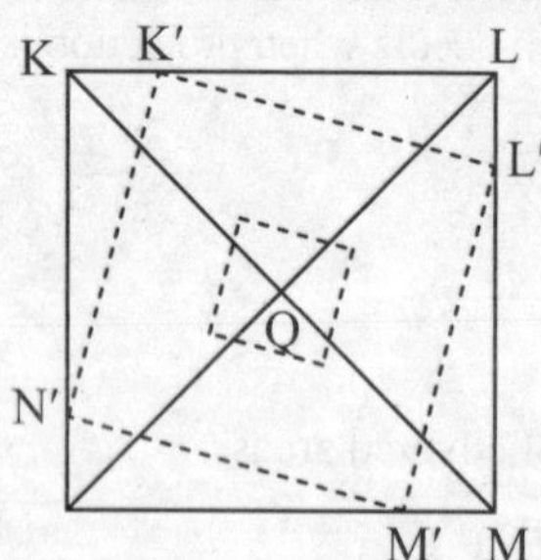

The speed of each person along the line joining his initial position and O will be $v\cos 45^\circ = \frac{v}{\sqrt{2}}$ to reach the centre.

The distance covered by each person is $d\cos 45^\circ = \frac{d}{\sqrt{2}}$

The four persons will meet at the centre of the square O after time

$$t = \frac{\text{Distance}}{\text{Speed}}$$

$$\therefore t = \frac{\frac{d}{\sqrt{2}}}{\frac{v}{\sqrt{2}}} = \frac{d}{v}$$

38. (4) $x = A\sin\omega t$

$y = 2A\cos\omega t$

$\Rightarrow \frac{x^2}{A^2} + \frac{y^2}{4A^2} = 1$

Also when time increases, X increases, Y decreases. Hence motion of particle will be elliptical clockwise.

39. (1) Time for hitting

$= \frac{300\cos 37^\circ}{700\cos 37^\circ} = \frac{3}{7}\text{sec}$

Height fallen by B

$= \frac{1}{2}gt^2 = 0.9\text{m} = 90\text{ cm.}$

40. (2) At any instant 't'

$v_x = 2 - t$

$v_y = 1 + 3t$

when $v_x = 0, t = 2$

$v_y = 7$ m/s

41. (3) For range

$y = 0 \Rightarrow t = 0, \frac{8}{5}$

$R = 6\left(\frac{8}{5}\right) = 9.6\,\text{m}$

42. (1)

$$t = \sqrt{\frac{2(H-h)}{g}} + \sqrt{\frac{2h}{g}} = \sqrt{\frac{2}{g}}\left(\sqrt{h} + \sqrt{H-h}\right)$$

43. (3) $(v_x)_{avg} = v\cos\theta$

$(v_y)_{avg} = \frac{0 + v\sin\theta}{2}$

$$v_{avg} = \sqrt{(v_x)_{av}^2 + (v_y)_{av}^2} = \sqrt{v^2\cos^2\theta + \frac{v^2\sin^2\theta}{4}}$$

$$= \frac{v}{2}\sqrt{1 + 3\cos^2\theta}$$

44. (4) There will be no impact on maximum height & time of flight.

Range will increase by $\frac{1}{2}\left(\frac{g}{4}\right)T^2$

as $H = \frac{1}{2}g\left(\frac{T}{2}\right)^2$

So, new range = R + H

45. (3) $v = \sqrt{v_x^2 + v_y^2} = \sqrt{a^2 + b^2t^2}$

46. (4) Separation between two bodies after T seconds $= \frac{1}{2}gT^2$

Relative velocity = gT

so time when separation becomes 'L' is

$$t = \frac{L - \frac{1}{2}gT^2}{gT}$$

47. **(2)** Components of velocity at t = 1s, are

$v_x = u\cos 60° = 5$ m/s

and $v_y = u\sin 60° + (-10)\times(1)$

$= (5\sqrt{3}-10)$ m/s

$\Rightarrow |v_y| = |10-5\sqrt{3}|$ m/s

Now, angle made by the velocity vector with horizontal

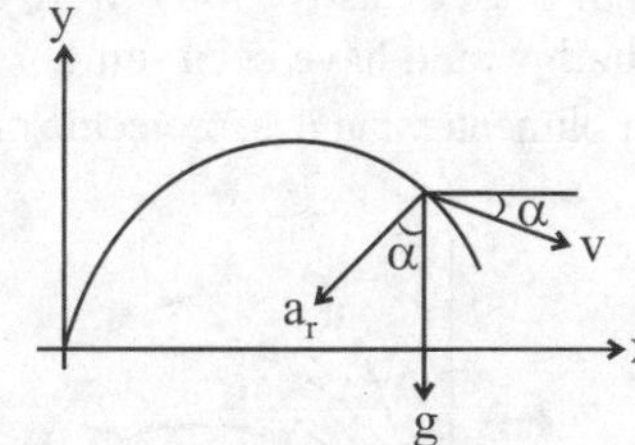

$$|\tan\alpha| = \left|\frac{v_y}{v_x}\right| = \frac{|10-5\sqrt{3}|}{5}$$

$\Rightarrow \tan\alpha = |2-\sqrt{3}|$

or $\alpha = 15°$

$$a_n = \frac{v^2}{R} = g\cos\alpha$$

$$R = \frac{v^2}{g\cos\alpha}$$

Substituting the values, we get

$R = 2.8$ m

48. **(2)** This is possible only when 'A' & 'B' are moving in a circular path & is located at the two ends of diameter 'D'.

So, time to reach their initial position $= \dfrac{\pi D}{v}$

49. **(1)** $s = \dfrac{1}{10}(2\pi r) = \dfrac{1}{2}(0.5)t^2$

$$\Rightarrow t = \sqrt{\frac{4}{10}(2\pi r)}$$

$$a_{CP} = \frac{v^2}{r} = \frac{a^2t^2}{r} = \frac{\left(\frac{1}{4}\right)(0.4)(2\pi r)}{r}$$

$a_{CP} = 0.2\pi$ and $a_T = 0.5$

$a_{Net} = \sqrt{a_{CP}^2 + a_T^2} \approx 0.8\ \text{m/s}^2$

50. **(2)** $v_{av} = \dfrac{\sqrt{2}R}{t}$

$$t = \sqrt{\frac{2\left(\pi/2\right)}{\pi/4}} = 2\text{s}$$

$v_{av} = 1$ m/s

51. **(4)** In the given interval the slope of v-t graph (i.e., acceleration) neither constant nor uniform. Therefore options (1), (2) and (3) are incorrect but (4) is correct.

52. **(2)** Given, initial position of particle

$\vec{r}_0 = (2\hat{i}+4\hat{j})$ m.

Initial velocity of particle, $\vec{u} = (5\hat{i}+4\hat{j})$ m/s

Acceleration of particle, $\vec{a} = (4\hat{i}+4\hat{j})\ \text{m/s}^2$

Position of particle at time t is, $\vec{r} = r_0 + ut + \dfrac{1}{2}at^2$

At t = 2s, position of particle is,

$$\vec{r} = (2\hat{i}+4\hat{j}) + (5\hat{i}+4\hat{j})\times 2 + \frac{1}{2}\left(4\hat{i}+\frac{1}{2}\right)at^2$$

or $\vec{r} = (2+10+8)\hat{i} + (4+8+8)\hat{j}$

$\Rightarrow \vec{r} = (20\hat{i}+20\hat{j})$ m

∴ Distance of particle from origin is,

$|\vec{r}| = \sqrt{(20)^2+(20)^2} = 20\sqrt{2}$ m

53. **(1)** $\sqrt{4^2+(4x-2)^2+2^2} = 2\sqrt{1^2+x^2+3^2}$

$5 + (2x-1)^2 = 10 + x^2$

$x = -\dfrac{2}{3}$

54. **(1)** $\vec{a} = 3\hat{i}$, $\vec{b} = 2\sqrt{3}\hat{i} + 2\hat{j}$

$\vec{c} = p(3\hat{i}) + q(2\sqrt{3}\hat{i}+2\hat{j})$

$= (3p + q2\sqrt{3})\hat{i} + 2q\hat{j}$

$|\vec{c}| = \sqrt{(3p+2q\sqrt{3})^2 - (2q)^2} = 10$

Only option (1) satisfying it.

55. **(2)** Let $\vec{A} = 2\hat{i}+3\hat{j}+\hat{k}$ and $\vec{B} = \hat{i}-\hat{j}+2\hat{k}$ unit vector perpendicular to both $\vec{A}$ and $\vec{B}$ is

$$n = \frac{\vec{A}\times\vec{B}}{|\vec{A}\times\vec{B}|}$$

$$\vec{A}\times\vec{B} = \begin{vmatrix} \hat{i} & \hat{j} & \hat{k} \\ 2 & 3 & 1 \\ 1 & -1 & 2 \end{vmatrix}$$

$= 7\hat{i} - 3\hat{j} - 5\hat{k}$

$\therefore |\vec{A}\times\vec{B}| = \sqrt{7^2+(-3)^2+(-5)^2} = \sqrt{83}$ unit

Therefore unit vector perpendicular to both A and B is

$$\therefore \hat{n} = \frac{1}{\sqrt{83}}(7\hat{i}-3\hat{j}-5\hat{k})$$

56. **(3)** $\vec{R}_1 = \vec{A}+\vec{B}$, $\vec{R}_2 = \vec{A}-\vec{B}$

$$R_1^2 + R_2^2 = \left(\frac{\sqrt{A^2+B^2}}{A^2+B^2+2AB\cos\theta}\right)^2 + \left(\frac{\sqrt{A^2+B^2}}{A^2+B^2-2AB\cos\theta}\right)^2$$

$= 2(A^2+B^2)$

57. (1) $\vec{r} = at\hat{i} - bt^2\hat{j}$

$x = at$ and $y = -bt^2$

$$\Rightarrow t = \left(\frac{x}{a}\right) \text{ and } y = -b\left(\frac{x}{a}\right)^2 \Rightarrow y = \frac{-b}{a^2}x^2$$

58. (1) $\omega = \frac{u}{\sqrt{10}}(3\hat{j} - \hat{k})$

$$\vec{r} = (4.1)\hat{i} + (-2-3)\hat{j} + (1+1)\hat{k}$$

$$\vec{r} = 3\hat{i} + 5\hat{j} + 2\hat{k}$$

$$\vec{v} = \vec{\omega} \times \vec{r}$$

$$\therefore \vec{v} = \frac{4}{\sqrt{10}}\begin{vmatrix} \hat{i} & \hat{j} & \hat{k} \\ 0 & 3 & -1 \\ 3 & -5 & 2 \end{vmatrix}$$

$$\vec{v} = \frac{4}{\sqrt{10}}(\hat{i} - 3\hat{j} - 9\hat{k})$$

59. (1) OA = OC

$\overrightarrow{OA} + \overrightarrow{OC}$ is along $\overrightarrow{OB}$, and its magnitude is

$2R\cos 45 = R\sqrt{2}$

$(\overrightarrow{OA} + \overrightarrow{OC}) + \overrightarrow{OB}$ is along $\overrightarrow{OB}$ and its magnitude is

$\sqrt{2}R + R = R(\sqrt{2} + 1)$

Statement Based MCQs

1. **(3)** When an object is projected with some velocity (v) at an angle (θ) with the horizontal then there is only one force acting on the object due to gravity in vertical (downward) direction.

   The horizontal component of velocity will remain same because there is no force in horizontal direction.

2. **(3)** If two vectors $(\vec{A} = \vec{B})$ are equal then magnitude and direction of both vectors will same.

   If negative vectors $(\vec{A} = -\vec{B})$ are equal the magnitude will same and direction will be opposite.

3. **(1)** In the presence of air resistance (drag force) the range, maximum height, speed at which the projectile strikes the ground will decreases whereas as the angle with which the projectile stripes ground and time of flight increases.

4. **(2)** The velocity on reaching the ground would have two components (v_x and v_y) and will not equal to thrown velocity (horizontally). Vertical component of velocity changes due to gravity effect.

5. **(2)** In uniform circular motion, the magnitude of velocity and acceleration remains same, but the direction of velocity and acceleration changes. Also, the centripetal acceleration is given by $a = \omega^2 r$.

Assertion & Reason MCQs

1. **(1)** Large angular displacement is scalar as the surface gets plane with increasing angular displacement.

2. **(1)** Minimum time taken to cross the river is measured using the component of boat's velocity along the width of river which is perpendicular to the river's velocity. Hence, increasing or decreasing river's velocity does not affect the minimum time to cross the river.

3. **(1)** In both cases, initial velocity in vertical direction is zero and acceleration acting in vertical direction is acceleration due to gravity.

4. **(1)** In this case one of the time is increased and the other is decreased just because the direction of resistive force has reversed. If the resistive force were acting in the same direction, both would have been same.

5. **(3)** For complimentary angles, projectile ranges remain same.

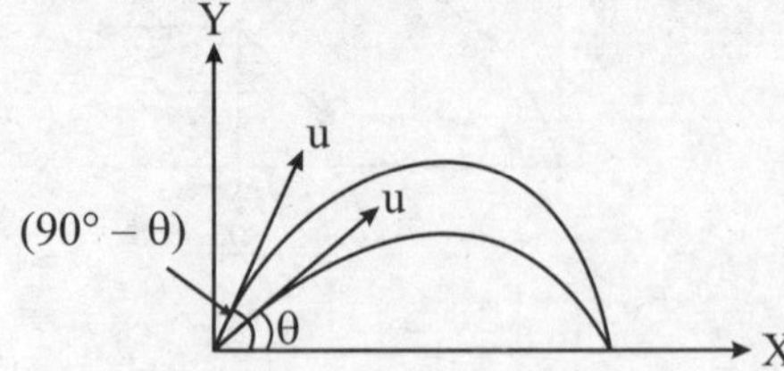

Match the Column MCQs

1. **(1)** $y = ax - bx^2$

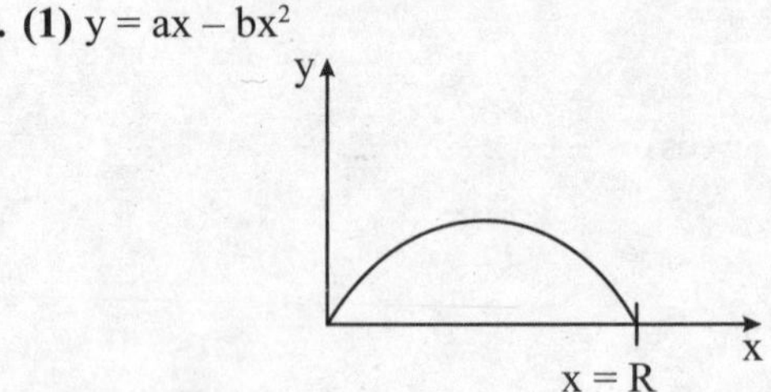

When x = R,

$y = 0 = ax - bx^2$

$x(a - bx) = 0$

$x = \frac{a}{b} = R$ (D → p)

At y = H

$v_y = 0 = \frac{dy}{dt}$ and $x = \frac{R}{2} = \frac{a}{2b}$

$\therefore y = a \times \frac{a}{2b} - b \times \frac{a^2}{4b^2} = \frac{a^2}{4b}$ (C → q)

Time of flight $= 2\sqrt{\frac{2y}{g}} = 2\sqrt{\frac{2}{g} \times \frac{a^2}{4b}} = \sqrt{\frac{2a^2}{bg}}$ (B → s)

Horizontal velocity $= \frac{\text{Range}}{\text{time}}$

$= \frac{\frac{a}{b}}{\sqrt{\frac{2a^2}{bg}}} = \sqrt{\frac{g}{2b}}$ (A → r)

2. **(3)** For minimum distance:

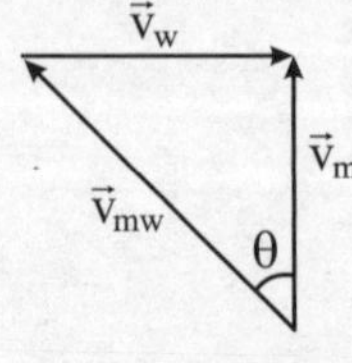

$\Rightarrow \vec{v}_m \perp \vec{v}_w$ and $\theta = \sin^{-1}\left(\frac{v_w}{v_{mw}}\right)$

For minimum time

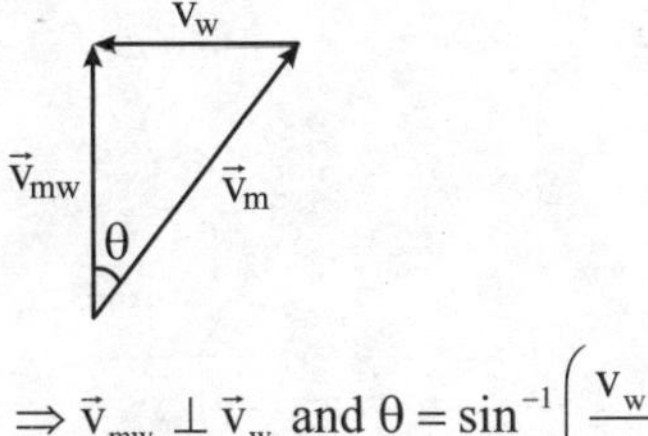

$\Rightarrow \vec{v}_{mw} \perp \vec{v}_w$ and $\theta = \sin^{-1}\left(\frac{v_w}{v_m}\right)$

3. (1) Initial velocity:

$\vec{u} = u\cos\theta\,\hat{i} + u\sin\theta\,\hat{j}$

$\vec{u} = 60\frac{\sqrt{3}}{2}\hat{i} + 60\left(\frac{1}{2}\right)\hat{j} = 30\sqrt{3}\,\hat{i} + 30\,\hat{j}$ (A → r)

Velocity after 3 s:

$\vec{v} = u\cos\theta\,\hat{i} + (u\sin\theta - gt)\,\hat{j}$

$= 30\sqrt{3}\,\hat{i} + (30 - 10\times 3)\,\hat{j} = 30\sqrt{3}\,\hat{i}$ (B → s)

Displacement after 2 s:

$\vec{s} = u\cos\theta t\,\hat{i} + \left(u\sin\theta t - \frac{1}{2}gt^2\right)\hat{j}$

$= 30\sqrt{3}\times 2i + \left(30\times 2 - \frac{1}{2}\times 10\times(2)^2\right)\hat{j}$

$= 60\sqrt{3}\,\hat{i} + 40\,\hat{j}$ (C → p)

Velocity after 2 s

$\vec{v} = u\cos\theta\,\hat{i} + (u\sin\theta - gt)\,\hat{j}$

$= 30\sqrt{3}\,i + (30 - 10\times 2)\,\hat{j}$

$= 30\sqrt{3}\,i + 10\hat{j}$ (D → q)

4. (4) $T = \frac{2u_y}{g} \Rightarrow T \propto u_y$ (A → q)

Height, $H = \frac{u_y^2}{2g} \Rightarrow H \propto u_y^2$ (B → r)

Range, $R = u_x T$

$R \propto T$ and $R \propto u_x$ (C → q)

Projection angle, $\tan\theta = \frac{u_y}{u_x} \Rightarrow \tan\theta \propto u_y$

By doubling u_y, tan θ will become two times, not θ. (D → s)

NCERT Exemplar MCQs

1. (2) Given,

$\vec{A} = \hat{i} + \hat{j}$

$\vec{B} = \hat{i} - \hat{j}$

We know that

$\vec{A}.\vec{B} = |\vec{A}||\vec{B}|\cos\theta$

$\Rightarrow (\hat{i}+\hat{j}).(\hat{i}-\hat{j}) = (\sqrt{1+1})(\sqrt{1+1})\cos\theta$

where θ is the angle between A and B

$\Rightarrow \quad \cos\theta = \frac{1-0+0-1}{\sqrt{2}\sqrt{2}} = \frac{0}{2} = 0 \Rightarrow \quad \theta = 90°$

2. (4) A scalar quantity is independent of direction. Hence has the same value for observers with different orientations of the axes.

3. (2) Clearly from the diagram, $\vec{u} = a\hat{i} + b\hat{j}$

As u is in the first quadrant. Hence both components a and b will be positive.

For $\vec{v} = p\hat{i} + q\hat{j}$ as it is in positive x-direction and located downward hence x-component p will be positive and y-component q will be negative.

4. (2) Let r makes an angle θ with positive X-axis component of r along X-axis

$r_x = |r|\cos\theta$

$(r_x)_{maximum} = |r|(\cos\theta)_{maximum}$

$c = |r|\cos 0° = |r|$ (∵ cosθ is maximum θ = 0°)

As θ = 0°, r is along positive X-axis.

5. (3) We know that

where θ is angle of projection

Given, θ = 15° and R = 50 m

Range, $R = \frac{u^2\sin 2\theta}{g}$

Putting all the given values in the formula, we get

$\Rightarrow R = 50\text{m} = \frac{u^2\sin(2\times 15°)}{g}$

$\Rightarrow 50\times g = u^2\sin 30° = u^2\times\frac{1}{2}$

$\Rightarrow 50\times g\times 2 = u^2$

$\Rightarrow u^2 = 50\times 9.8\times 2 = 100\times 9.8 = 980$

$\Rightarrow u^2 = \sqrt{980} = \sqrt{49\times 20} = 7\times 2\times\sqrt{5}$ m/s

= 14 × 2.23 m/s = 31.304 m/s

For $\theta = 45°; R = \frac{u^2\sin 2\times 45°}{g} = \frac{u^2}{g}$

(∵ sin 90° = 1)

$\Rightarrow \quad R = \frac{(14\sqrt{5})^2}{g} = \frac{14\times 14\times 5}{9.8} = 100$ m

6. (2) We know that impulse, I = F. Δt = Δp, where F is force, Δt is time duration and Δp is change in momentum. As Δp is a vector quantity, hence impulse is also a vector quantity. Sometimes area can also be treated as vector.

7. (4) We know that

speed, $v_0 = \frac{\text{Total distance travelled}}{\text{Time taken}}$

Hence, total distance travelled = Path length

= (speed) × time taken

8. (3) As given motion is two dimensional motion and given that instantaneous speed v_0 is positive constant. Acceleration is rate of change of velocity (instantaneous speed) hence it will also be in the plane of motion.

NEET & AIIMS Past Year MCQs

1. (4) Let:

$x = A\cos\omega t,\ y = 2A\cos\omega t \Rightarrow y = 2x$

Straight-line motion along $y = 2x$, with both coordinates undergoing SHM.

Velocity

$$\vec{v} = \frac{d\vec{r}}{dt} = -\left(\hat{i}+2\hat{j}\right)A\omega\sin\omega t$$

$$a = \frac{d^2r}{dt^2}$$

$$= -(1+2\hat{j})\omega^2 A\cos\omega t$$

Acceleration $\vec{a} = -\omega^2\vec{r}$, indicating Simple Harmonic Motion (SHM).

2. (4) Initial velocity $= -v\hat{j}$

Final velocity $= v\hat{i}$

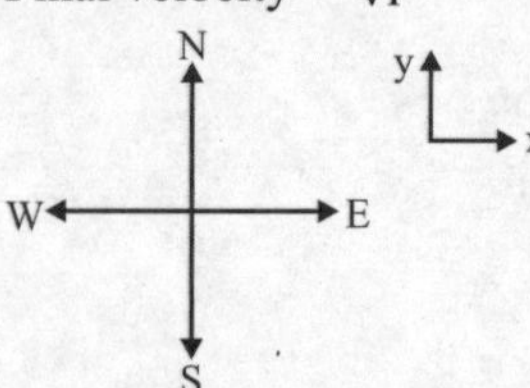

Change in velocity $= v\hat{i} - \left(-v\hat{j}\right) = v\left(\hat{i}+\hat{j}\right)$

Momentum gain is along $\hat{i}+\hat{j}$

$\Rightarrow$ Force experienced is along $\hat{i}+\hat{j}$

$\Rightarrow$ Force experienced is in North-East direction.

3. (4) $h_{max} = \frac{u^2\sin^2\theta}{2g} = \frac{280\times280}{2\times9.8}\times\frac{1}{4} = 1000\text{ m}$

4. (3) At highest point only horizontal component of velocity remains some.

$\because$ angle with the vertical is 60°, therefore angle with horizontal is 30°

$\Rightarrow u_x = u\cos\theta = 10\cos 30°$

$= 5\sqrt{3}\text{ ms}^{-1}$

5. (3) $T = \frac{2\pi R}{V}$

$$V = \frac{2\pi R}{T}$$

$$H = \frac{u^2\sin^2\theta}{2g}$$

$$4R = \frac{4\pi^2R^2\sin^2\theta}{T^2 2g}$$

$$\sin^2\theta = \frac{8RT^2g}{4\pi^2R^2}$$

$$\sin\theta = \sqrt{\frac{2T^2g}{\pi^2R}}$$

$$\theta = \sin^{-1}\left(\frac{2T^2g}{\pi^2R}\right)^{1/2}$$

6. (3) $u = 0$, $a = 5\text{ m/s}^2$, $t = 4$ sec

$\because v = u + at$

$v = 0 + 5\times4 = 20$ m/s

$v_x = 20$ m/sec

$v_y = u_y + a_y\Delta t$

$= 10\times2\ (\because \Delta t = 2\text{ sec})$

$v_{net} = 20\sqrt{2}$ m/sec and $a_{net} = 10\text{ m/sec}^2$

7. (1) $t = 2\frac{u\sin\theta}{g} \Rightarrow \frac{u\sin\theta}{g} = \frac{5}{2}$

$$H_m = \frac{u^2\sin^2\theta}{2g}\times\frac{g}{g} = \frac{g}{2}\frac{u^2\sin^2\theta}{g^2} = \frac{g}{2}\times\frac{25}{4}$$

$H_m = 31.25$ m

8. (3) Displacement along x-axis $= \frac{R}{2} = \frac{u^2\sin2\theta}{2g}$

Displacement along y-axis $= H = \frac{u^2\sin^2\theta}{2g}$

$$\therefore\ v_{avg} = \frac{\text{Net displacement}}{\text{Time}} = \frac{\sqrt{H^2+\frac{R^2}{4}}}{T/2}$$

Here $T = \frac{2u\sin\theta}{g}$

Substituting the given values in above equation

$$V_{avg} = \frac{v}{2}\sqrt{1+3\cos^2\theta}$$

9. (1) **10. (2)** **11. (4)**

Laws of Motion

NCERT Topic-wise MCQs

1. (3) Newton's first law of motion defines the inertia of body. It states that everybody has a tendency to remain in its state due to its inertia.

2. (4) Types of inertia are: Inertia of rest, Inertia of motion, and Inertia of direction.

3. (4) Horizontal velocity of ball and person are same so both will cover equal horizontal distance in a given interval of time so the ball falls exactly in the hand which threw it up.

4. (3) When bus suddenly applies brakes, lower part the man comes to rest whereas the upper part tends to remain in motion. These all are effects of Inertia.

5. (4) Newton's first law of motion.

6. (2) After a jerk, impulsive tension exceeds the breaking stress in string BC. So string breaks before the jerk wave can reach to string AB.

7. (3) $t = \frac{v}{a}\ \therefore t \propto \frac{1}{a}$ $\left[\begin{array}{l} F = ma \\ F = \text{same} \\ m \propto \frac{1}{a}\end{array}\right]$

$t \propto \frac{1}{a} \propto m$

$t \propto m$

$\frac{t_1}{t_2} = \frac{m_1}{m_2} = \frac{8}{10} = \frac{4}{5}$

8. (4) $F = m\left(\frac{dv}{dt}\right)$ (second law of motion)

$F = 150\left(\frac{8}{0.2}\right) = 150 \times 40 = 6000\text{ N}$

$\therefore$ To stop the car, 6000 N is required in opposite direction.

9. (3) $v = u + at = 0 + 3 \times 1 = 3\text{ ms}^{-1}$

$\Delta p = p_2 - p_1 = 2 \times 3 = 6\text{ kg ms}^{-1}$.

10. (1) Force exerted by bullets on the gun, $F = \frac{\Delta p}{\Delta t} = \frac{nm(v-u)}{\Delta t}$

$= nmv$ (as $u = 0$ and $\Delta t = 1\text{s}$)

11. (2) $u = 9 \times \frac{5}{18} = 2.5\text{m/s}$, $v = 0$, $t = 2 \times 60\text{sec}$

Resistance Force $= \frac{m(v-u)}{t} = \frac{2000(0-2.5)}{2 \times 60}$

Resistance = – 41.7 N (–ve sign shows that resistance occurs in opposite direction of motion.)

12. (4) Range of resultant of F_1 and F_2 varies between (3 + 5) = 8 N and (5 – 3) = 2N. It means for some value of angle (θ), resultant 6 can be obtained. So, the resultant of 3 N, 5N and 6N may be zero and the forces may be in equilibrium.

13. (1)

2 cos 45°, 1 cos 45°, 1 N, 2 N, 45°, 45°, 2 sin 45°, P, 90°, F_1, 1 sin 45°, F_2; y, x, T, 60°, Tsin60°, Tcos60°, T_1, 2kg - wt

Applying equilibrium conditions,

$\Sigma F_X = 0$

$\Rightarrow F_1 + 1\sin 45^\circ - 2\sin 45^\circ = 0$

or $F_1 = 2\sin 45^\circ - 1\sin 45^\circ$

$= \frac{2}{\sqrt{2}} - \frac{1}{\sqrt{2}} = \frac{1}{\sqrt{2}}\text{ N}$

14. (3) $T = \frac{F(L-x)}{L} \Rightarrow T = \frac{5(20-5)}{20} = \frac{75}{20} = 3.75\text{N}$

15. (4) After breaking the rope the lift undergoes free fall, hence acceleration becomes g.

$N = m(g - a) \Rightarrow N = m(g - g) = 0$

16. (1) $T = m(g - a)$

$T = mg\left(1 - \frac{a}{g}\right)$

$\frac{mg}{3} = mg\left(1 - \frac{a}{g}\right) \Rightarrow \frac{1}{3} = 1 - \frac{a}{g}$

$\Rightarrow \frac{a}{g} = 1 - \frac{1}{3} = \frac{2}{3} \Rightarrow a = 6.67\text{ m/s}^2$

17. (3) Forward force = 500 N

Backward force = 120 N

$\therefore$ Net force = 500 – 120 = 380 N

As, $F = ma$

$a = \frac{380}{1000} = 0.38\text{ m/s}^2$

Now $v = u + at$

$v = 15 + 0.38 \times 10 = 18.8\text{ m/s}^2$

18. (2) Let acceleration of monkey be 'a' m/s² upward.

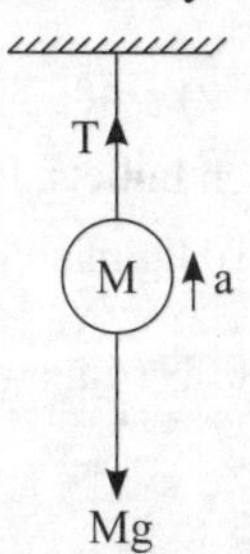

$T - mg = ma \Rightarrow T = m(g + a) = 40\ (10 + a)$

$\Rightarrow T = 400 + 40a$ ...(*i*)

Rope breaks if $T > 600\text{N}$.

$\Rightarrow 400 + 40a > 600$

$\Rightarrow a > 5\text{ m/s}^2$

So, string breaks if acceleration of monkey is any value more than 5 ms⁻².

19. (1) $T \sin 60^\circ = 2\text{kg wt}$

$T \times \frac{\sqrt{3}}{2} = 2\text{ kg wt}$

$T = \frac{4}{\sqrt{3}}\text{kgwt}$

$T_1 = T\cos 60^\circ \Rightarrow T_1 = \frac{4}{\sqrt{3}}\frac{\text{kgwt}}{2} = \frac{2}{\sqrt{3}}\text{kgwt}$

20. (3) Impulse = Area between force and time graph and it is maximum for graph (iii) and (iv). Which is equal to 0.5×10^{-3} Ns.

21. (3) Impulse = Force × Time

$= 50 \times 10^{-5} \times 3$

$= 1.5 \times 10^{-3}$ Ns

22. (1) Using $v^2 = u^2 + 2as$

$0 = (150)^2 + 2 \times a \times 0.06$

$\Rightarrow a = \frac{-22500}{0.12}$ (–ve sign indicates that acceleration is in opposite direction)

$F = 10^{-2} \times \frac{22500}{0.12} = 1875\text{N}$

23. (1) Change in Linear Momentum = Impulse = F × t
= Area under F – t graph
= (– 2 × 2) + 1 × (6 – 2) = – 4 + 4 = 0

24. (1) $F = \frac{mv-(-mv)}{t} = \frac{2mv}{t} = \frac{2\times200\times10^{-3}\times30}{0.01}$

F = 1200 N

25. (4) $F = \frac{dp}{dt} \Rightarrow t = \frac{P}{F} = \frac{135}{500} = 0.27\,\text{sec}$

26. (4) Δp = Area under F–t graph

$= \frac{1}{2}\times20\,N\times4\,s = 40\,Ns$

27. (2) $m_g v_g = m_b v_b$

$v_g = \frac{m_b v_b}{m_g} = \frac{20\times10^{-3}\times100}{2} = 1\,m/s$

28. (1) Apparent weight
= m(g + a) = 40(10 + 2) = 480 N = 48 kg wt.

29. (1) Momentum of each bullets, P = Mv
= 50 × 10^{-3} × 200 = 10 kg m/s

$F = \frac{nP}{t} = \frac{10\times10}{1} = 100N\,(\text{time} = 1\,\text{sec})$

Force = ma

$a = \frac{100}{2800} = 0.0357 = 3.6\ cm/s^2$

30. (4) We know that, $\vec{I} = \Delta\vec{P} = \vec{P}_f - \vec{P}_i$

$P = M(\sqrt{2gh})$

$M = 0.1$ kg
Hence,

$I = \Delta P = 0.1\left(\sqrt{2\times9.8\times5} - \left(-\sqrt{2\times9.8\times10}\right)\right)$

$= 0.1\left(14 + 7\sqrt{2}\right) \approx 2.39\ kg\ ms^{-1}$

31. (3) $v^2 - u^2 = 2as$
$0 - u^2 = 2 \times -5 \times 100$
$-u^2 = -2 \times 500 \Rightarrow u = \sqrt{1000} = 31.6$ m/s
Impulse = change in momentum
= m (v – u) = 0.1 (0 – 31.6) = – 3.16 Ns

32. (2) Impulse imparted to each billiard ball is equal to the change in momentum of the ball.
I = mv – mu = m(v – u) (v = –u rebound with same velocity)
I = m((–u) – u)

$I = -2mu = -2\times50\times10^{-3}\times40\times\frac{5}{18} = -1.11$ kg m/s

33. (2) $\vec{P}_i = Nmv\hat{i}$ and $\vec{P}_f = -Nmv\hat{i}$
where *N* is the number of balls that strike the wall

$\Delta\vec{P} = \vec{P}_f - \vec{P}_i = -2Nmv\hat{i} = -200\,Nmv\hat{i}$

$\vec{F}_{Total} = \frac{\Delta\vec{P}}{\Delta t} = -\frac{200mvt}{t}$

$|\vec{F}| = \frac{200mv}{t}$

34. (1) According to the formula

$a = \frac{m_2}{m_1+m_2}\times g = \frac{11\times10}{4+11} = \frac{110}{15} = 7.33 m/s^2$

35. (4) Acceleration of system,

$a = \frac{\text{Net pulling force}}{\text{Total mass}} = \frac{2mg - mg}{3m} = \frac{g}{3}$

New, from equation of motion of m

$T - mg = ma = \frac{mg}{3}$

$\therefore T = \frac{4mg}{3}$

For equilibrium of pulley,

T_{AB} = 2T + weight of pulley

$= \frac{8mg}{3} + 3mg = \frac{17mg}{3}$

36. (3) Forces on the three masses are shown in the figure.

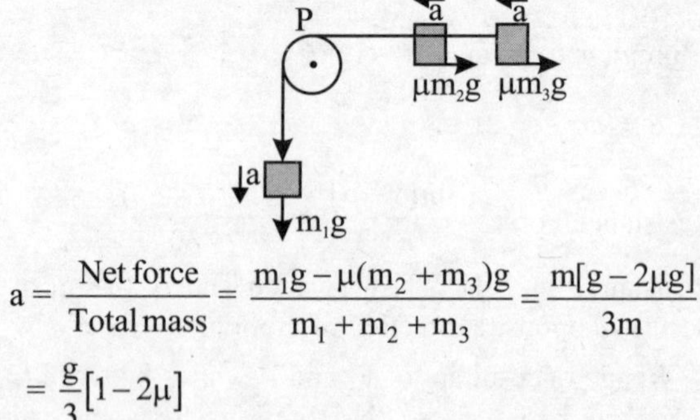

$a = \frac{\text{Net force}}{\text{Total mass}} = \frac{m_1 g - \mu(m_2+m_3)g}{m_1+m_2+m_3} = \frac{m[g-2\mu g]}{3m}$

$= \frac{g}{3}[1-2\mu]$

37. (3) Equations of motion are
for 2 kg block, $F - T_1 = 2a$ ...(*i*)
for 3 kg block, $T_1 - T_2 = 3a$ ...(*ii*)
For 5 kg block
$T_2 = 5a$ ...(*iii*)
putting the value of T_2 in eq. (*ii*)
$\therefore T_1 - 5a = 3a \Rightarrow T_1 = 8a$ [Putting in eq. (*i*)]
F – 8a = 2a
F = 10 N (a = 1 m/s²)

38. (1) $T = \left(\frac{2m_1m_2}{m_1+m_2}\right)g = \frac{2\times10\times15\times10}{25}$

$T = \frac{2\times2\times15\times10}{5} = 120N$

39. (1) v = constant ⇒ a = 0
$\therefore F_{net} = 0$
The net force on each block is zero.

40. (3) Free body diagram for bodies of masses m and $\sqrt{2}$m are shown below.

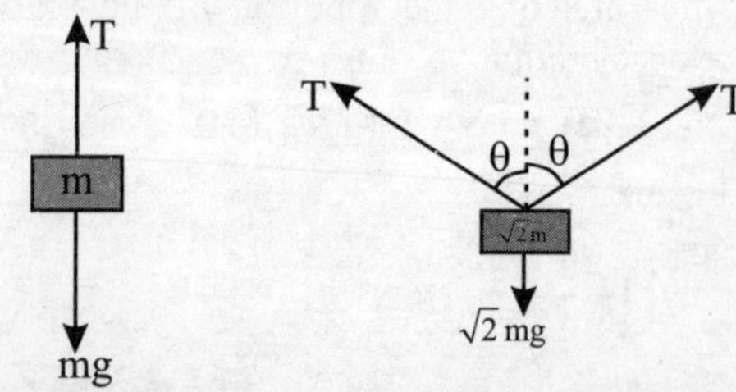

For equilibrium of mass m,

$T = mg$ ...(i)

For equilibrium of mass $\sqrt{2}$ m

$2T \cos\theta = \sqrt{2}\, mg$ ...(ii)

Dividing (ii) by (i), we get

$\cos\theta = \frac{1}{\sqrt{2}} \quad \therefore \quad \theta = 45°.$

41. **(4)** Acceleration of the system $a = \frac{F}{M+m}$ (common acceleration)

$$(M) \rightarrow \longleftarrow (m) \rightarrow F$$

Force on the block (M) = kx = Ma

$$T = \frac{M \times F}{M+m} \left(\text{putting } a = \frac{F}{m+M}\right)$$

42. **(4)** $m_2 g - T = m_2 a$ ...(i)

$T - m_1 g \sin 60° = m_1 a$ ...(ii)

Adding both equations

$m_2 g - m_1 g \sin 60° = m_2 a + m_1 a$

$$\frac{g(m_2 - m_1 \sin 60°)}{(m_2 + m_1)} = a$$

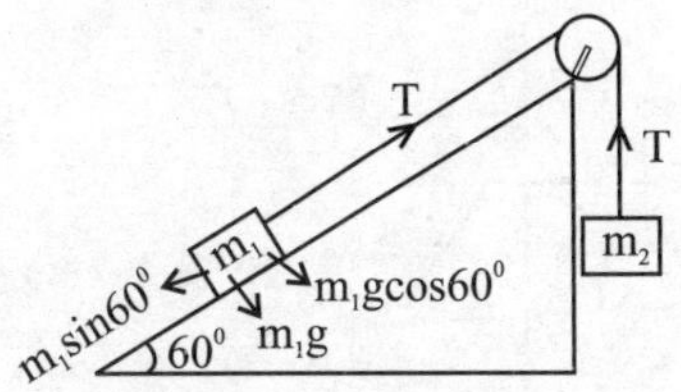

$$\frac{10\left(4 - 8 \times \frac{\sqrt{3}}{2}\right)}{(8+4)} = a$$

$$a = \frac{10 \times (-2.92)}{12} = -2.43 \text{ m/s}^2$$

–Ve sign indicates M_2 goes down.

43. **(1)** $T_2 - T_1 = M_2 a$ ...(i)

$T_1 = M_1 a$ ...(ii)

$$T_2 - T_1 = M_2 \times \frac{T_1}{M_1} (\text{from eq (ii)})$$

$$T_2 = \frac{T_1 M_2}{M_1} + T_1 \Rightarrow T_2 = T_1\left(1 + \frac{M_2}{M_1}\right)$$

$$\frac{T_1}{T_2} = \frac{M_1}{M_1 + M_2}$$

44. **(1)** Linear momentum is conserved because resultant of all external force acting on the system is zero.

45. **(4)** For the block of mass M to be in equilibrium, the tension in the string, T = Mg. Free body diagram of pulley is shown. Where, R_x and R_y are the components of force exerted by clamp on pulley

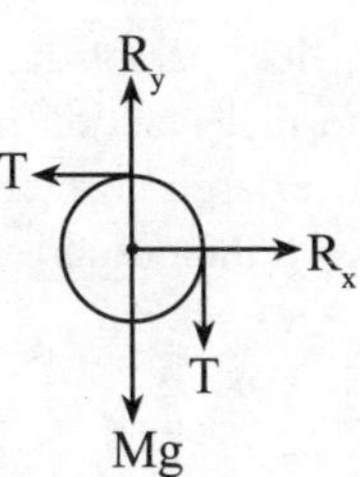

$R_x = T = Mg$

$R_y = Mg + T = (m + M)g$

$$R = \sqrt{(R_x)^2 + (R_y)^2} = \sqrt{(Mg + mg)^2 + (Mg)^2}$$

$$= \sqrt{(M+m)^2 + M^2}\, g$$

46. **(1)** Initial velocity of a ball = v

When ball strikes the wall normally it is reflected back, then final velocity = – v

Change in velocity $\Delta V = v - (-v) = 2v$

Force exerted by the ball on the wall is given by Newton's second law,

i.e., $F = ma = \frac{m\Delta v}{\Delta t} = \frac{(m2v)}{t} = \frac{2mv}{t}$

47. **(1)** Force required, $F = \frac{dp}{dt} = \frac{d(mv)}{dt} = v\left(\frac{dm}{dt}\right) = vM$

As velocity v is constant, hence

F = Mv newton

48. **(1)** In the frame of car

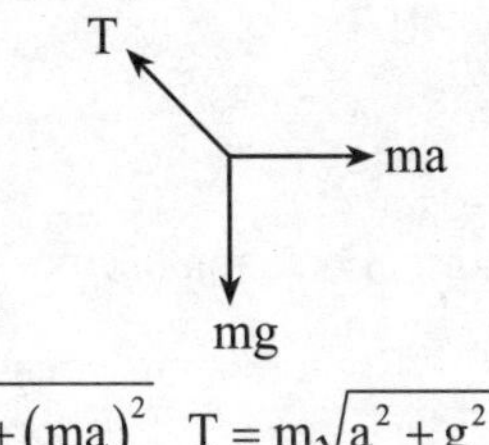

$$T = \sqrt{(mg)^2 + (ma)^2} \quad T = m\sqrt{a^2 + g^2}$$

49. **(3)** Tension in spring is nothing but spring force FBD of spring right end point is,

$$kx \longleftrightarrow 5N$$

Since spring is massless kx = 5N

50. **(4)** Let elongations of the springs X and Y be x and y respectively.

At equilibrium,

$K_1 Y = 8$ g ...(i)

$K_2 X = K_1 Y$ ...(ii)

From the above two equations, we get

$K_2 X = K_1 Y = 8g$

Hence, both springs will give the same reading

Hence, both springs will give the same reading

51. **(4)** After the string is cut,

For block (A)

$3\, ma_A = 4mg - 3mg$

$a_A = \frac{g}{3}$

For block (B)

$ma_B = mg \Rightarrow a_B = g$

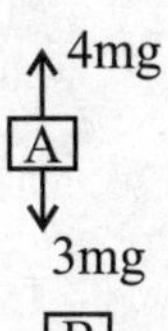

52. (2) For walking on ice, person has to take smaller steps to produce large normal reaction because on ice, friction is less, so to produce large normal reaction his legs must make small angle with the ground.

53. (2) $v^2 - u^2 = 2as$

$$a = \frac{(0^2)-(6)^2}{2\times 10} = \frac{-36}{20} = -1.8$$

$$a = \mu g \Rightarrow \mu = \frac{a}{g} = \frac{1.8}{10} \Rightarrow \mu = 0.18$$

54. (4) $(f_s)_{max} = \mu_s R = \mu_s mg = 0.4 \times 2 \times 9.8 = 7.8N$

Applied force is smaller than the friction force, then the block will not move, because static friction is self adjusting so, it will be equal to 2.5 N

55. (2) $\mu_s = \frac{\text{Length of chain hanging from the table}}{\text{length of chain lying on the table}}$

$$\mu_s = \frac{2/3}{1-\frac{2}{3}} = 2$$

56. (1) As we know that the relation

$$\mu = \frac{\ell}{1-\ell} \Rightarrow \mu - \mu\ell = \ell$$

$$\mu = \ell + \mu\ell \Rightarrow \ell = \left(\frac{\mu}{1+\mu}\right)$$

$$\therefore \ell = \left(\frac{0.50}{1+0.50}\right)\times 100$$

$$\ell = 33.3\%$$

57. (3) $\mu = \frac{F}{R}$ if, normal reaction becomes double then force also will be double. So, μ remains constant.

58. (4) We have, $N = mg\cos\theta$

Also, $F = ma = mg\sin\theta - \mu_K mg\cos\theta$

$\therefore a = g\sin\theta - \mu_K g\cos\theta$

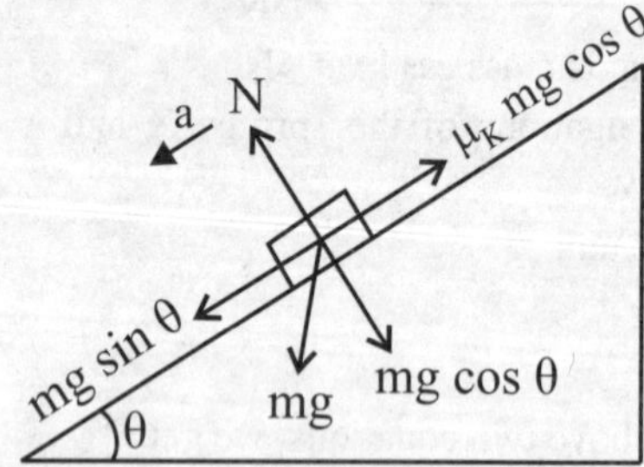

59. (1) Net pulling force $F_1 = 15g - 5g$

$F_1 = 10g = 100$ N

Net stopping force $F_2 = 0.2 \times 5 \times 10 = 10$ N

Since $F_1 > F_2$, therefore the system will move (anti-clockwise) with an acceleration.

$$a = \frac{F_1 - F_2}{15+5+5} = \frac{90}{25} = 3.6 \text{ m/s}^2$$

15 kg block

$15 \times 10 - T_2 = 15 \times a = 15 \times 3.6$

$\therefore T_2 = 96$ N

5 kg block

$T_1 - 5 \times 10 = 5 \times a = 5 \times 3.6$

$\therefore T_1 = 68$ N

$$\therefore \frac{T_1}{T_2} = \frac{68}{96} = \frac{17}{24}$$

60. (1) Since, the block is resting (not moving)

$\therefore f = mg\sin\theta \neq \mu_s mg\cos\theta$

$$\text{or } m = \frac{f}{g\sin\theta} = \frac{10}{10\times\sin 30°} = 2\text{kg}$$

61. (2) Acceleration down the inclined plane

$a = g(\sin\theta - \mu_k\cos\theta)$

$$a = 10\left(\frac{1}{\sqrt{2}} - \frac{0.71}{\sqrt{2}}\right) \Rightarrow a = \frac{10}{\sqrt{2}}(1-0.7) \Rightarrow a = 2.1 \text{ m/s}^2$$

62. (1) For vertical equilibrium of the block,

$R = mg + F\sin 60°$

$= \sqrt{3}g + \frac{\sqrt{3}}{2}F$

For horizontal equilirbium

$f = F\cos 60°$

For no motion of the block,

$f \geq F\cos 60°$

or $\mu R \geq F\cos 60°$

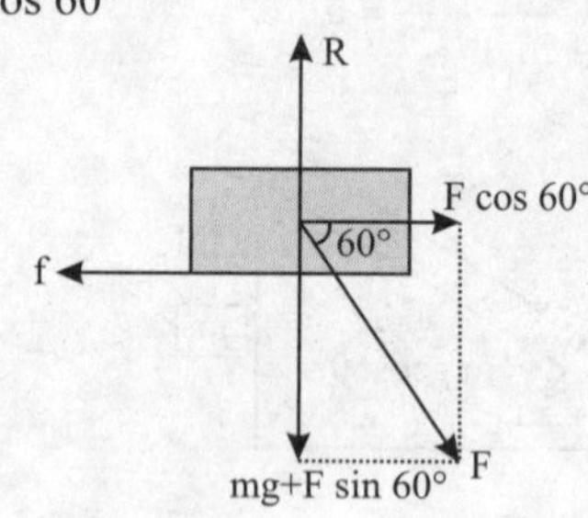

$$\text{or } \frac{1}{2\sqrt{3}}\left(\sqrt{3}g + \frac{\sqrt{3}}{2}F\right) \geq \frac{F}{2}$$

$$\text{or } g \geq \frac{F}{2}$$

or $F \leq 2g$ or $F_{max} = 20$ N.

63. (3) $s = ut + \frac{1}{2}at^2$

$$10 = \frac{1}{2}\times a\times(3)^2 \Rightarrow \frac{20}{9} = a$$

$\because a = g(\sin\theta - \mu\cos\theta)$

$$\frac{20}{9} = 9.8\left(\frac{1}{\sqrt{2}} - \mu\frac{1}{\sqrt{2}}\right) \Rightarrow \frac{20}{9} = \frac{9.8}{\sqrt{2}}(1-\mu)$$

$$1-\mu = \frac{20\times 1.41}{9\times 9.8} = \frac{28.2}{88.2} = 0.31$$

$\mu = 0.69$

64. (1) $N = mg = 40$ N

$\mu N = 0.2 \times 40 = 8N$

At t = 2 s, F = 4N

Since $F < \mu N$

$\therefore f = F = 4N$

65. (2) Different forces acting on the block are shown in the figure.

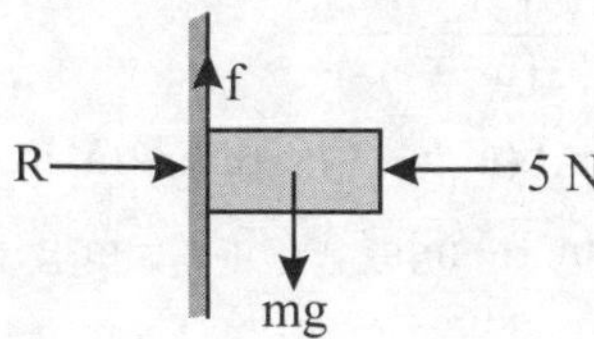

In equilibrium

Force of friction, $f = mg = 0.1 \times 9.8 \text{ N} = 0.98 \text{ N}$.

66. (1) For equilibrium,

$R = mg\cos\alpha$

$f = mg\sin\alpha$

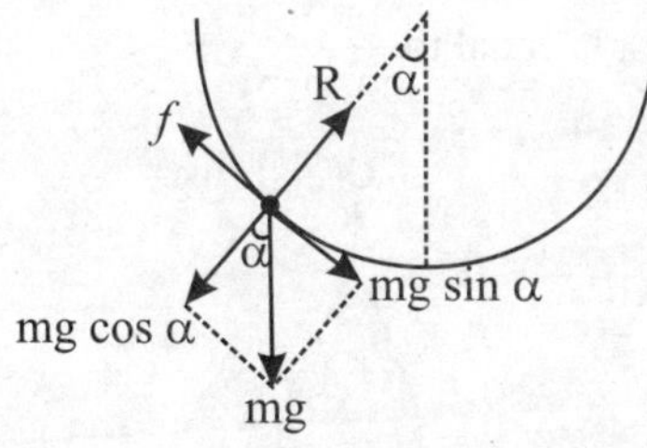

But for limiting case $f = \mu R = \mu mg\cos\alpha$

$\therefore \mu mg\cos\alpha = mg\sin\alpha$

or $\cot\alpha = \frac{1}{\mu} = \frac{1}{1/3} = 3$

67. (1) $\mu = \tan\theta \Rightarrow \theta = \tan^{-1}(\mu) = \tan^{-1}(1.73) = 60^\circ$

68. (3) Force of friction

$f = \mu N = \mu m\alpha$

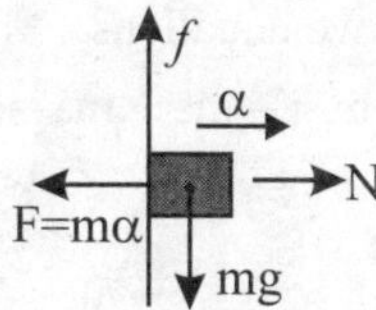

To prevent the block from falling

$\mu m\alpha \geq mg$

$\alpha \geq \frac{g}{\mu}$

69. (4)

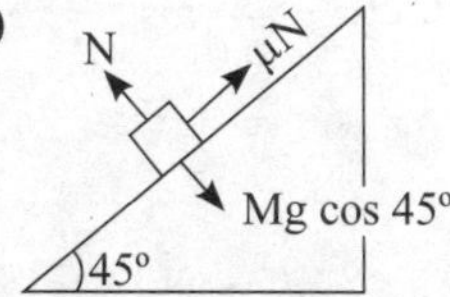

Criticle angle = 45°

$\therefore \mu = \tan 45^\circ = 1$

Friction force $= \mu N = 1 \times mg\cos 45^\circ$

$= 2 \times 9.8\cos 45^\circ = 19.6\cos 45^\circ$

70. (3) Critical angle is 30°, it means the block is just about to move.

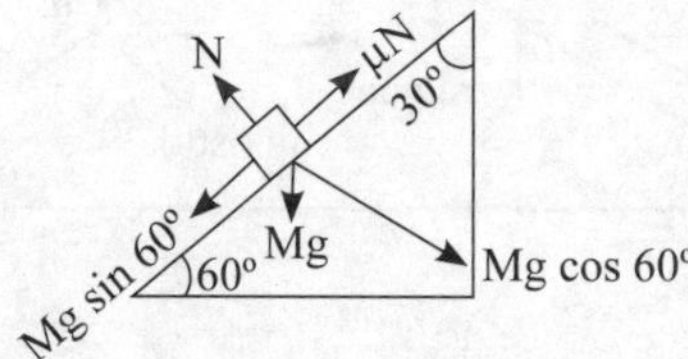

So, $Mg\sin 60^\circ = \mu Mg\cos 60^\circ$

$\mu = \tan 60^\circ = \sqrt{3}$

71. (1) $v = \sqrt{\mu rg} = \sqrt{1.5 \times 180 \times 10} = \sqrt{2700} = 51.96 \text{ m/s}$

72. (3) $v = 800 \times \frac{5}{18} = \frac{2000}{9} \text{ m/s}$

As we know that

$\tan\theta = \frac{v^2}{rg} \Rightarrow r = \frac{v^2}{g\tan\theta}$

$$r = \frac{\left(\frac{2000}{9}\right)^2}{\frac{1}{\sqrt{3}} \times 10} = 8.543 \times 10^3 \text{ m}$$

73. (4) Frictional force should be equal to the centripetal force required

$f = \frac{mv^2}{R} = \frac{2000 \times 15 \times 15}{20} = 22500$

As $f_{max} = \mu R = \mu mg$

$\mu = \frac{f}{mg} = \frac{22500}{2000 \times 10} = 1.125$

74. (1) If a body of mass m is moving in circle of radius r, then a force called centripetal force $F = \frac{mv^2}{r}$ act towards the centre and the acceleration $a_c = \frac{v^2}{r}$ called centripetal acceleration is present towards the centre.

75. (1) Tension in string provides the centripetal force.

76. (1) Gravitation force on the planet due to sun provide the centripetal force when a planet revolve around the sun.

77. (1) Centripetal force is provided by force of friction when a car is taking a circular turn on a horizontal road.

78. (3) As car is moving in a circle on a level road, then normal reaction balances the weight of the car.

79. (2)

80. (1) Here, v = 18km/h, R = 3m, $\mu_s = 0.1$

$\therefore$ Maximum velocity with which the cyclist should move to avoid slipping

$\sqrt{\mu_s gR} = \sqrt{0.1 \times 9.8 \times 3} = \sqrt{2.94} = 1.71 \text{ m/s}$

However, given in the question

$v = 18 \text{ km/h} = \frac{18 \times 1000}{60 \times 60} = 5 \text{ m/s}$

Here, $v > \sqrt{\mu_s gR}$.

So, cyclist will slip while taking the turn.

High Order Time Intensive MCQs

1. (1) $F = u\left(\frac{dm}{dt}\right) \Rightarrow \frac{dm}{dt} = \frac{F}{u} = \frac{210}{300} = 0.7 \text{ kg/s}$

2. (3) Initially due to upward acceleration apparent weight of the body increases but then it decreases due to decrease in gravity.

3. (1) $m = \frac{F}{a} = \frac{\sqrt{6^2 + 8^2 + 10^2}}{1} = \sqrt{200} = 10\sqrt{2} \text{ kg}$

4. (4) $F = ma = m\left(\dfrac{u^2 - v^2}{2s}\right) = \dfrac{20\times10^{-3}\times(250)^2}{2\times0.12}$

$\therefore\ F = 5.2\times10^3\ N$

5. (2) $a = \dfrac{5}{4+5}\times9.8 = \dfrac{49}{9} = 5.44\ m/s^2$

6. (3) $T_1 = \left(\dfrac{m_2+m_3}{m_1+m_2+m_3}\right)g = \dfrac{3+5}{2+3+5}\times10 = 8\ N$

7. (3) Impulse $= m(v_2 - v_1)$ ...(*i*)

Impulse = Area between the graph and time axis

$= \dfrac{1}{2}\times2\times4 + 2\times4 + \dfrac{1}{2}(4+2.5)\times0.5 + 2\times2.5$

$= 4 + 8 + 1.625 + 5 = 18.625$ ...(*ii*)

From (*i*) and (*ii*), $m(v_2 - v_1) = 18.625$

$\Rightarrow v_2 = \dfrac{18.625}{2} + 5 = 14.25\ m/s$

8. (3) Newton's first law defines force. Newton's second law gives us a measure of force. Impulse gives us the effect of force. Recoiling of gun is accounted for by Newton's third law.

9. (2) For equilibrium,

$2T\sin\theta = W$

$T = \dfrac{W}{2\sin\theta}$

θ lies between 0 to 90°

i.e., $\sin\theta < 1$

$\therefore\ T > W$

10. (2) $F = ma$ for a given body. If F is constant, then a is constant.

11. (4) Displacement in N[th] second

$S_n = u + \dfrac{1}{2}a(2n-1)$

$10 = u + \dfrac{1}{2}a(2\times1-1)$ ...(*i*)

$30 = u + \dfrac{1}{2}a(2\times2-1)$ ...(*ii*)

from (*i*) and (*ii*), we get

$a = 20\ m/s^2$

$F = ma = 0.2\times20 = 4\ N$

12. (1) Equations of motion are

$m_1g - T = m_1a$ ...(*i*)

$T - m_2g = m_2a$ ...(*ii*)

From eq (*i*) $T = -m_1a + m_1g$

Putting in eq (*ii*)

$-m_1a + m_1g - m_2g = m_2a$

$m_1g - m_1a = m_2a + m_2g$

$g(m_1 - m_2) = a(m_1 + m_2)$

Putting the value of a in eq (*i*) we get

$m_1g - T = \dfrac{m_1(m_1-m_2)g}{(m_1+m_2)}$

$m_1g - T = \dfrac{(m_1^2 - m_1m_2)g}{(m_1+m_2)}$

$\Rightarrow (m_1+m_2)(m_1g - T) = (m_1^2 - m_1m_2)g$

$\therefore\ m_1^2g - m_1T + m_1m_2g - m_2T = m_1^2g - m_1m_2g$

$m_1^2g - m_1T + m_2m_1g - m_2T - m_1^2g + m_1m_2g = 0$

$-T(m_1+m_2) + 2m_1m_2g = 0$

$\dfrac{2m_1m_2}{m_1+m_2}g = T$

Thrust on the pulley is 2T

$\therefore$ Thrust is equal to $\dfrac{4m_1m_2g}{m_1+m_2}$

13. (1)

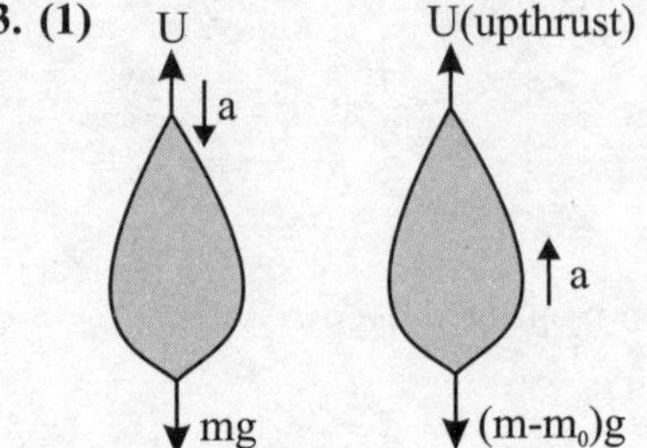

When the balloon descends down with acceleration a,

$mg - U = ma$

When the balloon moves up with acceleration a,

$U - (m - m_0)g = (m - m_0)a$

On adding the two equations,

$\Rightarrow mg - mg + m_0g = ma + ma - m_0a$

$\Rightarrow m_0 = \dfrac{2ma}{g+a}$

14. (1) Momentum carried by each bullet = Mv

$= 0.010\times500\ kg-ms^{-1} = 5\ kg-ms^{-1}$

Total momentum of 10 bullets = 5 × 10 = 50 Ns

Force due to 10 bullets = 50N

Hence, acceleration of the vehicle $= \dfrac{50}{200}ms^{-2} = 25cms^{-2}$

15. (1) $m_2g\sin60° - T = m_2a$ ...(*i*)

$T - m_1g\sin30° = m_1a$ ...(*ii*)

Adding both equations.

$\dfrac{m_2g\sin60° - m_1g\sin30°}{(m_1+m_2)} = a$

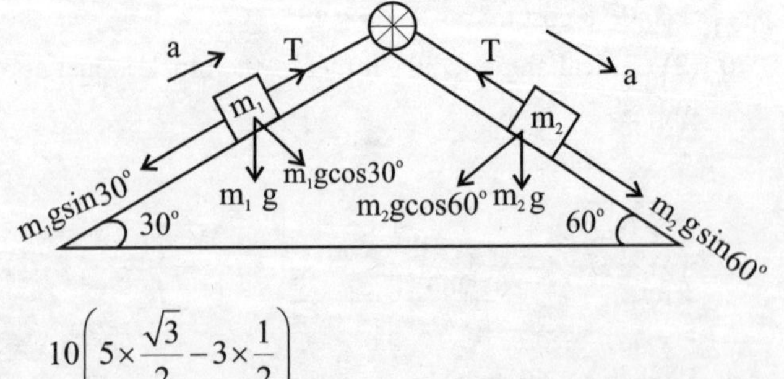

$= \dfrac{10\left(5\times\dfrac{\sqrt3}{2} - 3\times\dfrac{1}{2}\right)}{8}$

$$a = \frac{10(2.5\times1.73-1.5)}{8}$$

$a = 3.53$ m/s^2

16. (2) $\tan\theta = \mu,\ \cos\theta = \frac{1}{\sqrt{\mu^2+1}}$

$h = r - r\cos\theta = r[1-\cos\theta]$

$$= r\left[1-\frac{1}{\sqrt{\mu^2+1}}\right]$$

17. (2) The change in momentum is only due to the change in direction of the perpendicular components.

Hence, Δp = OQ sin 30° – (– OP sin 30°)

= 2 mv sin 30°

Its time rate will appear in the form of average force acting on the wall.

$\therefore$ F × t = 2 mv sin 30°

Or $F = \frac{2mv\sin30°}{t}$

Given, m = 0.5 kg, v = 12 ms^{-1}, t = 0.25 s, θ = 30°

Hence, $F = \frac{2\times0.5\times12\times\sin30°}{0.25} = 24$N

18. (4) Force on disc due to change in momentum of bullets is balanced by weight of the disc.

i.e., 2m'vn = mg

$$\therefore\ v = \frac{mg}{2m'n} = \frac{100\times980}{2\times5\times10} = 980\,\text{cms}^{-1}$$

19. (2) Impulse I = change in momentum

$\Delta p = m\Delta v$

Given, m = 0.1 kg,

$$\Delta v = v_f - v_i = 0 - \frac{4}{2} = -2\text{ms}^{-1}$$

$$I = 0.1\times(-2) = -0.2\,\text{kg-ms}^{-1}$$

20. (4) Using Newton's second law, we have

$F_{Thrust} - mg = ma$

$$V\frac{dm}{dt} - mg = ma$$

$$\therefore\ \frac{dm}{dt}\times800 = 5000(10+20)$$

$$\therefore\ \frac{dm}{dt} = \frac{5000\times30}{800} = 187.5\ \text{kgs}^{-1}$$

21. (1) Given, initial velocity = v

Final velocity = 0

Coefficient of friction = μ

For a constant acceleration,

v – u = at

Here, 0 – v = – (μg) t $\quad\left(\text{As, } a = \frac{\mu mg}{m}\right)$

Negative sign indicates retardation

$$\therefore\ t = \frac{v}{\mu g}$$

22. (1) When friction is absent, we have

$a_1 = g\sin\theta$

$S = \frac{1}{2}a_1t_1^2$ ...(*i*)

When friction is present, then

$a_2 = g\sin\theta - \mu_k g\cos\theta$

$S = \frac{1}{2}a_2t_2^2$ ...(*ii*)

From Eqs (*i*) and (*ii*) we get

$$\frac{1}{2}a_1t_1^2 = \frac{1}{2}a_2t_2^2$$

$$\Rightarrow g\sin\theta t_1^2 = [g\sin\theta - \mu g\cos\theta]\times n^2t_1^2$$

$$\Rightarrow \sin45° = (\sin45° - \mu\cos45°)\times n^2$$

$$\Rightarrow \mu = 1 - \frac{1}{n^2}$$

23. (4) $s = \frac{1}{2}at^2 \Rightarrow a = \frac{2s}{t^2} = \frac{2\times0.6}{(4)^2} = \frac{3}{40}$

Acceleration of the blocks are given by

$$a = \left(\frac{M-m}{M+m}\right)g \Rightarrow \frac{a}{g} = \frac{M-m}{M+m} \Rightarrow \frac{3}{40\times10} = \frac{M-m}{M+m}$$

3M + 3m = 400M – 400m

$$\Rightarrow \frac{m}{M} = \frac{397}{403}.$$

24. (4) Let the mass of the block B is m.

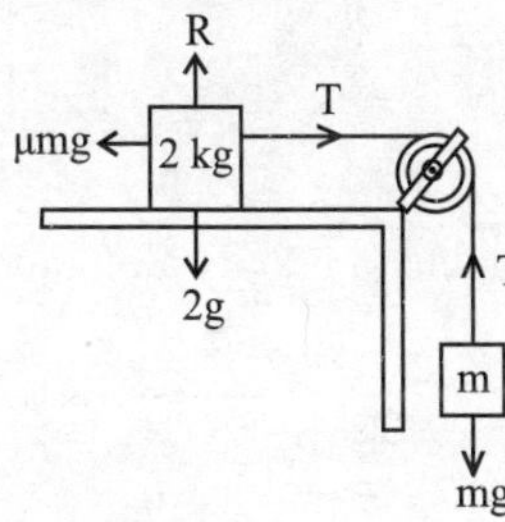

In equilibrium, T – mg = 0

$\Rightarrow$ T = mg ...(*i*)

If blocks do not move, then

$T = f_s$

for limiting case, $f_s = \mu_s R$

$\Rightarrow T = \mu_s R = \mu_s\,2g$ ...(*ii*)

Thus, from Eqs. (*i*) and (*ii*), we have

$mg = \mu_s 2g \Rightarrow m = \mu_s 2$

m = 0.2 × 2 = 0.4 kg

25. (2)

As the pulley is massless, the equation of motion is

$$F - 2T = 0 \Rightarrow T = \frac{F}{2}$$

The acceleration of the block is, $a = \frac{T-f}{m} = \frac{\frac{F}{2}-f}{m}.$

26. (1) If the same wedge is made rough, then time taken by the block to come down becomes n times. The coefficient of friction is

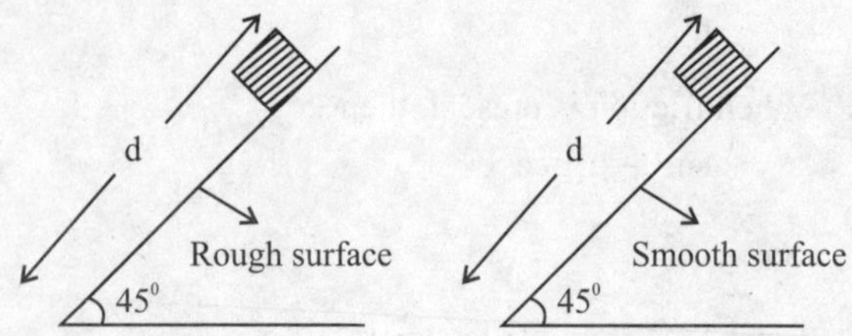

$$S=\frac{1}{2}at^2 \Rightarrow t \propto \frac{1}{\sqrt{a}}$$

$$\frac{t_2}{t_1}=\sqrt{\frac{a_1}{a_2}} \Rightarrow \frac{t}{2t}=\sqrt{\frac{g(\sin\theta-\mu\cos\theta)}{g\sin\theta}}$$

$$\Rightarrow \sin\theta = 4\sin\theta - 4\mu\cos\theta$$

$$\Rightarrow \mu=\frac{3}{4}\tan\theta=\frac{3}{4}$$

27. (1) $y=\frac{x^3}{6}$

$\therefore \tan\theta=\frac{dy}{dx}=\frac{3x^2}{6}=\frac{x^2}{2}$

In the limiting case of equilibrium,

$\mu = \tan\theta$

$0.5=\frac{x^2}{2}$

$\Rightarrow \quad x=\pm 1$

Then, $y=\frac{1}{6}$.

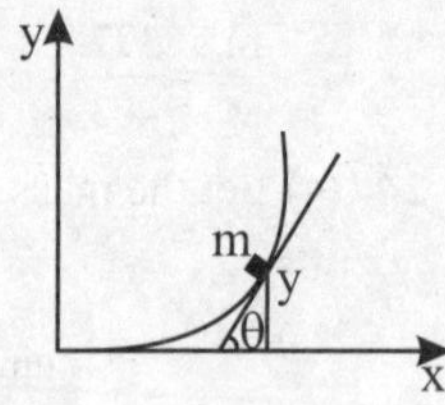

28. (3) Acceleration due to gravity along inclined plane $= g\sin\theta$

$\therefore$ Time taken, $t=\sqrt{\frac{2s}{a}}=\sqrt{\frac{2\ell}{g\sin\theta}}$

But $\sin\theta=\frac{h}{\ell}\Rightarrow \ell=\frac{h}{\sin\theta}$

Hence, $t=\sqrt{\frac{2}{g\sin\theta}\times\frac{h}{\sin\theta}}$

$$t=\frac{1}{\sin\theta}\sqrt{\frac{2h}{g}}$$

29. (3) In the frame of a pseudo force acts on the object. For the object to remain at rest relative to the incline, we have

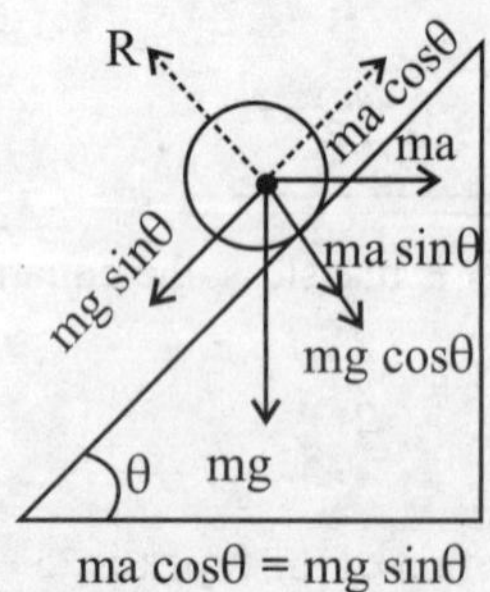

For equilibrium of body on the inclined plane,

$$ma\cos\theta = mg\sin\theta,\ \sin\theta=\frac{1}{\ell}$$

$$\Rightarrow \quad a=g\tan\theta=g\frac{1}{\sqrt{\ell^2-1}} \text{ or } a=\frac{g}{\sqrt{\ell^2-1}}$$

30. (1) The free body diagram shows the various forces acting on the block. When plane is frictionless, then acceleration, $a = g\sin\theta$.

Let time taken is t_1. Since, body starts from rest, initial velocity $u = 0$.

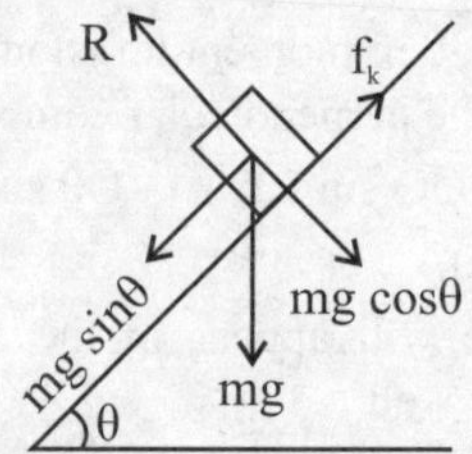

$\therefore \quad s=ut+\frac{1}{2}at_1^2 \Rightarrow s=\frac{1}{2}g\sin\theta t_1^2$

$\Rightarrow \quad t_1=\sqrt{\frac{2s}{g\sin\theta}}$

When plane is made rough, then acceleration $a = g(\sin\theta - \mu\cos\theta)$, where μ is coefficient of friction.

$\therefore \quad t_2=\sqrt{\frac{2s}{g(\sin\theta-\mu\cos\theta)}}=2\sqrt{\frac{2s}{g\sin\theta}}\ [\because t_2=2t_1]$

$\Rightarrow \frac{1}{\sin\theta-\mu\cos\theta}=\frac{4}{\sin\theta}\Rightarrow \sin\theta=4\sin\theta-4\mu\cos\theta$

$4\mu\cos\theta=3\sin\theta\Rightarrow\mu=\frac{3}{4}\tan\theta$

31. (3) For the block on incline, $mg\sin\theta - T = ma$ ...(*i*)

For the block on ground

$T = ma$ ...(*ii*)

$\Rightarrow mg\sin\theta = 2ma$

$\Rightarrow a=\frac{g\sin\theta}{2}$

$\Rightarrow T=\frac{mg\sin\theta}{2}$

32. (3) Equation of motion for the block A : $kx = ma$

$\Rightarrow a=\frac{T}{m}=\frac{kx}{m}$

For the block B : $F - kx = ma'$

$\Rightarrow a'=\frac{F-kx}{m}$

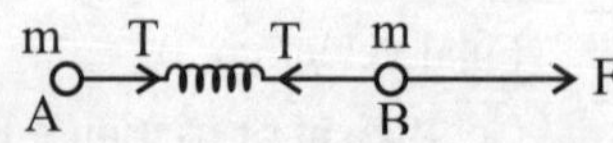

$\Rightarrow$ The relative acceleration

$a_r=|a'-a|=\frac{F-2kx}{m}$

33. (2) $Mg - T = Ma$ ...(*i*)

$2T-mg=\frac{ma}{2}$ ...(*ii*)

Adding eqn. (*i*) and eqn. (*ii*),

$$(2M-m)g = a\left(2M+\frac{m}{2}\right)$$

$$a = \frac{(2M-m)2g}{4M+m}$$

34. (1) Let x be the elongation in the spring, as the pulley is at rest, therefore, $kx = 2T$ & $T = 10g$

$$\Rightarrow x = \frac{20\times10}{1000} = 0.2 \text{ m}$$

35. (2)

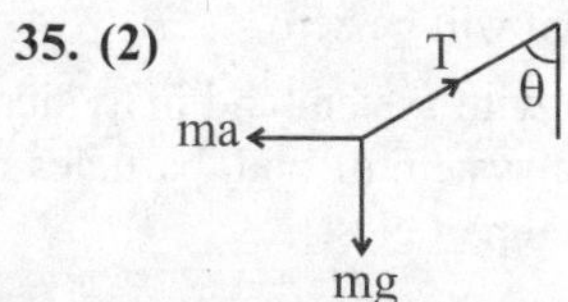

$T\cos\theta = mg$

& $T\sin\theta = ma$

$a = g\tan\theta = 10\tan 37^0$

$= 10\,(3/4) \Rightarrow 7.5 \text{ m/s}^2$

36. (1) From FBD of pulley

$F = 2T \Rightarrow T = 30$

$$a_1 = \frac{30-2g}{2}\uparrow \quad a_2 = \frac{4g-30}{4}\downarrow$$

37. (1)

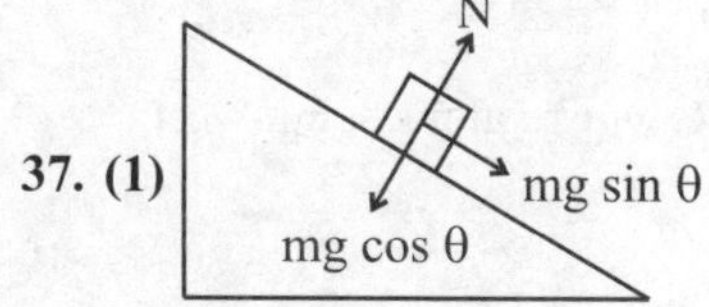

$mg\sin\theta = ma$

$\therefore a = g\sin\theta = 5 \text{ m/s}^2$

38. (1) $g\sin\theta = \mu g\cos\theta$

$\Rightarrow \mu = \tan\theta$

When block is projected up

$a = g\sin\theta + \mu g\cos\theta = 2g\sin\theta$

$$d_{stop} = \frac{v_0^2}{2a} = \frac{v_0^2}{2\times2g\sin\theta} = \frac{v_0^2}{4g\sin\theta}$$

39. (3) $F_{net} = ma = 60 \times 1 = 60.$

40. (4) Common acceleration possible for both blocks $= \frac{10}{6+4} = 1 \text{ m/s}^2$

Maximum possible acceleration of 6kg block moved by friction $a_{max} = \frac{\mu(4g)}{6} = \frac{4}{3} \text{ m/s}^2$

since $a_{max} > a_{common}$

So, both blocks will move together

41. (1) For no slipping $\tan\theta = \frac{v^2}{Rg} = \frac{h}{b}$

42. (3)

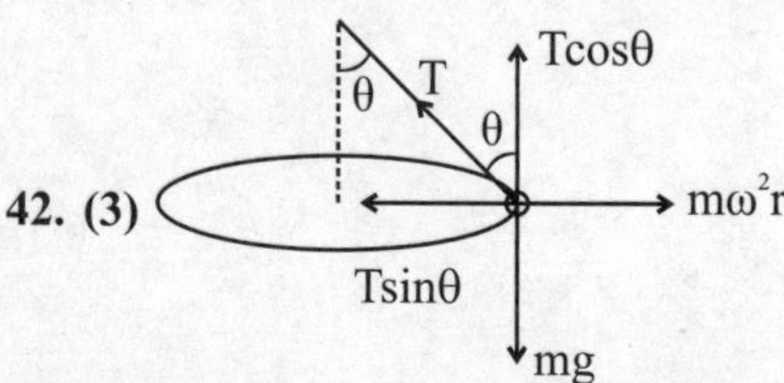

$\therefore T\sin\theta = m\omega^2 r$ ...(*i*)

$T\cos\theta = mg$ ...(*ii*)

Dividing eqn. (*i*) by (*ii*)

$$\tan\theta = \frac{\omega^2 r}{g}$$

43. (3) Third force to be applied will be equal in magnitude to the resultant forces and would point in a direction opposite to the resultant force.

44. (2) $a = \frac{m_1 - m_2}{m_1 + m_2}g$

$\because s = ut + 1/2\,at^2$ and $u = 0$

$$\Rightarrow t = \sqrt{\frac{2d}{a}} = \sqrt{\frac{2d(m_1+m_2)}{(m_1-m_2)g}}$$

45. (4) $F = \frac{nmv}{t} \Rightarrow n = \frac{F\times t}{m\times v} = \frac{144\times1}{0.04\times1200} = 3$

$\therefore n = 3$ bullets

46. (1) Impulse = Change in momentum

$F \times t = m(v-u).1$

$$F = \frac{m(u-v)}{t}$$

$$= \frac{0.150(0-20)}{0.1} = -30 \text{ N.}$$

47. (1) Let *a* be the acceleration of the lift.

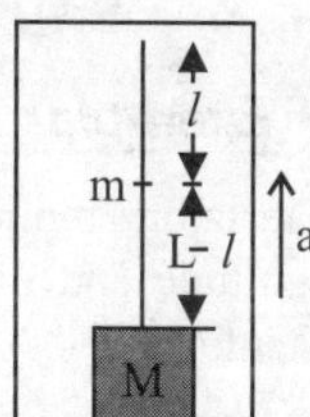

Mass of lower portion of string $= \frac{m}{L}(L-l)$

$$\therefore T - Mg - \frac{mg}{L}(L-l) = \left(M + \frac{m}{L}(L-l)\right)a$$

$$\therefore a = \frac{T}{M+m-\frac{ml}{L}} - g$$

48. (4) $a = g(\sin\theta - \mu\cos\theta)$

$9.8(\sin 45° - 0.5\cos 45°)$

$$= 9.8\left(\frac{1}{\sqrt{2}} - \frac{1}{2}\cdot\frac{1}{\sqrt{2}}\right) = \frac{9.8}{2\sqrt{2}} \text{ ms}^{-2}.$$

49. (1) Limiting frictional force $= \mu R = \mu\,(mg + Q\cos\theta)$

Horizontal push Force $= P + Q\sin\theta$

In equilibrium condition, all forces are balance equal

$\mu\,(mg + Q\cos\theta) = P + Q\sin\theta$

$$\mu = \frac{(P+Q\sin\theta)}{(mg+Q\cos\theta)}$$

Statement Based MCQs

1. **(3)** The tension is greater than weight of bird because the vertical component of the tension will be equal to the weight of the bird i.e., T cos θ = W cos θ < 1 i.e., T > W.
2. **(3)** Statement I is true and statement II is false.
3. **(3)** Statement I is true and statement II is false.
4. **(3)** Statement I is true and statement II is false.
5. **(4)** Statement I is false and statement II is true.

Assertion & Reason MCQs

1. **(1)** In rolling friction, surfaces in contact get momentarily deformed a little, and this results in a finite area of contact between the sphere and the surface and hence motion is opposed by component of contact force parallel to surface. Hence, rolling sphere on horizontal surface eventually stops.
2. **(1)** It is very difficult to produce slipping motion compared to rolling. In bicycle when brakes are on, the cycles can only do the slipping action and not rolling.
3. **(2)** By 3rd law of motion the pull of the teams is same.
4. **(1)** Rolling friction is always less than sliding friction as the surface area in contact is very less which reduces the friction. So providing ball bearing in motor's shaft minimizes friction and they increases its output power.
5. **(1)** Here the impulse is sharp so the brick is broken with bare hands.

Match the Column MCQs

1. **(4)** 2. **(3)** 3. **(1)** 4. **(1)**

NCERT Exemplar MCQs

1. **(3)** In a uniform translatory motion, all parts of the ball have the same magnitude and direction of velocity and this velocity is constant.

2. **(2)** We known that $F = \frac{dp}{dt}$

Given that meter scale is moving with uniform velocity. hence, dp = 0

Force = F = 0.

As all parts of the scale are moving with same velocity. So, the rod is not rotating. Hence, torque will also be zero.

3. **(3)** $\Delta\vec{p}$ = Change in momentum

= Final momentum – Initial momentum

$= m\vec{v} - m\vec{u}$

$= m(\vec{v}-\vec{u}) = (0.15)[-(3\hat{i}+4\hat{j})-(3\hat{i}+4\hat{j})]$

$= (0.15)[-6\hat{i}-8\hat{j}]$

$= -[0.15\times 6\hat{i}+0.15\times 8\hat{j}]$

$= -[0.9\hat{i}+1.20\hat{j}]$

$\Delta\vec{p} = -[0.9\hat{i}+1.2\hat{j}]$

4. **(3)** By previous solution $\Delta\vec{p} = -(0.9\hat{i}+1.2\hat{j})$

$\text{Magnitude} = |\Delta\vec{p}| = \sqrt{(0.9)^2+(1.2)^2}$

$= \sqrt{(0.81+1.44)} = 1.5\, kg-ms^{-1}$

5. **(4)** We know that for a system $F_{ext} = \frac{dp}{dt}$ (from Newton's second law)

If $F_{ext} = 0$, $dp = 0 \Rightarrow p = \text{constant}$

Hence, momentum of a system will remain conserved if external force on the system is zero.

In case of collision between particles equal and opposite forces will act on individual particles by Newton's third law.

Hence, total force on the system will be zero.

Note : We should not confuse with system and individual particles. As total force on the system of both particles is zero but force acts on individual particles.

6. **(3)** Let $\overrightarrow{OA} = \vec{p}_1$ = Initial momentum of player northward

$\overrightarrow{AB} = \vec{p}_2$ = Final momentum of player towards west.

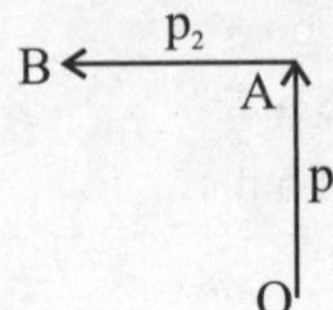

Change in momentum $= \vec{p}_2 - \vec{p}_1$

$= \overrightarrow{AB} - \overrightarrow{OA} = \overrightarrow{AB} + \left(-\overrightarrow{OA}\right)$

$\Rightarrow$ Clearly, resultant $\overrightarrow{AR}$ will be along south-west.

7. **(1)** $x(t) = pt + qt^2 + rt^3$

$v = \frac{dx}{dt} = p + 2qt + 3rt^2$

$a = \frac{dv}{dt} = 0 + 2q + 6rt$

at t = 2s; $a = 2q + 6 \times 2 \times r = 2q + 12r$

$= 2 \times 4 + 12 \times 5 = 68\ m/s^2$

Force, F = ma = 2 × 68 = 136N

8. **(2)** Given, mass = m = 5 kg

Acting force $F = (-3\hat{i}+4\hat{j})N$

Initial velocity at t = 0, $\vec{u} = (6\hat{i}-12\hat{j})m/s$

$\vec{a} = \frac{F}{m} = \left(-\frac{3\hat{i}}{5}+\frac{4\hat{i}}{5}\right)m/s^2$

As final velocity is along Y-axis only, its x-component must be zero.

From v = u + at, for X-component only, $0 = 6\hat{i} - \frac{3\hat{i}}{5}t$

$t = \frac{5\times 6}{3} = 10\ s$

9. **(2)** Given, mass of the car = m

As car starts from rest, u = 0

Velocity acquired along east $= v\hat{i}$

Duration, t = 2s.

We know that

v = u + at

$\Rightarrow \quad v\hat{i} = 0 + a\times 2 \Rightarrow \quad a = \frac{v}{2}\hat{i}$

Force, $F = ma = \frac{mv}{2}\hat{i}$

Hence, force acting on the car is $\frac{mv}{2}$ towards east. As external force on the system is only friction. Hence, the force on m is by friction. Force by engine is internal force.

10. (1,2,4) Given,

x = 0 for t < 0 s.

$x = 0;$ for $t > \frac{1}{4}s$

for $0 < t < \frac{1}{4}s$, X = A sin4πt

For, $0 < t < \frac{1}{4}s \quad v(t) = \frac{dx}{dt} = 4\pi A\cos 4\pi t$

$a(t) = \frac{dv(t)}{dt} = -16\pi^2 A\sin 4\pi t$

At $t = \frac{1}{8}s, a(t) = -16\pi^2\ A\sin 4\pi \times \frac{1}{8} = -16\pi^2 A$

$F = ma(t) = -16\pi^2 A \times m = -16\pi^2 mA$

Impulse = Change in linear momentum

$= F \times t = (-16\pi^2 Am) \times \frac{1}{4} = -4\pi^2 Am$

The impulse (change in linear momentum) At t = 0 is same as, $t = \frac{1}{4}s$.

Clearly, force depends upon t. Hence, force is not constant.

11. (1, 2, 4) Consider the diagram. Frictional force on B(f_1) and frictional force on A (f_2) will be as shown.

Let A and B are moving together

$$a_{common} = \frac{F - f_1}{m_A + m_B} = \frac{F - 0.1 \times 3mg}{(m/2) + m} = \frac{20F - 3mg}{30m}$$

For A, $f_2 = \frac{m}{2}a_{common}$

The force (F) will be maximum when,

$f_2 = \mu\frac{m}{2}g$

$$\frac{\mu mg}{2} = \frac{20F - 3mg}{30m} \times \frac{m}{2}$$

$$\Rightarrow \frac{0.2mg}{2} = \frac{20F - 3mg}{60}$$

$\Rightarrow F = 0.45\,mg$

⇒ Hence, maximum force upto which bodies will move together is $F_{max} = 0.45$ mg

1. Hence, for F = 0.25 mg < F_{max}, bodies will move together.

2. For F = 0.5 mg > F_{max}, body A will slip with respect to B.

3. $(f_1)_{max} = \mu(m_A + m_B)g = (0.1) \times \frac{3}{2}m \times g = 0.15$ mg

$(f_2)_{max} = \mu m_A g = (0.2) \times \left(\frac{m}{2}\right)(g) = 0.1$ mg

Hence, minimum force required for movement of the system (A + B). F must overcome (f_1)max. Thus,

$F > (f_1)_{max}$

⇒ F > 0.15 mg

4. Given, force F = 0.1 mg < F_{min}

Hence, the bodies will be at rest.

12. (1, 4) Let m_1 moves up the plane.

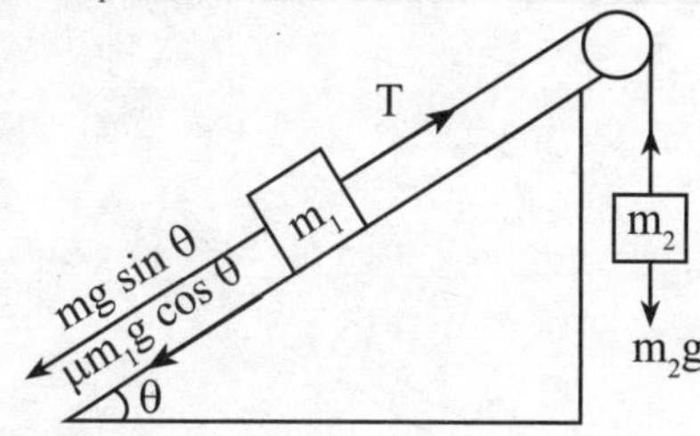

$m_2g - (m_1g \sin\theta + f) > 0$

$\Rightarrow m_2g - (m_1g \sin\theta + \mu\, m_1g \cos\theta) > 0$

$\Rightarrow m_2 > m_1 (\sin\theta + \mu \cos\theta)$

Hence, option (1) is correct.

Let m_1 moves down the plane, in this case f acts up the plane.

Hence, $m_1 g\sin\theta - f > m_2 g$

$\Rightarrow \quad m_1 g\sin\theta - \mu m_1 g\cos\theta > m_2 g$

$\Rightarrow \quad m_1(\sin\theta - \mu\cos\theta) > m_2$

$\Rightarrow \quad m_2 < m_1(\sin\theta - \mu\cos\theta)$

Hence, option (4) is correct.

13. (2, 3) Let A moves up the plane frictional force on A will be downward as shown.

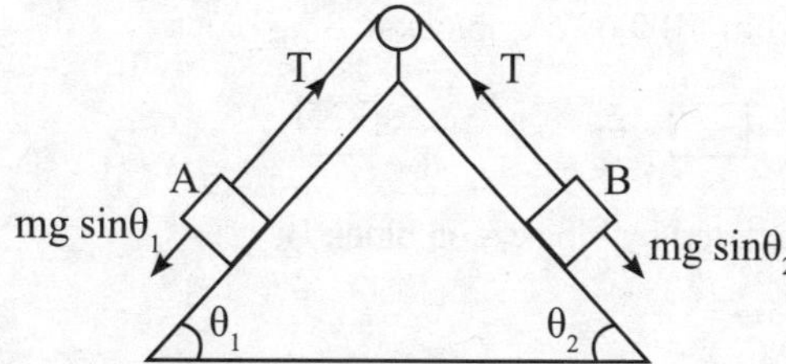

When A just starts moving up

$mg\sin\theta_1 + f = mg\sin\theta_2$

$\Rightarrow \quad mg\sin\theta_1 + \mu mg\cos\theta_1 = mg\sin\theta_2$

$$\Rightarrow \quad \mu = \frac{\sin\theta_2 - \sin\theta_1}{\cos\theta_1}$$

When A moves upwards

$f = mg\sin\theta_2 - mg\sin\theta_1 > 0$

$\sin\theta_2 > \sin\theta_1 \Rightarrow \theta_2 > \theta_1$

14. (3, 4) Given,

$m_1 = m_2 = 50(g) = \frac{50}{1000}kg = \frac{1}{20}kg$

Initial velocity $(u) = u_1 = u_2 = 5m/s$

Final velocity $(v) = v_1 = v_2 = -5 m/s$

Time duration of collision = 10^{-3} s.

Change in linear momentum = m (v – u)

$= \frac{1}{20}[-5-5] = -0.5$ N-s.

$$\text{Force} = \frac{\text{Impulse}}{\text{Time}} = \frac{\text{Change in momentum}}{10^{-3}\text{s}}$$

$$= \frac{0.5}{10^{-13}} = 500\text{N}$$

Impulse and force are opposite in directions.

NEET & AIIMS Past Year MCQs

1. (4) Mass m = kg,
Inclination θ = 30°,
Tension T = 30 N,
g = 10 m/s².

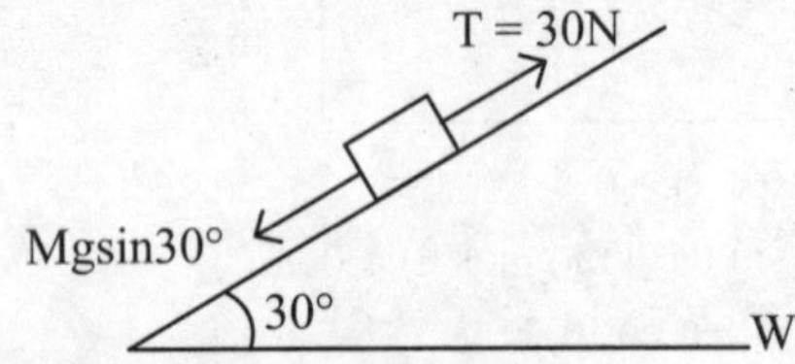

Net force along the incline:

$T - \text{mg}\sin\theta = \text{ma}$

Substitute:

$30 - 5 \times 10 \times \sin 30° = 5a$

$30 - 25 = 5a \Rightarrow a = 1$ m/s²

2. (1) In connected motion,

Common acceleration $a = \frac{F_{net}}{M_{total}}$

$a = \frac{10}{5} = 2$ m/s²

Now, from FBD of 2 kg block & 3 kg block

$\xrightarrow{a}$ [2] $\leftarrow F_{AB}$ with $F = 10$ N; $\xrightarrow{a}$ [3] with F_{BA}

Force exerted by block A on block B,

$F_{BA} = (3)a$

$F_{BA} = 3 \times 2 = 6$ N

3. (1) In horizontal circular motion, the tension (T) in the string provides the centripetal force required to keep the bob moving in a circle. I.e.,

$T = m\omega^2 r$

For constant m and r, the tension is proportional to the square of the angular velocity:

$T \propto \omega^2$

Initial State:

Tension T corresponds to ω = 10 rpm.

After Quadrupling Tension (T' = 4T):

Since $\omega \propto \sqrt{T}$, the new angular velocity is:

$\omega' = \sqrt{4} \times \omega = 2 \times 10 = 20\,rpm$

4. (2) A particle moving with uniform speed in a circular path has varying velocity as the direction of motion changes continuously. The magnitude of velocity remains constant. As the direction of centripetal acceleration is changing, the particle has varying acceleration.

5. (4) $T = M\omega^2 r$
$T \propto \omega^2$
T becomes 4 times

6. (4) N = mg and f = ma
$f \le \mu N \Rightarrow a \le \mu g$
$\Rightarrow a \le 1.5$ ms^{-2}
or $a_{max} = 1.5$ ms^{-2}

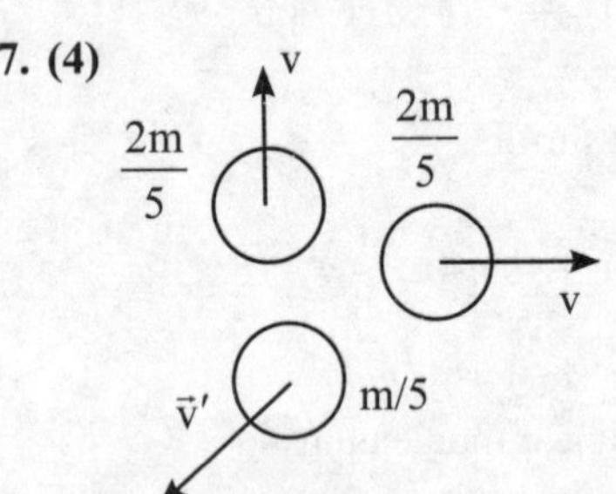

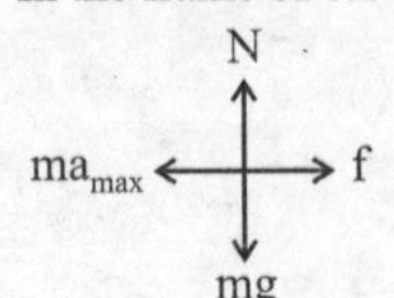

7. (4)

As the shell was initially at rest, therefore initial momentum is zero.

Using conservation of momentum:

Initial momentum = Final momentum

$$0 = \frac{2m}{5}(v\hat{i}) + \frac{2m}{5}(v\hat{i}) + \frac{m}{5}\vec{v}'$$

$$\Rightarrow \vec{v}' = -2v\hat{i} - 2v\hat{j}$$

$$\Rightarrow \vec{v}' = \sqrt{(2v)^2 + (2v)^2} = 2\sqrt{2}\,v$$

8. (1) $$\frac{S_n}{S_{n+1}} = \frac{\frac{a}{2}(2n-1)}{\frac{a}{2}(2(n+1)-1)} = \frac{2n-1}{2n+2-1} = \frac{2n-1}{2n+1}$$

9. (4) N – mg = ma
N - (50 × 9.8) = (50 × 0.49)
N = 514.5 Newton

10. (4)

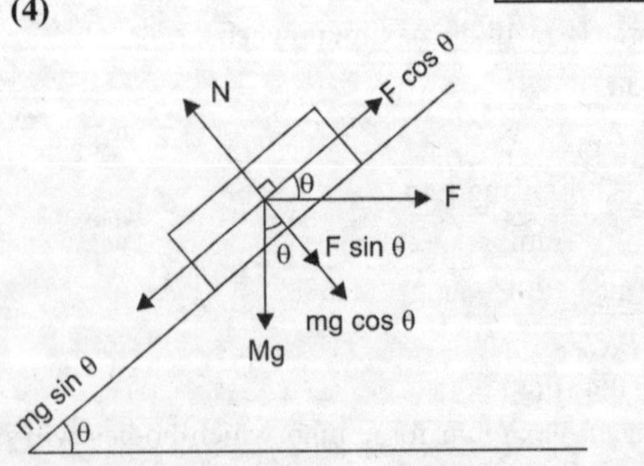

N = mg cos θ + F sin θ

11. (4)

12. (1)

Work, Energy and Power

NCERT Topic-wise MCQs

1. (3) Since, displacement of wall is zero.

Therefore, $W = \vec{F}\cdot\vec{S} = 0$.

2. (4) $F = ma \Rightarrow a = \frac{10}{10} = 1 m/s^2$

$$s_5 = ut + \frac{1}{2}at^2 \Rightarrow s_5 = 0 + \frac{1}{2}\times 1\times(5)^2 = 12.5m$$

$W_5 = 10 \times 12.5 = 125$ J

And

$$s_{5th} = u + \frac{a}{2}(2n-1) \Rightarrow s_{5th} = 0 + \frac{1}{2}(2\times 5 - 1) = 4.5\,m$$

W_{5th} = Force × displacement = 10 × 4.5 = 45 J

3. (2) $T = m(g - a)$ (downward)

$$T = m\left(g - \frac{g}{4}\right) = \frac{3mg}{4}$$

∴ work done by the cord

$$= \vec{F}.\vec{s}\cos\theta = \frac{3mg}{4}\times d\cos 180°$$

$$W = \frac{-3mgd}{4}J$$

4. (3) mg sinθ, mg cosθ, 30°, mg

$W = mg\sin\theta\times s = mg \sin 30° \times s$

$$W = 200\times\frac{1}{2}\times 10 = 10^3 J = 1\,kJ$$

5. (3) $W = \vec{F}.\vec{S}$

$$= (2\hat{i}+3\hat{j}+4\hat{k})(3\hat{i}+2\hat{j}+5\hat{k})$$

$= 6 + 6 + 20 = 32$ J

6. (2) $W = \vec{F}.\vec{S} = (\hat{i}+2\hat{j}+\hat{k})(4\hat{i}+\hat{j}+7\hat{k}) = 13J$

7. (2) $W = F\times\frac{1}{2}at^2\left[\text{from } s = ut + \frac{1}{2}at^2\right]$

$$\Rightarrow W = F\left[\frac{1}{2}\left(\frac{F}{M}\right)t^2\right] = \frac{F^2t^2}{2m} = \frac{25\times(1)^2}{2\times 15} = \frac{5}{6}J$$

8. (1) Work done = area under curve and displacement axis

= 1 × 10 – 1 × 10 + 1 × 10 = 10 J

9. (2) Gravitational force is a conservattive force and work done against does not depend on the path.

10. (3) $W = (mg \sin \theta)s$

11. (3) $W = \frac{MgL}{2}$ $h = \frac{L}{2}$, as the centre of mass is at a height of L/2 from the floor.

12. (4) $W = \int_{x_1}^{x_2} F\cdot dx = \int_0^5 Kxdx = \left[\frac{Kx^2}{2}\right]_0^5 = 12.5\,K$ joule

= 0.0125 K kJ

13. (3) $W = Fs\cos\theta = 60\times 100\times\cos 60° = \frac{6000}{2} = 3000J$

Wages for 15 J work = 10 paise = 0.1 rupee

Wages for 3000 J of work is $\frac{0.1\times 3000}{15}$ = Rs. 20

14. (1) $W = W_{12} + W_{23} + W_{34} + W_{45}$ (W = area under graph)

$$= 10\times 1 + \left[10\times 1 + \frac{5\times 1}{2}\right] + \frac{15\times 1}{2} - \frac{15\times 1}{2}$$

= 10 + 10 + 2.5 + 7.5 – 7.5 = 22.5 J

15. (4) W = Force × displacement

W = mg × d = 0.5 × 10 × 0.5 = 2.5 J

16. (1) $W = \int Fdx$

$$W = \int_0^2 (x + x^3)dx$$

$$W = \left[\frac{x^2}{2} + \frac{x^4}{4}\right]_0^2 \Rightarrow W = \left[\frac{4}{2} + \frac{16}{4}\right] - [0+0] = 6\,J$$

17. (2) $W = Mgh = \frac{1}{10}\times 9.8\times 100\times\frac{1}{1000} = 0.098$ J

18. (3) $W = \vec{F}.\vec{S}$, $\vec{F} = (6\hat{i} - 8\hat{j})$ and $\vec{S} = (4\hat{i} + 6\hat{j})$

19. (3) $W = FS\cos\theta$ where θ = 60° with the vertical and 30° with the horizontal

20. (2) W = Area of $F - S$ graph $= \frac{1}{2}\times b\times h = \frac{1}{2}\times 12\times 60 = 360\,J$

21. (4) Work done in stretching a string to obtain an extension ℓ is

$$W_1 = \frac{1}{2}K\ell^2$$

Similarly, work done in stretching a string to obtain extension ℓ_1 is

$$W_2 = \frac{1}{2}K\ell_1^2$$

∴ Work done in second stretching will be.

$$W = W_2 - W_1 = \frac{1}{2}K(\ell_1^2 - \ell^2)$$

22. (1) Friction always oppose motion. A body does work against friction at the expense of its kinetic energy.

23. (3) When a body is falling freely under the action of gravity alone in vacuum, its total mechanical energy (KE + PE) remains constant during the fall.

24. (3) As both balls are galling through same height therefore they possess same velocity.

but KE ∝ m (If v = constant)

$$\therefore \frac{(KE)_1}{(KE)_2} = \frac{m_1}{m_2} = \frac{2}{4} = \frac{1}{2}$$

25. (2) m_B v_B 950 g

Initial K.E. of system $= \frac{1}{2}m_Bv_B^2$.

By the law of conservation of linear momentum

$m_B v_B + 0 = m_{sys.} \times v_{sys.}$

$\Rightarrow v_{sys.} = \frac{50 \times 10}{50 + 950} = 0.5\,m/s$

Fractional loss in K.E.$= \frac{\frac{1}{2}m_B v_B{}^2 - \frac{1}{2}m_{sys.} v^2{}_{sys.}}{\frac{1}{2}m_B v_B{}^2}$

By substituting mB = 50 × 10^{-3} kg, v_B = 10 m/s

$m_{sys.}$ = 1 kg, v_s = 0.5 m/s we get

Fractional loss = $\frac{95}{100} = 95\%$

26. **(1)** Work = Force × Displacement (length)

If unit of force and length be increased by four times then the unit of energy will increase by 4 × 4 times.

27. **(1)** Initial energy of body K.E = $\frac{1}{2}mv^2 = \frac{1}{2} \times 4 \times 40 \times 40$

= 3200 J

At highest point, energy possessed by the body in the form of potential energy = mgh = 4 × 10 × 30 = 1200 J

Energy lost due to air friction is

$\Delta E = E_i - E_f \Rightarrow \Delta E = 3200 - 1200 \Rightarrow \Delta E = 2000J$

28. **(4)** Kinetic energy energy given by K = 1/2(mv^2)

since there is no force along the horizontal velocity along it remains same velocity along horizontal = v sin 30° = v/2

so KE at topmost point = K/4.

29. **(4)** Given $KE = \frac{1}{2}mv^2 = \frac{1}{2}m(gt)^2 \quad [\because v = gt]$

$\frac{K_1}{K_2} = \frac{t_1^2}{t_2^2 - t_1^2}$ Here $t_1 = 5$ sec; $t_2 = 10$ sec

$= \frac{(5)^2}{(10^2) - (5)^2} = \frac{25}{100 - 25} = \frac{25}{75}$

$K_1 : K_2 = 1:3$

30. **(1)** ΔK.E = ΔP.E. between A & B

= mgL(1 – cos 60°)

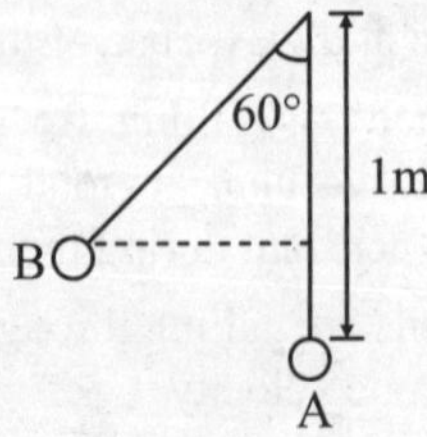

$= 10^{-1} \times 9.8 \times \left(1 - \frac{1}{2}\right) = 0.49\,J$

31. **(3)** As the body start from rest

$\therefore u = 0$

$\therefore v = 0 + at$ and $KE = \frac{1}{2}mv^2$

$\therefore K.E \propto t^2$

32. **(2)** $U = \frac{1}{2}kx^2 \Rightarrow U \propto x^2 \Rightarrow \frac{U_1}{U_2} = \left(\frac{2}{6}\right)^2$

$\therefore U_2 = U_1 \times \frac{36}{4} = 4 \times 9 = 36\,J$

33. **(1)** Total kinetic energy converted into potential energy.

$\frac{1}{2}mv^2 = \frac{1}{2}kx^2 \quad \Rightarrow \quad \sqrt{\frac{mv^2}{k}} = x$

$\sqrt{\frac{2 \times 18 \times 18}{1000}} = x \Rightarrow x = \sqrt{\frac{648}{1000}} = 0.80$ m

34. **(2)** x is the fraction of length hanging from the table.

$x = \frac{hanging\ length}{total\ length} = \frac{70}{300} = 0.23$

work done in pulling the chain

$w = \frac{MgLx^2}{2} = \frac{4 \times 10 \times 3 \times (0.23)^2}{2} = 3.174\,J$

35. **(2)** According to question,

$K \cdot E_A = K \cdot E_B$

$\frac{1}{2}m_A v_A{}^2 = \frac{1}{2}m_B v_B{}^2 \quad \Rightarrow \quad \frac{v_A}{v_B} = \sqrt{\frac{m_B}{m_A}} = \sqrt{\frac{5}{20}} = \frac{1}{2}$

$\therefore \frac{F\Delta t_A}{F\Delta t_B} = \frac{m_A \Delta v_A}{m_B \Delta v_B} \Rightarrow \frac{\Delta t_A}{\Delta t_B} = \frac{20}{5} \times \frac{1}{2} = 2$

36. **(2)** Using conservation of energy, kinetic energy at the time of throw = Potential energy at the maximum height.

$\frac{(KE)_A}{(KE)_B} = \frac{m_A g h_A}{m_B g h_B} = \frac{m}{2m} = \frac{1}{2}$

37. **(3)** $|\vec{v}| = \sqrt{v_1^2 + v_2^2}$

$|\vec{v}| = \sqrt{2^2 + 6^2} = \sqrt{40}$ m/s

$KE = \frac{1}{2}mv^2 = \frac{1}{2} \times 0.15 \times (\sqrt{40})^2 = 3J$

38. **(4)** Work done = ΔKE

Work done = $(KE)_f - (KE)_i$

Work done $= 0 - \frac{1}{2} \times 2 \times 5^2$

Work done = –25 J

Negative sign shows that the direction of force is opposite to displacement.

39. **(1)** Change in K.E. = work done

⇒ K.E. = F.x ⇒ 0 – 6 = 10 × s × (–1) ⇒ s = 0.6 m = $\frac{3}{5}$ m

40. **(4)** By work energy theorem,

$W_{Trunk} = \Delta K$

$\Rightarrow -F \times 5 = \frac{-K}{4} \Rightarrow F = \frac{K}{20}$

Now, again

$W_{Trunk} = \Delta K$

$\Rightarrow -F \times S = \frac{-3K}{4} \Rightarrow S = \frac{3K}{4F} = \frac{3}{4} \times 20 = 15$ cm

41. (2) As we know that $E = \frac{P^2}{2m} \Rightarrow E \propto \frac{1}{m}$

$\therefore \frac{E_1}{E_2} = \frac{m_2}{m_1} = \frac{6m}{3m} = \frac{2}{1}$

42. (1) $P = \sqrt{2mK.E}$

$\therefore \frac{P_1}{P_2} = \sqrt{\frac{2m_1 K.E_1}{2m_2 K.E_2}} = \sqrt{\frac{2\times3m\times27}{2\times m\times1}} = \sqrt{\frac{81}{1}} = \frac{9}{1}$

43. (4) If there is no air drag $H = \frac{u^2}{2g}$

$\therefore H = \frac{20\times20}{2\times10} = 20\,m$

Due to air drag acting on the ball, height achieved by ball is 10 m .

Loss in Energy $\Delta U = mg(h_2 - h_1)$

$= 2\times10(20-10) = 200\ J$

44. (1) Volume of tank = 10 × 10 × 10 ⇒ 1000 kg/m^3

Mass = Density × Volume = 10^3 m^3 × 1000 kg/m^3 = 10^6 kg

now, P.E. = mgh

[here height = 5m from the mid point of tank]

PE = 10^6 × 10 × 5 = 5 × 10^7 joule

45. (1) $\frac{1}{2}mv^2 = mg(h_1 - h_2)$

$v^2 = 2g(6.3 - 3.8)$

$v = \sqrt{2\times9.8\times2.5} = \sqrt{49} = 7$ m/s

46. (3) Energy of the spring, U $= \frac{1}{2}Kx^2$, where U = 2J, x = 0.04 m

and $F = Kx$

47. (2) $E = \frac{1}{2}Kx^2$ and force constant = F/x

48. (4) $E = \frac{P^2}{2m} = \frac{(Ft)^2}{2m} \quad \left\{F = \frac{P}{t}\right\}$

$E = \frac{F^2t^2}{2m}$

49. (2) $v^2 - u^2 = 2as$

$\Rightarrow \left(\frac{2u}{3}\right)^2 - u^2 = 2as_1$

$\Rightarrow \frac{4}{9}u^2 - u^2 = 2as_1$

$\Rightarrow -\frac{5u^2}{18s_1} = a$

Now: $v^2 - u^2 = 2as$

$0 - \left(\frac{2u}{3}\right)^2 = 2\times\left(\frac{-5u^2}{18s_1}\right)\times s_2$

$\Rightarrow -\frac{4u^2}{9} = \frac{-2\times5u^2}{18s_1}\times s_2 \Rightarrow \frac{4\times s_1}{5} = s_2$

$\Rightarrow s_2 = 3.2\,cm$

50. (3) By work-energy theorem,

$-F(3x) = 0 - \frac{mv^2}{2}$

$\Rightarrow Fx = \frac{mv^2}{6}$

Now, again by work-energy theorem,

$-F(nn) = 0 - \frac{m}{2}(3v)^2 \Rightarrow n = 27$

51. (3) Using conservation of momentum, 1 × 4 + 0 = 0 + 2 × v ⇒ v = 2

$\Delta KE = \frac{1}{2}m_1u_1^2 - \frac{1}{2}m_2v_2^2$

$= \frac{1}{2}\times1\times(4)^2 - \frac{1}{2}\times2\times(2)^2 = 4\,J$

52. (1) Decrease in P.E = Gain in K.E energies.

$mg(h_1 - h_2) = k_f - k_i \quad (k_i = 0)$

$\Rightarrow \frac{m_1}{m_2} = \frac{k_{f_1}}{k_{f_2}} \Rightarrow \frac{1}{3} = \frac{k_{f_1}}{k_{f_2}}$

53. (4) Work energy theorem is valid in the presence of all types of forces.

54. (3) Initial K.E of the bullet $= \frac{1}{2}mu^2 = \frac{1}{2}\times0.1\times(250)^2$

$= \frac{1}{2}\times0.1\times250\times250 = 3125$ J

Final K.E of bullet $\frac{20}{100}\times3125 = 625$ (20% given)

$k_f = 625$ J

$\frac{1}{2}mv_f^2 = 625 \Rightarrow v_f = \sqrt{\frac{625\times2}{0.1}} = 111.8\ m/s$

$\Rightarrow v_f \approx 112$ m/s

55. (1) K.E acquired by the body after falling is equal to the gain in P.E.

$\frac{1}{2}mv^2 = mgh \Rightarrow \frac{P^2}{2m} = mgh$

$\Rightarrow \frac{P}{\sqrt{2gh}} = m$

56. (1) Potential energy increases and kinetic energy decreases with height.

57. (2) By work energy theorem

$W_g = \Delta K$

$W_g = 0 - \frac{1}{2}mv^2 = \frac{-1}{2}\times0.1\times5^2 = -1.25\,J$

58. **(2)** Work done in stretching the spring initially by 5cm,

$$W_1 = \frac{1}{2}k \times x_1^2 = \frac{1}{2} \times 5 \times 10^3 \times (5 \times 10^{-2})^2 = 6.25 J$$

Now, work done in stretching the spring by 10 cm i.e., 5 cm + 5 cm

$$W_2 = \frac{1}{2}k(x_1 + x_2)^2$$

$$W = \frac{1}{2} \times 5 \times 10^3 (5 \times 10^{-2} + 5 \times 10^{-2})^2 = 25 J$$

Net work done = $W_2 - W_1 = 25 - 6.25 = 18.75$ N m

59. **(2)** Tension $T = mg + \frac{mv^2}{r} = mg + \frac{m}{l}[2gl(1-\cos\theta)]$

$= mg + 2mg(1 - \cos\theta) = mg + \frac{2mg}{2} = 2mg$

$= 2(50 \times 10^{-3})(10) = 1N$

60. **(3)** At maximum height, kinetic energy converts into potential energy, i.e.,

$U_{max} = 490$ J

suppose at height h′, potential energy becomes half.

$$\Rightarrow mgh' = \frac{U_{max}}{2}$$

$$h' = \frac{490}{2 \times 2 \times 9.8} = 12.5 m$$

61. **(2)** Down force on elevator is

F = mg + f

= (1800 kg × 10 ms^{-2}) + 4000 N = 22000 N

The motor must supply enough power to balance this force. hence,

p = Fv = (22000 N)(2 ms^{-1})

= 44000 W = 44 kW

62. **(3)** Work done W = potential energy of water = mgh

Therefore, W = 1000 × 9.8 × 10

Power = $\frac{W}{t}$ (1000 × 9.8 × 10)/980 = 100 W = 0.1 kW

63. **(1)** Power, p = $\vec{F} \cdot \vec{v} = (4\hat{i} + \hat{j} - 2\hat{k}) \cdot (2\hat{i} + 2\hat{j} + 3\hat{k})$

= (8 + 2 – 6)W = 4W

64. **(1)** Volume of blood discharge/ beat = 80 ml

= 80 × 10^{-6} m^{-3}

Pressure = P = 0.1 m of Hg

$P = 0.1 \times 13.6 \times 10^3 \times 9.8 \quad (P = h\rho g)$

$$\frac{\text{work done}}{\text{beat}} = PV$$

Total work done in 75 beat = 75 × PV

$$Power = \frac{work}{time}$$

$$P = \frac{0.1 \times 13.6 \times 10^3 \times 9.8 \times 80 \times 10^{-6} \times 75}{60}$$

$$P = \frac{79968 \times 10^{-3}}{60} = 1.33 \text{ watt}$$

65. **(4)** $P = \vec{F}.\vec{v} = ma \times at = ma^2t$

$$= m\left(\frac{v_1}{t_1}\right)^2 t = \frac{m{v_1}^2 t}{t_1^2} \qquad \left(a = \frac{v}{t}\right)$$

66. **(1)** work done in raising water = mgh.

∴ W = (volume × density) × gh (ρ = 1000 kg/m³)

W = 10 × 1000 × 10 × 20 = 2 × 10^6 J

$$\text{Power} = \frac{\text{work}}{\text{time}} = \frac{2 \times 10^6}{6 \times 60} = 0.0055 \times 10^6 = 5.5 \text{ kW}.$$

$$\therefore \text{ Efficiency} = \frac{5.5}{20} \times 100 = 27.5\%$$

67. **(3)** $P = \frac{W}{t} = \frac{mgh}{t} \Rightarrow t = \frac{mgh}{P} = \frac{200 \times 10 \times 50}{20 \times 1000} = 5 \sec$

68. **(3)** Power = $\vec{F} \cdot \vec{v} = (10\hat{i} + 10\hat{j} + 20\hat{k}) \cdot (5\hat{i} - 3\hat{j} + 6\hat{k})$

= 50 – 30 + 120 = 140 W

69. **(3)** Power = $\vec{F} \cdot \vec{v}$

$P = (4\vec{i} - 2\vec{j} + \vec{k}) \cdot (5\vec{i} - 10\vec{j} + 6\vec{k}) = 20 + 20 + 6 = 46$ W

70. **(1)** $W = Pt$ where t = 365 days × 24 hours

71. **(1)** ∵ Power is given by

∵ P = Fv where F is force and v is velocity

P = mg sin θ × v and $\sin\theta = \frac{2}{25}$

72. **(4)** ∵ $P = \frac{W}{t}$

$$P = \frac{(m+M)gh}{t} \Rightarrow M = \frac{Pt}{gh} - m = 50 \, kg$$

73. **(2)** $P = \frac{W}{t} = \frac{\Delta K}{t} = \frac{1}{2} \times \frac{m(v^2 - u^2)}{t}$

$$\Rightarrow P = \frac{2.05 \times 10^6((35)^2 - (5)^2)}{2 \times 15 \times 60} = 1.36 \text{MW}$$

74. **(2)** $P = \frac{W}{t} = \frac{F \cdot S}{t}$

$$= \frac{(3\hat{i} + 4\hat{j} + 5\hat{k}).(2\hat{i} + 3\hat{j} + 4\hat{k})}{5} = \frac{6 + 12 + 20}{5}$$

P = 7.6 Watt

75. **(3)** Output power $= \frac{mgh}{t} = \frac{200 \times 9.8 \times 20}{10} = 3920$ watt

$$\therefore \eta = \frac{P_o}{P_i} \Rightarrow P_i = \frac{P_o}{\eta} = \frac{3920 \times 100}{80} = 4900 \text{ watt}$$

76. **(1)** Power generated = Energy/sec

According to question,

$$P = \frac{mgh}{2} \times \frac{1}{t} = \frac{1}{2} \times \frac{9 \times 10^5 \times 10 \times 60}{60 \times 60}$$

∴ $P = 7.5 \times 10^4 = 75000$ Watt

then number of 100 Watt bulb can be light is $\frac{75000}{100}$

Bulb = 750

77. (4) $P = \frac{mgh}{t} = \frac{(\rho \times V)gh}{t} \Rightarrow \frac{P \times t}{\rho \times gh} = V$

$$V = \frac{5 \times 10^3 \times 5 \times 60}{10^3 \times 10 \times 50} \Rightarrow V = 3\text{m}^3 = 3000\,\text{L}$$

78. (3) $\text{mgh} = \frac{1}{2}\text{mv}^2$ [at the lowest point $v = \sqrt{5gR}$]

$\Rightarrow h = \frac{5R}{2}$

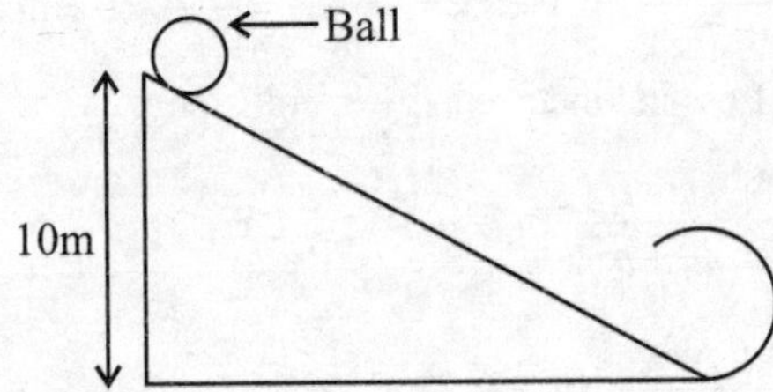

$\therefore R = \frac{2}{5}h$

R = 4m

79. (3) At the lowest position $v = \sqrt{5gR}$

Centripetal acceleration $a_c = \frac{v^2}{R} = \frac{\left(\sqrt{5gR}\right)^2}{R} = 5g$

80. (4) $F_c = T - mg$

$T = F_c + mg \Rightarrow T = mw^2r + mg$

$\therefore$ Tension in the string, T is

$T = 0.5 \times (4)^2 \times 1 + 0.5 \times 10$

= 13N

81. (2) $P = \vec{F}.\vec{V} = FV\cos 0° = 4500 \times 2 = 9$ kW

82. (3) Useful work done/sec.

$$P_{out} = \left(\frac{\text{mass}}{\text{sec}}\right) \times gh = \frac{0.5 \times 1000 \times 9.8 \times 30}{60}$$

Power of the engine $\frac{P_{out}}{\eta}$

$$= \frac{0.5 \times 1000 \times 9.8 \times 30}{60} \times \frac{100}{70} = 3500 \text{ watt.}$$

83. (3) $T_1 = \frac{mv^2}{r} - mg$

$T_2 = \frac{mv^2}{r} + mg$

Then $T_2 - T_1 = 2mg$

84. (4) Tension at lowest point is given by

$$T = \frac{mv^2}{r} + mg = mr\omega^2 + mg$$

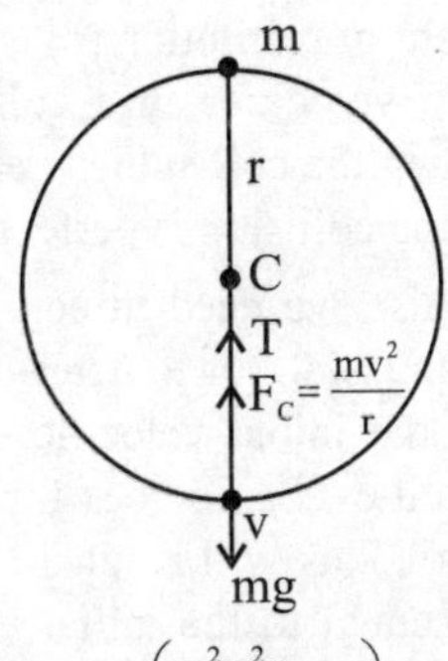

$$= mr\left(\frac{2\pi n}{60}\right)^2 + mg = m\left(\frac{\pi^2 n^2 r}{900} + g\right)$$

85. (2) The body describing a vertical circle,

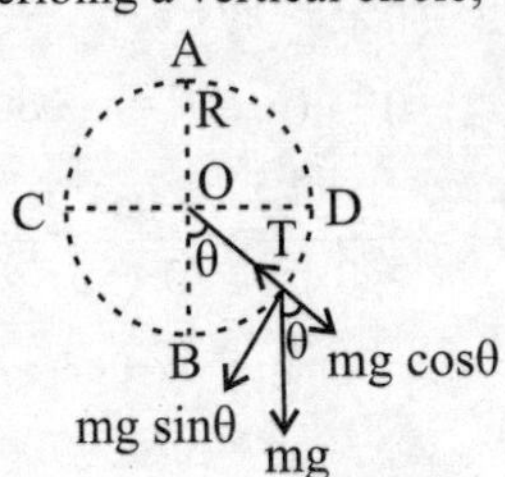

$$\therefore T - mg\cos\theta = \frac{mv^2}{R} \Rightarrow T = mg\cos\theta + \frac{mv^2}{R}$$

Tension is maximum when cos θ = 1 and velocity is maximum. Both conditions are satisfied at θ = 0° i.e., at lowest point B.

86. (3) Tension in the string when it makes an angle θ with the vertical

$$T = \frac{mv^2}{r} + mg\cos\theta$$

When the string is horizontal, θ = 90°

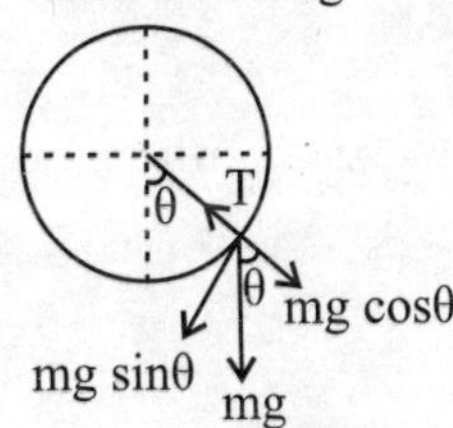

and cos 90° = 0

$\therefore \quad T = \frac{mv^2}{r} + mg \times 0$

$$= \frac{mv^2}{r} = \frac{0.6 \times (5)^2}{0.75} = 20\text{N}$$

87. (2) $T = \frac{mv^2}{r} + mg\cos\theta$

For θ = 30°,

$$T_1 = \frac{mv^2}{r} + mg\cos 30° = \frac{mv^2}{r} + \frac{\sqrt{3}mg}{2} \quad \text{... }(i)$$

For θ = 60°,

$$T_2 = \frac{mv^2}{r} + mg\cos 60° = \frac{mv^2}{r} + \frac{mg}{2} \quad \text{... }(ii)$$

From equation (*i*) and (*ii*), we can say

$\therefore \quad T_1 > T_2$

88. (1) The coefficient of restitution (e) is the ratio of the final to initial relative velocities in a collision system. If the value of e = 1, then the collision is perfectly elastic. If it is equal to 0, then the collision is perfectly inelastic.

89. (1) In this case also we need to conserve the momentum and write the expression for e. After doing so we will find that the sum of the initial velocities is equal to the sum of the final velocities. Same goes for the difference (with slight alterations). This will result in the final velocity of one body being equal to the initial velocity of the other. Hence, the velocities will get exchanged.

90. (3) conservation of momentum

$$m_A u_A + m_B u_B = m_A v_A + m_B v_B$$

$m_A u_A + m_B u_B = 0$ ($v_A = 0 = v_B$ both comes to rest)

$$u_B = -\frac{m_A u_A}{m_B} = -\frac{0.2\times 3}{0.15} = \frac{-0.6}{0.15} = -4 \text{ m/s}$$

91. (4) $m_1 = 10$ kg $\quad u_1 = 12$ m/s

$m_2 = 6$ kg $\quad u_2 = 8$ m/s

$$\therefore \quad v_1 = \frac{(m_1 - m_2)}{m_1 + m_2}u_1 + \frac{2m_2 u_2}{m_1 + m_2}$$

$$v_1 = \left(\frac{10-6}{10+6}\right)12 + \frac{2\times 6\times 8}{10+6} = \frac{144}{16} = 9 \text{ m/s}$$

$$v_2 = \left(\frac{m_2 - m_1}{m_1 + m_2}\right)u_2 + \left(\frac{2m_1 u_1}{m_1 + m_2}\right)$$

$$v_2 = \left(\frac{6-10}{16}\right)8 + \frac{(2\times 10\times 12)}{16} = \frac{-32+240}{16} = \frac{208}{16}$$

$v_2 = 13$ *m/s*

92. (2) According to the relation $e^{2n} = \frac{h_f}{h_i}$ where n is the number of collision

$$h = e^{2n} h_0$$

n = 1 after 1st rebound

$$h = (0.6)^2 \times 3 = 0.36 \times 3 = 1.08\, m$$

93. (2)

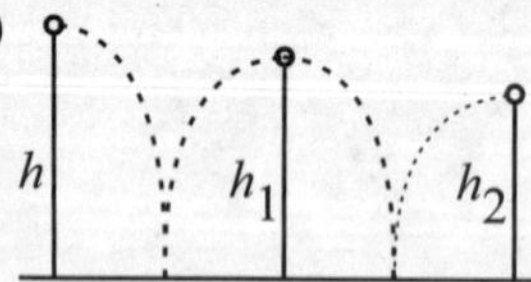

Let the particle rebounds to a height h_n after nth collision, then

$$h_n = e^{2n} h$$

$$H = h + 2h_1 + 2h_2 + 2h_3 + 2h_4 + ...$$

$$= h + 2he^2 + 2he^4 + 2he^6 + 2he^8 + ...$$

$$= h + 2h(e^2 + e^4 + e^6 + e^8 + ...) \quad \left\{\text{sum of } G.P = \left(\frac{a}{1+r}\right)\right\}$$

$$= h + 2h\left(\frac{e^2}{1-e^2}\right) = h\left[1 + \frac{2e^2}{1-e^2}\right] = h\left[\frac{1+e^2}{1-e^2}\right]$$

94. (1) Fractional decrease in kinetic energy of neutron

$$= 1 - \left[\frac{m_1 - m_2}{m_1 + m_2}\right]^2$$

As mass of neutron m_1 = m and mass of deuteron m_2 = 2m

$$\Rightarrow \text{fractional decrease in K.E.} = 1 - \left[\frac{1-2}{1+2}\right]^2 = 1 - \frac{1}{9} = \frac{8}{9}$$

95. (3) $m_1 v_1 + m_2 v_2 = (m_1 + m_2)v$

$$\frac{2\times 3 - 5\times 4}{7} = v \Rightarrow v = \frac{6-20}{7} = \frac{-14}{7} = -2 \text{ m/s}$$

96. (2) Loss in kinetic energy in collision is

$$= \frac{1}{2}\left(\frac{m_1 m_2}{m_1 + m_2}\right)(u_1 - u_2)^2(1-e^2)$$

$$= \frac{1}{2}\left(\frac{ww}{w+w}\right)(u_1 - u_2)^2(1-e^2) = \frac{w^2}{4w}(u_1 - u_2)^2$$

[$\because$ e = 0 in perfectly inelastic collision]

$$= \frac{w}{4}(u_1 - u_2)^2$$

97. (2) $v_1 = \frac{(m_1 - m_2)u_1 + 2m_2 u_2}{m_1 + m_2} \quad \{u_2 = 0\}$

$$v_1 = \frac{(m_1 - m_2)u_1}{m_1 + m_2} \Rightarrow \frac{u}{4} = \frac{(8 - m_2)u}{8 + m_2} \quad \left[\text{Given } v_1 = \frac{u}{4}\right]$$

$$8 + m_2 = 4(8 - m_2) \Rightarrow 8 + m_2 = 32 - 4m_2$$

$$\Rightarrow 5m_2 = 32 - 8$$

$$m_2 = \frac{24}{5} = 4.8 \text{ kg}$$

98. (4) Let v_1 and v_2 be the velocities of the spheres after collision,

Using conservation of momentum

$$m_1 u_1 + m_2 u_2 = m_1 v_1 + m_2 v_2$$

$$mu + 0 = m(v_1 + v_2)$$

$\therefore \; u = v_1 + v_2$ ... (1)

As the collision is inelastic

By $e = \frac{v_2 - v_1}{u_1 - u_2} = \frac{v_2 - v_1}{u - 0} \Rightarrow \therefore v_2 - v_1 = eu$ ... (2)

On putting the value of u in eq. (2)

$$\frac{v_2 - v_1}{e} = v_1 + v_2 \Rightarrow v_2 - v_1 = ev_1 + ev_2$$

$$v_2 - ev_2 = ev_1 + v_1 \Rightarrow v_2(1-e) = v_1(1+e)$$

$$\frac{v_1}{v_2} = \frac{1-e}{1+e}$$

99. (4) During inelastic collision of two particles, K.E after collision is not equal to kinetic energy before collision. Here, K.E appears in other forms. In some cases $(K.E)_{final} < (K.E)_{initial}$ such as when initial K.E is converted into internal energy of the product (as heat, elastic or excitation energy) while in other cases $(K.E)_{final} > (K.E)_{initial}$. Such as when internal energy stored in the colliding particles is released.

For example, when we fire a pistol, the hammer hits the bullet with relatively small initial kinetic energy but the kinetic energy of the hammer and bullet after collision is considerably greater than the initial kinetic energy.

100. (1) According to the conservation of linear momentum,

$$m_1u_1 + m_2u_2 = m_1v_1 + m_2v_2$$

$$500 + 1500 = 600 + 150\,v_2$$

$$v_2 = \frac{1400}{150} = 9.33$$

$$e = \frac{v_2 - v_1}{u_1 - u_2}$$

$u_1 = 5$ m/s $\quad u_1 = 10$ m/s

$v_1 = 6$ m/s $\quad v_2 = 9.3$ m/s

$$\therefore\; e = \frac{v_2 - v_1}{u_1 - u_2} = -\frac{(v_1 - v_2)}{u_1 - u_2} = -\frac{(6-9.3)}{5-10} = \frac{-3.3}{-5} = 0.66$$

101. (4)

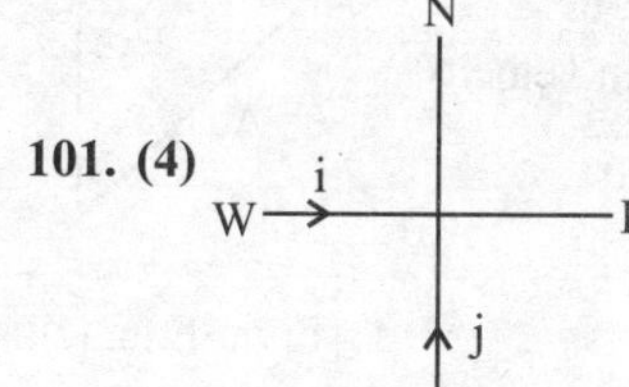

From law of conservation of momentum we have,

$m v\hat{i} + m v\hat{j} = 2mv'$ [v' is the final velocity]

$$\Rightarrow\; |2mv'| = m\left|v\hat{i} + v\hat{j}\right|$$

$$\Rightarrow\; 2mv' = m\sqrt{v^2 + v^2}$$

$$\Rightarrow\; v' = \frac{v\sqrt{2}}{2} = \frac{v}{\sqrt{2}}$$

102. (2) From equation of motion,

$v^2 = u^2 + 2gh$

Given, h = 4.9 m, g = 9.8 ms^{-2}, u = 0

Coefficient of restitution (e) of an object is a unique fractional value representing the ratio of relative velocities before and after an impact.

$$e = \frac{v'}{v}$$

$$\therefore\; v' = \frac{3}{4} \times 9.8 \;(\because v = \sqrt{2gh} \Rightarrow v = \sqrt{2 \times 9.8 \times 4.9} = 9.8)$$

Time taken from first bounce to second bounce is

$$t = \frac{2v'}{g}$$

$$t = 2 \times \frac{3}{4} \times 9.8 \times \frac{1}{9.8} = 1.5\text{s}$$

High Order Time Intensive MCQs

1. (3) When the block moves vertically downward with acceleration $\frac{g}{4}$, then tension in the cord

$$T = M\left(g - \frac{g}{4}\right) = \frac{3}{4}Mg$$

Work done by the cord = $Fs\cos\theta$

$$= Td\cos(180^\circ) = -\left(\frac{3Mg}{4}\right) \times d = -3Mg\frac{d}{4}$$

2. (3) Momentum = mass × velocity

$$u = \frac{p}{m} = \frac{5\text{kg ms}^{-1}}{4\text{kg}} = 2\text{ms}^{-1}$$

$$\text{Acceleration} = \frac{\text{Force}}{\text{Mass}}$$

$$a = \frac{0.2\text{N}}{4\text{kg}} = 0.05\text{ms}^{-2}$$

Distance travelled by the body in 10 s is

$$d = ut + \frac{1}{2}at^2$$

$$= (2\text{ms}^{-1})(10\text{s}) + \frac{1}{2} + (0.05\text{ms}^{-2})(10\text{s})^2$$

$$= 20\text{ m} + 2.5\text{ m} = 22.5\text{ m}$$

Work done, W = Fd = (0.2 N)(22.5 m) = 4.5 J

According to work-energy theorem,

Increase in kinetic energy = Work done = 4.5 J

3. (1) $F = \frac{C}{v}$, where C is constant of proportionality

$$\Rightarrow ma = \frac{C}{v} \text{ or } m\frac{dv}{dt} = \frac{C}{v}$$

$$vdv = \frac{Cdt}{m}$$

$$\frac{v^2}{2} = \frac{Ct}{m} \text{ or } \frac{1}{2}mv^2 = Ct$$

KE $\propto$ t

4. (3) Potential energy of the spring when stretched through a distance x,

$$U = \frac{1}{2}kx^2 = 10\text{J}$$

When x becomes 2x, the potential energy will be

$$U' = \frac{1}{2}k(2x)^2 = 4 \times 10 = 40\text{J}$$

$\therefore$ work done = U` – U = 40 – 10 = 30J

5. (2) Situation is shown in figure. When the mass m falls vertically on the spring, then spring is compressed by distance d.

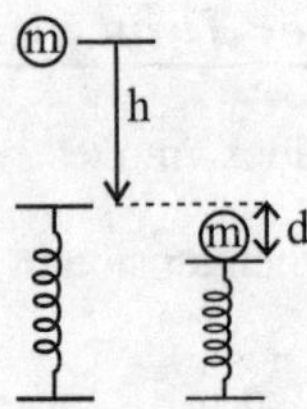

Hence, net work done in the process is

W = potential energy stored in the spring – loss of potential energy of mass

$= mg(h+d) - \frac{1}{2}kd^2$

6. (2) In vertical circular motion with uniform speed, tension is not constant.

- At the top, tension and gravity both act towards the center.

$$T_{top} = \frac{mv^2}{r} - mg \text{ (minimum)}$$

- At the bottom, tension opposes gravity:

$$T_{botom} = \frac{mv^2}{r} + mg \text{ (maximum)}$$

Hence, tension is minimum at the highest position.

7. (3) Area under F - t graph = change in momentum

$\Rightarrow p = \sqrt{2mk}$

$\therefore KE = \frac{p^2}{2m} = \frac{A^2}{2m}$ (A = net area of F - t graph)

$$= \frac{\left\{\left(\frac{4\times3}{2}\right) - \left(\frac{1.5\times2}{2}\right)\right\}^2}{2\times2} = 5.06 \text{ J}$$

8. (1) Block rises to height, h so speed of block after bullet gets embedded, $v = \sqrt{2gh}$

Now, applying conservation of momentum.

$$mu = (M+m)v \Rightarrow u = \left(\frac{M+m}{m}\right)\sqrt{2gh}$$

9. (2) To climb a height h, the boy utilizes potential energy = mgh

In order to climb, he will use the efficient energy.

Energy of one bread = 21 kJ = 21×10^3 J

Efficiency of boy = 28%

Hence, energy consumed by boy

$$= \frac{28}{100} \times 21000 = 5880 \text{J} \quad ...(i)$$

From law of conservation of energy, this energy is utilized in giving potential energy mgh, where g is acceleration due to gravity.

$\therefore mgh = 40 \times 9.8 \times h$ ...(*ii*)

Equation Eqs. (*i*) and (*ii*), we have

$\Rightarrow 40 \times 9.8 \times h = 5880$

$$\Rightarrow h = \frac{5880}{40\times9.8} = 15\text{m}$$

10. (1) When stone is at its lowest position, it has only kinetic energy, given by

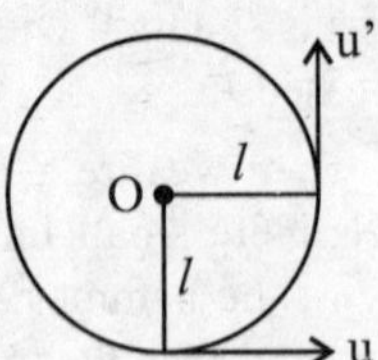

$$E_1 = K = \frac{1}{2}mu^2$$

At the horizontal position, the total energy is given by

$$E_2 = U + K = \frac{1}{2}mu'^2 + mg\ell$$

According to conservation of energy,

$E_1 = E_2$

$$\frac{1}{2}mu^2 = \frac{1}{2}mu'^2 + mg\ell$$

$$\Rightarrow \frac{1}{2}mu'^2 = \frac{1}{2}mu^2 - mgl$$

$$\Rightarrow u'^2 = u^2 - 2g\ell \Rightarrow u' = \sqrt{u^2 - 2g\ell} \quad ...(i)$$

The velocities are shown below

So, the magnitude of change in velocity

$$|\Delta u| = |u| = \sqrt{u'^2 + u^2 + 2u'u\cos 90°}$$

$$|\Delta u| = \sqrt{u'^2 + u^2} = \sqrt{2u^2 - 2gl}$$

[From Equ. (*i*)]

11. (1) According to work-energy theorem.

Work done = change in kinetic energy of the body

$\therefore W = K_f - K_i$

$$\int F.dx = K_f - \frac{1}{2}mv_i^2$$

$$\text{Or } \int_{x=20}^{x=30} (-0.1)x\,dx = K_f - 500$$

$$(-0.1)\left[\frac{x^2}{2}\right]_{x=20}^{x=30} = K_f - 500$$

$$(-0.1)\left[\frac{(30)^2}{2} - \frac{(20)^2}{2}\right] = K_f - 500$$

Or $K_f - 500 = -0.1\ (450 - 200)$

Or $K_f - 500 = -25$

$\therefore K_f = 500 - 25 = 475$ J

12. (2) Let a body of mass m initially at rest attains velocity v after a time t as shown in the figure.

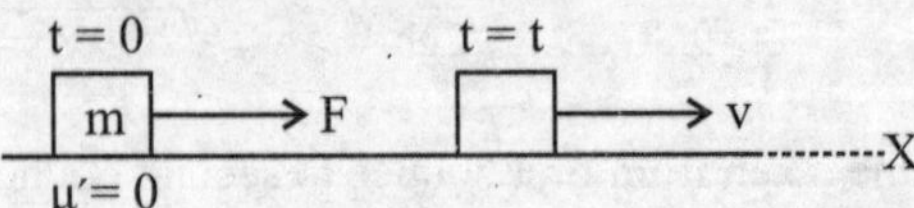

The constant force, F produces constant acceleration a.

Power delivered to the block by the force F at any time t is

$P = (F)(v)$ ...(*i*)

Using equation of motion along x

$v = u + at$

$\Rightarrow v = 0 + at$

$\Rightarrow v = at$ ...(ii)

From Equ. (i) and (ii), we get

$P = (F)(at) = (ma)(at)$ [$\because F = ma$]

$= (ma^2)t$

$\Rightarrow P \propto t$ [$\because ma^2 = \text{constant}$]

13. (4) The power of delivered to one bullet, $P = F \times v$

$P = mv \times v = mv^2 = 2K$

For n bullets power = 2nK

$$\text{Power} = \frac{\text{Kinetic energy}}{\text{Time}}$$

$$\text{For n bullets power} = \frac{n \cdot \frac{1}{2}mv^2}{t}$$

$$= nK \qquad \left(\because K = \frac{1}{2}mv^2, t = 1s\right)$$

14. (2) As, $P_0 = Fv = mav$

$$= \left(m\frac{dv}{dt}\right)v$$

$P_0 dt = mvdv$

Integrating both sides, we get

$$\int P_0 dt = \int mv\,dv$$

$$P_0 t = \frac{mv^2}{2} \Rightarrow v^2 = \frac{2P_0 t}{m} \Rightarrow v \propto t^{1/2}$$

15. (3) $P = m\frac{dv}{dt}v \Rightarrow vdv = \frac{p}{m}dt$

integrating both sides

$$\int vdv = \int \frac{P}{m}dt \Rightarrow \frac{v^2}{2} = \frac{Pt}{m}$$

$$\Rightarrow v = \sqrt{\frac{2P}{m}t} \Rightarrow \frac{dx}{dt} = \sqrt{\frac{2P}{m}t}$$

Again integrating both sides

$$\int dx = \sqrt{\frac{2P}{m}}\int t^{1/2}dt,$$

$$x = \sqrt{\frac{2P}{m}}\frac{t^{3/2}}{3/2}$$

$$\Rightarrow x \propto t^{3/2}$$

16. (3) If the motor pumps water (density = ρ) continuously through a pipe of area of cross-section A with velocity v, then mass flowing out per second.

$m = Av\rho$ ...(i)

Rate of increase of kinetic energy

$$= \frac{1}{2}mv^2 = \frac{1}{2}(Av\rho)v^2 = \frac{1}{2}A\rho v^3 \quad \text{... (ii)}$$

Mass m, flowing out per sec, can be increased to m′ by increasing v to v′, then power increases from P to P′.

$$\frac{P'}{P} = \frac{\frac{1}{2}A\rho v'^3}{\frac{1}{2}A\rho v^3} \text{ or } \frac{P'}{P} = \left(\frac{v'}{v}\right)^3$$

$$\text{Now, } \frac{m'}{m} = \frac{A\rho v'}{A\rho v} = \frac{v'}{v}$$

As, $m' = nm$, $v' = nv$

$$\therefore \frac{P'}{P} = n^3 \Rightarrow P' = n^3 P$$

17. (3) Centripetal acceleration, $a_c = \frac{v^2}{r} = k^2rt^2$

$\Rightarrow \quad v^2 = k^2r^2t^2$

kinetic energy is given by

$$K = \frac{1}{2}mv^2 = \frac{1}{2}mk^2r^2t^2$$

According to work-energy theorem, change in kinetic energy equal to work done.

$$\therefore \quad W = \frac{1}{2}mk^2r^2t^2$$

Thus, power delivered to the particle

$$P = \frac{dW}{dt} = mk^2r^2t$$

18. (1) The situation of collision is given as below:

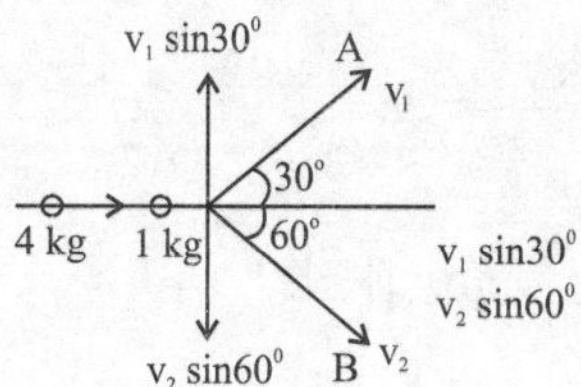

Along y axis, momentum is zero $m_1v_1 \sin 30° = m_2v_2 \sin 60°$

Hence, $4v_1 \sin 30° = v_2 \sin 60°$

$$\text{Or } \frac{v_1}{v_2} = \frac{\sqrt{3}}{4}$$

19. (3) In an inelastic collision, kinetic energy is not conserved but momentum remains conserved.

$m_1 = 4\text{kg}$, $m_1 = 4$ kg, $u_1 = 12$ ms^{-1}; $m_2 = 6$ kg, $u_2 = 0$

$m_1 = 4\text{kg}$, $m_2 = 6\text{kg}$

$u_1 = 12\ \text{ms}^{-1}$, $u_2 = 0$

Momentum before collision = Momentum after collision

$m_1u_1 + m_2u_2 = m_1v_1 + m_2v_2$

$\Rightarrow 4 \times 12 = (4 + 6)v$

$\Rightarrow v = 4.8\ \text{ms}^{-1}$

$$\text{Total kinetic energy before collision} = \frac{1}{2}m_1u_1^2$$

$$= \frac{1}{2} \times 4 \times (12)^2 = 288\,J$$

$$\text{Total kinetic energy after collision} = \frac{1}{2}(m_1 + m_2)v^2$$

$= \frac{1}{2}(4+6)(4.8)^2 = 115.2\text{J}$

∴ Loss in kinetic energy of system = 288 J – 115.2 J

= 172.8 J

20. (1) $W = \int \vec{F}.d\vec{s} = \int_{(0,0)}^{(1,1)} (2x dx + 2dy)$

$= \int_0^1 2x\,dx + \int_0^1 2\,dy = [x^2]_0^1 + [2y]_0^1$

= 3J

21. (4) $W = \int \vec{F}.d\vec{s} = K\int \frac{1}{s} ds = K(\log s)$

22. (2) Kinetic energy acquired by the body

= Force applied on it × Distance covered by the body

K.E. = F × d

If F and d both are same, then K.E. acquired by the body will be same.

23. (3) In inelastic collision between two bodies, total linear momentum is always conserved.

24. (2) $F = -\frac{dU}{dx} = 2Bx \Rightarrow F \propto \frac{1}{m}$

25. (3) $P = Fv = m\left(\frac{dv}{dt}\right)v \Rightarrow \frac{P}{m}dt = vdv$

$\Rightarrow \frac{P}{m}\times t = \frac{v^2}{2} \Rightarrow v = \left(\frac{2P}{m}\right)^{1/2}(t)^{1/2}$

Now $s = \int v dt = \int\left(\frac{2P}{m}\right)^{1/2}(t)^{1/2}$

$\therefore s = \left(\frac{2P}{m}\right)^{1/2}\left[\frac{2t^{3/2}}{3}\right] \Rightarrow s \propto t^{3/2}$

26. (4) $P = \frac{mgh}{t}$

$m = \frac{3000\,W \times 60\,s}{10\,ms^{-1} \times 10m} = 1200\,kg = 1200\,litre$

27. (1) $\Rightarrow F = \frac{P}{V}$

At the time of maximum velocity $\frac{dv}{dt} = 0 \Rightarrow a = 0$

Using II law of motion

$F - r = ma$

$\Rightarrow F = r \Rightarrow \frac{P}{V_{max}} = r \Rightarrow V_{max} = \frac{P}{r}$

28. (3) $\frac{v^2}{r} = K^2rt^2$

$\Rightarrow v = Krt$

$\Rightarrow \frac{dv}{dt} = Kr$

Centripetal force will not supply the power and power by the tangential force

$P = m\frac{dv}{dt}v$ ($\because P = Fv$ therefore $P = (ma)v$)

$\therefore m(Kr)Krt = mK^2r^2t$

29. (2) The situation at impact is shown $\sin\theta = \frac{r}{2r} = \frac{1}{2}$

Using principle of conservation of linear momentum we have

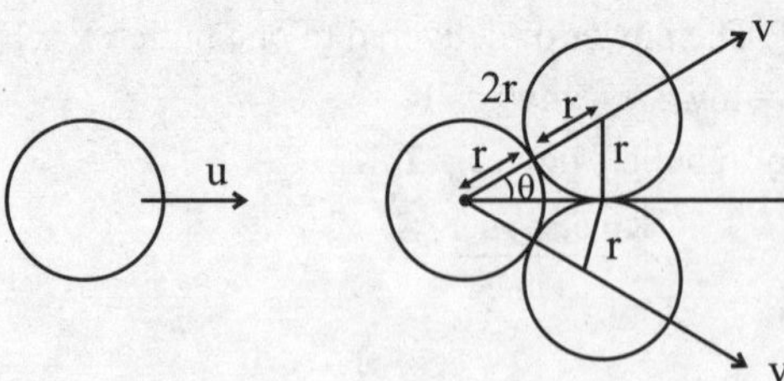

mu = mv cos 30° + mv cos 30°

mu = 2 mv cos 30°

$v = \frac{u}{2\cos 30°} = \frac{u}{2\times\frac{\sqrt{3}}{2}} = \frac{u}{\sqrt{3}}$

By definition of coefficient of restitution,

$e = \frac{\text{relative velocity of separation}}{\text{relative velocity of approach}}$

$\therefore e = \frac{v - 0}{u\cos 30° - 0} = \frac{v}{u\times\frac{\sqrt{3}}{2}} = \frac{\frac{u}{\sqrt{3}}}{u\sqrt{3}/2} = \frac{2}{3}$

30. (1) For U_{min} $\frac{d}{dx}U = 0 \Rightarrow x(x^2 - 1) = 0$

U is min at $x = \pm 1$

$\therefore U_{min} = \frac{1}{4} - \frac{1}{2} = -\frac{1}{4}$

$K_{max} + U_{min} = E_{Total}$

$K_{max} - \frac{1}{4} = 2 \Rightarrow K_{max} = \frac{9}{4}$

$\frac{1}{2}mv^2 = \frac{9}{4} \Rightarrow v^2 = \frac{9}{2}$

$v = \frac{3}{\sqrt{2}}$

31. (2) Relative to Belt

$0 - \frac{1}{2}mv_0^2 = -\mu mg(L)$

$\Rightarrow v_0 = \sqrt{2\mu gL}$

32. (2) Collision is perfectly elastic so both momentum and K.E. are conserved

$= \frac{1}{2}m_2 v_2^2 = \frac{1}{2}\times 2.5\,v_2^2 = 5$ joule

Therefore, $v_2 = 2ms^{-1}$

$$v_2 = \frac{2m_1}{m_1 + m_2}u_1$$

$$2 = 2\times\frac{5}{5+2.5}u_1$$

$$u_1 = \frac{15}{10} = \frac{3}{2}ms^{-1}$$

KE of the 1st body $= \frac{1}{2}m_1 u_1^2 = \frac{1}{2}\times 5\times(1.5)^2$ J

= 5.62 J

33. (1) Power $= p\frac{dV}{dt}$

where V is volume $= h\rho g\frac{dV}{dt}$

$$= 120\times10^{-3}\times13.6\times10^3\times10\times\frac{5\times10^{-3}}{60} = 1.36 \text{ W}$$

34. (3) $a = \frac{v}{t_1}$

velocity at time

$$v = at = \frac{vt}{t_1}$$

$$KE = W = \frac{1}{2}m\left[\frac{vt}{t_1}\right]^2$$

35. (4) $m\frac{dv}{dt} = Kv^{-2}$

$$\int m\left(v^2dv\right) = \int Kdt$$

$$m\left(\frac{v^3}{3}\right) = Kt$$

$$\frac{1}{2}mv^2 = \frac{3}{2}\frac{Kt}{v}$$

36. (1) $W_s = \Delta k$

$$-\frac{1}{2}kx^2 = \frac{1}{2}m\left(\frac{v}{2}\right)^2 - \frac{1}{2}mv^2$$

$$\frac{1}{2}kx^2 = \frac{3}{8}mv^2 \quad \Rightarrow k = \frac{3}{4}\frac{mv^2}{x^2}$$

37. (2) Force to lift

kx = 2g = 20

$$\Rightarrow \quad x = \frac{1}{2}$$

Now, $\frac{1}{2}mv^2 - mgh + \frac{1}{2}kx^2 = 0$

$$\Rightarrow \quad v^2 = 8 \quad \Rightarrow \quad v = 2\sqrt{2}$$

38. (1) As, $\frac{1}{2}mv^2 = \alpha x^2 \Rightarrow \frac{mv^2}{R} = \frac{2\alpha x^2}{R} = F_n$

Ae $\quad v = x\sqrt{\frac{2\alpha}{m}}$

$$a = \frac{dv}{dt} = \sqrt{\frac{2\alpha}{m}}\frac{dx}{dt}$$

$$\Rightarrow \quad F_t = ma = m\sqrt{\frac{2\alpha}{m}}\cdot\sqrt{\frac{2\alpha x^2}{m}}$$

$$\Rightarrow \quad F = \sqrt{F_n^2 + F_t^2} \Rightarrow F = 2\alpha x\sqrt{1+\frac{x^2}{R^2}}$$

39. (3) $W = Fl\sin\theta$

40. (4) $W = \Delta k = \frac{1}{2}mv^2$

$\Rightarrow W \propto v^2$ graph will be parabolic ($y^2 = 4ax$)

Statement Based MCQs

1. **(1)** Conservative forces do not contribute to the change in mechanical energy as when only these forces are in a closed system, the sum of changes in kinetic & potential energy is 0. Hence, work done (total) equals change in K.E. & P.E. is equal to negative of work done by conservative force when object undergoes a displacement from state 1 to state 2.
2. **(3)** Statement I is correct but Statement II is incorrect.
3. **(3)** Different SI units are adopted for same quantity in different domains of study in order to express quantities conveniently. So it does not matters what type of energy it is.
4. **(3)** Statement I is correct but Statement II is incorrect.
5. **(1)** Both statements are true.

Assertion & Reason MCQs

1. **(2)** Both are correct.
2. **(1)** Because the centripetal force is perpendicular to the velocity at all times. And power is the dot product of force & velocity. Hence, power is zero in circular motion.
3. **(4)** A body can have energy without momentum i.e. (potential energy). But body cannot have momentum without energy
4. **(2)** Reason is true but is not explaining the assertion.
5. **(3)** K.E can be increased by mutual attraction or repulsion forces for electrostatic or gravitation forces, without the application of any external force. However, reason is false.

Match the Columns MCQs

1. (1) **2. (1)** **3. (3)** **4. (2)** **5. (1)**

NCERT Exemplar MCQs

1. **(4)** Given, h = 1.5 m, v = 1 m/s, m = 10 kg, g = 10 ms^{-2}

From conservation of mechanical energy.

$(P.E)_i + (K.E)_i = (PE)_f + (K.E)_f$

$$\Rightarrow mgh + \frac{1}{2}mv^2 = 0 + (KE)_f$$

$\Rightarrow (KE)_f = mgh + \frac{1}{2}mv^2$

$\Rightarrow (KE)_f = 10\times10\times1.5 + \frac{1}{2}\times10\times(1)^2$

= 150 + 5 = 155 J

Note **:** We should be careful about the reference taken for P.E., it may or may not be the ground.

2. **(2)** First velocity of the iron sphere increases and after sometime becomes constant, called terminal velocity. Hence, accordingly first K.E increases and then becomes constant which is best represented by 2.

3. **(4)** When the man is squatting on the ground he is tilted somewhat, hence he also has to balance frictional force besides his weight in this case.

R = Reaction force = friction + mg

$\Rightarrow \quad R > mg$

When the man gets straight up in that case friction = 0

⇒ Reaction force = mg

4. **(3)** Here, work is done by the frictional force on the cycle and is equal to 200 × 10 = – 2000 J.

As the road is not moving, hence, work done by the cycle on the road = zero.

Note **:** We should be aware that here the energy of bicyclist lost during the motion is lost due to friction in the form of heat.

5. **(3)** As the body is falling freely under gravity, the potential energy decreases and kinetic energy increases but total mechanical energy (P.E + K.E) of the body and earth system will be constant as external force on the system is zero

6. **(3)** As the given tracks are frictionless, hence mechanical energy will be conserved. As both the tracks having same height, h.

From conservation of mechanical energy, $\frac{1}{2}mv^2 = mgh$

(for both tracks I and II)

Hence, speed is same for both stones. For stone I, a_1 = acceleration along inclined plane = g sin θ_1

Similarly, for stone II : a_2 = g sin θ_2 as $\theta_2 > \theta_1$ hence, $a_2 > a_1$.

And both length for track I is also less hence, stone II reaches earlier than stone I.

7. **(2)** Given $v = ax^{3/2}$

m = 0.5 kg, a = 5 $m^{-1/2}$ s^{-1}, work done (W) = ?

We know that

Acceleration $a_0 = \frac{dv}{dt} = v\frac{dv}{dx} = ax^{3/2}\frac{d}{dx}(ax^{3/2})$

$= ax^{3/2}\times a\times\frac{3}{2}\times x^{1/2} = \frac{3}{2}a^2x^2$

Now, Force $= ma_0 = m\frac{3}{2}a^2x^2$

Work done $= \int_{x=0}^{x=2} Fdx = \int_0^2 \frac{3}{2}ma^2x^2dx$

$= \frac{3}{2}ma^2\times[x^3/3]_0^2$

$= \frac{1}{2}ma^2\times8 = \frac{1}{2}\times(0.5)\times(25)\times8 = 50$ J

8. **(2)** Given, power = constant

We know that power (P)

$P = \frac{dW}{dt} = \frac{F.ds}{dt} = \frac{Fds}{dt}$ ($\because$ body is moving unidirectionally)

Hence, F.ds = Fds cos 0°

$P = \frac{Fds}{dt}$ = constant ($\because$ P = constant by question)

Now, writing dimensions

[F] [v] = constant

$\Rightarrow [MLT^{-2}][LT^{-1}]$ = constant

$\Rightarrow L^2T^{-3}$ = constant

$\Rightarrow L \propto T^{3/2} \Rightarrow$ Displacement (d) $\propto t^{3/2}$

9. **(3)** When a pendulum oscillates in air, it will lose energy continuously in overcoming resistance due to air. Therefore, total mechanical energy of the pendulum decreases continuously with time.

The variation is correctly represented by curve 3.

10. **(1)** Given,

Mass = m = 5 kg

Radius = 1 m = R

Revolution per minute ω = 300 rev/min

= (300 × 2π) rad/min

= (300 × 2 × 3.14) rad/60 s

$= \frac{300\times2\times3.14}{60}$ rad/s $= 10\pi$ rad/s

⇒ Linear speed = v = ωR

$= \left(\frac{300\times2\pi}{60}\right)(1) = 10\,\pi$ m/s

$KE = \frac{1}{2}mv^2 = \frac{1}{2}\times5\times(10\pi)^2$

$= 100\pi^2\times5\times\frac{1}{2} = 250\,\pi^2$ J

NEET & AIIMS Past Year MCQs

1. **(3)** Given:

Potential energy $V = \frac{Ax^2}{\sqrt{x+B}}$.

For dimensional consistency:

$[V] = [\text{Energy}] = ML^2T^{-2}$

Denominator $\sqrt{x+B}$

$\Rightarrow [B] = [x] = L^{1/2}$.

Thus:

$$[A]=\frac{[V]\sqrt{L^{1/2}}}{L^2}=ML^{1/2}T^{-2}$$

$$\frac{A^2}{B}=\frac{\left(ML^{1/2}T^{-2}\right)^2}{L^{1/2}}=M^2L^{1/2}T^{-4}$$

2. (4) Before hitting the ground,
$K.E_1 = mg \times 10$
Just after hitting the ground

$$K.E_2 = K.E_1 - 0.5\ K.E_1 = \frac{K.E_1}{2}$$

$$\therefore K.E_2 = mgh_2$$

$$\frac{K\cdot E_1}{2} = mgh_2 \Rightarrow \frac{mg\times 10}{2} = mgh_2$$

$h_2 = 5$ m

3. (3) Work done:

$$W=\vec{F}\cdot\Delta\vec{x}=(-2)(3)+(3)(0)=-6\,J.$$

From work energy theorem:

$$K_f = K_i + W = 10\text{ J} - 6\text{ J} = 4\text{ J}.$$

4. (3) $s = 2t - 1$

$$v=\frac{ds}{dt}=\frac{d}{dt}(2t-1)=2 \text{ (SI unit)}$$

Instantaneous power, $P=\vec{F}.\vec{v}$

$= 5 \times 2 = 10$ (SI unit)

5. (4) Before collision:

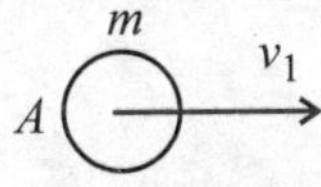

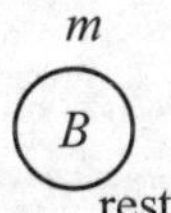

After collision:

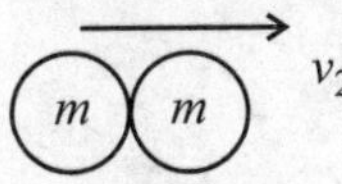

Now, from conservations of momentum,
Momentum before collision = Momentum after collision.

$$mv_1 + (0) = 2m\,v_2$$

$$mv_1 = 2mv_2$$

$$\frac{v_1}{v_2}=\frac{2}{1}$$

$$\therefore\ v_1 : v_2 = 2 : 1$$

6. (1) Potential energy stored in the spring $=\frac{1}{2}kx^2$

Now $\frac{1}{2}k(2)^2 = U$

$\&\frac{1}{2}k(8)^2 = U'$ (say)

$$\Rightarrow U'=\frac{64}{4}U=16\,U$$

7. (2) $\frac{1}{2}m\left(\frac{u}{3}\right)^2-\frac{1}{2}mu^2=-F_R\times 24$

$$0-\frac{1}{2}mu^2=-F_R\times d$$

$$\frac{\frac{1}{2}mu^2}{\frac{1}{2}mu^2\times\frac{8}{9}}=\frac{d}{24}$$

$$d=24\times\frac{9}{8}=27\text{ cm}$$

8. (4) As the velocity is constant, therefore net upward force is equal to the net downward force

$\Rightarrow F = W + f$

$= 20000 + 3000 = 23000$ N

As, Power $=\vec{F}\cdot\vec{v}$

$= 23000 \times 1.5 = 34500$ W

9. (2) Energy = Power × time

$= 100 \times 10^3 \times 3600$

$= 36 \times 10^7$ J

10. (1) $E = mgh$

$$P_{input}=\frac{mgh}{t}$$

$$=\frac{15\times10\times60}{1}=9000=9\text{kW}$$

Power lost $= 0.9 \times 10^3$

$P_{output} = 9 \times 10^3 - 0.9 \times 10^3 = 8.1$ kW

11. (3) Body initially at rest u = 0

$v^2 - u^2 = 2as$

$v^2 = 2as$

$$KE=\frac{1}{2}mv^2$$

$$KE=\frac{1}{2}m\times 2as$$

(KE $\propto$ S)

12. (4) **13. (1)** **14. (3)**

System of Particles and Rotational Motion

NCERT Topic-wise MCQs

1. (2) The mass distribution of the body determines the position of centre of mass. Therefore, centre of mass is independent of coordinate system.

2. (3) $x_{CM} = 0$ because masses also lies only along z-axis.

$$Z_{CM}=\frac{m_1z_1+m_2z_2}{m_1+m_2}=\frac{9\times5+11\times9}{9+11}$$

$$=\frac{45+99}{20}=\frac{144}{20}=7.2\text{ m}$$

3. (2) If mass is non-uniformly distributed, then centre of mass of ring may lie anywhere between centre and circumference (not on the centre). Hence $0 < a \le r$.

4. (3)

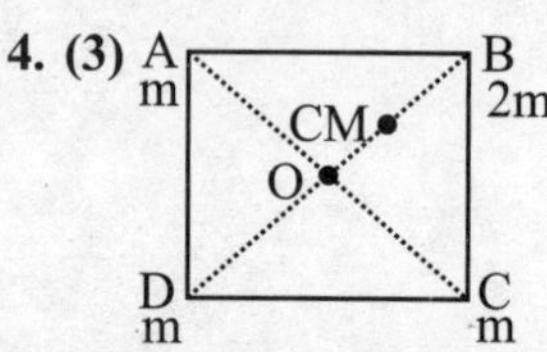

If all the masses were same, the centre of mass would have been at O but as the mass at B is 2 m, so the centre of mass of the system will shift towards B. So, centre of mass will be on the line OB.

5. (4) Mg is the only external force acting on the system

$\therefore$ Centre of mass will continue moving vertically downwards and there is not any relative shift of the center of mass of the system (B + C).

6. (1) Let radius of complete disc is R and that of small disc is R_1. Also, let centre of mass now shifts to O_2 at a distance x_{CM} from original centre.

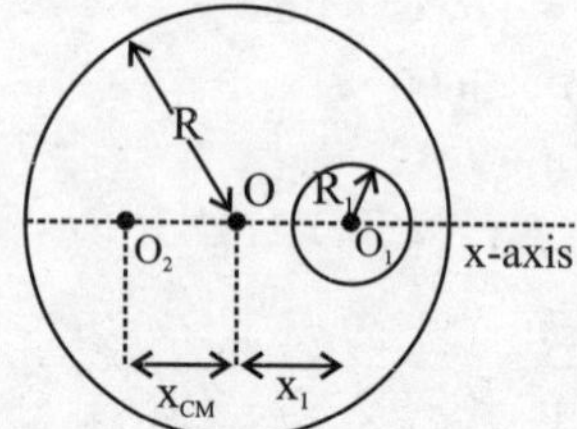

The position of new centre of mass is given by

$$x_{CM} = \frac{0 - \sigma\pi R_1^2 x_1}{\sigma\pi R^2 - \sigma\pi R_1^2}, \text{ where, } \sigma = \text{mass per unit area}$$

Here, R = 6 cm, R_1 = 2 cm, x_1 = 3.2 cm

$$\text{Hence, } x_{CM} = \frac{-\sigma \times \pi (2)^2 \times 3.2}{\sigma \times \pi \times (6)^2 - \sigma \times \pi \times (2)^2}$$

$$= -\frac{12.8\pi}{32\pi} = -0.4\text{cm}$$

Negative sign indicates the shift is towards left.

7. (1) Let mass per unit area of the plate be σ.

Mass of horizontal portion = 8 × 2 σ = 16 σ

Mass of vertical portion = 6 × 2 σ = 12 σ

The centres O_1 and O_2 of these portions lie at distances 1 m and 2 + 3 = 5 m from the point O. The CM of the T-shaped plate will lie at distance y from the point O which is given by

$$y = \frac{16\sigma \times 1 + 12\sigma \times 5}{16\sigma + 12\sigma} = \frac{76\sigma}{28\sigma} = 2.71\,\text{m}$$

8. (1) $x_{CM} = \dfrac{m_1 x_1 + m_2 x_2}{m_1 + m_2} = \dfrac{12 \times 0 + 16 \times 1.13}{12 + 16} = 0.6457\,\text{Å}$

9. (2) Coordinates of initial centre of mass,

$(x_{CM}, y_{CM}, z_{CM}) = (3, 3, 3)$

$\Rightarrow 1 \times x_1 + 2 \times x_2 + 3 \times x_3 = (1 + 2 + 3)3 = 18$ ...(*i*)

When centre of mass shifts to (1, 1, 1)

$$1 = \frac{1 \times x_1 + 2 \times x_2 + 3 \times x_3 + 4x_4}{(1+2+3+4)} \quad ...(ii)$$

Solving eq. (*i*) and (*ii*), we get

$4x_4 = 10 - 18 \Rightarrow x_4 = -2$

Similarly, $y_4 = -2$, $z_4 = -2$

The fourth particle must be placed at the point (–2, –2, –2).

10. (2) $\vec{V}cm = \dfrac{m_1\vec{v}_1 + m_2\vec{v}_2}{m_1 + m_2} = \dfrac{300(15\hat{i}) + 700(4\hat{i} + 6\hat{j})}{300 + 700}$

$$\vec{V}cm = \frac{4500\hat{i} + 2800\hat{i} + 4200\hat{j}}{1000} = 7.3\hat{i} + 4.2\hat{j}$$

11. (1) Let the tension in the string be T.

$T - m_2 g = m_2 a$ ...(*i*)

$m_1 g - T = m_1 a$ ...(*ii*)

from eq. (*i*) and (*ii*), we get

$$a = \left(\frac{m_1 - m_2}{m_1 + m_2}\right) g \quad ...(iii)$$

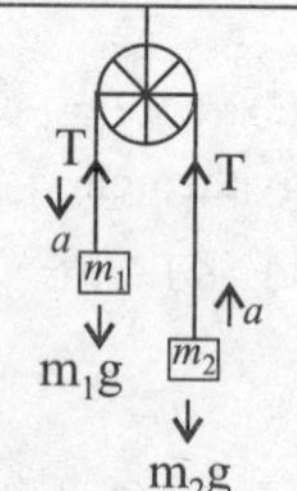

Acceleration of the centre of mass is

$$\vec{a}_{cm} = \frac{m_1\vec{a}_1 + m_2\vec{a}_2}{m_1 + m_2}$$

$(a_1 = a \;\&\; a_2 = -a)$

$$a_{cm} = \frac{m_1 a - m_2 a}{m_1 + m_2} = \left(\frac{m_1 - m_2}{m_1 + m_2}\right) a \quad ...(iv)$$

Put the value of a from equ. (*iii*)

$$a_{cm} = \left(\frac{m_1 - m_2}{m_1 + m_2}\right)^2 g$$

12. (1) $a_{cm} = \left(\dfrac{m_1 - m_2}{m_1 + m_2}\right)^2 g \Rightarrow a_{cm} = \left(\dfrac{20 - 10}{30}\right)^2 \times 9.8$

$$= \frac{9.8}{9} = 1.088\,\text{m/s}^2$$

13. (2) $\vec{a}_{cm} = \dfrac{m_1\vec{a}_1 + m_2\vec{a}_2}{m_1 + m_2} = \dfrac{0 + ma}{2m} = \dfrac{a}{2} \quad [\because m_1 = m_2 = m]$

14. (1) Velocity of centre of mass

$$V_{cm} = \frac{m_1 v_1 + m_2 v_2 + m_3 v_3}{m_1 + m_2 + m_3}$$

$$= \frac{20 \times 10\hat{i} + 30 \times 10\hat{j} + 50 \times 10\hat{k}}{100} = 2\hat{i} + 3\hat{j} + 5\hat{k}$$

15. (4) Here $m_1 = m_2 = m$; $v_1 = 2v$ and $v_2 = -v$

$$\therefore \; V_{CM} = \frac{m_1 v_1 + m_2 v_2}{m + m} = \frac{m \times 2v + m(-v)}{m + m} = \frac{v}{2}$$

16. (2) $\omega_0 = \dfrac{2\pi\times1200}{60}\text{rad s}^{-1} = 40\pi\,\text{rad s}^{-1}$

$\omega = \dfrac{2\pi\times3120}{60}\text{rad s}^{-1} = 104\pi\,\text{rad s}^{-1}$

$\therefore$ Angular acceleration, $\alpha = \dfrac{\omega-\omega_0}{t} = \dfrac{104\pi-40\pi}{16}$

$= 4\pi$ rad s^{-2}

17. (1) $v = 150\,\text{rpm} = \dfrac{150}{60}\text{rps} = \dfrac{5}{2}\text{rps}$

$v = 2\pi v = 2\pi\times\dfrac{5}{2} = 5\pi\,\text{rad s}^{-2}$

$\alpha = \dfrac{\omega-\omega_0}{t} = \dfrac{5\pi-0}{10} = \dfrac{\pi}{2}\text{rad s}^{-2}$

$\therefore$ Torque, $\tau = I\alpha = 6\times\dfrac{\pi}{2} = 3\pi\text{Nm}$

18. (4) Torque due to central force is zero.

$\therefore \tau = \dfrac{dL}{dt} = 0$

$\Rightarrow$ L = constant

19. (1) Here, $\vec{r} = i + 2\hat{j} - \hat{k},\ \vec{p} = 3\hat{i} + 4\hat{j} - 2\hat{k}$

$$\vec{L} = \vec{r}\times\vec{p} = \begin{vmatrix} \hat{i} & \hat{j} & \hat{k} \\ 1 & 2 & -1 \\ 3 & 4 & -2 \end{vmatrix}$$

$= \hat{i}(-4+4) + \hat{j}(-3+2) + \hat{k}(4-6) = 0\hat{i} - 1\hat{j} - 2\hat{k}$

$\vec{L}$ has components along –y axis and –z axis but it has no component along the x-axis. The angular momentum $\vec{L}$ is perpendicular to x-axis

20. (4) For translational equilibrium, $\sum\vec{F} = 0$

A-r

For rotational equilibrium, $\sum\vec{\tau} = 0$

B-s

Moment of inertia of a body = Mk^2

C-p

Torque is required to produce angular acceleration.

D-q

21. (2) $\omega = \omega_0 + \alpha t \Rightarrow \alpha = \dfrac{\omega-\omega_0}{t} = \dfrac{200-150}{18} = \dfrac{50}{18}\dfrac{\text{rotations}}{\text{s}^2}$

$\theta = \omega_0 t + \dfrac{1}{2}\alpha t^2$

$= 150\times18 + \dfrac{1}{2}\times\dfrac{50}{18}\times18\times18 = 3150$ rotations

22. (2) As we know that, $\alpha = \dfrac{\omega}{t} = \dfrac{2\times\pi\times\dfrac{680}{60}}{9} = 2.52\pi\,\text{rad s}^{-2}$

23. (4) Angular velocity is an axial vector, therefore its direction is along the axis of rotation

24. (2) Moment of inertia of disc about the diameter

$I = \dfrac{MR^2}{4} = \dfrac{1}{4}\times200\times10^{-3}\times(5\times10^{-2})^2$

$I = \dfrac{0.2\times25\times10^{-4}}{4} = \dfrac{5\times10^{-4}}{4}$

$\therefore L = I\omega = \dfrac{5\times10^{-4}}{4}\times15 = \dfrac{75}{4}\times10^{-4} = 18.75\times10^{-4}$ Js

25. (3) $L = mvr = 5\times10^{-6}\times2\times10^{3}\times60\times10^{-2}$

$= 6\times10^{-3}\,\text{kg m}^2\text{s}^{-1}$

26. (2) $I = MR^2 = 5\times(0.6)^2 = 5\times0.36 = 1.8\,\text{kg m}^2$

Torque = 80 × 0.6 = 48 Nm

$\tau = I\alpha \quad \therefore \alpha = \dfrac{\tau}{I} = \dfrac{48}{1.8} = 26.6\ \text{rad s}^{-2}$

27. (4) $\omega_0 = 0\quad \omega = 2\times\pi\times\dfrac{200}{60} = \dfrac{20\pi}{3}$ rad/sec

angular acceleration $\alpha = \dfrac{\omega-\omega_0}{t} = \dfrac{\dfrac{20\pi}{3}-0}{20} = \dfrac{\pi}{3}$rad/s^2

$\therefore\ \tau = I\alpha = 10\times\dfrac{\pi}{3} = 10.47\,\pi\,\text{Ns}$

28. (1) $\tau = I\alpha \Rightarrow \alpha = \dfrac{\tau}{I} == \dfrac{2000}{300} = \dfrac{20}{3}$rad/s^2

$\omega = \omega_0 + \alpha t = 0 + \dfrac{20}{3}\times5 = \dfrac{100}{3} = 33.3$ rad/s^2

29. (1) As we know that $\tau = \dfrac{dL}{dt} = \dfrac{7L-3L}{4} = \dfrac{4L}{4} = L$

30. (3) As we know that, $\omega = \omega_0 + \alpha t$

$\therefore\ \dfrac{\omega-\omega_0}{t} = \alpha \Rightarrow \alpha = \dfrac{0-5}{t} = \dfrac{-5}{t}\text{rad s}^{-2}$

$\therefore\ \tau = I\alpha \Rightarrow 9\times10^4 = 9\times10^5\times\dfrac{5}{t}$

$\Rightarrow$ t = 50 sec

31. (4) Angular momentum L = mvr = mωr × r = mωr^2

$\therefore\ \dfrac{\omega_2}{\omega_1} = \left(\dfrac{r_1}{r_2}\right)^2 \Rightarrow \omega_2 = \left(\dfrac{1}{5}\right)^2\times50 = \dfrac{50}{25} = 2\,\text{rad/s}$

32. (2) Centripetal force, $F = \dfrac{mv^2}{r}$

$\therefore F = \dfrac{m}{r}\times\left(\dfrac{L}{mr}\right)^2 = \dfrac{mL^2}{rm^2r^2} = \dfrac{L^2}{r^3m}\quad (L = mvr)$

33. (3) $I : \dfrac{1}{2}MR^2,\ \ m = v\times\rho = A\times t\times\rho = \pi R^2\times t\times\rho$

$\therefore\ I = \dfrac{1}{2}(\pi R^2\times t\times\rho)R^2 \propto R^4$

$\therefore\ I \propto R^4 \Rightarrow \dfrac{I_1}{I_2} = \left(\dfrac{R_1}{R_2}\right)^4 = \left(\dfrac{0.1}{0.4}\right)^4 = \dfrac{1}{256}$

34. (1) $I = Mk^2 \Rightarrow \sqrt{\dfrac{I}{M}} = K = \sqrt{\dfrac{200}{15}} = 3.65\,\text{m}$

35. (2) Moment of inertia of ring about its diameter

$$I = \frac{Mr^2}{2}$$

According to the parallel axis theorem

$$I = \frac{Mr^2}{2} + Mr^2 \quad I = \frac{3}{2}Mr^2$$

36. (4) Radius of gyration of a body depends on the mass distribution and axis of rotation.

37. (4) For top sphere $I_1 = \frac{MR^2}{2}$

For other two spheres

$$I_2 = I_3 = \frac{MR^2}{2} + MR^2 = \frac{3}{2}MR^2 \text{ (Parallel axis theorem)}$$

Total moment of inertia of the system

$$I = I_1 + I_2 + I_3 = \frac{MR^2}{2} + 3MR^2 = \frac{7}{2}MR^2$$

38. (2) $I = \frac{M}{2}r^2$

$$\therefore \frac{M_1}{M_2} = \frac{I_1}{I_2} \times \left(\frac{R_2}{R_1}\right)^2 = \frac{1}{2} \times \left(\frac{1}{2}\right)^2 = \frac{1}{8}$$

39. (2)

mass of cavity = M $\begin{bmatrix} \pi R^2 \to 9M \\ \pi\frac{R^2}{9} \to M \end{bmatrix}$

MI of Big Disc $= \frac{9MR^2}{2}$

MI of Cavity $= \frac{MR^2}{18}$

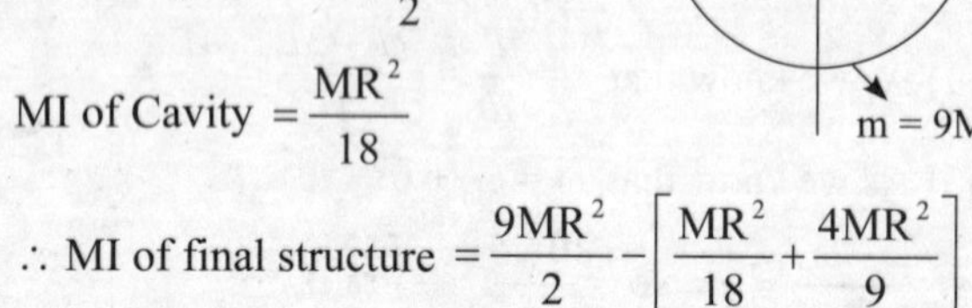

$$\therefore \text{ MI of final structure} = \frac{9MR^2}{2} - \left[\frac{MR^2}{18} + \frac{4MR^2}{9}\right]$$

$$= \frac{9MR^2}{2} - \left[\frac{MR^2 + 8MR^2}{18}\right] = 4MR^2$$

40. (3) Torque about O = 0

$$120\left(\frac{L}{2}\cos 60^\circ\right) - N_2 L = 0$$

$N_2 = 30$ N

41. (1)

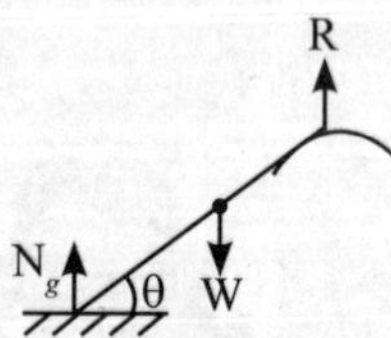

Balancing torque about pt of contact on ground:

$$R(L\cos\theta) = W\left(\frac{L}{2}\cos\theta\right)$$

$$\Rightarrow R = \frac{W}{2}$$

42. (4) $I = MR^2$

taking log on both sides

$\log I = \log M + 2\log R$

Differentiating it, we get

$$\frac{dI}{I} = 0 + 2\frac{dR}{R}$$

$$\frac{dI}{I} = 2\frac{dR}{R}$$

$$\therefore \frac{dI \times 100}{I} = 2 \times \frac{dR}{R} \times 100$$

$\Rightarrow$% increase in moment of inertia = 2 × 1% = 2 %

43. (1) Moment of inertia of system = $I_1 + I_2 + I_3$

$$I_2 = I_3 = \frac{ML^2}{3}$$

(moment of inertia of rod about an axis at one end and perpendicular to the rod)

$$I = 0 + \frac{ML^2}{3} + \frac{ML^2}{3} = \frac{2ML^2}{3}$$

44. (2) Moment of inertia of system $I = I_A + I_B + I_C$

$$I = 0 + \frac{ML^2}{3} + ML^2 = \frac{4}{3}ML^2$$

($I_c = ML^2$, by parallel axis theorem)

45. (4) Wire is bent into circular ring, then

$$l = 2\pi R \Rightarrow R = \frac{l}{2\pi}$$

$\therefore$ Moment of inertia of ring about its own axis

$$\therefore I = MR^2 = M\left(\frac{l}{2\pi}\right)^2 = \frac{Ml^2}{4\pi^2}$$

46. (1) By theorem of parallel axis

$$I = I_{cm} + M(4R)^2 = \frac{2}{5}MR^2 + 16MR^2$$

$$= \frac{2MR^2 + 80MR^2}{5} = \frac{82MR^2}{5}$$

$$\Rightarrow MK^2 = \frac{82MR^2}{5} \Rightarrow K = R\sqrt{\frac{82}{5}}$$

47. (4) $I_{disc} = \frac{Ma^2}{2}$, $I_{ring} = Ma^2$

$$I_{\text{square lamina}} = M\left[\frac{l^2 + b^2}{12}\right] = \frac{M\left((2a)^2 + (2a)^2\right)}{12} = \frac{2}{3}Ma^2$$

$$I_{rod} = \left[\frac{M(2a)^2}{4 \times 12} + \frac{M(2a)^2}{4}\right] = \frac{13}{12}Ma^2$$

$$\Rightarrow I_{square} = \frac{13Ma^2}{12} \times 4 = \frac{13}{3}Ma^2$$

48. (1) $(\tau) = \frac{dL}{dt}$

Given, $dL = 4A_0 - A_0 = 3A_0$, $dt = 4$ s

$$\Rightarrow \quad \tau = \frac{3}{4}A_0$$

49. (2)

50. **(2)** $K = \frac{L^2}{2I}$

so, $K \propto \frac{1}{I}$ {∵ L is same}

as, $I_1 > I_2$ so, $K_2 > K_1$

51. **(2)** Applying conservation of angular momentum :-

$I_1\omega = (I_1 + I_2)\omega'$

so, $\omega' = \frac{I_1\omega}{(I_1 + I_2)}$

52. **(2)** Using conservation of angular momentum, $I_1\omega_1 = I_2\omega_2$. On losing atmosphere, moment of inertia of earth decreases so, ω increases and time period T decreases

$\left\{\because T = \frac{2\pi}{\omega}\right\}$

53. **(3)** Kinetic energy of rotation is given by

$E = \frac{L^2}{2I}$

$\therefore \frac{E_A}{E_B} = \frac{L_A^2}{I_A}\frac{I_B}{L_B^2} \Rightarrow \frac{L_A}{L_B} = 5$

54. **(1)** Angular momentum remains constant, swimmer decreases his moment of inertia by forming a loop which results in increase in angular velocity.

55. **(1)** $\tau = I\alpha$

$\frac{ML^2}{3}\alpha = \frac{MgL}{2} \Rightarrow \alpha = \frac{3g}{2L}$ rad/sec^2

56. **(4)** For a system of particles,

$\frac{d\vec{P}}{dt} = \vec{F}_{ext}, \frac{d\vec{L}}{dt} = \vec{\tau}_{ext}$

57. **(3)** If $\tau_{ext} = 0$, then

$\frac{d\vec{L}}{dt} = 0$

$\Rightarrow \vec{L}$ = constant

58. **(4)** $\tau = \frac{\Delta L}{\Delta t} = \frac{2}{5} = 0.4$ N-m

59. **(3)** Direction of angular momentum is along $\vec{r} \times \vec{v}$.

For the given motion $\vec{r} \times \vec{v}$ and thus $\vec{L}$ will be along the axis.

60. **(2)** As the gymnast pulls his hands inwards, the radius of gyration decreases and hence moment of inertia decreases.

61. **(3)** $\because \vec{\tau} = \frac{d\vec{L}}{dt}$ and external torque about the axis = 0

$\Rightarrow \vec{L}$ = constant.

62. **(1)** $I_1\omega_1 = I_2\omega_2$

$\frac{I_1}{T_1} = \frac{I_2}{T_2}$ $\left[\omega = \frac{2\pi}{T}\right]$

$\frac{R^2}{24} = \frac{(R/2)^2}{T_2}$ $\left[I = \frac{2}{5}MR^2\right]$

$T_2 = 6$ hours

63. **(1)** We use, $I\alpha = \tau$

The torque, $\tau = FR$

$= 25 \times 0.20 = 5.0$ N-m

$I = \frac{MR^2}{2} = \frac{20.0 \times (0.2)^2}{2} = 0.4\,\text{kgm}^2$

$\alpha = \frac{5.0 \text{ N-m}}{0.4 \text{ kg-m}^2} = 12.50$ rad/s^2

64. **(2)** Work done by the force in unwinding 2 m of the cord

25 N × 2 m = 50 J

65. **(3)** Let ω be the final angular velocity.

The wheel starts from rest,

$\omega^2 = \omega_0^2 + 2\alpha\theta,$

The angular displacement $\theta = \frac{\text{Length of unwound string}}{\text{Radius of wheel}}$

$\Rightarrow \theta = \frac{2}{0.2} = 10$ rad

$\omega^2 = 2 \times 12.5 \times 10.0 = 250$ (rad/s)^2

$\therefore$ KE gained $= \frac{1}{2} \times 0.4 \times 250 = 50\,\text{J}$ $\left(\because I = 0.4\,\text{kg m}^2\right)$

66. **(1)** Given, M = 3 kg

R = 40 cm = 0.40 m, F = 30 N

Moment of inertia of the hollow cylinder about its axis of symmetry

$I = MR^2 = 3 \times (0.40)^2 = 0.48$ kg -m^2

Torque acting on the cylinder by the applied force,

$\tau = R \times F = 30 \times 0.40 = 12$ N-m

Angular acceleration produced in cylinder,

$\alpha = \frac{\tau}{I} = \frac{12}{0.48} = 25\,\text{rad/s}^2$

Linear acceleration, $a = R\alpha$

$= 0.40 \times 25 = 10.0$ m/s^2

67. **(3)** KE $= \frac{1}{2} I\omega^2$

$360 = \frac{1}{2} I (30)^2 \Rightarrow I = 0.8$ kg m^2

68. **(1)** Angular momentum is given by $\vec{L} = \vec{r} \times \vec{P}$

69. **(3)**

$$\text{ratio} = \frac{K_{rotation}}{K_{total}} = \frac{\frac{1}{2}I\omega^2}{\frac{1}{2}I\omega^2\left(1 + \frac{mR^2}{I}\right)}$$

$$= \frac{1}{1 + \frac{mR^2}{I}}$$

For disc,

$$\text{Ratio} = \frac{1}{1 + \frac{mR^2}{\left(\frac{mR^2}{2}\right)}} = \frac{1}{3}$$

Solid sphere

$$\text{Ratio} = \frac{1}{1+\frac{mR^2}{\frac{2}{5}mR^2}} = \frac{2}{7}$$

Ring

$$\text{Ratio} = \frac{1}{1+\frac{mR^2}{mR^2}} = \frac{1}{2}$$

$$\text{Hollow sphere ratio} = \frac{1}{1+\frac{mR^2}{\frac{2}{3}mR^2}} = \frac{2}{5}$$

70. (2) $\vec{L} = -\vec{r}\times\vec{p} = m(\vec{r}\times\vec{v})$

As the particle has decreasing linear speed $(|\vec{v}|)$, so $|\vec{L}|$ also decreases i.e., $\vec{L}$ is not conserved. Option (1) is not correct.

Direction of $(\vec{r}\times\vec{v})$ remains same, hence the direction of $\vec{L}$ remains the same. Option (2) is correct.

As the particle is confined to rotate in a circular path, it cannot spiral towards the centre. Option (3) is incorrect.

As the particle has a decreasing speed, it must have a transverse acceleration in addition to the centripetal acceleration. So its acceleration makes some angle with the radius. Option (4) is incorrect.

71. (4) Rotational K.E of ball $= \frac{1}{2}I\omega^2$

$$= \frac{1}{2}\times\frac{2}{5}MR^2\times\omega^2 = \frac{1}{5}\times 5\times(5\times10^{-2})^2\times(50)^2 = 6.25\text{ J}$$

72. (1)

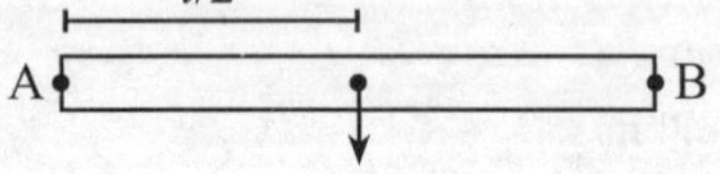

Torque, $\tau_A = mg\times\frac{\ell}{2}$ and $I_A = \frac{m\ell^2}{3}$

as, $\tau_A = I_A\times\alpha$ or $\alpha = \frac{\tau_A}{I_A} = \frac{\frac{mg\ell}{2}}{\frac{m\ell^2}{3}} \Rightarrow \alpha = \frac{3g}{2\ell}$

73. (4)

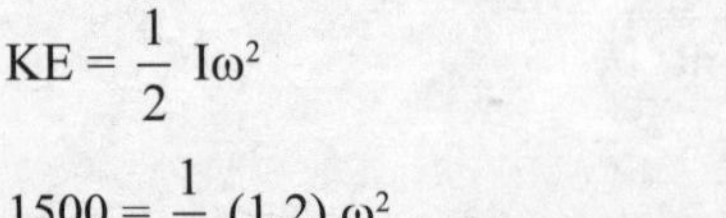

Normal reaction, N provides the centripetal acceleration to COM.

$N = m\omega^2\frac{\ell}{2}$

74. (2) $KE = \frac{1}{2}I\omega^2$

$\Rightarrow 1500 = \frac{1}{2}(1.2)\omega^2$

$\omega = \sqrt{\frac{3000}{1.2}} \Rightarrow \omega_f - \omega_i = \alpha\tau$

$\Rightarrow \sqrt{\frac{3000}{1.2}} = 25t \Rightarrow t = 2\text{ s}$

75. (3) Given : $I = 2\text{kg}-\text{m}^2$, $\omega_0 = \frac{60}{60}\times 2\pi$ rad/s, $\omega = 0$, $t = 60$s

The torque required to stop the wheel's rotation is

$$\tau = I\alpha = I\left(\frac{\omega_0-\omega}{t}\right) \Rightarrow \tau = \frac{2\times2\pi\times60}{60\times60} = \frac{\pi}{15}\text{ N-m}$$

76. (3) Using conservation of angular momentum $I_1\omega_1 = I_2\omega_2$ As person is coming towards axis, it's distance from axis decreases. So moment of inertia also decreases.

From conservation of angular momentum :-

as I decreases so, ω increases

77. (2) Before After

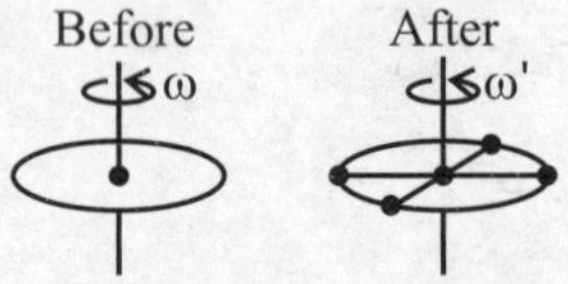

Applying conservation of angular momentum :-

$I_1\omega = I_2\omega' \Rightarrow Mr^2\times\omega = (Mr^2+4mr^2)\omega'$

so, $\omega' = \frac{Mr^2\omega}{r^2(M+4m)} = \frac{M\omega}{M+4m}$

78. (4) A sphere can not roll on a smooth surface (horizontal or inclined) because there is no backward force (friction force) to prevent its slipping. Rolling friction can act parallel or antiparallel to the direction of motion according to the condition of pulling and pushing a spherical or cylindrical body.

79. (4) Acceleration of center of mass of mass of rolling body depends on only angle of inclination.

Time taken depends on length of inclined plane and angle of inclination.

80. (4)

81. (1) For ring, $I = MR^2$

Acceleration of a body (pure) rolling on the inclined plane, $a = \frac{g\sin\theta}{1+\frac{I}{MR^2}}$

$$a = \frac{g\sin 30°}{1+\frac{MR^2}{MR^2}} \Rightarrow a = \frac{g}{4}$$

82. (3)

83. (3) The translational kinetic energy of the solid sphere

$K_t = \frac{1}{2}Mv^2$ ... i ... (*i*)

The rotational kinetic energy

$K_r = \frac{1}{2}I\omega^2 = \frac{1}{2}\times\frac{2}{5}MR^2\times\frac{v^2}{R^2}$

$\Rightarrow K_r = \frac{1}{5}Mv^2$ ... (ii)

On dividing Eq. (ii) by Eq. (i), we have

$\frac{K_r}{K} = \frac{1/5Mv^2}{1/2Mv^2} \Rightarrow \frac{K_r}{K} = \frac{2}{5}$

84. (2) Total kinetic energy $= \frac{1}{2}mv^2\left(1+\frac{K^2}{R^2}\right)$ ... (i)

Rotational kinetic energy $= \frac{1}{2}mv^2\left(\frac{K^2}{R^2}\right)$ ... (ii)

from eq. (i) & (ii) we get,

$$\therefore \frac{E_{Total}}{E_R} = \frac{\frac{1}{2}mv^2\left(1+\frac{K^2}{R^2}\right)}{\frac{1}{2}mv^2\left(\frac{K^2}{R^2}\right)} = \frac{1+\frac{K^2}{R^2}}{\frac{K^2}{R^2}}$$

$$= \frac{1+\frac{1}{2}}{\frac{1}{2}} = 3 \qquad \left(For\ disc\ \frac{K^2}{R^2} = \frac{1}{2}\right)$$

85. (4) Rotational kinetic energy $= \frac{2}{3}$ translational kinetic energy (given)

$\frac{1}{2}mv^2\left(\frac{K^2}{R^2}\right) = \frac{2}{3}\times\frac{1}{2}mv^2$

$\frac{K^2}{R^2} = \frac{2}{3}$

Therefore, body may be hollow sphere or spherical shell.

86. (1)

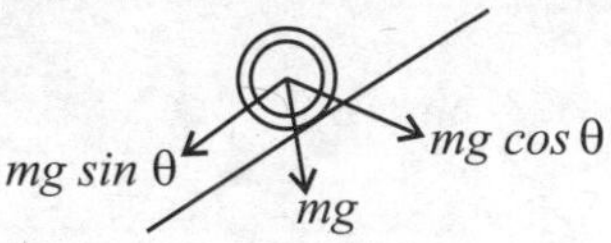

mg sin θ, mg cos θ, mg

$$a = \frac{g\sin\theta}{1+\frac{K^2}{R^2}} = \frac{g\times\frac{1}{2}}{1+1} = \frac{g}{2\times 2} = \frac{g}{4}$$

87. (2) KE $= \frac{1}{2}$ mV² $+ \frac{1}{2}$I ω² $= \frac{1}{2}$mV² $+ \frac{1}{2}$mR² $\left(\frac{V}{R}\right)^2$

KE = mV² = 4 J

88. (1) Total energy of disc $= \frac{1}{2}mu^2 + \frac{1}{2}I\omega^2$

$$= \frac{1}{2}mu^2 + \frac{1}{2}\frac{mR^2}{2}\times\frac{u^2}{R^2} = \frac{2mu^2+mu^2}{4} = \frac{3mu^2}{4}$$

Total energy gets converted into P.E.

$mgh = \frac{3mu^2}{4}$

$\therefore h = \frac{3u^2}{4g}$

89. (3)

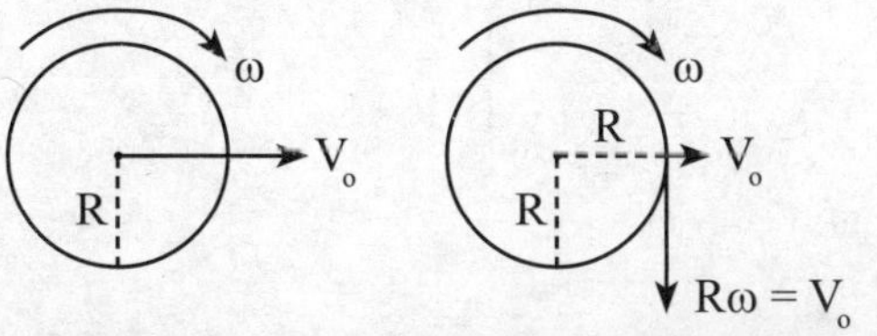

ω, V_o, R; ω, R, V_o, $R\omega = V_o$

P will have two velocities, one due to translation and one due to rotation.

$V_{net} = \sqrt{V_0^2 + V_0^2} = \sqrt{2}V_o$

90. (2)

$v_P = 2R\omega$, $v_Q = R\omega$

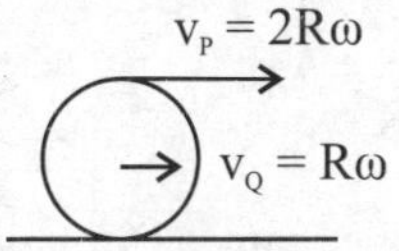

For pure rolling $v_Q = R\omega$, $v_P = v_Q + R\omega$, $v_P = 2v_Q$

91. (1) Given, $\frac{K_{Rotation}}{K_{Total}}\times 100 = 50\%$

$$\Rightarrow \frac{\frac{K^2}{R^2}}{1+\frac{K^2}{R^2}}\times 100 = 50 \Rightarrow \frac{\frac{K^2}{R^2}}{1+\frac{K^2}{R^2}} = \frac{1}{2}$$

so, $\frac{K^2}{R^2} = 1$, so it is a ring.

92. (1) Total KE $= \frac{1}{2}$ MV² $+ \frac{1}{2}$ Iω² $= \frac{3}{4}$ MV²

Δ KE $= \frac{3}{4}$ M (4V² – V²) $= \frac{9}{4}$ MV²

93. (1) $a = \frac{g\sin\theta}{1+\left(\frac{K^2}{R^2}\right)}$

K (solid sphere) $= \sqrt{\frac{2}{5}}$ R

K (solid cylinder) $= \frac{R}{\sqrt{2}}$

$\frac{a_{sphere}}{a_{cylinder}} = \frac{15}{14}$

94. (1) Kinetic energy of ring,

$K = \frac{1}{2}mv^2\left(1+\frac{K^2}{R^2}\right)$

$K = \frac{1}{2}0.4\times\left(\frac{10}{100}\right)^2\times(1+1) \Rightarrow K = 4\times 10^{-3}$ J $\left[\therefore \frac{K^2}{R^2} = 1\right]$

95. (4) As the inclined plane is smooth, the sphere can never roll rather it will just slip down.

Hence, the angular momentum remains conserved about any point on a line parallel to the inclined plane and passing through the centre of the ball because the only forces, Mg & N, acts at centre.

96. (1)

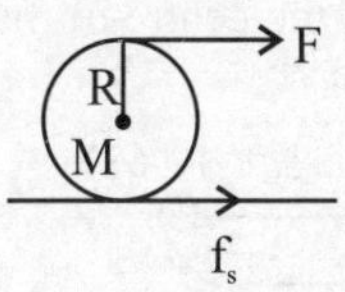

F, R, M, f_s

If F is applied at a height above $\left(1+\frac{K^2}{R^2}\right)$R from ground then friction is forward .

For height lower than $\left(1+\frac{K^2}{R^2}\right)$R, friction is backwards.

97. (2) The kinetic energy of a rolling body is the sum of kinetic energy of translation $\left(\frac{1}{2}Mv_{cm}^2\right)$ and kinetic energy of rotation

So, total KE of rolling motion.

$$K = K_{rot} + \frac{1}{2}M\,v_{CM}^2 \Rightarrow K_{rot} = K - \frac{Mv_{CM}^2}{2}$$

98. (4) Total kinetic energy is the sum of rotational and translational kinetic energy

$$K.E = K.E_r + K.E_T = \frac{1}{2}I\omega^2 + \frac{1}{2}mv^2$$

High Order Time Intensive MCQs

1. (2) Moment of inertia of rod about the axis, $I_1 = \frac{ML^2}{12}$

Moment of inertia of the 2 masses

$$I_2 = m\left(\frac{L}{2}\right)^2 + m\left(\frac{L}{2}\right)^2 = \frac{mL^2}{2}$$

$$\therefore I = I_1 + I_2 = \frac{ML^2}{12} + \frac{mL^2}{2} = \frac{(M+6m)L^2}{12}$$

2. (4) Using parallel axis theorem,

$I_{X'Y'} = I_{XY} + MR^2 = 2 + (1)(2)^2 = 6$ kg-m²

3. (1) One ring has axis perpendicular to its plane and other ring has axis along its diameter.

$$I = MR^2 + \frac{1}{2}MR^2 = \frac{3}{2}MR^2$$

4. (3) From conservation angular momentum,

$0 = I\omega - mvr \Rightarrow I\omega = mvr \Rightarrow 200 \times \omega = 50 \times 2 \times 1$

Or $\omega = \frac{1}{2}$ rad/s

Linear velocity of a point on the rim of the platform, $v = r\omega = 1$ m/s

Time taken, $t = \frac{2\pi r}{1-(-1)} = \frac{2\pi r}{2} = \pi r = 2\pi$ s

5. (2) Maximum height $H = \frac{v^2 \sin^2\theta}{2g}$

$$\Rightarrow \frac{v^2\sin^2 45°}{2g} = \frac{v^2}{4g}$$

At highest point, momentum $= p = mv\cos 45° = \frac{mv}{\sqrt{2}}$

Angular momentum $= H \times p = \frac{mv}{\sqrt{2}} \times \frac{v^2}{4g} = \frac{mv^3}{4\sqrt{2}g}$

6. (3) As there is no external force, the centre of mass of the system does not shift.

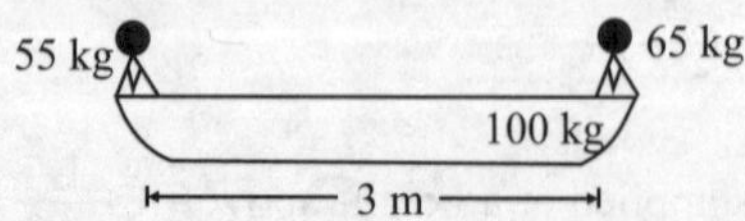

7. (2) We know that,

$\vec{L} = \vec{r} \times \vec{p}, L = r\,mv\sin\phi(-\hat{k})$

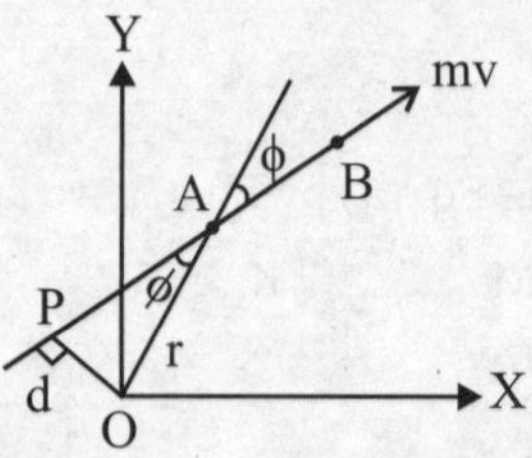

Therefore, the magnitude of L is mvr sinϕ = mvd

Where, d = r sinϕ is the distance of closed approach of the particle to the origin. As d is same for both the positions of the particle, hence, $L_A = L_B$.

8. (3) $I = \frac{2}{5}MR^2$ for a solid sphere,

Given, $R_1 = R, R_2 = \frac{R}{n}$

Using conservation of angular momentum

$$\therefore \frac{2}{5}MR^2\omega = \frac{2}{5}M\left(\frac{R}{n}\right)^2 \times \omega_2$$

$$\Rightarrow \omega_2 = n^2\omega$$

9. (2) Since, $\tau = I\alpha$

So, $\tau = \left(\frac{m(2\ell)^2}{12}\right)\left(\frac{\omega}{t}\right) = \left(\frac{m\ell^2\omega}{3t}\right)$

10. (1) Momentum of the particles = Mass × Velocity

$= 5 \times (3\sqrt{2}) = 15\sqrt{2}$

Slope of the line = tan θ = 1 ⇒ θ = 45°

The direction of momentum is given by slope of line θ = 45°

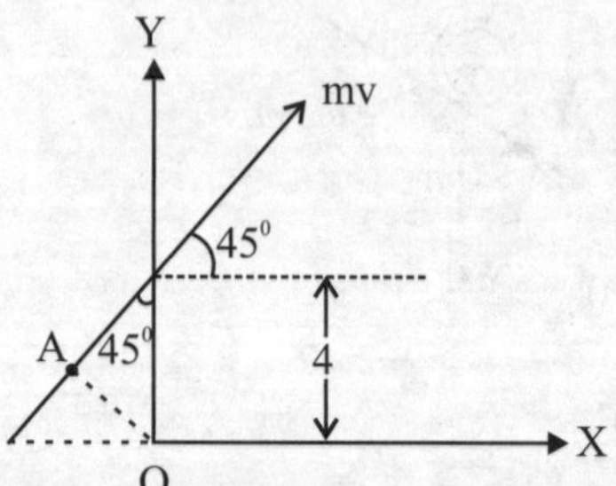

Intercept of straight line = 4

Length of the perpendicular from the origin OA on the straight line $= 4\sin 45° = \frac{4}{\sqrt{2}} = 2\sqrt{2}$

Angular momentum = momentum × perpendicular length

$= 15\sqrt{2} \times 2\sqrt{2} = 60$ units

11. (2) The torque exerted on the disc is given by

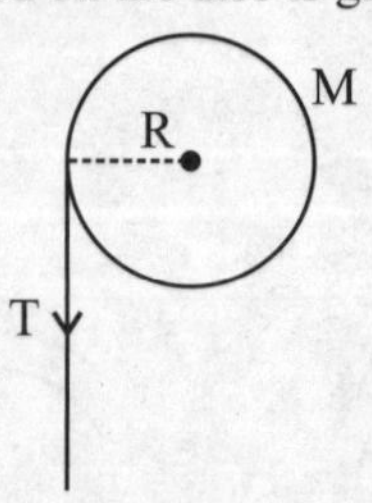

$\tau = TR$ ... (*i*)

$\tau = I\alpha$ ... (*ii*)

From Eqs. (*i*) and (*ii*), we get

$I\alpha = TR$

$\Rightarrow \quad \alpha = \frac{TR}{I} = \frac{2TR}{MR^2} \Rightarrow \alpha = \frac{2T}{MR}$

12. (3)

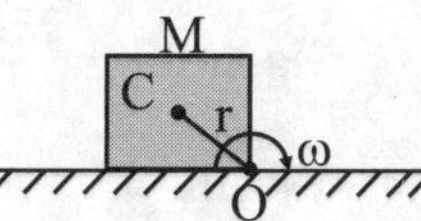

Clearly, $(2r)^2 = a^2 + a^2$

$\Rightarrow r = \frac{a}{\sqrt{2}}$

As the net torque about O is zero, angular momentum is conserved.

$L_i = L_f$

$Mv\left(\frac{a}{2}\right) = I_0\omega \Rightarrow \frac{Mva}{2} = (I_c + Mr^2)\omega$

or, $\frac{Mva}{2} = \left[\frac{Ma^2}{6} + M\left(\frac{a^2}{2}\right)\right]\omega$

$\therefore \omega = \frac{3v}{4a} = \frac{3 \times 2}{4 \times 0.30} = 5 \text{ rad s}^{-1}$

13. (1) Using conservation of energy (there is no loss of energy due to friction). The potential energy lost by the body in rolling down the inclined plane (mgh) = kinetic energy gained.

$K = \frac{1}{2}mv^2\left(1 + \frac{k^2}{R^2}\right)$, where v is the final velocity of the centre of mass of the body. Equating K and mgh.

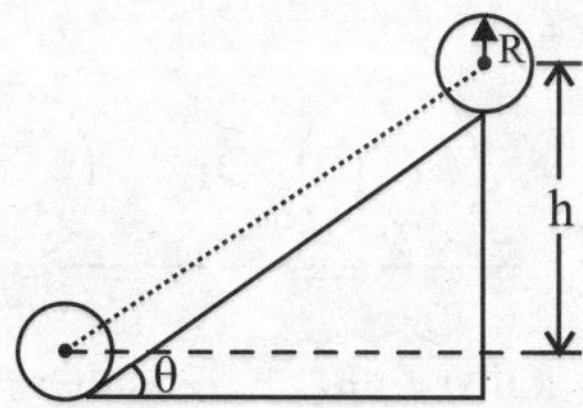

$mgh = \frac{1}{2}mv^2\left(1 + \frac{k^2}{R^2}\right) \quad$ or $\quad v^2 = \left(\frac{2gh}{1 + k^2/R^2}\right)$

Note: Velocity v is independent of the mass of the rolling body

For a ring, $k^2 = R^2$ [$\because I = MR^2$ and $I = Mk^2$]

$v_{ring} = \sqrt{\frac{2gh}{1+1}} = \sqrt{gh}$

For a solid cylinder, $k^2 = R^2/2$ [$\because I = 1/2\ MR^2$ and $I = Mk^2$]

$v_{disc} = \sqrt{\frac{2gh}{1+1/2}} = \sqrt{\frac{4gh}{3}}$

For a solid sphere, $k^2 = 2R^2/5$. [$\because I = 2/5\ MR^2$ and $I = Mk^2$]

$v_{sphere} = \sqrt{\frac{2gh}{1+2/5}} = \sqrt{\frac{10gh}{7}}$

From the results obtained, it is clear that among the three bodies the sphere has the greatest and the ring has the least velocity of the centre of mass at the bottom of the inclined plane.

14. (2) Kinetic energy of rotation, $K = \frac{1}{2}I\omega^2$

Where, $\omega = \frac{v}{r}$, r is radius of circle.

$\therefore \quad K = \frac{1}{2}I\left(\frac{v^2}{r^2}\right)$

$\Rightarrow \frac{K_2}{K_1} = \frac{r_1^2}{r_2^2} = \left(\frac{r_1}{r_2}\right)^2 \quad$ ($\because$ I and v is constant)

15. (3) For a rolling a solid cylinder, acceleration,

$a_{CM} = \frac{2g}{3}\sin\theta$

Now, $F_{net} = Ma_{CM}$

$\Rightarrow Mg\sin\theta - f = \frac{2Mg\sin\theta}{3} \Rightarrow f = \frac{Mg\sin\theta}{3}$

For no slipping,

$f \leq \mu_s N$

$\Rightarrow \frac{Mg\sin\theta}{3} < \mu_c Mg\cos\theta \Rightarrow \tan\theta \leq 3\mu_s$

16. (2) For cylinder,

$\frac{K^2}{R^2} = \frac{1}{2}$

So acceleration,

$a = \frac{g\sin\theta}{1 + \frac{K^2}{R}} = \frac{g\sin 60°}{1 + \frac{1}{2}} = \frac{g}{\sqrt{3}}$

17. (3) Time of descent on an inclined plane

$t_r = \frac{1}{\sin\theta}\sqrt{\frac{2h}{g}\left(1 + \frac{K^2}{R^2}\right)}$

which is independent of mass. Since radii of the two spheres are equal hence, both will reach at the same time.

18. (4) From equation of motion, $v^2 = u^2 + 2gh$

$(r\omega)^2 = 2gh$

$\Rightarrow r^2 = \frac{2 \times 10 \times 3}{(2\sqrt{2})^2}$

$\therefore \quad r = \frac{\sqrt{60}}{2\sqrt{2}} \approx \sqrt{7.5}$

19. (4) Let I be the M.I. of the square lamina about an axis through O and perpendicular to its plane. Then by perpendicular axis theorem,

$I_{AC} + I_{BD} = I$

or $I_{AC} = I/2 \quad [\because I_{AC} = I_{BD}]$

Again by perpendicular axis theorem,

$I_{EF} + I_{GH} = I$

or $I_{EF} = I/2 \quad [\because I_{EF} = I_{GH}]$

Hence, $I_{AC} = I_{EF}$

20. (2) Acceleration of a rolling body down an inclined plane is

$$a = \frac{g\sin\theta}{1+\frac{k^2}{R^2}}$$

For solid cylinder, $k^2 = \frac{R^2}{2}\left[\because I = \frac{MR^2}{2} \text{ and } I = Mk^2\right]$

$$\therefore\ a = \frac{g\sin\theta}{1+\frac{1}{2}} = \frac{2}{3}g\sin\theta$$

21. (3) Acceleration of the solid sphere when it rolls without slipping down an inclined plane is

$$a = \frac{g\sin\theta}{1+\frac{I}{MR^2}} \quad ...(i)$$

For a solid sphere, $I = \frac{2}{5}MR^2$

$$\therefore\ a = \frac{g\sin\theta}{1+\frac{2}{5}} = \frac{5}{7}g\sin\theta$$

Acceleration of the same sphere when it slides without friction down the same inclined plane is

$a' = g\sin\theta \quad ...(ii)$

Divide (*ii*) by (*i*), we get

$$\frac{a'}{a} = \frac{7}{5} \text{ or } a' = \frac{7}{5}a$$

22. (4) Acceleration of centre of mass

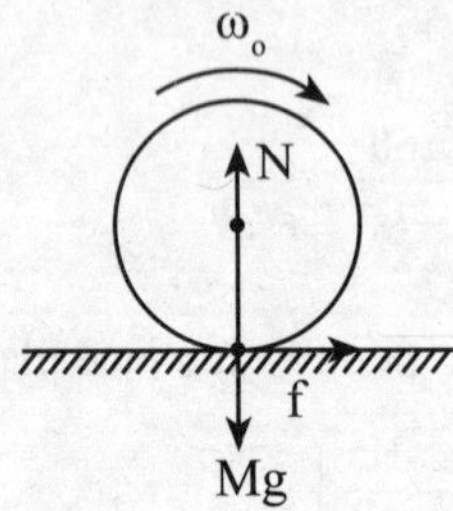

$$a_{CM} = \frac{f_k}{M} = \frac{\mu_k Mg}{M} = \mu_k g \quad ...(i)$$

Angular retardation produced by the torque

$$\alpha = \frac{\tau}{I} = \frac{f_k r}{I} = \frac{\mu_k Mgr}{I} = \frac{\mu_k g}{r} \quad ...(ii)$$

$V_{CM} = u_{CM} + a_{CM}t$

$\therefore\ V_{CM} = 0 + \mu_k gt$

$\omega = \omega_0 + \alpha t$

$\therefore\ \omega = \omega_0 - \frac{\mu_k gt}{r}$ (Using (*ii*))

For rolling without slipping

$V_{CM} = r\omega$

$$\frac{\mu_k gt}{r} = \omega_0 - \frac{\mu_k gt}{r} \Rightarrow t = \frac{r\omega_0}{2\mu_k g}$$

23. (1)

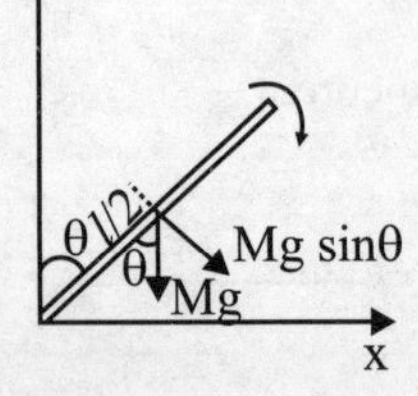

$$I = \frac{Ml^2}{3}$$

$$\tau = Mg\sin\theta.\frac{l}{2}$$

$\tau = I\alpha$

$$\therefore\ Mg\sin\theta.\frac{l}{2} = \frac{Ml^2}{3}.\alpha \Rightarrow \alpha = \frac{3g}{2l}\sin\theta$$

24. (4) $W = \Delta K$

$$W = \frac{1}{2}I\omega^2$$

$$= \frac{1}{2}\frac{MR^2}{2}\times(2\pi f)^2 = \frac{MR^2}{4}\times 4\pi^2\times(10)^2$$

$$= 1\times(0.4)^2\times 10\times 100 = 160$$

25. (4) In the frame of A, point B is moving in a circle with angular speed ω.

$\Rightarrow \vec{v_B} = \vec{v_A} + \vec{\omega}\times\overrightarrow{AB}$

$\Rightarrow \omega = 0.5\text{rad/s}$

Substituting,

$\vec{V_A} = -20j$

$\vec{V_B} = 30j$

Velocity vector of the center C of the rod satisfies the following equation,

$\vec{v_C} = \vec{v_A} + \vec{\omega}\times\overrightarrow{AC}$

$\Rightarrow \vec{v_C} = -20\hat{j} + 0.5\hat{k}\times 50\hat{i} = 5\hat{j}\text{ cm/s}$

26. (3) By conservation of angular momentum about the point of contact,

$I\omega_0 = mvr + I\omega$

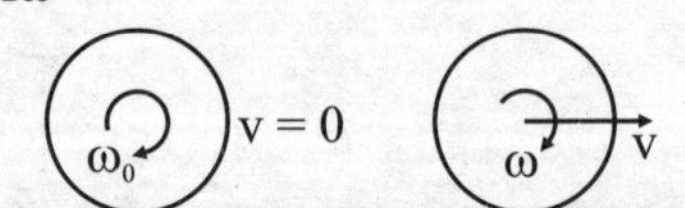

$$\Rightarrow mr^2\omega_0 = mvr + mr^2.\frac{v}{r} \Rightarrow r^2\omega_0 = 2vr$$

$$\therefore\ v = \frac{r\omega_0}{2}$$

27. (2) $a = R\alpha$

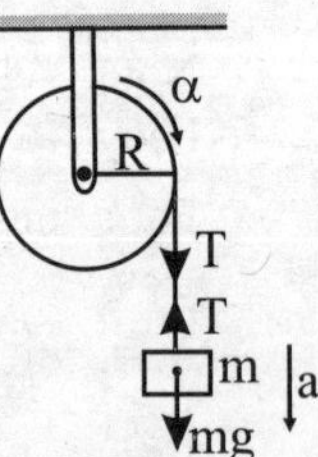

$mg - T = ma$

$\&\ T\times R = mR^2\alpha = mR^2\times\frac{a}{R}$

$\Rightarrow T = ma$

$\therefore\ mg - ma = ma \Rightarrow a = \frac{g}{2}$

28. (1) As the particle is moving in x-y plane,

$z = 0$ and $p_z = 0$

$$\vec{L} = \vec{r}\times\vec{p} = \begin{vmatrix} \hat{i} & \hat{j} & \hat{k} \\ x & y & 0 \\ p_x & p_y & 0 \end{vmatrix}$$

$$= \hat{k}(xp_y - yp_x)$$

But $x = vt$, $y = b$, $p_x = mv$, $p_y = 0$.

$\therefore \ \vec{L} = \hat{k}(vt \times 0 - bmv) = -mvb\hat{k}$

29. **(1)** By the law of conservation of energy

K.E = P.E

$$\frac{1}{2}I\omega^2 = mgh$$

$$\frac{1}{2}\left(\frac{ml^2}{3}\right)\omega^2 = mgh$$

$$h = \frac{l^2\omega^2}{6g}$$

30. **(3)** By conservation of angular momentum,

$$\frac{MR^2}{2}\omega = \left(\frac{MR^2}{2} - mR^2\right)\omega' + mR^2\omega$$

$$\Rightarrow \omega' = \frac{\left(MR^2 - 2mR^2\right)\omega}{\left(MR^2 - 2mR^2\right)} \Rightarrow \omega' = \omega$$

31. **(3)** $v_{CM} = \dfrac{m_1v_1 + m_2v_2}{m_1 + m_2} = \dfrac{10 \times 14 + 4 \times 0}{4 + 10} = 10\,ms^{-1}$

32. **(2)** For the circular motion of com:

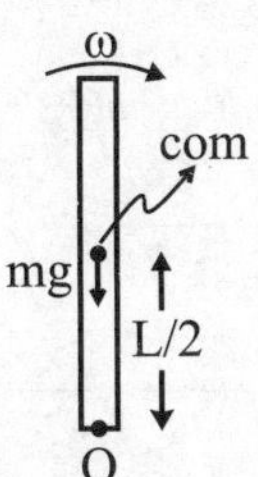

$$mg = m\left(\frac{L}{2}\right)\omega^2$$

$$\Rightarrow \omega = \sqrt{\frac{2g}{L}}$$

33. **(1)** For pure rolling of sphere on the plank, the point of contact of sphere should have velocity equal to the of the plank.

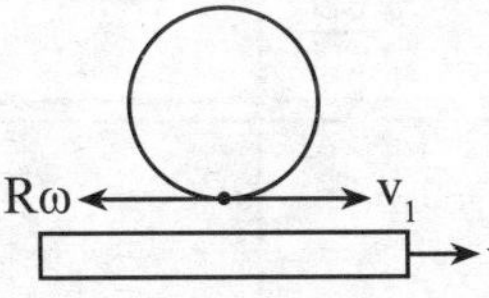

$\Rightarrow v_1 - R\omega = v$

$\Rightarrow v_1 - v = R\omega$

Statement Based MCQs

1. **(1)** Both statements are true.
2. **(1)** Both statements are true.
3. **(3)** Torque = Force × momentum

   So, larger the arm smaller the force needed to open the screw.
4. **(4)** MOI of rod about an axis passing through its centre $= \dfrac{ml^2}{12}$

   Since, gravitational force acts on COM, thus, $\tau_g = 0$ about COM.
5. **(3)** If there were only one propeller in the helicopter, the helicopter itself, would have turned in opposite direction of the direction of propeller due to conservation of angular momentum. Thus, two propellers provide helicopter a steady movement.

Assertion & Reason MCQs

1. **(1)** This is so because the top has greater linear velocity, i.e., it is moving faster with respect to bottom of wheel hence reason correctly explains the assertion.
2. **(3)** When the person is standing higher on the ladder the torque about bottom most point of the ladder is larger.
3. **(3)** Raw egg has more moment of inertia than a hard boiled egg so it rotates slower.
4. **(1)** The rivers deposit the sediments farther from the axis. So MI is increased & ω decreases.
5. **(4)** If there is a component of force acting along the motion of body then friction may act on it.

Match the Columns MCQs

1. **(1)** Let the sphere of mass M and radius R be struck horizontally at a height h from the floor, as shown in the figure.

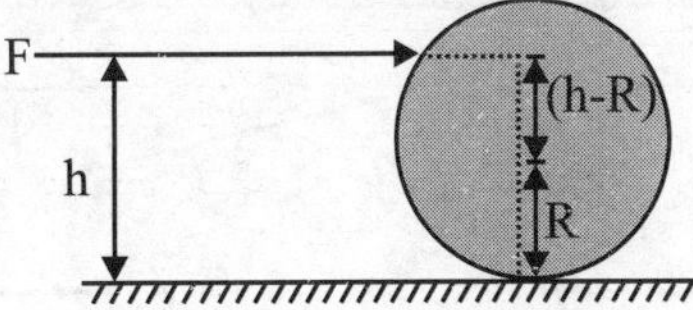

The sphere will roll without slipping when $\omega = \dfrac{v}{R}$.

Impulse, $J = MV$

Angular impulse about centre,

$J(h - R) = I\omega$

$$\Rightarrow Mv(h - R) = \left(\frac{2}{5}MR^2\right)\left(\frac{v}{R}\right)$$

$$Mv(h - R) = \frac{2}{5}MvR$$

$$h - R = \frac{2}{5}R \ \text{ or } \ h = \frac{7}{5}R$$

$\therefore$ The sphere will roll without slipping with a constant velocity and no loss of energy when

$$h = \frac{7}{5}R.$$

$\therefore$ D – (i)

Torque due to applied force about centre of mass,

$\tau = F(h - R)$

If $h = R$, $\tau = 0$, sphere will have only translational motion. It would lose energy by friction.

$\therefore$ B – (iv)

If $h > R \Rightarrow \tau = +ve$ sphere will spin clockwise.

$\therefore$ C – (ii)

Again, the sphere will spin anticlockwise, when τ is negative, i.e., $h < R$.

$\therefore$ A – (iii)

2. **(2)** Torque is equivalent to the force, Moment of Inertia is equivalent to the mass, Angular momentum is equivalent to the Linear Momentum and Angular velocity is equivalent to the Linear velocity.

3. **(1)** **4.** **(2)**

NCERT Exemplar MCQs

1. **(4)** We know that angular acceleration

$\alpha = \frac{d\omega}{dt}$, given ω = constant, where ω is angular velocity of the disc

$\Rightarrow \quad \alpha = \frac{d\omega}{dt} = \frac{0}{dt} = 0$

Hence, angular acceleration is zero.

2. **(2)** In the given diagrams, when the small piece Q is removed and glued to the centre of the plate, the mass comes closer to the z-axis. Hence, moment of inertia decreases.

3. **(3)** Consider the adjacent diagram, there is a line shown in the figure drawn along the diagonal. Initially, centre of mass of the system was at the dotted line diagonally away from the centre of the plate.

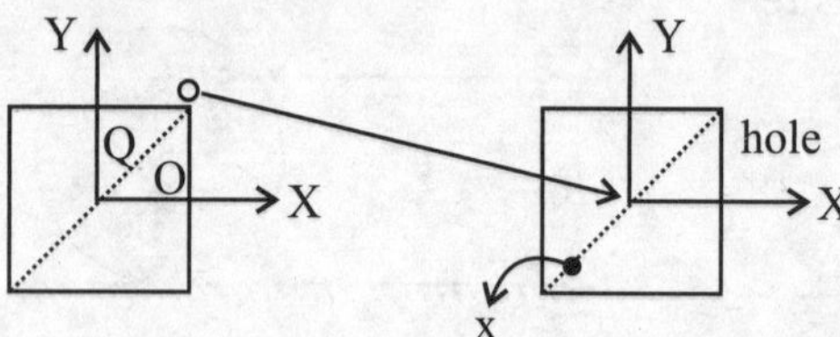

When mass is removed, it will be on the same line but shifted away from the centre and below (3rd quadrant). Position of C.M. is shown by x in the diagram.

4. **(1)** As no external torque acts on the system, angular momentum should be conserved. Hence I ω = constant ...(*i*)

$\Rightarrow I\omega = \frac{I}{2} \times \omega_2$ (as mass reduced to half, hence, moment of inertia also reduced to half)

$\Rightarrow \omega_2 = 2\omega$

NEET & AIIMS Past Year MCQs

1. **(1)**

Angular Speed (ω):

ω_1(second hand) $= \frac{2\pi}{60}$ rad/s (completes 2π in 60s).

ω_2 (minute hand) $= \frac{2\pi}{3600}$ rad/s (completes 2π in 3600s).

ω_3 (hour hand) $= \frac{2\pi}{43200}$ rad/s (completes 2π in 43200s).

Angular Distance in 1 Minute (x):

$x = \omega \times 60$ (since 1 minute = 60s).

$x_1 = 2\pi$ rad (second hand completes full circle).

$x_2 = \frac{2\pi}{60}$ rad (minute hand moves $\frac{1}{60}$th of a circle).

$x_3 = \frac{2\pi}{720}$ rad (hour hand moves $\frac{1}{720}$th of a circle).

For the second hand:

$$\frac{\omega_1}{x_1} = \frac{\frac{2\pi}{60}}{2\pi} = \frac{1}{60}$$

For the minute hand:

$$\frac{\omega_2}{x_2} = \frac{\frac{2\pi}{3600}}{\frac{2\pi}{60}} = \frac{1}{60}$$

For the hour hand:

$$\frac{\omega_3}{x_3} = \frac{\frac{2\pi}{43200}}{\frac{2\pi}{720}} = \frac{1}{60}$$

All ratios equal $\frac{1}{60}$ so $k = \frac{1}{60}$ is constant.

Hence, options 1 is correct.

2. **(4)** Moment of inertia about XY

$I_{XY} = I_{CM} + mR^2$

For a solid sphere: $I_{CM} = \frac{2}{5}mR^2$

Thus:

$I_{XY} = \frac{2}{5}mR^2 + mR^2 = \frac{7}{5}mR^2$

But $I_{XY} = mk^2$:

$\frac{7}{5}mR^2 = m(5)^2$

$R^2 = \frac{25 \times 5}{7}$

$R = \frac{5\sqrt{5}}{\sqrt{7}} = \frac{5x}{\sqrt{7}}$

Thus, $x = \sqrt{5}$.

3. **(3)**

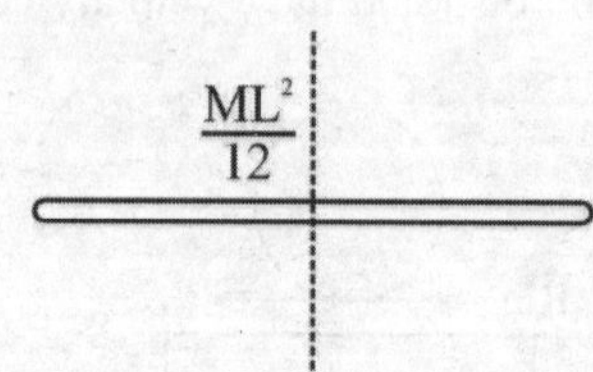

$\frac{ML^2}{12} = 2400 \Rightarrow \frac{400L^2}{12} = 2400$

$\Rightarrow L^2 = \frac{2400 \times 12}{400}$

$\Rightarrow L^2 = 72$

$\Rightarrow L = \sqrt{72}$

$\Rightarrow L = 8.5\ cm$

4. **(4)** In case of rolling, the velocity of the point of contact with the surface is zero.

Also, the velocity of the point at the highest point of the rolling body is twice the velocity of COM of the body.

So, point P moves faster than point Q.

5. **(1)** The angular acceleration direction is given along angular velocity or opposite to angular velocity depending upon whether angular velocity magnitude is increasing or decreasing and this direction remains along the axis of circular motion.

6. (None)
Radius of gyration of a solid surface,

$$K_S = \sqrt{\frac{2}{5}}R$$

Radius of gyration of a hollow surface,

$$K_H = \sqrt{\frac{2}{3}}R$$

$$\Rightarrow \frac{K_S}{K_H} = \sqrt{\frac{3}{5}} = \frac{\sqrt{3}}{\sqrt{5}}$$

7. (3) $\omega_0 = 1200 \text{ rpm} = \frac{1200}{60} \times 2\pi \text{ rad/s}$

and $\omega = 3120 \text{ rpm} = \frac{3120}{60} \times 2\pi \text{ rad/s}$

$\omega = \omega_0 + \alpha t$

$\alpha = \frac{\omega - \omega_0}{t}$

$\alpha = 4\pi \text{ rad/s}^2$

8. (3) Moment of inertia of the disc about an axis perpendicular to the disc and passing through the centre

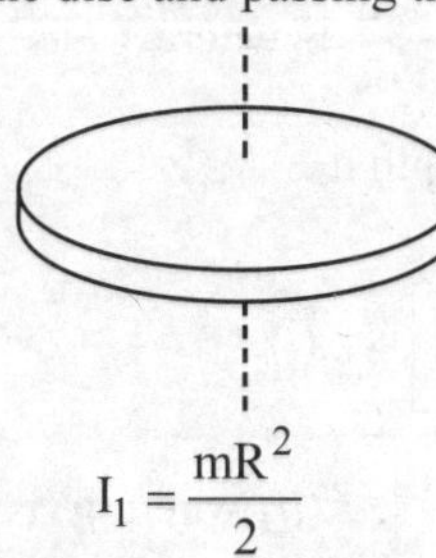

$$I_1 = \frac{mR^2}{2}$$

Moment of inertia of the disc about an axis in the plane of the disc and passing through the centre

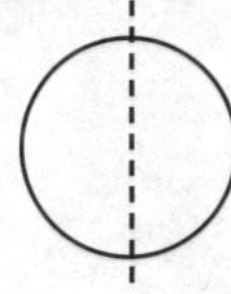

$$I_2 = \frac{mR^2}{4}$$

$$\Rightarrow \frac{k_1}{k_2} = \sqrt{\frac{I_1}{I_2}} = \sqrt{\frac{mR^2/2}{mR^2/4}} = \sqrt{2}:1$$

9. (3) 10 kg at (0, 0), CM at 10 m, 20 kg at (10, 0); X_{CM}

$$\because X_{CM} = \frac{m_1x_1 + m_2x_2}{m_1 + m_2} \Rightarrow X_{CM} = \frac{0 + 20 \times 10}{20 + 10} = \frac{20}{3} \text{m}$$

10. (3) By balancing torque

$2g \times 20 = 0.5g \times 60 + mg \times 120$

$m = \frac{1}{12} \text{kg}$

11. (2) Fact that C.O.M. of half disc is at distance of $\frac{4a}{3\pi}$ from center.

12. (1) 13. (1) 14. (1) 15. (1) 16. (4)

Gravitation

NCERT Topic-wise MCQs

1. (4) Only Archimedes uplift will change as it is dependent on acceleration due to gravity.

2. (4) $F = \frac{Gm_1m_2}{r^2}$ and $F' = \frac{Gm_1m_2}{(3r)^2} = \frac{F}{9}$

$$\% \text{ decrease} = \frac{\left(\frac{F}{9} - F\right) \times 100}{F} = -\frac{8}{9} \times 100 = -89\%$$

3. (2) $F = \frac{G M_E m}{R_E^2}$

$mg = \frac{G M_E m}{R_E^2}$

$g = \frac{G M_E}{R_E^2}$

$M_E = \frac{gR_E^2}{G}$

4. (3) $F = \frac{GM \times 5M}{(12R)^2} \Rightarrow a_{small} = \frac{5GM}{(12R)^2}$

$a_{large} = \frac{GM}{(12R)^2}$

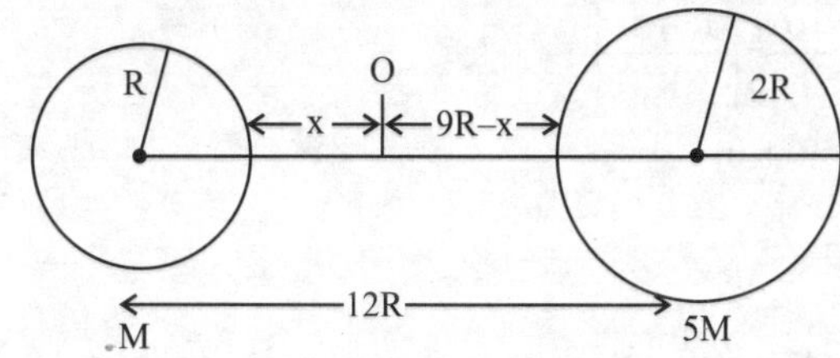

$$x = \frac{1}{2} \frac{G\,5M\,t^2}{(12\,R)^2} \quad ...(i)$$

$$9R - x = \frac{1}{2} \frac{GM\,t^2}{(12\,R)^2} \quad ...(ii)$$

$$\frac{x}{9R - x} = 5 \Rightarrow x = 7.5\,R$$

5. (4) The force of gravitation will be same as F because gravitational force depend on the mass of the body and distance between them and it is independent of the medium between the particles.

6. (2) Acceleration due to gravity at the equator is

$g_e = g - R\omega^2$

$\therefore mg_e = mg - mR\omega^2$

or $\frac{3}{5}mg = mg - mR\omega^2 \quad \left[\because mg_e = \frac{3}{5}mg\right]$

$$\therefore \omega = \sqrt{\frac{2g}{5R}} = \sqrt{\frac{2 \times 9.8}{5 \times 6400 \times 10^3}} = 7.8 \times 10^{-4} \text{ rad s}^{-1}.$$

7. (3) $\dfrac{g_e}{g_P}=\dfrac{G_e\frac{4}{3}\pi R_e^3\rho R_p^2}{G_P\frac{4}{3}\pi R_P^3\rho R_e^2}=\dfrac{G_eR_e}{G_pR_p}$

$\dfrac{R_P}{R_e}=\dfrac{1}{2}\quad[\because G_P=2G_e \text{ and } g_e=g_P]$

8. (1) Here h = 32 km, R = 6400 km

As h << R, so

$g_h=g\left(1-\dfrac{2h}{R}\right)=g-\dfrac{2gh}{R}$

or $g-g_h=\dfrac{2gh}{R}$

Percent decrease in weight

$=\dfrac{mg-mg_h}{mg}\times 100=\dfrac{g-g_h}{g}\times 100$

$=\dfrac{2gh}{g\times R}\times 100=\dfrac{2h}{R}\times 100=\dfrac{2\times 32}{6400}\times 100=1\%$

9. (1) Weight of body of mass "m" at pole = mg

$\because g'=g-\omega^2R\cos\lambda$

where ω is the angular velocity of the earth.

$\omega=0$ at pole

hence, $g'=g$ will not be changed

10. (3) $\dfrac{g}{g'}=\left[\dfrac{R+h}{R}\right]^2$

$\dfrac{16}{1}=\left[\dfrac{R+h}{R}\right]^2$

$\dfrac{R+h}{R}=4$

$h=3R$

11. (4) $g=\dfrac{4}{3}\pi\rho GR$ at earth surface

$g'=\dfrac{4}{3}\pi\rho G(R-d)$ at a depth d inside earth

$\therefore g'=g\left[1-\dfrac{d}{R}\right]$

12. (2) $\dfrac{g}{g'}=\left[\dfrac{R+h}{R}\right]^2$

$\because g'=\dfrac{g}{100}$

$\Rightarrow \dfrac{(R+h)^2}{R^2}=100$

$\Rightarrow R+h=10R\Rightarrow h=9R$

13. (3) $g'=g-\dfrac{gh}{R}$

compare this with straight line equation

$y=C+Mx$

$C=+g,\ M=-\dfrac{g}{R}$

14. (1) For R comparable to h

$\dfrac{g}{g'}=\left[\dfrac{R+h}{R}\right]^2$

$4=\left[\dfrac{R+h}{R}\right]^2$

$R+h=2R$

$h=R$

15. (2) $\dfrac{g'}{g}=\dfrac{R^2}{(R+h)^2}=\dfrac{1}{2}$

$R+h=\sqrt{2}R$

$h=R\left[\sqrt{2}-1\right]\Rightarrow 0.414R=0.414\times 6400=2649.9$ km

$\simeq 2650$ km

16. (1) $g=\dfrac{GM}{r^2}=\dfrac{G}{r^2}\times\dfrac{4}{3}\pi r^3 d=\dfrac{4}{3}\pi r d G\Rightarrow g\propto dr$

$\dfrac{g_1}{g_2}=\dfrac{d_1r_1}{d_2r_2}$

17. (1) $g''=g\left[1-\dfrac{d}{R}\right]$, where g'' is the acceleration due to gravity at a depth d

for d = 2h

$g''=g\left[1-\dfrac{2h}{R}\right]$ ...(*i*)

$g'=g\left[1-\dfrac{2h}{R}\right]$ ...(*ii*) where g' is the acceleration due to gravity of a body at height h from the surface of earth

$\Rightarrow \dfrac{g''}{g'}=\dfrac{1}{1}$

18. (3) $\dfrac{4}{100}g=g\left(\dfrac{R}{R+h}\right)^2$

$\dfrac{2}{10}=\dfrac{R}{R+h}$

$\therefore h=4R=4\times 6400=25{,}600$ km.

19. (4) We know acceleration due to gravity $g=\dfrac{Gm}{R^2}$

According to question acceleration due to gravity is same on both planets

$\therefore g_p=g_e$

$\dfrac{GM_p}{R_p^2}=\dfrac{GM_e}{R_e^2}$

$\dfrac{4}{3}\dfrac{\pi R_p^3}{R_p^2}\rho_p=\dfrac{4}{3}\pi\dfrac{R_e^3}{R_e^2}\rho_e$

$R_p\rho_p=R_e\rho_e$

$\because \rho_P=2\rho_e$

$R_p=\dfrac{R_e}{2}=\dfrac{R}{2}$

20. (1) $g_A = 9g_B$

$h_A = \frac{v^2}{2g_A}$ similarly $h_B = \frac{v^2}{2g_B}$

$\frac{h_A}{h_B} = \frac{g_B}{g_A} = \frac{1}{9}$

$h_B = 9h_A$

$\therefore h = 2m$

$h_B = 18$ m

21. (2) Here R = 6400 km, d = 16 km

$g_d = g\left(1-\frac{d}{R}\right) = g\left(1-\frac{16}{6400}\right) = \frac{399}{400}g$

$\therefore g - g_d = g - \frac{399}{400}g = \frac{1}{400}g$

The percentage decrease in weight of the body

$= \frac{mg - mg_d}{mg} \times 100 = \frac{g - g_d}{g} \times 100$

$= \frac{(1/400)g}{g} \times 100 = 0.25\%$

22. (1) Let h be the height at which 'g' is same as that at depth d. Now

$g_h = g_d$

or $g\left(1-\frac{2h}{R}\right) = g\left(1-\frac{d}{R}\right)$

or $\frac{2h}{R} = \frac{d}{R}$

$\therefore h = \frac{d}{2} = \frac{80}{2} = 40\,\text{km}$

23. (2) Acceleration due to $g' = g - \omega^2 R_e$

If the rotational speed is increased then acceleration due to gravity (g′) decrease this will cause weight of body to decrease.

24. (3) $g = \frac{4}{3}\pi\rho GR$ $g \propto dR$

(ρ = d given in the problem)

25. (3) At latitude λ, $g_\lambda = g - R\omega^2 \cos^2 \lambda$

At equator, $\lambda = 0°$, so $g_e = g - R\omega^2$

When $g_e = 0$, we have,

$\omega = \sqrt{\frac{g}{R}} = \sqrt{\frac{10}{6400\times10^3}} = 1.25\times10^{-3}\,\text{rad s}^{-1}$

26. (4) PE at surface $= \frac{-GMm}{R}$

PE at height (h = 10R) $= \frac{-GMm}{(R+h)} = -\frac{GMm}{11R}$

$\Delta PE = \frac{-GMm}{11R} - \left[-\frac{GMm}{R}\right] = \frac{10\,GMm}{11R}$

27. (2) Change in gravitational potential energy $\Delta U = U_f - U_i$

$\Delta U = \frac{-GMm}{R+2R} - \left[\frac{-GMm}{R}\right]$

$= \frac{GMm}{R} - \frac{GMm}{3R} = \frac{2}{3}mgR$

28. (1) Conservation of energy

$\frac{1}{2}mv^2 = \frac{1}{2}m\left[(20)^2 - v_e^2\right]$

where $v_e = 8\sqrt{2}\,\text{km/s}$

$\Rightarrow v = 16.5$ Km/s

29. (2) Inside spherical shell, E = 0

Outside shell $E \propto \frac{1}{r}$

30. (1) $v_e = \sqrt{\frac{2\,GM}{R}} = \sqrt{\frac{2G}{R} \times \frac{4}{3}\pi R^3 \rho}$

$v_e \propto R$

$\frac{v_A}{v_B} = \frac{R_A}{R_B} = \frac{2}{1}$

31. (2) Escape velocity of a body from the surface of the earth is given by $v_e = \sqrt{\frac{2\,GM_e}{R_e}}$

$M_p = \frac{M_e}{2}, R_p = \frac{R_e}{4}$

$v_e' = \sqrt{\frac{2GM_e \times 4}{2R_e}} = \sqrt{2}v_e$

32. (1) Gravitational potential energy at the surface of earth

$U_e = \frac{-GMm}{R}$,

Gravitational potential energy at a height nR above the surface of earth.

$U_h = \frac{-GMm}{R+nR} = \frac{-GMm}{R(n+1)}$

$\Delta U = U_h - U_e$

$= \frac{GMm}{R}\left[1 - \frac{1}{n+1}\right]$

$= \left[\frac{n}{n+1}\right] mgR$

33. (2) Energy required $= \frac{-GMm}{(3R)} - \left[\frac{-GMm}{(2R)}\right] = \frac{GMm}{6R}$

34. (2) Binding energy $= -E = \frac{GMm}{2r}$

$g = \frac{GM}{R^2}$

$BE = \frac{gmR^2}{2r}$

35. (3) $v_e = \sqrt{\frac{2GM_e}{R_e}}$

$M' = 8M, R' = 2R$

$v'_e = \sqrt{\frac{2G(8M)}{2R}}$

$v'_e = 2v_e$

36. (2) If ρ is the average density of the earth, then mass of the the earth,

$M_e = \frac{4}{3}\pi R_e^3 \rho$

Escape velocity on the earth,

$v_e = \sqrt{\frac{2GM_e}{R_e}} = \sqrt{\frac{2G}{R_e} \times \frac{4}{3}\pi R_e^3 \rho} = R_e\sqrt{\frac{8}{3}G\pi\rho}$

Similarly, escape velocity on the planet,

$v_p = R_P\sqrt{\frac{8}{3}G\pi\rho}$

$\therefore \frac{v_P}{v_e} = \frac{R_P}{R_e}$

But $R_P = 2R_e$

$\therefore v_P = 2v_e$

37. (4) Energy at the surface of earth = Energy at a height h above that surface of earth

$\frac{-GMm}{R} + \frac{1}{2}m\left(\frac{v_e}{3}\right)^2 = \frac{-GMm}{R+h}$

where $v_e = \sqrt{\frac{2GM}{R}}$

$\Rightarrow h = \frac{R}{8}$

38. (3) Potential energy = 2 × (Total energy) = $2E_0$

Because gravitational potential energy, $U = -\frac{GMm}{r}$

Total energy, $E_0 = -\frac{GMm}{2r}$

39. (1) $v_0 = \sqrt{\frac{gR^2}{R+R}} = \sqrt{\frac{gR}{2}} = \sqrt{\frac{9.8 \times 6400 \times 10^3}{2}} ms^{-1}$

$= 5600\ ms^{-1} = 5.6\ km\ s^{-1}$.

40. (1) According to question TE = PE

$\frac{-1}{2}\frac{GM_e m}{(R+H)} = \frac{-GM_e m}{4R}$

$4R = 2R + 2H$

$H = R$

41. (1) $v_{satellite} \propto \frac{1}{\sqrt{R}}$

$\frac{v_A}{v_B} = \sqrt{\frac{R_B}{R_A}}$

$\frac{3v}{v_B} = \sqrt{\frac{R}{4R}}$,

$v_B = 6v$

42. (1) Near the surface of the spherical planet acceleration due to gravity

$g_P = \frac{Gm_P}{R_P^2}$

Here, $R_P = \frac{D_P}{2}$ and $m_P = M_P$

$g_P = \frac{4GM_P}{D_P^2}$

43. (2) $T^2 \propto R^3$

$\left(\frac{T}{D}\right)^2 = \left(\frac{L_2}{L_1}\right)^3$

$T = D\left[\frac{L_2}{L_1}\right]^{3/2}$

44. (2) $mv_A \times OA = mv_B \times OB$

$\frac{v_B}{v_A} = \frac{OA}{OB} = x$

45. (1) Orbital velocity near earth's surface, $v_0 = \sqrt{gR_e}$

Escape velocity, $v_e = \sqrt{2gR_e} = 1.414\sqrt{gR_e}$

Additional velocity required

$= v_e - v_0 = (1.414 - 1)\sqrt{gR_e}$

$= 0.414\sqrt{9.8 \times 6400 \times 10^3} = 3.278 \times 10^3\ ms^{-1}$

$= 3.278$ km/s

46. (3) Angular momentum is conserved

$v_1 d_1 = v_2 d_2$

47. (2) Using conservation of angular momentum

$mv_1 r_1 = mv_2 r_2$

$v_1 r_1 = v_2 r_2 \Rightarrow \frac{v_1}{v_2} = \frac{r_2}{r_1}$

So, $\frac{K_1}{K_2} = \left(\frac{v_1}{v_2}\right)^2 = \left(\frac{r^2}{r_1}\right)^2$

High Order Time Intensive MCQs

1. (2) For motion of planet in circular orbit

Centripetal force = Gravitational force

$mr\omega^2 = \frac{GmM}{r^{5/2}}$

$\Rightarrow r^{7/2}\omega^2 = GM$

$\Rightarrow r^{7/2} = \frac{GM}{\omega^2} = \frac{GM}{(2\pi/T)^2} = \frac{T^2 \times GM}{4\pi^2}$

$\Rightarrow T^2 \propto r^{7/2}$

2. (3) Gravitational force of attraction between the particles.

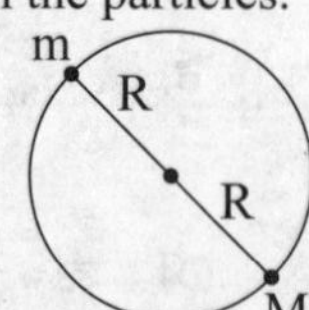

$F = \frac{Gm^2}{4R^2}$

$F = mR\omega^2 = \frac{mv^2}{R}$

$\therefore \frac{Gm^2}{4R^2} = \frac{mv^2}{R}$

$v = \sqrt{\frac{GM}{4R}} = \frac{1}{2}\sqrt{\frac{GM}{R}}$

3. (2) $F = \frac{GM}{r^2} \Rightarrow F \propto r^{-2}$

But if $F \propto r^{-5/2}$

Potential energy for a conservative force is defined as:

$U = -\int F.dr$

$= \int r^{-5/2} dr = r^{-3/2}$

$\therefore U \propto r^{-3/2}$

4. (2) Angular velocity $\omega = \frac{v}{R}$

$$\Rightarrow \omega = \frac{v_e}{R} = \frac{\sqrt{\frac{2GM}{R}}}{R} = \sqrt{\frac{2GM}{R^3}} \quad \left[\because v_e = \sqrt{\frac{2GM}{R}}\right]$$

5. (2) Let a small element dy at a distance y from the mass m be taken , then force due to this element

Gravitational force on the mass due to rod,

$$F = \int_x^{x+L} \frac{GMdym}{Ly^2}$$

$$= \frac{GMm}{L}\left[\frac{L}{x} - \frac{L}{x+L}\right] = \frac{GMm}{x(x+L)}$$

6. (4) Applying law of conservation of energy for asteroid at a distance 10 R_e and at earth's surface, $K_i + U_i = K_f + U_f$ ...(*i*)

Now, $K_i = \frac{1}{2}mv_i^2$ and $U_i = -\frac{GM_em}{10R_e}$

$K_f = \frac{1}{2}mv_f^2$

and $U_f = -\frac{GM_em}{R_e}$

$$\frac{1}{2}mv_i^2 - \frac{GM_em}{10R_e} = \frac{1}{2}mv_f^2 - \frac{GM_em}{R_e}$$

$$\frac{1}{2}mv_i^2 + \frac{GM_em}{R_e}\left[-\frac{1}{10}+1\right] = \frac{1}{2}mv_f^2$$

$$\Rightarrow v_i^2 + \frac{2GM_e}{R_e}\left[\frac{-1}{10}+1\right] = v_f^2$$

7. (3) $U_i + K_i = U_f + K_f$

$$-\frac{GMm}{R} + \frac{1}{2}m(2v_e)^2 = 0 + \frac{1}{2}mv^2$$

[$\because$ Gravitational potential energy $U = \frac{-GMm}{R}$]

$$\text{Or } -\frac{GM}{R} + 2v_e^2 = \frac{1}{2}v^2$$

$$\text{Or } \frac{2GM}{R} + \frac{8GM}{R} = v^2$$

$$\text{Or } v = \sqrt{\frac{6GM}{R}} = \sqrt{3\left(\frac{2GM}{R}\right)}$$

$$= \sqrt{3(2gR)} = \sqrt{3}v_e$$

8. (3) For black hole even lights cannot escape from the surface of the star.

Since, $v_{esc} = \sqrt{\frac{2Gm}{r}}$

Putting $v_{esc} = c$

Then, $\sqrt{\frac{2Gm}{r}} \geq c$

9. (2) $dF = \frac{Gm(A+Bx^2)dx}{x^2}$

$$F = Gm\int_a^{a+L}\left(\frac{A+Bx^2}{x^2}\right)dx$$

$$F = Gm\int_a^{a+L}\left(\frac{A}{x^2}+B\right)dx$$

$$= Gm\left[A\left(\frac{1}{a}-\frac{1}{a+L}\right)\right] + BL$$

10. (4) $g' = g - R\omega^2\cos^2\lambda$

Given $\lambda = 45° \Rightarrow g' = g - \frac{R\omega^2}{2}$

When Earth stops $\omega = 0$

$\therefore g' = g$

Increases in weight of body $= \frac{mR\omega^2}{2}$

11. (2) For the earth, T_1 = 1 year, r_1 = 1 AU
For the planet Venus, T_2 = ? r_2 = 0.72 AU

$$\therefore \quad T_2^2 = T_1^2\left[\frac{r_2}{r_1}\right]^3 = 1^2 \times \left[\frac{0.72}{1}\right]^3 = 0.37 \text{ yr}$$

or $T_2 = \sqrt{0.37} = 0.61\text{ yr} \simeq 223 \text{ days}$

12. (1) Gravitational Force = Necessary centripetal force required

$$\frac{mv^2}{R} = \frac{GM_em}{R^n} \quad \text{or} \quad v = \left(\frac{GM_e}{R^{n-1}}\right)^{1/2}$$

$$T = \frac{2\pi R}{v} = 2\pi R\left(\frac{R^{n-1}}{GM_e}\right)^{1/2}$$

$$= 2\pi\left(\frac{R^2R^{n-1}}{GM_e}\right)^{1/2} = 2\pi\frac{R^{\frac{(n+1)}{2}}}{(GM_e)^{1/2}} \quad \therefore T \propto R^{(n+1)/2}$$

13. (1) Area Velocity $= \frac{dA}{dt} = \frac{L}{2m} = \frac{mvr}{2m}$

$$= \frac{vr}{2} = \frac{r}{2}\sqrt{\frac{GM}{r}} = \frac{1}{2}\sqrt{GMr}$$

$$\frac{dA}{dt} \propto r^{1/2}$$

14. (4) Using Law of conservation of Energy

$$\frac{-GMm}{r}=\frac{1}{2}mv^2-\frac{GMm}{R}$$

$$\Rightarrow \frac{v^2}{2}=\frac{GM}{R}-\frac{GM}{r}=GM\left(\frac{r-R}{rR}\right)$$

$$\frac{v^2}{2}=gR\left(\frac{r-R}{r}\right) \qquad \left(\because \frac{GM}{R^2}=g\right)$$

$$v=\sqrt{2gR(r-R)/r}$$

15. (4) Intensity $I=\frac{GMr}{R^3}=\frac{Gr}{R^3}\left(\frac{4}{3}\pi R^3 J\right)=\frac{4\pi GrJ}{3}$

16. (1) Gravitational potential energy = – GMm/r

Orbital velocity, $v_0=\sqrt{\frac{GM}{R+h}}$

$$E_f=\frac{1}{2}mv_0^2-\frac{GMm}{3R}=\frac{1}{2}m\frac{GM}{3R}-\frac{GMm}{3R}$$

$$=-\frac{GMm}{6R}$$

$$\Rightarrow E_i=-\frac{GMm}{R}+K \Rightarrow E_i=E_f \Rightarrow K=\frac{5GMm}{6R}$$

$\therefore$ minimum required energy $=K=\frac{5GMm}{6R}$

17. (3) According to Kepler's 3rd Law $T^2 \propto r^3$

$$\left(\frac{T}{T'}\right)^2=\left(\frac{r}{r'}\right)^3 \Rightarrow \frac{25}{(T')^2}=\frac{r^3}{64r^3}$$

$\Rightarrow T'=40$ h

18. (4) The system will be bound where, total energy is –ve

For all the points magnitude of $E_p > E_k$

So total energy is negative. Therefore the system is a bound system for all the points.

19. (4) Potential energy at the surface of earth is

$U_{Earth}=-mgR$,

Energy required to move the satellite to infinity is

$U_{space}=U$

$\therefore E=mgR=10^3\times10\times6400\times10^3\,J=6.4\times10^{10}\,J$

20. (2) mg = 10 and m = 1 kg $\Rightarrow$ g = 10 m/s²

$$g'=g\frac{R^2}{(R+h)^2}=10\times\frac{R^2}{(3R/2)^2}=\frac{40}{9}$$

Pull on satellite $=m'g'=200\times\frac{40}{9}=889N$

21. (3) Gravitational field intensity will become zero at one point between Earth and Mars. Hence graph C.

22. (3) $g=\frac{GM_e}{R_e^{\,2}}$ when earth is shrinked

$g'=\frac{GM_e}{r^2}$ but $r=R_e$ so $g'=g$

23. (4) Above the surface of earth

$$\frac{\Delta g}{g}\times100=\frac{-2h}{R_e}$$

& below the surface of earth

$$\frac{\Delta g}{g}\times100=\frac{-h}{R_e}$$

24. (2) $W=-m\,\Delta v=-m(\vec{E}.d\vec{r})$

$\vec{E}$ is gravitational field

$d\vec{r}=5\hat{i}+4\hat{j}$

$W=-10(10+8)=-180$ J

25. (2) $\frac{1}{2}\mu v_{rel}^2=\frac{Gm_1m_2}{r}$

$$\frac{1}{2}\frac{m_1m_2}{m_1+m_2}v_{rel}^2=\frac{Gm_1m_2}{r}$$

$$v_{rel}=\left[\frac{2G(m_1+m_2)}{r}\right]^{1/2}$$

26. (1) $v_e=\sqrt{\frac{2GM}{R}}$

M = volume × density $=\frac{4}{3}\pi R^3\rho$

so, $v_e=\sqrt{\frac{8}{3}\pi\rho GR^2}$

$v_e \propto R$ if ρ = constant

Since, radius becomes twice. So, escape velocity becomes twice i.e., 2 × 11 = 22 km/s

27. (3) $m=\frac{4}{3}\pi R^3\rho_1$

$$m'=\frac{4}{3}\pi(2R)^3\rho_2-\frac{4}{3}\pi R^3\rho_2+\frac{4}{3}\pi R^3\rho_1$$

$$=\frac{28}{3}\pi R^3\rho_2+\frac{4}{3}\pi R^3\rho_1$$

$$m'=\frac{4}{3}\pi R^3(7\rho_2+\rho_1)$$

According to the question $F_1=F_2$

$$\Rightarrow \frac{m}{R^2}=\frac{m'}{4R^2}$$

$$m=\frac{m'}{4}$$

$m'=4m$

$7\rho_2+\rho_1=4\rho_1$

$$\Rightarrow \frac{\rho_1}{\rho_2}=\frac{7}{3}$$

28. (2) $W=U_f-U_i$

$$W=-\frac{3Gm^2}{\ell}$$

29. (1) Using conservation of angular momentum

$mv_A r_A = mv_B r_B$

$\Rightarrow v_A = v_B \dfrac{r_B}{r_A}$

$\Rightarrow v_A = v\left(\dfrac{r_B}{r_A}\right) \quad [\because v_B = v]$

As v_A is orbital velocity.

$v_{A(rA)} = \sqrt{\dfrac{GM}{r_A}}$ and $v_{A(1.2\,r_A)} = \sqrt{\dfrac{GM}{1.2r_A}}$

$\Rightarrow \dfrac{v_{A(r_A)}}{v_{A(1.2\,r_A)}} = \sqrt{\dfrac{1.2r_A}{r_A}}$

$\Rightarrow v_{A(1.2\,r_A)} = \dfrac{vr_B}{r_A} \times \dfrac{1}{\sqrt{1.2}}$

30. (1) $T = \dfrac{2\pi a}{v} = \dfrac{2\pi a}{\sqrt{\dfrac{GM}{a}}} \Rightarrow T^2 = \dfrac{4\pi^2 a^3}{GM}$

$\Rightarrow a = \left[\dfrac{GMT^2}{4\pi^2}\right]^{1/3}$

length of major axis $= 2a = 2\left(\dfrac{GM}{4\pi^2}\right)^{1/3} T^{2/3}$

31. (1) $v_a = \dfrac{dA}{dt} = \dfrac{1}{2} rv_0 = \dfrac{1}{2} r\sqrt{\dfrac{GM}{r}}$

$4v_a^2 = GMr$

$4v_a^2 = gR^2(R + h)$

$h = \dfrac{4v_a^2}{gR^2} - R$

32. (2) $T = 2\pi\sqrt{\dfrac{r^3}{GM_e}} = 2\pi\sqrt{\dfrac{r^3}{G\rho\dfrac{4}{3}\pi r^3}}$

$T = 2\pi\sqrt{\dfrac{1}{4G\rho\,\pi/3}}$

$T = 2\pi\sqrt{\dfrac{1}{4G\dfrac{\pi}{2}\times 10^4\left(\dfrac{\pi}{3}\right)}} = \sqrt{\dfrac{6}{6.67\times 10^{-11}\times 10^4}} = 3000$

33. (3) Sphere $v = \dfrac{-GM}{2R^3}(3R^2 - r^2) r \le R$

$v = \dfrac{-GM}{r}$ for $r > R$

Shell : $v = \dfrac{-GM}{R}$ for $r \le R$

$v = \dfrac{-GM}{r}$ for $r > R$

field sphere $E_g = \dfrac{GM}{R^3} . r$ for $r \le R$

$E_g = \dfrac{GM}{r^2}$ for $r > R$

Shell : E_g = zero for $r < R$

$E_g = \dfrac{GM}{r^2}$ for $r \ge R$

34. (3) $\dfrac{1}{2}Mv^2 - \dfrac{GM_e M}{R} = -\dfrac{GM_e M}{R+h}$

$\Rightarrow \dfrac{v^2}{2} = \dfrac{GM_e}{R} - \dfrac{GM_e}{R+h}$

$\Rightarrow \dfrac{GM_e}{R+h} = \dfrac{GM_e}{R} - \dfrac{v^2}{2}$

$\Rightarrow \dfrac{1}{R+h} = \dfrac{1}{R} - \dfrac{v^2}{2gR^2}$

$\dfrac{1}{R+h} = \dfrac{1}{R}\left[1 - \dfrac{v^2}{2gR}\right]$

$R + h = \dfrac{R}{1 - \dfrac{v^2}{2gR}}$

$h = \dfrac{R}{\left[\dfrac{2gR}{v^2} - 1\right]}$

Statement Based MCQs

1. (1) Both statements are true.

$\dfrac{GMm}{r^2} = \dfrac{mv^2}{r}$

$= \dfrac{m}{r} \times \left(\dfrac{2\pi r}{T}\right)^2 = \dfrac{4\pi^2 mr}{T^2}$

or $T^2 = \left(\dfrac{4\pi^2}{GM}\right) r^3$

$\therefore T^2 \propto r^3,$

2. (2) Both the statements are false. Acceleration due to gravity is more at the poles than at the equator.

3. (2) Both statements are false.

4. (1) Both statements are true.

5. (3) Statement I is true statement II is false. As $v_{esc} = \sqrt{2gR}$. Hence escape velocity on Moon is less than that on the earth.

Assertion & Reason MCQs

1. (2) Both assertion and reason are true but the reason is not a correct explanation of the assertion.

2. (1) Escape velocity of planet is independent of mass of body

$v_{escape} = \sqrt{2gR}$

3. **(1)** Assertion & Reason both are true and reason is the correct explanation of the assertion.

4. **(2)** Both assertion and reason are true but the reason is not a correct explanation of the assertion.

5. **(2)** According to kepler's second law, the angular momentum of the planet is constant.

Match the Columns MCQs

1. **(1)** $\frac{GM}{y} = x$ or $GM = xy$

(a) $E = \frac{GM}{(2y)^2} = \frac{xy}{4y^2} = \frac{x}{4y}$

(b) $|V| = \frac{GM}{y^3}\left(1.5\,y^2 - 0.5\frac{y^2}{4}\right)$

$= \left(\frac{xy}{y^3}\right)\left(\frac{11}{8}y^2\right) = \frac{11}{8}x$

(c) $E = \frac{GM}{y^3}\left(\frac{y}{2}\right) = \frac{xy}{y^3}\frac{y}{2} = \frac{x}{2y}$

(d) $|V| = \frac{GM}{2y} = \frac{xy}{2y} = \frac{x}{2}$

2. **(2)** (a) $W = \Delta U = \frac{mgh}{1+\frac{h}{R}}$

Put = R, $W = \frac{1}{2}\,mgR$

(b) $KE = \frac{GMm}{2r} = \frac{GMm}{4R} = \frac{1}{4}mgR$ (as $GM = gR^2$)

(c) $E = \frac{GMm}{2r}$

$E_1 = -\frac{GMm}{2(2R)} = -\frac{GMm}{4R} = -\frac{1}{4}mgR$

$E_2 = -\frac{GMm}{2(3R)} = -\frac{1}{6}mgR$

$E_1 \sim E = \frac{1}{12}\,mgR$

(d) $KE = \Delta U = \frac{mgH}{1+\frac{h}{R}}$

Putting h = R, $KE = \frac{1}{2}\,mgR$

3. **(1)** (a)-(ii, iii); (b)-(i); (c)-(i); (d)-(iv)

4. **(3)** (a) I side thin spherical shell:

$V = \frac{-GM}{r}$

$E = 0$

(b) Centre of solid sphere:

$V(r) = \frac{-Gm(3R^2 - r^2)}{2R^3}$

$\therefore V(o) = -\frac{3}{2}\frac{GM}{R}$

$E(r) = \frac{GM}{r^2}$ $(\because r < R)$

$\therefore E(o) = 0$

(c) In case of inside solid sphere:

Inside solid mass distribution $(\rho) \neq 0$

Hence, gravitational potential $V(o) \neq 0$

Gravitational Intensity $E(o) \neq 0$

(d) m o—r—P—r—o m

Gravitational potential at centre P

$V(P) = -\frac{Gm}{r} - \frac{GM}{r} = \frac{-2Gm}{r}$

$E(P) = 0$

5. **(1)**

Velocity	→	**Path**
$v = \sqrt{Gm/r}$	→	Circular
$\sqrt{Gm/r} < v < \sqrt{2Gm/r}$	→	Elliptical
$v = \sqrt{2Gm/r}$	→	Parabolic
$v > \sqrt{2Gm/r}$	→	Hyperbolic

NCERT Exemplar MCQs

1. **(4)** If we assume the Earth as a sphere of uniform density, then it can be treated as point mass placed at its centre. In this case acceleration due to gravity g = 0, at the centre.

It is not so, if the Earth is considered as a sphere of non-uniform density, in that case value of g will be different at different points and cannot be zero at any point.

2. **(3)** As observed from the Earth, the Sun appears to move in an approximate circular orbit. The gravitational force of attraction between the Earth and the Sun always follows inverse square law.

Due to relative motion between the Earth and mercury, the orbit of mercury, as observed from the Earth will not be approximately circular, since the major gravitational force on mercury is due to the Sun.

3. **(1)** As the Earth is revolving around the Sun in a circular motion due to gravitational attraction. The force of attraction will be of radial nature i.e., angle between position vector r and force F is zero. So, torque $= |\tau| = |r \times F| = rF\sin 0° = 0$

4. **(3)** As the total energy of the Earth satellite bounded system is negative $\left(\frac{-GM}{2a}\right)$, where, a is radius of the satellite and M is mass of the Earth.

Due to the viscous force acting on satellite, energy decreases continuously and radius of the orbit or height decreases gradually.

5. **(2)** As observed from the Sun, two types of forces are acting on the Moon one is due to gravitational attraction between the Sun and the Moon and the other is due to gravitational attraction between the Earth and the Moon. Hence, total force on the Moon is not central.

6. (4) Asteroids are also being acted upon by central gravitational forces, hence they are moving in circular orbits like planets and obey Kepler's laws.

7. (4) Gravitational mass of proton is equivalent to its inertial mass and is independent of presence neighboring heavy objects.

8. (3) Force on B due to A = $F_{BA} = \frac{FMm}{(BA)^2}$ towards BA

Force on B due to $C = F_{BC} = \frac{GMm}{(BC)^2}$ towards BC

As (BC) = 2AB

$$\Rightarrow \quad F_{BC} = \frac{GMm}{(2AB)^2} = \frac{GMm}{4(AB)^2} < F_{BA}$$

Hence, m will move towards BA (i.e., 2M)

9. (4) When the Earth is closest to the Sun, speed of the Earth is maximum, hence, KE is maximum. When the Earth is farthest from th Sun speed is minimum hence, KE is minimum but never zero and negative.

This variation is correctly represented by option (4).

NEET & AIIMS Past Year MCQs

1. (2) $g = \frac{GM}{R^2} = 9.8 \text{ m/s}^2$ ... (*i*)

Now for Planet,

$M' = \frac{M}{10}, D' = \frac{D}{2}$ [D is diameter of earth]

$\therefore R' = \frac{R}{2}$

Now, $g' = \frac{GM'}{(R')^2} = \frac{GM \times 4}{10 \times R^2} = \frac{4}{10} \times \frac{GM}{R^2}$

Substitute value of $\frac{GM}{R^2}$ from equation (*i*)

$g' = 0.4 \times 9.8 \text{ m/s}^2 = 3.92 \text{ m/s}^2$

2. (1) Given:

Mass m = 100 kg,

At point A: Height $h = 2R_E$,

At point B: Height $h = \frac{3R_E}{2}$,

$g = 10 \text{ m/s}^2$.

Gravitational acceleration at height *h*:

$$g' = g\left(\frac{R_E}{R_E + h}\right)^2$$

At A ($h = 2R_E$):

$$g_A = \frac{g}{9}$$

At B $\left(h = \frac{3R_E}{2}\right)$:

$$g_B = g\left(\frac{R_E}{R_E + \frac{3R_E}{2}}\right)^2 = \frac{4g}{25}$$

Change in weight:

$$\Delta W = m(g_B - g_A) = 100\left(\frac{4 \times 10}{25} - \frac{10}{9}\right) = 49 \text{ N}$$

3. (3) Given:

Escape velocity for Earth = *v*.

For the planet:

Mass (M_p) = $9M_e$ (where M_e is Earth's mass),

Radius (R_p) = $16\,R_e$ (where R_e is Earth's radius).

Escape velocity,

$$v_e = \sqrt{\frac{2GM}{R}}$$

Therefore,

$$\frac{(v_e)_p}{(v_e)_e} = \sqrt{\frac{M_p}{R_p} \times \frac{R_e}{M_e}}$$

$$\frac{(v_e)_p}{v} = \sqrt{\frac{9M_e}{16R_e} \times \frac{R_e}{M_e}} = \sqrt{\frac{9}{16}} = \frac{3}{4}$$

$$(v_e)_p = \frac{3}{4}v$$

4. (3) $(T.E)_{initial}$ = PE of satellite

$$= \frac{-GMm}{R}$$

$(T.E)_{final}$ = P.E of satellite + K.E of satellite

$$= \frac{-GMm}{3R} + \frac{1}{2}m\left(\sqrt{\frac{GM}{3R}}\right)^2 = \frac{-GMm}{6R}$$

Minimum energy required

$$= \frac{-GMm}{6R} + \frac{GMm}{R}$$

$$= \frac{-GMm + 6GMm}{6R} = \frac{5GMm}{6R}$$

5. (4)

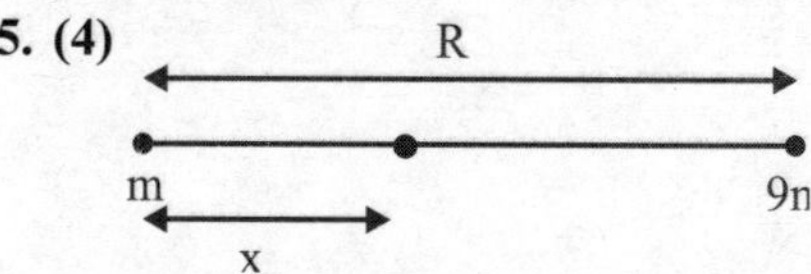

Let the gravitational field is zero at a distance x from the mass m.

$$\frac{Gm}{x^2} = \frac{G9m}{(R-x)^2}$$

$$\Rightarrow R - x = 3x \text{ or } x = \frac{R}{4}$$

Gravitational potential at $\frac{R}{4}$

$$= -\frac{Gm}{\frac{R}{4}} - \frac{G9m}{\frac{3R}{4}} = -\frac{4Gm}{R} - \frac{12Gm}{R} = -\frac{16Gm}{R}$$

6. (3) Time period of satellite

$$T = 2\pi\sqrt{\frac{R^3}{GM}}$$

$$= 2\pi\sqrt{\frac{R^3}{Gd\frac{4}{3}\pi R^3}}$$

$$\Rightarrow T = \sqrt{\frac{3\pi}{Gd}}$$

$$\Rightarrow \frac{3\pi}{Gd} = T^2$$

7. (3) Gravitational field intensity is

$$E_g = \frac{F}{m} = \frac{3}{60\times10^{-3}} = 50\,N/kg$$

8. (3) Gravitational constant = $[M^{-1}L^3T^{-2}]$

Gravitational potential energy = $[ML^2T^{-2}]$

Gravitational potential = $[L^2T^{-2}]$

Gravitational intensity = $[LT^{-2}]$

9. (3) Required figure,

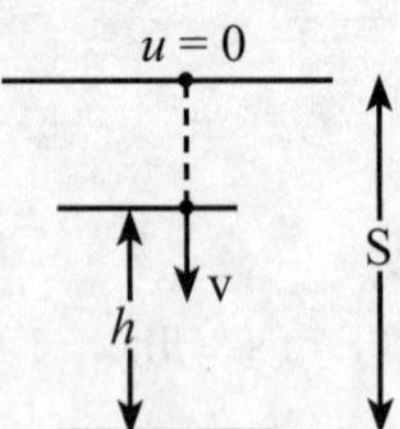

Particle is released, i.e., its initial velocity is zero, $u = 0$ m/s.

Let the height at any instant be 'h'.

Total energy remains conserved at any point.

$\Rightarrow K.E + P.E = E_{Total}$

$\Rightarrow 3\,P.E + P.E = E_{Total}$ [According to the question]

$\Rightarrow 4\,P.E = E_{Total}$

$\Rightarrow 4(mgh) = mgS \Rightarrow h = \frac{S}{4}$

Velocity of a particle under free fall is given by;

$$v = \sqrt{2gh}$$

$$v = \sqrt{2g\left(S - \frac{S}{4}\right)} = \sqrt{\frac{3gS}{2}}$$

10. (3) Escape velocity from the earth's surface is,

$$v = \sqrt{\frac{2GM}{R}} \Rightarrow v = \sqrt{\frac{2G}{R}\times\frac{4}{3}\pi R^3\rho}$$

$$\Rightarrow v = \sqrt{\frac{8\pi G\rho}{3}R^2} \Rightarrow v = R\sqrt{\frac{8\pi G\rho}{3}} \Rightarrow v \propto R \Rightarrow v_p = 4v_e$$

11. (3) Given, $v = Kv_e$

where, v_e = escape velocity = $\sqrt{\frac{2GM}{R}}$

Let maximum height reaches above the surface of earth be 'h'.

Total energy remains conserved at each point. So, at maximum height.

Now, $\frac{1}{2}mv^2 - \frac{GmM}{R} = 0 - \frac{GmM}{(R+h)}$

[$\because$ at max^{-m} height, velocity becomes zero.]

$$\Rightarrow \frac{1}{2}mv^2 - \frac{GmM}{R} = -\frac{GmM}{(R+h)}$$

$$\Rightarrow \frac{1}{2}m(Kv_e)^2 = -\frac{GmM}{(R+h)} + \frac{GmM}{R}$$

$$\Rightarrow \frac{1}{2}mK^2v_e^2 = GmM\left[\frac{1}{R} - \frac{1}{(R+h)}\right]$$

$$\Rightarrow \frac{1}{2}K^2\left(\sqrt{\frac{2GM}{R}}\right)^2 = GM\left[\frac{R+h-R}{R(R+h)}\right]$$

$$\Rightarrow \frac{1}{2}K^2\left(\frac{2GM}{R}\right) = GM\left[\frac{h}{R(R+h)}\right]$$

$$\Rightarrow K^2 = \frac{h}{(R+h)}$$

$$\Rightarrow RK^2 + hK^2 = h \Rightarrow RK^2 = h - hK^2 = h(1-K^2)$$

$$\Rightarrow h = \frac{RK^2}{1-K^2}$$

12. (2) $2a = r_1 + r_2$

$$a = \frac{r_1 + r_2}{2}$$

$$T = \frac{2\pi(a)^{3/2}}{\sqrt{Gm}} = \frac{2\pi\left(\frac{r_1+r_2}{2}\right)^{3/2}}{\sqrt{Gm}}$$

13. (1) Gravitational potential energy on the surface

$$U_s = \frac{GMm}{R} = mgR \quad ...(i)$$

Gravitational potential energy at height h

$$U_h = \frac{-GMm}{R+h}$$

$$= \frac{-GMm}{R+nR} = \frac{-GMm}{R(1+n)} \quad (\because h = nR)$$

$$U_h = \frac{-mgR}{1+n} \quad ...(ii)$$

Change in potential energy

$$\Delta U = U_h - U_s$$

$$= -\frac{mgR}{1+n} - mgR$$

$$= mgR\left[\frac{-1}{1+n} - 1\right]$$

14. (3)

15. (1)

Mechanical Properties of Solids

NCERT Topic-wise MCQs

1. (3) Given, radius of the rod, $r = 100$ mm $= 100 \times 10^{-3}$ m,

Length of the rod, $l = 10$ cm $= 10 \times 10^{-2}$ m

and force, $F_n = 100$ N

Longitudinal stress (developed in the rod)

$$= \frac{F_n}{A} = \frac{100}{\pi\left(100\times10^{-3}\right)^2} = 3182 \text{ Nm}^{-2}$$

2. (1) Young's modulus of a wire does not depend on the dimensions of the wire.

3. (3) $\because Y = \frac{F\times L}{A\times\Delta L} \Rightarrow F = \frac{YA\Delta L}{L}$

$$\therefore F = \frac{11.9\times10^{12}\times5\times10^{-6}\times5\times10^{-4}}{4}$$

$F = 7.4 \times 10^{+3}$N

4. (2) Increase in length,

$$\Delta L = \frac{FL}{YA} = \frac{FL}{Y\pi R^2}$$

As F, L and ΔL are same hence,

YR^2 = constant

$$\therefore 2.0\times10^{11}R_S^2 = 1.0\times10^{11}R_B^2$$

$$\Rightarrow R_S = \frac{R_B}{\sqrt{2}}$$

5. (4) Stress $= \frac{\text{force}}{\text{Area}}$

$$\Rightarrow \text{Stress} \propto \frac{1}{r^2}$$

$$\Rightarrow \frac{S_B}{S_A} = \frac{r_A^2}{r_B^2} = (3)^2 = 9S_A$$

$$\Rightarrow S_B = 9S_A$$

6. (4) $Y = \frac{F\times L}{A\times\Delta L} = \frac{500\times10\times0.7}{10\times10^{-6}\times0.5\times10^{-2}}$

$$= \frac{3500}{5\times10^{-8}} = 7\times10^{10}\text{ N/m}^2$$

7. (2) $F = \frac{YA\Delta L}{L}$ where, $A = \frac{\pi D^2}{4}$ $\Rightarrow F \propto \frac{r^2}{L}$

$$\therefore \frac{F_B}{F_A} = \frac{r_B^2}{(r_A)^2}\times\frac{L_A}{L_B} = \left(\frac{1}{5}\right)^2\times\frac{3}{4}$$

$$\Rightarrow \frac{F_B}{F_A} = \frac{3}{4}\times\frac{1}{25} = 0.03$$

8. (2) $Y = \frac{\text{Stress}}{\text{Strain}}$

or $Y \propto \frac{1}{\text{Strain}}$ (for same stress say σ)

$(\text{strain})_Q < (\text{strain})_P$

$\Rightarrow Y_Q > Y_P$

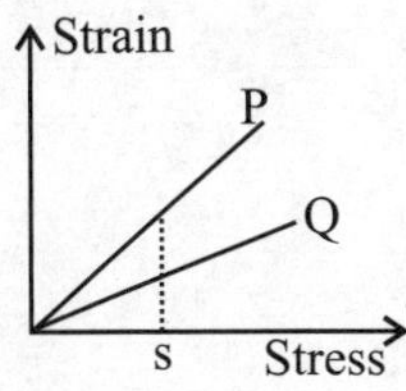

So, P is more ductile than Q.

9. (3) $\because$ Elongation, $l = \frac{FL}{AY} = \frac{FL}{\pi R^2 Y}$

New elongation, $l' = \frac{4F\times L}{\pi(4R)^2\times Y} = \frac{FL}{4\pi R^2 Y}$

$$\Rightarrow l' = \frac{l}{4}$$

10. (3) Breaking Stress $= \frac{\text{Force}}{\text{Area}}$

$\because$ wire is same, so breaking stress remains same

As $F \propto A$

$$\therefore \frac{F_1}{F_2} = \frac{A_1}{A_2} = \frac{\pi D_1^2}{4}\times\frac{4}{\pi D_2^2} = \left(\frac{D_1}{D_2}\right)^2 = \frac{1}{4}$$

$$\therefore F_2 = 4\times30 = 120\text{ kg wt}$$

11. (1) $\Delta l = \frac{Fl}{AY} = \frac{Fl}{\left(\frac{\pi d^2}{4}\right)Y}$

$$\Rightarrow (\Delta l) \propto \frac{l}{d^2}$$

Now, $\frac{l}{d^2}$ is maximum for option (1).

12. (1) $\because Y = \frac{F}{\pi r^2}\times\frac{L}{\Delta L}$

Given, Y and F are same

$$\Rightarrow \frac{L_1}{r_1^2\Delta L_1} = \frac{L_2}{r_2^2\Delta L_2}$$

$$\therefore \frac{\Delta L_1}{\Delta L_2} = \frac{r_2^2}{r_1^2}\times\frac{L_1}{L_2} \Rightarrow \frac{\Delta L_1}{\Delta L_2} = \left(\frac{D_2}{D_1}\right)^2\frac{L_1}{L_2}$$

$$\therefore \frac{\Delta L_1}{\Delta L_2} = \left(\frac{3}{2}\right)^2\times\frac{4}{3} = \frac{3}{1}$$

13. (2) Stress $= \frac{F}{A} = \frac{Mg}{A}$

$$= \frac{A\times L\times\rho\times g}{A} = L\rho g$$

Stress = LDg. $[\because \rho = D]$

$$Strain = \frac{\Delta L}{L} = \frac{l}{L/2} = \frac{2l}{L}$$

$$Y = \frac{LDg}{2l}\times L \Rightarrow l = \frac{L^2Dg}{2Y}$$

14. (1) $Y=\frac{F\times L}{A\times \Delta L}$

$\Rightarrow \Delta L \propto \frac{1}{r^2}$

$\therefore \frac{\Delta L_1}{\Delta L_2}=\left(\frac{r_2}{r_1}\right)^2 \Rightarrow \Delta L_2=\Delta L_1\times\left(\frac{r_1}{r_2}\right)^2 \quad \{\because r_1=2r_2\}$

$=5\times 4=20$ mm

15. (2) As we know $Y=\frac{FL}{A\times\Delta L}$

$F=\frac{YA\times\Delta L}{L}=\frac{YA(2\pi R-2\pi r)}{2\pi r}$

$\therefore \ F=YA\left(\frac{R-r}{r}\right)$

16. (1) Given: $A=10^{-6}$ m², $L=1$ m

$\therefore Y=\frac{FL}{A\Delta\ell}=\frac{20\times1}{10^{-6}\times10^{-4}}=20\times10^{10}=2\times10^{11}\text{Nm}^{-2}$

17. (2) PV = constant

Differentiate both sides,

PdV + VdP = 0.

$P=\frac{dP\times V}{-dV}=E$

$\therefore$ Isothermal elasticity $=\frac{-dP\times V}{dV}=P$

18. (4) $B=\frac{P}{\frac{-\Delta V}{V}}\Rightarrow B=\frac{h\rho g}{0.4\times10^{-2}}$

$h=\frac{9\times10^8\times0.4\times10^{-2}}{10^3\times10}=36\times10^1\text{m}$

19. (3) Each side of cube is L so $V=L^3$

$\frac{dV}{V}=3\frac{dL}{L}$

$\frac{dV}{V}\times100=3\frac{dL}{L}\times100$

$\therefore \ \frac{dV}{V}\times100=3\times3\%=9\%$

20. (4) $B=\frac{P}{\frac{\Delta V}{V}}\Rightarrow\frac{1}{B}=\frac{\left(\frac{\Delta V}{V}\right)}{P}$

$\Rightarrow C=\frac{\left(\frac{\Delta V}{V}\right)}{P}$

21. (3) $B=\frac{P}{\left(\frac{\Delta V}{V}\right)}$ for a rigid body $\Delta V=0$.

$\therefore$ B = ∞ (infinity)

22. (4) $\rho=\frac{M}{V}$ or $\rho\propto\frac{1}{V}\ \therefore\ \frac{\rho'}{\rho}=\frac{V}{V'}=\frac{V}{V+\Delta V}$

or $\rho'=\rho\left(\frac{V}{V+\Delta V}\right)=\rho\left(\frac{1}{1+\Delta V/V}\right)$

$\because B=\frac{-\Delta P}{\Delta V/V}$, so $\rho'=\rho\left(\frac{1}{1-\Delta P/B}\right)$

$=11.4\left[\frac{1}{1-\left(2.0\times10^8/8.0\times10^9\right)}\right]\approx11.7\,\text{g/cm}^3$

23. (4) $\text{B}=\frac{\text{Normal stress}}{\text{Volume strain}}=\left|\frac{\text{p}}{-\left(\frac{\Delta\text{V}}{\text{V}}\right)}\right|=\left|-\frac{\text{pV}}{\Delta\text{V}}\right|$

$\text{p}=\text{B}\frac{\Delta\text{V}}{\text{V}}=\text{B}\left(\frac{\Delta\text{V}}{\text{V}}\right)$

24. (2) $B=\frac{P}{\frac{\Delta V}{V}}$,

P = hρg = 300 × 10³ × 10 = 3 × 10⁶ N/m²

and, $\frac{\Delta V}{V}=0.3\%=\frac{0.3}{100}=3\times10^{-3}$

$\therefore\ B=\frac{3\times10^6}{3\times10^{-3}}=10^9\,\text{N/m}^2$

25. (1) $B=\frac{-VdP}{dV}=4(2-1.01)\times10^5=(0.99)4\times10^5$

$\therefore B=3.96\times10^5\,\text{N/m}^2$

26. (4) $\frac{\Delta V}{V}=\frac{0.08}{100}=8\times10^{-4},\ P=500\ atm$

$\therefore\ B=\frac{P}{\frac{\Delta V}{V}}=\frac{500\times0.76\times13.6\times10^3\times9.8}{8\times10^{-4}}$

$B=6330.8\times10^7\ \Rightarrow\ B=6.3\times10^{10}\,\text{N/m}^2$

27. (2) P = hρg = 5000 × 10³ × 9.8

$\Rightarrow\frac{\Delta V}{V}=\frac{P}{B}=\frac{5000\times10^3\times9.8}{3\times10^9}$

$\therefore\ \frac{\Delta V}{V}=16333.3\times10^{-6}=1.6\times10^{-2}$

$\frac{\Delta V}{V}\times100=1.6\%$

28. (2) Side of cube = 8 cm = 8 × 10⁻² m

Force on upper face = 10 × 10³ N

Displacement = 0.5 × 10⁻³ N

Area of upper face = (8 × 10⁻²)² = 64 × 10⁻⁴ m²

Displacement = 0.5 × 10⁻³ m

$\therefore\ G=\frac{F\times L}{A\times\Delta L}=\frac{10^4\times8\times10^{-2}}{64\times10^{-4}\times0.5\times10^{-3}}=2.5\times10^8\,\text{N/m}^2$

29. (1) $\therefore\ G = \dfrac{F \times L}{A \times \Delta L}$

Here, ΔL is the displacement of the upper edge of the slab due to tangential force, F.

$\therefore\ \Delta L = \dfrac{F \times L}{A \times G} = \dfrac{15 \times 10^4 \times 1}{1 \times 5.6 \times 10^9} \approx 2.7 \times 10^{-5}\,\text{m}$

30. (3) Angle of twist at free end

$= 30° = \dfrac{30}{180} \times \pi \text{ rad} = \dfrac{\pi}{6}\text{rad}$

Displacement of the free surface,

$\Delta L = \dfrac{2\pi r}{2\pi} \times \dfrac{\pi}{6} = \dfrac{\pi r}{6} = \dfrac{\pi \times 0.4}{6}\,\text{cm}$

Angle of shear or shearing strain $= \dfrac{\Delta L}{L}$

$= \dfrac{\pi \times 0.4}{6 \times 100}\text{rad.} = \dfrac{\pi \times 0.4}{6 \times 100} \times \dfrac{180}{\pi}\text{ degree} = 0.12°$

31. (3) According to the relation

$Y = 2G(1+\sigma)$

$\therefore\ Y > G$

32. (4) As we know that $Y = 2G(1+\sigma)$

$6G = 2G(1+\sigma)$

$\therefore\ \sigma = 2$

$Y = 3B(1-2\sigma) \Rightarrow B = \dfrac{Y}{-9}$

33. (2) Practical values of poisson ratio for a metal lies between 0 to 0.5.

34. (1) $\because\ Y = 3B(1-2\sigma) \quad \Rightarrow \sigma = 0.39$

35. (2) $\dfrac{9}{Y} = \dfrac{3}{G} + \dfrac{1}{B} \qquad (\because\ Y = 3B)$

$\therefore\ G = \dfrac{3}{2}B = 1.5\text{ B}$

36. (3) $\dfrac{\Delta V}{V} = (1-2\sigma)\dfrac{\Delta L}{L} \Rightarrow \dfrac{\Delta V}{V} = (1 - 2 \times 0.30)3 \times 10^{-4}$

$\therefore\ \dfrac{\Delta V}{V} = 0.4 \times 3 \times 10^{-4} = 1.2 \times 10^{-4}$

$\therefore\ \dfrac{\Delta V}{V} \times 100 = 1.2 \times 10^{-4} \times 100 = 0.012\%$

High Order Time Intensive MCQs

1. (2) When a metal wire is stretched by a load the fractional change in the transverse length is proportional to the fractional change in the longitudinal length.

Poisson's ratio $\sigma = \dfrac{-\Delta r}{r} \times \dfrac{l}{\Delta l}$

$\dfrac{\Delta r}{r} = -\sigma\dfrac{\Delta l}{l}$ ....(1) Volume = Area × length

$\because\ V = \pi r^2 l$

$\therefore\ \log V = \log \pi + 2 \log r + \log l$

$\dfrac{\Delta V}{V} = 2\dfrac{\Delta r}{r} + \dfrac{\Delta l}{l}$

$\Rightarrow \dfrac{\Delta V}{V} = 2\left(\dfrac{-\sigma \Delta l}{l}\right) + \dfrac{\Delta l}{l}$ (from equation 1)

$\Rightarrow \dfrac{\Delta V}{V} = \dfrac{\Delta l}{l}(1-2\sigma)$

(Poisson's ratio is constant for metal or material)

$\Rightarrow \dfrac{\Delta V}{V} \propto \dfrac{\Delta l}{l}$

2. (2) Stress $= \dfrac{\text{Force}}{\text{Area}} = \dfrac{mg}{A} = \dfrac{V\rho g}{A} = \dfrac{LA\rho g}{A} = L\rho g$

$L = \dfrac{\text{stress}}{\rho g} = \dfrac{6 \times 10^6}{3 \times 10^3 \times 10}$

L = 200 m

3. (2) Given, angle of shear = 30°

and change in length, ΔL = 250 cm = 2.5 m

$\therefore$ Shear strain, $\tan\phi = \dfrac{\Delta L}{L} \Rightarrow \tan 30° = \dfrac{2.5}{L}$

$L = \dfrac{2.5}{\tan 30°} = \dfrac{2.5}{0.577} = 4.33\,\text{m}$

Volume, $V = L^3 = 81.18\text{ m}^3$

4. (1) Let A is the area of cross section and l is the original length of the metallic wire.

$\because\ Y = \dfrac{T_1 l}{A(l_1 - l)}$

and $Y = \dfrac{T_2 l}{A(l_2 - l)}$

$\Rightarrow \dfrac{T_1 l}{A(l_1 - l)} = \dfrac{T_2 l}{A(l_2 - l)} \Rightarrow l = \dfrac{T_2 l_1 - T_1 l_2}{T_2 - T_1}$

5. (2) $\dfrac{dV}{V} = (1-2\sigma)\dfrac{dL}{L} = \left(1 - 2 \times \dfrac{1}{2}\right)3 \times 10^{-2}$

$\therefore\ \dfrac{dV}{V} \times 100 = 0$

6. (1) Let A be the area of cross-section of the rod.

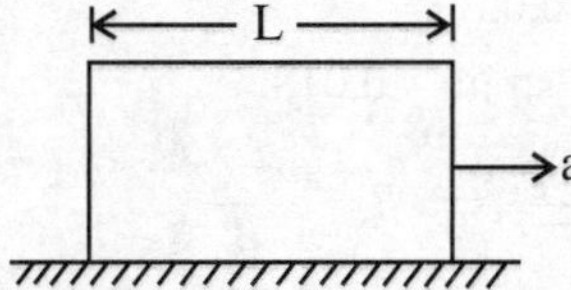

FBD of rod at mid-point:

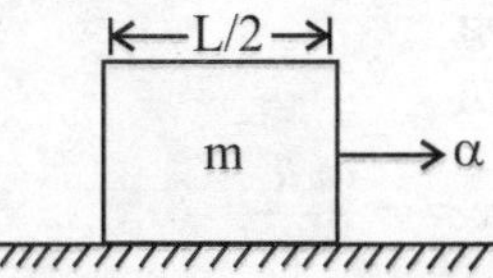

Mass m = Volume × density

$= \left(\dfrac{L}{2}.A\right)\rho$

$\therefore\ T = m\alpha = \left(\dfrac{L}{2}A\rho\alpha\right)$

$\therefore$ Stress $= \dfrac{T}{A} = \dfrac{1}{2}\rho\alpha L$

7. **(3)** For the first wire, elongation $l_1 = \frac{FL}{\pi R^2 Y}$

For the second wire, elongation $l_2 = \frac{F \times 2L}{4\pi R^2 Y}$

$\therefore \frac{l_1}{l_2} = 2$

8. **(1)** Tensile stress $= \frac{\text{Force}}{\text{Area}}$

$= \frac{F\cos\theta}{\frac{A}{\cos\theta}} = \frac{F\cos^2\theta}{A}$

9. **(2)** Bulk modulus, $B = \frac{PV}{\Delta V}$

$= \frac{P \times \frac{4}{3}\pi r^3}{\Delta V}$

$= 4\pi \times 3 \times 10^3$ atm

10. **(4)** Let Young's modulus of each wire is Y and each wire is stretched by length Δl.

$\because F = \frac{YA\Delta l}{l}$

$\Rightarrow \frac{F_A}{F_B} = \frac{A_A}{A_B} \times \frac{l_B}{l_A} = \frac{8}{1}$

11. **(4)** Stress $= \frac{\text{Weight}}{\text{Area}} = \frac{9^3 \times W_0}{9^2 \times A_0} = 9\left(\frac{W_0}{A_0}\right)$

Hence, the stress increases by a factor of 9.

12. **(3)** Breaking stress does not depend upon the length of the cable. Hence, maximum load will be same.

13. **(1)**

14. **(1)** Poisson ratio $= \frac{\text{Lateral strain}}{\text{Longitudinal strain}}$

$0.6 = \frac{\text{Lateral strain}}{0.03}$

$\Rightarrow$ Lateral strain = 0.018

15. **(2)** $B = \frac{dp}{\left(\frac{dV}{V}\right)} \Rightarrow B = \frac{mg/A}{\frac{dV}{V}}$

$\Rightarrow \frac{dV}{V} = \frac{mg}{AB}$ ... (*i*)

For a sphere $V = \frac{4\pi R^3}{3}$

$\Rightarrow dV = \frac{4\pi}{3}(3R^2 dR)$ ... (*ii*)

$\Rightarrow \frac{dV}{V} = \frac{3dR}{R}$

From eqn (*i*) and (*ii*)

$\frac{dR}{R} = \frac{mg}{3AB}$

16. **(2)** $F = F_1 + F_2$

$\frac{Y(2A)e}{\ell} = \frac{Y_1 Ae}{\ell} + \frac{Y_2 Ae}{\ell}$

$2Y_{eq} = Y_1 + Y_2$

$\Rightarrow Y_{eq} = \frac{Y_1 + Y_2}{2}$

Statement Based MCQs

1. **(1)** Both statements are true.
2. **(3)** Statement-I is true but statement-II is false.
3. **(2)** Both statements are false.
4. **(4)** Statement-I is false and statement-II is true.
5. **(1)** Both statements are true.

Assertion & Reason MCQs

1. **(1)** Both assertion & reason are true and reason is the correct explanation of the assertion.
2. **(3)** Breaking force $\propto$ area of cross-section
3. **(2)** Both assertion and reason are true but reason is not the correct explanation of the assertion.
4. **(1)** Both assertion and reason are true and reason is the correct explanation of assertion.
5. **(1)** Both assertion and reason are true and reason is the correct explanation of assertion.

Match the Columns MCQs

1. **(1)** Longitudinal stress causes the volume to change and generally the shape is not effected.

Shear stress causes the shape of an object to change but the volume remains the same.

Volumetric stress generally does not cause the shape to change but volume changes.

Tensile stress also causes the volume to change and does not effect the shape.

2. **(2)** Young' modulus, Bulk modulus and modulus of rigidity all depend on nature of material and temperature.

Volume depends on temperature because density depends on temperature.

3. **(3)** All are definitions.

NCERT Exemplar MCQs

1. **(2)** Here η = shearing stress/shearing strain and since liquids cannot sustain shearing stress, so $\eta = 0$.

2. **(4)** We know that,

Breaking stress $= \frac{\text{Breaking force}}{\text{Area of cross section}}$

Even when length of the wire changes, area of cross-section remains same.

Hence, breaking force will be same when length changes.

3. **(4)** With increase in temperature, length of a wire increases as $L_t = L_0(1+\alpha\Delta T)$

Where ΔT is change in the temperature, L_0 is original length, α is coefficient of linear expansion and L_t is length at temperature T.

$$\Rightarrow \Delta L = L_t - L_0 = L_0\,\alpha\,\Delta T$$

$$\therefore Y = \frac{\text{Stress}}{\text{Strain}} = \frac{FL_0}{A\times\Delta L} = \frac{FL_0}{AL_0\,\alpha\,\Delta T} \Rightarrow Y \propto \frac{1}{\Delta T}$$

When temperature increases Δ T increases, hence, Y decreases.

4. **(3)** Let the spring is stretched by applying a load to its free end. Therefore, the length and shape of the spring changes. Any change in length corresponds to longitudinal strain and change in shape corresponds to shearing strain.

5. **(2)** We know that Young's modulus

$$(Y) = \frac{\text{Stress}}{\text{Strain}} = \frac{F/A}{\Delta L/L} = \frac{FL}{A\Delta L}$$

$$= \frac{F}{\pi(D/2)^2}\times\frac{L}{\Delta L} = \frac{4FL}{\pi D^2\Delta L}$$

$$\Rightarrow D^2 = \frac{4FL}{\pi\Delta LY}$$

$$\Rightarrow D = \sqrt{\frac{4FL}{\pi\Delta LY}}$$

As F and $\frac{L}{\Delta L}$ are constants.

Hence, $D \propto \sqrt{\frac{1}{Y}}$

$$\Rightarrow \frac{D_{copper}}{D_{iron}} = \sqrt{\frac{Y_{iron}}{Y_{copper}}}$$

6. **(3)** Tension is least for case b.

NEET & AIIMS Past Year MCQs

1. **(3)**

$$(\text{Stress})_{max} = \text{Young's modulus} \times (\text{Strain})_{max}$$

$$\frac{8\times10^8}{2\times10^{11}} = (\Delta\ell)_{max}$$

$4 \times 10^{-3} = (\Delta\ell)_{max}$

$(\Delta\ell)_{max} = 4$ mm

2. **(3)** Stress $= \frac{F}{A} = \frac{T}{A} = \frac{W}{A}$

3. **(4)** As a spring is stretched its shape changes therefore shear modulus is used.

and $Y_{copper} < Y_{steel}$

Therefore, A is true but R is false.

4. **(2)**

Mechanical Properties of Fluids

NCERT Topic-wise MCQs

1. **(1)** Pressure at the bottom = $P_o + h\rho g$

Pressure at half the depth = $P_o + \frac{h}{2}\rho g$

According to the question, pressure at half depth of the lake = 1/4 (of the pressure at the bottom).

$$\therefore P_o + \frac{h}{2}\rho g = \frac{1}{4}(P_o + h\rho g)$$

$$4P_o + 2h\rho g = P_o + h\rho g \Rightarrow 3P_o = -h\rho g \Rightarrow \frac{-3P_o}{\rho g} = h$$

$$h = -\left(\frac{3\times10^5}{10^3\times9.8}\right) = -0.306\times10^2 = -30.6 \text{ m}$$

(–ve sign indicates depth but only magnitude is considered)

2. **(3)** Force acting on the base of a uniformly tapering vessel filled with a liquid = P × A = hρg × A

$\therefore F = 0.9 \times 1000 \times 9.8 \times 4 \times 10^{-3}$

$\therefore F = 35.28 \times 10^{-3} \times 10^3 = 35.28$ N

3. **(1)** Let V is the volume of ball and m is the mass of ball then, for equilibrium condition

$mg = (m_1 + m_2)g$

$$V\rho_3 g = \frac{V}{2}\rho_1 g + \frac{V}{2}\rho_2 g$$

$$\rho_3 = \frac{\rho_1 + \rho_2}{2}$$

Now, we can say that density of ball is intermediate or average of ρ_1 & ρ_2 but from diagram it is clear that $\rho_2 > \rho_1$ so $\rho_2 > \rho_3 > \rho_1$.

4. **(4)** As temperature remains constant throughout motion.

$P_1V_1 = P_2V_2$

$V(P_o + h\rho g) = P_o \times 2V$

$$h\rho g = P_o \Rightarrow h = \frac{P_o}{\rho g} = \frac{75\times13.6\times9.8}{\frac{13.6}{11}\times9.8} = 6.8\,\text{m}$$

5. **(3)** The mercury column will maintain its original height.

$$\cos 30^\circ = \frac{h}{\ell}$$

$$\ell = \frac{h}{\cos 30^\circ} = \frac{76}{\frac{\sqrt{3}}{2}} = \frac{152}{\sqrt{3}}$$

$\ell = 87.75\,\text{cm}$

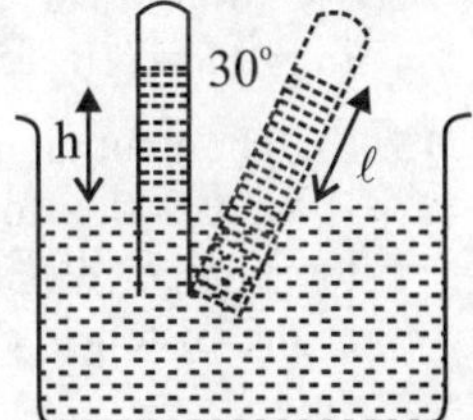

6. **(2)** Pressure on the swimmer at a depth of 40 m below the surface

$P = P_o + h\rho g$

$P = 1.01 \times 10^5 + 40 \times 10^3 \times 10 = 5.01 \times 10^5\ \text{Nm}^{-2}$

7. **(1)** On increasing the temperature of mercury its density will decrease. Hence, level of mercury in barometer tube will increase.

8. **(1)** Net Force = Pressure difference × Area.

$= (10 - 5)\ 10^5 \times 80 = 4 \times 10^7$ N

9. (2) $\Delta h'\rho_{Hg}g = h\rho_{air}g$

$$h = \frac{\Delta h'\rho_{Hg}g}{\rho_{air}g} = (90-60)10^{-2}\times10^{+4}$$

$h = 30 \times 10^2 = 3000 \approx 3$ km

10. (1) $\frac{V_{out}}{V} = \left(1-\frac{\rho}{\sigma}\right) = \frac{1000-700}{1000} = \frac{300}{1000}$

$$\frac{V_{out}}{V}\times100 = \frac{300}{1000}\times100 = 30\%$$

11. (3) Net up-thrust force (buoyant force) acting on the cylinder

$= \rho_1 l_1 Ag + \rho_2 l_1 Ag$

It implies that buoyant force acting on the cylinder is free from the depth of the liquid above cylinder (h).

Hence no change in the submergence if we increase the height of fluid-1 above fluid-2

12. (4) Weight of water displaced = weight of liquid displaced

$$\frac{1}{3}V\rho_w g = \frac{1}{2}V\rho_o g \Rightarrow \rho_0 = \frac{2}{3}\rho_w$$

$$\text{Specific gravity of liquid} = \frac{\rho_0}{\rho_w} = \frac{\frac{2}{3}\rho_w}{\rho_w} = \frac{2}{3}$$

13. (2) Let v be volume of sphere in the upper liquid and that in lower liquid is (V – v)

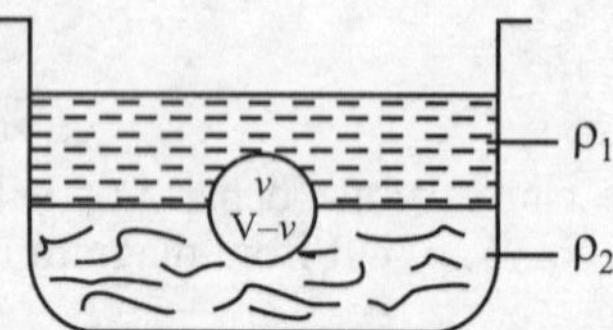

Using Archimedes principle $V\rho g = v\rho_1 + (V - v)\rho_2$

$V\rho g = v\rho_1 + V\rho_2 - v\rho_2$ or $V(\rho - \rho_2) = v(\rho_1 - \rho_2)$

$$\frac{v}{V} = \frac{\rho-\rho_2}{\rho_1-\rho_2} = \frac{\rho_2-\rho}{\rho_2-\rho_1}$$

Choice (2) is correct.

14. (2) According to Archimedes principle

weight of load = weight of water displaced by boat

$m \times 9.8 = (10 \times 5 \times 0.05) \times 9.8 \times 10^3 \Rightarrow m = 2500$ kg

15. (3) The output force should have a magnitude of $F_2 = 20{,}000$ N to supports the car. Initially the piston and the plunger are at the same height, so the fluid pressure at the piston is equal to the fluid pressure at the plunger $\frac{F_1}{A_1} = \frac{F_2}{A_2}$ where A_1 is the area of the piston and A_2 is the area of the plunger.

Therefore, the magnutude of the force $F_1 = F_2\left(\frac{A_1}{A_2}\right)$

Using $A = \pi r^2$ for the circular areas of the piston and plunger. We find that,

$$F_1 = F_2\left(\frac{A_1}{A_2}\right) = F_2\frac{\pi r_1^2}{\pi r_2^2} = 20000\left(\frac{0.01}{0.2}\right)^2 = 50\text{N}$$

16. (2) For a floating body,

Weight = force of upthrust $\Rightarrow V_B\rho_B\cdot g = V_{in}\cdot\rho g$

$$\Rightarrow V.\rho_B = \frac{4}{5}V.\rho_W$$

$$\Rightarrow \rho_B = \frac{4}{5}10^3 = 0.8\times10^3\,\text{kg/m}^3.$$

17. (2) Equating pressure at BD (line of constant pressure)

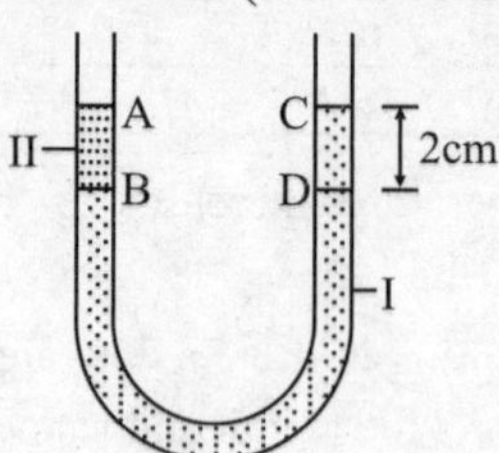

$\therefore 2 \times \rho' g = 2 \times 1.1 \times g$

$\rho' = 1.1$

Choice (2) is correct.

18. (1) System is falling under gravity due to which there is no upthrust to be applied by liquid.

19. (2) By equation of continuity, $A_1v_1 = A_2v_2$.

$\therefore \pi r^2 v = \pi(2r)^2 v'$

$$v' = \frac{v}{4}$$

20. (2) Again by equation of continuity, $A_1 v_1 = A_2 v_2$

$\pi r_1^2 v_1 = \pi r_2^2 v_2$

$$\frac{v_2}{v_1} = \frac{r_1^2}{r_2^2} = \frac{2^2}{1^2} = \frac{4}{1}; \quad \frac{v_2}{5} = 4,\ v_2 = 20\ \text{ms}^{-1}$$

21. (4) From equation of continuity

$V_B\,\pi R_B^2 = \pi(4R)^2\,5$

$V_B = 80$ m/s

22. (2) By equation of continuity $a_1 v_1 = a_2 v_2$

$\pi r_1^2 v_1 = \pi r_2^2 v_2$

$$\frac{v_1}{v_2} = \left(\frac{r_2}{r_1}\right)^2 = \frac{4}{9}$$

23. (2) According to the equation of continuity

$A_{en} \times V_{en} = A_{ex} V_{ex}$

$$\frac{V_{en}}{V_{ex}} = \frac{A_{ex}}{A_{en}} = \left(\frac{r_{ex}}{r_{en}}\right)^2 = \left(\frac{4}{5}\right)^2 = \frac{16}{25}$$

24. (2) By equation of continuity,

$A_1 V_1 = A_2 V_2 =$ constant

If area decreases then velocity increases.

25. (1) According to the equation of continuity

$AV = A_1V_1 + A_2V_2 + A_3V_3 + A_4V_4$

$\Rightarrow 70 \times 10^{-4} \times V$

$= 10\times10^{-4}\times5 + 10\times10^{-4}\times5 + 20\times10^{-4}\times2 + 40\times10^{-4}\times1$

$\Rightarrow 70 \times 10^{-4}\,V = 10^{-4}\,(50 + 50 + 40 + 40)$

$$\Rightarrow 70V = 180 \Rightarrow V = \frac{180}{70} = 2.57\,\text{m/s}$$

26. (4) Pressure at B is due to the water column AB and pressure at A which is applied by flowing liquid but $V_A > V_B$.

27. **(3)** Using equation of continuity in the pipe at the entry and exit.

$$\pi\left(\frac{D_1}{2}\right)^2 V_1 = \pi\left(\frac{D_2}{2}\right)^2 V_2$$

$$\Rightarrow \left(\frac{D_1}{D_2}\right)^2 V_1 = V_2$$

$$V_2 = \left(\frac{4}{2}\right)^2 \times 20 = 4\times 20 = 80\text{m/s}$$

28. **(2)** Apply Bernoulli's Theorem

$$\frac{P_1 - P_2}{\rho} = \frac{1}{2}\left(v_2^2 - v_1^2\right)$$

$$\frac{(5.1)\times 10^{-1}\times 1\times 1000}{1} = \frac{1}{2}\left(v_2^2 - 2^2\right)$$

$v_2^2 = 1020 + 4 = 1024$

$v_2 = 32$ cm/s

29. **(3)** According to the relation $t = \frac{A}{A_o}\sqrt{\frac{2}{g}H} - \frac{A}{A_o}\sqrt{\frac{2H'}{g}}$

$$t = \frac{A}{A_o}\sqrt{\frac{2}{g}}\times\left(\sqrt{H} - \sqrt{H'}\right)$$

$$t \propto \left(\sqrt{H} - \sqrt{H'}\right)$$

$$\frac{t_1}{t_2} = \frac{\sqrt{H} - \sqrt{\frac{H}{2}}}{\sqrt{\frac{H}{2}} - \sqrt{0}} = \frac{1 - \frac{1}{\sqrt{2}}}{\frac{1}{\sqrt{2}}} = \sqrt{2} - 1$$

30. **(4)** In a narrow constriction, velocity increases, pressure decreases. Pressure at B is low, Pressure in A and C is same. Choice (4) is correct.

31. **(3)** From Bernoullis equation $P_2 - P_1 = \frac{1}{2}\rho\left(v_1^2 - v_2^2\right)$

Upward force on the wing, $F = (P_2 - P_1)A$, where A is area of the wing

$$F = \frac{1}{2}\rho\left(v_1^2 - v_2^2\right)A = \frac{1}{2}\times 1.3\left(70^2 - 63^2\right)\times 2.5 = 1.5\times 10^3\text{ N.}$$

32. **(2)** Air speeds are different near the upper and lower surfaces of the wings which creates a pressure difference. Force due to this pressure difference supports the weight of the aeroplane.

33. **(2)** Streamline flow is applicable for the streamlined flow of incompressible fluids.

34. **(2)** Pressure head $= \frac{P}{\rho g}$; velocity head $= \frac{V^2}{2g}$

$$\frac{V^2}{2g} = \frac{2P}{\rho g} \Rightarrow V^2 = \frac{4P}{\rho}$$

$$V^2 = \frac{4\times 75\times 10^{-2}\times 13.6\times 10^3\times 9.8}{10^3} = 39984\times 10^{-2} = 399$$

$V \approx 20$ m/s

35. **(3)** Using equation of continuity,

$$a_1 v_1 = a_2 v_2 \Rightarrow v_2 = \frac{a_1 v_1}{a_2} = \frac{5\times 10^{-2}\times 5}{2\times 10^{-2}} = 12.5\text{m/s}$$

Now, using Bernoulli's equation,

$$P_1 + \frac{1}{2}\rho v_1^2 = P_2 + \frac{1}{2}\rho v_2^2 \Rightarrow P_2 = P_1 + \frac{1}{2}\rho\left(v_1^2 - v_2^2\right)$$

$$P_2 = 8\times 10^4 + \frac{1}{2}\times 10^3(25 - 156.25) = 146\text{ kPa}$$

36. **(4)** $R = 2\sqrt{HH'} = 2\times\sqrt{\frac{1}{\sqrt{3}}\times 1.71} = 2\text{m}$

37. **(2)** Velocity of discharge at orifice

$= \sqrt{2gh} = \sqrt{2\times 9.8\times 19.6} = 19.6\text{m/s.}$

38. **(4)** There is no upward or downward force acting on the ball due to air for a spinning, the air is dragged along a spinning ball. This result in dynamic shift the ball experiences are termed as magnus effect.

39. **(1)** As the two motor boats are moving in the same direction, therefore velocity of water current between them increases. This decreases the pressure between the motor boats and hence they are pulled towards each other.

40. **(3)** Energy remains conserved in Bernoulli's theorem.

41. **(1)** $v = \sqrt{2gh}$

Therefore, v is independent of radius.

42. **(1)** According to Bernoulli's Theorem

$$P + \frac{1}{2}dv^2 = \text{constant}$$

At sides the velocity is larger ($\because v = R\omega$) so the pressure is lower. Since the pressure at a given horizontal level must be equal, the liquid rises at the sides to compensate for the drop of pressure.

$$\therefore\ \frac{1}{2}dv^2 = dgh,$$

$$\text{or } h\,dg = \frac{1}{2}d\times R^2\omega^2$$

Choice (1) is correct.

43. **(3)** $t = \frac{A}{A_o}\sqrt{\frac{2H}{g}} = \frac{16\times 10^{-4}}{1\times 10^{-4}}\sqrt{\frac{2\times 32\times 10^{-2}}{9.8}}$

$$t = \frac{16\times 8\times 10^{-1}}{\sqrt{9.8}} = \frac{12.8}{3.13} = 4.08\text{sec}$$

44. **(1)** Eight drops coalesce to form one drop of radius R.

Volume of Big drop $= \frac{4}{3}\pi R^3$ ...(*i*)

Volume of each smaller drop $= \frac{4}{3}\pi r^3$ ... (*ii*)

from equation (*i*) & (*ii*) we get

$$\frac{4}{3}\pi R^3 = 8\times\frac{4}{3}\pi r^3 \Rightarrow R = 2r$$

Terminal speed v is related to radius r as,

$V_r \propto r^2 \Rightarrow \frac{V_r}{V_R} = \left(\frac{r}{2r}\right)^2 = \frac{1}{4}$

$\Rightarrow V_R = 4 \times 12 = 48 \text{ cm s}^{-1}$

45. **(4)** Formula for rate of flow of liquid through a tube of radius r (Poiseuille's relation) is $V = \frac{\pi}{8}\frac{Pr^4}{\eta l}$

For a tube of radius a and Pressure P, rate of flow Q is given as $Q = \frac{\pi}{8}\frac{Pr^4}{\eta l}$, if radius of tube is $\frac{a}{2}$ and pressure 2P, then $Q = \frac{\pi}{8}\frac{(2P)\left(\frac{a}{2}\right)^4}{\eta l} = \frac{1}{8}Q$

46. **(3)** In case of laminar flow the upper layer pulls if forward while lower layer pulls if backward.

47. **(2)** With the rise of temperature force of attraction between liquid decreases. While in case of gases, with rise of temperature collision increases in gases.

48. **(4)** Poiseuille's formula $V = \frac{\pi}{8}\frac{Pr^4}{\eta \ell} \Rightarrow \eta = \frac{\pi}{8}\frac{Pr^4}{V\ell}$

$\therefore \eta \propto r^4$

$\because$ Slight change in the value of radius will causes the more change in coefficient of viscosity.

49. **(1)** Velocity gradient or strain rate in liquid

$\eta = \frac{\text{shearing stress}}{\text{strain rate}}$

$\Rightarrow \text{shearing stress} = \eta \times \text{strain rate}$

$\text{Shearing stress} = 10^{-3} \times \frac{5}{20} = 2.5 \times 10^{-4} \text{ N/m}^2$

50. **(2)** Terminal velocity $(v) = \frac{2}{9}\frac{r^2(\rho-\sigma)g}{\eta}$

$$v = \frac{2r^2\rho\left(1-\frac{\sigma}{\rho}\right)g}{9\eta} = \frac{2r^3\rho\left(1-\frac{\sigma}{\rho}\right)g}{9r\eta}$$

$$v \propto \frac{m}{r} \quad \left[\because \left(m = \frac{4}{3}\pi r^3\rho\right) \Rightarrow \left(m \propto r^3\rho\right)\right]$$

51. **(2)** Due to increase in temperature, viscosity of oil decreases, surface tension also decreases but surface tension doesn't consider for lubrication property of oil/liquid.

52. **(2)** Value of coefficient of viscosity is very small as compared to coefficient of friction.

53. **(3)** Viscous drag and upthrust acts upward and weight acts downward for constant velocity (terminal velocity). Viscous drag to liquid and upthrust should balance weight of the ball.

54. **(4)** $V_T = \frac{2}{9}\frac{r^2(\rho-\sigma)g}{\eta}$

ρ = density of sphere

σ = density of liquid

$\therefore V \propto (\rho - \sigma) \Rightarrow \frac{V_g}{V_s} = \frac{\rho_g - \sigma}{\rho_s - \sigma}$

$$\frac{0.5}{V_s} = \frac{19.5-3}{10.5-3} = \frac{16.5}{7.5} \Rightarrow V_s = \frac{0.5 \times 7.5}{16.5} = 0.22 \text{ m/s}$$

55. **(2)** $\eta = 0.09 \text{ kgm}^{-1}\text{s}^{-1}$, $dv = 5 \text{m/s}$, $dx = 3 \times 10^{-3}$ m, $A = 0.2 \text{ m}^2$

$$F = \eta\frac{Adv}{dx} = \frac{0.09 \times 0.2 \times 5}{3 \times 10^{-3}} = 30\text{N}$$

56. **(2)** According to stokes's law, F = $6\pi\eta rv$ when ball is released its velocity increases first, after some time it becomes constant because viscous drag and upthrust balances weight of the body.

57. **(3)** We have two small drops having radii in the ratio 4 : 3.

P = mv

Terminal Velocity $v \propto r^2 \Rightarrow \frac{v_1}{v_2} = \left(\frac{r_1}{r_2}\right)^2 = \frac{16}{9}$

$\text{mass} = \frac{4}{3}\pi r^3\rho \Rightarrow m \propto r^3$

$\therefore$ Momentums are in ratio

$$\frac{P_1}{P_2} = \frac{r_1^3}{r_2^3} \times \frac{v_1}{v_2} = \left(\frac{4}{3}\right)^3 \times \frac{16}{9} = \frac{1024}{243} \approx 4:1$$

58. **(4)** Required force $F = \eta A\frac{\Delta v}{\Delta x}$

$$= \frac{1.0 \times 100 \times 10^{-4} \times (7 \times 10^{-2})}{10^{-3}} = 0.7\text{N}$$

59. **(2)** $F = S \times 2\pi r = 72 \times 10^{-3} \times 2 \times 3.14 \times 10 \times 10^{-2} = 4.5 \times 10^{-2}$N

60. **(2)** Work done $W = T \times \Delta A$

$= T \times 2 \times [4\pi(5)^2 - 4\pi(1)^2]$

$= 30 \times 8\pi \times 24$ ergs

$= 5760\ \pi$ ergs.

61. **(3)** Volume remains constant. Let R and r be the radius of bigger and smaller drops respectively.

$$\frac{4}{3}\pi R^3 = 10^3 \times \frac{4}{3}\pi r^3 \Rightarrow r = \frac{R}{10} = \frac{2 \times 10^{-3}}{10} = 2 \times 10^{-4}\text{m}$$

Work done in breaking drop into 1000 drops.

$W = S \times (10^3 \times 4\pi r^2 - 4\pi R^2) = S \times 4\pi\,(10^3 \times r^2 - R^2)$

$W = 72 \times 10^{-3} \times 4 \times 3.14\,[10^3 \times (2 \times 10^{-4})^2 - (2 \times 10^{-3})^2]$

$W = 0.904 \times 4 \times 10^{-5}\,(1 - 0.1) = 3.25 \times 10^{-5}$J

62. **(4)** Antiseptics have low surface tension due to which at normal temperature its surface tension reduces and spreads over the surface easily.

63. **(1)** Due to adhesive force, because this type of force of attraction is present in different molecules.

64. **(4)** Soaps decreases the surface tension of water so as to increase wetting ability of water which helps in cleaning clothes.

65. **(2)** Density of oil is lower than that of water that's why oil spreads on water surface.

66. **(2)** When small piece of camphor is drop on clean surface of water then they dissolve in water. The surface tension of solution is relatively less than the pure water. Because the shape of camphor pieces is irregular therefore one part of camphor dissolve more than the other side. Therefore the pieces of camphor pulls towards the area of excess surface tension. Where these pieces go the same incident occurs. Therefore it moves on the surface of water.

67. (4) Surface energy = $S \times \Delta A = 2 \times 10 \times 0.05 = 1$ J (Because film has two surfaces)

68. (4) $r = \dfrac{r_1 r_2}{r_2 - r_1} = \infty$ (radii are equal $r_1 = r_2$)

69. (1) As we know that $W = T \times \Delta A \Rightarrow T = \dfrac{W}{\Delta A}$

$$T = \frac{8\times10^{-2}}{2\times\left[(20\times8)10^{-4} - (11\times7)10^{-4}\right]}$$

$$= 0.048\times10^{2} = 4.8 \text{N/m}$$

70. (3) $R = n^{1/3} r \Rightarrow R = 3r$

$$\frac{\text{Energy of big drop}}{\text{Energy of small drop}} = \frac{T\times4\pi R^2}{T\times4\pi r^2}$$

$$\frac{R^2}{r^2} = \frac{(3r)^2}{r^2} = 9$$

71. (3) Work done = $T \times 2 \times 4\pi R^2$

$$\therefore \frac{w}{w'} = \frac{R^2}{(2R)^2} \Rightarrow w' = 4w$$

72. (3) Due to the presence of impurities cohesive force decreases because attraction among the particle of liquid or substance decreases due to which surface tension decreases. Surface tension depends on temperature also.

73. (4) Excess pressure = h dg.

$$\therefore \frac{2T}{r} = hdg;\ h = \frac{2T}{rdg}$$

74. (1) Surface tension of a liquid becomes zero at the critical temperature.

75. (2)

$$h = \frac{2S\cos\theta}{\rho rg} = \frac{2\times7.4\times10^{-3}\times\cos0^\circ}{10^3\times10^{-3}\times9.8} = 1.51\times10^{-3} = 1.51\text{mm}$$

76. (1) $h = \dfrac{2S\cos\theta}{r\rho g} \Rightarrow r = \dfrac{2S\cos\theta}{h\rho g}$

$$D = \frac{4S\cos\theta}{h\rho g}$$

$$\therefore D = \frac{4\times540\times10^{-3}\times\cos120^\circ}{-1.21\times10^{-2}\times13.6\times10^3\times9.8} = 0.06\text{mm}$$

77. (3) Because force of adhesion and cohesion considered.

78. (1) Liquid will rise in a tube if angle is acute due to capillary action.

79. (4) $h \propto \dfrac{1}{r}\quad \therefore \dfrac{r_1}{r_2} = \dfrac{h_2}{h_1} = \dfrac{5}{4}$

80. (3) $r = \dfrac{2S\cos\theta}{h\rho g} \Rightarrow D = \dfrac{4S\cos\theta}{h\rho g} = \dfrac{4S\cos\theta\rho_w}{h\times\rho_e\times g}$

$$\therefore \frac{h_1}{h_2} = \frac{S_1}{S_2}\times\frac{\rho_{\ell_2}}{\rho_{\ell_1}} = \frac{70}{40}\times\frac{0.6}{0.8} = \frac{21}{16}$$

81. (3) When wax is coated, angle of contact increases, so liquid level fall i.e, $(h_1 > h_2)$

82. (2) $h = \dfrac{2S\cos\theta}{r\rho g} \Rightarrow 0 = \dfrac{2S\cos\theta}{r\rho g}$

$\therefore \cos\theta = 0^\circ \Rightarrow \theta = 90^\circ$

83. (1) $h = \dfrac{2S\cos\theta}{\rho rg} \Rightarrow hr = \dfrac{2S\cos\theta}{\rho g} \Rightarrow hr = \text{constant}$

84. (1) The air through the horizontal tube will decrease the pressure and more liquid will be pushed into the capillary tube.

85. (3) $P \propto \dfrac{1}{r}$ Hence pressure decrease with time. Slope is parabolic downward.

86. (4) Excess pressure inside spherical drop $P = \dfrac{2S}{R}$

$$P_1 = \frac{2T}{R_1},\ P_2 = \frac{2T}{R_2}$$

Now,

$$\frac{2T}{R_1} = 4\times\frac{2T}{R_2} \Rightarrow R_2 = 4R_1$$

$$\frac{m_1}{m_2} = \frac{V_1\rho}{V_2\rho} = \frac{4/3\pi R_1^3\rho}{4/3\pi R_2^3\rho} = \frac{R_1^3}{64R_1^3} = \frac{1}{64}$$

87. (2) Excess pressures $(\Delta P_1)P_i - P_o = 1.06 - 1 = 0.06$ atm

$(\Delta P_2)P_i - P_o = 1.08 - 1 = 0.08$ atm

$$\Delta P = \frac{4S}{R} \Rightarrow R = \frac{4S}{\Delta P} \Rightarrow \frac{R_1}{R_2} = \frac{\Delta P_2}{\Delta P_1}$$

$$\frac{R_1}{R_2} = \frac{0.08}{0.06} = \frac{8}{6} = \frac{4}{3}$$

$$\therefore \frac{V_1}{V_2} = \left(\frac{R_1}{R_2}\right)^3 = \left(\frac{4}{3}\right)^3 = \frac{64}{27}$$

88. (4) According to the relation, $P_1V_1 = P_2V_2$

$$(H+h)\rho g\times\frac{4}{3}\pi r^3 = H\rho g\times\frac{4}{3}\pi(3r)^3$$

$$(H+h)r^3 = 27Hr^3 \Rightarrow h = 26H$$

89. (4) $P = h\rho g = 2\times10^{-3}\times0.8\times10^3\times9.8$ (sp. gravity $= \dfrac{\rho_\ell}{P_w}$)

P = 15.68 Pa

For soap bubble. $P = \dfrac{4S}{R} \Rightarrow 15.68 = \dfrac{4\times S}{2\times10^{-2}}$

$\therefore S = 7.84\times10^{-2}$ N/m

High Order Time Intensive MCQs

1. (2) Time taken by water to reach the foot

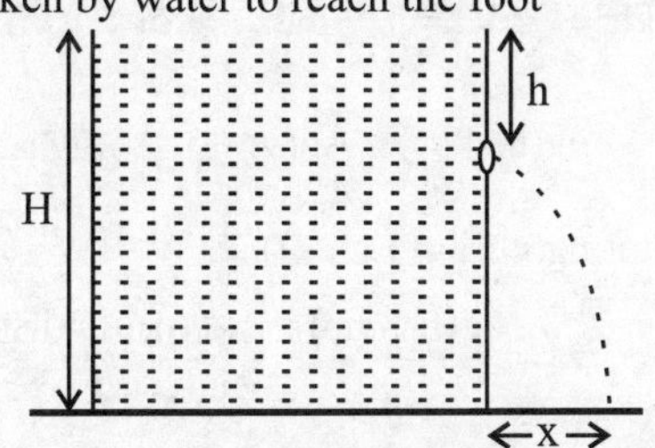

$t = \sqrt{\frac{2(H-h)}{g}}$

Velocity of water coming out of hole $v = \sqrt{2gh}$

$\therefore$ Emerging stream of water strike the foot at, $x = v \times t$

$= \sqrt{2gh} \times \sqrt{\frac{2(H-h)}{g}} = 2\sqrt{h(H-h)}$

2. **(3)** The height h of a liquid risen in a capillary tube is given by:

$h = \frac{2\,T\cos\theta}{r\rho g} \propto \frac{1}{r}$

When the radius of capillary tube is doubled the height becomes half

$\therefore\ h' = \frac{h}{2}$ and $r' = 2r$

Mass of liquid is $M = \pi r^2 h$

$M' = \pi(2r)^2 \times \frac{h}{2} = 2M$

3. **(1)** Using Bernoulli's theorem,

$p_1 + \frac{1}{2}\rho v_1^2 = p_2 + \frac{1}{2}\rho v_2^2$ ...(*i*)

Here, $p_1 = \rho_m g h_1 = 13600 \times 9.8 \times 10^{-2}$

$p_2 = 13600 \times 9.8 \times h$

$\rho = 1000$ kg-m^{-3}

$v_1 = 35 \times 10^{-2}$ ms^{-1}, $v_2 = 65 \times 10^{-2}$ ms^{-1}

$\therefore$ From Equ. (*i*), we get

$\Rightarrow 13600 \times 9.8 \times 10^{-2} + \frac{1}{2} \times 1000 \times (0.35)^2$

$= 13600 \times 9.8 \times h + \frac{1}{2} \times 1000 \times (0.65)^2$

After solving, h = 0.89 cm of Hg

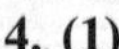

4. **(1)**

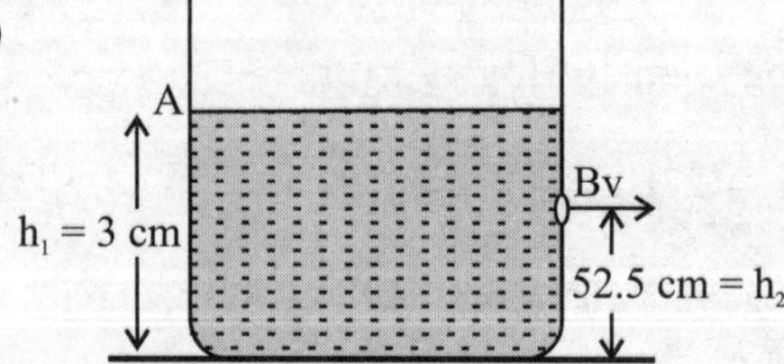

From equation of continuity,

$av = AV \Rightarrow \quad V = \frac{av}{A}$ ...(*i*)

Using Bernoulli's theorem,

Energy per unit volume at point A = energy per unit volume at point B

$p_0 + \rho g h_1 + \frac{1}{2}\rho V^2 = p_0 + \rho g h_2 + \frac{1}{2}\rho v^2$

$\Rightarrow \quad \rho g(h_1 - h_2) = \frac{1}{2}\rho(v^2 - V^2)$

Substituting v from Equ. (*i*)

$\Rightarrow \quad v^2 = \frac{2g(h_1 - h_2)}{1 - \left(\frac{a}{A}\right)^2} = \frac{2 \times 10 \times (3 - 0.525)}{1 - (0.1)^2} = 50(\text{m/s})^2$

5. **(3)** We know that terminal velocity is given by $v = \frac{2}{9}\frac{r^2\rho g}{\eta}$

$\Rightarrow \quad \eta = \frac{2}{9} \cdot \frac{r^2 \rho g}{v}$

$= \frac{2}{9}\frac{(1\times10^{-2})^2 \times (1.5\times10^3) \times 9.8}{0.25 \times 10^{-2}} = 130\,\text{Pas}$

6. **(2)** From equation of continuity,

$(a_1 v_1)_{tube} = (a_2 v_2)_{holes}$

$(8\times10^{-4}) \times \left(\frac{0.15}{60}\right) = (40 \times 10^{-8}) \times v$

$\Rightarrow \quad v = \frac{8\times10^{-4} \times 0.15}{40\times10^{-8} \times 60} = 5\,\text{ms}^{-1}$

7. **(1)** Let height of water column in the tank be h.

Total pressure (p) = atmospheric pressure (p_0) + pressure due to water column in tank (p′)

$\therefore p' = p - p_0 = 3 - 1 = 2$ atm $= 2 \times 10^5$ N/m^2

$p' = h\rho g$

$h\rho g = 2 \times 10^5 \Rightarrow h = 20$ m

Hence, velocity of water coming from hole, i.e. velocity of efflux is

$v = \sqrt{2gh} = \sqrt{2 \times 10 \times 20} = \sqrt{400}$ ms^{-1}

8. **(3)** $\therefore$ Upward lift = pressure difference × area of wing

$[\because F = p \times A]$

From Bernoulli's theorem,

$p_1 + \frac{1}{2}\rho v_1^2 = p_2 + \frac{1}{2}\rho v_2^2$

Pressure difference is

$p_2 - p_1 = \frac{1}{2}\rho(v_1^2 - v_2^2)$

upward lift $= (p_2 - p_1) \times A = \frac{1}{2}\rho(v_1^2 - v_2^2)A$

9. **(1)** Given, $v_2 = 2v$, $v_1 = v$

$\therefore\ p + \frac{1}{2}\rho v^2 = p' + \frac{1}{2}\rho(2v)^2$

$\Rightarrow \quad p' = p - \frac{3}{2}\rho v^2$

10. **(4)** There is no effect of rotation on levels of water in two limbs. This is because each limb is under vertical pressure due to atmosphere only (p_0). Hence, water level remain same in both sections.

11. **(3)** Let, thickness of layer be x.

So, volume V = Area × x

$V = A \times x \ (\because x = 2r)$

$\Rightarrow \quad 2r = \frac{V}{A} \Rightarrow r = \frac{V}{2A}$ ...(*i*)

And $\Delta p = \frac{T}{r}$

We know that $F = \Delta p \times A$

$= \frac{T}{r} \times A$

$F = \frac{T}{\left(\frac{V}{2A}\right)} \times A$ [from Eq. (*i*)]

$$T = \frac{F \times V}{2A^2}$$

Where, $F = 16 \times 10^5$ dyne, $V = 0.04\ cm^3$,

$A = 20cm^2 \Rightarrow T = \frac{16 \times 10^5 \times 0.04}{2 \times 20^2} = 80$ dyne/cm

12. (2) $V = \frac{P\pi r^4}{8\eta l}; V' = \frac{P\pi(2r)^4}{8\eta(4l)}$

Dividing both the equation $V' = 4V, \theta = 60°$

13. (2) Excess pressure $= \frac{2T}{r}$

$= \frac{2 \times 70 \times 10^{-3}}{1 \times 10^{-3}} = 140 Nm^{-2}$

14. (3) Excess pressure inside the air bubble is given by.

$$\Delta P = P_i - P_o = \frac{2T}{r}$$

$$P_i = P_o + \frac{2T}{r}$$

$$= 1.013 \times 10^5 + \frac{2 \times 7.2 \times 10^{-2}}{0.1 \times 10^{-3}}$$

$= 1.013 \times 10^5 + 0.0144 \times 10^5$
$= 1.027 \times 10^5\ Nm^{-2}$

15. (3) If the length of the capillary tube is less, the liquid comes upto the height of the capillary tube and stops there. According to equation for capillary rise

$$h = \frac{2T \cos\theta}{r\rho g}$$

Then valve of (angle of contact) keeps changing until the equation is balanced.
i.e., meniscus adjust itself so that the water does not spill.

16. (4) Surface tension × circumference = force

$\Rightarrow$ Circumference $= \frac{F}{T} = 12.5 \times 10^{-2}m$

17. (3) $h \propto \frac{1}{r}$

$$\frac{h_2}{h_1} = \frac{r_1}{r_2} = \frac{D_1}{D_2}$$

As $D_2 = \frac{1}{2} D_1$

$\Rightarrow h_2 = 2h$

18. (1) From Boyle's law, PV = constant

$\therefore\ P_1V_1 = P_2V_2$

Here, $p_1 = (h + l), V_1 = \frac{4}{3}\pi r^3$

$p_2 = l, V_2 = \frac{4}{3}\pi(3r)^3$

$\therefore\ (h + l)\frac{4}{3}\pi r^3 = l \times \frac{4}{3}\pi(3r)^3$

or $h + l = 27l$

$\therefore\ h = 26l$

19. (1) Energy needed = Increment in surface energy
= (surface energy of n small drops) – (surface energy of one big drop)
$= n4\pi r^2T - 4\pi R^2T = 4\pi T\ (nr^2 - R^2)$

20. (2) Velocity gradient $= \frac{0.5}{\frac{2.5}{2} \times 10^{-2}}$

As force on the plate due to viscosity is from upper as well as lower portion of the oil, equal from each part,

Then, $F = 2\eta A \frac{dv}{dz} \Rightarrow 1 = 2 \times \eta \times (0.5) \frac{0.5}{1.25 \times 10^{-2}}$

$\Rightarrow \eta = 2.5 \times 10^{-2}$ kg – sec/m²

21. (1) Weight of cylinder = upthrust due to both liquids

$$V \times D \times g = \left(\frac{A}{5} \cdot \frac{3}{4} L\right) \times d \times g + \left(\frac{A}{5} \cdot \frac{L}{4}\right) \times 2d \times g$$

$$\Rightarrow \left(\frac{A}{5} \cdot L\right) D \cdot g = \frac{ALdg}{4} \Rightarrow \frac{D}{5} = \frac{d}{4} \therefore D = \frac{5}{4} d$$

22. (1) Pressure at A : $P_A = P_{atm} - \rho_1\ gl \sin\theta$
Pressure at B : $P_B = P_{atm} + \rho_2\ gh$
But P_B is also equal to $P_B = P_A + \rho_3\ gl \sin\theta$
Hence, $P_{atm} + \rho_2\ gh = P_A + \rho_3\ gl \sin\theta$
$\Rightarrow P_{atm} + \rho_2\ gh = P_{atm} - \rho_1\ gl \sin\theta + \rho_3\ gl \sin\theta$

$\Rightarrow \sin\theta = \frac{\rho_2 h}{(\rho_3 - \rho_1)\ell}$.

23. (4) $p_2 - p_1 = (d_2 - d_1)\rho g = 3.03 \times 10^6$

$\Rightarrow d_2 - d_1 = \frac{3.03 \times 10^6}{\rho g}$

$\therefore\ d_2 - d_1 = 303m \approx 300$ m

24. (4) As we know that, $V = \frac{P}{R_{eff}}, R = \frac{8\eta\ell}{\pi r^4} = \frac{8\eta^2\ell}{\pi\left(\frac{r}{2}\right)^2}$

$$R' = \frac{8\eta\ell \times 2}{\pi\left(\frac{r}{2}\right)^4} = \frac{8\eta\ell \times 32}{\pi r^4}$$

$\therefore R = 32R$

$$V = \frac{P}{R + 32R} = \frac{P}{33R} = \frac{V}{33}$$

25. (1) In equilibrium, the pressure of liquid at the same level must be equal. Considering pressure at level D in both arms of U-tube. Pressure of h cm of oil + pressure of (20 – h) cm of mercury = pressure of 20 cm of carbon tetrachloride.
$h \times 0.9 \times g + (20 - h) \times 13.6 \times g = 20 \times 1.6 \times g$
$\Rightarrow 0.9h + 272 - 13.6h = 32$
$\Rightarrow 12.7h = 240$

$\Rightarrow h = \frac{240}{12.7} = 18.9$ cm.

26. (2) Apply Bernoulli's theorem between two ends

$$P_0 + \rho\int_3^5 x\omega^2 dx = P_0 + \frac{1}{2}\rho v^2 \Rightarrow \frac{\rho\omega^2}{2}[5^2 - 3^2] = \frac{1}{2}\rho v^2$$

$\Rightarrow\ v = 2m/s$

27. (2) Hemispherical soap bubble has large radius of curvature, so it has less pressure and hence air flows from 1 (high pressure) towards 2 (at low pressure).

28. (2) $W = \Delta k$

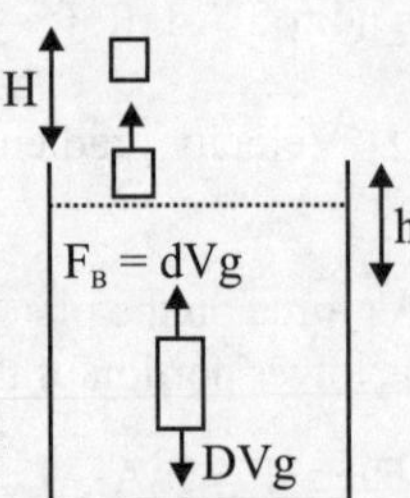

$$(d-D)Vgh = \frac{1}{2}mv^2$$

$$(d-D)Vgh = \frac{1}{2}(DV)v^2$$

$$v^2 = 2gh\left(\frac{d-D}{D}\right)$$

To reach maximum height from the free surface, this velocity

$v^2 = 2gH$

$$2gh\left(\frac{d-D}{D}\right) = 2gH$$

$$H = h\left(\frac{d}{D} - 1\right)$$

29. (3) Work done = T × increase in area (both sides)

$$= 30 \times 2 \times 4\pi\left(\frac{2^2}{\pi} - \frac{1^2}{\pi}\right) = 720 \text{ erg}$$

30. (4) The moment coin falls into the water, the weight of the block decreases, as a result the block rises up and therefore *l* decreases.

Further, with the rise of the block, the volume of displaced water also decreases, as a result the water level in the tank falls and thus h also decreases.

31. (3) $V_1 = V_2 \Rightarrow a_1 v_1 = a_2 v_2$

$L^2\sqrt{2gh} = \pi r^2\sqrt{12gh}$

$$r^2 = \frac{L^2\sqrt{2gh}}{\pi\sqrt{2gh} \times \sqrt{6}} \Rightarrow r^2 = \frac{L^2}{\pi\sqrt{6}}$$

32. (3) According to the question.

$V_o(P_o + h\rho g) = P_o V \Rightarrow P_o V_o + V_o d\rho g = P_o V$

$$\frac{P_o V_o + V_o d\rho g}{P_o} = V \Rightarrow V = V_o\left(1 + \frac{d\rho g}{P_o}\right)$$

$$V = V_o\left[1 + \frac{d\rho g}{h\rho g}\right]$$

$$\Rightarrow V = V_o\left[1 + \frac{d\rho_w}{h\rho_{Hg}}\right]$$

$$V = 100\left(1 + \frac{5000 \times 1}{70 \times 13.6}\right) = 625.2 \text{cm}^3$$

33. (3) Velocity of ball before entering into the water

$V = \sqrt{2gh} = \sqrt{2 \times 10 \times 5} = 10 \text{m/s}$

Due to the upthrust the ball motion will retard.

upthrust (F) = ma

Apparent weight = ma

$$\therefore \left(\frac{\rho - \sigma}{\rho}\right)g = a \Rightarrow \left(\frac{5-1}{5}\right) \times 10 = 8 \text{m/s}^2$$

$v^2 - u^2 = 2ah$

Due to upthrust after reaching its maximum depth final velocity becomes zero

$(0)^2 - (V)^2 = 2as$

$-100 = 2 \times 8 \times h$

$h = -6.25$ m (–ve indicates depth)

34. (4) Let r be the radius of common interface

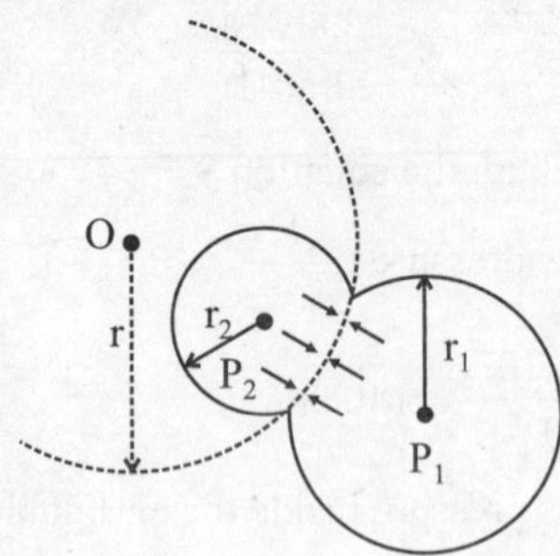

Pressure difference

$$P_1 - P = \frac{4T}{r_1} \quad \text{... (i)}$$

$$P_2 - P = \frac{4T}{r_2} \quad \text{... (ii)}$$

Eq (*ii*) – Eq (*i*) gives,

$$P_2 - P_1 = 4T\left(\frac{1}{r_2} - \frac{1}{r_1}\right)$$

But $P_2 - P_1 = \dfrac{4T}{R}$

$$\therefore \frac{4T}{R} = 4T\left(\frac{1}{r_2} - \frac{1}{r_1}\right)$$

$$\frac{1}{r} = \frac{1}{r_2} - \frac{1}{r_1} = \frac{r_1 - r_2}{r_2 r_1} \Rightarrow r = \frac{r_2 r_1}{r_1 - r_2}$$

35. (3) Net force on the free surface of the liquid in equilibrium (from accelerated frame) should be perpendicular to it. Forces on a water particle P on the free surface have been shown in the figure. In the figure ma is the pseudo force.

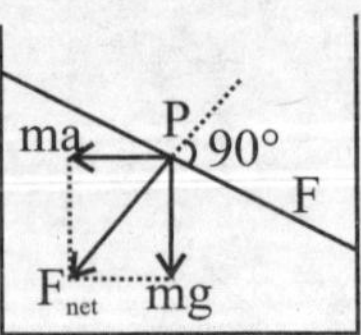

36. (3) Volume of the water flowing per second.

$$V = \frac{20}{60} = \frac{1}{3} \times 10^{-3} \text{m}^3\text{s}^{-1} \quad \left(\begin{array}{l} 1\text{m}^3 = 1000\text{L} \\ 1\text{L} = 10^{-3}\text{m} \end{array}\right)$$

(a) Area $= \pi r^2 = 3.14 \times \left(\frac{10}{2} \times 10^{-2}\right)^2 = 3.14 \times 25 \times 10^{-4}$

As we know that $V = av$ $\left(\begin{array}{l} \text{rate of flow} = \frac{\text{volume}}{t} \\ \frac{\text{vol}}{t} = \frac{\text{Area} \times \text{length}}{t} \\ V = a \times v \\ v = \text{velocity} \end{array}\right)$

$$\frac{1}{3}\times 10^{-3} = 78.5\times 10^{-4}\times v$$

$$v = \frac{10^{-3}\times 10^{4}}{3\times 78.5} = \frac{10}{235.5} = 0.0424 \text{m/s}$$

37. (4) Let h be the height at which water is filled in the vessel.

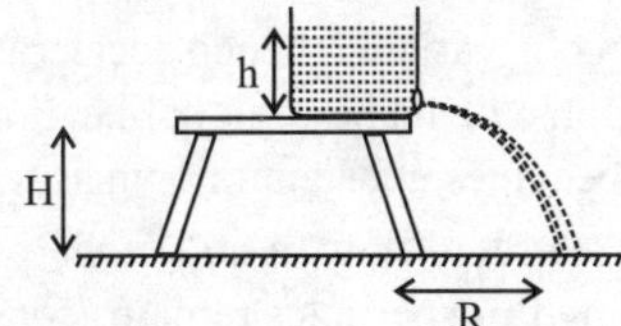

Velocity of efflux is

$$V_{eff} = \sqrt{2gh}$$

$$H = ut + \frac{1}{2}gt^2$$

$$H = 0 + \frac{1}{2}gt^2 \Rightarrow t = \sqrt{\frac{2H}{g}}$$

Horizontal range R = v × t

$$R = \sqrt{2gh}\times\sqrt{\frac{2H}{g}}$$

$$\therefore\ h = \frac{R^2}{4H}$$

38. (1) $V = a_1a_2\sqrt{\frac{2gh}{a_1^2 - a_2^2}}$ {venturimeter}

$a_1 = \pi\,(40\times 10^{-2})^2 = \pi\times 1600\times 10^{-4} = \pi\times 16\times 10^{-2}\ m^2$

$a_2 = \pi\,(5\times 10^{-2})^2 = \pi\times 25\times 10^{-4} m^2$

$$V = \pi^2\times 16\times 10^{-2}\times 25\times 10^{-4}\sqrt{\frac{2\times 9.8\times 120\times 10^{-2}}{(\pi\times 0.16)^2 - (\pi\times 0.0025)^2}}$$

$$V = \pi^2\times 400\times 10^{-6}\sqrt{\frac{1372\times 10^{-2}}{\pi^2\times 256\times 10^{-4}}}$$

$V = 3\times 10^{-2}\ m^3/sec$

39. (1) The container must be accelerated with acceleration g.
From Bernoulli's theorem

$$p_0 + H\rho g + \frac{1}{2}\rho v_1^2 = p_0 + (H-h)\rho g + \frac{1}{2}\rho v_2^2$$

Solving above equation, we get

$$\frac{1}{2}\rho v_1^2 = -h\rho g + \frac{1}{2}\rho v_2^2$$

$$\Rightarrow v_1^2 + 2gh = v_2^2$$

$$v_1^2 - v_2^2 = -2gh \quad ...(i)$$

If u = 0 (free fall), $v^2 - u^2 = 2gh \Rightarrow v^2 = 2gh$

So, from Equ. (i) $v_2^2 = 0 \Rightarrow v_2 = 0$

Water will not come out.

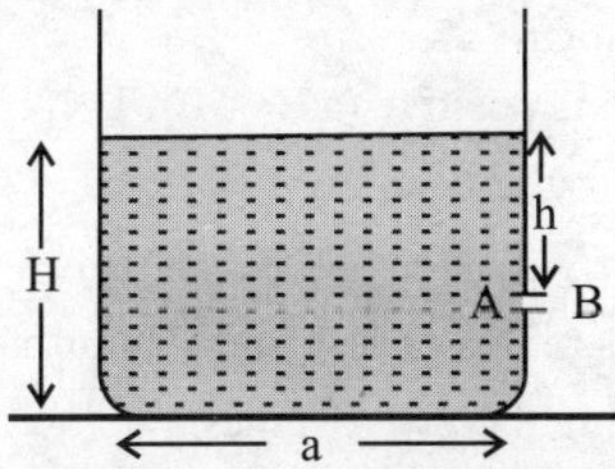

When vessel is accelerated down with an accelerating g (free fall), then pseudo acceleration g will be vertically upwards and effective value of g is zero. Hence, water will not flow.

40. (2)

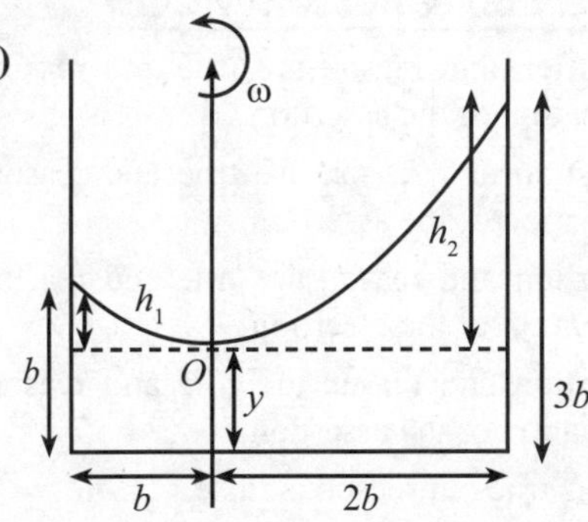

$h_2 = 3b - y,\ h_1 = b - y,\ r_2 = 2b,\ r_1 = b$

$h_2 - h_1 = 2b$

$$h = \frac{r^2\omega^2}{2g}$$

So $h_1 = (b-y) = \frac{\omega^2 r_1^2}{2g}$

$$h_2 = (3b - y) = \frac{\omega^2 r_2^2}{2g}$$

$$h_2 - h_1 = 2b = \frac{\omega^2}{2g}(r_2^2 - r_1^2) = \frac{\omega^2}{2g}(3b^2),\ \omega = 2\sqrt{\frac{g}{3b}}$$

41. (4)

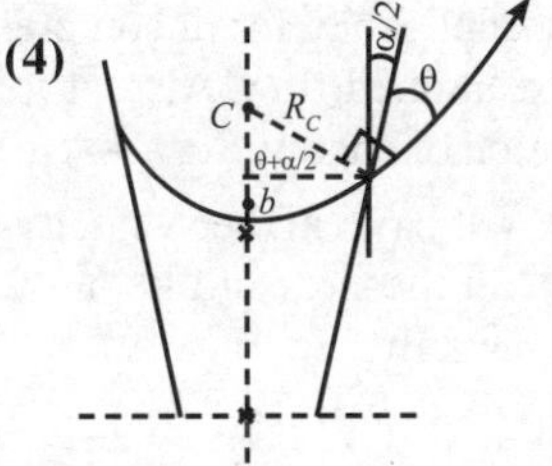

Using geometry: $\frac{b}{R_e} = \cos\left(\theta + \frac{\alpha}{2}\right)$

Using Pressure method: $P_0 - \frac{2S}{R_e} + h\rho g = P_0$

$$\Rightarrow h = \frac{2S}{R_e\rho g} = \frac{2S}{b\rho g}\cos(\theta + \alpha/2)$$

Statement Based MCQs

1. **(4)** Statement-I is false but statement-II is true.
2. **(1)** Since the water trapped between slides has a lower pressure as compared to atmospheric. The outside pressure of air tries to push slides towards each other. Also, there is adhesive force acting between glass and water which is responsible for sticking up of slides.
3. **(1)** It is so because the smaller bubble has smaller radii hence larger excess pressure inside and larger bubble has smaller pressure. Therefore, when they are connected by capillary tube air flows from smaller bubble to larger one till pressure become equal.

4. (1) A fish possess streamlined body which help to cut its way through water

5. (3) Statement-I is true Statement-II is false.

Assertion & Reason MCQs

1. (1) Both assertion and reason are true and reason is the correct explanation of the assertion.

2. (1) Both assertion and reason are true and reason is the correct explanation of the assertion.

3. (1) Both assertion and reason are true and reason is the correct explanation of the assertion.

4. (1) Both assertion and reason are true and reason is the correct explanation of the assertion.

5. (3) Assertion is true but reason is false.

Match the Column MCQs

1. (2) Kinetic energy is 1/2 mv^2. Hence,

$\frac{v^2}{2g} = \frac{\left(\frac{1}{2}mv^2\right)}{mg}$ is energy per unit weight.

$\frac{\rho v^2}{2} = \frac{\left(\frac{1}{2}mv^2\right)}{m\rho}$ is energy per unit volume and

$\frac{v^2}{2} = \frac{\left(\frac{1}{2}mv^2\right)}{m}$ is energy per unit mass.

2. (1) Bernoulli's theorem is applicable for incompressible, non-viscous liquids with streamlined flow. Also, it is based on the principal of conservation of energy. So, a $\rightarrow$ (iii), (iv).

A ball moving with a spin will have higher velocity of air above the ball than below. So the pressure below will be more than that above, causing a lift. So, b $\rightarrow$ (v).

Based on the principle of Barnoulli's theorem, artificial high pressure created pushes the paint through the gun. So, c $\rightarrow$ (ii).

Streamlined flow is possible in only non-viscous medium. So, d $\rightarrow$ (iii).

As the stream passes through smaller cross section or narrower pipers, the velocity increases causing the pressure to fall. So, d $\rightarrow$ (i).

3. (3) Magnus effect talks of difference in pressure caused due to difference in velocity to lift a structure. Lifting of asbestos roofs during cyclonic storms is based on this phenomenon.

Viscous force is wet friction and brings loss in internal energy.

At the same level in the liquid in a container, the pressure is the same according to Pascal's law.

Hydraulic machines use the idea of Pascal's law to tranfer equal pressure in all directions from the free surface where it is applied.

4. (2) $F_2 - F_1$ upthrust

= weight of cylinder in equilibrium or floating condition. From liquid-1, horizontal pair of forces cancel out. So, net force = 0.

NCERT Exemplar MCQs

1. (3) When the pebble is falling through the viscous oil, the viscous force is

$F = 6\pi\eta rv$

Where r is radius of the pebble, v is instantaneous speed, η is coefficient of viscosity.

As the force is variable, hence acceleration is also variable. So, v-f graph will not be straight line. First velocity increases and then becomes constant known as terminal velocity.

2. (4) In a streamline flow at any given point, the velocity of each passing fluid particles remains constant. If we consider a cross-sectional area, then a point on the area cannot have different velocities at the same time, hence two streamlines of flow cannot cross each other.

3. (2) As we know for a streamline flow of a liquid, velocity of each particle at a particular cross-section is constant, because Av = constant (law of continuity) between two cross-section of a tube of flow.

4. (1) Consider the diagram where an ideal fluid is flowing through a pipe.

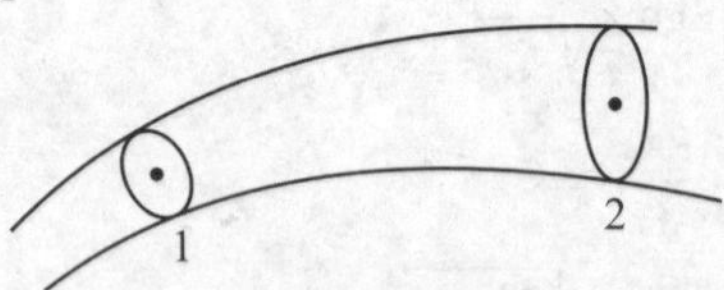

As given

d_1 = Diameter at 1st point is 2.5.

d_2 = Diameter at 2nd point is 3.75.

Applying equation of continuity for cross-sections A_1 and A_2.

$\Rightarrow A_1 v_1 = A_2 v_2$

$$\Rightarrow \frac{v_1}{v_2} = \frac{A_2}{A_2} = \frac{\pi(r_2^2)}{\pi(r_1^2)} = \left(\frac{r_2}{r_1}\right)^2 \Rightarrow \frac{v_1}{v_2} = \frac{9}{4}$$

5. (3) According to the question, the observed meniscus is of convex figure shape, which is only possible when angle of contact is obtuse. Hence, the combination will be mercury-glass (140°)

NEET & AIIMS Past Year MCQs

1. (3) According to Bernoulli's theorem:

$P + \frac{1}{2}\rho v^2 + \rho gh =$ constant

Applying Bernoulli's principle at points *X* and *Y*:

At $X \Rightarrow P + K_1 + \rho g(0)$

At $Y \Rightarrow P + K_2 + \rho g(h)$

Therefore,

$P + K_1 = P + K_2 + \rho gh$

$K_1 = K_2 + \rho gh$

Since ρgh is positive (as $h > 0$), Therefore, $K_1 > K_2$.

2. (3) For circular disc,

$F_{excess} = 2\pi r T$

where *r* is radius of disc and *T* is surface tension.

Substituting the given values,

$F_{excess} = 2\pi \times 4.5 \times 10^{-2} \times 0.07$

$= 2 \times \frac{22}{7} \times 4.5 \times 10^{-2} \times 7 \times 10^{-2}$

$= 198 \times 10^{-4}$ N = 19.8 mN

3. (4) Surface energy of bubble = 2 × charge in surface area × surface tension

$= 8\pi R^2 \times T = 8 \times 3.142 \times 4 \times 10^{-4} \times 3 \times 10^{-2}$ J $= 30.1 \times 10^{-5}$ J

4. (3) Factual (theory based)

5. (3) Initially the speed of the ball is zero, then it increases & after some time becomes constant due to the viscous liquid.

⇒ Acceleration of the ball first decreases and after some time becomes zero.

6. (2) Total pressure inside a soap bubble is given by

$P = P_0 + \frac{4T}{R}$

⇒ As the bubble expands R increases and P decreases

7. (4) Archimedes upward thrust will be absent for a fluid, if there were no gravity.

8. (2)

9. (3)

10. (1)

11. (1)

Thermal Properties of Matter

NCERT Topic-wise MCQs

1. (2) According to the formula:

$\frac{C}{5} = \frac{F-32}{9} \Rightarrow \frac{-183}{5} = \frac{F-32}{9}$

$F = -297.4 \simeq -297°F$

2. (3) Absolute zero is the lowest temperature possible. At absolute zero there is no motion and no heat, all the molecular motion ceases but the electronic motion continues to occur. Absolute zero cannot be achieved.

3. (2) The error in the reading will be

Δl = (scale reading) (α) (ΔT)

$= (60)(1.2 \times 10^{-5})(27° - 15°)$

= 0.00864 cm

4. (1) $\alpha = \frac{\Delta L}{L_o \Delta T} \Rightarrow \Delta T = \frac{\Delta L}{L_o \times \alpha} = \frac{3}{100 \times 2 \times 10^{-5}}$

$\Delta T = 1.5 \times 10^3 = 1500°C$

5. (3) $V_t = V_o(1 + \gamma\Delta T) \Rightarrow V_o = 4 \times 10^3\ m^3$

$\alpha = \frac{\gamma}{3} \Rightarrow \gamma = 3\alpha = 3 \times 18 \times 10^{-6}$

$\therefore V_t = V_o(1 + \gamma\Delta T)$

$\Rightarrow V_t = 4 \times 10^{-3}(1 + 3 \times 18 \times 10^{-6} \times 10^2)$

$V = 4 \times 10^{-3}(1 + 5.4 \times 10^{-3})$

$V = 4.002 \times 10^{-3}\ m^3$

6. (2) $\frac{\Delta A}{A_o \Delta T} = \beta = \frac{2}{4000 \times 100} = 5 \times 10^{-6}\ °C^{-1}$

$\alpha = \frac{\beta}{2} = 2.5 \times 10^{-6}/°C$

7. (4) $\frac{\Delta T}{T} = \frac{1}{2}\alpha\Delta\theta = \frac{1}{2} \times 3 \times 10^{-5} \times 10 = 15 \times 10^{-5}$

% increase $= \frac{\Delta T}{T} \times 100 = 15 \times 10^{-5} \times 100 = 15 \times 10^{-3} = 0.015\%$

8. (2) $C = \frac{5}{9}(F - 32) = \frac{5}{9}F - \frac{160}{9}$.

Slope of line $AB = \frac{5}{9}$

9. (2) $\frac{X-L}{U-L} = \frac{C}{100}$

$\frac{59-5}{95-5} = \frac{C}{100} \Rightarrow \frac{54}{90} = \frac{C}{100} \Rightarrow C = 60°$

10. (1) Given that coefficient of volumetric expansion is $5 \times 10^{-5}/°C$

$\alpha = \frac{\gamma}{3} = \frac{5 \times 10^{-5}}{3} = 1.6 \times 10^{-5}/°C$

11. (3) Actual area of cross section

$= (100 \times 15) = 15 \times 10^{-2}\ m^2 = 1500\ cm^2$

$A = A_o(1 + \beta\Delta T)$

$A = A_o + A_o\beta\Delta T \Rightarrow \frac{\Delta A}{A_o} = \beta\Delta T \Rightarrow \frac{\Delta A}{A_o \Delta T} = \beta$

$\beta = 2\alpha = \frac{2}{15 \times 10^2 \times 10^2} \Rightarrow \alpha = \frac{1}{15 \times 10^4} = 0.066 \times 10^{-4}$

$\alpha = 6.6 \times 10^{-6}/°C$

12. (4) Let the mass remained be x

$\gamma = \frac{\Delta V}{V \times \Delta T}$ $\{m = \rho V \Rightarrow \Delta m = \rho\Delta V\}$

$\gamma = \frac{\Delta m}{m \times \Delta T} = \frac{x/50}{x \times 80} \Rightarrow \gamma = \frac{1}{4000} = 0.25 \times 10^{-3}$

$\gamma = 2.5 \times 10^{-4}/°C$

13. (3) $\frac{X-L}{U-L} = \frac{C}{100}$

$\frac{62-(-10)}{110-(-10)} = \frac{C}{100} \Rightarrow C = 60°C$

14. (1) Moment of inertia of rod, $I = \frac{M\ell^2}{12}$

$dI = \frac{M}{12} 2\ell d\ell$

On differentiating w.r.t I, we get

$\Rightarrow \frac{dI}{I} = 2\frac{d\ell}{\ell} = 2\alpha\Delta t \left[\because \frac{dl}{l} = \alpha\Delta t\right]$

15. (4) On heating a body, the distance between any two points of it increases. So if we consider two particle to be diametrically opposite on either holes diameter will increase with increase in temperature.

16. **(2)** Heat lost by water = Heat gained by ice

$m_w s_w (T_2 - T_1) = m_i L + m_i s_w (T_1 - T_{mix})$

$0.30 \times 4186 (70 - 10) = 0.20 \times 4186 (10 - 0) + 0.2 L$

$\Rightarrow 75348 = 0.20L + 8372$

$\Rightarrow \frac{75348 - 8372}{0.20} = L$

$\therefore L = 3.34 \times 10^5 \text{ Jkg}^{-1}$

17. **(1)** $80 \times 1 \times (30 - 0) = x \times 80 \Rightarrow x = 30$ gm

18. **(3)** $Q = mc\Delta T \Rightarrow \frac{Q}{mc} = \Delta T$

$c = \infty, \Delta T = 0$

19. **(3)** $m = \rho v$

$\frac{m_1}{m_2} = \frac{v_1}{v_2} \quad \left(Q = mc\Delta T, c \propto \frac{1}{m}\right)$

$\frac{m_1}{m_2} = \frac{(4/3)\pi(r_1)^3}{(4/3)\pi(r_2)^3} = \left(\frac{1}{4}\right)^3 = \frac{1}{64}$

$\therefore \frac{c_1}{c_2} = \frac{m_2}{m_1} = 64$

20. **(1)** Let x kg of ice melt when 1 kg of steam at 100°C is used. Steam first gets converted into water at 100°C and then water is brought to 0°C, whereas ice at 0°C melts becomes water at 0°C.

Heat lost by steam at 100°C to form water at 0°C

$= (1 \times (540 \times 10^3) + 1 \times 1000 \times 100)$ cal

Heat gained by ice at 0°C to become water at 0°C

$= x \times (1000 \times 80)$

using, Heat gained = heat lost,

$(540 + 100) \times 10^3 = x \times 80 \times 10^3 \Rightarrow x = \frac{640}{80} = 8$ kg

Choice (1) is correct.

21. **(2)** The amount of heat absorbed or rejected by the substance to change its temperature by one unit is called heat capacity.

22. **(3)** $Q = nc_p\Delta T \qquad (R = 2\text{cal/k mol}^{-1})$

$Q = 8 \times \frac{5R}{2} \times 60 = 2400$ cal

23. **(2)** At the time of boiling $c = \infty$ because $\Delta T = 0$

24. **(3)** Because water drops that sprinkled in room starts absorbing heat or energy of room for evaporation so room temperature slightly reduces.

25. **(4)** Coefficient of thermal conductivity is an intrinsic property. It depends on the nature of the material.

26. **(2)** According to the relation $K_{eq} = \frac{2k_1k_2}{k_1 + k_2}$

$k_{eq} = \frac{2 \times 4 \times 6}{4 + 6} = \frac{48}{10} = 4.8$

27. **(4)** As hot air has lesser density than the colder air. So if rises above, the freezer is kept at the top section so that cold air being dense sinks down and the warm air is forced to rise up and it gets cold in the freezer.

28. **(4)** It takes place by radiation because there is no disturbance of particles.

29. **(1)** In case (i) $Q = 20 = \frac{KA(100 - 0)4}{2d}$ ... (*i*)

In case (ii) $Q = 20 = \frac{K\,2\,A(100 - 0)t'}{d}$ ... (*ii*)

[In 2nd case area becomes double and thickness is half].

eq^n (*ii*) ÷ eq^n (*i*) gives,

t' = 1 min.

Choice (1) is correct.

30. **(4)** According to the relations.

In series combination $K_S = \frac{2k_1k_2}{k_1 + k_2}$

In parallel combination $k_P = \frac{k_1 + k_2}{2}$

$\frac{k_S}{k_P} = \frac{\frac{2k_1k_2}{k_1 + k_2}}{\frac{k_1 + k_2}{2}} = \frac{4k_1k_2}{(k_1 + k_2)^2}$

31. **(1)**

A	B	C

Given, $K_B = K_A/2$

And $K_B = 3K_C$

$\therefore K_C = K_A/6$

Rods are in series form, so

$\frac{L}{K} = \frac{\ell_1}{K_A} + \frac{\ell_2}{K_B} + \frac{\ell_3}{K_C} \quad (\because \ell = \ell_1 = \ell_2 = \ell_3)$

$\frac{3\ell}{K} = \frac{\ell}{K_A} + \frac{\ell}{K_A/2} + \frac{\ell}{K_A/6}$

Or $\frac{3\ell}{K} = \frac{9\ell}{K_A}$ or $K = \frac{K_A}{3}$

32. **(2)** In parallel combination, the flow of heat is the sum of the individual.

$H_1 = H_2 + H_3$

Let us consider T is the temperature of junction

$\frac{4KA(150 - T)}{x} = \frac{2KA(T - 20)}{x} + \frac{3KA(T - 80^\circ)}{x}$

$4(150 - T) = 2(T - 20) + 3(T - 80^\circ)$

$600 - 4T = 2T - 40 + 3T - 240$

$600 + 280 = 9T \Rightarrow T = \frac{880}{9} = 97.7^\circ C$

33. **(2)** In satellite, weightlessness condition exist and no gravitational pull is present whereas, base of the connection is the difference in weight and upthrust.

34. **(3)** $\frac{dQ}{dt} = L\left(\frac{dm}{dt}\right)$ or $\frac{\text{Temperature difference}}{\text{Thermal resistance}} = L\left(\frac{dm}{dt}\right)$

or $\frac{dm}{dt} \propto \frac{1}{\text{Thermal resis tan ce}} \Rightarrow q \propto \frac{1}{R}$

In the first case rods are in parallel and thermal resistance is $\frac{R}{2}$ while in second case rods are in series and thermal resistance is 2R.

$$\frac{q_1}{q_2} = \frac{2R}{R/2} = \frac{4}{1}$$

35. (1)

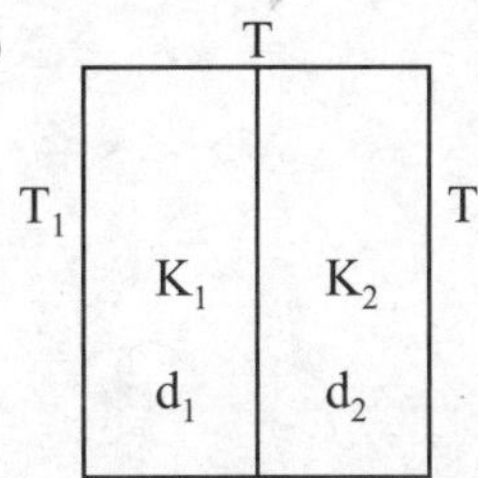

$$Q_1 = Q_2 \Rightarrow \frac{K_1A(T_1 - T)t}{d_1} = \frac{K_2A(T - T_2)t}{d_2}$$

or $(K_1 T_1 - K_1 T)d_2 = (K_2 T - K_2 T_2)d_1$

$K_1 T_1 d_2 + K_2 T_2 d_1 = (K_2Td_1 + K_1 d_2 T)$

or $T = \frac{K_1T_1d_2 + K_2T_2d_1}{K_2d_1 + K_1d_2}$

Choice (1) is correct.

36. (2) $Q_1 + Q_2 = Q$ ...(*i*)

$$Q_1 = \frac{K_1(\pi R^2)(\theta_1 - \theta_2)t}{d} \quad ...(ii)$$

$$Q_2 = \frac{K_2\left(\pi (3R)^2 - \pi R^2\right)(\theta_1 - \theta_2)t}{d} \quad ... (iii)$$

$$Q = \frac{K\left[\pi \times (3R)^2\right](\theta_1 - \theta_2)t}{d} \quad ... (iv)$$

Putting in (*i*) and solving, we get

$$K = \frac{(K_1 + 8K_2)}{9}$$

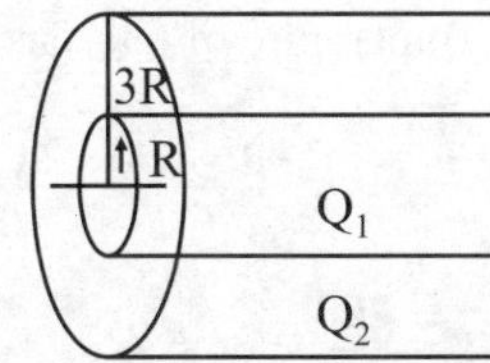

37. (4) $\frac{dQ}{dt} = -KA\frac{d\theta}{dx}$; when $K = \infty$, $\frac{d\theta}{dx} = 0$

i.e. θ is independent of x, i.e. constant or uniform.

38. (1) Temperature of interface,

$$\theta = \frac{K_1\theta_1\ell_2 + K_2\theta_2\ell_1}{K_1\ell_2 + K_2\ell_1}$$

Given, $K_{Cu} = 9Ks$,

So, if $K_s = K_1 = K$,

Then $K_{Cu} = K_2 = 9K$

$$\Rightarrow \quad \theta = \frac{9K \times 100 \times 6 + K \times 0 \times 18}{9K \times 6 + K \times 18}$$

$$\theta = \frac{5400K}{72K} = 75°C$$

39. (4) Rate of heat flow is $\frac{Q}{t} = \frac{KA(\theta_1 - \theta_2)}{\ell}$

For rate of heat flow to be the same, $\frac{Q_1}{t} = \frac{Q_2}{t}$

$$\therefore \quad \frac{K_1A_1(\theta_1 - \theta_2)}{\ell} = \frac{K_2A_2(\theta_1 - \theta_2)}{\ell}$$

$K_1A_1 = K_2A_2$ or $\frac{A_1}{A_2} = \frac{K_2}{K_1}$

High Order Time Intensive MCQs

1. (2) Thermal capacity per unit volume is $= \frac{mc}{V} = \rho c$

∴ required ratio of thermal capacities is

$$\frac{\rho_1c_1}{\rho_2c_2} = \frac{5}{6} \times \frac{3}{5} = \frac{1}{2}$$

Choice (2) is correct.

2. (3) Initial length of rod A = ℓ_1

Initial length of rod B = ℓ_2

Linear coefficient of rod A = α_1

Linear coefficient of rod B = α_2

If temperature of combined rod increased by t°C, then increase in length of rod A is given by

$$\Delta\ell_1 = \frac{\ell_1}{\alpha_1 t}$$

Increase in length of II rod is given by

$$\Delta\ell_2 = \frac{\ell_2}{\alpha_2 t} \text{ and } \Delta\ell_1 = \Delta\ell_2$$

$$\therefore \quad \frac{\ell_1}{\alpha_1 t} = \frac{\ell_2}{\alpha_2 t}$$

$$\Rightarrow \quad \frac{\alpha_1}{\alpha_2} = \frac{\ell_1}{\ell_2} \Rightarrow \quad \frac{\alpha_1}{\alpha_2} + 1 = \frac{\ell_1}{\ell_2} + 1$$

$$\Rightarrow \quad \frac{\alpha_1 + \alpha_2}{\alpha_2} = \frac{\ell_1 + \ell_2}{\ell_2} \Rightarrow \frac{\alpha_2}{\alpha_1 + \alpha_2} = \frac{\ell_2}{\ell_1 + \ell_2}$$

3. (4) Given, m = 50 kg, v = 5 ms^{-1}, L = 80 cal/g,

1 cal = 4.2 J

Heat lost, i.e, Q_1 = work done = change in kinetic energy = K

i.e., $\frac{1}{2}mv^2 = \frac{1}{2} \times 50 \times 5^2 = 625J$

and heat gained i.e.,

$Q_2 = m' \times L = 80 \times m' \times 4.2$

heat lost = heat gained

$\therefore Q_1 = Q_2$

$625 = m' \times 8 \times 42$

$m' = 1.86$ g

4. (3) $\because \; Q = \frac{KA\Delta Tt}{x}$

$$\therefore mL = \frac{KA(\Delta T)t}{x} \rightarrow m = \frac{92 \times 10^{-3} \times 100 \times 180}{2 \times 9 \times 10^4}$$

$m = 92 \times 10^{-4}$ kg

5. **(3)** Let the final temperature of mixture be T°C

Heat lost by water,

$H_1 = 19 \times 1\,(30 - T) = 570 - 19\,T$

Heat taken by ice, $H_2 = ms_i\Delta T + mL + ms_wT$

$H_2 = 5 \times 0.5 \times 20 + 5 \times 80 + 5T$

Using principle of calorimetry

$H_1 = H_2$

$\Rightarrow 5 \times (0.5)20 + 5 \times 80 + 5T = 570 - 19T$

$\Rightarrow 24T = 570 - 450 = 120$

$\Rightarrow T = 5°C$

6. **(2)** Heat energy required to just melt the bullet

$Q = Q_1 + Q_2$

Here, $Q_1 = ms\Delta T = m \times 125 \times (327 - 27) = 3.75 \times 10^4$ m

$Q_2 = mL = m \times 2.5 \times 10^4 = 2.5\times 10^4$ m

$\therefore Q = 3.75 \times 10^4 \text{ m} + 2.5 \times 10^4$ m

$Q = 6.25 \times 10^4$ m

Let v be the speed of bullet, then 50% of $\frac{1}{2}mv^2$ should be equal to Q.

$\therefore 50\% \times \frac{1}{2}mv^2 = Q$

$\therefore 0.5 \times \frac{1}{2}mv^2 = 6.25 \times 10^4 m \Rightarrow v = \sqrt{\frac{6.25 \times 10^4 \times 2}{0.5}} = 500 ms^{-1}$

$= 1800$ km h^{-1}

7. **(1)** Initial mass of ice block, m = 42 kg

Initial velocity of ice block, u = 4 ms^{-1}

Final velocity of ice block, v = 0

Let retardation be a. Using 3rd equation of motion, we get

$v^2 = u^2 - 2as$

$\Rightarrow \quad as = \frac{u^2 - v^2}{2} = \frac{(4)^2 - (0)^2}{2} = 8 m^2 s^{-2}$

Heat generated = Work done against friction

$Q = m \times a \times s \ [\because W = F \times s]$

$Q = 42 \times 8$

$Q = 336$ J

Let M kg of ice melts due to the heat generated

So, Q = ML

$\Rightarrow M = \frac{Q}{L} = \frac{336}{3.36 \times 10^5}$ kg

$M = 1 \times 10^{-3}$ kg

8. **(3)**

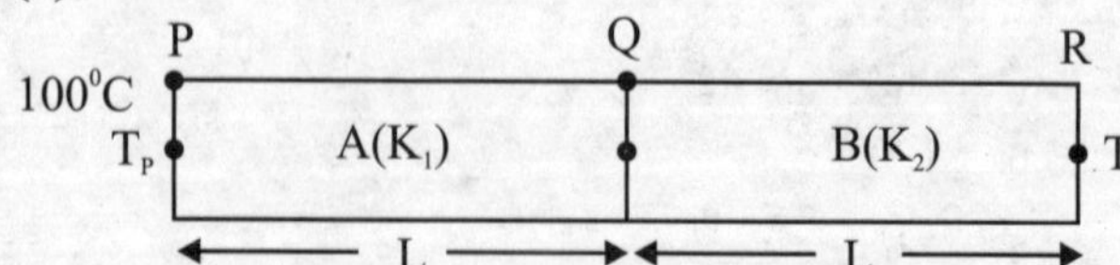

Rate of heat flow through rod will be given by

$\left(\frac{dQ}{dT}\right)_{PQ} = \left(\frac{dQ}{dT}\right)_{PR}$

$\left(\frac{100 - T_Q}{\frac{L}{K_1A}}\right) = \left(\frac{T_P - T_R}{\frac{L}{K_1A} + \frac{L}{K_2A}}\right)$

Using $K_1 = \frac{K_2}{2}$

or $K_2 = 2K_1$

$T_P - T_R = 12°C$

$\frac{(100 - T_Q)}{\frac{1}{K_1}} = \frac{12}{\frac{1}{K_1} + \frac{1}{2K_1}}$

$(100 - T_Q) = \frac{2 \times 12}{3} = 8$

$T_P - T_Q = 8°C$

9. **(3)** $\frac{Q}{t} = \frac{KA\Delta T}{x} = \frac{\Delta T}{R}\left[R = \frac{\rho\ell}{A} = \frac{1}{K}\frac{x}{A}\right]$

$\frac{Q}{t} = \frac{\Delta T}{R}$ ...(*i*)

As the Rods B and C are in Parallel combinations

$\frac{1}{R_P} = \frac{1}{R} + \frac{1}{R} = \frac{2}{R}$

$\therefore R_P = \frac{R}{2}$

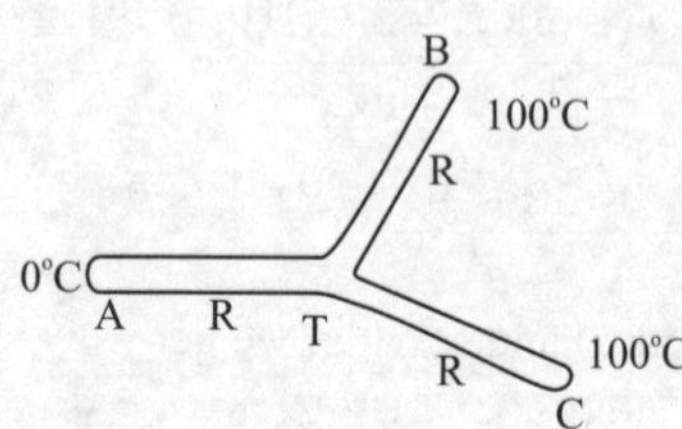

In series combination flow of heat is same

Let T be the temperature of interface.

$\frac{T - 0}{R} = \frac{100 - T}{R/2} \Rightarrow T - 0 = 200 - 2T$

$3T = 200 \Rightarrow T = \frac{200}{3} = 66.6°C$

10. **(2)** Heat capacity of beaker is 20 gm means that its water equivalent is 20 gm. Heat gained by water and its container is (200 + 20) × (θ – 20) where θ is the final temperature.

Heat lost by hot water = 440 × 1 × (92 – θ)

using, heat gained = heat lost

220(θ – 20) = 440 × (92 – θ)

or θ – 20 = 184 – 2 θ

3θ = 204,

θ = 68°C

11. **(3)** If V is the volume of spherical iron ball, then volume of ice melted is $\frac{V}{2}$

Heat lost by iron ball = $MC\,\theta = V \times \rho_{Fe} \times C \times \theta$

$= V \times 7675 \times 504\,\theta$ J ... (*i*)

Heat gained by ice = $mL = \frac{V}{2} \times 921 \times 336 \times 10^3$ J ... (*ii*)

equating the two values or using heat gained = heat lost

$$\frac{V}{2} \times 921 \times 336 \times 10^3 = V \times 7675 \times 504\,\theta$$

$$\theta = \frac{921 \times 336 \times 10^3}{2 \times 7675 \times 504} = 40°C$$

12. (4) Heat released to convert x gm of steam at 100°C to water at 100°C is x × 540 cal.

If y gm of ice is converted from 0°C to water at 100°C it requires heat y × 80 + y × 1 × 100 = 180 y

$$x \times 540 = 180\,y \quad \text{or} \quad \frac{y}{x} = \frac{540}{180} = \frac{3}{1}$$

13. (1) If heat produced is equally shared then half of it shared between the two

$$ms\Delta T = \frac{1}{2}\left(\frac{1}{2}mv^2\right)$$

$$\Delta T = \frac{v^2}{s \times 4} = \frac{400 \times 400}{125 \times 4} = 320°C$$

14. (3) Heat liberated $\frac{dQ}{dt} \propto R^3$ and also $\frac{dQ}{dt} = R^2T^4$

Or $T^4 \propto R$

Thus $\frac{T_1^4}{T_2^4} = \frac{R}{R/2}$

Or $T_2 = \left(\frac{1}{2}\right)^{1/4} T_1$ or $T_2 = \frac{T_1}{1.19}$

That is, temperature decreases by a factor of 1.19.

15. (2) Applying Newton's law

In case (1)

$$K_1\left(T - T_{r_1}\right) = K_2\left(T_{r_2} - T_{out_1}\right)$$

And in case (2)

$$K_1\left(T - T_{r_2}\right) = K_2\left(T_{r_2} - T_{out_2}\right)$$

Dividing these equations

$$\frac{T - T_{r_1}}{T - T_{r_2}} = \frac{T_{r_2} - T_{out_1}}{T_{r_2} - T_{out_2}} \Rightarrow \frac{T - 20}{T - 10} = \frac{20 - (-20)}{10 - (-40)}$$

or $T = 60°C$

16. (1) m = mass of water, f = fraction which freezes

L_1 = latent heat of vaporization

L_2 = latent heat of fusion ∴ $L_1 = 7L_2$

Mass of water frozen = mf

Heat lost by freezing water = mfL_2

Mass of vapour formed = m(1 – f)

Heat gained by vapour = m (1 – f) L_1

$mfL_2 = m(1 - f) \times 7L_2$

f = 7 – 7f or f = 7/8

17. (1) 800 calories of heat raise the temperature of the substance from 0°C to 80°C.

∴ 800 = m (1000 × 0.5) × 80

(∵ specific heat = relative sp. heat x sp. heat of water)

or m = 0.02 kg

18. (3) Let I stand for the iron ball and B stand for the brass plate. $L_I = 6$ cm and $L_I - L_B = 0.001$ cm at t = 30°C. Since the brass plate expands uniformly, the hole must expand in the same proportion. Then heating both the ball and the plate leads to increases in the diameters of the ball and the hole, with the hole increasing more, since $\alpha_B > \alpha_I$. We require $\Delta L_B - \Delta L_I = 0.001$ cm. $\Delta L_B = \alpha_B L_B \Delta t$; $\Delta L_I = \alpha_I L_I \Delta t$. We can approximate L_B in this formula by 6 cm $= L_I$. Then

$\Delta L_B - \Delta L_I = (\alpha_B - \alpha_I)L_I \Delta t = 0.001$ cm

On solving, Δt = 23.8°C, and finally t = 30°C + 23.8°C = 53.8°C

19. (2) It is given that the difference in length of the two rods is always 5 cm. Thus the expansion in both the rods must be same for all temperatures. Thus we can say that at all temperature differences, we have

$$\Delta L_{Cu} = \Delta L_{steel}$$

or $\alpha_{Cu}\ell_1\Delta t = \alpha_{st}\ell_2\Delta t$ [If ℓ_1 and ℓ_2 are the initial lengths of Cu and steel rods]

or $\alpha_{Cu}\ell_1 = \alpha_{st}\ell_2$

or $1.7\ell_1 = 1.1\ell_2$ ...(*i*)

It is given that $\ell_2 - \ell_1 = 5$cm ...(*ii*)

$$\left(\frac{1.7}{1.1} - 1\right)\ell_1 = 5\text{cm}$$

or $\ell_1 = \frac{5 \times 1.1}{0.6} = 9.17$cm

Now from equation (*ii*) $\ell_2 = 14.17$ cm

20. (1) Thermal resistance $R = \frac{\ell}{kA} = \frac{\ell}{k\left(\pi r^2\right)}$

or $R = \frac{2}{401 \times \pi \times \left(10^{-2}\right)^2} = 15.9\,K/W$

21. (3) Temperature gradient

$$= \frac{0 - 100}{2} = -50K/m = -50°C/m$$

22. (1)

100°C —3K— T (H →); from T: H_1 through 2K to 50°C; H_2 through K to 0°C

Rate of flow of heat,

$H = H_1 + H_2$

Using $H = \frac{KA(T_1 - T_2)}{L}$

$$\frac{3K(T - 100)A}{L} = \frac{2K(50 - T)A}{L} + \frac{K(0 - T)A}{L}$$

3(T – 100) = 2(50 – T) + (0 – T)

$6T = 400$

$T = \frac{400}{6} = \frac{200}{3}°C$

23. (3) $E_P = A_P e \sigma T_1^4$

$E_Q = A_Q e \sigma T_2^4$

$$\therefore \frac{E_P}{E_Q} = \frac{4\pi(8)^2 \times e\sigma(273+127)^4}{4\pi(2)^2 \times e\sigma(273+527)^4}$$

$$= \frac{64}{4} \times \frac{(400)^4}{(800)^4} = \frac{1}{1}$$

Choice (3) is correct.

24. (1) The bulb which is painted black will absorb more heat, air on the side x will be heated up, it expands and will push the level towards y. Therefore, the level of alcohol in limb x falls while that in limb y rises. Choice (1) is correct.

25. (4) According to Kirchoff's law good emitter is a good absorber

$\therefore \in_y > \in_x$ and $A_y > A_x$

26. (4) For slabs in series, we have;

$R_{eq} = R_1 + R_2$

i.e., $\frac{5x}{K_{eq}A} = \frac{4x}{2KA} + \frac{x}{KA}$

$K_{eq} = \frac{5K}{3}$

Now, in steady state, rate of heat transfer through the slab

$$= \frac{K_{eq}A(T_2 - T_1)}{5x} = \left(\frac{A(T_2 - T_1)K}{x}\right)f$$

Putting the value of $K_{eq.}$ we get;

$f = \frac{1}{3}$

27. (2) When the sheet is heated both the holes d_1 and d_2, increase in size because thermal expansion of an isotropic solid is very similar to a true photographic enlargement.

28. (1) $\frac{E_2}{E_1} = \frac{(273+127)^4}{(273+27)^4} = \frac{(400)^4}{(300)^4} = \frac{4^4}{3^4} = \frac{256}{81}$

Choice (1) is correct.

29. (1) Rotational kinetic energy of the sphere

$KE = \frac{1}{2}I\omega^2$ where $I = \frac{2}{5}Mr^2$

$$= \frac{1}{2} \times \frac{2}{5}Mr^2\omega^2 = \frac{1}{5}Mr^2(2\pi f)^2 = \frac{4}{5}\pi^2r^2Mf^2$$

Since half of the KE is converted into heat,

$$\therefore A = \frac{1}{2}\left[\frac{4}{5}\pi^2r^2Mf^2\right] = M.S.\Delta t$$

$$M.S.\Delta t = \frac{2}{5}\pi^2r^2Mf^2$$

$$\therefore \Delta t = \frac{2}{5S}\pi^2r^2f^2$$

30. (1) Mass = volume × density

$\therefore M_0 = V_0 d_0 = V_t d_t$

$\frac{V_t}{V_0} = \frac{d_0}{d_t}$, but $V_t = V_0(1 + gt)$

$\therefore 1 + \gamma t = \frac{d_0}{d_t}$, $\gamma = \frac{d_0 - d_t}{td_t}$

$\gamma = 3\alpha = \frac{d_0 - d_t}{td_t}$

$\alpha = \frac{d_0 - d_t}{3t(d_t)} = \frac{10 - 9.7}{3 \times 100 \times 9.7} = 10^{-4}$

31. (1) According to the question

Loss in energy = $mg(h_1 - h)$

$= 0.2 \times 10(15 - 5) = 20$ J

$10 = ms\Delta T \Rightarrow 10 = 0.2 \times 386.4 \times \Delta T$

$\Rightarrow \Delta T = 0.129°C$

Statement Based MCQs

1. **(3)** Statement-I is True but the Statement-II is False.
2. **(1)** Both Statement-I and Statement-II are True.
3. **(4)** Statement-I is false but Statement-II is true.
4. **(1)** Both Statement-I and Statement-II are True.

Assertion & Reason MCQs

1. **(3)** Assertion is true but reason is false.
2. **(1)** Wool and air are bad conductors of heat. Hence, enclose and maintain envelope of heat around body.
3. **(1)** As skate blades are sharp, area is very small due to which pressure becomes high which in turn decreases melting point of ice.
4. **(3)** Assertion is true but reason is false.
5. **(1)** Dental materials used for filling must be matched in their thermal expansion properties to those of tooth enamel (Otherwise consuming hot coffee or cold ice cream would be quite painful).

Match the Columns MCQs

1. (2) Fraction of volume submerged, $f = \frac{V_2}{V_1} = \frac{\rho_1}{\rho_2}$

After increasing the temperature

$$f' = \frac{V_2'}{V_1'} = \frac{\rho_1(1 - \gamma_1\Delta T)}{\rho_2(1 - \gamma_2\Delta T)} > f \text{ (because } \gamma_2 > \gamma_1)$$

If $\gamma_2 = \gamma_1$, then $f' = f$

If $\gamma_2 < \gamma_1$, then $f' < f$; it means fraction of volume submerged decreases and solid lifts up.

2. (4) When P and Q are mixed

$ms(T - 20) = (2m)s(40 - T) \Rightarrow T = \frac{100}{3} = 33.3°C$

When P and R are mixed $ms(T - 20) = (3m)s(60 - T)$ $\Rightarrow T = 50°C$

When Q and R are mixed (2m)s(T – 40) = (3m)s(60 – T)

$\Rightarrow$ T = 52°C

When P, Q and R are mixed

ms(T – 20) + (2m)s(T – 40) = (3m)s(60 – T)

$\Rightarrow$ T = 46.67°C

3. (1)

(i) Rate at which heat is radiated from the body

$= Q_r$ J/s

$= e\sigma AT_1^4 = 0.55 \times 5.67 \times 10^{-8} \times 1.5 \times (323)^4$ J/s

= 509 W

(ii) Rate at which heat radiation is absorbed by the body

$= Q_a$ J/s

$= e\sigma AT_2^4 = 0.55 \times 5.67 \times 10^{-8} \times 1.5 \times (295)^4$ J/s

= 354 W

(iii) Rate at which net radiation is emitted by the body

$= Q_n$ J/s $= Q_r - Q_a$ = (509 – 354)W = 155 W

NCERT Exemplar MCQs

1. (4) As $\alpha_{Al} > \alpha_{steal}$, aluminum will expand more. So, it should have larger radius of curvature. Hence, aluminum will be on convex side.

2. (2) As the rod is heated, it expands. No external torque is acting on the system so angular momentum should be conserved.

L = Angular momentum = Iω = constant

$\Rightarrow \quad I_1\omega_1 = I_2\omega_2$

Due to expansion of the rod $I_2 > I_1$

$\Rightarrow \quad \frac{\omega_2}{\omega_1} = \frac{I_1}{I_2} < 1$

$\Rightarrow \quad \omega_2 < \omega_1$

So, angular velocity (speed of rotation) decreases.

3. (1) Let volume of the sphere is V and ρ is its density, then we can write buoyant force F = Vρg (g = acceleration due to gravity)

$\Rightarrow \quad F \propto \rho$

(∵ V and g are almost constant)

$\Rightarrow \quad \frac{F_{4°C}}{F_{0°C}} = \frac{\rho_{4°C}}{\rho_{0°C}} > 1 \quad (\because \rho_{4°C} > \rho_{0°C})$

$\Rightarrow \quad F_{4°C} > F_{0°C}$

Hence, buoyancy will be less in water at 0°C than that in water at 4°C.

4. (1) As the temperature is increased length of the pendulum increases. We know that time period of pendulum

$T = 2\pi\sqrt{\frac{L}{g}}$

$\Rightarrow \quad T \propto \sqrt{L}$

As L increases, so, time period (T) also increases.

5. (1) We know that as temperature increases, vibration of molecules about their mean position increases. Hence, kinetic energy associated with random motion of molecules increases.

6. (4) Let the radius of the sphere is R. As the temperature increases radius of the sphere increases.

Original volume $V_0 = \frac{4}{3}\pi R^3$

Coefficient of linear expansion = α

∴ Coefficient of volume expansion = 3 α

$\therefore \frac{1}{V}\frac{dV}{dT} = 3\alpha \Rightarrow dV = 3V\alpha dT = 4\pi R^3 \alpha \Delta T$

= Increase in the volume

NEET & AIIMS Past Year MCQs

1. (4) Thermal stress = thermal strain × Y

$\frac{F}{A} = (\alpha\,\Delta\,T)\,Y$

$\frac{F}{10^{-3}} = (10^{-5}) \times 100 \times 0.5 \times 10^{11}$

F = 50 × 10³ N

2. (2) Water is used as a coolant due to its high specific heat capacity, which allows it to absorb large amounts of heat without a significant temperature rise.

3. (2) Foam is an insulator, it keeps the room cooler, not hotter.

Assertion False, Reason True → Option (2)

4. (2) **5. (1)** **6. (2)**

Thermodynamics

NCERT Topic-wise MCQs

1. (2) First law of thermodynamics is essentially a law of conservation of energy.

2. (3) For cyclic process, as the final and initial states are same so change in internal energy is zero.

ΔU = '0'

According to Ist law of thermodynamics

W = ΔQ

$2200 - 825 - 1100 + W_4 = 5960 - 5585 - 2980 + 3645$

$W_4 = 765$ J

3. (1) dQ = dU + dW

2 × 1000 × 4.2 = dU + 500

dU = 7900 J

4. (4) $\frac{dQ}{dW} = \frac{dQ}{dQ - dU}$

$\frac{dW}{dQ} = \frac{dQ - dU}{dQ} \Rightarrow 1 - \frac{dU}{dQ}$

$\frac{dW}{dQ} = 1 - \frac{nC_V\Delta T}{nC_P\Delta T}$

$\frac{dW}{dQ} = 1 - \frac{1}{\gamma}$

5. (1) As ΔU = ΔQ – ΔW = ΔQ – PdV

$= \frac{540 - 1.013 \times 10^5 (1670) \times 10^{-6}}{4.2} \approx 500$ calorie

6. (2) As volume decreases, work is done on the gas and so it is –ve.

$W = 50\,[4 - 10] = -300$ Joule

Heat supplied, $Q = 100$ Joule

$Q = \Delta U + W$ (Ist Law of thermodynamics)

$\Delta U = Q - W = 100 - (-300) = 400$ Joule

7. (4) Since, volume = constant

$\Delta U = nC_v\Delta T \qquad \left(\because C_v = \dfrac{R}{\gamma - 1}\right)$

$= \dfrac{3}{2}\times 2\times 8.31\times 20 = 498.6 \text{ Joule} \approx 499 \text{ J}$

8. (3) Work done = $P\Delta V$, where, P is pressure and ΔV is change in volume.

$\Delta V = 0$

$W = 0$

9. (4) $W = +30$ joule

$\Delta Q = 40$ Joule

from Ist law of thermodynamics

$\Delta U = \Delta Q - \Delta W = 40 - 30 = 10$ Joule

10. (3) Using Ist law of thermodynamics

$\Delta Q = \Delta U + \Delta W$

ΔU is a state function, so $\Delta Q - \Delta W$ is not a path function

11. (4) Using Ist law of thermodynamic

$\Delta Q = \Delta U + \Delta W$

$+150 = \Delta U + 110$

$\Delta U = 40$ Joule

12. (3) Work done during each cycle = Area enclosed under P – V graph

$= \dfrac{1}{2}(2P - P)(3V - V) = PV$

13. (3) Pressure, volume, temperature and mass

14. (3) In a cyclic process the initial and final states are same hence, $\Delta U = 0$, therefore thermal energy added to the system must be equal to the work done during the cyclic process $\Delta U = 0$ or $E = 0$.

15. (1) W = area under P – V graph

$= \dfrac{1}{2}[(3-1)+(2-1)]\times(2\times 10^2 - 10^2)$

= 150 Joule

16. (1) Internal energy change is path independent function

$\Delta U_1 = \Delta U_2$

$Q_1 - W_1 = Q_2 - W_2$

17. (1) $W = P\Delta V$ = Area of triangle

$= \dfrac{1}{2}3P.2V = 3\,PV$

18. (2) In a cyclic process $\Delta U = 0$

$\Delta Q = \Delta W$

19. (2) For adiabatic process $TV^{\gamma-1}$ = constant

For mono atomic gas $\gamma = 5/3$

$\therefore\ T_1V_1^{\left(\frac{5}{3}-1\right)} = T_2V_2^{\left(\frac{5}{3}-1\right)}$

or $T_2 = T_1\left(\dfrac{V_1}{V_2}\right)^{2/3}$

$= T_1\times\left(\dfrac{27}{8}\right)^{2/3} = T_1\times\left(\dfrac{3}{2}\right)^2 = \dfrac{9}{4}T_1$

$T_1 = 27 + 273 = 300$ K

$\therefore\ T_2 = 300\times\dfrac{9}{4}\text{K} = 675\text{K} = 402°\text{C}$

Increase in temperature is $402 - 27 = 375°$C.

Choice (2) is correct.

20. (1) There is no heat exchange in Adiabatic process.

21. (3) $W = PdV \Rightarrow \dfrac{W}{n} = \int_{V_1}^{V_2}\dfrac{RT}{V}dV = RT\log_e\dfrac{V_2}{V_1}$

22. (4) $W_{isothermal} = 2.303\,nRT\log\dfrac{V_2}{V_1}$

[$\therefore$ Initial volume = V_1, final volume = V_2]

$= 2.303\times 10\times 8.3\times 600\log\dfrac{10}{100}$

$= -11.4\times 10^4$ Joule

23. (1) $PV^\gamma = K$ putting $P = \dfrac{nRT}{V}$

$TV^{\gamma-1} = K$

24. (1) The gas expands so temperature of the system decrease but no flow of heat takes place.

25. (1) Internal energy $\Delta U = nC_v\Delta T$ [Internal energy of a system is a function of temperature any change in temperature leads to change in internal energy]

$\Delta T = 0$ for isothermal process

26. (4) $B = -\dfrac{\Delta P}{\Delta V/V}$

for isobaric process $\Delta P = 0 \Rightarrow B = 0$

27. (2) Work done = Area under P – V graph

Isobaric curve enclosed maximum area

28. (2) Adiabatic process $P^{1-\gamma}T^\gamma = K$

general equation $P \propto T^{-\gamma/1-\gamma}$

given equation $P \rightarrow T^C$

Comparing the powers of T in the above equations.

$\dfrac{-\gamma}{1-\gamma} = C$

$C = \dfrac{-\frac{5}{3}}{1-\frac{5}{3}} = \dfrac{\frac{-5}{3}}{\frac{-2}{3}} \qquad \left[\because \gamma \text{ for monoatomic} = \dfrac{5}{3}\right]$

$C = \dfrac{5}{2}$

29. (4) $P_1^{1-\gamma}T_1^{\gamma} = P_2^{1-\gamma}T_2^{\gamma}$

$$\left[\frac{P_1}{P_2}\right]^{1-\gamma} = \left[\frac{T_2}{T_1}\right]^{\gamma}$$

$$\left[\frac{8P_1}{P_1}\right]^{1-\frac{5}{3}} = \left[\frac{T_2}{300}\right]^{5/3}$$

$$\left[\frac{1}{8}\right]^{2/3} = \left[\frac{T_2}{300}\right]^{5/3}$$

$T_2 = 131$ K

or $T_2 = -142°C$

30. (2) An isobaric process is one in which volume and temperature of system may change but $\Delta P = 0$.

31. (3) $PV^{\gamma} = P'V'^{\gamma} = \text{constant}$

$$PV^{3/2} = P'\left(\frac{V}{4}\right)^{3/2};\ P' = P(4)^{3/2} = 8P$$

32. (2) Using $C_P - C_V = R$

divide the above equation by C_V

$$\frac{C_P}{C_V} - 1 = \frac{R}{C_V}$$

$$\gamma - 1 = \frac{R}{C_V}$$

$$C_V = \frac{R}{\gamma - 1}$$

33. (1) Gas cylinder suddenly exploding is an irreversible adiabatic change, therefore work done against expansion reduces the temperature of the gas cylinder.

34. (1) Thermo-flask system undergoes a change under the condition that no exchange of heat takes place between the system and the surroundings, then such a process is called adiabatic process $\Delta Q = 0$. But work is done on the system by shaking it against viscous force, so the temperature of the system will increase due to increase in internal energy.

High Order Time Intensive MCQs

1. (2) Net work done by the gas = Area enclosed in P-V curve, i.e., area of ΔABC

$$= \frac{1}{2} \times 4 \times 10^{-3} \times 2 \times 10^{5} = 4 \times 10^{2} = 400 \text{ J}$$

2. (1) Here, process is cyclic so change in internal energy is zero. We know that work done by gas is given by $W = P\Delta V$

Total work done by gas is

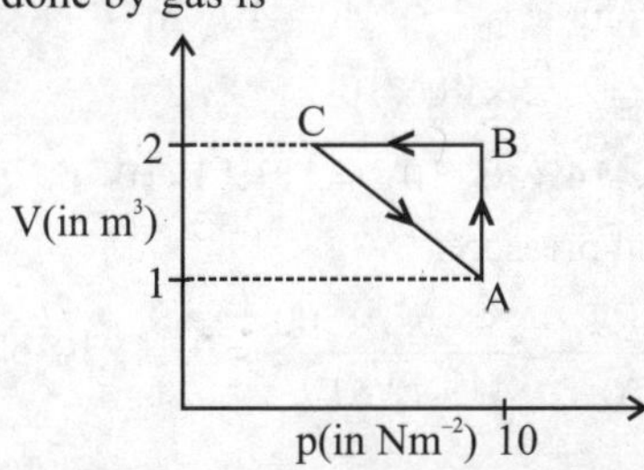

$W = W_{AB} + W_{BC} + W_{CA}$ ...(*i*)

First law of thermodynamics,

$\Delta Q = \Delta U + W$ ...(*ii*)

Where, ΔQ is heat supplied, ΔU is the change in internal energy which is zero and W is the work done.

$\therefore W_{AB} = P\Delta V = 10(2 - 1) = 10$ J

$W_{BC} = P\Delta V = 2 \times 0 = 0$

Hence, Eqs. (*i*) and (*ii*) will be written as

$5 = 0 + (10 + 0 + W_{CA}) \Rightarrow W_{CA} = -5$J

Work done by the gas in the process $C \rightarrow A$, $W_{CA} = -5$ J

3. (1) For all process 1, 2 and 3

Change in internal energy,

i.e., $\Delta U = U_B - U_A$

$\therefore \Delta U_1 = \Delta U_2 = \Delta U_3$ ...(*i*)

Now, $\Delta Q = \Delta U + \Delta W$ ...(*ii*)

ΔW = work done by the gas = area under P-V curve, area (1) > area (2) > area (3) $\Rightarrow \Delta W_1 > \Delta W_2 > \Delta W_3$

From equation (*ii*)

we have

$\therefore \Delta Q_1 > \Delta Q_2 > \Delta Q_3$

4. (1) From the first law of thermodynamics

$\Delta U = U_f - U_i = Q - W$ ...(*i*)

Where Q is the heat absorbed by the system and W is the work done by the system.

For the path iaf; we have

$Q = +50$ cal, $W = +20$ cal

$\therefore \Delta U = Q - W = 50 - 20 = 30$ cal

Since internal energy of the system depends only on temperature and no other factor. Therefore ΔU between i to f is always 30 cal whatever path be adopted.

For the path ibf,

$Q = +36$ cal and $\Delta U = 30$ cal

$\therefore W = Q - \Delta U = 36 - 30 = 6$ cal

5. (2) In adiabatic process, slope of P – V graph

$$\frac{dP}{dV} = \frac{-\gamma P}{V}$$

(Slope 2) > (slope 1) $(\because \gamma_2 > \gamma_1)$

Plot 1 → O_2 ($\gamma = 1.4$)

Plot 2 → He ($\gamma = 1.67$)

6. (2) $\Delta Q = Q_1 + Q_2 + Q_3 + Q_4$

$= 6000 - 5500 - 3000 + 3500$

$\Delta Q = 1000$ J

Total work done by ideal gas i.e.,

$\Delta W = W_1 + W_2 + W_3 + W_4$

$= 2500 - 1000 - 1200 + x = 300 + x$

For cyclic process,

$U_f = U_i$

$\Rightarrow \Delta U = U_f - U_i = 0$

So, from first law of thermodynamics,

i.e, $\Delta Q = \Delta U + \Delta W$

$\Delta Q = \Delta W \quad (\therefore \Delta U = 0)$

We have, $1000 = (300 + x)$

i.e., $x = 1000 - 300 = 700$ J

Efficiency of a cycle is defined as $\eta = \frac{\text{Work done}}{\text{heat Input}}$

$= \frac{\Delta W}{Q_1 + Q_4} = \frac{\Delta Q}{(Q_1 + Q_4)}$

$= \frac{1000}{(6000+3500)} = \frac{1000}{9500} \times 100 = 0.105 \times 100 = 10.5\%$

7. (3) $W_{AB} = -P_0V_0$

$W_{BC} = 0$

and $W_{CD} = 4P_0V_0$

$\therefore W_{ABCD} = -P_0V_0 + 0 + 4P_0V_0 = 3P_0V_0$

8. (3) For isothermal expansion, temperature remains constant.

$\Rightarrow PV = \text{constant}$

i.e., $PV = P' \times 2V \qquad [\because V = 2V]$

$P' = \frac{P}{2}$

For adiabatic expansion, $PV^\gamma = \text{constant}$

$\Rightarrow \frac{P}{2}[2V]^{5/3} = P''[16V]^{5/3}$

$\frac{P}{2}[2V]^{5/3} = P''[16V]^{5/3} \qquad \left[\because \gamma = \frac{5}{3}\right]$

$P'' = \frac{P}{2}\left[\frac{2V}{16V}\right]^{5/3} = \frac{P}{2}\left[\frac{1}{8}\right]^{5/3}$

$P'' = \frac{P}{64}$

9. (3) In a cyclic process, $\Delta U = 0$

$\Delta Q = \Delta U + \Delta W = 0 + \Delta W = \Delta W$

= work done in one cycle

= area of closed curve = πr^2

$= \pi\left(\frac{20}{2}\right)^2$ (k Pa) litre

$= \pi(100)\times 10^3 \frac{N}{m^2} \times 10^{-3} m^3 = 100\,\pi$ J.

10. (4) According to question, $P \propto T^3$

$PV = nRT$ and $P \propto T^3$

$P \propto (PV)^3$

$\Rightarrow P^2V^3 = \text{constant}$

$\Rightarrow PV^{3/2} = \text{constant}$

comparing the above equation with

$PV^\gamma = \text{constant}$

i.e., $\gamma = \frac{3}{2}$

11. (4) Isothermal P – V curve is rectangular hyperbola & isobaric, curve will be straight line parallel to V axis.

12. (4) For cyclic process, change in internal energy is zero

$\Rightarrow \Delta U = 0$

$80 + Q = 60 + (-30) \Rightarrow Q = -50$ cal

13. (4) $T_A = \frac{4P_0V_0}{nR}, T_B = \frac{12P_0V_0}{nR}$

$\Delta T = \frac{8P_0V_0}{nR}$

$\Delta U = nC_V\Delta T = \frac{3}{2}nR\Delta T$

$\Delta U = 12\,P_0V_0$

ΔW = Area under P-V curve = $3\,P_0V_0$

$\Delta Q = \Delta U + \Delta W = 15\,P_0V_0$

14. (1) $\%\eta = \frac{W}{Q_{supplied}} \times 100;\; W = 12kJ \qquad [Q_{supplied} = 30\text{ kJ}]$

15. (1) From ideal gas equation

$PV = nRT$

initial pressure $P = \frac{nRT}{V} = \frac{2\times 8.3\times 300}{20\times 10^{-3}}$

$= 2.49\times 10^5\,N/m^2$

When volume of the gas isdoubled at constant pressure, its temperature is also doubled. This process is shown on P-V curve by line AB. The gas then cools to temperature T adiabatically. This is shown by curve BC. The whole process is represented by the curve ABC.

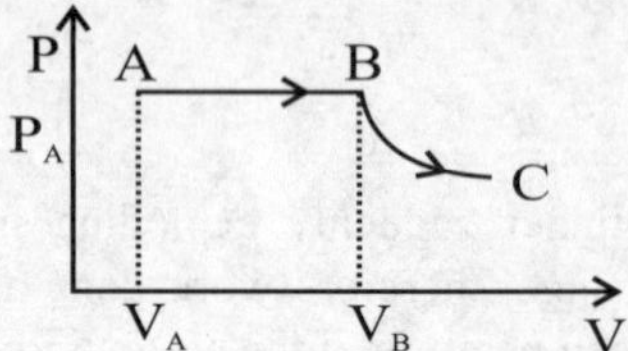

At point B, pressure

$P_B = P_A = 2.49\times 10^5\,N/m^2$.

Volume $V_B = 2V_A = 40\times 10^{-3}\,m^3$,

Temperature $T_B = \frac{2V_AT_A}{V_A} = 600K$.

For adiabatic process $TV^{\gamma-1} = \text{constant}$

$\Rightarrow T_BV_B^{\gamma-1} = T_CV_C^{(\gamma-1)}$

$\therefore \left(\frac{V_C}{V_B}\right)^{\gamma-1} = \frac{T_B}{T_C} = \frac{600}{300} = 2$

$\therefore \frac{V_C}{V_B} = 2^{1/(\gamma-1)} = 2^{3/2}$

Final volume = $2^{3/2} \times 40 \times 10^{-3}\,m^3$

$\therefore V_C = 2\sqrt{2} \times 40 \times 10^{-3}$

$= 2\times 1.414\times 40\times 10^{-3} = 113.13\times 10^{-3}\,m^3$

16. (1) Final pressure

$P_C = \frac{nRT_C}{V_C} = \frac{2\times 8.3\times 300}{113.13\times 10^{-3}} = 0.44\times 10^5\,N/m^2$

17. (4) The work done by gas in isobaric process AB

$= 2.49 \times 10^5 \times (40-20) \times 10^{-3} = 4980\text{J}$

The work done by gas during adiabatic process BC

$$W_2 = \frac{nR}{1-\gamma}[T_2 - T_1] = \frac{2 \times 8.3}{1-\left(\frac{5}{3}\right)}[300-600]$$

$\Rightarrow W_2 = 7470\text{J}.$

$\therefore$ Net work done by gas $W = W_1 + W_2$

$= 4980 + 7470 = 12450$ J

18. (4) As the process is isobaric

$\Delta W = \int PdV = P[V_F - V_i]$ $\because$ P is constant

$= 1.01 \times 10^6 [1671 - 1] = 1686.7 \times 10^6 \text{erg}$

$= 168.67$ Joule $\left[\because 1 \text{ erg} = 10^{-7}\text{J}\right]$

19. (1) $\Delta Q = mL = 1 \times 540 \text{ cal} = 2268$ J, [1 cal = 4.2J]

From 1st law of thermodynamics

$\Delta Q = \Delta U + \Delta W$

so, $\Delta U = \Delta Q - \Delta W = (2268 - 168.67)\text{J} = 2099.33$ J

20. (3) Change in internal energy depends only on change in temperature since internal energy is a function of state only i.e., $dU = nC_v dT$.

In adiabatic process, $dQ = 0$,

Hence, $dU + dW = 0$

$\Rightarrow dU = -dW$

i.e., magnitude of change in internal energy is equal to magnitude of work done.

21. (1) For a gas, $PV = \mu RT = \frac{m}{R}RT$

For graph A, $PV = \frac{m}{M}RT$

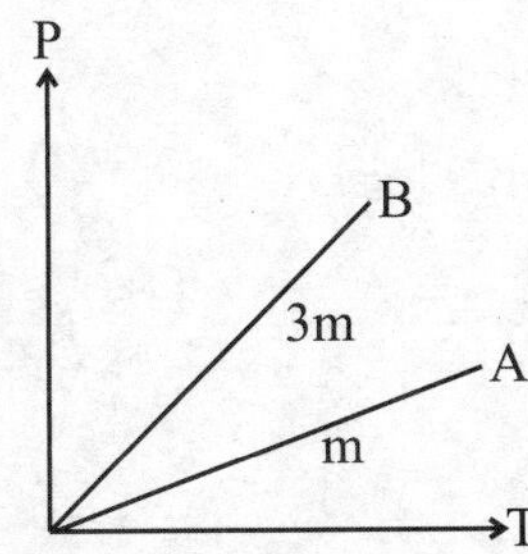

Slope of graph A, $\left(\frac{P}{T}\right) = \frac{m}{M}\frac{R}{V}$ ...(*i*)

For graph B, $PV = \frac{3m}{M}RT$

Slope of graph B, $\left(\frac{P}{T}\right) = \frac{3m}{M}\frac{R}{V}$ ...(*ii*)

$$\frac{\text{Slope of curve B}}{\text{slopeofcurveA}} = \frac{\frac{3m}{M}\frac{R}{V}}{\frac{m}{M}\frac{R}{V}} = \frac{3}{1}$$

Statement Based MCQs

1. **(1)** Both the statements are true.
2. **(2)** Pressure is constant in isobaric process.
3. **(1)** When a gas is allowed to expand, the force due to displacement and pressure are in the same direction.

   So, $\theta = 0$

   $\therefore$ Work done = Fs cos θ = Fs = +ve
4. **(4)** Statement-I is false but statement-II is true.
5. **(3)** Statement-I is true but statement-II is false.

Assertion & Reason MCQs

1. **(3)** Isothermal process is thermodynamic process in which the temperature of a system remains constant, this is due to slow process.
2. **(1)** Both assertion & reason are true and reason is the correct explanation of the assertion.
3. **(1)** Both assertion & reason are true and reason is the correct explanation of the assertion.
4. **(1)** In this process $\Delta P = 0$. The heat transfer to the system does work, but also changes the internal energy of the system.
5. **(2)** Both assertion and reason are true but reason is not the correct explanation of the assertion.

Match the Columns MCQs

1. (1) $W_{iaf} + \Delta U_{if} = H_{iaf}$

$25 + \Delta U_{ij} = 60$

$\Delta U_{if} = 35$

and $W_{ibf} + \Delta U_{if} = H_{ibf}$

$W_{ibf} + 35 = 42$

$W_{ibf} = 7$ cal

(a) $W_{ibf} = 7$ cal

(b) $H_{if} = W_{fi} + \Delta U_{fi} = 13 + 35 = 48$ cal

(c) $\Delta U_{if} = U_f - U_i$

$\Rightarrow U_f = 35 + 10 = 45$ cal

(d) $H_{ib} = W_{ibf} + \Delta U_{ib} = 7 + (22 - 10) = 19$ cal

2. (1) (a) $dQ = dU + dW$

Here $dQ = \frac{400 \times 4184}{1000} = 1674$ J

$dW = +300$ J. Since the work is done by the system,

$dU = dQ - dW = 1374$ J

(b) $\Delta U = \Delta Q - \Delta W = (200)(4.184) - (-320) = 1157$ (approx).

(c) $\Delta U = (\Delta Q - \Delta W) = (-800)(4.184) - 0 = -3347\text{J}$

3. (4) a-(i); b-(ii); c-(iv); d-(iii)

NCERT Exemplar MCQs

1. (1) Amount of sweat evaporated/minute

$$= \frac{\text{Sweat produced / minute}}{\text{Number of calories required for evaporation / kg}}$$

$= \frac{\text{Amount of heat produced per minute in jogging}}{\text{Latent heat (in cal/ kg)}}$

$= \frac{14.5\times10^3}{580\times10^3} = \frac{145}{580} = 0.25\text{kg}$

2. **(3)** According to the question given that PV = constant

Hence, we can say that the gas is going through an isothermal process.

Clearly, from the graph that between process 1 and 2 temperature is constant and the gas expands and pressure decreases i.e., $P_2 < P_1$ which corresponds to diagram (iii).

3. **(2)** Consider the P-V diagram given in the question.

Work done in the process ABCD = area of rectangle ABCDA

$= (AB) \times BC = (3V_0 - V_0) \times (2P_0 - P_0)$

$= 2V_0 \times P_0 = 2P_0V_0$

As the process is going anti-clockwise, hence there is a net compression in the gas.

So, work done by the gas $= -2P_0V_0$.

4. **(1)** Consider the P-V diagram shown for the container A (isothermal) and for container B (adiabatic).

Both the process involving compression of the gas.

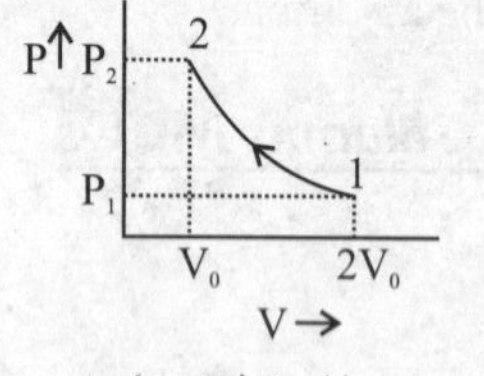

(container A)

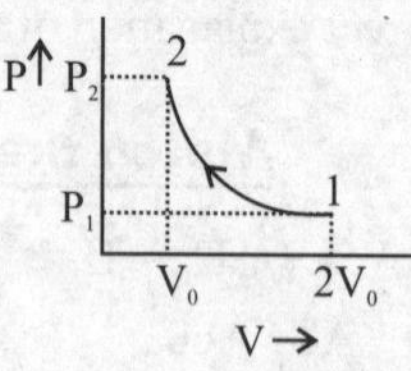

(container B)

For isothermal process (gas A) (during $1 \rightarrow 2$)

$P_1V_1 = P_2V_2$

$\Rightarrow P_0(2V_0) = P_2(V_0)$

$\Rightarrow P_2 = 2P_0$

For adiabatic process, (gas B) (during $1 \rightarrow 2$)

$P_1V_1^\gamma = P_2V_2^\gamma \Rightarrow P_0(2V_0)^\gamma = P_2(V_0)^\gamma$

$$\Rightarrow P_2 = \left(\frac{2V_0}{V_0}\right)^\gamma P_0 = (2)^\gamma P_0$$

Hence, $\frac{(P_2)B}{(P_2)A}$ = ratio of final pressure $\frac{(2)^\gamma P_0}{2P_0} = 2^{\gamma-1}$

where, γ is ratio of specific heat capacities for the gas.

5. **(2)** Let the equilibrium temperature of the system is T

Let us assume that $T_1, T_2 < T < T_3$.

According to question, there is no net heat loss to the surroundings.

Heat lost by M_3 = Heat gained by M_1 + Heat gained by M_2

$\Rightarrow M_3s(T_3 - T) = M_1s(T - T_1) + M_2s(T - T_2)$

(where, s is specific heat of the copper material)

$\Rightarrow T[M_1 + M_2 + M_3] = M_3T_3 + M_1T_1 + M_2T_2$

$$\Rightarrow T = \frac{M_1T_1 + M_2T_2 + M_3T_3}{M_1 + M_2 + M_3}$$

NEET & AIIMS Past Year MCQs

1. **(4)** The equilibrium state of a thermodynamic system is described by:

Pressure (A)

Temperature (C)

Volume (D)

Total heat (B) and work done (E) are path-dependent, not state variables.

2. **(3)** As volume is constant along path *bc*, hence the work done, $W = \int PdV = 0$

3. **(2)** $\eta = 1 - \frac{T_{sink}}{T_{source}} = 0.5$

$\frac{T_{sink}}{T_{source}} = 0.5$

$\Rightarrow T_{sink} = \frac{1}{2}\times(273+327) = \frac{1}{2}\times600 = 300\text{ K} = 27°\text{C}$

4. **(3)** For the curve 4 pressure is constant, so this is an isobaric process.

For the curve 1, volume is constant, so it is isochoric process. Between curves 3 and 2, curve 2 is steeper, so it is adiabatic and 3 is isothermal.

Note: We should be careful while deciding between isothermal and adiabatic curves because these curves look similar.

5. **(3)** $\Delta Q = \Delta u + \Delta w$

m = 1 gm

L_v = 540 cal/gm

$\Delta Q = 1 \times 540 = 540$

$540 = \Delta u + P\Delta v$

$540 = \Delta u + 10^5 \times (1671.1) \times 10^{-6}$

$540 = \Delta u + 167$

$\Delta u = 540 - 167$

=373 cal

6. **(3)** Efficiency, $\eta = 1 - \frac{T_2}{T_1}$ where T_1 = Temp. of source

T_2 = Temp. of Sink

$\Rightarrow \frac{T_2}{T_1} = 1 - \eta = 1 - \frac{1}{6} = \frac{5}{6}$ ...(*i*)

case 2 $\frac{T_2 + 62}{T_1} = 1 - \left(2\times\frac{1}{6}\right) = \frac{2}{3} \Rightarrow \frac{T_2}{T_1} - \frac{6_2}{T_1} = \frac{2}{3}$

$\Rightarrow \frac{5}{6} - \frac{62}{T_1} = \frac{2}{3} \Rightarrow \frac{62}{T_1} = \frac{1}{6} \Rightarrow T_1 = 372\text{k}$

$= 372 - 273 = 99°\text{C}$

Kinetic Theory

NCERT Topic-wise MCQs

1. **(3)** $PV = \mu RT$

$P_{H_2} = \frac{m}{M_{H_2}}\frac{RT}{V}$, $P_{He} = \frac{m}{M_{He}}\frac{RT}{V}$

$\therefore \frac{P_{H_2}}{P_{He}} = \frac{M_{He}}{M_{H_2}} = \frac{4\times10^{-3}}{2\times10^{-3}} = 2$

2. (3) Slope $\propto \frac{1}{P}$

$\tan\theta_2 > \tan\theta_1$

$\frac{1}{P_2} > \frac{1}{P_1}$

$P_1 > P_2$

3. (1) Gases cannot be liquefied above critical temperature however large the pressure may be.

4. (3) $\frac{P_1V_1}{T_1} = \frac{P_2V_2}{T_2}$

$V_2 = \frac{P_1V_1T_2}{P_2T_1}$

$V_2 = \frac{1\times 500\times(273-3)}{0.5\times(273+27)} = 900\,m^3$

5. (1) Air inside a closed container is saturated with vapour, so if temperature remains constant, the saturated vapour pressure does not change, Also, pressure will be twice, when volume is halved.

6. (1) $PV = nRT = \frac{m}{M}RT$

$\frac{P}{m/V} = \frac{RT}{M}$

$\frac{P}{\rho} \propto T$

$\therefore T_1 > T_2$

7. (4) $P_{mix} = P_1 + P_2$

$P_{mix} = \frac{n_1RT}{V} + \frac{n_2RT}{V}$

$10 = \frac{8\,RT}{32\,V} + \frac{7\,RT}{28\,V} \Rightarrow 10 = \frac{RT}{2V}$

$\Rightarrow \frac{RT}{V} = 20$

If O_2 is absorbed, pressure exerted by N_2

$P_{N_2} = \frac{7}{28}\frac{RT}{V} = \frac{20}{4} = 5$ atm

8. (4) $\because PV = nRT$

$\Rightarrow PV = \frac{m}{M}RT$

$T \propto \frac{1}{m}$

$\frac{T_1}{T_2} = \frac{m_2}{m_1}$ $\left\{m_2 = m - \frac{m}{4} \Rightarrow m_2 = \frac{3}{4}m\right\}$

$T_2 = \frac{4}{3}\times(273+60)$

$T_2 = 444$ K or 171°C

9. (4) At constant temp.

PV= constant

$P_1V_1 = P_2V_2$ $\left\{V_2 = V_1 - \frac{10V_1}{100} \Rightarrow 0.90V_1\right\}$

$P_2 = \frac{P_1V_1}{V_2}$ $\frac{P_2}{P_1} = \frac{V_1}{V_2} = \frac{100}{90}$

$\%\left(\frac{P_2 - P_1}{P_1}\right) = \frac{1}{9}\times 100 \Rightarrow 11.1\%$

10. (1) $\frac{PV}{T}$ = constant ; V = rms

$\therefore \frac{P}{T}$ = constant $\Rightarrow P\times\frac{100.4}{100}\times\frac{1}{T+1}$

T = 250 K

$P_2 = P_1 + P_1$ of $4\% = \frac{100.1P_1}{1000}$

$\frac{P_1}{P_2} = \frac{T_1}{T_2}$

$\frac{1000P_1}{1004P_1} = \frac{T}{T+1}$

$1004T = 1000T + 1000$

$4T = 1000$

T = 250 K

11. (3) $PV = \mu RT$

$T = \frac{PV}{\mu R}$

$T_1 = \frac{PV_1}{\mu R}$ ... (*i*)

$T_2 = \frac{PV_2}{\mu R}$ ... (*ii*)

Dividing Eq (*i*) and (*ii*)

$\frac{T_1}{T_2} = \frac{V_1}{V_2}$ or $T \propto V$

From graph $V_1 < V_2 : T_1 < T_2$ or $T \propto V$

12. (4) $P_{avg} = \frac{P+(-P)}{2} = 0$ average momentum is always zero.

It does not depend on any physical parameter

$\therefore$ For any temperature mass and volume, the average momentum of the gas will be zero

13. (4) $\frac{P_1V_1}{T_1} = \frac{P_{mix}\,V_{mix}}{T_{mix}}$

$T_1 = T_{mix} = T$ and $V_1 = V + V$, $V_{mix} = V$

$\frac{P(2V)}{T} = \frac{P_{mix}\,V}{T}$

$P_{mix} = 2P$

14. (3) Number of moles (n) $= \frac{m}{M}$

$n = \frac{14}{28} = 0.5$

$Q_{constant\,p} = n\,C_P\,\Delta T$

$= \frac{1}{2}\times\frac{7}{2}R\times 40 = 70R$

15. (1) Since, Brownian motion explains the collision between particle and their scattering. It is an evidence in support of kinetic theory of gases which is based on movement of particles.

16. (2) $V = \text{constant} \Rightarrow \frac{P_1}{M_1T_1} = \frac{P_2}{M_2T_2}$

Final mass $\Rightarrow \frac{P_1}{M_1T_1} = \frac{P_2}{M_2T_2}$

$M_2 = \frac{P/2}{300} \times \frac{6g \times 400}{P} = 4g$

Leak out mass $= M_1 - M_2 = 6 - 4 = 2g$

17. (1) PV = nRT

$\frac{n}{V} = \frac{P}{RT}$ (constant)

At constant pressure and temperature, both balloons will contain equal no. of molecules per unit volume

18. (3) $n = \frac{PV}{RT}$

n = 1.98

$n = \frac{m}{M} = 1.98$

$m = 1.98M \Rightarrow m = 1.98 \times 28 \approx 56g$

19. (4) $\frac{PV}{T} = K$

$\frac{P_1V_1}{T_1} = \frac{P_2V_2}{T_2}$ $\left\{\because V = \frac{m}{d}\right\}$

$\frac{P_1m}{T_1d_1} = \frac{P_2m}{T_2d_2} \Rightarrow \frac{P_1}{d_1T_1} = \frac{P_2}{d_2T_2}$

20. (3) Mean free path, $\lambda = \frac{1}{\sqrt{2}\,\pi n d^2}$, (d = 2r)

21. (3) Oxygen is diatomic. It has 5 degrees of freedom as heat is given to O_2, C_V increases from $\frac{5R}{2}$ to $\frac{7R}{2}$. Hence internal energy increases In such cases molecules have an additional degree of freedom due to their vibrational motion

22. (2) $C_V = \frac{fR}{2}$, degree of freedom for diatomic gas and polyatomic gas at room temperature is f but at higher temperature their degree of freedom increases by one due to vibrational motion of molecule

23. (1)

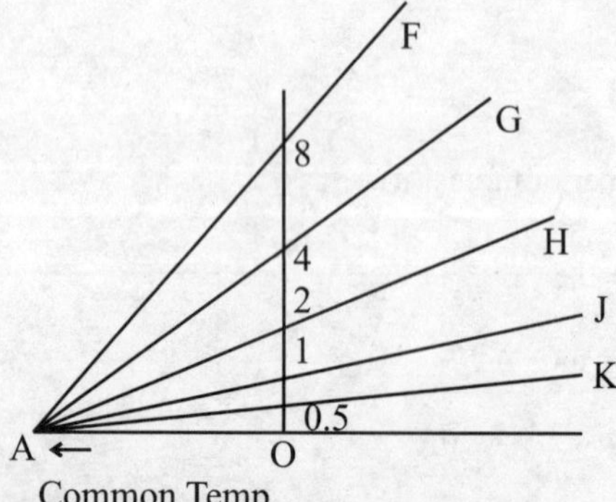

$PV = \mu RT = \frac{m}{M}RT$

$V = \frac{m}{M}\frac{RT}{P}$

Slope of V – T-graph is $\frac{m}{M}\frac{R}{P}$

For line H, slope $\frac{2}{OA} = \frac{m}{M}\frac{R}{P}$

If m → 2m; P → P/2

Slope becomes

$\frac{2m}{M}\frac{R}{\left(\frac{P}{2}\right)} = 4 \times \frac{mR}{MP}$

$= 4$ times the slope of line $H = 4 \times \frac{2}{OA} = \frac{8}{OA}$

Corresponds to line F.

24. (2) $\because C_V = \frac{f}{2}R, C_P = \left(\frac{f}{2}+1\right)R$

For mono atomic, f = 3

$C_V = \frac{3}{2}R, C_P = \left(\frac{3}{2}+1\right)R = \frac{5}{2}R$

From, $C_P - C_V = R$

$\frac{5}{2}R - \frac{3}{2}R = R$

$r = \frac{C_P}{C_V} = \frac{5/2}{3/2} = \frac{5}{3}$

25. (1) Mean free path, $\lambda = \frac{1}{\sqrt{2}\,\pi n d^2}$

n = number density (number of molecules per unit volume)

when n is reduced, λ increases.

26. (3) $Q = nC_P\Delta T$

$1163.4 = 1 \times \frac{7}{2} \times 8.31 \times \Delta T$

$\Delta T = 40$ K

27. (1) Use Mayer's equation

$C_P - C_V = R$

$C_V = \frac{7}{2}R - R = \frac{5}{2}R$

$\gamma = \frac{C_P}{C_V} = \frac{\frac{7}{2}R}{\frac{5}{2}R} \Rightarrow \gamma = \frac{7}{5}$

28. (4) Let, Specific heat at constant volume is (C_V) and degree of freedom is f, then

$C_V = \frac{fR}{2} = \frac{6R}{2}$ $\therefore f = 6$

$C_V = 3R$

29. (3) For one degree of freedom, energy = $\frac{1}{2}kT$ and energy is equally shared for each of n degrees of freedom.

$\therefore$ Energy per molecule is = $\frac{n\,kT}{2}$

30. (4) $C_P - C_V = R$

$$\frac{C_P}{C_P} - \frac{C_V}{C_P} = \frac{R}{C_P}$$

$$1 - \frac{1}{\gamma} = \frac{R}{C_P}$$

$$C_P = \frac{\gamma R}{\gamma - 1}$$

31. (3) $v_{rms} = \sqrt{\frac{3RT}{M}}$

Gas is compressed isothermally, so T remains constant and v_{rms} will remains same

32. (3) $v_{rms} = \sqrt{\frac{(1)^2 + 3^2 + 5^2 + 7^2}{4}} = \sqrt{21}$

$= 4.57$ Km/sec

$$v_{avg} = \frac{1+3+5+7}{4} = 4 \text{ km/s}$$

$v_{rms} - v_{avg} = 0.57$ km/sec

33. (1) $v_{rms} = \sqrt{\frac{3\,RT}{M}}$

$T \propto M$ ($\because v_{rms}$ and R are constant)

$$\frac{T_{O_2}}{T_{H_2}} = \frac{M_{O_2}}{M_{H_2}} = \frac{32}{2} = 16$$

$T_{O2} = 16 \times 300 = 4800$ K

34. (4) Average velocity of the molecule of an ideal gas is zero because molecules possess all sorts of velocities in all direction whose vector sum is zero.

35. (2) $\frac{v_{rms_1}}{v_{rms_2}} = \sqrt{\frac{T_1}{T_2}} = \sqrt{\frac{120}{480}} = \frac{1}{2}$

$$V = \frac{1}{2}V_2 \Rightarrow V_2 = 2V_1 \Rightarrow V_2 = 2V$$

36. (4) $T_m = \frac{n_1T_1 + n_2T_2}{n_1 + n_2}$

$$= \frac{10 \times 10 + 20 \times 20}{10 + 20} = \frac{500}{3} = 16.6\,°C$$

37. (4) $v_{rms} = \sqrt{\frac{3\,RT}{M}}$

v_{rms} and R are constants

$$\frac{T}{M} = \text{constant or } \frac{T_1}{M_1} = \frac{T_2}{M_2} \Rightarrow \frac{T_1}{T_2} = \frac{M_1}{M_2}$$

$$\frac{T}{2} = \frac{273 + 47}{32} \Rightarrow \frac{T}{2} = \frac{320}{32} = 10$$

T = 20 K

38. (4) Root mean square velocity does not depend upon pressure. Hence, the rms velocity remains same

39. (4) $v_{rms} = \sqrt{\frac{3\,RT}{M}} \propto \sqrt{\frac{1}{M}}$

$$\frac{(v_{rms})_{O_3}}{(v_{rms})_{O_2}} = \sqrt{\frac{M_{O_2}}{M_{O_3}}} = \sqrt{\frac{32}{48}} = \frac{\sqrt{2}}{\sqrt{3}}$$

40. (4) $Q = \frac{f}{2}\mu R\Delta T$

f = 3 for He (mono atomic gas)

$$Q = \frac{3}{2} \times \frac{1}{4} \times K_b N_a \Delta T \qquad \therefore K_b = \frac{R}{N_a}$$

$$Q = \frac{3}{8}K_b N_a (T_2 - T_1)$$

41. (3) γ (diatomic) = $\frac{7}{5}$

$W = Q - \Delta U = nC_p\,\Delta T - nC_V\Delta T$

$W = n(C_P - C_V)\Delta T$

$$\frac{Q}{W} = \frac{C_P}{C_P - C_V}$$

$$C_P = \gamma C_V \text{ thus, } \frac{Q}{W} = \frac{\gamma}{\gamma - 1}$$

$$\therefore \frac{Q}{W} = \frac{(7/5)}{(7/5) - 1} = \frac{7}{2}$$

42. (4) Kinetic energy (per mol) $= \frac{f}{2}RT$

$$K = \frac{3}{2}RT$$

Note; when nothing said about the atomicity of gas, goes for translational kinetic energy

$$K = \frac{3}{2} \times 8.31 \times 273 \quad \Rightarrow K = 3.4 \times 10^3 \text{ Joule}$$

43. (2) Pressure exerted by the gas

$$P = \frac{1}{3}\rho v_2 \Rightarrow \frac{1}{3}\frac{m}{V}v^2 \text{ (multiply and divide RHS by 2.)}$$

$$P = \frac{2}{3V} \times \left(\frac{1}{2}mv^2\right) \Rightarrow P = \frac{2}{3}E, \text{ (For unit volume)}$$

44. (2) From non inertial frame of reference. An pseudo force acts opposite to real acceleration on the rear face, as a result the pressure in the front side would be lowered

45. (3) $K = \frac{1}{2}mv_{rms}^2$

$$v_{rms} = \sqrt{\frac{3KT}{M}}$$

$$v_{rms} \propto \sqrt{T}$$

$$\frac{K_2}{K_1} = \left(\sqrt{\frac{T_2}{T_1}}\right)^2 \Rightarrow K_2 = K_1\frac{T_2}{T_1}$$

$$K_2 = K_1\frac{600}{300} \Rightarrow K_2 = 2K_1$$

46. (1) At a given temperature (T) all the ideal gas molecules no matter what their masses, have the same average translational kinetic energy

$$E = \frac{3}{2}KT$$

so E does not depend upon density

$$\frac{E_1}{E_2} = \frac{1}{1}$$

High Order Time Intensive MCQs

1. (4) Given, $v_{O_2} = \frac{1}{2} v_{H_2}$

$$v_{rms} = \sqrt{\frac{3RT}{M}}$$

where, M = molecular weight

$$\therefore \sqrt{\frac{3RT}{32}} = \frac{1}{2}\sqrt{\frac{3R\times 273}{2}}$$

$$\Rightarrow \frac{T}{32} = \frac{273}{8}$$

or $T = 1092\text{ K} = 1092 - 273 = 819°\text{C}$

2. (1) As we know that,

$$v_{rms} \propto \sqrt{T}$$

Hence, $\frac{v_{rms(1)}}{v_{rms(2)}} = \sqrt{\frac{T_1}{T_2}}$

$$\Rightarrow \frac{400}{v_s} = \frac{\sqrt{27+273}}{\sqrt{227+273}} = \sqrt{\frac{300}{500}}$$

$$\Rightarrow v_s = \sqrt{\frac{500}{300}} \times 400 = 516.39\text{ ms}^{-1}$$

3. (2) $V_{Sound} = \sqrt{\frac{\gamma P}{\rho}};\ 330 = \sqrt{\frac{\gamma \times 1.013\times 10^5}{1.3}}$

$$\gamma = \frac{(330)^2 \times 1.3}{1.013\times 10^5} \cong 1.3 \cong \frac{7}{5}$$

Also $\gamma = 1 + \frac{2}{f}$ or $\frac{7}{5} = 1 + \frac{2}{f};\ f = 5$

4. (4) Using ideal gas equation, $PV = nRT$

$$\Rightarrow P_{H_2} = \frac{m}{M_{H_2}}\cdot\frac{RT}{V} \text{ and } P_{He} = \frac{m}{M_{He}}\cdot\frac{RT}{V}$$

$$\therefore \frac{P_{H_2}}{P_{He}} = \frac{M_{He}}{M_{H_2}} = \frac{4\times 10^{-3}}{2\times 10^{-3}} = 2 \Rightarrow P_{H_2} = 2P_{He}$$

5. (4) Average kinetic energy, $E = \frac{3}{2}kT \Rightarrow E \propto T$

Thus, average kinetic energy of a gas molecule is directly proportional to the absolute temperature of gas.

6. (3) Kinetic energy of ideal gas depends only on its temperature. Hence, it remains constant whether its pressure is increased or decreased.

7. (2) We know, $E = \frac{3}{2}nkT$

where, k = Boltzmann's constant

using energy conservation,

$$\therefore \frac{3}{2}n_1kT_1 + \frac{3}{2}n_2kT_2 = \frac{3}{2}(n_1+n_2)kT$$

$$\Rightarrow T = \frac{n_1T_1 + n_2T_2}{(n_1+n_2)}$$

8. (2) $U \propto T$

$$\frac{U_i}{U_f} = \frac{T_i}{T_f} \Rightarrow \frac{300}{U_f} = \frac{T}{3T} \Rightarrow U_f = 900\text{ J}$$

$$v \propto \sqrt{T}$$

$$\frac{v_i}{v_f} = \sqrt{\frac{T_i}{T_f}} \Rightarrow \frac{400}{v_f} = \sqrt{\frac{1}{3}} \Rightarrow v_f = 400\sqrt{3}\text{ m/sec}$$

9. (3) We know, PV = nRT

For oxygen $P \times V = 1 \times R \times T$

$P_1 \times V = 1 \times R \times 2T$ [for Helium]

$P_1 = 2P$

So, correct option is (3)

10. (2) $\frac{(Q)_V}{(Q)_P} = \frac{C_V}{C_P} = \frac{1}{\gamma} = \frac{3}{5} \Rightarrow Q_v = \frac{3}{5}Q_P \Rightarrow Q_v = \frac{3}{5}\times 200$

$\therefore (Q)_V = 120\text{ J}$

11. (2) $U = \frac{n_1f_1RT}{2} + \frac{n_2f_2RT}{2}$

$$= \frac{8\times 5RT}{2} + \frac{2\times 3RT}{2} = 23RT$$

12. (4) In Cylinder A piston is movable so it is isobaric process. In cylinder B piston is fixed. So, it is adiabatic process. As same amount of heat is given to two cylinder.

$$\Rightarrow Q_A = Q_B$$

$$nCp\Delta T_A = nCv\Delta T_B$$

$$n\frac{7}{2}\Delta T_A = n\frac{5}{2}T_0$$

$$\Delta T_A = \frac{5}{7}T_0$$

13. (2) $v_{rms} = \sqrt{\frac{3kT}{m}}$l; $v_{mp} = \sqrt{\frac{2kT}{m}}$

$$\frac{v_{rms}}{v_{mp}} = \sqrt{\frac{3}{2}} \text{ or } v_{rms} = \sqrt{\frac{3}{2}}v_{mp}$$

14. (2) $\frac{3}{2}nR\Delta T = 100$

$$\Delta Q = \Delta U + \Delta W$$

For isobaric $\Delta Q = n\frac{3}{2}R\Delta T + Rn\Delta T$

$$= \frac{5}{2}(nR\Delta T)\quad \Delta Q = \frac{5}{3}\left(\frac{3}{2}nR\Delta T\right) = \frac{500}{3}$$

$$\Delta Q = \Delta U = \frac{500}{3}$$

15. (2)

16. (2) For hydrogen $v_H = \sqrt{\frac{3RT}{M_H}}$ ...(i)

For oxygen $v_O = \sqrt{\frac{3RT}{M_O}}$ ...(ii)

Given $v_H = v_1$ and $v_O = v_2$ at same temperature

$\therefore \quad \frac{v_1}{v_2} = \sqrt{\frac{M_O}{M_H}}$

$M_O = 32, M_H = 2$

$\therefore \quad \frac{v_1}{v_2} = \sqrt{\frac{32}{2}} = \sqrt{16} = 4$

17. (2) According to the given problems

VP^2 = constant

From the gas law

PV = nRT

$\Rightarrow \left(\frac{k}{\sqrt{V}}\right)V = nRT \Rightarrow \sqrt{V} = \left(\frac{nR}{K}\right)T$

$\Rightarrow \sqrt{V} \propto T$

$\therefore \quad \sqrt{\frac{V_1}{V_2}} = \left(\frac{T_1}{T_2}\right)$, i.e., $\sqrt{\frac{V}{2V}} = \frac{T}{T'}$

$\Rightarrow T' = \sqrt{2}\,T$

18. (3) Apply $PV = \mu RT = \frac{m}{M}RT$

$20 \times V = \frac{m}{M} \times R \times 300$ ... (i)

$P' \times V = \frac{(m/2)}{M} \times R \times 350$ ... (ii)

Eq. (ii) ÷ (i)

$P' = \frac{140}{12} = 11.67$ atm.

19. (4) $V_{rms} = \sqrt{\frac{3KT}{m}} = \sqrt{\frac{3RT}{M}}$ If M is less V_{rms} is more

Thus H_2 is the fastest. Choice (4) is correct.

20. (2) Gauge pressure is measured above atmospheric pressure; to use the ideal gas law, we convert it to absolute pressure by adding 1 atm.

Given:

$V = 0.03m^3, T_1 = 300K, T_2 = 290K$

Initial pressure $P_1 = 15 + 1 = 16$ atm Final pressure $P_2 = 11 + 1 = 12$ atm

Convert to Pa: 1atm = 1.013×10^5 Pa

Using $n = \frac{PV}{RT}$:

$$n_1 = \frac{16 \times 1.013 \times 10^5 \times 0.03}{8.31 \times 300} = 19.52\text{mol},$$

$$n_2 = \frac{12 \times 1.013 \times 10^5 \times 0.03}{8.31 \times 290} = 15.13\text{mol},$$

Oxygen withdrawn

$= (19.52 - 15.13) \times 32 = \boxed{140.5g} = \boxed{0.140kg}$

21. (4) The specific heat at constant pressure (C_p) is the amount of heat required to raise the temperature of one gram through 1°C when the pressure of the gas is kept constant. Again, the mean kinetic energy per molecule (3/2)kT depends only upon temperature. Clearly both the specific heats at constant pressure and mean kinetic energy are depending on the temperature which is again same for the two gases.

22. (1) Dalton's law of partial pressure gives total pressure P + P = 2P

Statement Based MCQs

1. **(1)** Both Statement-I and Statement-II are True.
2. **(4)** Statement-I is false but Statement-II is true
3. **(1)** Statement-I and Statement-II both are correct.
4. **(3)** Statement-I is true but Statement-II is false.
5. **(1)** Statement-I and Statement-II both are correct.

Assertion & Reason MCQs

1. **(3)** Assertion is true but Reason is false.
2. **(1)** Both Assertion and Reason are true and Reason is the correct explanation of the Assertion.
3. **(2)** Both Assertion and Reason are true but Reason is not the correct explanation of the Assertion.
4. **(2)** Both assertion and reason are true but reason is not the correct explanation of the assertion.
5. **(4)** Assertion is false but Reason is true.

Match the Columns MCQs

1. (3)

(a) Average speed $\rightarrow \sqrt{\frac{8RT}{\pi M}}$

(b) R.M.S speed $\rightarrow \sqrt{\frac{3RT}{M}}$

(c) Most probable speed $\rightarrow \sqrt{\frac{2RT}{M}}$

(d) Mean square speed $\rightarrow \frac{3RT}{M}$

2. (2)

(a) $\frac{C_P}{C_V} = \gamma + \frac{2}{n}$

(b) Energy associated with each degrees of freedom $\rightarrow \frac{kT}{2}$

(c) Pressure $\rightarrow \frac{nRT}{V}$ (n = no. of atoms/volume)

(d) Gay Lussac's law, $P \propto T$

NCERT Exemplar MCQs

1. **(2)** The motion of the vessel as a whole does not effect the relative motion of the gas molecules with respect to the walls of the vessel. Hence, pressure of the gas inside the vessel, as observed by us, on the ground remains the same.
2. **(4)** In an ideal gas, when a molecule collides elastically with a wall, the momentum transferred to each molecule will be twice the magnitude of its normal momentum.
   For the face EFGH, it transfers only half of that.
3. **(2)** Boyle's law is applicable when temperature is constant
   i.e., pV = nRT =constant
   $\Rightarrow$ pV = constant (at constant temperature)
   i.e., $p \propto \frac{1}{V}$
   [where , p = pressure, V= volume]
   So, this process can be called as isothermal process.
4. **(3)** Consider the diagram where an ideal gas is contained in a cylinder, having a piston of mass M. Friction is absent.
   The pressure inside the gas will be
   $p = p_a + Mg/A$
   where, p_a = atmospheric pressure
   A = area of cross-section of the piston.
   Mg = weight of piston
   Hence, p = constant.
   as $pv = nRT \Rightarrow$ volume (V) increases at constant pressure.
5. **(4)** When the molecules breaks into atoms, the number of moles would become twice.
   Now, by ideal gas equation
   P = Pressure of gas,
   n = Number of moles
   R = Gas constant,
   T = Temperature
   pV = nRT
   As volume (V) of the container is constant.
   As gases breaks, number of moles becomes twice of initial, so $n_2 = 2n_1$
   So, $p \propto nT$
   $$\Rightarrow \frac{p_2}{p_1} = \frac{n_2T_2}{n_1T_1} = \frac{(2n_1)(3000)}{n_1(3000)} = 20 \Rightarrow p_2 = 20p_1$$
   Hence, final pressure of the gas would be 20 times the pressure initially.
6. **(2)** For a function f(v), the number of molecules n = f(v), which are having speeds between v and v dv.
   For each function $f_1(v)$ and $f_2(v)$, n will be different, hence each function $f_1(v)$and $f_2(v)$ will obey the Maxwell's distribution law separately.
7. **(4)** We know for an ideal gas, pV = nRT (Ideal gas equation)
   $$n = \frac{pV}{RT}$$
   As number of moles of gas remains fixed, hence, we can write
   $$\frac{p_1V_1}{RT_1} = \frac{p_2V_2}{RT_2} \Rightarrow \frac{(p)(V)(1.1T)}{(1.05)V(T)}$$
   $[p_1 = p, V_2 = 1.05V \text{ and } T_2 = 1.1T]$
   $$= p \times \left(\frac{1.1}{1.05}\right) = p\,(1.0476) \approx 1.05\,p$$
   Hence, final pressure p_2 lies between p and 1.1p.

NEET & AIIMS Past Year MCQs

1. **(2)** According to Charle's law V $\propto$ T

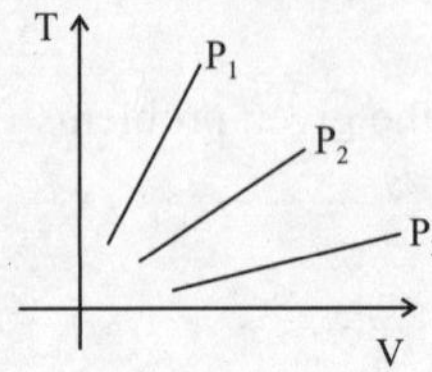

   $P_1 > P_2 > P_3$
2. **(3)** Given: $\gamma = \frac{C_p}{C_v}$
   For polyatomic gas:
   $C_v = (3 + f)R,\ C_p = (4 + f)R$
   $$\gamma = \frac{4+f}{3+f} \Rightarrow f = \frac{4-3\gamma}{\gamma-1}$$
3. **(3)** Ti = –50°C = 223 K
   $V_{rms} \propto \sqrt{T}$
   As v_{rms} increased by 3 times
   So $(v_{rms})_f = 4(v_{rms})_{initial}$
   $T_f = 16\,T_i = 16 \times 223 = 3568$ K
   $T_f = (3568 - 273)°C = 3295°C$
4. **(1)** Volume V = (no. of moles) (22.4 litre)
   $$\Rightarrow V = \frac{\text{Total mass of the molecules}}{\text{molar mass}}(22.4 \times 10^{-3}\,m^3)$$
   $$\Rightarrow V = \frac{4.5 \times 10^3}{18} \times 22.4 \times 10^{-3}\,m^3 = 5.6\,m^3$$
5. **(2)**
6. **(3)** $\frac{V_H}{V_o} = \sqrt{\frac{T_H}{T_O}}$
   $$= \sqrt{\frac{300}{675}} = \sqrt{\frac{4}{9}} = \frac{2}{3}$$
   $$\frac{V_H}{V_O} = \frac{2}{3} \Rightarrow 3V_H = 2V_O$$
7. **(2)** For monoatomic gas, $\gamma = \frac{C_p}{C_v} = \frac{5}{3}$
   We know that,

$\Delta Q = \mu C_p \Delta T$
$\Delta U = \mu C_v \Delta T$

$$\frac{\Delta U}{\Delta Q} = \frac{Cv}{Cp} = \frac{3}{5}$$

i.e. fraction of heat energy to increase the internal energy be 3/5

8. (4) Mole Fraction of O_2

$$= \frac{64/32}{64/32 + \frac{60}{20}} = \frac{2}{2+3} = \frac{2}{5}$$

$\therefore$ Partial Pressure of $O_2 = P_T \times$ Mole fraction

$$= 30 \times \frac{2}{5} = 12 \text{ bar}$$

9. (2) **10. (1)** **11. (3)**

Oscillations

NCERT Topic-wise MCQs

1. (4) Instantaneous displacement is

$x = 5\cos(2\pi t + \pi/4)$

put t = 1

$$x = 5\cos\left[2\pi + \frac{\pi}{4}\right] \qquad [\therefore \cos(2\pi + \theta) = \cos\theta]$$

$$\Rightarrow x = 5\cos\frac{\pi}{4} = \frac{5}{\sqrt{2}} \text{m}$$

2. (4) For SHM, $\frac{d^2y}{dt^2} \propto -y$

If $y = A\sin^2 \omega t$

$$\frac{dy}{dt} = (2A\sin\omega t\cos\omega t)\omega = A\omega\sin 2\omega t$$

$$\frac{d^2y}{dt^2} = 2A\omega^2\cos 2\omega t$$

i.e., $\frac{d^2y}{dt^2}$ is not proportional to $-y$.

3. (2) $y = a\sin\frac{2\pi}{T}t$

$$y = \frac{a}{2}$$

$$\frac{a}{2} = a\sin\frac{2\pi}{T}t$$

$$\frac{2\pi}{T}t = \frac{\pi}{6}$$

$$t = \frac{T}{12} = \frac{6}{12} = \frac{1}{2} \text{ sec}$$

4. (3) Phase difference $= \left(\omega t + \frac{3\pi}{6}\right) - \left(\omega t + \frac{\pi}{6}\right)$

$$\Rightarrow \omega t + \frac{3\pi}{6} - \omega t - \frac{\pi}{6} = \frac{\pi}{3}$$

5. (2) T = time period = 1 sec

Angular frequency $= \frac{2\pi}{T} = 2\pi$

Equation of motion of block

$N - mg = ma$

at extreme position

$a = -\omega^2 A$

separation will begin

when N = 0 i.e., (a = g)

$a = \omega^2 A = g$

$$\Rightarrow A = \frac{g}{\omega^2} = \frac{10}{4\pi^2}$$

$\Rightarrow A = 0.25$ m

6. (3) $v^2 = \omega^2(A^2 - x^2)$

$v^2 = \omega^2A^2 - \omega^2x^2$

$v^2 + \omega^2x^2 = \omega^2A^2$ [dividing whole equation by ω^2A^2]

$$\frac{v^2}{\omega^2A^2} + \frac{x^2}{A^2} = 1$$

It is an ellipse equation

7. (3) $y = A\sin\omega t - B\cos\omega t$

let $A = a\cos\theta$, $B = a\sin\theta$

$a^2 = A^2 + B^2$

$a = \sqrt{A^2 + B^2}$

8. (1) $\because x = A\sin\omega t$

$\Rightarrow v = A\omega\cos\omega t$

$$2 = A\frac{2\pi}{16}\cos\frac{2\pi}{16} \times 2$$

$$A = \frac{16\sqrt{2}}{\pi} = 7.2 \text{ m}$$

9. (4) Maximum acceleration of the particle executing SHM

$a_{max} = \omega^2 x$

$$x = \frac{a_{max}}{\omega^2}$$

$$x = \frac{7.5}{(3.5)^2} = 0.61 \text{ m}$$

10. (1) $\frac{V_1}{V_2} = \frac{\omega\sqrt{A^2 - Y_1^2}}{\omega\sqrt{A^2 - Y_2^2}}$

$$\frac{10}{7} = \frac{\sqrt{A^2 - 9}}{\sqrt{A^2 - 16}}$$

we get A = 4.76 cm $\simeq$ 4.8 cm

Length of path = 2A = 9.5 cm $\simeq$ 10 cm

11. (4) $x = 4(\cos\pi t + \sin\pi t)$

$$= 4\sqrt{2}\left[\frac{1}{\sqrt{2}}\cos\pi t + \frac{1}{\sqrt{2}}\sin\pi t\right]$$

$$= 4\sqrt{2}\left[\sin\frac{\pi}{4}\cos\pi t + \cos\frac{\pi}{4}\sin\pi t\right] = 4\sqrt{2}\sin\left(\pi t + \frac{\pi}{4}\right).$$

Standard equation of displacement is,

$x = A\sin(\omega t + \phi)$

Comparing the given equation with standard equation.

$A = 4\sqrt{2}$

12. (2) From $x = 0$ to $x = \frac{A}{2}$, Phase change = 30°

$\text{time} = \frac{T}{12} = 2\,\text{sec}$

13. (1) Time taken to go from x = 0 to x = A/2 is

T_1 and $T_1 = \frac{T}{12}$

Time taken to go from x = A/2 to x = A is $T_2 = T/6$

$\therefore T_1 < T_2$.

14. (3) Amplitude = 5m

time period will be

$T = \frac{2\pi}{\omega}$

$T = \frac{2\pi}{\pi}$

T = 2 sec.

15. (2) Acceleration of particle executing SHM at a displacement *x* from the mean position is given by, $= -k(x + a)$.

16. (3) $x = A\cos\omega t$

$x = A/2$

we get $\omega t = 60°$

$$\frac{KE}{PE} = \frac{\frac{1}{2}m\omega^2A^2\sin^2\omega t}{\frac{1}{2}m\omega^2A^2\cos^2\omega t} = \tan^2\omega t = \tan^2 60° = 3$$

17. (4) $T_1 = 2\pi\sqrt{\frac{M}{k}}$ ... (*i*)

When a mass m is placed on mass M, the new system is of mass = (M + m) attached to the spring. New time period of oscillation.

$T_2 = 2\pi\sqrt{\frac{(m+M)}{k}}$ ... (*ii*)

Consider v_1 is the velocity of mass M passing through mean position and v_2 velocity of mass (m + M) passing through mean position.

Using, law of conservation of linear momentum

$Mv_1 = (m + M)v_2$

$M(A_1\omega_1) = (m + M)(A_2\omega_2)$ $\quad (\because v_1 = A_1\omega_1 \text{ and } v_2 = A_2\omega_2)$

or $\frac{A_1}{A_2} = \frac{(m+M)}{M}\frac{\omega_2}{\omega_1}$

$= \left(\frac{m+M}{M}\right) \times \frac{T_1}{T_2}$ $\quad \left(\because \omega_1 = \frac{2\pi}{T_1} \text{ and } \omega_2 = \frac{2\pi}{T_2}\right)$

$\frac{A_1}{A_2} = \sqrt{\frac{m+M}{M}}$ [Using eqns (*i*) and (*ii*)]

18. (4) Let amplitude of S.H.M be A. At mean position the velocity of particle is maximum, so $\omega = \frac{V_{max}}{a} = \frac{16}{4} = 4$

$V = \omega\sqrt{A^2 - y^2} \Rightarrow 8\sqrt{3} = 4\sqrt{16 - y^2} \Rightarrow 2\sqrt{3} = \sqrt{16 - y^2}$

$\Rightarrow 12 = 16 - y^2 \Rightarrow y^2 = 4$ (squaring both side)

$\Rightarrow y = 2\text{cm}$

19. (4) Let the amplitude of the oscillation be A

For displacement $y = \frac{\sqrt{3}A}{2}$

$PE = \frac{1}{2}m\omega^2y^2 \Rightarrow \frac{1}{2}m\omega^2\left[\frac{\sqrt{3}A}{2}\right]^2$

$KE = \frac{1}{2}m\omega^2(A^2 - y^2) = \frac{1}{2}m\omega^2\left(A^2 - \left[\frac{\sqrt{3}A}{2}\right]^2\right)$

$$\therefore \frac{PE}{KE} = \frac{\frac{1}{2}m\omega^2 \times \frac{3A^2}{4}}{\frac{1}{2}m\omega^2\frac{A^2}{4}} = \frac{3}{1}$$

20. (1) $F = -m\omega^2x$, amplitude = 10 cm = 0.1 m

on comparing this with $F = -kx$

$k = m\omega^2$

$\omega = \sqrt{\frac{K}{m}} = \sqrt{\frac{1000}{10}} = 10\,\text{rad/s}$

$a_{max} = \omega^2x = (10)^2 \times \frac{10}{100} = 10\,\text{m/s}^2$

21. (4) Given, KE = PE

$\frac{1}{2}m\omega^2(A^2 - x^2) = \frac{1}{2}m\omega^2x^2$

$A^2 - x^2 = x^2$

$2x^2 = A^2$

$x = \frac{A}{\sqrt{2}}$

22. (1) $\because v = \omega\sqrt{A^2 - x^2}$

at x = A/2

$v = \omega \times \frac{\sqrt{3}A}{2}$

$v = \frac{\sqrt{3}}{2}v_0$

23. (3) Potential energy of spring

$U = \frac{1}{2}kx^2$

$U' = \frac{1}{2}K(nx)^2 = n^2\left[\frac{1}{2}Kx^2\right]$

$U' = n^2U$

24. (2) For 1st situation

$U = \frac{1}{2}K(10 \times 10^{-2})^2$ ... (*i*)

For 2nd situation

$$U' = \frac{1}{2}K\left(20\times10^{-2}\right)^2 \quad \text{... (ii)}$$

From (*i*) and (*ii*)

$U' = 4U$

25. (2) Potential energy, $U = \frac{1}{2}Kx^2$

$2U = kx^2 \quad [\because F = -kx]$

$\Rightarrow 2U = -Fx$

$\Rightarrow \frac{2U}{F} = -x \Rightarrow \frac{2U}{F} + x = 0$

26. (1) Time period $T = \frac{2\pi}{\omega} = \frac{2\pi}{2\pi}$

$2\pi t + \frac{\pi}{3} = \pi$

$2\pi t = \frac{2\pi}{3}$

$\therefore t = \frac{1}{3} = 0.33$ sec

27. (1) Initially t = 0

$y = 10\sin\frac{\pi}{3}$

$y = 10\times\frac{\sqrt{3}}{2} = 5\sqrt{3}$ m

$v = \frac{dy}{dt}$

$v = 60\cos(6t + \pi/3)$

At t = 0

$v = 60\times\cos\frac{\pi}{3}$

v = 30 m/s

28. (2) $y = A\sin(\omega t + \theta)$

A = 6 cm

m = 1 kg

$\omega = 100$ rad/sec

$KE_{max} = \frac{1}{2}m\omega^2A^2$

$KE_{max} = \frac{1}{2}\times1\times(100)^2\times\left(6\times10^{-2}\right)^2 = 18$ J

29. (3) The average acceleration of a particle performing SHM over one complete oscillation is zero. Because acceleration is a function of sine and average value of sin for 1 complete oscillation = 0

30. (3) Equivalent force constant of three springs,

$$k_{eq.} = \frac{k_1k_2}{k_1+k_2} = \frac{k\times2k}{3k} = \frac{2k}{3}$$

$$\therefore \text{Time period, } T = 2\pi\sqrt{\frac{m}{k_{eq.}}} = 2\pi\sqrt{\frac{m}{2k/3}} = 2\pi\sqrt{\frac{3m}{2k}}.$$

31. (3) $V_{max} = a_{max}$

$\omega A = \omega^2 A$

$\omega = 1$

$\frac{2\pi}{T} = 1$

$T = 2\pi$

T = 6.28 sec

32. (3) For free fall $g_{eff} = 0$, so frequency of oscillation will be zero

$\because f = \frac{1}{2\pi}\sqrt{\frac{g_{eff}}{\lambda}} \Rightarrow f = 0$

33. (4) Acceleration $= \frac{d^2x}{dt^2}$

In given equation

$2a + 32x = 0$

$a = -16x$

general eqn of SHM

$a = -\omega^2x$

we get

$\omega^2 = 16$

$\omega = 4$

$\frac{2\pi}{T} = 4$

$T = \frac{\pi}{2}$ sec

34. (2) $\frac{T_e}{T_P} = \sqrt{\frac{g_P}{g_e}}$

$\frac{2}{T_P} = \sqrt{\frac{1}{2}}$

$T_P = 2\sqrt{2}$ sec

35. (4) A spring-mass system has frequency

$\omega(t) = \sqrt{\frac{k}{m(t)}}$

As mass $m(t)$ decreases (ice melts), $\omega(t)$ increases.

Since no external demaping is mentioned, mechanical energy remains constant. But with decreasing mass, the same total energy gets distributed over a lighter mass, which leads to increase in amplitude (kinetic energy gets converted more easily into displacement).

36. (2) g_{eff} (in water) $= g\left[1-\frac{\sigma}{\rho}\right]$

$\frac{T}{T'} = \sqrt{\frac{g'}{g}} = \sqrt{\frac{g\left[1-\frac{\sigma}{\rho}\right]}{g}}$ ($\sigma = 1$ for water)

$T' = \sqrt{2}\,T \Rightarrow \frac{1}{\sqrt{2}} = \sqrt{1-\frac{1}{\rho}}$

$\rho = 2$

37. (2) When a bigger pendulum of time period $T_1 = 5T/4$ complete one vibration, the smaller pendulum will complete 5/4 vibration. It means the smaller pendulum will be leading, the bigger pendulum by phase T/4 sec $= \frac{\pi}{2}$ rad = 90°

38. (3) Time period of simple pendulum $= 2\pi\sqrt{\frac{L}{g}}$

$$\Rightarrow \frac{T_1}{T_2} = \sqrt{\frac{g_2}{g_1}}$$

$$g_1 = g,\ g_2 = \frac{g}{\left[1+\frac{h}{R}\right]^2}$$

h = R, we get

$$g_2 = \frac{g}{4}$$

$$\frac{T_1}{T_2} = \sqrt{\frac{g}{4g}} = \frac{1}{2} \Rightarrow \frac{T_2}{T_1} = \frac{2}{1}$$

39. (1) liquid density $= \frac{\rho}{g}$

upthrust $= V\frac{\rho}{8}g$

mg = Vρg

$$F_{net} = V\rho g - V\frac{\rho}{8}g$$

$$F_{net} = \frac{7}{8}V\rho g$$

effective value of g, $g = \frac{7}{8}g$

$$T' = 2\pi\sqrt{\frac{\ell}{g'}} \Rightarrow T' = 2\pi\sqrt{\frac{8\ell}{7g}} = \sqrt{\frac{8}{7}}T$$

40. (3) $F = -\frac{dU}{dx} = -8\sin 2x$

$$a = \frac{F}{m} = -8\sin 2x \qquad (\because m = 1\text{ kg})$$

For small oscillations, $\sin 2x \approx 2x$

i.e., $a = -16x$

Since, $a \propto -x$

the oscillations are SHM in nature.

$$\therefore T = 2\pi\sqrt{\left|\frac{m}{k}\right|} = 2\pi\sqrt{\frac{1}{16}} = \frac{\pi}{2}\text{sec.}$$

41. (3) Again, since the mass is decreasing slowly,

$$\omega(t) = \sqrt{\frac{k}{m(t)}} \Rightarrow \text{increases with time}$$

But if the system is losing mass through a hole, and that mass falls away, it may carry away energy. So mechanical energy is not conserved, and decreases slowly.

42. (2) $F = -m\omega^2 x$ so $\omega^2 = \frac{13.5}{1.5} = 9$

$$\Rightarrow \omega = 3$$

$$T = \frac{2\pi}{3}$$

43. (1) Its acceleration is g sin θ, since vehicle is accelerating a pseudo force mg sin θ will act on bob of the pendulum which cancel the sin θ component of weight of the bob. $g_{eff} = g\cos\alpha$

$$T = 2\pi\sqrt{\frac{L}{g_{eff}}} = 2\pi\sqrt{\frac{L}{g\cos\alpha}}$$

44. (4) Velocity of S.H.M. at mean position = Aω

$$\Rightarrow 12 = A\omega \Rightarrow 12 = 2\times\omega \Rightarrow \omega = \frac{12}{2} = 6$$

$$\text{Frequency} = \frac{\omega}{2\pi}$$

$$f = \frac{6}{2\pi} = \frac{3}{\pi}\quad f = \frac{3}{\pi}$$

High Order Time Intensive MCQs

1. (4) The time period of a pendulum in air,

$$T = 2\pi\sqrt{\frac{l}{g}} \qquad \text{... (i)}$$

l being the length of simple pendulum.

In liquid, effective weight of sphere

Apparent weight w' = weight of bob in air – upthrust

Apparent weight $w' = (\rho - \rho')Vg$

where, ρ' = density of sphere $= \frac{\rho}{10}$

where, ρ = density of liquid.

$$\because w' = mg_{eff} = V\rho g_{eff}$$

$$\therefore \quad g_{eff} = \left(\frac{\rho - \rho/10}{\rho}\right)g = \frac{9}{10}g$$

Thus, $T' = 2\pi\sqrt{\frac{l}{g_{eff}}} = 2\pi\sqrt{\frac{l}{\frac{9}{10}g}}$

$$\frac{T'}{1} = \sqrt{\frac{10}{9}} \Rightarrow T' = \sqrt{\frac{10}{9}}T$$

2. (4) Time period $T = 2\pi\sqrt{\frac{m}{K}} = 2\times\frac{22}{7}\sqrt{\frac{0.1}{100}} = 0.198$ sec.

$$\text{time} = \frac{2}{3}T = \frac{2}{3}\times 0.198 = 0.132\text{ sec.}$$

3. (4) Frictional force ≥ Restoring force

$$\mu mg \ge KA$$

$$\frac{\mu g}{A} \ge \frac{K}{m}$$

$$f = \frac{1}{2\pi}\sqrt{\frac{K}{m}}$$

$$f = \frac{1}{2\pi}\sqrt{\frac{\mu g}{A}}$$

4. (4) The reduced mass,

$$\mu = \frac{(m)(m)}{m+m} = \frac{m}{2}$$

The given system is equivalent to a system of a particle of mass m/2 connected to a spring of stiffness k rigidly.

The required period of oscillation,

$$T = 2\pi\sqrt{\frac{\mu}{k}} = 2\pi\sqrt{\frac{m/2}{k}} = \pi\sqrt{\frac{2m}{k}}$$

5. **(2)** Given, potential energy $V(x) = A(1\text{-}\cos px)$.

The force experienced by the particle is given as

$$F = -\frac{dV}{dx} = -Ap\sin px$$

For small oscillations, we have

$F \approx -Ap^2x$ [$\because$ $\sin px \approx px$]

Hence, the acceleration would be given by

$$a = \frac{F}{m} = -\frac{Ap^2}{m}x$$

Also, we know that

$$a = \frac{F}{m} = -\omega^2 x$$

$$\Rightarrow \omega = \sqrt{\frac{Ap^2}{m}} \Rightarrow T = \frac{2\pi}{\omega} = 2\pi\sqrt{\frac{m}{Ap^2}}$$

6. **(1)** As Young's modulus Y is given by

$$Y = \frac{\text{Stress}}{\text{Strain}} = \frac{mg/A}{l/L} \quad (\because mg = kl)$$

$$\Rightarrow mg = \frac{YAl}{L}$$

So, $kl = \frac{YAl}{L} \Rightarrow k = \frac{YA}{L}$

Now, frequency is given by

$$n = \frac{1}{2\pi}\sqrt{\frac{k}{m}} = \frac{1}{2\pi}\sqrt{\left(\frac{YA}{mL}\right)}$$

7. **(1)** Displacement of damped oscillator is given by

$x = x_m e^{-bt/2m} \sin(\omega' t + \phi)$,

where, ω' = angular frequency of damped oscillator

$$\omega_0 = \sqrt{\frac{k}{m}}$$

$$\omega' = \sqrt{\omega_0^2 - \left(\frac{b}{2m}\right)^2} = \sqrt{\frac{k}{m} - \frac{b^2}{4m^2}}$$

8. **(4)** Speed of particle executing SHM is given by

$v = \omega\sqrt{a^2 - x^2}$ and $\omega = \frac{2\pi}{T}$

where, x = displacement = $a/2$

$$\Rightarrow v = \frac{2\pi}{T}\sqrt{a^2 - \left(\frac{a}{2}\right)^2} \quad \Rightarrow \quad v = \frac{2\pi}{T}\left(\frac{\sqrt{3}a}{2}\right) = \frac{\sqrt{3}\pi a}{T}$$

9. **(1)** The angular frequency after removing the mass m_1 is given by,

$$\omega = \sqrt{\frac{k}{m_2}}$$

This is because as m_1 is removed, the system will oscillate with the same frequency as the system with only mass m_2.

10. **(2)** Let block is displaced through x meter, then weight of the water displaced or upthrust (upwards) $= -Ax\rho g$ [$\because f = mg = (Ax\rho)g$]

where, A is area of cross-section of the block and ρ is its density. This must be equal to force applied, where m is mass of the block and a is acceleration.

$\therefore ma = -Ax\rho g$

$$\Rightarrow \quad a = -\frac{A\rho g}{m}x = -\omega^2 x$$

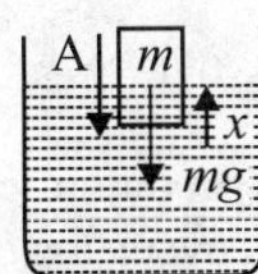

This is the equation of simple harmonic motion.

Time period of oscillation,

$$T = \frac{2\pi}{\omega} = 2\pi\sqrt{\frac{m}{A\rho g}} \Rightarrow T \propto \frac{1}{\sqrt{A}}$$

$$\left[\because T = 2\pi\sqrt{\frac{\text{displacement}}{\text{acceleration}}}\right]$$

11. **(1)** The frequency of K.E. is twice that of a particle executing S.H.M.

12. **(2)** μ = reduced mass

$$\mu = \frac{m_1 m_2}{m_1 + m_2} = 2\text{kg}$$

$$T = 2\pi\sqrt{\frac{\mu}{k}} = \frac{\pi}{10}\text{ sec}$$

$$A \propto \frac{1}{m} \quad \frac{A_3}{A_6} = \frac{6}{3} = \frac{2}{1}$$

$\therefore A_3 = 4\text{cm}$

and $A_6 = 2$ cm

13. **(3)** Radius of circular path in the horizontal plane ΔOAB, $r = l\sin\theta$

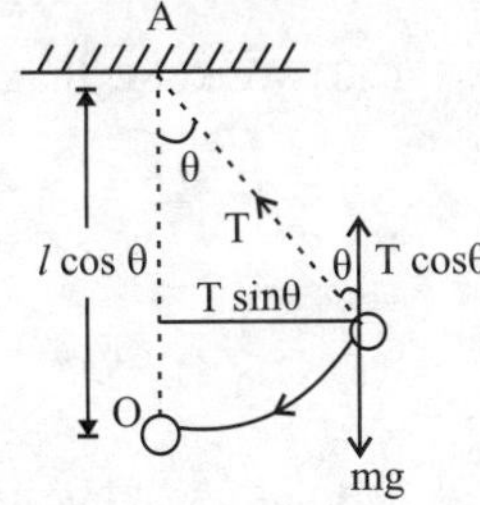

Resolving T along the vertical and horizontal directions, we get

$T\cos\theta = Mg$ ...(*i*)

$T\sin\theta = mr\omega^2 = M(l\sin\theta)\,\omega^2$

$\Rightarrow \quad T = Ml\omega^2$ ...(*ii*)

On dividing Eq. (*ii*) by Eq. (*i*), we get

$$\frac{1}{\cos\theta} = \frac{l\omega^2}{g} \Rightarrow \omega^2 = \frac{g}{l\cos\theta} \quad \Rightarrow t = \frac{2\pi}{\omega} = 2\pi\sqrt{\frac{l\cos\theta}{g}}$$

14. **(3)** Time period, $T = 2\pi\sqrt{\frac{M}{k}}$ ... (*i*)

and $T' = 2\pi\sqrt{\frac{M+m}{k}} \Rightarrow \frac{5T}{3} = 2\pi\sqrt{\frac{M+m}{k}}$ ... (*ii*)

On dividing Eq. (*i*) by Eq. (*ii*), we have

$$\frac{3}{5} = \sqrt{\frac{M}{M+m}}$$

After squaring $\frac{9}{25} = \frac{M}{M+m}$

$16M = 9m$

$$\frac{m}{M} = \frac{16}{9}$$

15. (4) The simple pendulum at angular amplitude θ_0 is shown in figure.

Maximum tension in the string is

$$T_{max} = mg + \frac{mv^2}{l} \quad ...(i)$$

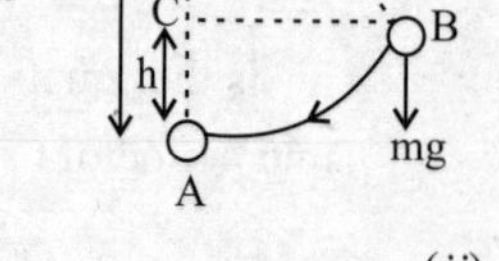

When bob of the pendulum comes form A to B, it covers a vertical distance h

$$\therefore \quad \cos\theta_0 = \frac{l-h}{l}$$

$$\Rightarrow h = l(1 - \cos\theta_0) \quad ... (ii)$$

Also, during B to A, potential energy of bob converts into kinetic energy i.e., $mgh = \frac{1}{2}mv^2$

$$\therefore \quad v = \sqrt{2gh} \quad ... (iii)$$

Thus, using Eqs. (*i*), (*ii*) and (*iii*), we obtain

$$T_{max} = mg + \frac{2mg}{l} l(1 - \cos\theta_0)$$

$$= mg + 2mg\,(1 - \cos\theta_0)$$

$$= mg\,(3 - 2\cos\theta_0)$$

16. (4) Effective spring constant, $k = 2k$

Frequency of oscillation, $f = \frac{1}{2\pi}\sqrt{\frac{2k}{m}}$

When one spring is removed, then spring factor = k

New frequency of oscillation will be constant

$$f' = \frac{1}{2\pi}\sqrt{\frac{k}{m}} = \frac{f}{\sqrt{2}}$$

17. (3) We know that,

$$T = 2\pi\sqrt{\frac{m}{k}}$$

$$\frac{\ell_1}{\ell_2} = \frac{1}{3}, \ell_1 + \ell_2 = \ell \quad k_1\frac{\ell}{4} = k_2\left(\frac{3\ell}{4}\right)$$

When spring is cut into ratio 1 : 3, the new time constant is

$$k' = \frac{4}{3}k$$

$$T' = \frac{\sqrt{3}}{2}T$$

18. (4) The displacement of particle, executing SHM,

$$y = 5\sin\left(4t + \frac{\pi}{3}\right) \quad ... (i)$$

Velocity of particle,

$$\left(\frac{dy}{dt}\right) = \frac{5d}{dt}\sin\left(4t + \frac{\pi}{3}\right) = 5\cos\left(4t + \frac{\pi}{3}\right)\cdot 4$$

$$= 20\cos\left(4t + \frac{\pi}{3}\right)$$

Velocity at $t = \left(\frac{T}{4}\right)$

$$\left(\frac{dy}{dt}\right)_{t=\frac{T}{4}} = 20\cos\left(4 \times \frac{T}{4} + \frac{\pi}{3}\right)$$

$$\text{or} \quad v = 20\cos\left(T + \frac{\pi}{3}\right) \quad ... (ii)$$

Comparing the given equation with standard equation of SHM $y = a\sin(\omega t + \phi)$,

$$\text{As, } \omega = \frac{2\pi}{T} \quad \text{or} \quad T = \frac{2\pi}{\omega}$$

$$T = \frac{2\pi}{4} \quad \text{or} \quad T = \left(\frac{\pi}{2}\right)$$

Now, putting value of T in eqn. (*ii*), we get;

$$v = 20\cos\left(\frac{\pi}{2} + \frac{\pi}{3}\right) = -20\sin\frac{\pi}{3}$$

$$= -20 \times \frac{\sqrt{3}}{2} = -10 \times \sqrt{3}$$

The kinetic energy of particle,

$$KE = \frac{1}{2}mv^2 = \frac{1}{2} \times 2 \times 10^{-3} \times (-10\sqrt{3})^2 = 0.3 \text{ J.}$$

19. (1) $\because x = A\cos\omega t$

$$\frac{A}{2} = A\cos\frac{2\pi}{T}t \Rightarrow \frac{\pi}{3} = \frac{2\pi t}{T} \Rightarrow t = \frac{T}{6}$$

20. (2) $l_1 + l_2 = l$

$l_2 = nl_1$

$$\ell_1 = \frac{\ell}{n+1}$$

$Kl = K_1 l_1$

$K_1 = (n + 1)\,K$

$$T = 2\pi\sqrt{\frac{m}{K_1}} = 2\pi\sqrt{\frac{m}{(n+1)K}}$$

21. (3) Since $\omega(t) = \sqrt{k/m(t)}$, and mass decreases gradually, the frequency increases non-linearly (inverse square-root dependence), giving a gradually increasing curve that flattens over time.

22. (4) In Ist sec. $\quad x = 0$ to $x = \frac{A}{\sqrt{2}}$

$$\Rightarrow d_1 = \frac{A}{\sqrt{2}}$$

In 2nd sec. $\quad x = \frac{A}{\sqrt{2}}$ to $x = A$

$$d_2 = A - \frac{A}{\sqrt{2}}$$

$$\therefore \frac{d_1}{d_2} = \frac{\frac{A}{\sqrt{2}}}{A - \frac{A}{\sqrt{2}}} = \sqrt{2} + 1$$

23. (4) $\frac{T_1}{T_2}=\sqrt{\frac{g_2}{g_1}}$ $g_2=g+\frac{g}{4}$

and $g_1 = g$

$\Rightarrow \frac{T_1}{T_2}=\sqrt{\frac{5}{4}}$

Given, $T_1=\sqrt{5}$

$\Rightarrow T_2 = 2$ sec

24. (1) $T\propto\frac{1}{\sqrt{K}}$ Let the equivalent spring constant in the three cases are K_1, K_2 and K_3

$\Rightarrow K_1 = K,\ K_2=\frac{K}{2}\ \&\ K_3 = 2K$

25. (2) $v=\omega\sqrt{A^2-x^2};\ v^2=\omega^2\left(A^2-\frac{A^2}{16}\right)$

$v^2=\omega^2A^2\times\frac{15}{16};\ K.E.=\frac{1}{2}mv^2$

$\therefore$ Fraction of $K.E.\rightarrow 15/16$

26. (2) $\because F = kA$ and $F = mg \Rightarrow A=\frac{mg}{K}=0.25$ m

27. (3) $k = k_1 + k_2$

$\Rightarrow \omega^2=\omega_1^2+\omega_2^2$

$\Rightarrow \frac{1}{T^2}=\frac{1}{T_1^2}+\frac{1}{T_2^2}$

$\frac{1}{T^2}=\frac{25}{144}\Rightarrow T=2.4$ s

28. (4) At mean position F = 0, v → max. Acceleration is max at extremes

Statement Based MCQs

1. **(1)** Both the Statements are true.

   Tension in the string $T = mg\cos\theta+\frac{mv^2}{r}$.

   Hence tension and velocity are maximum at mean position, when $\theta = 0$.

2. **(1)** Both the Statements are true.

   Total energy of SHM = K.E of particle + P.E of particle

   $=\frac{1}{2}mw^2A^2$

3. **(3)** Statement-I is true but Statement-II is false. At mean position, the displacement of the particle is zero.

   As $a\ \alpha - x$

   $\therefore a_{mean}=0$

   Time-period of simple pendulum $T=2\pi\sqrt{\frac{l}{g}}$, where g is the acceleration due to gravity.

4. **(2)** Both the Statements are false.

Assertion & Reason MCQs

1. **(1)** Both Assertion and Reason are true but Reason is the correct explanation of the Assertion.
2. **(2)** Acceleration is in opposite direction to the displacement. At the extreme position velocity of bob is zero but acceleration is maximum.
3. **(2)** $K.E=\frac{1}{2}mw^2(A^2-y^2)$

   $P.E=\frac{1}{2}mw^2y^2$

   For K.E = P.E

   $\Rightarrow a^2 = 2y^2$

   $\Rightarrow y=\frac{a}{\sqrt{2}}$

4. **(4)** Assertion is false but Reason is true.
5. **(2)** The motion of earth around the sun repeats itself in a regular pattern, making it periodic. As it is not a to and fro motion, so not oscillatory motion

Match the Columns MCQs

1. **(2)** A constant force and a constant torque affect only the mean position. In third case, as the block falls on mean position, the mean position is not affected.

   In a car a constant pseudo force will act which will affect only the mean position.

2. **(1)** Position and velocity both can be positive and negative. If at mean position, PE is zero, then at any other position, PE is positive. Total energy is always constant.
3. **(4)**

   (i) Linear combination Linear of two SHMs will be an SHM if the individual SHMs have equal frequencies, their magnitudes may be different.

   (ii) $y = A\sin\omega_1 t + A\sin(\omega_2 t+\phi)$

   If $\omega_1=\omega=\omega,\ \phi=\pi/2$

   $y = A\sin\omega t + A\cos\omega t$

   Its amplitude will be $\sqrt{2}A$.

   (iii) Time period of a pendulum of infinite length is $T=2\pi\sqrt{\frac{R}{g}}$

   (iv) The above will be the maximum value of time period of an oscillating pendulum.

4. **(3)** $\sin\omega t-\cos\omega t=\sqrt{2}\sin\left(\omega t-\frac{\pi}{4}\right)$

   $\therefore T=\frac{2\pi}{\omega}$

   $\Rightarrow$ It is SHM with time period $\frac{2\pi}{\omega}$

   $\cos\left(\frac{\pi}{4}-2\omega t\right)$ represents SHM with time period

   $T=\frac{2\pi}{2\omega}=\frac{\pi}{\omega}$

$1 + \omega t + \omega^3 t^3$ is non periodic

$\sin^3\omega t = \frac{1}{4}[3\sin\omega t - \sin 3\omega t]$ is periodic but not SHM.

5. (2) For parallel combination of springs

$k_{eq} = k_1 + k_2$

For series combination of springs

$k_{eq} = \frac{k_1 k_2}{k_1 + k_2}$

For option A, $k_{eq} = \frac{k \times k}{k + k} = \frac{k}{2}$

$\therefore T = 2\pi\sqrt{\frac{M}{k_{eq}}} = 2\pi\sqrt{\frac{2M}{k}}$

For option B, $k_{eq} = k + k = 2k$

$\therefore T = 2\pi\sqrt{\frac{M}{k_{eq}}} = 2\pi\sqrt{\frac{M}{2k}}$

For option C, $k_{eq} = \frac{k}{2} + k = \frac{3k}{2}$

$\therefore T = 2\pi\sqrt{\frac{2M}{3k}}$

For option D, $k_{eq} = \frac{k}{2} + \frac{k}{2} = k$

$\therefore T = 2\pi\sqrt{\frac{M}{k}}$

NCERT Exemplar MCQs

1. (2) Given, $y = 3\cos\left(\frac{\pi}{4} - 2\omega t\right)$

Velocity of the particle

$$v = \frac{dy}{dt} = \frac{d}{dt}\left[3\cos\left(\frac{\pi}{4} - 2\omega t\right)\right]$$

$$= 3 - (2\omega)\left[-\sin\left(\frac{\pi}{4} - 2\omega t\right)\right]$$

$$= 6\omega\sin\left(\frac{\pi}{4} - 2\omega t\right)$$

As acceleration, $a = \frac{dv}{dt} = \frac{d}{dt}\left[6\omega\sin\left(\frac{\pi}{4} - 2\omega t\right)\right]$

$$= 6\omega \times (-2\omega)\cos\left(\frac{\pi}{4} - 2\omega t\right)$$

$$= -12\omega^2\cos\left(\frac{\pi}{4} - 2\omega t\right)$$

$$= -4\omega^2\left[3\cos\left(\frac{\pi}{4} - 2\omega t\right)\right]$$

$\Rightarrow \quad a = -4\omega^2 y$

$\Rightarrow \quad$ As acceleration, $a \propto -y$

Hence, due to negative sign motion is SHM.

Clearly, from the equation

$\omega' = 2\omega$

[$\because$ Standard equation $y = a\cos(\omega t + \phi)$]

$\Rightarrow \quad \frac{2\pi}{T'} = 2\omega \Rightarrow T' = \frac{2\pi}{2\omega} = \frac{\pi}{\omega}$

[and given equation $y = 3\cos\left(-2\omega t + \frac{\pi}{4}\right)$]

So, motion is S.H.M with period $\frac{\pi}{\omega}$.

2. (2) Given equation of motion is

$y = \sin^3\omega t = (3\sin\omega t - 4\sin 3\omega t)/4$

$\left[\because \sin 3\theta = 3\sin\theta - 4\sin^3\theta\right]$

$\Rightarrow \quad \frac{dy}{dt} = \left[\frac{d}{dt}(3\sin\omega t) - \frac{d}{dt}(4\sin 3\omega t)\right]/4$

$\Rightarrow \quad 4\frac{dy}{dt} = 3\omega\cos\omega t - 4\times\left[3\omega\cos 3\omega t\right]$

$\Rightarrow \quad 4\times\frac{d^2y}{dt^2} = \left(-3\omega^2\sin\omega t + 12\omega\sin 3\omega t\right).3\omega$

$\Rightarrow \quad \frac{d^2y}{dt^2} = -\frac{3\omega^2\sin\omega t + 36\omega^2\sin 3\omega t}{4}$

$\Rightarrow \quad \frac{d^2y}{dt^2}$ is not proportional to y.

Hence, motion is not S.H.M.

As the expression is involving sine function, hence it will be periodic.

3. (4) For motion to be S.H.M acceleration of the particle must be proportional to negative of displacement.

i.e., $a \propto -(y \text{ or } x)$

We should be clear that y has to be linear.

4. (3) Consider the diagram in which a liquid column oscillates. In this case, restoring force acts on the liquid due to gravity. Acceleration of the liquid column, can be calculated in terms of restoring force.

Restoring force, f = Weight of liquid column of height 2 y

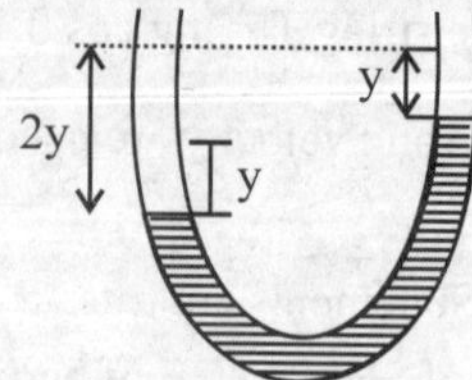

$\Rightarrow \quad f = -(A \times 2y + \rho) \times g = -2A\rho g y$

$\Rightarrow \quad f \propto -y \Rightarrow$ Motion is SHM with forces constant $k = 2A\rho g$.

$\Rightarrow$ Time period

$$T = 2\pi\sqrt{\frac{m}{k}} = 2\pi\sqrt{\frac{A \times 2h \times \rho}{2A\rho g}} = 2\pi\sqrt{\frac{h}{g}}$$

$T = 2\pi\sqrt{\frac{l}{g'}}$, where $l = h$

Which is independent of the density of the liquid.

5. (3) Given, x = a cos ωt ... (*i*)

y = a sin ωt ... (*ii*)

Squaring and adding Eqs. (*i*) and (*ii*),

$x^2 + y^2 = a^2(\cos^2\omega t + \sin^2\omega t) = a^2$

$\Rightarrow x^2 + y^2 = a^2$

$[\because \cos^2 \omega t + \sin^2 \omega t = 1]$

This is the equation. of a circle

Clearly, the locus is a circle of constant radius a.

6. (4) According to the question, the displacement

Y= a sin ωt + b cos ωt

Let a = A sinϕ and b = A cosϕ

Now, $a^2 + b^2 = A^2 \sin^2 \phi + A^2 \cos^2 \phi$

$\Rightarrow A = \sqrt{a^2 + b^2}$

Thus, $y = A \sin(\omega t + \phi)$

$$\frac{dy}{dt} = A\omega \cos(\omega t + \phi)$$

$$\frac{d^2y}{dt^2} = -A\omega^2 \sin(\omega t + \phi) = -Ay\omega^2 = (-A\omega^2)y$$

$$\Rightarrow \quad \frac{d^2y}{dt^2} \propto (-y)$$

Hence, it is an equation of S.H.M with amplitude $A = \sqrt{a^2 + b^2}$.

7. (2) According to the question, A is given a transverse displacement.

Through the elastic support the disturbance is transferred to all the pendulums.

A and C are having same length, hence they will be in resonance, because their time period of oscillation.

$T = 2\pi\sqrt{\frac{l}{g}}$ and hence, frequency is same.

So, amplitude of A and C will be maximum.

***Note*:** In this problem, we have assumed that the support is perfectly elastic and there is no damping Hence, oscillation is considered as undamped.

8. (1) Let angular velocity of the particle executing circular motion is ω and when it is at Q makes an angle θ as shown in the diagram.

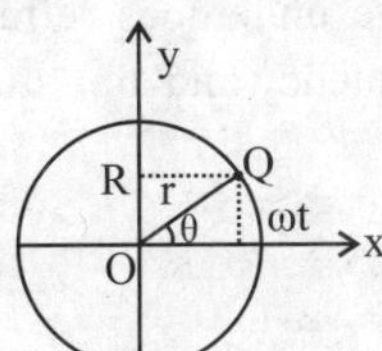

Clearly, θ = ωt

Now, we can write OR = OQ cos (90 – θ)

= OQsinθ = OQsinωt = r sin ωt $[\because OQ = r]$

$\Rightarrow \quad x = r \sin \omega t = B \sin \omega t$

$[\because r = B]$

$$x = B\sin\frac{2\pi}{T}t = B\sin\left(\frac{2\pi}{30}t\right)$$

Clearly, this equation represents S.H.M.

9. (3) As the given equation is

X = a cos $(\alpha t)^2$ is a cosine function. Hence, it is an oscillatory motion.

Now, putting t + T in place of t

$x(t + T) = a \cos[\alpha(t + T)]^2$ $\quad [\because x(t) = a\cos(\alpha t)^2]$

$= a\cos\left[\alpha t^2 + \alpha T^2 + 2\alpha t T\right] \neq x(t)$

where, T is supposed as period of the function ω(t).

Hence, it is not periodic.

10. (1) Let equation of an S.H.M is represented by

y = a sinωt

$$v = \frac{dy}{dt} = a\omega \cos \omega t$$

$\Rightarrow \quad (v)_{max} = a\omega = 30$ ... (*i*)

$$\text{Acceleration, } (A) = \frac{dx^2}{dt^2} = -a\omega^2 \sin \omega t$$

$A_{max} = \omega^2 a = 60$

Eqs. (*i*) and (*ii*), we get,

$\Rightarrow \omega(\omega a) = 60$

$\Rightarrow \omega(30) = 60$

$\Rightarrow \quad \omega = 2$ rad/s

$\Rightarrow \quad \frac{2\pi}{T} = 2$ rad/s $\Rightarrow T = \pi$ sec

NEET & AIIMS Past Year MCQs

1. (3) For a particle in SHM with amplitude *A*:

Total energy $E = \frac{1}{2}kA^2$.

When *KE* = *PE*:

$$\frac{1}{2}kx^2 = \frac{1}{2}\left(\frac{1}{2}kA^2\right)$$

$$x^2 = \frac{A^2}{2} \Rightarrow x = \frac{A}{\sqrt{2}}$$

2. (4) $m' = 3m$

$$l' = \frac{1}{2}l$$

$$T = 2\pi\sqrt{\frac{l}{g}}$$

$$T' = 2\pi\sqrt{\frac{l'}{g}} = 2\pi\sqrt{\frac{l}{2g}}$$

$$T' = \left(\frac{1}{\sqrt{2}}\right)2\pi\sqrt{\frac{l}{g}} = \frac{1}{\sqrt{2}}T$$

$$T' = \frac{x}{2}T$$

$$\frac{1}{\sqrt{2}}T = \frac{x}{2}T$$

$$x = \frac{2}{\sqrt{2}} = \sqrt{2}$$

3. (4) Keeping the battery connected

V remains same

$C = \frac{\varepsilon_0 A}{d}$, d is reduced

$\therefore$ C increases

From

Q = CV

As C↑ Q↑ $\Rightarrow$ A is correct

$E = \frac{1}{2}CV^2$ As C↑ E increases

$\Rightarrow$ B is incorrect

Capacitance increases $\Rightarrow$ C is correct

Product of charge and voltage increases

$\Rightarrow$ E is correct.

4. (1) From x – t graph,

A = 1, T = 8

$\Rightarrow \omega = \frac{2\pi}{T} \Rightarrow \omega = \frac{\pi}{4}$

at t = 2, x = 1

$a = -\omega^2 x \Rightarrow a = \frac{-\pi^2}{16} \times 1$

$\Rightarrow a = \frac{-\pi^2}{16} m/s^2$

5. (2) Let the time period of pendulums having longer length and shorter length arc T_ℓ and T_s.

$\Rightarrow n(T)_\ell = (n + 1)T_s$

$(n)2\pi\sqrt{\frac{1.21}{g}} = (n+1)2\pi\sqrt{\frac{1}{g}}$

(n)(1.1) = (n + 1)

n = 10

No. of oscillation of smaller one = n + 1 = 11

6. (1) $X = A \sin(\omega t) = A \sin(2\pi nt)$

$U = \frac{1}{2}kx^2 = \frac{1}{2}KA^2 \left[\frac{1-\cos(2\pi(2n)t)}{2}\right]$

Hence frequency of U (Potential energy) = 2n

7. (3) $F = kx \Rightarrow K = \frac{1000}{5} = 200 N/m$

As $T = 2\pi\sqrt{\frac{M}{K}} \Rightarrow T = \frac{2\pi}{10} = 0.628s$

8. (2) $E = \frac{1}{2}m\omega^2 A^2$

$= \frac{1}{2} \times 1 \times 25 \times 10^6 \times 4 \times 10^{-4}$

$= 50 \times 10^2 = 5$ kJ

9. (4) Apply conservation of energy at B & C

$K_B + P_B = K_C + P_C$

$0 + mgl(1 - \cos\theta_0) = \frac{1}{2}mu^2 + 0$

$u^2 = 2gl(1 - \cos\theta_0)$ ... (*i*)

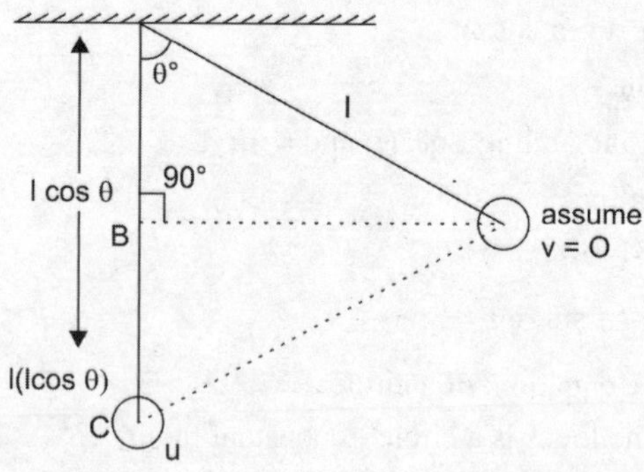

at point C

$T - mg = \frac{mu^2}{l}$

$T = mg + \frac{m}{\ell} \times 2g\ell(1 - \cos\theta_0)$

$T = mg + 2mg\left[1 - 1 + \frac{\theta_0^2}{2}\right]$

[Since θ_0 is small]

$\Rightarrow T = mg[1 + \theta_0^2]$

10. (1) T = $T = 2\pi\sqrt{\frac{M}{K}}$ ...(*i*)

$T'' = \frac{5}{4}T = 2\pi\sqrt{\frac{M+m}{K}}$ ...(*ii*)

$\frac{5}{4} \times 2\pi\sqrt{\frac{M}{K}} = 2\pi\sqrt{\frac{M+m}{K}}$

Squaring both sides $\frac{25}{16}\frac{M}{K} = \frac{M+m}{K}$

25M = 16M + 16M 9 M = 16m

$\frac{m}{M} = \frac{9}{16}$

11. (4) Loss in K.E = gain in P.E = $\frac{1}{2}Kx^2$

12. (1)

Waves

NCERT Topic-wise MCQs

1. (4) When a wave undergoes refraction, then its velocity changes, but frequency and amplitude remains unchanged.

2. (3)

3. (1) $v = n\lambda$ [$\therefore$ n = frequency]

25 = n × 100

n = 0.25

$T = \frac{1}{n} = \frac{1}{0.25} = 4$ sec

4. (2) $\omega = 100$, K = 20

velocity of waves = $\frac{\omega}{K} = \frac{100}{20} = 5$ m/s

5. (3) Wave travelling along y axis is given by $x = A\sin(ky - \omega t)$

6. (4) $(v_p)_{max} = A\omega = A \times 2\pi\nu$

wave velocity $v = \frac{\omega}{k} = \lambda\nu$

As $(v_p)_{max} = 3\nu$

$A \times 2\pi\nu = 3\lambda\nu \Rightarrow \lambda = \frac{2\pi A}{3}$

7. (1) Given, $y = A\sin(100\pi t - 3x)$

Equation of a progressive wave is given by

$y = A\sin(wt - kx) \Rightarrow K = 3$

$\Rightarrow \lambda = \frac{2\pi}{3} \quad \because K = \frac{2\pi}{\lambda}$

$\phi = \frac{\pi}{3}$

$\frac{2\pi}{\lambda}x = \frac{\pi}{3}$

$x = \frac{\pi}{9}\text{m}\left[\text{put } \lambda = \frac{2\pi}{3}\right]$

8. (1) $v_P = \frac{dy}{dt}$ = slope of wave ⇒ slope at A and B is zero.

Their velocity at A and B same. Distance between A and B is λ/2.

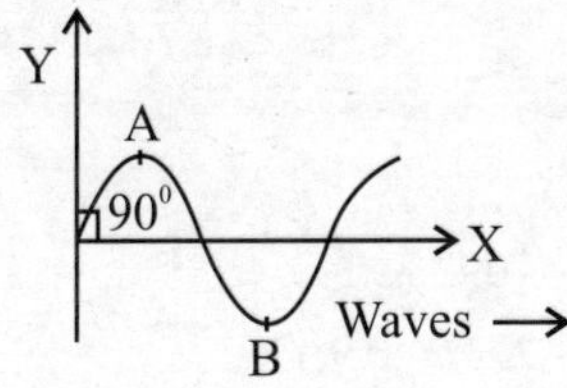

9. (1) $(v_p)_{max} = A\omega$

$v_w = \frac{\omega}{K}$

$\frac{(v_p)_{max}}{v_w} = KA$

10. (4) Velocity of a progressive wave $= \frac{\omega}{K} = \frac{8\pi/2}{\frac{\pi}{2\times 8}}$

= 64 cm/sec

11. (1) Phase difference $= \frac{2\pi}{\lambda}$ × (Path difference)

$\Rightarrow \quad \frac{\pi}{3} = \frac{2\pi}{\lambda}$ × (Path difference)

$\Rightarrow \quad \text{path difference} = \frac{\lambda}{6}$

12. (4) $\because \phi = \frac{2\pi}{\lambda} \times x = \frac{2\pi}{\lambda} \times \frac{7\lambda}{2} = 7\pi$

13. (3) Frequency of waves $n = \frac{54}{60}$ per second

$\lambda = 10$ m

$\therefore v = n\lambda = \frac{54}{60} \times 10 = 9$ m / sec.

14. (3) $v = \frac{d}{t} = \frac{20}{0.5} = 40$ m/s

15. (1) Liner mass density $(\mu) = \frac{M}{L} = \frac{10\times10^{-3}}{1} = 10^{-2}$

Speed of wave $= \sqrt{\frac{T}{\mu}} = \sqrt{\frac{100}{10^{-2}}} = 100$ m/s

16. (4) $f = \frac{1}{2\pi}\sqrt{\frac{T}{\mu}}$ ... (*i*)

Now $f' = 3f$

$3f = \frac{1}{2\pi}\sqrt{\frac{T'}{\mu}}$ ... (*ii*)

from (*i*) and (*ii*)

$T' = 3^2 T = 9T$

17. (1) Distance travelled by the wave in four round trips 32 m

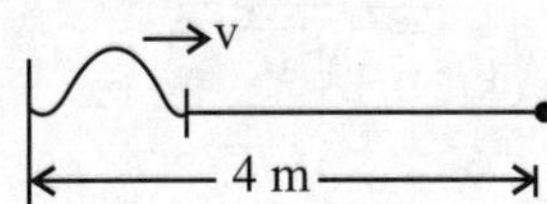

$t = \frac{d}{v} \Rightarrow v = \frac{d}{t} = \frac{32}{0.8} = 40\text{ m / s}$

$v = \sqrt{\frac{T}{\mu}}$

$\Rightarrow T = \mu v^2$

$T = \frac{0.2}{4} \times (40)^2 = \frac{2\times16\times10}{4}$

T = 80 N

18. (2) $v = \sqrt{\frac{T}{\mu}} = \sqrt{\frac{T}{\rho A}} \Rightarrow v \propto \frac{1}{\sqrt{A}} \Rightarrow \frac{v_1}{v_2} = \frac{2}{1}$

where, ρ = density

A = Area of cross-section

μ = ρA = mass of unit length of the string

19. (1) $v = \frac{dy}{dt} = a\left(\frac{2\pi}{\lambda}\right)v\cos\left(\frac{2\pi}{\lambda}(vt - x)\right)$

20. (3) Velocity of wave in a string is given by

$v = \sqrt{\frac{T}{\mu}}$

Where T is tension and m is mass per unit length

$\frac{v_1}{v_2} = \sqrt{\frac{T_1}{T_2}}$ (mass is same)

$T_1 = mg = 5$ g, $\quad T_2 = (15 + 5)g = 20$ g

$\Rightarrow \frac{v_1}{v_2} = \sqrt{\frac{5}{20}} = \sqrt{\frac{1}{4}} = \frac{1}{2}$

$\Rightarrow v_2 = 2v_1$.

Since frequency remains same

$\lambda_2 = 2\lambda_1 = 2 \times 0.08 = 0.16$ m.

21. (3) Frequency is given by:

$$\nu = \frac{1}{2L}\sqrt{\frac{T}{\mu}} \quad \therefore \nu_1 = \frac{1}{2l}\sqrt{\frac{T}{\mu}}$$

$$\nu_2 = \frac{1}{4l}\sqrt{\frac{4T}{\mu}} \quad \therefore \frac{\nu_1}{\nu_2} = 1:1$$

22. (1) Frequency of vibration is given by:

$$f = \frac{n}{2l}\sqrt{\frac{T}{m}} \quad \left(\text{where } v = \sqrt{\frac{T}{m}}\right)$$

$$\therefore f = \frac{nv}{2l} = \frac{5\times 20}{2\times 10} = 5 \text{ Hz.}$$

23. (4) According to question;

$$\frac{2}{2l_1}\sqrt{\frac{T}{\mu_1}} = \frac{3}{2l_2}\sqrt{\frac{T}{\mu_2}}$$

$$\Rightarrow \frac{l_1}{l_2} = \frac{2}{3}\sqrt{\frac{\mu_2}{\mu_1}} = \frac{2}{3}\sqrt{\frac{a_2\rho}{a_1\rho}}$$

$$\Rightarrow \frac{l_1}{l_2} = \frac{2}{3}\sqrt{\frac{r_2^2}{r_1^2}} = \frac{2}{3}\sqrt{\left(\frac{1}{2}\right)^2} \Rightarrow \frac{l_1}{l_2} = \frac{1}{3}.$$

24. (2) $v = \sqrt{xg}$ where 'x' is the distance from lower end, so on moving up velocity also increases.

25. (1) At constant temperature, velocity is independent of Pressure.

26. (2) $v_1 = \sqrt{\frac{\gamma RT}{M}}, v_2 = \sqrt{\frac{3RT}{M}}$

So $v_2 > v_1$ As $\gamma_{max} = 1.67$

27. (2) $\frac{I_1}{I_2} = \frac{4}{4} = \left(\frac{a_1}{a_2}\right)^2 \quad \therefore \frac{a_1}{a_2} = \frac{2}{1}$

$\frac{a_1 + a_2}{a_1 - a_2} = \frac{3}{1}$, So $\frac{I_{max}}{I_{min}} = \frac{9}{1}$

$\therefore L = 10\log\frac{I_{max}}{I_{min}} = 10 \log 9 = 20 \log 3$

28. (2) Sound travel in a gaseous medium as a longitudinal wave which consists of compressions and rarefactions.

29. (4) Sound waves having wavelength greater than that of audible sound has frequency less than that of audible sound are called infra-sonic.

30. (1) Wavelength of ultrasonic waves in air is of the order of 5×10^{-5} cm.

31. (3) Audible range of frequency is 20 Hz to 20 kHz

32. (2) As the sound waves are mechanical waves they require medium for propagation.

33. (2) Intensity after passing through one slab

$$I' = \left[I - \frac{20}{100}\times I\right] = \left[I - \frac{I}{5}\right] = \frac{4I}{5}$$

So, intensity after passingh through two slabs

$$I'' = \left[I' - \frac{20}{100}\times I'\right] = \frac{4I'}{5} = \frac{16I}{25}$$

$$\therefore \% \text{ decrease} = \left[\frac{\left(I - \frac{16I}{25}\right)}{I}\right]\times 100 = 36\%$$

34. (3) $|B| = \frac{\Delta P}{\Delta V / V} = \frac{100\times 10^3}{5\times 10^{-3}/100} = 2\times 10^9$

Speed of sound $V = \sqrt{\frac{B}{d}} = \sqrt{\frac{2\times 10^9}{10^3}}$

$= 1.4 \times 10^3 = 1400$ m/s

35. (2) $\rho_{mix} = \frac{m_{O_2} + m_{H_2}}{V_{O_2} + V_{H_2}} = \frac{32+2}{V+V} = \frac{34}{2V} = \frac{17}{V}$

$\rho_{H_2} = \frac{2}{V}$

velocity $\propto \frac{1}{\rho}$

$$\frac{v_{mix}}{v_{H_2}} = \sqrt{\frac{\rho_{H_2}}{\rho_{mix}}} = \sqrt{\frac{2/v}{17/v}} \Rightarrow \sqrt{\frac{2}{17}} \Rightarrow \frac{v_{mix}}{v_{H_2}} = \sqrt{\frac{2}{17}}$$

36. (1) When a sound wave propagates in a gaseous medium pressure varies as an adiabatic process.

37. (2) Velocity $\propto \sqrt{\text{Temperature}}$

$\frac{2v}{v} = \sqrt{\frac{T_2}{T_1}}$

$T_2 = 4T_1$

$T_2 = 4 \times 273 = 1092$ K

or $T_2 = 1092 - 273 = 819°C$

38. (4) Amplitude $\propto \sqrt{\text{Intensity}}$, Intensity $\propto \frac{1}{(\text{Distance})^2}$

$\therefore \frac{A_P}{A_Q} = \frac{r_Q}{r_P} = \frac{9}{4}$

39. (1) At night, amount of carbon dioxide in atomsphere increases which raises the density of atmosphere as intensity is directly proportional to density, intensity of sound is more at night.

40. (1) $v' = 250 \pm 4 = 254$ or 246

$v'' = 2v'$ and $v'' = 513 \pm 5 = 518$ or 508

As 508 = 2(254)

So, Required frequency = 254 Hz.

41. (4) The number of beats produced per second

$v = v_1 - v_2$

42. (3) $n_A = 258$ Hz, $n_B = 262$ Hz

If it produces X beats with 258Hz and 2x with 262 Hz

262 – (258 – x) = 2x

262 – 258 = x

x = 4

unknown frequency = 254 Hz

43. **(4)** Let f_1 be frequency of tuning fork and f_2 be frequency of piano string

Now, $f_1 - f_2 = 4HZ$

$f_1 = 512$ Hz, f_2 has either has 516 Hz or 508 Hz

On increasing tension frequency f_1 is increased and beat frequency decreases. So, f_2 is 508 Hz before increasing tension.

44. **(2)** Speed = 360 rev/min $\Rightarrow \frac{360}{60} \Rightarrow$ 6rev/sec

Frequency = 6 × 60 = 360 Hz

45. **(3)** General equation of waves

$y_0 = a \sin \omega t$

$\omega_1 = 2000\pi \quad 2\pi f_1 = 2000\pi$

$\omega_2 = 2008\pi \quad 2\pi f_2 = 2008\pi$

$f_1 = 1000$ Hz $\quad f_2 = 1004$ Hz

beats = $f_2 - f_1$

= 1004 – 1000 = 4

46. **(3)** Beats produced by 1st & 2nd wave

= 1 beat & will be heard at = 1s

Beat produced by 2nd & 3rd wave

= 1 beat & will be heard at = 1s

Beat produced by 1st & 3rd wave

= 2 beats & will be heard at, 1s.

So, Maximum beats heard in 1 sec. = 2 Beats

47. **(4)** Time interval between two successive beats

$$T = \frac{1}{\text{Beat frequency}} = \frac{1}{f_1 - f_2}$$

48. **(2)** Number of beats A and B, $n_1 = 5$

Number of beats B and C, $n_2 = 3$

No of beats between A and C = 5 – 3 = 2

49. **(3)** $a_{res} = \sqrt{a_1^2 + (a_2 - a_3)^2}$

50. **(3)** The frequency of A, $v_A = v + \frac{2}{100}v$

and the frequency of B, $v_B = v - \frac{3}{100}v$

According to the question,

$$v_A - v_B = 6 \Rightarrow \left(v + \frac{2}{100}v\right) - \left(v - \frac{3}{100}v\right) = 6$$

$$\Rightarrow \frac{5}{100}v = 6 \Rightarrow v = \frac{600}{5} = 120\text{Hz}$$

The frequency of A

$$v_A = \left(v + \frac{2}{100}v\right) = 120 + \frac{2}{100} \times 120 = 122.4\text{Hz}$$

51. **(4)** Let the frequency of tuning fork C is x Hz.

The frequency of tuning fork $A \Rightarrow x + \frac{1.5}{100}x = 1.015x$

The frequency of tuning fork $B \Rightarrow x - \frac{2.5}{100}x = 0.975x$

Beat frequency of tuning fork A and B is

$1.015x - 0.975x = 0.04x$

Thus, $0.04x = 12 \Rightarrow x = \frac{12}{0.04} = 300\text{Hz}$

52. **(1)** $x = v$ ...(*i*)

$x - 85 = \frac{v(1.5)}{2}$ ...(*ii*)

$\Rightarrow x = 340$m,

$\therefore v = 340$ m/s

53. **(4)** Fundament frequency (f)

$$= \frac{v_s}{4\ell} = \frac{340}{4 \times 85 \times 10^{-2}} = \frac{340}{340 \times 10^{-2}}$$

f = 100 Hz

only odd harmonics are produced

100 Hz, 300 Hz, 500 Hz, 700 Hz, 900 Hz, 1100 Hz

54. **(2)** Fundamental frequency of open organ pipe is

$$f_0 = \frac{v_s}{2\ell_0} \Rightarrow \ell_0 = \frac{v_s}{2f_0} \Rightarrow \ell_0 = \frac{10}{3}\text{ cm}$$

55. **(4)** $n = \frac{1}{2\ell}\sqrt{\frac{T}{m}}$ ... (*i*)

$\frac{n}{2} = \frac{1}{2\ell}\sqrt{\frac{T'}{m}}$ ... (*ii*)

by (*i*) & (*ii*)

$\frac{T'}{T} = \frac{1}{4}$

Subtract both side from 1

$1 - \frac{T'}{T} = 1 - \frac{1}{4}$

$\frac{T - T'}{T} = \frac{3}{4}$

56. **(3)** In the case of a travelling wave, the reflection at a rigid boundary will take place with a phase change of π radian.

57. **(1)** In the fundamental mode

$n = \frac{v}{4\ell} \Rightarrow \frac{\ell}{t \times 4\ell} \Rightarrow \frac{1}{4t}$

Given $t = 10^{-2}$

$\Rightarrow n = \frac{100}{4} = 25$

58. **(1)** $f_0 = \frac{v}{2\ell} = \frac{333}{2 \times 33.3 \times 10^{-2}} = 500$ Hz

frequency for 5th overtone

$\phi = 6f_0 \Rightarrow \phi = 6 \times 500 = 3000$ Hz

59. **(3)** For an open organ pipe

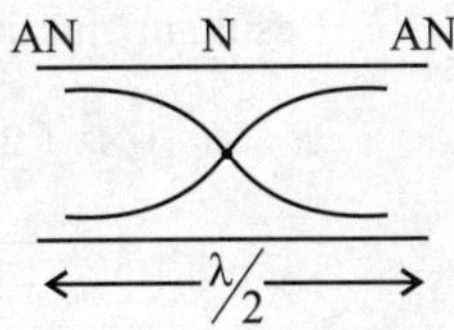

frequency $f = n\frac{v}{2\ell}$

f = 480 Hz, ℓ = 1m, v = 330 m/s

$$n = \frac{f(2\ell)}{v} = \frac{480 \times 2 \times 1}{330} \approx 3$$

60. (1) $\frac{\lambda}{2}+\frac{\lambda}{2}=1.21\ \text{Å}$

$\lambda = 1.21\text{Å}$

61. (1) $f_1 = \frac{1}{2\ell}\sqrt{\frac{T}{\mu}}$

$f_1 \propto \sqrt{T}$

$\frac{f_1}{f_2}=\sqrt{\frac{T_1}{T_2}} \Rightarrow \frac{450}{900}=\sqrt{\frac{9}{T_2}}$

$T_2 = 36$ kg wt

62. (4)

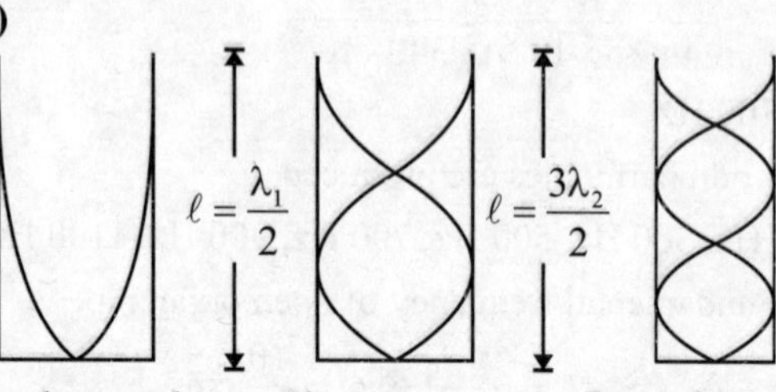

Fundamental tone First overtone Second overtone

$\ell=\frac{\lambda_1}{2} \Rightarrow \lambda_1 = 2\ell$, $\ell=\frac{3\lambda_2}{2} \Rightarrow \lambda_2=\frac{2\ell}{3}$

$\ell=\frac{5\lambda_3}{2} \Rightarrow \lambda_3=\frac{2\ell}{5}$

$\therefore\ \lambda_1:\lambda_2:\lambda_3 = 2\ell:\frac{2\ell}{3}:\frac{2\ell}{5}=1:\frac{1}{3}:\frac{1}{5}$

High Order Time Intensive MCQs

1. (1) Path difference = $(\pi r - 2r)$

$= (\pi-2)r = n\lambda \Rightarrow \lambda = \frac{(\pi-2)r}{n}$

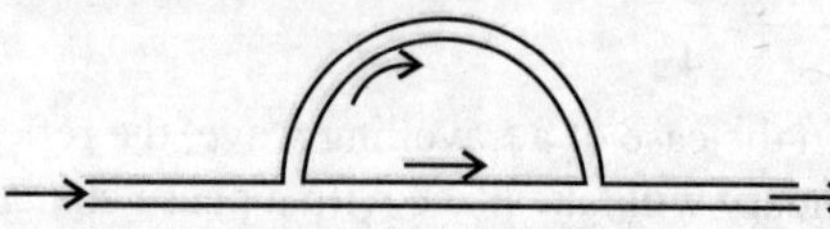

where, $v = f\times\lambda$

f = frequency

$\Rightarrow\quad f=\frac{v}{\lambda} \Rightarrow f=\left[\frac{v}{(\pi-2)r}\right]n$

So, multiple integral $=\left[\frac{v}{(\pi-2)r}\right]$

2. (4) Let $y_1 = A\sin(\omega t)$

and $y_2 = A\sin\left(\omega t+\frac{\pi}{2}\right)$

Resultant amplitude of two waves

$R = A^2 + A^2 + 2A^2\cos\left(\frac{\pi}{2}\right)$

$\Rightarrow R=\sqrt{2A^2+2A^2\times 0} \Rightarrow R=\sqrt{2A^2} \Rightarrow R=\sqrt{2}A$

However, both will have the same frequency on superimposing.

3. (1) We know that,

$5 = n_1 - n_2 = \frac{v}{4}\left[\frac{1}{l_1}-\frac{1}{l_2}\right]$

$v = 300$ m/s $\Rightarrow \frac{1}{l_1}-\frac{1}{l_2}=\frac{20}{300\times100}=\frac{1}{1500}$

$\frac{1}{l_2}=\frac{1}{10}-\frac{1}{1500}\Rightarrow\frac{149}{1500}\Rightarrow l_2=\frac{1500}{149}=10.06\text{cm}$

4. (2) Frequency of first string

$f_1=\frac{1}{2l_1}\sqrt{\frac{T}{\mu}}=\frac{1}{2\times51.6\times10^{-2}}\sqrt{\frac{20}{10^{-3}}}=137.03\text{Hz}$

Similarly, frequency of second string

$f_2=\frac{1}{2\times49.1\times10^{-2}}\sqrt{\frac{20}{10^{-3}}}=144.01$

Number of beats = difference of frequency of two strings.

$= f_2 - f_1 = 144 - 137 = 7$ beats

5. (1) Given, $T = 20$ N, $A = 0.2\ \text{mm}^2 = 0.2\times10^{-6}\ \text{m}^2$,

$\rho = 7.5\ \text{g cm}^{-3} = 7.5\times10^3\ \text{kg m}^{-3}$

Mass per unit length of the wire, $\mu = A\rho$

$= 0.2\times10^{-6}\times7.5\times10^3 = 1.5\times10^{-3}\ \text{kg m}^{-1}$

when a string is plucked at the mid-point, it vibrates with fundamental frequency.

Fundamental frequency of the stretched string is:

$f=\frac{1}{2L}\sqrt{\frac{T}{\mu}}=\frac{1}{2L}\sqrt{\frac{20}{1.5\times10^{-3}}}$

Speed of transverse wave, $v = f(2L)$

or $\quad v=\frac{1}{2L}\sqrt{\frac{20}{1.5\times10^{-3}}}\times2L=116\ \text{ms}^{-1}$.

6. (2) $v=\frac{dy}{dt}=y_0\cos\left[2\pi\left(ft-\frac{x}{\lambda}\right)\right]\times2\pi f$

$=2\pi y_0\cos\left[2\pi\left(ft-\frac{x}{\lambda}\right)\right]$

The particle velocity is maximum, when

$\cos\left[2\pi\left(ft-\frac{x}{\lambda}\right)\right]=1$

$\therefore\quad v_{max} = 2\pi f y_0 \qquad \text{... }(i)$

We know that $y = a\sin(\omega t - kx)$. The wave velocity v is given by:

$v=\frac{\omega}{K}=\frac{2\pi f}{2\pi/\lambda}=f\lambda \qquad \text{... }(ii)$

Given that, $v_{max} = 4v$

$\therefore\quad 2\pi f y_0 = 4f\lambda$ or $\lambda=\frac{\pi y_0}{2}$.

7. (1) As the wave is going along positive x axis, so option (2) and (3) are wrong.

Moreover, A = 3 mm

Speed of the wave, $v = \frac{d}{t} = \frac{6\times10^{-2}}{4\times10^{-3}} = 15\ m/s$

$k = \frac{2\pi}{\lambda} = \frac{2\pi}{0.40} = 16 rad m^{-1}$

$\omega = kv = 16 rad/m \times 15\ m/s = 2.4 \times 10^2\ rad/s$

$\therefore$ option (1) is correct.

8. (1) $\frac{b}{a} = \frac{d}{c}$ (as velocity is same)

9. (4) $n = \frac{v}{4L_1}$ and $n = \frac{3v}{4L_2}$

$\therefore v = 4nL_1$ and $3v = 4nL_2$

or $2v = 4n(L_2 - L_1)$ or $v = 2n(L_2 - L_1)$

10. (2) Linear mass density

$\mu = \frac{M}{L} = 3 \times 10^{-3}$ Kg/m

Speed of Transverse wave

$v = \sqrt{\frac{T}{\mu}} = \sqrt{\frac{T}{3\times10^{-3}}} \Rightarrow v = \frac{\omega}{K} = \sqrt{\frac{T}{3\times10^{-3}}}$

$\frac{1.5}{60} = \sqrt{\frac{T}{3\times10^{-3}}} \Rightarrow T = 3\times10^{-3}\times\frac{1}{1600}$

$= 0.187 \times 10^{-5}$ N

11. (2) $\lambda = \frac{100}{25} = 4\,m$

$v = \sqrt{\frac{T}{\mu}} = \sqrt{\frac{100}{4/100}} = \frac{100}{2} = 50 m/s$

$v = \frac{\omega}{K} \Rightarrow 50 = \frac{\omega}{2\pi/\lambda}$

$\left\{\omega = 50\times\frac{2\pi}{4} = 25\pi\right\}$

Since at t = 0 and at x = 0 particle is going down. So $\phi = \pi$.

$\therefore$ Equation becomes

$y = A\sin(\omega t - Kx + \pi)$

$v_{max} = A\omega$

$\frac{5}{100} = A\times25\pi$

$A = \frac{1}{500\pi} m$

$\left\{y = \frac{1}{500\pi}\sin\left(25\pi t - \frac{\pi}{2}x + \pi\right)\right\}$

12. (4) $I \propto a^2n^2$

$\frac{I_A}{I_B} = \frac{A_1^2\omega_1^2}{A_2^2\omega_2^2} = \left(\frac{2}{1}\right)^2\times\left(\frac{1}{2}\right)^2 = 1:1$

13. (1) $v_1 = v_2$

$\frac{2\pi\times50}{2\pi\times2} = \frac{2\pi\times100}{2\pi a} \Rightarrow \frac{I_1}{I_2} = \frac{A_1^2\omega_1^2}{A_2^2\omega_2^2}$

$\frac{50}{2} = \frac{100}{a} \qquad = \frac{2^2}{4^2}\times\frac{(2\pi\times50)}{(2\pi\times100)^2}$

$\therefore a = 4 \qquad = \frac{1}{4}\times\frac{1}{4}$

$\Rightarrow \left\{\frac{I_1}{I_2} = \frac{1}{16}\right\}$

14. (1) $\frac{I_1}{I_2} = \frac{\omega_1^2}{\omega_2^2}\cdot\frac{A_1^2}{A_2^2}$

$\because \omega_2 = 2\omega_1$ and $A_1 = 2A_2 \Rightarrow \frac{I_1}{I_2} = 1:1$

15. (2) Let intensity due to a single person = I

then $10\log I/I_0 = 60$

Also, Intensity due to 8 person = 8I

$\therefore$ Final decibel level

$= 10\log\left(\frac{8I}{I_0}\right) = 10\left(\log\frac{I}{I_0} + \log 8\right)$

$= 60 + 10\log 8 = 60 + 30\log 2$

$= 60 + 30(0.3010) = 69$

16. (2) $y = 5\sin\frac{\pi x}{3}\cos 40\pi t$

Comparing with the standard equation of stationary wave

$y = A\sin kx\cos\omega t$

we get; $k = \frac{\pi}{3} = \frac{2\pi}{\lambda}$

$\therefore \lambda = 6$ cm

Hence, the separation between two adjacent nodes

$= \frac{\lambda}{2} = 3$ cm.

17. (1) $f = \frac{5}{2l}\sqrt{\frac{9g}{\mu}} = \frac{3}{2l}\sqrt{\frac{Mg}{\mu}}$

$\therefore M = 25$ kg

18. (4) $v = 350$ m/s $\qquad A = 350$ Hz

$\therefore \lambda = 1$ m

$AP - BP = 25\,cm = \frac{1}{4}m. = \frac{\lambda}{4}$

$\therefore \quad \Delta x = \frac{\lambda}{4}$

$\frac{\Delta x}{\lambda} = \frac{\Delta\phi}{2\pi} \Rightarrow \Delta\phi = \frac{2\pi}{\lambda}\times\Delta x$

$\left\{\Delta\phi = \frac{2\pi}{1}\times\frac{1}{4} = \frac{\pi}{2}\right\}$

$A_{net} = \sqrt{a_1^2 + a_2^2 + 2a_1a_2\cos(\Delta\phi)} = \sqrt{0.3^2 + 0.4^2 + 0}$

$A_{net} = 0.5$mm

19. (3) $\frac{v_1}{v_2} = \sqrt{\frac{T_1}{T_2}\frac{\rho_2}{\rho_1}}$

As $T_1 = T_2$ and $\rho_2 = 2\rho_1$

$\Rightarrow \frac{v_1}{v_2} = \frac{\sqrt{2}}{1}$

20. (4) $n_1\sqrt{T_1} = n_2\sqrt{T_2} \Rightarrow 6\sqrt{36} = 4\sqrt{T_2} \Rightarrow T_2 = 81N$

21. (2) Fundamental frequency (closed pipe)

$n_c = \frac{v}{4\ell}$

fundamental frequency (open pipe)

$n_0 = \frac{v}{2\ell}$

number of beats = 2

$\frac{v}{2\ell} - \frac{v}{4\ell} = 2$

$v = 8\ell$ ... (*i*)

when length of open pipe is halved and closed pipe is doubled

$\frac{v}{\ell} - \frac{v}{8\ell} = n$

$\frac{7v}{8\ell} = n$

$n = \frac{7}{8\ell} \times 8\ell$ [from (*i*)]

$n = 7$

22. (2) $n_p = \frac{v}{4L}, n_q = \frac{v}{2L}, n_r = \frac{2v}{2L}, n_s = \frac{3v}{4L}$

$\therefore n_p : n_q : n_r : n_s = 1 : 2 : 4 : 3$

23. (3) $L = 10\log_{10}\frac{I}{I_0}$ decibel

$L' = 10\log_{10}\frac{2I}{I_0}$ decibel

$L' - L = 10\left[\log_{10}\frac{2I}{I_0} - \log_{10}\frac{I}{I_0}\right]$

$= 10\log_{10}2 = 10 \times 0.3010 = 3.01$ dB

24. (2) Using $\frac{P}{4\pi R^2} = I \Rightarrow \frac{\frac{P}{4\pi(D-50)^2}}{\frac{P}{4\pi D^2}} = \frac{2}{1} \Rightarrow \frac{D^2}{(D-50)^2} = 2$

$\Rightarrow D = \sqrt{2}(D-50) \Rightarrow (\sqrt{2}-1)D = 50\sqrt{2} \Rightarrow D = \frac{50\sqrt{2}}{\sqrt{2}-1}$

25. (3) Using the diagram

$v_{max} = A\omega = 5$

$\because \omega = \frac{2\pi}{T} \Rightarrow \omega = \frac{2\pi}{4} \Rightarrow A = \frac{10}{\pi}$

Statement Based MCQs

1. **(3)** The distance between two consecutive compressions or rarefaction is called the wavelength of the wave.
2. **(1)** Both the Statements are true.
3. **(4)** Statement-I is false but Statement-II is true.
4. **(4)** In water sounds travel faster than in air because in water, particles are much closer and they can transmit vibration energy from one particle to next particle more quickly.
5. **(1)** For open organ pipe $f_o = \frac{v}{2L}$

For closed organ pipe $f_c = \frac{v}{4L}$

Assertion & Reason MCQs

1. **(2)** Due to the vibrations of molecules of medium, sound waves are produced. In vacuum, there is no matter in space so, sound waves cannot propagate in vacuum.
2. **(3)** The presence of moisture in air lowers the density of air.

As velocity of sound in air = $\sqrt{\frac{\gamma P}{d}}$ [d = density]

3. **(2)** Both Assertion and Reason are true but Reason is not the correct explanation of the Assertion.
4. **(2)** Both Assertion and Reason are true but Reason is not the correct explanation of the Assertion.
5. **(2)** Sound waves having frequency range above 20,000 Hz are called ultrasonics and below 20 Hz are called infrasonics.

Match the Column MCQs

1. **(1)** Second overtone is shown in the figure.

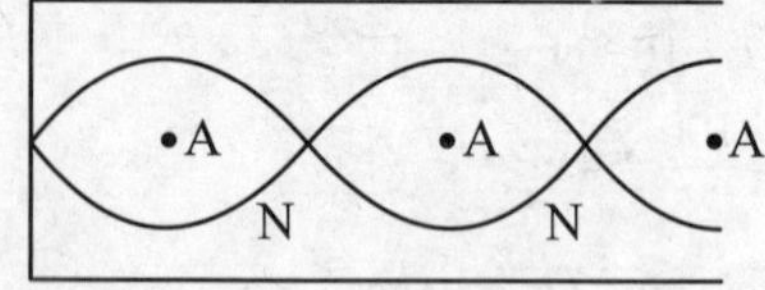

Displacement wave

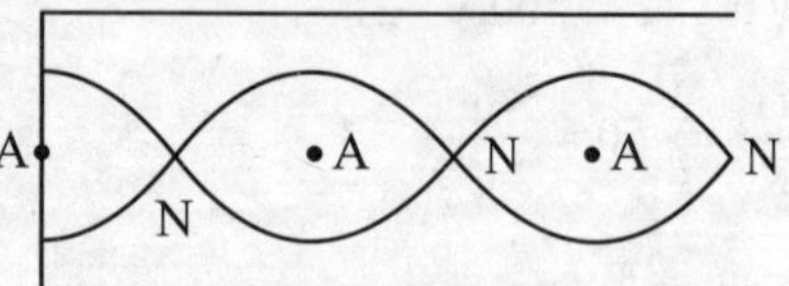

Pressure wave

Distance between nearest node and antinode is 2/5

2. **(1)** $\nu_1 = 1310$ Hz, $\nu_2 = 1834$ Hz, $\nu_3 = 2358$ Hz

$\frac{\nu_2}{\nu_1} = \frac{7}{5}; \frac{\nu_3}{\nu_1} = \frac{9}{5}$

$\nu_1 : \nu_2 : \nu_3 = 5 : 7 : 9$

(This corresponds to a pipe closed at one end)

Fundamental frequency $n_0 = \frac{2358}{9} = 262$ Hz

Frequency of the first overtone $n_1 = 3n_0 = 786$ Hz

Frequency of the fifth overtone = $5n_0 = 1310$ Hz

For fundamental frequency

$\lambda_0/4 = l$

$\lambda_0 = 4l$

$n_0 = \frac{v}{4l}$

$\Rightarrow l = \frac{v}{4n_0} = \frac{340}{4 \times 262} = 32.4 \times 10^{-2}$ m

Length of the pipe = 32.4 cm.

3. (3) In string $v = \sqrt{T/\mu} = 320$ m/s

Open pipe and string fixed at both ends.

$$f = \frac{nv}{2L} = 320, 640, 960, ...$$

Closed pipe and stirng free end

$$f = (2n-1)\frac{v}{4L} = 160, 480, 800, ...$$

4. (2) (a) Pitch is related to frequency,

(b) Quality is related to wave form.

(c) Loudness depends on intensity.

(d) Nodes are the position of zero amplitude.

5. (4) (a) $y = \sin(5x - 4t) + 3\cos(4t - 5x + \pi/6)$

This is combination of two wave moving in same direction with same angular frequency but in different phases. Hence, super position of these two will produce traveling wave.

(b) $y = 10\cos\left(t - \frac{x}{330}\right) \times \sin(100)\left(t - \frac{x}{330}\right)$

This is combination of two waves having different frequencies hence, it will produce beats.

(c) $y = 10\sin(2\pi x - 120t) + 10\cos(120t + 2\pi x)$

This is combination of two waves having same frequency but moving in opposite direction hence it is a standing wave.

(d) $y = 10\sin(2\pi x - 120t) + 8\cos\left(118t - \frac{59}{30}\pi x\right)$

Here, the frequencies at combining waves are different hence, beats will be produced.

NCERT Exemplar MCQs

1. (2) Water waves produced by a motorboat sailing in water are both longitudinal and transverse, because the waves produce transverse as well as lateral vibrations in the particles of the medium.

2. (3) Let the frequency in the first medium is ν and in the second medium is ν'.

Frequency remains same in both the medium

$$\text{So}, v = v' \Rightarrow \frac{v}{\lambda} = \frac{v'}{\lambda'} \Rightarrow \lambda' = \left(\frac{v'}{v}\right)\lambda$$

λ and λ', v and v' are wavelengths and speeds in first and second medium respectively.

$$\text{So}, \lambda' = \left(\frac{2v}{v}\right)\lambda = 2\lambda$$

3. (3) Due to presence of moisture, density of air decreases.

We know that speed of sound in air is given by

$$v = \sqrt{\frac{\gamma p}{\rho}}$$

For air γ and p are constants.

$$v \propto \frac{1}{\sqrt{\rho}}, \text{ where } \rho \text{ is density of air.}$$

where ρ_1 is density of dry air and ρ_2 is density of moist air.

$$\text{As } \rho_2 < \rho_1 = \frac{v_2}{v_1} > 1 \Rightarrow v_2 > v_1$$

Hence, speed of sound wave in air increases with increase in humidity.

4. (3) Speed of sound wave in a medium $v \propto \sqrt{T}$ (where T is temperature of the medium). Clearly, when temperature changes speed also changes.

As, $v = \nu\lambda$

where ν is frequency and λ is wavelength.

Frequency (ν) remains fixed

$\Rightarrow v \propto \lambda$ or $\lambda \propto v$

As frequency does not change, so wavelength (λ) changes.

5. (2) Propagation of longitudinal waves through a medium leads to transmission of energy through the medium without matter being transmitted.

6. (3) When mechanical transverse wave propagates through a medium, the constituent of the medium oscillate perpendicular to wave motion causing change in shape. That is each, element of the medium is subjected to shearing stress. Solids and strings have shear modulus. That is why, sustain shearing stress.

Fluids have no shape of, their own, they yield n shearing stress. This is why transverse waves are possible in solids and strings but not in fluids.

7. (4) Due to compression and rarefactions density of the medium (air) changes. At compressed regions density is maximum and at rarefactions density is minimum.

As density is changing, so Boyle's law is not obeyed.

Bulk modulus remains same.

The time of compression and rarefaction is too small i.e., we can assume adiabatic process and hence no transfer of heat.

8. (2) Amplitude of reflected wave

$$A_r = \frac{2}{3} \times A_i = \frac{2}{3} \times 0.6 = 0.4 \text{ units}$$

Given equation of incident wave

$$y_i = 0.6\sin 2\left(t - \frac{x}{2}\right)$$

Equation of reflected wave is

$$y_r = A_r \sin 2\pi\left(t + \frac{x}{2} + \pi\right) \text{ [∵ At denser medium}$$

phase changes by π]

The positive sign is due to reversal of direction of propagation

$$\text{So, } y_r = -0.4\sin 2\pi\left(t + \frac{x}{2}\right) \quad [\because \sin(\pi + \theta = -\sin\theta)]$$

9. (2) Mass m = 2.5 kg

μ = mass per unit length

$= \frac{m}{l} = \frac{2.5kg}{20} = \frac{1.25}{10} = 0.125kg/m$

Speed $v = \sqrt{\frac{T}{\mu}} = \sqrt{\frac{200}{0.125}}$

[speed of transversed waves in any string]

$l = v \times t \Rightarrow 20 = \sqrt{\frac{200}{0.125}} \times t$

$\Rightarrow \quad t = 20 \times \sqrt{\frac{125}{2\times10^5}} = 20 \times \sqrt{\frac{25\times5}{2\times10^5}}$

$= 20 \times \sqrt{25 \times \frac{1}{0.4\times10^5}} = 20\times5\sqrt{\frac{1}{0.4\times10^4}} = \frac{25\times5}{2\times10^2}$

$t = \frac{1}{2} = 0.5s$

NEET & AIIMS Past Year MCQs

1. (4) Wave equation:

$y = C\sin\frac{2\pi}{\lambda}(at - x)$

Compare with standard form $y = A\sin(\omega t - kx)$:

$\omega = \frac{2\pi a}{\lambda},\ k = \frac{2\pi}{\lambda}$

Frequency $f = \frac{\omega}{2\pi} = \frac{a}{\lambda}$.

2. (3) Fundamental harmonic frequency open pipe

$= \frac{v}{2L} = v_1$ (say)

Fundamental harmonic frequency of closed pipe $= \frac{v}{4L}$

$= v_2$ (say)

$\Rightarrow \frac{v_1}{v_2} = \frac{\frac{v}{2L}}{\frac{v}{4L}} = 2:1$

3. (4) Velocity of transverse wave in a string is

$v = \sqrt{\frac{T}{\mu}}$

T is tension in the string and μ is mass per unit length of the string.

$\Rightarrow v \propto \sqrt{T} \Rightarrow \frac{v_i}{v_f} = \sqrt{\frac{T_i}{T_f}}$

$\Rightarrow \frac{v_i}{v_f} = \sqrt{\frac{T}{2T}} \Rightarrow \frac{v_i}{v_f} = \frac{1}{\sqrt{2}}$

4. (1) Given, wavelength $\lambda = L$

Frequency $= \nu$

$\therefore$ Velocity of wave, $v = \upsilon\lambda = \upsilon L$

5. (2) Given, $\lambda_1 = 50cm = 0.5m$, $\lambda_2 = 0.51m$

$f_1 - f_2 = 12 \Rightarrow \frac{v}{\lambda_1} - \frac{v}{\lambda_2} = 12$

$\Rightarrow v\left(\frac{1}{0.5} - \frac{1}{0.51}\right) = 12$

$\Rightarrow 0.0392\ v = 12 \Rightarrow v = 306$ m/sec

CLASS-XII-Explanations

Electric Charges and Fields

NCERT Topic-wise MCQs

1. (4) Net force on the electron will be zero along the diagonal AC. It will move towards D due to attraction by the proton.

2. (1) The force between 4q and q; $F_1 = \frac{1}{4\pi\varepsilon_0}\cdot\frac{4q\times q}{\lambda^2}$

The force between Q and q; $F_2 = \frac{1}{4\pi\varepsilon_0}\cdot\frac{Q\times q}{(\lambda/2)^2}$

We want $F_1 + F_2 = 0$ or $\frac{4q^2}{\lambda^2} = -\frac{4Qq}{\lambda^2} \Rightarrow Q = -q$

3. (1) $q_1 = q_2 = 2 \times 10–8C$, r = 1m

Tension in the string will be equal to the force between the two charges.

According to coulomb's law,

$$F = \frac{Kq_1q_2}{r^2}$$

$$= \frac{9\times10^9\times(2\times10^{-8})^2}{(1)^2}$$

$= 3.6 \times 10^{-6}N$

4. (2) Let the distance between the charge q and Q be x. System is in equilibrium then net force on Q is zero.

∴ Total force on Q at B is

$$\frac{KQq}{x^2}+\frac{KQQ}{(2x)^2}=0$$

(Q at A, q at middle, Q at B; distances x and x)

$$\frac{KQq}{x^2}=\frac{-KQ^2}{4x^2}$$

$$q=\frac{-Q}{4}$$

5. (3) According to quantisation of charge,

q = ne

$$n=\frac{q}{e}=\frac{1\times10^{-7}}{1.6\times10^{-19}}=6.25\times10^{11}$$

6. (1) As we know the relation

Dielectric constant K is defined as

$$\frac{F}{F_m}=K$$

$$\Rightarrow F_m=\frac{F}{5}$$

7. (3) When there is no additional charge $F = \frac{K(+10)(-5)}{R^2}$

$$F=\frac{-50K}{R^2} \quad ... (i)$$

when +2 μC charge is added to both spheres.

Charge on 1st sphere = 10 μC + 2 μC = 12 μC

Charge on 2nd sphere = –5 μC + 2 μC = –3 μC

New force will be $F' = \frac{K(+12)(-3)}{R^2} = -\frac{36K}{R^2}$ ...(*ii*)

From (*i*) and (*ii*)

$$\therefore \frac{F'}{F} = -\frac{36K}{R^2}\times\frac{R^2}{-50K}=\frac{36}{50}\Rightarrow\frac{F'}{F}=\frac{18}{25}\Rightarrow F'=\frac{18F}{25}$$

8. (4) let the triangle be ABC as follow:-

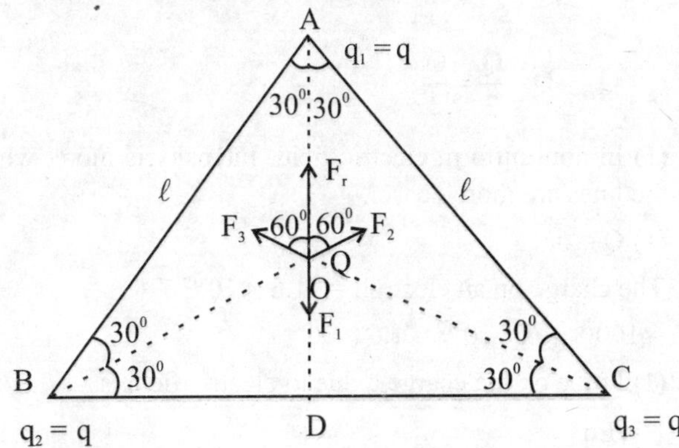

$$\frac{AD}{AB}=\cos 30°$$

$$AD=\frac{\ell\sqrt{3}}{2}$$

Distance AO of centroid from A is $\frac{2}{3}AD$

$$=\frac{2}{3}\times\frac{\ell\sqrt{3}}{2}=\frac{\ell}{\sqrt{3}}$$

Similarly BO and CO are equal to $\frac{\ell}{\sqrt{3}}$

Force on Q at O due to charge q_1 placed at A

$$F_1=\frac{kQq}{(\ell/\sqrt{3})^2}=\frac{3KQq}{\ell^2} \text{ along AO}$$

Similarly $F_2=\frac{3KQq}{\ell^2}$ along OB

$F_3=\frac{3KQq}{\ell^2}$ along CO

Angle between F_2 and F_3 is 120°

According to parallelogram law of vector addition:-

$$F_r = \sqrt{F_2^2 + F_3^2 + 2F_2F_3\cos 120^\circ}$$

Let, $F_1 = F_2 = F_3 = F$

$$F_r = \sqrt{F^2 + F^2 + 2F^2\left(\frac{-1}{2}\right)} = F$$

Force due to charge q at A is equal and opposite to the resultant force F_1. So the force experienced is Zero.

Trick:- The charge is at the centre of the symmetrical system so the net force on it will be zero.

9. (4) $Q_1 + Q_2 = Q$ ...(*i*)

$Q_2 = Q - Q_1$ ...(*ii*)

$$F = \frac{KQ_1(Q-Q_1)}{R^2} \quad ...(iii)$$

For F to be maximum $\frac{dF}{dQ_1} = 0$

$$\frac{1}{R^2}[KQ - 2KQ_1] = 0$$

$$\Rightarrow Q_1 = \frac{Q}{2}$$

$$Q_2 = Q - \frac{Q}{2} = \frac{Q}{2}$$

10. (1) In non-uniform electric field. Intensity is more, where the lines are more denser.

11. (1) Q = ne

The charge on an electron = -1.6×10^{-19}

$= 100 \times 1.6 \times 10^{-19} = 1.6 \times 10^{-17}$

12. (1) Force on the charge q due to electric field is

F = Eq

Both of them have charge and electric field is also same so the force on them will be same.

13. (1)

14. (2) Electric field at the centre is not zero in option (2), due to lack of symmetry.

15. (1) If on 2q, force due to q is to the left and force due to –3q is to the right

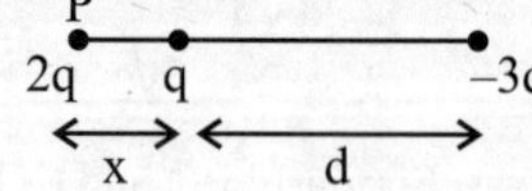

$$\therefore \frac{2q^2}{4\pi\varepsilon_o x^2} = \frac{6q^2}{4\pi\varepsilon_o (d+x)^2}$$

$\therefore (d + x)^2 = 3x^2$

$$\Rightarrow x = \frac{d}{2} \pm \frac{\sqrt{3}d}{2}$$

(–ve sign is unacceptable)

$$x = \frac{d}{2} + \frac{\sqrt{3}d}{2} = \frac{d}{2}\left(1+\sqrt{3}\right) \text{ to the left of q}$$

16. (2) Electrostatic field lines start from positive charge and end at a negative charge.

17. (1) The Particle thrown against electric field experiences retardation which is equal to

$$a = \frac{-qE}{m}$$

as we know $v^2 - u^2 = 2as$

$$0 - u^2 = 2\frac{-qE}{m} \times s$$

$$s = \frac{mu^2}{2qE}$$

18. (2) $\because 0.5e = \frac{1}{2}mv_x^2 \Rightarrow v_x = \sqrt{\frac{e}{m}}$

Along x axis, $L = v_x t = \sqrt{\frac{e}{m}}t$ ...(*i*)

Along y axis $v_y = \frac{eE}{m}t$ ...(*ii*)

$$\frac{v_y}{L} = E\sqrt{\frac{e}{m}} = Ev_x$$

$$\because \tan\theta = \frac{v_y}{v_x}$$

$$\Rightarrow \tan\theta = E \times L = 10 \times 0.1 = 1$$

$$\Rightarrow \theta = 45^\circ$$

19. (3) We know that

$F_1 = qE = m_1a_1$ ...(*i*)

$F_2 = q\,E = m_2a_2$ ...(*ii*)

From eq (*i*) & (*ii*) we get

$$\frac{a_1}{a_2} = \frac{F_1/m_1}{F_2/m_2} = \frac{F_1.m_2}{F_2.m_1}$$

$\because F_1 = F_2$

$$\frac{a_1}{a_2} = \frac{m_2}{m_1} = \frac{1}{0.5} = 2$$

$$\frac{a_1}{a_2} = 2$$

20. (4) The electric field intensity at a point in a vacuum is equal to force experienced by a unit positive charge placed at that point.

21. (3)

22. (3) When two spheres are joined by a wire charge flows till electric potential on both the spheres become equal.

$V_1 = V_2$

$$\frac{kq_1}{R_1} = \frac{kq_2}{R_2} \Rightarrow \frac{q_1}{q_2} = \frac{R_1}{R_2} \quad ...(i)$$

$$\therefore \frac{E_1}{E_2} = \frac{kq_1}{R_1^2} \times \frac{R_2^2}{kq_2} = \frac{q_1}{q_2} \times \frac{R_2^2}{R_1^2}$$

Put value of $\frac{q_1}{q_2}$ {from eqn (i)}

$$\therefore \frac{E_1}{E_2} = \frac{R_1}{R_2} \times \frac{R_2^2}{R_1^2} = \frac{R_2}{R_1}$$

$$\frac{E_1}{E_2} = \frac{R_2}{R_1}$$

23. (2) We have, $E = \frac{8m}{e}$ v/m

time to come out, $t = \frac{1}{2}$ sec

$$V_y = at = \left(\frac{eE}{m}\right)t = \left(\frac{e}{m} \times \frac{8m}{e}\right) \times \frac{1}{2} = 4\text{m/s}$$

So, $\tan\theta = \frac{V_y}{V_x} = \frac{4}{2} \Rightarrow \theta = \tan^{-1} 2$

24. (2) $E = \frac{Kq}{r^2}$

$$5 \times 10^5 = \frac{9 \times 10^9 \times q}{(3)^2}$$

$$\Rightarrow q = 5 \times 10^{-4} C = 500\ \mu C$$

25. (2)

26. (3) According to Gauss's Law,

$$\oint \vec{E} \cdot \vec{ds} = \frac{q}{\varepsilon_0} \quad ...(i)$$

Cube has six equal faces, so electric flux pass through each would be equal.

Then total flux $= 6(E \times A) = \frac{q}{\varepsilon_0}$

So, net electric flux passing through each face of the cube is given by

$$E \times A = \frac{q}{6\varepsilon_0}$$

27. (3) By Gauss's law $\phi = \frac{q}{\varepsilon_0}$

$q = 1C \quad \therefore \phi = \frac{1}{\varepsilon_0} = (\varepsilon_0)^{-1}$

28. (4) When charged ball is placed at the center of another ball, it induces opposite charge on the inner surface by the induction. Further same nature of charge will be induced on the outer surface of the shell. Hence the total charge on the surface of the shell is $-100e - 20e = -120e$

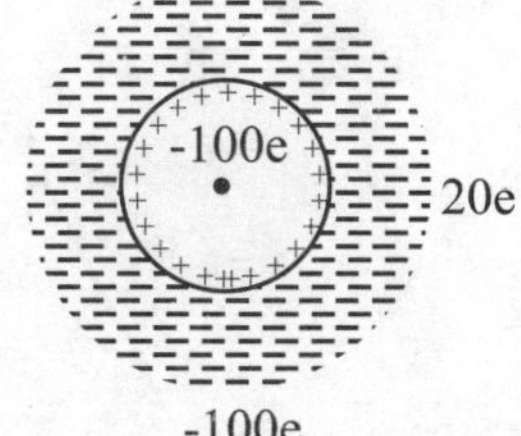

29. (1) Because all charge are present on the outer surface of the shell. Hence, no electric lines are present inside the shell so electric field is absent there.

30. (4) Due to parallel sheets electric field is

$E \neq 0$ | $\vec{E}$ $\vec{E}$ | $E = 0$ | $E \neq 0$

$$E = \left(\frac{\sigma}{2\varepsilon_o}\right) + \left(\frac{-\sigma}{2\varepsilon_o}\right) = 0$$

Because both electric fields are in opposite direction.

31. (3) Integral $\int \vec{E}.\vec{dS}$ gives flux through a surface.

So, the total flux through the surface is zero.

32. (3) Total flux passing through the cube is

$$\phi = \frac{q}{\varepsilon_o}$$

A cube has 6 faces so flux through one face is given by

$$\Rightarrow \phi = \frac{q}{6\varepsilon_o}$$

33. (2) Gauss law can't be valid if the inverse square law were not exactly true.

34. (1) Total charge enclosed $(q_{en}) = 4q - 2q$

$\Rightarrow q_{en} = 2q$

According to Gauss theorem

$$\phi = \frac{q_{en}}{\varepsilon_o}$$

$$\phi = \frac{2q}{\varepsilon_o}$$

35. (1) Gauss law is valid for closed loops.

36. (2) Electric lines of forces are not passing through it so flux is zero and electric lines of force are perpendicular to the areal plane of disc so according to the relation it will again be zero.

$\phi = E.A \cos\theta$

$\phi = E.A \cos 90°$

$\phi = 0$

37. (3) Total charge into the shell is zero because a dipole consists of equal amount of negative and positive charge.

So, $q_{net} = 5(-q + q) = 0$

$$\phi = \frac{q}{\varepsilon_o} = \text{zero}$$

38. (4) 8 cubes are required to enclose that charge

So, from gauss law total electric flux through all the 8 cubes

$$\phi = \frac{q_{en}}{\varepsilon_o}$$

Electric flux through each cube

$$\Rightarrow \phi = \frac{2q}{8\varepsilon_o} = \frac{q}{4\varepsilon_o}$$

$$\therefore \phi = \frac{q}{4\varepsilon_o}$$

39. (2) $\phi = EA\cos\theta$

$\phi = (8\hat{i} + 8\hat{j} + \hat{k})(10\hat{i})\cos 0°$

$\phi = 80\ Nm^2C^{-1}$

40. (3) $\phi = \frac{q}{\varepsilon_o}$

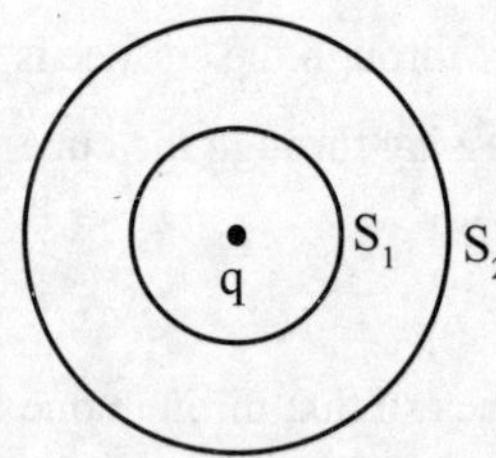

Let ϕ_1 and ϕ_2 be the electric flux through the two spheres S_1 and S_2.

$$\Rightarrow \frac{\phi_1}{\phi_2} = \frac{q_1}{\varepsilon_o} \times \frac{\varepsilon_o}{q_2} = \frac{q_1}{q_2}$$

$$\frac{\phi_1}{\phi_2} = \frac{q}{q} = 1$$

41. (2) Electric field due to a line of charge is r

$$E = \frac{\lambda}{2\pi\varepsilon_0 r}$$

$$\because E = E_{+\lambda} + E_{-\lambda}$$

$$\therefore E = \frac{2\lambda}{2\pi\varepsilon_o R} \times 2 \qquad \left[\because r = \frac{R}{2}\right]$$

$$E = \frac{2\lambda}{\pi\varepsilon_o R}$$

42. (4) From gauss law $\phi_0 = \frac{q}{\varepsilon_o}$

This is the net flux coming out of the cube.

Since a cube has 6 sides. So electric flux through any face is

$$\phi = \frac{\phi_0}{6} = \frac{q}{6\varepsilon_o} \Rightarrow \frac{4\pi q}{6(4\pi\varepsilon_o)}$$

Amount of flux passing through the two faces

$$= 2\phi = \frac{4\pi q}{3(4\pi\varepsilon_o)}$$

43. (4) Electric flux through any closed surface is equal to the net charge inside the surface divided by ε_0

Therefore $\phi = \frac{q}{\varepsilon_o}$

Let q_1 be the charge, due to which flux ϕ is entering the surface (this charge will be –ve)

$q_1 = -\phi_1\,\varepsilon_0$

Let $+q_2$ be the charge due to which flux ϕ_2 is leaving the surface

$$\phi_2 = \frac{q_2}{\varepsilon_o}$$

$q_2 = \phi_2\,\varepsilon_0$

So, electric charge inside the surface

$= q_2 + q_1 = \varepsilon_0\,\phi_2 - \varepsilon_0\phi_1$

$\Rightarrow q_1 + q_2 = \varepsilon_0\,(\phi_2 - \phi_1)$

44. (2) $\phi = \vec{E}.d\vec{s}$

$$\phi = \left(4\hat{i} + 4\hat{j} + 4\hat{k}\right) \cdot \left(5\hat{i}\right) = 20 \text{ units}$$

45. (4) $E = \frac{2Kp}{r^3}$

Now the potential energy of a dipole is given by

$U = -p\,E\cos(x)$

Substituting E and x = 0 since it is co axial

$$U = \frac{-2Kpp}{r^3}$$

$$F = \frac{-dU}{dr}$$

$$F = \frac{6Kpp}{r^4}$$

46. (1) Dipole moment and electric field are opposite to each other on equatorial line. Hence the angle between them is 180 degrees.

47. (2)

48. (2) $q = \pm 4\mu C$

$2a = 5 \times 10^{-3} m$

$|\vec{p}| = q \times 2a$

$= 4 \times 10^{-6} \times 5 \times 10^{-3}$

$|\vec{p}| = 2 \times 10^{-8}$ C-m

49. (3) $qc = 0.1\ \mu C = 10–7C$

$2a = 2cm = 2 \times 10^{-2} m$

$E = 10^5 N/C$, $\tau = ?$

As we know : $\tau = p \times E \sin\theta$

For max τ, $\sin\theta = 1$

$\tau = 2 \times 10^{-4} N - m$

50. (1) $p_{net} = \sqrt{p_1^2 + p_1^2 + 2p_1p_2\cos\theta}$

$p_1 = p_2 = p$

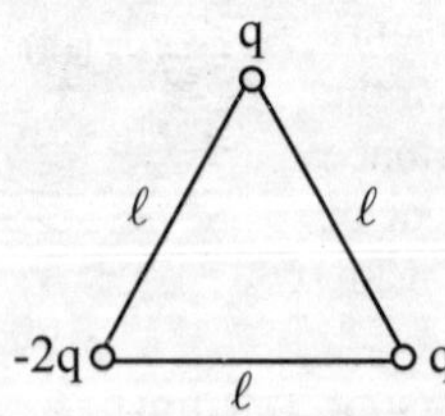

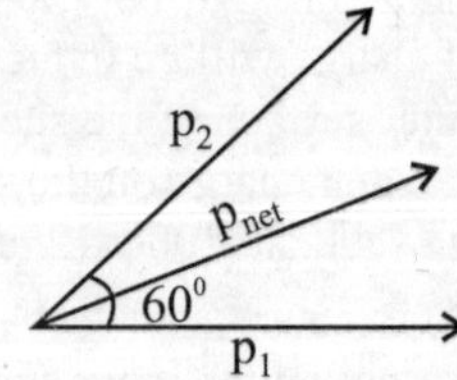

$$p_{net} = \sqrt{p^2 + p^2 + 2p^2\cos 60^o}$$

$$p_{net} = \sqrt{3}p = \sqrt{3}q\ell$$

51. (3) $|\vec{p}| = q \times 2a = 8 \times 10^{-6} \times 5 \times 10^{-3}$

$|\vec{p}| = 40 \times 10^{-9} = 4 \times 10^{-8} C - m$

52. (4) $E = \frac{p}{4\pi\varepsilon_o r^3}\sqrt{3\cos^2\theta + 1}$

$$E = \frac{4 \times 10^{-8} \times 9 \times 10^9}{0.5 \times 0.5 \times 0.5} \times \sqrt{3(0.5)^2 + 1}$$

$= 3.8 \times 10^3$ N/C

53. (4) $p = q \times 2a$

Distance between the center is a

$p = q \times a$

$$a = \frac{7\times10^{-30}}{1.6\times10^{-19}} = 4.375\times10^{-11}\,m$$

54. (1) Torque, $\tau = pE\sin\theta$

$$12\times10^{-20} = p\times10^{7}\times\frac{\sqrt{3}}{2}$$

$\therefore p = 1.4\times10^{-26}\,C-m$

55. (2) $\tau = \vec{p}\times\vec{E}$

$\tau = pE\sin\theta$

$\tau = q\times2\ell\times E\times\sin\theta$

$$q = \frac{\tau}{2\ell\times E\times\sin\theta} \Rightarrow q = \frac{15\times10^{2}}{2\times5\times10^{-2}\times10^{7}\times\frac{1}{\sqrt{2}}}$$

$q = 2.121\times10^{-3} = 2.121\ mC \approx 2\ mC$

56. (3) $W = pE(\cos\theta_1 - \cos\theta_2)$

$W = pE(\cos0^\circ - \cos180^\circ)$

$W = pE(1-(-1)) = 2pE$

High Order Time Intensive MCQs

1. (2) For the **Case-I:**

In equilibrium

$T\cos\theta = mg$ ...(i)

$T\sin\theta = \frac{kq^2}{r^2}$ ...(ii)

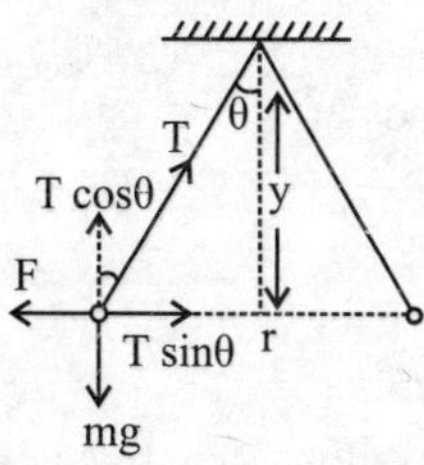

From Eqn. (i) and (ii), we get

$$\tan\theta = \frac{kq^2}{r^2mg} \Rightarrow \frac{r}{2y} = \frac{kq^2}{r^2mg}$$

$$\Rightarrow \frac{r^3}{2y} = \frac{kq^2}{mg} \quad ...(iii)$$

For **Case-II:**

$$\Rightarrow \frac{(r')^3}{2\frac{y}{2}} = \frac{kq^2}{mg} \Rightarrow \frac{(r')^3}{y} = \frac{kq^2}{mg} \quad ...(iv)$$

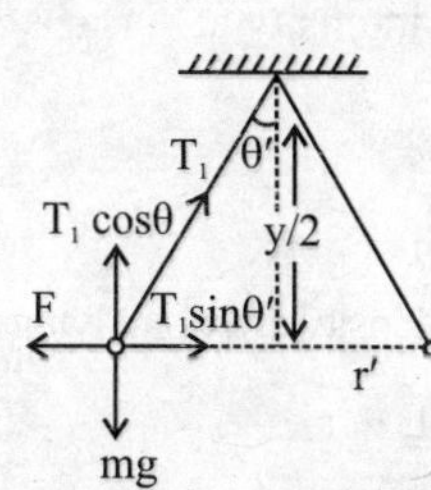

From Eqs. (iii) and (iv), we get

$$\frac{(r')^3}{y} = \frac{r^3}{2y} \Rightarrow r' = \frac{r}{\sqrt[3]{2}}$$

2. (4) We know that,

$$\text{Charge density} \propto \frac{1}{(\text{Radius of curvature})^2} \quad \left[\because \sigma = \frac{q}{4\pi r^2}\right]$$

$\therefore \sigma_2 > \sigma_1 > \sigma_3$

3. (3) $F = \frac{1}{4\pi\varepsilon_0}\frac{q_1q_2}{r^2}$

4. (4) $\frac{14.4\times10^{-19}}{1.6\times10^{-19}} = 9$

5. (2) Deutron (e, 2m), α-particle (2e, 4m)

$F_1 = eE,\ F_2 = 2eE$

$$a_1 = \frac{eE}{2m},\ a_2 = \frac{2eE}{4m} = \frac{eE}{2m}$$

$F_1 \neq F_2,\ a_1 = a_2$

6. (2) The surface charge densities on the outer surface of the shells will be equal then charges on the shells are proportional to the squares of their radii. Hence the charge distribution will be as shown.

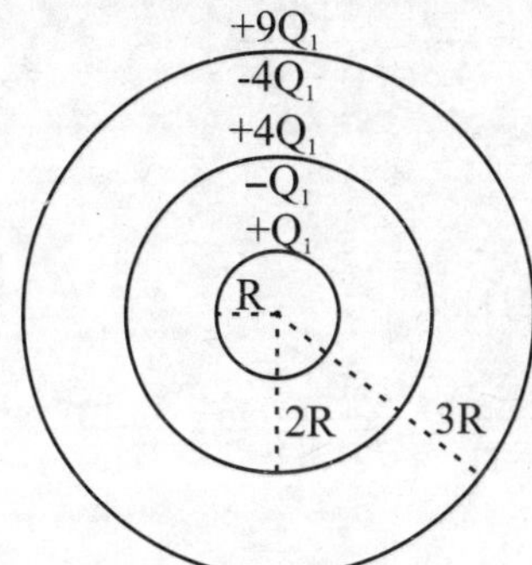

$Q_2 = 4Q_1 - Q_1 = 3Q_1$

$Q_3 = 9Q_1 - 4Q_1 = 5Q_1$

$Q_1 : Q_2 : Q_3 := 1 : 3 : 5.$

7. (1) Relative permittivily is given by

$$\varepsilon_r = \frac{F_a}{F_{in}} = \frac{mg}{(m-m_1)g} = \frac{\rho V}{(\rho-\sigma)V} = \frac{\rho}{(\rho-\sigma)}$$

8. (3) $F = qE = ma$

$$\Rightarrow a = \frac{qE}{m} = \frac{1.6\times10^{-19}\times1\times10^{3}}{9\times10^{-31}} = \frac{1.6\times10^{15}}{9}\,ms^{-2}$$

Given, $u = 5\times10^6$ m/s and $v = 0$

$\therefore$ From $v^2 = u^2 - 2as \Rightarrow s = \frac{u^2}{2a}$

$$\therefore \text{Distance, } s = \frac{(5\times10^6)^2\times9}{2\times1.6\times10^{15}} = 7\ cm$$

9. (1) The flux through a closed surface depends only on the enclosed charge and is independent of size and shape of the surface.

10. (1) The electric intensity outside a charged sphere,

$$E=\frac{kq}{r^2}=\frac{k(\sigma A)}{r^2}$$

$$=\left(\frac{1}{4\pi\varepsilon_0}\right)\frac{\sigma 4\pi R^2}{r^2}=\frac{\sigma R^2}{\varepsilon_0 r^2}$$

11. (1) $T=2\pi\sqrt{\frac{\ell}{g_{eff}}}=2\pi\sqrt{\frac{\ell}{g+\frac{qE}{m}}}$

Hence time period decreases and frequency increases.

12. (3) Electric field intensity at a particular point is directly proportional to the number of electric line of forces crossing per unit area.

$\Rightarrow E_A=E_C$ ...(*i*)

As, density of electric line of force in region *A* and *C* is more as compared to B.

$E_A>E_B$ ...(*ii*)

From Eqs., (*i*) and (*ii*), we get

$E_A=E_C>E_B$

13. (2) Let electric field be zero at distance x from sphere a.

Then, $E_1=E_2$

$$\frac{1}{4\pi\epsilon_0}\frac{10}{x^2}=\frac{1}{4\pi\epsilon_0}\frac{20}{(r-x)^2}$$

$(r-x)^2=2x^2$

$r-x=\sqrt{2}x$

$r=\left(\sqrt{2}+1\right)x$

$$x=\frac{r}{(\sqrt{2}+1)}=\frac{80}{(\sqrt{2}+1)}\text{ cm}=33.137\text{ cm}$$

14. (2) $E=\frac{\lambda}{(2\pi\varepsilon_0 r)}=\frac{2\lambda}{(4\pi\varepsilon_0 r)}$

$E=18\times10^4$ N/C

$\frac{1}{(4\pi\varepsilon_0)}=9\times10^9$ Nm^2/C^2

Linear charge density $\lambda=E.\frac{4\pi\varepsilon_0 r}{2}$

$$=\frac{(18\times10^4\text{ N/C})(2)}{((9\times10^9\text{ Nm}^2/\text{C}^2)\times2)}$$

$=2\times10^{-5}$ C/m

15. (2) At the center net electric field will be due to only that charge which does not has charge on the opposite corner.

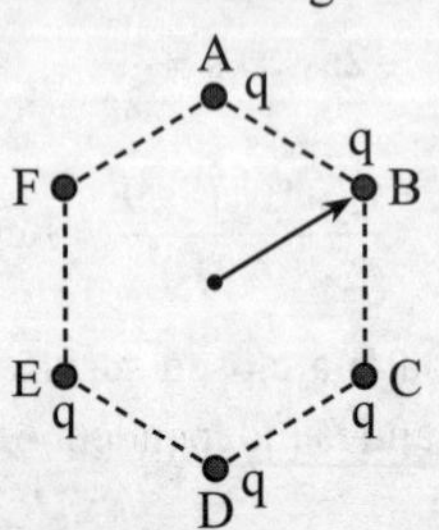

$$E_{net}=\frac{1}{4\pi\varepsilon_0}\frac{q}{x^2}=\frac{1}{4\pi\varepsilon_0}\frac{q}{(2a)^2}$$

$$=\frac{1}{16\pi\varepsilon_0}\frac{q}{a^2}$$

16. (2) Flux = E × Projected area

$=E\pi R^2$

17. (1) As the loop is lying in y-z plane,

So, area vector, $\vec{A}=10\hat{i}$

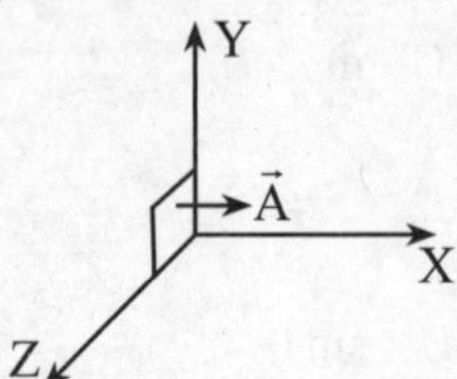

Electric flux, $\phi=\vec{E}\cdot\vec{A}$

$=(10\times10)+(3\times0)+(4\times0)\Rightarrow\phi=100$ units

18. (3)

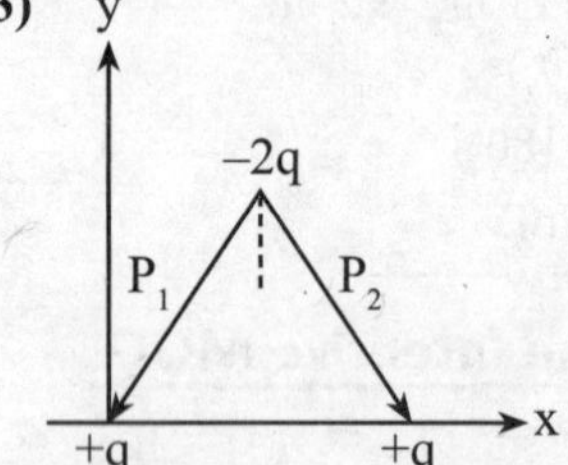

Dipole moment is given by,

P = charge × distance between two charges

So for two pairs of dipoles we can write,

$|P_1|=q(\ell)$

$|P_2|=q(\ell)$

|Resultant| = 2P cos 30°

$$2q\ell\left(\frac{\sqrt{3}}{2}\right)=\sqrt{3}q\ell$$

So, $\vec{P}_{Resultant}=-\sqrt{3}q\ell\hat{j}$

As the resultant is along negative y-axis direction.

19. (2) $E=\frac{1}{4\pi\varepsilon_0}\frac{2p}{r^3}$

20. (4) The electric field intensity at a point lying outside the sphere (non-conduction) is

$$E=\frac{1}{4\pi\varepsilon_0}\frac{q}{r^2}$$

where, *r* is the distance of that point from centre of sphere.

$\therefore\quad E\propto\frac{1}{r^2}$ ...(*i*)

The electric field intensity at surface of sphere,

$E=\frac{q}{4\pi\varepsilon_0 R^2}$ ...(*ii*)

where R, radius of sphere.

The electric field intensity at a point lying inside the sphere is

$E=\frac{qr}{4\pi\varepsilon_0 R^2}$ or $E\propto r$ ...(*iii*)

Also, at the centre of sphere r = 0. Hence, E = 0. The graphical distribution is shown below.

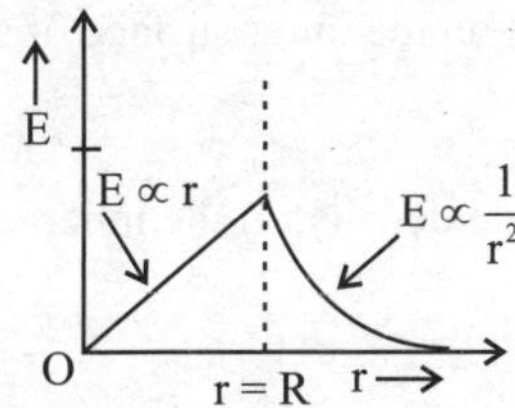

21. (4)

P – +Q, U –, +R, +T, S –, O, $\vec{E}$

Field as O due to P and S are equal and opposite and cancel out. Fields at O due to Q and T also cancel out. But the fields at O due to R and U are in same direction. The net field at O is $2\vec{E}$

22. (1) Gauss' law states that the net electric flux through any closed surface is equal to the net charge inside the surface divided by ε_0.

i.e. $\phi_{total} = \dfrac{q}{\varepsilon_0}$

Let electric flux linked with surfaces A, B and C be ϕ_A, ϕ_B and ϕ_C respectively

Since, $\phi_C = \phi_A$

total $= \phi_A + \phi_B + \phi_C$

$$\therefore \quad 2\phi_A + \phi_B = \phi_{total} = \frac{q}{\varepsilon_0} \text{ or } \phi_A = \frac{1}{2}\left(\frac{q}{\varepsilon_0} - \phi_B\right)$$

$$\text{Hence,} \quad \phi_A = \frac{1}{2}\left(\frac{q}{\varepsilon_0} - \phi\right) \quad (\text{give, } \phi_B = \phi)$$

23. (2) The fields due to the three sheets at point P will point in the negative Z-direction.

Total electric field at pont P is

$$\vec{E} = \frac{\sigma}{2\varepsilon_0}(-\hat{k}) + \frac{2\sigma}{2\varepsilon_0}(-\hat{k}) + \frac{\sigma}{2\varepsilon_0}(-\hat{k})$$

$$= -\frac{2\sigma}{\varepsilon_0}\hat{k}.$$

24. (1) ∵ Centripetal force = Electrostatic Force

$$\frac{mv^2}{r} = \frac{1}{4\pi\varepsilon_0}\frac{q_1q_2}{r^2}$$

$$\Rightarrow v = \sqrt{\frac{q_1q_2}{4\pi\varepsilon_0 mr}}$$

25. (3) Period of revolution: $T = \dfrac{2\pi r}{v} = \dfrac{2\pi r}{\sqrt{q_1q_2}}\sqrt{4\pi\varepsilon_0 rm}$

$$T = 4\sqrt{\frac{\pi^3\varepsilon_0 r^3 m}{q_1q_2}}$$

26. (3) q • ←— 2ℓ —→ • q (charge q displaced by x)

When the charge is displaced by an amount x. Force on it

$$F = \frac{Kq^2}{(\ell+x)^2} - \frac{Kq^2}{(\ell-x)^2} = \frac{Kq^2}{\left[\ell^2 - x^2\right]^2}(-4\ell x)$$

$$\text{or } F = -\frac{4K\ell xq^2}{\left(\ell^2 - x^2\right)^2} = -\frac{\ell xq^2}{\pi\epsilon_0\left(\ell^2 - x^2\right)^2}$$

When x is very small

$$F = -\frac{q^2}{\pi\epsilon_0\ell^3}x \quad \text{or} \quad a = -\frac{q^2}{\pi\epsilon_0 m\ell^3}x$$

$$\omega^2 = \frac{q^2}{\pi\varepsilon_0 m\ell^3} \quad \& \quad T = \frac{2\pi}{\omega}$$

$$\omega = \frac{q}{\sqrt{\pi\varepsilon_0 m\ell^3}}$$

$$T = 2\pi\sqrt{\frac{\pi\varepsilon_0 m\ell^3}{q^2}}$$

27. (3) ∵ Sharp surfaces have less area A and surface charge density $\sigma = \dfrac{q}{A}$

∴ Sharp surfaces will have high charge density.

Surface at A is sharpest.

As radius of C is less than B. So, C is more sharp than B.

$\therefore \sigma_A > \sigma_C > \sigma_B > \sigma_D$

28. (4) Charge = 0 (A) + q (B) → q/2 (A) q/2 (B)

q/2 (A) + q (C) → 3q/4 (A) 3q/4 (C)

B (q/2) ——— r ——— C (3q/4)

$$\therefore F = \frac{1}{4\pi\varepsilon_0}\frac{q.q}{r^2} \quad \text{(Initial force between B and C)}$$

New force between B and C,

$$F' = \frac{1}{4\pi\varepsilon_0}\frac{\left(\frac{q}{2}\right)\left(\frac{3q}{4}\right)}{r^2} = \frac{3}{8}F$$

29. (3)

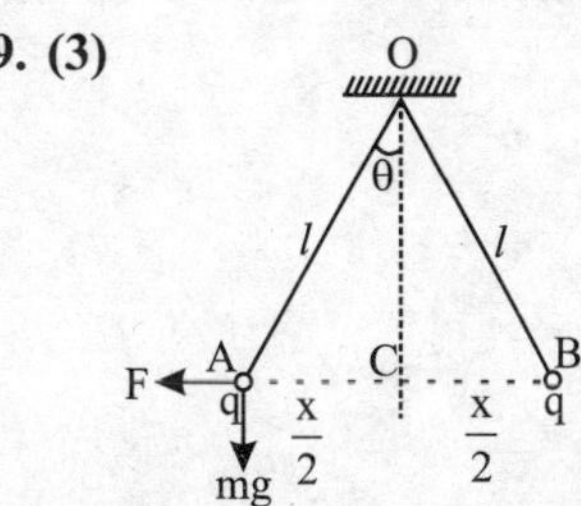

From ΔACO of forces,

$$\frac{F}{AC} = \frac{mg}{OC}$$

$$\frac{kq^2}{x^2(x/2)} = \frac{mg}{\sqrt{l^2 - \left(\frac{x}{2}\right)^2}}$$

$$\frac{2kq^2}{x^3} = \frac{mg}{l} \qquad \text{(x is very small)}$$

$$\therefore\ q^2 = \frac{mg}{2kl}.x^3$$

$$\Rightarrow q^2 \propto x^3 \Rightarrow q \propto x^{3/2}$$

$$\Rightarrow \frac{dq}{dt} \propto \frac{3}{2}x^{\frac{1}{2}}.\frac{dx}{dt} \Rightarrow \frac{dq}{dt} \propto \frac{3}{2}x^{\frac{1}{2}}.v$$

As $\frac{dq}{dt}$ is constant for both spheres, so

$$v \propto \frac{1}{x^{\frac{1}{2}}} \Rightarrow v \propto x^{-\frac{1}{2}}$$

30. (2) We have, $q'_1 = q'_2 = \left(\frac{q_1 + q_2}{2}\right) = \left(\frac{10-20}{2}\right) = -5\mu C$

$$\therefore\ \frac{F_1}{F_2} = \frac{q_1 q_2}{q'_1 q'_2} = \frac{(10)(-20)}{(-5)\times(-5)} = -\frac{8}{1}$$

31. (3) Total flux enclosed by surface $= \frac{Q_{inclosed}}{\varepsilon_0}$

$$= \frac{1-7-4+10+2-5-3+6}{\varepsilon_0} \times 10^{-6} = 0$$

32. (4) At the centre of the ring, E = 0 when a positive charge q > 0 is displaced away from the centre in the plane of the ring, say to the right, force of repulsion on q, due to charge on right half increases and due to charge on left half decreases. Therefore, charge q is pushed back towards the centre. So option a. is correct.

When charge q is negative (q < 0), force is of attraction, Therefore, charge q displaced to the right continues moving to the right till it hits the ring.

Along the axis of the ring, at a distance r from the centre,

$$E = \frac{Qr}{4\pi\varepsilon_0 (r^2 + a^2)^{3/2}}$$

If charge q is negative (q < 0), it will perform SHM for small displacement along the axis.

33. (2) When a charge +q is placed at the centre of spherical cavity as shown in figure.

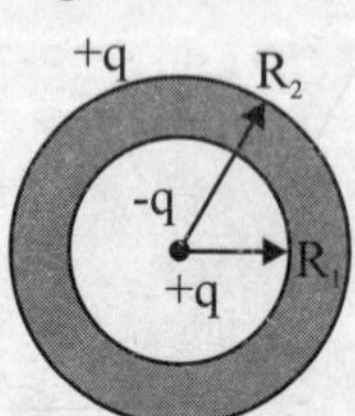

Charge induced on the inner surface of shell $= -q$ ...(*i*)

Charge induced on the outer surface of shell $= +q$ ...(*ii*)

∴ Surface charge density on the inner surface $= \frac{-q}{4\pi R_1^2}$

34. (1) By Gauss's law,

$$\phi = \frac{q}{\varepsilon_0}$$

or $q = \phi\varepsilon_0$
$= 1.05 \times 8.854 \times 10^{-12}$ C
$= 9.27 \times 10^{-12}$ C

35. (4) As the field lies in the plane of the surface, it is perpendicular to the normal to this surface.

Flux, $\phi = \vec{E}.\vec{S} = ES\cos 90° = 0$

36. (3)

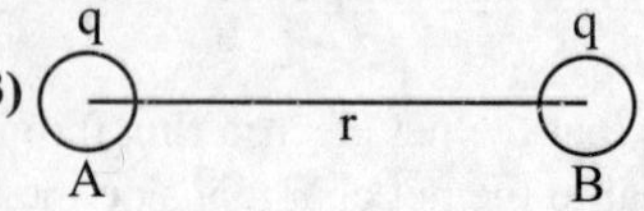

Original force of repulsion, $F = k\frac{qq}{r^2}$

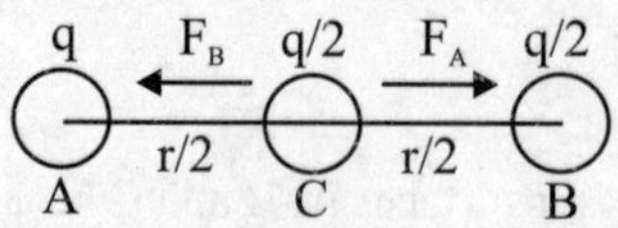

Net force on C,

$$F_C = F_A - F_B = k\left[\frac{q \times \frac{q}{2}}{\left(\frac{r}{2}\right)^2} - \frac{\frac{q}{2} \times \frac{q}{2}}{\left(\frac{r}{2}\right)^2}\right]$$

$$F_C = k\frac{qq}{r^2}[2-1] = F.$$

37. (3) Let n_e be the number of electrons inside the body. Then

$q = n_e e^- + n_p e^+$

$-2 \times 10^{-6} = -n_e \times 1.6 \times 10^{-19} + 2.5 \times 10^{13} \times 1.6 \times 10^{-19}$

$\therefore\quad n_e = 3.75 \times 10^{13}$

38. (1) As $P = \frac{F}{A}$ (P : Pressure, F : Force, A: Area)

$$\Rightarrow F = PA = \frac{\sigma^2}{2\varepsilon_0}\pi R^2$$

$$\Rightarrow F \propto \frac{\sigma^2 R^2}{\varepsilon_0}$$

39. (2) $E_q + E_{3q}$ is along PA

$E_{2q} + E_{4q}$ is along PB

∴ E_{net} is along the direction of CB.

40. (1) Surface charge density $(\sigma) = \frac{\text{Charge}}{\text{Surface area}}$

So, surface charge density on the inner surface

$$\sigma_{inner} = \frac{-2Q}{4\pi b^2}$$

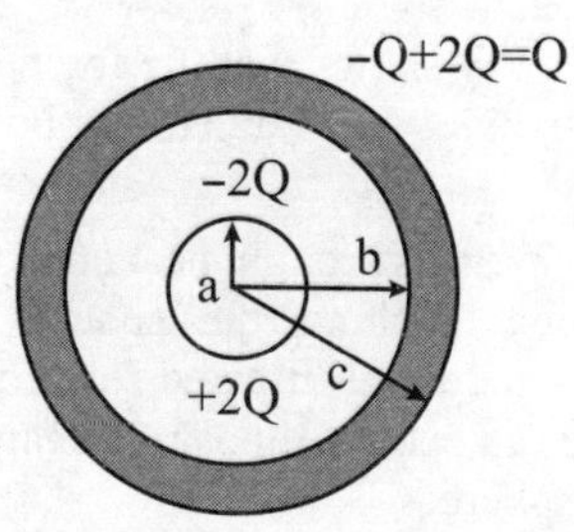

and surface charge density on the outer surface

$$\sigma_{outer} = \frac{Q}{4\pi c^2}$$

41. (1)

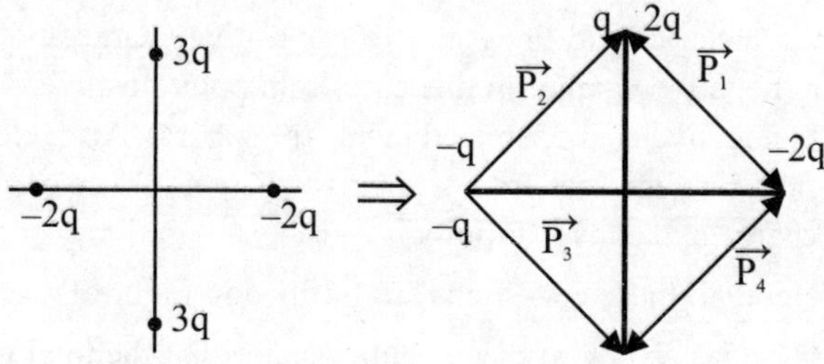

Here $\quad p_x = 0$

and $\quad p_y = 2qa$

$\therefore\ |\vec{p}_2| = |\vec{p}_3|$

& $\ |\vec{p}_1| = |\vec{p}_4|$

$\Rightarrow\ \vec{p}_{net} = 2qa\hat{j}$

42. (1) Coulomb's force on each bob due to other will have equal magnitude and mass of both the bob is same. So deflection in both spring will be same.

Statement Based MCQs

1. **(1)** The charges gets induced on the conductor plate due to polarization which creates electric field in the surroundings due to which the point charges will experience a force.
2. **(3)** Surface energy is dissipated in the form of heat during sharing of charges between different bodies. Charge is conserved quantity. Energy conservation never fails.
3. **(4)** $\vec{E}$ decreases at different rates when going away from a point charge or a dipole. Tangent at any point of electric line of force shows direction of electric field.
4. **(1)** As, surface densities are equal,

$\sigma_1 = \sigma_2$

$$\therefore\ \frac{q_1}{4\pi r_1^2} = \frac{q_2}{4\pi r_2^2} \Rightarrow \frac{q_1}{q_2} = \frac{r_1^2}{r_2^2} \qquad ...(i)$$

Where q_1, q_2 and r_1, r_2 are charges and radii of two spheres.

Now, ratio of electric fields

$$\frac{E_1}{E_2} = \frac{kq_1}{r_1^2} \times \frac{r_2^2}{kq_2} = \frac{q_1}{q_2} \times \frac{r_2^2}{r_1^2}$$

Using equation (*i*)

$$\frac{E_1}{E_2} = 1 \Rightarrow E_1 = E_2$$

5. **(1)** When dielectric (K) is placed between two charges then force,

$$F_m = \frac{F_a}{k}$$

Force reduces for dielectric medium.

For conducting medium, K = ∞

$$\therefore\ F_m = \frac{F_a}{\infty} = 0$$

Assertion & Reason MCQs

1. **(1)** Charge is invariant as it is independent of frame of reference. It is a fixed quantity.
2. **(1)** The end of the rod is made pointed which have very less radius hence it has more surface charge density.
3. **(1)** As a conductor has no electric field inside it so it can be used as a shield to block the electric field.
4. **(1)** Both assertion and reason are correct and reason is the correct explanation of the assertion.
5. **(4)** The charges are only at the surface of a conductor, the charge enclosed in the Gaussian surface in the cavity is zero. Hence the field is zero.

Match the Columns MCQs

1. (2)

(a) 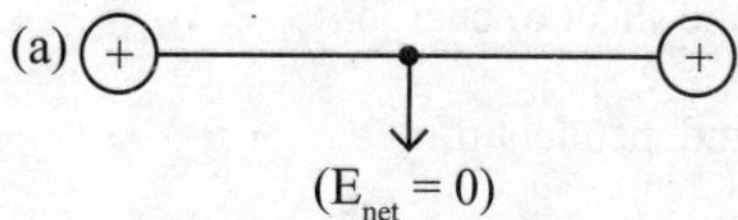

($E_{net} = 0$)

As we go closer to either of the charges, then E will increase.

So, option (i, iii, iv) are correct.

(b) Inside the dipole, when we will go closer to the either of charge, E will increase.

And when we go x > a then system will behave as dipole and upon increasing x, E will decrease.

(c) It is very clear the at origin, E = 0 and upon increasing x, E will decrease.

(d) (ii, iv)

2. (2)

(a) Initially, the potential difference exist between both shells, so positive charge is flow from high to low potential.

Very system wants to acquire minimum potential energy if possible for stability. So charge flown to achieve it.

(b) As explained in [a], charge flow does not depends on the size of sphere.

(c) Charge flow through wire until the potential becomes same for both shells.

(d) Potential is same everywhere inside a conducting shell. So no charge is flow through connecting wire, so no heat is produced.

3. (1)

(a) As charge at B is removed. $\vec{E}_C$ and $\vec{E}_F$ cancelled each other.

$\vec{E}_E$ and $\vec{E}_D$ are at 60°. Hence $|\vec{E}_{net}| = \sqrt{3}E$

(b) As charge at C is removed. $\vec{E}_B$ and $\vec{E}_E$ cancelled each other.

$\vec{E}_F$ and $\vec{E}_D$ are at 120°. Hence $|\vec{E}_{net}| = E$

(c) As charge at D is removed. $\vec{E}_F$ and $\vec{E}_C$ cancelled each other. $\vec{E}_B$ and $\vec{E}_E$ cancelled each other. Hence $|\vec{E}_{net}| = 0$

(d) As charge at B and C are removed. $\vec{E}_F$ and $\vec{E}_D$ at 120°. $\vec{E}_E$ is also in resultant direction. Hence $|\vec{E}_{net}| = 2E$

4. (2) For a point charge Q: $E = \frac{1}{4\pi\varepsilon_0}\frac{Q}{d^2} \Rightarrow E \propto \frac{1}{d^2}$

For point on the axis of a dipole:

$$E = \frac{1}{4\pi\varepsilon_0}\cdot\frac{2pd}{(d^2-\ell^2)^2} \simeq \frac{1}{4\pi\varepsilon_0}\cdot\frac{2(Q\times 2\ell)}{d^3}$$

$$\Rightarrow E \propto \frac{1}{d^3}$$

For an infinite line charge:

$$E = \frac{\lambda}{2\pi\varepsilon_0 d} \Rightarrow E \propto \frac{1}{d}$$

For an infinite plane sheet of charge:

$$E = \frac{\sigma}{2\varepsilon_0} \Rightarrow \text{E is independent of d}$$

NCERT Exemplar MCQs

1. (1) The net force on q_1 by q_2 and q_3 is along the +x -direction, so nature of force between q_1, q_2 and q_3 is attractive. This can be represent by the figure given below.

The attractive force between these charges states that q_1 is a negative charge (since, q_2 and q_3 are positive).

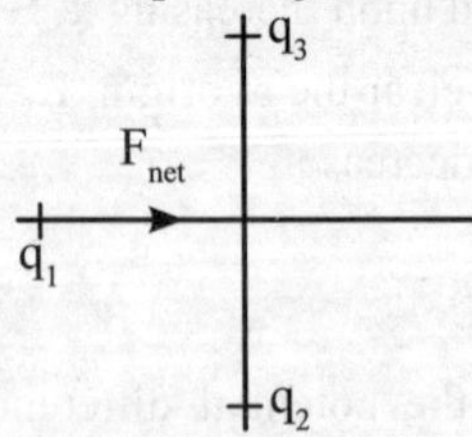

Thus, nature of force between q_1 and newly introduced charge Q (positive) is attractive and net force on q_1 by q_2, q_3 and Q are along the same direction as given in the diagram below.

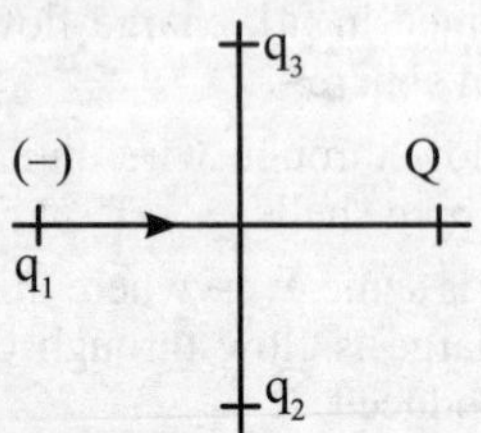

The figure given above clearly shows that the force on q_1 shall increase along the positive x-axis due to the positive charge Q.

2. (1) When a positive point charge is brought near an isolated conducting sphere without touching the sphere, then the free electrons in the sphere are attracted towards the positive charge. This leaves an excess of positive charge on the rear (right) surface of sphere.

Both kinds of charges are bound in the metal sphere and cannot escape. They, therefore, reside on the surface.

Thus, the left surface of sphere has an excess of negative charge and the right surface of sphere has an excess of positive charge.

An electric field lines start from positive charge and ends at negative charge (in this case from point positive charge to negative charge created inside the sphere). Also, electric field line emerges from a positive charge, in case of single charge and ends at infinity.

Here, all these conditions are fulfilled in Figure 1.

3. (4) Gauss's law of electrostatics states that the total of the electric flux out of a closed surface is equal to the charge enclosed divided by the permitivity i.e., $\phi = \frac{Q}{\varepsilon_0}$.

Thus, electric flux through a surface doesn't depend on the shape, size or area of a surface but it depends on the number of charges enclosed by the surface.

So, here in this question, all the figures have same electric flux as all of them has single positive charge.

4. (2) According to Gauss's law, the term q on the right side of the equation $\int_S E.dS = \frac{q}{\varepsilon_0}$, includes the sum of all charges enclosed by the surface.

The charges may be located anywhere inside the surface, if the surface is so chosen that there are some charges inside and some outside, the electric field on the left side of equation is due to all the charges, both inside and outside S.

So, E on LHS of the above equation will have a contribution from all charges while q on the RHS will have a contribution from q_2 and q_4 only.

5. (3) The space between the electric field lines is increasing, here from left to right and its characteristics states that, strength of electric field decreases with the increase in the space between electric field lines. As a result force on charges also decreases from left to right.

Thus, the force on charge –q is greater than force on charge + q in turn dipole will experience a force towards left.

6. (1) When a point positive charge brought near an isolated conducting plane, some negative charge develop on the surface of the plane towards the charge and an equal positive charge develop on opposite side of the plane. This process is called charging by induction.

7. **(1)** When the point is situated at a point on diameter away from the centre of hemisphere charged uniformly positively, the electric field is perpendicular to the diameter. The component of electric intensity parallel to the diameter cancel out.

NEET & AIIMS Past Year MCQs

1. **(4)** Given:
   Side of cube (a) = 5×10^{-2} m,
   Charge (Q) = 6 μC = 6×10^{-6} *C*.
   Surface area:
   A cube has 6 faces, so:
   $A = 6a^2 = 6 \times (5 \times 10^{-2})^2 = 6 \times 25 \times 10^{-4} = 150 \times 10^{-4}\ m^2$
   Surface charge density (σ):
   $$\sigma = \frac{Q}{A} = \frac{6\times10^{-6}}{150\times10^{-4}} = \frac{6}{150}\times10^{-2}$$
   $$= 0.04\times10^{-2} = 0.4\times10^{-3}\,C\,m^{-2}$$
2. **(1)**

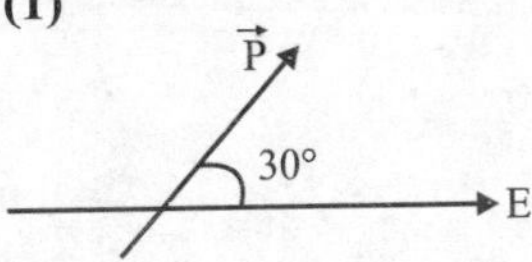

$$\vec{\tau} = \vec{P}\times\vec{E}$$
$$|\vec{\tau}| = PE\sin\theta \Rightarrow 4 = q \times 2a \times E\sin 30^\circ$$
$$q = \frac{4}{(2\times10^{-2})\times2\times10^{5}\times\left(\frac{1}{2}\right)} = 2 \times 10^{-3}\text{ C} = 2\text{ mC}$$

3. **(2)** $\oint \vec{E}\cdot d\vec{s} = 0 \Rightarrow \phi_{net} = \phi_{in} - \phi_{out} = 0 \Rightarrow \phi_{in} = \phi_{out}$
4. **(3)** Electric field intensity due to a dipole at a large distance is given by
   $$E \propto \frac{P}{R^3} \Rightarrow E \propto \frac{1}{R^3}$$
5. **(1)**
6. **(1)** Curl of any conservative field is zero.
7. **(1)** Both assertion and reason are true and reason is the correct explanation of assertion.
8. **(2)** In a non-uniform electric field
   $$\text{Net force, } F = q\,\Delta E = q\left(\frac{\Delta E}{\Delta x}\right)\Delta x$$
   $$= (q\Delta x)\left(\frac{\Delta E}{\Delta x}\right) = p\times1 = p$$

Electrostatic Potential and Capacitance

NCERT Topic-wise MCQs

1. **(1)** The potential due to smaller sphere is given by

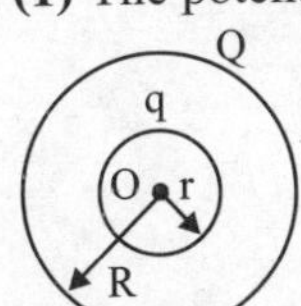

$$V_1 = \frac{1}{4\pi\varepsilon_0}\cdot\frac{q}{r} + \frac{1}{4\pi\varepsilon_0}\cdot\frac{Q}{R} \quad \ldots(i)$$

The potential due to bigger sphere is given by

$$V_2 = \frac{1}{4\pi\varepsilon_0}\cdot\frac{Q}{R} + \frac{1}{4\pi\varepsilon_0}\cdot\frac{q}{R} \quad \ldots(ii)$$

So, the potential difference between the spheres

$$V = V_1 - V_2$$

or $V = \frac{1}{4\pi\varepsilon_0}\cdot\frac{q}{r} + \frac{1}{4\pi\varepsilon_0}\cdot\frac{Q}{R} - \frac{1}{4\pi\varepsilon_0}\cdot\frac{Q}{R} - \frac{1}{4\pi\varepsilon_0}\cdot\frac{q}{R}$

$$V = \frac{1}{4\pi\varepsilon_0}\cdot\frac{q}{r} - \frac{1}{4\pi\varepsilon_0}\cdot\frac{q}{R} = \frac{1}{4\pi\varepsilon_0}\left(\frac{q}{r} - \frac{q}{R}\right)$$

2. **(3)** As we know that electric field and electric potential are related as:
   $$E = \frac{-dV}{dx}$$
   $$E = \frac{-d(3x^2+5)}{dx} = -(6x+0)$$
   $$E = -6x$$
   $$E_{(-2,\,1,\,0)} = -6(-2) = +12\text{ Vm}^{-1}$$
3. **(3)** $W = -pE(\cos\theta_2 - \cos\theta_1)$
   $= pE(\cos\theta_1 - \cos\theta_2)$
   $= 2 \times 10^{-12} \times 3 \times (\cos 60^\circ - \cos 180^\circ)$
   $W = 9 \times 10^{-12}$ J
4. **(4)** Electrostatic force is a conservative force so work done in carrying an electric charge Q, once around a circle of radius R will be zero.
5. **(1)** K.E = qV = $1.6 \times 10^{-19} \times 100 = 1.6 \times 10^{-17}$J.
6. **(4)** Electrostatic potential energy of the given system is given as.
   $$U = \frac{1}{4\pi\varepsilon_o}\times\frac{1}{a}(Qq+Qq+q^2)$$
   $2Qq + q^2 = 0 \qquad [\because U = 0]$
   $2Qq = -q^2$
   $$\therefore\ Q = \frac{-q}{2}$$
7. **(1)** Surface charge densities are given as
   $$\sigma = \frac{Q}{A} = \frac{Q}{4\pi R^2}$$
   As $\sigma_1 = \sigma_2$
   $$\Rightarrow \frac{Q_1}{4\pi\varepsilon_o R_1^2} = \frac{Q_2}{4\pi\varepsilon_o R_2^2} \qquad [\because Q_1 = Q_2 = Q]$$
   $$\frac{V_1}{V_2} = \frac{R_1}{R_2} \qquad \left[\because V = \frac{Q}{4\pi\in_0 R}\right]$$

8. (4) As we know

$$E = \frac{-dV}{dx}$$

$$E = \frac{-d(8x^3)}{dx} = -(16x^2)$$

$$E_{(0,-3,1)} = -16 \times (0)^2 = 0$$

9. (4) $V_B - V_A = \frac{W}{q},\ V_A = 10V$ (Given)

$$V - 10 = \frac{20}{4} \Rightarrow V = 15V$$

10. (3) Volume of 8 drops = volume of the bigger drop

$$\frac{4}{3}\pi r^3 \times 8 = \frac{4}{3}\pi R^3 \Rightarrow 2r = R$$

According to the charge conservation $8q = Q$

Potential of one small drop $V' = \frac{q}{4\pi\varepsilon_o r}$

Similarly potential of big drop $V = \frac{Q}{4\pi\varepsilon_o R}$

$$\frac{V'}{V} = \frac{q}{Q} \times \frac{R}{r} \Rightarrow \frac{V'}{20} = \frac{q}{8q} \times \frac{2r}{r}$$

$$V = \frac{20}{4} = 5V$$

Short trick:

$V = n^{2/3}V'$

$20 = (8)^{2/3}V'$

$20 = 4V'$

$V' = 5$ volt

11. (4) As electric potential inside the sphere is equal to the potential at its surface.

Electric potential at center = Electric potential on the surface.

12. (2) Electric potential at surface

$$V = \frac{1}{4\pi\varepsilon_o}\frac{q}{R}$$

$$V = \frac{9\times10^9 \times 50\times1.6\times10^{-19}}{9\times10^{-15}} = 8\times10^6 V$$

13. (1) $V_B - V_A = \frac{W}{q_o}$

$$\Delta V = \frac{40}{500} = 0.08V$$

14. (2) $\Delta U = U_2 - U_1$

$$\Delta U = \frac{q_1q_2}{4\pi\varepsilon_o r_2} - \frac{q_1q_2}{4\pi\varepsilon_o r_1}$$

$$\therefore \Delta U = \frac{q_1q_2}{4\pi\varepsilon_o}\left[\frac{1}{0.4} - \frac{1}{1}\right]$$

$$\Delta U = 18 \times 15 \times 10^{-12} \times 9 \times 10^9 (1.5)$$

Work done = Change in Potential energy

$W = \Delta U = 3.6 \times 10^{-6}$ J

15. (4) $W = q\,\Delta V$

$W = 150 \times 2e\,(V_B - V_A)$

$\therefore\ W = 150 \times 2 \times 1.6 \times 10^{-19} (-5 - 18)$

$W = 1.1 \times 10^{-15}$ J

16. (3) $U = U_1 + U_2 + U_3$

$$U = \frac{K(q)(-2q)}{a} + \frac{K(-2q)(q)}{a} + \frac{K(q)(q)}{2a}$$

$$U = \frac{-4Kq^2}{a} + \frac{Kq^2}{2a}$$

$$U = \frac{Kq^2}{a}\left[-4 + \frac{1}{2}\right]$$

$$\Rightarrow U = \frac{-7Kq^2}{2a}$$

$$U = \frac{-7q^2}{8\pi\varepsilon_o a}$$

17. (1) Due to unlike charges potential at a point midway between them is zero.

$q_1 = q,\ q_2 = -q$, Distance = 5d

$$V_A = \frac{k(q)}{2.5d} + \frac{k(-q)}{2.5d}$$

$V_A = 0$ volt

(Mid Point)
+q A −q
2.5d 2.5d

18. (3) Potential Energy = $-pE\cos\theta$

$= -17 \times 10^{-8} \times 2 \times 10^5 \times \cos 30^o$

$= -2.9 \times 10^{-2}$ J

19. (1) Total electric potential at point C is

$V_{net} = V_{AC} + V_{BC}$

$$V_{net} = \frac{Kq}{r} + \frac{Kq}{r}$$

$$V_{net} = \frac{2Kq}{r}$$

$$V_{net} = \frac{2\times9\times10^9\times10\times10^{-6}}{0.5m} = 36\times10^4 V$$

B 10μC
0.5m 0.5m
A 10μC C
0.5m

20. (2) Let $r_1 = 20$ cm, $r_2 = 5$ cm

Given $q_1 = -8\mu C,\ q_2 = +4\mu C,\ q = +3\mu C$

Potential at A

$$V_A = \frac{Kq_1}{r_1} + \frac{Kq_2}{r_2} \Rightarrow V_A = K\left[\frac{-8\times10^{-6}}{20\times10^{-2}} + \frac{4\times10^{-6}}{5\times10^{-2}}\right]$$

$V_A = K \times 10^{-4}[-0.4 + 0.8]$

$V_A = 0.4 \times K \times 10^{-4}$ V

Potential at B, $V_B = \frac{Kq_2}{r_1} + \frac{Kq_1}{r_2}$

$$V_B = K\left[\frac{4\times10^{-6}}{20\times10^{-2}} - \frac{8\times10^{-6}}{5\times10^{-2}}\right]$$

$V_B = K \times 10^{-4}\,[0.2 - 1.6]$

$V_B = -1.4\,K \times 10^{-4}\,V$

Charge is moved from B to A then, work done is

$W = q\,(V_A - V_B)$

$W = 3 \times 10^{-6}\,((0.4K \times 10^{-4}) - (-1.4\,K \times 10^{-4}))$

$\therefore\; W = 3 \times 10^{-6} \times 1.8 \times 9 \times 10^{9} \times 10^{-4}$

$W = 48.6 \times 10^{-1} = 4.86\,J \approx 4.9\,J$

21. (3) Electric potential at point P

$$V_P = \frac{KQ}{r} + \frac{Kq}{r}$$

$$V_P = \frac{KQ}{\frac{R}{2}} + \frac{Kq}{\frac{R}{2}}$$

$$V_P = \frac{2K}{R}(10-2)10^{-6}$$

$$V_p = \frac{2\times9\times10^{9}\times8\times10^{-6}}{R} = \frac{144\times10^{3}}{R}\text{ volt}$$

$$V_P = \frac{1.44\times10^{5}\text{ volt}}{R}$$

22. (2) q_1 (Proton) = e, q_2 (α – Particle) = 2e

$$P.E = \frac{kq_1q_2}{r} \Rightarrow P.E = \frac{9\times10^{9}\times1.6\times10^{-19}\times2\times1.6\times10^{-19}}{2\times10^{-10}}$$

$$P.E = \frac{4.8\times9}{2}\times10^{-19} = 21.6 \times 10^{-19}J \simeq 2.2 \times 10^{-18}J$$

23. (2) Capacitance of a parallel plate capacitor having a metallic plate of thickness t is

$$C = \frac{\varepsilon_o A}{d-t}$$

$\because$ Metallic plate is thin, so t = 0

$$\therefore\; C = \frac{\varepsilon_o A}{d-0} = \frac{\varepsilon_o A}{d}$$

24. (1) $C = \frac{\varepsilon_o A}{d}$

$$C \propto \frac{1}{d}$$

Capacity of capacitor depends upon the shape, size and seperation between the plates.

25. (2) $\frac{C_1}{C_o} = \frac{\varepsilon_o A_1}{d_1} \times \frac{d}{\varepsilon_o A}$ $\quad\therefore$ where $A_1 = 2A$

$d_1 = 3d$

$$\frac{C_1}{C_o} = \frac{2A}{3d} \times \frac{d}{A} = \frac{2}{3} \Rightarrow C_1 = 60 \times \frac{2}{3} = 40\mu F$$

26. (4) $Q = CV \;\Rightarrow Q = 500 \times 10^{-6} \times 10 = 5000 \times 10^{-6}C$

The time interval required to charge the capacitor to $5000 \times 10^{-6}C$ will be equal to the one required for producing potential difference of 10V.

$$\text{Time } t = \frac{\text{Total charge}}{\text{Rate of charging}} = \frac{5000\times10^{-6}}{125\times10^{-6}} = 40\text{sec}$$

27. (1) For the two capacitors connected in parallel,

$V_1 = V_2$

or $\frac{C_1}{q_1} = \frac{C_2}{q_2}$

or $\frac{q_1}{q_2} = \frac{C_1}{C_2}$.

28. (1) As we know, Electric field = $\frac{\text{Voltage}}{\text{Separation}}$

$$\therefore\; E = \frac{V}{d} = \frac{100}{10^{-3}} = 10^{5} = 100000\text{ V/m}$$

29. (1) Potential difference = 10 – (–10) = 20

From the relation

$q = CV$

$$C = \frac{q}{V} = \frac{40}{20} = 2\,F$$

30. (3) $C = \frac{\varepsilon_0 KA}{d}$

$$C' = \frac{\varepsilon_0(2K)A}{d/2} = \frac{4\varepsilon_0 KA}{d} = 4C$$

31. (4) When separation increases capacitance decreases.

$$C \propto \frac{1}{d} \Rightarrow \text{and } V = \frac{q}{C}$$

$\therefore$ V increases.

32. (2) Capacity of an isolated sphere is $4\pi\varepsilon_o R$

Therefore, capacity $\propto$ R

33. (3) $C = 4\pi\varepsilon_o R$

$$\therefore C = \frac{2}{9\times10^{9}} = 0.22\times10^{-9}$$

$C = 2.2 \times 10^{-10}\,F$

34. (1) $C_{AB} = \frac{C_1(C_2+C_3)}{C_1+C_2+C_3} = \frac{10(10+20)}{10+10+20} = \frac{300}{40} = 7.5\,\mu F$

35. (1) $C_{AB} = C_o + C_{med.}$

$$= \frac{\varepsilon_0\frac{A}{2}}{d} + \frac{\varepsilon_0 K\frac{A}{2}}{d}$$

$$= \frac{\varepsilon_0 A}{2d}[1+K]$$

$$C_{AB} = \frac{C}{2}(1+K)$$

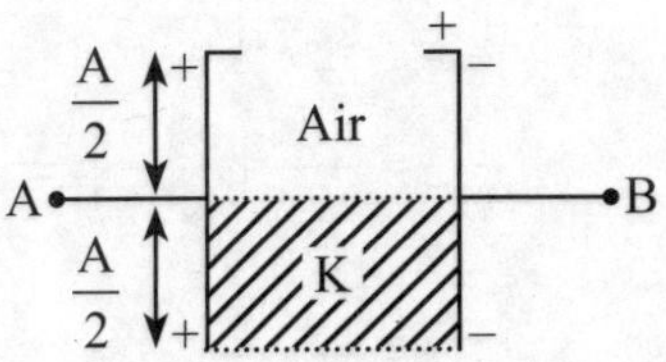

36. (2) Charge on the capacitor,

$$Q = \frac{q_1-q_2}{2} = \frac{10-(-2)}{2} = 6\,\mu C$$

$$C = \frac{Q}{V} = \frac{6\times10^{-6}}{12}F = 0.5\,\mu F.$$

37. (2) $V = Ed$

As E remains the same, so V increases as distance increases.

38. (4) When a dielectric slab is introduced between the plates, the new capacitance will be

$$C_1 = \frac{\varepsilon_o A}{d - t\left(1 - \frac{1}{K}\right)}$$

$$\therefore \frac{C_1}{C} = \frac{\varepsilon_o A}{d - t\left(1 - \frac{1}{K}\right)} \times \frac{d}{\varepsilon_o A}$$

$$\therefore \frac{C_1}{C} = \frac{d}{d - t\left(1 - \frac{1}{K}\right)}$$

$$\frac{C_1}{C} = \frac{5\times10^{-3}}{5\times10^{-3} - 2\times10^{-3}\left(1 - \frac{1}{3}\right)}$$

$$C_1 = \frac{5\times 20}{3.67} = 27.2\mu F$$

39. (1) $P = \frac{W}{t} \Rightarrow P = \frac{1}{2}\frac{CV^2}{t}$

$$P = \frac{80\times10^{-6}\times10^6}{2\times3\times10^{-3}} = 13.3\times10^3 \text{ watt}$$

$P = 13.3$ kW

40. (1) $\Delta E = \frac{1}{2}CV_2^2 - \frac{1}{2}CV_1^2$

$$\therefore \Delta E = \frac{1}{2}C\left(V_2^2 - V_1^2\right)$$

$$\therefore \Delta E = \frac{1}{2}\times12\times10^{-6}\left((40)^2 - (20)^2\right)$$

$\Delta E = 6 \times 10^{-6}\,(1600 - 400)$

$\Delta E = 6 \times 10^{-6}\,(1200)$

$\Delta E = 72 \times 10^{-4}\text{ J} \simeq 7.2 \times 10^{-3}\text{ J}$

41. (3) This is a series combination because there is separation charges and in parallel combination area changes.

$$\frac{1}{C_{eq}} = \frac{1}{C_1} + \frac{1}{C_2} \Rightarrow \frac{1}{C_{eq}} = \frac{d_1}{K_1\varepsilon_o A} + \frac{d_2}{K_2\varepsilon_o A}$$

$$\therefore \frac{1}{C_{eq}} = \frac{d}{2\varepsilon_o A}\left[\frac{1}{2} + \frac{1}{5}\right]$$

$$\therefore \frac{1}{C_{eq}} = \frac{d}{2\varepsilon_o A}\left(\frac{5+2}{10}\right) \Rightarrow \frac{1}{C_{eq}} = \frac{7d}{20\varepsilon_o A}$$

$$\therefore C_{eq} = \frac{20\varepsilon_o A}{7d}$$

42. (2) $C_m = \dfrac{\varepsilon_o A}{d - t + \frac{t}{K}}$

$$C_0 = \frac{\varepsilon_o A}{d - t} \quad (K = \infty \text{ for metals})$$

$$C_0 = \frac{\varepsilon_o A}{d} \quad \Rightarrow \frac{C_m}{C_0} = \frac{\varepsilon_o A}{d - t} \times \frac{d}{\varepsilon_o A} = \frac{d}{d - t}$$

$$\frac{C_m}{C_0} = \frac{8}{8-2} = \frac{8}{6} \Rightarrow C_m = 25\times\frac{4}{3} = 33.3\mu F$$

43. (2) Area is divided by dielectrics so, this is parallel combination.

$$\therefore C_1 = C_m + C_o$$

$$C_1 = \frac{K\varepsilon_o A}{3d} + \frac{2\varepsilon_o A}{3d}$$

$$C_1 = \frac{\varepsilon_o A}{3d}(K+2) \Rightarrow C_1 = \frac{C}{3}(K+2)$$

44. (3) $C_1 = \dfrac{K_1\varepsilon_o A}{d/2},\; C_2 = \dfrac{K_2\varepsilon_o A}{d/2}$

$$C_1 = \frac{2K_1\varepsilon_o A}{d},\; C_2 = \frac{2K_2\varepsilon_o A}{d}$$

$$\frac{1}{C_{eq}} = \frac{1}{C_1} + \frac{1}{C_2} \Rightarrow \frac{1}{C_{eq}} = \frac{1}{\frac{2K_1\varepsilon_o A}{d}} + \frac{1}{\frac{2K_2\varepsilon_o A}{d}}$$

$$\frac{1}{C_{eq}} = \frac{d}{2K_1\varepsilon_o A} + \frac{d}{2K_2\varepsilon_o A} \Rightarrow \therefore \frac{1}{C_{eq}} = \frac{d}{2\varepsilon_o A}\left(\frac{K_1 + K_2}{K_1K_2}\right)$$

$$C_{eq} = \frac{2\varepsilon_o A}{d}\left(\frac{K_1K_2}{K_1 + K_2}\right)$$

45. (3) From the relation:

$E = \frac{1}{2}CV^2$ (Heat produced in the form of energy)

$$E = \frac{1}{2}\times2\times10^{-6}\times10^4$$

$E = 0.01$ J

46. (4) Doubling the separation of plates will halve the value of capacitance equal to 1 pF. Now dielectric insertion will increase the capacitance K times.

47. (4) $C_{air} = \dfrac{\varepsilon_o A}{d}$ with dielectric slab

$$C_1 = \frac{\varepsilon_o A}{\left(d - t + \frac{t}{K}\right)}$$

$$\text{Given } C_1 = \frac{4}{3}C = \frac{4}{3}\frac{\varepsilon_o A}{d} \Rightarrow \frac{\varepsilon_o A}{d - t + \frac{t}{K}} = \frac{4}{3}\frac{\varepsilon_o A}{d}$$

$$K = \frac{4t}{4t - d} = \frac{4(d/2)}{4[d-2] - d} = 2$$

$K = 2$

48. (1) The capacitor with air has capacitance,

$$C_1 = \frac{\varepsilon_0}{d}\left(\frac{3A}{4}\right) = \frac{3\varepsilon_0 A}{4d}$$

Similarly, the capacitor with slab of K dielectric constant K has capacitance,

$$C_2 = \frac{\varepsilon_0 K}{d}\left(\frac{A}{4}\right) = \frac{\varepsilon_0 AK}{4d}$$

Since C_1 and C_2 are in parallel

$$C_{net} = C_1 + C_2 = \frac{3\varepsilon_0 A}{4d} + \frac{\varepsilon_0 AK}{4d} = \frac{\varepsilon_0 A}{d}\left(\frac{3}{4} + \frac{K}{4}\right)$$

$$= \frac{C}{4}(K+3)$$

49. (2) Heat produced in a wire is equal to the energy stored in capacitor.

$$H = \frac{1}{2}CV^2 = \frac{1}{2}\times(2\times10^{-6})\times(200)^2 = 4\times10^{-2}\text{ J}$$

50. (2) According to question, $U_0 = \frac{1}{2}CV^2$

where, $U_0 = \frac{1}{2}\frac{\varepsilon_0 A}{d}V^2$ $\left(\because C = \frac{\varepsilon_0 A}{d}\right)$

Now, new potential energy, $U = \frac{1}{2}C_n V^2$

$U = \frac{1}{2}\frac{K\varepsilon_0 A}{d}V^2$ $\left(\therefore \text{new capacitance}, C_n = \frac{K\varepsilon_0 A}{d}\right)$

$U = U_0 K$

51. (1) Equivalent capacitance $\frac{1}{C_{eq}} = \frac{1}{C_1} + \frac{1}{C_2} + \frac{1}{C_3} + \frac{1}{C_4}$

$$\frac{1}{C_{eq}} = \frac{1}{4} + \frac{1}{6} + \frac{1}{5} + \frac{1}{3}$$

$$\frac{1}{C_{eq}} = \frac{15+10+12+20}{60} = \frac{57}{60}$$

$C_{eq} = 1.05\ \mu\text{F}$

$q = C_{eq}V$

$\therefore q = 1.05\times10^{-6}\times150 = 157.5\times10^{-6}$

$q = 1.5\times10^{-4}\text{C}$

$\therefore V = \frac{q}{C} = \frac{1.5\times10^{-4}}{5\times10^{-6}} = 0.3\times10^2$

$V = 30$ volt

52. (2) For the capacitance between P and R C_1, C_2, C_3 are in series but they form parallel combination with C_4 and C_5 which are together form series combination

$\therefore C_1 = C_2 = C_3 = C_4 = C_5 = C$

Effective capacitance between P and R

$$C_{PR} = \frac{C}{3} + \frac{C}{2} = \frac{5C}{6}$$

Similarly between P and Q

$$C_{PQ} = C + \frac{C}{4} = \frac{5C}{4}$$

$$\therefore \frac{C_{PR}}{C_{PQ}} = \frac{5C}{6}\times\frac{4}{5C} = \frac{4}{6}$$

$$\frac{C_{PR}}{C_{PQ}} = \frac{2}{3}$$

53. (1) It forms 3 parallel plate capacitor which are joined in parallel

$C_P = C_1 + C_2 + C_3$

$$C_P = \frac{\varepsilon_0 A}{d} + \frac{\varepsilon_0 A}{d} + \frac{\varepsilon_0 A}{d}$$

$$C_P = \frac{3\varepsilon_0 A}{d}$$

54. (3) Both capacitors are joined in series combination in which charge remain same in all capacitors

So, charge on C_1 = charge on C_2

$\therefore C_1\Delta V = C_2\Delta V$

$\therefore C_1(V_A - V_\Delta) = C_2(V_\Delta - V_B)$

$C_1(V_1 - V_\Delta) = C_2(V_\Delta - V_2)$

$\therefore C_1V_1 - C_1V_\Delta = C_2V_\Delta - C_2V_2$

$\therefore C_1V_1 + C_2V_2 = C_2V_\Delta + C_1V_\Delta$

$$V_\Delta = \frac{C_1V_1 + C_2V_2}{C_1 + C_2}$$

It means there is a common potential at point O.

55. (1) Let C be the capacitance of infinite ladder

$$\frac{1}{C'} = \frac{1}{C} + \frac{1}{5} = \frac{5+C}{5C}$$

$$C' = \frac{5C}{5+C}$$

$C'' = 2 + C'$

$$C'' = 2 + \frac{5C}{5+C}$$

$C'' = C$ (Addition of one or more 5 μF and 2 μF do not change the total capacitance. So, overall capacitance should be C)

$$C = \frac{10+2C+5C}{5+C}$$

$5C + C^2 = 10 + 7C$

$C^2 = 10 + 2C$

$C^2 - 2C - 10 = 0$

$$D = \frac{-b\pm\sqrt{b^2-4ac}}{2a}$$

$$\therefore C = \frac{2\pm\sqrt{44}}{2}$$

$$C = \frac{2+6.6}{2},\ C = \frac{2-6.6}{2}$$

$C = \frac{8.6}{2}$, $C = \frac{-4.6}{2}$

$C = -2.3$, $C = 4.3$

–ve value neglected

So, total capacitance is 4.3 μF

56. (4) $C_P = 2 + 5 + 8 = 15\ \mu F$

$$\frac{1}{C_S} = \frac{1}{2} + \frac{1}{5} + \frac{1}{8} = \frac{20+8+5}{40}$$

$$C_S = \frac{40}{33}$$

$$\frac{C_P}{C_S} = \frac{15}{40} \times 33 \approx \frac{12}{1}$$

57. (2) $C_{AE} = \frac{C}{3} = \frac{3}{3} = 1\mu F$

$$C_{EF} = \frac{C}{3} = \frac{3}{3} = 1\mu F$$

$$C_{GF} = \frac{C}{2} = \frac{3}{2} = 1.5\mu F$$

$$C_{GB} = \frac{C}{2} = \frac{3}{2} = 1.5\mu F$$

All are joined in parallel

$C = C_{AE} + C_{EF} + C_{FG} + C_{GB}$

$C = 1 + 1 + 1.5 + 1.5$

$C = 5\ \mu F$

58. (1) $C_P = C_1 + C_2 = 18 + 18 = 36\ \mu F$

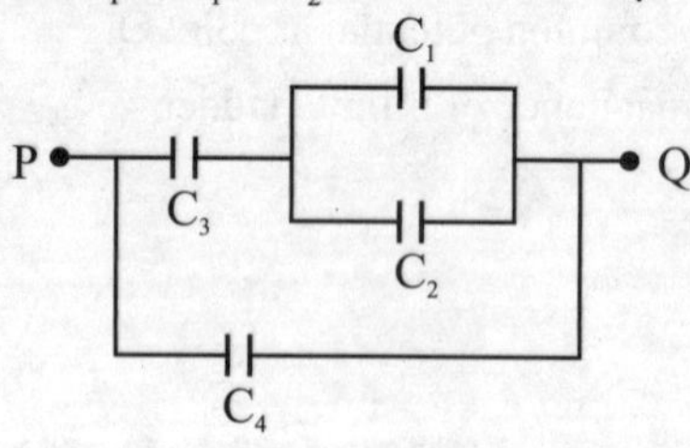

$$\frac{1}{C_s} = \frac{1}{36} + \frac{1}{18} = \frac{1+2}{36}$$

$C_s = 12\ \mu F$

$C_P = 12 + 18 = 30\ \mu F$

59. (3) The given circuit is as follows

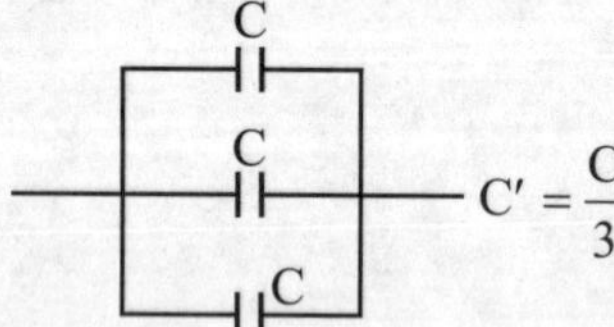

60. (4)

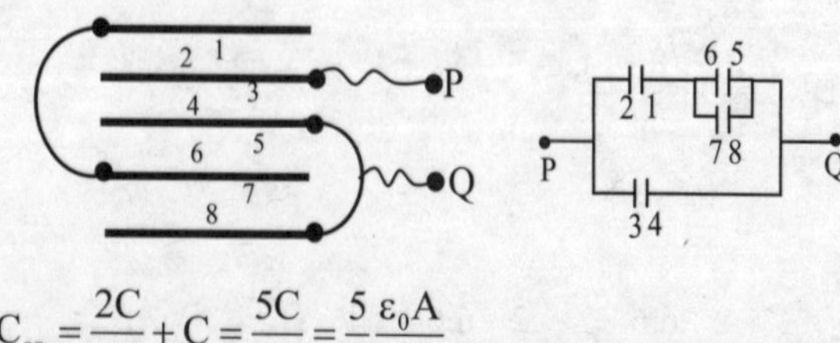

$$C_{eq} = \frac{2C}{3} + C = \frac{5C}{3} = \frac{5}{3}\frac{\varepsilon_0 A}{d}$$

61. (3)

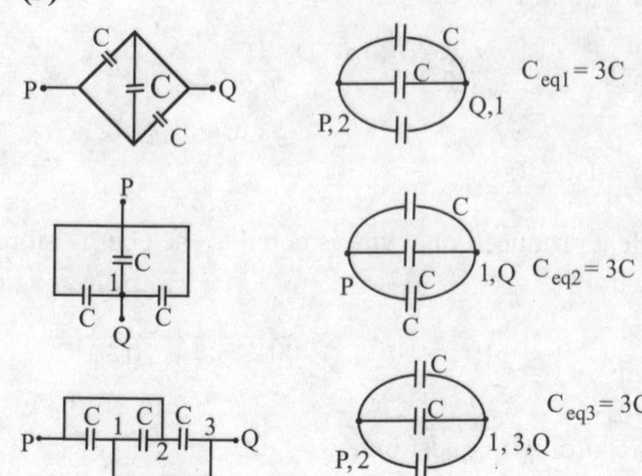

$\therefore C_{eq1} : C_{eq2} : C_{eq3} = 1 : 1 : 1$

62. (2) $Q = C_{eq}.V$

63. (1) Common potential $= \frac{C_1V_1 - C_2V_2}{C_1 + C_2}$

64. (1) Common potential $= \frac{C_1V_1 + C_2V_2}{C_1 + C_2} = 100\ V$

Charge flown = 3(300 – 100) = 600 μC

65. (3) Net charge on the isolated system = 0.

The capacitors of 2 μF and 3 μF are in parallel.

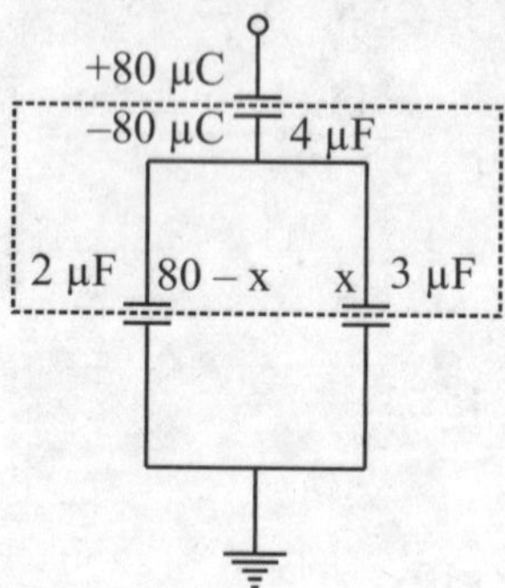

P.D. across each of them must be same.

$$\frac{80 - x}{2} = \frac{x}{3}$$

$\therefore x = +48\ \mu C$

66. (4) Energy loss $= \frac{1}{2}\frac{C_1C_2}{C_1 + C_2}(V_1 - V_2)^2$

Energy loss $= \frac{1}{2}\left(\frac{5}{2}\right)(15)^2$

Energy loss = 281.25 J

67. (2) C_2 capacitor is uncharged capacitor that means potential difference across it is zero because capacity can't be zero.

$C = \frac{\varepsilon_0 A}{d}$, $q = CV$

then potential difference will be

$$V' = \frac{C_1V_1 + C_2V_2}{C_1 + C_2}$$

$V_2 = 0$, $V_1 = V$

then, $V' = \dfrac{C_1V + C_2(0)}{C_1 + C_2}$

$$V' = \left(\frac{C_1}{C_1 + C_2}\right)V$$

68. (4)

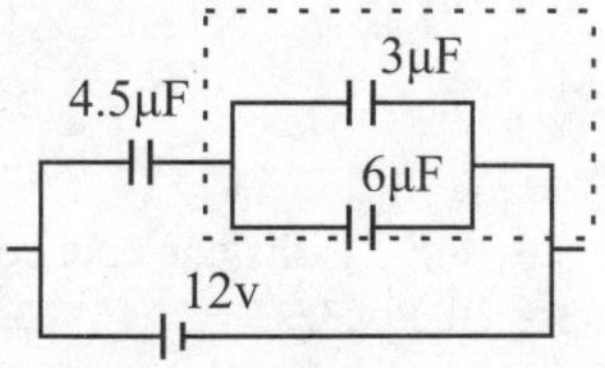

4.5μF, 3μF, 6μF, 12v

$C_{eq} = 3 + 6 = 9$ μF

Then capacitor 4.5 μF becomes is series with C_{eq}. So, potential difference across 4.5 μF capacitor.

$$V = \left(\frac{9}{4.5+9}\right)12 = 8V \quad \left[\because V_1 = \frac{C_2V}{C_1 + C_2}\right]$$

69. (4) Equivalent capacitance of three capacitors in series,

$$\frac{1}{C} = \frac{1}{C_1} + \frac{1}{C_2} + \frac{1}{C_3} \Rightarrow \frac{1}{C} = \frac{1}{3} + \frac{1}{6} + \frac{1}{6}$$

$$\Rightarrow C = \frac{6}{4} = 1.5\mu F$$

The charge of this circuit, $q = CV = 1.5 \times 120$

$= 180$ μC

The potential difference across the 3 μF

$q = CV,\ V = \dfrac{q}{C} = \dfrac{180}{3} = 60\,V$ (charge in same in all capacitor)

70. (1) Given, $C_1 = 10$ pF $= 10 \times 10^{-12}$F

$C_2 = 20$ pF $= 20 \times 10^{-12}$ F

$V_1 = 200$ V, $V_2 = 100$ V

Using $\Rightarrow q_1 = V_1C_1$

$q_2 = V_2C_2$

Common potential of capacitors

$$V = \frac{q_1 + q_2}{C_1 + C_2} = \frac{V_1C_1 + V_2C_2}{C_1 + C_2}$$

$$= \frac{200 \times 10 \times 10^{-12} + 100 \times 20 \times 10^{-12}}{10 \times 10^{-12} + 20 \times 10^{-12}} = \frac{4000}{30} = 133.3V$$

71. (3) Let the radius of each small drop is r and the radius of big drop is R.

So, the volume of 64 small drops = the volume of big drop

i.e. $64 \times \dfrac{4}{3}\pi r^3 = \dfrac{4}{3}\pi R^3 \Rightarrow 64r^3 = R^3$ ...(*i*)

$\Rightarrow 4r = R$

$\Rightarrow R = 4r$

Now, the capacitance of small drop

$C_1 = 4\pi\varepsilon_0 r$ ...(*ii*)

and the capacitance of big drop is

$C_2 = 4\pi\varepsilon_0 R$ $(R = 4r)$

$C_2 = 4\pi\varepsilon_0(4r)$

$\Rightarrow C_2 = 16\pi\varepsilon_0 r$ ...(*iii*)

$$\frac{C_2}{C_1} = \frac{4}{1} \Rightarrow C_2 : C_1 = 4:1$$

72. (3) Charge Q_1 will get divided on C_2 & C_3

$\therefore\ Q_1 = Q_2 + Q_3$

$V_1 + V_2$ or $V_1 + V_3 = V$ because $V_2 = V_3$

since capacitors are in parallel.

73. (3) $\Delta U = U_i - U_f$

$$= \frac{1}{2}C(V_1^2 + V_2^2) - \frac{1}{2}(2C)\left(\frac{V_1 + V_2}{2}\right)^2$$

$$= \frac{1}{4}C(V_1 - V_2)^2$$

High Order Time Intensive MCQs

1. (4) Loss of energy $= \dfrac{1}{2}\dfrac{C_1C_2}{(C_1 + C_2)}(V_1 - V_2)^2$

$$= \frac{1}{2} \times \frac{5 \times 5(2000 - 1000)^2}{(5+5)} = 1.25 \times 10^6 \text{ J}$$

2. (1) In the given figure capacitors 3 μF, 3 μF and 3 μF (between R and S) are in series. If Q is the charge on each of these capacitors, then

$$30 = \frac{Q}{3\mu} + \frac{Q}{3\mu} + \frac{Q}{3\mu}$$

Q = 30 μC

Potential difference between R and S

$$V = \frac{Q}{C} = \frac{30\mu}{3\mu}\text{ volt}$$

V = 10 volt

3. (3) We know that, qE = mg

$\because$ Electric field, $E = \dfrac{Q}{\varepsilon_0 A}$

$\therefore\ \dfrac{qQ}{\varepsilon_0 A} = mg$ or $q = \dfrac{\varepsilon_0 Amg}{Q}$

$$= \frac{8.85 \times 10^{-12} \times 2 \times 10^{-2} \times 2.5 \times 10^{-7} \times 10}{5 \times 10^{-7}}$$

$= 8.85 \times 10^{-13}$C

4. (4) Initial energy, $U_i = \dfrac{1}{2}C_0V^2$

Final energy, $U_f = \dfrac{1}{2}(KC_0)\left(\dfrac{V}{K}\right)^2$

Change in energy $U_f - U_i = \dfrac{1}{2}C_0V^2\left(\dfrac{1}{K} - 1\right)$

Note that this change of energy is negative, i.e. there is a decrease of energy.

5. **(3)** The two condensers in the circuit are in parallel connection, hence

$$C' = C + \frac{C}{2} = \frac{3C}{2}$$

The work done in charging the equivalent capacitor is stored in the form of potential energy.

Hence, $W = U = \frac{1}{2}C'V^2$

$$= \frac{1}{2}\left(\frac{3C}{2}\right)V^2 = \frac{3}{4}CV^2$$

6. **(2)** As insulator plate is passed between the plates of the capacitor, its capacity increases first and then decreases as the plate slips out. As a result, positive charge on plate *A* increases first and then decreases, hence current in outer circuit flows form *B* to *A* and then form *A* to *B*.

7. **(3)** $\because C = \frac{\varepsilon_0 A}{d-t}$ (for metal plate of thickness 't' inserted between plates of capacitor)

If t is very less, then

$$C = \frac{\varepsilon_0 A}{d}$$

So, capacity remains unchanged.

8. **(2)** Initially, the capacitance of capacitor

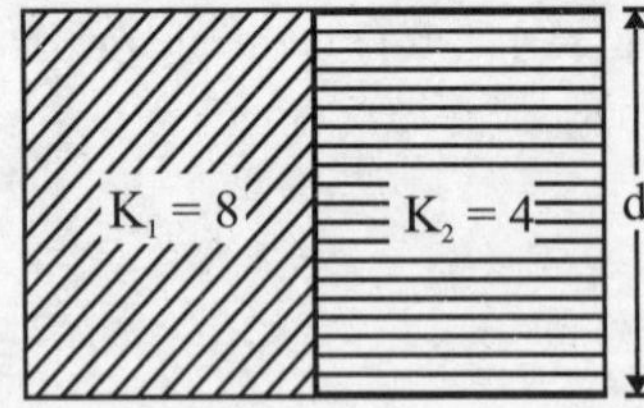

$$C = \frac{\varepsilon_0 A}{d}$$

Where, A is area of each plate and d is the separation between the plates.

After filling with dielectrics, we have two capacitors of capacitance.

$$C_1 = \frac{K_1\varepsilon_0(A/2)}{d} = \frac{8}{2}\frac{\varepsilon_0 A}{d} = \frac{4\varepsilon_0 A}{d} = 4C$$

and $C_2 = \frac{K_2\varepsilon_0(A/2)}{d} = \frac{4}{2}\frac{\varepsilon_0 A}{d} = \frac{2\varepsilon_0 A}{d} = 2C$

Hence, their equivalent capacitance is

$C_{eq} = C_1 + C_2 = 4C + 2C = 6C$

i.e. new capacitance will be six times the original.

9. **(1)** When S_1 and S_3 are closed, the capacitors C_1 and C_2 are in series arrangement. So, the potential difference developed across capacitors are in the inverse ratio of their capacitances.

$$\therefore \quad \frac{V_1'}{V_2'} = \frac{C_2}{C_1} = \frac{3\text{pF}}{2\text{pF}} = \frac{3}{2}$$

and $V_1' + V_2' = V_1 + V_2 = 30 + 20 = 50\text{ V}$

We get, $V_1' = V_1 = 30\text{ V}$ and $V_2' = V_2 = 20\text{ V}$

10. **(1)** Capacitance of the parallel plate capacitor is $C = \frac{\varepsilon_0 A}{d}$

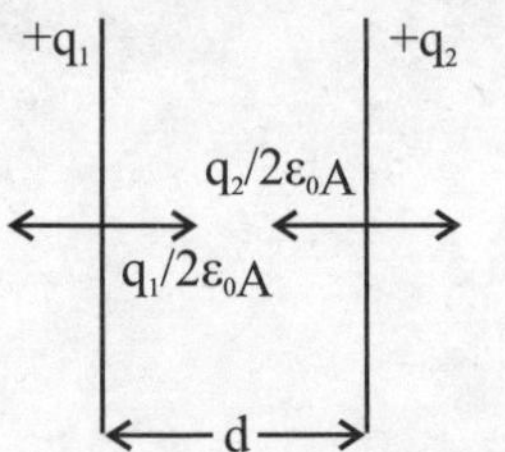

Where, A is area of plates and d is distance between them.
Potential difference across the plates,

V = Ed

Here, $E = \frac{(q_1 - q_2)}{2\varepsilon_0 A}$

$$V = \frac{(q_1 - q_2)d}{2\varepsilon_0 A} = \frac{q_1 - q_2}{2C}$$

11. **(4)** **Case-I:** When the capacitors are joined in series

$$U_{\text{series}} = \frac{1}{2}\frac{C_1}{n_1}(4V)^2$$

Case-II: When the capacitors are joined in parallel

$$U_{\text{parallel}} = \frac{1}{2}(n_2C_2)V^2$$

$U_{\text{series}} = U_{\text{parallel}}$ (Given)

or $\frac{1}{2}\frac{C_1}{n_1}(4V)^2 = \frac{1}{2}(n_2C_2)V^2 \Rightarrow C_2 = \frac{16C_1}{n_1n_2}$

12. **(3)** The capacitance of a parallel plate capacitor in the absence of the dielectric is,

$$C_0 = \frac{\varepsilon_0 A}{d} \quad ...(i)$$

where A is the area of each plate and d is the distance between them.

The capacitance of a parallel plate capacitor in the presence of dielectric slab of thickness t and dielectric constant K, is

$$C = \frac{\varepsilon_0 A}{(d-t)+\frac{t}{K}} = \frac{\varepsilon_0 A}{\left(d - \frac{3}{4}d\right) + \frac{3d}{4K}}$$

$$= \frac{\varepsilon_0 A}{\frac{d}{4} + \frac{3d}{4K}} = \frac{4K\varepsilon_0 A}{d(k+3)} \quad ...(ii)$$

Dividing eqn. (*ii*) by eqn. (*i*), we get

$$\frac{C}{C_0} = \frac{4K\varepsilon_0 A}{d(K+3)} \times \frac{d}{\varepsilon_0 A} = \frac{4K}{K+3}$$

13. **(4)** Doubling the separation of plates will halve the value of capacitance equal to 1 pF. Now dielectric insertion will increase the capacitance K times.

14. **(4)** As battery is disconnected, charge remains constant in the work process.

Work done = final potential energy – initial potential energy

$$= \frac{Q^2}{2C_f} - \frac{Q^2}{2C_i} = \frac{Q^2}{2}\left(\frac{1}{C_f} - \frac{1}{C_i}\right)$$

Where $Q = CV = \frac{A\varepsilon_o V}{d}$

and $C_i = \frac{A\varepsilon_o}{d}$, $C_f = \frac{A\varepsilon_o}{2d}$

Work done $= \frac{Q^2}{2}\left(\frac{2d}{\varepsilon_0 A} - \frac{d}{\varepsilon_0 A}\right) = \frac{Q^2 d}{2\varepsilon_0 A} = \frac{\varepsilon_0 A V^2}{2d}$

15. (3) When K_1 is closed

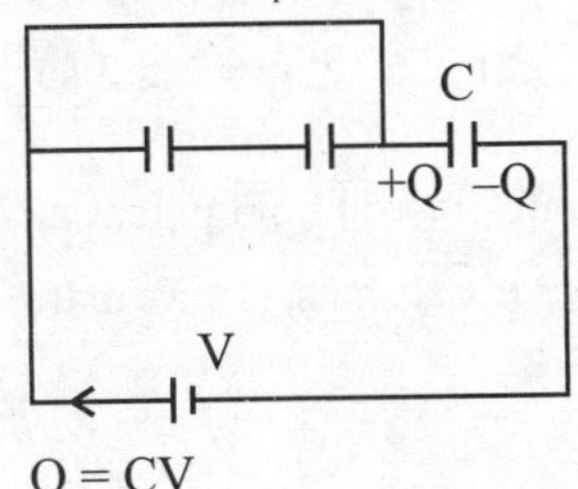

$Q = CV$

Energy $U_i = \frac{1}{2}CV^2$

When K_2 is also closed

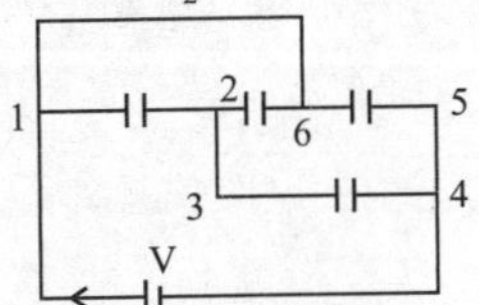

Equivalent circuit

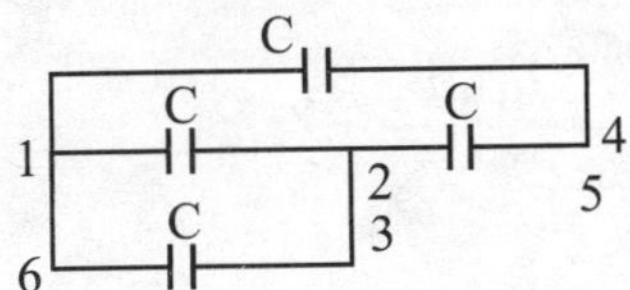

$C_{eq} = C + \frac{2C}{3} = \frac{5C}{3}$

Energy $U_f = \frac{1}{2} \times \frac{5C}{3} V^2 = \frac{5}{6}CV^2$

Charge supplied by battery after closing K_2

$= \frac{5}{3}CV - CV = \frac{2}{3}CV$

Energy supplied by battery $= U_f - U_i + \Delta H$

$$\frac{2}{3}CV^2 = \frac{5}{6}CV^2 - \frac{1}{2}CV^2 + \Delta H$$

$$\therefore \quad \Delta H = \frac{1}{3}CV^2$$

16. (4) Infinite

17. (2) $C_1 = 3\mu F$ $\quad V_1 = 300V$

$C_2 = 5\mu F$ $\quad V_2 = 500V$

$$V_C = \frac{C_1V_1 + C_2V_2}{C_1 + C_2}$$

$$V_C = \frac{3 \times 300 + 5 \times 500}{3+5} = \frac{900 + 2500}{8} = \frac{3400}{8} = 425V$$

18. (1) Q is constant

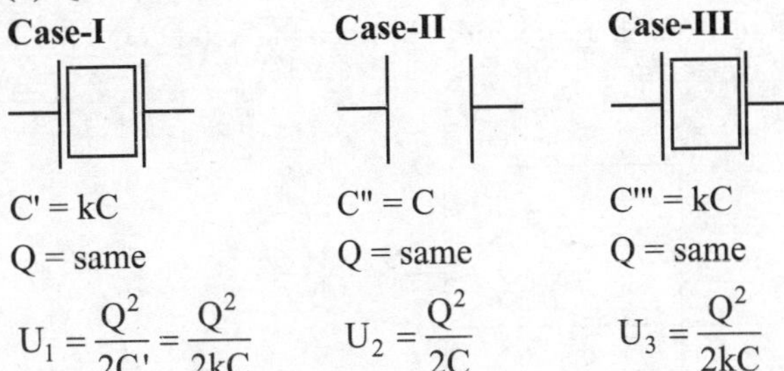

Case-I	Case-II	Case-III
$C' = kC$	$C'' = C$	$C''' = kC$
Q = same	Q = same	Q = same
$U_1 = \frac{Q^2}{2C'} = \frac{Q^2}{2kC}$	$U_2 = \frac{Q^2}{2C}$	$U_3 = \frac{Q^2}{2kC}$

Work done $= U_3 - U_1 = 0$

19. (3)

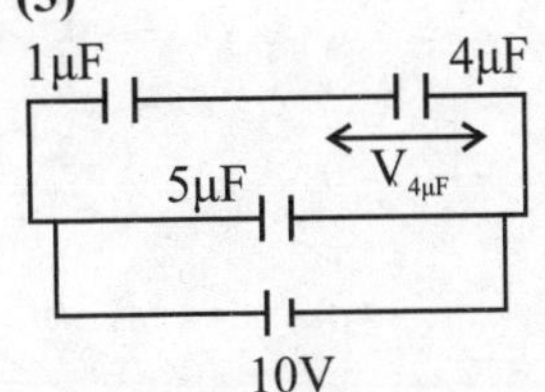

2 μF & 2 μF are reduced to 1 capacitor of 1 μF and 3 μF & 1 μF is reduced to 4 μF.

$\therefore V_{4_{\mu F}} = \frac{10 \times 1}{5} = 2V = V1_{\mu F}$

(potential difference across 4 μF capacitor is calculated by voltage division Rule)

$\therefore V_{5_{\mu F}} = 10V$

(potential difference across 5 μF is same as Battery)

$\therefore \frac{V_{1_{\mu F}}}{V_{5_{\mu F}}} = \frac{2}{10} = \frac{1}{5}$

20. (3)

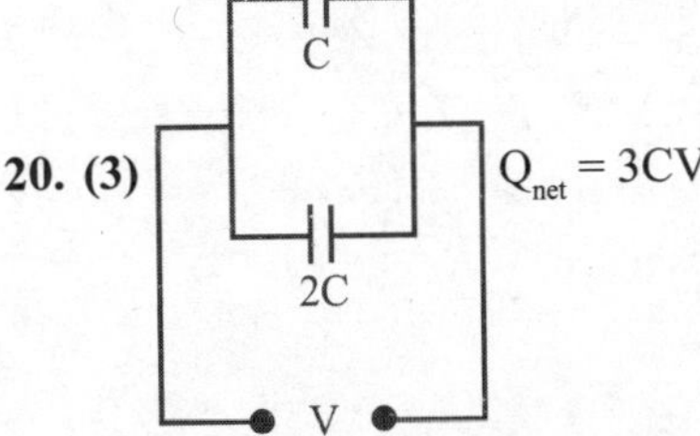

As the battery is removed therefore, Q remains constant and $C' = KC$

Let potential difference V'

$Q'_{net} = KCV' + 2CV' = (K + 2)CV'$

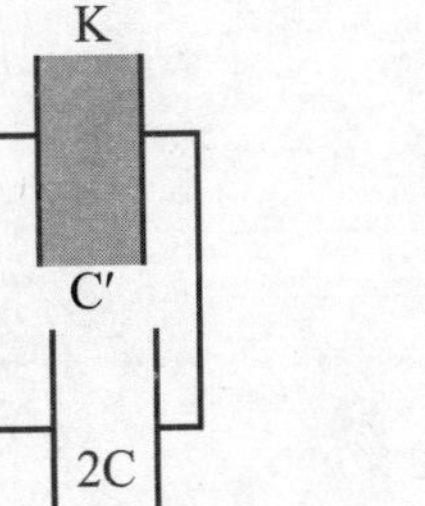

Charge conservation $Q_{net} = Q'_{net}$

$3CV = (K + 2)\,CV'$

$V' = \frac{3V}{K+2}$

21. (4) C_3 is short circuited and C_1, C_2 are in parallel.

22. (4) Note here potential has been talked about and not the capacitance.

23. (1)

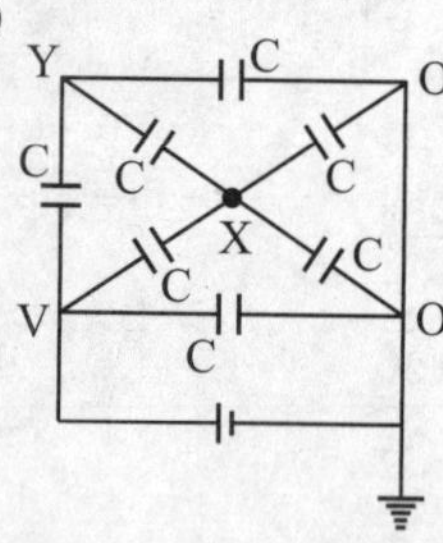

$$(X-0)C+(X-0)C+(X-V)C+(X-Y)C=0$$

$$4X-Y=V \quad ...(i)$$

$$(Y-0)C+(Y-X)C+(Y-V)C=0$$

$$3Y-X=V \quad ...(ii)$$

On solving

$$11X=4V$$

$$X=\frac{4V}{11}$$

24. (3) Field exist only between the spheres in the annular gap.

25. (1) $C=4\pi\varepsilon_0 R \quad \therefore d=2R=\frac{2C}{4\pi\varepsilon_0}$

26. (4) $U_i=\frac{1}{2}C_1V^2=\frac{1}{2}\times 2\times V^2=V^2$

After connecting S to 2,

$$V_{common}=\frac{C_1V_1+C_2V_2}{C_1+C_2}=\frac{2V+8\times 0}{2+8}$$

$$V_{common}=\frac{V}{5}$$

$$U_f=\frac{1}{2}(2+8)\left(\frac{V}{5}\right)^2=\frac{V^2}{5}$$

Percentage energy dissipated $=\frac{U_i-U_f}{U_i}\times 100$

$$=\frac{V^2-\frac{V^2}{5}}{V^2}\times 100=80\%$$

27. (1) As is clear from figure

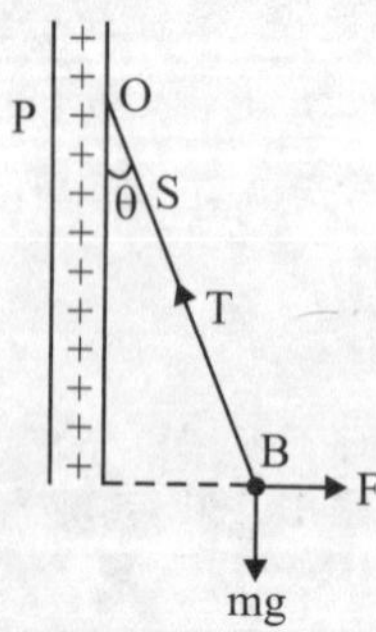

$$\frac{F}{AB}=\frac{mg}{OA}$$

$$\frac{F}{mg}=\frac{AB}{OA}=\tan\theta$$

$$\frac{q\sigma}{2\varepsilon_0(mg)}=\tan\theta$$

$$\therefore \sigma\propto\tan\theta$$

28. (1) $\because$ K.E. = QV

$$\frac{1}{2}m(v_1^2-v_2^2)=QV$$

$$\frac{1}{2}\times 10^{-3}[v_1^2-(0.2)^2]=10^{-8}(600-0)\Rightarrow v_1=22.8\,\text{cms}^{-1}$$

29. (1) For a uniform electric field, the equipotential surfaces are placed at right angles to the lines of forces. Therefore, only two points have the same electric field potential.

30. (2) Potential inside the sphere will be same as that on its surface i.e.

$$V=V_{surface}=K\frac{q}{10}\text{volt}$$

$$V_{out}=K\frac{q}{15}\text{volt}$$

$$\therefore \quad \frac{V_{out}}{V}=\frac{2}{3}\Rightarrow V_{out}=\frac{2}{3}V$$

31. (3) Potential energy of the system

$$\frac{-kqQ}{x}-\frac{kQq}{x}+\frac{kq^2}{2x}=0$$

$$\Rightarrow \quad \frac{-4kqQ+kq^2}{2x}=0\Rightarrow kq^2=4kQq$$

$$\Rightarrow \quad \frac{q}{Q}=4$$

32. (1) When charge q_3 is at C, then its potential energy is

$$U_C=\frac{1}{4\pi\varepsilon_0}\left(\frac{q_1q_3}{0.4}+\frac{q_2q_3}{0.5}\right)$$

When charge q_3 is at D, then

$$U_D=\frac{1}{4\pi\varepsilon_0}\left(\frac{q_1q_3}{0.4}+\frac{q_2q_3}{0.1}\right)$$

Hence, change in potential energy.

$$\Delta U=U_D-U_C$$

$$=\frac{1}{4\pi\varepsilon_0}\left(\frac{q_2q_3}{0.1}-\frac{q_2q_3}{0.5}\right)=\frac{q_3q_2}{4\pi\varepsilon_0}\left(\frac{1}{0.1}-\frac{1}{0.5}\right)$$

$$=\frac{q_3q_2}{4\pi\varepsilon_0}(8)$$

as $\Delta U=\frac{q_3K}{4\pi\varepsilon_0}$

$$\Rightarrow K=8q_2$$

33. (1) $C=\frac{\varepsilon_0(\pi R^2)}{x}=4\pi\varepsilon_0 R \quad \Rightarrow x=R/4$

34. (1) Capacitors of 4μF are in series and the equivalent of them is in parallel with 2μF capacitor.

35. (1) Maximum capacitance is obtained in series combination and minimum capacitance is obtained in parallel combination.

$C_{max} = C_{series} = 1 + 2 + 3 = 6\mu F$

$C_{min} = C_{parallel} = \frac{6}{6+3+2} = \frac{6}{11}\mu F$

36. (2) Energy density $U = \frac{1}{2}\varepsilon_o E^2$ and $E = \frac{V}{d}$

$\therefore\ U = \frac{1}{2}\varepsilon_0 \frac{V^2}{d^2} = \frac{8.8\times10^{-12}\times(400)^2}{2\times(2\times10^{-3})^2}$

$U = 0.176\ Jm^{-3}$

37. (3) Potential ratio,

$\frac{V_1}{V_2} = \frac{q_1}{q_2}\times\frac{r_2}{r_1} = \frac{1.5\times10^{-8}}{0.3\times10^{-7}}\times\frac{2\times10^{-2}}{1\times10^{-2}}$

$V_1 = V_2$

As the potential of two spheres are same hence there will be no flow of charge.

38. (1) $\because\ C = K\varepsilon_0\frac{A}{d}$, where k is dielectric medium

39. (2) Isolated capacitor $\Rightarrow$ Q = constant

separation d increase $\Rightarrow$ C = decrease

$\because$ Q = CV $\Rightarrow$ V = increase

40. (4) $C_{AB} = C_1 + C_2 + C_3 = 1 + 2 + 3 = 6\mu F$

41. (4) $\frac{1}{C_1} = \frac{1}{3}+\frac{1}{3}+\frac{1}{3}$

$\Rightarrow C_1 = 1\ \mu F,\ C_2 = 2 + 1 = 3\ \mu F$

$C_{eq} = C_1 = 1\ \mu F.$

42. (3) $C' = \frac{\varepsilon_0\varepsilon_r A}{d} \Rightarrow C' = 6C$

As charge is constant then

$U = \frac{q^2}{2C} \Rightarrow U' = \frac{q^2}{2\times C'} = \frac{q^2}{2\times 6C} = \frac{U}{6}$

43. (2) Energy supplied by battery $U = C_{net}V^2$

(double of stored energy)

$C_{net} = \{C_1 \text{ series } C_2\} \parallel C_3$

$C_{net} = \frac{6}{3+1} + 4 = \frac{11}{2}\mu F$

$\therefore\ U = \frac{11}{2}\times 2 = 11\mu J = 11\times10^{-6} J$

44. (2) Charge in capacitors spheres will be same. Let V_1 and V_2 be the potential difference for the two capacitors then

$V_1 + V_2 = V = 100$ ...(*i*)

$q_1 = q_2$ and $C_1 = C$; $C_2 = 4C$

$CV_1 = 4CV_2$

$V_1 = 4V_2$ ...(*ii*)

On solving equation (*i*) and (*ii*)

$V_1 = 80$ V and $V_2 = 20$V

45. (1) After filling the space with oil.

$C' = \varepsilon_r C$ and $Q' = Q\varepsilon_r$

$\varepsilon_r = \frac{Q'}{Q} = \frac{100}{40} = 2.5$

46. (1) Initially when switch is open

R E C $q_1 = CE$

Finally when switch is closed

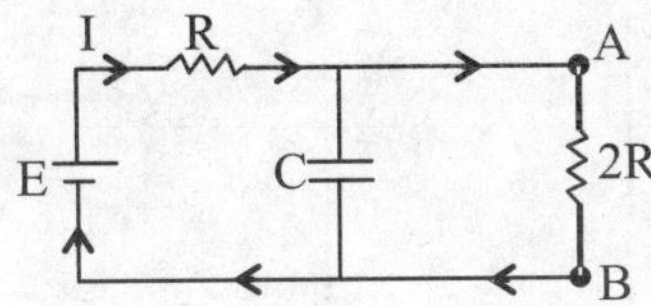

$I = \frac{E}{3R};\ \therefore\ V_A - V_B = I\,2R = (E/3R)\times 2R = 2E/3$

$\therefore$ Charge on capacitor = CV; $q_2 = \frac{C2E}{3} = \left(\frac{2}{3}\right)CE$

$\therefore\ \frac{q_1}{q_2} = \frac{CE}{(2/3)CE} = 3:2$

47. (3) $C = \varepsilon_0\frac{A}{d}$

So increasing area and decreasing distance between plates results in increased capacitance.

48. (2) $U_0 = \frac{1}{2}CV^2$

New energy, $U' = \frac{1}{2}C'V^2$

$C' = CK^2$

$U = \frac{1}{2}CV^2K^2 = U_0K^2$

49. (2) Without copper $C = \frac{\varepsilon_0 A}{d}$

with copper $C' = \frac{\varepsilon_0 A}{d - t\left(1 - \frac{1}{\varepsilon_r}\right)}$

for copper $\varepsilon_r = \infty$ and $t = \frac{d}{2}$

Hence $C' = \frac{\varepsilon_0 A}{d - \frac{d}{2}} = 2\frac{\varepsilon_0 A}{d} \Rightarrow \frac{C'}{C} = 2$

50. (3) Potential same at both spheres $V_1 = V_2$

$\therefore\ \frac{kQ_1}{R_1} = \frac{kQ_2}{R_2} \Rightarrow \frac{Q_1}{Q_2} = \frac{R_1}{R_2}$

surface charge density $\sigma = \frac{Q}{4\pi R^2}$

$$\therefore \frac{\sigma_1}{\sigma_2} = \frac{Q_1}{4\pi R_1^2} \times \frac{4\pi R_2^2}{Q_2} = \frac{Q_1}{Q_2} \times \left(\frac{R_2}{R_1}\right)^2 = \left(\frac{R_1}{R_2}\right)\left(\frac{R_2}{R_1}\right)^2$$

$$= \frac{R_2}{R_1} = \frac{20}{10} = \frac{2}{1}$$

51. **(2)** Charge across C_2, $Q_2 = \frac{6}{11} \times C$

Charge across C_4

$Q_4 = 4C$

$$\Rightarrow \frac{Q_2}{Q_4} = \frac{\frac{6}{11}C}{4C} = \frac{3}{22}$$

52. **(1)**

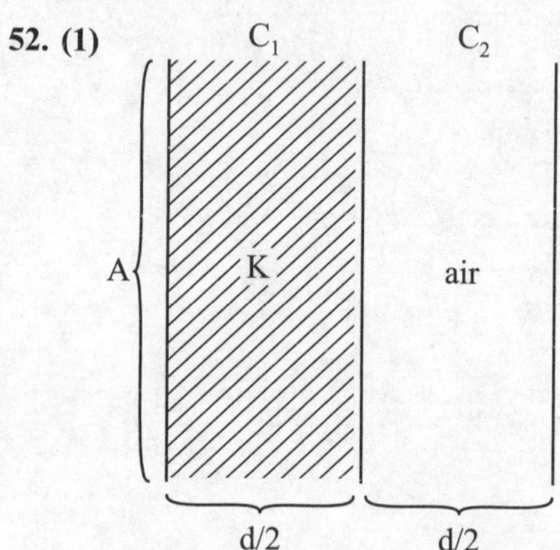

$C = \frac{\varepsilon_0 A}{d}$ $C_1 = \frac{\varepsilon_0 KA}{d/2} = \frac{2\varepsilon_0 KA}{d}$

Eq. $C_2 = \frac{\varepsilon_0 A}{d/2} = 2\left(\frac{\varepsilon_0 A}{d}\right)$

C_1 & C_2 are in series combination

Hence

$$\frac{1}{C_s} = \frac{1}{C_1} + \frac{1}{C_2} = \frac{d}{2\varepsilon_0 A}\left(\frac{1}{K} + \frac{1}{1}\right)$$

$$\Rightarrow \frac{1}{C_s} = \frac{d}{2\varepsilon_0 A}\left(\frac{K+1}{K}\right)$$

$$C_s = \frac{2\varepsilon_0 A}{d}\left(\frac{K}{K+1}\right) \quad C_s = \frac{2CK}{K+1}$$

53. **(3)** When length x of the dielectric slab is placed between the plates of the capacitors, then

$$C_1 = \frac{K\varepsilon_0 tx}{d} \qquad \left(\because C = K\frac{\varepsilon_0 A}{d}\right)$$

and similarly $C_2 = \frac{\varepsilon_0 t(L-x)}{d}$

$\therefore$ Net capacitance, $C_{net} = C_1 + C_2 = \frac{\varepsilon_0 t}{d}[L + x(K-1)]$

Thus, energy stored in the capacitor,

$$U = \frac{1}{2}CV^2 = \frac{1}{2}C_{net}V^2 = \frac{\varepsilon_0 tV^2}{2d}[L + x(K-1)]$$

Statement Based MCQs

1. **(4)** According to statement

Charge density (σ) is same is false but charge same.

Charge density, $\sigma = \frac{Q}{A}$

In case of charge distribute uniformly a conductor has to be equipotential under static condition.

2. **(1)** If two protons are brought near one another then work has to be done against electrostatic froce because of like charges. This work done stored as potential energy of the system.

Magnitude of charge on both the plates of capacitor will be same because all electrons lost from one plate will reach to another plate.

3. **(3)** The bubble on giving charge rather expands so as to compensate for the increased potential energy due to uniform distribution of charge on surface.

4. **(1)**

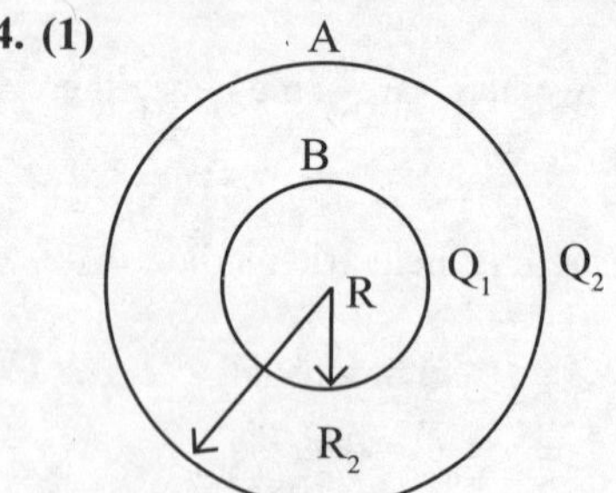

Potential difference

$$V_0 - V_1 = \frac{KQ_1}{R_1} - \frac{KQ_1}{R_2}$$

5. **(1)** $C = \frac{\varepsilon_0 A}{d}$

Where d = distance between them

Capacitance of the capacitor is inversely proportional to distance between them therefore, as capacitance increases, the distance between plates decrease.

Assertion & Reason MCQs

1. **(1)** When they are connected by a wire then the potential will become equal inside the sphere. Thus only charge on the bigger sphere are left and inner charge density gets redistributed so as to make equal potential everywhere inside the bigger sphere

2. **(1)** When the outer shell is earthed then only outer surface gets zero potential.

3. **(1)** As charge gets transferred from one plate to another, hence charge remains conserved so, the charges must be same on both the plates.

4. **(2)** A charge conductor may initially have a charged particle inside it but since $\vec{E}$ field inside it is always zero so it gets redistributed on the outer surface.

5. **(2)** During takeoff and landing, the friction between tyres and the run way may induced charges on the tyres. Due to conducting nature of the tyre, the charge so collected is conducted to the ground and electrical sparking is avoided.

Match the Columns MCQs

1. (4)

(a) The electrostatic potential due to a point charge varies as inversely proportional to the distance of the charge. It is represented as:

$$V \propto \frac{1}{r}$$

(b) The potential due to a dipole depends on the magnitude of the position vector and the dipole moment vector. It is given by

$$V = \frac{\hat{p}.\hat{r}}{4\pi\varepsilon_0 r^2}$$

$$V \propto \frac{1}{r^2}$$

(c) Potential difference between any two points on equatorial surface is zero.

(d) Electric field inside a conductor is zero so the potential at any point is constant.

2. (1)

(a) $C_i = \frac{4\times4}{4+4} = 2\mu F$

$$C_f = \frac{C_1C_2}{C_1+C_2} = \frac{8\times2}{8+2} = 1.6\mu F$$

$q = CV$

Since, total capacity is decreasing. Hence, charge on both capacitors will decrease.

(b) $U_2 = \frac{1}{2}\frac{q^2}{C}$

q has become $\frac{1.6}{2}$ or 0.8 times but C is halved. Hence, U_2 will increase.

(c) $V_2 = \frac{q}{C}$

q has become 0.8 times but C is halved. Hence, V_2 will increase.

(d) $E_2 = \frac{V_2}{d}$ or $E_2 \propto V_2$

3. (2) $\int dV = -\int E\cdot dr$

$$\therefore\ V_A - V_B = -\int_B^A E\cdot dr$$

4. (3) Volume of big drop = n × Volume of small drop

$$\therefore\ \frac{4}{3}\pi R^3 = n\times\frac{4}{3}\pi r^3$$

$$\Rightarrow\ R = n^{1/3}r$$

Capacitance of big drop, $C = 4\pi\varepsilon_0 R = 4\pi\varepsilon_0 n^{1/3}r = n^{1/3}C'$

Potential of big drop, $V = \frac{1}{4\pi\varepsilon_0}.\frac{Q}{R} = \frac{1}{4\pi\varepsilon_0}.\frac{nq}{n^{1/3}r}$

$$= \frac{n}{n^{1/3}}\left(\frac{1}{4\pi\varepsilon_0}\times\frac{q}{r}\right) = n^{2/3}V'$$

5. (3) V_1 —|C_1|— V —|C_2|— V_2

$$V = \frac{C_1V_1 + C_2V_2}{C_1+C_2}$$

$$\frac{V-V_1}{C_1} = \frac{V_2-V}{C_2}$$

$$\text{Total energy} = \frac{1}{2}C_1(V-V_1)^2 + \frac{1}{2}C_2(V_1-V)^2$$

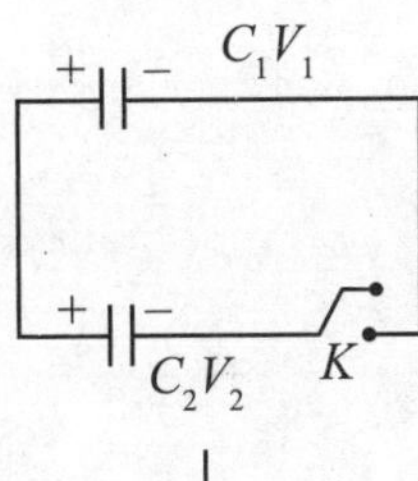

↓

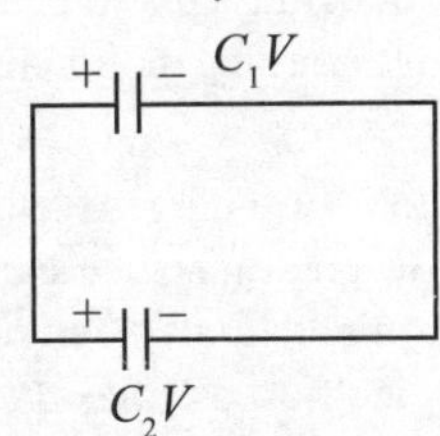

$$V = \frac{C_1V_1 + C_2V_2}{C_1+C_2}$$

$$\Delta V = \frac{1}{2}\frac{C_1C_2}{C_1+C_2}(V_1-V_2)^2$$

$$\frac{V-V_1}{C_1} = \frac{V_2-V}{C_2}$$

NCERT Exemplar MCQs

1. (4) Initially capacitor gets charges so no current will flow through the branch of capacitor.

Current flows through 2Ω resistance from left to right, is given by

$$I = \frac{V}{R+r} = \frac{2.5V}{2+0.5} = 1A$$

The potential difference across 2Ω resistance $V = IR = 1\times 2V = 2V$

Since, capacitor is in parallel with 2Ω resistance, so it also has 2V potential difference across it.

The charge on capacitor

$q = CV = (2\ \mu F)\times 2V = 8\mu C$

Note: The potential difference across 2Ω resistance solely occurs across capacitor as no potential drop occurs across 10 Ω resistance.

2. (3) The direction of electric field is always perpendicular to one equipotential surface maintained at high electrostatic potential to other equipotential surface maintained at low electrostatic potential.

The positively charged particle experiences electrostatic force along the direction of electric field i.e., from high electrostatic potential to low electrostatic potential. Thus, the work is done by the electric field on the positive charge, hence electrostatic potential energy of the positive charge decreases.

3. **(3)** The work done by a electrostatic force is given by $W_{12} = q(V_2 - V_1)$. Here initial and final potentials are same in all three cases and same charge is moved, so work done is same in all three cases.

4. **(3)** In this problem, the electric field intensity E and electric potential V are related as

$$E = -\frac{dV}{dr}$$

Electric field intensity E = 0 suggest that $\frac{dV}{dr} = 0$

This imply that V = constant.

Thus, E = 0 inside the charged conducting sphere causes, the same electrostatic potential 100V at any point inside the sphere.

5. **(1)** In this problem, the collection of charges. whose total sum is not zero, with regard to great distance can be considered as a point charge. The equipotential due to point charge are spherical in shape as electric potential due to point charge q is given by

$$V = k_e \frac{q}{r}$$

This suggest that electric potentials due to point charge is same for all equidistant points. The locus of these equidistant points, which are at same potential, form spherical surface.

6. **(3)** The capacitance of parallel plate capacitor filled with dielectric block has thickness d_1 and dielectric constant K_1 is given by

$$C_1 = \frac{K_1 \varepsilon_0 A}{d_1}$$

Similarly, capacitance of parallel plate capacitor filled with dielectric block has thickness d_2 and dielectric constant K_2 is given by

$$C_2 = \frac{K_2 \varepsilon_0 A}{d_2}$$

Since, the two capacitors are in series combination, the equivalent capacitance is given by

$$\frac{1}{C} = \frac{1}{C_1} + \frac{1}{C_2}$$

$$\text{Or } C = \frac{C_1 C_2}{C_1 + C_2} = \frac{\frac{K_1 \varepsilon_0 A}{d_1} \frac{K_2 \varepsilon_0 A}{d_2}}{\frac{K_1 \varepsilon_0 A}{d_1} + \frac{K_2 \varepsilon_0 A}{d_2}} = \frac{K_1 K_2 \varepsilon_0 A}{K_1 d_2 + K_2 d_1}$$

But the equivalent capacitances is given by

$$C = \frac{K \varepsilon_0 A}{d_1 + d_2}$$

On comparing, we have

$$K = \frac{K_1 K_2 (d_1 + d_2)}{K_1 d_2 + K_2 d_1}$$

NEET & AIIMS Past Year MCQs

1. **(4)** Given:

Charge $q = 4 \times 10^{-7}$ C,

Distance r = 9 cm = 0.09 m,

$$\frac{1}{4\pi\varepsilon_0} = 9 \times 10^9 \, N m^2 / C^2.$$

Electric potential:

$$V = \frac{kq}{r} = \frac{9 \times 10^9 \times 4 \times 10^{-7}}{0.09} = 4 \times 10^4 \text{ V}$$

2. **(1)** Formula of potential is given by $V = \frac{Kp\cos\theta}{r^2}$

For axial point θ = 0° or 180°

$$V = \pm\frac{Kp}{r^2} = \pm\frac{1}{4\pi\varepsilon_0} \times \frac{p}{r^2}$$

$$= \pm\frac{9 \times 10^9 \times 4 \times 10^{-6}}{4} = \pm 9 \times 10^3$$

3. **(2)** For spherical shell

$V_{centre} = V_{surface}$

Difference in potential is zero.

4. **(2)** Capacitance C depends on:

Geometry (area A, distance d): $C = \frac{\varepsilon_0 A}{d}$.

It is independent of charge q and potential V.

5. **(3)** Balanced Wheatstone bridge

Equivalent circuit

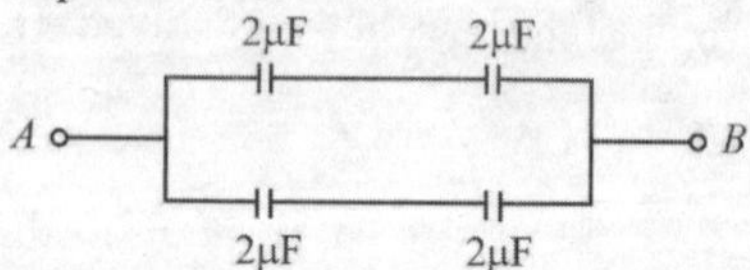

6. **(1)** The electrostatic energy stored in a capacitor is given by:

$$U = \frac{1}{2} CV^2$$

Given:

Capacitance C = 12 pF = 12×10^{-12} F

Voltage V = 50 V

Substitute the values:

$$U = \frac{1}{2} \times 12 \times 10^{-12} \times (50)^2$$

$U = 6 \times 10^{-12} \times 2500$

$U = 15 \times 10^{-9}$ J = 15 nJ

7. **(2)** 3μF and 3μF in parallel

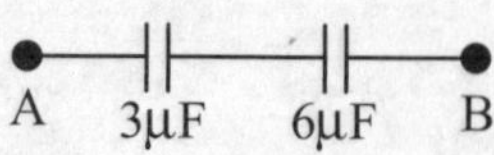

$\Rightarrow C_{eqv} = 3 + 3 = 6\mu F$

Now, $C'_{eqv} = \frac{3 \times 6}{9} = 2\mu F$

8. (2) $V_P = V_q + V_{-q}$

$$= \left(\frac{Kq}{5-3} + \frac{K(-q)}{5+3}\right) \times 10^2$$

$$= \left(\frac{Kq}{2} - \frac{Kq}{8}\right) \times 10^2 = \left(\frac{3Kq}{8}\right) \times 10^2$$

9. (3) Potential V is constant

$$\Rightarrow \frac{1}{4\pi\varepsilon_0}\frac{q}{R} = \text{constant}$$

$$\therefore \ V \propto \frac{1}{R}$$

10. (4) Electric field lines arc always perpendicular to the equipotential surface.

11. (1)

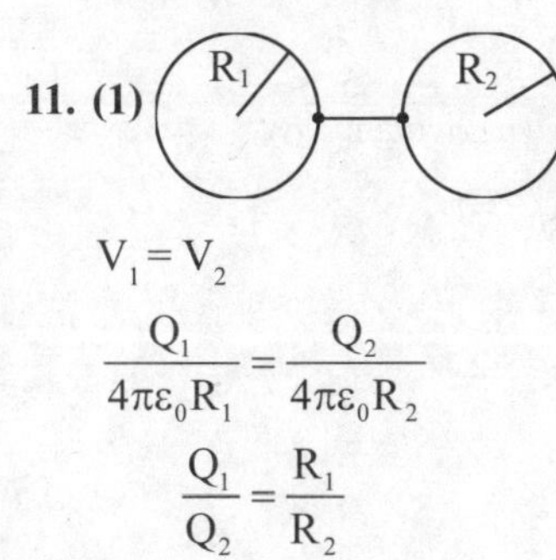

$V_1 = V_2$

$$\frac{Q_1}{4\pi\varepsilon_0 R_1} = \frac{Q_2}{4\pi\varepsilon_0 R_2}$$

$$\frac{Q_1}{Q_2} = \frac{R_1}{R_2}$$

$$\frac{\sigma_1}{\sigma_2} = \frac{Q_1}{4\pi R_1^2} \times \frac{4\pi R_2^2}{Q_2}$$

$$= \frac{Q_1}{Q_2} \times \left(\frac{R_2}{R_1}\right)^2 = \frac{R_1}{R_2} \times \left(\frac{R_2}{R_1}\right)^2$$

$$\Rightarrow \frac{\sigma_1}{\sigma_2} = \frac{R_2}{R_1}$$

12. (1)

$C_{AB} = 2C$

13. (2) Energy = Energy density × volume

$$= \frac{1}{2}\varepsilon_0 E^2 Ad$$

14. (3) The radius of the bigger drop will be

$$\frac{4}{3}\pi R^3 = 27 \times \frac{4}{3}\pi r^3$$

$R = 3r$

The potential of the small drop

$$= \frac{KQ}{r} = V$$

The potential of the big drop

$Q = nq = 27q$

$$V' = \frac{KQ}{R} = \frac{K \times 27q}{3r} = 9V = 9 \times 220 = 1980 \text{ V}$$

15. (1) Both assertion and reason are true and reason is the correct explanation of assertion.

16. (1) Equipotential surfaces are surfaces of constant scalar potential. They are used to visualize an n-dimensional scalar potential function in (n – 1) dimensional space. The gradient of the potential, denoting the direction of greatest increase in perpendicular to the surfaces. Hence, equipotential surfaces associated with an electric field which is increasing in magnitude along the x-direction are planes parallel to yz-plane

17. (2)

18. (3)

19. (2)

Current Electricity

NCERT Topic-wise MCQs

1. (2) According to the relation

$\sigma E = J$, where σ is conductivity and is a constant for specific material

$E \propto J$

2. (1) $I = \frac{dq}{dt} = 3t^2 + 2t + 5$

$\therefore \ dq = (3t^2 + 2t + 5)dt$

$$\int_0^q dq = \int_0^2 (3t^2 + 2t + 5)dt$$

$$q = \left[\frac{3t^3}{3} + \frac{2t^2}{2} + 5t\right]_0^2$$

$$q = [t^3 + t^2 + 5t]_0^2$$

$q = 8 + 4 + 10$

$q = 22$ C

3. (1) At constant temperature, resistance of conductor remains constant, so graph between V and I is linear.

∴ Option (1) is correct.

4. (1)

(a) $\frac{\rho\ell}{A}$: This represents the formula for resistance R of a material, where ρ is the resistivity, (ℓ) is the length, and (A) is the cross-sectional area. So, (a) matches with (iii).

(b) $\frac{RA}{\ell}$: This is resistivity ρ which is derived from the formula of resistance. So, it matches (iv).

(c) $\frac{1}{\rho}$: This is the formula for conductivity (σ), which is the reciprocal of resistivity. So, (c) matches with (ii) Conductivity.

(d) V = IR: This is Ohm's Law, which states that the voltage (V) across a conductor is equal to the product of the current (I) flowing through it and its resistance (R). So, (d) matches with (i)

5. (4) For two conductors of the same material and the same cross-sectional area, the resistance is proportional to their lengths.

$R \propto \ell$

$\frac{R_2}{R_1} = \frac{\ell_2}{\ell_1} = \frac{20}{10} = 2$

$R_2 = 2 \times R_1$

$R_2 = 2 \times 100$ ohms

$R_2 = 200$ ohms

6. (2) Volume = $A \times \ell = 3$

$A = \frac{3}{\ell}$

$R = \rho\frac{\ell}{A} \Rightarrow \rho \times \frac{\ell}{3} \times \ell = 3$

$\Rightarrow \rho \times \ell^2 = 9$

$\therefore \ell = \frac{3}{\sqrt{\rho}}$

7. (4) Mobility, $\mu = \frac{v_d}{E}$

Given that, V = 200V, $v_d = 0.5ms^{-1}$ and $\ell = 2$ m

As we know: $E = \frac{V}{\ell} \Rightarrow \mu = \frac{v_d}{V} \times l$

$\mu = \frac{0.5}{200} \times 2 = 5 \times 10^{-3} m^2V^{-1}s^{-1}$

8. (4) $q = I \times t = 5 \times 60 = 300$ C

9. (2) $R \propto \frac{\ell}{A} \propto \frac{\ell}{r^2}$

$\frac{R_1}{R_2} = \frac{\ell_1}{\ell_2} \times \left(\frac{d_2}{d_1}\right)^2 = \frac{L}{4L}\left(\frac{2d}{d}\right)^2 = 1$

$\therefore R_1 = R_2 = R$

10. (1) Since $R \propto \ell^2$

$\therefore \frac{\Delta R}{R} = \frac{2\Delta\ell}{\ell}$

If the length is 10% increased, then resistance also increases almost by 20%.

Hence, new resistance, R = 10 + 20% of 10 Ω

$R = 10 + \frac{20}{100} \times 10 = 12\,\Omega$

R = 12 ohm

11. (2) Net current $I = I_e + I_p$

$I = \frac{nq^+}{t} + \frac{nq^-}{t}$

$I = 2.9 \times 10^{18} \times 1.6 \times 10^{-19} + 1.2 \times 10^{18} \times 1.6 \times 10^{-19}$

I = 0.66A

12. (1) Volume before compression = volume after compression

$\pi r^2 L = \pi(nr)^2 \times L'$

$\Rightarrow L' = \frac{L}{n^2}$

$\therefore R' = \frac{\rho\frac{L}{n^2}}{\pi(nr)^2} \Rightarrow R' = \rho\frac{L}{\pi r^2} \times \frac{1}{n^4} = \frac{R}{n^4}$

13. (3) As we know:

$I = qv \Rightarrow 1.6 \times 10^{-19} \times 6.8 \times 10^{16} = 11 \times 10^{-3}$A

$I = 1.1 \times 10^{-2}$ A

14. (4) We know that, $R_t - R_0 = \alpha R_0 t$

$\therefore R_{50} - R_0 = \alpha R_0 \times 50$ ...(*i*)

and $R_{100} - R_0 = \alpha R_0 \times 100$ ...(*ii*)

On dividing equation (*i*) by (*ii*), we get

$\frac{R_{50} - R_0}{R_{100} - R_0} = \frac{50}{100} \Rightarrow \frac{5 - R_0}{6 - R_0} = \frac{1}{2}$

$\Rightarrow 10 - 2R_0 = 6 - R_0 \Rightarrow R_0 = 4\Omega$

15. (3) By using $v_d = \frac{I}{neA}$

$= \frac{100}{10^{28} \times 1.6 \times 10^{-19} \times \frac{\pi}{4} \times (0.02)^2} = \frac{10^2}{10^9 \times 0.4 \times 3.14 \times 0.0004}$

$v_d = 2 \times 10^{-4}$m/s

16. (1) Drift velocity

or $v_d = \frac{eE}{m}\tau = \frac{e\tau V}{mL}$ $\left[\because E = \frac{V}{L}\right]$

$v_d \propto V$

$\therefore \frac{v_{d_1}}{v_{d_2}} = \frac{V_1}{V_2}$

$\frac{v_{d_1}}{v_{d_2}} = \frac{1}{2}$

17. (2) Conductance = $G = \frac{1}{R} = \frac{A}{\rho\ell}$

$\therefore G \propto \frac{1}{\ell}$

So, conductance will be halved.

18. (2) For a limited range of temperature, the graph between resistivity and temperature is a straight line for a material like nichrome because it is an alloy.

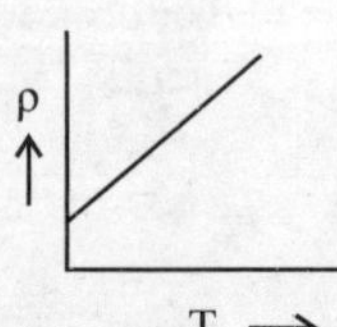

19. (2) $R = \rho\dfrac{\ell}{A}$

Volume V of the wire is,

$V = A\ell$

By differentiation, $0 = \ell dA + Ad\ell \quad \Rightarrow -\ell dA = Ad\ell \quad ...(i)$

Differentiating, $R = \rho\dfrac{\ell}{A},\ dR = \dfrac{\rho(Ad\ell - \ell dA)}{A^2} \quad ...(ii)$

By putting value of $-\ell dA$ in equation (*ii*), we get

$dR = \dfrac{\rho 2Ad\ell}{A^2}$

$dR = \dfrac{2\rho d\ell}{A}$

Dividing both sides by R and putting $R = \rho\dfrac{\ell}{A}$ in RHS.

$\dfrac{dR}{R} = \dfrac{2d\ell}{\ell}$

So, $\dfrac{dR}{R}\% = 2\times\dfrac{d\ell}{\ell}\%$

$\dfrac{dR}{R}\% = 2\times 0.1\%$

Hence, $\dfrac{dR}{R}\% = 0.2\%$

20. (1) In parallel circuits, equivalent resistance (Req) is given by

$\dfrac{1}{R_{eq}} = \dfrac{1}{R} + \dfrac{1}{R} + \dfrac{1}{R} + ...\text{N times}$

$\Rightarrow \dfrac{1}{R_{eq}} = \dfrac{1}{R}(1+1+...\text{N times}) = \dfrac{1}{R}\times N$

$\Rightarrow R_{eq} = \dfrac{R}{N}$

21. (3) In series circuits, equivalent resistance (R_{eq}) is given by

$R_{eq} = R + R + R...\text{n times}$

$\Rightarrow R_{eq} = R(1 + 1 + ...\text{ n times}) = nR$

22. (3) Lowest resistance will be obtained when connected in parallel

$\dfrac{1}{R_p} = \dfrac{1}{R_1} + \dfrac{1}{R_2} + \dfrac{1}{R_3} .... \dfrac{1}{R_n}$

$\dfrac{1}{R_p} = \dfrac{1}{\frac{1}{10}} + \dfrac{1}{\frac{1}{10}} + \dfrac{1}{\frac{1}{10}} + \dfrac{1}{\frac{1}{10}} + \dfrac{1}{\frac{1}{10}} + \dfrac{1}{\frac{1}{10}} + \dfrac{1}{\frac{1}{10}} + \dfrac{1}{\frac{1}{10}} + \dfrac{1}{\frac{1}{10}} + \dfrac{1}{\frac{1}{10}}$

$\dfrac{1}{R_p} = 100 \Rightarrow R_p = \dfrac{1}{100}\text{ ohm}$

23. (4) When a metallic wire of resistance 12Ω is bent to form a square, the equivalent circuit diagram is as shown in figure.

Resistance of arm, ABC = 3 + 3 = 6 Ω

Similarly, resistance for arm, ADC = 3 + 3 = 6 Ω

As the arms ADC and ABC are in parallel

Their effective resistance is $\dfrac{6\times 6}{6+6} = 3\Omega$

24. (3) $V_P - V_Q = R_{eq}\times I = \left[\dfrac{6}{3} + \dfrac{12\times 6}{12+6}\right]\times\dfrac{1}{2} = (2+4)\dfrac{1}{2} = 3V$

25. (4) 2Ω and 2Ω in parallel gives = $\dfrac{2\times 2}{2+2} = 1\Omega$

∴ Total resistance = 1 + 2 + 1 + 4 = 8Ω

∴ Current = $\dfrac{\text{emf}}{\text{resistance}} = \dfrac{4}{8} = 0.5A$

26. (2)

∴ $R_{ab} = R + \dfrac{R}{3} = \dfrac{4R}{3}$

27. (4) After cutting of wire, resistance of each part will be $\dfrac{R}{n}$ and all n parts are joined into parellel combination then effective R resistance will be $\dfrac{R}{n^2}$.

28. (4) 6 Ω and 6 Ω are in series so effective resistance is 12 Ω which is in parallel with 3 Ω so

$\dfrac{1}{R} = \dfrac{1}{3} + \dfrac{1}{12} = \dfrac{15}{36}$

$R = \dfrac{36}{15}$

$I = \dfrac{V}{R} = \dfrac{4.8\times 15}{36} = 2A$

29. (1) According to Kirchhoff's first law,

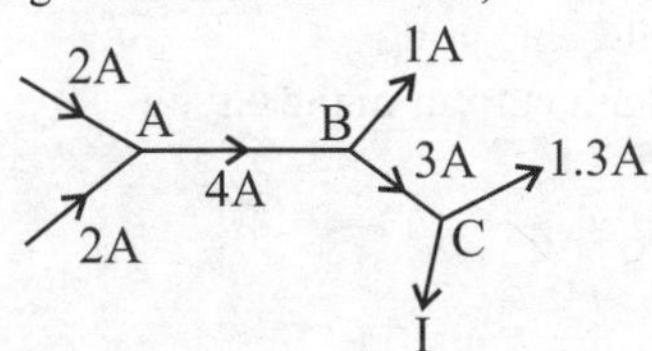

At junction A, $I_{AB} = 2 + 2 = 4A$

At junction C, $I = I_{BC} - 1.3 = 3 - 1.3 = 1.7A$

30. (2) In series, $I_1 = \dfrac{2E}{2+2r} = \dfrac{E}{1+r}$

In parallel, $I_2 = \dfrac{E}{2+\frac{r}{2}} = \dfrac{2E}{4+r}$

Since $I_1 = I_2$

$$\Rightarrow \frac{E}{1+r} = \frac{2E}{4+r} \Rightarrow r = 2\Omega$$

31. (1) Let m be the number of rows

Let n be the number of cells in a row

Then $m \times n = 100$ ...(*i*)

Also condition of maximum current is

$$R = \frac{nr}{m}$$

$$25 = \frac{1 \times n}{m} \Rightarrow n = 25m \quad ...(i)$$

On solving (*i*) and (*ii*), we get $m = 2$

32. (3) By using, $I = \dfrac{E}{R+r}$

$$0.5 = \frac{E}{11+r}$$

$$E = 5.5 + 0.5r \quad ...(i)$$

$$\text{and } 0.9 = \frac{E}{5+r}$$

$$\Rightarrow E = 4.5 + 0.9r \quad ...(ii)$$

On solving (*i*) and (*ii*) we get

$r = 2.5$ ohm

33. (3) Let X be the equivalent resistance between A and B. The network consist of infinite units of two resistors of 1Ω, 1Ω. The addition of one more such unit across AB will not affect total resistance. Th network will then appear as shown below:

$$R_{PQ} = 1 + \frac{1 \times X}{1+X} = R_{AB}$$

$$\text{or } 1 + \frac{X}{1+X} = X$$

$$\text{or } X^2 - X - 1 = 0$$

$$\text{or } X = \frac{1+\sqrt{1+4}}{2} = \frac{1+\sqrt{5}}{2}\Omega$$

34. (3) Total cells $= m \times n = 24$ ...(*i*)

For maximum current in the circuit

$$R = \frac{mr}{n}$$

$$\Rightarrow 3 = \frac{m}{n} \times \frac{1}{2} \Rightarrow m = 6n \quad ...(ii)$$

On solving equation (*i*) and (*ii*) we get

$n = 2, m = 12$

35. (4) Reading of voltmeter,

$$E_{eq} = \frac{E_1 r_2 + E_2 r_1}{r_1 + r_2} = \frac{18 \times 1 + 12 \times 2}{1+2}$$

$$= \frac{18+24}{3} = \frac{42}{3} = 14V$$

36. (4) $I = \dfrac{2E}{R + R_1 + R_2}$

For cell (2), $E = V + IR_2 = 0 + IR_2$

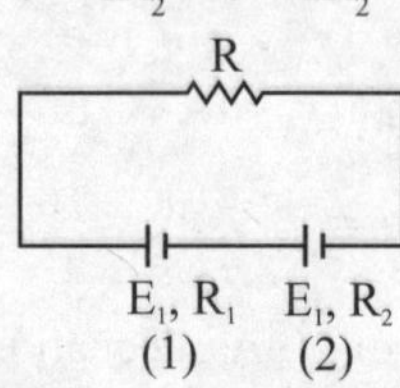

$$I = \frac{2I}{R+R_1+R_2} \times R_2 \Rightarrow R = R_2 - R_1$$

37. (2) Initially when no current flows, the terminal voltage will be emf so $\varepsilon = 40V$. The internal resistance of battery is given by

$$r = \left(\frac{E}{V} - 1\right)R = \left(\frac{40}{30} - 1\right)9 = \frac{9 \times 10}{30} = 3\,\text{ohm}$$

38. (2) Let the current in the circuit be, $I = \dfrac{E}{R+r}$ ...(*i*)

Across the cell, $V = E - Ir$ ...(*ii*)

$$V = E - \frac{E}{R+r} \times r \quad \text{[From equation (}i\text{)]}$$

$$\therefore V(R+r) = ER$$

$$r = \left(\frac{E-V}{V}\right)R$$

39. (3) Power: $P = \dfrac{V^2}{R}$

For small variation in the value of power after differentiation:

$$\frac{\Delta P}{P} \times 100 = \frac{2 \times \Delta V}{V} \times 100\%$$

$$= 2 \times 2.5 = 5\%$$

$\therefore$ Power decreased by 5%.

40. (3) $R_2 = R$

$R_1 = 5\Omega$, $P = 30W$

In parallel combination

$$P = \frac{V^2}{R_1} + \frac{V^2}{R_2}$$

$$P - \frac{V^2}{R_1} = \frac{V^2}{R_2}$$

$$30 - \frac{10^2}{5} = \frac{10^2}{R_2} \Rightarrow \frac{100}{R_2} = 30 - 20 = 10$$

$\therefore R_2 = 10\Omega$

41. (2) Resistance of 25W bulb

$$R_1 = \frac{V^2}{P} = \frac{(220)^2}{25} = 1936\Omega$$

Resistance of 100W bulb

$$R_2 = \frac{V^2}{P} = \frac{(220)^2}{100} = 484\Omega$$

Series resistance $= R_1 + R_2 = 1936 + 484 = 2420\Omega$

Current through series combination will be,

$$I = \frac{440}{2420} = \frac{2}{11}$$

$$V_2 = R_2 \times I = 484 \times \frac{2}{11} = 88V$$

$$V_1 = R_1 \times I = 1936 \times \frac{2}{11} = 352V$$

Thus the bulb of 25W will be fused because it can tolerate only 220V while the voltage across it is 352V.

42. (1) Let, each bulbs are rated at V volts, men.

$$\frac{P_{40}}{P_{100}} = \frac{V^2}{R_1} \times \frac{R_2}{V^2} \Rightarrow \frac{40}{100} = \frac{R_2}{R_1} \Rightarrow R_2 = \frac{2}{5}R_1$$

When bulbs are connected in series (same current flow across the bulbs):

$$\frac{P_{40}}{P_{100}} = \frac{I^2R_1}{I^2R_2} = \frac{R_1}{R_2} = \frac{5}{2}$$

Hence, 40W bulb glow brighter as P_{40} is 2.5 times P_{100}.

When bulbs are connected in parallel (Same voltage across the bulbs)

$$\frac{P_{40}}{P_{100}} = \frac{V^2}{R_1} \times \frac{R_2}{V^2} = \frac{R_2}{R_1} = \frac{2}{5}$$

Hence, 100W bulb glow brighter as P_{100} is 2.5 times p_{40}.

43. (3) $P = \frac{V^2}{R} \Rightarrow R = \frac{V^2}{P} \Rightarrow R \propto V^2$

$$\frac{R_1}{R_2} = \left(\frac{200}{300}\right)^2 = \frac{4}{9}$$

When connected in series, potential drop and power consumed is

$P = I^2R$ and $V = IR$

$P \propto R$ and $V \propto R$

$$\therefore \frac{P_1}{P_2} = \frac{R_1}{R_2} = \frac{4}{9}, \frac{V_1}{V_2} = \frac{4}{9}$$

44. (2) Heat produced in each resistor

$$H = \frac{V^2}{R}t$$

Heat H_S produced in resistors in series.

$$H_S = \frac{V^2}{2R}T$$

As the heat produced is same in both the cases

$\Rightarrow H = H_S$

$$\Rightarrow \frac{V^2}{R}t = \frac{V^2}{2R}T$$

$\Rightarrow T = 2t = 2(10) = 20$ minutes

45. (1) For I[st] filament

$$H = \frac{V^2}{R_1}t_1 \Rightarrow R_1 = \frac{V^2}{H}t_1 \quad ...(i)$$

For 2nd filament

$$H = \frac{V^2}{R_2}t_2 \Rightarrow R_2 = \frac{V^2}{H}t_2 \quad ...(ii)$$

When placed in parallel

$$H = \frac{V^2}{R_P}t_P \Rightarrow R_P = \frac{V^2}{H}t_P \quad ...(iii)$$

From eq. (*i*), (*ii*), & (*iii*) we get

$$\frac{1}{R_p} = \frac{1}{R_1} + \frac{1}{R_2}$$

$$\frac{H}{V^2t_p} = \frac{H}{V^2t_1} + \frac{H}{V^2t_2}$$

$$\frac{1}{t_p} = \frac{1}{t_1} + \frac{1}{t_2} \Rightarrow \frac{1}{10} + \frac{1}{15} = \frac{1}{6}$$

$t_p = 6$ min

46. (3) $R_s = R + 2R = 3R$

$$R_p = \frac{R \times 2R}{R + 2R} = \frac{2R^2}{3R} = \frac{2}{3}R$$

$$\frac{H_s}{H_p} = \frac{V^2}{R_s} \times \frac{R_p}{V^2} = \frac{2R}{3 \times 3R} = 2:9$$

47. (3) Resistance of each part = R/n

Effective resistance = $\frac{R}{n^2}$ $\left[R_p = \frac{R}{n \times n}\right]$

$$P \propto \frac{1}{R} \Rightarrow \frac{P'}{P} = \frac{R}{R/n^2} \Rightarrow P' = n^2P$$

48. (2) Initially, $\frac{l_1}{100 - l_1} = \frac{X}{R} = \frac{2}{3}$

$$\therefore l_1 = \frac{3}{5} \times 100 = 40 \text{ cm}$$

Finally, $\frac{l_2}{100 - l_2} = \frac{X}{R'} = \frac{12}{8} = \frac{3}{2}$

$$\therefore l_2 = \frac{3}{5} \times 100 = 60 \text{ cm}$$

$\therefore$ J is displaced by $l_2 - l_1 = 20$ cm

49. (3) For the balance point of meterbridge.

Case - I, when balancing length is 55 cm

$$\frac{P}{Q} = \frac{\ell_1}{100 - \ell_1} \Rightarrow \frac{3}{Q} = \frac{55}{45}$$

$$Q = \frac{45}{55} \times 3 \quad ...(i)$$

Case - II When an unknown resistance x is involved

$$\frac{P + x}{Q} = \frac{\ell_1}{100 - \ell_1}$$

$$\frac{3 + x}{Q} = \frac{75}{25} \Rightarrow 3 + x = \frac{75}{25} \times Q \quad ...(ii)$$

By putting the value of Q from equation (*i*) in equation (*ii*)

$$3+x=\frac{75}{25}\times\frac{45\times3}{55}$$

$$x=\frac{81}{11}-3=\frac{81-33}{11}=\frac{48}{11}\Omega$$

50. (3) Potential gradient = $\left(\frac{E}{R+r}\right)\frac{r}{L}$

$$=\frac{10\times3}{(3+3)\times5}=\frac{30}{6\times5}=1Vm^{-1}$$

V = 10 mV/cm

51. (4) $S=\left(\frac{100-\ell}{\ell}\right)R$

Initially, $30=\left[\frac{100-\ell}{\ell}\right]\times10\Rightarrow\ell=25cm$

Finally, $10=\left[\frac{100-\ell}{\ell}\right]\times30\Rightarrow\ell=75cm$

So, shift in null point/balance point = 50 cm

52. (1) The ratio $\frac{AC}{CB}$ will remain unchanged.

High Order Time Intensive MCQs

1. (3) $\frac{1}{R}=\frac{1}{R_1}+\frac{1}{R_2}$

$$\Rightarrow\frac{\sigma2A}{\ell}=\frac{\sigma_1A}{\ell}+\frac{\sigma_2A}{\ell}$$

$$\Rightarrow2\sigma=\sigma_1+\sigma_2=10+15=25$$

$$\Rightarrow\sigma=\frac{25}{2}=12.5\Omega^{-1}m^{-1}$$

2. (1) A shunt resistor in an ammeter is used to increase the range of the ammeter.

The shunt provides a parallel resistance path to allow excess current to bypass the ammeter's internal coil, enabling the ammeter to measure currents beyond its normal range without getting damaged.

3. (2) Since coil and shunt are in parallel.

Therefore, $R_{eff}=\frac{100\times10}{100+10}=\frac{100}{11}=9.1$

4. (2) The current 'I' through the circuit according to Ohm's Law when a battery with electromotive force (emf) 'E' and internal resistance 'Ri' is connected to a resistor of resistance 'R' can be calculated using the formula: $\frac{E}{R_i+R}$

5. (3) When bulbs are in series

$$P=\frac{V^2}{3R}\Rightarrow\frac{V^2}{R}=3P$$

When bulbs are connected in parallel

$$P'=\frac{V^2}{\frac{R}{3}}=\frac{3V^2}{R}=3(3P)=9P$$

6. (2) As $R\propto l$, so the length increases n-fold. Also, since the volume of the wire remains constant, so the area of the wire decreases n-fold.

Hence, the resistance increases to n^2R.

7. (4)

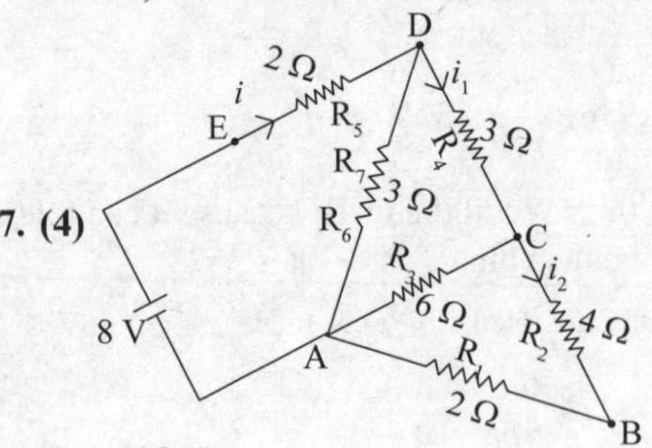

$R_{eq}=4\Omega$

$i=\frac{8}{4}=2A\quad i_1=\frac{2\times3}{3+6}=\frac{2}{3}A$

$i_2=\frac{2/3}{2}=\frac{1}{3}A$

8. (2) $R_{eff}=\frac{4}{5}+\frac{6}{5}=2\Omega$

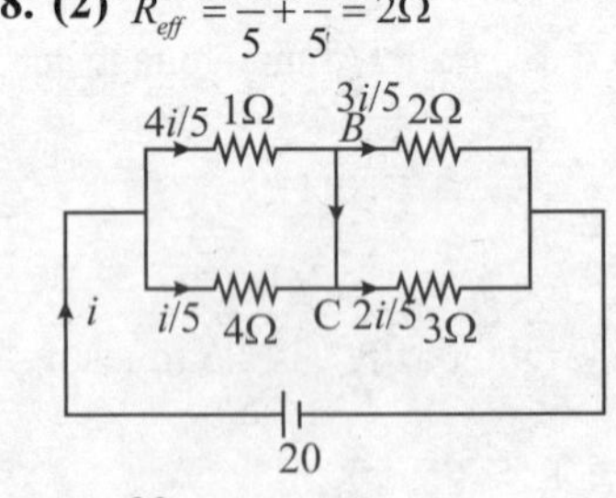

$i=\frac{20}{2}=10A$

$I=\frac{4i}{5}-\frac{3i}{5}=+\frac{i}{5}=2A$

9. (3) From the graph, part CD has negative slope, thus it has negative dynamic resistance.

10. (4) Resistance of wire, $R=\rho\frac{l}{A}$

$\therefore\quad R\propto l$,

Let l_1 and l_2 be the lengths of the wires having resistances R_1 and R_2

Here, $R_0=R_1+R_2=12\,\Omega$ ...(*i*)

and $\frac{R_1\times R_2}{R_1+R_2}=\frac{8}{3}\Omega$ ($\because R_1$ and R_2 are in parallel)

Putting, $R_1+R_2=12\,\Omega$ in above equation, we get

$R_1R_2=32\,\Omega$ ...(*ii*)

After solving Eqs. (*i*) and (*ii*), we get $R_1=4\,\Omega$ and $R_2=8\,\Omega$

$\because\frac{R_1}{R_2}=\frac{\ell_1}{\ell_2}$

Hence, $\frac{l_1}{l_2}=\frac{1}{2}$

11. (1) As, $R=10$ cm $=0.1$ m

$\therefore$ Circumference of circle $=2\pi R$

$\Rightarrow R'=$ Total resistance $=2\pi R(12)$

$$\frac{1}{R_{eq}} = \frac{1}{R'/2} + \frac{1}{R'/2} = \frac{4}{R'} \Rightarrow R_{eq} = \frac{R'}{4}$$

$\Rightarrow R_{eq} = 6\pi\,(0.1) = 0.6\,\pi\Omega$

12. **(2)** Applying Kirchhoff's law in loop 1, we get, $500I + 100I = 12$

So, $I = \frac{12\times10^{-2}}{6} = 2\times10^{-2}$ A

Hence, $V_B = IR = 100(2\times10^{-2}) = 2V$

If the cell V_2 has a potential of 2V, potential difference across the resistance R and V_2 would be equal. Hence no current would flow through V_2.

13. **(4)** Let the emf of cell be E and internal resistance be r.

Then, $I_1 = \frac{E}{(r+R_1)}$ and $I_2 = \frac{E}{(r+R_2)}$

On dividing, we get

$\frac{I_1}{I_2} = \frac{(r+R_2)}{(r+R_1)}$ or $r = \frac{I_2R_2 - I_1R_1}{I_1 - I_2}$

14. **(2)** $\left[\frac{dI}{dV}\right]_1 = \frac{1}{R_1} = \tan\theta_1$

$\left[\frac{dI}{dV}\right]_2 = \frac{1}{R_2} = \tan\theta_2$

$\theta_1 > \theta_2 \Rightarrow \tan\theta_1 > \tan\theta_2$

$R_2 > R_1$

Hence, $T_2 > T_1$ [$\because R \propto T$]

15. **(3)** $V_P - V_Q = E + I(r + R) = 2 + 3(2 + 4) = 20$ V.

16. **(3)** Emf, $E = I(nr + r)$

$E = Ir(n + 1)$

But potential difference = Inr

$$\Rightarrow \frac{\text{Potential difference}}{\text{Emf}} = \frac{n}{n+1}$$

17. **(2)** For the circuit given in question,

$V_A - (6\times2) - 12 - (9\times2) + 4 - (5\times2) = V_B$

or $V_A - 12 - 12 - 18 + 4 - 10 = V_B$

or $V_A - V_B = 48$ V

18. **(2)** There are two cases

Cases-I: For balanced point of meter bridge, $\frac{5}{R} = \frac{l_1}{100 - l_1}$ ...(*i*)

Cases-II: When R is shunted with equal resistance i.e. R

$$\frac{1}{R'} = \frac{1}{R} + \frac{1}{R} \Rightarrow R' = R/2$$

$$\therefore \frac{5}{R/2} = \frac{1.6l_1}{100 - 1.6l_1} \quad ...(ii)$$

From Eqs. (*i*) and (*ii*), we get

$$2\left(\frac{l_1}{100 - l_1}\right) = \frac{1.6l_1}{100 - 1.6l_1} \Rightarrow l_1 = 25\text{cm}$$

From Eq. (*i*), $\frac{5}{R} = \frac{25}{75} \Rightarrow R = 15\Omega$

19. **(4)** As the circuit is open, no current flows through it.

Potential difference between X and Y = emf of the battery = 120 V.

20. **(1)** Equivalent resistance of 3 Ω and 6 Ω connected in parallel is

$R' = \frac{3\times6}{3+6} = 2\Omega$

Total resistance = 2 + R

$I = \frac{V}{R} \Rightarrow 2 = \frac{6}{2+R}$

$\Rightarrow 2 + R = 3$

$\therefore R = 1\,\Omega$

21. **(1)**

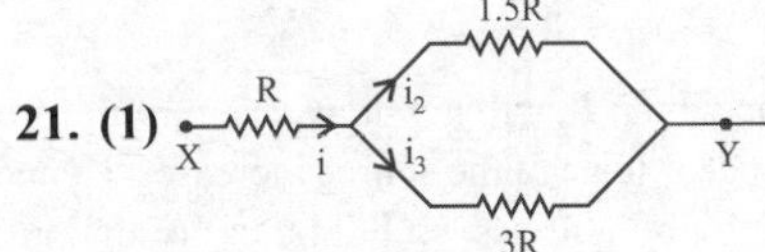

Here, $V_2 = V_3$

i.e. $i_2 \times 1.5R = 3R \times i_3$ ...(*i*)

and $i_2 + i_3 = i$ ...(*ii*)

On solving (*i*) and (*ii*)

$\Rightarrow i_2 = \frac{2i}{3}$ and $i_3 = \frac{i}{3}$

Now, $V_1 = iR$

$V_2 = \frac{2i}{3}\times1.5R = iR$

and $V_3 = \frac{i}{3}\times3R = iR$ i.e. $V_1 = V_2 = V_3$

22. **(3)** Total current drawn from the battery,

$i = \frac{E}{R+r} = \frac{6}{100+0} = 0.06$A

Resistance of 50 cm wire is

$R' = \frac{\rho l'}{A} = \left(\frac{\rho}{A}\right)l' = \left(\frac{R}{l}\right)l'$

$R = \frac{100}{300}\times50 = \frac{50}{3}\Omega$

Hence, the potential difference between two points on the wires separated by a distance 50 cm is

$V = iR' = 0.06\times\frac{50}{3} = 1$V

23. **(4)** Since due to wrong connection of each cell, the total emf is reduced by 2ε, so then for wrong connection of three cells the total emf will be reduced to (nε–6ε) whereas the total or equivalent internal resistance of the cell combination will be nr.

24. **(3)** Resistance of first bulb,

$R_1 = \frac{V_1^2}{P_1} = \frac{250\times250}{100} = 625\Omega$

Resistance of second bulb,

$$R_2 = \frac{V_2^2}{P_2} = \frac{200 \times 200}{100} = 400\Omega$$

In a series circuit, potential divides in proportion to resistance. Potential drop across bulb B_2 is

$$\therefore V_2' = \frac{R_2}{R_1 + R_2} \times V = \frac{400}{625 + 400} \times 250\text{ V} = 97.56\text{ V} \approx 98\text{ V}.$$

25. (1) According to Kirchhoff's first rule, current in arm

DC = 3 − 1 = 2 A

Potential difference across 2 Ω = 2 × 2 = 4 V

As point A is at zero potential, so potential of point C is $V_C = 2V$

Now, $V_D - V_C = 4 \Rightarrow V_D - 2 = 4 \Rightarrow V_D = 6\text{ V}$

$\because V_D - V_B = 4 \Rightarrow V_B = V_D - 4 = 6 - 4 = 2V$

26. (3) When key, k is opened, bulb B_2 will not draw any current form the source, so that terminal voltage of the source increases. Hence, power consumed by the bulb increases, so brightness of the bulb becomes more. The brightness of bulb B_1 decreases.

27. (1) Let R_1 and R_2 be the resistances of the bulbs.

Power consumed, $P = \frac{V^2}{R}$

So, the power consumed by the first bulb is

$$P_1 = \frac{V^2}{R_1} \quad ...(i)$$

and power consumed by the second bulb is

$$P_2 = \frac{V^2}{R_2} \quad ...(ii)$$

If these two bulbs are combined in series. So, the total resistance of the combination is

$R = R_1 + R_2$

Hence, the power consumed in combination is

$$P = \frac{V^2}{(R_1 + R_2)} \quad ...(iii)$$

Now, from Eqs. (*i*) and (*ii*)

$$\Rightarrow \frac{1}{P_1} + \frac{1}{P_2} = \frac{1}{\frac{V^2}{R_1}} + \frac{1}{\frac{V^2}{R_2}} = \frac{R_1}{V^2} + \frac{R_2}{V^2} = \frac{(R_1 + R_2)}{V^2}$$

Form Eq. (*iii*),

$$\frac{1}{P_1} + \frac{1}{P_2} = \frac{1}{P} \Rightarrow \frac{P_2 + P_1}{P_1 P_2} = \frac{1}{P}$$

$$\Rightarrow P = \frac{P_1 P_2}{P_1 + P_2}$$

28. (4)

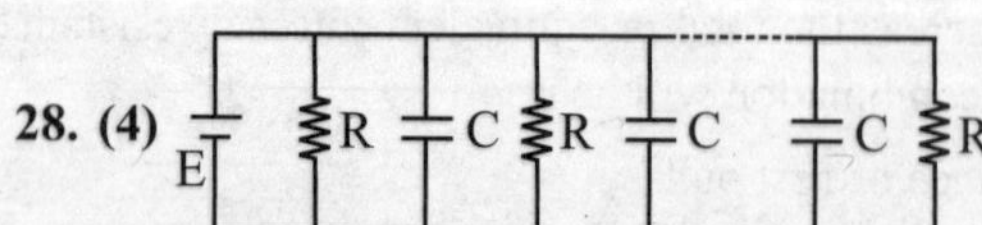

As all the capacitors are connected in parallel, therefore potential difference across each capacitor is E.

$$\therefore Q_{first} = Q_{last} = CE$$

$$\text{Ratio} = \frac{Q_{first}}{Q_{last}} = 1$$

29. (2) In steady state the potential drop across 'R' will be more than that across R/2. So $V_X < V_Y$.

Hence A will have -ve charge.

30. (4)

If a battery supplies some amount of current. It also receives same amount of current so,

$I - I_1 + I_2 = I$

$\Rightarrow I_2 = I_1$

31. (1) Let the resistances of both segments be R_1 and R_2

then, $R_1 = \frac{\rho\ell}{A'}$ where $A'\ell = A\frac{\ell}{2} \Rightarrow A' = \frac{A}{2}$

so, $R_1 = \frac{2\rho\ell}{A} = 2R$

$$R_2 = \frac{\rho \ell/4}{2A} = \frac{R}{8}$$

Effective resistance, $R_{eq} = R_1 + R_2 = \frac{17R}{8}$

32. (3) $\frac{\varepsilon_1 + \varepsilon_2}{r_1 + r_2 + R} = 2$ and $\frac{\varepsilon_1 - \varepsilon_2}{r_1 + r_2 + R} = 1$

$$\Rightarrow \frac{\varepsilon_1 + \varepsilon_2}{\varepsilon_1 - \varepsilon_2} = \frac{2}{1} \text{ or } \frac{\varepsilon_1}{\varepsilon_2} = \frac{3}{1}$$

33. (1) Current flowing through the resistor will be same throughout & $J \propto \frac{1}{A}$

34. (3) If $V_B - V_C = 12$ then potential drop across 2Ω resistor between B and C is 2V so, current flow from B to C $= \frac{2}{2} = 1\text{A}$

By Junction rule, $I_1 = 2 + 1 = 3\text{A}$.

35. (1) When switch S is closed, $R_{eff} = 2R$ (R is resistance of one bulb)

$$P = \frac{V^2}{2R}$$

When switch S is open, $R_{eff} = 3R$

$$P' = \frac{V^2}{3R} \Rightarrow \frac{P'}{P} = \frac{2}{3}$$

36. (3) Internal resistance of the cell,

$$r = \left(\frac{l_1 - l_2}{l_2}\right) = \left(\frac{55 - 50}{50}\right) \times 10\Omega = 1\Omega$$

37. (1)

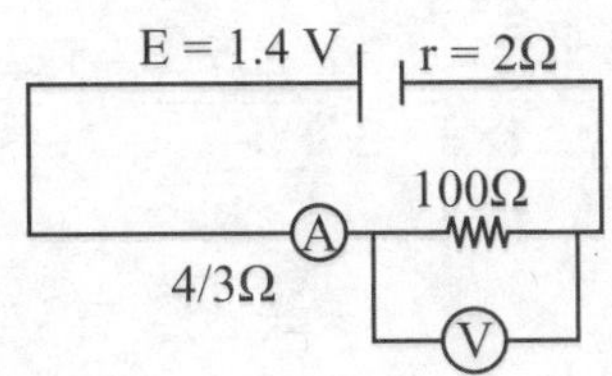

Total resistance in the circuit $= \left[2+\frac{4}{3}+\frac{100R_V}{100+R_V}\right]$

$$I=\frac{\text{Emf}}{\text{Total resistance}}$$

$$\Rightarrow 0.02A=\frac{1.4V}{\left[2+\frac{4}{3}+\frac{100R_V}{100+R_V}\right]\Omega}$$

$\Rightarrow R_v = 200\ \Omega$

38. (4)

Total resistance in the circuit $= \left[2+\frac{4}{3}+\frac{100R_V}{100+R_V}\right]$

$$I=\frac{\text{Emf}}{\text{Total resistance}} \Rightarrow 0.02A=\frac{1.4V}{\left[2+\frac{4}{3}+\frac{100R_V}{100+R_V}\right]\Omega}$$

$\Rightarrow R_v = 200\ \Omega$

Potential difference across the voltmeter

$$=0.02A\left[\frac{100\times 200}{100+200}\right]=1.33V.$$

Voltmeter reading = 1.10 V

$\Rightarrow$ Error = 1.33 – 1.10 = 0.23 V.

39. (1) Arrangement is shown in the figure.

$$\frac{X}{Y}=\frac{60}{40}=\frac{3}{2} \quad ...(i)$$

When X is shunted then resistance in the left gap becomes

$$X'=\frac{X.\frac{X}{2}}{X+\frac{X}{2}}=\frac{X}{3} \quad ...(ii)$$

Now from equation (*i*) and (*ii*)

$$\frac{\left(\frac{X}{3}\right)}{Y}=\frac{l}{(100-l)} \Rightarrow \frac{1}{3}\times\frac{3}{2}=\frac{l}{(100-l)} \Rightarrow l = 33.3cm$$

$\therefore$ Shift = 60 – 33.3 = 26.7 cm.

40. (4) As current is rate of flow of charge in the direction in which positive charge will move, the current due to electron will be $i_e = \frac{n_e q_e}{t} = 3 \times 10^{18} \times 1.6 \times 10^{-19} = 0.48$ A

(Opposite to the motion of electrons, i.e. right to left)

Current due to protons $i_p = \frac{n_p q_p}{t} = 2 \times 10^{18} \times 1.6 \times 10^{-19} =$ 0.32 A (Right to left) so total $I = i_e + i_p = 0.48 + 0.32 = 0.80$ A (Right to left)

41. (3) As the equivalent resistance in the two arms of the resistances are equal. Therefore, equal current will flow through each of them.

Reading of $A_2 = \frac{1}{4}A$

Hence reading of $A_1 = \frac{1}{2}$ amp. and reading of A_2 =1/4 amp.

42. (3) Equal current (of 1 A) will be distributed in the two branches.

$V_D - V_A = 2 \times 1 = 2$ V

and $V_D - V_B = 3 \times 1 = 3$ V

$\therefore V_A - V_B = (V_D - V_B) - (V_D - V_A) = 1$ V

43. (3)

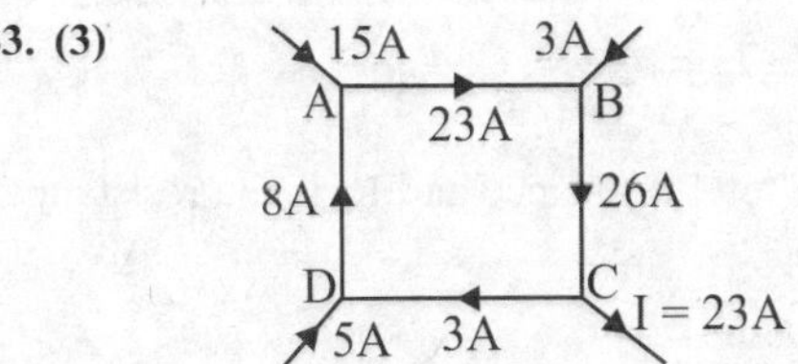

Applying Kirchhoff's first law at junction A, B, C, D

At A, $i_{AB} = 15 + 8 = 23$ A

At B, $i_{BC} = 23 + 3 = 26$ A

At D, $i_{CD} = 8A - 5A = 3A$

At C, $i_{CD} + I = i_{BC}$

or $3 + I = 26$

$\therefore I = 23$ A

44. (1) In series i is same, $H = i^2Rt$. Therefore, $H \propto R$.

45. (2) In steady state capacitor will be in open circuit.

$\Rightarrow R_{eff} = 1.2 + 2.8 = 4\Omega$

$$i=\frac{V}{R_{eff}}=1.5A$$

$$i_1=\frac{3}{5}(i)$$

$$i_1=\frac{3}{5}\times 1.5=0.9A$$

46. (1) $R_{eq}=\frac{R\theta(2\pi-\theta)}{4\pi^2}$

Putting $\theta = 45° = \frac{\pi}{4}$ rad and $R = 15\Omega$

$$R_{eq}=\frac{15\times\frac{\pi}{4}\times\left(2\pi-\frac{\pi}{4}\right)}{4\pi^2}=\frac{\frac{15\pi}{4}\left(\frac{7\pi}{4}\right)}{4\pi^2}$$

$$=\frac{105}{64}\Omega=1.64\Omega$$

47. (1) The given figure is equivalent to

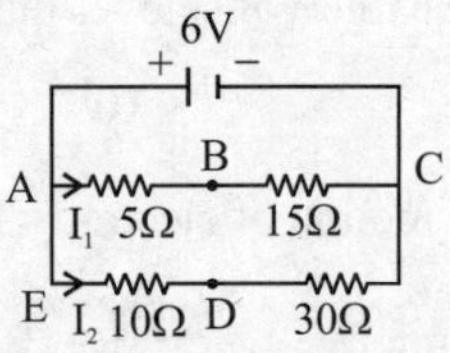

$$I_1 = \frac{6}{20} = \frac{3}{10};\ I_2 = \frac{6}{40} = \frac{3}{20}A$$

$$V_B = V_A - I_1 \times 5 = 6 - \frac{3}{10} \times 5 = 6 - \frac{3}{2} = \frac{9}{2}V$$

$$V_D = V_C - I_2 \times 10 = 6 - \frac{3}{20} \times 10 = 6 - \frac{3}{2} = \frac{9}{2}V$$

$$\therefore V_B - V_D = 0$$

48. (3) If one had considered a solid cylinder of radius b, one can suppose that it is made of two concentric cylinders of radius a and the outer part, joined along the length concentrically one inside the other.

If I_a and I_x are the currents flowing through the inner and outer cylinders

$$\therefore I_{total} = I_b = I_a + I_x \Rightarrow \frac{V}{R_b} = \frac{V}{R_a} + \frac{V}{R_x}$$

Where R_b is the total resistance and R_x is the resistance of the tubular part.

$$\therefore \frac{1}{R_x} = \frac{1}{R_b} - \frac{1}{R_a}$$

$$\text{But } R_a = \frac{\rho l}{\pi a^2} \text{ and } R_b = \frac{\rho l}{\pi b^2}$$

$$\therefore \frac{1}{R_x} = \frac{\pi}{\rho l}(b^2 - a^2)$$

$$\therefore R_x = \frac{\rho l}{\pi(b^2 - a^2)}$$

49. (2) As $q = \alpha t - \beta t^2$

$$i = \frac{dq}{dt} = \alpha - 2\beta t$$

$$\text{For } i = 0,\ t = \frac{\alpha}{2\beta}$$

$\therefore$ Current flow from $t = 0$ to $t = \frac{\alpha}{2\beta}$

$$\Rightarrow H = \int_0^t i^2 R dt \text{ on solving we get}$$

$$= R\int_0^{\alpha/2\beta} (\alpha - 2\beta t)^2 dt = R\int_0^{\alpha/2\beta} (\alpha^2 + 4\beta^2 t^2 - 4\alpha\beta t) dt$$

$$H = \frac{\alpha^3 R}{6\beta}$$

50. (2) For potential difference between B and D to be zero,

$$\frac{P}{Q} = \frac{R}{S}$$

$$\frac{6+15}{\frac{8x}{8+x}+3} = \frac{15 + \frac{6\times 6}{6+6}}{4 + \frac{4\times 4}{4+4}} \Rightarrow \frac{21}{\frac{8x}{8+x}+3} = \frac{18}{6} = 3$$

$$\frac{8x}{8+x} + 3 = 7$$

$$8x = 32 + 4x$$

$$4x = 32 \Rightarrow x = 8\ \Omega.$$

51. (3) $R \propto \frac{l}{A} = \frac{l^2\rho}{Al\rho} = \frac{l^2\rho}{m}$

For same m, $R \propto l^2$

$$\therefore \quad P \propto \frac{1}{R} \propto \frac{1}{l^2}$$

$$\frac{P_A}{P_B} = \frac{l_B^2}{l_A^2} = \frac{4}{1}$$

$$P_A = 4 \times 5 = 20\ W$$

Statement Based MCQs

1. **(1)** Manganin has high conductivity and its temperature coefficient of resistance is almost zero that is why its resistivity does not change with variation in temperature. Hence, it is used as standard resistance coil.
2. **(3)** When we close circuit, electric field sets up in entire closed circuit instantly with speed of electromagnetic wave which cause electron drive at every portion of circuit. Due to this current set up in entire circuit instantly. The drift velocity in a metallic wire is not that high,
3. **(1)** Since in a conductor, electrons are in random motion and hence conductor has average thermal velocity of electrons as zero. Therefore, no net current flows through it and it does not experience any magnetic force.
4. **(2)** Bulb is never fused in switched off condition. It only fuses when excess current passes through it.
5. **(4)** Since, metal has moving or conducting electrons and electrolytes has ions to conduct electricity, and ions have less mobility than electrons so metal conducts electricity more than electrolyte.

Assertion & Reason MCQs

1. **(1)** When electrons are accelerated by an electric field they gain kinetic energy and then collide with ions present in the lattice. They transfer energy and the energy is lost in the form of heat.
2. **(1)** Nichrome has high resistivity and high melting point, so it can operate at very high temperature and does not burn.
3. **(1)** Electricity chooses the path of least resistance and the water present inside a tree is a much better conductor than wood.
4. **(1)** For case I: $V_A - Ir - E = V_B \quad \Rightarrow \quad V_A - V_B = E + Ir$
   For case II: $V_A - E + Ir = V_B \quad \Rightarrow \quad V_A - V_B = E - Ir.$
5. **(3)** Heat dissipated in the electric bulb filament is due to high resistance of filament wire where as in a line wire there is very less resistance so as to transfer very high power. Reason is false as the melting point of the filament must be very high else it cannot glow or last for much time.

Match the Columns MCQs

1. **(2)** (a) (ii, iii)

   Electrical conductivity of conductor depends on nature of conductor and temperature.

   (b) (i, ii, iii)

   Conductance of conductor (σ = neμ) dimensions, temperature and nature of conductor.

   (c) (iii, iv)

   For a given conductor of given dimensions and at a given temperature current density (J = σ neμE) depends on electric field strength.

   (d) (i, ii, iii)

   For a given potential difference applied across a conductor of given length, current in it will depends on dimensions, temperature and nature of conductor.

2. **(3)** Let current through various A branches is as shown in the figure.

   $i_1 + i_2 + i_3 = 0$ ...(*i*)

   $V_A - V_B = V_D - V_C = V_E - V_F$

   $1 - li_1 = 2 - 21_2 = 3 - Ri_3$ ...(*ii*)

   From Eqs. (*i*) and (*ii*), we get

   $$\frac{2-1+i_1}{2} + \frac{3-1+i_1}{R} = 0 \text{ or } i_2 = \frac{-(R+4)}{3R+2}$$

   Again from Eqs. (*i*) and (*ii*), we get

   $$\frac{2i_2 - 1}{i} + i_2 + \frac{1+2i_2}{R} = 0 \text{ or } i_2 = \frac{R-1}{3R+2}$$

   Now $i_3 = -i_1 - i_2 = \dfrac{5}{3R+2}$

   (a) Current i_3 cannot be zero.

   (b) Current i_1 cannot be zero.

   (c) $i_2 = 0 \Rightarrow R = 1\Omega$

   (d) $i_3 = \dfrac{5}{8}$ or $\dfrac{5}{3R+2} = \dfrac{5}{8}$ or $R = 2\,\Omega$

3. **(1)** (a) Average velocity of free electrons in metals when $E = 0$, is zero, in the absence of electric field, the electron move in zig-zag direction so net displacement is zero, hence average velocity of free electrons in zero.

   (b) Random velocity of free electron = 10^5 m/s.

   $(E = 0)$

   (c) Drift velocity (v_d) = 10^{-5} m/s.

   $(E \neq 0)$

NCERT Exemplar MCQs

1. **(2)** Current per unit area (taken normal to the current), I/A, is called current density and is denoted by J. The SI unit of the current density is A/m^2. The current density is also directed along E and is also a vector and the relationship is given by J = σE

   The J changes due to electric field produced by charges accumulated on the surface of wire.

   Note: As the current progresses along the wire, the direction of J (current density) changes in an exact manner, while the current I remain unaffected. The agent that is essentially responsible for this.

2. **(1)** Given by, $\varepsilon_{eq} = \dfrac{\varepsilon_2 r_1 + \varepsilon_1 r_2}{r_1 + r_2}$

   This suggest that the equivalent emf ε_{eq} of the two cells is given by the equivalent emf of this combination is

   $\varepsilon_1 < \varepsilon_{eq} < \varepsilon_2$

3. **(3)** The percentage error in R can be minimized by adjusting the balance point near the middle of the bridge, i.e., when l_1 is close to 50 cm. This requires a suitable choice of S.

   $$\frac{R}{S} = \frac{l_1}{100 - l_1}$$

   Since here, R : S = 2.9 : 97.1 imply that the S is nearly 33 times to that of R. In order to make this ratio 1 : 1, it is necessary to reduce the value of S nearly $\dfrac{1}{33}$ times, i.e., nearly 3 Ω.

4. **(1)** The resistance of wire is given by

   $$R = \rho \frac{l}{A}$$

   For greater value of R, ℓ must be higher and A should be lower and it is possible only when the battery is connected across 1 cm × $\left(\dfrac{1}{2}\right)$ cm (area of cross-section A).

5. **(1)** The relationship between current and drift speed is given by I = ne Avd

   Here, I is the current and v_d is the drift velocity.

   So, $I \propto v_d$

   Thus, only drift velocity determines the current in a conductor.

NEET & AIIMS Past Year MCQs

1. **(2)** The current I is related to drift velocity v_d by:

   $I = nAev_d$

   For a wire of diameter d, the cross-sectional area $A = \dfrac{\pi d^2}{4}$.

   Given:

   Initial current I_1 = 100 mA, diameter d, drift velocity v.

   New current I_2 = 200 mA, diameter $\dfrac{d}{2}$, drift velocity v'.

   Since $I \propto d^2 v_d$:

   $$\frac{I_1}{I_2} = \frac{d^2 v}{\left(\frac{d}{2}\right)^2 v'}$$

   $$\frac{100}{200} = \frac{d^2 v}{\frac{d^2}{4} v'}$$

   $$\frac{1}{2} - \frac{4v}{v'}$$

   $v' = 8v$

2. (2) Given:

Voltage $V = 200 \pm 4V$

Current $I = 20 \pm 0.2A$

Resistance $R = \frac{V}{I} = \frac{200}{20} = 10\Omega$

Relative uncertainty:

$\frac{\Delta R}{R} = \frac{\Delta V}{V} + \frac{\Delta I}{I} = \frac{4}{200} + \frac{0.2}{20}$

$= 0.02 + 0.01 = 0.03$

Absolute uncertainty:

$\Delta R = 0.03 \times 10 = 0.3\Omega$

Thus, resistance $R = (10 \pm 0.3)\Omega$.

3. (3) At steady state, capacitor → open circuit.

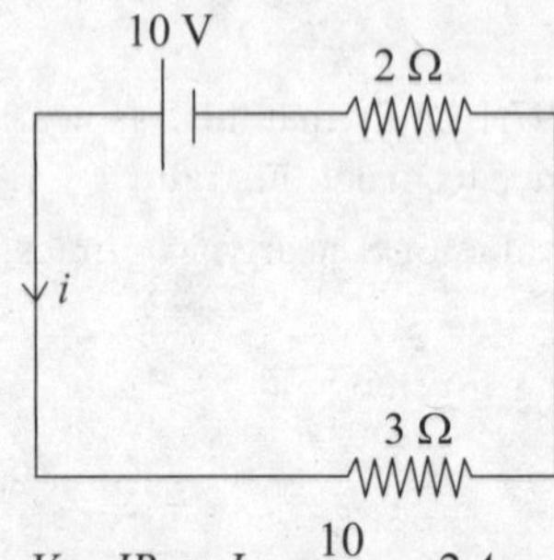

$V = IR \Rightarrow I = \frac{10}{2+3} = 2A$

4. (1) Original resistance:

$R_0 = 10\ \Omega$

Wire stretched to double length $2l$, then:

$R \propto l^2 \Rightarrow R_1 = \left(\frac{2l}{l}\right)^2 \times R_0 = 4R_0 = 40\Omega$

This wire is now a circle. Resistance across any diameter = two halves in parallel:

$R_{eq} = \frac{20 \times 20}{20+20} = \frac{400}{40} = 10\Omega$

5. (4) Resistance of each part = 10Ω

First five parts equivalent = 50Ω

Next five parts equivalent = 2Ω

Again for series combination $R_{eq} = 52\Omega$

6. (1)

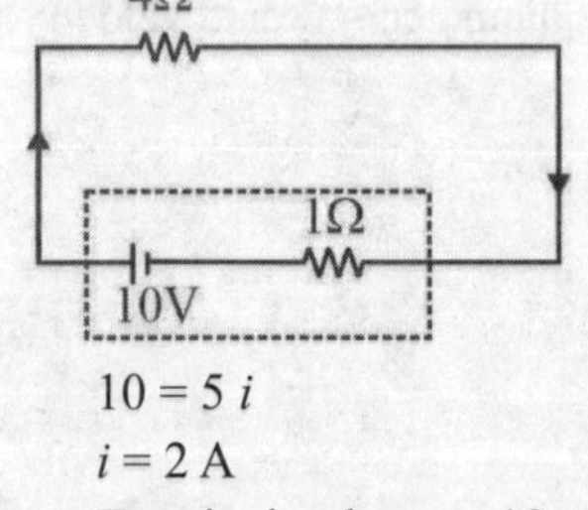

$10 = 5i$

$i = 2$ A

Terminal voltage = 10 – 1(2)

= 8 V

7. (4) Given:

Resistances: 8Ω, 12Ω, and uniform wire $AB = 40$ cm.

Let x be the distance from A.

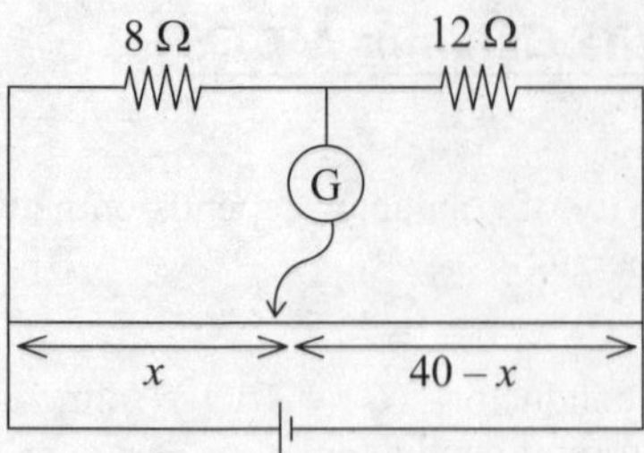

For null point:

$\frac{8}{x} = \frac{12}{40-x} \Rightarrow x = 16\text{cm from } A$

Distance from $B = 40 - 16 = 24$ cm.

8. (3) In option 3 if the diode is considered non ideal with resistance 10Ω then the bridge can be balanced.

9. (4) $P_1 = 1\text{kW}$, $P_2 = 2\text{kW}$

<u>For parallel combination</u>

$P_{Parallel} = P_1 + P_2 = 1 + 2 = 3\text{kW}$

<u>For series combination</u>

$\frac{1}{P_{series}} = \frac{1}{P_1} + \frac{1}{P_2} = 1 + \frac{1}{2}$

$P_{series} = \frac{2}{3}\text{kW}$

$\frac{P_{parallel}}{P_{series}} = \frac{3}{2} \times 3 = \frac{9}{2}$

$\frac{P_{series}}{P_{parallel}} = \frac{2}{9}$

10. (3) The circuit can be redrawn as an equivalent circuit given below

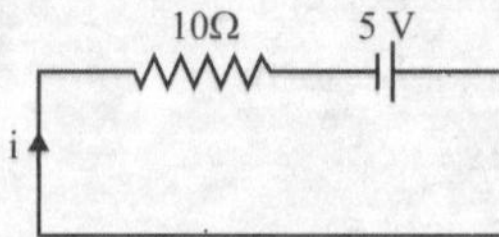

$\Rightarrow i = \frac{5}{10} = 0.5$ A

11. (4) No current through 'G'

So potential difference across R is 2V

$i = \frac{8}{400} \Rightarrow R = \left(\frac{2}{8} \times 400\right) = 100\ \Omega$

12. (4) $R = 22 \times 10^3 \pm 5\%$

First band = Red

2nd band = Red

3rd band = Orange (Multiplier)

4th band = Gold (Tolerance)

13. (4) $R = R_0(1 + \alpha\Delta T)$

$\Rightarrow 6.8 = 2[1 + \alpha \times (80 - 0)]$

$\Rightarrow \alpha = \frac{3.4 - 1}{80} = 0.03 = 3 \times 10^{-2}\ °C^{-1}$

14. (3) $i_{series} = \frac{E}{R_{series}} \Rightarrow i_{series} = \frac{E}{10R}$

$i_{parallel} = \frac{E}{R_{Parallel}} = \frac{E}{R/10}$

$i_{parallel} = n \times i_{series}$

$\Rightarrow \frac{10E}{R} = \frac{nE}{10R} \Rightarrow n = 100$

15. (4) For conductors, temperature coefficient of resistance a is positive, therefore resistance of the conductor rises with temperature and for semiconductor a is negative, therefore its resistance decreases with temperature.

16. (1) Radius of wire $= \frac{10^{-2}}{\sqrt{\pi}}$

Cross sectional area $A = \pi r^2 = 10^{-4}\ m^2$

$\because J = \frac{1}{\rho} \times E \Rightarrow J = \frac{10 \times 10}{10 \times 10^{-4}} = 10^5\ A/m^2$

17. (3)

100Ω

V

200Ω

$P = \frac{V^2}{R}$

As both resistors are in parallel so potential drop (V) across both the resistors are same.

$\Rightarrow P \propto \frac{1}{R} \Rightarrow \frac{P_1}{P_2} = \frac{R_2}{R_1} = \frac{200}{100} = \frac{2}{1} = 2:1$

18. (3) $R_{eq} = R/4 = 0.25 \Rightarrow R = 1\Omega$

Series: $R_{eq} = 4R = 4\Omega$

19. (4) $J = \frac{I}{A} = nev_d = \frac{ne^2\tau}{m}E = \sigma E = \frac{E}{\rho}$

$v_d = \frac{eE}{m}I; \rho = \frac{m}{ne^2\tau}$ or $\rho = \frac{E}{J}, j = nev_d$

$\tau = \frac{m}{ne^2\rho}$

20. (1) $\frac{i_2}{i_3} = \frac{r_3}{r_2}$

$\Rightarrow \frac{i_1 + i_3}{i_3} = \frac{r_3}{r_2}$

$\Rightarrow \frac{i_3}{i_2} = \frac{r_2}{r_2 + r_3}$

21. (3) $S = \frac{I_g.G}{I - I_g}$

$S = \frac{5 \times 10^{-7} \times 50}{5 - (5 \times 10^{-7})} = 5 \times 10^{-6}\Omega$

22. (4) Let $V_A = 0$ so $V_B = 0$

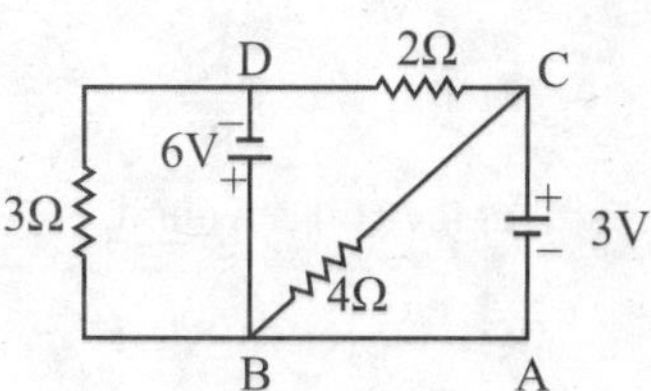

$V_C = 3$volt $V_D = .\ 6$volt

So, $V_C\ .\ V_D = IR$

$9 = I \times 2 \quad I = \frac{9}{2}$ Amp.

23. (2) Let is be the current flowing through ammeter and I the total current. So, a current $i - ia$ will flow through shunt resistance. Potential difference across ammeter and shunt resistance is same,

i.e. $ia \times R = (i - ia) \times S$

$S = \frac{ia\ R}{i - ia}$

Given, $ia = 100$ A, $i = 750$ A, $R = 13\ \Omega$

Hence, $S = \frac{100 \times 13}{750 - 100} = 2\ \Omega$

24. (3) **25. (3)** **26. (4)** **27. (3)** **28. (4)**
29. (3)

Moving Charges and Magnetism

NCERT Topic-wise MCQs

1. (2) The magnetic field at a distance d from an infinitely long straight current-carrying wire is given by the expression:

$B = \frac{\mu_0 i}{2\pi d}$

Here, μ_0 is the permeability of free space, i is the current, and d is the perpendicular distance from the wire to the point where the magnetic field is being measured.

2. (2) The magnetic field due to a current element can be calculated using Biot-Savart's Law.

3. (2) The magnetic field outside a long straight current-carrying conductor varies with distance r as $B \propto \frac{1}{r}$. This is described by Ampère's circuital law and the Biot-Savart Law for a long, straight wire, where the magnetic field at a distance r from an infinitely long straight wire carrying current I is given by $B = \frac{\mu_0 I}{2\pi r}$.

4. (3) Circumference of circle $= 2\pi R$

$L = 2\pi R$

$R = \frac{L}{2\pi}$

Magnetic moment $(\mu) = IA = I \times \frac{L^2}{4\pi^2} \times \pi = \frac{IL^2}{4\pi}$

5. (1) Net force acting on electron is zero

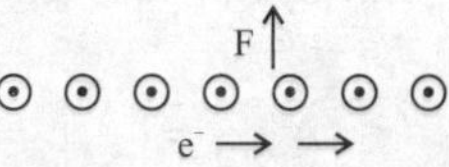

$F = q(\vec{v} \times \vec{B})$

$F = q\ (vB) \sin 0°$

$F = 0$

So, work done

$W = F.S = 0$

6. (2) $B_{centre} = \frac{\mu_0 NI}{2R}$

As the coil rewound,

$$R' = \frac{R}{3}, N' = 3, N = 1$$

$$\frac{B_0}{B'} = \frac{\mu_0 NI}{2R} \times \frac{2R'}{\mu_0 N'I} = \frac{1 \times R}{R \times 3 \times 3} = \frac{1}{9}$$

$$B' = 9B_0$$

7. (4) $|\vec{B}| = \sqrt{B_1^2 + B_2^2}$

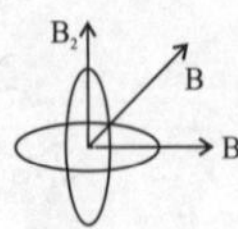

$$|\vec{B}| = \frac{\mu_o NI}{2R}\sqrt{1+3} = \frac{\mu_o NI \times 2}{2R}$$

$$|\vec{B}| = \frac{\mu_o NI}{R}$$

8. (2) Magnetic field at a distance r from the straight wire.

$$B = \frac{\mu_o i}{2\pi r}$$

$$\therefore \quad B \propto \frac{1}{r}$$

$$\frac{B}{B'} = \frac{20}{5} \Rightarrow B' = \frac{B}{4}$$

9. (4) For equilibrium of wire

$$F = mg \Rightarrow BIL = mg$$

$$B = \frac{mg}{IL} = \frac{\left(\frac{200}{1000}\right)10}{2 \times 1.5} = \frac{2}{2 \times 1.5} = \frac{10}{15}$$

$$B = \frac{2}{3} \Rightarrow B = 0.66T$$

10. (1) $r_1 : r_2 = 1 : 2$ and $B_1 : B_2 = 1 : 3$ (flux density)

$$B = \frac{\mu_o}{4\pi}\frac{2\pi nI}{R} \Rightarrow I \propto BR$$

$$\frac{I_1}{I_2} = \frac{B_1}{B_2}\frac{R_1}{R_2} = \frac{1 \times 1}{3 \times 2} = \frac{1}{6}$$

$$\frac{I_1}{I_2} = \frac{1}{6}$$

11. (2) $F = BIL$

$$B = \frac{F}{IL} \Rightarrow \frac{[MLT^{-2}]}{A \times L} = [MT^{-2}A^{-1}]$$

12. (3) Suppose length of each wire is ℓ

$$A_{square} = \left(\frac{\ell}{4}\right)^2 = \frac{\ell^2}{16}$$

$\because$ Magnetic moment

$$M = IA$$

$$\frac{M_{square}}{M_{circle}} = \frac{A_{square}}{A_{circle}}$$

$$\frac{M_{square}}{M_{circle}} = \frac{\ell^2}{16} \times \frac{4\pi}{\ell^2} = \frac{\pi}{4}$$

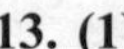

13. (1)

$$\frac{B_{centre}}{B_{axis}} = \left[1 + \frac{x^2}{r^2}\right]^{3/2}$$

$$\frac{B}{54} = \left[1 + \left(\frac{4}{3}\right)^2\right]^{3/2} = \left(\frac{25}{9}\right)^{3/2} = \left(\frac{5}{3}\right)^3$$

$$\frac{B}{54} = \frac{125}{27}$$

$$B = \frac{125 \times 54}{27}$$

$$B = 250\ \mu T$$

14. (1) $B = \frac{I d\ell \sin\theta}{r^2} \times \frac{\mu_0}{4\pi}$

$$B \propto I$$

15. (1) According to Maxwell's right hand thumb rule the direction of magnetic field at a point above the conductor is towards north.

16. (1) Magnetic field at the ends of solenoid is half of the field present inside.

$$B_{end} = \frac{B_{inside}}{2}$$

$$B_{end} = \frac{\mu_o nI}{2}$$

17. (4) The radius r of the path of a charged particle moving in a magnetic field is given by the formula:

$$r = \frac{mv}{qB}$$

where:

m is the mass of the particle,

v is the velocity of the particle,

q is the charge of the particle,

B is the magnetic field strength.

18. (2) Radius of circular path

$$R = \frac{mv}{qB} = \frac{p}{qB} = \frac{\sqrt{2m(KE)}}{qB} = \frac{\sqrt{2mqV}}{qB}$$

$$R \propto \sqrt{m}$$

$$m \propto R^2$$

mass ratio of X and Y is

$$\frac{m_1}{m_2} = \left(\frac{R_1}{R_2}\right)$$

19. (3) The total force (F) on a charged particle with charge (q), moving with velocity (v) in the presence of both electric field (E) and magnetic field (B) is known as the Lorentz force.

20. **(1)** As we know that

$F = ma$

$qvB \sin 90^\circ = ma$

$$a = \frac{qvB}{m} \Rightarrow a = \frac{1.6\times10^{-19}\times2\times3.4\times10^{7}}{1.67\times10^{-27}}$$

$a = 6.5 \times 10^{15}$ m/sec^2

21. **(1)** A moving charge will gain energy due to the electric field, while magnetic field energy remains constant.

22. **(3)** $F = \frac{\mu_0}{4\pi}\frac{2I_1I_2}{r} = 10^{-3}$ N

When both current will doubled then

$$F' = \frac{\mu_0}{4\pi}\frac{2\times(2I_1\times2I_2)}{r}$$

$= 4 \times F$

$F' = 4 \times 10^{-3}$N

23. **(3)** $M = IA$

$M = I \times \pi r^2$

24. **(1)** $F = BIL \sin\theta$

$7.5 = 2 \times 5 \times 1.5 \sin\theta$

$$\frac{7.5}{15} = \sin\theta \Rightarrow \frac{1}{2} = \sin\theta \Rightarrow \theta = 30^\circ$$

25. **(1)** Proton has positive charge on it. Therefore two beams of protons parallel to each other will produce current in the same direction resulting in magnetic attraction.

26. **(4)** $\tau_{max} = NABI$

$$= I\times1\times(\pi r^2)B \qquad \left[\begin{aligned}2\pi r &= L\\ \therefore r &= \frac{L}{2\pi}\end{aligned}\right]$$

$$\tau_{max} = \pi I\left(\frac{L}{2\pi}\right)^2 B$$

$$\tau_{max} = \frac{L^2IB}{4\pi}$$

27. **(2)** Magnetic field by the wire is perpendicular to the motion of electron and it is given by

$$B = \frac{\mu_o}{4\pi}\times\frac{2I}{r} \Rightarrow B = 10^{-7}\times\frac{2\times5}{0.1}$$

$B = 10^{-5}$ Wb/m^2

Hence force on electron

$F = qvB$

$F = (1.6 \times 10^{-19}) \times 5 \times 10^6 \times 10^{-5}$

$F = 8 \times 10^{-18}$N

28. **(4)** Kinetic energy of a particle in magnetic field remains constant and it is

$K = qV$

$K \propto q$ [V = const]

$\therefore K_p : K_d : K_\alpha - q_p : q_d : q_\alpha$

$K_p : K_d : K_\alpha = 1 : 1 : 2$

29. **(2)** $\vec{a}.\vec{B} = 0$

$(\alpha\hat{i} - 4\hat{j}).(2\hat{i} + 3\hat{j}) = 0$

$\alpha(2) + (-4)(3) = 0$

$$\alpha = \frac{12}{2} \Rightarrow \alpha = 6$$

30. **(3)** $N_{per\,sec} = \frac{1}{T} = \frac{v}{2\pi R}$

Current through circular orbit

$$i = \frac{q}{T} = \frac{vq}{2\pi R}$$

Magnetic moment, $\mu = IA$

$$= \frac{vq}{2\pi R}\times\pi R^2 = \frac{vqR}{2}$$

31. **(4)** Velocity component which is perpendicular to magnetic field provide circular motion to charge particle and horizontal component which is parallel or along the magnetic field will provide it straight line motion. So the combined motion will form a helix.

32. **(2)** $F = qvB\sin\theta$

$F = 1.6 \times 10^{-19} \times 2 \times 10^7 \times 1.5 \times \sin 30^\circ$

$$F = 1.6 \times 10^{-12} \times 2 \times 1.5 \times \frac{1}{2}$$

$F = 2.4 \times 10^{-12}$N

33. **(1)** $qvB = \frac{mv^2}{r}$

$$r = \frac{mv}{Bq} = \sqrt{\frac{2mE}{q^2B^2}} \qquad \left(E = \frac{B^2q^2r^2}{2m}\right)$$

$$r_p = \sqrt{\frac{2mE}{e^2B^2}},\ r_d = \sqrt{\frac{2(2m)E}{e^2B^2}},\ r_\alpha = \sqrt{\frac{2\times4m\times E}{4e^2\times B^2}}$$

Hence. $r_p : r_d : r_\alpha = 1 : \sqrt{2} : 1$

34. **(2)** Charged particle enters in magnetic field perpendicularly it moves in circular path. The required centripetal force is provided by magnetic force.

$$r = \frac{mv}{Bq}$$

Kinetic energy of the particle

$$K = \frac{1}{2}mv^2 \Rightarrow mv = \sqrt{2mK}$$

$$r = \frac{\sqrt{2mK}}{qB} \Rightarrow r \propto \sqrt{m}$$

$$\frac{r_e}{r_p} = \sqrt{\frac{m_e}{m_p}}$$

So, $m_e < m_p$

$r_e < r_p$

Trajectory of electron is more curved and trajectory of proton is less curved.

35. (3) The cyclotron equation is $r = \frac{mv}{qB_0}$

When charged particle should not hit the screen

$$\Rightarrow \frac{mv}{qB_0} < d, \therefore V_{max} = \frac{qB_0 d}{m}$$

36. (1) $R = \frac{\text{Current sensitivity}}{\text{Voltage sensitivity}} = \frac{100}{25} = 4\Omega$

37. (3) To use a galvanometer as an ammeter, a low-resistance, known as a shunt resistor, must be connected in parallel with the galvanometer. This allows most of the current to bypass the galvanometer, preventing it from being damaged by the high currents it is measuring.

38. (2) If current is I then current passes through the galvanometer is 0.1% of $I = \frac{I}{1000}$ So, current passes through shunt $= I - \frac{I}{1000} = \frac{999}{1000} I$

39. (1) Case-(I): $A_1 = \frac{\pi D^2}{4}, n_1 = n, \theta = \theta_1$

Case-(II): $A_2 = \frac{\pi}{4}\left(\frac{D}{4}\right)^2, n_2 = 10n, \theta = \theta_2$

$\because$ Current is same,

$$I = \frac{K\theta_1}{n_1BA_1} = \frac{K\theta_2}{n_2BA_2} \Rightarrow \theta_2 = \theta_1 \times \frac{n_2A_2}{n_1A_1}$$

$\theta_2 = \theta \times 10 \times \frac{1}{16} \Rightarrow \theta_2 = \frac{5}{8}$ times the original

40. (4) $K = \frac{I}{\theta} = \frac{V}{R\theta}$

$$K = \frac{1}{(900+100)100} = \frac{1}{10^5}$$

$K = 10^{-5}$ A/division

41. (2) According to the relation:

$$\theta = \frac{NIAB}{C}$$

$\theta \propto N$

42. (4) $\tau = MB\sin\theta$

$\tau_{max} = NIAB \quad (\because \sin\theta = 1)$

43. (4) Current sensitivity, $\frac{\theta}{I} = \frac{NBA}{C}$

$$= \frac{100 \times 5 \times 10^{-4}}{10^{-8}} = 5\,\text{rad}/\mu\,\text{amp}$$

44. (1) $\sigma_i = \frac{\theta}{i} = \frac{\theta \times G}{i \times G}$

$\sigma_i = \sigma_v G \quad \left(\sigma_v = \frac{\theta}{i \times G}\right)$

$\sigma_v = \frac{\sigma_i}{G}$

45. (2) The potential difference

$V_{ab} = (V_a - V_b)$ is the same for the path.

$I_gG = (I - I_g)S$

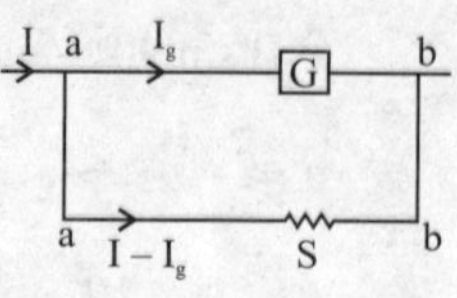

$I_g(G + S) = IS$

$\frac{I_g}{I} = \frac{S}{G+S}$

The fraction of current passing through shunt $= \frac{I - I_g}{I}$

$$= 1 - \frac{I_g}{I} = 1 - \frac{S}{S+G} = \frac{G}{S+G} = \frac{8}{2+8} = 0.8\text{ A}$$

46. (1) Current flowing through galvanometer

$I_g = 16\mu A/\text{div} \times 30\text{ div} = 480\mu A$

The value of resistance in series to the galvanometer is

$$R = \frac{V}{I_g} - G = \frac{3}{480 \times 10^{-6}} - 0 = \frac{3 \times 10^5}{48} = \frac{10^5}{16}$$

$R = 6.25k\Omega$

47. (1) For a voltmeter with full scale reading

$V = I_g(G + R)$

$$R = \frac{V}{I_g} - G \Rightarrow R = \frac{12}{0.1} - 2 = 120 - 2 = 118\Omega$$

48. (1) Potential across G and S is same therefore

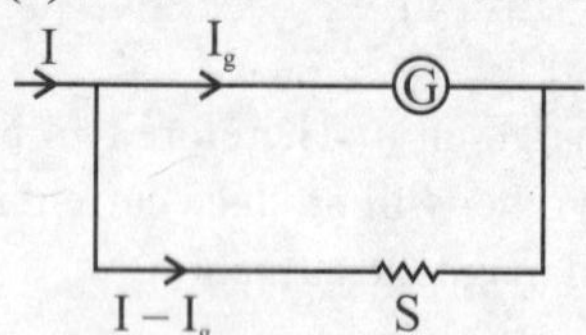

$I_g \times G = (I - I_g) \times S$

$$S = \frac{I_gG}{(I - I_g)} \Rightarrow S = \frac{0.0002 \times 400}{3 - 0.0002} = 0.027\Omega$$

49. (4) Let the current for one division of galvanometer be i

Current through the galvanometer $I_g = 10i$

Current through the ammeter $I = 50i$

$$S = \frac{I_gG}{(I - I_g)} \Rightarrow G = S\left(\frac{I - I_g}{I_g}\right)$$

$$G = 12 \times \left[\frac{50i - 10i}{10i}\right] \Rightarrow = 12 \times \left(\frac{40i}{10i}\right) = 48\ \Omega$$

$G = 48\ \Omega$

High Order Time Intensive MCQs

1. (3) Magnetic field,

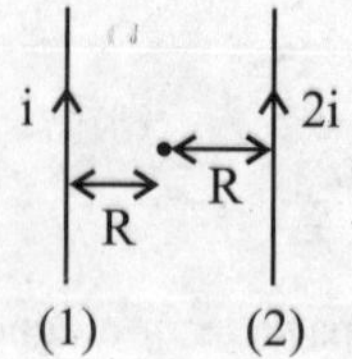

$$B_1 = \frac{\mu_0}{4\pi}\frac{2i}{R} \text{ and } B_2 = \frac{\mu_0}{4\pi}\frac{2.2i}{R}$$

$$B_2 - B_1 = \frac{\mu_0}{4\pi}\frac{2i}{R} = B_1 = B \qquad (\because B_1 = B)$$

Now $i_2 = 0$, $B_{net} = B$

2. (1)

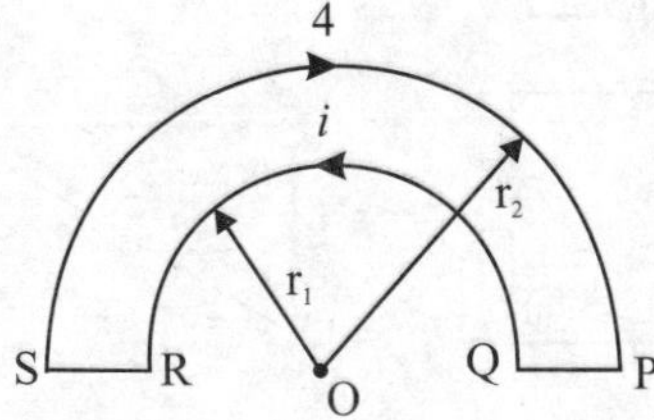

$B_{PQ} = B_{RA} = 0$

$B_{QR} = \frac{\mu_0}{4} \cdot \frac{i}{r_1}$ (vertically upward)

$B_{SP} = \frac{\mu_0}{4} \cdot \frac{i}{r_2}$ (vertically downward)

So, $B_{net} = B_{QR} - B_{SP} = \frac{\mu_0}{4}\left(\frac{1}{r_1} - \frac{1}{r_2}\right)i$

3. (2) The magnetic field due to the straight wire at a distance r is given by

$$B = \frac{\mu_0 i}{2\pi r} \Rightarrow \text{or } B \propto \frac{1}{r} \text{ or } \frac{B_2}{B_1} = \frac{r_1}{r_2}$$

$$\therefore \quad \frac{B_2}{B} = \frac{5}{25} = \frac{1}{5} \text{ or } B_2 = \frac{B}{5}$$

4. (4) Magnetic field due to a long current carrying wire, at r [i.e. $r \geq a$]

$$B = \frac{\mu_0}{2\pi}\frac{I}{r} = \frac{\mu_0 I}{2\pi r} \quad \left[B \propto \frac{1}{r}\right]$$

For $r < a$

Let current (I') flow through wire

$$I' = \left(\frac{I}{\pi a^2}\right) \times \pi (r)^2$$

$$I' = \frac{Ir^2}{a^2}$$

$$B = \frac{\mu_0 I'}{2\pi r} = \frac{\mu_0 I r^2}{2\pi a^2 r} = \frac{\mu_0 I r}{2\pi a^2}$$

$B \propto r$

5. (2) $B = B_1 + B_2 + B_3 + B_4$

$$B = 4\left[\frac{\mu_0}{4\pi} \cdot \frac{I}{a}(\sin\phi_1 + \sin\phi_2)\right]$$

$$= 4 \times \frac{\mu_0}{4\pi} \frac{I}{(l/2)}\left[\frac{1}{\sqrt{2}} + \frac{1}{\sqrt{2}}\right]$$

Here, $a = \frac{l}{2}, \phi_1 = \phi_2 = 45°$

$$\therefore \quad B = 4 \times \frac{16}{\sqrt{2}}\left[\frac{\mu_0}{4\pi}\frac{I}{l}\right] = 8\sqrt{2}\left(\frac{\mu_0}{4\pi}\frac{I}{l}\right)$$

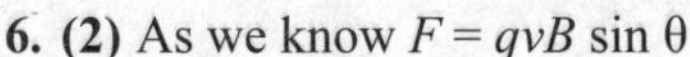

6. (2) As we know $F = qvB \sin\theta$

If $\theta = 90°$

Then, $F = qvB$ ...(*i*)

Also, $F = \frac{mv^2}{r}$ ...(*ii*)

From Eqs. (*i*) and (*ii*), we get

$$qvB = \frac{mv^2}{r}, r = \frac{mv}{qB}$$

$$\therefore E = \frac{1}{2}mv^2, v = \sqrt{\frac{2E}{m}}$$

$$\therefore r = \frac{m\sqrt{\frac{2E}{m}}}{qB} \Rightarrow r = \frac{\sqrt{2Em}}{qB}$$

7. (2) Radius of circular path,

$$R = \frac{\sqrt{2m(K)}}{qB}$$

So, $q \propto \sqrt{m(K)}$

Hence, $\frac{e}{2e} = \sqrt{\frac{(m_p)(1\text{MeV})}{(4m_e)(\text{K})}} \Rightarrow \frac{1}{4} = \frac{1}{4\text{K}}$

Energy of α- particle, $K = 1$ MeV

8. (2) We know that

$B = \frac{\mu_0 I}{4\pi r}\theta$ due to the arc

∴ Total magnetic field induction at O. due to current through the three arc is

$$B = \frac{\mu_0 I}{4\pi}\theta\left[\frac{1}{r} - \frac{1}{2r} + \frac{1}{3r}\right]$$

$$B = \frac{5\mu_0 I\theta}{24\pi r}$$

9. (1) $B_O = B_{CG} + B_{DEF}$

$$= \frac{\mu_0}{4\pi}\frac{2I}{R} + \frac{\mu_0}{4\pi}\frac{2\pi I}{R}$$

$$B_o = \frac{\mu_o I}{2R}\left[\frac{1}{\pi} + 1\right]$$

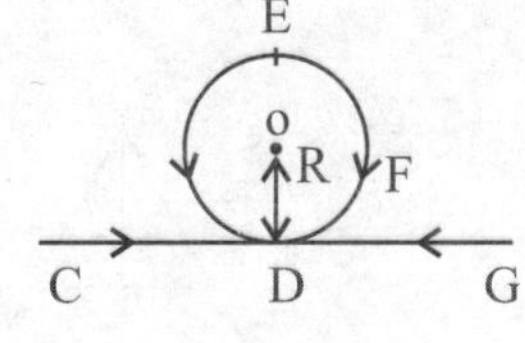

10. (3) $\vec{F}_E = q\vec{E}$ and $\vec{F}_B = q\left(\vec{V} \times \vec{B}\right)$

Net force on particle must be zero i.e.

$q\vec{E} + q\vec{V} \times \vec{B} = 0$

Possible cases are

(i) $\vec{E}$ & $\vec{B} = 0$

(ii) $\vec{V} \times \vec{B} = 0, \vec{E} = 0$

(iii) $q\vec{E} = -q\vec{V} \times \vec{B}$

$\vec{E} \neq 0$ & $\vec{B} \neq 0$

11. (3) Wires placed close to each other carrying current in the same direction attract and wires placed far apart carries current in opposite direction and repel each other.

The magnitude of attraction is greater than magnitude of repulsion. Therefore, loop moves towards the wire.

12. (2) Forces, F_1 and F_2 acting on the coil are equal in magnitude and opposite in directions. As the forces F_1 and F_2 have the same line of action, their resultant effect on the coil is zero.

$\tau = NiBA \sin\theta$

$\tau_{max} = NiBA\ (\theta = 90^0)$

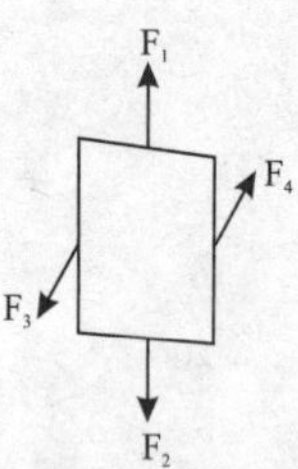

The two forces F_3 and F_4 are equal in magnitude and opposite in direction. As the two forces have different lines of action, they constitute a torque. Thus, if the force on one arm of the loop is F, the net force on the remaining three arms of the loop is – F.

13. (4) Mutual force between conductors A and C,

$$F_1 = \frac{\mu_0}{2\pi}\frac{I_1 I_2 l}{r}$$

$$= \frac{\mu_0}{2\pi}\frac{2\times3\times1}{0.05}$$

$= 2.40 \times 10^{-5}$N [towards A]

Mutual force between conductors B and C

$$= \frac{\mu_0}{2\pi}\frac{4\times3\times1}{0.08} = 3\times10^{-5}\text{ N [towards B]}$$

Hence, the resultant force experienced by C

$F = F_2 - F_1$

$= (3 - 2.4) \times 10^{-5}$ N

$= 0.6 \times 10^{-5}$ N

14. (2) I = 10 A, θ = 45°, B = 0.2 T

$\therefore\ \vec{F} = I\vec{l}\times\vec{B} \Rightarrow F = IlB\sin\theta$

$$\therefore\ \frac{F}{l} = IB\sin 45^0 = 10\times0.2\times\frac{1}{\sqrt{2}} = \frac{2}{\sqrt{2}}\text{Nm}^{-1}$$

15. (4)

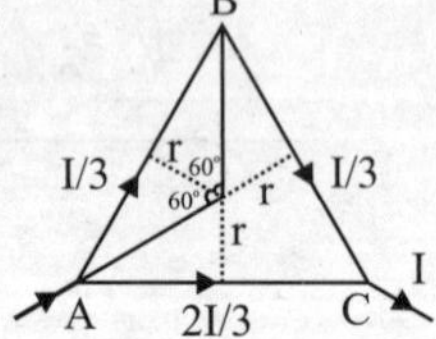

$$B_{AB} = \frac{\mu_0}{4\pi r}\left(\frac{I}{3}\right)[\sin 60° + \sin 60°]\otimes$$

$$B_{BC} = \frac{\mu_0}{4\pi r}\left(\frac{I}{3}\right)[\sin 60° + \sin 60°]\otimes$$

$$B_{AC} = \frac{\mu_0}{4\pi r}\left(\frac{2I}{3}\right)[\sin 60° + \sin 60°]\odot$$

$\therefore$ So B_{net} at O = zero

16. (1) Magnetic field, $B_1 = \dfrac{\mu_0 i}{4\pi R}, \odot$

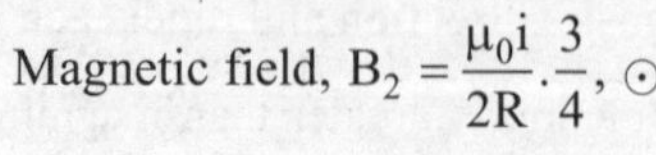
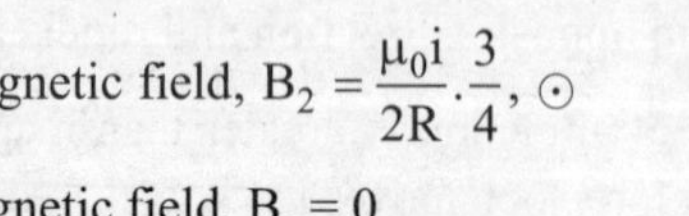

Magnetic field, $B_2 = \dfrac{\mu_0 i}{2R}.\dfrac{3}{4}, \odot$

Magnetic field, $B_3 = 0$

Magnetic field at the centre, $B_O = B_1 + B_2$

$$= \frac{\mu_0 i}{4R}\left(\frac{1}{\pi}+\frac{3}{2}\right), \odot$$

17. (3)

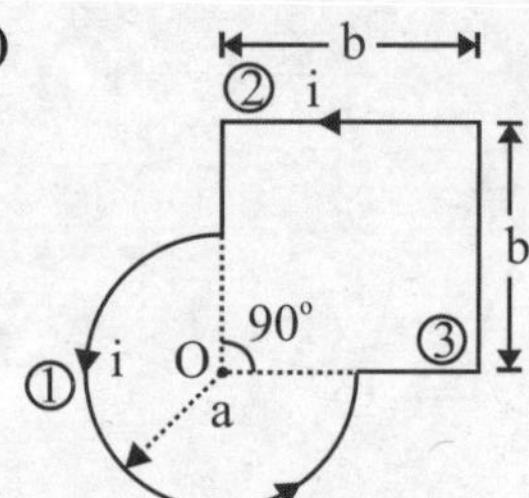

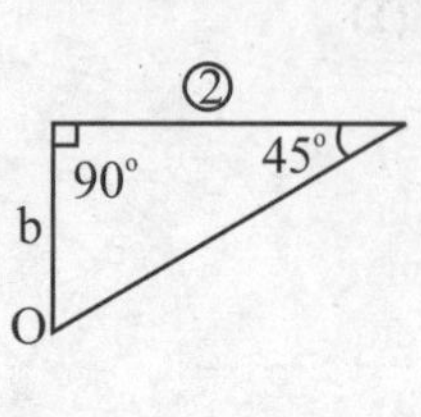

$$\text{Magnetic field, } B_1 = \frac{\mu_0 i}{2a}.\frac{3}{4} = \frac{\mu_0 i}{2a}.\frac{3}{4} = \frac{3\mu_0 i}{8a}\ \odot$$

$$\text{Magnetic field, } B_2 = \frac{\mu_0 i}{4\pi b}(\cos 90° + \cos 45°)$$

$$= \frac{\mu_0 i}{4\sqrt{2}\pi b} = B_3\ \odot$$

Magnetic field at the centre O,

$$B_O = B_1 + (B_2 + B_3) = \frac{3\mu_0 i}{8a} + \frac{\mu_0 i}{2\sqrt{2}\pi b}, \odot$$

18. (1) Current density, $J = \dfrac{I}{\pi a^2}$

From ampere's circuital law, $\oint B\cdot dl = \mu.I_{enclosed}$

For r < a, $B \times 2\pi r = \mu_0 \times J \times \pi r^2$

$$\Rightarrow\ B = \frac{\mu_0 I}{\pi a^2}\times\frac{r}{2}$$

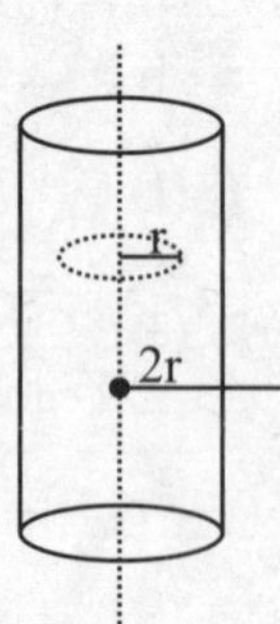

$$\text{At}\quad r = \frac{a}{2},\ B_1 = \frac{\mu_0 I}{4\pi a}$$

$$\text{For } r > a,\ B\times 2\pi r = \mu_0 I \Rightarrow B = \frac{\mu_0 I}{2\pi r}$$

$$\text{At } r = 2a,\quad B_2 = \frac{\mu_0 i}{4\pi a}$$

$$\text{So, } \frac{B_1}{B_2} = 1$$

19. (2) By $B = \dfrac{\mu_0}{4\pi}\dfrac{I}{r}(\theta_R)$

$$B_1 = \frac{\mu_0 I}{4\pi R_1}\times\frac{\pi}{2} = \frac{\mu_0 I}{8R_1}\otimes$$

$$B_2 = \frac{\mu_0 I}{4\pi R_2}\times\frac{\pi}{2} = \frac{\mu_0 I}{8R_2}\otimes$$

$B_{net} = B_1 + B_2$

$$B_{net} = \frac{\mu_0 I}{8}\left(\frac{1}{R_1}+\frac{1}{R_2}\right)$$

20. (2) Using Ampere's law

$\oint \vec{B}\cdot d\vec{\ell} = \mu_0 I_{inside}$

$\because\ I_{inside} = 0$

$\therefore\ \vec{B} = 0$

Statement Based MCQs

1. **(3)** An electric charge produces an electric field and moving charges produces magnetic field. Therefore, a moving charge produces both fields. The magnitude of charge remains constant on changing the velocity of particle.
2. **(4)** A moving charge produces magnetic as will as electric field. The line integral of the magnetic field along any closed path in free space equal to μ_0 times net current.

   $\oint \vec{B}\cdot d\vec{\ell} = \mu_0 I$
3. **(2)** A solenoid tends to contract when current is passed through it. Two wires carrying current in the same direction attract each other due to magnetic force.
4. **(1)** When a charge is projected in a magnetic field it experiences a force which is in a perpendicular direction to its motion. Therefore no work is done on the particle and since its speed remains unchanged so is its kinetic energy remains constant.
5. **(1)** $qvB = \frac{mv^2}{r}$

   $r = \frac{mv}{qB}$

Assertion & Reason MCQs

1. **(1)** Low resistance of the ammeter ensures that main current is least affected in the circuit.
2. **(3)**

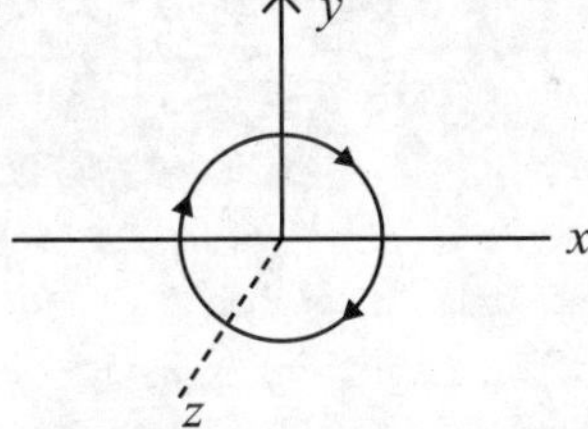

The direction of magnetic flux depends of direction of current in the conductor.

The direction defined by right hand thumb rule.

3. **(2)** $T = \frac{2\pi M}{qB}$, mass of α-particles is 4 times of proton.

   Charge on α-particles is 2 times of proton.

   So, $T_{new} = \frac{2\times\pi\times 4M}{2a\times B} = 2T$
4. **(2)** Only current flowing through the Ampere's loop is considered.

Match the Column MCQs

1. **(2)** $\vec{F} = q(\vec{E} + \vec{v}\times\vec{B})$

   $= q\vec{E} + q(\vec{v}\times\vec{B})$

   The direction of $(\vec{v}\times\vec{B})$ magnetic force define by right hand thumb rule.

   and when the sign of the charge (q) charge of the particle is from positive to negative (or vice versa).

   The direction of magnetic force also reverses.

2. **(4)** $\vec{F} = I\vec{\ell}\times\vec{B}$

   $\vec{\tau} = \vec{\mu}\times\vec{B}$

   $= I\vec{A}\times\vec{B}$

   $P.E = -\vec{\mu}.\vec{B}$

   (a) $\vec{F} = I\left(\int_a^b d\vec{s}\right)\times\vec{B}$

   $\oint d\vec{s} = 0$ the net magnetic force on a closed loop is $\vec{F} = 0$

   $P.E = -\vec{\mu}.\vec{B}$

   Here, direction of $\vec{\mu}$ and $\vec{B}$ is opposite.

   $= -\mu B\cos\theta \qquad (\theta = 180°)$

   $P.E = -\mu B\cos(180°)$

   $P.E. = \mu B$

   (b) $\vec{F} = 0 \ \left(\oint ds = 0\right)$ that is (i)

   $\vec{\tau} = \vec{\mu}\times\vec{B}$

   Here, in the case, $\theta = 90°$

   $= |\mu|\,|B|\sin\theta$

   So, $\sin\theta$ is max τ is max i.e. (ii)

   (c) $\vec{F} = 0 \ \left(\oint ds = 0\right)$ i.e. (i)

   $P.E = -\vec{\mu}.\vec{B}$ in this case $\theta = 0°$

   $= -|\vec{\mu}|\,|\vec{B}|\cos\theta$ i.e. (iii)

   (d) $\vec{F} = 0 \ \left(\oint ds = 0\right)$ i.e. (i)

   $P.E = -\vec{\mu}.\vec{B}$

   $= -|\vec{\mu}|\,|\vec{B}|\cos\theta$

   angle between $\vec{\mu}$ and $\vec{B}$ is $\theta > 90°$

   So, $\cos\theta < 1$

   So, P.E is positive i.e. (iv)

3. **(2)**

   (a) If a charged particle at rest experiences a force, then this should be electrostatic force. So electric field should be present, magnetic field may or may not present.

   (b) This possible if (i) both fields are zero (ii) both fields are not zero and their forces balance (iii) $E = 0$, $B \neq 0$ and velocity is parallel to magnetic field.

   But if only E is there, then velocity cannot remain same.

   (c) Speed can vary only if electric field is present. Particle will go undeviated with varying speed if v, E and B all are in same direction. B may also be zero.

   (d) For helical path, B should be present, E may or may not be.

4. **(1)**

   (a) Velocity of the particle may be constant, if forces of electric and magnetic fields balance each other. Then, path of particle will be straight line. Also, path of particle may be helical if magnetic and electric fields are in same direction. But path of particle cannot be circular. Path can be circular if only magnetic field is present, or if some other force is present which can cancel the effect of electric field.

(b) Here, all the possibilities are possible depending upon the combinations of the three fields.

(c) This situation is similar to part (a).

(d) In a uniform electric field, path can be only straight line or parabolic.

5. (4)

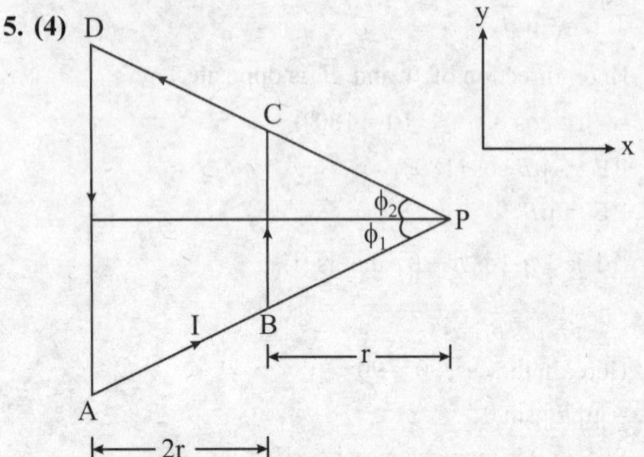

At P, magnetic field due to AB and CD will be zero.

$$\vec{B}_{BC} = \frac{\mu_0}{4\pi}\frac{I}{r}(\sin\phi_1 + \sin\phi_2)(\hat{k})$$

$$\vec{B}_{DA} = \frac{\mu_0}{4\pi}\frac{I}{3r}(\sin\phi_1 + \sin\phi_2)(\hat{k})$$

Net field at P:

$$\vec{B} = \vec{B}_{BC} + \vec{B}_{DA} = -\frac{2\mu_0}{4\pi}\frac{I}{3r}(\sin\phi_1 + \sin\phi_2)\hat{k}$$

Hence, we can find the answers given below.

(ii) speed remain the same inside the dees because of the magnetic field but direction undergoes change continuously.

NCERT Exemplar MCQs

1. (4) For given pitch d correspond to charge particle, we have

$$\frac{q}{m} = \frac{2\pi m v\cos\theta}{dB} = \text{constant}$$

Since, charged particles traverse identical helical paths in a completely opposite sense in a uniform magnetic field $\vec{B}$, $\frac{q}{m}$ for two particles should be same and of opposite sign.

Therefore,

$$\left(\frac{e}{m}\right)_1 + \left(\frac{e}{m}\right)_2 = 0$$

2. (1) In Biot-Savart's law, magnetic field $\vec{B} = id\vec{\ell}\times\vec{r}$ and $id\ell$ due to flow of electron is in opposite direction of v and by direction of cross product of two vectors $\vec{B}\perp\vec{v}$.

3. (1) The direction of magnetic moment of circular loop of radius R placed in the x–y plane is along z-direction and given by M = I (πr^2), when half of the loop with x > 0 is now bent so that it now lies in the y-z plane, the magnitudes of magnetic moment of each semicircular loop of radius R lie in the x-y plane and the y-z plane is M′ = I (πr^2)/4 and the direction of magnetic moments are along z-direction and x-direction respectively. Their resultant

$$M_{net} = \sqrt{M'^2 + M'^2} = \sqrt{2}M' = \sqrt{2}I(\pi r^2)/4$$

So, $M_{net} < M$ or M diminishes

4. (4) Magnetic Lorentz force on electron is projected with uniform velocity along the axis of a current carrying long solenoid.

F = – evB sin 180° = 0 (θ = 0°)

As magnetic field and velocity are parallel. The electron will continue to move with uniform velocity along the axis of the solenoid.

5. (4) The rotation of the loop by 30° about an axis perpendicular to its plane make no change in the angle made by axis of the loop with the direction of magnetic field, therefore, the work done to rotate the loop is zero.

Note: The work done to rotate the loop in magnetic field W = MB ($\cos\theta_1 - \cos\theta_2$). (where signs are as usual)

NEET & AIIMS Past Year MCQs

1. (1) Magnetic field $B = \frac{\mu_0 NI}{2r}$

$$= \frac{4\pi\times10^{-7}\times100\times7}{2\times100\times10^{-2}} = 0.0044\,\text{T} = 4.4\,\text{mT}$$

2. (4) M = mL

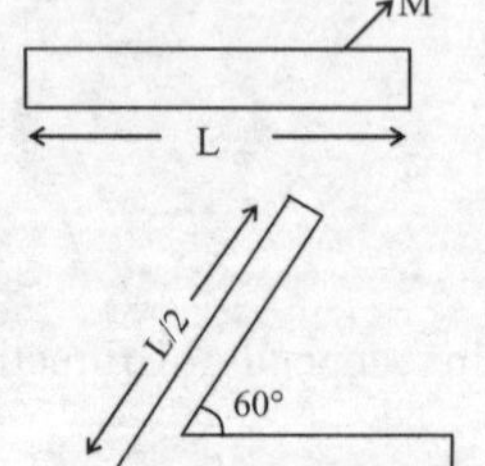

⇒

Magnetic pole strength remains the same

$$M_{new} = mL_{eff} = \frac{mL}{2} = \frac{M}{2}$$

(leffective is distance between ends)

3. (4) $\vec{B}_P = \vec{B}_{upper\ wire} \otimes + \vec{B}_{semi-circle} \odot + \vec{B}_{lower-wire} \otimes$

$$B_P = -\frac{\mu_0 i}{4\pi R} + \frac{\mu_0 i}{4R} - \frac{\mu_0 i}{4\pi R}$$

$$= \frac{\mu_0 i}{4R}\left(1 - \frac{2}{\pi}\right) \text{ pointing away from the page}$$

4. (4) $|\vec{F}| = |I(\vec{L}\times\vec{B})| = |I[L\hat{i}\times(2\hat{i}+3\hat{j}-4\hat{k})]| = 5\,IL$

5. (4) According to Biot Savart's law

$$d\vec{B} = \frac{\mu_0\left(Id\vec{\ell}\times\vec{r}\right)}{4\pi r^3}$$

Therefore, the expression for magnetic field depends on current carrying element $Id\vec{\ell}$, which is a vector quantity, therefore, statement-I is correct and statement-II is wrong because charge q is not a vector source.

6. (3) Magnetic field inside a long solenoid is given by

$$B = \mu_0 \frac{N}{\ell} i$$

$$\therefore\ B = 4\pi \times 10^{-7} \times \frac{100}{10^{-3}} \times 1 = 12.56 \times 10^{-2}\,T$$

7. (4) Inside the wire $B \propto r$ and outside the wire $B \propto \frac{1}{r}$.

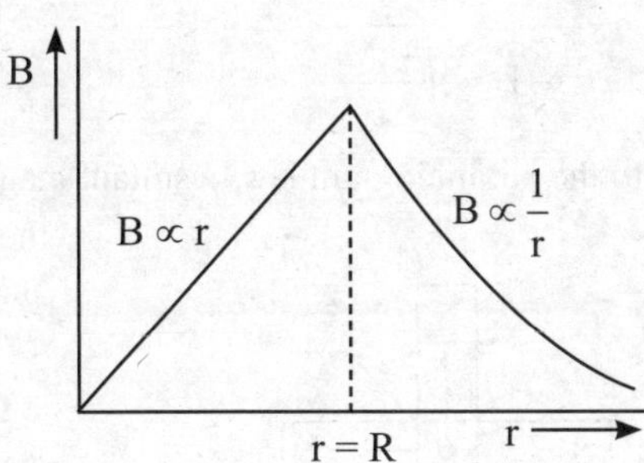

8. (3)

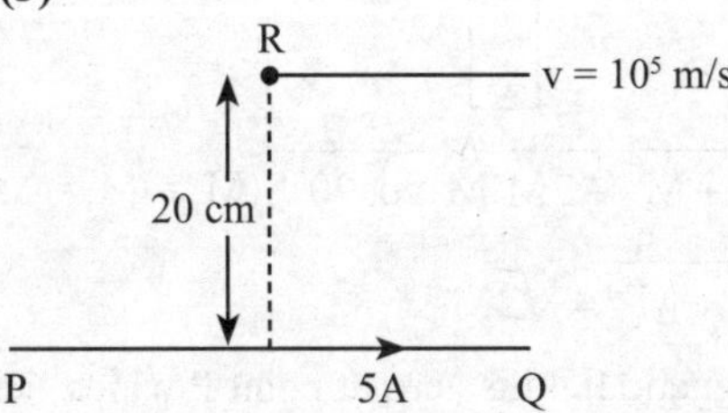

A current carrying wire will produce a closed magnetic field lines around itself. So, applying

Biot-savart law for infinite straight current-carrying wire and the magnitude of magnetic field at point r due to infinite wire will be given as follows:

$$B = \frac{\mu_0 I}{2\pi r} \quad ...(i)$$

Given,

$I = 5A$

$r = 20\text{ cm} = 0.2\text{ m}$

Now, putting in eqn (*i*);

$$|\vec{B}| = \frac{4\pi \times 10^{-7} \times 5}{2\pi \times 0.2\text{ m}} = 5 \times 10^{-6}\,T$$

$\vec{B} = 5 \times 10^{-6}\,T$ [i.e., directed out of the plane]

The force experienced by electron due to magnetic field produced by current carrying wire is given as below:

$$F = q(\vec{v} \times \vec{B})$$

Since, $\vec{v} \perp \vec{B}$

i.e., $F = qvB \sin\theta$

$F = qvB \sin 90°$

$F = qvB$

$F = 1.6 \times 10^{-19} \times 10^5 \times 5 \times 10^{-6}$

$= 16 \times 5 \times 10^{-21} = 80 \times 10^{-21}\,N$

$F = 8 \times 10^{-20}\,N$

9. (2) For current carrying cable (i.e., cylindrical) magnetic field are as follows:

Inside (r < R)

$$B = \left(\frac{\mu_0 I}{2\pi R^2}\right) r \text{ i.e., } B_{inside} \propto r$$

Outside ($r \geq R$)

$$B = \frac{\mu_0 I}{2\pi r} \text{ i.e., } B_{outside} \propto \frac{1}{r}$$

Both relations gives plot of magnetic field (B) with respect of distance from the axis of the cable as follow:

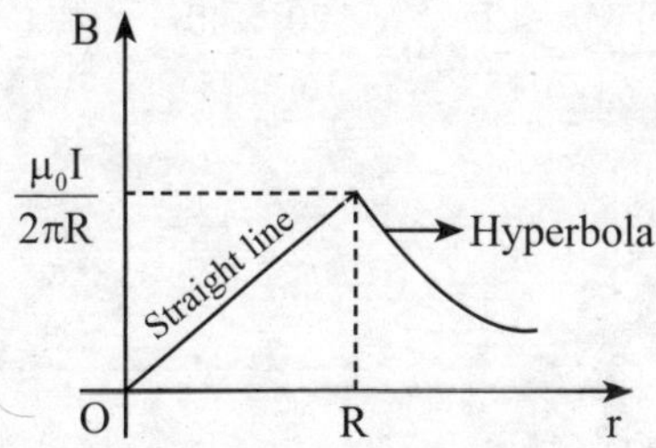

10. (1) $\because\ \vec{F} \perp \vec{B}$

$\therefore\ \vec{F}.\vec{B} = 0$

Hence, option (1) is the correct answer.

11. (2)

12. (1)

13. (2)

Magnetism and Matter

NCERT Topic-wise MCQs

1. (1) When the wire is bent in the form of a semicircle of radius r, then

$L = \pi r = 3.14r$

$r = \frac{L}{3.14} = \frac{31.4}{3.14} = 10\text{cm}$

Distance between the two ends of the wire

$2r = 20\text{cm} = 0.2\text{m}$

$M = m \times 2r = 0.2 \times 0.2$

$M = 0.04\text{ Am}^2$

2. (4) As we know,

$$\vec{B} = \frac{\mu_o 2M}{4\pi r^3}$$

$$= \frac{10^{-7} \times 2 \times 50}{(0.2)^3} = \frac{10^{-7} \times 100}{8 \times 10^{-3}}$$

$\vec{B} = 1.25 \times 10^{-3}\,T$

3. (3) Here, $M = 4.8 \times 10^{-2}\text{ Am}^2$

$\theta = 30°$, $B = 3 \times 10^{-2}T$

$\because\ \tau = MB \sin\theta$

$\tau = 4.8 \times 10^{-2} \times 3 \times 10^{-2} \sin 30°$

$\tau = 7.2 \times 10^{-4}\text{ N–m}$

4. (2) M = NIA

$M = NI(\pi R^2)$

$= 50 \times 12\pi(0.2)^2 = 75.4 Am^2$

5. (2) We know that

$W = MB(\cos\theta_1 - \cos\theta_2)$

$\theta_1 = 0^\circ,\ \theta_2 = 180^\circ$

$W = MB(\cos 0^\circ - \cos 180^\circ)$

$W = 2MB$

6. (1) H = ni

$$i = \frac{4\times10^3\times12\times10^{-2}}{60} = \frac{12}{15}\times10 = \frac{120}{15} = \frac{40}{5} = 8\,A$$

7. (4)

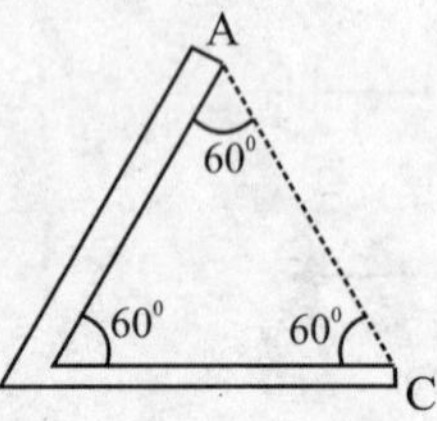

Let ABC be the magnet of length 2L.

∵ AB = BC

∴ ∠BAC = ∠BCA = 60°

∴ ABC is an equalteral triangle.

On bending the magnet, the length of the magnet

AC = AB = BC = L

8. (1) As, $\tau = MB\sin\theta$

$\tau = (m \times 2l) \times 2 \times 10^{-5}\sin 30^\circ$

$$\tau = 15\times10\times10^{-2}\times2\times10^{-5}\times\frac{1}{2}$$

$\tau = 1.5 \times 10^{-5}$Nm

9. (4) $B_1 = \frac{\mu_o}{4\pi}\frac{2M}{x^3}$ and $B_2 = \frac{\mu_o}{4\pi}\frac{M}{y^3}$

$B_1 = B_2$

$$\frac{\mu_o}{4\pi}\frac{2M}{x^3} = \frac{\mu_o}{4\pi}\times\frac{M}{y^3} \Rightarrow \left(\frac{x}{y}\right)^3 = 2 \Rightarrow \frac{x}{y} = 2^{1/3}$$

10. (3) Here, $\theta = 30^\circ$, B = 0.35 T, $\tau = 4.5 \times 10^{-2}$ N m

then, $M = \frac{\tau}{B\sin\theta} = \frac{4.5\times10^{-2}}{0.35\times\sin30^\circ}$

$$M = \frac{4.5\times10^{-2}}{0.35\times\frac{1}{2}} = \frac{2\times4.5\times10^{-2}}{0.35};\ M = 0.26\ J\,T^{-1}$$

11. (1) $W = MB(\cos\theta_1 - \cos\theta_2)$

$= MB(\cos0^\circ - \cos60^\circ)$

$$= MB\left(1-\frac{1}{2}\right) = \frac{MB}{2}$$

and $\tau = MB\sin\theta = MB.\sin 60^\circ = MB\frac{\sqrt{3}}{2}$

$$\therefore \tau = \left(\frac{MB}{2}\right)\sqrt{3} \Rightarrow \tau = \sqrt{3}\,W$$

12. (1) $M_{net} = \sqrt{M^2 + M^2 + 2M^2\cos120^\circ}$

$$= M\sqrt{2\left(1-\frac{1}{2}\right)} = M.$$

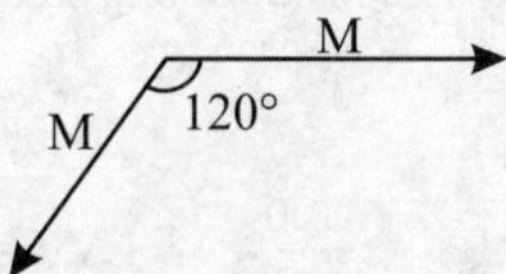

13. (3) Magnetic moment of circular loop carrying current

$M = IA = I(\pi R^2)$

$$M = I\pi\left(\frac{L}{2\pi}\right)^2 \Rightarrow L = \sqrt{\frac{4\pi M}{I}}$$

14. (2) According to the parallelogram law, resultant magnetic moment will be

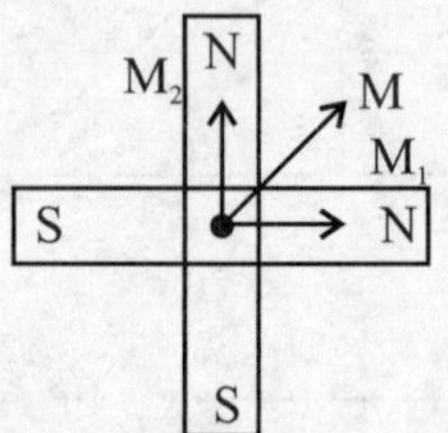

$$M = \sqrt{M_1^2 + M_2^2 + 2M_1M_2\cos90^\circ}\ \ (M_1 = M_2 = M)$$

$$M = \sqrt{M^2 + M^2} = \sqrt{2}M$$

15. (2) Let the magnetic field is zero at point P which lies at a distance x from the magnetic pole having pole strength of 10Am.

$$\frac{\mu_o}{4\pi}.\frac{10}{x^2} = \frac{\mu_o}{4\pi}.\frac{40}{(30-x)^2}$$

$$\frac{1}{x^2} = \frac{4}{(30-x)^2}$$

$$\frac{1}{x} = \frac{2}{30-x}$$

$\Rightarrow$ x = 10 cm

Therefore, from the stronger pole distance of the null point is 20 cm

16. (4) $\tau = MB$ (maximum value of $\sin\theta$ is 1)

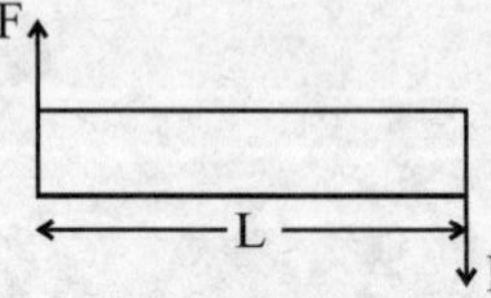

F × L = MB

$$F = \frac{M}{L}\times B$$

$$6\times10^{-4}=\frac{3}{L}\times2\times10^{-5}$$

$L = 0.1$ m

17. (1) Both points lies on axial line

$$B\propto\frac{1}{d^3},\ \frac{B_A}{B_B}=\left(\frac{d_B}{d_A}\right)^3$$

$$\frac{B_A}{B_B}=\left(\frac{48}{24}\right)^3=(2)^3\Rightarrow\ \frac{B_A}{B_B}=8$$

18. (2) Potential energy, $U=-\vec{M}\cdot\vec{B}=0.25\text{ T},\ \theta=30^\circ$

(i) $U = -MB\cos 180^\circ = MB$

(ii) $U = -MB\cos 90^\circ = 0$

(iii) $U = -MB\cos\theta$

(iv) $U = +MB\cos\theta$

19. (3) $B = \mu H$

$B = 8\times10^{-3}\times160$

$= 1.28$ Wb/m^2

20. (3) $\mu=\frac{B}{H}=\frac{\phi}{AH}$

$$=\frac{2.4\times10^{-5}}{0.2\times10^{-4}\times16\times10^{2}}=\frac{1.2}{16}\times10^{-2}$$

$= 0.75\times10^{-3}$ Wb/Am

$$\mu_r=\frac{0.75\times10^{-3}}{4\pi\times10^{-7}}\Rightarrow\mu_r=\frac{75}{4\pi}\times10^{2}=5.9713376$$

$\mu_r = 597$

$\chi_m = \mu_r - 1$

$\chi_m = 597 - 1 \Rightarrow 596$

21. (2) Intensity of magnetisation, $(I)=\frac{M}{V}$

$$I=\frac{M}{\text{mass / Density}}$$

$$\text{density}=5\text{g / cm}^3=\frac{5\times10^{-3}\text{Kg}}{(10^{-2})^3}=5\times10^3\text{kg / m}^3$$

$$I=\frac{6\times10^{-7}\times5\times10^{3}}{10^{-3}}=3\text{Am}^{-1}$$

22. (3) As we know $\chi_m=\frac{I}{B}$

$I = \chi_m B \Rightarrow 3\times10^{-4}\times4\times10^{-4} = 12\times10^{-8}$ Am^{-1}

23. (1) Magnetising field $H = 2\times10^3$ A/m

Magnetic field $B = 8\pi T$

$\because\ B = \mu H$

$\therefore\ B = \mu_0\mu_r H$

So, relative permeability

$$\mu_r=\frac{B}{\mu_0 H}$$

$$\mu_r=\frac{8\pi}{4\pi\times10^{-7}\times2\times10^{3}}$$

$\mu_r = 10^4$

24. (2) Relationship between magnetic susceptibility χ_m and relative permeability μ_r is

$\mu_r = 1+\chi_m$ or $\chi_m = \mu_r - 1$

25. (1) According to Curie's law

$$\chi=\frac{C}{T}\ \Rightarrow\chi\propto\frac{1}{T}.$$

i.e., magnetic susceptibility is inversely proportonal to temperature hence graph (a) is the best representation of this relation.

26. (3) As $\chi_m=\frac{C}{T}\quad\therefore\ \frac{\chi_m}{\chi_{m'}}=\frac{T'}{T}$

$$T'=\frac{\chi_m}{\chi_{m'}}\times T=\frac{1.2\times10^{-5}}{1.44\times10^{-5}}\times300=250\text{K}$$

27. (3) $\chi_m=\frac{I}{H}=\frac{\text{Am}^{-1}}{\text{Am}^{-1}}$ = no units & dimensions.

28. (4) $ui = -MB(\cos\pi)$

$= MB$

$u_f = -MB$

$w = u_f - u_i$

$w = -2MB$

29. (2) $\mu_r = 1+\chi_m \Rightarrow 1+5500 = 5501$

$\mu = \mu_0\mu_r = 4\pi\times10^{-7}\times5501$

$\mu = 6.9\times10^{-3}$ TmA^{-1}

30. (2) $\mu=\frac{B}{H}=\frac{1.6}{1000}=1.6\times10^{-3}\text{TmA}^{-1}$

$$\mu_r=\frac{\mu}{\mu_o}=\frac{1.6\times10^{-3}}{4\pi\times10^{-7}}=\frac{1.6\times10^{-3}}{4\times3.14\times10^{-7}}=1.3\times10^{3}$$

$\chi_m \Rightarrow \mu_r - 1 \approx 1.3\times10^3$

$\mu_r = 1.3\times10^3$

31. (2) Susceptibility χ_m of a paramagnetic substance is given by

$$\chi_m=\frac{C}{T}\ \Rightarrow\chi_m\propto\frac{1}{T}\ \text{(Temperature dependent)}$$

32. (4) Net magnetic induction $B = B_0 + B_m = \mu_0 H + \mu_0 M$

33. (1) In elements or materials all the electrons they contain have opposite spin so they cancel each other. But some elements or material contain domains of electron i.e., ferromagnetic and paramagnetic material. All the domains are randomly oriented in absence of electric field but in the presence of electric field they are oriented along electric field and provide magnetic effect.

34. (3) With rise in temperature, the magnetism of magnet falls and at critical temperature it becomes zero.

$$\chi=\frac{C}{T}$$

35. (3) Since, $\chi_m\propto\frac{1}{T}$

$$\Rightarrow\frac{\chi_2}{\chi_1}=\frac{T_1}{T_2}$$

$$\frac{\chi_m}{0.0060}=\frac{273-73}{273-173}=2\ \Rightarrow\chi_m=0.0120$$

36. **(1)** Since, $\mu_r = 1 + \chi_m$

$6000 = 1 + \chi_m$

$\chi_m = 5999$

37. **(4)** Magnetic susceptibility of paramagnetic materials are positive but small positive.

38. **(4)** Chromium (Cr) is a ferromagnetic substance.

High Order Time Intensive MCQs

1. **(1)** Work done,

$W = -MB(\cos\theta_2 - \cos\theta_1)$

$W = -1.5 \times 0.22(-1 - 1)$

$W = 0.66$ J

Torque,

$\tau = MB \sin\theta$

$\tau = 1.5 \times 0.22 \sin 180°$

$\tau = 0$

2. **(2)** Work done, $W = MB(\cos\theta_1 - \cos\theta_2)$

Amount of work done to rotate the bar magnet through 60°

$W_2 = MB(\cos 0° - \cos 60°)$

$= MB(1 - 1/2) = \frac{MB}{2}$

Amount of work done to rotate the bar magnet through 90°

$W_1 = MB(\cos 0° - \cos 90°)$

$W_1 = MB$

Given

$W_1 = nW_2$

$MB = n \times \frac{MB}{2} \Rightarrow n = 2$

3. **(1)** Work done in rotating a magnet (from angle 0 to θ) is given by

$W = MH(\cos 0° - \cos\theta) = MH(1 - \cos\theta)$

4. **(2)** $W = MB(\cos\theta_1 - \cos\theta_2)$

As $W = \sqrt{3}\ J$

$\sqrt{3} = MB(\cos 0° - \cos 60°)$

$= MB\left(1 - \frac{1}{2}\right)$

$\Rightarrow MB = 2\sqrt{3}$ J

Torque $\tau = MB \sin 60° = (2\sqrt{3})\left(\frac{\sqrt{3}}{2}\right) = 3$ J

5. **(2)** Magnetic needle kept in a non-uniform magnetic field experiences both a force and torque.

6. **(4)** Magnet is cut equally

$M = \frac{m}{2} \times \ell = \frac{M_1}{2}$

On putting both parts ⊥

$M_2 = \sqrt{\frac{(M_1)^2}{(2)^2} + \frac{(M_1)^2}{(2)^2}} = \frac{M_1}{\sqrt{2}}$

$\frac{M_1}{M_2} = \sqrt{2}$

7. **(2)** The magnetic susceptibility of ferromagnetic material is large and positive.

8. **(2)** For a paramagnetic substance, the intensity of magnetization of a material is directly proportional to the magnetic intensity.

As $I \propto H \Rightarrow I = \chi H$ ...(*i*)

$\chi \propto \frac{1}{T} \propto \chi = \frac{C}{T}$ ...(*ii*) (Curie's law)

where T = absolute temperature,

From Eqs. (*i*) and (*ii*), we get $I \propto \frac{1}{T}$.

9. **(2)** Magnetic field at center due to orbiting charge:

$B = \frac{\mu_0 ev}{4\pi r^2}$

Area of the orbit:

$A = \pi r^2$

Magnetic flux:

$\Phi = B \cdot A = \left(\frac{\mu_0 ev}{4\pi r^2}\right)(\pi r^2) = \frac{\mu_0 evr}{4\pi}$

10. **(4)**

F, 5cm, 30°, F, 5 sin 30°

Deflecting couple

K = 2 [mB] [perpendicular distance from axis of rotation]

$= 2\left[2 \times 0.32 \times 10^{-4} \times 5 \times \frac{1}{2} \times 10^{-2}\right]$

$= 0.32 \times 10^{-5} = 32 \times 10^{-7}$ N-m

11. **(3)** $T \propto \frac{1}{\sqrt{B_H}} = \frac{1}{\sqrt{B\cos\theta}}$

$\frac{T_1}{T_2} = \sqrt{\frac{B_2 \cos\theta_2}{B_1 \cos\theta_1}}$

$\frac{B_1}{B_2} = \frac{T_2^2 \cos\theta_2}{T_1^2 \cos\theta_1} = \left(\frac{3}{2}\right)^2 \times \frac{\cos 60^0}{\cos 30^0} \Rightarrow \frac{B_1}{B_2} = \frac{9}{4\sqrt{3}}$

12. **(3)** $M_1 = M_2 = \frac{M}{2};\ M_R = \sqrt{M_1^2 + M_2^2}$

13. **(3)** For part P:

$M_p = m \times 2r$ where r is radius

$= m \times 2 \times \frac{l}{\pi}$ $[\because l = \pi r]$

$\therefore M_P = \frac{M}{\pi}$

For part Q: $M_Q = \frac{M}{2}$

$\therefore M_R = M_P + M_Q$

$M_R = \frac{M(2+\pi)}{2\pi}$

14. (3) $M' = \frac{2M\sin\left(\frac{\theta}{2}\right)}{\theta}$ and $\theta = \frac{\pi}{3}$ $\therefore$ $M' = \frac{3H}{\pi}$

Percentage change in the magnetic moment

$= \frac{M'-M}{M}\times 100$

15. (4) F = mB and M = m × 2l $\Rightarrow F = \frac{M}{2l}\times B$

$\therefore 2l = \frac{M\times B}{F}$

∴ Length of the magnet (2*l*) = 0.1 m

16. (3) $B = \frac{\mu_0}{4\pi}\frac{m}{d^2}, F = \frac{\mu_0}{4\pi}\frac{m_1 m_2}{(2d)^2}$

and $m_1 = m_2 = m$ $\therefore F = \frac{mB}{4}$

17. (4) $B = \frac{\mu_0}{4\pi}\frac{m}{d^2}, F = \frac{\mu_0}{4\pi}\frac{m\times m}{d^2}$

$\therefore B = \frac{F}{m}$

Resultant magnetic field

$B_R = \sqrt{B^2 + B^2 + 2B^2\cos 60} = \sqrt{3}\,B$

$B_R = \sqrt{3}\frac{F}{m}$

18. (1) $C\times(180° - 45°) = M_1 B\sin 45°$

$C\times(180° - 30°) = M_2 B\sin 30°$

Dividing both the equations gives $\frac{M_1}{M_2} = \frac{9}{10\sqrt{2}}$

19. (3) For circular loop ($l = 2\pi r$),

$M_{cir} = IA = I\times\pi r^2 = I\pi\left(\frac{l}{2\pi}\right)^2$

For square loop,

$M_{sq} = I\times\left(\frac{l}{4}\right)^2 \therefore \frac{M_{sq}}{M_{cir}} = \frac{I\times\left(\frac{l}{4}\right)^2}{I\pi\left(\frac{l}{2\pi}\right)^2} = \frac{\pi}{4}$

20. (4)

2d

d

Magnetic field B between the two magnets is

$\left[B_{axial} = \frac{\mu_0}{4\pi}\frac{2m}{d^3}\ \&\ B_{eq} = \frac{\mu_0}{4\pi}\frac{m}{d^3}\right]$

$B = \sqrt{B_{axial}^2 + B_{eq}^2}$

21. (3) Here, $\vec{M} = 30\hat{j}\,A\,m^2$ and $\vec{B} = (2\hat{i} + 5\hat{j})\,T$

Since $\vec{\tau} = \vec{M}\times\vec{B}$

$= 30\hat{j}\times(2\hat{i} + 5\hat{j}) = 60\hat{j}\times\hat{i} + 150\hat{j}\times\hat{j} = 60(-\hat{k}) + 150\times 0$

$= -60\hat{k}\,N\,m$ $[\because \hat{j}\times\hat{i} = -\hat{k}$ and $\hat{j}\times\hat{j} = 0)$

22. (2) Resultant magnetic moment due to the combination

$B = \sqrt{M^2 + M^2} = M\sqrt{2}$

$\therefore B = \frac{\mu_0}{4\pi}\cdot\frac{2(M\sqrt{2})}{d^3}$

As the point P lies along the axial line of the combination of the magnets.

23. (3) $B_H = B_{e_1} = \frac{\mu_0 2M}{4\pi d_1^3}$; $B_H = B_{e_2} = \frac{\mu_0 M}{4\pi d_2^3}$

24. (3) Here, $\theta = 60°$, $B_1 = 1.2\times 10^{-2}$ T

$\theta_1 = 30°$ and $\theta_2 = 60° - 30° = 30°$

In stable equilibrium, torques due to two fields must be balanced i.e. $\tau_1 = \tau_2$

$\Rightarrow MB_1\sin\theta_1 = MB_2\sin\theta_2$

or $B_2 = B_1\frac{\sin\theta_1}{\sin\theta_2}$

$= B_1\frac{\sin 30°}{\sin 30°} = B_1 = 1.2\times 10^{-2}$ T

$\vec{B}_2$, n, 30°, $\vec{B}_1$, S

25. (2) $\frac{M_1}{M_1 - M_2} = \left(\frac{d_1}{d_2}\right)^3$

26. (1) $I = \frac{M}{V} = \frac{n\times \text{magnetic moment of each molecule}}{\text{Volume}}$

27. (4) $\tan\delta = \frac{B_V}{B_H}$ and

$\tan\delta^1 = \frac{B_V^1}{B_H^1} = \frac{B_V}{B_H\cos 45°} \Rightarrow \tan\delta^1 = \frac{\tan\delta}{\cos 45°} = \sqrt{2}\tan\delta$

Statement Based MCQs

1. (1) At geomagnetic poles horizontal component of Earth's magnetic field is zero. Thus, the needle experience no force.

2. (1) The true geographic north-south direction is inclined at an angle with magnetic north-south direction. The south of needle points towards the magnetic north pole. Magnetic makes angle with geographic meridian (11.3°).

3. (3) Because lines of force prefer to pass through or concentrate in the iron walls of the box. Also, μ for iron is high.

4. (1) The earth has only vertical component of its magnetic field at the magnetic poles. Since compass needle is only free to rotate in horizontal plane where H = 0, hence compass needle may stay in any direction.

The dip needle rotates in a vertical plane and angle of dip at poles is 90°, so it will stand vertical.

5. (1) Since magnetic levitation is a method by which an object is suspended with no support other than magnetic field. Magnetic force is used to counter the effects of gravitational force.

Assertion & Reason MCQs

1. **(1)** At the equator vertical component of earth's magnetic field is zero. Hence, the needle will not be deflected anywhere in vertical direction, also a freely suspended magnet will be horizontal at equator.
2. **(1)** A solenoid is analogous to bar magnet.
3. **(1)** As the angle between velocity and magnetic field is zero. Therefore, no force present acts on it.
4. **(4)** When magnet is brought near iron nails, magnetic induction produces poles in the iron which are attracted to the magnet. Therefore, the iron piece will align itself in the direction of magnetic field. Hence, not only translatory force but a torque will also act on it. Field due to magnet is uniform.
5. **(1)** At high magnetic fields there is maximum magnetization of ferromagnetic materials. And at very low temperature the ferromagnetic material cannot be transformed into paramagnetic materials. Therefore, it would be difficult to break the domains under low temperature and high fields.

Match the Columns MCQs

1. **(4)**

(a) Mobiles, T.V., music system, computers and many more electronic devices are made up of magnetic materials and magnet in it. When we bring external magnets near these electronic devices, magnets will interfere with magnetic component of devices so magnets must always be kept away from electronic devices.

(b) Horse shoe magnet is a bar magnet of U-shaped.

(c) Bar magnet should be kept in pairs with their opposite pole on the same side separated by a non-magnetic material (wood).

(d) Magnet loses its magnetic power called demagnetisation. It is due to heating, hammering, A.C. Current field etc.

2. **(2)**

(a) Magnetic moment (M) = IA.

Unit : $A-m^2$

(b) Pole strength $= \frac{M}{L} = \frac{IA}{L}$

Unit : A – m

(c) Intensity of magnetisation $= \frac{M}{V} = \frac{IA}{V}$

Unit : $\frac{A-m^2}{m^3} = A-m^{-1}$

(d) Magnetic Flux (ϕ) = B.A

Unit : $\frac{\text{Weber}}{m^2}.m^2 \Rightarrow \text{Weber}$

3. **(2)** The ratio of magnitude of intensity of magnetisation to that of magnetising field strength is called magnetic susceptibility. It is scalar with no units and dimensions.

Material	Magnetic susceptibility
Diamagnetic	Small and negative
Paramagnetic	Small and positive
Ferromagnetic	High and positive.

4. **(3)**

Quantity	Formula	Dimension
Pole strength	$m = \frac{M}{I}$	[LA]
Intensity of magnetisation	$I = \frac{M}{V}$	$[L^{-1} A]$
Magnetic moment	M = IA	$[L^2 A]$
Relative permeability	$\mu_r = \frac{\mu}{\mu_0}$	No-dimension

NCERT Exemplar MCQs

1. **(2)** Here, $M_1 = 8\ A\ M^{-1}$, $B_1 = 0.6$ T, $T_1 = 4$ K, $B_2 = 0.2$ T, $T_2 = 16$ K

Then, for paramagnetic materials,

Magnetisation (M) $= C\frac{B}{T}$(Curie's law)

Now, in the first case, $M_1 = \frac{CB_1}{T_1}$ ... (*i*)

and in the second case, $M_2 = \frac{CB_2}{T_2}$ ... (*ii*)

Dividing equation (*i*) by (*ii*), we get

$$\frac{M_1}{M_2} = \frac{B_1}{B_2}.\frac{T_1}{T_2} \Rightarrow \frac{M_1}{M_2} = \frac{0.6}{0.2} \times \frac{16}{4};\ \frac{8}{M_2} = 3\times 4$$

$$M_2 = \frac{8}{12} = \frac{2}{3}\ A\ m^{-1}$$

2. **(1)** For the earth's magnetism, the magnetic field lines of the earth resemble that of a hypothetical magnetic dipole located at the centre of the earth.

The axis of the dipole does not coincide with the axis of rotation of the earth but is presently tilted by approximately 11.3° with respect to the later.

Hence, the declination varies between 11.3° W to 11.3° E.

3. **(2)** As Gauss' law states, $\int_s E.ds = \frac{q}{\varepsilon_0}$ for electrostatic field. It does not contradict for electrostatic fields as the electric field lines do not form continuous closed path.

According to Gauss' law in magnetic field.

$$\oint_s B.ds = 0$$

It contradicts for magnetic field, because there is a magnetic field inside the solenoid and no field outside the solenoid carrying current but the magnetic field lines form the closed path.

NEET & AIIMS Past Year MCQs

1. **(4)** If sheet is magnetic, then force is required to hold the sheet because magnetic pole exert force on sheet
   As sheet is conducting, eddy current formed which oppose the way by which it is produced, so forced is required.
   As sheet is non conducting & non polar, so no magnetization occur in sheet hence no force is required.

2. **(3)**

Material	Magnetic susceptibility (χ)
Diamagnetic	-10^{-5} to 10^{-9}
Paramagnetic	10^{-5} to 10^{-3}
Ferromagnetic	>> 1
Non–magnetic	0

3. **(2)** $\oint_s \vec{B}\cdot d\vec{A} = 0$ (No monopole exist)

4. **(4)** For equilateral triangle

$$x = \sqrt{a^2 - \frac{a^2}{4}} = \sqrt{\frac{3a^2}{4}} = \sqrt{\frac{3a^2}{4}} = \frac{\sqrt{3}}{2}a$$

$$A_1 = \frac{1}{2}\times a \times \frac{\sqrt{3}}{2}a \Rightarrow A_1 = \frac{\sqrt{3}}{4}a^2$$

∴ Magnetic dipole moment

$$\mu_1 = N_1 I A_1 \Rightarrow \mu_1 = \frac{4I\sqrt{3}}{4}a^2$$

$$\mu_1 = \sqrt{3}\,Ia^2$$

For square

$$A_2 = a^2 \Rightarrow \mu_2 = N_2 I A_2 = 3\times I\times a^2$$

$$\mu_2 = 3Ia^2$$

5. **(1)** H_2O has $\mu_r < 1$. It do not allow magnetic field lines of forces through it. Hence, it is diamagnetic

6. **(1)** Magnetic lines of force are continuous and closed. It originates from one pole and terminates at another poles. Magnetic monopoles do not exist

7. **(3)** 8. **(2)** 9. **(1)** 10. **(4)**

Electromagnetic Induction

NCERT Topic-wise MCQs

1. **(1)** According to the Faraday's law of EMI

$$e = \frac{\phi_2 - \phi_1}{t} = \frac{(4B_0 - B_0)}{t}A_0$$

$$e = \frac{3B_0 A_0}{t}$$

2. **(4)** $|e| = N\left(\frac{\Delta B}{\Delta t}\right)A\cos\theta$

$$= 500 \times 1 \times (10^{-1})^2 \cos 0°$$

$$|e| = 5 \text{ volt}$$

3. **(2)** $e = -N\frac{(B_2 - B_1)}{\Delta t}A\cos\theta$

$$= \frac{-500\times(0-0.1)\times 100\times 10^{-4}\cos 0°}{0.1} = 5 \text{ volt}$$

4. **(2)** $e = -N\frac{(B_2 - B_1)}{\Delta t}A\cos\theta$

$$= \frac{-50(0.35-0.10)\times\pi(3\times 10^{-2})^2\times\cos 0°}{2\times 10^{-3}} = 17.7 \text{ volt}$$

5. **(2)** Amount of charge flowing through the circuit is given by

$$\Delta Q = \frac{\Delta\phi}{R}$$

$$= \frac{(10-2)}{2} = 4C$$

6. **(1)** $e = -N\frac{(B_2 - B_1)}{t}A\cos\theta$

$$0.1 = \frac{-50\left[0-\left(2\times 10^{-2}\right)\right]\times 100\times 10^{-4}\times\cos 0°}{t}$$

$$\Rightarrow t = \frac{10^{-2}}{10^{-1}} = 0.1 \text{ sec}$$

7. **(4)** $|e| = \frac{d\phi}{dt} = \frac{d(3t^2+4t+9)}{dt} = 6t + 4$

$= (6 \times 2) + 4$ ($t = 2s$, given)

$e = 16$ volt

8. **(3)** It is based on the law of conservation of momentum.

9. **(1)** $E = \frac{-d\phi}{dt}$ & $E = IR$

As, $I = \frac{dq}{dt}$

$$\Rightarrow \left(\frac{dq}{dt}\right)R = \frac{-d\phi}{dt}$$

$$\Rightarrow dq = \frac{20}{R} = \frac{20}{20}$$

$\therefore Q = 1\,C.$

10. **(1)** As we know, $\phi = BA$

Change in flux, $d\phi = 0.05\,(101–100)10^{-4} = 5 \times 10^{-6}$ Wb

Now, $dq = \frac{d\phi}{R}$

$$dq = \frac{5\times 10^{-6}}{2} = 2.5\times 10^{-6}C$$

11. **(1)** A bar magnet on falling downward through a long hollow metallic cylinder produces eddy current due to change in magnetic flux and the motion of the bar magnet is retarded, and with time motion ultimately stops because retarding force becomes equal to its weight. Hence, deceleration becomes zero.

12. (3) $F = Bvq$, $\phi = BA$

$$B = \frac{F}{vq} \Rightarrow \phi = \frac{F \times A}{vq} \Rightarrow \left(\frac{[MLT^{-2}][L^2]}{[LT^{-1}][AT]}\right)$$

$\phi = [ML^2T^{-2}A^{-1}]$

13. (4) $e = BAv = B.A \times \frac{\omega}{2\pi} = B \times \pi R^2 \times \frac{\omega}{2\pi}$

$$e = \frac{1}{2}BR^2\omega = \frac{1}{2} \times 0.05 \times (2)^2 \times 60 = 6\,V$$

14. (3) Given: $\phi = 6t^2 + 7t + 1$

The induced emf is given by, $e = \frac{d\phi}{dt}$

or $e = \frac{d}{dt}(6t^2) + \frac{d}{dt}(7t) + \frac{d}{dt}(1)$

$= 12t + 7$

$e_{t=2} = 12 \times 2 + 7$

$= 24 + 7 = 31$ mV

15. (1) $e = \frac{-d\phi}{dt} = \frac{-BA\cos 180° \times N}{t}$

$$= \frac{-0.4 \times 10^{-4} \times 500 \times 10^{-4} \times (-1) \times 1000}{1/10} = 0.02\,V$$

16. (1) As, $I = \frac{e}{R} = \frac{d\phi}{Rdt}$ and $Idt = \frac{d\phi}{R} \Rightarrow \Delta q = \frac{\Delta\phi}{R}$

If coil contain n turns then, $\Delta q = \frac{n\Delta\phi}{R}$

$\therefore \Delta q = \frac{n\Delta\phi}{R} = \frac{1}{7} \times (1.35 - 0.79) = 0.08\,C$

17. (2) $|e| = \frac{d\phi}{dt} = \frac{d}{dt}(\pi r^2 B) = \pi B \frac{d}{dt}(r^2)$

$= \pi B 2r\left(\frac{dr}{dt}\right) = 2\pi r\,B\left(\frac{dr}{dt}\right)$

18. (3) As we know, $e = Blv$

$\therefore B = \frac{e}{\ell v} = \frac{25}{10 \times 5} = 0.5\,T$

19. (4) If on changing current through the coil, the emf induced in the coil is e, then by Faraday's 2nd law, we have

$e = -L\frac{\Delta i}{\Delta t}$

Given, $L = 5H \frac{\Delta i}{\Delta t} = -2$ A/s (decreasing)

$\therefore e = -5 \times (-2) = +10\,V$

20. (1) As we know,

$e = L\frac{dI}{dt} \Rightarrow 0.20 = \frac{L[5-(-5)]}{0.20}$

$L = 4 \times 10^{-3}\,H$

21. (2) As we know the relation, $L = \frac{\mu_o N^2 A}{\ell}$

$\therefore L \propto N^2$

$\frac{L_1}{L_2} = \left(\frac{N_1}{N_2}\right)^2 \Rightarrow \frac{108}{L_2} = \left(\frac{600}{500}\right)^2$

$\therefore L_2 = \frac{25}{36} \times 108 = 25 \times 3 = 75\,MH$

22. (2) Since $e = \omega B\pi r^2 \sin\omega t$

Emf is zero when $\omega t = n\pi$

i.e. when $\omega t = 0, \pi, 2\pi$... etc

So induced emf changes direction after every half rotation.

23. (3) $e_{max} = NBA\omega$

$= 50 \times 80 \times 10^{-4} \times 0.05 \times 2000 \times \frac{2\pi}{60} = \frac{4\pi}{3}V$

24. (3) $\because e = Blv$

$\therefore e \propto v \propto gt$ $(v = u + gt)$

25. (4) As we know

$|e| = \frac{d\phi}{dt} = B.Av$

$|e| = B \times \pi r^2 \times \frac{\omega}{2\pi} = \frac{Br^2\omega}{2}$

$r = 0.5$ m, $\omega = 4\pi$ rad/s and $B = 0.4 \times 10^{-4}$ T

$|e| = 0.4 \times 10^{-4} \times (0.5)^2 \times 2\pi = 6.28 \times 10^{-5}\,V$

26. (2) Energy stored in an inductor is given by

$E = \frac{1}{2}LI^2$

$E = \frac{1}{2} \times 50 \times 10^{-3} \times (2)^2 = 0.1$ Joule

27. (3) When flux completely linked with each other the coefficient of coupling becomes unity, $K = \frac{M}{\sqrt{L_1 L_2}}$

28. (2) $\tau_L = \frac{L}{R} = \frac{1}{2}s$

$i = i_0(1 - e^{-t/\tau_L})$

$\therefore \frac{i_0}{i} = \frac{1}{1 - e^{-t/\tau_L}}$

Substituting, t = 1s and $\tau_L = \frac{1}{2}s$

We get, $\frac{i_0}{i} = \frac{e^2}{e^2 - 1}$

29. (4) $\varepsilon = L\frac{dI}{dt} = 50 \times 10^{-3} \times \frac{1}{0.1} = 0.5V$

30. (2) In steady state current passing through the solenoid

$I = \frac{E}{R} = \frac{10}{10} = 1A$

31. (4) According to Fleming's right hand rule AB, DC are parallel to the direction of motion.

32. (4) Self inductance of a solenoid is given by

$$L = \frac{\mu_0 N^2 A}{l}$$

$\therefore L \propto N^2$

$$L' = \left(\frac{4N}{N}\right)^2 L \Rightarrow L' = 16L$$

33. (2) $\because$ Energy stored in an inductor is, $U = \frac{1}{2}LI^2$

$\therefore U \propto I^2$

$$U' = \left(\frac{I}{2I}\right)^2 U \Rightarrow U' = \frac{U}{4}$$

34. (4) Given $\Delta I = 20$ A, $\Delta t = 0.006$ s

$$\therefore \frac{\Delta I}{\Delta t} = \frac{20}{0.006}$$

$$\therefore E = .09 \times \frac{20}{.006} = \frac{180}{.6} = 300 \text{ V}$$

35. (1) Mutual inductance between coils is.

$$M = K\sqrt{L_1 L_2}$$

$$M = 1 \times \sqrt{2 \times 10^{-3} \times 8 \times 10^{-3}} = \sqrt{16 \times 10^{-6}} = 4 \text{ mH}$$

36. (3) $L = \frac{\mu_0 N^2 A}{l}$

$$L = \frac{4\pi \times 10^{-7}(10^3)^2 \times 10^{-3}}{1} = 1.256\text{mH}$$

37. (4) $\because N\phi = LI$

$$\therefore \frac{N\phi}{I} = L$$

$$\Rightarrow L = \frac{100 \times 10^{-5}}{5} = 20 \times 10^{-5} = 0.2\text{mH}$$

38. (4) If in time 't', the rod turns by the angle 'θ', the area generated by the rotation of the rod will be:

$$A = \left(\frac{1}{2}l\right) \times (l\theta) \Rightarrow A = \frac{1}{2}l^2\theta$$

So, the flux linked with the area generated by the rotation of the rod:

$$\phi = B\left(\frac{1}{2}l^2\theta\right) \Rightarrow \phi = B\left(\frac{1}{2}l^2\omega t\right)$$

The induced EMF is:

$$E = \frac{1}{2}\frac{d(Bl^2\omega t)}{dt} \Rightarrow E = \frac{1}{2}(Bl^2\omega)$$

39. (3) $U = \frac{1}{2}LI^2$

$$I = \frac{V}{R} = \frac{10}{2} = 5A \Rightarrow U = \frac{1}{2} \times 2 \times (5)^2 = 25\text{Joules}$$

40. (2) We knows that $L = \frac{\mu_0 N^2 A}{\ell} \Rightarrow \frac{\mu_0 N^2 \pi r^2}{\ell}$

$$\frac{L_2}{L_1} = \left(\frac{N_2}{N_1}\right)^2 \times \left(\frac{r_2}{r_1}\right)^2 \times \left(\frac{\ell_1}{\ell_2}\right) = (1)^2 \times \left(\frac{2}{1}\right)^2 \times \frac{1}{2} = 2:1$$

$$\frac{L_2}{L_1} = \frac{2}{1}$$

41. (1) As $L_0 = \frac{\mu_0 N^2 A}{\ell}, L = \frac{\mu N^2 A}{\ell}$ $(L \propto \mu)$

Relative magnetic permeability,

$$\mu_r = \frac{\mu}{\mu_0} \Rightarrow \mu_r = \frac{L}{L_0} = \frac{16}{0.04} = 400$$

42. (3) $v = 1800$ km/h $= 1800 \times \frac{5}{18} = 500\text{m/s}$

$l = 25$ m, $B = 5 \times 10^{-4}$ T, $\delta = 30^\circ$

$e = B_V l v$ (B_V is the vertical component of the earth's field).

$e = (B \sin\delta)\, lv$ $\{\because B_V = B\sin\delta\}$

$e = 5 \times 10^{-4} \sin 30^\circ \times 25 \times 500 = 3.1$ volt

43. (1) Here, $\ell = 1$m, $\omega = 400$ rad s^{-1}, $B = 0.5$T

For rotation of a rod of length 'ℓ' about its end.

$$e = \frac{B\ell^2\omega}{2} \Rightarrow e = \frac{0.5 \times 1^2 \times 400}{2} = 100 \text{ V}$$

44. (2) As we know that

$e = NBA\omega$

$\therefore e \propto \omega$

Hence induced e.m.f is doubled

45. (4) $N = 10$, $A = 12 \times 12\text{cm}^2 = 144 \times 10^{-4}$cm

$B = 0.025$ Wb/m^2, $\omega = ?$

$e_{max} = 20$mV

As we know, $e_{max} = NBA\omega$

$$\therefore \omega = \frac{e_{max}}{NBA} = \frac{20 \times 10^{-3}}{10 \times 0.025 \times 144 \times 10^{-4}}$$

$$\omega = 5.55 \text{ rad/s}$$

46. (2) $V = 200$V ; $r = 10\ \Omega$ and $R = 100\ \Omega$

$\therefore$ Total resistance, $R_T = 10 + 100 = 110\ \Omega$

$$I = \frac{V}{R_T} = \frac{220}{110} = 2A$$

$P = I^2 R = 4 \times 100 = 400$W

47. (2) As we know

$e = NAB\omega$

$$e = 1 \times (20 \times 10^{-2})^2 \times 500 \times 10^{-4} \times 2\pi \times \frac{240}{60}$$

$e = 400 \times 10^{-4} \times 500 \times 10^{-4} \times 2\pi \times 4$

$e = 5.02 \times 10^{-2}$ V

48. (2) $e = BNA\,\omega$

So, $e \propto \omega$

$$\text{New speed} = \frac{120}{100} \times 1500 = 1800 \text{ rpm}$$

49. (4) All of these are application of eddy current.

50. (3) Electric generator is based on the principle of electromagnetic induction.

High Order Time Intensive MCQs

1. (1) Induced emf of coil is given by, $E = \left|-\frac{d\phi}{dt}\right|_t$

$$E = \left|-\frac{d\phi}{dt}\right|_{t=2} = |100\,t|_{t=2} = 200\text{ V}$$

Current in the coil, $i = \frac{E}{R} = \frac{200}{400} = \frac{1}{2} = 0.5\text{A}$

2. (2) Current in coil is given by, $I = \left|\frac{1}{R}\frac{d\phi}{dt}\right| \Rightarrow |d\phi| = |IR\,dt|$

Change in flux = (area of triangle) × R

$$= \left(\frac{1}{2} \times 4 \times 0.1\right) \times 10 = 2\text{ Wb}$$

3. (4) In steady state, whole current passes through the inductor.

4. (1) The induced emf is given by

$$e = -\frac{d\phi}{dt} \quad ...(i)$$

and the magnetic flux is given as

$\phi = BA\cos\theta$

(Here, θ is the angle between the magnetic field B and area vector of ring A)

So, $\theta = 0° \Rightarrow \phi = BA\cos 0° \Rightarrow \phi = BA$ ...(*ii*)

From Eqs. (*i*) and (*ii*), we get

$$e = -\frac{d}{dt}(BA)$$

So, $e = -\frac{d}{dt}[(B_0 + \alpha t)A]$ (given; $B = B_o + \alpha t$)

$$\Rightarrow = -A\frac{d}{dt}[B_0 + \alpha t] \Rightarrow = -A[0 + \alpha] = -A\alpha$$

But $A = \pi r^2$ is the area of ring, here r is the radius of ring.

So, $e = -\pi r^2\alpha \Rightarrow e = -\pi\alpha r^2$

5. (1) $\tau_L = \frac{L}{R} = 2\text{ s}$

$$I_0 = \frac{E}{R} = 3\text{A},\ t = 2\text{s}$$

$$i =_0 (1 - e^{-t/\tau_L})$$

6. (4) The induced emf will oppose the motion of the magnet. The direction of induced current will be anti-clockwise (applying right hand rule).

7. (2) Magnetic flux linked with each turn of the coil

$\phi_1 = BA$ (as coil is perpendicular to B)

On rotating the coil through 180°,

Magnetic flux linked with the coil, $\phi_2 = -BA$

∴ Change in magnetic flux

$\Delta\phi = \phi_2 - \phi_1 = -2BA$

So, emf induced in the coil

$$e = -N\frac{\Delta\phi}{\Delta t} = \frac{2NBA}{\Delta t}$$

where, N is the number of turns in the coil.

The current in the coil is given by:

$$i = \frac{e}{R} = \frac{2NBA}{R\Delta t}$$

$$= \frac{2 \times 1200 \times 4 \times 10^{-4} \times 500 \times 10^{-4}}{20 \times 0.1}$$

$= 24 \times 10^{-3}\text{ A} = 24\text{ mA}$

8. (3) Initially the inductance will oppose the current which tries to flow through the inductance. But 10 Ω and 20 Ω can conduct. The current will be

According to ohm's law

$$I = \frac{V}{R} = \frac{2}{30} = \frac{1}{15}\text{A}$$

9. (1) → i —L— i →

Flux through a inductor of inductance carrying current is given by:

$\phi = Li$

Given, $L = 40$ mH, $\Delta t = 4$ms, $\Delta i = 11 - 1 = 10$A

So, $|\text{emf}| = \frac{\Delta\phi}{\Delta t} = L\frac{\Delta i}{\Delta t} = (40\text{mH})\frac{(10)}{4\text{ ms}}$

∴ emf = 10 × 10 = 100 V

10. (4) $e = -L\frac{di}{dt} = -L\frac{(-2-2)}{0.05}$

$$8 = L\frac{(4)}{0.05} \Rightarrow L = \frac{8 \times 0.05}{4} = 0.1\text{H}$$

11. (1) Amount of magnetic flux linked with the inductor is

$\phi = Li$

The emf induced in the inductor is

$$e = -\frac{d\phi}{dt} = -\frac{d}{dt}(Li) = -L\frac{di}{dt}$$

$$\Rightarrow |e| = L\frac{di}{dt}$$

Here, induced current $= \frac{V}{R} = \frac{10}{5} = 2$ A

Circuit switches off in 1 ms

$\Rightarrow dt = 1 \times 10^{-3}$s

∴ Induced emf in the inductor is

$$|e| = 10 \times \frac{2}{1 \times 10^{-3}} = 2 \times 10^4\text{ V}$$

12. (4) The self- inductance of the solenoid is given by

$$L = \frac{\mu_0 N^2 A}{l}$$

Where N is the number of turns, A is area of cross section and l is length of the solenoid

$$\because\ N' = 2N \Rightarrow L' = \frac{\mu_0 (2N)^2 A}{l} = \frac{4\mu_0 N^2 A}{l} = 4L$$

13. (1) To calculate the emf we can imagine a closed loop by connecting the centre with any point on the circumference, say X with a resistor.

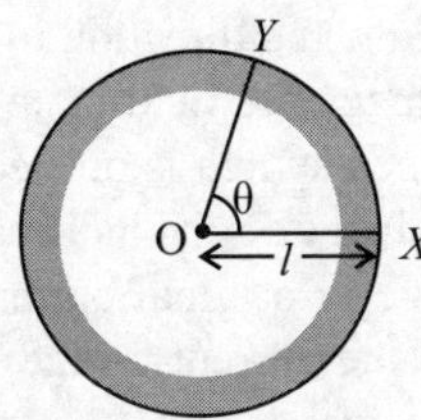

Potential difference across the resistor is then equal to the induced emf. It arises due to separation of charges.
$e = B \times$ (rate of change of area of loop)
If θ is the angle between the rod and l the radius of circle at X at time t, the area of the arc formed by the rod and radius is

$$\text{Area}(OXY) = \frac{1}{2} l^2 \theta$$

where, l is radius of the circle.

$$\therefore e = B \times \frac{d}{dt}\left(\frac{1}{2} l^2 \theta\right) \Rightarrow e = \frac{1}{2} B l^2 \frac{d\theta}{dt} = \frac{1}{2} B l^2 \omega \quad \left(\because \omega = \frac{d\theta}{dt}\right)$$

14. (4) $e = Bvl = 4 \times 10^{-5} \times 90 \times 35 = 0.126$ V

15. (4) $e = Bvl \sin\theta = 0.1 \times 10 \times 4 \sin 30° \Rightarrow e = 2$ V

16. (4) Coefficient of mutual inductance is given by

$$M = \frac{\phi}{I_p} = \frac{\mu_0 \mu_r N_1 N_2 A I_p}{l I_p} = \frac{\mu_0 \mu_r N_1 N_2 A}{l} \quad (I_p = \text{current in primary coil})$$

17. (1) Just after pressing the key K, the inductor offers infinite resistance.

$$I_1 = \frac{10}{6+8} = \frac{5}{7}\text{A}$$

18. (2) At t = 0, L_1 and L_2 will produce infinite resistance.

19. (2) $\because$ e = Bvl

$$i = \frac{e}{R} = \frac{Bvl}{R} \Rightarrow F = ilB = \frac{B^2 l^2 v}{R}$$

$$P = F \cdot v = \frac{B^2 l^2 v^2}{R} \Rightarrow P = \frac{(1 \times 1 \times 20)^2}{0.2} = 2000 \text{ W} = 2 \text{ kW}$$

20. (2) The magnetic energy stored in the solenoid
$E = 1/2 \times L \times I^2$
$E = 1/2 \times 5 \times 10^2 = 250$ Joule

21. (2) Mutual inductance, $M = \frac{N_s \phi}{I}$

$$= \frac{12 \times 2.5 \times 10^{-4}}{2} = 15 \times 10^{-4}\text{H}$$

22. (2) By short-circuiting the battery, net resistance across inductor is R/2 (R and R in parallel).

$$\therefore \tau_{net} = \frac{L}{R_{net}} = \frac{2L}{R}$$

23. (3) $e = -L\left(\frac{di}{dt}\right) = -L$(Slope of i-t graph

Initially, slope = 0 $\Rightarrow e = 0$

Then in remaining two regions slopes are constant but of opposite signs. Hence, induced emfs are constant but of opposite signs.

24. (4)

25. (1) At time t, angle rotated by loop is $\theta = \omega t$. This is also the angle between B and S. Then

$\phi = BS \cos\theta$

$= Bb^2 \cos \omega t$

$$e = \left|\frac{d\phi}{dt}\right| = b^2 B\omega \sin \omega t$$

26. (4) Here, area $A = \left(\frac{10}{100}\right)^2 = \frac{1}{100}\text{m}^2$

$$e = NA\frac{dB}{dt} = 500 \times \frac{1}{100} \times 1V = 5V.$$

27. (4) $\therefore$ In parallel, $L_p = \frac{L_1 L_2}{L_1 + L_2}$

$\Rightarrow (L_1 + L_2)1.5 = L_1 L_2$

and in series, $L_S = L_1 + L_2 = 8$

$\therefore\ L_1 L_2 = 12$

$(L_1 - L_2)^2 = (L_1 + L_2)^2 - 4L_1 L_2$

$L_1 - L_2 = \sqrt{(8)^2 - 4 \times 12} = 4H$

28. (4) As the loop remains inside the region of magnetic field, no current will be induced.

29. (1) Area of square loop, S = 10 cm × 10 cm = 10^{-2} m^2
Initial magnetic flux linked with the loop,

$$\phi_{B_1} = B_1 S \cos\phi = 0.1 \times 10^{-2} \times \cos 45° = \frac{10^{-3}}{\sqrt{2}}\text{Wb}$$

Final magnetic flux linked with the loop,
$\phi_{B_2} = 0$ Wb $\quad [\because B_2 = 0]$
The induced emf in the loop,

$$\varepsilon = -\frac{d\phi}{dt} = -\frac{(\phi_2 - \phi_1)}{t}$$

$$= -\frac{\left(0 - \frac{10^{-3}}{\sqrt{2}}\right)}{0.7} = \frac{10^{-3}}{0.7 \times \sqrt{2}} \cong 10^{-3}\text{V}$$

The induced current in the loop,

$$I = \frac{\varepsilon}{R} = \frac{10^{-3}\text{V}}{1\Omega} = 10^{-3}\text{A} = 1.0\text{mA}$$

30. (2) $e = -L\frac{di}{dt} = -4.0 \times 10^{-3} \times \frac{250 \times 10^{-3}}{0.1} = -1\text{mV}$

31. (3) According to Faraday's law of electromagnetic induction,

$$E_{induced} = \frac{-\Delta\phi}{\Delta t} = -\frac{B(A_f - A_i)}{\Delta t}$$

Let r be the radius of circle; then side of square formed

$$= \frac{2\pi r}{4} = \frac{\pi r}{2}$$

Change in area of the loop

$$= A_i - A_f = \pi r^2 - \left(\frac{\pi r}{2}\right)^2 = \frac{\pi(4-\pi)r^2}{4}$$

Hence average emf induced $= \frac{\pi(4-\pi)r^2}{4} \cdot \frac{B}{t}$

$$= \frac{\pi(4-\pi)\times(0.1)^2\times 100}{4\times 0.1} = 6.75 \text{ volt.}$$

32. **(3)** $e = L\frac{dI}{dt}$

$$3.28 = L\times\frac{(10-2)}{0.1}$$

$\therefore L = 0.04$ H

33. **(3)** Four resistance form a balanced Wheatstone bridge So no current will flow through the inductor. Hence, stored energy in coil is zero.

Statement Based MCQs

1. **(2)** An electric motor converts electrical energy to mechanical energy.

2. **(3)** If two coils of inductance L_1 and L_2 are joined together, then their mutual inductance is given by

$$M = k\sqrt{L_1 L_2} \quad ...(i)$$

It is clear from the relation, if self-inductance of primary and secondary coil are doubled the mutual inductance of the coils will be doubled.

3. **(3)** Mutual inductance depends only on size, number of turns, relative position and orientation of 2 coils.

4. **(1)**

- We know that when we move a conductor in a magnetic field, then an emf is induced across its end. The emf induced is given by the equation,

  emf $= Bv\ell$

- Also, we know that if a coil lies in a magnetic field and if the magnetic field is changed, the flux linked with the coil also changes and hence an emf is induced in the coil according to the equation,

  emf $= -\frac{d\phi}{dt}$

Assertion & Reason MCQs

1. **(2)** Both Assertion and Reason are True but Reason is not a correct explanation of the Assertion.

2. **(1)** Both Assertion and Reason are True and the Reason is a correct explanation of the Assertion.

3. **(3)** Induced emf is proportional to rate of change of magnetic flux. And as the magnetic field is equal for both the loops, induced emf is same. Induced current depends on resistance. Thus, current will not be same in both the loops. It will be more in the copper compared to aluminum loop.

4. **(1)** Current induced in the solenoid offers retarding force to the falling magnet, so, its acceleration is reduced.

5. **(2)** From Lenz's law, the direction of induced emf is such that the induced magnetic field oppose change in original magnetic flux. When switch is opened, the sudden drop in the magnetic field in the circuit induces an emf in a direction that oppose change in original current. This induced emf causes a spark as the current bridges the air gap between the poles of the switch.

Matching Type MCQs

1. **(1)** Steady state current through inductor,

$$i_0 = \frac{9}{3} = 3A$$

Now, this current decays exponentially across inductor and two resistors.

$$\tau_L = \frac{L}{R} = \frac{9}{6+3} = 1\,s$$

$t_{1/2} = (\ln 2)\,\tau_L = (\ln 2)$s

Given time is half-life time. Hence, current will remain 1.5 A.

$$i = i_0 e^{-t/\tau_L} = 3e^{-t}$$

$$\left(-\frac{di}{dt}\right) = 3e^{-t}$$

In the beginning $\left(-\frac{di}{dt}\right) = 3$ A/s

After one half-life time $\left(-\frac{di}{dt}\right) = 1.5$ A/s

(a) $V_L = L\left(\frac{-di}{dt}\right) = 9\times 1.5 = 13.5$ V

(b) $V_{3\Omega} = iR = 1.5 \times 3 = 4.5$ V

(c) $V_{6\Omega} = iR = 1.5 \times 6 = 9$ V

(d) $V_{bc} = V_L - V_{3\Omega} = 9$V

2. **(2)** (a) $e = \frac{d\phi}{dt} = 2$V

(b) $i = \frac{e}{R} = 1$ A = constant

(c) $\Delta q = i\Delta t = 1 \times 2 = 2$ C

(d) $H = i^2R\Delta t = (1)^2\,(2)\,(2) = 4$J

3. **(3)** (a) If current is increased, ⊗ magnetic field passing through loop will increase. So, induced current will produce ⊙ magnetic field. Hence, induced current is anti-clockwise.

Now, i and I currents in PQ are in opposite directions. Hence, they will repel each other.

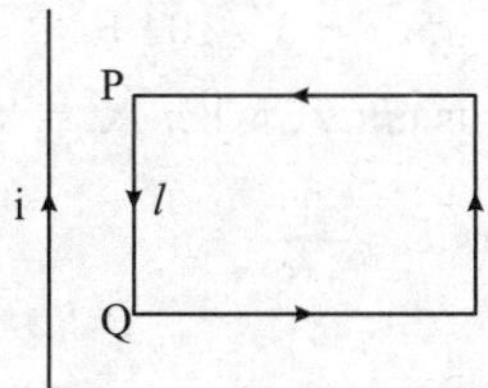

4. (4) (a) At 1s, ⊗ flux is increasing. So, induced current will produce ⊙ magnetic field or induced current is anticlockwise.

$$i=\frac{(d\phi/dt)}{R}=\frac{5}{10}=\frac{1}{2}A$$

NCERT Exemplar MCQs

1. (3) Here, $\vec{A}=L^2\hat{k}$ and $\vec{B}=B_0\left(2\hat{i}+3\hat{j}+4\hat{k}\right)T$

$$\phi=\vec{B}\cdot\vec{A}=B_0\left(2\hat{i}+3\hat{j}+4\hat{k}\right)\cdot L^2\hat{k}=4B_0L^2\ \text{Wb}$$

2. (2) Also, the magnetic flux linked with uniform surface of area A in uniform magnetic field is given by

$$\phi=\vec{B}\cdot\vec{A}$$

$$A=A_1+A_2=\left(L^2\hat{k}+L^2\hat{i}\right)$$

and $\phi=\vec{B}\cdot\vec{A}=B_0\left(\hat{i}+\hat{k}\right)\cdot\left(L^2\hat{k}+L^2\hat{i}\right)=2B_0L^2\text{Wb}$

3. (2) When cylindrical bar magnet is rotated about its axis, no change in flux linked with the circuit takes place, consequently no EMF is induced and hence, no current flows through the ammeter A.

4. (4) Coil A must be carrying a constant current in counter clockwise direction. Because of that, when a moves towards B, current induced in B is counterclockwise direction as per Lenz's law. The current in B would stop when A stops moving.

5. (1)

6. (2) The self-inductance of a solenoid of cross-sectional area A and length l, having n turns per unit length, filled the inside of the solenoid with a material of relative permeability (e.g., soft iron, which has a high value of relative permeability) is given by

$$L=\mu_r\mu_o n^2A\ell$$

Where, $5\times n=N/l$

$$\Rightarrow L=\frac{\mu_r\mu_0N^2A}{l}$$

$$\therefore\ L\propto A \text{ and } L\propto\frac{1}{\ell}$$

Note: The capacitance, resistance, self and mutual inductance depends on the geometry of the devices as well as permittivity/permeability of the medium.

NEET & AIIMS Past Year MCQs

1. (3) Lenz Law

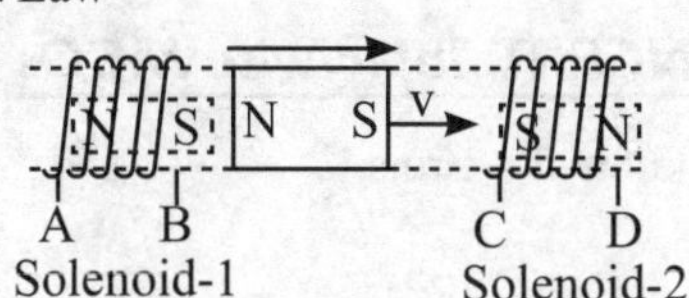

2. (1) Magnetic energy stored in an inductor $=\frac{1}{2}LI^2$

$$=\frac{1}{2}\times4\times10^{-6}\times(2)^2=8\ \mu J$$

3. (3) B = 0.5 T

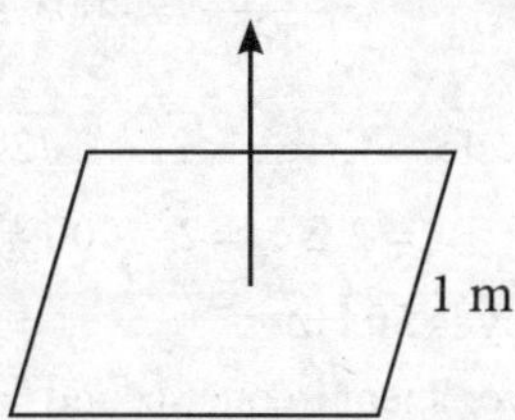

As the angle between the magnetic field and normal to the plane of the coil is zero

$\Rightarrow\theta=0°$

$\because\ \phi=B.A.\cos\theta=0.5\times(1)\times1$

$\phi=0.5$ Wb

4. (4) Maximum emf induced in a circular coil having N turns, area A and angular speed ω is given by

$$E_{max}=NBA\omega$$

∴ Maximum current induced in the coil is given by

$$i_{max}=\frac{E_{max}}{R}\Rightarrow i_{max}=\frac{NBA\omega}{R}$$

$$i_{max}=\frac{1000\times2\times10^{-5}\times\pi(10^2)\times2}{12.56}$$

$$i_{max}=1A$$

5. (3)

$$M=\frac{N_2\phi_2}{I_1}$$

$$M=\frac{N_2B_1A_2}{I_1}$$

$$M\propto B_1A_2$$

$$M\propto\frac{\mu_0I_1}{2\pi R_1}\times\pi R_2^2$$

$$M\propto\frac{R_2^2}{R_1}$$

6. (1) $W=\Delta U$

$$W=\frac{1}{2}LI_2^2-\frac{1}{2}LI_1^2$$

$$=\frac{1}{2}L\left(I_2^2-I_1^2\right)$$

7. (2)

8. (1)

Alternating Current

NCERT Topic-wise MCQs

1. (2) As we know $P = I^2rms$

$$P = \left(\frac{I_P}{\sqrt{2}}\right)^2 \times R = \frac{I_P^{\ 2}}{2}R$$

2. (1) Here, $R = 100\Omega$, $E_v = 200V$, $n = 50$ Hz

Average power/cycle $= E_v I_v = E_v \times \frac{E_v}{R}$

$$= \frac{(200)^2}{100} = 400 \text{ watt}$$

Peak power / cycle $= E_0 I_0 = \sqrt{2}E_v \times \sqrt{2}I_v$

$= 2E_vI_v = 2 \times 400 = 800$ W

As the frequency is 50 Hz

$\therefore$ Average power spent in one second = 400 × 50

$\therefore$ Energy spent in one minute = 400 × 50 × 60 = 12×10^5 J

3. (4) As we know, phase angle, $\phi = 90^0$

$\therefore$ power = VI cos ϕ

$P = 0$

4. (3) $X_L = \omega L = 300 \times 800 \times 10^{-3} = 240\Omega$

$$Z = \sqrt{R^2 + X_L^{\ 2}} = \sqrt{(10)^2 + (240)^2} = 240.2\Omega$$

Power factor, $\cos\phi = \frac{R}{Z} = \frac{10}{240.2} = 0.0416$

5. (4) Power factor, $\cos\phi = \frac{R}{Z}$

$$Z = \sqrt{R^2 + X_L^{\ 2}} = \sqrt{R^2 + (\omega L)^2}$$

$$Z = \sqrt{(30)^2 + (100 \times 400 \times 10^{-3})^2} = \sqrt{900 + 1600} = \sqrt{2500}$$

$Z = 50\ \Omega$

$$\therefore \cos\phi = \frac{R}{Z} = \frac{30}{50} = 0.6$$

6. (3) As we know,

$C = ?$, $R = 10\Omega$, $\cos\phi = 0.5$, $E_v = 80V$, $\nu = 100$ Hz

As $\cos\phi = \frac{R}{Z} \Rightarrow Z = \frac{R}{\cos\phi} = \frac{10}{0.5} = 20$

$\therefore R^2 + X_C^{\ 2} = Z^2$

$$X_C = \sqrt{Z^2 - R^2} = \sqrt{(20)^2 - (10)^2} = \sqrt{400 - 100} = \sqrt{300}$$

$X_C = 10\sqrt{3}$

$$\frac{1}{\omega C} = 10\sqrt{3} \Rightarrow C = \frac{1}{\omega \times 10\sqrt{3}} = \frac{1}{2 \times \pi \times 100 \times 10\sqrt{3}}$$

$C = 9.2 \times 10^{-5}$ F

7. (4) As we know, $P \propto \cos\phi$

Power factor angle ϕ is phase angle between V & I and the phase angle between V & I is not given.

8. (3) $L = ?$, $C = 5\mu F = 5 \times 10^{-6}$ F, $R = 10\Omega$, $\nu = 50$ Hz

Power factor is unity, when $X_L = X_C$

$$\omega L = \frac{1}{\omega C} \text{ or } L = \frac{1}{\omega^2 C}$$

$$L = \frac{1}{(2\pi \times 50)^2 \times 5 \times 10^{-6}} \Rightarrow L = 2.026 \text{ H}$$

9. (4) Power dissipated in pure resistor will be $\frac{(V_{rms})^2}{R}$

$$P = \frac{(30)^2}{100} = \frac{900}{100} = 9W$$

10. (4) Average power, $P = V_{rms} I_{rms} \cos\phi$

Here, $Z = \sqrt{R^2 + (X_L - X_C)^2}$, $\cos\phi = \frac{R}{Z}$

But, $I_{rms} = \frac{V_{rms}}{Z}$.

$$\therefore P = V_{rms}^2 \frac{R}{Z^2}$$

$$\therefore P = V_{rms}^2 \frac{R}{\{R^2 + (X_L - X_C)^2\}} = \frac{\varepsilon^2 R}{\left[R^2 + \left(\omega L - \frac{1}{\omega C}\right)^2\right]}$$

11. (2) As we know

$$\cos\phi = \frac{R}{Z} = \frac{R}{\sqrt{R^2 + (X_L - X_C)^2}}$$

If, $X_L = X_C$

$\cos\phi = 1$ (max value)

Hence, power delivered will be maximum

where $X_L = X_C \Rightarrow \omega L = \frac{1}{\omega C}$

12. (2) At resonance LCR circuit behaves as purely resistive circuit. For purely resistive circuit power factor is 1.

13. (2) As we know,

$$\cos\phi = \frac{R}{Z} = \frac{R}{\sqrt{R^2 + \omega^2 L^2}}$$

$$= \frac{12}{\sqrt{(12)^2 + 4 \times \pi^2 \times (60)^2 \times (0.1)^2}}$$

$\cos\phi = 0.30$

14. (2) $\cos\phi = \frac{R}{Z} = \frac{R}{\sqrt{R^2 + \omega^2 L^2}} = \frac{5}{\sqrt{25 + (50)^2 \times (0.1)^2}}$

$$\cos\phi = \frac{5}{\sqrt{25 + 25}} = \frac{5}{5\sqrt{2}} = \frac{1}{\sqrt{2}}$$

$$\cos\phi = \frac{1}{\sqrt{2}}$$

$$\therefore \phi = \frac{\pi}{4}$$

15. (4) Average power in ac circuit is given by:

$P = V_{rms} I_{rms} \cos\phi$

For pure capacitor circuit, $\phi = 90^0$

so, P = 0

16. (3) As we know,

The effective voltage is, $V_{rms} = \frac{V_0}{\sqrt{2}}$

$V_{rms} = \frac{423}{\sqrt{2}} = 299\,V$

17. (3) As we know

$V_{rms} = \frac{V_0}{\sqrt{2}}$

(Peak) $V_0 = \sqrt{2} \times V_{rms} = 1.414 \times 220$

$V_0 = 311$ volts

18. (4) Time taken by the current to reach the maximum value

$t = \frac{T}{4} = \frac{1}{4v} = \frac{1}{4 \times 50} = 5 \times 10^{-3}$ second

$\therefore I_0 = I_{rms}\sqrt{2} \quad \left(I_{rms} = \frac{I_0}{\sqrt{2}}\right)$

$I_0 = 10\sqrt{2}$

$I_0 = 14.14$ A

19. (3) This is according to the relation

$I_{rms} = \sqrt{\frac{i_1^2 + i_2^2}{2}} = \frac{(i_1^2 + i_2^2)^{1/2}}{\sqrt{2}}$

20. (2) $\because E_{rms} = \frac{E_0}{\sqrt{2}} = \frac{120}{\sqrt{2}} = 84.85$ V

21. (3) Phase difference, $\Delta\phi = \phi_2 - \phi_1 = \frac{\pi}{6} - \left(\frac{-\pi}{6}\right) = \frac{\pi}{3}$

As the current is ahead in phase.

$\therefore$ Current leads the voltage by 60^0.

22. (1) The mean square voltage,

$$\bar{V}^2 = \frac{1}{t}\int_0^t V^2 dt = \frac{1}{t}\int_0^1 V_0^2 t^3 dt$$

$$= \frac{V_0^2}{t}\left[\frac{t^4}{4}\right]_0^1 = \left[\frac{V_0^2 t^3}{4}\right]_0^1 = \frac{V_0^2}{4}$$

$\therefore \; V_{rms} = \sqrt{\bar{V}^2} = \frac{V_0}{2}$

23. (1) rms value of current

$$I_{rms} = \sqrt{I_0^2 + \frac{I_1^2}{2}} = \sqrt{4^2 + \frac{(3)^2}{2}} = 4.52 \text{ unit}$$

24. (3) Mean value or average value of voltage $= \frac{2E_0}{\pi}$

25. (4) $I_V - 5A$, $I_0 = \sqrt{2} \times I_{r.m.s} = 1.414 \times 5 = 7.07A$

From $I = I_0 \sin \omega t$

$I = 7.07 \sin 2\pi \times 50 \times \frac{1}{600}$

$= 7.07 \sin\frac{\pi}{3} = 7.07 \times \frac{1}{2} = 3.535A$

26. (2) Here, $C = 10\,\mu F = 10^{-5}F$, $V = 25$ volt

$L = 100$ mH $= 10^{-1}H$, $I_0 = ?$

Maximum energy in inductor = Maximum energy in condensor

$\frac{1}{2}LI^2 = \frac{1}{2}CV^2 \Rightarrow I = V\sqrt{\frac{C}{L}}$

$= 25\sqrt{\frac{10^{-5}}{10^{-1}}} = 25 \times 10^{-2}$ A

$I_0 = 0.25A$

27. (2) $C = ?$ $L = 101.4\mu H = 101.4 \times 10^{-6}H$

$\nu = 1$ MHz $= 10^6$ Hz

As, $\nu = \frac{1}{2\pi\sqrt{LC}}$

$\therefore C = \frac{1}{4\pi^2\nu^2 L}$

$C = \frac{1 \times 7 \times 7}{4 \times 22 \times 22 \times 10^{12} \times 101.4 \times 10^{-6}}$

$= 2.496 \times 10^{-10}$ farad

$C = 249.6$ pF

28. (1) Here, $L = 100$ mH $= 10^{-1}$ H

$C = 25\mu F = 25 \times 10^{-6}$ F

$R = 15\Omega$, $E_V = 120$ V, $\nu = 50$ Hz

At resonance,

$\nu = \frac{1}{2 \times \pi\sqrt{LC}} = \frac{1 \times 7}{2 \times 22 \times \sqrt{10^{-1} \times 25 \times 10^{-6}}}$

$\nu = \frac{7 \times 10^3}{44 \times 1.58} = 100.7\,Hz$

29. (1) $I_V = 1.5$ mA $= 1.5 \times 10^{-3}$ A

$\omega = 300$ rad/sec, $R = 10\,k\Omega = 10^4\,\Omega$

$C = 0.50\,\mu F = 0.50 \times 10^{-6}$ F

$X_C = \frac{1}{\omega C} = \frac{1}{300 \times 0.5 \times 10^{-6}} = 6.67 \times 10^3\,\Omega$

$Z = \sqrt{R^2 + X_C^2} = \sqrt{(10^4)^2 + (6.67 \times 10^3)^2}$

$Z = 1.2 \times 10^4\,\Omega$

RMS voltage across the capacitor $= I_V X_C$

$= 1.5 \times 10^{-3} \times 6.67 \times 10^3 = 10V$

30. (1) Here, $R = 10\Omega$, $E_V = 100V$, $\nu = 50$ Hz, $\theta = \frac{\pi}{4}$ rad

$L = ?$

$\tan\theta = \frac{X_L}{R} = \frac{2\pi\nu L}{R}$

$L = \frac{R\tan\theta}{2\pi\nu} = \frac{10 \times \tan\pi/4}{2 \times 3.14 \times 50} = 0.032\,H$

31. (3) $R = 300\Omega$, $L = 1H$, $\nu = \frac{300}{2\pi}$ Hz

$$\therefore \tan\phi = \frac{X_L}{R} = \frac{2\pi\nu L}{R}$$

$$\therefore \tan\phi = \frac{2\pi\times\frac{300}{2\pi}\times 1}{300} = 1$$

$\phi = 45^0$

32. (2) As we know that,

$$\omega = \frac{1}{\sqrt{LC}} \Rightarrow \therefore \nu = \frac{1}{2\pi\sqrt{LC}}$$

$(\nu) \propto (LC)^{-1/2}$

33. (4) $Z = \sqrt{R^2 + X_L^2} = \sqrt{R^2 + (2\pi Lf)^2}$

$$= \sqrt{(30)^2 + \left(2\pi\times 50\times\frac{0.4}{\pi}\right)^2} = \sqrt{900+1600} = \sqrt{2500}$$

$Z = 50\Omega$

$$I = \frac{V}{Z} = \frac{200}{50} = 4A$$

34. (1) Here, $C = 3\times 10^{-6}$ F, $E_V = 220V$, $\nu = 50$ Hz

$I_V = ?$ $E_0 = ?$

$$I_V = \frac{E_V}{X_C} = \frac{E_V}{\frac{1}{\omega C}} = E_V\times\omega C$$

$$I_V = E_V\, 2\pi\nu C$$

$$I_V = 220\times 2\times\frac{22}{7}\times 50\times 3\times 10^{-6} = 0.207A$$

$$\therefore E_0 = \sqrt{2}\,E_v = 1.41\times 220 = 311.1V$$

35. (2) As we know that

$$X_L = \omega L = 2\pi\nu L = \frac{E_0}{I_0}$$

$$X_L = \frac{\sqrt{2}\times E_{rms}}{I_0} = \frac{\sqrt{2}\times 200}{0.9}$$

$$\Rightarrow 2\times\pi\times\nu\times 1 = \frac{\sqrt{2}\times 200}{0.9}$$

$$\Rightarrow \nu = \frac{\sqrt{2}\times 200}{0.9\times 2\pi} = 50Hz$$

$\nu = 50$ Hz

36. (3) $I_o = 0.25A$, $\nu = 60Hz$, $L = 2H$

$$V_L = I_v X_L = \frac{I_o}{\sqrt{2}}X_L = \frac{I_o}{\sqrt{2}}\omega L$$

$$V_L = \frac{I_o 2\pi\nu L}{\sqrt{2}} = \frac{0.25}{\sqrt{2}}\times 2\times\frac{22}{7}\times 60\times 2$$

$V_L = 133.4V$

37. (4) Here, $P = 100$ W, $E_v = 220$ V,

$I_v = ?$

$$I_v = \frac{P}{E_v} = \frac{100}{220} = 0.45A$$

38. (3) $R = 100\Omega$, $E_v = 110V$

$$E_0 = \sqrt{2}E_v = 1.414\times 110 = 155.5V$$

$\therefore$ Potential difference over a half cycle

$$E_v = \frac{2E_0}{\pi} = \frac{2\times 155.5}{3.14}$$

$E_v = 99.1V$

39. (3) Here, $L = 5H$, $R = 20\Omega$, $V_o = 100$ volt

When current reaches its final value, role of L vanishes.

Maximum current, $I_o = \frac{V_o}{R} = \frac{100}{20} = 5A$

$\therefore$ Energy stored in magnetic field $= \frac{1}{2}LI^2$

$$E = \frac{1}{2}\times 5\times(5)^2 = 62.5J$$

$$E = \frac{62.5}{4.2}\text{ cal}$$

$E = 14.8$ calorie

40. (4) $L = 0.4$ mH $= 0.4\times 10^{-3}$H

$C = 400$ pF $= 4\times 10^{-10}$F

$\lambda = ?$

$$\nu = \frac{1}{2\pi\sqrt{LC}}$$

$$\nu = \frac{1}{2\times 3.14\times\sqrt{0.4\times 10^{-3}\times 4\times 10^{-10}}}$$

$$\nu = \frac{10^7}{6.28\times 4}\text{ Hz}$$

If c is velocity of em waves, then

$$\lambda = \frac{c}{\nu} = \frac{3\times 10^8}{10^7/6.28\times 4}$$

$\lambda = 6.28\times 10\times 4\times 3$

$\lambda = 753.6$ m

41. (2) Here, $C = 100\ \mu F = 10^{-4}F$, $R = 50\Omega$, $L = 0.5H$

$E_v = 110V$, $\nu = 50Hz$

$X_L = \omega L = 2\pi\nu L = 2\times 3.14\times 50\times 0.5 = 157\Omega$

$$X_C = \frac{1}{\omega C} = \frac{1}{2\times 3.14\times 50\times 10^{-4}}$$

$X_C = 31.85\ \Omega$

$$Z = \sqrt{R^2 + (X_L - X_C)^2}$$

$$Z = \sqrt{(50)^2 + (157-31.85)^2} = 134.77\Omega$$

$$\therefore I_v = \frac{E_v}{Z} = \frac{110}{134.77} = 0.816A$$

42. (1) $L = ?, E_v = 200V, \nu = 50Hz, \quad I_o = 0.9A$

As, $I_o = \frac{E_o}{X_L} = \frac{\sqrt{2}E_V}{2\pi\nu L}$

$\therefore L = \frac{\sqrt{2}E_V}{2\pi\nu I_o} = \frac{1.414 \times 200}{2 \times 3.14 \times 50 \times 0.9} \simeq 1H$

43. (3) Here V = 80V, I = 10A, $E_v = 100V$

Then, $Z = \frac{E_v}{I_v} = \frac{100}{10} = 10\Omega$

As $R^2 + X_L^2 = Z^2$

$\therefore X_L^2 = Z^2 - R^2 = 10^2 - 8^2 = 36 \Rightarrow X_L = 6\ \Omega$

Now, $X_L = \omega L = 2\pi\nu L$

$\therefore\ 2\pi\nu L = 6$

$L = \frac{6}{2 \times 3.14 \times 50} = 1.9 \times 10^{-2} H$

44. (2) $\nu = ?, L = 0.5H, X_L = 1000\Omega$

As, $X_L = \omega L = 2\pi\nu L$

$\nu = \frac{X_L}{2\pi L} = \frac{1000}{2 \times 3.14 \times 0.5}$

$\nu = 318.5$ Hz

45. (1)

46. (1) $E_v = 100\sqrt{2}\,V$, $E_0 = ?$, $E_m = ?$

$E_0 = \sqrt{2}E_v = \sqrt{2}\left(100\sqrt{2}\right) = 200\,V$

$E_m = \frac{2E_0}{\pi} = \frac{2 \times 200}{3.14} = 127.4\,V$

47. (1) For L – C oscillation, $T = 2\pi\sqrt{LC}$

From one extreme to another when electric field is maximum, magnetic field is zero. When electric field is zero magnetic field is maximum.

At time, $t = \frac{T}{4}$, energy stored is completely magnetic

$\therefore t = \frac{T}{4} = \frac{2\pi\sqrt{LC}}{4} = \frac{\pi\sqrt{LC}}{2}$

$\therefore t = \frac{\pi\sqrt{20 \times 10^{-3} \times 50 \times 10^{-6}}}{2} = \frac{\pi\sqrt{10^{-6}}}{2}$

t = 1.57 ms

48. (2) As we know,

$X_C = \frac{1}{2\pi\nu C}$

for d.c $\nu = 0$

$X_C = \frac{1}{2 \times \pi \times 0 \times C} = \infty$

$X_C = \infty$

49. (3) $Z = \sqrt{R^2 + X_L^2} = \sqrt{R^2 + (2\pi\nu L)^2}$

$Z = \sqrt{(40)^2 + 4\pi^2 \times (50)^2 \times (95.5 \times 10^{-3})^2} = 50\Omega$

50. (3) $X_L = 2 \times \pi \times \nu \times L$

$= 2 \times \pi \times \frac{1}{\pi} \times 50 = 2 \times 50 = 100\Omega$

51. (2) $\cos\phi = \frac{R}{Z} = \frac{R}{\sqrt{(R)^2 + \omega^2 L^2}}$

$= \frac{200}{\sqrt{200^2 + (200)^2 \times (1)^2}}$

$\cos\phi = \frac{1}{\sqrt{2}} \Rightarrow \phi = 45^0$

52. (1) $V = \sqrt{V_R^2 + (V_L - V_C)^2} = \sqrt{(40)^2 + (60 - 30)^2} = 50V$

53. (1) Frequency of LC oscillation $= \frac{1}{2\pi\sqrt{LC}}$

$\frac{f_1}{f_2} = \frac{1}{\sqrt{L_1C_1}}\sqrt{L_2C_2} = \left(\frac{L_2C_2}{L_1C_1}\right)^{1/2}$

$\frac{f_1}{f_2} = \left(\frac{2L \times 4C}{L \times C}\right)^{1/2} = (8)^{1/2}$

$\frac{f_1}{f_2} = 2\sqrt{2} \Rightarrow f_2 = \frac{f_1}{2\sqrt{2}} \qquad (f_1 = f)$

$f_2 = \frac{f}{2\sqrt{2}}$

54. (2) $I = \frac{V}{R} = \frac{100}{1000} = 0.1A$

In resonance condition,

$V = V_R, V_L = V_C = \frac{I}{\omega C} = \frac{0.1}{200 \times 2 \times 10^{-6}} = 250V$

55. (3) Here, $X_L = \omega L = 100 \times 1 = 100\Omega$

$X_C = \frac{1}{\omega C}$

$= \frac{1}{100 \times 20 \times 10^{-6}} = 500\Omega$

Impedance of the a.c. source,

$Z = \sqrt{(X_L - X_C)^2 + R^2}$

$\sqrt{(100 - 500)^2 + 300^2}$

$= 500\ \Omega$

$= \frac{V_{rms}}{Z} = \frac{50}{500} = \frac{1}{10} A$

= voltage across capacitor

$= X_C\, i_{rms}$ [RMS]

$= 500 \times 0.1 = 50V$

56. (1) For imparting maximum power

$X_L = X_C \Rightarrow \omega L = \frac{1}{\omega C}$

$C = \frac{1}{\omega^2 L} = \frac{1}{(2\pi\nu)^2 L} = \frac{1}{(100 \times \pi)^2 \times 10}$

$C = 1 \times 10^{-6} = 1\mu F$

57. (1) At resonance, V_L and V_C are equal in magnitude but have phase difference of 180^0 relative to each other.

$\therefore V_{LC} = V_L - V_C = 0$

Hence, voltameter V_2 read 0 volt

58. (2) Here, $L = 30$ mH $= 30 \times 10^{-3}$ H, $V_{rms} = 220V$

Frequency of a.c source, $\nu = 50$ Hz

The inductive reactance is $X_L = 2\pi\nu L$

$X_L = 2 \times 3.14 \times 50 \times 30 \times 10^{-3}$

$X_L = 9.42\Omega$

The r.m.s current in the coil is

$$I_{r.m.s} = \frac{V_{rms}}{X_L} = \frac{220}{9.42} = 23.4A$$

59. (3) Given $\lambda = 300$ m, $c = 3 \times 10^8$ m/s

$C = 2.4 \times 10^{-6}$ F

$$\nu = \frac{c}{\lambda} = \frac{3\times10^8}{300} = 10^6\,\text{Hz}$$

Resonant frequency

$$\nu = \frac{1}{2\times\pi\times\sqrt{LC}} \Rightarrow \sqrt{LC} = \frac{1}{2\times\pi\times\nu}$$

$$LC = \frac{1}{4\times\pi^2\times\nu^2} \Rightarrow L = \frac{1}{4\times\pi^2\left(10^6\right)^2\times2.4\times10^{-6}}$$

$L = 10^{-8}$ H

60. (1) Charge on the capacitor

$q_0 = q_0 \sin \omega t$

or $I_0 = \omega q_0 =$ maximum current

$$\text{Now, } \omega = \frac{1}{\sqrt{LC}} = \frac{1}{\sqrt{10^{-9}}} = \left(10^9\right)^{1/2}$$

$$\therefore I_0 = \left(10^9\right)^{1/2}\times\left(1\times10^{-6}\right) = \sqrt{1000}\,\text{mA}$$

61. (1) Here, Resistance, $R = 3\Omega$

Inductive reflectance, $X_L = 10\Omega$

Capacitive reactance, $X_C = 14\Omega$

The impedence of the series LCR circuit is:

$$Z = \sqrt{R^2 + \left(X_C - X_L\right)^2} = \sqrt{(3)^2 + (14-10)^2}$$

$Z = 5\Omega$

62. (1) Current through the bulb, $I = \frac{P}{V} = \frac{60}{10} = 6A$

$V = \sqrt{V_R^{\,2} + V_L^{\,2}}$

$(100)^2 = (10)^2 + V_L^{\,2}$

$V_L = 99.5$ V

Also, $V_L = I \times X_L = I \times (2\pi fL)$

$99.5 = 6 \times 2 \times 3.14 \times 50 \times L$

$L = 0.052$ H

63. (2) $$I_{rms} = \frac{V_{rms}}{\sqrt{R^2 + \frac{1}{\omega^2C^2}}}$$

When ω increases, I_{rms} increases so the bulb glows brighter.

64. (3) Current through the inductor, $I_L = \frac{E}{X_L}$,

$\therefore I_L = \frac{90}{30} = 3\,A$,

Current through the capacitor, $I_C = \frac{E}{X_C}$,

$\therefore I_C = \frac{90}{20} = 4.5\,A$

[As the phase diff between the currents through the capacitor and the inductor is 180°]

Net current through circuit $I = I_C - I_L$

$I = 1.5$ A

$$\therefore Z = \frac{V}{I} = \frac{90}{1.5} = 60\,\Omega$$

65. (2) At $t = 0$, phase of the voltage is zero, while phase of the current is $\frac{-\pi}{2}$ i.e. voltage leads by $\frac{\pi}{2}$

66. (4) Current through the circuit is given by $I = \frac{V}{Z}$,

$$\text{where, } Z = \sqrt{R^2 + \left(2\pi fL - \frac{1}{2\pi fC}\right)^2}$$

From the above equation at $f = 0$, $Z = \infty$

As frequency increases Z decreases therefore the current, I increases

When, $f = \frac{1}{2\pi\sqrt{LC}}$ (resonant frequency) Z becomes minimum & I becomes maximum

For $f > \frac{1}{2\pi\sqrt{LC}} \Rightarrow$ Z starts increasing and current I goes on decreasing

67. (3) Resistance of the circuit,

$R = R_1 + R_2 = 40\ \Omega + 40\ \Omega = 80\ \Omega$

Impedance of the circuit,

$$Z = \sqrt{R^2 + (X_L - X_C)^2} = \sqrt{(80)^2 + (100-40)^2}$$

$$= \sqrt{(80)^2 + (60)^2} = 100\Omega$$

Power factor, $\cos\phi = \frac{R}{Z} = \frac{80}{100} = 0.8$

68. (2) $V_L = 16$ V, $V_R = 20$ V

In the inductor, the voltage across the resistor V_R lags by $\frac{\pi}{2}$

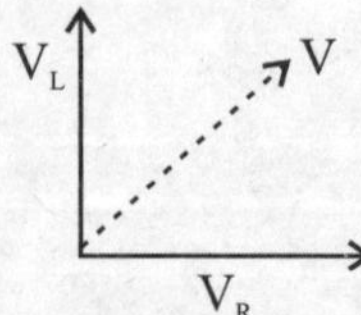

Therefore, total potential difference of the source is

$$\therefore V = \sqrt{V_L^{\,2} + V_R^{\,2}} = \sqrt{256 + 400} = 25.6\,V$$

$V = 25.6$ Volt

69. (1) In series LCR circuit the phase angle ϕ is given by-

$$\tan\phi = \frac{X_L - X_C}{R}$$

When $X_L > X_C$, $\tan\phi$ is (+ve). Therefore, the phase difference is (+ve)

70. (2) The mechanical equivalent of spring constant in LC oscillating circuit is $K = \frac{1}{C}$

71. (4) Restoring force is provided by inductor.

72. (1) $I_p = ?\ E_p = 220V, E_s = 22\ V, R_s = 220\Omega$

$$\therefore\ I_s = \frac{E_s}{R_s} = \frac{22}{220} = 0.1A$$

$$As \frac{I_p}{I_s} = \frac{E_s}{E_p} \Rightarrow \therefore I_p = \frac{E_s}{E_p} \times I_s = \frac{22 \times 0.1}{220} = 0.01A$$

73. (4) $N = 10, A = 12 \times 12cm^2 = 144 \times 10^{-4}cm$

$B = 0.025\ Wb/m^2$, $e_o = 20mV, \omega = ?$

As we know, $e = NBA\omega$

$$\therefore \omega = \frac{e}{NBA} = \frac{20 \times 10^{-3}}{10 \times 144 \times 10^{-4} \times 0.025}$$

$$\omega = 5.55\ rad / s^{-1}$$

74. (2) $V = 200V$; $r = 10\ \Omega$ and $R = 100$

$\therefore$ Total resistance, $R_T = 10 + 100 = 110\Omega$

$$I = \frac{V}{R_T} = \frac{220}{110} = 2A$$

$P = I^2 R = 4 \times 100 = 400W$

75. (2) 5000 V to be converted to 240V

$\therefore$ This is a step down transformer.

$$\because \frac{V_P}{V_S} = \frac{n_P}{n_S} \Rightarrow \frac{5000}{240} = \frac{n_p}{n_s}$$

$\therefore$ Primary has 20.8 times the number of turns in the secondary.

76. (2) As we know, $e = NAB\omega$

$$e = 1 \times (20 \times 10^{-2})^2 \times 500 \times 10^{-4} \times 2\pi \times \frac{240}{60}$$

$e = 400 \times 10^{-4} \times 500 \times 10^{-4} \times 2\pi \times 4$

$e = 5.02 \times 10^{-2}\ V$

77. (3) According to the transformation ratio

$$\frac{N_s}{N_p} = \frac{I_p}{I_s} \Rightarrow \frac{3}{2} = \frac{3}{I_s} \Rightarrow I_s = 2$$

$I_s = 2A$

78. (1) $\frac{N_s}{N_p} = \frac{V_s}{V_p} \Rightarrow \frac{200}{100} = \frac{V_s}{120}$

$\therefore V_s = 240V$

$$\because \frac{V_s}{V_p} = \frac{I_p}{I_s} \Rightarrow \frac{240}{120} = \frac{10}{I_s}$$

$I_s = 5A$

79. (4) A transformer does not work on d.c. Therefore, voltage across the secondary will be zero.

80. (3) $E_p = 2300\ V,\ n_p = 4000, E_s = 230V, n_s = ?$

Assume 100% efficiency

$$\frac{n_s}{n_p} = \frac{E_s}{E_p} = \frac{230}{2300} = 0.1$$

$$\therefore n_s = \frac{n_p}{10} = \frac{4000}{10}$$

$n_s = 400$

81. (1) As, here source match to the load then consider, there is no loss of energy.

$\therefore$ Output power = Input power

$\Rightarrow I_s^2 R_s = I_p^2 R_p$

$$\frac{I_s}{I_p} = \sqrt{\frac{R_p}{R_s}} = \sqrt{\frac{9000}{10}} = 30$$

$$\text{Also,} \frac{I_s}{I_p} = \frac{n_p}{n_s} = 30 \Rightarrow \frac{n_s}{n_p} = \frac{1}{30}$$

82. (1) As we know that, $\eta = \frac{\text{Output power}}{\text{Input power}}$

$$\eta = \frac{E_s I_s}{I_p E_p} \Rightarrow \frac{80}{100} = \frac{200 \times I_s}{4 \times 10^3}$$

$$I_s = \frac{80}{100} \times \frac{4 \times 10^3}{200} = 16A$$

Also $E_p I_p = 4kW$

$$I_p = \frac{4 \times 10^3}{100} = 40$$

$\therefore I_p = 40A, I_s = 16A$

83. (3) $\eta = \frac{\text{output power}}{\text{Input power}}$

$$\frac{88}{100} = \frac{P_0}{P_i} = \frac{880}{P_i}$$

$\Rightarrow P_i = 1000W \Rightarrow V_p I_p = 1000 \Rightarrow 2200 \times I_p = 1000$

$$\Rightarrow I_P = \frac{10}{22} = 0.4545A$$

$\therefore I_p = 0.4545\ A$

84. (2) In a step down transformer, voltage is reduced and current is increased

85. (3) As we know, $\eta = \frac{\text{Output power}}{\text{Input power}}$

$\because$ Losses are neglected in the case of 100% efficiency

$\therefore \eta = 1$

then, $P_o = P_i$

86. (4) Input power = 4 kW, Efficiency is 90%

$\therefore$ output power = 0.9×4 kW = 3.6 kW

3.6 kW = $I^2R = 36 \times R$

$$R = \frac{3.6kW}{36} = \frac{3.6 \times 10^3}{36} = 10^2\Omega = 100\Omega$$

87. (1) Secondary voltage will be zero because transformer works only on a.c. and the battery give dc.

High Order Time Intensive MCQs

1. (3) $I_0 = 150$ A and $V_0 = 150$ V and $\phi = 60^0$

$$P = \frac{1}{2}V_0 I_0 \cos\phi = 0.5\times150\times150\times\cos 60° = 5625 \text{ W}.$$

2. (3) $\tan\theta = \dfrac{X_L - X_C}{R}$

where, $X_L - X_C$ is the net reactance of the circuit.

$$\Rightarrow \quad \theta = \tan^{-1}\frac{X_L - X_C}{R}$$

$$\Rightarrow \quad \theta = \tan^{-1}\left(\frac{\sqrt{3}R}{R}\right) \Rightarrow \theta = \tan^{-1}(\sqrt{3}) = 60° = \frac{\pi}{3}$$

3. (3) Phase difference is given by

$$\tan\phi = \frac{X_L - X_C}{R} \Rightarrow \tan\frac{\pi}{3} = \frac{X_L - X_C}{R}$$

when L is removed, $\tan\dfrac{\pi}{3} = \dfrac{X_C}{R} = \sqrt{3}$

$$\therefore \quad X_C = \sqrt{3}R \quad \ldots(i)$$

Similarly, when C is removed

$$\tan\frac{\pi}{3} = \frac{X_L}{R} = \sqrt{3} \Rightarrow X_L = \sqrt{3}R \quad \ldots(ii)$$

$$\Rightarrow \quad X_C = X_L \quad \ldots(iii)$$

Now, $\theta = 0 \Rightarrow \theta = 0°$

$\therefore$ Power factor, $\cos\theta = \cos 0° = 1$

4. (3) Power factor of AC circuit is given by

$$\cos\phi = \frac{R}{Z} \quad \ldots(i)$$

where, R is resistance employed and Z the impedance of the circuit.

$$\cos\phi = \frac{R}{\sqrt{R^2 + (X_L - X_C)^2}} \quad (\because Z = \sqrt{R^2 + (X_L - X_C)^2})$$

Given, $R = 8\ \Omega$, $X_L = 31\Omega$ and $X_C = 25\ \Omega$

$$\cos\phi = \frac{8}{\sqrt{(8)^2 + (31-25)^2}} = \frac{8}{\sqrt{64+36}} = \frac{8}{10} = 0.80$$

5. (1) $\tan\phi = \dfrac{X_L}{R} = \dfrac{L\omega}{R}$

$R = X_L = 3\Omega \qquad (X_L = \omega L)$

$$\Rightarrow \tan\phi = \frac{3\Omega}{3\Omega} \Rightarrow \tan\phi = 1$$

$$\therefore \phi = \tan^{-1}(1) \Rightarrow \phi = 45° \Rightarrow \phi = \frac{\pi}{4} \text{ rad}$$

6. (1) Let the applied voltage be V volt.

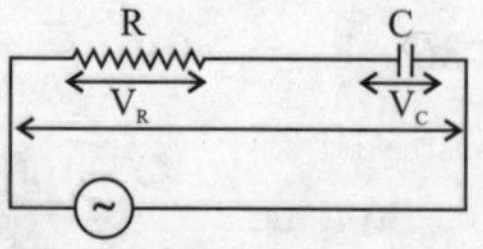

$$V = \sqrt{V_R^2 + V_C^2} = \sqrt{(12)^2 + (5)^2}$$
$$= \sqrt{144+25} = \sqrt{169} = 13\text{V}$$

7. (4) Reactance of capacitance, $X_C = \dfrac{1}{\omega C}$

$$\therefore \quad X = \frac{1}{\omega C}$$

In the given case, $X_C = \dfrac{1}{(2\omega)(2C)} = \dfrac{1}{4}\left(\dfrac{1}{\omega C}\right) = \dfrac{X}{4}$

8. (2) $P = Vi\cos\phi = V\left(\dfrac{V}{Z}\right)\left(\dfrac{R}{Z}\right)$ $(\because V = iZ \text{ and } \cos\phi = R/Z)$

$$= \frac{V^2 R}{Z^2} = \frac{V^2 R}{(R^2 + \omega^2 L^2)}$$

9. (1) Let the applied voltage be V volt.

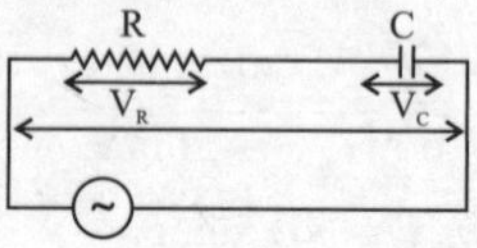

$$V = \sqrt{V_R^2 + V_C^2} = \sqrt{(12)^2 + (5)^2}$$
$$= \sqrt{144+25} = \sqrt{169} = 13\text{V}$$

10. (1) $\tan\phi = \dfrac{\dfrac{1}{\omega C} - \omega L}{R}$

The current leads the voltage by 45^0.

$\Rightarrow \phi = 45°$

$$\therefore \tan 45° = \frac{\frac{1}{\omega C} - \omega L}{R} \Rightarrow \therefore C = \frac{1}{\omega(\omega L + R)} = \frac{1}{2\pi f(2\pi f L + R)}$$

11. (1) $<i> = \dfrac{\int_{T/8}^{T/4} i\,dt}{\int_{T/8}^{T/4} dt} = \dfrac{\int_{T/8}^{T/4} 5\sin\omega t\, dt}{\int_{T/8}^{T/4} dt}$

$$= \frac{\frac{5}{\omega}[-\cos\omega t]_{T/8}^{T/4}}{\left[\frac{T}{4} - \frac{T}{8}\right]}$$

$$= \frac{5}{2\pi}\times T\times\frac{8}{T}\left[-\cos\left(\frac{2\pi}{T}\times\frac{T}{4}\right) + \cos\left(\frac{2\pi}{T}\times\frac{T}{8}\right)\right]$$

$$= \frac{5}{2\pi}\times 8\left[-\cos\frac{\pi}{2} + \cos\left(\frac{\pi}{4}\right)\right]$$

$$= \frac{5}{2\pi}\times 8\left[\frac{1}{\sqrt{2}}\right] = \frac{5\times8}{2\sqrt{2}\pi}$$

$$= \frac{10}{\pi}\sqrt{2}\text{A}$$

12. (1) $<i> = \dfrac{\int_o^{T/4} i\,dt}{\int_o^{T/4} dt}$

13. (1) $i = \dfrac{(V_0)_R}{R}$, $V_L = (V_0)_L \sin\left(\omega t + \dfrac{\pi}{2}\right)$ *and* $(V_0)_L = X_L i$

14. (1) $I_{rms} = \dfrac{E_{rms}}{R} = 2A; V_{rms} = I_{rms}(X_L - X_C) = 0$

$\therefore$ Circuit is at resonance

15. (2) $I = \dfrac{E}{Z} = \dfrac{E}{\sqrt{R^2 + \left[2\pi f L - \dfrac{1}{2\pi f C}\right]^2}}$

In both cases, the impedance is infinite. Hence, current flowing in the circuit is zero.

16. (1) $I_{r.m.s.} = \dfrac{V_{r.m.s.}}{Z} = \dfrac{V_{r.m.s.}}{R} = \dfrac{100}{50} = 2amp$

$V = \sqrt{V_R^2 + (V_L - V_C)^2} = V_R [\because V_L = V_C]$

17. (3) $n = \dfrac{1}{2\pi\sqrt{LC}}; \dfrac{n_1}{n_2} = \sqrt{\dfrac{C_2}{C_1}} = \sqrt{\dfrac{kC}{C}};$

$\dfrac{n_1}{n_2} = \sqrt{k}$

18. (2) Reading of voltmeter in rms value

$E = \dfrac{14.14}{\sqrt{2}} = 10V$

19. (3) $I = \dfrac{E}{Z}$, $Z = 220\Omega$

$\therefore I = \dfrac{220}{Z} = 1A$

$Z^2 = R^2 + X_L^2$

$\therefore X_L = \sqrt{Z^2 - R^2} = 214.24$

$V_L = I X_L$

$= 214.24$ V

20. (2) $V = 120 \sin 100\pi t \cos 100\pi t$

$= 60 \times 2 \sin 100\pi t \cdot \cos 100\pi t$

$= 60 \sin 200\pi t$

$\therefore \omega = 200\pi$

$f = \dfrac{\omega}{2\pi} = \dfrac{200\pi}{2\pi} = 100$

21. (1) Energy $E = 1/2LI^2$

Change of current from minimum to maximum. $= 2 - 0 = 2A$.

Energy spent

$= \dfrac{1}{2}LI^2 = \dfrac{1}{2} \times 5 \times 4$

$= 10\,J$

22. (3) $V = V_0$ for $0 \le t \le \dfrac{T}{2}$

$V = 0$ for $\dfrac{T}{2} \le t \le T$

$$V_{rms} = \left[\frac{\int_0^T V^2 dt}{\int_0^T dt}\right]^{1/2} = \left[\frac{\int_0^{T/2} V_0^2 dt + \int_{T/2}^{T} (0) dt}{\int_0^T dt}\right]^{1/2}$$

$$= \left[\frac{V_0^2}{T}(t)_0^{T/2}\right]^{1/2} = \left[\frac{V_0^2}{T}\left(\frac{T}{2}\right)\right]^{1/2} = \frac{V_0}{\sqrt{2}}$$

23. (1) Efficiency, $\eta = \dfrac{P_{out}}{P_{in}}$

$\Rightarrow \quad \eta = \dfrac{V_s I_s}{V_p I_p} \Rightarrow \dfrac{80}{100} = \dfrac{200 \times I_s}{2000} \Rightarrow I_s = 8\,A$

Also, $V_p I_p = 2000 \Rightarrow I_p = \dfrac{2000}{100} = 20\,A$

24. (3) If I_p and I_s be the currents in the primary and secondary coils at any instant and the energy losses be zero, then

Power in secondary coil = Power in primary coil

$V_s \times I_s = V_p \times I_p \qquad (I_p = 6A)$

$\Rightarrow \quad \dfrac{I_p}{I_s} = \dfrac{V_s}{V_p} = \dfrac{N_s}{N_p}$ = transformer ratio

Given, $\dfrac{N_p}{N_s} = \dfrac{2}{3}, \dfrac{N_p}{N_s} = \dfrac{2}{3}, I_p = 6A \Rightarrow I_s = \dfrac{N_p}{N_s} I_p = \dfrac{2}{3} \times 6 = 4A$

25. (4) Since power factor has to be made 1.

$\therefore$ Effective capacitance should be increased thus connecting in parallel.

$\because \cos\phi = 1$

$\therefore \phi = 0$

$I\omega L = \dfrac{I}{\omega(C + C')}$ Resonance condition

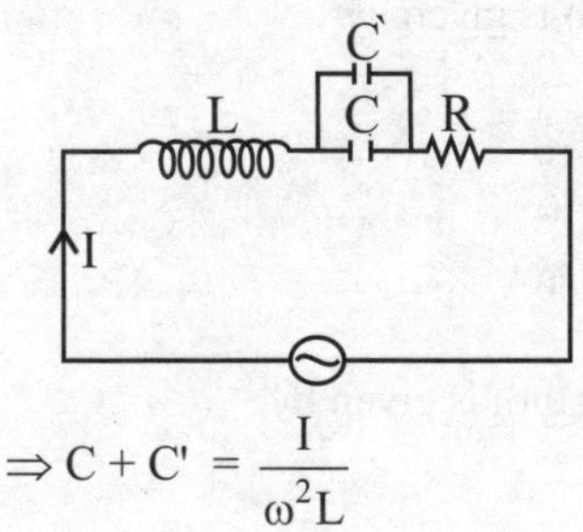

$\Rightarrow C + C' = \dfrac{I}{\omega^2 L}$

$\therefore C' = \dfrac{I}{\omega^2 L} - C$

$\therefore C' = \dfrac{1 - \omega^2 LC}{\omega^2 L}$ in parallel.

26. (3) $\%\eta = \dfrac{P_{out}}{P_{in}} \times 100 = \dfrac{V_s I_s}{V_p I_p} \times 100$

$50 = \dfrac{2200 \times I_s}{220 \times 5} \times 100 \Rightarrow I_s = 0.25A$

27. (2) Voltage across primary coil

$V_p = \dfrac{d\phi}{dt} = \dfrac{d}{dt}(\phi_0 + 4t) = 4V$ (as ϕ_0 = constant)

Also, we have

$N_p = 50$ and $N_s = 1500$

From relation,

$$\frac{V_S}{V_p} = \frac{N_s}{N_p} \Rightarrow V_s = V_p \frac{N_s}{N_p} = 4\left(\frac{1500}{50}\right) = 120\,V$$

28. (4) As the A.C source / a.v source complete its cycle in T time as (+) half cycle and (–ve) half

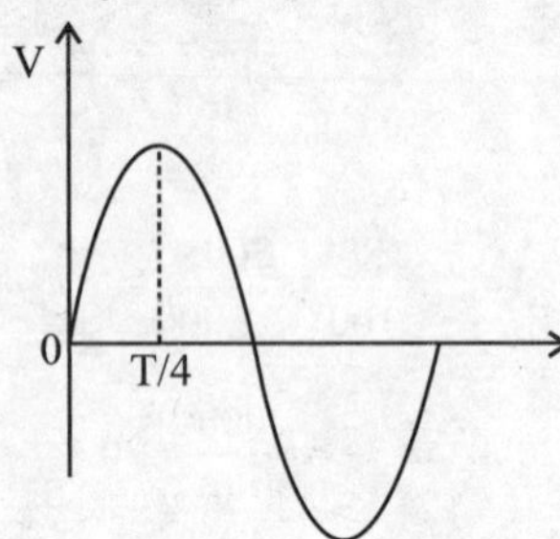

$$\therefore \text{ Required time } t = \frac{T}{4} = \frac{1}{4\times 50} = \frac{1}{200}$$

$t = 5 \times 10^{-3}$ second

29. (3) $E = 141 \sin(628\,t)$

$$E_{rms} = \frac{E_0}{\sqrt{2}} = \frac{141}{1.41} = 100V$$

and $2\pi f = 628$

$$f = \frac{628}{6.28} = 100\,Hz$$

30. (2) We have,

$$\frac{N_s}{N_p} = \frac{I_p}{I_s}, \frac{10}{500} = \frac{I_p}{I_s} \Rightarrow \frac{I_p}{I_s} = \frac{1}{50}$$

$$\Rightarrow I_s = 50 I_p$$

This condition is satisfied only with option (2).

31. (1) Transformation ratio is given by

$$\frac{V_s}{V_p} = \frac{N_s}{N_p} \Rightarrow V_s = \frac{V_p \times N_s}{N_p}$$

$$V_S = \frac{220 \times 40000}{200} = 44000\text{ V}$$

Potential difference per turn is given by

$$\frac{V_s}{N_s} = \frac{44000}{40000} = 1.1V$$

32. (1) Short trick:

$$V_{rms} = \sqrt{\frac{(V_1)^2 + (V_2)^3 + (V_3)^2}{\text{Number of loops}}}$$

$$V_{rms} = \sqrt{\frac{(10)^2 + (10)^2 + (10)^2}{3}} = 10V$$

Statement Based MCQs

1. (1) An inductor is a low pass filter for D.C. and blocks A.C. currents. As it has zero resistance for D.C. currents therefore inductor acts as perfect conductor for D.C.

2. (3) The capacitive reactance limits the amplitude of the current in a purely capacitive circuit in the same way as the resistnace limits the current in a purely resistive circuit. i.e., $I_0 = \varepsilon_0/X_C$

3. (1) 120 V rms turn out to be about 170 V peak to ground. Hence, alternating voltage is more than it appears.

4. (3) The inductive reactance limits the amplitude of current in a purely inductive circuit in the same way as the resistance limits the current in a purely resistive circuit. i.e., $I_0 = \varepsilon_0/X_L$

5. (1) Both of the given statements are correct.

Assertion & Reason MCQs

1. (1) Direct current doesn't have variable magnetic field induced through the primary winding of transformer. Thus, only constant magnetic field will act on secondary and shall not induce any voltage across the terminals of the secondary. Reason is also true.

2. (3) KVL rule can also be applied in A.C. circuits as well as in D.C. Since in an L.C. circuit

$$V^2 = V_C^2 + V_R^2$$

$$V_C^2 = V^2 - V_R^2$$

$$= 10^2 - 8^2 = 10^2 - 64 = 100 - 64$$

$$V_C = \sqrt{36} = 6V$$

Thus, the reason is false

3. (3) Like direct current, an alternating current also produces magnetic field. But the magnitude and direction of the field goes on changing continuously with time.

4. (2) Both assertion and reason are correct but reason is not the correct explanation of the assertion.

Match the Columns MCQs

1. (1) (a) For the circuit containing C only, $\phi = 90°$

$$\text{Current, } I = \frac{V\angle 0^0}{X_C} = \frac{V\angle 0^0}{\left(\frac{1}{j\omega c}\right)} = V\omega C\angle 90^0$$

(b) For the circuit containing L only, $\phi = 90°$

$$\text{Current, } I = \frac{V\angle 0^0}{jX_L} = \frac{V}{\omega L}\angle -90^0$$

(c) For the circuit containing R only.

$$\text{Current, } I = \frac{V\angle 0^0}{R} = \frac{V}{R}\angle 0^0$$

(d) For the circuit containing R and C only.

$$I = \frac{V\angle 0^0}{Z\angle\phi} = \frac{V}{\sqrt{R^2 + X_C^2}}\angle \tan^{-1}\left(\frac{X_C}{R}\right)$$

2. (1) (a) $\phi = 0°$ between voltage function and current function.

(b) $I = I_0 \sin(\omega t - 90°)$

i.e. $\phi = 90°$ and voltage function leads the current function.

(c) Current function leads the voltage function.
So, $X_C > X_L$

(d) Voltage function leads the current function.
So, $X_L > X_C$

3. (3) • $X_L = \omega L$

$$X_L = 2\pi \times 50 \times \frac{100}{\pi} \times 10^{-3} = 10\Omega$$

$$X_C = \frac{1}{\omega C}$$

$$X_C = \frac{1}{2\pi \times 50 \times \frac{100}{\pi} \times 10^{-6}} \quad 100\ \Omega$$

- Impedance; $Z = \sqrt{R^2 + (X_L - X_C)^2}$

$$Z = \sqrt{100^2 + (10-100)^2} = 134.5\ \Omega$$

$$i_{rms} = \frac{V}{Z} = \frac{220}{134.5} = 1.63\ A$$

- RMS value of voltage drop across capacitor;
$V_C = i_{rms}.X_C = 1.63 \times 100 = 163$ V
- Average power dissipated in the resistor is;
$P_{avg} = i^2_{rms} R = (1.63)^2 \times 100 = 265.7$ W
- Average power dissipated in inductor and capacitor would be zero.

NCERT Exemplar MCQs

1. (2) Given, $v = 50$ Hz, $I_{rms} = 5$A

$$t = \frac{1}{300} s$$

We have to find I (t)

I_0 = Peak value = $\sqrt{2} I_{rms} = \sqrt{2} \times 5$

$= 5\sqrt{2} A$

$$I = I_0 \sin \omega t = 5\sqrt{2} \sin 2\pi vt = 5\sqrt{2} \sin 2\pi \times 50 \times \frac{1}{300}$$

$$= 5\sqrt{2} \sin \frac{\pi}{3} = 5\sqrt{2} \times \frac{\sqrt{3}}{2} = 5\sqrt{3/2} A$$

2. (3) For delivering maximum power from the generator to the load, total internal reactance must be equal to conjugate of total external reactance.

Hence, $X_{int} = X_{ext}$

$\Rightarrow X_g = -X_L$

$\Rightarrow X_L = -X_g$

3. (3) The voltmeter connected to AC mains reads mean value ($< v^2 >$) and is calibrated in such a way that it gives value of $\sqrt{< v^2 >}$.

4. (2) We know that resonant frequency in an L-C-R circuit is given by $v_0 = \dfrac{1}{2\pi\sqrt{LC}}$

Now to reduce v_0 either we can increase L or we can increase C.

To increase capacitance, we must connect another capacitor parallel to the first.

5. (3) Quality factor (Q) of an L-C-R circuit is given by,

$$Q = \frac{1}{R}\sqrt{\frac{L}{C}}$$

Where R is resistance, L is inductance and C is capacitance of the circuit. To make Q high, R should be low, L should be high and C should be low.

These conditions are best satisfied by the values given in option (3).

Note: We should be careful while writing formula for quality factor, because we are considering series L-C-R circuit.

6. (3) Given, $X_L = 1\ \Omega$, $R = 2\ \Omega$

$E_{rms} = 6$ V, $P_{av} = ?$

Average power dissipated in the circuit

$$P_{av} = E_{rms} I_{rms} \cos\phi$$

$$I_{rms} = \frac{I_0}{\sqrt{2}} = \frac{E_{rms}}{Z}$$

$$Z = \sqrt{R^2 + X_L^2} = \sqrt{4+1} = \sqrt{5}$$

$$I_{rms} = \frac{6}{\sqrt{5}} A \Rightarrow \cos\phi = \frac{R}{Z} = \frac{2}{\sqrt{5}}$$

$$P_{av} = 6 \times \frac{6}{\sqrt{5}} \times \frac{2}{\sqrt{5}} \quad \text{[from equation (}i\text{)]}$$

$$= \frac{72}{\sqrt{5}\sqrt{5}} = \frac{72}{5} = 14.4 \text{ W}$$

7. (1) Secondary voltage $V_s = 24$ W,

Power associated with secondary, $P_s = 12$ W

$$I_S = \frac{P_S}{V_S} = \frac{12}{24} = \frac{1}{2} A = 0.5A$$

Peak value of the current in the secondary coil

$$I_0 = I_S\sqrt{2}$$

$$= (0.5)(1.414) = 0.707 = \frac{1}{\sqrt{2}} A$$

NEET & AIIMS Past Year MCQs

1. (4) The voltage V_L leads the current by $\frac{\pi}{2}$, so:

$$V_L = V_0 \sin\left(\omega t + \frac{\pi}{2}\right)$$

this simplifies to:

$V_L = V_0 \cos(\omega t)$

The peak voltage V_0 is given by:

$V_0 = I_0 X_L$

where $X_L = \omega L$

$\therefore V_0 = I_0 \omega L$

Substitute V_0 back into the expression for V_L:

$V_L = I_0 \omega L \cos(\omega t)$

2. (2) Given:

$Q = Q_0 e^{-Rt/2L}$

At $Q = 0.5\,Q_0$:

$0.5 = e^{-Rt/2L} \Rightarrow t = \dfrac{2L\ln 2}{R}$

Substitute $R = 1.5\ \Omega$, $L = 12$ mH,

$\ln 2 = 0.693$:

$$t = \frac{2\times 12\times 10^{-3}\times 0.693}{1.5} = 11.09\,ms$$

3. (1) Given:

Secondary voltage $V_s = 11000\ V$

Power $P = 88\ W$

For a transformer, power in primary = power in secondary (ignoring losses):

$V_p I_p = V_s I_s$

Current in secondary:

$$I_s = \frac{P}{V_s} = \frac{88}{11000} = 8\times 10^{-3}\ A = 8\,mA$$

4. (4) $V_{rms} = 210$ volt,

$V_0 = 210\sqrt{2}$ volt

$V_0 = I_0 X_C$

$V_0 = I_0 \times \dfrac{1}{\omega C}$

$$210\sqrt{2} = I_0 \times \frac{1}{2\pi\times 50\times 10\times 10^{-6}}$$

$I_0 = 210\sqrt{2}\times 2\pi\times 500\times 10^{-6}$

$I_0 = 0.932$A

5. (4) $\dfrac{N_P}{N_S} = \dfrac{1}{2}$ (given)

We know that transformation ratio (r) is $r = \dfrac{N_S}{N_P} = \dfrac{V_S}{V_P}$

$\therefore \dfrac{V_S}{V_P} = \dfrac{2}{1}$

$V_S : V_P = 2 : 1$

6. (1) For resonance frequency

$\omega L = \dfrac{1}{\omega C}$

$$\Rightarrow \omega = \frac{1}{\sqrt{LC}} = \frac{1}{\sqrt{10\times 10^{-3}\times 1\times 10^{-6}}} = \frac{1}{\sqrt{10^{-8}}} = 10^4\ \text{rad/sec}$$

$$f = \left(\frac{\omega}{2\pi}\right) = \frac{10^4}{2\pi} = 1.59\ \text{kHz}$$

7. (2) For transformer

$$\left(\frac{i_p}{i_s}\right) = \left(\frac{V_s}{V_P}\right) = \left(\frac{N_s}{N_P}\right)$$

For bulb, Vs = 12 V

$$i_s = \left(\frac{60}{12}\right) = 5\ A \Rightarrow \frac{i_p}{5} = \left(\frac{12}{220}\right)$$

$$i_P = \frac{60}{220} = \left(\frac{3}{11}\right) = 0.27\ A$$

8. (4) $\omega = 2\pi f \Rightarrow \omega = 100\pi$

$$Z = \sqrt{R^2 + (X_L - X_C)^2} = \sqrt{10^2 + \left(\omega L - \frac{1}{\omega C}\right)^2}$$

$$= \sqrt{100 + \left(100\pi\times\frac{50}{\pi}\times 10^{-3} - \frac{1}{100\pi\times\frac{10^3}{\pi}\times 10^{-6}}\right)^2}$$

$$= \sqrt{100 + (5-10)^2} = \sqrt{100+25}$$

$Z = 5\sqrt{5}\,\Omega$

9. (4) Peak or maximum is $\sqrt{2}$ times the rms voltages in ac.

10. (3) Given, $\omega = 100$

$\because \omega = 2\pi\nu \Rightarrow \nu = \dfrac{\omega}{2\pi} = \dfrac{50}{\pi}$ Hz

Resonance frequency

$$\nu_0 = \frac{1}{2\pi\sqrt{LC}} = \frac{1}{2\pi}\sqrt{\frac{1}{10\times 10\times 10^{-6}}} = \frac{50}{\pi}\ \text{Hz}$$

11. (4) $q = CV$

$\dfrac{dq}{dt} = \dfrac{CdV}{dt}$

$I_1 = C(V_0\omega\cos\omega t) = V_0\omega C\cos\omega t$

12. (3) $I_0 = 10\sqrt{2}$A

$$I_{rms} = \frac{I_0}{\sqrt{2}} = 10\text{A} \Rightarrow V_{rms} = \sqrt{V_R^2 + (V_L - V_C)^2}$$

$$= \sqrt{(40)^2 + (40-10)^2} = 50\text{V}$$

$$Z = \frac{V_{rms}}{I_{rms}} = \frac{50\text{V}}{10\text{V}} = 5\Omega$$

13. (4) $\eta = 100\%$

$P_{in} = P_{out}$

$V_P I_P = V_S I_S$

$220 \times I_P = 44$

$$I_P = \frac{44}{220} = \frac{1}{5}\text{A} = 0.2\text{A}$$

14. (2) $Q = \dfrac{\omega}{\Delta\omega} = \dfrac{\omega L}{R} \Rightarrow \Delta\omega = R/L = \dfrac{40}{5} = 8\text{rad/sec}$

$$\omega_0 = \frac{1}{\sqrt{LC}} = \frac{1}{\sqrt{5\times 80\times 10^{-6}}} = 50\,\text{rad/sec}.$$

$$\omega_1 = \omega_0 - \frac{\Delta\omega}{2} = 46\text{rad/sec}$$

$$\omega_2 = \omega_0 + \frac{\Delta\omega}{2} = 54\,\text{rad/sec}$$

15. (2)

16. (1)

Electromagnetic Waves

NCERT Topic-wise MCQs

1. (1) $I_D = 2A$, $C = 1\mu F = 10^{-6}$ F

As we know that,

$$I_D = C\frac{dV}{dt} \Rightarrow \frac{I_D}{C} = \frac{dV}{dt}$$

$$\frac{2}{10^{-6}} = \frac{dV}{dt} \Rightarrow \frac{dV}{dt} = 2\times10^6\,Vs^{-1}$$

2. (2) $\because \lambda = 40m = 40 \times 10^9$ nm

$$\therefore E(eV) = \frac{1240}{\lambda(nm)} = \frac{1240}{40\times10^2} = 3.1\times10^{-8}\text{ eV}$$

3. (3) According to the relation

$$E = \frac{hc}{\lambda} = \frac{6.6\times10^{-34}\times3\times10^8}{3000\times10^{-10}}$$

$E = 6.6 \times 10^{-19}$ J

$$E = \frac{6.6\times10^{-19}}{60\times60} = 1.83\times10^{-22}\text{ Wh}$$

4. (4) $B_0 = 5 \times 10^{-9}$ T

q = Charge on α-particle = 2 × e

$= 3.2 \times 10^{-19}$ C

Maximum electric field $E_0 = cB_0$

$E_0 = (3 \times 10^8) \times (5 \times 10^{-9})$

$= 1.5\ Vm^{-1}$

Maximum force on α-particle due to electric field

$F_E = q \times E_0$

$= 3.2 \times 10^{-19} \times 1.5$

$F_E = 4.80 \times 10^{-19}$ N

Maximum force on α-particle due to magnetic field

$F_B = qvB$

$= 3.2 \times 10^{-19} \times 3 \times 10^8 \times 5 \times 10^{-9}$

$F_B = 4.80 \times 10^{-19}$ N

5. (1) Pressure exerted by reflected wave on the surface

$$P = \frac{2I}{c} = \frac{2\times1}{3\times10^8} = 0.667\times10^{-8}\,N/m^2$$

$= 6.67 \times 10^{-9}\ N/m^2$

6. (4) Pressure exerted by absorbed wave on the surface

$$P = \frac{I}{c}$$

$\because F = PA$

$$\therefore F = \frac{IA}{c} = \frac{6\times12}{3\times10^8} = 24\times10^{-8}\,N$$

7. (1) According to the relation, $I_D = \varepsilon_0\frac{EA}{t} = \varepsilon_0\frac{\left(\frac{V}{d}\right)A}{t}$

$$I_D = \frac{\varepsilon_0 VA}{dt} = \frac{8.85\times10^{-12}\times400\times(60\times10^{-4})}{10^{-6}\times2\times10^{-3}}$$

$I_D = 1.062 \times 10^{-2}$ A

8. (2) As we know

$$B = \frac{E}{c} = \frac{6.3}{3\times10^8} = 2.1\times10^{-8}\,T$$

As $\vec{E}$ is along the y-direction and wave is travelling along x-direction, therefore $\vec{B}$ is along z-direction because in propagation of electromagnetic wave E & B fields are ⊥ to each other.

$$\vec{B} = 2.1\times10^{-8}\hat{k}\,T$$

9. (4) Flow of energy per unit area per unit time in electromagnetic wave is called poynting vector

$$\vec{S} = (\vec{E} \times \vec{B})/\mu_0$$

10. (3) $E = 11$ keV $= 11 \times 10^3 \times 1.6 \times 10^{-19}$ J

$$11\times10^{-16}\times1.6 = \frac{hc}{\lambda}$$

$$\lambda = \frac{6.6\times10^{-34}\times3\times10^8}{1.6\times11\times10^{-16}} = 1.125\times10^{-10}\,m$$

$\lambda = 1.125$ Å

This wavelength lies in the region of X-ray

11. (3) Maxell added displacement current to Ampere's law, resulting in the Ampere-Maxwell law, ensuring magnetic fields can be generated even without conduction current (e.g., in a capacitor gap).

12. (3) Displacement current arises due to time-varying electric field, such as when a capacitor is charging or discharging.

13. (1)

14. (3) If $\vec{E}$ components are not perpendicular to $\vec{B}$ still electromagnetic waves can pass through this region.

15. (1)

16. (1) Electromagnetic waves transfer energy & momentum on hitting the object or material.

17. (2) If magnetic monopole exists, then the magnetic flux over a closed path will not be equal to zero.

18. (1) Frequency range for television broadcasting is 30-300 MHz.

19. (1) Displacement current,

$$I_d = \varepsilon_0\frac{d\phi_E}{dt}$$

$$\therefore \frac{d\phi}{dt} = \frac{I_d}{\varepsilon_0} = \frac{2\times10^{-7}}{8.85\times10^{-12}} = 2.26\times10^4\text{ Wb/s.}$$

20. (3) Amplitude of oscillating magnetic field

$$B_0 = \frac{E_0}{c} = \frac{480}{3\times10^8} = 1.6\times10^{-6}\,Wb/m^2$$

21. (3) Refractive index of medium is:

$$\mu = \frac{c}{v} \quad \text{where, } c = \frac{1}{\sqrt{\mu_0 K_0}}$$

and $v = \frac{1}{\sqrt{\mu_0 K_0 \mu_r K_r}}$

$\therefore \mu = \frac{1/\sqrt{\mu_0 K_0}}{1/\sqrt{\mu_0 K_0 \mu_r K_r}} = \sqrt{\mu_r K_r}$

Given $\mu_r = \mu_0$ and $K_r = K_0$ then

$\mu = \sqrt{\mu_0 K_0}$

22. (2) $\mu = \frac{c}{v} = \frac{1/\sqrt{\mu_0 \varepsilon_0}}{1/\sqrt{\mu\varepsilon}} = \sqrt{\frac{\mu\varepsilon}{\mu_o \varepsilon_o}}$

23. (3) According to Maxwell's hypothesis, a changing electric field gives rise to magnetic field.

24. (1) Electric energy stored in capacitor

$U = \frac{1}{2}\frac{q^2}{C} = \frac{1}{2}\frac{q^2}{\varepsilon_0 A/d} = \frac{q^2 d}{2\varepsilon_0 A}$

$\therefore$ Mean electric energy density between the plates of charged capacitor

$\frac{U}{Ad} = \frac{1}{2}\frac{q^2}{\varepsilon_0 A^2}$

25. (3) $n = \frac{x}{\lambda} = \frac{10^{-3}}{4\times10^{-7}} = 0.25\times10^4$

26. (2) Intensity $= \frac{P}{A} = \frac{3.9\times10^{26}}{4\pi r^2}$

$I = \frac{3.9\times10^{26}}{4\times\frac{22}{7}\times(6.96\times10^8)^2} = \frac{3.9\times10^{26}\times7}{4\times22\times48.44\times10^{16}}$

$= \frac{27.3}{4262.72}\times10_{10}$

$I = 6.4 \times 10^7 \ W/m^2$

27. (1) The speed of electromagnetic wave in a medium is-

$v = \frac{1}{\sqrt{\mu\varepsilon}} = \frac{1}{\sqrt{\mu_0\varepsilon_0\mu_r\varepsilon_r}} = \frac{c}{\sqrt{\mu_r\varepsilon_r}}$

$v = \frac{3\times10^8}{\sqrt{4\times2.25}} = 1\times10^8 \ m/s$

28. (4) $B_0 = \frac{E_0}{c} = \frac{60}{3\times10^8} = 2\times10^{-7} T$

29. (1) Cosmic rays are not electromagnetic waves because electromagnetic waves are produced by accelerating charge particle.

30. (3) The stationary or standing wave is a superposition of two waves of equal magnitude / amplitude and frequency moving in opposite direction to each other. In the case of stationary waves, there is no transfer of energy.

31. (1) SONAR emitts ultrasonic waves. Ultrasonic waves have frequency range above audible range, we can't hear them.

32. (1) Wavelength of microwave is more than that of infrared.

33. (2) Infrared rays or heat rays used for heating thermopiles to produce electricity.

34. (3) They are all electromagnetic waves so they ravel with same speed in vacuum.

35. (1) The frequency order of electromagnetic rays are γ-rays > X-rays > UV rays

A > B > C

36. (4) $\lambda = 3$ mm $= 3 \times 10^{-3}$ m ; $E_0 = 66 \ V \ m^{-1}$

$B_0 = \frac{E_0}{c} = \frac{66}{3\times10^8} = 2.2\times10^{-7} T$

As electromagnetic wave is propagating along x-axis and electric field variation is along y-direction so the magnetic field variation is along z-direction.

Using the relation for harmonic wave

$E_y = E_0 \cos\frac{2\pi}{\lambda}(ct - x)$

$E_y = E_0 \cos\frac{2\pi c}{\lambda}(t - x/c)$

$\therefore E_y = 66\cos\frac{2\pi\times3\times10^8}{3\times10^{-3}}(t - x/c)$

$E_y = 66 \cos 2\pi \times 10^{11}(t - x/c)$

and $B_z = B_0 \cos\frac{2\pi c}{\lambda}(t - x/c)$

$= 2.2 \times 10^{-7} \cos 2\pi \times 10^{11} (t - x/c)$

Thus option (4) is correct.

37. (1) Direction of magnetic field vector will be in the direction of $\hat{K}\times\hat{E}$ magnitude of magnetic field $B = \frac{E}{C} = \frac{K}{\omega}E$

$\Rightarrow \vec{B} = \frac{1}{\omega}(\vec{K}\times\vec{E})$

38. (3) On comparing with equation, $E = E_0 \sin(kx - \omega t)$, we get $\omega = 10 \times 10^{10}$ rad/s $k = 500$

$\therefore$ Velocity, $v = \frac{\text{coefficient of time}}{\text{coefficiont of } x} = \frac{\omega}{k}$

$\Rightarrow v = \frac{10\times10^{10}}{500} = 2\times10^8 = \frac{2\times c}{3}$

39. (3) EM wave is in direction $\rightarrow \frac{\hat{i}+\hat{j}}{\sqrt{2}}$

Electric field is in direction $\rightarrow \hat{k}$

$\vec{E}\times\vec{B} \rightarrow$ direction of propagation of EM wave

40. (2) IR radiation have heating effects because it is produced by heating bodies or molecules.

41. (4) 6×10^{-7} m lies in visible spectrum which ranges from 4×10^{-7} to 8×10^{-7} m.

42. (3) α-rays is nothing but nucleus of helium moving at higher speed.

43. (3) Heat radiation is the another name of infrared radiation and it is a type of electromagnetic wave which travels with the speed of light 3×10^8 m/s.

44. (1) In electromagnetic wave, the electric field $(\vec{E})$ and magnetic field $(\vec{B})$ are perpendicular to each other and also perpendicular to the direction of propagation of wave. In the given question, the electromagnetic wave is propagating along +z axis. $\vec{E} \times \vec{B}$ gives the direction of propagation of wave, so

$\vec{E} = E_0 \hat{i}$ and $\vec{B} = B_0 \hat{j}$

45. (4) $\vec{E} = \hat{i}\, 40 \cos\omega\left(t - \frac{z}{c}\right)$

$\vec{v}$ is along +z direction

$\vec{E}$ is along +x direction

$\Rightarrow |\vec{B}| = \frac{E}{c}$ and its direction will be along +y.

$\left[\text{As } \vec{C} \parallel (\vec{E} \times \vec{B})\right]$

$\therefore \vec{B} = \hat{j}\frac{40}{c}\cos\omega\left(t - \frac{z}{c}\right)$

46. (3) $\lambda_m > \lambda_v > \lambda_x$

47. (4) Refractive index, $n = \frac{c}{v}$

or $\quad v = \frac{c}{n} = \frac{3\times10^8}{1.5} = 2\times10^8 \text{ m/s}$

Here, $\mu = 5 \times 10^{-7}$ μ m^{-1}

$$v = \frac{1}{\sqrt{\mu \in}} = \frac{1}{\sqrt{\mu \in_0 \in_r}}$$

$$\therefore \quad \in_r = \frac{1}{v^2 \mu \in_0}$$

$$= \frac{1}{(2\times10^8)^2 \times (5\times10^{-7}) \times (8.85\times10^{-12})}$$

$= 5.63 \approx 6.$

48. (4) γ-rays are produced when radioactive decay of radioactive substance occurs.

49. (3) Radar system operated on microwaves help us to detect the speed of tennis ball or cricket ball in motion.

High Order Time Intensive MCQs

1. (2) Angular frequency,

$$\omega = 2\pi v = \frac{2\pi c}{\lambda} = \frac{2\pi \times 3\times10^8}{6\times10^{-3}} = \pi\times10^{11} \text{ rad/s}$$

The electric field along y-axis is given by

$$E_y = E_0 \sin\omega(t - x/c) = 33\sin\pi\times10^{11}(t - x/c)$$

2. (3) Total power = Electromagnetic flux × area

$= 10^4 \times (10 \times 10) = 10^6$ W

3. (2) Let, E = energy falling on the surface per second

$= 10$ J

Momentum of photons $p = \frac{h}{\lambda} = \frac{h}{(c/v)} = \frac{hv}{c} = \frac{E}{c}$

On reflection, change in momentum per second

$$= 2p = \frac{2E}{c} = \frac{2\times10}{3\times10^8} = 6.7\times10^{-8}\text{N}$$

4. (3) A force and a pressure are exerted on the surface when a surface intercepts electromagnetic radiation as the surface is non-reflecting, so it completely absorbs the radiation and in such case the force is

$$F = \frac{IA}{c}$$

The radiation pressure is the force per unit area

$$p = \frac{F}{A} = \frac{I}{c}$$

5. (1) When frequency of microwaves match with the resonant frequency of the water molecules then food items with water molecules are effectively heated by microwaves.

6. (2) The wavelength of radio waves is largest. So, the radio waves undergo largest diffraction.

7. (3) Since surface charge density σ increases at a constant rate, the electric field $E = \frac{\sigma}{\varepsilon_0}$ between the plates also increases linearly with time. This changing electric field gives rise to a displacement current, which in turn produces a magnetic field as per the Ampere Maxwell law:

$$\oint \vec{B}\cdot d\vec{l} = \mu_0\varepsilon_0 \frac{d\Phi_E}{dt}$$

For a circular loop of radius r, $\Phi_E = E \cdot \pi r^2$, so

$$B \cdot 2\pi r = \mu_0 \pi r^2 \frac{d\sigma}{dt} \Rightarrow B = \frac{\mu_0 r}{2}\cdot\frac{d\sigma}{dt}$$

Thus, $B \propto r$ – the magnetic field increases linearly with radial distance and is maximum at the plate edges.

8. (2) The temperature of air decreases as one goes up form the earth's surface, because the earth radiates in the infrared region which is the main heat source and temperature drops as we go away from it.

9. (4) Here, $\vec{B} = 1.2\times10^{-8}\hat{k}$T

The magnitude of $\vec{E}$ is

$E = Bc = (1.2 \times 10^{-8}\text{ T})(3 \times 10^8 \text{ m s}^{-1})$

$= 3.6 \text{ V m}^{-1}$

$\vec{B}$ is along Z-direction and the wave propagates along X-direction. Therefore $\vec{E}$ should be in a direction perpendicular to both X and Z axes. Using vector algebra $\vec{E}\times\vec{B}$ should be along X-direction.

Since $(+\hat{j})\times(+\hat{k}) = \hat{i}$

$\therefore$ $\vec{E}$ is along the Y-direction.

Thus, $\vec{E} = 3.6\hat{j}\text{ Vm}^{-1}$.

10. (1) The frequency of γ-rays is more than that of X-rays and frequency of X-rays is more than that of ultraviolet rays.

$\therefore$ b > a > c

11. (3) Here, power of bulb = 100 W

As intensity, $I = \frac{\text{Power of visible light}}{\text{area}}$

$$= \frac{100\times6/100}{4\pi(8)^2} = 7.46\times10^{-3}\text{ Wm}^{-2}$$

12. **(1)** Here, B = 12 × 10^{-8} sin (1.20 × 10^7 z –3.6 × 10^{15} t) T

Comparing it with, $B = B_0 \sin(kz - \omega t)$,

we have $B_0 = 12 \times 10^{-8}$T

$$\therefore I_{av} = \frac{1}{2}\frac{B_0^2 c}{\mu_0}$$

$$\frac{1}{2} \times \frac{(12\times10^{-8})^2 \times 3\times10^8}{4\pi\times10^{-7}}$$

$$= 1.71 \text{ W m}^{-2}$$

13. **(2)** $E_y = 2.5 \frac{N}{C} \times$

$$\cos\left[\left(2\pi\times10^6 \frac{rad}{m}\right)t - \left(\pi\times10^{-2}\frac{rad}{s}\right)x\right]$$

$E_z = 0, E_x = 0$

The wave is moving in the positive direction of x.

This is in the form $E_y = E_0(\omega t - kx)$

$\omega = 2\pi \times 10^6$

$2\pi\nu = 2\pi \times 10^6 \qquad \Rightarrow \nu = 10^6$ Hz

$k = \pi \times 10^{-2}$

$$\Rightarrow \lambda = \frac{2\pi}{\pi\times10^{-2}} = 2\times10^2 = 200 \text{ m}$$

14. **(3)** Here, $B_y = 2\times10^{-7} \sin(0.5 \times 10^3 x + 1.5 \times 10^{11} t)$

Comparing it with, $B_y = B_0 \sin(kx + \omega t)$, we have

$\omega = 1.5 \times 10^{11}$;

$$\nu = \frac{\omega}{2\pi} = \frac{1\cdot5\times10^{11}}{2\times3\cdot14} = 2\cdot39\times10^{10} \text{ Hz}$$

Range of microwave is 10^9 Hz to 10^{12} Hz. Hence the wave is a microwave.

15. **(1)** Intensity. $I = \frac{\text{energy / time}}{\text{area}} = \frac{\text{power}}{\text{area}}$

$$= \frac{(100\times2\cdot5/100)}{4\pi(3)^2} = 0\cdot022 \text{ W / m}^2$$

16. **(4)**

(i) Audio information can be transmitted in the from of optical waves through optical fibres.

(ii) Very hot objects emit not only infrared radiation but also emit visible radiation

(iii) Humans can detect infrared radiations by feeling the heat on the portion of body exposed by radiations.

Therefore, all the given statements are not completely correct. Hence the option (4) is correct.

Statement Based MCQs

1. **(1)** The induced electric field has different properties than electric field produced by static charges. As we know induced $\vec{E}$ forms a closed loop and induced $\vec{E}$ is always parallel to its displacement will do a non zero work, unlike conservative forces where work done is zero.

2. **(1)** The atmosphere envelop the earth and is responsible for green house effect to maintain essential warmth on earth.

3. **(1)** Both the statements are correct.

4. **(2)** Microwaves are the electromagnetic waves of wavelength of the order of a few millimetres, which is less than those of T.V. signals. On account of smaller wavelength, the microwaves can be transmitted as beam signals in a particular direction and are much better than radio waves because microwaves do not spread or bend around the corners of any obstacle coming in their way. Therefore microwaves are better carriers of signals than radiowaves. Moreover, em waves do not requie a material medium for propagation.

5. **(3)** Cones in the retina of the human eye are most sensitive for radiation of wavelength λ = 5600 × 10^{-10} m. Therefore, its frequency

$$\nu = \frac{c}{\lambda} = \frac{3\times10^8}{5600\times10^{-10}} = 5.36\times10^{14} \text{Hz.}$$

Hence, eye is most sensitive for the light of frequency 5.36 × 10^{14} Hz.

Assertion & Reason MCQs

1. **(2)** An accelerating charge particle produces time varying electric field which in turn will produce time varying magnetic field which again produces time varying electric field and that is how EM wave is propagated.

2. **(1)** Both Assertion and Reason are True and the Reason is a correct explanation of the Assertion.

3. **(3)** The differential equation of electric field and magnetic field is given as,

$$\frac{\partial^2 t}{\partial x^2} = \mu_o\varepsilon_o\left(\frac{\partial^2 E}{\partial t^2}\right) \text{ and } \left(\frac{\partial^2 B}{\partial x^2}\right) = \mu_o\varepsilon_o\left(\frac{\partial^2 B}{\partial t^2}\right)$$

The equation of em waves and standing waves are different.

4. **(4)** The free electrons in metal are well displaced causing sparks; are accumulated and may lead to shocks.

5. **(1)** Both Assertion and Reason are True and the Reason is a correct explanation of the Assertion.

Match the Columns MCQs

1. **(3)**

(a) Hindrance = $\frac{E}{H}$

(b) Poynting vector, $\vec{S} = \frac{1}{\mu_o}(\vec{E}\times\vec{H})$

(c) Energy density, $U = \frac{1}{2}\varepsilon_o E^2$

(d) $\frac{|\vec{B}|}{|\vec{E}|} = \frac{1}{c}, \quad \left(\because \frac{E}{B} = c\right)$

2. **(2)** (a) The troposphere is the lowest layer of our atmosphere. Starting at ground level to about 12 km above the earth surface.

(d) The stratosphere extends from the top of troposphere to about 50 km above the earth surface.

(b) Above the stratosphere is the mesosphere it extends upwards to a height of about 80 km above the earth surface.

(c) The ionosphere is not a distinct layer like the others mentioned above. Ionosphere is found above the mesosphere.

3. (3) We know, $n = \sqrt{\mu_r \varepsilon_r}$ .... (*i*)

(a) $\varepsilon_r = 1, \mu_r = 2$

$n = \sqrt{\varepsilon_r \mu_r} = \sqrt{1 \times 2} = \sqrt{2}$

(b) $\varepsilon_r = 2, \mu_r = 1$

$n = \sqrt{\varepsilon_r \mu_r} = \sqrt{2}$

(c) $\varepsilon_r = 2, \mu_r = 2$

$n = \sqrt{2 \times 2} = \sqrt{4} = 2$

(d) $\varepsilon_r = 1, \mu_r = 3$

$n = \sqrt{1 \times 3} = \sqrt{3}$

4. (2)

(a) Infrared waves used to treat muscle tension and also used in remote control etc.

(b) Radio waves used for broadcasting.

(c) X-rays is used to detect fracture of bones.

(d) Ultraviolet rays absorbed by the ozone layer of the atmosphere.

NCERT Exemplar MCQs

1. (3) Given, energy required to dissociate a carbon monoxide molecule into carbon and oxygen atoms E = 11 eV

We know that, E = hν, where h = 6.62×10^{-34} J-s

ν = frequency

$\Rightarrow$ 11eV = hν

$$\Rightarrow \nu = \frac{11 \times 1.6 \times 10^{-19}}{6.62 \times 10^{-34}} \text{J}$$

$$\Rightarrow \quad v = \frac{11 \times 1.6 \times 10^{-19}}{6.62 \times 10^{-34}} \text{J}$$

$= 2.65 \times 10^{15}$ Hz

This frequency radiation belongs to ultraviolet region.

2. (2) When a wave is reflected from denser medium, then the type of wave doesn't change but only its phase changes by 180° or π radian.

Thus, for the reflected wave $\hat{z} = -\hat{z}, \hat{i} = -\hat{i}$ and additional phase of π in the incident wave.

Given, here the incident electromagnetic wave is,

$$E = E_0 \hat{i} \cos(kz - \omega t)$$

The reflected electromagnetic wave is given by

$$E_r = E_0(-\hat{i})\cos[k(-z) - \omega t) + \pi]$$

$$= -E_0 \hat{i} \cos\left[-(kz + \omega t) + \pi\right]$$

$$= -E_0 \hat{i} \cos\left[-(k + \omega t)\right] = E_0 \hat{i} \cos(kz + \omega t)$$

3. (2) Given, energy flux $\phi = 20\text{W/ cm}^2$

Area, $A = 30\text{cm}^2$

Time, t = 30 min = 30×60 s

Now, total energy falling on the surface in time t is,

$U = \phi At = 20 \times 30 \times (30 \times 60)$ J

Momentum of the incident light $= \frac{u}{c}$

$$= \frac{20 \times 30 \times (30 \times 60)}{3 \times 10^8}$$

$$= 36 \times 10^{-4} \text{kg-ms}^{-1}$$

Momentum of the reflected light = 0

Momentum delivered to the surface

$= 36 \times 10^{-4} - 0 = 36 \times 10^{-4}$ kg-ms^{-1}

4. (1) We know that, $E = \frac{P}{A}$

$E \propto P$

$$\frac{E'}{E} = \frac{P'}{P} = \frac{50}{100} = \frac{1}{2}$$

$\Rightarrow E' = E/2$

5. (4) The direction of propagation of electromagnetic wave is perpendicular to both electric field vector $\vec{E}$ and magnetic field vector $\vec{B}$, *i.e.*, in the direction of $\vec{E} \times \vec{B}$.

6. (1, 4) Here, in electromagnetic wave, electric field vector is given as,

$$\vec{E} = \left(E_1 \hat{i} + E_2 \hat{j}\right)\cos(kz - \omega t)$$

In electromagnetic wave, the associated magnetic field vector,

$$\vec{B} = \frac{\vec{E}}{c} = \frac{E_1 \hat{i} + E_2 \hat{j}}{c}\cos(kz - \omega t)$$

Also, E and B are perpendicular to each other and the propagation of electromagnetic wave is perpendicular to E as well as B, so the given electromagnetic wave is plane polarised.

NEET & AIIMS Past Year MCQs

1. (3) Given:

$$\frac{\mu_r}{\varepsilon_r} = \frac{1}{4}$$

The ratio of electric field *E* to magnetic field *H* is:

$$\frac{E}{H} = \sqrt{\frac{\mu}{\varepsilon}} = \sqrt{\frac{\mu_0 \mu_r}{\varepsilon_0 \varepsilon_r}} = \sqrt{\frac{\mu_0}{\varepsilon_0}}\sqrt{\frac{\mu_r}{\varepsilon_r}}$$

Given $\sqrt{\frac{\mu_0}{\varepsilon_0}} = 120\pi$ and $\frac{\mu_r}{\varepsilon_r} = \frac{1}{4}$:

$$\frac{E}{H} = 120\pi \times \frac{1}{2} = 60\pi$$

2. (2) Order of EM waves by wavelength (shortest to longest):

γ–rays < X-rays < UV < Visible < IR < Microwaves

Gamma rays have the smallest wavelength

3. **(4)** As capacitor is being charged the displacement current is

$$I = \varepsilon_0 \frac{d(\phi)_E}{dt}$$

$$I = \varepsilon_0 \frac{d}{dt}(EA) \quad I = \varepsilon_0 \frac{d}{dt}\left(\frac{V}{d}A\right)$$

$$I = \frac{\varepsilon_0 A}{d}\frac{d}{dt}\left(\frac{Q}{C}\right)$$

$$I = \frac{dQ}{dt} = I_{conduction}$$

4. **(2)** EM waves do not originate from charges moving with uniform speed.

5. **(4)** Capacitive reactance $= \frac{1}{\omega C} = X_c$ (say)

on decreasing the operating frequency ω reduces

As X_c is inversely proportional to ω the value of X_c increase

$\therefore I_C = I_D = \frac{V_o}{X_c}$

As X_c increases, therefore displacement current Id decreases.

6. **(4)** $B = \frac{E}{C} = \frac{48}{3\times10^8} = 16\times10^{-8} = 1.6\times10^{-7}$ T

7. **(1)** Refractive index $n = \sqrt{\epsilon_r u_r}$

However $v = \frac{c}{n} \Rightarrow v = \left(\frac{c}{\sqrt{\epsilon_r \mu_r}}\right)$

8. **(1)** (A) Wavelength of Radio wave ≈ 10^2 m

(B) Wavelength of Microwave ≈ 10^{-2} m

(C) Wavelength of Infrared radiations ≈ 10^{-4} m

(D) Wavelength of X-ray ≈ Å = 10^{-10} m

9. **(1)** Given,

Direction of propagation of electromagnetic waves is along x-axis. (i.e., i)

Direction of propagation of electromagnetic wave is given by $\vec{E}\times\vec{B}$.

i.e., $\vec{V} \parallel (\vec{E}\times\vec{B})$

Now, we need to check all options one by one as follows:

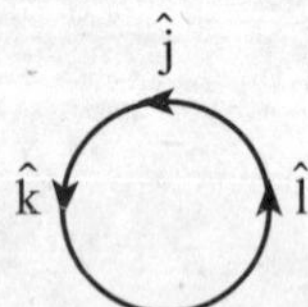

Here, when electric field $(\vec{E})$ is directed along $(-\hat{j}+\hat{k})$ and magnetic field $(\vec{B})$ is directed along $(-\vec{j}-\vec{k})$, then the propagation of electromagnetic wave will be in $\hat{i}$ direction i.e., in x-axis.

10. **(2)**

11. **(3)** **12.** **(2)**

13. **(1)** An electromagnetic wave carries energy and momentum. Since it carries, momentum, it also exerts pressure called radiation pressure.

14. **(2)** Electromagnetic waves are those waves in which there are sinusoidal variation of electric and magnetic field vectors at right angles to each other as well as at right angles to the direction of wave propagation. They are self-sustaining oscillations of electric and magnetic fields in free space or vacuum.

Ray Optics and Optical Instruments

NCERT Topic-wise MCQs

1. **(2)** Using mirror formula

$$\frac{1}{v}+\frac{1}{u}=\frac{1}{f}$$

$$\frac{1}{v}-\frac{1}{30}=\frac{-1}{20}$$

$\Rightarrow v = -60$ cm

$\Rightarrow m = \frac{-v}{u} = -2$

2. **(1)**

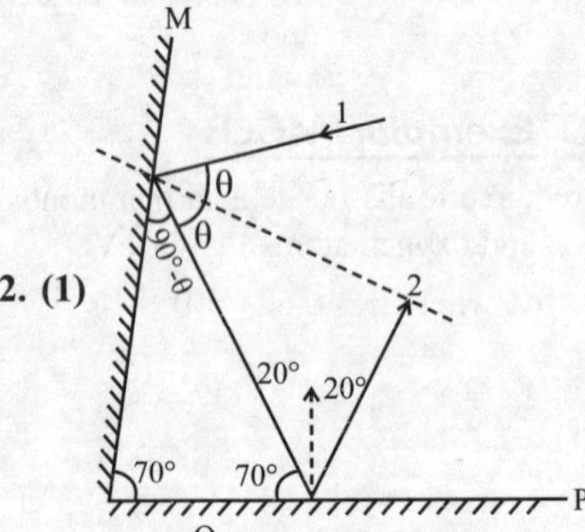

$90° - \theta + 70° + 70° = 180°$

$\therefore \theta = 50°$

3. **(2)** Total no. of image formed $= \frac{360°}{\theta} = \frac{360°}{72°} = 5$

Since 5 is an odd number.

4. **(2)** Image formed by plane mirror is virtual, erect and of equal size.

5. **(1)** $\therefore \frac{1}{v}+\frac{1}{u}=\frac{1}{f}$

$\Rightarrow \frac{1}{v}+\frac{1}{30}=\frac{1}{20}$

$\Rightarrow \frac{1}{v}=\frac{3-2}{60} \Rightarrow v = 60$ cm

6. **(3)** Using mirror formula,

$$\frac{1}{f}=\frac{1}{v}+\frac{1}{u}$$

$$\frac{1}{-10}=\frac{1}{v}+\frac{1}{-15}$$

$\Rightarrow v = -30$ cm.

$|\text{Axial magnification}| = \frac{dv}{du} = \frac{v^2}{u^2} = 4$

Given, du = 2 mm

$\Rightarrow$ dv = amplitude of image = 4 × 2 = 8 mm.

7. (1) $m = +\frac{1}{n} = -\frac{v}{u} \Rightarrow v = -\frac{u}{n}$

From mirror formula

$\frac{1}{f} = \frac{1}{-\frac{u}{n}} + \frac{1}{u} \Rightarrow u = -(n-1)f$

8. (1) Different angles are as shown in fig. In triangle ABC,

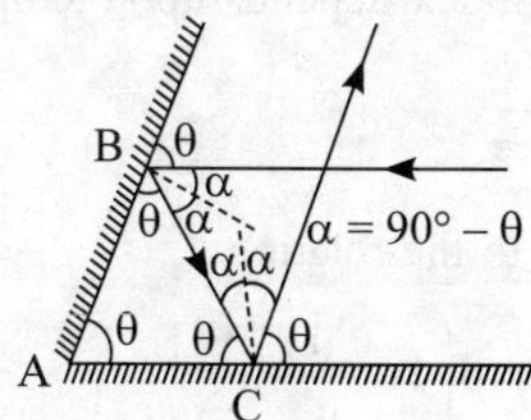

$\theta + \theta + \theta = 180°$

$\theta = 60°$

9. (1) Here, we have $n = \frac{360°}{\theta} = \frac{360°}{90°} = 4$

Number of images = (n – 1) = (4 – 1) = 3

10. (2) Distance between the object and the image = 0.5 + 0.5 = 1 m

11. (2) $f = -\frac{R}{2} = -18$ cm (concave mirror)

Given, virtual image and m = 3

$\Rightarrow m = \frac{f}{f-u}$ or $3 = \frac{-18}{-18-u}$

u = – 12 cm

12. (3) The ray diagram for the situation is as shown below

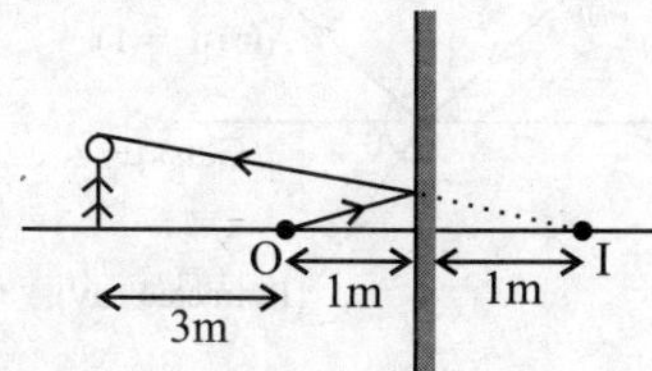

Required distance = 5m.

13. (4) In going from air to water, rays of light bend towards normal. Therefore, a real, inverted image is formed at a point between C and O.

14. (3) In case of a concave mirror, when we move the object from infinity to focus of the miror, (i.e. u decreases from ∞ to f), the image moves from focus to ∞ (i.e. v increases from f to ∞). Therefore, the graph looks like the curve shown in (3).

15. (1) Linear magnification,

$m = \frac{I}{O} = \frac{f}{f+u} = \frac{10}{10-25} \Rightarrow m = \frac{-10}{15} = \frac{-2}{3}$

Areal magnification, $m^2 = \frac{4}{9}$

Area of image $= \frac{4}{9} \times$ Area of object $= \frac{4}{9} \times (3)^2 = 4\ \text{cm}^2$

16. (4) A ray coming parallel to principal axis after reflection passes through the focus.

17. (3) In the graph at P,

v = u, which is possible only when u = 2*f*.

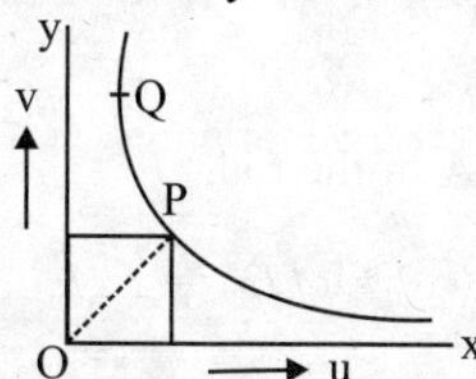

At any other point Q above P, v > 2 *f*.

18. (1) Here, $u = -1.5f, f = -f$

$\frac{1}{f} = \frac{1}{v} + \frac{1}{u}$

We have,

$\frac{1}{(-f)} = \frac{1}{v} + \frac{1}{(-1.5f)}$

$\frac{1}{v} = \frac{1}{-3f}$ or $v = -3f$

Now, magnification, $m = \frac{I}{O} = \frac{v}{-u}$

Here, $O = +2.5$ cm

$\therefore \quad \frac{I}{2.5} = \frac{-3f}{-(-1.5f)}$ or $I = -5$ cm

19. (3) For concave mirror

u = –u, f = – f

using mirror formula

$v = \frac{uf}{u-f}$

Length of image L = | v | – | f |

$= \frac{uf}{u-f} - f$

Solving we get $L = \frac{f^2}{u-f}$

20. (3) Using mirror formula, $\frac{1}{v} + \frac{1}{u} = \frac{1}{f}$

Now, $\frac{du}{dt}\left(\frac{-1}{u^2}\right) = \frac{1}{v^2}\left(\frac{dv}{dt}\right)$ [∵ f is constant]

$\frac{du}{dt} = 4$ (towards mirror), v = – 30 cm, u = –20

$\therefore \frac{-4}{20^2} = \frac{1}{30^2} \times \frac{dv}{dt}$

So $\frac{dv}{dt} = -9$ cm/s

–ve so away from the mirror.

21. (2) As one end of magnified image touches an end of the rod, therefore this end of rod must be at the centre of curvature of mirror.

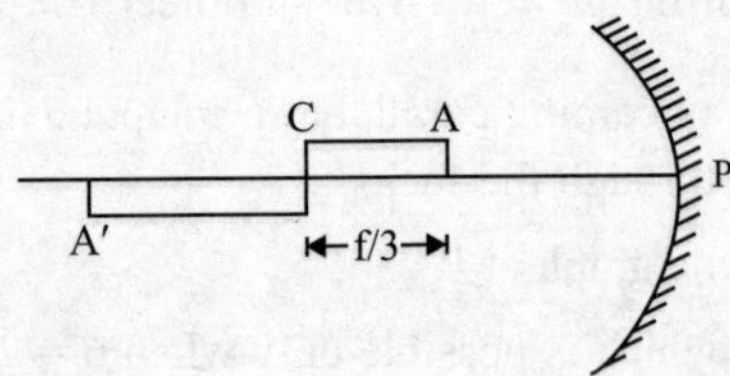

$\because$ CP = 2f

For the end A of the rod,

$$u = -PA = -(2f - f/3) = -\frac{5f}{3}$$

From $\frac{1}{v} + \frac{1}{u} = \frac{1}{f}$

$$\frac{1}{v} = \frac{1}{f} - \frac{1}{u} = \frac{1}{-f} + \frac{3}{5f} = \frac{-2}{5f}$$

$v = -5f/2$

$\therefore$ length of image = CA' = $\frac{5}{2}f - 2f = f/2$

22. (2) Here, h = 4 m. r = ?, μ = 5/3, For total internal reflection,

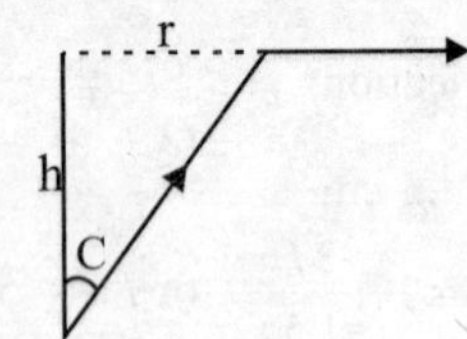

$$\mu = \frac{1}{\sin C} = \frac{\sqrt{h^2 + r^2}}{r}$$

$\Rightarrow h^2 + r^2 = \mu^2 r^2$

$$r = \frac{h}{\sqrt{\mu^2 - 1}} = \frac{4}{\sqrt{\frac{25}{9} - 1}} = 3\text{m}$$

23. (1) As we know,

$$\mu = \frac{c}{v} \qquad (v = n\lambda)$$

$$\therefore \mu = \frac{3 \times 10^8}{4 \times 10^{14} \times 5 \times 10^{-7}} = 1.5$$

24. (3) Let the water should be filled up to the height of x so that the bottom of container rises upto the height $(21 - x)$

$$\mu = \frac{\text{real depth}}{\text{apparent depth}}$$

$$\frac{4}{3} = \frac{x}{21 - x}$$

21 cm (21 – x) (21 – x) x

$4(21 - x) = 3x$

$x = 12$ cm

25. (3) As we know,

$$\lambda \propto \frac{1}{\mu}$$

$$\frac{\lambda_{air}}{\lambda_g} = \frac{\mu_g}{\mu_{air}} = \frac{\mu}{1} \Rightarrow \frac{\lambda_1}{\lambda_2} = \frac{\mu}{1}$$

26. (2) $v \propto \lambda \Rightarrow \frac{V_\omega}{V_a} = \frac{\lambda_\omega}{\lambda_a}$

$$\therefore V_\omega = \frac{\lambda_\omega}{\lambda_a} \times V_a = \frac{4500}{6000} \times 3 \times 10^8$$

$V_\omega = 2.25 \times 10^8$ m/s

27. (1) Refractive index depends upon temperature as

$$\mu \propto \frac{1}{\text{temperature}}$$

28. (4) According to the relation

$$\text{Apparent rize} = d\left(1 - \frac{1}{{}_a\mu_w}\right) = 12 \times \left(1 - \frac{3}{4}\right) = \frac{12 \times 1}{4} = 3\,\text{cm}$$

29. (4) Apparent depth of the bottom $(y) = \frac{\text{real depth }(x)}{\mu}$

$$y = \frac{H}{4n_1} + \frac{H}{4n_2} + \frac{H}{4n_3} + \frac{H}{4n_4}$$

$$y = \frac{H}{4}\left(\frac{1}{n_1} + \frac{1}{n_2} + \frac{1}{n_3} + \frac{1}{n_4}\right)$$

Note: "Apparent depth can be algebraically add".

30. (3) Here, i = 45°

Applying snell's law at air-glass, surface

$${}^a\mu_g = \frac{\sin i}{\sin r'}$$

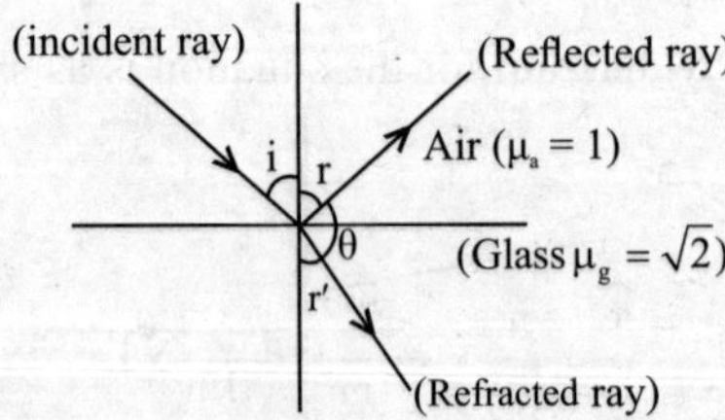

$$\frac{\mu_g}{\mu_a} = \frac{\sin i}{\sin r'} \Rightarrow \frac{\mu_a \sin i}{\mu_g} = \sin r'$$

$$\frac{1 \times \sin i}{\sqrt{2}} = \sin r' \Rightarrow \frac{1}{\sqrt{2}}\sin 45^0 = \sin r'$$

$$\frac{1}{\sqrt{2}} \times \frac{1}{\sqrt{2}} = \sin r' \Rightarrow \frac{1}{2} = \sin r'$$

$r' = 30°$

From figure,

$\theta + r + r' = 180°$ $\qquad (\because \angle i = \angle r)$

$\theta = 180° - 75°$

$\theta = 105°$

31. (4) From face (1)

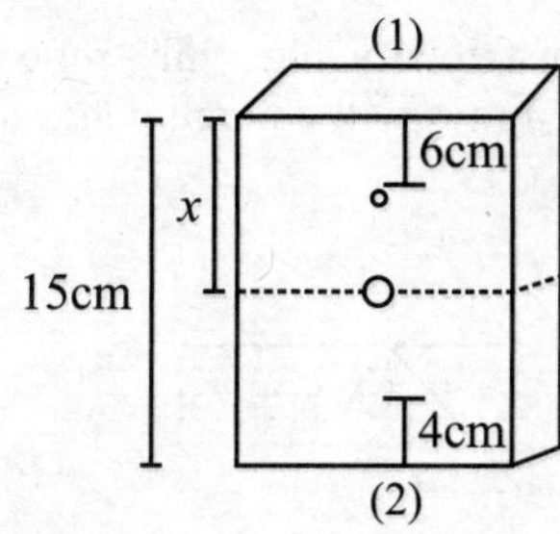

$$\mu = \frac{x}{6} \quad \ldots(i)$$

From face (2)

$$\mu = \frac{15-x}{4} \quad \ldots(ii)$$

By comparing eq. (*i*) & (*ii*)

$$\frac{x}{6} = \frac{15-x}{4} \Rightarrow 4x = 90 - 6x$$

$$10x = 90$$

$$x = 9$$

$$\mu = \frac{x}{6} = \frac{9}{6} = \frac{3}{2} = 1.5$$

$\mu = 1.5$

Short trick:

$$\mu = \frac{\textit{real depth}}{\textit{Apparent depth}}$$

"Apparent depth can be algebrically added

$$\mu = \frac{15}{6+4} = \frac{15}{10} = 1.5$$

$\mu = 1.5$

32. (3) ${}^2\mu_1 \times {}^3\mu_2 \times {}^4\mu_3 = \frac{\mu_1}{\mu_2} \times \frac{\mu_2}{\mu_3} \times \frac{\mu_3}{\mu_4}$

$${}^2\mu_1 \times {}^3\mu_2 \times {}^4\mu_3 = \frac{\mu_1}{\mu_4}$$

$${}^2\mu_1 \times {}^3\mu_2 \times {}^4\mu_3 = {}^4\mu_1 = \frac{1}{{}^1\mu_4}$$

33. (4) $c = \frac{1}{\sqrt{\mu_0 \varepsilon_0}}$ $, v = \frac{1}{\sqrt{\mu\varepsilon}}$

$$\frac{c}{v} = \left(\sqrt{\frac{\mu\varepsilon}{\mu_0\varepsilon_0}}\right)$$

$$\frac{c^2}{v^2} = \frac{\mu}{\mu_0} \times \frac{\varepsilon}{\varepsilon_0} \Rightarrow \frac{\varepsilon}{\varepsilon_0} = \frac{\mu_0}{\mu} \times \frac{c^2}{v^2} \quad \ldots(i)$$

$$\therefore 1.5 = \frac{c}{v}$$

$$v = \frac{3\times10^8}{1.5} = 2\times10^8 \text{ [Put in } (i)]$$

$$\frac{\varepsilon}{\varepsilon_0} = \frac{4\times\pi\times10^{-7}}{5\times10^{-7}} \times \frac{9\times10^{16}}{4\times10^{16}}; \ \varepsilon_r = \frac{3.14\times9}{5} = 5.8$$

$\varepsilon_r \approx 6$

34. (3)

$$\tan\theta_c = \frac{R}{8}, \quad \sin\theta_c = \frac{1}{\mu} = \frac{3}{5} \Rightarrow \tan\theta_c = \frac{3}{4}$$

R = 6 cm

35. (3) ${}^a\mu_w = \frac{4}{3}$, ${}^a\mu_g = \frac{3}{2}$

$${}^w\mu_g = \frac{{}^a\mu_g}{{}^a\mu_w} = \frac{3/2}{4/3} = \frac{9}{8}; \ \frac{\sin i}{\sin r} = {}^w\mu_g = \frac{9}{8}$$

$$\sin r = \frac{8}{9} \times \frac{1}{2} \Rightarrow \sin r = \frac{4}{9} \Rightarrow r = 26.38°$$

36. (3) According to the question

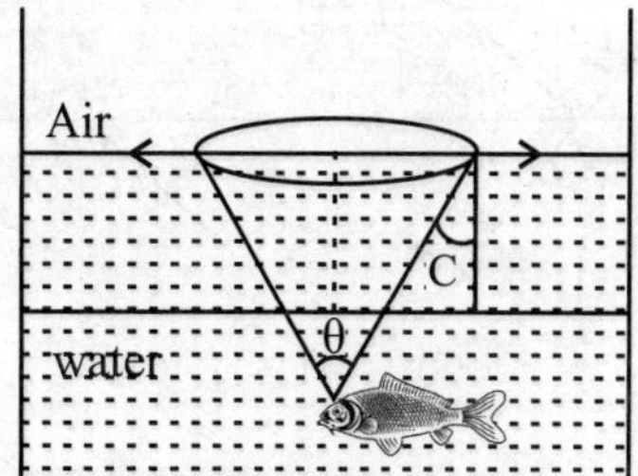

θ = C + C (as by the figure)

θ = 2C ; θ = 2 × 49° ⇒ θ = 98°

37. (3)

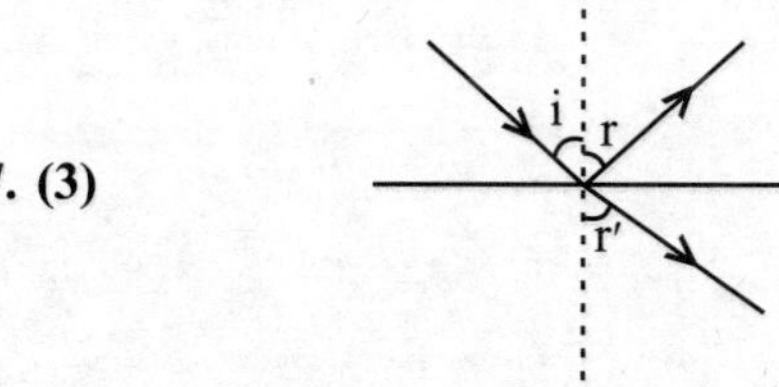

$${}^D\mu_R = \frac{\sin i}{\sin r'} \Rightarrow {}^R\mu_D = \frac{\sin r'}{\sin i}$$

$$\frac{1}{\sin C} = \frac{\sin r'}{\sin i}$$

$$\sin C = \frac{\sin i}{\sin(90 - r)} = \frac{\sin i}{\cos r} \ (\angle i = \angle r)$$

∴ sin C = tan i

C = $\sin^{-1}$ (tan i)

38. (3) One wave means one wavelength. So, if t is the thickness of slab, λ the wavelength and n the number of waves, then

$$n\lambda = t \Rightarrow \lambda = \frac{t}{n}$$

or $\lambda \propto \frac{1}{n}$ (as t is same) or $\frac{\lambda_1}{\lambda_2} = \frac{n_2}{n_1}$

Now, refractive index of second medium w.r.t first medium is

$${}_1\mu_2 = \frac{\lambda_1}{\lambda_2} = \frac{n_2}{n_1}$$

39. (3) As we know,

$$\mu = \frac{1}{\sin C}\ ;\ \sin C = \frac{1}{{}^w\mu_g}$$

$$C = \sin^{-1}\left(\frac{1}{{}^w\mu_g}\right);\ C = \sin^{-1}\left(\frac{8}{9}\right)$$

Alternate method:

applying Snell's law

$${}^g\mu_w = \frac{\sin i}{\sin r} \quad i = C,\ r = 90°$$

$$\frac{\mu_w}{\mu_g} = \frac{\sin C}{\sin 90°};\ \frac{4 \times 2}{3 \times 3} = \sin C$$

$$C = \sin^{-1}\frac{8}{9}$$

40. (1)

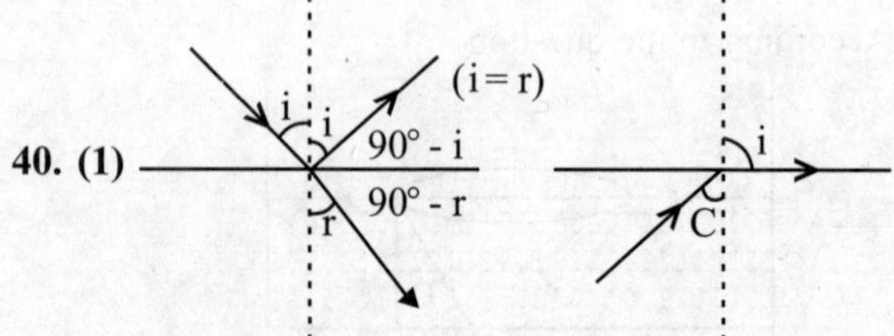

$90° - i + 90° - r = 90°;\ -(i + r) = 90° - 180°$

$i + r = 90°$

$$\mu = \frac{\sin i}{\sin r};\ \mu = \frac{\sin i}{\sin(90 - i)}$$

$$\mu = \frac{\sin i}{\cos i} = \tan i\ ;$$

$$\because \sin C = \frac{1}{\mu}$$

$$\sin C = \frac{1}{\tan i};$$

$\Rightarrow C = \sin^{-1}(\cot i)$

41. (3) $${}^a\mu_g = \frac{1}{\sin\theta} \Rightarrow \mu = \frac{1}{\sin\theta} \quad ...(i)$$

From the Snell's law

$$\mu = \frac{\sin i}{\sin r} = \frac{\sin\theta}{\sin r}$$

$$\sin r = \frac{\sin\theta}{\mu} \quad ...(ii)$$

Put the value of $\sin\theta$ in equation (*ii*)

$$\therefore \sin r = \frac{1}{\mu^2} \Rightarrow r = \sin^{-1}\left(\frac{1}{\mu^2}\right)$$

42. (1) As, $C = \sin^{-1}\left(\frac{1}{\mu}\right), \mu \propto \frac{1}{\lambda}$

$\xrightarrow[\text{[VIBGYOR]}]{\lambda}$ increase

So, $\lambda_R > \lambda_O > \lambda_Y$

If green is just totally internally reflected then yellow, orange and red rays will emerge out.

43. (2)

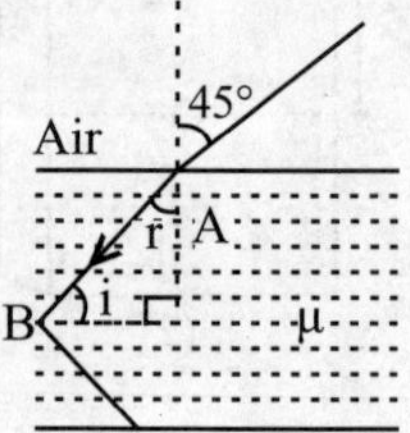

Using snell's law at the air water interface

$$\mu = \frac{\sin 45°}{\sin r}$$

$$\Rightarrow \sin r = \frac{\sin 45°}{\mu}$$

$$\therefore \sin r = \frac{1}{\mu\sqrt{2}}$$

At point B for total internal reflection

$$\sin i = \frac{1}{\mu}$$

$i = 90° - r$

$$\therefore \sin(90° - r) = \frac{1}{\mu} \Rightarrow \cos r \frac{1}{\mu} \quad ...(i)$$

$$\text{Now, } \cos r = \sqrt{1 - \sin^2 r} = \sqrt{1 - \frac{1}{2\mu^2}} = \sqrt{\frac{2\mu^2 - 1}{2\mu^2}} \quad ...(ii)$$

Putting the value of cos r in equation (*ii*) and squaring both sides

$$\frac{1}{\mu} = \sqrt{\frac{2\mu^2 - 1}{2\mu^2}} \Rightarrow \frac{1}{\mu^2} = \frac{2\mu^2 - 1}{2\mu^2}$$

$$2 = 2\mu^2 - 1 \Rightarrow \frac{3}{2} = \mu^2 \Rightarrow \mu = \sqrt{\frac{3}{2}}$$

44. (2) Critical angle C is equal to incident angle if ray is reflected normally

$\Rightarrow C = 90°$

45. (3) Apparent shift of bottom position of beaker in water is

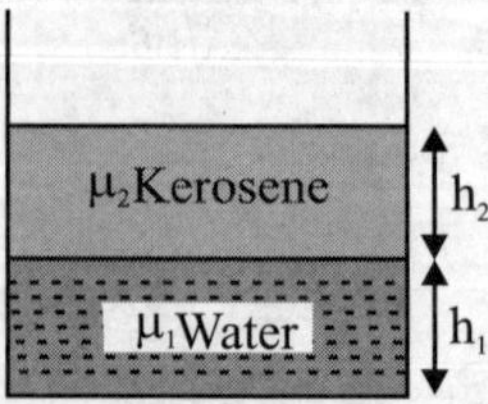

$$x_1 = h_1 - \frac{h_1}{\mu_1} = h_1\left(1 - \frac{1}{\mu_1}\right)$$

Apparent shift of bottom position in kerosene is

$$x_2 = h_2 - \frac{h_2}{\mu_2} = h_2\left(1 - \frac{1}{\mu_2}\right)$$

$$\text{Total shift} = x_1 + x_2 = h_1\left(1 - \frac{1}{\mu_1}\right) + h_2\left(1 - \frac{1}{\mu_2}\right)$$

46. (2) In vacuum c = d/t

In medium $v = \frac{5d}{T}$;

$\because \mu = \frac{c}{v} = \frac{T}{5t}$

Also, $\sin C = \frac{1}{\mu}$ (C is critical angle)

$\therefore C = \sin^{-1}\left(\frac{5t}{T}\right)$

47. (3) deviation in red ray will be least because of highest wavelength out of seven color rays.

48 (1) Here, $\mu > \sqrt{2} \Rightarrow \frac{1}{\mu} < \frac{1}{\sqrt{2}}$

$\Rightarrow \sin C < \sin 45°$

$\Rightarrow C < 45°$

They ray striking a glass-air surface from inside the prism will be reflected back if angle of incidence is not less than 45°.

90° deviation is produced if the ray strikes on one of the sides normally as shown.

Here, I = 45° and hence reflection takes place.

49. (2) Dispersive power depends only on material of prism.

50. (1) We know, $\mu = \frac{\sin\left(\frac{A+\delta m}{2}\right)}{\sin A/2}$

$$\frac{c}{v} = \frac{\sin\left(\frac{A+\delta m}{2}\right)}{\sin A/2} \Rightarrow v = \frac{3\times10^8 \sin 30°}{\sin 45°}$$

$\Rightarrow v = 2.12 \times 10^8$ m/s

51. (2) As we know $\omega = \frac{\mu_b - \mu_r}{\mu - 1}$ [μ=mean refractive index]

$$\mu = \frac{\mu_b + \mu_r}{2} = \frac{1.664+1.644}{2} = 1.654$$

$$\omega = \frac{\mu_b - \mu_r}{\mu - 1}$$

$$\omega = \frac{1.664-1.644}{1.654-1} = 0.0305$$

52. (2) According to the relation

$\delta = (\mu - 1)A \Rightarrow 5 = (\mu - 1)A$

$5 = (1.5 - 1)A$

$A = 10°$

53. (4) $A + \delta = i + e$

$60° + 55° = 15° + e$;

$\therefore e = 115 - 15 = 100°$

54. (1) According to the relation

$$\mu = \frac{\sin\left(\frac{A+\delta m}{2}\right)}{\sin A/2} = \frac{\sin\left(\frac{60°+60°}{2}\right)}{\sin\frac{60°}{2}}$$

$$\mu = \frac{\sin 60°}{\sin 30°} = \frac{\sqrt{3}}{2}\times\frac{2}{1} = \sqrt{3}$$

$\mu = \sqrt{3}$

55. (1) i = 60°, A = 30°, δ = 30°

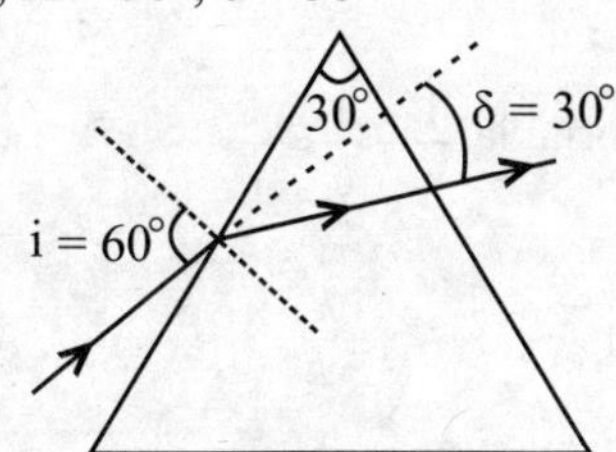

As we known

$i + e = A + \delta$

$e = A + \delta - i$

$= A + \delta - i = 30° + 30° - 60°$

$e = 0°, r_2 = 0°$

As, $r_1 + r_2 = A$

$\therefore r_1 = 30°$

$$\mu = \frac{\sin i}{\sin r} = \frac{\sin 60°}{\sin 30°} = \sqrt{3}$$

$\mu = 1.732$

56. (1) According to the lens maker formula

$$\frac{1}{f} = (\mu - 1)\left[\frac{1}{R_1} - \frac{1}{R_2}\right]$$

$$\frac{1}{f} = \frac{1}{2}\left[\frac{2}{40}\right] \Rightarrow f = 40\text{cm}$$

57. (4) $\frac{1}{F} = \frac{1}{f_1} + \frac{1}{f_2} \Rightarrow \frac{1}{80} = \frac{1}{20} + \frac{1}{f_2}$

$$\frac{1}{f_2} = \frac{1-4}{80}$$

$$\Rightarrow f_2 = -\frac{80}{3}\text{cm}$$

$\therefore$ power of second lens

$$P_2 = \frac{100}{f_2} = \frac{100}{\frac{-80}{3}} = -3.75\text{D}$$

58. (2) According to the relation

$$\frac{1}{f} = \frac{1}{f_1} + \frac{1}{f_2}$$

$$f = \frac{f_1 f_2}{f_1 + f_2} = \frac{10\times(-10)}{10+(-10)} = \frac{-100}{0} = \infty$$

59. (1) $\frac{1}{f} = \frac{1}{v} - \frac{1}{u} = \frac{1}{90} + \frac{1}{45}$

$\frac{1}{f} = \frac{3}{90}$

f = 30 cm

$\frac{I}{O} = \frac{v}{u}$

$\frac{I}{5} = \frac{90}{-45} \Rightarrow I = -10\text{cm}$

60. (4) Here u = mf

Using lens formula $\frac{1}{v} - \frac{1}{u} = \frac{1}{f}$

$\Rightarrow v = \frac{mf}{m+1}$

Now magnification = $\frac{v}{u}$

$\Rightarrow \mu = \frac{mf}{m+1} \times \frac{1}{mf}$

$\mu = \frac{1}{m+1}$

61. (2) According to displacement method

$O = \sqrt{I_1 I_2}$

$O = \sqrt{4 \times 16} = 8\text{cm}$

62. (1) Here, $\frac{1}{f} = (\mu - 1)\left(\frac{1}{R_1} - \frac{1}{R_2}\right)$

Taking $R_1 = R$ and $R_2 = -R$,

$$\frac{1}{f} = (\mu - 1)\frac{2}{R} \quad \text{...}(i)$$

When the lens is cut into two halves along XOX', and f' is focal length of each half, then

$$\frac{1}{f'} = (\mu - 1)\left(\frac{1}{R_1} - \frac{1}{R_2}\right) = (\mu - 1)\frac{2}{R} \quad \text{...}(ii)$$

From (*i*) and (*ii*), $f' = f$

When the lens is cut into two halves along YOY' then each half behaves as a plano convex lens. If f'' is focal length of each half, then

$$1/f'' = (\mu - 1)\left(\frac{1}{R} - \frac{1}{\infty}\right) = \frac{\mu - 1}{R} \quad \text{...}(iii)$$

From (*ii*) and (*iii*)

$\frac{1}{f'} = \frac{2}{f''}$ or $f'' = 2f' = 2f.$

63. (2) Case (*i*), u = –25 cm, f = 20 cm; v = ?

$\because \frac{1}{f} = -\frac{1}{u} + \frac{1}{v}$

$\Rightarrow \frac{1}{v} = \frac{1}{20} - \frac{1}{25} = \frac{1}{100} \Rightarrow v = 100 \text{ cm}$

$m_{25} = \frac{v}{u} = \frac{100}{-25} = -4$

Case (ii), u = – 50 cm ; f = 20 cm, v = ?

$\Rightarrow \frac{1}{v} = \frac{1}{20} - \frac{1}{50} = \frac{3}{100}$

or $v = \frac{100}{3}$ cm

$m_{50} = \frac{(100/3)}{-50} = \frac{-2}{3} \Rightarrow \frac{m_{25}}{m_{50}} = \frac{-4}{-2/3} = 6$

64. (2) Full image will be formed of less intensity.

65. (3) Here, R_1 = + 20 cm ; R_2 = – 20 cm

Using lens maker's formula

$\frac{1}{f} = (\mu - 1)\left(\frac{1}{R_1} - \frac{1}{R_2}\right) = \left(\frac{3}{2} - 1\right)\left(\frac{1}{20} - \frac{1}{-20}\right) = \frac{1}{20}$

f = 20 cm

Here, u = – 30 cm, f = 20 cm, v = ?

Using lens formula, we have

$\frac{1}{f} = \frac{1}{v} - \frac{1}{u}$

or $\frac{1}{20} = \frac{1}{v} - \frac{1}{-30} = \frac{1}{v} + \frac{1}{30}$

or $\frac{1}{v} = \frac{1}{20} - \frac{1}{30} = \frac{1}{60}$ or v = 60 cm.

Magnification, $m = \frac{v}{u} = \frac{h_i}{h_0}$

$\frac{60\,\text{cm}}{-30\,\text{cm}} = \frac{h_i}{2\,\text{cm}}$

or $h_i = -4$ cm

It means, the image is real, inverted and of height 4 cm. The path of rays is shown in the given figure.

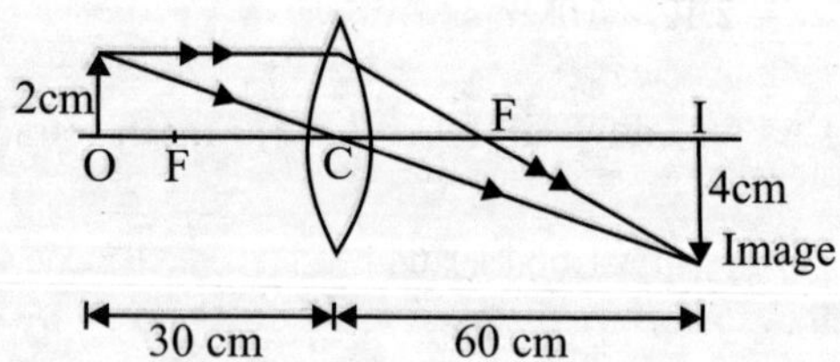

66. (1) $\frac{1}{f} = \left(\frac{n}{1} - 1\right)\left(\frac{1}{R_1} - \frac{1}{R_2}\right)$

$\frac{1}{f_l} = \left(\frac{n}{n'} - 1\right)\left(\frac{1}{R_1} - \frac{1}{R_2}\right)$

$\frac{f_l}{f} = \frac{(n-1)n'}{n - n'}$ or $f_l = \frac{-fn'(n-1)}{n' - n}$

67. (2) Using lens maker's formula

$\frac{1}{f} = (\mu - 1)\left(\frac{1}{R_1} - \frac{1}{R_2}\right)$

For first plano convex lens.

$$\frac{1}{f_1} = (1.5-1)\left(\frac{1}{14} - \frac{1}{\infty}\right) = \frac{1}{28}$$

For second plano convex lens.

$$\frac{1}{f_2} = (1.2-1)\left(\frac{1}{\infty} - \frac{1}{-14}\right) = \frac{1}{70}$$

If f is combined focal length of bi-convex lens, then

$$\frac{1}{f} = \frac{1}{f_1} + \frac{1}{f_2} = \frac{1}{28} + \frac{1}{70} = \frac{5+2}{140} = \frac{1}{20}$$

$f = 20$ cm

For this lens, u = – 40 cm, v = ?

From $\frac{1}{v} - \frac{1}{u} = \frac{1}{f}$

$$\frac{1}{v} = \frac{1}{f} + \frac{1}{u} = \frac{1}{20} - \frac{1}{40} = \frac{1}{40}$$

v = 40 cm

68. (4) According to the lens maker formula

$$\frac{1}{f} = (\mu-1)\left[\frac{1}{R_1} - \frac{1}{R_2}\right]$$

$$\frac{1}{f} = (\mu-1)\left[\frac{2}{R}\right] \quad [R_1 = R_2 = R]$$

$$f = \frac{R}{2(\mu-1)} \quad ...(i)$$

After cutting the lens as shown below convex lens becomes plano-convex lens whose $R_1 = R$, $R_2 = \infty$

$$\therefore \frac{1}{f'} = (\mu-1)\left[\frac{1}{R} - \frac{1}{\infty}\right] \Rightarrow \frac{1}{f'} = (\mu-1)\left[\frac{1}{R}\right]$$

$$f' = \frac{R}{(\mu-1)} \quad ...(ii)$$

From equation (*i*) and (*ii*) we get

$$\frac{f'}{f} = \frac{R}{\mu-1} \times \frac{(\mu-1)\times 2}{R} \Rightarrow f' = 2f$$

Short Note : (1) If lens is cut perpendicular to its principal axis. Its focal length becomes twice the previous focal length.

(2) If a lens is cut along its principal axis its focal length remain same because nature of the lens remains as converging

$R_1 = R$, $R_2 = -R$

69. (4) Focal length depends upon as

$$\frac{1}{f} = (\mu-1)\left(\frac{1}{R_1} - \frac{1}{R_2}\right)$$

Three materials are used to make the lens so there should be three focal lengths then three images will be produced.

70. (1) By using the relation; $\frac{\mu_2}{v} - \frac{\mu_1}{u} = \frac{\mu_2 - \mu_1}{R}$

$$\frac{1.5}{v} - \frac{1}{(-15)} = \frac{(1.5-1)}{+30}$$

v = – 30 cm [–ve sign indicates that image is formed on same side of object (left side)]

71. (1) Here, u = – 10 cm, R = – 5 cm,

$\mu_2 = 1.5$, $\mu_1 = 1$, v = ?

In the present case, refraction occurs from denser to rarer medium.

$$\therefore \quad \frac{\mu_2}{-u} + \frac{\mu_1}{v} = \frac{\mu_1 - \mu_2}{R}$$

$$\frac{1.5}{10} + \frac{1}{v} = \frac{1-1.5}{-5} = \frac{1}{10}$$

$$\text{or} \quad \frac{1}{v} = \frac{1}{10} - \frac{3}{20} = -\frac{1}{20}$$

$\therefore$ v = – 20 cm.

72. (4) For real image m = –ve

m = –2

$\therefore$ As we know $m = \frac{f}{u+f}$

$$-2 = \frac{f}{u+f} \Rightarrow -2 = \frac{20}{u+20}$$

$\Rightarrow$ –2u + (– 40) = 20

u = – 30 cm

73. (1) As $P_{eq} = \Sigma P_i$

[beccuse all lenses are identical]

$\therefore 5P = 25D$

$\Rightarrow P = 5D$

$$\Rightarrow \frac{1}{f} = 5 \Rightarrow f = \frac{1}{5}m = 20cm$$

74. (3) L_1 will converge the parallel ray at its focus. L_2 will render it parallel only when the refracted ray starts from the focus of L_2

$\therefore$ Distance between L_1 L_2 = $F_1 + F_2$

75. (3) By covering the aperture of diameter d/2, focal length of lens is not affected. Area reduces by 1/4th. So does the intensity.

$\therefore$ New focal length = f, and

New intensity = $I - \frac{I}{4} = \frac{3I}{4}$

76. (3) $$\frac{1}{f} = (\mu-1)\left(\frac{1}{R_1} - \frac{1}{R_2}\right) = (1.5-1)\left(\frac{1}{15} - \frac{1}{20}\right)$$

$$= \frac{1}{2} \times \frac{1}{60} = \frac{1}{120} \Rightarrow f = 120 \text{ cm.}$$

77. (3) Here, $m = \frac{f_o}{f_e} = 9 \quad \therefore \quad f_o = 9f_e \quad ...(i)$

When adjusted for parallel rays, the distance between the objective and eye piece is

$f_o + f_e = 20$ cm

Using (i), $9f_e + f_e = 20$ or $f_e = 2$ cm

From (i), $f_o = 9 \times 2$ cm $= 18$ cm.

78. (1) $\frac{f_o}{f_e} = 10$

$f_o + f_e = 1.1$

$\therefore f_o = 100$ cm and $f_e = 10$ cm

Final magnification $= f_o\left(\frac{1}{D}+\frac{1}{f_e}\right) = 100\left[\frac{1}{25}+\frac{1}{10}\right] = 14$

79. (1) $MP = \frac{f_0}{f_e} = \frac{200}{5} = 40$

80. (1) (i) Given, $f_0 + f_e = 105$ and $f_0 = 20\ f_e$

$\therefore 20\ f_e + f_e = 105 \Rightarrow f_e = 5$ cm

$f_0 = 20 \times 5 = 100$ cm

Hence, $m = \frac{f_0}{f_e} = \frac{100}{5} = 20.$

81. (2) In compound microscope objective forms real image while eye piece forms virtual image.

82. (4) $f_o = 1.5$ cm, $f_e = 6.25$ cm,

$u_o = -2$ cm, $v_e = -D = -25$cm

For objective lens $\frac{1}{f_o} = \frac{1}{v_o} - \frac{1}{u_o}$

$\frac{1}{1.5} = \frac{1}{v_o} - \frac{1}{-2} \Rightarrow \frac{1}{v_o} = \frac{1}{1.5} - \frac{1}{2}$ or $v_o = 6$ cm

For eye piece $\frac{1}{f_e} = \frac{1}{v_e} - \frac{1}{u_e}$

$\frac{1}{6.25} = \frac{1}{-25} - \frac{1}{-u_e} \Rightarrow \frac{1}{u_e} = \frac{1}{6.25} + \frac{1}{25} = \frac{4}{25} + \frac{1}{25} = \frac{1}{5}$

$u_e = 5$ cm,

Length of tube $= L = v_o + u_e$

$= 6.0$ cm $+ 5.0$ cm, $L = 11$ cm

83. (4) Magnification $= 5 = \frac{f_o}{f_e}$

Tube length $= 36 = f_o + f_e$

$\Rightarrow f_o = 30$ cm

$f_e = 6$ cm

84. (1) M.P. $= m_o \times m_e$

$32 = m_o \times 4$

$\therefore m_o = 8$

85. (3) $\Delta\theta = \frac{1.22\lambda}{a} = \frac{1.22 \times 5 \times 10^{-7}}{2} = 0.31 \times 10^{-6}$ rad

86. (1) $\frac{1}{v_o} - \frac{1}{-1.1} = \frac{1}{+1} \Rightarrow v_o = +11$ cm

$\therefore MP = \frac{v_0}{u_o}\left(1+\frac{D}{f_e}\right) = \frac{11}{1.1}\left(1+\frac{25}{5}\right) = 60$

87. (2) Microscope is required to magnify the object which is too small. So, spectacles should be removed while working with microscope.

88. (4) Magnification of compound microscope

$m = \frac{-v_o}{u_o} \times \frac{D}{u_e}$

$m_1 = m_o \times m_e$; $m = m_1 \times m_2$

89. (4) Maximum magnification $= 1 + \frac{D}{f}$

$m = 1 + \frac{25}{2.5}$

$m = 1 + 10 = 11$; $m = 11$

90. (1) For objective lens $\frac{1}{f} = \frac{1}{v_o} - \frac{1}{u_o}$

$\frac{1}{v_o} = \frac{1}{f_o} + \frac{1}{u_o} = \frac{1}{4} + \frac{1}{-5}$

$\frac{1}{v_o} = \frac{1}{20} \Rightarrow v_o = 20$ cm

Now, $m = \left|\frac{v_0}{u_0}\right|\left(1+\frac{D}{f_e}\right)$

$m = \frac{20}{5}\left[1+\frac{20}{10}\right] = 4(3) = 12$

91. (2) We know that $m = \frac{f_o}{f_e}$

$m = \frac{\text{Angle subtended by the image}}{\text{Angle subtended by the object}}$

$\therefore \frac{f_o}{f_e} = \frac{\beta}{\alpha}$

$\beta = \frac{f_o \times \alpha}{f_e} = \frac{60 \times 3°}{5} = 36°$

$\beta = 36°$

92. (1) $f_o = 1$ cm $\quad f_e = 2.5$ cm

$u_o = -1.2$ cm $\quad L = ?$

$\frac{1}{v_o} = \frac{1}{f_o} + \frac{1}{u_o}$

$\Rightarrow \frac{1}{v_o} = \frac{0.2}{1.2} \Rightarrow v_o = 6$ cm

$\therefore L = v_o + f_e \Rightarrow L = 6 + 2.5 = 8.5$ cm

$L = 8.5$ cm

93. (3) $\frac{1}{f_o} = \frac{1}{v_o} - \frac{1}{u_o} \Rightarrow \frac{1}{v_o} = \frac{1}{f_o} + \frac{1}{u_o}$

$\frac{1}{v_o} = \frac{5}{150}$

$\Rightarrow v_o = 30$ cm

Again $\frac{1}{f_e} = \frac{1}{v_e} - \frac{1}{u_e}$

$$\frac{1}{u_e}=\frac{1}{v_e}-\frac{1}{f_e}=-\frac{1}{25}-\frac{1}{2.5}=\frac{-11}{25}$$

$u_e = -2.27$ cm

Length of the telescope, $L = |v_o| + |u_e|$

$L = 30 + 2.27$

$L \approx 32$ cm

94. (4) As we know, length of microscope tube

$L = v_o + u_e$

$$\frac{1}{f_o}=\frac{1}{v_o}-\frac{1}{u_o} \Rightarrow v_o=\frac{u_o f_o}{(u_o+f_o)} \quad ...(i)$$

$$\frac{1}{f_e}=\frac{1}{v_e}-\frac{1}{u_e}$$ [for eye piece, least distance of distinct vision is 25cm shown as D]

$$\frac{1}{f_e}=\frac{1}{D}-\frac{1}{u_e}$$

$$u_e=\frac{f_e D}{f_e-D} \quad ...(ii)$$

from (*i*) & (*ii*), we get

$$\therefore L=\frac{u_o f_o}{(u_o+f_o)}+\frac{f_e D}{f_e-D}$$

$$=\frac{-2\times1.5}{(-2+1.5)}+\frac{6.25\times(-25)}{6.25-(-25)}=\frac{3}{0.5}+\frac{156.25}{31.25}$$

$L = 11$ cm

95. (3) As we know the relation, $m=\frac{LD}{f_o f_e}$

$$m=\frac{20\times25}{0.5\times1}$$

$m = 1000$

96. (2) When final image is at d = 25 cm ;

$$m=\frac{-f_o}{f_e}\left(1+\frac{f_e}{d}\right)=\frac{-20}{5}\left(1+\frac{5}{25}\right)=-4.8$$

97. (3) $f_o=\frac{1}{1.25}=0.8,\ f_e=\frac{1}{-20}=-0.05$

$\therefore |L| = |f_o| + |f_e|$ [for telescope]

$|L| = 0.8 - 0.05;$

$|L| = 0.75$ m = 75cm

$$|m|=\frac{|f_o|}{|f_e|}=\frac{0.8}{0.05}=16$$

98. (1) Here, $f_o = 15$ cm, $f_e = 1$ cm $=\frac{1}{100}$ m

Diameter of image of moon formed by objective lens

$$=\alpha f_o=\frac{3.42\times10^6}{3.8\times10^8}\times15\text{m}=0.135\text{m}$$

= 13.5 cm.

High Order Time Intensive MCQs

1. (2) Here, real depth of mark, x = 3 cm

apparent depth of mark, y = ?

refractive index, $\mu = 1.5$

As $\mu=\frac{x}{y}$

$$\therefore y=\frac{x}{\mu}=\frac{3}{1.5}=2\text{cm}$$

Distance through which mark appears to be raised $= x - y = 3 - 2 = 1$ cm.

2. (3)

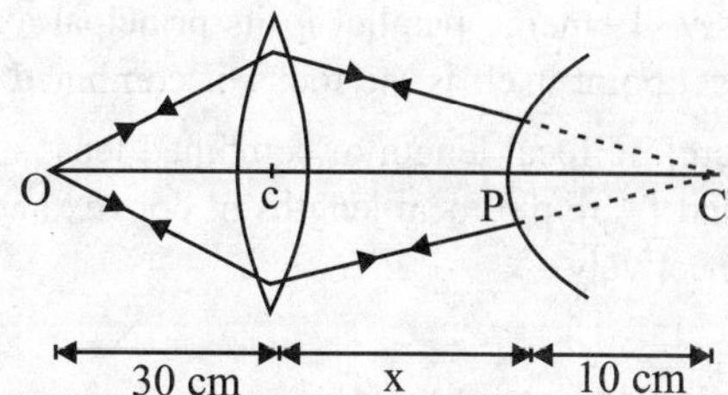

Here, u = OC = – 30 cm, f = 20 cm.

$$\frac{1}{v}=\frac{1}{f}+\frac{1}{u}=\frac{1}{20}-\frac{1}{30}=\frac{1}{60}$$

v = 60 cm.

To obtain upright image of O at O itself,

$v = OC_1 = x + 10$

$\therefore x = v - 10 = 60 - 10 = 50$ cm

3. (2) Using mirror formula

$$\frac{1}{f}=\frac{1}{v}+\frac{1}{u} \quad ...(i)$$

On differentiating Eq. (*i*), we get $0=-\frac{1}{v^2}dv-\frac{1}{u^2}du$

$$dv=-\frac{v^2}{u^2}\times b \quad (du=b) \quad ...(ii)$$

From Eq. (*i*),

$$\frac{1}{v}=\frac{1}{f}-\frac{1}{u}=\frac{u-f}{fu} \text{ or } \frac{u}{v}=\frac{u-f}{f}$$

$$\frac{v}{u}=\frac{f}{u-f} \quad ...(iii)$$

Now from Eqs. (*ii*) and (*iii*).

$$dv=-\left(\frac{f}{u-f}\right)^2 b$$

Therefore, size of image $=\left(\frac{f}{u-f}\right)^2 b$

4. (1) According to Snell's law,

$$\mu=\frac{\sin i}{\sin r}$$

Given, $i=2r \Rightarrow \mu=\frac{\sin 2r}{\sin r}$

$$\Rightarrow \quad \mu=\frac{2\sin r\cos r}{\sin r} \Rightarrow \cos^{-1}\frac{\mu}{2}=r$$

$$\Rightarrow \quad i=2\cos^{-1}\frac{\mu}{2}$$

5. (4) If C is critical angle then, $\sin C = \frac{v_1}{v_2}$

$$C = \sin^{-1}\left(\frac{v_1}{v_2}\right) = \sin^{-1}\left(\frac{1.8\times10^8}{2.4\times10^8}\right) = \sin^{-1}\left(\frac{3}{4}\right)$$

6. (2) For total internal reflection, $i > c$

$$\Rightarrow \sin i > \sin c \Rightarrow \sin 45° > \frac{1}{\mu} \Rightarrow \mu > \sqrt{2} \Rightarrow \mu > 1.4$$

7. (2) Since the image of the pin is formed at the position of the pin itself, therefore light rays incident on the plane mirror normally which means that the rays falling over the lens will emerge parallel to its principal axis. Hence, the object point itself is the focus of combined lens.

Therefore focal length of combined lens is F = 40 cm, If f_1 and f_2 are the focal lengths of convex and concave lens respectively.

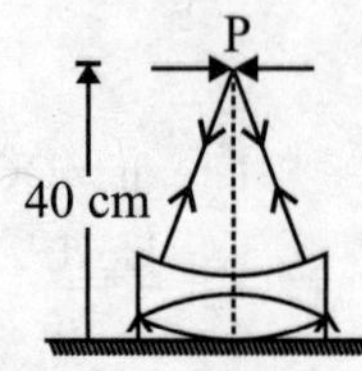

$$\frac{1}{F} = \frac{1}{f_1} + \frac{1}{f_2}$$

$$\frac{1}{f_2} = \frac{1}{F} - \frac{1}{f_1}$$

$$f_2 = \frac{Ff_1}{f_1 - F} = \frac{40\times10}{10-40} = -\frac{40}{3}\text{cm}$$

Negative sign indicates concave nature of the lens.

8. (1) As shown in figure, a light ray from the coin will not emerge out of liquid if, $i > c$ therefore, minimum radius R corresponds to $i = c$.

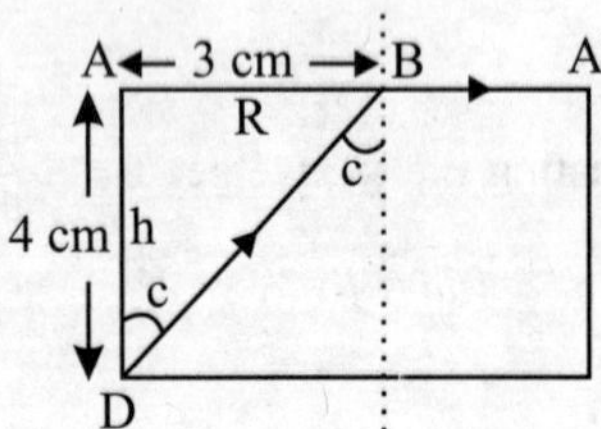

In $\Delta DAB \frac{R}{h} = \tan c \Rightarrow R = h\tan c \Rightarrow R = \frac{h}{\sqrt{\mu^2 - 1}}$

Given, R = 3 cm and h = 4 cm

$$\therefore \quad \frac{3}{4} = \frac{1}{\sqrt{\mu^2-1}} \Rightarrow \mu^2 = \frac{25}{9} \Rightarrow \mu = \frac{5}{3}$$

But $\mu = \frac{c}{v} \Rightarrow v = \frac{c}{\mu} \Rightarrow v = \frac{c}{\mu} = \frac{3\times10^8}{5/3}$

$$= 1.8\times10^8 \text{ ms}^{-1}$$

9. (2)

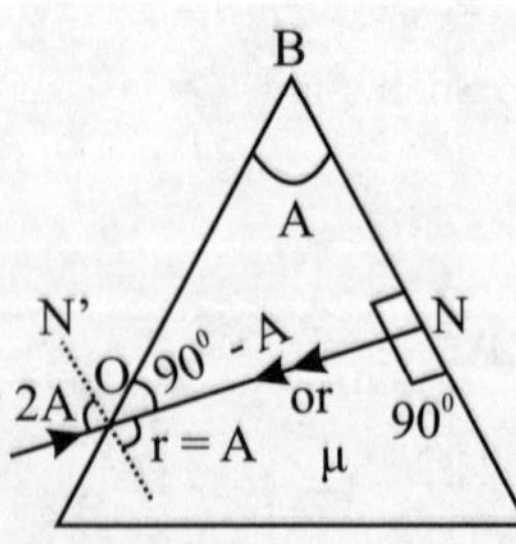

$\angle BON = 90° - A$

So, $\angle r = 90° - (90° - A) \Rightarrow \angle r = A$

By Snell's law, $\frac{\sin i}{\sin r} = \mu \Rightarrow \frac{\sin(2A)}{\sin(A)} = \mu$

$$\frac{2\sin A\cos A}{\sin A} = \mu \Rightarrow \mu = 2\cos A$$

10. (3) Focal length of the combination $\frac{1}{f} = \frac{1}{f_1} + \frac{1}{f_2}$ ...(*i*)

We have, $f_1 = \frac{R}{(\mu_1 - 1)}$ and $f_2 = \frac{-R}{(\mu_2 - 1)}$

On putting these values in Eq. (*i*), we get

$$\frac{1}{f} = \frac{(\mu_1-1)}{R} - \frac{(\mu_2-1)}{R} \quad \Rightarrow \quad f = \frac{R}{\mu_1 - \mu_2}$$

11. (3) There will be no image at position ± 45 cm from mirror A.

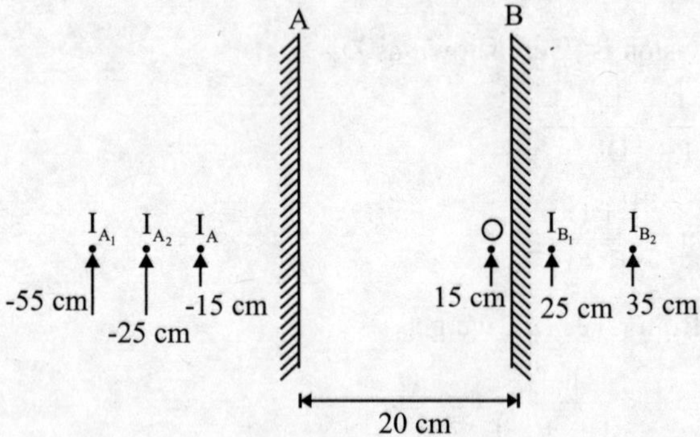

12. (4) After silvering the plane surface, planoconvex lens behaves as a concave mirror of focal length

$$\frac{1}{f'} = \frac{2}{f_{lens}}$$

but f' = 0.2 m

$\Rightarrow f_{lens} = 2f' = 2 \times 0.2 = 0.4$ m

Now, from lens maker's formula,

$$\frac{1}{f_{lens}} = (\mu - 1)\left(\frac{1}{R_1} - \frac{1}{R_2}\right)$$

$$\therefore \quad \frac{1}{0.4} = (1.5-1)\times\frac{1}{R_1} \qquad (\because R_2 = \infty)$$

$\Rightarrow R_1 = 0.5 \times 0.4 = 0.2$ m

13. (3) Required semi vertical angle = Critical angle i_C

$$= \sin^{-1}\frac{1}{\mu} = \sin^{-1}\left(\frac{3}{4}\right).$$

14. (4) $n = \frac{\text{Real depth}}{\text{Apparent depth}}$

$$n_1 = \frac{6}{4} = \frac{3}{2} = 1.5$$

$$\Rightarrow \frac{n_1}{u} - \frac{n_2}{v} = \frac{n_1 \sim n_2}{R} \Rightarrow \frac{1.5}{6} - \frac{4}{17} = \frac{1.5-1}{R} \Rightarrow R = 34 \text{ cm}$$

15. (1) $\frac{\text{Distance of sun's image}}{\text{Distance of sun}}$

$$= \frac{\text{Distance of image from the lens}}{\text{Distance of sun from the lens}}$$

$$\frac{x}{1.39\times10^9}=\frac{10^{-1}}{1.5\times10^{11}}$$

$$x=\frac{1.39\times10^9\times10^{-1}}{1.5\times10^{11}}$$

$x = 9.2 \times 10^{-4}$ m

16. **(1)** $\frac{1}{f}=(n-1)\left(\frac{1}{R_1}-\frac{1}{R_2}\right)=(1.5-1)\left[\frac{1}{(+20)}-\frac{1}{(-20)}\right]$

$\Rightarrow f=+20\,\text{cm}$

Incident rays travelling parallel to the axis of lens will converge at its principal focus.

$\therefore L = +20$ cm

17. **(4)** The focal length of the lens,

$$\frac{1}{f}=(\mu-1)\left(\frac{1}{R_1}-\frac{1}{R_2}\right)\Rightarrow\frac{1}{f}=(1.5-1)\left(\frac{-1}{20}-\frac{1}{-30}\right)$$

$$\Rightarrow \frac{1}{f}=0.5\left(\frac{-30+20}{600}\right)$$

$$\Rightarrow f=-120\text{ cm}$$

18. **(4)** For a thin prism, angle of minimum deviation is given by

$\delta = (\mu - 1)A$

where, μ is refractive index of the prism and A the angle of prism. For dispersion without deviation

$\delta_1=\delta_2 \Rightarrow (\mu_1-1)A_1=(\mu_2-1)A_2$

$$\Rightarrow A_2=\frac{(\mu_1-1)}{(\mu_2-1)}A_1$$

Given, $\mu_1 = 1.54$, $A_1 = 4°$ and $\mu_2 = 1.72$

$$\Rightarrow A_2=\frac{(1.54-1)}{(1.72-1)}\times4=3°$$

19. **(4)** $A = 30°$ $\mu=1.414=\sqrt{2}$ $i_2=0°$

$r_2 = 0°$ (as the refracted ray retraces its path)

$r_1 + r_2 = A$

$r_1 = A = 30°$

$$\mu=\frac{\sin i_1}{\sin r_1}\Rightarrow\sqrt{2}=\frac{\sin i_1}{\sin 30°}\Rightarrow\sin i_1=\frac{1}{\sqrt{2}}\Rightarrow i_1=45°$$

20. **(3)** $A=60°, r=A/2=30°$

using Snell's law

$$\frac{\mu_g}{\mu_i}=\frac{\sin i}{\sin 30°}$$

$$\Rightarrow \sin i=\frac{\sqrt{3}}{2}\Rightarrow i=60°$$

21. **(4)** The focal length of the plano-convex lens is 20 cm with the plane surface being silvered.

So, the net power of the lens is given by $P=\frac{1}{f}\times2=\frac{2}{f}$

Thus, the net focal length is, $\frac{f}{2}=\frac{20}{2}=10\text{cm}$

22. **(4)** $\mu=\frac{\sin i_1}{\sin r_1}=\frac{\sin i_2}{\sin r_2}$

23. **(3)** Dispersive power, $\omega=\frac{\delta_B-\delta_R}{\delta}$

For 1st prism $\omega_1=\frac{\delta_B-\delta_R}{\delta}=\frac{12°-8°}{10°}=\frac{2}{5}$

where, $\delta=\frac{\delta_B+\delta_R}{2}$

similarly, for 2nd prism

$$\omega_2=\frac{14°-10°}{12°}=\frac{1}{3}\Rightarrow\frac{\omega_1}{\omega_2}=\frac{2}{5}\times\frac{3}{1}=\frac{6}{5}$$

24. **(2)** Equation of object, $x = y$

Equation of image

$x = -y$

$x + y = 0$

25. **(4)**

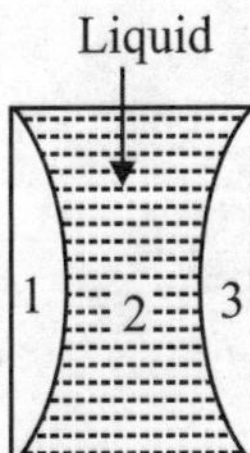

By lens maker's formula

$$\frac{1}{f_1}=(1.5-1)\left(\frac{1}{-\infty}-\frac{1}{-12}\right)=0.5\times\frac{1}{12}=\frac{1}{24}$$

$$\frac{1}{f_2}=(\mu-1)\left(\frac{1}{-12}-\frac{1}{12}\right)=-\frac{(\mu-1)}{6}$$

$$\frac{1}{f_3}=(1.5-1)\left(\frac{1}{12}-\frac{1}{\infty}\right)=\frac{1}{24}$$

Also, $F = -60$ cm

$$\frac{1}{F}=\frac{1}{f_1}+\frac{1}{f_2}+\frac{1}{f_3}$$

$$\therefore \frac{1}{-60}=\frac{1}{24}-\frac{(\mu-1)}{6}+\frac{1}{24}=-\frac{(\mu-1)}{6}+\frac{1}{12}$$

or $\frac{(\mu-1)}{6}=\frac{1}{12}+\frac{1}{60}$; or $\frac{(\mu-1)}{6}=\frac{5+1}{60}=\frac{6}{60}$

or $\frac{(\mu-1)}{6}=\frac{36}{60}=0.6$ or $\mu = 0.6 + 1 = 1.6$

26. **(2)** Let thickness of B = t

$\therefore$ thickness of C = 2t

and thickness of A = t + 2t = 3t

If λ_0 is wavelength in air, then as number of waves in A = no. of waves in B + no. of waves in C.

$$\therefore \frac{3t}{\lambda_0/\mu_0}=\frac{t}{\lambda_0/\mu_1}+\frac{2t}{\lambda_0/\mu_2}$$

or, $3\mu_0 = \mu_1 + 2\mu_2$

or $\mu_1 = 3\mu_0 - 2\mu_2 = 3 \times 1.4 - 2 \times 1.2 = 1.8$

27. (4) $d_{APP} = \frac{d_1}{\mu_1} + \frac{d_2}{\mu_2} = \frac{d + d_2}{v_{eff}} \Rightarrow v_{eff} = \frac{d_1 + d_2}{\frac{d_1}{\mu_1} + \frac{d_2}{\mu_2}}$

28. (1) $\frac{D_1}{D_2} = \frac{f_1}{f_2} \Rightarrow \frac{10}{D_2} = \frac{15}{20} \Rightarrow D_2 = \frac{40}{3}\text{cm}$

29. (3)

30. (2) No. of ref $= \frac{2\sqrt{3}}{\frac{0.2}{\sqrt{3}}} = 30$

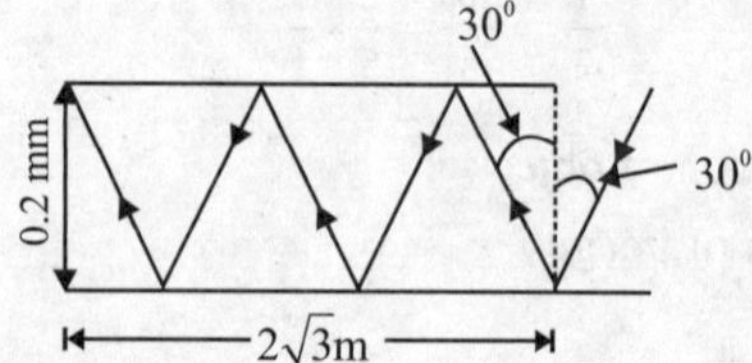

31. (3) Let $f_1 = f_2 = f$

$d = \frac{2f}{3} = 12\,\text{cm}$

$\therefore f = 18$ cm

As $\frac{1}{F} = \frac{1}{f_1} + \frac{1}{f_2} - \frac{d}{f_1 f_2}$

$\therefore \frac{1}{F} = \frac{1}{18} + \frac{1}{18} - \frac{12}{18\times18} = \frac{1}{9} - \frac{1}{27} = \frac{2}{27}$, $F = \frac{27}{2} = 13.5\,\text{cm}$

32. (1) Time, $t = \frac{\text{Thickness of glass}}{\text{Speed of light in glass}} = \frac{d}{c/\mu}$

$= \frac{\mu d}{c} = \frac{3\times4\times10^{-3}}{3\times10^{8}} = 4\times10^{-11}\text{s.}$

33. (4) From lens makers formula

$\frac{1}{f} = \left(\frac{n_2}{n_1} - 1\right)\left(\frac{1}{R_1} - \frac{1}{R_2}\right)$

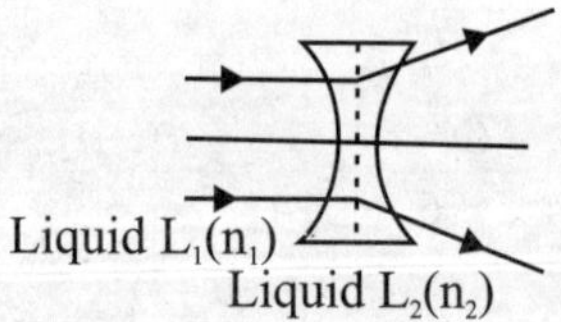

For a concave lens, R_1 is negative and R_2 is positive

$\therefore \left(\frac{1}{R_1} - \frac{1}{R_2}\right)$ is negative $\Rightarrow f$ will be negative only when $n_2 > n_1$

$\therefore$ lens must be filled with L_2 and placed in L_1 as shown is the above figure.

34. (4) From $\frac{1}{v} - \frac{1}{u} = \frac{1}{f}$

$\frac{u}{v} - 1 = \frac{u}{f}$

$\frac{1}{m} - 1 = \frac{u}{f}$ or $\frac{1}{m} = \left(\frac{1}{f}\right)u + 1$

which is the equation of a st line whose slope

$\frac{1}{f} = \frac{b}{c}$

$\therefore f = \frac{c}{b}$

35. (2) $\frac{\mu_2}{v} - \frac{\mu_1}{-u} = \frac{\mu_2 - \mu_1}{R}$

or $\frac{1.5}{x} + \frac{1}{x} = \frac{1.5 - 1}{R} \Rightarrow x = \frac{2.5R}{0.5} = 5R$

36. (1) Applying Snell's law at point A,

$\mu_1 \sin\theta_1 = \mu_2 \sin\theta_2$ ...(*i*)

But $\theta_2 = 90° - \theta_c$

$\therefore \cos\theta_2 = \sin\theta_c = \frac{\mu_1}{\mu_2}$ ...(*ii*)

Elimination of θ_2 between (*i*) and (*ii*), we get

$\therefore \sin\theta_1 = \sqrt{\left(\frac{\mu_2}{\mu_1}\right)^2 - 1}$

37. (1) Here, $R_1 = -R$, $R_2 = R$, $\mu_2 = 1.5 = \frac{3}{2}$

$\mu_1 = 1.75 = \frac{7}{4}$

$\frac{1}{f} = \left(\frac{\mu_2}{\mu_1} - 1\right)\left(\frac{1}{R_1} - \frac{1}{R_2}\right)$

$\frac{1}{f} = \left(\frac{3/2}{7/4} - 1\right)\left(\frac{1}{-R} - \frac{1}{R}\right) = \frac{-1}{7}\left(\frac{-2}{R}\right)$

$\therefore f = +\frac{7R}{2} = +3.5R$

$\therefore$ Concave lens will behave as a convergent lens of focal length 3.5 R.

38. (3) Given, $v_1 = 1.5 \times 10^8$ m/s ; $v_2 = 2.0 \times 10^8$ m/s

Refractive index for medium M_1 is

$\mu_1 = \frac{c}{v_1} = \frac{3\times10^8}{1.5\times10^8} = 2$

Refractive index for medium M_2 is

$\mu_2 = \frac{c}{v_2} = \frac{3\times10^8}{2.0\times10^8} = \frac{3}{2}$

If i is the angle of incidence and C is the critical angle then for total intenal reflection

$\sin i \geq \sin C$ But $\sin C = \mu_2/\mu_1$

$\therefore \sin i \geq \frac{\mu_2}{\mu_1} \geq \frac{3/2}{2}$ or $i \geq \sin^{-1}\left(\frac{3}{4}\right)$

39. (2) $\mu = \frac{t_1 + t_2}{\frac{t_1}{\mu_1} + \frac{t_2}{\mu_2}} = \frac{10 + 15}{\frac{10}{4/3} + \frac{15}{3/2}}$

$= \frac{25}{7.5 + 10} = 1.43$

40. (1) For total internal reflection,

$i > i_C$

$\Rightarrow \quad \sin i > \sin i_C$

$\Rightarrow \quad \sin 45° > \frac{1}{\mu} \Rightarrow \mu > \sqrt{2} \Rightarrow \mu > 1.414$

For green and violet colours, $\mu > 1.414$, they will suffer total internal reflection. But red colour will be refracted.

41. (4) $\sin i_c = \frac{1}{\mu} = \frac{\sqrt{3}}{2}$

$\sin r = \sin(90° - i_c) = \cos i_c = \frac{1}{2}$

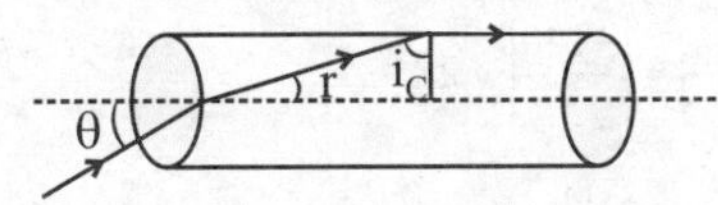

$\frac{\sin\theta}{\sin r} = \frac{\mu_2}{\mu_1} = \frac{2}{\sqrt{3}}; \ \sin\theta = \frac{2}{\sqrt{3}}\sin r = \frac{2}{\sqrt{3}}\cdot\frac{1}{2} = \frac{1}{\sqrt{3}}$

$\therefore \ \theta = \sin^{-1}\left(\frac{1}{\sqrt{3}}\right).$

Statement Based MCQs

1. **(2)** A microscope cannot work as a telescope by interchanging its two lenses. On the other hand, in a telescope, the objective has much larger focal length than the eyepiece.
2. **(4)** Velocity of light in a glass slab is different for different colours of light. There is no dispersion when a ray of white light emerges from a rectangular glass slab.
3. **(4)** As light travels through many layers of earth's atmosphere, light of star is bent many times and in random direction, which results in twinkling of star. In the case of looming phenomena light from a ship reaches us in such a way that the ship appears to be inverted. It is due to total internal reflection.
4. **(1)** Cone shaped cells in retina are responsible for detection of colours.

Assertion & Reason MCQs

1. **(2)** Light is dispersed by water droplets present in air to form a colored pattern called rainbow.

   Intensity of seattering is inversely propertinala to the wavelength of light.
2. **(1)** Both Assertion and Reason are True and the Reason is a correct explanation of the Assertion.
3. **(2)** Focal length of mirror does not depend on wavelength, frequency as well as refractive index. It is fixed for a given curvature of the spherical mirror. Reflection of light from more denser medium result the phase change of π.
4. **(1)** The nature of lens depends on refractive indices of material of the lens and the surrounding medium.

Match the Columns MCQs

1. (1) (a) $m = -2 \Rightarrow$ An enlarged real image formed by concave mirror.

(b) $m = \frac{-1}{2} \Rightarrow$ A diminished real image formed by concave mirror.

(c) $m = +2 \Rightarrow$ An enlarged virtual image formed by concave mirror.

(d) $m = +\frac{1}{2} \Rightarrow$ A diminished virtual image produced by convex mirror.

2. (1) Image is inverted and real, shows concave mirror. As the image is on the opposite side of principal axis, the mirror is concave.

Plane mirrors shows same size of image as object.

3. (1) (a) and (b)

$v = \frac{\mu_2}{\mu_1}u$

$\Rightarrow |v| < |u|$ as $\mu_1 > \mu_2$

i.e., v and u are of same sign.

Or they are on same side of plane surface.

From plane surface, if object is real, image is virtual and vice-versa.

(c) and (d)

$v = \frac{\mu_1}{\mu_2}u \Rightarrow |v| > |u|$

Other explanations are same.

4. (2) $\frac{1}{v} - \frac{1}{u} = \frac{1}{f} \Rightarrow \frac{1}{u} = -\frac{1}{f} + \frac{1}{v}$

$= \frac{f - v}{vf} \Rightarrow u = \frac{vf}{f - v}$

$m = \frac{v}{u} = \frac{v(f - v)}{vf} = \frac{f - v}{f} = 1 - \frac{v}{f}$

$m = \frac{-1}{f}\cdot v + 1$

Slope = inverse of focal length

Intercept an y-axis = 1

Intercept an x-axis = focal length

NCERT Exemplar MCQs

1. (1) Since, deviation

$\delta = (\mu - 1)A = (1.5 - 1) \times 5° = 2.5°$

By geometry, angle of refraction by first surface is 5°. Using snell's law for prism,

$\mu_1 i = \mu_2 r$

$1 \times \theta = 1.5 \times 5$

$\theta = 7.5°$

2. (4) Since $v \propto \lambda$, the light of red colour is of highest wavelength and therefore of highest speed. Therefore, after travelling through the slab, the red colour emerge first.

3. **(3)** When an object approaches a convergent lens from the left of the lens with a uniform speed of 5 m/s, the image moves away from the lens with a non-uniform acceleration.
4. **(2)** A passenger in an aeroplane may see a primary and a secondary rainbow like concentric circles.
5. **(3)** According to VIBGYOR. among all given sources of light, the blue light has smallest wavelength. According to Cauchy relationship, smaller the wavelength higher the refractive index and consequently smaller the critical angle.

   So, corresponding to blue colour, the critical angle is least which facilitates total internal reflection for the beam of blue light. The beam of green light would also undergo total internal reflection.
6. **(3)** Here, R = 20 cm, $\mu = 1.5$, on substituting the values in $f = \frac{R}{\mu - 1} = \frac{20}{1.5 - 1} = 40$ cm of converging nature as $f > 0$.

   Therefore, lens act as a convex lens irrespective of the side on which the object lies.
7. **(2)** The phenomenon involved in the reflection of radiowaves by ionosphere is similar to total internal reflection of light in air during a mirage i.e., angle of incidence is greater than critical angle.
8. **(2)** The PQ ray of light passes through focus F and incident on the concave mirror, after reflection, should become parallel to the principal axis and shown by ray-2 in the figure.
9. **(2)** Here, light ray goes from optically rarer medium (air) to optically denser (turpentine), then it bends towards the normal, i.e., $i > r$ whereas when it goes from to optically denser medium turpentine to rarer medium water. Then it bends away the normal, i.e., $i < r$.
10. **(4)** The speed of the image of the car would appear to increase as the distance between the cars decreases.
11. **(1)** The negative refractive index of material shows reflected ray will be in clockwise direction from normal.

NEET & AIIMS Past Year MCQs

1. **(3)** Statement I: Correct. Image formation requires regular reflection/refraction (e.g., mirrors/lenses).

   Statement II: Correct. Object color depends on incident light's constituent colors.
2. **(4)**

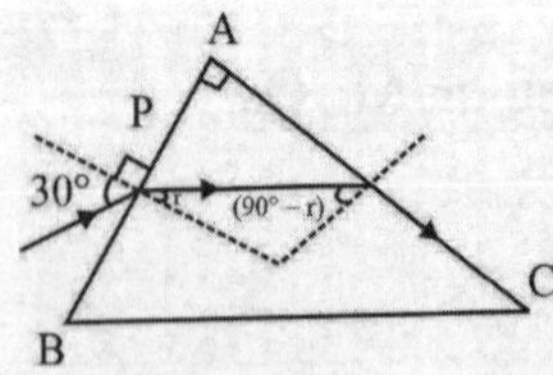

For surface *AB*

$1 \sin 30° = \mu \sin r$ ... (*i*)

For surface *AC*

$\mu \sin (90° - r) = 1$ ... (*ii*)

$\Rightarrow \mu \cos r = 1$

$\Rightarrow \cos r = \frac{1}{\mu}$

From (*i*) and (*ii*)

$\Rightarrow \frac{1}{2} = \mu\sqrt{1 - \frac{1}{\mu^2}}$

$\Rightarrow \frac{1}{4} = \mu^2\left(1 - \frac{1}{\mu^2}\right)$

$\Rightarrow \frac{1}{4} = \mu^2 - 1 \Rightarrow \mu^2 = 1 + \frac{1}{4}$

$\Rightarrow \mu = \frac{\sqrt{5}}{2}$

3. **(4)** $f_0 = 140$ cm, $f_e = 5$ cm

   For normal adjustment

   $M = \left|\frac{-f_0}{f_e}\right| = \left|\frac{-140}{5}\right| = 28$
4. **(1)** Speed of light in air is

   $C = \frac{x}{t_1}$

   Speed of light in another denser medium.

   $C_2 = \frac{10x}{t_2} \Rightarrow \mu = \frac{C}{C_2} = \frac{x}{t_1} \times \frac{t_2}{10x} \Rightarrow \mu = \frac{t_2}{10t_1}$

   For total internal reflection

   $\sin C = \frac{1}{\mu} \Rightarrow \sin C = \frac{10t_1}{t_2}$

   $C = \sin^{-1}\left(\frac{10t_1}{t_2}\right)$
5. **(4)** $\frac{1}{f_1} = (1.6 - 1)\left(\frac{1}{\infty} - \frac{1}{20}\right) = -\frac{3}{100}$

   $\frac{1}{f_2} = (1.5 - 1)\left(\frac{1}{20} - \frac{1}{-20}\right) = \frac{1}{20}$

   $\frac{1}{f_3} = \frac{1}{f_1} = -\frac{3}{100}$

   $\frac{1}{f_{net}} = \frac{1}{f_1} + \frac{1}{f_2} + \frac{1}{f_3} = -\frac{3}{50} + \frac{1}{20}$

   $\Rightarrow f_{net} = -100$ cm
6. **(1)** $F_{convex} = f$

   $F_{concave} = -f$

   $\frac{1}{f_{comb}} = \frac{1}{F_{convex}} + \frac{1}{F_{concave}} = \frac{1}{f} - \frac{1}{f}$

   $\Rightarrow f_{comb} = \infty$
7. **(4)** $R_1 = R_2 = 20$ cm = 0.2 m and $\mu = 1.5$

   $R_1 = +20$ cm $\quad R_2 = -20$ cm

   As, $P = \frac{1}{f} = (\mu - 1)\left(\frac{1}{R_1} - \frac{1}{R_2}\right)$

   $P = \left(\frac{3}{2} - 1\right)\left(\frac{1}{0.2} + \frac{1}{0.2}\right)$

   $P = +5D$

8. (3)

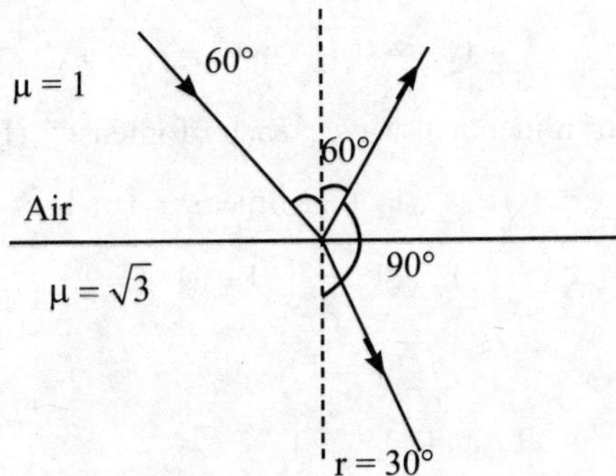

As the reflected and refracted rays are perpendicular therefore angle of incidence is equal to Brewster's angle.

$\Rightarrow \tan i_P = \mu = \sqrt{3} \Rightarrow i_p = 60° = i$

9. (3) $\mu = \dfrac{c}{u} \Rightarrow u \propto \dfrac{1}{\mu}$

$\because \sin i_c = \dfrac{\mu_R}{\mu_D} = \dfrac{v_d}{v_r} = \dfrac{1.5}{2} = \dfrac{3}{4} \Rightarrow i_c = \sin^{-1}\left(\dfrac{3}{4}\right)$

10. (4) $MP = \dfrac{f_0}{f_e}$

$R.P. = \dfrac{A}{1.22\lambda}$

Large aperture of the objective lens provides better resolution

∴ Good quality of image is formed and also it gathers more light.

11. (1) $d = f' - f_1 = 20 - 5 = 15\text{cm}$

12. (4) $r_1 + r_2 = A = 30°$

$r_2 = 30° (r_1 = 0°)$

from snell's law

$\sqrt{3} \sin r_2 = 1 \times \sin e$

$\Rightarrow \angle e = 60°$

13. (3) First for image formation from lens

$u = -60$ cm

$f = +30$cm

$$\Rightarrow v = \frac{uf}{u+f} = \frac{-60 \times 30}{-60+30} = 60cm$$

this real image formed by lens acts as virtual object for mirror

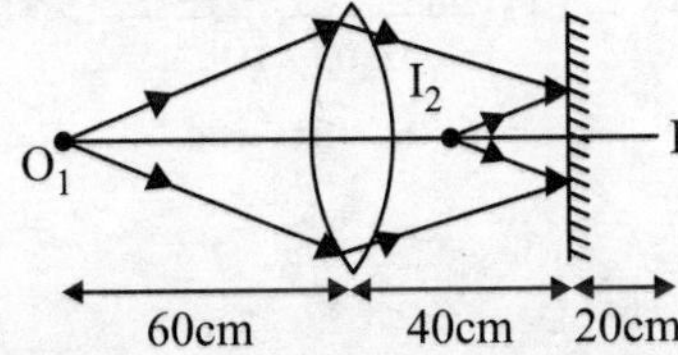

Real image from plane mirror is formed 20cm in front of mirror, hence at 20cm distance from lens, Now for second refraction from lens,

$u = -20$cm

$f = +30$cm

$$v = \frac{uf}{u+f} = \frac{-20 \times 30}{-20+30} = 60cm$$

So, final virtual image is 60cm from lens, or 20cm behind mirror

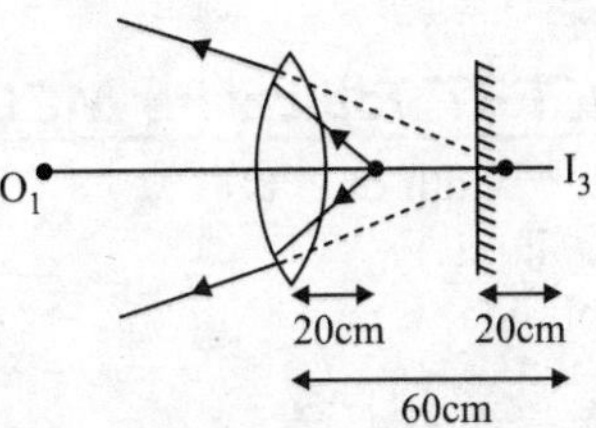

14. (4) Magnifying power of an astronomical telescope is given by

$$I_m l = \frac{t_o}{f_e}\left(1 + \frac{F_e}{D}\right)$$

Where t_o is focal length of objective, f_e is focal length of eye piece, D is least distance of distinct vision. From this formula we observe that

Magnifying power $\propto \dfrac{1}{\text{Focal of eye-piece}}$

Hence, to produce the largest magnification the focal length of eye-piece must be smallest.

Hence, $f_e = 15$ cm

15. (1) $^r\mu_d = \dfrac{v_r}{v_d}$

$^r\mu_d = \dfrac{2v}{v} = 2$

$u = \dfrac{1}{\sin_c}$

$\sin_c = \dfrac{1}{\mu} = \dfrac{1}{2}$

$c = 30$

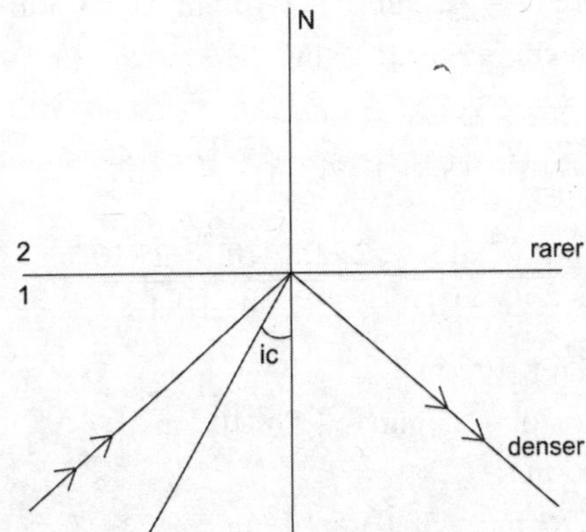

16. (2) $f_0 = 200$cm $\quad f_e = 4$cm

at normal adjustment ⇒ strainless vision

17. (4) ∵ Deviation, $\delta_{mean} = 0$

$\Rightarrow \delta_1 + \delta_2 = 0 \Rightarrow (\mu_1 - 1)A_1 + (\mu_2 - 1)A_2 = 0$

$$\Rightarrow A_2 = \frac{(\mu_1 - 1)A}{(\mu_2 - 1)} = \frac{(1.54-1)\times 4}{(1.72-1)} = 3°$$

18. (1) In simple microscope object is kept in between optical center and focus of the lens so it forms erect, enlarged and virtual image.

19. (3)

Wave Optics

NCERT Topic-wise MCQs

1. (1) Intensity ∝ width of slits

width ∝ (amplitude)^2

$\therefore I \propto (\text{Amplitude})^2$

$$\therefore \frac{I_1}{I_2} = \frac{a^2}{b^2} = \frac{1}{4} \Rightarrow \frac{1}{2} = \frac{a}{b}$$

So, width ∝ intensity

or b = 2a

$$\therefore \frac{I_{max}}{I_{min}} = \frac{(a+b)^2}{(a-b)^2} = \left(\frac{a+2a}{a-2a}\right)^2 = \left(\frac{3a}{-a}\right)^2 = \frac{9}{1}$$

$$\therefore \frac{I_{max}}{I_{min}} = \frac{9}{1}$$

2. (3) we know, $\beta = \frac{\lambda D}{d} = \frac{5900 \times 10^{-10} \times 3 \times 10^{-2}}{3 \times 10^{-3}}$

$\beta = 5.9 \times 10^{-6}$ m

$\beta = 5.9$ μm

3. (3) $\beta = \frac{\lambda D}{d}, \beta' = \frac{\lambda' D}{d}$

$$\therefore \frac{\beta'}{\beta} = \frac{\lambda' D}{d} \times \frac{d}{\lambda D} = \frac{\lambda'}{\lambda} = \frac{1}{\mu}$$

$$\beta' = \frac{\beta}{\mu} = \frac{2}{1.33} = 1.5 \text{ mm}$$

4. (2) As we know R = (a + b). This is the case of constructive interference because both are in same phase.

5. (1) Here, d = 0.2 mm = 2×10^{-4} m, D = 1.5 m

x = 1.8 cm = 1.8×10^{-2} m

n = 4, λ = ?

For dark fringes, $x = (2n-1)\frac{\lambda D}{2d}$

or $\lambda = \frac{2xd}{(2n-1)D} = \frac{2 \times 1.8 \times 10^{-2} \times 2 \times 10^{-4}}{(2 \times 4 - 1) \times 1.5}$

$\lambda = 6.86 \times 10^{-7}$ m

6. (2) Here, d = 1.5 mm = 1.5×10^{-3} m

D = 2.5 m

λ = 589.3 nm = 589.3×10^{-9} m

Fringe width, $\beta = \frac{\lambda D}{d} = \frac{589.3 \times 10^{-9} \times 2.5}{1.5 \times 10^{-3}} = 982.17 \times 10^{-6}$ m

Number of fringes $= \frac{x}{\beta} = \frac{4.9 \times 10^{-3}}{982.17 \times 10^{-6}} = 4.99 \approx 5$

7. (4) For 3rd bright fringe, path difference = 3 λ = 15000 Å and phase difference = 3 (2 π) = 6 π radian.

8. (2) Huygen's principle is a geometrical method to find a wavefront.

9. (3) Intensity of each of n waves = I_0

When interference is due to coherent sources.

$I = I_1 + I_2 + 2\sqrt{I_1 I_2} \cos\phi$

$I_{max} = I_1 + I_2 + 2\sqrt{I_1 I_2} \cos 0° = (\sqrt{I_1} + \sqrt{I_2})^2$

For n identical waves, each of intensity (I_0)

$I_{max} = (\sqrt{I_0} + \sqrt{I_0} + \ldots n \text{ times})^2 = (n\sqrt{I_0})^2 = n^2 I_0$

10. (3) $I_{max} = \left(\sqrt{I_1} + \sqrt{I_2}\right)^2 = \left(\sqrt{I} + \sqrt{4I}\right)^2 = 9I$

$I_{min} = \left(\sqrt{I} - \sqrt{4I}\right)^2 = I$

$I_{max} = 9I$ and $I_{min} = I$

11. (3) $A^2 = a_1^2 + a_2^2 + 2a_1 a_2 \cos\delta$

$\because a_1 = a_2 = a$

$\therefore A^2 = 2a^2(1 + \cos\delta)$

$\Rightarrow A^2 \propto \cos^2\frac{\delta}{2}$ $\quad \left[\because 1 + \cos\delta = 2\cos^2\frac{\delta}{2}\right]$

$\because I \propto A^2$

$\therefore I \propto A^2 \propto \cos^2\frac{\delta}{2}$

Then $I \propto \cos^2 \delta/2$

12. (2) Total intensity at any point is given by

$I = kR^2 = k(a^2 + b^2 + 2ab\cos\phi)$

when b = a, $I = k(a^2 + a^2 + 2a^2\cos\phi) = 2ka^2(1 + \cos\phi)$

At the centre of a bright fringe,

$\phi = 0°$ ∴ $I_1 = 2ka^2(1 + \cos 0°) = 4ka^2$

Distance between two fringes = β, which is proportional to wavelength (λ)

Now $\frac{\lambda}{4}$ corresponds to a phase diff. $= \frac{2\pi}{4} = \frac{\pi}{2}$

$\therefore I_2 = 2ka^2(1 + \cos\pi/2) = 2ka^2$

$$\therefore \frac{I_1}{I_2} = \frac{4ka^2}{2ka^2} = \frac{2}{1}$$

13. (2) Wavelength of H_α line, λ = 6563 Å = 6563×10^{-10} m

Red shift observed in star, (λ' – λ) = 15 Å = 15×10^{-10} m

Let the velocity of the star with which it is receding away from the earth be v.

∴ Red shift relation, $\lambda' - \lambda = \frac{v}{c}\lambda$

$$\therefore v = \frac{c}{\lambda} \times (\lambda' - \lambda) = \frac{3 \times 10^8 \times 15 \times 10^{-10}}{6563 \times 10^{-10}}$$

$= 6.86 \times 10^5$ ms^{-1}

14. (4) $\frac{I_1}{I_2} = \beta = \frac{a^2}{b^2}$ $\quad \therefore \frac{a}{b} = \sqrt{\beta}$

$I_{max} = (a + b)^2$

$I_{min} = (a - b)^2$

$$\frac{I_{max} - I_{min}}{I_{max} + I_{min}} = \frac{(a+b)^2 - (a-b)^2}{(a+b)^2 + (a-b)^2} = \frac{4ab}{2(a^2 + b^2)}$$

$$= \frac{2ab}{(a^2 + b^2)} = \frac{2b.b\sqrt{\beta}}{b^2 + \beta b^2} = \frac{2\sqrt{\beta}}{1 + \beta}$$

15. (3) Here, $\frac{I_1}{I_2} = \frac{a^2}{b^2} = \frac{1}{9}$ or $\frac{a}{b} = \frac{1}{3} \Rightarrow b = 3a$

(a) When the sources are incoherent, no interference occurs. Resultant intensity at any point is the sum of the intensities of individual waves, i.e., $I = I_1 + I_2 = k(a^2 + b^2)$

(b) When the sources are coherent, and constant phase diff. $\phi = 60°$, then

$I' = K(a^2 + b^2 + 2ab\cos\phi)$

$= K(a^2 + b^2 + 2ab\cos 60°)$

$= K(a^2 + b^2 + ab)$

$$\therefore \frac{I}{I'} = \frac{K(a^2+b^2)}{K(a^2+b^2+ab)} = \frac{K(a^2+9a^2)}{K(a^2+9a^2+3a^2)} = \frac{10a^2}{13a^2} = \frac{10}{13}$$

16. (3) $\beta = \frac{\lambda D}{d} = \frac{5\times10^{-7}\times2}{10^{-3}} = 10^{-3}\,m$

$\beta = 1\,mm$

17. (4) The phase difference between the light waves reaching the third bright fringe (n = 3) from the central bright fringe will be = $2n\pi$

$= 2 \times 3 \times \pi = 6\pi$ (case of constructive interference)

18. (1) $\beta = \frac{\lambda D}{d} = \frac{\beta_1}{\beta_2} = \left(\frac{D_1}{D_2}\right)\left(\frac{\lambda_1}{\lambda_2}\right)\left(\frac{d_2}{d_1}\right) \Rightarrow \frac{D_1}{D_2} = \frac{4}{1}$

19. (1) From $I_R = 4I_0\cos^2\frac{\phi}{2} = \frac{1}{2}(4I_0)$

$\cos^2\frac{\phi}{2} = \frac{1}{2}, \quad \cos\frac{\phi}{2} = \frac{1}{\sqrt{2}}; \quad \frac{\phi}{2} = 45°, \phi = 90°$

Minimum distance from central maximum corresponding to $\phi = \pi/2$ would be : $\frac{\phi}{2\pi}\times\beta$

$$x = \frac{\pi/2}{2\pi} = \left(\frac{\lambda D}{d}\right) = \frac{1}{4}\times\frac{500\times10^{-9}\times1}{1\times10^{-3}}$$

$x = 125 \times 10^{-6}$ m

$x = 1.25 \times 10^{-4}$ m.

20. (2) Angular fringe width

$\theta = \frac{\lambda}{d}, \quad \theta \propto \lambda$

$\lambda_w = \frac{\lambda_{air}}{\mu_w}$, so $\theta_w = \frac{\theta_{air}}{\mu_w} = \frac{0.20}{4/3} = 0.15°$

21. (4) Distance of nth dark fringe from central fringe

$x_n = \frac{(2n-1)\lambda D}{2d}$

$x_2 = \frac{(2\times2-1)\lambda D}{2d} = \frac{3\lambda D}{2d}$

$10^{-3} = \frac{3\times\lambda\times1}{2\times0.9\times10^{-3}} \Rightarrow \lambda = 6\times10^{-5}\,cm$

22. (1) $$\frac{I_{max}}{I_{min}} = \left[\frac{\sqrt{\frac{I_1}{I_2}}+1}{\sqrt{\frac{I_1}{I_2}}-1}\right]^2 = \left(\frac{\sqrt{2}+1}{\sqrt{2}-1}\right)^2$$

$$= \left(\frac{1.414+1}{1.414-1}\right)^2 = \left(\frac{2.414}{0.414}\right)^2 = 33.99 \simeq 34$$

23. (3) When path diff. is zero, phase diff. $\phi = 0$

$\therefore$ from $I = k[I_1 + I_2 + 2\sqrt{I_1I_2}\cos\phi]$

$I_X = k[a^2 + b^2 + 2\sqrt{a^2b^2}\cos0°] = k[a^2 + b^2 + 2ab]$

At Y, path diff. $= \frac{\lambda}{4}$; phase diff $= \frac{2\pi}{4} = \frac{\pi}{2}$ rad

$\therefore \quad I_Y = k[a^2 + b^2 + 2\sqrt{a^2b^2}\cos\pi/2] = k[a^2 + b^2]$

$$\frac{I_X}{I_Y} = \frac{k[a^2+b^2+2ab]}{k[a^2+b^2]} = \frac{(a+b)^2}{a^2+b^2}$$

24. (1) If thin film appears dark

$2\mu t\cos r = n\lambda$ [for normal incidence r = 0°]

$2\mu t = n\lambda \Rightarrow t = \frac{n\lambda}{2\mu}$

$t_{min} = \frac{5890\times10^{-10}}{2\times1} = 2.945\times10^{-7}\,m$

25. (4) $\beta = \frac{\lambda D}{d} \xrightarrow{\text{VIBGYOR}}, \lambda$ increases

$\lambda_R > \lambda_G > \lambda_B$

so $\beta_R > \beta_G > \beta_B \Rightarrow \beta \propto \lambda$

26. (4) Distance of the nth bright fringe from the center

$x_n = \frac{n\lambda D}{d}$

$x_3 = \frac{3\times6000\times10^{-10}\times2.5}{0.5\times10^{-3}} = 9\times10^{-3}\,m$

$x_3 = 9$ mm

27. (1) For dark fringe

$x = \frac{(2n-1)\lambda D}{2d}$

$$\lambda = \frac{2xd}{(2n-1)D} \Rightarrow \frac{2\times1.8\times10^{-2}\times2\times10^{-4}}{(2\times4-1)\times1.5} = \frac{7.2\times10^{-6}}{7\times1.5}$$

$\lambda = 6.86 \times 10^{-7}$ m

28. (2) Frequency, $\nu = 6\times10^{14}\,s^{-1}$

$\lambda = c/\nu = \frac{3\times10^8}{6\times10^{14}} = 5\times10^{-7}\,m$

$\beta = 0.75mm = 0.75 \times 10^{-3}m$

$D = 1.5$ m

$d = \frac{\lambda D}{\beta} = \frac{5\times10^{-7}\times1.5}{0.75\times10^{-3}} = 10^{-3}\,m$

$d = 10^{-3}$m

29. (2) $I = 4I_o \cos^2 \phi/2$

$$\frac{(4I_o)}{2} = 4I_o \cos^2 \phi/2 \quad \text{[Intensity is half of the maximum]}$$

$$\therefore \frac{1}{\sqrt{2}} = \cos \phi/2$$

$90° = \phi$

Minimum distance from central maximum corresponding to = 90° is

$$x = \frac{\phi}{2\pi} \times \beta \Rightarrow \frac{\pi/2}{2\pi} \times \frac{\lambda D}{d}$$

$$x = \frac{1}{4} \times \frac{500 \times 10^{-9} \times 1}{1 \times 10^{-3}}$$

$x = 1.25 \times 10^{-4}$m

30. (1) Here $\frac{I}{I_0} = \frac{3}{4}$ (given)

$$\Rightarrow \cos^2\left(\frac{\phi}{2}\right) = \frac{3}{4} \qquad \left(\because I = I_0 \cos^2\left(\frac{\phi}{2}\right)\right)$$

$$\text{or } \cos\frac{\phi}{2} = \frac{\sqrt{3}}{2} \text{ or } \phi = 60° = \frac{\pi}{3}$$

Phase difference, $\phi = \frac{2\pi}{\lambda} \times$ path difference

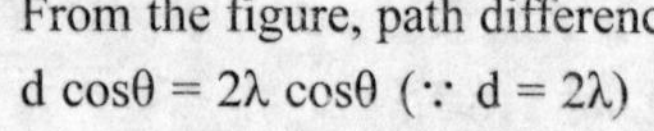

From the figure, path difference

$d \cos\theta = 2\lambda \cos\theta$ ($\because d = 2\lambda$)

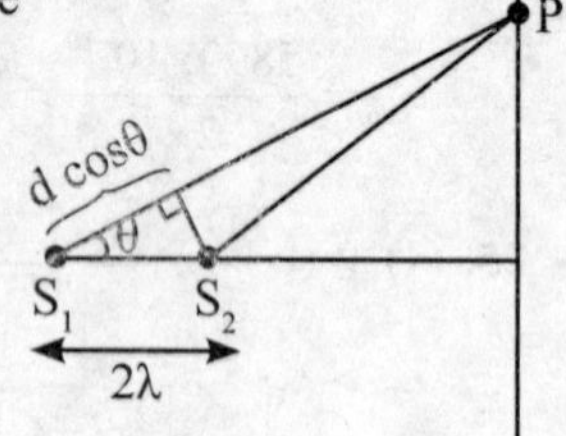

$$\therefore \frac{\pi}{3} = \frac{2\pi}{\lambda} 2\lambda \cos\theta$$

$$\therefore \cos\theta = \frac{1}{12}$$

$$\therefore \theta = \cos^{-1}\left(\frac{1}{12}\right)$$

31. (4) Phase difference $\phi = \frac{2\pi}{\lambda} \times \text{path difference}$

$$= \frac{2\pi}{\lambda} \times \frac{\lambda}{6} = \frac{\pi}{3} = 60°$$

Intensity, $I = I_o \cos^2(\phi/2)$

$$\frac{I}{I_o} = \cos^2(30°) \Rightarrow \frac{I}{I_o} = \left(\sqrt{3}/2\right)^2 = 3/4 = 0.75$$

32. (1) Angular fringe width of first minima

$$\theta = \frac{2(2n-1)\lambda}{2d} = \frac{(2n-1)\lambda}{d}$$

$$\theta = \frac{(2 \times 1 - 1) \times 4.8 \times 10^{-7}}{0.6 \times 10^{-3}} = 8 \times 10^{-4} \text{ rad}$$

Angular width = 8×10^{-4} rad

33. (3) Average of I_{max} and I_{min} is given as

$$\frac{I_{max} + I_{min}}{2} = \frac{(a_1 + a_2)^2 + (a_1 - a_2)^2}{2} = \frac{2(a_1^2 + a_2^2)}{2}$$

$$\frac{I_{max} + I_{min}}{2} = a_1^2 + a_2^2 = I_1 + I_2$$

Hence, during phenomenon of interference energy is conserved but it is redistributed.

34. (4) $n_1 \lambda_1 = n_2 \lambda_2 \Rightarrow$ Position of fringe $= \frac{nD\lambda}{d}$

$$\Rightarrow \frac{\lambda_1}{\lambda_2} = \frac{n_2}{n_1} = \frac{6000}{\lambda_2} = \frac{24}{16}$$

$$\lambda_2 = \frac{6000 \times 16}{n_2} = \frac{96000}{24} = 4 \times 10^3 = 4000 \text{ Å}$$

35. (4) Total phase difference = $\phi_1 + \phi_2$

where $\phi_1 = 54° = 54 \times \frac{\pi}{180} = 0.3\pi$ radian,

$$\phi_2 = \frac{2\pi}{\lambda} \times 1.5\lambda = 3\pi \text{ radian}$$

$\therefore \phi_1 + \phi_2 = 0.3\pi + 3\pi = 3.3\pi$ radian.

36. (1) $Z_F = \frac{a^2}{\lambda} = \frac{(2 \times 10^{-3})^2}{600 \times 10^{-9}} = 6.67$m

37. (4) $a = 4\text{cm} = 4 \times 10^{-2}$ m

$\lambda = 2 \text{ cm} = 2 \times 10^{-2}$ m

Angular spread of central maximum (2θ) is

$a \sin\theta = \lambda n$

$$\sin\theta = \frac{2 \times 10^{-2}}{4 \times 10^{-2}} \Rightarrow \sin\theta = \frac{1}{2} \Rightarrow \theta = 30°$$

Angular spread = $2\theta = 2 \times 30° = 60°$

38. (3) Resolving Power (R.P.) $= \frac{1}{d} = \frac{2\mu \sin\theta}{\lambda}$

$$\text{R.P.} = \frac{2 \times 1 \times \sin 30°}{600 \times 10^{-9}} = \frac{10}{6} \times 10^6 = 1.67 \times 10^6$$

39. (1) As we know, $Z_f = \frac{a^2}{\lambda} = \frac{(3 \times 10^{-3})^2}{500 \times 10^{-9}}$

$$Z_f = \frac{9 \times 10^{-6}}{5 \times 10^{-7}} = \frac{1.8}{10^{-1}}$$

$Z_f = 18$ m

40. (3) $d = \frac{\lambda}{2\mu \sin\theta}$

$$= \frac{5000 \times 10^{-10}}{2 \times 1 \times \sin 30°} \qquad (\because 2\theta = 60° \Rightarrow \theta = 30°)$$

$d = 5 \times 10^{-7}$

$d = 0.5$ μm

41. (3) For the first minima $d \sin\theta = \lambda$

$$\sin\theta = \frac{\lambda}{d} \Rightarrow \frac{5000 \times 10^{-10}}{0.001 \times 10^{-3}} = \frac{5 \times 10^{-7}}{10^{-6}}$$

$$\sin\theta = \frac{1}{2} \Rightarrow \theta = 30°$$

42. (2) Condition for the minima = $d \sin\theta = n\lambda$

If angle is small, then $\sin\theta = \theta$

$\therefore d\theta = \lambda$

Half angular width $2\theta = \frac{\lambda}{d}$

full angular width $\omega = 2\theta = \frac{2\lambda}{d}$

Also $\omega' = \frac{2\lambda'}{d}$

$\therefore \frac{\lambda'}{\lambda} = \frac{\omega'}{\omega}$ or $\lambda' = \frac{\lambda\omega'}{\omega} \Rightarrow 6000 \times 0.7 = 4200\,\mathring{A}$

43. (3) From coherent source on superposition gives two beams interference pattern, But when their vibrations are perpendicular to each other interference will not be possible.

44. (4) Width of central bright fringe $= \frac{2\lambda D}{d}$

$= \frac{2 \times 500 \times 10^{-9} \times 80 \times 10^{-2}}{0.2 \times 10^{-3}}$

$= 4 \times 10^{-3}$ m

Width of central bright fringe = 4mm

45. (2) Most of the light is diffracted between the first order minima on either side of central maximum, for which

$\sin\theta = \pm\frac{\lambda}{d} = \pm\frac{450 \times 10^{-9}}{0.2 \times 10^{-3}} = \pm 2.25 \times 10^{-3}$

As $\sin\theta$ is small, therefore, $\sin\theta \approx \theta$

$\therefore \quad \theta = \pm 2.25 \times 10^{-3}$ radian.

Angular divergence $= 2\theta = 4.5 \times 10^{-3}$ rad.

46. (4) Relation for the maximum

$\frac{(2n+1)\lambda}{2} = a\sin\theta$

$\frac{(2 \times 1+1)\lambda}{2} = a\sin\theta \qquad [\because n = 1]$

$\frac{3\lambda}{2} = a\sin\theta$

47. (3) Distance between the first dark fringe on either side of the central bright fringe

$x = \frac{2\lambda D}{d}$

$\therefore x = \frac{600 \times 10^{-9} \times 2 \times 2}{10^{-3}}$

$x = 24 \times 10^{-4}\text{m} \Rightarrow 2.4$ mm

48. (2) Distance $= \frac{2\lambda D}{d} = \frac{2 \times 6500 \times 10^{-10} \times 1.8}{0.5 \times 10^{-3}}$

Distance $= 4.68 \times 10^{-3}\text{m} = 4.68$ mm

49. (3) Most of the light is diffracted between the first order minima on either side of central maximum.

for which $\sin\theta = \frac{\lambda}{d} = \frac{450 \times 10^{-9}}{0.2 \times 10^{-3}}$

$= 5 \times 45 \times 10^{-5} = 225 \times 10^{-5}$

$\therefore$ As $\sin\theta$ is small, therefore $\sin\theta \approx \theta$

$\theta = 2.25 \times 10^{-3}$rad

50. (2) Colored lines are observed due to diffraction.

51. (3) When the speed of the electrons is increased the wavelength decreases.

So, the angular width decreases as the angular width is directly proportional to the wavelength.

52. (2) Angular width $= \frac{2\lambda}{d} = \frac{2 \times 12000 \times 10^{-9}}{5 \times 10^{-3}} = \frac{24}{5} \times 10^{-3}$

Angular width $= 4.8 \times 10^{-3}$ rad

53. (1) Here, $\lambda = ?$

Limit of resolution, $d = 5 \times 10^{-7}$ m

$2\theta = 90°$ or $\theta = 45°$ $\mu = 1$

As $d = \frac{\lambda}{2\mu\sin\theta}$

$\therefore \lambda = 2\mu d\sin\theta = 2 \times 1 \times 5 \times 10^{-7} \sin 45°$

$\lambda = 10^{-6} \times \frac{1}{\sqrt{2}} = 0.707 \times 10^{-6}$m

54. (2) According to Brewster law. $\mu = \tan i_p$

$1.732 = \tan i_p$

$\tan^{-1}(1.732) = i_p$

$60° = i_p$

55. (1) As $\tan i_p = \tan^{-1}(\mu) = \tan^{-1}(\sqrt{3})$

$\therefore i_p = 60°$

as $r = 90° - i_p \Rightarrow r = 90° - 60° = 30°$

56. (2) When the ordinary light passes through the polarizer it makes the light particle to vibrate in one direction.

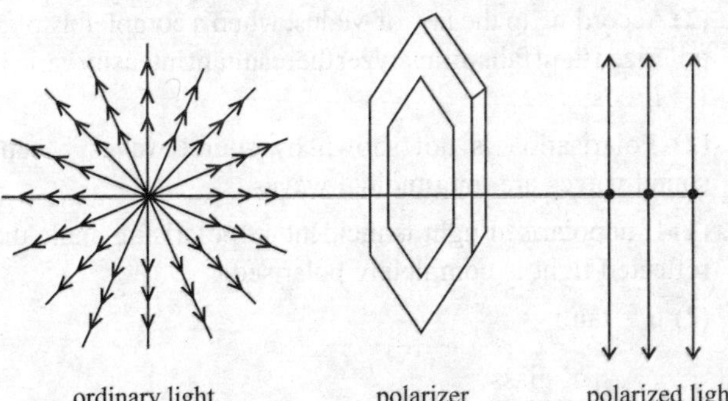

57. (2) According to the law of Malus.

$I = I_o \cos^2\theta$

$\frac{I_o}{2} = I_o\cos^2\theta \Rightarrow \frac{1}{\sqrt{2}} = \cos\theta$

$\theta = 45°$

58. (4) By using $\mu = \tan i_p$

$\mu = \tan 60° = \sqrt{3}$

So critical angle,

$C = \sin^{-1}\left(\frac{1}{\mu}\right) \Rightarrow C = \sin^{-1}\left(\frac{1}{\sqrt{3}}\right)$

59. (3) From Brewster's law

$\mu = \tan i_p$

$\frac{c}{v} = \tan 60° = \sqrt{3}$

$v = \frac{c}{\sqrt{3}} = \frac{3\times10^8}{\sqrt{3}}$

$v = \sqrt{3}\times10^8 \,m/s$

60. (3) If an unpolarized light is converted into plane polarised light by passing through a Polaroid, its intensity becomes half.

61. (2) According to the Brewster $\mu = \tan i_p$

$\mu = \tan(53.74°)$

$\mu = \sqrt{2}$

$\therefore \sqrt{2} = \frac{\sin i}{\sin r} \Rightarrow \sqrt{2} = \frac{\sin 45°}{\sin r}$

$\sin r = \frac{1}{\sqrt{2}}\times\frac{1}{\sqrt{2}} = \frac{1}{2}$

$r = \sin^{-1}\left(\frac{1}{2}\right)$

$r = 30°$

62. (2) Intensity of light after passing through the first sheet

$I_1 = \frac{I}{2}$

Intensity of light after passing through the second sheet

$I_2 = I_1\cos^2(45°) = \frac{I}{4}$

After passing through n^{th} sheet

$I_n = \frac{I}{2^n} = \frac{I}{64} \Rightarrow n = 6$

63. (2) According to the law of Malus, when a completely plane polarized light falls on analyzer the resultant intensity varies as $I = I_0 \cos^2\theta$.

64. (2) Polarisation is not shown by sound waves because sound waves are longitudinal waves.

65. (1) If unpolarised light is incident at polarising angle then reflected light is completely polarised.

66. (2) $\mu = \tan i_p$

${}^w\mu_g = \frac{\mu \text{ of glass}}{\mu \text{ of water}} = \tan i_p$

μ of glass = μ of water × $\tan i_p$

$= \frac{4}{3}\times\tan 51°$

$= \frac{4}{3}\times1.2349 = 1.647$

67. (4) The new amplitude will be

$A' = A\cos 60° = A/2$

68. (1) After passing through the first polaroid, intensity of the light is halved $= \frac{256}{2}$

Polaroid P_2 and P_3 will make intensity cos2(60o) and cos2(30o) times respectively.

⇒Intensity of light point O

$= \frac{256}{2}\times\frac{1}{4}\times\left(\frac{\sqrt{3}}{2}\right)^2 = 24$

High Order Time Intensive MCQs

1. (3) For net intensity,

$I' = 4I_0\cos^2\frac{\phi}{2}$

Case I $K = 4I_0\cos^2\left(\frac{2\pi}{2}\right)$ $\left(\because \phi = \frac{2\pi}{\lambda}\times\lambda\right)$

$\Rightarrow K = 4I_0$

Case II $K' = 4I_0\cos^2\left(\frac{\pi/2}{2}\right)$ $\left(\because \phi = \frac{2\pi}{\lambda}\times\frac{\lambda}{4}\right)$

$\Rightarrow K' = K/2$

2. (1) According to given question,

$\frac{I_{max}}{I_{min}} = \frac{(a+b)^2}{(a-b)^2}$

$\frac{3}{1} = \frac{(a+b)^2}{(a-b)^2}$

$\therefore \frac{a}{b} = \frac{\sqrt{3}+1}{\sqrt{3}-1}$

3. (1) Here, $\lambda_r = 660$ nm ; $\lambda' = ?$

For diffraction minima, $a\sin\theta = n\lambda$, $\sin\theta = \frac{n\lambda}{a}$

For first minima of red light, $\sin\theta = \frac{1\lambda_r}{a}$

For diffraction maxima, $a\sin\theta = (2n+1)\frac{\lambda}{2}$

$\therefore$ for first maxima of λ', $a\sin\theta' = \frac{3\lambda'}{2}$; $\sin\theta' = \frac{3\lambda'}{2a}$;

As the two coincide, therefore, $\sin\theta' = \sin\theta$

$\frac{3\lambda'}{2} = \frac{\lambda_r}{a}$ or $\lambda' = \frac{2}{3}\lambda_r$ or $\lambda' = \frac{2}{3}(660) = 440\,\text{nm}$

4. (2) Fringe width, $\beta = \frac{D\lambda}{d} \Rightarrow \beta \propto \lambda$

If the Young's double slit experiment is performed in water, wavelength λ decreases, so fringe width also decreases.

5. (4) Phase difference, $\phi = \frac{2\pi}{\lambda}\times$ path difference

$= \frac{2\pi}{\lambda}\times\frac{\lambda}{6} = \frac{\pi}{3} = 60°$

Intensity,

$I = I_0\cos^2\left(\frac{\phi}{2}\right) \Rightarrow \frac{I}{I_0} = \cos^2(30°) = \left(\frac{\sqrt{3}}{2}\right)^2 = 0.75$

6. (4) $\beta = \frac{\lambda D}{d} = \frac{600\times10^{-9}}{1\times10^{-3}} = 12\times10^{-4}\,\text{m}$

So, the distance between the first dark fringes on either side of the central bright fringe is

$x = 2\beta = 2\times12\times10^{-4}\,\text{m} = 2.4\,\text{mm}$

7. (3) Two sources should have the same wavelength, nearly the same amplitude and have a constant phase difference. For interference pattern to remain sustained.

8. (4) The rays of light from two coherent sources superimpose each other on the screen forming alternate maxima and minima. For two sources of equal intensities the intensity of the maxima is four times the individual intensities. If two non-coherent sources superimpose, there will be no maxima and minima, instead the intensity will be $\left(\frac{I_0}{4}+\frac{I_0}{4}\right)=\frac{I_0}{2}$ throughout.

9. (2) The distance traversed by light in a medium of refractive index μ in time t is given by:

$$d = vt$$

where, v is velocity of light in the medium. The distance traversed by light in a vacuum in this time, d' = ct

$$d' = c.\frac{d}{v} \qquad \text{[from Eq. (i)]}$$

$$d' = d\frac{c}{v} = \mu d \quad ...(ii) \qquad \left(\text{since } \mu = \frac{c}{v}\right)$$

Here, optical path for first ray = n_1L_1

Optical path for second ray = n_2L_2

Path difference = $n_1L_1 - n_2L_2$

Now-phase difference = $\frac{2\pi}{\lambda}\times$ path difference

$$\Delta\phi = \frac{2\pi}{\lambda}\times(n_1L_1 - n_2L_2) = \frac{2\pi}{\lambda}(n_1L_1 - n_2L_2)$$

10. (4) For destructive interference

Path difference = $S_2P - S_1P = (2n-1)\frac{\lambda}{2}$

where n = 1, 2, 3, ... etc.

For, n = 6

$$S_2P - S_1P = 11\frac{\lambda}{2}$$

11. (3) For the first minima, $d \sin\theta = \lambda$

$$\Rightarrow \sin\theta = \frac{\lambda}{d} \Rightarrow \theta = \sin^{-1}\left[\frac{5000\times10^{-10}}{0.001\times10^{-3}}\right] = \sin^{-1}(0.5) = 30°$$

12. (4) The phase difference (ϕ) between the wavelets from the top edges and th⇒e bottom edges of the slit is

$$\phi = \frac{2\pi}{\lambda}(d\sin\theta)$$

where, d is the slit width.

The first minima of the diffraction pattern occurs at $\sin\theta = \frac{\lambda}{d}$

$$\text{So, } \phi = \frac{2\pi}{\lambda}\left(d\times\frac{\lambda}{d}\right) = 2\pi$$

13. (3) For n^{th} secondary minima,

Path difference = a sin $\theta_n = n\lambda$

For n^{th} secondary maxima

path difference = $a\sin\theta_n = (2n+1)\frac{\lambda}{2}$

For 1st minima, λ = 5500 Å, $\theta_n = 30°$

$$a \sin 30° = \lambda \qquad ...(i)$$

For 1st secondary maxima, path difference = a sin θ_n

$$= (2+1)\frac{\lambda}{2} \qquad ...(ii)$$

Dividing Eq. (*i*) by Eq. (*ii*), we get

$$\frac{\frac{1}{2}}{\sin\theta_n} = \frac{2}{3} \Rightarrow \sin\theta_n = \frac{3}{4} \Rightarrow \theta_n = \sin^{-1}\left(\frac{3}{4}\right)$$

14. (3) In single slit diffraction experiment, width of central maxima, $y = \frac{2\lambda D}{d}$

$$\Rightarrow \quad y_1 = \frac{2\lambda_1 D}{d_1} \text{ and } y_2 = \frac{2\lambda_2 D}{d_2}$$

$$\therefore \quad \frac{y_1}{y_2} = \frac{\lambda_1}{\lambda_2}\times\frac{d_2}{d_1} = \frac{400}{600}\times\frac{d/2}{d} = \frac{1}{3}$$

$$\Rightarrow y_2 = 3y_1$$

15. (4) Intensity reflected along AB, I_1 = I/4

Intensity transmitted along CA′ = 3 I/4

Intensity reflected at CA′ = $\frac{1}{4}\times\frac{3I}{4} = 3I/16$

Intensity reflected at A' = $\frac{1}{4}\times\frac{3I}{16} = \frac{3I}{64}$

∴ Intensity transmitted along

$$A'B' = I_2 = \frac{3I}{16} - \frac{3I}{64} = \frac{9I}{64}$$

$$\frac{I_1}{I_2} = \frac{a^2}{b^2} = \frac{I}{4}\times\frac{64}{9I} = \frac{16}{9}$$

$$\therefore \quad \frac{a}{b} = \frac{4}{3}$$

$$\frac{I_{max}}{I_{min}} = \frac{(a+b)^2}{(a-b)^2} = \frac{(4+3)^2}{(4-3)^2} = \frac{49}{1}$$

16. (1) Amplitudes A_1 and A_2 are added as vectors. Angle between these vectors is the phase difference $(\beta_1 - \beta_2)$ between them.

$$\therefore R = \sqrt{A_1^2 + A_2^2 + 2A_1A_2\cos(\beta_1 - \beta_2)}$$

17. (2) As 4th bright fringe of unknown wavelength (λ) coincides with 3rd bright fringe of known wavelength (λ)

$$\therefore \quad \frac{4\lambda' D}{d} = \frac{3\lambda D}{d}$$

$$\text{or } \lambda' = \frac{3}{4}\lambda = \frac{3}{4}\times 590 = 442.5\,\text{nm}$$

18. (2) For coherent sources, $I_1 = 4\,I_0$

For incoherent sources, $I_2 = I_0 + I_0 + 2\,I_0$

$$\therefore \quad \frac{I_1}{I_2} = \frac{4I_0}{2I_0} = \frac{2}{1}$$

19. (1) Path difference $=\left(d+\frac{b^2}{2d}\right)-d=\frac{b^2}{2d}$

For missing wavelengths, $(2n-1)\frac{\lambda}{2}=\frac{b^2}{2d}$

If $n=1, \frac{\lambda}{2}=\frac{b^2}{d}$

If $n=2, 3\times\frac{\lambda}{2}=\frac{b^2}{2d}$ or $\lambda=\frac{b^2}{3d}$

20. (3) Using $x=\frac{\lambda D}{d}$

$x_2-x_1=(3\lambda-\lambda)\frac{D}{d}=\frac{2\lambda D}{d}$

$d=\frac{2\lambda D}{x_2-x_1}=\frac{2\times6000\times10^{-10}\times0.5}{3\times10^{-3}}$

$d = 2\times10^{-4}$ m = 0.2 mm.

21. (4) Here, $\lambda = 6\times10^{-7}$ m, α = 0.4 mm = 4×10^{-4} m.

f = 0.8 m

Linear diameter of second maximum,

$2x'=\frac{2(2n+1)\lambda f}{2a}=\frac{2\times5\times6\times10^{-7}\times0.8}{2\times4\times10^{-4}}$

$= 6\times10^{-3}$ m = 6 mm

22. (2) Here, α = 0.25 mm = 0.25×10^{-3} m

λ = 5890 Å = 5890×10^{-10} m

Angular separation between central maximum and first order maximum

$\theta_1=\frac{3\lambda}{2a}=\frac{3\times5890\times10^{-10}}{2\times0.25\times10^{-3}}=3.53\times10^{-3}$ radian

23. (3) $x=n\lambda\frac{D}{d}=(n+1)\lambda'\frac{D}{d}$

$\therefore n\times12000=(n+1)\times10000$

or $n=5$ and $x=n\lambda\frac{D}{d}$

$\therefore d=\frac{n\lambda D}{x}=\frac{5\times12000\times10^{-10}\times2}{6\times10^{-3}}$ m

$= 2\times10^{-3}$ m = 2 mm

24. (4) $\theta_1=\frac{\lambda_1}{d}, \theta_2=\frac{\lambda_2}{d}$

$\therefore \frac{\theta_1}{\theta_2}=\frac{\lambda_1}{\lambda_2}\quad \therefore \lambda_2=\lambda_1\times\frac{\theta_2}{\theta_1}$

or $\lambda_2=5890\times\frac{0.22}{0.20}=6479\,\mathring{A}$

25. (3) Distance between central maximum and 10th maximum = 5 – 2 = 3 cm.

When YDSE apparatus is immersed in liquid of $\mu=1.5=3/2$

λ becomes $\frac{2}{3}$ and fringe width also becomes $\frac{2}{3}$.

The position of central max remains unchanged i.e., It $y = 2$ cm but 10th maximum shrinks to $y=2+\frac{2}{3}(3)=4$cm

26. (4) Here, $I_1 = I_2 = 1\ Wm^{-2}$

At 3rd vertex of equilateral triangle, the waves from two sources reach in phase.

$\therefore I_{max}=I_1+I_2+2\sqrt{I_1I_2}\cos0°$

$= 1 + 1 + 2\times1\times1 = 4\ Wm^{-2}$.

Statement Based MCQs

1. (1) Fringe width is inversely proportional to distance between sources. When coherent sources are placed very close, fringe width is very large and single fringe may occupy whole field of view and pattern would not be detected.

2. (1) When a Polaroid is rotated in the path of unpolarized light, the intensity of light transmitted from the Polaroid remains undiminished (because unpolarized light contains waves vibrating in all possible planes with equal probability).

However, when the Polaroid is rotated in the path of plane polarised light, its intensity will vary from maximum (when the vibrations of the plane polarised light are parallel to the axis of the Polaroid) to minimum (when the direction of the vibrations becomes perpendicular to the axis of the crystal). Thus using a Polaroid we can easily verify that whether the light is polarised or not.

3. (4) The maximum and minimum intensity obtained is in accordance with conservation of energy.

4. (1) The wave front originating from a point source is spherical in shape. It is because, the locus of all such points, which are equidistant from the point source, is a sphere.

5. (1) Light suffers diffraction through the small slit of muslin cloth which is comparable to its wavelength of light, to form a colored spectrum.

Assertion & Reason MCQs

1. (2) Diffraction takes place for all types of waves mechanical or non mechanical transverse or longitudinal. However its effects are practical only if wavelength of radiation is comparable to the dimensions of diffracting device, i.e., $\lambda < d$.

2. (3) This is because of destructive interference. The destructive interference of two coherent light sources may give dark fringe.

3. (4) The thin film is produced in upper slit, the shifting of fringe pattern will be in upper direction,

And minima is independent of slit widths, minima will be dark in any case.

4. (1) Interference is due to superimposition of light wave from two coherent sources. Interference pattern will disappears when one of slit is closed

5. (1) A wavefront is the locus of points having the same phase. A wavefront is composed of all points where the light reaches from the source in the same time.

Match the Columns MCQs

1. (2)(a) $I_{max}=(\sqrt{I_1}+\sqrt{I_2})^2\Rightarrow I_1=I_0$ and $I_2>I_0$

$\therefore\quad I_{max}>4I_0$

(b) When one slit is closed, fringe pattern will disappear.

$I_{max} = I_{min} = I_0$

(c) When both slits are made incoherent, again fringe pattern will disappear.

$I_{max} = I_{min} = I_1 + I_1 = 2I_0$

(d) Glass slab will only shift the fringe pattern.

2. (1) $I = 4I_0 \cos^2\left(\frac{\phi}{2}\right)$

(a) $\phi = 60° \Rightarrow I = 3I_0$

(b) $\phi = 90° \Rightarrow I = 2I_0$

(c) $\phi = 0° \Rightarrow I = 4I_0$

(d) $\phi = 120° \Rightarrow I = I_0$

3. (3) $Y \rightarrow$ maxima

$y \rightarrow$ minima

Y_6, y_6, Y_5, y_5, Y_4, y_4, Y_3, y_3, Y_2, y_2, Y_1, y_1, ω

4. (1) (a) Inclined rays and slab both will shift the fringe pattern upwards. So, zero order maxima will definitely lie about O.

(b) Inclined rays will shift the fringe pattern upwards, but slab will shift the fringe pattern downwards. Hence, zero order maxima may lie above O, below O or at O.

(c) Same explanations can be given for this option.

(d)

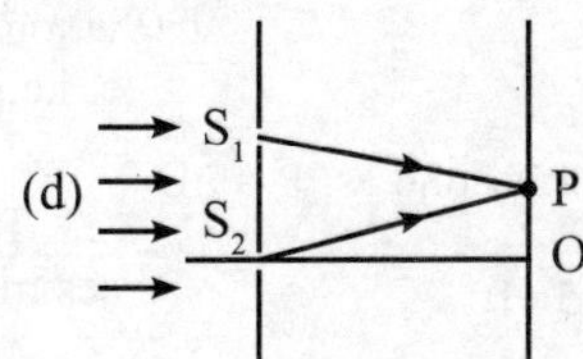

NCERT Exemplar MCQs

1. (3) Consider the diagram, the light beam incident from air to the glass slab at Brewster's angle (i_p). The incident ray is unpolarised and is represented by dot (.).

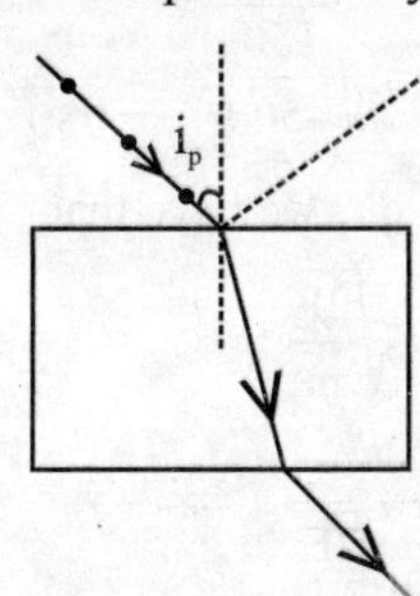

The reflected light is plane polarised represented by arrows. As the emergent ray is unpolarised, hence intensity cannot be zero when passed through polaroid.

2. (1) Given, width of the slit 10^4Å $= 10^4 \times 10^{-10}$ m $= 10^{-6}$ m $= 1\mu$m

Wavelength of (visible) sunlight varies from 4000Å to 8000Å.

As the width of slit is comparable to that of wavelength, hence diffraction occurs with maxima at centre. So, at the centre all colours appear, i.e., mixing of colours form white patch at the centre.

3. (3) For the interference pattern to be formed on the screen, the sources should be coherent and emits lights of same frequency and wavelength.

In a Young's double-slit experiment, when one of the holes is covered by a red filter and another by a blue filter. In this case due to filtration only red and blue lights are present. In YDSE monochromatic light is used for the formation of fringes on the screen. Hence, in this case there shall be no interference fringes.

4. (4) According to question, there is a hole at point P_2. From Huygen's principle, wave will propagates from the sources S_1 and S_2. Each point on the screen will acts as secondary sources of wavelets.

Now, there is a hole at point P_2 (minima). The hole will act as a source of fresh light for the slits S_3 and S_4.

Therefore, there will be a regular two slit pattern on the second screen.

NEET & AIIMS Past Year MCQs

1. (4) Fringe width $\beta = \frac{\lambda D}{d}$.

$\lambda = 600 \times 10^{-9}$ m, $D = 1\ m$, $d = 1.5 \times 10^{-3}\ m$

$$\beta = \frac{600 \times 10^{-9} \times 1}{1.5 \times 10^{-3}} = 4 \times 10^{-4}\ m$$

2. (3) Interference requires coherent sources (same frequency and constant phase difference).

Only waves $A(y = \alpha \sin\omega t)$ and $C(y = \alpha \sin(\omega t - \phi))$ meet this condition.

3. (1) Unpolarized light intensity I_0 passes through:

Polaroid A$\rightarrow$ Intensity $I_1 = \frac{I_0}{2}$.

Polaroid B (60° to A) $\rightarrow$ Intensity I_B

$= I_1 \cos^2 60° = \frac{I_0}{8}$

Polaroid C(45° to B) $\rightarrow$ Intensity

$I_C = I_B \cos^2 45° = \frac{I_0}{16}$.

4. (1) When we use white light in place of monochromatic light, the central fringe appears white, while the other fringes display different colours. This is because the central fringe for all colours is formed at central point, so it appears white after mixing up. As fringes of other colours fall at different positions, we see fringes of all colours on either side of the central bright fringe.

5. (2) At brewster angle, only the reflected light is completely polarised, but the refracted light is partially polarised.

6. (4) $\Delta\theta = \frac{\lambda}{d}$

$\Delta\theta$ is proportional to λ but independent of D.

7. (1) $\because\ y = (n\lambda)\left(\frac{D}{d}\right)$

$n_1\lambda_1\frac{D}{d} = n_2\lambda_2\frac{D}{d} \Rightarrow n_1\lambda_1 = n_2\lambda_2$

$(8)\ (600\text{ nm}) = n_2(400\text{ nm})$

$n_2 = 12$

8. (1) After passing through first polarizer $I = \frac{I_0}{2}$

After second polarizer $I = \frac{I_0}{2}\cos^2(30°) = \frac{3}{8}I_0$

After third polarizer $I = \frac{3I_0}{8}\cos^2(60°) = \frac{3I_0}{32}$

So final intensity $= \frac{3I_0}{32}$

9. (4) $I = I_0\cos^2\left(\frac{\phi}{2}\right)$

$\frac{I_0}{2} = I_0\cos^2\left(\frac{\phi}{2}\right), \phi = \frac{\pi}{2}$

$\frac{2\pi}{\lambda}.\Delta x = \frac{\pi}{2}$

$\Delta x = \frac{\lambda}{4}$

10. (2) Path difference

$a\sin\theta = \frac{5\lambda}{2}$

$a\tan\theta = \frac{5\lambda}{2}$

$a.\frac{y}{D} = \frac{5\lambda}{2}$

$\lambda = \frac{2ay}{5D} = \frac{2\times 0.7\times 10^{-3}\times 2\times 10^{-3}}{5\times 1}$

$= \frac{2.8}{5}\times 10^{-6}\text{ meter}$

$= \frac{28}{5}\times 10^{-7} = 5600\ \text{Å}$

Dual Nature of Radiation and Matter

NCERT Topic-wise MCQs

1. (3) For an electron, $\lambda = \frac{h}{\sqrt{2mK}}$

$\therefore \lambda \propto \frac{1}{\sqrt{K}}$

$\therefore \frac{\lambda'}{\lambda} = \frac{\sqrt{K}}{\sqrt{3K}} \Rightarrow \lambda' = \frac{\lambda}{\sqrt{3}}$

2. (1) As $\lambda = \frac{h}{\sqrt{2mE}}$

$\therefore E = \frac{h^2}{2m\lambda^2} \Rightarrow \lambda$ is same for all so,

$\therefore E \propto \frac{1}{m}$.

Hence energy will be maximum for particle with lesser mass.

3. (4) $\lambda = \frac{h}{\sqrt{2mqV}}$ or $\lambda \propto \frac{1}{\sqrt{mq}}$

$\therefore\ \lambda_e : \lambda_p : \lambda_\alpha = \frac{1}{\sqrt{m_e q_e}} : \frac{1}{\sqrt{m_p q_p}} : \frac{1}{\sqrt{m_\alpha q_\alpha}}$

$\because m_e > m_p > m_\alpha$.

$\therefore \lambda_e > \lambda_p > \lambda_\alpha$.

4. (3) $E = 3.3 \times 10^{-20}$ J, $h = 6.6 \times 10^{-34}$ Js

$\therefore\ \nu = \frac{E}{h} \Rightarrow \nu = \frac{3.3\times 10^{-20}}{6.6\times 10^{-34}} = \frac{1\times 10^{14}}{2} = 5 \times 10^{13}$ Hz

5. (4) Energy of each photon, $E = \frac{hc}{\lambda} = \frac{6.63\times 10^{-34}\times 3\times 10^8}{6000\times 10^{-10}}$

$E = 3.31 \times 10^{-19}$ J

No. of photons emitted per second as, $P = nE$

$n = \frac{P}{E} = \frac{25}{3.31\times 10^{-19}} = 7.55\times 10^{19}$

6. (2) As we know, $\lambda = \frac{h}{\sqrt{2mK}}$, $(K = eV_0 \Rightarrow K \propto V_0)$

$\lambda \propto \frac{1}{\sqrt{V_0}}$

$\frac{\lambda_2}{\lambda_1} = \sqrt{\frac{V_1}{V_2}} \Rightarrow \lambda_2 = \sqrt{\frac{150}{4500}}\ \text{Å} = \frac{1}{\sqrt{30}}\ \text{Å}$

7. (4) We have, $E = qV$, we have that

$E = \frac{1}{2}mv^2 \Rightarrow v = \sqrt{\frac{2E}{m}}$

as, $\lambda = \frac{h}{mv} = \frac{h}{m\sqrt{\frac{2E}{m}}} \Rightarrow \lambda = \frac{h}{\sqrt{2mqV}}$ ...(*i*)

For electron, $\lambda_e = \dfrac{h}{\sqrt{2m_e qV}}$ ...(*ii*)

For proton $\lambda_p = \dfrac{h}{\sqrt{2m_p qV}}$

$\therefore \quad \lambda_p = \dfrac{h}{\sqrt{2m_p q \cdot 4V}} \quad (\because V = 4V)$ ...(*iii*)

On dividing Eq. (*ii*) by Eq. (*iii*), we get

$$\frac{\lambda_e}{\lambda_p} = 2\sqrt{\frac{m_p}{m_e}}$$

8. (1) As we know, $\lambda = \dfrac{h}{\sqrt{2mE}}$ where E is kinetic energy

Energy of an atom at absolute temperature, $E = \dfrac{3}{2}k_B T$

where, k_B = Boltzmann constant.

$$\lambda = \frac{h}{\sqrt{2m \times \frac{3}{2}k_B T}}$$

$$\therefore \lambda = \frac{h}{\sqrt{3mk_B T}}$$

9. (2) $n = \dfrac{P}{E} \Rightarrow n = \dfrac{10 \times 1000}{6.6 \times 10^{-34} \times (880 \times 1000)}$

$= \dfrac{1}{6.6 \times 88 \times 10^{-34}}$

$n = 1.72 \times 10^{31}$

10. (4) Energy falling per second on the mirror

$$E = \frac{nhc}{\lambda}$$

Momentum of the incident light per sec.

$$P = \frac{E}{c} = \frac{nh}{\lambda}$$

Momentum of the incident light per second normal to the plane mirror $P_1 = P \cos 60^\circ$

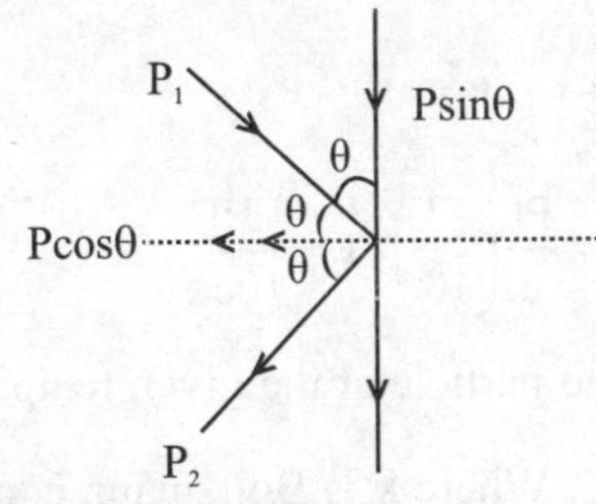

Force on mirror = $P_2 - (-P_1)$

$\therefore \; F = 2P \cos 60^\circ = \dfrac{2 \times nh}{\lambda} \cos 60^\circ$

$F = \dfrac{nh}{\lambda} \Rightarrow \dfrac{2 \times 10^{19} \times 6.6 \times 10^{-34}}{660 \times 10^{-9}}$

$F = 2 \times 10^{-8}$ N

11. (3) Gain in K.E = charge × potential difference.

$= 1e \times 1V = 1eV \; \{\because W = q \times v\}$

12. (2) $\lambda = \dfrac{h}{\sqrt{2meV}}$ ... (*i*)

$\therefore \lambda' = \dfrac{h}{\sqrt{2MeV}}$ ... (*ii*)

On dividing (*ii*) eq. from (*i*) we get

$$\frac{\lambda'}{\lambda} = \frac{h}{\sqrt{2eVM}} \times \frac{\sqrt{2meV}}{h}$$

$$\frac{\lambda'}{\lambda} = \sqrt{\frac{m}{M}}$$

$$\lambda' = \lambda\sqrt{\frac{m}{M}}$$

13. (1) Photons move with velocity of light and possess energy hv. Therefore it also exerts pressure.

14. (1) Energy of photon, $E = hv = \dfrac{hc}{\lambda_p}$

$$\lambda_p = \frac{hc}{E}$$

Wavelength of electron, $\lambda_e = \dfrac{h}{\sqrt{2mE}}$

$$\therefore \frac{\lambda_p}{\lambda_e} = \frac{hc}{E} \times \frac{\sqrt{2mE}}{h} = c\sqrt{\frac{2m}{E}}$$

15. (4) As, $\lambda = \dfrac{h}{\sqrt{2mqV}}$

$$\therefore \lambda \propto \frac{1}{\sqrt{mq}}$$

$$\therefore \frac{\lambda_p}{\lambda_\alpha} = \frac{\sqrt{m_\alpha q_\alpha}}{\sqrt{m_p q_p}}$$

$$\frac{\lambda_p}{\lambda_\alpha} = \sqrt{\frac{4m_p \times 2e}{m_p \times e}} = \sqrt{8}$$

16. (2) de-Broglie wavelength of a particle in terms of kinetic energy K is given by

$$\lambda = \frac{h}{\sqrt{2mK}},$$

when the energy becomes K/4

$$\lambda = \frac{h}{\sqrt{2m\frac{K}{4}}} = \frac{2h}{\sqrt{2mK}} = 2\lambda$$

17. (1) Let n_1 & n_2 be the no. of photons of X-rays and visible light emitted from the two source.

$\therefore \dfrac{n_1 hc}{\lambda_1} = \dfrac{n_2 hc}{\lambda_2}$ (power same)

$\therefore \dfrac{n_1}{\lambda_1} = \dfrac{n_2}{\lambda_2}$ or $\dfrac{n_1}{n_2} = \dfrac{\lambda_1}{\lambda_2} = \dfrac{1}{500}$

18. (2) $\lambda = \frac{h}{\sqrt{2mK}} \Rightarrow \lambda \propto \frac{1}{\sqrt{K}}$

$$\therefore \frac{\lambda_1}{\lambda_2} = \sqrt{\frac{K_2}{K_1}} = \sqrt{\frac{16}{1}} = \frac{4}{1}$$

$$\therefore \frac{\lambda_1}{\lambda_2} = \frac{4}{1}$$

% change in the de-Broglie wavelength =

$$\left(\frac{\lambda_1 - \lambda_2}{\lambda_1}\right)\times 100 = \left(1 - \frac{\lambda_2}{\lambda_1}\right)\times 100 = \left(1 - \frac{1}{4}\right)\times 100 = 75\%$$

19. (2) At, $\nu = 0$, $E_k = h \times 0 - \phi_0 = -\phi_0$

or $\phi_0 = -E_k = -(-4 \text{ eV}) = 4$ eV.

20. (2)

21. (1) $K_{particle} = \frac{1}{2}mv^2 \left[\lambda = \frac{h}{mv}\right]$

$$K_{particle} = \frac{1}{2}\left(\frac{h}{\lambda v}\right).v^2 \Rightarrow K_{particle} = \frac{vh}{2\lambda}$$

$$K_{photon} = \frac{hc}{\lambda} \Rightarrow \therefore \frac{K_{particle}}{K_{photon}} = \frac{v}{2c} = \frac{2.25\times10^8}{2\times3\times10^8} = \frac{3}{8}$$

22. (2) $\lambda = \frac{h}{\sqrt{2mE}}$

$$\Rightarrow E = \frac{h^2}{2m\lambda^2} = \frac{(6.6\times10^{-34})^2}{2\times9.1\times10^{-31}\times(0.3\times10^{-9})^2} = 2.65\times10^{-18}\text{ J}$$

$$E = \frac{2.65\times10^{-18}}{1.6\times10^{-19}} = 16.8\text{eV}$$

23. (2) $\because E = \frac{hc}{\lambda} - W$ ...(*i*)

$$4E = \frac{hc}{\frac{\lambda}{3}} - W \quad ...(ii)$$

Solving (*i*) and (*ii*) we get

$$W = \frac{hc}{3\lambda}$$

24. (1) $\lambda_{neutron} \propto \frac{1}{\sqrt{T}} \Rightarrow \frac{\lambda_1}{\lambda_2} = \sqrt{\frac{T_2}{T_1}}$

$$\therefore \frac{\lambda_1}{\lambda_2} = \sqrt{\frac{273+927}{273+27}} = \sqrt{\frac{1200}{300}} = \sqrt{4} = 2$$

$$\Rightarrow \lambda_2 = \frac{\lambda_1}{2} = \frac{\lambda}{2}$$

25. (4) de Broglie wavelength, $\lambda = \frac{h}{p}$

For electron $\lambda = \frac{h}{p_e}$, for proton $\lambda_p = \frac{h}{p_p}$

As $\lambda_e = \lambda_p$ given

$\Rightarrow p_e = p_p$

26. (1) For elastic collision,

$p_{before\ collsision} = p_{after\ collision}$

$$mv = mv_A + \frac{m}{2}v_B$$

$2v = 2v_A + v_B$ ...(*i*)

Now, coefficient of restitution,

$$e = \frac{v_B - v_A}{u_A - u_B}$$

Here, $u_B = 0$ (Particle at rest) and for elastic collision e = 1

$\therefore 1 = \frac{v_B - v_A}{v} \Rightarrow v = v_B - v_A$ ...(*ii*)

From Eq. (*i*) and Eq. (*ii*),

$$v_A = \frac{v}{3} \text{ and } v_B = \frac{4v}{3}$$

$$\text{Hence, } \frac{\lambda_A}{\lambda_B} = \frac{\left(\frac{h}{mv_A}\right)}{\left(\frac{h}{\frac{m}{2}v_B}\right)} = \frac{v_B}{2v_A} = \frac{4/3}{2/3} = 2$$

27. (4) $\lambda_{min} = \frac{hc}{eV}$

$$\log(\lambda_{min}) = \log\left(\frac{hc}{e}\right) - \log V$$

y = c – mx.

So, the required graph is given in option (4).

28. (3) Let E be the energy of the incident light of power P falling on surface for time t then.

E = Pt

Momentum of the incident light, $P_1 = \frac{E}{c} = \frac{Pt}{c}$

Momentum of the reflected light, $P_2 = \frac{30}{100}\frac{E}{c}$

Force on the surface = $\frac{\text{Change in momentum}}{\text{Time}}$

$$F = \frac{P_2 - (-P_1)}{t} = \frac{P_2 + P_1}{t} = \left(\frac{30}{100}\frac{E}{c} + \frac{E}{c}\right)/t$$

$$F = \frac{E}{c}\left(\frac{30}{100} + 1\right)/t$$

$$F = \frac{E}{ct}\left(\frac{13}{10}\right) = \frac{Pt}{c}\left(\frac{13}{10}\right)/t = \frac{13P}{10c}$$

29. (4) K.E. of the particle at the given temp. T is given by

$\frac{1}{2}mv^2 = \frac{3}{2}kT$, Where k is Boltzmann constant

or $mv = \sqrt{3mkT}$

De-Broglie wavelength, $\lambda = \frac{h}{mv} = \frac{h}{\sqrt{2mkT}}$

i.e., $\lambda \propto \frac{1}{\sqrt{mT}}$

$$\therefore \frac{\lambda_H}{\lambda_{He}} = \sqrt{\frac{m_{He}T_{He}}{m_H T_H}} = \sqrt{\frac{4m}{2m}\times\frac{(273+127)K}{(273+27)K}} = \sqrt{\frac{8}{3}}$$

30. (1) $K_{max} = E - W = \frac{hc}{\lambda} - W$

Here, W = work function

Similarly $K'_{max} = \frac{hc}{(3\lambda/4)} - W$

$= \frac{4}{3}\frac{hc}{\lambda} - W = \frac{4}{3}\left[\frac{hc}{\lambda} - W\right] + \frac{W}{3}$

$= \frac{4}{3}K_{max} + \frac{W}{3}$

$\Rightarrow K'_{max} > \frac{4}{3}K_{max}$

or $v' > \left(\sqrt{\frac{4}{3}}\right)v$

31. (1) From the relation : K.E = $h\nu - \phi_o$

$\therefore$ K.E = $h\nu - h\nu_o$

Thus $\phi_0 = h\nu_o = \frac{hc}{\lambda_o}$

$= \frac{6.6\times10^{-34}\times3\times10^{8}}{6500\times10^{-10}} = 0.003\times10^{-16}\,J = \frac{0.003\times10^{-16}}{1.6\times10^{-19}} \approx 2eV$

32. (2) K.E = eV_o

K.E = e × 1.5V = 1.5eV

33. (2) Photo current in Photoelectric effect depends on the intensity of the incident light and it is independent of the frequency of incident light.

34. (3) From Einstein's photoelectric equation, we have

$eV_0 = hc\left[\frac{1}{\lambda} - \frac{1}{\lambda_0}\right]$ ...(*i*)

$\frac{eV_0}{6} = hc\left[\frac{1}{3\lambda} - \frac{1}{\lambda_0}\right]$ ...(*ii*)

On dividing Eq. (*i*) by Eq. (*ii*), we get

$$6 = \frac{\left(\frac{1}{\lambda} - \frac{1}{\lambda_0}\right)}{\left(\frac{1}{3\lambda} - \frac{1}{\lambda_0}\right)}$$

$\frac{6}{3\lambda} - \frac{6}{\lambda_0} = \frac{1}{\lambda} - \frac{1}{\lambda_0} \Rightarrow \frac{1}{\lambda} = \frac{5}{\lambda_0}$

$\Rightarrow \frac{\lambda_0}{\lambda} = 5 \Rightarrow \lambda_0 = 5\lambda$

35. (1) Work function, $\phi_o = h\nu_o$

Here ν_o is the threshold frequency.

So, $\phi_o \propto \nu_o$, Hence Pt > Al > K

36. (1) The photo-current increases linearly with increase in intensity of light but it is independent of frequency. The stopping potential is independent of intensity of light but increases linearly with increase in frequency of light waves.

37. (2) The maximum kinetic energy of the emitted electron is given by.

$K_{max} = h\nu - \phi_o = h(4\nu - \nu) = h\,(3\nu)$

$K_{max} = 3h\nu$

38. (3) Here, $eV_o = \frac{hc}{\lambda_o} - W_o$

$0.5e = \frac{hc}{6\times10^{-7}} - W_o$

$0.5 = \frac{hc}{e\times6\times10^{-7}} - \frac{W_o}{e}$ ...(*i*)

Similarly, $1.5 = \frac{h}{e}\left(\frac{c}{4\times10^{-7}}\right) - \frac{W_o}{e}$ ...(*ii*)

From eq (*i*) & (*ii*)

$1 = \frac{hc}{e\times10^{-7}}\left[\frac{1}{4} - \frac{1}{6}\right] = \frac{hc}{e\times10^{-7}}\left(\frac{3-2}{12}\right)$

$\frac{h}{e} = \frac{12\times10^{-7}}{3\times10^{8}} = 4\times10^{-15}\,Vs$

39. (1) $V_s e = E - \phi_o$

$V_s = \frac{(25-7)}{e}eV$

$V_s = 18$ V

40. (3) $\frac{E_{K_1}}{E_{K_2}} = \frac{h\nu_1 - \phi_1}{h\nu_2 - \phi_2} = \frac{1-0.5}{3-0.5} = \frac{0.5}{2.5} = \frac{1}{5}$

41. (4) $E_k = \frac{hc}{\lambda} - \phi_o$

$E_k = \left(\frac{6.6\times10^{-34}\times3\times10^{8}}{300\times10^{-9}\times1.6\times10^{-19}} - 3.2\right)eV$

$= 4.125 - 3.2 = 0.925$ eV

42. (1) $\Delta E = hc\left[\frac{1}{\lambda_1} - \frac{1}{\lambda_2}\right] = \frac{hc(\lambda_2 - \lambda_1)}{\lambda_1\lambda_2}$

$= \frac{6.62\times10^{-34}\times3\times10^{8}\times(5000-2500)\times10^{-10}}{2500\times5000\times10^{-20}}$

$\Delta E = 3.96\times10^{-19}$ J

43. (1) As by the relation

$eV_o = h(\nu - \nu_o)$

$V_o = \frac{h(\nu - \nu_o)}{e}$

$= \frac{6.6\times10^{-34}\times(8.2-3.3)\times10^{14}}{1.6\times10^{-19}}$

$= 2.27 \approx 2V$

44. (1) According to Einstein's photoelectric equation.

$h\nu = h\nu_o + \frac{1}{2}mv^2_{max}$

$\frac{1}{2}mv^2_{max} = h\nu - h\nu$

According to the question

$$\frac{1}{2}m\left(8\times10^{6}\right)^{2}=h\left(5\nu_{o}-\nu_{o}\right) \quad ...(i)$$

$$\frac{1}{2}mv_{max}^{2}=h\left(2\nu_{o}-\nu_{o}\right) \quad ...(ii)$$

Dividing equation (*i*) by (*ii*), we get

$$\frac{\left(8\times10^{6}\right)^{2}}{v_{max}^{2}}=\frac{4\nu_{o}}{\nu_{o}}\Rightarrow v_{max}^{2}=\frac{\left(8\times10^{6}\right)^{2}}{4}$$

$$v_{max}^{2}=\frac{\left(64\times10^{12}\right)}{4}\Rightarrow v_{max}=\frac{\left(8\times10^{6}\right)}{2}=4\times10^{6}\,m/s$$

45. (1) As electron describes a circular path of radius r in the magnetic field therefore.

$$\frac{mv^{2}}{r}=evB$$

$$r=\frac{mv}{eB}=\frac{P}{eB}=\frac{\sqrt{2mK}}{eB} \quad ...(i)$$

By the Einstein equation.

$$K=E-\phi_{0} \quad ...(ii)$$

By putting the value of K in eq (*i*)

$$\therefore r=\frac{\sqrt{2m\left(E-\phi_{o}\right)}}{eB}$$

46. (4) According to photo electric equation

$$K=\frac{hc}{\lambda}-\frac{hc}{\lambda_{o}}=hc\left(\frac{1}{\lambda}-\frac{1}{\lambda_{o}}\right)$$

$$K=hc\left[\frac{\lambda_{o}-\lambda}{\lambda\lambda_{o}}\right]$$

47. (4) Tungsten filament is coated with barium oxide.

48. (1) Velocity of photons of different frequencies in the vacuum is same 3×10^{8}m/s because it is the energy packets of light.

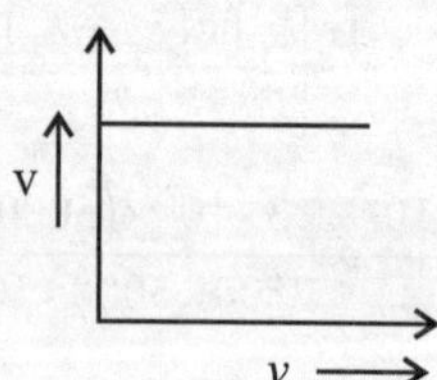

49. (3) $W_{o}=\frac{6.6\times10^{-34}\times3\times10^{8}}{1.6\times10^{-19}\times\lambda_{o}}=\frac{12375}{\lambda_{o}}\times10^{-10}$

$$=\frac{12375}{4.2}\times10^{-10}=2946\text{Å}$$

50. (3) $W_{o}=\frac{hc}{\lambda_{o}}$

$$\therefore W_{o}\propto\frac{1}{\lambda_{o}}$$

$$\frac{W_{o(T)}}{W_{o(Na)}}=\frac{\lambda_{o(Na)}}{\lambda_{o(T)}}\Rightarrow\lambda_{o(T)}=\frac{\lambda_{o(Na)}\times W_{o(Na)}}{W_{o(T)}}$$

$$\lambda_{o(T)}=\frac{5460\times2.3}{4.5}=2791\,\text{Å}$$

51. (1) Energy of incident light, $E=\frac{12375}{2000}=6.18\text{eV}$

According to relation $E=W_{o}+eV_{o}$

$$V_{o}=\frac{E-W_{o}}{e}$$

$$V_{o}=\frac{(6.18-5.01)\text{eV}}{e}=1.17\text{V}\approx1.2\text{V}$$

52. (3) $\lambda_{0}=\frac{hc}{W_{0}}=\frac{6.63\times10^{-34}\times3\times10^{8}}{4.0\times1.6\times10^{-19}}\text{m}$

$=310\times10^{-9}$ m $=310$ nm.

53. (1) $\frac{1}{2}mv_{1}^{2}=hf_{1}-\phi_{o}$ ...(*i*)

$$\frac{1}{2}mv_{2}^{2}=hf_{2}-\phi_{o} \quad ...(ii)$$

Subtracting both equations, we have

$$\frac{1}{2}m\left(v_{1}^{2}-v_{2}^{2}\right)=h\left(f_{1}-f_{2}\right)$$

$$v_{1}^{2}-v_{2}^{2}=\frac{2h}{m}\left(f_{1}-f_{2}\right)$$

54. (3) $V_{o}=\frac{hc}{\lambda}-\phi\Rightarrow\frac{1}{2}mv^{2}=q.V_{o}$

$$V_{o}=\frac{mv^{2}}{2q}\Rightarrow V_{o}=\frac{v^{2}}{2\times\frac{q}{m}}$$

$$V_{o}=\frac{1\times\left(1.2\times10^{6}\right)^{2}}{2\times1.8\times10^{11}}\Rightarrow V_{o}=\frac{1.44\times10^{12}}{3.6\times10^{11}}$$

$V_{o}=4$

55. (3) $W_{o}=\frac{hc}{\lambda_{o}}=\frac{6.625\times10^{-34}\times3\times10^{8}}{5000\times10^{-10}}=4\times10^{-19}$ Joules

56. (1) $W_{o}=\frac{hc}{\lambda_{o}}$

Energy of the radiation of wavelength 2000Å is greater than the energy of 3000Å hence we know

$$eV_{o}=\frac{hc}{\lambda}-\frac{hc}{\lambda_{o}}$$

$\because\lambda<\lambda_{o}$

$\therefore$ electron will be emitted.

57. (1) $K_{max}=eV_{o}=e\times2V=2eV$

[$\because$ kinetic energy = charge × potential difference]

58. (2) Einstein's equation

$(KE)_{max}=h\nu-\phi_{o}$

For first condition

$$0.5=E-\phi_{o} \quad ...(i)$$

For second condition

$$0.8=\left[E+\frac{E\times20}{100}\right]-\phi_{o}$$

$$0.8=1.2\,E-\phi_{o} \quad ...(ii)$$

$0.3 = 0.2E \Rightarrow E = \frac{3}{2} = 1.5\text{eV}$

From equation (*i*)

$\phi_o = E - 0.5 = 1.5 - 0.5 = 1\text{eV}$

59. (2) Photoelectric equation, $h\nu - h\nu_o = K_{max}$

$h(\nu_1 - \nu_o) = K_1$ ...(*i*)

and $h(\nu_2 - \nu_o) = K_2$ ...(*ii*)

from eq (*i*) & (*ii*) we get

$$\frac{\nu_1 - \nu_o}{\nu_2 - \nu_o} = \frac{1}{n} \Rightarrow n\nu_1 - n\nu_o = \nu_2 - \nu_o$$

$$\Rightarrow \frac{(n\nu_1 - \nu_2)}{(n-1)} = \nu_o$$

60. (2) $\because \frac{1}{2}mv^2_{max} = h\nu - h\nu_0$

$$v_{max} = \sqrt{\frac{2h(\nu - \nu_o)}{m}}$$

$$v_{max} = V_o = \sqrt{\frac{2h(2\nu_o - \nu_o)}{m}} \quad \text{(given, } \nu = 2\nu_o\text{)}$$

$$v_{max} = V_o = \sqrt{\frac{2h\nu_o}{m}} \quad ...(i)$$

Again, $\nu = 5\nu_o$

$$v_{max} = \sqrt{\frac{2h(5\nu_o - \nu_o)}{m}} = \sqrt{\frac{2h(4\nu_o)}{m}} = 2V_o$$

61. (2) Energy corresponding to 2000Å

$$E = \frac{12375}{2000}\text{eV} = 6.2\text{eV}$$

Maximum kinetic energy.

$K = h\nu - h\nu_o$

$K = E - W_o \Rightarrow K_{max} = (6.2 - 5.01)\text{ eV} = 1.19\text{eV}$

Now, $K_{max} = \frac{1}{2}mv^2_{max}$

$$1.19 \times 1.6 \times 10^{-19} = \frac{1}{2} \times 9.1 \times 10^{-31} \times v^2_{max}$$

$$v^2_{max} = \frac{1.19 \times 10^{-19} \times 1.6 \times 2}{9.1 \times 10^{-31}} = 0.418 \times 10^{12} = 41.8 \times 10^{10}$$

$v_{max} = 6.46 \times 10^5$ m/s

62. (2) $E - W_o = \frac{1}{2}mv^2$

$$2W_o - W_o = \frac{1}{2}mv_1^2 \quad ...(i)$$

$$5W_o - W_o = \frac{1}{2}mv_2^2 \quad ...(ii)$$

Therefore, from the equation (*i*) & (*ii*) we get

$$\frac{2W_o - W_o}{5W_o - W_o} = \frac{v_1^2}{v_2^2} = \frac{W_o}{4W_o} = \left(\frac{v_1}{v_2}\right)^2 \Rightarrow \therefore \frac{1}{2} = \frac{v_1}{v_2}$$

63. (2) As, $eV_0 = \frac{hc}{\lambda} - \phi_0$; so $2.50\ V_0 e = \frac{hc}{\lambda} - \phi_0$ ...(*i*)

and $V_0 e = \frac{hc}{1.5\lambda} - \phi_0$ ...(*ii*)

From (*i*) ; $eV_0 = \frac{hc}{2.5\lambda} - \frac{\phi_0}{2.5}$ ...(*iii*)

From (*ii*) and (*iii*),

$$\frac{hc}{1.5\lambda} - \phi_0 = \frac{hc}{2.5\lambda} - \frac{\phi_0}{2.5}$$

$$\frac{hc}{\lambda}\left[\frac{1}{1.5} - \frac{1}{2.5}\right] = \phi_0\left(1 - \frac{1}{2.5}\right) = \frac{1.5}{2.5}\phi_0 = \frac{3}{5}\phi_0$$

or $\frac{hc}{\lambda}\left[\frac{10}{15} - \frac{10}{25}\right] = \frac{3}{5}\frac{hc}{\lambda_0}$ or $\frac{hc}{\lambda}\left[\frac{4}{15}\right] = \frac{3}{5}\frac{hc}{\lambda_0}$

or $\lambda_0 = \frac{9}{4}\lambda = 2.25\ \lambda$.

64. (4) Photon-electron collision is an example of elastic collision because energy conservation takes place i.e. energy of photon absorbs by the electron to emit within 10^{-9} sec time lapse.

65. (4) As we know, $\frac{K.E_1}{K.E_2} = \frac{h(\nu) - \phi_o}{h(2\nu) - \phi_o} = \frac{\frac{1}{2}mv^2}{\frac{1}{2}m(2v)^2}$

$$\frac{1}{4} = \frac{h\nu - \phi_o}{2h\nu - \phi} \Rightarrow 4h\nu - 4\phi_o = 2h\nu - \phi_o \Rightarrow 2h\nu = 3\phi_o$$

$$\therefore \frac{2h\nu}{3} = \phi_o$$

66. (3) From photoelectric equation

$$E = h\nu = \frac{hc}{\lambda}$$

$$E = \frac{6.6 \times 10^{-34} \times 3 \times 10^8}{5000 \times 10^{-10}} = 3.96 \times 10^{-19}\text{ J}$$

$$E = \frac{3.96 \times 10^{-19}}{1.6 \times 10^{-19}} = 2.48\text{eV}$$

Hence, $E_K = 2.48 - 1.90 = 0.58\text{eV}$

67. (2) According to the Planck's quantum law

$$E = nh\nu = n\left(\frac{hc}{\lambda}\right)$$

$$\frac{E}{t} = \frac{n}{t}\left(\frac{hc}{\lambda}\right) \Rightarrow 10^{-7} = \left(\frac{n}{t}\right)\frac{6.6 \times 10^{-34} \times 3 \times 10^8}{5000 \times 10^{-10}}$$

$$\frac{n}{t} = \frac{5000 \times 10^{-10} \times 10^{-7}}{6.6 \times 10^{-34} \times 3 \times 10^8} = 2.5 \times 10^{11}\text{ sec}^{-1}$$

68. (3) Specific charge of electrons

$$\frac{e}{m} = 1.8 \times 10^{11}\text{ Ckg}^{-1}$$

Maximum kinetic energy of photoelectrons

$\frac{1}{2}mv^2_{max} = eV_s$ [$\therefore$ work function = 0 for K_{max}]

$eV_s = \frac{1}{2}mv^2 \Rightarrow V_s = \frac{mv^2}{2e}$

$V_s = \frac{(1.2\times10^6)^2}{2\times1.8\times10^{11}} = 0.4\times10 = 4V$

High Order Time Intensive MCQs

1. (3) Here, $\phi_0 = 6.2$ eV, $V_0 = 5$ V

As, $eV_0 = \frac{hc}{\lambda} - \phi_0$ or $\frac{hc}{\lambda} = eV_0 + \phi_0$

$= e \times 5\text{ V} + 6.2\text{ eV} = 11.2\text{ eV}$

or $\lambda = \frac{hc}{11.2\text{ eV}} = \frac{12400\text{ eV Å}}{11.2\text{ eV}} = 1107\text{Å}$

This wavelength is in ultraviolet region.

2. (2) $qvB = mv^2/r$ or $mv = qBr$

de-Broglie wavelength, $\lambda = \frac{h}{mv} = \frac{h}{qBr}$

so $\lambda \propto \frac{1}{qr}$

$\therefore \frac{\lambda_\alpha}{\lambda_p} = \frac{q_p r_p}{q_\alpha r_\alpha} = \frac{e}{2e}\times\frac{r}{r} = \frac{1}{2}$

3. (3) Since, the values of v is not specified, therefore it is not possible to say that which metal has comparatively greater or lesser work function.

4. (1) Let energy given to photon and electron be E

$\lambda_p = \frac{hc}{E}$

$\lambda_e = \frac{h}{\sqrt{2m_e E}} \Rightarrow \frac{\lambda_p}{\lambda_e} = c\sqrt{\frac{2m}{E}} > 1 \Rightarrow \lambda_p > \lambda_e$

5. (1) Kinetic energy,

$K = \phi - \phi_0$

Here, $K_1 = 1 - 0.5 = 0.5$ eV

$K_2 = 2.5 - 0.5 = 2$ eV

$\therefore \frac{K_1}{K_2} = \frac{0.5}{2} = \frac{1}{4}$ or $\frac{v_1^2}{v_2^2} = \frac{1}{4}$ or $\frac{v_1}{v_2} = \sqrt{\frac{1}{4}} = \frac{1}{2}$

6. (1) The de-Broglie wavelength, of an electron in terms of potential difference V is $\lambda = \frac{1.227}{\sqrt{V}}$ nm

7. (2) $\lambda_e = \frac{h}{mv}$ and $\lambda_p = \frac{hc}{E_p}$ (E_p = energy of photon)

Given, $\lambda_e = \lambda_p$

$\Rightarrow \frac{h}{mv} = \frac{hc}{E_p} \Rightarrow \frac{h}{P_e} = \frac{hc}{E_P}$ ($P_e = mv$ = momentum of electron)

or $\frac{E_P}{P_e} = c$

8. (1) According to Einstein photoelectric equation,

$K_A = \frac{hc}{\lambda_A} - \phi \Rightarrow 2K_A = \frac{2hc}{\lambda_A} - 2\phi$

or $2K_A + \phi = \frac{2hc}{\lambda_A} - \phi$

Given $\lambda_A = 2\lambda_B$

$\lambda_B = \lambda_A/2$

$K_B = \frac{2hc}{\lambda_A} - \phi \Rightarrow 2K_A + \phi = K_B \Rightarrow K_A < \frac{K_B}{2}$

9. (3) As, $(K)_{max} = h\nu - \phi_0 = h(\nu - \nu_0)$ (K_{max} = maximum kinetic energy)

$\frac{1}{2}mv^2_{max} = h[2\nu - \nu] = h\nu$ [$\because \nu_0 = \nu$]

$\Rightarrow v^2_{max} = \frac{2h\nu}{m} \Rightarrow v_{max} = \sqrt{\frac{2h\nu}{m}}$

10. (1) Efficient power, $P = \frac{EN}{t}$

$P = \frac{N}{t}\times\frac{hc}{\lambda} = 200\times0.25 = 50$

$\frac{N}{t} = 50\times\frac{\lambda}{hc} \Rightarrow \frac{N}{t} = \frac{50\times0.6\times10^{-6}}{6.6\times10^{-34}\times3\times10^8} = 1.5\times10^{20}$

11. (4) $eV_0 = \frac{hc}{\lambda_0} - W_0$ and $eV' = \frac{hc}{2\lambda_0} - W_0$

Subtracting them, we have

$e(V_0 - V') = \frac{hc}{\lambda_0}\left[1 - \frac{1}{2}\right] = \frac{hc}{2\lambda_0}$

or $V' = V_0 - \frac{hc}{2e\lambda_0}$

12. (1) Energy radiated as visible light $= \frac{5}{100}\times100 = 5\text{ J/s}$

Let N be the number of photons emitted per second. Then, $Nh\nu = E = 5$

$\therefore N = \frac{5\lambda}{hc} = \frac{5\times5.6\times10^{-7}}{(6.62\times10^{-34})(3\times10^8)}$ $\left[\because E = \frac{hc}{\lambda}\right]$

$= 1.4\times10^{19}$

13. (2) By using equation,

$eV = hc\left(\frac{1}{\lambda} - \frac{1}{\lambda_0}\right)$ ...(*i*)

and $\frac{eV}{3} = hc\left(\frac{1}{2\lambda} - \frac{1}{\lambda_0}\right)$ ...(*ii*)

Dividing Eq. (*i*) by Eq. (*ii*), we get

$3 = \frac{\left(\frac{1}{\lambda} - \frac{1}{\lambda_0}\right)}{\left(\frac{1}{2\lambda} - \frac{1}{\lambda_0}\right)}$ or $3\left(\frac{1}{2\lambda} - \frac{1}{\lambda_0}\right) = \left(\frac{1}{\lambda} - \frac{1}{\lambda_0}\right)$

or $\frac{3}{2\lambda} - \frac{1}{\lambda} = \frac{3}{\lambda_0} - \frac{1}{\lambda_0}$ or $\frac{1}{2\lambda} = \frac{2}{\lambda_0} \Rightarrow \lambda_0 = 4\lambda$

14. (3) Energy of photoelectron is given by equation,

$$E=\frac{hc}{\lambda}-W \Rightarrow \frac{hc}{\lambda}=E+W \quad \ldots(i)$$

where, W is the work function for the metal surface (constant).

$$\Rightarrow \frac{hc}{\lambda'}=2E+W \quad \ldots(ii)$$

Diving Eq. (*i*) by Eq. (*ii*), we get

$$\frac{\lambda'}{\lambda}=\frac{E+W}{2E+W} \Rightarrow \frac{\lambda'}{\lambda}=\frac{(E+W)}{2\left(E+\frac{W}{2}\right)}$$

$$\therefore \quad \frac{\lambda'}{\lambda}>\frac{1}{2} \text{ or } \lambda'>\frac{\lambda}{2} \Rightarrow \lambda>\lambda'>\frac{\lambda}{2}$$

15. (2) Maximum kinetic energy, $K_{max}=hv-\phi_0=hv-hv_0$
$=hf-hf_0=h(f-f_0)$

16. (4) $\lambda=\frac{h}{\sqrt{2mE}}$ or $E=\frac{h^2}{2m\lambda^2}$

$$\therefore \ E_1=\frac{h^2}{2m\lambda_1^2} \text{ and } E_2=\frac{h^2}{2m\lambda_2^2}$$

$$\therefore \ E_2-E_1=\frac{h^2}{2m}\left[\frac{1}{\lambda_2^2}-\frac{1}{\lambda_1^2}\right]$$

$$=\frac{(6.6\times10^{-34})^2}{2\times9.1\times10^{-31}}\left[\frac{1}{(0.5\times10^{-9})^2}-\frac{1}{(2\times10^{-9})^2}\right]J$$

$$=\frac{(6.6\times10^{-34})^2}{2\times9.1\times10^{-31}\times10^{-18}}\times\frac{3.75}{1.6\times10^{-19}}eV=5.6\ eV.$$

17. (2) Kinetic Energy, $\frac{1}{2}mv_1^2=2hv_0-hv_0=hv_0$

and $\frac{1}{2}mv_2^2=3hv_0-hv_0=2hv_0$

$$\therefore \quad \frac{1}{2}mv_2^2=2\left[\frac{1}{2}mv_1^2\right] \Rightarrow v_2^2=2v_1^2$$

or $v_2=\sqrt{2}v_1=\sqrt{2}\times2\times10^6 \Rightarrow v_{12}=2\sqrt{2}\times10^6\ ms^{-1}$

18. (1) Intensity of light is inversely proportional to square of distance.

i.e. $I\propto\frac{1}{r^2}$ or $\frac{I_1}{I_2}=\frac{(r_2)^2}{(r_1)^2}$

$$I_1=\frac{(1)^2}{(0.5)^2}I_2$$

$I_1=4I_2$

Now, since number of photoelectrons emitted per second is directly proportional to intensity, so number of electrons emitted would decrease by a factor of 4.

19. (3) According to photoelectric equation,

$$eV=\frac{hc}{\lambda}-W_0$$

$$0.5e=\frac{hc}{6\times10^{-7}}-W_0$$

$$\Rightarrow \quad 0.5=\frac{h}{e}\times\frac{c}{6\times10^{-7}}-\frac{W_0}{e} \quad \ldots(i)$$

Similarly, $1.5=\frac{h}{e}\times\frac{c}{4\times10^{-7}}-\frac{W_0}{e} \quad \ldots(ii)$

From Eq. (*i*) and (*ii*),

$$1=\frac{h}{e}\times\frac{c}{10^{-7}}\left(\frac{1}{4}-\frac{1}{6}\right) \Rightarrow \frac{h}{e}=\frac{12\times10^{-7}}{3\times10^8}=4\times10^{-15}\ V\text{-}s$$

20. (3) When a particle of kinetic energy E goes to a region where it has now a potential energy V, then its new kinetic energy, E′ = E – V, because the energy V has been taken out of its K.E., similar to a body going to a height losing K.E. and gaining P.E.

De-Broglie wavelength, $\lambda=\frac{h}{\sqrt{2mE}}$

$$\lambda'=\frac{h}{\sqrt{2mE'}}=\frac{h}{\sqrt{2m(E-V)}}$$

$$=\frac{h}{\sqrt{2mE(1-V/E)}}=\frac{\lambda}{\sqrt{(1-V/E)}}.$$

21. (2) Work function of a metal is given by

$$W=hv_0=\frac{hc}{\lambda_0}$$

or $W_0\propto\frac{1}{\lambda_0}$

$$\therefore \frac{W_0}{W'_0}=\frac{\lambda'_0}{\lambda_0} \text{ or } \frac{W_0}{(W_0/2)}=\frac{\lambda'_0}{\lambda_0} \Rightarrow \lambda'_0=2\lambda_0$$

22. (2) According to Einstein's photoelectric equation

$$\frac{hc}{\lambda}=\phi_0+KE_{max}$$

where ϕ_0 is a work function of a metal.

$$\therefore \ K_1=\frac{hc}{\lambda_1}-\phi_0 \quad \ldots(i)$$

$$K_2=\frac{hc}{\lambda_2}-\phi_0 \quad \ldots(ii)$$

or $K_1-K_2=hc\left[\frac{1}{\lambda_1}-\frac{1}{\lambda_2}\right]$

$$=hc\left[\frac{1}{3\lambda_2}-\frac{1}{\lambda_2}\right]=\frac{-2hc}{3\lambda_2} \quad \text{(Given } \lambda_1=3\lambda_2\text{)}$$

$$=-\frac{2}{3}(K_2+\phi_2) \quad \text{(Using (ii)}$$

or $K_1=K_2-\frac{2}{3}K_2-\frac{2}{3}\phi_0=\frac{K_2}{3}-\frac{2}{3}\phi_0$

or $K_1<\frac{K_2}{3}$

23. (4) Einstein's photoelectric equation can be written as

$hf = K + \phi$

but $K = eV_s$, V_s being stopping potential

$\therefore\ hf = eV_s + \phi$ ...(i)

So, $hf_1 = eV_1 + \phi$ ...(ii)

and $hf_2 = eV_2 + \phi$

Subtraction Eq. (ii) from Eq. (i) we get

$h(f_1 - f_2) = e(V_1 - V_2)$

or $eV_2 = eV_1 - h(f_1 - f_2)$

or $V_2 = V_1 + \frac{h}{e}(f_2 - f_1)$

24. (4) Using Einstein's photoelectric equation as

$\frac{1}{2}mv^2 = hv - \phi$

$\Rightarrow \frac{1}{2}m \times (4\times10^6)^2 = 2hv_0 - hv_0$ ...(i)

Similarly, $\frac{1}{2}m \times v^2 = 5hv_0 - hv_0$ ...(ii)

Dividing Eq. (ii) by Eq. (i), we get

$\frac{v^2}{(4\times10^6)^2} = \frac{4hv_0}{hv_0}$

$\Rightarrow v^2 = 4 \times 16 \times 10^{12}$

$\Rightarrow v^2 = 64 \times 10^{12}$

$v = 8 \times 10^6\ ms^{-1}$

25. (1) Using Einstein's photoelectric equation

$hv = \phi_0 + \frac{1}{2}mv^2$

The intercept of KE-v graph on KE axis gives the value of work function.

Hence, work function $\phi_0 = 1.5$ eV

26. (2) $\phi = \frac{hc}{\lambda_0} = \frac{12400}{5420}$ eV $= 2.28$ eV

27. (3) $KE = 2E_0 - E_0 = E_0$

$\lambda_1 = \frac{h}{\sqrt{2mE_0}}$

when, $KE = 2E_0$ (for x > 1)

$\lambda_2 = \frac{h}{\sqrt{2m(2E_0)}} \Rightarrow \frac{\lambda_1}{\lambda_2} = \sqrt{2}$

28. (1) $k = \frac{2\pi}{5\times10^{-7}} = 2\pi/\lambda$

$\Rightarrow \lambda = 5 \times 10^{-7}$ m = 5000 Å ...(i)

$\frac{hc}{\lambda} - \phi = (KE)_{max} = eV_0$ ...(ii)

For photon, substituting the given values,

$E = \frac{hc}{\lambda} = \frac{12375\ eV}{\lambda}$ [Given]

or, $\frac{hc}{\lambda} = \frac{12375}{5000}$ eV [using Eq. (i)] ...(iii)

Now, substituting the values from Eq. (iii) in Eq. (ii), we get

$\frac{12375}{5000}$ eV − 2 eV = eV_0

$\Rightarrow 2.475$ eV − 2 eV = eV_0

or $V_0 = 2.475$ V − 2V

$\Rightarrow V_0 \approx 0.48$ V

29. (1) For photoelectric emission to take place, the wavelength of incident light should be less than the threshold wavelength for that metal. If λ_m is the threshold wavelength, then work function of metal

$\phi_0 = \frac{hc}{\lambda_m}$ or $\lambda_m = \frac{hc}{\phi_0}$

For metal p, $\lambda_{m_p} = \frac{1240\ eV\ nm}{2.0\ eV} = 620$ nm

For metal q, $\lambda_{m_q} = \frac{1240\ eV\ nm}{2.5\ eV} = 496$ nm

For metal r, $\lambda_{m_r} = \frac{1240\ eV\ nm}{3.0\ eV} = 413.3$ nm

Wavelength in the incident beam are 550 nm, 450 nm and 350 nm. Therefore, the photoelectric emission is possible by light of wavelength 350 nm (being shorter than λ_m) from metals p, q and r. For light of wavelength 450 nm (< 620 nm) and 496 nm) photoelectric emission is possible from metals p and q. For light of wavelength 550 nm (< 620 nm), the photoelectric emission is possible from metal p.

Current (I) ∝ Intensity, where, Intensity = Nhν of photoelectrons. Therefore, I is maximum for p metal, less for q metal and least for r metal. Therefore, graph (1) is correct.

30. (3) Given, threshold wavelength, $\lambda_0 = 380$ nm

Wavelength of incident light, $\lambda = 260$ nm

Using Einstein's relation of photoelectric effect,

$(KE)_{max} = eV_0 = h\nu - h\nu_0$ ...(i)

But $h\nu = E = \frac{1237}{\lambda_0\,(nm)}$ eV

$\therefore\ E_0 = \frac{1237}{\lambda_0\,(nm)\,eV}$ ...(ii)

From Eq. (i) and (ii), we get

$(KE)_{max} = E - E_0 = \left(\frac{1237}{\lambda} - \frac{1237}{\lambda_0}\right)$ eV

$= 1237\left[\frac{1}{\lambda} - \frac{1}{\lambda_0}\right]$ eV (λ in nm) ...(iii)

By putting values of λ and λ_0 in Eq. (*iii*), we get

$$(KE)_{max} = 1237\left(\frac{1}{260} - \frac{1}{380}\right) eV \Rightarrow (KE)_{max} = 1.5 \text{ eV}.$$

31. **(1)** $h\nu = W_0 + K_{max} = W_0 + eV_0 = (6.2 + 5)$ eV = 11.2 eV

Wavelength of incident radiation in Å,

$$\lambda = \frac{12400}{11.2} = 1107 \text{ Å (UV-region)}$$

32. **(2)** We know that, stopping potential increases with increase in frequency. Therefore from figure, stopping potential $V_a = V_b$ but V_a or $V_b < V_c$.

Hence, $f_a = f_b$ but f_a or $f_b < f_c$.

Also, $I_a < I_b$ but $I_b = I_c$.

Hence, correct option is (2), according to which $f_a = f_b$ and $I_a < I_b$.

33. **(4)** Radius of the circular path of a charged particle in a magnetic field is given by

$$R = \frac{mv}{Bq}$$

or $mv = RBq$

Here, $R = 0.83$ cm $= 0.83 \times 10^{-2}$ m

$B = 0.25$ Wb m^{-2}

$q = 2e = 2 \times 1.6 \times 10^{-19}$C

$\therefore mv = (0.83 \times 10^{-2})(0.25)(2 \times 1.6 \times 10^{-19})$

de Broglie wavelength,

$$\lambda = \frac{h}{mv} = \frac{6.6\times10^{-34}}{0.83\times10^{-2}\times0.25\times2\times1.6\times10^{-19}} = 0.01 \text{ Å}$$

34. **(2)** For electron,

$$\lambda_e = \frac{h}{m_e v_e} = \frac{h}{m_e(c/100)} = \frac{100h}{m_e c} \quad \ldots(i)$$

$$E_e = \frac{1}{2}m_e v_e^2 \text{ or } m_e v_e = \sqrt{2E_e m_e}$$

$$\lambda_e = \frac{h}{m_e v_e} = \frac{h}{\sqrt{2m_e E_e}} \text{ or } E_e = \frac{h^2}{2\lambda_e^2 m_e} \quad \ldots(ii)$$

For photon

$$E_e = \frac{hc}{\lambda_p} = \frac{hc}{2\lambda_e} \quad (\because \lambda_p = 2\lambda_e \text{ (given)})$$

$$\therefore \frac{E_p}{E_e} = \frac{hc}{2\lambda_e} \times \frac{2\lambda_e^2 m_e}{h^2} = \frac{\lambda_e m_e c}{h}$$

$$= \frac{100h}{m_e c} \times \frac{m_e c}{h} = 100 \quad \text{[Using } (i)\text{]}$$

$$\therefore \frac{E_e}{E_p} = \frac{1}{100} = 10^{-2}$$

Thus option (2) is correct.

35. **(2)** $\dfrac{hc}{\lambda} - \phi = \dfrac{1}{2}mv^2$

$$\sqrt{2\left(\frac{hc - \lambda\phi}{\lambda m}\right)} = v$$

36. **(2)** $\dfrac{hc}{\lambda} = \phi + eV$

$$\frac{hc}{2\lambda} = \phi + \frac{eV}{3}; \quad \frac{hc}{\lambda} - \phi = \frac{3hc}{2\lambda} - 3\phi \Rightarrow 2\phi = \frac{hc}{2\lambda};$$

$$\phi = \frac{hc}{4\lambda} \Rightarrow \frac{hc}{\lambda_0} = \frac{hc}{4\lambda}$$

$\lambda_0 = 4\lambda$

37. **(4)** Given, $\lambda_1 = 300$ nm

$\lambda_2 = 400$ nm

$$\frac{hc}{e} = 1240 \text{ nmV}$$

Using Einstein equation for photoelectric effect,

$E = h\nu = \phi + eV_0$ ...(*i*)

(here, ϕ is work function of the metal and V_0 is stopping potential)

For λ_1 wavelength's wave,

$E_1 = h\nu_1 = \phi + eV_{01}$

or $\dfrac{hc}{\lambda_1} = \phi + eV_{01}$ ...(*ii*)

Similarly, $\dfrac{hc}{\lambda_2} = \phi + eV_{02}$ ...(*iii*)

From Eqs. (*ii*) and (*iii*), we get

$$hc\left[\frac{1}{\lambda_1} - \frac{1}{\lambda_2}\right] = e(V_{01} - V_{02}) \text{ or } \frac{hc}{e}\left[\frac{1}{\lambda_1} - \frac{1}{\lambda_2}\right] = \Delta V$$

By using given values,

$$\Delta V = 1240\left[\frac{1}{300} - \frac{1}{400}\right]\frac{\text{nmV}}{\text{nm}} = 1240 \times \frac{1}{1200} \text{V}$$

$$\Rightarrow \Delta V = 1.03 \approx 1\text{V}$$

38. **(2)** As, $\lambda_1 = \dfrac{h}{\sqrt{2m_p E}}, \lambda_2 = \dfrac{hc}{E}$

$$\therefore \frac{\lambda_1}{\lambda_2} = \frac{h/\sqrt{2m_p E}}{hc/E} = \frac{\sqrt{E}}{\sqrt{2m_p c^2}}$$

$$\Rightarrow \frac{\lambda_1}{\lambda_2} \propto \sqrt{E} \Rightarrow \frac{\lambda_1}{\lambda_2} \propto E^{1/2}$$

39. **(3)** By conservation of linear momentum,

$p_1 + p_2 = 0$ or $p_1 = -p_2$ or $|p_1| = |p_2|$

$$\therefore \frac{\lambda_2}{\lambda_1} = \frac{|p_1|}{|p_2|} = 1$$

40. **(2)** $E = W_0 + K_{max}$ ...(*i*)

$\Rightarrow hf = W_A + K_A$

and $2hf = W_B + K_B = 2W_A + K_B$ ...(*ii*)

$$\left(\because \frac{W_A}{W_B} = \frac{1}{2}\right)$$

Dividing equation (*i*) by (*ii*)

$$\frac{1}{2} = \frac{W_A + K_A}{2W_A + K_B} \Rightarrow \frac{K_A}{K_B} = \frac{1}{2}$$

41. (1) By using, $\frac{hc}{\lambda} = W_0 + \frac{1}{2}mv^2$

$$\Rightarrow \frac{hc}{400\times10^{-9}} = W_0 + \frac{1}{2}mv^2 \quad ...(i)$$

and $\frac{hc}{250\times10^{-9}} = W_0 + \frac{1}{2}m(2v)^2 \quad ...(ii)$

On solving (*i*) and (*ii*)

$$\frac{1}{2}mv^2 = \frac{hc}{3}\left[\frac{1}{250\times10^{-9}} - \frac{1}{400\times10^{-9}}\right] \quad ...(iii)$$

From equations (*i*) and (*iii*), $W_0 = 2hc \times 10^6$ J

42. (3) Given, $\lambda_p = \lambda_e$ or $\frac{h}{\sqrt{2m_pq_pV_p}} = \frac{h}{\sqrt{2m_eq_eV}}$

or $m_p\, q_p\, V_p = m_e\, q_e\, V$

or $V_p = \left(\frac{m_e}{m_p}\right)\left(\frac{q_e}{q_p}\right)V$

$$= \frac{1}{1840}\times1\times V = \frac{V}{1840} \text{ volt}$$

43. (3) de-Broglie Wavelength, $\lambda = \frac{h}{mv}$

Fringe width, $\beta = \frac{D\lambda}{d}$

As the velocity v of the electrons increases, their wavelength λ decreases and hence fringe width β decreases.

44. (3) $\lambda = \frac{h}{p} \quad ...(i)$

$$\left(\lambda + \frac{0.25}{100}\lambda\right) = \frac{h}{(p-p_0)}$$

or $\frac{100.25\,\lambda}{100} = \frac{h}{p-p_0} \quad ...(ii)$

From (*i*) and (*ii*),

$$\therefore \frac{100.25}{100} = \frac{p}{p-p_0}$$

On solving, $p = 401\, p_0$.

45. (2) Given, $\lambda_1 = 2\lambda$, $\lambda_2 = \lambda$

$$E_1 = \frac{hc}{\lambda_1} - \phi_0 \quad ...(i)$$

and $E_2 = \frac{hc}{\lambda_2} - \phi_0$

or $\frac{hc}{\lambda_2} = (E_2 + \phi_0) \quad ...(ii)$

$$\therefore E_1 - E_2 = hc\left[\frac{1}{\lambda_1} - \frac{1}{\lambda_2}\right] = hc\left[\frac{1}{2\lambda} - \frac{1}{\lambda}\right] = -\frac{hc}{2\lambda}$$

$$= -\frac{1}{2}(E_2 + \phi_0) \quad \text{from } (ii)$$

or $E_1 = E_2 - \frac{1}{2}E_2 - \frac{1}{2}\phi_0 = \frac{E_2}{2} - \frac{1}{2}\phi_0$

or $E_1 < \frac{E_2}{2}$.

46. (3) Max. KE $= \frac{1}{2}mv_{max}^2 = eV_s$.

or $v_{max} = \sqrt{\frac{2eV_s}{m}} = \sqrt{2\times1.76\times10^{11}\times0.71}$

$= 5 \times 10^5$ ms^{-1} = 500 km s^{-1}

Statement Based MCQs

1. **(1)** The electrons in metals occupies different energy levels, thus have different minimum energy required to be ejected from atom. Also electron might not be emitted from metal surface, rather it could be within the metal due to which it suffers collision and hence energy losses.

2. **(1)** The moving mass m of a photon is given by

$E = mc^2 = hc$

$\therefore m = \frac{E}{c^2} = \frac{h\nu}{c^2}$

Momentum, $p = mc = \frac{h\nu}{c} = \frac{h}{\lambda}$.

3. **(3)** For an incident photon of given energy, velocity of photoelectron ejected from near the surface is larger than that coming from the interior of the metal because less energy is required to eject an electron from the surface than from the interior. The velocity of ejected electron may not be zero.

4. **(1)** A photocell works on the principle of photoelectric emission. It is also called an electric eye.

5. **(1)** $\lambda = \frac{h}{mv}$

For constant v, $\lambda \propto \frac{1}{m}$

λ is significantly measurable only in case of microscopic or sub-microscopic particles.

Assertion & Reason MCQs

1. **(2)** Photoelectric current is directly proportional to intensity of incident radiation as each photon ejects one electron constituting photocurrent.
2. **(1)** Frequency does not change as energy associated with a photon is fixed and does not depends upon medium in which they propagate. On changing medium the speed as well as wavelength does change but frequency remains unchanged.
3. **(1)** Both Assertion and Reason are True and the Reason is a correct explanation of the Assertion.
4. **(2)** Both Assertion and Reason are True but Reason is not the correct explanation of the Assertion.
5. **(2)** Both Assertion and Reason are True but Reason is not a correct explanation of the Assertion.

Match the Columns MCQs

1. **(3)** A-(ii); B-(i); C-(iii); D-(iv)

$K_{max} = hf - W$

$\therefore K_{max}$ versus f graph is a straight line of slope h and intercept – W.

$eV_0 = hf - W$

$\therefore V_0 = \left(\frac{h}{e}\right)f - \frac{W}{e}$

$\therefore V_0$ versus f graph is again a straight line of slope $\frac{h}{e}$ and intercept $\frac{-W}{e}$.

2. **(3)** (a)-(iii); (b)-(i, ii); (c)-(i, ii), (d)-(iv)

(a) If intensity changes, then number of photons/time incident on the metal surf ace change and hence number of photoelectrons liberated change, so saturation photocurrent changes. Stopping potential and K_{max} will remain the same.

(b) From $eV_0 = hf - \phi$ and $K_{max} = hf - \phi$

If f changes, then V_0 and K_{max} change.

(c) From $eV_0 = hf - \phi$ and $K_{max} = hf - \phi$·

If target material changes, then ϕ changes, then V_0 and K_{max} change.

(d) If we change the potential difference between emitter and collector, then time taken for electrons to eject changes.

3. **(3)** (a) The Franck-Hertz experiment was the first electrical measurement to clearly show the Quantum (discrete) nature of atoms. The experiment confirmed the quantum theory which predicted that electrons occupied only discrete energy levels.

(b) Photoelectric effect experiment, supports the particle theory of light.

(c) The Davision and Germer experiment showed that, electron beams can undergo diffraction when passed through the atomic crystal. This implies the wave nature of electrons as waves can exhibit interference and diffraction.

NCERT Exemplar MCQs

1. **(4)** Velocity of a body falling from a height H is given by

$v = \sqrt{2gh}$

We know that de-Broglie wavelength

$$\lambda = \frac{h}{mv} = \frac{h}{m\sqrt{2gH}} \Rightarrow \lambda = \frac{h}{m\sqrt{2g}\sqrt{H}}$$

Here, $\frac{h}{m\sqrt{2g}}$ is a constant say 'K'.

So, $\lambda = K\frac{1}{\sqrt{H}} \Rightarrow \lambda \propto \frac{1}{\sqrt{H}} \Rightarrow \lambda \propto (H)^{\frac{-1}{2}}$

2. **(2)** Given in the question,

Energy of a photon,

$E = 1$ MeV

$\Rightarrow E = 10^6$ eV

Now, hc = 1240 eV-nm

Now, $E = \frac{hc}{\lambda}$

$$\lambda = \frac{hc}{E} = \frac{1240 \text{ eV-nm}}{10^6 \text{eV}} = 1.24 \times 10^{-3} \text{ nm}$$

3. **(4)** When a beam of electrons of energy E_0 is incident on a metal surface kept in an evacuated chamber, electrons can be emitted with maximum energy E_0 (due to elastic collision) and with any energy less than E_0, when part of incident energy of electron is used in liberating the electrons from the surface of metal.

4. **(2)** We know that the relation between λ and K is given by

$v = \frac{h}{\sqrt{2mK}}$

Here, for the given value of energy $\frac{h}{\sqrt{2K}}$ is a constant.

Thus, $\lambda \propto \frac{1}{\sqrt{m}}$

$\therefore \quad \lambda_p : \lambda_n : \lambda_e : \lambda_\alpha = \frac{1}{\sqrt{m_p}} : \frac{1}{\sqrt{m_n}} : \frac{1}{\sqrt{m_e}} : \frac{1}{\sqrt{m_\alpha}}$

Since, $m_p = m_n$, hence $\lambda_p = \lambda_n$

As, $m_\alpha > m_p$, therefore $\lambda_\alpha < \lambda_p$

As, $m_e < m_n$, therefore $\lambda_e > \lambda_n$

Hence, $\lambda_\alpha < \lambda_p = \lambda_n < \lambda_e$

5. **(1)** Given, $\vec{v} = v_0\hat{i} \Rightarrow B = B_0\hat{j}$

Force on moving electron due to magnetic field is,

$F = -e(\vec{v} \times \vec{B})$

$F = -e\left[v_0\hat{i} \times B_0\hat{j}\right] \Rightarrow F = -ev_0B_0\hat{k}$

As this force is perpendicular to v and B, so the magnitude of v will not change, i.e., momentum (= mv) will remain constant in magnitude. Hence, de-Broglie wavelength $\lambda = \frac{h}{mv}$ remains constant.

6. **(1)** Initial de-Broglie wavelength of electron, $\lambda_0 = \frac{h}{mv_0}$

Force on electron in electric field,

$F = -eE = -e[-E_0\hat{i}] = eE_0\hat{i}$

Acceleration of electron, $a = \frac{F}{m} = \frac{eE_0\hat{i}}{m}$

Velocity of electron after time t,

$$v = v_0\hat{i} + \left(\frac{eE_0\hat{i}}{m}\right)t = \left(v_0 + \frac{eE_0}{m}t\right)\hat{i} = v_0\left(1 + \frac{eE_0}{mv_0}t\right)\hat{i}$$

de-Broglie wavelength associated with electron at time t is

$$\lambda = \frac{h}{mv} = \frac{h}{m\left[v_0\left(1+\frac{eE_0}{mv_0}t\right)\right]} = \frac{\lambda_0}{\left[1+\frac{eE_0}{mv_0}t\right]}$$

$$\left[\therefore \lambda_0 = \frac{h}{mv_0}\right]$$

7. (3) Initial de-Broglie wavelength of electron,

$$\lambda_0 = \frac{h}{mv_0}$$

Force on electron in electric field,

$$F = -eE = eE_0\hat{j}$$

Acceleration of electron,

$$a = \frac{F}{m} = \frac{eE_0\hat{j}}{m}$$

It is acting along negative y-axis.

The initial velocity of electron along x-axis, $v_{x0} = v_0\hat{i}$.

Initial velocity of electron along Y-axis, $v_{y0} = 0$.

Velocity of electron after time t along x-axis, $v_x = v_0\hat{i}$

($\because$ There is no electron along x-axis.)

Velocity of electron after time t along y-axis,

$$v_y = 0 + \left(-\frac{eE_0}{m}\hat{j}\right)t = -\frac{eE_0}{m}t\hat{j}$$

$$\Rightarrow |v| = \sqrt{v_x^2 + v_y^2} = v_0\sqrt{1+\frac{e^2E_0^2t^2}{m^2v_0^2}}$$

de-Broglie wavelength,

$$\lambda' = \frac{h}{mv} = \frac{h}{mv_0\sqrt{1+e^2E_0^2t^2/(m^2v_0^2)}} = \frac{\lambda_0}{\sqrt{1+e^2E_0^2t^2/m^2v_0^2}}$$

NEET & AIIMS Past Year MCQs

1. (3) The maximum kinetic energy of the ejected photoelectron is:

$$(K.E.)_{max} = \frac{hc}{\lambda} - \phi$$

Given $\lambda = \frac{hc}{e}$, substitute:

$$(K.E.)_{max} = \frac{hc}{\frac{hc}{e}} - \phi = e - \phi$$

2. (1) de Broglie wavelength $\left(\lambda = \frac{h}{p}\right)$ depends on momentum ($p = \sqrt{2mE_k}$), where $E_k = qV$ (same potential difference for both particles).

For the same $V, \lambda \propto \frac{1}{\sqrt{mq}}$.

Alpha particle ($m_\alpha \gg m_e, q_\alpha = 2e$) vs. electron ($q_e = e$):

$$\frac{\lambda_\alpha}{\lambda_e} = \sqrt{\frac{m_e q_e}{m_\alpha q_\alpha}} \ll 1 \Rightarrow \lambda_e > \lambda_\alpha.$$

3. (2)

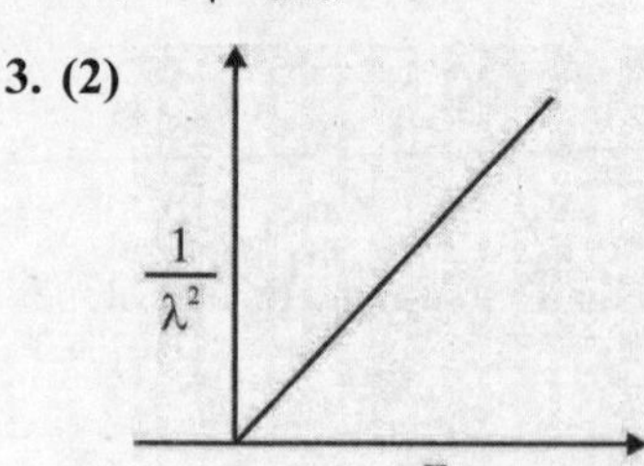

$$\lambda = \frac{h}{p} = \frac{h}{\sqrt{2mE}}$$

$$\lambda = \frac{h}{\sqrt{2mE}}$$

$$\lambda^2 = \frac{h^2}{2mE}$$

$$E \propto \frac{1}{\lambda^2}$$

4. (4) Energy $E = h\nu$

Velocity = $C = \nu\lambda$

$$\lambda = \frac{c}{\nu}$$

$$P = \frac{h}{\lambda} = \frac{h}{\frac{c}{\nu}} = \frac{h\nu}{c}$$

$F_{ext} = 0$, So momentum conserved energy conserved.

5. (3) eV = Energy of electron

for minimum wavelength, maximum loss of energy

$$eV = \left(\frac{hc}{\lambda}\right)$$

$$\lambda \propto \left(\frac{1}{V}\right)$$

6. (2) Incident energy = 2.20 eV

If $\phi < 2.20$ eV electron will emit.

$\phi > 2.20$ eVNo electron emission

Only caesium will emit electron

7. (1) According to de-Broglie's hypothesis

$$\lambda = \frac{h}{p} \Rightarrow \lambda \propto \frac{1}{P}$$

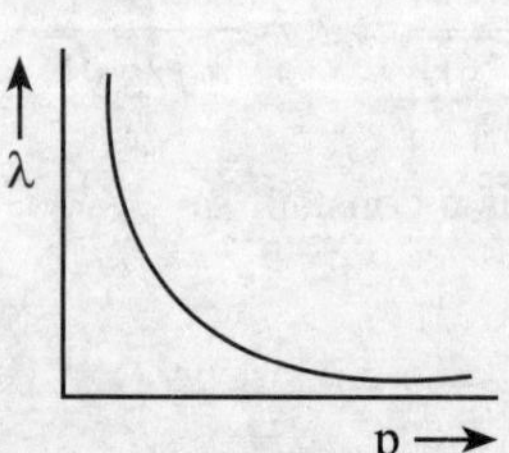

8. (None) Using the equation

$eV = h\nu - \phi$

$$\frac{eV_s}{2} = \frac{h\nu}{2} - \phi$$

This equation cannot be solved using given data.

9. (2) $\frac{hc}{\lambda} = K_{max} + \phi$[given ϕ is negligible]

So, $\frac{hc}{\lambda} = K_{max}$

$$\lambda_d = \frac{h}{\sqrt{2mK_{max}}} \Rightarrow K_{max} = \frac{h^2}{2m\lambda_d^2}$$

$$\left(\frac{hc}{\lambda}\right) = \frac{h^2}{2m\lambda_d^2} \Rightarrow \lambda = \left(\frac{2mc}{h}\right)\lambda_d^2$$

10. (1) In solar cell bond breakage becomes at depletion region.

11. (4) $\lambda = \frac{h}{mv}$

$\lambda \alpha \frac{1}{m}$ $\quad m_{\alpha} > m_p > m_n > m_{\beta}$

λ is highest for β particle

12. (1) de Broglie wavelength, $\lambda = \frac{h}{p} = \frac{h}{mv}$

$$\therefore \frac{\lambda_p}{\lambda_e} = \frac{m_e V_e}{m_p V_p} \Rightarrow m_p = m_e\left(\frac{v_e}{v_p}\right) \times \left(\frac{\lambda_e}{\lambda_p}\right)$$

Given, $\frac{v_e}{v_p} = \frac{1}{3}, \frac{\lambda_e}{\lambda_p} = \frac{1}{1.814 \times 10^{-4}}$

$m_e = 9.1 \times 10^{-31}$ kg

$$\therefore m_p = \frac{9.1 \times 10^{-31}}{3 \times 1.814 \times 10^{-4}} = 1.672 \times 10^{-27}\,kg$$

$\therefore$ The particle is a neutron.

13. (2) Energy corresponding to 2000 Å

$E = \frac{12375}{2000}$ ev = 6.2 eV

Maximum kinetic energy

$K = h\nu - h\nu_0$

$k = E - W$

$= (6.2 - 5.01)$ eV = 1.19 eV

Now, $\frac{1}{2} \times 9.1 \times 10^{-31} \times v_{max}^2$

$= 1.19 \times 1.6 \times 10^{-19}$

Or

$$v_{max}^2 = \frac{1.19 \times 1.6 \times 10^{-19} \times 2}{9.1 \times 10^{-31}}$$

Or $4.18 \times 10^{12} = 41.8 \times 10^{10}$

$v_{max} = 6.46 \times 10^5$ m/s

14. (3)

Atoms

NCERT Topic-wise MCQs

1. (1) The number of scattered particles detected will be maximum at the angle of scattering $\theta = 0°$ and minimum at $\theta = 180°$ because only 1 out of 10000 particles are perfectly reflected back.

2. (3) Impact parameter is the perpendicular distance of the initial velocity vector of the α-particle close to the nucleus and trajectory of α-particle depends on impact parameter.

3. (2) $$E = \frac{(Ze)(2e)}{4\pi\varepsilon_o r} = \frac{9 \times 10^9 \times 79 \times 2(1.6 \times 10^{-19})^2}{41.3 \times 10^{-15}}$$

$$= \frac{9 \times 79 \times 3.2 \times 1.6 \times 10^{-14}}{41.3}\,J$$

$$= 8.814 \times 10^{-13}\,\text{Joule} = \frac{8.814 \times 10^{-13}}{1.6 \times 10^{-19}}$$

E = 5.51 MeV

4. (2) $$E = \frac{(Ze)(e)}{4\pi\varepsilon_o r} = \frac{9 \times 10^9 \times 79 \times (1.6 \times 10^{-19})^2}{r \times 1.6 \times 10^{-13}}$$

$r = 2.3 \times 10^{-14}$m

5. (2) $$\frac{1}{\lambda_{32}} = R\left(\frac{1}{2^2} - \frac{1}{3^2}\right) = \frac{5R}{36}$$

$$\therefore \lambda_{32} = \frac{36}{5R}$$

6. (3) Centripetal force is provided by the electrostatic force of attraction between the revolving electrons and the nucleus in an atom.

$F_e = F_c$

$$\therefore \frac{mv^2}{r} = \frac{e^2}{4\pi\varepsilon_o r^2}$$

or $$v^2 = \frac{e^2}{4\pi\varepsilon_o rm} \Rightarrow v = \frac{e}{\sqrt{4\pi\varepsilon_o mr}}$$

7. (3) $E = \frac{q_1 q_2}{4\pi\varepsilon_o r}$, where E is energy

8. (2) At zero impact parameter α-particles rebound back ($\theta = \pi$) and suffers large scattering.

9. (2) At the distance of closest approach, Initial K.E. of α-particle

= P.E. of α-particle and target nucleus

or $$\frac{1}{2}mv^2 = \frac{1}{4\pi\varepsilon_0} \cdot \frac{Ze \cdot 2e}{r_0}$$

or $$r_0 = \frac{1}{4\pi\varepsilon_0} \cdot \frac{Ze \cdot 2e}{\frac{1}{2}mv^2}$$ i.e., $r \propto \frac{1}{m}$

10. (4) $$\frac{\text{Volume of atom}}{\text{Volume of nucleus}} = \frac{\frac{4}{3}\pi(10^{-10})^3}{\frac{4}{3}\pi(10^{-15})^3} = 10^{15}$$

11. (4) All the statements are correct.

12. (1) Bismuth $\left({}^{214}_{83}\text{Be}\right)$ is used to produce alpha particle beam.

13. (1) At the distance of closed approach, Initial K.E. of the particle = P.E. of the particle and the massive nucleus

$$\frac{1}{2}mv^2 = \frac{1}{4\pi\varepsilon_0}\cdot\frac{Ze\times e}{r_0}$$

$$\text{or } r_0 = \frac{Ze^2}{2\pi\varepsilon_0 mv^2}$$

14. (4) The atomic electrons being so light, do not effect the alpha particles.

15. (3) Rutherford's atomic model could account for the positive charge central core atom.

16. (1) In emission spectra an atomic gas or vapour is excited at low pressure by passing an electric current through them, it emits radiations of specific wavelength.

17. (1) In 1885, the first spectral series were observed by Johann Jakob Balmer, This series is called the Balmer series.

18. (4) H_δ line has smallest wavelength, Hence The frequency of this line will be maximum because $\nu \propto \frac{1}{\lambda}$.

19. (2) Balmer series falls under visible region.

20. (1) For Lyman series

$$\nu = Rc\left[\frac{1}{(1)^2}-\frac{1}{n^2}\right],\ n = 2,3,4....$$

For the series limit for Lyman series n = ∞

$$\therefore \nu_1 = Rc\left[\frac{1}{(1)^2}-\frac{1}{\infty^2}\right] = Rc \quad ...(i) \qquad ...(i)$$

For the first line of Lyman series, n = 2

$$\therefore \nu_2 = Rc\left[\frac{1}{(1)^2}-\frac{1}{(2)^2}\right] = \frac{3}{4}Rc \qquad ...(ii)$$

For Balmer series

$$\therefore \nu = Rc\left[\frac{1}{(2)^2}-\frac{1}{n^2}\right]\ n = 3,4,5...$$

For the series limit of Balmer series, n = ∞
Similarly as eq. (1)

$$\therefore \nu_3 = Rc\left[\frac{1}{(2)^2}-\frac{1}{(\infty)^2}\right] = \frac{Rc}{4} \qquad ...(iii)$$

From eq (*i*), (*ii*), & (*iii*) we get

$\nu_1 = \nu_2 + \nu_3 \quad \Rightarrow \nu_1 - \nu_2 = \nu_3$

21. (3) For the first line $n_1 = 2$, $n_2 = 3$

For the second line $n_1 = 2$, $n_2 = 4$

$$\frac{\lambda_1}{\lambda_2} = \frac{\left[\frac{1}{(2)^2}-\frac{1}{(4)^2}\right]}{\left[\frac{1}{2^2}-\frac{1}{3^2}\right]} = \frac{3/16}{5/36} = \frac{3}{16}\times\frac{36}{5}$$

$$\frac{\lambda_1}{\lambda_2} = \frac{27}{20}$$

$$\lambda_2 = \frac{20}{27}\times\lambda_1 = \frac{20}{27}\times 6561 = 4860\,\text{Å}$$

22. (4) $\nu = \frac{c}{\lambda}$ ...(*i*)

$$\frac{1}{\lambda} = (Z)^2 R\left[\frac{1}{n_1^2}-\frac{1}{n_2^2}\right]$$

$$\frac{1}{\lambda} = (1)^2 R\left[\frac{1}{9}-\frac{1}{4}\right]$$

$$\frac{1}{\lambda} = R\left[\frac{4-9}{36}\right]$$

$$\frac{1}{\lambda} = R\left[\frac{-5}{36}\right]$$

By putting the value of $\frac{1}{\lambda}$ in eq. (*i*)

$$\nu = \frac{-5Rc}{36}$$ [–ve sign indicates emission]

23. (4) $\frac{1}{\lambda} = R\left[\frac{1}{n_1^2}-\frac{1}{n_2^2}\right] = R\times K,\ \left[K = \frac{1}{n_1^2}-\frac{1}{n_2^2}\right]$

For λ_{max}, K should be minimum

which corresponds to $n_1 = 3$, $n_2 = 4$, so transition takes place between n = 4 and n = 3 states

24. (4) Number of spectral lines obtained due to transition of electrons from n^{th} orbit to lower orbit is given by

$$N = \frac{n(n-1)}{2}$$

Case-I: $6 = \frac{n_1(n_1-1)}{2}$

$\Rightarrow n_1^2 - n_1 - 12 = 0$

$\Rightarrow (n_1 - 4)(n_1 + 3) = 0$

$\Rightarrow n_1 = 4$

Case-II: $3 = \frac{n_2(n_2-1)}{2}$

$\Rightarrow n_2^2 - n_2 - 6 = 0 \Rightarrow (n_2 + 2)(n_2 - 3) = 0$

$\Rightarrow n_2 = 3$

$$\therefore \nu \propto \frac{1}{n} \Rightarrow \frac{\nu_1}{\nu_2} = \frac{n_2}{n_1} = \frac{3}{4}$$

25. (4) For the first line of Lyman series, $n_1 = 1$, $n_2 = 2$

$$\therefore \frac{1}{\lambda_1} = R\left[\frac{1}{(1)^2}-\frac{1}{(2)^2}\right] = R\left[1-\frac{1}{4}\right] \Rightarrow \frac{3R}{4} \qquad ...(i)$$

For the first line of Paschen series, $n_1 = 3$ and $n_2 = 4$

$$\frac{1}{\lambda_2} = R\left[\frac{1}{(3)^2}-\frac{1}{(4)^2}\right] = R\left[\frac{1}{9}-\frac{1}{16}\right] \Rightarrow \frac{7R}{144} \qquad ...(ii)$$

On dividing equation (*i*) and (*ii*) we get

$$\frac{\lambda_1}{\lambda_2}=\frac{7R}{144}\times\frac{4}{3R}=\frac{7}{108}$$

$$\therefore \frac{\lambda_1}{\lambda_2}=\frac{7}{108}$$

26. (3) In Bohr's atomic model, $T \propto n^3$

27. (2) $E=\frac{-13.6}{n^2}=-3.4 \quad \therefore n=2$

Angular momentum of electron

$$=\frac{nh}{2\pi}=\frac{2\times h}{2\times\pi}=\frac{h}{\pi}$$

$$L=\frac{h}{\pi}$$

28. (1) According to the Bohr's theory, the angular momentum of electrons in the orbits are integral multiples of $\frac{h}{2\pi}$. Angular momentum can not be in fractions.

29. (1) In the nth orbit, let r_n = radius and v_n = speed of electron.

Time period, $T_n=\frac{2\pi r_n}{v_n}\propto\frac{r_n}{v_n}$

Now, $r_n \propto n^2$ and $v_n \propto \frac{1}{n}$

$\therefore \frac{r_n}{v_n}\propto n^3$ or $T_n \propto n^3$

Here, $8=\left(\frac{n_1}{n_2}\right)^3 \Rightarrow \frac{n_1}{n_2}=2 \Rightarrow n_1=2n_2$

30. (4) By absorbing 10.2 eV of energy, an electron jumps from ground level (n = 1) to the first excited state (n = 2).

∴ Increase in angular momentum

$$=\frac{2h}{2\pi}-\frac{h}{2\pi}=\frac{6.6\times10^{-34}}{2\times3.14}=1.05\times10^{-34}\text{ Js}$$

31. (3) In hydrogen atom, the lowest orbit corresponds to minimum energy because it is also in bounded state, i.e., negative energy.

32. (2) Radius of nth orbit in hydrogen like atoms is

$r_n=\frac{a_o n^2}{Z}$ (Where a_o is the Bohr's radius)

For hydrogen atom, $Z=1$

$\therefore r_1=a_0$ (n = 1 for ground state)

For Be^{3+}, $Z=4 \therefore r_n=\frac{a_o n^2}{4}$

According to given problem

$r_1=r_n$

$a_o=\frac{n^2 a_o}{4}$

n = 2

33. (2) $E_n=\frac{Z^2}{n^2}E_0$

For H atom : Z = 1 and n = 2

$\therefore E_2=\frac{1}{4}E_0$

For He atom : Z = 2 and n = 3

$\therefore E_3=\frac{2^2}{3^2}E_0=\frac{4}{9}E_0$

Hence $\frac{E_3}{E_2}=\frac{\frac{4}{9}E_0}{\frac{1}{4}E_0}=\frac{16}{9}$

or $E_3=\frac{16}{9}E_2=\frac{16}{9}E$

34. (4) According to Bohr's model $mvr=\frac{nh}{2\pi}$

$$\Rightarrow n=\frac{2\pi mvr}{h}=\frac{2\times22\times6\times10^{24}\times3\times10^4\times1.5\times10^{11}}{7\times6.6\times10^{-34}}$$

$\therefore n=2.57\times10^{74}$

35. (4) The circumference of the orbit must be an integral multiple of de Broglie wavelength λ, i.e., $2\pi r=n\lambda$.

36. (1) In the ground state of hydrogen atom, suppose, a_o = Bohr radius.

$T=\frac{2\pi a_o}{v_o}$

$\therefore$ current created, $I=\frac{e}{T}=\frac{ev_o}{2\pi a_o}$

37. (4) $E=\frac{-13.6}{n^2}$ ev

The energy of photon $=E_3-E_2$

$$\frac{-13.6}{(3)^2}-\left(\frac{-13.6}{(2)^2}\right)=13.6\left[-\frac{1}{9}+\frac{1}{4}\right]$$

$$=13.6\left[\frac{-4+9}{36}\right]=13.6\times\frac{5}{36}$$

E = 1.9 eV

38. (2) ∵ K.E. = –total energy

K.E. = 3.4 eV = $3.4\times1.6\times10^{-19}$ J

$$\because \lambda=\frac{h}{P}=\frac{h}{\sqrt{2mE}}=\frac{6.6\times10^{-34}}{\sqrt{2\times9.1\times10^{-31}\times3.4\times1.6\times10^{-19}}}$$

$\lambda=6.6\times10^{-10}$ m

39. (1) K.E. of an electron in nth orbit: $K_n\propto\frac{1}{n^2}$ and

P.E. of an electron in nth orbit: $U_n\propto\frac{1}{n^2}$

∴ When an electron passes from state n = 2 to n = 1

$\frac{K_2}{K_1}=\frac{1^2}{2^2}=\frac{1}{4}$ or $K_1=4K_2$

$\frac{U_2}{U_1}=\frac{1^2}{2^2}=\frac{1}{4}$ or $U_1=4U_2$

choice (1) is correct.

40. (4) $r = \frac{a_o n^2}{Z} \Rightarrow \frac{0.53\times10^{-10}}{4} = \frac{a_0 n^2}{Z}$

$\frac{0.53\times10^{-10}}{4} = \frac{(1)^2}{Z}\times0.53\times10^{-10}$

$Z = 4$

∴ It must be Be^{3+}

41. (3) $T_e = \frac{2\pi r}{v}$

$T_e = \frac{4\varepsilon_o^2 n^3 h^3}{me^4}$

Now, current, $I = \frac{e}{T_e} \Rightarrow \frac{e}{\frac{4\varepsilon_o^2 n^3 h^3}{me^4}} = \frac{me^5}{4\varepsilon_o^2 n^3 h^3}$

$I \propto e^5$

42. (1) The electron acts as a current loop with frequency $f = \frac{v}{2\pi r}$, so the current is $I = ef = \frac{ev}{2\pi r}$. The magnetic field at the center of the loop is $B = \frac{\mu_o I}{2r} = \frac{\mu_o ev}{4\pi r^2}$. Using the quantization $mvr = \frac{nh}{2\pi}$, we get $v = \frac{nh}{2\pi mr}$, so $B = \frac{\mu_0 enh}{8\pi^2 mr^3}$.

The flux through the orbit (area πr^2) is $\phi = B\cdot\pi r^2 = \frac{\mu_0 enh}{8\pi mr}$.

Substitute $r = \frac{n^2h^2\epsilon_0}{\pi me^2}$: $\phi = \frac{\mu_0 enh}{8\pi m}\cdot\frac{\pi me^2}{n^2h^2\epsilon_0} = \frac{\mu_0 e^3}{8nh\epsilon_0}$.

43. (1) Centripetal force = Electrostatic attraction

$\frac{mv^2}{a_0} = \frac{1}{4\pi\varepsilon_0}\cdot\frac{e^2}{a_0^2} \quad \therefore v = \frac{e}{\sqrt{4\pi\varepsilon_0 a_0 m}}$.

44. (3) $v = \frac{c}{137}$

$\frac{v}{c} = \frac{1}{137}$

45. (2) Total energy of electron is always negative in bounded state. But K.E is equal in magnitude it can never be negative.

46. (4) According to the de-Broglie relation.

$2\pi r = n\lambda$

For first orbit n = 1

$2\pi r = \lambda$

47. (4) As we know that $v \propto \frac{1}{n}$

$\frac{v_5}{v_2} = \frac{2}{5} \Rightarrow \frac{v_5}{v} = \frac{2}{5}$

$v_5 = \frac{2}{5}v$

48. (2) $(E_C - E_B) + (E_B - E_A) = E_C - E_A$

$\Delta E_{CB} + \Delta E_{BA} = \Delta E_{CA}$

As we know,

$\Delta E_n = \frac{hc}{\lambda_n}$ then,

$\frac{hc}{\lambda_1} + \frac{hc}{\lambda_2} = \frac{hc}{\lambda_3}$

$\frac{1}{\lambda_1} + \frac{1}{\lambda_2} = \frac{1}{\lambda_3} \Rightarrow \lambda_3 = \frac{\lambda_1\lambda_2}{\lambda_1+\lambda_2}$

49. (1) Ionisation energy = $(Z)^2hcR$

$= (3)^2hcR = 9hcR$

50. (2) In hydrogen atom electron moves with constant centripetal acceleration.

$a_c = \frac{v^2}{r} = \frac{(2.18\times10^6)^2}{0.528\times10^{-10}} = 9\times10^{22}\,m/s^2$

51. (2) P.E = 2 × T.E

= 2 × –13.6 eV

P.E = –27.2 eV

52. (1) $\frac{1}{\lambda_{max}} = R\left[\frac{1}{(1)^2} - \frac{1}{(2)^2}\right] \Rightarrow \lambda_{max} = \frac{4}{3R} \approx 1213\,\mathring{A}$

$\frac{1}{\lambda_{min}} = R\left[\frac{1}{(1)^2} - \frac{1}{\infty}\right] \Rightarrow \lambda_{min} = \frac{1}{R} = 910\,\mathring{A}$

53. (1) $r \propto n^2$, $A = \pi r^2$

$\frac{A_1}{A_o} = \frac{\pi(r_1)^2}{\pi(r_o)^2} = \frac{(n_2)^4}{(n_1)^4} = \frac{16}{1}$

54. (2) $E_n = \frac{13.6\times Z^2}{n^2}$

For the first excited state n = 2 and for Li^{2+}, Z = 3

$E_n = \frac{13.6\times9}{4} = 30.6\,eV$

55. (3) $E_n = -\frac{13.6}{n^2}eV$

$\therefore \Delta E = E_\infty - E_2 = 0 + \frac{13.6}{2^2} = 3.4\,eV$

56. (2) As we know that, $E_n = \frac{-13.6Z^2}{n^2}eV$

$E_1 = \frac{-13.6\times(2)^2}{(1)^2} = -54.4eV$

57. (3) $\frac{1}{\lambda} = RZ^2\left[\frac{1}{n_1^2} - \frac{1}{n_2^2}\right]$

$\lambda \propto \frac{1}{Z^2}$, $\lambda_{Li^{++}} : \lambda_{He^+} : \lambda_H = 4:9:36$

58. (3) For an electron to remain orbiting around the nucleus, the angular momentum (L) should be an integral multiple of $\frac{h}{2\pi}$

$$mvr = \frac{nh}{2\pi}$$

59. (2) $T = \frac{4\varepsilon_o^2 n^3 h^3}{mZ^2 e^4}$

$\therefore T \propto n^3$

Hence, frequency $\propto n^{-3}$

High Order Time Intensive MCQs

1. (2) Energy of the first time in Lyman series

$$E = 13.6 \times \left(1 - \frac{1}{4}\right) eV = 10.2 \text{ eV}$$

velocity of atoms $v = \frac{p}{m} = \frac{E}{c} \cdot \frac{1}{m}$

$$= \frac{10.2 \times 1.6 \times 10^{-19}}{3 \times 10^8 \times 1.67 \times 10^{-27}} \text{ m/s}$$

$= 3.25$ m/s

2. (1) Here, $E_5 - E_1 = \frac{hc}{\lambda}$

$$\frac{Rhc}{25} - Rhc = \frac{hc}{\lambda}$$

$$\frac{24}{25} R = \frac{1}{\lambda}$$

$$P = \frac{h}{\lambda} \Rightarrow v = \frac{h}{m\lambda} = \frac{24 \text{ Rh}}{25 \text{ m}}$$

3. (3) We know that, frequency

$$v = \frac{c}{\lambda} = c.R\left(\frac{1}{n_1^2} - \frac{1}{n_2^2}\right)$$

$$= 3 \times 10^8 \times 10^7 \left(\frac{1}{2^2} - \frac{1}{4^2}\right) = \frac{9}{16} \times 10^{15} \text{ Hz}$$

4. (2) For transition from 4E to E,

$$(4E - E) = \frac{hc}{\lambda_1} \Rightarrow \lambda_1 = \frac{hc}{3E} \quad ...(i)$$

For transition from $\frac{7}{3}$ E to E

$$\left(\frac{7}{3}E - E\right) = \frac{hc}{\lambda_2} \Rightarrow \lambda_2 = \frac{3hc}{4E} \quad ...(ii)$$

From Eqs. (*i*) and (*ii*), we get $\frac{\lambda_1}{\lambda_2} = \frac{4}{9}$

5. (2) When the electron jumps from n = 3 to n = 1 orbit,

$$v = Rc\left[\frac{1}{1^2} - \frac{1}{3^2}\right]$$

or $2.7 \times 10^{15} = Rc \times \frac{8}{9}$

When the electron jumps from n = 4 to n = 1 orbit,

$$v' = Rc\left[\frac{1}{1^2} - \frac{1}{4^2}\right] = Rc \times \frac{15}{16}$$

$$\therefore \frac{v'}{2.7 \times 10^{15}} = \frac{15}{16} \times \frac{9}{8}$$

or $v' = \frac{15}{16} \times \frac{9}{8} \times 2.7 \times 10^{15}$

$= 2.8 \times 10^{15}$ Hz.

6. (3) Absorption spectrum involves excitation of ground level to any higher level. The spectral lines 1, 2, and 3 (which involve ground level) emission will occur in the absorption spectrum.

7. (1) The wavelength can be given as

$$\frac{1}{\lambda} = Z^2 R\left(\frac{1}{n_1^2} - \frac{1}{n_2^2}\right) \quad ...(i)$$

For λ to be shortest, $\left(\frac{1}{n_1^2} - \frac{1}{n_2^2}\right)$ should be maximum

so, $n_2 = \infty$ and $n_1 = 1$

Now, from Eq. (*i*) $\frac{1}{\lambda} = R\left(\frac{1}{1}\right) \Rightarrow \lambda = \frac{1}{R} = \frac{1}{10^7} m \simeq 1000 \text{ Å}$

8. (3) When transition is from upper state to lower state, the energy is evolved. Therefore, options (3) and (4) are practicable.

In option (3), $E = -Rch\left(\frac{1}{1^2} - \frac{1}{2^2}\right) = -\frac{3}{4} Rch$

In option (4), $E = -Rch\left(\frac{1}{2^2} - \frac{1}{6^2}\right) = -\frac{2}{9} Rch$

Thus, maximum energy is evolved in option (3), when transition takes place between n = 2 to n = 1.

9. (1) Energy of emitted radiation, $E = Rhc\left(\frac{1}{n_1^2} - \frac{1}{n_2^2}\right)$

$$\Rightarrow E_{(4\to3)} = Rhc\left(\frac{1}{3^2} - \frac{1}{4^2}\right) = Rhc\left(\frac{7}{9 \times 16}\right) = 0.05Rhc$$

$$\Rightarrow E_{(4\to2)} = Rhc\left(\frac{1}{2^2} - \frac{1}{4^2}\right) = Rhc\left(\frac{3}{16}\right) = 0.2Rhc$$

$$\Rightarrow E_{(2\to1)} = Rhc\left[\frac{1}{(1)^2} - \frac{1}{(2)^2}\right] = Rhc\left(\frac{3}{4}\right) = 0.75Rhc$$

$$\Rightarrow E_{(1\to4)} = Rhc\left[\frac{1}{(4)^2} - \frac{1}{(1)^2}\right] = -\frac{8}{9} Rhc = -0.9Rhc$$

10. (2) $\frac{1}{\lambda_1} = R\left(\frac{1}{2^2} - \frac{1}{3^2}\right) = R\left(\frac{5}{36}\right)$

$$\frac{1}{\lambda_2} = R\left(\frac{1}{2^2} - \frac{1}{4^2}\right) = R\left(\frac{12}{16 \times 4}\right) = \frac{3R}{16}$$

$$\frac{\lambda_2}{\lambda_1} = \frac{16}{3} \times \frac{5}{36} = \frac{20}{27}$$

Given, $\lambda_1 = \lambda$

$$\lambda_2 = \frac{20}{27} \lambda$$

11. (1) Let n_1 be initial state of electron. Then $E_1 = -\frac{13.6}{n_1^2}eV$

Here $E_1 = -0.85$ eV, therefore $-0.85 = \frac{13.6}{n_1^2} \Rightarrow n_1 = 4$

Let n_2 be the final excitation state of the electron. Since excitation energy is always measured with respect to the ground state, therefore.

$\Delta E = 13.6\left[1-\frac{1}{n_2^2}\right]$ here $\Delta E = 10.2$ eV, therefore,

$10.2 = 13.6\left[1-\frac{1}{n_2^2}\right] \Rightarrow n_2 = 2$

12. (3) Energy of an electron in the ground state of an atom (Bohr's hydrogen like atom) is given as

$E = -13.6\ Z^2$ eV

(Z = atomic number of the atom

$\Rightarrow E_{ionisation} = 13.6\ Z^2$

$\Rightarrow \frac{(E)_H}{(E_{ion})_{Li}} = \left(\frac{Z_H}{Z_{Li}}\right)^2 = \left(\frac{1}{3}\right)^2 = \frac{1}{9}$

13. (3) By using $E = -\frac{13.6}{n^2}eV$ (for H_2 atom)

$\Rightarrow -0.544 = -\frac{13.6}{n^2} \Rightarrow n^2 = 25 \quad n = 5$

$\therefore$ Angular mometum $= n\frac{h}{2\pi} = \frac{5h}{2\pi}$

14. (3) $\therefore$ Orbiting speed of an electron $v \propto \frac{1}{n}$

Frequency $\nu \propto \frac{1}{n^3}$ and radius $r \propto n^2$

15. (4) $2E - E = \frac{hc}{\lambda} \Rightarrow E = \frac{hc}{\lambda}$

$\frac{4E}{3} - E = \frac{hc}{\lambda'} \Rightarrow \frac{E}{3} = \frac{hc}{\lambda'} \therefore \frac{\lambda'}{\lambda} = 3$

$\Rightarrow \lambda' = 3\lambda$

16. (4) $E_n = -13.6\frac{Z^2}{n^2}eV$

Required energy for said transition

$\Delta E = E_3 - E_1 = 13.6Z^2\left[\frac{1}{1^2}-\frac{1}{3^2}\right]$

$\Rightarrow \Delta E = 13.6\times 3^2\left[\frac{8}{9}\right] = 108.8eV$

$\Rightarrow \Delta E = 108.8 \times 1.6 \times 10^{-19}$ J

Now $\Delta E = \frac{hc}{\lambda} = 108.8\times 1.6\times 10^{-19}$

$\Rightarrow \lambda = \frac{6.6\times 10^{-34}\times 3\times 10^8}{108.8\times 1.6\times 10^{-19}}$

$= 0.11374 \times 10^{-7}$ m $= 113.74$ Å

17. (2) By using $\frac{1}{\lambda} = RZ^2\left[\frac{1}{n_1^2}-\frac{1}{n_2^2}\right]$

For hydrogen atom $\frac{1}{(\lambda_{min})_H} = R\left[\frac{1}{1^2}-\frac{1}{\infty}\right] = R$

$\Rightarrow (\lambda_{min})_H = \frac{1}{R}$ ...(*i*)

For hydrogen like atom

$\left(\frac{1}{\lambda_{min}}\right)_{atom} = RZ^2\left(\frac{1}{2^2}-\frac{1}{\infty}\right)$

$(\lambda_{min})_{atom} = \frac{4}{RZ^2}$ ...(*ii*)

From equation (*i*) and (*ii*)

$\Rightarrow Z = 2 = \frac{1}{R} = \frac{4}{RZ^2}$

18. (3) For first Balmer line,

$\bar{v} = \frac{1}{\lambda} = R\left(\frac{1}{2^2}-\frac{1}{3^2}\right) = \frac{5R}{36}$

19. (3) The hydrogen atom before the transition was at rest. Therefore from conservation of momentum.

$p_{H-atom} = p_{photon} = \frac{E_{radiated}}{c} = \frac{13.6\left(\frac{1}{n_1^2}-\frac{1}{n_2^2}\right)eV}{c}$

$1.6\times 10^{-27}\times v = \frac{13.6\left(\frac{1}{1^2}-\frac{1}{5^2}\right)\times 1.6\times 10^{-19}}{3\times 10^8}$

$v \simeq 4$ m /s

20. (3) We have, $\lambda = \frac{hc}{\Delta E}$

$\therefore \frac{\lambda_1}{\lambda_2} = \frac{hc/\Delta E_1}{hc/\Delta E_2} = \frac{\Delta E_2}{\Delta E_1}$

$= \frac{\left(\frac{4}{3}E - E\right)}{2E - E} = \frac{1}{3}$

21. (1) Kinetic energy $K \propto Z^2$

$\therefore \frac{K_H}{K_{He^+}} = \left(\frac{1}{2}\right)^2 = \frac{1}{4}$

22. (3) As, for conservative fields $\mathbf{F} = -\left(\frac{dU}{dr}\right)$

$\therefore$ Magnitude of force on particle is

$\Rightarrow F = \frac{dU}{dr} = \frac{d}{dr}\left(\frac{1}{2}kr^2\right) \Rightarrow F = kr$

This force is acting like centripetal force.

$\therefore \frac{mv^2}{r} = kr$ ... (*i*)

So, for nth orbit,

$\Rightarrow v_n^2 = kr_n^2/m$

$$\Rightarrow \frac{n^2h^2}{4\pi^2 r^2} = mkr_n^2 \qquad \left[\because v_n = \frac{nh}{2\pi mr}\right]$$

Therefore, $r_n^4 \propto n^2$

$\Rightarrow r_n^2 \propto n$

So, $r_n \propto \sqrt{n}$

Energy of particle is

$$E_n = PE + KE = \frac{1}{2}kr_n^2 + \frac{1}{2}mv_n^2$$

$$= \frac{1}{2}kr_n^2 + \frac{1}{2}kr_n^2 \qquad \text{[Using Eq. (i)]}$$

$= kr_n^2$

So, energy $E_n \propto r_n^2$

$\Rightarrow E_n \propto n$

Statement Based MCQs

1. (2) Electric conduction in a gas takes place under low pressure. Statement II is also false.

2. (1) This is because when the visible light from sun's surface passes through atmosphere, some of light at particular wavelength is absorbed by atoms and ions and so is missing in the spectrum we see.

3. (1) Since Bohr's model considers nucleus massive in comparison to electrons and both Hydrogen and deuterium consists of one electron only, so Bohr's model can't differentiate between them.

4. (1) Both the statements are true.

5. (1) We know that principal quantum number represent the main energy level or energy shell. Since each energy level is associated with a definite amount of energy, this quantum number determines to a large extent energy of an electron. It also determines the average distance of an electron around the nucleus.

Assertion & Reason MCQs

1. (1) The conclusion drawn was that almost all space inside atom is empty as α particles weren't deflected so they went straight.

2. (1) Both assertion and reason are true and the reason is the correct explanation of the assertion.

For Balmer series,

$$\frac{1}{\lambda_{max}} = R\left(\frac{1}{2^2} - \frac{1}{3^2}\right) = \frac{5R}{36}$$

$$\therefore \lambda_{max} = \frac{26}{5R} = \frac{36}{5 \times 1.097 \times 10^7} = 6563 \text{ Å}$$

$$\frac{1}{\lambda_{min}} = R\left(\frac{1}{2^2} - \frac{1}{\infty^2}\right) = \frac{R}{4}$$

$$\lambda_{min} = \frac{4}{R} = \frac{4}{1.097 \times 10^7} = 3646 \text{ Å}$$

Both of these wavelengths lie in the visible region of the spectrum.

3. (2) Both Assertion & Reason are correct but Reason is not the correct explanation of the assertion.

4. (2) Both Assertion & Reason are correct but Reason is not the correct explanation of the assertion.

5. (2) According to classical physics, all accelerating charged particle radiate electromagnetic radiation. So moving electrons will also radiate energy. If we see the atomic structure we find that electrons revolve around the nucleus in some particular orbits. Bohr termed these orbits as the stationary orbits this is because the electrons do not radiate energy as long as they are moving in these orbits. This is one of Bohr's postulates. This postulate is based on the fact that if the moving electrons radiate thereby losing energy, they have got a chance to finally fall back onto the nucleus and the atom will collapse.

Match the Column MCQs

1. (2) For Bohr's Model

- Velocity of electron in n^{th} orbit $(v_n) = \frac{Z}{n}\frac{e^2}{2\varepsilon_o h}$
- Total Energy of electron in n^{th} orbit

$$E_n = \frac{-13.6Z^2}{n^2} eV$$

- Radius of nth orbit $(r_n) = \frac{n^2}{Z}\frac{\varepsilon_o h^2}{\pi me^2}$

Hence, $v_n \alpha \frac{Z}{n}$, $\quad E_n \alpha \frac{Z^2}{n^2}$, $\quad r_n \alpha \frac{n^2}{Z}$

2. (1) Lyman series → Electron jumps from higher state to first orbit (n = 1)

Balmer series → Electron jumps from higher state to second orbit (n = 2)

Paschen series → Electron jumps from higher state to third orbit (n = 3)

Pfund series → Electron jumps from higher state to fifth orbit (n = 5)

3. (4) By Bohr's Model

$P_E = 2\ T_E = -2\ K_E$

If $|T_E| = a$ (magnitude)

$T_E = -a$

$P_E = -2a$

$K_E = a$

NCERT Exemplar MCQs

1. (3) The atomic number of lithium is 3, therefore, the radius of Li^{++} ion in its ground state, on the basis of Bohr's model, will be about $\frac{1}{3}$ times to that of Bohr radius.

Therefore, the radius of lithium ion is near $\frac{53}{3} \approx 18$ pm.

2. (3) When one decides to work in a frame of reference where the electron is at rest, the given expression is not true as it forms the non- inertial frame of reference.

3. (1) The simple Bohr model cannot be directly applied to calculate the energy levels of an atom with many electrons. This is because of the electrons not being subject to a central force.

4. (1) In the simple Bohr model, only the magnitude of angular momentum is kept equal to some integral multiple of $\frac{h}{2\pi}$, where, h is Planck's constant and thus, the Bohr model gives incorrect values of angular momentum.

5. (1) In the molecules, nuclear force between the nuclei of the two atoms is not important because nuclear forces are short-ranged.

6. (1) The total energy associated with the two H-atoms in the ground state collide inelastically

$= 2 \times (13.6 \text{ eV}) = 27.2 \text{ eV}$

The maximum amount by which their combined kinetic energy is reduced when any one of them goes into first excited state after the inelastic collision.

The total energy associated with the two H-atoms after the collision

$$= \left(\frac{13.6}{2^2}\right) + (13.6) = 17.0 \text{ eV}$$

Therefore, maximum loss of their combined kinetic energy $= 27.2 - 17.0 = 10.2$ eV

7. (1) A set of atoms in an excited state decays in general to any of the states with lower energy.

NEET & AIIMS Past Year MCQs

1. (4) Since,

$$\lambda \propto \frac{1}{\Delta E}$$

Smaller energy difference → Larger wavelength

Hence:

$\lambda_2 > \lambda_1, \lambda_2 = 2\lambda_3$

2. (2) Electronic transitions to $n = 4$ correspond to the Brackett series in hydrogen.

Brackett Series

3. (1) Statement I is correct and II is incorrect.

Atoms are electrically neutral as they have equal number of positive and negative charges. Also, atoms of most elements are stable and they emit characteristics spectrum of their own.

4. (4)

$$\frac{1}{\lambda} = RZ^2\left(\frac{1}{n_1^2} - \frac{1}{n_2^2}\right)$$

$$\lambda = \frac{1}{RZ^2}\left(\frac{n_1^2 n_2^2}{n_2^2 - n_1^2}\right)$$

For H, $Z = 1$ So,

$$\lambda = 91.2\left(\frac{n_1^2 n_2^2}{n_2^2 - n_1^2}\right) nm$$

For A $\quad \lambda = 656.3$

For B $\quad \lambda = 486.1$

For C $\quad \lambda = 434.1$

For D $\quad \lambda = 410.2$

5. (3) $\because \frac{1}{\lambda} = R\left(\frac{1}{n_2^2} - \frac{1}{n_1^2}\right) \Rightarrow \frac{1}{\lambda} = R\left(\frac{1}{2^2}\right) \Rightarrow \lambda = \frac{4}{R}$

$$\frac{1}{\lambda'} = R\left(\frac{1}{4^2}\right) \Rightarrow \lambda' = \frac{16}{R} \Rightarrow \lambda' = 4\lambda$$

6. (1) $r_n = a_0 n_2$

$r_1 = a_0 = 5.3 \times 10^{-11}$m

$r_3 = a_0(3)2 = 5.3 \times 10^{-11} \times 9 = 4.77$Å

7. (1) For the First excited state n = 2

$$T_1 = -13.6\frac{z^2}{n^2} = -\frac{13.6}{4}\text{eV}$$

For the second excited state

$$T_2 = -13.6\frac{z^2}{n^2} = -\frac{13.6}{9}\text{eV}$$

$\Rightarrow T_1 : T_2 = 9 : 4$

8. (1) For maximum wave length of Balmer series

$n_2 = 3$; $n_1 = 2$

$$\frac{1}{\lambda} = R.(1)^2\left[\frac{1}{4} - \frac{1}{9}\right]$$

On solving $\lambda = 6566.4$ Å

9. (2) The wavelength of Balmer series is given by:

$$\frac{1}{\lambda} = R\left(\frac{1}{(2)^2} - \frac{1}{(n)^2}\right), \text{ n = 3, 4, 5------}$$

$$\frac{1}{\lambda_{max}} = R\left(\frac{1}{(2)^2} - \frac{1}{(3)^2}\right) \quad R\left[\frac{9-4}{9 \times 4}\right] = R\left[\frac{5}{36}\right]$$

$$\lambda_{max} = \frac{36}{5R} = \frac{36}{5 \times 1.097 \times 10^7} = 6563 \text{ Å}$$

and $\frac{1}{\lambda_{min}} = R\left[\frac{1}{(2)^2} - \frac{1}{(60)^2}\right]$

$$\lambda_{min} = \frac{4}{R} = \frac{4}{1.097 \times 10^7} = 3646 \text{ Å}$$

Hence it lies in the visible region.

Nuclei

NCERT Topic-wise MCQs

1. (1) $\frac{R_1}{R_2} = \left(\frac{A_1}{A_2}\right)^{\frac{1}{3}} = \left(\frac{216}{64}\right)^{\frac{1}{3}} = \frac{6}{4} = 3:2$

2. (1) Number of atoms in 14 g of ${}^{14}_{6}C$

= Avogadro's number $= 6 \times 10^{23}$.

Number of neutrons in each atom = 14 – 6 = 8

∴ Total number of neutrons $6 \times 10^{23} \times 8 = 48 \times 10^{23}$.

3. (1) The mass of nucleus formed is always less than the sum of the masses of the constituents protons and neutrons, i.e., $m < (A-Z)m_n + Zm_p$

4. (2) Density of uranium nucleus $\approx 10^{17}$ kg m^{-3}

5. (1) No. of protons = 2 + 2 + 6 + 2 = 6 = 18

No. of Neutrons = 40 – 18 = 22

6. (4) $\because \frac{r(Te)}{r(Al)} = \left(\frac{125}{27}\right)^{\frac{1}{3}}$

$\therefore r(Te) = \frac{5}{3} r(Al) = \frac{5}{3} \times 3.6 = 6.0\,\text{fm}$

7. (2) Let nucleus be $_ZX^A$

$\therefore \frac{R_x}{RC_s} = \left(\frac{A}{189}\right)^{1/3}$

$\therefore \frac{1}{3} = \left(\frac{A}{189}\right)^{1/3}$

$A = \frac{189}{27} = 7$

8. (3) $\Delta m = 2 \times 1.007825 + 2 \times 1.008665 - 4.002800$

$= 4.03298 - 4.002800 = 0.03018$ amu

Binding energy = $\Delta m \times 931.5$ Mev

$= 0.03018 \times 931.5 = 28.1$ Mev

9. (4) $_1H^2 + {_1H^2} \rightarrow 2He^4 + \Delta E$

The binding energy per nucleon of a deuterium = 1.1 MeV

$\therefore$ Total binding energy = 2 × 1.1 = 2.2 MeV

The binding energy per nucleon of a helium nuclei = 7 MeV

$\therefore$ Total binding energy = 4 × 7 = 28 MeV

Hence, energy released

$\Delta E = (28 - 2 \times 2.2)\text{MeV} = 23.6$ MeV

10. (4) For nuclei having A > 56 binding energy per nucleon gradually decreases.

11. (2) Energy released,

$= 2$ B.E. $({}^4_2\text{He})$ – B.E. $({}^1_1H + {}^7_3\text{Li})$

$= 2 \times 4 \times 7.06 - [0 + 7 \times 5.60]$

$= 56.48 - 39.20 = 17.28$ MeV.

12. (1) Number of protons = 20

Number of neutrons = 40 – 20 = 20

Total mass of 20 protons and 20 Neutrons

$= 20\,m_p + 20m_n = 20\,(m_p + m_m)$

$= 20(1.007825 + 1.008665)$

$= 40.3298$

Mass defect, $\Delta m = 40.3298 - 39.962589$

$= 0.367211$u

Total B.E = 0.367211 × 931

= 341.873441 MeV

$\text{B.E/N} = \frac{341.873441}{41} = 8.34\ \text{MeV/N}$

13. (3) B.E = E = mc^2 = $(21 \times 10^{-3})(3 \times 10^8)^2$

$= 21 \times 10^{-3} \times 9 \times 10^{16}$

$E = 1.89 \times 10^{15}$ J

14. (2) $E_b + E_c > E_a$

15. (2) $P = \frac{nE}{t} \Rightarrow 1000 = \frac{n \times 200 \times 10^6 \times 1.6 \times 10^{-19}}{t}$

$\frac{n}{t} = 3.125 \times 10^{13}$

16. (2) $\frac{\text{Energy}}{\text{Fission}} = 200\,\text{Mev} = 200 \times 10^6 \times 1.6 \times 10^{-19}\,\text{J}$

$\text{Fission rate} = \frac{5}{200 \times 10^6 \times 1.6 \times 10^{-19}} = 1.56 \times 10^{11}\,\text{fission/sec}$

17. (4) Energy release is the fission of one nucleus = 200 MeV

$200 \times 1.6 \times 10^{-19} \times 10^6$ J $= 3.2 \times 10^{-11}$ J

P = 16kW = 16×10^3 watt

Now, number of nuclei required per second.

$n = \frac{P}{E} = \frac{16 \times 10^3}{3.2 \times 10^{-11}} = 5 \times 10^{14}$

18. (4) $E = \Delta mc^2 = 1 \times (3 \times 10^8)^2 = 9 \times 10^{16}$ J

$E = \frac{9 \times 10^{16}}{1.6 \times 10^{-19}} = 5.625 \times 10^{35}\,\text{eV} = 5.625 \times 10^{29}\,\text{MeV}$

19. (2) $_0n^1 \rightarrow {_1H^1} + {_{-1}e^o} + \bar{v} + Q$

$\Delta m = m_n - m_p - m_e$

$= \left(1.6785 \times 10^{-27} - 1.6725 \times 10^{-27} - 9 \times 10^{-31}\right)\text{kg}$

$= 9 \times 10^{-31}\,\text{kg}$

Energy = $9 \times 10^{-31} \times (3 \times 10^8)^2 = 0.511$ Mev

nearly equal to 0.73 Mev

20. (2) As we know that, E = mc^2

$\frac{E_2}{E_1} = \frac{m\left(\frac{2}{3}c\right)^2}{mc^2}$ [mass remain same]

$\frac{E_2}{E_1} = \frac{4c^2}{qc^2} = \frac{4}{9}$

21. (4) Nuclear forces are short range forces.

22. (3) Neutron is unstable outside a nucleus.

23. (1) Nuclear force is charge independent it also acts between two neutrons above the distance 15 Fermi forces get weak.

24. (3) B.E. = $\Delta m \times c^2 = (m_o - 8m_p - 8m_n)c^2$

25. (2) Nuclear force is independent of charge.

26. (2) As we know that, $N = N_0\left(\frac{1}{2}\right)^n$

$10^4 = 8 \times 10^4 \times \left(\frac{1}{2}\right)^{\frac{t}{T}}$

$10^4 = 8 \times 10^4 \times \left(\frac{1}{2}\right)^{t/3}$

$$\frac{1}{8}=\left(\frac{1}{2}\right)^{t/3}$$

$$\left(\frac{1}{2}\right)^{3}=\left(\frac{1}{2}\right)^{t/3}$$

$$\frac{t}{3}=3 \Rightarrow t=9 \text{ year}$$

27. (1) By the relation $N=N_o\left(\frac{1}{2}\right)^n$

$$\therefore \frac{N}{N_o}=\left(\frac{1}{2}\right)^{\frac{2700}{300}} \Rightarrow \frac{N}{N_p}=\left(\frac{1}{2}\right)^9$$

$$\therefore \frac{N}{N_o}=\frac{1}{512}$$

28. (4) Average life is $\frac{1}{\lambda}=\frac{2000}{0.693}=2886$

29. (3) $N=N_0\left[\frac{1}{2}\right]^{t/T}$

$$N=\left(1-\frac{7}{8}\right)N_0$$

$$\frac{N_0}{8}=N_0\left[\frac{1}{2}\right]^{t/5}$$

$$\left(\frac{1}{2}\right)^3=\left[\frac{1}{2}\right]^{t/5} \Rightarrow t=15 \text{days.}$$

30. (4) Half life of a substance doesn't depends upon amount, temperature and pressure. It depends upon the nature of the substance.

31. (2) β - decay from nuclei is based on this process only.

32. (1) Fraction of material = $1-\frac{N}{N_o}$

$$=1-\left(\frac{1}{2}\right)^{\frac{t}{T_{1/2}}}=1-\left(\frac{1}{2}\right)^{\frac{20}{5}}=1-\frac{1}{16}=\frac{15}{16}=93.75\%$$

33. (3) Half life $T_{1/2}=\frac{0.693}{\lambda}=\frac{0.693}{1.07\times10^{-4}}=6476 \text{ years}$

34. (2) ${}^{22}_{10}Ne \rightarrow 2\,{}^{4}_{2}He+{}^{14}_{6}X$

The unknown nucleus is ${}^{14}_{6}C$

35. (2) Half-life $T_{1/2}=\frac{\ln 2}{\lambda}=\frac{0.693}{\lambda}$

Mean life, $\tau=\frac{1}{\lambda}$

36. (2) $A=A_o\left[\frac{1}{2}\right]^{\frac{t}{T_{1/2}}}$

$$\frac{A}{A_o}=\left[\frac{1}{2}\right]^{27/3}=\left[\frac{1}{2}\right]^9=\frac{1}{512}$$

$$\frac{A}{A_o}=\frac{1}{512}$$

37. (3) $(T_{1/2})_X=(T)_Y$

$$\frac{0.693}{\lambda_X}=\frac{1}{\lambda_Y}$$

or $\lambda_X=0.693\,\lambda_Y$ or $\lambda_X<\lambda_Y$

Rate of decay = λN

Initially, both samples have the same number of atoms(N), but $\lambda_Y>\lambda_X$, so Y will decay at a faster rate than X.

38. (4) $A=A_o\left[\frac{1}{2}\right]^{\frac{t}{T_{1/2}}}$

$$5\times10^{-6}=64\times10^{-5}\left(\frac{1}{2}\right)^{t/3}$$

$$\frac{1}{128}=\left(\frac{1}{2}\right)^{t/3} \Rightarrow t=21 \text{days}$$

39. (1) $\frac{N}{N_o}=\left(\frac{1}{2}\right)^n \Rightarrow \frac{1}{8}=\left(\frac{1}{2}\right)^n$

n = 3

$\therefore$ Now $t=n\times T_{1/2}=3\times3.8=11.4$ days

40. (4) A given radioactive nucleus may decay at any time after t = 0.

41. (3) $\frac{3}{4}$th active decay takes place in time

$$t=2\left(T_{1/2}\right) \Rightarrow \frac{3}{4}=2\left(T_{1/2}\right) \Rightarrow T_{1/2}=\frac{3}{8}\text{sec}$$

42. (3) By using $N=N_oe^{-\lambda t}$ and average life time $t=\frac{1}{\lambda}$

$$\text{so, } N=N_oe^{-\lambda t} \Rightarrow N=N_oe^{-\lambda\times\frac{1}{\lambda}}$$

$$N=\frac{N_o}{e} \Rightarrow \frac{N}{N_o}=e^{-1}$$

Now disintegrated fraction = $1-\frac{N}{N_o}=1-\frac{1}{e}=\frac{e-1}{e}$

43. (4)

44. (2) Activity decreases from 6000 dps to 3000 dps in 140 days. It indicates that the half life = 140 days.

In 280 days (or two half lives), the activity reduces to 1/4th of the initial activity.

$\therefore$ $R_0=4R=4\times6000$

= 24000 dps.

45. (3) Mean life = $\frac{1}{\lambda}=\frac{1}{1.5\times10^{-9}}=6.67\times10^8$ sec

46. (3) Remaining substance = $N=N_o\left(\frac{1}{2}\right)^n$

$$N=10\times\left(\frac{1}{2}\right)^{\frac{20}{15}}$$

$$N=\frac{10}{(2)^{4/3}}=\frac{10}{2.51}=3.96$$

$\therefore$ Decayed , material = 10 – 3.96 = 6.04 g

47. (2) $\frac{R}{R_0} = \left(\frac{1}{2}\right)^n$

or $\frac{1}{64} = \left(\frac{1}{2}\right)^n$

$\therefore$ n = 6 = number of half lives

$t = n \times T_{1/2}$

= 6 × 2 = 12 h.

48. (3) Since, average life = 1.44 × half life.

and average life = 5 years

$\therefore$ half life = 3.47 years.

49. (2) $N = N_o\left(\frac{1}{2}\right)^{\frac{t}{t_{1/2}}}$

No. of atoms at t = 2hr, $N_1 = 8\times10^{10}\left[\frac{1}{2}\right]^{\frac{2}{1}} = 2\times10^{10}$

No. of atoms at t = 4hr

$N_2 = 8\times10^{10}\left[\frac{1}{2}\right]^{\frac{4}{1}} = \frac{1}{2}\times10^{10}$

$\therefore$ No. of atoms decayed in given duration

$= \left(2-\frac{1}{2}\right)\times10^{10} = 1.5\times10^{10}$

50. (1) $\frac{N}{N_o} = \frac{12.5}{100}$, t = ?, T = 139 days

As $\frac{N}{N_o} = \left(\frac{1}{2}\right)^n \quad \therefore \frac{125}{1000} = \left(\frac{1}{2}\right)^n$

$\left(\frac{1}{8}\right) = \left(\frac{1}{2}\right)^n$

$\therefore$ n = 3

$\Rightarrow t = 3\times139 = 417$ days

51. (4) Number of half lives of of P in 18 hours $= \frac{18}{9} = 6$

Number of half lives of Q in 18 hours $= \frac{18}{9} = 2$

$\therefore$ Number of nuclei left undecayed

$N_1 = N_o\left[\frac{1}{2}\right]^6 = \frac{N_o}{64}$

$N_2 = N_o\left[\frac{1}{2}\right]^2 = \frac{N_o}{4}$

The ratio of their rates of disintegration is

$\frac{R_1}{R_2} = \frac{\lambda_1 N_1}{\lambda_2 N_2} = \frac{T_2}{T_1}\left[\frac{N_1}{N_2}\right] = \frac{9}{3}\times\frac{N_o\times4}{64\times N_o} = \frac{3}{16}$

52. (1) Here R = 8.0 mCi

= 8.0 × 10^{-3} × 3.7 × 10^{10} dis s^{-1}

= 29.6 × 10^7 dis s^{-1}

$T_{1/2}$ = 5.3 years = 5.3 × 3.16 × 10^7 s

But $R = \lambda N = \frac{0.693}{T_{1/2}}\cdot N$

$\therefore N = \frac{RT_{1/2}}{0.693} = \frac{29.6\times10^7\times5.3\times3.16\times10^7}{0.693}$

= 7.15 × 10^{16} atoms

53. (1) T = 4.5 × 10^9 years

T = 4.5 × 10^9 × 365 × 24 × 60 × 60

$\lambda = \frac{0.693}{T}s^{-1}$

Number of atoms in one gram of $_{92}U^{238}$ is

$N = \frac{6.03\times10^{23}}{238}$

$\frac{dN}{dt} = \lambda N \Rightarrow \frac{dN}{dt} = \frac{0.693\times6.03\times10^{23}}{238\times4.5\times10^9\times365\times24\times60\times60}$

$\frac{dN}{dt} = 2.34\times10^4 s^{-1}$

54. (2) $N = \frac{6.023\times10^{23}}{226}$

$\lambda = \frac{0.693}{T} = \frac{0.693}{1620\times3.16\times10^7}$

$\therefore$ No. of disintegration per second $\frac{dN}{dt} = \lambda N$

$= \frac{0.693}{1620\times3.16\times10^7}\times\frac{6.023\times10^{23}}{226}$ = 3.16 × $10^{10}s^{-1}$

≈ 3 × 10^{10} s^{-1}

55. (2) $\left|\frac{dN}{dt}\right| = \lambda N \Rightarrow \left|\frac{dN}{dt}\right| \propto N$

56. (1) $_8O^{16} +_1 H^2 \rightarrow_z X^A +_2 He^4$

Conservation of mass number:

A + 4 = 16 + 2 = 18, A = 18 – 4 = 14

Conservation of charge number

Z + 2 = 8 + 1 = 9

Z = 9 – 2 = 7

$\therefore$ Element = $_7N^{14}$

57. (3) Decay constant for the processes are

$\lambda_1 = \frac{0.693}{t_1}, \quad \lambda_2 = \frac{0.693}{t_2}$

The probability that an active nucleus decays by the first process in small time dt is $\lambda_1$dt.

Similarly for second decay. The probability that it decays either by first process or second process is ($\lambda_1$dt + $\lambda_2$dt).

If effective decay constant is λ this probability is also equal to λ dt

$\therefore \lambda dt = \lambda_1 dt + \lambda_2 dt$

$\lambda = \lambda_1 + \lambda_2$

$\therefore \frac{0.693}{t} = \frac{0.693}{t_1} + \frac{0.693}{t_2}$

$\frac{1}{t} = \frac{1}{t_1} + \frac{1}{t_2}$

58. (1) Short trick/ note : penetrating power is inversely proportional to mass.

$\therefore$ penetrating power $\propto \frac{1}{\text{mass}}$

59. (1) β - emission from the nucleus is always accompanied with an anti-neutrino. The β^- decay is.

$$n \rightarrow +\bar{e} + \bar{u} \rightarrow \text{anti-neutrino}$$

$\downarrow$

electron

60. (4) Actual process is

$${}^{238}_{92}U \xrightarrow{\alpha} {}^{234}_{90}X \xrightarrow{\beta^-} {}^{234}_{91}Y$$

So, Z = 91. A = 234

61. (3) Atomic number decreases by 10 and on. of α - particle is 5.

62. (1) The process is

$${}_{92}U^{238} \xrightarrow{\alpha} {}_{90}Z^{234} \xrightarrow{\beta} {}_{91}Y^{234} \xrightarrow{e} {}_{92}U^{234}$$

$\therefore$ A= 234, B = 90, C = 234. D = 91. e = β

63. (1) 180 – 4 – 0 – 4 – 0 = 172 Mass number

72 – 2 + 1 – 2 + 0 = 69 Atomic number

64. (4) Hydrogen is combined to form helium nucleus.

65. (2) ${}_7N^{17} + 2He^4 \rightarrow {}_8O^{17} + {}_1H^1$

This is the reaction of formation of isotope of oxygen

66. (1) It is also used because it is most reactive in number reaction.

67. (2) It stops the high energy neutrons.

High Order Time Intensive MCQs

1. (2) Binding energy = 0.042 × 931 = 39.10 MeV

$\therefore$ Binding energy per nucleon $= \frac{39.10}{7} = 5.58 \simeq 5.6\text{MeV}$

2. (3) Loss of kinetic energy,

$$\frac{1}{2}m_1u_1^2 - \frac{1}{2}m_1v_1^2 = \frac{75}{100} \times \frac{1}{2}m_1u_1^2$$

$$\Rightarrow u_1^2 - v_1^2 = \frac{3}{4}u_1^2 \Rightarrow v_1 = \frac{1}{2}u_1 \quad \text{...(i)}$$

$$v_1 = \frac{(m_2 - m_1)u_1}{(m_1 + m_2)} \quad \text{...(ii)}$$

$$\frac{1}{2}u_1 = \frac{(m_2 - m_1)u_1}{(m_1 + m_2)}$$

$m_2 = 3m_1 = 3m$

3. (3) For each α-emission, 2 protons and 2 neutrons are lost. For each positron emission, 1 proton is lost and 1 neutron is gained

$n_p = Z - 2 \times 3 - 2 \times 1 = Z - 8$

$n_n = (A - Z) - 2 \times 3 + 2 \times 1 = A - Z - 4$

$$\therefore \frac{n_n}{n_p} = \frac{A - Z - 4}{Z - 8}$$

4. (4) Binding energy of neutron, E = $\Delta m \times 931$ MeV

= (13.00335 – 13) × 931 MeV = 3.12MeV

5. (1) $e = 3E_b - E_a \Rightarrow E_a + e = 3E_b$

6. (1) Total mass of the initial particles

m_i = 1.007825 + 7.016004 = 8.023829 u and the total mass final particles

m_f = 2 × 4.002603 = 8.005206

$\Delta m = m_i - m_f$ = 8.023829 – 8.005206 = 0.018623 u

The Q – value is given by

$Q = (\Delta m)c^2$ = 0.018623 × 931.5

= 17.35 MeV

7. (4) By conservation of momentum,

$$m_1v_1 = m_2v_2 \quad \text{or} \quad \frac{m_1}{m_2} = \frac{v_1}{v_2} = \frac{1}{2}$$

$$\text{As} \quad m = \frac{4}{3}\pi r^3 \rho \quad \therefore \quad m \propto r^3 \quad \text{or} \quad r \propto m^{1/3}$$

$$\therefore \frac{r_1}{r_2} = \left(\frac{m_1}{m_2}\right)^{1/3} = \left(\frac{1}{2}\right)^{1/3} = 1 : 2^{1/3}$$

8. (2) Binding energy of $({}_1^2H + {}_1^3H)$ = a + b

Binding energy of ${}_2^4He$ = c

In a nuclear reaction, the resultant nucleus is more stable than the rectants. Hence, binding energy of ${}_2^4He$ will be more than that of $({}_1^2H + {}_1^3H)$

Thus, energy released = charge in binding energy

= c – (a + b) = c – a – b

9. (3) X : Y = 1 : 7

$$\frac{m_x}{m_y} = \frac{1}{7} \Rightarrow 7m_x = m_y$$

Let the initial total mass is m.

$\Rightarrow m_x + m_y = m$

$$\frac{m_y}{7} + m_y = m \Rightarrow \frac{8m_y}{7} = m \Rightarrow m_y = \frac{7}{8}m$$

Only $\frac{1}{8}$ part remains left undecayed

Number of half-lives involved are three,

$$1 \xrightarrow{T_{1/2}} \frac{1}{2} \xrightarrow{T_{1/2}} \frac{1}{4} \xrightarrow{T_{1/2}} \frac{1}{8}$$

So, time taken to becomes $\frac{1}{8}$ unstable part

$= 3 \times T_{1/2} = 3 \times 1.43 \times 10^9 = 4.29 \times 10^9$ yr

10. (2) As $\frac{N}{N_0} = \left(\frac{1}{2}\right)^n$

$$\text{As,} \quad \frac{N}{N_0} = \left(\frac{1}{2}\right)^3 = \frac{1}{8}$$

Number of half-lives = 3

$$\Rightarrow T = 20\text{yr} \Rightarrow T = \frac{t}{n} \Rightarrow t = T \times n = 20 \times 3 = 60\text{yr}$$

11. **(4)** 40 g $\xrightarrow[\text{half-life}]{20s}$ 20g $\xrightarrow{20s}$ 10g

160g $\xrightarrow[10s]{}$ 80g $\xrightarrow{10s}$ 40g $\xrightarrow{10s}$ 20g $\xrightarrow{10s}$ 10g

So, after 40s, A_1 and A_2 become equal

12. **(2)** We know that, $N = N_0 e^{-\lambda t}$

For A, $N_A = N_0 e^{-5\lambda t}$

For B, $N_B = N_0 e^{-\lambda t}$

and $\frac{N_A}{N_B} = \frac{1}{e^{4\lambda t}}$

$\Rightarrow \frac{1}{e^2} = \frac{1}{e^{4\lambda t}}$

$t = \frac{1}{2\lambda}$

13. **(1)** Momentum of photon, $P = \frac{h\nu}{c}$

$$E = \frac{P^2}{2M} = \frac{\left(\frac{h\nu}{c}\right)^2}{2M} = \frac{h^2\nu^2}{2Mc^2}$$

14. **(4)** Half-life, $\tau_{1/2} = \frac{0.693}{\lambda}$

Activity, $I_1 = N_1 \lambda$ and $I_2 = N_2 \lambda$

Where, λ is disintegration constant.

$\therefore\ (I_1 - I_2) = (N_1 - N_2)\frac{0.693}{\tau_{1/2}}$

$(N_1 - N_2) \propto (I_1 - I_2)\, \tau_{1/2}$

15. **(4)** Fraction remains after n half-lives,

$$\frac{N}{N_0} = \left(\frac{1}{2}\right)^n = \left(\frac{1}{2}\right)^{t/T}$$

$$\frac{N}{N_0} = \left(\frac{1}{2}\right)^{5/T} = \frac{1}{e} \qquad \left(\because N = \frac{N_0}{e}\right)$$

Taking log on both sides, we get

$$\log 1 - \log e = \frac{5}{T}\log\frac{1}{2}$$

$$-1 = \frac{5}{T}(-\log 2)$$

$\Rightarrow T = 5 \log_e 2$

Now, let t′ be the time after which the activity reduces to half

$$\left(\frac{1}{2}\right) = \left(\frac{1}{2}\right)^{t'/5\log_e 2} \Rightarrow t' = 5\log_e 2$$

16. **(3)** By Moseley's law, $\sqrt{v} = a(Z-b)$ or $v = a^2(Z-b)^2$

Comparing with the equation of a parabola, $y^2 = 4ax$, it conforms to graph (3).

17. **(1)** As $\frac{N}{N_0} = \left(\frac{1}{2}\right)^n$

where, number of half lives, $n = \frac{t}{T}$

whose T is the half life period

for sample X

$$\frac{1}{16} = \left(\frac{1}{2}\right)^{8/T_X} \text{ or } \left(\frac{1}{2}\right)^4 = \left(\frac{1}{2}\right)^{8/T_X}$$

$$\Rightarrow 4 = \frac{8}{T_X} \qquad \ldots(i)$$

For sample Y

$$\left(\frac{1}{256}\right) = \left(\frac{1}{2}\right)^{8/T_Y}$$

$$\text{or } \left(\frac{1}{2}\right)^8 = \left(\frac{1}{2}\right)^{8/T_Y}$$

$$8 = \frac{8}{T_Y} \qquad \ldots(ii)$$

Dividing (*i*) by (*ii*), we get

$$\frac{4}{8} = \frac{8}{T_X} \times \frac{T_Y}{8} \Rightarrow \frac{1}{2} = \frac{T_Y}{T_X} \text{ or } \frac{T_X}{T_Y} = \frac{2}{1}$$

18. **(1)** Let the number of fissions per second be n. Energy released per second

$= n \times 200\text{ MeV} = n \times 200 \times 1.6 \times 10^{-13}$ J

Energy required per second = power × time

$= 1\text{ kW} \times 1\text{ s} = 1000$ J

$\therefore n \times 200 \times 1.6 \times 10^{-13} = 1000$

$$\text{or } n = \frac{1000}{3.2\times10^{-11}} = \frac{10}{3.2}\times10^{13} = 3.125\times10^{13}$$

19. **(3)** $\frac{N}{N_0} = e^{-\lambda t}$

$$1 - \frac{1}{3} = e^{-\lambda t_1} \text{ and } 1 - \frac{2}{3} = e^{-\lambda t_2}$$

$$\Rightarrow \frac{e^{-\lambda t_1}}{e^{-\lambda t_2}} = 2 \Rightarrow t_2 - t_1 = \frac{\ln 2}{\lambda} = T_{1/2} = 20\text{ min}$$

20. **(3)** Here, $\lambda = 0.3465\text{ day}^{-1}$

t = 4 days

$$T_{1/2} = \frac{0.693}{\lambda} = \frac{0.693}{0.3465} = 2\text{ days}$$

$$\therefore n = \frac{t}{T_{1/2}} = \frac{4}{2} = 2$$

Hence, sample left undecayed after a period of 4 days.

$$\because \frac{N}{N_0} = \left(\frac{1}{2}\right)^n \Rightarrow \frac{N}{N_0} = \left(\frac{1}{2}\right)^2 = \left(\frac{1}{4}\right) = 25\%$$

$\therefore$ Sample decayed = 75%

21. **(3)** If R_0 be the initial activity of the sample, then

$R_1 = R_0 e^{\lambda t_1}$ and $R_2 = R_0 e^{-\lambda t_2}$

Where $\lambda = \frac{1}{T}\left(\because \text{Mean life} = T = \frac{1}{\lambda}\right)$

$$\Rightarrow \frac{R_2}{R_1} = \frac{e^{-\lambda t_2}}{e^{-\lambda t_1}} = e^{\lambda(t_1 - t_2)} \Rightarrow R_2 = R_1 \exp\left(\frac{t_1 - t_2}{T}\right)$$

Statement Based MCQs

1. **(1)** An α-particle have 2 units of positive charge. Therefore the atomic no. of the daughter nucleus decrease by 2 on α-decay.
2. **(3)** It is so because electrons do not initially resides inside a nucleus, infact electron is produced by decaying of nucleus inside a nucleus. And this process results in β-rays emission so electrons from nuclear origin are termed as β-particles.
3. **(1)** Since fusion reactions requires a very high temperatures so energy released from fission reaction can be used to initiate the fusion reactions.
4. **(1)** The necessary conditions for nuclear fusion reaction are (1) High temperature (2) High pressure
5. **(1)** It is so because when mass number increases then number of protons also increases due to which repulsion increase which further decrease the binding energy per nucleon. Also, number of nucleon increases which decreases B.E per nucleon value.

Assertion & Reason MCQs

1. **(1)** Both Assertion and Reason are True and the Reason is a correct explanation of the Assertion.
2. **(3)** Stability of nucleus depends on average binding energy per nucleon. Iron has highest B.E. per nucleon so it is more stable than uranium, but it does not have higher binding energy than U^{235} nucleus. So Reason is false.
3. **(1)** This is because hydrogen nucleus contains only one nucleon.
4. **(2)** Both Assertion and Reason are True and the Reason is not correct explanation of the Assertion.
5. **(3)** Assertion is true but the reason is false. Electrons are emitted as β-particles when a neutron decays into a proton during a radioactive disintegration.

Match the Columns MCQs

1. **(4)** P will continuously decrease, but R will increase. (P + Q) will continuously decrease as R is formed only from P and Q.

   (R + Q) = Total – P

   P is continously decreasing. So, (R + Q) will continuously increase.
2. **(2)** $E_1 = [2m({}_1H^2) - m({}_1H^3) - m({}_1H^1)] \times 931.5$ MeV

   $= 4$ MeV

   $E_2 = [-m({}_2H^4) - m({}_0n^1) + m({}_1H^3) + m_1({}_1H^2)] \times 931.5$ meV

   $= 17.6$ MeV

   $E_3 = [-m({}_2He^3) - m({}_0n^1) + 2m({}_1H^2) \times 931.5$ meV $= 3.3$ MeV

   $E_4 = [m({}_2He^3) - m({}_1n^2) + m({}_2He^4) + m_1(H^1)] \times 931.5$ meV

   $= 18.3$ MeV
3. **(3)**

   (a) In α-decay, mass number decrease by 4 and atomic number decrease by 2.

   (b) In β^+-decay, mass number remain unchanged while atomic number decrease by 1.

   (c) In fission, parent nucleus break into almost two equal fragments.

   (d) In proton emission both mass number and atomic number decrease by 1.
4. **(1)**

   (a) Mass of products formed is less than the original mass of the system in α-decay, β-decay, nuclear fission and nuclear fusion.

   (b) Binding energy per nucleon increases in α- decay, β-decay, nuclear fission and nuclear fission.

   (c) Mass number is conserved in all possibilities.

   (d) Charge number is conserved in all possibilities.

NCERT Exemplar MCQs

1. **(3)** Radioactivity is a process due to which a radioactive material spontaneously decays. In half-life (t =1 yr) of the material on the average half the number of atoms will decay.

   Therefore, the containers will in general have different number of atoms of the material, but their average will be approximately 5000.
2. **(2)** Given, $F = \frac{GMm}{r^2}$

   M = effective mass of hydrogen atom

   $=$ mass of electron + mass of proton $- \frac{B^2}{C}$

   where B is BE of hydrogen atom = 13.6 eV.
3. **(1)** Let the nucleus is ${}_zX^A$. β^+ decay is represented as

   $${}_zX^A \rightarrow {}_{z-1}Y^A + {}_{+1}e^0 + \upsilon + Q_2$$

   $$\therefore\ Q_2 = \left[m\left({}_zX^A\right) - m\left({}_{z-1}Y^A\right) - m_e\right]c^2$$

   $$= \left[m\left({}_zX^A\right) + zm_e - m\left({}_{z-1}Y^A\right) - (z-1)m_e - 2m_e\right]c^2$$

   $$= [m({}_zX^A) - m({}_{z-1}Y^A) - 2m_e]c^2$$

   $$= (M_x - M_y - 2m_e)c^2$$

   β^-decay is represented as

   $$= {}_zX^A \rightarrow {}_{z+1}A^Y + {}_{-1}e^0 + \overline{v} + Q_1$$

   $$Q_1 = \left[m\left({}_zX^A\right) - m\left({}_{z+1}Y^A\right) - m_e\right]c^2$$

   $$= \left[m\left({}_zX^A\right) + zm_e - m\left({}_{z+1}Y^A\right) - (z+1)m_e\right]c^2$$

   $$= [m({}_zX^A) - m({}_{z-1}Y^A)]c^2$$

   $$= (M_x - M_y)c^2$$
4. **(1)** Tritium (${}_1H^3$) contains 1 proton and 2 neutrons. A neutron decays as $n \longrightarrow p + \overline{e} + \overline{v}$, the nucleus may have 2 protons and one neutron, i.e., tritium may transform into ${}_2He^3$ (2 protons and 1 neutron).

Tritium energy is less than that of $_2He^3$ nucleus, i.e., transformation is not allowed energetically.

5. **(2)** Stable heavy nuclei have more neutrons than protons. This is because electrostatic force between protons is repulsive, which may reduce stability.

6. **(2)** According to the question, the moderator used, has light nuclei (like proton). When protons undergo perfectly elastic collision with the neutron emitted their velocities are exchanged, i.e., neutrons come to rest and protons move with the velocity of neutrons.

Heavy nuclei will not serve the purpose because elastic collisions of neutrons with heavy nuclei will not slow them down.

NEET & AIIMS Past Year MCQs

1. **(1)** A: Slow neutrons cause fission in ^{235}U more effectively than fast neutrons. (True)
B: α-rays are helium nuclei. (True)
C: β-rays are fast-moving electrons/positrons. (True)
D: γ-rays have shorter wavelengths than X-rays. (False)

2. **(1)** $\frac{N}{N_0} = \frac{1}{16} = \left(\frac{1}{2}\right)^{\frac{t}{T_{1/2}}}$

$t = 4T_{1/2} = 20 \times 4 = 80$ minutes

3. **(4)** $^{22}_{11}Na \rightarrow X + e^+ + \nu$
Above equation is of β^+ decay
Therefore the atomic number decreases by 1 and mass number remain unchanged
$^{22}_{11}Na \rightarrow ^{22}_{10}Ne + e^+ + \nu$

4. **(4)** Nuclear Radius : R is proportional to $R_0(A)^{1/3}$

$$\Rightarrow \frac{R(125)}{R(64)} = \frac{125^{1/3}}{64^{1/3}} = \frac{5}{4}$$

5. **(3)** $A^{240} \longrightarrow B^{120} + C^{120}$
Given,
Binding energy per nucleon of A, $E_A = 7.6$ MeV
$F_B = 8.5$ MeV
and $E_C = 8.5$ MeV
Total gain in Binding energy is given as below;
Q = Binding energy of products – Binding energy of rectants
$= [(A_B \times F_B + A_C \times E_C) - (A_A \times E_A)]$
$= [(120 \times 8.5 + 120 \times 8.5) - (240 \times 7.6)]$
$= [2 \times 120 \times 8.5 - 240 \times 7.6] = [2040 - 1824]$
Q = 216 MeV

6. **(1)** For a radioactive nuclide,

$$N = N_0\left(\frac{1}{2}\right)^{\frac{t}{T_{\frac{1}{2}}}} \Rightarrow \frac{N}{N_0} = \left(\frac{1}{2}\right)^{\frac{150}{100}}$$

$$\Rightarrow \frac{N}{N_0} = \left(\frac{1}{2}\right)^{\frac{3}{2}} \Rightarrow \frac{N}{N_0} = \frac{1}{2\sqrt{2}}$$

7. **(2)** $^A_ZX \xrightarrow{B^+} {}_{Z-1}B \xrightarrow{\alpha} {}_{Z-3}C \xrightarrow{B^-} {}_{Z-2}D$
In B^+ decay atomic number decreases by 1.
In B^- decay atomic number increases by 1.
In α decay atomic number decreases by 2.

8. **(2)** $\lambda_{min} = \frac{12400}{V_{(\text{in volt})}} \mathring{A} = \frac{12400}{15 \times 10^3} \mathring{A} = 0.82\,\mathring{A}$

9. **(4)** $^{15}X_7 + ^4He_2 \longrightarrow ^{18}Y_8 + ^1H_1$
Nuclear nuclei reaction follow conservation of atomic number and mass number.

10. **(1)** Continuous energy statges

11. **(2)** For Lyman series $\frac{1}{\lambda} = R\left[\frac{1}{1^2} - \frac{1}{n^2}\right]$
where n = 2, 3 .....

Semiconductor Electronics

NCERT Topic-wise MCQs

1. **(1)** Mean free path, $d = 4 \times 10^{-8}$ m, E = ?
Energy of electron = 2 eV
F = eE
Work done on clectron when it moves through distance d = eEd. This work done is equal to the energy transfered to the electron, so
∴ eEd = 2 eV
Ed = 2 Volt

$$E = \frac{2}{d}\text{Volt/m} = \frac{2}{4 \times 10^{-8}} = 5 \times 10^7 \text{ V/m}$$

2. **(3)** Number of density of electron, $n_e = 6 \times 10^{19}$ m^{-3}
Volume of sample, V = 1 cm × 1 cm × 2mm
$= 10^{-2}$ m × 10^{-2}m × 2 × 10^{-3}m = 2 × 10^{-7}m^3
∴ No. of electrons in the given sample = $n_e \times V$
$= 6 \times 10^{19} \times 2 \times 10^{-7} = 12 \times 10^{12}$

3. **(2)** $\rho = 0.50$ Ωm
$\mu_e = 0.39$ m^{2}V^{-1}s^{-1}
$\mu_h = 0.11$ m^{2}V^{-1}s^{-1}
Resistivity of intrinsic semiconductor is

$$\frac{1}{\rho} = e(\mu_e n_i + n_i\mu_h)$$

Where, $n_i = \frac{1}{\rho e(\mu_e + \mu_h)}$

$$n_i = \frac{1}{(0.5)(1.6 \times 10^{-19})(0.39 + 0.11)} = 2.5 \times 10^{19}\text{ m}^{-3}$$

4. **(2)** More energy is required by an electron to jump from valence band to conduction band.

5. **(4)** $n_i = 1.5 \times 10^{16}$ m^{-3}
$n_h = 4.5 \times 10^{22}$ m^{-3}
Now, $n_h n_e = n_i^2$

$$\therefore n_e = \frac{n_i^2}{n_h} = \frac{1.5 \times 1.5 \times 10^{32}}{4.5 \times 10^{22}} = 5 \times 10^9\text{ m}^{-3}$$

6. (1) Safe limit of temperature for germanium and silicon is 80°C to 200°C.

7. (2) Electron and hole concentration in a semiconductor in thermal equilibrium is

$n_e n_h = n_i^2$

$$\text{or } n_h = \frac{n_i^2}{n_e} = \frac{(1.5\times10^{16})^2}{5\times10^{22}} = 4.5\times10^9\,m^{-3}$$

8. (2) If all the donor states in n-type semiconductor are filled, the number of electrons in donor states will increase. Due to it the charge density in donor states will become more than one.

9. (3) Energy gap, $E_g = \frac{hc}{\lambda} = \frac{12400\text{eV}\,\mathring{A}}{0.72\text{eV}} = 17222\,\mathring{A}$

10. (3) At 0K semiconductor behaves as an insulator.

11. (4) Drift velocity, $v_d = \frac{I}{nAe}$, *i.e.*, $v_d \propto \frac{I}{n}$

$$\therefore \frac{(v_d)_e}{(v_d)_h} = \frac{I_e}{I_h}\times\frac{n_h}{n_e} = \frac{7}{4}\times\frac{5}{7} = \frac{5}{4}$$

12. (3) Because with rise in temperature, resistance of semiconductor decreases, hence overall resistance of the circuit decreases, which in turn increases the current in the circuit.

13. (1) At room temperature, some covalent bonds break and semiconductor behaves slightly as a conductor.

14. (1) Phosphorus is pentavalent impurity and for pentavalent impurity atoms $n_e >> n_h$.

15. (2) When a strong current passes through the semiconductor it heats up the crystal and covalent bonds are broken. Hence because of excess number of free electrons it behaves like a conductor.

16. (2) Formation of energy bands in solids are due to Pauli's exclusion principle.

17. (2) Resistance will increase as Resistance of a semiconductor $\propto \frac{1}{\text{Temperature}}$

18. (2) In insulators, the forbidden energy gap is very large, in case of semiconductor it is moderate and in conductors the energy gap is zero. Hence, $EG_3 > EG_1 > EG_2$

19. (4) Resistance of conductors (Cu) decreases with decrease in temperature while that of semi- conductors (Ge) increases with decrease in temperature.

20. (2)

Conduction Band (CB)

1 eV Impurity level

Valence Band (VB)

In a N-type semiconductor, there is one extra electron left, after completing the octet. This electron has very small ionisation energy and so, it lies very close to the conduction band. So, the donor level lies closely below the bottom of the conduction band.

21. (4) Extrinsic semiconductor is neutral because the total number of positive and negative charges are equal.

22. (1) The mobility of an electron in the conduction band is more then the mobility of hole in the valence band.

23. (4) $\frac{1}{\rho} = en_e\mu_e \Rightarrow \rho = \frac{1}{1.6\times10^{-19}\times10^{13}\times1200}$

$= 520.9\ \Omega$ cm

24. (3) For a germanium crystal,

$E_g = 0.7\text{ eV} = 0.7\times1.6\times10^{-19}\text{ J} = 1.12\times10^{-19}\text{ J}.$

25. (3) N-type semiconductors are neutral because neutral atoms are added during doping.

26. (1) $I = I_e + I_h = en_eA\upsilon_e + en_hA\upsilon_h = eA(n_e\mu_e + n_h\mu_h)E$

$J = \frac{I}{A} = en_i(\mu_e + \mu_h)E.$

27. (2) If the temperature is increased in a semi-conductor then the number of charge carrier will increase but average drift speed will decrease.

28. (1) Due to doping the conductivity of the semiconductor will increase

29. (3) $E = \frac{hc}{\lambda} \Rightarrow \lambda = \frac{hc}{E}$

30. (1) Phosphorus (P) when added as an impurity into silicon produces n-type semi conductor.

31. (1) Decreases exponentially with increasing band gap.

The probability of electrons to be found in the conduction band of an intrinsic semiconductor.

$P(E) = \frac{1}{1+e^{\frac{(E-E_F)}{kT}}}$; where k = Boltzmann's constant

Hence, at a finite temperature, the probability decreases exponentially with increasing band gap.

32. (1) As $n_h > n_e$, the semiconductor is of p-type.

33. (2) In semiconductor, no covalent bond breaks at 0 K, so, no electron is free at 0 K. At room temperature some covalent bonds break, so, some electrons and holes become free. More covalent bonds break at high temperature and so, number of free electrons increases at high temperature.

34. (3) Si act as an insulator at absolute zero.

35. (3) Increase of temperature increases resistance of conductor but decreases resistance of semiconductor.

36. (4)

37. (2) In p-type semiconductor, the number density of holes in the valence is greater than the number density of electrons in the conduction band. Therefore, the conduction band has a smaller number of electrons than the holes in the valence band.

In the given figure, the valence band has more holes as compared to the number of electrons in the conduction band. Therefore, the material is most likely a p-type semiconductor.

38. (2) The majority charge carriers in an n-type semiconductor are electrons.

39. (2) Energy band gap of carbon is more than the energy band gap of silicon.

40. (1) Effective voltage in the circuit, V = 4.5 – 0.5 = 4V

Current in the circuit, $I = \frac{V}{R} = \frac{4}{100} = 0.04A$

$I = 0.04A = 40mA$

41. (3) In an unbiased p-n junction the diffusion of charge carriers across the junction take from higher concentration to lower concentration.

42. (3) The depletion region created at the junction is devoid of free charge carriers.

43. (3) Diode is reversed biased, so only drift current due to minority carriers which is 20μA will flow

Potential drop across Resistor,

$V = 15\Omega \times 20\mu A = 300\ \mu v = 0.0003V$

Potential difference across the diode = 4 – 0.0003 = 3.99 = 4V

44. (1) p-n Junction diode is forward biased when p terminal is at high potential with respect to n.

45. (3) Input, $V_{rms} = 20V$

Peak value of input voltage, $V_m = \sqrt{2}V_{rms}$

$= \sqrt{2} \times 20 = 28.28V$

Since the transformer has a transformation ratio 1 : 2 the maximum value of output voltage of the transformer applied the diode will be

$V_o = 2\ V_{in} = 2 \times 28.28V$

The dc voltage in the rectified output

$$= \frac{2V_o}{\pi} = \frac{2 \times 2 \times 28.28}{\frac{22}{7}} = 36V$$

46. (1) A maximum permissible current is

$$I_{z_{max}} = \frac{P}{V_Z} = \frac{364 \times 10^{-3}}{9.1} = 40mA$$

47. (3) Diode/circuit is reverse bias then effective voltage across diode is = 3 – 1 = 2V

$\therefore$ Current across the diode $= \frac{2}{100} = 20mA$

48. (1) Depletion layer consists of mainly stationary ions.

49. (2) The unidirectional flow of current through a p-n junction enables it to be used as a rectifier.

50. (3) After a large reverse voltage in p-n junction diode, a huge current flows in the reverse direction suddenly.

51. (3) At junction a potential barrier / depletion layer is formed, with n-side at higher potential than p-side . Therefore there is an electric field at the junction directed from the n-side to p-side.

E
p ← n
• → Holes
∘ → Electrons

52. (1) This circuit is reversed bias.

53. (3) The reverse current is mainly due to drift of charges.

54. (2) Reverse saturation current is due to minority charge carriers. When temperature rises by 100°C, the semiconductor loses the property of controlled conduction due to the breaking of many covalent bonds.

55. (1) $E = \frac{V}{d} = \frac{0.5}{5 \times 10^{-7}} = 10^6\ V/m$

56. (2) Zener breakdown can occur in heavily doped diodes. In lightly doped diodes the necessary voltage is higher, and avalanche multiplication is then the chief process involved.

57. (2) Reverse resistance $= \frac{\Delta V}{\Delta I} = \frac{1}{0.5 \times 10^{-6}} = 2 \times 10^{+6}\Omega$

58. (2) As the output voltage obtained in a half wave rectifier circuit has a single variation in one cycle of ac voltage, hence the fundamental frequency in the ripple of output voltage would be = 50Hz

59. (1) The detection occurs only when the energy of incident photon greater than or equal to the energy bond gap.

$\frac{hc}{\lambda} = 2.5eV$

$\therefore \lambda = \frac{hc}{2.5eV} = \frac{1240\,eV - nm}{2.5\,eV} = 496\,nm \approx 5000Å$

60. (1)

61. (2)

62. (1) In the given circuit in forward biased condition the voltage across R is V.

63. (2) Reverse bias raises the potential barrier when applied to a junction diode.

64. (1) Electric field is zero in the middle of the depletion layer of a reverse biased P-N junction

65. (4) Potential barrier is formed at the P-N junction due to the excess of positive and negative charge at the junction.

66. (2) For detecting intensity of light, a photodiode in reverse bias is used.

67. (2) When a Zener diode is operated in the reverse breakdown region, the voltage across it remains practically constant (equal to V_Z) for a large change in the reverse current.

68. (2) A diode conducts only during the positive half cycle of a.c. Accordingly, the capacitor charges and discharges. During charging, the p.d. across capacitor

$= 100 \times \sqrt{2} = 141V$

69. (4) The barrier potential depends on type of semiconductor (for Si, V_b = 0.7 volt & for Ge, V_b = 0.3 volt), amount of doping and also on the temperature.

70. (3) When p-side of p-n junction is given more positive voltage as compared to n-side, then junction diode becomes forward biased.

71. (4) The term LED is abbreviated as 'Light Emitting Diode'. It is forward biased p-n junction which emits spontaneous radiation. Current in the circuit = 10 mA = 10×10^{-3} A and voltage in the circuit = 6 – 2 = 4V. From ohm's law,

$\therefore R = \frac{V}{I} = \frac{4}{10 \times 10^{-3}} = 400\Omega$

72. (1) Rectification is the process of conversion of a.c into d.c.

73. (2) $I = \dfrac{(3.2-3)V}{100\Omega} = 2\times10^{-3}\,A$

74. (2)

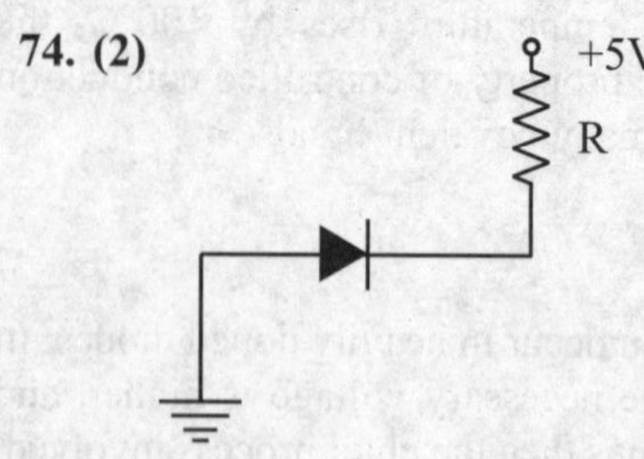

As the n-type of the diode is at the higher potential therefore, the diode is reversed biased.

∴ Option (2) is correct.

75. (3) $Y = \overline{\overline{A}.\overline{B}} = \overline{\overline{A}} + \overline{\overline{B}} \Rightarrow Y = A + B$

76. (1) Truth table of this logic circuit

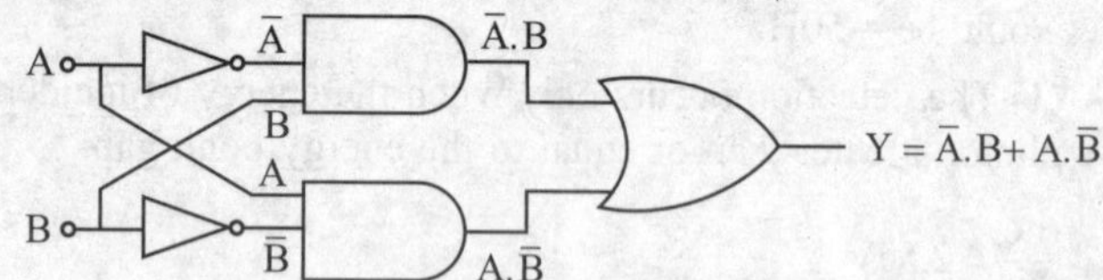

A	B	$\overline{A}$	$\overline{B}$	$\overline{A}.B$	$A.\overline{B}$	$Y = \overline{A}.B + A.\overline{B}$
0	0	1	1	0	0	0
0	1	1	0	1	0	1
1	0	0	1	0	1	1
1	1	0	0	0	0	0

77. (3)

A	B	C = A.B	$D = \overline{A}B$	E = C + D
0	0	0	0	0
0	1	0	1	1
1	0	0	0	0
1	1	1	0	1

78. (2) Two gates are required as follow

A, B → $\overline{A.B}$ → Y

$Y = \overline{\overline{AB}.\overline{AB}} = A{\cdot}B$

79. (1) $Y = \left(\overline{A+B}\right).\left(\overline{AB}\right) = \left(\overline{A}.\overline{B}\right).\left(\overline{A}+\overline{B}\right) = \left(\overline{A}\overline{A}\right).\overline{B} + \overline{A}\left(\overline{B}.\overline{B}\right)$

$Y = \overline{A}.\overline{B} + \overline{A}\overline{B} = \overline{A}\overline{B}$

A	B	Y
0	0	1
1	0	0
0	1	0
1	1	0

80. (4) 'NOR' gates are considered as universal gates because all the gates like AND, OR, NOT can be obtained by using only NOR gates.

81. (2) A → $\overline{A}$, B → $\overline{B}$, → $\overline{\overline{A}+\overline{B}}$ → Y

$Y = \overline{\overline{A}+\overline{B}} = \overline{\overline{A}}.\overline{\overline{B}} = A.B$ (AND gate)

82. (1) OR gate

A	B	Y
0	0	0
0	1	1
1	0	1
1	1	1

Y will be 1 when either of the inputs or both the inputs are 1.

83. (1) $y = A + \overline{A}.B = A(1+B) + \overline{A}.B \quad [\because 1+B=1]$

$= A + \left(A+\overline{A}\right)B = A + 1.B = A + B$

which is for OR gate

It means, one OR gate is required for the above design.

84. (4) Any digital circuit can be realised by repetitive use of NOR gates or NAND gates. NOR gates or NAND gates are called universal gates because by using these gates we can realise other basic gates like OR, AND and NOT gates.

85. (4) It is clear from the given logic circuit that the output Y is low when both the inputs are high otherwise it is high. Thus logic circuit is NAND gate.

High Order Time Intensive MCQs

1. (4) Number density of atoms in silicon specimen

$= 5 \times 10^{28}$ atom/m^3 = 5×10^{22} atom/cm^3

Since 1 atom of indium is doped in 5×10^7 silicon atoms, so total number of indium atom doped per cm^3 of silicon will be

$n = \dfrac{5\times10^{22}}{5\times10^{7}} = 1.0\times10^{15}$ atom/cm^3

2. (2) During the positive half cycle of input, the upper diode gets forward biased and conducts current while lower diode 1 gets reverse biased and does not conduct current. During the negative half cycle of input, diode 1 gets forward biased and conducts current, So the contributions from diode 1are B, D type, Hence option (2) is correct.

3. (1) Voltage on resistor, $R_S = V_i - V_L$

$\Rightarrow (I_L + nI_L)R_S = V_i - V_L$

$\therefore R_S = \dfrac{(V_i - V_L)}{(n+1)I_L}$

4. (3) The forward biased p-n junction does not offer any resistance.

$\therefore R_{AB} = \dfrac{10\times10}{10+10} = 5\,k\Omega$

Total resistance,

R = 10 + 5 = 15 kΩ

Current in the circuit,

$$I = \frac{V}{R} = \frac{30V}{15 \times 10^3} A = 2 \times 10^{-3} A$$

Current through each arm $= \frac{I}{2} = 10^{-3}$ A

$\therefore V_{AB} = 10 \times 10^3 \times 10^{-3} = 10$ V.

5. (2) In circuit 1, first p-n junction is forward bisaed and second is reverse biased. Their potential drops cannot be equal. In circuits 2 and 3, the two p-n junctions are connected in series, they carry equal (forward or reverse) currents. Their potential drops are equal.

6. (2) The initial current, $I_i = 0$

V_i —— 150Ω —— ▷| —— +3V

and the final current, $I_f = \frac{6-3}{150} = \frac{3}{150} = 0.02$ A

So, change in current, $\Delta I = I_f - I_i = 0.02$ A = 20 mA

7. (4) For $V_i < 0$, the diode is reverse biased and offers infinite resistance.

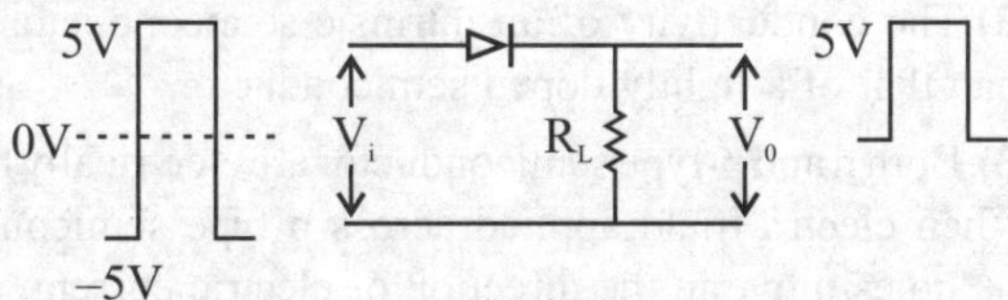

For $V_i > 0$, the junction diode is forward biased. The amplitude of the input signal is 5 V. In the negative part of signal, the diode doesn't conduct.

8. (3) Reverse voltage across the zener diode,

$$V_L = 20 \times \frac{2}{3} = 13.333 \text{ V}$$

$\Rightarrow V_L > V_z(12V)$

The voltage across 2 kΩ resistor = 12 V (since applied voltage is more than breakdown voltage for Zener diode)

Therefore, current passing through 2 kΩ resistor $= \frac{V_Z}{R_L}$

$$= \frac{12}{2 \times 10^3} = 6 \times 10^{-3} A = 6 \text{ mA}$$

9. (4) As A = 0, B = 1 $\Rightarrow Y = \overline{A.B} = \overline{1.0} = 1 = NAND$

10. (4) In common emitter configuration.

$$I_C = \frac{\Delta V_C}{R_L} = \frac{0.5}{500} = 1 \text{mA}$$

$$\beta = \frac{\alpha}{1-\alpha} = \frac{0.96}{0.04} = 24 \quad (\alpha = 0.96)$$

$$\because \beta = \frac{I_C}{I_B} \Rightarrow I_B = \frac{I_C}{\beta} = \frac{1}{24} \text{ mA}$$

11. (2) Here, $V = V_{CE} + I_C R_C \Rightarrow 15 = 7 + I_C \times 2 \times 10^3$

$I_C = 4$ mA ; β = 100

$$\because \beta - \frac{I_C}{I_B} \Rightarrow I_B = \frac{4}{100} = 0.04 \text{mA}$$

12. (1) CE amplifier causes a phase difference of π.

$$A_v = \frac{V_0}{V_i}$$

$$\therefore V_0 = A_\upsilon V_i = 150 \times 2\cos\left(15t + \frac{\pi}{3} + \pi\right)$$

$$= 300 \cos\left(15t + \frac{4\pi}{3}\right)$$

13. (4) The boolean expression of this gate is

$$Y = \overline{(A+B).(A+B)} = \overline{(A+B)} + \overline{(A+B)} = \overline{(A+B)}$$

which is for NOR gate.

14. (2) Output of the given combination is,

$Y = (A + B).\ C$

The truth table for the given combination is

A	B	C	Y = (A + B).C
0	0	0	0
0	0	1	0
0	1	0	0
0	1	1	1
1	0	0	0
1	0	1	1
1	1	0	0
1	1	1	1

Hence, $A = 1$, $B = 0$ and $C = 1$.

15. (3) Hence, for $A = 0$, $B = 0$ and $C = 1$

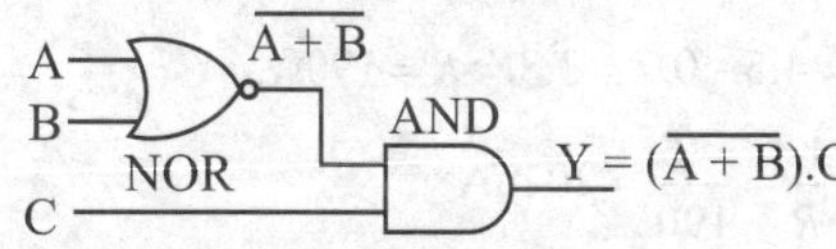

16. (4) It is diode-diode logic ckt of OR gate

17. (1) The given V-I characteristic is for a solar cell.

When I = 0, V = emf

⇒ A represents open circuit voltage

When V = 0, current = I

⇒ B represents short circuit voltage.

18. (4) When the inputs of a AND gate are shorted, then there are two conditions arise:

(i) When the input is 0, then the output is 0.

(ii) When the input is 1, then the output is 1.

Hence, by using above two conditions, the output is 0 for the given logic circuit. When A = 0, B = 1

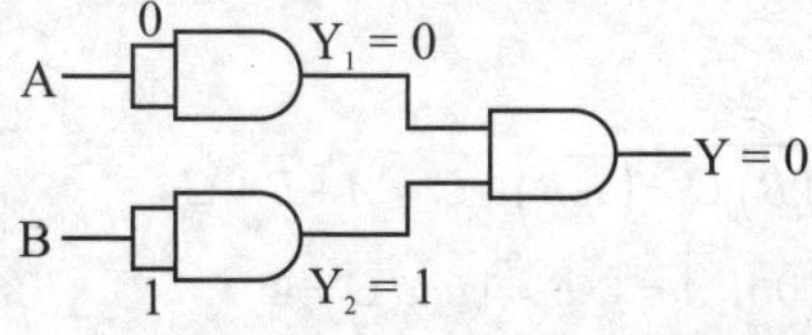

19. (2)

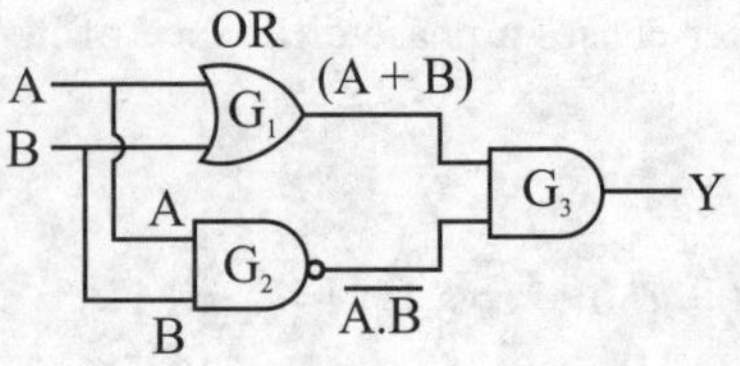

Output, $Y = (A+B)\overline{AB}$ (De-Morgan's theorem)

$= A\overline{AB} + \overline{AB}B$

$= A(\overline{A}+\overline{B}) + (\overline{A}+\overline{B})B$

$= A\overline{A} + A\overline{B} + \overline{A}B + B\overline{B}$

$= 0 + A\overline{B} + \overline{A}B + 0 = \overline{A}B + A\overline{B} = A \oplus B$

20. (4)

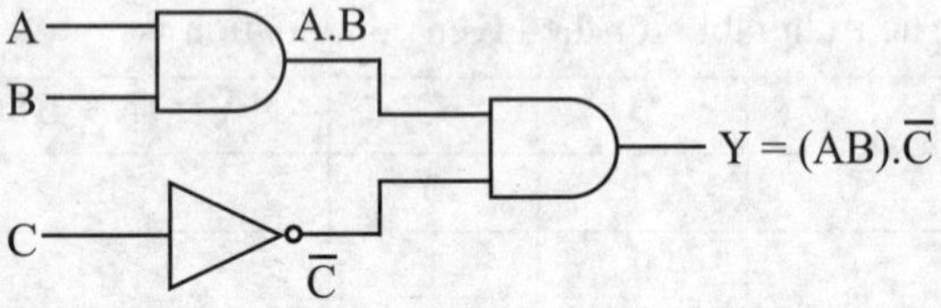

The output of gate-1 $Y_1 = A.\ B$

Thus, output will be 1, if $A = 1$, $B = 1$ and $C = 0$

21. (2) If positive terminal is connected to A, D_1 is in forward bias and D_2 is in reverse bias. So 20Ω is ineffective.

$\therefore i = \frac{V}{R} = \frac{3}{10} = 0.3A$

22. (1) $R = \frac{V}{i} = \frac{1.5-0.5}{5\times10^{-3}} = \frac{1}{5\times10^{-3}}$

$= 0.2\times10^3 = 2\times10^2\,\Omega$

23. (1) $V = 4.5 - 0.7 = 3.8V; R = 190\Omega,$

$i = \frac{V}{R} = \frac{3.8}{190} = 20$ mA

24. (2) $X = \overline{\overline{A.B}} = \overline{\overline{A}+\overline{B}} = A + B$. It is OR gate

25. (1) In a common-base arrangement, the current amplification factor.

$\alpha = \frac{I_C}{I_E} \Rightarrow I_C = \alpha I_E = 0.88\times1 = 0.88mA.$

and $I_B = I_E - I_C$

26. (2) $\beta = \frac{I_C}{I_B} \Rightarrow I_C = \beta I_B = 40\times25\times10^{-6} = 1000\,\mu A$

$I_E = I_B + I_C = 25 + 1000 = 1025\mu A = 1.025\,\text{mA}$

27. (3) $\beta = \frac{\alpha}{1-\alpha}$, But $\beta = \left(\frac{\delta I_C}{\delta I_B}\right)_{V_{CE}}$

$\Rightarrow \delta I_C = \beta.\delta I_B = 9\times4mA = 36mA$

28. (4) Here

$Y = \overline{(\overline{A.B}).\overline{C}} = (A.B) + C = \overline{A} + \overline{B} + C$

$\therefore Y = 0$ if $A = 1$, $B = 1$ and $C = 0$

29. (1) Logic of A is 1010, Logic of B is 1001

Truth table for the combination of gates is

$\overline{A}$	$\overline{B}$	Y
0	0	0
1	1	1
0	1	1
1	0	0

Logic of Y is 0111, which is option (1).

30. (3) An oscillator is an amplifier with positive feedback.

31. (3) For an OR gate if the input are A and B, then the output is A + B

As all the gates are OR gate

$Y_1 = A + D, Y_2 = A + D + B,$

$Y = A + D + B + C$

Statement Based MCQs

1. (1) In an unbiased p-n Junction, holes diffuse from p region to n region while electrons diffuse from n region to p region due to difference in concentration of hole and electrons on either side of junction

2. (3) The conductivity of an intrinsic semiconductor is less than that of a lightly doped semiconductor.

3. (3) Both n and p-type semiconductor are electrically neutral. When electric field applied across p type semiconductor, the holes move in the direction of electric current.

4. (2) Both statements are false.

5. (3) $\beta = \frac{\alpha}{1-\alpha} = \frac{0.98}{1-0.98} = 49$

Voltage gain $= \beta\frac{R_0}{R_i} = 49\times10^3$

Also, $\alpha = \frac{\beta}{1+\beta}$ and $\alpha \neq \frac{\beta}{1-\beta}$.

Assertion & Reason MCQs

1. (2) Both A & R are true but R not correctly explaining A.

2. (1) They can be used to implement any circuitry. All basic gates OR, AND, NOT can be derived from universal gates (NOR & NAND).

3. (1) Both Assertion and Reason are True and the Reason is a correct explanation of the Assertion.

Match The Column MCQs

1. (4)

(a) P-Type semiconductor is formed when trivalent (IIIrd-group) doping is used.

Intrinsic Semiconductor + Trivalent Doping (IIIrd-group) → P-Type Semiconductor

(b) N-Type semiconductor is formed when pentavalent (V[th]-group) doping is used.

Intrinsic Semiconductor + Pentavalent Doping (V^{th}-group) $\rightarrow$ N-Type Semiconductor

(c) Intrinsic semiconductor is pure semiconductor. (No-doping)

(d) Drift velocity (v_d) = Mobility (μ) $\times$ Electric field ($\vec{E}$)

2. (3)

(a,b) A PN junction diode is formed when P-Type and N-Type material joined together by a bonding force.

Diode symbol $\Rightarrow$

Forward-bias diode $\Rightarrow$ + –

Reverse-bias diode $\Rightarrow$ – +

(c) LED (Light emitting diode) is one of the special diode. Generally used material for fabrication is GaAs.

Symbol $\Rightarrow$

(d) Zener diode is voltage regular device, works in reverse bias.

Symbol $\Rightarrow$

3. (2) • Form factor is defined as the ratio of RMS value and AVERAGE value.

$$\text{Form Factor (F.F.)} = \frac{\text{RMS-Value}}{\text{Avg-Value}}$$

$\Rightarrow$ For Half wave rectifier;

$$(\text{F.F.})_{HWR} = \frac{V_m/2}{V_m/\pi} = \frac{\pi}{2} = 1.57$$

$\Rightarrow$ For Half wave rectifier;

$$(\text{F.F.})_{FWR} = \frac{V_m/2}{2V_m/\pi} = \frac{\pi}{2\sqrt{2}} = 1.11$$

• Ripple factor measures the percentage of ac component in the rectified output.

$$\text{Ripple factor } (\gamma) = \sqrt{(\text{F.F.})^2 - 1}$$

$\Rightarrow$ For Half wave rectifier;

$$(\gamma)_{HWR} = \sqrt{(1.57)^2 - 1} = 1.21$$

$\Rightarrow$ For Half wave rectifier;

$$(\gamma)_{FWR} = \sqrt{(1.11)^2 - 1} = 0.48$$

NCERT Exemplar MCQs

1. (4) The conductivity of a semiconductor increases with increase in temperature because the number density of current carriers increases, relaxation time decreases but effect of decrease in relaxation time is much less than increase in number density.

2. (2) When p-n junction is forward biased, it opposes the potential barrier junction. When p-n junction is reverse biased, it supports the potential barrier junction, resulting increase in potential barrier across the junction.

3. (2) In the given circuit p-side of p-n junction, D_1 is connected to lower voltage and n-side of p-n to higher voltage.

Thus D_1 is reverse biased.

The p-side of p-n junction D_2 is at higher potential and n-side of D_2 is at lower potential. Therefore D_2 is forward biased.

Hence, current flows through the junction B to A.

4. (4) As p-n junction conducts during positive half cycle only, the diode connected here will work in positive half cycle. Potential difference across C = peak voltage of the given AC voltage = $V_0 = V_{rms}\sqrt{2} = 220\sqrt{2}$V.

5. (2) The concept of hole describes the lack of an electron at a position where one could exist in an atom or atomic lattice. If an electron is excited into a higher state, it leaves a hole in its old state.

Thus, hole can be defined as a vacancy created when an electron leaves a covalent bond.

6. (2) Consider the figure given here, suppose 'V' is the potential difference between A and B. r_1 = 5kΩ and r_2 = 5kΩ are resistance in series connection.

Then, $V - 0.3 = [(r_1 + r_2)] \times [(0.2 \times 10^{-3})]$ [$\because$ V = ir]

$= [(5 + 5)\ 10^3] \times (0.2 \times 10^{-3})$

$= 10 \times 10^3 \times 0.2 \times 10^{-3} = 2$

$V = 2 + 0.3 = 2.3$ V

7. (3) Here, $C = A.B$ and $D = \bar{A}.B$

$E = C + D = (A.B) + (\bar{A}.B)$

The truth table of this arrangement of gates can be given by

A	B	$\bar{A}$	C = A.B	D = $\bar{A}$B	E = (C + D)
0	0	1	0	0	0
0	1	1	0	1	1
1	0	0	0	0	0
1	1	0	1	0	1

NEET & AIIMS Past Year MCQs

1. (4) The I-V characteristics of a solar cell show:

Negative current (reverse bias) for $V < 0$.

Positive current (forward bias) for $V > 0$.

2. (2) OR gate $\rightarrow$ Output is 1 if any input is 1

NOT $\rightarrow$ Inverts the out put

$\rightarrow$ Thi s is NOR Gate

3. (3)

The circuit is a combination of NAND and NOT gates.

Output $Y = \overline{\overline{A \cdot B}} = A + B$.

4. (3) The $I - V$ characteristics of solar cell is drawn in the fourth quadrant of the coordinate axes. This is because a solar cell does not draw current but supplies the same to the load.

In reverse bias condition, the positive terminal is connected to the n-type and the negative terminal is connected to the *p*-type. Hence, majority charge carriers are swept into terminals and minority charge carriers are repelled by the terminals. Thus, a very small drift current flows through the diode due to minority charge carriers of both *n*-side and *p*-side.

5. (1) Output for given options.

(1) $\overline{B}$	(2) B	(3) $A.B+\overline{A}$	(4) $A.\overline{B}+\overline{A}$
1	0	1	1
0	1	1	1
1	0	0	1
0	1	1	0

So option (1) is correct.

6. (2)

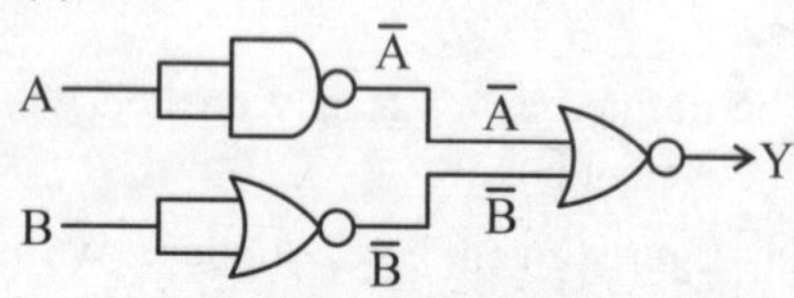

$Y = \overline{\overline{A}+\overline{B}} = \overline{\overline{A}}.\overline{\overline{B}}$ (AND gate)

7. (4) Theory based

Capacitor in parallel removes the ac ripple from the rectified output.

8. (2) Factual (theory based)

9. (3)

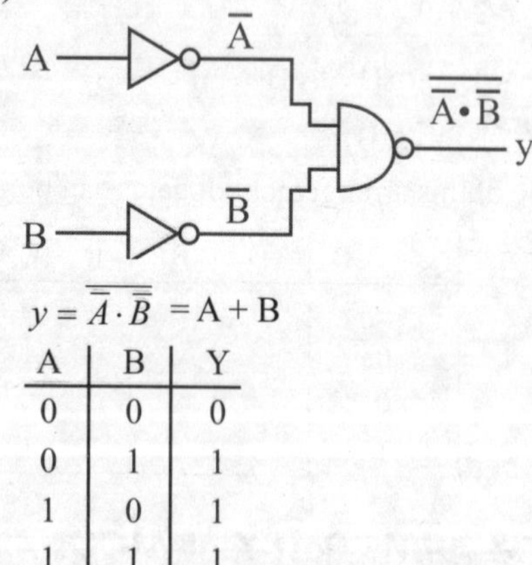

$y = \overline{\overline{A}\cdot\overline{B}} = A + B$

A	B	Y
0	0	0
0	1	1
1	0	1
1	1	1

10. (4) In the case of half wave rectifier

Input frequency = Output frequency

$\Rightarrow$ Output frequency = 60 Hz

11. (1) In (a) & (c) circuits, both the junctions are in similar biasing conditions so they equal resistances. Since both are in series, therefore equal potential will drop across them.

12. (4) $C = \overline{\overline{A\cdot B}.\overline{\overline{A}\cdot B}}$

using De-Morgan Theorem

$C = \overline{\overline{A\cdot B + \overline{A}\cdot B}} \Rightarrow C = B(A+\overline{A}) = \overline{\overline{B}}$

Therefore

A	B	C
0	0	1
0	1	0
1	0	1
1	1	0

13. (2) In N type semiconductor majority charge carriersare e^- and P type semiconductor majority chargecarriers are holes.

$I = neAV_d = neA(\mu E)$

$\mu_e > \mu_h \Rightarrow I_e > I_h$

14. (3) $Y = AB + \overline{BC}$

A	B	C	Y
0	0	1	0 + 1 = 1
1	0	1	0 + 1 = 1
0	1	0	0 + 1 = 1
0	0	1	0 + 1 = 1

15. (1) Current in zero diode is 5 times.

So total current drawn from battery = 6 mA + 30 mA = 36 mA

Potential difference Across R = 24 volt

So V = IR; $24 = 36 \times 10^{-3}$ R; R = 2000/3 Ω

16. (4) $Y = \left[\overline{(\overline{A.A}).(\overline{B.B})}\right] = \left[\overline{\overline{A}.\overline{B}}\right]$

$= \overline{\overline{A}} + \overline{\overline{B}}$

$= A + B$

17. (2) The p-side of p-n junction is taken at higher potential than n-side so; p-n junction is forward biased. Taking its resistance to be zero and applying Ohm's law

$$I = \frac{V}{R} = \frac{5-2}{300} = 10^{-2}\ A$$

$= 10^{-2} \times 10^{3}$ mA = 10 mA

18. (1) *n*-type semiconductor is formed when group 13 in doped with group 14. 3 electron of group 13 make bond with 3 electron of group 14 and 4th electron of group 14 is free but the atom is electrically neutral. So, n-type semiconductor is neutral.

19. (1)

A, B → or → A + B; B → $\overline{B}$; A + B and $\overline{B}$ → and → D

$II = (A + B)\,\overline{B}$

20. (1) Power = V × I

$364 \times 10^{-3} = \frac{91}{10} I$

$I = 40 \times 10^{-3}$ A

I = 40 mA

21. (1)

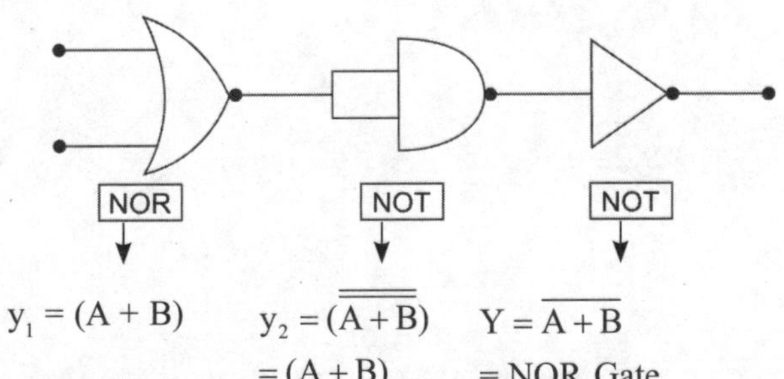

$y_1 = (A + B)$ $y_2 = \overline{\overline{(A+B)}} = (A+B)$ $Y = \overline{A+B} \equiv$ NOR Gate

Experiment Skills in Physics

NCERT Topic-wise MCQs

1. **(3)** 2mm
2. **(1)** If the zeros of vernier scale and main scale do not coincide when the two jaws are in close contact without any gap
3. **(1)** The thin rod projecting at the back
4. **(2)** 0.1mm
5. **(1)** 0.05mm
6. **(1)** The difference between one main scale division and one vernier scale division
7. **(3)** 0.529 cm
8. **(1)** Accuracy increase
9. **(4)** Zero
10. **(2)** 0.2%
11. **(2)** $T_1 > T_2$
12. **(2)** Yield point
13. **(3)**

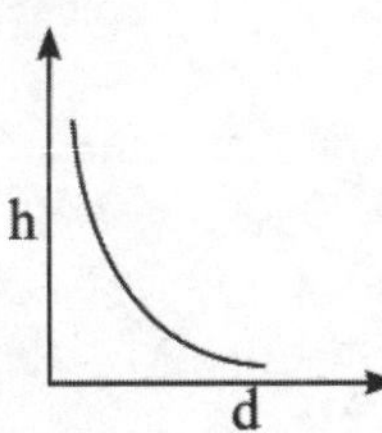

14. **(2)** 6.85 Ω
15. **(1)** High specific heat and low temperature coefficient of resistance
16. **(3)** One
17. **(2)** Compare the resistance
18. **(2)** Thick copper strips
19. **(1)** Both instruments will be damaged for flow of large current
20. **(1)** To change the resistance of the circuit
21. **(4)** Both (2) and (3)
22. **(4)** All of the above
23. **(4)** All of these
24. **(4)** All of the above
25. **(3)** $\left(\frac{n+1}{n}\right)F$
26. **(2)** smaller than focal length of eye piece
27. **(4)** at least distance of distinct of vision from eye piece
28. **(4)** None of above
29. **(2)** When the p type section is connected to the positive terminal and the n type section to the negative terminal of the battery the diode is said to be reverse biased.
30. **(3)** When a p-n diode is reverse biased, it offers high resistance. Forward biasing of p-n junction diode offers low resistance. In the case of an ideal p-n junction, resistance is zero. So, the ohmmeter can measure zero resistance.
31. **(2)** When a p-n diode is reverse biased, its depletion region is widened. This is because the voltage at the cathode will be higher than at the anode. This has the effect of increasing or decreasing the effective resistance of the junction itself allowing or blocking current flow through the diode.

High Order Time Intensive MCQs

1. **(3)** 6.93mm
2. **(3)**
3. **(3)** Percentage error $= \frac{1}{100} \times 100\% = 1\%$
4. **(4)** Closing of doors and window of the room
5. **(3)** Square of amplitude
6. **(2)** Balance point is near 50cm mark
7. **(2)** Voltmeter in parallel and ammeter in series
8. **(1)** The voltmeter reads the emf of the source and the ammeter reading is zero
9. **(4)** The current required to produce a deflection of one division
10. **(4)** Determination of refractive index of glass slab
11. **(2)** $P = Q = R = X$
12. **(1)** A straight line
13. **(4)** Gold
14. **(3)** By decreasing the area of cross section of the potentiometer wire
15. **(2)** 5A div^{-1}

Statement Based MCQs

1. **(3)** Statement-I is correct and Statement-II is incorrect.
2. **(1)** Statement-I and Statement-II both are correct.
3. **(4)** Statement-I is incorrect but Statement-II is correct.
4. **(3)** Statement-I is correct and Statement-II is incorrect.

Assertion & Reason MCQs

1. **(1)** If both Assertion and Reason are True and the Reason is the correct explanation of the Assertion.
2. **(3)** If Assertion is True but the Reason is False.
3. **(4)** If Assertion is False but the Reason is True.

Match The Column MCQs

1. (1) **2. (4)** **3. (4)** **4. (2)**

NEET Solved Papper

Year

2025

NEET Solved Paper

1. Consider a water tank shown in the figure. It has one wall at $x = L$ and can be taken to be very wide in the z direction. When filled with a liquid of surface tension S and density ρ, the liquid surface makes angle θ_0 ($\theta_0 << 1$) with the x-axis at $x = L$. If $y(x)$ is the height of the surface then the equation for $y(x)$ is:

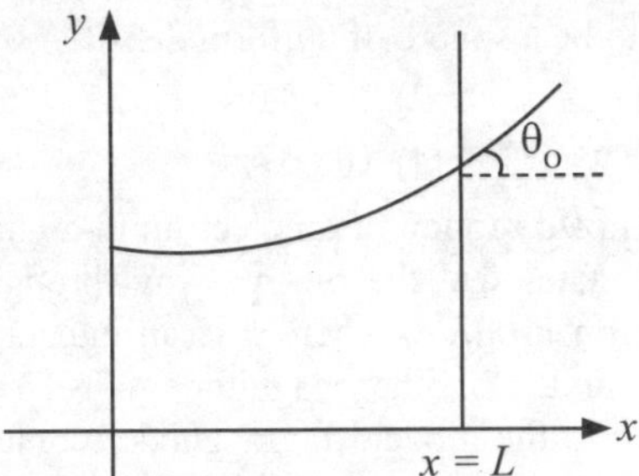

(take $\theta(x) = \sin\theta(x) = \tan\theta(x) = \frac{dy}{dx}$, g is the acceleration due to graity)

1) $\frac{d^2y}{dx^2} = \frac{\rho g}{S}x$ 2) $\frac{d^2y}{dx^2} = \frac{\rho g}{S}y$

3) $\frac{d^2y}{dx^2} = \sqrt{\frac{\rho g}{S}}$ 4) $\frac{dy}{dx} = \sqrt{\frac{\rho g}{S}}x$

2. A microscope has an objective of focal length 2 cm, eyepiece of focal length 4 cm and the tube length of 40 cm. If the distance of distinct vision of eye is 25 cm; the magnification in the microscope is:

1) 100 2) 125

3) 150 4) 250

3. An electron (mass 9×10^{-31} kg and charge 1.6×10^{-19} C) moving with speed c/100 (c = speed of light) is injected into a magnetic field $\vec{B}$ of magnitude 9×10^{-4} T perpendicular to its direction of motion. We wish to apply an uniform electric field $\vec{E}$ together with the magnetic field so that the electron does not deflect from its path. Then (speed of light $c = 3 \times 10^8$ ms^{-1}).

1) $\vec{E}$ is perpendicular to $\vec{B}$ and its magnitude is 27×10^4 V m^{-1}

2) $\vec{E}$ is perpendicular to $\vec{B}$ and its magnitude is 27×10^2 V m^{-1}

3) $\vec{E}$ is parallel to $\vec{B}$ and its magnitude is 27×10^2 V m^{-1}

4) $\vec{E}$ is parallel to $\vec{B}$ and its magnitude is 27×10^4 V m^{-1}

4. There are two inclined surfaces of equal length (L) and same angle of inclination 45° with the horizonal. One of them is rough and the other is perfectly smooth. A given body takes 2 times as much time to slide down on rough surface than on the smooth surface. The coefficient of kinetic friction (μ_k) between the object and the rough surface is close to:

1) 0.25 2) 0.40

3) 0.5 4) 0.75

5. The kinetic energies of two similar cars A and B are 100 J and 225 J respectively. On applying breaks, car A stops after 1000 m and car B stops after 1500 m. If F_A and F_B are the forces applied by the breaks on cars A and B, respectively, then the ratio F_A/F_B is:

1) $\frac{3}{2}$ 2) $\frac{2}{3}$ 3) $\frac{1}{3}$ 4) $\frac{1}{2}$

6. The current passing through the battery in the given circuit, is:

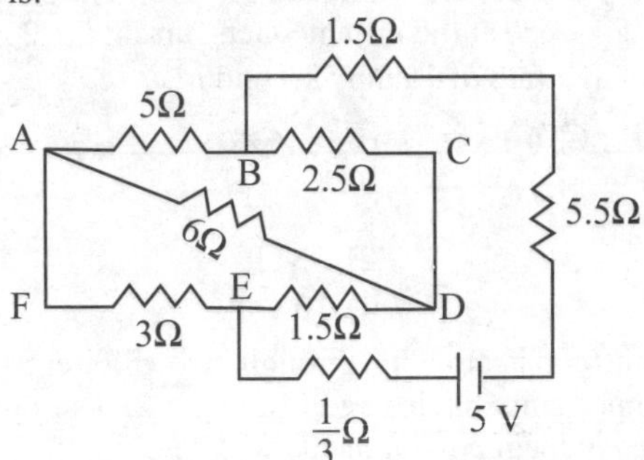

1) 2.0 A 2) 0.5 A

3) 2.5 A 4) 1.5 A

7. A bob of heavy mass m is suspended by a light string of length l. The bob is given a horizontal velocity v_0 as shown in figure. If the string gets slack at some point P making an angle θ from the horizontal, the ratio of the speed v of the bob at point P to its initial speed v_0 is:

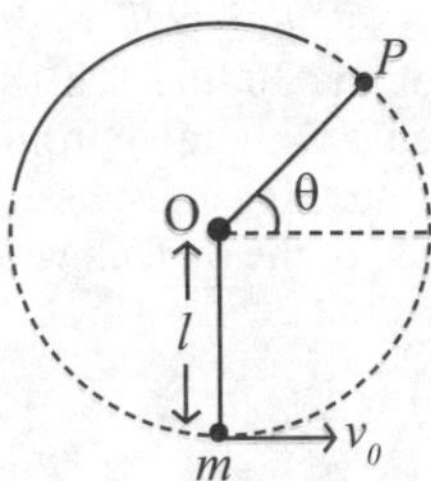

1) $(\sin\theta)^{1/2}$ 2) $\left(\frac{1}{2+3\sin\theta}\right)^{1/2}$

3) $\left(\frac{\cos\theta}{2+3\sin\theta}\right)^{1/2}$ 4) $\left(\frac{\sin\theta}{2+3\sin\theta}\right)^{1/2}$

8. The output (Y) of the given logic implementation is similar to the output of an/a ________ gate.

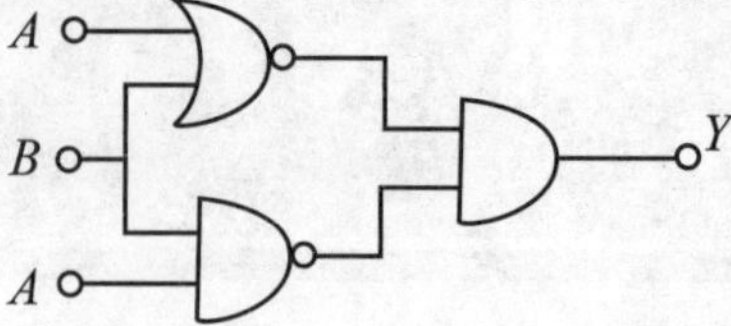

1) AND 2) NAND
3) OR 4) NOR

9. The electric field in a plane electromagnetic wave is given by
$E_z = 60 \cos (5x + 1.5 \times 10^9 t)\ V/m$.
Then expression for the corresponding magnetic field is (here subscripts denote the direction of the field):
1) $B_y = 2 \times 10^{-7} \cos (5x + 1.5 \times 10^9 t)T$
2) $B_x = 2 \times 10^{-7} \cos (5x + 1.5 \times 10^9 t)T$
3) $B_z = 60 \cos (5x + 1.5 \times 10^9 t)T$
4) $B_y = 60 \sin (5x + 1.5 \times 10^9 t)T$

10. A ball of mass 0.5 kg is dropped from a height of 40 m. The ball hits the ground and rises to a height of 10 m. The impulse imparted to the ball during its collision with the ground is (Take $g = 9.8\ m/s^2$)
1) 21 NS 2) 7 NS
3) 0 4) 84 NS

11. AB is a part of an electrical circuit (see figure). The potential difference "$V_A - V_B$", at the instant when current $i = 2\ A$ and is increasing at a rate of 1 amp / second is:

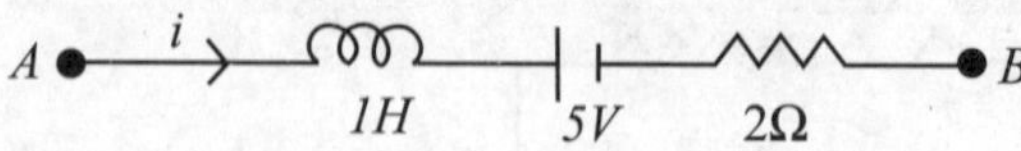

1) 5 volt 2) 6 volt
3) 9 volt 4) 10 volt

12. A 2 amp current is flowing through two different small circular copper coils having radii ratio 1 : 2. The ratio of their respective magnetic moments will be:
1) 1 : 4 2) 1 : 2
3) 2 : 1 4) 4 : 1

13. In a certain camera, a combination of four similar thin convex lenses are arranged axially in contact. Then the power of the combination and the total magnification in comparison to the power (p) and magnification (m) for each lens will be, respectively–
1) $4p$ and $4m$ 2) p^4 and $4m$
3) $4p$ and m^4 4) p^4 and m^4

14. An oxygen cylinder of volume 30 litre has 18.20 moles of oxygen. After some oxygen is withdrawn from the cylinder, its gauge pressure drops to 11 atmospheric pressure at temperature 27°C. The mass of the oxygen withdrawn from the cylinder is nearly equal to:
[Given, $R = \frac{100}{12}$ J mol^{-1} K^{-1}, and molecualr mass of $O_2 = 32$,
1) 0.125 kg 2) 0.144 kg
3) 0.116 kg 4) 0.156 kg

15. In some appropriate units, time (t) and position (x) relation of a moving particle is given by $t = x^2 + x$. The acceleration of the particle is:
1) $-\frac{2}{(x+2)^3}$ 2) $-\frac{2}{(2x+1)^3}$
3) $+\frac{2}{(x+1)^3}$ 4) $+\frac{2}{2x+1}$

16. To an ac power supply of 220 V at 50 Hz, a resistor of 20 Ω, a capacitor of reactance 25Ω and an inductor of reactance 45Ω are connected in series. The corresponding current in the circuit and the phase angle between the current and the voltage is, respectively:
1) 7.8 A and 30° 2) 7.8 A and 45°
3) 15.6 A and 30° 4) 15.6 A and 45°

17. The Sun rotates around its centre once in 27 days. What will be the period of revolution if the Sun were to expand to twice its present radius without any external influence? Assume the Sun to be a sphere of uniform density.
1) 100 days 2) 105 days
3) 115 days 4) 108 days

18. A model for quantized motion of an electron in a uniform magnetic field B states that the flux passing through the orbit of the electron in $n(h/e)$ where n is an integer, h is Plank's constant and e is the magnitude of electron's charge. According to the model, the magnetic moment of an electron in its lowest energy state will be (m is the mass of the electron).
1) $\frac{he}{\pi m}$ 2) $\frac{he}{2\pi m}$
3) $\frac{heB}{\pi m}$ 4) $\frac{heB}{2\pi m}$

19. Three identical heat conducting rods are connected in series as shown in the figure. The rods on the sides have thermal conductivity $2K$ while that in the middle has thermal conductivity K. The left end of the combination is maintained at temperature 37 and the right end at T. The rods are thermally insulated from outside. In steady state, temperature at the left junction is T_1 and that at the right junction is T_2. The ratio T_1/T_2 is:

3T	2K	K	2K	T
		T_1	T_2	

1) $\frac{3}{2}$ 2) $\frac{4}{3}$
3) $\frac{5}{3}$ 4) $\frac{5}{4}$

20. The plates of a parallel plate capacitor are separated by d. Two slabs of different dielectric constant K_1 and K_2 with thickness $\frac{3}{8}d$ and $\frac{d}{2}$, respectively are inserted in the capacitor. Due to this, the capacitance becomes two times larger than when there is nothing between the plates.
If $K_1 = 1.25\ K_2$, the value of K_1 is:
1) 2.66 2) 2.33
3) 1.60 4) 1.33

21. Two cities X and Y are connected by a regular bus service with a bus leaving in either direction every T min. A girl is driving scooty with a speed of 60 km/h in the direction X to Y notices that a bus goes past her every 30 minutes in the direction of her motion, and every 10 minutes in the opposite direction. Choose the correct option for the period T of the bus service and the speed (assumed constant) of the buses.

1) 9 min, 40 km/h 2) 25 min, 100 km/h
3) 10 min, 90 km/h 4) 15 min, 120 km/h

22. A uniform rod of mass 20 kg and length 5 m leans against a smooth vertical wall making an angle of 60° with it. The other end rests on a rough horizontal floor. The friction force that the floor exerts on the rod is:

(take $g = 10$ m/s^2)

1) 100 N 2) $100\sqrt{3}$ N 3) 200 N 4) $200\sqrt{3}$ N

23. In an oscillating spring mass system, a spring is connected to a box filled with sand. As the box oscillates, sand leaks slowly out of the box vertically so that the average frequency $\omega(t)$ and average amplitude $A(t)$ of the system change with time t. Which one of the following options schematically depicts these changes correctly?

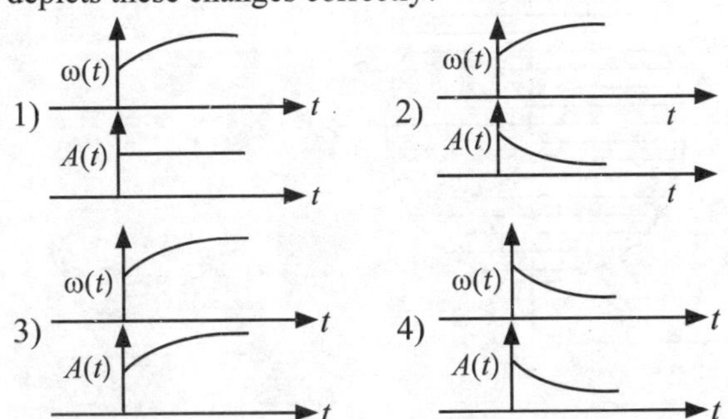

24. A balloon is made of a material of surface tension S and its inflation outlet (from where gas is filled in it has small area A. It is filled with a gas of density ρ and takes a spherical shape of radius R. When the gas is allowed to flow freely out of it, its radius r changes from R to 0 (zero) in time T. If the speed $v(r)$ of gas coming out of the balloon depends on r as r^a and $T \propto S^\alpha A^\beta \rho^\gamma R^\delta$ then:

1) $a = \frac{1}{2}, \alpha = \frac{1}{2}, \beta = -1, \gamma = +1, \delta = \frac{3}{2}$

2) $a = -\frac{1}{2}, \alpha = -\frac{1}{2}, \beta = -1, \gamma = -\frac{1}{2}, \delta = \frac{5}{2}$

3) $a = -\frac{1}{2}, \alpha = -\frac{1}{2}, \beta = -1, \gamma = \frac{1}{2}, \delta = \frac{7}{2}$

4) $a = \frac{1}{2}, \alpha = \frac{1}{2}, \beta = -\frac{1}{2}, \gamma = \frac{1}{2}, \delta = \frac{7}{2}$

25. Consider the diameter of a spherical object being measured with the help of a Vernier callipers. Suppose its 10 Vernier Scale Divisions (V.S.D.) are equal to its 9 Main Scale Divisions (M.S.D.). The least division in the M.S. is 0.1 cm and the zero of V.S. is at $x = 0.1$ cm when the jaws of Vernier callipers are closed.

If the main scale reading for the diameter is M = 5 cm and the number of coinciding vernier division is 8, the measured diameter after zero error correction, is:

1) 5.18 cm 2) 5.08 cm 3) 4.98 cm 4) 5.00 cm

26. A parallel plate capacitor made of circular plates is being charged such that the surface charge density on its plates is increasing at a constant rate with time. The magnetic field arising due to displacement current is:

1) zero at all places
2) constant between the plates and zero
3) non-zero everywhere with maximum at the imaginary cylindrical surface connecting peripheries of the plates.
4) zero between the plates and non-zero outside.

27. An unpolarized light beam travelling in air is incident on a medium of refractive index 1.73 at Brewster's angle. Then–

1) reflected light is completely polarized and the angle of reflection is close to 60°.
2) reflected light is partially polarized and the angle of reflection is close to 30°.
3) both reflected and transmitted light are perfectly polarized with angles of reflection and refraction close to 60° and 30°, respectively.
4) transmitted light is completely poarized with angle of refraction close to 30°.

28. Two identical charged conducting spheres A and B have their centres separated by a certain distance. Charge on each sphere is q and the force of repulsion between them is F. A third identical uncharged conducting sphere is brought in contact with sphere A first and then with B and finally removed from both. New force of repulsion between spheres A and B (Radii of A and B are negligible compared to the distance of separation so that for calculating force between them they can be considered as point charges) is best given as:

1) $\frac{3F}{5}$ 2) $\frac{2F}{3}$ 3) $\frac{F}{2}$ 4) $\frac{3F}{8}$

29. A container has two chambers of volumes $V_1 = 2$ litres and $V_2 = 3$ litres separated by a partition made of a thermal insulator. The chambers contains $n_1 = 5$ and $n_2 = 4$ moles of ideal gas at pressures $p_1 = 1$ atm and $p_2 = 2$ atm, respectively. When the partition is removed, the mixture attains an equilibrium pressure of :

1) 1.3 atm 2) 1.6 atm 3) 1.4 atm 4) 1.8 atm

30. A particle of mass m is moving around the origin with a constant force F pulling it towards the origin. If Bohr model is used to describe its motion, the radius r of the nth orbit and the particle's speed v in the orbit depend on n as:

1) $r \propto n^{1/3}$; $v \propto n^{1/3}$ 2) $r \propto n^{1/3}$; $v \propto n^{2/3}$
3) $r \propto n^{2/3}$; $v \propto n^{1/3}$ 4) $r \propto n^{4/3}$; $v \propto n^{-1/3}$

31. The radius of Martian orbit around the Sun is about 4 times the radius of the orbit of Mercury. The Martian year is 687 Earth days. Then which of the following is the length of 1 year on Mercury?

1) 88 earth days 2) 225 earth days
3) 172 earth days 4) 124 earth days

32. A body weighs 48 N on the surface of the earth. The gravitational force experienced by the body due to the earth at a height equal to one-third the radius of the earth from its surface is:

1) 16 N 2) 27 N 3) 32 N 4) 36 N

33. A wire of resistance R is cut into 8 equal pieces. From these pieces two equivalent resistances are made by adding four of these together in parallel. Then these two sets are added in series. The net effective resistance of the combination is:

1) $\frac{R}{64}$ 2) $\frac{R}{32}$ 3) $\frac{R}{16}$ 4) $\frac{R}{8}$

34. De-Broglie wavelength of an electron orbiting in the $n = 2$ state of hydrogen atom is close to

(Given Bohr radius = 0.052 *nm*)

1) 0.067 *nm* 2) 0.67 *nm*

3) 1.67 *nm* 4) 2.67 *nm*

35. An electric dipole with dipole moment 5×10^{-6} cm is aligned with the direction of a uniform electric field of magnitude 4×10^5 N/C. The dipole is then rotated through an angle of 60° with respect to the electric field. The change in the potential energy of the dipole is:

1) 0.8 J 2) 1.0 J

3) 1.2 J 4) 1.5 J

36. A constant voltage of 50 V is maintained between the points A and B of the circuit shown in the figure. The current through the branch CD of the circuit is:

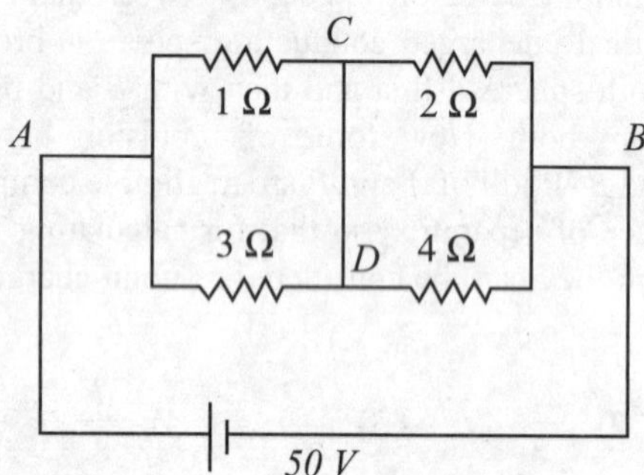

1) 1.5 *A* 2) 2.0 *A* 3) 2.5 *A* 4) 3.0 *A*

37. A photon and an electron (mass m) have the same energy E. The ratio $(\lambda_{photon}/\lambda_{electron})$ of their de Broglie wavelengths is (c is the speed of light).

1) $\sqrt{\frac{E}{2m}}$ 2) $c\sqrt{2mE}$ 3) $c\sqrt{\frac{2m}{E}}$ 4) $\frac{1}{c}\sqrt{\frac{E}{2m}}$

38. Which of the following options represent the variation of photoelectric current with property of light shown on the x-axis?

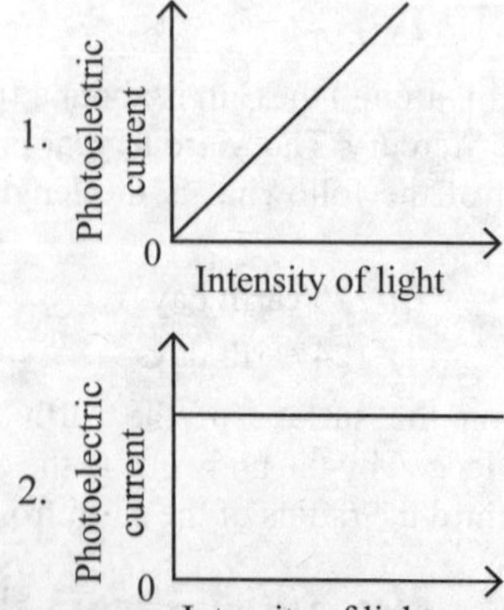

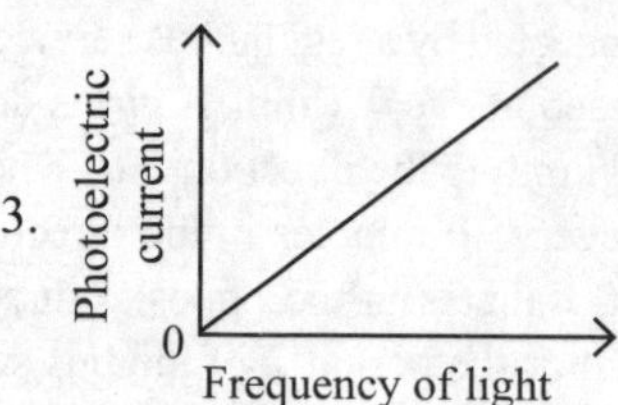

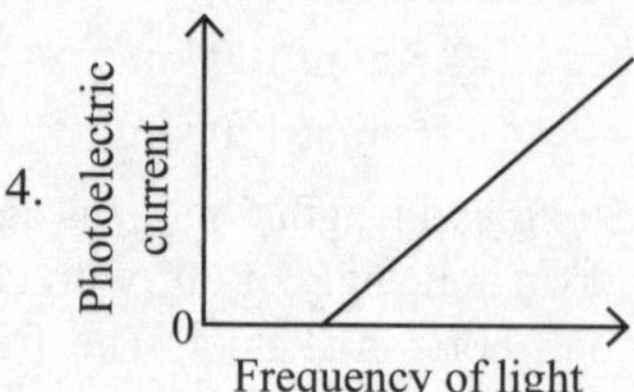

1) 1 only 2) 1 and 3

3) 1 and 4 4) 2 and 4

39. A sphere of radius R is cut from a larger solid sphere of radius $2R$ as shown in the figure. The ratio of the moment of inertia of the smaller sphere to that of the rest part of the sphere about the Y-axis is:

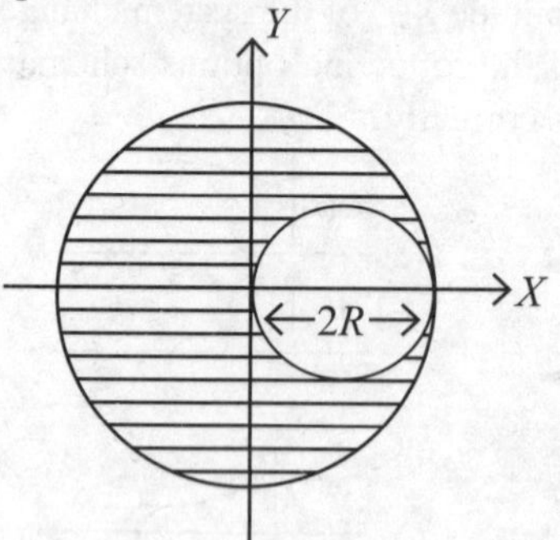

1) $\frac{7}{8}$ 2) $\frac{7}{40}$ 3) $\frac{7}{57}$ 4) $\frac{7}{64}$

40. A full wave rectifier circuit with diodes (D_1) and (D_2) is shown in the figure. If input supply voltage V_{in} = 220 sin (100 π t) volt, then at t = 15 *msec*.

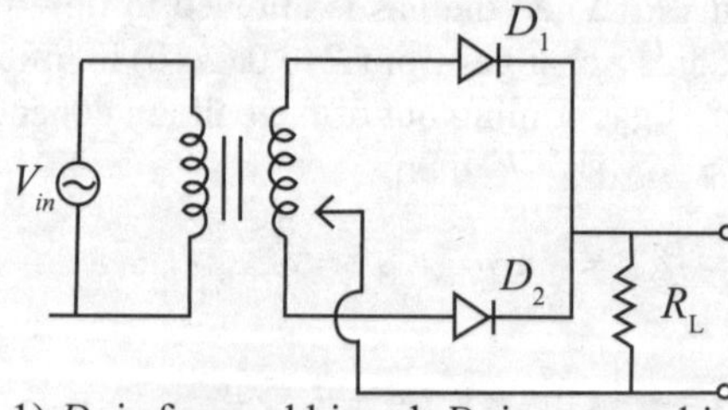

1) D_1 is forward biased, D_2 is reverse biased.

2) D_1 is reverse biased, D_2 is forward biased.

3) D_1 and D_2 both are forward biased.

4) D_1 and D_2 both are reverse biased.

41. Two gases A and B are filled at the same pressure in separate cylinders with movable pistons of radius r_A and r_B, respectively. ON supplying an equal amount of heat to both the systems reversibly under constant pressure, the pistons of gas A and B are displaced by 16 cm and 9 cm, respectively. If the change in their internal energy is the same, then the ratio $\frac{r_A}{r_B}$ is equal to:

1) $\frac{4}{3}$ 2) $\frac{3}{4}$ 3) $\frac{2}{\sqrt{3}}$ 4) $\frac{\sqrt{3}}{2}$

42. A physical quantity P is related to four observations a, b, c and d as follows:

$P = a^3b^2 / c\sqrt{d}$

The percentage errors of measurement in a, b, c and d are 1%, 3%, 2% and 4% respectively. The percentage error in the quantity P is:

1) 10% 2) 2%

3) 13% 4) 15%

43. The intensity of transmitted light when a polarioid sheet, placed between two crossed polaroids at 22.5° from the polarization axis of one of the polaroid, is (I_0 is the intensity of polarised light after passing through the first polaroid):

1) $\frac{I_0}{2}$ 2) $\frac{I_0}{4}$ 3) $\frac{I_0}{8}$ 4) $\frac{I_0}{16}$

44. Two identical point masses P and Q, suspended from two separate massless springs of spring constants k_1 and k_2, respectively, oscillate vertically. If their maximum speeds are the same, the ratio (A_Q/A_P) of the amplitude A_Q of mass Q to the amplitude A_P of mass P is:

1) $\frac{k_2}{k_1}$ 2) $\frac{k_1}{k_2}$ 3) $\sqrt{\frac{k_2}{k_1}}$ 4) $\sqrt{\frac{k_1}{k_2}}$

45. A pipe open at both ends has a fundamental frequency f in air. The pipe is now dipped vertically in a water drum to half of its length. The fundamental frequency of the air column is now equal to:

1) $\frac{f}{2}$ 2) f 3) $\frac{3f}{2}$ 4) $2f$

Answer Key

1	2	3	4	5	6	7	8	9	10	11	12	13	14	15	16	17
2	2	2	4	2	2	4	4	1	1	4	1	3	3	2	2	4
18	19	20	21	22	23	24	25	26	27	28	29	30	31	32	33	34
2	3	1	4	2	2	3	3	3	1	4	2	3	1	2	3	2
35	36	37	38	39	40	41	42	43	44	45						
2	2	3	1	3	2	2	3	3	4	2						

Explanations

1. (2) (NCERT XI, Part-II, Ch-Mechanical properties of fluids)

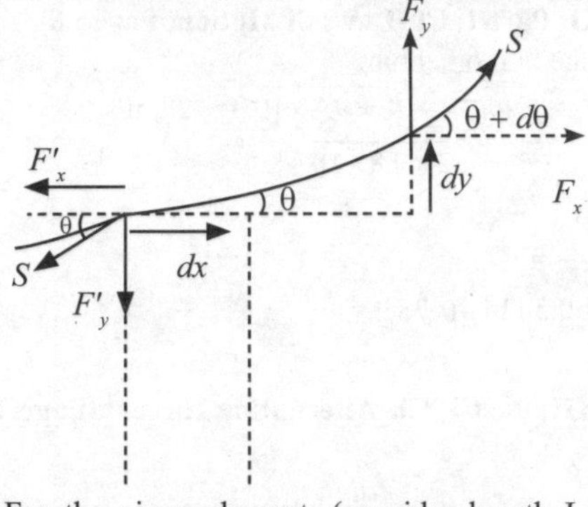

For the given element, (consider length L in Z direction) Net force in upward direction = Weight (S sin(θ + dθ) − S sinθ) L = mg

Angle is small $\therefore$ sinθ ≈ θ

$$\Rightarrow \frac{d\theta}{ydx} = \frac{\rho g}{S} \quad \dots (i)$$

$$\tan\theta = \frac{dy}{dx}$$

$$\Rightarrow \sec^2\theta \frac{d\theta}{dx} = \frac{d^2y}{dx^2} \quad \dots (ii)$$

Put $d\theta$ from (ii) in (i) take $\cos\theta \approx 1$, we get

$$\Rightarrow \frac{1}{4} = 1 - \mu$$

2. (2) (NCERT XII, Part-II, Ch-Ray Optics and Optical Instruments, Page-240)

$f_0 = 2$ cm $f_e = 4$ cm

L = 40 cm D = 25 cm

When image is formed at infinity

$$\text{Magnification} = \frac{L}{f_0} \cdot \frac{D}{f_e}$$

$$= \frac{40}{2} \times \frac{25}{4}$$

= 125

3. (2) (NCERT XII, Part-I, Ch-Moving Charges and Magnetism, Page-109)

A particle goes undeflected when Magnetic force = Electric force

qE = qVB sin90

$$E = \frac{c}{100} \times 9 \times 10^{-4} \sin 90$$

$E = 27 \times 10^2$ V/m and

$\vec{E} = \vec{B} \times \vec{V}$

4. (4) (NCERT XI, Part-I, Ch-Laws of motion, Page-61)

On smooth surface acceleration on a_1 = g sinθ & on rough surface $a_2 = g\sin\theta - \mu g\cos\theta$

$$L = \frac{1}{2}a_1t_1^2 = \frac{1}{2}a_2t_2^2$$

$$\Rightarrow \left(\frac{t_S}{t_R}\right)^2 = \frac{a_2}{a_1}$$

$$\Rightarrow \left(\frac{1}{2}\right)^2 = \frac{g\sin\theta - \mu g\cos\theta}{g\sin\theta}$$

$$\Rightarrow \frac{1}{4} = \frac{\sin 45^\circ - \mu\cos 45^\circ}{\sin 45^\circ}$$

$$\Rightarrow \frac{1}{4} = 1 - \mu \Rightarrow \mu = 1 - \frac{1}{4} = \frac{3}{4} = 0.75$$

5. (2) (NCERT XI, Part-I, Ch-Work Energy and Power, Page-73)

Work done by forces = change in kinetic energy

$$\frac{(\Delta)KE)_A}{(\Delta)KE)_B} = \frac{-F_A\Delta r_A}{-F_B\Delta r_B}$$

$$\frac{-100}{-225} = \frac{-F_A(1000)}{-F_B(1500)}$$

$$\Rightarrow \frac{F_A}{F_B} = \frac{2}{3}$$

6. (2) (NCERT XII, Part-I, Ch-Current Electricity, Page-97)

Diagram is a balanced wheatstone bridge so remove 6Ω

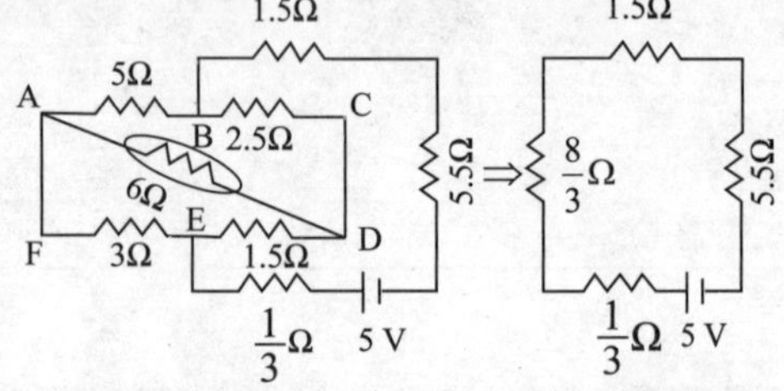

$$R_{eq} = \frac{1}{3} + \frac{8}{3} + 1.5 + 5.5$$

$$i = \frac{V}{R_{eq}} = \frac{5}{\frac{1}{3} + \frac{8}{3} + 1.5 + 5.5} = \frac{5}{10} = 0.5A$$

7. (4) (NCERT XI, Part-I, Ch-Work Energy and Power, Page-79)

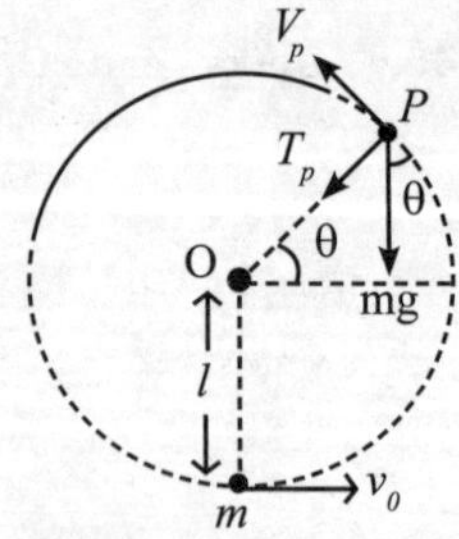

At pt P

Component of *mg* along *PO* = centripetal force

$$Tp + mg\sin\theta = \frac{mv_p^2}{\ell} \text{ (as } T_p = 0)$$

$$mg\sin\theta = \frac{mv_p^2}{\ell} \Rightarrow mv_p^2 = mg\ell\sin\theta \quad \ldots(i)$$

From mechanical energy conservation

$$\frac{1}{2}mv_0^2 = mg\ell(1+\sin\theta) + \frac{1}{2}mv_p^2 \quad \ldots(ii)$$

From (*i*) & (*ii*)

$$\frac{1}{2}mv_0^2 = mg\ell(1+\sin\theta) + \frac{1}{2}mg\ell\sin\theta$$

$$\Rightarrow v_0 = \sqrt{2g\ell + 3g\ell\sin\theta} \quad \ldots(iii)$$

$$v_p = \sqrt{g\ell\sin\theta}$$

$$v_0 = \sqrt{2g\ell + 3g\ell\sin\theta}$$

$$\frac{v_p}{v_0} = \sqrt{\frac{\sin\theta}{2 + 3\sin\theta}}$$

8. (4) (NCERT XII, Part-II, Ch-Semiconductor Electronics)

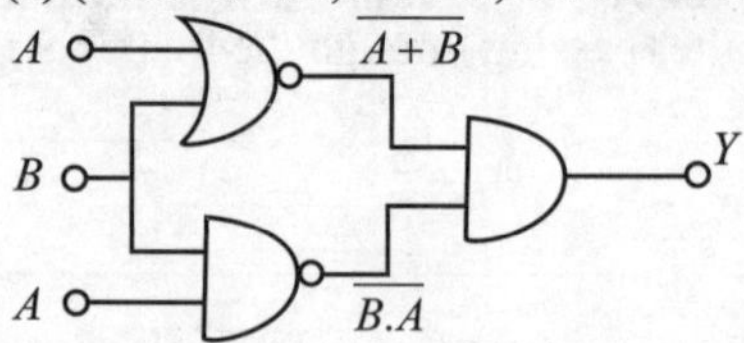

& the output of NOR gate is $\overline{A+B}$ the output of NAND gate is $\overline{AB}$

Output of AND gate *Y* is $Y = \overline{A+B}.\overline{A.B}$

$$y \Rightarrow \overline{A}.\overline{B}.(\overline{A} + \overline{B})$$

$$\Rightarrow \overline{A}.\overline{B}.\overline{A} + \overline{A}.\overline{B}.\overline{B}$$

$$\Rightarrow \overline{A}.\overline{B} + \overline{A}.\overline{B}$$

$$\Rightarrow \overline{A}.\overline{B} = \overline{(A+B)}$$

Behaves as NOR gate

9. (1) (NCERT XII, Part-I, Ch-Electromagnetic waves, Page-207)

Velocity of wave is given by

$$v = \frac{\omega}{k} = \frac{1.5\times10^9}{5} = 3 \times 10^8 \text{ m/s}$$

$$B_0 = \frac{\textit{Electric Field } (E_0)}{\textit{Speed } (v)} = \frac{60}{3\times10^8}$$

$= 20 \times 10^{-8}$ m/s $= 2 \times 10^{-7}$ T

$\overline{E} \times \overline{B} = \overline{C}$

$B_y = 2 \times 10^{-7} \cos(5x + 1.5 \times 10^9 t)$T

10. (1) (NCERT XI, Part-I, Ch-Laws of Motion, Page-55)

Impulse = change in momentum

Initial velocity $-\sqrt{2gh_i} = -\sqrt{2\times9.8\times10} = -28$ m/s

Final velocity $\sqrt{2gh_i} = \sqrt{2\times9.8\times10}$

= 14 m/s

Impulse $= m\vec{v}_f - m\vec{v}_i$

$= m[\vec{v}_f - \vec{v}_i] = 0.5\,[14 - (-28)]$

= 21 NS

11. (4) (NCERT XII, Part-I, Ch-Alternating current, Page-181)

Given data:

Current i = 2 A, increasing at $\frac{di}{dt} = 1\ A/s$.

Inductor L = 1H.

Battery voltage = 5 V.

Resistor R = 2Ω.

The voltage drop across the resistor is given by Ohm's Law:

$V_R = i \times R = 2\,A \times 2\Omega = 4V$

The voltage drop across the inductor is given by:

$$V_L = L\frac{di}{dt} = 1\text{ H} \times 1\ A/s = 1\ V$$

$$V_A - V_B = V_{battery} + V_L + V_R = 5 + 1 + 4 = 10V.$$

12. (1) (NCERT XII, Part-I, Ch-Moving charge and magnetism, Page-124)

Given:
Radii ratio: $r_1 : r_2 = 1 : 2$
Current: $I_1 = I_2 = 2\ A$
Area of the first coil (A_1):

$$A_1 = \pi r_1^2$$

Area of the second coil (A_2):

$$A_1 = \pi r_1^2$$

Magnetic moment of the first coil (μ_1):

$$\mu_1 = I \cdot A_1 = 2 \cdot \pi r_1^2$$

Magnetic moment of the second coil (μ_2):

$$\mu_2 = I \cdot A_2 = 2 \cdot \pi r_2^2$$

Therefore, $\frac{\mu_1}{\mu_2} = \frac{2\pi r_1^2}{2\pi r_2^2} = \frac{r_1^2}{r_2^2}$

$$\frac{\mu_1}{\mu_2} = \left(\frac{1}{2}\right)^2 = \frac{1}{4}$$

13. (3) (NCERT XII, Part-II, Ch-Ray optics, Page-238)

Since there are four identical lenses, each with power *p*, the total power is:
$P = p + p + p + p = 4p$
If each lens has magnification m, the total magnification for four lenses is:
$M = m \times m \times m \times m = m^4$

14. (3) (NCERT XI, Part-II, Ch-Kinetic theory of gases, Page-246)

Using the ideal gas law:
$$n_f = \frac{P_f V}{RT}$$
Substituting the given values:
$$= n_f = \frac{12 \times 1.01 \times 10^5 \times 30 \times 10^{-3}}{\left(\frac{100}{12}\right) \times 300}$$
= 14.544 moles

Moles Withdrawn

$\Delta n = n_i - n_f = 18.20 - 14.544 = 3.656$ moles

Mass Withdrawn
Converting moles to mass:
Mass = Δn × Molecular Mass = 3.656 × 32g = 116.992 g = 0.116 kg

15. (2) (NCERT XI, Part-I, Ch-Motion in straight line, Page-15)

Given: $t = x^2 + x$.
Differentiate $t = x^2 + x$ with respect to t:
$$1 = (2x+1)\frac{dx}{dt} \Rightarrow \frac{dx}{dt} = \frac{1}{2x+1}$$

Differentiate $\frac{dx}{dt}$ to find acceleration:

$$a = \frac{d}{dt}\left(\frac{1}{2x+1}\right) = \frac{-2\frac{dx}{dt}}{(2x+1)^2}$$

$$= \frac{-2\left(\frac{1}{2x+1}\right)}{(2x+1)^2} = \frac{-2}{(2x+1)^3}$$

16. (2) (NCERT XII, Part-I, Ch-Alternating current, Page-187)

Net Reactance X:
$X = X_L - X_C = 45\Omega - 25\Omega = 20\Omega$
Impedance Z:
$$Z = \sqrt{R^2 + X^2} = \sqrt{20^2 + 20^2}$$
$$= \sqrt{400 + 400} = \sqrt{800} = 20\sqrt{2}\,\Omega$$

Current I:
$$I = \frac{V_{r.m.s}}{Z} = \frac{220}{20\sqrt{2}} = \frac{11}{\sqrt{2}} \approx 7.8\text{ A}$$

Phase Angle ϕ: $\tan\phi = \frac{X}{R} = \frac{20}{20} = 1$
$\phi = 45°$

17. (4) (NCERT XI, Part-I, Ch-Systems of particles and rotational motion, Page-122)

Given Data:
Initial period of rotation, $T_1 = 27$ days.
Initial radius, $R_1 = R$.
Final radius after expansion, $R_2 = 2R$.
Conservation of Angular Momentum:
$I_1\omega_1 = I_2\omega_2$

For a sphere of uniform density, the moment of inertia is $I = \frac{2}{5} MR^2$, and angular velocity $\omega = \frac{2\pi}{T}$.

Substitute Moments of Inertia and Angular Velocities:

$$\left(\frac{2}{5}MR_1^2\right)\left(\frac{2\pi}{T_1}\right) = \left(\frac{2}{5}MR_2^2\right)\left(\frac{2\pi}{T_2}\right)$$

$$\frac{R_1^2}{T_1} = \frac{R_2^2}{T_2}$$

$$T_2 = T_1\left(\frac{R_2}{R_1}\right)^2 = 27 \times \left(\frac{2R}{R}\right)^2$$
$= 27 \times 4 = 108$ days

18. (2) (NCERT XII, Part-I, Ch-Moving charge and magnetism, Page-124)

The magnetic flux Φ through the electron's orbit is quantized as:
$$\Phi = n\left(\frac{h}{e}\right)$$
where n is an integer, h is Planck's constant, and e is the electron charge.
The magnetic flux through a circular orbit of radius r is:
$\Phi = B \cdot \pi r^2$
$$B\pi r^2 = \frac{h}{e} \Rightarrow r^2 = \frac{h}{eB\pi}$$
when charge moves in external field
then $r = \frac{mv}{qB} \Rightarrow \frac{v}{r} = \frac{eB}{m}$
The magnetic moment μ for a current loop is:

$\mu = I.\ A = IA = I(\pi r^2)$
$$= \left(\frac{ev}{2\pi r}\right)(\pi r^2)$$

$$= \frac{e}{2\pi}\left(\frac{eB}{m}\right)\pi\left(\frac{h}{B\pi e}\right) = \frac{eh}{2\pi m}$$

19. (3) (NCERT XI, Part-II, Ch-Thermal properties of matter, Page-216)

Thermal Resistances:

Left & Right rods (conductivity $2K$): $R = \frac{L}{2K\,A}$

Middle rod (conductivity K): $R = \frac{L}{K\,A}$

In steady state, heat flow Q is the same through all rods.

- Left rod: $Q = \frac{3T - T_1}{L/(2K\,A)} = 2K\,A(3T - T_1)/L$
- Middle rod: $Q = \frac{3T - T_2}{L/(K\,A)} = K\,A(T_1 - T_2)/L$
- Right rod: $Q = \frac{T_2 - T}{L/(2K\,A)} = 2K\,A(T_2 - T)/L$

Equate and Simply:

- From left and middle rods:
  $2(3T - T_1) = T_1 - T_2$
  $\Rightarrow 6T + T_2 = 3T_1$ ... (*i*)
- From middle and right rods:
  $T_1 - T_2 = 2(T_2 - T)$
  $\Rightarrow T_1 + 2T = 3T_2$ ... (*ii*)
- From (*ii*) equation: $T_1 = 3T_2 - 2T$
- Substitute into (*i*) equation:
  $6T + T_2 = 3(3T_2 - 2T)$

$\Rightarrow T_2 = \frac{3}{2}T$

$\Rightarrow T_2 = \frac{5}{2}T$

Final Ratio: $\frac{T_1}{T_2} = \frac{5/2}{3/2} = \frac{5}{3}$

20. (1) (NCERT XII, Part-I, Ch-Electrostatic potential and capacitance, Page-69)

Given:

Total separation d

Thickness of slabs: $\frac{3}{8}d$ (with K_1) and $\frac{d}{2}$ (with K_2)

Remaining air gap: $d - \frac{3}{8}d - \frac{d}{2} = \frac{d}{8}$ (with $K = 1$)

Capacitance doubles: $C_{new} = 2C_0$, where $C_0 = \frac{\epsilon_0 A}{d}$

Relation: $K_1 = 1.25\,K_2$

Effective Capacitance Formula:

$$\frac{1}{C_{new}} = \frac{\frac{3}{8}d}{K_1 \epsilon_0 A} + \frac{\frac{d}{2}}{K_2 \epsilon_0 A} + \frac{\frac{d}{8}}{\epsilon_0 A}$$

$$\frac{1}{2C_0} = \frac{3d}{8K_1 \epsilon_0 A} + \frac{d}{2K_2 \epsilon_0 A} + \frac{d}{8\epsilon_0 A}$$

$$\frac{d}{2\epsilon_0 A} = \frac{3d}{8K_1 \epsilon_0 A} + \frac{d}{2K_2 \epsilon_0 A} + \frac{d}{8\epsilon_0 A}$$

$$\frac{1}{2} = \frac{3}{8K_1} + \frac{1}{2K_2} + \frac{1}{8}\;\frac{3}{8} = \frac{3}{8K_1} + \frac{1}{2K_2}$$

Substitute $K_1 = 1.25\,K_2$:

$$\frac{3}{8} = \frac{3}{10K_2} + \frac{1}{2K_2} = \frac{8}{10K_2}$$

$\Rightarrow K_2 = 2.133$

$\Rightarrow K_1 = 1.25 \times 2.133 \approx 2.66$

21. (4) (NCERT XI, Part-I, Ch-Motion in a straight line, Page-14)

Let the bus speed be v km/h and time interval T minutes.

In the same direction, relative speed = $v - 60$; time = 30 min $\rightarrow$ distance = $\frac{(v-60)\times 30}{60}$

In the opposite direction, relative speed = $v + 60$; time = 10 min $\rightarrow$ distance = $\frac{(v+60)\times 10}{60}$

Both distances are equal. So, $v = 120$

Inter-Bus spacing (each min) $= v.\left(\frac{T}{60}\right) km$

This will be equal to both distances.

$\Rightarrow \frac{vT}{60} = \frac{(v-60)\times 30}{60} \Rightarrow T = 15\,\text{min}$

22. (2) (NCERT XI, Part-I, Ch-System of particles and rotational motion, Page-109)

Rod is in equilibrium — apply torque about the base:

$200 \times \frac{5}{2} \cos 60° = N \times 5 \sin 60°$

$\Rightarrow N = 100\,N$

Friction balances vertical weight:

$f = N \tan 60° = 100 \times \sqrt{3} = 100\sqrt{3}\,N$

Answer: (2) $100\sqrt{3}\,N$

23. (2) (NCERT XI, Part-II, Ch-Oscillations, Page-262)

In the spring-mass system, as sand leaks out, the mass m(t) decreases over time.

- **Frequency ($\omega(t)$):** The angular frequency is $\omega = \sqrt{\frac{k}{m}}$, where k is the spring constant. As m decreases, $\omega(t)$ increases because it's inversely proportional to the square root of mass.
- **Amplitude ($A(t)$):** The amplitude decreases because the leaking sand carries away energy. In a realistic scenario, the system loses energy over time, reducing the amplitude of oscillation, as shown by the downward trend in $A(t)$.

24. (3) (NCERT XI, Part-II, Ch-Mechanical properties of solids)

The speed of gas coming out depends on pressure due to surface tension(S). $\Delta P \sim \frac{S}{R}$.

Using Bernoulli's principle, $\frac{1}{2}\rho v^2 \sim \frac{S}{R}$

$\Rightarrow v \sim \left(\frac{S}{\rho R}\right)^{1/2}$, so $v \propto r^{-1/2}$,

Hence $a = -\frac{1}{2}$.

Using dimensional analysis on 'T' $T \propto S^{\alpha} A^{\beta} \rho^{\gamma} R^{\delta}$,

we find $\alpha = -\frac{1}{2},\ \beta = -1,\ \gamma = \frac{1}{2},\ \delta = \frac{7}{2}$

25. (3) (NCERT XI, Part-I, Ch-Units and Measurements)

Given: 10 V.S.D. = 9 M.S.D.,

So 1 V.S.D. = 0.09 cm and Least Count = 0.01 cm.

Main scale reading = 5 cm;

Vernier reading = 8 × 0.01 = 0.08 cm.

Total reading = 5 + 0.08 = 5.08 cm.

Zero error = +0.1 cm

$\Rightarrow$ Corrected reading = 5.08 – 0.1 = 4.98 cm.

26. (3) (NCERT XII, Part-I, Ch-Electromagnetic waves, Page-202)

When a capacitor is being charged, the changing electric field between its plates creates a displacement current, which produces a magnetic field, just like a real current.
According to **Ampere-Maxwell Law:**

$$\oint \vec{B} \cdot d\vec{l} = \mu_0 \epsilon_0 \frac{d\Phi_E}{dt}$$

Here, the changing electric flux $\frac{d\Phi_E}{dt}$ is due to increasing surface charge density.
For a circualr capacitor, the magnetic field is non-zero everywhere, and is maximum on the cylindrical surface joining the edges (peripheries) of the plates where the electric flux charges the most.
Hence, correct answer is (3).

27. (1) (NCERT XII, Part-II, Ch-Wave optics, Page-269)

At Brewster's angle(θ_B), the reflected light is completely polarized perpendicular to the plane of incidence.
Refractive index = 1.73,
So $\theta_B = \tan^{-1}(1.73) \approx 60°$
Hence, angle of reflection = 60°, and reflected light is completely polarized.

28. (4) (NCERT XII, Part-I, Ch-Electric charges and fields, Page-7)

Initial force: $F = \frac{1}{4\pi\varepsilon_0} \cdot \frac{q^2}{r^2}$

When uncharged sphere touches A: A and C each get $q/2$.
Then C touches B: total charge = $q/2 + q = 3q/2$, so B and C get $3q/4$ each.
Now, A has $q/2$, B has $3q/4$

New force = $\frac{1}{4\pi\varepsilon_0} \cdot \frac{(q/2)(3q/4)}{r^2} = \frac{3F}{8}$

29. (2) (NCERT XI, Part-II, Ch-Thermal properties of matter, Page-204)

When two ideal gases mix in an insulated container, total pressure at equilibrium is found using:

$$P_{\text{final}} = \frac{P_1V_1 + P_2V_2}{V_1 + V_2}$$

Given:
$P_1 = 1$ atm, $V_1 = 2L$
$P_2 = 2$ atm, $V_2 = 3L$

$$P_{\text{final}} = \frac{(1)(2)+(2)(3)}{2+3} = \frac{2+6}{5} = \frac{8}{5} = 1.6 \text{ atm}$$

30. (3) (NCERT XII, Part-II, Ch-Atoms, Page-297)

Force balance: $F = \frac{mv^2}{r} \Rightarrow v \propto \sqrt{r}$

Bohr's quantization: $mvr = n\hbar \Rightarrow v \propto \frac{n}{r}$

Combine: $v \propto \sqrt{r}$, $v \propto \frac{n}{r}$

$\Rightarrow \frac{n}{r} \propto \sqrt{r}$
$\Rightarrow n \propto r^{3/2}$
$\Rightarrow r \propto n^{2/3}$
Then $v \propto \sqrt{r} \propto n^{1/3}$

31. (1) (NCERT XI, Part-I, Ch-Gravitation, Page-129)

Use Kepler's third law: $\frac{T^2}{R^3}$ = constant.

So, $\frac{T_{\text{Mars}}^2}{R_{\text{Mars}}^3} = \frac{T_{\text{Mercury}}^2}{R_{\text{Mercury}}^3}$.

Given: $R_{\text{Mars}} = 4 \times R_{\text{Mercury}}$, $T_{\text{Mars}} = 687$ days.

$\therefore \frac{687^2}{(4R_{\text{Mercury}})} = \frac{T_{\text{Mercury}}^2}{R_{\text{Mercury}}^3}$

$\Rightarrow T_{\text{Mercury}}^2 = \frac{687^2}{64}$.

$\Rightarrow T_{\text{Mercury}} = \frac{687}{8} = 85.875 \approx 86$ days.

Closest option to 86 is 88 Earth days (real value of Mercury's year is 88 days.

32. (2) (NCERT XI, Part-I, Ch-Gravitation, Page-133)

Gravitational force varies with height: $F = \frac{mgR^2}{(R+h)^2}$, where R is Earth's radius, $h = \frac{R}{3}$.
At surface ($h = 0$), $F = mg = 48\ N$.

At $h = \frac{R}{3}$, distance from center = $R + \frac{R}{3} = \frac{4R}{3}$.

New force: $F' = 48 \times \left(\frac{R}{\frac{4R}{3}}\right)^2$

$= 48 \times \left(\frac{3}{4}\right)^2 = 48 \times \frac{9}{16} = 27\ N.$

33. (3) (NCERT XII, Part-I, Ch-Current electricity, Page-84)

Resistance of one piece: $R/8$.

Four pieces in parallel: Equivalent resistance = $\frac{R/8}{4} = \frac{R}{32}$.

Two such sets in series: Net resistance = $2 \times \frac{R}{32} = \frac{R}{16}$.

34. (2) (NCERT XII, Part-II, Ch-Atom, Page-299)

Bohr radius $a_0 = 0.052$ nm. Radius for $n = 2$: $r = a_0n^2 = 0.052 \times 4 = 0.208$ nm.

Velocity in n^{th} orbit: $v = \frac{2.18 \times 10^6}{n}$ m/s, so for $n = 2$, $v = 1.09 \times 10^6$ m/s.

De-Broglie wavelength: $\lambda = \frac{h}{mv}$. Using $h = 6.63 \times 10^{-34}$ J. s, $m = 9.1 \times 10^{-31}$ kg.

$$\lambda = \frac{6.63 \times 10^{-34}}{9.1 \times 10^{-31} \times 1.09 \times 10^6} \approx 6.68 \times 10^{-10} \text{ m}$$

= 0.668 nm.

35. (2) (NCERT XII, Part-II, Ch-Electrostatic Potential and capacitance, Page-60)

Potential energy of a dipole: $U = -pE \cos\theta$. Initially ($\theta = 0°$), $U_i = -pE$. At $\theta = 60°$, $U_f = -pE \cos 60° = -pE \times 0.5$.
Change in potential energy: $\Delta U = U_f - U_i = -0.5pE - (-pE) = 0.5pE$.
So $\Delta U = 0.5 \times (5 \times 10^{-6}) \times (4 \times 10^5) = 1.0$ J.

36. (2) (NCERT XII, Part-I, Ch-Current electricity, Page-98)

Refer the circuit diagram below:

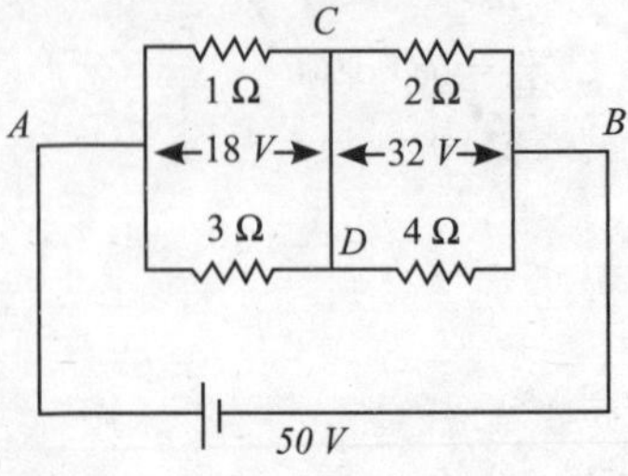

Current through 1Ω and 3Ω is 18 A and 6 A respectively.

Current through 2Ω and 4Ω is 16 A and 4 A respectively.

By Kirchhoff's current law: current through CD = 2A

37. (3) (NCERT XII, Part-II, Ch-Dual nature of radiation and matter, Page-285)

Photon energy: $E = pc$, so $p_{\text{photon}} = \dfrac{E}{c}$.

De Broglie wavelength: $\lambda_{\text{photon}} = \dfrac{h}{p_{\text{photon}}} = \dfrac{hc}{E}$.

Electron energy: $E = \dfrac{p_{electron}}{2m}$, so $p_{\text{electron}} = \sqrt{2mE}$. De Broglie wavelength: $\lambda_{\text{electron}} = \dfrac{h}{p_{electron}} = \dfrac{h}{\sqrt{2mE}}$.

Ratio: $\dfrac{\lambda_{photon}}{\lambda_{electron}} = \dfrac{\frac{hc}{p}}{\frac{h}{\sqrt{2mE}}} = \dfrac{c}{E} \times \sqrt{2mE} = c\sqrt{\dfrac{2m}{E}}$.

38. (1) (NCERT XII, Part-II, Ch-Dual nature of radiation and matter, Page-278)

Photoelectric current depends on light intensity (linear increase) and frequency (current exists above threshold frequency).

Graph A: Current vs. intensity (linear) → Correct.

Graph C and D: Incorrect, Graph should be constant above threshold.

39. (3) (NCERT XI, Part-I, Ch-Systems of particles and rotational motion, Page-115)

Moment of inertia of a solid sphere about an axis through its center: $I = \dfrac{2}{5} MR^2$.

Smaller sphere (radius R): Mass $M_1 = \rho\dfrac{4}{3}\pi R^3$, $I_1 = \dfrac{2}{5} M_1R^2$.

Larger sphere (radius $2R$): Mass M_2

$= \rho\dfrac{4}{3}\pi(2R)^3 = 8M_1$. Remaining mass

$= 8M_1 - M_1 = 7M_1$.

Use parallel axis theorem for both spheres about Y-axis (distance R from center for smaller sphere, adjusted for remaining part).

$I_{\text{smaller}} = \dfrac{2}{5}M_1R^2 + M_1R^2 = \dfrac{7}{5} M_1R^2$.

$I_{\text{remaining}} = \dfrac{2}{5}(8M_1)(2R)^2 - \dfrac{7}{5}M_1R^2$

$= \dfrac{57}{5}M_1R^2$

Therefore, required ratio $= \dfrac{I_{\text{smaller}}}{I_{\text{remaining}}} = \dfrac{7}{57}$

40. (2) (NCERT XII, Part-II, Ch-Semiconductor electronics: Materials, devices and simple circuits, Page-339)

Input: $V_{\text{in}} = 220 \sin(100\pi t)$. Frequency: $\omega = 100\pi$, so $T = \dfrac{2\pi}{\omega}$ $= 0.02\text{s} = 20$ msec.

At $t = 15$ msec: $\theta = 100\pi \times 0.015 = 1.5\pi = 270°$. $V_{\text{in}} = 220 \sin(270°) = -220\ V$.

Full-wave rectifier: When V_{in} is negative, D_1 (top) is reverse biased, D_2 (bottom) is forward biased, allowing current through the load.

41. (2) (NCERT XI, Part-II, Ch-Thermodynamics, Page-230)

Work done at constant pressure:

$W = P\Delta V$. $\Delta V_A = \pi r_A^2 \times 16$, $\Delta V_B = \pi r_B^2 \times 9$.

Heat supplied: $Q = \Delta U + W$.

Since $\Delta U_A = \Delta U_B$, $Q_A = Q_B$, so $W_A = W_B$.

$P\pi r_A^2 \times 16 = P\pi r_B^2 \times 9$.

$\Rightarrow \left(\dfrac{r_A}{r_B}\right)^2 = \dfrac{9}{16} \Rightarrow \dfrac{r_A}{r_B} = \dfrac{3}{4}$.

42. (3) (NCERT XI, Part-I, Ch-Unit and measurement, Page-6)

For $P = a^3b^2/cd^{1/2}$, the percentage error is:

$\dfrac{\Delta P}{P} \times 100 = 3\left(\dfrac{\Delta a}{a}\right) + 2\left(\dfrac{\Delta b}{b}\right) + \left(\dfrac{\Delta c}{c}\right) + \dfrac{1}{2}\left(\dfrac{\Delta d}{d}\right)$.

$= 3 \times 1\% + 2 \times 3\% + 2\% + \dfrac{1}{2} \times 4\%$

$= 3\% + 6\% + 2\% + 2\% = 13\%$.

43. (3) (NCERT XII, Part-II, Ch-Wave optics, Page-270-271)

1. After the first polaroid, intensity = I_0. The second polaroid (middle) is at $\theta = 22.5°$ to the first: $I_1 = I_0 \cos^2 \theta$.

The third polaroid is crossed with the first (90° apart), so it's at $90° - \theta$ to the second. $I_2 = I_1 \cos^2 (90° - \theta)$

$I_2 = I_0 \cos^2 \theta \cos^2 (90° - \theta)$

$= \dfrac{I_0}{4}\sin^2(2\theta) = \dfrac{I_0}{4}\sin^2(2\times 22.5) = \dfrac{I_0}{8}$

44. (4) (NCERT XI, Part-II, Ch-oscillation, Page-267)

For SHM, maximum speed: $v_{\max} = A\omega$. *For P*: $\omega p = \sqrt{\dfrac{k_1}{m}}$, so $v_{p,\max} = Ap\sqrt{\dfrac{k_1}{m}}$. For Q: $v_{Q,\max} = A_Q\sqrt{\dfrac{k_2}{m}}$.

Given $v_{p,max} = v_{Q,max}$: $Ap\sqrt{\dfrac{k_1}{m}} = A_Q\sqrt{\dfrac{k_2}{m}}$.

So, $\dfrac{A_Q}{A_P} = \sqrt{\dfrac{k_1}{k_2}}$.

45. (2) (NCERT XI, Part-II, Ch-Waves, Page-292)

Open pipe: Fundamental frequency $f = \dfrac{v}{2L}$, where L is the length of the pipe.

Dipped halfway: Air column length $= \dfrac{L}{2}$, now closed at one end (water end). For a closed pipe, fundamental frequency $= \dfrac{v}{4\left(\frac{L}{2}\right)} = \dfrac{v}{2L} = f$.

New frequency $= f$.